40th European Photovoltaic Solar Energy Conference and Exhibition (EU PVSEC 2023)

Lisbon, Portugal
18-22 September 2023

Volume 1 of 3

ISBN: 979-8-3313-1553-5

Printed from e-media with permission by:

Curran Associates, Inc.
57 Morehouse Lane
Red Hook, NY 12571

proceedings.com

Some format issues inherent in the e-media version may also appear in this print version.

Printed with permission by Curran Associates, Inc. (2025)

For permission requests, please contact WIP – Renewable Energies
at the address below.

WIP – Renewable Energies
Sylvensteinstr. 2
81369 Munchen
Germany

Phone: +49 89 72012735
Fax: +49 89 72012791

wip@wip-munich.de

Additional copies of this publication are available from:

Curran Associates, Inc.
57 Morehouse Lane
Red Hook, NY 12571 USA
Phone: 845-758-0400
Fax: 845-758-2633
Email: curran@proceedings.com
Web: www.proceedings.com

TABLE OF CONTENTS

VOLUME 1

VOLUME 2

VOLUME 3

40th European Photovoltaic Solar Energy Conference and Exhibition

---

Proceedings of the International Conference

18 September – 22 September 2023

Edited by:

J.M. ALMEIDA SERRA
University of Lisbon
Faculty of Sciences
Portugal

R. KENNY
European Commission
Joint Research Centre
Italy

J. BERGMILLER
WIP Renewable Energies
Germany

J. DE GREGORIO
WIP Renewable Energies
Germany

Photos at:

Coordination of the Technical Programme:

European Commission Joint Research Centre
Via E. Fermi 1
21020 Ispra (VA)
Italy

Institutional Support:

European Commission

WCRE – World Council for Renewable Energy

Supporting Organisations:

APVI – Australian PV Institute

EREF – European Renewable Energies Federation

ESA – European Space Agency

EUFORES –The European Forum for Renewable Energy Sources

EUREC – The Association of European Renewable Energy Research Centres

ETIP PV – European Technology & Innovation Platform PV

GÜNDER – The Turkish Solar Energy Society

IEA PVPS – International Energy Agency, Photovoltaic Power Systems Programme

INES – National Solar Energy Institute

NUS /SERIS – National University of Singapore / Solar Energy Research Institute of Singapore

ODTÜ-GÜNAM – Center for Solar Energy Research and Applications

REN21 – Renewable Energy Policy Network for the 21st Century

Institutional PV Industry Cooperation:

SolarPower Europe

ESMC – European Solar Manufacturing Council

Supporting Associations:

ABENS – Associação Brasileira de Energia Solar

ALER – Associação Lusófona de Energías Renováveis

ARE – Aliance for Rural Electrification

AREA – African Renewable Energy Alliance

ASOM – Aliance for Solar Mobility

CABA – Continental Automated Buildings Association

ESIA – European Solar PV Industry Aliance

VDMA Photovoltaic Equipment

Supporting Educational Institutions:
Loughborough University

EU PVSEC 2023 realised by:

WIP Renewable Energies
Sylvensteinstr. 2, 81369 München, Germany
Tel: +49 89 720 12 735, Fax: +49 89 720 12 791
Email: pv.conference@wip-munich.de
www.eupvsec.org
www.wip-munich.de

Proceedings produced and published by:

WIP Renewable Energies
Sylvensteinstr. 2, 81369 München, Germany
Tel: +49 89 720 12 735, Fax: +49 89 720 12 791
Email: pv.conference@wip-munich.de
www.eupvsec.org
www.wip-munich.de

ISSN 2196-0992

40th EUROPEAN PHOTOVOLTAIC SOLAR ENERGY CONFERENCE AND EXHIBITION
18 SEPTEMBER – 22 SEPTEMBER 2023

EU PVSEC 2023 COMMITTEES

INTERNATIONAL SCIENTIFIC ADVISORY COMMITTEE (ISAC)

Chair

P. Szymanski, European Commission Joint Research Centre, Director of Energy, Transport and Climate, Petten, The Netherlands

Committee Members

P. Frankl, Head of the Renewable Energy Division, International Energy Agency, France
M. Getsiou, European Commission, DG RTD, Brussels, Belgium
S.W. Glunz, Head of Division "Solar Cells – Development and Characterization", Fraunhofer ISE, Freiburg, Germany
R. Kenny, European Commission Joint Research Centre, Directorate for Energy and Transport and Climate, Ispra, Italy
P. Malbranche, General Director, CEA INES, France
S. Nowak, Managing Director of NET Nowak Energy & Technology, St. Ursen, Switzerland
W.C. Sinke, Principal Scientist Solar Energy, TNO Energy Transition, The Netherlands
M. Topič, Chairman ETIP PV, Head of Laboratory of Photovoltaics and Optoelectronics of the University of Ljubljana, Slovenia
P. Verlinden, Director at Amrock, Visiting Professor at Sun Yat-Sen University, Guangzhou, China
J. Bergmiller, Managing Director Events, WIP Renewable Energies, Munich, Germany
J. de Gregorio, Head of Unit, Scientific Services and Cooperation, WIP Renewable Energies, Munich, Germany

CONFERENCE EXECUTIVE COMMITTEE

Conference General Chair

J.M. Almeida Serra, University of Lisbon- FCiencias.ID, Portugal

Technical Programme Chair

R. Kenny, European Commission Joint Research Centre, Directorate for Energy and Transport and Climate, Italy

Committee Members

W.C. Sinke, Principal Scientist Solar Energy, TNO Energy Transition, The Netherlands
S. Nowak, Managing Director of NET Nowak Energy & Technology, St. Ursen, Switzerland
M. Topič, Chairman ETIP PV, Head of Laboratory of Photovoltaics and Optoelectronics of the University of Ljubljana, Slovenia
V. Bermudez Benito, QEERI, Qatar
E. Voroshazi, CEA / INES, France
H. Ossenbrink, Band Gap, Germany
J. Bergmiller, Managing Director Events, WIP Renewable Energies, Munich, Germany
J. de Gregorio, Head of Unit, Scientific Services and Cooperation, WIP Renewable Energies, Munich, Germany

2023 SCIENTIFIC COMMITTEE

Programme Technical Chair

R. Kenny (Technical Chair), European Commission, Joint Research Centre, Italy

Topic Chairs

Topic 1: Silicon Materials and Cells
S. Glunz, Fraunhofer ISE, Germany

Topic 2: Evolving and Emerging Technologies: Tandems; Thin Film absorbers; III-V; New Materials and Concepts; Advanced Modelling
S. Veenstra, TNO Energy Transition, The Netherlands

Topic 3: Photovoltaic Modules and BoS Components
E. Voroshazi, CEA / INES, France

Topic 4: PV Systems Engineering, Integrated/Applied PV
D. Moser, Eurac Research, Italy

Topic 5: Energy Transition – Integration, Storage, Sustainability, Policy, Economics, Energy Poverty, Society
N. Della-Valle, European Commission JRC, Italy

Topic Organisers and Paper Review Experts

Topic 1: Silicon Materials and Cells
D. Bredemeier, ISFH, Germany
S. Riepe, Fraunhofer ISE, Germany
S. W. Glunz, Fraunhofer ISE, Germany
G. Hahn, University of Konstanz, Germany
K. Ding, Forschungszentrum Jülich, Germany
D. Muñoz, CEA / INES, France
P. Roca i Cabarrocas, CNRS-LPICM, France
A. W. Weeber, TNO Energy Transition, The Netherlands
M. Topic, University of Ljubljana, Slovenia
K. Bothe, ISFH, Germany
P. Fath, RCT-Solutions, Germany
S. Peters, Hanwha Q CELLS, Germany

M. P. Bellmann, STIFTELSEN SINTEF, Norway
M. Benmansour, CEA / INES, France
T. Bruton, Progress in Photovoltaics, United Kingdom
A. Ciesla, UNSW Australia, Australia
C. Hagendorf, Fraunhofer CSP, Germany
X. Yu, Zhejiang University, China
R. Brendel, ISFH, Germany
S. Dubois, CEA / INES, France
T. Dullweber, ISFH, Germany
M. Ernst, ANU, Australia
M. Hermle, Fraunhofer ISE, Germany
J. Horzel, Fraunhofer ISE, Germany
R. C. G. Naber, Hanwha Q CELLS, The Netherlands
W. Nemeth, NREL, USA
R. Preu, Fraunhofer ISE, Germany

B. Terheiden, University of Konstanz, Germany
R. Turan, METU, Türkiye
W. Dauksher, Arizona State University, USA
P. Delli Veneri, ENEA, Italy
W. Favre, CEA / INES, France
T. Matsui, AIST, Japan
J. Meier, Meier Technologies, Switzerland
Y. Ohshita, Toyota Technological Institute, Japan
J. Rentsch, Fraunhofer ISE, Germany
J. Schmidt, ISFH, Germany
M. Wright, University of Oxford, UK
J. Zhao, CSEM, Switzerland
A. Augusto, Arizona State University, USA
N. Azkona, UPV/EHU, Spain
M. di Sabatino, NTNU, Norway
F. Ferrazza, ENI, Italy
D. MacDonald, ANU, Australia
A. Otaegi, UPV/EHU, Spain
M. C. Schubert, Fraunhofer ISE, Germany
L. Serenelli, ENEA, Italy
A. Yamada, Tokyo Institute of Technology, Japan
D. L. Bätzner, Meyer Burger Research AG, Switzerland
D. Brunner, RENA Technologies
A. Danel, CEA / INES, France
C. Gerardi, ENEL Green Power, Italy
A. Metz, Germany
H. J. Nonnenmacher, Meyer Burger, Germany
P. J. Verlinden, AMROCK, Australia
Q. Wang, Zhejiang Jinko Solar, China
W. Wu, Tongwei Solar (Meishan), China
W. Zhang, Gonda Electronic Technology, China

Topic 2: Evolving and Emerging Technologies: Tandems; Thin Film absorbers; III-V; New Materials and Concepts; Advanced Modelling

J. C. Goldschmidt, University of Marburg, Germany
I. Gordon, imec, Belgium
W. Tress, Zurich University of Applied Sciences, Switzerland
S. Veenstra, TNO Energy Transition, The Netherlands
M. Edoff, Uppsala University, Sweden
G. Siefer, Fraunhofer ISE, Germany
A. N. Tiwari, EMPA, Switzerland
A. Martí Vega, UPM, Spain
J. J. Poortmans, imec, Belgium
T. Magorian-Friedlmeier, ZSW, Germany
I. Ramiro, UPM, Spain

S. Albrecht, HZB, Germany
S. Berson, CEA / INES, France
A. Bruno, Nanyang Technological University, Singapore
P. Carroy, CEA / INES, France
C. Case, Oxford PV, UK
G. Coletti, TNO Energy Transition, The Netherlands
K. Makita, AIST, Japan
M. Morales-Masis, University of Twente, The Netherlands
U. W. Paetzold, KIT Karlsruher Institut für Technologie, Germany
H. Sivaramakrishnan Radhakrishnan, imec, Belgium
M. J. Brites, LNEG, Portugal
S. Cros, CEA / INES, France
S. Hayase, The University of Electro-Communications, Japan
S. Huang, Macquarie University, Australia
M. Khenkin, HZB, Germany
N. Kyranaki, CEA / INES, France
C. F. Lin, National Taiwan University, Taiwan
B. Mihaylov, European Commission JRC, Italy
M. Provost, IPVF, France
Y. Saygili, ODTÜ-GÜNAM, Türkiye
L. Vesce, University of Rome "Tor Vergata", Italy
N. Barreau, University of Nantes, France
G. Brammertz, imec, Belgium
T. Dalibor, Avancis, Germany
D. Dimova-Malinovska, Bulgarian Academy of Sciences, Bulgaria
G. Flamand, imec, Belgium
C. Hages, University of Florida, USA
S. Ishizuka, AIST, Japan
R. R. King, Arizona State University, USA
A. Meeder, ZSW, Germany
R. Po, ENI, Italy
A. Redinger, University of Luxembourg, Luxembourg
A. Romeo, University of Verona, Italy
V. Sittinger, Fraunhofer IST, Germany
M. Theelen, TNO/Solliance, The Netherlands
G. Timò, RSE, Italy
J. P. Connolly, GeePs-CentraleSupelec, France
M. A. Hernandez-Fenollosa, UPV, Spain
K. Hummelen, University of Groningen, The Netherlands
J. P. Kleider, CNRS/GeePs, France
D. König, UNSW Australia, Australia
I. Konovalov, University of Applied Sciences Jena, Germany
Y. Okada, University of Tokyo, Japan
F. Roca, ENEA, Italy
M. Rusu, HZB, Germany
C. Becker, HZB, Germany
N. Ekins-Daukes, UNSW Australia, Australia
D. Kuciauskas, NREL, USA
C. Lévy-Clément, CNRS, France
P. Pearce, UNSW Australia, Australia
R. I. Schropp, Utrecht University, The Netherlands
T. Tayagaki, AIST, Japan

Topic 3: Photovoltaic Modules and BoS Components

E. Voroshazi, CEA / INES, France
V. Bermudez Benito, QEERI, Qatar
T. Barnes, NREL, USA
R. Gottschalg, Fraunhofer CSP, Germany
W. Herrmann, TÜV Rheinland Energy, Germany
G. Friesen, SUPSI, Switzerland
C. Buerhop-Lutz, HI ERN, Germany
J. Moschner, KU Leuven, Belgium

V. Barth, CEA / INES, France
A. Faes, CSEM, Switzerland
P. Kratzert, SOLARWATT, Germany
M. Mittag, Fraunhofer ISE, Germany
M. Á. Muñoz-García, UPM, Spain
H. Nagel, Fraunhofer ISE, Germany
S. M. Pietralunga, CNR – ISMAC, Italy
T. Timofte, ISC Konstanz, Germany
G. Beaucarne, Dow Silicones, Belgium
T. Bejat, CEA, France
N. Bosco, NREL, USA
C. Camus, LayTec AG, Germany
E. Fokuhl, Fraunhofer ISE, Germany
U. Jahn, VDE Renewables, Germany
E. Krassowski, CE Cell Engineering, Germany
S. Lindig, Eurac Research, Italy
S. Oktik, Kadir Has University Central Campus, Türkiye
G. Oreski, PCCL, Austria
M. Pander, Fraunhofer CSP, Germany
T. Sample, European Commission JRC, Italy
S. Bordihn, ISFH, Germany
A. J. Carr, TNO Energy Transition, The Netherlands
M. Despeisse, CSEM, Switzerland
J. Govaerts, imec, Belgium
Y. Hishikawa, AIST, Japan

J. Lopez-Garcia, QEERI, Qatar
T. Lyubenova, Joint Research Centre, Italy
C. Monokroussos, TÜV Rheinland (Shanghai), China
H. Nussbaumer, Zurich University of Applied Sciences, Switzerland
M. Paverttoni, SERIS, Singapore
S. Riechelmann, PTB, Germany
J. Roy, IIT Kharagpur, India
R. Sinton, Sinton Instruments, USA
C. Ulbrich, PVcomB, Germany
P. Bauwens, Ghent University, Belgium
N. M. Pearsall, Northumbria University, UK

Topic 4: PV Systems Engineering, Integrated/Applied PV

A. M. Gracia Amillo, Fraunhofer ISE, Germany
W. M. van Sark, Utrecht University, The Netherlands
F. P. Baumgartner, Zurich University of Applied Sciences, Switzerland
K. Lappalainen, Tampere University, Finland
J. M. Almeida Serra, University of Lisbon - FCiencias.ID, Portugal
D. Moser, Eurac Research, Italy
G. Mütter, Alternative Energy Solutions, Austria
F. Frontini, SUPSI, Switzerland
A. Scognamiglio, ENEA, Italy
T. Reindl, NUS, Singapore
I. Tsanakas, CEA, France
R. M. E. Valckenborg, TNO Energy Transition, The Netherlands
G. Adinolfi, ENEA, Italy
P.-J. Alet, CSEM, Switzerland
I. Antón Hernández, UPM, Spain
S. Taylor, European Space Agency, The Netherlands

A. Alcañiz, Delft University of Technology, The Netherlands
P. Blanc, MINES ParisTech, France
V. Lara-Fanego, Solargis, Slovakia
A. Louwen, Eurac Research, Italy
A. Martinez Fernandez, European Commission JRC, Italy
T. Oozeki, AIST, Japan
C. Protogeropoulos, EEPS, Greece
J. Remund, Meteotest, Switzerland
M. Sengupta, NREL, USA
M. Zehner, Rosenheim Technical University of Applied Sciences, Germany
C. Alonso-Tristán, UBU, Spain
M. Carbone, ENEL Green Power, Italy
P. Darez, 350 Renewables, Chile
C. W. Hansen, Sandia National Laboratories, USA
C. Nyman, Soleco, Finland
R. Einhaus, ZSW, Germany
P. Hacke, NRE, USA
A. Heimsath, Fraunhofer ISE, Germany
J. Lin, PV Guider, Taiwan
O. Mayer, Bayern Innovativ, Germany
A. Migan-Dubois, GeePs, France
S. Rodrigues, Forschungszentrum Jülich, (HI-ERN), Germany
J. S. Stein, Sandia National Laboratories, USA
D. Stellbogen, ZSW, Germany
P. Suratkar, PV Diagnostics, India
M. Theristis, Sandia National Laboratories, USA
A. Virtuani, CSEM, Switzerland
S. Aden, HZB, Germany
P. Alonso, BWP Baumanagement, Germany
Y. B. Assoa, CEA, France
P. Bonomo, SUPSI, Switzerland
V. D'Ambrosio, University of Naples Federico II, Italy
G. C. Eder, OFI, Austria
T. Reijenga, BEAR-iD, The Netherlands
E. Román Medina, Fundación Tecnalia, Spain
L. H. Slooff-Hoek, TNO Energy Transition, The Netherlands
F. Colucci, ENEA, Italy
M. Di Blasi, Enel Green Power, Italy
M. Dörenkämper, TNO Energy Transition, The Netherlands
M. Heinrich, Fraunhofer ISE, Germany
B. Lim, ISFH, Germany
B. Newman, Lightyear, The Netherlands
R. Peibst, ISFH, Germany
A. Reinders, Eindhoven University of Technology, The Netherlands
S. Schindele, BayWa R.e., Germany
R. Schlatmann, HZB, Germany
T. Tanahashi, AIST, Japan
G. Barchi, Eurac Research, Italy
R. Bründlinger, AIT, Austria
V. Efthymiou, University of Cyprus, Cyprus
K. Araki, University of Miyazaki, Japan
R. Campesato, CESI, Italy
V. Khorenko, Azur Space Solar Power, Germany
K. Nishioka, University of Miyazaki, Japan

Topic 5: Energy Transition – Integration, Storage, Sustainability, Policy, Economics, Energy Poverty, Society

R. Pestana, R&D NESTER, Portugal
M. Victoria, Aarhus University, Denmark
I. Weiss, WIP Renewable Energies, Germany
M. de Wild-Scholten, SmartGreenScans, The Netherlands
K. Wambach, Wambach-Consulting, Germany
M. Getsiou, European Commission DG RTD, Belgium
P. Malbranche, International Solar Aliance, India
S. Nowak, NET Nowak Energy & Technology, Switzerland
C. Breyer, Lappeenranta University of Technology, Finland
I. Kaizuka, RTS Corporation, Japan
G. Masson, Becquerel Institute, Belgium
S. Caneva, WIP Renewable Energies, Germany
N. Della-Valle, European Commission JRC, Italy

M. C. Brito, University of Lisbon - Faculty of Science, Portugal
F. Carigiet, Zurich University of Applied Sciences, Switzerland
J. S. da Costa Fernandes, University of Applied Sciences Offenburg, Germany
B. Gaiddon, HESPUL, France
J. M. Klein, CEA / INES, France
P. Lalanne, HydroClapeyron, France
A. C. Neves Foles, University of Évora, Portugal
M. Aleman, Becquerel Institute, Belgium
A. Anctil, Michigan State University, USA
S. Arancón, Plug and Play, Spain
S. Capaccioli, ETA - Florence Renewable Energies, Italy
V. Fthenakis, Columbia University, USA
G. Heath, NREL, USA
K. Komoto, Mizuho Research & Technologies, Japan
W. Palitzsch, LuxChemtech, Germany
E. Parnham, DuPont, United Kingdom
S. Guastella, RSE, Italy
H. Ossenbrink, Band Gap, Germany
D. Polverini, European Commission DG GROWTH, Belgium
M. Ray, IIT Kharagpur, India
N. Taylor, European Commission JRC, Italy
R. Blanchard, Loughborough University, United Kingdom
S. De Iuliis, ENEA, Italy
T. Haarberg, BNW-Energy AS, Germany
A. Nayfeh, Khalifa University, United Arab Emirates

E. Vartiainen, Fortum Growth, Finland
G. Ruggieri, Insubria University, Italy
S. Tay En Rong, National University of Singapore, Singapore
Kaizuka, RTS Corporation, Japan

Awards Coordinators

Student Awards Coordinator
A.H.M. Smets, Delft University of Technology, The Netherlands

Poster Awards Coordinator
R. Kenny (Technical Chair), European Commission Joint Research Centre, Directorate for Energy and Transport and Climate, Italy

SUBJECT INDEX

Topic Code

1 Silicon Materials and Cells
2 Thin Films and New Concepts
3 Photovoltaic Modules and BoS Components
4 PV Systems Engineering, Integrated/Applied PV
5 PV in the Energy Transition

Session Type

P = Plenary Session
O = Oral Session
V = Visual Session

Day Codes

A = Monday, 23 September 2023
B = Tuesday, 24 September 2023
C = Wednesday, 25 September 2023
D = Thursday, 26 September 2023
E = Friday, 27 September 2023

e.g. 1AO.4 ⇒ 1= Silicon Materials and Cells, A=Monday, O=Oral session, 4=Session 4

TABLE OF CONTENTS OF EU PVSEC 2023 PROCEEDINGS PAPERS

Visual SESSION 1CV.3 High Temperature Routes for Si Cells / Heterojunction Solar Cells / Recent Advances in Silicon Solar Cell Characterisation

Visual SESSION 2BV.2 Perovskite Photovoltaics

Oral SESSION 2CO.1 High-Efficiency Pervoskite Silicon Tandems

Plenary SESSION 2CP.3

Visual SESSION 2CV.2 Compound and Organic Semiconductors

Oral SESSION 2DO.6 Chalcogenide Solar Cells, Devices and Back Contacts

Oral SESSION 2DO.8 CIGS and Related Materials, Device Concepts and Stability

Oral SESSION 2DO.9 III-V and Related Compound Semiconductors

Visual SESSION 2DV.1 Perovskite-based Tandem Devices

Visual SESSION 3AV.1 PV Module Design and Manufacturing / Electronic Systems for BOS

Visual SESSION 3AV.2 Module Reliability

Visual SESSION 3AV.3 PV Module Performance

Oral SESSION 3BO.11 Reliability in Harsh Environments

Oral SESSION 3BO.14 Reliability and Degradation of Module Packaging

Oral SESSION 3BO.15 Durability Testing, Modelling, and In-Situ Characterisation

Oral SESSION 3CO.10 Optical and Electrical Characterisation of PV Modules

Oral SESSION 3DO.19 Encapsulants, Light-Weight and Eco-Designed Modules

Oral SESSION 3EO.1 Innovations in PV Modules and BOS Components

Oral SESSION 4AO.7 Field Performance - Imaging, Temperature, Climate

Oral SESSION 4BO.7 Operation of PV Systems in Specific Contexts

Visual SESSION 4BV.3 Integrated PV and Emerging Synergistic Applications

Visual SESSION 4BV.4 PV and Buildings / Managing Local Fluctuations with Storage / Concentrators; Space Applications

Oral SESSION 4CO.8 Solar Forecasting

Oral SESSION 4CO.9 Solar Resource Assessment and Modelling

Oral SESSION 4DO.1 PV System Design and Optimisation

Oral SESSION 4DO.12 Design and Optimisation of Bifacial PV Systems

Oral SESSION 4DO.2 Agrivoltaics: Advances in Design, Modelling and Performance Analysis

Oral SESSION 4DO.3 Floating PV

Oral SESSION 4DO.4 Transport / Vehicle Integrated Photovoltaics (VIPV)

Oral SESSION 4DO.5 "PV Everywhere": Infrastructure Integrated Photovoltaics (I2PV) and New Applications

Visual SESSION 4DV.4 Solar Resource Assessment, Modelling and Forecasting / PV System Engineering

Oral SESSION 5DO.11 Broadening the PV Markets: Tools and Results for Better Analysis, Modelling and Forecasting

Oral SESSION 5DO.14 Circularity and Recycled Material Supply of PV

Visual SESSION 5DV.3 PV Deployment: Realising the Full Potential of PV in Various Markets and World Regions / Markets, Cost, Economics of PV / Energy

FOREWORD

The European Photovoltaic Solar Energy Conference and Exhibition (EU PVSEC) is the World's leading and most renowned forum for PV research and development and the biggest conference on PV solar energy. In 2023, celebrating its 40th edition, the EU PVSEC was the central meeting and exchanging point for global PV experts from research, development, and industry.

The EU PVSEC 2023 took place from 18 – 22 September 2023 in Lisbon, Portugal, was a resounding success providing diverse and state of the art research findings and the key message of the pivotal role of photovoltaics in mitigating climate change and its importance in the energy transition. Building on a rapidly expanding production and deployment rate and 1.5 TW of global installed capacity expected by the end of 2023, the PV community discussed, covering the complete value chain, and from an innovative perspective, the pathways and challenges towards becoming a major, sustainable, circular, clean renewable energy source in the coming decades.

Combining Conference and Exhibition, this year, more than 1850 participants from 63 countries attended this highly inspiring platform delivering over 1050 presentations. It provided a vital forum for information exchange in the field of PV solar energy research, demonstration and applications. In the exhibition 71 companies from all parts of the world welcomed visitors and presented their products and services.

The EU PVSEC 2023 Proceedings give a comprehensive overview of the photovoltaic solar sector, its current status and future prospects in science, research, innovation, development and deployment on almost 3,900 pages. Selection for inclusion in the conference was made by paper review experts and topic organisers (see the listing on pages 010002-001-004), to whom we express our sincere gratitude for their comprehensive review work and overall contribution to the success of the conference. In addition to the 326 submitted papers, the proceedings include 102 presentations (slides) shown during the plenary and oral presentations as well as 156 poster files of the visual presentations. In total this amounts to 584 publications.

In a new publication tool updated this year, the EU PVSEC 2023 maintains its commitment to providing quick and open access to high quality scientific results. This is a powerful source for targeted and quick information search and retrieval, enabling you to search by topic, keywords, paper title, DOI, author, or organization. We are sure that these new features will help to simplify the use and exploit the full potential of this extensive fountain of knowledge.

The Conference Proceedings are published as downloadable files and are also fully accessible online. A DOI code (Digital Object Identifier) has been assigned to each paper. This ensures unequivocal and permanent identification and full citability. The EU PVSEC 2023 papers can be viewed and downloaded in a full free open access from the Proceedings website https://userarea.eupvsec.org/proceedings.

We are confident that these Proceedings will play an important role in providing a comprehensive overview of the current actors and activities in the global PV sector and that they will disseminate information on the state-of-the-art of technologies and applications. This can generate further research, add momentum to innovation and promote interest in PV worldwide.

We would like to cordially thank all authors and participants of the EU PVSEC 2023 for their contributions and look forward to welcoming you in Vienna, Austria from 23 – 27 September 2024 at the EU PVSEC 2024, the 41st European Photovoltaic Solar Energy Conference and Exhibition

The Editors

THE EFFECT OF HYDROGEN ON BORON-OXYGEN-RELATED DEGRADATION AND REGENERATION IN CRYSTALLINE SILICON

Rune Søndenå[1,a], Philip M. Weiser[2], Eduard Monakhov[2], and Halvard Haug[1]
[1]Institute for Energy Technology, Norway
[2]Centre for Materials Science and Nanotechnology, Department of Physics, University of Oslo, Norway
[a]Rune.Sondena@ife.no

ABSTRACT: The role of hydrogen on boron-oxygen related light induced degradation and regeneration of the minority carrier lifetime in silicon wafers under illuminated annealing has been studied using a combination of low temperature infrared spectroscopy and carrier lifetime measurements. The localized vibrational mode of the hydrogen-passivated boron acceptor complex (H-B) is detected by transmission measurements through a stack of 1×1 cm^2 p-type monocrystalline silicon wafer pieces polished to optical quality along one set of parallel edges. Boron-oxygen related degradation, with the characteristic fast and slow decay, is demonstrated in the wafers during illumination at room temperature. With the presence of hydrogen, the initial minority carrier lifetime is somewhat reduced and the extent of the fast decay during the first few minutes of illumination decreases. A subsequent illuminated annealing process is used to regenerate much of the initial carrier lifetimes. The concentration of H-B complexes decreases by about 16% during light induced degradation at room temperature, whereas a stronger decrease is seen during the illuminated annealing step. Unintentional in-diffusion of a small amount of hydrogen into the wafer during the surface passivation process makes it difficult to determine if the presence of hydrogen is required for regeneration to occur. Accelerated recovery of the degraded lifetime is, however, observed with an increased H-B concentration.
Keywords: Crystalline silicon, degradation, hydrogen

1 INTRODUCTION

The minority charge carrier lifetime (here: lifetime) in silicon wafers containing both boron and oxygen is known to degrade considerably when exposed to light.[1] This effect is named light induced degradation (LID), although charge injection through a bias voltage passing current through the material will also generate these recombination-active defects. Boron-oxygen related light induced degradation (BO-LID) in *p*-type Czochralski (Cz) silicon wafers may reduce the lifetime by as much as 90%[2]. BO-LID is characterized by its initial fast decay during the first few minutes of illumination followed by a slower decay until lifetime saturation.[3] Charge injection at an elevated temperature, e.g. illuminated annealing,[4], [5] can be used to recover much of the initial lifetime in a so-called regeneration process[6], [7]. Other defects and impurities in silicon are also sensitive to light, e.g. the FeB-pair, and the much investigated light and elevated temperature induced degradation (LeTID)[8], [9]. LeTID also results in an initial degradation followed by a regeneration of the lifetime with prolonged illumination at an elevated temperature[10], [11]. The introduction of hydrogen (H) into the bulk silicon, either by contact firing with a hydrogen-rich anti reflection coating (ARC) present or by plasma injection, is a requirement for the activation of the LeTID recombination centres[12], [13]. Unlike the defect complex causing LeTID, formation of recombination active BO-defects has not been linked to the presence of hydrogen. However, hydrogen has a demonstrated effect on the regeneration rates[14], [15]. Both Münzer, and Nampalli *et al.* have reported that the lifetime in cells and wafers containing no hydrogen could not be regenerated following the degradation[16], [17]. Walter and Schmidt have, however, reported that the presence of hydrogen is not required for regeneration[15]. By measuring the regeneration rate in wafers with different hydrogen content and extrapolating towards low hydrogen concentrations, Fattah et al. propose that some regeneration can be expected even without H present[18]. Different regeneration mechanisms for the fast and slow decay, where only regeneration of the fast forming recombination centres require hydrogen, have been proposed by Voronkov et al.[19]. During contact firing hydrogen is introduced into the wafers and cells mainly as H_2 molecules, but also a fraction of hydrogen-boron (H-B) pairs[20], [21]. Splitting of the H_2 dimers with a subsequent formation of hydrogen-acceptor-pairs during dark annealing has been used to quantify the hydrogen content in silicon wafers[22], [23]. A mechanism with passivation of recombination active BO-complexes using hydrogen in its neutral state (H^0) has been proposed[24], [25]. During the illuminated annealing regeneration process, it is hypothesized that H-B pairs are split, releasing H^0 to passivate the BO-related defect(s).

Hydrogen-related defects in bulk silicon can be studied via their localized vibration modes (LVMs) using infrared (IR) absorption spectroscopy[26], [27]. The LVMs of defect complexes of intrinsic defects, transition metals, acceptors, donors, and other impurities such as carbon, together with hydrogen in silicon are well studied[28]–[34]. A method to detect H-related point defects, including H-B pairs, in multicrystalline silicon wafers using low temperature Fourier Transform-IR (FT-IR) spectroscopy was recently proposed[20], [21], [35].

In the present work, we combine quasi steady-state photoconductance (QssPC) measurements and low temperature FT-IR spectroscopy to study changes in the lifetime and the H-B pair concentration, respectively, at different stages throughout the degradation and regeneration cycle of a *p*-type Cz-Si wafer. The importance of hydrogen on the BO-related degradation and regeneration will be discussed.

2 EXPERIMENTAL DETAILS

Commercially available 6-inch *p*-type (B-doped) Cz-Si wafers with a resistivity of about 2.2 Ω-cm have been investigated in this study. Wafers taken from the middle section of a Cz-Si crystal were damage etched in an HNA solution (hydrofluoric, nitric, and acetic acid), followed by an emitter in-diffusion in a $POCl_3$ tube furnace. Hydrogenation is performed on half the wafers by depositing a hydrogen-rich SiN_x anti-reflective coating (ARC) on both sides using plasma enhanced chemical vapour deposition (PECVD) at 400 °C, followed by a simulated contact firing process in a belt furnace (without metal pastes present). The ARC was then removed by dilute HF, and the emitters etched away from all wafers in a new HNA solution. All wafers were then passivated using a stack of amorphous silicon capped with silicon nitride (a-Si:H/a-SiNx:H) at a temperature of 230 °C, also using PECVD. A surface recombination velocity (SRV) of less than 5 cm/s is routinely obtained using this process[36].

QssPC measurements were performed at room temperature (RT) using a Sinton WCT-120TS. Carrier lifetimes have been extracted at an injection level of $0.1 \times p_0$, where p_0 is the equilibrium hole concentration. Degradation and regeneration of the lifetimes in segments from both fired/hydrogenated and non-fired reference wafers were induced first using low intensity LED illumination (ca. 5 mW/cm^2) at RT, followed by illuminated annealing (ca. 70 mW/cm^2 at 150 °C). This setup and procedure has previously been used to study BO-LID and LeTID degradation in multicrystalline silicon wafers[10], [11]. Lifetime degradation and regeneration have been modelled using sigmoidal fits of the form

$$\tau(t) = \tau_0 + (\tau_\infty - \tau_0) \cdot \frac{t^n}{(k^n + t^n)}, \qquad (1)$$

where k is a time constant corresponding to the time-to-half-max, n is a coordination factor, and τ_0 and τ_∞ are the initial and the final lifetime, respectively. Expressions based on first order kinetics provided fits of considerably poorer quality, and therefore were not considered further. Photoluminescence (PL) images have been obtained using a LIS-R1 setup from BTI[37]. Wafer segments halted at different stages of the degradation-regeneration cycle, as shown in Figure 1, were investigated with respect to the hydrogen complexes using FT-IR spectroscopy.

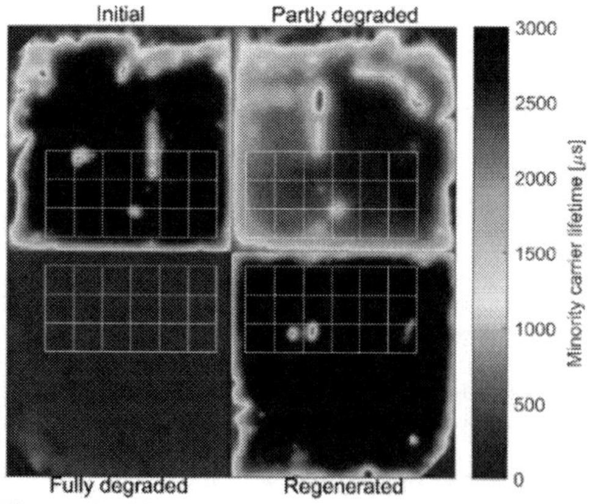

Figure 1: PL-images of the four segments of the hydrogenated wafer halted at various stages of processing. The 1×1 cm^2 wafer samples are indicated. Similar samples are made from the reference wafer.

Infrared transmittance measurements are made by sending the beam through the longitudinal direction of a stack of 1×1 cm^2 wafer samples. [35], [38] The sample stacks were cooled to 5.0 K in a Janis PTSHI-950-5 low vibration pulse tube cryostat, and IR transmittance measurements were made using a Bruker IFS 125HR FT spectrometer equipped with a globar light source, a KBr beamsplitter, and a liquid-nitrogen-cooled InSb detector. The empty sample holder served as the background for all measurements. Absorption coefficient spectra (α) were calculated from the transmittance (T) spectra using the standard equation, $T = \exp(-\alpha d)$, where d is the thickness of the sample stack. The hydrogen-boron passivated acceptor complex has been studied using its LVM at 1904 cm^{-1} in wafer stacks at different points of the degradation-regeneration process.

3 RESULTS AND DISCUSSION

Figure 2 shows the degradation and the subsequent regeneration of the lifetimes in both hydrogenated and reference Cz-Si wafers under illumination at RT and at 150 °C, respectively. Comparable lifetimes are expected as they are almost neighbouring wafers from the middle of a Cz-Si crystal. All wafers were etched and diffused together, and the processing of the wafers differ only by the hydrogenation. A simulated contact firing with an ARC present was performed on half of the wafers. These hydrogenated wafers show a slightly lower initial lifetime, τ_0, than the reference wafers. The extent of the fast decay during the first few minutes is also somewhat reduced, resulting in a higher lifetime in the partly degraded state for the hydrogenated wafers. Both the hydrogenated and the reference wafers degrade to a lifetime of about 250 µs after 96 hours of illumination at RT. It has been suggested that the defects responsible for BO-LID is caused by activation of two different latent recombination centres. The thermal history may shift the thermal equilibrium between these two centres, causing the differences observed in the degradation curves, as the amount of recombination sites causing the fast decay can be modulated with pre-degradation annealing steps [39], [40]. Comparable timescales for the decay and similar lifetimes in the fully degraded wafers indicate that hydrogen a minimal or no role in the degradation process.

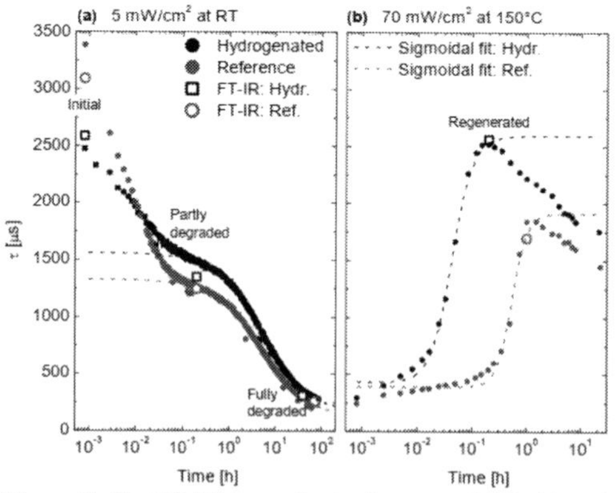

Figure 2: QssPC-lifetimes for hydrogenated (black) and reference (red) p-type Cz wafers during (a) illumination at RT, and (b) illuminated annealing. Dark annealing at 200 °C for 20 minutes was performed immediately before the automated degradation measurements. Sigmoidal fits based on the rate equation (Eq. 1) are shown as dashed lines[2].

Table 1: Time constants for the degradation and regeneration curves using Eq. 1. Standard errors from the fitting, when using n as a free parameter, are included.

	k_{Ref} [min]	k_{Hyd} [min]
Fast BO decay	0.19±0.01	0.48±0.07
Slow BO decay	311±6	294±4
Regeneration	32±2	2.5±0.1

Much of the initial lifetimes can be recovered in both hydrogenated and reference wafers during illuminated annealing. The initial lifetime in the hydrogenated wafer is completely recovered, whereas about 55% of τ_0 is regenerated in the reference wafer. Importantly, the regeneration rate is considerably higher in the hydrogenated, than in the reference wafers. Time constants for the fast and slow decays as well as regeneration are shown in Table I. The regeneration rate in the hydrogenated wafer is one order of magnitude faster than in the reference wafer. Sperber *et al.* have demonstrated an increase in the SRV of hydrogen-rich dielectric layers upon extended illuminated annealing, potentially explaining the secondary degradation after the maximum regenerated lifetime has been reached[41], [42]. The onset of the passivation layer degradation before full regeneration is reached may explain the lower fraction of the initial lifetime regenerated in the wafer with the lower regeneration rate. Another potential explanation is depletion of hydrogen.

The H-B complex has been investigated using FT-IR spectroscopy in wafer segments halted at different stages in the degradation-regeneration curve. The lifetimes in the different stages, as shown in Figure 1, are marked with open symbols in Figure 2. Figure 3(a) shows the IR absorption coefficient spectra of the H-B defect's LVM in the hydrogenated wafer at four different stages of the degradation-regeneration process. The corresponding concentrations shown in Figure 3(b) are calculated using calibration factors from the literature[43]. The initial H-B concentration decreases by 13% and 16% after illumination at RT for 0.12 h (partly degraded) and 96 h (fully degraded), respectively. After 12 minutes of illuminated annealing, which regenerates practically all the initial lifetime, the H-B content is decreased to about 35% of the initial concentration. Very small H-B concentrations (≤1.6e13 cm^{-3}) were detected in the reference samples, presumably due to in-diffusion from the low-temperature surface passivation.[35]

A mechanism where H^0, generated by dissolution of H-B pairs during illuminated annealing, passivating the recombination active defect complexes causing BO-LID has been proposed[24], [25]. Our results show that there is indeed a considerable reduction in the H-B concentration coinciding with the regeneration process in the hydrogenated wafer, supporting this theory. It is worth commenting that the reference wafer also is processed with a low-temperature hydrogen-rich passivation layer prior to the illuminated annealing. This passivation layer is a potential source for unintentional hydrogen in-diffusion. Thus, we cannot rule out that the small amount of hydrogen detected is sufficient to cause regeneration of the lifetime in the reference samples. Different initial hydrogen levels may even explain the differences in the regeneration rates.

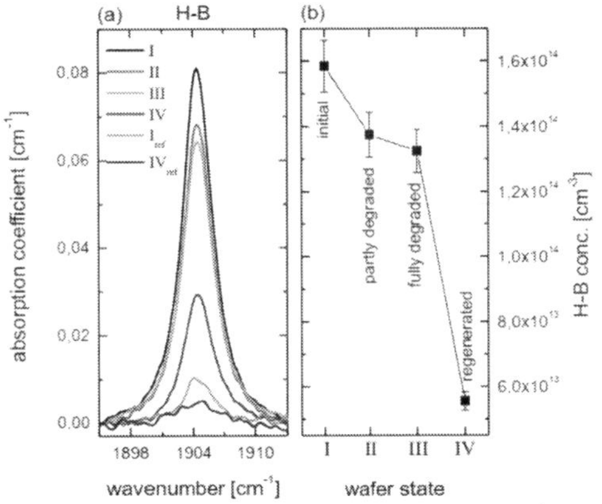

Figure 3: **(a)** IR absorption spectra of the 1904 cm^{-1} H-B LVM. **(b)** The H-B concentration in each wafer state as calculated from the integrated absorption coefficient for the H-B LVM.

4 SUMMARY

Low temperature FT-IR spectroscopy has been used to study the role of hydrogen on the BO-related light induced degradation and regeneration. The main difference with the presence of hydrogen in the bulk silicon is a considerable increase in the regeneration rate (by one order of magnitude) during illuminated annealing. Regeneration is also observed in the reference wafer, albeit slower and only to about 55% of the initial lifetime.
Changes in the H-B concentration throughout the degradation-regeneration cycle has been monitored through its LVM. An initial H-B concentration of 1.58e14 cm^{-3} was measured in the hydrogenated wafer. H-B concentrations decreased to ~84% and ~35% of the initial concentration after RT illumination and illuminated annealing, respectively. Thus, H-B may act as a source of H during the regeneration process.

5 ACKNOWLEDGMENTS

Funding for this work was provided by the Norwegian Research Council through the Research Center for Sustainable Solar Cell Technology (FME SUSOLTECH, #257639). The data that support the findings of this study are available from the corresponding author upon reasonable request.

[1] T. Niewelt, J. Schön, W. Warta, S. W. Glunz, and M. C. Schubert, *IEEE J. Photovolt.*, 7 (2017) 383–398.
[2] T. U. Nærland, H. Haug, H. Angelskår, R. Søndenå, E. S. Marstein, and L. Arnberg, *IEEE J. Photovolt.*, 3 (2013) 1265–1270.
[3] K. Bothe and J. Schmidt, *Appl. Phys. Lett.*, 87 (2005) 262108.
[4] A. Herguth, G. Scubert, M. Kaes, and G. Hahn, in *IEEE 4th World Conference on Photovoltaic Energy Conversion*, Waikoloa, Hawaii, (2006) 940–943.

[5] B. Hallam, A. Herguth, P. Hamer, N. Nampalli, S. Wilking, M. Abbott, S. Wenham, and G. Hahn, *Appl. Sci.*, 8, (2018) 10.

[6] A. Herguth, G. Schubert, M. Kaes, and G. Hahn, *Prog. Photovolt.*, 16 (2008) 135–140.

[7] A. Herguth and G. Hahn, *J. Appl. Phys.*, 108 (2010) 114509.

[8] D. Macdonald, J. Tan, and T. Trupke, *J. Appl. Phys.*, 103 (2008) 73710.

[9] K. Ramspeck, S. Zimmermann, H. Nagel, A. Metz, Y. Gassenbauer, B. Birkmann, and A. Seidl, in *27th EUPVSEC*, Frankfurt, Germany (2012) 861-865.

[10] R. Søndenå and M. S. Wiig, *J. Appl. Phys.*, 125 (2019) 85701.

[11] R. Søndenå, H. Haug, C. C. You, J. Zhu, and M. S. Wiig, *AIP Conf. Proc.*, 2147 (2019) 140010.

[12] C. Vargas, K. Kim, G. Coletti, D. Payne, C. Chan, S. Wenham, and Z. Hameiri, *IEEE J. Photovolt.*, 8, (2018) 413–420.

[13] M. A. Jensen, A. Zuschlag, D. Skorka, S. Wieghold, A. Morishige, G. Hahn, and T. Buonassisi, *J. Appl. Phys.*, 124 (2018) 85701.

[14] S. Wilking, A. Herguth, and G. Hahn, *Energy Procedia*, 38 (2013) 642–648.

[15] D. C. Walter and J. Schmidt, *Sol. Energy Mater. Sol. Cells*, 158 (2016) 91–97.

[16] K. A. Münzer, in *24th EUPVSEC*, Hamburg, Germany, WIP (2009) 1558–1561.

[17] N. Nampalli, B. J. Hallam, C. E. Chan, M. D. Abbott, and S. R. Wenham, *IEEE J. Photovolt.*, 5 (2015) 1580–1585.

[18] T. O. Abdul Fattah, V. P. Markevich, J. A. T. De Guzman, J. Coutinho, S. B. Lastovskii, I. D. Hawkins, I. F. Crowe, M. P. Halsall, and A. R. Peaker, *Phys. Status Solidi A*, 219 (2022) 2200176.

[19] V. Voronkov and R. Falster, *Phys. Status Solidi B*, 253 (2016) 1721–1728.

[20] R. Søndenå, P. M. Weiser, and E. V. Monakhov, *AIP Conf. Proc.*, 2487 (2021) 130012.

[21] R. Søndenå, P. M. Weiser, F. Mosel, N. Aßmann, P.-A. Hansen, and E. Monakhov, *AIP Conf. Proc.*, 2826 (2023) 110008.

[22] V. V. Voronkov and R. Falster, *Phys. Status Solidi B*, 254 (2017) 1600779.

[23] D. C. Walter, D. Bredemeier, R. Falster, V. V. Voronkov, and J. Schmidt, *Sol. Energy Mater. Sol. Cells*, 200 (2019) 109970.

[24] M. Gläser and D. Lausch, *Energy Procedia*, 77 (2015) 592–598.

[25] S. Wilking, C. Beckh, S. Ebert, A. Herguth, and G. Hahn, *Sol. Energy Mater. Sol. Cells*, 131 (2014) 2–8.

[26] M. Stavola, *Identification of Defects in Semiconductors, Volume 51B*. Boston, MA: Academic Press, 1998.

[27] M. Stavola and W. B. Fowler, *J. Appl. Phys.*, 123 (2018) 161561.

[28] A. Nakanishi, N. Fukata, and M. Suezawa, *Phys. B Condens. Matter*, 308–310 (2001) 216–219.

[29] J. Mullins, V. P. Markevich, M. P. Halsall, and A. R. Peaker, *Phys. Status Solidi A*, 216 (2019) 1800611.

[30] M. Stavola, S. J. Pearton, J. Lopata, and W. C. Dautremont-Smith, *Appl. Phys. Lett.*, 50 (1987) 1086–1088.

[31] K. Bergman, M. Stavola, S. J. Pearton, and J. Lopata, *Phys. Rev. B*, 37 (1988) 2770–2773.

[32] J.-F. Zheng and M. Stavola, *Phys Rev Lett*, 76 (1996) 1154–1157.

[33] C. Peng, H. Zhang, M. Stavola, V. Yelundur, A. Rohatgi, L. Carnel, M. Seacrist, and J. Kalejs, *J. Appl. Phys.*, 109 (2011) 53517.

[34] B. Hourahine, R. Jones, S. Öberg, P. R. Briddon, V. P. Markevich, R. C. Newman, J. Hermannsson, M. Kleverman, J. L. Lindström, L. I. Murin, N. Fukata, and M. Suezawa, *Phys. B Condens. Matter*, 308–310 (2001) 197–201.

[35] P. M. Weiser, E. V. Monakhov, H. Haug, M. S. Wiig, and R. Søndenå, *J. Appl. Phys.*, 127 (2020) 065703.

[36] H. Haug, R. Søndenå, M. S. Wiig, and E. S. Marstein, *Energy Procedia*, 124 (2017) 47–52.

[37] T. Trupke, R. A. Bardos, M. C. Schubert, and W. Warta, *Appl. Phys. Lett.*, 89 (2006) 44107.

[38] F. Wolny, A. Krause, A. Oehlke, and M. Wagner, *Energy Procedia*, 92 (2016) 274–277.

[39] M. Kim, M. Abbott, N. Nampalli, S. Wenham, B. Stefani, and B. Hallam, *J. Appl. Phys.*, 121 (2017) 053106.

[40] J. Schmidt, K. Bothe, V. V. Voronkov, and R. Falster, *Phys. Status Solidi B*, 257 (2020) 1900167.

[41] D. Sperber, A. Heilemann, A. Herguth, and G. Hahn, *IEEE J. Photovolt.*, 7 (2017) 463–470.

[42] D. Sperber, A. Graf, D. Skorka, A. Herguth, and G. Hahn, *IEEE J. Photovolt.*, 7 (2017) 1627–1634.

[43] S. A. McQuaid, R. C. Newman, J. H. Tucker, E. C. Lightowlers, R. A. A. Kubiak, and M. Goulding, *Appl. Phys. Lett.*, 58 (1991) 2933–2935.

Universität Konstanz

Impact of Hydrogen in Ga-doped Silicon on Maximum LeTID Defect Density

R. Zerfaß, J. Simon, A. Herguth, G. Hahn

40th EU PVSEC, Lisbon, September 18, 2023

40th European Photovoltaic Solar Energy Conference and Exhibition

Introduction

Light- and elevated Temperature Induced Degradation (LeTID) :

- Efficiency limiting degradation phenomenon in c-Si
- Root cause still unknown
- Dependent on hydrogen released by SiN_x:H passivation layer during firing [1-3]

Goals

- Direct measurement of different hydrogen configurations in Ga-doped c-Si
- Correlation between hydrogen content and maximum LeTID defect density
- Examination of the GaH-pair dynamics during degradation
- Determination of the activation energy of the degradation step

[1] J.Simon et al., Solar Energy Materials and Solar Cells 260, 112456, 2023
[2] J. Schmidt et al., IEEE Journal of Photovoltaics 9(6), 1497, 2019
[3] U. Varshney et al., IEEE Journal of Photovoltaics 11(1), 65, 2021

Sample Preparation / Processing

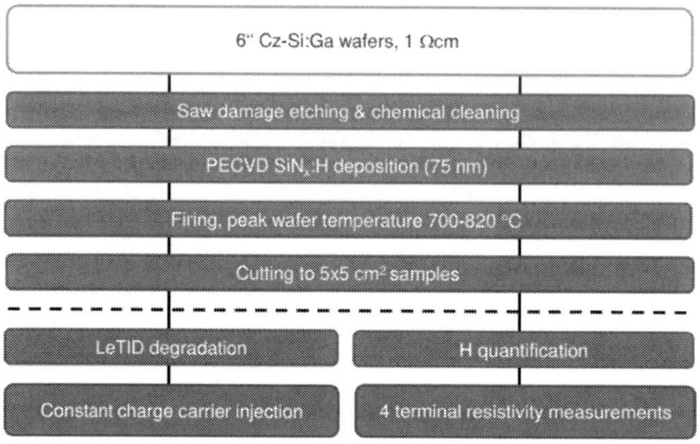

Lifetime Degradation

Lifetime τ_{eff} measurement using PCD method

Lifetime equivalent defect density: $\Delta N_{leq} = \frac{1}{\tau_{eff}(0)} - \frac{1}{\tau_{eff}(t)}$

Degradation conditions

τ_{eff} changes during degradation

- Constant illumination → change in injection level Δn
- Defect dynamics dependent on injection level

Solution: adjust illumination intensity during degradation (constant injection) [4]

- Constant charge carrier injection of $\Delta n = 10^{15}$ cm^{-3}
- Temperature 120°C, results comparable to 80°C [5]

[4] A. Graf et al., AIP Conference Proceedings 2147, 140003, 2019
[5] F. Maischner et al, Solar Energy Materials and Solar Cells 260, 112451, 2023

Hydrogen Measurements

Measurement method

- 4-terminal resistivity measurements [6]
- $-\Delta p = \Delta[\mathrm{GaH}] \propto \Delta\sigma$

$2\,\mathrm{Ga}^- + \mathrm{H}_2 + 2\,h^+ \rightleftharpoons 2\,\mathrm{GaH}^0$

- Formation: 180°C, darkness
- Dissociation: 300°C, 2 suns

→ Distinction between initial concentration $[H_2]_{init}$ and $[GaH]_{init}$ possible

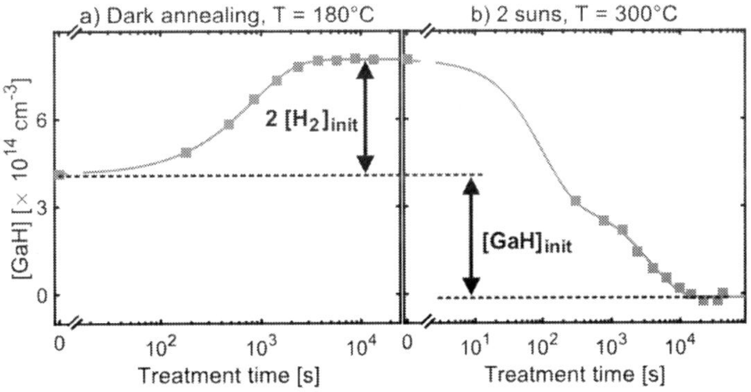

[6] A. Herguth and C. Winter, IEEE Journal of Photovoltaics 11(4), 1059-1068, 2021

Correlation between LeTID and Hydrogen

Influence of $[H_2]_{init}$ and $[GaH]_{init}$ on LeTID

- $[GaH]_{init}$: linear trend with offset from origin
  - → GaH pairs do not take part in LeTID formation
  - → direct involvement of GaH in LeTID unlikely
- $[H_2]_{init}$: linear trend without offset

→ $[H_2]_{init} \propto \Delta N_{leq,\,max}$

H_2 directly(?) taking part in LeTID formation

→ Ratio of $[H_2]_{init}$ to $[GaH]_{init}$ important?

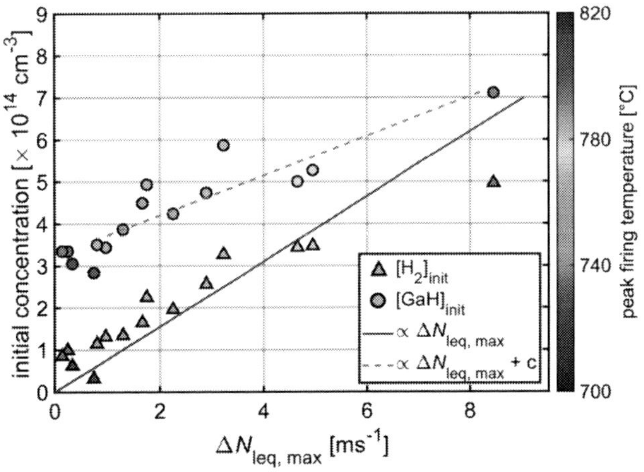

Ratio between $[H_2]_{init}$ and $[GaH]_{init}$

Impact of firing T and total [H] on $[GaH]_{init}$ and $[H_2]_{init}$

Fraction f of $[GaH]_{init}$ dependent on total H concentration and firing profile

- $f = \frac{[GaH]_{init}}{[H]}$
- Limit for large [H] may be dependent on exact firing profile

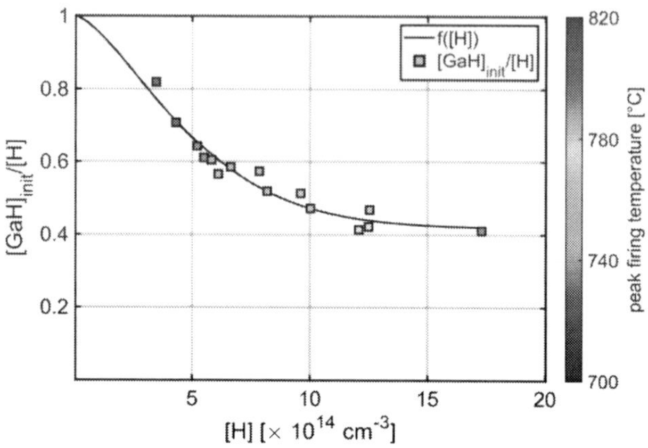

Ratio between $[H_2]_{init}$ and $[GaH]_{init}$

Impact of firing T and total [H] on $[GaH]_{init}$ and $[H_2]_{init}$

Fraction f of $[GaH]_{init}$ dependent on total H concentration and firing profile

- $f = \frac{[GaH]_{init}}{[H]}$
- Limit for large [H] may be dependent on exact firing profile

Comparison with B-doped Si [7,8]

- Higher fraction of $[GaH]_{init}$ (lower $[H_2]_{init}$) in Ga-doped Si

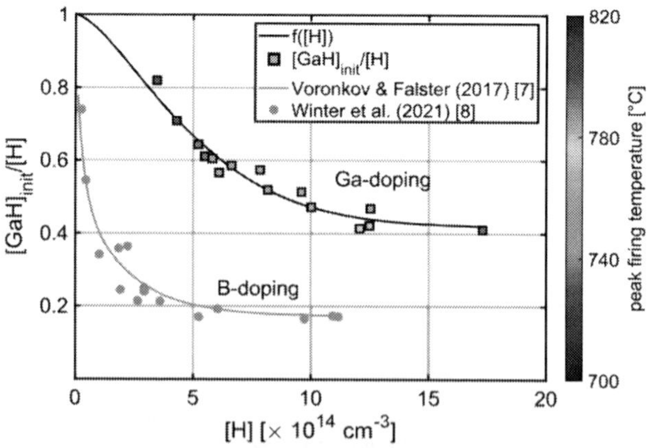

Less LeTID due to less $[H_2]_{init}$ compared to B-doped Si

[7] V. Voronkov, R. Falster, Physica Status Solidi B 254(6) 1600779, 2017
[8] C. Winter et al., Physica Status Solidi A 218, 2100220, 2021

Correlation between LeTID and Hydrogen

Fixed relation between different hydrogen configurations

- $f([\mathrm{H}]) = \frac{[\mathrm{GaH}]_{\mathrm{init}}}{[\mathrm{H}]}$
- $[\mathrm{H}] = [\mathrm{GaH}]_{\mathrm{init}} + 2\,[\mathrm{H}_2]_{\mathrm{init}}$

Also:

- $[\mathrm{H}_2]_{\mathrm{init}} \propto \Delta N_{\mathrm{leq,\,max}}$

→ Correlation between $[\mathrm{GaH}]_{\mathrm{init}}$ and $\Delta N_{\mathrm{leq,\,max}}$

- $[\mathrm{GaH}] \propto \frac{f \cdot \Delta N_{\mathrm{leq,\,max}}}{(1-f)}$

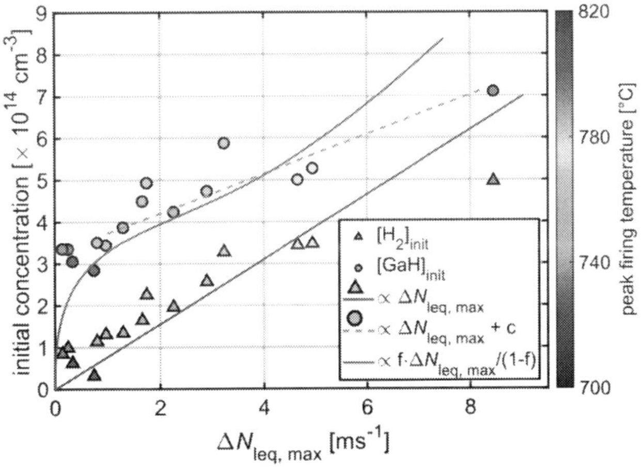

$[\mathrm{GaH}]_{\mathrm{init}}$ trend can be explained as a consequence of $[\mathrm{H}_2]_{\mathrm{init}} \propto \Delta N_{\mathrm{leq,\,max}}$

Variation of Degradation Conditions

Degradation at 120 °C and different constant injections

- Higher injection: faster and less degradation
- Slow GaH-pair formation process, not relevant during degradation
- No net release of hydrogen from GaH-pairs

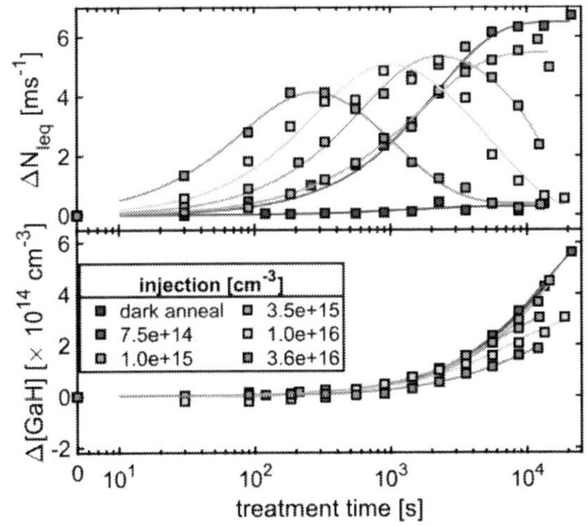

Variation of Degradation Conditions

Degradation at 120 °C and different constant injections

- Higher injection: faster and less degradation
- Slow GaH-pair formation process, not relevant during degradation
- No net release of hydrogen from GaH-pairs

Degradation at different T and $\Delta n = 10^{15}$ cm^{-3}

- Faster at higher T
- Also no release of $[H_2]$ from GaH-pairs

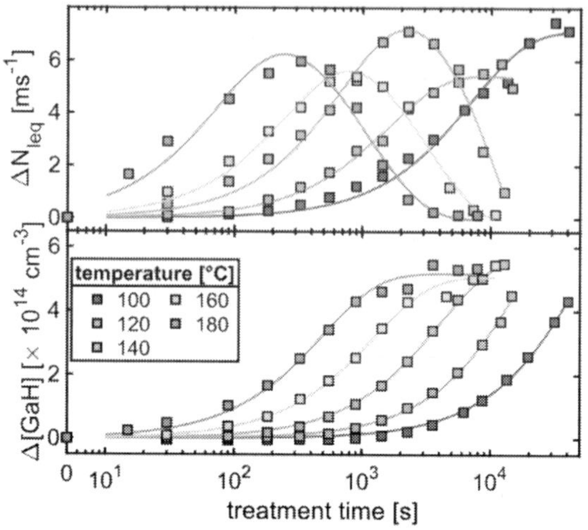

GaH dissociation irrelevant for LeTID kinetics

→ Maximize of $[GaH]_{init}$ as possible LeTID mitigation strategy

Activation Energy of Degradation Process

Activation energy at $\Delta n = 10^{15}$ cm^{-3}

Arrhenius equation: $R_{\deg}(T) = R_0 \cdot \exp\left(-\frac{E_A}{k_B T}\right)$

→ $E_A = 0.76(17)$ eV

Other values from literature

- 0.58(4) eV [9]
- 0.74 eV [10]
- 0.96(4) eV and 0.90(5) eV [11]

→ degradation not under constant injection

→ E_A may also be dependent on Δn

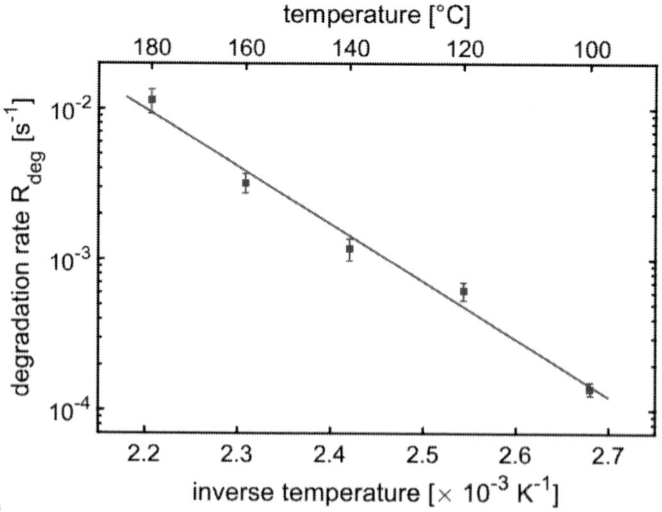

[9] M. Winter et al., IEEE Journal of Photovoltaics 11(4), 866, 2021
[10] S. Jafari et al., Solar Energy Materials and Solar Cells 251, 112121, 2023
[11] D. Lin et al., Solar Energy 225, 407-411, 2021

Conclusions

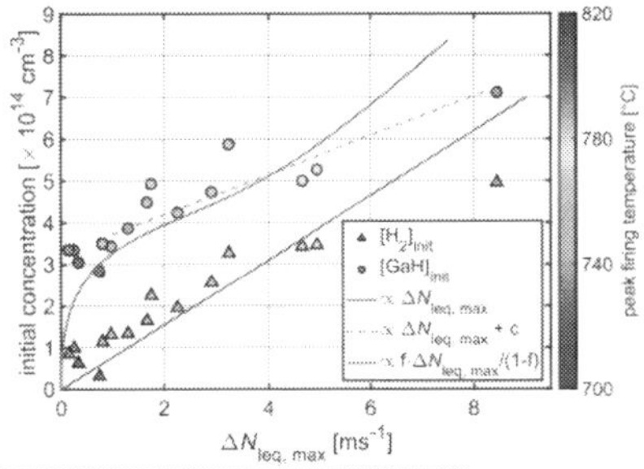

H_2 as precursor for LeTID formation
- $[H_2]_{init} \propto \Delta N_{leq,\,max}$
→ Behavior of $[GaH]_{init}(\Delta N_{leq,\,max})$ as a consequence

Less LeTID due to less $[H_2]_{init}$ in Ga-doped Si compared to B-doping

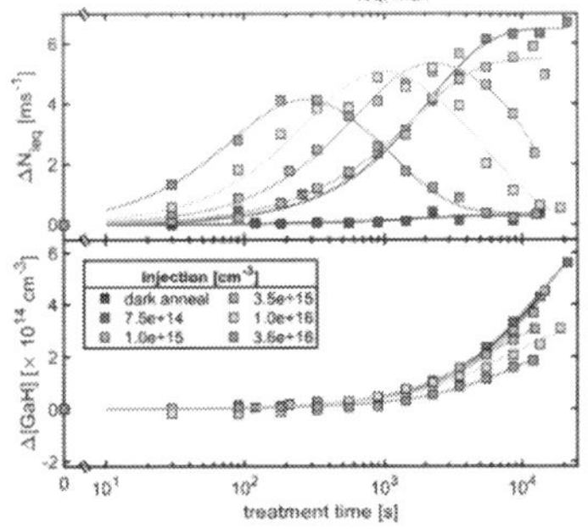

GaH dissociation irrelevant for LeTID kinetics
→ Maximation of $[GaH]_{init}$ as possible mitigation strategy

Activation Energy under constant Δn: $E_A = 0.76(17)$ eV

Thank you for your Attention!

Part of this work was funded by BMWK under contract numbers 03EE0152A. The content is the responsibility of the authors.

Supported by:

Federal Ministry for Economic Affairs and Climate Action

on the basis of a decision by the German Bundestag

Ruben Zerfaß
Department of Physics, Photovoltaics Division
ruben.zerfass@uni-konstanz.de

Paper published in SolarRRL: doi.org/10.1002/solr.202300501

24.2 %-EFFICIENT POLO BACK JUNCTION SOLAR CELL WITH INDUSTRIAL PECVD ALO$_X$/SIN$_Y$ PASSIVATION

B. Min[1], V. Mertens[1], Y. Larionova[1], T. Pernau[2], H. Haverkamp[2], T. Dullweber[1], R. Peibst[1] and R. Brendel[1, 3]

[1]Institute for Solar Energy Research Hamelin (ISFH), Am Ohrberg 1, 31860 Emmerthal, Germany
[2]centrotherm international AG, Württemberger Str. 31, 89143 Blaubeuren, Germany
[3]Dep. Solar Energy, Inst. Solid-State Physics, Leibniz University of Hannover, Appelstr. 2, 30167 Hannover, Germany

ABSTRACT: In this work, we transfer the cleaning and passivation steps for fabricating our p-type back junction (BJ) solar cells featuring passivating polysilicon on oxide (POLO) rear contacts from lab scale to industrial processes and equipment. We therefore develop an industrial PECVD AlO_x/SiN_y passivation stack with excellent passivation quality on the undiffused textured p-type crystalline silicon surface. In addition, we replace the RCA clean with an industrial cleaning sequence using HCl/O_3. We integrate the developed industrial cleaning and passivation processes into our current POLO BJ cell fabrication flow, resulting in a cell performance which is at the same level as the reference which we fabricate with a lab-scale processes. For the industrial process we determine a surface recombination current density J_{0s} of 4 fA/cm² for the passivated undiffused and textured surface from comparing measurements on lifetime samples with numerical device simulations. This value is on the same level as that of the reference sample passivated with an ALD AlO_x layer capped by a PECVD SiN_y layer. The novel PECVD AlO_x/SiN_y passivation stack is successfully integrated into our current cell fabrication process: Our best POLO BJ solar cell has an efficiency of 24.2 % with an open-circuit voltage of 725 mV on a cell area of 244.5 cm² using p-type Ga-doped Czochralski Si, as independently confirmed by ISFH CalTeC. To our knowledge, this is the first report on high quality industrial cleaning and dielectric passivation on textured and undiffused p-type crystalline silicon surfaces.
Keywords: p-type, passivating contacts, PECVD, aluminum oxide, surface passivation

1 INTRODUCTION

Our p-type back junction (BJ) solar cell concept featuring passivating polysilicon on oxide (POLO) rear contacts achieves high efficiencies while keeping the process complexity and the production cost at the level of the mainstream PERC concept. As shown in Figure 1, one of the unique features of the POLO BJ concept is the undiffused textured front surface passivated with an AlO_x/SiN_y dielectric stack. The absence of a front surface diffusion requires highly efficient cleaning and passivation. We so far applied an RCA clean and an AlO_x layer from atomic layer deposition (ALD) [1].

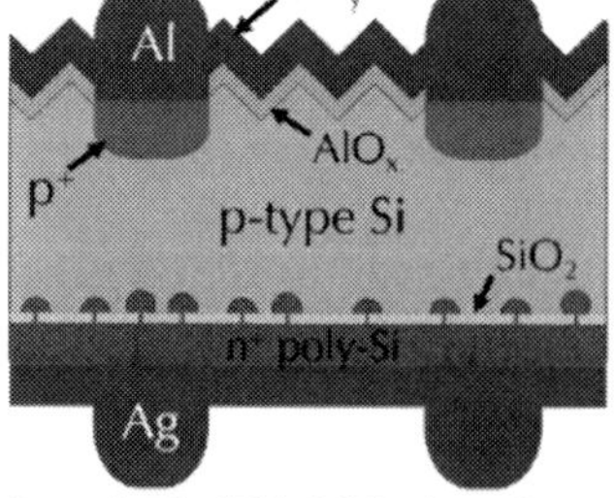

Figure 1: Schematic of a POLO BJ solar cell.

In this work, however, we replace the expensive RCA clean with an industrial cleaning sequence using HCl/O_3. In addition, we develop an AlO_x/SiN_y passivation stack using an industrial direct plasma-enhanced chemical vapor deposition (PECVD) system instead of ALD. The surface recombination current density of the passivated undiffused and textured surface is carefully determined by comparing measurements in lifetime samples with numerical device simulations. The novel PECVD AlO_x/SiN_y passivation stack is successfully integrated into our current cell fabrication process, reaching a cell efficiency of 24.2 % with an open-circuit voltage of 725 mV on a p-type Czochralski (Cz) Si wafer as independently confirmed by ISFH CalTeC.

Due to the page constraints, this conference paper is limited to the essential results of our work. More details on the determination of J_{0s} as well as the solar cell results are submitted for publication to *Progress in Photovoltaics*.

2 EXPERIMENTAL

We use 156.75 mm×156.75 mm (M2) gallium-doped Cz silicon wafers to develop a PECVD AlO_x/SiN_y passivation stack fabricated with industrial direct plasma-enhanced CVD system (cPLASMA-LAB 2600, centrotherm). For the cleaning step, we use the Batchlab (Product type) tool from RENA. The surface passivation quality is analyzed with lifetime measurements (photoconductance decay method and photoluminescence) and numerical device simulations. We fabricate POLO BJ solar cells to validate the developed PECVD AlO_x/SiN_y passivation stack in our current process flow. More details about the process flows and the advantages of our cell concept are presented in our previous works [2, 3].

3 RESULTS

3.1 A surface recombination current density of 4 fA/cm² on undiffused textured surface passivated with PECVD AlO_x/SiN_y dielectric stacks

The circles in Figure 2 show the measured effective carrier lifetimes τ_{eff} of a textured symmetric lifetime sample passivated with an optimized AlO_x/SiN_y dielectric stack fabricated with an industrial PECVD system. Within a single deposition process, we deposited a 5 nm-thick AlO_x layer and a 100 nm-thick SiN_y layer. Typically, the surface recombination current density J_{0s} can be determined with the well-known slope method from Kane and Swanson if the base of the sample is in high injection [4]. However, the base resistivities of our samples are between 0.73 and 0.9 Ω cm, hence the base is not completely in high injection during illumination with the flash lamp from the Sinton lifetime tester. Therefore, we first reproduce the measured τ_{eff} with numerical device modeling, using the software SENTAURUS DEVICE and the most recent device models as well as silicon parameters [5, 6]. The optical generation is calculated with 3D ray tracing software SUNRAYS [7]. We then extract the J_{0s} from the device simulation using the general definition of J_0 [8]. We obtain sufficiently low J_{0s} values of 4 fA/cm^2 on the textured surface.

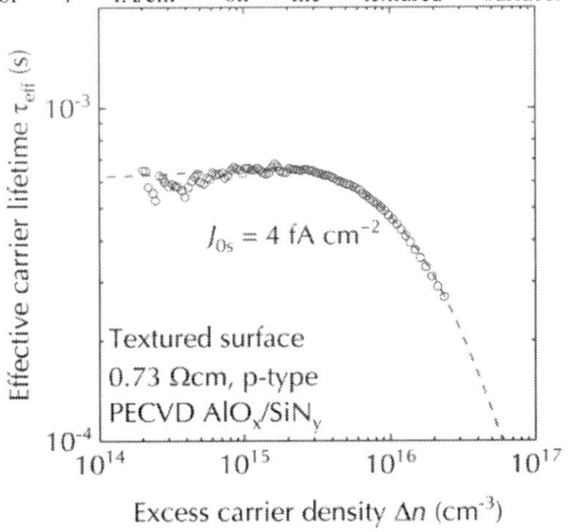

Figure 2: Measured effective carrier lifetime τ_{eff} (circles) of symmetric lifetime samples with PECVD AlO_x/SiN_y dielectric stack on textured surface. The simulated τ_{eff} curves are depicted as dashed lines.

3.2 Process integration for POLO back junction solar cells

Figure 3 shows the *IV* parameters of four groups of POLO BJ solar cells measured with the LOANA tool from pvtools. We applied a contacting scheme that neglects the resistance of the metal grid [9]. The group "black" is fabricated with lab-scale processes (RCA clean and ALD AlO_x). The groups "green", "blue" and "red" are fabricated with industrial processes by applying the cleaning with HCl/O_3 and the PECVD AlO_x/SiN_y passivation stack. We applied two different Al pastes for the front grid. While the group "black" and "green" are fabricated with the Al paste A, the groups "blue" and "red" are fabricated with the Al paste B.

By comparing the group 1 and 2, it can be shown that the industrial cleaning process and the PECVD AlO_x/SiN_y passivation stack provide the same level of cell performance as the reference which is fabricated with RCA clean and ALD AlO_x layer. The industrial processes lead to even higher cell efficiencies, mainly due to improved short-circuit current density (J_{sc}) and fill factor (*FF*). Notably the groups "green", "blue", and "red" have, on purpose, slightly thinner SiN_y layers at the cell front side than the group "black". This explains the improvement in short-circuit current density (J_{sc}) values in the group "green" compared to the group 1 due to reduced surface reflection losses. On the other hand, a thinner SiN_y layer can increase the Al spiking at the cell front side [2], which may be the reason for the slightly reduced open-circuit voltage (V_{oc}) values of the group "green" compared to the group "black". Detailed investigations on the Al spiking are currently in progress.

Further improvements are achieved by applying the Al paste B as shown in the group "blue". The improvement in the V_{oc} and J_{sc} values indicates the reduction of the recombination losses at Al front contacts. Finally, we reduce the recombination losses at the Ag contacts by adjusting the SiN_y layer thickness at the rear as shown in group 4.

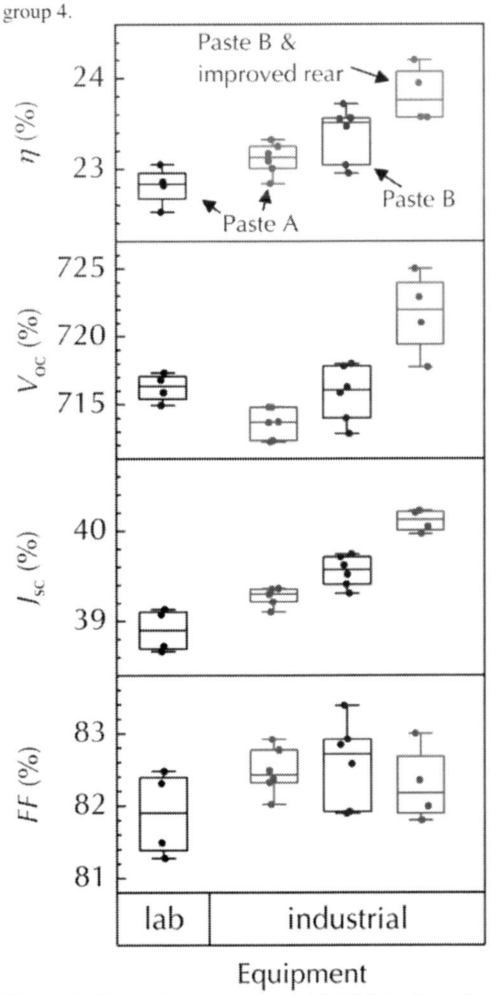

Figure 3: Comparison between the POLO BJ cell results fabricated with lab-scale and industrial equipment.

The details about those Al pastes will be the subject of a dedicated publication at a later date. Our 24.2 % efficient champion cell has a V_{oc} of 725 mV with a J_{sc} of 40.2 mA cm^{-2} and an *FF* of 83.0 % as independently confirmed by ISFH CalTeC.

4 CONCLUSION

We successfully transfer the cleaning and passivation steps of POLO BJ solar cells from lab-scale to industrial processes and equipment. We demonstrate for the first time that the ALD AlO_x is not necessarily needed to achieve a high passivation quality on undiffused textured surfaces of p-type Cz silicon. As the best values, we achieve a surface recombination current density of 4 fA/cm² for the passivated undiffused and textured surface and a cell efficiency of 24.2 % with an open-circuit voltage of 725 mV on a p-type Ga-doped Cz Si wafer. Our results demonstrate the feasibility to clean and passivate the textured and undiffused surfaces of p-type c-Si wafers with industrial processes. This also opens a path for the industrialization of novel cell concepts such as POLO BJ and POLO IBC.

5 ACKNOWLEDGEMENTS

The authors thank to M. Pollmann, B. Gehring, T. Brendemuehl, T. Neubert and D. Sylla for processing solar cells (all ISFH), M. Dhamrin and K. Tsuji from Toyo Aluminium K. K. for their supports regarding Al pastes. This work was financially supported by the state of Lower Saxony and the German Federal Ministry for Economic Affairs and Climate Action (BMWK) under contact number 03EE1150A (APOLON).

6 REFERENCES

[1] B. Min, N. Wehmeier, T. Brendemuehl, F. Haase, Y. Larionova, L. Nasebandt, H. Schulte-Huxel, R. Peibst, R. Brendel. "716 mV Open-Circuit Voltage with Fully Screen-Printed p-Type Back Junction Solar Cells Featuring an Aluminum Front Grid and a Passivating Polysilicon on Oxide Contact at the Rear Side", Solar RRL 5, 2000703 (2021).

[2] B. Min, N. Wehmeier, T. Brendemuehl, A. Merkle, F. Haase, Y. Larionova, L. David, H. Schulte-Huxel, R. Peibst, R. Brendel. "A 22.3% Efficient p-Type Back Junction Solar Cell with an Al-Printed Front-Side Grid and a Passivating n+-Type Polysilicon on Oxide Contact at the Rear Side", Solar RRL 2000435 (2020).

[3] B. Min, L. Nasebandt, C. Hollemann, D. Bredemeier, L. Thiemann, T. Brendemühl, K. Bothe, R. Peibst, R. Brendel. "23.1 %-efficient POLO back junction solar cell", Proc. of the 8th World Conference on Photovoltaic Energy Conversion, Milan, Italy, 2022, pp. 107 - 109.

[4] D. Kane, R. Swanson. "Measurement of the emitter saturation current by a contactless photoconductivity decay method", Proc. of the 18th IEEE Photovoltaic Specialist Conference, New York, 1985, pp. 578-583.

[5] P.P. Altermatt. "Models for numerical device simulations of crystalline silicon solar cells—a review", J. Comput. Electron. 10, 314-330 (2011).

[6] A. Richter, S.W. Glunz, F. Werner, J. Schmidt, A. Cuevas. "Improved quantitative description of Auger recombination in crystalline silicon", Phy. Rev. B 86, 165202 (2012).

[7] R. Brendel. "SUNRAYS: a versatile ray tracing program for the photovoltaic community", Proc. of the 12th European Photovoltaic Solar Energy Conference, Amsterdam, The Netherlands, 1994, pp. 1339–1342.

[8] J.O. Schumacher, P.P. Altermatt, G. Heiser, A.G. Aberle. "Application of an improved band-gap narrowing model to the numerical simulation of recombination properties of phosphorus-doped silicon emitters", Sol. Energ. Mat. Sol. 65, 95-103 (2001).

[9] K. Bothe, C. Kruse, D. Hinken. "Contacting of busbarless solar cells for accurate I-V measurements", Proc. of the 37th European Photovoltaic Solar Energy Conference and Exhibition, Online event, 2020.

40th European Photovoltaic Solar Energy Conference and Exhibition

This presentation was selected by the Sc. Committee of the EU PVSEC 2023 for submission of a full paper to one of the EU PVSEC's collaborating peer-reviewed journals.

POLY-SI (N) REMOVAL FOR TOPCON SOLAR CELLS

K. Krieg, S. Mack, M. Zimmer, J. Vollmer[1], T. Dannenberg[1], D. Brunner[1]
Fraunhofer Institute for Solar Energy Systems, Heidenhofstr. 2, 79110 Freiburg, Germany
[1]RENA Technologies GmbH, Höhenweg 1, 78148 Gütenbach, Germany
katrin.krieg@ise.fraunhofer.de

ABSTRACT: This work gives results on wet chemical parasitical polysilicon removal for tunnel oxide passivated contact (TOPCon) n-type silicon solar cells. The poly-Si removal is important to fabricate TOPCon cells with low reverse bias junction leakage current density ($j_{rev,2}$). TOPCon solar cells with in-situ n-doped LPCVD and PECVD poly-Si layers were fabricated with wet chemical inline and wet chemical batch poly-Si removal processes. Our results show that a wet chemical alkaline batch removal for parasitical PECVD poly-Si is possible using the anneal oxide from N_2 anneal process to protect the functional poly-Si on rear. This batch process led to an efficiency of 23.4% solar cell with in-situ doped PECVD poly-Si.
Keywords: PECVD/LPCVD/batch cluster etching/inline etching/wet chemical

1 INTRODUCTION

Tunnel oxide passivated contact (TOPCon) solar cells are in focus of photovoltaics (PV) industry and research. The deposition of the poly-Si layer is a key aspect for TOPCon solar cells. In this paper we will use the expression poly-Si for both as deposited and annealed silicon layers and specify the status if needed. There are several options how to fabricate a TOPCon silicon solar cell [1, 2]. The TOPCon layer stack on the rear side of the solar cell consists of a tunnel oxide and a poly-Si(n) layer. It is needed for carrier selectivity and usually is deposited by low pressure chemical vapor deposition (LPCVD), or plasma enhanced chemical vapor deposition (PECVD). Up to now the TOPCon technology at Fraunhofer ISE PV-TEC was based on in-situ phosphorus doped LPCVD layers [3]. The ITRPV 2022 roadmap suggests that transition from LPCVD to PECVD poly-Si deposition will take place in the upcoming years [4]. The doping of the PECVD TOPCon layers usually is done in-situ, i.e. during deposition, because the deposition rate for in-situ doped poly-Si is as high as for intrinsic poly-Si. All the deposition processes, LPCVD and PECVD, lead to a parasitical wraparound [1]. The wraparound to the front side of the solar cell is complete for LPCVD, whereas for PECVD it is located only at the wafer edge (Fig.1).

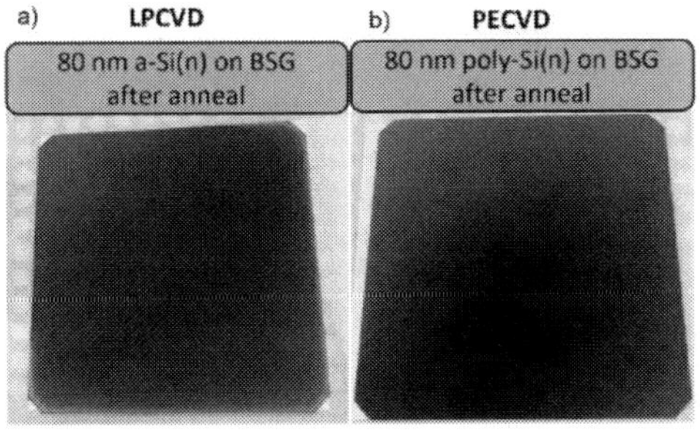

Fig. 1: Photograph of the front side with poly-Si on boron silicate glass (BSG) for (a) LPCVD poly-Si as complete and for (b) PECVD only at edges (mostly BSG visible) after anneal.

For solar cells, a high-quality edge isolation needs to be ensured, and this is in particular true for processes with a parasitic polysilicon deposition on the front side and the wafer edge [5]. There are ideas to avoid the wrap-around completely e.g., by using masked deposition in case of physical vapor deposition (PVD) [6] or PECVD [7].

There are several options to remove the parasitical poly-Si, but in this article, we solely focus on wet chemical processes for the poly-Si removal wrap-around since wet chemical etching is successfully used to remove the parasitic emitter on the rear side of the solar cell [8]. In general, the polysilicon removal can be realized in hot diluted potassium hydroxide (KOH) as an inline process or as a batch process. A batch process combines an inline single-side oxide etching tool and a batch tool with KOH/additive for selective etching of the parasitical exposed poly-Si, while the rear side with the functional poly-Si must be protected (Fig. 2).

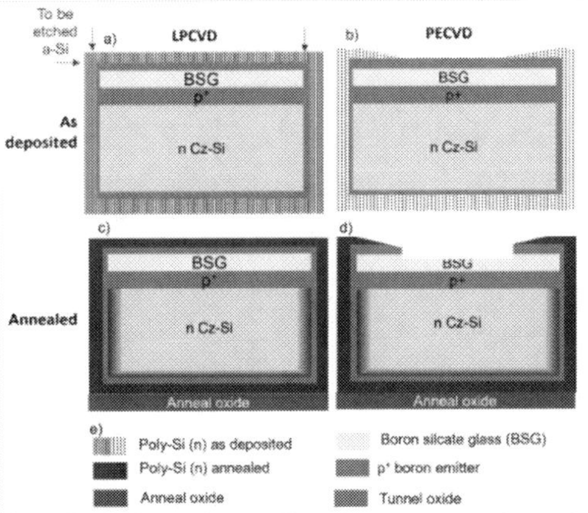

Fig. 2: Cell scheme for in-situ-doped poly-Si just before the parasitical poly-Si etch on the textured front side for LPCVD poly-Si (a) as deposited and (c) annealed and PECVD poly-Si (b) as deposited and (d) annealed. For c) and d) the oxide layer on front and edge was already removed in a single side HF etch.

In order to avoid an additional process step, we suggest using the anneal oxide that is on the wafer after the mandatory thermal anneal process in the TOPCon sequence in which the functional polysilicon is formed. Usually, the atmosphere is nitrogen during the anneal process. Is the oxide, that is formed during unloading the wafers after the N_2-anneal process, chemically stable enough to allow for a poly-Si removal process in a batch process?

The aim of this work is to give an overview on wet chemical process technology for parasitical poly-Si removal for TOPCon silicon solar cells. Therefore,

- Both in-situ doped LPCVD or PECVD poly-Si are investigated.
- The N_2-anneal step was performed right after poly-silicon deposition in order to produce an oxide on the rear functional poly-Si. The anneal step after poly-Si removal was used in case of the LPCVD reference.
- The parasitical poly-Si was removed selectively from the front side of the solar cell by either inline or batch wet chemical etching.
- J-V and reverse current density were measured for finished TOPCon solar cells.

2 EXPERIMENTAL

TOPCon solar cells were manufactured at Fraunhofer PV-TEC on 156.75 x 156.75 mm² n-type Cz-Si wafers with a resistivity of 1.4 Ωcm. The group size was 10 wafers. The wafers were prepared with alkaline texturing, boron diffusion, a single side boron silicate (BSG) etch in HF/HCl and batch alkaline edge isolation process followed by ozone cleaning (Fig. 3). The TOPCon stack consists of a thin thermal tunnel oxide and 80 nm in-situ n-doped poly-Si layer formed either by LPCVD or by PECVD. The mandatory annealing step for transforming the a-Si layer into the desired final poly-Si was performed directly after poly-Si(n) deposition, whereas only for the LPCVD reference cells it was done after the parasitical polysilicon removal, as this represents our current standard, see Fig. 3.

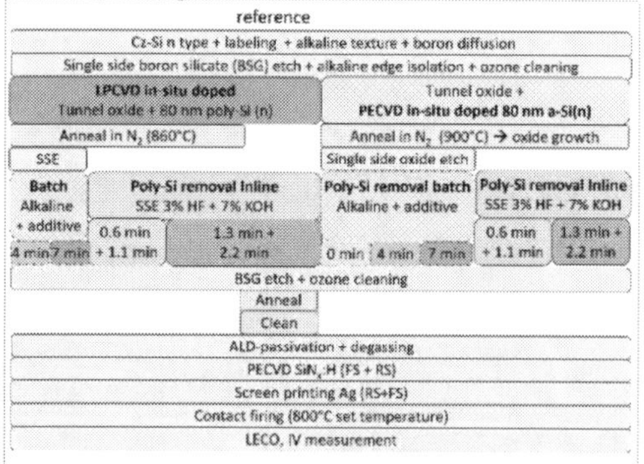

Fig. 3: TOPCon solar cell process for in-situ n-doped LPCVD and PECVD poly-Si with batch and inline poly-Si removal processes after anneal. For the reference the anneal was done after poly-Si etch. FS front side, RS rear side.

The poly-Si removal was performed either in a wet chemical batch or inline tool. To perform a poly-Si removal in a batch tool, the front side oxide layer on top of the poly-Si was removed in a single side etching tool applying a 13 wt.% HF/13% wt.% HCl solution until the wrap-around was hydrophobic at 2 m/min (corresponds to a 0.5 min etching time). The batch poly-Si removal was performed for 4 and 7 min in 3% KOH including the additive CellEtch from ICB GmbH. The inline polysilicon etching was varied at speeds of 1 and 2 m/min in a 3% HF bath with a length of 1 m, followed by an alkaline bath with 7% KOH at 70°C and a bath length of 2.2 m. After the polysilicon removal, the BSG etch and subsequent cleaning were performed directly in adjacent modules within the inline machine, whereas for the batch groups this was done in the batch machine, respectively. For the reference group the annealing process was done after the poly-Si etch (after BSG etch, Fig.3).

The wafers were cleaned in an ozone cleaning sequence and passivated by ALD Al_2O_3 and subsequent PECVD silicon nitride deposition. The bifacial metallization pattern was applied by conventional flat-bed screen print technology, the set temperature in the fast firing furnace was 800°C. The solar cells were treated by laser enhanced contact optimization (LECO) [10] and measured at an industrial flash cell tester from Halm under standard test condition, including short time thermo-graphy. The given cell tester data are for grid touch measurement on a non-reflective black chuck. The reverse bias current density $j_{rev,2}$ was measured at -12 V.

3 RESULTS

3.1 Shunt behavior of TOPCon solar cells

First, we will have a look at thermography images of TOPCon solar cells, keeping in mind that an ideal thermography picture would be homogeneously black. Without a poly-Si removal process and thus also no edge treatment, we find increased local recombination, indicated by bright edges (see Fig. 5g). In case of inline poly-Si removal processes, a 1.1 min KOH etching time is not sufficient to remove the PECVD and LPCVD poly-Si completely (Fig. 5a+c). This might be due to an insufficient HF process prior to KOH etching, as a present oxide layer on the front poly-Si layer might prevent or at least retard a high-quality removal of the front poly-Si layer. The groups with shiny edges are not considered further in the following.

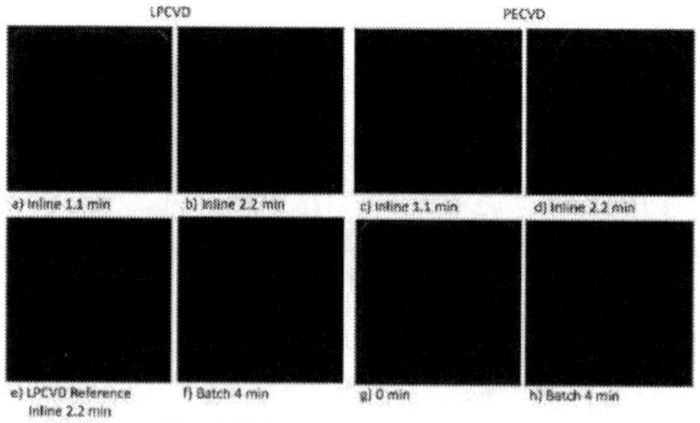

Fig. 5: Thermography pictures for TOPCon solar cells (156 x 156 mm²) with LPCVD and PECVD poly-Si(n) with an alkaline poly-Si removal by inline in a+c+e) 1.1 min, b+d) in 2.2 min, g) without and d+h) batch alkaline etching.

In case of inline etching and a KOH etching time of 2.2 min all bright lines vanish (see Fig. 5 b and d). This leads to a desired low a reverse current density $j_{rev,2}$ = 0.3 mA/cm² for PECVD and LPCVD (see Fig. 6). The LPCVD reference with poly-Si etch after deposition showed a higher $j_{rev,2}$ = 1 mA/cm² compared to the groups with anneal after deposition. The 4 min alkaline batch poly-Si removal process results also in a low $j_{rev,2}$ = 0.1 mA/cm² for PECVD and 0.3 mA/cm² for LPCVD poly-Si. A 7 min alkaline batch process reduced this even further to $j_{rev,2}$ = 0.04 mA/cm² for PECVD and $j_{rev,2}$ = 0.02 mA/cm² for LPCVD (Fig. 6).

10.4229/EUPVSEC2023/1AO.5.4
020005-002

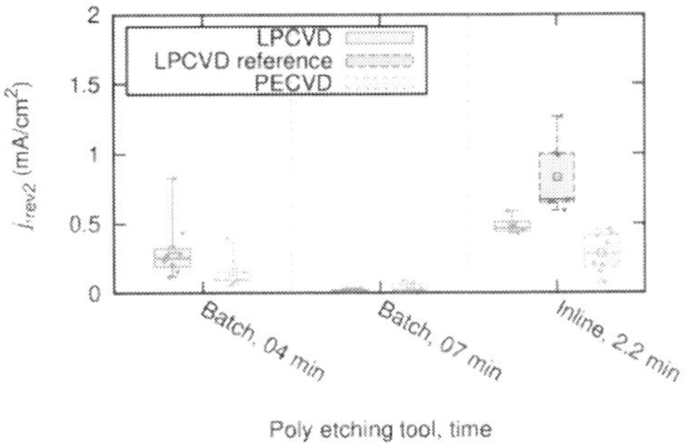

Fig. 6: $j_{rev,2}$ for TOPCon solar cells (n-type Cz-Si, M2) with PECVD and LPCVD poly-Si and for alkaline wet chemical batch and inline poly-Si removal process performed after annealing. For the LPCVD reference the anneal was done after poly-Si etch.

3.1 J-V for TOPCon solar cells

The TOPCon solar cell J-V results for fill factor FF and Voc are shown in Fig. 7, measured on a black chuck. Only groups with a $j_{rev,2}$ < 2 mA/cm² are considered. First, we look at FF. When comparing PECVD poly-Si to LPCVD poly-Si with anneal right after poly-Si deposition, a -0.2%abs lower FF and +7 mV higher Voc are found for PECVD poly-Si. This higher passivation quality might be due to a higher hydrogen content in the PECVD layer, which can improve interface passivation. A 7 min batch KOH etching time for poly-Si results in a 2 mV lower Voc in our experiment compared to a 4 min etching time for PECVD and LPCVD poly-Si.

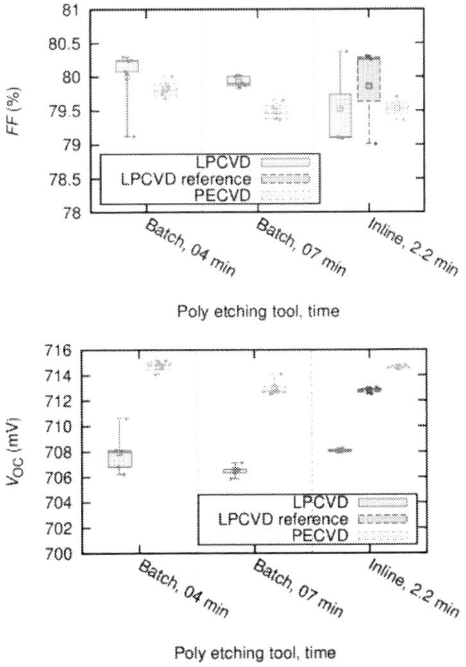

Fig. 7: Fill factor (*FF*) and V_{oc} results for TOPCon solar cells (n-type Cz-Si, M2) with LPCVD and PECVD poly-Si with batch and inline poly-Si removal processes after anneal. For the reference the anneal was done after poly-Si etch.

For further analysis, we will examine photoluminescence (PL) images of the solar cells in Fig. 8. They show black stripes for samples with poly-Si removal in an alkaline batch bath (shown for LPCVD in Fig. 8, left). All samples underwent annealing immediately following the poly-Si deposition. This effect has nothing to do with the front side of the cell. In confocal images the black stripes resulting from batch etched poly-Si are observed on the rear side of cell as areas where the poly-Si has been etched away (Fig. 8, right). These stripes originate from the necessary downholders in the inline dryer used during the single side oxide etch process which inadvertently damaged the thin anneal oxide on the rear side of the sample. Consequently, the damaged anneal oxide could not withstand the poly-Si etching during batch etching process leading to localized poly-Si etchin and adversely impacting the quality of the rear-side passivation (Fig. 8). This issue can be avoided in an inline process, due to a different tool set-up.

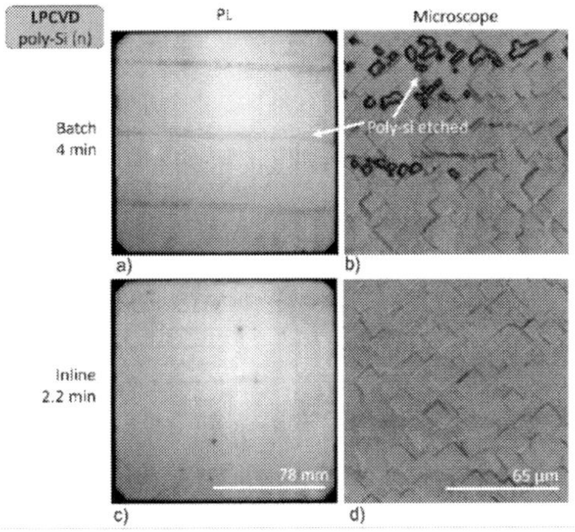

Fig. 8: PL images (left, 156x156 mm²) and confocal laser intensity microscope picture (right) for TOPCon solar cells with LPCVD poly-Si wrap-around removed (a+b) in 4 min in a batch alkaline bath and c+d) in 2.2 min in line. Undamaged poly-Si is transparent.

j_{sc} values are highest for PECVD poly-Si with inline poly-Si removal and lowest for PECVD poly-Si with 7 min batch poly-Si removal (Fig. 9). The higher Voc for PECVD poly-Si solar cells compared to LPCVD poly-Si could be transferred to cell efficiency. The cell efficiency for cells with PECVD poly-Si was 23.3% compared to 23.2% for LPCVD poly-Si.

Altogether, the highest cell efficiency in this experiment was measured with 23.4% for PECVD poly-Si with anneal after poly-Si deposition and a poly-Si batch removal process for 4 min. This was as high as the reference TOPCon sequence with LPCVD poly-Si with wet-chemical inline poly-Si removal before anneal. The solar cell efficiency of 23.2% for LPCVD poly-Si with anneal after deposition is lower than the LPCVD reference, which might be a result of the local etching of poly-Si layer on the rear side, as shown in Fig 8.

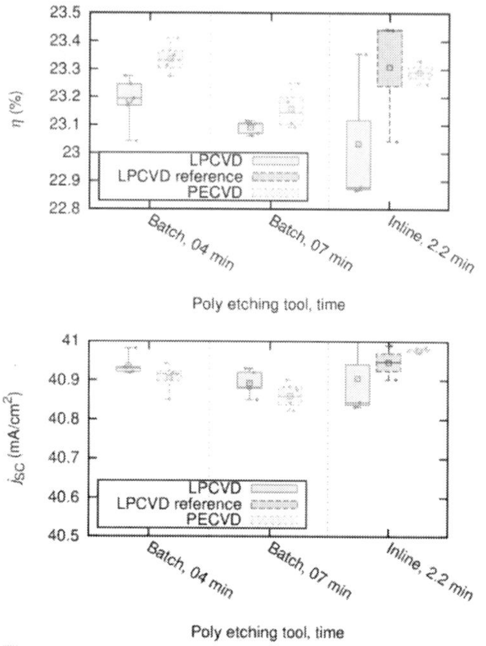

Fig. 9: Efficiency and j_{sc} results for TOPCon solar cells (n-type Cz-Si, M2) with LPCVD and PECVD poly-Si with batch and inline poly-Si removal processes after anneal. For the reference the anneal was done after poly-Si etch.

4 CONCLUSIONS

TOPCon silicon solar cells were prepared with PECVD and LPCVD n-doped poly-Si followed by annealing after poly-Si deposition. Additionally, TOPCon cells with annealing after parasitical LPCVD poly-Si removal were fabricated. For parasitical poly-Si removal, a comparison included both inline and batch cluster wet chemical processes with two different etching times, respectively. The short PECVD sequence with anneal right after poly-Si deposition yielded the highest efficiency of 23.4% and lowest $j_{rev,2}$ for a batch parasitical polysilicon removal process. For protection of the rear side only the anneal oxide was used. A higher KOH etching time reduces the V_{oc} due to unwanted etching of functional poly-Si on the rear side. The inline wet chemical parasitical poly-Si removal does not damage the functional polysilicon on rear.

The poly-Si etch process still has to be optimized. The use of an anneal oxide etch is essential. A solar cell process sequence should be studied with the poly-Si removal after PECVD poly-Si deposition, in order to perform the anneal process without BSG. Ideally, the chemical stability of the anneal oxide should be increased in order to be able to extend the KOH batch poly-Si removal time, or alternatively other bath compositions could be tested.

5 ACKNOWLEDGMENTS

The authors acknowledge that part of this work was supported by funding from the German Federal Ministry for Economic Affairs and Climate Action within the research project "Taler" (contract no. 03EE1021B).

5 REFERENCES

1. S.W. Glunz, B. Steinhauser, J.-I. Polzin, Ch. Luderer, B. Grübel, T. Niewelt, et al., Prog Photovolt Res Appl. 31, 4 (2021)
2. X. Zhang, R. Dumbrell, W. Li, M. Xu, D. Yan, J. Jin, Z. Wang, P. Zheng, C. Liu, J. Yang Prog Photovolt Res Appl., 31 (2023)
3. S. Mack, B. Kafle, Ch. Tessmann, K. Krieg, S. Bashardoust, E. Lohmüller, U. Belledin, P. Saint-Cast, H. Höffler, D. Ourinson, T. Fellmeth, B. Steinhauser, S. Kluska, A. Wolf, in Proceedings of the 8th World Conference on Photovoltaic Energy Conversion, Milano, Italy, (2022)
4. M. Fischer, M. Woodhouse, P. Baliozian, J. Trube, International technology roadmap for Photovoltaic (ITR PV) 2022 Results 12th edition (2023)
5. Q. Wang, W. Wu, D. Chen, L. Yuan, S. Yang, Y. Sun, et al., Solar Energy 211 (2020)
6. E. Schneiderloechner, T. Dietsch, J. Hoss, J. Linke, J.-I. Polzin, S. Mack, H. Nagel, V. Linss, in Proceedings IEEE-PVSC, San Juan, Puerto Rico (2023)
7. M. Stöhr, B.Beier, J.Aprojanz, A.Vogt, N.Ambrosius, R.Brendel, T.Dullweber in Proceedings of the 37th European Photocoltacis Solar Energy Conference and Exhibition, online (2020)
8. T. Dannenberg, J. Vollmer, M. Passig, C. Scheiwe, D. Brunner, A. Pediaditakis, U. Jäger, I. Wang, W. Xie, S. Xu, J. Wu, H. Kühnlein, K. Krieg, C. Teßmann, M. Zimmer, Sol. RRL 7 (2023)
9. B. Kafle, S. Mack, Ch. Teßmann, S. Bashardoust, L. Clochard, E. Duffy, et al. Sol. RRL 6 (2022)
10. T. Fellmeth, H. Höffler, S. Mack, E. Krassowski, K. Krieg, B. Kafle, J. Greulich, Prog Photovolt, Res Appl. 30 (2022)

10.4229/EUPVSEC2023/1AO.5.4
020005-004

DEVELOPMENT AND PROCESS OPTIMIZATION OF A P-IBC SOLAR CELL WITH PECVD DEPOSITED PASSIVATED CONTACTS

L. Rachdi[1], J. Lossen[1], D. Rudolph[1], Y.P. Sharma[1], L.J. Koduvelikulathu[1]
J.I Polzin[2], S. Schmidt[2], A. Wolf[2] and T. Pernau[3]
[1]ISC Konstanz e.V., Rudolf-Diesel-Str. 15, 78467 Konstanz, Germany
[2]Fraunhofer Institute for Solar Energy Systems ISE, Heidenhofstr. 2, 79110 Freiburg, Germany
[3]Centrotherm International AG, Württemberger Str. 31, 89143 Blaubeuren, Germany

ABSTRACT: In this research, we present the design and optimization of an interdigitated back contact (IBC) solar cell. The cell utilizes a cost-effective, commonly used substrate material, a Cz-Si Ga-doped wafer. The proposed processing techniques are all well-known from high-volume Passivated Emitter and Rear Cell (PERC) production, allowing for efficient utilization of existing production tools. The solar cell design features a front floating emitter and rear passivated contacts, with an in-situ doped a-Si layer deposited by plasma enhanced chemical vapor deposition (PECVD). Our process optimization efforts presented in this work have resulted in a current champion cell with a conversion efficiency of 22.94%.
Keywords: interdigitated back contact, IBC, TOPCon, passivated contacts

1 INTRODUCTION

In recent years, the photovoltaic industry has seen a dominant use of Passivated Emitter and Rear Cell (PERC) technology in silicon solar cell production [1]. Despite this, Interdigitated Back Contact (IBC) solar cells have been shown to offer significant advantages, including better light management with higher Jsc values, compared to PERC cells [2]. However, the industry's reliance on PERC production equipment presents a challenge for a widespread adoption of the IBC technology.

One potential solution is to select an IBC technology with high tool reutilization, which would allow for a technology transition while maximizing usage of existing equipment [3]. Another key enabler for high efficiency is the use of passivating contacts, also known as tunnel oxide passivating contact (TOPCon)-technology [4]. The combination of a thin oxide and a polysilicon layer offers excellent passivation properties and simple process requirements, making them a key enabler for devices with high efficiency.

Currently, low pressure chemical vapor deposition (LPCVD) is the dominant technology used for polysilicon deposition in the PV industry. However, plasma enhanced chemical vapor deposition (PECVD) offers many advantages over LPCVD, including a higher deposition rate and the ability to deposit single sided, which enhances its industrial compatibility [4]. These benefits make PECVD a cost-effective alternative for depositing passivating contacts and a potential solution for high efficiency solar cells while minimizing the cost.

In this study, we present our advances in the development of a Front floating Emitter- Interdigitated Back Contact (FE-pIBC) solar cell architecture on p-type Ga-doped mono Cz-substrates with a-Si(n) deposited utilizing a Centrotherm cPlasma PECVD tool.

2 SAMPLE PREPARATION

2.1 Solar cells

The experiments are conducted on commercial M2-sized Ga-doped p-type Cz-Si wafers with a resistivity ranging from 1 to 2 Ω.cm. Figure 1 shows the applied process flow. The p-type wafers undergo an alkaline texturing process, followed by a single-side chemical etching and cleaning procedure. Subsequently, a thin interfacial oxide layer is thermally grown in a tube furnace and an in-situ doped a-Si(n) layer is deposited on the rear side using PECVD. The wafers are then subject to thermal annealing and phosphorous (n+) diffusion in a $POCl_3$-tube furnace to create the front floating emitter (FFE) region. The aforementioned processes are carried out at Fraunhofer ISE and the samples are then shipped to ISC Konstanz for further processing.

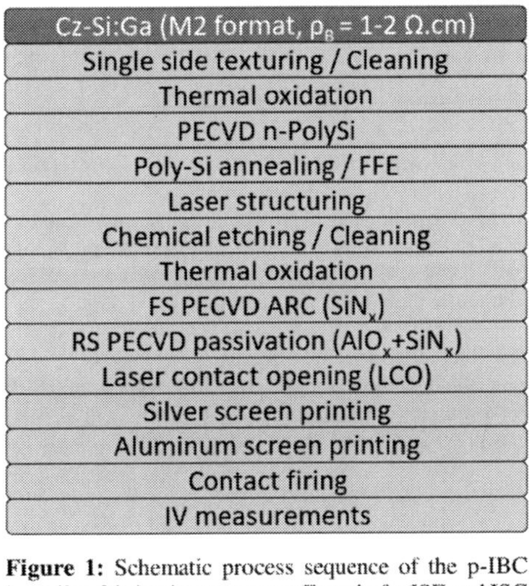

Figure 1: Schematic process sequence of the p-IBC base-line fabrication process at Fraunhofer ISE and ISC Konstanz

At ISC Konstanz, the interdigitated base and emitter region pattern is defined using laser ablation utilizing the phosphosilicate glass (PSG) layer as a mask, followed by a chemical removal of laser damage, PSG removal and wet chemical cleaning. After a thermal oxidation process, a passivation stack of AlO_x-SiN_y is deposited on the rear side and a SiN_x layer on the front side using PECVD.

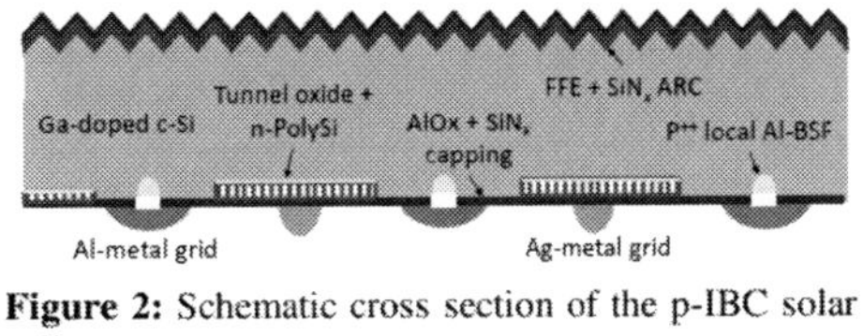

Figure 2: Schematic cross section of the p-IBC solar cell

Laser contact openings of the AlO_x-SiN_y passivation in the base region are carried out and subsequently aluminum base contacts are printed, similarly to the PERC process base contact formation. Silver contacts are established on the poly-Si emitter region using screen printing, followed by a standard co-firing process. The contacts are realized using a metallization pattern with 9 busbars per polarity. A schematic cross section of the finalized solar cell is shown in Figure 2.

2.2 Lifetime samples

In addition to the fabrication of solar cells, a set of carrier lifetime samples is prepared to assess the passivation quality across distinct regions. For the evaluation of the n-poly and base regions passivation, symmetrical flat samples are produced. These samples incorporated n-poly on both sides and were subject to laser ablation following a test structure configuration, with large area ablated test fields. After removal of the laser damage, PSG removal and wet chemical cleaning, thermal oxidation and PECVD AlO_x+SiN_y deposition on both sides forms the passivation layer stack.

Furthermore, cell precursors are generated, following the same laser ablation pattern, with the intention of assessing cells implied open circuit voltage (iV_{oc}) potential without metallization.

Both lifetime structures are subjected to fast firing in a conveyor belt furnace using the same temperature profile as the p-IBC solar cells. Subsequently, the samples underwent characterization using quasi-steady-state photo conductance (QSSPC) and photoluminescence imaging (PL).

3 OPTIMIZATION OF VARIOUS PARAMETERS

Several process parameter optimizations were carried out at various steps in the fabrication process. This paper will showcase a selection of optimization instances, without aiming to provide an exhaustive list.

3.1 Rear side surface morphology

Initially, the FE-pIBC cells featured a textured front side and a flat rear side, achieved by a first alkaline texturing batch process and subsequent single sided etching in acidic solution to remove the rear side texture. However, the diminished popularity of the latter method in numerous modern factories due to the HNO_3 wastewater and nitrogen oxides produced, has led us to explore substitute surface etching techniques for the rear side. In the current process flow, a sacrificial layer enables a single sided texturing process with a saw damage etched (SDE) rear side.

In Figure 3, fill factor (FF) values obtained by current voltage (IV) measurements are compared for groups with different rear side surface morphologies. The first reference group features the initial structure with an acidic polished rear surface. The second group features an SDE rear side and textured front side, as described above. Finally, the third group features a textured front and rear side. Additionally, two variants of silver finger pastes were subjected to testing.

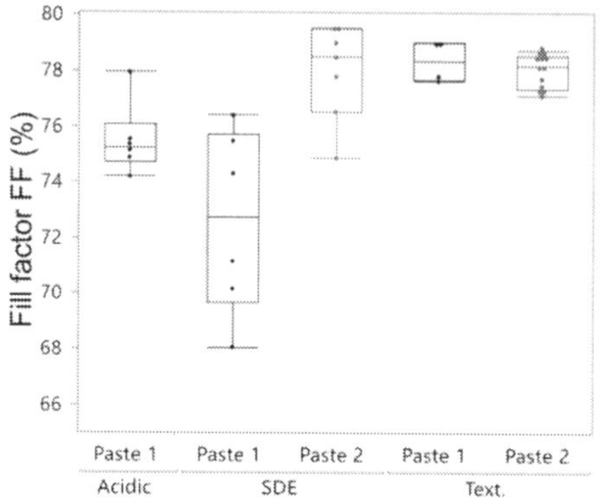

Figure 3: Fill factor results of FE-pIBC solar cells with different silver finger pastes and rear side morphology

In Figure 3, we can see that the FF values highly depend on the rear side surface morphology. Using the Ag paste 1, the FF improves when transitioning from an acidic polished rear side to a textured one, whereas for an SDE surface a lower FF is observed, which we attribute to differences in contact formation. The initial FF drop on SDE surface could be recovered with a new paste (Paste 2 in Figure 3) and highest efficiencies are obtained by combining the latter to SDE rear surface.

3.2 Laser structuring

Laser patterning experiments are conducted testing different sets of laser parameters both on test structures and on solar cells. Planar test structures with an equivalent PSG thickness are fabricated by the deposition of poly-Si(n) layers and subsequent $POCl_3$-diffusion. After laser ablation, the samples were chemically etched in alkaline solution and characterized by laser scanning microscopy (LSM). An example for such laser parameters variations, in this case different power settings, is depicted in Figure 4.

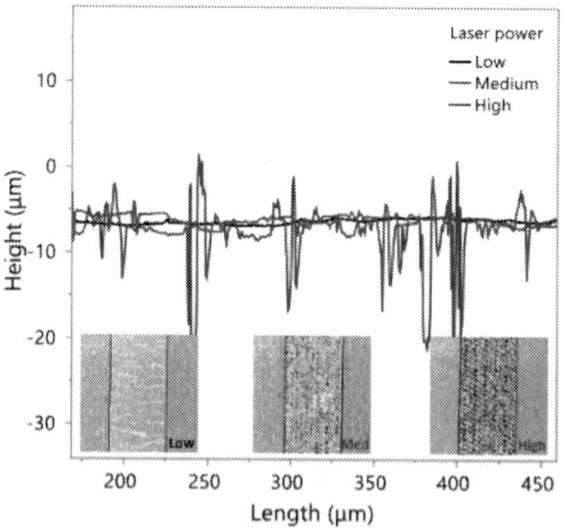

Figure 4: Height profile after laser ablation using different laser powers and subsequent alkaline etching measured by LSM

It is interesting to see, that when increasing the laser power to a certain extent, overlap spikes begin to appear. An optimum of laser parameters should be found to be able to etch the n-poly layer while mitigating the formation of these overlap spikes, which have the risk to introduce both passivation and contact-related defects.

3.3 FFE formation and poly-Si annealing

The $POCl_3$ tube furnace process following the a-Si deposition plays an important role by serving dual purposes: First, the high temperature facilitates the crystallization of the n-poly layer, and second, engendering the formation of a lightly doped front floating emitter on the front surface of the FE-pIBC solar cell. The developed $POCl_3$ process recipes enable the formation of lightly doped FFE profiles with high sheet resistance values of $R_{sh} \approx 1000$ Ω/sq., despite the high temperature required for fully crystallizing the silicon layer in the same process. A test diffusion with a full load of 500 HF-dipped wafers shows a uniform doping result over the process boat and wafer surface, confirming the industrial feasibility of the process. Figure 5 depicts doping profiles of the FFE observed for different process parameters.

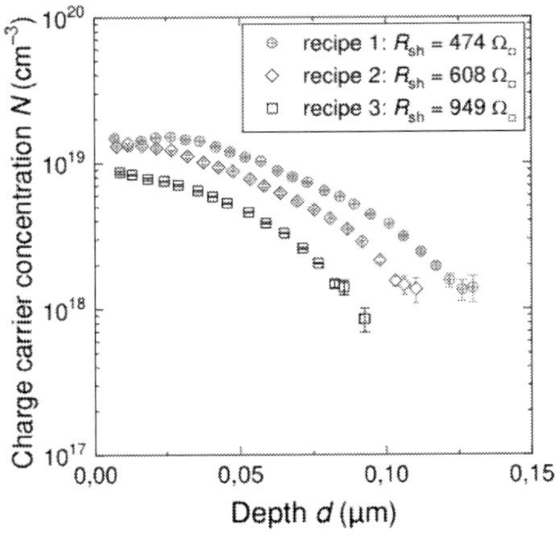

Figure 5: FFE doping profiles of different diffusion recipes determined by electrochemical capacitance voltage (ECV) measurements on alkaline textured surfaces. The profiles are scaled to match the local sheet resistance determined by eddy current measurements.

For two different diffusion recipes investigations are conducted on both lifetime samples and solar cells. The results shown in Figure 6 demonstrate that a FFE with a higher sheet resistance notably enhances the implied open-circuit voltage iV_{oc} of test structures that feature a FFE front side and a poly-Si passivated rear side. This improved implied iV_{oc} subsequently contributes to an overall elevation in the V_{oc} at the solar cell level, as evident in Figure 6.

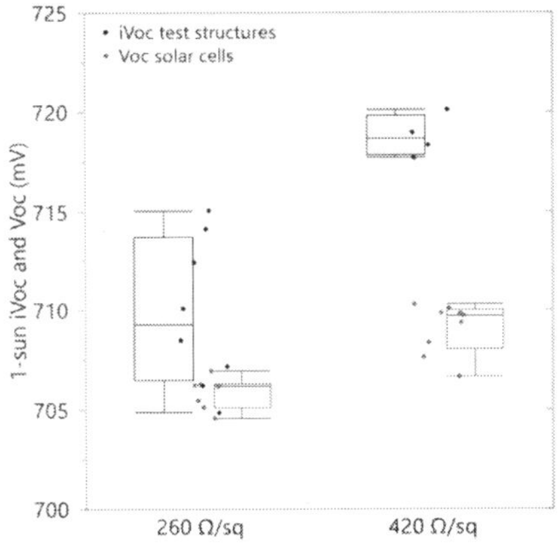

Figure 6: Influence of the front floating emitter diffusion process on iV_{oc} and V_{oc} obtained by quasi steady state photoconductance (QSSPC) and IV measurements, respectively.

3.4 Laser contact opening

In pursuit of enhancing the fill factor, experimentation is conducted involving two distinct laser contact opening geometries. Other printing parameters, including the specific aluminum paste employed and the fast firing temperature profile, are maintained constant.

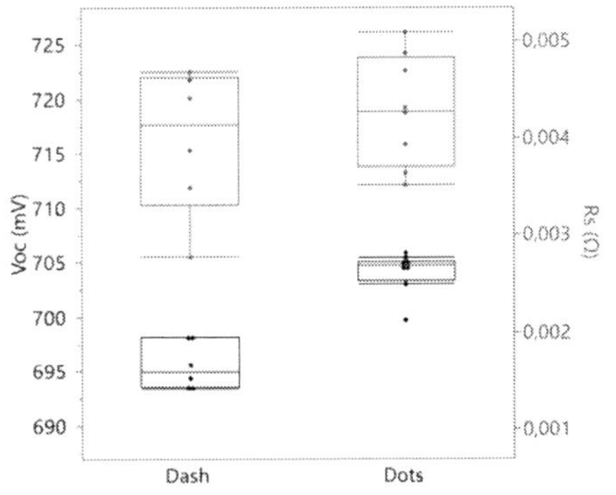

Figure 7: Open-circuit voltage V_{oc} (black, left scale) and series resistance R_s (red right scale) results obtained by IV measurements on solar cells with dash and dots shaped laser contact openings

While the implementation of dash openings succeeds in slightly diminishing the series resistance, a higher fraction of metallization results in a significant reduction in V_{oc}, a trend visualized in Figure 7. This decrease in V_{oc} leads to a decline in efficiency, not compensated by higher FF

4 RESULTS

4.1 Lifetime results

Lifetime samples prior to metallization are fabricated in order to evaluate the iV_{oc} potential within distinct regions of the solar cell. Two lifetime sample configurations are fabricated:

- Symmetrical flat samples with n-poly on both sides and laser ablated squares on one side following the PL image in Figure 8. The n-poly and the ablated area are passivated by thermal oxide + PECVD AlO_x+SiN_x layers
- Asymmetrical samples (cell precursors) following the same FE-pIBC architecture prior to the metallization printing.

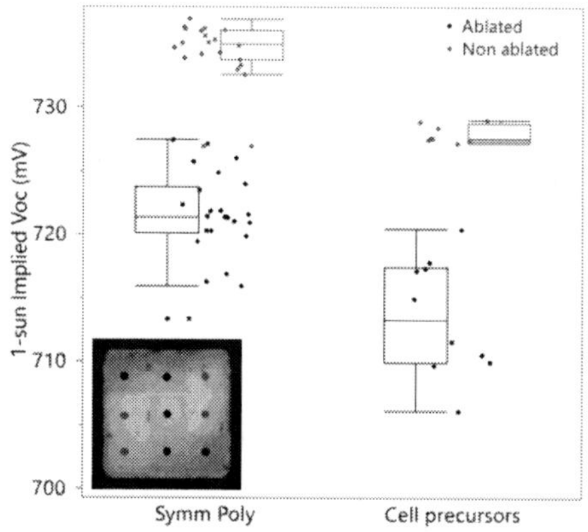

Figure 8: Implied open-circuit voltage iV_{oc} results obtained by QSSPC on different samples architectures. The PL image in the corner shows a typical result for the symmetrical n-poly-sample structure with ablated (black data) and non-ablated (red-data) regions

Apparent from Figure 8, the poly-Si passivation (non ablated area) yields a higher passivation quality compared to the ablated and dielectrically passivated regions. This is confirmed on cell precursers, where

We obtained an iV_{oc} of 715mV and 730mV in the ablated and non-ablated area, respectively.

4.2 Champion cell

After implementing the learnings from a series of optimization procedures, the current-voltage (IV) characteristics of our most exemplary solar cell are succinctly outlined in Figure 9.

The champion solar cell achieves an efficiency of 22.94%, as ascertained through measurements conducted at Fraunhofer ISE CalLab PV Cells. We are currently in the process of planning further optimization strategies with the objective of further efficiency increase.

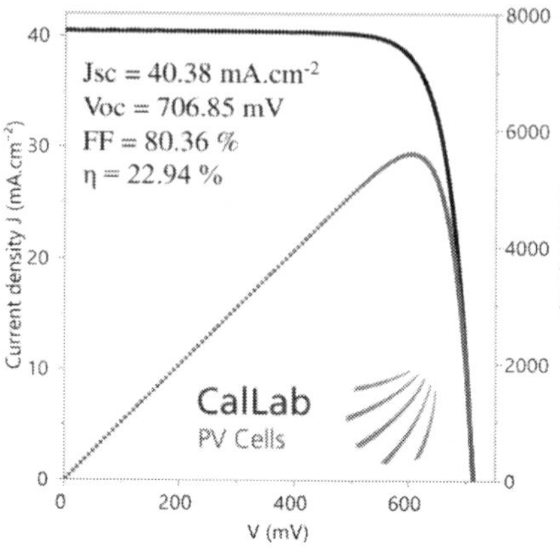

Figure 9: IV curves and characteristics of our champion p-IBC cell measured by Fraunhofer ISE CalLab PV Cells

5 SUMMARY

This manuscript summarizes our efforts concerning the design and optimization of a p-type front floating emitter- interdigitated back contact (FE-pIBC) solar cell, employing passivation through plasma-enhanced chemical vapor deposition (PECVD) n-poly Si layers. The study showcases the feasibility of fabricating FE-pIBC passivated contacts, guided by a process flow emphasizing high reusability of PERC related equipment. Notably, the process sequence harmonizes seamlessly with prevalent high-temperature processing utilized in solar cell fabrication. The front-side floating emitter and poly annealing are merged into a single step. Systematic optimization allowed us to reach an efficiency of 22.94%. Ongoing refinements and adjustments fuel our optimism in achieving further increase in efficiency.

ACKNOWLEDGMENTS

The project POPEI is financially supported by Bundesministerium für Wirtschaft und Klimaschutz (BMWK) Germany, with grant agreement No. 03EE1102.

6 REFERENCES

[1] International Technology Roadmap for Photovoltaic (ITRPV) Roadmap 2022
[2] R. Kopecek et al., "ZEBRA technology: low cost bifacial IBC solar cells in mass production with efficiency exceeding 23.5%," 2020 47th IEEE Photovoltaic Specialists Conference (PVSC), Calgary, AB, Canada, 2020, pp. 1008-1012
[3] B. Kafle et al., "TOPCon – Technology options for cost efficient industrial manufacturing", Solar Energy Material & Solar cells, 227, 2021
[4] F. Feldmann et.al, "Industrial TOPCon solar cells realized by PECVD tube process", 37th European Photovoltaic Solar Energy Conference and Exhibition, EU PVSEC 2020

40th European Photovoltaic Solar Energy Conference and Exhibition

This presentation was selected by the Sc. Committee of the EU PVSEC 2023 for submission of a full paper to one of the EU PVSEC's collaborating peer-reviewed journals.

FORMATION OF INTERFACIAL SILICON OXIDE BY MAGNETRON SPUTTERING: TOWARDS PVD IN-LINE PROCESSING OF TUNNEL OXIDE PASSIVATING CONTACTS

Volker Linß[1,a)], Henning Nagel[2)], Tina Dietsch[1)], Winfried Wolke[2)], Sebastian Mack[2)], Sebastian Gatz[1)]
Martin Bivour[2)], Eric Schneiderlöchner[1)]
[1)] VON ARDENNE GmbH, Am Hahnweg 8, D-01328 Dresden
[2)] Fraunhofer Institute for Solar Energy Systems ISE, Heidenhofstraße 2, D-79110 Freiburg
[a)] Linss.Volker@VONARDENNE.com

ABSTRACT: The TOPCon approach (tunnel oxide passivated contact) has a higher solar cell efficiency potential compared to PERC. Therefore, TOPCon is the next step towards industrial high throughput mass production of silicon solar cells. All processes of the solar cell production must be cost-optimized. Sputter deposition is a well-known scalable large-area high-throughput technique and has been shown to be suitable for in-situ-doped poly-Si(n) deposition. This contribution focuses on the creation of the interfacial oxide of the TOPCon structure by magnetron sputtering that can be combined with the sputter deposition of the in-situ doped silicon in a single tool to further reduce cost. Magnetron sputter deposition of silicon oxide is a well-established large-area technology in the industrial glass coating industry. The process had to be adapted to produce an ultra-thin, high-quality interfacial oxide that fits the needs of the TOPCon structure. Precursors of solar cells were used to create the interfacial oxide by reactive DC magnetron sputter deposition from a silicon target, using an in-line sputter tool. To test the quality of the interfacial oxide it was covered with an in-situ doped aSi layer grown by PECVD followed by annealing, hydrogen passivation, printing, and a firing step. The resulting efficiency of the so produced solar cell is compared to cells with different interfacial oxides and doped poly-silicon layers. It is shown that the reactive sputter process is capable to produce the interfacial oxide in high quality suitable for high volume manufacturing.
Keywords: TOPCon, interfacial oxide, magnetron sputtering, solar cell efficiency

1 INTRODUCTION

The TOPCon approach (tunnel oxide passivated contact, [1]) has a high cell efficiency potential compared to PERC and therefore is the next step towards industrial high throughput mass production of silicon solar cells. Process routes with the lowest possible number of process steps and preferred scalable processes have to be used for a cost effective, competitive high-volume production of silicon solar cells. Sputter deposition is an easily scalable single-sided deposition method available in industrial tool platforms [2] and proven over years in thin film deposition for instance in the large area glass coating [3]. For the TOPCon approach, the industrial-size sputter deposition has already been demonstrated to work for the sputter deposition of the in-situ-doped aSi(n) layer [4] with solar cell efficiencies equal to the LPCVD reference process. In the meantime, a further development of the sputter process of the aSi(n) layer applied on the solar cell architecture shown in Fig. 1 yielded an increase of the so far reported efficiency of 23.4 % to 23.9 %, see Fig. 2.

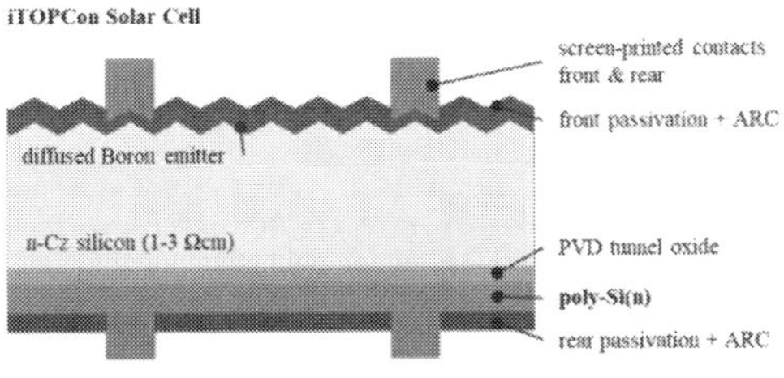

Figure 1: Cell structure of i-TOPCon solar cells produced on M2-n-Cz wafers

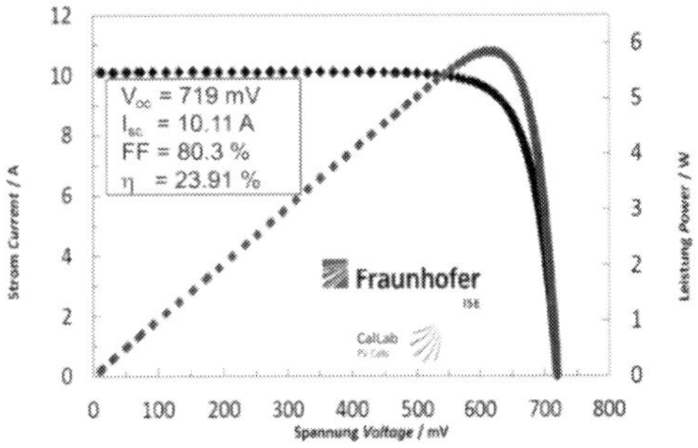

Figure 2: Calibrated measurement of an i-TOPCon solar cell with wet chemical interfacial oxide and in situ-doped magnetron sputtered aSi(n)

It will further reduce cost if the formation of the interfacial oxide and the following deposition of the aSi(n) layer are combined in a single inline sputter tool. Plasma assisted methods for the formation of the interfacial oxide and the deposition of the aSi(n) layer are already in industrial use (e.g. plasma oxidation & plasma assisted in-situ-doping "POPAID") [5]. However, the most common methods for the tunnel oxide (or interfacial oxide) layer formation are wet chemical oxidation, ozone oxidation (wet and dry), thermal oxidation or oxidation by PECVD [6]. The cleaned surface of the wafer is oxidized, that means the interfacial oxide is grown into the wafer surface. This can be done by PVD methods as well. Another approach is to deposit a new interfacial SiOx layer on top of the cleaned silicon wafer surface by a scalable deposition method like magnetron sputtering [7]. It is the aim of this paper to demonstrate the suitability of an industrial sputter process for the formation of the interfacial oxide for the use in high-quality i-TOPCon solar cells.

2 EXPERIMENTAL SECTION

Fig. 3 presents the overall process flow used for the production of TOPCon solar cells using n-type M2 Cz wafers with different combination of interfacial oxide and aSi(n). In the following we will focus on the interfacial oxide creation by magnetron sputtering and the deposition methods of the aSi(n) layer.

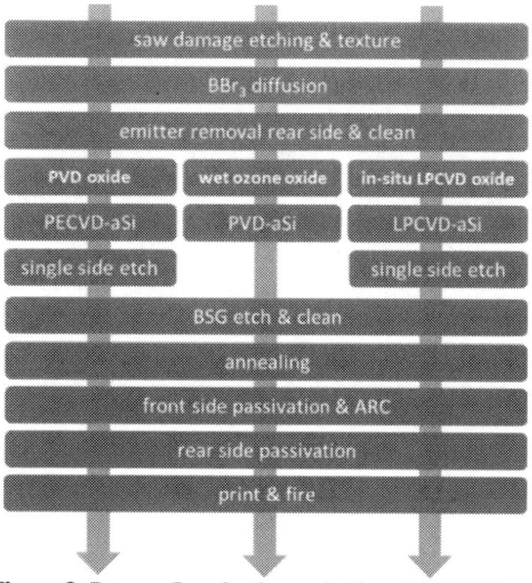

Figure 3: Process flow for the production of i-TOPCon solar cells with different interfacial oxide / aSi(n) layer combinations

2.1 Interfacial oxide preparation by magnetron sputtering

A single tube magnetron sputter process was used for the preparation of the interfacial oxide by PVD. Fig. 4 shows a sketch of the used compartment of the inline tool "SCALA" (VON ARDENNE GmbH) installed at Fraunhofer ISE.

Figure 4: Sketch of used compartment equipped with a single tube target of the inline sputter tool "SCALA"

The material of the target was pure silicon and it was sputtered in an oxygen-containing atmosphere. First, the reactive sputter process was characterized for different argon/oxygen compositions by voltage-power curves and then a suitable working point was chosen for the formation of the interfacial oxide. The thickness of the oxide layers was measured by spectral ellipsometry.

2.2 Deposition of the doped silicon layer

The deposition of the aSi(n) to get the doped poly silicon after annealing was done in three different ways, all of them using in situ doping. The deposition of the amorphous silicon layer was done by PECVD (140 nm) for the PVD oxide wafers and by PVD (140 nm), and LPCVD (80 nm) for the two reference groups. The PECVD and LPCVD deposition were carried out at Fraunhofer ISE while the PVD deposition was carried out at VON ARDENNE [4]. Annealing temperatures between 860 °C and 930 °C were chosen to elaborate an optimal diffusion of the phosphorus dopant through the interfacial oxide.

3 RESULTS AND DISCUSSION

The results section will present basic process properties and possibilities to adjust the thickness and properties of the PVD-grown interfacial oxide and the application in solar cells in comparison to the above-mentioned reference solar cells.

3.1 Sputter process characterization and growth of the interfacial oxide

In Figure 5 the voltage power-dependance for a given mixture of the process gases argon and oxygen is shown. The reactive sputtering mode in the low power region can be clearly distinguished from the metallic mode as known for sputtering silicon oxide layers. The working point for the experiments was set to <1 kW to be in a stable processing condition.

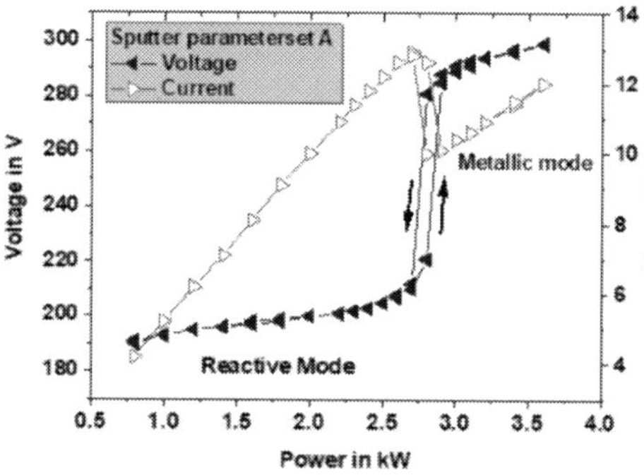

Figure 5: Example for a voltage-power dependence of the reactive SiOx sputter process at a given gas composition Ar/O_2

To tune in the oxide thickness within the range from 1.2 nm to 2 nm oxygen flow, carrier speed and processing temperature have been varied. As expected, lower carrier speeds as well as a higher oxygen flow do provide thicker SiOx layers. The temperature variation between 150°C and 350°C did not show a significant influence on the thickness of the silicon oxide layer grown onto the wafer surface. Figure 6 shows the dependance of the thickness of the SiOx layer on the argon content in the sputter gas and the process duration, that itself depends on the carrier velocity, given in arbitrary units.

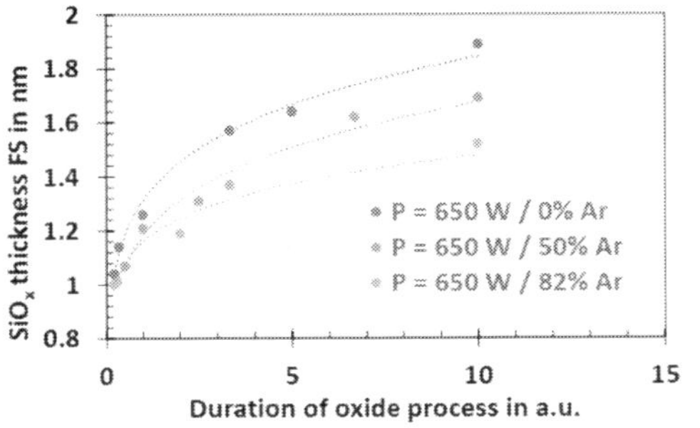

Figure 6: Measured interfacial oxide thickness grown at different sputter gas compositions Ar/O_2 at constant sputter power in dependence on the process duration

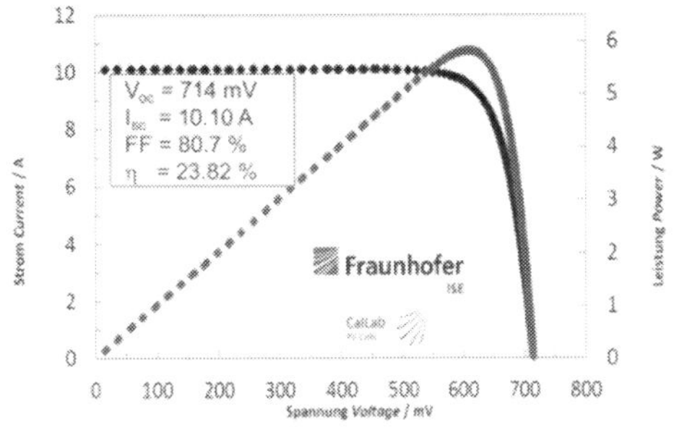

Figure 8: Calibrated measurement of an i-TOPCon solar cell with sputter-process created interfacial oxide and PECVD-deposited aSi(n)

3.2 Application of sputtered SiOx layers in solar cells

To evaluate the developed oxide growth processing regime i-TOPCon solar cells have been produced according to the sequence given in Figure 3. Five different PVD oxides have been processed with varying thickness of 1.1 nm, 1.3 nm, and 1.6 nm after removal of a native oxide layer from the wafer surface in diluted HF (1 %). Then a 140 nm thick aSi(n) layer has been deposited using PECVD. The anneal step to form the poly-Si layer has been performed in a tube furnace at 870 °C, 900 °C, and 930 °C using a plateau time of ten minutes.

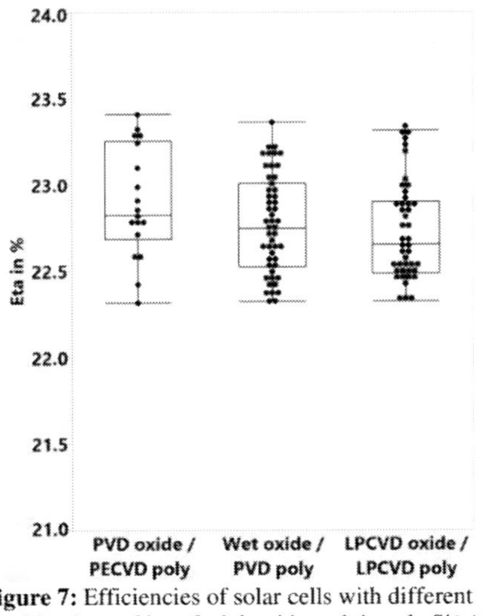

Figure 7: Efficiencies of solar cells with different combinations of interfacial oxide and doped aSi(n)

Figure 7 shows the efficiency distribution of the PVD oxide group in comparison to the two reference solar cell groups. It can be seen that the developed PVD SiOx layer is perfectly suited to produce high quality i-TOPCon solar cells at the median efficiency level of the references. The best PVD-oxide solar cell has been measured at ISE CalLab PV Cells as shown in Figure 8. A promising efficiency of 23.8% has been demonstrated. As the FF leaves some room for improvement, we expect that future iTOPCon solar cells with PVD processing of the interfacial oxide as well as the in-situ doped aSi(n) layer will reach 24% and beyond.

4 CONCLUSIONS

It has been shown that the interfacial oxide of i-TOPCon solar cells can be prepared using a reactive magnetron sputter process. This process can be run in inline mode and is scalable for high-throughput production. The properties of the sputter-deposited interfacial oxide can be adjusted by the sputter process parameters. The resulting interfacial oxide was used in solar cells and yielded a high efficiency comparable or even better than the same type of reference solar cells with different interfacial oxides / poly silicon layers. The combination of inline prepared interfacial oxide and inline deposited in situ doped aSi(n) by magnetron sputter processes is enabling a lean process flow for the rear passivating contact and the high manufacturing capacity of VON ARDENNE's sputter equipment of up to 1.2 GW per tool will help to further reduce manufacturing cost.

ACKNOWLEDGEMENT

The authors like to thank Uwe Graupner (VA) for assistance in numerous film deposition experiments and all colleagues at Fraunhofer ISE who contributed to this work. This work was funded by the German Federal Ministry for Economic Affairs and Energy within the project "SpuTOPiA" under contract number 03EE1068A (VA) and 03EE1068C (ISE).

REFERENCES

[1] F. Feldmann, M. Bivour, C. Reichel, M. Hermle, S.W. Glunz, "A Passivated Rear Contact for High-Efficiency n-Type Si Solar Cells Enabling High Voc's and FF>82 %", Proceedings of 28th EU-PVSEC, (2013) 988–992.
[2] B. Kafle, B. S. Goraya, S. Mack, F. Feldmann, S. Nold, J. Rentsch, "TOPCon – Technology options for cost efficient industrial manufacturing", Solar Energy Materials & Solar Cells 227 (2021) 111100.
[3] G. Bräuer, "Large area glass coating", Surface&Coatings Technology 112 (1999) 358.
[4] V. Linss, T. Dietsch, J. Baumann, U. Graupner, J. Hoß, J. Linke, J. Lossen, J.-I. Polzin, S. Mack, H. Nagel, M. Bivour, E. Schneiderlöchner, "Single sided high throughput sputter process technology for in-situ doped n-type amorphous silicon layers for high efficiency TOPCon solar cells", Proceedings of 8th

World Conference on Photovoltaic Energy Conversion 2022, p. 202-207.
[5] M. Jiayu, "The POPAID Route of Producing TOPCon for Utility Application", TaiyangNews Virtual Conference 2022.
[6] H. Yousuf, M.Q. Khokhar, M.A. Zahid, M. Rabelo, S. Kim, D.P. Pham, Y. Kim, J. Yi, "Tunnel Oxide Deposition Techniques and Their Parametric Influence on Nano-Scaled SiOx Layer of TOPCon Solar Cell: A Review", energies 15 (2022) 5753.
[7] C. Allebé, A. Descoeudres, P. Wyss, P. Boillat, N. Pernès, B. Paviet-Salomon, C. Ballif, "Sputtering for the formation of Si-based passivating contacts", 50th IEEE PVSC (2023), submitted to IEEE Journal of Photovoltaics.

FABRICATION OF THIN SILICON-HETEROJUNCTION SOLAR CELLS FOR BENDABLE PEROVSKITE/C-SI TANDEM SOLAR CELLS

Kimihiko Saito, Hirotaka Shishido, Ryousuke Ishikawa
Tokyo City University, Advanced Research Laboratories
1-28-1 Tamatsutsumi Setagaya-ku Tokyo 158-8557, Japan
Phone +81-3-5707-2769 e-mail: kisaitou@tcu.ac.jp

ABSTRACT: We conducted the development of silicon heterojunction (SHJ) solar cells towards the fabrication of bifacial bendable perovskite (PVK)/SHJ tandem solar cells. By depositing a silicon nitride film onto the surface of a 300 μm-thick CZ Si substrate and using it as a KOH etching mask, we produced a substrate with a thickness of approximately 60 μm, featuring a mirrored front surface and a rear textured surface. Using this substrate, the power conversion efficiencies of 19.80 % and 18.87 % for the front and rear emitter structure cells were achieved, respectively. To mitigate the decrease in bottom cell current caused by thinning of the Si substrate, we fabricated p- and n-type films with refractive indices of 2.7 to 2.9 in the range of 800 to 1100 nm, which should contribute to reduce reflection at the interface between top and bottom cell. Furthermore, we introduced micro texture (μTex) with a height of around <1 μm on the light-incident side surface, confirming an increase in short-circuit current due to the enhancement of the long-wavelength sensitivity of the SHJ cell. Additionally, we verified that PVK single cells formed on Si substrates with these μTex operated without any issues.
Keywords: Si heterojunction, perovskite, texture, tandem solar cell

1 INTRODUCTION

In recent years, there has been a demand for lightweight and bendable high-efficiency solar cells to be installed on buildings in energy-intensive urban areas. To address this demand, the development of perovskite (PVK) solar cells [1,2] and thin silicon heterojunction (SHJ) solar cells [3~5] has been progressing. On the other hand, tandem solar cells incorporating monolithic two-terminal PVK/SHJ solar cells have achieved the power conversion efficiencies (PCEs) exceeding 30 % [6~8]. Furthermore, regarding the PVK/SHJ tandem solar cells, to mitigate the tendency for the bottom cell current to become smaller compared to the top cell current, a double-sided light-receiving configuration has been proposed [9]. Considering these, we have been studying the development of bifacial bendable PVK/thin SHJ tandem solar cells. In this study, we report on the element technology development of thin SHJ solar cells for the PVK/SHJ tandem cell efficiency enhancement.

2 EXPERIMENTS

2.1 Substrate fabrication

On one side of a 5 cm square n-type CZ (100) crystalline silicon (c-Si) substrate (thickness of 300 ± 20 μm and a resistivity of 1 to 10 Ωcm) with silicon nitride (SiN_x) films deposited on both sides, the removal of the SiN_x film leaving a 5 mm width from the edge was carried out using a YVO_4 laser. Subsequently, a coating of an alkali-resistant resist (ProTEK B3) on the SiN_x film on the other side was applied to block the pinholes in it. The substrate was immersed in a 13 wt% KOH solution at 90 °C and Si etching was carried out in the area where the SiN_x film was removed to achieve a thickness of approximately 60 μm (etching rate ~2.4 μm/min). Then, a pyramid texture was formed on the etched Si surface by immersing the substrate in an aqueous solution containing 4 wt% KOH and additives for texture formation (0.75 wt% TT72C13 and 15 wt% TK81 from Hayashi Pure Chemical) at 70 °C for 15 minutes. Subsequently, the substrate was immersed in a 5 % HF solution to remove the ProTEK film and SiN_x film. Afterward, an organic contamination was removed by a sulfuric acid-hydrogen peroxide mixture (SPM) cleaning, resulting in the production of a substrate with a mirrored front surface/textured back surface (M/T substrate). Next, to remove the potential damage layer on the mirrored surface that might have been introduced during SiN_x film formation, isotropic etching of the Si substrate surface was carried out by immersing the substrate in a solution consisting of a mixture of 46-48 % HF and 61 % HNO_3 in a volumetric ratio of 1:40 for 30 seconds. Further cleaning processes, including SC2, were employed, followed by the removal of the oxide layer using 5 % HF, after which chemical vapor deposition (CVD) was performed.

Regarding the fabrication of a substrate with a micro-textured front surface/textured back surface (μT/T substrate), after the SPM cleaning in the above-mentioned process, the M/T substrate was immersed in an aqueous solution containing 3.1 wt% KOH and additives for texture formation (0.75 wt% TT72C13 and 19 wt% TK81) at 65 °C for 15 minutes to form micro texture (μTex) on the front surface side. The subsequent steps after the HF and HNO_3 solution treatment were the same as previously described.

The substrate thickness was measured by a digital thickness gauge (Mitutoyo ID-S112S) and by cross-sectional observations using scanning electron microscopy (SEM, HITACHI SU8230).

2.2 Solar cell fabrication and evaluation

In the case of PVK/SHJ tandem solar cells, there are two options depending on whether the hole transport layer (HTL) or the electron transport layer (ETL) is placed on the light-incident side of the PVK top cell. Consequently, two different device structures were fabricated for the SHJ cell, namely the front emitter (FE) structure and the rear emitter (RE) structure, as illustrated in Fig. 1. The film thicknesses indicated in Fig.1 represent the thicknesses achieved when deposited on a flat substrate. The p, i and n layers based on hydrogenated amorphous silicon (a-Si:H), hydrogenated amorphous silicon oxide (a-SiO_x:H) and on hydrogenated microcrystalline silicon oxide (μc-SiO_y:H) material were deposited by plasma enhanced CVD (PECVD, SAMCO PD-2203L) with an excitation frequency of 40.68 MHz. The deposition conditions of the

10.4229/EUPVSEC2023/1AO.6.6

intrinsic a-Si:H/a-SiO_x:H bilayer films for c-Si surface passivation are detailed in reference [10]. The deposition conditions for the p-layer and n-layer are as shown in Table I. After deposition of the p, i and n layers, the in-plane distribution of effective minority carrier lifetime (τ_{eff}) was measured using microwave photoconductivity decay (μ-PCD) method (SEMILAB, WT-2000PVN). The electrodes on the light-incident side, composed of a 1 cm² sized indium tin oxide (ITO) film and a silver (Ag) grid electrode, and on the back side, composed of ITO and Ag electrode, were formed by sputtering and evaporation, respectively. Current-voltage (IV) characteristics were measured under AM1.5 filtered light with an intensity of 100 mW/cm² using a 1 cm² aperture mask attached to the light-incident side electrode. External quantum efficiency (EQE) measurements were conducted similarly to the IV measurements by attaching a 1 cm² aperture mask to the light-incident side electrode.

Furthermore, the wavelength dispersion of the complex refractive index of the p and n layers was estimated by analyzing spectroscopic ellipsometry (SE) data obtained by using an ex-situ ellipsometer (J. A. Woollam, M-2000) at room temperature and under atmospheric conditions with three angles (65°, 70°, and 75°) through analysis with CompleteEASE (ver.6.68b).

Table I: Deposition conditions of conventional and low *n* p and n layers

Film	Gas flow rate (sccm)	Pressure (Pa)	RF power (mW/cm²)	$T_{substrate}$ (°C)
conventional p μc-SiO_x:H	$SiH_4/H_2/B_2H_6/CO_2$=2/340/1/0.5	90	79	200
conventional n a-Si:H	$SiH_4/H_2/PH_3$=5/95/5	60	53	200
low *n* p μc-SiO_x:H	$SiH_4/H_2/B_2H_6/CO_2$=2/340/1/1	90	79	160
low *n* n μc-SiO_x:H	$SiH_4/H_2/PH_3/CO_2$=2/98/8/2	60	53	200

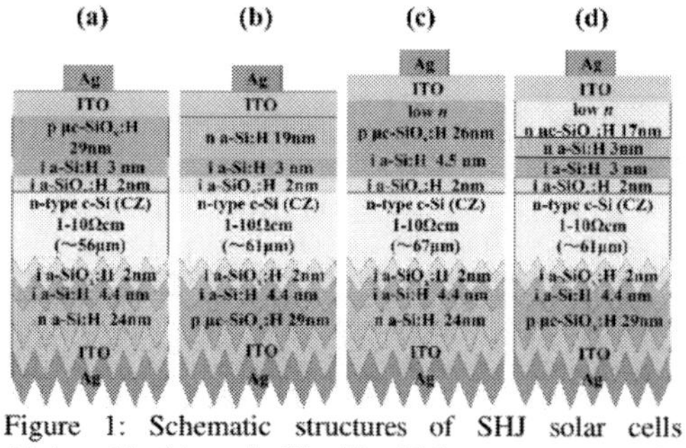

Figure 1: Schematic structures of SHJ solar cells fabricated in this study. The film thicknesses represent the thicknesses achieved when deposited on a flat substrate.

3 RESULTS AND DISCUSSIONS

3.1 Solar cell results on thin M/T substrates

Generally, PVK solar cells are fabricated using coating methods such as spin-coating. Therefore, when using PVK solar cells as the top cell in tandem configurations, it's desirable to keep the surface of the bottom cell flat. On the other hand, when c-Si solar cells are used as the bottom cell, it is advantageous to introduce a textured structure on the cell surface for enhancing current output. In other words, for the SHJ bottom cell used in PVK/SHJ tandem solar cells, a substrate with a mirrored surface on the front side and a textured surface on the back side is preferred.

Figure 2 (a) and (b) shows the appearance of the mirrored front surface and the textured backside, respectively, of the M/T substrate obtained by the fabrication process in this study, with a thickness of approximately 60 μm in the region subjected to Si etching. Figure 2 (c) displays the appearance of the curved cells fabricated on this substrate which is attached to a flexible plastic film after removing its thick edge portion. Figure 3 (a) and (b) shows the μPCD results of the FE structure cell with a substrate thickness of approximately 56 μm and of the RE structure cell with a substrate thickness of approximately 61 μm, respectively. The corresponding IV characteristics with the cell structures as shown in Fig. 1 (a) and (b) are presented in Table II. (For further discussion, EQE data is presented in Fig. 9 (c).) τ_{eff} of 1 to 2 msec were obtained in the thin region of the substrate with a thickness around 60 μm. The thickness-dependent τ_{eff} correlation between this thin region and the edge portion, which has an approximate thickness of 300 μm, suggests a low surface recombination velocity of <1 cm/sec. Due to this pronounced surface passivation performance and the reduced substrate thickness, both FE and RE structures achieved a high open-circuit voltage (V_{oc}) of approximately 740 mV, resulting in the PCEs of 19.80 % for the FE structure cell and of 18.87 % for the RE structure cell.

Table II: IV characteristics of thin SHJ solar cells obtained in this study using a 1 cm² aperture mask

sub.	sub. thickness (μm)	cell structure	front doped layer	V_{oc} (V)	I_{sc} (mA)	FF	PCE (%)
M/T	56	FE	conv.	742.0	33.27	0.802	19.80
M/T	61	RE	conv.	737.1	31.83	0.804	18.87
M/T	67	FE	low *n*	726.5	34.47	0.789	19.77
M/T	69	RE	low *n*	739.0	33.91	0.802	20.11
μT/T	57	FE	conv.	738.0	35.94	0.794	21.06

Figure 2: Appearance of a 60 μm-thick M/T substrate obtained in this study. (a) mirrored front surface, which reflects the ceiling, (b) textured backside, in which texture is formed in a thinned 4 cm² area, (c) thinned portion of the substrate, on which four solar cells were fabricated, is attached to a plastic film

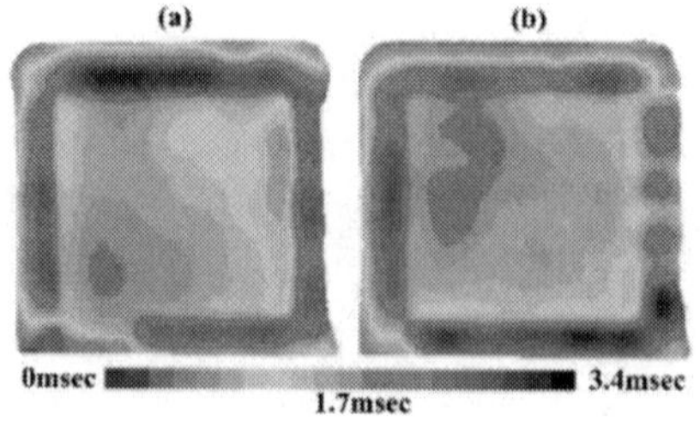

Figure 3: μPCD results of (a) FE and (b) RE structure cells using the M/T substrates.

3.2 Reduction of refractive index (*n*) of p and n layers

In order to increase the current output from the bottom cell in the tandem structure, it's necessary to reduce the reflection at the interface between top and bottom cell. To achieve this, it's desirable to insert a λ/4 film with a refractive index (*n*) intermediate to the respective *n* values of the PVK and c-Si layers to the interface between the top and bottom cell. Hence, for the doped layer on the light-incident side of the SHJ bottom cell, a film with a lowered *n* of around 2.8 and with a thickness of approximately 100 nm has been proposed, and the efficacy of this film has been practically demonstrated [11,12]. Furthermore, it is considered that deploying p^+.layer/n^+.layer or n^+.layer /p^+.layer tunnel recombination junction layers between the top and bottom cell using the low *n* thin-film Si p and n layers is more advantageous for minimizing reflection at the interface between top and bottom cell, as opposed to utilizing transparent conductive films like ITO with a *n* of around 2.0 as a recombination junction layer [13]. Therefore, in this study, an investigation into the reduction of *n* for the p-layer and for the n-layer was conducted, and the cell evaluation was carried out by applying it to the FE and RE structure, respectively.

For the p-layer, optimization of substrate temperature, CO_2 flow rate and of B_2H_6 flow rate were carried out. For the n-layer, the conventional a-Si:H film was replaced with a μc-SiO_x:H film by adjusting the SiH_4/H_2 flow ratio and by introducing CO_2 into the deposition gases, and additionally the PH_3 flow rate was further adjusted to reduce resistivity. Figure 4 shows a cross-sectional transmission electron microscopy (TEM) image of multilayer films composed of the μc-SiO_x:H p-layer and of the passivation bilayer formed on a c-Si (100) substrate under the same conditions as the cell fabrication described in Table I and in the reference [10]. As shown in this figure, the μc-SiO_x:H p-layer forms a film where the microcrystalline phase grows conically from the passivation layer side, resulting in a film with a changing ratio of amorphous and microcrystalline phases along the film thickness direction. This means that the film quality of the microcrystalline doped film is strongly influenced by the underlying layer. Therefore, for the SE evaluation substrate, a substrate similar to the cell fabrication was provided, wherein a passivation bilayer was formed on the surface of an n-type c-Si(100) substrate after removing the surface oxide layer. Moreover, based on our prior investigations, we have observed that depositing the doping layer under conditions that enhance the influence of hydrogen radicals during deposition, thereby accelerating the crystallization process, leads to a significant reduction in passivation performance. This reduction is probably due to the thinning or crystallization of the underlying passivation layer caused by the hydrogen radical impact on the passivating layer during the initial stages of doping layer deposition [14,15]. Therefore, in this SE evaluation, the thickness of the a-Si:H i-layer in the underlying passivation bilayer was varied while observing its impact on τ_{eff}. Additionaly, the SE analysis was carried out in the range of 300 to 1200 nm using a model that takes into account the thickness-dependent ratio of crystalline to amorphous phases in the doped layer. Table I presents the optimized deposition conditions for the low *n* p and n layers, and Fig.5 displays the results of the wavelength dispersion of *n* for both conventional and low *n*-modified p and n layers. The mean squared error (MSE) values in these SE analyses were less than 5. For this low *n* p-layer deposition, varying the underlying i a-Si:H film thickness from 3 nm to 4.5 nm did not result in significant changes in τ_{eff}. However, in this evaliation, we used an i a-Si:H film with a thickness of 4.5 nm. On the other hand, for the evaluation of the low *n* n-layer deposition, we utilized a substrate with an underlying i a-Si:H film thickness of 6 nm. As shown in Fig. 5, we obtained the films that reduced the value of *n* in the range from 800 to 1100 nm to approximately 2.9 for the p-layer and 2.7 ~ 2.8 for the n-layer. These values are within the optimal range for minimizing the reflection at the interface between the top and bottom cells. In addition, while the conventional p-layer exhibited different *n* values at the top and bottom of the film due to the differences in the crystalline phase/amorphous phase ratio along the thickness direction, the low *n*-modified μc-SiO_x:H p and n layers exhibited consistent *n* values at the top and bottom. This indicates that the low *n*-modified μc-SiO_x:H p and n layers form a homogeneous film in the thickness direction from the early stages of the film growth just above the passivation layer. The SE analysis showed almost constant values of about 43% and 13% of the crystalline component from the bottom to the top of the film for both the p-layer and n-layer films, respectively.

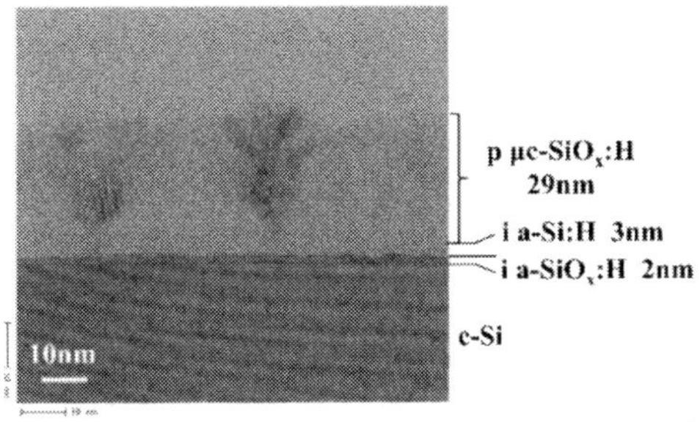

Figure 4: Cross-sectional TEM image of the sample equivalent to the front side of an FE structure cell

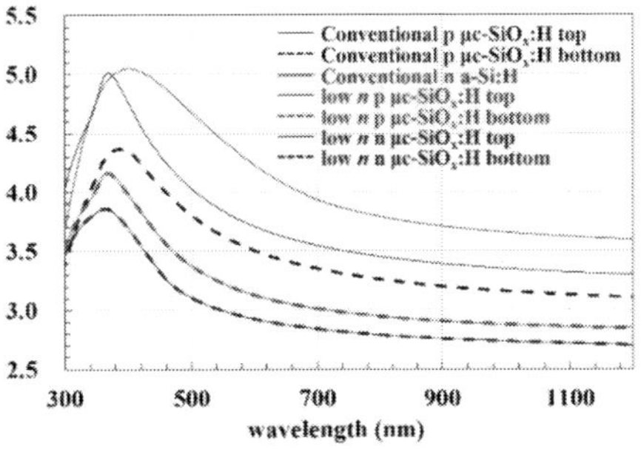

Figure 5: Wavelength dispersion of *n* for both conventional and low *n*-modified p and n layers.

The cells employing the optimized low *n* p and n layers were evaluated using the cell structure shown in Fig. 2 (c) and (d), respectively. For the FE structure cells using the low *n* p-layer, the same structure as used for the SE evaluation was employed, with the a-Si:H i-layer thickness of the underlying passivation bilayer set at 4.5 nm. For the RE structure cells using the low *n* n-layer, the cell fabrication was performed using a three-layer underlying structure of n a-Si:H (3 nm)/ i a-Si:H (3 nm)/ i a-SiOx:H (2 nm), with the a-Si:H n-layer being deposited under the conditions indicated in Table I. The IV characteristics of cells incorporating these low *n* doped layers are presented in Table II. As demonstrated by these results, with respect to the characteristics of the FE cell utilizing a low *n* p-layer, although there was a slight decrease in V_{oc} and fill factor (FF) compared to the cell utilizing the conventional p-layer, no significant decrease was observed. Additionally, for the RE cell incorporating a low *n* n-layer, it was confirmed that both V_{oc} and FF remained nearly unchanged compared to the cell utilizing the conventional n-layer.

Moreover, TiO_2 films formed by thermal atomic layer deposition (ALD-TiO_2) have been reported to form good ohmic contacts with p-type c-Si [16]. Furthermore, ALD-TiO_2/p^+-Si layers have been reported to form low-resistance and well-ohmic junctions, and tandem cells using this junction between the PVK top cell and the c-Si bottom cell have been demonstrated [17]. Namely, if the formation of low-resistance ohmic contacts between this low *n* p-layer and ALD-TiO_2 films can be confirmed, it would enable the creation of a further simplified structure tandem cell, in which the bottom cell's p-layer and the ETL of the top cell are directly connected, avoiding an increase in the reflection at the interface between top and bottom cell. Therefore, we investigated the contact resistance between the aforementioned low *n* p-layer and ALD-TiO_2 film using the Cox-Strack method [18]. The analysis of contact resistance in multilayer films using the Cox-Strack method is noted to be challenging due to variations in current spreading depending on the film's resistance [19,20]. However, as mentioned above, the film quality of the μc-SiO_x:H p-layer is strongly influenced by the underlying layer. Therefore, for the devices under this investigation, a structure resembling the cell configuration as shown in Fig. 6 (a) was used. In this structure, a passivation bilayer was formed on one side of a low-resistance p-type c-Si (100) substrate (0.01-0.001 Ωcm) after the removal of the surface oxide on the substrate using hydrofluoric acid. This was followed by the deposition of a low *n* μc-SiO_x:H p-layer on top of this bilayer. Then, a μc-Si:H p-layer were formed on the opposite side of the substrate. In this investigation, the thickness of the a-Si:H i-layer was set to 3 nm to avoid the influence of the high bulk resistance associated with this intrinsic layer. The deposition of the TiO_2 film on the p-layer was carried out with a sequence of introducing $Ti[N(CH_3)_2]_4$ (TDMAT) and H_2O alternately into the chamber, repeated 600 times to achieve a film thickness of 20 nm, at a substrate heater temperature of 250 °C. Finally, the Ag electrode with the pattern shown in Fig. 6 (a) was formed on the TiO_2 surface, and a full-area Ag electrode was also created on the reverse side. An example of the IV curve for this device structure is presented in Fig. 6 (b). From the IV characteristics measured with electrodes of various areas, the contact resistance of the entire device structure was estimated to be 0.25 Ωcm^2. However, due to the presence of multiple layers and interfaces, in order to estimate the contact resistance of the ALD-TiO_2/low *n* μc-SiO_x:H p-layer, it is necessary to obtain the resistances of other layers and interfaces. Therefore, another structure was fabricated to evaluate the resistances of multiple layers and of interfaces using the Cox-Strack method, where the low *n* p-layer in the Fig. 6 (a) configuration was substituted with the conventional p-layer (29 nm) and a dual-layer electrode pattern of Ag/ITO was directly deposited onto the p-layer without the formation of the TiO_2 film. The obtained contact resistance across this entire device structure was determined to be 0.31 Ωcm^2. Furthermore, for the samples fabricated on non-alkali glass substrates, the Ag/ALD-TiO_2 contact resistance, ALD-TiO_2 film conductivity, ITO/conventional μc-SiO_x:H p-layer contact resistance, conventional μc-SiO_x:H p-layer film conductivity, low *n* μc-SiO_x:H p-layer film conductivity, backside Ag/μc-Si:H p-layer contact resistance, and μc-Si:H p-layer film conductivity obtained through the transfer length method (TLM) test structures were 0.002 Ωcm^2, 4.95 S/cm, 0.025 Ωcm^2, 0.13 S/cm, 0.15 S/cm, 0.003 Ωcm^2, and 1.98 S/cm, respectively. From these results, it is inferred that the overall contact resistance of the device, as obtained in this study, is mainly attributed to the bulk resistance of the intrinsic passivation layer, namely, the contact resistance of the ALD-TiO_2/low *n* μc-SiO_x:H p-layer can be considered negligible. This suggests that the low *n* μc-SiO_x:H p-layer studied here can directly contact the ALD-TiO_2 film in the tandem cell configurations.

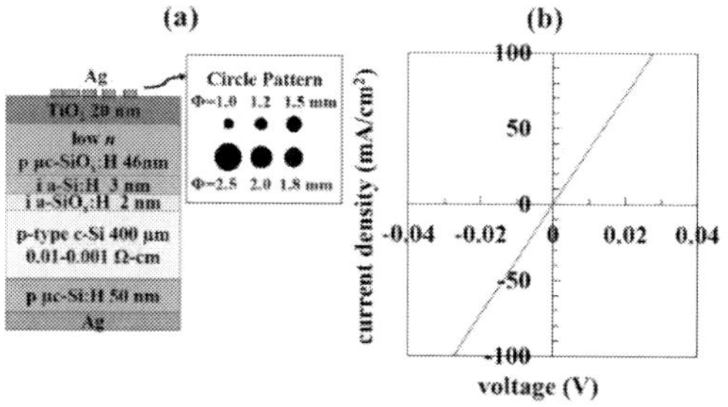

Figure 6: (a) Sample structure and electrode patterns for contact resistance measurement using Cox-Strack method, (b) IV characteristics of the device structure shown in this figure (a)

3.3 Introduction of μTex on the light-incident side

To mitigate the reduction in output current from the bottom cell caused by thinning, it is desirable to introduce texture on its light-incident side. On the other hand, since the thickness of the PVK film formed by spin coating is generally 1 μm or less, the surface roughness of the bottom cell substrate should ideally be 1 μm or less [21~23]. To achieve this, we attempted to create micro-scale texturing using the KOH solution by adjusting the etching temperature and the amount of additives commonly used in commercial c-Si solar cell texturing as denoted in the "Experiments" section. Furthermore, we carried out an additional isotropic etching on the textured Si surface using the mixed solution of HF and HNO_3 (HF/HNO_3 treatment) for 30 seconds as same as what is done on the M/T substrates. Figure 7 shows surface and cross-sectional SEM images of the fabricated textured substrate. Since a solution containing less than 1% tetramethylammonium hydroxide (TMAH) was employed for substrate cleaning, the image of the substrate just before CVD process is also presented. As shown in the figures, the HF/HNO_3 treatment resulted in a reduction in texture size overall, yielding a height of approximately less than 1 μm. Additionally, rounding of the pyramid tops and valleys is observed [24], as well as the emergence of more flattened areas. This flattening is presumed to occur due to the removal of the "nano" pyramids formed during the KOH texture etching process. Figure 8 (a) shows the reflectance spectra of these μTex substrates formed on c-Si substrates

Figure 7: SEM images of surface and cross-section of μTex substrates at the stages of (a) post-texture etching, (b) post-HF/HNO_3 treatment, (c) pre-CVD process

with a thickness of 300 μm. It also includes the reflectance of the mirror surface and the conventional texture used on the backside of the cells as a reference. As shown in this figure, the reflectance just after KOH texture etching is nearly equivalent to that of the conventional texture. However, after the HF/HNO_3 treatment, the reflectance increases by approximately 5 %, attributed to the rounding of pyramid tops and valleys and to the emergence of flattened surface areas, as observed in SEM images. On the contrary, based on the SEM images in Fig. 7(b) and (c), as well as the reflectance results in Fig. 8(a), the cleaning with a solution containing TMAH is considered to have a negligible impact on the μTex characteristics. Figure 8 (b) shows the reflectance distribution at 900 nm obtained from measurements at multiple points on several μTex substrates just after KOH texture etching and in a state ready for CVD process. As shown in this figure, the HF/HNO_3 treatment causes a significant increase in reflectance distribution. This phenomenon might originate from non-uniform formation of "nano" pyramids within the plane during the KOH texture etching stage, which are then eliminated through the HF/HNO_3 treatment, resulting in an enlargement of flattened surface area and in an enhancement of the reflectance. It is noted that the impact of this μTex formation on the backside texture has been confirmed to be minimal through SEM observations, reflectance measurements, and cell characterization.

As for cell characteristics, Fig. 9 (a) and (b) display the schematic diagram of the FE cell structure using the μT/T substrate with a thickness of approximately 57 μm, and the cross-sectional SEM image of the fabricated cell, respectively. The results of its IV characteristics are presented in Table II. Furthermore, Fig. 9 (c) presents its EQE spectrum of its cell along with the FE and RE structure cells with the M/T substrate as described in the "Solar cell results on thin M/T substrates" section. In the cell fabricated utilizing μT/T substrate, an asymmetric surface texture is formed on the light-incident side and the

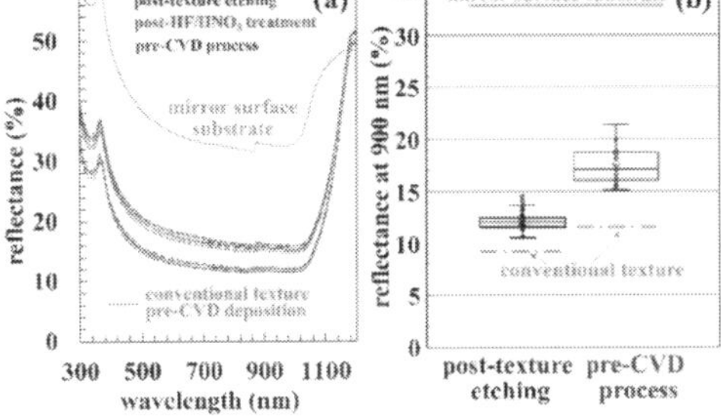

Figure 8: (a) Reflection spectra measured at five different locations on the same μTex substrates at the stage of post-etching (black line), post-HF/HNO_3 treatment (red line) and pre-CVD process (blue line). The types of solid and dashed lines are varied based on the measurement positions in the substrate. (b) Reflectance distribution at 900 nm obtained from measurements at multiple points on several μTex substrates at the stage of post-etching and pre-CVD process.

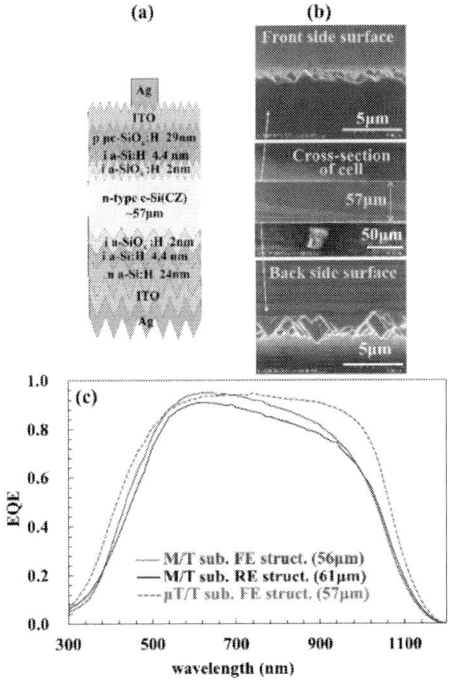

Figure 9: (a) Schematic diagram of the FE cell structure using the μT/T substrate. (b) Cross-sectional SEM image of the fabricated cell with the structure in this figure (a). (c) EQE spectra of the fabricated cell with the structure in this figure (a) along with the FE and RE structure cells with the M/T substrate as indicated in the "Solar cell results on thin M/T substrates" section.

rear side. By introducing the μTex studied in this investigation on the front side, we confirmed an increase in the short circuit current (I_{sc}) due to the enhancement of long-wavelength sensitivity, which is crucial for the bottom cell of the tandem solar cell, without significantly decreasing V_{oc} and FF. In terms of the effects of HF/HNO_3

treatment on cell characteristics, based on the current evaluation stage results using the FE structure cells, it has been observed that there is no change in I_{sc} when comparing between cells with HF/HNO_3 treatment and those without. However, V_{oc} decreases without HF/HNO_3 treatment. Namely, in the μTex and CVD conditions studied here, it is suggested that HF/HNO_3 treatment, which leads to rounding of pyramid tops and valleys as well as an increase in flat areas, can improve V_{oc}, while on the other hand, there may be little significant impact on I_{sc} achieved by the introduction of μTex. However, with regard to the fact that I_{sc} remains unchanged despite the increase in flat regions due to HF/HNO_3 treatment, further meticulous investigation is required, making it a subject for future research.

Now, while the improvement in long-wavelength sensitivity through the formation of surface μTex has been confirmed, it is meaningless if practical PVK cells cannot be fabricated on top of it. Therefore, we formed an 80 nm-thick ITO back electrode with a width of 8 mm on a 30 nm-thick SiO_2-deposited μTex-formed c-Si wafer, followed by UV/O_3 treatment of its surface, and then deposited a self-assembled monolayer (SAM) using [2-(3,6-Dimethoxy-9*H*-carbazol-9-yl) ethyl] phosphonic Acid (MeO-2PACz) through spin-coating. Subsequently, a $Cs_{0.05}(FA_{0.77}MA_{0.23})_{0.95}Pb(I_{0.77}Br_{0.23})_3$ composition PVK film [25] was formed by spin-coating by using anisole as an antisolvent. On its surface, a 20 nm-thick C_{60} film was deposited via evaporation, followed by the deposition of a 20 nm-thick SnO_2 film through ALD. A 90 nm-thick ITO film with a width of 3.5 mm was then sputter-deposited to form a cross with the ITO electrode on the backside. Finally, a 100 nm-thick MgF_2 film was deposited by evaporation. Additionally, for reference, a PVK single cell was fabricated on a flat glass substrate using the same procedure. The cell measurements were carried out by attaching a 3 mm square aperture mask to the region where the ITO electrodes intersected. Figure 10 (a) displays the cross-sectional SEM image of the PVK single cell formed on the μTex substrate. As clearly observed in this figure, the PVK film adequately covers the textured surface in the spin-coating process studied in this work. Figure 10 (b) shows the IV characteristics of the fabricated PVK cells. Both the μTex substrate and the glass substrate exhibit reduced FF due to significant series resistance. This can be attributed to the distance of approximately 5 mm on the front side and of approximately 5-10 mm on the back side between the 3 mm square area exposed to the light and the ITO film in contact with the probes used for measuring the IV characteristics. However, apart from this aspect, both substrates exhibit short-circuit current density (J_{sc}) of over 20 mA/cm^2 and V_{oc} of approximately 1.1 V, with little difference between them. Namely, it was confirmed that using the μTex substrate studied in this work, comparable solar cell characteristics to those fabricated on flat substrates can be achieved without short-circuiting between the pyramid tips and the surface electrode.

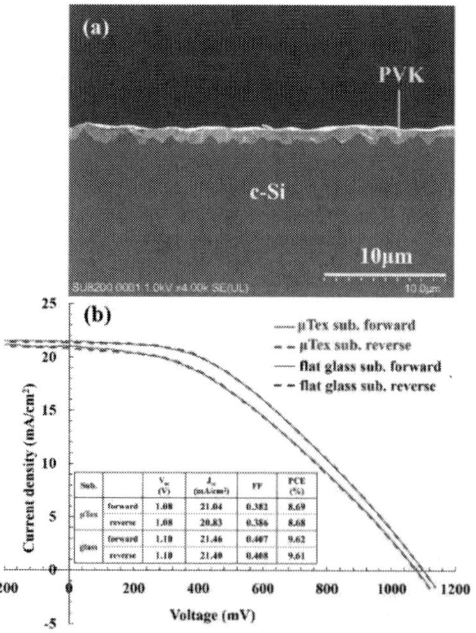

Sub.		V_{oc} (V)	J_{sc} (mA/cm²)	FF	PCE (%)
μTex	forward	1.08	21.04	0.382	8.69
	reverse	1.08	20.83	0.386	8.68
glass	forward	1.10	21.46	0.407	9.62
	reverse	1.10	21.40	0.408	9.61

Figure 10: (a) Cross-sectional SEM image of the PVK single cell formed on the μTex substrate. (b) IV characteristics of the fabricated PVK cells on the μTex and flat glass substrates. "forward" and "reverse" signify voltage sweeping in measurement to be boosting and lowering, respectively.

4 SUMMARY

We carried out the fabrication of SHJ cells using n-type CZ Si substrates for the development of bifacial bendable PVK/SHJ tandem solar cells. Utilizing a 300 μm-thick Si substrate with SiN_x films deposited on both sides, we fabricated the M/T substrate with a thickness of approximately 60 μm by removing the SiN_x film on one side using a YVO_4 laser and then by performing substrate etching and surface texture etching using KOH. Using these substrates, we fabricated solar cell cells with FE and RE structures, resulting in the PCEs of 19.80 % and 18.87 %, respectively, with V_{oc} reaching approximately 740 mV. With the aim of enhancing the output current of the bottom cell, we fabricated p-type and n-type films with n of 2.7 to 2.9 in the range of 800 to 1100 nm to reduce reflection at the top cell/bottom cell interface. In terms of solar cell characteristics, we also confirmed that utilizing these low n p and n layers does not lead to significant degradation. This low n p-layer was verified to form a low-resistance ohmic contact with the ALD-TiO_2. Namely, it is suggested that the direct integration between the SHJ bottom cell with FE structure and the PVK top cell using the ALD-TiO_2 as the ETL is acceptable without the optically disadvantageous recombination junction layer using a transparent conductive film. Furthermore, by employing KOH texture etching with the use of commercial additives, followed by HF/HNO_3 treatment, the μTex with a height of approximately <1 μm were introduced on the light-incident side surface. This led to an increase in I_{sc} due to the enhancement of long-wavelength sensitivity in the SHJ cell. In addition, the PVK single cell fabricated on these μTex formed on Si substrates provided the verifications that the PVK film effectively covered the μTex without any issues, as confirmed by the cross-sectional SEM images, and that the cell operated normally without short-circuiting.

ACKNOWLEDGEMENT

This work was supported by New Energy and Industrial Technology Development Organization (NEDO) under the project code JPNP20015.

REFERENCES

[1] H. S. Jung, G. S. Han, N-G. Park and M. J. Ko, Joule, 3 (2019) 1850

[2] G. Tang and F. Yan, Nano Today, 39 (2021) 101155

[3] H. Sai, H. Umishio and T. Matui, Solar RRL, 5 (2021) 2100634

[4] M. Yang, Presentation at 33rd International Photovoltaic Science and Engineering Conference (2022) MoO-31b-01

[5] W. Liu, Y. Liu, Z. Yang, C. Xu, X. Li, S. Huang, J. Shi, J. Du, A. Han, Y. Yang, G. Xu, J. Yu, J. Ling, J. Peng, L. Yu, B. Ding, Y. Gao, K. Jiang, Z. Li, Y. Yang, Z. Li, S. Lan, H. Fu, B. Fan, Y. Fu, W. He, F. Li, X. Song, Y. Zhou, Q. Shi, G. Wang, L. Guo, J. Kang, X. Yang, D. Li, Z. Wang, J. Li, S. Thoroddsen, R. Cai, F. Wei, G. Xing, Y. Xie, X. Liu, L. Zhang, F. Meng, Z. Di and Z. Liu, Nature 617 (2023) 717

[6] X. Y. Chin, D. Turkay, J. A. Steele, S. Tabean, S. Eswara, M. Mensi, P. Fiala, C. M. Wolff, A. Paracchino, K. Artuk, D. Jacobs, Q. Guesnay, F. Sahli, G. Andreatta, M. Boccard, Q. Jeangros and C. Ballif, Science, 381 (2023) 381

[7] Best Research-Cell Efficiency Chart | Photovoltaic Research | NREL

[8] https://www.linkedin.com/posts/erkan-aydin-1a02a92a_solar-photovoltaics-tandempv-activity 7069197510673620993-MD9j/

[9] R. Asadpour, R. V. K. Chavali, M. R. Khan and M. A. Alan, Appl. Phys. Lett. 106 (2015) 243902

[10]K. Saito, T. Takamura, Y. Ichikawa and M. Konagai, 36th European Photovoltaic Solar Energy Conference proceedings (2019) 522

[11]R. Santbergen, R. Mishima, T. Meguro, M. Hino, H. Uzu, J. Blanker, K. Yamamoto and M. Zeman, Optics Express, 24 (2016) A1289

[12]L. Mazzarella, M. Werth, K. Jager, M. Jost, L. Korte, S. Albrecht, R. Schlatmann and B. Stannowski, Optics Express, 26 (2018) A487

[13]C. McDonald, H. Sai, V. Svrcek, A. Kogo, T. Miyadera, T. N. Murakami, M. Chikamatsu, Y Yoshida and T. Matui, ACS Appl. Mater. Interfaces, 14 (2022) 33505

[14]W. Liu, L. Zhang, R. Chen, F. Meng, W. Guo, J. Bao and Z. Liu, J. Appl. Phys., 120 (2016) 175301

[15]S. Sriaman, S. Agarwal, E. S. Aydil and D. Maroudas, Nature, 418 (2002) 62

[16]T. Matui, M. Bivour, M. Hermle and H. Sai, ACS Appl. Mater. Interfaces, 12 (2020) 49777

[17]H. Shen, S. T. Omelchenco, D. A. Jacobs, S. Yalamanchili, Y. Wan, D. Yan, P. Phang, T. Duong, Y. Wu, Y. Yin, C. Samundsett, J. Peng, N. Wu, T. P. White, G. G. Andersson, N. S. Lewis and K. R. Catchpole, Sci. Adv. 4 (2018) eaau9711

[18]R. H. Cox and H. Strack, Solid State Electron, 10 (1967) 1213

[19]M. Rijnbach, R. J. E. Hueting, M. Stodolny, G. Janssen, J. Melskens and J. Schmitz, IEEE Transactions on Electron Devices, 67 (2020) 1757

[20]B. Wijngaarden, J. Yang and J. Schmitz, Solar Energy Materials and Solar Cells, 246 (2022) 111909

[21]B. Chen, Z. J. Yu, S. Manzoor, S. Wang, W. Weigand, Z. Yu, G. yang, Z. Ni, X. Dai, Z. C. Holman and J. Huang, Joule, 4 (2020) 850

[22]P. Tockhorn, J. Sutter, A. Cruz, P. Wagner, K. Jager, D. Yoo, F. Lang, M. Grischek, B. Li, J. Li, O. Shargaieva, E. Unger, A. Al-Ashouri, E. Kohnen, M. Stolterfoht, D. Neher, R. Schlatmann, B. Rech, B. Stannowski, S. Albrecht and C. Becker, Nature Nanotechnology, 17 (2022) 1214

[23]A. Harter, S. Mariotti, L. Korte, R. Schlatmann, S. Albrecht and B. Stannowski, Prog. Photovolt Res Appl. 31 (2023) 813

[24]F. Meng, J. Liu, L. Shen, J. Shi, A. Han, L. Zhang, Y. Liu, J. Yu, J. Zhang, R. Zhou and Z. Liu, Front. Energy, 11 (2017) 78

[25]A. Al-Ashouri, E. Kohnen, B. Li, A. Magomedov, H. Hempel, P. Caprioglio, J. A. Marquez, A. B. M. Vilches, E. Kasparavicius, J. A. Smith, N. Phung, D. Menzel, M. Grischeck, L. Kegelmann, D. Skroblin, C. Gollwitzer, T. Malinauskas, M. Jost, G. Matic, B. Rech, R. Schlatmann, M. Topic, L. Korte, A. Abate, B. Stannowski, D. Neher, M. Stolterfoht, T. Unold, V. Getautis, S. Albrecht, Science, 370 (2020) 1300

020010-001

40th European Photovoltaic Solar Energy Conference and Exhibition

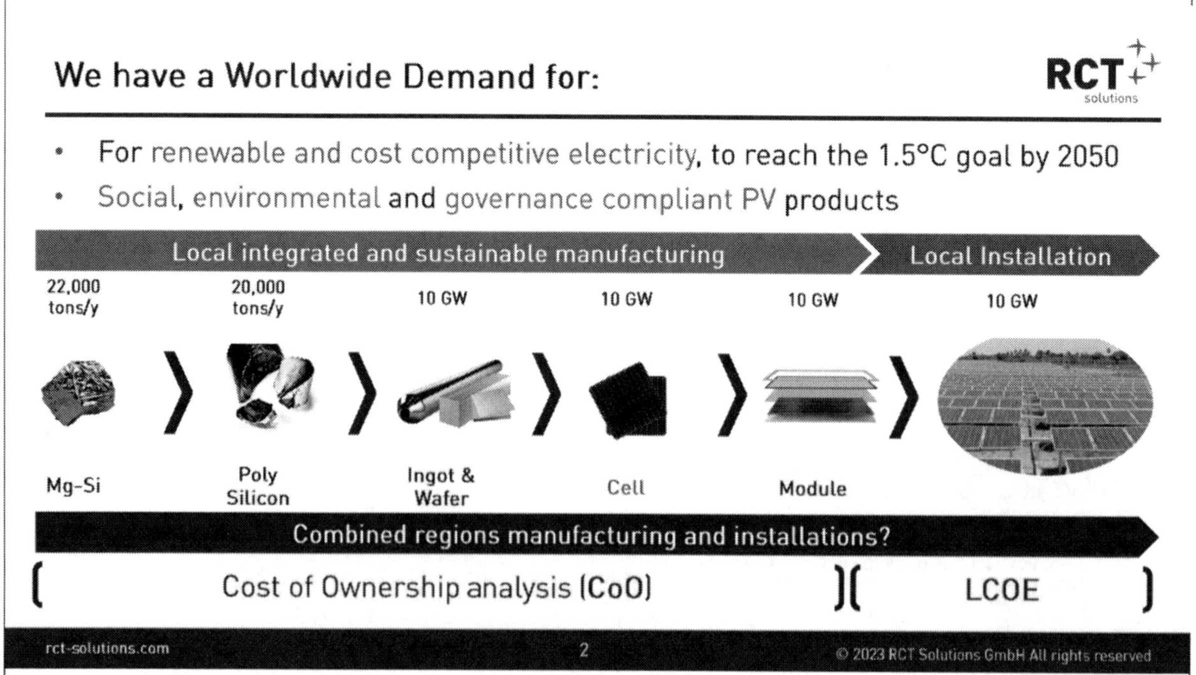

020010-002

What to select from the Solar Cell Zoo?

RCT solutions

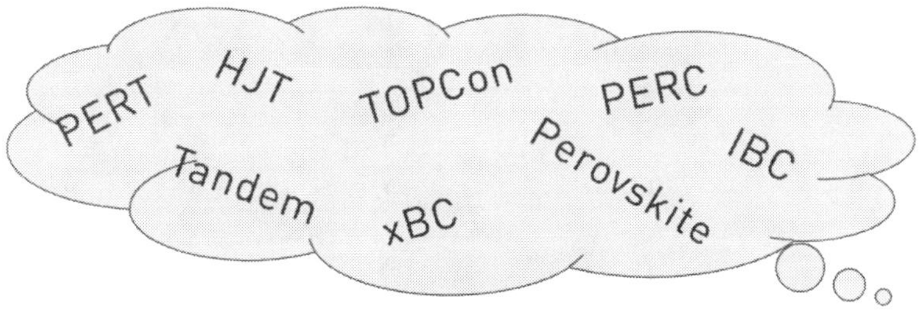

Economic analysis required!

1. Technologies
Locations
2. CoO
3. LCOE
Summary

The „Engine" of PV – the Solar Cell

RCT solutions

Technology	PERC SE	TOPCon SE	HJT	TBC
Description	Passivated Emitter and Rear Cell with Selective Emitter	Tunnel Oxide Passivated Contact Solar Cell with Selected Emitter	Silicon Hetero Junction Solar Cell Technology	TOPCon Back Contact Solar Cell with partial Cu metallization
Production Capacity	300+ GW	300+ GW	20+ GW	(5+ GW)
Cell Eff. (Fab-Theoretical)	23.2% → 24.5%*	24.5% → 27.1%*	25.0% → 28.5%*	25.5% → 29.1%**

Cell Structure

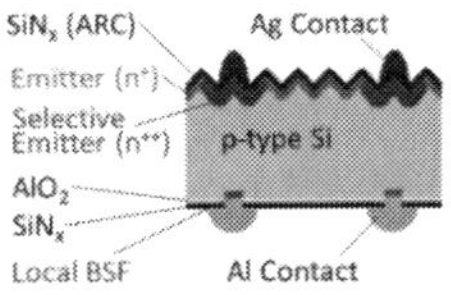

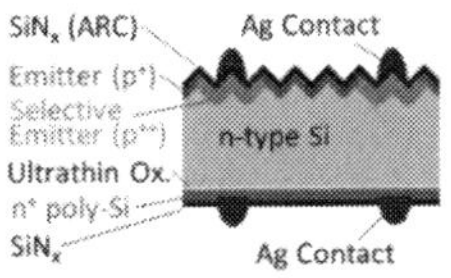

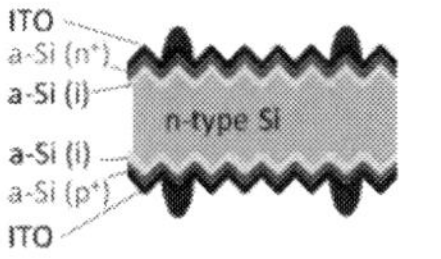

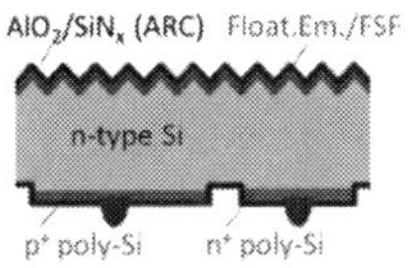

*- Solar Energy Materials & Solar Cells 231 (2021) 111291, **- Solar Energy Materials & Solar Cells 238 (2022) 111560*

020010-005

40th European Photovoltaic Solar Energy Conference and Exhibition

Selected Locations for CoO and LCOE

RCT solutions

rct-solutions.com 6

020010-006

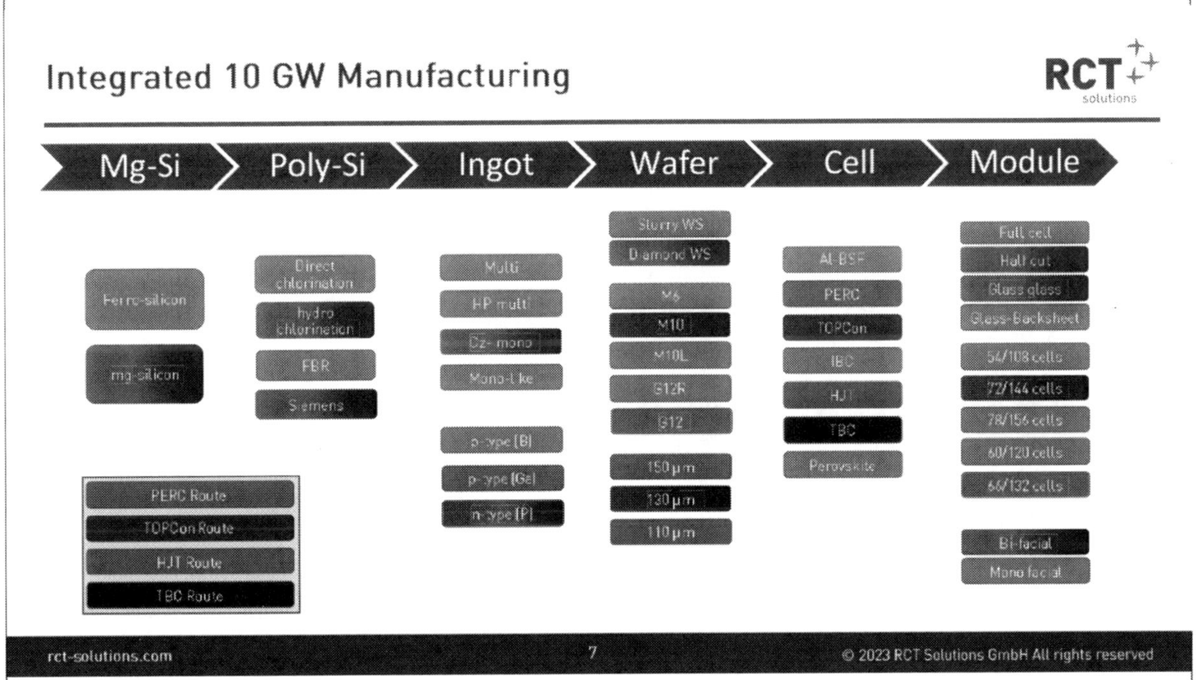

020010-007

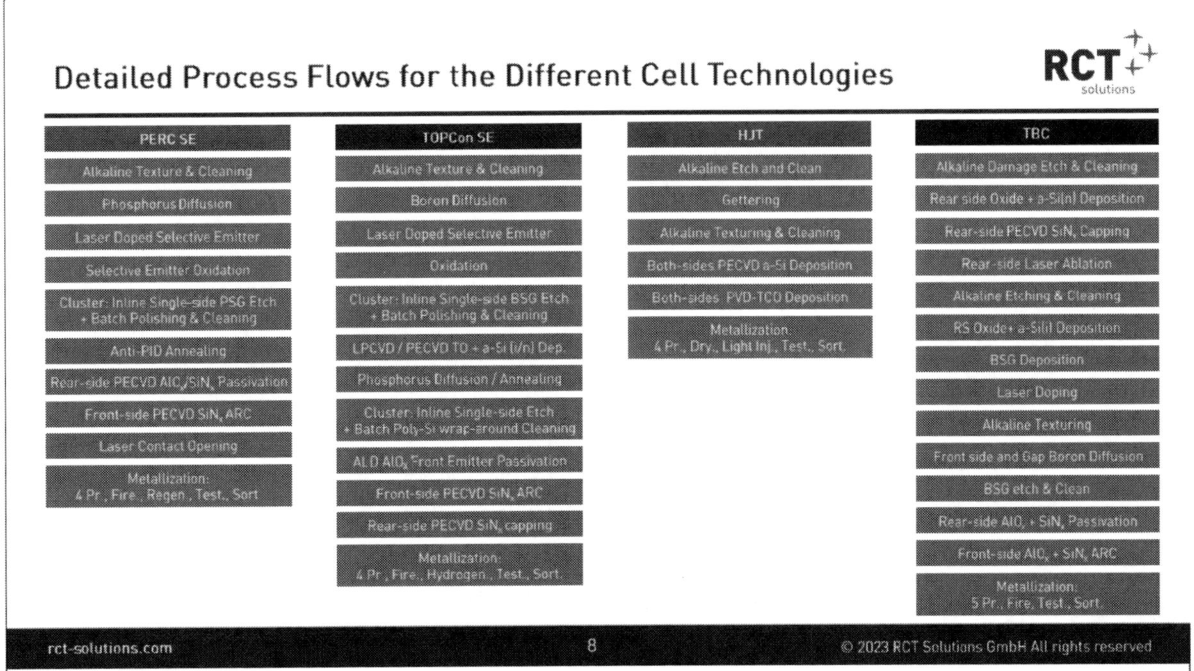

020010-008

Component Wise COO – 10 GW Factory Breakdown

RCT solutions

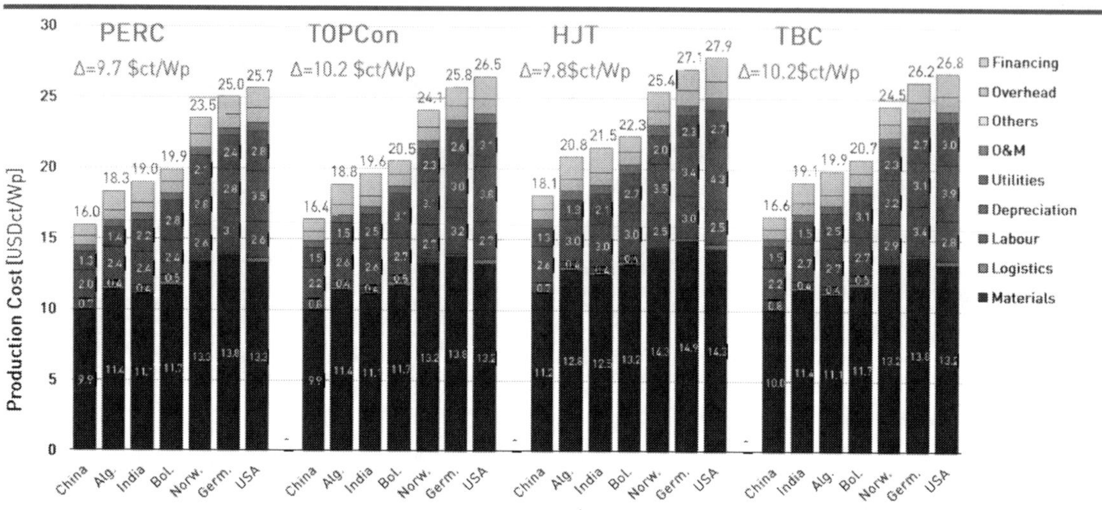

40th European Photovoltaic Solar Energy Conference and Exhibition

Location Wise COO – 10 GW Factory Breakdown

RCT solutions

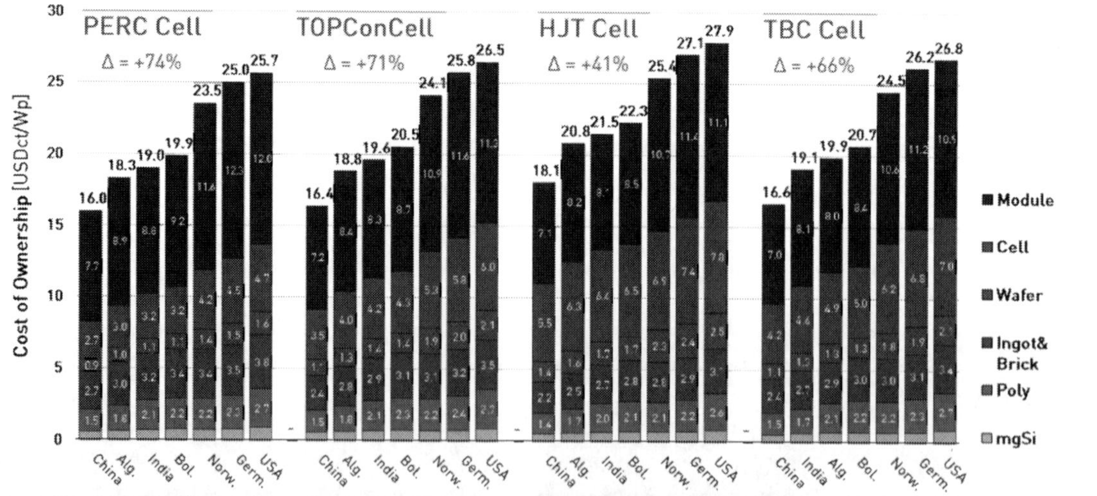

Technology **Related Cost** Differences

RCT solutions

Module – Nearly identical cost
- Differences caused by module power ($/Wp!) and module BOM

Cell – Responsible for most difference in CoO
- Ag content (highest for HJT)
- Equipment CAPEX (amount, type of equipment (thermal))

Poly-Si, Ingot and Wafer – Nearly identical cost
- Higher Poly-Si material quality for n-type is balanced by thinner wafer

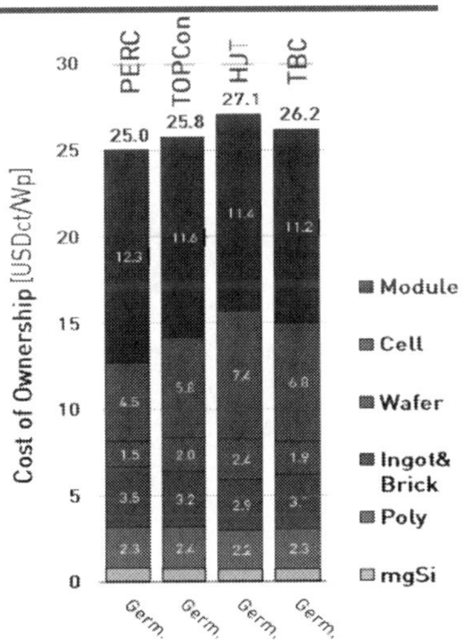

Location **Related Cost** Differences

RCT solutions

OPEX:

- Material pricing (up to 30%):
  - Shipment costs for imported materials
  - Economy of Scale in local markets
  - Import taxes (e.g. glass and other module materials in Germany)
- Labor cost (up to a factor of 8)
- Electricity cost (up to a factor of 2)

CAPEX/Financing:

- Equipment (up to 40%)
  - Country of origin
  - Import taxes (e.g. Chinese equipment in US)
- Building (> 100%)
  - Min. factor 2 between Chinese and European building

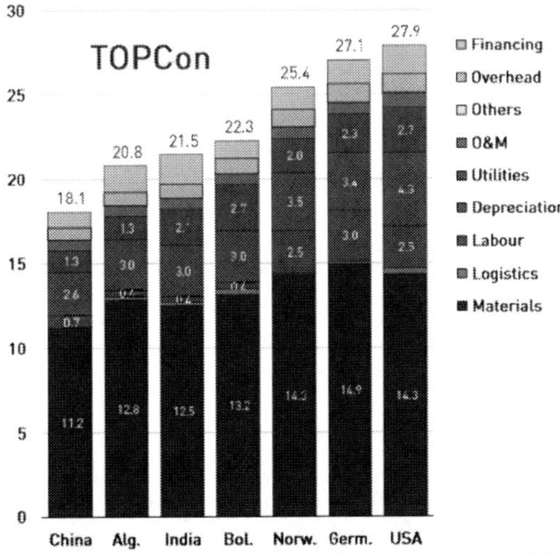

Selected Locations for CoO and LCOE

RCT solutions

Albedo for all locations will be set to 20%

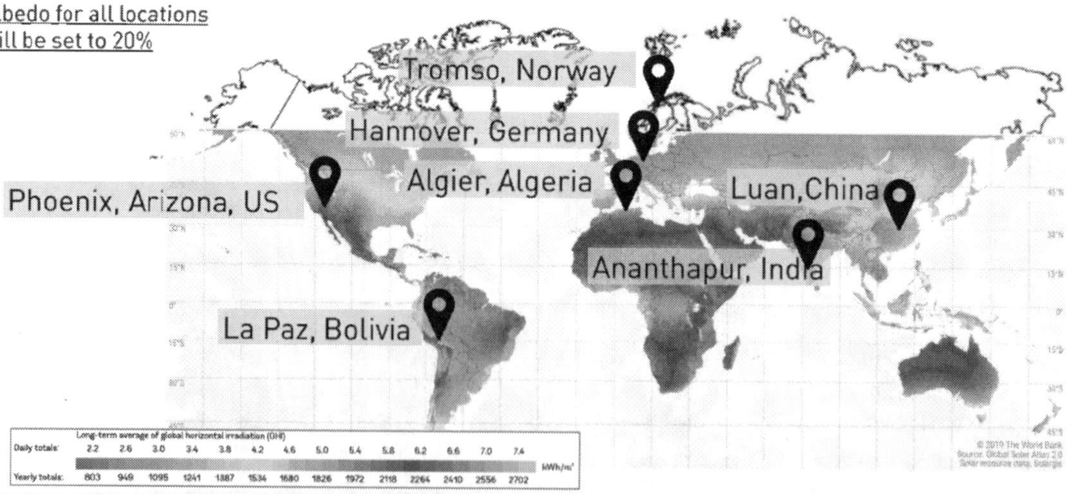

Levelized Cost of Electricity Impacted by Cell Parameters

RCT solutions

I_0: Investment costs

- Module Price
- BOS
  - BOS "power related"
  - BOS "area related" (higher module efficiency benefits)
- Land costs (higher module efficiency benefits)

$$LCOE = \frac{I_0 + \sum_{t=1}^{n} \frac{A_t}{(1+i)^t}}{\sum_{t=1}^{n} \frac{M_{t,el}}{(1+i)^t}}$$

M_{tel}: Energy Yield: produced amount in electricity in kWh

- <u>Prop. GHI (site dependent)</u>
- Thermal loss (Indicated by the temperature coefficient)
- Bifacial gain
- Degradation loss
- Others

At: Annual total cost (O&M)

- Minor for PV power plants (1%)

wacc: Weighted average cost of capital

- Taken worldwide with 6% or country dependent

n: Economic lifetime (taken as 25 years)

- Reliable product are required

Higher Temperature Coefficient Increases Energy Yield

RCT solutions

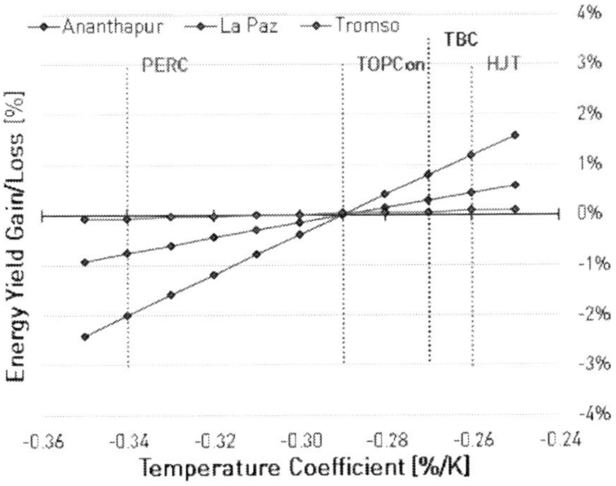

Simulation by PV-Syst and meteo data of Meteonorm 8.1, NREL, NASA and PVGIS

- Efficiency of PV modules reduces under high operating temperatures, this is indicated by temperature coefficient
- PERC cells have the lowest temp. coefficient, highest losses
- HJT cells have the highest temperature coefficient, lowest losses
- Yield gain between PERC and HJT:
  - 3.0% for Ananthapur, India
  - 1.3% for La Paz in Bolivia
  - Negligible for Tromso, Norway

Higher Bifaciality Factor Increases Energy Yield

RCT solutions

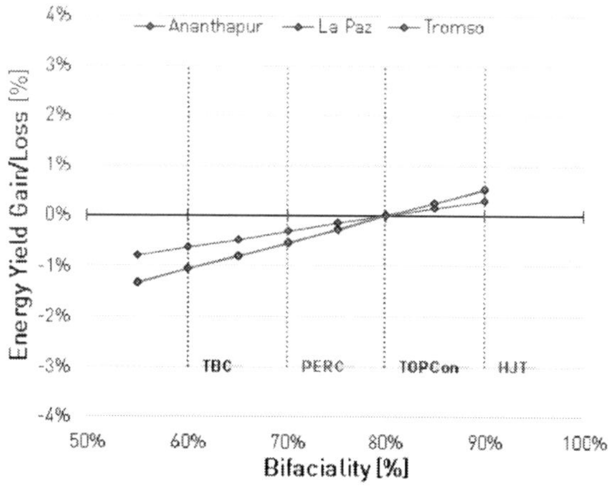

- HJT have the highest bifaciality, whereas TBC cells have the lowest, since both contacts are on rear-side
- Yield gain between PERC and HJT with an Albedo of 20%:
  - 1.9% for Ananthapur and Tromso
  - 1.1% for La Paz in Bolivia
  - Diffused light contribution as root cause

Result temperature coefficient and bifaciality:

- Norway: Bifaciality more important
- Bolivia: Balanced influence
- India: Temp. coefficient more important

Simulation by PV-Syst and meteo data of Meteonorm 8.1, NREL, NASA and PVGIS

rct-solutions.com 17

020010-017

KPI's of Different Cell Technologies for Energy Yield

RCT solutions

Technology	PERC SE	TOPCon SE	HJT	TBC
Wafer Size	G12 (440.96 cm²)	M10 (330.15cm²)	G12 half-cut (220.48 cm²)	M10 (330.15cm²)
Cell Efficiency	23.2%	**24.5%**	**25.0%**	25.5%
P_{mpp} Temperature Coefficient	-0.34 %/°C	**-0.29 %/°C**	-0.26 %/°C	**-0.27 %/°C**
Bifaciality +-5%	**70%**	**80%**	90%	60%
First Year Degradation	2%	1%	1%	1%
Second Year Ongoing Degradation	0.45%	0.40%	0.40%	0.40%
Comments	• No significant difference in low light intensities expected • Second year ongoing degradation mostly dependent on module packaging • Only PERC affected by LID			

rct-solutions.com 18

020010-018

System Costs for LCOE (Horizontal Single-Axis Tracker 10 MW)

RCT solutions

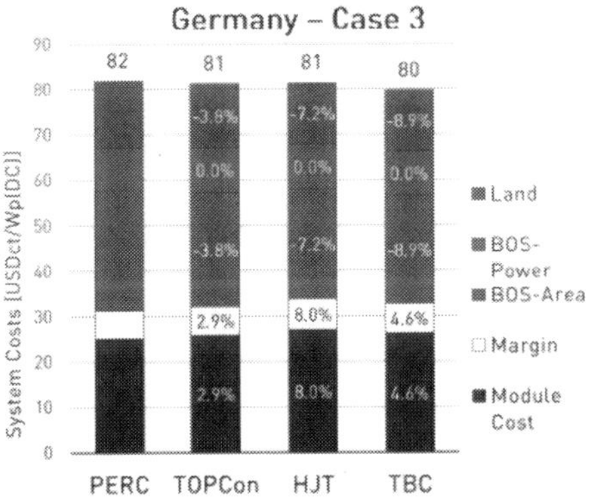

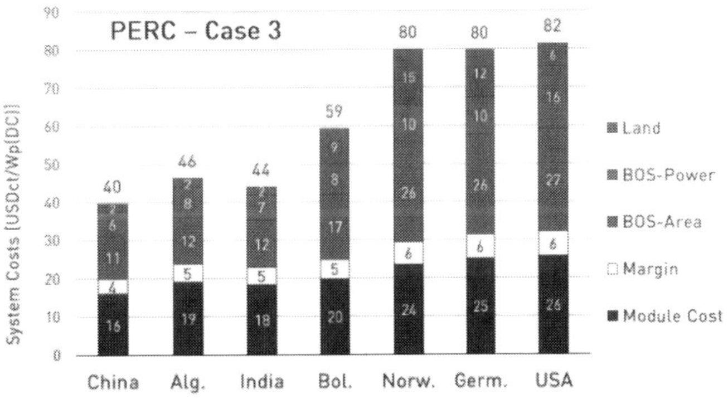

Installation Case 1: **Module from lowest-cost location + Generic BOS + Generic Financing**

Installation Case 2: **Module from lowest-cost location + Local BOS + Local Financing**

Installation Case 3: **Local produced Module + Local BOS + Local Financing**

40th European Photovoltaic Solar Energy Conference and Exhibition

Installation Case 1: **Module from Lowest-Cost Location + Generic BOS + Generic Financing** (HSAT 10 MW)

RCT solutions

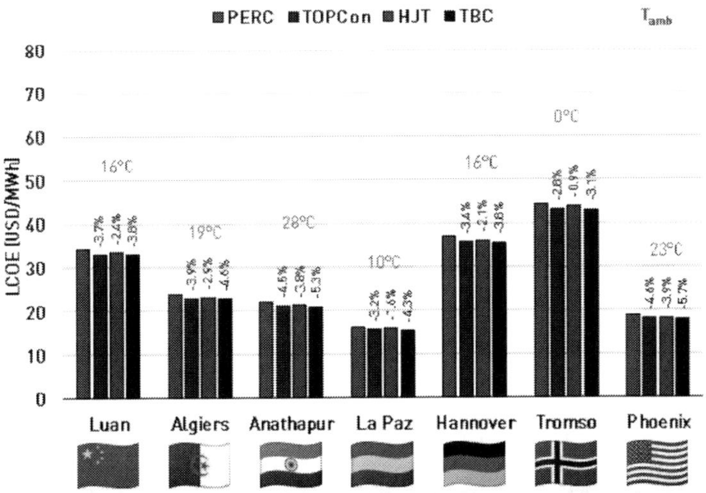

LCOE comparison (PERC as reference)

- Lowest LCOE in La Paz, Bolivia
- All n-type based system with lower LCOE
- TBC with lowest LCOE in all cases, followed by TOPCon as more established technology
- Increasing advantage with higher temperature

Generic Data: Land Costs 4 USD/m²; WACC 6.0% - Pitch optimized according economical feasible

Installation Case 2: **Module from Lowest-Cost Location + Local BOS + Local Financing** (HSAT 10 MW)

RCT solutions

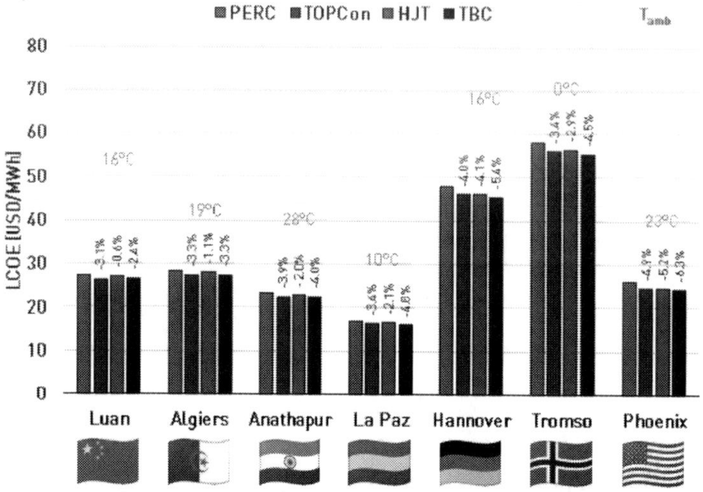

- Lowest LCOE in La Paz, Bolivia – no change in order
- Increase in countries with high BOS cost
- HJT undercuts or comes closer to in high BOS cost countries and benefits from lower wacc in e.g. USA and Germany

Installation Case 3: **Local produced Module + Local BOS + Local Financing** (HSAT 10MW)

RCT solutions

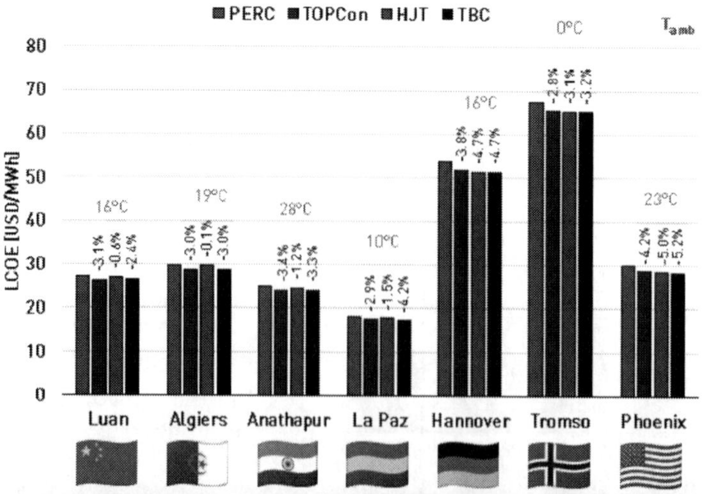

- Lowest LCOE still in La Paz, Bolivia
- Increase in LCOE differences in countries with highest module manufacturing cost (Europe, US)
- HJT and TBC outperforms in high system cost countries
- TOPCon and TBC lead in the lower system cost countries
- Also influenced highly by wacc

Case 2 and 3: Local Manufacturing of Modules vs. Importing from Lowest-Cost Location (HSAT)

RCT solutions

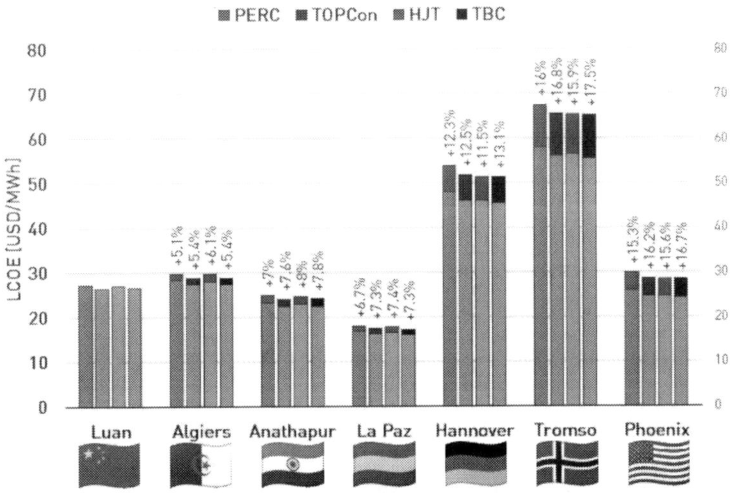

What are the impacts of higher local module manufacturing costs (up to 50%) on LCOE?

- In low system cost countries, around 5-8 % higher LCOE
- In Germany 12 -13% higher LCOE
- 15 -17% higher LCOE in Norway and US

Government and policy support can be one measure to reduce the financing cost to gap the difference in LCOE

40th European Photovoltaic Solar Energy Conference and Exhibition

Summary

RCT solutions

10 GW Manufacturing of solar modules at different locations: CoO $\Delta_{avg.}$ = 10ct/Wp (US → China)

- Location differences - material, labor, electricity and wacc

10 GW Manufacturing of four different solar cell technologies: CoO $\Delta_{avg.}$ = 10% (PERC → HJT)

- Cell technology differences - Ag consumption and equipment CAPEX

Differences in LCOE for 10 MW Project:

- Energy yield – Bifaciality and Temperature coefficient up to 4% gain compared to PERC
- n-type solar cell structures, with differences in-between of 2%, show better LCOE than out phasing PERC technology
- Low System Cost + High wacc → TOPCon and TBC lowest LCOE (e.g. India and Algeria 3-4% lower)
- High System Cost + Low wacc → TBC and HJT lowest LCOE (e.g. Europe and US 3-5% lower)
- TBC applicable on both markets

PV enables lowest LCOE at serval location options worldwide to enable clean electricity/hydrogen production

Does it make sense to manufacture and install at the same Location?

50% - 60% higher cost → 15% to 17% higher LCOE

RCT solutions

A competitive production cost structure and high energy yield can reduce the impact between manufacturing in the lowest cost and in the own country on LCOE to a minimum of 5-7%

Institutional support, such as financing support can counterbalance the manufacturing cost.

Investing to ensure :
Environment goals,
Ethnical and sustainable supply
Resilience trough overall margin capturing
Local job and value creation and
Self-reliance on energy need.

ENVIRONMENT **SOCIAL** **GOVERNANCE**

rct-solutions.com 25

020010-025

40th European Photovoltaic Solar Energy Conference and Exhibition

RCT Solutions GmbH
Line-Eid-Strasse 1
D-78467 Konstanz, Germany

Phone +49 7531 58470 12
info@rct-solutions.com
http://www.rct-solutions.com

Regd. HRB 708952,
Executive Board: Dr. Peter Fath

Thank you

Authors Contact: julian.reichle@rct-solutions.com

020010-026

THE UNIVERSITY OF MELBOURNE

UNSW

Helmholtz Zentrum Berlin

TU/e

EU PVSEC 2023

Electron contact interlayers for low-temperature-processed crystalline silicon solar cells

Speaker: J. Jesús Ibarra Michel

19/09/2023

Jesus Ibarra Michel et al., Prog. Photovolt. (2023), Under Review

Intro | Results | Conc

Introduction and motivation

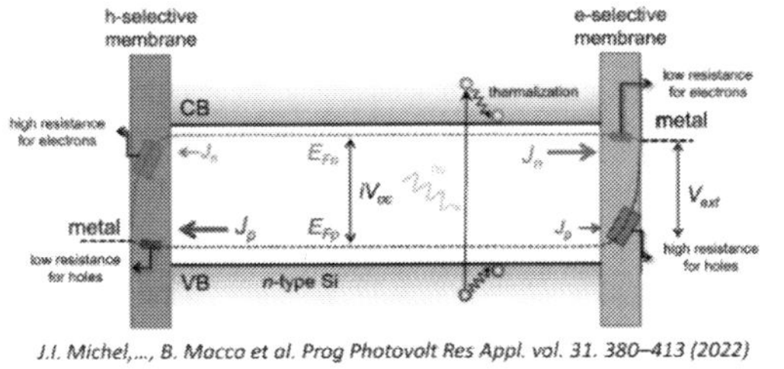

J.I. Michel,..., B. Macco et al. Prog Photovolt Res Appl. vol. 31. 380–413 (2022)

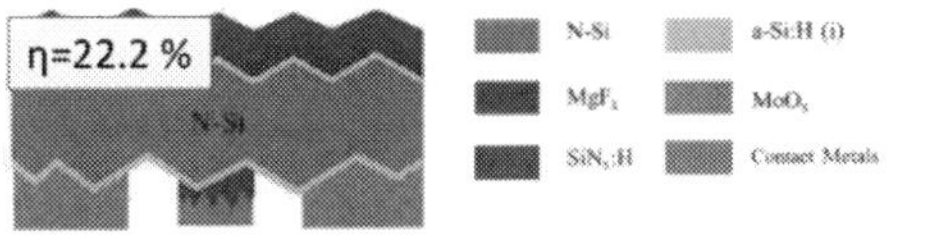

AIP Conf. Proc. 1999, 040025 (2018)

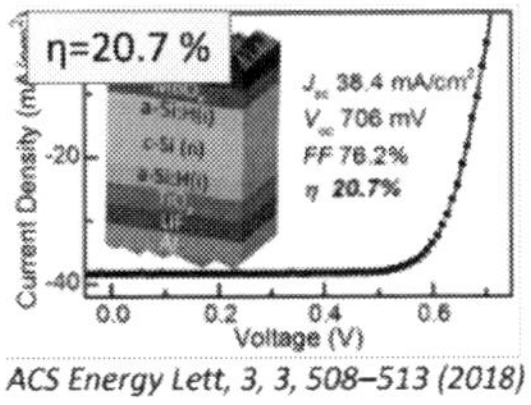

ACS Energy Lett, 3, 3, 508–513 (2018)

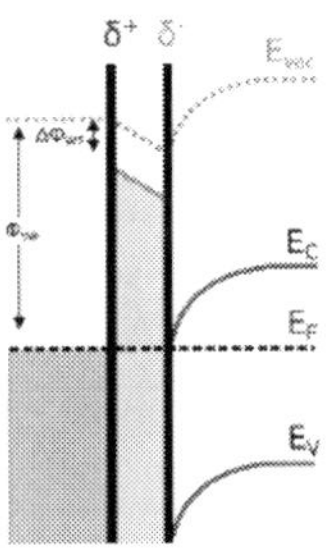

- Thin interlayers have been used as electron-selective films
- Physical origin:
  - Fermi-level depinning
  - Interface dipole (charge exchange and Pauli repulsion)
- Downwards band bending at the c-Si surface will promote electron accumulation

2

Intro Results Conc

Results and discussion

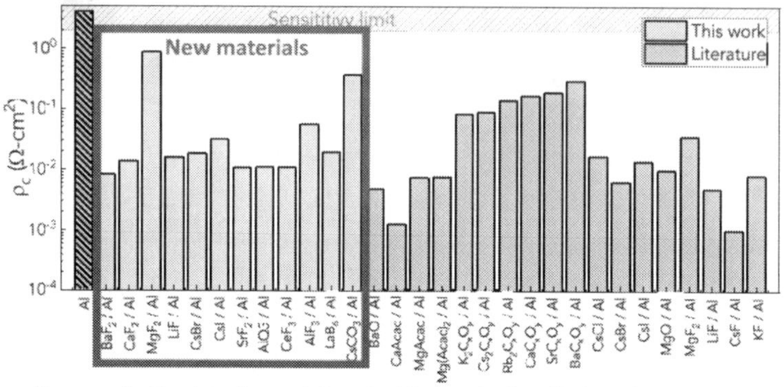

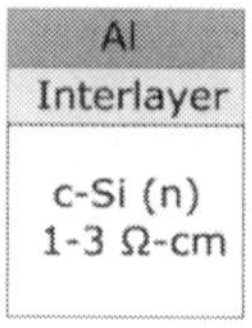

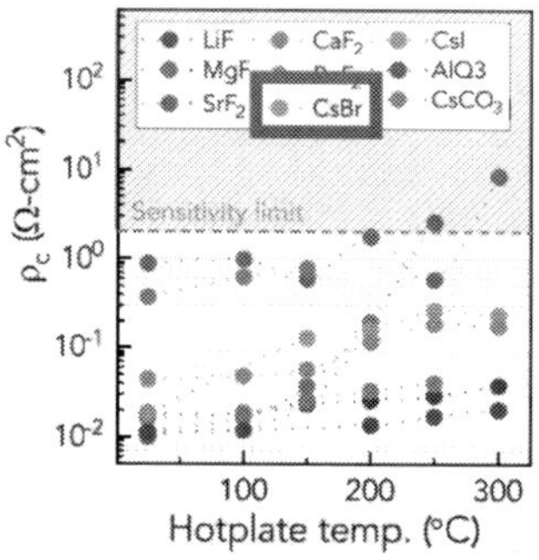

- New e-selective interlayers (<5 nm) added to the 'pool' of e-selective contacts on c-Si.
- Some of these materials are less thermally stable than others.
  - Remember back-end processing (150-200 °C)
- Set focus on CsBr (improved performance with pass. interlayers)

3

40th European Photovoltaic Solar Energy Conference and Exhibition

Intro Results Conc

Results and discussion

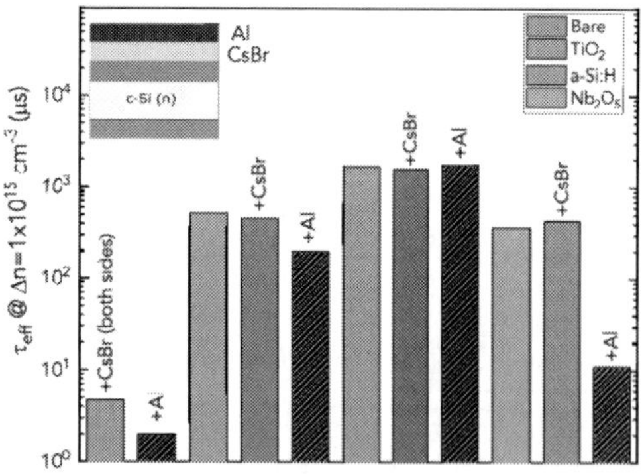

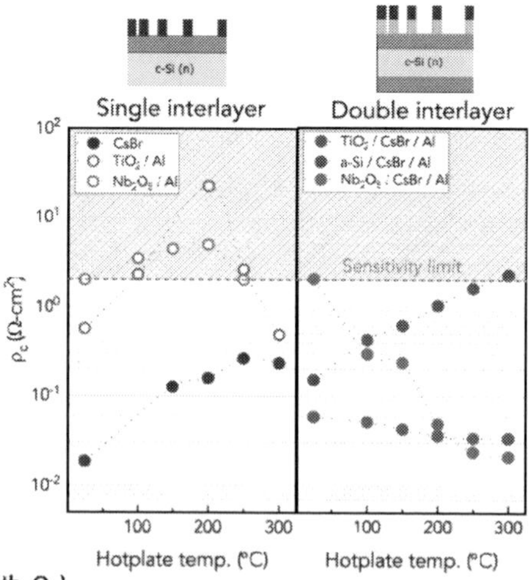

- Different passivating interlayers were tested (TiO_2, a-Si:H, Nb_2O_5)
- QSSPC measurement at each deposition step
  - Metal deposition causes surface passivation damage
- Specific contact resistivity as a function of hotplate anneal
  - Combining CsBr with either TiO_2 or Nb_2O_5 yields a ρ_c low and stable

4

Intro | Results | Conc

Results and discussion

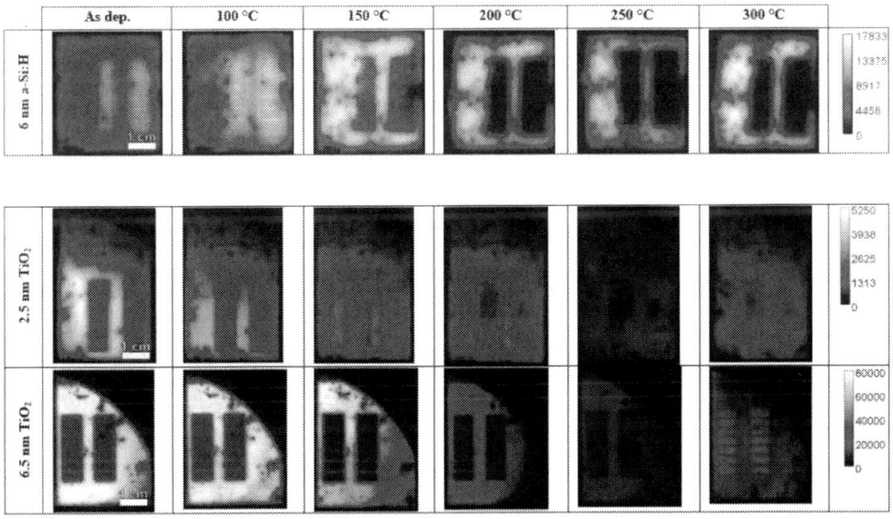

- Photoluminescence measurements revealed metalized regions with higher SRH recombination
- For the case of TiO_2, improvement in metal regions was observed at temperature >250 °C.

5

Intro | Results | Conc

Results and discussion

- Annealing at a temperature higher than 250 °C yields a modest improvement in τ_{eff} and ρ_c
- Ultra-violet photoelectron spectroscopy (UPS) showed a lower secondary-electron cutoff after vacuum annealing at 250 °C

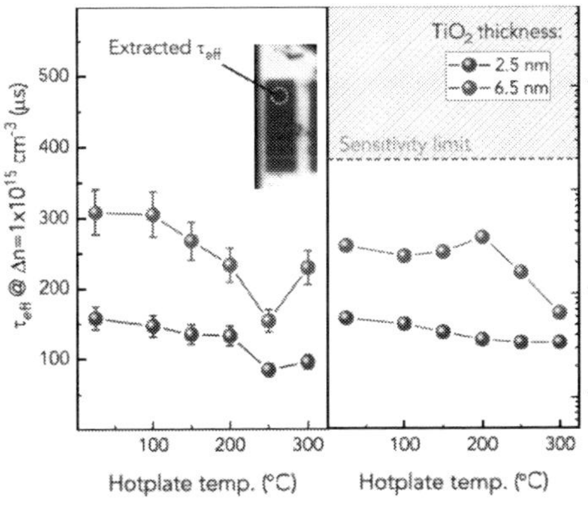

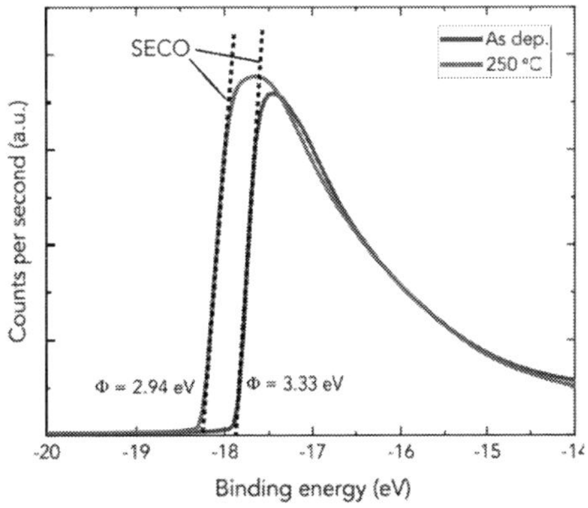

6

Intro Results Conc

Results and discussion

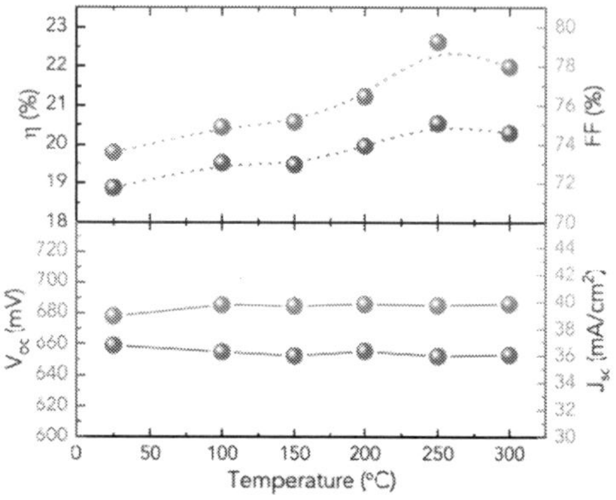

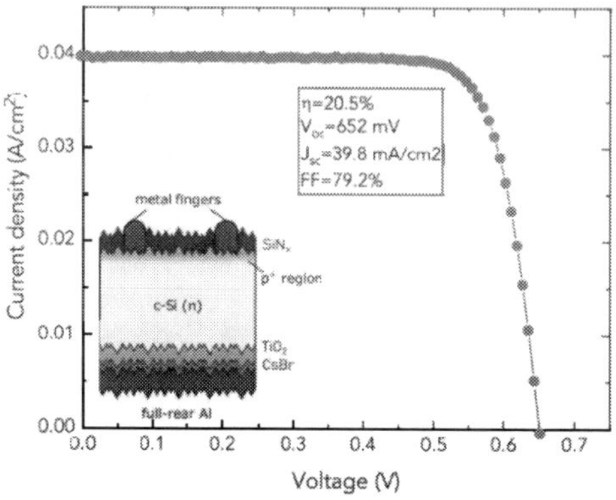

- Test device showed an efficiency of 20.5%, after annealing at 250 °C.
- The Voc and Jsc showed stability up to 300 °C
- The improvement in efficiency comes clearly from an improvement in FF
  - Front fingers sintering
  - Slight Improvement of ρ_c at the rear

7

020011-007

Intro Results Conc

Conclusions

- We have demonstrated new materials as electron-selective contacts, adding to the pool of metal compounds replacing n^+ doped Si
- Experimentation with CsBr in combination with other passivating layers shows its promising electrical performance in τ_{eff} and ρ_c structures
- Best performance was observed on our TiO_2/CsBr/Al after annealing at 250 °C. A device fabricated with a full-area rear contact based on TiO_2/CsBr has yielded an efficiency of 20.5% (with promising thermal stability)
- **Future work?** Lookout for a suitable TCO compatible with these electron contacts and, in general, still on the quest of finding a material that can simultaneously passivate the c-Si surface and provide e-selectivity.

8

020011-008

¡Gracias!

Process optimizations for edge passivation and high efficiency shingle heterojunction cells compatible with industry

Mickaël ALBARIC
Benoit MARTEL
Samuel HARRISON
Saliha EDOUARD

INSTITUT NATIONAL DE L'ENERGIE SOLAIRE

EU PVSEC 2023 – Lisbon – 1BO.1.6 – 19/09/2023

40th European Photovoltaic Solar Energy Conference and Exhibition

Outline

1. **Context and objectives of the study**
2. **45°oriented wafers in industrial environment**
3. **Edge passivation**
   * Reduction of the thickness of the edges passivation layer
   * Effect of the Qtime
4. **Conclusion and Perpectives**

40th European Photovoltaic Solar Energy Conference and Exhibition

1. Context

020012-003

40th European Photovoltaic Solar Energy Conference and Exhibition

Half cells module from Longi

Shingled solar module (CEA - 2022)

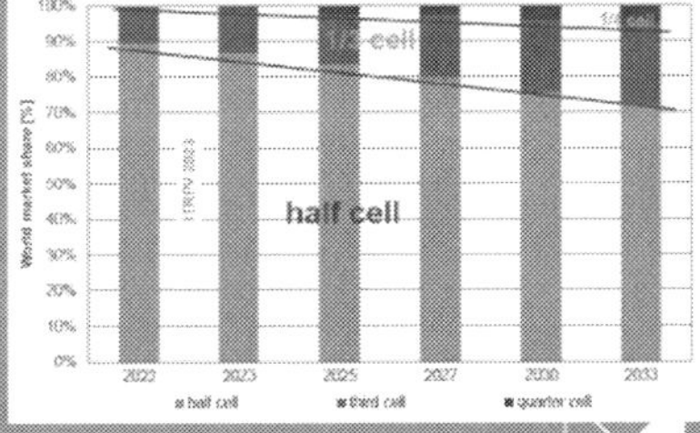

ITRPV 2023: full cell disappeared for size >= M10

Context

Evolution of the PV panel:

- Mainstream : half cells solar modules architecture
- Increase PV panel power (+10%)[1]: reducing the resistive losses
- Shingled solar modules are also proposed by manufacturers – High performance & improved aesthetics [2]

[1] *Mittag, et al. (2017), 44th IEEE PVSC, Washington, D.C.*

[2] *Harrison et al. (2022), 39th European Photovoltaic Solar Energy Conference and Exhibition*

020012-004

Half cells module from Longi

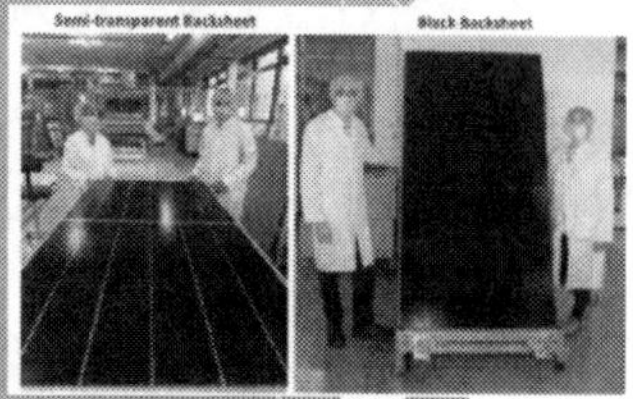

Shingled solar module (CEA - 2022)

Prototype kit to "solarize" electric vehicles (CEA - 2022)

Colored PV modules BIPV (CEA - 2020)

Context

Evolution of the PV panel:

- Mainstream : half cells solar modules architecture
- Increase PV panel power (+10%)[1]: reducing the resistive losses
- Shingled solar modules are also proposed by manufacturers – High performance & improved aesthetics [2]
- Good match between new demands for PV (BIPV, VIPV) and minimum technology alteration

Drawbacks:

- Cell cutting → Cell efficiency losses in the range of ~0.2%$_{abs.}$ for half cells and ~1%$_{abs.}$ for shingles[2]
- Expected electrical gain can be partially offset by the cell cutting ...

[1] *Mittag, et al. (2017), 44th IEEE PVSC, Washington, D.C.*

[2] *Harrison et al. (2022), 39th European Photovoltaic Solar Energy Conference and Exhibition*

[3] *Giglia, et al. (2020), 37th European Photovoltaic Solar Energy Conference and Exhibition*

020012-005

Context: cut edge

- **Smooth cut edge mandatory for repassivation**

020012-006

Context: cut edge

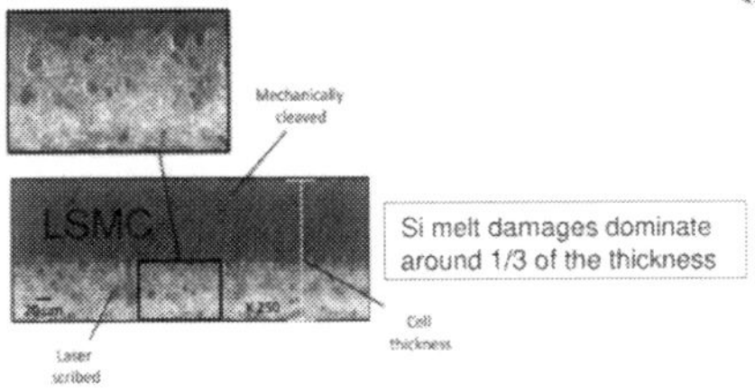

SEM images of cell edge separated by conventional laser scribe

- **Smooth cut edge mandatory for repassivation**
- "Conventional" cell cutting technologies [4, 5]
  - Laser scribe & mechanical cleaving (LSMC)

[4] *Baliozian et al. (2020), IEEE journal of photovoltaics vol. 10, no. 2.*
[5] *Lohmüller et al. (2023), PV magazine webinar*

40th European Photovoltaic Solar Energy Conference and Exhibition

Context: cut edge

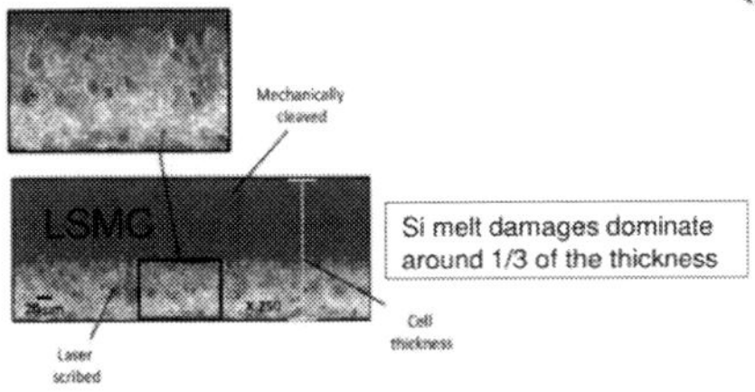

SEM images of cell edge separated by conventional laser scribe

- **Smooth cut edge mandatory for repassivation**
- "Conventional" cell cutting technologies [4, 5]
  - Laser scribe & mechanical cleaving (LSMC)
- "Low-damage" cell cutting technologies [4, 5]
  - Themal laser separtion (TLS[6])
  - Laser direct cleaving (LDC[7])
  - 45° rotated silicon brick for wafering (CEA - INES)

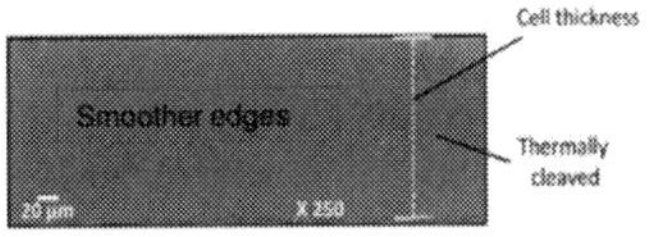

SEM images of cell edge separated by laser scribe + thermomechanical cleave (TLS)

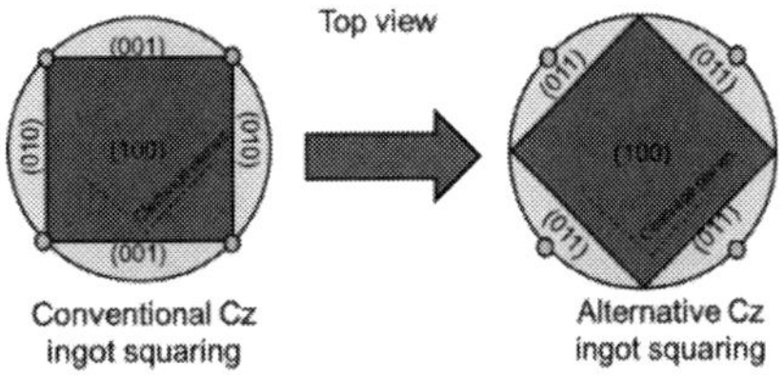

[4] *Baliozian et al. (2020), IEEE journal of photovoltaics vol. 10, no. 2.*
[5] *Lohmüller et al. (2023), PV magazine webinar*
[6] *3D Micromac*
[7] *Innolas*

Context: cut edge

- **Smooth cut edge mandatory for repassivation**
- "Conventional" cell cutting technologies [4, 5]
  - Laser scribe & mechanical cleaving (LSMC)
- "Low-damage" cell cutting technologies [4, 5]
  - Themal laser separtion (TLS[6])
  - Laser direct cleaving (LDC[7])
  - 45° rotated silicon brick for wafering (CEA - INES)
    - Simple & cost effective
    - Purely mechanical cleavage
    - No modification of the cell process
    - No laser needed

Is the 45° rotated wafers compliant with production lines?

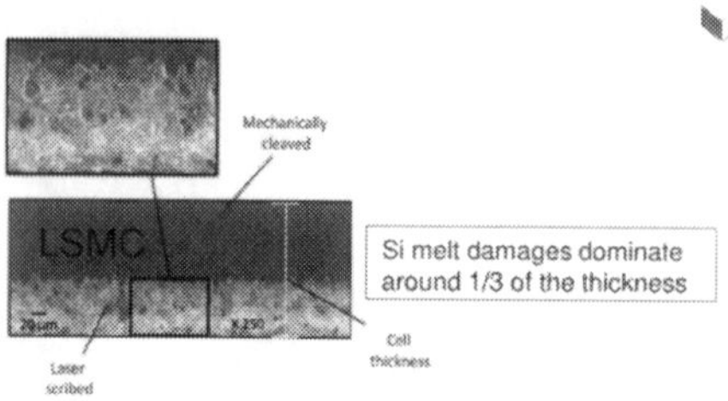

SEM images of cell edge separated by conventional laser scribe

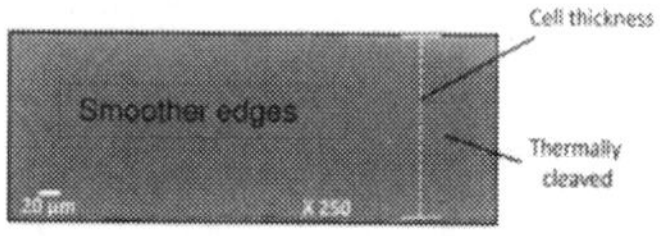

SEM images of cell edge separated by laser scribe + thermomechanical cleave (TLS)

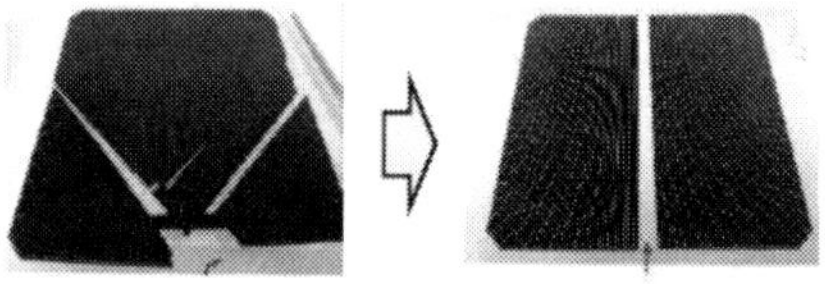
Mechanical Cleavage plane aligned with metal pattern!

[4] Baliozian et al. (2020), IEEE journal of photovoltaics vol. 10, no. 2.
[5] Lohmüller et al. (2023), PV magazine webinar
[6] 3D Micromac
[7] Innolas

Context : Edge passivation solutions

- **Process the cell after wafer separation: not investigated in this presentation**
- **Edge passivation after cell separation @ "Labscale"**
  - Teams have successfully developed edge passivation solutions based on ALD/AlO_x
  - Excellent results observed on SHJ[8,9,12] and TOPCon[10,11]
- **Industrial interests**
  - Singulus[13]: Generis PET tool

[8] S. Harrison et al., EU-PVSEC, 2021
[9] B.Martel et al, Solar Energy Materials & Solar Cells 250, 2023
[10] F. Dhainaut et al., EPJ Photovoltaics 14, 2023
[11] P. Baliozian et al., IEEE J. Photovoltaics, 2020
[12] A. Münzer et al., IEEE J. Photovoltaics, 2021
[13] Singulus, press release, 2023

Context : Edge passivation solutions

- **Process the cell after wafer separation: not investigated in this presentation**
- **Edge passivation after cell separation @ "Labscale"**
  - Teams have successfully developed edge passivation solutions based on ALD/AlO_x
  - Excellent results observed on SHJ[8,9,12] and TOPCon[10,11]
- **Industrial interests**
  - Singulus[13]: Generis PET tool
- **Edge passivation protocol for SHJ applied at CEA to 45° rotated wafers**

ALD deposition on full wafer

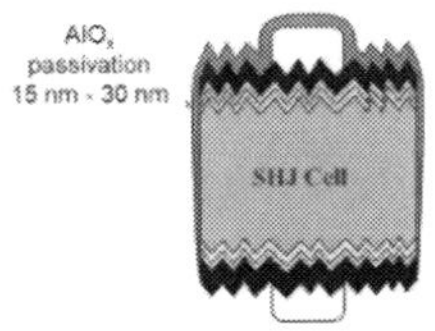

Light Soaking Anneal of both: Cell & AlO_x layer

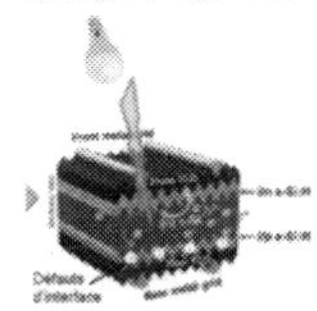

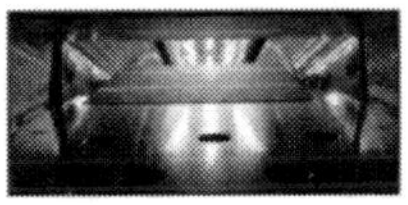

AMAT oral presentation 1BO.4.4 Tuesday afternoon

- High Irradiance intensity
- Controlled light wavelength (diodes)
- Controlled T° on the cell (chuck)

Excellent edge repassivation achieved on cut-cells → Up to ~80% of efficiency recovery

[8] S. Harrison et al., EU-PVSEC, 2021
[9] B. Martel et al, Solar Energy Materials & Solar Cells 250, 2023
[10] F. Dhainaut et al., EPJ Photovoltaics 14, 2023
[11] P. Baliozian et al., IEEE J. Photovoltaics, 2020
[12] A. Münzer et al., IEEE J. Photovoltaics, 2021
[13] Singulus, press release, 2023

Questions to be answered to assess the potential for a large-scale use:

- 45° rotated wafers in industrial environment ?
- Throughput and cost: **Minimum passivation layer thickness ?**
- Flexibility: **Qtime between the separation and the passivation step ?**

2. 45°wafer in industrial environment

45° rotaded wafers: High performance & industry compatibility

Integration at CEA on the SHJ industrial pilot line

Compare to ref. wafers, 45° rotated wafers:

- Similar high efficiencies
- Similar breakage rate

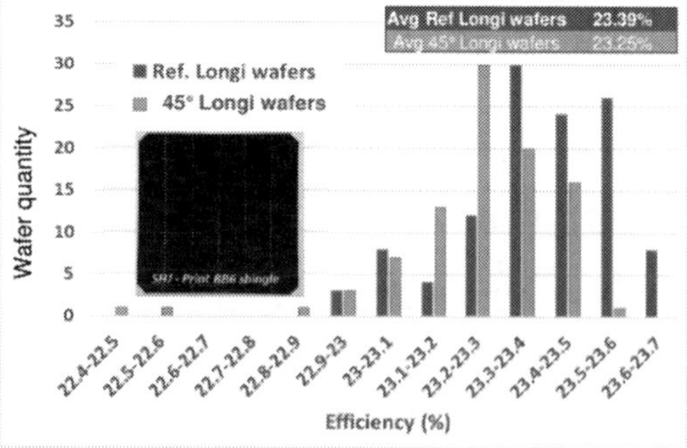

Albaric et al, CSSC-11 (2022)

SHJ industrial pilot line - CEA

Observed shingle cut losses

Separation path	FF losses (%)	Efficiency losses (%)
Laser	2.1	1.0
TLS	2.5	1.0
45° wafers	2.6	1.1

Edge passivation solutions needed to recover the losses !

45° rotaded wafers: High performance & industry compatibility

kalyon·PV IBC4EU

- B-Doped Cz ingot sent to CEA for 45° squaring (45DR) and grinding at G1 scale
- Wafering and PERC cell processing on the Kalyon production line
  - Two groups of wafers used: 45DR and standard (STD) for reference (~200 wafers each)
  - 175 µm thick wafers (thicker wafers for a first trial)
- Promising results for the 45DR
  - No problem observed during wafering and cell processing
  - Similar behavior of the 45DR compare to the standard

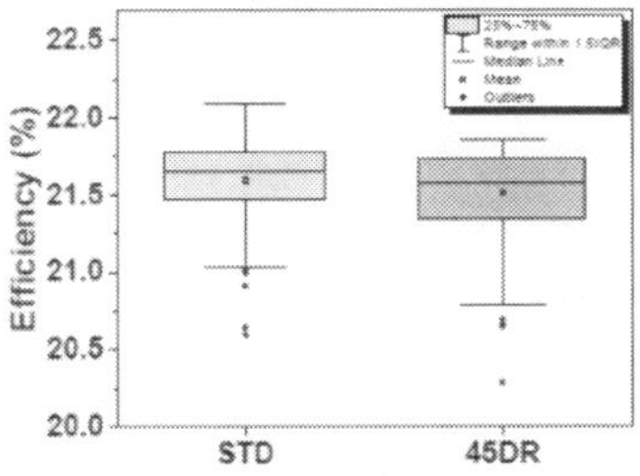

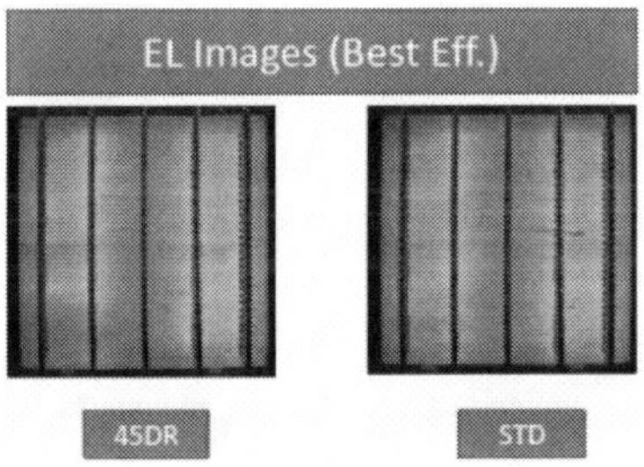

cea

3. Edge passivation

cea

Reduction of the thickness of the layer

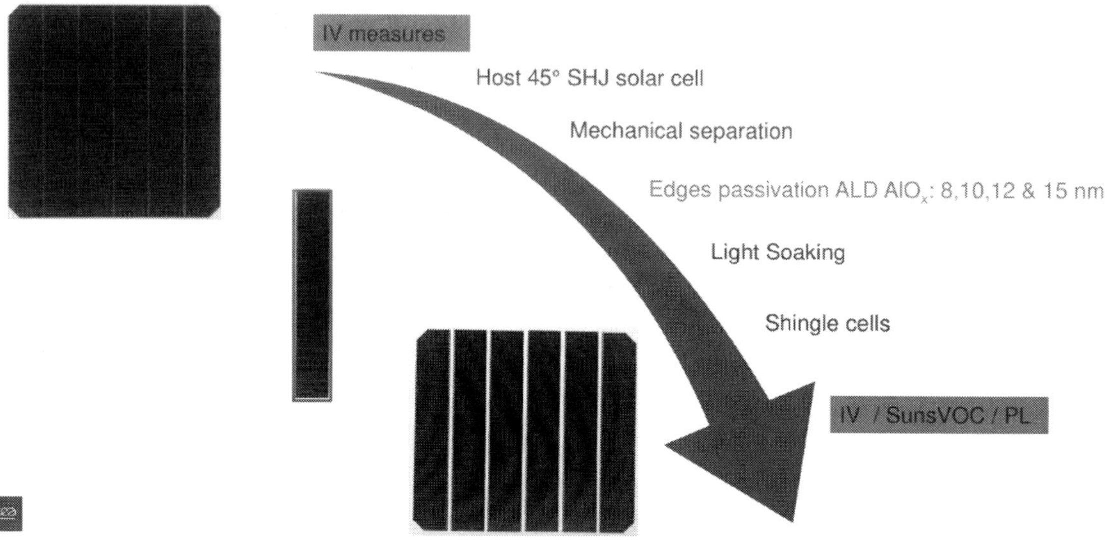

40th European Photovoltaic Solar Energy Conference and Exhibition

Reduction of the thickness of the layer

- Host SHJ cells @ 22.8% - 23%
- Separation → 1.1%$_{abs}$ efficiency losses
- 4 host SHJ cells per configuration → 24 shingles per configuration

Reduction of the thickness of the layer

- Host SHJ cells @ 22.8% - 23%
- Separation → 1.1%$_{abs}$ efficiency losses
- 4 host SHJ cells per configuration → 24 shingles per configuration

- Variability results observed due to sample handling issues

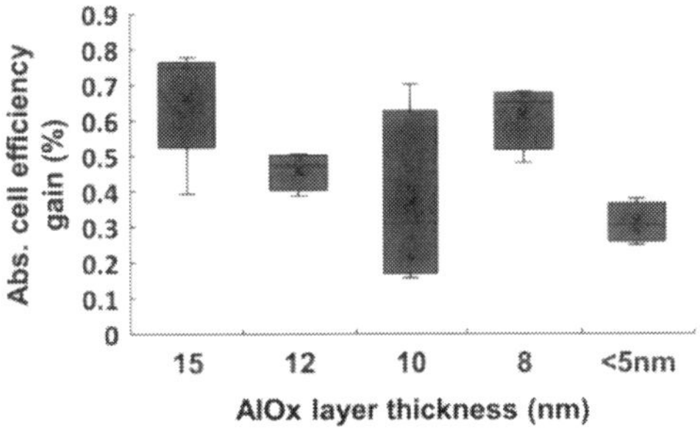

- Cell efficiency gain observed after edge passivation whatever the AlO_x layer thickness
- No significant difference in maximum cell efficiency gains for thicknesses between 8 nm and 15 nm

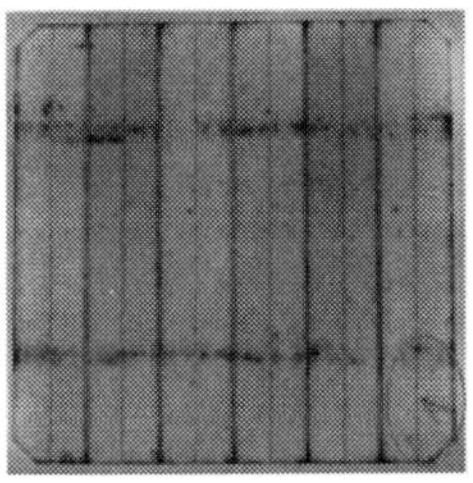

PL : handling crack

Qtime effect

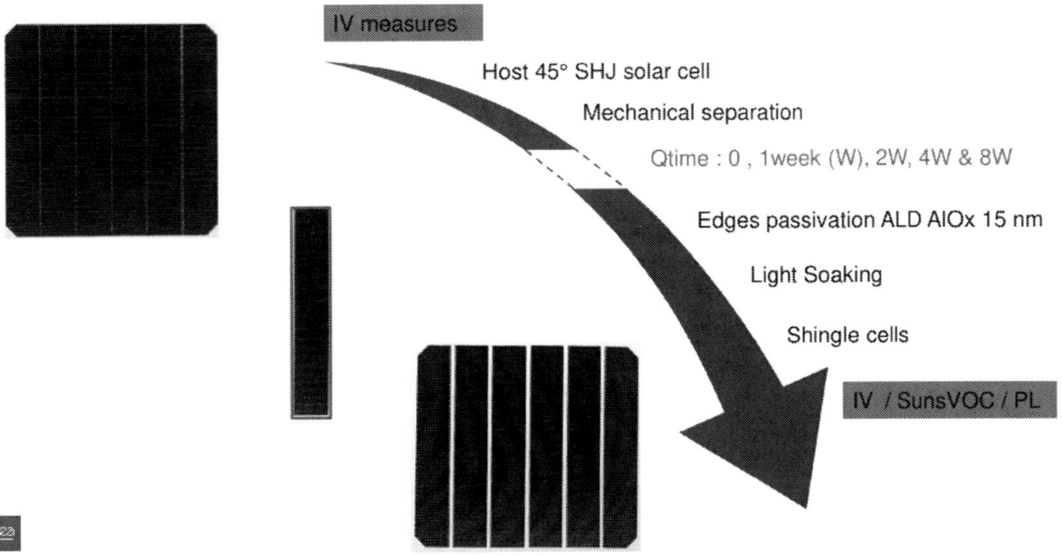

Qtime effect

- Host SHJ cells @ 22.9%
- Separation → 1.2%$_{abs}$ efficiency losses
- 2 host SHJ cells per configuration → 12 shingles per configuration

40th European Photovoltaic Solar Energy Conference and Exhibition

Qtime effect

- Host SHJ cells @ 22.9%
- Separation → 1.2%$_{abs}$ efficiency losses
- 2 host SHJ cells per configuration → 12 shingles per configuration

- Variability of the edge passivation process
- Limited efficiency losses after edges passivation : 0.1% - 0.3%

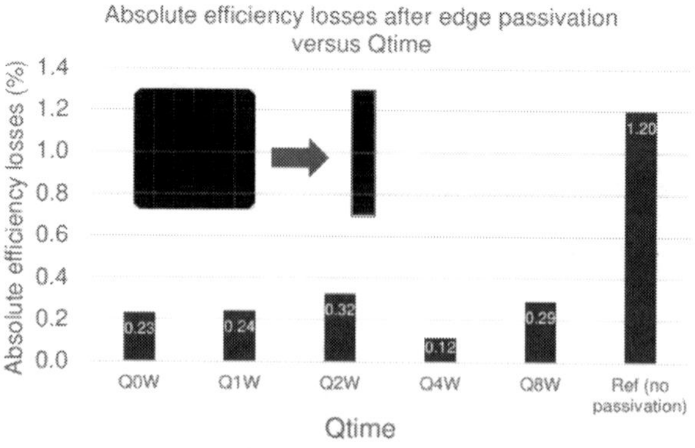

Qtime effect

- Host SHJ cells @ 22.9%
- Separation → 1.2%$_{abs}$ efficiency losses
- 2 host SHJ cells per configuration → 12 shingles per configuration

- Variability of the edge passivation process
- Limited efficiency losses after edges passivation : 0.1% - 0.3%

- No noticeable effect of the Qtime between cleavage and passivation up to 8 weeks

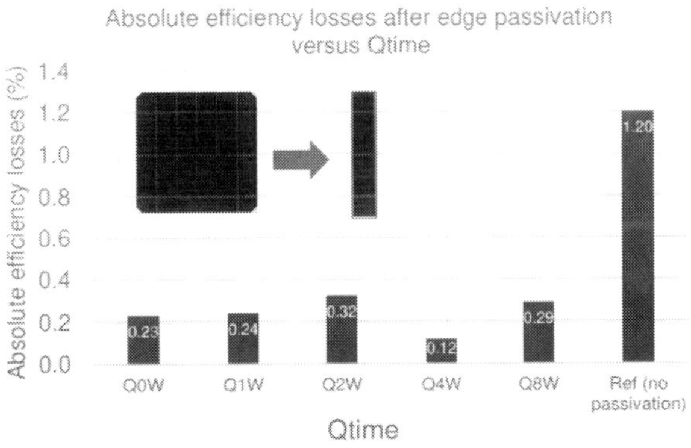

40th European Photovoltaic Solar Energy Conference and Exhibition

Conclusions

- **45° oriented wafers compatible with industrial environment cell process**
  - Evaluation on the CEA pilot line environment & SHJ cells process
  - Evaluation on the Kalyon production line: wafering and cell processing
  - No specific issues with the breakage rate

- **For SHJ cells processed from 45° oriented wafers (mechanical cleavage)**
  - Possible reduction of the thickness of the ALD edge passivation layer until 8 nm
  - No effect of the Qtime on the edge passivation up to 8 weeks

Perspectives

- Edge passivation solution will continue in the frame of the IBC4EU project:
  - Further developments for edge passivation solutions will continue at industrial level with Kalyon and Norsun
  - At larger wafer size (M10)
  - On the full industrial chain : cell / module

cea liten

Thank you

Mickael ALBARIC
CEA-Liten, Grenoble, France
liten.cea.fr
mickael.albaric@cea.fr
T. + 33 (0)4 79 79 20 25

IBC4EU

Acknowledgement

Funded by the European Union, under the Horizon Europe programme, Grant agreement number 101084259. Views and opinions expressed are, however, those of the author(s) only and do not necessarily reflect those of the European Union or European Climate, Infrastructure and Environment Executive Agency. Neither the European Union nor the granting authority can be held responsible for them

40th European Photovoltaic Solar Energy Conference and Exhibition

Call for Abstracts: Submit your latest results for SiliconPV and nPV by **November 4!**

SiliconPV 2024
14th International Conference on Crystalline Silicon Photovoltaics

npworkshop Chambéry 2024

bifiPV 2024 workshop

Advanced technologies, materials and concepts for crystalline Si solar cells and modules

April 15 – 19, 2024 | Chambéry, France

Three PV Events in One Week – Save the Date!

SiliconPV 2024: April 15 – 17
www.siliconpv.com

nPV Workshop 2024: April 17 – 18
www.2024.npv-workshop.com

bifiPV Workshop 2024: April 18 – 19
www.bifipv-workshop.com

Venue: Chambéry, France

Registration will be open online in January.

© Grand Chambéry Alpes Tourisme © Centre de Congrès Le Manège

Conference Host: cea, ines INSTITUT NATIONAL DE L'ENERGIE SOLAIRE

Conference Organizer: conexio pse

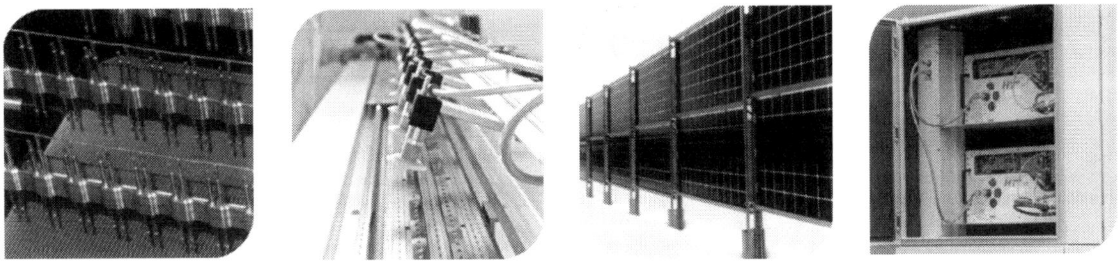

New approaches to edge passivation of laser cut PERC solar cells

ISC research for a sunny future

1BO.2.2, Oral Presentation

19th September 2023

Akash Thukaram
Project Engineer
halm elektronik GmbH

halm.

EU PVSEC 2023

40th European Photovoltaic Solar Energy Conference and Exhibition

Contents

- Motivation and goals
- Literature review
- Investigated methods for edge passivation
  1) Laser scribing and *simultaneous* Al doping
  2) Laser scribing and *subsequent* Al doping
- Results and discussions
- Conclusions

Motivation and goals

- Trend towards larger wafer formats
- ↑Area ⟶ ↑short circuit current at cell level
- Subcells - Half cells, shingles
- Cutting techniques - ****LSMC***, TLS, LIC etc.
- Edges are not passivated ⟶ high recombination of charge carriers
- Optimization of laser process is a key step
- **Aim**: Solutions to passivate the laser cut edges

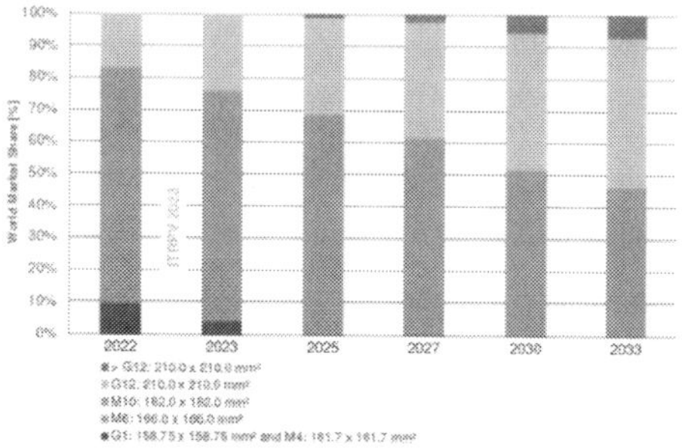

World market share of Si wafer formats [1]

* LSMC: Laser scribe & mechanical cleavage, TLS: Thermal laser separation, LIC: Laser induced cutting

[1] ITRPV VDMA 2023

40th European Photovoltaic Solar Energy Conference and Exhibition

Literature review

- Removal of emitter by creating passivation isolation trenches [2]
- Creation of surface field effect at the edges [3]
- Wet chemical treatment of edges [4]
- Passivated Edge Technology (PET) [5]
- PET applied to TOPCon shingle cells [6]
- Edge passivation with a fluorinated polymer [7]

[2] J. Zhao, A. Wang, P. P. Altermatt, and G. Zhang, 2000
[3] W. P. Mulligan, A. Terao, D. D. Smith, P. J. Verlinden, and R. M. Swanson, 2000
[4] P. P. Altermatt, G. Heiser, and M. A. Green, 1996
[5] P. Baliozian et al., 2020
[6] E. Lohmüller, P. Baliozian , L. Gutmann , et al., 2023
[7] D. Tune, F. Buchholz and A. Halm, 2020

Method 1 – *simultaneous* doping

- **Laser scribing and *simultaneous* Al doping**
  - ➢ Standard p-type PERC cells
  - ➢ Two step laser process: Al melting + standard laser scribing
  - ➢ Variations of laser parameters for Al melting
  - ➢ Cell scribed into 9 square parts

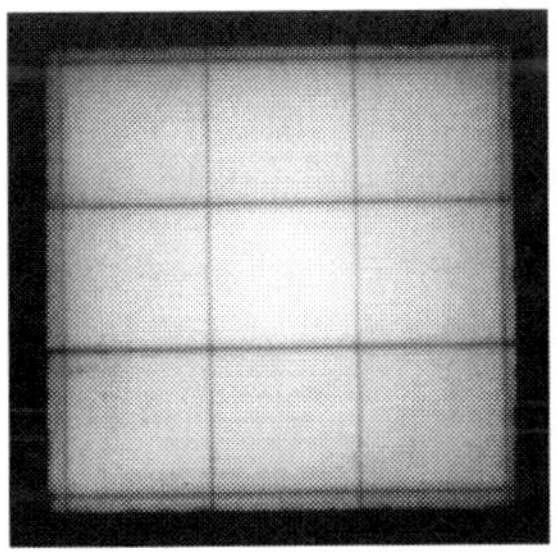

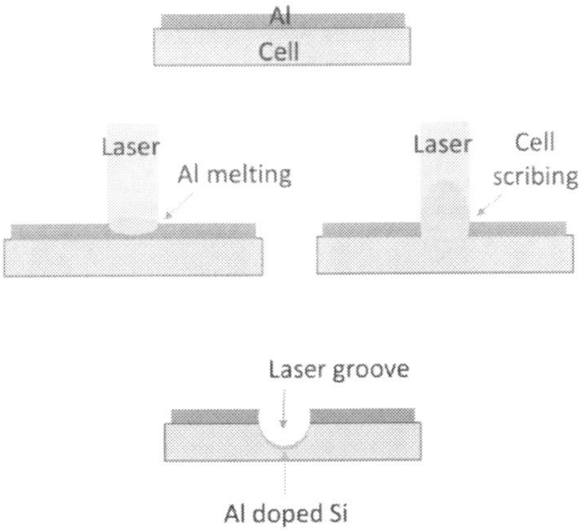

Laser parameter evaluation

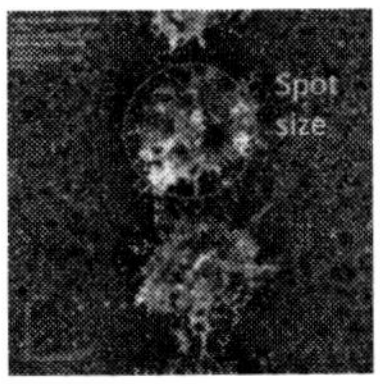

20kHz

50kHz

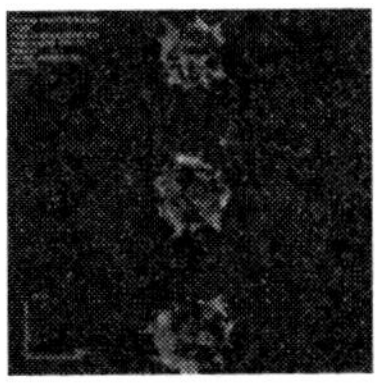

80kHz

Table 1: Laser parameters for Al melting

Set	f_{rep} [kHz]	Pulse overlap [%]	Measured spot diameter [µm]	Repetitions
1	20	95	96	0
2	20	50	96	0
3	50	95	72	0
4	50	50	72	0
5	80	95	60	0
6	80	50	60	0
7	80	95	60	10
8	80	50	60	10

Standard laser parameters for cell scribing:

- ➢ Frequency - 30 kHz
- ➢ Cutting speed - 100 mm/s
- ➢ Line repeat (no. of passes) - 2

I–V characterization

(a)

(b)

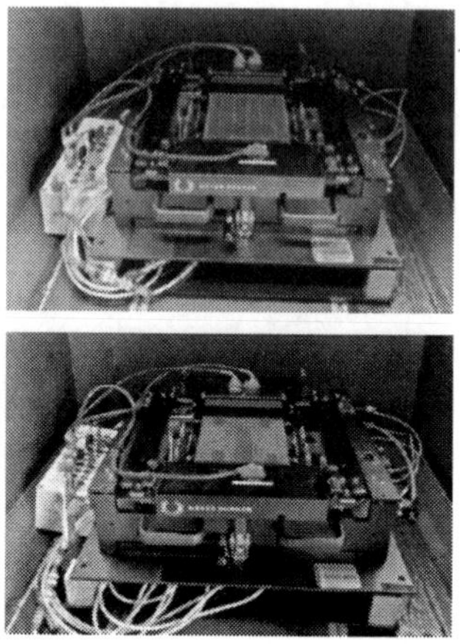

Grid touch setup for (a) Full cell & (b) Cleaved cell

I–V measurements comparison – 1 sun

Delta values between initial IV measurements & IV measurements after cell cleavage.

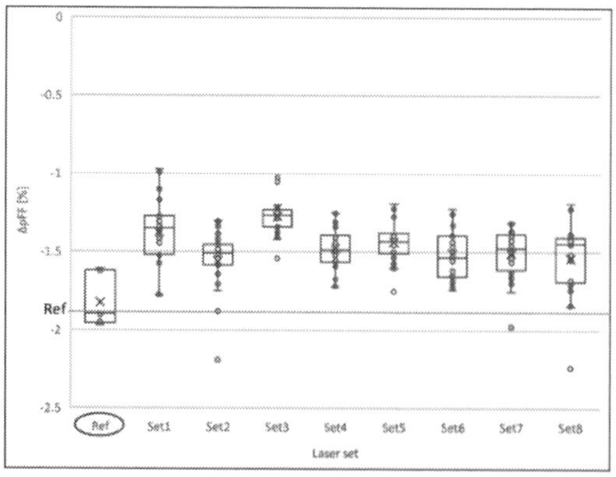

ΔpFF [%]

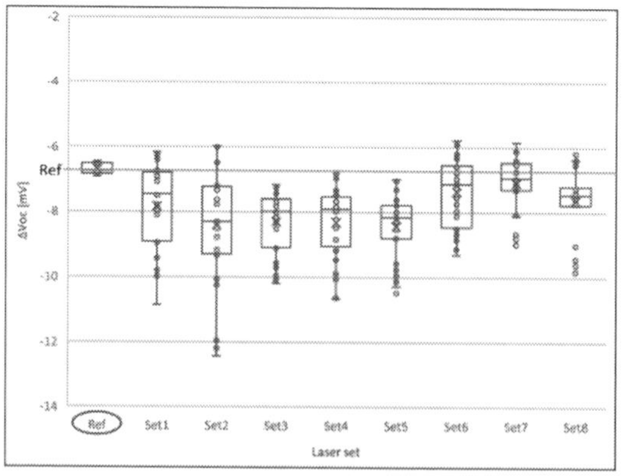

ΔV_{OC} [mV]

I–V measurements comparison – 1 sun

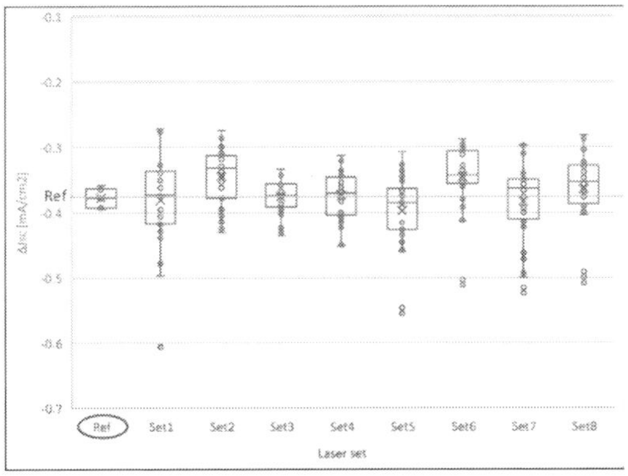

ΔJ_{SC} [mA/cm²]

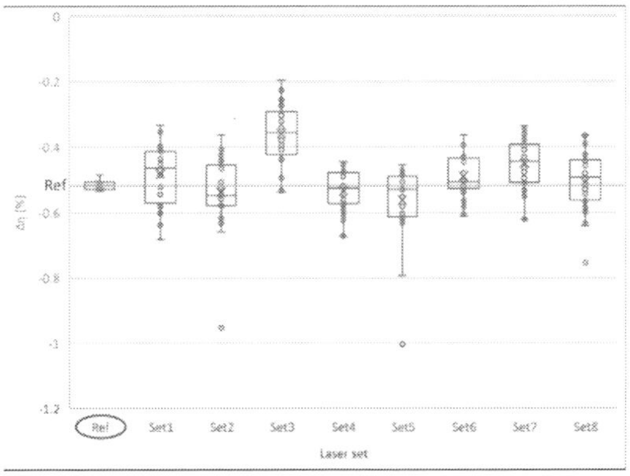

$\Delta\eta$ [%]

I–V measurements comparison – 0.3 sun (cleaved)

Absolute values measured at 0.3 sun after cell cleavage.

- FF [%]

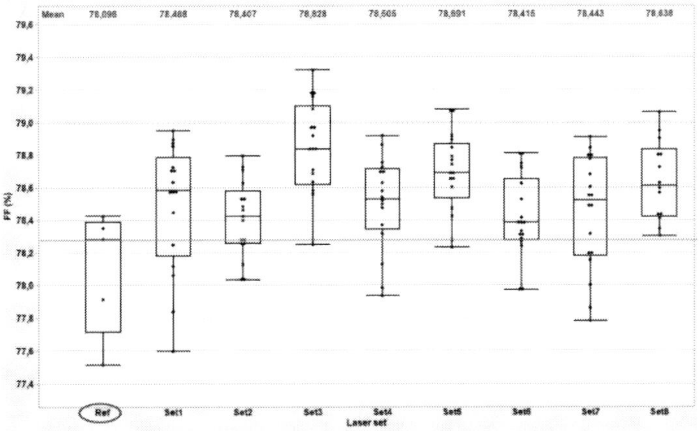

I–V measurements comparison – 0.3 sun (cleaved)

- V_{OC} [V]

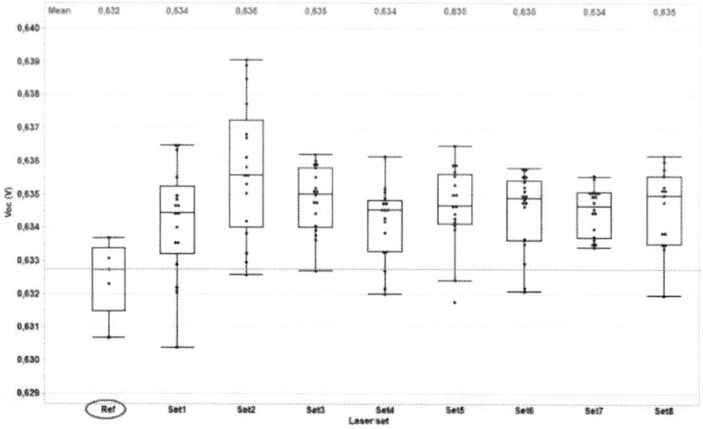

I–V measurements comparison – 0.3 sun (cleaved)

- J_{SC} [mA/cm^2]

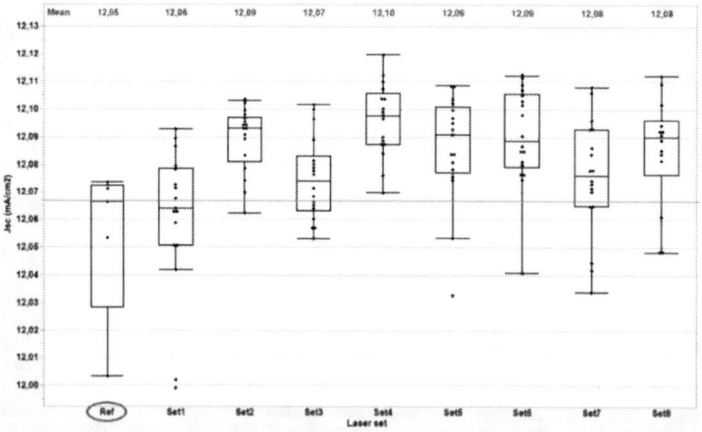

IV measurements comparison – 0.3 sun (cleaved)

- η [%]

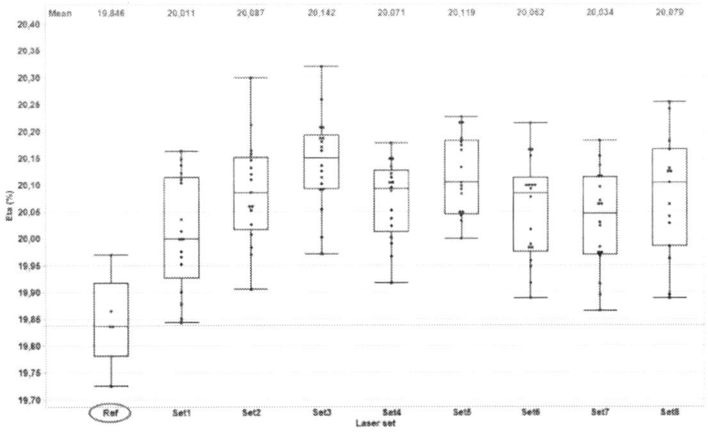

40th European Photovoltaic Solar Energy Conference and Exhibition

Photoluminescence measurements (cleaved cells)

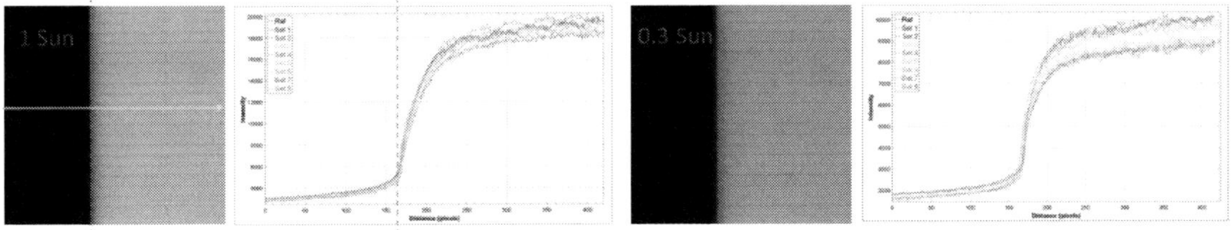

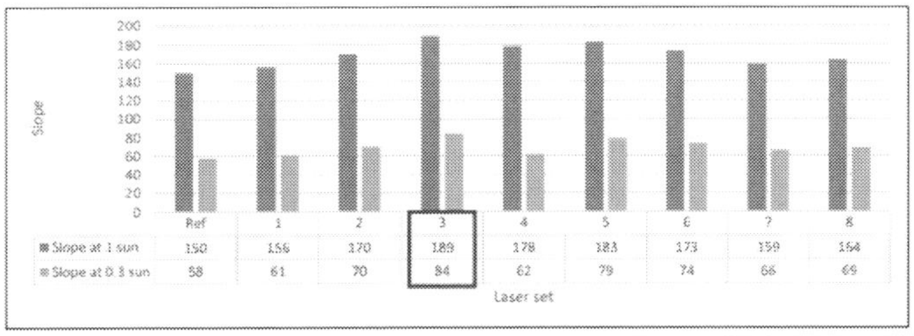

Method 1 – *simultaneous* doping

Summary:

- Improvement in pFF of cleaved cells ----> + 0.2 to 0.5 $\%_{abs,\ 1\ sun}$
- Improvement in V_{OC} ----> + 2 to 4 $mV_{abs,\ 0.3\ sun}$
- Improvement in efficiency ----> 0.1 $\%_{abs,\ 1\ sun}$, 0.3 $\%_{abs,\ 0.3\ sun}$

40th European Photovoltaic Solar Energy Conference and Exhibition

Method 2 – *subsequent* doping

- **Laser scribing and *subsequent* Al doping**
  - ➢ Laser scribing of PERC precursors prior to metallization
  - ➢ Variations of standard scribing recipe
  - ➢ Standard printing process & firing
  - ➢ Cells were cut into 6 stripes

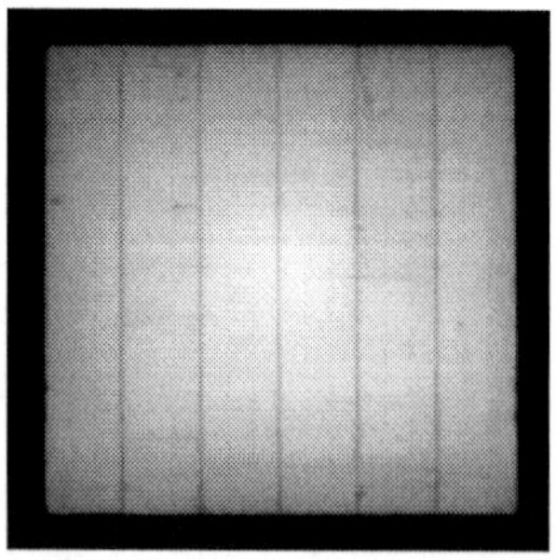

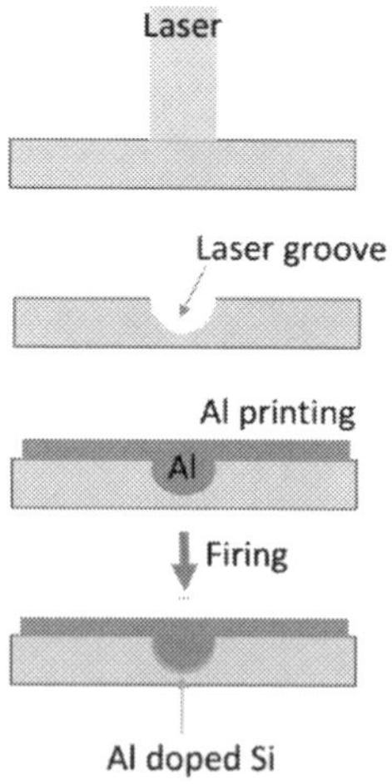

Method 2 – *subsequent* doping

Group	Type of laser processing
Group 1	Standard – single scribe
Group 2	Standard – two scribes
Group 3	Two scribes – 20µm offset
Group 4	Three scribes – 20µm offset
Reference	No laser processing prior metallization

Table 2: Laser scribing variations

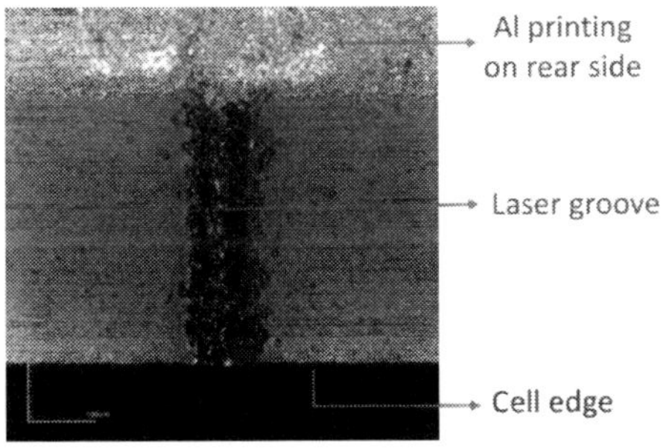

I–V measurements comparison – 1 sun (cleaved)

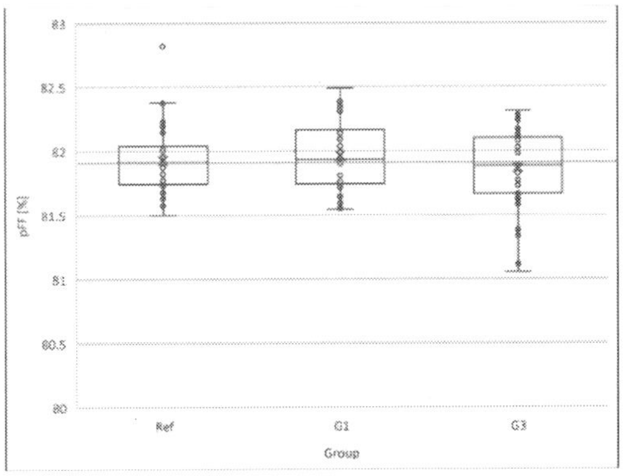

pFF [%]

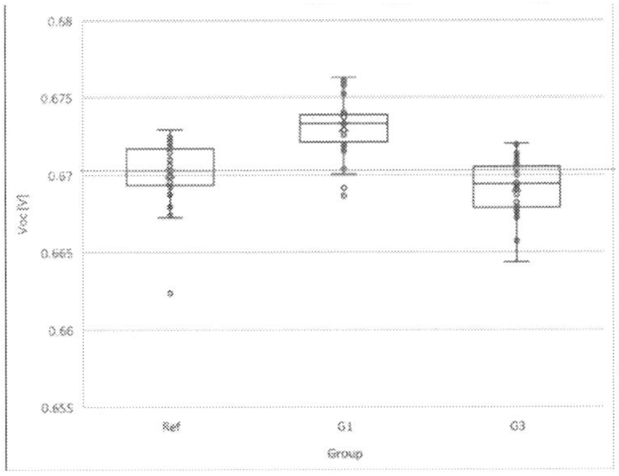

V_{OC} [V]

I–V measurements comparison – 1 sun (cleaved)

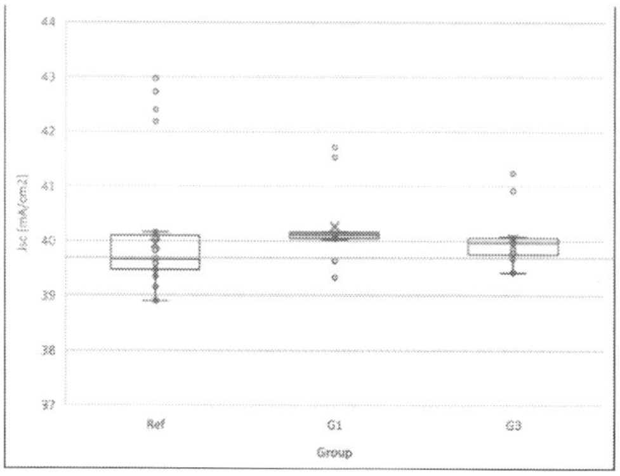

J_{SC} [mA/cm^2]

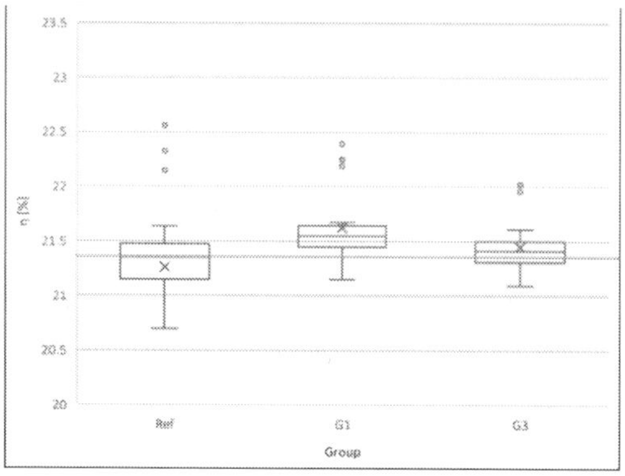

η [%]

I–V measurements comparison - 0.3 Sun (cleaved)

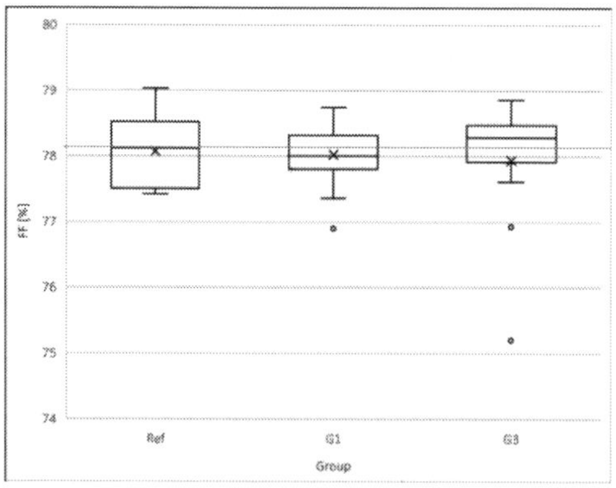

FF [%]

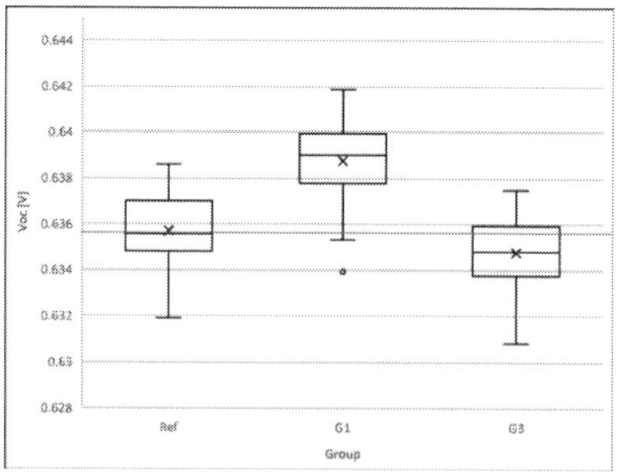

V_{OC} [V]

I–V measurements comparison - 0.3 Sun (cleaved)

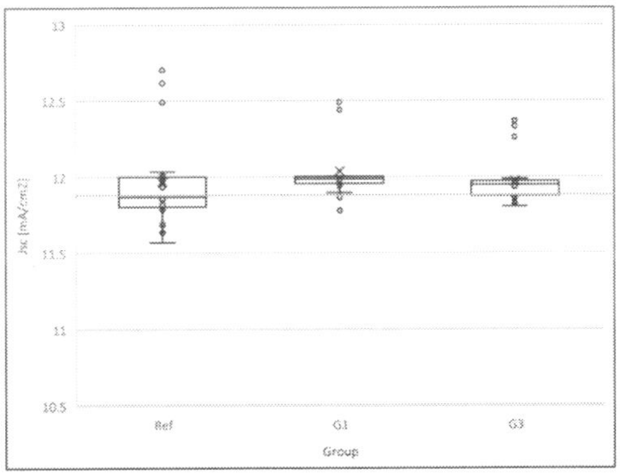

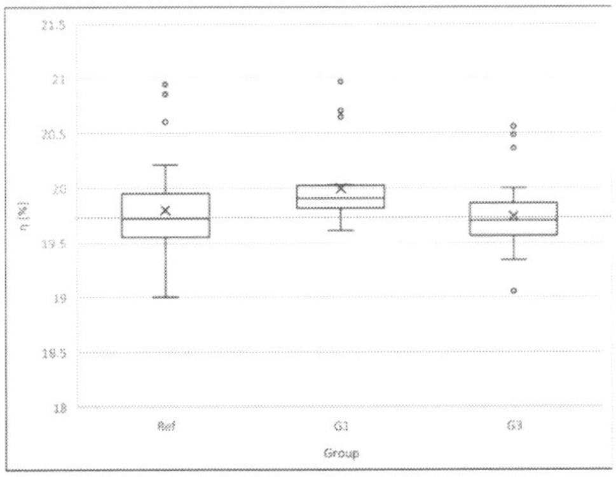

J_{SC} [mA/cm^2]   η [%]

Photoluminescence measurements (cleaved cells)

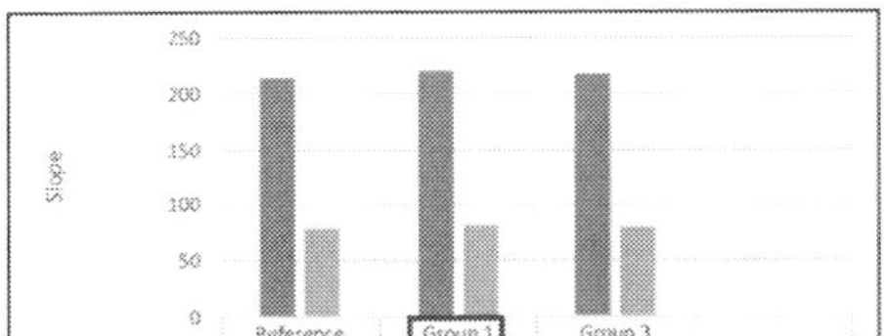

	Reference	Group 1	Group 3
Slope at 1 sun	215	221	218
Slope at 0.3 sun	79	82	80

Group

Method 2 – *subsequent* doping

Summary:

- Improvement in V_{OC} ----> + 3 $mV_{abs,\ 1\ sun}$, + 3 $mV_{abs,\ 0.3\ sun}$
- Improvement in efficiency ----> + 0.2 $\%_{abs,\ 1\ sun}$, + 0.2 $\%_{abs,\ 0.3\ sun}$

40th European Photovoltaic Solar Energy Conference and Exhibition

Conclusions

		Method 1 – simultaneous doping	Method 2 – subsequent doping
1	Passivation improvement seen in I-V?	pFF ++ V_{OC} ~ J_{SC} ~ η +	pFF ~ V_{OC} ++ J_{SC} + η ++
2	Obstacles	No obstacles	• Finding optimal laser parameters • Adjusting screen printing process

1BO.2.2

ISC research for a sunny future

Thank you

For the attention

Fraunhofer CSP

EU-PVSEC 2023

Rapid Inline Quantum Efficiency Measurements for an Accelerated Failure Diagnostics

M. Turek, M. Meusel, S. Lange, A. Müller, D. Adner, C. Hagendorf

Fraunhofer Center for Silicon Photovoltaics CSP, Halle, Germany

Supported by: Federal Ministry for Economic Affairs and Climate Action on the basis of a decision by the German Bundestag

Funded by the European Union

PILATUS

020014-001

40th European Photovoltaic Solar Energy Conference and Exhibition

Rapid Inline Quantum Efficiency Measurements and High-Resolution Diagnostics

Overview of the presentation

1. **Introduction and objective: Early detection of cell process variations**

2. **Experimental setup & data acquisition**
   - Detection of process fluctuations: Rapid inline EQE
   - Pre-selection of relevant solar cells: Statistical data analysis
   - Microstructural solar cell investigation: TEM+EDX

MK4-platform @ Fraunhofer CSP

3. **Result:** nm-scale analysis (structure & composition) of layer stacks for "outlier" cells

4. **Conclusion: Improved cell process monitoring**

„TEM" – transmission electron microscopy, „EDX" - energy-dispersive X-ray spectroscopy

020014-002

INCREASING CHALLENGES FOR QUALITY CONTROL OF NEXT GENERATION SOLAR CELLS
Statistical microstructure workflow required

Technology trend: Solar cells, i.e. their layer stacks, becoming more complex

General objective:
Reliable detection and root cause analysis of "hidden" process instabilities

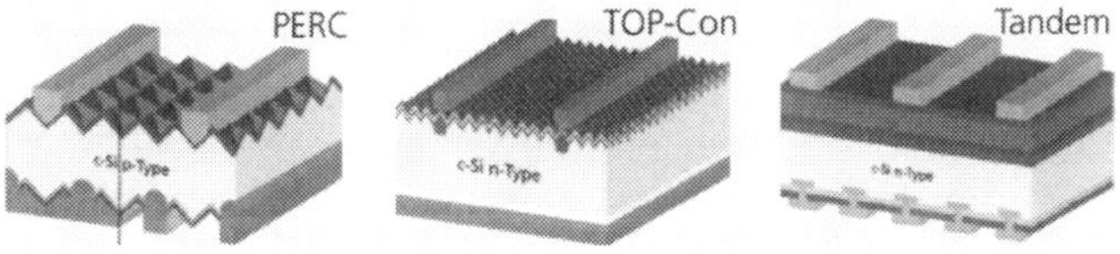

40th European Photovoltaic Solar Energy Conference and Exhibition

INCREASING CHALLENGES FOR QUALITY CONTROL OF NEXT GENERATION SOLAR CELLS
Statistical microstructure workflow required

Technology trend: Solar cells, i.e. their layer stacks, becoming more complex

General objective:
Reliable detection and root cause analysis of "hidden" process instabilities

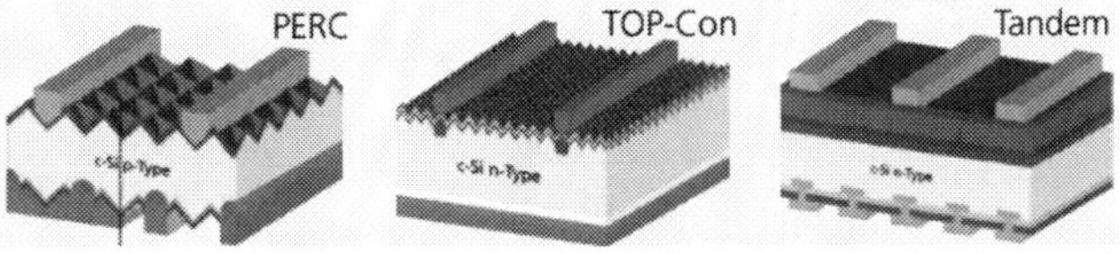

Production data – cell performance
- New inline-characterization approaches, e.g. rapid-EQE
- Automated, statistical data analysis for quality control

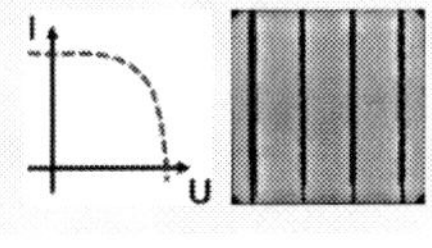

INCREASING CHALLENGES FOR QUALITY CONTROL OF NEXT GENERATION SOLAR CELLS

Statistical microstructure workflow required

Technology trend: Solar cells, i.e. their layer stacks, becoming more complex

General objective:
Reliable detection and root cause analysis of "hidden" process instabilities

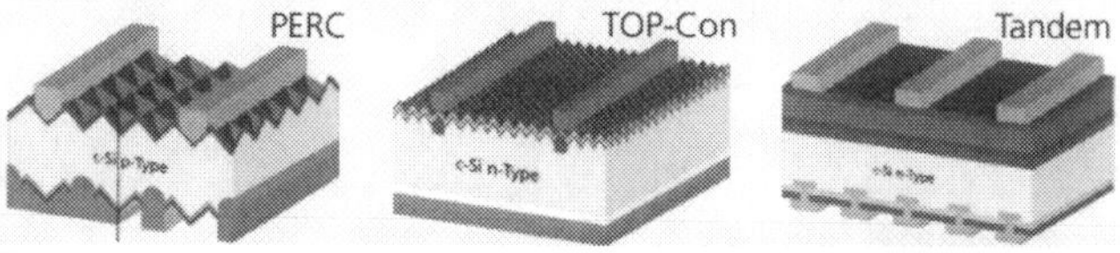

Production data – cell performance

- New inline-characterization approaches, e.g. rapid-EQE
- Automated, statistical data analysis for quality control

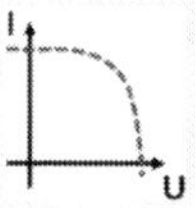

Advanced nm-scale lab-diagnostics

- Sample selection & complex target preparation
- Result: Layer thickness and composition (nm-scale)

INCREASING CHALLENGES FOR QUALITY CONTROL OF NEXT GENERATION SOLAR CELLS

Statistical microstructure workflow required

Technology trend: Solar cells, i.e. their layer stacks, becoming more complex

General objective:
Reliable detection and root cause analysis of "hidden" process instabilities

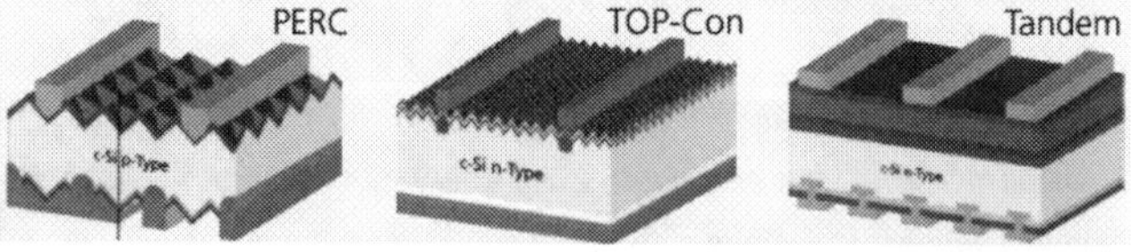

Production data – cell performance

- New inline-characterization approaches, e.g. rapid-EQE
- Automated, statistical data analysis for quality control

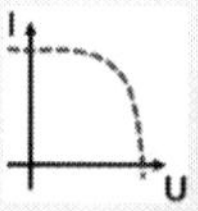

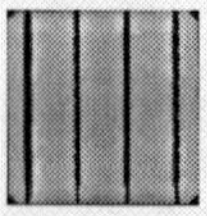

⇔

Advanced nm-scale lab-diagnostics

- Sample selection & complex target preparation
- Result: Layer thickness and composition

Challenge: Linking cell performance, e.g. production data, with high-resolution nm-scale lab-methods

Statistical Microstructure Workflow

General concept and key elements

Approach @ Fraunhofer CSP: Three key elements combined in one workflow

(I) Advanced inline measurements, e.g. LED sun simulators – rapid-EQE[1,2] ⇨ **(II)** Data analysis on larger cell groups (MK4) ⇨ **(III)** Pre-selection of relevant cells and nm-scale diagnostics

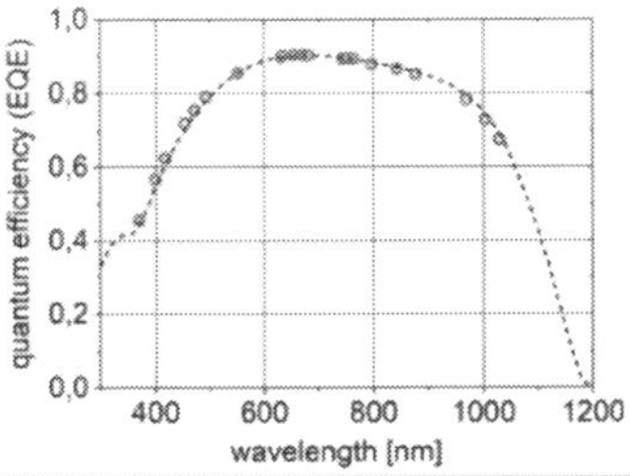

[1] T. Luka et al, Ener. Proc. **77**, p113 (2015); [2] M. Turek et al, Solar Ener Mat & Solar Cells **194**, p 142 (2019)

40th European Photovoltaic Solar Energy Conference and Exhibition

Statistical Microstructure Workflow – Challenges and Implementation

Lab-to-fab platform MK4: Advanced inline-characterization and data analysis

Advanced production data analysis @ CSP
- Production-like MK4 platform: testing **500+ cells**
- Inline processing time: **< 1 sec / cell**

High resolution nm-scale diagnostics @ CSP
- TEM Hitachi HF5000: testing **single samples**
- Preparation TEM-lamella + TEM-analysis: **> 1 day / cell**

MK4-platform

Solar cell stack

⇔

Hitachi TEM HF5000

TEM-lamella

Statistical Microstructure Workflow – Challenges and Implementation

Lab-to-fab platform MK4: Advanced inline-characterization and data analysis

Advanced production data analysis @ CSP

- Production-like MK4 platform: testing 500+ cells
- Inline processing time: < 1 sec / cell

High resolution nm-scale diagnostics @ CSP

- TEM Hitachi HF5000: testing single samples
- Preparation TEM-lamella + TEM-analysis: > 1 day / cell

MK4-platform

Solar cell stack

Hitachi TEM HF5000

TEM-lamella

Major challenge: Reliable identification of relevant and suitable cells out of large batches → rapid inline EQE

„TEM" – transmission electron microscopy, „EDX" - energy-dispersive X-ray spectroscopy

Experimental Approach: Inline-EQE Using LED Sun Simulators

Comparison of two cell batches: Inline measurements and data analysis

Step 1: Lab-to-fab MK4

Batches B1 & B2: **500 cells** each

→ STC I-V & rapid-EQE[1,2]

- STC
- STC @ 75%
- I_{sc} @ 370 nm
- I_{sc} @ 415 nm
- I_{sc} @ 760 nm
- I_{sc} @ 875 nm
- I_{sc} @ 1030 nm

[1] T. Luka et al, Ener. Proc. **77**, p113 (2015); [2] M. Turek et al, Solar Ener Mat & Solar Cells **194**, p 142 (2019)

EXPERIMENTAL APPROACH: INLINE-EQE USING LED SUN SIMULATORS

Comparison of two cell batches: Inline measurements and data analysis

Step 1: Lab-to-fab MK4
Batches B1 & B2: **500 cells** each
→ STC I-V & rapid-EQE[1,2]

- STC
- STC @ 75%
- I_{sc} @ 370 nm
- I_{sc} @ 415 nm
- I_{sc} @ 760 nm
- I_{sc} @ 875 nm
- I_{sc} @ 1030 nm

Time series: I_{sc}-values B1

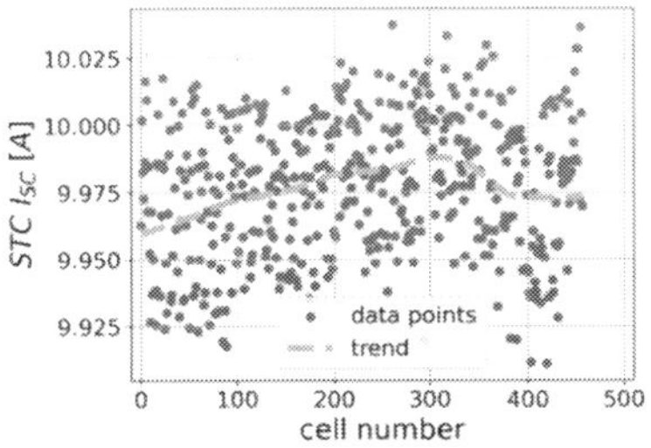

Statistical data analytics & cell selection

[1] T. Luka et al, Ener. Proc. **77**, p113 (2015); [2] M. Turek et al, Solar Ener Mat & Solar Cells **194**, p 142 (2019)

EXPERIMENTAL APPROACH: INLINE-EQE USING LED SUN SIMULATORS

Comparison of two cell batches: Inline measurements and data analysis

Step 1: Lab-to-fab MK4
Batches B1 & B2: 500 cells each
→ STC I-V & rapid-EQE[1,2]

⇨ **Step 2: Lab tools**
Selected cells: **3x B1a** & **3x B1b**
→ full-EQE & reflectance

⇨ **Step 3: nm-scale analytics**
Präparation **1x B1a** & **1x B1b** cell
→ ToF-SIMS, FIB, TEM-EDX

- STC
- STC @ 75%
- I_{sc} @ 370 nm
- I_{sc} @ 415 nm
- I_{sc} @ 760 nm
- I_{sc} @ 875 nm
- I_{sc} @ 1030 nm

Time series: I_{sc}-values B1

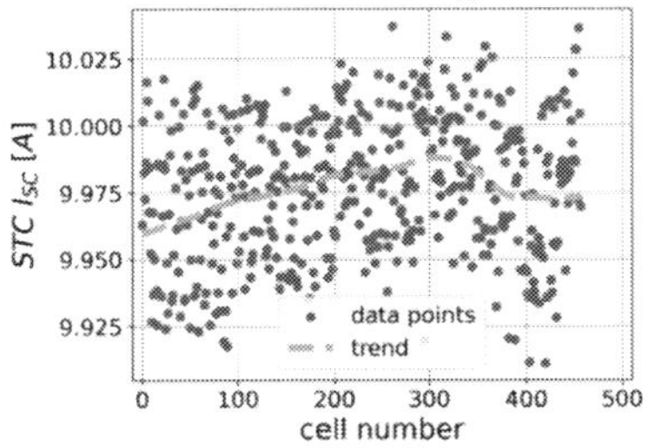

Statistical data analytics & cell selection

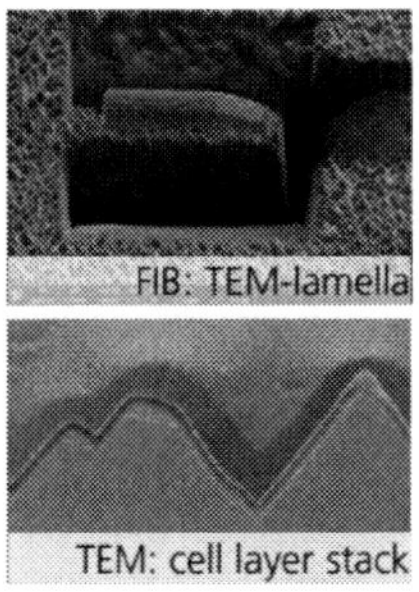

[1] T. Luka et al, Ener. Proc. **77**, p113 (2015); [2] M. Turek et al, Solar Ener Mat & Solar Cells **194**, p 142 (2019)

RESULTS – LAB-TO-FAB DATA ACQUISITION (STEP 1): 500+ CELLS

Data pre-processing: Trend correction needed for assessment of process fluctuations

Raw data: Stable STC I-V measurement but drift in some rapid-EQE values

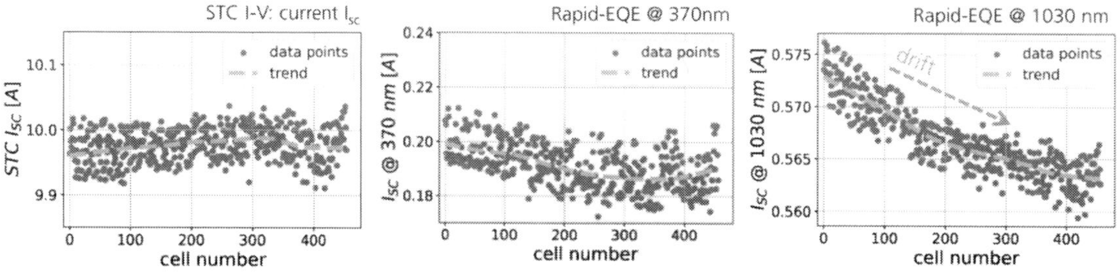

Result: Trend correction required to isolate measurement drifts from process fluctuations

40th European Photovoltaic Solar Energy Conference and Exhibition

RESULTS – LAB-TO-FAB DATA ANALYSIS (STEP 1): 500+ CELLS

Batch comparison: STC-measurements not sensitive enough for clear distinction B1 vs. B2

Statistical distribution: Two batches **B1** & **B2** appear **similar** w.r.t. **STC-value** distributions

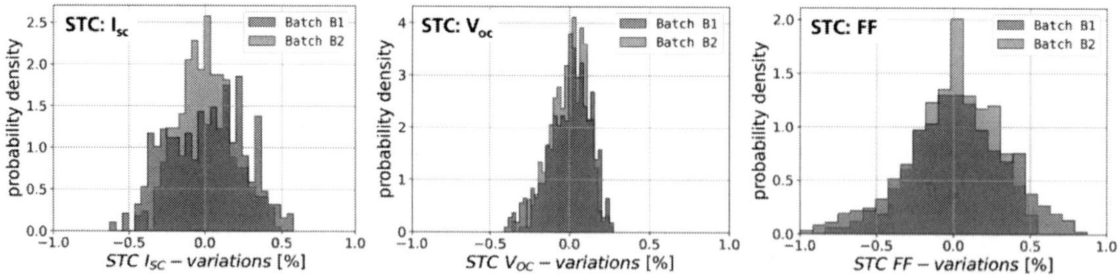

Result: Data analysis beyond STC (current I_{sc}, voltage V_{oc}, FF) required → rapid-EQE

RESULTS – LAB-TO-FAB DATA ANALYSIS (STEP 1): 500+ CELLS

Batch comparison: Short-wavelength EQE shows largest process fluctuations

Distribution rapid-EQE values: Large relative **fluctuations in 370nm-EQE**, more pronounced for **batch B1**

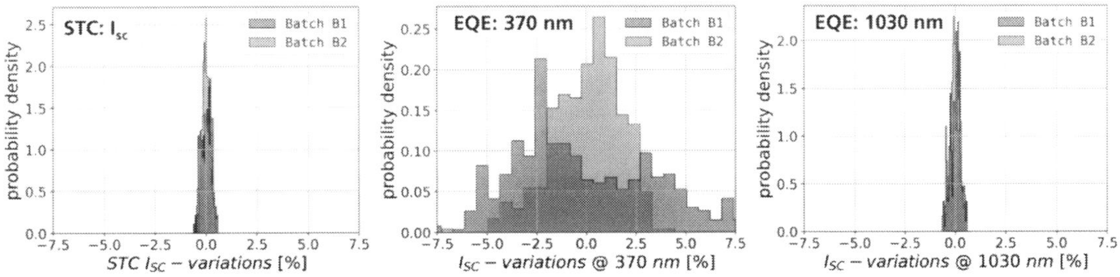

Result: Front-side cell process for batch B1 with larger fluctuations than other processes

40th European Photovoltaic Solar Energy Conference and Exhibition

RESULTS – LAB-TO-FAB DATA ANALYSIS (STEP 1): 500+ CELLS

Batch comparison: Short-wavelength EQE shows largest process fluctuations

Distribution rapid-EQE values: Large relative **fluctuations in 370nm-EQE**, more pronounced for **batch B1**

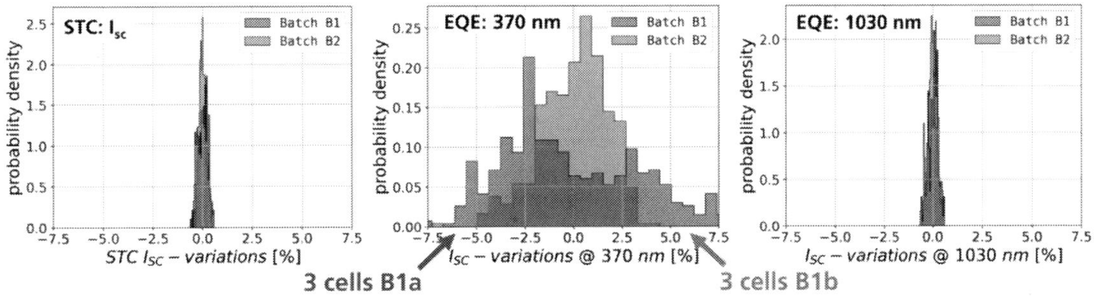

Result: Front-side cell process for batch B1 with larger fluctuations than other processes

> **Verification** using lab-tools on few **pre-selected cells**: **B1a** (lower 370nm-EQE) & **B1b** (larger 370nm-EQE)

Results – Verification and Cell Selection with Lab Tools (Step 2: Six cells)

Inline-EQE results clearly confirmed by offline measurements

EQE and reflectance: Clear differences between **B1a** (lower 370nm-EQE) and **B1b** (larger 370nm-EQE)

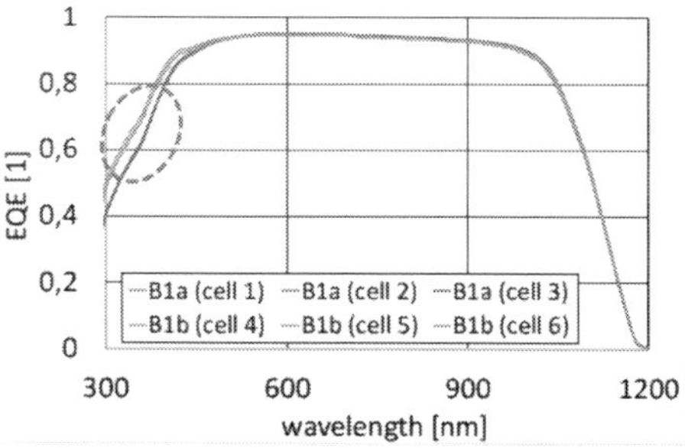

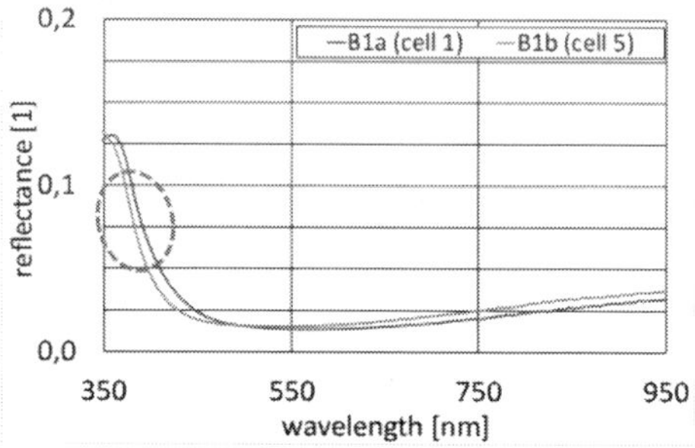

Result: Rapid-EQE confirmed → Process instabilities at front side layer stack, optical properties affected

40th European Photovoltaic Solar Energy Conference and Exhibition

Results – Verification and Cell Selection with Lab Tools (Step 2: Six cells)

Inline-EQE results clearly confirmed by offline measurements

EQE and reflectance: Clear differences between **B1a** (lower 370nm-EQE) and **B1b** (larger 370nm-EQE)

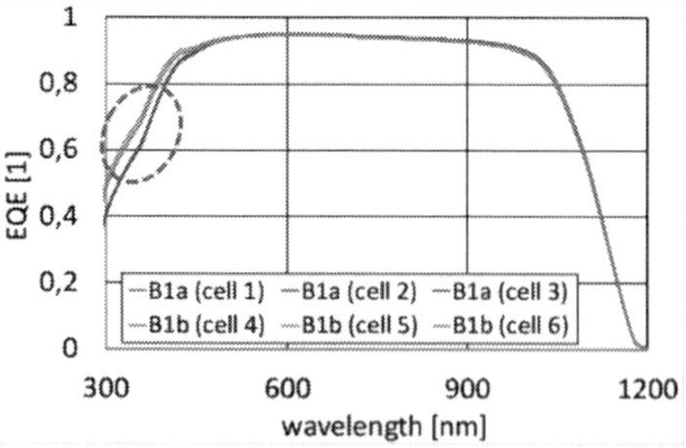

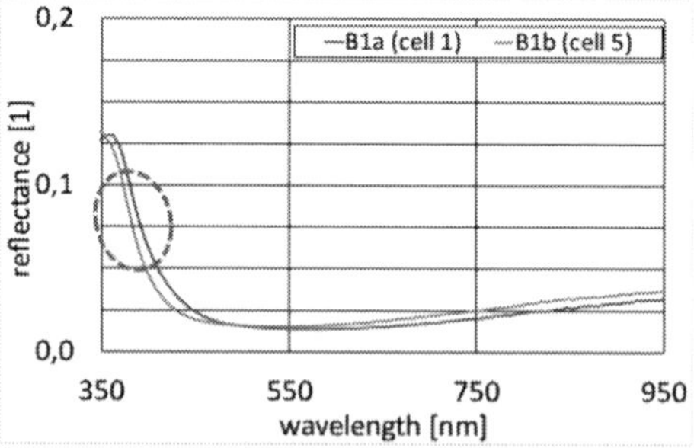

Result: Rapid-EQE confirmed → Process instabilities at front side layer stack, optical properties affected

➢ **Root cause analysis: layer thickness or differences in SiN-composition?** (not accessible by reflectance)

Results – nm-Scale Analytics (Step 3: Two Cells)

Sample preparation and STEM/EDX on layers and interfaces: nm-differences in layer thickness

TEM-analysis: Comparable stack composition but nm-differences in layer thickness

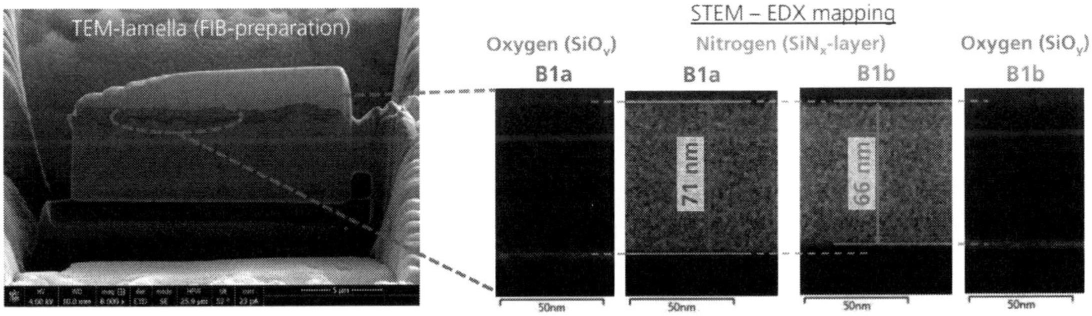

Result: Stack SiO_y (4nm) – SiN_x (B1a: 71 nm ↔ B1b: 66 nm) – SiO_y (5nm); but identical composition

„FIB" – focused ion beam, „STEM" – scanning transmission electron microscopy, „EDX" - energy-dispersive X-ray spectroscopy

Fraunhofer CSP

Results – nm-Scale Analytics (Step 3: Two Cells)

Sample preparation and STEM/EDX on layers and interfaces: nm-differences in layer thickness

TEM-analysis: Comparable stack composition but nm-differences in layer thickness

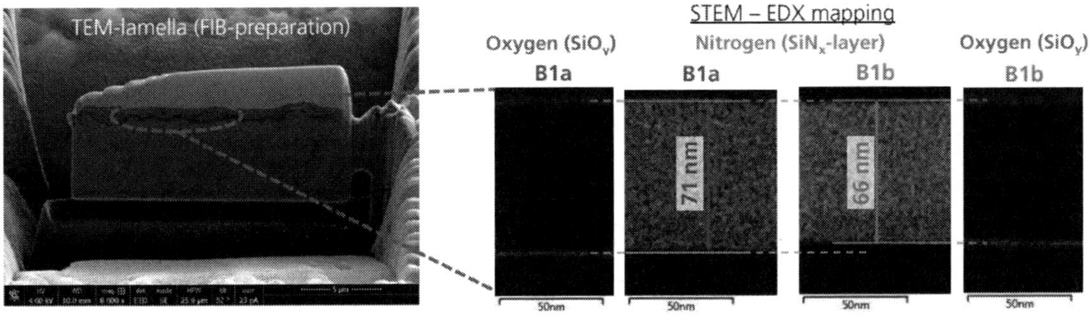

Result: Stack SiO_y (4nm) – SiN_x (B1a: 71 nm ↔ **B1b: 66 nm**) – SiO_y (5nm); but identical composition

➢ **Differences in SiN_x layer thickness explain variations in optical properties and quantum efficiency**

„FIB" – focused ion beam, „STEM" – scanning transmission electron microscopy, „EDX" - energy-dispersive X-ray spectroscopy

Fraunhofer CSP

SUMMARY: RAPID INLINE EQE FOR ACCELERATED FAILURE DIAGNOSTICS

Early detection of cell process fluctuations combined with root-cause analysis

Three key-elements of statistical microstructure workflow:

1. Detection of "hidden" process variations → rapid inline EQE
2. Pre-selection of outlier cells → lab-to-fab platform MK4 + data analysis
3. Root cause analysis → High-resolution nm-scale diagnostics: TEM/EDX on layer stack

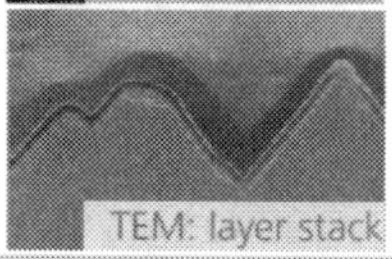

Result: Minor process variations in layer thickness and/or composition can be identified using the statistical microstructure workflow

Page 10 15.09.2023 - public information - Fraunhofer CSP

020014-021

40th European Photovoltaic Solar Energy Conference and Exhibition

Fraunhofer CSP

Thank you for your attention!

Marko Turek
Tel. +49 345 5589 5121
marko.turek@csp.fraunhofer.de

Fraunhofer CSP
Otto-Eißfeldt- Straße 12
06120 Halle (Saale)
www.csp.fraunhofer.de

Funded by the Federal Ministry for Economic Affairs and Climate Action within the "OptiLearn"-project (FKZ 03EE1108) and by the European Union within the "PILATUS" project.

Funded by the European Union

Supported by: Federal Ministry for Economic Affairs and Climate Action on the basis of a decision by the German Bundestag

PILATUS

020014-022

STUDY OF DIFFERENT INTERFACIAL OXIDES FOR INDUSTRIAL N-POLY SI PASSIVATION ON POLO IBC SOLAR CELLS

Verena Mertens[1], Silke Dorn[1], Jonathan Langlois[1], Maximilian Stöhr[1], Yevgeniya Larionova[1], Rolf Brendel[1,2], Norbert Ambrosius[3], Aaron Vogt[3], Thomas Pernau[4], Helge Haverkamp[4], Thorsten Dullweber[1]

[1]Institute for Solar Energy Research Hamelin (ISFH), Am Ohrberg 1, 31860 Emmerthal, Germany
[2]Institute of Solid-State Physics, Leibniz Universität Hannover, Appelstrasse 2, 30167 Hannover, Germany
[3]LPKF Laser & Electronics AG, Osteriede 7, 30827 Garbsen, Germany
[4]centrotherm international AG, Württemberger Str. 31, 89143 Blaubeuren, Germany

Corresponding author: Verena Mertens
E-mail: mertens@isfh.de, Phone: +49-5151-999315

ABSTRACT: We present investigations on in situ grown PECVD interfacial oxides for PECVD n-poly Si. In this study we investigate the impact of thickness variation of PECVD oxide layers and compare two different variants of PECVD oxides, grown in N_2O plasma (SiO_xN_y) or O_2 plasma (SiO_x) with oxygen as environmentally friendly alternative as oxidant. For these experiments we use an industrial centrotherm tool for PECVD deposition. We perform our study on POLO IBC solar cells using p-type silicon wafers in M2 format and with our masking process, using LIDE processed (Laser Induced Deep Etching) glass shadow masks for structured deposition of PECVD n-amorphous silicon to form the IBC pattern on the rear side of the solar cells. The PECVD oxides are formed in situ through the glass shadow mask directly before the deposition of PECVD-n-amorphous Si. The POLO IBC cell efficiencies show that the PECVD oxides SiO_xN_y and SiO_x with optimized thickness outperform our wet chemical SiO_x with maximum efficiency of 23.8% and 23.7%, respectively. These data demonstrate the feasibility to deposit a high-quality in situ PECVD interfacial oxide for optimum surface passivation and current transport of passivated contacts at the same time.

1 INTRODUCTION

The PECVD deposition of n-poly Si receives currently more and more interest as it is a one-sided deposition process which is suitable for various solar cell concepts like Topcon, POLO BJ or POLO IBC.

The POLO IBC solar cell concept targets the implementation of a passivating contact in a solar cell structure and at the same time highly reusing processing steps from PERC+ technology like p-type silicon wafers, AlO_x/SiN_x passivation, rear side polishing, aluminum finger BSF contacts [1, 2]. Our processing sequence uses structured deposition of PECVD-interfacial oxide plus n-amorphous silicon through a glass shadow mask to create the rear side IBC layout [3]. The shadow masks are manufactured with the LIDE process (Laser Induced Deep Etching) developed by LPKF Laser & Electronics. The solar cell schematic and the structured deposition of the IBC layout are sketched in Figure 1.

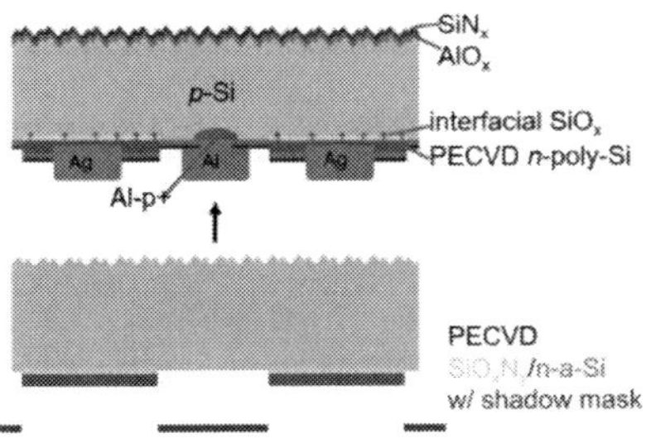

Figure 1: Schematic of the structured deposition of PECVD SiO_xN_y and n-a-Si through a glass shadow mask

As starting point, a process for PECVD in situ grown interfacial oxide plus n-amorphous silicon passivation was developed on a lab-type single wafer tool [4]. We implemented the poly silicon layer in the POLO IBC solar cell process with glass mask and obtained a median implied open-circuit voltage of 738 mV on solar cells without metallization [6] and 23.0% solar cell efficiency [5]. The poly Si process was then transferred from the lab tool to an industrial centrotherm furnace [5].

The data presented in this paper deal with a study on the PECVD interfacial oxide. We show results from a fine-tuning experiment of the interfacial oxide grown by N_2O plasma SiO_xN_y and PECVD-oxide grown in O_2 plasma, SiO_x. We show solar cell results for M2 sized POLO IBC solar cells and discuss the impact of interfacial oxide thickness and precursor gas on current-voltage parameters.

2 EXPERIMENTAL

We use 0.9 Ohm*cm Ga-doped p-type material for our solar cells. We texture the wafer and perform an acidic rear polishing process followed by an alkaline post-treatment in order to remove the pyramids on the rear side of the wafers. After applying an HCl/ ozone clean, we perform our in-situ interfacial oxide variations: we grow a wet chemical interfacial oxide from DI water/ ozone bath immersion for 30 minutes, and deposit afterwards a PECVD n-a-Si (n-type doped amorphous silicon) layer with a thickness of around 150 nm. Alternatively, we use an industrial centrotherm c.plasma PECVD-tool to in situ grow the interfacial oxide followed by a 150 nm n-a-Si layer in a single PECVD process. We grow the PECVD interfacial oxides either from N_2O (SiO_xN_y) or from oxygen plasma (SiO_x): In order to achieve a structured deposition for the IBC solar cells' rear side we grow or deposit all PECVD layers through a glass shadow mask fabricated by LPKF Laser and Electronics. We vary the interfacial oxide thickness by variation of the plasma immersion time. The resulting PECVD oxide thicknesses used in this experiment are in the range of 1 to 2 nm. However, as the glass mask reduces the oxide growth rate and also the deposition rate of the *n*-a-Si, we cannot experimentally assess the absolute oxide thickness due to the lack of feasible measurement techniques which are

Table 1: SiO_xN_y experiment: variation of SiO_xN_y thickenss and comparison to a wet chemical SiO_x

A1	A2	A3	A4
Wet chem SiO_x	SiO_xN_y thin	SiO_xN_y medium	SiO_xN_y thick
PECVD *n*-a-Si, ~150 nm			
Anneal in N_2, 850°C			

Table 2: SiO_x experiment: variation of SiO_x thickness and poly Si-anneal temperature, reference is the SiO_xN_y with medium thickness.

B1	B2	B3	B4	B5	B6	B7	B8
SiO_xN_y med	SiO_x thin			SiO_x medium			SiO_x thick
PECVD n-poly Si, ~150 nm							
Anneal in N_2, T (°C):							
850	800	825	850	800	825	850	850

applicable to roughly only several 100 µm wide IBC fingers. Therefore, the oxides are named here thin, medium, thick. The n-a-Si thickness is adjusted by deposition time to 150 nm through the glass mask as verified by SEM measurements. The subsequent anneal is done at 850°C for the SiO_xN_y. For the SiO_x we also perform an *n*-poly Si anneal temperature variation at 800°C, 825°C and 850°C as detailed in **Table 2**.

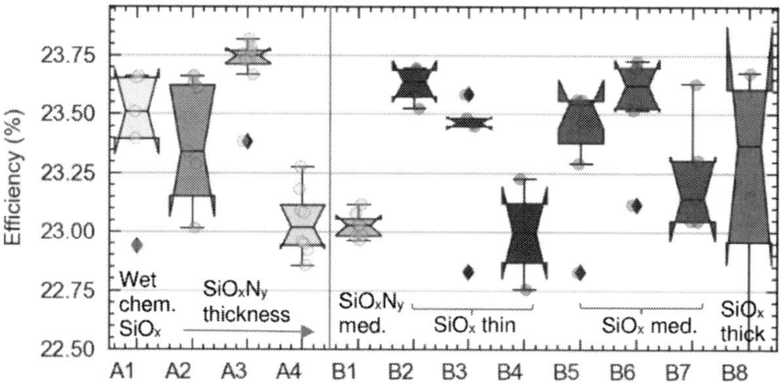

v

Figure 1: Solar cell efficiencies for POLO IBC cells corresponding to experimental plan A and B: variation of interfacial oxide thickness for PECVD oxides grown from N_2O (A2 to A4, B1) and O_2 plasma (B2 to B8) compared to a wet chemical interfacial oxide (A1).

Table 3: Summary of solar cell results for the SiO_xN_y and SiO_x experiment: efficiency eta, short-circuit current density J_{sc}, open-circuit voltage V_{oc}, fill factor *FF*, pseudo fill factor p*FF* and series resistance R_s. Here the median values for the different groups are given.

		Anneal in N_2 (°C)	Eta (%)	J_{sc} (mA/cm²)	V_{oc} (mV)	*FF* (%)	p*FF* (%)	R_s (Ohm*cm²)
A1	Wet chem. SiO_x	850	23.5	40.6	715	80.6	84.2	0.74
A2	SiO_xN_y thin	850	23.3	40.4	716	81.0	84.0	0.54
A3	SiO_xN_y medium	850	23.8	40.8	721	80.8	84.4	0.74
A4	SiO_xN_y thick	850	23.0	40.6	719	78.9	84.1	1.09
B1	SiO_xN_y medium	850	23.0	40.3	715	79.8	84.2	0.93
B2	SiO_x thin	800	23.6	40.5	720	80.8	84.4	0.76
B3	SiO_x thin	825	23.5	40.6	716	80.7	84.1	0.69
B4	SiO_x thin	850	23.0	40.4	704	81.0	84.1	0.64
B5	SiO_x medium	800	23.6	40.4	721	80.7	84.3	0.76
B6	SiO_x medium	825	23.6	40.7	718	80.9	84.3	0.69
B7	SiO_x medium	850	23.1	40.5	709	81.1	84.2	0.65
B8	SiO_x thick	850	23.4	40.3	714	81.0	84.1	0.65

10.4229/EUPVSEC2023/1BO.3.3
020015-002

Afterwards we passivate the front side of our solar cells using ALD AlO_x from a FastALD tool from Solay Tec combined with an outgassing step at 600°C for 30 minutes and a SiN_x layer from a SiNA tool from Roth and Rau. The rear side of our solar cells is passivated with PECVD-AlO_x/ SiN_x from a centrotherm PECVD tool. For the base contacts we us a picosecond laser to create laser contact openings. We screen print our wafers using a commercial aluminum paste from Toyo Aluminium for the base contacts and a commercial Ag paste for the contact to the *n*-poly Si. After contact formation in a fast firing process we measure the IV characteristics of our solar cells on an IBC compatible measurement chuck developed together with PV tools in a Loana system.

3 RESULTS AND DISCUSSION

The POLO IBC cell efficiencies obtained in the SiO_xN_y experiment and the SiO_x experiment are summarized in Figure 1. All results are summarized in Table 3.
Looking first at the results for A1 to A4, the data show a clear impact of the interfacial oxide thickness on efficiency for SiO_xN_y (A1 to A4). With optimized thickness as in group A3, the PECVD SiO_xN_y with 23.7% efficiency outperforms our wet chemical SiO_x (A1) which yields 23.5%. However, a slightly lower anneal temperature for the wet chemical oxide might decrease this difference.
In the SiO_x experiment with group B1 to B8, the reference group B1 shows decreased efficiency compared to the identical group A3 of the SiO_xN_y experiment due to a preconditioning issue experienced in the B-experiment.
Therefore, we compare our SiO_x results to the best group of the SiO_xN_y: The variation of poly Si anneal temperature for the different SiO_x thicknesses shows, that in principle, the same optimum is found either by variation of oxide thickness or variation of anneal temperature and that the optimum anneal T shifts to higher values for increasing SiO_x thickness. For the thin SiO_x the optimum anneal temperature is 800°C whereas for medium SiO_x 825°C, both groups reach a median efficiency of ~23.6%.
Comparing SiO_xN_y and SiO_x at optimum process conditions, both PECVD interfacial oxides give quite similar efficiencies of 23.7% and 23.6% with slight advantage for the SiO_xN_y.
In **Table 3** we provide all IV parameters for this experimental study. The largest impact of oxide thickness can be seen on the open-circuit voltage and the series resistance of the resulting POLO IBC solar cells. For all groups a test of optimum firing temperature was performed, so that impact of insufficient firing on series resistance can be excluded.
The series resistance increases for increasing SiO_xN_y thicknesses, in A2 to A4 and SiO_x thickness in B2 to B8. For the latter, the additional variation of poly Si anneal temperature adds a slight decrease in series resistance R_s with increasing anneal temperature. The open-circuit voltage exhibits an optimum for SiO_xN_y groups A2 to A4 and decreases for increasing anneal temperature for the SiO_x groups B2 to B8.
We speculate that this is due to an optimum of in diffusion of phosphorus into the wafer and oxide break up, leading to highest passivation level.
For the SiO_x experiment the specific range chosen for poly Si anneal variation and oxide thickness seems to be the root cause that no optimum is observed here.

In order to determine the impact of metallization on the open-circuit voltage V_{oc} and the pseudo fill factor *pFF* of our solar cells, we compare these data to the respective implied values iV_{oc} and *iFF* of lifetime measurements on solar cells without LCO and metallization. These data are determined by injection dependent photoconductance calibrated photoluminescence imaging (PCPLI) [8].
The PCPLI calibration procedure is performed on a both sides textured sample with the front side passivation of our solar cell deposited on both sides of the wafer. The results are given in Table 4.
The implied V_{oc} differs by roughly 20 mV from the V_{oc} of the solar cells which corresponds to a calculated $J_{0,metal}$ contribution of ~15 fA/cm^2 for our solar cells. The implied fill factor and the pseudo fill factor are in the same range. Referring to Green's formula, at the current level of V_{oc} and J_{sc} of our solar cells, an improvement to ~85.0% is theoretically possible [7].

Table 4: Comparison of implied open-circuit voltage iV_{oc} and implied fill factor *iFF* determined by PCPLI for solar cells without metallization with V_{oc} and *FF* of full solar cells with metallization for the SiO_x experiment of Table 2. Given are median values.

Gr	Var	iV_{oc} (mV)	*iFF* (%)	V_{oc} (mV)	p*FF* (%)
B1	SiO_xN_y medium	737	84.4	715	84.2
B2	SiO_x thin, 800°C	741	84.2	720	84.4
B3	SiO_x thin, 825°C	736	84.5	716	84.1
B4	SiO_x thin, 850°C	726	84.2	704	84.1
B5	SiO_x medium, 800°C	745	84.9	721	84.3
B6	SiO_x medium, 825°C	738	84.6	716	84.3
B7	SiO_x medium, 850°C	729	84.3	709	84.2
B8	SiO_x thick, 850°C	735	84.4	714	84.1

4 SUMMARY AND CONCLUSIONS

We tested two different oxidants for a PECVD interfacial oxide grown in situ together with a PECVD-*n*-a-Si on POLO IBC solar cells using our structured PECVD deposition process through a glass shadow mask and optimized the oxide thickness for both variants. For both PECVD interfacial oxides SiO_xN_y from N_2O plasma and SiO_x from oxygen plasma, which is a more environmentally friendly alternative, we get similar median efficiencies at optimum oxide thickness of 23.7% and 23.6% respectively and at maximum 23.8% and 23.7%. These solar cells outperform our wet chemical interfacial oxide.
For SiO_x we showed that either an oxide thickness variation or an anneal temperature variation can lead to

optimum efficiency results.
Comparison of solar cells without metallization with the results of full solar cells for the SiO_x experiment shows a decrease of ~20 mV by our metallization.

5 ACKNOWLEDGEMENTS

This work is funded by the German Federal Ministry for Economic Affairs and Climate Action under grant FKZ 0324294C (UltraPERC), 0324274B (Genesis), 03EE1102E (Popei), by the state of Lower Saxony, Germany and the European Union's Horizon Europe research and innovation program under grant agreement No.101084259, project IBC4EU.

6 REFERENCES

[1] EE. Bende et al., 7th Metallization and Interconnection Workshop, Konstanz, Germany, 2017.

[2] F. Haase et al., Proc. 46th IEEE PVSC, Chicago, IL, USA, 2019, 2200-2206.

[3] T. Dullweber et al., presented at SNEC, Shanghai, China, 2021.

[4] M. Stöhr et al., ACS Appl. Energy Mater., 4 (5), 4646–4653 (2021).

[5] T. Dullweber et al., Proc. 8th WCPEC, Milano, Italy, 2022, 35-39.

[6] V. Mertens et al., Proc. 38th EUPVSEC, 2021, 135-139.

[7] M. Green, Solar Cells 7, 337 – 340 (1982-1983).

[8] S. Herlufsen et al., phys. stat. sol. (RRL) 2, 245-247. https://doi.org/10.1002/pssr.200802192

DEMONSTRATION OF AN INDUSTRIAL CONCEPT FOR THE METALISATION AND INTERCONNECTION OF P-IBC CELLS USING ALUMINIUM FOIL BY LASER PROCESSES

Gernot Emanuel[1], Oliver John[1], Jan Paschen[1], Matheus Melati Menegassi[1], Andreas Nägele[1], Jan Lossen[2], Tobias Messmer[2], Thomas Pernau[3], Jonas D. Huyeng[1], Jan Nekarda[1]
[1]Fraunhofer Institute for Solar Energy Systems ISE, Heidenhofstraße 2, 79110 Freiburg, Germany
[2]International Solar Energy Research Center Konstanz e.V.,Rudolf-Diesel-Straße 15, 78467 Konstanz, Germany
[3]centrotherm international AG, Württemberger Str. 31, 89143 Blaubeuren, Germany
Corresponding author: gernot.emanuel@ise.fraunhofer.de

ABSTRACT: This paper presents a metallisation concept for p-IBC solar cell based on FoilMet® technology. Necessary adjustments to the process flow are identified. The partly substitution of screen printed fingers, bus bars and interconnecting pads using one piece of aluminium foil (Al-foil), can drastically reduce silver and aluminium consumption. Furthermore, this opens up several possibilities to easily interconnect these solar cells to strings. The contacting of the p-base is performed by welding the foil onto screen printed and fired aluminium contacts, as already shown for a standard PERC structure [1–3]. For the n-regions a similar process was used on locally printed and fired screen printed Ag structure respectively. To improve the adhesion the foil is bonded to the surrounding passivation layer. To avoid shunting of both polarities, the foil is subsequently separated by means of a further laser process. It is shown that this can be achieved without inflicting damage by applying a defined pattern of an insulating layer underneath the lasered area. All relevant laser processes are proven by applying the processes on p-Cz solar cells.
Keywords: Solar cell, back contact, laser processing

1 INTRODUCTION

In the current mass production of PV cells, n-type TOPCon cells represent the second largest part, next to the PERC technology [10]. Furthermore, in addition to heterojunction and tandem cell technology, IBC cells are also expected to further establish on the market and a cell efficiency potential of 25% has been demonstrated [4]. For the latter, the two projects POPEI and SPINAT are investigating concepts and potentials for an industrial implementation on a p-type substrate. The basic idea in "POPEI" is to use already existing technologies in p-type PERC production lines with few adjustments, to establish a cell process flow for p-IBC cells, based on standard screen printed contact grid. The project SPINAT aims to replace parts of this conventional metallisation by using aluminium foil instead, which is joined and structured by means of laser radiation.

2 AIM AND APPROACH

The main objective of this work is to adapt the FoilMet® technology for the use in contacting and interconnecting p-IBC cells, as well as to reduce the amount of Ag and Al paste and to enable an innovative cell interconnection with further material saving options. This is to be achieved by using a thin piece of aluminium foil with a thickness of less than 12µm in combination with Laser based bonding, welding and foil separation processes. This paper explains the complete approach in detail and describes the development of the different laser processes and their successful evaluation on solar cell level.

2.1 p-IBC production processes

Starting from a PERC solar cell production based on a p-type c-Si wafer, Figure 1 show the necessary adaptations of the process flow for the production of FoilMet® metallised p-IBC cells. Processes and technologies are shown in dark green for which process adaptation is required, but for which the same equipment can be used in principle. Additional process steps are shown in light green. This includes the necessary TOPCon-layer (Tunnel Oxid Passivating Contact) deposition [5]. No adaptation for the FoilMet® metallisation is required in the "Frontend" process flow.

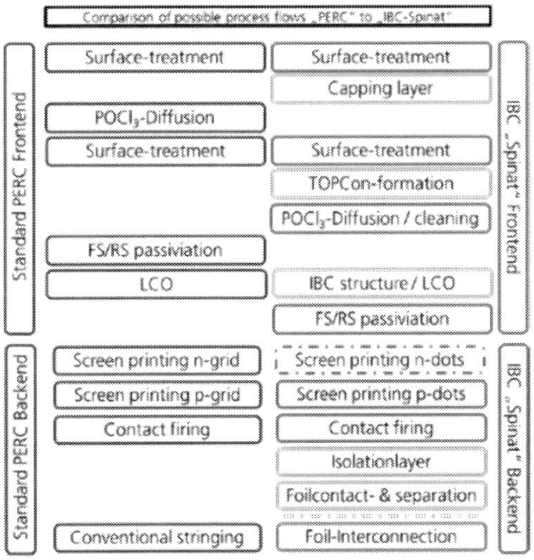

Figure 1: Comparison of the basic process flows for the production of a PERC solar cells (in red) and p-IBC cells. In dark green are processes where only adaptations are required, in light green are additional processes.

At the backend process flow, an additional step is required, namely the application of an insulation layer. This will be discussed in more detail below. The necessary steps of applying the Al-foil, connecting and adhering the foil, as well as separating it, can all be accomplished within a single piece of equipment.

2.2 p-IBC POPEI cell structure

A p-IBC cell features contacts for both polarities on the rear side. A cross-section is shown in Figure 2 The contacts are realized with a standard screen printing (SP) technology. The cell base is contacted with an aluminium paste analogous to a PERC solar cell, which forms a back surface field during the co-firing step. The TOPCon layer at the rear side of the p-IBC cell (see right side in Figure 3), forms the n-emitter and is contacted with a fired-through Ag paste. The front side has a "Free-Floating-Emitter" (FFE) with a passivating anti reflection layer on top.

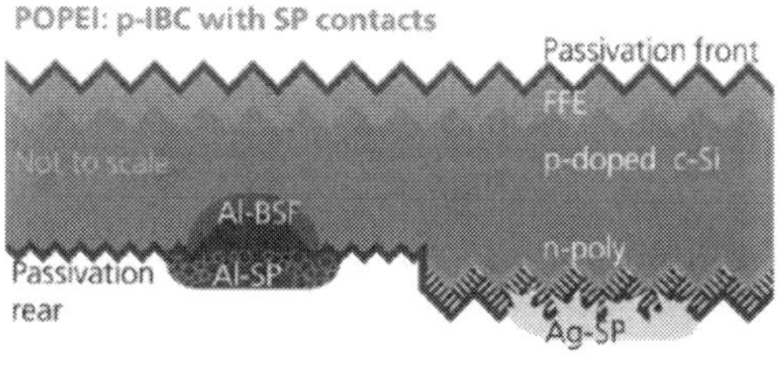

Figure 3: p-IBC cell structure developed in the POPEI project, with the screen-printed (SP) contacts on the rear. The base contact is shown on the left and the emitter contact is shown on the right.

2.3 Concept for FoilMet® metallisation of p-IBC

Transitioning from a conventional screen printed Interdigitated Back Contact (IBC) structure (Figure 3, top) to FoilMet® metallisation requires several adjustments which are illustrated step by step in Figure 3.

By contacting the p- and n- areas with a single piece of Al-foil, it is possible to omit the busbars of both polarities (Figure 3, 2nd row).

As a second optimization, the contact fingers can be interrupted in regions where the Al-foil assumes the contact role. This is because screen printed structures are solely responsible for extracting charge carriers from underlying regions and no longer need to laterally transport the current, as this task is now handled by the aluminium foil. Consequently, a "dot-dash" structure is formed as shown in Figure 3, 3rd row.

As the current flow in the transport direction intensifies, an appropriate geometry in the foil layout must be considered. In the current concept, this is accomplished by utilizing a progressively enlarging triangle that aligns with the direction of the current flow (Figure 3, bottom). To prevent electrical short circuits, the contact fingers in the emitter and base areas must be covered with an insulating layer (depicted in green). Additionally, the insulating layer (represented by the green lines in the "triangular structure") is employed to segregate the two areas and shield the cell from laser damage during the separation step. This insulating layer can be realised via a screen printed solder resist. The separation of the foil is achieved through a "laser foil separation" (LFS) process, ensuring electrical isolation.

2.4 Laser processes for FoilMet® metallisation

Figure 4 illustrates the applied laser processes involved in the production of a p-IBC FoilMet® cell. These processes encompass three distinct types of laser process. The laser metal bonding (LMB) [6] process, which is utilized to securely adhere the foil to the passivating layer. The laser metal welding (LMW) process on screen printed structures. This process has been previously demonstrated by J. Paschen et al. for aluminium fingers [1]. The laser foil separation (LFS) process, as expounded upon in section 2.3, to separate the Al foil within the active cell area.

Figure 2: (Top) Basic structure of two IBC rear sides with emitter metallisation (orange) and base metallisation (blue) and standard cell connector strings on the busbars. (2nd row) the IBC structure without bus bars. (3rd row) the adapted IBC structure for FoilMet® metallisation and (Bottom) with isolation layer in green.

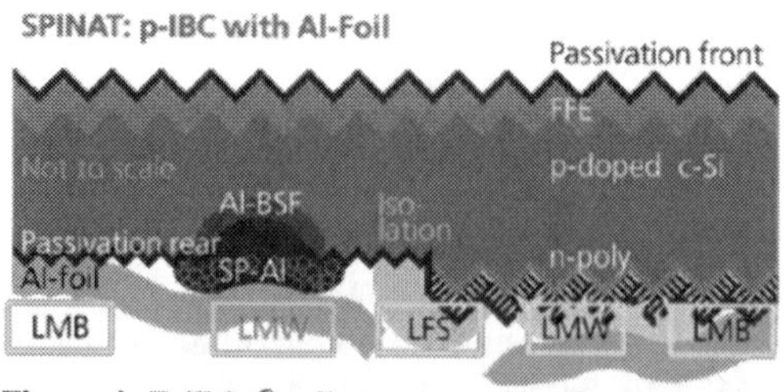

Figure 4: FoilMet® cell structure with all relevant laser processes. Abbreviations are explained in the text. For the laser processes with a yellow frame, the prove for contacting a p-IBC cell are demonstrated in this work.

3 EXERIMENTAL SETUP

The verification of the feasibility of the required laser processes is shown in Figure 4 and was conducted on two distinct test structures.

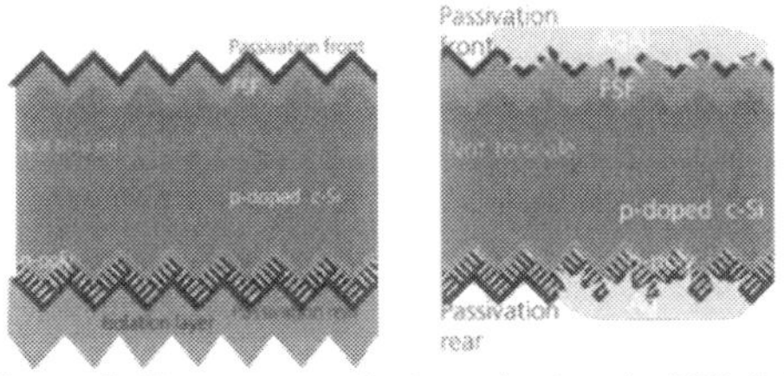

Figure 5: Test structure for investigating the LFS (left). The insulation layer (in green) on the rear side of the TOPCon layer is applied on the entire surface. TOPCoRE cell with an identical rear side as in the SPINAT p-IBC cell (right). The front side is realised with a boron FSF, which can be contacted by screen printing (AgAl).

3.1 Laser foil separation process

To develop the laser foil separation process (LFS), samples with a similar structure to our p-IBC cell, but without any electrical contacts and only featuring the kind of rear side stack, which appears where the LFS process is performed later in cell processing (see Figure 5 left). This structure involved the production of p-IBC lifetime samples without metallisation. The backside of the structure featured a surface-filling TOPCon layer, which was identical to the emitter regions of the p-IBC cell. Two different deposition technologies for the TOPCon layer (LPCVD / PECVD) were employed in two separate groups. Following this, the test structures were coated with a full-area isolation layer on top of the TOPCon layer.

The determination of parameters for the laser foil separation (LFS) process relies on two essential criteria. Firstly, it is crucial that the foil is completely separated, ensuring successful isolation. Secondly, the process must not result in significant damage to the underlying TOPCon layer. To assess potential laser-induced damage, initial measurements of the test samples (Figure 5 left) were conducted using modulated photoluminescence (mod-PL) equipment [7]. Subsequently, the sample was positioned with its front side on a vacuum chuck of a laser device, with the Al foil placed over it and the air between the foil and the sample evacuated. Using this setup, a randomized 10x10 field pattern with various laser parameters was applied to different samples. An industrial low-power picosecond infrared (IR) fiber laser served as the laser source. Following the laser processes, another mod-PL measurement was performed, and the resulting images were analysed accordingly and the iV_{oc} value was determined. By comparing the images before and after the laser processes, the extent of damage caused by different laser parameters could be evaluated (see Figure 7).

It is important to note that the lasered foils were not bonded to the test samples, allowing for the subsequent removal of the foils after the laser processes. This enables the assessment of separation quality for each laser parameter using transmitted light microscopy (see Figure 6).

3.2 Laser metal bonding & welding on TOPCoRE cells

To join the Al-foil to the emitter area of the p-IBC cell, the combination of the laser metal welding (LMW) process to bond the Al foil to the screen printed Ag structure, and laser metal bonding (LMB) process to attach the Al foil to the passivation layer are applied. In order to validate the combination of these both processes four criteria have to be met. Firstly, the foil should adhere strongly enough to the rear side of the cell to prevent detachment. Moreover, the process must not cause damage to the passivation layer (2), Ag fingers (3), or the underlying emitter structure beneath the passivation layer (4).

TOPCoRE (Tunnel Oxid Passivating Rear Emitter) cells [8]were utilized (see Figure 5 right). The rear side of the TOPCoRE cell is identical to the emitter region of the p-IBC cell. On the front side, a screen-printed and contactable front surface field (FSF) was applied. This is relevant for a I-V measurement but does not affect the laser processes under investigation. The TOPCoRE cells (Figure 5 right) were placed with their front side on a laser vacuum chuck. Subsequently, the Al foil was placed over the cell, following a similar procedure as in the LFS experiments, and the air between the foil and the cell was evacuated. A "multi-dash" laser pattern, comprising four fields with three dashes each, approximately 500µm long, arranged orthogonally, was applied to the foil, covering the entirety of the rear side of the cell. In regions without screen-printed structures, the foil was bonded to the passivation layer through the laser process. In areas with metallisation, a welding process takes place between the Al foil and the Ag finger [9].

The adhesion force between the foil and the rear side of the cell was determined using a 90° peel-off test, which is not discussed here, as only processes with sufficient adhesion were utilized. To analyse potential damage to the passivation layer and the emitter structure, modulated photoluminescence (mod-PL) measurements were conducted before and after the laser processes on the respective cells. Additionally, cells with a pure screen-printed grid (without bus bars) were manufactured. To demonstrate the effectiveness of the FoilMet® metallisation on these structures, cells of this type were fabricated using the process mentioned. I-V measurements were performed on both cell types using a commercially cell tester.

4 RESULTS

The results obtained with the described experimental setup are presented below, along with one possible concept to implement the FoilMet® technology in a p-IBC cell.

4.1 Proof of Laser Foil Separation on TOPCon layer

Figure 6 presents transmitted light images of the samples described in section 3.1. The first two lines, indicated by green frames, depict successful separation of the foil. The grey regions represent the ablated area, while the black regions represent the Al foil. The process of the intermediate state of separation (Figure 6 middle) cannot be employed due to insufficient process stability, which hinders the achievement of electrically isolated polarities throughout the entire structure. The red frames in the lower section of Figure 6. illustrate isolated areas that have been penetrated but are not contiguous. In this instance, the isolation has failed and therefore the polarities would not have been separated.

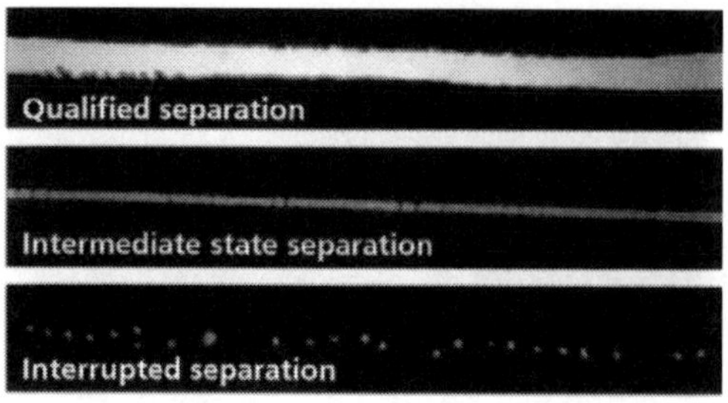

Figure 6: Microscope transmitted light images that were analyzed via Pythonscript. From qualified separation (top) to an interrupted separation (bottom)

Only processes with laser parameters that lead to a "qualified separation" can be used for solar cell processing.

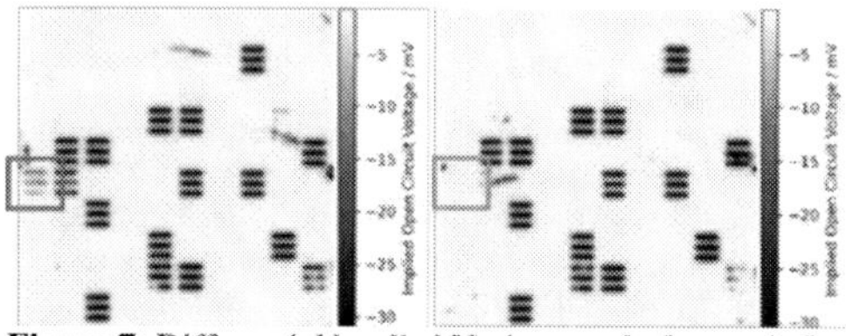

Figure 7: Differential implied V_{oc} images, before and after LFS. The visible fields correspond to the 10x10 laser parameter variation performed. The green and orange rectangles correspond to the marked parameters in Figure 8.

In order to assess the potential damage caused by the LFS processes on the underlying cell structure beneath the isolation layer, the relevant laser parameters were evaluated using delta iV_{oc} images. Figure 8 illustrates the amount of damage observed for different laser parameters, as indicated by the 10x10 field pattern mentioned earlier. In Figure 8 the subplots arranged over the y-axis the laser power is varied, over the x-axis the pulse pitch. Each Graph shows the number of passes on x-axes versus the drop in V_{oc} after the LFS process on y-axes. Symbols in grey are parameters not having fully separated the foil. The shape and the colour of the symbols represent the variation in TOPCon-layer and isolation layer thickness. The framed areas in both Figure 8 and Figure 9 represent parameter sets that enable complete separation of the foil while minimizing or even eliminating damage to the sample. For the insulating layer with a thickness of 5µm, it is evident that an iV_{oc} value of less than 10mV can be achieved, indicating minimal damage. If the thickness of the insulating layer is doubled to 10µm, damage can be reduced further. This observation holds true for both the TOPCon layers produced using PECVD and LPCVD deposition techniques.

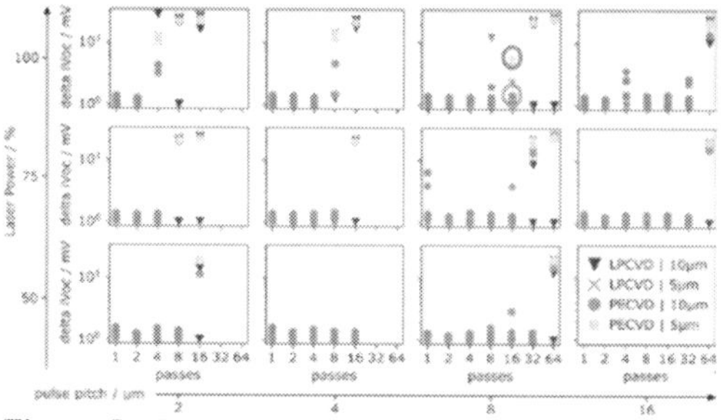

Figure 8: Overview of the laser parameter variation, applied on PECVD and LPCVD TOPCon-layers with two different thickness of insulating layers.

4.2 Proof of Laser Metal Welding & Bonding on TOPCoRE cells

iV_{oc} images resulting from the analysis of the mod-PL measurement, indicate, that the laser processes lead to damage in the form of speckles but results in a similar iV_{oc} (Figure 9).

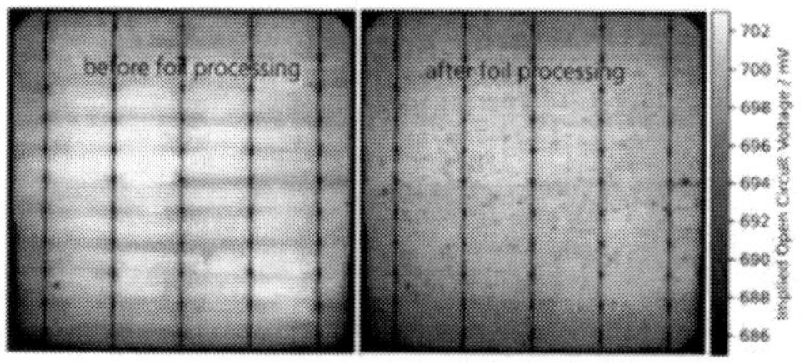

Figure 9: mod-PL images before (left) and after (right) processing the cells with FoilMet® metallisation.

To investigate any potential influence of the laser process on cells, the screen-printed references and the cells produced using the FoilMet® technique were subjected to measurements using the commercial *I-V* cell tester. Table I presents the data for the best-performing cells. Upon comparing the two cell types, it is evident that both the efficiency, with 21.5%, and the short circuit current density, with 39.0 mA/cm², are identical. The open circuit voltage can also be considered practically identical within the limits of measurement accuracy. However, a slight decrease in the fill factor (FF) was observed. The exact cause of this effect has not been fully explored and will require further investigation. In summary, cell performance is not significantly affected by the FoilMet process.

Table I: *I-V* measurements performed with a commercially available cell tester

	V_{oc}	J_{sc}	η	FF
	(mV)	(mA/cm²)	(%)	(%)
SP reference	698.8	39.0	21.5	79.1
FoilMet® cell	701.2	39.0	21.5	78.5

4.3 Concept for the FoilMet® metallisation on p-IBC cells

Based on the findings in the previous section, a concept for FoilMet® metallisation of p-IBC cells has been developed. Beginning with a cell featuring the modified IBC structure that includes the insulation layer

(Figure 3, bottom), the Al foil is applied uniformly over the cell(Figure 10). Leveraging the established laser processes ("Foilcontact & separation" from Figure 1,), the Al foil can be bonded to the cell, welded to the contacts and precisely separated (Figure 10, laser path in red).

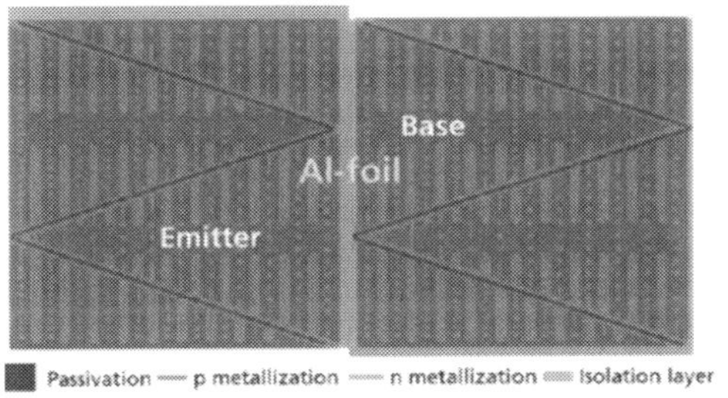

Figure 10: Concept to realise FoilMet® metallisation on p-IBC cells with LFS (in red) and an interconnection of cells to fabricate a string.

Furthermore, Al foil loosely extended over the cell border can be utilized to establish various FoilMet® interconnections possibilities between cells, as illustrated by the "Foil-interconnect" process in Figure 1.
Implementing this concept, an initial estimation indicates that the potential for reducing silver usage, considering the potential modifications in the IBC layout, amounts to approximately 45% for an p-IBC cell in total. Around 30% of this reduction is attributed to the elimination of busbars and pads. The remaining 15% reduction is achieved through the implementation of partially interrupted fingers.

5 CONCLUSION AND OUTLOOK

5.1 Conclusion

In this study a successful application of all essential laser processes for contacting p-IBC cells using FoilMet® metallisation technology at cell level was presented. The laser processes (LMB / LMW) demonstrated excellent quality, achieving equal efficiency to screen-printed references. An efficiency of 21.5% was achieved on TOPCoRE solar cells. Furthermore, a novel concept for the industrial production of FoilMet® p-IBC solar cellsvia the application of an enhanced IBC geometry was developed. This concept offers significant potential for reducing silver usage, with an estimated total reduction potential of 45%. Overall, this work contributes to advancing the field of p-IBC solar cell technology and highlights the feasibility and benefits of implementing the FoilMet® metallisation approach.

5.2 Outlook

Having successfully demonstrated the necessary laser processes for FoilMet® metallisation of p-IBC cells and developed a comprehensive concept, the next phase involves fabricating p-IBC cells and utilizing them to produce a FoilMet®-IBC string. To further minimize the usage of Ag screen printing paste, it is imperative to simulate and subsequently implement these results for reducing the amount of Ag paste within the cells.

In order to completely eliminate the need for silver paste in the emitter regions, the development of a Laser Metal Contact (LMC) process is necessary. This process aims to establish direct contact between the TOPCon-Emitter region and the Al foil by the using the FoilMet® metallisation technology (see Figure 11).

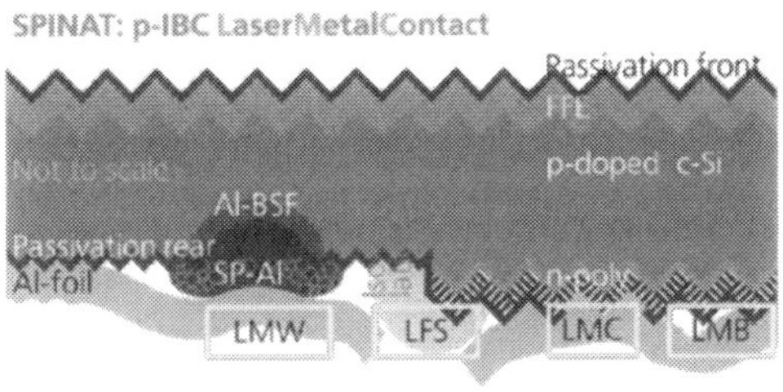

Figure 11: Direct FoilMet® metallisation of the TOPCon layer (emitter) with the Laser-Metal-Contact (LMC) process

5 ACKNOWLEDGMENTS

The authors acknowledge that part of this work was supported by funding from the German Federal Ministry for Economic Affairs and Climate Action within the research project "POPEI" (contract no. 03EE1102) and "SPINAT" (contract no. 03EE1124)

7 REFERENCES

[1] J. Paschen, G. Emanuel, A. A. Brand, T. Fellmeth, A. De Rose, O. John, and J.-F. Nekarda, "FoilMet® - Connect: A new rear metallization upgrade for PERC and other cell concepts," *AIP Conf. Proc.*, vol. 2156, p. 20004, 2019.

[2] J.-F. Nekarda, M. Graf, A. Rodofili, A. Wolf, and R. Preu, "Laser-based foil rear side metallization for crystalline silicon solar cells," *Proc. of SPIE*, vol. 8473, 847307-847307-14, 2012.

[3] M. Graf, J.-F. Nekarda, D. Eberlein, N. Wöhrle, R. Preu, R. Böhme, and T. Große, "Progress in laser-based foil metallization for industrial PERC solar cells," in *29th European Photovoltaic Solar Energy Conference and Exhibition*, Amsterdam, The Netherlands, 2014, pp. 532–535.

[4] F. Haase, C. Hollemann, S. Schafer, J. Krugener, R. Brendel, and R. Peibst, "Transferring the Record p-type Si POLO-IBC Cell Technology Towards an Industrial Level," in *46th IEEE Photovoltaic Specialists Conference (PVSC)*, Chicago, IL, USA, 2019, pp. 2200–2206.

[5] F. Feldmann, B. Steinhauser, V. Arya, A. Büchler, A. A. Brand, S. Kluska, M. Hermle, and S. W. Glunz, "Evaluation of TOPCon technology on large area solar cells," in *33rd European Photovoltaic Solar Energy Conference and Exhibition*, Amsterdam, The Netherlands, 2017, pp. 465–467.

[6] O. John, J. Paschen, A. De Rose, B. Steinhauser, G. Emanuel, A. A. Brand, and J.-F. Nekarda, "Laser Metal Bonding (LMB) - low impact joining of thin aluminum foil to silicon and silicon nitride surfaces," *Procedia CIRP*, vol. 94, pp. 863–868, 2020.

[7] H. Höffler, F. Schindler, A. Brand, D. Herrmann, R. Eberle, R. Post, A. Kessel, J. M. Greulich, and

M. C. Schubert, "Review and Recent Development in Combining Photoluminescence- and Electroluminescence Imaging with Carrier Lifetime Measurements via Modulated Photoluminescence at variable Temparatures," in *37th European Photovoltaic Solar Energy Conference and Exhibition*, Online, 2020, 264 - 276.

[8] A. Richter, R. Müller, J. Benick, F. Feldmann, B. Steinhauser, C. Reichel, A. Fell, M. Bivour, M. Hermle, and S. W. Glunz, "Design rules for high-efficiency both-sides-contacted silicon solar cells with balanced charge carrier transport and recombination losses," *Nat. Energy*, vol. 6, no. 4, pp. 429–438, 2021.

[9] J. Paschen, P. Baliozian, O. John, E. Lohmüller, T. Rößler, and J.-F. Nekarda, "FoilMet ® - Interconnect: Busbarless, electrically conductive adhesive-free, and solder-free aluminum interconnection for modules with shingled solar cells," *Prog Photovolt Res Appl.*, vol. 30, no. 8, pp. 889–898, 2022.

[10] VDMA, ITRPV_14th_edition_presentation_PVCellTech

40th European Photovoltaic Solar Energy Conference and Exhibition

This presentation was selected by the Sc. Committee of the EU PVSEC 2023 for submission of a full paper to one of the EU PVSEC's collaborating peer-reviewed journals.

ALTERNATIVE METALLIZATION: SCREEN PRINTED NICKEL CONTACTS WITH GRAPHENE LAYER IN BETWEEN

Berkeli Akgayev[1] Aysegul Sezgin[2], Mucahit Yilmaz[3] and Veysel Unsur[3,4] *

[1] Department of Energy Systems Engineering, Necmettin Erbakan University, Konya, Türkiye

[2] Department of Nano Science and Nano Engineering, Necmettin Erbakan University, Konya, Türkiye

[3] Department of Fundamental Sciences, Necmettin Erbakan University, Konya, Türkiye

[4] Center of Solar Energy Research and Applications, ODTÜ-GÜNAM, Ankara, Türkiye

*Corresponding author: vunsur@erbakan.edu.tr, veysel.unsur@odtugunam.org

ABSTRACT: This research investigates the potential of nickel (Ni) metal paste as an alternative to traditional silver contacts in crystalline silicon (Si) solar cells. The primary objective is to achieve cost reductions while maintaining or enhancing cell efficiency. A distinctive feature of this approach is the introduction of a graphene interlayer between the nickel and silicon, designed to circumvent the challenges associated with Schottky contact formation. This graphene layer ensures an ohmic or near-ohmic contact, enhancing the cell's performance. The study further explores the properties and composition of the Ni metal paste, emphasizing the roles of glass frit and organic vehicle. The selected glass frit demonstrates a transition temperature conducive to optimal etching behavior, while the organic vehicle ensures the paste's printability. Preliminary findings indicate that the Ni contacts achieve a contact resistance of approximately 5.2 mΩ.cm^2, and light current-voltage measurements suggest an efficiency rate surpassing 21%.

Keywords: silicon solar cells, metallization, screen printing, Ni contacts

1 INTRODUCTION

Traditionally, screen-printed silver (Ag) has been the preferred material for contacts owing to its superior conductivity and ease of soldering onto silicon (Si). However, the high cost of silver, which constitutes nearly 40% of the total cell cost [1] due to limited availability, has been a significant drawback. Numerous research efforts have been directed towards finding a cost-effective alternative, with a focus on metals like nickel (Ni) and copper (Cu) [2–5], which can be applied through electro- or light-induced plating methods.

In the realm of photovoltaic applications, Cu and Ni are often used in tandem, since copper tends to diffuse into silicon even at ambient temperatures, thereby compromising the p-n junction. Typically, nickel serves as a foundational layer on the silicon, with copper plated onto it to establish a conductive yet non-diffusive contact for the solar cells. However, this two-step process, coupled with the plating procedure, escalates the overall costs due to increased processing steps and substantial consumable usage.

Moreover, the prevalent use of metals forms a Schottky contact [6] with silicon instead of an ohmic contact, resulting in high contact resistivity at the metal/silicon interface, which adversely affects the overall efficiency. For the Ni/Cu contacts, relying on the random formation of nickel silicate (NiSi) between Ni and Si to make an ohmic contact makes the process more complicated. Therefore, there is a pressing need for innovative approaches that can replace silver without diminishing the solar cell's efficiency.

In this study, we propose a groundbreaking technique that employs screen-printed nickel in lieu of silver for silicon wafer-based solar cells, potentially reducing the metallization cost per cell by over 70%. Given nickel's lower conductivity compared to silver and the potential formation of a Schottky contact at the metal-semiconductor junction, we suggest introducing a graphene layer between the nickel and n-type silicon (emitter). This intervention is anticipated to lower the Schottky barrier height, enabling the contact to function similarly to an ohmic contact [7,8].

Screen-printable metal pastes are typically composed of metal powder, glass frit, and an organic vehicle, each playing a crucial role during the metallization and contact co-firing stages. The metal powder forms the contacts, fingers, and busbars, facilitating carrier transport out of the cells, with its structure influencing gridline and contact resistances [9–11]. Meanwhile, the glass frit, containing various metal oxides, etches through the anti-reflection coating (ARC) layer, usually a silicon nitride film approximately 70 nm thick, facilitating contact between the metal bulk and the silicon emitter without causing damage or shunting issues [12–15]. The organic vehicle, comprising solvents, binders, surfactants, and rheological agents, not only governs the paste's printability but also determines the aspect ratio and stability of the final contacts [16–18].

2 EXPERIMENTAL DETAILS

In this research, custom-made glass frit and organic vehicle were developed to cater to the specific needs of the nickel paste formulation. To pinpoint the optimal glass frit, we utilized Thermogravimetric Analysis (TGA) and Differential Scanning Calorimetry (DSC) measurements, facilitated by equipment from Seteram Labsys Evo.

The components were uniformly mixed using a three-roll milling process (Exakt50I), and the viscosity of the resultant paste was assessed using a Brookefield-DVNext rheometer, aiding in the selection of the suitable nickel metal paste.

Simultaneously, 50 mm x 50 mm monolayer graphene sheets were generated through Chemical Vapor Deposition (CVD) on copper foils and later transferred onto textured silicon wafers with the assistance of photoresist. Subsequently, these pastes were screen printed onto M2 size PERC cells (156.75mm x 156.75mm), which were pre-coated with a graphene layer on the upper surface, followed by a contact co-firing procedure in a Rapid Thermal Process (RTP) furnace.

The final stage of the experiment involved evaluating the solar cells' performance through various assessments including electrical output parameters, light current-voltage (I-V) measurements, and analyzing the Transmission Line Model (TLM) contact and finger resistances.

3 RESULTS AND DISCUSSION

Initially, a detailed analysis of the metal paste components, namely the glass frit and the organic vehicle, was conducted to ascertain the printability of the Ni metal paste and the etching capability of the glass frit on the solar cells' ARC layer. As depicted in Figure 1a, there are no crystal peaks in XRD patterns implying that the glass frit is in completely amorphous state. The DSC curve in Fig 1b illustrates the transition temperature (Tg) of the prepared glass frit, a critical factor influencing the etching behavior. A lower Tg initiates early softening during contact co-firing, providing ample time for etching, though with a risk of shunting the p-n junction. The chosen glass frit exhibits a Tg between 260°C and 300°C, aligning with softening temperatures near 500°C, where a brief plateau in the firing profile ensures sufficient time for ARC layer etching and establishing contact between the Ni bulk and the Si emitter.

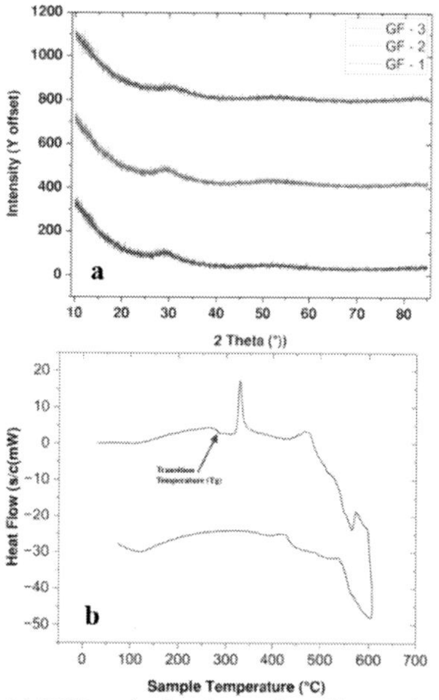

Fig. 1: (a) XRD peaks for chosen glass frits proving that frits are in amorphous state **(b)** DSC analysis that the transition temperature for the glass frit used is 285°C.

Examining the second component, the organic vehicle, Observations indicate that maintaining a contact angle below 30° ensures a uniform and printable dispersion of the paste components. Additionally, the viscosity of the organic vehicles, influencing the printability and final contact aspect ratio, ranged from 320 Pa·s to 480 Pa·s, confirming satisfactory printability. Figure 2 presents the SEM image of the chosen Ni paste evenly spread on a wafer and subjected to a two-minute drying process under infrared illumination. The images illustrate the consistency in particle size and distribution, as evidenced in Figure 2, which depicts the homogeneous distribution of metal powder, glass frit, and binders. Notably, the mean metal particle dimension stands at 1.5μm, a critical attribute for a paste, facilitating smoother particle sintering during the firing phase. This characteristic is instrumental in reducing porosity and consequently diminishing gridline resistance.

Fig. 2: SEM images of the dispersed Ni paste and particle distribution

Regarding the incorporation of the graphene layer on the top surface of the cells, Figure 3 delineates the outcomes of Raman spectroscopy analyses conducted at various locations on identical cells. The findings confirm the successful transfer of graphene onto the solar cells, with distinct G and 2D peaks manifestly visible, attesting to the high quality of the graphene layer. Despite the satisfactory quality of the graphene, localized defect peaks appearing around 1350 cm-1 were noted, potentially arising during the transfer process of the graphene layer.

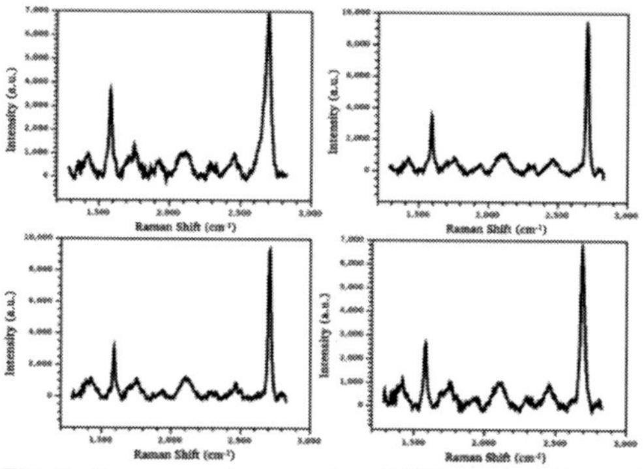

Fig. 3: Raman spectrum peaks of different locations of transferred graphene

Following the formulation of the Ni metal paste and the successful transfer of graphene onto the Si wafers, the front Ni contacts were meticulously screen printed onto the 50mm x 50mm solar cells. These were then subjected to a firing process in an IR belt furnace, the temperature profile of which is illustrated in Fig. 4a. The firing profile adopted for the Ni paste mirrors that of the silver paste, owing to the utilization of compatible metal oxides. To facilitate the enhanced sintering of Ni metal particles, the organic vehicle incorporated suitable surfactants. Furthermore, the glass frit's etching capability on the ARC layer was optimized to approximately 450°C, ensuring that the thermal dynamics of the Ni contacts remained analogous to their silver counterparts. Delving deeper into the contact specifics, Figure 4b presents the contact resistance values observed across various samples, with an encouraging average nearing 5.2 mΩ.cm², showcasing a competitive edge compared to silver-based contacts.

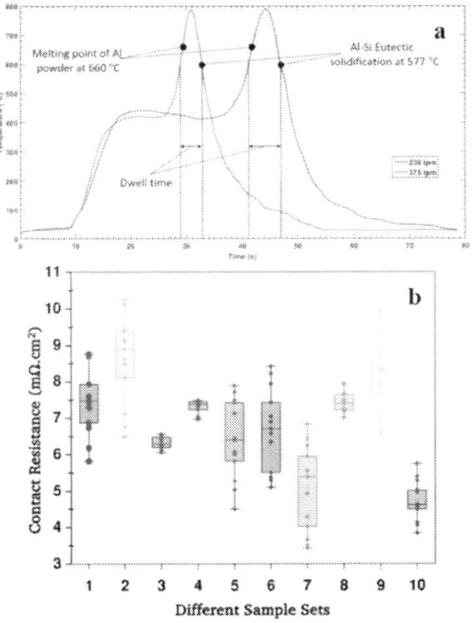

Fig. 4. (a) Firing temperature profile from the IR conveyor belt furnace **(b)** TLM measurements of different Ni contacts

Additionally, the light current-voltage (I-V) measurements for cells outfitted with Ni contacts, revealing promising results. The data indicates an open circuit voltage (VOC) of 660mV, a short circuit current density (JSC) of 39.11 mA/cm^2, and a fill factor (FF) of 81.4%. These metrics culminate in an efficiency rate of approximately 21%, affirming that the contact formation process is devoid of any significant issues.

4 CONCLUSIONS

This research unveils a pioneering nickel metal paste that can be screen printed, offering a viable alternative to the traditionally utilized silver contacts, thereby addressing their associated cost limitations. Historically, nickel contacts have been employed solely as a foundational layer upon which copper contacts are plated, serving as a protective barrier preventing the aggressive diffusion of copper into the silicon substrate. However, this dual printing and/or plating approach introduces additional steps and materials to the manufacturing process, rendering it less desirable. Despite nickel's tendency to form a Schottky contact with silicon, the introduction of a graphene interlayer serves to facilitate a bond between the nickel and graphene, effectively reducing the metal's work function and enabling the acquisition of ohmic contacts.

Acknowledgment
The authors gratefully acknowledge the support by the Scientific and Technological Research Council of Turkiye (TUBITAK) under grant no 120E474.

References

[1] A. Kumar, M. Bieri, T. Reindl, A.G. Aberle, Energy Procedia. 130 (2017) 43–49.

[2] S.W. Glunz, M. Aleman, J. Bartsch, N. Bay, K. Bayer, R. Bergander, A. Filipovic, S. Greil, A. Grohe, M. Horteis, A. Knorz, M. Menko, A. Mette, D. Pysch, V. Radtke, P. Richter, D. Rudolph, T. Rublack, C. Schetter, D. Schmidt, O. Schultz, R. Woehl, 33rd IEEE Photovoltaic Specialists Conference, 2008: pp. 1–4.

[3] V. Unsur, T. Klein, M.F.A.M. van Hest, M. Al Jassim, A. Ebong, Progress in Photovoltaics: Research and Applications. 27 (2019) 453–459.

[4] D. Rudolph, R. Farneda, T. Timofte, A. Halm, N. Chen, J. Libal, F. Buchholz, I. Rosen, M. Grouchko, O. Shochet, , AIP Conference Proceedings. 2709 (2022) 020006.

[5] A. Ebong, S. Huneycutt, S. Grempels, K. Ankireddy, R. Dharmadasa, T. Druffel, 2021 IEEE 48th Photovoltaic Specialists Conference (PVSC), 2021: pp. 1417–1420.

[6] W. Schottky, Z. Physik. 118 (1942) 539–592.

[7] K.-E. Byun, H.-J. Chung, J. Lee, H. Yang, H.J. Song, J. Heo, D.H. Seo, S. Park, S.W. Hwang, I. Yoo, K. Kim, Nano Lett. 13 (2013) 4001–4005.

[8] M.-H. Lee, Y. Cho, K.-E. Byun, K.W. Shin, S.-G. Nam, C. Kim, H. Kim, S.-A. Han, S.-W. Kim, H.-J. Shin, S. Park, Nano Lett. 18 (2018) 4878–4884.

[9] V. Unsur, A. Chowdhury, N. Chen, A. Ebong, 2015 IEEE 42nd Photovoltaic Specialist Conference (PVSC), 2015: pp. 1–5.

[10] K. Batchu, A. Ebong, V. Unsur, 12th International Conference on High-Capacity Optical Networks and Enabling/Emerging Technologies (HONET), 2015: pp. 1–4.

[11] M.M. Hilali, K. Nakayashiki, C. Khadilkar, R.C. Reedy, A. Rohatgi, A. Shaikh, S. Kim, S. Sridharan, J. Electrochem. Soc. 153 (2005) A5.

[12] K. Ren, V. Unsur, A. Chowdhury, Y. Zhang, A. Ebong, 2018 IEEE 7th World Conference on Photovoltaic Energy Conversion (WCPEC)(A Joint Conference of 45th IEEE PVSC, 28th PVSEC & 34th EU PVSEC), IEEE, 2018: pp. 1051–1054.

[13] Y. Kim, J. Huh, H. Kim, Solar Energy Materials and Solar Cells. 185 (2018) 97–103.

[14] M.M. Hilali, S. Sridharan, C. Khadilkar, A. Shaikh, A. Rohatgi, S. Kim, J. Electron. Mater. 35 (2006) 2041–2047.

[15] Y. Tai, G. Zheng, H. Wang, H. Wang, J. Bai, RSC Advances. 5 (2015) 92515–92521.

[16] L. Song, C. Chen, T. GUO, W. Zhang, EP2590177A2, 2013.

[17] J. Qin, S. Bai, W. Zhang, Z. Liu, H. Wang, Circuit World. 42 (2016) 77–83.

[18] T. Nie, C. Cui, S. Ma, H. Wang, ACS Appl. Electron. Mater. 3 (2021) 861–871.

STABILITY OF MINIMODULES WITH COPPER PLATED HETEROJUNCTION CELLS

AND PATTERN TRANSFER PRINTING OF COPPER PARTICLE PASTE FOR SEED-GRID FORMATION

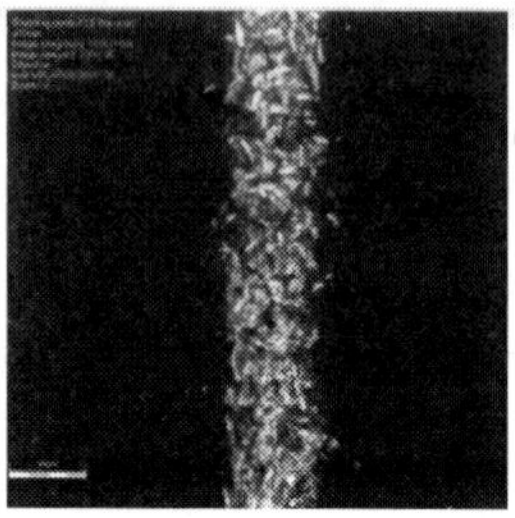

Agata Lachowicz[1], Nicolas Badel[1], Alexis Barrou[1], Vincent Barth[2], Samuel Harrison[2], Nicola Frasson[3], Marco Galiazzo[3], Natali Cohen[4], Eyal Cohen[4], Tal Marcu[4], Dor Dror[4], Jun Zhao[1], Bertrand Paviet-Salomon[1], Christophe Ballif[1]

[1]CSEM, Switzerland / [2]CEA INES, France / [3]AMAT, Italy / [4]DR UTILIGHT, Israel

40th EU PVSEC
Lisbon, 19.09.2023

EU PVSEC

MOTIVATION & OUTLINES

- High Ag consumption for HJT [1]
- Process complexity

3-STEP PROCESS

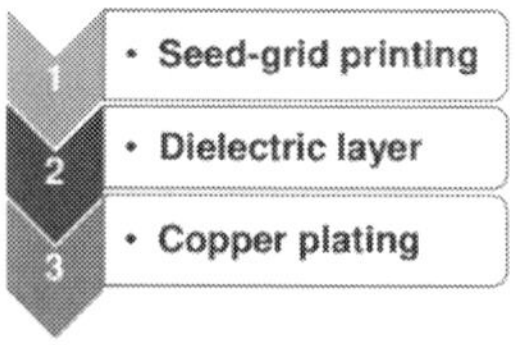

STABILITY OF MINI-MODULES
with **Smartwire** interconnection and **Shingles**

NARROW COPPER LINES
using pattern-transfer-printing of copper paste

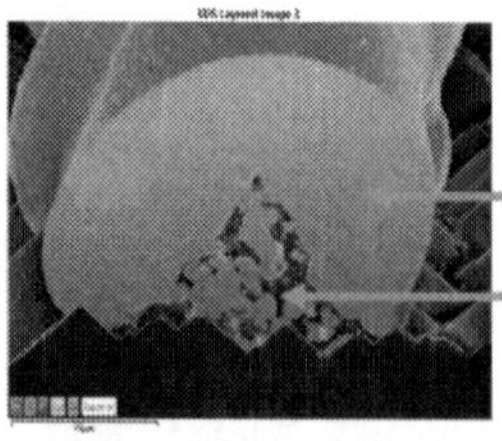

1 J. Reichle, RCT Solutions, this conference

csem

3-STEP PROCESS

STABILITY OF MINI-MODULES

Smart wires

Shingles

NARROW COPPER LINES

Pattern-transfer printing PTP

40th European Photovoltaic Solar Energy Conference and Exhibition

SELF-ALIGNED COPPER ELECTRODEPOSITION

- 3-step-process:
  1. **Seed-grid printing on (thin) TCO and curing (200°C)**
  2. **Dielectric layer deposition over the entire wafer surface**
  3. **Copper electrodeposition**

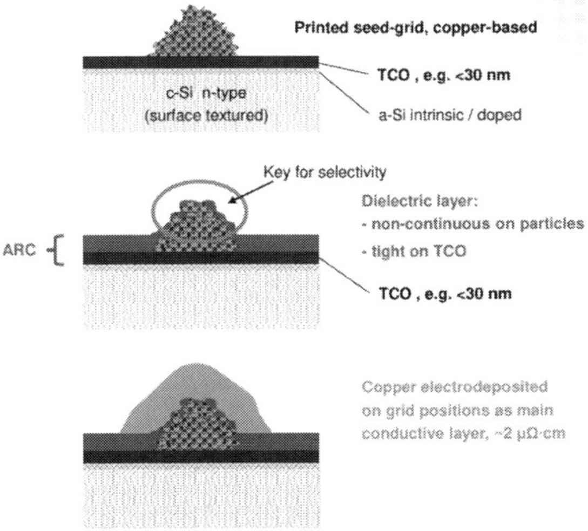

- Printing is already part of the production
- A stack of thin ITO and a dielectric layer is one of the strategies followed for Indium reduction [1,2]
- Only plating to add

4 EU PVSEC 2023 Stability of minimodules with copper plated HJT cells / PTP

1 A. Cruz et al., "Optoelectrical analysis of TCO+Silicon oxide double layers", SOLMAT, 2022
2 S. Janke et al., "Approaches for SHJ cells with low or no In content", Silicon PV, 2022

:: csem

CU ELECTRODEPOSITED ON PRINTED PASTE: SCC

- Copper pastes for screen-printing in development for PV – goal: narrow lines / only small interruptions acceptable
- Silver-coated-copper paste (SCC) used for seed-grid formation, knotless screen

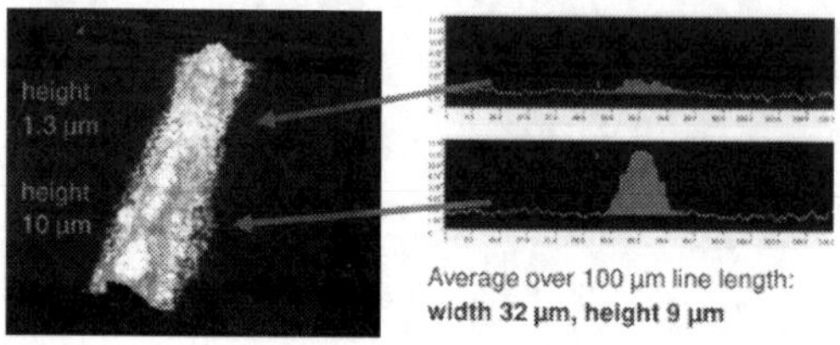

Average over 100 µm line length:
width 32 µm, height 9 µm

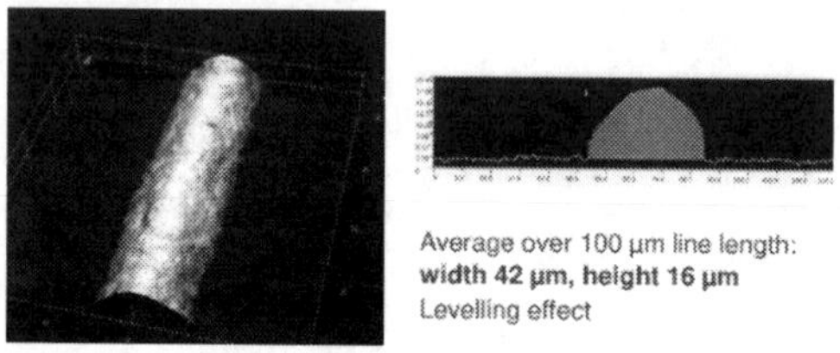

Average over 100 µm line length:
width 42 µm, height 16 µm
Levelling effect

Best result so far with SCC paste seed-grid, M6 high quality industrial precursor[1]

M6 industrial HJT precursor	Area [cm²]	Voc [mV]	Jsc [mA/cm²]	FF [%]	Eff [%]
SCC-seed-grid / dielectric layer / electrodeposited copper, 9BB-layout	274.15	742.4	39.9	81.7	24.2

Internal measurement with 9 bars on a gold-chuck, after calibration with a certified M6 cell; cell as plated / without any additional processes

5 EU PVSEC 2023 Stability of minimodules with copper plated HJT cells / PTP

1 A. Lachowicz et al., Self-aligned Copper Electrodeposition", Proc. WCPEC-8, 2022

csem

3-STEP PROCESS

STABILITY OF MINI-MODULES
Modules w. Smartwire interconnection

NARROW COPPER LINES
Pattern-transfer printing PTP

MINIMODULES WITH SMARTWIRE INTERCONNECTION: TC

- **Glass-glass modules**
  Polyolefin encapsulant
  One ½-cell, industrial M6 precursors
  Wires: with In-free low melting point alloy and supporting foil
- Seed-grid
  Silver-coated-copper paste (SCC) only
- With copper
  plated on SCC paste, same paste and same print as the reference, dielectric layer as plating mask, copper with capping layer
- Degradation PTC:
  SCC-reference: -1.4% after 690 cycles
  With copper: -0.2% after 690 cycles

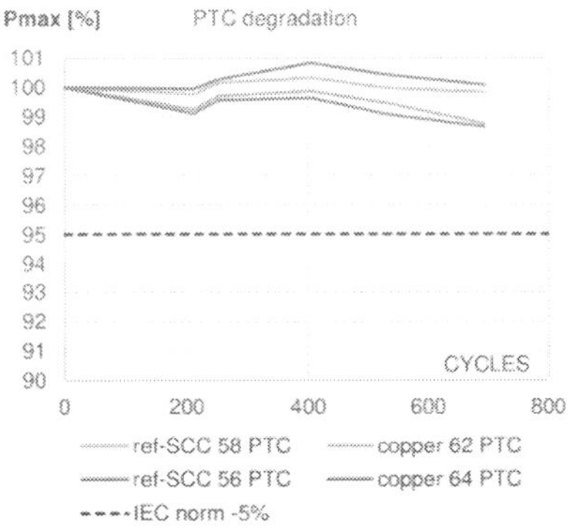

Orange/brown lines: modules with copper

Grey lines: SCC seed-grid only

:: csem

MINIMODULES WITH SMARTWIRE INTERCONNECTION: DH

- Strong DH degradation of modules with silver-coated-copper paste
  Mainly in fill factor
- The used paste not previously tested with module materials used in this experiment (wire supporting foil)
- In general:
  Good module stability has been proven for SCC paste by the industry
- Better stability w. dielectric layer and copper
- **5% degradation of modules with <u>dielectric layer</u> and copper after 2700 h**
  To re-test / with another materials

Dashed lines: Storage control

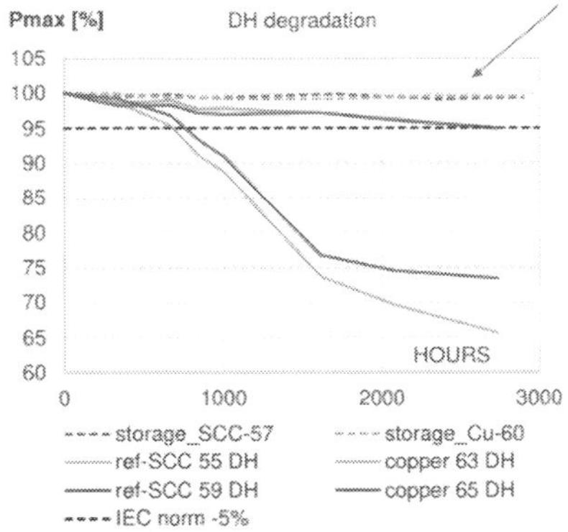

Orange/brown lines: modules with copper

Grey lines: SCC seed-grid only

:: csem

MINIMODULES WITH SMARTWIRE INTERCONNECTION: DH/EL

- EL: damp-heat degradation starts at the edges

Seed-grid
SCC paste

Seed-grid
CC paste
+ dielectric layer
+ plated Cu

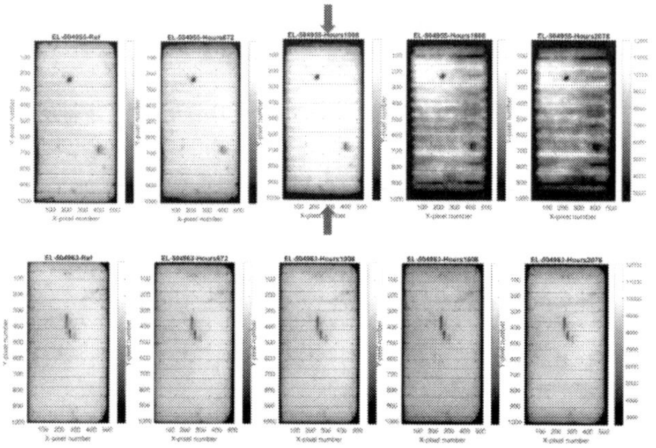

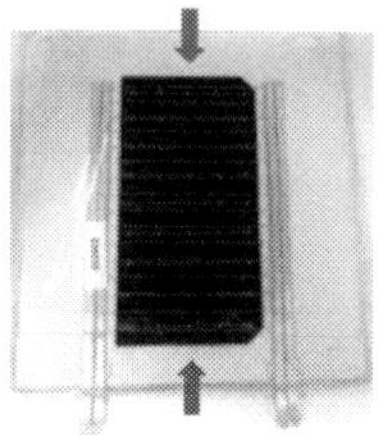

- 1-cell mini-modules: worst case for testing DH degradation? Humidity ingress from unsealed module edges close to the cell

5% degradation of 1/2-cell-modules with dielectric layer + copper

MINIMODULES WITH SMARTWIRES: PID

- Modules after storage and PTC have been re-used for PID test

- Conditions:
  85°C/85% rel. humidity (like DH)
  +1000V or -1000V
  1x IEC norm = 96 h

- **STABLE**
  Tested 2x 96 h

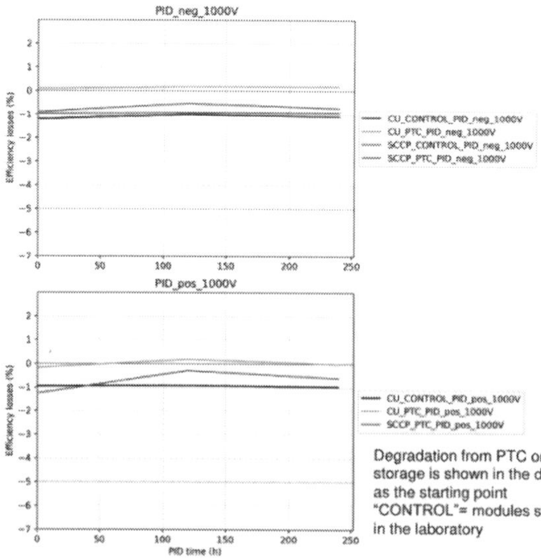

Degradation from PTC or storage is shown in the diagrams as the starting point "CONTROL"= modules stored in the laboratory

3-STEP PROCESS

STABILITY OF MINI-MODULES
Shingle modules

NARROW COPPER LINES
Pattern-transfer printing PTP

SHINGLE MODULES

- Based on M2 precursors fabricated at **CEA INES**
  Seed-grid: Ag paste with reduced laydown, single print, lower line aspect ratio
  Layout with 6 shingles and additional pads for contacting for copper electrodeposition
- Plating at **CSEM**
  Cu thickness ~20 µm
- Shingle separation and stringing on industrial equipment developed by **AMAT** Italy
  6 shingles, corresponding to one M2 cell
  ECA wet deposit per shingle ~3.5 mg
- Module fabrication at **CEA INES**
  Glass-glass
  Polyolefin encapsulant

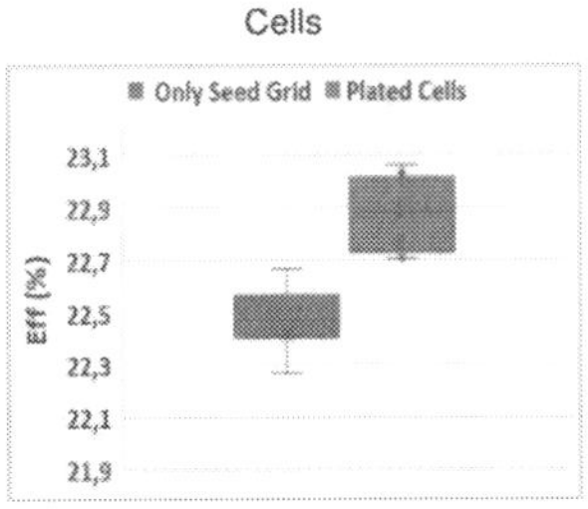

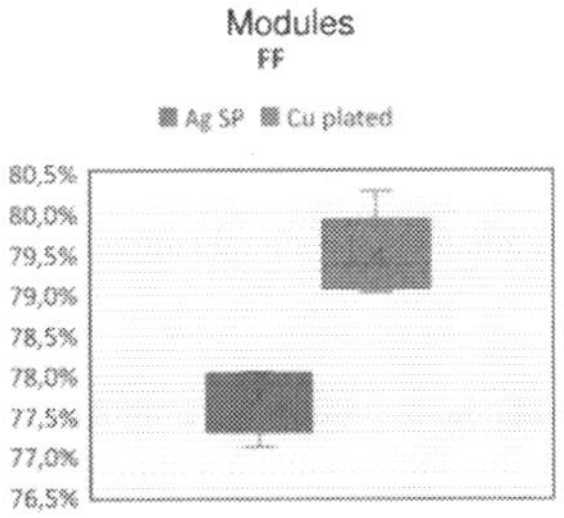

Line resistance	as printed (Ag paste)	w. electrodeposited copper
per shingle width 25 mm	0.43 Ω	0.07 Ω

Measurement with Kelvin probes on contacting pads over the width of one shingle tile

:: csem

SHINGLE MODULES: EL / Pmax

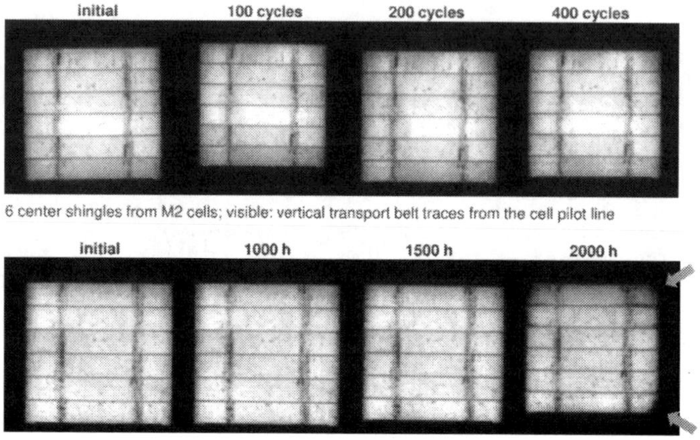

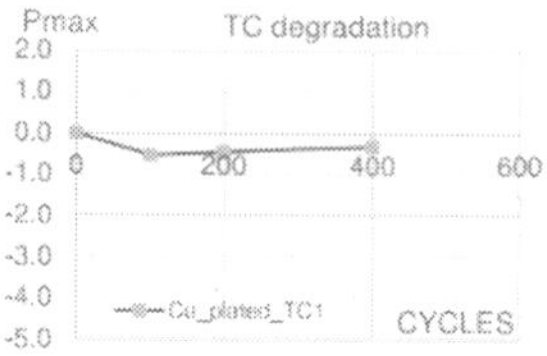

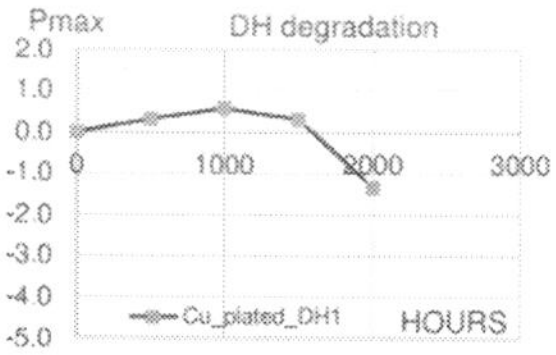

Loss in FF: >2%

13 EU PVSEC 2023 Stability of minimodules with copper plated HJT cells / PTP

:: csem

020018-013

3-STEP PROCESS

STABILITY OF MINI-MODULES
Shingle modules

NARROW COPPER LINES
Pattern-transfer-printing PTP

020018-014

PATTERN TRANSFER PRINTING PTP™

1. Filling stage
   Filling of transparent film with trenches.
   A metal blade pushes the paste inside the trenches and cleans the surfaces between the trenches from excessive paste residue.

2. Transfer stage
   The film is placed above the wafer with a minor distance gap (non-contact printing) and the transfer occurs via laser.
   The filled trenches are scanned by a CW IR high-power laser that drives the paste out and causes the deposition of a whole line onto the wafer.

   Time for the laser transfer: <1s/M10 wafer.

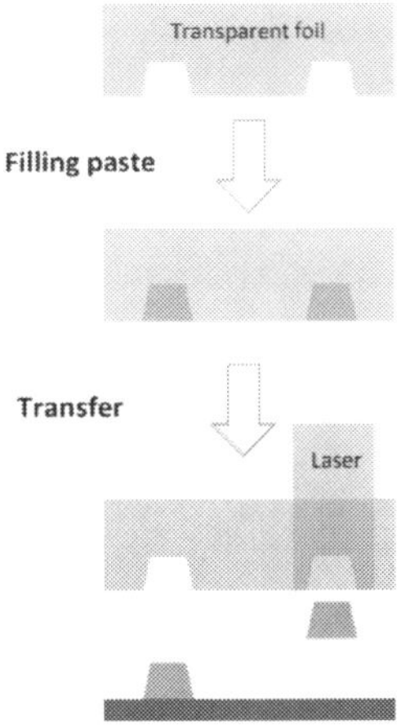

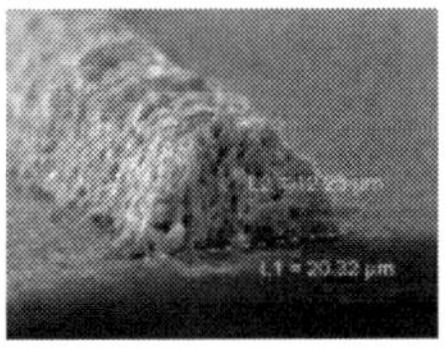

DR Utilight :: csem

PATTERN TRANSFER PRINTING FOR SEED-GRID FORMATION

- SCC and Cu paste from Namics Corp
  Both cured at 200°C / air
- PTP trenches: trapezoid
  width 15/10 µm, height 20 µm

Narrow lines
No interruptions
Good adhesion (tape test)

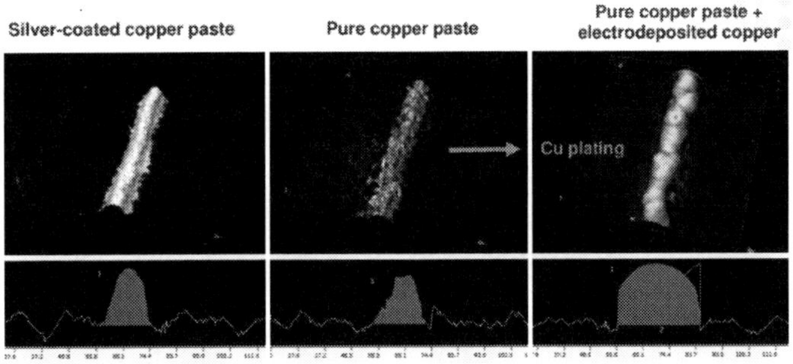

	Silver-coated copper paste	Pure copper paste	Pure copper paste + electrodeposited copper
Line width	**18.9 µm**	**21.5 µm**	**25.0 µm**
Line height	10.3 µm	9.3 µm	17.6 µm
CS area	116 µm²	102 µm²	335 µm²

- Comparison resistance measured over 26 mm distance (on pads)
- Contact resistance measured with screen-printed pattern: 1.5 mΩ·cm²

Pad to pad distance 26 mm	SCC paste	Cu paste	Cu paste + electrodeposited Cu
Ω	1.5	5.1	0.18

:: csem

PATTERN TRANSFER PRINTING FOR SEED-GRID FORMATION

- Pure copper paste
- Plating mask: dielectric layer
- Copper electrodeposition

Line width	26.1 µm
Line height	21.9 µm

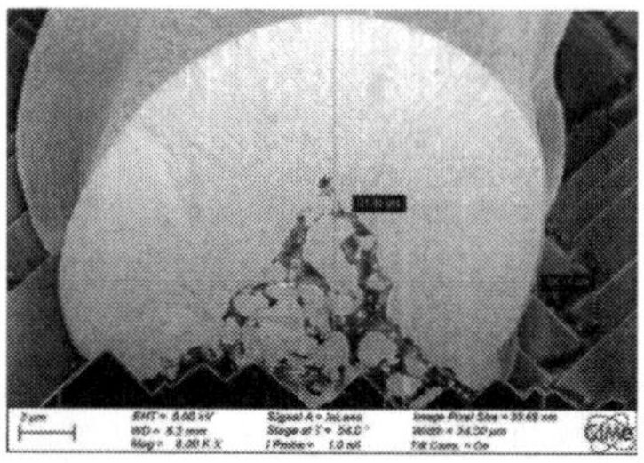

EDX map

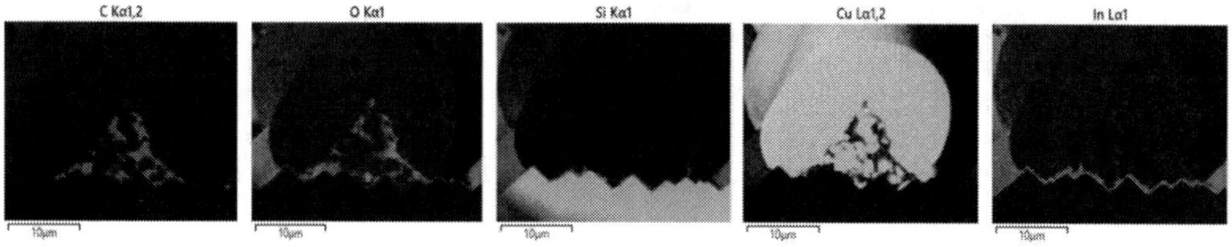

IV RESULTS PTP-SCC AND Cu-PASTE / Cu-PASTE PLATED

SCC-PTP	**PTP with silver-coated-copper paste**
Cu-PTP	**PTP with (uncoated) copper particle paste**
Cu & plated	**Cu electrodeposited on Cu paste**

- First test on older M2 industrial HJT precursors
  DR Utilight: PTP front
  CSEM: SP back, 6 narrow BBs w. contacting pads
- Extremely low FF with Cu paste alone
- With electrodeposited copper:
  FF and efficiency close to SCC reference
  Result of best cells after modification
  of pre-treatment / deoxidation before copper plating
- Next / optimization:
  - Test on M10 precursors started
  - Further FF improvement / pre-treatment

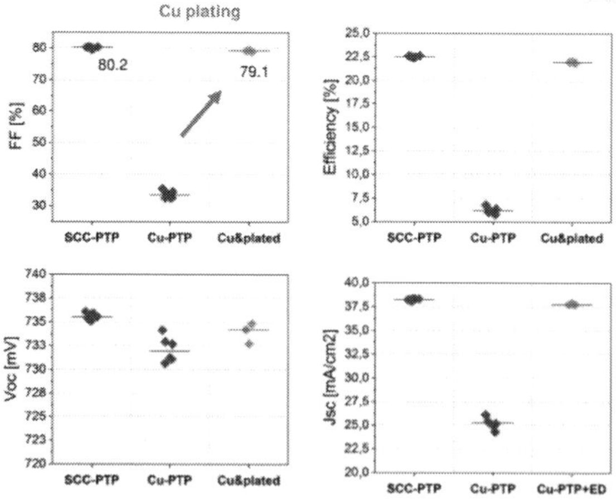

Measurement with 6 pogo-bars on a gold chuck

SUMMARY AND OUTLOOK

- Industrially viable 3-step process
- Modules
  - Very good stability in TC and PID
  - DH to improve: another materials / edge sealing
- Narrow lines with PTP of Cu-Paste
  - Cell performance close to SCC reference
  - FF / pretreatment further to improve
  - Next: PTP on M10 wafers

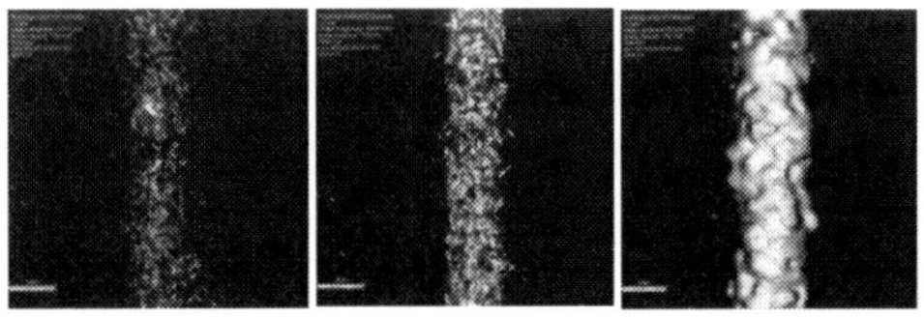

40th European Photovoltaic Solar Energy Conference and Exhibition

THANK YOU VERY MUCH FOR YOUR ATTENTION!

We thank the companies DR UTILIGHT and NAMICS for fruitful collaboration
We thank the companies ATOTECH and DUPONT for providing us plating chemistry

This work has been partially carried out within the European projects HighLite (GA 857793) and Resilex (GA 101058583)

HighLite RESiLEX Funded by the European Union

Contact: agata.lachowicz@csem.ch

:: csem
FACING THE CHALLENGES OF OUR TIME

PROGRESSES ON THE DESIGN OF A TOOL FOR THE POST-TREATMENT OF HJT CELLS: FROM PROTOTYPE TO IN-LINE INDUSTRIAL TOOL

Alessandro Voltan[1], Francesco Dalla Torre[1], Jordi Veirman[2], Jean-Sebastien Caron[2], David Pelletier[2], Mickael Albaric[2], Pedro Jeronimo[2], Anne-Sophie Ozanne[2], M. Sciuto[3]
[1]Applied Materials, Via Postumia Ovest, 244, 31048, Olmi di San Biagio di Callalta, TV, Italy
[2]Univ Grenoble Alpes, CEA, LITEN, DTS, INES, F-38000, Grenoble, France
[3]Enel Green Power, Contrada Blocco Torrazze, Zona Industriale, 95121, Catania, Italy
Mail address: Alessandro_voltan@amat.com

ABSTRACT: High intensity light exposure with controlled heating is known to be a method for performance enhancement of amorphous-crystalline silicon heterojunction (HJT) solar cells and modules [1,2]. The related efficiency enhancement is correlated to a simultaneous defect recovery and series resistance improvements, the physical origin of which is still under investigation. Recently, with the aim to integrate such intense illumination post-treatment in the cell manufacturing sequence, multiple processes involving intense illumination densities have been developed [1]. The commercially available tools for in-line post-treatment are generally based on Xenon lamps (longer and energy-intensive processes) or on Laser-based sources (high cost of ownership).

In this work, we present a fast post-treatment process for HJT solar cells, using an LED-based oven jointly designed by Applied Materials (AMAT, Italy) and CEA/LITEN (France) and in collaboration with Enel Green Power (EGP, Italy) [3]. The described oven will be capable to achieve a maximum illumination power of 50kW/m^2 with multiple wavelengths illumination. The expected exposure time will be shorter than 20s, minimizing the overall oven footprint and related power consumption.

We report here an overview of our latest achievements including data collected during the process tuning at laboratory level (CEA/LITEN) and results obtained during the validation of final oven design. Currently, up to +0.6% efficiency gain was achieved, regardless of the cell thickness. Testing of HJT cells with n-doped nanocrystalline silicon oxide layers at the cell front side, as a replacement of the historical n-doped amorphous silicon layers, revealed that the same post-treatment process can be successfully extended to new generation HJT cells.

Keywords: Heterojunction, Light Soaking, In-line post processing

1 INTRODUCTION

The behavior of HJT cells under Light Soaking (LS) and the related defect recovery is still actively investigated in literature. With our intensive illumination approach, the associated conversion efficiency gain can be harvested only if the temperature does not exceed the cell damage threshold temperature (<250°C) even in localized regions (Fig. 1). Furthermore, it has been demonstrated that recovery mechanisms can simultaneously take place at different positions into the solar cells, from surface passivation (incl. on the backside) to the cell front side where the TCO can be improved [4]. Based on these requirements, the development of a dedicated LED-based post-treatment oven was initiated with the aim to i) reduce as much as possible the exposure time and the process length ii) achieve a uniform illumination with intensities up to 50kW/m2 iii) reduce as much as possible the power consumption iv) keep the cells in a safe temperature range <250°C v) include multiple light wavelengths for spectrum tuning.

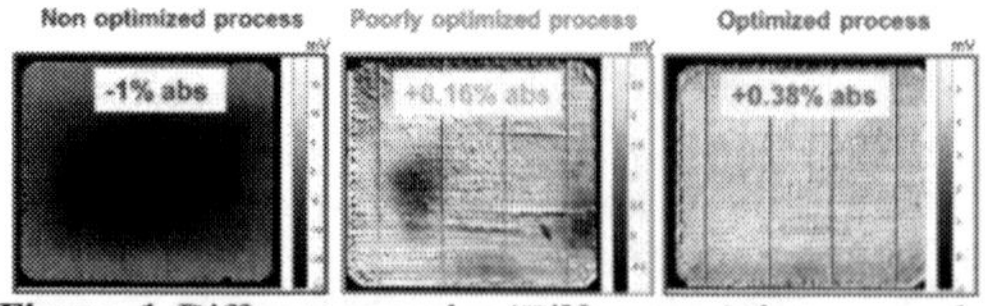

Figure. 1. Different scenarios (□iVoc maps) that occurred during process optimization: temperature profile during treatment well above (left), slightly and locally above (center), or safely below (right) the threshold acceptable temperature value.

As a first step, a stand-alone LED-based prototype, in which the cell is static and manually loaded, has been developed by Applied Materials (AMAT) for feasibility study and post-treatment development at CEA/LITEN. Once this step finalized, the second phase was dedicated to the design of the final in-line post-treatment oven, which was validated through thermal simulations and separate component testing. In the following, the main results are reported.

2 EXPERIMENTAL RESULTS AT LAB SCALE

Investigation about fast light soaking process feasibility has been performed at CEA/LITEN using the first static lamp prototype provided by Applied Materials and M2 cells both from EGP (Enel Green Power) and CEA/LITEN.

Static, water cooled LEDs lamp has been designed to overcome 50kW/m^2 light intensity with multiple wavelengths (Fig. 2). In order to test different illumination recipes both light intensity and spectral distribution were independently programmable and tunable.

Figure. 2: static LEDs lamp

Multiple DOE tests have been performed to identify the proper combination of process parameters suited for high volume manufacturing and fast in-line processing.

As preliminary result, it was demonstrated that >+0.3%abs

efficiency gains was achievable using an optimized illumination time-profile and wavelength mix with 12s exposure time, without thermal degradation (Fig. 3) on M2 cells.

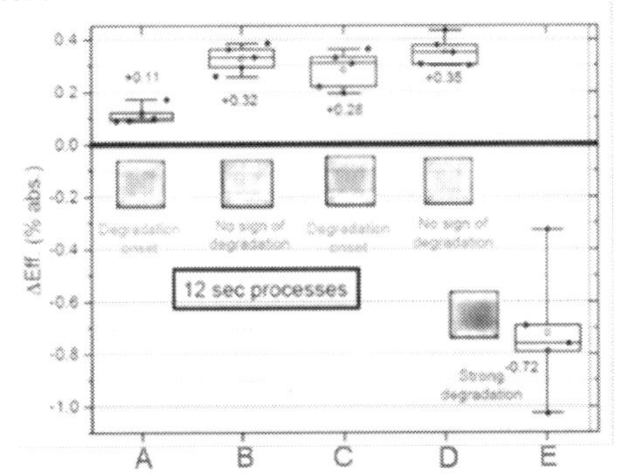

Figure. 3: comparison of different post-treatment recipes. Best performances achieved with recipe D (12s exposure time)

The best efficiency gains were achieved with a 3 illumination-step recipe with precise control of exposure time and temperature distribution/evolution. The transposition of this recipe to HJT cells with decreasing thicknesses (160µm, 130µm and 100µm), showed similar behaviors.

Efficiency gain stability was also monitored in this phase. After having obtained an efficiency gain of +0.5% with fast treatment, a maximum stabilized loss of -0.1% was measured after 15 days regardless of the storing conditions (N2 vs Dark), Fig. 4. Similar behavior was found on cells exposed to longer light soaking time (180s).

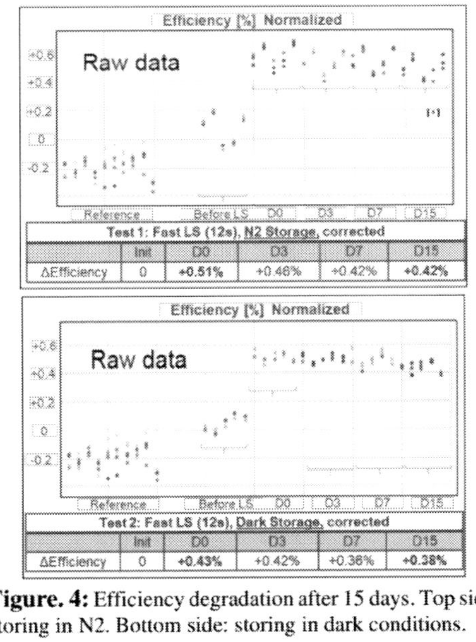

Figure. 4: Efficiency degradation after 15 days. Top side: storing in N2. Bottom side: storing in dark conditions.

In order to improve the optoelectrical cell properties, the HJT community is progressively shifting to doped n-doped nanocrystalline silicon oxide layers at the cell front side, as a replacement of the original n-doped amorphous silicon layers. The benefits are enhanced short-circuit currents (Isc) (more transparency), and fill factors (FF) (improved carrier transport). To assess the post-treatment benefits on such cells, 4 batches featuring variations in the nc-SiO(n) layers were also processed at CEA/LITEN, along with a reference batch with standard a-Si:H (n) layer using the static LEDs lamp prototype. The expected initial advantages in FF and Isc were observed with nc-SiO(n) cells outperforming the reference a-Si:H(n) population by 0.25 to 0.40% abs. Cells from each batch then received an industrial intense illumination treatment. The changes in photovoltaic parameters before/after the treatment are shown in Fig.5. The response of nc-SiO(n) cells appears to be significantly greater than that of reference cells, with larger FF and Voc improvements (origins under investigation). These results highlight the promising compatibility of the developed post-treatments with future commercial HJT cells, as such nc-SiO(n) layers are being/will be implemented in production.

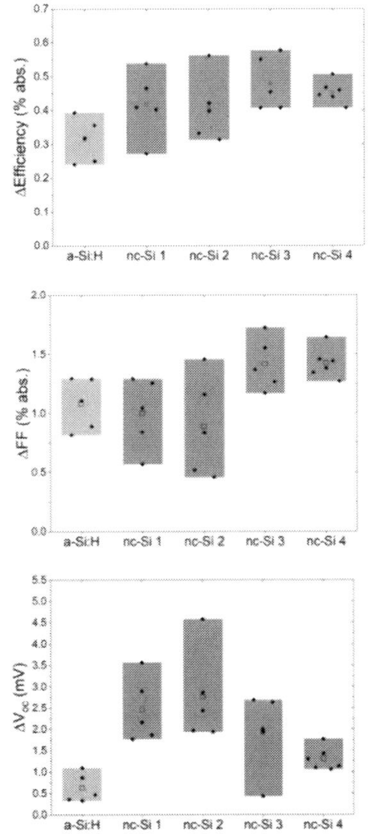

Figure. 5: Efficiency, FF and V_{oc} improvements measured after the industrial intense illumination treatment for 4 variations in nc-SiO(n) layers and the reference a-Si:H process

3 EQUIPMENT DEVELOPMENT

Process data collected with the first static lamp prototype have then been used to design the final oven able to replicate the same illumination profiles and thermal conditions in dynamic in-line mode (Fig. 6), now under testing and validation. The distribution of LEDs has been defined in order to have less than 10% non-uniformity

10.4229/EUPVSEC2023/1BO.4.4
020019-002

light distribution. After optical design and simulation, uniformity mapping has been verified on stand-alone oven mock-up. Same "small sized" oven has also been used for testing of critical components (stress test) and lifetime assessment. In particular, LEDs stability was measured, obtaining that light variation during warm-up (20s) is <2% at maximum intensity that is considered acceptable for the process.

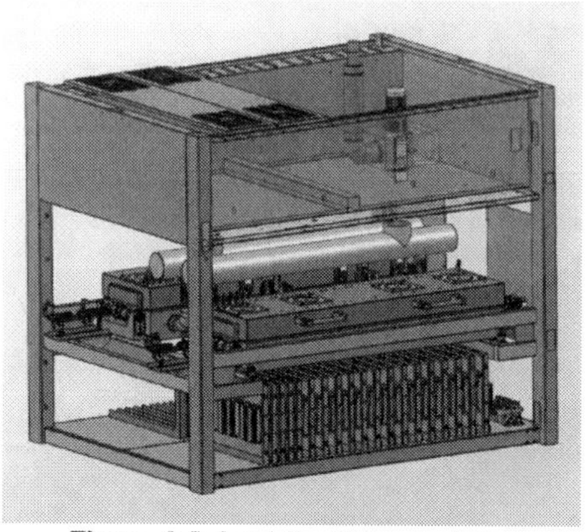

Figure. 6: Industrialized LS oven design

One key difference between the two tool configurations (static prototype used for feasibility study and in-line dynamic oven) is the possible non-uniformity of post-treatment across the cell in the in-line tool. Local temperature overshoots could occur when transitioning across zones of different illuminations, due to heat pile-up at the exit side of the cell (i.e. entering the tool last). To verify if such overshoots do occur, a 3D time-resolved numerical simulations was performed in order to reconstruct the thermal profile experienced by the cell (G12) throughout post-treatment. To this end, a physical model of the cell involving conduction, convection and radiation was developed at CEA and the energy equation is solved by ANSYS Fluent. Fig. 7 shows the spatially-resolved field of the maximum temperature deviation (in °C) from the average maximum temperature experienced during the entire post-treatment. The computation was performed for two thicknesses, 160µm (mainstream) and 100µm (prospective value).

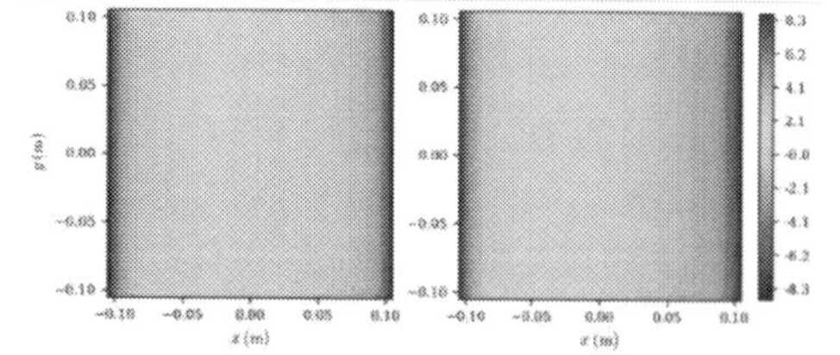

Figure. 7: 3D thermal simulation output: imaging of the maximum temperature deviation from the average maximum as calculated for the entire post-treatment. The cell motion is from left to right.

Fig.7 reveals indeed an overshoot in both cases at the exiting cell side (left), as this side experiences heating not only from the illumination but also from lateral heat conduction from the cell part already under illumination for a while. The spatial inhomogeneity in the maximum temperature is predicted to remain below 10°C and appears to be very little sensitive to the cell thickness. Thus, the transitioning between different illumination zones should not generate unwanted hot spots at specific areas, strengthening the robustness of the in-line process.

4 VALIDATIONS OF INDUSTRIAL TOOL

The overall length of the final oven (Fig. 8) is less than 3m with $(0.25\times2)m^2$ irradiated area, suited for HJT cells with size up to $(210\times210)mm^2$ with LEDs grouped into independently tunable zones. Temperature of the cells will be precisely monitored by means of 3 pyrometers embedded into the thermal chamber and kept below the threshold limit by a patent pending cooling system, working in closed loop control mode.

Figure. 8: LS oven under testing and validation

Industrialized LS oven is currently under testing and validation with the aim to replicate results obtained at lab level and to verify thermal simulations reported in Fig. 7. In fig. 9 it is reported a thermal profile acquired after recipe optimization. The overall process length, including cooling, is 20s and the 2 curves reported in Fig. 9 have been measured with thermocouples placed on edge (position 1) and center (position 2) of M2 cell during the advancement inside the light soaking oven.
It has been demonstrated that i) thermal behavior obtained in static mode can be replicated in dynamic condition ii) lateral temperature variation is within 10°C in good agreement with previous simulations iii) peak T can be achieved in few seconds without overshooting.

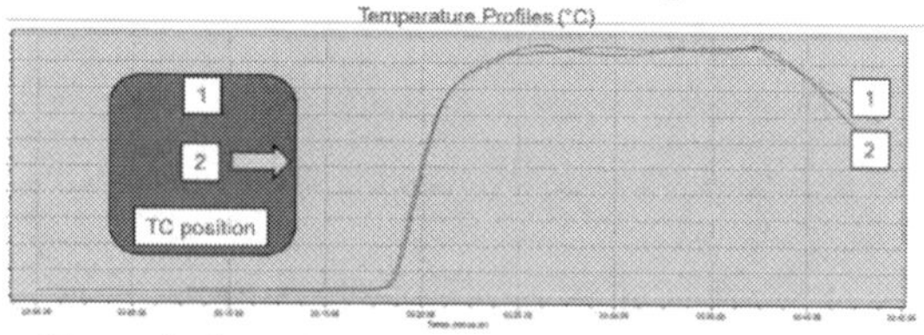

Figure. 9: thermal profiles acquired during 20s Light Soaking process in dynamic mode

After thermal characterization, a first test has been performed using a batch of 10 HJT cells, M2 size. This wafer format was used in order to replicate same tests done using static prototype.
All the 10 cells have been characterized before and after fast light soaking process including both IV measurements and Photoluminescence analysis.
After a first light soaking process, a second LS run was performed on same cells in order to evaluate possible

additional improvements.
Figure 10 shows the Photoluminescence images collected on the same M2 cell before and after Light soaking.
After LS, no signs of Temperature overshoots are visible, confirming simulations reported in Fig. 7.

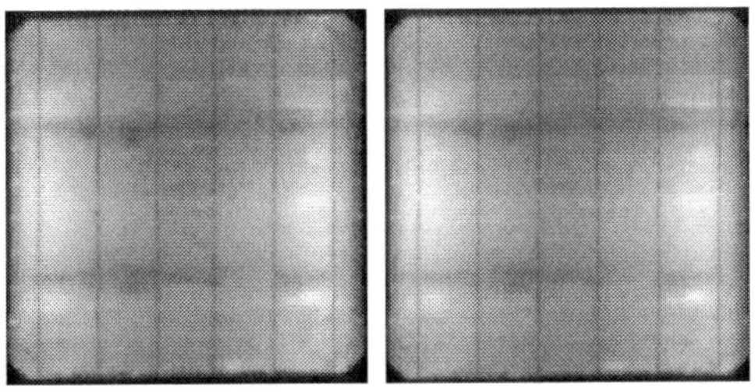

a) b)

Figure. 10: Photoluminescence image on same cell. a) before LS b) after LS.

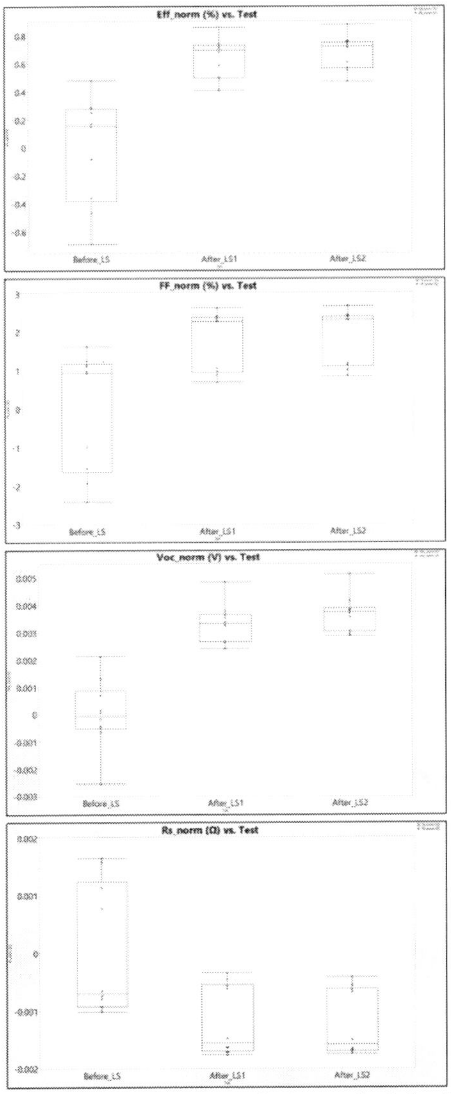

Figure. 11: IV measurements normalized to avg. data before LS.

IV measurements are reported in Figure 11 and summarized in Table I. All data are normalized to average data before LS.
After first light soaking process, an average +0.6%abs efficiency gain has been obtained, well aligned with lab data. Improvements are mainly due to Voc (3.3mV), FF (1.77%) and Rs (-1.2mOhm). Isc improvement is negligible.
After second Light soaking step, no additional performance improvement has been achieved.

Table I: IV measurements normalized to average data before LS.

IV measurements: Δ between average results					
	Efficiency [%]	FF [%]	Voc [mV]	Isc [A]	Rs [mOhm]
Before_LS	0.00	0.00	0.00	0.00	0.00
After_LS1-Before_LS	0.64	1.77	3.32	0.01	-1.19
After_LS2-AfterLS1	0.03	0.08	0.34	0.00	-0.04

5 CONCLUSIONS

Preliminary tests performed with the industrialized Light Soaking oven has shown that the same process obtained in static mode at lab scale can be replicated with in-line/dynamic conditions. The latest lab tests with static prototype have demonstrated that up to +0.6% efficiency gains can be achieved with short exposure time (12s) , with limited loss at module level. Similar results have been also obtained on M2 cells with industrialized equipment. A promising transposition to future HJT cells with nc-SiO(n) layers has also been demonstrated at laboratory scale. The newly developed LED-based oven will be ready for installation at customer site by end of 2023.

References

[1] J. Veirman et al., "Towards an industrial in-line solution for efficient post-treatment of silicon heterojunction solar cells", Solar Energy Materials and Solar Cells, Volume 245, 2022, 111867

[2] M. Wright et al., "High-Intensity Illuminated Annealing of Industrial SHJ Solar Cells: A Pilot Study," in IEEE Journal of Photovoltaics, vol. 12, no. 1, pp. 267-273, Jan. 2022.

[3] A. Voltan, "Fast Light Soaking for inline cell efficiency improvement", presented at the 5th International workshop on Silicon Heterojunction solar cells, November 2022, Aix-les-Bains, France

[4] J. Veirman et al., AIP Conference Proceedings 2487, 020017 (2022)

Gas-Immersion Laser-Doped Silicon Tunnel Junctions for Multi-junction Solar Cells

G Gaspar[1], FC Serra[1], JM Serra[1], **K Lobato[1,*]**, F Geml[2], G Hahn[2], J Kern[3], M Müller[3],L Vines[4], A Fave[5]

[1]**Instituto *Dom Luis*, Faculty of Sciences, University of Lisbon**
[2]Department of Physics, University of Konstanz
[3]Institut für Angewandte Physik, Technische Universität Bergakademie Freiberg
[4]Department of Physics, Centre for Materials Science and Nanotechnology, University of Oslo
[5]Université de Lyon, INSA Lyon, CNRS

Contents

N-I-P configuration - n layer is deposited first onto substrate

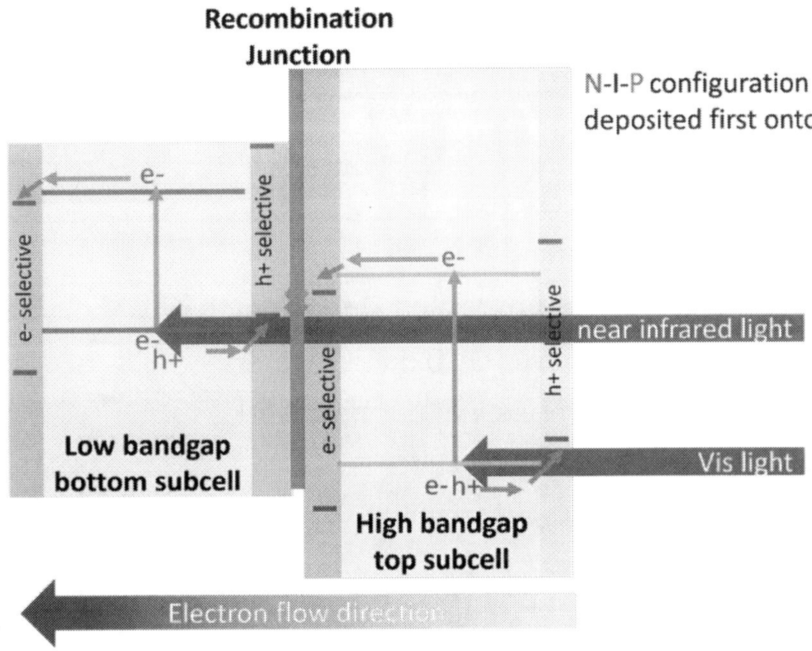

40th European Photovoltaic Solar Energy Conference and Exhibition

- For monolithic perovskite on silicon tandem devices, two strategies can be employed for the recombination junction
  - Transparent Conducting Oxide (ITO, InOx)
  - Tunnel Junction
- Each has their advantages and disadvantages.
  A good review about this topic can be found here: de Bastiani et al., Materials Horizons 7 (2020) 2791–2809

Contents

40th European Photovoltaic Solar Energy Conference and Exhibition

- A tunnelling Esaki diode requires
    - highly (degenerately) doped layers
    - with abrupt changes in doping profiles
- Conduction and valence bands become energetically aligned
- Band bending is such that the depletion region is narrow enough to allow tunnelling

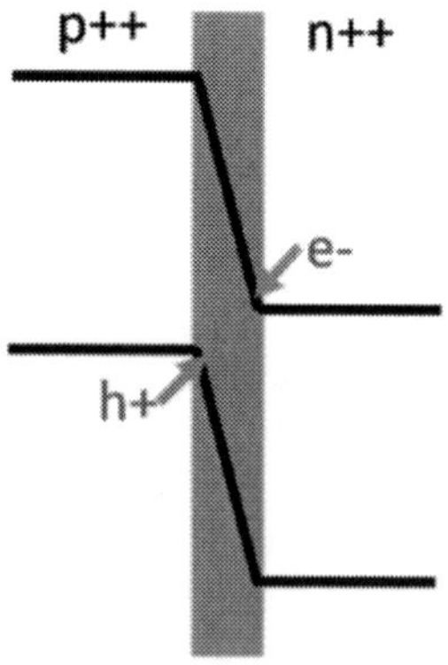

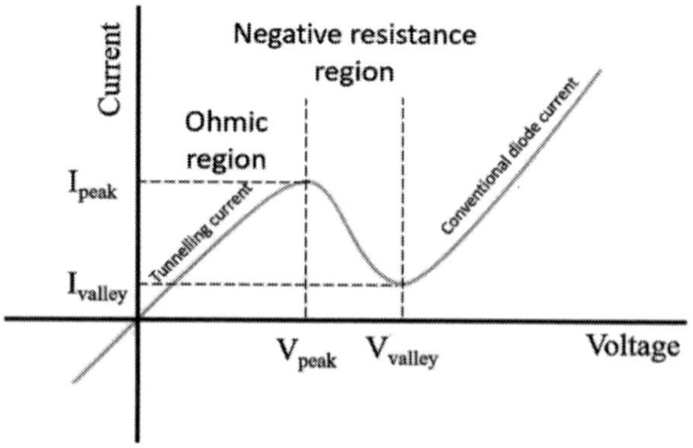

Contents

- GILD process
    1. Substrate is immersed in atmosphere containing dopant source

1. Immersion

- GILD process
  1. Substrate is immersed in atmosphere containing dopant source
  2. **Pulsed laser (tens of ns) melts a thin surface layer**

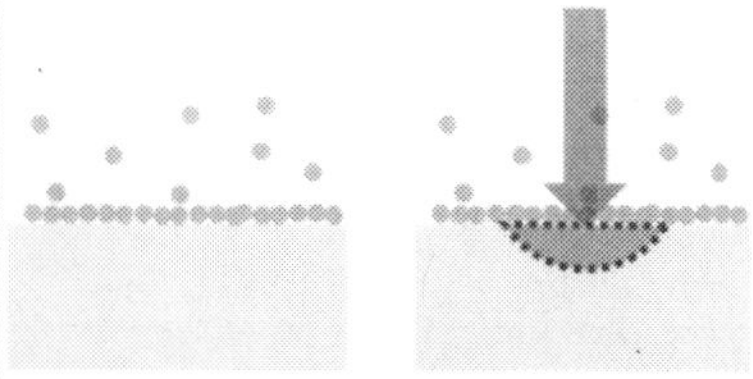

1. Immersion

2. Laser pulse & melting

40th European Photovoltaic Solar Energy Conference and Exhibition

- GILD process
  1. Substrate is immersed in atmosphere containing dopant source
  2. Pulsed laser (tens of ns) melts a thin surface layer
  3. **Dopant adsorbed at sample surface dissolves into melt and rapidly diffuses within melt region**

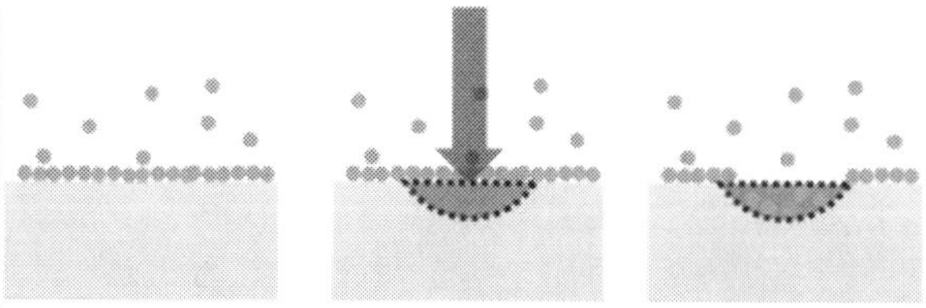

1. Immersion

2. Laser pulse & melting

3. Dopant diffusion in melt

- GILD process
  1. Substrate is immersed in atmosphere containing dopant source
  2. Pulsed laser (10s of ns) melts a thin surface layer
  3. Dopant adsorbed at sample surface dissolves into melt and rapidly diffuses within melt region
  4. **Melt recrystallizes**

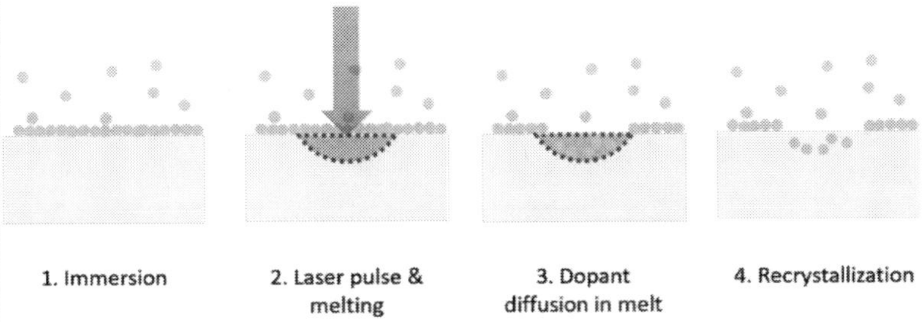

40th European Photovoltaic Solar Energy Conference and Exhibition

- GILD process
  1. Substrate is immersed in atmosphere containing dopant source
  2. Pulsed laser (10s of ns) melts a thin surface layer
  3. Dopant adsorbed at sample surface dissolves into melt and rapidly diffuses within melt region
  4. Melt recrystallizes
  5. **Dopant adsorption**

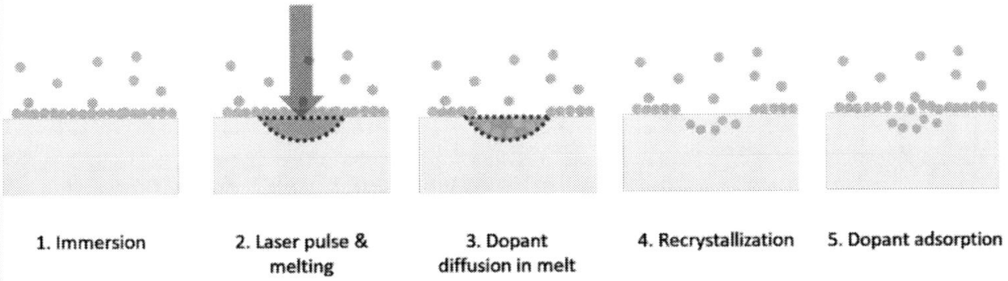

- GILD process
  1. Substrate is immersed in atmosphere containing dopant source
  2. Pulsed laser (10s of ns) melts a thin surface layer
  3. Dopant adsorbed at sample surface dissolves into melt and rapidly diffuses within melt region
  4. Melt recrystallizes
  5. Dopant adsorption
  6. **Laser fires further along**

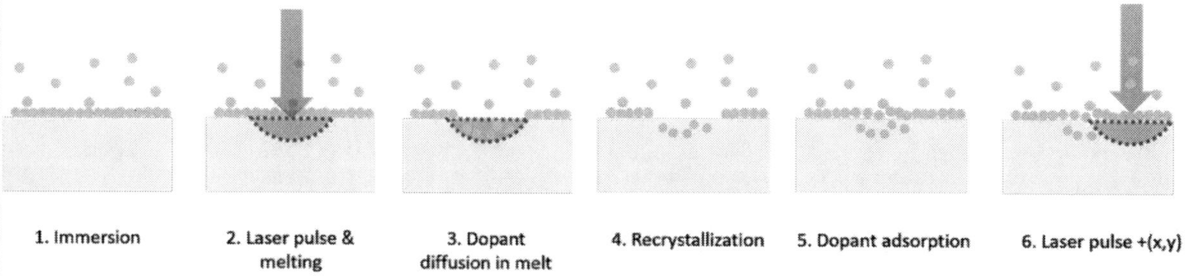

40th European Photovoltaic Solar Energy Conference and Exhibition

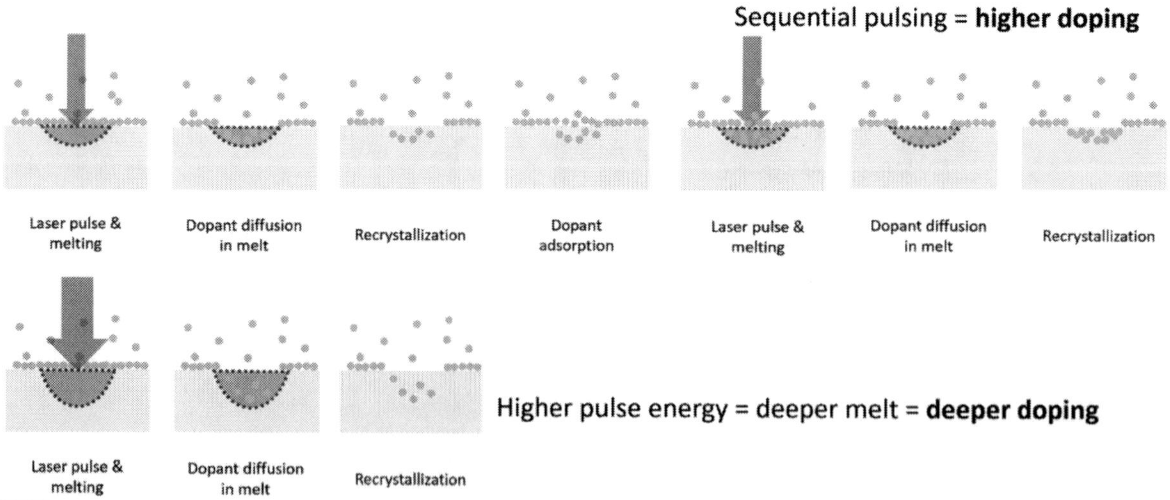

- GILD permits
  - Very high doping 10^{21} cm^{-3}
  - Abrupt doping profiles < 10 nm / decade
  - Control depth of doping
  - Dopants are electrically active
  - Material recrystallizes epitaxially
- **Ideal for Tunnel Junction**

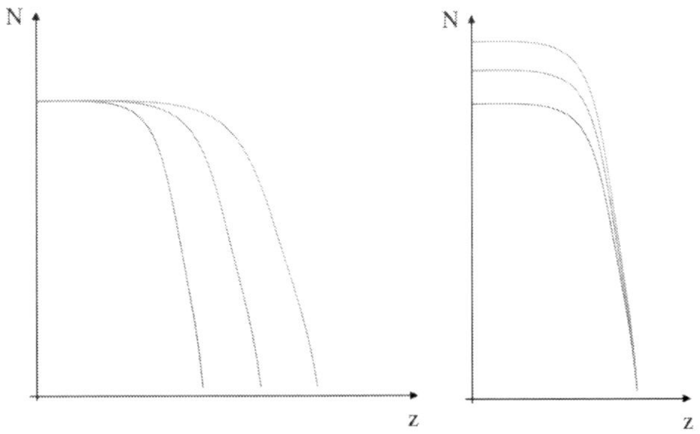

40th European Photovoltaic Solar Energy Conference and Exhibition

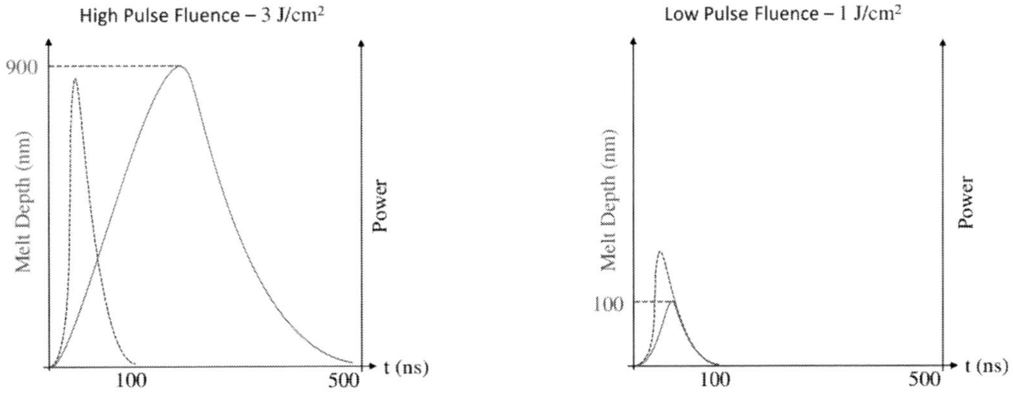

E. Landi et al., *Numerical Simulation of the Gas Immersion Laser Doping (GILD) Process in Silicon*, IEEE Transactions on Computer-Aided Design of Integrated Circuits and Systems. 7 (1988) 205–214. https://doi.org/10.1109/43.3150.

Contents

Laser rastering properties:
- Distance between scan lines: 10 µm;
- **Distance between laser spots: *d* = 1 µm;**
- **Number of rastering sequences: *n* = 6 and 9 times;**
- 5×10 mm² areas were laser scanned;
- 0.5 to 10 cm²/min of laser processed area.

Laser pulse:
- Energy: 0.04 mJ;
- Repetition frequency: 500 kHz;
- Beam diameter: 20 µm.

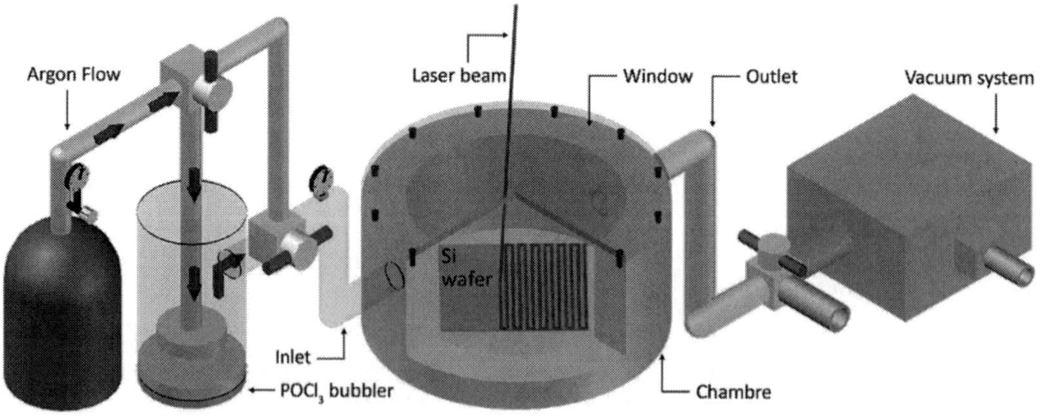

The initial idea

- Start with p++ emitter on c-Si n-type

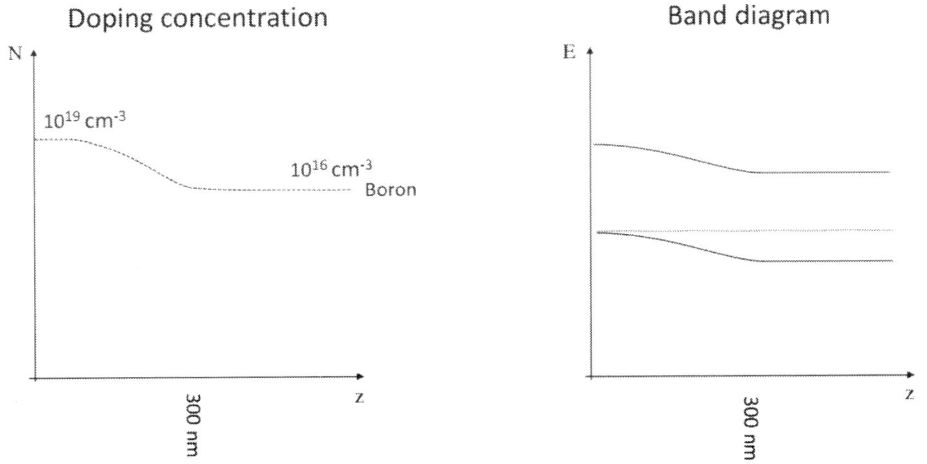

40th European Photovoltaic Solar Energy Conference and Exhibition

- Perform GILD to ***try*** to form Tunnel Junction

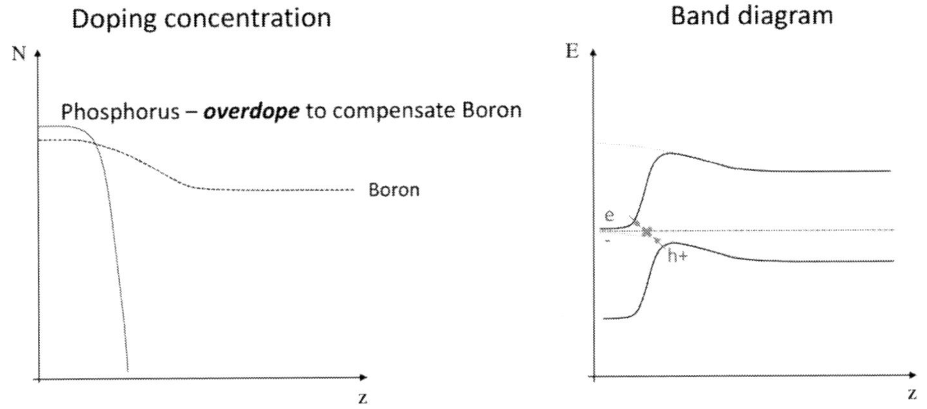

Contents

- Starting with Si p+ wafer with a p++ emitter

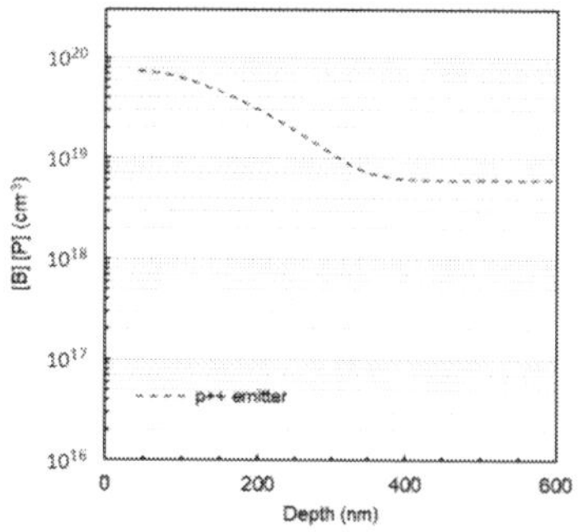

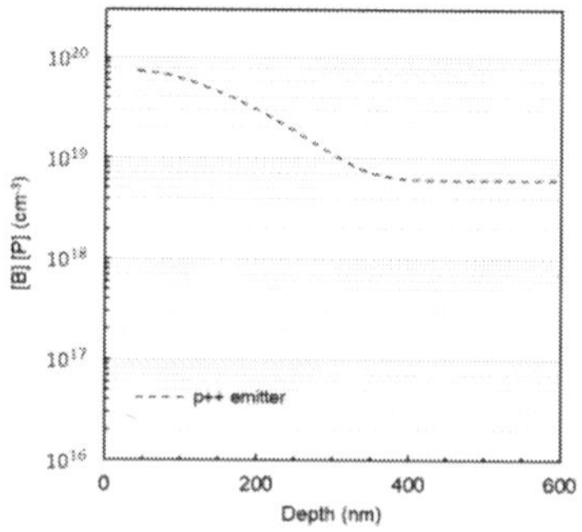

- Perform GILD to n++ dope with Phosphorous

Aggressive laser processing

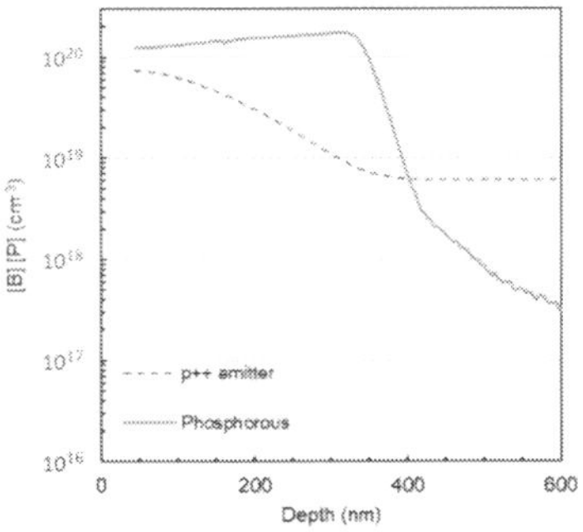

Mild laser processing

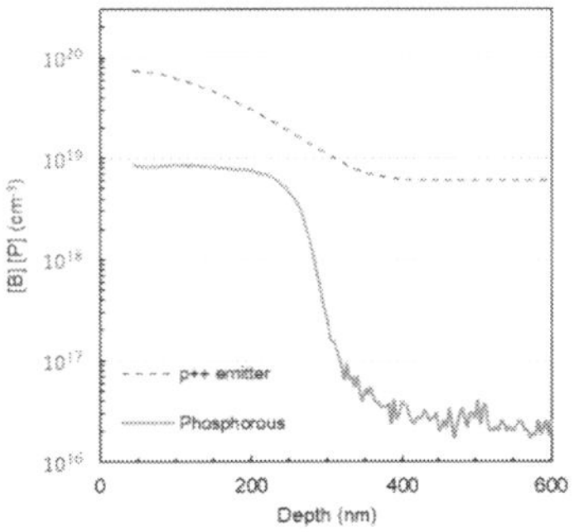

Doping concentration determined by Secondary Ion Mass Spectroscopy (SIMS)

- However, what about the underlying p++ Boron doping profile?

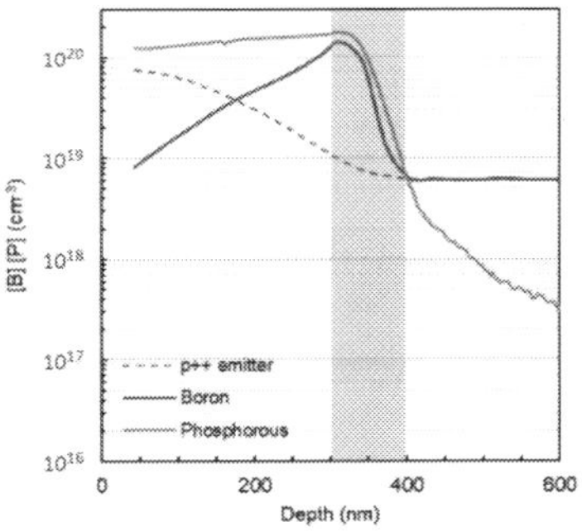

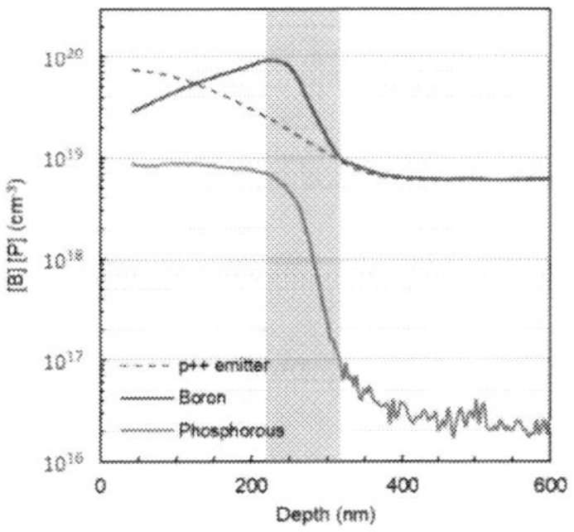

- Boron undergoes *pile-up* with the concentration building up at the maximum melt depth and decreasing at the surface.
- This was obviously a problem! Concentration profiles will track each other.

- Boron *pile-up* – occurs because the segregation (partition) coefficient under these conditions $k_p > 1$.
  - i.e., Boron *prefers* to be in the solid rather than liquid phase
  - Well know phenomenon – PC Lill et al., Materials. 10 (2017) 189.

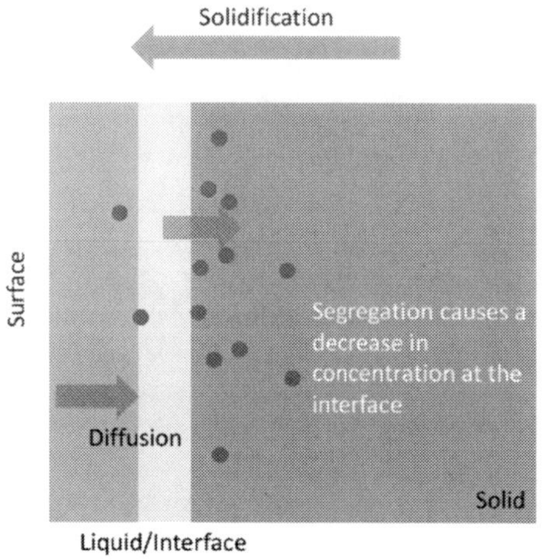

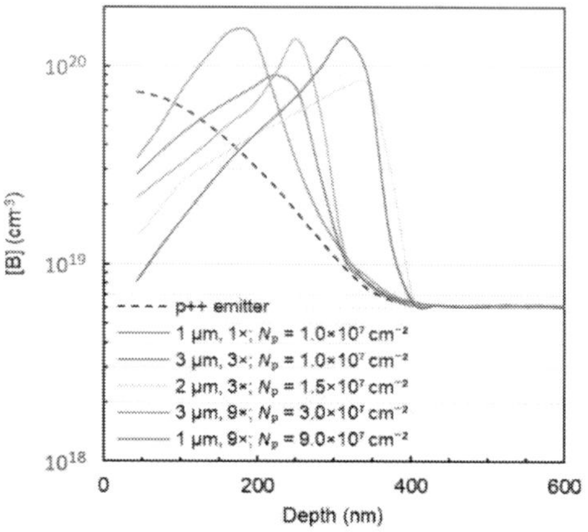

Contents

- Start, again, with p++ emitter on c-Si n-type

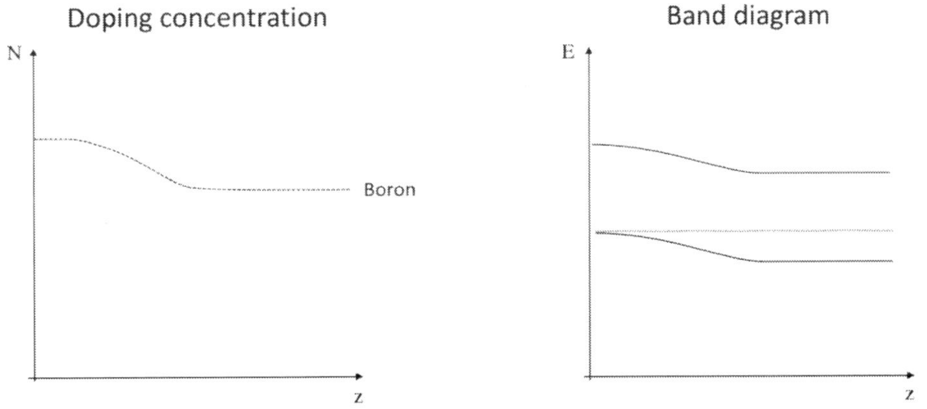

40th European Photovoltaic Solar Energy Conference and Exhibition

- Then, firstly, perform laser dopant *pile-up* (LDP)

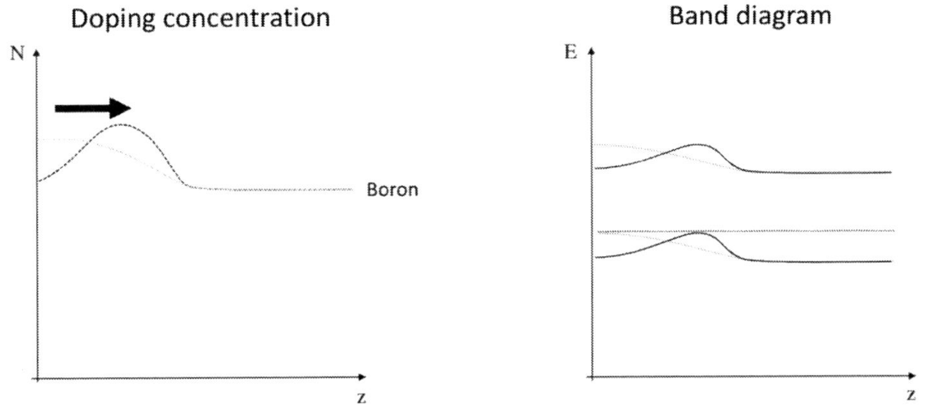

- Followed, secondly, perform GILD ensuring melt depth is shallower

2-steps: Laser dopant pileup + **Gas Immersion Laser Doping**

Doping concentration

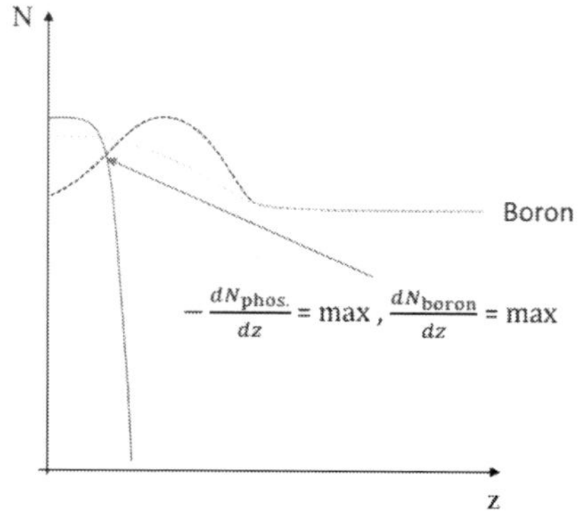

Band diagram

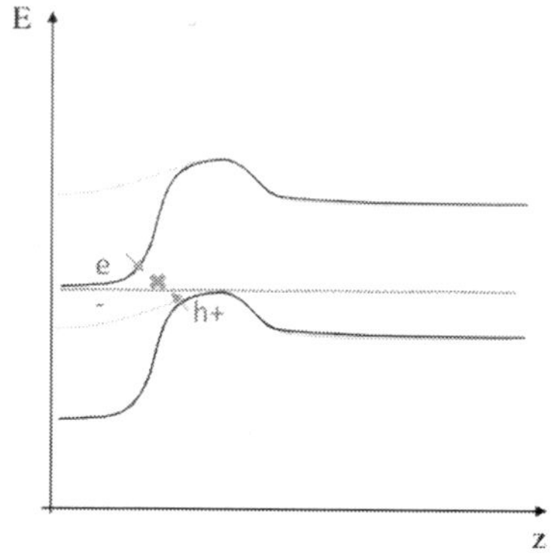

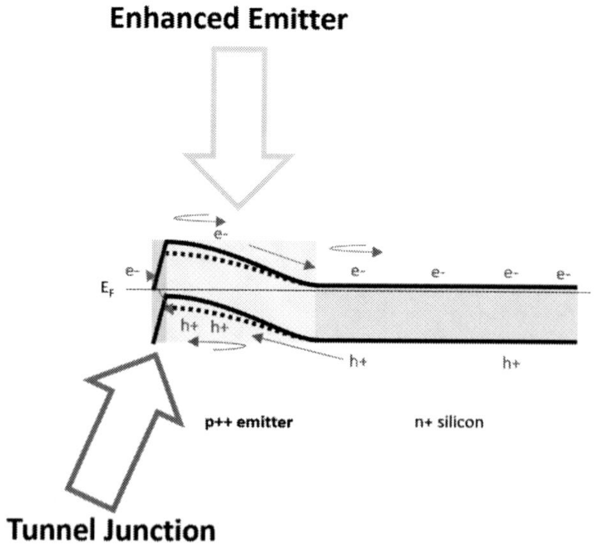

LPD+GILD

- Secondary Ion Mass Spectroscopy (SIMS) results

Aggressive GILD

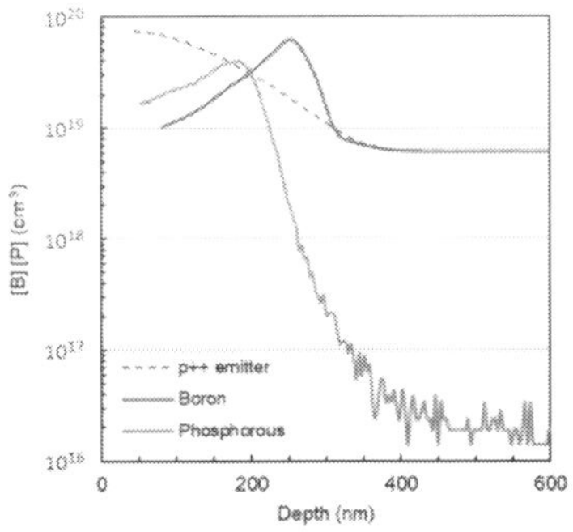

Mild GILD

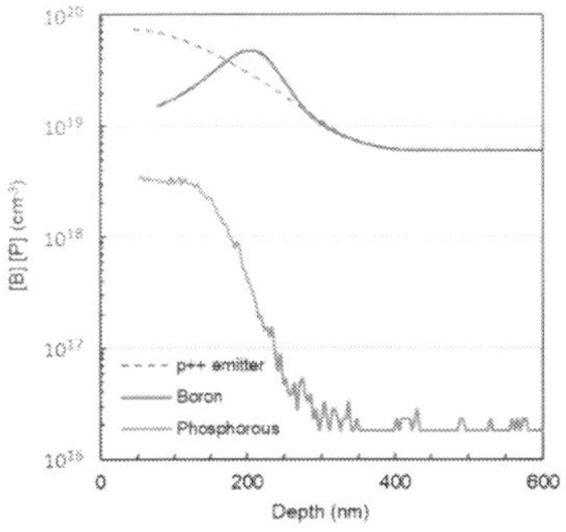

- Clear that process is not homogenous – as such SIMS results are all over the place...
- However, ECV proved useful.

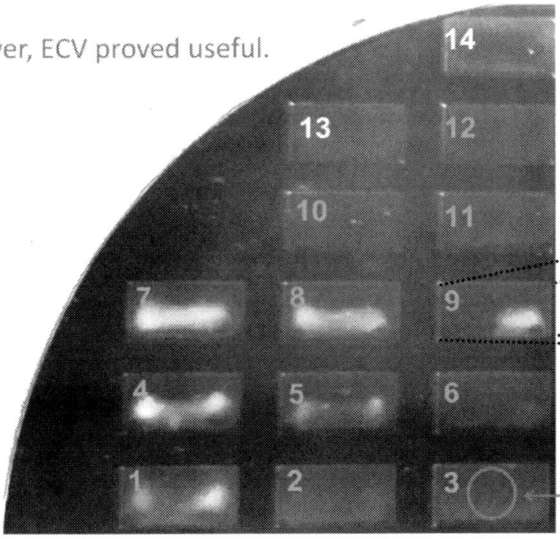

Image of ¼ wafer processed by GILD+LDP (1-12) and LDP only (13-14).

All concurrent areas were processed in identical fashion.

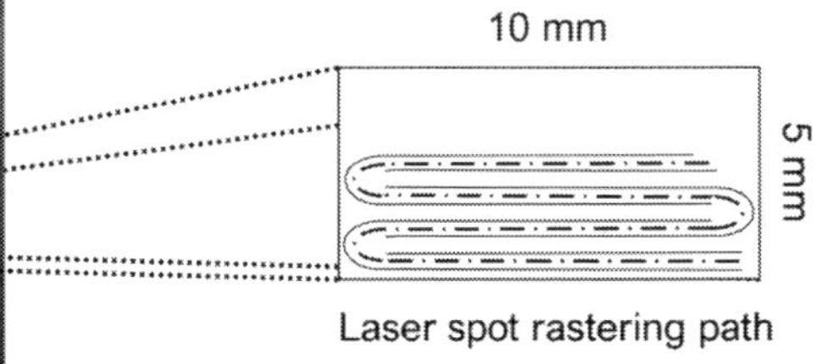

ECV measurement area within laser processed areas: ⌀3.5 mm

LDP step

- ECV results show that LDP alters p++ dopant profile as expected.

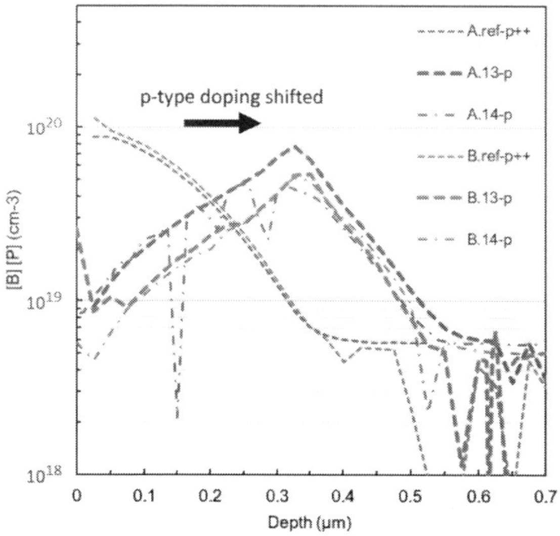

Two identical wafers, A and B, processed in identical fashion.

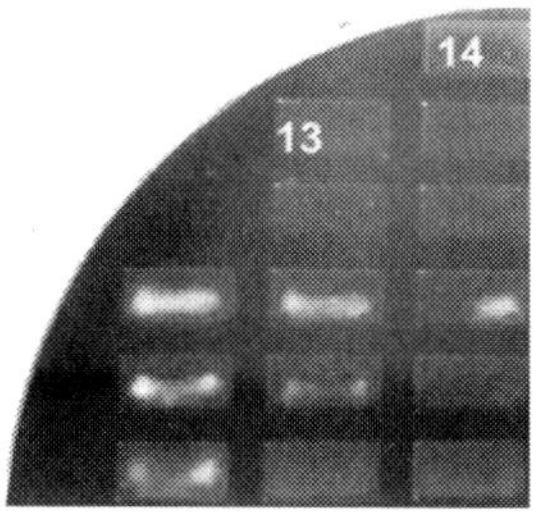

40th European Photovoltaic Solar Energy Conference and Exhibition

GILD step

- ECV results show that GILD successfully introduces n++ phosphorous doping.
- However, results are extremely varied.

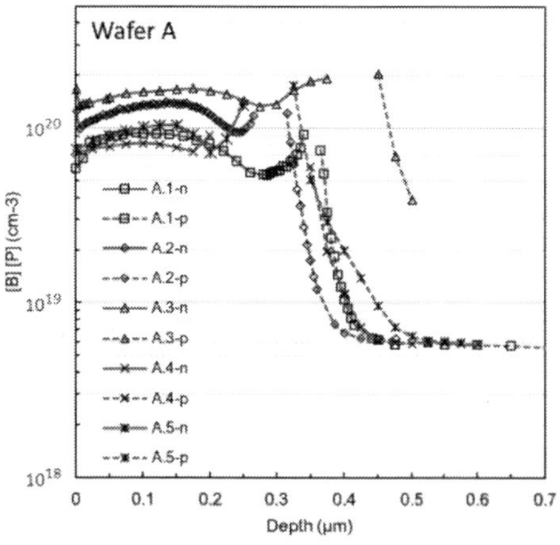

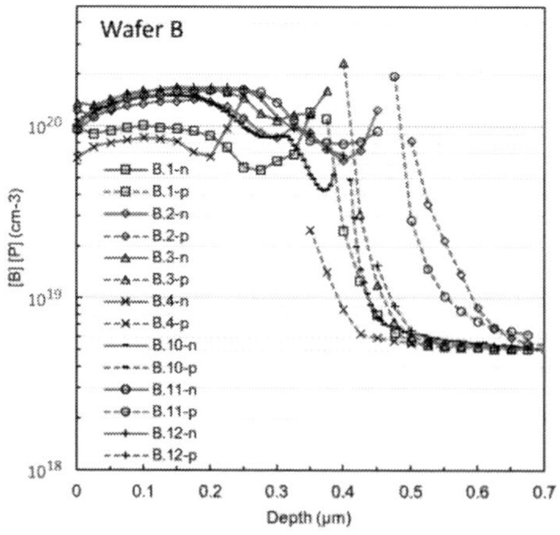

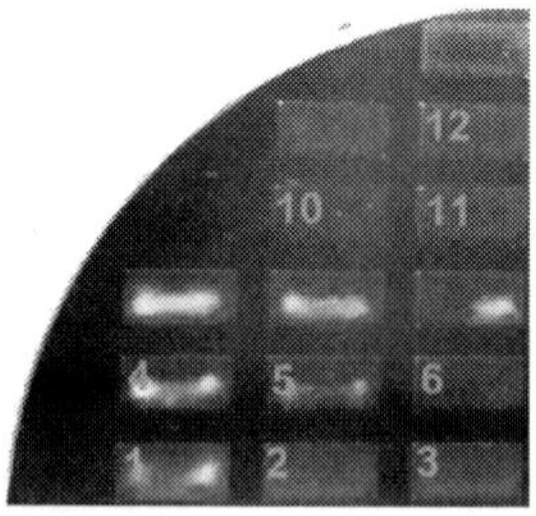

- If we examine by sample *position*, there now appears to be some measure of *reproducibility*.
- Suggests that laser processing is very sensitive to thermal behaviour of wafer during laser processing?

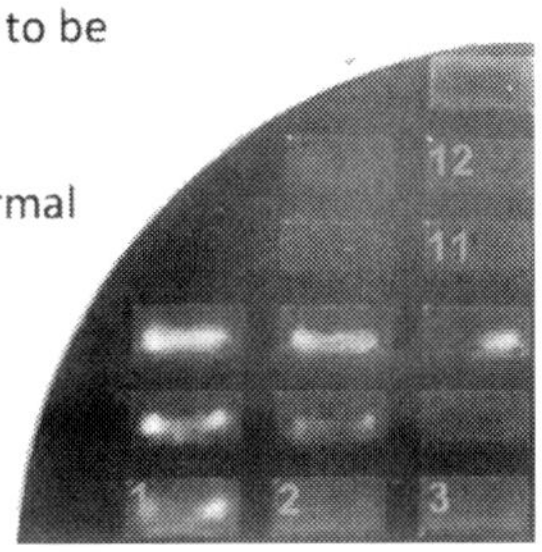

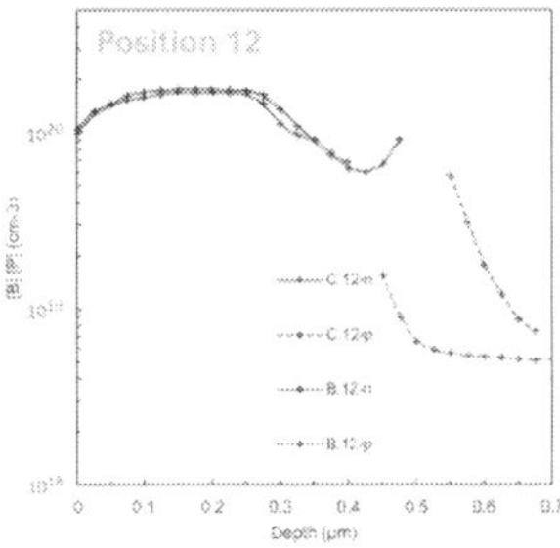

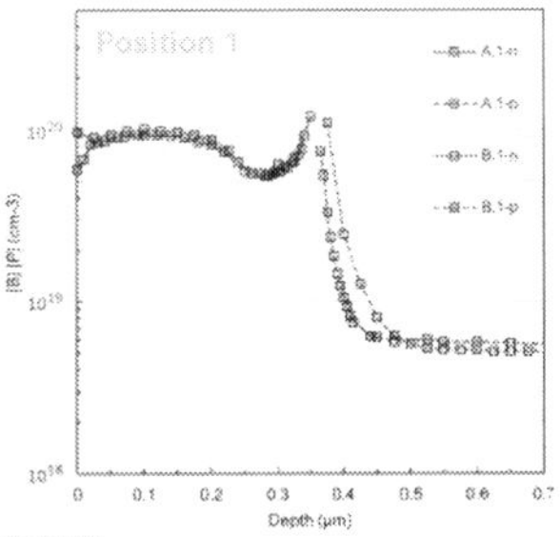

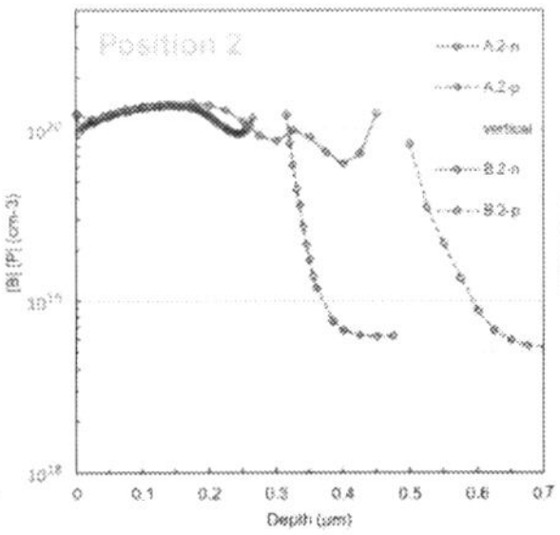

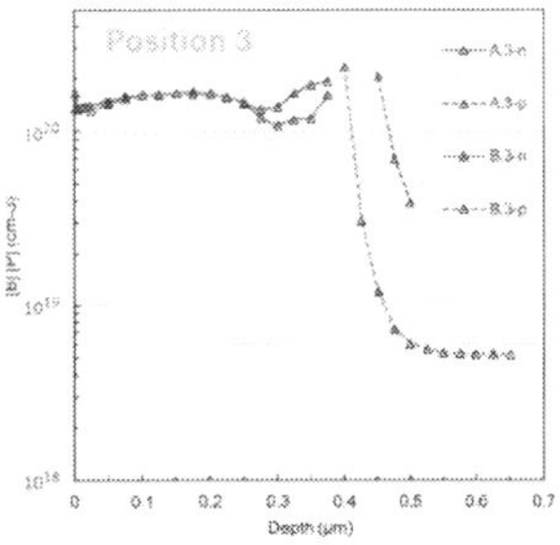

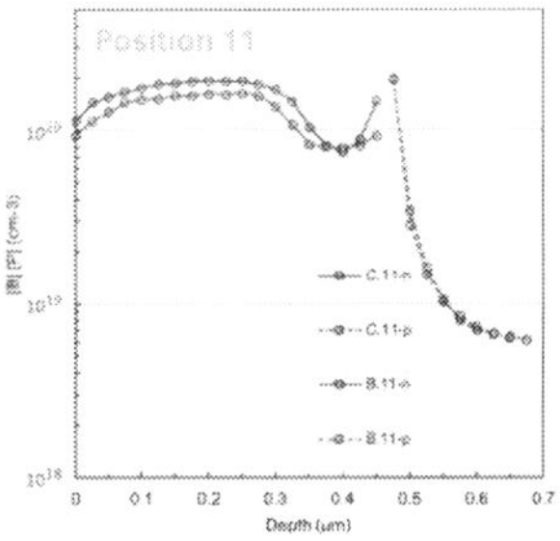

ECV oddities

- This effect is always observed (was removed from previous data)
- Unsure if real or artifact

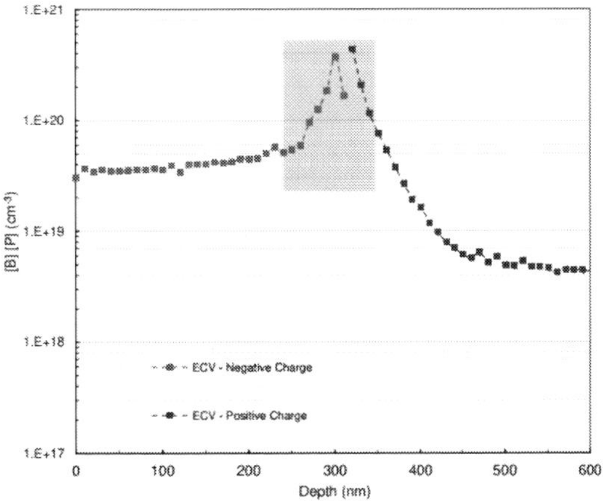

And what about oxygen?

- It too is incorporated – origin is $POCl_3$

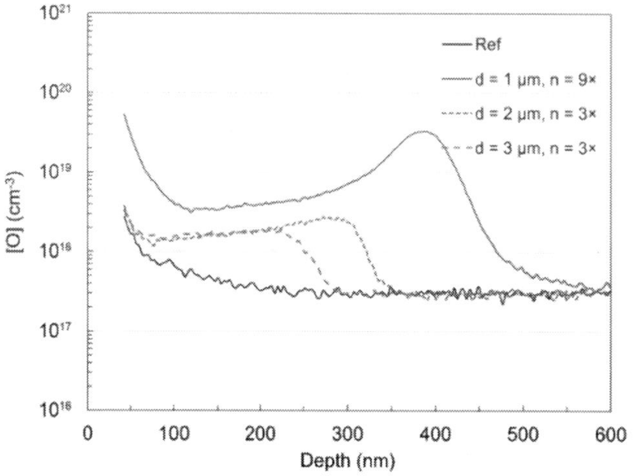

SIMS evaluation

40th European Photovoltaic Solar Energy Conference and Exhibition

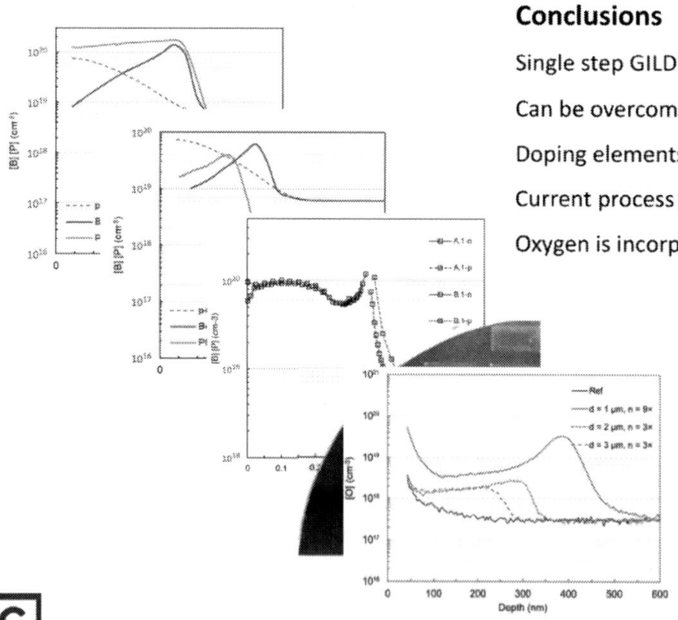

Conclusions

Single step GILD is unlikely to result in desired doping profiles

Can be overcome by 2-step process

Doping elements are electrically active (ECV)

Current process does not result in homogenous samples

Oxygen is incorporated from the dopant atmosphere

Gas-Immersion Laser-Doped Silicon Tunnel Junctions for Multi-junction Solar Cells

Laser doping @ Lisbon

Ciências ULisboa

Prof João A. Serra   Dr Filipe C. Serra   Dr Guilherme Gaspar   Prof Killian Lobato

ECV @ Konstanz and Lyon

Dr Fabian Geml   Prof Giso Hahn

Prof Alain Fave

SIMS @ Oslo

Prof Lasse Vines

Modelling @ Freiburg

Prof Matthias Müller   Jonas Kern

Universität Konstanz

INSA INSTITUT NATIONAL DES SCIENCES APPLIQUÉES LYON

UiO : University of Oslo

BERGAKADEMIE TECHNISCHE UNIVERSITÄT FREIBERG

1BO.4 Gas-Immersion Laser-Doped Silicon Tunnel Junctions for Multi-junction Solar Cells – 40th EU PVSEC, 19th September 2023

39

020020-039

Impact of the Front Side Multi Anti-Reflection Coating layer on PERC Solar Cells

Cheng-Wen Kuo, Ta-Ming Kuan, Yung-Chih Li, Chun-Wei Lee, Wei-Lo Chueh, Li-Guo Wu, Shih-Chieh Lin and Cheng-Yeh Yu
TSEC Corporation, No.85, Gaungfu N. Rd., Hsin-Chu 30351, Taiwan.

ABSTRACT: In this work, we demonstrated that the p-type PERC solar cells can also be improved the efficiency based on front side $SiO_x/SiO_yN_x/SiN_x$ multi anti-reflection coating (ARC) layer structure. Improved the electrical properties are obtained by depositing the SiO_x layer after the SiO_yN_x/SiN_x stack layer. Considering the follow-up visual color and production capacity, there is an ARC design with the best conditions, and its photocurrent and open circuit voltage are both improved. Compared to the results of the reference (REF) solar cell, the conversion efficiency was imprŏved by about 0.11%. Finally, the relative quality tests of the optimized ARC PERC solar cells are also in compliance with IEC specifications.
Keywords: PERC, Multi-ARC, SiO

1 INTRODUCTION

As the need for net-zero emissions accelerates, the energy trend is to use renewable energy to replace traditional fossil fuels. Many research institutions have conducted research on renewable energy from the aspects of its sources and its environmental impact [1]. The solar energy is one of the renewable energy sources. It is both a free energy source and environmentally friendly. Converting solar energy into electricity using photovoltaics has been found to reduce CO2 emissions [2].

Photovoltaic(PV) converts solar energy into electrical energy based on the PV effect, a process that generates a voltage (direct current, DC) between two different semiconductor materials when exposed to sunlight [3].

Solar cell efficiency loss can occur through a variety of mechanisms, including reduced efficiency due to reflection at the interface between different media.

The high reflectance values result from the high refractive index of the solar cell substrate, which results in asignificant percentage of the incident solar radiation being reflected at the interface between the solar cell surface and air. Therefore, this radiation does not contribute to the carrier pair generation process, thereby reducing the efficiency of the solar cell [1,4].

For mass production of solar cells, more efficient solutions must be found to limit photons lost at the front of the cell. The idea is to make the deposition of one or more ntireflective layers[5].

In PVs, the refractive index of the amorphous hydrogenated form of SiN (SiN_x:H) has been found to be ideal because it leads to very low absorption in the solar spectral range and has excellent passivation properties on the silicon surface and within the silicon bulk in the case of mono and multi-crystalline material.Plasma-enhanced chemical vapor deposition (PECVD) is currently one of the preferred deposition methods for SiN thin films due to its lower thermal budget than other CVD processes and the high hydrogen content in the layers [6].

The amorphous hydrogenated form of SiON dielectric (SiO_xN_y:H) shares some similarities with SiN which can be deposited by PECVD at low temperature, it exhibits low absorbance and an adjustable refractive index from below 1.5 to as high as 3.0 by hanging the precursor gas flow ratio during the deposition process[7].

Therefore, by adjusting the optical index and thickness of the SiON films, these properties are suitable for antireflection multilayer coating structures for solar cells. Previous literature [8] also reported the anti-reflection coating ARC on crystalline silicon solar cell plays an important role not only on optical absorption but also on PID sensitivity. Refractive index of SiN_x ARC is reported to influence the PID sensitivity.

In this work, we introduce novel $SiO_x/SiO_yN_x/SiN_x$ in the ARC layer to improve the optical properties of silicon solar cells. It can be found that the wavelength can be improved, and the photocurrent can also be increased. Finally, the potential-induced degradation (PID) test of PERC solar cells also complies with the IEC TS 62804-1 specification, which is also similar to SiO_yN_x/SiN_x PERC solar cell degradation.

2 EXPERIMENTAL

Figure 1 showed the PERC+ structure in this work. The M6 size (166.0 mm x 166.0 mm) bifacial PERC solar cells were fabricated on 155 μm thick p-type boron doped Cz-silicon wafers with a resistivity of 0.4~1.1 Ω-cm.

The process sequence in this work was shown in Figure 2. Chemical wet etching methods were used for saw damage removal and surface texturing, which include both acidic etching for multi-crystalline silicon wafers and alkaline etching for mono-crystalline silicon wafers. The solution composition of acidic etching was nitric acid (HNO3) : hydrofluoric acid (HF) : deionized water (DI) = 1 : 2 : 1.5, and the solution composition of alkaline etching was potassium hydroxide (KOH) : deionized water (DI) = 1 : 50. Conventional phosphorous diffusion was employed by furnace to form the n+ emitters with a sheet resistance of 130 - 140 ohm/sq. The diffusion process also improved to uniformity for this work.

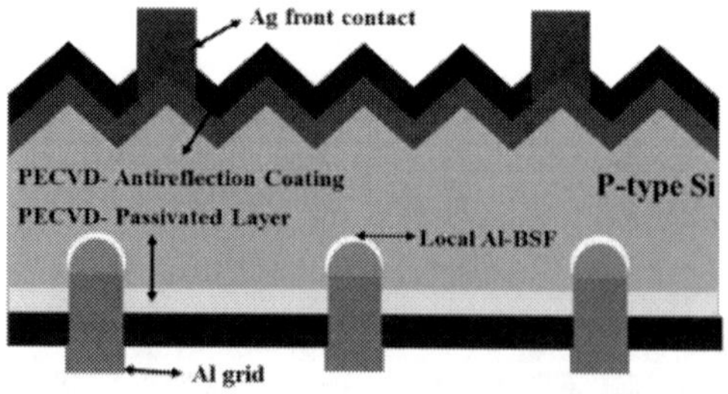

Figure 1: PERC+ structure as investigated in this work.

The phosphosilicate glass (PSG) was then removed by HF-contained solutions, formed edge isolation in the meantime. Then, the rear surface is polished which removes 3-5 μm silicon in this work, in order to reduce the

10.4229/EUPVSEC2023/1BV.5.
020021-001

roughness and its impact on the effective charge carrier lifetime [9]. In order to decrease PID effect and increase efficiency, we optimized the CVD process parameter to improve. An antireflection coating (ARC) layer of silicon nitride (SiN_x: H) and/or $SiO_x/SiO_xN_y/SiN_x$ multi ARC was deposited by plasma enhanced chemical vapor deposition (PECVD) with thickness of 90 nm. The rear side passivation was carried out on rear side also using a PECVD [10]. The Fig 3 shown that the thickness of rear side passivation thickness was about 120~140 nm. Afterward, the rear side laser opening ratio is also optimized. All PERC solar cells have the laser-ablated opening widths in the rear-side passivation stacks which were about 30 μm, as shown in Fig 4. The rear Al finger width is 150 μm in this work. The detail parameters were listed in Table 1. The front side silver and rear side aluminum metallization was formed by screen printing technique and followed by co-firing process, which resulted in formation of front silver grid contact and local back-surface-field. Finally, the current-voltage measurements of solar cells under AM 1.5 illumination were performed.

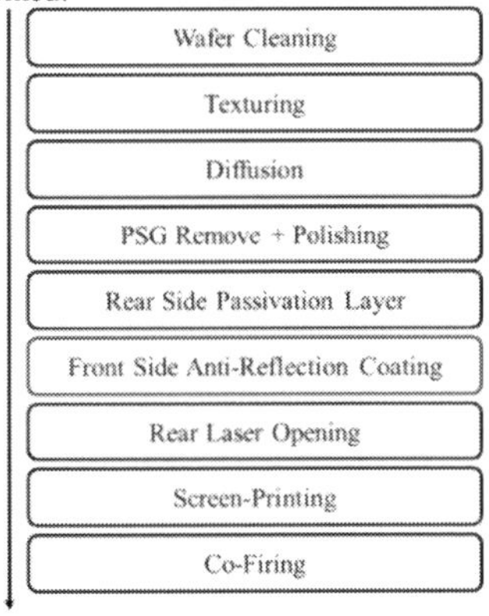

Figure 2: Process flow for the fabrication of PERC solar cell as developed in this work

Figure 3: The scanning electron microscope (SEM) image of rear side passivation layer for this work.

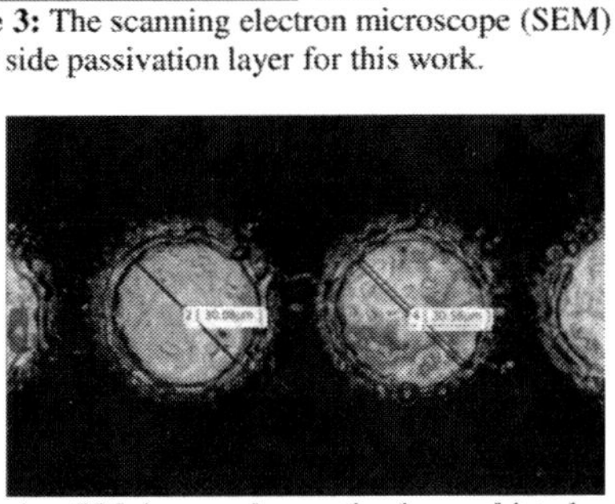

Figure 4 : OM image shown the laser ablated opening width in this work.

3 EXPERIMENTAL RESULTS AND DISCUSSION

We have performed an optical reflectance analysis for the different ARC layer, as shown in Figure 5. The distinct reduction of the reflectance at long wavelengths (>780 nm) is apparent for the optimized ARC layers compared with the baseline REF. the front average reflection of short wavelengths is about 2.46% and 2.18%, respectively.

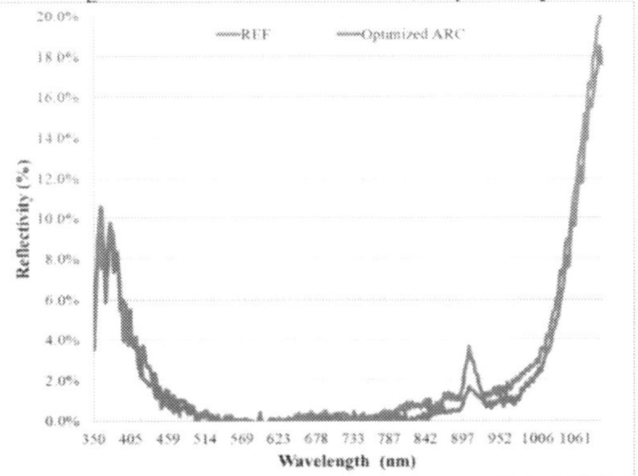

Figure 5: Measured reflectivity for the optimized ARC (red) PERC solar cell and Ref (blue) PERC solar cell.

Table 1 shows the average electrical performance of REF and optimized ARC PERC solar cells. The average efficiency of optimized ARC PERC solar cell was 22.98%, the average efficiency of 22.87% was obtained for the REF PERC solar cells. It shows that the photocurrent can also be increased by about 43mA and 0.11% the absolute gain in conversion efficiency can be obtained by introducing $SiO_x/SiO_yN_x/SiN_x$ as an ARC instead of multi SiN_x layer. Figure 6 shows SEM images of the different front side ARC thickness for REF and optimized ARC PERC solar cell. It can be found that the ARC thickness is similar to the process setting. It means that the use of optimized ARC layers can improve the optical characteristics, however, the appearance color need to be considered.

Table 1: The average electrical results of M6 REF and optimized ARC PERC cells.

	EFF (%)	FF (%)	V_{OC} (V)	I_{SC} (A)
REF PERC	22.87	80.14	0.6892	11.349
Optimized ARC PERC	22.98	79.94	0.6918	11.392

On the other hand, the ARC layer structure of PERC solar cells has an effect on PID [8], the potential-induced degradation (PID) test of the solar cell also conforms to the IEC TS 62804-1 specification and was used in this experiment. The REF and optimized ARC PERC solar cells also passed PID test which condition was indicated to be 85 degree Celsius and 85% RH in climatic chamber apply -1000 V for 96 hours as shown in Figure 7. Observed the power loss < 5% and no EL darkened area were found after PID test for the REF and optimized ARC PERC solar cells. Besides, the environmental test of the damp heating (DH) 1000 hours and the thermal cycle 200 cycle for REF and optimized ARC PERC solar cells were shown in Fig. 8 and 9. The DH 1000 test shows that decay rate < 3% and TC 200 test show that the decay rate < 1%. It could be found that there is not much difference in the change of ARC when the screen printing conditions are the same from the

10.4229/EUPVSEC2023/1BV.5.
020021-002

DH 1000 and TC 200 test. It also shows that the experimental results are effective.

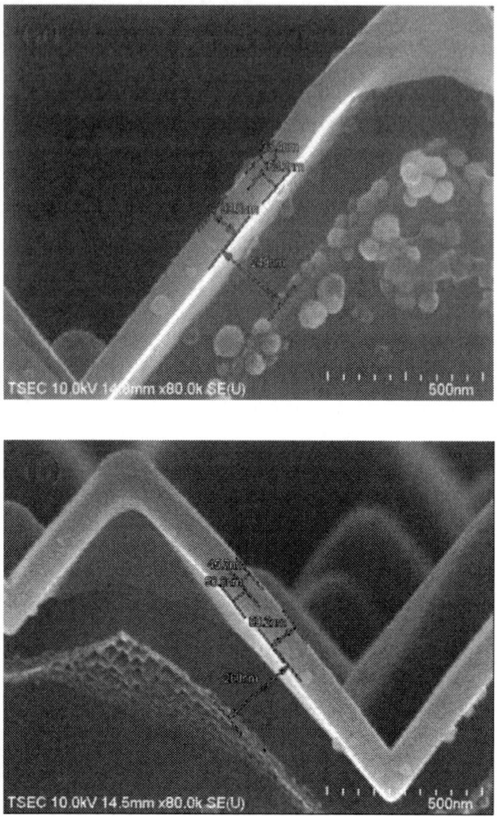

Figure 6: Scanning electron microscope images of rear side passivation thickness for the (a) REF, (b) Optimized ARC PERC

	Pre-PID		Post-PID		EL		Decay Rate(%) (Spec<5%)
Sample	Rsh	Pmpp	Rsh	Pmpp	Pre-PID	Post-PID	△Pmpp
REF-1	281.9	5.9845	92.9	5.7508			3.91%
REF-2	286.9	5.8610	213.7	5.8043			0.97%
REF-3	506.0	5.8105	441.9	5.6204			3.27%
Optimized ARC-1	1025.4	6.1018	825.8	5.8861			3.53%
Optimized ARC-2	1415.3	6.0212	530.6	5.9841			0.62%
Optimized ARC-3	546.4	6.0294	272.1	5.9456			1.39%

Figure 7: Summary of power loss and EL images of pre-PID and post-PID of REF and optimized ARC PERC solar cells. PID test condition is 85 degree Celsius / 85% RH biased -1000V for 96 hours.

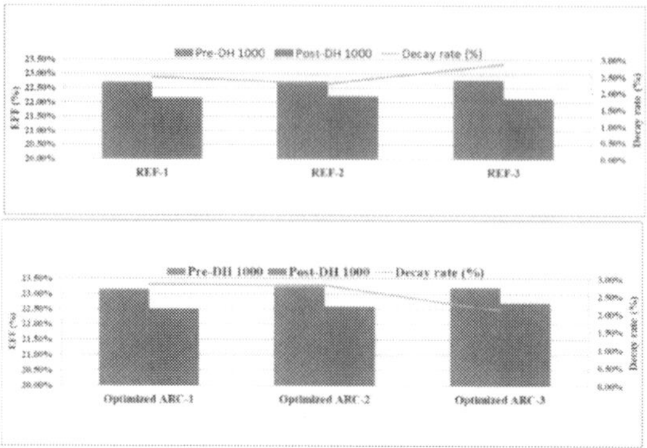

Figure 8: DH 1000 test for REF and Optimized ARC PERC

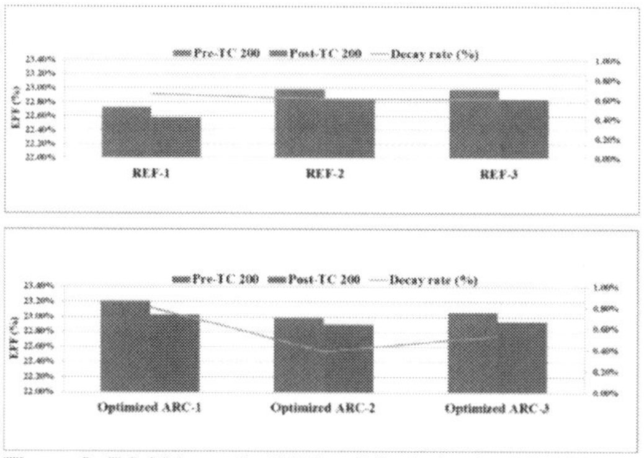

Figure 9: TC 200 test for REF and Optimized ARC PERC

4 CONCLUSIONS

In this work, we have demonstrated that the p-type PERC solar cells could be improved front side performance based on front side $SiO_x/SiO_yN_x/SiN_x$ ARC layer structure. Improved ARC layer was obtained by depositing the SiO_x layer after the SiO_yN_x/SiN_x layer. The distinct reduction of the reflectance at long wavelengths (>780nm) is apparent for the optimized ARC layers compared with the baseline REF. About 0.11% gain in conversion efficiency was obtained compared to the result of reference solar cell. Finally, PID test of REF and optimized ARC PERC solar cell is also in compliance with IEC TS 62804-1 specifications and, the power loss of both is similar in this experiment.

5 REFERENCES

[1] H. J. El-Khozondar, R.J. El-Khozondar, R. Al Afif, and C. Pfeifer, "Modified Solar cells with antireflection coatings" Int. J. Thermofluids Vol. 11, pp. 100103, 2021.

[2] C.-.H. Park, Y.-J. Ko, J.-.H. Kim, H. Hong, "Greenhouse gas reduction Effect of solar energy systems applicable to 'high-rise apartment housing structures in South Korea", *Energies*, vol. 13, pp. 1, 2020.

[3] N. Khordehgah, A. Żabnieńska-Góra, and H. Jouhara, "Analytical modelling of a photovoltaics-thermal technology combined with thermal and electrical storage systems", Renew. Energy, Vol. 165, pp. 350, 2021.

[4] O.V. Semenova, V.A. Yuzova, T.N. Patrusheva, F.F.

Merkushev, M.Y. Railko, and S.A. Podorozhnyak, "Antireflection and protective films for silicon solar cells", IOP Conf. Ser.: Mater. Sci. Eng., Vol. 66, pp. 012049, 2014.

[5] A. Dieye, N. Mbengue, O. A. Niasse, F. Dia, M. Diagne, and B. Ba, "Influence of Multi Anti – Reflective Layers on the Silicon Solar Cells", Journal of Materials Science and Surface Engineering, Vol. 5, pp. 729, 2017.

[6] J. Dupuis, E. Fourmond, J. Lelièvre, D. Ballutaud, M. Lemiti, "Impact of PECVD SiON stoichiometry and post-annealing on the silicon surface passivation." *Thin Solid Films*, vol.516, pp.6954, 2008.

[7] J. Dupuis, E. Fourmond, D. Ballutaud, N. Bererd, M. Lemiti, "Optical and structural properties of silicon oxynitride deposited by plasma enhanced chemical vapor deposition" *Thin Solid Films*, vol. 519, pp.1325, 2010.

[8] C. Zhou, J. Zhu, S.E. Foss, H. Haug, Ø. Nordseth, E.S. Marstein, W. Wang, "SiOyNx/SiNx Stack Anti-reflection Coating with PID-resistance for Crystalline Silicon Solar Cells," Energy Procedia, vol. 77 pp.434, 2015.

[9] E. Urrejola, R.Petres, J. Glatz-Reichenbach, K. Peter, E. Wefringhaus, H. Plagwitz, G. Schubert., "High Efficiency Industrial PERC Solar Cells with all PECVD-Based Rear Surface Passivation" 26th EUPVSEC, pp. 2233, 2011.

[10] C.-W. Kuo, T.-M. Kuan, W.-L. Chueh, L.-G. Wu, S.-C. Lin, C.-Y. Yu, "Investigation of the Rear Side Passivation Layer on Bifacial PERC Solar Cells" 38th EUPVSEC, pp. 288, 2021

INVESTIGATION OF PROCESS PARAMETERS ON THE DISTRIBUTION OF THE SPECIFIC RESISTIVITY IN GA-DOPED CZ-CRYSTALS

F. Mosel[1], K.Hess[1], M. Trempa[2], J. Friedrich[2]
[1] PVA Crystal Growing Sytems GmbH, Im Westpark 10-12, 35435 Wettenberg, Germany
[2]Fraunhofer-Institut für Integrierte Systeme IISB, Schottkystraße 10, 91058 Erlangen, Germany
e-mail: frank.mosel@pvatepla.com
tel: +49 64168690-125, fax: +49 64168690-822

ABSTRACT: Light induced degradation (LID) which is based on a defect state of a boron atom with two oxygen atoms in Czochralski-grown crystals (BO-LID) reduces the bulk lifetime and hence the energy conversion efficiency of solar cells upon illumination. This boron-oxygen complex forms during the cooling process of the crystal. Gallium has replaced boron as doping element because it shows negligible BO-related degradation. However, as a doping element, gallium has two severe disadvantages. Firstly, a small distribution coefficient resulting in a highly inhomogeneous resistivity distribution, which means a significant reduction in the yield of substrate material in a specified resistivity range. Secondly, the hardly reproducible adjustment of the specific resistivity in Ga-doped Cz-crystals due to evaporation phenomena of gallium species. We have examined the influence of Cz-process parameters on the evaporation behavior of gallium during the crystal growth of silicon crystals in the resistivity range of interest for solar cells. We will show here our main results.
Keywords: resistivity distribution, gallium-doping, LID, Czochralski process, evaporation

1 INTRODUCTION

The fact that p-type solar cells were only produced on boron-doped wafers until recently is due to the favorable incorporation behavior of boron during the growth of silicon crystals by the Czochralski (Cz) technique. The boron distribution coefficient of k=0.8 means that boron is incorporated into the solidifying crystal with almost the same concentration as the boron concentration in the melt. Thus, a relatively uniform dopant distribution over the crystal length can be achieved. However, a disadvantage of the solar cells whose base is boron-doped is the light-induced degradation (LID). The mechanism that triggers the degradation is a boron-oxygen correlated defect. During the cooling process of the crystal, a substitutionally incorporated boron atom (B_S) combines with a mobile oxygen dimer (O_{2i}). In its ground state, this complex is a deep donor [1], which acts as a trap for minority carriers and not as a recombination center. Under the influence of light, which is equivalent to charge carrier injection, this B_SO_{2i} complex binds an electron and transforms to one of two possible shallow acceptor states. These two acceptor states result in rapid (FRC: fast forming recombination center) and slow (SRC: slow forming recombination center) formation of the BO complex that induces LID. Since the simultaneous presence of boron and oxygen is a prerequisite for the BO complex, there are various possibilities to reduce or avoid LID. Floating zone grown silicon mono crystals are almost free of oxygen. This production method is expensive compared to Cz-technology and is limited in terms of crystal diameter. An alternative growth technology is the use of a magnetic field during the Cz-process (MCz method). A constant magnetic field dampens the melt convection and thus reduces the oxygen incorporation into the crystal. Yet, this method also requires a high investment for the superconducting magnet.

The simplest method is the substitution of boron by another group III element. Since neither aluminum nor indium are alternative doping elements, gallium is the only possible candidate. Both former mentioned elements also form A_SO_2 complexes (A is either B, Al, Ga, In), of which the AlO complex shows strong recombination activity. Indium has an ionization energy of Ev +0.16eV, i.e. at room temperature only about 50% of the substitutionally incorporated indium is ionized. The non-ionized indium acts as a recombination path [2]. This leaves Ga as the only alternative for substituting boron in Cz monocrystals.

We have determined that, in contrast to boron-doped crystals, the expected resistivity distributions in gallium-doped crystals did not match those measured on the grown crystals. Fig.1 shows the axial resistivity distribution of a boron doped crystal as expected before growth (dashed green line) and measured on the surface of the crystal after growth (data). With the exception of the first two measuring points, there is a good accordance. The two outliers at the beginning of the boron-doped crystal are due to compensation of thermal donors, which often occurs at the beginning of the crystal. The red line is obtained by a least squares fit to the data. Fig.2 shows the same plot for a gallium-doped crystal. A significant deviation from the expected (green line) to the measured data is evident.

To clarify this situation we investigated the evaporation behavior of gallium during the Cz-process. We compared the results of electrical characterization with the chemical content of Ga-doped samples, measured by means ICP-MS[1].

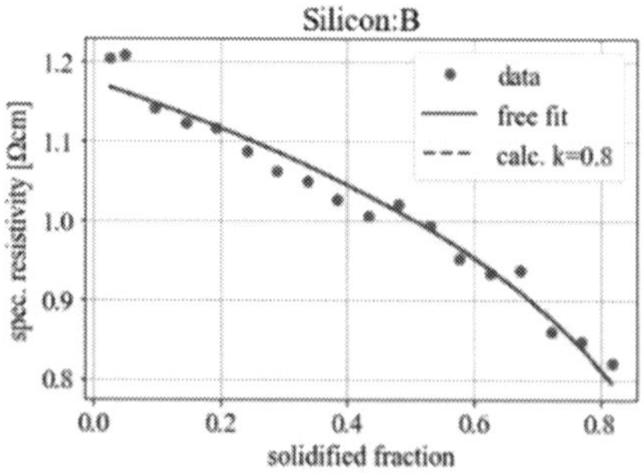

Figure 1: Axial distribution of the specific resistivity in a boron-doped Cz-crystal

[1] Fraunhofer-Center für Silizium-Photovoltaik CSP

10.4229/EUPVSEC2023/1BV.5.3
020024-001

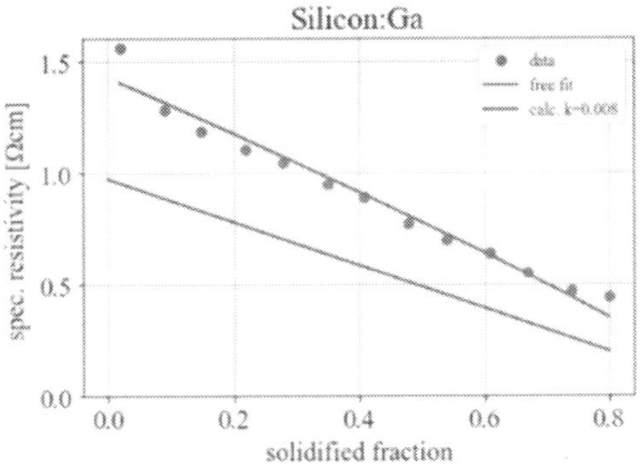

Figure 2: Axial distribution of the specific resistivity in a gallium-doped Cz-crystal

2 EXPERIMENTAL APPROACH

2.1 Overview

The crystal growth experiments presented in this paper were performed in a Cz-puller from PVA Crystal Growing Systems GmbH. A growth configuration for 26-inch crucibles was used. All samples examined in this work were grown from solar-grade polysilicon.

In the first step, the evaporation behavior of gallium during the melting phase of the charged silicon was examined. It turned out that under the test conditions during the melting phase there was no noticeable loss of Ga from the melt. In further crystal growth experiments, the evaporation was investigated under different growth conditions mainly with respect to the inert gas flow and furnace pressure. The electronic properties in the samples measured by means of Hall-effect were compared with the chemical contents of the doping elements. The results are summarized in Chapter 2.4. For completeness, we also studied the evaporation rates of aluminum and indium under standard process conditions.

2.2 Evaporation behavior of Ga during the melting phase

Two different doping methods were used to investigate the evaporation behavior of the weighed quantity of gallium during the melting phase of the polysilicon. The gallium was added to the silicon charge both in a closed silicon container (Fig.3) and unprotected as beads (Fig.4). Fig.5 shows the silicon container with its cap.

Figure 3: Ga embedded in a closed silicon container surrounded by silicon chunks

Figure 4: Ga beads surrounded by silicon chunks

Fig.6 shows the closed Si-container just before the charge was completely melted.

Figure 5: Silicon container

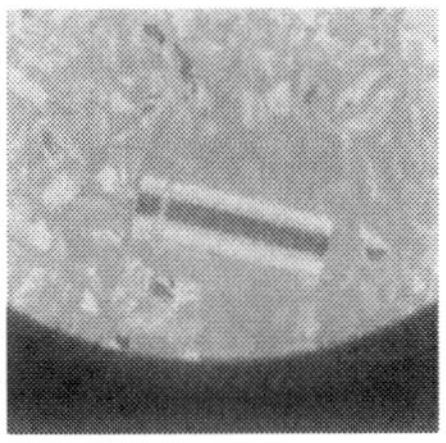

Figure 6: Silicon container during melting phase

The initial process conditions were identical for both experiments. 40 kg of polysilicon were melted, the calculated initial concentration of gallium in the melt was $4.2*10^{18}$ at/cm^3. The ratio of the free melt surface A to the melt volume V was A/V=0.18 cm^{-1}.The Ar flow rate was 30 l/min, the furnace pressure 20 mbar. The melting phase lasted 5 hours, the time gap of the melt until the sample was withdrawn from the melt was 2.5 hours. Fig 7 shows an example of the crystallized sample used for characterization.

Figure 7: Si-sample and 2 € coin

The expected Ga-concentration in the sample without evaporation was $3.4*10^{16}$ at/cm^3, assuming a segregation coefficient of k=0.008. The results of the ICP measurements of the Ga-content in the two samples are given in Tab.I. Both measurements gave identical values within the framework of the standard deviation (SD). At this point it can be stated that the melting phase seems not to be a critical process phase concerning the loss of gallium which is in contradiction to the work of Hoshikawa et al. [3].

Table I: ICP-MS results of experiment chap. 2.2

	atoms/cm^3 sample Fig.3	atoms/cm^3 sample Fig.4
Gallium	$3.98*10^{16}$	$4.00*10^{16}$
SD	$1.39*10^{15}$	$1.35*10^{15}$
LOD	$2.08*10^{11}$	$1.84*10^{11}$

2.3 Evaporation behavior under process conditions:

The evaporation of volatile elements is represented by a first-order kinetic reaction, i.e. by eq.(1)

$$-\frac{dC}{C} = \gamma \frac{A}{V} dt \quad or \quad ln\left(\frac{C_t}{C_0}\right) = -\gamma \frac{A}{V} t \qquad (1)$$

where C_t, C_0, γ, A, V, t denote the actual doping element concentration [at/cm^3], initial doping element concentration [at/cm³], evaporation rate constant [cm/s], free melt surface [cm²], melt volume [cm³] and process time [s], respectively.

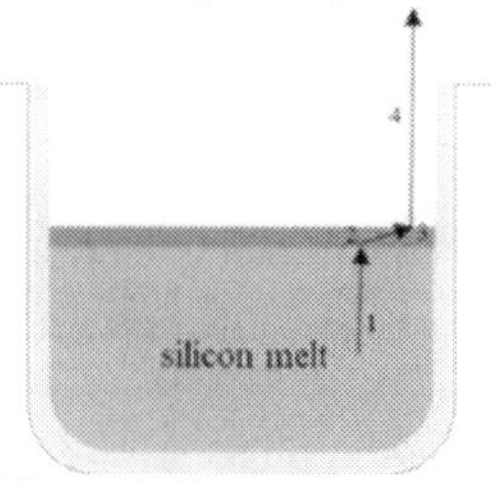

Figure 8: Sketched sequence of evaporation

Evaporation takes place in four sequential steps, which are outlined in Fig.8: 1) transport of the doping element or the oxygen compound of the doping element in the melt volume to the boundary layer, 2) transport through the melt/gas boundary layer, 3) physical evaporation process, 4) removal through the gas phase. The time dependence of the evaporation process results from the four sub steps. The transport in the melt volume is dominated by the convection phenomena in the melt volume. The transport in the boundary layer is determined by convection and diffusion of the solute. The chemical evaporation process can be described by the Hertz-Knudsen equation [4]. The transport through the gas phase is essentially influenced by the gas flow and chamber pressure.

It seems practically impossible to describe the totality of the mentioned transport processes of evaporation as a function of the crystal growth process in a manageable model. Therefore, an attempt was made to provide a tool for predicting the dopant incorporation and the resulting specific resistivity by means of numerical fit calculations on measured data of the specific resistivity (2.3.2).

2.3.1 Determination of the evaporation rate using a 2-point measurement method

For the measurement of the evaporation rates two samples were grown according to fig.9. After the first sample a second one was withdrawn from the melt after a time gap between 14 and 22 hours. Then the final crystal growth process followed.

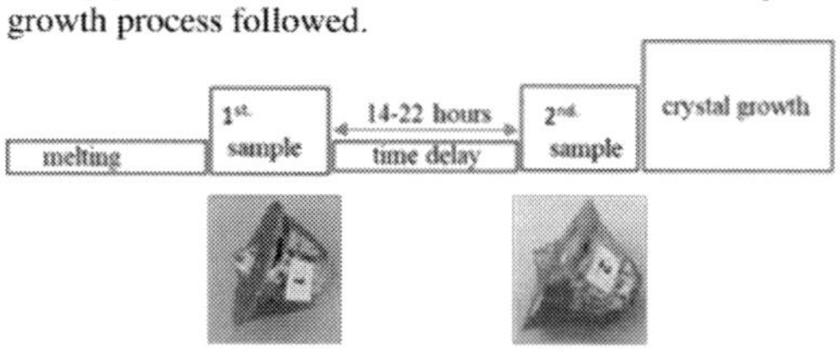

Figure 9: Time schedule of the experiments

The main process conditions for the different evaporation experiments are given in the legend of Fig.10. The diagram shows the evaporation rates determined for gallium in silicon melt under the applied process conditions plotted versus the initial Ga concentration in the melt. With the exception of the experiments under increased and reduced argon pressure an average evaporation coefficient of 3-4*10^{-5} cm/s for Ga under the applied process conditions is applicable. However, the values determined from two measurements can only be evaluated as static reference points. Fig.11 is intended to illustrate the pressure dependence of the evaporation coefficient of gallium in a silicon melt. At this point, it should be expressly noted that the silicon melt was hold in a silica crucible as a source of oxygen.

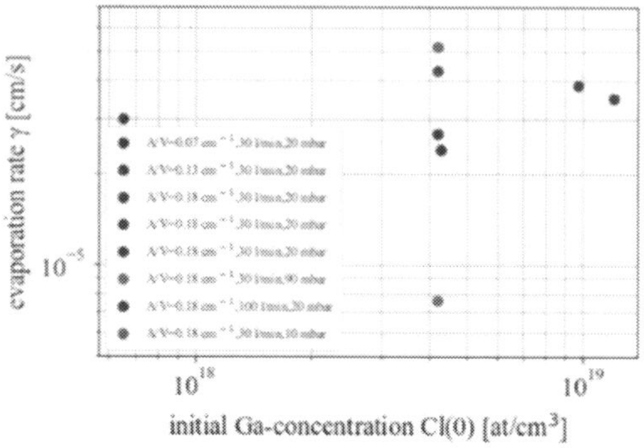

Figure 10: Overview of the evaporation coefficients γ [cm/s] of Ga in silicon melt according equation (1)

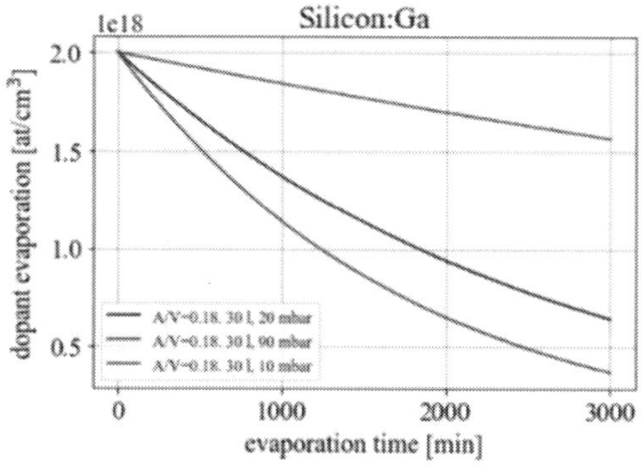

Figure 11: Evaporation of Ga in silicon melt for different gas pressures (Ar) in the process chamber, the blue line represents a mean value of 3.5*10^{-5} cm/s.

For the sake of completeness, the evaporation rates of aluminum and indium in a silicon melt were also determined and compared with the evaporation rate of gallium. Fig.12 shows the calculated loss of the different elements versus the time of the melt in the silica crucible.

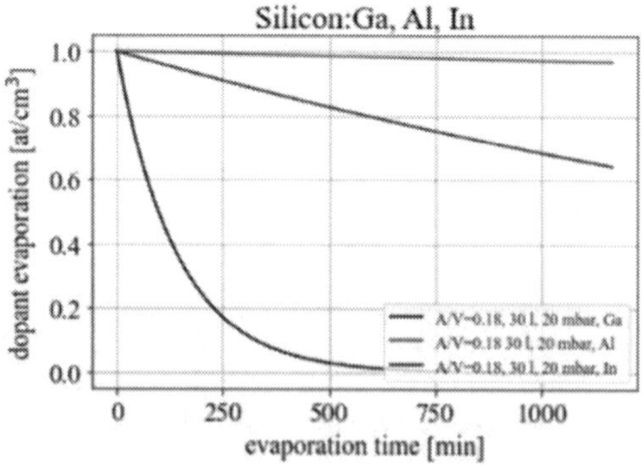

Figure 12: Estimated evaporation of group III elements in silicon melt versus time

2.3.2 Determination of the evaporation rate by least squares fitting of the extended Scheil equation

The axial dopant distribution C_S in a Cz-crystal can be

described by the directional solidification model known as Scheil equation eq.(2). The corresponding concentration of the dopant in the melt is given by eq.(4).

$$C_S = kC_l(0)(1-g)^{(k-1)} \quad (2)$$

The relationship between the concentrations in the crystal C_S and in the melt volume C_l is given by the segregation coefficient k: eq.(3).

$$\frac{C_S}{C_l} = k \quad (3)$$

$$C_l = C_l(0)(1-g)^{(k-1)} \quad (4)$$

The model describes the segregation of dopants that only take part in the melt-crystal phase transition (closed system). This assumption is entirely justified for dopants that have a negligible vapor pressure in the silicon melt. Strictly speaking, the Scheil equation applies in a conservative, closed melt/crystal system without exchange of the dissolved doping element with the environment. This idealized situation cannot be applied to volatile doping elements, which show a noticeable evaporation behavior under process conditions.
Nevertheless, there are various approaches in literature [5,6] to describe the evaporation behavior of volatile dopants by extending the Scheil equation. In our case we have expanded the exponent in the equation by a term γ_{fit}/v eq.(5). Here γ_{fit} is an expression describing the evaporation coefficient and v is the growth rate. This modified Scheil equation, takes into account the evaporation through a modified evaporation coefficient γ_{fit} and the process dynamics through the growth rate v.

$$C_l = C_l(0)(1-g)^{\left(k-1+\frac{\gamma_{fit}}{v}\right)} \quad (5)$$

The evaporation rate is influenced by a large number of process parameters, such as the concentration of the doping element in the melt, the process gas conditions (pressure and flow), the oxygen concentration in the melt and thus also the hydrodynamic conditions and the quality of the crucible, the ratio of free melt surface to melt volume A/V and many other secondary process conditions, all dependent on the nature of the Cz-process.
Fig.13 - Fig.15 show the axial resistivity distributions for some Ga-doped silicon crystals grown under identical process gas conditions. The main growth parameters are given in Tab.II. With the exception of Fig.15, the expected resistivity distributions shown as green lines are significantly lower than the data measured on the surface of the as grown crystal (Fig.13 and Fig.14). This means that the amount of gallium was not sufficient to achieve the target resistivity. The deviations were mainly caused by the evaporation of gallium from the silicon melt in the silica crucible.

Table II: Main growth parameters of crystal growth experiments, shown in Fig.13 – Fig.15

Si:Ga	$C_l(0)$ [at/cm3]	target resistvity [Ωcm] at start of body	drift mobility μ_D [cm^2/(Vs)], used in eq.(6)	A/V at process start	process gas
Fig.13	$1.8*10^{18}$	1.00	410	0.07	30 l/min, 20 mbar
Fig.14	$4.2*10^{18}$	0.50	410	0.13	30 l/min, 20 mbar
Fig.15	$4.4*10^{18}$	0.50	350	0.1	30 l/min, 20 mbar

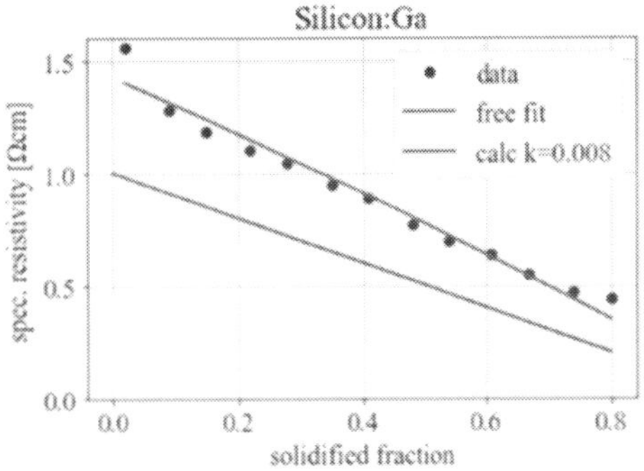

Figure 13: Axial distribution of the specific resistivity in a Ga-doped silicon crystal versus the solidified fraction

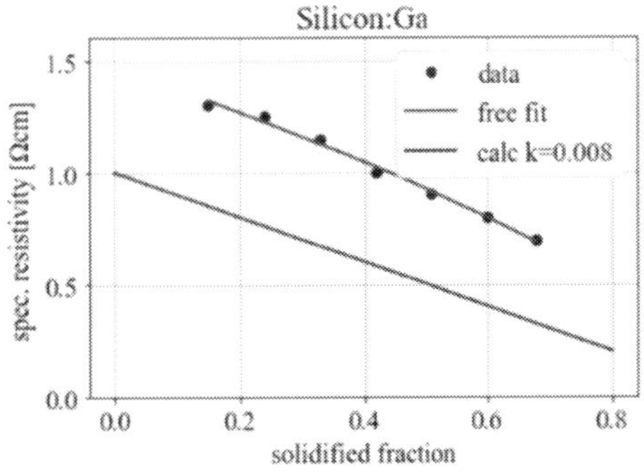

Figure 14: Axial distribution of the specific resistivity in a Ga-doped silicon crystal versus the solidified fraction

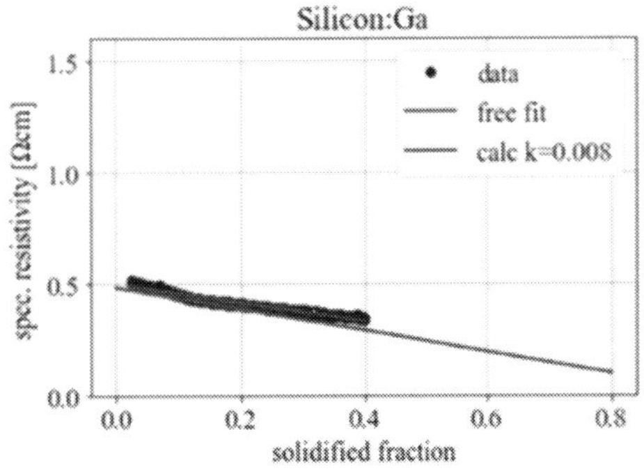

Figure 15: Axial distribution of the specific resistivity in a Ga-doped silicon crystal versus the solidified fraction

From the axial resistivity distributions (Fig.13 - Fig.15), the corresponding concentrations in the melt C_l were estimated according to eq.(6) with the drift mobilities given in Tab.II.

$$C_l = \left(\frac{1}{e\,\mu_D\,\rho\,k}\right) \quad (6)$$

with e: elementary charge, μ_D: drift mobility, ρ: specific resistivity, k: segregation coefficient.

Figs. 16 – 18 show the gallium concentrations in the melt volume corresponding to the solidified fraction. The green curves show the concentrations according to Scheil eq.(4) without evaporation and the weighed amount of dopant

Cl(0) at the start of solidification. The red concentration curves were fitted to the measured data (blue symbols) according to the extended Scheil equation eq.(5). The fit parameters are the initial concentrations in the melt at the beginning of the body phase Cl(0) and the term (γfit/v), where v is the growth rate of 1 mm/min corresponding to the applied average pulling speed.

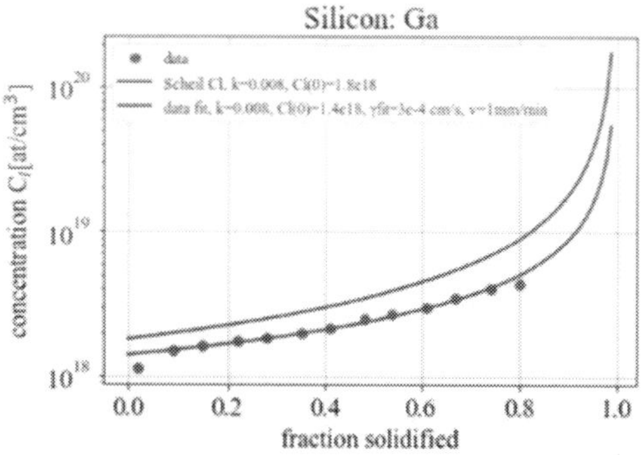

Figure 16: Gallium concentration in the melt versus the solidified fraction

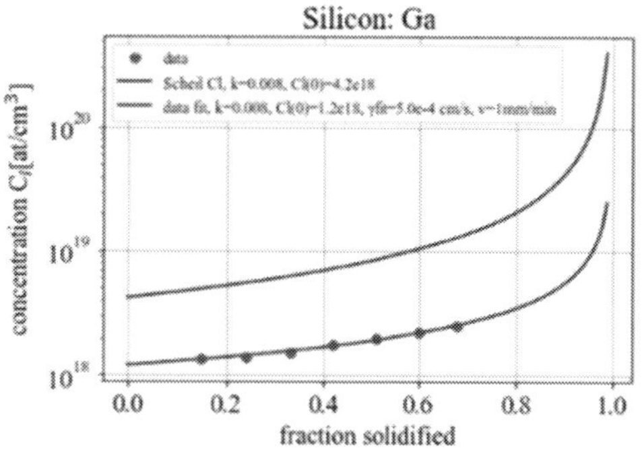

Figure 17: Gallium concentration in the melt versus the solidified fraction

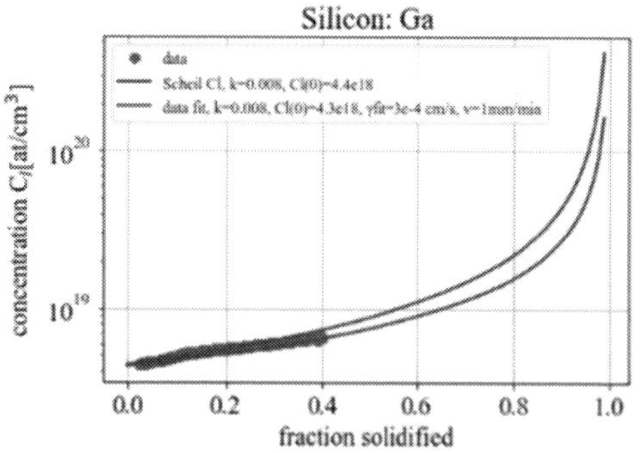

Figure 18: Gallium concentration in the melt versus the solidified fraction, the data points correspond to 1200 as cut wafers

The deviation of the calculated Ga-concentrations without the effect of evaporation (green curves) and the fitted concentrations (red curves) according to eq.(5) is a result of the duration of the time after the melting process and the final start of the body phase. During this period, the volatile dopant may evaporate, depending on the applied process conditions.

The gallium loss due to evaporation can be estimated according to eq.(7) with $\lambda = \gamma$ A/V:

$$C(t) = C(0)e^{-\lambda t} \quad (7)$$

Here, t is the time that elapses after the melting phase to the start of the final body phase. Tab.II summarizes the data for the calculation of the concentration losses according to eq.(7). An average evaporation coefficient of $3.5*10^{-5}$ was applied. The estimated values for the examples of Fig.16 and Fig.18 agree well with the initial values from the fit calculations (see legend). The initial concentration of Fig.17 is significantly below the calculated value, which may be due to the unfavorable A/V ratio.

The evaporation coefficients γ_{fit} obtained from the least square fit calculations (see legend in Figs 16-18) are an order of magnitude higher than the measured evaporation rates. They have to be regarded as pure fitting parameters of the measured values for the evaluation of the totality of the process parameters.

Table III: Data used for calculation of eq. 7

Si:Ga	C(0) [at/cm^3]	A/V [cm^{-1}]	γ [cm/s]	$\lambda=\gamma$A/V [s^{-1}]	t [s]	C(t) [at/cm^3]
Fig.16	$1.8*10^{18}$	0.07	$3.5*10^{-5}$	$2.5*10^{-6}$	87000	$1.5*10^{18}$
Fig.17	$4.2*10^{18}$	0.18	$3.5*10^{-5}$	$6.3*10^{-6}$	70000	$2.7*10^{18}$
Fig.18	$4.4*10^{18}$	0.1	$3.5*10^{-5}$	$3.5*10^{-6}$	36000	$3.9*10^{18}$

2.4 Electrical characterization of the investigated samples

For electrical characterization, Hall effect samples were prepared adjacent to the samples for chemical analysis. For the evaluation of the measurement a Hall constant of r = 0.72 was applied, which allows a correct determination of the drift mobility and carrier concentration [7].

$$p = \frac{r}{qR_H} \quad (8)$$

The Hall coefficient R_H is measured experimentally, p is the free hole concentration and r is the hall factor.

Tab.IV summarizes the results of the Hall effect measurements and the chemical content of gallium in the examined samples.

Table IV: Comparison of electronic transport properties with measured gallium concentrations

Ga-conc. [at/cm^3]	4.00E+16	2.72E+16	3.36E+16	2.55E+16
p-conc. [at/cm^3]	3.00E+16	2.52E+16	3.00E+16	2.00E+16
μ drift [cm^2/Vs]	354	360	413	440
μ Hall = $R_H\sigma$ [cm^2/Vs]	255	259	297	317
[p]/[Ga]$_{chem}$	0.77	0.91	0.89	0.78

2.5 Discussion of the measurement results (ICP-MS and Hall effect)

We have calculated the temperature dependent free carrier concentration by numerical solution of the charge-balance equation eq.(9) in weakly compensated p-type silicon for some group III elements.

$$n + N_A^- = p + N_D^+ \quad (9)$$

with n free electron concentration in the conduction band, p free hole concentration in the valence band ,N_A^- concentration of ionized acceptors, N_D^+ concentration of ionized donors.

10.4229/EUPVSEC2023/1BV.5.3
020024-005

For the calculation of the different group III elements, an acceptor concentration of $N_A=2*10^{16}$ at/cm³ and a donor concentration of $N_D=5*10^{14}$ at/cm³ were assumed. Tab.V summarizes the energy levels for the group III elements used in the charge-balance equation and the calculated degree of ionization. Fig.19 shows the calculated temperature dependent hole concentration for the different elements. The degree of ionization measured in the samples by means of Hall effect and ICP-MS is given in Tab.IV. The ratio $p/[Ga]_{chem}$ includes compensation effects that could not be taken into account in the free hole concentration [p]. The phosphorus and oxygen concentrations were therefore also measured on comparable sample material. With the very low oxygen concentration in the analyzed samples ($3\text{-}4*10^{17}$ at/cm³), no thermal donors acting as compensation centers should be present. The phosphorus concentration was in the range of 10^{14} at/cm³ and should therefore only contribute insignificantly to the compensation.
Since the energy level of the gallium acceptor in Si is somewhat deeper in the band gap than the boron acceptor, only about 94 % of the substitutionally incorporated Ga has emitted holes into the valence band at 300 K (Tab. V). This fact cannot explain the degree of ionization $[p]/[Ga]_{chem}$ estimated in the examined samples (Tab.IV). It can be stated that in the studied concentration range of gallium in Cz-silicon monocrystals, a significant fraction is not ionized which is confirmed by Hoshikawa et al. [8]. In order to make reliable statements, a systematic investigation on more samples would be necessary. The unionized Ga-atoms should occupy interstitial sites, as they are not electrically active. The extent to which they influence the drift mobility as neutral-impurity scatter centers is unclear.
The fact that only an undefined fraction of the gallium is electrically active affects the determination of the dopant concentration from the measured resistivity data according to eq.(6). Thus, two effects overlap that make a reproducible adjustment of the specific resistivity in Ga-doped Cz-crystals difficult, namely the evaporation of the gallium and the undefined degree of ionization.

Table V: Energy level and degree of ionization of group III elements at room temperature

acceptor	energy level [eV]	degree of ionization @ 300 K [%]
boron	0.045	98
aluminum	0.067	95
gallium	0.072	94
indium	0.16	49

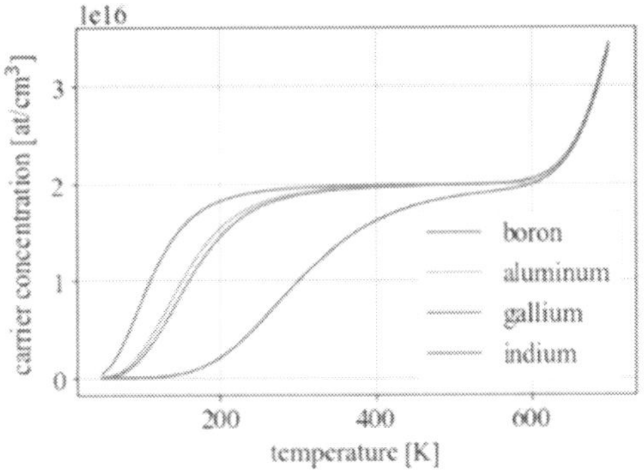

Figure 19: Temperature dependent hole concentration for the group III elements

3 SUMMARY

- Literature values for evaporation rates during the Czochralski process are hardly available and given without sufficient background information.
- During the melting process, only insignificant amount of gallium seems to evaporate.
- Evaporation rates were determined under different process conditions but without a growing crystal.
- The process pressure has a noticeable influence on the evaporation behavior of gallium.
- Since noticeable evaporation of gallium was measured in silicon melt and during growth, participation of oxygen from the silica crucible seems to be involved in the evaporation process.
- The evaporation behavior was determined using an approximate equation with two fit parameters for the different process conditions.
- Under the process conditions studied in this paper, a noticeable fraction of the gallium is apparently incorporated electrically inactive on interstitial site.

4 APPENDIX

Eq.(9b) shows the charge-balance equation of a semiconductor with an acceptor and a donor impurity in detail.

$$N_C exp\left(\frac{E_F - E_C}{kT}\right) + N_A \frac{1}{1 + 4exp\left(\frac{E_A - E_F}{kT}\right)} = N_V exp\left(\frac{E_V - E_F}{kT}\right) + N_D \frac{1}{1 + 2exp(E_F - E_D)} \quad (9b)$$

$N_{C,V}$: effective density of states in the conduction and valence band, $E_{C,V}$: energy of the conduction and valence band edge, E_F: Fermi energy, N_A,N_D: concentration of acceptor and donor atoms, k: Boltzmann-constant, T: absolute temperature.

4 ACKNOWLEDGEMENTS
This work was supported by the German Federal Ministry for Economy Affairs and Climate Action under contract number 03EE1126B

5 REFERENCES

[1] Joyce Ann T. De Guzmann et al., Defect Reactions Responsible for Boron-Oxygen Degradation in Crystalline Silicon Photovoltaics, Proceedings of the 37th EU PVSEC (2020), 145

[2] Joyce Ann T. De Guzmann et al., Indium-Doped Silicon for Solar Cells – Light-Induced Degradation and Deep-Level Traps, Phys. Status Solidi A, **2021**, 218, 2100108

[3] Takeshi Hoshikawa, Toshinori Taishi, Shuji Oishi, Keigo Hoshikawa, Investigation of methods for doping CZ silicon with gallium, Journal of Crystal Growth, 275 (2005) e2141 – e2145

[4] Jafar Safarian, Merete Tangstad, Vacuum refining of molten silicon, Metallurgical and Materials Transactions B, Vol. 43B, 2012, 1427-1445

[5] M. Porrini, R. Scala and V.V. Voronkov, Behavior of Volatile Dopants (P, Sb) in Czochralski Silicon Growth, Journal of Crystal Growth 460, 13-15 (2017).

[6] Zhensheng Liu, Torbjörn Carlberg, A Model for Dopant Concentration in Czochralski Silicon Melts, J. Electrochem. Soc., Vol. 140, No. 7, July 1993, 2052-2057

[7] Blood-Orton "the electrical characterization of semiconductors: majority carriers and electron states", 1992; Fig. 3.14, p. 127; ISBN 0-12-528627-9

[8] Takeshi Hoshikawa et al., Relationship between Gallium Concentration and resistivity of Gallium-doped Czochralski Silicon crystals: Investigation of a conversion curve, Japanese Journal of Applied Physics, Vol. 47, No. 12 2008, pp 8691 – 8695

[9] S.M. Sze, Semiconductor devices, ISBN 0-471-33372-7

[10] Frank Szmulowicz, Calculation oft he mobility and the Hall factor for doped p-type silicon, Physical Review B **34,6** (1986) 4031-4047

CHARACTERIZATION AND HYDROMETALLURGICAL TREATMENT OF INDUSTRIAL SI-KEF

Author(s)

Tinotenda Mubaiwa[1], [2]Tyke Naas, Jafar Safarian[1]

Company / Institute(s)

[1]Norwegian University of Science and Technology, [1,2]The Norwegian Research Center for Sustainable Solar Cell Technology, [2]REC Solar Norway AS

Address(es)

tinotenda.mubaiwa@ntnu.no

ABSTRACT: The current study aims to characterize and purify Kerf Loss Waste (KLW) to solar grade. The use of acid leaching has been chosen as an initial purification step to remove metallic impurities, and different acids and conditions were tested. However, before focusing on the hydrometallurgical treatment, the first step is to extensively characterize the industrial Si-kerf from different sources. Several characterization techniques of scanning electron microscopy/ energy dispersive x-ray spectroscopy (SEM/EDS), X-ray diffraction (XRD), Brunauer-Emmett-Teller (BET), infrared absorption and thermal conductivity (LECO), particle size distribution (PSD), and glow discharge mass spectrometry (GDMS), have been applied. It was found that Si-kerf, with moisture content in the range 20-50%, is a complex material containing Si particles smaller than 2μm, and contains fine polymer, oxides, and metallic particles. Different types of acids were used for leaching the impurities and it was found that it is possible to purify Si-kerf for further recycling into the PV silicon value chain for Si crystal growth. 1M HCl solution was found to be the best leaching agent with 90.4% efficiency, followed by H_2SO_4 with 88.1% efficiency and finally HNO_3 with 83.1% efficiency for the removal of Ni.

Keywords: Silicon kerf, acid leaching, characterization, solar grade silicon

1 INTRODUCTION

Over the last couple of decades, photovoltaics has grown to become a dominant source of renewable energy, with silicon being by far the most dominant PV material [1]. Solar energy plays a pivotal role in reducing the global carbon emissions [3,4]. Solar grade silicon feedstock is produced via two main routes: chemical and pyrometallurgical. These processes begin by production of metallurgical grade silicon [5]. To minimize cost, as well as reduce environmental pollution, recycling of waste from solar silicon production can be a viable option. Approximately 94% of solar cell production in the PV industry is taken up by crystalline silicon, and production of silicon wafers takes up a significant part of the manufacturing costs. The main process to manufacture crystalline silicon is the so-called Siemens process [6, 8]. During the multi-wire sawing process for producing silicon wafers, more than 40% is lost in the form of kerf loss waste (KLW). The main constituents of this KLW include high purity silicon with a surface oxide layer, polymer particles from the sacrificial mounting beam, adhesive, cutting additive, and traces of metal from the slicing wire [6, 7]. It is therefore apparent that various techniques should be employed to recover the silicon and thus make the production process circular. Acid leaching can be used as a purification step before melting, in the recycling of silicon kerf. The aim is to target and eliminate specific impurities, especially metallic imputrities like Al, Ni, Fe, etc. In acid leaching there are basically two phases, namely stable phase, which is unaffected by the acid, and the impurity-containing phase, which is attacked by the acid. The commonly used acids for industrial solid waste treatment include HCl and H_2SO_4, among others. Some researchers have combined acid leaching with other metallurgical processes in the purification of Si-kerf from diamond wire slicing. Wang et al (2008) combined acid leaching, induction melting and directional solidification and obtained solar grade silicon. In 2019 Kong et al used acid leaching, agglomeration, and inductive melting to produce 4N silicon ingots. Yang et al also combined acid leaching and high temperature deoxidation and produced 5N silicon to be used in Si_3N_4 crucibles. [10-12] [21,22]

2 EXPERIMENTAL METHODOLOGIES

2.1 Characterization and leaching

Solar cell silicon is a high purity material which therefore requires very sensitive analytical methods for quantification of the different elements. In the present study several characterization techniques were employed. These are; scanning electron microscopy/energy dispersive X-ray spectroscopy (SEM/EDS), X-ray diffraction (XRD), physisorption of inert gas by the Brunauer-Emmett-Teller (BET) method, infrared absorption and thermal conductivity to measure combustion gases within the kerf sample using the infrared absorption and thermal conductivity (LECO) method, particle size distribution (PSD) analysis, inductively coupled plasma mass spectrometry (ICP-MS), and glow discharge mass spectrometry (GDMS). In this study there were five (5) industrial silicon kerf samples from different suppliers, and we sought to compare the similarities and differences of different properties, with the aim of coming up with standardized treatment methods. Before characterization the samples were dried in an oven at 80°C for 24 hours to remove moisture. A selected sample (REC1) was leached with 1M concentration of HCl, H_2SO_4 and HNO_3 acids at 40°C for 1hour using a jacketed glass reactor with a mechanical stirrer. The silicon particles were then separated using a lab scale centrifuge at 4000rpm. The solid: liquid ratio in leaching was 1:10. The solid residue was washed by distilled water on a hot plate two times and dried in an oven at 80°C for 24 hours. After drying the solid was crushed and analyzed by GDMS.

3 RESULTS AND DISCUSSION

3.1 Microstructural analysis

The as received kerf was dried and analyzed by SEM (Zeiss ULTRA 55 is a scanning electron microscope (SEM) with a field emission electron gun (FEG)). The morphology is shown in Figure 1.

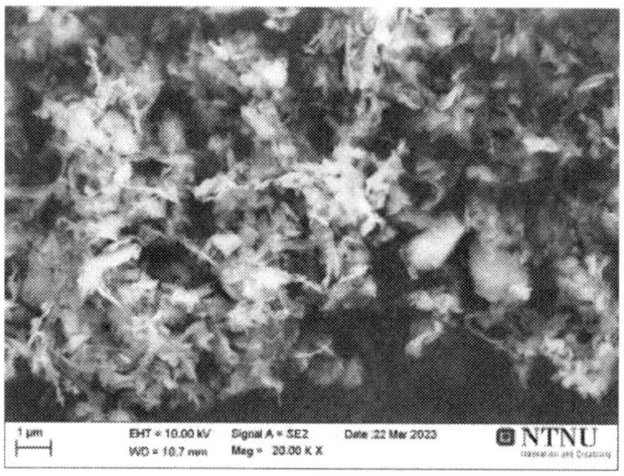

Figure 1: SEM image of as received kerf afer drying

The image shows that the kerf particles are tiny flakes of loosely agglomerated material, irregular in shape. Element mapping by EDS shows the distribution of the main impurities on one selected sample (Figure 2). The main impurities (for the leaching treatment) of interest are Al, Ni and Fe.

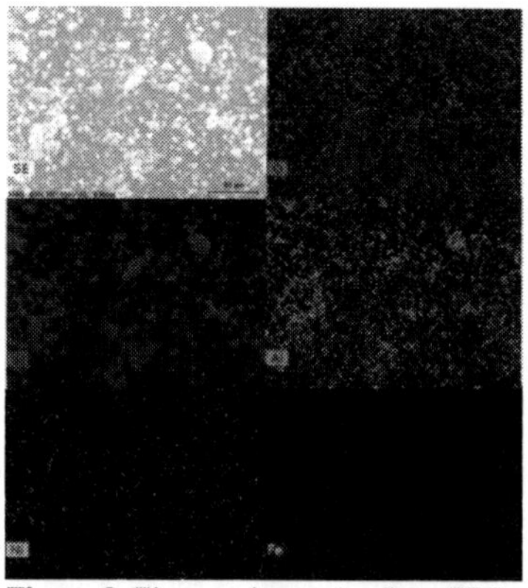

Figure 2: Elemental mapping from EDS

Key requirements of KLW are low content of boron (target <0.08 ppmw), phosphorous (target <0.3 ppmw) and metals. (Ni and Fe tend to dominate.) As can be seen from Table 1 and Table 2, the impurities in KLW differ from source to source. However, the order of magnitude is quite comparable for most elements, showing that the general composition is quite similar across the different suppliers. One of the samples (REC5), however, is analyzed with 0.98wt% Al, which is quite different from the other 4 samples. The sample is also the oldest and the high amount of Al can be attributed to the use of a slicing beam containing $Al(OH)_3$ filler.

3.2 Elemental analysis

Table 1: Impurity content from ICP-OES/MS (ppmw).

	REC1	REC2	REC3	REC4	REC5
Al	6.3	126.6	165.5	114.3	9800
B	0.02	0.02	0.03	0.09	0.25
Ca	84.2	44.7	53.0	22.4	14.6
Cr	< 0.7	<0.7	1.0	<0.7	<0.7
Cu	<0.5	<0.5	<0.5	<0.5	0.7
Fe	6.5	25.1	50.3	18.3	2.5
Ga	0.9	1.5	1.7	1.5	2.6
Mg	27	2.1	16.4	1.6	2.7
Mn	<0.4	<0.4	0.4	0.5	<0.4
Ni	236	246	83.8	257	114
P	<0.6	1.01	3.85	0.9	1.02
Ti	<0.4	1.1	3.4	1.0	3.4
V	<0.4	<0.4	<0.4	<0.4	<0.4

Non-metallic impurities (C, O, N) were analyzed by the infrared combustion (LECO) method. The carbon content of the final, purified Solar Grade Silicon is critical for the Cz ingot pulling process. The acceptable C content in solar cell silicon is approximately < 5ppmw, preferably lower. In KLW, the carbon content (mainly coming from the slicing beam) is in the percent range. Efficient carbon removal steps are thus a necessary part of a process to upgrade KLW to Solar Grade Silicon. The analyzed oxygen content is presumably mainly coming from surface SiO_2 on the silicon particles, and thus represents a yield loss. Oxygen analysis values as low as possible are desirable. Table 2 shows analysis for C, O and N.

Table 2: C, O and N composition from LECO

Sample	Carbon (wt%)	Oxygen (wt%)	Nitrogen (wt%)
REC1	2,62	4,65	0,134
REC2	4,16	4,47	0,137
REC3	2,42	4,36	0,122
REC4	1,39	3,71	0,125
REC5	0,951	5,66	0,109

3.3 X-Ray Diffraction (XRD)

The XRD analysis in Figure 3 indicates single phase crystalline Si in all the samples. The peaks occur at ~ 28.5, 47.4, 56.2, 69.1 and 76.2 theta degrees.

10.4229/EUPVSEC2023/1BV.5.4
020026-002

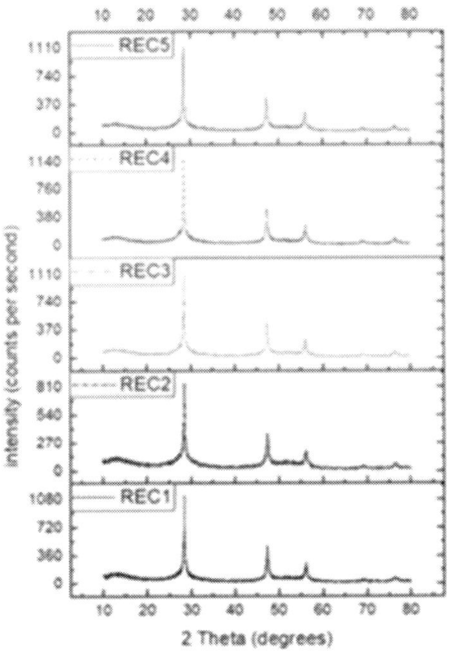

Figure 3: XRD analysis

3.4 Particle Size Distribution (PSD)

The particle size distribution of 4 of the 5 samples used showed a consistent particle size of 1-18 μm, with an average size (D50) of 1-2 μm. However, one of the samples (REC2) showed a completely different particle size range, up to ~100 μm and an average size of ~11 μm as shown in Figure 4. BET surface area measurements also showed that sample 2 had almost half of the surface area compared to other samples. Most of the samples have particle sizes within the expected range for silicon kerf. The difference in one of the samples could be a result of different cutting methods from one of the suppliers.

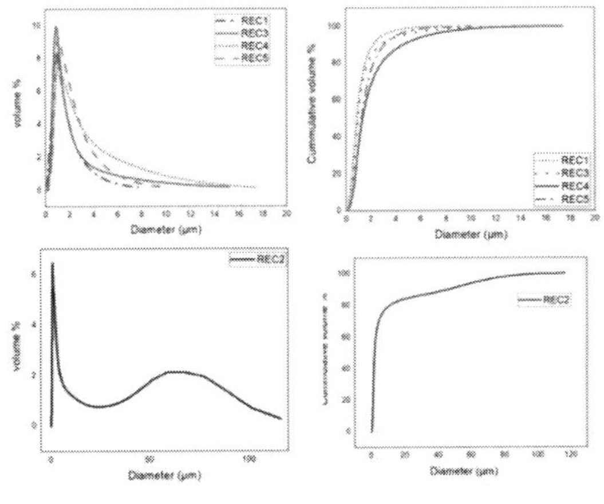

Figure 4: Particle size distribution for different samples.

3.5 BET surface area

The surface area can give an indication of the quality of the cutting process during wafering. The results in Table 3 show a surface area range of ~30-40 m^2/g for 4 of the 5 samples, whilst one sample (REC2) had close to half this value. This sample also behaved differently when particle size distribution was analyzed, showing differences in the cutting processes depending on supplier.

Table 3: The measured BET surface area of the kerf samples.

Sample	BET surface area (m^2/g)
REC1	39.8
REC2	17.7
REC3	41.7
REC4	35.7
REC5	30.4

3.6 Leaching using different acids

After leaching with 3 different acids, the Ni (the most abundant impurity for most of the samples) and P content was measured using GDMS and the results in Table 4 show that leaching by HCl is more efficient than H_2SO_4 and HNO_3.

Table 4: GDMS analysis before and after leaching.

	Initial (ppmw)	After HNO_3 (ppmw)	After H_2SO_4 (ppmw)	After HCl (ppmw)
Ni	249	42.0	29.6	22.8
P	8.1	1.3	0.5	2.5

4 Conclusions

It can be concluded that the morphology of Si-kerf shows loosely agglomerated irregular material from the cutting process, with an average size of around 2μm. From ICP/GDMS it can be concluded that Ni is the main metallic impurity, except for the older sample (REC5) which had Al as the main metallic impurity. The particle size distribution shows an average of 1- 2 μm for most of the sample except REC 2 which has an average particle size of around 11 μm. The BET surface area averages 30 - 40 m^2/g for most of the samples except for REC2 which had around 17 m^2/g. XRD analysis shows the typical diffraction peaks for crystalline Si. Leaching results show that HCl > H_2SO_4 > HNO_3.

5 ACKNOWLEDGEMENTS

This research was funded by Research Centre for Sustainable Solar Cell Technology (FME SuSolTech) co-sponsored by the Norwegian Research Council and industry partners, grant by project number 257639.

6 References

1. Parida, B., Iniyan, S. and Goic, R. (2011) 'A review of solar photovoltaic technologies', Renewable and Sustainable Energy Reviews, 15(3), pp. 1625–1636. Available at: https://doi.org/10.1016/j.rser.2010.11.032.
2. Braga, A.F.B. et al. (2008) 'New processes for the production of solar- grade polycrystalline silicon: A review', Solar Energy Materials and Solar Cells, 92(4), pp. 418–424. Available at: https://doi.org/10.1016/j.solmat.2007.10.003.
3. Obeidat, F. (2018) 'A comprehensive review of future photovoltaic systems', Solar Energy, 163(January), pp. 545–551. Available at: https://doi.org/10.1016/j.solener.2018.01.050.
4. Kabir, E. et al. (2018) 'Solar energy: Potential and future prospects', Renewable and Sustainable Energy Reviews, 82(September 2016), pp. 894–900. Available at: https://doi.org/10.1016/j.rser.2017.09.094.
5. Zhang, et al. (2013) 'Development of Solar Grade Silicon (SoG-Si) Feedstock by Recycling SoG-Si Wastes DOE award ED-EE0000575 DE- PS36-09GO99003 (CFDA number: 81 . 087): Photovoltaic Supply Chain and Cross-Cutting Technologies: Topic 1: Proof of Concept / Feasibility Ass', 99003, pp. 1–42.
6. Li, J. et al. (2021) 'Progress in recovery and recycling of kerf loss silicon waste in photovoltaic industry', Separation and Purification Technology, 254(March 2020), p. 117581. Available at: https://doi.org/10.1016/j.seppur.2020.117581.
7. Hachichi, K. et al. (2018) 'Silicon Recovery from Kerf Slurry Waste: A Review of Current Status and Perspective', Silicon, 10(4), pp. 1579–1589. Available at: https://doi.org/10.1007/s12633-017-9642-x.
8. Huang, L. et al. (2019) 'A metallurgical route to upgrade silicon kerf derived from diamond-wire slicing process', Vacuum, 163(December 2018), pp. 164–171. Available at: https://doi.org/10.1016/j.vacuum.2019.02.019.
9. Bhuvanesh, N.S.P. (1995) 'Acid Leaching', 5(1094), p. 260.
10. Wang, T.Y. et al. (2008) 'A novel approach for recycling of kerf loss silicon from cutting slurry waste for solar cell applications', Journal of Crystal Growth, 310(15), pp. 3403–3406. Available at: https://doi.org/10.1016/j.jcrysgro.2008.04.031.
11. Kong, J. et al. (2019) 'An Economical Approach for the Recycling of High-Purity Silicon from Diamond-Wire Saw Kerf Slurry Waste', Silicon, 11(1), pp. 367–376. Available at: https://doi.org/10.1007/s12633-018-9889-x.
12. Yang, H.L. et al. (2019) 'Recycling and reuse of kerf-loss silicon from diamond wire sawing for photovoltaic industry', Waste Management, 84, pp. 204–210. Available at: https://doi.org/10.1016/j.wasman.2018.11.045.
13. Chigondo, F. (2018) 'From Metallurgical-Grade to Solar-Grade Silicon: An Overview', Silicon, 10(3), pp. 789–798. Available at: https://doi.org/10.1007/s12633-016-9532-7.
14. Cowern, N.E.B. (2012) Silicon-based photovoltaic solar cells, Functional Materials for Sustainable Energy Applications. Woodhead Publishing Limited. Available at: https://doi.org/10.1533/9780857096371.1.1.
15. Xakalashe, B.S. and Tangstad, M. (2011) 'Silicon processing: from quartz to crystalline silicon solar cells', Southern African Pyrometallurgy International Conference, (March), pp. 1–18.
16. Jäger-Waldau, A. (2012) Overview of the global PV industry, Comprehensive Renewable Energy. Elsevier Ltd. Available at: https://doi.org/10.1016/B978-0-08-087872-0.00110-4.
17. Faraji, F. et al. (2022) 'Kinetics of leaching: A review', Reviews in Chemical Engineering, 38(2), pp. 113–148. Available at: https://doi.org/10.1515/revce-2019-0073.
18. Wang, T.Y. et al. (2008) 'A novel approach for recycling of kerf loss silicon from cutting slurry waste for solar cell applications', Journal of Crystal Growth, 310(15), pp. 3403–3406. Available at: https://doi.org/10.1016/j.jcrysgro.2008.04.031.
19. Terje Halvorsen, Moen, M. and Mørk, K. (2012) 'Cabriss Recycling of Si-Kerf From PV', Psychology Applied to Work: An Introduction to Industrial and Organizational Psychology, Tenth Edition Paul, 53(9), pp. 1689–1699.
20. Yang, H.L. et al. (2019) 'Recycling and reuse of kerf-loss silicon from diamond wire sawing for photovoltaic industry', Waste Management, 84, pp. 204–210. Available at: https://doi.org/10.1016/j.wasman.2018.11.045.
21. Kong, J. et al. (2021) 'Recycling high-purity silicon from diamond-wire saw kerf slurry waste by vacuum refining process', Journal of Cleaner Production, 286. Available at: https://doi.org/10.1016/j.jclepro.2020.124979.
22. Yang, S. et al. (2019) 'Kinetic mechanism of aluminum removal from diamond wire saw powder in HCl solution', Journal of Hazardous Materials, 368(October 2018), pp. 1–9. Available at: https://doi.org/10.1016/j.jhazmat.2019.01.020.19.
23. Gijbels, R. and Bogaerts, A. (1997) 'Recent trends in solid mass spectrometry: GDMS and other methods', Fresenius' Journal of Analytical Chemistry, 359(4–5), pp. 326–330. Available at: https://doi.org/10.1007/s002160050581.

INFLUENCE OF DIFFERENT ALUMINUM CONCENTRATIONS IN CZ-SI WAFERS ON LETID

Melanie Mehler[1], Annika Zuschlag[1], Matthias Trempa[2], Thomas Buck[3], Giso Hahn[1]

[1]University of Konstanz, Department of Physics, 78457 Konstanz, Germany
[2]Fraunhofer-Institute for Integrated Systems and Device Technology, 91058 Erlangen, Germany
[3]International Solar Energy Research Center Konstanz e.V., Rudolf-Diesel-Straße 15, 78467 Konstanz, Germany

ABSTRACT: In this work the influence of different Al concentrations in Cz-Si wafers and the bonding of acceptors with H is investigated. For this purpose, the long-term behavior of the excess charge carrier lifetime is examined for Cz-Si wafers with Al concentrations in the range of ~10^{13}-10^{16} cm^{-3}. By varying the Al concentration in the Si bulk via changing the doping procedure during crystal growth, samples can be processed containing either more or less Al atoms compared to the overall H concentration in the wafer. At higher Al concentrations in the Si bulk, a significant delay of the degradation and regeneration behavior can be observed. The effect of the Al-induced delay can be shown not only for lifetime samples, but also for PERC solar cells processed from the same Cz-Si material. By using AlO_x passivation layers with varying thickness as a barrier layer for H, the H content in the Si bulk can be reduced after firing of AlO_x/SiN_y:H stacks. When comparing the lifetime equivalent defect densities ΔN_{leq} of the AlO_x/SiN_y:H passivated samples with the SiN_y:H passivated samples, a delay in degradation can be observed at higher Al concentrations and low H content in the Si bulk. A possible explanation for the Al-induced delay could be due to the expected stronger binding of H with Al compared to B. From the results of the B-H and Al-H measurements, it can also be suggested that the release of H from the acceptor contributes to the degradation. AlH pairs do not dissolve in the dark like BH. This behavior could also be observed on Ga-doped wafers. The results allow further insights into the origin and kinetics of LeTID.

Keywords: Degradation, Hydrogen, Dopants

1 INTRODUCTION

Degradation phenomena such as light and elevated temperature induced degradation (LeTID) can reduce the efficiency of Si solar cells. It could be shown that the strength of LeTID degradation of Cz-Si PERC solar cells correlates with the Al concentration in the Cz wafer [1,2]. Thus, an influence of Al atoms on the LeTID effect can be concluded. The aim in this work is to investigate the impact of Al dopant and its bonding with H on the degradation and regeneration kinetics. For this purpose, complementary to previous work [2], different Al concentrations in the range of ~10^{13}-10^{16} cm^{-3} in the Cz-Si wafers are investigated. By using AlO_x layers deposited via atomic layer deposition (ALD) in different thicknesses, which act as a diffusion barrier for H [3], the H content in the Si bulk can be changed. Using Cz-Si wafers differing in Al concentration, samples can be investigated whose H content is higher or lower compared to the Al dopant concentration. This opens up the possibility to study the influence of Al dopants in combination with H-acceptor pair formation on LeTID kinetics.

2 EXPERIMENTAL

For the degradation experiments ~1 Ωcm B-doped, two different B+Al co-doped, and 'pure' Al doped Cz-Si wafers (referred to as B-Reference, B+Al(1), B+Al(2) and 'pure' Al, respectively) serve as base materials. Tab. 1 shows the secondary ion mass spectrometry (SIMS) measurement results of the dopant concentrations for the respective samples.

After the saw damage and a cleaning step, the samples are gettered using a $POCl_3$ diffusion (55 Ω/sq). The resulting emitter is removed in an etching step and the samples are cleaned again. Afterwards, both sides of the samples are deposited with 0-25 nm AlO_x from ALD at

Table I: B and Al concentrations in the investigated Cz-Si materials (obtained via SIMS analysis)

	B conc. [at/cm^3]	Al conc. [at/cm^3]
B-Reference	1.15-1.2·10^{16}	3-5·10^{13}
B+Al(1)	0.9-1.1·10^{16}	2.8-3.4·10^{14}
B+Al(2)	0.9-1.0·10^{16}	4.7-6.0·10^{15}
Al	2.5-3·10^{13}	1.4-1.7·10^{16}

300°C, followed by deposition of 75 nm plasma-enhanced chemical vapor deposition (PECVD) SiN_y:H. The samples are fired at $T_{sample,peak}$=810°C and then degraded at 80°C and 0.9(1) suns (same degradation parameters as for the PERC cells). For the measurement under constant injection, the laser setup of [4] was used. The samples were degraded at 220°C and at a constant injection of Δn=1.0·10^{16} cm^{-3}. The effective minority charge carrier lifetime τ_{eff} is determined using photoconductance decay (PCD) at 30°C and evaluated at Δn=0.1 p_0, with p_0 being the doping density. For better comparison of degradation and regeneration behavior from the different samples, the lifetime equivalent defect density ΔN_{leq} is calculated [5]. For determination of BH and/or AlH pairs via resistivity measurements [6], Al is evaporated on the samples after firing and electrically contacted by LFCs (laser fired contacts). For the formation of acceptor-H pairs, the resistivity samples were treated at 220°C in the dark, and for dissolution at 220°C and 2.2(2) suns. The resistance measurement is performed at 25°C. The process flow for the lifetime and resistivity samples is shown in Fig. 1.

The Cz-Si PERC solar cells have been fabricated according to an industrial-type solar cell process from the same base materials. The front side is passivated with silicon nitride SiN_y:H and the rear side with silicon oxy-nitride layer SiO_xN_y. The solar cell parameters are determined at 25°C using a flasher from H.A.L.M. For a better comparison, the difference ΔV_{OC} of the measured values to the initial value is considered.

10.4229/EUPVSEC2023/1BV.5.10
020027-001

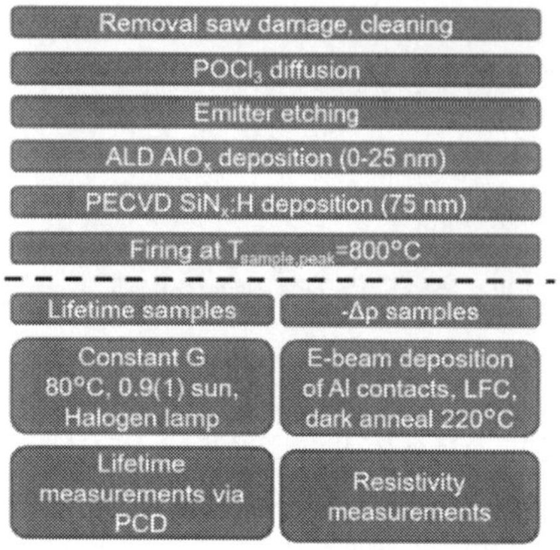

Figure 1: Process diagram for the lifetime and resitivity samples.

3 RESULTS AND DISCUSSION

Fig. 2 shows the resulting τ_{eff} and ΔN_{leq} values for the different samples, passivated with SiN_y:H layers only on both sides. Under these degradation conditions, no regeneration occurs for the 'pure' Al sample within 900 h. The initial τ_{eff} values of the B-Reference and B+Al(1) are comparable, whereas the B+Al(2) and 'pure' Al samples show significantly lower initial τ_{eff}. The reduction in τ_{eff} at high Al concentrations could be (partly) attributed to Al-O complexes [7].

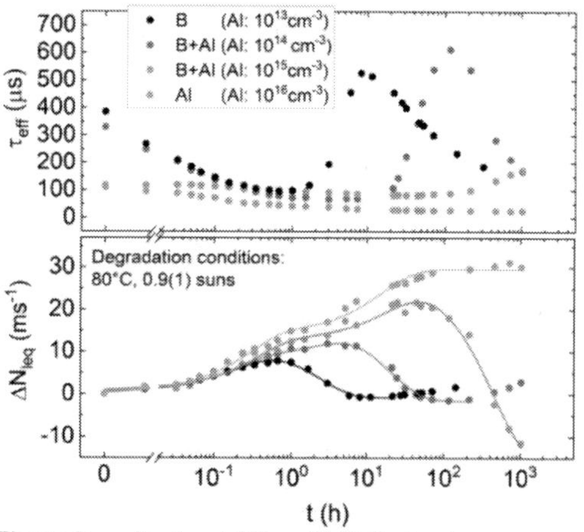

Figure 2: τ_{eff} (top) and ΔN_{leq} with fit (bottom) over accumulated time for treatment at 80°C, 0.9 suns illumination for the Cz-Si materials with different Al concentrations.

With higher Al concentration, the maximum defect density increases and a delay in degradation and regeneration kinetics is observed. The resulting time constants confirm the delay of degradation and regeneration at higher Al concentrations although the absolute values should be handled with care as injection changes during treatment time. Except for the B-Reference, two exponential functions are used for fitting the degradation. A shoulder can be seen during degradation for the materials B+Al(1), B+Al(2) and 'pure' Al, indicating a two-step degradation. The effect of the Al-induced delay for degradation and regeneration could also be observed at PERC cell level, see Fig. 3.

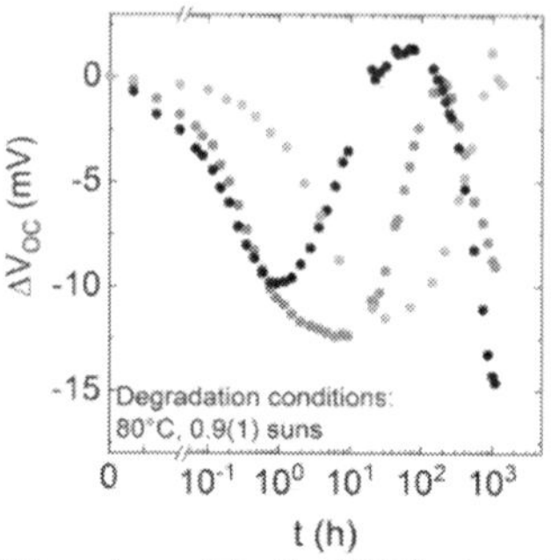

Figure 3: ΔV_{oc} values of the Cz-PERC solar cells over accumulated time for treatment at 80°C, 0.9 suns illumination for the Cz-Si materials with different Al concentrations.

Since τ_{eff} is different for the various samples, a measurement under constant injection is more reasonable. Fig. 4 shows the resulting τ_{eff} and ΔN_{leq} values for the B-Reference, B+Al(1) and 'pure' Al. The degradation temperature was increased to 120°C to analyze the regeneration from 'pure' Al. Also here, a higher Al concentration leads to delayed kinetics. Compared to the measurement with constant illumination, 'pure' Al shows a lower $\Delta N_{leq,max}$ and stronger regeneration compared to the other two materials, maybe indicating that directly after firing a significant amount of LeTID defects is already present. The degradation kinetics have also changed, since only one exponential function is sufficient to fit the degradation. This change in degradation kinetics at constant injection could also be shown in [2].

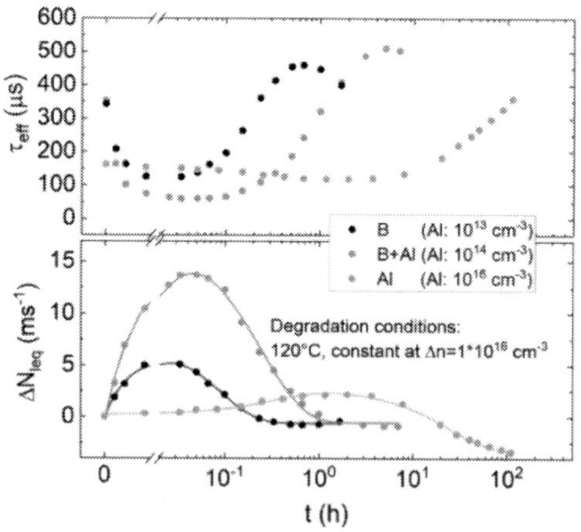

Figure 4: τ_{eff} (top) and ΔN_{leq} with fit (bottom) over accumulated time for treatment at 120°C and at constant injection for the Cz-Si materials with different Al concentrations.

Fig. 5 shows for the B-Reference, B+Al(1) and 'pure' Al the ΔN_{leq} values for the samples passivated with ALD AlO_x / PECVD SiN_y:H followed by firing. The samples in Fig. 5 are from another run but were processed in the same way as the samples shown in Fig. 2 Since the samples were not in the same SiN_y:H deposition and not fired on the same day, the SiN_y:H passivated samples differ from those in Fig. 2. As expected, the samples with AlO_x/SiN_y:H layers show lower ΔN_{leq} values than the samples with SiN_y:H layer only. For the 'pure' Al sample, the maximum ΔN_{leq} could be lowered from 20 ms^{-1} to below 5 ms^{-1}. Compared to the SiN_y:H passivated sample, the degradation time constants of the AlO_x/SiN_y:H passivated B+Al(1) samples show comparable values, whereas the Al samples with an additional AlO_x layer show a more significant delay in degradation. Thus, a higher Al concentration and a lower H content in the Si bulk lead to a delay in degradation and regeneration. Again, a shoulder during degradation can be seen in the B+Al(1) samples. A possible explanation for the 2-step degradation in the B+Al(1) material could be that the bonding of H to Al is stronger compared to B, which results in different "release times" of H.

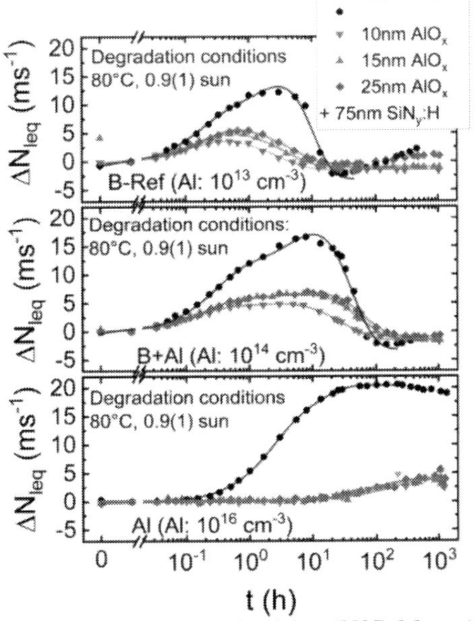

Figure 5: ΔN_{leq} over accumulated time (80°C, 0.9 suns) for samples with AlO_x layers of different thickness plus an additional 75 nm SiN_y:H layer for the differently Al doped samples. The data of the B-Reference was already shown in [8].

To explain the delay in degradation and regeneration, the role of H is examined in more detail. Fig. 6 shows for the B-Reference and 'pure' Al the change in hole concentration $-\Delta p$ of the SiN_y:H samples. As can be seen, the $-\Delta p$ values show a formation and dissolution of BH pairs. But the pairs do not completely dissolve in the dark, and the resistance can also increase again, for example, due to thermal donors [9]. The degradation temperature for the B-Reference sample is higher compared to Al in order to accelerate the dynamics. In [10] the formation and dissolution of BH pairs between 160-220°C could be shown. In comparison, Al in Fig. 6 shows no dissolution of the AlH pairs at a degradation temperature of 180°C. A similar behavior has also been observed in Ga doped Cz-Si wafers [10].

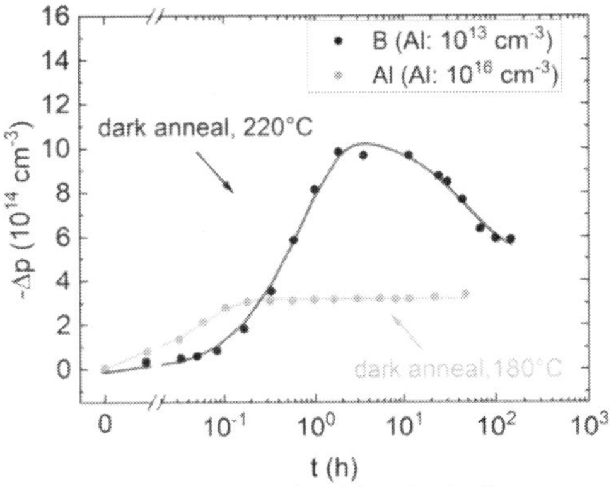

Figure 6: $-\Delta p$ over accumulated time for the B-Reference and 'pure' Al samples passivated with 75 nm SiN_y:H. The 'pure' Al sample was degraded at 180°C and the B-Reference sample at 220°C in the dark

Assuming that the AlH bond is stronger than the BH bond and the degradation and/or regeneration is also influenced by the dissolution of the acceptor-H complex, this can be an explanation for the observed delayed LeTID kinetics. To resolve the acceptor-H pairs, the samples were treated under light and elevated temperature, analogous to [11]. Fig. 7 shows the formation and dissolution of the acceptor H pairs for the B-Reference and the two B+Al co-doped samples.

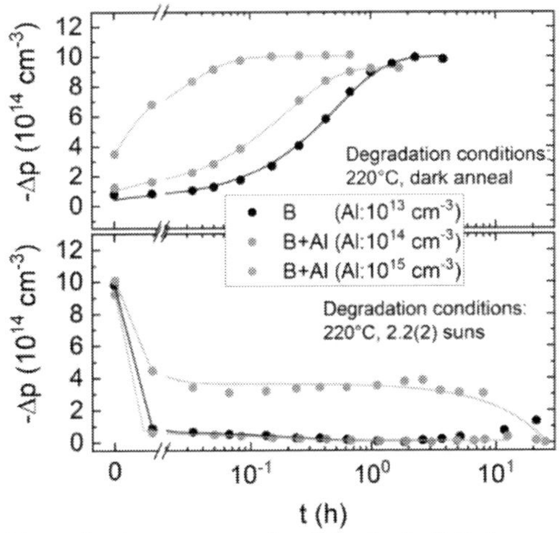

Figure 7: $-\Delta p$ over accumulated time for the B-Reference and the two B+Al co-doped samples passivated with 75 nm SiN_y:H. For the formation of the acceptor-H pairs, the samples were degraded at 220°C in the dark (top) and for the dissolution at 220°C and 2.2 suns.

The B-Reference and B+Al(1) ([Al]: 10^{14} cm^{-3}) contain less Al, and B+Al(2) ([Al]: 10^{15} cm^{-3}) similar or even more Al compared to the overall amount of H. The formation of acceptor-H pairs is faster for the B+Al co-doped samples than for the B reference. The dissolution of the acceptor-H pairs for B+Al(2) ([Al]: 10^{15} cm^{-3}) proceeds via 2 steps compared to the other samples. Thus, a higher Al concentration seems to accelerate the formation of the acceptor-H pairs and the dissolution of the pairs occurs more slowly.

From the resistance measurements, the initial $[H_2]$

dimers and the concentration of initial acceptor-H pairs can be determined according to [12]. Fig. 8 shows the initial concentrations of the respective species for the B-Reference and the two B+Al co-doped materials. As expected, the concentration of initial acceptor-H pairs increases with higher Al concentration, and the concentration of the initial [H_2] dimers decreases. Due to the higher activation energy for the dissolution of AlH-pairs, H dissolves more slowly from Al, which is a possible cause of the delay in LeTID kinetics.

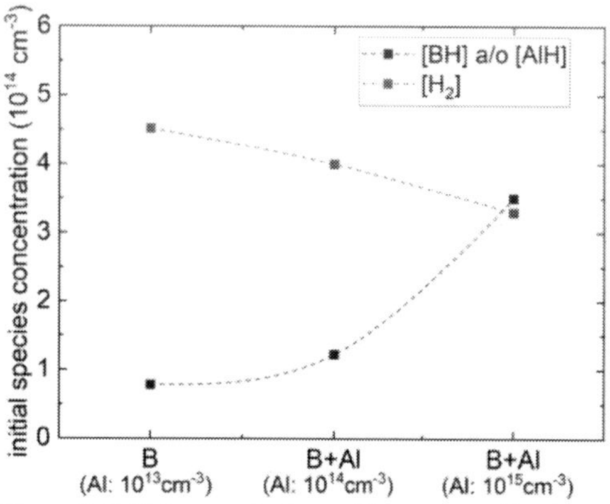

Figure 8: Initial concentrations of the acceptor-H pairs and [H_2] dimers for the B-Reference and the two B+Al co-doped materials.

4 CONCLUSIONS

A higher Al concentration in the Si bulk leads to delayed LeTID degradation and regeneration kinetics. This behavior could be observed under both constant generation and constant injection, as well as on PERC solar cell level.

In addition, using AlO_x barrier layers has shown that higher [Al] and lower [H] leads to delayed kinetics.

Resistance measurements in the dark have shown no dissociation of AlH pairs, unlike for BH pairs. Thus, LeTID degradation and/or regeneration seem to be influenced by the dissociation of acceptor-H pairs. As expected, the B+Al co-doped samples show more initial acceptor-H pairs than the B-Reference. This could be explained by trapping of H at Al, and its release happens more slowly as compared to trapping at B due to the higher activation energy.

5 ACKNOWLEDGEMENTS

Part of this work was financially supported by the German Federal Ministry for Economic Affairs and Climate Action (FKZ 03EE1051). The content is the responsibility of the authors.

6 REFERENCES

[1] M. Wagner et al., Solar Energy Materials and Solar Cells 187 (2018) 176-188.
[2] M. Mehler et al., Physica Status Solidi A 218(22) (2021) 2100603.
[3] A. Schmid et al., IEEE Journal of Photovoltaics 11(4) (2021) 967-973.
[4] A. Graf et al., AIP Conference Proceedings 2147 (2019) 140003
[5] A. Herguth, IEEE Journal of Photovoltaics 9(5) (2019) 1182-1194.
[6] A. Herguth, C. Winter, IEEE Journal of Photovoltaics 11(4) (2021) 1059-1068.
[7] J. Schmidt et al., IEEE Journal of Photovoltaics 9(6) (2019) 1497-1503.
[8] M. Mehler et al., in Proc. 8th WCPEC, (2022) 142-144.
[9] C. Winter et al., Physica Status Solidi A 218(23) (2021) 2100220.
[10] Y. Acker et al., Physica Status Solidi A 219(17) (2022) 220014.
[11] J. Simon et al., in Proc. 8th WCPEC, (2022) 19-21.
[12] J. Simon et al., Solar Energy Materials and Solar Cells 260 (2023) 112456.

MINORITY CARRIER TRAPPING EFFECTS IN MONO-CRYSTALLINE SILICON

Tarek O. Abdul Fattah[1], Vladimir P. Markevich[1], Stanislau B. Lastovskii[2], Nikolay V. Abrosimov[3], Matthew P. Halsall[1], Anthony R. Peaker[1]
[1] Photon Science Institute and Department of EEE, University of Manchester, Manchester M13 9PL, United Kingdom
[2] Scientific-Practical Materials Research Center of the National Academy of Sciences of Belarus, Minsk 220072, Belarus
[3]Leibniz-Institut für Kristallzüchtung (IKZ), Max-Born-Straȝe 2, 12489 Berlin, Germany
e-mail: tarek.abdulfattah@postgrad.manchester.ac.uk

ABSTRACT: The available literature on persistent photoconductivity (PPC) and minority carrier trapping effects in silicon crystals is considered based on the importance of these phenomena to the analysis of both, minority carrier lifetime measurements and solar cell performance. It is shown what kind of information about the electronic properties of carrier traps can be extracted from studies of PPC kinetics and supporting electrical measurements. We show the characteristic equations used to describe the trapping effect, how the cases of single-level traps and defects with negative-U properties can be distinguished and how to account for the presence of minority carrier traps in photoconductivity-based recombination lifetime measurements. The results of our recent studies of the minority carrier trapping effects in n- and p-type Si crystals, which were grown by either Czochralski or float-zone techniques and doped with different group-III and group-V impurities during growth are presented. The cases of traps with solid identifications of their structures and electronic properties are elucidated and origins of the observed PPC effects are discussed.
Keywords: persistent photoconductivity, minority carrier traps, carrier capture rates, recombination lifetime

1 INTRODUCTION

Minority carrier traps in semiconductor materials can be defined as the defects having charge states with high ratios of capture rates of minority and majority carriers, in contrast to recombination centers, which have significant capture rates of both types of charge carriers. In other words, the probability of capturing both types of carriers is high for a recombination center but is much lower for one type of carriers in the case of traps. The presence of minority carrier traps with deep energy levels in the band gap of a semiconductor material results in the appearance of several effects, such as persistent photoconductivity (PPC) or long-time decays of injection-induced junction capacitance. Deep-level traps with concentrations of the order of 10^{12}-10^{14} cm^{-3} can significantly affect the dynamics of the photo-excited charge carriers and the interpretation of minority carrier recombination lifetime measurements in semiconductor materials by PC-based techniques, which are frequently used in photovoltaic (PV) industry [1]. Therefore, information about electronic properties and origins of the deep level traps is needed for i) correct measurements of recombination lifetime and ii) evaluation of the influence of these traps on solar cell performance.

In the 1950s, Hornbeck and Haynes [2,3] developed a model for explaining PPC effects, which were observed in n- and p-type silicon and germanium crystals. The model explained the PPC effects by temporary trapping and subsequent emission of trapped minority carriers from single deep energy levels with extremely high ratio of capture cross sections of minority and majority carriers. However, no meaningful physical reasons of the existence of such high capture cross section ratios and suppression of carrier emission to a majority carrier band were given. Modifications to the Hornbeck-Haynes (HH) model have then been proposed and used in several studies to explain PPC and carrier trapping effects in mono- and multi-crystalline Si crystals [4-7]. Recently, Jafari and Hameiri [8] have implemented the HH model for explaining the effects of firing and laser annealing on the minority carrier trapping phenomena in p-type multi-crystalline Si materials and linking these effects to light and elevated temperature induced degradation (LeTID) of solar cells. In some cases, however, the HH model and its modifications cannot explain all the details of PPC-related effects, particularly, when the defects under consideration have negative-U properties [9]. These cases usually involve the occurrence of significant structural transformations of defects upon trapping or emission of charge carriers. Examples of such cases are the three first species of oxygen-related thermal double donors (O-TDDs) [10-13] and defects incorporating a substitutional acceptor atom (A) and the oxygen dimer (O_2) in silicon [14].

A lot of information about the electronic properties of traps can be obtained from studies of the temperature dependencies of PPC kinetics in samples with different charge carrier concentrations. The analysis of the temperature dependencies of characteristic time (or rate) of the PPC decays can be used to distinguish if the corresponding traps are associated with either single-level defects or with centers possessing negative-U properties. Furthermore, the analysis of the occurrence of traps in Si crystals with different doping concentrations and residual impurities can be used to obtain information about the chemical constituents of the traps.

It is argued in the present work that in the case of negative-U defects, the PPC decays after injection-induced minority carrier trapping are related to the capture of two majority carriers accompanied by a configurational transformation of a defect. The equations, which describe non-equilibrium occupancy statistics of the negative-U defects (and linked changes in the concentrations of free charge carriers) will be analyzed and compared with the equations describing the capture/emission processes for single-level traps. It is shown how to distinguish minority carrier trapping defects of different types. Harmful and beneficial effects of minority carrier traps on the performance of Si solar cells are considered, and some links between the traps and the injection-induced mechanisms of degradation of B-doped Si solar cells are discussed.

2 EXPERIMENTAL DETAILS

Schottky barrier, n^+-p-p^+, and p^+-n-n^+ diodes made from mono-crystalline Si wafers, grown by either the Czochralski (Cz) or float-zone (Fz) techniques and doped with either group-III (B, Al, Ga, or In) impurities or group-V (P) during growth, have been studied. Current-voltage (*I-V*) and Capacitance-Voltage (*C-V*) measurements have been used to assess the quality of the diodes and determine the uncompensated doping impurity concentrations and the width of the probed depletion regions. Emission and capture of carriers by the traps have been measured with the use of deep-level transient spectroscopy (DLTS) and high-resolution Laplace DLTS techniques [15,16]. We have also carried out measurements of changes in total bias capacitance of the diodes, which can be considered as a measure of changes in majority carrier concentration in the samples, after minority carrier injection pulses and subsequent flooding of the depletion regions with majority carriers (either with electrons or holes). Such measurements can be considered as analogues of photoconductivity measurements for bulk silicon samples.

3 EXPERIMENTAL RESULTS

3.1 Peculiarities of non-equilibrium occupancy statistics for some specific traps responsible for PPC

In many cases, the PPC effects in semiconductors are related to defects with negative-*U* properties [9]. Fig. 1 shows the configuration-coordinate diagram (CCD) for an amphoteric defect with negative-*U* properties in a p-type semiconductor. The defect has three charge states and exhibits bistability, i.e. it can exist in different stable atomic configurations for the same charge state (neutral charge state in the considered case). The non-equilibrium occupancy statistics for such defects have been proposed in Ref. [12] and further developed in Refs. [9] and [17].

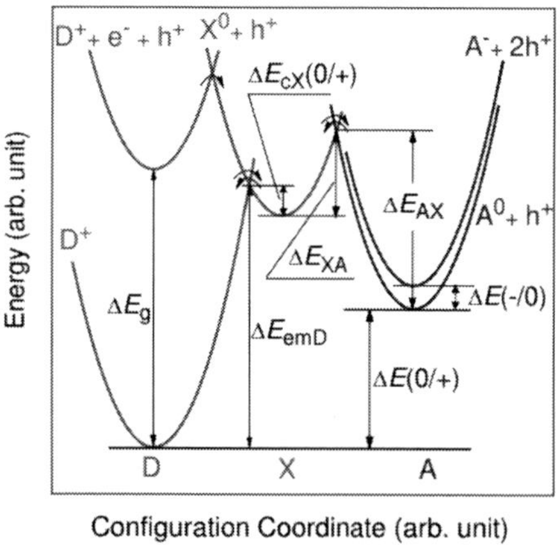

Fig. 1. Configuration-coordinate diagram of an amphoteric defect with negative-*U* properties in a p-type semiconductor.

Under equilibrium conditions, the defect can be in either D^+ or A^- states depending on the position of the Fermi level in the material, and the probability of occupation of the metastable X^0 and A^0 states is negligible [9]. In the positively charged minimum energy configuration (D^+), the defect is usually an effective trap for electrons. A capture of an electron in this charge state, which can be induced by minority carrier injection pulses, can result in the transformation of the defect into the metastable X^0 state and further to the metastable A^0 and stable A^- states under favorable conditions. The back transition to the minimum energy D^+ state, after the end of minority carrier injection phase, requires the capture of two holes and surmounting the E_{AX} energy barrier to achieve the $A^0 \rightarrow X^0$ structural transformation. If the barrier is high, which could occur in many cases, the back transition to the equilibrium occupancy of the D^+ state after the end of an injection pulse could be a long process causing PPC. In the following we will briefly consider the peculiarities in temperature and carrier concentration dependencies of the transitions rates.

It was found that for small deviations from the equilibrium occupancy of the D^+ state and when the concentration of the trap is much lower than the concentration of majority carriers ($N_T << p$), the changes in the density of the D^+ state at a constant temperature can be expressed by the following equation [9,17]:

$$\Delta[D^+](t) = \Delta[D^+]_0 exp(-t/\tau), \quad (1)$$

where is $\Delta[D^+]_0$ is the magnitude of the initial deviation of D^+ state density from an equilibrium value, and τ is the characteristic time of the process. The characteristic time depends on rates of transitions between the states as [17]:

$$\tau^{-1} = \frac{\omega_{XA} e_D + c_X p \omega_{AX} f_A}{\omega_{XA} + c_X p} \quad (2)$$

$$e_{D(A)} = c_p N_v exp(-\Delta E_{em}/kT) \quad (2.1)$$

$$c_{X(D)} = \sigma_0 v_{th} exp\left(-\Delta E_{cX(D)}/kT\right) \quad (2.2)$$

$$\omega_{XA(AX)} = \left(\omega_{XA(AX)}\right)_0 exp\left(-\Delta E_{XA(AX)}/kT\right) \quad (2.3)$$

where $e_{D(A)}$ and $c_{X(D)}$ are the emission and capture coefficients, and $\omega_{XA(AX)}$ are the rates of a structural transformation without a charge state change. f_A is the occupancy function for the defect in the configuration with a shallow *E*(-/0) acceptor level defined as:

$$f_A = [A^0]/\{[A^0] + [A^-]\} = \{1 + exp[(E_F - E_A(-/0))/kT]\}^{-1} = \{1 + (N_v/p)exp[-\left(E_V + E_A(-/0)\right)/kT]\}^{-1} \quad (2.4)$$

Equation (2) can be expressed in a common way as:

$$\tau^{-1} = e^{eff} + c_p^{eff} p \quad (3)$$

with

$$e^{eff} = e_D\left(1 + c_{pX} p/\omega_{XA}\right)^{-1} \quad (4)$$

$$c_p^{eff} = c_{pX} \omega_{AX} \omega_{XA}^{-1} f_A \left(1 + c_{pX} p/\omega_{XA}\right)^{-1} \quad (5)$$

An analysis of Eqs. (2)-(5) indicates that several terms with their specific activation energies and power dependence on the free carrier concentration can be dominant in the temperature dependencies of τ^{-1} values in respect to the Fermi level position relative to positions of the *E*(0/+), *E*(-/0) and *E*(-/+) levels and $c_{pX} p/\omega_{XA}$ ratio.

The e^{eff} term is the dominant term in Eq. (3) when the Fermi level is above the *E*(-/+) occupancy level of a negative-*U* defect in a p-type semiconductor. Under these conditions, $\tau^{-1} \cong e^{eff}$. When $c_{pX} p \ll \omega_{XA}$, $\tau^{-1} \cong e_D$, so the temperature dependence of τ^{-1} can be described by a commonly used equation for single hole emission. For the cases of $c_{pX} p \gg \omega_{XA}$, $\tau^{-1} \cong e_D \omega_{XA}/c_{pX} p$, and the transformation rate (actually the rate of occupancy of the D^+ state with holes) is inversely proportional to the free hole concentration. Such an unusual dependence of τ^{-1} versus p has been observed for a complex consisting of a substitutional acceptor atom and the oxygen dimer in silicon [14,18,19].

The capture term is dominant in Eq. (3) when $E_F < E(-/+)$. In this case, $\tau^{-1} \cong c_p^{eff} p$ and up to four specific terms with different values of activation energies and power dependencies on p can occur depending on the Fermi level position relative to the $E(-/0)$ level and $c_{pX}p/\omega_{XA}$ ratio. For a number of defects, e.g., bistable thermal double donors in Si and Ge crystals, under certain conditions, $\tau^{-1}{\sim}n^2$ dependencies have been observed [12,13]. Such dependencies clearly show a two units change in charge state of a defect, so indicating its negative-U properties.

From the analysis of temperature dependencies of transition rates between configurations of a negative-U defect, which can be monitored by measurements of changes in either free carrier concentration or density of a specific defect state in semiconductor crystals with different $p(n)$, it is possible to elucidate the electronic structure of the defect and determine practically all its energy differences and transformation barriers. This has been done for a number of defects in Si and Ge crystals [12-14,20].

It should be noted that the Eqs. (2)-(5) are also valid for bistable defects with a single energy level in the gap or with energy levels with positive U. In such case, for a donor type defect in p-type semiconductor, the A^- state in Fig. 1 does not exists and the term f_A in the Eqs. (2) and (5) is eliminated.

3.2 PPC and minority carrier trapping effects in n-type Czochralski-grown Si crystals

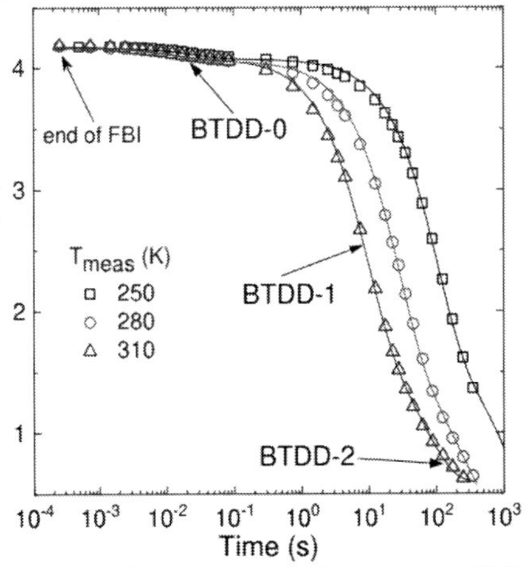

Fig. 2. Decays of compensated diode capacitance ($\Delta C = C_d - C_b$) of a p^+-n-n^+ diode at 250, 280 and 300 K after the end of a forward current injection pulse (FBI) (I_p = 0.5 A/cm^2, t_p = 100 ms). The solid lines are calculated for a three-exponent decay process with least-square fitting values of ΔC_n (n from 1 to 3) and characteristic decay times, τ_n. A backing off capacitor of C_b = 200 pF was used to compensate the capacitance of the diode studied (C_d). All the measurements were carried out on a p^+-n-n^+ diode from 5 Ω·cm n-type Cz-Si material, which was annealed at 350 °C for 10 hours for generation of TDDs.

As it has already been mentioned, the PPC effects in n-type Cz-Si crystals were observed and described about 70 years ago by Haynes and Hornbeck [3]. The authors argued that the PPC effects could be related to temporary trapping of holes in the volume of n-type silicon by two different traps (referred to as α and β traps in [3]) with single deep energy levels in the lower part of the Si gap. However, in 1980s some strong arguments were presented that the PPC effects in n-type Cz-Si crystals are associated with structural transformations with charge state changes of three first species of oxygen-related thermal double donors, which were found to be defects with negative-U properties [10-12]. It was found that PPC effects in n-type Cz-Si could not be related to a hole emission from a deep-level hole trap as it was suggested by Haynes and Hornbeck [3] but was caused by the subsequent capture of two electrons by shallow donor states of O-TDDs followed by a transformation into a deep donor state [10-12]. The non-equilibrium occupancy statistics for such a case was developed [12], and from the combined analysis of PPC, equilibrium occupancy statistics and infrared absorption measurements, the configuration-coordinate diagrams of all three bistable O-TDDs (BTDDs) species were constructed. It is important to note that the PPC decay measurements in Ref. [12] were carried out over a wide temperature range from 150 K to 350 K for n-type Cz-Si crystals with a range of electron concentrations from about $2{\times}10^{13}$ cm^{-3} to about $5{\times}10^{15}$ cm^{-3}. All the obtained experimental results were described well by the proposed negative-U defect model.

It should be noted that the negative-U model of PPC in n-type Cz-Si crystals is still not accepted by all researchers in the field, and some recent studies attempted to describe PPC in n-type Cz-Si with the use of the Hornbeck-Haynes model with some modifications [6,21,22].

We have recently observed two emission signals related to electron emission from a deep donor state of the BTDD-0 and BTDD-1 centers in the DLTS spectra of diodes on n-type Cz-Si crystals [13]. Emission and capture characteristics of the defects have been investigated and a confirmation of the relation of minority-carrier-injection-induced long-time conductivity (or diode capacitance) decays in these samples with the BTDD defects has been obtained [13]. Fig. 2 shows the kinetics of decay of diode capacitance {$\Delta C(t)$}, which is proportional to free carrier concentration in the bulk volume, to an equilibrium value for a p^+-n-n^+ diode, which was subjected to a heat treatment at 350 °C for 10 h for generation of O-related TDDs. The decays were recorded at 250, 280, and 300 K after the end of minority-carrier injection forward current pulse. There are three well separated steps in the decay curves, which can be associated with different BTDD species. The solid lines in Fig. 2 are calculated for multi-exponential decay processes with least-square fitting of exponents parameters. The fitting parameters were the magnitudes of the ΔC steps and characteristic decay times, τ.

Fig. 3 shows temperature dependencies of τ^{-1} values for BTDD-1 measured in two diodes with different concentrations of free electrons. It is found that the decay rates in the low-temperature range {capture region where the second term in Eq. (3) dominates} are proportional to the squared concentration of free electrons and the rates are not dependent on n in the high-temperature range {emission region with the dominance of the first term in

Eq. (3)}. The occurence of the $\tau^{-1}{\sim}n^2$ dependence indicates unambiguously that the considered change in excess carrier concentration is related to capture of two electrons by a defect with negative-U properties [9], and confirms the results on electronic properties of BTDDs reported in Refs. [10-13]. The structural models of O-related BTDDs, which are consistent with the experimental results obtained, were elucidated by ab-initio calculations [23,24]. It was argued that the deep-donor – shallow-donor transitions of the BTDDS are related to reversible transformations of oxygen atoms from twofold to threefold bonding configurations in [110] aligned O_n chains.

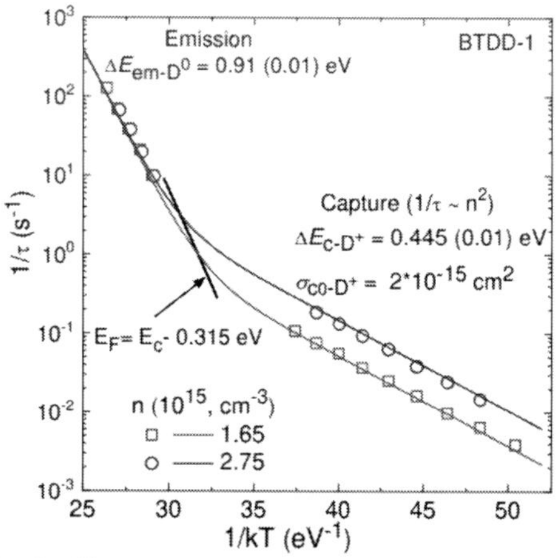

Fig. 3. Temperature dependencies of effective capture rates (injection-induced excess capacitance decays) for BTDD-1 defect in Cz-Si samples with different free electron concentrations. The values of free electron concentration in the samples studied are given in the graphs.

3.3 PPC and minority carrier trapping effects in p-type Cz-Si crystals

As for the minority carrier traps in p-type Cz-Si crystals, Haynes and Hornbeck reported long (minutes) two-stage decays of conductivity at room temperature in B-doped Si crystals after the application of short light pulses [2]. The authors interpreted the observed effects as being related to temporary trapping and emission of injected electrons by two different ("shallow" and "deep") traps in the volume of p-type silicon: From an analysis of the conductivity decays, the energy level of the "shallow" trap was determined as E_c-0.57 eV, while the energy level of the "deep" trap was placed at E_c-0.79 eV [2].

Recently the PPC effects in B-doped Cz-Si have been linked with the reversible transformations of a defect consisting of a substitutional boron atom and an oxygen dimer (B_sO_2) from a configuration with a deep donor state into a configuration with a shallow acceptor state [25-27]. The order of the deep donor and shallow acceptor states in the Si gap is found to be reverse, so the B_sO_2 defect is a center with negative-U properties. It has been argued that the deep-shallow transformations of the B_sO_2 defect are associated with the structural changes of the oxygen dimer between square and staggered configurations [25,26]. It has been further shown that the complexes of the oxygen dimer with other group-III impurity atoms (Al, Ga, and In) in Si have electronic properties similar to that of the B_sO_2 defect and cause PPC [14,18,19,27].

Fig. 4 shows decays of normalized values of excess-capacitance induced by minority-carrier-injection {$\Delta C(t)/\Delta C_{max}$} recorded at 290 K for the diodes on Si materials doped with either Al, B, or In impurities. It can be seen that the long-time decays in all the samples can be described by mono-exponential transients.

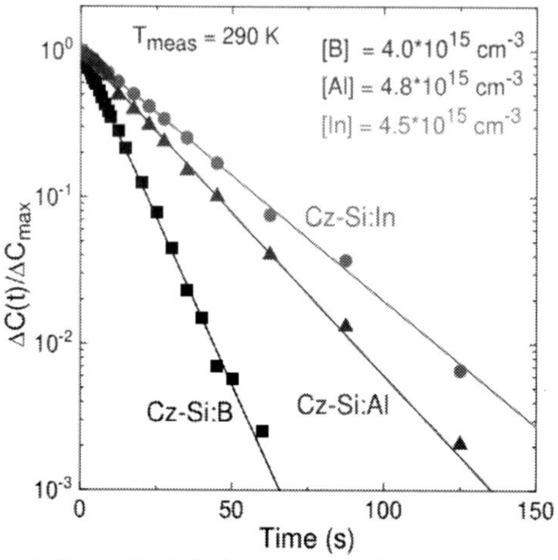

Fig. 4. Normalized diode capacitance decays recorded at 290 K on as-fabricated Schottky diodes from Cz-Si materials doped with either Al, B or In impurity atoms during growth. Solid lines are calculated with the use of an equation for the mono-exponential decay with a decay rate value as a fitting parameter.

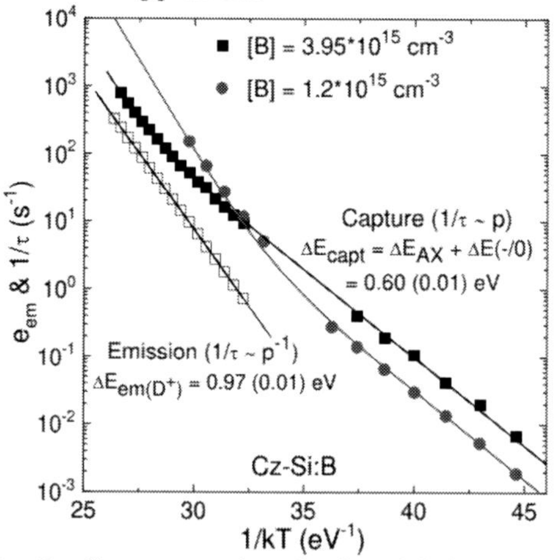

Fig. 5. Temperature dependencies of hole emission (empty symbols) and capture (filled symbols) rates for the B_sO_2 defect. The values have been derived from an analysis of capacitance measurements described above on n^+-p-p^+ diodes from 3 and 10 Ω·cm Cz-Si:B materials, which were subjected to heat-treatments at 375 °C for 10 hours. Solid lines were calculated with the use of Eqs. (3-5) and values of energy barriers as fitting parameters. Concentrations of uncompensated shallow acceptors determined from capacitance-voltage measurements were (3.95±0.1)×10^{15} cm^{-3} and (1.2±0.1)×10^{15} cm^{-3} in diodes from 3 and 10 Ω·cm materials, respectively.

Fig. 5 shows temperature dependencies of τ^{-1} values measured for two diodes on Cz-Si:B samples with

10.4229/EUPVSEC2023/1BV.5.12

different concentrations of free electrons. It is found that the decay rates in the low-temperature range {capture region where the second term in Eq. (3) dominates} are linearly proportional to the concentration of free holes and are proportional to p^{-1} in the high-temperature range {emission region with the dominance of the first term in Eq. (3)}. The observed features can be explained assuming that the acceptor-O_2 complexes in p-type Si are negative-U defects with the CCD diagram as that shown in Fig. 1 and with $c_{pX}p \gg \omega_{XA}$. The values of the $E(-/+)$ occupancy levels, ΔE_{em} and ΔE_{AX} barriers for the acceptor-O_2 defects have been determined in Refs. [14] and [27].

It has been found that prolonged exposure of B-doped Cz-Si crystals to minority carrier injection results in the disappearance of PPC and in degradation of recombination lifetime of excess carrier (BO-LID degradation) [25,28]. It has been argued that this effect is related to the transformation of B_sO_2 into another shallow acceptor configuration, which has strong recombination activity [25]. Such transformation has not been observed for the complexes of O_2 with other than B acceptor atoms [14,18,19,29]. A hypothesis has been proposed that the dominant recombination process associated with the BO-LID degradation is related to trap-assisted Auger recombination via a shallow acceptor state of B_sO_2 [25].

3.4 Detection of a bistable hole trap FZ-Si:Ga crystals

With the recent industrial trend of using Ga-doped Si wafers instead of B-doped ones for the production of silicon solar cells, it is important to expand the knowledge on recombination active defects and minority carrier traps in Ga-doped silicon material grown by Cz and FZ techniques. We have recently observed PPC at room temperature and detected a DLTS emission signal due to a defect responsible for it in Ga-doped FZ-Si crystals [19].

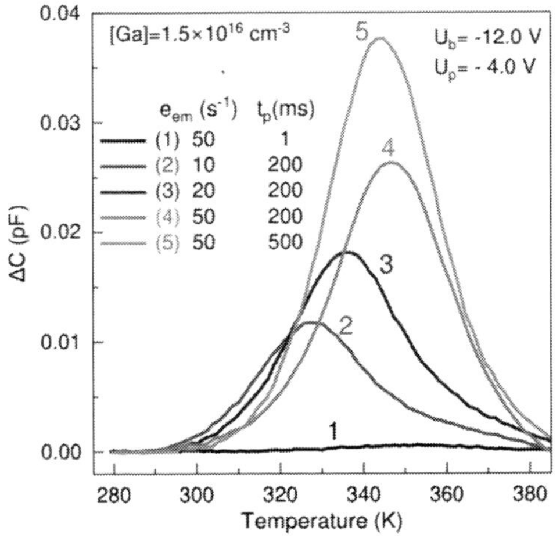

Fig. 6. Conventional DLTS spectra recorded in the temperature range 280-385 K on a diode fabricated on FZ-Si:Ga material with [Ga] = 1.5×10^{16} cm^{-3}. Measurement conditions, which were used for recoding the spectra, are given in the graph.

Fig. 6 shows conventional DLTS spectra recorded with different emission rate windows and filling pulse lengths for a Schottky diode on a FZ-Si:Ga sample. It can be seen that magnitude of the detected emission peak, labelled as the H_{345K} trap in the following, strongly depends on both the rate window and filling pulse length. The H_{345K} signal has negligible magnitude in the DLTS spectra recorded with the application of filling pulses shorter than 10 ms. These findings indicate that capture cross section of holes by the defect is temperature dependent because of the existence of high energy barrier for capture. We have carried out measurements of temperature dependencies of hole emission and capture rates by the H_{345K} trap in the samples with different hole concentrations.

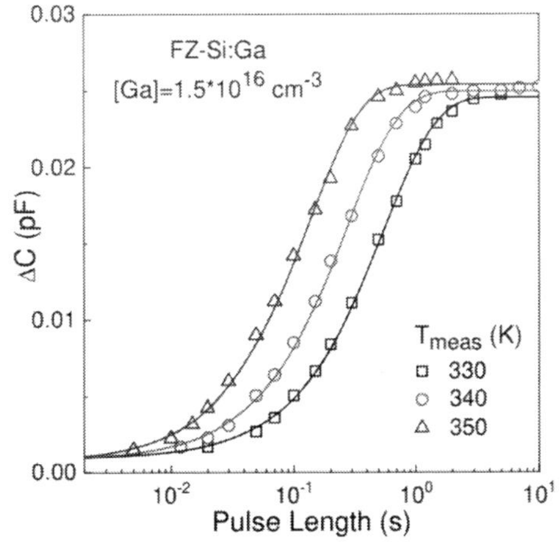

Fig. 7. Changes in magnitude of the L-DLTS signal due to the H_{345K} trap upon changes in the filling pulse length for a Schottky diode on Fz-Si:Ga sample with [Ga] = 1.5×10^{16} cm^{-3}. The solid lines are calculated for a mono-exponential growth process with least-square fitting values of ΔC_{max} and characteristic growth rates.

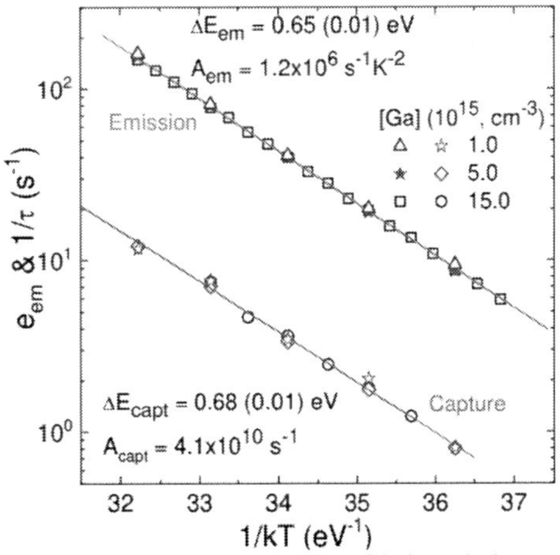

Fig. 8. Temperature dependencies of hole emission and capture rates for the H_{345K} trap detected in the DLTS spectra recorded for samples from different FZ-Si:Ga wafers. The values of activation energies for hole emission and capture processes derived from the presented Arrhenius plots are given in the graph.

Fig. 7 shows changes in hole population of the trap upon changes in filling pulse length in the temperature range 330-350 K in one of the Fz-Si:Ga samples. It can be seen that the capture rate of hole is temperature dependent, and the maximum achievable ΔC magnitude is nearly the same at different temperatures. Fig. 8 shows Arrhenius plots of emission and occupancy rates of the H_{345K} trap in three FZ-Si:Ga samples with different hole concentrations. An analysis of these temperature dependencies reveals some very peculiar features. Particularly, i) emission rates were significantly higher than hole capture rates, with the activation energy for hole capture being higher that the activation energy for hole emission; ii) hole occupancy rates do not depend on hole concentration. These observations can be explained assuming that the defect has two charge states (one energy level), it is bistable in one charge state, and it has $c_{pX}p \gg \omega_{XA}$ (Fig. 1). In this case the effective capture coefficient (Eq. 3) will be $c_p^{eff}p = \omega_{AX}$ with $\Delta E_{AX} = 0.68 \pm 0.01\ eV$, and the emission rate from DLTS measurements is $e_D = c_pN_v exp(-\Delta E_{em}/kT)$ with $\Delta E_{em} = 0.65 \pm 0.01\ eV$. Based on the obtained results a CC diagram of the defect has been constructed in Ref. [19]. The CCD for the H_{345K} trap resembles that shown in Fig.1 without the A^- state.

The H_{345K} trap has not been observed in FZ-Si:B crystals, which we have studied [19]. It is likely that Ga atoms are involved in the formation of the defect either directly or indirectly.

4 DISCUSSION AND CONCLUSIONS

According to the definition of minority carrier traps in semiconductors, they are not effective recombination centers and, therefore, seem to be not very harmful for solar cells. However, they still capture charge carriers, which finally recombine, although with slow rates. This means the trapped carriers do not contribute to photovoltaic current, and thus, the presence of traps reduces efficiency of solar cells slightly.

The analysis of the charge carrier dynamics associated with the oxygen-related bistable thermal double donors indicates that these defects behave as minority carrier trapping centers in n-type Cz-Si crystals but not as effective recombination centers. However, it has been found that the presence of TDDs results in the degradation of efficiency of hetero-junction solar cells with their base from n-type Cz-Si [21,30]. It has been argued that in n-type Si crystals containing high concentration of TDDs, these defects are not fully ionized in the shallow donor configuration at room temperature and effective minority carrier recombination can occur through the $E(+/+2)$ level of TDDs at about E_c-0.15 eV [13].

Strong arguments have been presented that the PPC effects in Czochralski-grown p-type Si crystals doped with different group-III impurities are associated with the complexes of substitutional acceptor atoms with the oxygen dimer. Ab-initio calculations have shown that that the deep-shallow transformations of the acceptor-O_2 defects are related to the structural changes of the oxygen dimer between square and staggered configurations in the vicinity of substitutional group-III atoms [14,25].

The two cases of PPC effects in CZ-Si samples, which are considered above, show an important role of oxygen atoms in the appearance of the PPC phenomenon. It appears that the reversible transformations of oxygen atoms from twofold to threefold bonding configurations in silicon lattice result in significant changes in electronic properties of the defects, which incorporate a few oxygen atoms. These changes are further reflected in changes of electrical properties of Si materials. The detected changes can be long-living because of the existence of rather high energy barriers for the atomic reconfigurations of the O atoms. This brings back to attention the importance of avoiding the incorporation of high concentrations of oxygen and/or any processing conditions leading to the formation/activation of oxygen-related defects in Si.

The defect observed in FZ-Si:Ga material shows very unusual electronic properties with some peculiar features in emission and capture rates of holes. These features can be explained assuming that the defect has one energy level, it is bistable in one charge state, and there is an energy barrier of 0.68 eV for a transformation of the defect from one configuration to another. This defect has not been observed in FZ-Si:B samples, which we have studied [19]. The origin of this defect remains an open question that needs further investigations.

All of these case studies illustrate the importance of good understanding of electronic properties of deep-level traps in order to accurately calculate charge carrier dynamics upon injection of minority carriers and predict their effects on solar cell performance.

ACKNOWLEDGEMENTS

The work was done with funding from the UK EPSRC under contract No EP/TO25131/1.

REFERENCES

[1] R. Sinton, A. Cuevas, Appl. Phys. Let. 69 (1996) 2510.
[2] J. A. Hornbeck, J. R. Haynes, Phys. Rev. 94 (1955) 311.
[3] J. R. Haynes, J. A. Hornbeck, Phys. Rev. 100 (1955) 606.
[4] D. Macdonald, A. Cuevas, Appl. Phys. Lett. 74 (1999) 1710.
[5] J. Schmidt, K. Bothe, R. Hezel, Appl. Phys. Lett. 80 (2002) 4395.
[6] Yu Hu, S. Shøn, Ø. Nielsen, E. J. Øvrelid, L. Arnberg, J. Appl. Phys. 111 (2012) 053101.
[7] Y. Zhu, M. K. Juhl, G. Coletti, Z. Hameiri, IEEE J. Photovoltaics 9 (2019) 652.
[8] S. Jafari, Z. Hameiri, Sol. En. Mat. & Solar Cells 232 (2021) 111341.
[9] J. Coutinho, V. P. Markevich, A. R. Peaker, J. Phys.: Condens. Matter 32 (2020) 323001.
[10] L. F. Makarenko, V. P. Markevich, L. I. Murin, Sov. Phys. Semicond. 19 (1985) 1192.
[11] Ya. I. Latushko, V.P. Markevich, L. F. Makarenko, L. I. Murin, Phys. Status Solidi A 93 (1986) K181.
[12] L. F. Makarenko, L. I. Murin, Phys. Status Solidi B 145 (1988) 241.
[13] V.P. Markevich, M. Vaqueiro-Contreras, S. B. Lastovskii, L. I. Murin, M. P. Halsall, A. R. Peaker, J. Appl. Phys. 124 (2018) 225703.
[14] J. A. T. De Guzman, et al., J. Appl. Phys. 130 (2021) 254703.
[15] L. Dobaczewski, K. Bonde Nielsen, A. R. Peaker, J. Appl. Phys. 96 (2004) 4689.
[16] A. R. Peaker, V. P. Markevich, J. Coutinho, J. Appl. Phys. 123 (2018) 161559.
[17] V. P. Markevich, L. I. Murin, T. Sekiguchi, M. Suezawa, Mater. Sci. Forum 258-263 (1997) 217.
[18] J. A. T. De Guzman et al., Phys. Status Solidi A 218 (2021) 2100108.
[19] T. O. Abdul Fattah, et al., AIP Conf. Proc. 2826 (2023) 110001.
[20] V. V. Litvinov, G. V. Palchick, V. I. Urenev, Phys. Status. Solidi A 108 (1988) 311.

[21] M. Tomassini, et al., J. Appl. Phys. 119 (2016) 084508.
[22] M. Siriwardhana, Y. Zhu, Z. Hameiri, D. Macdonald, F. Rougieux, IEEE J. Photovoltaics 11 (2021) 273.
[23] M. Pesola, Y. J. Lee, J. von Boehm, M. Kaukonen, R. M. Nieminen, Phys. Rev. Lett. 84 (2000) 5343.
[24] J. Coutinho, et al., Phys. Rev. Lett. 87 (2001) 235501.
[25] M. Vaqueiro-Contreras, et al., J. Appl. Phys. 125 (2019) 185704.
[26] V. P. Markevich, et al., Phys. Status Solidi A 216 (2019) 1900315.
[27] J. A. T. De Guzman, et al., in Proc. of the 47th IEEE PV Specialist Conference (2020) pp. 1013-1018.
[28] S. Jafari, et al., ACS Appl. Mater. & Interfaces 13 (2021) 6140.
[29] A. R. Meyer, et al., ACS Appl. Energy Mater. 5 (2022) 13161.
[30] J. Li, X. Yu, S. Yuan, L. Yang, Z. Liu, D. Yang, Sol. En. Mat. & Solar Cells 179 (2018) 17.

DEVELOPMENT OF LOW-COST HIGH-EFFICIENT AND RELIABLE UMG PV CELLS

Carlos del Cañizo[1*], David Fuertes Marrón[1], Nerea Dasilva-Villanueva[1], Luis Jaime Caballero[1], Bülent Arikan[2], Hasan Hüseyin Canar[2,7], Rasit Turan[2,3], Hasan Asav[2,7], Ahmet Emin Keçeci[2,7], Guillermo Sánchez Plaza[4], Laura Méndez[5], Manuel Funes[5], Eduardo Forniés[5], Daniel Garraín[6]

[1] Instituto de Energía Solar, ETSI Telecomunicación, Universidad Politécnica de Madrid, Spain
[2] ODTÜ-GÜNAM (Center for Solar Energy Research and Applications), Turkey
[3]Middle East Technical University, Department of Physics, Turkey
[4]Nanophotonics Technology Center – Universidad Politécnica de Valencia, Valencia, Spain
[5]Aurinka PV Group, Rivas-Vaciamadrid, Spain
[6]CIEMAT, Madrid, Spain
[7]Micro and Nanotechnology, Middle East Technical University, Ankara, Turkey

* Corresponding author: Instituto de Energía Solar. ETSI Telecomunicación, Avenida Complutense 30, 20840 Madrid, Spain, carlos.canizo@upm.es, phone: +34-910672521

ABSTRACT: Upgraded Metallurgical Silicon (UMG) has the potential to reduce the cost of Photovoltaic (PV) technology, while at the same time improving the environmental profile of solar-grade silicon in terms of energy payback time and CO_2 emissions. In this contribution we report recent work made in the framework of a European SOLAR-ERA.NET project (CHEER-UP) to demonstrate that UMG multicrystalline silicon is a competitive alternative to polysilicon for the production of high efficiency solar cells. Defect engineering techniques based on phosphorus diffusion gettering have been explored to improve the bulk quality of UMG, reaching bulk lifetimes >300 µs, with record values of 1.1 ms. Black silicon has been implemented, designing this texturing process to make it compatible with gettering and surface passivation effects and obtaining reflectivities in the range of 12% after smoothing. Passivated Emitter and Rear Cell (PERC) has been the cell architecture used to assess the efficiency of UMG, conveniently tailoring an industrial manufacturing process to the specificities of the material, and TOPCon solar cell processing key steps have been tested on the material to further evaluate the potential of the material beyond PERC. Degradation mechanisms related to light exposure and operation temperature have been shown not to be significant in UMG PERC solar cells when a regeneration step is implemented. The project has resulted in the achievement of 20,1% UMG multicrystalline PERC solar cells, an efficiency target that is accompanied by a reduction in the environmental impact of silicon technology.

Keywords: Feedstock. Upgraded Metallurgical Silicon. Multi-crystalline.

1 INTRODUCTION

Upgraded MetallurGical (UMG) silicon, presented as an alternative to conventional polysilicon, experienced very relevant advances in its development in the 2000s and 2010s, and reached commercialization for some time. Even if those commercial initiatives could not keep up with the steep growth of the Photovoltaic (PV) market, it still makes sense to explore the potential of UMG, as it offers reduced CAPEX and thus faster capacity expansions, competitive manufacturing costs and more benign environmental impact.

This paper reports on the progress in the development of UMG achieved under the framework of a SOLAR-ERA.NET project, CHEER-UP, that has run from February 2020 to May 2023 with the goal of fabricating UMG multi-PERC solar cells and assessing its potential for advanced cell architectures [1].

A review of the specificities of the UMG material produced by Ferroglobe is first presented. Then, results in the optimization of key processing steps are reported, followed by the description of the PERC solar cell process and the results obtained. Preliminary works for the manufacturing of UMG TOPCore solar cells follow. Finally, main findings from a life cycle assessment of the UMG technology are shared.

2 THE UMG ROUTE BY FERROSOLAR

The feedstock production process at Ferroglobe in Spain is sketched in figure 1 and has been described in detail elsewhere [2] [3]. The resulting UMG-Si product is a compensated material, with boron concentration below 0.2 ppmw and phosphorus concentration in the range of 0.1–0.3 ppmw. Total metal concentration is below 0.5 ppmw.

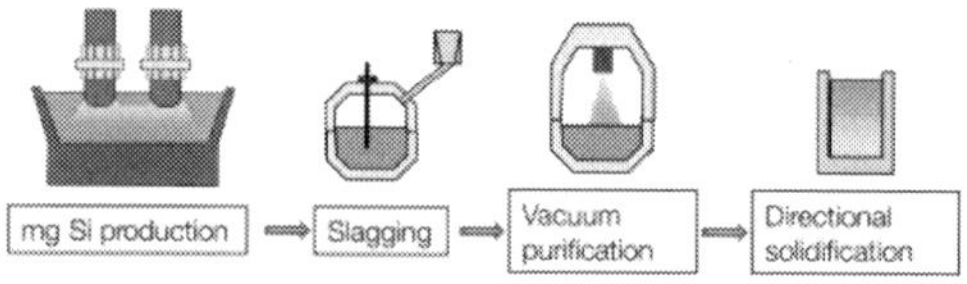

Figure 1: Main steps in the production of UMG at Ferroglobe.

High Performance Multicrystalline ingots were grown at Pillar in Ukraine with this silicon feedstock, adding small quantities of gallium and phosphorus to control de variation of resistivity along the ingot. Ingots were further cut into 25 bricks with a square area of 156.75 × 156.75 mm² and later sliced into wafers.

The wafers further processed in the CHEER-UP project come from a variety of ingots and bricks, and their resistivities lie in the 1-1.5 Ωcm range.

3 OPTIMIZING KEY PROCESSING STEPS IN SOLAR CELL PROCESSING

3.1 Enhancement of carrier lifetime by P diffusion

gettering (PDG)

To effectively remove impurities from the bulk of the UMG-Si wafers, a phosphorous diffusion gettering (PDG) step is introduced as a first step in cell processing [4].

A test was done with randomly selected wafers that were cut into 52 x 52 mm^2 samples. They were firstly characterized with the Sinton Photoconductance Decay (PCD) tool, and then chemically etched and cleaned before the thermal process. After PDG, the diffused emitter was removed by etching and PCD measurements were conducted again.

The impact of sequential PDG processes carried out under different conditions (duration and temperatures) can be seen in figure 2. Best results for the first gettering process are systematically obtained at 780°C, resulting in lifetimes upgraded to an average level of 300 µs. Further improvements can be observed after a second PDG process, reaching an outstanding maximum lifetime of 722 µs.

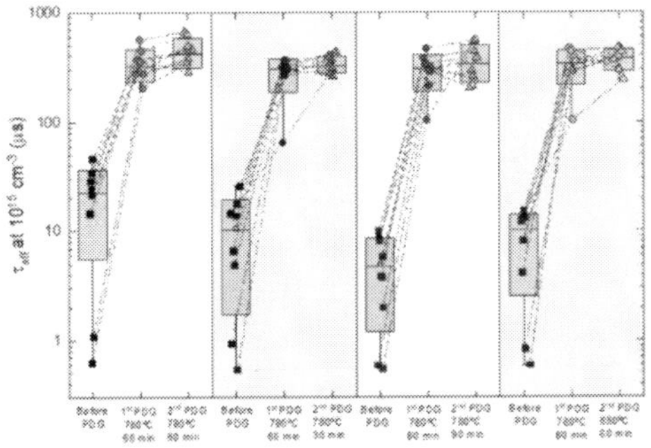

Figure 2: Effective lifetime variation after several PDG conditions for UMG-Si samples passivated with Iodine-Ethanol (IE).

3.2 Enhancement of rear surface passivation with dielectric stacks

In order to determine the appropriate alumina thickness for the rear passivation structure, 5, 10 and 15 nm alumina layers were symmetrically deposited onto UMG wafers, followed by the symmetrical deposition of a 120 nm hydrogenated amorphous silicon nitride coating layer. The sample group without alumina layer and with hydrogenated amorphous silicon oxynitride layer, an alternative passivation layer for the rear of the PERC cell, were also considered [4].

The results shown in figure 3 indicate that 5 nm thickness is the most appropriate as average lifetimes of 300 µs and implied open-circuit voltages (iV_{oc}) of 709 mV are obtained.

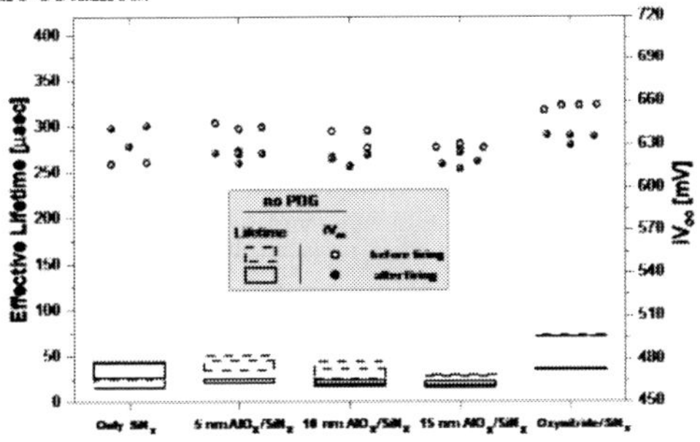

(a)

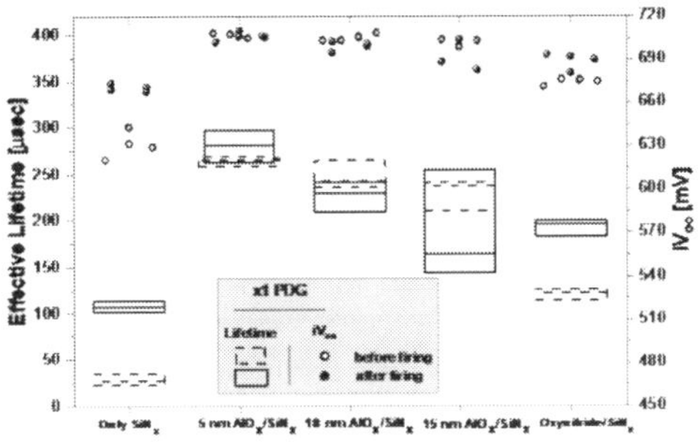

(b)

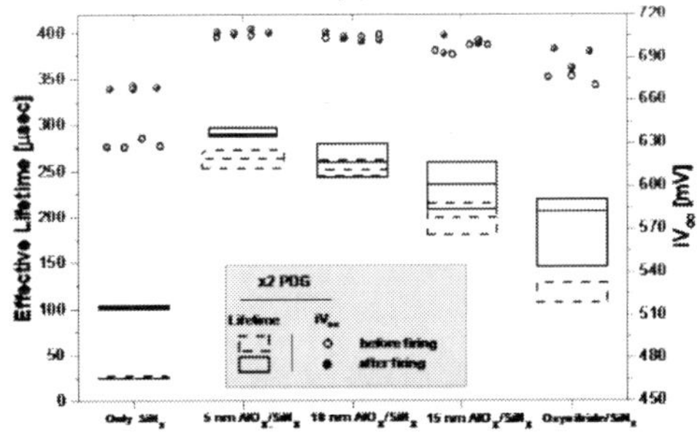

(c)

Figure 3: Effective lifetime (left axis) and iVoc (right axis) of symmetrically passivated UMG wafers. (a) Wafers without pregettering. (b) Wafers with one pregettering. (c) Wafers with two sequential pregetterings.

3.3 Reduction of reflectivity with metal assisted etching + smoothing

The surface nanotexturing of multi-UMG wafers was performed by means of metal assisted etching (MAE) in an industrial line at Parla Solar in Turkey. After the MAE process, surface smoothing by HNO_3/HF/DI water is done in order to obtain the most suitable surface for emitter formation and passivation.

Reflectance measurements before and after smoothing on visually different grains are shown in figure 4, where a certain increase in the surface reflection can be observed after smoothing. The average reflectance value was determined to be in the range of 11.5-12%.

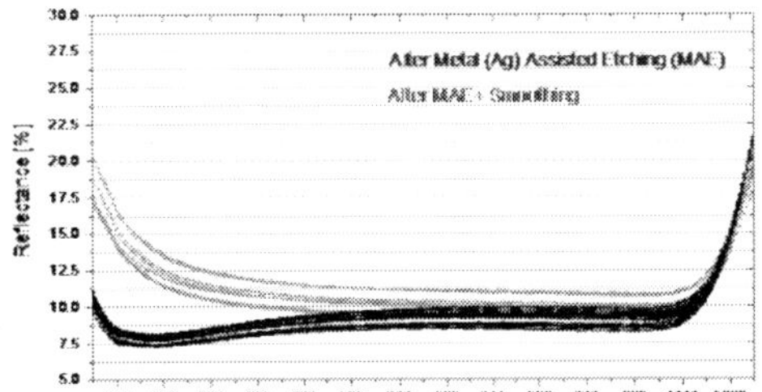

Figure 4: Reflectance of different grains in UMG wafers after MAE (in black) and after MAE + Smoothing (in brown).

4 SOLAR CELL DEVELOPMENT

4.1 Implementing PERC solar cells

A schematic representation of the PERC solar cell process for multi-UMG, involving twelve key production

10.4229/EUPVSEC2023/1BV.5.14
020031-002

steps, is given in figure 5(a), in which the learnings described in section 3 have been incorporated. All production steps were carried out at ODTU-GÜNAM pilot line in Turkey. The front and rear images of a UMG PERC cell, once the production process has been completed, are presented in figure 5(b).

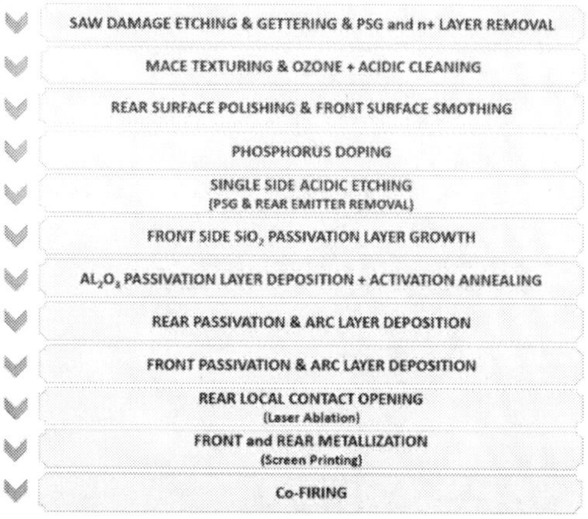

(a)

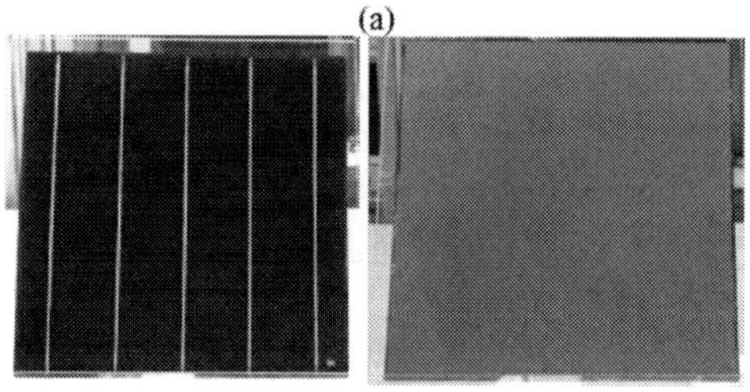

(b)

Figure 5. (a) PERC solar cell process designed for industrial scale production of multi-UMG PERC solar cells. (b) Front and rear side view of a M2 UMG PERC solar cell.

Two different approaches are followed for the front side passivation and ARC, labelled "2-stack front passivation" (formed by a silicon oxynitride layer on top of a silicon nitride layer) and "4-stack front passivation" (consisting of two silicon oxynitride layers of different refraction index and thickness, and two silicon nitride layers also with different refraction indexes).

The graphs showing the short-circuit current density (J_{SC}), open-circuit voltage (V_{OC}), fill Factor (FF), and cell efficiency distributions of both the 2-stack and 4-stack front passivated cells are given in figure 6.

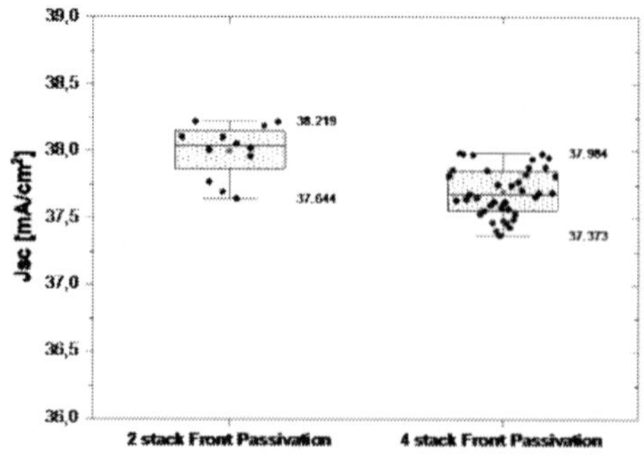

(a)

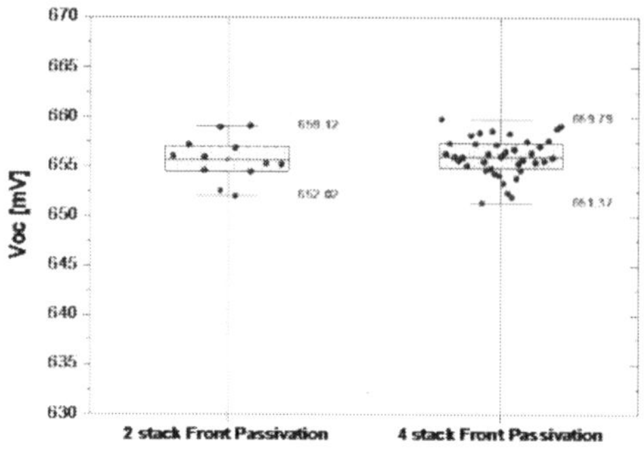

(b)

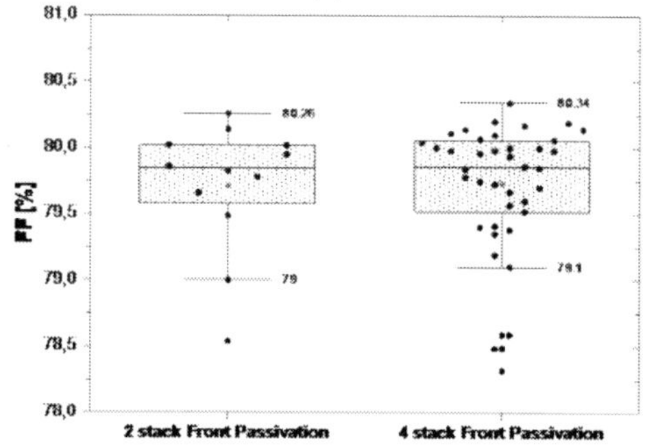

(c)

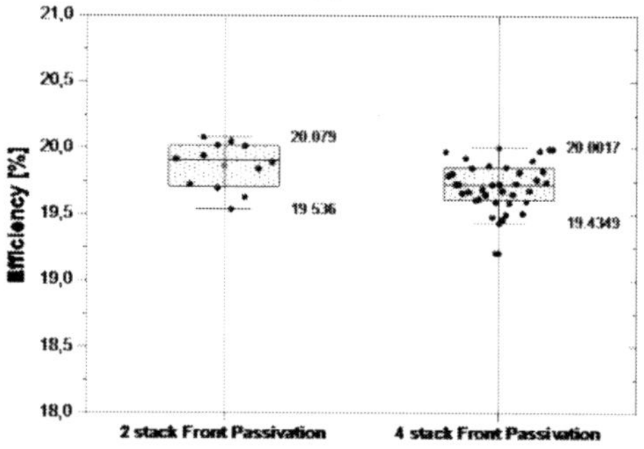

(d)

Figure 6. Distribution of solar cell performance parameters measured under Standard Test Conditions for both the 2-stack and the 4-stack front passivation options. (a) Short-circuit current density (J_{SC}), (b) open-circuit voltage (V_{OC}), (c) fill Factor (*FF*), and (d) cell efficiency.

V_{OC} values range between 651-660 mV, with similar results for both front-passivation structures. J_{SC} values of 38.2 mA/cm^2 for the 2-layer front passivation structure and 38.0 mA/cm^2 for the 4-layer structure have been reached. The fill factor values reach a maximum of 80.34%, with a wide distribution, attributed to the fact that the silver pastes used in the front surface metallization are developed for conventional pyramidal textured surfaces, not optimized for MAE natotextured ones. This caused discontinuity and narrowing of the fingers in some of the cells. It is foreseen that this problem, which is not directly related to mc-UMG material quality, can be eliminated by using new generation silver pastes and masks adapted for nanotextured surfaces.

Efficiency values of 20% were achieved with both front passivation approaches, with highest value of 20.1%. This value can be surpassed with the inclusion of new generation production materials specific for nanotextured multicrystalline wafers, and with the use of new generation production approaches such as selective emitter, floating busbar and large wafer formats.

4.2 Assessing UMG potential for TOPCon architectures

Tests have been done at Fraunhofer ISE to assess the feasibility of UMG for TOPCon cell architectures, in particular a TOPCon rear emitter structure called TOPCoRE [6].

Bulk lifetimes after key processing steps (PDG, TOPCon anneal and boron diffusion steps) have been measured by calibrated Photoluminiscence (PL), confirming that values in the hundreds microseconds range are obtained in most of the area of the processed wafers, surpassing locally the millisecond in some cases.

It has been confirmed that high passivation levels are reached with Al_2O_3/SiN_x stacks on Reactive Ion Etching RIE-textured UMG samples, measuring implied V_{oc} >700 mV in good areas of the processed wafers.

Based on the injection-dependent PL results, device simulations have been implemented to give the spatial efficiency distribution that can be expected from the material quality following the ELBA methodology [8]. As shown in Figure 7, a global efficiency higher than 22% is within reach with these multicrystalline UMG-Si substrates.

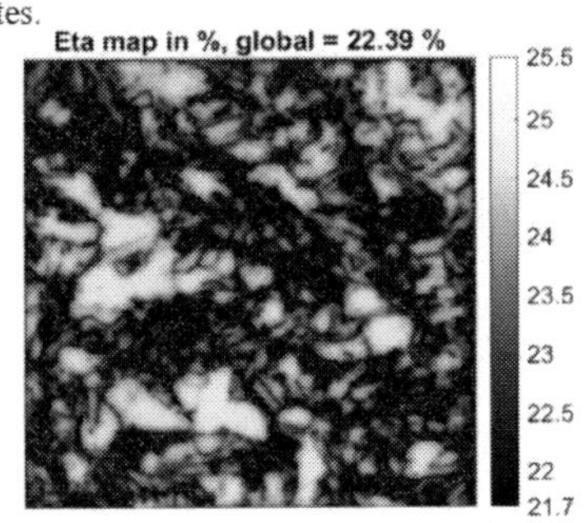

Figure 7: Estimated efficiency of UMG-Si solar cells according to an ELBA analysis for a TOPCore structure. Taken from [3].

A solar cell processing batch is currently underway to confirm these promising results.

5 LIFE CYCLE ASSESSMENT OF UMG TECHNOLOGY

To evaluate the differences in the potential environmental impact between the UMG cells and those based on standard polysilicon, a process-based Life-Cycle Analysis (LCA) has been performed, using the Simapro 9.0 software with Ecoinvent 3.5 as database for all the background data [5].

Different electricity mixes are considered, as the electric power input has a decisive role in the overall impact of the PV value chain [8].

A first analysis has been done for a BSF technology, covering the whole value chain, from metallurgical silicon manufacturing to electricity generation by a ground-mounted multicrystalline PV system with a nominal capacity of 100 MWp installed in the South of Spain. The details can be found in [9] and corroborate the reduced environmental impact when using Ferro's UMG-Si instead of conventional polysilicon. For example, for the Spanish electricity mix, UMG-Si can lead to a reduction over 20% of climate change related emissions (12 vs. 15 gCO_{2eq}/kWhe) and over 25% of the Energy Pay Back Time of the electricity generated (0.52 vs. 0.68 years) during the expected lifetime of PV system, considered 30 years.

Additional studies have been performed to assess the environmental impact of PERC and TOPCon structures, in this case focusing only on cell technology. A selection of results is presented in Figure 8.

In Figure 8(a) the climate change emissions of a state-of-the-art polysilicon PERC technology are compared to those for PERC and TOPCore cell technologies based on UMG-Si developed in the CHEER-UP project. Four different geographical locations are represented, with very different carbon intensity in their electricity mix. Cells made of polysilicon show values between 0.6-0.28 $kgCO_2$/Wp, while for UMG cells they are 0.32-0.13 $kgCO_2$/Wp. Manufacturing cells using China (CN), United States (US) and European Union (EU) electricity mixes imply 54%, 48% and 16% increase of climate change emissions, respectively, as compared to the manufacture under the Spanish (ES) mix.

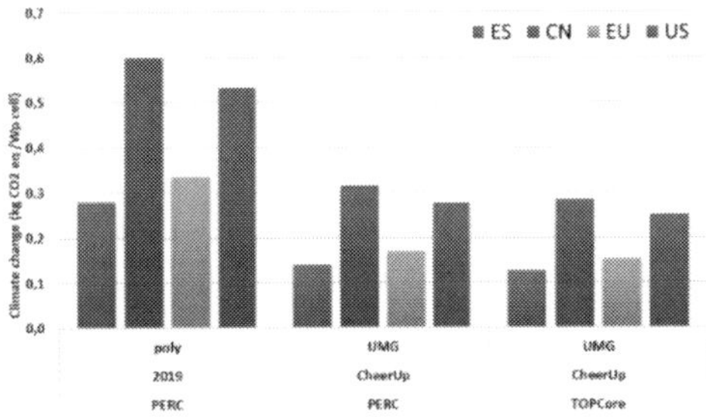

(a)

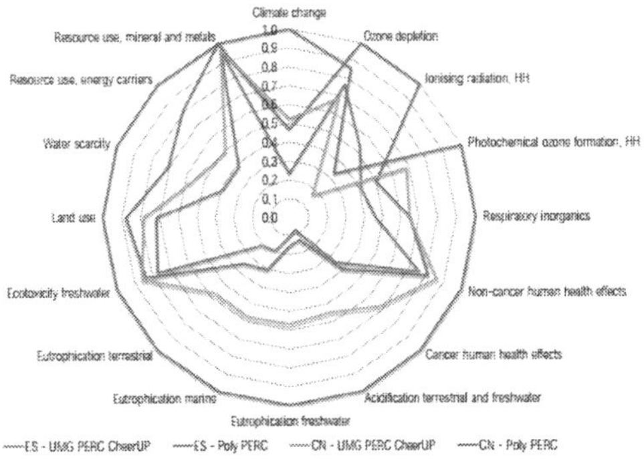

(b)

Figure 8: (a) Climate change emissions in $kgCO_2$eq per cell W_p of conventional polysilicon PERC technology and PERC and TOPCore UMG-Si technologies developed within the CHEER-UP project, for four electricity mixes: Spain (ES), China (CN), European Union (EU) and United States (US). (b) Normalized environmental impacts of polysilicon and UMG PERC technologies, comparing the manufacture in Spain and China.

The trends shown by normalized values of all Environmental Footprint categories analysed for PERC architectures in ES and CH can be seen in Figure 8(b). The strong influence of the location is clearly seen, as the predominant presence of fossil fuels, specifically coal, in the Chinese mix has a strong influence on the results. As an example, it is foreseen that, for PERC architecture, UMG-Si leads approximately to a 50% reduction in terms of climate change emissions for both ES and CN mixes. However, for Acidification – Terrestrial and Fresh Water category, the reduction is 45% for ES mix, and 76% for CN mix, thereby highlighting the relevance of the silicon feedstock employed for PV-manufacturing when the energy mix counts with low shares of renewables energies.

5 CONCLUSIONS

Through the optimization of gettering, texturing and surface passivation strategies and the tailoring of a solar cell process to the specificities of the material, the CHEER-UP project has resulted in the achievement of 20,1% UMG multi PERC solar cells, an efficiency target that is accompanied by important reductions in the fabrication cost and the environmental impact of silicon technology.

As a means to assess the feedstock potential for advanced cell architectures, TOPCon solar cell processing key steps have been tested on UMG wafers, indicating that efficiencies beyond 22% are within reach.

For a given energy mix, the environmental impacts of UMG-Si-based cells can be ~50% lower than those based on conventional polysilicon.

Therefore, Upgraded Metallurgical Silicon can be the basis of a low-CAPEX, low-cost and environmental-friendly alternative supply of solar silicon.

6 ACKNOWLEDGEMENTS

This work is part of the R&D SOLAR-ERA.NET Cofund 2 project CHEER-UP, funded by the Spanish MCIN/ AEI/10.13039/501100011033/ through projects PCI2019-111834-2 and PCI2019-111903-2, by the Spanish Centro para el Desarrollo Tecnológico Industrial (CDTI), and by the Turkish TÜBITAK through Project 219M029. MCIN/AEI/ 10.13039/501100011033 is also acknowledged for financial support through the GREASE project (PID2020-113533RB-C31).

The involvement of Fraunhofer ISE's Group Innovative Cleanroom Technologies lead by Dr. Jan Benick is greatly acknowledged.

7 REFERENCES

[1] https://www.era-learn.eu/network-information/networks/solar-cofund-2/solar-cofund-2-joint-call/low-cost-high-efficient-and-reliable-umg-pv-cells-cheer-up

[2] E. Forniés *et al.*, Mass Production Test of Solar Cells and Modules Made of 100% UMG Silicon. 20.76% Record Efficiency, Energies. 12 (2019) 1495. doi:10.3390/en12081495.

[3] C. del Cañizo *et al.*, Progress in the development of upgraded metallurgical silicon solar cells for industrial production, 8th World Conf. on PV Energy Conv. (2022).

[4] N. Dasilva-Villanueva *et al.*, High carrier lifetimes in UMG multicrystalline wafers after P- diffusion compatible with high-efficiency cell structures", Solar Energy Materials&Solar Cells 250 (2023) 112080.

[5] G. Wernet *et al.*, The ecoinvent database version 3 (part I): overview and methodology, Int. J. Life Cycle Assess., vol. 21, n.o 9, pp. 1218-1230, sep. 2016, doi: 10.1007/s11367-016-1087-8.

[6] A. Ritcher *et al.*, Design rules for high-efficiency both-sides contacted silicon solar cells with balanced charge carrier transport and recombination losses, Nature Energy 6 (2021), 429-438, doi:10.1038/s41560-021-00805-w

[7] B. Michl *et al.*, Efficiency limiting bulk recombination in multicrystalline silicon solar cells, Solar Ener. Mat. Solar Cells 98 (2012) 441–447, doi:10.1016/j.solmat.2011.11.047.[8] D. Yue, F. You, S. B. Darling, Domestic and overseas manufacturing scenarios of silicon-based photovoltaics: Life cycle energy and environmental comparative analysis, Sol. Energy, vol. 105, pp. 669-678, jul. 2014, doi: 10.1016/j.solener.2014.04.008.

[9] L. Méndez *et al.*, Upgraded metallurgical grade silicon and polysilicon for solar electricity production: A comparative life cycle assessment, Sci. Total Environ., vol. 789, p. 147969, oct. 2021, doi: 10.1016/j.scitotenv.2021.147969.

40th European Photovoltaic Solar Energy Conference and Exhibition

This presentation was selected by the Sc. Committee of the EU PVSEC 2023 for submission of a full paper to one of the EU PVSEC's collaborating peer-reviewed journals.

ASSESSING THE POTENTIAL OF TOPCON SOLAR CELLS ARCHITECTURE USING INDUSTRIAL N-TYPE CAST-MONO SILICON MATERIAL

B. Bazer-Bachi[1*], P. Saint-Cast[2], J. Posada[3], S. Williatte[1], C. Tessmann[2], R. Bodeux[3], S. Mack[2] and G. Goaer[1]
[1]Photowatt, 33 rue Saint-Honoré, 38300 Bourgoin-Jallieu, France
[2]Fraunhofer Institute for Solar Energy Systems (ISE), Heidenhofstrasse 2, 79110 Freiburg, Germany
[3]EDF R&D - Institut photovoltaïque d'Ile de France (IPVF), 18 Boulevard Thomas Gobert, 91120 Palaiseau, France
*b.bazerbachi@photowatt.com

ABSTRACT: Cast-mono silicon material is interesting for its lower carbon footprint compared to Cz monocrystalline silicon. However, solar cells fabricated using cast-mono (CM) silicon show lower performances. In this work, two routes to make cast-mono silicon advantageous over Cz silicon are considered.
The first route is to further reduce carbon footprint of cast-mono silicon, by using Upgraded Metallurgical Grade silicon (UMG-Si) feedstock instead of Solar Grade silicon (SoG-Si) feedstock. TOPCon solar cells are fabricated using both feedstocks, and cast-mono growing technology, using industrial-type furnaces. Laboratory studies show that UMG-Si can result in efficiencies higher than solar cells made of SoG-Si when feeding the material to a CM crystallization process. But when compared to Cz, CM-UMG-Si TOPCon solar cells results are still 0.5%$_{abs}$ lower.
The second route is to take advantage of the TOPCon passivation layer ability to getter metallic impurities, and thus improve the quality of cast-mono material. Several TOPCon sequences are tested and their effect on the passivation properties of the layer are studied. In the end, solar cells are fabricated and again, UMG-Si solar cells show better results than SoG-Si solar cells, with efficiency up to 22.65%, independently confirmed.
Keywords: Cast-mono, SoG, UMG, TopCon

1 INTRODUCTION

Cast-mono material is a hybrid material between Cz monocrystalline silicon and cast grown multicrystalline silicon[1]. During the ingot crystallization, the seeds deposited at the bottom of the crucible allow a controlled growth with the same crystal orientation as the seeds, usually (100). Compared to Cz material, cast-mono wafers contain more metallic impurities mainly originating from the crucible and crystal defects such as dislocations appearing during the growth. Such drawback leads to lower overall performance of the cast-mono material. This reduced performance is counterbalanced by lower production costs and lower emissions of greenhouse gases [2].

The replacement of Solar Grade silicon (SoG-Si) by Upgraded Metallurgical Grade silicon (UMG-Si) is a solution to further reduce the carbon footprint, and production costs [3]. However, as UMG-Si shows higher impurity content, the resulting efficiency might be lower than using conventional SoG-Si.

This study aims to compare the use of these two silicon materials, using n-type silicon wafers produced in an industrial-size furnace. To evaluate the potential of each material, TOPCon solar cells are fabricated, as it is state-of-the-art architecture, and it shows high gettering potential [4].

The second part of this work aims to test several TOPCon processes, as the gettering potential differs, depending on the type of process used [5].

2 EXPERIMENTAL DETAILS

The wafers used in this study originate from two 650 kg industrial size Phosphorus-doped ingots, grown either using SoG-Si feedstock or using UMG-Si feedstock. The 15.7 cm x 15.7 cm N-type cast-mono Si wafers produced from these ingots show resistivities of 1.6 – 1.9 Ω.cm and 0.6 – 0.9 Ω.cm for cast-mono SoG-Si and cast-mono UMG-Si respectively. N-type Cz wafers, with resistivity of 0.9 – 1.1 Ω.cm are added as references. As the CM wafers come from two different ingots, the dislocation growth within the ingot differs. Moreover, for each ingot (CM-SoG and CM-UMG), wafers are coming from 2 different bricks. In each brick, the dislocation growth varies so the position and rate of dislocation within the wafer are different. Bricks 1 and 2 are associated to the UMG-Si ingot and Bricks 3 and 4 are linked to SoG-Si ingot.

TOPCon solar cells are fabricated using process similar as the process presented in [6]. The electrical properties of the cells are characterized by current voltage measurements in light illumination at 25 °C (AM1.5 global spectrum). Photoluminescence imaging are carried out on the cells.

Symmetrical lifetime samples are also fabricated. In order to take into account all the high temperature process steps, the samples first undergo the boron diffusion (after texturing and cleaning). After boron diffusion, the *p+* layer is removed and the TOPCon sequence is applied on both sides to all samples, varying several parameters. Finally, the samples are passivated and fired. QSSPC measurements determine the $iV_{oc,max}$ which corresponds to the maximum implied V_{oc} achievable, as the boron diffused layer is removed beforehand.

3 COMPARISON OF SOLAR CELLS RESULTS FROM CAST-MONO AND CZ WAFERS

The first part of this study aims to compare solar cells fabricated from cast-mono wafers to Cz material, with silicon from two different materials: SoG-Si and UMG-Si. Standard process is used for the TOPCon sequence for the solar cells, as described in part 2.

Table I shows the solar cells results obtained for the 3 types of wafers i.e. Cz-Si, cast-mono SoG-Si (CM-SoG) and cast-mono UMG-Si (CM-UMG). The efficiency difference is 0.51%$_{abs}$ between Cz-Si and and best CM-Si efficiencies. The efficiency gain for Cz-Si is due to an increase of all I/V parameters. Contrary to Cz-Si, cast-mono silicon, both from UMG-Si and SoG-Si, contain dislocation clusters and metallic impurities lowering bulk lifetime and affecting the solar cells parameters [7], [8]. The maximum efficiencies are 22.25% and 22.52% for CM-UMG and Cz-Si respectively.

10.4229/EUPVSEC2023/1BV.5.15
020032-001

Si material	V_{oc} (mV)	J_{sc} (mA/cm²)	*FF* (%)	*η* (%)	*pFF-FF* (%)	*FF₀-pFF* (%)
Cz-Si	695.8	40.28	80.1	22.45	3.4	1.1
Cast-mono UMG-Si	689.8	40.14	79.2	21.94	3.2	2.1
Cast-mono SoG-Si	684.5	40.22	77.7	21.40	3.7	3.0

Table I : I/V parameters of the TOPCon solar cells fabricated from different material type (Cz-Si, CM-SoG and CM-UMG). The results are averaged out of 6-9 cells with optimized firing temperature. These measurements are in-house measurements, the front contact is obtained by GridTouch, the rear contact by a non conductive dark chuck.

Surprisingly, the efficiency of CM-UMG is 0.5%$_{abs}$ higher than the efficiency of CM-SoG. This gain is partly explained by the lower resistivity of UMG-Si, resulting in higher V_{oc} and *FF*. However, the loss in *FF* is not only due to higher series resistance, resulting in higher *pFF-FF* for CM-SoG compared to CM-UMG. A higher loss in FF_0-*PFF* (2.1% for CM-UMG and 3% for CM-SoG) is also observed.

In order to understand the lower results obtained for SoG-Si, one has to consider the spatial repartition of the dislocations within the wafers, as dislocations create shunt paths within the space charge region [9]. Figure 1 shows PL images of cells for the three studied materials. For CM-SoG and CM-UMG, two PL images are shown, corresponding to the 4 different groups of cells described in the experimental part, with different dislocations location and concentration, representative of the other cells.

Cz	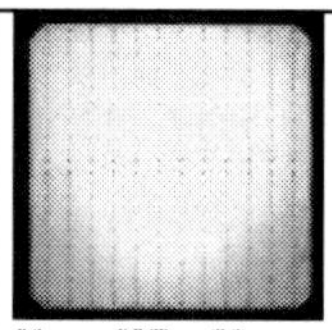 V_{oc} = 697 mV FF_0-*PFF* = 1.1%	
CM-UMG	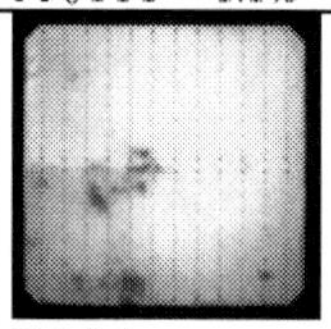 Brick 1 V_{oc} = 693 mV FF_0-*PFF* = 1.7 %	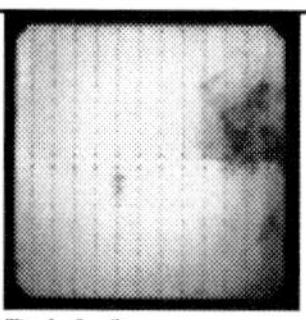 Brick 2 V_{oc} = 691 mV FF_0-*PFF* = 2.0%
CM-SoG	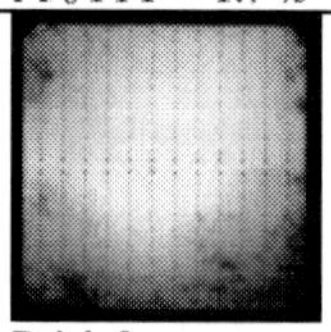 Brick 3 V_{oc} = 686 mV FF_0-*PFF* = 2.7%	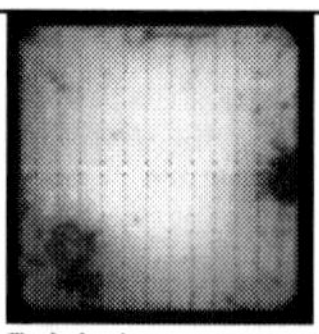 Brick 4 V_{oc} = 685 mV FF_0-*PFF* = 3.3%

Figure 1: PL images of representative cells, for three materials, Cz, CM-SoG and CM-UMG. For each cell, the corresponding V_{oc} and FF_0-*PFF* values are depicted.

Figure 1 also shows that when the dislocation rate is very low, as for Brick 1, V_{oc} and FF_0-*PFF* are close to Cz results. As the dislocation rate increases, as for Brick 4, FF_0-*PFF* increases and V_{oc} decreases. It turns out that the wafers from SoG material (Bricks 3 and 4) have a highest amount of dislocations, resulting in higher losses both for V_{oc} and *FF*.

Moreover, a measurement of $iV_{oc,\,max}$ on symmetrical samples passivated with TopCon layers can be conducted on dislocation free regions of the samples with QSSPC. On dislocation free regions, $iV_{oc,max}$ are 704 mV and 703 mV respectively for CM-UMG and CM-SOG. The value for Cz is 710 mV. The difference between Cz and CM-Si is then partly due to the metallic impurities, which might be removed using efficient gettering.

4 OPTIMIZATION OF THE TOPCON PROCESS

As shown in the previous section, there is a gap between the Cz and CM efficiencies, explained by the presence of dislocation and metallic impurities. However, the TOPCon layer generates a strong impurity gettering effect, removing metallic impurities such as iron [10]. Even though the authors conducted their experiment on *p*-type substrate, one can assume that a positive effect of enhanced gettering should be reached on *n*-type substrate, as *n*-type material also well reacts to phosphorus gettering [8], [11].

The second part of this work thus deals with the optimization of the TOPCon process to enhance the impurity gettering. In order to test the ability of the TOPCon layer to getter, several TOPCon process steps are modified: the type of tunnel oxide (chemical or thermal), the doping process (in-situ or ex-situ), the poly-Si layer thickness (low or high) and the anneal temperature (low or high).

4.1 Symmetrical lifetime samples results

The experiment deals with the evaluation of $iV_{oc,max}$ for the different TOPCon sequences, as described above. Figure 2 shows the variation of the $iV_{oc,max}$, for the three types of material, i.e. Cz-Si, CM-UMG and CM-SoG and for all the TOPCon processes tested. The standard sequence corresponding to a thermal oxide with a thin poly Si, an in-situ doping and a low temperature thermal anneal, is shown on the left-hand side of the graph (Thin poly-Si).

In order to take into account, the different spatial position of the dislocations within the sample (as discussed in the previous section), the measurements are done in 5 different positions (center and 4 corners). For Cz-Si, $iV_{oc,max}$ is mainly dominated by the surface passivation quality. The $iV_{oc,max}$ variations for Cz samples thus show the impact of the TOPCon process changes. The CM-Si $iV_{oc,max}$ results follow the same trend as Cz-Si results. Even though some parameters such as chemical oxide, high temperature or thicker poly-Si have more potential for enabling external gettering, $iV_{oc,max}$ is dominated by the surface passivation quality.

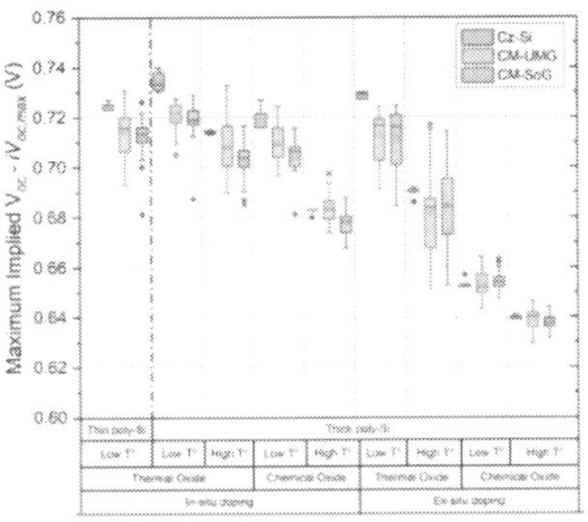

Figure 2: Variation of the $iV_{oc,max}$ as a function of the material type and on the recipe parameters : thin or thick poly-Si, in-situ or ex-situ doping, thermal or chemical oxide and high or low temperature. The values shown are mean values of 4-8 samples.

4.2 Solar cell results

Following the lifetime samples measurements, solar cells are fabricated using the two optimum TOPCon sequences in addition to the standard process used in Section 3. The results for the standard process are the same as the ones presented in Table I. Ex-situ doping resulted in lifetime inhomogeneities visible on PL image and was not used for solar cells fabrication. The solar cell fabrication thus focuses on the best results obtained for in-situ doping, defined as:

- Standard: Thermal Oxide + Thin poly-Si + In-situ Doping + Low Temperature anneal
- Thermal Oxide: Thermal Oxide + Thick poly-Si + In-situ Doping + Low Temperature anneal
- Chemical Oxide: Chemical Oxide + Thick poly-Si + In-situ Doping + Low Temperature anneal.

Table II shows the solar cells I/V results for the three selected recipes and the two CM-Si materials. Table II also gives details about the *FF* losses, showing *pFF-FF* and *FF0-PFF* values for the three TOPCon recipes and the two CM-Si materials.

Similarly to the lifetime samples, the efficiency is higher using a thermal oxide associated with a thicker poly-Si layer, compared to the Standard recipe. The highest efficiency results from an enhancement of V_{oc} (as already observed in previous section) and *FF*.

Thicker poly-Si leads to reduce the series resistance (lower *pFF-FF*) thanks to the higher conductivity of the thicker poly-Si layer and/or a lower contact resistance between Ag paste and poly-Si. The lower FF_0-*PFF* shows that the thicker TOPCon layer also reduces the recombination within the SCR. It was shown previously that a thicker layer suppresses recombination at the metal contacts [12]. There could be also an effect of an enhanced gettering of impurities, thanks to thicker layer, but the two effects can't be decoupled for now.

Using chemical oxide instead of thermal oxide gives lower efficiency, mainly due to reduced V_{oc} even though *FF* increases. The reduction of V_{oc} is explained by the reduced passivation quality of the TopCon layer using chemical oxide, as already stated in the previous section. The *FF* increase is both due to a reduction of FF_0-*PFF* and *pFF-F*. Using chemical oxide results in the formation of pinholes which facilitate the conduction [13]. The presence of pinholes enables faster gettering [5], leading to a further reduction of the SCR recombination resulting in lower FF_0-*PFF* than thermal oxide.

For both Thermal and Chemical Oxide, the thicker polySi layer leads to lower J_{sc}, due to a higher free carrier absorption of the thicker polySi layer [14].

The trend observed in Table II is the same as in Table I. CM-UMG shows better results that CM-SoG because of dislocation on both V_{oc} and *FF*.

Finally, the two champion cells using CM-Si and including a thicker polySi layer with a thermal oxide (Thermal Oxide in Table II) where independently measured at Fraunhofer ISE CalLab in Germany, resulting in efficiency of 22.45% and 22.65% for CM-SoG and CM-UMG respectively (CalLab measurement condition: 12 front bus bars contact on the front side and rear reflective conducting chuck, AM1.5, 1000W/m²).

5 CONCLUSION

This study showed that using UMG-Si as feedstock for fabricating cast-mono TOPCon solar cells is a promising option, as the efficiency is only 0.5 $\%_{abs}$ lower than Cz-Si. The harmful effect of dislocation rate was shown qualitatively, and is especially detrimental for the *FF*, as dislocations create recombination paths within the space charge region. This explains the weaker results obtained with SoG-Si, as the dislocation rate is slightly higher.

Optimizing the TOPCon sequences leads to an additional efficiency gain of +0.4 $\%_{abs}$, using a thicker polySi layer coupled with thermal oxide. This enhancement is induced by the V_{oc} and the *FF* rise, thanks to a better surface passivation, a reduced metal recombination and probably to a more effective gettering.

	TopCon process	V_{oc} **(mV)**	J_{sc} **(mA/cm²)**	***FF* (%)**	***η* (%)**	***pFF-FF* (%)**	FF_0***-pFF* (%)**
CM-SoG	Standard	684,5	40,22	77,7	21,40	3,7	3,0
	Thermal	690,5	40,09	78,7	21,78	3,5	2,4
	Chemical	680,1	40,06	79,3	21,59	2,9	2,1
CM-UMG	Standard	689,8	40,14	79,2	21,94	3,2	2,1
	Thermal	693,0	40,05	79,9	22,18	2,7	2,0
	Chemical	683,2	40,02	80,3	21,96	2,5	1,6

Table II : I/V parameters of the TOPCon solar cells fabricated as a function of the TopCon sequence, for CM-UMG and SoG-UMG The results are averaged out of 6-9 cells with optimized firing temperature. These measurements are in-house

10.4229/EUPVSEC2023/1BV.5.15

For chemical oxide, the better gettering induced by the presence of pinholes does not counterbalance the degradation of the surface passivation, due to the detrimental effect of the pinholes, leading to a decrease of the efficiency.

6 REFERENCES

[1] N. Stoddard *et al.*, « Casting Single Crystal Silicon: Novel Defect Profiles from BP Solar's Mono2 TM Wafers », in *Gettering and Defect Engineering in Semiconductor Technology XII*, in Solid State Phenomena, vol. 131. Trans Tech Publications Ltd, janv. 2008, p. 1-8. doi: 10.4028/www.scientific.net/SSP.131-133.1.

[2] Riepe, Stephan *et al.*, « Cast-mono silicon wafers for a sustainable PV market growth », présenté à 37th European Photovoltaic Solar Energy Conference and Exhibition, 2020, p. 121-125. doi: 10.4229/EUPVSEC20202020-2AO.4.3.

[3] L. Méndez, E. Forniés, D. Garrain, A. P. Vázquez, A. Souto, et T. Vlasenko, « Upgraded metallurgical grade silicon and polysilicon for solar electricity production: A comparative life cycle assessment », *Science of The Total Environment*, vol. 789, p. 147969, 2021, doi: https://doi.org/10.1016/j.scitotenv.2021.147969.

[4] Desrues, Thibaut *et al.*, « Ultra-thin Poly-Si/SiOx Passivating Contacts Integration for High Efficiency Solar Cells on n-type Cast Mono Silicon Wafers », *AIP Conference Proceedings*, 2021.

[5] Z. Yang *et al.*, « Impurity Gettering in Polycrystalline-Silicon Based Passivating Contacts—The Role of Oxide Stoichiometry and Pinholes », *Advanced Energy Materials*, vol. 12, n° 24, p. 2103773, 2022, doi: https://doi.org/10.1002/aenm.202103773.

[6] S. Mack *et al.*, « Status and Perspective of Industrial TOPCon Solar Cell Development at Fraunhofer ISE », présenté à Proceedings of the 8th WCPEC, Milano, Italy, 2022.

[7] C. Liu *et al.*, « Industrial TOPCon solar cells on n-type quasi-mono Si wafers with efficiencies above 23% », *Solar Energy Materials and Solar Cells*, vol. 215, p. 110690, 2020, doi: https://doi.org/10.1016/j.solmat.2020.110690.

[8] H. C. Sio *et al.*, « The electrical properties of high performance multicrystalline silicon and mono-like silicon: Material limitations and cell potential », *Solar Energy Materials and Solar Cells*, vol. 201, p. 110059, 2019, doi: https://doi.org/10.1016/j.solmat.2019.110059.

[9] O. Breitenstein, J. P. Rakotoniaina, M. H. Al Rifai, et M. Werner, « Shunt types in crystalline silicon solar cells », *Progress in Photovoltaics: Research and Applications*, vol. 12, n° 7, p. 529-538, 2004, doi: 10.1002/pip.544.

[10] A. Liu *et al.*, « Direct Observation of the Impurity Gettering Layers in Polysilicon-Based Passivating Contacts for Silicon Solar Cells », *ACS Applied Energy Materials*, vol. 1, n° 5, p. 2275-2282, 2018, doi: 10.1021/acsaem.8b00367.

[11] H. Park *et al.*, « Effect of the Phosphorus Gettering on Si Heterojunction Solar Cells », *International Journal of Photoenergy*, vol. 2012, p. 794876, févr. 2012, doi: 10.1155/2012/794876.

[12] F. Feldmann *et al.*, « Large Area TOPCon Cells Realized by a PECVD Tube Process », sept. 2019. doi: 10.4229/EUPVSEC20192019-2EO.1.4.

[13] K. Lancaster, S. Großer, F. Feldmann, V. Naumann, et C. Hagendorf, « Study of Pinhole Conductivity at Passivated Carrier-selected Contacts of Silicon Solar Cells », *Energy Procedia*, vol. 92, p. 116-121, 2016, doi: https://doi.org/10.1016/j.egypro.2016.07.040.

[14] F. Feldmann, M. Nicolai, R. Müller, C. Reichel, et M. Hermle, « Optical and electrical characterization of poly-Si/SiOx contacts and their implications on solar cell design », *Energy Procedia*, vol. 124, p. 31-37, 2017, doi: https://doi.org/10.1016/j.egypro.2017.09.336.

DEVELOPMENT BY USING FUNCTIONAL GRAIN BOUNDARY OF N-TYPE MONO-CAST SILICON FOR SOLAR CELLS

[1]Hitoshi Matsuo*, [1]Hideyoshi Tanabe, [2]Noritaka Usami
[1]Kyocera Corporation, Japan, [2]Graduate School of Engineering Nagoya University, Japan
*hitoshi.matsuo.hs@kyocera.jp

ABSTRACT: We have developed a crystal growth technology for growing high-quality n-type mono-cast silicon using a one-way solidification furnace for polycrystalline silicon for solar cells. Because it is difficult to obtain large monocrystals to cover the entire surface of a unidirectional solidification furnace, plural monocrystals must be used as the seed. However, there was a problem that dislocation clusters were generated from the seed junctions. We report a growth method of a high-quality mono-cast with few dislocation clusters. The crystal orientation and growth conditions were optimized by implementing this countermeasure with a simple method. By reducing the stress in the block by making the solid-liquid interface shape convex above and by forming an original functional grain boundary based on the SMART method of Nagoya University, we could grow high-quality mono-cast crystals of G5 size.

1 INTRODUCTION

Monocrystalline silicon solar cells based on the CZ method are currently the mainstream of crystalline silicon solar cells. Further cost reductions are needed to achieve costs comparable to those of commercial power sources in terms of larger areas and higher costs.

The growth of mono-cast silicon using seed crystals and a unidirectional solidification furnace is very cost-effective. It has advantages such as the ability to fabricate wafers with low oxygen concentration, which is a characteristic of the casting method [1], and the ease of expanding the area of wafers. It is also reported that the output is equivalent to that of a single-crystal silicon solar cell [2]. However, dislocation clusters generated from seed crystals reduce conversion efficiency compared to monocrystalline silicon solar cells [3]. Dislocation clusters generated from seed crystal junctions are particularly problematic, and the key to growing high-quality quasi-single crystals is how to suppress dislocation clusters generated from seed crystal junctions. We propose the growth of high-quality G5 size mono-cast crystal growth by forming unique functional grain boundaries based on the SMART approach (Seed Manipulation for ARtificially controlled defect Technique) [4-6] developed at Nagoya University.

2 EXPREMENTAL AND CALCULATION METOD

2.1 Experimental method

Seed crystals arranged so that the crystal growth direction is <100> were set in the crucible, and Si feedstock was set on the seed crystals so that the total weight of silicon was 400 kg. Si feedstock and a part of seed crystals were melted to the desired amount and then solidified so they gradually solidified from the bottom. The crucible can be moved to allow heat removal from the bottom. After confirming that the silicon was solidified entirely, it was cooled and solidified gradually. The solidified block was a G5-size n-type pseudo monocrystalline block measuring 830 mm in length and width and 250 mm in height. A block of approximately 160 mm was cut from the block, sliced to a thickness of 200 μm, and evaluated.

2.2 Calculation method

Numerical calculations were performed to understand the furnace's internal stress and temperature distribution. For the numerical calculations, global simulation technology was used to trace the furnace's conditions from melting, solidifying, and cooling of silicon. [7]

3 RESULTS AND DISCUSSION

3.1 Stress reduction by optimization of solid-liquid interface shape

In general, dislocations tend to occur more easily when the solid-liquid interface shape is close to being sub-convex. Therefore, we attempted to optimize the conditions for controlling the solid-liquid interface shape to be upper convex by numerical analysis. First, the validity of the calculations used in this study was verified: Fig. 1 shows a graph comparing the results of the numerical calculations with the dislocation density (EPD) measured from an actual block. The experimental and calculated data were in close agreement. Fig. 2 shows the solid-liquid interface geometry and dislocation density calculated from the temperature distribution in the furnace using numerical analysis. Under the initial temperature conditions, there was a downward convex trend from the middle to the latter half of solidification. Therefore, the solid-liquid interface was adjusted by the heater conditions. As a result, the upper convexity of the solid-liquid interface was increased. As a result, the area ratio of dislocation density below 10000 1/cm^2 was improved from 23% to 36%.

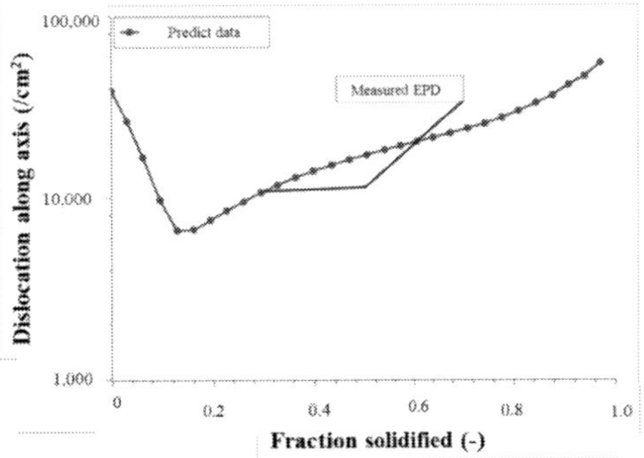

Figure 1: Comparison of between experimental and calculated results

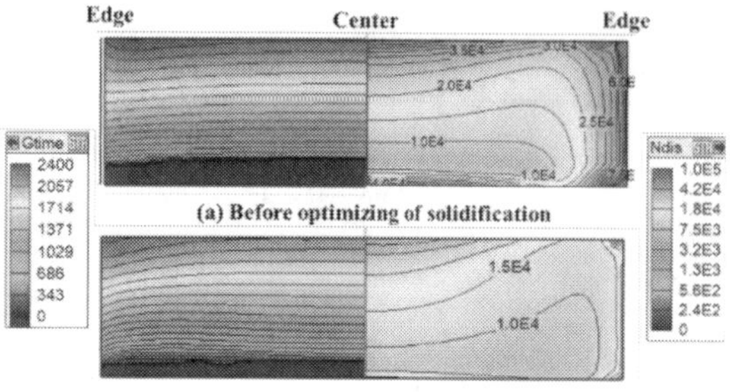

(b) After optimizing of solidification

Figure 2: Numerically adjusted solid-liquid interface and dislocation density distribution. The left figure shows the solid-liquid interface, and the right figure shows the dislocation density distribution.

3.2 Improvement of the rate of dislocation clustering by introducing functional grain boundaries

Dislocations could not be counteracted. To improve dislocation, we developed and introduced our functional grain boundary based on the SMART method developed at Nagoya University. The functional grain boundary developed by Nagoya University is characterized by a two kind of layer structure, one layer that generates dislocations and the other that dislocation blocking, while the functional grain boundary we have developed has a simple structure with only one kind of layer to enable high-quality crystal growth even for large G5 silicon crystals. The orientation difference between the functional grain boundaries was set in the 36° to 44° relative to the lateral (100) plane. The degree of crystal lattice curvature with and without functional grain boundaries is shown in Fig. 3. It can be seen that the lattice curvature is smaller when functional grain boundaries are introduced. This result suggests that the functional grain boundaries we have developed may mitigate the interstitial stresses.

Pseudo-single crystal silicon was prepared under these conditions. The results of the photoluminescence measurement of the ingot head are shown in Fig. 4. Together with this, the results of the lifetime measurement are shown in Fig. 5. The crystal quality is improved with temperature distribution and functional grain boundary formation. Fig. 6 shows the results of evaluating the area ratio of dislocation clusters in the entire block using photoluminescence images. It can be seen that the

(a) X-ray topographic image (b) Difference between the respective lattice angles

Figure 3: Measurements results of X-ray topographic. (a) shows an image of the X-ray topograph, and (b) is the difference in lattice angles taken along the arrows shown in (a). The blue and red arrows correspond to the red and blue plots in the graph.

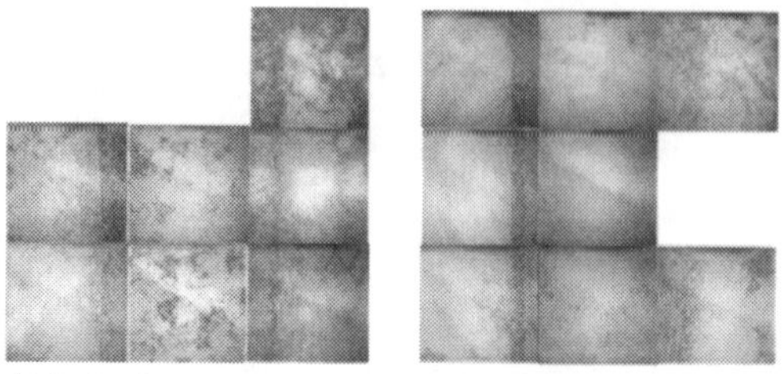

(a) Before improvement (b) After improvement

Figure 4: Photoluminescence images of the cross-section sliced silicon wafer at the top of the ingot.

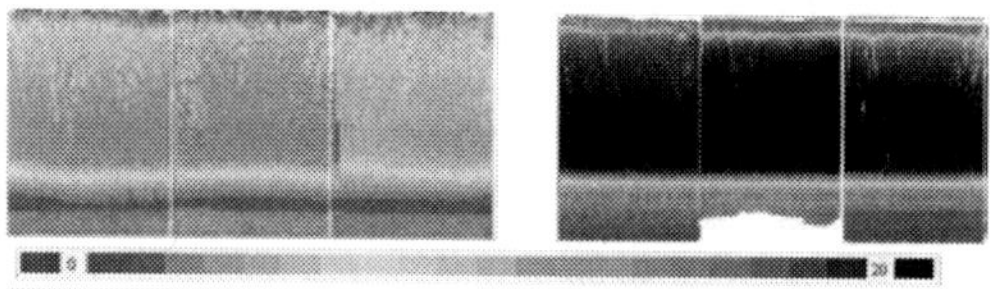

(a) Before improvement (b) After improvement

Figure 5: The results of Lifetime measurement of the cross-section in the center of the ingot only at the inner corner

improved crystals have high quality data. In addition, the spread of dislocation clusters was successfully suppressed by confining the generated dislocation clusters to narrow functional grain boundaries. This method reduced the dislocations in the entire G5 block, as shown in Fig. 6. In our evaluation, the area ratio of dislocation clusters, which is considered to have little effect on the conversion efficiency of solar cells, is 3%. The area ratio of dislocation clusters in the entire ingot evaluated this time was within 3%, and we could grow single crystals of good quality. By optimizing the temperature conditions in the furnace and functional grain boundaries, which is our original technology, we succeeded in growing high-quality mono-casts with few dislocation clusters.

Solar cells were processed using the prepared wafers and evaluated. Fig. 7(a) shows the results of n-type solar cells with HIT structure, and Fig. 7(b) shows the conversion efficiency of p-type solar cells with PERC process. The p-type mono-cast wafers had higher average and maximum solar cell conversion efficiencies than the CZ wafers. The conversion efficiency of mono-cast wafers was higher than that of CZ wafers

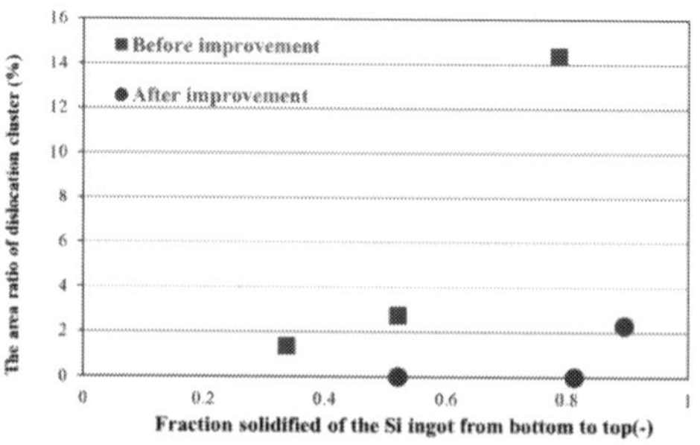

Figure 6: The area ratio of dislocation clusters. The fraction solidified "0" represents the bottom, and "1" represents the top of the ingot.

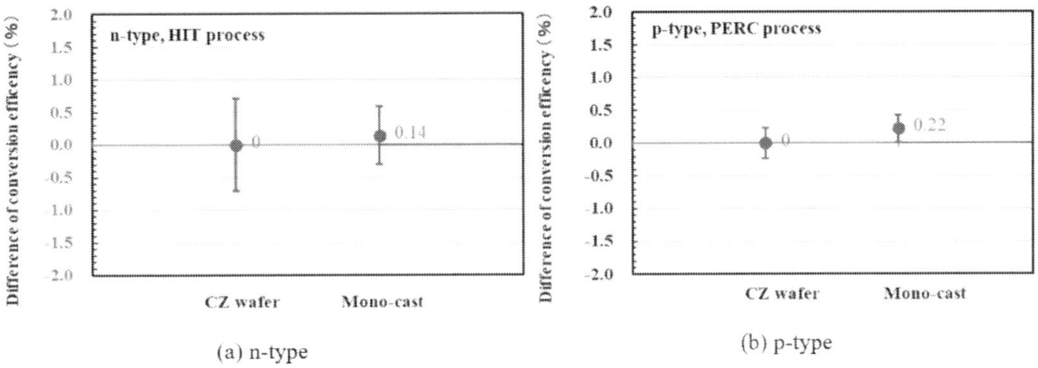

Figure 7: The difference in conversion efficiency from CZ wafer. Solid circles represent averaging of each conversion efficiency.

because mono-cast wafers had lower lattice curvature and were of higher quality than CZ wafers.

In the future, these results can be utilized to contribute to the cost reduction of silicon wafers.

4 CONCLUSION

We worked on developing crystal growth technology to grow high-quality n-type mono-cast silicon using a one-way solidification furnace for polycrystalline silicon for solar cells. By reducing the stress in the block by making the solid-liquid interface shape convex and by forming unique functional grain boundaries based on Nagoya University's SMART method, we have achieved the growth of high-quality G5-size mono-cast crystals. The company intends to apply this technology to larger sizes and automatic control to contribute to the cost reduction of silicon wafers.

5 ACKNOWLEDGMENTS

This article is based on results obtained from a project, JPNP15003, subsidized by the New Energy and Industrial Technology Development Organization (NEDO).

6 REFERENCE

[1] H. Matsuo et al., J. Crystal Growth 310 (2008) 4666.
[2] F. Jay et al., Sol. Energy Mater. Sol. Cells 130 (2014) 690.
[3] D. Hu et al., Sol. Energy Mater. Sol. Cells 140 (2015) 121.
[4] I. Takahashi, et al., Appl. Phys. Express 8 (2015) 105501.
[5] K. Kutsukake, et al., Appl. Phys. Express 6 (2013) 25505.
[6] P. Krenckel, et al., Crystals 11 (2021) 90.
[7] X. Liu, et al., ACS Omega 2022, 7 (2022) 6665.

INVESTIGATION OF CONTACT FORMATION ON LOCALLY LASER-DOPED SURFACES ON A MICROSCOPIC SCALE

Daniel Wurmbrand, Heiko Plagwitz, Frank Huster, Giso Hahn, Barbara Terheiden
University of Konstanz, Department of Physics, 78457 Konstanz, Germany
Daniel.Wurmbrand@uni-konstanz.de Tel: +49 7531 885295

ABSTRACT: In this work we present a detailed investigation of the contact formation of screen-printed silver paste on locally laser-doped silicon surfaces by Kelvin probe force microscopy (KPFM) and scanning electron microscopy (SEM). Although laser doping has been studied many times in the past, there is a lack of systematic studies that shed light on the correlation of laser-induced surface modifications and the associated electrical contactability of the surface, on a microscopic scale. The use of phosphosilicate glass (PSG) deposited by atmospheric pressure chemical vapor deposition (APCVD) as a dopant source and a nanosecond pulsed laser, which can be operated with both a Gaussian and a flat-top beam profile, enables the generation of locally doped structures. Contacting by means of screen-printed silver paste and subsequent evaluation by transfer length method (TLM) shows that, depending on the selected laser parameters and pulse overlap, specific contact resistances can be obtained that vary by more than an order of magnitude. By using the aforementioned KPFM and SEM analysis methods, we show that laser pulse overlap is a key aspect affecting the extent of laser induced surface damage and ultimately silver crystallite growth.
Keywords: Laser doping, local doping, contact formation, APCVD

1 INTRODUCTION

Laser doping is with more than 80% the main process to create selective emitters in PERC (Passivated Emitter and Rear Cell) solar cells which are in turn the main cell product with above 80% in 2022 [1]. A sound overview on laser doping in its different aspects is given by Vaqueiro-Contreras et al. [2]. Vaqueiro-Contreras et al. showed that laser doping can be implemented besides in PERC also in PERT (Passivated Emitter and Rear Totally diffused) and TOPCon (Tunnel Oxide Passivated Contact) solar cell concepts [2].

In contrast to selective emitters, the application of laser doping for the generation of only locally doped areas has only rarely been investigated. Heilig et al. studied the different doping profiles from dopant containing silicate glasses that were used first as a dopant source and then treated locally with a laser, as is the common procedure to generate selective emitters, with dopant concentrations from "fresh" P containing silicate glasses (PSG, BSG) [3]. They also studied the difference of doping from $POCl_3$ formed PSG to PSG from APCVD (atmospheric pressure chemical vapor deposition). Using APCVD for PSG formation allows to create the desired locally laser doped areas, this is not possible with $POCl_3$ based PSG.

In this laser doping process, the laser radiation is transmitted almost unhindered by the dopant containing glass layer and absorbed near the surface in the silicon. In the area hit by the laser pulse, the silicon is melted which leads to local heating of the overlying glass layer. Liquid phase diffusion of the released dopant atoms takes place in the liquefied silicon. Finally, the molten silicon solidifies with the dopant atoms embedded. Since this process takes place on the nanosecond scale, defect formation during recrystallization of the silicon surface cannot be completely avoided under certain conditions [4]. Other studies investigated laser damage regarding the damage to the pyramid textured wafer surface and its effects on the efficiency of selective emitter solar cells [5].

The microscopic Ag contact formation on these laser-doped areas was not investigated up to now. Therefore, a systematic microscopic investigation by Kelvin probe force microscopy (KPFM) and scanning electron microscopy (SEM) of locally laser-doped surfaces with focus on Ag crystallite growth in dependence of laser process parameters is presented and related to specific contact resistivities ρ_C determined by the transfer length method (TLM).

2 EXPERIMENTAL

Figure *1* shows the process flow of sample preparation and subsequent analysis method. Different silicon substrates were used according to the different characterization methods. Low resistivity n-/p-type Czochralski (Cz) Si substrate (0.8 & 1.0 Ωcm, respectively) were textured using an aqueous alkaline etch solution based on KOH with alcohol-based additives.

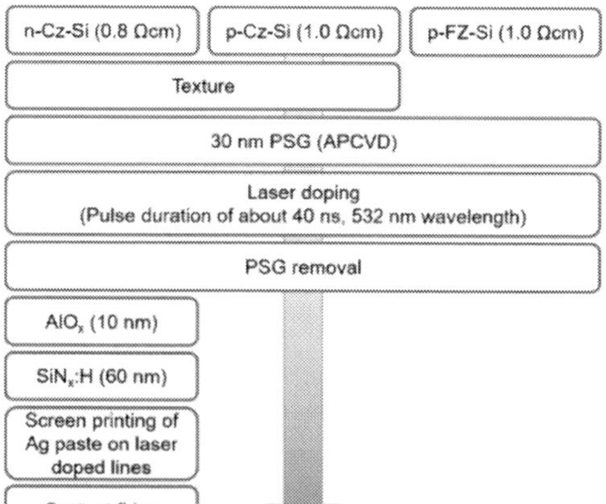

Figure 1: Process flow diagram of sample preparation

For the KPFM samples p-type float-zone (FZ) silicon substrates (1.0 Ωcm) have been used without alkaline texture to avoid roughness-related effects on the measurement of the surface potential. All samples receive a deposition of 30 nm PSG with a phosphorus content of about 8.0-8.5 at% by APCVD. For laser doping a pulsed

10.4229/EUPVSEC2023/1BV.5.37

nanosecond laser (532 nm wavelength, ca. 40 ns pulse duration) with gaussian (G) or flat top (FT) pulse profile is used to create 120 µm wide laser doped lines (TLM and KPFM samples) or 2 x 2 cm² fields with full-area doping to determine doping profiles by electrochemical capacitance-voltage measurements (ECV). The remaining PSG layer and residues from the laser process were removed in diluted HF. Samples for TLM examination receive a typical passivation layer stack consisting of 10 nm AlO_x (APCVD) and 60 nm SiN_x:H (PECVD). 40 µm wide silver contacts are screen-printed onto the 120 µm wide doped lines and sintered in a fast firing process in a belt furnace. The silver contact fingers are centered on the doped lines. To account for the current flow through the base substrate, the evaluation follows the Eidelloth approximation [6]. A simulation-based correction also accounts for the local high-low junctions below the contacts. Details of this evaluation method will be published elsewhere [7].

3 RESULTS AND DISCUSSION

3.1 Influence of laser parameters on contact resistivities

Figure 2 shows measured specific contact resistivities as a function of the distances between pulse centers, for both the Gaussian beam profile and the flat-top profile.

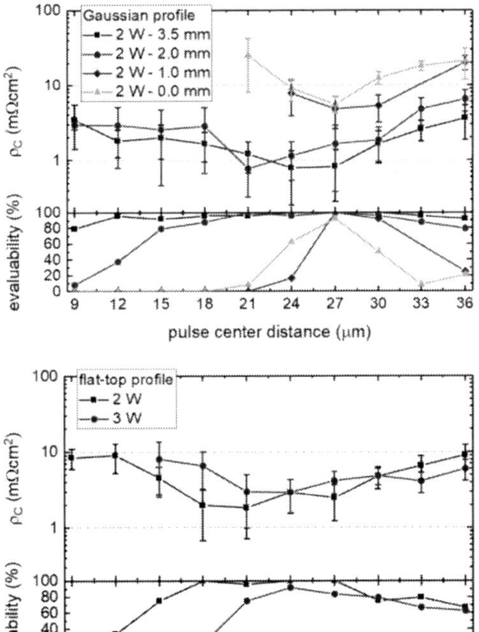

Figure 2: Specific contact resistivity plotted against laser pulse distance for a Gaussian laser profile with a variation of the laser focus (upper graph) and flat-top beam profile with a variation in laser power (lower graph).

With both beam profiles a minimum of the specific contact resistivity is reached at a pulse distance of 21-27 µm. Also, the evaluability of the TLM measurements, shown in the lower part of the graphs, has a maximum at the optimum pulse spacings. The evaluability indicates the percentage of measured structures whose linear fit of the TLM evaluation reaches an R^2 value greater than 0.99 and can thus be considered to be reliably and homogeneously contacted. Accordingly, higher contact resistances are obtained for very large and also for small pulse overlaps. For both Gaussian and flat-top laser pulse, the achieved contact resistivities tend to be lower for the lower fluences, i.e. lower power for flat top and larger defocus for Gaussian profiles. For the Gaussian beam profile, already a defocus of 2.0 mm leads to higher average contact resistances compared to a defocus of 3.5 mm and a reduced evaluability for low point distances. For 1.0 and 0.0 mm defocus this effect becomes so strong that up to 21 µm point spacing almost no measurement shows a linear correlation within the TLM evaluation and thus allows an evaluation. Likewise, the contact resistances for the larger pulse distances are significantly increased.

This already gives a hint that the surface damage induced by the laser pulses has an impact on the contactability of the laser-doped surfaces, because the resulting doping profiles do not fully account for the observed differences as will be shown in the next section.

3.2 Doping profiles

Figure 3 shows the doping profiles that are obtained using different laser parameters.

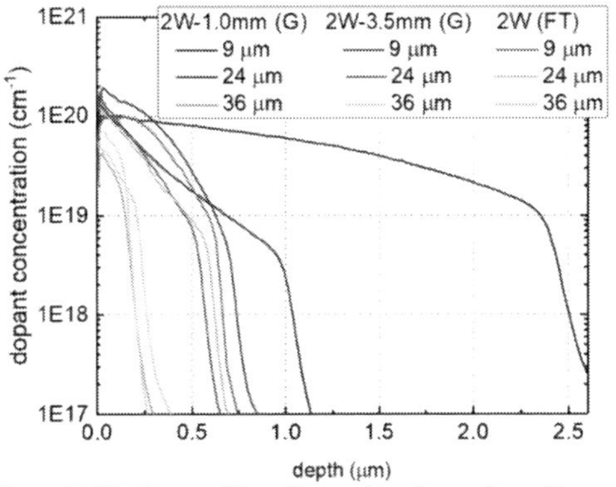

Figure 3: Doping profiles of laser doped samples with different laser parameters.

Green and blue lines indicate doping profiles generated by Gaussian laser pulses with 2 W and 1.0 or 3.5 mm defocus, respectively. In brown, doping profiles obtained at 2 W power with the flat top laser pulses are shown. For each set of laser parameters, the doping profile at 9, 24 & 36 µm pulse spacing is shown, in order to represent the minimum and maximum pulse spacing (9 and 36 µm), as well as the mostly optimal pulse spacing (24 µm) with respect to ρ_C. Indeed, the shallow doping profiles with a much lower surface concentration at 36 µm pulse spacing could explain the increase in contact resistivity at large pulse spacing. However, at 9 µm pulse spacing, profiles with similarly high surface concentrations of about 1×10^{20} cm^{-1} compared to those at 24 µm, are obtained. For the highly focused Gaussian laser pulses (1 mm defocus), a very deep doping profile is obtained at the small pulse spacing of 9 µm due to locally multiple melting events. Interestingly, according to Figure 2, no significant contact formation takes place with these

10.4229/EUPVSEC2023/1BV.5.37
020035-002

laser parameters and pulse spacings, indicating that the achieved doping levels play a minor role at low pulse spacings. Thus, we conclude that the different doping profiles cannot be the main reason for the shown dependence of the contact resistivity on the pulse spacing.

3.3 Kelvin probe force microscopy

To understand what happens on the silicon surface during laser doping, we perform KPFM imaging of laser-doped smooth FZ-Si samples. In this AFM mode the surface potential of a sample is measured locally and can therefore depict the local dopant concentration. Accordingly, the KPFM images depicted in Figure *4* show the local dopant concentration of the upper right edge of laser-doped lines with pulse spacing of 36 µm (a), 24 µm (b), and 9 µm (c), using the flat-top beam profile at 2 W. Especially in Figure *4*a the nearly square shapes of the areas doped by single laser pulses are clearly visible. Within their upper half there is an area of stronger contrast surrounded by artifacts indicating a heavily doped but also potentially damaged area created by excessive laser fluence. It is circled in green for more clarity. The brighter artifacts originate from surface structures and thus indicate the melting edge. In addition, elliptical secondary maxima can be seen next to the main spot. Comparing the images a-c with the specific contact resistances depicted in Figure *2*, it can be deduced that no significant contact formation occurs when these heavily melted areas of highest fluence overlap in such a way that they form a continuous surface.

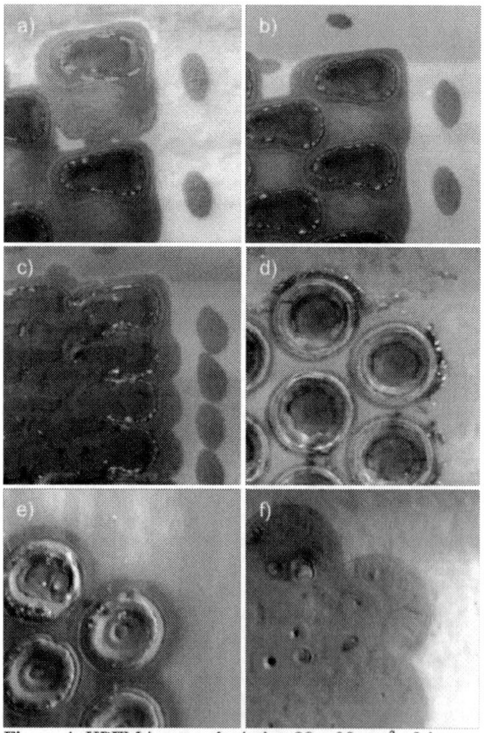

Figure 4: KPFM images depicting 90 x 90 µm^2 of the upper right edge of laser doped lines with point distances of 36, 24 & 9 µm (a-c) on a FZ-Si substrate created by the flat-top profile at 2 W and with a fix pulse distance at 36 µm, and 1, 2 & 3.5 mm (d-f) defocus created by the Gaussian profile.

If the pulse spacing is increased, areas with lower doping and less surface damage are formed between these strongly damaged areas, which allows for lower contact resistances (cf. Figure 2b). However, as the pulse spacing continues to increase, the contact resistivity rises again. This seems to be the case when the damaged but highly doped areas are no longer in contact. Therefore, multiple melting processes no longer occur in the regions of increased peak fluence and the obtained doping profiles start to show reduced surface concentrations.

Figure 4d-f instead shows laser doped sites created with the Gaussian profile at a constant pulse spacing of 36 µm but an increasing defocus of 1.0, 2.0 & 3.5 mm. Figure *4*f shows several laser spots with 3.5 mm defocus, forming an almost undamaged, uniformly doped surface with a dark KPFM contrast that corresponds to the low doping concentration shown in Figure *3* (bright green line). That same contrast can be found in Figure *4*e, surrounding melted and thus potentially damaged areas in the center of the laser doped spots produced with 2 mm defocus. These moderately doped areas are still forming a continuum in this case, whereas they vanish with 1 mm defocus as shown in Figure 4d, leaving only unconnected highly doped and highly damaged single spots. This phenomenon explains the similar behavior of ρ_C with respect to the pulse spacing at defocusing values of 2.0 and 3.5 mm, despite the significant difference in the laser damaged area. With a defocus of 1 mm and decreasing pulse spacing, only small areas of moderate doping are obtained, which would quickly disappear with increasing pulse overlap in favor of a fully damaged surface. This explains the drop in contactability between 21 and 24 µm pulse spacing in Figure 2.

3.4 Scanning electron microscopy

The conclusion that contacting takes place mainly in the areas between the damaged surfaces can be confirmed with the SEM images shown in Figure *5*. These images show the silicon surface after laser doping with flat top beam profile at 2 W, and subsequent screen-printing of a silver contact that has finally been removed with dilute HF. The area formerly covered with silver is on the right side of the image, whereas the left side shows the part of the laser-doped area that extended laterally beyond the silver finger. In this area the SEM images reveal the strong change of the surface structure in the heavily damaged areas by melting and recrystallisation, within the green circle in Figure 5a. In the intermediate areas pyramid-like structures are found which show a small change in topology only at the pyramid tips. In Figure 5b, at higher magnification, it can be seen that the pyramids are densely covered with silver crystallites in the lower half of the image, while almost no crystallites are found in the more damaged area in the upper half of the image.

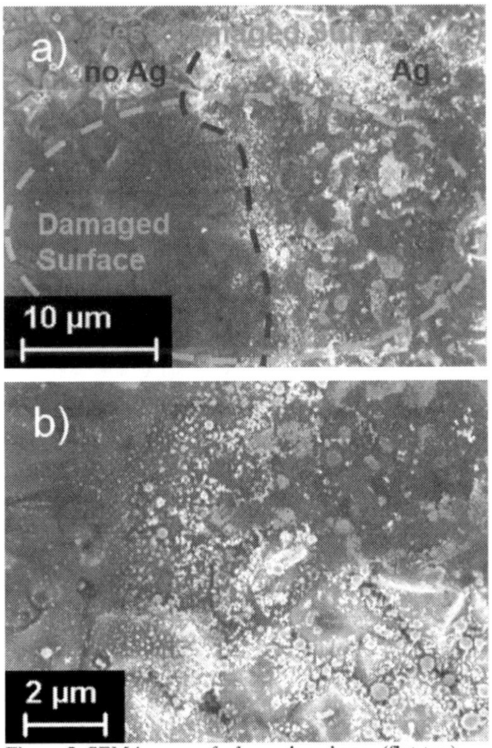

Figure 5: SEM images of a laser-doped area (flat-top) after removal of the screen-printed silver contact finger by HF. The area formerly covered by the silver finger is located right side of the image.

4 SUMMARY

In summary, high peak fluences during laser doping in particular lead to local damage and modification of the silicon surface, which in extreme cases can prevent the formation of silver crystallites during contacting by screen-printed silver paste. This occurs mainly at very small pulse spacing, which causes full-surface damage and thus prevents silver crystallite growth and contacting at all. Accordingly, an optimal combination of laser fluence and pulse spacing simultaneously provides a sufficiently high doping level and interconnected doped but undamaged areas between the pulse centers or other areas with higher damage due to excessive peak fluence, thus creating an environment in which silver crystallites can form.

5 ACKNOWLEDGEMENTS

Part of this work was supported by the German BMWK under contract 03EE1018B. The content of the publication is the responsibility of the authors.

REFERENCES

[1] International technology roadmap for photovoltaics, 14th edition, 2023
[2] M. Vaqueiro-Contreras et al., IEEE Journal of Photovoltaics 13(3) (2023) 373.
[3] M. Heilig et al., AIP Conference Proceedings 2147 (2019) 070004.
[4] K. Ohmer et al., IEEE Journal of Photovoltaics 1(2) (2011) 183.
[5] M. Kim et al., Solar Energy Materials and Solar Cells 132 (2015) 215.
[6] S. Eidelloth and R. Brendel, IEEE Electron Device Letters 35(1) (2014) 9.
[7] D. Wurmbrand, to be published

NEW PROTOTYPE IPL DEVICE WITH FULL AREA ILLUMINATION OF SHJ SOLAR CELLS FOR DRYING AND CURING

Daniel Ourinson[1], Denis Erath[1], Maurice Lion[2], Matthias Drews[3], Sebastian Tepner[3], Lukas Folcarelli[1], Andreas Lorenz[1], Florian Clement[1]
[1]Fraunhofer Institute for Solar Energy Systems ISE, Heidenhofstraße 2, 79110 Freiburg, Germany
[2]Botest Systems GmbH, Unterwittbacher Straße 28, 97892 Kreuzwertheim, Germany
[3]Asys GmbH, Benzstraße 10, 89160 Dornstadt, Germany
daniel.ourinson@ise.fraunhofer.de, denis.erath@ise.fraunhofer.de, mlion@botest.com, Matthias.Drews@asys-group.com, Sebastian.Tepner@asys-group.com, lukas.folcarelli@ise.fraunhofer.de, andreas.lorenz@ise.fraunhofer.de, florian.clement@ise.fraunhofer.de

ABSTRACT: In this work, a new prototype of an intense pulsed light (IPL) device by the company *Botest Systems* is presented. This device is used for drying and curing silicon heterojunction (SHJ) solar cells in the range of milliseconds, while conventional oven drying and curing (in furnaces) is in the range of seconds to minutes. Therefore, this IPL-based alternative can significantly increase the solar cell throughput. The IPL drying and curing process achieves power conversion efficiencies similar to those of the conventional method. Prior to this work, Schube et al. [1] have already demonstrated similar results using a different IPL device. Compared to that device, the one used in the present work is capable of illuminating larger areas, allowing M2 size solar cells to be processed at once. In the future, the illumination area can be upgraded to current standard wafer formats such as G12. This device has the potential to be integrated into an industrial SHJ production line.
Keywords: intense pulsed light, flash light anneal, silicon heterojunction solar cells, drying, curing

1 INTRODUCTION

The recent ITRPV report predicts a significant increase in PV market share for silicon heterojunction (SHJ) solar cells [2]. The drying and curing of the paste is part of the SHJ manufacturing process route. In 2020, Schube et al. [1] successfully demonstrated that industrial SHJ solar cells dried and curing using a so-called "intense pulsed light" (IPL) process [3] had similar power conversion efficiencies (η) to conventionally dried and cured samples (i.e., in a furnace). An IPL device contains flash lamps that illuminate a sample with a high power density flash whose spectrum is mainly in the visible range. Hence, this spectrum can be absorbed by a silicon solar cells, which heats the sample for the duration of the flash, typically in the range of milliseconds. Thus, the IPL method allows for very short drying and curing processes, compared to conventional drying and curing times, which are typically in the range of a seconds to minutes [3]. Thus, the IPL process has the potential to increase the throughput of solar cells and reduce the cycle time for manufacturing a solar cell. However, in Ref. [1], an IPL device is used that only allows illumination of an area of 5x7 cm^2. In contrast, the present work presents a novel prototype IPL device by the company *Botest Systems* [4] that can illuminate an area of 16x16 cm^2, allowing a solar cell of M2 size to be fully illuminated at once. Upgrading to fully illuminate larger solar cells (up to G12 size) is estimated to be relatively straightforward. Thus, such a device has the potential to be integrated into an industrial SHJ production line. The aim of the present work is to demonstrate at least similar η for industrial SHJ solar cells dried and cured with the novel IPL device as for conventionally dried and cured samples.

2 METHODOLOGY

Figure 1 shows images of the *Botest* IPL device. The centerpiece of this device are two xenon flash lamps that illuminate the sample. The lamps are powered by two driver racks and are located in a mirror-coated chamber to ensure a uniform illumination of the sample. To remove volatile organics during paste drying, compressed dry air is introduced into the sample chamber and exhausted at the opposite side of the chamber. The sample is placed on a height-adjustable chuck. The following parameters can be set via a software: lamp current (I_{LAMP}), pulse duration (t_P), pulse number (n_P), and pulse frequency (f_P). The power density of the lamp is controlled via I_{LAMP}. The power density on the sample is regulated by I_{LAMP} and the distance (d) between the lamps and the sample.

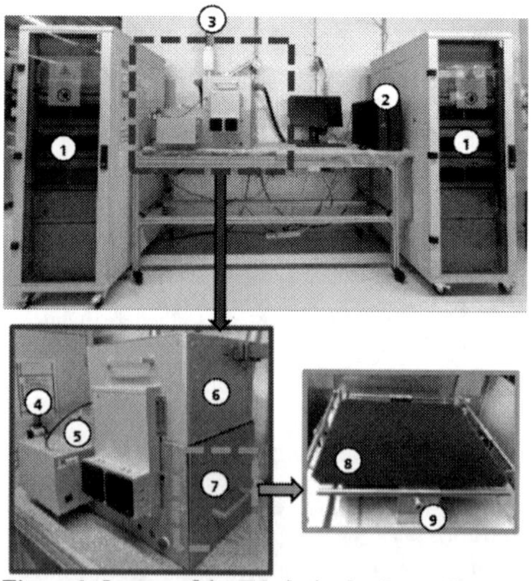

Figure 1: Images of the IPL device by *Botest Systems*. 1) Two lamp driver racks. 2) PC with software to set I_{LAMP}, t_P, n_P, f_P. 3) Exhaust. 4) Compressed dry air. 5) Chamber with photodiode for power density measurement. 6) Mirror-coated chamber with two xenon flash lamps. 7) Process chamber. 8) Sample. 9) Height-adjustable chuck.

10.4229/EUPVSEC2023/1BV.5.38
020036-001

In this work, only the drying and curing process is varied. The rest of the SHJ process steps are kept similar and conventional. Industrial M2 size SHJ precursors are metallized on the rear and front side by conventional silver pastes.

In the first step, the implied open circuit voltage (iV_{OC}) is measured for all samples before metallization via modulated photoluminescence [5]. Only the samples with similar iV_{OC} before metallization and curing are selected for further processing, since different initial lifetime levels may affect the lifetime gain due to curing. In this context, it is ensured that the lifetime of the solar cell batch used is improved by the curing process. Here, the conventional drying and curing are conducted in a conveyor belt furnace at a temperature of $T = 200$ °C and a process time of $t = 1$ min, respectively, $t = 10$ min. In a second step, the goal is to find the best IPL parameter set for the curing process by sweeping I_{LAMP}, t_P and d, and comparing the resulting current-voltage (I-V) parameters with conventionally cured solar cells, while keeping the drying step conventional for both groups. In a third step, the solar cells are completely dried and cured using the IPL method and compared with the conventionally processed samples in terms of the I-V results. To find the best parameter set for the IPL drying process, a parameter sweep is conducted again, taking the best parameter set for the IPL curing process as the initial condition. Tape tests are performed to ensure proper drying conditions.

Table 1: Different drying and curing conditions used in this work. "I_{LAMP} (RS 1)" contains I_{LAMP} after rear side printing. "I_{LAMP} (FS 1)" contains I_{LAMP} of the first flash after front side printing. "I_{LAMP} (FS 2)" contains I_{LAMP} of the second flash after front side printing. "Conv" stands for the conventional method.

Name	Definition	I_{LAMP} (RS 1)	I_{LAMP} (FS 1)	I_{LAMP} (FS 2)
Conv	Conventional drying and curing	-	-	-
Conv/ IPL_1	Conventional drying, IPL curing	-	I_{OPT}	-
IPL_2	IPL drying and curing	I_{OPT}	I_{OPT}	I_{OPT}
IPL_3		I_{OPT}	I_{OPT} - 5 A	I_{OPT}
IPL_4		I_{OPT}	I_{OPT}	I_{OPT} + 5 A
IPL_5		I_{OPT}	I_{OPT} + 4 A	-

3 RESULTS AND DISCUSSION

It is positive to note that no technical problems, such as deflagration of the solvent, have occurred during processing. The best parameters for the IPL curing process that give the highest η are: $I_{LAMP} = I_{OPT}$, $t_P = 200$ ms and $d = 1$ cm. This parameter set results in a similar η to conventionally cured samples (see Figure 2). Conveniently, the best parameter set for the IPL drying process is found to be the same as for the IPL curing process. Regarding the best sequence for the IPL route, the first IPL flash is conducted on the sample with its rear side facing the lamps after printing the rear side grid, followed by a second and a third IPL flash conducted on the sample with its front side facing the lamps after printing the front side grid (corresponding to "IPL3" in Table 1 and Figure 2). The first IPL flash acts primarily as the rear side drying step, the second flash as the front side drying and the third flash as the curing step, although both drying and curing can and probably do occur during the same flash. Tape tests indicate proper drying conditions for the chosen IPL parameter set.

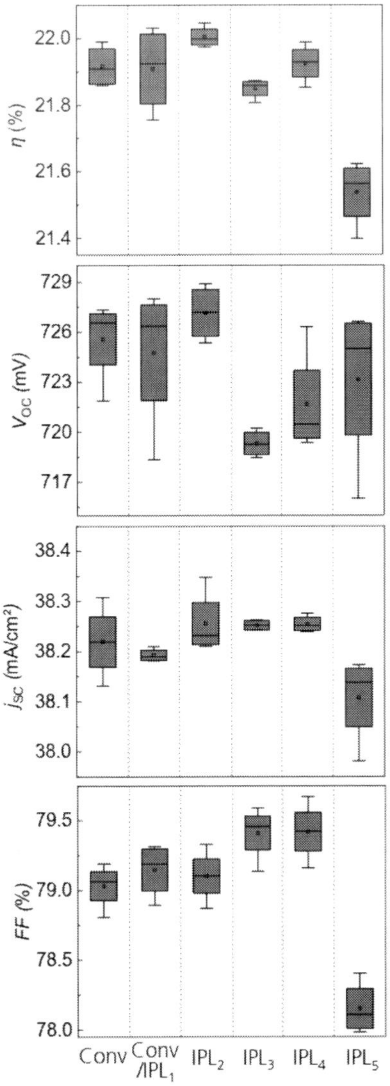

Figure 2: I-V results of the IPL-dried and -cured SHJ solar cells (boxplots 3-6) compared to conventionally dried and IPL-cured (boxplot 2) as well as conventionally dried and cured samples (boxplot 1). Here, η, the open circuit voltage (V_{OC}), the short circuit current density (j_{SC}) and the fill factor (FF) are displayed. The different methods of drying and curing shown in the x-axis are described in Table 1.

As a next step, the solar cell interconnection and module integration should be investigated for IPL-processed SHJ solar cells. Further, Cu pastes should be investigated by IPL-drying and curing, as well. Finally, further technologies like perovskite-tandem devices should be investigated.

4 ACKNOWLEDGEMENTS

This work was supported in part by the German Federal Ministry for Economic Affairs and Climate Action within the research project "Rock-It" (0324306B). Fraunhofer ISE would like to thank all co-workers involved in this contribution.

5 REFERENCES

[1] J. Schube, L. Tutsch, T. Fellmeth, F. Feldmann, M. Weil, A. Harter, J.-I. Polzin, M. Bivour, R. Keding, and S. W. Glunz, "Intense pulsed light in back end processing of solar cells with passivating contacts based on amorphous or polycrystalline silicon layers," *Solar Energy Materials and Solar Cells*, vol. 216, p. 110711, 2020.

[2] VDMA, "International Technology Roadmap for Photovoltaic (ITRPV): 2022 Results," 2023.

[3] T. Druffel, R. Dharmadasa, B. W. Lavery, and K. Ankireddy, "Intense pulsed light processing for photovoltaic manufacturing," *Solar Energy Materials and Solar Cells*, vol. 174, pp. 359–369, 2018.

[4] *ASYS Group: The ASYS Group*. [Online] Available: https://www.asys-group.com/en/company/the-asys-group. Accessed on: Jan. 29 2023.

[5] Johannes A. Giesecke and Wilhelm Warta, "Microsecond carrier lifetime measurements in silicon via quasi-steady-state photoluminescence," *Progress in Photovoltaics*, vol. 20, pp. 238–245, 2012.

PLASMA TEXTURING OF MULTIPLE KERF-LESS SEPARATED ULTRA-THIN SILICON LAYERS

Alena Okhorzina[1], Richard Schalinski[1,2], Norbert Bernhard[1], Stefan L. Schweizer[2]

[1]Department of Electrical, Mechanical and Industrial Engineering, Anhalt University of Applied Sciences, Bernburger Str. 55, 06366 Koethen, Germany
[2]Institute of Physics, Martin Luther University Halle-Wittenberg, 06099 Halle, Germany

ABSTRACT: Kerf-less separated ultra-thin silicon layers are an interesting future perspective for ultra-thin silicon solar cells, but difficult to texture by the classical wet-chemical processes, because the initial micro-roughness of the surface due to the sawing process is missing. In this investigation it is demonstrated that multiple kerf-less separated ultra-thin silicon layers can be textured by a mask-less plasma etching process. The silicon layers were separated by the so-called millefeuille process [1] in which modulated photo-electrochemically etched macro-pores grow in a subsequent tempering process which leads to separation of several ultra-thin silicon layers in one single step. Afterwards they were plasma-textured in a laboratory plasma etch system in which by superposition of a capacitively and inductively coupled plasma the two plasma parameters ion density and ion energy could be separately controlled to a certain extent. It was shown that plasma texturing is possible for these ultra-thin films.

1 INTRODUCTION

The current industrial main stream process of the separation of ingots to slices (wafering) for crystalline silicon solar cells is diamond-wire sawing. This process causes a micro-rough and damaged surface, which afterwards is usually wet-chemically etched in order to remove the saw damage and create effective light-trapping structures for reflectivity reduction and increasing the effective mean path length of the incident light. Since the relative material loss of sawing increases with decreasing wafer thicknesses, and also the process itself is limited, if wafer thicknesses become too low due to mechanical stability reasons, several attempts in investigating so-called kerf-less separation techniques (i.e. without loss of Si material) have been made.

Besides hydrogen implantation (SmartCut [5]), deposition of a ductile metal (SlimCut [6]), the photo-electrochemical growth of pores (PorousCut [7, 8]) is a promising technique. By Hernandez and co-workers it was shown that the PorousCut technique could be run in such a way that several layers could by separated at the same time [1], a process which could improve the productivity in case of a possible future volume production. Since the layered structure resembles puff pastry (in French millefeuille) the authors called it millefeuille process.

In this process – as in many other kerf-less separation processes – the surfaces of the separated layers are very smooth and difficult to texture wet-chemically. Therefore the target of this research project was to investigate whether plasma etching would be a suited method to texture these ultra-thin silicon layers such effectively that they could be used for photovoltaic applications.

2 EXPERIMENTAL

For this investigation the millefeuille process of Hernandez and co-workers [1] was adopted (see figure 1). This process is based on photo-electrochemical etching (PECE) [2], but uses a modulation of voltage and photocurrent to create periodically wide and narrow pores. In a subsequent anneal step under hydrogen or inert gas atmosphere the pores re-arrange by Si diffusion. Under appropriate conditions the wide pores grow together and several thin silicon layers are separated.

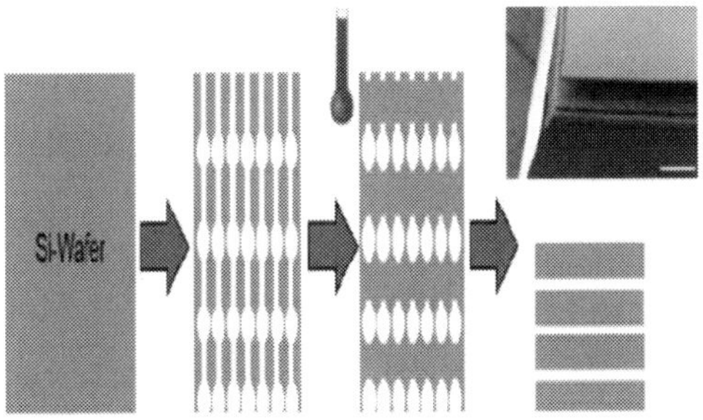

Figure1: Schematic process flow of the millefeuille process which creates several ultrathin smooth Si layers in one single process sequence. Insert top right from [1], length of scale bar is 100 µm.

These thin layers with rather smooth surface are then transferred (using an adhesive plasma resistant and chemically stable Kapton tape) to a plasma etch system and textured by a mask-less plasma etch process (the so-called black-silicon process [3, 9, 10]). In this investigation we focus on the plasma etching process. The technological details of the adapted millefeuille process will be described in another paper which is in preparation. An Oxford Instruments laboratory plasma etch system with capacitive and inductive plasma coupling was used for the texturing process. Both coupling modes CCP (capacitively coupled plasma) and ICP (inductively coupled plasma) can be either used independently or in superimposed mode.

Capacitive coupling always generates a self-bias of several ten to several hundred volts due the much higher electron mobility in comparison to the ion mobility. This self-bias creates a certain ion bombardment to the etched surface which is urgently necessary for pattern transfer in reactive ion etching (as used in microelectronics manufacturing). On the other side, this ion bombardment

10.4229/EUPVSEC2023/1BV.5.39

can create plasma damage near the surface of the sample under treatment and possibly enhance surface recombination velocities in solar cells.

In an inductively coupled plasma there is no self-bias, only the always present slight positive potential of a few volts of the plasma relative to grounded chamber walls and substrate holder. Since here (as usual for solar cell texturing) we use a mask-less self-patterning process, the ion bombardment is not essentially necessary, if a moderate aspect ratio of the textured structure (similar to that in wet chemical texturing) is sufficient. For higher aspect ratios a certain extent of ion bombardment is also necessary in mask-less texturing. Here a combined CCP and ICP process is advantageous over a pure CCP process, because at a still productive etch rate the plasma damage can be minimized, since ion density (and therefore also chemical radical density and etch speed) can be controlled independently of the ion energy (which is determined by the self-bias and therefore CCP power).

The advantage is that the reflectivity of the silicon surface is more reduced than in a pure ICP process which is already visible with the naked eye, as can be seen in figure 2 where a polished mono-crystalline c-Si reference wafer was etched with a pure ICP process vs. a combined ICP and CCP process. The diameter of roughly 5 cm corresponds to the diameter of the ceramic tube through which the coil couples the HF field to the plasma. A coil with a bigger diameter in a possible industrial etch machine could create also a larger uniformly etched area.

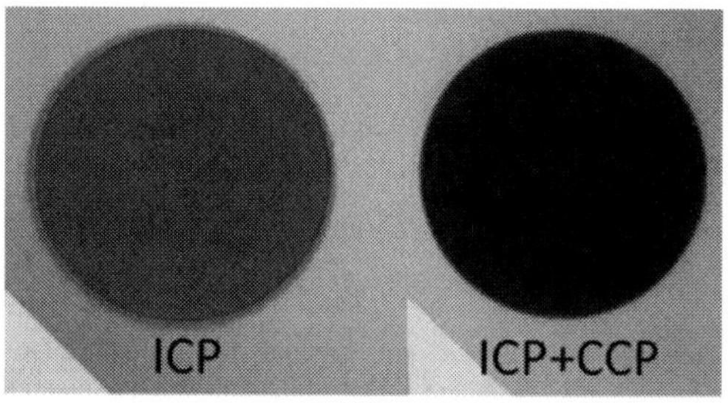

Figure 2: Comparison of the reflection reduction of a pure ICP (inductively coupled plasma) process with a process of superposed ICP and CCP (capacitively coupled plasma) on a polished mono-crystalline Si reference wafer.

In a pure CCP process the self-bias (determining the ion energy of bombarding ions) and the plasma density (determining the amount of ions and chemically etching radicals) are inherently connected to each other, by applying a superposition of CCP and ICP both parameters can be independently changed, opening a greater window of possible process results and especially aspect ratios of the texturing. A high CCP contribution gives very high aspect ratios (almost needle-like structures), a fair balance between CCP and ICP gives moderate aspect ratios (similar to wet chemical etching) [3, 4, 10, 11, 12]. For a pure CCP process a conformal deposition of a passivation layer might be a challenge, too. Therefore in this investigation we focused on combined CCP and ICP, and pure ICP processes. For both a SF_6/O_2 chemistry in the range of a few 10 µbar pressure and substrate temperature slightly below room temperature at a frequency of 13.56 MHz was used. For the pure ICP process a much higher HF power than for the combined process was needed. The fluorine radicals created in the plasma are the etching agents, whereas oxygen provides by oxidation of Si a certain random-induced sidewall and surface passivation to create the texturing. Before starting experiments on the millefeuille samples a process optimization on crystalline Si reference wafers was performed.

3 RESULTS AND DISCUSSION

After optimization on Si reference wafers, the plasma etch processes were applied to thin foils created by the millefeuille process. In order to generate the multiple thin layers with typical thicknesses in the range of 5 to 20 µm several process optimization steps, especially of the tempering step were necessary. Details of this optimization will be described in another publication. A separation of up to four to six layers at the same time was reached, although the separation and layer quality was not yet perfect across larger areas. Main problems were remaining pores in some layers (figure 3) and also thin connection bridges below neighbouring layers (figure 4). But all the same, the principal feasibility of the process was shown.

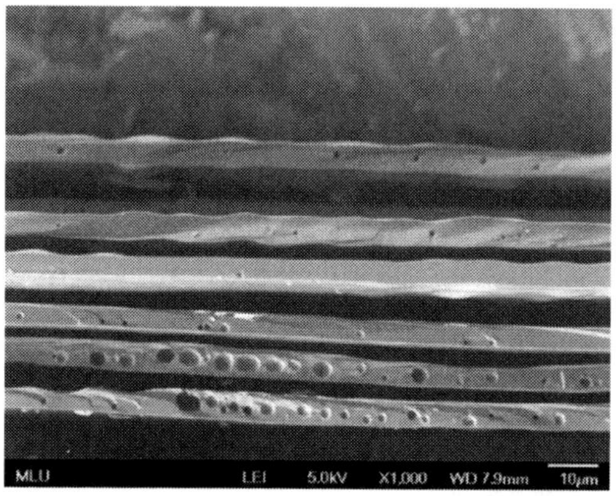

Figure 3: SEM cross section of several separated layers of a few µm thickness, but with remaining pores in some layers.

Figure 5 shows one of the best results, the separation of several layers as it was reached after several optimization steps of the annealing step, the process parameters of which turned out to be very critical for a good separation and good layer quality. As also visible in figure 5, the surfaces of the layers are rather smooth, which means that a wet-chemical process which needs an initial roughness as points of etch attack would not work well. But it is known that plasma etching works almost independent of crystallographic orientation and also for perfectly smooth wafers [3, 4]. Therefore, plasma etch processes which were already widely investigated on wafers of normal thickness with promising performance [3, 4, 10, 11, 12] were applied to the thin millefeuille layers. For this purpose, a special sample holder for the plasma etch chamber was designed, by which the millefeuille layers were transferred on a tape (Kapton) to the plasma etch chamber.

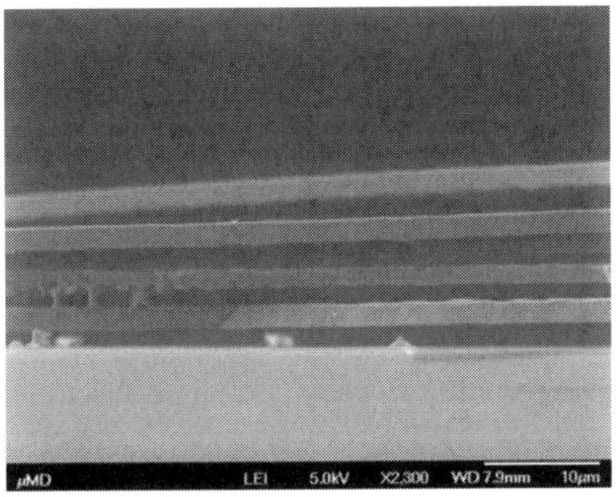

Figure 4: SEM cross section of four created layers, but with incomplete layer separation between the 3rd and 4th layer (from top).

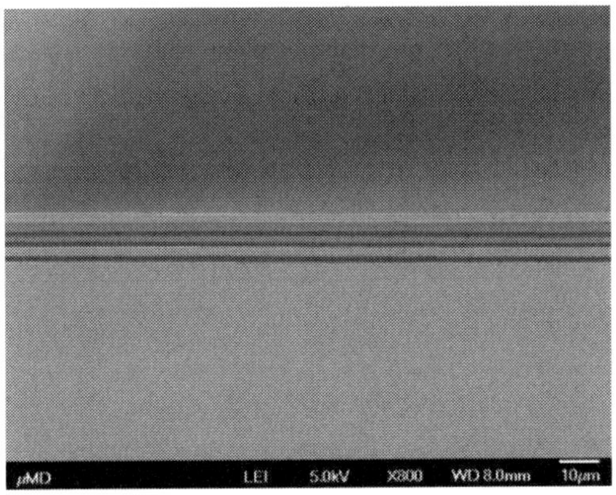

Figure 5: SEM cross section of a multiple layer generated by the millefeuille process. A good layer separation and a good lateral homogeneity is visible.

With a combined ICP and CCP process an aspect ratio of more than one with typical sizes in the range of 100 to 200 nm could be reached, a texturing which is known to act as a very effective light trapping structure, as studied previously on thicker layers [3, 4]. Figure 6 shows the SEM cross section of such a textured sample with ICP and CCP excitation superimposed. Such a texturing with moderate aspect ratio was chosen, because this is a good compromise between reflectivity reduction and a reasonable charge carrier lifetime after surface passivation [4]. Figure 7 shows the same etch recipe on another sample to illustrate the size ratio of the texturing to the whole sample thickness.

Samples were also etched with a pure ICP plasma as found after the initial optimization process. Although on crystalline Si reference wafers a clear reflectivity reduction was observed (cf. figure 2), in the SEM pictures of such treated millefeuille layers only very slight texturing was observed (figure 8). This might be due to the fact that the texturing was so fine with such low aspect ratios that it could not be resolved well in the SEM pictures. This issue is open for further clarification.

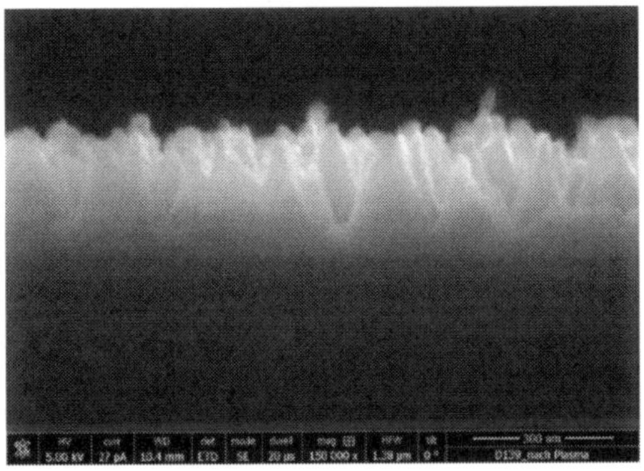

Figure 6: SEM cross section of the plasma textured surface (ICP + CCP process). Clearly visible the reflection reducing surface topography with structures in the range 100 to 150 nm width and roughly 200 nm height.

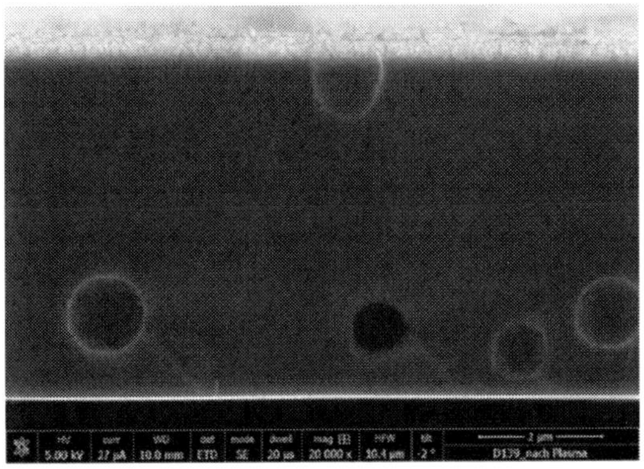

Figure 7: SEM cross section of the plasma texturing (ICP + CCP process) in comparison to total layer thickness of about 5 µm, and with some remaining pores.

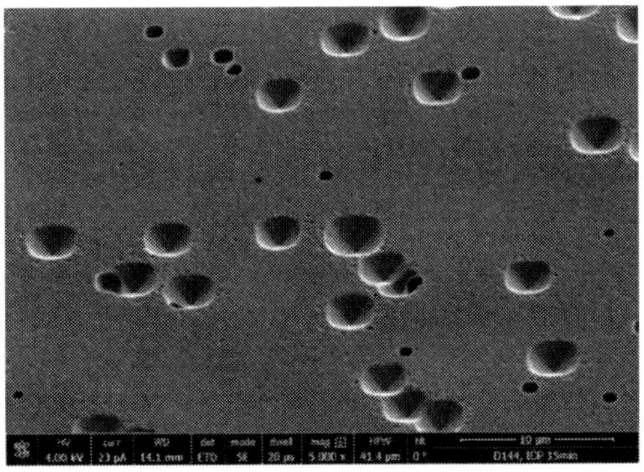

Figure 8: SEM view inclined of top of sample textured with pure ICP recipe. The big craters are probably remaining pores. Only a slight texturing of the area between the pores is visible.

An alternative or complementary explanation could be

the fact that the pure ICP etching process for this setup and process parameters requires a significant stabilization time of the process, and the etch rate in the middle of the sample may be less than at the edges. This was noticeable during process optimization on a polished mono-crystalline Si wafer. Unfortunately we could not prepare an SEM image of the edge of the sample (where we would had expected a slightly stronger texturing), as the outer borders of the sample were badly damaged during transportation. This topic needs further clarification, too.

4 CONCLUSIONS

It was shown that multiple kerf-less separated layers from the millefeuille process can be textured by a combined capacitively (CCP) and inductively coupled plasma (ICP) etch process. The big advantage of plasma etching over wet chemical etching is that the process works also for very smooth surfaces as created by the millefeuille process or other kerf-less separation techniques. In addition, by balancing the CCP to ICP excitation mode the aspect ratio of the textured structures can be controlled. The challenge for a possible future industrial application is certainly the handling of the very thin foils in an automatic transfer process from the anneal furnace where the multiple separated layers are created, to the plasma etch chamber. In this investigation only small areas in the range of one cm^2 were transferred. The millefeuille process itself also needs still optimization with respect to unwanted remaining pores and incomplete separation of neighboring layers. But the principal feasibility of both the separation process and the plasma texturing was shown.

Acknowledgement: Funding by the Investitionsbank Sachsen-Anhalt and the European Union within the EFRE program (funding no. ZS/2016/12/83145) is gratefully acknowledged.

References:

[1] D. Hernandez, T. Trifonov, M. Garin, and R. Alcubilla., Appl. Phys. Lett. **102**, 172102 (2013)
[2] T. Geppert, S.L. Schweizer, U. Gösele, and R.B. Wehrspohn., Appl. Phys. A **84**, 237-242 (2006)
[3] M. Gaudig, J. Hirsch, Th. Schneider, A.N. Sprafke, J. Ziegler, N. Bernhard, and R.B. Wehrspohn, J. Vac. Sci. Technol. A 33, 05E132 (2015)
[4] J. Hirsch, M. Gaudig, N. Bernhard, and D. Lausch, Applied Surface Science, 374, 252 – 256 (2016)
[5] F. Henley, S. Kang, Z. Liu, L. Tian, J. Wang and Y.-L. Chow, Proc. 34th IEEE Photovoltaic Specialists Conference (2009)
[6] F. Dross, A. Milhe, J. Robbelein, I. Gordon, P.-O. Bouchard, G. Beaucarne and J. Poortmans, Proc. 33rd IEEE Photovoltaic Specialists Conference (2008)
[7] C.S. Solanki, R.R. Bilyalov, J. Poortmans, J. Nijs and R. Mertens, Sol. En. Mat. and Solar Cells **83**, 101-113 (2004)
[8] R. Brendel und M. Ernst, Phys. Stat. Sol. RRL **4**, pp. 40-42 (2010)
[9] H. Jansen, M. de Boer, R. Legtenberg, M. Ellwenspoek, J. Micromech. Microeng. **5**,, 115- 120 (1995)
[10] M. Gaudig, J. Hirsch, J. Ziegler, T. Schneider, M. Werner, A. Sprafke, N. Bernhard; R. B. Wehrspohn, Proc. 29th EU PVSEC, 885 (2014)
[11] J. Hirsch, M. Gaudig, M. Gläser, S. Großer, N. Bernhard, D. Lausch, Proc. 31st EU PVSEC 662 (2015)
[12] J. Hirsch, M. Gaudig, M. Gläser, M. Werner, S. Großer, N. Bernhard, D. Lausch, Proc. 32nd EU PVSEC 722 (2016)

40th European Photovoltaic Solar Energy Conference and Exhibition

This presentation was selected by the Sc. Committee of the EU PVSEC 2023 for submission of a full paper to one of the EU PVSEC's collaborating peer-reviewed journals.

A SUSTAINABLE AND FAST SILICON TEXTURIZATION PROCESS DRIVEN BY LIGNOSULFONATES

Rocío Barrio (CIEMAT, rocio.barrio@ciemat.es,);
Ignacio Torres (CIEMAT, ignacio.torres@ciemat.es); Nieves González (CIEMAT, nieves.gonzalez@ciemat.es);
Guillermo Godoy (CIEMAT, ggodoy@ucm.es); Ross Ellis (BORREGARD; ross.ellis@borregaard.com)
CIEMAT, Avda. Complutense 40, 28040 Madrid, Spain
BORREGARD, Hjalmar Wessels vei 6, 1721 Sarpsborg, Norway

ABSTRACT: High-efficiency solar cells based on crystalline silicon continue to be the leading technology in the photovoltaic (PV) market. To achieve an energy model based on renewable energies, a large volume of PV installations is expected in the coming years. To make the PV industry more sustainable and oriented towards a circular economy, minimizing waste and promoting recycling is imperative. In search of this goal, we propose an innovative, cost-effective, and environmentally sustainable method for texturing silicon wafers. We explore the utilization of a solution containing NaOH and lignosulfonates (LS) without isopropyl alcohol (IPA). Lignin and lignosulfonates are byproducts of the paper industry that can be purified and modified to create valuable performance additives for use in the photovoltaic sector.

Our preliminary results with LS are very promising, producing highly homogeneous pyramidal surfaces with reflectance comparable to the traditional methods. We have observed that the process could be further improved with the addition of small quantities of additives to the solution. Overall, the proposed process is faster and less toxic that procedures with IPA and supports the sustainability and the circular economy in the PV industry.

Keywords: texturization, silicon, lignosulfonates, circular economy, sustainability

1 INTRODUCTION

Climate change is one of the most profound challenges we urgently face globally. Addressing this objective will require substantial reductions in carbon emissions across all productive sectors, along with a rapid transition to a net-zero carbon emissions energy system based on renewable energy. The European Commission had set ambitious targets for renewable energy deployment, including photovoltaic (PV) installations, as part of their broader energy and climate goals. According to the integrated National Energy and Climate Plan (NECP) [1] of the different European country members, nearly 600 GW of PV installation are planned by the end of this decade, with Spain expected to contribute 76 GW to this total [2]. Due to this spectacular growing of the PV installations in the coming years, it is crucial to make the photovoltaic industry more sustainable and focused on the circular economy, and to implement measures in the mass production of solar cells that minimize waste and promote the use of byproducts and the recycling of their components.

An example of these improvements in the production of more sustainable solar cells is addressed in the presentation of this work at EU PVSEC 2023 held in Lisbon. It is summarized in Figure 1, representing the continous flow of waste from paper industry, which, once treated and purified can be incorporated into texturing process in a circular economy. The center of this loop illustrates the possible photovoltaic modules produced after the texturing process that we propose. These processes will be detailed further in the continuation of this text.

One of the key steps in the fabrication of high-efficiency silicon solar cells (c-Si) is the controlled pyramid texturing of silicon substrates to minimize reflection with a subsequent passivation strategy to reduce recombination losses. Conventional texturization processes are based on anisotropic etching from aqueous solutions of NaOH or KOH, deionized water (DIW) and isopropyl alcohol (IPA), the latter acting as a wetting agent [3]. However, IPA entails serious concerns, due to IPA being a toxic, volatile (that limits the process temperature or to maintain a constant concentration of the solution), and relatively expensive compound. In order to replace it, we explore using a more sustainable solution containing NaOH and lignosulfonates for texturizing crystalline silicon wafers [4].

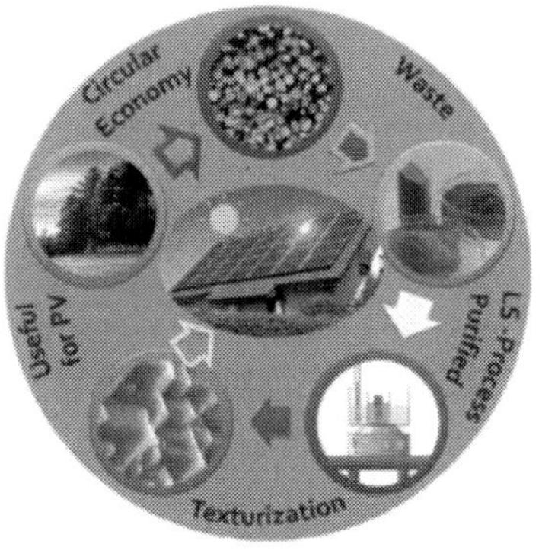

Figure 1. Explanatory Chart: circular and sustainable silicon wafer-based solar cell manufacturing process with lignosulfonates.

Lignin and lignosulfonates are waste products of the pulp and paper industry that are commonly incinerated or sold as crude mixtures that can be purified and modified to generate valuable performance additives that can be used in a variety of industries. It is important to note that approximately 70 million tons of lignin are produced annually. Thus, its utilization for other purposes is highly beneficial from an environmental perspective, even more so if it can be employed in the manufacturing of PV modules as part of a renewable energy transition. By replacing IPA with lignosulfonates, we would be reducing a major contaminant of the PV industry.

The purpose of this study is to evaluate this cost-effective texturing technique using lignosulfonates and an alkali solution to create adequate pyramidal structures on crystalline silicon wafers with size next to 10 microns. The aim is to enhance the optical trapping and overall performance of solar cells while ensuring a higher level of sustainability and reduced environmental impact compared to conventional texturing methods that rely on IPA. As texturizing is a sensitive process, it is very important to use lignosulfonates with low levels of impurity and tight specifications to ensure that the process runs consistently.

This research work is the result of the current collaboration between a Norwegian company, Borregaard and a Spanish public Research Centre, CIEMAT (Centro de Investigaciones Energéticas, Medioambientales y Tecnológicas). Borregaard is a world leader in biorefining and lignin polymer technology, offering a wide range of purified and modified lignosulfonate products. The lignin biopolymers are produced from a single wood type at the biorefinery with strict specifications. Borregaard purifies and modifies the crude lignin materials to produce a range of specialized lignin biopolymers with low levels of impurity and consistent and specific molecular character to enhance their properties. CIEMAT is a technological research center focused on applied research on energy and environment. CIEMAT, in this work, is testing the texturization process using the lignosulfonates provided by Borregard, and is involved in developing and characterizing materials and devices for photovoltaic solar cells.

2 EXPERIMENTAL DETAILS

2.1 Samples and texturing process

Our experiments focused on the study of lignosulfonates to texture silicon wafers. We examined lignosulfonate concentration ranging from 5 ppm to 3200 ppm and evaluated the effect of different etching temperatures and times. In addition, we explored the effect of using additives together with the lignosulfonates to improve texture quality, to control the size of the obtained pyramids and to enhance solution reproducibility and 'shelf-life'.

All the experiments were conducted using small pieces (< 6 cm^2) of (100)-oriented n-type monocrystalline silicon wafers (c-Si) with a thickness of 280 µm. The processes were carried out in a small beaker containing the specified solutions. The beaker was placed in a water bath, and the temperature of this bath was controlled with a hot-plate and monitored with a temperature probe. The solutions were agitated using a magnetic stirrer and the beaker was covered to prevent water evaporation. Once the etching time had elapsed, samples were removed from the solution and rinsed with DIW to halt the chemical reaction.

2.2. Methods and characterization

To characterize the morphology of the surfaces, we individually checked all the textured samples under an optical microscope and took SEM images (Scanning Electron Microscope) of a few selected samples. Lastly, we measured the reflectance of all samples using a Perkin Elmer spectrophotometer equipped with a 6 cm integrating sphere, and the average reflectance was estimated from 400 to 1100 nm to analyze the light-trapping effect of the resulting surfaces.

After the texturing process, samples were cleaned with a standard cleaning process (SC also called RCA1+RCA2) [5], and a final HF etching. Finally, high quality intrinsic and n-doped hydrogenated amorphous silicon thin layers (a-Si:H(i) and a-Si:H(n) respectively), were symmetrically deposited by Plasma Enhanced Chemical Vapour Deposition (PECVD) on both sides of the surfaces (Figure 2) to reduce recombination losses and assure long minority carrier lifetimes. This was done on both, polished and textured wafers, to analyze the quality of the a-Si:H/c-Si interface after texturing. To characterize the passivation of these interface and the validity of the texturing process, the implied open-circuit voltage (i-Voc at 1 sun) and the effective minority-carrier lifetime (τ at $N_D = 10^{+15}$ cm^{-3}) was extracted using the quasi-steady-state photoconductance (QSSPC) technique developed by Sinton and Cuevas with a commercial setup (Sinton Consulting WCT-120) [6].

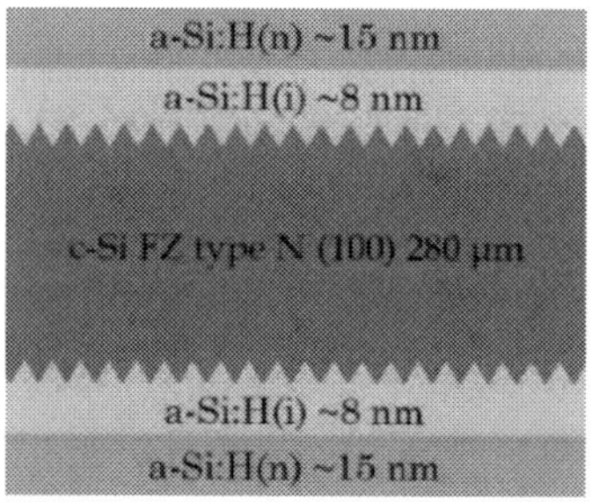

Figure 2. Schematic of the a-Si:H(n)/a-Si:H(i)//c-Si heterostructures used to assess the passivation quality and the quality os the textured wafers.

3 RESULTS AND DISCUSSION

The texturisation process of silicon wafers depends strongly on the etching time, the temperature and the concentration of the alkaline solution. In a previous work we optimized these parameters for conventional alkaline solutions including IPA. [3]. In this proposal, the incorporation of lignosulfonates into the solution as a wetting agent instead of IPA involved investigating various concentrations (ranging from 5 to 3200 ppm), along with adjustments in etching times (ranging from 10 to 60 minutes) and process temperature (ranging from 70 to 90 ºC). These parameters were systematically varied in order to find their optimun values. The concentration of NaOH was 5% (w/w) for all the experiments.

For most of the conditions tested, optical microscope images revealed a very homogeneous texturization when using NaOH and lignosulfonates over the whole area, with no shiny parts indicating incomplete texturing. The addition of small amounts of lignosulfonate (<3200 ppm) appears to accelerate the etching process drastically, with processing times as little as 10 minutes. In our standard IPA process at 80°C [3], processing times less than 45-60 minutes resulted in incomplete texturing. Thus, processes involving lignosulfonates are markedly faster. However, the resulting pyramids are smaller than samples etched with IPA and similar concentration of NaOH.

A more concise analysis of the morphology was

carried out using SEM images, revealing that the base dimensions of the obtained pyramids ranged from <1 µm to approximately 4-5 µm across a wide range of tested lignosulfonate concentrations (Figure 3-A). Larger pyramids were not observed with extended etching times or using higher concentrations of lignosulfonates in the solution. However, when additives, specifically sodium carbonate (Na_2CO_3) and para-toluene sulfonic acid (PTS), were introduced in conjuction with lignosulfonates, it led to the formation of larger pyramids with bases reaching approximatly 10 µm (Figure 3-B) [7]. The addition of PTS alone as an additive led to the formation of smaller pyramids on the surfaces (Figure 3-C), whereas the incorporation of only Na_2CO_3 as an additive did not induced any noticeable morphological changes in the pyramid size (Figure 3-D). Therefore, it appears necessary to incorporate both additives to effectively control pyramid size and achieve suitable surface homogenization on the wafers. To achieve this goal required a comprehensive study of the concentration of both additives and etching times to ensure uniformity and prevent the formation of deteriorated pyramids or shiny zones.

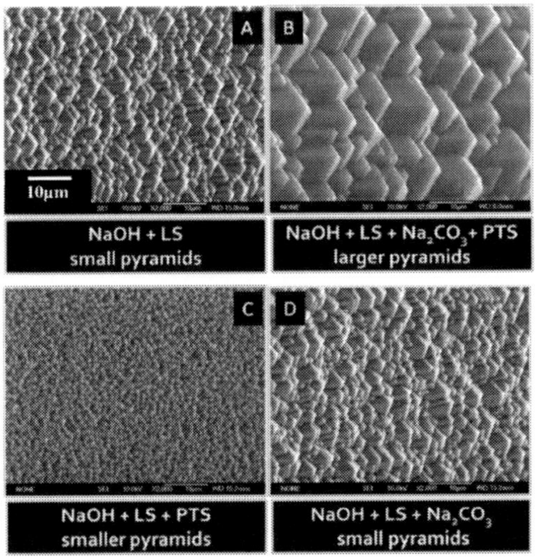

Figure 3. SEM images of silicon wafer textured using a solution of 5% NaOH and 400 ppm of lignosulfonates (LS) at 80 °C. (A) Without additives, (B) with both additives, (C) using only PTS as an additive, and (D) only Na_2CO_3 as an additive. The scale of the four SEM-images is the same for all of them, indicating a value of 10 µm.

Regarding the optical behavior (*Figure 4*), the hemispherical reflectance from 400 nm to 1100 nm (spectral range of interest for photovoltaic cells based on silicon wafers) was reduced from an initial average value of 40% (in the case of polished wafers) to values between 11 and 20% on average (for textured wafers), depending on the concentration of the additives.

The best results, both morphologically and optically, were achieved with the following optimized solution and etching conditions: 5% NaOH, 400 ppm of lignosulfonates, 3% Na_2CO_3 and 5 % PTS, at 80 °C during 20 minutes. These results are comparable to values obtained from samples etched using the standard process with IPA, resulting in an average reflectance of approximately 11-13% (Figure 4). The reflectance value is expected to decrease to an average of 3% with the deposition of an antireflective coating on the textured surface. These optical results confirm that the wafer texturing process with lignosulfonates can effectively enhance the optical trapping of incident photons, thereby improving the efficiency of silicon-based solar cells.

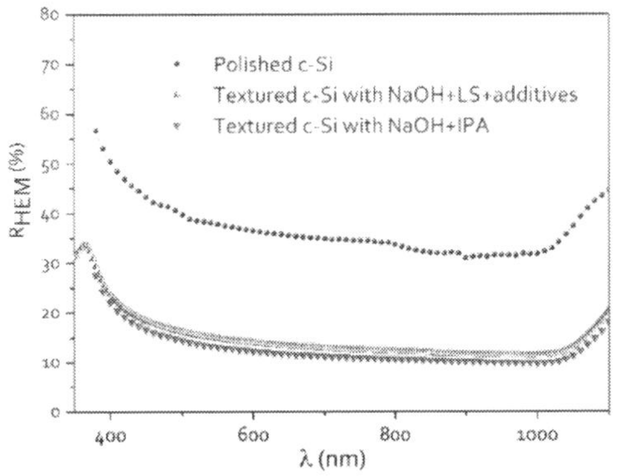

Figure 4. Hemispherical reflectance spectra before and after texturization with NaOH and Lignosulfonates (our proposal in this work) and with NaOH and IPA (conventional process), in comparison with a polished wafer.

Upon optimizing the texturing process with lignosulfonates, it is imperative to analyze the influence of this process on the passivation of the textured wafers. After cleaning treatments, the layers of amorphous silicon were deposited by PECVD as schematically shown in Figure 2 in a textured and in a polished wafer simultaneously. The passivation results were highly satisfactory, with lifetimes approaching 2 ms and open-circuit voltages exceeding 700 mV (Table I).

Table I. Lifetime of minority carriers and implied open circuit voltage of wafers with no standard cleaning process (polished wafer) and after standard cleaning (polished and textured wafers) with the same a-Si:H thin layers.

c-Si wafer	τ (µs)	i-V_{oc} (mV)
Polished	8.450	736
Polished + SC	4.179	716
Textured + SC	1.920	709

According to these results, the texturing procedure is a viable option in the fabrication of monocrystalline silicon solar cells, particularly in HIT (Heterojunction with Intrinsic Thin layers) solar cells [8]. However, the differences in τ and iVoc values measured on the textured and polished wafers after cleaning, and the values achieved without cleaning, suggest the need to optimize the standard cleaning process. The polished wafers have already been cleaned previously through standard cleaning processes by the wafer manufacturer.

4 CONCLUSIONS

The texturization process driven by lignosulfonates for (100)-oriented monocrystalline silicon wafers was achieved though collaboration between Borregard and CIEMAT. The results are highly promising, thanks to its remarkable homogeneity, rapid etching speed and favorable optical trapping achieved. Furthermore, the high minority carrier lifetime attained after passivation indicates that a high quality surface is maintained after texturization. It appears that the use of additives plays a crucial role in controlling pyramid size.

Based on our investigations thus far, we can confidently assert that lignosulfonates have the potential to replace IPA, simplifying and expediting the texturization process. With only a small concentration of lignosulfonates in the solution, this approach also offers a more sustainable alternative to traditional methods. It holds great appeal for adoption on a larger scale within the photovoltaic industry, especially for silicon-based technology.

We believe that this research could be of significant interest to manufacturers seeking to incorporate sustainable solutions into their silicon photovoltaic cell production chain and contribute to the pursuit of an optimal circular economy. That includes reducing waste, conserving and ensuring the avalability of resources in the long term, and minimizing the carbon footprint compared to traditional manufacturing methods.

5 ACKNOWLEDGEMENTS

The authors would like to thank the CNME for SEM measurements, and the student G. Godoy for the development of his Master's Degree Project on this topic.

6 REFERENCES

[1] M. B. Nieto-Morone, M. C. Alonso-García, F. G. Rosillo, J. D. Santos, and M. A. Muñoz-García, "State and prospects of photovoltaic module waste generation in China, USA, and selected countries in Europe and South America," *Sustain. Energy Fuels*, vol. 7, no. 9, pp. 2163–2177, 2023, doi: 10.1039/d2se01685k.

[2] Ministerio para la transición ecológica y el reto demográfico, "Borrador para la actualización del PNIEC 2023-2030," 2023.

[3] R. Barrio, N. González, J. Cárabe, and J. J. Gandía, "Optimisation of NaOH texturisation process of silicon wafers for heterojunction solar-cells applications," *Sol. Energy*, vol. 86, no. 3, 2012, doi: 10.1016/j.solener.2011.12.013.

[4] O. K. Karimov, G. A. Teptereva, I. A. Chetvertneva, E. M. Movsumzade, and E. K. Karimov, "The structure of lignosulfonates for production of carbon catalyst support," *IOP Conf. Ser. Earth Environ. Sci.*, vol. 839, no. 2, 2021, doi: 10.1088/1755-1315/839/2/022086.

[5] S. Cells, "Surface Cleaning and Passivation Technologies for the Fabrication of High-Efficiency Silicon Heterojunction," 2023.

[6] A. Cuevas and D. Macdonald, "Measuring and interpreting the lifetime of silicon wafers," *Sol. Energy*, vol. 76, no. 1–3, pp. 255–262, 2004, doi: 10.1016/j.solener.2003.07.033.

[7] Y. Zhang *et al.*, "A novel additive for rapid and uniform texturing on high-efficiency monocrystalline silicon solar cells," *Sol. Energy Mater. Sol. Cells*, vol. 222, no. January, p. 110947, 2021, doi: 10.1016/j.solmat.2020.110947.

[8] M. Taguchi *et al.*, "24 . 7 % Record Efficiency HIT Solar Cell on," *IEEE J. Photovolt*, vol. 4, no. 1, pp. 96–99, 2014.

40th European Photovoltaic Solar Energy Conference and Exhibition

This presentation was selected by the Sc. Committee of the EU PVSEC 2023 for submission of a full paper to one of the EU PVSEC's collaborating peer-reviewed journals.

CHALLENGES AND ADVANTAGES OF CUT SOLAR CELLS FOR SHINGLING AND HALF-CELL MODULES

Jonas D. Huyeng, Elmar Lohmüller, Torsten Rößler, Behnaz Shabanzadeh,
Christian Reichel, Julian Weber, Marc Hofmann, Daniel von Kutzleben,
Najwa Abdel Latif, Achim Kraft, Holger Neuhaus, Florian Clement, Ralf Preu
Fraunhofer Institute for Solar Energy Systems ISE, Heidenhofstr. 2, 79110 Freiburg, Germany
Corresponding author: jonas.huyeng@ise.fraunhofer.de

ABSTRACT: This work discusses challenges and advantages of cut solar cells, as used for shingling and half-cell photovoltaic modules. Cut cells have generally lower current output and allow reduced ohmic losses at the module level. Experimental results are collected, combining industrial blue wafers with different cell layouts, which are then implemented into a combined simulation method up to module level. These simulations accurately compare host cell and cut cell performance. Flexible methods like GridTouch® for current-voltage (*I-V*) measurements can provide quick results but can also lead to overestimation of host cell performance, resulting in large cell-to-module (CTM) losses, if not correctly interpreted. In addition, unpassivated edges introduce new losses affecting fill factor and open-circuit voltage. The passivated edge technology (PET) yields *I-V* results close to an ideal edge without recombination. Module performances are compared, highlighting the highest efficiency in a shingled module with PET.
Keywords: shingling, edge passivation, passivated edge technology, PET, solar cell simulation, module simulation

1 INTRODUCTION AND MOTIVATION

Photovoltaic (PV) modules with half-cut cells have become state of the art in the industry today [1]. Compared to full-cell modules, ohmic losses are reduced through lower generated current.

Alternative module configurations, such as shingling, have also gained attention due to their potential for further enhancing power density [2–5]. Shingling involves overlapping cut solar cells (typically 1/5th or 1/6th of a full cell), known as shingle cells, enabling the reduction of inactive area and increasing active cell area within a given module size [6, 7].

However, the process of cutting cells introduces challenges, particularly cutting losses due to unpassivated edges at the separation path. While these losses are present in half cut cells, they are compensated by the advantages at the module level and are seldomly discussed.

In this work, we expand previous findings presented by von Kutzleben *et al.* [8], comparing half-cell and shingling modules. For our study, we have created experimental results of half-cell and shingling host cells and cut cells, fabricated on industrial solar cell precursors. We demonstrate the ability to accurately simulate host cell and cut cell performance (even for unknown precursor properties) using Gridmaster [9] modelling. Also, we illustrate the critical influence of grid-neglecting contacting during current-voltage (*I-V*) measurements, which can be easily demonstrated using Gridmaster+ [10]. Additionally, we analyze the impact of cutting-induced losses through experimental results and Gridmaster+ simulations.

Our experimental realization initially shows lower performance for shingling host cells due to a more conservative metallization layout. However, through module performance simulations using SmartCalc.Module [11], we illustrate that shingling modules with edge passivation can achieve higher efficiencies on the same module area.

2 EXPERIMENTAL AND SIMULATION

2.1 Experimental details

The following flow chart illustrates the experimental realization of two types of solar cell hosts, based on the same industrial precursor, for a comparison of the different concepts. In addition, one group of shingle cells was also further processed to receive the patented "Passivated Edge Technology (PET)" [12, 13] treatment for edge passivation.

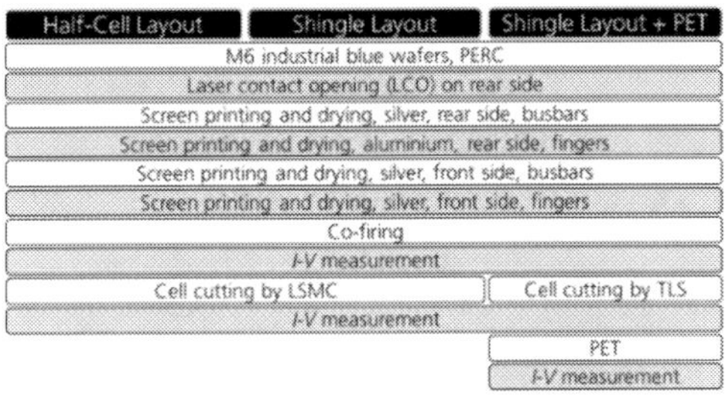

Figure 1: Flow chart of the experimental groups. LSMC: "Laser scribe and mechanical cleave", TLS: "Thermal laser separation", PET: "Passivated edge technology".

The precursor wafer format is M6 (edge length 166 mm). All processes were performed at Fraunhofer ISE PV-TEC [14]. The half-cell layout is a multi-busbar configuration with nine busbars. The shingle layout is a 1/6-cut layout, utilizing the whole wafer. In this work, only the center strips are discussed for simplicity. Finger grid and busbars were applied in dual print, *i.e.*, separate printing steps. Further details are given in Table I.

Table I: Nominal metallization grid parameters for the different layouts.

		Half cell	Shingle
Front busbar width	(mm)	0.1	0.5
Front pad width	(mm)	1.0	-
Busbar pitch / shingle width	(mm)	16.8	27.7
Front finger screen opening	(µm)	24	33
Number of front fingers		2 × 67	128
Rear finger screen opening	(µm)	150	200
Number of rear fingers		2 × 83	166

The two grid layouts differ slightly to account for the different current flow in multi busbar and shingle solar cells. They have been kept mostly similar, to allow for comparability. Yet, accounting for the grid shading on host level (+ 2.8 % more shading), the shingle layout is at a disadvantage from the start, due to some conservative design choices, which we have already improved on in follow-up experiments (not discussed in this work).

The finished solar cells were measured on an industrial *I-V* tester, either by directly contacting the finger grid, resulting in grid-neglecting results, or by a pin array placed on the busbars, including the finger grid resistance in the results.

Photographs of the finished cells are shown in Figure 2.

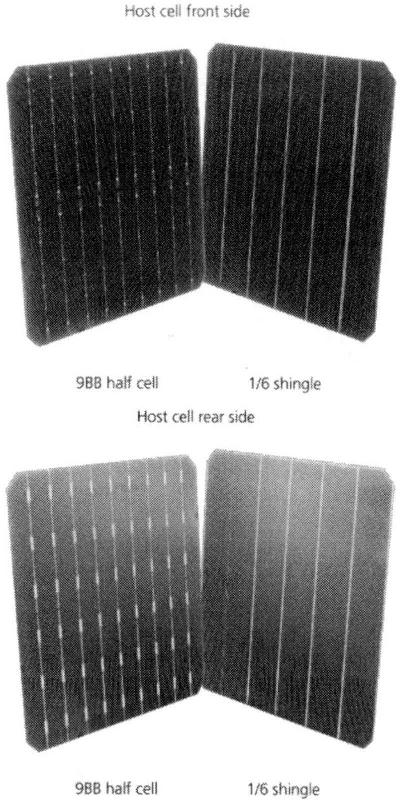

Figure 2: Photographs of finished host solar cell, based on the same industrial M6 PERC precursor, with different grid layouts. ©Fraunhofer ISE.

2.2 Simulation tools

For solar cell simulation, we used the expanded Gridmaster [9] implementation "Gridmaster+", which now runs on MatLab or Python. It includes an empirical parametrization for edge recombination. [15], the option to simulate grid-neglecting *I-V* measurements as done by the GridTouch® contacting scheme [16] and can also simulate monolithic tandem solar cell configurations (not used in this work) [10].

As the precursor properties are unknown, the input parameters have been varied within reasonable ranges, to achieve good agreement between experiments and simulations on host level.

The final solar cell *I-V* parameters relevant for module integration were used as an input for SmartCalc.Module [11] to simulate the module performance of full-cell, half-cell, and shingle modules. For this, the same glass size was assumed, but for shingle modules, additional shingle cells were added to benefit from the tighter packaging due to the overlapping shingle cells. All parameters of the bill of materials (BOM) have been kept the same between the three different layouts (full-cell, half-cell, shingle) and are based on state-of-the-art commercial materials.

3 RESULTS

The experimental and simulated results on cell and module level are presented in this section. The discussion of the results is presented in the next section.

3.1 Solar cell results (experimental and simulation)

The cells were measured in the host (uncut) state and after cutting. The resulting *I-V* parameters are given in Figure 3 and 4 as black symbols.

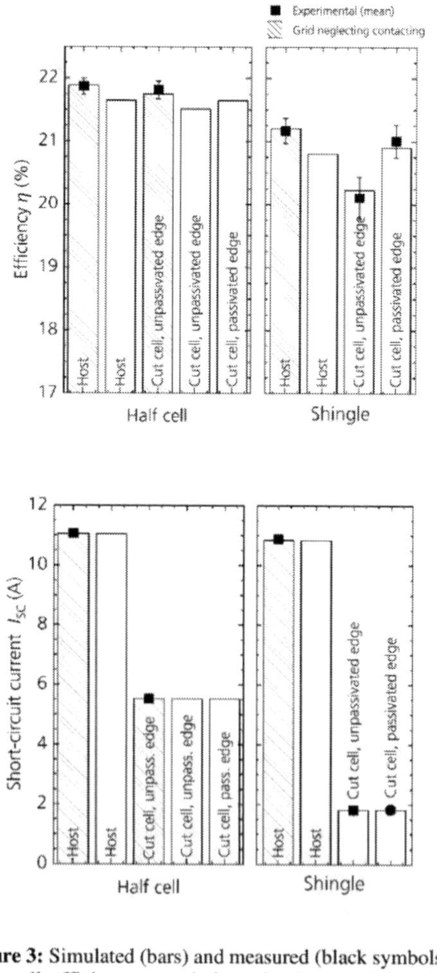

Figure 3: Simulated (bars) and measured (black symbols) solar cell efficiency η and short-circuit current I_{SC}, for different grid layouts in different stages.

While Figure 3 shows the solar cell efficiency η and short-circuit current I_{SC}, Figure 4 shows open-circuit voltage V_{OC} and fill factor FF. The standard deviation of the experimental groups is given by the error bars. The labelled bars represent the simulated values by Gridmaster+, in the configuration given by the label. The assumed systematic error of the simulations is within 5 $\%_{rel}$.

The grid-neglecting contacting is highlighted by a dashed filling of the bar. This provides a quick method to measure very different layouts, but it is important to note that the values determined by this should not be compared to module results, where the grid resistance plays a role.

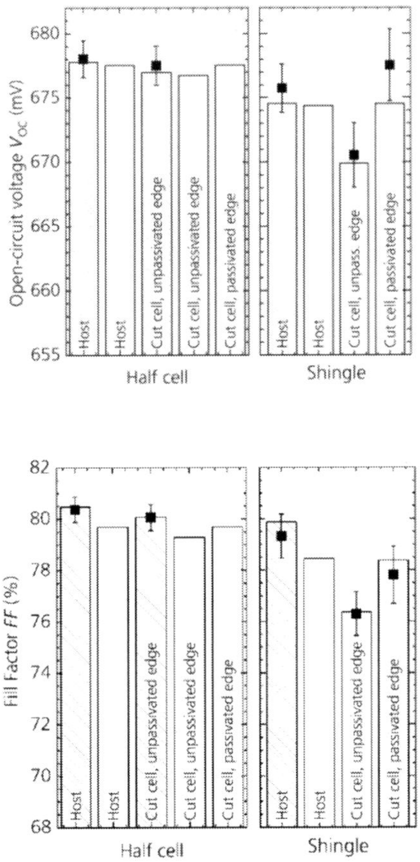

Figure 4: Simulated (bars) and measured (black symbols) open-circuit voltage V_{OC} and fill factor FF for different grid layouts in different stages.

3.2 Module simulation results

The solar cell parameters were used as an input to simulate five different modules, with comparable size and BOM. The results are given in Figure 5.

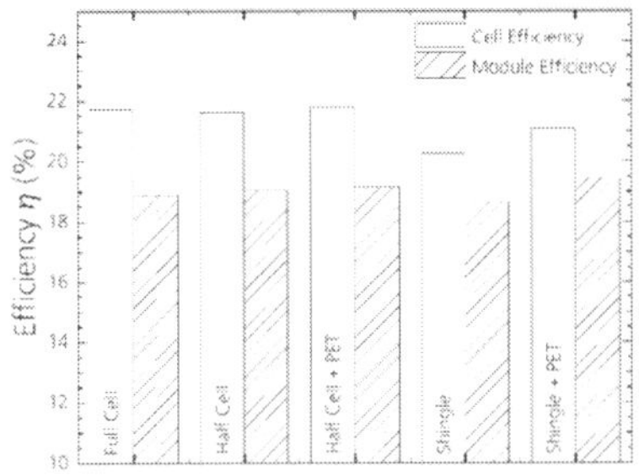

Figure 5: Simulated solar cell and module efficiency, for different cut size cells, with and without additional edge passivation (PET) after cutting.

4 DISCUSSIONS

The implemented shingle layout results in less efficient host solar cells, due to the conservative layout with increased shading. However, the cut stripes from these hosts will be overlapped in the shingling process, eliminating the busbar shading, which is calculated in SmartCalc.Module and leads to an improved cell-to-module (CTM) ratio. Therefore, the comparison on solar cell host value can be seen to have limited relevance.

The advantage of cut cells can be seen in the systematic lowering of the I_{SC}, which reduces the ohmic losses on module level and therefore also increases CTM.

The major challenge of cut cells can be seen in the V_{OC}, which is significantly lower after cutting, due to the unpassivated edges. This effect can be easily switched off in the simulation, given the performance of an ideal cut cell with no additional recombination on the cut edges. Interestingly, applying PET to the third experimental group recovered a significant amount of the cutting-induced losses.

The fill factor is also affected by the cutting and drops significantly. However, it is important to ensure a fair comparison, considering the contacting scheme in the *I-V* measurement: As can be seen, the host cells were measured using grid-neglecting contacting for simplicity. When comparing this grid-neglecting result to the busbar contacting by pins used for the cut shingle measurements, the drop is significantly higher than comparing it, more accurately, to the (simulated) host measurement with pins.

The pin contacting is more relevant for the module performance, as the current flow is more comparable to the current flow in the module [17]. Referencing to the grid-neglecting *I-V* measurements for CTM analysis would result in much higher losses.

It is noteworthy that the effect of edge recombination is also visible for half-cut cells. As they only have one unpassivated edge after cutting and a smaller edge-to-area ratio, the effect is much less pronounced than in the case of the inner shingle strips (shown here) with two unpassivated edges.

These effects also translate to module level. As expected, the half-cell configuration benefits from the lower ohmic losses and has a lower CTM loss than the full cell module.

Von Kutzleben *et al.* [8] have already shown that a shingle module can outperform a half-cell module of the same glass size, if the host level of shingle and half-cut cells is very close. In this study, the shingle design was not yet fully optimized (see above), resulting in a significant difference between both layouts on cell level.

On module level, both cut cell configurations benefit from the additional edge passivation in terms of the simulated module efficiency. Due to the lower CTM losses for shingle modules and due to the higher gain from edge passivation, the shingle module with PET is predicted to have the highest module efficiency in the end.

5 CONCLUSIONS

Three sets of cut solar cells have been fabricated based on the same industrial precursor to include minimal variation of the blue wafer properties. Between the groups the metallization layout was varied, to compare half cells, shingle cells and shingle cells with Passivated Edge Technology (PET). As the shingling approach hides the busbars in the module, comparing shingle host cells and half-cell host cells is questionable.

Additional simulations we used to check the results and to address the deviations originating from different *I-V* contacting schemes. It is important to clarify these conditions to make reasonable comparisons between different cell states.

When simulating module performance, it is demonstrated that a lower host cell efficiency does not necessarily translate into a lower module performance.

All cut cells benefit from edge passivation, which directly translates into a gain in module efficiency. This gain is smaller for half-cut cells than for shingles, as the latter are also more negatively affected from the cutting.

With the boost by PET, shingled solar modules can outperform full-cell and half-cell configurations on comparable bill of materials, due to a higher power density enabled by the shingling approach.

6 ACKNOWLEDGEMENTS

We thank our colleagues at Fraunhofer ISE who supported the presented experiments and simulations, especially Tobias Fellmeth, Puzant Baliozian, Gernot Emanuel, Diana Witt, Marc Retzlaff, Milad Salimi Sabet, Felix Maischner, and Alexander Krieg.

This work was supported by the Federal Ministry for Economic Affairs and Climate Action (BMWK) through the project "GutenMorgen" (FkZ: 03EE1101A) and funding is gratefully acknowledged.

7 REFERENCES

[1] ITRPV, "International Technology Roadmap for Photovoltaic (ITRPV): 2023 Results," VDMA, 2023.

[2] P. Baliozian, N. Klasen, N. Wöhrle, C. Kutter, H. Stolzenburg, A. Münzer, P. Saint-Cast, M. Mittag, E. Lohmüller, T. Fellmeth, M. Al-Akash, A. Kraft, M. Heinrich, A. Richter, A. Fell, A. Spribille, H. Neuhaus, and R. Preu, "PERC-based shingled solar cells and modules at Fraunhofer ISE," *Photovoltaics International*, no. 43, 129-145, 2019.

[3] D. Tune, T. Rössler, G. Oreski, C. Carrière, C. Kaiser, T. Timofte, N. Klasen, J. Weber, B. Jäckel, S. Großer, M. Pander, M. Turek, M. Galiazzo, M. I. Devoto, T. Fischer, I. Ullmann, D. Rudolph, and A. Halm, "The Sun is Rising on Conductive Adhesives," *Photovoltaics International*, vol. 47, pp. 57–76, 2022.

[4] H. Schulte-Huxel, A. Kraft, T. Roessler, and A. De Rose, "Chapter 10: Module interconnection technologies," in *Silicon Solar Cell Metallization and Module Technology*, T. Dullweber and L. Tous, Eds.: IET, 2021, pp. 435–489.

[5] V. Nikitina, D. Reinwand, T. Fellmeth, S. Nevendu Mont, D. von Kutzleben, T. Roessler, and D. H. Neuhaus, "Application of SHJ and TOPCon Shingle Cells in Full Format and Integrated Modules," (eng), 2022.

[6] N. Wöhrle, T. Fellmeth, E. Lohmüller, A. Fell, J. Greulich, R. Preu, T. Fellmeth, P. Baliozian, A. Fell, and R. Preu, "The SPEER solar cell – simulation study of shingled PERC technology based stripe cells," in *33rd European Photovoltaic Solar Energy Conference and Exhibition*, Amsterdam, The Netherlands, 2017.

[7] N. Wöhrle, E. Lohmüller, M. Mittag, A. Moldovan, P. Baliozian, T. Fellmeth, K. Krauss, A. Kraft, and R. Preu, "Solar cell demand for bifacial and singulated-cell module architectures," *Photovoltaics International*, vol. 36, pp. 48–62, 2017.

[8] D. von Kutzleben, T. Roessler, M. Mittag, J. Weber, S. Sigdel, N. Klasen, P. Zahn, A. Kraft, and H. Neuhaus, "Development of Shingle Matrix Technology for Integrated PV Applications," (eng), 2022.

[9] T. Fellmeth, F. Clement, and D. Biro, "Analytical modeling of industrial-related silicon solar cells," *IEEE J. Photovoltaics*, vol. 4, no. 1, pp. 504–513, 2014.

[10] B. Shabanzadeh, "Modeling of Photovoltaic Modules by Evaluating Silicon and Tandem Solar Cells," Master thesis;, Albert-Ludwigs-University, Freiburg im Breisgau.

[11] M. Mittag, "Systematic PV-module optimization with the cell-to-module (CTM) analysis software "SmartCalc.CTM"," *Photovoltaics International*, vol. 36, pp. 97–104, 2017.

[12] E. Lohmüller, P. Baliozian, M. Hofmann, L. Gutmann, L. Kniffki, A. Richter, J. Geng, L. Wang, R. Dunbar, and A. Lepert, "Thermal Laser Separation and Edge Passivation for TOPCon Shingle Solar Cells," in *SiliconPV 2023*, Delft, The Netherlands, 2023.

[13] P. Baliozian, M. Al-Akash, E. Lohmüller, A. Richter, T. Fellmeth, A. Münzer, N. Wöhrle, P. Saint-Cast, H. Stolzenburg, A. Spribille, and R. Preu, "Postmetallization "passivated edge technology" for separated silicon solar cells," *IEEE J. Photovoltaics*, vol. 10, no. 2, pp. 390–397, 2020.

[14] R. Preu, J. Rentsch, S. Rein, F. Clement, J. F. Nekarda, P. Saint-Cast, S. Lohmüller, E. Lohmüller, J. Greulich, N. Wöhrle, S. Mack, A. Wolf, S. Pingel, A. Steinmetz, P. Baliozian, B. S. Goraya, S. Nold, S. W. Glunz, and A. W. Bett, "Lessons Learned from 25 Years Production Technology Research & Development," (eng), 2022.

[15] A. Fell, J. Schön, M. Müller, N. Wöhrle, M. C. Schubert, and S. W. Glunz, "Modeling Edge Recombination in Silicon Solar Cells," *IEEE J. Photovoltaics*, vol. 8, no. 2, pp. 428–434, 2018.
[16] N. Bassi, C. Clerc, Y. Pelet, J. Hiller, V. Fakhfouri, C. Droz, M. Despeisse, J. Levrat, A. Faes, D. Bätzner, and P. Papet, "Grid^{TOUCH}: Innovative solution for accurate IV measurement of busbarless cells in production and Laboratory Environments," in *29th European Photovoltaic Solar Energy Conference and Exhibition*, Amsterdam, The Netherlands, 2014, pp. 1180–1185.
[17] T. Rößler, D. von Kutzleben, N. Klasen, V. Nikitina, P. Baliozian, A. Münzer, E. Fokuhl, and A. Kraft, "Progress in shingle interconnection based on electrically conductive adhesives at Fraunhofer ISE," in *PROCEEDINGS OF THE 10TH WORKSHOP ON METALLIZATION AND INTERCONNECTION FOR CRYSTALLINE SILICON SOLAR CELLS*, Genk, Belgium, 2022, p. 20012.

10.4229/EUPVSEC2023/1BV.5.42

NUMERICAL AND EXPERIMENTAL INVESTIGATION OF SUSTAINABLE RINSING PROCESSES FOR SILICON SOLAR CELL MANUFACTURING

Lena Emmer (née Mohr), Geron Isele, Katrin Krieg, Martin Zimmer
Fraunhofer Institute for Solar Energy Systems, Heidenhofstr. 2, 79110 Freiburg
lena.emmer@ise.fraunhofer.de

ABSTRACT: In this work, potential improvements for the rinsing process of Si wafers are identified. A numerical model was developed which describes the concentration distribution near the wafer surface of potassium hydroxide (KOH) as a function of time, temperature, volume flow, wafer size and gap distance between the wafers. The simulation results showed that a temperature increase from 25°C to 45°C reduces the KOH concentration at the wafer surface by 35%. An increase of wafer size from M2 to M12 led to an increase in rinsing time and water consumption of 10%. In contrast to constant low-flow rinsing, the combination of a phase without flow and subsequent high flow saved 35 s of process time. The simulation results were transferred to experimental rinse recipes. With regards to the flow conditions two different inlet pipes, an industrial standard and an optimized pipe are compared using three different recipes. In the experimental comparison of the rinsing scenarios, the final conductance was reduced by half for a recipe of two minutes without flow and three minutes high flow using the optimized inflow pipe, compared to the five-minute low flow recipe using the standard inflow pipe.
Keywords: Wet-chemical processes, Rinsing, Water consumption, Modelling, CFD Simulation

1 INTRODUCTION

Silicon solar cells are produced in a series of different process steps in which wet-chemical etching and rinsing steps have a considerable influence on the quality of the wafers [1,2]. The requirements for optimal flow conditions over each individual wafer surface are very high since the rinsing process must stop the reaction from the previous process step and the process solution must be rinsed off to prevent carryover into further basins. To increase the throughput in the industrial production of solar cells, wafers are processed in parallel batch processes in which two, four or six carriers with 100 wafers each are dipped in a circulated process bath. The most common method for rinsing wafers is overflow rinsing [3]. In this process, pure deionized water flows around the wafers from bottom to top in a batch process. Batch-type processes have been studied in several publications, the mass transfer and the kinetic effects [4,5], as well as the water motion [6–8]. In [9] the following adjustments were identified for water reduction during the rinse process after hydrochloric acid (HCL)-containing process: Reducing the distances between the wafers in the basin from each other (a), reducing the flow velocity of the purge flow (b), change of flow direction from upward to downward (c) and reduction of carryover of process solution (d).

For this study the concentration decrease of the process solution in a rinsing bath should be simulated to better understand the rinsing processes and thus save water in the future. A numerical model is developed which describes the concentration distribution of potassium hydroxide (KOH) as a function of temperature and volume flow at the wafer surface. Bath changes can be considered as well as the gap distance between wafers in a carrier and the wafer size. The knowledge gained from these simulations is transferred to rinsing recipes in an industrial process plant by means of experimental tests. The aim is to identify potential for improvement in the rinsing of wafers and to implement resource-saving rinsing recipes in the future.

2 APPROACH

Potential improvements for the rinsing process of Si wafers are identified regarding water consumption and rinsing times with the aim to identify potential for improvement in the rinsing of wafers and to implement resource-saving rinsing recipes in the future. During wet chemical process rinsing steps respectively rinse recipes are required to stop the reaction at the wafer surface from previous chemical process steps as saw damage edge or texturing. Since the concentration distribution directly at the wafer surface in the process basins is not accessible experimentally, the investigations are carried out by means of numerical simulation with COMSOL Multiphysics. A numerical model is developed which describes the concentration distribution of potassium hydroxide (KOH) as a function of time, temperature, volume flow and means of a bath change. With respect to larger wafer formats and higher throughputs, the effect of these larger wafer formats on the rinsing performance is investigated.

Based on the simulation results, rinsing recipe adjustments were carried out, in order to experimentally investigate at an industrial process plant. A rinsing bath of an industrial process plant consist of two inlet pipes, a perforated plate and two carriers with 100 wafer each (see Fig.1). Regarding to the flow conditions two different inlet pipes, an industrial standard pipe and an optimized pipe [10] are compared using three different recipes. The target values of the study were the measured final conductivity and the rinsing time, resulting together with the flow rate in the water consumption.

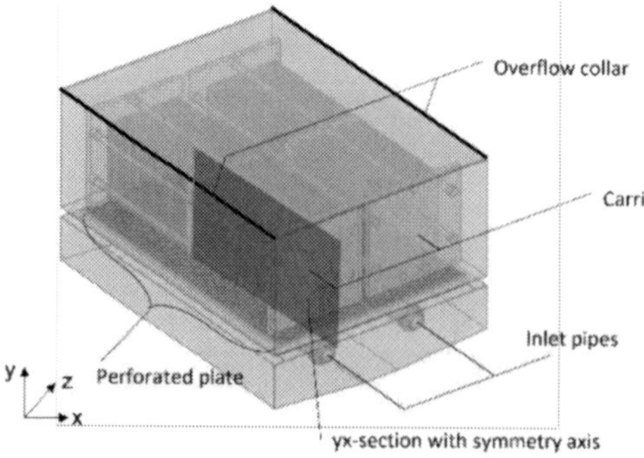

Figure 1: Industrial process basin, modified taken from [5].

10.4229/EUPVSEC2023/1BV.5.44
020041-001

3 EXPERIMENTAL SETUP

3.1 2D simulation model

In the first step a 5-wafer model is built to verify the boundary conditions for plausibility and to investigate the influence of temperature on the concentration decrease in the basin. According to the yx-section shown in Fig.1 a 100-wafer model (Fig. 2, top) is created and reduced to a 5-wafer model (Fig. 2, down). The findings are then transferred to the 100-wafer model and the effects of the number of wafers and the wafer format on the rinsing performance are observed.

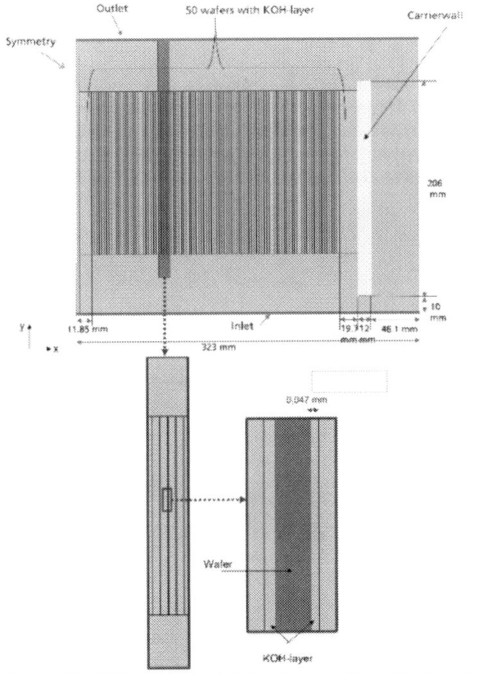

Figure 2: 100-wafer model (top), as well as the 5-wafer model (bottom left) and a detail of the 5-wafer model (bottom right).

To reduce the computational effort, the model is reduced according to its symmetry and therefore just 50 wafers must be simulated. In Fig. 1 and Fig. 2, the symmetry is marked in orange. When symmetry is chosen $v \times N = 0$, where v is the velocity field and N is the normal unit vector, it is assumed that the boundary extends indefinitely. There are no penetration conditions, and the velocity is not set to zero [11] as it is the case when the normal boundary condition is chosen. The inlet for the 100-wafer model and the 5-wafer model is the lower edge and is marked in green. Here the flow velocity is varied according to low flow (3.25 mm/s) or high flow (6.5 mm/s) conditions. The upper edge is defined as outlet, a pressure boundary condition of 0 Pa is assumed compared to the reference pressure of 1 atm. The side wall of the carrier with a thickness of 12 mm is on the right of the wafers, shown in Fig. 2. The wafers are inserted as sections with a thickness of 0.16 mm. On the left and right side of the wafers are narrow domains representing the KOH-layer with a thickness of 0.047 mm (see detail Fig.2, bottom right), carried over into the bath by the previous process step. The starting KOH concentration is assumed with 25 g/l and therefore corresponds to the concentration of the texture process. The addition of organic additives is not considered in the simulation. The area below the perforated plate (Fig. 1) is not included. A homogeneous inflow is assumed.

A laminar flow was assumed for the calculation of both, high and low flow. The Reynolds number (Re) provides information on whether a flow is laminar or turbulent. It was assumed that the rinsing process corresponds to a flow around plates, with twice the plate length as the characteristic length, the flow velocities for low and high flow and the density and viscosity of water at 20°C. With $Re_{(low\ flow)} = 1142$ and $Re_{(high\ flow)} = 2283$ are the Reynolds numbers below the critical Reynolds number $Re_c \approx 500000$ [12] for flow over a flat plate and therefore laminar flow is assumed.

To calculate mass transfer based on diffusion and convection, the transport of diluted species interface was used in addition to the laminar interface of COMSOL Multiphysics. For this purpose, the material constants and thus also the diffusion coefficient of KOH are selected with $3.7 \cdot 10^{-9}$ m^2/s in the domains in which KOH is located.

The heat-transfer interface is used to map the heat exchange between the KOH-layer and the rinsing medium. It enables the calculation of heat distributions. More detailed information about the heat-transfer interface can be found in [13].

As initial value a temperature of 80°C in the KOH-layer and therefore the temperature of the alkaline texture is assumed, while the remaining domains have the initial temperature of 20°C. The calculated heat distribution is coupled with the mass transfer and flow calculation, which determines the influence of the temperature distribution on the concentration decrease in the KOH-layer.

At given KOH concentrations, the deviation of the heat transfer related material constants of KOH and the rinsing water is very small. Therefore, the thermal conductivity and capacity of water is used for the entire model. Only the heat resulting from previous alkaline texture in the KOH-layer is considered, and the stored heat in the wafer itself is neglected.

To investigate the influence of a heated rinsing water, a further calculation is carried out. However, since COMSOL specifies the diffusion coefficients for KOH and water isothermally, the coefficients must be adjusted. In this simulation a temperature in the entire domain of 45°C is specified. This means a temperature increase of 25°C and therefore an approximately doubled diffusion coefficient for KOH in water.

The 5-wafer model is locally discretized with an unstructured mesh in the inlet and outlet regions and with a structured mesh in the KOH-layer and the region between the wafers. A coarser mesh with approx. 240000 elements can be used to calculate the flow and a finer mesh with approx. $3.7 \cdot 10^6$ elements to calculate the mass transport. The number of mesh elements and thus the calculation time is increased for the calculations of the 100-wafer model.

First, the influence of the temperature is investigated using the 5-wafer model. The heat input from the texture process through the KOH-layers into the rinser is simulated. The second simulation represents the mass transfer when the total basin temperature is increased. Both calculations are performed without a laminar flow for 60 s.

Furthermore, the influence of the wafer gab distance as number of wafers in the rinsing bath over 100 s is investigated. For this purpose, the width of the wafer gap and the number of wafers is varied using low flow conditions.

- 50 wafers with a gap width of 9.70 mm
- 100 wafers with a gap width of 4.76 mm
- 200 wafers with a gap width of 2.20 mm

In addition, the effect of the wafer format on the rinsing performance was investigated. For this purpose, the two wafer formats M2 and M12 were simulated with a low flow rate for 100 s.

To investigate the influence of different rinsing steps, the following recipes are simulated with the 100-wafer model:

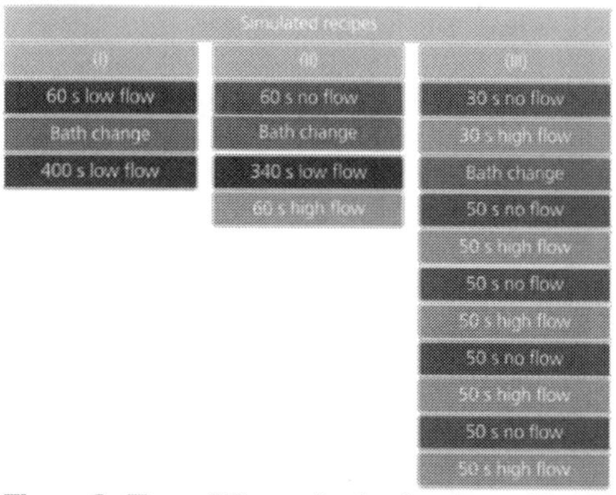

Figure 3: Three different simulated recipes with 460 s rinsing time each.

The bath change after one minute represents the quick dump process and allows a significant increase in the concentration decrease, as the wafers again enter a clean environment with a higher concentration gradient. Previous simulations showed that the KOH-layer remains near the wafer surface during rinsing and does not diffuse far into the wafer gap. Therefore, a large part of the rinse water passes through the wafer gap without KOH loading. It is also known from the simulations that in the initial phase the strength of the rinsing flow does not influence the decrease of the concentration in the KOH-layer. Therefore, the modified recipes include phases during which there is no flow in the basin, so that KOH diffuses towards the center of the wafer gap. Subsequent high-flow phases are intended to ensure faster removal of KOH.

All flow simulations are calculated stationary, the solution is transferred to the mass transport simulation, which is calculated time dependent. In the first seconds the calculation is done in steps of 0.1 s and from the first seconds on in steps of 1 s. For evaluation, the mean KOH concentration in the layers of the middle wafer is examined over time.

2.2 Experiments

Based on the numerical results, rinse recipe adjustments will be carried out, which will be experimentally investigated at an industrial process plant. Two different inlet pipes, an industrial standard pipe and an optimized pipe [10] regarding to the flow conditions are compared using three different recipes, see Fig. 3.

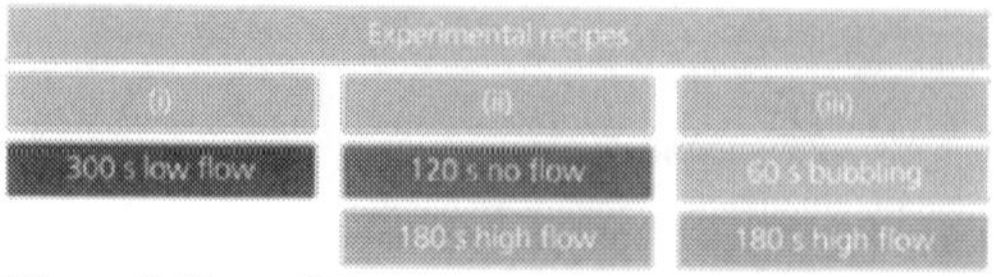

Figure 5: Three different experimental recipes.

A process basin is filled with 25 g/l KOH, two carriers loaded with 100 multicrystalline wafers are placed in the basin for five minutes. Afterwards the carriers are placed in the rinse. The recipes shown in Fig.3 are carried out and repeated three times. The final conductivity and the rinsing time are measured, resulting together with the flow rate in the water consumption.

3 Results

3.1 2D simulation results

In Fig. 5 the temperature in the 5-wafer model is shown. The initial temperature of the rinsing water is 20°C (purple) and the initial temperature of the KOH-layer is 80°C (white), shown in the cutout after 0.1 seconds. The temperature difference after five seconds is 3°C.

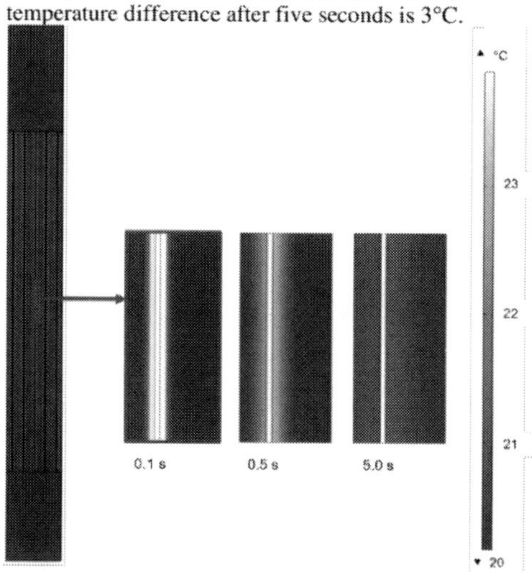

Figure 4: Simulated time dependent heat distribution from KOH-layer (left) and details on the middle wafer at seconds 0.1, 0.5 and 5 (from left to right).

It can be concluded that the influence of the temperature on the mass transport can only have a decisive influence in the initial period. At the beginning of the simulations, the diffusive mass transport dominates, and the flow does not contribute to the concentration decrease in the KOH-layer. Therefore, the influence of the temperature distribution on the flow is neglected and not further investigated.

Fig. 6 shows the result of the simulated decrease in KOH concentration after 20 s at a rinse water temperature of 45°C and 20°C.

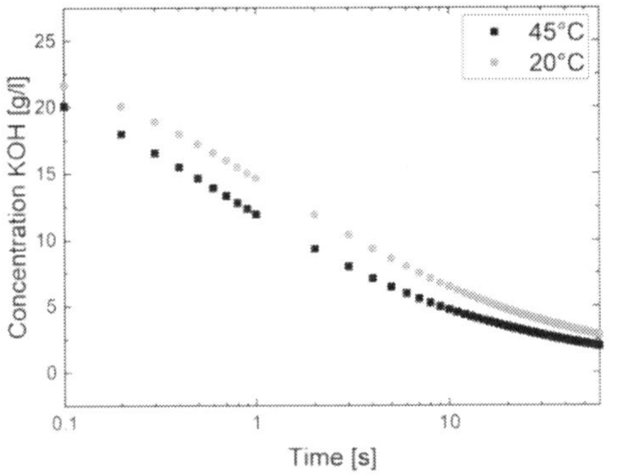

Figure 6: Simulated comparison of the decrease in KOH concentration near the wafer surface at a rinse water and KOH-layer temperature of 45°C and 20°C without flow in the first 60 seconds.

Both concentrations curves are based on diffusion processes since the simulation was carried out without flow. The figure shows a faster decrease in concentration at a water temperature of 45°C with an average concentration reduction by 35%.

Doubling the number of wafers from 100 (orange) to 200 (blue) in a process basin showed no significant effect on the concentration at the wafer surface during rinsing, as shown in Fig 7.

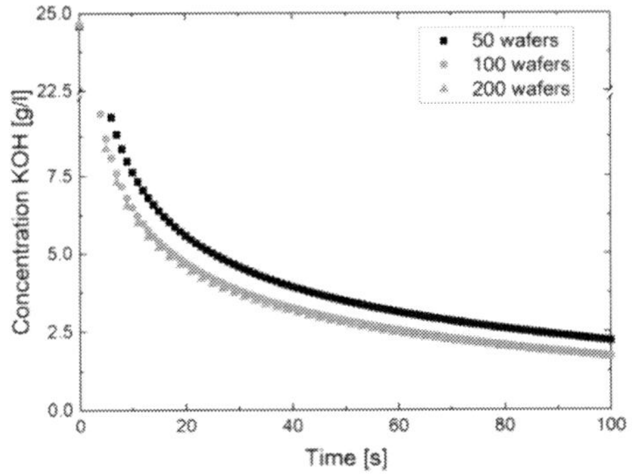

Figure 7: Simulated concentration near the wafer surface with different number of wafers per carrier, 200 wafer (blue), 100 wafer (orange) and 50 wafer (black) at low flow condition.

Although 50 wafers bring less KOH into the bath, and the flow resistance caused by the number of wafers is the lowest, the curve with 50 wafers shows a poorer decrease in the concentration. This is due to the laminar flow profile in the wafer gap for 50, 100 and 200 wafers, which decreases according to the number of wafers and the KOH diffusion near the wafer surface. The maximum flow velocity decreases as the number of wafers increases. Fig. 8 shows the flow velocities for 50, 100 and 200 wafers in the wafer gap near the wafer surface on the left y-axis. Also, the local concentration at 0 and 50 s near the wafer surface is shown on the right y-axis. The mass transport simulation is based on KOH diffusion without flow and is independent of the carrier loading. At t=0 s (green, closed symbols) the entire amount of KOH is within 0.05 mm from the wafer surface. After 50 s (green, open symbols), the diffusion of KOH concentration is within 0.7 mm from the wafer surface.

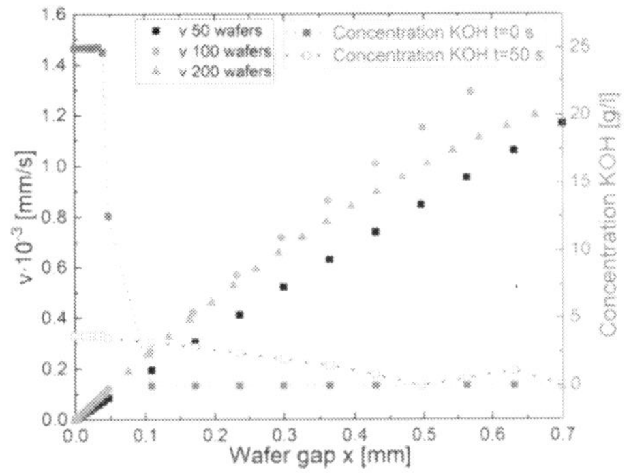

Figure 8: Simulated flow velocity v in mm/s for 50 (black), 100 (orange) and 200 (blue) wafers at low flow conditions and the KOH concentration in g/l due to diffusion without flow after 0 and 50 seconds over the wafer gap x in mm. The wafer is located at x=0 mm.

Within this marginal area, the velocity (v) for 200 wafers (blue) and 100 wafers (orange) increases more steeply than for 50 wafers (black). The velocity increase for 200 wafers is lower from x=0.35 mm compared to the 100 wafers curve since the wafer gap is significantly narrower for 200 wafers and the maximum of the laminar flow profile is therefore closer to the wafer surface. Because most of the KOH is within the first 0.4 mm, the lower curve of 200 wafers and its lower velocity has no effect on the decrease in concentration in the KOH-layer over the time. The higher velocities of the 200 and 100 wafer curves in the area between $0 < x > 0.3$ mm are decisive for the better concentration decrease. When doubling the number of wafers from 100 to 200 wafers in the basin, the concentration decreases in the KOH-layer directly at the wafer surface remains unchanged. This would allow an increase in throughput by a factor of 2, while the rinsing result, water consumption and rinsing time remain constant.

When comparing the two wafer formats M2 and M12, the wafer gaps remain the same for both simulations. After 100 s rinsing time, the average concentration for M12 with a value of 1.94 g/l KOH is just 10 % above the value for M2 with 1.77 g/l. While the diffusive mass transport of both calculations is the same, the convective mass transport in relation to the wafer surface is lower for the larger format.

When comparing the three simulated rinsing recipes, the water consumption is the same, because the increased water consumption during high flow phases is compensated by the phases without flow. The final value of the average KOH concentration for recipe (II) is 24.5% below recipe (I), see Fig. 9. The final concentration of recipe (III) is below the concentration of recipe (I) and (II). The final concentration of rinsing recipe (I) is reached 35 s earlier with recipe (III). Due to the change from no flow to high flow every 50 s the curve has a slight oscillatory shape. The final value of the average KOH concentration for recipe (III) is at least 73 % below recipe (I).

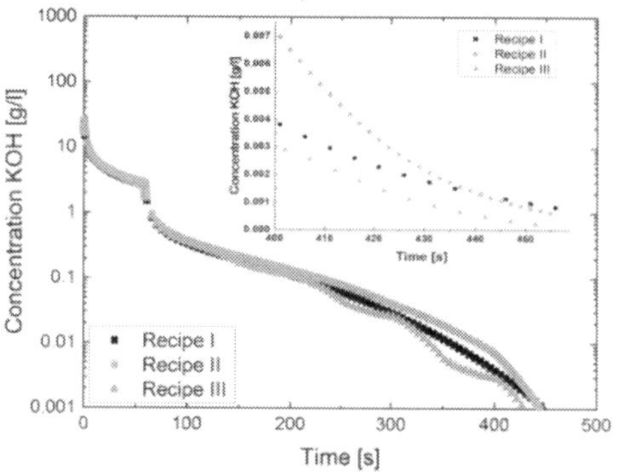

Figure 9: Simulated mean KOH concentration near the wafer surface of the 100-wafer model over rinsing time for three different recipes in the time range 0-460 s and detailed 400-460 s.

3.2 Experiments results

The simulations showed that the KOH-layer remains near the wafer surface during rinsing and does not diffuse far into the wafer gap, hence a large part of the rinse water passes through the wafer gap without KOH loading. Therefore, a recipe without flow is investigated.

In the experimental comparison of the rinsing scenarios, the optimized pipe showed an improvement in terms of rinsing performance since the final mean conductivity is below the results of the standard pipe, see Fig.10. For all three recipes the accumulated water consumption is 300 L.

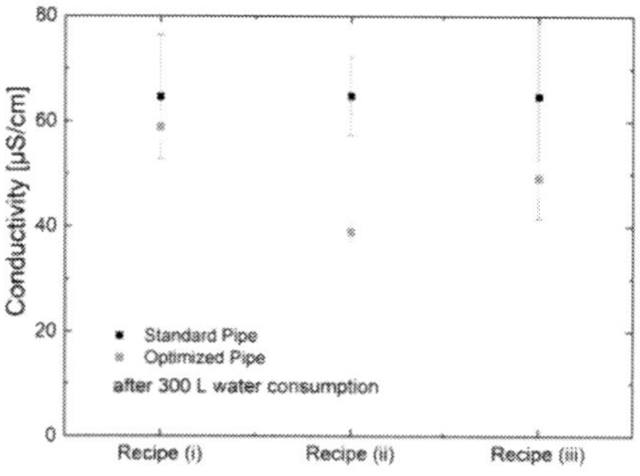

Figure 10: Mean final conductivity and standard deviation (n=3) for different recipes with standard pipe (black) and optimized pipe (orange) after 300 L of water consumption.

Compared to the standard pipe, the results of the optimized pipe also show the lower variance in the final conductivity for all recipes, resulting in more reproducible rinsing results. The final conductivity was reduced by half for recipe (ii) and the optimized inflow pipe, compared to the five-minute low flow recipe (i) and the standard inflow pipe. Recipe (ii) shows the lowest variance in the final conductivity. Therefore, the conductivity for recipe (ii) over the accumulated water consumption is shown in Fig. 11 for the standard (black) and optimized pipe (orange). In each case, the mean value with the standard deviation is shown as error band.

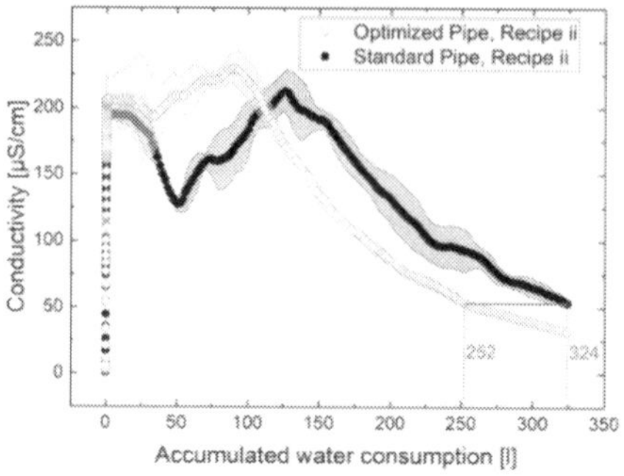

Figure 11: Conductivity and standard deviation (n=3) for recipe ii over the accumulated water consumption. At approx. 54 µS/cm the water consumption of the optimized pipe (black) is 252 l and 324 l of the standard pipe resulting in a water reduction of 72 l.

A lower standard deviation is shown within the experiments with the optimized pipe. At the end of the experiments, the curve of the standard pipe is above that of the optimized pipe. If the final conductivity value of the standard pipe is used for this recipe, the optimized pipe saves 72 l of water per run compared to the standard pipe, which corresponds to a water saving of approx. 25% per run.

4 CONCLUSION

With the use of flow simulations, it is possible to gain a deeper understanding of the rinsing processes. A temperature of 45°C reduces the average KOH concentration at the wafer surface by 35% compared to 25°C. No significant decrease in the removal of KOH near the wafer surface is found by doubling the number of wafers from 100 to 200 wafers. Larger wafer formats do not pose major challenges in terms of rinsing effects. To optimize the rinsing recipes for a sustainable rinsing, it is important to allow diffusion of KOH into the wafer gap; periods without rinsing flow are ideal for this purpose. In the experimental comparison of the rinsing scenarios, the final conductivity was reduced by half for a recipe of two minutes without flow and three minutes high flow using the optimized inflow pipe, compared to the five-minute low flow recipe using the standard inflow pipe.

ACKNOWLEDGEMENTS

This work was funded by the German Federal Ministry for Economic Affairs and Climate Action within the research projects "CHEPHREN" (contract no. 03EE1103B).

REFERENCES

1. N. Johan, M. Mohamad Shahimin, and S. Shaari, IEEE Student Conference on Research and Development (SCOReD) (2010).
2. W. Kern, ed., *Handbook of semiconductor wafer cleaning technology* (Noyes Publications, Park Ridge, N.J, 1993).

3. C.-Y. Chen, B. Panigrahi, K.-S. Chong, W.-H. Li, Y.-L. Liu, and T.-Y. Lu, J. Fluids Eng. **140**, 199 (2018).
4. M. S. Kulkarni and H. F. Erk, J. Electrochem. Soc. **147**, 176 (2000).
5. L. Mohr, T. Dannenberg, A. Moldovan, M. Zimmer, and C. Müller, Ind. Eng. Chem. Res., 17680 (2020).
6. H. Habuka, S. Kobayashi, M. Kato, T. Takeuchi, and M. Aihara, J. Electrochem. Soc. **151** (2004).
7. S. Singhal, B. Elkhatib, J. Stuber, S. V. Sreenivasan, and O. A. Ezekoye, IEEE Trans. Semicond. Manufact. **22**, 399 (2009).
8. L. Mohr, T. Krick, M. Zimmer, T. Röder, and C. Müller, IEEE Trans. Semicond. Manufact. **32**, 334 (2019).
9. O. Raccurt, F. Tardif, L. Kerber, T. Lardin, and T. Vareine, Journal of Micromechanics and Microengineering **13**, 442 (2003).
10. L. Mohr, *Numerische und experimentelle Untersuchung der Strömung sowie der chemischen Reaktionen in einem Prozessbecken für die Herstellung von Silizium-Solarzellen.* Dissertation, ISBN 978-3-8396-1764-9 (2021).
11. COMSOL 5.4, ed., *CFD Module User's Guide* (2018).
12. C. P. Kothandaraman, *Fundamentals of heat and mass transfer*, Rev. 3. ed. (New Age International (P) Ltd., Publishers, New Delhi, 2006).
13. COMSOL 5.2a, ed., *COMSOL Multiphysics 5.2*, COMSOL Reference Manual (2016).

LOCALIZED BIFACIAL PASSIVATED CONTACTS SOLAR CELL WITH 23.5% EFFICIENCY

Shih-Peng Hsu[1], Chao-Ping Huang[1], Kuang-Chieh Lai[2], Chi-Chun Li[2], Bo-Cheng Kung[1], Sung-Yu Chen[1], Ming-Tsun Kuo[1], Han-Chen Chang[1,*]

[1]Photovoltaic Technology Division, Green Energy and Environment Research Laboratories, Industrial Technology Research Institute, Tainan, Taiwan

[2]Green Energy Technology R&D Section Technology Development Division, Motech Industries Inc., Tainan, Taiwan

* Email of Han-Chen Chang：YESSEY@itri.org.tw

ABSTRACT: Recently, silicon solar cells with tunnel oxide passivated contact (TOPCon) structure have demonstrated the feasibility of industrial production and received great attention. In this work, we present a localized bifacial passivated contacts (L-biPC) solar cell which with the selectively p-type TOPCon and full n-type TOPCon fabricated under the front electrode and at the rear side, respectively. The device design aims to reduce metal-induced defects and parasitic absorption of sunlight caused by the polysilicon layer. The best conversion efficiency of 23.5% is achieved with the cell area of 100 cm^2. Simulation result further reveals that the open circuit voltage, short circuit current, and cell efficiency of the optimized L-biPC solar cell are 732 mV, 41.1 mA/cm^2, and 25.1%, respectively.
Keywords: Tunnel oxide passivated contact, simulation, bifacial passivated contact

1 INTRODUCTION

As the market demands higher and higher efficiency solar cells, the p-type silicon solar cells encounter increasing difficulties in improving its efficiency. The n-type silicon solar cell has the advantages of high carrier lifetime and no light induced degradation (LID), which is beneficial to improve the efficiency of solar cells. There are three types of high-efficiency solar cells based on n-type silicon wafers: TOPCon solar cells, heterojunction with Intrinsic Thin layer (HIT) solar cells and interdigitated back contact (IBC) solar cells. The n-type cells fabricated according to these three structures can achieve a conversion efficiency of more than 23%. Compared to HIT and IBC solar cells, TOPCon solar cells are more compatible with the current process, and have potential for low-cost production. According to ITRPV annual report, the market share of TOPCon solar cells will increase to more than 50% in the next five years as shown in Figure 1 [1]. There are two reasons for the phenomenon: First, the cost of n-type wafer is gradually lower than that of p-type wafer. The second is the gradual maturity of n-type solar cell technologies.

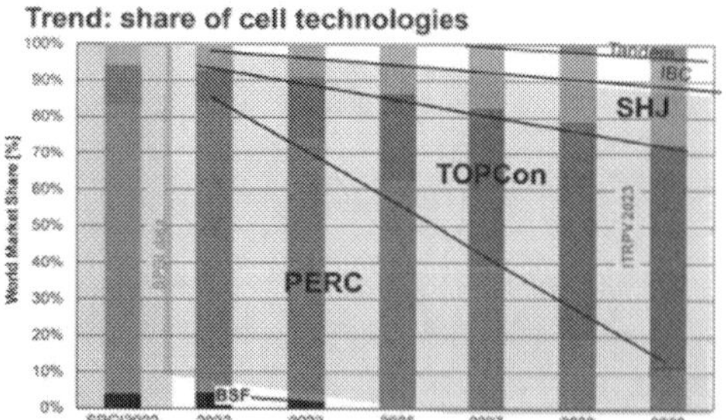

Figure 1：Market share of c-Si PV cell technologies.

Tunnel oxide passivated contact is a technique based on the principle of selective carrier passivation to reduce surface recombination and improve electrode contact. Bifacial passivated contact (biPC) solar cells take TOPCon technology a step further by applying it to both the front and back sides of the cell, achieving a theoretical efficiency of up to 28.7%, which is closer to the theoretical maximum efficiency of 29.43% for crystalline silicon solar cells. However, using heavily doped polycrystalline silicon on the front side of the solar cell results in increased absorption losses. Therefore, there are currently two methods to implement bifacial passivated contacts: (1) Thinning the heavily doped polycrystalline silicon layer on the front side and reducing absorption losses by using wide bandgap passivated contact materials with better transparency (full area biPC, F-biPC). (2) Using locally doped polycrystalline silicon under the electrodes to enhance passivation effects (L-biPC). Here, we will present the key processes and simulation results for L-biPC.

2 EXPERIMENTAL METHODS

In order to further assess the efficiency potential of high-efficiency solar cells, the renowned German solar research institute ISFH conducted simulation calculations based on the concept of carrier selectivity for various types of solar cell structures [2], as shown in Figure 2.

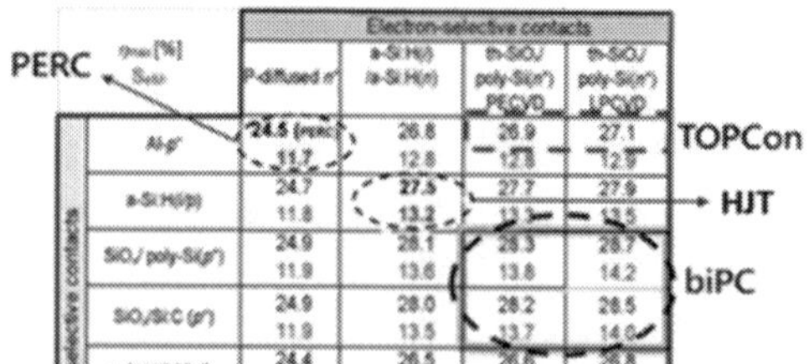

Figure 2：Theoretical efficiencies of different passivated contact solar cells. (ISFH, Silicon PV 2019)

It can be observed that the theoretical efficiency of PERC solar cells is 24.5%, while that of TOPCon solar cells can reach 27.1%. Notably, for solar cells with bifacial passivated contact structures (passivated contacts on both sides) exhibit even better efficiency limits (28.2%~28.7%), significantly higher than HJT's 27.5%, and closer to the theoretical limit of crystalline silicon solar cells at 29.4% [3]. However, when heavily doped polycrystalline silicon is used on the front side of solar cells, the optical bandgap is similar to that of the silicon wafer, leading to absorption, as shown in Figure 3 [4]. This results in relatively larger absorption losses.

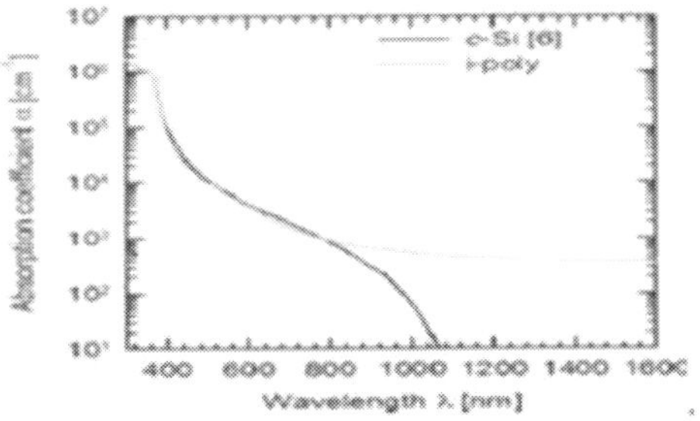

Figure 3 : Absorption in polycrystalline silicon and silicon wafer.

In this study, we used Wafer Ray Tracer simulation (PV Lighthouse) to obtain the current density of various film layers at different thicknesses, as shown in Figure 4. When the thickness of the front p^+ poly-Si exceeds 20 nm, each additional 20 nm leads to a degradation in current density of approximately 0.3~0.5 mA/cm^2. As for the thickness of the backside n^+ poly-Si and SiN_X, the impact on light absorption is not significant.

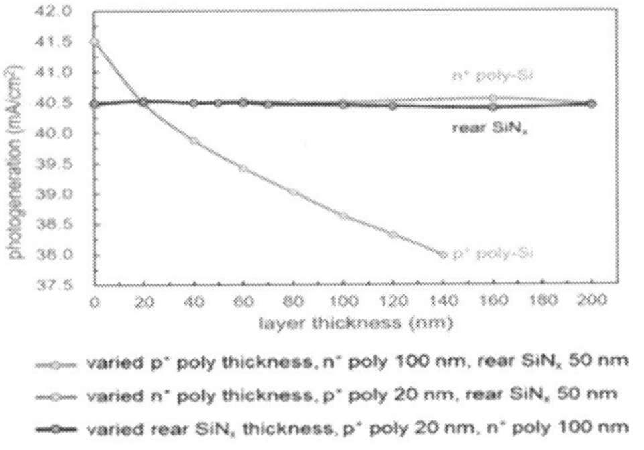

Figure 4 : Simulation of the current density at different thicknesses of each layer.

Therefore, to achieve biPC solar cells with high efficiency, besides using more transparent wide-bandgap passivated contact materials like silicon carbide (SiC_X) to reduce absorption losses, thinning the heavily doped polycrystalline silicon on the front side or selectively doping polycrystalline silicon under the electrodes (as shown in Figure 5) can significantly mitigate the front-side absorption issue, thereby increasing the overall conversion efficiency.

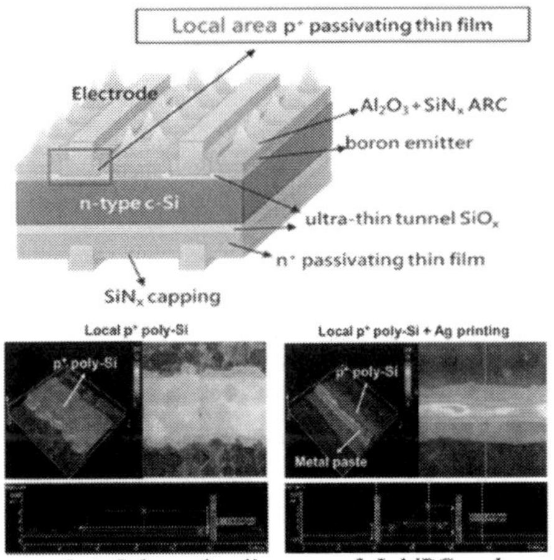

Figure 5 : Schematic diagram of L-biPC solar cell Structure and local area captured by confocal microscopy Imaging.

The process of L-biPC solar cell is shown in Figure 6. First, the as-cut wafer was polished by an alkali etching to remove surface defects. After cleaning the wafer by hydrochloric acid-hydrogen peroxide-water mixture (HPM) and hydrofluoric acid (HF), a passivation layer formed by tunnel silicon oxide and poly silicon was deposited on the front side of the solar cell. The front locally doped polycrystalline silicon area was defined by mask paste, and then the silicon wafer surface textured and rear TOPCon structured were performed. Then, the anti-reflection coating (ARC) layers were respectively deposited on the front and rear side of the solar cell. Finally, silver paste was screen printed, the co-firing process was carried out after screen printing processes. During the co-firing process, the silver paste will fire through the passivation layer to form an ohmic contact.

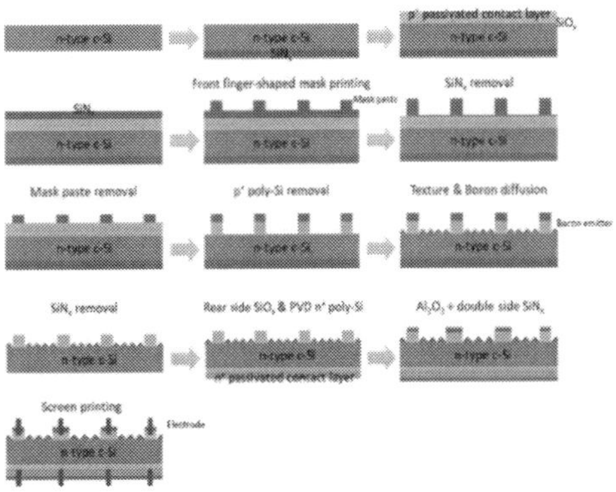

Figure 6 : The processes flow of L-biPC solar cell.

3 RESULTS AND DISSUSION

In addition to the thickness of the front-side passivation layer, the quality of the wafer is another crucial factor affecting solar cell efficiency. Figure 7 shows the optimization values for the overall bulk defect state density and resistivity of an n-type wafer simulated using AFORS-HET, a one-dimensional device simulation software developed by the Helmholtz-Zentrum Berlin.

10.4229/EUPVSEC2023/1CV.3.1
020042-002

When N_{tr} = 6×10⁹ cm⁻³/eV for the wafer and the sheet resistance is 0.5 Ω-cm, the optimal efficiency can reach 23.9%. Under the same film layer structure, for wafers with poor quality and high resistance, such as $N_{tr} = 8\times10^{10}$ cm^{-3}/eV, the efficiency drops to only 20.1%. Therefore, when substrate defect density is high, it is essential to choose wafers with low resistance. However, as defect density decreases, the efficiency has narrow differences between wafers with different resistances. Nevertheless, it is crucial to note that wafers with the same N_{tr} but different doping concentrations can result in varying carrier lifetimes.

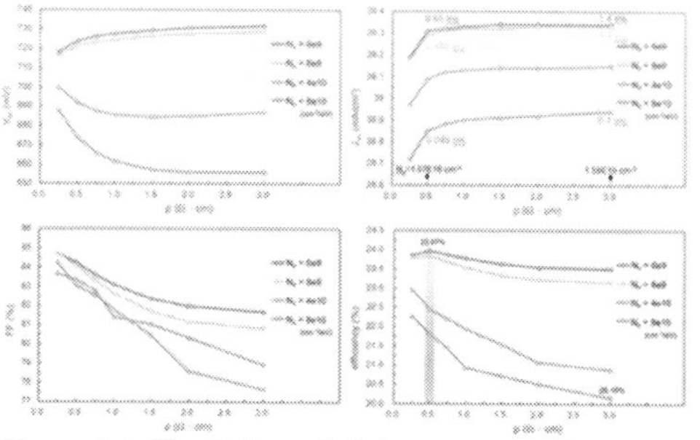

Figure 7 : Simulation of defect state density N_{tr} and resistivity ρ for n-type wafers using AFORS-HET.

Figure 8 presents the simulation results of the electrical properties concerning the depth and concentration gradient of the boron tail under boron doping in the front p⁺ poly-Si. The horizontal axis represents the variation in concentration within the depth. There are three concentration gradients, namely $N_a=5\times10^{19}$ cm^{-3} to 1×10^{18} cm^{-3}, $N_a=1\times10^{19}$ cm^{-3} to 1×10^{18} cm^{-3} and $N_a=2\times10^{18}$ cm^{-3} to 1×10^{17} cm^{-3}. The results indicate that the boron tail depth is another crucial parameter that needs optimization to enhance solar cell efficiency. The effect of boron tail doping concentration, on the other hand, is not as significant as doping depth. When the concentration gradient is 80 nm with N_a at 1×10^{18} cm^{-3}, the simulated solar cell efficiency can reach 24.2%. Furthermore, as the boron tail depth exceeds 80 nm, the efficiency continues to increase. However, it remains to be explored whether this in-diffusion profile can be achieved in actual manufacturing processes. Although increasing annealing temperature and duration time can extend the boron tail depth [5-6], the potential impacts on the device are still unknown.

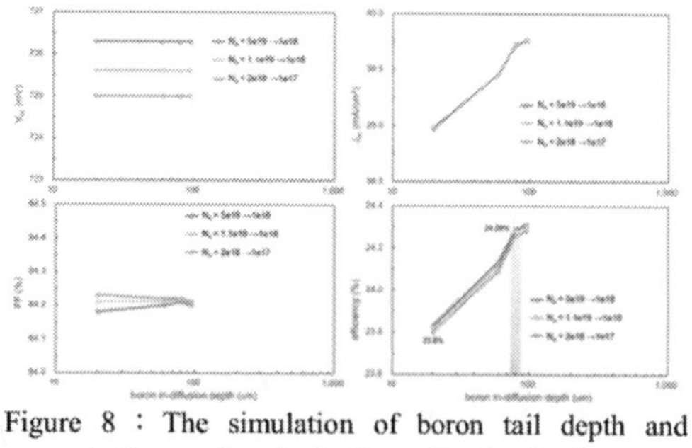

Figure 8 : The simulation of boron tail depth and concentration gradient in the front side p⁺ poly-Si.

After completing the optimization of the front side simulation parameters, simulations of the rear side structure are conducted. Figure 9 presents the simulation results of the impact of different rear side n⁺ poly-Si thicknesses and doping concentrations (N_d) on the electrical characteristics of the solar cell. It shows that in conditions where Shockley-Read-Hall recombination (SHR) is not severe, the doping concentration of the rear side n⁺ poly-Si has a significant effect on Jsc (short-circuit current density) and cell efficiency. Increasing the doping concentration is beneficial for electrical performance. When the doping concentration is increased from 5×10^{19} cm^{-3} to 4×10^{20} cm^{-3}, Jsc significantly improves from 36.5 mA/cm² to over 39.5 mA/cm². Additionally, it can be observed that the optimal thickness for the n⁺ poly-Si layer is approximately between 100 nm and 120 nm, efficiency begins to drop when the thickness exceeds 120 nm.

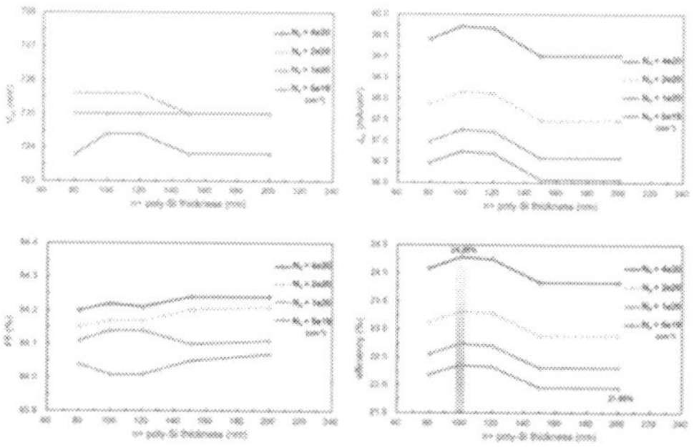

Figure 9 : The impact of different rear side n⁺ poly-Si thicknesses and doping concentrations on the electrical characteristics of the solar cell.

Finally, utilizing the wafer ray tracer, the optimal thin-film thickness obtained from AFORS-HET simulations is used to derive the corresponding optical properties, which are then incorporated into Quokka 2 for efficiency simulations. Due to the absorption issue of the front side p⁺ poly-Si, the efficiency can only reach between 24% and 25%. However, if we switch to a selective passivated contact structure with locally doped polycrystalline silicon under the electrodes with a resistivity of 1 Ω-cm, carrier lifetime of 7000 μs, a thickness of 160 μm for the n-type wafer, a 20 nm local p⁺ poly-Si layer on the front side with a 0.7 μm boron emitter, and a rear side with SiO_x of 1.2 nm and n⁺ poly-Si of 100 nm, we obtain the simulated electrical characteristics of the L-biPC solar cell as shown in Figure 10: Voc=732 mV, Jsc=41.1 mA/cm², F.F.=83%, and the cell efficiency reaches 25.1%. Based on these conditions, the solar cells we produce can achieve an efficiency of 23.5%, as shown in Figure 11.

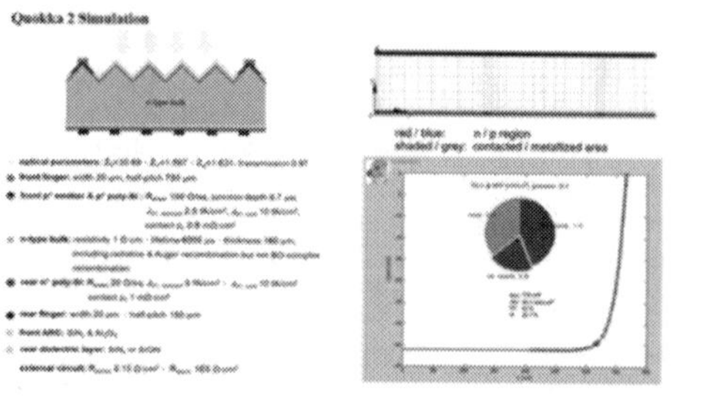

Figure 10 : Efficiency simulations in Quokka 2 using the optimal thin-film thickness obtained from AFORS-HET simulations.

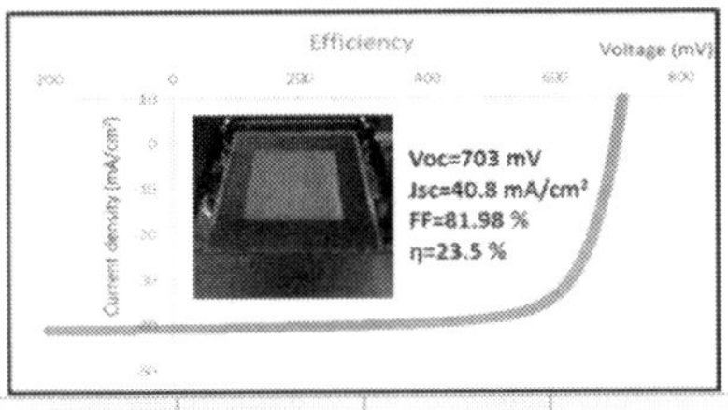

Voc (mV)	J_{sc} (mA/cm^2)	F.F. (%)	Eff. (%)
703	40.8	81.98	23.5

Figure 11 : L-biPC solar cell efficiency.

4 CONCLUSION

The Industrial Technology Research Institute (ITRI) has developed various materials and structures suitable for passivated contact applications, with the most prominent being the TOPCon solar cell, based on a SiO_x/poly-Si bilayer stack structure. This passivated contact structure typically involves the growth of ultra-thin oxide layers through chemical or thermal oxidation, providing passivation at the single-crystal silicon/oxide interface to reduce interface defect density. It serves as a barrier/protective layer for the diffusion of dopants from polycrystalline silicon into the silicon wafer while still allowing charges to drift from single-crystal silicon to polycrystalline silicon, achieving the desired carrier selectivity.

With the support of the Bureau of Energy, Ministry of Economic Affairs, Motech Industries Inc. and ITRI have collaborated to develop TOPCon solar cells and modules, which have already entered mass production and are actively expanding production capacity.

ITRI is further developing high efficiency biPC solar cells based on the TOPCon structure, which can be upgraded from the current mainstream PERC and TOPCon production lines. Through simulation, L-biPC solar cells achieve electrical characteristics with Voc=732 mV, Jsc=41.1 mA/cm2, F.F.=83%, and a cell efficiency of 25.1%, showing significant future potential. And based on the simulation conditions, the solar cells we produce can achieve an efficiency of 23.5%.

5 ACKNOWLEDGEMENTS

The financial support provided by Bureau of Energy, Ministry of Economic Affairs, R.O.C. (Grant No. 112-S0102) is gratefully acknowledged.

6 REFERENCE

[1] International Technology Roadmap for Photovoltaic (ITRPV) Results 2023 incl. maturity repot, 14th edition.
[2] J. Schmidt, R. Peibst, R. Brendel " Surface passivation of crystalline silicon solar cells: Past, present and future", Silicon PV (2019).
[3] A. Richter, M. Hermle, S. W. Glunz " Reassessment of the limiting efficiency for crystalline silicon solar cells", IEEE Journal of Photovoltaics, 3, 1184 (2013).
[4] S. Reiter, et al., "Parasitic absorption in polycrystalline Si-layers for carrier-selective front junctions", Energy Procedia, 92, 199 (2016).
[5] E. Hande, et al., "Study of screen printed metallization for polysilicon based passivating contacts", Energy Procedia, 124, 851 (2017).
[6] F. Feldmann, et al., "Studying dopant diffusion from Poly-Si passivating contacts", Solar Energy Materials and Solar Cells, 200 (2019).

ALL-ALUMINIUM SCREEN-PRINTED POLO BACK JUNCTION SOLAR CELLS

B. Min[1], F. Haase[1], T. Brendemuehl[1], K. Tsuji[2], M. Dhamrin[2, 3], R. Peibst[1] and R. Brendel[1, 4]

[1]Institute for Solar Energy Research Hamelin (ISFH), Am Ohrberg 1, 31860 Emmerthal, Germany
[2]Toyo Aluminium K.K., 341-14 Higashiyama, Ohtani, Hino-Cho, Gamo-Gun, Shiga 529-1608, Japan
[3]Graduate School of Engineering, Osaka University 1-1 Yamadaoka, Suita 565-0871, Japan
[4]Dep. Solar Energy, Inst. Solid-State Physics, Leibniz University of Hannover, Appelstr. 2, 30167 Hannover, Germany

ABSTRACT: In this work, we present our current development status of all-aluminum screen-printed p-type back junction (BJ) solar cells featuring passivating polysilicon on oxide (POLO) rear contact. We achieve a contact resistivity with a median of 2 mΩcm^2 between screen-printed Al and poly-Si, measured with the transfer-length-method (TLM). The laser contact opening process is essential to achieve such low contact resistivity since it changes the surface morphology of poly-Si layer beneath AlO_x/SiN_y dielectric stack and improves the contact formation. Our investigation of the co-firing of the front and rear Al grids in a single firing step shows that a restriction of the peak firing temperature to around 700 °C is needed to avoid a strong deterioration of POLO contacts caused by screen-printed Al paste at the cell rear side. Since this temperature range is not sufficiently high to form the p$^+$-type layer beneath the front Al fingers properly, we first fire the front Al grids by a peak firing temperature of 810 °C. Subsequently, we screen-print the rear Al grid and fire the samples in a second firing step with a peak firing temperature range between 660 °C and 740 °C. We achieve a V_{oc} of 700 mV with the best all-aluminum screen-printed POLO BJ solar cell fabricated with a gallium-doped Cz silicon wafer.
Keywords: aluminum, silver-free, passivating contacts

1 INTRODUCTION

Current commercial silicon solar cells rely on a screen-printed silver metallization on both sides in case of SHJ (silicon heterojunction) and TOPCon (tunnel oxide passivated contact) solar cells, or at least on the front side like PERC (passivated emitter and rear cell) solar cells. However, silver is the most critical metal posing significant demand risk if PV production expands to serve TW-scale installation capacity in next years for fulfilling the Paris Climate Agreement [1]. Hence, there is a strong need to reduce the silver consumption in industrial silicon solar cells. The most suitable way for the current PV industry would be to keep the well-established screen-printing technology and just replace the expensive Ag paste with an alternative metal paste, such as Al paste. Al is highly abundant and it is already applied in the PV production, in example for the rear side metallization of the PERC solar cells.

Our p-type back junction (BJ) solar cell concept featuring passivating polysilicon on oxide (POLO) rear contact is one of the most favorable concept for realizing all-aluminum screen-printed metallization [2, 3]. As shown in Figure 1, the lateral hole conductivity is provided by the wafer, hence it allows to place 60 µm-wide Al fingers at the front side with a pitch of 2 mm to reduce shadowing losses. This would be not possible by front junction cell concepts such as PERC or TOPCon, since they need a small finger pitch around 1 mm due to a high sheet resistance above 150 Ω/□, hence narrow finger width beneath 35 µm is necessary which is difficult with currently commercially available Al pastes due to lager particle sizes of Al compared to Ag. By contacting passivating POLO contacts at the rear with screen-printed Ag, we recently achieve 24.2 % with our cell concept [4].

In this work, we fabricate all-aluminum screen-printed POLO BJ solar cells by replacing the Ag grid with an Al grid for contacting phophorus-doped poly-Si layer at the rear side. An innovative Al paste is applied to form contacts to n$^+$-type poly-Si using special Al alloy powder (TTC series from TOYO Aluminium K.K.) [5]. We first investigate the contact formation between phosphorus-doped poly-Si layer and screen-printed Al with transfer length method (TLM) and scannig electron microscope (SEM). We then investigate the impact of different firing methods on cell performance by varing the peak firing temperature between 660 °C and 740 °C.

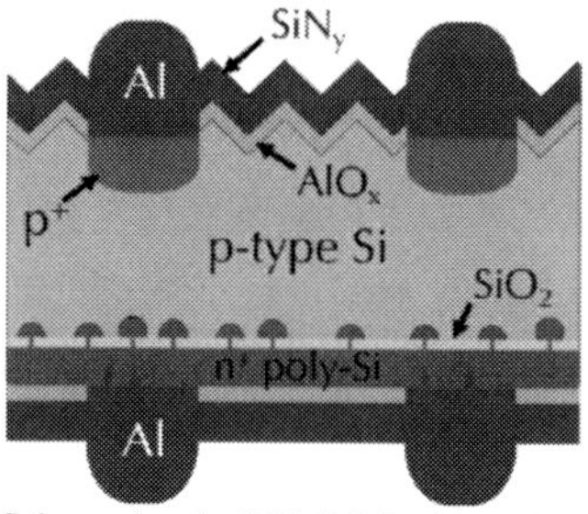

Figure 1: Schematic of a POLO BJ solar cell with front and rear Al grids

2 EXPERIMENTAL

We investigate the contact formation with test structures for TLM measurements by screen-printing Al paste on bare poly-Si and on AlO_x/SiN_y stack on top of poly-Si after opening the dielectric stack with a pico-second laser system.

For the fabrication of POLO BJ solar cells, we use 156.75 mm×156.75 mm (M2) gallium-doped Cz silicon wafers. After a saw damage etch and a cleaning step, all samples receive of a wet-chemical interfacial oxide grown in an ozonized wafer. We then deposit insitu P-doped poly-Si layers on both sides using low-pressure chemical vapor deposition furnace. A 30 minutes long oxidation step at 880 °C serves for the formation of POLO junctions. Simultaneously, a 130 nm thick oxide layer grows on top of the poly-Si layer which will protect the poly-Si layer during the later texture step. After a single side removal of the oxide layer, the samples receive a texture step followed by a clean step whereby the protective oxide layer on top of the poly-Si layer at rear is removed. The front and rear

10.4229/EUPVSEC2023/1CV.3.2
020044-001

sides are passivated with an AlO_x/SiN_y stack deposited with a plasma-enhanced chemical vapor deposition furnace. After a laser contact opening (LCO) step on both sides, we print front and rear Al grid. In the first batch, we fire the solar cells in a single firing step. In the second batch, we apply two different firing steps for the front and rear Al grids.

3 RESULTS

3.1 Impact of the laser contact opening process on the contact formation between screen-printed Al and phosphorus-doped poly-Si

We measure the contact resistivity between screen-printed Al and poly-Si by different peak firing temperature to find the optimum. We apply the TTC-series from Toyo Aluminum K.K. which do not fire through the dielectric stack. For our POLO BJ cell as shown in Figure 1, we use the laser contact opening (LCO) process on both sides. In order to evaluate the impact of the LCO process on the contact formation, we fabricate TLM test structures with and without AlO_x/SiN_y stack on top of the poly-Si. For the sample with AlO_x/SiN_y we open the dielectric stack with a pico-second laser system before the screen-printing step. The sample without AlO_x/SiN_y stack receive a dip in a HF solution to remove the native oxide before the screen-printing step. As shown in Figure 2, the samples with AlO_x/SiN_y stack lead to significantly lower contact resistivities than that without AlO_x/SiN_y stack for all considered peak firing temperatures.

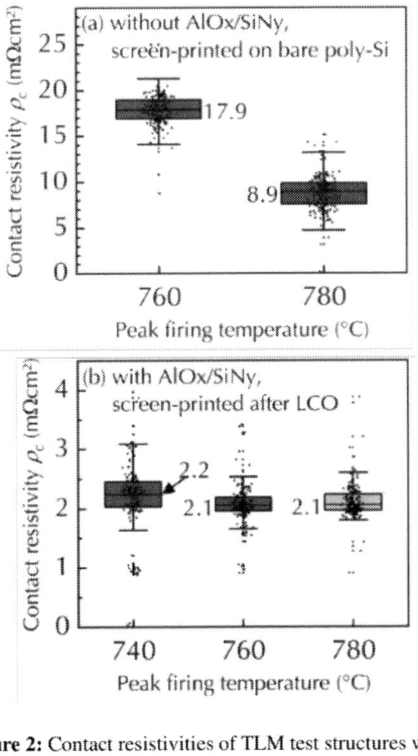

Figure 2: Contact resistivities of TLM test structures with screen-printed Al on (a) bare poly-Si and (b) AlO_x/SiN_y on top of poly-Si. The AlO_x/SiN_y stack is locally opened with a pico-second laser system.

We than remove the screen-printed Al wet-chemically and investigate the contacted area with scanning electron microscope (SEM). As shown in Figure 4, the poly-Si layer is mostly intact, indicating that Al is not alloying into the poly-Si on in large areas (in contrast to "normal PERC-like" Al pastes). However, a few grains of the poly-Si are removed in the Al etch as well, in particular for the sample shown in (a) w./o. AlO_x/SiN_y. We therefore cannot exclude the formation of Al p^+ alloyed regions on the scale of a few 100 nm.

The LCO process increases the surface roughness. We therefore speculate that the LCO process creates defects which improve the contact formation between screen-printed Al and poly-Si.

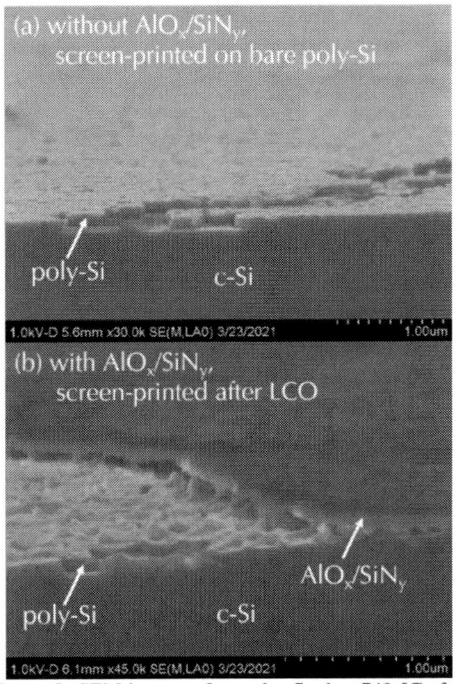

Figure 3: SEM images of samples fired at 740 °C after removal of screen-printed Al on (a) bare poly-Si and (b) AlO_x/SiN_y after LCO process.

3.2 Co-firing of front and rear Al grids

In the first batch, we co-fire the front and rear Al grids in a single firing step. Figure 3 shows the *IV* parameters of busbarless POLO BJ solar cells as a function of peak firing temperature, measured in-house with the LOANA tool from pvtools. For both peak firing temperatures of 740 °C and 760 °C, the median *FF* is beneath 40 %, mainly caused by a low value of R_{shunt} less than 100 Ωcm². By decreasing the peak firing temperature to 700 °C, the efficiency, *FF* and R_{shunt} increase, resulting in the best cell with an efficiency of 17.9 % with a V_{oc} of 675 mV and a *FF* of 68 %. The increased R_{shunt} indicates that lower peak firing temperature reduces the deterioration of POLO contact at the rear. But such low temperature leads to a poor formation of p^+-type layers beneath Al contacts at the front side, as we observe with SEM at multiple contacts (not shown here). Therefore, we conclude to separate the firing steps for the front and rear Al grids.

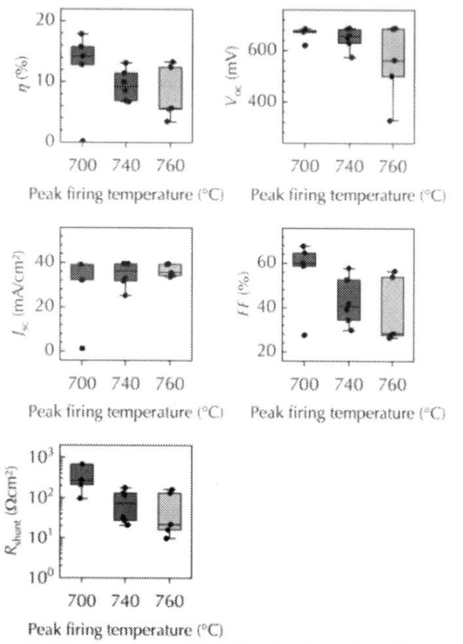

Figure 4: *IV* parameters of POLO BJ solar cells with co-fired front and rear Al grids as a function of peak firing temperature.

3.3 Separated firing of front and rear Al grids

Figure 6 shows the *IV* parameters of the second batch with separately fired front and rear Al grids. After screen-printing the front grid, we fire all samples at a peak firing temperature of 810 °C. We then screen-print the rear grid and fire the samples at temperatures between 660 °C and 740 °C. The firing of the front grid at 810 °C improves the p^+-type layer formation beneath the front Al contacts, resulting in V_{oc} of 700 mV, if a peak firing temperature of 680 °C is applied in the second firing step. The shunt resistance R_{shunt} increases by decreasing peak firing temperature during the second firing step for the rear Al grid. However, for the lowest peak firing temperature of 660 °C, the median *FF* is only 26 %, indicating the poor contact formation between screen-printed Al and poly-Si at the rear.

It is important to note that all POLO BJ solar cells of the second batch show strong inhomogeneity in EL images as shown exemplary in Figure 5, indicating a high resistive losses at Al contacts. We did not observe such inhomogeneity in the first batch. Since the front Al contacts are fired at 810 °C, it is likely to conclude that the rear Al contacts are not formed properly. Since both sides receive the LCO process before the screen-printing step of the front grid, it is possible that an oxide layer grows on top of poly-Si during the first firing step that may hinder the contact formation between screen-printed Al and poly-Si at the rear. In such cases, the uncertainty in the measured *FF* can be especially high, since our busbarless solar cells are measured with the grid-resistance neglecting contacting scheme [6]. In order to quantify the measurement uncertainty caused by contacting inhomogeneous busbarless solar cells, we fabricate a third batch of POLO BJ solar cells with front and rear Al fingers, and 9 Ag busbars at both sides. The best cell in the third batch shows an efficiency of 19 % with a V_{oc} of 689 mV and a *FF* of 67 %.

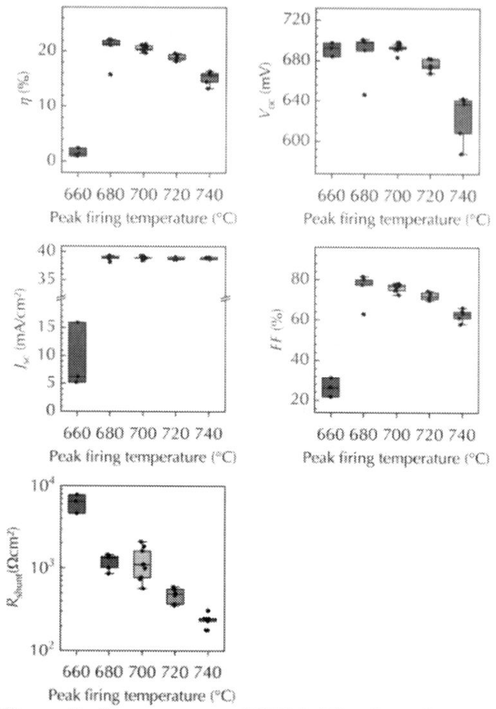

Figure 5: *IV* parameters of POLO BJ solar cells with separately fired front and rear Al grids as a function of peak firing temperature for the rear Al grid.

The reduced V_{oc} may be explained by additional recombination losses beneath the unoptimized Ag busbars. The low *FF* of the 9 busbar POLO BJ cells shows that the uncertainty in the measured *FF* of the busbarless POLO BJ solar cells in the second batch is considerably high, while the V_{oc} values are on comparable level for both busbarless and 9 busbar POLO BJ solar cells. We therefore explicitly emphasize that the *FF* and hence the efficiency in Figure 5 possess high measurement uncertainties and should be treated with caution.

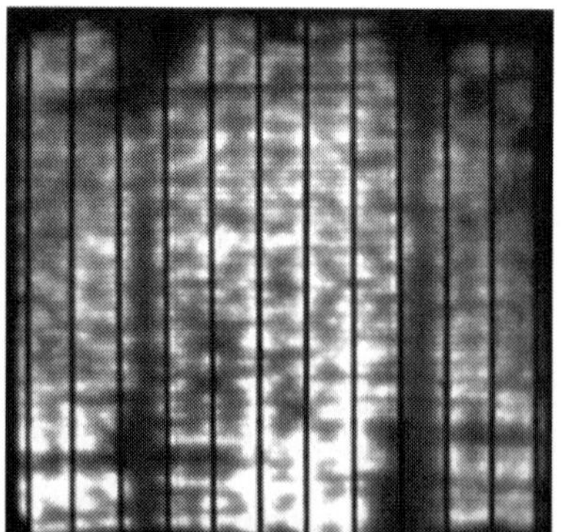

Figure 6: EL image of a POLO BJ solar cell with separately fired front and rear Al grids

4 CONCLUSION

We demonstrate experimentally the feasibility to

realize all-aluminum screen-printed p-type back junction solar cells featuring a passivating POLO contact. Our results show the possibility to reduce the Ag consumption of crystalline solar cells significantly. We show that the laser contact opening process reduces the contact resistivities between screen-printed Al and phosphorus-doped poly-Si significantly due to changed surface morphology of the poly-Si layer. Our best all-aluminum screen-printed POLO BJ cell has a V_{oc} of 700 mV on a p-type Cz silicon wafer. This is a very promising starting point to realize Ag-free crystalline silicon solar cells for the forthcoming multi-terawatt era of photovoltaics.

[1] B. Hallam, M. Kim, Y. Zhang, L. Wang, A. Lennon, P. Verlinden, P.P. Altermatt, P.R. Dias. "The silver learning curve for photovoltaics and projected silver demand for net-zero emissions by 2050", Prog. Photovoltaics 31, 598-606 (2023).

[2] R. Brendel, C. Kruse, A. Merkle, H. Schulte-Huxel, F. Haase, R. Peibst. "Screening Carrier Selective Contact Combinations for Novel Crystalline Si Cell Structures", Proc. of the 35th European Photovoltaic Solar Energy Conference and Exhibition, Brussels, Belgium, 2018, pp. 39 - 46.

[3] B. Min, N. Wehmeier, T. Brendemuehl, F. Haase, Y. Larionova, L. Nasebandt, H. Schulte-Huxel, R. Peibst, R. Brendel. "716 mV Open-Circuit Voltage with Fully Screen-Printed p-Type Back Junction Solar Cells Featuring an Aluminum Front Grid and a Passivating Polysilicon on Oxide Contact at the Rear Side", Solar RRL 5, 2000703 (2021).

[4] B. Min, V. Mertens, Y. Larionova, T. Pernau, H. Haverkamp, T. Dullweber, R. Peibst, B. R. "24.2 %-efficient POLO back junction solar cell with industrial PECVD AlO_x/SiN_y passivation", Proc. of the 40th European Photovoltaic Solar Energy Conference and Exhibition, Lisbon, Portugal, 2023.

[5] S. Suzuki, Peng Z., Tsuji K., Nakahara M., Morishita N., Kuroki T., Minde A., Dhamrin M., B. T. "Screen-Printed Aluminium Contacts on n^+-Doped Silicon", Proc. of the 37th European Photovoltaic Solar Energy Conference and Exhibition, Online, 2020, pp. 542 - 545.

[6] K. Bothe, C. Kruse, D. Hinken. "Contacting of busbarless solar cells for accurate I-V measurements", Proc. of the 37th European Photovoltaic Solar Energy Conference and Exhibition, Online event, 2020.

PROGRESS IN DEVELOPMENT OF POLYZEBRA IBC SOLAR CELLS

Jonathan Linke, Florian Buchholz, Saman Sharbaf, Jan Hoß, Valentin Mihailetchi, Jan Lossen, Radovan Kopecek
International Solar Energy Research Center Konstanz e.V.
Rudolf-Diesel-Str. 15, D-78467 Konstanz, Germany
Corresponding author: Jonathan Linke, +49 7531 36 183 362, jonathan.linke@isc-konstanz.de

ABSTRACT: The polyZEBRA IBC solar cell concept features both polarities as passivating poly-Si/SiO_x contacts on the rear side. The cells are fabricated on industrially established equipment only. In particular, both polarities are contacted in the same screen-printing step. In the case of optimal passivating contacts, a diminishing iV_{oc}-to-V_{oc} loss is expected. Recent progress in the polyZEBRA development strongly reduced to this iV_{oc}-to-V_{oc} loss from >10mV to only a few mV by applying an optimized rear side AlO_x passivation capping. Another progress was the reduction of the base resistivity as simulations predicted a fill factor gain of 2.6%. However, in practice only a gain of 1% could be achieved, meaning that the fill factor is limited by an effect not considered in the simulation. Bringing together these optimizations and assuming literature-based high passivation values for the individual regions of the solar cells reveals a simulated cell efficiency potential of >25%.

1 INTRODUCTION

The polyZEBRA IBC solar cell concept features both polarities as passivating poly-Si/SiO_x contacts on the rear side, while emitter and base are separated by a p-diffused gap region. Patterning of the emitter and base are achieved by laser processes, namely laser-induced SiN_x mask ablation for patterning of the base and laser-induced activation of dopants for local etch-stability of the emitter region. Details about the process flow can be found in [1,2].

The polyZEBRA IBC solar cell concept bases on the industrially-established ZEBRA [3] process. As such, existing ZEBRA production lines may be upgraded to polyZEBRA with only a few additional fabrication steps. In particular, the metallization scheme is not altered which enables to transfer the recently developed Cu-ZEBRA concept [4] also to polyZEBRA in order to reduce silver consumption. So far, 23.4% cell efficiency has been achieved on full-area M6 wafers with manual wafer handling in the lab and >24% is expected still this year. These numbers are expected to be significantly increased in an automated production line.

In this contribution, recent progress in the development of polyZEBRA cells is presented. First, based on the passivating contact nature, a very low iV_{oc}-to-V_{oc} loss is expected. This was demonstrated by an improved capping of the rear side AlO_x passivation. Second, simulations predicted a fill factor gain of 2.6% by changing from high to low base resistivity, which was tested experimentally. And third, bringing together these optimizations and literature-based passivation values for the emitter, base and gap/front side regions yields a simulated efficiency potential of >25% for the polyZEBRA cells.

2 EXPERIMENTAL

polyZEBRA solar cells were fabricated according to the process flow described in [1]. Additionally, test structures were fabricated that followed the exact same process flow as the cells, but laser patterning was adjusted so that on each wafer three 4x4 cm² squares for each region (emitter, base and gap/front side) were produced. In this way, each region may be characterized individually without limitations of the IBC pattern on µm-scale.

For testing the improved rear side AlO_x capping, one cell group received the reference capping A, which is a stack of SiO_xN_y/SiN_z. The second group received the improved capping B, which is a single SiN_x layer and thicker than capping A.

The influence of the base resistivity was investigated by using high resistivity 20 Ωcm base material for one group and low resistivity 1.4 Ωcm base material for the second group.

Simulations were performed using Quokka3 [5]. The baseline simulation used experimental data as input values and yields cell efficiencies within the error bars of the experimental cell efficiencies. The efficiency potential was estimated by gradually replacing the experimental input values with literature values, as depicted in Table 1.

Cell characterization was done with a conventional Halm IV flasher. The passivation of cell precursors and the test structures was measured by quasi steady-state photoconductance decay (QSSPC) using a Sinton lifetime tester.

3 RESULTS AND DISCUSSION

3.1 Improved rear AlO_x capping

In Fig. 1, a summary of the IV-measurements is shown of both cell groups with the reference capping A and the improved capping B. The iV_{oc} values were measured on cell precursors, which were fired but not metallized. It is striking, that the iV_{oc}-to-V_{oc} loss is reduced from ~15 mV for capping A to only a few mV for capping B. At the same time, the pFF-to-FF loss is increased from ~2%$_{abs}$ for capping A to ~3%$_{abs}$ for capping B.

Both observations are compatible with the following reasoning: Using the thinner capping A, the paste spikes through the interfacial oxide into the base and causes significant recombination, but enables a rather low series resistance, as one part of the carriers can avoid the interfacial oxide by entering the metal directly from the base. This low series resistance represented by the low difference of pFF and FF. Changing to the thicker capping B, the spiking of the paste through the interfacial oxide and thus the high recombination is suppressed, which limits the iV_{oc}-to-V_{oc} loss to a few mV. However, as the interfacial oxide is assumed to be intact, all carriers have to pass the interfacial oxide. This additional resistivity contribution causes a higher series resistance compared to capping A, represented by the larger difference between pFF and FF.

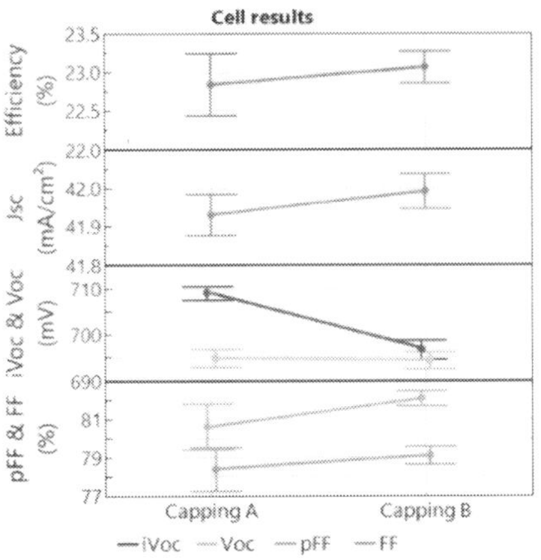

Figure 1: Cell result summary of both the reference capping A and the improved capping B.

Interestingly, even though the series resistance seems to be increase with capping B, the absolute value of the FF is ~0.5%$_{abs}$ higher than for capping A due to a an ~1.5%$_{abs}$ higher pFF, which eventually also increases the mean value of the cell efficiency of capping B compared to capping A. This higher pFF is also consistent with the hypothesis of a suppressed paste spiking through the interfacial oxide.

A second observation is the iV_{oc} loss for capping B compared to capping A of >10 mV. QSSPC measurements of the test structures reveals that this loss is solely attributed to a loss in p$^+$ poly-Si passivation as shown in Fig. 2.

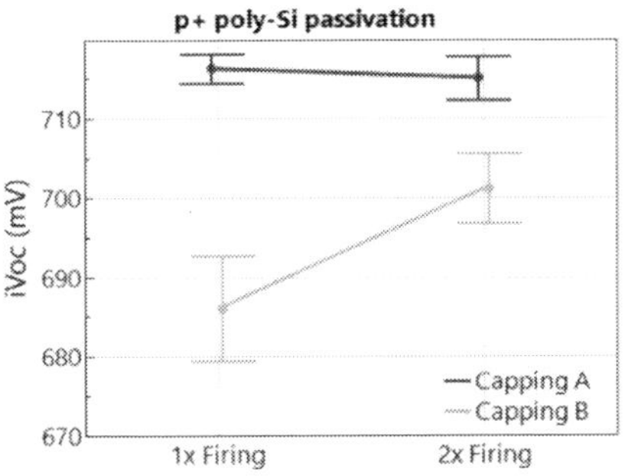

Figure 2: iV_{oc} of the p$^+$ poly-Si emitter region measured on the test structures for both cappings. Measurements were done on the same wafers after 1st and 2nd firing with same parameters.

At a first glance it is not clear, why the passivation should be worse by only changing the capping layer. A hint for the underlying mechanism was found by subjecting the wafers to a second firing step with identical parameters, in which the passivation significantly improved. A possible explanation would be that the doubled thermal budget provides more energy and time for the hydrogen atoms to diffuse to the interface below the p$^+$ poly-Si, where they may saturate defects. In [6] it was reported, that different properties of SiN_x layers change the mechanics of hydrogen in-diffusion and effusion if subjected to a fast firing step. Thus, it is possible that the microstructure of capping B prevents sufficient hydrogen supply, while for the same thermal budget the microstructure of capping A enables the hydrogen to saturate a sufficient number of defects.

This conclusion opens up an interesting route for optimization of the microstructure of the rear AlO_x capping while maintaining the parameters of the firing step, which were optimized for contact formation. In this way both the sufficient hydrogen supply from capping A and the very low iV_{oc}-to-V_{oc} loss of capping B would be combined.

3.2 Lower base resistivity

In this section, the results of the comparison of high and low base resistivity material for polyZEBRA cells is presented. Based on the experience with standard ZEBRA cells, the polyZEBRA development was initiated on medium to high base resistivity material >10 Ωcm. However, for IBC cells usually rather low base resistivity ~2 Ωcm material is recommended as result of a trade-off between bulk lifetime and internal resistance [7]. Considering a change from high (20 Ωcm) to low (1.4 Ωcm) base resistivity for polyZEBRA cells, Quokka3 simulations predicted a loss in J_{sc} of ~0.3 mA/cm², but a significant gain in fill factor of ~2.6%$_{abs}$ from the additional lateral conductivity in the bulk. According to this simulation, the fill factor gain overcompensated the J_{sc} loss, yielding a 0.6%$_{abs}$ gain in cell efficiency.

These predictions have been tested experimentally. The summary of the IV-measurement cell results are shown in Fig. 3. When changing from high to low base resistivity, a J_{sc} loss of ~0.3 mA/cm² can be seen, which is in line with the predicted behavior. However, the gain in fill factor of ~1%$_{abs}$ is less then half of the predicted value. This means, that the fill factor is limited by some other effect like transport through the poly-Si/SiO_x structures to the metal contacts. Nevertheless, also this lower gain in fill factor overcompensates the J_{sc} loss and results in a mean cell efficiency gain of 0.1%$_{abs}$.

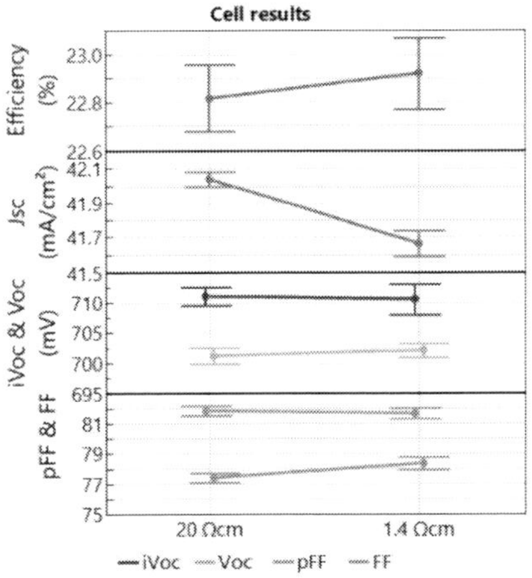

Figure 3: Cell result summary of both the high and low base resistivity groups.

One could expect, that a lower bulk lifetime of the low base resistivity material could negatively affect the pFF.

Indeed, the mean value is slightly reduced, but this variation lies within the statistical uncertainty. So this effect seems to play a minor role at least for this specific tested material.

3.3 Cell efficiency potential

The cell efficiency potential for the polyZEBRA concept was estimated by using Quokka3 simulations. First, all experimentally accessible input parameters were fixed for the baseline simulation. These include in particular the base resistivity ρ_{base}, the surface recombinations $J_{0,pass}$ and sheet resistances R_{sheet} of emitter/base/gap/front side regions, the contact resistivities ρ_c of emitter/base region and the rear pattern geometrics, i.e. width of emitter/base/gap region W and cell pitch p. All parameters that are not experimentally accessible were roughly estimated and adjusted, so that the resulting simulated cell efficiency roughly matches the experimental values. In particular, the recombination below the metal contacts $J_{0,met}$ was varied in this way, as it is difficult to determine experimentally the rather low values achieved with passivating poly-Si/SiO_x contacts.

Based on the baseline simulation, improvements expected in the near future were gradually implemented in the simulation in order to show an optimization path towards high efficiency cells. First, the lower base resistivity and the capping B were implemented according to the results presented in the previous sections. The next step is the assumption to solve the loss in p$^+$ poly-Si passivation of capping B by an further optimized capping. As next step an optimized rear pattern was implemented. And lastly, the surface recombination values were set to the values reported in the literature [8] and the metal recombination values to 1 fA/cm² representing ideal passivating contacts. The simulation parameters that were changed during each gradual improvement are summarized in Table 1.

Table I: Simulation parameters used for gradual improvements.

Improvement	Altered input values	Old values	New values
Low base resistivity	ρ_{base}	20 Ωcm	1.4 Ωcm
Capping B	$J_{0,met,emi}$ $J_{0,pass,emi}$	400 fA/cm² 20 fA/cm²	50 fA/cm² 40 fA/cm²
Optimized capping	$J_{0,pass,emi}$	40 fA/cm²	10 fA/cm²
Optimized rear pattern	Cell pitch	1080 µm	800 µm
Optimized gap/front side passivation	$J_{0,pass,gap}$	25 fA/cm²	5 fA/cm² [8]
Optimized poly-Si passivation	$J_{0,met,emi/base}$ $J_{0,pass,emi}$ $J_{0,pass,base}$	50 fA/cm² 10 fA/cm² 5 fA/cm²	1 fA/cm² 5 fA/cm² [8] 1 fA/cm² [8]

Figure 4 shows the simulated polyZEBRA cell efficiency gain for each gradual improvement. The baseline of ~23% is consistent with the experimental data from the reference groups "Capping A" and "High base resistivity" from the previous sections. As discussed before, changing to low base resistivity predicted an efficiency gain of 0.6%$_{abs}$. This was not yet confirmed experimentally, but is expected to be as soon as the other effect limiting the fill factor will be identified and solved. Applying the capping B does not change the simulated efficiency significantly as the lower $J_{0,met}$ and the higher $J_{0,pass}$ of the p$^+$ poly-Si cancel out. The latter is solved by the optimized capping and so the efficiency increases. The optimization of the rear pattern enables ~24% cell efficiency and eventually using the optimum literature recombination values of all regions of the cell, the simulated cell efficiency exceeds the 25%.

It should be noted, that the baseline efficiency is based on an in-house lab-scale cell process with no automation. It is expected to gain ~0.5%$_{abs}$ in cell efficiency in a future production line so that the efficiency potential might even be higher.

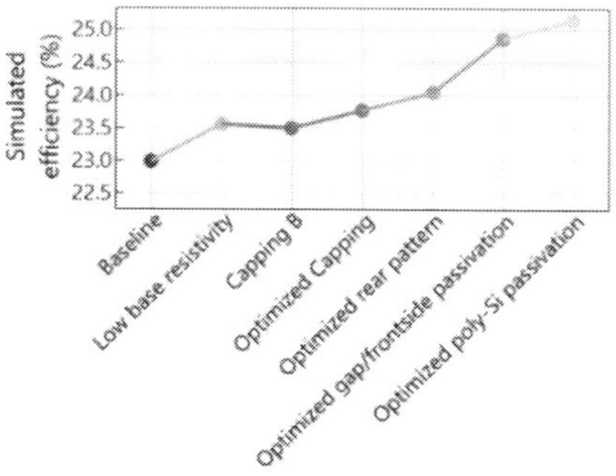

Figure 4: Simulated cell efficiency by gradual improvements according to the input values in Table 1.

4 CONCLUSION

In this contribution, recent progress in the polyZEBRA IBC cell development was reported. It was shown, that an optimization of the rear AlO_x capping enables to minimize the iV_{oc}-to-V_{oc} loss as well as the recombination below the p$^+$ poly-Si. Furthermore, switching from high to low base resistivity material causes a slight J_{sc} loss, which is overcompensated by a fill factor gain. Bringing these two findings and further optimizations together, simulations reveal an efficiency potential of >25% for the polyZEBRA cell concept.

5 ACKNOWLEDGEMENTS

This work was supported by the project IBC4EU and has received funding from the European Union's Horizon Europe research and innovation programme under grant agreement No.101084259.

6 REFERENCES

[1] J. Linke *et al.*, "Fully Passivating Contact IBC Solar Cells Using Laser Processing," in *Proc. of 8th World Conference on Photovoltaic Energy Conversion WCPEC*, Milan, 2022. doi: 10.4229/WCPEC-82022-1CV.2.11.

[2] F. Buchholz *et al.*, "Local Passivating Contacts from Laser Doped p+ Polysilicon," in *Proc. of 38th European Photovoltaic Solar Energy Conference and Exhibition EUPVSEC*, Online, 2021. doi: 10.4229/EUPVSEC20212021-2BO.11.3.

[3] R. Kopecek *et al.*, "Interdigitated Back Contact Technology as Final Evolution for Industrial Crystalline Single-Junction Silicon Solar Cell," *Solar*, vol. 3, pp. 1–14, 2023, doi: 10.3390/solar3010001.

[4] N. Chen *et al.*, "Thermal Stable High-Efficiency Copper Screen Printed Back Contact Solar Cells," *Solar RRL*, vol. 7, no. 2, p. 2200874, 2022, doi: 10.1002/solr.202200874.

[5] A. Fell, "Quokka3." 2023.

[6] D. Bredemeier, D. C. Walter, R. Heller, and J. Schmidt, "Impact of Silicon Nitride Film Properties on Hydrogen In-Diffusion into Crystalline Silicon," in *Proc. of 36th European Photovoltaic Solar Energy Conference and Exhibition EUPVSEC*, Marseille, 2019, pp. 112–115. doi: 10.4229/EUPVSEC20192019-2AO.4.4.

[7] P. Verlinden, "Interdigitated Back Contact Solar Cells," in *Photovoltaic Solar Energy*, A. Reinders, P. Verlinden, W. van Sark, and A. Freundlich, Eds., Chichester, UK: John Wiley & Sons, Ltd, 2017, pp. 92–103. doi: 10.1002/9781118927496.ch10.

[8] J. Schmidt, R. Peibst, and R. Brendel, "Surface passivation of crystalline silicon solar cells: Present and future," *Sol. Energy Mater. Sol. Cells*, vol. 187, pp. 39–54, 2018, doi: 10.1016/j.solmat.2018.06.047.

REAR-JUNCTION TUNNEL OXIDE PASSIVATED CONTACT SILICON SOLAR CELLS WITH SELECTIVE FRONT SURFACE FIELD

Vaibhav V. Kuruganti*[1], Thomas Buck[1], Sebastian Veerman[1], Christian Ebert[2], Sven Seren[2], Radovan Kopecek[1], Valentin D. Mihailetchi[1]
[1] ISC Konstanz e.V., Rudolf-Diesel-Str. 15, 78467 Konstanz, Germany
[2] SCHMID, Robert-Bosch str. 32-36, 72250 Freudenstadt, Germany
*vaibhav.kuruganti@isc-konstanz.de

ABSTRACT: Tunnel oxide passivated contacts (TOPCon) solar cells are a promising technology for the next generation of c-Si solar cells. In this study we explore the efficiency potential and optimize the fabrication process of a TOPCon front-back contact (FBC) n-type cell concept with a rear junction (RJ) topology, i.e., the cell has a APCVD p^+ poly-Si rear emitter and a $POCl_3$ diffused n^+ front surface field (FSF). Additionally, the FSF is selectively laser doped to form an n^{++} region under the metal contacts whereas the n^+ region between contact fingers is optimized solely for best surface passivation. The optimization of the FSF processes including laser doping and the deposition parameters of the TOPCon RJ were optimized independently. The APCVD deposition parameters were optimized to obtain the "in-situ annealed" p^+ poly-Si layers with excellent electrical and passivation quality. Subsequently, the learnings from these individually optimized front FSF and rear side TOPCon junction processes will be integrated for fabricating high efficiency TOPCon RJ solar cells. Also, a Quokka 3 modelling of the performance potential of this cell concept based on realistic parameters from the experimental data showed conversion efficiency exceeding 24% on large wafers using industrial processes.

1 INTRODUCTION

Impressive high conversion efficiency in crystalline silicon (c-Si) solar cells have been achieved in mass-production in the past few year [1][2], which resulted from a steady improvement in production technology and implementation of some advance technologies such as passivating contacts. Among passivating contacts approaches, two technologies are dominating the market; a low temperature approach based on hydrogenated amorphous silicon (a-Si:H) used in silicon heterojunction cell concepts (SHJ) [3], or a high temperature approach based on polycrystalline silicon thin films (also known as TOPCon - tunnel oxide passivating contact) [4]. The application of these passivating contacts technology led to conversion efficiencies up to 26.7% for an interdigitated back contact (IBC) cell architecture [5], and up to 26% for a front and back contact (FBC) cell topology [6].

The TOPCon design is based on the principle of separating the carrier extraction zone from the carrier generation zone using a thin interfacial silicon dioxide layer and highly doped polycrystalline silicon (SiO_2/poly-Si) layer [7]. The first report on the development of TOPCon solar cells was published in 2013 by Fraunhofer ISE [4] in which an ultra-thin SiO_2 layer along with a phosphorous-doped poly-Si layer at metal-semiconductor interface was used to significantly reduce the surface recombination at the rear side of n-type crystalline silicon solar cells. Ever since the inception of TOPCon solar cell concept, phosphorous-doped poly silicon layer was used due to historic reasons, excellent passivation quality and low contact resistivity [8].

In this work, we present TOPCon front-back contact (FBC) n-type cell concept with a rear junction (RJ) topology, i.e., the cell has a p^+ poly-Si rear emitter and a diffused n^+ front surface field (FSF). Some of the advantages of this cell concept over the traditional TOPCon front junction n-type cell concept topology includes:

1) Lower the thermal budget of the process sequence due to lack of high temperature boron diffusion.
2) Because of the RJ configuration, very lightly doped FSF can be used in the non-metallized areas thereby decreasing the parasitic absorption and improving the passivation quality.
3) Selective n^{++} FSF laser doping under the front contacts is performed using the well-known PERC technology.
4) Reducing the silver consumption by decreasing the metal fraction (by increasing the finger pitch) on the front side.
5) Using the novel concept of in-situ annealing of APCVD p^+ poly-Si layer during APCVD deposition, thereby omitting the need for an additional high temperature step for crystallization and dopant activation.

2 EXPERIMENT

The cross-section view of the RJ-TOPCon solar cell is depicted in Figure 1, while the corresponding process sequence of the TOPCon RJ solar cell is shown in Figure 2. This study was carried on industrial type Cz-Si n-type M2 wafer format. All the process equipment used was industrial or industrial-like, such as Centrotherm tube furnaces for interface oxide growth and $POCl_3$ diffusion. APCVD deposition of p^+ poly-Si was carried out on SCHMID APCVD system. The wet-bench processes were carried out on RENA inline and batch tools. Local laser doping for selective FSF was performed in an industrially feasible high throughput tool. The PECVD SiN_x deposition was carried out on Centrotherm PECVD system and firing in a Centrotherm fast firing furnace.

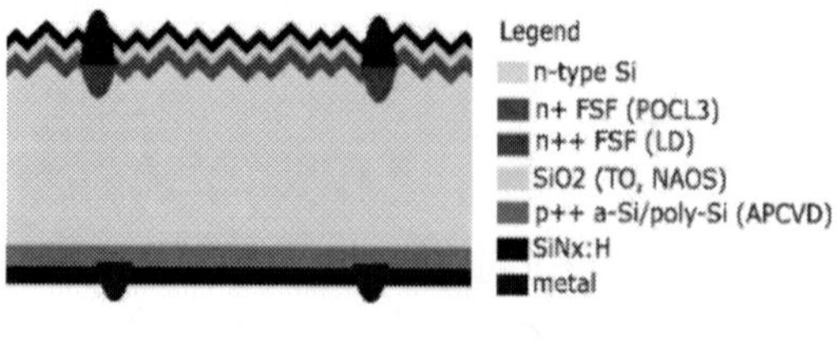

Figure 1: A schematic cross-section view of the resulting TOPCon RJ solar cell.

10.4229/EUPVSEC2023/1CV.3.5
020047-001

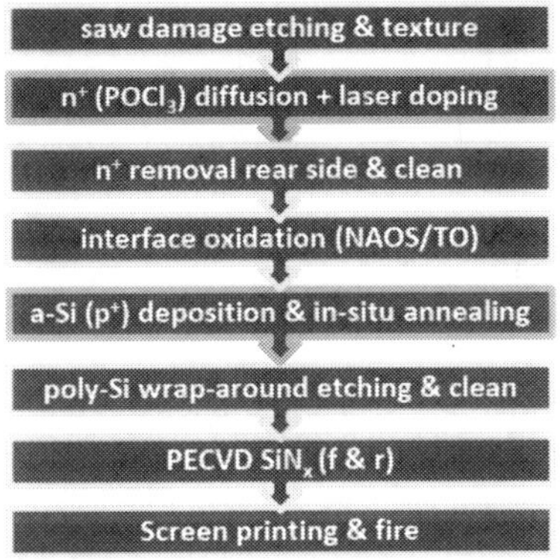

Figure 2: Process sequence of TOPCon RJ with APCVD deposited and annealed p^+ a-Si emitter.

3 RESULTS

3.1 n^+ $POCl_3$ FSF OPTIMIZATION

Figure 3 shows the typical carrier profile of the FSF n^+ $POCl_3$ diffusion process and their corresponding sheet resistance are depicted in the legend. A wide range of $POCl_3$ diffusion recipes without the subsequent high drive-in step were studied in this work. The $POCl_3$ diffusion recipes were optimized by varying the deposition temperature and deposition time with the aim to achieve a very shallow doping profile in order to minimize the parasitic absorption [9].

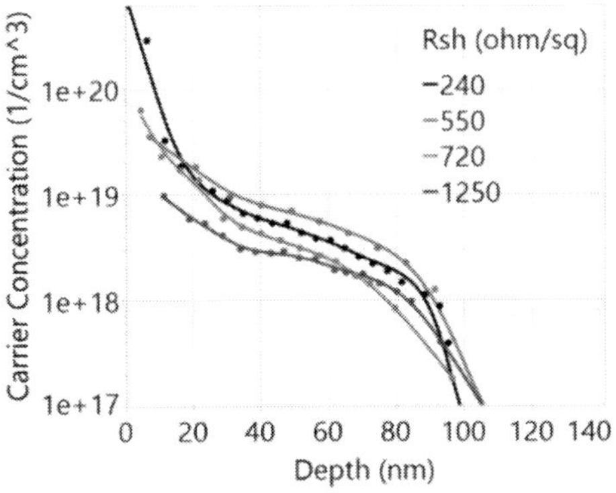

Figure 3: Active dopant concentration profiles of the n^+ FSF regions, determined from ECV measurements, for different $POCl_3$ diffusion recipes.

The diffusions were optimized to lower the surface concentration in order to obtain better passivation quality. The Figure 4 depicts how the dark saturation current density (J_{0pas}) decreases with increasing the sheet resistance and the obtained results are in good agreement with literature [10]. Due to the excellent passivation quality 1250 Ω/sq $POCl_3$ FSF was shortlisted for fabrication of TOPCon RJ solar cells.

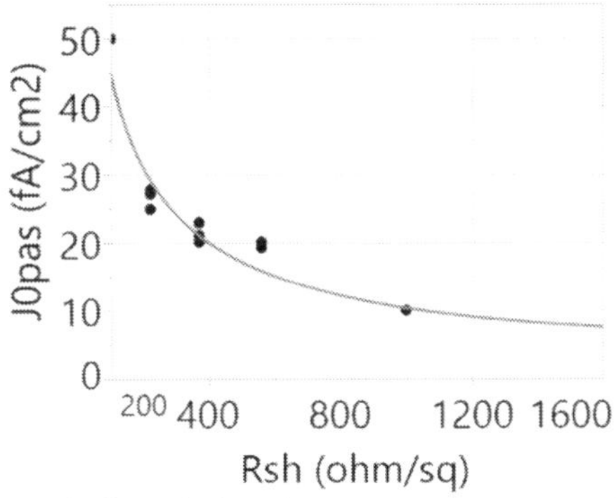

Figure 4: J_{0pas} of n^+ regions measured on symmetric diffused/passivated lifetime samples. The line represents a guide for the eye.

3.2 n^{++} $POCl_3$ SELECTIVE FSF OPTIMIZATION

Laser doping was performed to obtain the highly conductive n^{++} FSF layers selectively under the metal contacts. From Figure 5 it can be observed that the sheet resistance decreases with increasing laser fluence. Already at a fluence of 2 J/cm² we observe the sheet resistance decreases by one order of magnitude and with subsequent increase in the laser fluence the sheet resistance remains constant.

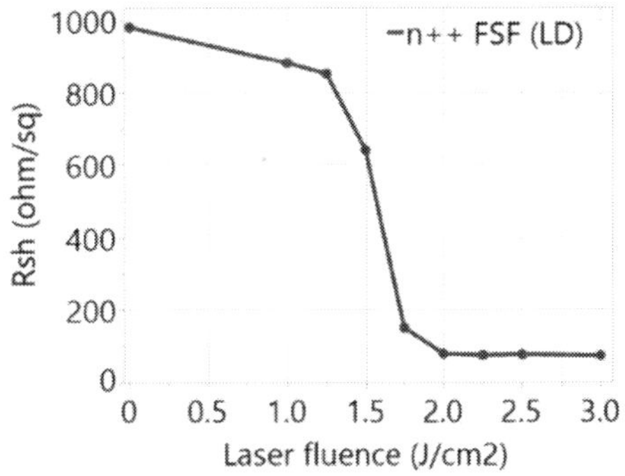

Figure 5: Influence of laser fluence (H_p) on the sheet resistance (R_{sh}).

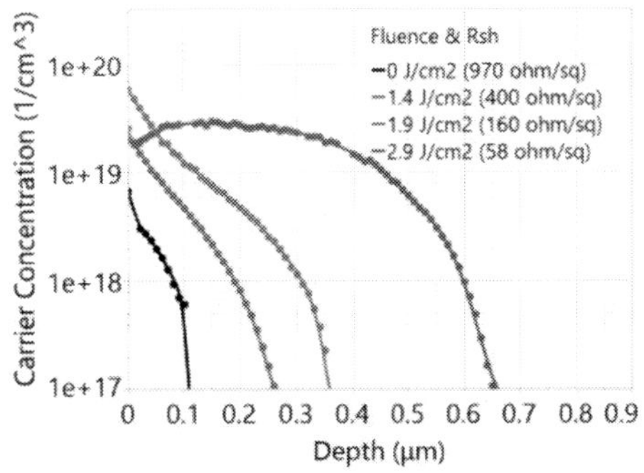

Figure 6: Active dopant concentration profiles of the n^{++} selective FSF regions, determined from ECV measurements, for different laser fluence.

Figure 6 shows the corresponding doping profiles of the laser doped regions. We observed that for laser fluence below 2 J/cm^2 the surface concentration increases from 8× 10^{18} to 7×10^{19} cm^{-3} with a slight increase in the junction depth. For higher laser fluence, such as 3 J/cm^2, the surface concentration decreases due to evaporation and a substantial increase in the junction depth is observed from 100 nm to almost 700 nm. In order to obtain minimal laser induced damages while maintaining highly conductive n^{++} FSF regions, the fluence of 2 J/cm^2 was shortlisted for fabrication of TOPCon RJ solar cells.

4.2 p+ poly-Si EMITTER OPTIMIZATION

In this work APCVD technology was used for the deposition of p$^+$ poly-Si layer. We develop a concept of in-situ annealing in which the deposited p$^+$ a-Si layer is annealed already during the APCVD deposition process, thereby omitting the need for an additional high temperature annealing [11]. In order to compare the electrical properties and passivation properties of in-situ annealed layer, we additional anneal half of the samples again in the quartz tube furnace at a temperature of 950°C for 30 minutes and are referred to as ex-situ annealed samples in this work. Figure 7 shows the three different temperature profiles studied in this work for the deposition of p$^+$ poly-Si. It can be categorized as three zones, namely: Zone-I as pre-deposition zones, Zone-2 as deposition zones and Zone-3 as post deposition zones. For all the profiles, the set temperatures in pre-deposition and deposition zone were kept constant while only the post deposition set temperatures were varied as shown schematically in Figure 7.

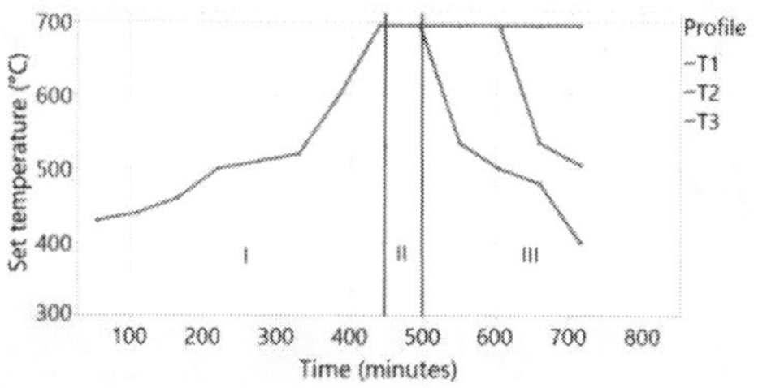

Figure 7: The T1, T2 and T3 temperature profiles (see legend) of APCVD p$^+$ a-Si layers studied in this work. The zone I, II and III represent pre-deposition zones, deposition zone and post-deposition zones respectively.

Figure 8 shows the measured sheet resistance as a function of different temperature profile used for APCVD p$^+$ poly-Si layer deposition for both in-situ and ex-situ annealed samples. The additional extensive annealing in the quartz tube furnace for 30 minutes at 950 °C for the ex-situ annealed layers leads to better crystallization and dopant activation yielding high-conductive and homogeneous layers in comparison to in-situ annealed layers. Progressing from T1 to T3 temperature profiles, the overall thermal budget during the deposition increases and it strongly influences the in-situ annealed layers when compared to ex-situ annealed layers. We obtain highly non-conductive layers with mean R_{sh} of 4000 Ω/sq at the T1 temperature profile to very conductive 195 Ω/sq layers with the T3 temperature profile in the case of in-situ annealed layers. The substantial decrease in the R_{sh} explains the influence of the temperature in the post-deposition zones of the APCVD system on the deposited layer's crystallinity and electrical properties.

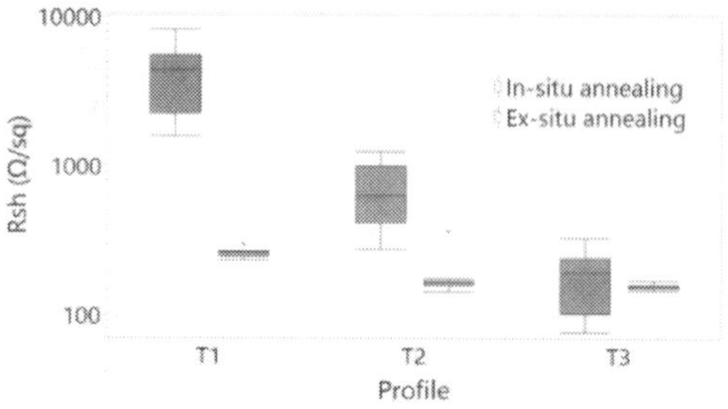

Figure 8: Sheet resistance (R_{sh}) of in-situ and ex-situ annealed p$^+$ poly-Si layers as a function of deposition temperature profiles T1, T2, and T3 respectively.

Figure 9 compares the measured iV_{oc} of the samples with in-situ and ex-situ annealing as a function of the APCVD deposition temperature profiles T1, T2, and T3 respectively. Excellent passivation quality is demonstrated for the p$^+$ poly-Si layers on symmetrical lifetime samples. It can be observed that ex-situ annealed p$^+$ poly-Si layers show better passivation homogenuity across the wafer in comparison to in-situ annealed p$^+$ poly-Si layers. Also the influence of the APCVD deposition temperature profile has a strong effect on passivation quality in-situ annealed p$^+$ poly-Si as compared to ex-situ annealed p$^+$ poly-Si. Consequently, we obtained the same passivation quality with iV_{oc} of 715–720 mV after fast-firing for both in-situ and ex-situ annealed poly-Si layers deposited with T3 temperature profiles.

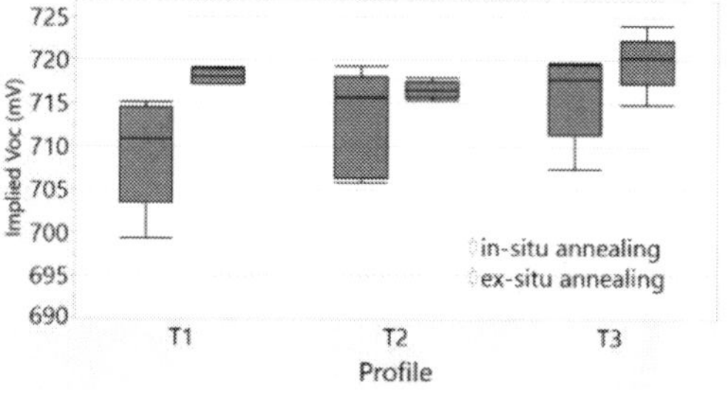

Figure 9: Implied open circuit voltage (iV_{oc}) of in-situ and ex-situ annealed poly-Si(p$^+$) layers after fast firing as a function of deposition temperature profiles T1, T2, and T3 respectively.

We have demonstrated comparable electrical and passivation quality of p$^+$ poly-Si layers deposited with T3 temperature profile for both in-situ and ex-situ annealed poly-Si layers. Due to the lack of additional high temperature step and still maintain good electrical and passivation quality, we aim to implement in-situ annealed p$^+$ poly-Si deposited with T3 temperature profile for our TOPCon RJ solar cells.

4.3 OPTIMIZED CELL MODELLING

The individual optimized processes for the front selective FSF and rear TOPCon p$^+$ emitter are currently integrated in a process flow to fabricate our TOPCon RJ n-type FBC cells on large area (M2) wafers. In the Figure 10, a Quokka3 modelling of the performance potential of this cell concept is presented based on realistic input parameters from the experimental data. In these simulations we show the dependence of this TOPCon RJ cells on front side metallization grid, FSF R_{sh}, and wafer resistivity. More than 24% conversion efficiency can realistically be expected on large wafers using industrial processes from this cell concept.

Table I: Experimental best obtained electrical parameters.

Parameter	Value/units
Front selective FSF: J_{0pa}/J_{0met}	10 / 250 fA/cm^2
Rear poly-Si emitter: J_{0pa}/J_{0met}	6 / < 50 fA/cm^2
FSF R_{sh} (n^+/n^{++}):	1000 / 80 Ω/sq
Emitter R_{sh} (p^+):	250 Ω/sq

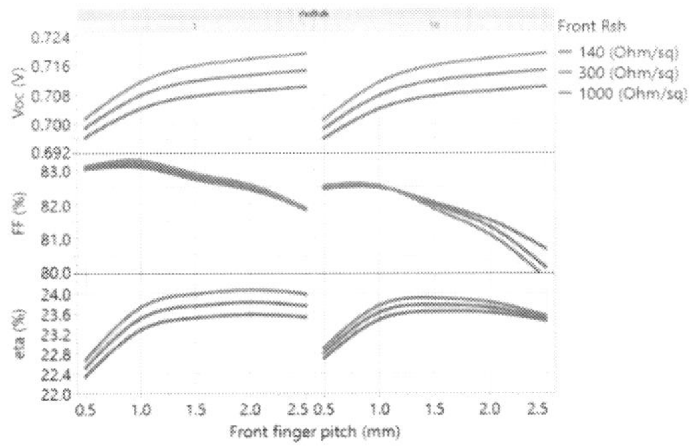

Figure 10: IV parameters of a TOPCon RJ cell simulated by Quokka3 as a function of front side finger pitch and FSF R_{sh} under the passivation area. Input parameters taken from experimental data or assumed based on literature reported data [6].

4.3 SUMMARY AND OUTLOOK

In this work, we present results of process optimization of our n-type TOPCon RJ solar cell concept. For the fabrication of the TOPCon RJ solar cells, we present the optimizations performed for selecting the suitable $POCl_3$ diffsuion recipe which provides less parasitic absorption and good passivation quality. The laser doping settings were optimized to obtain highly doped n^{++} c-Si layer under the metal contacts with minimal laser induced damage. The deposition parameters of the APCVD p^+ poly-Si layer were optimised such that no additonal high temperature annealing step was needed for the crystallization and dopant activation. Excellent electrical and passivation quality of the novel in-situ annealed p^+ poly-Si layers was demonstrated. Incorporation of the individual optimized process parameters in Quokka3 simulation model demonstrated efficiency exceeding 24% for this cell concept. Currently, work is underway to intergrate all the individual leanrings into fabrication process of TOPCon RJ n-type FBC cells on large area (M2) wafers.

8 REFERENCES

[1] D. Chen *et al.*, "24.58% total area efficiency of screen-printed, large area industrial silicon solar cells with the tunnel oxide passivated contacts (i-TOPCon) design," *Sol. Energy Mater. Sol. Cells*, vol. 206, no. October 2019, pp. 1–8, 2020, doi: 10.1016/j.solmat.2019.110258.

[2] Y. Chen *et al.*, "Mass production of industrial tunnel oxide passivated contacts (i-TOPCon) silicon solar cells with average efficiency over 23% and modules over 345 W," *Prog. Photovoltaics Res. Appl.*, vol. 27, no. 10, pp. 827–834, Oct. 2019, doi: 10.1002/PIP.3180.

[3] J. Haschke, O. Dupré, M. Boccard, and C. Ballif, "Silicon heterojunction solar cells: Recent technological development and practical aspects - from lab to industry," *Sol. Energy Mater. Sol. Cells*, vol. 187, pp. 140–153, Dec. 2018, doi: 10.1016/J.SOLMAT.2018.07.018.

[4] F. Feldmann, M. Bivour, C. Reichel, M. Hermle, and S. W. Glunz, "Passivated rear contacts for high-efficiency n-type Si solar cells providing high interface passivation quality and excellent transport characteristics," *Sol. Energy Mater. Sol. Cells*, vol. 120, no. PART A, pp. 270–274, Jan. 2014, doi: 10.1016/J.SOLMAT.2013.09.017.

[5] K. Yamamoto, K. Yoshikawa, H. Uzu, and D. Adachi, "High-efficiency heterojunction crystalline Si solar cells," *Jpn. J. Appl. Phys.*, vol. 57, no. 8, p. 08RB20, Aug. 2018, doi: 10.7567/JJAP.57.08RB20/XML.

[6] A. Richter *et al.*, "Design rules for high-efficiency both-sides-contacted silicon solar cells with balanced charge carrier transport and recombination losses," *Nat. Energy*, vol. 6, no. 4, pp. 429–438, 2021, doi: 10.1038/s41560-021-00805-w.

[7] B. Kafle, B. S. Goraya, S. Mack, F. Feldmann, S. Nold, and J. Rentsch, "TOPCon – Technology options for cost efficient industrial manufacturing," *Sol. Energy Mater. Sol. Cells*, vol. 227, no. February, p. 111100, 2021, doi: 10.1016/j.solmat.2021.111100.

[8] A. Chaudhary *et al.*, "Influence of Polysilicon Thickness on Properties of Screen-Printed Silver Paste Metallized Silicon Oxide/Polysilicon Passivated Contacts," *Phys. Status Solidi Appl. Mater. Sci.*, vol. 218, no. 18, 2021, doi: 10.1002/pssa.202100243.

[9] C. Battaglia, A. Cuevas, and S. De Wolf, "High-efficiency crystalline silicon solar cells: status and perspectives," *Energy Environ. Sci.*, vol. 9, no. 5, pp. 1552–1576, May 2016, doi: 10.1039/C5EE03380B.

[10] M. J. Kerr, J. Schmidt, A. Cuevas, and J. H. Bultman, "Surface recombination velocity of phosphorus-diffused silicon solar cell emitters passivated with plasma enhanced chemical vapor deposited silicon nitride and thermal silicon oxide," *J. Appl. Phys.*, vol. 89, no. 7, pp. 3821–3826, 2001, doi: 10.1063/1.1350633.

[11] Kuruganti, V. V. et al., "In-situ annealing of boron-doped amorphous silicon layers using APCVD technology" (Manuscript submitted for publication)

IMPROVED P⁺POLY-Si/SiOx PASSIVATING QUALITY ENABLED BY TUNING *EX-SITU* BBr3 DIFFUSION AND HYDROGENATION RECIPES

Yiğit Mert Kaplan[1,2]*, Berkay Uygun[1,2], Hasan Hüseyin Canar[1,2], Emine Hande Çiftpınar[1], Gökhan Altıner[1,2], Raşit Turan[1,2,3], Hisham Nasser[1]**

[1] ODTÜ-GÜNAM, Middle East Technical University, 06800 Ankara, Türkiye
[2] Micro and Nanotechnology Graduate Program of Natural and Applied Sciences, Middle East Technical University (METU), 06800 Ankara, Türkiye
[3] Department of Physics, Middle East Technical University (METU), 06800 Ankara, Türkiye

*mert.kaplan@odtugunam.org
**nasser@metu.edu.tr | hisham.nasser@odtugunam.org

ABSTRACT: Realizing high-quality passivation of boron-doped poly-Si on SiO_x passivating contacts (p-TOPCon) is one of the main challenges in the development of TOPCon solar cells. To overcome this challenge for *ex-situ* doped p-TOPCon, processes such as BBr_3 doping, and hydrogenation must be carefully optimized to reach high passivation quality on textured and polished c-Si surfaces. In this work, TOPCon structures (thin SiO_x with undoped poly-Si) with different SiO_x and poly-Si thicknesses were prepared by LPCVD on p-type Cz wafers. *Ex-situ* boron doping was realized by an atmospheric pressure diffusion furnace with two different boron doping recipes. The recipes were developed by changing the deposition and drive-in process parameters. For the hydrogenation step, p-TOPCon symmetrically passivated p-type wafers were coated by two and three hydrogenation stacks comprising ALD-AlO_x/PECVD-SiN_x and AlO_x/SiN_x/AlO_x, respectively. To further improve the passivation quality, the deposition temperature of SiN_x varied. Additionally, the passivation parameters were investigated after the high temperature firing step. The highest iV_{OC} of 710 mV ($J_{0,pass}$ = 6.5 fA/cm^2) was obtained from 100 nm poly-Si/1.6 nm SiO_x on SDE, and 675 mV ($J_{0,pass}$ = 22.3 fA/cm^2) is obtained from 150 nm poly-Si/1.6 nm SiO_x on textured p-type Cz wafers.
Keywords: passivating contacts, p-TOPCon, hydrogenation, *ex-situ* BBr_3 doping

1 INTRODUCTION

Recently, tunnel oxide passivating contact (TOPCon) structure has demonstrated promising advancements in c-Si solar cell performance. It has already been reported that TOPCon solar cells have attained an impressive efficiency value exceeding 26% [1]. Therefore, given the potential outcomes and developments, TOPCon-based solar cells are expected to dominate the PV industry. The passivation quality of hole-selective passivating contacts (p-TOPCon) is lower than that of the electron-selective counterpart (n-TOPCon) [2]. So far, substantial efforts have been invested to bring p-TOPCon to the same level as that achieved by n-TOPCon on polished and textured c-Si wafers [3]. While the inferior passivation quality of p-TOPCon structures is not fully understood, it can be mainly attributed to the boron doping profile in the poly-Si, boron segregation at the interfacial oxide [3], chemical passivation of the interfacial oxide, and the requirement of proper hydrogenation stacks [4]. Therefore, to achieve p-TOPCon structures with superior passivation quality, the above-mentioned points need to be carefully addressed.

In this work, we study the influence of varying the *ex-situ* doping recipe, tunnel oxide thickness, hydrogenation stacks parameters, and poly-Si thickness on the passivation quality of p-TOPCon on both alkaline textured and chemically polished Cz p-type c-Si wafers. The passivation parameters were also studied after the conventional fast firing (FFO) processing step, usually utilized to realize industrial screen-printed metallization.

2 DESIGN OF EXPERIMENT

The design of the experiment is given in Figure 1. In this work, p-type c-Si (156.75 x 156.75 mm^2, Cz, 1.2—1.5 Ω.cm) textured (TXT) and saw-damage etched (SDE) wafers were utilized.

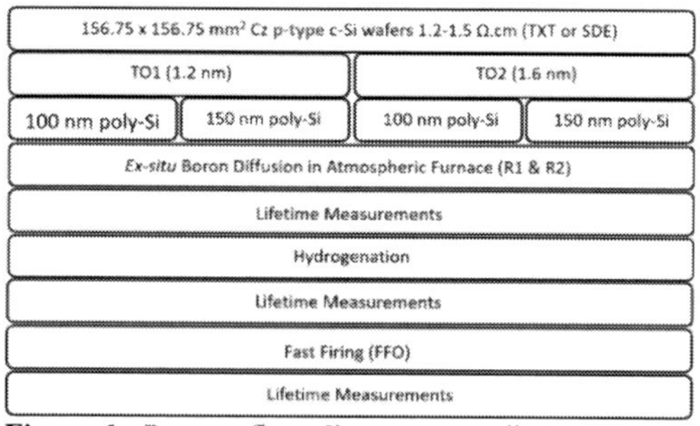

Figure 1: Process flow diagram to realize p-TOPCon structures on textured and chemically SDE c-Si wafers.

Low pressure chemical vapor deposition (LPCVD) was used to deposit both SiO_x and undoped poly-Si. Interfacial SiO_x thickness was fixed at either 1.2 nm (TO1) or 1.6 nm (TO2) while undoped poly-Si thickness was fixed at 100 nm or 150 nm. The thicknesses of these layers were measured by spectroscopic ellipsometry on SDE wafers. The poly-Si thickness on TXT surfaces is expected to be thinner by a factor of ~0.7 than the intended thicknesses (100 nm and 150 nm). Doping processes were realized by *ex-situ* boron diffusion using BBr_3 liquid precursor in an atmospheric diffusion furnace.

Two different boron doping recipes (R1 & R2) were implemented in the process. The recipes differ in duration, temperature of deposition, and temperature of drive-in. R1

10.4229/EUPVSEC2023/1CV.3.12
020048-001

has a deposition temperature of 850°C and 900°C for drive-in, while R2 has 850°C for both deposition and drive-in steps. For the hydrogenation, two different stacks were investigated. The first one consists of AlO_x/SiN_x (termed as two-stacks), and the other consists of $AlO_x/SiN_x/AlO_x$ (termed as three-stacks). SiN_x (refractive index fixed at n = 2.04) was prepared by PECVD at two different deposition temperatures (380°C and 450°C). AlO_x was deposited using a high throughput spatial ALD tool.

Finally, symmetrically coated samples with different hydrogenation stacks experienced FFO at various peak temperatures (750°C, 800°C, 850°C, 900°C, and 925°C). After each step, lifetime measurements were collected using a Sinton WCT-120TS lifetime tester.

3 RESULTS AND CONCLUSIONS

3.1 Sheet Resistance

The dependence of sheet resistance (R_{sheet}) of p-TOPCon on the doping recipe and the poly-Si thickness is given in Figure 2. P-TOPCon prepared using R1 exhibit R_{sheet} of 44 Ω/square and 53 Ω/square for 100 nm and 150 nm, respectively. The R_{sheet} values obtained for R2 recipes are 60 Ω/square and 62 Ω/square for 100 nm and 150 nm, respectively. The lower R_{sheet} values of p-TOPCon prepared using R1 are attributed to the higher process temperature compared to R2 recipe.

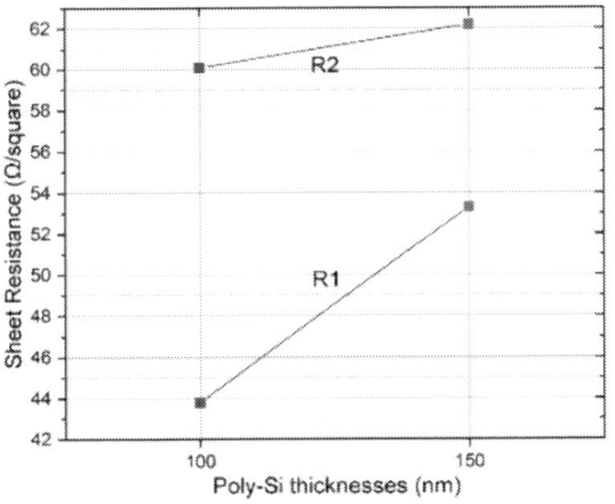

Figure. 2: Sheet resistance values of 100 nm and 150 nm thick poly-Si layers with respect to R1 and R2 doping recipes.

3.2 Dopant Profile

To assess the active boron concentration, electrochemical capacitance voltage (ECV) measurements were conducted (Figure 3 and Figure 4). ECV results show that the samples with a thicker poly-Si layer have a higher average boron concentration within the poly-Si layer compared to the thinner poly-Si layer. In the case of R1, the sample with 100 nm poly-Si/1.6 nm SiO_x has an average boron concentration of $6.58x10^{19}$ cm^{-3} in the poly-Si layer, while the sample with 150 nm poly-Si/1.6 nm SiO_x has $6.82x10^{19}$ cm^{-3}. Similarly, the same trend is observed for samples prepared using R2 recipe, where the 100 nm poly-Si/1.6 nm SiO_x sample has a concentration of $1.04x10^{20}$ cm^{-3}, and the 150 nm poly-Si/1.6 nm SiO_x sample has $1.32x10^{20}$ cm^{-3} in the poly-Si layer. These higher boron concentrations within the poly-Si layer compared to the samples with 1.2 nm SiO_x can be attributed to the improved field-effect passivation and selectivity of the TOPCon structure, resulting in better passivation quality [5], [6].

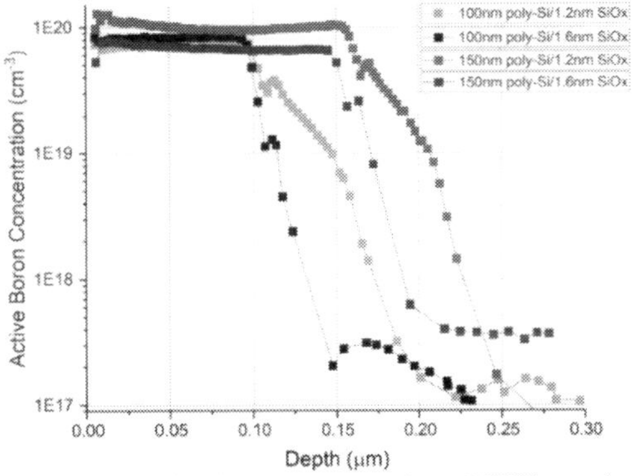

Figure 3: Active boron concentration of TXT samples with R2 boron doping recipe.

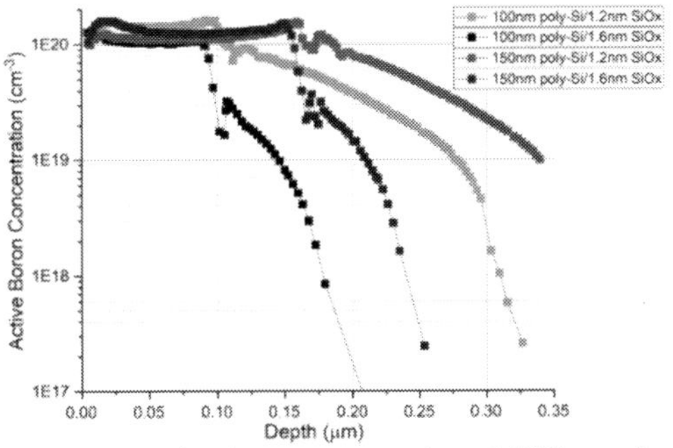

Figure 4: Active boron concentration of SDE samples with R1 boron doping recipe.

ECV spectra clearly demonstrate that samples with a 1.2 nm SiO_x layer are ineffective in blocking boron in-diffusion into the c-Si wafer when compared to 1.6 nm SiO_x counterpart. When we investigate these two cases in terms of SiO_x thicknesses and the $J_{0,\,pass}$ parameter, the sample with thinner SiO_x (R2/150 nm poly-Si/TO1/SDE) results in a $J_{0,pass}$ of 23.65 fA/cm^2, while the thicker SiO_x (R2/150 nm poly-Si/TO2/SDE) has a $J_{0,pass}$ of 6.40 fA/cm^2. It's worth noting that the hydrogenation and firing parameters were kept the same. Therefore, it can be concluded that the deeper boron in-diffusion in the c-Si results in a more pronounced Auger recombination when compared to the shallower boron in-diffusion case, which is also supported by [7]. Consequently, the passivation quality of the p-TOPCon structure can be correlated with the dopant concentration in poly-Si and the boron in-diffusion tail in the c-Si, as has been proposed in the literature [8].

3.3 Passivation Results

Two- and three-stacks hydrogenation configurations were investigated to ensure an efficient hydrogen supply to achieve the best passivation quality for the p-TOPCon structure.

Figures 5-9 show the iV_{OC} results of the p-TOPCon at each processing step. In all cases, p-TOPCon with 1.2 nm SiO_x exhibit lower passivation quality compared to these with 1.6 nm SiO_x. These results can be attributed to the diffusion blocking capability of the SiO_x, and of course, its chemical passivation.

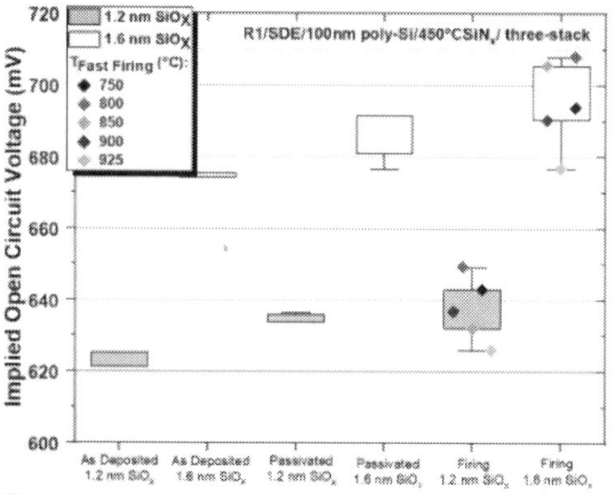

Figure 5: Variation in iV_{OC} values of 100 nm poly-Si with TO1 (grey) and TO2 (white) doped with R1 on SDE surfaces and coated by three-stacks with SiN_x deposited at 450°C.

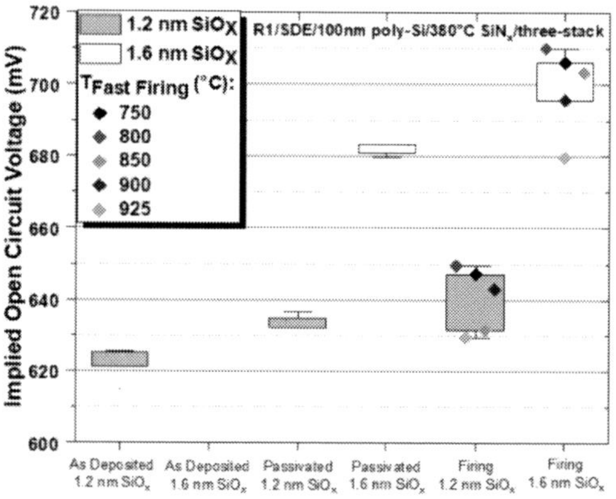

Figure 6: Variation in iV_{OC} values of 100 nm poly-Si with TO1 (grey) and TO2 (white) doped with R1 recipe on SDE surfaces and coated by three-stacks with SiN_x deposited at 380°C.

For all configurations, FFO at 800°C resulted in the highest iV_{OC}. Among p-TOPCon on SDE c-Si, the highest passivation parameters were achieved by R1/100 nm poly-Si/1.6 nm SiO_x with three-stacks hydrogenation as shown in Figure 5 (iV_{OC} = 710 mV, $J_{0,\,pass}$ = 6.5 fA/cm^2). On the other hand, for the p-TOPCon/ on textured c-Si, the highest passivation parameters were achieved by R2/150 nm poly-Si/1.6 nm SiO_x (Figure 9) with iV_{OC} = 675 mV, $J_{0,\,pass}$ = 22.3 fA/cm^2).

When FFO temperature is increased beyond 800°C, passivation quality starts to degrade. This degradation can be attributed to the redistribution of hydrogen atoms, i.e., hydrogen effusion [4][9]–[11].

To understand the influence of SiN_x deposition temperature, SiN_x was prepared at 380°C or 450°C. Changing the deposition temperature has no pronounced influence on the passivation parameters of R1/SDE/100 nm poly-Si/1.6 nm SiO_x/three-stack (Figure 5 and Figure 6). The passivation structures with SiN_x deposited at 450°C resulted in an iV_{OC} of 708 mV while that with SiN_x deposited at 380°C resulted in an iV_{OC} of 710 mV at the same FFO temperature. Similar behavior is observed for p-TOPCon on TXT c-Si (Figure 7 and Figure 8). Two-stacks hydrogenation configuration resulted in the highest passivation parameters (675 mV, 22.3 fA/cm^2) of p-TOPCon on TXT surfaces. This can be attributed to the more pronounced Si-H bonding in SiN_x deposited at 380°C [12]

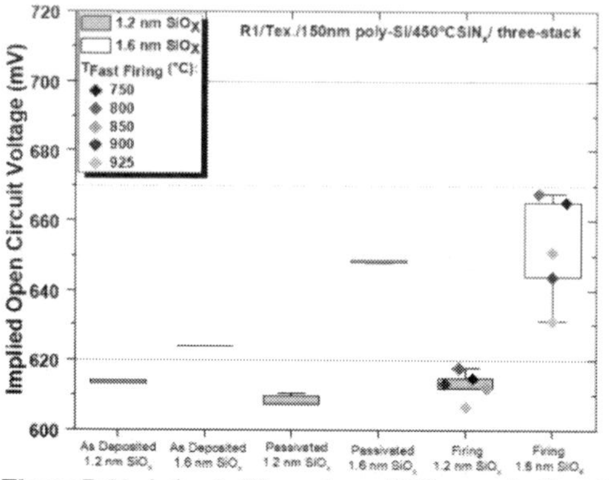

Figure 7: Variation in iV_{OC} values of 150 nm poly-Si with TO1 (grey) and TO2 (white) doped with R1 recipe on TXT surfaces and coated by three-stacks with SiNx deposited at 450°C.

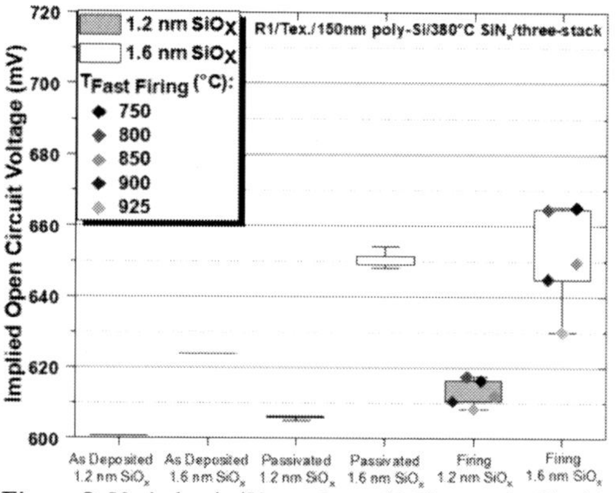

Figure 8: Variation in iV_{OC} values of 150 nm poly-Si with TO1 (grey) and TO2 (white) doped with R1 recipe on TXT surfaces and coated by three-stacks with SiNx deposited at 380°C.

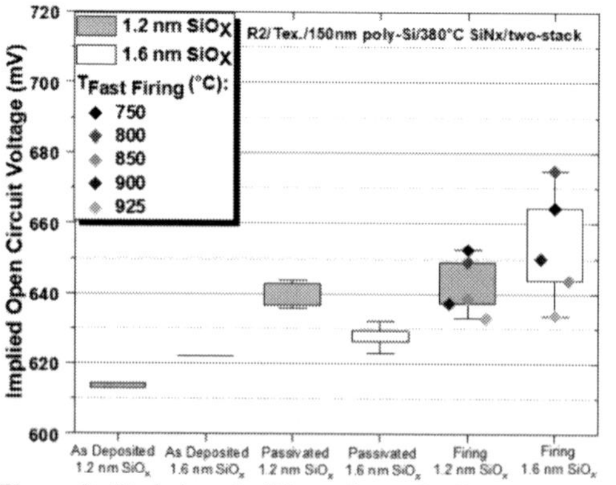

Figure 9: Variations in iV_{OC} values of 150 nm poly-Si with TO1 (grey) and TO2 (white) doped with R2 recipe on TXT surfaces and coated by two-stacks with SiN_x deposited at 380°C.

10.4229/EUPVSEC2023/1CV.3.12
020048-003

In conclusion, it is found that R1/100 nm poly-Si/1.6 nm SiO_x/three-stacks on SDE c-Si exhibits the highest passivation parameters with iV_{OC} of up to 710 mV and $J_{0,pass}$ of 6.5 mA/cm^2. On the other hand, the R2/TXT/150 nm poly-Si/1.6 nm SiO_x/three-stacks exhibits a maximum iV_{OC} of only ~675 mV and $J_{0,pass}$ of 22.3 fA/cm^2 which demonstrates promising results for further studies.

REFERENCES

[1] A. Richter *et al.*, "Design rules for high-efficiency both-sides-contacted silicon solar cells with balanced charge carrier transport and recombination losses," *Nat Energy*, vol. 6, no. 4, pp. 429–438, Apr. 2021, doi: 10.1038/s41560-021-00805-w.

[2] G. Kaur, Z. Xin, R. Sridharan, A. Danner, and R. Stangl, "Engineering aluminum oxide/polysilicon hole selective passivated contacts for high efficiency solar cells," *Solar Energy Materials and Solar Cells*, vol. 218, Dec. 2020, doi: 10.1016/j.solmat.2020.110758.

[3] D. L. Young *et al.*, "Gallium-Doped Poly-Si:Ga/SiO2 Passivated Emitters to n-Cz Wafers with IVoc >730 mV," *IEEE J Photovolt*, vol. 7, no. 6, pp. 1640–1645, Nov. 2017, doi: 10.1109/JPHOTOV.2017.2748422.

[4] M. K. Stodolny *et al.*, "NOVEL SCHEMES OF P + POLYSI HYDROGENATION IMPLEMENTED IN INDUSTRIAL 6'' BIFACIAL FRONT-AND-REAR PASSIVATING CONTACTS SOLAR CELLS," 2018. doi: 10.4229/35thEUPVSEC20182018-2CO.10.3.

[5] H. Steinkemper, F. Feldmann, M. Bivour, and M. Hermle, "Numerical Simulation of Carrier-Selective Electron Contacts Featuring Tunnel Oxides," *IEEE J Photovolt*, vol. 5, no. 5, pp. 1348–1356, Sep. 2015, doi: 10.1109/JPHOTOV.2015.2455346.

[6] G. Yang *et al.*, "Will SiOx-pinholes for SiOx/poly-Si passivating contact enhance the passivation quality?," *Solar Energy Materials and Solar Cells*, vol. 252, Apr. 2023, doi: 10.1016/j.solmat.2023.112200.

[7] M. Firat *et al.*, "Local Enhancement of Dopant Diffusion from Polycrystalline Silicon Passivating Contacts," *ACS Appl Mater Interfaces*, vol. 14, no. 15, pp. 17975–17986, Apr. 2022, doi: 10.1021/acsami.2c01801.

[8] S. Mack, D. Herrmann, M. Lenes, M. Renes, and A. Wolf, "Progress in p-type Tunnel Oxide-Passivated Contact Solar Cells with Screen-Printed Contacts," *Solar RRL*, vol. 5, no. 5, May 2021, doi: 10.1002/solr.202100152.

[9] B. Steinhauser, F. Feldmann, D. Ourinson, H. Nagel, T. Fellmeth, and M. Hermle, "On the Influence of the SiNx Composition on the Firing Stability of Poly-Si/SiNx Stacks," *Physica Status Solidi (A) Applications and Materials Science*, vol. 217, no. 21, Nov. 2020, doi: 10.1002/pssa.202000333.

[10] J. I. Polzin, B. Hammann, T. Niewelt, W. Kwapil, M. Hermle, and F. Feldmann, "Thermal activation of hydrogen for defect passivation in poly-Si based passivating contacts," *Solar Energy Materials and Solar Cells*, vol. 230, Sep. 2021, doi: 10.1016/j.solmat.2021.111267.

[11] C. Madumelu *et al.*, "Assessing the stability of p+ and n+ polysilicon passivating contacts with various capping layers on p-type wafers," *Solar Energy Materials and Solar Cells*, vol. 253, May 2023, doi: 10.1016/j.solmat.2023.112245.

[12] H. Mäckel and R. Lüdemann, "Detailed study of the composition of hydrogenated SiN x layers for high-quality silicon surface passivation," *J Appl Phys*, vol. 92, no. 5, pp. 2602–2609, Sep. 2002, doi: 10.1063/1.1495529.

APPLICATION POTENTIAL OF LASER CRYSTALLIZATION FOR USE IN TOPCON SILICON SOLAR CELLS

Mohammed Jaber[1,2], Stefan Haas[1,3], Maurice Nuys[1], Frank Pennartz[1], Alain Doumit[1], Uwe Rau[1,2]
[1] IEK-5 Photovoltaik, Forschungszentrum Julich, Leo-Brandt Straße, 52425 Jülich, Germany
[2] Electrical Engineering, Information Technology and Computer Engineering, RWTH Aachen University, Aachen, Germany
[3] Department of Aerospace Engineering, FH Aachen University of Applied Sciences, 52066 Aachen, Germany
m.jaber@fz-juelich.de

ABSTRACT: Passivating contacts based on poly-Si/SiO_2 contact systems offer high efficiency potential for silicon based solar cells due to their excellent surface passivation and carrier selectivity. Typically, the poly-silicon layers are realized by deposition of amorphous silicon and subsequent high temperature furnace annealing. Disadvantages of furnace annealing are the lack of spatial resolution of the crystallization as well as high energy consumption. Here an alternative method of crystallization and dopant activation of highly-doped n-type hydrogenated amorphous silicon layers (a-Si:H) by pulsed-laser processing is investigated and the potential of the process for application in tunnel oxide passivated contact (TOPCon) silicon solar cells is analyzed. The laser process is investigated in symmetrical lifetime samples consisting on both sides of an ultra-thin chemically grown tunnel oxide and a poly-Si(n) layer, prepared by hot-wire chemical vapor deposition (HWCVD) of a-Si:H(n) followed by crystallization. Crystallization of a-Si:H has been done by furnace annealing (reference) or spatially resolved pulsed-laser processing. Here, dependency of passivation quality on laser wavelength (355 nm and 532 nm) laser pulse energy, spatial pulse overlaps as well as number of process repetitions was investigated. For laser processed samples, Raman backscattering indicates a high degree of crystallinity (X_C) with an asymmetric TO peak centered at 517.5 cm^{-1}. Quasi-steady-state photoconductance (QSSPC) measurements show for the sample with highest lifetime implied open-circuit voltage (iV_{OC}) up to 701 mV and 876 µs effective carrier lifetime.
Keywords: Passivated Contacts, Laser Crystallization, Dopant Activation, TOPCon, a-Si:H

1 AIM AND APPROACH

In this study, the potential of laser crystallization as a cost-effective alternative to furnace annealing for the crystallization of a-Si:H will be presented. In addition, the quality of passivation of the n-type contacts in the TOPCon solar cell structure will be evaluated.

Sample fabrication begins with wet chemical oxidation with ozone water at 30℃ for 10 minutes after RF cleaning, producing tunnel oxide in the range of 1.5 nm on both wafer surfaces. Subsequently, a-Si:H(n) is prepared by HWCVD with thickness between 11-26 nm as measured by spectroscopy ellipsometry (SE). After a-Si:H deposition, the process chain is divided into two routes, the conventional route based on furnace annealing and the novel method based on laser crystallization. In the conventional method, the sample is high temperature annealed in the furnace at 820℃ for 30 minutes to convert a-Si:H into poly-Si. In the parallel route based on laser crystallization, the samples are treated locally on both sides with two different laser wavelengths (355 and 532 nm). After crystallization, the process chains merge. All samples undergo hydrogenation treatment by deposition of SiN_X on both sides of the sample, followed by forming gas annealing treatment at 400℃ for 30 min.

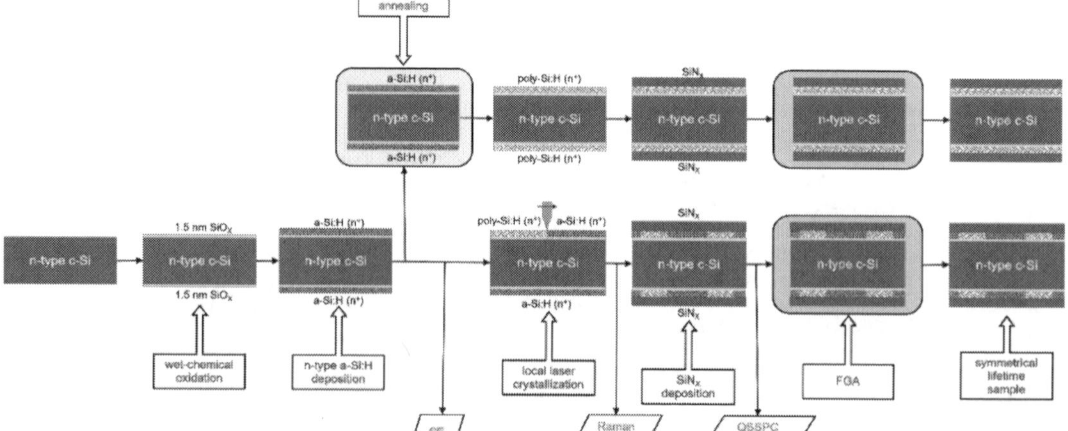

Figure 1 Schematic of the processes chain

2 SCIENTIFIC INNOVATION AND RELEVANCE

Increasing solar cell efficiency involves more complex processes but simplifying the fabrication process makes highly efficient solar cells more appealing for industrial applications. The project aims to achieve low-temperature and free-mask processes. My work specifically focuses on replacing high-temperature treatment with laser treatment for back contacts (IBC) and tunnel oxide passivated contact (TOPCon) solar cell structure fabrication. Laser treatment is used to crystallize the thin-film selective contact layer and activate the dopant. This treated layer serves as a conductive and passivation layer on the back side of the solar cells.

3 RESULTS AND CONCLUSIONS

The conversion of a-Si:H into poly-Si was proved by the Raman spectrum of the silicon thin-film before and after thermal annealing. Figure 2 (a) shows the spectrum of the thin-film silicon layer before and after oven annealing. The black line shows an amorphous silicon structure with peak around 483 cm^{-1}. Red spectrum shows a strong centered peak at 522 cm^{-1}. Figure 2 (b) represents of fitting of poly-Si, the strong peak at 522 cm^{-1} represents the crystalline peak, and the polycrystalline peak at 508 cm^{-1}. The small shoulder at the left lower part of the spectrum represents the amorphous portion with a peak at 472 cm^{-1}. High-temperature treated sample shows good passivation quality after the hydrogenation process by the SiN_x layer and FGA. 1.2 ms minority carrier lifetime measured by QSSPC.

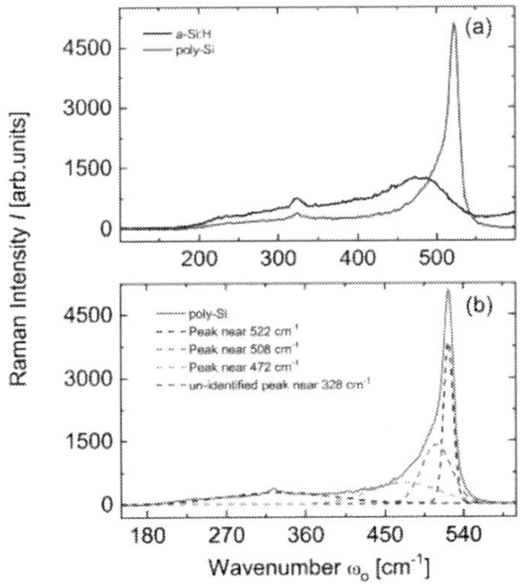

Figure 2 Raman spectrum of furnace annealed sample before and after

Two distinct laser wavelengths, namely 355 nm and 532 nm, were employed to investigate the depth of the crystallization process and its impact on the passivation quality of the TOPCon stake. Furthermore, two different Raman excitation wavelengths, specifically 320 nm and 532 nm, were utilized to examine both the bulk and surface of the crystallized layer. The obtained results, as illustrated in Figure 3, demonstrate a sequence of UV-laser energy characterized by 320 nm Raman. These findings highlight the feasibility of achieving crystallization in a-Si:H through the utilization of a pulsed laser that delivers a fluence exceeding 267.7 mJ/cm^{-2}. Additionally, the 532 nm Raman (Green-Raman) analysis confirms that the depth of crystallization extends to the bulk of a-Si:H in proximity to the SiOx/c-Si interface.

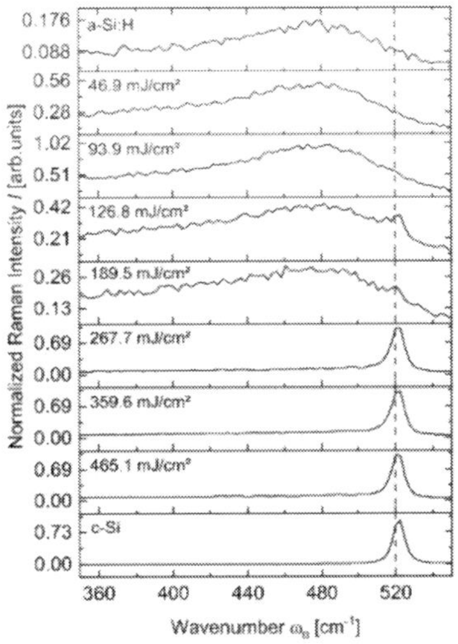

Figure 3 UV Raman for single spot setup by UV laser

The crystallized polysilicon layer, formed using a UV laser, exhibits tensile stress as a result of the differential expansion coefficients among the layers in the TOPCon stack. This stress can potentially cause damage to the oxide layer at the interface, leading to a loss of its functionality. Raman spectra analysis of the poly-Si clearly demonstrates a redshift of the TO peak towards a lower wavenumber of 517 cm^{-1}. In order to release the stress, it is recommended to preheat the sample both before and during the laser treatment, ensuring that the temperature remains below the crystallization temperature of the silicon. In Figure 3, the peak position of the TO mode is depicted, showing a shift towards 520 cm^{-1} as the heat plate temperature increases.

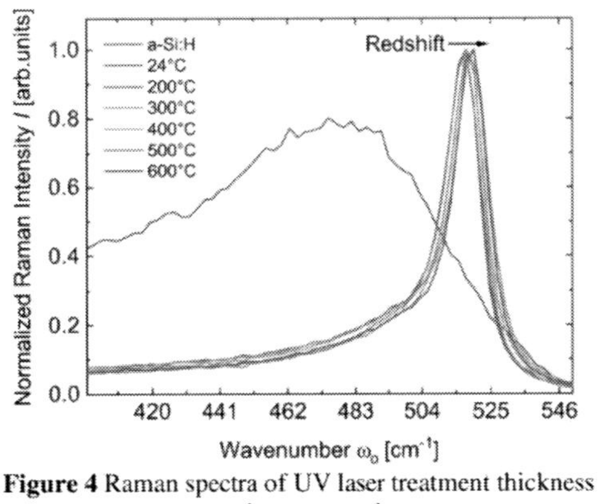

Figure 4 Raman spectra of UV laser treatment thickness series measured

10.4229/EUPVSEC2023/1CV.3.18
020049-002

Figure 5 displays the lifetime map for a series of laser fluence (F) after the hydrogenation process using SiN_x:H and FGA as determined by Photoluminescence Imaging (PL). It is observed that increasing the laser fluence (F) beyond 194 mJ/cm^{-2} results in a decrease in passivation quality. The highest achieved lifetime is 859 µs, which corresponds to a laser fluence of 130 mJ/cm^{-2}, indicating the best outcome. The best laser crystallization parameters are exported to large-scale treatment. Figure 6 shows large-scale laser treatment with four sets of laser parameters. The PL mapping lifetime clearly proves the ability to provide a homogenous crystallization process for precursor a-Si:H layer for TOPCon solar cell structure.

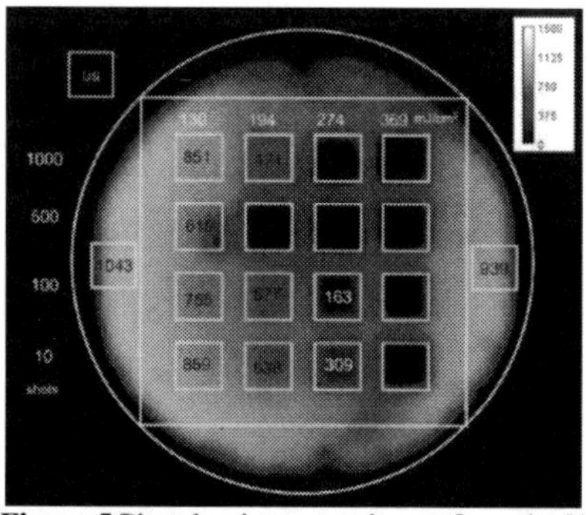

Figure 5 Photoluminescence image for poly-Si crystalized by UV laser source.

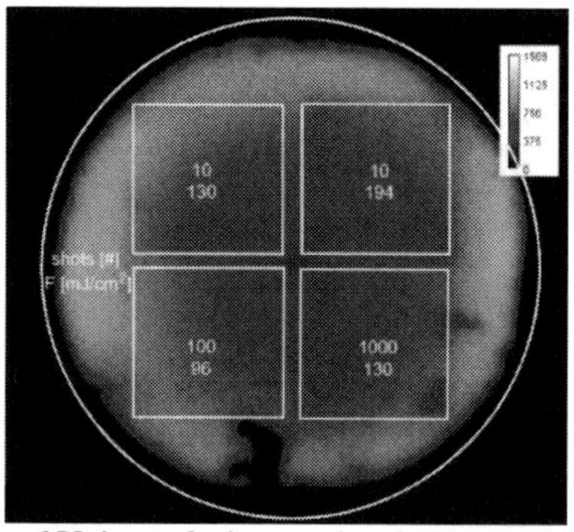

Figure 6 PL image for large-scale poly-Si crystalized by UV laser source.

10.4229/EUPVSEC2023/1CV.3.18
020049-003

MICRODOPING OF LAYERS FOR HIGH EFFICIENCY SILICON HETEROJUNCTION SOLAR CELLS

I. Voicu Vulcanean[1], J. Seif[2], S. Pingel[1], I. Koc[1], M. Bivour[1], A. Steinmetz[1]
[1]Fraunhofer Institute for Solar Energy Systems ISE, Germany
[2]Idener, Spain

ABSTRACT: Within this paper two approaches have been attempted to improve the interfaces of silicon heterojunction devices, either intrinsic amorphous silicon a-Si:H(i) to doped layer a-Si:H(n) or a-Si:H(i) layer to silicon oxide and nanocrystalline layer nc-Si:H (n or p). For the conventional a-Si:H based device, microdoping was implemented in the i2 layer of the a-Si:H(i) layer stack, aiming to increase the fill factor. An improvement in fill factor of up to 0.5%$_{abs.}$ was shown by introducing microdoping into a fraction of the i2 layer, resulting in an improved efficiency reaching 23.4% compared to the undoped reference at 23.2%. For "next generation" SHJ devices based on nc-Si:H, microdoping was implemented in the interfacial silicon oxide layer. For nc-Si:H(n) the fill factor showed a gain of up to 0.3%$_{abs.}$ compared with the undoped group, with the best cell reaching 22.9% efficiency. When introducing microdoping in the silicon oxide layer for nc-Si:H(p), a similar trend is observed as an increase in *FF* of ~0.3%$_{abs.}$ compared with undoped group drives the efficiency for the best cell at 23%. The two microdoping approaches decreased the series resistance, positively affecting the fill factor.
Keywords: silicon heterojunction solar cell, a-Si, nanocrystalline, microdoping, fill factor

1 INTRODUCTION

The progress in achieving high efficiency solar cells with silicon heterojunction (SHJ) devices has been substantial in recent years, with record energy conversion efficiencies reaching 26.81% [1]. The technology is known to exhibit high open-circuit voltages (V_{oc}), enabled by the excellent passivation properties of the hydrogenated amorphous silicon layers (a-Si:H) [2]. The layers are deposited in stacks with intrinsic (i) followed by n-doped amorphous layer for the front, and intrinsic (i) followed by p-doped amorphous layer for the rear. As next step, sputtering of transparent conductive oxide (TCO) is introduced for antireflection and lateral charge transport [3][4]. The amorphous silicon layers tend to limit the cell performance due to fill factor (*FF*) limitation, induced by high series resistance (R_s) [5]. Therefore, to further improve the device, a key factor in achieving higher efficiencies is to improve the *FF*. Besides improving the a-Si:H layer stack, nanocrystalline silicon layers are a viable solution to improve both optical and transport properties in SHJ solar cells. These layers exhibit lower absorption at shorter wavelengths, enabling higher short-circuit current densities (J_{sc}) [6]. In addition to this, improved transport properties and more efficient doping compared with a-Si:H, results in improved contacts [6]. In order to get the benefits of the nc-Si:H layers, several requirements have to be met: a) obtaining a fast nucleation of the nc-Si:H layer on top of the intrinsic a-Si:H layer; b) no damage to the passivation layer during nc-Si:H deposition; c) high crystallinity of the nc-Si:H layer to benefit from higher doping efficiency and transparency [7]. It has been shown that silicon oxide layers (SiO_x) grown prior to the nc-Si:H material, play an important role in helping to accelerate the nucleation of the nc-Si:H layers [5], thus obtaining thinner layers and subsequently improving the J_{sc}.

2 EXPERIMENTAL

The samples in this study were based on n-type monocrystalline Czochralski (Cz) M2 silicon wafers with a thickness of 160 µm (after texturing) and base resistivity of 1 Ωcm. The as-cut wafers have been subjected to saw damage removal, texturing and ozone-based cleaning followed by HF dip to remove the native oxide prior the deposition of the a-Si:H layers. The a-Si:H and nc-Si:H layers have been deposited by PECVD at 13.56 MHz, using SiH_4, H_2, PH_3 (5% in H_2), TMB (2% in H_2) and additional CO_2 (for SiO_x) as process gasses. The PECVD processes for the a-Si:H layers were conducted at 200°C, while for the nc-Si:H layers a temperature of 140°C was used.

Adding dopant gases in the mix for the i2 layer deposition changes the deposition rate of the standard process. The deposition rate of the microdoped layers was determined depositing layers on flat float-zone (FZ) silicon samples and characterizing them by means of spectroscopic ellipsometry. The deposition time was then adjusted to reach the same thickness as for the undoped reference. The samples were finalized by sputtered TCO followed by industrial screen-printing metallization. The final parameters of the manufactured SHJ cells are analyzed by I-V measurements under standard testing conditions (GridTouch, full area measurement, black non-reflecting chuck). All cells are processed in a rear emitter configuration (p-doped layer at the rear).

2.1 Microdoping the i2 layer of the a-Si:H(i) layer stack

A first approach was implemented for the amorphous layers for SHJ by microdoping fractions of the total thickness of the i2 layer (Figure 1). The dopant concentration and the overall thickness of the i2 layer were kept constant, only the doped layer thickness of the initial undoped i2 layer was changed in ¼ steps from 0% (undoped ref.) to 100% (completely doped i2 layer). A trade-off between maintaining the passivation and the generated current on a high enough level versus the potential improvement in the series resistance and the fill factor can be expected, as the potential improvement in the carrier transport from adding extra doping in the i2 layer can come at the cost of degradation in passivation and optical losses by decreasing the transparency of the layer.

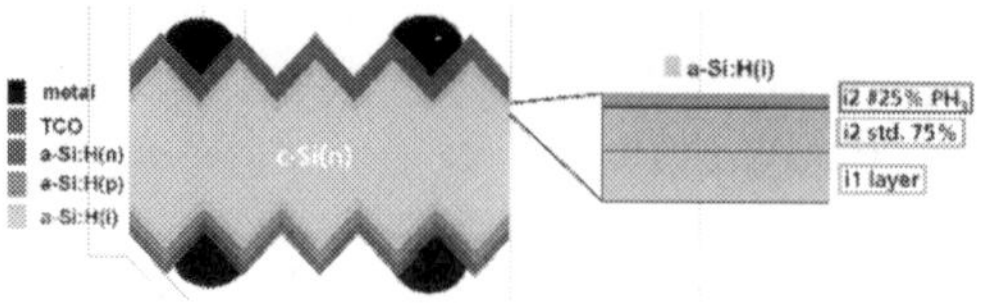

Figure 1: Schematic illustration of the SHJ solar cell structure with microdoping 25% of the thickness of the i2 layer.

2.2 Microdoping the interfacial SiO_x layer for nc-Si:H (n or p) devices

Following the second approach, the interfacial SiO_x seed-layers for subsequent nc-Si:H (either n or p) depositions were microdoped (Figure 2 & 3). The SiO_x layer can be grown via a vacuum break after a-Si:H deposition or by using a PECVD-based plasma treatment [8], as it was implemented for the experiments included in this paper. Before the implementation of the dopant in the SiO_x layer, variations in the SiO_x layer thickness have been conducted in order to optimize the process. The plasma treatment time has been varied for a CO_2 concentration of 5%. The thickness of the nc-Si:H doped layers was approximately 25 nm.

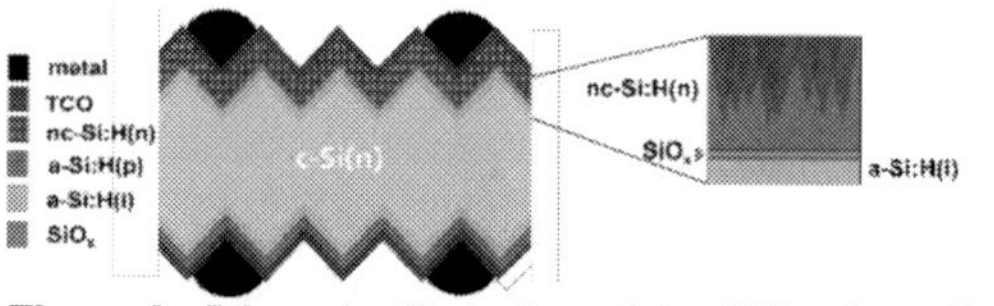

Figure 2: Schematic illustration of the SHJ solar cell structure with nc-Si:H(n).

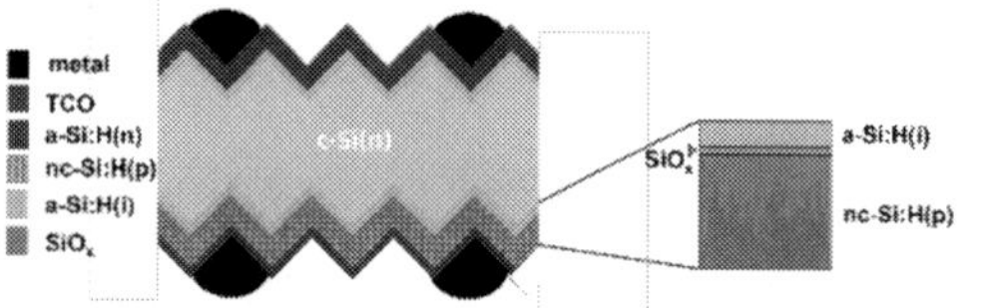

Figure 3: Schematic illustration of the SHJ solar cell structure with nc-Si:H(p).

3 RESULTS AND DISCUSSION

3.1 Microdoping the i2 for i/n layer stack

The results of implementing microdoping into the i2 layer are summarized in Figure 4. Clear trends can be seen in all cell parameters with increasing the fraction (#xx%) of the doped i2 layer. Strong impact on passivation is noticed with a decreasing V_{oc} with increasing the thickness fraction of the doped layer. While for the #25% group the loss in V_{oc} is below 1 mV, for the #100% group a V_{oc} drop of ~4 mV is observed compared to the undoped reference.

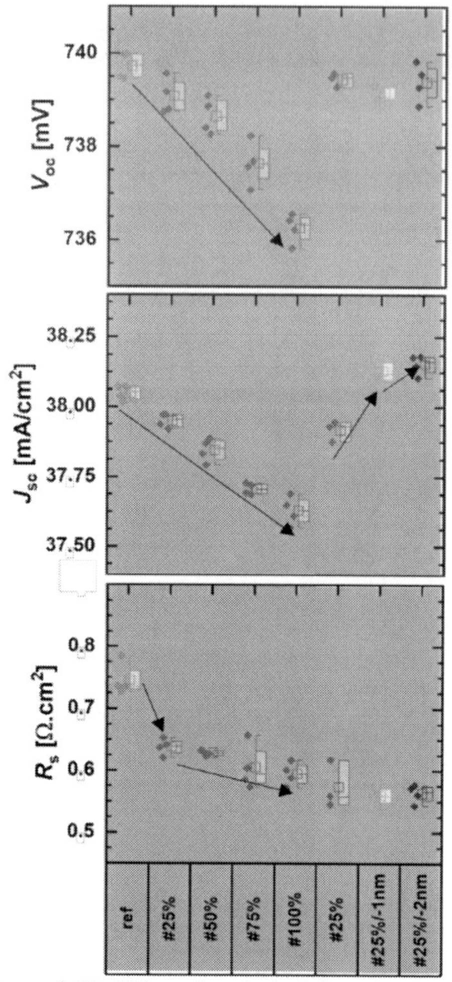

Figure 4:(A) I-V results obtained by microdoping fractions of the i2 layer thickness, increasing in ¼ steps from 0% (ref) to 100%.

10.4229/EUPVSEC2023/1CV.3.26
020051-002

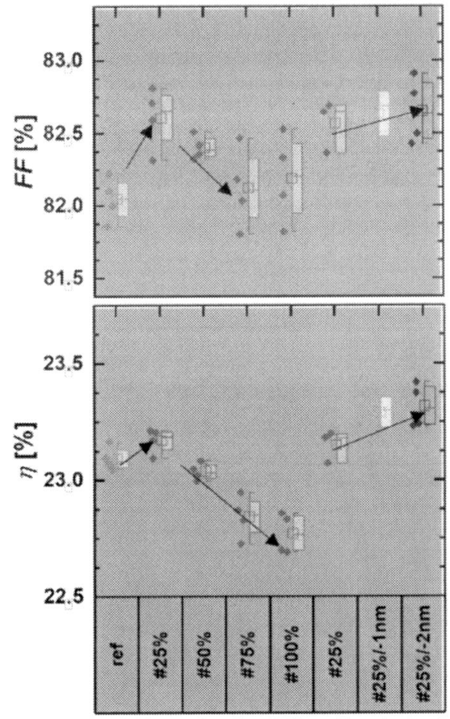

Figure 4:(B) I-V results for SHJ cells obtained by microdoping fractions of the i2 layer, increasing in ¼ steps from 0% (ref) to 100%.

Likely due to parasitic light absorption from the additional dopant included in the i2 layer, a downward trend is observed for the J_{sc}, recording losses of up to 0.5 mA/cm^2 from undoped to 100% doped layer. However, due to improved carrier transport at the interface and a drop in R_s, the FF increases above the undoped reference for all groups that have been subjected to microdoping. This values up to 0.5%$_{abs.}$ gain compared to the undoped reference, reaching a FF of 82.8% for the #25% group. This large boost in FF leads to roughly 0.1%$_{abs.}$ increase in efficiency. While doping larger fractions of the intrinsic layer shows lower R_s values, it is not enough to compensate for the degradation in passivation observed as decreasing V_{oc} and the optical losses indicated by lower J_{sc}.
To offset the losses due to microdoping of the i2 layer, the focus on the second experiment was to improve the J_{sc}. Therefore, the thickness of the standard n-doped layer was reduced. The process sequence for #25% group from the previous experiment is used as a reference. As both groups with thinner n-layer show an increase in J_{sc} of ~0.2 mA/cm^2 above the reference, we demonstrated that this approach has been successfully implemented. The V_{oc} didn't show major changes with all groups being in their standard deviation range. This is also the case for the FF. The J_{sc} improvement, together with a mean FF of ~82.6%, resulted in efficiencies of up to 23.4%.

3.2 Microdoping the SiO_x seed-layers for nc-Si:H(n)

Implementing plasma treatment time variation for interfacial SiO_x layers followed by the subsequent nc-Si:H(n) deposition leads to the results summarized in Figure 5.

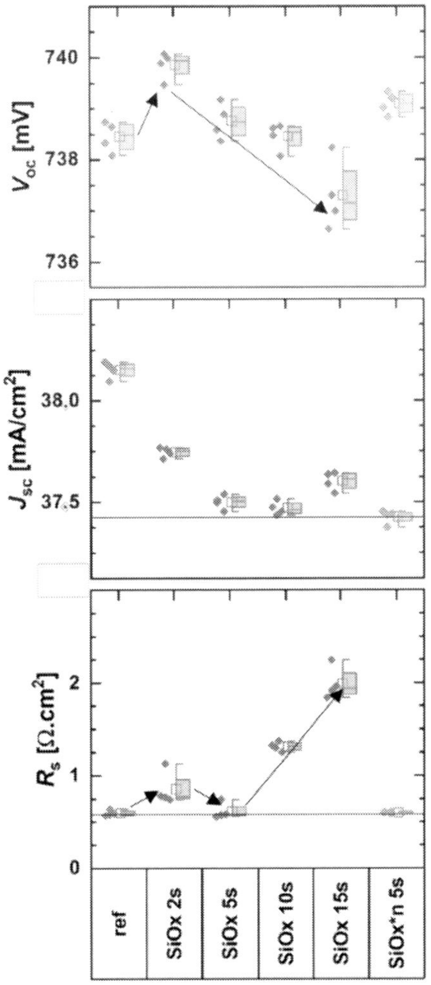

Figure 5:(A) I-V results for SHJ cells obtained with nc-Si:H(n) layers at the front, with variation in plasma treatment time, without and with microdoping for the SiO_x layers present at the a-Si:H(i) interface.

An optimum between sufficiently thin SiO_x layer for efficient charge carrier transport and sufficiently thick SiO_x for adequate nucleation must be found. Ellipsometry results show a layer thickness between 0.4 nm at 2 s and 2 nm at 15 s. For the short plasma treatment time of 2 and 5 seconds the FF is on the same level as the reference. Beyond 5 seconds the charge carriers are not efficiently extracted anymore and a drop in FF and overall loss in conversion efficiency is observed, mainly due to an increase in R_s, up to 2 Ωcm^2. Based on mean reflection values (data not shown), the nc-Si:H(n) groups have shown lower reflection compared with the a-Si:H reference, however this is not enough to compensate the losses of ~20% in the short wavelength range. Additionally, a thicker nc-Si:H(n) layer by a factor of 3 compared to the a-Si:H(n) reference leads to a lower J_{sc}. The group with 5 s is chosen for microdoping due to better FF and a higher mean efficiency. Doping the SiO_x layer by small amounts with PH_3 (indicated as SiO_x*n) maintains low R_s and improves the *FF* up to 82.8%.

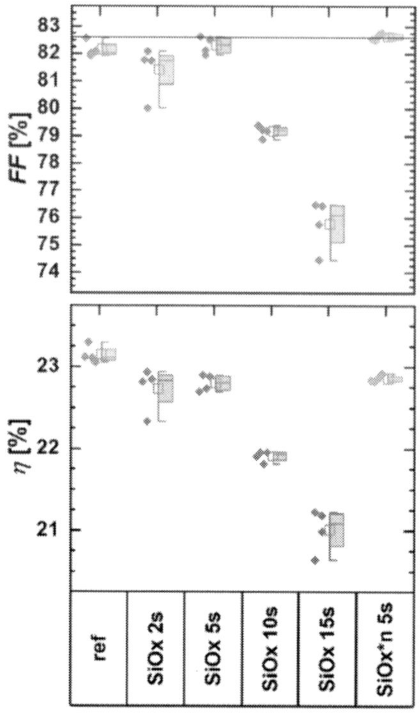

Figure 5:(B) I-V results obtained with nc-Si:H(n) layers at the front, with variation in deposition time for the SiOx layers present at the a-Si:H(i) interface.

Similar as for microdoping the i2 of the a-Si:H(i) layer stack due to transparency issues in the oxide layer as dopant is added, the SiO_x*n 5 s group shows lower J_{sc} values as the SiOx 5 s group, dropping below 37.5 mA/cm^2. Due to poor *FF*, the groups with long oxidation time fall in the range of 21 to 22% in efficiency. The efficiency for the nc-Si:H(n) group with short plasma treatment is close to 23%, while the SiO_x*n group being only slightly better due to improved *FF*.

3.3 Microdoping the SiO_x seed-layers for nc-Si:H(p)

Implementing nc-Si:H(p) on the rear side of the SHJ cells (see Figure 3) leads to the results summarized in Figure 6. To evaluate the benefits of nc-Si:H(p) layers for bi-facial applications, the cells were measured from the front (FS - green symbols) and rear (RS - orange symbols) side. Based on the results from the nc-Si:H(n) experiment, the SiO_x with a plasma treatment time of 5 seconds was chosen as a starting point, followed by a group with added dopant in the oxide layer (indicated as SiO_x*p) and the last group (indicated as SiO_x**p) for which we increased the doping concentration. The difference in J_{sc} between the FS and RS by more than 2 mA/cm^2 for all groups is caused by a lower reflection on the front side. The slightly better J_{sc} for the nc-Si:H(p) films from the rear side measurement can be attributed to a better IR response. Controversial trends can be seen for the J_{sc} where the heavy doped group (SiOx**p) gains ~0.1 mA/cm² for the FS. However, for the RS measurement a downward trend in J_{sc} can be seen as we increase the doping concentration. The SiO_x*p group achieves a similar V_{oc} (739-740 mV- data not shown), in comparison to the undoped group and the a-Si:H reference. A downward trend can be observed for the R_s (rear side measurement), as the series resistance drops from the undoped SiO_x group to the doped SiO_x (p* and p**) groups by ~0.1 Ωcm^2. As for microdoping of the SiO_x*n, an improvement of ~0.3%$_{abs.}$ in *FF* is reached for the SiO_x p** group (RS). These gains bring the SiO_x**p group to a mean efficiency of 22.7%, with the champion cell reaching a conversion efficiency of 23%.

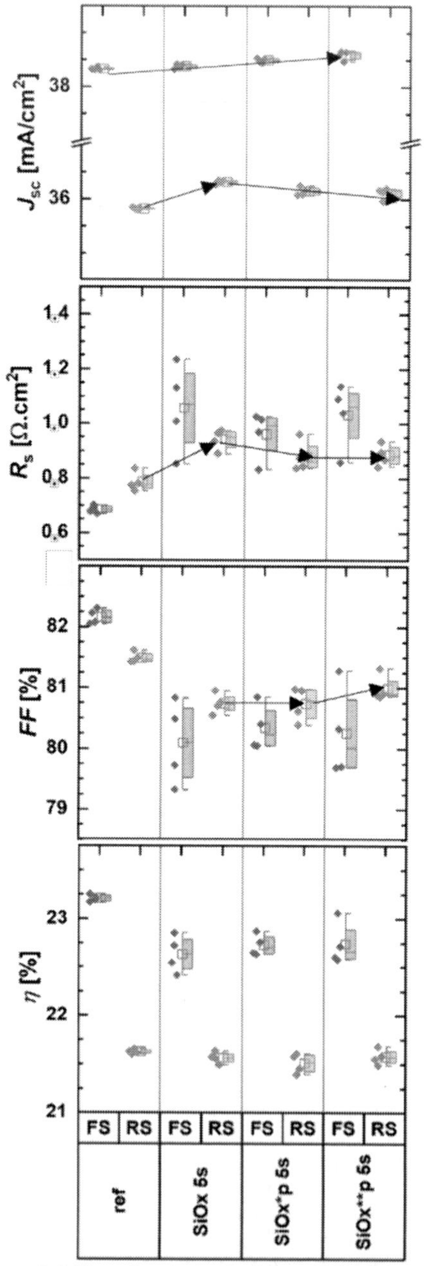

Figure 6: I-V results obtained with nc-Si:H(p) layers at the rear, with microdoping variation for the SiO_x layers present at the a-Si:H(i) interface. Front side (FS) and rear side (RS) measurements are shown.

10.4229/EUPVSEC2023/1CV.3.26

4 CONCLUSION

In this paper we showed a promising way to improve the interface of the SHJ layers, and subsequently the *FF* via microdoping. For a classic a-Si:H based SHJ device, doping a fraction of the total thickness of the i2 layer showed a drop in J_{sc} and Voc but these losses were compensated by improved carrier transport via decreasing the R_s, which resulted in a *FF* gain of up to 0.5%$_{abs.}$ for the #25% group, improving the efficiency up to 23.4% compared to the undoped reference 23.2%. A similar approach has been used for "next generation" SHJ devices based on nc-Si:H by doping the interfacial SiO_x for nc-Si:H (n or p). We observed an improvement by lowering the R_s, resulting in a positive trend for the *FF*, for both front and rear layers.

For nc-Si:H(n) with the SiOx*n group, the *FF* showed a gain of up to 0.3%$_{abs.}$ compared with the undoped group, with the best cell reaching 22.9% efficiency. Introducing microdoping in the SiO_x for nc-Si:H(p), a similar trend is observed as an increase in *FF* of ~0.3%$_{abs.}$ compared with undoped group drives the efficiency for the best cell at 23%. Reducing the J_{sc} gap of nc-Si:H based solar cell device compared to the a-Si:H will be a focus point for further optimization and will be addressed by reducing the thickness of the doped nc-Si:H layers.

FUNDING

This work was supported by the German Federal Ministry for Economic Affairs and Climate Action BMWK within the research project "Utility4Indium" under contract no. 03EE1127D.

ACKNOWLEDGEMENT

The authors would like to thank all colleagues from the PV division contributing to this work for sample preparation and characterization.

REFERENCES

[1] LONGi Green Energy Technology Co., Ltd. (LONGi). At 26.81%, LONGi sets a new world record efficiency for silicon solar cells. Xi'an, China; 19.11.2022.

[2] Duan W, Lambertz A, Bittkau K, Qiu D, Qiu K, Rau U, Ding K. A route towards high-efficiency silicon heterojunction solar cells. Progress in Photovoltaics. 2022;30:384–92. doi:10.1002/pip.3493.

[3] Mazzarella L, Morales-Vilches AB, Korte L, Schlatmann R, Stannowski B. Ultra-thin nanocrystalline n-type silicon oxide front contact layers for rear-emitter silicon heterojunction solar cells. Solar Energy Materials and Solar Cells. 2018;179:386–91. doi:10.1016/j.solmat.2018.01.034.

[4] Antognini, Luca; Paratte, Vincent; Haschke, Jan; Cattin, et al. Influence of the Dopant Gas Precursor in P-Type Nanocrystalline Silicon Layers on the Performance of Front Junction Heterojunction Solar Cells. IEEE J. Photovoltaics. 10.1109/JPHOTOV.2021.3074072

[5] Boccard M, Monnard R, Antognini L, Ballif C. Silicon oxide treatment to promote crystallinity of p-type microcrystalline layers for silicon heterojunction solar cells. In: SiliconPV2018. p. 40003. doi:10.1063/1.5049266.

[6] Nogay G, Seif JP, Riesen Y, Tomasi A, Jeangros Q, Wyrsch N, et al. Nanocrystalline Silicon Carrier Collectors for Silicon Heterojunction Solar Cells and Impact on Low-Temperature Device Characteristics. IEEE J. Photovoltaics. 2016;6:1654–62. doi:10.1109/JPHOTOV.2016.2604574.

[7] Seif JP, Descoeudres A, Nogay G, Hanni S, Nicolas SM de, Holm N, et al. Strategies for Doped Nanocrystalline Silicon Integration in Silicon Heterojunction Solar Cells. IEEE J. Photovoltaics. 2016;6:1132–40. doi:10.1109/JPHOTOV.2016.2571619.

[8] Steinmetz A, Seif J, Koc I, Voicu Vulcanean I, Kurt D, Pingel S, Bivour M. Nanocrystalline Silicon Layer for the Application in Silicon Heterojunction Solar Cells. SiliconPV 2023 Conference. To be published.

LITHIUM-DOPED NICKEL OXIDE GROWN BY DIFFERENT PVD METHODS FOR HOLE-SELECTIVE CONTACTS IN SILICON-BASED HETEROJUNCTIONS

F. Menchini[1], S. Rakhshani[1], L. Serenelli[1], L. Martini[1], E.Salza[1], P. Mangiapane[1], M. Izzi[1], A. Latini[2], M. Tucci[1]
[1]ENEA, TERIN Department, via Anguillarese 301, 00123 Roma, Italy
[2]Sapenza University, Chemistry Department, Roma, Italy

ABSTRACT: Carrier-selective contacts are well known to be one of the keys to obtain cost effective high-efficiency solar cells. In silicon-based heterojunctions the contacts are usually obtained by doped hydrogenated amorphous-silicon (a-Si:H) layers, which however introduce important absorption losses and require expensive production processes. Many candidate materials have been identified to substitute a-Si:H layers, but their full exploitation is frequently limited by the constrain of being resistant to thermal treatments up to at least 250°C. Among them, doped nickel oxide seems to be ideal to form a hole-selective contact thanks to the alignment of its valence band with that of silicon. In this work we grew films of lithium-doped nickel oxide (NiOx:Li) by different PVD techniques and tested their behavior in hole-selective contacts in silicon heterojunctions. The J-V characteristics of the proof-of-concept devices demonstrate the possibility to replace p-type a-Si:H film with NiOx:Li, although further improvements are needed to optimize the performances and solve the issue related to silicon lifetime degradation.
Keywords: selective contact, nickel oxide, heterojunction

1 INTRODUCTION

In the first half of years 2010s the historical power conversion efficiency record for crystalline silicon solar cells, held by UNSW, was reached and overcome by Panasonic amorphous silicon/crystalline silicon heterojunctions (HJs) [1]. This finally established the potential of HJ technology, which currently still hits records by different PV players. At the time of writing this work LONGI settled an efficiency of 27.6%, which is close to the theoretical limit of 29.43% [2, 3].

The main keys to the HJ success are the silicon surface passivation with very thin hydrogenated amorphous-silicon (a-Si:H) layers and the carrier-selective contacts, as obtained by doped a-Si:H layers, which are able to induce the inversion layer (p) and then the p-n junction inside the n-doped crystalline silicon (n-c-Si) absorber and to form an effective selective base contact able to collect the electrons and repulse the holes by simply using the band alignment mismatch at the a-Si:H/c-Si interface. Indeed, if the p-n junction is fully formed within the high-lifetime silicon bulk, the SHR recombination at midgap when the Fermi level crosses the p-n junction is strongly reduced, thus resulting in excellent open-circuit voltages. However, such doped a-Si:H layers are at the same time the limit of HJ cells, because they introduce important absorption losses and require expensive production processes due to the toxic gases needed for their doping. These drawbacks have recently urged researchers to test new alternative materials to be introduced in the hole- and electron-selective stacks as carrier-selective contacts [4]. Many candidates have been identified to substitute a-Si:H and successfully employed [5]. Among them, nickel oxide (NiOx) is theoretically characterized by a good alignment of its valence band with that of silicon, which confers it optimal characteristics as hole-extracting layer as confirmed by theoretical calculations [6]. On the other side, a doping of NiOx is needed in order to push its Fermi level close to the valence band and induce a deep inversion layer in the n-type doped c-Si [7]. Nevertheless, other calculations [8] illustrated that, due to the interactions of NiOx with a-Si:H, a thin SiOx layer could be formed, which could worsen the transport properties also because of a valence band energy modulation. Then it was shown that it is necessary to greatly increase the NiOx doping to enhance the carrier selectivity. Nonetheless, only a few experimental results have been presented on the topic yet [9, 10].

One of the crucial parameters to obtain working devices is the defect density at the c-Si/NiOx interface. A simulation contained in [11] evidenced that a defect density below 2×10^{10} cm^{-2} is needed to avoid undesirable barriers against carrier collection. These findings establish the necessity of a good passivation of the c-Si surface to improve the V_{OC}. Once the surface passivation is ensured, doping is the main parameter to adjust. Different elements (not only monovalent) can be used as dopants for NiOx, among which Li^+ is recommended for the purpose due to its ionic radius (0.135 Å) which is very similar to that of Ni^{2+} (0.145 Å).

In this work we illustrate the application of lithium-doped nickel oxide, at various concentrations, to the aim of obtaining a hole selective contact for n-type silicon. In order to avoid sputtering damages, we chose a PVD approach, employing e-beam and thermal evaporation. The material was initially characterized by its optical and electrical properties, and then used as emitter in HJ devices production on n-type c-Si wafers.

2 EXPERIMENTAL

We deposited films of lithium-doped nickel oxide (NiOx:Li) starting from calcined precursors containing different dopant concentrations. Firstly, $Ni(NO_3)_2{\cdot}6H_2O$ was calcinated at 800°C to obtain NiOx. Then, to obtain the NiOx:Li doped samples, the proper amount of $Li(NO)_3$ was added to NiOx, pressed in a pellet and heated up to 1000°C for 3 hours. Two doping concentrations were considered, 3% and 5% at.

E-beam evaporation was conducted in a Balzer system at room temperature, with a deposition rate of 0.1 nm/sec, a base pressure of $6\text{-}8\times10^{-6}$ mbar and an accelerating voltage of the electron gun of 3.5 kV. The substrates, positioned on a rotating carrier, were not intentionally heated during the deposition. The thickness of the films was controlled by a quartz oscillator during the growth process.

Thermal evaporation was obtained by heating the precursors in an alumina crucible inserted in a tungsten heater with 0.1-0.2 Å/s deposition rate at a 5×10^{-6} mbar base pressure.

10.4229/EUPVSEC2023/1CV.3.28
020052-001

The films, deposited on soda lime glasses, were optically characterized with the aid of a Perkin Elmer Lambda 950 spectrophotometer, measuring reflectance and transmittance and extracting the optical gap (Eg) from the Tauc plot. The dopant activation energy was calculated by current measurements at various temperatures in a cryogenic system.

In order to test the performances of the NiOx:Li films inside a device, n-type 1-5 Ω·cm textured c-Si wafers were used as substrates, which were passivated on both sides by a 5.5 nm-thick layer of amorphous hydrogenated silicon oxide (a-SiOx:H) deposited via PECVD [12]. This latter material equals the good passivation properties of amorphous silicon, but with an enhanced Eg that favours short-wavelength light absorption. Different NiOx:Li layers were deposited on one side of the c-Si substrates and then covered by an 80 nm-thick layer of Indium Tin Oxide (ITO) to form the emitter contact. The base contact was fabricated in two ways: one set of samples had an unpassivated ohmic contact obtained by means of a thick film of InGa eutectic, the other one contained a 18 nm-thick layer of n-type a-Si:H covered by an 80 nm-thick layer of ITO and a full metal contact of screen printed Ag, subsequently annealed at 180°C for 10 minutes to improve TCO lateral conductivity and transparency and recover sputtering damage.

Selected samples were characterised by Glow Discharge Optical Emission Spectroscopy (GDOES) in a Horiba Jobin Yvon GD Profiler 2 spectrometer. This technique provides an in-depth chemical composition of the film by etching the sample with an Ar plasma and analysing the plasma radiation emitted from the etched material, which wavelength can be correlated to the excited element. The spectra show the signal variation for each element vs the etching time, which is proportional to the etched thickness of material. The devices were characterized by light AM1.5G and dark current density-voltage (J-V) measurements, as well as by external quantum efficiency (EQE) measurements.

3 RESULTS

3.1 Optical characterization

The e-beam evaporated NiOx:Li films deposited on glass, undoped and with 3% and 5% doping, were optically characterized. The absorption coefficient (α) comparison, shown in Fig. 1A, evidences some differences introduced by the doping atoms influencing the microstructural properties of material, and in turn their optical properties. It is expected that by increasing the Li doping concentration the absorption increases [13], however, no clear trend was observed in this case. It is interesting to notice the difference between absorption coefficients of the various Ni oxides, both doped and undoped, and the p-type doped a-Si:H, reported in the same graph as a grey dashed line, confirming the higher transparency of NiOx:Li with respect to the amorphous silicon counterpart. Then the band gap energy (Eg) of the material was extrapolated from the Tauc plot shown in Fig. 1B. The bandgap of all films is around 3.7 eV, with a slightly higher value for the doped films, in accordance with other literature values [14, 15].

3.2 Electrical characterization

In order to estimate the lateral conductivity of the films, two ITO pads were deposited on the samples to ensure the contact, which was proven to be ohmic, and a current-voltage (I-V) characteristic was collected on the as-deposited sample and after a thermal annealing at 180°C for 30 minutes. The results are reported in Fig. 2A as a function of Li doping.

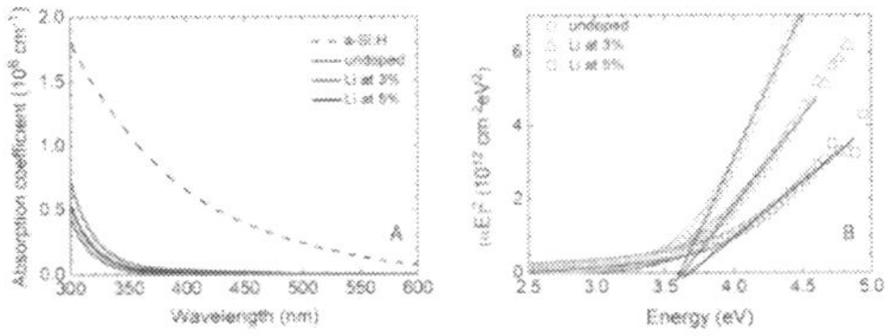

Figure 1: Optical characterization of NiOx:Li films with Li doping at 0% (undoped), 3% and 5% at. concentrations. A: absorption coefficient, B: Tauc plots for the same NiOx film as in A, with the extrapolation of the linear part indicating the Eg. In Fig. 1A the absorption coefficient of a-Si:H is reported for comparison

Figure 2B reports the trend of the activation energy as a function of Li doping as evaluated from I-V measurements as a function of temperature.

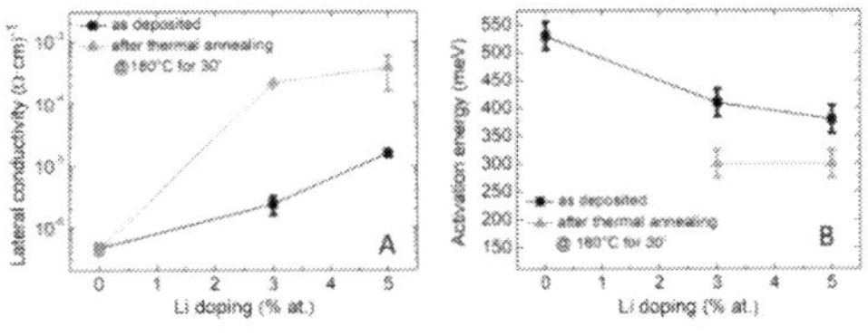

Figure 2: A: lateral conductivity and B: activation energy of NiOx:Li films with different Li doping before (blue square symbols) and after (orange triangle symbols) a thermal annealing treatment of 180°C for 30'

As the Li doping increases the conductivity of the films increases from 4.7×10^{-7} (Ω·cm)$^{-1}$ in the undoped sample to 1.8×10^{-3} (Ω·cm)$^{-1}$ in the highly-doped sample, while in the same range of Li doping the activation energy drops from 530 mV to 140 mV. Both results are in accordance with the increase of doping: an increased concentration of Li in the material induces the substitution of more Ni^{2+} lattice cations, forming valence vacancies, i.e. free holes that increase the p-type conductivity of the material. The activation energy is consequently reduced because of the approaching of the Fermi level to the valence band. After thermal annealing the conductivity of the samples increase, and the activation energy consequently reduces. The lower the activation energy value is, the more the Fermi level of NiOx:Li is shifted towards the valence band edge, and a better band alignment is achieved. Thus, a compromise between doping concentration and activation energy and usage of a buffer layer with lower Eg can be investigated.

3.3 Devices

3.3.1 E-beam evaporation

Based on these premises, several devices were fabricated in order to test the performances of the NiOx:Li films as hole extractors, as illustrated in the experimental section. 20 nm-thick films were deposited on passivated c-Si substrates contacted by a-Si:H and ITO con the back side, as described in the Experimental section and depicted inside Fig. 3A. The light and dark J-V characteristics and external quantum efficiency of the samples were measured, and the best results, corresponding to the sample containing a Li doping of 5% at., are shown in Fig. 3.

The J-V characteristic (Fig. 3A) shows a V_{OC} of 324 mV, a FF of 37.1% and a Jsc of 35.4 mA/cm^2, confirmed by the external quantum efficiency reported in Fig. 3B. The fact that the light and dark characteristics do not cross each other (lower inset in Fig. 3A) is an indication that no V_{OC} drop due to a valence band offset is present in the cell, as explained in [16].

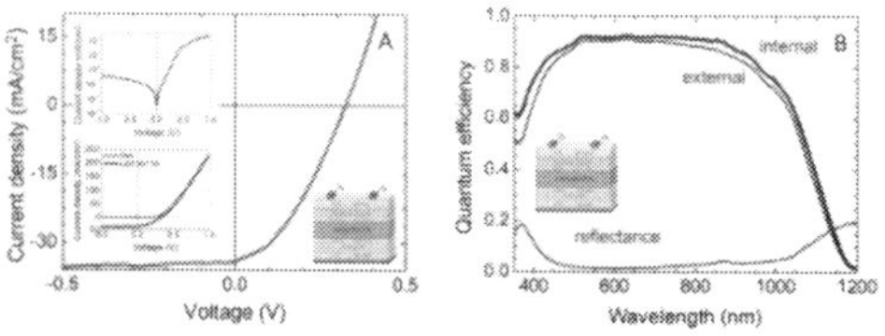

Figure 3: A: light and dark (upper inset) J-V characteristics and B: quantum efficiency of the best performing cell, containing a NiOx:Li film with Li at 5% at. The lower inset in A shows the superposition of dark and light characteristics, showing that the two curves do not cross each other

However, the overall features of the cell are still scarce. Besides the non-optimised front and back contacts and thickness of the NiOx:Li film, the main reason for the low performances of the cell was connected to the low value of the V_{OC}; this was certainly limited due to small area of the cells (which were obtained by cleaving larger wafer [17]) and by the very low lifetime of the structure. In fact the initial lifetime of the passivated substrate (above 1 ms) was strongly deteriorated after the NiOx:Li film deposition, when it reached 100 µs, and not even the thermal treatment was able to restore it. We theorised that this could be a consequence of the e-beam deposition technique, during which x-rays are emitted from the electronic gun [18].

To conclude, the exposed experiments constitute a proof-of-concept for Si-based HJ cell containing a NiOx:Li hole selective layer, which shows that the material is well suited for such application and could be even further optimised in terms of doping. However, the e-beam deposition technique turned out not to be suitable for depositing materials over passivated c-Si substrates, because of the remarkable and permanent damages to the passivation, which leaded to very low V_{OC} values.

3.3.2 Thermal evaporation

To overcome this problem, we explored the possibility to deposit the NiOx:Li films by thermal evaporation. The technique does not imply any emission of radiation, but presents some complications due to the high evaporation temperatures of the source material: in particular, NiOx has a fusion temperature of around 2000°C. The evaporation material was constituted by the same pellets already used for e-beam evaporation, while different evaporation crucibles were tested. Indeed, not all the crucible materials could be suitable to sustain high electrical power to ensure the high temperature needed to evaporate the pellets. At the same time, it is difficult to find a material for crucible which, once heated at such high temperatures, does not evaporate itself, thus contaminating the deposited film. A cylindrical alumina crucible was chosen due to its high temperature resistance, and was inserted in a tungsten heater. The deposition rate was kept very low to avoid excessive thermal stress. Starting from the best results obtained from e-beam deposition, we deposited 10 nm-thick 5%-doped NiOx:Li films over a passivated c-Si substrate contacted on the back by InGa, as described in the Experimental section and reported in the scheme in Fig. 4B. The light J-V characteristics of the sample are reported in Fig. 4A and show a typical p-n junction behaviour. The relatively high Jsc of 32.3 mA/cm^2 confirms the high transparency of the window layer, and the FF (nearly 55%) higher than the previously reported ones confirms the effectiveness of the doping. The cell behaviour is compromised mostly in terms of V_{OC} (335 mV), which is anyway slightly higher than that of the corresponding cell obtained by e-beam evaporation: the latter one had also the back side passivated, while in this case the ohmic back contact is simply made by InGa, as evident from the EQE in Fig. 4B. The dark characteristic (inset in Fig. 4A) shows a high recombination behaviour evident from the high current density in reverse bias and low forward bias conditions. The initial lifetime of the passivated substrate of 1.5 ms dropped to a few tens of µs after the deposition of the emitter, and could not be recovered after the thermal annealing. In this case, even if the damaging x-rays connected to e-beam evaporation was no more an issue, some other source of damage for the passivation should have been present. Probably a diffusion of some metal may have happened during deposition, coming from the heater-crucible system, as they were heated at very high temperatures.

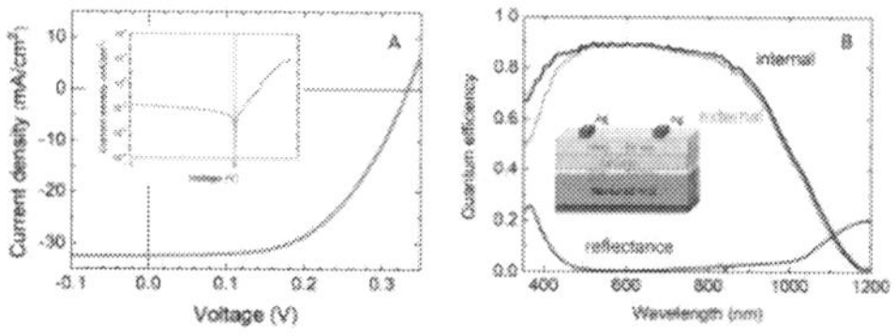

Figure 4: A: light and dark (inset) J-V characteristics and B: quantum efficiency of a cell containing a thermally-evaporated NiOx:Li film with 5% at. Li doping

To check this hypothesis the sample was examined with the aid of GDOES, and the results are shown in Fig. 5. In the first part of the profile, besides the expected signals form Ni and Li (and O), also a contribution from Al is clearly visible, revealing that a metallic contamination is coming from the alumina crucible (we previously checked that this contamination was not present in the pellets). The metal contamination is also suggested by the unexpectedly high conductivity of the sample (around 10^{-2} Ω·cm), which is many orders of magnitude higher than the corresponding value found for the e-beam evaporated sample (Fig. 2B). Obviously, these contaminations are not acceptable, as they are completely uncontrolled, especially after a thermal treatment, so they must be avoided a priori. Moreover, they could also be at the basis of the unexpected damage of the substrate after deposition.

Thus, since NiOx:Li appears to be a promising candidate to form a hole selective contact to silicon, an

10.4229/EUPVSEC2023/1CV.3.28
020052-003

alternative deposition method should be found. A viable hypothesis could be the evaporation of metallic nickel in oxygen atmosphere to form NiOx. In this case the doping could be obtained by co-evaporating a source containing the doping element.

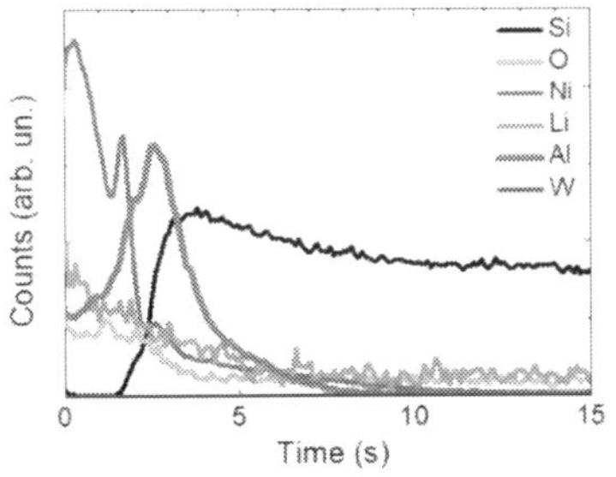

Figure 5: GDOES profiles of NiOx:Li films deposited on a c-Si/a-SiOx:H substrate by thermal evaporation from an alumina crucible in a W heater.

4 CONCLUSIONS

NiOx is a promising material to form an emitter on HJ solar cells on n-type doped c-Si wafer. However, there is the necessity of a strong doping in order to obtain good results. Li was used to dope NiOx and conductivity values of 10^{-4} $(\Omega\cdot cm)^{-1}$ and activation energies around 300 mV were measured. The introduction of a passivating layer allowed the production of working devices, even with low performances and efficiencies around 5%, mainly connected to the evaporation techniques of the ceramic NiOx:Li precursors. In particular, the e-beam evaporation technique was demonstrated not to be compatible with passivated substrates, due to the unavoidable damage caused by x-rays. The more versatile thermal evaporation technique was also tested, but after many and different attempts it was found out that the used NiOx:Li sintered precursor required too much power to be evaporated, thus causing contaminations coming from the crucible. The reduced lifetimes after deposition suggested both a damage induced by diffused species and a band misalignment due to low NiOx:Li doping. However, the material continues being interesting, so that the challenge for its optimisation still remains open. Thus, since the evaporation from a ceramic precursor seems to produce inevitable drawbacks, an alternative approach could be the co-evaporation of metallic nickel and dopant precursor in oxygen atmosphere, which could reduce the necessity for high temperatures.

5 ACKNOWLEDGEMENT

This work was funded under the National Recovery and Resilience Plan (NRRP), Mission 4 Component 2 Investment 1.3 - Call for tender No. 1561 of 11.10.2022 of Ministero dell'Università e della Ricerca (MUR); funded by the European Union – NextGenerationEU. Project code PE0000021, CUP I83C22001800006, Project title "Network 4 Energy Sustainable Transition – NEST" and by the Italian Ministry of Environment and Energy Security in the framework of the Operating Agreement with ENEA for Research on the Electric System.

6 REFERENCES

[1] https://www.nrel.gov/pv/cell-efficiency.html, 2023.
[2] M.A. Green, E.D. Dunlop, G. Siefer, et al. Solar cell efficiency tables (Version 61), Progress in Photovoltaics: Research and Applications 31:1 (2023) 3-16
[3] A. Richter, M. Hermle, S.W. Glunz, IEEE Journal of Photovoltaics 3:4 (2013) 1184-1191
[4] J. Bullock, M. Hettick, J. Geissbühler, et al., Nature Energy 1 (2016) 15031
[5] Y. Wang, S. T. Zhang, L. Li, et al., EcoMat 5:2 (2022) e12292
[6] R. Islam and K. C. Saraswat, Proceedings IEEE 40th Photovoltaics Specialists Conference PVSC, 2014, pp. 285-289
[7] H. Imran, T. M. Abdolkader, IEEE Transactions in Electron Devices 63:9 (2016) 3584-3590
[8] M. Boccard, R. Woods-Robinson, A. Fioretti, et al., Proceedings of the 37th EU PVSEC 2020, online, 2BO.3.3, 21 pages
[9] H. Hsu, N. F. Wang, Y. Z. Tsai, et al., Thin Solid Films 573 (2014) 159-163
[10] X. Yang, W. Liu, J. Chen, et al., Applied Physics Letters 112 (2018) 173904
[11] F. Menchini, M. L. Grilli, T. Dikonimos, et al., Physica Status Solidi C 13:10-12 (2016) 1006-1010
[12] L. Serenelli, L. Martini, L. Imbimbo et al., Applied Surface Science 392 (2017) 430-440
[13] W. Chia-Ching, Y. Cheng-Fu, Nanoscale Research Letters 8:1 (2013) 33
[14] J. Arunodaya and T. Sahoo, Materials Research Express 7 (2020) 016405
[15] D. Paul Joseph, M. Saravanan, B. Muthuraaman, et al., Nanotechnology 19 (2008) 485707
[16] L. Serenelli, L. Martini, F. Menchini, et al., Progress in Photovoltaics: Research and Applications 29:7 (2021) 876-884
[17] L. Serenelli, L. Martini, F. Menchini, et al., Solar Energy 258 (2023) 2-7
[18] T. Rahman, A. To, M.E. Pollard, et al., Progress in Photovoltaics: Research and Applications 23 (2018) 38-47

EPITAXIALLY GROWN CRYSTALLINE SILICON AS ELECTRON SELECTIVE CONTACT LAYER FOR CRYSTALLINE GERMANIUM TPV CELLS

M. Gamel[1,a)], G. López[1], T. Jawhari[2], A. J. Olivares[3], P. Roca i Cabarrocas[3], M. Garín[4] and I. Martín[1]

[1]Departament d'Enginyeria Electrònica, Universitat Politècnica de Catalunya, Barcelona, Spain.
[2]Centres Científics i Tecnològics de la Universitat de Barcelona (CCiTUB), Barcelona, Spain.
[3]Lab. de Physique des Interfaces et des Couches Minces (UMR 7647 CNRS), Ecole Polytechnique, Palaiseau, France.
[4]Department of Engineering, Universitat de Vic—Universitat Central de Catalunya (UVIC-UCC), Vic, Spain.
a) Use this author for all correspondence: Phone.: +34 93 405 41 93, e-mail: mansur.gamel@upc.edu

ABSTRACT: Crystalline germanium has been proposed as a cost-effective absorber for the fabrication of thermophotovoltaic (TPV) cells which require, among other technologies, the development of electron-selective contacts. In this work, we explore the deposition process of n-type high-quality crystalline silicon (c-Si(n)) on p-type crystalline germanium (c-Ge) substrates using Plasma-Enhanced Chemical Vapor Deposition (PECVD) to create c-Si(n)/c-Ge(p) heterojunctions that function as electron-selective contacts. Our analysis includes material characterization using Raman spectroscopy, X-ray diffraction, and Spectroscopic Ellipsometry measurements, confirming the crystalline nature of the epitaxially grown silicon layers. The interface quality of the heterojunction is studied by measuring the effective lifetime. The results demonstrate that the addition of a thin intrinsic amorphous silicon layer at the interface improves its passivation. Additional improvement is obtained after depositing a thin n-type amorphous layer on top of the c-Si(n), achieving a surface recombination velocity of 334 cm/s. This better passivation positively impacts both the open circuit voltage and the fill factor of 1x1 cm^2 finished devices, leading to an efficiency of 3.38%. The results suggest the potential of c-Si/c-Ge heterojunction for cost-effective c-Ge cells, with room for efficiency improvements through further optimization.

Keywords: epitaxial crystalline silicon, thermophotovoltaic, crystalline germanium, PECVD.

1 INTRODUCTION

Crystalline germanium (c-Ge) has a strong absorption up to 2 μm wavelength, which has been traditionally utilized in tandem and multijunction solar cells to boost their efficiency by harvesting infrared radiations [1]. Recently, c-Ge cells have been proposed as an affordable alternative to low-bandgap III-V semiconductors for thermophotovoltaic (TPV) applications, where the cell converts infrared thermal radiation into electricity [2-5]. Numerous efforts have been devoted to develop efficient carrier-selective contacts in c-Ge, since they are crucial in obtaining high-performance photovoltaic cells by separating the carriers photogenerated at the absorber with minimum recombination losses. Some examples are the homojunction-based carrier-selective contacts using thermal diffusion from spin-on dopant [2][3] or the application of laser firing process [4][5]. On the other hand, wide band gap semiconductors as carrier-selective contacts, forming heterojunction structures, offer potential advantages. In particular, the use of electron and hole selective contacts with an appropriate band alignment to the c-Ge absorber can lead to a significant reduction in carrier recombination. For this reason, several heterojunctions have been developed on c-Ge, including MoO_x [6], hydrogenated amorphous silicon (a-Si:H) [7][8], InGaP [5], and microcrystalline silicon [9], among others.

Recently, ultrathin c-Ge epitaxial films were deposited on crystalline silicon (c-Si) wafers, acting as an effective buffer layer in tandem cells with comparable performance to the III-V layers growth on c-Ge substrates in terms of mismatching and crystallinity [10]. Simple Plasma-Enhanced Chemical Vapor Deposition (PECVD) was utilized in the process, opening a new perspective for low-cost epitaxial films, which have better doping efficiency, higher conductivity, higher mobility, and lower optical absorption than their amorphous or microcrystalline counterparts. Moreover, PECVD has been demonstrated to produce high-quality epitaxial Si on GaAs, which has similar lattice parameter as c-Ge [11].

Therefore, this study aims to investigate epitaxial c-Si(n) grown by PECVD on p-type c-Ge wafers in order to develop high-quality electron-selective contacts. Firstly, we discuss the development and characterization of the c-Si(n) film on p-type c-Ge substrate. Secondly, we detailed strategies to improve the surface passivation of the c-Si(n)/c-Ge(p) heterojunction. Finally, devices were completed for further analysis of the interface quality. The findings of this study provide valuable guidance for developing highly efficient c-Ge TPV cells based on cost-effective and scalable manufacturing processes.

2 METHODOLOGY

This study utilizes 4-inch p-type c-Ge (100) wafers with a resistivity, doping, and thickness of 1.2 Ω·cm, 2×10^{15} cm^{-3}, and 175 ± 15 μm, respectively. The c-Ge wafers are cleaned to obtain a H-terminated surface by dipping them into diluted hydrofluoric acid (HF, 1%) for one minute, followed by 30 seconds dip in DI water and 3 minutes dip in hydrochloric (HCl 32%):H_2O water at 1:1 ratio. Firstly, we process the rear side of the wafers by depositing the dielectric stack (see the inset in Fig. 5), which consists of a H_2 plasma treatment followed by ~2 nm intrinsic amorphous silicon carbide (a-SiC_x:H(i)) layer deposited by PECVD at 400 °C. ALD is then used to deposit 50 nm aluminum oxide (Al_2O_3) layer at 200°C. Finally, a 45 nm thick, close to the stoichiometric amorphous silicon carbide film (a-SiC) is deposited by PECVD at 370 °C. A post-annealing treatment at 360 °C is performed to fully activate the fixed charge of the

10.4229/EUPVSEC2023/1CV.3.29
020053-001

Al_2O_3 layer, which is responsible for the field-effect surface passivation. This dielectric stack has resulted in the state-of-the-art passivation of the c-Ge wafers with a surface recombination velocity of 9 cm/s (see reference [12] for further details).

On the front side of the c-Ge wafer, the c-Si(n) thin film is deposited using a standard RF (13.56 MHz) PECVD reactor [13]. The H_2, SiH_4, and PH_3 flow rates are optimized as 350 sccm, 3 sccm, and 1 sccm, respectively. The optimum deposition temperature is set to 175 °C, and the electrode power was 12 W, with a pressure of 1,444 Torr. The developed c-Si film on the c-Ge is characterized by spectroscopic ellipsometry, Raman spectroscopy, and X-ray diffraction (XRD) measurements.

To get information about surface recombination at the c-Si(n)/c-Ge(p) interface, we use Sinton-WCT120 photoconductance tool to determine the dependence of effective lifetime (τ_{eff}) on excess carrier density. Considering negligible bulk recombination, i.e., bulk lifetime (τ_b) tending to infinity, and assuming that the rear side is well passivated by the dielectric stack a-SiC_x:H(i)/Al_2O_3/a-SiC with S_{rear} = 9 cm/s, the front surface recombination (S_{front}) between the c-Si(n)/c-Ge(p) wafer can be calculated using the following expression:

$$\frac{1}{\tau_{eff}} = \frac{1}{\tau_b} + \frac{S_{front}}{w} + \frac{S_{rear}}{w} \Rightarrow S_{front} = \left(\frac{w}{\tau_{eff}} - S_{rear}\right) \qquad (1)$$

where w is the thickness of the wafer.

The heterojunction was studied under different configurations by introducing a thin a-Si:H film in the interface and on the highly doped c-Si(n). Finally, 1x1 cm^2 photovoltaic devices are completed by sputtering 80 nm indium tin oxide (ITO) film on the c-Si(n) layer using Magnetron system. After the ITO deposition, ~1 µm silver (Ag) metal grid was evaporated on the ITO surface. Finally, a ~1 µm Al mirror is evaporated at the rear side of the structure, followed by laser firing to create the local hole selective contact with a fixed processing area of 5%. Afterwards, these devices are characterized by measuring current density-voltage (*J-V*) curves under air mass 1.5 global (AM1.5G) conditions.

3 RESULTS AND DISCUSSION

3.1 Characterization of the c-Si(n) film

The structure of the deposited c-Si(n) film has been analyzed at room temperature using Raman spectroscopy. Fig. 1 shows the Raman spectroscopy measurements of the c-Si films deposited on the c-Ge wafer using the developed process. We can observe sharp Raman peaks in ~ 300 and 520 cm^{-1}, which can be attributed to c-Ge and c-Si, respectively [14]. In particular, the sharp peak at ~520 cm^{-1} indicates a dominant crystalline phase of silicon with negligible presence of amorphous phase whose Raman response is located at 480 cm^{-1}. Additionally, the absence of a characteristic peak at 400 cm^{-1} for Si–Ge phonon mode illustrates the lack of Si–Ge mixing material at the interface [15].

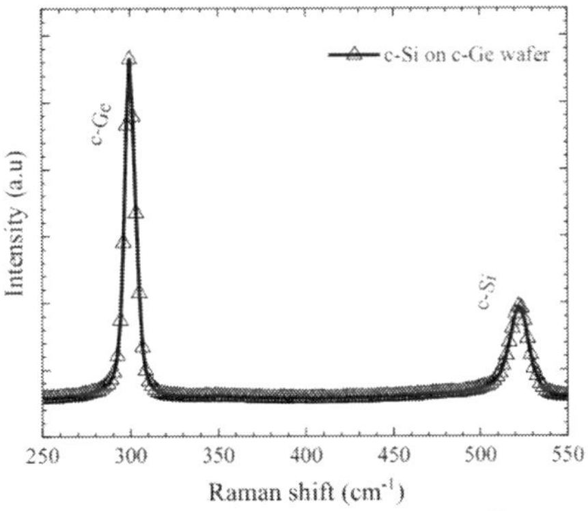

Figure 1: Raman spectroscopy of the c-Si(n) film on c-Ge(p) wafers.

To confirm the crystalline quality of the silicon film, additional characterization is conducted using Spectroscopic Ellipsometry and XRD measurements. Regarding Spectroscopic Ellipsometry, Fig. 2 shows the real and imaginary parts of the pseudo-dielectric function <ε_r> and <ε_i> of the c-Si(n) thin film deposited on a c-Ge wafer. The experimental data is excellently fitted by considering a silicon layer of 117 nm consisting of 100% large grain c-Si and a top layer of ~5 nm with 72% of c-Si, 13% of a-Si, and 15% of voids, which can be interpreted as roughness on top of the c-Si layer. This result indicates the deposition of a high-quality c-Si(n) layer directly on the Ge wafer without forming any interfacial mixed Si-Ge layer, which agrees with the Raman spectroscopy analysis.

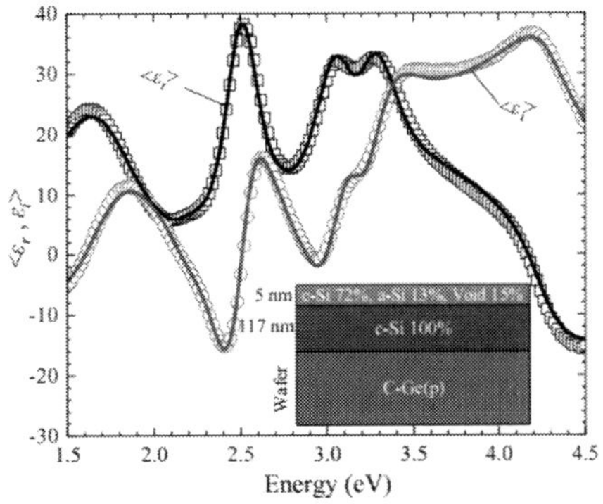

Figure 2: The real <ε_r> and imaginary <ε_i> part of the pseudo-dielectric function of c-Si(n) thin film deposited on a c-Ge wafer, where the symbol represents the experimental data while the line is the theoretical curve. The inset shows the layers used in the model to fit the experimental data with 117 nm c-Si bulk and 5 nm roughness.

Regarding XRD, Fig. 3 shows the XRD pattern of the c-Si(n) film on the c-Ge(p) wafer compared to the bare c-Ge wafer. The bare c-Ge wafer has a diffraction peak at a specific angle (2 theta) of 63.26°, representing the (100) orientation. After c-Si deposition, the c-Ge peak can still be observed, but two new peaks appear at

10.4229/EUPVSEC2023/1CV.3.29
020053-002

44.24° and 21.38°. Estimated by Bragg's Law, these two peaks can be attributed to c-Si planes of (111) and (100), respectively. This result again indicates no Si-Ge intermixing at the interface [10]. More importantly, the XRD spectrum exhibits sharp peaks, which also points to a high crystalline quality of the involved films, agreeing with the ellipsometry results.

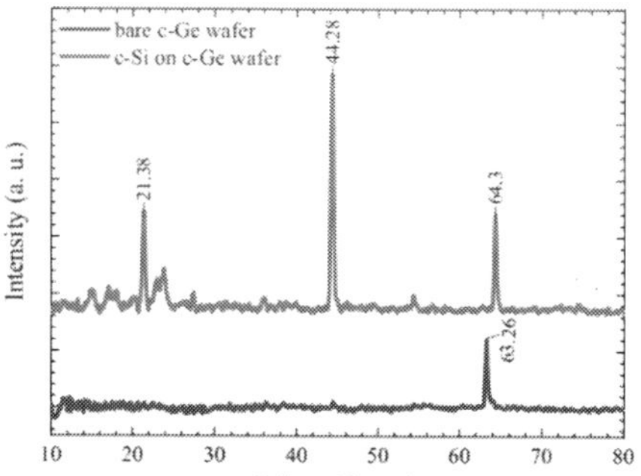

Figure 3: XRD measurements of the c-Si(n) on c-Ge substrate (red line) and the bare c-Ge wafer (black line).

3.2 Surface passivation

The passivation quality of the heterojunction is characterized by quasi-steady-state photoconductance lifetime measurements, using a Sinton WCT-120, as described in the experimental section. Fig. 4 shows the effective lifetime as a function of cumulative annealing of 10 minutes in forming gas (N_2/H_2) starting at 225 °C up to 350 °C. For the c-Si(n)/c-Ge(p) heterojunction, a low effective lifetime of 7 µs is measured in the as-deposited state, which improves to 14 µs after annealing at 300 °C. This short effective lifetime can most likely be attributed to the c-Si(n)/c-Ge(p) interface, which has a high defect density due to the mismatch in lattice parameters between c-Si (a = 0.5431 nm) and c-Ge (a = 0.56575 nm). In order to reduce the interface dangling bonds, a thin a-Si:H(i) layer (~ 1nm) between the interface c-Si(n)/c-Ge(p) was introduced, resulting in a significant enhancement in the effective lifetime, as shown in Fig. 4. Suggesting an enhanced chemical passivation achieved by the presence of the intrinsic a-Si:H(i) layer. Therefore, a ~5 nm n-type a-Si:H film was deposited on highly doped c-Si(n) to improve the effective lifetime, obtaining a maximum effective lifetime in the range of 50 µs. This improvement is linked to the passivation of the ITO/c-Si(n) interface. Given the high quality of the c-Si(n) layer, minority carrier diffusion lengths longer than the layer thickness (~100 nm) are expected, which makes the ITO/c-Si(n) interface play a role in the recombination. Consequently, introducing a thin a-Si:H(n) layer can offer moderate passivation, effectively achieving a surface recombination velocity of 334 cm/s (calculated by equation 1) for the whole electron selective contact.

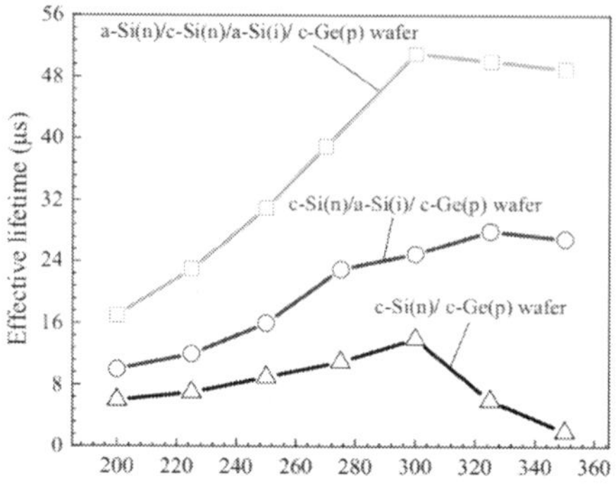

Figure 4: Effective lifetime measurements of three different configurations with c-Si(n) on c-Ge substrate as a function of the annealing temperature. For all samples, the rear side was coated by the dielectric stack a-SiC_x:H(i)/Al_2O_3/a-SiC with a rear surface recombination velocities of 9 cm/s.

3.3 Device Characterization

Solar cell devices have been fabricated with three different configurations, as shown in the inset of Fig. 5 and as explained in detail in the methodology section. The current density-voltage (*J-V*) curves of the devices measured under AM1.5G illumination are shown in Fig. 5. The output performance of the cells followed a similar pattern to the passivation findings. Devices with c-Si(n)/c-Ge(p) heterojunction produce the lowest cell performance due to high surface recombination formed by a mismatch in the lattice characteristics of c-Si(n) and c-Ge(p). This mismatch resulted in a high interface state density and hampered overall device performance. The addition of the thin a-Si:H(i) film between the interface c-Si(n)/c-Ge(p) is critical in lowering the interface state density, resulting in an increase in cell performance, notably in improving the open-circuit voltage (V_{oc}) and fill factor (*FF*). Moreover, further improvement in the V_{oc} and *FF* is observed by depositing the a-Si:H(n) film on the c-Si(n). This additional step contributed significantly to the overall enhancement of device performance, which results in 3.38% (V_{oc} = 186.3 mV, J_{sc} = 40.1 mA/cm^2, *FF* = 45.33%).

Although there is room for improvement (for example, reference [7] where a 5.26 % efficiency is reported with a-Si:H(n)/cGe(p) heterojunction), our findings highlight the potential of the utilization of high-quality c-Si(n) layers and the importance of a-Si:H(i) and a-Si:H(n) thin films at the interfaces to mitigate interface state density and improve the performance of the c-Ge cell. The results presented here pave the way for developing advanced Ge TPV cell technologies with higher conversion efficiencies and improved performance characteristics.

10.4229/EUPVSEC2023/1CV.3.29
020053-003

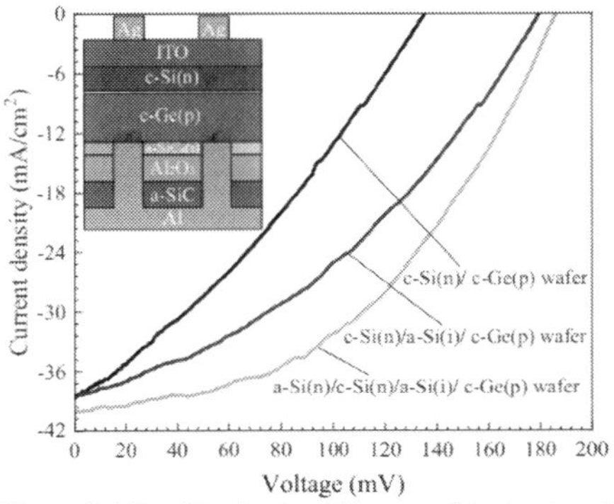

Figure 5: 1 Sun-illumination J-V curve of the developed c-Ge cell. A cross-section of the device structure is included in the inset.

4 CONCLUSIONS

In summary, we have developed high-quality crystalline c-Si(n) layers on c-Ge(p) substrates. The crystalline quality of the silicon layers is confirmed by Raman spectroscopy, Spectroscopic Ellipsometry, and XRD characterization. Furthermore, the c-Si(n)/c-Ge(p) heterojunction solar cells were studied. It was found that the presence of the thin intrinsic a-Si:H(i) layer at the interface is beneficial to improve interface passivation, and further improvement is obtained when a thin a-Si:H(n) layer is deposited on c-Si(n). This better passivation positively impacts the V_{oc} and FF in finished devices. The reported efficiencies show room for improvement, but the results reported hereby suggest the potential of c-Si(n)/c-Ge(p) heterojunction for cost-effective c-Ge cells.

ACKNOWLEDGEMENT
This work is part of projects PID2020-115719RB-C21 (GETPV), TED2021-131778B (TROPIC), and PID2020-116719RB-C44 (MaterOne) funded by MCIN/AEI/10.13039/501100011033. Project TED2021-131778B (TROPIC) is also funded by European Union "NextGenerationEU"/PRTR.

REFERENCES

[1] S. P. Tobin, S. M. Vernon, C. Bajgar, V. E. Haven, L. M. Geoffroy and D. R. Lillington, IEEE Electron Device Lett., Vol. 9 (1988) 2–4.
[2] N. E. Posthuma, J. Van Der Heide, G. Flamand, and J. Poortmans, IEEE Trans. Electron Devices, Vol. 54 (2007) 1210–1215.
[3] E.U. Onyegam, D. Sarkar, M. Hilali, S. Saha, R.A. Rao, L. Mathew, D. Jawarani, J. Mantey, M. Ainom, R. Garcia, W. James, S.K. Banerjee, Sol. Energy Mater. Sol. Cells 111 (2013) 206–211.
[4] J. van der Heide, N. E. Posthuma, G. Flamand, W. Geens, and J. Poortmans, Sol. Energy Mater. Sol. Cells 93 (2009) 1810–1816.
[5] J. Fernández, F. Dimroth, E. Oliva, M. Hermle, and A. W. Bett, AIP Conf. Proc., Vol. 890 (2007) 190–197.
[6] A. Alcañiz, G. López, I. Martín, A. Jiménez, A. Datas, E. Calle, E. Ros, L.G. Gerling, C. Voz, C. del Cañizo, R. Alcubilla, Sol. Energy 181 (2019) 357–360.
[7] S. Nakano, Y. Takeuchi, T. Kaneko, and M. Kondo, in Journal of Non-Crystalline Solids 358 (2012) 2249–2252.
[8] T. Kaneko and M. Kondo, Jpn. J. Appl. Phys. 50 (2011) 23–26.
[9] B. Hekmatshoar, D. Shahrjerdi, S. W. Bedell and D. K. Sadana, 2012 38th IEEE Photovoltaic Specialists Conference, Vol. 1 (2012) 001590-001593.
[10] M. Ghosh, P. Bulkin, F. Silva, Erik V. Johnson, I. Florea, D. Funes-Hernando, A. Tanguy, C. Renard, N. Vaissiere, J. Decobert, I. García, I. Rey-Stolle, P. Roca i Cabarrocas, Sol. Energy Mater. Sol. Cells 236 (2022) 111535.
[11] R. Cariou, W. Chen, J-L. Maurice, J. Yu, G. Patriarche, O. Mauguin, L. Largeau, J. Decobert, P. Roca i Cabarrocas, Sci Rep 6 (2016) 25674 .
[12] I. Martín, G. López, M. Garín, C. Voz, P. Ortega, and J. Puigdollers, Surfaces and Interfaces 31 (2022) 102070.
[13] P. Roca i Cabarrocas, J. B. Chévrier, J. Huc, A. Lloret, J. Y. Parey, and J. P. M. Schmitt, J. Vac. Sci. Technol. A9 (1991) 2331–2341.
[14] W. S. Yoo, K. Kang, T. Ueda, T. Ishigaki, H. Nishigaki, N. Hasuike, H. Harima, M. Yoshimoto, C. S. Tan, ECS J. Solid State Sci. Technol., 4 (2) (2015), P9–P15.
[15] R. Cariou, J. Tang, N. Ramay, R. Ruggeri, and P. Roca i Cabarrocas, Sol. Energy Mater. Sol. Cells 134 (2015) 0927–0248.

10.4229/EUPVSEC2023/1CV.3.29
020053-004

40th European Photovoltaic Solar Energy Conference and Exhibition

This presentation was selected by the Sc. Committee of the EU PVSEC 2023 for submission of a full paper to one of the EU PVSEC's collaborating peer-reviewed journals.

INVESTIGATING THE PASSIVATION POTENTIAL OF THERMAL EVAPORATION GROWN VANADIUM SUB-OXIDE FILMS FOR SILICON HETEROJUNCTION SOLAR CELLS

Rajesh Kanakala[#1], Narendra Bandaru[#,1,2], Rajesh Maurya[1], Jatindra Kumar Rath[1,2,*]

[1]Department of Physics, Indian Institute of Technology Madras, Chennai, 600036, Indias
[2]DSEHC, Department of Physics, Indian Institute of Technology Madras, Chennai, 600036, India

Corresponding author *: jkr@iitm.ac.in
# - Equal contribution

ABSTRACT: In recent years, transition metal oxides as carrier-selective contacts for silicon have gained a lot of research interest in the photovoltaic community as promising candidates to replace doped amorphous Si layers in Silicon heterojunction solar cells. In this work, thickness-dependent passivation of thermally evaporated vanadium oxide (V_2O_x) on n-type crystalline Silicon (n-Si) was investigated by lifetime studies using photoconductivity decay (PCD). The structural, optical, and electrical characteristics of the deposited films were examined using a wide range of characterization techniques, such as AFM, GIXRD, UV/Visible analysis, Hall measurements, TLM, temperature-dependent conductivity, minority carrier lifetime measurements using SINTON photoconductivity decay (PCD), and lifetime mapping by SEMILAB µ-PCD. The minority carrier lifetime values ranged from 64.07±3.39 µs to 207.52±4.30 µs for film thickness varied from 1.5 nm to 4.5 nm respectively, indicating that thicker films will show enhanced passivation. It is suggested that the formation of a sub-stoichiometric SiO_x interlayer as a result of the bonding of silicon with oxygen may be responsible for surface passivation. It is hypothesized that the silicon sub-oxide layer's fixed charges add to the field effect passivation mechanism. With increasing film thickness, the activation energy values reduced from 0.51 eV to 0.47 eV, indicating improved charge carrier transfer. From the results, a 4.5 nm-thick V_2O_x film has good optoelectronic properties, with a minority lifetime of 207.52±4.30 µs, an implied V_{oc} of 651±1 mV, an electron concentration of 2.62×10^{13} cm^{-3}, mobility of 40 cm^2/V-s, and optical transmittance of above 90%.
Keywords: transition metal oxide, vanadium oxide, passivation, thermal evaporation

1 INTRODUCTION

Recent years have seen a substantial increase in interest in the integration of vanadium pentoxide (V_2O_5) layers in solar cells, particularly in the context of silicon heterojunction (SHJ) solar cells [1-4]. Usage of hydrogenated amorphous/nano-crystalline silicon (a-Si:H/nc-Si:H) with crystalline silicon (c-Si) in SHJ solar cells has emerged as a key photovoltaic innovation. These thin film stacks benefit from the outstanding surface passivation provided by intrinsic a-Si:H layers, particularly those with thicknesses below 5 nm [5-7]. Novel approaches have been explored, such as the use of dopant-free transition metal oxides (TMOs) as carrier selective materials, to enhance SHJ solar cells' performance and streamline the manufacturing procedure [8, 9]. V_2O_5 has drawn much interest among TMOs because of its wide band gap (Eg > 2.5 eV) and a greater work function (≈ 6.9 eV) than TMOs such as tungsten oxide (WO_3) and molybdenum oxide (MoO_3) [10-12].

In this study, we deposited V_2O_x films with precisely regulated thicknesses using thermal evaporation deposition. A well-established, reliable, and cost-effective method for depositing thin films is thermal evaporation. It has advantages including simplicity of usage, environmental friendliness, and the ability to create high-quality films [13, 14]. In a previous work, we found that changing the film thickness from 5 nm to 20 nm, peaking at 10 nm, caused a variation in lifetime values ranging from 220 microseconds to 293 microseconds [15]. However, its thickness makes it difficult to extract or tunnel the carriers from c-Si. The goal of this research is to identify an ultrathin layer with which a high carrier lifetime can be achieved. By systematically changing the thickness of the V_2O_x films, we investigate the initial growth, conformity, evolution, and impact on a variety of properties, including structural, optical, electrical, and surface passivation characteristics.

2 EXPERIMENTAL PROCEDURES

V_2O_x thin films were deposited using thermal evaporation in an ultra-high vacuum environment at room temperature on cleaned n-type c-Si (111) Cz wafers (resistivity of 2 cm and thickness of 160 m; commercial wafer) and Corning (Eagle XG) glass substrates. The source material, a Sigma Aldrich 99.99% pure V_2O_5 powder, was evaporated using a tungsten boat. The pressure inside the chamber was maintained at 5×10^{-5} mbar. The c-Si wafer was subjected to saw damage removal in KOH solution for 5 min. Then, the saw damage removed samples were cleaned with standard RCA cleaning and dipped in 2% HF before loading into the chamber. Glass was cleaned with soap solution, then rinsed with deionized water, and then sonicated in IPA for 10 min for the deposition of V_2O_x. V_2O_5 powder (from Sigma Aldrich, 99.99% pure) was evaporated using a tungsten boat at a processing pressure of 5×10^{-5} mbar. To study and characterise the deposited films, a number of samples have been fabricated using both wafer and glass substrates of varied sizes. After the deposition of V_2O_x films, the samples were subjected to various characterization like UV/Visible, Grazing Incident X-ray Diffraction (GIXRD), Atomic Force Microscopy (AFM), hall measurement, temperature-dependent conductivity, minority carrier lifetime measurements with SINTON photoconductivity decay (PCD) and lifetime mapping by SEMILAB µ-PCD to study the structural and optoelectronic properties.

The thickness and topography of these films were determined via AFM analysis, which was performed using the Park NX10 AFM apparatus in a non-contact mode. The structural study of the films was carried out using GIXRD patterns from an X-ray diffractometer (Model: Smart Lab Diffractometer, Rigaku). The GIXRD measurements were conducted with a 2 range of 10-70° and a 0.5° optimum grazing angle. Cu K_α radiation with a wavelength of 0.15418 nm, a step size of 0.12°, and a scan speed of 4 deg/min was used for the measurements. The optical characteristics of the films were examined using a UV/Visible spectrometer (JASCO V-570). The electrical parameters of the films, including electrical resistivity (ρ), carrier concentration (n_e), and carrier mobility (μ_e), were measured using a Hall Effect measuring device (Model: Ecopia, HMS 3000). The studies were conducted in the Vander-Pauw configuration with a current range of -1 mA to +1 mA at ambient temperature. For a further examination of the samples' electrical characteristics, temperature-dependent conductivity measurements for activation energy (AE) were used. Temperature-dependent conductivity measurements for AE were investigated using an experimental setup that included a home-built vacuum system that could sustain a high vacuum environment up to 1×10^{-6} mbar, a Keithley 617 Multimeter for voltage sourcing and current measurement, and a LakeShore 331 Temperature Controller. The temperature dependence of dark conductivity measurements was performed on samples having a surface area of 1.5 x 2.5 cm^2 and silver electrodes positioned 1 mm apart to determine AE. After the samples were placed within the high vacuum chamber, a fixed voltage of 50 V was given. The samples were then maintained for an hour at 220 °C after being gradually heated. The temperature-dependent current was measured at an interval of 1 °C during the cool-down process at a cooling rate of 0.5 °C/min. Minority carrier lifetimes (τ_{eff}) of V_2O_x deposited Si wafers were measured using a photo-conductance decay (PCD) system (WCT-120, Sinton instrument). The aim of this research was to investigate how V_2O_x thickness affected τ_{eff}. This was accomplished using a symmetric device structure made up of V_2O_x (1.5 nm to 4.5 nm)/c-Si/V_2O_x (1.5 nm to 4.5 nm). The τ_{eff} was mapped across the whole device structure on a full-size wafer using the µ-PCD method in the SEMILAB instrument (WT 2000). The analysis was conducted using a laser with a pulse energy of 0.103 µJ, a wavelength of 905±10 nm, and a frequency of 2.5 kHz. The τ_{eff} values for the various thicknesses of V_2O_x in the device structure mentioned above are derived using this method.

3 RESULTS AND DISCUSSION

Using tapping mode in AFM, it was estimated that the thicknesses of the V_2O_x films on corning (EAGLE XG) glass substrates were around 1.5 nm, 2.5 nm, 3.5 nm, and 4.5 nm. GIXRD measurements of these films on corning glass substrates show a broad peak at 2θ = 22.25° and with no other peaks corresponding to crystalline vanadium oxide (VO_2, V_2O_3, V_2O_5) indicating that the films are sub-oxides and in the amorphous phase (see Fig. 1) [16-20]. The amorphous structure of the films is advantageous for device applications since it offers layer flexibility and substrate compatibility. From surface topography studies of the V_2O_x films on commercial c-Si wafers using AFM, the root mean square (RMS) roughness values increase from 0.4 nm to 0.7 nm with the increase in film thickness indicating that the films are smooth (see Fig. 2). From Fig. 3, Uv/Visible spectroscopy results of all the films show that the films are highly transparent with the optical transmittance greater than 90% in the visible range. These results show that the films can be used in SHJ as carrier-selective contacts to reduce parasitic absorption.

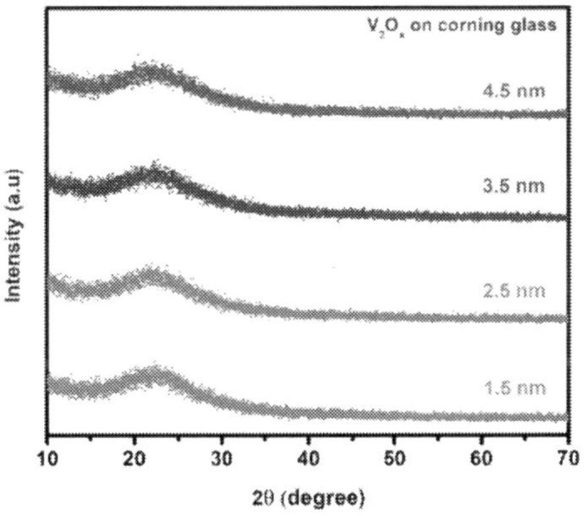

Fig. 1: The XRD pattern V_2O_x films with different thickness deposited on Corning glass

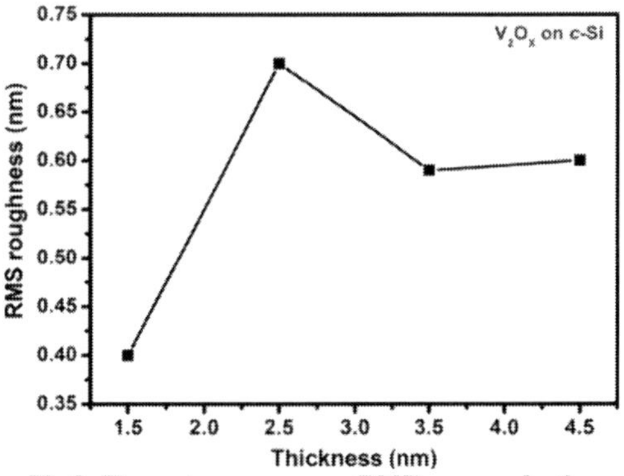

Fig 2: The root mean squares (RMS) average for the V_2O_x films.

Hall measurement confirmed that all the deposited V2Ox films are n-type with electron concentration (n_e) decreasing from 1.06×10^{15} cm^{-3} to 2.62×10^{13} cm^{-3} with the increase in film thickness while the mobility (μ_e) is increased slightly (i.e., from 37.65 cm^2/V-s to 40.00 cm^2/V-s). As a result, the resistivity of the film is increased from 1.50×10^2 Ω-cm to 5.95×10^3 Ω-cm (see Fig. 4). To

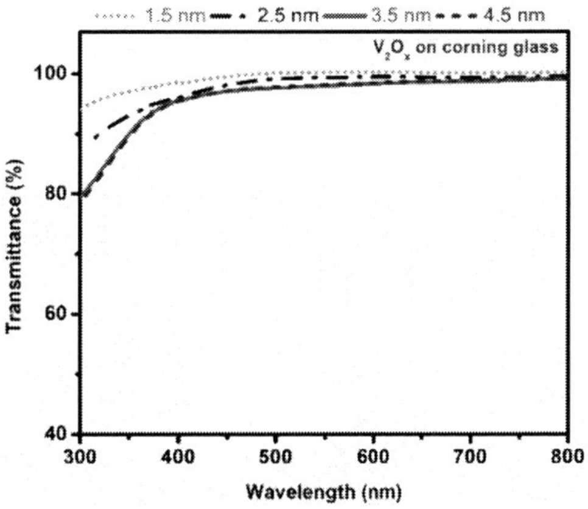

Fig 3: Transmittance Spectra for V_2O_x films

understand these results, the growth of the V_2O_x films was modeled as an amorphous material growth e.g., a-Si:H. From this model, it is predicted that the initial growth structure of these films was disordered and differed from the ideal vanadium-to-oxygen ratio, resulting in sub-

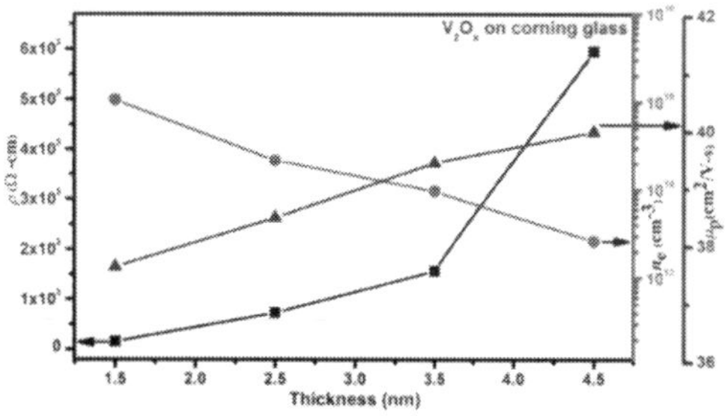

Fig 4: Resistivity, carrier concentration and mobility of V_2O_x films

stoichiometry during this incubation phase. As the growth proceeds films get more orderly which can be seen in improved carrier mobility. The decreased carrier concentration exhibits this gradual increase in stoichiometry. This inherent tendency, which also contributes to increased structural order and decreased defect density, allows for higher carrier mobility and more efficient charge transport.

Fig. 5 shows the activation energy values of the V_2O_x films. Despite a decrease in carrier concentration, the AE values declined from 0.514 eV to 0.474 eV as the film thickness increased from 1.5 nm to 4.5 nm. This is because

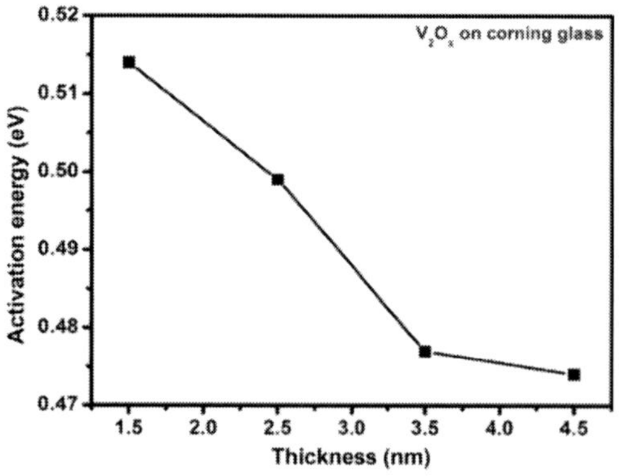

Fig 5 : Effect of thickness variation of V2Ox thin films on activation energy.

the AE of thin films depends mostly on interface band bending rather than bulk material [21]. A decrease in AE implies an increase in work function, which is a curial factor hole extraction. High work function is a necessary characteristic of a material to be a hole-selective contact, these results confirm that 4.5 nm film can extract the hole from c-Si easily.

To study the surface passivation by V_2O_x films, four samples were made by depositing V_2O_x on both sides of the n-type c-Si with thicknesses 1.5 nm, 2.5 nm, 3.5 nm, and 4.5 nm. Fig. 6 shows the minority carrier lifetime (τ_{eff}) measurement with different thicknesses of V_2O_x films. From Fig. 6, the τ_{eff} values increase from 64.07±3.39 µs to 207.52±4.30 µs with increase in thickness. Fig. 7 shows how implied V_{oc} (iVoc) and τ_{eff} vary with an increase in V_2O_x thickness. A lifetime of 207.52±4.30 µs was achieved with 4.5 nm thick V_2O_x film with an iVoc of

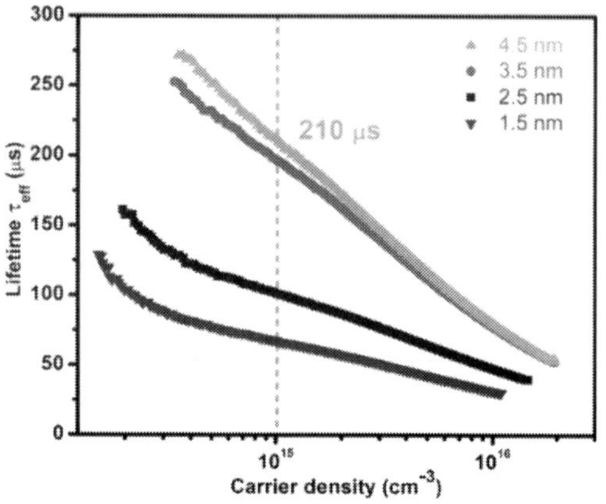

Fig 6: Minority carrier lifetime of V_2O_x films with different thickness

651±1 mV. This iVoc of 651±1 mV was achieved without any post-deposition and hydrogen treatment. A post-deposition treatment could enhance the passivation quality and achieve the iVoc above 700 mV. These results emphasize that V_2O_x can be used as a hole transport and passivating contact for c-Si.

The lifetime mapping of τ_{eff} measured by µ-PCD for different thickness V_2O_x films is shown in Figures 8a–d. The average τ_{eff} values are increasing from 62.35 µs to 139.7 µs with the increase in

V_2O_x thickness from 1.5 nm to 4.5 nm. The observed surface passivation is due to the formation of a sub-stoichiometric SiO_x interlayer, which results from silicon's reaction with oxygen in the V_2O_x layers. It is believed that the fixed charges in the silicon sub-oxide layers have an

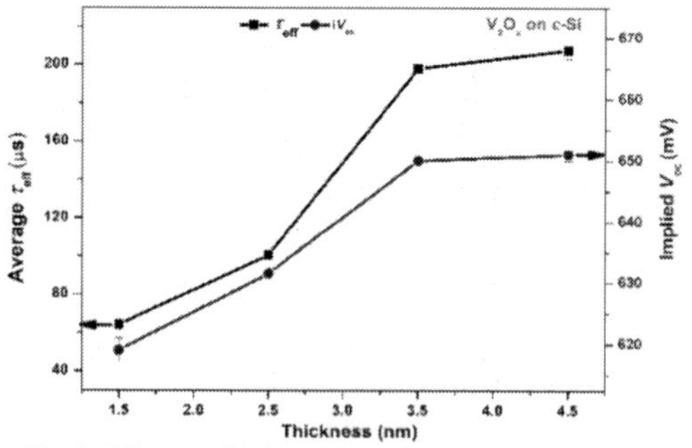

Fig 7: Effective lifetimes (τ_{eff}) and implied Voc (iV_{oc}) dependence of of the V_2O_x thin films

impact on the field-effect passivation [22]. A comprehensive examination of the experimental data reveals the structural and optoelectronic properties of the V2Ox films, and the findings demonstrate the relationships between electrical conductivity, surface passivation effects, layer thickness, and structural properties.

4 CONCLUSIONS

10.4229/EUPVSEC2023/1CV.3.33
020054-003

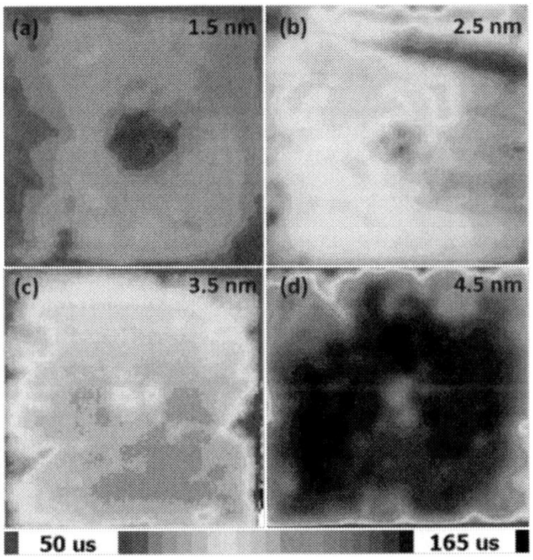

Fig 8: μ-PCD minority carrier lifetime mapping of V_2O_x films deposited on both sides of c-Si

The findings of this work highlight the impact of film thickness on the surface passivation and electrical properties of V_2O_x thin films. The electrical resistance of V_2O_x film increases with layer thickness, indicating a decrease in carrier concentration. The mobility of the carrier has slightly increased with the increase in thickness. The activation energy of the films decreases with layer thickness, indicating enhanced charge carrier transport due to high work function. The studied surface passivation effects indicate that V_2O_x films have the potential to be effective passivation layers. These results underline how important it is to optimise the film thickness in V_2O_x-based devices in order to achieve the necessary electrical and surface passivation properties. Results of the research work showed that a lifetime of 210 μs with an implied Voc of 651 mV was achieved for the V_2Ox layer with a thickness 4.5 nm on an n-type Cz(111) Si commercial wafer without any post-deposition treatments, using a simple physical deposition (thermal evaporation). The obtained results provide valuable information for designing and producing V_2O_x films for heterojunction silicon solar cells by elucidating the relationship between film thickness and film properties.

5 ACKNOWLEDGMENTS

The Department of Science and Technology (DST), Government of India, funded the DST IITM Solar Energy Harnessing Centre under grant no. DST/TMD/SERI/HUB/1(C), which provided financial assistance for the project.

6 DECLARATIONS

Conflict of interest: There is no conflict of interest, according to the authors.

7 REFERENCES

[1] J. Ibarra Michel, J. Dréon, M. Boccard et al., Carrier-selective contacts using metal compounds for crystalline silicon solar cells. Progress in Photovoltaics: Research and Applications, 31(4), pp.380-413 (2023).

[2] J. Melskens, B.W. Van de Loo, B. Macco et al., Passivating contacts for crystalline silicon solar cells: From concepts and materials to prospects. IEEE Journal of Photovoltaics, 8(2), pp.373-388 (2018).

[3] X. Yang, H. Xu, W. Liu et al., Atomic layer deposition of vanadium oxide as hole-selective contact for crystalline silicon solar cells. Advanced Electronic Materials, 6(8), p.2000467 (2020).

[4] G. Du, L. Li, H. Zhu, et al., High-performance hole-selective V2OX/SiOX/NiOX contact for crystalline silicon solar cells. EcoMat, 4(3), p.e12175 (2022).

[5] H.A. Gatz, J.K. Rath, M.A Verheijen et al., Silicon heterojunction solar cell passivation in combination with nanocrystalline silicon oxide emitters. physica status solidi (a), 213(7), pp.1932-1936 (2016).

[6] H. Sai, H.J. Hsu, P.W. Chen, et al., Intrinsic amorphous silicon bilayers for effective surface passivation in silicon heterojunction solar cells: A comparative study of interfacial layers. physica status solidi (a), 218(9), p.2000743 (2021).

[7] H. Lin, M. Yang, X. Ru et al., Silicon heterojunction solar cells with up to 26.81% efficiency achieved by electrically optimized nanocrystalline-silicon hole contact layers. Nature Energy, pp.1-11 (2023).

[8] 1Y. Wang, S.T. Zhang, L. Li et al., Dopant-free passivating contacts for crystalline silicon solar cells: Progress and prospects. EcoMat, 5(2), p.e12292 (2023).

[9] C. Yu, S. Xu, J. Yao et al., Recent advances in and new perspectives on crystalline silicon solar cells with carrier-selective passivation contacts. Crystals, 8(11), p.430 (2018).

[10] Li. Du, L. Li, X. Yang et al., Improved V2OX passivating contact for p-type crystalline silicon solar cells by oxygen vacancy modulation with a SiOX tunnel layer. Advanced Materials Interfaces, 8(22), p.2100989 (2021).

[11] S.Q. Hussain, K. Mallem, Y.J. Kim et al., Ambient annealing influence on surface passivation and stoichiometric analysis of molybdenum oxide layer for carrier selective contact solar cells. Materials Science in Semiconductor Processing, 91, pp.267-274 (2019).

[12] G. Terán-Escobar, J. Pampel, J.M. Caicedo et al., Low-temperature, solution-processed, layered V 2 O 5 hydrate as the hole-transport layer for stable organic solar cells. Energy & Environmental Science, 6(10), pp.3088-3098 (2013).

[13] W. Wu, J. Bao, X. Jia et al., Dopant-free back contact silicon heterojunction solar cells employing transition metal oxide emitters. physica status solidi (RRL)–Rapid Research Letters, 10(9), pp.662-667 (2016).

[14] G. Masmitjà, P. Ortega, J. Puigdollers et al., Interdigitated back-contacted crystalline silicon solar cells with low-temperature dopant-free

selective contacts. Journal of Materials Chemistry A, 6(9), pp.3977-3985 (2018).
[15] B. Narendra, K. Rajesh, M. Ramakrishna et al., The structural, optical, and electrical properties of thermal evaporation-deposited V2OX films for use in silicon heterojunction solar cells. Journal of Materials Science: Materials in Electronics, 34(12), p.1015 (2023).
[16] N. Fieldhouse, S.M. Pursel, R. Carey et al., Vanadium oxide thin films for bolometric applications deposited by reactive pulsed dc sputtering. Journal of Vacuum Science & Technology A, 27(4), pp.951-955 (2009).
[17] R.M. Öksüzoğlu, P. Bilgiç, M. Yıldırım et al., Influence of post-annealing on electrical, structural and optical properties of vanadium oxide thin films. Optics & Laser Technology, 48, pp.102-109 (2013).
[18] M. Benmoussa, E. Ibnouelghazi, A. Bennouna, Structural, electrical and optical properties of sputtered vanadium pentoxide thin films. Thin solid films, 265(1-2), pp.22-28 (1995).
[19] Y. Ke, S. Wang, G. Liu et al., Vanadium dioxide: The multistimuli responsive material and its applications. Small, 14(39), p.1802025 (2018).
[20] R.K. Jain, A. Khanna, Structural, optical and electrical properties of crystalline V2O5 films deposited by thermal evaporation and effects of temperature on UV–vis and Raman spectra. Optik, 144, pp.271-280 (2017).
[21] Ye. Ya-Gu James, W. A. Anderson, and Y. S. Tsuo. Thickness dependence of photoelectrical properties of intrinsic amorphous silicon. Solar cells, 23(3-4), pp.191-199 (1988).
[22] L.G. Gerling, G. Masmitja, P. Ortega et al., Passivating/hole-selective contacts based on V2O5/SiOx stacks deposited at ambient temperature. Energy Procedia, 124, pp.584-592 (2017).

DOPANT-FREE BASE CONTACTS FOR SILICON HETEROJUNCTIONS

F. Menchini, L. Martini, L. Serenelli, E. Salza, M. Izzi, M. Tucci
ENEA, Casaccia Research Centre
Via Anguillarese 301, 00123 Roma, Italy

ABSTRACT: The search for an efficient ohmic contact for n-type crystalline silicon-based heterojunctions, alternative to amorphous silicon, has long been sought but has not yet come to a convincing solution. Some materials are currently used to improve electron extraction in Si-based devices, but they are still limited by a low resistance to thermal treatments. In this work we investigated MgF_2 and MgF_2/ZnS films covered by Ti as base contacts for n-type crystalline silicon. The structures were characterised by measuring their J-V characteristics and the minority carrier lifetime variations after deposition of the layers on the passivated substrates. We found out that the introduction of an ultrathin film of ZnS helps enhancing the linearity of the characteristic and then the effectiveness of the ohmic contact, while simultaneously better preserving the silicon surface passivation. The device with a 1.5 nm-thick ZnS film kept its ohmic nature even after a thermal annealing treatment at 250°C. Further experiments are ongoing to replace the full metal Ti contact with a transparent TCO layer.
Keywords: heterojunction solar cells, dopant-free selective contacts, MgF_2, ZnS.

1 INTRODUCTION

Silicon-based heterojunctions (HJ) are currently the most favorable technology for solar cells. Recent developments are aiming at the substitution of the commonly-used doped hydrogenated amorphous silicon (a-Si:H) films used in the selective contacts by alternative materials, in order to reduce the optical losses introduced by the highly-doped a-Si:H layers and to avoid the use of dangerous and toxic gases that are needed for industrial safety procedures. For this reason, the new investigated materials are usually referred to as "dopant-free" carrier-selective layers [1]. In particular, the search for an efficient ohmic contact for n-type crystalline silicon (n-c-Si) alternative to amorphous silicon has not yet come to a convincing solution, notwithstanding the numerous suggestions [2]. Actually, the performances of such alternative solutions still do not overcome those of "doped" contacts, because the former heterostructures are generally mainly limited by low endurance to thermal treatments which should simulate typical screen-printable paste curing or Transparent Conductive Oxide (TCO) annealing steps in the manufacturing process, that cannot be avoided during the production process. To date, no effective undoped ohmic contact for the n-side of c-Si that is not damaged by annealing has been individuated yet, nor with a full-area metallic film neither with a transparent TCO, which is the long-chased solution for bifacial devices.

Recent results have reported highly efficient heterostructure solar cells with dopant-free selective contacts employing a metal compound as part of the electron-transport layer (ETL), namely LiF, in conjunction with an Al electrode [1, 3-5]. Efficiencies above 20% have been claimed on complete cells, but no investigation has been made on the sole electron contact. In all cases, no thermal annealing has been performed after the deposition of the ETL, suggesting that the structure had underwent some degradation after heating. Only a few authors [6, 7] have shown that an interlayer of TiO_2 is able to improve the passivation stability of the cells, most likely thanks to a decoupling of the passivating layer from the LiF/Al layers. Another metal fluoride, MgF_2, has been successfully applied in conjunction with Al in HJ selective contacts, but again no data about temperature resistance have been reported [8]. Finally, no significant results have been produced for thermally-annealed transparent selective contacts employing metal compounds and TCOs.

The major function of the metal-compound films in the selective contacts is that of avoiding the formation of gap states at the c-Si/metal interface, thus depinning the Fermi level from midgap. In addition to this, the work function (WF) of the outer metallic contact plays a fundamental role, as it determines the barrier height at the interface. For example, it has been calculated [9] that for n-type c-Si a WF below 4.2 eV is needed for the TCO to form an ohmic contact with no barrier to the transport of electrons. In this sense Al, with a WF of 4.08 eV, as suggested in [10], should be a suitable candidate as outer electrode, even though different WF values are reported for Al depending on its crystallinity grade and deposition conditions. However, the lack of results on thermally annealed structures reinforces the suspect that Al diffusion towards the c-Si is responsible for the poor performances of the cells.

In this work we explored the possibility of substituting Al with Ti, another low-WF metal [11]. We investigated the properties of electron-selective contacts containing MgF_2 and ZnS thin films deposited by room-temperature thermal evaporation, a technique that has the advantages of being versatile, low-cost, scalable and high-throughput. We showed the formation of ohmic contacts with a metallic Ti outer electrode, and the conditions for which the contacts can withstand thermal treatments. The analyses were performed on "stand alone" base contacts, as explained in the Experimental section, and not on full cells as mainly done in the literature, in order to characterise and optimise the bare base contact. We finally performed numerical calculations to simulate the possibility of substituting the metallic contact with a TCO for future applications in bifacial devices.

2 EXPERIMENTAL

Flat, FZ, n-type crystalline silicon (n-c-Si) wafers of 3-5 Ω·cm resistivity and 250 μm thickness were used as substrates. The wafers were passivated by amorphous hydrogenated silicon oxide (a-SiOx:H) 5 nm/thick thin films deposited by plasma enhanced chemical vapour deposition at 300°C [12, 13].

10.4229/EUPVSEC2023/1CV.3.35

The selective contacts were deposited on one side of the passivated substrates by thermal evaporation at a base pressure of 10^{-6} mbar from precursor pellets placed in a Mo boat. Two sets of samples were produced to investigate different base contacts. The first set of samples contained a 1 nm-thick MgF_2 film, while a second set contained an additional overlayer of 0.7 or 1.5 nm-thick ZnS. All samples were then covered by 50 nm-thick Ti films to enhance lateral conductivity. The structures are sketched in Fig. 1.

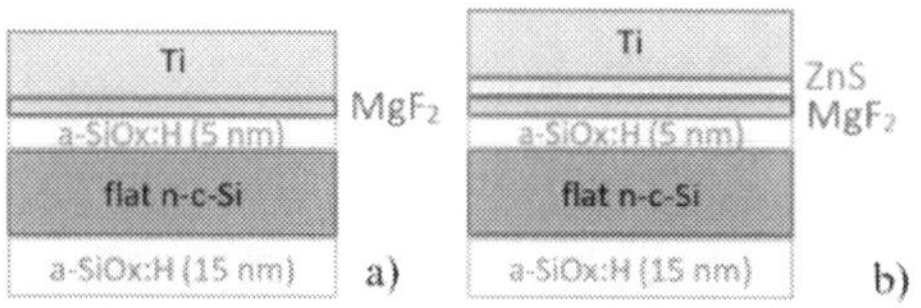

Figure 1: Schematic section of the investigated samples. The n-c-Si wafer is passivated by a-SiOx:H and the electron-selective contact is formed by a) an MgF_2 film or b) an MgF_2/ZnS bilayer covered by a Ti film. The rear-side passivation is removed and replaced by an InGa eutectic thick film to ensure an ohmic contact for J-V measurements

The minority carrier lifetime (τ) was evaluated by quasi-steady-state photoconductance measurements performed with a Sinton 100 instrument. The nature of the base contacts was evaluated by transverse current density vs voltage (J-V) measurements after acidic removal of the rear side passivation and application of a thick film of InGa eutectic directly on the c-Si surface, which ensures a perfect and low resistive ohmic contact, verified each time by a coplanar current-voltage measurement.

A thermal annealing treatment was performed on selected samples in a furnace in N_2 flux with fast (2°C/s) heating rate at 250°C for 10 minutes.

A final proof-of-concept cell with structure TCO/MoOx (10 nm)/a-SiOx:H (5 nm)/n-c-Si/a-SiOx:H(5 nm)/MgF_2 (1 nm)/ZnS (1.5 nm)/TCO was fabricated, containing all dopant-free contacts, in order to compare its optical properties to those of a conventional HJ cell based on doped amorphous silicon. Reflectance measurements were performed in a commercial Perkin Elmer λ950S double-beam spectrophotometer equipped with an integrating sphere.

Numerical simulations were performed by a software described in [14].

3 RESULTS AND DISCUSSION

The minority carrier lifetime of each structure can be considered as an indicator of recombination losses, thus it represents the most important reference during the fabrication process. All the investigated samples had a good starting level of passivation right after the a-SiOx:H deposition, corresponding to a 2-3 ms lifetime value.

When a MgF_2/Ti bilayer was deposited over the passivated c-Si wafer, as depicted in Fig. 1a, the lifetime of the heterostructure was slightly increased to 3.7 ms from the starting value of 3.1 ms, confirming the overall compatibility of the process, and in particular of the MgF_2 layer deposition, with the underlying a-SiOx:H passivating film. Figure 2 shows the J-V characteristics of the sample contacted on the back surface by InGa as described in the Experimental section. The quasi-linear trend of the J-V characteristic reveals the ohmic nature of the contact, confirming that no barrier to electron transport is formed inside the heterostructure and proving that Ti has a sufficiently low work function to avoid any undesired depletion within the c-Si. This latter fact was confirmed by indirect measurements (as described in [9]) that allowed WF estimation at the inner interface with the MgF_2 layer and returned a value lower than 4.1 eV.

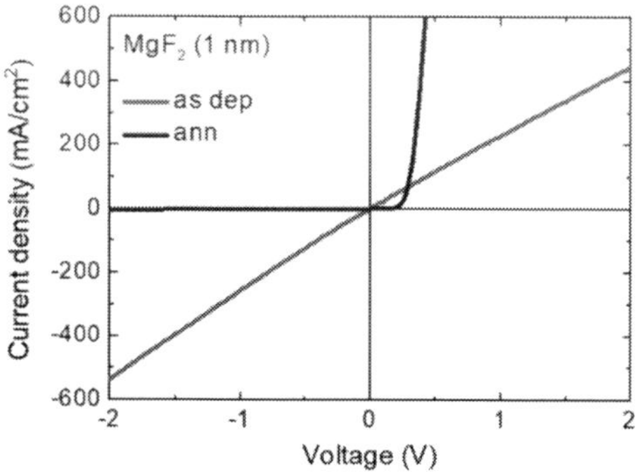

Figure 2: Dark J-V characteristics of the heterostructure n-c-Si/MgF_2(1 nm)/Ti after deposition (as dep) and after a thermal annealing at 250°C (ann). The contact is ohmic after deposition, but rectifies after annealing

Unfortunately, after thermal annealing at 250°C the lifetime of the structure dropped to 200 μs and the contact became rectifying (see Fig. 2). We believe that the damages to passivation were caused by Ti diffusion from the outer contact towards the c-Si/a-SiOx:H interface: indeed the lifetime of a twin structure without Ti capping was preserved and even increased of about 20% after thermal annealing, proving that the damage is linked to the presence of Ti. The same diffusion could have caused the formation of junction similar to a Schottky, with consequent rectification of the contact. Another hypothesis could be the modification of the work function of Ti deposited over MgF_2, due for example to the polar nature of MgF_2 [2]. This could have caused the shift of the Fermi level position toward midgap, thus inducing increased recombination that reflects in the rectifying behaviour of the contact. In the first case a barrier to metal diffusion could be beneficial, in the second case the introduction of a material able to pin the Fermi level below the conduction band could help electron transport.

ZnS is known to be able to form an electron-selective contact on n-c-Si due to its low electron affinity (3.9 eV) and to a possible slight n-type conductivity due to S vacancies or Zn interstitials [15]. These characteristics encouraged us to introduce a thin layer of ZnS between MgF_2 and Ti (heterostructure depicted in Fig. 1a). We considered two thicknesses for the ZnS layer: 0.7 nm and 1.5 nm, so to exploit the potential conductivity of the ZnS layer for transverse carrier transport and limiting the tunnelling only to the MgF_2 layer. We observed an ohmic behaviour of the contact for both thicknesses, as evidenced by the linear J-V characteristic shown in Figs. 3a and 3b. Moreover samples with the thinner (0.7 nm) ZnS layer showed a slight reduction of τ (of about 15%) after deposition, while the passivation was not further damaged after the thermal annealing, even if the contact became rectifying. Nevertheless this was a positive hint that ZnS had somehow acted as a physical barrier to the hypothesised Ti diffusion. When the thickness of the ZnS layer was increased to 1.5 nm the lifetime was doubled just after deposition and further increased by 20% after the thermal treatment. Remarkably, after thermal annealing not only the lifetime was enhanced, but the contact was still ohmic. This was the evidence of the fabrication of an undoped electron-selective contact resistant to thermal annealing up to 250°C.

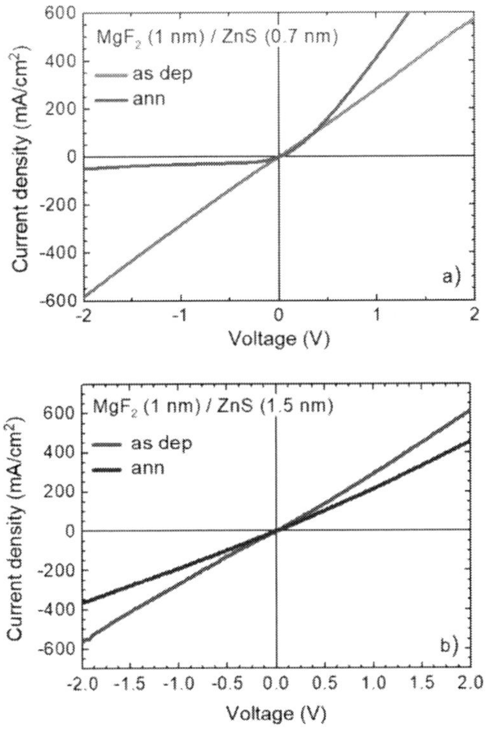

Figure 3: Dark J-V characteristics of the heterostructures a) n-c-Si/MgF_2(1 nm)/ZnS(0.7 nm) and b) n-c-Si/MgF_2(1 nm)/ZnS(1.5 nm) after deposition (as dep) and after a thermal annealing at 250°C (ann). Both heterostructures behave as an ohmic contact after deposition, while only structure b) remains ohmic after the thermal annealing

Different metals could be tested for the outer contact among those with a WF around 4.1 eV. It is not obvious, in fact, that the lowest-WF material should work better [16]. We tested Mg and Mn but the results were not encouraging.

Future applications in bifacial devices require transparent contacts, in which the metallic layer has to be substituted by a TCO film. To our knowledge, no transparent selective contacts for electrons that can withstand a thermal treatment at T>200°C have been reported yet. We have performed several preliminary experiments with different TCOs but with not encouraging results. With the aim of better understanding the transport mechanisms inside the structures, we performed numerical simulations of the dark band diagrams of different structures terminated with a TCO. The parameters used for the n-c-Si substrate (1 Ω·cm resistivity) and the a-SiOx:H passivating layers were kept fixed as they are well known, and MgF_2 was depicted as a thin insulating layer with a bandgap of about 12 eV. The electron affinity (Χ) of ZnS was varied between 3.9 and 4.1 eV to simulate possible differences from the deposited material and values found in the literature, while the work function (Φ) of the TCO was varied between 4.0 and 4.2 eV. Table I reports the overall results inferred from the simulations, indicating which kind of contact is obtained. It was found out that for a 1.5 nm thickness of the ZnS layer all combinations of parameters bring to the formation of an ohmic contact. For this reason in the simulations we also varied the thickness of the ZnS layer between 1.5 and 6 nm, so to account for possible errors in thickness estimation and experimental difficulties in producing too thin layers.

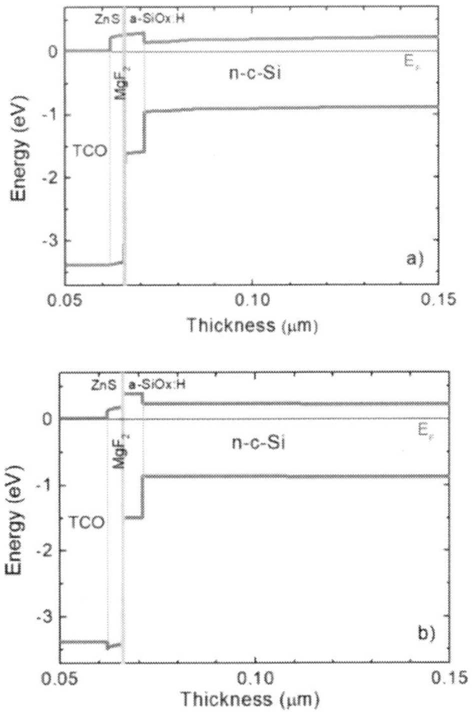

Figure 4: Simulated band diagrams for the heterostructure n-c-Si/a-SiOx:H(5 nm)/MgF_2(1 nm)/ZnS(3 nm) a) with X_{ZnS}=3.9 eV and Φ_{TCO}=4.1 eV and b) with X_{ZnS}=4.1 eV and Φ_{TCO}=4.2 eV. In case a) the c-Si surface is in accumulation and the contact is ohmic, in case b) the c-Si surface is depleted and the contact is rectifying

The results showed that not all combinations allowed favourable electron extraction. Two representative band diagrams for the structure n-c-Si/MgF_2(1 nm)/ZnS(3 nm)/TCO obtained from the simulation are reported in Fig. 4. Figure 4a refers to the situation in which the electron affinities of ZnS and TCO are 3.9 and 4.1 eV, respectively. It can be seen that the n-c-Si surface is in accumulation and allows the transport of electrons towards the TCO. However, if the TCO's electron affinity and the ZnS thickness are increased, the formation of the ohmic contact is not always ensured, as can be seen from Table I. As already mentioned, in the case of a ZnS film with 1.5 nm thickness there are no unfavourable conditions among

10.4229/EUPVSEC2023/1CV.3.35
020055-003

those investigated, while if the thickness of the ZnS film is increased some options are not acceptable. For example, the structure n-c-Si/MgF_2(1 nm)/ZnS(3 nm)/TCO where the electron affinities of ZnS and TCO are 4.1 and 4.2 eV, respectively, is represented in the band diagram in Fig. 4b. The downward shifting of the Fermi level brings to a depletion of the c-Si surface and to the formation of a barrier. If the ZnS thickness is increased, the situation gets even worse, because all the band bending is induced in the ZnS and not in the c-Si. However even in this condition unfavourable to charge transport, the Fermi level would still be far from midgap, preserving passivation.

Table I: Nature of the electron transport in the heterostructure base contact as deduced from numerical calculations with different valued for the electron affinity (X) of ZnS and work function (Φ) of the TCO

ZnS thickness = 1.5 nm

X_{ZnS} / Φ_{TCO}	3.9	4.0	4.1
4.0	ohmic	ohmic	ohmic
4.1	ohmic	ohmic	ohmic
4.2	ohmic	ohmic	ohmic

ZnS thickness = 3 nm

X_{ZnS} / Φ_{TCO}	3.9	4.0	4.1
4.0	ohmic	ohmic	ohmic
4.1	ohmic	ohmic	ohmic
4.2	ohmic	ohmic	rectifying

ZnS thickness = 6 nm

X_{ZnS} / Φ_{TCO}	3.9	4.0	4.1
4.0	ohmic	ohmic	ohmic
4.1	ohmic	ohmic	ohmic
4.2	rectifying	rectifying	rectifying

The situation represented in Fig. 4a could be also applied to our structures, with a ZnS layer of 1.5 nm, WF_{ZnS}=3.9 eV and Φ_{Ti}<4.1 eV, explaining the formation of an ohmic contact for the electrons. On the basis of calculations, however, also the device with the 0.7 nm-thick ZnS layer should allow the transport of electrons to the Ti electrode. Probably the reason of the failure of the device after annealing has to be sought in an incomplete coverage of the surface during ZnS deposition, due to the very low thickness of the films, which could be even more critical when moving to textured substrates. Fortunately, the simulations show that a certain tolerance in the thickness of the ZnS layer is possible.

As an additional remark, our n-c-Si substrates had a resistivity of 3-5 Ω·cm; a higher doping would obviously help lowering the contact resistivity and reduce the possibility to form a Schottky junction.

Finally, we tested the heterostructure from the optical point of view. We deposited a proof-of-concept complete cell containing all dopant-free selective contacts, TCO/MoOx/a-SiOx:H/n-c-Si/a-SiOx:H/MgF_2/ZnS/TCO, with adequate thicknesses of the layers. Its reflectance from the electron-contact side is compared to that of a conventional HJ cell based on doped amorphous silicon in Fig. 5. The reflectance of the "dopant-free" device is lower than that of the corresponding traditional one on the whole wavelength spectrum, confirming the outstanding optical performances of our device that would help maximising the current in complete cells.

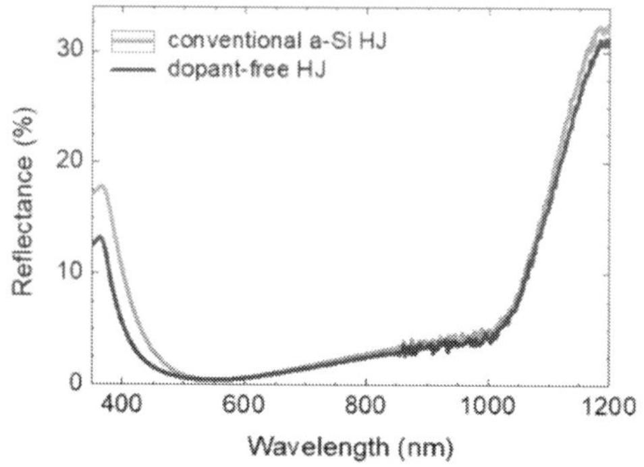

Figure 5: Comparison between the reflectances measured from the electron-selective side of a conventional HJ based on doped amorphous silicon and a "dopant-free" heterostructure in which the electron selective side is composed by a MgF_2/ZnS/TCO stack. The "dopant-free" device shows a lower reflectance the whole wavelength spectrum, confirming the high transparency of the contact

4 CONCLUSIONS

An ohmic contact for electrons in selective contact HJ solar cell based on n-type doped c-Si wafer resistant to thermal annealing up to 250°C was fabricated with a MgF_2/ZnS bilayer and a metallic Ti contact. We found out that a minimum thickness of the ZnS layer is needed to ensure the ohmic nature of the contact and preserve wafer passivation even after the thermal treatment. Numerical simulations limit the acceptable parameters for TCOs substituting the Ti contact for applications in bifacial devices.

ACKNOWLEDGEMENTS

This work was funded under the National Recovery and Resilience Plan (NRRP), Mission 4 Component 2 Investment 1.3 - Call for tender No. 1561 of 11.10.2022 of Ministero dell'Università e della Ricerca (MUR) by the European Union – NextGenerationEU. Project code PE0000021, Concession Decree No. 1561 of 11.10.2022 adopted by Ministero dell'Università e della Ricerca (MUR), CUP I83C22001800006, Project title "Network 4 Energy Sustainable Transition – NEST" and by the Italian Ministry of Environment and Energy Security in the framework of the Operating Agreement with ENEA for Research on the Electric System.

REFERENCES

[1] J. Bullock, M. Hettick, J. Geissbühler, A. J. Ong, T. Allen, C. M. Sutter-Fella, T. Chen, H. Ota, E. W. Schaler, S. De Wolf, C. Ballif, A. Cuevas, A. Javey, Nature Energy 1 (2016) 15031

[2] J. Ibarra Michel, J. Dreon, M. Boccard, J. Bullock, B. Macco, Progress in Photovoltaics: Research and Applications 31 (2023) 380-413

[3] L. Fang, S.J. Baik, S. Lim, S. Yoo, K.S. Lim, IEEE Transactions on Electron Devices 58:9 (2011) 3048-3052

[4] J.H. Yang, H.H. Jung, J. Seo, K. D. Kim, D. H. Kim, D. C. Lim, S. G. Park, J.W. Kang, M. Song, M. S. Choi, J. D. Kwon, K. S. Nam, Y. Jeong, S. H. Kwon, Y. C. Park, Y. C. Kang, K. B. Chung, C. S. Kim, K. S. Lim, S. Y. Ryu, Journal of Physical Chemistry C 117:45 (2013) 23459-23468

[5] W. Lin, Solar RRL 5:5 (2021) 2000771

[6] J. Bullock, Y. Wan, Z. Xu, S. Essig, M. Hettick, H. Wang, W. Ji, M. Boccard, A. Cuevas, C. Ballif, A. Javey, ACS Energy Letters 3:3 (2018) 508-513

[7] J. Bullock, Y. Wan, M. Hettick, X. Zhaoran, S. P. Phang, D. Yan, H. Wang, W. Ji, C. Samundsett, Z. Hameiri, D. Macdonald, A. Cuevas A. Javey, Advanced Energy Materials 9:9 (2019) 1803367

[8] Y. Wan, C. Samundsett, J. Bullock, T. Allen, M. Hettick, D. Yan, P. Zheng, X. Zhang, J. Cui, J. A. McKeon, A. Javey, A. Cuevas, ACS Applied Materials and Interfaces 8:23 (2016) 14671-14677

[9] L. Martini, L. Serenelli, F. Menchini, M. Izzi, M. Tucci, Progress in Photovoltaics: Research and Applications 28 (2020) 307-320

[10] P. A. Tipler, R. A. Llewellyn, Modern Physics, 3rd Ed., W.H. Freeman, 1999

[11] H. L. Skriver, N. M. Rosengaard, Physical Review B 46:11 (1992) 7157-7168

[12] J. P. Seif, A. Descoeudres, M. Filipič, F. Smole, M. Topič, Z. C. Holman, S. De Wolf, C. Ballif, Journal of Applied Physics 115 (2014) 024502

[13] L. Serenelli, L. Martini, L. Imbimbo, R. Asquini, F. Menchini, M. Izzi, M. Tucci, Applied Surface Science 392 (2017) 430-440

[14] D. Caputo, U. Forghieri, F. Palma, Journal of Applied Physics 82 (1997) 733-741

[15] X. Yang, B. Chen, J. Chen, Y. Zhang, W. Liu, Y. Sun, Materials Science in Semiconductor Processing 74 (2018) 309-312

[16] W. Wu, W. Lin, S. Zhong, et al., Physica Status Solidi Rapid Research Letters 14:4 (2020) 1900688

SELECTIVE LASER ABLATION OF TRANSITION METAL OXIDE THIN FILMS FOR SHJ SOLAR CELLS

C. Munoz-Garcia[a], D. Canteli[a], S. Lauzurica[a], M. Morales[a], C. Molpeceres[a], T. Tom[b], J.M. Asensi[b], J. Bertomeu[b], Eloi Ros[c], P. Ortega[c], J.M. López-González[c], C. Voz[c]
[a] Centro Láser. Universidad Politécnica de Madrid (UPM). C/ Alan Turing 1, 28031, Madrid, Spain
[b] Universitat de Barcelona (UB). C/ Gran Via de les Corts Catalanes, 585, 08007, Barcelona, Spain
[c] Universitat Politècnica de Catalunya (UPC). C/ Jordi Girona 1-3, 08034, Barcelona, Spain
cristina.munozg@upm.es

ABSTRACT: An alternative to the use of amorphous silicon (a-Si:H) in silicon heterojunction (SHJ) solar cells are transition metal oxide (TMO) films. TMOs act as passivators and as selective contacts, i.e. hole transport layers (HTL) or electron transport layers (ETL). Due to their chemical properties these oxides act as transparent contacts with semiconducting properties. Three TMOs, molybdenum oxide, vanadium oxide, and tungsten oxide, show good properties for use as p-type front contacts for n-type crystalline silicon heterojunction solar cells. In this work we show a complete study of the laser removal of WOx, VOx and MoOx films deposited on crystalline silicon using laser sources of three different wavelengths (355 nm, 532 nm, and 1064 nm) and in two temporal regimes in pulse width (picoseconds and nanoseconds). We measure the fluence threshold to remove TMO film, and the threshold to induce damage in the substrate, allowing us to calculate the parametric window of the process. We also studied the relati on between crater dimensions and ablation parameters. Complete removal of the three different TMO films has been achieved without damage to the silicon substrate. The results of diode isolation and their electrical characteristics are also included, showing the quality of the laser isolation processes.
Keywords: Transition Metal oxides (TMO), laser ablation, fluence threshold, silicon solar cells

1 INTRODUCTION

Silicon Heterojunction (SHJ) devices were developed to enhance the efficiency of first-generation silicon solar cells by incorporating thin layers of hydrogenated amorphous silicon (a:Si-H) onto crystalline silicon (c-Si) wafers. This technology offers the advantages of being easily doped, very good passivating properties, and low-temperature processing, although heterojunction emitters exhibit high sheet resistance [1]. Various strategies have been proposed to improve efficiency, including the Heterojunction with Intrinsic Thin Layer [2,3] and or Tunnel Oxide Passivated Contact (TOPcon) [4]. Another approach involves using Transition Metal Oxides (TMOs) as hole or electron transport layers (HTL or ETL) to serve as selective contacts, simplifying solar cell fabrication. Three transition metal oxides -molybdenum oxide (MoO_3), vanadium oxide (V_2O_5), and tungsten oxide (WO_3) have been investigated to be used as front hole-selective contacts for n-type crystalline silicon heterojunction solar cells [5, 6, 7]. These oxides, particularly thermally evaporated non-stoichiometric MoOx, VOx, and WOx films, possess a high work function (up to 6.9 eV) and a wide energy band gaps (Egap > 3 eV). Their transparency and semiconductive properties result from oxygen vacancy defects [8].

However, the ability to selectively remove the oxide film without damaging the substrate is critical to the production of electronic devices using these materials, and laser systems appear to be a good option for this due to their high precision and versatility.

Laser scribing involves irradiating the surface of a material to induce heating, melting and evaporation. Understanding thin film laser scribing requires analysis of two key parameters: laser wavelength selection, which affects process efficiency and the potential for selective removal of the top layer based on the absorption coefficients of the layers [9], and laser pulse fluence, which is critical for complete removal of the top layer without damaging the underlying substrate. Appropriate pulse repetition frequency and overlapping are also important for multi-pulse processing [10]. IN this short of concept is the damage fluence threshold, defined as the minimum fluence needed to induce visible damage in the film [11]. This concept applies also to oxide film removal and silicon substrate damage thresholds, and the parametric window, range between oxide removal and substrate damage, of the process, i.e., the range of working parameters that allow the complete removal of the oxide layer without damaging the substrate, is defined by these two threshold fluences.

In this study, the laser scribing process is optimized for three transition metal oxides—WOx, VOx, and MoOx—of significant interest for their application as HTL layers on silicon. The main objective is to investigate how laser pulses of different wavelengths and durations interact with these oxide films.

2 MATERIAL AND METHODS

2.1 Laser sources

Laser ablation experiments are carried out in the nanosecond and picosecond regimes with three wavelengths using different laser sources: For nanosecond pulses Nd:YVO4 DPSS sources emitting at 355 nm (HIPPO, Spectra-Physics), at 532 nm (Explorer, from Spectra-Physics), and at 1064 nm (Yb-fibre laser RFL-P20QE). In the picosecond range a Nd: YVO4 DPSS laser source (Lumera Super Rapid - HE) emitingt at 355 nm, 532 nm and 1064 nm. All laser sources work at TEM00 mode.

2.2 Sample preparation

To study of the ablation process, three different oxide thin-films -MoOx, VOx, WOx- were deposited on crystalline silicon. First, polished n-type Si wafers were dipped in dilute hydrofluoric acid (1%) to remove any native silicon oxide from their surface. Then, the different TMO layers were thermally sublimated in vacuum (<10^{-5} mbar) from a tantalum boat using powdered V_2O_5, MoO_3 and WO_3 precursors (from Sigma Aldrich) with more than

99.99% purity. The deposition rate was regulated at around 0.2 Å/s with a quartz micro-balance. The thin layers grown on the substrates at room temperature result clearly sub-oxidized, which turns out to be positive for their hole-selective behavior.

2.3. Ablation thresholds. Measurement and characterization techniques.

To establish the parametric windows it is important to know the ablation thresholds, that can be easily measured from the relation between pulse fluence and crater diameter generated in the sample according to Liu's methodology [11].

$$D^2 = 2 \cdot w_0^2 ln\left[\frac{\phi_0}{\phi_{th}}\right] \qquad (1)$$

A confocal microscope (Sensofar PL µ2300) was used to measure the ablation profile and perform morphological characterization.

2.4 Laser isolation of TMO diodes

Once the parametric windows for the selective ablation of the TMO films were defined, we evaluated the viability of the process by the characterization of diodes with and without laser isolation. The diodes were deposited on n-type crystalline silicon (c-Si) substrates. The back (cathode) side featured a reference ohmic contact made of doped amorphous silicon. On the front (anode) side a TMO film was used as hole-selective contact, and a metallic electrode structure was used, consisting of a thin nickel layer (10 nm) with a high work function, covered by a 200 nm aluminium layer to facilitate electrical characterization. To evaluate the performance of these devices, the current-voltage (I-V) characteristics were measured, using a high-precision Keithley 2410 sourcemeter.

3 RESULTS AND DISCUSSION

In all samples, confocal microscopy reveals that nanosecond pulses at wavelengths of 355 nm and 532 nm induce craters with a clean edge, whereas 1064 nm pulses generate more thermal affection. On the contrary, picosecond pulses produced craters with a larger affected zone, probably due to the higher beam radius (See figure 1).

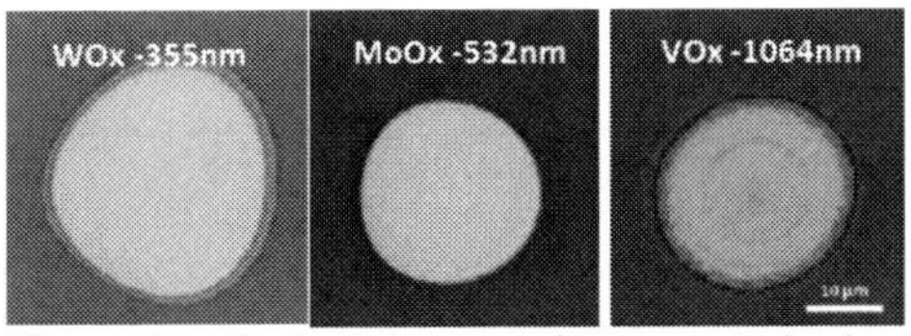

Figure 1: Confocal images of craters obtained with nanosecond pulses at 355 nm, 532 nm and 1064 nm.

Figure 2 shows confocal microscopy images of the VOx sample ablated with the nanosecond laser system at 532 nm. As can be seen, an adequate pulse energy allows the selective removal of the TMO layer, although higher energies can damage the silicon substrate.

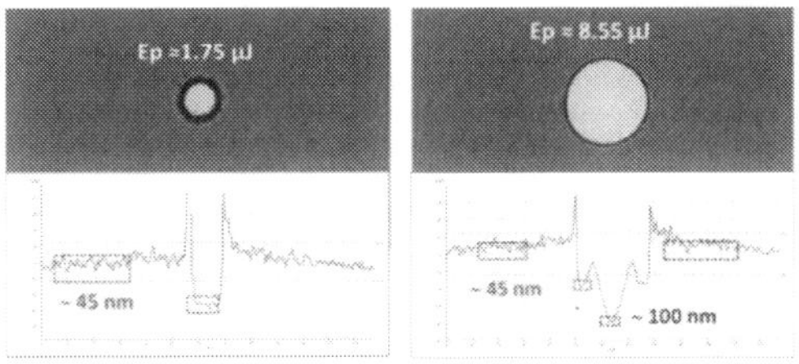

Figure 2: Confocal microscope images and profiles of craters generated in a VOx film at two different pulse energies, with a ns, 532 nm laser source.

In order to calculate the thresholds, the figure 3 shows an example of the relation between the square of the crater diameter and the logarithm of maximum laser pulse fluence (ϕ_0). In all oxides, the behaviour is the same, allowing the calculation of the beam radius (ω_0). As the ω_0 value does not depend on the oxide, an average value of the different ω_0 values obtained for the three oxides has been used for the fluence calculations. For the nanosecond laser systems, the obtained beam radius (ω_0) values are 18.7 ± 0.7 µm, 14.2 ± 0.8 µm and 20.6 ± 0.1 µm for 355 nm, 532 nm, and 1064 nm wavelength, respectively. For the picoseconds laser system, the beam radii are higher : 33.1 ± 0.5 µm (at 355 nm), 42.2 ± 0.8 µm (at 532 nm) and 62 ± 1 µm (at 1064 nm).

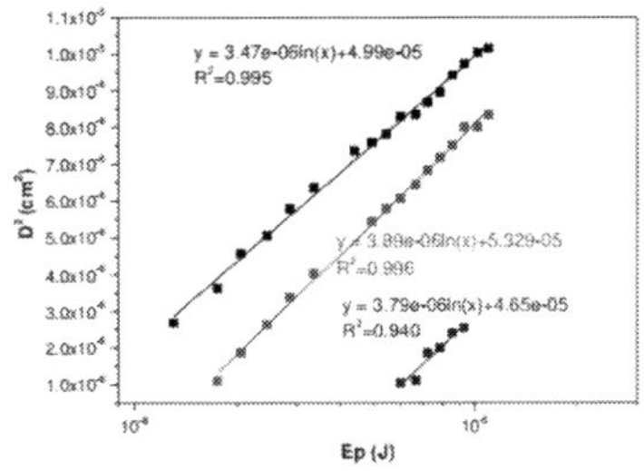

Figure 3: Squared diameter of craters in VOx plotted against laser pulse energy, at 532 nm, ns pulses. The different lines correspond to damage in the oxide film (black), total film removal (red) and substrate damage (blue).

Figures 4 and 5 show the threshold fluence values for nanosecond and picosecond pulse regimes at the different wavelengths for three different oxides. Working with ns pulses, a given wavelength leads to very similar threshold fluence values for oxide removal for the three different oxides (red columns, Figure 3): 0.15 ± 0.02 J/cm² for 355 nm, 0.34 ± 0.08 J/cm² for 532 nm and 4.9 ± 0.4 J/cm² for 1064 nm. The ablation threshold obtained at 1064 nm is more than an order of magnitude higher than at the other wavelengths, probably due to the lower absorption of the material at this wavelength and, to a lesser extent, to the longer pulse duration (120 ns at 1064 nm versus 12 ns and 15 ns at 355 nm and 532 nm, respectively).

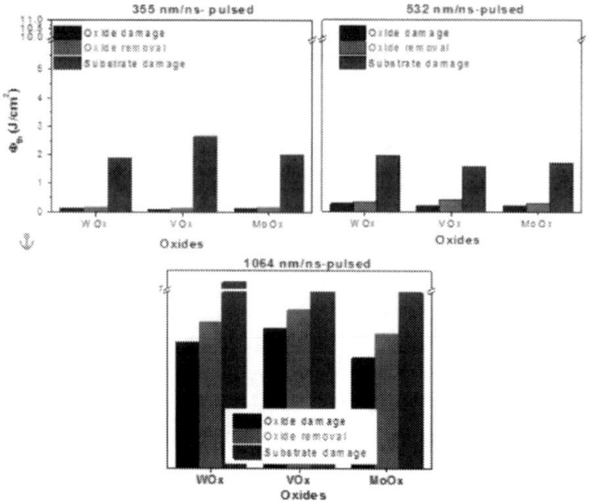

Figure 4: Threshold fluence for the three oxides studied in this work with nanoseconds laser pulses at different wavelengths.

In contrast, when working with picosecond (ps) pulses, the use of infrared (IR) produces no substrate damage in any of the studied oxides, regardless the pulse energy used. Moreover, in the case of WOx, at 532 nm wavelength, there is no substrate damage either. The fluence thresholds for oxide removal obtained are 0.06 ± 0.02 J/cm^2 for 355 nm, 0.16 ± 0.02 J/cm^2 for 532 nm and 0.63 ± 0.11 J/cm^2 for 1064 nm. These values are lower than the obtained in ns pulses due to the lower duration of pulse -8ps in all wavelength-, as a longer laser pulse allows a higher thermal diffusion into the material, being needed more energy to remove the oxide layer.

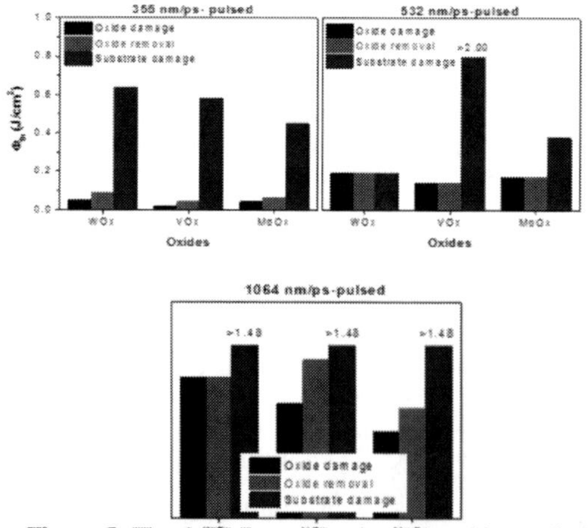

Figure 5: Threshold fluence for the three oxides studied in this work with picoseconds laser pulses at different wavelengths.

Table I shows the average parametric windows for the three studied oxides at different wavelengths using both nanosecond (ns) and picosecond (ps) laser pulses. In the ns regime, the parametric windows indicating the range between oxide removal and substrate damage, are approximately 2.0 J/cm² for 355 nm, 1.5 J/cm² for 532 nm, and roughly 5.5 J/cm² for 1064 nm. In the ps regime, it is possible to remove the oxide layer for all three oxides at 355 nm wavelength. At 532 nm, a parametric window is only found in the case of MoOx. Moreover, when employing infrared (IR) wavelengths, the highest pulse fluence values used do not induce damage to the silicon substrate. In summary, excluding the 532 nm picosecond pulses for WOx, all laser systems allow the removal of VOx, WOx and MoOx films without damaging the substrate.

Table I: Parametric windows of oxide removal process for ns and ps pulses at 355 nm, 532 nm and 1064 nm.

		Parametric Window (J/cm^2)		
		355 nm	**532 nm**	**1064 nm**
WOx	**ns-pulsed**	0.17 – 1.87	0.13 – 1.98	4.93 – 10.36
	ps-pulsed	0.09 – 0.64	0.19 – 0.19	0.65 – >1.48
VOx	**ns-pulsed**	0.13 – 2.65	0.43 – 1.59	5.39 – 9.53
	ps-pulsed	0.04 – 0.58	0.14 – >2.00	0.73 – >1.48
MoOx	**ns-pulsed**	0.16 – 2.00	0.29 – 1.72	4.57 – 9.33

Finally, the effectiveness of this approach has been studied by the isolation of diodes of the three TMOs. The isolation was performed with the nanosecond, 532 nm system, as it generates craters with good morphology, and a wide parametric window for all three studied oxides.

To prevent damage to the substrate, a parameterisation of the isolation process centred on the overlap between laser pulses was carried out. Figure 6 shows a MoOx diode that has been isolated by laser ablation at 0.25 mm from the contact edge. The homogeneity and continuity of the process can be observed, with no damage to the substrate.

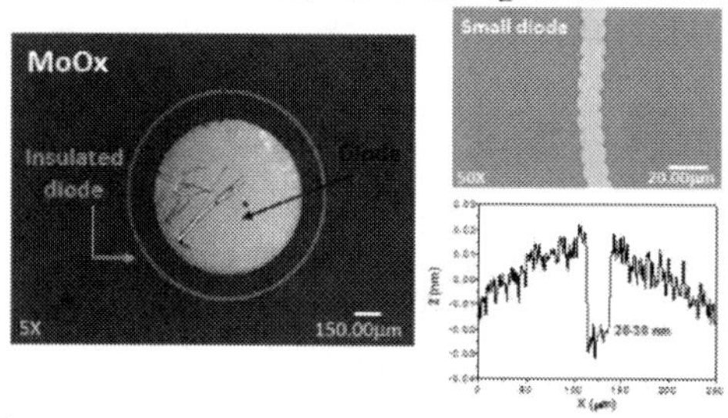

Figure 6: MoOx diode isolated by laser ablation (532 nm, ns-pulsed regime) (left). Confocal image and profile of ablated spot on the MoOx film (right).

The electrical characteristics of the diodes with the three different oxides before and after the laser isolation process were compared (See figure 7). A significant enhancement is noticeable for all cases, characterized by a reduction in the reverse current density of two orders of magnitude. The high work function of transition metal oxides induces an inversion layer on the c-Si surface [12]. This results in a leakage current spreading in the region surrounding the metal contact. This unintended effect can be mitigated by employing laser ablation to isolate the active region. The method developed here could be applied to define more complex patterns, such as interdigitated contacts or planar device architectures.

10.4229/EUPVSEC2023/1CV.3.37
020056-003

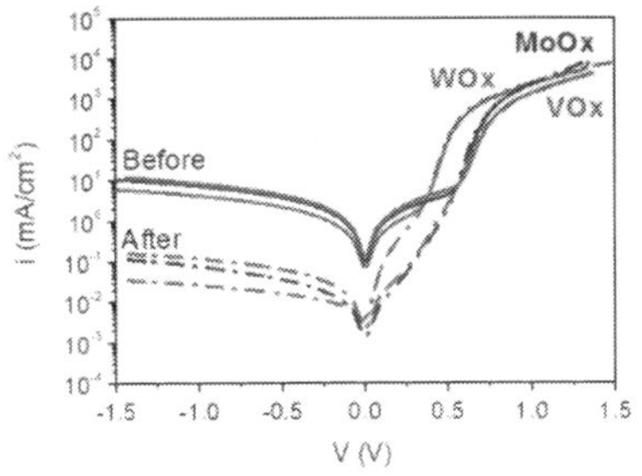

Figure 7: Dark JV curves for oxide diodes before and after isolation.

4 CONCLUSIONS

In summary, we have studied the influence of laser ablation on three different oxides - WOx, VOx and MoOx - deposited by evaporation on silicon substrates. The study was performed for laser wavelengths of 355 nm, 532 nm and 1064 nm, combined with pulse durations of nanoseconds and picoseconds. The ablation behavior of the three oxides remains consistent under laser irradiation, regardless of the employed wavelength the temporal regime considered. The study revealed that using longer wavelengths results in rougher crater edges, whereas shorter wavelengths yield cleaner edges. To prevent substrate damage, we observed that the best choice is to use laser sources in the nanosecond-regime, as they offer a broader range of parameter values leading to an effective oxide film removal. Moreover, it was observed that using nanosecond pulses, along with shorter wavelengths in the ultraviolet (UV) and visible (VIS) range, produces lower thermal effect. The isolation of diodes using a ns pulsed laser at 532 nm led to homogeneous and continuous laser grooves, and a remarkable reduction—up to two orders of magnitude—in the reverse current density for all three distinct TMOs.

5 REFERENCES

[1] Y. Hamakawa, K. Fujimoto, K. Okuda, Y. Kashima, S. Nonomura, and H. Okamoto, 'New types of high efficiency solar cells based on a-Si', *Appl. Phys. Lett.*, vol. 43, no. 7, pp. 644–646, 1983, doi: 10.1063/1.94462.

[2] A. Descoeudres *et al.*, 'Improved amorphous/crystalline silicon interface passivation by hydrogen plasma treatment', *Appl. Phys. Lett.*, vol. 99, no. 12, Sep. 2011, doi: 10.1063/1.3641899.

[3] M. Taguchi, '"Improvement of the conversion efficiency of polycrystalline silicon thin film solar cell," ', *Tech. Dig. Fifth Int. Photovolt. Sci. Eng. Conf.*, pp. 689–692, 1990.

[4] F. Feldmann, M. Bivour, C. Reichel, M. Hermle, and S. W. Glunz, 'Passivated rear contacts for high-efficiency n-type Si solar cells providing high interface passivation quality and excellent transport characteristics', *Sol. Energy Mater. Sol. Cells*, vol. 120, no. PART A, pp. 270–274, 2014, doi: 10.1016/j.solmat.2013.09.017.

[5] J. Meyer, S. Hamwi, M. Kröger, W. Kowalsky, T. Riedl, and A. Kahn, 'Transition metal oxides for organic electronics: Energetics, device physics and applications', *Advanced Materials*, vol. 24, no. 40. pp. 5408–5427, Oct. 23, 2012, doi: 10.1002/adma.201201630.

[6] C. Battaglia *et al.*, 'Silicon heterojunction solar cell with passivated hole selective MoO x contact', *Appl. Phys. Lett.*, vol. 104, no. 11, Mar. 2014, doi: 10.1063/1.4868880.

[7] F. Feldmann *et al.*, 'High and Low Work Function Materials for Passivated Contacts', in *Energy Procedia*, 2015, vol. 77, pp. 263–270, doi: 10.1016/j.egypro.2015.07.037.

[8] L. G. Gerling *et al.*, 'Transition metal oxides as hole-selective contacts in silicon heterojunctions solar cells', *Sol. Energy Mater. Sol. Cells*, vol. 145, pp. 109–115, Feb. 2016, doi: 10.1016/j.solmat.2015.08.028.

[9] S. Rung, A. Christiansen, and R. Hellmann, "Influence of film thickness on laser ablation threshold of transparent conducting oxide thin-films," *Appl. Surf. Sci.*, vol. 305, pp. 347–351, 2014.

[10] [L. Torrisi, A. Borrielli, and D. Margarone, 'Study on the ablation threshold induced by pulsed lasers at different wavelengths', *Nucl. Instruments Methods Phys. Res. Sect. B Beam Interact. with Mater. Atoms*, vol. 255, no. 2, pp. 373–379, 2007, doi: 10.1016/j.nimb.2006.12.144.

[11] J. M. Liu, 'Simple technique for measurements of pulsed Gaussian-beam spot sizes', *Opt. Lett.*, vol. 7, no. 5, p. 196, 1982, doi: 10.1364/ol.7.000196.

[12] J. Bullock, A. Cuevas, T. Allen, and C. Battaglia, "Molybdenum oxide MoOx : A versatile hole contact for silicon solar cells," *Appl. Phys. Lett.*, vol. 105, no. 23, p. 232109, Dec. 2014.

6 ACKNOWLEDGEMENTS

Partial financial support has been provided by the Spanish Ministry of Science and Innovation under the projects CHENOC (ENE2016–78933-C4–1-R and ENE2016–78933-C4–4-R) and SCALED (PID2019–109215RB-C41 and PID2019–109215RB-C44).

OPTIMIZING AN N-PERT SOLAR CELL TO ATACAMA DESERT SOLAR SPECTRUM USING A GENETIC ALGORITHM

Pablo Ferrada[1]*, Aitor Marzo[2], Benjamin Ivorra[3], Miriam Ruiz Ferrández[4,5], Francisco Araya[6], Emilio Ruiz Reina[7]
[1]University of Antofagasta, Centro de Desarrollo Energético Antofagasta, Angamos 601, 1240000 Antofagasta, Chile
[2]Department of Optics, University of Granada, 18071, Granada, Spain
[3]Interdisciplinary Mathematics Institute (IMI), MOMAT Research Group & Department of Applied Mathematics and Mathematical Analysis, Complutense University of Madrid, Plaza de las Ciencias 3, 28040 Madrid, Spain
[4]MOMAT Research Group, Complutense University of Madrid, 28040 Madrid, Spain
[5]Supercomputing-Algorithms Research Group, University of Almería, 04120 Almería, Spain
[6]CEA-INES, 50 Avenue du Lac Léman, 73375 Le Bourget-du-Lac, France
[7]Department of Applied Physics II, Escuela de Ingenierías Industriales, Campus de Teatinos, University of Málaga, C/Doctor Ortiz Ramos, 29071 Málaga, Spain

ABSTRACT: The spectral distribution of solar radiation in the Atacama Desert is different from the global standard, exhibiting high irradiation values. As the response of photovoltaic (PV) technologies depends on the spectrum, it is needed to consider local conditions and type of technology to optimize PV devices. The reason is that solar cells are usually conceived for best performance under standard testing conditions (STC). In this work, we obtained geometrical and doping parameters which maximize the power of an n-type bifacial passivated emitter and rear totally diffused solar cell (n-PERT). Six control variables were used: the thicknesses of cell, emitter, and back surface field, as well as doping concentration of emitter, base, and back surface field. This optimization was carried out for the representative solar spectrum of Atacama Desert (air mass, AM1.08) and the standard reference AM1.5 through a genetic algorithm. For validation, the calculated current-voltage parameters of the cell at STC were compared with experimental measurements. The relative difference between computed and measured efficiencies were below 1% under STC. It came out that decreasing the thickness of all layers and increasing doping can lead to a relative increment of 5.4% in the cell efficiency under AM 1.08.

1 INTRODUCTION

From the point of view of solar resource, the Atacama Desert stands out worldwide for having the highest levels of solar irradiation in the world, with annual values that exceed up to 2600 kWh/m^2 [1–3], of which 175 kWh/m^2 corresponds to incident solar irradiation within the ultraviolet (UV) spectral range, 280–400 nm [4].

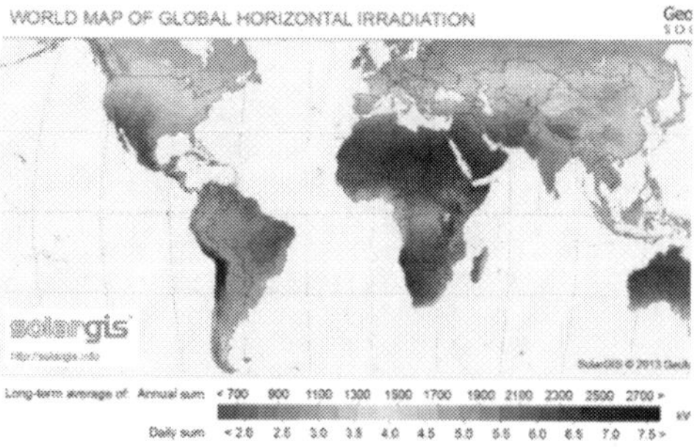

FIG. 1. Global Horizontal Irradiation map from Solar GIS [5].

The solar resource and the solar spectrum are relevant aspects for photovoltaic (PV) technologies because their production depends on the amount of incident solar radiation and on its spectral distribution, as PV cells response dependents on the photon energy (E_{ph}=h c/λ).

Photovoltaic cells and modules are certified at standard testing conditions (STC). The STC means that they are tested in an indoor controlled environment at 25°C and at 1000 W/m^2 with lamps that try to simulate the shape of the standard reference solar spectrum [6]. The American Society for Testing and Materials and research centers of the United States (U.S.) [7] was the entity supporting the definition of the reference spectrum norm G173-03. Under this scheme, two main spectra were defined as the standards for direct normal and global tilted (at 37°) spectral irradiances [8] with an Air Mass of 1.5 (AM1.5). For the definition of these spectra, the corresponding atmospheric conditions are an average for the United States (U.S.) over one year. The standard spectral irradiances were calculated considering the atmospheric composition and using the Simple Model of the Atmospheric Radiative Transfer of Sunshine (SMARTS) [9,10].

Using a similar methodology, which is based on atmospheric conditions and SMARTS, spectra associated to different locations can be calculated. A representative spectrum for the Atacama Desert region was obtained from the available long-term databases of local atmospheric parameters (AOD, PW, O3, among others) [11]. That is, the mean direct and global spectral irradiances for the area of interest were estimated. It is remarkable that 5.6% of Reference Spectrum (AM1.5G) is in the ultraviolet (UV) spectral range for the Tilted Global Irradiance (GTI). Conversely, due to the low presence of ozone (O3), the Atacama Desert presents a blueshift in the shape of the local solar spectrum which means that 7.7% of its energy is in the UV range. This result can lead to implications in the operation of PV technologies in outdoor conditions. Moreover, degradation issues may be enhanced.

The differences between the spectra mainly refer to intensity and UV content. In this way, authors found arguments to define a Desert Label for Atacama Desert [4]. This outcome highlights the necessity of determining solar spectrum for specific locations, which can have implications in the definition of measurement standards and procedures for PV module durability tests.

10.4229/EUPVSEC2023/1CV.3.54
020058-001

As the solar spectrum varies during the day, PV modules rarely operate at the optimum. In fact, the PV device shows losses when operating at non-standard conditions. Losses have been investigated in [12], where authors explain that the mismatch between solar spectrum and the mono-energetic absorption of a single bandgap leads to non-absorption of photons with energy below the bandgap. Additionally, the mismatch between excited carriers and phonons causes thermalization. Provided that the mean GTI of Atacama Desert (AM1.08) has a higher intensity and UV content, one can state that thermalization losses can be even larger compared to the case at STC. To this, recombination phenomena is temperature dependent [13].

Today, many of the industrial crystalline silicon (c-Si) solar cells are bifacial, having the advantage of photon absorption from front and rear sides. According to the ITRPV [14], relevant technologies in the past and present are the surface field (BSF), passivated emitter and rear cell (PERC), passivated emitter and rear locally diffused solar cell (PERL), passivated emitter and rear totally diffused solar cell (PERT) and Si heterojunction (SHJ). In particular, n-PERT bifacial solar cell achieved efficiencies of 22% with a bifaciality of 95% [15].

The aim of this work is to find optimal parameters of a n-PERT solar cell (see cell structure in the inset of FIG. 2), when illuminated with a representative Atacama Desert solar spectrum (AM1.08).

2 METHODS

The methodology consists of two steps: 1) Validation and 2) optimization. For validation (1), empirical data was collected, a numerical model was developed using COMSOL Multiphysics and experimental results were compared with those obtained through simulations. For optimization (2), selected parameters of the solar cell were optimized to the mean Atacama Spectrum (AM1.08). See spectra in FIG. 2. See more details in [16].

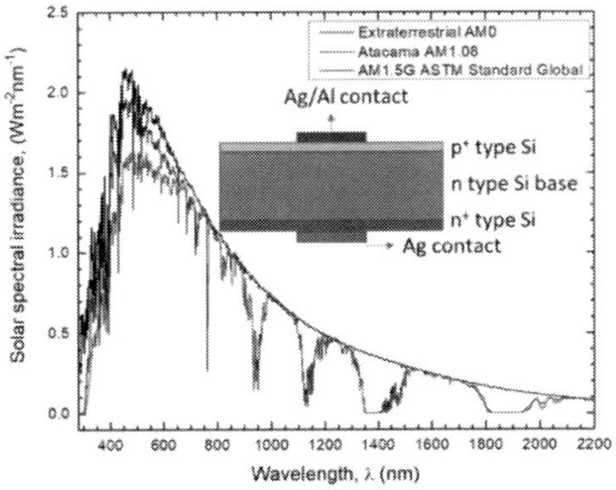

FIG. 2. Extraterrestrial AM0, Atacama Desert AM1.08 and reference AM1.5G solar spectrum. A sketch of the cell is in the inset.

2.1 Model validation

This stage required the fabrication of n-PERT solar cells, characterization, and results analysis. See the experimental details in the published work of Ferrada et al [17]. Authors reported geometry, doping level and output current-voltage characteristics. For the model of the present work, the emitter and BSF doping profiles were simulated as gaussian functions, where the argument is the depth. Regarding metal fraction (f_{met}) at the front and rear side of the cell, $f_{met,f} = 0.052$ and $f_{met,r} = 0.099$ were considered. This value is defined as the ratio of the covered area by the metallization to the cell area. The SRH carrier lifetime values were obtained from [18,19]. See Table I.

Table I: Input parameters obtained through measurements.

Name	Value	Description
d_{cell}	180 µm	Solar cell thickness
d_E	0.65 µm	Emitter depth
d_{BSF}	0.45 µm	Thickness of BSF
N_E	2.4×10^{19} cm^{-3}	Emitter surface conc.
N_B	8.4×10^{14} cm^{-3}	Base doping
N_{BSF}	6.2×10^{19} cm^{-3}	BSF surface conc.
τ_n	1.5 ms	SRH Carrier lifetime
τ_p	1.5 ms	SRH Carrier lifetime

Additionally, the spectral reflectance (R_λ) of the metallized solar cell was determined using a Perkin Elmer 950 Spectrophotometer (FIG. 3) [20].

A one-dimensional (1D) solar cell model was was developed in the Semiconductor Module of COMSOL Multiphysics v6 and tested using the AM1.5g. The current-voltage curve (JV) was calculated, leading to the electrical outputs such as the short circuit current density (J_{sc}), open circuit voltage (V_{oc}), fill factor (FF), power (P_{mpp}) and efficiency (η). The mpp stands for maximum power point. Thus, simulated JV values were compared with the measured JV values. Results and discussion section.

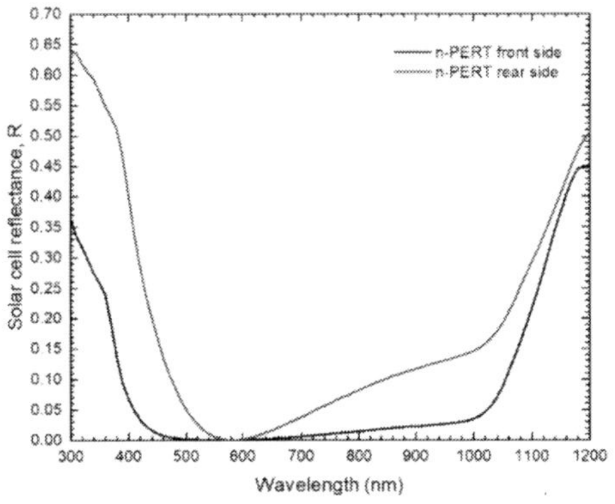

FIG. 3. Spectral reflectance of the n-PERT solar cell.

To identify the minimum N° of elements, ensuring quality of results, a mesh independence study was carried out (see FIG. 4, where each electrical output parameter was normalized to those using the largest number of mesh elements). For more than 600 mesh elements, the electrical response closely match those computed with the extremely fine mesh. It turned out that a fine mesh was possible to use.

10.4229/EUPVSEC2023/1CV.3.54
020058-002

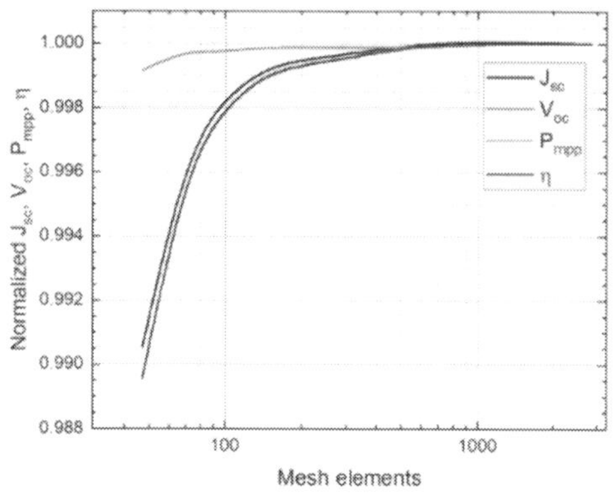

FIG. 4. Mesh independence study of the main output JV parameters as function of the mesh elements for a 1D model.

The next task in this stage was to detect any possible differences between the experimental results and the simulations. There are mainly two sources of discrepancies between model an experiment:

(1) The model assumes flat surfaces. Therefore, a correction factor was used to match J_{sc}. However, the larger surface area leads to a larger saturation current density J_0 due to recombination. Researchers [19] recognize this issue and defined correction factors (f_{corr}) for different solar cells. To compute J_{sc} at STC conditions, $f = 1.17$ for the front side of the n-PERT was used, where $f_{corr} = J_{sc,meas}/J_{sc,cal}$. It is the ratio of the measured short circuit current density ($J_{sc,meas}$) to the calculated value ($J_{sc,cal}$).

(2) The model does not consider metal induced recombination. Therefore, V_{oc} can be higher than the experimental result. This discrepancies were determined in [21]. A reduction in the open circuit voltage ΔV_{oc} of n-PERT cells was studied by varying the metal fraction f_{met} and measuring the dark saturation current density at metal/semiconductor interfaces. In this work, the V_{oc} decreases 45 mV due to the front side metallization and 35 mV due to the rear side metallization. On the other hand, the effect of series and shunt resistance (R_{ser} and R_{shunt}) can be observed in the shape of the JV curve, and therefore, on the FF and P_{mpp} [22]. Based on the that work [21], metal induced recombination can be implemented after the model is solved, by subtracting ΔV_{oc} for a certain metal fraction. Moreover, ohmic losses can affect P_{mpp} and FF. Thus, we used $P_{mpp} = P_{mpp,0} - R_{ser} I_{mpp}^2$, where $P_{mpp,0}$ is the power without resistance effects. For the fill factor, $FF = FF_0(1 - r_{ser})$, where FF_0 is the fill factor without resistive effects, r_{ser} is the normalized series resistance to the characteristic resistance ($r_{ser} = R_{ser}/R_{CH}$ and $R_{CH} = V_{oc}/I_{sc}$) [23].

2.2 Optimization

The first stage consists of defining the objective function, control parameters, constraints, and output variable. Methodologically, the COMSOL model was converted as a Matlab script which returns the output power as function of the control parameters. Six optimization parameters were used to maximize the maximum power of the solar cell. The parameters were the emitter (d_E), the cell (d_{cell}), and the back surface field (d_{BSF}) thickness as well as the emitter (N_E), the base (N_B) and the BSF (N_{BSF}) doping concentration. A genetic algorithm (GA) was used to find the optimum parameters. The details of the GA are described in [24]. See FIG. 5.

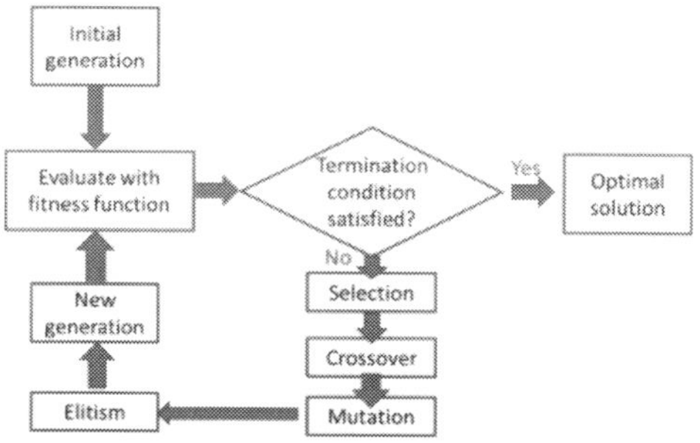

FIG. 5. General procedure of the genetic algorithm (GA).

For the first step in the GA, we randomly create an initial generation of individuals in the search space. As iterations increase, called generations, sub processes are repeated. These sub processes are the selection, crossover, mutation, and elitism. After completing this process, a next generation is obtained. When it is finished, we reach the generation containing the approximated parameters of the solar cell. Using the parameters found through the GA, the optimized solar cell performance was calculated and compared to the JV response without optimization. That is, under AM1.08 and AM1.5.

4 RESULTS

4.1. Measurements and Simulation at Standard Conditions

The measurements and simulations at STC were performed to validate the model and shown. Thus, a direct comparison between the measured and computed JV parameters is presented in Table II and III. For this purpose, a set of 6 solar cells were measured by illuminating the front and rear side, separately. The corresponding standard deviation is indicated in Table 2. The area of the measured cells was 244.3 cm² and the series resistance was 0.45 Ωcm² from experimental measurements.

According to the values in Table II and III, we efine the relative differences ΔX for the parameter X as $\Delta X = (X_{calc} - X_{meas})/X_{meas}$, where X_{calc} and X_{meas} stand for calculated and measured, respectively. The J_{sc} and V_{oc} exhibit relative differences between computed and measured values below 1% for the front side and below 2% for the rear side. The power P_{mpp} presents the largest difference between measurement and calculation; that is 8% for the front and 7% for the rear side. Nevertheless, the fill factor FF and efficiency keep relative differences below 1% for both sides. To sum up, the error of the simulated JV parameters is at most 2%, except for the power, which reach values up to 8% when illuminated with the AM1.5G reference solar spectrum. The discrepancy of the 8% for power can be attributed to the following argument. The power already takes series resistance though $P_{mpp} = P_{mpp,0} - R_{ser} I_{mpp}{}^2$ into account. Here, the corresponding error δP_{mpp} uses the value of $R_{ser} = 0.45\ \Omega\text{cm}^2$, and the propagation of the short circuit current density error, δJ_{sc}. Thus, the propagation error can reach 16%, which provides insights to understand the discrepancy of 8% with the measurement of the maximum power.

Table II: Current-voltage JV measurements at STC.

	JV Measurement	
Parameter	Front	Rear
J_{sc} (mA/cm^2)	39.2 ± 0.03	34.6 ± 0.03
V_{oc} (mV)	653.1 ± 2	649.7 ± 2
P_{mpp} (W)	4.9 ± 0.02	4.3 ± 0.09
FF (%)	78.3 ± 0.2	78.2 ± 0.16
η (%)	20 ± 0.08	17.6 ± 0.1

Table III: Current-voltage JV calculations at STC.

	JV simulation	
Parameter	Front	Rear
J_{sc} (mA/cm^2)	39.2	34.2
V_{oc} (mV)	646.4	654.6
P_{mpp} (W)	5.3	4.6
FF (%)	78.7	78.7
η (%)	20.0	18.0

4.2. Optimal Solar Cell Parameters

The initial values and range for the selected control parameters parameters are summarized in Table IV; whereas the parameters of the GA are found in Table V. As stated in the methodology, the optimal cell parameters were found for front side illumination at the AM1.5G standard and the AM1.08.

Table IV. Thickness and doping parameters of the cell.

Parameters	Initial Values	Range
d_E (nm)	650	50–750
d_{cell} (nm)	180 × 103	150 × 103–200 × 103
d_{BSF} (nm)	450	50–750
N_E (cm^{-3})	2.44 × 10^{19}	1 × 10^{19}–1 × 10^{20}
N_B (cm^{-3})	8.44 × 10^{14}	1 × 10^{14}–5 × 10^{15}
N_{BSF} (cm^{-3})	6.16 × 10^{19}	1 × 10^{19}–5 × 10^{20}

Table V. Parameters of the genetic algorithm (GA).

GA Parameters	Value
Population size	70
Generations	110
Stopping criterium	10
Mutation probability	10%

FIG. 5 shows the simulated JV characteristic curve of the the optimized (full line) and non-optimized (dotted curve) n-PERT solar cell at the AM1.5G (black) and AM1.08 (red) solar spectra. It is seen that the J_{sc} of the optimized solar cell is increased with respect to that of the non-optimized cell, while keeping the V_{oc} nearly at the same value, for both spectra. It follows that the power output is increased. The outcome means that the global optimization, carried out through a genetic algorithm, succeed in finding solar cell parameters that enhance the maximum output power out though the current density and reducing recombination.

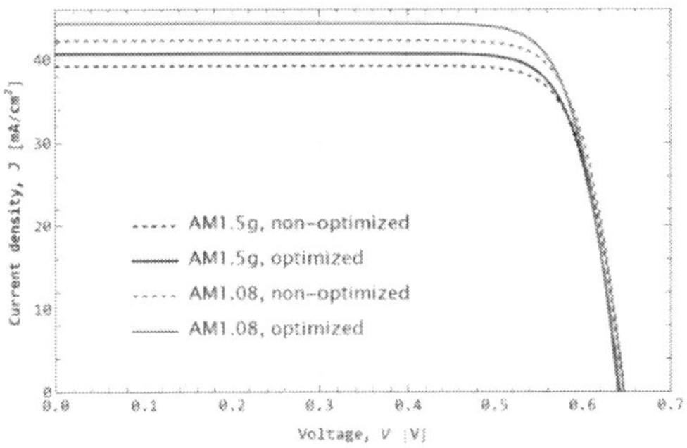

FIG. 6. Current–voltage curve of the optimized and non-optimized n-PERT solar cell at the AM1.5G and AM1.08 spectra.

The relative increment after optimization was calculated as $100(X_{opt} - X_{non\text{-}opt})/X_{non\text{-}opt}$, where X_{opt} and $X_{non\text{-}opt}$ refer to optimal and non-optimal values. By illiminating with AM1.5 and AM.108, optimized cells exhibit increments in all parameters with the exception of the V_{oc}. The largest improvements were obtained under the AM1.08 Atacama spectrum, as there was an increment of 5.7% in the maximum power and of 5.4% in efficiency. This enhanced performance was expected since solar cells are optimized to STC.

The geometrical and doping parameters for the optimized solar cell under both spectra in comparison with the experimental values of the n-PERT device are shown in Table VI.

Table VI: Summary of initial and optimized solar cell parameters.

Parameter	Exp. Values	AM1.5G	AM1.08
d_E (nm)	650	200.2	201
d_{cell} (nm)	180×10^3	154.2 × 10^3	165.1 × 10^3
d_{BSF} (nm)	450	330	250
N_E (cm^{-3})	2.44×10^{19}	9.89 × 10^{19}	9.36 × 10^{19}
N_B (cm^{-3})	8.44×10^{14}	9.83 × 10^{14}	9.81 × 10^{14}
N_{BSF} (cm^{-3})	6.16×10^{19}	3.87 × 10^{20}	4.92 × 10^{20}

According to Table VI, the emitter doping concentration is required to be slightly lower at AM1.08 compared to that at the standard spectrum. This modification in the design can reduce recombination caused by the high-power density spectrum in Atacama Desert. A larger intensity implies a larger number of photons impinging the solar cell, leading to an increase in the the excess minority carrier density and Auger recombination [25] at the front side. Hence, reducing the doping concentration allows to maintain recombination low enough. Further, BSF thickness is required to be thinner and heavier doped at AM1.08 compared to that at STC. See more details in the discussion provided in [16].

5 CONCLUSIONS

A passivated emitter and rear totally diffused (n-PERT) crystalline silicon solar cell was modeled with COMSOL Multiphysics v6. Based on this model and a genetic algorithm (GA) implemented in MATLAB, the maximum output power was optimized. For this goal, the control variables were six cell parameters such as the emitter (d_E), the cell (d_{cell}), and the back surface field BSF (d_{BSF}) as well as the doping concentration of each layer, that is, N_E, N_{cell}, and N_{BSF}. The optimization of cell power was carried out under two different spectra, a representative spectrum of the Atacama Desert (AM1.08) and the standard AM1.5. Provided that the Atacama and reference spectra differ, the optimal parameters of the n-PERT solar cell resulted to be specific for each illumination. For AM1.08 illumination, the BSF is required to be thinner and lightly doped compared to values under AM1.5 spectrum. In addition, emitter doping needs to be heavier compared to the doping under the AM1.5 spectrum. To sum up, the optimized solar cell under the AM1.08 spectrum resulted in a gain of +4.9% for the short circuit current density J_{sc}, +5.7% for the power P_{mpp}, and +5.4% for the efficiency η, with a loss in the open circuit voltage V_{oc} of 1%. These results highlight the importance of considering local solar spectrum in solar cell design as the performance can be improved.

ACKNOWLEDGEMENTS

We acknowledge ANID/FONDAP/15110019 "Solar Energy Research Center" SERC-Chil and the Spanish Ministry of Science and Innovation under Projects PID2019-106337GB-I00, and PID2021-126805OB-I00 (HELIOSUN project). A. Marzo thanks for the Ramon y Cajal contract (RYC2021-031958-I), funded by the Spanish Ministry of Science and Innovation MCIN/AEI/10.13039/501100011033 and by the European Union "NextGenerationEU/PRTR.

In addition, the authors gratefully acknowledge the technical and academic support of Multiphysics Modeling School (MMS) (https://www.multiphysics.uma.es/home).

REFERENCES

[1] R.A. Escobar, C. Cortés, A. Pino, M. Salgado, E.B. Pereira, F.R. Martins, J. Boland, J.M. Cardemil, Estimating the potential for solar energy utilization in Chile by satellite-derived data and ground station measurements, Sol. Energy. 121 (2015) 139–151. doi:10.1016/j.solener.2015.08.034.

[2] R.A. Escobar, C. Cortés, A. Pino, E.B. Pereira, F.R. Martins, J.M. Cardemil, Solar energy resource assessment in Chile: Satellite estimation and ground station measurements, Renew. Energy. 71 (2014) 324–332. doi:10.1016/J.RENENE.2014.05.013.

[3] R. Rondanelli, A. Molina, M. Falvey, R. Rondanelli, A. Molina, M. Falvey, The Atacama Surface Solar Maximum, Bull. Am. Meteorol. Soc. 96 (2015) 405–418. doi:10.1175/BAMS-D-13-00175.1.

[4] J.F. Lelièvre, R. Couderc, N. Pinochet, L. Sicot, D. Munoz, R. Kopecek, P. Ferrada, A. Marzo, D. Olivares, F. Valencia, E. Urrejola, Desert label development for improved reliability and durability of photovoltaic modules in harsh desert conditions, Sol. Energy Mater. Sol. Cells. 236 (2022). doi:10.1016/j.solmat.2021.111508.

[5] Solar GIS: Global Horizontal Irradiation, (2022). https://solargis.com/es/maps-and-gis-data/download/world (accessed April 20, 2023).

[6] A. Gallo, A. Marzo, E. Fuentealba, E. Alonso, High flux solar simulators for concentrated solar thermal research: A review, Renew. Sustain. Energy Rev. 77 (2017) 1385–1402. doi:10.1016/j.rser.2017.01.056.

[7] ASTM, ASTM International - Standards Worldwide, (2017). https://www.astm.org/ (accessed February 15, 2022).

[8] D.R. Myers, K. Emery, C. Gueymard, Revising and Validating Spectral Irradiance Reference Standards for Photovoltaic Performance, in: ASES/ASME Sol. 2002 Conf., 2002: pp. 567–574. doi:https://doi.org/10.1115/1.1638784.

[9] C.A. Gueymard, Parameterized transmittance model for direct beam and circumsolar spectral irradiance, Sol. Energy. 71 (2001) 325–346. doi:10.1016/S0038-092X(01)00054-8.

[10] C. Gueymard, SMARTS: Simple Model of the Atmospheric Radiative Transfer of Sunshine | Grid Modernization | NREL, 1995. https://www.nrel.gov/grid/solar-resource/smarts.html (accessed March 23, 2019).

[11] A. Marzo, P. Ferrada, F. Beiza, P. Besson, J. Alonso-Montesinos, J. Ballestrín, R. Román, C. Portillo, R. Escobar, E. Fuentealba, Standard or local solar spectrum? Implications for solar technologies studies in the Atacama desert, Renew. Energy. (2018). doi:10.1016/j.renene.2018.05.039.

[12] L.C.H. and N.J. Ekins-Daukes, Fundamental losses in solar cells, Prog. Photovolt Res. Appl. 19 (2010) 286–293. doi:10.1002/pip.

[13] O. Dupré, R. Vaillon, M.A. Green, Physics of the temperature coefficients of solar cells, Sol. Energy Mater. Sol. Cells. 140 (2015) 92–100. doi:10.1016/j.solmat.2015.03.025.

[14] ITRPV 2021, International Technology Roadmap of Photovoltaicos (ITRPV), 2021.

[15] R. Kopecek, J. Libal, Bifacial photovoltaics 2021: Status, opportunities and challenges, Energies. 14 (2021). doi:10.3390/en14082076.

[16] P. Ferrada, A. Marzo, M.R. Ferrández, E.R. Reina, B. Ivorra, J. Correa-Puerta, V. del Campo, Optimization of N-PERT Solar Cell under Atacama Desert Solar Spectrum, Nanomaterials. 12 (2022). doi:10.3390/nano12203554.

[17] P. Ferrada, D. Rudolph, C. Portillo, A. Adrian, J. Correa-Puerta, R. Sierpe, V. del Campo, M. Flores, T.P. Corrales, R. Henríquez, M.J. Kogan, J. Lossen, Interface analysis of Ag/n-type Si contacts in n-type PERT solar cells, Prog. Photovoltaics Res. Appl. (2020) pip.3242. doi:10.1002/pip.3242.

[18] A. Edler, Development of bifacial n-type solar cells for industrial application, Universität Konstanz, 2014. http://nbn-resolving.de/urn:nbn:de:bsz:352-275017.

[19] A. Fell, K.R. McIntosh, P.P. Altermatt, G.J.M. Janssen, R. Stangl, A. Ho-Baillie, H. Steinkemper, J. Greulich, M. Muller, B. Min, K.C. Fong, M. Hermle, I.G. Romijn, M.D. Abbott, Input Parameters for the Simulation of Silicon Solar Cells in 2014, IEEE J. Photovoltaics. 5 (2015) 1250–1263.

doi:10.1109/JPHOTOV.2015.2430016.

[20] P. Ferrada, A. Marzo, E. Cabrera, H. Chu, J. Rabanal, D. Diaz-almeida, A. Schneider, R. Kopecek, Potential for photogenerated current for silicon based photovoltaic modules in the Atacama Desert, Sol. Energy. 144 (2017) 580–593. doi:10.1016/j.solener.2017.01.053.

[21] A. Edler, V.D. Mihailetchi, L.J. Koduvelikulathu, C. Comparotto, R. Kopecek, R. Harney, Metallization–induced recombination losses of bifacial silicon solar cells, Prog. Photovoltaics Res. Appl. 23 (2015) 620–627. doi:10.1002/PIP.2479.

[22] L.H. Slooff, B. Brockholz, W.J.H. Verhees, S.C. Veenstra, E.M. Cobussen-Pool, J.M. Kroon, E.E. Bende, Determination of the intrinsic diode parameters of polymer solar cells, Energy Procedia. 31 (2012) 11–20. doi:10.1016/j.egypro.2012.11.159.

[23] M.A. Green, Solar cell fill factors: General graph and empirical expressions, Solid State Electron. 24 (1981) 788–789. doi:10.1016/0038-1101(81)90062-9.

[24] S. Katoch, S.S. Chauhan, V. Kumar, A review on genetic algorithm: past, present, and future, Multimed. Tools Appl. 80 (2021) 8091–8126. doi:10.1007/s11042-020-10139-6.

[25] R.A. Sinton, R.M. Swanson, Recombination in Highly Injected Silicon, IEEE Trans. Electron Devices. 34 (1987) 1380–1389. doi:10.1109/T-ED.1987.23095.

Minimum bandgap criteria for carrier selective layers in Si solar cells

Nithin Chatterji[1,2], and Pradeep R. Nair[2]
[1] Dept. Of Electronics Engineering, Sardar Vallabhbhai National Institute of Technology, Surat, India,
[2] Dept. Of Electrical Engineering, Indian Institute of Technology Bombay, India
nithinc@eced.svnit.ac.in, prnair@ee.iitb.ac.in

ABSTRACT: Silicon solar cell technology use layers, having high doping or asymmetrical band offset, for selective collection of photo-generated carriers from the bulk active layer material. As such a variety of materials having different parameters like band gaps, band alignment with the bulk photoactive material, dielectric constant etc., are investigated as carrier selective layers with silicon. The bandgap of these materials in addition to regulating the amount of light coupled to the bulk photoactive layer also controls the transport and hence the collection of the photo-generated carriers. The parasitic absorption concerns would demand a carrier selective layer bandgap to be very large with proper band alignment for one photo-generated carrier. However, ignoring the parasitic absorption concerns, the theoretical limits on the minimum bandgap criteria for the selective layer is not well understood. In this regard, we investigate the minimum band gap criteria for a selective layer in silicon based carrier selective solar cells. Our results indicate that the blocking barrier detrimentally affects the device when it is less than 0.6eV and that the minimum bandgap for a material to act as perfect carrier selective layer for silicon technology is around 1.4eV. Finally, materials are compared based on their band alignment with Si and hence our results are useful for the selection of appropriate materials for silicon based carrier selective solar cells.

Keywords: photovoltaics, numerical modeling, device physics

1 INTRODUCTION

Selective collection of photo-generated carriers at respective contacts is a fundamental challenge in Photovoltaics. Typically, carrier selection in solar cells is achieved by using heavy doping and/or by using materials having asymmetrical band alignment with the bulk photo-active material, facilitating the transport of only one type of carrier. Expectedly, such a material should not offer any barrier for collecting the desired photo-generated carrier while ensuring that the other type of carrier is effectively blocked. This is clearly evident for the case of TiO_2_- a well-known electron selective material employed in Si-based carrier selective (CS) solar cells [1]–[3]. The conduction band of[TiO_2 aligns well with that of Si thus ensuring efficient collection of electrons while the 1.2eV offset in the valence band ensures that holes are effectively blocked. As such, a variety of materials [1]-[14] with different band alignment and bandgaps are used as carrier selective CS layer with Si.

2 MODEL SYSTEM

As mentioned above variety of materials are tried as CS layers, however, it is a surprise that the theoretical limits on the band gap criteria of CS layers are still not clearly elucidated in literature – which we address in this study. Specifically, (a) for Si, what would be the minimum band gap of a CS layer, (b) Does such a limiting band gap depend on the type of carriers being collected – electron vs. hole, (c) How would the interface properties (like traps, fixed charges, etc.) affect the minimum band gap criteria?. In this manuscript, we address the above aspects comprehensively through detailed numerical simulations [15].

Band gap of CS layers influence the performance of a solar cell in two key aspects – (i) parasitic absorption by the CS layer and hence the net photo-carrier generation in the active layer, (ii) influence on carrier collection through the band alignment, interface properties, etc. Here, we mainly focus on item (ii) listed above and identify the minimum band gap criteria from a carrier collection perspective. As such the results are directly valid when the selective layer is extremely thin or for interdigitated back contact (IBC) solar cells.

Fig. 1 shows the equilibrium band alignment of typical Si-based CS solar cells, used for numerical simulations in this report. In Si-based CS solar cells the Electron selective (ESL) allows the transport of electrons from photoactive material to the contacts and block holes. Here ΔE_c is the band discontinuity for electrons in the transport band (conduction band) and E_{bh} is the barrier for holes from the photoactive material to ESL. Similarly in the hole selective (HSL), the discontinuity for holes in the transport band (valence band) is ΔE_v and the barrier for electrons is E_{be}. Further, ΔE_c and ΔE_v are assumed to be positive if there is a barrier for the corresponding carrier (i.e., electron and hole, respectively) from photo-active material to SL and ΔE_c, ΔE_v is assumed to be negative if the band discontinuity at the interface do not hinder the collection of carriers to the SL. Note that under this description the blocking barriers, E_{be} and E_{bh} , are always positive.

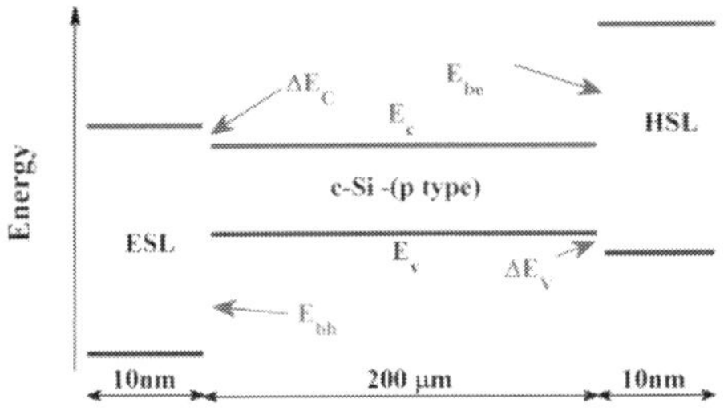

Fig. 1: Energy band alignment of various layers in silicon based CS solar cells used for Numerical simulations.

To simplify the analysis we assume that the interface traps are uniform, the contacts are ohmic and the transport is over-the-barrier. Further, uniform generation rate was assumed in Si corresponding to 1 sun. We note that the effect of the transport band offset and the interface trap density on the device performance and the temperature coefficient of silicon based CS solar cells was recently reported [16], [17]. In this work, we consider the effect of blocking barrier as well to estimate the minimum band gap criteria for CS layers.

3 RESULTS AND CONCLUSIONS

Fig. 2 shows the effect of ESL band alignment with silicon on CS solar cell performance parameters with no interface traps. Part (a) indicates that (i) as expected, V_{oc} improves with increase of barrier for holes, E_{bh}, at ESL/Si interface due to reduction in recombination in ESL, (ii) E_{bh} is more critical for V_{oc} when the band bending in Si near ESL is reduced - due to the lower work function difference between the two layers when transport band offset ΔE_c is negative, and (iii) when $E_{bh} > 0.4$eV, V_{oc} is invariant with ΔE_c due to the reduction of hole recombination in ESL. Fig. 2b shows that J_{sc} remains invariant with band alignment due to the negligible hole transfer from Si to ESL because of the large band bending in Si near the pn junction between ESL and Si at short circuit conditions irrespective of the band alignment. Part (c) shows that the FF follows the trends of V_{oc} for most values of ΔE_c [18]; however for $\Delta E_c = 0.5$eV, FF is affected by the reduction in over-the-barrier transport of electrons from Si to ESL [16]. Fig. 2d shows that the efficiency follows the trends of V_{oc} except at $\Delta E_c = 0.5$eV where FF limits it. Finally, a comparison between different materials used as ESL indicates that in the absence of interface traps these materials are perfect selective layers.

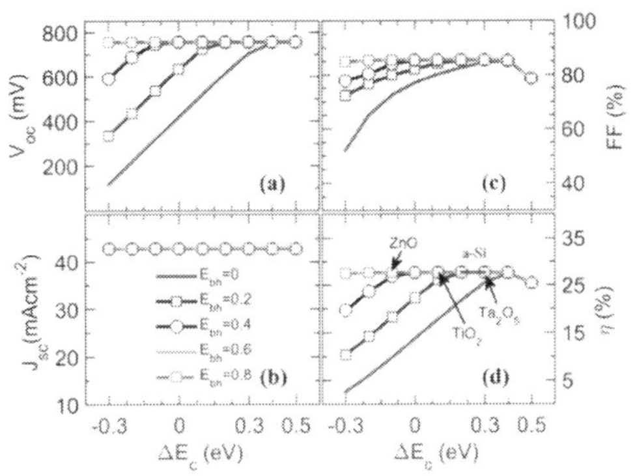

Fig. 2: The effect of ESL band alignment with silicon on CS solar cell performance parameters - (a) V_{oc}, (b) J_{sc}, (c) FF, and (d) efficiency without interface traps. The legends indicate the blocking barrier, E_{bh} and the x axis indicates the transport layer band discontinuity ΔE_c.

The interface at the ESL/Si interface was assumed to be without any traps in the last paragraph which is not a valid assumption. The interface recombination depends on the product of the carrier densities at the ESL/Si interface ($R_s \sim n_s p_s$) and is expected to significantly be affected by the work function of the ESL layer and hence on the transport band offset, E_c. In this regard, Fig. 3 shows the effect of interface traps ($D_{it} = 10^{12} cm^{-2} eV^{-1}$) on the solar cell performance parameters as the band alignment of ESL with Si is varied. Part (a) shows that (i) for $E_{bh} = 0$eV, the curve roughly follows the trends in Fig. 2, (ii) the effect of interface traps become more prominent and distinguishable as E_{bh} is increased, and (iii) for $E_{bh} > 0.4$eV, the enhanced interface recombination for low magnitudes of ΔE_c reduce the V_{oc} as reported elsewhere [2]. It was shown that for low magnitudes of ΔE_c the interface carrier densities (n_s, p_s), for the entire voltage range, are of approximately the same magnitude and leads to the increase in interface recombination ($R_{interface} \sim n_s p_s$, where n_s, p_s are interface electron and hole carrier densities). However, increase in the magnitude of ΔE_c (with doping remaining a constant) leads to a change in the build-in-potential and hence to the increase in one of the carriers and the reduction of the other, thus reducing the interface recombination and improvement in V_{oc}.

Part (b) of Fig, 3 shows that J_{sc} remains invariant with the band alignment (as in Fig. 2(b)). FF follows the trends of V_{oc} except at $\Delta E_c = 0.5$eV where over-the-barrier transport of electrons is limited (see Fig. 3c). Finally, part (d) indicates that the efficiency also follows the trends of V_{oc} until FF is limited by large ΔE_c and that the minimum band gap for a material to act as ideal ESL layer is around 1.4eV ($\Delta E_c = -0.3$eV, $E_{bh} = 0.6$eV. Additionally, based on their band alignment with Si, a-Si and Ta_2O_5 are better than ZnO and TiO_2 in the presence of interface traps [3] (based on the effect of band offset alone. Variation in dielectric constant and parasitic absorption are important factors to consider). As reported in [16], the optimal band offset in the conduction band is around 0.2-0.4 eV and hence our results are in agreement with the reported values for optimal transport band offset for ESL.

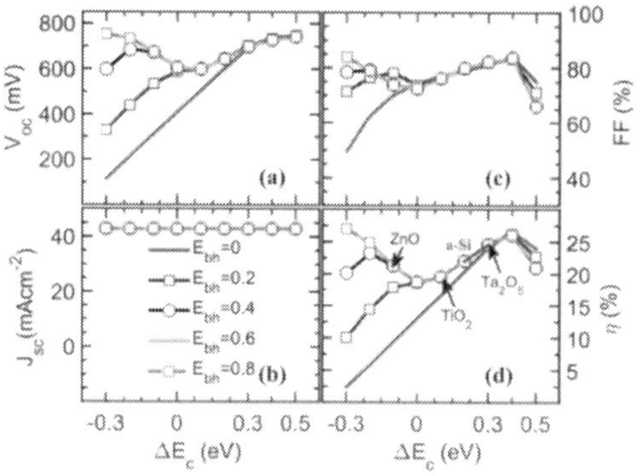

Fig. 3: The effect of trap density ($D_{it} = 10^{12} cm^{-2} eV^{-1}$) and band alignment of ESL with silicon on CS solar cell performance parameters - (a) V_{oc}, (b) J_{sc}, (c)FF, and (d) efficiency. For low E_{bb} the trends are similar to that in Fig. 1 for negative ΔE_c. The effect of interface traps becomes prominent as E_{bh} is increased.

The minimum band gap criteria for the selective layer in Si solar cells, considering electrostatics and carrier transport, were explored in detail using numerical simulations. Our results indicate that a blocking barrier, E_{bh}, of around 0.6eV is required for ESL layer to act as a perfect blocking layer in these device structures. Further, the minimum band gap for the ESL layers was found to be around 1.4eV Si solar cells. Additionally, HSL layers also showed similar results with the same minimum magnitudes for the blocking barrier E_{be} and bandgap (results are not shared here). Finally, our results provide a road map for selecting appropriate materials based on its alignment with the bulk material for Si solar cells and the insights could be extended to other technologies as well.

Acknowledgements: The authors acknowledge Center of Excellence in Nanoelectronics (CEN) and National Center for Photovoltaic Research and Education (NCPRE), IIT Bombay for computational facilities. NC acknowledges SVNIT Faculty seed grant.

References

[1] T. G. Allen et al., "A Low Resistance Calcium/Reduced Titania Passivated Contact for High Efficiency Crystalline Silicon Solar Cells," Adv. Energy Mater., vol. 7, no. 12, p. 1602606, 2017.

[2] S. Avasthi, W. E. McClain, G. Man, A. Kahn, J. Schwartz, and J. C. Sturm, "Hole-blocking titanium-oxide/silicon heterojunction and its application to photovoltaics," Appl. Phys. Lett., vol. 102, no. 20, p. 203901, May 2013.

[3] K. A. Nagamatsu et al., "Titanium dioxide/silicon hole-blocking selective contact to enable double-heterojunction crystalline silicon-based solar cell," Appl. Phys. Lett., vol. 106, no. 12, p. 123906, Mar. 2015.

[4] K. Masuko et al., "Achievement of More Than 25% Conversion Efficiency With Crystalline Silicon Heterojunction Solar Cell," IEEE J. Photovoltaics, vol. 4, no. 6, pp. 1433–1435, Nov. 2014.

[5] F. Feldmann, M. Bivour, C. Reichel, M. Hermle, and S. W. Glunz, "Passivated rear contacts for high-efficiency n-type Si solar cells providing high interface passivation quality and excellent transport characteristics," Sol. Energy Mater. Sol. Cells, vol. 120, no. Part A, pp. 270–274, 2014.

[6] J. Melskens, B. W. H. van de Loo, B. Macco, L. E. Black, S. Smit, and W. M. M. Kessels, "Passivating Contacts for Crystalline Silicon Solar Cells: From Concepts and Materials to Prospects," IEEE J. Photovoltaics, vol. 8, no. 2, pp. 373–388, Mar. 2018.

[7] J. Bullock *et al.*, "23% efficient n-type crystalline silicon solar cells with passivated partial rear contacts," in *2018 IEEE 7th World Conference on Photovoltaic Energy Conversion (WCPEC) (A Joint Conference of 45th IEEE PVSC, 28th PVSEC 34th EU PVSEC)*,

[8] S. Acharyya et al., "Dopant-free materials for carrier-selective passivating contact solar cells: A review," Surfaces and Interfaces, vol. 28, p. 101687, 2022.

[9] J. Bullock *et al.*, "Efficient silicon solar cells with dopant-free asymmetric heterocontacts," *Nat. Energy*, vol. 1, p. 15031, Jan. 2016.

[10] D. Zielke, A. Pazidis, F. Werner, and J. Schmidt, "Organic-silicon heterojunction solar cells on n-type silicon wafers: The BackPEDOT concept," *Sol. Energy Mater. Sol. Cells*, vol. 131, pp. 110–116, 2014.

[11] B. Macco, M. F. J. Vos, N. F. W. Thissen, A. A. Bol, and W. M. M. Kessels, "Low-temperature atomic layer deposition of MoOx for silicon heterojunction solar cells," *Phys. status solidi – Rapid Res. Lett.*, vol. 9, no. 7, pp. 393–396, 2015.

[12] M. Bivour, J. Temmler, H. Steinkemper, and M. Hermle, "Molybdenum and tungsten oxide: High work function wide band gap contact materials for hole selective contacts of silicon solar cells," *Sol. Energy Mater. Sol. Cells*, vol. 142, pp. 34–41, 2015.

[13] J. Geissbühler *et al.*, "22.5% efficient silicon heterojunction solar cell with molybdenum oxide hole collector," *Appl. Phys. Lett.*, vol. 107, no. 8, p. 81601, Aug. 2015.

[14] L. G. Gerling *et al.*, "Transition metal oxides as hole-selective contacts in silicon heterojunctions solar cells," *Sol. Energy Mater. Sol. Cells*, vol. 145, pp. 109–115, 2016.

[15] "Synopsys." Synopsys TCAD, 2016.

[16] N. Chatterji, A. Antony, and P. R. Nair, "Interface-Dependent Efficiency Tradeoff in Si-Based Carrier-Selective Solar Cells," *IEEE Trans. Electron Devices*, vol. 65, no. 6, pp. 2509–2516, Jun. 2018.

[17] N. Chatterji, A. Antony, and P. R. Nair, "Temperature Coefficient of Silicon-Based Carrier Selective Solar Cells," *IEEE J. Photovoltaics*, 2019.

[18] M. A. Green, "Semiempirical Expressions for Solar Cell Fill Factors," *Electron. Lett.*, vol. 21, no. 12, pp. 529–530, 1985.

10.4229/EUPVSEC2023/1CV.3.56

ACCURATE CRYSTALLITE SIZE DETERMINATION IN POLY-SI/SIO$_X$ PASSIVATING CONTACTS

Tobias Okker[1], Raphael Glatthaar[1], Frank Huster[1], Sven Seren[2], Thomas Pernau[3], Giso Hahn[1], Barbara Terheiden[1]

[1]University of Konstanz, Universitätsstr. 10, 78464 Konstanz, Germany
[2]Schmid Group, Robert-Bosch-Str. 32-36, 72250 Freudenstadt, Germany
[3]Centrotherm international AG, Württemberger Str. 31, 89143 Blaubeuren, Germany

ABSTRACT: In this work the crystallite size of phosphorus doped polycrystalline-silicon ((n)poly-Si) layers for passivating contacts is determined by X-ray diffraction (XRD) measurements and by scanning electron microscope (SEM) imaging, where a novel preparation technique is used to visualize grain boundaries. For XRD evaluation, it is shown that grazing incidence XRD (GIXRD) measurements lead to signal enhancement for thin poly-Si layers (<300 nm) compared to the Bragg-Brentano configuration. In particular, at low incidence angles of GIXRD measurements, peak broadening caused by an enlarged beam footprint makes crystallite size determination more difficult, especially when using only Scherrer's equation. Thus, crystallite size analysis via Scherrer equation is compared to a Rietveld refinement procedure using a fundamental parameter approach considering instrument and strain broadening, yielding different results. The novel HF-SEM method confirms results obtained through the Rietveld refinement. Key differences in crystallite shape are visible when comparing poly-Si deposited by plasma enhanced chemical vapor deposition (PECVD) and by atmospheric pressure chemical vapor deposition (APCVD), whereas for both techniques the crystallite height agrees well with XRD crystallite size. Comparing both deposition techniques, the crystallite height for APCVD is comparable to the layer thickness, which is different compared to PECVD.

Keywords: poly-Si crystallite size, XRD, SEM

1 INTRODUCTION

With the increasing interest and drive for industrial implementation of TOPCon solar cell concepts in recent years, it is of utmost importance to control the properties of these contacts [1,2]. The doped poly-Si layer exhibits the main impact on the quality of the passivating contact junction, which is located on a thin interfacial SiO_x layer [3]. The poly-Si crystallite size is one of the most important properties of the layer, since it influences the conductivity [4,5] and also the dopant activation and thus the passivation of the junction [6]. In numerous investigations the crystallite size was determined predominantly by XRD, but also by transmission electron microscopy (TEM), and Raman measurements [8-10]. XRD has proven itself as a reliable technique to determine crystallite sizes for which there are two prevalent measurement modes used. The first is the Bragg-Brentano method (here referred to as 2θ-ω mode), where the incidence and detector angle are varied simultaneously [11,12]. This is used for a lot of different applications due to the high-resolution and high intensity without monochromatization or parallelization. The second method is called gracing-incidence-XRD (GIXRD), where a constant low incidence angle is used which is often used to investigate thin films to allow for an increased interaction volume of the X-ray and the poly-Si [13]. Besides the measurement modes, the evaluation of the measurements usually is of higher importance when it comes to the determination of the crystallite size D. This is often done directly via the Scherrer equation for an individual peak at the diffraction angle Θ with

$$D = \frac{K\lambda}{\beta \cos\theta},$$

where K is a factor related to the shape of the crystallites, λ the X-ray wavelength, and β the width of the diffraction peak [14]. However, this neglects additional peak broadening effects from the instrument itself, the strain within the layer, different crystallite size distributions, and other broadening effects [4]. An evaluation that considers all these factors is the Rietveld refinement [15] which is compared to the Scherrer equation approach in this work. As mentioned before, TEM images are also commonly used to determine the poly-Si crystallite size, where additionally the distribution of crystallite sizes can be extracted. In contrast, XRD measurements do not allow accurate conclusions on the size distributions [16]. However, TEM requires time-intensive sample preparation and it is not feasible for large numbers of samples. The other possible imaging microscopy method is scanning electron microscopy (SEM). However, with standard examination of the sample's cross-section, the spatial and in particular the contrast resolution is too small for the grain boundaries of the poly-Si crystallites. Here, we present a preparation method in which the grain boundaries are etched and thus made visible after ion milling of the sample cross-section. In addition, we compare with this method PECVD and APCVD poly-Si regarding crystallite size.

2 EXPERIMENTAL

2.1 Processing

For this work, 150 µm thick n-type Cz-Si wafers with a base doping of 5 Ωcm were saw damage etched with KOH and cleaned in an ozone and piranha solution. Afterwards a 1.7 nm thick thermal oxide was grown in a tube furnace. This was followed by a single sided deposition of phosphorous doped amorphous silicon (a-Si) using an APCVD tool from SCHMID. Multiple drive throughs between 1 and 5 were applied, where each individual layer shows a nominal thickness of 50 nm. Additionally, a PECVD tool by Centrotherm was used to deposit a 140 nm thick (n)a-Si layer on top of the oxide. After that the silicon layers were crystallized in a diffusion furnace at 920°C for 30 min in N_2 atmosphere. Preparation for SEM imaging was done via ion-milling for a few minutes, followed by an HF dip in 2.5% diluted hydrofluoric acid (HF) for 1 s. The HF step presumably

10.4229/EUPVSEC2023/1CV.3.58

starts to etch the crystallite boundaries at a higher speed than the poly-Si crystallites themselves, increasing the imaging contrast between grain and grain boundary. Crystallite sizes then were determined by investigating multiple images and extracting the crystallite dimensions (length, height and area) by a conventional image editing tool.

2.2 XRD measurements

XRD measurements were conducted using a Bruker D8 Advance in a 2θ range of 20°-70°, using a step width of 0.005°. An exemplary measurement is shown in Fig. 1. A primary divergence slit of 0.2 mm and a 2.5° axial Soller slit was used on the detector side.

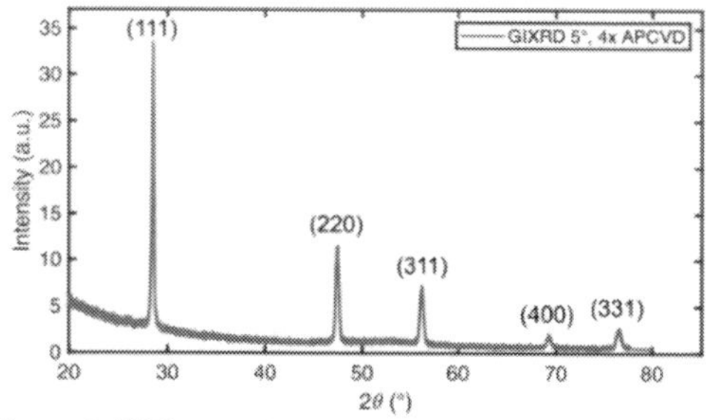

Figure 1: XRD exemplary measurement of a poly-Si layer consisting of 4 drive throughs in the APCVD the corresponding poly-Si peaks are marked.

For usage of the Scherrer equation for determination of the crystallite size, the DIFFRAC.EVA software by Bruker was used to analyze the (111) peak at 2θ=28.5°. Importantly, the integral breadth was used as a measure of the peak width instead of the often-used full width at half maximum (FWHM) which can overestimate the crystallite size [4]. For Rietveld refinement the commercial software Topas was used [17]. The instrumental contribution was modelled via the fundamental parameter approach (FPA) [18], instead of using a reference sample. The refinement includes a polynomial background subtraction, accounting for surface roughness after Pitschke et al. [19] and a zero-error peak shift. Starting with Fd-3m:2 crystal structure, lattice constant, scale factor, and crystallite size were refined. The profile was fitted with a pseudo-Voigt function, whereas for the crystallite size only the Lorentzian part was employed. The fit accuracy was evaluated with R-factors to ensure a comparability between measurements. For a detailed description of the R-factors we refer to [15] or [11]. Furthermore, strain is often not considered when evaluating crystallite size. To quickly estimate the magnitude of strain, Williamson-Hall plots can be used, where strain broadening $\beta_{\text{strain}} = \epsilon \sin\theta$ and crystallite size broadening $\beta_{\text{crystal}} = \frac{K\lambda}{D\cos\theta}$ are added to obtain the measured peak width [20,21]

$$\beta_{\text{tot}} = \beta_{\text{strain}} + \beta_{\text{crystal}} \rightarrow \beta_{\text{tot}}\cos(\theta) = \epsilon\sin(\theta) + \frac{K\lambda}{D}.$$

This uses the fact that strain and crystallite size have different angular dependencies. To obtain the strain one can graphically solve this equation for strain and crystallite size, as is shown in Fig. 2. This is only a rough estimation, as there are only at maximum 5 poly-Si peaks available for graphical determination. Williamson-Hall-plots show that strain cannot be neglected and are thus considered through a Gaussian contribution in the Topas analysis [13]. Possible sources of strain within the poly-Si are the doping concentration, as it was found that a higher phosphorous concentration can increase strain [22, 23]. Furthermore, annealing and the cooling down step can induce additional stress in the layer [24].

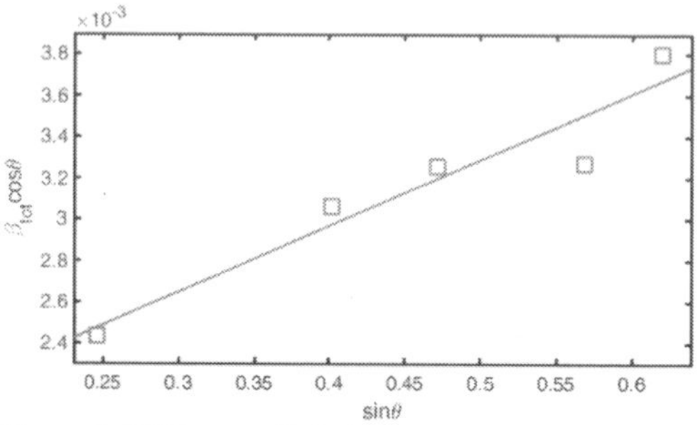

Figure 2: Williamson-Hall plot for the in Fig. 1 shown measurement, yielding an estimation for the crystallite size of D≈78 nm and ε≈0.007.

3 RESULTS

3.1 XRD – Measurement modes

In Fig. 3 XRD profiles of a sample with 5 drive throughs (DT) in the APCVD tool are shown for different measurement modes at the (111) peak. As all measurements were taken with the same detection parameters, one can clearly see that the signal strength is enhanced for the GIXRD measurements which allows for a decreased measurement time compared to the 2θ-ω mode. This is true for all investigated layers with thicknesses <250 nm and is a big advantage of the method as the illuminated sample volume is increased. Especially for even thinner layers (<100 nm), the 2θ-ω mode is more and more unsuitable due to the lacking signal height. Moreover, XRD peaks stemming from the Cz-Si wafer background can interfere with the poly-Si signal in the 2θ-ω mode which is the case for the (400) peak at 70°. However, more importantly it is clearly visible that for a lower incidence angle such as for 1° the XRD peaks broaden immensely due to the beam footprint on to the sample [13]. We found that this broadening is magnified by a larger divergence slit. There have been attempts to consider the beam footprint broadening $G_{hkl}(\omega)$ in Rietveld refinements, as it can be calculated as $G_{hkl}(\omega) = \frac{180}{\pi}\frac{f\sin(2\theta-\omega)}{R\sin(\omega)}$ with the primary slit width f and goniometer radius R [13,25]. This leads to $G_{111}(1°)$ =1.94 and $G_{111}(5°)$ =0.37 and shows the drastic broadening at lower angles. As the peak width is the way to determine the crystallite size from XRD measurements, this broadening presents a hinderance. This broadening is not accounted for in the Topas analysis. Thus, it is not possible to determine the crystallite size for GIXRD 1° measurements via Rietveld refinement with the current Bruker setup available at University of Konstanz. By using a small divergence slit we minimized the broadening effect for GIXRD 5° measurements leading to almost negligible beam footprints. It was found for the samples used here that GIXRD measurements at 5° lead to less peak broadening and thus present a good compromise between signal strength and peak broadening.

10.4229/EUPVSEC2023/1CV.3.58
020062-002

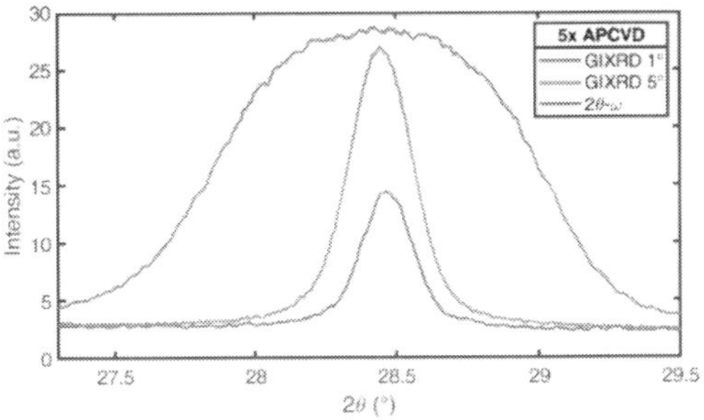

Figure 3: Different XRD measurement modes of the 250 nm thick poly-Si layer accomplished by driving five times through the APCVD System.

3.2 XRD – Analysis of poly-Si layers

The crystallite size stemming from different evaluations and measurement modes for APCVD and PECVD poly-Si are shown in Fig. 4. It is apparent that GIXRD measurements at 1° lead to a rather small crystallite size due to the peak broadening through the beam footprint using the Scherrer equation. On the other hand, GIXRD at 5° produces larger values for crystallite size, while the 2θ-ω mode yields marginally larger crystallite sizes. This points to the fact that, as discussed above, there is still a small peak broadening effect due to beam divergence remaining for most of the samples investigated in this study. However, it has to be pointed out that for the 2θ-ω mode, the signal strength of the 50 nm thick poly-Si layer is reduced immensely compared to GIXRD measurements and requires a longer measurement time to produce valid diffraction peaks. The largest crystallite sizes are obtained from the Rietveld refinement using the FPA approach. Here, the GIXRD and 2θ-ω modes largely agree with one another, leading to crystallite sizes between 50-60 nm. Of course, this crystallite size corresponds to the layers scattering perpendicular to the surface. This is important to keep in mind, as the crystallite shape and distribution is not easily obtainable through XRD measurements.

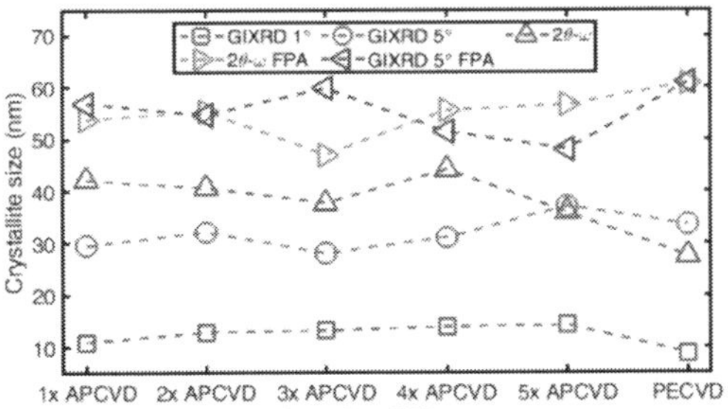

Figure 4: Crystallite sizes for different measurement modes obtained by applying the Scherrer equation and Rietveld Refinement using FPA.

3.3 HF-SEM method

To gain more information about the specific crystallite size distribution within the poly-Si layers and to compare the different evaluation methods, imaging techniques are considered next. For one of the here used APCVD layers, TEM investigations were made. In [26] the poly-Si/SiO_x structure was analyzed on an atomic scale, allowing the resolution of individual crystallites. Moreover, the crystallite height is strongly correlated to the layer thickness. In Fig. 5 SEM images are shown after sample preparation for a 4 DT APCVD layer. As already pointed out in [26], for APCVD the crystallites show a cuboidal shape which are usually wider than high. Furthermore, one striking difference is observable when comparing APCVD and PECVD layers in Fig. 5. Additionally, the crystallite height is limited by the thickness of the individual drive-through layers itself. On the contrary, PECVD poly-Si layers have a less ordered structure, where the crystallites are ordered in no discernible way.

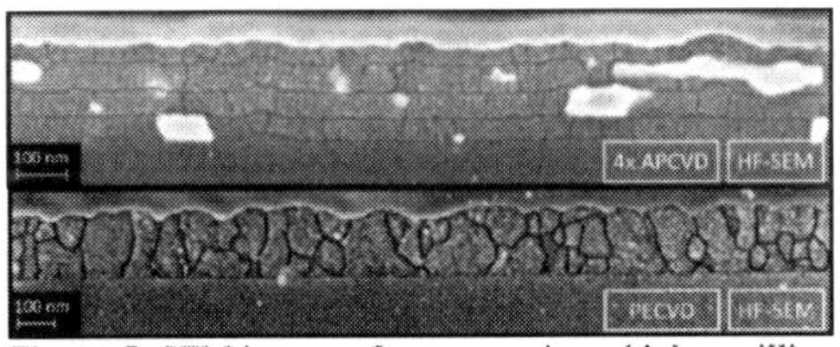

Figure 5: SEM images after preparation with ion-milling and HF for a (top) 4 DT APCVD stack and a (bottom) PECVD layer.

Imaging techniques also allow for determination of a size distribution, as is shown in Fig. 6 where a PECVD and a 5 DT APCVD sample are compared. To obtain the histogram, the crystallite height is shown in Fig. 6. While the distribution is quite narrow around the layer thickness for the APCVD layer, for PECVD layers there is a much wider distribution of crystallite heights which agrees with the expectation from the images in Fig. 5.

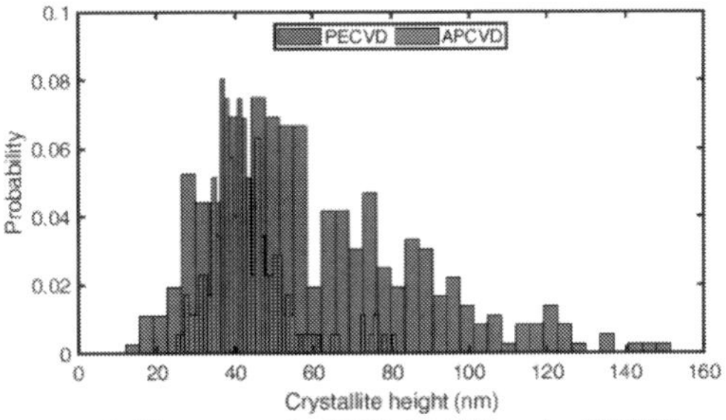

Figure 6: Histogram of a 5 DT APCVD and a PECVD layer, where the occurrence of a respective crystallite height is shown.

Comparing the HF-SEM method with the XRD results in Fig. 7 for the Rietveld refinement, one can see that they largely agree with each other. Of course, neither method is without errors when determining crystallite size. For XRD minimizing the R-factors ensures comparability between the measurements. However, XRD crystallite size mostly presents an estimate which is why no error could be determined. In the case of SEM analysis there are error sources originating from the determination process by hand. These can be related to the resolution of the SEM images as this may be a limiting factor. For some crystallites the boundaries itself also are not able to clearly be identified due to lacking contrast. By averaging multiple images this error can be reduced, but cannot be avoided completely. This should be kept in mind when looking at data from SEM images.

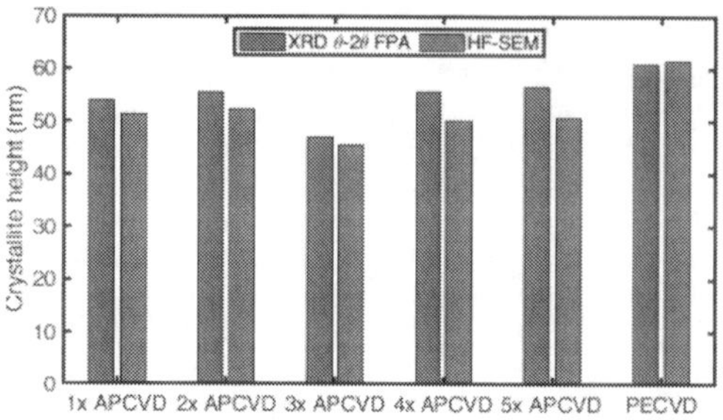

Figure 7: Comparison of XRD crystallite size with the results obtained through analysis of HF-SEM images for APCVD and PECVD layers.

4 CONCLUSIONS

We showed the influence of GIXRD and 2θ-ω measurements on the peak profiles of crystallized APCVD and PECVD (n)poly-Si layers. The peak broadening effects at lower incidence angles can lead to a miscalculation in the crystallite size determination.

Furthermore, a comparison between Scherrer equation and Rietveld refinement using the fundamental parameter approach was made that showed a larger crystallite size for the Rietveld refinement procedure and thus a more reasonable value.

When comparing HF-SEM and XRD results we found that they agree with each other in the case of an evaluation using Rietveld refinements.

The newly developed HF-SEM method was applied for PECVD and APCVD poly-Si layers and provided results comparable to those of TEM images. This technique revealed the different shape and order of crystallites between APCVD and PECVD poly-Si which was visible in crystallite height distributions.

Further investigations by the HF-SEM method on structural properties of differently deposited amorphous or polycrystalline silicon layers will give a deep insight in the dependence of electrical properties of passivating contacts on deposition conditions.

5 AKNOWLEDGEMENTS

Part of this work was financially supported by the German Federal Ministry for Economic Affairs and Climate Action (FKZ 03EE1106B).

6 REFERENCES

[1] D.K. Ghosh, S. Bose, G. Das, S. Acharyya, A. Nandi, S. Mukhopadhyay, A. Sengupta, Surfaces and Interfaces 30 (2022) 101917.
[2] S.W. Glunz, B. Steinhauser, J.-I. Polzin, C. Luderer, B. Grübel, T. Niewelt, A.M.O.M. Okasha, M. Bories, H. Nagel, K. Krieg, F. Feldmann, A. Richter, M. Bivour, M. Hermle, Progress in Photovoltaics: Research and Applications 31 (2021) 341.
[3] T.G. Allen, J. Bullock, X. Yang, A. Javey, S.D. Wolf, Nature Energy 4 (2019) 914.
[4] C. Weidenthaler, Nanoscale 3 (2011) 792.
[5] T. Okker, R. Glatthaar, S. Seren, G. Hahn, B. Terheiden, AIP Conference Proceedings 2826 (2023), 020005.
[6] J.Y.W. Seto, Journal of Applied Physics 46 (1975) 5247.
[7] P.F. Chou, in 2012 IEEE International Reliability Physics Symposium (IRPS), 2012.
[8] J.P. Colinge, E. Demoulin, F. Delannay, M. Lobet, J.M. Temerson, Journal of The Electrochemical Society 128 (1981) 2009.
[9] K. Juraić, D. Gracin, I. Djerdj, A. Lausi, M. Čeh, D. Balzar, Nuclear Instruments and Methods in Physics Research Section B: Beam Interactions with Materials and Atoms 284 (2012) 78.
[10] J.M. Westra, V. Vavruňková, P. Šutta, R.A.C.M.M. van Swaaij, M. Zeman, Energy Procedia 2 (2010) 235.
[11] J. Epp, in Materials Characterization Using Nondestructive Evaluation (NDE) Methods, Elsevier, 2016, 81.
[12] J. Narvaez-Monroy, E. Villalobos-Portillo, L. Fuentes-Cobas, L. Fuentes-Montero, Journal of Physics: Conference Series 1723 (2021) 012047.
[13] D. Simeone, G. Baldinozzi, D. Gosset, S.L. Caer, J.-F. Bérar, Thin Solid Films 530 (2013) 9.
[14] J.I. Langford, A.J.C. Wilson, Journal of Applied Crystallography 11 (1978) 102.
[15] R.E. Dinnebier, A. Leineweber, J.S.O. Evans, Rietveld Refinement, De Gruyter, 2018.
[16] B.B. He, Powder Diffraction 37 (2022) 76.
[17] A.A. Coelho, J. Evans, I. Evans, A. Kern, S. Parsons, Powder Diffraction 26 (2011) 22.
[18] A.L. Ortiz, F.L. Cumbrera, F. Sánchez-Bajo, F. Guiberteau, R. Caruso, Journal of the European Ceramic Society 20 (2000) 1845.
[19] W. Pitschke, H. Hermann, N. Mattern, Powder Diffraction 8 (1993) 74.
[20] D. Nath, F. Singh, R. Das, Materials Chemistry and Physics 239 (2020) 122021.
[21] A.K. Zak, W.H.A. Majid, M.E. Abrishami, R. Yousefi, Solid State Sciences 13 (2011) 251.
[22] C.S. Lee, J.H. Lee, C.A. Choi, K. No, D.M. Wee, Journal of Micromechanics and Microengineering 9 (1999) 252.
[23] K.D. Weeks, S.G. Thomas, P. Dholabhai, J. Adams, Thin Solid Films 520 (2012) 3158.
[24] L. Tutsch, F. Feldmann, B. Macco, M. Bivour, E. Kessels, M. Hermle, IEEE Journal of Photovoltaics 10 (2020) 986.
[25] G. Abbondanza, A. Larsson, F. Carlá, E. Lundgren, G.S. Harlow, Journal of Applied Crystallography 54 (2021) 1140.
[26] R. Glatthaar, F.-P. Schmidt, A. Hammud, T. Lunkenbein, T. Okker, F. Huster, S. Seren, B.C. Greven, G. Hahn, B. Terheiden, Solar RRL (2023).

40th European Photovoltaic Solar Energy Conference and Exhibition

This presentation was selected by the Sc. Committee of the EU PVSEC 2023 for submission of a full paper to one of the EU PVSEC's collaborating peer-reviewed journals.

DEPTH DISTRIBUTION XPE REVEALS PHOSPHO-OXYGEN BONDS IN POLYSILICON FILMS OF TOPCON DEVICES

Z.Q.Ma[1,2], Y.L.Wang[1], Z.X.Lan[1], L.Zhao[3], F.Xu[1], F.Hong[1]
[1] SHU-SOEN's R&D Lab, Department of Physics, Shanghai University, Shanghai 200444, China
[2] Instrumental Analysis & Research Center, Shanghai University, Shanghai 200444, China
[3] School of Microelectronics, Shanghai University, Shanghai 200444, China
e-mail: zqma@shu.edu.cn

ABSTRACT: In this report, we show that the oxidative phosphorus derivatives may be formed in phosphorus (P)-doped polycrystalline silicon film (n^+-poly Si) in TOPCon (tunneling oxide passivation contact) device by using X-ray photoemission (XPE) coupled with depth profile of Ar ion milling, grazing incidence X-ray diffraction (GIXRD) pattern and energy dispersive X-ray spectroscopy (EDXS), as well as the thermodynamic analyses. The affirmation of the phosphorus oxides present in the polysilicon film is based on the discovery of P 2p 3/2 peak arranged in 129-138 eV region of binding energy (BE) in XPE, excluding plasmon loss peak (PLP) of Si 2p state. On the other side, the localized segregation of oxygen (O) and phosphorus (P) after 900 °C annealing has been revealed along the inside of the poly-Si film. As a criterion for the existence of a new chemical bond, the thermodynamic function and the minimum energy principle are used to confirm the oxides and derivatives of phosphorus through the interface of n-Si/SiO_x/ n^+-poly Si system.
Keywords: TOPCon device, poly Si film, phosphorus dopant, oxidized phosphide, XPE

1 INTRODUCTION

Currently, it has been widely recognized by the photovoltaic (PV) community that n-type silicon based tunneling-oxide passivating contact (TOPCon) solar cell is most promising route towards the next-generation passivating contact technologies [1–2], demonstrating highly efficient solar cells as well as remarkable efficiency potentials. In contrast with passivated emitter and rear cells (PERC) solar cell [3], TOPCon construction employs indirect contact of metallic electrodes with high concentration phosphorus-dopant polycrystalline silicon material, resulting in significantly reduced recombination of carriers at electrical contacts and elevated open-circuit voltage (V_{OC}) of the solar cellss. So far, p-TOPCon, n-TOPCon, and bi-facial TOPCon solar cells have been developed in the massive manufacture. Not surprisingly, bifacial TOPCon demonstrated substantial efficiency potential due to superior surface passivation and complete elimination of metal/c-Si contact. The limiting efficiency for bifacial TOPCon is 28.7 %. In other words, these technologies are capable of achieving higher conversion efficiencies, while the efficiency gap is insignificant [4]. Of course, these great progresses are beneficial from heavily phosphorus-doped polysilicon films, its functions more than just "passivation" and low-resistance contacts. Actually, the composited layers acts as a selectivity of photo-generated electron and enable the PV device achieve a highly power conversion efficiency (PCE). However, the microstructure (chemical configuration and physical phase) on the poly-Si film is still open for further investigation concerning on the behavior of passivation contact.

Due to the high concentration of phosphorus elements in n^+-poly Si film, (10^{20}-10^{21}) / cm^3 [5,6], and high temperature (~ 900 °C) crystallization process, making the diffusion and redistribution of phosphorus and internal oxygen elements, will change its microstructure (configuration and phase), leading to the favorable variation of optical and electric properties for the TOPCon device. However, in addition to phosphorus substitutional doping, what we are concerned is on the production of new components arisen from phosphorus and oxygen elements, which has not been reported.

2 EXPERIMENTAL IN DETAIL

2.1 Sample preparation

There are two types of samples prepared for the research. One is double-side-textured surface for the PV device; another is planar surface covered by ~ 9 nm SiO_2 for test. In any cases, *n*-type (100)-Cz Si wafers were cleaned with the standard RCA method. To form a front boron-doped emitter, boron diffusion at 860 °C with BBr_3 source was applied to n-Si substrate. Subsequently, a low temperature silicon oxide (~1.5 nm) were grown on both sides of the wafer with nitrous oxide (N_2O) source, followed by the in-situ growth of ~130 nm phosphorus-doped (n^+) poly-Si layer through low pressure chemical vapor deposition (LPCVD) on both sides, then an thermal annealing in a tube furnace at 900 °C was applied to activate phosphorus elements accompanied by a crystallization of amorphous to polycrystalline state. After removing n^+-doped polysilicon stuck on the edge by wet etching method, a rear SiN_x film was deposited via PECVD on the rear side to enhance the passivation effect and bifacial power generation. To reduce emitter recombination rate and minimize reflection loss, a stacked layer with 4-nm-thick ALD-Al_2O_3 and 75-nm-thick PECVD-SiN_x was coated on the textured surface. After finishing electrode on both sides, the conversion efficiency of the PV device can reach to 24.85%.

2.2 Characterization methods

In order to identify the chemical bond states within the heavily P-doped Si thin films, near normal angle X-ray photoemission (XPE) coupled with depth etching were carried out using (Thermo-fish Scientific) monochromated Al Kα X-ray (hν=1486.6 eV) source. The pass energy window of 200 eV and 40 eV for the wide-scan (survey) and core-level modes, respectively. No extra XPE signals were detected within the sensitivity of the XPE measurements except those from chemical elements that make up the chemical composition of P-doped Si thin films and bulk Si. All the binding energies (BE) were precisely determined within ± 0.3 eV using an XPE handbook. To precisely identify the chemical bond states of spin-orbit doublet peaks in P 2p and O 1s core-level spectra, peak fitting was performed using Shirley background subtraction and Gaussian-Lorentzian functions [7].

The Grazing incident X-ray diffraction (GIXRD) measurements were carried out by using an X-ray diffractometer (Rigaku D/MAX) in the θ-2θ configuration with a CuKα source (λ =1.5418 Å). All of the measurements were performed with a small scanning. The X-ray tube was operated at 40 kV and 150 mA. The Si (400) peak at 69.13° was excluded because too strong intensity. Additionally, a wide range of 25-65° for 2θ angle was scanned so that all the major diffraction peaks could be detected.

The lattice morphology as well as P and O elemental seggregation in the multilayers of n-Si / SiOx / n^+-poly Si films were examined by using high resolution transmission electron microscopy (HR-TEM, JEOL-2100F) at 200 keV, together with energy dispersive X-ray spectra (EDXS).

3 RESULTS AND DISCUSSION

3.1 XPE for P-O bonds

As noted in the symptoms explored in our previous investigation [8], for the case of the P 2p XPE spectra, as shown in Fig.1, the experimental collected spectra taken at 1000 seconds (around 75 nm, calibrated by etching rate) deep inside n^+-poly Si film and three fitted P 2p peaks for each one were observed in the scopes of 127.2 – 130.4, 130.1 – 135.4 and 133.2 - 138.4 eV, respectively, which were assigned to be the phosphided species ($P^{\delta-}$, $\delta < 5$) and oxidized species of phosphorus (P^{5+}) [9]. We found that the fitted peaks centered at 129.3, 132.4 and 135.3 eV should be ascribed to be P-Si (or P-P), P-O and PO_x bonds, respectively [10-12]. Especially, there was no more chemical shift of BE for the peaks as the phosphorus oxides were uniformly distributed in the n^+-poly Si film. This depth profile data were gained through time (s)-dependent Ar ion etching point-by-point for the 130 nm poly Si film of planar surface.

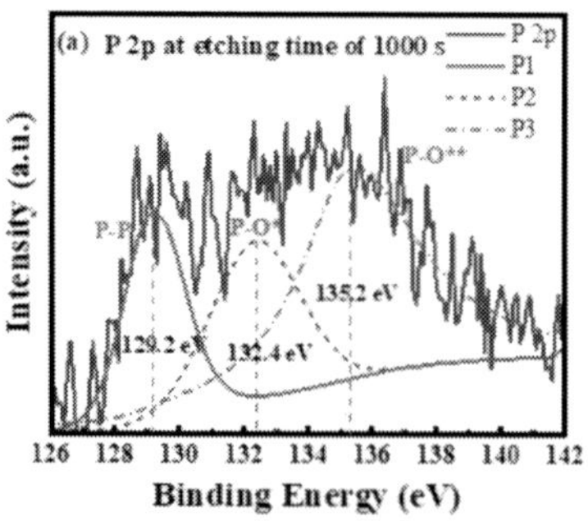

Fig.1 Fitted XPE spectra of P 2p core-level at 1000s.

3.2 GIXRD for crystallographic structure

The phase and components in the n^+-poly Si film were verified by GIXRD patterns, as demonstrated in Fig.2. Well-known crystallized silicon directions [(1 1 1), (2 2 0), (3 1 1)] are at 2θ = 28.4, 47.3, and 56.2°, respectively. Most of the crystallized silicon film using LPCVD and annealing processes at 850-910 °C grew in the direction of (111), which measured a full-width-at-half-maximum (FWHM) peak at (111) of about 0.28° (silicon substrate was 0.05°).

Grain sizes in the n^+-poly Si film films can be calculated using Scherrer's formula [13]:

$$D = k\,\lambda / \beta \cos\theta \qquad (1)$$

where D is the average crystallite size, λ (1.5406 Å) is the wave-length of the used X-ray, β is the full width at half maximum (FWHM) of the measured peak in radians and θ is the angle of diffraction. k is a constant which is determined by the geometry of the crystallites and taken as 0.9. The grain sizes of the films were obtained in the range of 46.2-55.4 nm.

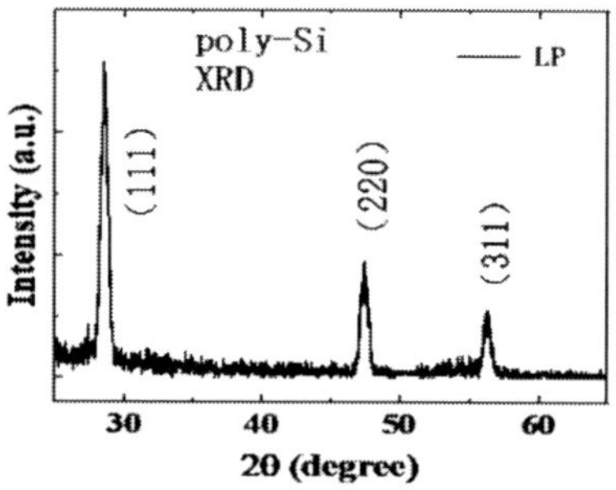

Fig.2 GIXRD of n^+-Poly Si film.

3.3 HR-TEM and EDXS for n^+-poly Si film

The atomic lattice morphology of n^+-poly Si film on SiO_x/c-Si substrate was performed by HR-TEM for textured Si wafer, as illustrated in Fig.3, with an interstaggered atomic distribution states within the interfacial regime. There was no clear atomic arrangement structure in this interface area, as displayed in Fig.3a, but a distinct dark-gray brightness resolution was present.

In Fig.3b, a further observation of amorphous SiO_x layer was existed with phinholes. As pinhole formation in the interfacial silicon oxide during the high temperature anneal was found to play a critical role for majority carrier transport (electron), these pinholes might also contribute to an increase of recombination in the n^+-poly Si/SiO_x structure, depending on the individual pinhole density [6,14]. Additionally, the pinholes were shown to grow with both temperature and duration up to the extreme case of the formation of a little large radius (5-7 nm) of three times the original oxide thickness and the higher pinhole densities [6, 15].

On the other side, as the diffusivity of phosphorus in the interfacial oxide is quite low [6], it is easier to control the in-diffusion of phosphorus through the oxide barrier to form a shallow dopant tail below the interfacial oxide, which seems to be one of the reasons for the present superiority of adopting phosphorus doped poly-Si/SiO_x contacts [17] in the advanced PV stracture.

However, besides the substitutional dopants inside the crystallites of n^+-poly Si film, those excess phosphorus might localize within the grain boundaries (GBs), where are the sites with lowest energies and compresive stress [18]. Thus, to explore the unfavourable situation of these superfluous phosphorus seems to be neccesary for the well

10.4229/EUPVSEC2023/1CV.3.59
020063-002

understanding of special opto-electronic property and passivation mechanism of n^+-poly Si film.

As a result, the presence of phosphorus decreases the crystallization temperature of the n^+-poly Si layers during growth. In addition, the ratio of phosphorus incorporated into the n^+-poly Si film decreases with increasing growth temperature because of the increase in the growth rate. The resistivity of the phosphorus-doped n^+-poly Si films decreases with increasing deposition temperature at the same phosphorus concentration, indicating that the use of a high growth temperature foster an enhancement in the activation of phosphorus in the n^+-poly Si films during growth [19].

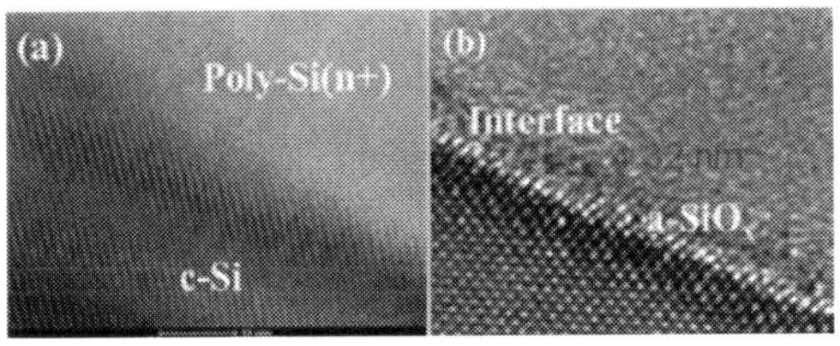

Fig.3 HR-TEM images for the textured c-Si/SiO_x/ n^+-poly Si wafer.

The elemental segregation phenomena of O and P in n^+-poly Si film for a random micron-pyramided interface of n^+-poly Si/SiO_x/n-Si wafer was shown in Fig.4. It was found that these islands-like segregations of O and P within the atomic networks of n^+-poly Si film might come from the non-uniform oxidation of n-Si substrate with random pyramid surface and the high concentration of O and P in n^+-poly Si film, which are indicated by Figs.4a and b, respectively. However, we did not observe the similar segregation of O and P elements in the n^+-poly Si film for the planar Si wafer. In this case, it may meant that the oxidized issue of Si at the random pyramid zone varied dependent on the surface morphologies of the c-Si substrate. Thus, the segregation of the phosphorus (P) and oxygen (O) elements within the n^+-poly Si film may be assigned to the atomic in-diffusion of both a high concentration of P and de-adsorbed O on the fractional surface of the pyramid structure of the n-Si substrate due to high-temperature annealing for a long time.

Furthermore, a strong interfacial segregation of dopants to the grain boundaries (GBs) was revealed in the annealed samples. The heterogeneous precipitation of O and P to the GBs was observed, as well as the clustering in the interior of the grains. Their shapes (loop, rod) strongly suggest that these clusters are the result of dopant segregation to extended defects [20]. Nanometer size oxygen clusters were also observed. They originate from the thermal relaxtation process through the oxide layer. On the other side, oxygen is incorporated in n^+-poly Si film during the fabrication process and is generally present in the as-grown material in a supersaturated state [21]. Hence, during thermal treatments it will tend to form clusters and precipitates. These oxide precipitates have both effects on the electrical and mechanical properties of n^+-poly Si film [22]. The possibility for internal gettering of other impurities by the oxide precipitates and defect clusters in the n^+-poly Si film is advantageous because it results in an increase in the minority carrier lifetime.

Additionally, the doping properties of a polysilicon material are different from those of monocrystalline silicon. Grain boundaries are areas with high density defects and unsaturated dangling bonds that have an important impact on both impurity diffusion and impurity distribution. At the grain boundary, the diffusion coefficient of the doped atom is significantly higher than that inside the grain, and the rapid diffusion of impurities along the grain boundary makes the impurity diffusion rate of the entire polysilicon increase significantly. Similarly, the distribution of impurities is also affected by grain boundaries. At the same temperature, the solid solubility of impurities on the grain boundary is usually higher than that inside the grain, and the segregation phenomenon of impurities occurs between the grain-grain boundary, and the segregation coefficient is usually less than 1 [23].

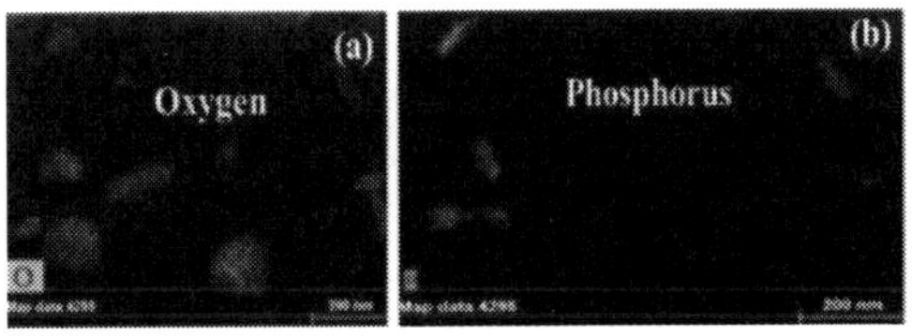

Fig.4 EDXS of P and O component's segregation in n^+-poly Si films.

4 THERMODYNAMIC ANALYSIS

It is well-known that the formation energies of the silica and the phosphorus oxide are about -9.0 and -15.0 eV, respectively [24, 25]. In addition, the mole Gibbs energy and enthalpy are about -856.0 and -910.0 kJ/mol for SiO_2, while the values are about -1348.0 and -1504.0 kJ/mol for P_2O_5 or P-O bonds [26, 27], under the equilibrium state (1 Atm., 298.15 K), respectively. On the other side, the oxidative phosphorus might be formed at the sites of the high concentration-doped phosphorus (10^{20} – 10^{21} atom / cm^3) [5] and the higher oxygen composition (10^{17}-10^{18} /cm^3) by in-diffusion, thermal relaxation and chemical absorption [28] in the grain boundaries (GBs) of n^+-poly Si film. Basing on the varition of thermodynamic parameters of silicon- and phosphorus- oxides, we can conjecture the possibility of the product of solid phase reaction during the processes of high-temperature annealing and atomic diffusion in the n^+-poly Si film afterward.

5 CONCLUSION

We have investigated the chemical bonding states in highly P-doped n^+-poly Si thin films grown on textured and planar SiOx/n-Si (100) substrates using X-ray photoemission (XPE). The fitted XPE P 2p spectra clearly manifested that the oxidized phosphorus present in the modes of bridging or non-bridging linking bonds within n^+-poly Si films, such as P-P, P-Si, P-O, O-P-O bonds. The elemental segregation phenomena of phosphorus (P) and oxygen (O) in the textured sample have been observed from EDXS. It is addressed that the excess oxygen or phosphorus elements are condensed in the grain boundaries, which may arise from in-diffusion at high temperature and compressive stress within grain boundaries. These results indicate that more P atoms are incorporated into the substitutional sites of the Si lattice with the increase of P concentrations. The fitted peaks at three characteristic binding energies of XPE spectra only for the heavily P-doped n$^+$-poly Si samples, which means a new composion produced. Thus, the experimental findings open up new doors for evaluating the chemical states of P-doped Si materials in fundamental researches as well as in photovoltaic industrial applications.

6 ACKNOWLEDGEMENTS

This work is mainly supported by the National Natural Science Foundation of China (No. 61874070, 61674099 and 61274067) and the R&D Foundation of the SHU-SOENs PV Joint Laboratory (No. SS-E0700601).

7 REFERENCES

[1] T.G. Allen, J. Bullock, X. Yang, A. Javey, S. De Wolf, Nature Energy 4 (2019) 914.
[2] M. Hermle, F. Feldmann, M. Bivour, J.C. Goldschmidt, S.W. Glunz, Appl. Phys. Rev. 7 (2020) 021305.
[3] R. Preu, E. Lohmüller, S. Lohmüller, P. Saint-Cast, J.M. Greulich, Appl. Phys. Rev. 7 (2020) 041315.
[4] J. Schmidt, R. Peibst, R. Brendel, Sol. Energy Mater. Sol. Cells 187 (2018) 39.
[5] Q. Wang, H. Peng, S. Gu, et al., Solar Energy Materials & Solar Cells 259 (2023) 112458.
[6] J. Linke, R. Glatthaar, F. Huster, et al., Solar Energy Materials & Solar Cells 246 (2022) 111890.
[7] D. Briggs, M.P. Seah, Practical Surface Analysis: By Auger and X-ray Photoelectron Spectroscopy, Wiley (1990) 572.
[8] Y.L.Wang, Z.X. Lan, F. Xu, et al., ACS Applied Electrical Materials, 4 (2022) 3947.
[9] A. W. Burns, A. F. Gaudette and M. E. Bussell, J. Catal., 260 (2008) 262.
[10] A. Thogersen, M. Syre, B. R. Olaisen and S. Diplas, Journal of Applied Physics 113 (2013) 044307.
[11] W.B Ying, Y Mizokawa, Y Kamiura, K Kawamoto, W.Y Yang, Appl. Surf. Sci. 181 (2001) 1.
[12] M. Lee, H.-Y. Ryu, E. Ko, D.-H. Ko, ACS Appl. Electron. Mater. 1 (2019) 288.
[13] A. Monshi, M. R. Foroughi, M. R. Monshi, *World Journal of Nano Science and Engineering* 2 (2012) 154.
[14] T. Aoyama, H. Tashiro, K. Suzuki, J. Electrochem. Soc. 146 (1999) 1879.
[15] G.T. Yang, R. Gram, P. Procel, Solar Energy Materials & Solar Cells 252 (2023) 112200.
[16] F. A. Trumbore, J. Bell Syst. Tech., 39 (1960) 205.
[17] J. Schmidt, R. Peibst, R. Brendel, Sol. Energy Mater. Sol. Cells 187 (2018) 39.
[18] S. Duguay, A. Colin, D. Mathiot, P Morin, D. Blavette, J. of Applied Physics 108 (2010) 034911.
[19] B. Kim, H. Jang, S. W. Kim, J. of Vacuum Science & Technology A 32, 031510 (2014).
[20] Peter Käshammer, Talid Sinno, J. of Applied Physics **118** (2015) 095301.
[21] M.Geddo, B.Pivac, A.Sassella, et al., J. of Applied Physics, 72 (1992)4313.
[22]J.M.Serra, R.Gamboa, A.M.Vallera, Mater. Sci. Engineer. B 36 (1996)73.
[23] S. Duguay, A. Colin, D. Mathiot, et al., J. of Applied Physics 108 (2010) 034911.
[24] https://materialsproject.org/#apps/interfacereactions
[25] H. L. Lukas, S. G. Fries, B. Sundman, Computational Thermodynamics, The Calphad Method, Cambridge University Press, Cambridge, 2007.
[26] In-Ho Jung, P. Hudon, J. Am. Ceram. Soc., 95 (2012) 3665.
[27] W. M. Haynes, Handbook of Chemistry and Physics (95th), 2014.
[28] R Angelucci, M Severi, S Solmi, L Baldi, Thin Solid Films 103, (1983) 275.

FULL SPECTRUM INLINE EXTERNAL QUANTUM EFFICIENCY MEASUREMENTS OF SOLAR CELLS IN LESS THAN 100 MS

Sven Wasmer*, Stephan Bergemann, Bernhard Klöter, Peter Fuß-Kailuweit
WAVELABS Solar Metrology Systems GmbH - Spinnereistr. 7, 04179 Leipzig, Germany
*s.wasmer@wavelabs.de

ABSTRACT: The spectrally-resolved external quantum efficiency is one of the most important characterization techniques of solar cells as it can help to understand the origins of optical and recombination losses [1]. Yet, due to long measurement acquisition times and usually complex equipment, the method is still constrained to laboratories. To bring it from lab to fab, we present the latest results of our method which we term "quasi External Quantum Efficiency" (*qEQE*) utilizing each of the 20 channels of an LED-based solar simulator, covering a total wavelength range from 375 nm to 1115 nm. A stability analysis of *qEQE* reveals that we can successfully reduce the total measurement time for a full spectrum down to well below 100 ms, one order of magnitude faster than reported for setups using for example Fourier optical systems [2]. As we directly calibrate to the known *EQE* of a reference solar cell, neither a spectrometer nor a sophisticated calculation procedure as done in Ref. [3] is needed. The comparison with a reference external quantum efficiency measurement of a full area (M6) Silicon heterojunction technology (HJT) solar cell results in a low mean absolute error of only $1.7\,\%_{abs}$ at the positions of the effective wavelengths of the 20 LEDs. We furthermore show the results of a large industrial dataset consisting of 20.000 PERC solar cells applying *qEQE* measurements. Using the blue and red response of the solar cells allowed us to attribute variations in the open-circuit voltage most probably to fluctuations of the bulk lifetime or the recombination at the rear side, while we could rule out front side effects. This demonstrates the added value of inline *qEQE* measurements to metrology in solar cell manufacturing, where it can assist in speeding up troubleshooting by narrowing down potential process issues. The method also allows to refine the overall loss analysis of the solar cells for a much more detailed picture needing just end-of-line data acquired by the cell tester.

Keywords: solar cells, characterization, inline metrology, quantum efficiency

1 INTRODUCTION

Measurements of the external quantum efficiency (*EQE*), although one of the most insightful characterization techniques regarding the loss analysis of solar cells [1], have not yet made the leap from lab to fab. Here, we show that full spectrum inline quasi-*EQE* measurements of solar cells are possible in less than 100 ms, one order of magnitude faster than reported for setups using e.g. Fourier optical systems [2]. As we directly calibrate to the known *EQE* of a reference solar cell at the wavelengths of the LEDs, neither a spectrometer nor a sophisticated calculation procedure as done in Ref. [3] is needed. At the hands of the data of 20.000 industrial PERC solar cells, we furthermore show how the *qEQE* method significantly extends and enhances the end-of-line metrology applied in solar cell testers during mass-production.

2 AIM AND APPROACH

Our aim is to implement measurements of the external quantum efficiency using LED-based solar simulators, where each of the single LED channels are switched on and off subsequently while recording the corresponding solar cell currents. A time resolved-graph of the internal monitor diode's intensity output can be seen in Figure 1. The LED channels, ascendingly sorted by wavelength, cover a total wavelength range from 375 nm to 1115 nm and are a mixture of narrow- and broadband ones. As the single LED channels provided by our SINUS-300 solar simulator induce significant cell currents, no bias light is needed. We refer to this method as "quasi External Quantum Efficiency" (*qEQE*), as by relating to the known *EQE* of a reference solar cell (here a small 4 cm² WPVS solar cell), we determine "effective" photon fluxes as calibration factors, inherently taking the non-linearity of the *EQE* (Ref. [3]) into account. The center or effective wavelengths of the 20 single channels are calculated in such a way that a strict monochromatic light source would need this wavelength λ_{center} to generate the same spectrally integrated photon ϕ_{tot} and energy flux Irr_{tot} of the spectrum of the single LED ($\lambda_{center}=hc\phi_{tot}/Irr_{tot}$). To allow for highly reproducible and repeatable measurements, it is vital to take the actual intensity and eventually technically induced variations into account for each individual LED channel.

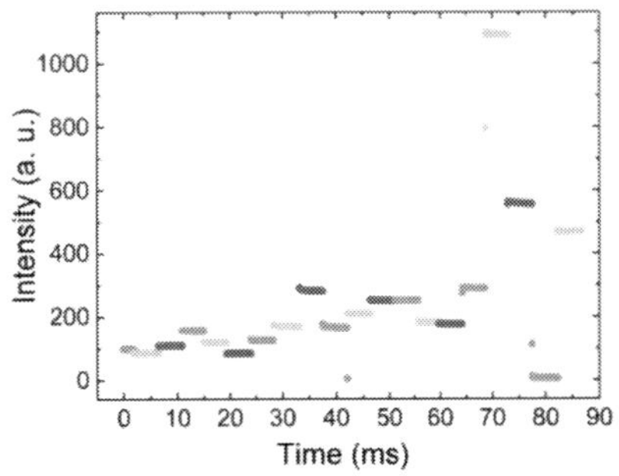

Figure 1: Measured time-resolved intensity of the internal intensity diode during an 88 ms *qEQE* recipe step, which works by subsequently switching each of the 20 LED channels of the SINUS-300 solar simulator on and off.

The intensity output of the SINUS-300's internal monitor cell's is used to ensure stable operation conditions throughout each measurement. We test our method regarding repeatability and dependence on the total acquisition time, compare it to the reference method based on a grating monochromator for a full area (M6) Silicon

10.4229/EUPVSEC2023/1CV.3.61
020064-001

heterojunction technology (HJT) solar cell and finally show as an application the analysis of 20.000 industrial PERC solar cells using our *qEQE* method.

3 RESULTS AND DISCUSSION

The results of the stability analysis regarding repeatability and acquisition time are shown in Figure 2. The individual LED channels are referenced to the longest measurement of 704 ms by plotting the relative deviation of the obtained cell current from the reference for each LED channel, instead of their absolute values. Each set measurement time was repeated around 30 times.

Concerning the total measurement duration variation, we find an increasing deviation from the reference line with decreasing acquisition times. No recurring systematic trend can be observed with respect to the LED channels. However, the relative deviations with respect to the 704 ms measurement stay well below 1%, which is acceptable in a production environment and defines an upper limit for the uncertainty of the measured local quantum efficiency for each LED channel. Furthermore, the repeatability depicted by the error bars for the 30 repetitions per data point remains even below 0.6%.

The results from Figure 2 demonstrate that a full spectrum *qEQE* measurement is feasible within less than 100 ms, while being independent of the total measurement time and having a high repeatability.

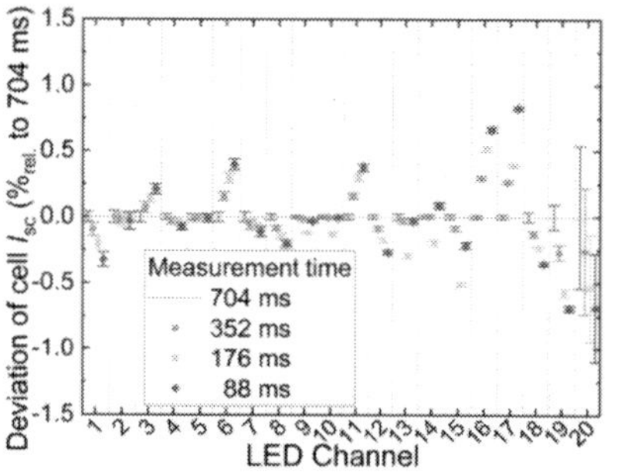

Figure 2: Results of the stability analysis regarding repeatability and acquisition time. Relative deviation of our method from the longest 704 ms measurement is below 1% for each of the 20 LED channels.

A comparison with the reference method based on a grating monochromator for a full area (M6) Silicon heterojunction technology (HJT) solar cell is shown in Figure 3. The *qEQE* values derived from our method within the given wavelength range recorded using a SINUS-300 LED solar simulator nicely follow the reference line taken with a *LOANA* system by *pv tools* and exhibit a low mean absolute error of just 1.7%$_{abs}$. Major deviations may be explained by the fact that we calibrate using a small 4 cm² WPVS solar cell, being affected by slight lateral inhomogeneities of the single LEDs. Calibrating to known *EQEs* of full-area reference cells should improve the accordance even further. For a detailed J_{sc} loss analysis, the values are smooth enough to be inter- and slightly extrapolated.

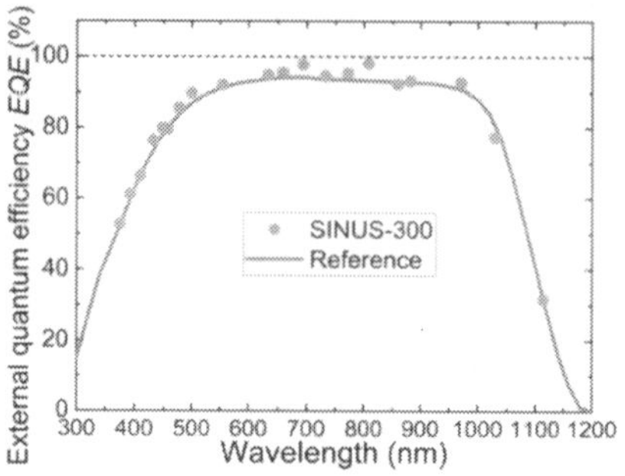

Figure 3: Our *qEQE* method for a full area M6-Si-HJT solar cell agrees very well with a reference measurement conducted using a *LOANA* system by *pv tools*.

While 100 ms measurement duration is impressively fast for a quantum efficiency measurement, this is still too long to integrate such a measurement into a production line, where overall cycle time of a complete measurement is critical. To address this practical consideration, we investigated a reduced *qEQE*, using only LED channels in the blue and red part of the spectrum. This allows to cut down measurement time to below 30 ms, which can be integrated much better into a regular solar cell production line. We performed these reduced *qEQE* measurements on a set of 20.000 industrial PERC solar cells to prove the feasibility for production.

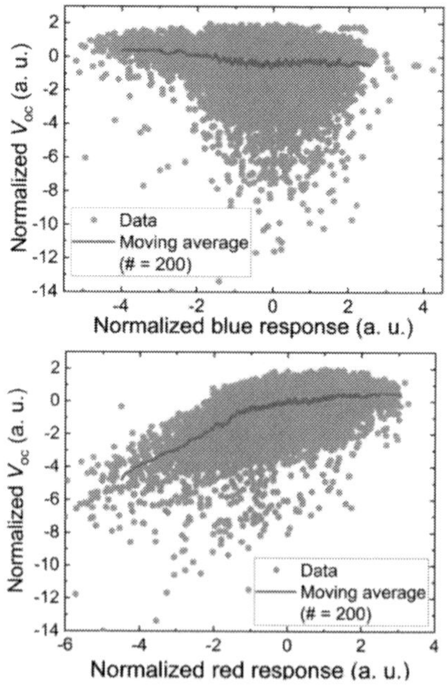

Figure 4: Raw data scatterplots of the open circuit voltage Voc versus (*top*) the blue and (*bottom*) the red response of 20.000 industrial PERC solar cells. Whereas the first shows no correlation, the second shows a clear trend, revealing that the front side is most likely not responsible for V_{oc} variations. All data are normalized to zero mean and unit variance due to confidentiality reasons.

The results of this experiment are shown in Figure 4, where we plot the variations of the open-circuit voltage V_{oc} of the solar cells with respect to the normalized responses to the red and the blue LED channels. Whereas V_{oc} shows almost no correlation with the blue response, an increasing trend can be found in the plot of the red response. The short wavelength light is predominantly absorbed in the emitter and the first micrometers within the Silicon base. Long wavelength absorption in silicon is much smaller, which allows the red part of the light to penetrate the whole solar cell.

Hence, from our example, it can be concluded that the emitter quality in this production line is stable enough to not affect V_{oc}. However, the bulk lifetime and/or the recombination at the rear side has a significant impact. This shows the added benefit of inline *qEQE* measurements to metrology in solar cell production, as sources of variations can be distinguished much more quickly and on a large statistical foundation. The method also allows to refine the overall loss analysis of solar cells for a much more detailed picture needing just end-of-line data acquired by the cell tester and allowing to characterize every single produced cell.

4 SUMMARY AND CONCLUSION

We introduce the first inline-ready approach to measure the full spectral external quantum efficiency of solar cells within less than 100 ms using LED-based solar simulators. We refer to it as "quasi External Quantum Efficiency" (*qEQE*). It is one order of magnitude faster than previous methods and significantly enhances the end-of-line metrology applied in solar cell testers during mass-production, as exemplarily shown at the hands of 20.000 PERC solar cells.

ACKNOWLEDGEMENT

The Fraunhofer CSP is gratefully acknowledged for the LOANA reference and HJT inline measurements within the project "OptiLearn" (03EE1108E).

The authors thank the Institute for Solar Energy Research Hamelin (ISFH) for valuable preliminary tests and evaluations within the project "P3T" (03EE1017E). Parts of the light engine firmware and of the SinusGUI's *qEQE* mode were also implemented and tested within the "P3T" project.

REFERENCES

[1] Fischer, B. *"Loss analysis of crystalline silicon solar cells using photoconductance and quantum efficiency measurements"*. Cuvillier Verlag, 2003.

[2] Melskens, J., Heirman, S. G. M., Koornneef, R. N., Elshinawy, M. A. A., & Schouten, M. *"Fast Optical Measurement System: Ultrafast external quantum efficiency measurements on silicon solar cells"*. Solar Energy Materials and Solar Cells, 173, 6-11, 2017.

[3] Turek, M., Sporleder, K., & Luka, T. *"Spectral characterization of solar cells and modules using LED-based solar simulators"*. Solar Energy Materials and Solar Cells, 194, 142-147, 2019.

IV MEASUREMENTS OF SHINGLE SOLAR CELLS

Alexander Krieg[1], Klaus Ramspeck[2], Nico Wöhrle[1], Philipp Kunze[1], Michael Rauer[1], Andreas Fell[1], Johannes Greulich[1], Stefan Rein[1]
[1]Fraunhofer Institute for Solar Energy Systems, Heidenhofstr. 2, D-79110 Freiburg, Germany
[2]halm elektronik GmbH, Friesstr. 20, D-60388 Frankfurt/Main, Germany

ABSTRACT:

Due to the large variety of contacting probes, the measured *IV* characteristics of solar cells differ considerably depending on the contacting used. In particular, the *FF* heavily depends on the selected contacting layout. The number and distribution of the contact points on the cell but also the shading by the measuring probes affect the *FF*. To be able to determine losses due to the cutting process from host to shingle cell and cell to module losses, the impact of two different contact units installed in an inline cell tester was evaluated. With PCBTOUCH contacting, the resistance losses of the front wires must be considered, otherwise the *FF* losses caused by cutting will be underestimated by $0.5\%_{abs.}$ because the resistance contribution of the wires in the host cell measurement is significantly stronger than in the shingle cell measurement. For the universal contact probes, series resistance losses caused by the probes must also be corrected, but these do not affect the *FF* losses due to the cutting process, but only the absolute value of the *FF*.

Keywords: *IV* measurement, shingle solar cells, contacting units, fill factor losses

1 INTRODUCTION

Shingle cell modules are a promising concept for integrated PV applications with strong area limitations and high aesthetic requirements, such as vehicle-integrated PV. An efficient development of shingle-specific cell process steps such as wafer-cutting and edge passivation requires a current-voltage *(IV)* measurement of individual shingle cells to be able to evaluate the positive or negative effects of the measures carried out. Compared to full-size solar cells, the *IV* measurements of shingle cells have some challenging boundary conditions that need to be considered. For example, the individual shingle cells on the host cells are not electrically connected to each other with grid fingers, which can lead to inhomogeneities in the voltage distribution on the cell if the cable resistances vary. Due to their small area, the shingle cells are also particularly sensitive to shadowing by the measuring probes, which makes precise alignment of the cells in the measuring chamber necessary. To be able to measure shingle cells precisely, the inline cell tester at Fraunhofer ISE was upgraded so that shingle cells can be automatically unloaded from a box, aligned and contacted in the measuring station and after the measurement sorted into one of six BIN classes (Figure 1).

Figure 1: Shingle cells during magazine unloading. The blowing nozzles, which allow the cells to be separated in the magazine, are located to the right and left of the cell stack.

2 AIM AND APPROACH

In industrial production, the *IV* characteristics of individual shingle cells are usually not measured and sorted after they have been separated. However, for process development and optimization of individual process steps, such as cutting or edge passivation, an *IV* measurement of individual shingle cells is essential to be able to evaluate the positive or negative effects of the measures carried out. Since the contacting of the solar cells is not standardized, there is a variety of options for contacting the host wafers and shingle cells [1]. The differences in the *IV* parameters and electroluminescence (EL) images for the different contacting configurations for host and shingle cells are evaluated in this work and the advantages and disadvantages for the different contacting units are pointed out.

2.1 Challenges in correct solar cell contacting

2.1.1 PCBTOUCH

During the measurement with production PCBTOUCH, which is based on GridTOUCH technology [2] , 30 load- and 5 sense wires are pressed onto the cell from above, while the cell is also contacted from the bottom with 30 load and 5 sense contact tracks located on a printed-circuit board (pcb). Solar cells without busbars can also be contacted with the PCBTOUCH unit. Due to the large number of load wires and the resulting small distance of approx. 5 mm between the wires, the finger resistance of the cell is almost negligible. However, the situation is different with the resistance of the wires on the front. Since these are kept thin with a diameter of 0.3 mm to minimize shading of the solar cell, there is a significant voltage drop along the wires. With the 4-wire measurement, the wire resistances are compensated from the source-measure unit up to the edge of the cell. On the cell, however, the load and sense wires are electrically connected by the cell's grid fingers and a voltage gradient occurs across the cell. This shows that the resistance of the wires is included in the measurement. The voltage curve for a section of a cell is shown below the EL image in Figure 2. To make the effect clearly visible, the PCBTOUCH wires were only

10.4229/EUPVSEC2023/1CV.3.62
020066-001

connected on the right side, by disconnecting the plug on the left side, whereas they are connected on both sides in standard configuration. Since an external voltage is applied to the cell during the EL measurement, the voltage on the right side of the cell is approx. 16 mV higher than on the left side for this measurement with a current of 10 A applied to the M2-sized solar cell. During the *IV* measurement, with current flowing out of the cell, the terminal would be on the side with the lower voltage.

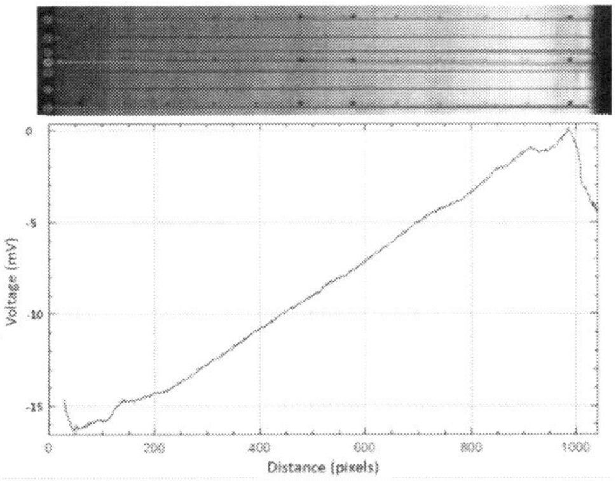

Figure 2: Top: EL image section of a solar cell that is contacted with the PCBTOUCH. The wires are only connected on the right side. Bottom: The voltage gradient was calculated from the grey value gradient of the EL image.

In production, the current and voltage connection on the PCBTOUCH takes place at both ends of the wires, which is equivalent to a parallel connection. The distance over which the current flows through the wires is halved so that the voltage drop across the cell is only a quarter as large as in the example shown above.

On the EL image in Figure 3 it is clearly visible that there is a higher gray value, and therefore a higher cell voltage between the load wires in which a sense wire runs over the cell. Since there is no current output via the sense wires from the measuring device during the EL measurement, the current must flow from the load wires via the busbars on the edge into the sense wires.

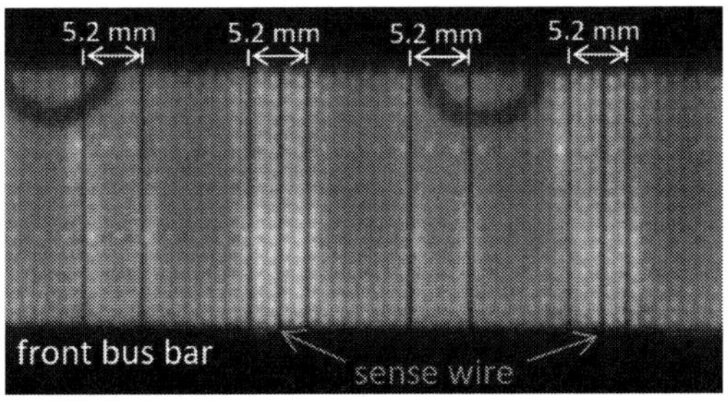

Figure 3: EL image of a shingle cell at a current of 10 A, which is 6 times I_{MPP} and clearly shows the series resistance effects.

Nevertheless, a correction of the wire resistance is necessary for a correct voltage measurement. As can be seen from the relevant literature [3], not the full series resistance of the wire is effective, but only one third of the resistance, because the current density to the external contacts increases continuously during an *IV* measurement. The resistance of the wires R_{wire} is calculated with:

$$R_{wire} = \frac{1}{3 \cdot n_{wire} \cdot n_{connection}} \cdot \rho_{wire} \cdot \frac{l_{wire}}{A_{wire}} \quad (1)$$

n_{wire} is the number of wires, $n_{connection}$ is the number of connection lines, l_{wire} is the length of the wires on the cell in which the current flows in direction of the connections. A_{wire} is the cross-section of the wire, which in case of the PCBTOUCH is 0.0707 mm² with a diameter of 0.3 mm. The specific resistance of a wire is 0.0792 Ω mm²/m and was determined using a 4-wire measurement on one wire with a length of 30 cm.

Since busbars or fingers on the cell that run parallel to the wires also contribute to the current transport and reduce the series resistance loss of the wires, their resistance R_{met} must also be considered when correcting the series resistance R_{wire} of the wires calculating the total resistance $R_{wire,cor}$ according to the formula of parallel interconnected ohmic resistances:

$$R_{wire,cor} = \frac{R_{wire} \cdot R_{met}}{R_{wire} + R_{met}} \quad (2)$$

Furthermore, care must be taken to ensure homogeneous contacting. The PCBTOUCH is equipped with 5 sense wires and each of the sense wires is surrounded by 6 parallel load wires. This leads to an asymmetric contacting when contacting host cells with 6 shingle cells. Since the individual shingle cells are not connected with each other via the metallization, the individual shingle solar cells are at different voltage potentials during the *IV* measurement, which can be clearly seen in the EL image in Figure 4 (top). The edge shingles are contacted with one less wire. This leads to an incorrect *FF* measurement. Contacting the host cells perpendicular to the busbars, which ensure an equipotential voltage distribution, is significantly more homogeneous as shown in Figure 4 (bottom).

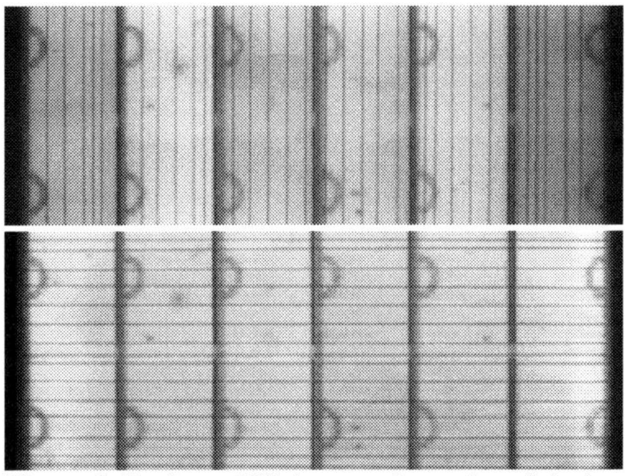

Figure 4: Sections of EL images of a section of the host cell ID 798 with 6 shingles contacted perpendicular to the fingers (top) and perpendicular to the busbars (bottom). The voltage distribution across the cell is significantly more homogeneous when contacting perpendicular to the busbars.

2.1.2 Universal contact probes (ucp)

With universal contact probes (ucp) [4], solar cells can also be contacted independently of the presence or location of busbars. This contacting scheme was used for the following experiments and the host and shingle cells were contacted in the center perpendicular to the fingers (Figure 5). With the ucp, the voltage measurement does not take place on the cell, but in the middle of the contact strip. As a result, the contact resistance between the ucp and the metallization of the cell and the line resistance of the ucp are included in the measurement. In addition, the shading of one probe, which has a shading width of approx. 5 mm and thus covers approx. 15% of the active shingle cell area, reduces the *FF*. To compensate the shading and to illuminate the unshaded part of the cell at 1000 W/m² (related to the whole cell area), the flash power must be increased, which leads to higher currents in the unshaded part of the cell and thus leads to higher series resistances in the grid fingers. For one strip, halm determined the additional series resistance contribution of 1.3 mΩ for PERC cells [5]. This value was used for the corrections in the experiments carried out.

When measuring host cells, clear signal fluctuations between the individual shingle cells have been observed in the EL image of solar cells contacted with ucp (Figure 5, top). Since the individual shingle cells are not electrically connected to each other via grid fingers and the contact strips are connected pairwise with individual cables to the measuring electronics, no compensating currents can flow between the contact strips. Slight differences in the cable resistances of the measuring stripes lead to uneven voltage distribution during the measurement. In this case, limiting resistors in the individual cables improved the homogeneity (Figure 5, bottom).

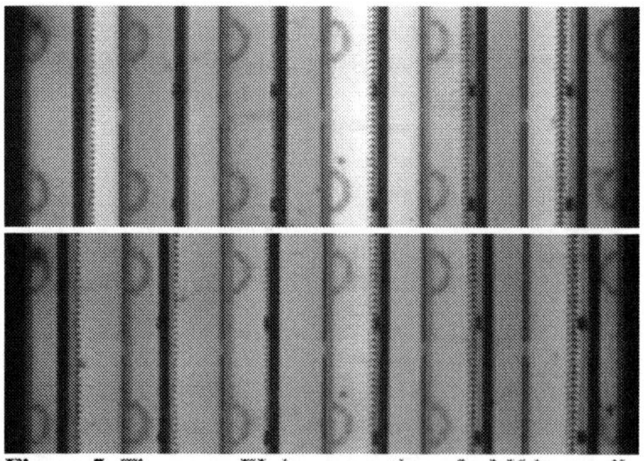

Figure 5: The upper EL image section of a M6 host cell shows clear signal fluctuations between the individual shingle cells. In the lower image, limiting resistors in the individual leads have improved the signal homogeneity.

2.1.4 Correction of the *FF* for comparability of the measurements with different contacting units

Since the current paths differ depending on the contact layout used, the finger resistance must be considered when comparing the results obtained with PCBTOUCH and ucp [6]:

$$\Delta FF = \frac{1}{12} \cdot \frac{I_{mpp}^{2}}{I_{sc} \cdot V_{oc}} \cdot GridRes \cdot N_{cont,1} \cdot \left(\frac{1}{N_{cont,1}^{2}} - \frac{1}{N_{cont,2}^{2}} \right) \quad (3)$$

$N_{cont,1}$ corresponds to the number of contact strips or wires of the measuring setup. $N_{cont,2}$ is the number of contacts to convert to. For the complete compensation of the finger resistances, $N_{cont,2} = \infty$ should be chosen. *GridRes* is the grid resistance between the contacts.

3 EXPERIMENTAL

3.1 Impact of the number of shingle cells on the *IV* measurements

In this experiment, it shall be shown that the influence of the series resistance of the PCBTOUCH wires is not negligible, and a correction of the wire resistance is necessary for the comparability of measurements on host and shingle cells based on different contact units.

Five shingle cells were used for the measurements, which have almost identical *IV* characteristics in the individual measurements.

In the first series of measurements, one shingle cell was measured first and then the number of separated shingle cells measured in parallel was increased to five. The cells were contacted with 20 load and 5 sense wires perpendicular to the busbar (Figure 6, top). In the second variant, the cells were contacted with 30 load and 5 sense wires perpendicular to the fingers. In this configuration, the fingers run longitudinally across the cell (Figure 6, bottom). In addition, in a series of measurements with these cells, the current and voltage terminals were only connected on one side of the wires.

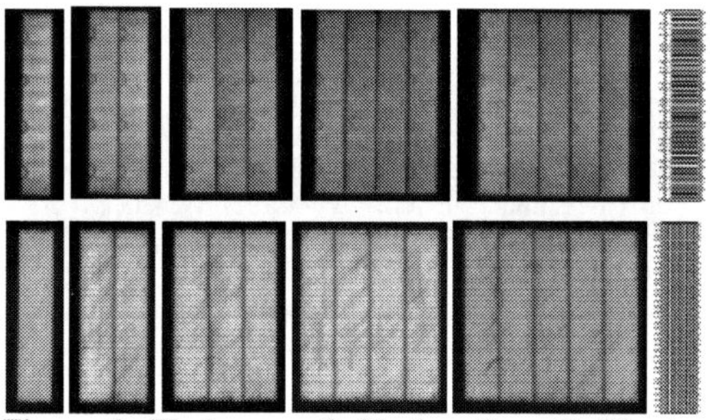

Figure 6: EL images of shingle cells with 25 wires running perpendicular to the busbars (top) and shingle cells with 35 wires running perpendicular to the fingers (bottom) during *IV* measurement. A contacting scheme is shown on the right side.

3.2 Determination of the due to the process for shingle separation

For *IV* measurements of 20 host cells, the Pasan PCBTOUCH unit with 30 load wires was used for contacting the cells perpendicular to the busbars. Furthermore, the halm ucp were used for central contacting each of the shingle cells of the host wafer in the middle perpendicular to the contact fingers.

After the measurement, some of the cells were separated using laser scribe and mechanical cleave (LSMC), the second part was separated using thermal laser separation (TLS). After cutting the measurements on 30 shingle cells are carried out with PCBTOUCH and ucp, respectively (Figure 7).

10.4229/EUPVSEC2023/1CV.3.62
020066-003

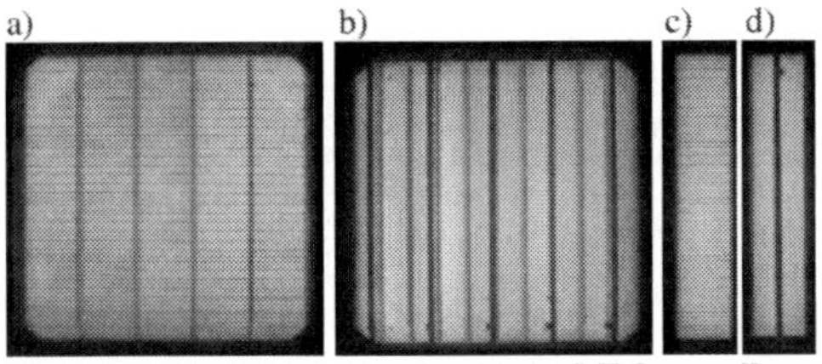

Figure 7: The EL images of (a-b) the M2 host cells and (c-d) the 1/5 shingle cells show the impact of contacting units. The host cells were contacted with (a) the PCBTOUCH unit perpendicular to the busbars and (b) the ucp stripes in the center of each of the shingle cells. The shingle cells were contacted using (c) the PCBTOUCH perpendicular to the busbar with 35 wires and (d) a centrally arranged ucp.

4 RESULTS

4.1 Impact of the number of shingle cells on the *IV* measurements

For all contacting configurations, the *FF* decreases with an increasing number of shingle cells measured in parallel (open symbols in Figure 8), although a constant *FF* would be expected. In the case of single-sided connection of the current and voltage wires, the drop in the *FF* is about 2% (abs.), i.e., a factor of 4 higher compared to the configuration with wires connected on both sides. If the series resistance contribution of the wires is corrected according to formula (1), the *FF* does not change with an increasing number of cells measured in parallel. In the case of the wires running perpendicular to the busbars, the resistance of the grid fingers was also considered.

Since the grid resistance $R_{met} = 5.42$ mΩ/cm is relatively high compared to the wire resistance $R_{wire} = 0.224$ mΩ/cm, the correction resistance $R_{wire,cor}$ is only 4% smaller than R_{wire}.

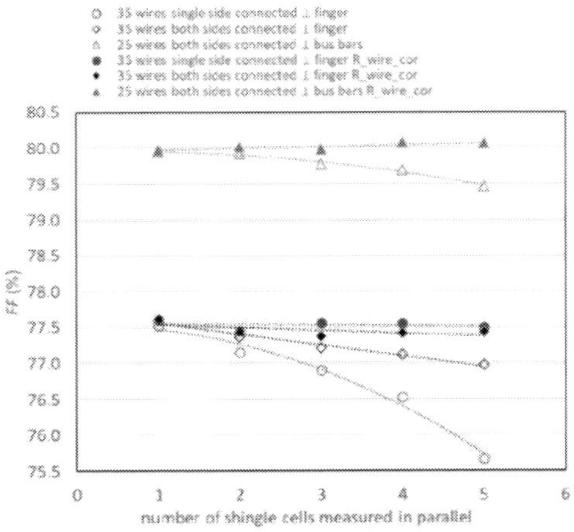

Figure 8: The measured *FF*s decrease in all test variants of contact configurations (see legend) with an increasing number of shingle cells measured in parallel (open symbols). If the series resistance contribution of the PCBTOUCH wires is considered (filled symbols), the *FF* remains almost at the same level with varying number of shingle cells measured in parallel.

Measuring perpendicular to the busbars with 25 wires as well as perpendicular to the fingers with one- and two-sided connection, the deviation of the corrected *FF* is below 0.1%$_{rel.}$ and thus in the range of measurement uncertainty for PCBTOUCH measurements.

This experiment confirms that the resistance of the PCBTOUCH wires has a significant impact on the *FF* determination of the cells and must be considered when comparing cells of different sizes. The conductivity of the backside PCB contact is high enough and does not have a significant influence on the *FF*.

4.2 Determination of the losses due to the process for shingle separation

Jsc differences between the host cells and separated shingle cells due to differences in the contacting units used can be minimized by an appropriate calibration of the measuring system by compensating different amounts of contact-probe-induced shading with an increased intensity of the measuring light. Remaining deviations in J_{sc}, which were less than 0.2 mA/cm² in all cases, were corrected in this experiment so that the J_{sc} mean value for each group is 39.7 mA/cm² and *Jsc* influence on the η is eliminated. Since no power is extracted from the cell under the U_{oc} conditions, U_{oc} measurements are not affected by series resistance losses. Note that U_{oc} deviations were negligible in this experiment. However, this is different with the *FF* determination (Figure 9). Without wire resistance correction, the PCBTOUCH measurements underestimate the *FF* loss of apparently 1.6%$_{abs.}$ due to the TLS cutting process by 0.5%$_{abs}$. Taking the wire resistance correction into account, the *FF* losses of 2.1%$_{abs.}$ determined with the PCBTOUCH unit are close to the *FF* losses of 1.9% determined with the ucp unit. In contrast to the PCBTOUCH, the *FF* losses determined with the ucp unit remain the same with and without resistance corrections, but the absolute *FF* level before and after shingle separation changes. If both the PCBTOUCH and ucp measurements are converted to the grid finger resistance free (and hence comparable) situation, ucp and PCBTOUCH results are at the same *FF* level.

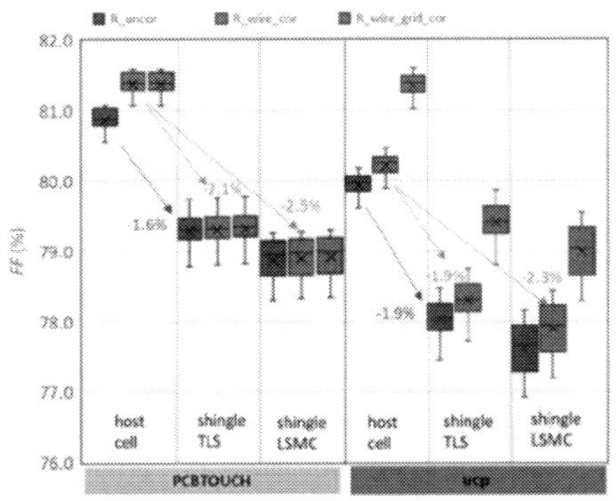

Figure 9: *IV* parameters of the host and shingle cells measured with (left) the PCBTOUCH unit and (right) the universal contact probes (ucp).

This is also reflected in the efficiency, which only differs in the 2nd decimal digit if the finger resistance is neglected. With both contact units, a 0.4% higher *FF* loss is determined for the LSMC process compared to the TLS process.

10.4229/EUPVSEC2023/1CV.3.62

Table 1: Cell efficiencies η of host and shingle cells only wire corrected $R_{\_wire\_cor}$ and wire and grid resistance corrected $R_{\_wire\_grid\_cor}$.

η (%)	Host	Shingle TLS	Shingle LSMC
PCBT $R_{\_wire\_cor}$	21.94	21.30	21.16
ucp $R_{\_wire\_cor}$	21.63	21.01	20.88
PCBT $R_{\_wire\_grid\_cor}$	21.94	21.30	21.16
ucp $R_{\_wire\_grid\_cor}$	21.92	21.30	21.17

4 CONCLUSIONS

To determine the cutting losses from shingle host cells to separated shingle cells precisely, both, the PCBTOUCH and the ucp unit, are suitable for contacting the solar cells. With PCBTOUCH contacting, the resistance losses of the front wires must be considered, otherwise the *FF* losses caused by cutting will be underestimated by $0.5\%_{abs}$ for the present situation. Series resistance losses must also be corrected for the ucp, but these do not affect the *FF* losses before and after cutting, but only the absolute value of the *FF* in both states. The *FF* losses determined with both contacting units differ by $0.2\%_{abs}$. Taking the series resistance of the cell´s contact fingers into account, the contacting-unit-induced differences in the *FF* and η become smaller than the measurement uncertainty.
The wire resistance correction method presented shows good results for the samples investigated. The load and sense wires are effectively shorted, which is important for the validity of the correction. The general case without short circuit of the wires is more complex though and the correction becomes more difficult.

5 ACKNOWLEDGMENTS

The authors acknowledge that part of this work was supported by funding from the German Federal Ministry for Economic Affairs and Climate Action within the research project "Guten Morgen" (contract no. 03EE1101A).

5 REFERENCES

[1] International Electrochemical Commission, IEC 60904: Photovoltaic devices – Part 1: Measurement of photovoltaic current-voltage characteristics, International Standard, Second Edition, 2006.
[2] Bassi, Nicolas, et al. "GridTouch: innovative solution for accurate *IV* measurement of busbarless cells in production and laboratory environments." Proc. of 29th European Photovoltaic Solar Energy Conference and Exhibition. EUPVSEC, 2014.
[3] Goetzberger, A.; Voß, B.; Knobloch, J. (1997): Sonnenenergie. Photovoltaik - Physik und Technologie der Solarzelle. Freiburg im Breisgau: B. G. Teubner.
[4] Ramspeck, K., et al. "Contacting New Solar Cell Metallization Layouts and Contact Quality Surveillance in Solar Cell Production"." Proceedings of the 36th European Photovoltaic Solar Energy Conference and Exhibition, Marseille, France. 2019.
[5] Waleska, P., et al. "Lifetime Analysis of Bifacial-Ready Contacting of Busbarless and Multibusbar Solar Cells for Industrial Mass Production" Proceedings of the 37th European Photovoltaic Solar Energy Conference and Exhibition, Lisbon, Portugal. 2020
[6] Rauer, Michael; Krieg, Alexander; Fell, Andreas; Pingel, Sebastian; Wöhrle, Nico; Greulich, Johannes M. et al. (2022): Assessing current-voltage measurements of busbarless solar cells. In: Sol. Energy Mater. Sol. Cells 248, S. 111988. DOI: 10.1016/j.solmat.2022.111988.

COMPARING MDP, QSSPC AND PL IMAGING FOR THE PRODUCTION CONTROL OF HETEROJUNCTION SOLAR CELLS

Christian Diestel[1], Saravana Kumar[1], Nadine Schüler[2], Marcel Leonhardt[3], Jonas Haunschild[1], Stefan Rein[1]
[1]Fraunhofer Institute for Solar Energy Systems ISE, Freiburg, Germany
[2]Freiberg Instruments, Freiberg, Germany
[3]Meyer Burger, Hohenstein-Ernstthal, Germany
christian.diestel@ise.fraunhofer.de , +49 761 4588 5052
Hans Bunte Straße 19, Geb. 7, 79108 Freiburg, Germany

ABSTRACT: Minority carrier lifetime is a valuable parameter for quality control of the early production steps in solar cell manufacturing. It is particularly powerful for heterojunction technology as passivation of the amorphous silicon (aSi) layers is fully established directly after PECVD without further activation steps (as e.g. in the PERC process). This allows sensitive quality assessment especially if lifetime is measured injection-dependently and/or spatially resolved. Quasi-steady-state photoconductance (QSSPC), photoluminescence (PL) imaging and microwave-detected photoconductivity (MDP) are fast and contactless techniques that allow lifetime measurements inline. Each technique has its own strengths and weaknesses regarding carrier injection resolution, spatial information, quantitative accuracy, and signal understanding. At Fraunhofer ISE, each of the systems is installed in our inline wafer inspection system FWIS which allows direct comparison of the results for each wafer on a statistical scale. We present here such a comparison of QSSPC, PL and MDP on a set of 4000 silicon heterojunction precursors. We observe strong correlation between the three techniques, indicating good qualitative agreement of the measured lifetimes. Despite MDP being designed for measurement of differential lifetime over a steady-state bias injection level, even high-power MDP measurements without bias injection agree quantitatively with QSSPC measured at a carrier injection level of 10^{16} cm^{-3}, suggesting that this may be taken as the effective carrier injection density of the MDP tool for the given setup. By calibrating the instruments against each other, we are thus able to gain spatially and injection-resolved lifetime data for each individual wafer in a production environment.
Keywords: Microwave-detected photoconductivity, carrier lifetime, silicon heterojunction solar cells

1 INTRODUCTION

The global photovoltaic power installation capacity is growing by 174 GWp/a and accelerating by 48 GWp/a^2 (data from end of 2021). 95% of this is c-Si technology [1]. Silicon single-junction cell efficiency is approaching its theoretical limit [2], and to gain further incremental improvements, individual process steps during production need to be fine-tuned efficiently. Improving overall production efficiency by upscaling, as well as by reducing waste and machine downtime during repairs, can additionally increase cost efficiency.

To meet these requirements during cell manufacturing, automatic inline characterisation of cell precursors between production steps is highly beneficial. The data gathered enables tight feedback loops for optimisation of individual processes and early detection of faulty wafers.

For many of the process steps, a variety of important quality parameters can be tested with inline characterisation tools. To assess the quality of the junction formation and surface passivation steps, for example, minority carrier lifetime is one such parameter: It can be used to predict the final cell performance without the influence of the grid and contacts, by calculating the implied open-circuit voltage (V_{oc}).

Resolving lifetime measurements spatially across the wafer surface as well as for different injection levels provides detailed information about the previous processes. Values measured at multiple injection levels show the relative strengths of the various recombination mechanisms: at low levels, defect-enabled Shockley-Read-Hall recombination dominates, whereas at high levels the interaction between excited carriers increases, which causes Auger recombination to dominate [3]. Peak lifetime values usually occur in an intermediate range of injection levels, in which radiative recombination has a higher contribution.

Lifetime maps measured across the surface of each wafer with high spatial resolution further allow for differentiating between different sources of defects, such as impurities introduced through handling or spatial inhomogeneities of deposition and diffusion processes.

In this work, we compare different techniques for the inline lifetime measurement of silicon heterojunction (SHJ) solar cell precursors. The aim is to demonstrate the agreement between the measured data from the different techniques on a wafer-by-wafer basis. The combination of these techniques can therefore provide spatially and injection-resolved lifetime data for each wafer, suitable for rapid, robust, reliable and meaningful inline quality control.

2 BACKGROUND

In the quasi-steady-state photoconductance (QSSPC) method, established by Sinton et al. in 1996 [4], free carriers are excited by a flash lamp of decaying intensity. In steady-state mode, the decay time of the flash intensity is long compared to the carrier lifetime, while it is short in transient mode. The change of inductance caused by the density of excess carriers is measured via an inductively coupled coil and converted to lifetime with prior knowledge of the doping density. As the measurement runs continuously during the intensity decay of the lamp, lifetime values at different injection levels are recorded, the injection-dependent lifetime curve providing insight into the type of limitations and performance at different operating conditions. Our IL-800 QSSPC tool from Sinton Instruments takes one measurement at the centre of each wafer, and it has an integration area with a diameter of about 8 cm due to the size of the coil and the flash lamp. Thus, this technique provides an averaged lifetime without spatial resolution. It is, however, a well-understood and widely used technique for contactless lifetime measurement.

Photoluminescence (PL) imaging [5,6] is an equally well understood method, capable of capturing high-resolution images of the radiative recombination intensity in a semiconductor due to concurrent laser excitation. Our inline PL tool from Meyer Burger creates images from on-the-fly line scans. As the illumination occurs only locally via the line laser, lateral carrier diffusion prevents us from knowing the local injection level at the detection point precisely. Nor can the brightness of the PL image be directly translated into lifetime values, as it depends not only on the excess minority carrier density but also on the net background doping and the optical properties of each sample. Nevertheless, a calibrated lifetime map can be generated by scaling each PL image to the lifetime value measured by means of QSSPC [7–10] or MDP [11] across their respective integration regions on the same wafer.

Finally, microwave-detected photoconductivity (MDP) [11] uses laser diodes to excite the sample to steady-state conditions. During and after this pulse of light, the photoconductivity is measured via its effect on the reflectance of microwaves in a resonant cavity. The lifetime can be inferred from the transient decay of photoconductivity after the light has been switched off. Our MDP-Linescan tool from Freiberg Instruments has an integration area with a diameter of 1.1 mm, and it performs a single line scan down the middle of each wafer as it passes underneath on the fly. The MDP data therefore contain more spatial information than those from QSSPC, but less than PL images. As with QSSPC, carrier lifetime can be calculated directly; but unlike QSSPC, the value is computed over the entire slope of the photoconductance decay curve. As we discuss later in Section 4.1, this makes it difficult to determine an injection level directly.

In summary, QSSPC gives high injection resolution but low spatial resolution, PL imaging is the other way around, and MDP is intermediate in both categories. The combination of these tools, if calibrated well, could deliver comprehensive lifetime data for each wafer with high resolution both spatially and in terms of injection levels.

The conventional way to calibrate different instruments against each other is to use reference samples. In our case, these would have to be pristine, uniform solar cell precursors without any defects, be it from handling, process inhomogeneities or other sources. Such samples take great effort to produce and can easily be damaged, so relying on them can be expensive. Our approach relies instead on comparing a large enough population of production samples to gain a representative statistical distribution of values for each instrument. The correlation of these distributions can then be used to see, in the first instance, whether two instruments are actually measuring the same physical property, and if so, to then calibrate one instrument against the other.

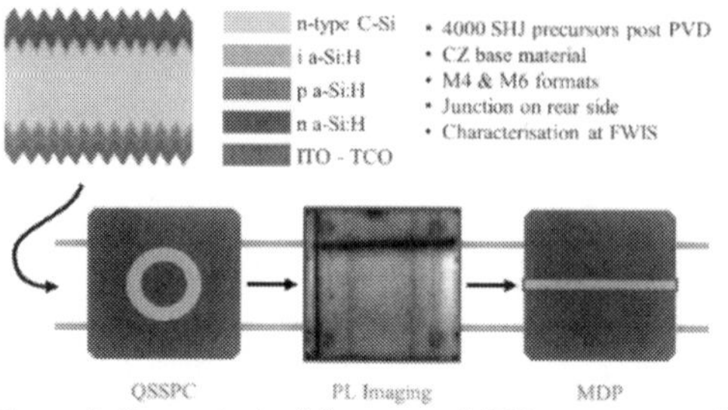

Figure 1: Layer stack of the textured SHJ precursors and process flow of the inline lifetime measurements. Wafers are placed onto the conveyor belt and automatically measured using QSSPC, PL imaging and MDP tools. The integration areas of the individual techniques are highlighted red in the sketch.

3 EXPERIMENTAL

Figure 1 shows the process flow of the inline lifetime measurements performed at our frontend wafer inspection system (FWIS) at Fraunhofer ISE; this system allows us to characterise solar cell precursors inline as it could be implemented in production. To gain the statistical data for this work, 4164 textured SHJ precursors were loaded onto the conveyor belt and passed by the various contactless characterisation tools. The stop-and-go motion of the belt facilitates both on-the-fly scanning and stationary measurements, depending on the requirements of each instrument.

Three of the instruments installed in the FWIS can be used for lifetime measurements: (i) the Sinton IL-800 QSSPC tool, (ii) the Meyer-Burger PL imaging tool, and (iii) the MDPlinescan tool from Freiberg Instruments. The first operates in stationary mode, while the latter two record their data on the fly. The data from each tool are initially collected by the respective control computer and then passed on to a central machine, where they are consolidated and processed. The areas of data acquisition are indicated in Figure 1 by the red outlines.

For the statistical comparison of the three methods of lifetime measurement, a representative lifetime value was calculated for each wafer and each method. Geometrical consistency was achieved by averaging sections of the PL image that corresponded to the integration area of each of the other two tools when comparing with them (an annulus for QSSPC and a strip for MDP).

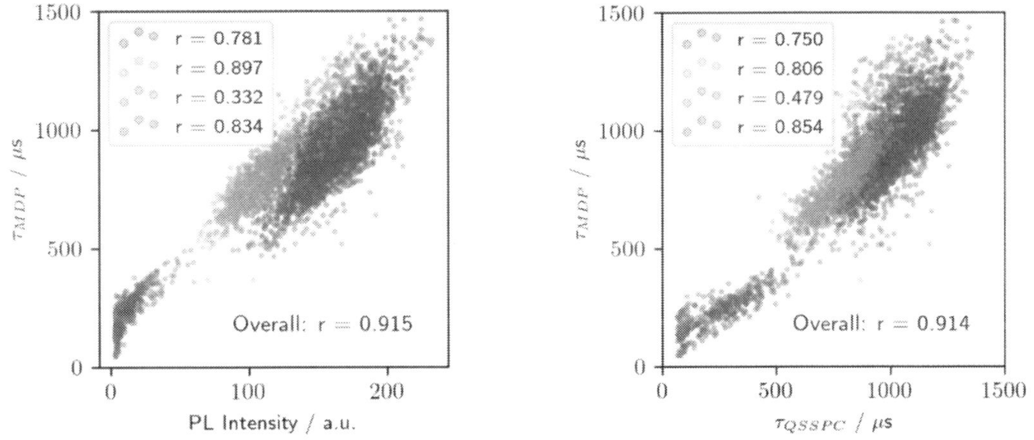

(a) (b)

Figure 2: MDP lifetime (averaged over the full trace) plotted against (a) mean PL intensity and (b) QSSPC lifetime at $1 \times 10^{16} cm^{-3}$ carrier density. The different coloured clusters correspond to the same groups as in Figure 3.

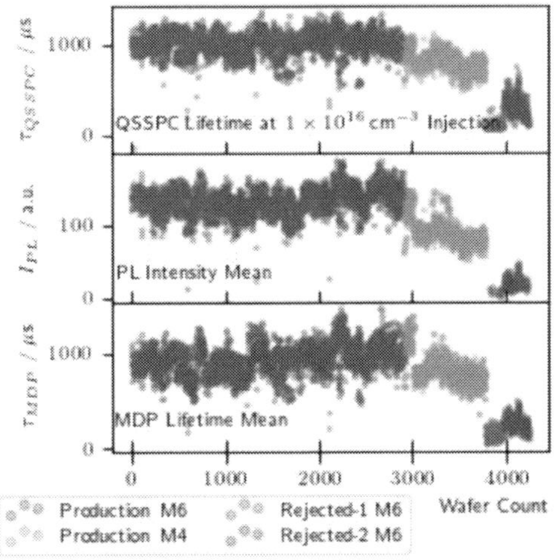

Figure 3: QSSPC (top), PL (middle) and MDP (bottom) data averaged for each wafer and plotted chronologically for 4164 SHJ precursors. Groups from different production batches are colour-coded.

4 RESULTS AND CONCLUSIONS

Figure 3 shows average values for the three lifetime tools plotted as a function of wafer count. The top subplot shows QSSPC lifetime in the central region at an injection level of $1 \times 10^{16} cm^{-3}$. The middle subplot shows PL image intensity averaged over the full surface of each wafer. The bottom subplot shows lifetime measured by MDP.

The wafers are sorted in no particular order. However, four groups of wafers can be distinguished by the different colours in the plot. The blue and orange groups were production-quality precursors of M6 and M4 sizes, respectively. The green and red groups were wafers from two different batches, rejected from the production line due to their low performance in previous inline characterisation.

It is already visible that all three datasets correlate well with each other, and that the QSSPC and MDP values are even in good absolute agreement. Further details can be seen by plotting the lifetimes against each other, as has been done in Figure 2. Here, we see that the different groups form distinguishable clusters, but that the correlations between MDP lifetime (on each y-axis) and the other two methods (on each x-axis) are generally good for groups with significant spread in both parameters, and excellent ($r > 0.9$) overall.

4.1 Injection levels

The level of excess carrier injection influences the measured lifetime substantially. QSSPC lifetime data for the best wafers range from 1 ms at $1 \times 10^{14} cm^{-3}$ and at $1 \times 10^{16} cm^{-3}$ to over 3 ms at $1 \times 10^{15} cm^{-3}$.

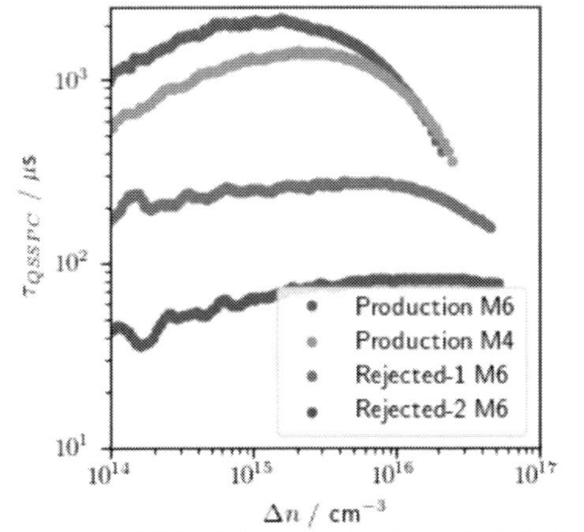

Figure 4: QSSPC lifetime vs. minority carrier injection level for the different production batches.

This can be seen in the plotted QSSPC lifetimes versus injection level for representative wafers from each production batch in Figure 4. Note that the strong oscillations at low Δn are artifacts of data processing and do not represent actual trends. Note also that the positions of positive and negative slope shift between wafers of different quality: at $\Delta n = 5 \times 10^{15} cm^{-3}$ the red and green curves are still rising while the orange and blue curves are falling.

Resolving the injection level of QSSPC measurements is possible because data are collected continuously during the brightness decay of the excitation lamp, and the gradient at each point of the decay curve can be determined to yield a separate lifetime value.

MDP, on the other hand, is designed as a differential method of lifetime measurement: A steady-state homogeneous injection level is to be set by a bias light, and the laser adds a perturbance to the carrier density. As long as this perturbance is small, the decay curve is linear in the semi-logarithmic plot, and so a fit over the entire range can yield accurate results. Even so, the so-obtained differential lifetimes would actually need to be integrated over all bias light intensity levels to calculate true absolute lifetimes.

Bias light series were not implemented in our inline MDP setup at the time of this experiment. Instead, a stronger laser pulse was used to enable noise-free measurement. However, this introduced a significant uncertainty in the lifetime and injection level as calculated from the entire decay slope for two reasons: Firstly due to the strong and wafer-specific injection dependence of lifetime; and secondly due to significant diffusion of carriers away from the excitation spot during the measurement, which effectively dampens the injection density as a function of time since the start of the transient.

We can estimate an upper limit on the steady-state injection level from the known power (500 mW), wavelength (925 nm) and spot size (1.1 mm) of the laser excitation; this gives us a value of $2.5 \times 10^{19} cm^{-3}$. However, in our current setup this is serves only as a rough guide.

Bias light series and integration of differential lifetimes will solve this problem and are currently in development, to be published in future works. In absence of this, our statistical survey can be used for a practical and empirical relation of MDP lifetime to injection level by plotting the MDP lifetime against the QSSPC lifetime at different injection levels, as has been done in Figure 5. The red, orange and yellow datasets correspond to injection levels of $1 \times 10^{15} cm^{-3}$, $5 \times 10^{15} cm^{-3}$, and $1 \times 10^{16} cm^{-3}$, respectively. As we can see, not only is the correlation best for $10^{16} cm^{-3}$, the datapoints also line up well with the 1:1 line, indicating good absolute correspondence of the values. This suggests that $10^{16} cm^{-3}$ may be treated as the effective injection level at which the lifetime is measured with MDP.

4.2 Region of interest selection

A PL image contains a brightness value of radiative recombination for each pixel, which corresponds to a resolution of about 150 µm on the sample. The MDP tool records a line scan down the centre of the wafer, with a resolution of about 1 cm. In order to compare these methods with each other, the different measurement locations must be considered.

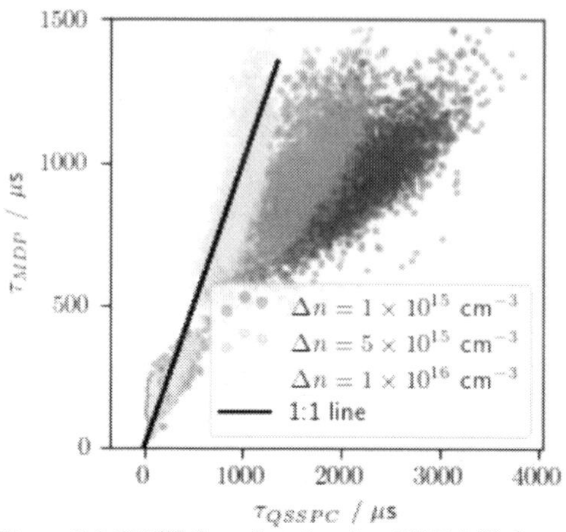

Figure 5: MDP lifetime plotted against QSSPC lifetime at different injection levels. The line of 1:1 proportionality is shown in black.

Figure 6 shows MDP lifetime averages plotted against different types of PL intensity averages. In the purple and red groups, the arithmetic mean *A* was computed over the whole wafer and a stripe corresponding to the MDP detection area, respectively. The correlation is better for the stripe subset ($r = 0.915$ vs 0.835), indicating on the one hand that matching the integration areas allows for a more precise comparison between tools. On the other hand, it shows that this narrow stripe is generally not a representative sample of the lifetime inhomogeneities across each wafer. This means that despite sampling a whole cross-section of the wafer surface, MDPlinescan is no replacement for PL imaging in terms of spatial resolution.

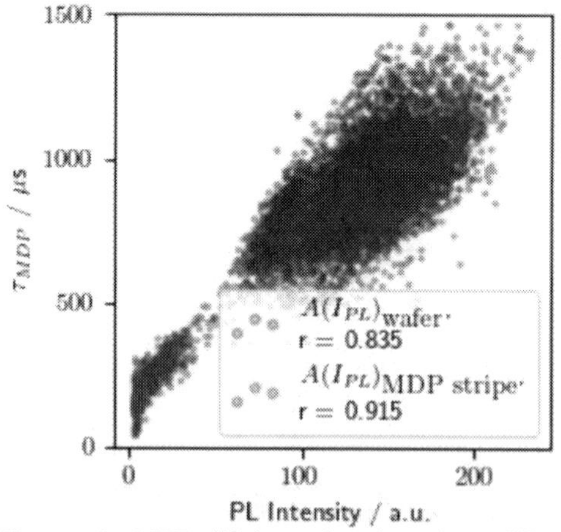

Figure 6: MDP lifetime plotted against different averagings of the PL intensity images. The blue and orange clouds, respectively, show the arithmetic mean A of each whole wafer, and a stripe matching the integration area of the MDP linescan.

4.3 Calibration of PL images

Our standard and most accurate method for converting PL images into lifetime maps is by calibrating the pixel values of each image against the QSSPC lifetime measured for that wafer. Being able to calibrate each wafer individually is one of the benefits of having both measurement devices installed as inline tools in the production line.

However, if space or budget allows for only one tool for lifetime measurement to be installed inline, lifetime maps can still be generated with acceptable accuracy using an inline PL and calibrating it via other methods.

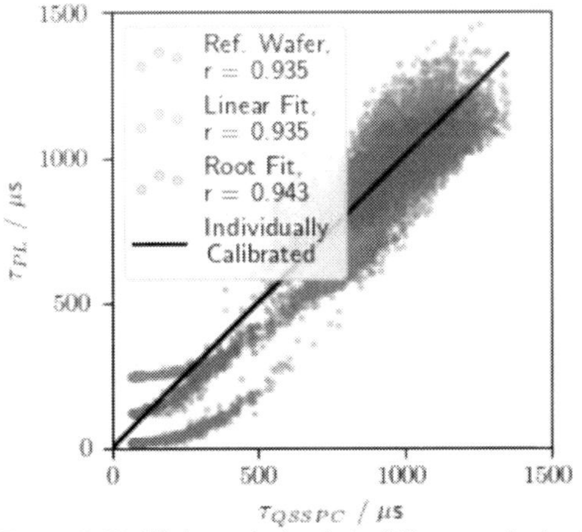

Figure 7: PL lifetime calculated via different methods of calibration against QSSPC.

Figure 7 shows mean PL lifetimes calibrated by different methods against the corresponding QSSPC lifetimes. The black line indicates the calibration against QSSPC for each individual wafer. By definition, this is the identity relation.

The blue-green dots arose from calibration against the QSSPC lifetime of a single reference wafer, picked from the centre of the correlation cloud of mean PL values versus QSSPC lifetime. With this simple method, a single representative wafer can be measured by offline QSSPC to generate a constant scaling factor for all PL measurements.

The olive-green dots show a two-parameter calibration using an affine linear fit ($\tau(I_{PL}) = A + B \cdot I_{PL}$) of the mapping of QSSPC lifetime to mean PL intensity. Although the correlation coefficients are the same, clearly this calibration yields more accurate mean PL lifetimes than the single-wafer calibration (blue dots), as it accounts for differences in signal offset from stray light or other sources. This method could be used in production if there is capacity to measure QSSPC not just on a single wafer but on a larger selection.

An even better match with the individual QSSPC values is achieved by fitting a square root function ($\tau(I_{PL}) = A + B \cdot \sqrt{I_{PL} - C}$) (grey dots). This is expected from the theoretical τ^2-dependence of PL intensity over large ranges of τ.

4.4 Conclusion

In conclusion, we have demonstrated that QSSPC, MDP and PL imaging are compatible methods for inline measurements of minority charge carrier lifetimes in silicon solar cell precursors. Their data, averaged over corresponding areas of each sample, correlate well with each other over a wide range of lifetime values. QSSPC directly offers absolute lifetime values with a high injection level resolution. However, it provides little spatial information. MDP, designed for measuring differential lifetime over a series of bias light intensities, can nevertheless provide a value that corresponds fairly well to absolute lifetime even from a single measurement with a powerful laser excitation, without using the bias light. We were able to show that an effective injection level may be inferred for the MDP measurements from the statistical correlation with QSSPC measurements, with a value of $10^{16} cm^{-3}$ in the case of our MDP setup. Compared to QSSPC, the MDP line scan provides more detailed spatial information.

When comparing methods of different integration geometries, we show that comparing only matching areal subsets of the data yields more precise results. PL imaging by itself yields high-resolution maps of relative carrier lifetime; the excellent correlation of mean PL intensity with QSSPC lifetime justifies calibration of PL images against QSSPC, which can yield highly resolved *absolute* lifetime maps. [7]

Finally, we demonstrated that although measuring each wafer with a reference method such as QSSPC yields the most accurate calibration of PL images, more cost-effective methods offer acceptable results as well: in order of accuracy, one may apply (a) a scale factor, (b) a linear function or (c) a square-root function to PL values. These require only (a) one typical reference wafer or (b,c) a selection of reference wafers to be measured by QSSPC.

In future work we intend to establish inline MDP bias light series in order to measure differential lifetime with accurate and precise knowledge of the corresponding injection level. Injection series of absolute lifetime may then be determined by integration and compared directly with those of QSSPC.

Furthermore, we aim to extend the statistical comparison to measurements of the finished solar cell parameters. This will help us to quantify the predictive power of precursor lifetime measurements by the different methods.

Acknowledgments: This work was funded by the German Federal Ministry for Economic Affairs and Climate Action within the project SALSA under contract number 03EE1096A.

5 References

1. Simon Philipps, *Photovoltaics Report*, <https://www.ise.fraunhofer.de/en/publications/studies/photovoltaics-report.html>.
2. VDMA, International Technology Roadmap for Photovoltaic (ITRPV), Results 2022 (2023).
3. S. Rein, *Lifetime spectroscopy. A method of defect characterization in silicon for photovoltaic applications* (Springer, Berlin, heidelberg, New York, 2005).
4. R. A. Sinton, A. Cuevas, and M. Stuckings, "Quasi-steady-state photoconductance, a new method for solar cell material and device characterization," in *Conference Record of the Twenty fifth IEEE Photovoltaic Specialists Conference, 1996. Hyatt Regency Crystal City, Washington, D.C., May 13-17, 1996* (Institute of Electrical and Electronics Engineers, New York, N.Y., Piscataway, N.J, 1996), p. 457.
5. T. Trupke, R. A. Bardos, M. C. Schubert, and W. Warta, Appl. Phys. Lett. **89**, 44107 (2006).

6. T. Trupke, B. Mitchell, J. W. Weber, W. McMillan, R. A. Bardos, and R. Kroeze, Energy Procedia **15**, 135 (2012).
7. D. Impera, M. Bivour, B. Michl, M. C. Schubert, and M. Hermle, Energy Procedia **27**, 280 (2012).
8. T. Trupke, R. A. Bardos, and J. Nyhus, "Photoluminescence characterization of silicon wafers and silicon solar cells," in *Proc. 18th Workshop on Crystalline Silicon Solar Cells & Modules* (2008).
9. S. Herlufsen, J. Schmidt, D. Hinken, K. Bothe, and R. Brendel, phys. stat. sol. (RRL) **2**, 245 (2008).
10. T. Trupke, J. Nyhus, and J. Haunschild, phys. stat. sol. (RRL) **5**, 131 (2011).
11. K. Dornich, N. Schüler, B. Berger, and J. R. Niklas, Materials Science and Engineering: B **178**, 676 (2013).

10.4229/EUPVSEC2023/1CV.3.66

40th European Photovoltaic Solar Energy Conference and Exhibition

This presentation was selected by the Sc. Committee of the EU PVSEC 2023 for submission of a full paper to one of the EU PVSEC's collaborating peer-reviewed journals.

OPTICAL DESIGN AND BANDGAP ENGINEERING IN ULTRATHIN SI/GE MULTIPLE QUANTUM WELLS SOLAR CELL

Hosni Meddeb*, Maximilian Götz-Köhler, Kai Gehrke, Martin Vehse
German Aerospace Center (DLR), Institute of Networked Energy Systems, Urban and Residential Technologies, Carl-von-Ossietzky-Str. 15, 26129 Oldenburg, Germany

ABSTRACT: Emerging photovoltaics (PV) enable multifunctional utilization in versatile applications beyond massive electricity production by conventional PV technologies. Novel functional materials and innovative device architectures are sought of to provide tunable solar cells with customized design capabilities. In this work, we present the optical design and bandgap engineering features in ultrathin hydrogenated amorphous Si/Ge multiple quantum well (MQW) solar cell integrated as a photonic nanocavity. Optically, the absorption resonance condition, the absorption edge and the resulting photocurrent outputs are strongly modulated by changing the period thicknesses, the numbers and the positions of QWs inside the deep-subwavelength nanophotonic resonator. Regarding the electronic aspect, the photovoltage and the carrier collection efficiency are strongly thickness-dependent due to the tuning of QW bandgap and QW/barrier band heterojunction band offset under quantum confinement effect. The established design rules along our study can be extended for the development of relevant multifunctional semitransparent PV devices as well as ultrathin PV based on quantum-confined nanostructures from different material systems.

Keywords: emerging PV technologies, ultrathin solar cell, bandgap engineering, nanophotonic design, optoelectronic modelling

1 INTRODUCTION

Unlike conventional solar cells, tunable photovoltaic (PV) technologies can afford diverse functionalities beyond solar energy harvesting, such as aesthetic appearance, visual comfort and thermal management. The adaptation of emerging PV to distributed applications involves adjustable tuning capabilities like bandgap, transparency, color characteristics [1,2]. This requires controlled manipulation of inherent properties of composing materials and customized design of the device architecture. The tuning approaches can be manifested within different regions of solar cell devices including photoactive materials, electrode layers and optical structures [1]. In particular, the specifications of the absorber material in terms of thickness, bandgap and optical properties are substantial for the control of the spectral response, the transparency aspect and the output characteristics of the relevant solar cell device [1,2]. By drastically thinning the absorber layer, significant visible light transmission can be achieved. Ultrathin solar cells feature reduced nanoabsorber thicknesses by several orders of magnitude relative to conventional PV technologies. Interestingly, the thickness reduction in low-dimensional absorber materials induces quantum confinement (QC) effect leading to size-dependent changes in the optoelectronic properties [3]. As one of the most successful integration of quantum confined nanostructures in PV, multiple quantum well (MQW) solar cells feature significant bandgap engineering abilities [4-6]. To enhance the optical absorption and maintain high photocurrent in ultrathin solar cells, efficient light trapping schemes are required. Among these approaches, optical nanocavities or nanophotonic resonators allow strong optical confinement into small spatial dimensions with thicknesses significantly smaller than the incident wavelength, due to multiple light passes and strong interference effects [7-9].

In this work, we report the optical design and bandgap engineering capabilities in ultrathin hydrogenated amorphous Si/Ge MQW solar cell featuring photonic nanocavity. By changing the QW period thicknesses and the positions of QWs inside the deep-subwavelength nanophotonic resonator, the spatial and spectral distributions of the optical field and the local absorption are strongly affected. This leads to a significant modulation of the absorption resonance condition, the absorption edge and the resulting photocurrent outputs. Due to quantum confinement effect, the change of MQW configurations with different individual QW periods alters both the bandgap energy and the band offset at the QW/barrier heterojunctions. This in turn controls the photovoltage as well as the electronic transport and collection in solar cells. These characteristics of different MQW solar cells devices were analyzed via a combination of optoelectronic modelling. It is expected that the established design rules along this study can be generalized for the development of derived multifunctional semitransparent PV devices as well as ultrathin PV based on quantum-confined nanostructures from different material systems.

2 EXPERIMENTAL METHODS

Solar cell devices were on commercial glass substrates with 1 µm-thick Aluminum-doped Zinc Oxide ZnO:Al films (AZO) as front contacts. The structures consisted of hydrogenated amorphous silicon and germanium multilayers with n-i-p superstrate configuration as (n-a-Si:H/i-a-Si:H/ MQW /i-a-Si:H/p-a-Si:H) with thicknesses of (7 nm / 3 nm / MQW / 3 nm / 7 nm). The transport layers for electrons (ETL) and for holes (HTL) as well as the buffer passivation layers on both sides forming the outer n/i and i/p regions were kept similar for all solar cell devices. In the intrinsic MQW region, the alternation between i-a-Si:H quantum barrier (QB) and i-a-Ge:H quantum well (QW) layers was changed to obtain different periodic stratified architectures, while keeping similar total thickness for i-a-Ge:H material in the range of 20 nm. By varying the deposition duration of i-a-Ge:H layers, different QW thicknesses L_{QW} were obtained . Therefore, the following combinations of QW thicknesses and numbers are adopted: (2 × 10 nm), (3 × 6.7 nm), (4 × 5 nm), (5 × 4 nm), (8 × 2.5 nm). The thicknesses of i-a-Si:H layers as QB interlayers separating for all MQW configurations were kept constant with L_{QB} of 1 nm. After the front contact opening through the glass substrate by laser scribing, 200 nm-thick silver (Ag) layers were deposited through shadow mask. as reflective back contacts to define cell core areas of 1 cm^2. For the fabrication of solar cell devices, DC magnetron sputtering technique was used for the front and back electrodes deposition, whereas low-temperature Plasma Enhanced Chemical Vapor Deposition (PECVD) method at 13.56 MHz was employed for the coating of semiconductor functional layers.

To assess the outputs of different solar cell devices, external quantum efficiency (EQE) and illuminated J-V measurements were performed at standard test conditions.

The experimental results of different MQW solar cells devices were analyzed via a combination of optoelectronic modelling approaches using wave-optic-based software package Scout/CODE and drift-diffusion numerical AFORSHET program.

3 RESULTS AND DISCUSSIONS

3.1 Synergy between photonic and quantum confinements in ultrathin Si/Ge MQW solar cells featuring photonic nanocavity

Figure 1-a exhibits the structure of superstrate n-i-p QW solar cell based on hydrogenated amorphous silicon (a-Si:H) and germanium (a-Ge:H) nanostructures integrated in subwavelength photonic nanocavity. The photoactive material based on optical absorber is in the order of 20 nm, while the total thickness (L) of the cell core remains below 50 nm which is significantly smaller than the incident optical wavelength (λ). In such device configuration, lossy a-Si:H/a-Ge:H multilayer with high and comparable refractive and extinction indices is embedded between transparent front AZO film and Ag metallic back reflector as two regions with lower refractive-indices. Such architecture with ultrathin and highly lossy films on opaque reflecting metal surfaces, forms an "asymmetric Fabry-Perot (F-P)" deep-subwavelength absorbing nanocavity. This enables one-dimensional interference of light waves through ultrathin nanoabsorbers and hence, the field build-up in the nanophotonic cavity maximizes the amount of light absorption within the photoactive materials [10].

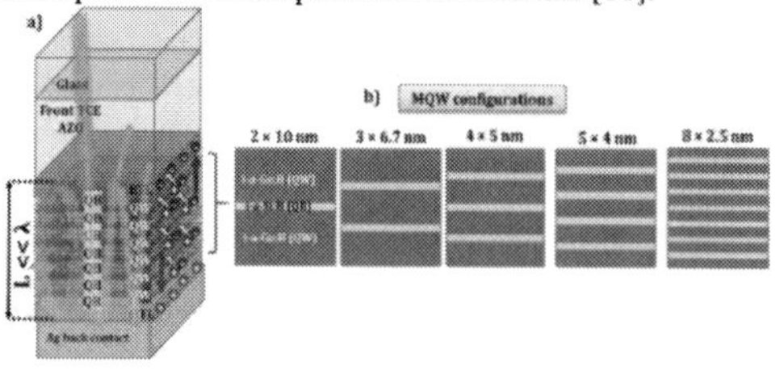

Figure 1: Synergy between optical and electronic confinements in ultrathin Si/Ge MQW solar cells. a) Schematic of n-i-p solar cell device based on a-Si:H (QB) / a-Ge:H (QW) nanostructures embedded as an absorbing optical nanocavity. The multiple passes of light inside the nanophotonic cavity are highlighted by yellow arrows. L is the total thickness of n-i-p multilayer forming the nanocavity length. b) Illustration of different periodic MQW configurations.

Figure 1-b shows different periodic MQW configurations based on a-Si:H (QB)/a-Ge:H (QW) nanostructures. By keeping a constant 1 nm-thick QB interlayer, the following QW periods are adopted: (2 × 10 nm), (3 × 6.7 nm), (4 × 5 nm), (5 × 4 nm), (8 × 2.5 nm). For different periodic MQW configurations, the energy band alignment consists of periods including narrow-bandgap QW sandwiched between two wide-bandgap QBs, and then, duplicated according to the number of QWs. Importantly, the reduction of a-Ge:H QW thickness induces a bandgap widening due an upward shift of the conduction band edge to higher discrete energy level [3]. Therefore, QC effects enable significant bandgap engineering in ultrathin MQW Si/Ge solar cell by altering both the QW bandgap Eg and the conduction band offset ΔE_C at the a-Si:H (QB)/ a-Ge:H (QW) heterojunction [11].

3.2 Experimental characteristics of different ultrathin MQW Si/Ge solar cells

The experimental results of solar cell devices with different MQW configurations are represented in Figure 2. In figure 2-a, it is clearly seen broader EQE are obtained for thicker QW with smaller bandgap. Thereafter, the absorption edge of the EQE spectra is shifted to short wavelength as QW thickness is reduced due to the bandgap widening under QC effect. This leads to a photocurrent loss at long wavelengths in MQW configuration with the thinnest QW (8 × 2.5 nm) compared to thicker QW counterparts with broader EQE profiles. Moreover, by reducing QW width, a slight deviation to short wavelengths can be noticed for the maximum EQE peak which characterize the absorption resonance. Over the spectral absorption range, comparable EQE level of about 60% is reached for all MQW configuration with a slight drop as QW thickness decreases. This is a clear indication for a collective contribution of MQW nanoabsorbers in different configurations to the total photocurrent.

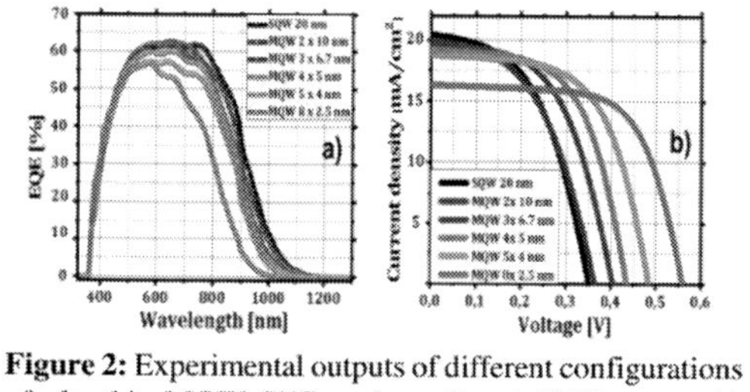

Figure 2: Experimental outputs of different configurations of ultrathin MQW Si/Ge solar cells. a) EQE spectra b) illuminated J-V curves.

Table I: Solar cell characteristics as a function of MQW configuration. J_{sc}: Short-circuit current, V_{oc}: Open-circuit voltage, FF: Fill factor and PCE: Photoconversion efficiency.

Cell configuration	J_{sc} [mA/cm^2]	V_{oc} [mV]	FF [%]	PCE [%]
SQW 20 nm	20.4	350	50	3.6
MQW 2×10 nm	19.9	363	47	3.39
MQW 3×6.7 nm	19.5	404	53	3.97
MQW 4×5 nm	19	435	56	4.45
MQW 5×4 nm	18.6	484	58	5.05
MQW 8×2.5 nm	16.5	560	62	5.5

10.4229/EUPVSEC2023/2AO.1.2
020069-002

Figure 2-b shows the J-V curves under illumination for different configurations of MQW solar cells. The corresponding performance characteristics are mentioned in table I. In accordance with EQE results, the short-circuit current (J_{SC}) values rise progressively from 16.5 mA/cm^2 up to 20.4 mA/cm^2 when QW thickness is increased from 2.5 nm to 20 nm. While, J_{sc} values remain comparable for large QW periods, a remarkable decrease is particularly noticed for the thinnest L_{QW} of 2.5 nm. The J_{sc} values increase in MQW configurations with wider QW periods due to additional absorption of low-energy photons in the lower band-gap QWs. However, an opposite trend is obtained for open-circuit voltage (V_{oc}) and Fill factor (FF) evolution where thinner QWs lead gradual increases. In combination, the best PCE above 5.5% is achieved with MQW (8 × 2.5 nm) corresponding to a relative enhancement of more than 70% relative to lowest PCE of about 3.4% for thicker QW counterparts.

These significant variations in the experimental results of different ultrathin MQW Si/Ge solar cells point out the importance of the understanding of optical design and bandgap engineering in our emerging PV technology. Therefore, in the following sections, the output characteristics of solar cell devices will be explained based on optoelectronic modelling.

3.3 Bandgap engineering in ultrathin Si/Ge MQW solar cells

In order the explain the photovoltage change for different MQW solar cell configurations, band diagrams at the corresponding V_{oc} conditions are computed with numerical simulation in figure 3. It is well known that the supplied V_{oc} voltage is dependent on the absorber bandgap E_g. In fact, the photovoltage of MQW solar cell device is directly linked difference of quasi-Fermi levels between electrons ($E_{F,n}$) and holes ($E_{F,p}$) determining the electrochemical potential of electrons/holes and the occupation of states between the conduction and valence bands. Hence, thinner QWs with larger bandgap yield higher V_{oc} values. This is induced by the large spatial splitting of the quasi-Fermi levels between $E_{F,n}$ (electrons) and levelled down $E_{F,p}$ (holes) due to bandgap widening under QC effects, as can be clearly observed when QW thickness decreases from 20 nm down to 2.5 nm. This implies a reduction of recombination between opposite electrons and holes leading to longer electron lifetime in MQW configurations with thinner QWs [3]. Wider QWs with deeper conduction band prolong the escape time of electrons, leading to higher recombination and lower V_{oc}.

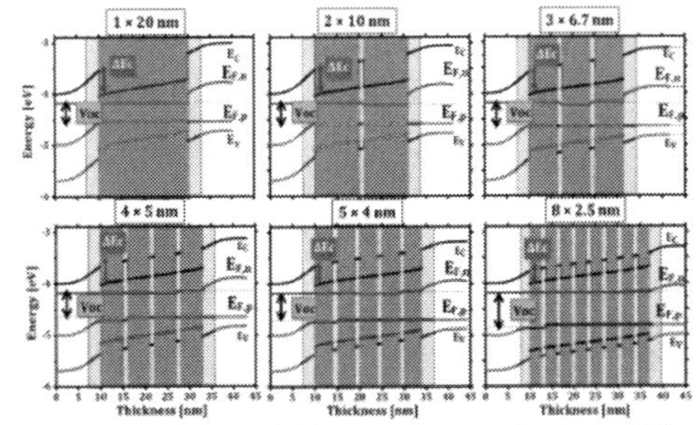

Figure 3: Energy band diagram at open circuit conditions for different n-i-p MQW configurations. The band gap energy of QW E_g and the QW/barrier conduction band offsets ΔE_c are highlighted with black and red arrows respectively.

Regarding the FF variation as a function of MQW configuration, this characteristic parameter is strongly affected by the carrier transport and the resistive losses at the junctions [12-15]. Mainly, the change of the conduction band discontinuity ΔE_c at the a-Si:H (QB)/ a-Ge:H (QW) heterojunctions under QC effect reduces the potential barrier height and improves the carrier collection by tunneling and thermal emission mechanisms [12-15]. This leads to a significant improvement in FF for thinner QW.

3.4 Optical design in ultrathin Si/Ge MQW solar cells

In the following parts, explain the change in photocurrent generation of different ultrathin MQW Si/Ge solar cells by optical modelling. Therefore, we investigate the impact of optical design of nanoabsorbers in terms of MQW configurations on the distribution of the optical field and the local absorption. Across the MQW solar cell device, the optical absorption (A) in different functional layers mainly depends on the amplitude of light electromagnetic field and on the optical refractive indices (n, k) of materials, given by [9]:

$$A\,(z,\lambda) = \frac{2\pi c\,\varepsilon}{\lambda}\,n(z,\lambda) \times k(z,\lambda)\,|E(z,\lambda)|^2 \quad (1)$$

where c is the speed of light, ε is the permittivity of free space and $|E(z,\lambda)|^2$ is the electric field amplitude as a function of depth z and wavelength λ.

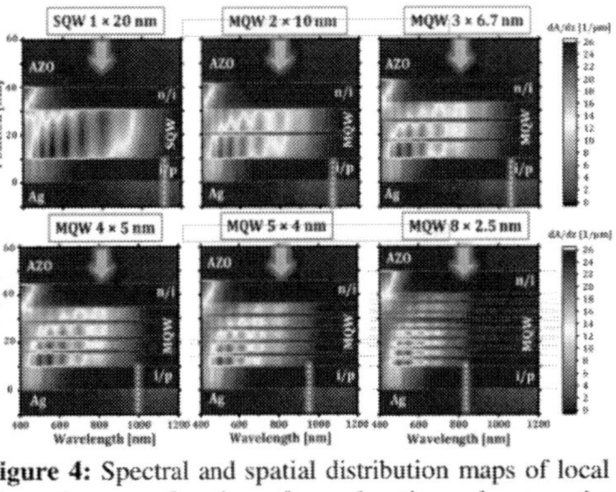

Figure 4: Spectral and spatial distribution maps of local absorption as a function of wavelengths and across the depth of ultrathin Si/Ge MQW solar cells with different configurations. The dashed white lines point out the interfaces between different functional layers. The corresponding absorption thresholds for each a-Ge:H QW thickness is highlighted with dashed yellow lines. Yellow arrows show the direction of the incident light.

Figure 4 depicts the distribution of local absorption as a function of wavelengths and across the depth of ultrathin Si/Ge MQW solar cells with different configurations. The regions with high absorption require a simultaneous combination of strong electric field intensity $|E(\lambda)|^2$ and lossy materials with higher product of refractive indices $n(\lambda) \times k(\lambda)$. In the spatial wavelength-dependent distributions, suitable spectral and spatial overlapping of electric field maxima with medium having high refractive indices in the photoactive MQW regions is desired to foster photocurrent generation [10,11]. In all MQW configurations, more intense and broader absorption in obtained for each a-Ge:H QW region due to higher product of optical coefficients $n(\lambda) \times k(\lambda)$ and lower bandgap energy E_g compared to other a-Si:H buffer or barrier functional layers. This implies dominant contributions of a-Ge:H QW regions with all different thicknesses in the generated photocurrent from the intrinsic photoactive absorbers of different MQW solar cells [10,11].

For the spectral distribution of the mapping plot as a function of light wavelengths, it is observed that the absorption thresholds of periodic MQW devices move towards longer wavelengths as the corresponding thicknesses of individual i-a-Ge:H QWs increase. This red-shift of the turn-on absorption in the local absorption distribution by increasing QWs thickness elucidates the related trend in the experimental EQE results [10,11].

Interestingly, along the depth of solar cell devices, the regions of QWs nanoabsorbers in the vicinity of the Ag back reflectors are characterized by the strongest local absorption in all the QW configurations. These high-absorption regions (red zones) are extended up to 30 nm away from the Ag/p-a-Si:H back edge of the nanophotonic cavity. The enhanced optical absorption is a consequential outcome of the strong resonance in the planar nanocavity due to the electric field build-up and interference effects. Then, a gradual decrease in the absorption intensity of all wavelengths is visualized over different stratified MQW configurations, when QWs position get further away from the back reflective contact. Therefore, lower absorption is noticed in the QW regions adjacent to the front side, especially, for the MQW configuration (8 × 2.5 nm) with thinner QW thickness/period and large number of QWs. The spatial extension of the nanophotonic cavity length maximizes the propagation phase shift, hinders the compensation of nontrivial total phase shift and weakens the light confinement capability by multiple passes recirculation. This in turn limits further improvement in photocurrent. To this end, by changing the thickness of QWs in different MQW configurations with the same total thickness for i-a-Ge:H absorbing material in the range of 20 nm, the values of photocurrent slightly change for large QW period, but, remarkably drop is especially observed for the thinnest L_{QW} of 2.5 nm. The main discrepancy is attributed to the spectral absorption where the high-absorption regions and the absorption threshold are restricted to shorter wavelengths by decreasing QWs thickness due QC effects.

Overall, MQW configuration with the thinnest periods is the best architecture yielding the highest open circuit voltage and fill factor values owing to bandgap engineering. However, the number of QW should be considered in the optical design for efficient enhancement of optical absorption and photogeneration.

4 CONCLUSIONS

In this paper, we study the optical design and the bandgap engineering implications in ultrathin Si/Ge multiple quantum well solar cells featuring photonic nanocavity.

By changing the period thicknesses, in the optical absirbing nanocavity, the absorption resonance condition, the absorption edge and the photocurrent output can be significantly modulated. Based on the optical modeling of the local absorption distributions in such Si/Ge MQW solar cells, the experimental photocurrent results were explained. Mainly, the variation of QW thicknesses and the optical bandgaps in each MQW configuration tunes the turn-on absorption threshold and the spectral extension of high-intensity absorption zones. Whereas, a gradual drop of the absorption intensities was observed as MQW region get away from the reflective back side of the device. Regarding the impact of bandgap engineering, it was demonstrated that the reduction of a-Ge:H QW thickness induces a widening of the QW bandgap E_g and a drop of the conduction band offset ΔE_C at the a-Si:H (QB)/ a-Ge:H (QW) heterojunctions. This leads to both improved Voc and FF electrical parameter of MQW configuration with thinner period. The best efficiency of about 5.5% was achieved for Si/Ge MQW solar cell with the thinnest QW period (8 × 2.5 nm), due to the optimum trade-off between all the optoelectronic characteristic outputs.

It is believed that the presented design rules along this study can be generalized for the optimization of relevant multifunctional semitransparent PV devices as well as ultrathin PV based on other quantum-confined with bandgap tuning capabilities.

5 ACKNOWLEDGMENTS

The authors acknowledge funding from the Energy branch of the German Aerospace Center (DLR). We would also like to thank M. Götz-Köhler, O. Banik, O. Sergeev, C. Koch, N. Osterthun, L. Busch, N. Neugebohrn, D. Berends and all the group colleagues at DLR Institute of Networked Energy Systems for the helpful inputs.

References

[1] H. Meddeb et al., Adv. Energy Mater. (2022); 12(28): 2200713.

[2] O. Almora et al, Adv. Energy Mater. (2023), 13, 2203313.

[3] H. Meddeb et al., Nano Energy (2020); 76, 105048.

[4] K. Barnham et al. J. of Applied Physics (1990); 67(7): 3490–3.

[5] R. France et al Joule (2022); 6(5): 1121–35.

[6] J. Liang et al. Energy Environ. Sci. (2022); 15(1): 296–310.

[7] M. Kats MA et al. Nature Mater (2013); 12(1): 20–4.

[8] M. Kats et al. Laser Photonic Rev. (2016), 10(5):735-749.

[9] K-T Lee et al. Advanced materials (2014); 26(36): 6324–8.

[10] H. Meddeb et al., IEEE J. Photovoltaics (2022); 12(3):1-11.

[11] H. Meddeb et al. Prog Photovolt Res Appl. 2022; 1–13.

[12] S. Ramey et al IEEE Trans Electron Dev. (2003), 50(5):1179-1188.

[13] H. Fujii et al. IEEE J. Photovoltaics (2014); 4(1): 237–43.

[14] K. Toprasertpong et al. Solar Energy Materials and Solar Cells (2018); 174: 146–56,

[15] K. Toprasertpong Phys. Rev. B (2017); 96(7).

2D FERROELECTRIC SEMICONDUCTORS WITH VARIABLE GAP FOR PHOTOVOLTAICS: FUNCTIONALIZED IN_2SE_3 NANOLAYERS

R. Minnings and A.I. Shkrebtii*
Ontario Tech University, Oshawa, ON, Canada
*Corresponding author: Anatoli.Chkrebtii@ontariotech.ca

ABSTRACT: Two-dimensional (2D) ferroelectric films, that possess strong out-of-plane dipole moments, are also affected by quantum confinement, and demonstrate a possibility of electron band gap tuning, are interesting as novel building blocks for photovoltaics, photodetectors, heterostructures, etc. The aim of this research is to investigate the microscopic mechanisms of the band gap tuning in semiconducting α-In_2Se_3 nanolayers and understand the interplay of ferroelectric polarization and quantum confinement on their electron band gap as well as the films' stability. (Unlike, *e.g.*, graphene, the primitive cell of this exotic In_2Se_3 system is five layers thick, forming the so-called quintuple layer (QL) nanofilm). By predicting macroscopic atomic structure-dependent electron and optical properties of In_2Se_3 2D nanofilms, which consist of several differently oriented QLs, we demonstrate that they are promising for applications in photovoltaic-based devices. The hydrogenation of the In_2Se_3 films, investigated with the molecular dynamics (MD), indicates on chemical stability of the material, which is also important for photovoltaics.
Keywords: Thin films, Ferroelectricity, Optical properties, Simulation, Nanoparticles

1 INTRODUCTION

Semiconducting thin films, used as a photoactive layer, are important components in photovoltaic technology (see, *e.g.*, [1] and Refs therein). Such films have to be sufficiently thick to maximize solar photon absorption. In photovoltaic devices, however, also thinner films (in nanoscale, down to one atom thick), are used [2,3]. 2D thin films, often composed of layered materials can offer the possibility of tunable electronic properties, particularly in terms of their bandgap, a critical parameter that determines its electronic and optical properties. Even though such films weakly absorb the incoming light, 2D materials have demonstrated optical and electronic properties, which are promising for next-generation solar cells [4,5]. This includes, for instance, nanoscale absorber layers, films for efficient carrier transport and collection, the top solar cell protective layer, efficient light emitters, photodetectors, and many more. In bulk materials, this bandgap is typically a fixed property, but in 2D thin films, quantum confinement and reduced dimensionality play a significant role.

Pristine and functionalized 2D semiconducting indium selenide α-In_2Se_3 nanofilms have recently attracted attention in both fundamental research (see, e.g., [6-8] and Refs. therein) and applications [3,9,10]. Contrary to the one-atom-thick flat graphene-like materials, the elemental 2D α-In_2Se_3 layer contains five alternating Se and In atoms. It forms a hexagonal cell (in the film's plane) with a quintuple layer (QL) structure with asymmetric Se-In-Se=In=Se tetra-octahedral configuration along the c-axis (perpendicular to the film). Two slightly different ground state structures, the *α* phases, labeled as FE-ZB' and FE-WZ', within the space group R3m, and the same tetra-octahedral bond configuration, (see Fig. 1 and [7]). Microscopic asymmetry of the QL film in the normal direction causes ferroelectricity, thus forming a strong dipole moment with direction, which can be switched by applying a constant external electric field [10]. In our notation, the out-of-plane polarization in Fig. 1 is down (↓) or "D". In Fig. 2 both up (↑) (notation "U") or down (↓) out-of-plane polarizations are shown.

When several QLs form nanofilms, each of the QLs is bonded by a weak van der Waals (vdW) force, the electron band gap of the nanolayers depends strongly not only on its thickness due to quantum confinement, but also, as we show, on the dipole orientation of each layer as well. Important to stress that the most common film growth techniques can be applied to deposit multi-QL In_2Se_3 nanolayers [11-13]. Additionally, hydrogenation is a versatile tool in photovoltaic technology that allows for the functionalization of grown nanofilms. Indeed, 2D semiconducting α-In_2Se_3 nanofilm's integration with a hydrogenated Si(111) surface, for solar cell applications [9] has been demonstrated.

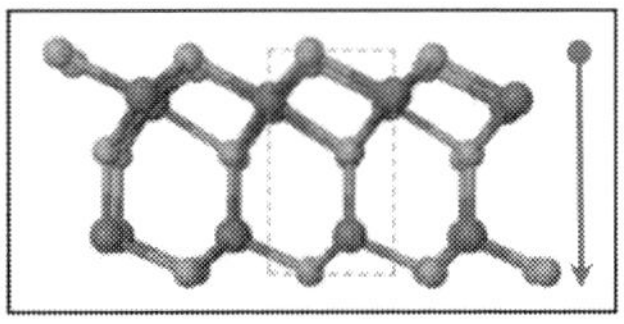

Figure 1: One quintuple layer (QL) of α-In_2Se_3 quasi two-dimensional nanofilm, side view. Yellow spheres correspond to Se atoms, brown spheres are In atoms. The blue dashed line outlines the primitive 2D unit cell. The figure clearly demonstrates that there is no symmetry plane along the QL, which is the source of out-of-plane polarization, as indicated by a blue arrow.

2 THE GOALS OF THE RESEARCH

Following the approach, we developed in [14] and [15], we further investigate from the first principles within Density Functional Theory (DFT) the impact of stacked 3QLs, and 4QLs with all possible combinations of out-of-plane polarization, either up (↑) or down (↓). The band structure and dielectric response of the 2D In_2Se_3 3QL and 4QL nanomaterials were modelled. All possible QLs combinations of out-of-plane polarization of each individual QL layer in the stack were considered, within an ABC stacking arrangement of each layer. The In_2Se_3 nanolayers are constructed from vertically stacked semiconducting quintuple layers (QLs), which also exhibit

10.4229/EUPVSEC2023/2AO.1.6
020070-001

spontaneous out-of-plane polarization (as well as in-plane polarization, not considered here).

We addressed the above-mentioned features theoretically, focusing on the interpretation of the simulated results for the energy gap and optical response in terms of the microscopic charge redistribution, related bonding, and their use in noninvasive linear optical characterization. In particular, we simulated from first principles, the structural, dynamic, electron, and optical properties of 2D pristine In_2Se_3-based nanolayers: one QL, trilayer, and four-QL film with different dipole moment orientations. Such a "numerical experiment" allowed us to theoretically interpret the films' optical response in terms of the microscopic atomic geometry, number of QLs, their vertical dipole orientation, and interlayer bonding. Theoretical optical spectra agree well with the experimental results available (see [6]).

We also predicted that the hydrogenation (at one side, both at the top and bottom of QL or several QLs) substantially modifies the charge density and related microscopic geometry, leading to the semiconducting or metallic behavior of the functionalized films. However, in [15] our "numerical hydrogenation experiment" with complete hydrogen coverage, indicated a possibility of top layer etching as well as forming two separate semiconducting InSe-H and InSe2-H nanofilms. During the next steps of the hydrogenation study, we carried out the finite temperature molecular dynamics (MD) and the total energy minimization. This indicated the instability of the hydrogenated In_2Se_3 system at nonzero temperature.

We previously carried out an experiment that demonstrated that α-In_2Se_3 possesses room-temperature out-of-plane ferroelectricity down to a thickness of a few layers [6], and this ferroelectricity is expected to persist down to a single QL. However, the presence of a strong dipole moment favors the formation of 2D domains with opposite dipole orientations which reduces the electrostatic energy; for instance, the calculated potential barrier for changing the dipole orientation (that is shift between Up and Down structures) is approximately only 0.07 eV [7]. Applying the electric field in the normal direction allows to reversibly switch preferential domain orientation.

The In_2Se_3 2D film also exhibits a thickness-dependent band gap [17], due to the quantum confinement, also allows for tuning of its dielectric function, and thus, optical properties [18]. This enables several useful optoelectronic and photovoltaic device properties.

In more detail, the basic In_2Se_3 layer consists of five alternating Se and In atoms that form the so-called quintuple layer (QL), as shown in Figure 1. Numerous structural variations of the QLs are possible, with the two most energetically stable layered structures belonging to the so-called alpha phase [7]. Both are semiconducting with an energy bandgap in the optical range.

The QLs can be stacked on top of each other (via several film deposition techniques) and we have previously modelled [14] a few QLs, stacked in an ABC stacking arrangement. We showed that when the stacked QL layers are arranged so that the net out-of-plane polarisation is zero, the band gap is not reduced as much relative to the bandgap for a single QL layer, as for the case when the stacked QL layers have net out of plane polarisation.

In this paper, we focus on the direct bandgap for all possible combinations of polarisation of each QL, in 3QL and 4QL this film structures. This has applications for band gap tuning in potential applications.

Since the experimentally observed symmetry for single and multiple QLs at room temperature is consistent with the R3m space group, we also discuss the optical characterization of α-In_2Se_3 as a promising ferroelectric material. Knowledge of how the spontaneous polarization changes with the thickness and microscopic arrangement of this material is crucial for engineering α-In_2Se_3 for useful device applications where band gap tuning can play an important role.

Previously [15] we discussed the hydrogenation of the In_2Se_3 QL. We showed that hydrogenating both sides of the single QL resulted in the single QL separating into two separate compounds In-Se-H and In-Se_2-H respectively. We also mentioned that we encountered instability in the molecular dynamics modelling of these and other hydrogenated compounds of In_2Se_3. In this paper, we have performed extensive finite temperature molecular dynamics for hydrogenated In_2Se_3 QL, and a double layer 2QL at approximately 700K. In this MD run, the bottom layer Se atoms were fixed in position and the top layer Se atoms each were bonded to a single H atom.

3. COMPUTATIONS AND EXPERIMENT

Two ground-state pristine structures of α-In_2Se_3 single QL with different polarization directions are used in the simulations, (see Figure 2) for zincblende-like (ZB') QL. These structurally optimized basic single QLs are used both as is, and to build stacked QLs in a hexagonal ABC stacking arrangement. To obtain a stacked arrangement (*i.e.,* multiple QL films) one QL is placed above the other (which coincides with the *c*-axis direction in the In_2Se_3 layered crystal, see Figure 2) with a pre-optimized vertical separation of 2.95Å, and a shift is made in the (*a,b* axis plane), to obtain the ABC stacking arrangement. In Figure 2 (left), the QL net polarization is down (negative c-axis direction). That polarisation can be reversed by taking the mirror image of the single QL in the *a,b* axis plane. In this way, stacked QLs can be constructed with each QL layer having an out-of-plane polarisation up (parallel to the c-axis) or down (anti-parallel to the c-axis), as desired.

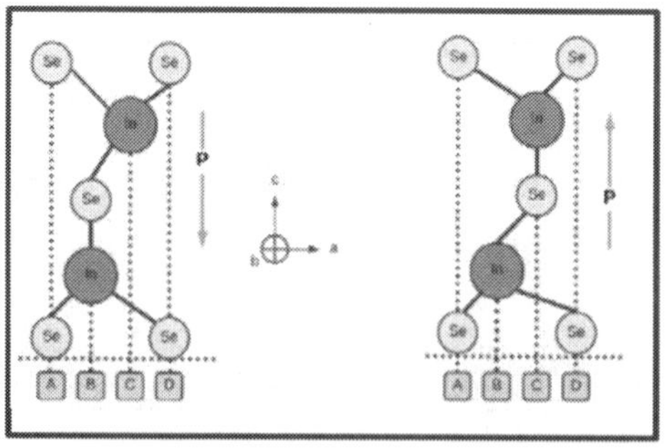

Figure 2: Two zincblende-like (ZB') α-In_2Se_3 atomic arrangements with opposite dipole moments. The first three layers from the bottom-up (at the left) or top-down (at the right) are identical. The grey arrows indicate the net polarisation direction for each quintuple layer.

10.4229/EUPVSEC2023/2AO.1.6
020070-002

Please note that a bi-layer with Up-Down polarisation (in a CVD process this would be laying down the first layer with Up polarisation) is not the same structure as a bi-layer with Down-Up polarisation (in a CVD process this would be laying down the first layer with Down polarisation). Think of it as $[U_1U_2][D_1D_2]$ in one case, where $[U_1U_2]$ is one QL with U polarisation where U_1 and U_2 represent a partial structure in the QL that combined male the full QL.

As previously mentioned, we started with an ABC QL stacking arrangement with an inter-QL separation of 2.95 Å. After the optimisation of the structure with a vdW correction, the optimized inter-QL distance was reduced to 2.6 Å. This separation is slightly different for different stacked configurations, which is within the expected numerical accuracy of the calculations, while the atomic structure of the stacked multi-QL systems, does not depend on the dipole orientation of the individual QLs within the stack. We performed such optimization for stacked 2 QLs, 3 QLs, 4 QLs, and the bulk. In all cases, different combinations of out-of-plane polarisation (c-axis direction, see Figure 2 for axis orientation) were used in each layer to obtain either net out-of-plane polarisation or zero polarisation for the stacked QL structure (note: net zero polarisation is only possible for an even number of QLs). For instance, for a "bilayer", that is a stacked structure with two QLs, four configurations are possible with different arrangements of the out-of-plane dipole orientation: up-up ↑↑, down-down ↓↓, up-down ↑↓, and down-up ↓↑. The first two up-up and down-down configurations are identical energetically.

It should be noted that DFT-LDA underestimates the band gap compared to the experimental values. This is a well-known problem and there are techniques to fix this, including the accurate many-particle GW formalism (see, *e.g.*, [20]) or the simple but reasonably accurate so-called "scissors operator" approach, which rigidly shifts the energy of the empty surface states up by a well-defined amount. It is important to note that despite the underestimation of the energy gap, DFT-LDA produces the correct electron wave functions for both filled and empty states. Therefore, we show here only DFT-LDA results, while the many-particle effects will be considered in the next phase of the research.

Experimentally, the In_2Se_3 samples were grown using Physical Vapor Deposition (PVD), as described in [6]. Images of the In_2Se_3 flakes grown, taken from [6], are shown in Figure 3.

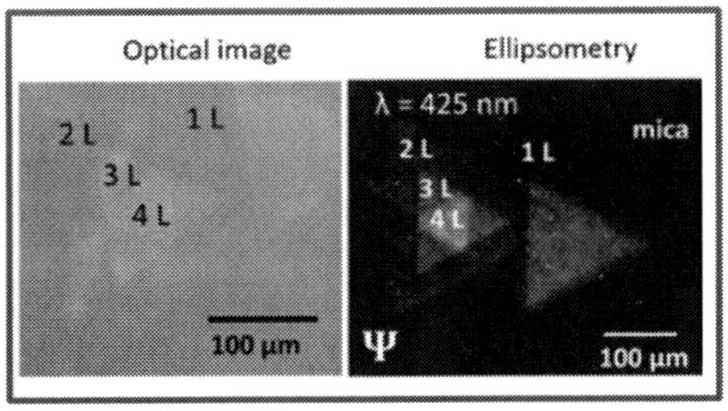

Figure 3: Optical and ellipsometry images of In_2Se_3 flakes, consisting of one, two, three, and four QLs.

4 RESULTS AND DISCUSSION

We have modelled from first principles the bandgaps of all possible combinations of 3QL and 4QL layers made from U or D polarizations, within an ABC stacking arrangement. This work is a continuation of [14]. The QL itself is modelled with two different structures, within the so-called α phases, labeled as FE-ZB' and FE-WZ', members of the space group R3m, and the same Se-In-Se=In=Se tetra-octahedral configuration along the c-axis in one QL layer, Figure 1. The stacked QLs interact between each QL layer with a weak Van der Walls (vdW) force. The stacking configuration used is ABC QL stacking with an optimized inter-quintuple layer separation of approximately 2.6 Å. This makes the theoretically calculated QL thickness of 9 Å in good agreement with the AFM data of 8.4 Å from [21]. The stacking configuration (ABC) and QL interlayer separation do not impact the net polarization of the resulting film, that is controlled by the sum of the net polarization of the individual QL layers in the stacking arrangement.

To have a consistent picture of the energy gap modification from one QL to four QL films, we compiled below in Table 1 from our previous results [14] for a select different number of stacked QLs.

Number of QL's	Net Polarization (Y/N)	Bandgap (eV)	Bandgap reduction relative to one QL
1	Y	1.12	N/A
2	Y	0.38	65.9%
2	N	0.81	27.5%
3	Y	0.41	63.4%
3	N (*partial*)	0.69	38.2%
4	Y	0.29	74.4%
4	N	0.59	46.6%
Bulk	N (*partial*)	0.31	72.1%

Table 1. The DFT-LDA bandgap, obtained from the calculated electron band structure, for the films with 1QL, 2 QLs, 3QLs, and 4QLs, and bulk, with net or zero-net out-of-plane polarisation. Net polarization: "Y" means that all layers have the same directions of the out-of-plane dipoles, "N" indicates on complete or at least partial compensation of the dipoles.

As mentioned earlier in this paper, a UD polarization cannot be made into a DU polarization by a 180-degree rotation. There are therefore many combinations of polarizations that are possible in 3QL and 4QL structures. In this paper, we explore the bandgaps for all of these combinations as described below. We calculated the optical response and bandgap for all possible multilayer 3QL and 4QL ABC stacking arrangements where each layer can have an Up or Down polarisation. In this paper, we focus on the band structure. For 3QLs for each of the WZ' and ZB' types there are 8 possible combinations of Up or Down polarisations in the 3 layers. For 4QLs there are 16 combinations of Up or Down polarisations, for each of the WZ' and ZB' types. In total, we have modelled 48 Up/Down polarisation combinations in 3QL and 4QL ABC stacked structures for WZ' and ZB'. We have not considered a mixture of WZ' and ZB' in the same stacked QL in this paper. Within a fixed number of QLs it is, therefore, possible to vary the bandgap by the combination of Up and Down polarisations in that structure.

Here we did not consider calculations with GW self-corrections for the optical response due to the computational cost. A correction [16] to the optical response and band gap has not been applied to the results (see [21] as an example) as the intent here is to show the trend in the value of the bandgap with different numbers of stacked QLs and the presence of vertical net or the absence of net out-of-plane polarisation in the stack. Therefore, these theoretical results, presented here, have no quasiparticle correction applied to them.

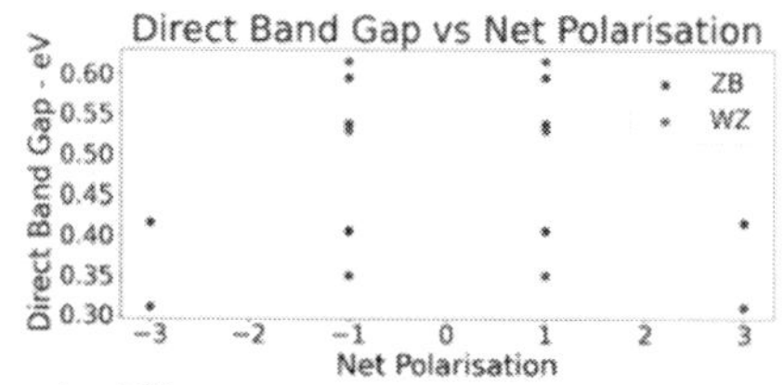

a) 3QL

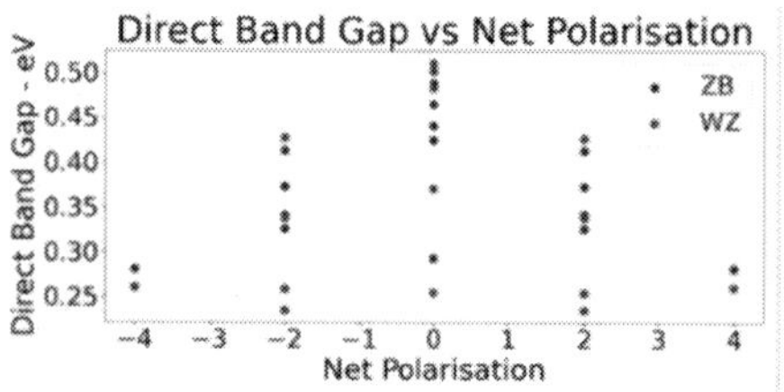

b) 4QL

Figure 4: The direct and indirect bandgap for WZ (red) and ZB (blue) structures. 3QL (a) and 4QL (b) show stacked layers with all possible combinations of Up (U) and Down (D) polarisations, ABC stacking arrangement.

The direct band gap *vs.* net polarisation in Figure 4 shows some interesting features. In Table 2 we only show the case for 4QLs where the net polarisation is exactly zero. For both WZ' and ZB' there appears to be a significant difference in the direct bandgap depending upon the up/down polarisation layer positioning in the 4QL. This difference is not caused by quantum confinement effects, as they are all 4QLs and within a possible small difference are the same. They all have net zero polarisation, so, on the face of it the difference in band gap should not be down to net polarisation effects. However, the lowest band gap for both WZ' and ZB' has the polarisation combination of UUDD (if using CVD to create the layers, they would be layered in the order UUDD), and the highest band gap with the polarization combination of DUDU. Given that the source of the electric field due to polarisation is from the dipole moments of each layer, a local electric field within the QLs will vary with location and layer due to these effective four electric field sources. An UUDD combination then could have stronger E -E-fields at certain points within the QL's than DUDU where the E-field may be canceled more generally within the structure due to the interleaving of different polarisations layer by layer.

QL Polarisations	Direct Band Gap - eV	Net Polarisation
ZB_D_U_U_D	0.4412	0
ZB_U_D_U_D	0.4662	0
ZB_U_U_D_D	0.2937	0
WZ_D_U_U_D	0.5033	0
WZ_U_D_U_D	0.4842	0
WZ_U_U_D_D	0.2558	0
ZB_D_D_U_U	0.4251	0
ZB_U_D_D_U	0.4413	0
ZB_D_U_D_U	0.4893	0
WZ_D_D_U_U	0.3712	0
WZ_U_D_D_U	0.5027	0
WZ_D_U_D_U	0.5114	0

Table 2: The electron band gap for the case for 4QLs only where the net polarisation is exactly zero. For both WZ' and ZB' there appears to be a significant difference in the direct bandgap depending upon the up/down polarisation.

Interestingly though the 4QL-DDUU combination has a significantly larger band gap than 4QL-UUDD, as shown in Table 3.

QL Polarisations	Direct Band Gap - eV
WZ-DU	0.7842
WZ-UD	0.7828
WZ-UU	0.3340
WZ-DD	0.3341
ZB-DU	0.7925
ZB-UD	0.7667
ZB-UU	0.4046
ZB-DD	

Table 3: Bandgaps for net zero non-zero polarisations in 2QL films.

If we look at the bi-layer (2QL), the results are in Table 3) we see that for WZ' or ZB' the UD and DU configurations have very similar bandgaps. Another item to note is that the WZ' structures have a greater range of band gaps than the ZB' in the 3QL and 4QL combinations. There is some interesting physics in the structure of the 4QLs impacting the direct bandgap. If we have a two-layer quintuple layer (QL) with opposite polarizations in each layer, the resulting electric field in the material is not constant and will vary with the z-direction (height): Each layer, due to its polarization, will create an electric field pointing in a certain direction. When these two layers are stacked on top of each other with opposite polarizations (say Up in the first layer and Down in the second), the electric field in the material will differ.

10.4229/EUPVSEC2023/2AO.1.6
020070-004

However, it is important to note that the specifics can be quite complex and depend on a range of factors such as the exact geometry, distance between the layers, the magnitude of the dipole moments, and the relative alignment of the dipoles. Therefore, a detailed computational model or experimental measurements would likely be necessary to accurately predict the electric field distribution in such a system.

Reduction of the optical gap for the parallel-aligned out-of-plane dipole moments can be attributed to a strong Franz-Keldysh-like effect, due to an internal dipole electric field (of the order of $5 \cdot 10^8$ V/m).

Finally, in [15] we looked at various hydrogenation structures of the In_2Se_3 QL. We found that during energy relaxation a double-sided Hydrogenated 1QL separates into two hydrogenated structures. In effect, double-sided hydrogenation appears to etch the structure. However, under final temperature molecular dynamics at various temperatures, the resulting two structures were found to be unstable.

Here, we hydrogenated just a single side of a 1QL and a 2QL structure, and ran molecular dynamics at 700K with the bottom Se atoms fixed and all the top Se with one H. During the molecular dynamics run in both cases the H atoms were stripped off the 1QL and the 2QL. (For the 2QL there was one H atom of four left bonded to a Se atom. "Manually removing" the last H atom from the Se and placing it near to one of the other stripped of H atoms and then relaxing shows the total energy of the system drop of about 1 Ry. Thus, unlike, *e.g.*, graphene films, the In_2Se_3 QL systems demonstrate chemical stability with respect to highly reactive atomic hydrogen.

4 CONCLUSIONS

We simulated atomic geometry, electron bands, and the linear optical response for pristine quasi-2D In_2Se_3 quintuple layer (QL) nanomaterials, as well as several QLs systems, bonded through a weak van der Waals (vdW) interaction. Within the DFT LDA approach, 48 direct band gaps for different polarization arrangements of ZB and WZ 3QL and 4QL thin films in an ABC stacking arrangement have been simulated. We also showed that even with the same net polarization of zero the 4QL combinations exhibit significantly different bandgaps.

Multi-QL films, which have been grown using PVD [11], demonstrated a good agreement with the theory for the linear optical response for several stacked In_2Se_3 QLs as well as for the bulk material. It was proven both theoretically and experimentally that due to the confinement effect, the optical gap can be tuned (reduced) when the QLs number increases.

We demonstrated theoretically that due to the ferroelectric nature of the In_2Se_3 nano-layers, the optical gap can further be reversibly tuned by the external field. For photovoltaics, the incorporated In_2Se_3 layers can be used to modify, e.g., not only the bandgap but also semiconductor surface conductivity by switching the films' dipole orientations.

The hydrogenation simulations using the finite temperature molecular dynamic, and the total energy minimization indicate that In_2Se_3 nanofilms are chemically stable, which is another attractive feature for application in photovoltaics.

5 REFERENCES

[1] T. D. Lee and A. U. Ebong, Renewable and Sustainable Energy Reviews 70, 1286 (2017).

[2] S. Das, D. Pandey, J. Thomas, and T. Roy, Adv Mater 31, 1802722 (2019).

[3] H. Li, W. Ren, G. Wang, L. Gao, R. Peng, H. Li, P. Zhang, M. Shafa, X. Tong, S. Luo, et al, J. Phys. D 49, 145108 (2016).

[4] E. Singh, K. S. Kim, G. Y. Yeom, and H. S. Nalwa, ACS Appl. Mater. Interfaces 9, 3223 (2017).

[5] D. Jariwala, A. R. Davoyan, J. Wong, and H. A. Atwater, ACS Photonics 4, 2962 (2017).

[6] Y. Cho, S. M. Anderson, B. S. Mendoza, S. Okano, N. Arzate, A. I. Shkrebtii, D. Wu, K. Lai, R. Carriles, D. R. T. Zahn, et al, Phys. Rev. Mater. 6, 034006 (2022).

[7] W. Ding, J. Zhu, Z. Wang, Y. Gao, D. Xiao, Y. Gu, Z., et all. Nature Communications 8, 14956 (2017).

[8] F. Lyu, Y. Sun, Q. Yang, B. Tang, M. Li, Z. Li, M. Sun, P. Gao, L. Ye, and Q. Chen, Nanotechnology 31, 315711 (2020).

[9] J. Wang, H. Yu, C. Hou, and J. Zhang, ACS Appl. Mater. Interfaces 12, 26543 (2020).

[10] D. Dutta, S. Mukherjee, M. Uzhansky, and E. Koren, npj 2D Materials and Applications 5, 81 (2021).

[11] Y. Cho, S. M. Anderson, B. S. Mendoza, S. Okano, N. Arzate, A. I. Shkrebtii, D. Wu, K. Lai, R. Carriles, D. R. T. Zahn, et al, Phys.Rev.Materials 6, 034006 (2022).

[12] V. D. Botcha, Y. Hong, Z. Huang, Z. Li, Q. Liu, J. Wu, Y. Lu, and X. Liu, J. Alloys Compounds 773, 698 (2019).

[13] M. Küpers, P. M. Konze, A. Meledin, J. Mayer, U. Englert, M. Wuttig, and R. Dronskowski, Inorg. Chem. 57, 11775 (2018).

[14] R. Minnings, A. I. Shkrebtii, G. Perinparajah, et al, Proceedings 37th European Photovoltaic Solar Energy Conference, Sep. 6, 2020, p. 51.

[15] A. I. S. R. Minnings G. Perinparajah, Proceedings 38th European Photovoltaic Solar Energy Conference and Exhibition, Sep 30 2021, p. 80.

[16] Y. Zhou, D. Wu, Y. Zhu, Y. Cho, Q. He, X. Yang, K. Herrera, Z. Chu, Y. Han, M. C. Downer, et al, Nano Letters 17, 5508 (2017).

[17] J. Quereda, R. Biele, G. Rubio-Bollinger, N. Agraït, R. D'Agosta, and A. Castellanos-Gomez, Advanced Optical Materials 4, 1939 (2016).

[18] B. Hwang, J. Hwang, J. K. Yoon, S. Lim, S. Kim, M. Lee, J. H. Kwon, H. Baek, D. Sung, G. Kim, et al, Scientific Reports 6, 31160 (2016).

[19] P. Giannozzi, O. Andreussi, T. Brumme, O. Bunau, M. B. Nardelli, M. Calandra, R. Car, C. Cavazzoni, D. Ceresoli, M. Cococcioni, et al, Journal of Physics: Condensed Matter 29, 465901 (2017).

[20] P. Cudazzo, C. Attaccalite, I. V. Tokatly, and A. Rubio, Phys.Rev.Lett. 104, 226804 (2010).

[21] Y. Cho, S. M. Anderson, B. S. Mendoza, S. Okano, R. Carriles, N. Arzate, A. I. Shkrebtii, D. Wu, K. Lai, D. R. T. Zahn, et al, "Opto-electronic properties of alpha-In2Se3: single-layer to bulk" https://arxiv.org/abs/1911.01784 .

10.4229/EUPVSEC2023/2AO.1.6
020070-005

HOT-CARRIER EFFECT AND INVERTED PYRAMID STRUCTURE TO ACHIEVE 4% EFFICIENCY FOR SOLAR ENERGY BELOW SI BANDGAP FROM METAL-SILICON JUNCTION SOLAR CELL

Hsin-Ting Lin[1], Chung-Han Chang[2], Ching-Fuh Lin[1, 2*]

1. Graduate school of Advanced Technology Program in Semiconductor Devices, Materials, and Hetero-integration, National Taiwan University, No. 1, Sec. 4, Roosevelt Road, Taipei, Taiwan, 10617 R.O.C.
2. Graduate Institute of Photonics and Optoelectronics, National Taiwan University, No. 1, Sec. 4, Roosevelt Road, Taipei 106319, Taiwan

Email:
1. R10k43005@ntu.edu.tw
2. 123564Haley@gmail.com
*lincf@ntu.edu.tw

ABSTRACT: Traditional monocrystalline silicon solar cells can only absorb light with wavelength under 1.1 μm in the solar spectrum. The rest of solar spectrum cannot be absorbed and is wasted. In this study, we propose a metal/silicon junction solar cell to utilize the hot carrier effect to break through the problem that single crystal silicon solar cells cannot absorb light with wavelength above 1.1 μm. The metal thin film is used to absorb photon energy to excite hot carriers across Schottky barrier, generate photocurrent, and maximize the use of sunlight spectrum. In addition, we successfully combine inverted pyramid structures on the silicon with the metal/silicon junction solar cell to enhance the light absorption, thus generating more hot carriers and higher photocurrent, and finally achieving the photovoltaic conversion efficiency of 4.779%.

Keywords: Ultra-thin metal films, Metal-semiconductor junction, Schottky energy barrier, Hot carrier effect, Mid-infrared light absorption.

1 INTRODUCTION

For the AM 1.5 solar spectrum, the main radiation energy is concentrated in wavelength between 0.15 and 4 μm. Ultraviolet regime, visible regime and infrared regime account for 5%, 52% and 43% respectively. Since the photon energy needs to be higher than the band gap of the semiconductor material in order to be absorbed, it determines the minimum energy (hv) of photons that can be absorbed by the material, which is equal to the maximum wavelength. Typically, the energy band gap ranges from 1 to 1.7 eV, which is equal to wavelength of 730-1240 nm. In other words, the range is from visible regime to near-infrared regime. Currently, absorption spectrum of commercial solar cells is mostly between 300-1100 nm. In the solar spectrum, the radiation energy below wavelength of 1.1 μm accounts for 80.43%, while the remaining energy accounts for 19.57% is wasted.

Therefore, in order to effectively utilize infrared light with wavelength above 1100 nm, some materials with energy band gap lower than 1.1 eV are used in solar cells as absorber layers. III-V and II-VI compound materials, such as GaSb, InAs, CIS or InGaAsSb compounds are mostly used. Although these materials have matured epitaxial technology and are used in high-efficiency tandem cells, there are still several drawbacks, such as the high cost of certain elemental materials and the high difficulty of epitaxial and fabrication processes, which are not conducive to large-scale production. On the other hand, almost all metals can absorb infrared light, but since metals have no band gap, there are still many challenges to overcome.

Here we overcome those challenges and make the inverted pyramid structure (IPS) on the silicon substrate to further improve the absorption and increase the efficiency of photovoltaic conversion efficiency of metal/silicon junction solar cell.

2 EXPERIMENTS

2.1 Inverted pyramid structure on silicon substrate

N-type double-side polished silicon wafer with a doping level of approximately 10^{15} cm^{-3} and a thickness of around 600 μm were employed in the fabrication of IPS. To ensure the wafers free from organic contaminants such as dust particles, grease, or silica gel, as well as the oxidation layer, a cleaning process involving buffer oxide etching (BOE) was employed.

Plasma-enhanced chemical vapor deposition (PECVD) was used to deposit a 600 nm thick silicon dioxide (SiO_2) layer on both sides of the Si wafer. Subsequently, using a photolithography process, the desired pattern for the inverted pyramid array was defined. The IPS pattern was generated by exposing and developing ENPI-307 negative photoresist. Following this, a reactive-ion etching (RIE) process was utilized to create the etching mask necessary for the subsequent wet etching process.

The samples were then immersed in a potassium hydroxide (KOH) solution to facilitate the formation of inverted pyramid shapes through anisotropic wet etching. Finally, any remaining SiO_2 was removed using BOE, completing the fabrication process of the IPS.

2.2 Device fabrication

To fabricate the metal-silicon junction solar cell, silicon wafers were initially cut into 2.5 cm x 2.5 cm squares and subjected to a standard cleaning process. Subsequently, a 9.5 nm thick layer of silver was deposited onto IPS silicon substrate through thermal evaporation, serving as the absorption layer and forming a metal-semiconductor junction structure. Transparent silver fingers were then evaporated onto this metal layer to act as the transparent electrode. On the reverse side of the substrate, aluminum (Al) was deposited to establish an ohmic contact with the silicon, serving as the back electrode, thus completing the metal-silicon junction solar cell.

Current density and voltage curves were obtained using a solar simulator connected to a Keithley 2400 under AM 1.5G illumination conditions with an intensity of 100 mW/cm2. Additionally, to evaluate the photovoltaic performance of the metal-silicon junction solar cell in the infrared spectral range, an optical filter was integrated into the measurement setup to block wavelengths below 1100 nm within the solar spectrum. The filter's transmittance was measured using a spectrophotometer, and the remaining transmitted radiation intensity was assessed using a power meter.

3 RESULTS AND DISCUSSION

We use the metal thin film as the absorber layer, which can absorb infrared light with wavelength above 1100 nm. Because of the lack of band gap in the metals, the excitation of hot carriers interacts with other carriers or phonons and redistribute the energy [1]. The electrons leaping to the high energy level will fall back to the low energy levels in less than one picosecond [2]. As a result, it is difficult to collect those electrons as photocurrents. In order to export the hot carriers quickly, the metal film should be thin enough to allow the hot carriers to pass through the metal layer before cooling and then cross the Schottky barrier between the metal and silicon junction to generate the photocurrent, as illustrated in Fig 1.

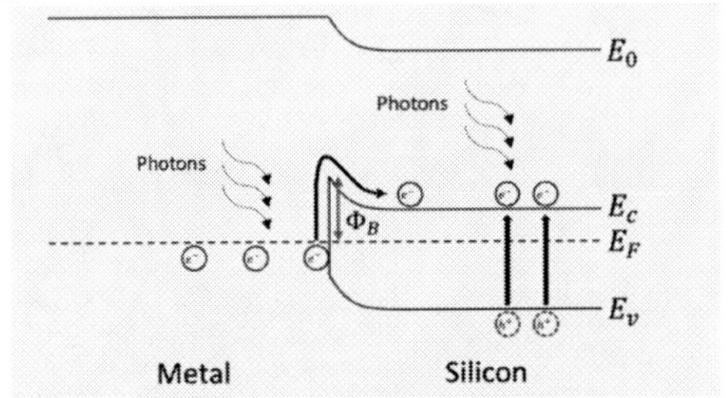

Figure 1: Hot carriers are excited after receiving photons and cross Schottky barrier (Φ_B) to generate photocurrent.

In this study, to ensure that the incident light received by the metal/silicon junction solar cell is in the wavelength that cannot be absorbed by the monocrystalline silicon solar cells, we used a 600 μm thick silicon wafer as a filter and measured the transmission spectrum by spectrophotometer to confirm that the wavelength below 1100 nm is effectively blocked, and the transmission rate above 1100 nm is 55 %. The residual light intensity measured via power meter is 13.85 mW/cm^2.

However, if the metal is too thin, large amount of infrared light will penetrate and the absorption rate will only be about 10%-30%. To overcome this, continuing our previous research [3], we combined with our group's research [4] and optimized the absorption of metal/silicon junction solar cells on silicon substrates with IPS. In Fig 2, confining the incident light in inverted pyramid cavity and thus enhancing light capture capability. IPS successfully increases the absorption rate of infrared light with wavelength above 1100 nm.

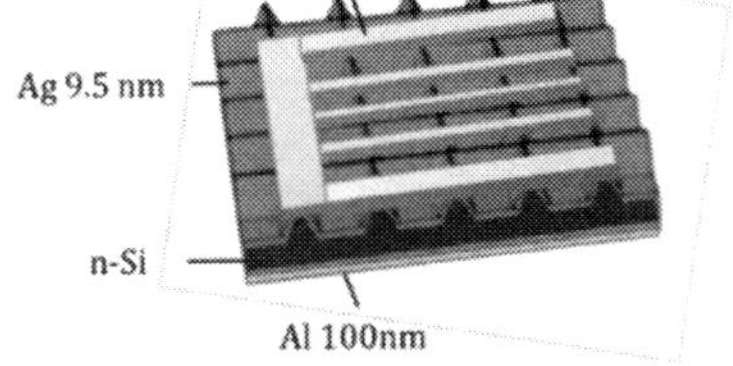

Figure 2: Structure of metal/silicon junction solar cell with IPS.

As shown in Fig 3, the photovoltaic conversion efficiency of the metal/silicon junction solar cell is 2.456% with a smooth-surface silicon substrate and 3.296% with a rough-surface silicon substrate, and 4.779% with IPS. The photovoltaic conversion efficiency is higher than metal/silicon junction solar cell with smooth surface silicon substrate.

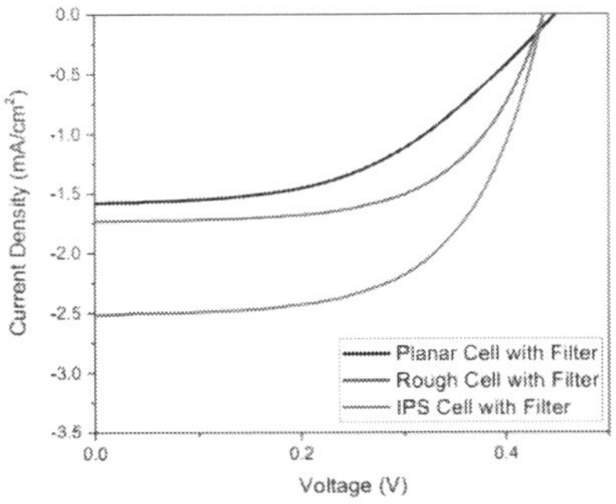

Device	V_{oc} (V)	J_{sc} (mA/cm^2)	R_s (Ohm)	R_{sh} (Ohm)	Fill Factor (%)	Cell eff. (%)
IPS Cell under Filter	0.4375	2.51	22.2574	3041.9006	60.17	4.779
Rough Cell under Filter	0.4374	1.7	22.2574	3041.9006	60.17	3.296
Planar Cell under Filter	0.4488	1.6	63.0699	2166.3491	48.09	2.456

Figure 3: Comparison of Photovoltaic characteristics of metal/silicon junction solar cell with smooth-surface silicon substrates, rough-surface silicon substrates, and IPS silicon substrates in infrared regime (>1.1 μm)

4 CONCLUSIONS

In this study, we achieved silicon-based solar cells with higher infrared conversion performance and a breakthrough in the collection and conversion of hot carriers into photocurrent by forming an Schottky barrier at the interface of metal/semiconductor. Furthermore, the photovoltaic conversion effieciency of metal/silicon junction solar cells can be increased by creating an inverted pyramide structure on the silicon. For infrared regime above 1.1 μm, the cell efficiency was measured at 4.779%, which is much higher than the photovoltaic conversion efficiency of metal/silicon junction solar cell with smooth-surface silicon substrate. There are still many parameters that can be investigated. Hence, the infrared conversion efficiency of the metal/silicon junction solar cell is expected to be even higher.

5 ACKNOWLEDGEMENT

The authors gratefully acknowledge Ministry of Science and Technology and National Science and Technology Council in Taiwan for supporting the research with the contract numbers: MOST 111-2221-E-002-080-MY3.

6 REFERENCE

[1] Ana Brown, Ravishankar Sundararaman, Prineha Narang, William Goddard, and Harry Atwater, " Nonradiative Plasmon Decay and Hot Carrier Dynamics: Effects of Phonons, Surfaces, and Geometry." ACS nano, 10, 957-966, 2016.

[2] Dirk Koenig, Casalenuovo Kristen, Yasuhiko Takeda, Gavin Conibeer, Jean-François Guillemoles, Robert-John Patterson, and Lunmei Huang, " Hot carrier solar

cells: Principles, materials and design." Physica E: Low-dimensional Systems and Nanostructures, 42, 2862-2866, 2010.

[3] Chung-Han Chang, Jia-Ci Jhou, and Ching-Fuh Lin, "Harvesting Solar Energy below Si Bandgap Using Metal-Silicon Junction through Hot Carrier Effect, "39th European Photovoltaic Solar Energy Conference and Exhibition (39th EU PVSEC), Milan, Italy, September 26-30 2022.

[4] Hong-Jhang Syu, Hung-Chieh Chuang, Meng-Jie Lin, Chun-Chung Cheng, Po-Jui Huang, and Ching-Fuh Lin, "Ultra-broadband photoresponse of localized surface plasmon resonance from Si-based pyramid structures." Photonics Research, 7, 1119-1126, 2019.

10.4229/EUPVSEC2023/2AO.2.2

MACHINE LEARNING TO PREDICT MODULE PERFORMANCE FROM CELL AND MODULE PARAMETERS

Hannes Wagner-Mohnsen[1,*], Sven Wasmer[1], Bernhard Klöter[1], Pietro P. Altermatt[2], and Marco Ernst[3]
[1]WAVELABS Solar Metrology Systems GmbH, Spinnereistr. 7, 04179 Leipzig, Germany
[2]Trina Solar Co Ltd, State Key Laboratory for Photovoltaic Science and Technology (SKL), No 2 Trina Road, Xinbei District, Changzhou, Jiangsu Province, China 213031
[3]The Australian National University, School of Engineering, Canberra, ACT 2600, Australia
*Corresponding author: h.wagner-mohnsen@wavelabs.de

ABSTRACT: Algorithms of artificial intelligence offer new ways to optimize solar modules in mass-production. We apply state-of-the-art algorithms combined with sophisticated device physics to identify the dominant parameters that affect solar module performance. We analyze solar cell parameters, three different binning strategies, and module parameters. All this data is fed into a machine learning model to reveal the underlying correlations between these parameters and the scattering of the modules power output.
In our first results presented here, two module specific parameters (ribbon shading and Rs) lead to the strongest scattering of module power, while solar cell parameters have a smaller influence. This suggests that the cells were effectively binned, but we still observe an overall impact from the binning strategy on these impacting factors. The results of our framework depend on the architecture of the solar cells and modules, the ranges of inputs for relevant cell and module parameters fed into the model, and the binning strategy. Therefore, our approach is fully flexible and can be applied to different production lines and scenarios.

Keywords: Machine Learning, Simulation, PERC, Module, SHAP

1 INTRODUCTION

The PV-industry is continuously growing with factories in the multi-GW range. Like in other mass-production environments mathematical concepts like SixSigma can be used for continuous process improvement or methodologies like LEAN to analyze workflows for cycle time reduction and elimination of waste. Other concepts including new developed techniques from artificial intelligence (AI) bear the potential to work along with these conventional methodologies and to further improve todays mass-production.

Part of AI is the concept of machine learning (ML), with the two sub-areas of supervised and unsupervised machine learning. In supervised ML, a previously trained model predicts an output from known inputs feed into the model. Instead in unsupervised ML a model finds hidden patterns in data. Recent work in this area with reference to solar industry has been published by [1- 4].

We will use supervised machine learning in this paper and combine it with sophisticated device modelling. Our goal is to show what actually limits a solar module in mass-production. Therefore, we include information from solar cells, the binning strategy and solar modules and feed all this data into a machine learning model. This model helps use to reveal the complicated interactions off all these parameters and to show what actually impacts the solar modules power output most.

2 MODELLING

2.1 Solar Cell Simulation

We simulate a set of 100,000 PERC solar cells from an industrial line, using the Quokka 2D [5] simulator. To describe the typical scattering of cell performance, we vary eight different device parameters within experimental boundaries observed three years ago, namely: wafer resistivity (0.6 – 1.4 Ωcm) and bulk lifetime (by iron concentration 3 – 5 x 10^{10} cm^{-3}), rear surface recombination velocity (10 – 20 cm/s), saturation currents of the emitter (30 – 40 fA/cm^2) and the BSF (250 – 350 fA/cm^2), front finger width (33 – 40 um, with coupled series resistance and shading), an optical scaling factor to take variations in front reflectivity, texture, and ARC into account, and the emitter sheet resistivity (140 – 160 Ω/sq). Based on these parameters, we simulate the IV-curves and the average PL-signal. The simulated IV-curves are used in the next step to simulate solar modules.

2.2 Solar Module Simulation

To demonstrate the general behavior of our AI analysis, we categorized the PERC solar cells only into six evenly spaced intervals, or bins, and repeated this process three times to establish bins for efficiency, V_{oc} and J_{sc}. From each bin, 120 cells each were randomly selected for the simulation of a total of 40,000 typical half-cell modules, using PVMismatch [6]. For each bin, the number of modules generated corresponded to the number of cells within the bin.

Six additional parameters relating to the module are varied: the distance between the cells along the string and between neighboring strings (in the x and y directions), the thickness of the anti-reflective coating (ARC) on the front glass of the module, a factor to account for shading from the ribbons, a cell-to-module (CTM) factor, and the module series resistance.

The effect of the first three factors on module power is estimated using SunSolve [7] and parameterized for our analysis. All parameters are assumed to directly affect only the module current, except for the module series resistance. The defined parameters are assumed to have normal distributions with their mean and standard deviation, such as the cell separation in the x- and y-direction, which are assumed to be (0.3 ± 0.04) cm and (0.2 ± 0.02) cm, respectively. For the ARC, a thickness variation of (110 ± 10) nm is assumed. The ribbon shading factor is 0.99 ± 0.0025, whereas we assume that the CTM factor is 0.99 at 22.5% cell efficiency and decreases linearly with increasing cell efficiency (0.01 per 1%). Lastly, the estimated module series resistance is (0.02 ± 0.005) Ω.

2.3 Machine Learning

The machine learning process is implemented in Python with the models from xgboost [8]. We trained models for each binning strategy and for two different scenarios. The first scenario used more accessible parameters from the solar cells, the second uses detailed parameters that might not be measurable inline during solar cell production. The accessible parameters scenario uses the mean value of cells V_{oc}, J_{sc}, FF and the average PL signal from the solar cells and the additional six parameters from the solar module. The detailed scenario uses the average value of the eight previously described parameters (cf. 2.1) together with the six parameters from the solar module. The data are fed each into a machine learning model as input parameters with solar modules power output as target, giving us six different trained machine learning models.

2.4 SHAP Analysis

Trained machine learning models are often black boxes that makes their interpretation difficult. In order to understand what are the main features that affect the output of a trained model, explainable machine learning techniques are required. This means for each input parameter of the trained machine learning model the effect to the output parameter has to be quantified. One technique to achieve this is given by SHAP values (SHapley Additive exPlanations) [9]. A SHAP value is the average marginal contribution of a feature considering all possible combinations. These values can be used to understand the importance of each feature and to explain the result of the model. We calculate these values in Python with the shap package.

3 RESULTS

3.1 Simulation results

Figure 1 shows the simulated PERC solar cells, with natural scattering of the IV-curve parameters, as know from a stable production environment. Figure 2 shows an example of the simulated solar modules, with cells binned by efficiency.

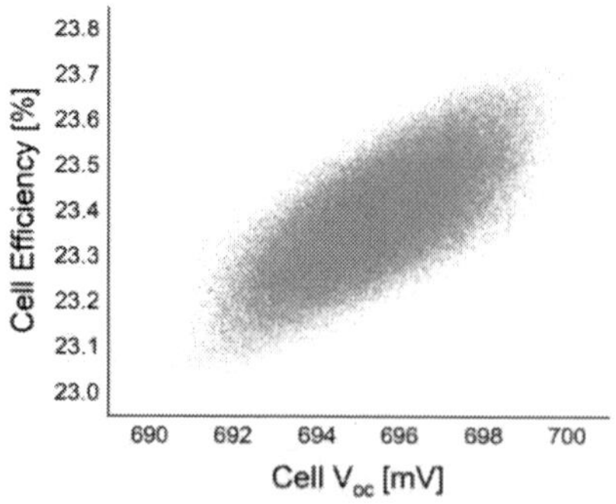

Figure 1: Simulated data for 100,000 PERC solar cells. These data are used with different binning strategies to simulate solar modules (cf. Fig. 2).

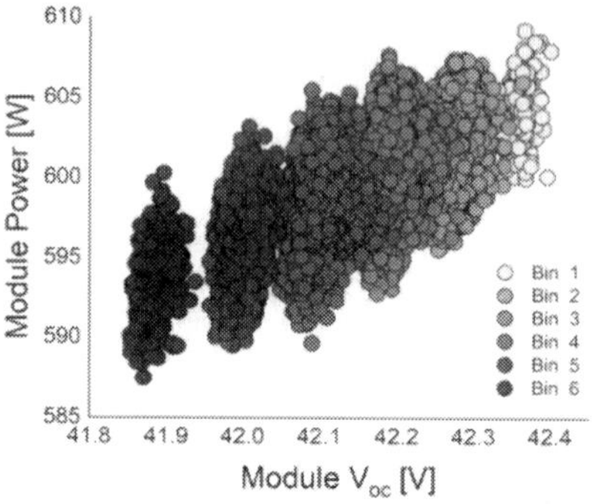

Figure 2: Simulated data for 40,000 half-cell modules. The modeled PERC solar cells were binned by efficiency (cf. Bin 1 to Bin 6).

3.2 Machine Learning

We fed the cell and module data into the machine learning model and got a very good agreement between actual model power output and the predicted power output from the trained machine learning model (cf. Fig 3). The RMSE between both is 0.0549 W. We obtain nearly identical results for the other trained machine learning models. This good agreement is mainly attributed to the simulation-based input data which does not contain any noise. We subsequently use the trained models to calculate the SHAP values.

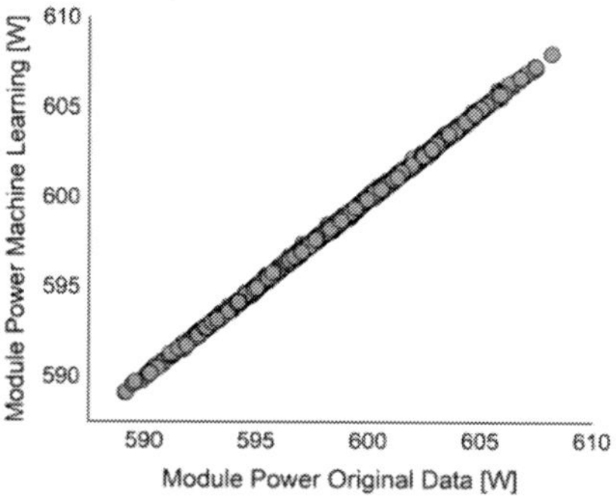

Figure 3: Results from machine learning process. Test data set from the machine learning model is plotted against actual power output from the simulation models.

3.3 SHAP analysis

Figures 4 and 5 show the calculated SHAP values for the accessible and the detailed parameter scenario for each binning strategy. Module parameters have the strongest impact on the module output power. Minor impacts are observed from the solar cell parameters. This suggests that the cells were effectively binned, but we still observe an overall impact in the SHAP values from the binning strategy itself on modules power output. This can be seen in the PL signal in the accessible parameter scenario, were the influence of the PL signal on modules power output is less pronounced with a current binning compared to a binning by V_{oc} or efficiency. In the detailed parameter scenario, it can be observed that the wafer's iron concentration has a greater influence on the module power output when binned solely by cell efficiency.

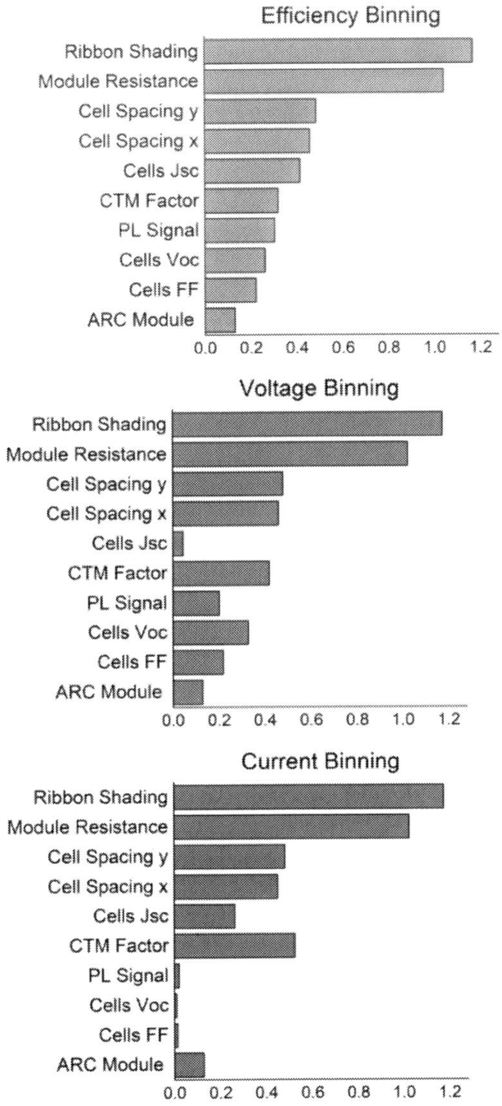

Figure 4: Results from SHAP analysis for the accessible parameter scenario for three different binning approaches.

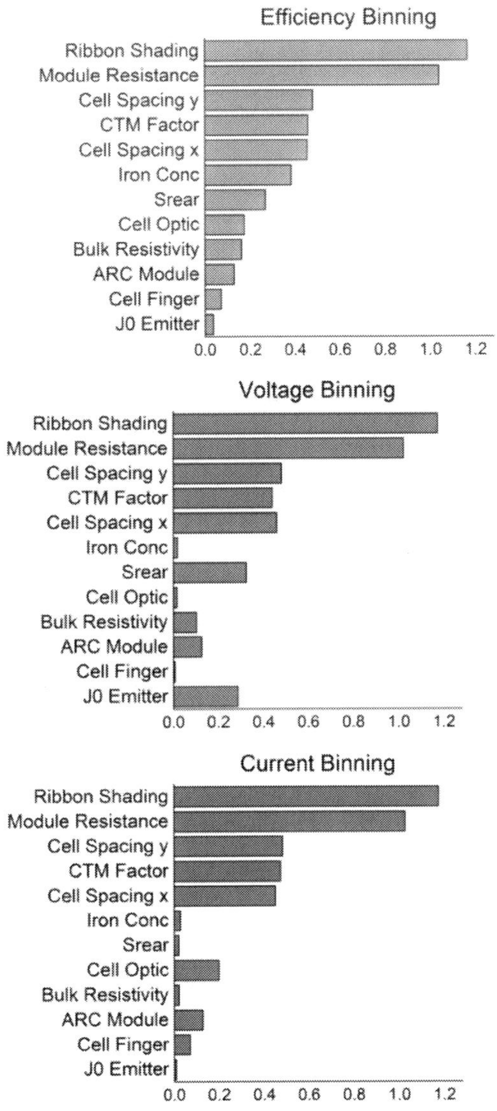

Figure 5: Results from SHAP analysis for the detailed parameter scenario for three different binning approaches.

4 DISCUSSION

Figures 4 and 5 show, that the main influences are from ribbon shading and modules resistance in all three binning strategies. An improvement of modules power output would thus mainly be achieved by optimizing these parameters. In both scenarios (accessible and detailed), the solar cell specific parameters are influenced by the binning strategy. For example, the iron concentration influences module power significantly more in the efficiency binning than in V_{oc} and J_{sc} binning (cf. Figure 5). This can be explained by the different sorting of the cells in the binning: the distribution of iron concentration in the efficiency binning is much more inhomogeneous and thus has a higher influence, compared to a V_{oc} or J_{sc} binning. These effects might be helpful in fine tuning the binning strategy.

5 CONCLUSIONS

We presented how a trained machine learning model in combination with explainable AI techniques (SHAP) is able to visualize the complex correlations between inputs and outputs for a multidimensional problem. By applying this technique to cell and module data we were able to show what influences solar modules strongest. In our case the module parameters have strongest impact, but for other cell types, binning strategies, or module architectures this will be different. An interesting finding for the cells, on the other hand, shows that the binning strategy can change the influence on cells impact on modules power. A binning by efficiency is much more sensitive against iron contamination of the cells, compared to a binning by V_{oc} of the cells. An optimum might be a mixture of different binning strategies.

Currently, we focus on additional impacting factors such as the effect of temperature and light intensity on modules power output. With these additional input parameters, we can analyze solar modules power output in the field.

REFERENCES

[1] H. Wagner-Mohnsen and P. P. Altermatt, "Machine Learning for Optimization of Mass-Produced Industrial Silicon Solar Cells," 2021 International Conference on Numerical Simulation of Optoelectronic Devices (NUSOD), Turin, Italy, 2021, pp. 51-52, doi: 10.1109/NUSOD52207.2021.9541457.

[2] H. Wagner-Mohnsen and P. P. Altermatt, "A Combined Numerical Modeling and Machine Learning Approach for Optimization of Mass-Produced Industrial Solar Cells," in IEEE Journal of Photovoltaics, vol. 10, no. 5, pp. 1441-1447, Sept. 2020, doi: 10.1109/JPHOTOV.2020.3004930.

[3] R. Jaiswal, M. Martínez-Ramón and T. Busani, "Recent Advances in Silicon Solar Cell Research Using Data Science-Based Learning," in IEEE Journal of Photovoltaics, vol. 13, no. 1, pp. 2-15, Jan. 2023, doi: 10.1109/JPHOTOV.2022.3221003.

[4] S. Datta, A. Baul, G. C. Sarker, P. K. Sadhu and D. R. Hodges, "A Comprehensive Review of the Application of Machine Learning in Fabrication and Implementation of Photovoltaic Systems," in IEEE Access, vol. 11, pp. 77750-77778, 2023, doi: 10.1109/ACCESS.2023.3298542.

[5] Quokka 2 - Version 2.2.5

[6] Mark Mikofski, Bennet Meyers, Chetan Chaudhari (2018). "PVMismatch Project: https://github.com/SunPower/PVMismatch". SunPower Corporation, Richmond, CA.

[7] SunSolve Power – Version 6.14.1

[8] Chen, T. & Guestrin, C., 2016. XGBoost: A Scalable Tree Boosting System. In Proceedings of the 22nd ACM SIGKDD International Conference on Knowledge Discovery and Data Mining. KDD '16. New York, NY, USA: ACM, pp. 785–794. Available at: http://doi.acm.org/10.1145/2939672.2939785.

[9] Scott M. Lundberg and Su-In Lee, "A unified approach to interpreting model predictions", NIPS'17: Proceedings of the 31st International Conference on Neural Information Processing Systems, December 2017, Pages 4768–4777.

40th European Photovoltaic Solar Energy Conference and Exhibition

This presentation was selected by the Sc. Committee of the EU PVSEC 2023 for submission of a full paper to one of the EU PVSEC's collaborating peer-reviewed journals.

RADIATIVE COOLING OF SOLAR CELLS WITH CEMENT-BASED MATERIALS

Matteo Cagnoni[a,1], Pietro Testa[a,2], Jorge S. Dolado[b,c,3] and Federica Cappelluti[a,4]

[a]Department of Electronics and Telecommunications, Politecnico di Torino, C.so Duca degli Abruzzi 24, 10129 Turin, Italy

[b]Centro de Fisica de Materiales, CSIC-UPV/EHU, Paseo Manuel de Lardizabal 5, San Sebastián 20018, Spain

[c]Donostia International Physics Center, Paseo Manuel de Lardizabal 4, San Sebastián 20018, Spain

[1]matteo.cagnoni@polito.it, [2]pietro.testa@polito.it, [3]j.dolado@ehu.eus, [4]federica.cappelluti@polito.it

ABSTRACT: Lowering the operating temperature of solar cells can increase efficiency and lifespan of these devices. Among the available thermal management options, radiative cooling shines because of its passive nature and systemic simplicity. Unfortunately, the applicability of most radiative coolers to industrial manufacturing is questioned by high cost or UV instability. We have recently shown that stable and cost-effective cement-based materials can be designed to be suitable radiative coolers for solar cells. However, we have done so in line with the literature, by modeling the solar cell in the radiative limit and neglecting the thermal contact resistance at the cell/cooler interface, which might actually be large enough to hinder heat transfer because of the poor adhesion properties of cement-based materials. In this work, we have generalized the model used to assess radiative coolers by incorporating solar cell non-radiative losses (Auger, Shockley-Read-Hall) and a thermal barrier between cell and cooler. The final model provides a description of the thermal behavior of a solar cell with radiative cooler closer to reality, while preserving the transparency of the detailed-balance approach, that has allowed us to better assess these systems and provide design guidelines, with focus on cement-based radiative coolers.

Keywords: solar cells, radiative cooling, cement-based materials, detailed-balance principle

1 INTRODUCTION

Radiative cooling has been drawing a lot of attention in the last decade as a thermal management solution for buildings and solar cells [1], after the experimental demonstration of sub-ambient temperatures under direct sunlight [2]. Radiative coolers are designed to dump excess heat directly into outer space by emitting thermal radiation within the atmosphere transparency window (AW) between 8 and 13 μm [3]. When thermally coupled to a solar cell, a radiative cooler can reduce the cell temperature by collecting the excess heat generated upon sunlight absorption and thermally radiating it out of the device [4]. This approach is extremely attractive because of potential efficiency and systemic simplicity, but its applicability beyond research is questioned by the reliance on expensive materials [5] or fabrication processes [6], or on organic polymers with potential UV instability [7].

Faced with this challenge, we have recently shown that, from an electromagnetic perspective, cement-based materials could be ideal candidates for highly effective radiative coolers of solar cells, fulfilling also economic and stability requirements for large-scale deployment [8]–[10]. In particular, by combining chemical kinetics, molecular mechanics, effective-medium theory and electromagnetic simulations, we have first explained how these materials can be equipped with optimal thermal emissivity by tuning their micro-structure. Then, we have shown how they can be exploited in a system like the one depicted in Figure 1 to reduce the temperature of standard silicon cells by up to 20 K, *de-facto* matching the performance of much more expensive meta-materials. The system in Figure 1 corresponds to the detailed-balance model commonly employed in the literature [11] to determine the operating temperature of a solar cell with a radiative cooler, by describing the energy exchange between cell, cooler, environment and end-user.

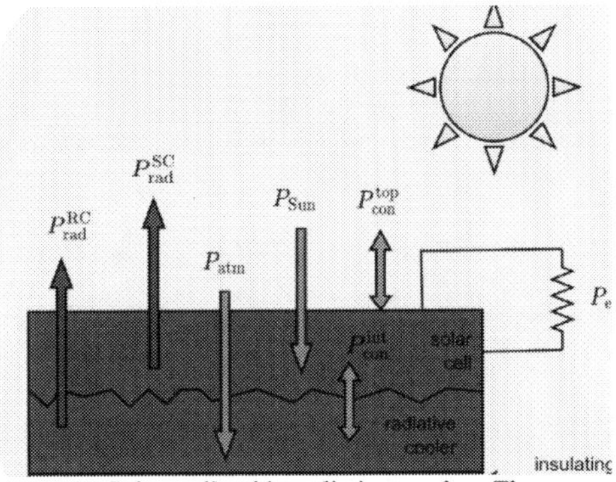

Figure 1: Solar cell with radiative cooler. The terms entering the corresponding detailed-balance model are shown.

Experimental realization of this concept would be technologically disruptive, since it could lead to efficiency relative gains up to 9% [12] and lifetime prolongation up to 4 times [13] at a very small additional cost, thanks to the reliance on cheap, available, stable and scalable materials.

In this work, we have extended the analysis done in [9] by including relevant non-idealities in the photovoltaic and thermal models, to obtain a more accurate evaluation of the cooling capability and derive useful guidelines at material/device level toward the practical realization of solar cells with cement-based radiative coolers.

First, the performance of radiative coolers is commonly assessed by describing the solar cell with the Shockley-Queisser model in the radiative limit [9], [11]. However, intrinsic and extrinsic non-radiative recombination mechanisms such as Auger [14] and Shockley-Read-Hall (SRH) [15] are often more prominent and strongly temperature dependent. By incorporating these terms into the description of the cell, we have obtained a thermal model for the whole system much closer to reality. Furthermore, by introducing these terms in a simple parametric form, we have eschewed the use of specific solutions and treated the problem on a rather general footing, which can provide the researchers with a broader view.

Second, the system is usually assumed to be isothermal in its stationary state, with a perfect thermal contact between the cell and the cooler. However, cement-based materials usually exhibit poor adhesion to surfaces [16], suggesting that the thermal resistance at the cell/cooler interface might be non-negligible; this might happen also in the case of good adhesion because of acoustic mismatch [17]. Unfortunately, a poor thermal contact is going to hinder heat transfer from the cell to the cooler, hence worsening the cooling performance. By extending the detailed-balance model to incorporate the effect of the thermal contact resistance at the cell/cooler interface, we have enabled a better assessment of the potential of cement-based radiative coolers and the definition of thermal contact requirements.

The extended detailed-balance model hereby developed should not be intended as applicable to cement-based radiative coolers only, but to any kind of radiative cooler, as long as the cooler-dependent quantities (*e.g.*, emissivity) are adapted to the specific case under consideration.

2 METHODS

In case of perfect thermal contact between cell and cooler, the net power density P_{net} ejected by the cell/cooler stack in terrestrial environment can be calculated as:

$$P_{\text{net}} = P_{\text{rad}}^{\text{SC}} - P_{\text{Sun}} + P_{\text{el}} + P_{\text{rad}}^{\text{RC}} - P_{\text{atm}} + P_{\text{con}}^{\text{top}}$$

where $P_{\text{rad}}^{\text{SC}}$ is the power density radiated by the solar cell, P_{Sun} the one absorbed from the Sun, P_{el} the one delivered to the end-user, $P_{\text{rad}}^{\text{RC}}$ the one radiated by the cooler, P_{atm} the one absorbed from the atmosphere, and $P_{\text{con}}^{\text{top}}$ the one exchanged with the environment by conduction and convection at the device top surface.

Explicit formulas to calculate these terms can be found in [9]. In particular, $P_{\text{el}} = \max_{V}\{J(V)\,V\}$ can be obtained as the power density at maximum-power-point, where the $J(V)$ relation is usually described in the radiative limit considered by the original Shockley-Queisser model [18].

In this work, we have first improved the description of the cell/cooler system by incorporating Auger and Shockley-Read-Hall non-radiative recombination into $J(V)$ according to the models developed in [14] and [15], respectively. Interestingly, by performing some algebraic manipulations, we could separate the dependence on temperature of these terms from the one on the solar cell band-gap E_g and the one on material- and technology-related parameters (*e.g.*, effective density-of-states, depletion region thickness). Then, we have incorporated the latter into a single parameter for each non-radiative recombination channel, denoted as β_{AUG} and β_{SRH}, which quantifies the T-normalized strength of the losses associated to the corresponding mechanism. More details on the math can be found in the Full Paper [19].

Finally, this approach has let us express P_{net} only as a function of the solar cell operating temperature T and the triplet $\left(E_g, \beta_{\text{AUG}}, \beta_{\text{SRH}}\right)$, given the emissivity of the radiative cooler and the environmental conditions. Then, one can determine the solar cell operating temperature T as a function of $\left(E_g, \beta_{\text{AUG}}, \beta_{\text{SRH}}\right)$ by solving $P_{\text{net}} = 0$ with respect to T.

As a second extension, we have introduced a finite thermal contact resistance at the cell/cooler interface, to empirically account for acoustic mismatch and poor adhesion properties, quantified by the heat transfer coefficient h_c^{int}. Because of this, solar cell and radiative cooler are not isothermal anymore, having temperature T_{SC} and T_{RC}, respectively. Their net power densities are coupled through $P_{\text{con}}^{\text{int}}\left(T_{\text{SC}}, T_{\text{RC}}\right) = h_c^{\text{int}}\left(T_{\text{SC}} - T_{\text{RC}}\right)$, as depicted in Figure 1. Their operating temperatures are again determined by the condition of zero net power density [20] according to the following system of equations, which we have used to study the impact of h_c^{int} on T_{SC} in the radiative limit:

$$\begin{cases} P_{\text{net}}^{\text{SC}}\left(T_{\text{SC}}, T_{\text{RC}}\right) = P_{\text{rad}}^{\text{SC}} - P_{\text{Sun}} + P_{\text{el}} + P_{\text{con}}^{\text{top}} + P_{\text{con}}^{\text{int}} = 0 \\ P_{\text{net}}^{\text{RC}}\left(T_{\text{SC}}, T_{\text{RC}}\right) = P_{\text{rad}}^{\text{RC}} - P_{\text{atm}} - P_{\text{con}}^{\text{int}} = 0 \end{cases}$$

3 RESULTS

Although the methods developed are applicable to any radiative cooler, we have hereby considered the best cement-paste from our previous work [9] to perform the numerical calculations, as discussed in more detail in [19].

First, we have calculated the temperature reduction provided by the cement-based radiative cooler to the solar cell as a function of band-gap and strength of the non-radiative recombination losses, by calculating and comparing the cell operating temperature with and without radiative cooler; we have assumed that cell and cooler are isothermal.

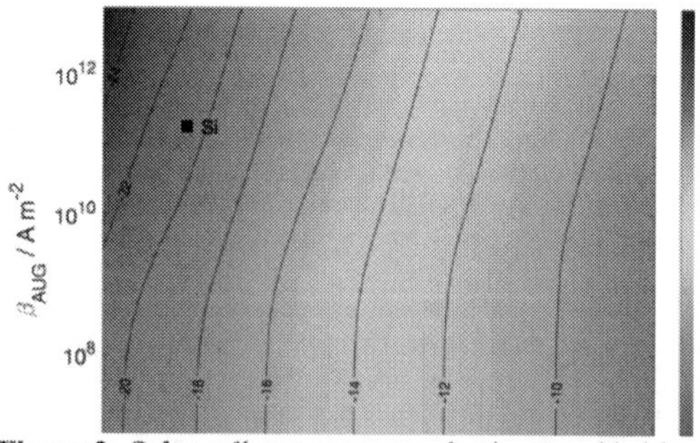

Figure 2: Solar cell temperature reduction provided by the considered cement-based radiative cooler as a function of band-gap and strength of the Auger recombination losses, for negligible SRH losses.

Figure 2 shows the results for the cases with negligible Shockley-Read-Hall recombination, *i.e.*, $\beta_{\text{SRH}} = 0$. Interestingly, the temperature reduction provided by the radiative cooler increases for larger values of β_{AUG}, suggesting that the cooling performance of the latter is underestimated in the radiative limit on the one hand, and that the presence of a radiative cooler is more beneficial for solar cells whose behavior departs more from the radiative limit. Clearly, the same conclusions can be reached concerning the efficiency gains provided by the radiative cooler, which are larger for stronger non-radiative recombination because of the larger temperature reduction attainable by the radiative cooler and the larger efficiency temperature coefficient due to increasing non-radiative recombination; this latter point is shown and discussed in more detail in [19]. We have observed a similar behavior for the cases with negligible Auger recombination, *i.e.*, $\beta_{AUG} = 0$, but $\beta_{SRH} > 0$ [19], suggesting that these conclusions are qualitatively independent from the non-radiative recombination channels considered.

Second, we have estimated the impact of a finite thermal contact resistance at the cell/cooler interface as a function of the associated heat transfer coefficient h_c^{int}. To minimize the number of parameters affecting the results and preserve a simple data visualization, we have performed this step by modeling the solar cell in the radiative limit.

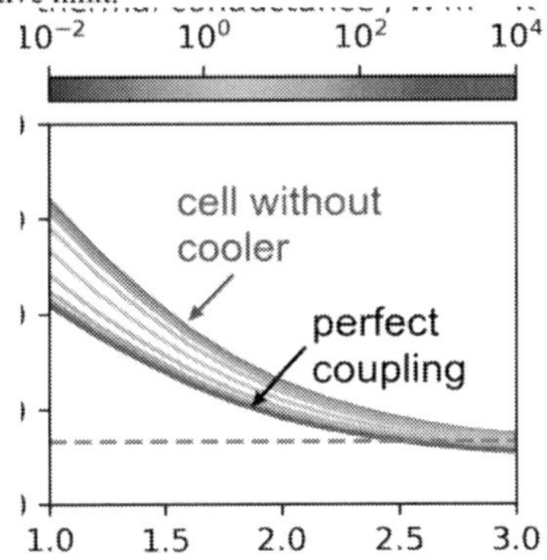

Figure 3: Solar cell operating temperature vs band-gap as a function of the heat transfer coefficient at the cell/cooler interface. The solar cell is modeled in the radiative limit.

Figure 3 shows the solar cell operating temperature T_{SC} as a function of its band-gap E_g for several h_c^{int} values. As expected, as the heat transfer coefficient decreases, the solar cell temperature approaches the one of the system without radiative cooler, suggesting that a too large thermal barrier completely inhibits the effect of the radiative cooler by decoupling it from the solar cell. Although not surprising, this model and the corresponding data allow one to quantify the temperature reduction that can be achieved as a function of the thermal contact resistance and define specifications for the values of the latter depending on the temperature reduction required for the application of interest. Therefore, it provides important design guidelines concerning a critical but often overlooked aspect of these systems. A more in dept discussion can be found in [19].

4 CONCLUSION

In conclusion, we have generalized the detailed-balance model commonly employed in the literature to assess the performance of radiative coolers by incorporating non-radiative recombination losses in the Shockley-Queisser model of the solar cell and by introducing an empirical finite heat transfer coefficient between cell and cooler to model an eventual non-negligible thermal contact resistance at their interface due to acoustic mismatch or poor adhesion properties.

Although general, we have applied this model to the specific case of cement-based materials, in particular to the best cement-based radiative cooler that we have designed in our previous work [9]. Our choice to focus on these materials has been driven by the disruptive potential of this solution due to its stability and low-cost, which are problems affecting current state-of-the-art radiative coolers.

This model extension and the corresponding simulations and data analysis have enabled us to obtain a description of the thermal behavior of a solar cell with radiative cooler closer to reality, while preserving the transparency of the detailed-balance approach, and to better assess these systems while providing design guidelines. Our results show that the beneficial impact of radiative cooling and the subsequent efficiency gain are relevant for all solar cell technologies, but more significant for devices that depart more from operation in the radiative limit. Moreover, some amount of non perfect-thermal coupling between radiative cooler and cell can be accommodated without hindering the cooling performance.

ACKNOWLEDGEMENTS

This project has received funding from the European Union's Horizon 2020 research and innovation program under grant agreement No. 964450.

REFERENCES

[1] M. Zeyghami, D. Y. Goswami, and E. Stefanakos, "A Review of Clear Sky Radiative Cooling Developments and Applications in Renewable Power Systems and Passive Building Cooling," Solar Energy Materials and Solar Cells, vol. 178, pp. 115–128, May 2018, doi: 10.1016/j.solmat.2018.01.015

[2] A. P. Raman, M. A. Anoma, L. Zhu, E. Rephaeli, and S. Fan, "Passive Radiative Cooling Below Ambient Air Temperature Under Direct Sunlight," Nature, vol. 515, no. 7528, pp. 540–544, Nov. 2014, doi: 10.1038/nature13883

[3] W. Li and S. Fan, "Radiative Cooling: Harvesting the Coldness of the Universe," Optics and Photonics News, vol. 30, no. 11, p. 32, Nov. 2019, doi: 10.1364/OPN.30.11.000032

[4] L. Zhu, A. Raman, K. X. Wang, M. A. Anoma, and S. Fan, "Radiative Cooling of Solar Cells," Optica, vol. 1, no. 1, p. 32, Jul. 2014, doi: 10.1364/OPTICA.1.000032

[5] M. A. Kecebas, M. P. Menguc, A. Kosar, and K. Sendur, "Passive Radiative Cooling Design with Broadband Optical Thin-Film Filters," Journal of Quantitative Spectroscopy and Radiative Transfer, vol. 198, pp. 179–186, Sep. 2017, doi: 10.1016/j.jqsrt.2017.03.046

[6] G. Perrakis, A. C. Tasolamprou, G. Kenanakis, E.

N. Economou, S. Tzortzakis, and M. Kafesaki, "Combined Nano and Micro Structuring for Enhanced Radiative Cooling and Efficiency of Photovoltaic Cells," Scientific Reports, vol. 11, no. 1, p. 11552, Dec. 2021, doi: 10.1038/s41598-021-91061-1

[7] J. Mandal et al., "Hierarchically Porous Polymer Coatings for Highly Efficient Passive Daytime Radiative Cooling," Science, vol. 362, no. 6412, pp. 315–319, Oct. 2018, doi: 10.1126/science.aat9513

[8] M. Cagnoni, A. Tibaldi, P. Testa, J. S. Dolado, and F. Cappelluti, "Passive Radiative Cooling of Solar Cells by Low-Cost and Scalable Metamaterials: Physical Simulation and Efficiency Limits," in Physics, Simulation, and Photonic Engineering of Photovoltaic Devices, San Francisco, US-CA: SPIE, Mar. 2022, p. 1199606. doi: 10.1117/12.2607489

[9] M. Cagnoni, A. Tibaldi, J. S. Dolado, and F. Cappelluti, "Cementitious Materials as Promising Radiative Coolers for Solar Cells," iScience, vol. 25, no. 11, p. 105320, Nov. 2022, doi: 10.1016/j.isci.2022.105320

[10] J. S. Dolado et al., "Radiative Cooling Properties of Portlandite and Tobermorite: Two Cementitious Minerals of Great Relevance in Concrete Science and Technology," ACS Applied Optical Materials, Jun. 2023, doi: 10.1021/acsaom.3c00082

[11] T. S. Safi and J. N. Munday, "Improving Photovoltaic Performance Through Radiative Cooling in Both Terrestrial and Extraterrestrial Environments," Optics Express, vol. 23, no. 19, p. A1120, Sep. 2015, doi: 10.1364/OE.23.0A1120

[12] E. Skoplaki and J. A. Palyvos, "On the Temperature Dependence of Photovoltaic Module Electrical Performance: A Review of Efficiency/Power Correlations," Solar Energy, vol. 83, no. 5, pp. 614–624, May 2009, doi: 10.1016/j.solener.2008.10.008

[13] O. Dupré, R. Vaillon, and M. A. Green, Thermal Behavior of Photovoltaic Devices: Physics and Engineering. Cham, CH: Springer, 2017.

[14] T. Tiedje, E. Yablonovitch, G. D. Cody, and B. G. Brooks, "Limiting Efficiency of Silicon Solar Cells," IEEE Transactions on Electron Devices, vol. 31, no. 5, pp. 711–716, May 1984, doi: 10.1109/T-ED.1984.21594

[15] A. Schenk, "A Model for the Field and Temperature Dependence of Shockley-Read-Hall Lifetimes in Silicon," Solid-State Electronics, vol. 35, no. 11, pp. 1585–1596, Nov. 1992, doi: 10.1016/0038-1101(92)90184-E

[16] M. Bohnet and F. Ullmann, Eds., Ullmann's Encyclopedia of Industrial Chemistry. Weinheim, DE: Wiley, 2003.

[17] E. T. Swartz and R. O. Pohl, "Thermal Resistance at Interfaces," Applied Physics Letters, vol. 51, no. 26, pp. 2200–2202, Dec. 1987, doi: 10.1063/1.98939

[18] W. Shockley and H. J. Queisser, "Detailed Balance Limit of Efficiency of p-n Junction Solar Cells," Journal of Applied Physics, vol. 32, no. 3, pp. 510–519, Mar. 1961, doi: 10.1063/1.1736034

[19] M. Cagnoni, P. Testa, J. S. Dolado, and F. Cappelluti, "Extended Detailed Balance Modeling Toward Solar Cells with Cement-Based Radiative Coolers," *Progress in Photovoltaics: Research and Applications*, Under Review.

[20] Y. A. Çengel and A. J. Ghajar, Heat and Mass Transfer: Fundamentals & Applications, 5th ed. New York, US-NY: McGraw-Hill, 2015.

HZB Helmholtz Zentrum Berlin

9 DIFFERENT ISOS PROTOCOLS FOR ACCELERATED AGING OF A PEROVSKITE SOLAR CELL: COMPARISON AND ANALYSIS OF INDOOR AND OUTDOOR TESTS

Ulas Erdil, Mark Khenkin, Quiterie Emery, Hans Köbler, Antonio Abate, Eugene A.Katz, Rutger Schlatmann, Carolin Ulbrich

HI-SCORE

Introduction

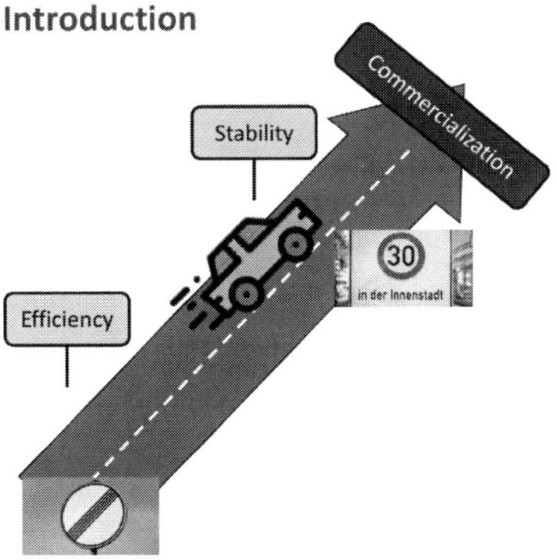

nature energy — CONSENSUS STATEMENT

OPEN

Consensus statement for stability assessment and reporting for perovskite photovoltaics based on ISOS procedures

Khenkin, M.V., Katz, E.A., Abate, A. *et al. Nat Energy* 5, 35–49 (2020)

International Summit on Organic PV Stability (ISOS)

https://www.rbb24.de/panorama/beitrag/2023/01
https://www.merkur.de/wirtschaft/

HZB Helmholtz Zentrum Berlin

40th European Photovoltaic Solar Energy Conference and Exhibition

Introduction

Aim: To reproduce and accelerate degradation mechanisms observed outdoor via indoor stability tests

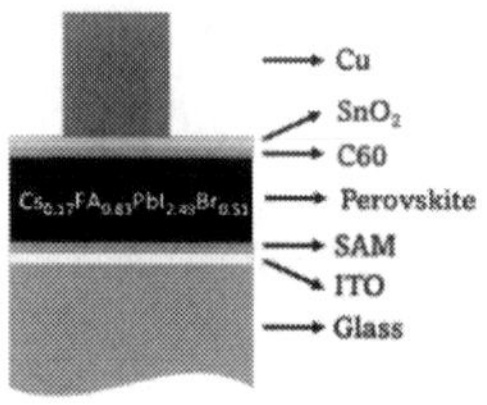

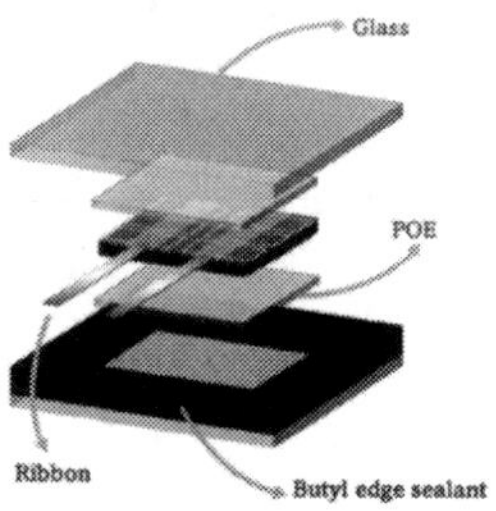

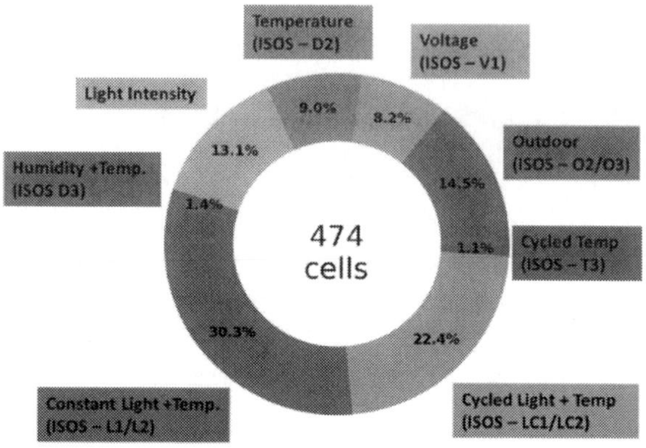

3

HZB Helmholtz Zentrum Berlin

020076-003

40th European Photovoltaic Solar Energy Conference and Exhibition

Outdoor - Berlin

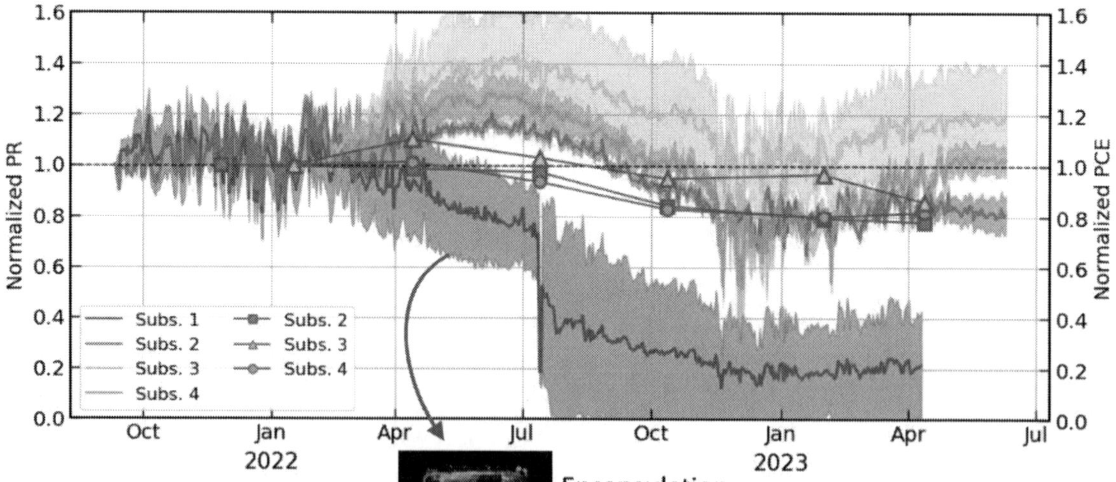

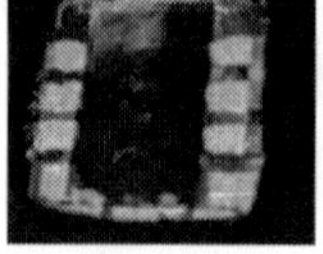

4

HZB Helmholtz Zentrum Berlin

020076-004

Cycled Temperature

Outdoor

Sde Boker, Israel

BGU
Ben-Gurion University of the Negev

Nicosia, Cyprus

University of Cyprus

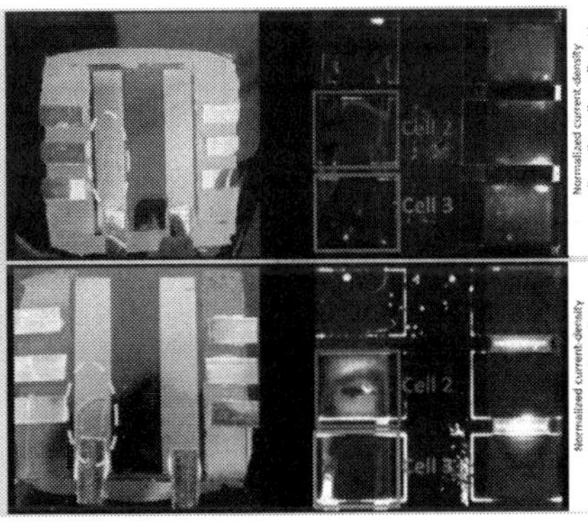

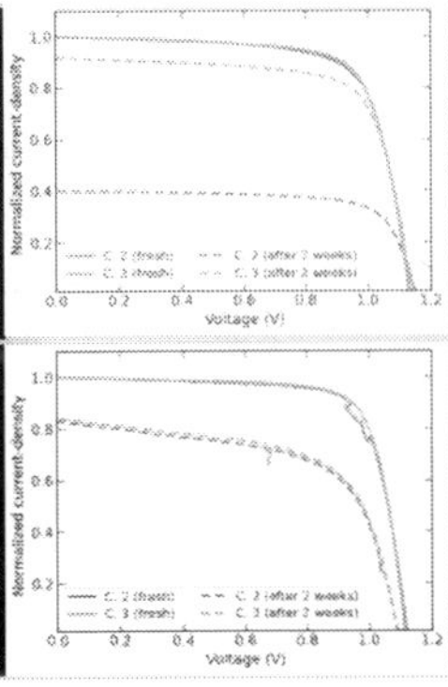

Delamination

- Not observed in Berlin(>1.5 years operation), but observed in Israel and Cyprus
- If it happens, it happens fast

HZB

40th European Photovoltaic Solar Energy Conference and Exhibition

Cycled Temperature

Outdoor

Sde Boker, Israel

BGU
Ben-Gurion University of the Negev

Nicosia, Cyprus

University of Cyprus

Indoor
(ISOS – T3)

-40 – 85 °C cycling profile

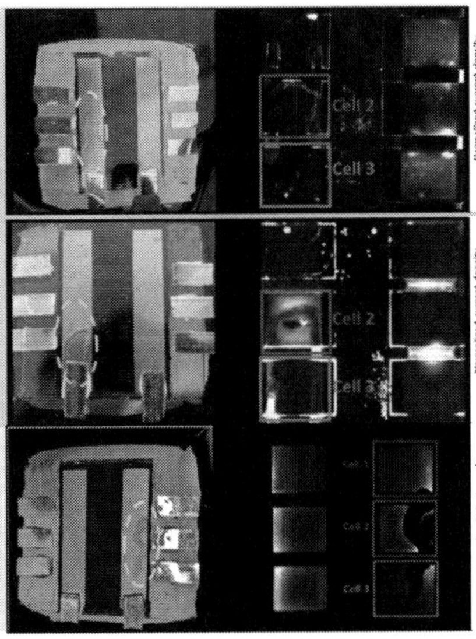

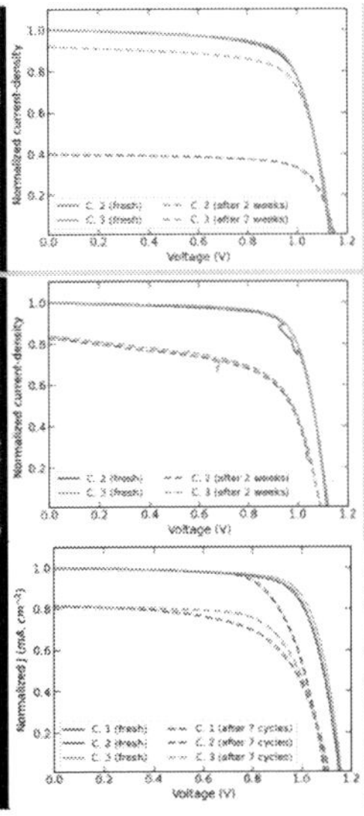

Delamination

- Not observed in Berlin(<1.5 years operation), but observed in Israel and Cyprus
- If delamination happens, it happens fast

HZB

Light intensity (not exactly on ISOS)

Outdoor (ISOS O-3)

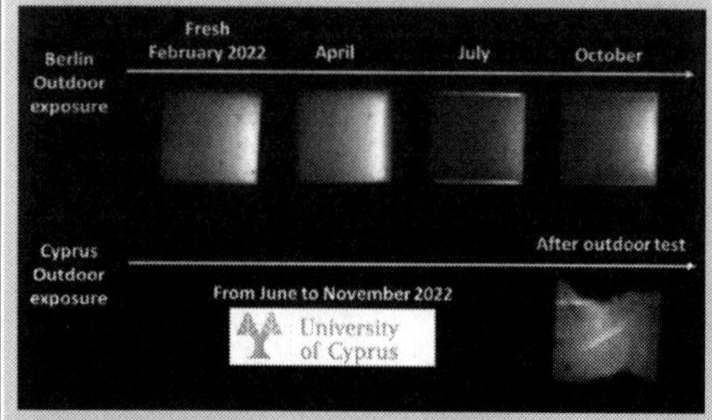

Indoor

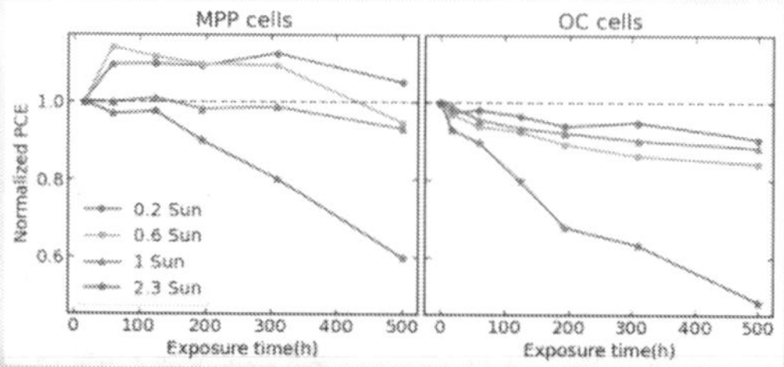

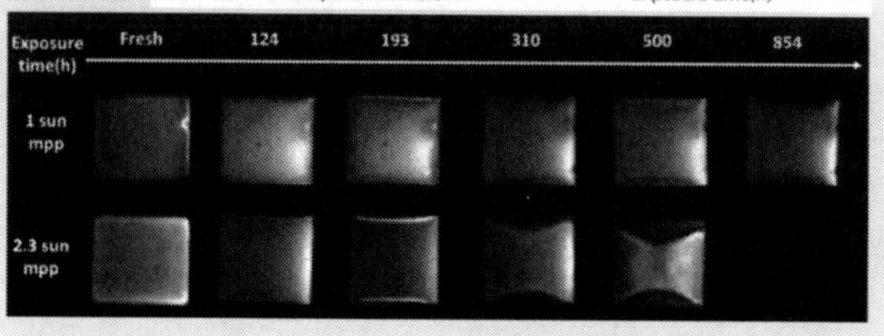

Edge Effects

- Has a characteristic pattern for MPP cells
- Can be accelerated by increasing light intensity

Cycled Light + Temperature

Outdoor

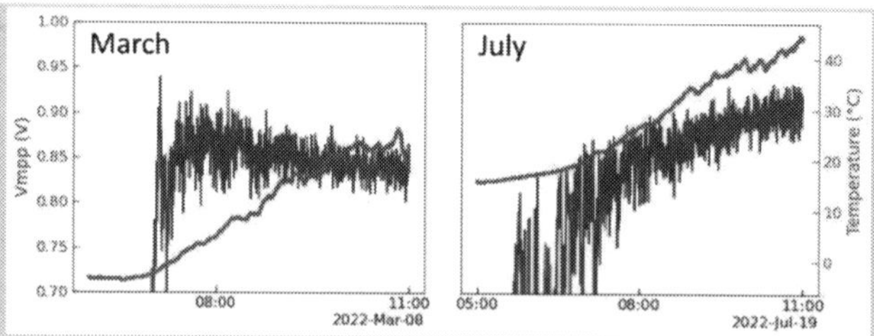

'Fatigue'

- Can be accelerated by increasing temperature during cycled experiments
- Not reversible

Cycled Light + Temperature

Outdoor

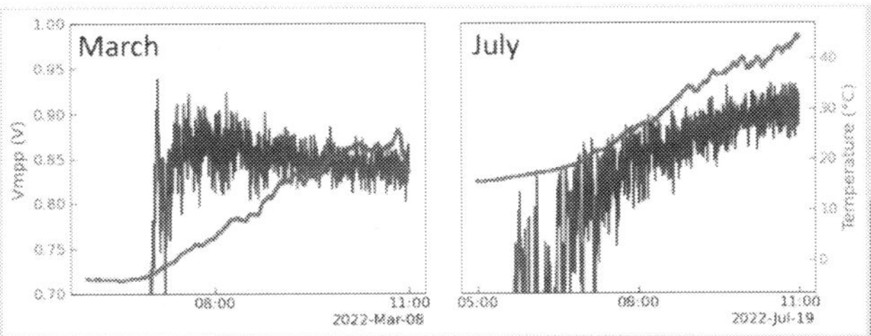

Indoor
(ISOS-LC1/LC2)

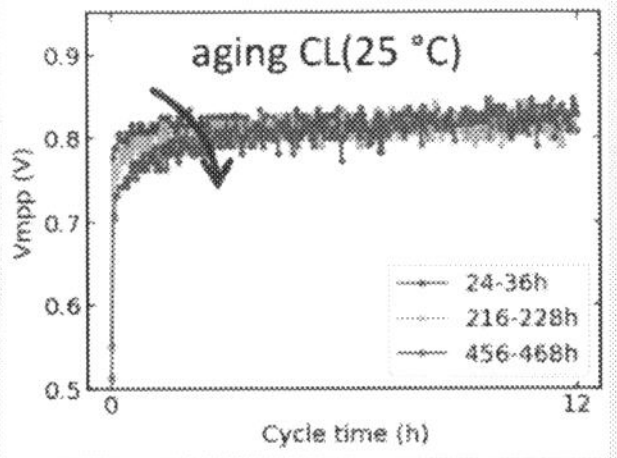

'Fatigue'

- Can be accelerated by increasing temperature during cycled experiments
- Not reversible

9

HZB

Cycled Light + Temperature

Outdoor

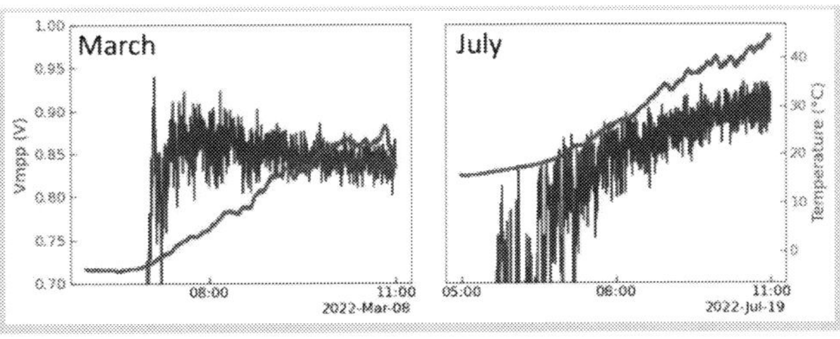

Indoor
(ISOS-LC1/LC2)

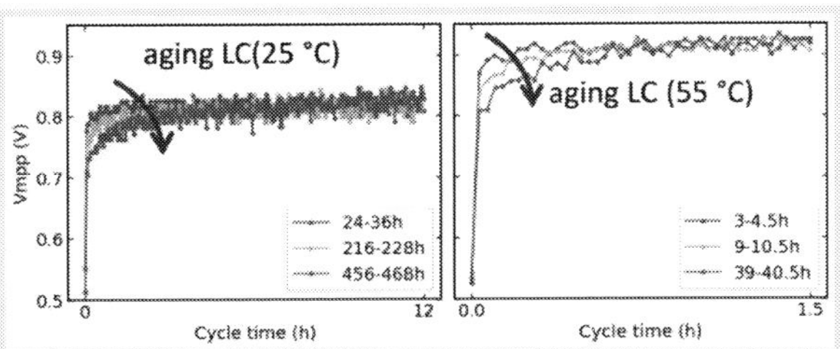

'Fatigue'

- Can be accelerated by increasing temperature during cycled experiments
- Not reversible

10

HZB

Voltage

Indoor
(ISOS-V1 but dark included)

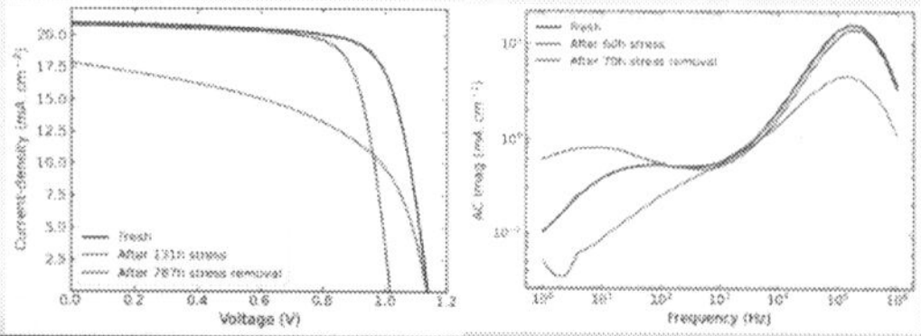

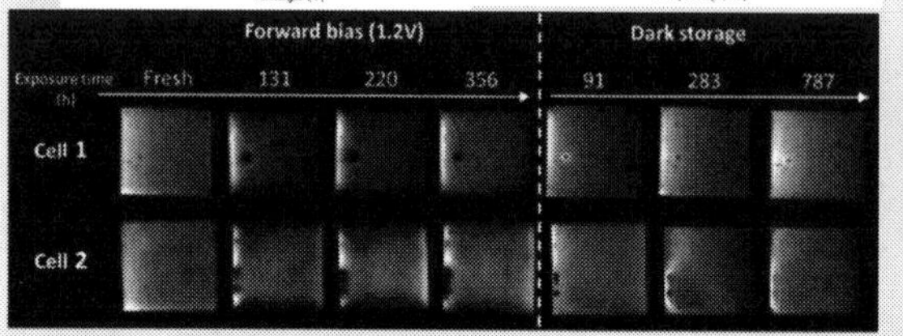

Outdoor

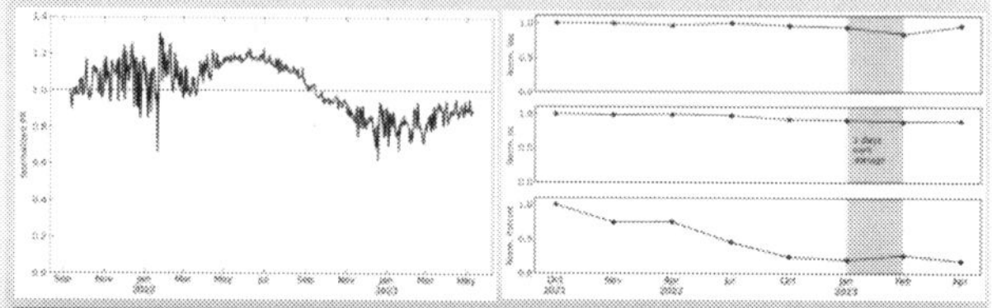

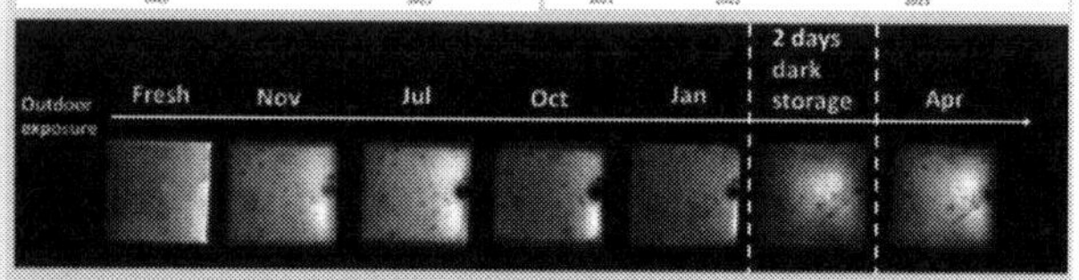

Ion-migration Induced Degradation

- Macroscopic defect formation but reversible
- However, irreversible trap formation

HZB Helmholtz Zentrum Berlin

40th European Photovoltaic Solar Energy Conference and Exhibition

Voltage

Photo-CELIV

Voc-Light intensity

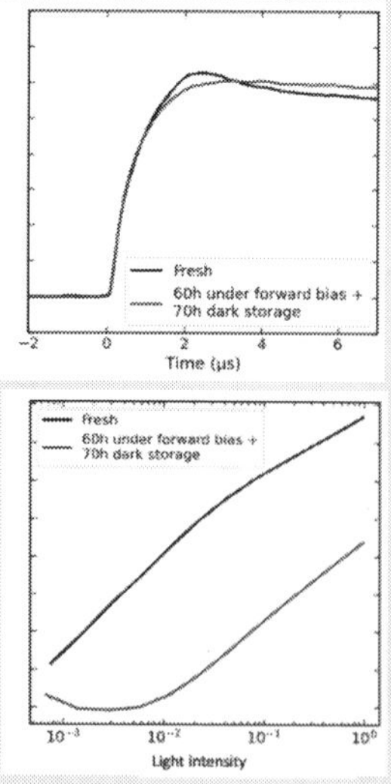

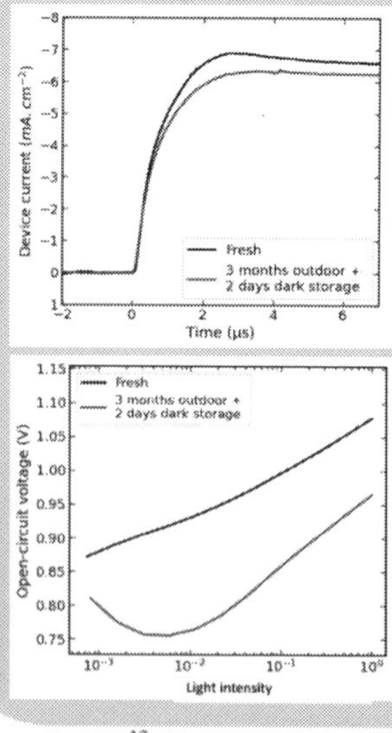

BGU
Ben-Gurion University of the Negev

Ion-migration Induced Degradation

- A simple test to resemble outdoor
- We don't even need light to simulate outdoor! (for these devices)

All icons were taken from Flaticon.com

HZB Helmholtz Zentrum Berlin

Summary

Outdoor Observed Degradation/Failure Modes	Possible Origin(s)	Reproduced Indoor Test
Moisture-Induced Degradation	Trace moisture, cracked encapsulation	None (but DH test is the best way to go)
Delamination	TEC mismatch but possibly residue stress assisted	Cycled Temperature
Edge Effects	Laser pattern (as specific site)	Light intensity
Fatigue Behaviour	Ion-migration mediated defect formation	Cycled light + Temperature
Ion-migration Induced Degradation	Ion-migration mediated defect formation	Voltage

Thank you!

HZB

Zurich University of Applied Sciences
zh aw School of Engineering
ICP Institute of Computational Physics

SOLARONIX

Schweizerische Eidgenossenschaft
Confédération suisse
Confederazione Svizzera
Confederaziun svizra
Innosuisse – Schweizerische Agentur für Innovationsförderung

FLUXiM
boosts your R&D - www.fluxim.com

Investigation of Time and Location Dependent Variations in Electroluminescence Images of Perovskite Solar Cells

EU PVSEC 40 2023

18th – 22nd of September

Ennio Comi[1], Sandra Jenatsch[2], Balthasar Blülle[2], Mattia Battaglia[1], Miguel Angel Torre Cachafeiro[1]
Christoph Kirsch[1], Roman Hiestand[2], Urs Aeberhard[2], Beat Ruhstaller[1,2] and Evelyne Knapp[1]

[1] Institute of Computational Physics, Zurich University of Applied Sciences, Technikumstr. 71, 8401 Winterthur, Switzerland

[2] Fluxim AG, Katharina-Sulzer-Platz 2, 8406 Winterthur, Switzerland

EU PVSEC
EU PVSEC 2023

Ennio Comi
Scientific Associate
ZHAW (ICP)

zh aw School of Engineering
ICP Institute of Computational Physics

Introduction – Transient EL

Transient EL imaging of a PSC at 1.2 V bias in a loop.

Our Measurement:

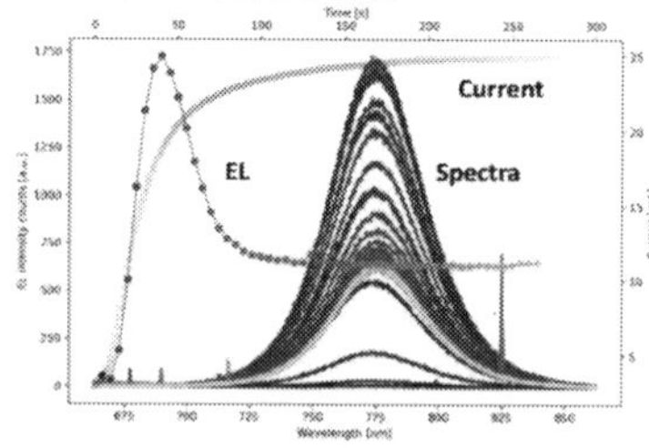

Transient EL (TrEL) intensity, spectrum and current of a mesoporous carbon PSC.

Literature [1]:

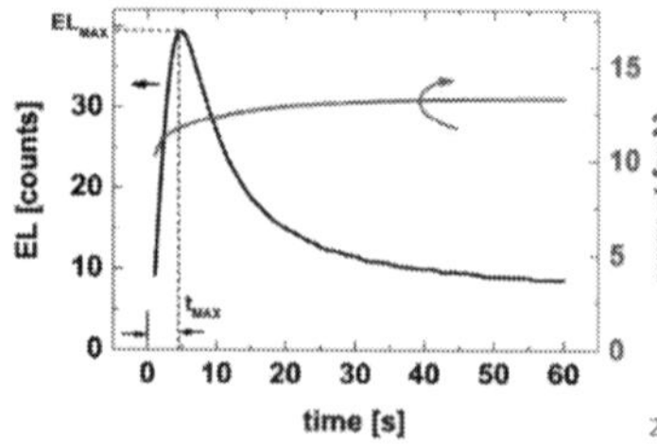

[1] M. Córdoba et al. "Electroluminescence transients and correlation with steady-state solar output in solution-prepared CH 3 NH 3 PbI 3 perovskite solar cells using different contact materials," J. Phys. D. Appl. Phys., vol. 53, no. 11, p. 115501, Mar. 2020.

2

Introduction/Outline

Goal:

Understand the temporal dynamics and local variations in transient electroluminescence (TrEL) measurements of hole transport layer-free carbon-based perovskite solar cells.

Methods:

- Electroluminescence imaging and spectrometry
- Transient 1D drift-diffusion simulations (Setfos)
- Large area 2D+1D FEM Simulation (Laoss)

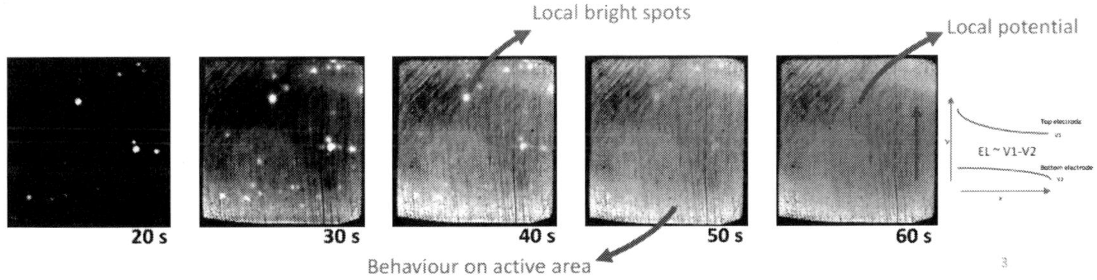

3

40th European Photovoltaic Solar Energy Conference and Exhibition

Measurement Setup & Solar Cells

Measurement Setup

- Basler ace PRO Monochrome
- 1.1" Sony IMX541 CMOS (20.2 MP)
- Near-infrared sensitivity up to 1000 nm
- Probing with NI PXIe SMU
- Spectrometer: Andor Kymera 191-B2_SIL

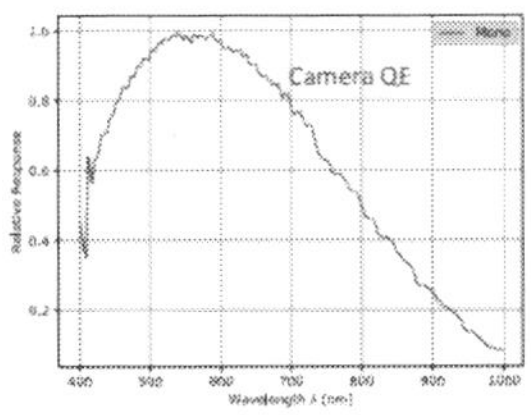

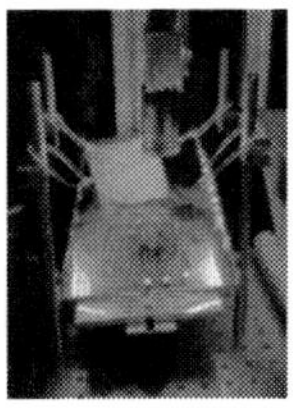

Perovskite Solar Cells

- Screen- & inkjet-printed mesoporous HTM-free carbon cell
- FTO/cTiO_2/mTiO_2/mZrO_2/ $MAPbI_3$/carbon structure
- ≈1.5 cm^2 active area
- Perovskite precursor infiltrated through mesoporous stack
- Fabricated by Solaronix SA in ambient conditions

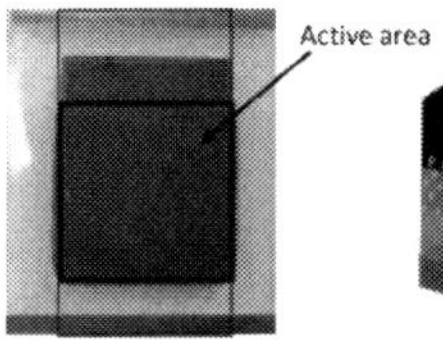

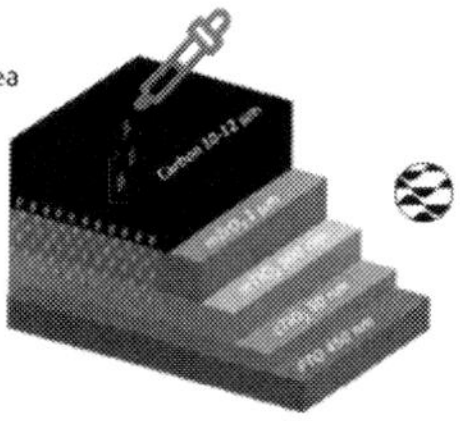

SOLARONIX

4

TrEL Measurement Results

→The active area behaves similar as reported in literature [1].

→ In some devices, local **bright spots** appear **early** and are **more intense** compared to the active area.

→ The EL signal decreases towards the center as the local potential drops.

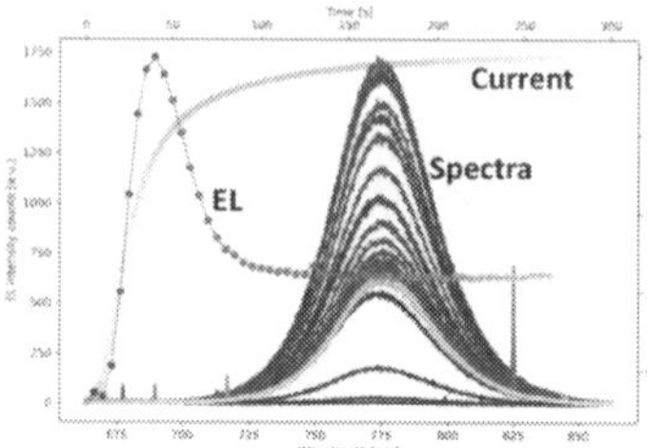

*Transient EL intensity, current and spectrum of an **encapsulated PSC** measured for 300 s at 1.2 V bias.*

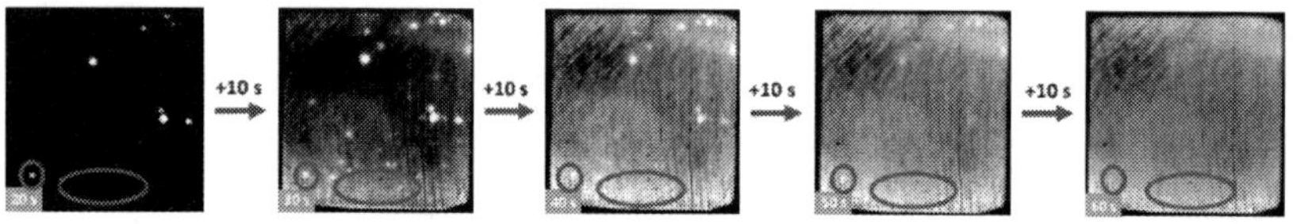

*EL images of an **unencapsulated PSC** taken at 1.2 V bias.*

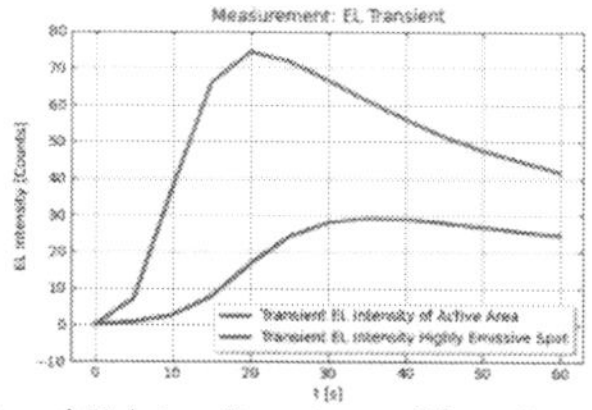

Local EL intensity extracted from images.

[1] M. Córdoba et al. "Electroluminescence transients and correlation with steady-state solar output in solution-prepared CH 3 NH 3 PbI 3 perovskite solar cells using different contact materials," J. Phys. D. Appl. Phys., vol. 53, no. 11, p. 115501, Mar. 2020.

5

40th European Photovoltaic Solar Energy Conference and Exhibition

Transient 1D Drift-Diffusion Model

- Incorporating **anions, cations** and **trap-assisted recombination** (SRH).
- Langevin **recombination coupled emission** model.
- Baseline parameters found with **steady-state and impedance measurements**.

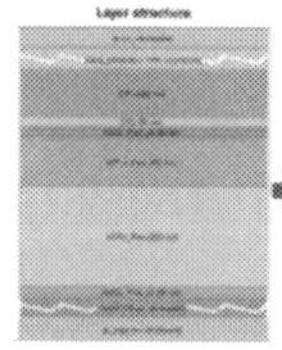

Parameter variation and fitting: Influence of **ion density and mobility** on transient luminance.

Parameter fitting for **bright spots** and normal **active area**. → Doubling ion density for bright spot.

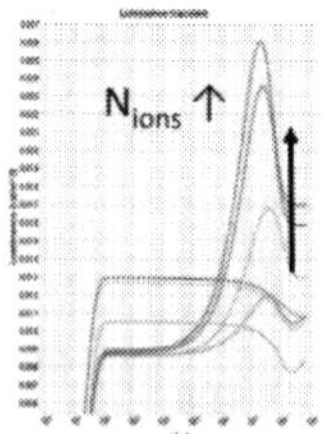

→ The ion density and mobility is crucial for the rise and fall of the EL signal and its timing.

→ Ion migration increases the non-radiative SRH recombination rate [2] and leads to the intensity drop at around 40 s.

→ Slow anions can also cause an increase in radiative recombination minutes after the intensity drop.

→ **Doubling the ion density** led to an increase in EL signal and an earlier rise and fall. → Bright spots

→ **The bright spots are probably originating from perovskite filled cavities in the mesoporous device structure.**

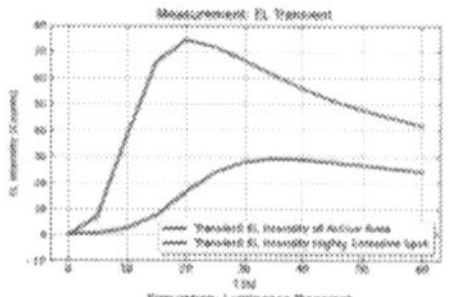

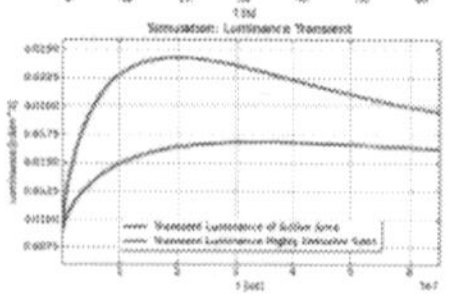

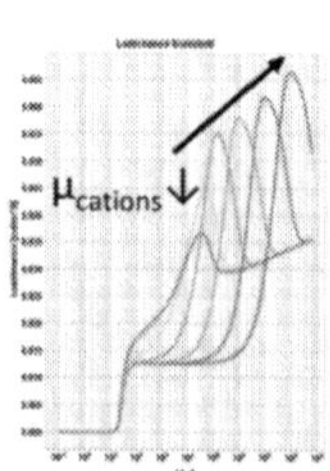

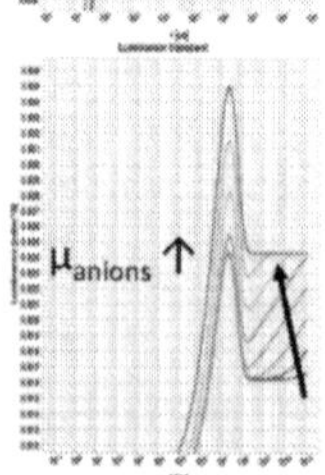

[2] D. Walter, A. Fell, Y. Wu, T. Duong, and K. Weber, "The Interaction of Ion Migration with Shockley-Read-Hall Recombination in the Bulk of Perovskite Solar Cells Explains Anomalous Voltage and Luminescence Transients," 2018 IEEE 7th World Conf. Photovolt. Energy Conversion, WCPEC 2018 - A Jt. Conf. 45th IEEE PVSC, 28th PVSEC 34th EU PVSEC, pp. 3518–3521, 2018.

6

Transient Large Area 2D+1D FEM Simulation

- To account for resistive losses caused by the sheet resistance of the electrodes, a **2D+1D FEM solver (Laoss)** has been used.
- The **quasi steady-state JVLs** (voltage-dependent current density and luminance characteristics) of the simulated bright spots and the active area are **used as 1D coupling laws.**

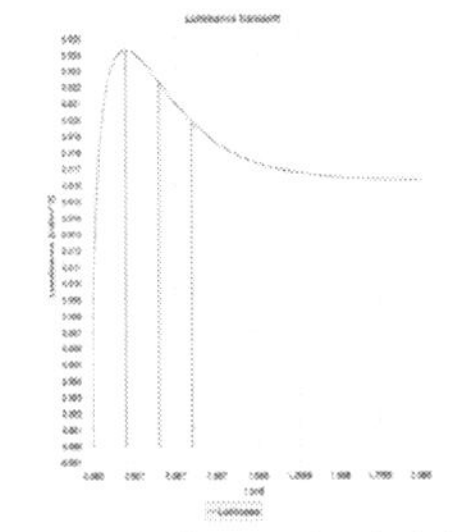

JVL extraction from transient simulation

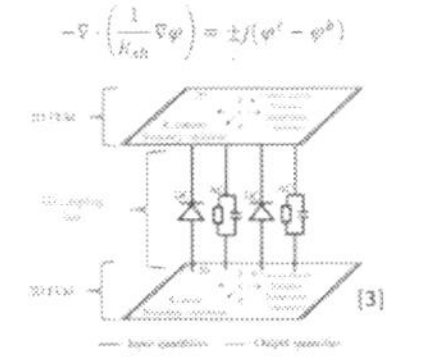

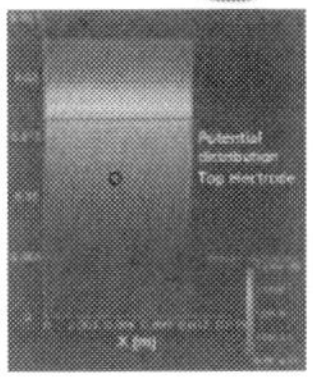

[3] E. L. Comi, E. Knapp et al. "Sinusoidal small-signal (AC) and steady-state (DC) analysis of large-area solar cells" Solar Energy Adv. 1, 100003 (2021)

7

40th European Photovoltaic Solar Energy Conference and Exhibition

Transient Large Area 2D+1D FEM Simulation

- To account for resistive losses caused by the sheet resistance of the electrodes, a **2D+1D FEM solver (Laoss)** has been used.
- The **quasi steady-state JVLs** (voltage-dependent current density and luminance characteristics) of the simulated bright spots and the active area are **used as 1D coupling laws.**
- A subdomain was defined in the center to represent a bright spot in the EL images.

→ This results in simulated **transient 2D luminance** images **considering resistive electrodes**, which cause local potential and luminance variations.

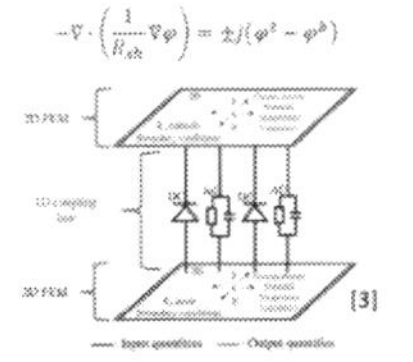

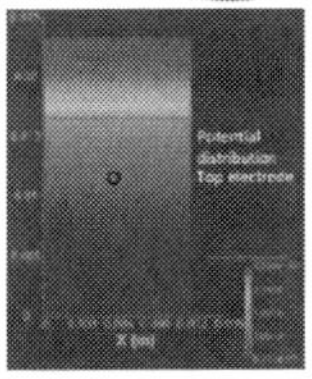

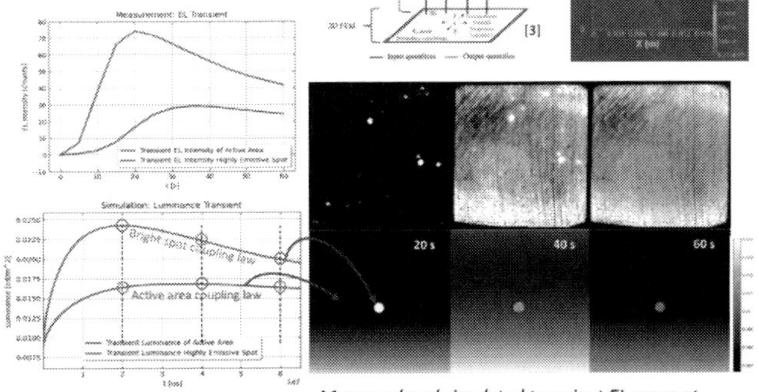

Measured and simulated transient EL maps at 1.2 V bias shown at 20, 40, and 60 s.

8

[3] E. L. Comi, E. Knapp et al. "Sinusoidal small-signal (AC) and steady-state (DC) analysis of large-area solar cells" Solar Energy Adv. 1, 100003 (2021)

What if we Measured for a Longer Period?

- A later increase of the EL signal has been predicted by the simulation when anion mobility is low.
- Our simulation did not predict the correct intensity of the late increase.

→ Changing material parameters over time? Injection barrier forming and reducing the current? Decreasing trap density? Increasing ion density?

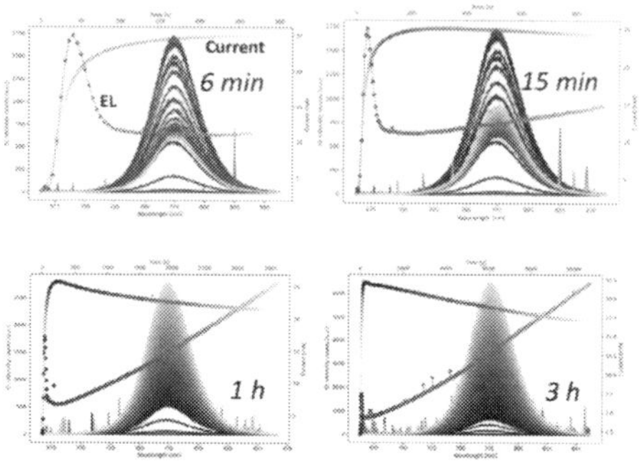

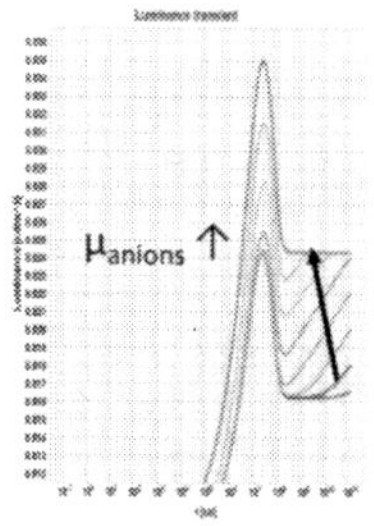

9

40th European Photovoltaic Solar Energy Conference and Exhibition

What if we Measured for a Longer Period?

- A later increase of the EL signal has been predicted by the simulation when anion mobility is low.
- Our simulation did not predict the correct intensity of the late increase.

→ Changing material parameters over time? Injection barrier forming and reducing the current? Decreasing trap density? Increasing ion density?

PSC without HTM:

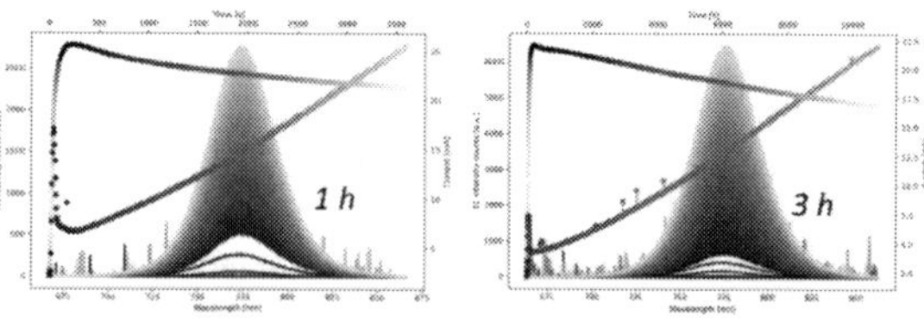

PSC with HTM:

Other devices with HTM do not show a late increase.
→ Anions moving into the HTM increasing interface recombination?

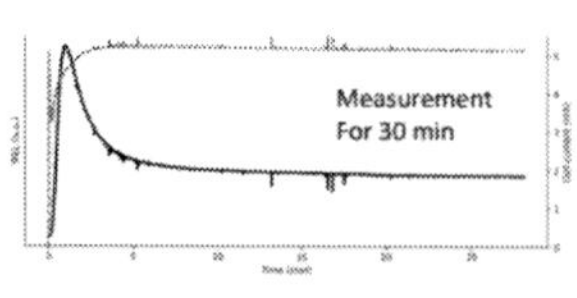

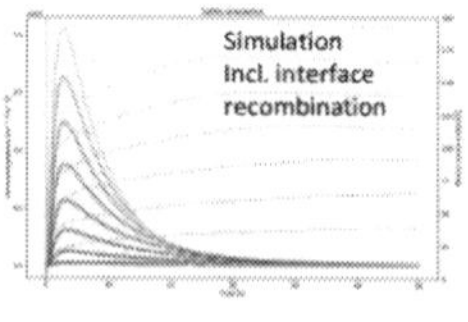

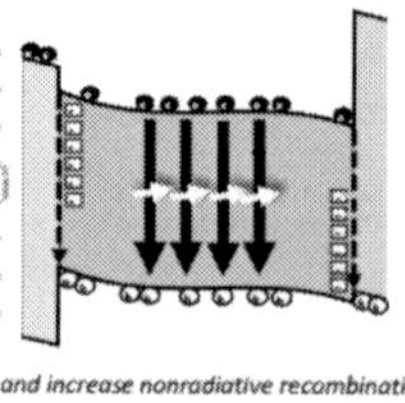

Miguel Angel Torre Cachafeiro
ZHAW ICP

Transient EL measurement (30 min) and simulation of a PSC with HTM. Anions are penetrating the HTL and increase nonradiative recombination → EL does not increase again.

10

Next Steps

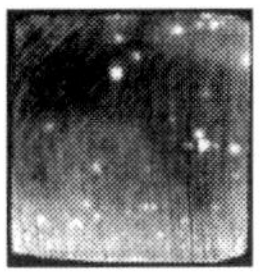

- Further investigation of late EL increase:
  →Measure devices with different fabrication methods / layer thicknesses (impact on TrEL?)
- Measure transient spectrum of such bright spots (different peak wavelength?).
- Device fabrication with intentionally placed cavities in the mesoporous structure.
  →Are perovskite filled cavities really the cause of these bright spots?

11

40th European Photovoltaic Solar Energy Conference and Exhibition

Conclusions

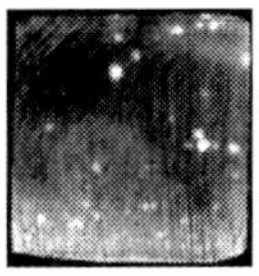

- We found an explanation for the transient EL dynamics in mesoporous HTM-free PSCs locally and globally based on fundamental charge carrier dynamics.
- The simulations predicted a late increase in EL signal. Validation with further measurements is outstanding.
- Mobile ionic charge carriers can affect long term stability of PSCs [4]. Understanding the origin of these bright spots could lead to an improvement in efficiency and stability.
- Dynamics of transient EL measurements can help to extract ion mobility and density.

[4] M. H. Futscher et al., "Quantification of ion migration in $CH_3NH_3PbI_3$ perovskite solar cells by transient capacitance measurements," pp. 1497–1503, 2019.

12

Thank you for your attention!

SOLARONIX

FLUXiM

boosts your R&D - www.fluxim.com

Booth D7

2CO.2.6

School of Engineering

ICP Institute of Computational Physics

Acknowledgement

Toby Meyer and Stephanie Narbey from Solaronix SA for providing the perovskite solar cells.

Funding support

Innosuisse project Advanced Imaging and Machine Learning for PV Quality Assurance (no. 58054.1 IP-EE)

13

Backup Slides

14

Potential Drop an EL Signal in PSCs

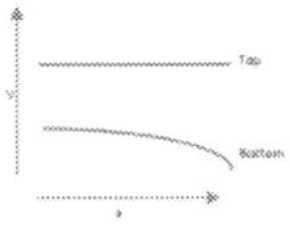

Top Electrode: **Gold** (~2 Ω/sq)
Bottom Electrode: **FTO on PET** (~60 Ω/sq)

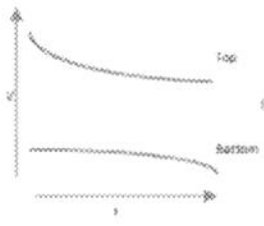

Top Electrode: **Carbon** (~15 Ω/sq)
Bottom Electrode: **FTO on glass** (~5 Ω/sq)

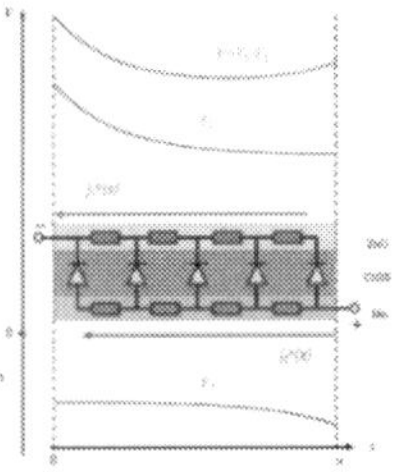

Top Contact

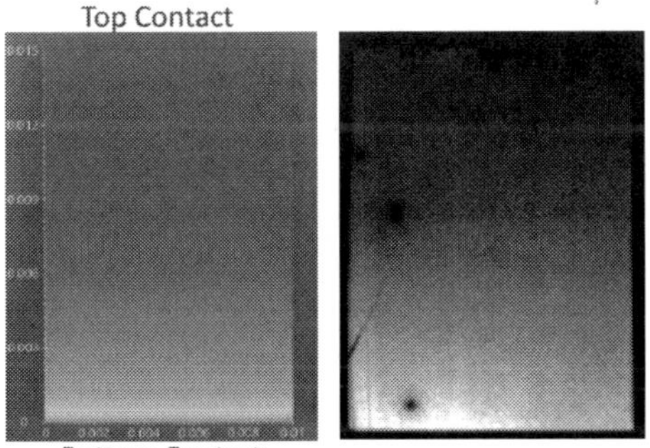

Bottom Contact

Simulated and measured EL image.

Top Contact

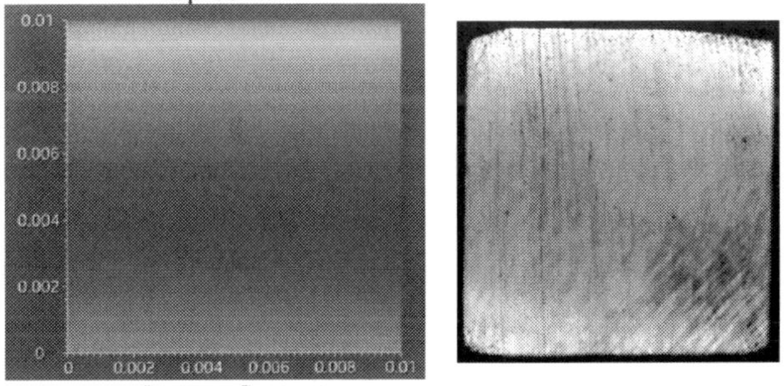

Bottom Contact

Simulated and measured EL image.

15

Quasi steady-state JVL from Setfos

- Short voltage ramp at 20, 40 and 60 s to extract JVL for local coupling law with fixed charge carrier distribution.

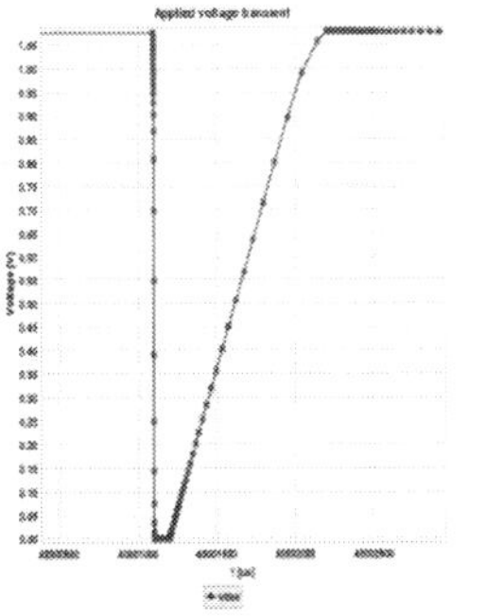

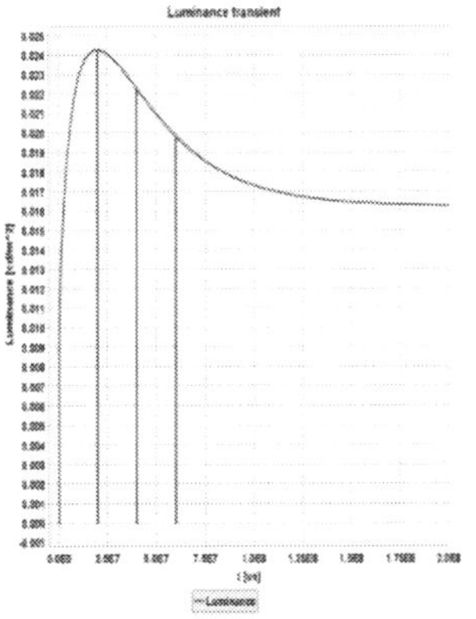

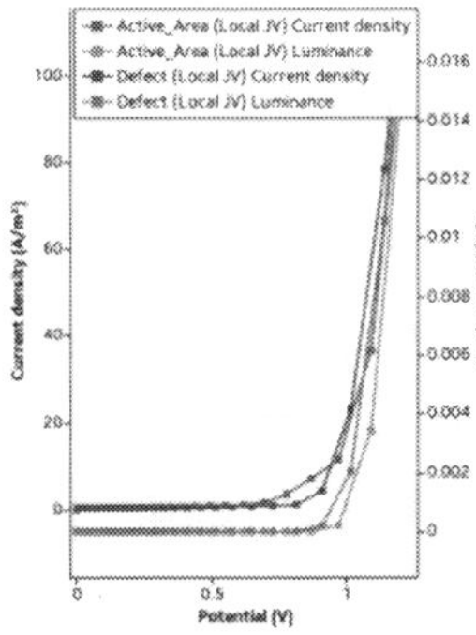

16

zh aw School of Engineering ICP Institute of Computational Physics

Simulation Parameters

Parameter	Compact TiO_2	Mesoporous TiO_2 (internal) / MAPI	Mesoporous TiO_2 / MAPI	Mesoporous ZrO_2 / MAPI	Mesoporous ZrO_2 (internal) / MAPI
Thickness	50 nm	50 nm	450 nm	950 nm	50 nm
HOMO	6 eV	5.4 eV	5.4 eV	5.4 eV	5.4 eV
LUMO	4 eV	3.9 eV	3.9 eV	3.9 eV	3.9 eV
Dielectric constant	60	6	6	6	6
DOS (HOMO & LUMO)	1e27 m^{-3}	1e19 m^{-3}	1e19 m^{-3}	1e19 m^{-3}	1e19 m^{-3}
Electron mobility	1e-3 cm^2/Vs	10 cm^2/Vs	10 cm^2/Vs	10 cm^2/Vs	10 cm^2/Vs
Hole mobility	1e-3 cm^2/Vs	10 cm^2/Vs	10 cm^2/Vs	10 cm^2/Vs	10 cm^2/Vs
Trap density (electrons)	0 cm^{-3}	3e17 cm^{-3}	3e17 cm^{-3}	3e17 cm^{-3}	3e17 cm^{-3}
Trap energy depth (electrons)	0 eV	0.5 eV	0.5 eV	0.5 eV	0.5 eV
Electron capture rate (electrons)	0 cm^3/s	3e-10 cm^3/s	3e-10 cm^3/s	3e-10 cm^3/s	3e-10 cm^3/s
Hole capture rate (electrons)	0 cm^3/s	3e-12 cm^3/s	3e-12 cm^3/s	3e-12 cm^3/s	3e-12 cm^3/s
Ion density (anions & cations)	0 m^{-3}	1.5e17 cm^{-3} / *3e17 cm^{-3}**	1.5e17 cm^{-3} / *3e17 cm^{-3}**	1.5e17 cm^{-3} / *3e17 cm^{-3}**	1.5e17 cm^{-3} / *3e17 cm^{-3}**
Anion mobility	0 cm^2/Vs	1e-19 cm^2/Vs (slow)	1e-19 cm^2/Vs (slow)	1e-19 cm^2/Vs (slow)	1e-19 cm^2/Vs (slow)
Cation mobility	0 cm^2/Vs	5e-10 cm^2/Vs	5e-10 cm^2/Vs	5e-10 cm^2/Vs	5e-10 cm^2/Vs

** Bright spot case*

Electrode Layer	Layer Glass (incoherent)	FTO	Carbon
Thickness	100 nm	450 nm	10 um
Work Function	-	4.2 eV	5.35 eV

Laoss Parameters	Sheet Resistance
Top electrode (Carbon)	15 Ohm/sq
Bottom electrode (FTO)	5 Ohm/sq

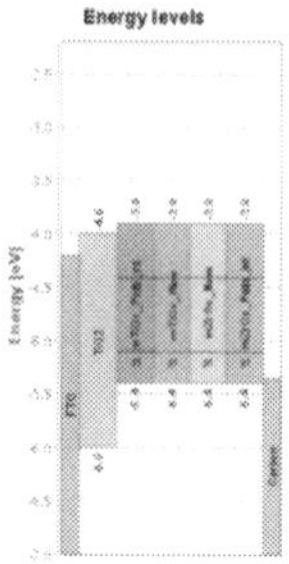

17

020077-017

zh aw School of Engineering ICP Institute of Computational Physics

TrEL Simulation with Lower Trap Density

- Trap density reduced from 3e17 cm^{-3} to 1e17 cm^{-3}
- Constant material parameters for transient simulation
- Late EL overshoot after initial EL drop appears
- Proportions of EL drop and overshoot is incorrect

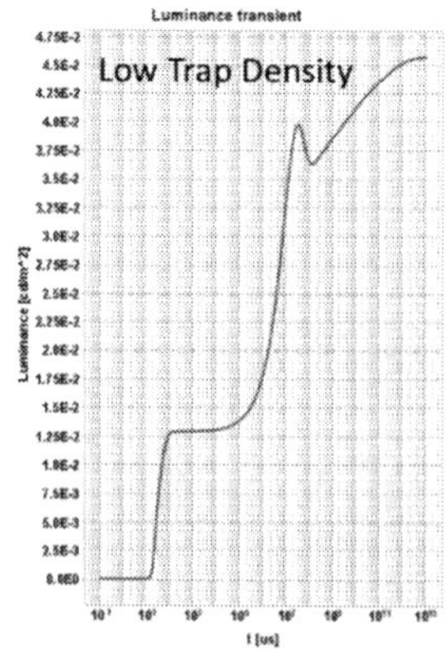

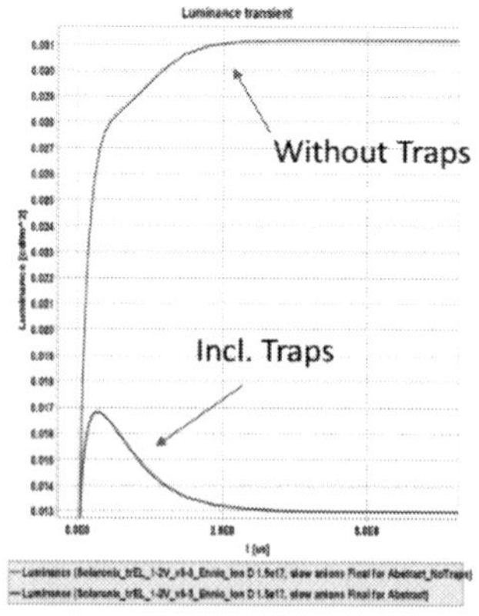

18

020077-018

PL Map of Emissive Spot (532 nm Laser)

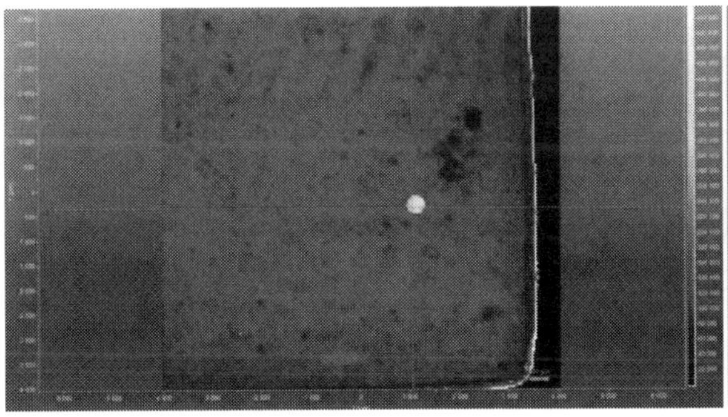

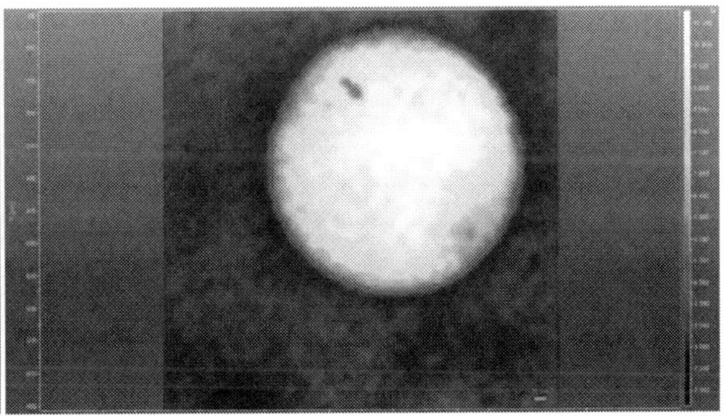

Spectra for active area and emissive Spot is the same. Just the intensity varies.

PL and EL of the Encapsulated Cell

Brightened by 80%

EL at 1.2 V after 60 s (Cell 312) ***before*** *applying a bias of 1.2 V for 3 hours.*

Delta PL (Voc-Isc) under green LED at 0.7 suns ***after*** *applying a bias of 1.2 V for 3 hours.*

EL at 1.2 V after 60 s (Cell 312) ***after*** *applying a bias of 1.2 V for 3 hours.*

Open PSC from recent Batch – Same Behaviour as Cell 312

After 1 min (Current 24.3 mA)

After 10 min (Current peak 26.9 mA)

After 1 h (Current 23.1 mA)

40th European Photovoltaic Solar Energy Conference and Exhibition

Cell with Bright Spots – DLIT Imaging

Amplitude

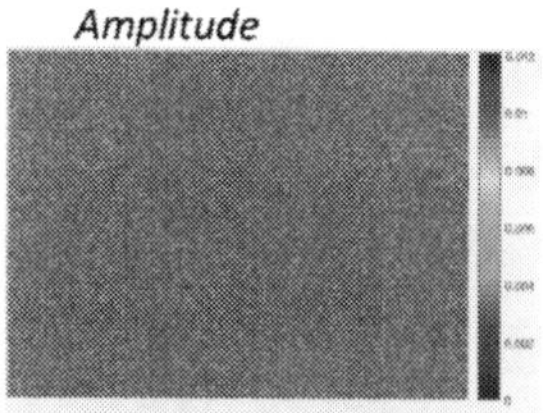

Phase

DLIT with 1.2 V pulses, 1 Hz

EL image mirrored

Bright spots also appear around shunts (probably due to a temperature increase).

The bright spots in this device probably did not originate from shunts increasing the temperature as they are not visible in a DLIT measurement.

Effect of Substrate Temperature on Structure, Morphology and Optical Properties of Sb_2Se_3 Thin Films Fabricated by Chemical-molecular Beam Deposition Method from Sb and Se precursors for solar cells

T.M. Razykov[a], A. Bosio[b], K.M. Kouchkarov[a], R. Khurramov[a*],
M.S. Tivanov[c], D.S. Bayko[c], A. Romeo[d], N. Romeo[b], N. Torabi[d], Sh.B.Utamuradova[e]

[a] Physical-Technical Institute, Chingiz Aytmatov Street 2B, Tashkent 100084, Uzbekistan
[b] University of Parma, G. P Usberti 7/A - 43124 Parma, Italy
[c] Faculty of Physics, Belarusian State University, 220030 Minsk, Belarus
[d] Universita' di Verona, Ca' Vignal 2- Strada Le Grazie 15, 37134 Verona, Italy
[e] Institute of Semiconductors Physics and Microelectronics, Yangi Olmazor Street 20, Tashkent 100057, Uzbekistan
*Corresponding author, e-mail: rkhurramov@yahoo.com

Abstract

Sb_2Se_3 thin films were obtained by chemical-molecular beam deposition on soda-lime glass from high purity Sb and Se precursors at 400 °C, 450 °C and 500 °C substrate temperature. By the exact control of the Sb/Se ratio, Sb_2Se_3 thin films with stoichiometric composition were obtained. The effect of substrate temperature on morphology, structure and optical properties of Sb_2Se_3 thin-films were studied by energy-dispersive X-ray microanalysis, X-ray diffraction, Raman spectroscopy, scanning electron microscopy and atomic force microscopy. Average diameters and lengths of Sb_2Se_3 rods deposited at different substrate temperature were the range of 0.5÷2 μm and 1÷4 μm respectively which was grown at different slope and compactness to the substrate. The optical bandgap of the films was determined from the transmission and reflection spectra and 1.16, 1.21 and 1.26 eV band gap energies were observed for 500, 450 and 400 °C substrate temperature Sb_2Se_3 thin films respectively.

Key words: Antimony triselenide, chemical molecular beam deposition, Raman spectroscopy.

1. Introduction

Over the last years, as the interest in Sb_2Se_3 chalcogenide thin films became stronger, researchers devoted to the synthesis of this compound have doubled in the recent decade. Studies showed that this binary compound has good electrochemical, optoelectric properties. The direct bandgap of this material for a single-junction solar element is between 1.1 and 1.3 eV [1] and strong absorption of sunlight $\alpha > 10^5$ cm^{-1} in the visible spectra [2], carrier mobility ≈ 10 cm^{2}V^{-1}s^{-1} for minority carriers and based on the transient absorption spectroscopy, a carrier lifetime ≈ 60 ns [3]. In addition, the elements that make up these materials have a relatively low cost, with amphibious Sb and Se with contents of 0.2 and 0.05 ppm (parts per million), respectively, resistance to external influences and nontoxicity [4]. According to the electrochemical properties, Sb_2Se_3 has been suggested as an anode material for lithium-ion batteries [5] and higher initial hydrogen storage capacity material [6]. Sb_2Se_3 binary compound is used in optical recording material [7], in thermoelectric devices [8], solar cells and photoelectrochemical cell [9] applications due to excellent optoelectronic properties. At present, the efficiency of thin film solar cells based on Sb_2Se_3 is 10.57% [10]. However, the theoretically calculated efficiency of Sb_2Se_3 solar cells can exceed 31% by the ideal Shockley-Queisser limit [11]. The efficiency of Sb_2Se_3 solar cell is highly dependent on the physical properties of the base layer as well as the growth method. Among Sb_2Se_3 thin-film growth technology, the most common are rapid vacuum thermal evaporation [12], ionic layer adsorption and reaction methods [13], magnetron sputtering [14], chemical bath deposition [15], pulsed laser deposition [16], but all these methods required additional processes to improve the physical properties, such as selenization and subsequent annealing (structural, morphological and optical) of the absorbing layer. In our method, we can obtain different compositions of thin films without additional procedures by changing the temperature of the precursors during growth. Recently, we described fabrication of Sb_2Se_3 films by chemical-molecular beam deposition (CMBD) method [17] and discussed their characteristics in [18]. Structure, morphology and optical properties of Sb_2Se_3 films fabricated at substrate temperatures of 400 °C, 450 °C and 500 °C are discussed in this paper.

2. Experimental

Sb_2Se_3 thin films were deposited on soda-lime glass (SLG). The SLG substrates were cleaned using detergent, deionized water, acetone, and ethanol in an ultrasonic cleaning bath and N_2 gas drying in sequence. Sb_2Se_3 films were deposited by CMBD processes using two different quartz crucibles each loaded with Sb (99.999% purity) or Se (99.99% purity) beads. Quartz crucibles were positioned to heat treatment from special temperature-controlled molybdenum heaters, and Sb and Se granules were melted and evaporated from the crucibles at atmospheric pressure with gaseous hydrogen at a flow rate of 20 cm^3 /min. The SLG substrates were placed at the top of reaction camera by molybdenum holder mask which was further detailed our previous works [17,18]. We put an additional thermocouple to the substrate for monitoring the exact temperature. At 400, 450 and 500 °C substrate temperatures Sb_2Se_3 thin films were obtained at Sb 900÷960 °C and Se 400÷500 °C with a 30 min evaporation time. We limited Sb_2Se_3 thin films substrate temperature from 400 °C to 500 °C since to keep away from thermal damage the SLG substrates (<540 °C) and poor quality crystallization of Sb_2Se_3 thin films at temperature below 400 °C. The reactor was naturally cooled down after film deposition. The 2÷3 μm thickness of the thin films was determined using a MIM-7 microscope.

10.4229/EUPVSEC2023/2BV.1.2
020078-001

The morphological, structural and optical properties of Sb_2Se_3 thin films were analysed by various techniques. The chemical composition of different substrate temperature samples were determined by energy dispersive X-ray (EDX) analysis using an Aztec Energy Advanced X-Max 80 (Oxford Instruments) spectrometer. The surface morphology, average grain size and topography of the films were investigated by using a LEO-1455 VP (Carl Zeiss) scanning electron microscope (SEM) and a Solver Nano (NT-MDT) atomic force microscope (AFM). The AFM instrument was used in the semi-contact mode, with a scanning probe of 10 nm tip radius, at a resonance frequency of 236 kHz. The crystal structure was performed by using an Ultima IV (Rigaku) X-ray diffractometer in the grazing incidence diffraction geometry at 1 degree of incident X-rays with CuKα radiation source (λ = 1.54178 Å). X-ray diffraction (XRD) analysis was carried out with the use of the database of the Joint Committee on Powder Diffraction Standard (JCPDS) and Crystallography Open Database (COD). Regarding structure analysis, Raman spectra were measured at room temperature, using a Nanofinder HE (LOTIS TII) confocal spectrometer with spectral resolution was better, than 2.5 cm^{-1}. A solid-state laser, with a radiation wavelength of 532 nm, was used to excite. The power of laser radiation incident on the samples was reduced to 60 µW, to avoid thermal damage. The size of the area, on which the laser radiation was focused on the surface of the samples, was about 0.7 µm in diameter. The optical properties of the films were acquired by a spectrophotometer Photon RT (Essent Optics) in the wavelength range of 400÷3000 nm with spectral resolution better than 5 nm using unpolarized light at room temperature.

3. Results and discussions

Firstly, we characterized the chemical compositions of samples by EDX analysis and the stoichiometric ratio Sb_2Se_3 films deposited at 400 °C, 450 °C and 500 °C substrate temperature were obtained. The surface morphology and grain size of rods were investigated by using SEM results. The structure of Sb_2Se_3 thin films is an array of randomly arranged rods (Fig. 1), which, depending on the temperature of the substrate. Average diameters and lengths of Sb_2Se_3 rods deposited at different substrate temperature were the range of 0.5÷2 µm and 1÷4 µm respectively which was grown at different slope and compactness to the substrate.

The crystal structure is formed by $(Sb_4Se_6)_n$ ribbons stacked in parallel in the [001] film depth direction through strong covalent Sb–Se bonds [19,20]. In general, the surface morphology of pure Sb_2Se_3 demonstrates from fine to coarse grain structures during grain growth as the substrate temperature increases.

Figure 1. Surface morphology of Sb_2Se_3 thin films deposited at different substrate temperatures

To further study the morphology of Sb_2Se_3 thin films, AFM measurements were performed. Fig. 2 shows the AFM images of Sb_2Se_3 thin films. All AFM parameters for the different substrate temperature Sb_2Se_3 thin films are given in Table 1. Studies conducted using AFM showed that the thin films have a highly developed surface relief. The value of the average surface roughness Sa is in the range of ~0.1÷0.4 µm. The average and root mean-square roughness of thin films surface increases with the increase of substrate temperature, while the surface skewness decreases from the characteristic values of the normal distribution (zero skewness reflects symmetrical height distribution) to a value of 0.44. These AFM results indicated that the grain size of thin films enlarges with the substrate temperature increases. Typically, when the substrate temperature increases, the critical radius of nucleus of grains increases and it is explained by Gibbs free energy [21].

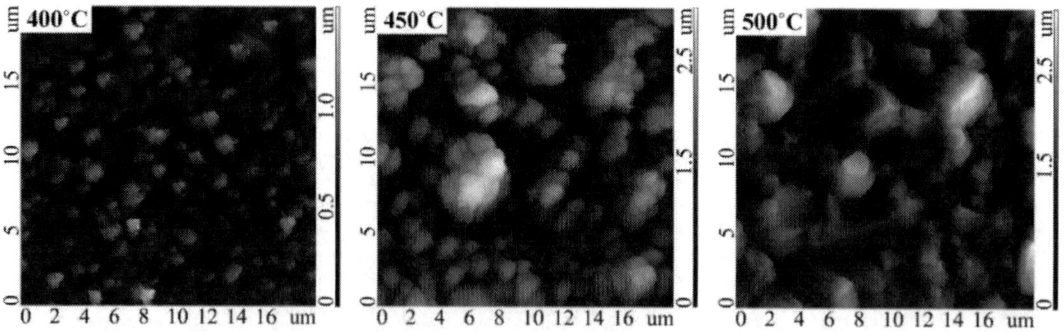

Figure 2. AFM images of Sb_2Se_3 thin films deposited at different substrate temperatures

10.4229/EUPVSEC2023/2BV.1.2
020078-002

Table 1. AFM parameters for the different substrate temperature Sb_2Se_3 thin films

$T_{substrate}$	400°C	450°C	500°C
Roughness average Sa, um	0.16	0.37	0.43
Root Mean Square Roughness Sq, um	0.22	0.48	0.55
Skewness Ssk	0.93	0.94	0.44
Kurtosis Ska	4.97	4.25	3.08

XRD analysis was carried out by Rigaku Ultima IV diffractometer. The results showed that all thin films have noticeable peaks that are in good agreement with the standard JCPDS map (No. 15–0861) of the Sb_2Se_3 orthorhombic phase. Thin films deposited at substrate temperatures of 450°C and 500°C crystallite peaks with the preferred orientation (hk1) are dominant, which are attributed to vertically oriented ribbons. Among the orientations (hk1), the preferred orientations (211) and (221) became more intense. (hk1)-Oriented grains, consisting of $(Sb_4Se_6)_n$ ribbons, have a certain angle of inclination with the substrate, which has the best carrier transport ability. At 400 °C substrate temperature, dominant peaks had (hk0) orientation (fig. 3). Orientation of crystallites of thin films is critically important for ensuring efficient charge transfer in a photoactive layer, since due to structural features, a strong directional anisotropy of electrooptical properties [22]. J. Zhou et al. found that in Sb_2Se_3 thin films, the high current part corresponds to the grain position of the film, while the low current part corresponds to the grain boundary, which indicates that carriers in Sb_2Se_3 thin films are mainly conducted through grains [23].

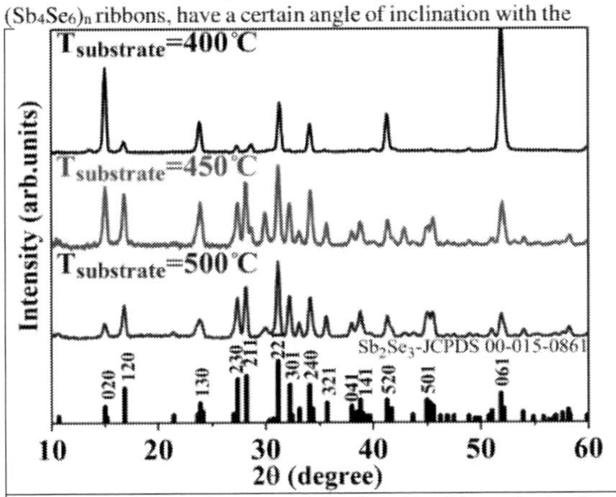

Figure 3. X-Ray Patterns of Sb_2Se_3 thin films obtained at different substrate temperatures

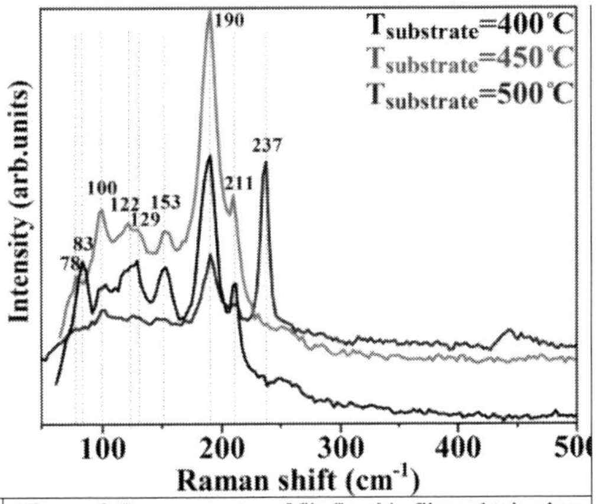

Figure 4. Raman spectra of Sb_2Se_3 thin films obtained at different substrate temperatures

Raman scattering spectra were measured (Fig. 4). to obtain additional information about the structural properties of Sb_2Se_3 films obtained by CMBD at different substrate temperatures with a stoichiometric composition of Sb/Se = 0.66. As can be seen from fig. 4, the Raman spectra for all films show peaks characteristic of Sb_2Se_3:78, 83, 100, 122, 129, 153, 190, 211, 237 cm^{-1} [24]. The peaks at 153, 237 cm^{-1}, which are characteristic of Sb and trigonal Se, are also observed. But, the Raman peak at about 153 cm^{-1} is assigned to the characteristic of both Sb and Sb_2Se_3 [25, 26]. Thus, the Raman results allow us to speak about the formation of the main phase Sb_2Se_3 and the secondary phase Sb, Se on the films. The emergency of secondary phases is a result of the flux deviations from appropriate proportion of Sb and Se contents from separate sources on the deposition process.

The absorption coefficient (α) of Sb_2Se_3 thin films was determined via transmission (T) and reflection (R) spectra (Fig. 5) [18]. As widely known, the absorber layer bandgap for band-to-band direct allowed transitions can be determined by the dependence of the absorption coefficient on the photon energy in the Tauc plot which is described in the equation as below:

$$(\alpha h\nu)^2 = A(h\nu - E_g) \quad (1)$$

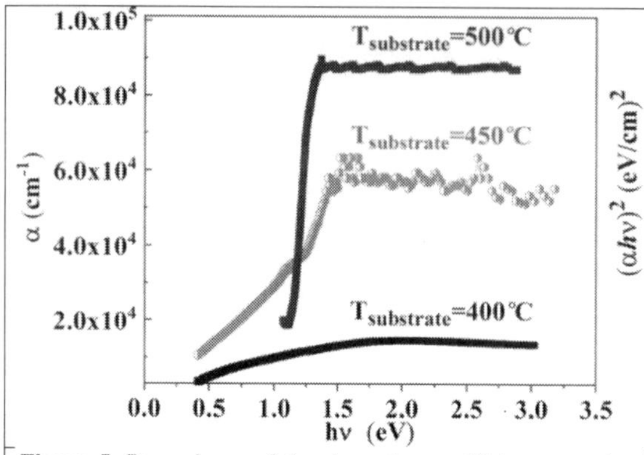

Figure 5. Dependence of the absorption coefficient α on the quantum energy of Sb_2Se_3 films obtained at different substrate temperatures.

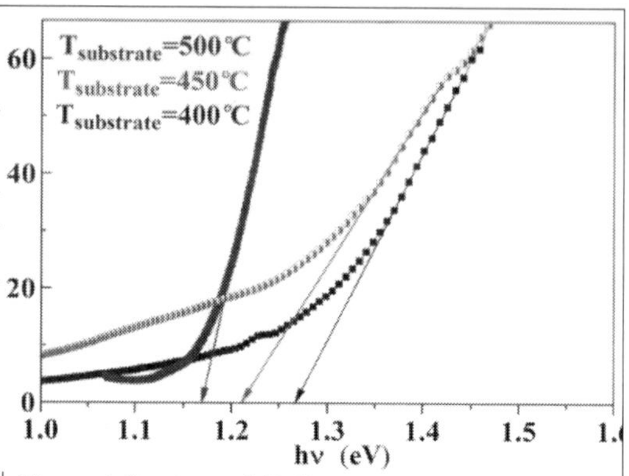

Figure 6. Band gap of Sb_2Se_3 films obtained at different substrate temperatures

1.16, 1.21 and 1.26 eV band gap energies were observed for 500, 450 and 400 °C substrate temperature Sb_2Se_3 thin films respectively (Fig. 6). These values of band gap are nearly similar with pure Sb_2Se_3 thin film (1.15 eV) [27]. A slight change in the band gap for three different substrate temperature Sb_2Se_3 thin films is an accordance with the change of thickness of the rods in relation to the bulk.

4. Conclusion

In this study, we have grown Sb_2Se_3 thin films with different substrate temperature from separate sources of Sb and Se by the CMBD method and investigated the effect of substrate temperature to the morphological, structural and optical properties of them as the absorber materials for the photovoltaic absorber applications. From the SEM and AFM analysis, it was observed that the size of grains on films and average surface roughness enlarged slightly with the increase of substrate temperature. Thin films deposited at substrate temperatures of 450°C and 500°C crystallite peaks with the preferred orientation (hk1) were dominant, which are attributed to vertically oriented ribbons, while at 400 °C substrate temperature, dominant peaks had (hk0) orientation. The Raman spectra for all films showed the peaks of Sb-Se bonding. An analysis of the dependency functions $(\alpha h\nu)^2 = f(h\nu)$ showed suitable direct bandgap for different substrate temperature Sb_2Se_3 thin films.

Acknowledgements

This work was supported by the Ministry of innovative development of the Republic of Uzbekistan (Grant № MRB-2021-540) and the State Committee on Science and Technology of the Republic of Belarus (Grant No: F21UZBG-022).

References

[1] K. Li, F. Li, C. Chen, P. Jiang, S. Lu, S. Wang, Y. Lu, G. Tu, J. Guo, L. Shui, Z. Liu, B. Song, J. Tang, One-dimensional Sb_2Se_3 enabling ultra-flexible solar cells and mini-modules for IoT applications, Nano energy 86 (2021) 106101.

[2] Y. Zhou, M. Leng, Z. Xia, J. Zhong, H. Song, X. Liu, B. Yang, J. Zhang, J. Chen, K. Zhou, J. Han, Y. Cheng, J. Tang, Solution-processed antimony selenide heterojunction solar cells, Adv. Energy Mater. 4 (8) (2014) 1301846, https://doi.org/10.1002/aenm.201301846.

[3] N. Spalatu, R. Krautmann, A. Katerski, E. Karber, R. Josepson, J. Hiie, I.O. Acik, M. Krunks, Screening and optimization of processing temperature for Sb_2Se_3 thin film growth protocol: Interrelation between grain structure, interface intermixing and solar cell performance, Sol. Energy Mater. Sol. Cells 225 (2021) 111045, https://doi.org/10.1016/j.solmat.2021.111045.

[4] K. Zeng, D.J. Xue, J. Tang, Antimony selenide thin-film solar cells. Semicond. Sci. Technol (2016). https://doi.org/10.1088/0268-1242/31/6/063001

[5] M.-Z. Xue, Z.-W. Fu, Pulsed laser deposited Sb_2Se_3 anode for lithium-ion batteries, Journal of Alloys and Compounds 458 (1) (2008) 351–356.

[6] J. Ma, Y. Wang, Y. Wang, Q. Chen, J. Lian, W. Zheng, Controlled synthesis of onedimensional Sb2Se3 nanostructures and their electrochemical properties, The Journal of Physical Chemistry C 113 (31) (2009) 13588–13592.

[7] Y. Nakane, N. Sato, H. Makinon, S. Miyaoka, Principle of laser recording mechanism by forming an alloy in the multilayer of thin metallic films, Proc. SPIE 0529, Optical Mass Data Storage I 76 (1985).

[8] H.C. Kim, T.S. Oh, D.-B. Hyun, Thermoelectric properties of the p-type Bi_2Te_3- Sb_2Te_3- Sb_2Se_3 alloys fabricated by mechanical alloying and hot pressing, Journal of Physics and Chemistry of Solids 61 (5) (2000) 743–749.

[9] B.R. Sankapal, C.D. Lokhande, Studies on photoelectrochemical (PEC) cell formed with SILAR deposited Bi_2Se_3 – Sb_2Se_3 multilayer thin films, Sol. Energy Mater. Sol. Cells 69 (1) (2001) 43–52.

[10] Y. Zhao, Sh. Wang, Ch. Li, B. Che, X. Chen, H. Chen, R. Tang, X. Wang, G. Chen, T. Wang, J. Gong, T. Chen, X. Xiao, J. Li, Regulating deposition kinetics via a novel additive-assisted chemical bath deposition technology enables fabrication of 10.57%- efficiency Sb_2Se_3 solar cells, Energy Environ. Sci. (2022), DOI: 10.1039/d2ee02261c.

[11] S. Rühle, Tabulated values of the Shockley-Queisser limit for single junction solar cells. Solar Energy 130, (2016) 139–147. https://doi.org/10.1016/j.solener.2016.02.015.

[12] L. Wang, D.B. Li, K. Li, C. Chen, H.X. Deng, L. Gao, Y. Zhao, F. Jiang, L. Li, F. Huang, Y. He, H. Song, G. Niu, J. Tang, Stable 6%-efficient Sb_2Se_3 solar cells with a ZnO buffer layer. Nat. Energy 2 (2017). https://doi.org/10.1038/ nenergy.2017.46.

[13] C.D. Lokhande, B.R. Sankapal, S.D. Sartale, H.M. Pathan, M. Giersig, V. Ganesan, A novel method for the deposition of nanocrystalline Bi_2Se_3, Sb_2Se_3 and Bi_2Se_3-

Sb_2Se_3 thin films SILAR, Applied surface science 182 (3) (2001) 413–417.

[14] Ma, C., Guo, H., Wang, X., Chen, Z., Cang, Q., Jia, X., Li, Y., Yuan, N., Ding, J., 2019. Fabrication of Sb_2Se_3 thin film solar cells by co-sputtering of Sb_2Se_3 and Se targets. Solar Energy 193, 275–282. https://doi.org/10.1016/j.solener.2019.09.046.

[15] Maghraoui-Meherzi, H., ben Nasr, T., Dachraoui, M., 2013. Synthesis, structure and optical properties of Sb2Se3. Mater. Sci. Semicond. Process 16, 179–184. https://doi. org/10.1016/j.mssp.2012.04.019.

[16] Mavlonov, A., Shukurov, A., Raziq, F., Wei, H., Kuchkarov, K., Ergashev, B., Razykov, T., Qiao, L., Structural and morphological properties of PLD Sb2Se3 thin films for use in solar cells. Solar Energy 208 (2020), 451–456. https://doi.org/10.1016/j. solener.2020.08.004.

[17] T.M. Razykov, A. Bosio, B.A. Ergashev, D. Isakov, R. Khurramov, K.M. Kouchkarov, M.A. Makhmudov, A. Romeo, N. Romeo, M.S. Tivanov, Sh.B. Utamuradova, D.S. Bayko, L.S. Lyashenko, O.V. Korolik, A.A. Mavlonov, Structural and optical properties of Sb_xSe_y thin films obtained by chemical molecular beam deposition method from Sb and Se precursors, Solar Energy 254 (2023) 67–72. https://doi.org/10.1016/j.solener.2023.03.010.

[18] T.M. Razykov, K.M. Kuchkarov, M.S. Tivanov, D.S. Bayko, L.S. Lyashenko, B.A. Ergashev, A. Mavlonov, A.N. Olimov, R. Khurramov, D.Z. Isakov, M. Pirimmatov, Characteristics of thin Sb2Se3 films obtained by the chemical molecular beam deposition method for thin-film solar cells, Thin Solid Films 774 (2023) 139844. https://doi.org/10.1016/j.tsf.2023.139844.

[19] D. Wang, C. Song, X. Fu, X. Li, Growth of one-dimensional Sb_2S_3 and Sb_2Se_3 crystals with straw-tied-like architectures, J. Cryst. Growth 281 (2005) 611–615. https://doi.org/10.1016/j.jcrysgro.2005.04.064.

[20] V.L. Deringer, R.P. Stoffel, M. Wuttig, R. Dronskowski, Vibrational properties and bonding nature of Sb_2Se_3 and their implications for chalcogenide materials, Chem. Sci. 6 (2015) 5255–5262.

[21] S.N. Park, S.Y. Kim, S.J. Lee, Sh.J. Sung, K.J. Yang, J.K. Kang, and D.H. Kim, Controlled synthesis of (hk1) preferentially oriented Sb_2Se_3 rod arrays by co-evaporation for photovoltaic applications, J. Material Chem. A (2019), https://doi.org/10.1039/c9ta08289a.

[22] H. Lei, J. Chen, Z. Tan, G. Fang, Review of Recent Progress in Antimony Chalcogenide-Based Solar Cells: Materials and Devices. Solar RRL (2019) https://doi.org/10.1002/solr.201900026.

[23] J. Zhou, H. Chen, X. Zhang, K. Chi, Y. Cai, Y. Cao, J. Pang, Substrate dependence on $(Sb_4Se_6)_n$ ribbon orientations of antimony selenide thin films: Morphology, carrier transport and photovoltaic performance, Journal of Alloys and Compounds 862 (2021) 158703. https://doi.org/10.1016/j.jallcom.2021.158703.

[24] V.F. Pedro, G. Maxim, A. Xavier, J. Tariq, P. Marcel, Multiwavelength excitation Raman scattering study of Sb_2Se_3 compound: fundamental vibrational properties and secondary phases detection. 2d Mater 6 (2019). https://doi.org/10.1088/2053-¬1583/ab4029

[25] A. Kumar, V. Kumar, A. Romeo, C. Wiemer and G. Mariotto, Raman spectroscopy and in situ XRD probing of the thermal decomposition of Sb_2Se_3 thin films, J. Phys. Chem. C 125 (36) (2021) 19858-19865, https://doi.org/10.1021/acs.jpcc.1c05047.

[26] A. Shongalova, M.R. Correia, J.P. Teixeira, J.P. Leitão, J.C. González, S. Ranjbar, S. Garud, B. Vermang, J.M.V. Cunha, P.M.P. Salomé, P.A. Fernandes, Growth of Sb_2Se_3 thin films by selenization of RF sputtered binary precursors. Sol. Energy Mater. Sol. Cells 187 (2018) 219–226, https://doi.org/10.1016/j.solmat.2018.08.003.

[27] C. Chen, W. Li, Y. Zhou, C. Chen, M. Luo, X. Liu, K. Zeng, B. Yang, C. Zhang, J. Han, J. Tang, Optical properties of amorphous and polycrystalline Sb_2Se_3 thin films prepared by thermal evaporation. Appl. Phys. Lett. 107 (2015) https://doi.org/ 10.1063/1.4927741.

GROWTH OF SB_2SE_3 BY IONIZED JET DEPOSITION FOR THIN FILM SOLAR CELLS APPLICATION

Narges Torabi, Elisa Artegiani, Gianpiero Tedeschi[+], Ikram Anefnaf, Simya Olavil Karay, Maryam Mukhtar, Alessandro Romeo*

LAPS-Laboratory for Photovoltaics and Solid State Physics, Department of Computer Science, University of Verona, Ca' Vignal 1, Strada Le Grazie 15, 37134 Verona, Italy.

[+]Noivion S.r.l., Piazza Manifattura,1, 38068, Rovereto (TN), Italy

*author for correspondence, e-mail address: alessandro.romeo@univr.it

ABSTRACT: The composition ratio of Sb_2Se_3 plays a key role in the thin film quality and defects formation. To address this problem different post-treatments and growth techniques have been applied. In this work, we study the growth of Sb_2Se_3 thin film by a novel vacuum-based method called ionized jet deposition (IJD). This technique is based on pulsed electron ablation that directs ionized gas through a metallic nozzle into the solid target to generate plasma with the same composition as the target. This method is therefore able to grow Sb_2Se_3 with the same stoichiometry as the source while using low substrate temperatures. Here Sb_2Se_3 thin film has been grown with this technique on CdS buffer layer for a superstrate configuration device, also the results are compared with Sb (deposited by IJD)

Keywords: Sb_2Se_3, Thin Films, Sb.

1 Introduction

Sb_2Se_3 has received intensive attention recently on account of its outstanding structural and optoelectronic properties like one dimensional (1D) crystal structure, suitable band gap (1.0-1.2 eV) [1], high absorption coefficient (>10^5 cm^{-1}), and low-toxic earth-abundant elements [2]. Its crystal structure consists of building blocks stacked along one direction and when the ribbons grow well-oriented, the recombination loss at the grain boundaries is minimized. Up to now the highest efficiency of 10% has been reported for Sb_2Se_3 grown by vapor deposition [3]. According to the literature, Sb_2Se_3 has some defects showing higher concentrations under Se-poor conditions. For instance, Sb_2Se_3 thin films grown by thermal evaporation reveal the presence of selenium vacancies because of the lower vapor pressure of Se [4]. However, this problem could be addressed through Se co-evaporation or post-selenization, as we have already demonstrated [5]. Moreover, it has been indicated that the stoichiometry ratio of Sb_2Se_3 can influence the orientation of the thin film [6].

Ionized jet deposition (IJD) technique is an improved version of the pulsed electron deposition, which favors the growth of the thin film with the same composition as the target from where is ablated. It is based on pulsed electron ablation that directs ionized gas through a metallic nozzle into the solid target to generate plasma. This technique consists of applying electron beam pulses at high voltage (up to 25 kV) with a duration of less than 1 µs to the target material, which is transformed into its plasma phase. The ablation keeps the stoichiometric composition of the plasma, and it is able to generate a high-power density of about 108 W/cm^2 at the target surface.

Different variables can be tuned to improve the properties of the deposited thin film such as electron pulse frequency, acceleration voltage, chamber pressure, and substrate temperature. Last but not least, it has the potential to deposit thin films on a large area by a small target.

In this paper we explore the possibility of preparing Sb2Se3-based solar cells by applying IJD on an Sb_2Se_3-target and we compare it with devices prepared by deposition by IJD of pure Sb followed by selenization.

2 Materials and Methods

2.1 Device fabrication

The devices have been fabricated on TEC12 glass, a soda lime glass coated by a bilayer of SnO_2:F/SnO_2. CdS has been applied as a buffer layer by chemical bath deposition (CBD). The substrate is immersed in a solution containing 15 ml of cadmium acetate, 10 ml of thiourea, and 32 ml of ammonia (20 %) in 200 ml of distilled water for 16 min at 60 °C.

Sb_2Se_3 is deposited, on the CdS buffer layer, by IJD, the chamber is pumped down to 10^{-7} mbar, and then a flux of argon is inserted in order to generate the above-mentioned plasma. The 8-cm in diameter Sb_2Se_3 target is then ablated by the plasma, generated by the electron pulses. The parameters that can be regulated in the electron pulse deposition are typically pulse frequency, acceleration voltage, and deposition pressure which was kept in the 10^{-3} mbar range. The substrate temperature has been fixed to 300 °C, a value that allows for good crystallization without selenium loss.

The back contact in our Sb_2Se_3-based devices is typically a layer of 50 nm-thick gold.

To analyze the properties of the absorber according to the process, we have monitored the effect of the above-mentioned process parameters. We propose here 3 different deposition conditions, shown in Table I, and we have studied the different physical properties according to the deposition parameters. Moreover, we have confronted these -as deposited- with Sb_2Se_3 post-treated ones at 300 °C in vacuum.

Table I: Process parameters for Sb_2Se_3 deposited by IJD.

Repetition rate	10Hz	20Hz	30Hz
Voltage (kV)	11,7	8.1	4
Current (mA)	5.6	8.1	8.5
Pressure (10^{-3} mbar)	1.26	1.43	1.4

A second approach used in this work is the IJD of pure Sb target. In order to obtain, in the end, the wished Sb_2Se_3

compound, the antimony layer has been treated under selenium atmosphere at 450 °C for one hour.

2.2 Physical and Electrical Characterization.

Morphological analysis was carried out by NT-MDT Solver Pro atomic force microscopy (AFM), the structural properties were monitored by X-ray diffraction (XRD) analysis with a Thermo ARL X'TRA powder diffractometer (in Bragg- Brentano geometry, equipped with a Cu-anode X-ray source (Kα, λ=1.5418 Å) and a Peltier Si (Li) cooled solid state detector).

3 ANALYSIS AND DISCUSSION

3.1 Sb_2Se_3 from Sb_2Se_3 target

The AFM images of Sb_2Se_3 thin films deposited at a substrate temperature of 300 °C with an acceleration voltage of 12 kV and a repetition rate of 10 Hz are shown in Figure 1. Pictures of as-deposited (left) and post-deposition annealed (PAT) at 300 °C in vacuum for one hour (right) thin films indicate that the grain size remarkably increases. The roughness root mean square (RMS) value, calculated from the AFM data, also increases from 10.4 nm to 21.3 nm.

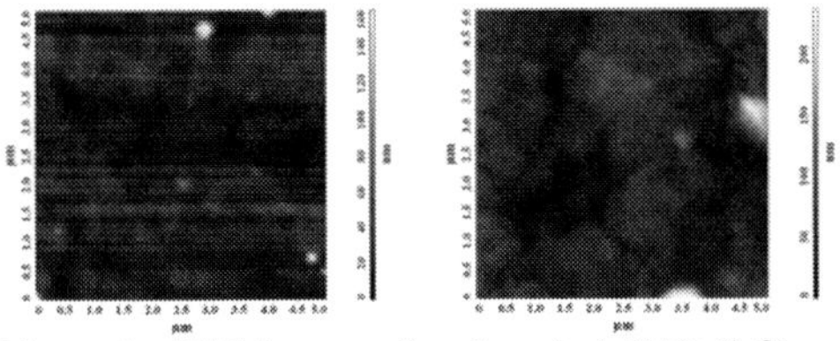

Figure 1: AFM images of as-deposited (AD) (left) and post-annealed (PAT) (right), deposited with a frequency rate of 10 Hz.

The deposition without heating the substrate results in an amorphous layer (not shown here), while, by increasing the temperature to 300 °C the film begins to crystallize. On the other hand, the crystallinity considerably improves by applying the PAT to the film. After the annealing treatment, the layer shows XRD peaks (see Figure 2) corresponding to Sb_2Se_3 -JCPDS card n. PDF 15 0861. The expected (hk1) peaks, with high intensity, are present, but some (hk0) planes (which are known to be detrimental for the carriers' transport) are also detected.

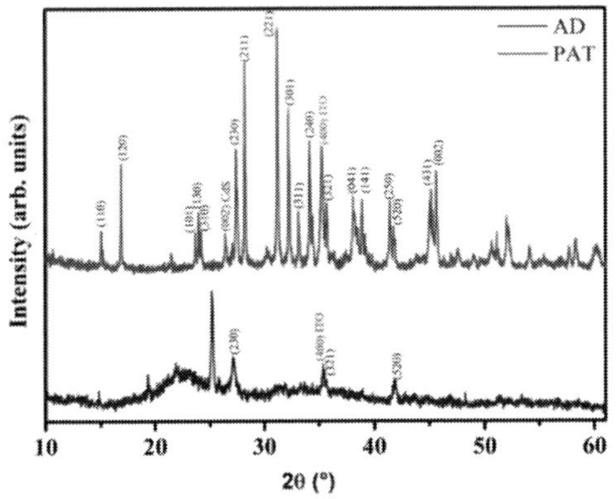

Figure 2: XRD pattern of as-deposited (AD) (bottom in blue) and post-deposition annealed (PAT) (top, in red), deposited with a frequency rate of 10 Hz.

Figure 3 illustrates the AFM pictures of Sb_2Se_3 thin films grown at a substrate temperature of 300 °C with acceleration voltage of 8 kV and repetition rate of 20 Hz. AFM images of as-deposited and post-annealed Sb_2Se_3 thin film show that for as-deposited layers -ordered and homogeneously distributed- nanograins are generated and that then, after annealing, the grains are clustered into larger agglomeration. Also, by post-annealing treatment, RMS roughness is enhanced from 11.42 nm to 18.42 nm.

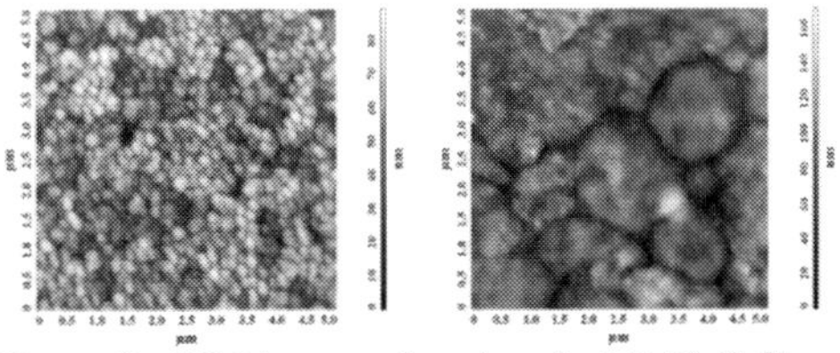

Figure 3: AFM images of as-deposited (AD) (left) and post-annealed (PAT) (right), deposited with a frequency rate of 20 Hz.

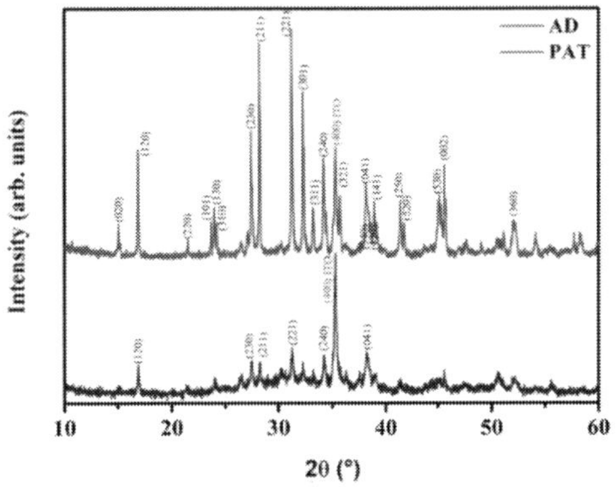

Figure 4: XRD pattern of as-deposited (AD) (bottom in blue) and post-deposition annealed (PAT) (top, in red) deposited with a frequency rate of 20 Hz.

With this higher repetition rate, the as-deposited thin film shows a larger number of XRD peaks corresponding to Sb_2Se_3 (see figure 4). After annealing at 300 °C in vacuum, the layer shows much more XRD peaks, suggesting that a second recrystallization has occurred. But the number of (hk0) reflections is also increased.

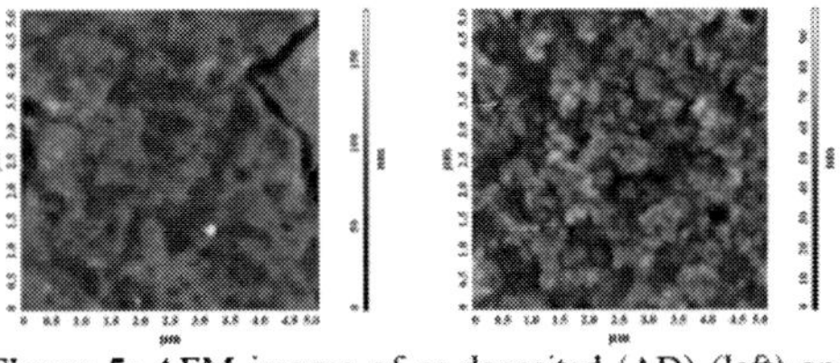

Figure 5: AFM images of as-deposited (AD) (left) and post-annealed (PAT) (right), deposited with a frequency rate of 30 Hz.

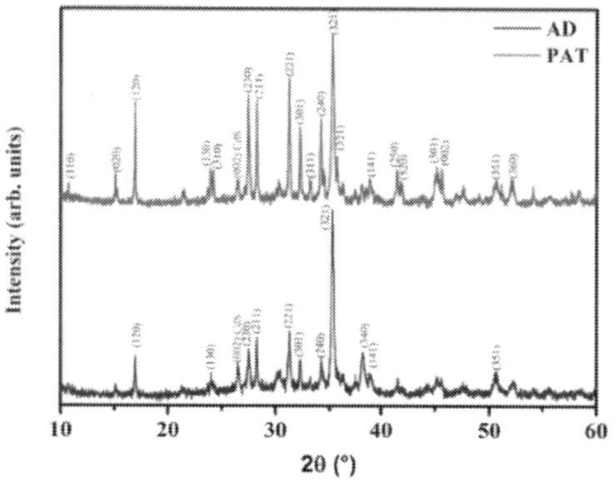

Figure 6: XRD pattern of as-deposited (AD) (bottom in blue) and post-deposition annealed (PAT) (top, in red), deposited with a frequency rate of 30 Hz.

The AFM and XRD analysis of samples deposited at a substrate temperature of 300 °C with an acceleration voltage of 4 kV and a repetition rate of 30 Hz are shown respectively in Figure 5 and 6. In this sample, more (hk1) planes are present -before and after post-annealing- compared to the previous ones. In addition, by comparing the AFM images of Figure 5 in the latter the grains have started to cluster. The RMS roughness of this sample is enhanced from 10.88 nm to 13.89 nm after post-annealing. The XRD patterns of the different samples suggest that by increasing the repetition rate the crystallinity of the as-deposited films actually improves with sharper peaks and an increase in (hk1) orientation.

3.2 Sb_2Se_3 from Sb target

Sb_2Se_3 thin films have also been prepared by Sb deposition by IJD followed by post-deposition treatment in selenium atmosphere and subsequently analyzed. The morphology changes drastically after the annealing in selenium atmosphere, in Figure 7 an AFM picture of the selenised antimony is shown, grains form clusters and the structure is effectively crystalline.

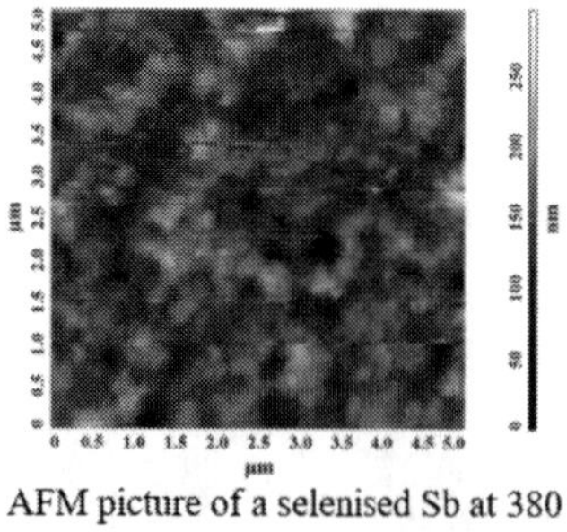

Figure 7: AFM picture of a selenised Sb at 380 °C.

Different selenization times and selenization temperatures have been applied. In Figure 8 we show the XRD patterns of Sb layer after a 10 min. or 15 min. post-deposition selenization ramp, from 350 °C to 380 °C.
As can be observed, the peaks of Sb_2Se_3 are present, demonstrating the transformation of the Sb layer. But what is interesting is that the main peaks are in the (h,k,1) orientation.

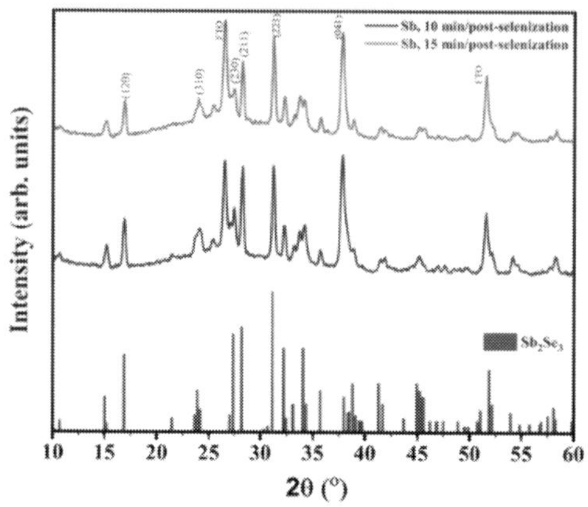

Figure 8: XRD pattern of selenium treated Sb layer, two treatment times are shown and compared to the powder reflections.

4 RESULTS AND CONCLUSIONS

Preliminary finished devices have been prepared and measured with a Keithley Source Meter 2420, under an AM 1.5 spectrum at 100 mW/cm2, using a LOT Quantum Design Europe solar simulator LS0306. The solar cells made with Sb_2Se_3 target have delivered very small current densities while the ones fabricated from the selenized Sb layer show promising efficiencies of around 2%. The possible explanation for these two behaviors could be attributed to the different orientations of the ribbons in the two cases. The Sb_2Se_3 target ablation still delivers different (hk0) orientations while the Sb-based ones have a few peaks but are mainly oriented in the (hk1) direction. In any case, the different frequency rate influences the orientation of the deposited layer as demonstrated by the comparison of the samples made at 10, 20 and 30 Hz. Possibly by a further investigation of the IJD deposition parameters this orientation can be driven into a higher efficiency device.

5 ACKNOWLEDGMENTS

This project has been funded by Cariverona Foundation in the project INSOBILD Ref. 2022.0091-CUP B33C22001760007.

6 REFERENCES

[1] F. Kosek, J. Tulka, and L. Štourač, "Optical, photoelectric and electric properties of single-crystalline Sb2Se3," *Czechoslov. J. Phys. B 1978 283*, vol. 28, no. 3, pp. 325–330, Mar. 1978.

[2] C. Chen *et al.*, "Optical properties of amorphous and polycrystalline Sb2Se3 thin films prepared by thermal evaporation," *Appl. Phys. Lett.*, vol. 107, no. 4, p. 043905, Jul. 2015.

[3] Z. Duan *et al.*, "Sb2Se3 Thin-Film Solar Cells Exceeding 10% Power Conversion Efficiency Enabled by Injection Vapor Deposition Technology," *Adv. Mater.*, vol. 34, no. 30, p. 2202969, Jul. 2022.

[4] X. Liu *et al.*, "Thermal Evaporation and

Characterization of Sb 2 Se 3 Thin Film for Substrate Sb 2 Se 3 /CdS Solar Cells," 2014.
[5] V. Kumar *et al.*, "Analysis of Se Co-evaporation and Post-selenization for Sb 2 Se 3-Based Solar Cells," 2021.
[6] K. J. Tiwari *et al.*, "Efficient Sb2Se3/CdS planar heterojunction solar cells in substrate configuration with (hk0) oriented Sb2Se3 thin films," *Sol. Energy Mater. Sol. Cells*, vol. 215, p. 110603, Sep. 2020.

EFFICIENT NARROWBAND PHOTOVOLTAIC OPTICAL ENERGY RECEIVERS

Igor Sokolovskyi[1], Vitaliy Kostylyov[1], Anatoliy Sachenko[1], and Anatoli Shkrebtii[2]*
[1] V. Lashkaryov Institute of Semiconductor Physics (ISP), NAS of Ukraine, Kyiv 03028, Ukraine
[2] Ontario Tech University, 2000 Simcoe St. N., Oshawa, ON, L1G 0C5, Canada
*Corresponding authors: anatoli.chkrebtii@ontariotechu.ca

ABSTRACT: The demand for wireless power transfer (WPT) systems is increasing with the growing need not only for portable electronic devices, electric vehicle charging, smartphones, etc. Long-range monochromatic laser/photovoltaic receiver systems that convert the monochromatic light by means of optical transmission of energy at a distance can be used, *e.g.*, (i) to power unmanned aerial vehicles (UAVs) or space vehicles in the direct line of sight of a ground station (allowing them to be in the air up to the technical limitations); (ii) to provide sufficient energy for effective use of ion engines [1-3], *etc*. In such WPT systems, solar cell (SC) based receivers can be effectively used, while they have to be optimized in terms of the transmitter power and its wavelength, as well as the SC design, thermal balance, and power propagation through the atmosphere. To evaluate the photovoltaic receiver efficiency, we modify a previously developed theoretical model of photoconversion in silicon solar cells [4] to optimize the SC receiver in terms of the incident light wavelength, incoming power, the receiver design, and its characteristics. As demonstrated, such a narrowband photoconverter can be at least twice as efficient (reaching up to 50%) compared to terrestrial solar cells.
Keywords: power transfer, narrow band, photoconversion efficiency

1 INTRODUCTION

Wireless optical power transmission (WOPT) systems with a photovoltaic cell as a receiver can be used for achieving long-range, high-power, and hazard-free power delivery in the air. The development of photovoltaic systems with an internal conversion of monochromatic light (similar to photoluminescent solar concentrators (LSCs) [5]) and systems of the optical transmission of energy at a distance, makes the question of the efficiency of converting monochromatic light into electrical power relevant. Such transmitter/receiver systems can be used, for instance, (i) to power unmanned aerial vehicles or space vehicles in the direct line of sight of a ground station or transmitter and allow them to be in the air up to the limit of their technical possibilities and (ii) to provide sufficient energy for effective use of ion engines [1]. To evaluate such efficiency, we modify the theoretical model of photoconversion in silicon solar cells, which we developed previously [4]. The solar cell efficiency for the transmitter spectral shape, close to monochromatic, can be tuned to the transmitter parameters such as the power density and wavelength of irradiated power.

Four main directions of wireless optical energy transfer can be proposed:

1. The light transfer through the solid transparent plate, similar to the delivery of power in the LSC photoconverters [5]. The photoconverters area is matched with the power source automatically. Optical energy is uniformly delivered to the edges of the plate.
2. Recharging or powering of mobile electronic devices on short distances [1,2].
3. Powering of unmanned aerial vehicles at medium distances in the atmosphere [3]. This requires using transmitter wavelength without significant absorption in the air and the match between the receiver's area and the beam's size.
4. Long-range energy transfer to spacecraft. The beam's size becomes crucial, but microgravity gives more possibilities for precise attenuation of the lenses for the laser.

2 MODEL

We modified previously developed theoretical tools for terrestrial solar cell optimization and applied the tools for the novel highly demanded wireless optical power transmission systems. The ability of such systems is seriously limited by the low efficiency of power conversion. Unlike standard AM1.5 conditions for SCs, the optimization for aerospace vehicle operation requires consideration of the adequate thermal balance between high incoming power density and radiative cooling for space, or air convection for UAV cooling. Power propagation through the atmosphere should also be included in addition to the photoconverter's parameters optimization.

Here we demonstrate a simplified version of the formalism without going into details of the recombination (our complete model includes, *e.g.*, such recombination parameters as bulk Shockley-Read-Hall lifetime and surface recombination velocity), and other parameters to be optimized. In this model the short-circuit current for the photovoltaic receiver can be written as:

$$J_{SC} = q \int d\lambda \, EQE(\lambda) \, I(\lambda), \qquad (1)$$

where q is the elementary charge, λ is the wavelength of the incident light, $I(\lambda)$ is its intensity, and $EQE(\lambda)$ is the external quantum efficiency of the photoconverter. For the case of close to monochromatic illumination from the transmitter, the spectral shape $I(\lambda)$ can be considered as a narrow band.

It is established that the experimental spectral dependencies of the external quantum efficiency $EQE(\lambda)$ in textured silicon SCs near the absorption edge are described by an empirical formula $EQE(\lambda)=[1+b/(4n_r(\lambda)^2 \cdot \alpha(\lambda)d)]^{-1}$, where $\alpha(\lambda)$ is the light absorption coefficient, d is the base thickness, n_r is the refractive index, and b is a non-dimensional coefficient, which characterizes the texturing quality. It is usually greater than 1, when b=1 the established empirical formula transforms into the well-known Yablonovitch formula.

10.4229/EUPVSEC2023/2BV.1.9
020081-001

The physical meaning of the parameter b is the ratio of the photon path length in the case of a perfectly randomized surface, to the photon path length in a particular sample with nonrandomized surface. This formula allows modelling and optimization the photoconversion efficiency of the SC in terms of base thickness d.

This self-consistent approach for textured Si solar cells efficiency η of has been validated experimentally. Dark and light I–V characteristics as well as light intensity dependence of the open-circuit voltage were measured for a number of highly efficient silicon p-n junction SCs with the photoconversion efficiency $\eta \geq 20\%$. The spectral dependencies of $EQE(\lambda)$ under AM1.5 conditions were also investigated experimentally.

For the textured silicon SCs considered, the experimental dependencies of $EQE(\lambda)$ were analyzed. It is shown that close to the absorption edge $EQE(\lambda)$ behavior can be accurately described by the above empirical formula with the parameter b, ranging from 1.6 to 4. Using the experimental values $EQE(\lambda)$ and the proposed formula, we calculated the dependence of the short-circuit current density JSC on the base thickness d for the textured SCs.

The light I-V characteristics for the rear contact solar cells under consideration were calculated using the expressions from [5]:

$$I(V) = I_{ph} - I_r(V) - \frac{V+IR_s}{R_{SH}}, \quad (2)$$

$$I_r(V) = qA_{SC}\left(\frac{d}{\tau_{eff}^b} + S_{00s}\left(1+\frac{\Delta n}{n_0}\right) + \frac{A_B}{A_{SC}}S_{sc}\right)\Delta n(V), \quad (3)$$

$$\tau_{eff}^b(n) = \left[\frac{1}{\tau_{SRH}} + \frac{1}{\tau_r(n)} + \frac{1}{\tau_{nr}(n)} + \frac{1}{\tau_{Auger}(n)}\right]^{-1}, \quad (4)$$

$$\Delta n(V) = -\frac{n_0}{2} + \sqrt{\frac{n_0^2}{4} + n_{i0}(T)^2 e^{\Delta E_g/kT}\left(exp\frac{q(V-IR_s)}{kT} - 1\right)}, \quad (5)$$

where $I(V)$ is the full current, I_{ph} is the photogeneration current, $I_r(V)$ recombination (dark) current, V is applied voltage, S_{00s} is initial surface recombination rate, A_{SC} and A_B are total and doped SC areas, R_s and R_{SH} are series and shunt resistances, $\tau_{eff}^b(n)$ is effective bulk lifetime, n_{i0} is the intrinsic concentration at low injection, and ΔE_g (n_0, Δn) is the magnitude of bandgap narrowing in Si.

The expression for the dark current according to (2) has the form:

$$I_D(V) = \frac{qA_{SC}d\ \Delta n}{\tau_{eff}(\Delta n)} - \frac{V-IR_s}{R_{SH}}. \quad (6)$$

The photogeneration current I_L dependence on the open circuit voltage V_{OC} can be found from (2) by putting $I = 0$:

$$I_L = \frac{A_{SC}qd}{\tau_{eff}(\Delta n_{OC})}\Delta n_{OC} - \frac{V_{OC}}{R_{sh}}, \quad (7)$$

with

$$\Delta n_{OC} = -\frac{n_0}{2} + \sqrt{\frac{n_0^2}{4} + n_{i0}^2 e^{\Delta E_g/kT}\left(e^{qV_{OC}/kT} - 1\right)}. \quad (8)$$

If we compare the expressions for Δn and Δn_{OC}, as well as the dark current (7) and photogeneration current (8), one sees that they coincide with each other, if we replace V by V_{OC} and I_D by I_L and also put R_S in (5) to zero.

This means that the solutions $I_D(V, R_{sh}, R_s)$ and $I_L(V_{OC}, R_{sh})$ are identical if $R_S = 0$. And $V_{OC}(\Delta n_{OC})$ can be found from (8):

$$V_{OC} = \frac{kT}{q}ln\left(\frac{(\Delta n_{OC}+n_0)\cdot\Delta n_{OC}}{n_{i0}(T)^2 e^{\Delta E_g/kT}} + 1\right). \quad (9)$$

Multiplying current $I(V)$ on the applied voltage V, we get the power $P(V)$, and maximizing this from $dP/dV = 0$ condition we find the voltage V_M at the point of maximum power. Substituting V_M in equation (2), we obtain the current at the maximum power point I_M. This allows calculating photoconversion efficiency η and the I–V fill-factor FF in the usual way. Note that in the approximation used, the short-circuit current I_{SC} value is a parameter.

On the other hand, the short-circuit current in the textured silicon SCs can be calculated if the external quantum efficiency of the photocurrent $EQE(\lambda)$ is known.

In the limiting case of zero spectrum width, the incoming power, entering the photoconverter, can be written as:

$$P = q\int J_{ld}(\lambda)\frac{hc}{\lambda}d\lambda = qJ_l\frac{hc}{\lambda_l}, \quad (10)$$

where $J_{ld}(\lambda)$ is the photon flux density, delivered from the transmitter, J_l is the integral of $J_{ld}(\lambda)$, λ_l is the wavelength of the monochromatic light, h is Planck's constant, and c is the speed of light. The photoconversion efficiency η can be written in the standard form:

$$\eta = \frac{qJ_mV_m}{P}, \quad (11)$$

where J_m and V_m are the current density and voltage of the photoconverter at the maximum power point.

3 RESULTS

To illustrate the theoretical formalism, we applied the model using parameters of Si-based thin film solar cells from the work of Sai et al. [6]. In particular, we used Si-based textured solar cells with different base thicknesses for the modeling. The resulting wavelength-dependent photoconversion efficiency for the monochromatic light of the three solar cells was considered. The incoming to the receiver power density was basically assumed to be 100 mW/cm[2], equal to the one for AM1.5 irradiation.

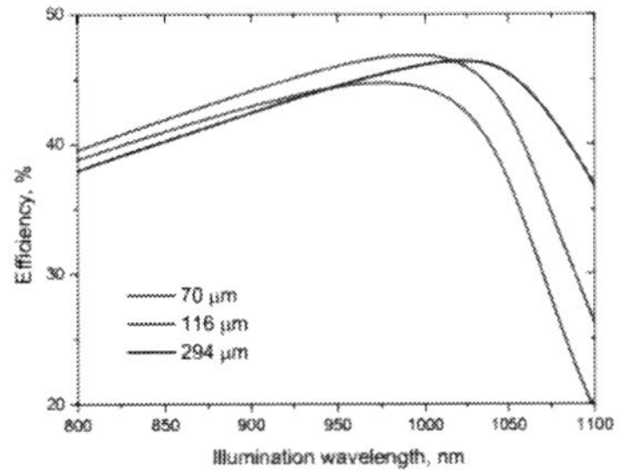

Figure 1. Photoconversion efficiency η of the photovoltaics receiver *vs.* wavelength of the incident light for three Si solar cell samples with different base thicknesses [3].

10.4229/EUPVSEC2023/2BV.1.9
020081-002

As shown, the efficiency of photoconversion for the receiver significantly exceeds that for the Si-based solar cells under sunlight. A properly optimized receiver for the incoming beam, close to monochromatic, practically does not suffer from the high photon thermalization loss. Indeed, the maximum efficiency is 46.9% for a sample with a base thickness of 116 nm. At the same time, the wavelength of monochromatic light, the thickness of the sample, and its other parameters are subject to optimization.

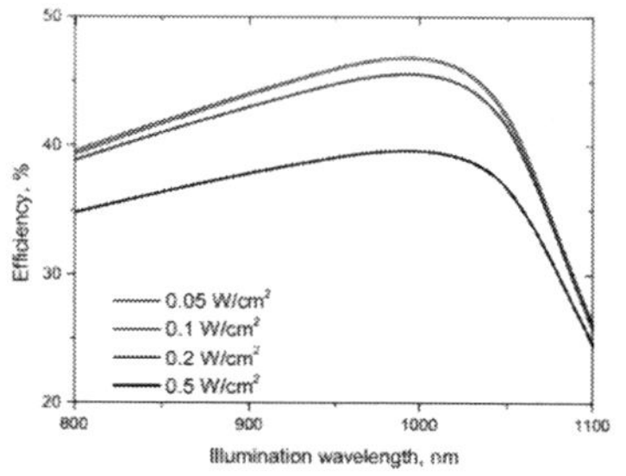

Figure 2. Photoconversion efficiency η of the photovoltaics receiver *vs.* wavelength of the incident light for Si solar cell samples with different illumination powers: 50, 100, and 200 mW/cm^2.

Regarding the possibility of tuning the wavelength of monochromatic incoming light, nanoscale structures can be used, in which the wavelength changes due to quantization when the size of the structure changes. Figure 1 shows the λ dependent photoconversion efficiency for the incoming power density of 100 mW/cm^2. As the calculations showed, the efficiency of photoconversion for this case significantly exceeds that for sunlight due to the minimization of losses on photon energy relaxation. Their maximum efficiency is 46.87% for a 116 μm thick sample, which is about twice its efficiency under solar illumination. Maximum η values for 100 mW/cm^2 illumination power confirm effective optimization of the solar cell's parameters. Figure 2 shows the wavelength-dependent efficiency η for different illumination power. One can see, that the optimal wavelength is practically independent from the input power density in the wide range of the parameter.

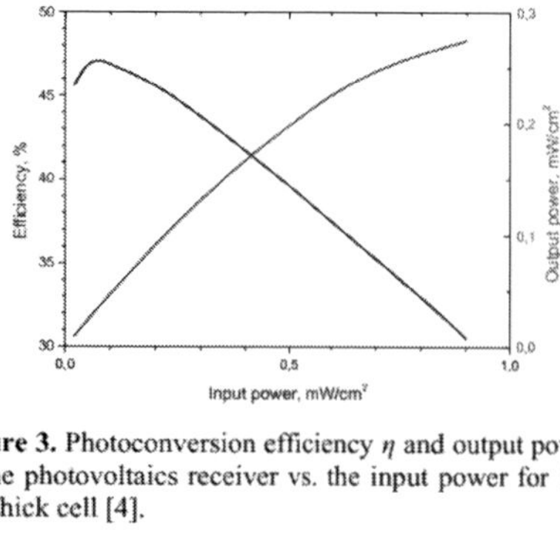

Figure 3. Photoconversion efficiency η and output power of the photovoltaics receiver vs. the input power for 116 μm thick cell [4].

Figure 3 shows the dependence of the photoconversion efficiency and, therefore, output power density on the input power density. Unlike classical solar cells, considered photoconverters can receive power in a wide range of densities.

The thickness dependence of photoconversion efficiency is shown at Figure 4. Base optical and recombination parameters of 116 μm thick cell from [6] were implemented.

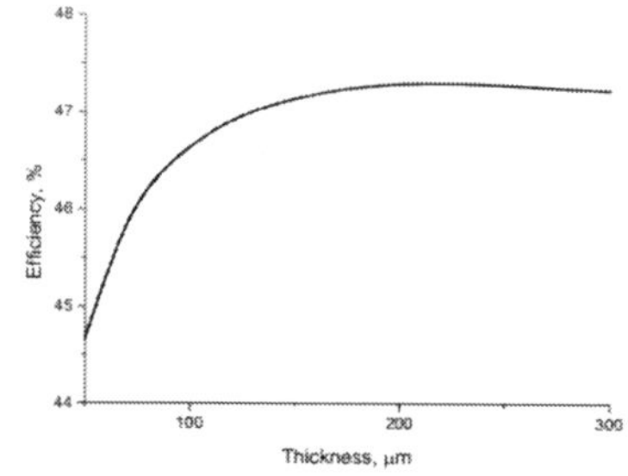

Figure 4. Thickness dependence of photoconversion efficiency for the cell with parameters of 116 μm thick solar cell [4].

At the same time, to fully optimize the WPT systems for the wide range of transmitter/receiver parameters, not only the wavelength of monochromatic light but also the thickness of the sample, its recombination (our comprehensive model includes such recombination parameters as bulk Shockley-Read-Hall lifetime, and surface recombination velocity), and other SCs characteristics, all are the subject of optimization.

4 CONCLUSIONS

A theoretical formalism for solar cell photoconversion efficiency, used as narrowband photoconverters in WPT receivers, is developed. As shown, the SCs efficiency can approach 50% for the optimized solar cells.

The balance between relaxation and recombination losses is shifted toward thicker elements for monochromatic photoconverters. Therefore, optical path-increasing texturization suites better, than just antireflection coating for this case and minimization of bulk recombination is crucial.

Unlike classical solar cells, considered photoconverters can receive power in a wide range of densities. Our calculations show a slow enough decrease in the photoconversion efficiency with the rise in the input power density.

5 REFERENCES

[1] Zhang, Zhuo, Jianwei Zhang, Yuxiang Gong, Yinli Zhou, Xing Zhang, Chao Chen, Hao Wu, Yongyi Chen, Li Qin, Yongqiang Ning, and Lijun Wang. "Long-Distance High-Power Wireless Optical Energy Transmission Based on VECSELs" Crystals, **12**, pp. 1475-1488 (2022).

10.4229/EUPVSEC2023/2BV.1.9
020081-003

[2] Nadeem Javed, Ngoc-Luu Nguyen, Syed Farhan Ali Naqvi, and Jinyong Ha, "Long-range wireless optical power transfer system using an EDFA," Opt. Express, **30**, 33767-33779 (2022).

[3] See https://biz.crast.net/laser-charged-chinese-drone-can-stay-in-the-air-indefinitely/

[4] A. V. Sachenko, V.P. Kostylyov V.M. Vlasiuk, I.O. Sokolovskyi, M. Evstigneev, A.I. Shkrebtii, D. Johnston, P. Michael, and T. Missimer, "Characterization and Optimization of Highly Efficient Silicon-Based Textured Solar Cells: Theory and Experiment," *Proceedings of 48th IEEE Photovoltaic Specialists Conference (PVSC48)*, pp. 0544-0550, 2021.

[5] M.R. Kulish, A.V. Sachenko, V.P. Kostylyov, I.O. Sokolovskyi, A.I. Shkrebtii. "Light Separation for the Lateral System of Luminescent Solar Concentrators", *Proceedings of 8th World Conference on Photovoltaic Energy Conversion (WCPEC8)*, pp. 412 – 414, 2022.

[6] Sai, H, Oku, T, Sato, Y, Tanabe, M, Matsui, T, Matsubara, K. Potential of very thin and high-efficiency silicon heterojunction solar cells. Prog. Photovolt. Res. Appl., **27**, pp. 1061– 1070 (2019).

10.4229/EUPVSEC2023/2BV.1.9
020081-004

SOME FEATURES OF CHARGE CARRIERS INJECTION IN NANOCOMPOSITE SOLAR CELLS BASED ON CLONIDINE/SI HETEROJUNCTION WITH PLASMONIC AU NANOPARTICLES

S.V. Mamykin, T.R. Barlas, V.R. Romanyuk, I.B. Mamontova, O.S. Kondratenko, N.V. Kotova, N.M. Roshchina, P.S. Smertenko
V. Lashkaryov Institute of Semiconductor Physics National Academy of Sciences of Ukraine
41 Nauky ave., Kyiv 03680, Ukraine, Tel: +38 (044) 5256546; E-mail: mamykin@isp.kiev.ua

ABSTRACT: A detailed analysis of the fine structure of the I-V characteristics of the clonidine/Si hybrids with plasmonic Au nanoparticles is carried out. The features of the injection processes were determined on the basis of the fine structure of the I-V characteristics as processed by the method of dimensionless sensitivity. Dimensionless sensitivity is the slope of I-V characteristic in log-log scale. Photoconverting heterostructures for research were fabricated by self-organization process at room temperature from aqueous solution of clonidine (CLO) in chemical bath. *n*-type Si (100) with 0.5-2.4 Ohm·cm resistivity have been used as substrate. Two sets of structures were selected for comparison: flat Si – flat Si + CLO – flat Si + CLO +Au and porous Si – porous Si + CLO – porous Si + CLO + Au. For both cases Au nanoparticles increase photocurrent with restriction at high bias even for contact restriction for flat substrate. In all cases the exponential behavior of I-V characteristics at low biases is perfect, but in case of flat substrate $I(V) \propto \exp(V^{1.5})$, while for porous one $I(V) \propto \exp(V)$. There are monomolecular recombination in dielectric at double injection (α=3) and monomolecular recombination in semiconductor (α=2). This allows choosing a suitable design for a solar cells.
Keywords: nanocomposite solar cell, clonidine-Si heterojunction, plasmonic Au nanoparticles, charge flow, dimensionless sensitivity.

1 INTRODUCTION

The search for new materials [1-3], architectures [4, 5], principles [6, 7], and physical phenomena [8, 9] to increase the efficiency of silicon-based organic/inorganic hybrids for solar cells application, as well as to reduce their cost, is continues. We applied several new approaches to solve these problems: (i) use of one of the inexpensive representatives of heterocyclic amines (clonidine), (ii) fabrication of thin film heterostructures at room temperatures, (iii) application of a simple chemical bath method, (iv) use of plasmonic Au nanoparticles, (v) application of dimensionless sensitivity for injection and recombination diagnostics. For the first time, a detailed analysis of the fine structure of the I-V characteristics was carried out in this work.

It is well known that injection of both types of charge carriers and their recombination are the main processes that determine the efficiency almost of all types of semiconductor devices [10]. Classic works on monopolar [11] and double [12] injections described theoretically and experimentally the base injection mechanisms as well as research approaches. It is the ability to inject a sufficient amount of charge carriers that is a factor in the separation of these charges during the operation of solar cells.

The aim of this study is to expand knowledge about injection mechanisms and processes, as well as to understand how to control these processes during various technological operations.

This analysis allows us to select the appropriate structure design for further application of these structures for solar cells.

2 SAMPLE PREPARATION AND EXPERIMENTAL

Photoconverting heterostructures were prepared by the method of self-organization at room temperature from an aqueous solution of clonidine (CLO) in a chemical bath [9].

Factory polished *n*-Si (100) with a resistivity of 0.5-2.4 Ohm·cm was used as a substrate. Nanostructured surface was obtained by etched in a solution of H_2O_2:HF:H_2O=1:2:10 at room temperature for t=1-3 min.

The contacts were the ohmic InGa-to-Si contact and the 50 nm thick Au top contact (S = 0.16 cm^2). Thus, two sets of photoconverting heterostructures were selected for comparison: flat Si – flat Si + CLO – flat Si + CLO +Au and porous Si – porous Si + CLO – porous Si + CLO + Au.

The forward and reverse I-V characteristics of the fabricated photoconverting heterostructures were measured in the dark and under illumination by standard technique.

3 METHODOLOGY DESCRIPTION

Previously [13] the injection approach was proposed for investigation of solar cell. It is based on the analysis of the fine structure of the I-V characteristics as processed by the method of dimensionless sensitivity [14, 15]. Dimensionless sensitivity is the slope of current-voltage characteristic in log-log scale:

$$DS \equiv \alpha = \frac{d\ln(I)}{d\ln(V)} = \frac{(dI/I)}{(dV/V)} = \frac{dI}{dV}\cdot\frac{V}{I} \quad (1)$$

We also define the value *DS of DS* as γ

$$\gamma = \frac{d\ln(\alpha)}{d\ln(V)} = \frac{(d\alpha/\alpha)}{(dV/V)} = \frac{d\alpha}{dV}\cdot\frac{V}{\alpha} \quad (2)$$

In fact, the α and γ values are the exponents in the expressions, respectively,

$$I(V) \propto V^{\alpha} \text{ and } I(V) \propto \exp\left(V^{\gamma}\right) \quad (3)$$

Table I: Recombination and injection regimes in terms of dimensionless sensitivity.

Value of DS	Mark in the Fig. 1	Recombination and injection regimes
$0<DS_{min}<1$	*7*	Regime of constant field between contacts
1	*2*	Ohm's law
1	*6*	Contact limited current
$1<DS<1.5$	*3*	Low injection
1.5	*10*	Double injection in semiconductor, bimolecular recombination of carriers
2	*9, 4*	Monopolar injection in semiconductor or dielectric without traps
2	*5*	Monopolar injection in semiconductor or dielectric with traps
2	*4, 5*	Double injection in ideal dielectric, monomolecular recombination of carriers
$DS_{max}>2$	*8*	Traps field limited (monopolar injection) or recombination limited (double injection)
3	-	Double injection in ideal dielectric, monomolecular recombination of carriers
4	-	Super high double injection level

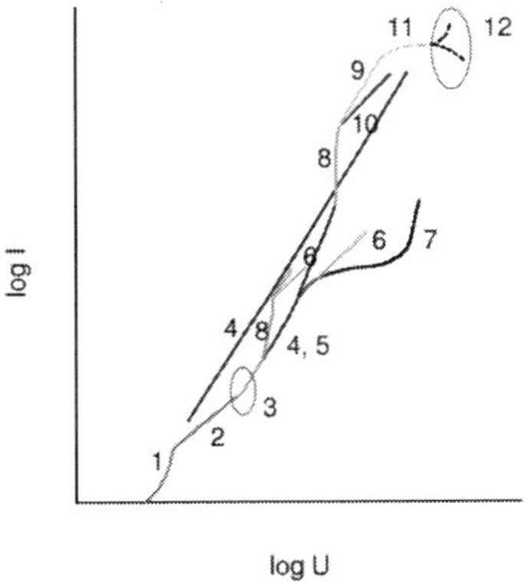

Figure 1: Peculiarities of injection and recombination regimes: (1) voltage drop on the junction, (2) Ohm's low, (3) low injection of charge carriers, (4) monopolar injection of charge carriers in semiconductor or dielectric, (5) double injection of charge carriers in ideal dielectric, bimolecular recombination of charge carriers, (6) contact limited current, (7) regime of constant field between contacts, (8) traps field limited (monopolar injection of charge carriers) or recombination limited (double injection of charge carriers), (9) double injection of charge carriers in semiconductor, monopolar recombination of charge carriers, (10) double injection of charge carriers in semiconductor, bimolecular recombination of charge carriers, (11) saturation of the junction of charge carriers, (12) Joule heating.

Regions of α and γ constancy can be found in experimental or theoretical I-V characteristic data processed using (1) and (2).

A detailed description of α values is given in [13, 14], where the main results of the injection classics are collected, for example, [10-12].

The values of dimensionless sensitivity α and their corresponding modes of recombination and injection are systematized in Table 1. The possible sequence of these injection modes can also be presented graphically as in [12] (see Fig. 1).

4 RESULTS AND DISCUSSION

Our previous research [9] presented clonidine/silicon hybrids with plasmonic Au nanoparticles. It was shown that the clonidine layer always contributes to the injection of charge carriers and increases the efficiency of the obtained solar cells, but the detailed features of the injection process were not analyzed. A detailed analysis of dark I-V characteristic of clonidine/silicon hybrids with plasmonic Au nanoparticles is carried out below.

The forward dark I-V characteristics for photoconverting heterostructures on the base of flat Si, flat Si + CLO and flat Si + CLO +Au are presented in Fig. 2.

As we can see from Fig.2 the exponential behavior at low biases is almost perfect: in usual plot DS=f(V) the I-V characteristic curves for flat Si (Fid. 2, a, b, e) with DS approximation α=28(V-0.055), and flat Si + CLO +Au (Fig. 2, a, d, g) with DS approximation α=35(V-0.05) have a small cut-off voltage of about 0.05 V, but only in the case (c) with approximation α=18 V gamma=1, while cases (b) and (c) have gamma 1.5, i.e. current is proportional to the exponent with gamma=1.5 $I(V)\propto \exp(V^{1.5})$. At the same time, in the studied range, the current in cases (b) and (d) is limited by monomolecular recombination (DS=2) and contact (DS=1), respectively, while in case (c) there is the current restriction (DS=3) by monomolecular recombination in dielectric.

So, from these experiments for flat Si substrate, we can see that a thin layer of 400 nm CLO increases the dark current, and Au nanoparticles further increase the dark current. At the same time, the behavior of the DS is different. The photoconverting heterostructure on flat Si exhibits the usual classical behavior with bimolecular recombination at high bias with DS=1.5, Fig. 2 a,b, CLO removes the current limitation due to recombination processes and shows double injection with DS=3, Fig. 2 a,c, whereas the Au nanoparticles layer leads to a current limitation at high voltages with a linear characteristic and DS=1, Fig. 2 a,d. Comparison of these three samples suggests that the flat Si+CLO sample is preferable in these configurations for further use in solar cells, primarily due to the direct exponential characteristic without cut-off voltage and the absence of current limitation at high voltages.

The reverse dark I-V characteristics of this set of samples are presented in Fig. 3. The feature of these curves is the absence of rectification on the flat Si sample (Fig. 3, b) while the other two samples have sublinearity (Fig. 3, c and d) with DS of approximately 0.7. For a classic rectifier, DS should be 0.5. Therefore, the I-V characteristics of these structures show that the flat

10.4229/EUPVSEC2023/2BV.1.12

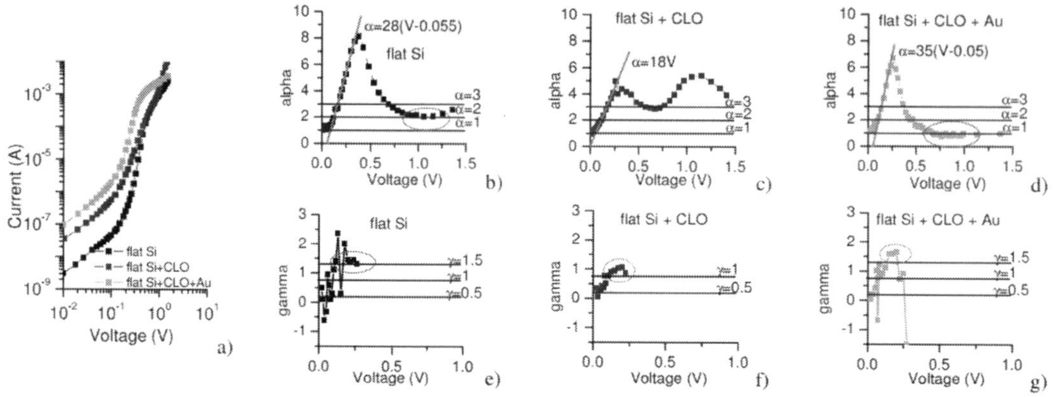

Figure 2: The dark forward I-V characteristics (*a*) and their DSs α (*b-d*) and γ (*e-g*) for the photoconverting heterostructures on the base of flat Si (*a, b, e*), flat Si + CLO (*a, c, f*) and flat Si + CLO +Au (*a, d, g*). The red circles highlight characteristic gamma values.

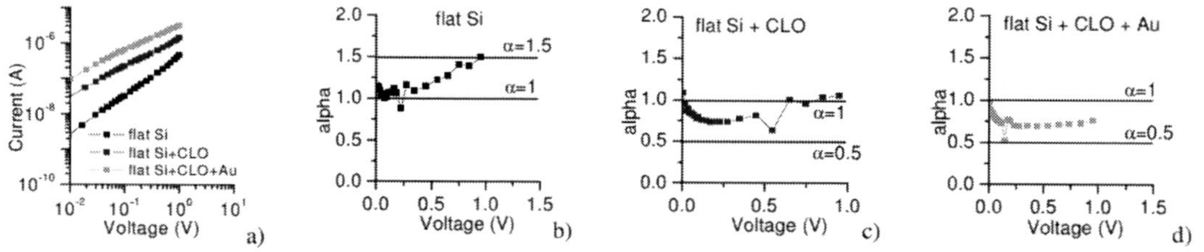

Figure 3: The dark reverse I-V characteristics (a) and their DSs α (*b-d*) for the photoconverting heterostructures on the base of flat Si (*a, b*), flat Si + CLO (*a, c*) and flat Si + CLO +Au (*a, d*).

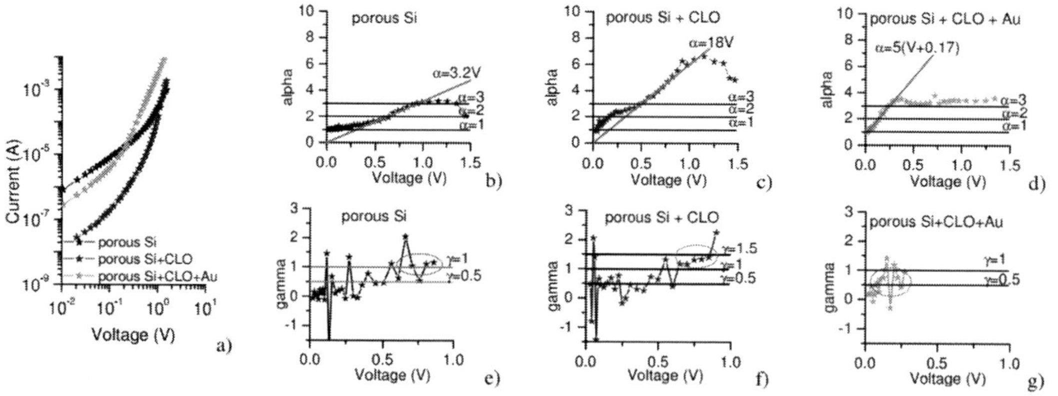

Figure 4: The dark forward I-V characteristics (a) and their DSs α (*b-d*) and gamma (*e-g*) for the photoconverting heterostructures on the base of porous Si (*a, b, e*), porous Si + CLO (*a, c, f*) and porous Si + CLO +Au (*a, d, g*). The red circles highlight characteristic gamma values.

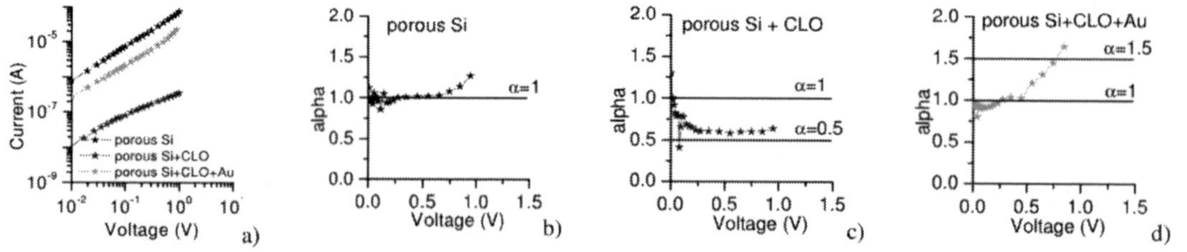

Figure 5: The dark reverse I-V characteristics (a) and their DSs α (*b-d*) for the photoconverting heterostructures on the base of porous Si (*a, b*), porous Si + CLO (*a, c*) and porous Si + CLO +Au (*a, d*).

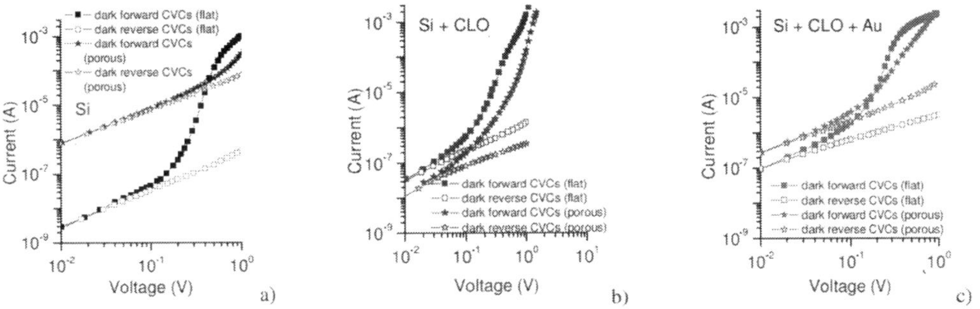

Figure 6: The dark forward and reverse I-V characteristics for the photoconverting heterostructures on the base of flat and porous Si (*a*), flat and porous Si + CLO (*b*) and flat and porous Si + CLO +Au (*c*).

Si+CLO and the flat Si+CLO+Au samples are close to the requirements of a good solar cell.

Fig. 4 demonstrates the forward dark I-V characteristics of the photoconverting heterostructures: porous Si – porous Si + CLO – porous Si + CLO +Au. Here the exponential behavior at low biases is also ideal and $I(V) \propto \exp(V)$. The DS approximations are as follows: α=3.2V, α=18V, α=5(V+0.17) for porous Si, Fig. 4b; porous Si + CLO, Fig. 4c; and porous Si + CLO +Au, Fig. 4d, respectively. For the first and second samples there is the lack of the cut-off voltage with gamma=1 (Fig. 4, b) and gamma between 1 and 1.5 (Fig. 4, c). For the third porous Si + CLO +Au sample (Fig. 4, d), the cut-off voltage is about 0.17 V, which means the presence of an additional barrier at the porous Si/CLO interface due to the presence of Au nanoparticles. In this case, gamma = 0.5, and the dependence of the current on the applied voltage is described by the square root exponent. In the investigated range, the current for the photoconverting heterostructures porous Si (Fig. 4, b) and porous Si + CLO +Au (Fig. 4, d) samples is restricted by monomolecular recombination in the dielectric at double injection (alpha=3), while for the porous Si + CLO + Au (Fig. 4, c) sample there is no current restriction (alpha>3). In addition, it should be noted that the CLO layer on the porous Si substrate reduces the dark current by a factor of 15.

The reverse dark I-V characteristics of this set of samples are presented in Fig. 5. Here, the behavior of I-V characteristics differs from that of flat Si. The peculiarity of these curves is the absence of rectification on the photoconverting heterostructures porous Si and porous Si + CLO + Au samples (Fig. 5, b and d, respectively) while the samples of photoconverting heterostructures porous Si + CLO has the sublinearity (Fig. 5, c) with the DS of approximately 0.6 which is necessary for a good solar cell.

The effect of porosity on the I-V characteristics in structures with CLO and Au nanoparticles should be considered separately. Fig. 6 shows the forward and reverse I-V characteristics for these samples. It is obvious that porosity increases the dark currents in both directions. At the same time, the porosity reduces the barrier in the forward direction and the exponent becomes flatter than for the flat sample, (Figs. 6a, 2b and 4b). For samples with the clonidine layer, the behavior is different: porosity reduces the dark current and the current restriction due to monomolecular recombination into the dielectric, when DS>>3, (Figs. 6b, 2c and 4c). The samples with the clonidine layer and Au nanoparticles show increase of dark forward I-V characteristic in samples with porosity. The porosity cancel the contact restriction with DS=1 and there is the monomolecular recombination into the dielectric, with DS=3, (Figs. 6c, 2d and 4d).

The reverse dark I-V characteristics actually repeat the behavior of forward I-V characteristics: porosity increase the current for Si substrate, decreases the current for Si with CLO layer and increases for Si with CLO layer and Au nanoparticles, Fig. 6a-c.

5 CONCLUSIONS

The dark I-V characteristic of two sets of photoconverting heterostructures: flat Si - flat Si + clonidine (CLO) - flat Si + CLO +Au and porous Si - porous Si + CLO - porous Si + CLO + Au were studied using dimensionless sensitivity to recognize fine features of their behavior. The CLO layer increased the dark current for flat substrates and decreased the current for porous substrates, while the Au nanoparticles increased the current in both cases. In this case, for the flat substrate, the classical current limitation by bimolecular recombination, with DS=1.5, was replaced by current limitation by the double injection into dielectric with monomolecular recombination, with DS=3, when adding the CLO layer and current limitation by the contact when applying Au nanoparticles, with DS=1. For the porous substrate, double injection in ideal dielectric with monomolecular recombination with DS=3 dominated in all cases except for structures with the CLO layer, where an exponentially increasing current occurs. In all cases the exponential behavior of I-V characteristics at low biases was ideal, but in the case of the flat substrate $I(V) \propto \exp(V^{1.5})$, while for porous one $I(V) \propto \exp(V)$. In terms of injection, the CLO layer always facilitated the injection of carriers, while the Au nanoparticles limited the current, especially in the case of the flat surface.

Regarding the reverse dark I-V characteristics, it should be noted that no rectification was observed on the photoconverting heterostructures on the base of flat Si, while the samples with the CLO layer and Au nanoparticles had the sublinearity with DS≈0.7. For the set of porous samples, only the porous Si + CLO samples showed sublinearity with a DS≈0.6, which is required for a good solar cell.

The effect of porosity on the forward and reverse I-V characteristics for Si samples and samples with Au nanoparticles was to increase the dark currents in both directions. For the samples with a clonidine layer, on the contrary, the porosity reduced the dark currents. The dominant mechanism of charge transfer was current limitation due to double injection into a dielectric with monomolecular recombination with DS=3.

6 REFERENCES

[1] A. Hagfeldt, M. Lira-Cantu, Applied Surface Science 482 (2019) 55.

[2] M. Lira-Cantú, Nature Energy 2 (2017) 17115.

[3] Z. Peng, N. Stingelin, H. Ade, J. Michels. Nature Reviews Materials 8 (2023) 439.

[4] M. Delgado, I. García, M. Hinojosa, C. Algora, Progress in Photovoltaics: Research and Applications 31 (2023) 555.

[5] K. Nirmal, W. Ren, A. Khot, D. Kang, T. Dongale, T. Kim, Advanced Materials 10 (2023) 2300433.

[6] M. Khenkin, E. Katz, A. Abate et al. Nature Energy 5 (2020) 35.

[7] J. Franeker, W. Voorthuijzen, H. Gorter et al., Solar Energy Materials & Solar Cells 117 (2013) 267.

[8] Jie Ouyang, Qi-Chao Lu, Sheng Shen and Shuang-Feng Yin, Nanomaterials 13 (2023) 1919.

[9] S. Mamykin, I. Mamontova, N. Kotova, O. Kondratenko, T. Barlas, V. Romanyuk, P. Smertenko, N. Roshchina. Physics and Chemistry of Solid State 21 (2020) 390.

[10] S.M. Sze, Y. Li, K.K. Ng. Physics of Semiconductor Devices, 4th Edition. Wiley, 2021.

[11] M. Lampert, P. Mark, Current Injection in Solids, Academic Press, New York & London, 1970.

[12] R. Baron, J.W. Mayer, Double injection in Semiconductors, in: Semiconductors and Semimetals, Academic Press, New York & London, Vol. 6 (1970) 201.

[13] R. Ciach, Yu.P. Dotsenko, V.V. Naumov, A.N. Shmyryeva, P.S. Smertenko, Solar Energy Materials & Solar Cells 76 (2003) 613.

[14] P. Smertenko, L. Fenenko, L. Brehmer, S. Schrader, Advances in Colloid and Interface Science 116 (2005) 255.

[15] R. Schifano, R. Dluzewski, W. Zajkowska et al., Applied Surface Science 552 (2021) 149067.

OPTIMIZATION OF BiOI/HTL HETEROJUNCTION FOR EFFICIENT CHARGE EXTRACTION FROM SOLAR CELL: FOR INDOOR LIGHT HARVESTING

Sarita Manjhi[1], Nithin Chatterji[1], Brajendra S. Sengar[2], Vivek Garg[1]
1 Department of Electronics Engineering, SVNIT, Surat-395007, India
2 Department of Electronics Engineering, NIT Srinagar-190006, India
*Email: vivekgarg0101@gmail.com

ABSTRACT: Indoor photovoltaics (IPVs) have piqued the interest of many because of their potential to power small and portable electronics and photonic devices. Lead free perovskite-inspired materials (PIMs) have recently attracted significant attention in photovoltaics research due to similarity in their electronic structure to high-performance lead-halide perovskites, but without the same toxicity limitations. This paper investigates one of the exemplary PIM, Bismuth Oxy-iodide (BiOI). While its bandgap is too large for single-junction solar cells, it is nearly perfect for indoor light harvesting. In order to explore the potential of BiOI in indoor environment, we first developed the baseline model of BiOI device (ITO/NiOx/BiOI/ZnO/Contact) stack taking experimental research paper as reference, which was able to achieve power conversion efficiency (PCE) of 4%. The performance of the proposed device is improved by investigating the effect of: (a) absorber thickness and defect density, (b) valence band offset (VBO) between hole transport layer (HTL) and absorber, and (c) interface (NiOx/BiOI) defect density. Further, series and shunt resistance of device is optimized. Additionally, performance of the optimized device is investigated under different WLED light intensities. Finally, after optimizing the device under WLED illumination, the best performance parameters achieved are: Jsc: 1.83mA/cm2 , Voc:1.33 V, FF: 85.91%, Eff: 40%. Moreover, the optimized device's performance under different indoor light sources such as halogen and CFL has been performed to estimate the performance under widely utilized lightning sources.
Keywords: three to five keywords in order of importance

1 INTRODUCTION

The market value and number of devices connected to the Internet of Things (IoT) are expected to grow at an exponential rate, driving up demand for indoor photovoltaics (IPV) to power autonomous devices. Lead free perovskite-inspired materials (PIMs) have recently attracted significant attention in photovoltaics research due to similarity in their electronic structure to high-performance lead-halide perovskites, but without the same toxicity limitations. The ability of PIMs to harvest indoor light, on the other hand, has yet to be considered. A theoretical maximum efficiency of ~62% can be achieved under both FL and WLED lighting with absorbers possessing a bandgap of 1.9 eV based on the detailed balance approach [2]. One of the promising material among PIMs is BiOI. Bandgap of BiOI is too large for single-junction solar cells, it is nearly perfect (~1.95 eV) for indoor light harvesting. The spectroscopically limited maximum efficiency (SLME) of the photovoltaic conversion efficiency of BiOI under indoor WLED is reported as approximately 45% [1, 2]. Thus, a large scope of improvement is available which motivates us to further explore the potential of BiOI device under indoor environment.

In this work, HTL/BiOI heterojunction is investigated for improved device performance by optimizing a) absorber layer, b) hole transport layer (HTL), c) HTL/absorber interface. Further, the optimized device performance is also investigated for the different indoor lightning sources and intensities to investigate the performance under real time scenarios.

2 DEVICE STRUCTURE AND BASELINE DEVELOPMENT

The layer configuration of the proposed device is: glass substrate/ITO/NiOx/BiOI/ZnO/Cr/Ag, where, ZnO is ETL and NiOx is HTL, indium tin oxide (ITO) and Cr/Ag are employed as transparent front contact and back contact respectively as shown in Fig. 1(a). The energy band alignment diagram of the presented solar cell structure is shown in Fig. 1(b). The simulation is done in SCAPS 1D using basic parameters derived from theoretical and experimental research papers as listed in Table 1. Simulation of the device structure has been carried out at illumination of WLED 1000 lux with incident power of 320 uWcm-2 .

From Fig 1 (c), we can observe that simulated device parameters are closely matching with the previously reported theoretical and experimental works.

(a)

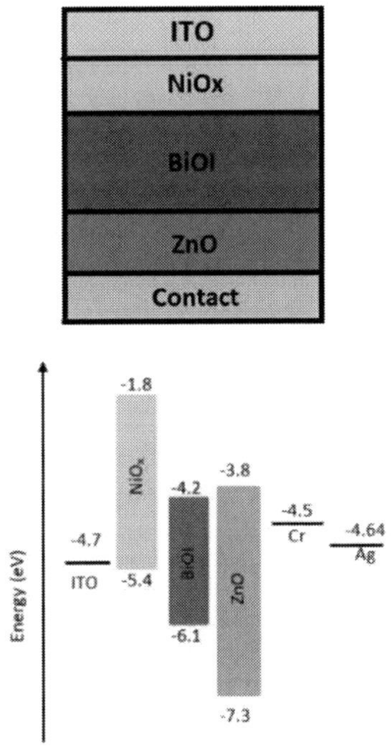

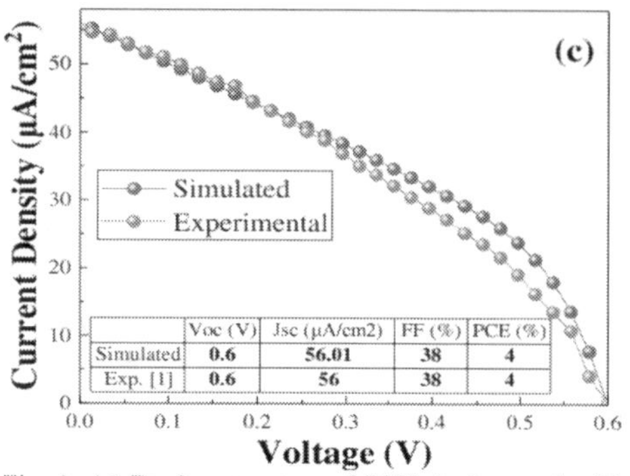

Fig. 1: (a) Device structure of BiOI device stack. (b) Band alignment of BiOI device stack (c) Experimental v/s simulated J-V curve of BiOI device (under 1000 lux WLED, 320 μW/cm2)

Table 1: Solar Cell parameters used for simulation of BiOI device structure in SCAPS

	NiOx (HTL) [7]	BiOI [2]	ZnO (ETL) [6]
Thickness (nm)	30	300	100
Bandgap(eV)	3.6	1.93	3.5
Electron affinity (eV)	1.8	4.2	3.8
Relative permittivity	10.7	45	9
CB density of states (cm^{-3})	1.0E+19	4.839E+19	2.2E+18
VB density of states (cm^{-3})	1.0E+19	1.329E+19	1.9E+19
Electron Thermal Velocity (cms^{-1})	1.0E+7	1.6E+7	1.0E+7
Hole Thermal Velocity (cms^{-1})	1.0E+7	1.02E+7	1.0E+7
Electron mobility	12	14	100
Hole mobility	25	7	25
Donor conc. (cm^{-3})	0	1.0E+15	1.0E+20
Acceptor conc. (cm^{-3})	1.0E+15	1.0E+15	0
Series resistance ($ohm.cm^2$)	1.22E+2		
Shunt resistance ($ohm.cm^2$)	1.67E+4		

3 RESULTS AND DISCUSSIONS

For optimization of WLED of 1000 lux intensity is taken as source and a standard power density of approximately 520 mWcm-2 is considered. The front and back contacts are taken as flat bands for initial optimization.

3.1 Optimization of absorber parameters

The first optimization is performed on the absorber layer (BiOI). Effect of variation in thickness and defect density is studied here. From Fig 2 (a). We can observe that PCE value of approximately 12.3% is achieved at thickness 1.4 um and defect density 1E+12 cm-3 for absorber.

The performance of device is directly affected by absorber layer thickness as it affects the diffusion length (lifetime) of charge carriers. If the absorber layer is too thick, the charge carriers may recombine before reaching the charge collecting electrodes, and if the absorber layer is too thin, it will not be able to absorb the incident radiation effectively, thus lowering the device's overall performance. The decrease in performance with increasing defect density is primarily due to an improvement in the recombination process, which results in charge carrier annihilation.

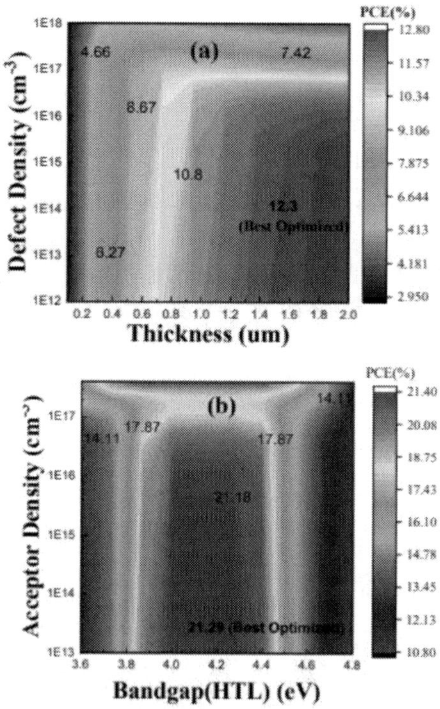

Fig. 2: (a) Variation of performance parameters of BiOI device with change in absorber layer thickness and defect density, (b) Variation in PCE of BiOI device with VBO and acceptor density of HTL

3.2 Optimization of HTL layer

The optimization of HTL layer is performed by studying variation in valence band offset ($VBO = \chi HTL - \chi perovskite + Eg.HTL - Eg.perovskite$) and carrier concentration. For varying VBO, bandgap is changed from 3.5 eV to 4.8 eV. From Fig. 2 (b), we can observe that device has constant performace for a range of HTL bandgap i.e. from 3.7 eV to 4.4 eV. Thus other potential HTL materials with valence band position in the range 5.5 eV to 6.2 eV can be explored for further improvement of device. The optimised bandgap obtained is 4.08 eV and acceptor density is 3E+13 cm-3 as we can observe from fig. 5. PCE at the optimised values is 21.29%.

3.3 Optimization of HTL/absorber interface

The defect density of interface NiOx / BiOI i.e. between HTL and absorber is varied from 1E+10 cm-2 to 1E+13 cm-2 . From Fig. 3 (a), it can be seen that PCE starts decreasing rapidly after 1E+10 cm-2 . Thus, the optimum value of interface defect density is chosen as 1E+10 cm-2 for which PCE achieved is 32.4%.

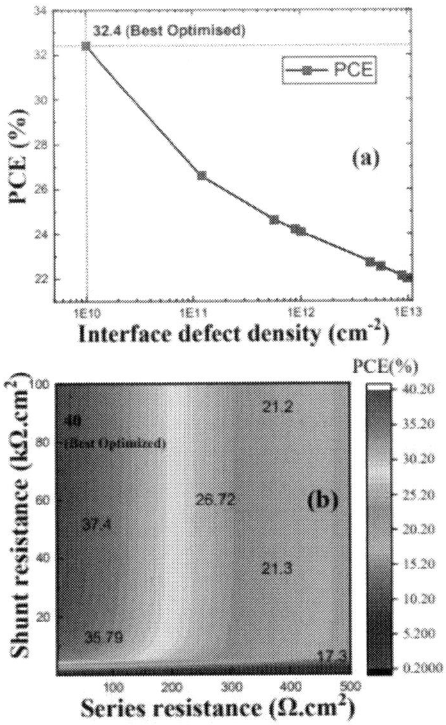

Fig. 3: (a) Variation in performance parameters of BiOI device with variation in defect density of NiOx/BiOI interface, (b) Variation in performance parameters of BiOI device with series and shunt resistance

3.4 Optimization of series and shunt resistance of device

Series resistance (Rs) is varied from 5 ohm.cm2 to 500 ohm.cm2 and shunt resistance (Rsh) from 100 to 105 ohm.cm2 , as shown in Fig 3 (b). The movement of current through the emitter and base of the solar cell causes series resistance; second, the contact resistance between the metal contact and the silicon; and finally, the resistance of the top and rear metal contacts. The main effect of series resistance is that it reduces the fill factor, though extremely high values may also reduce the short circuit current. Shunt resistance is mainly due to manufacturing defects. Low shunt resistance values affect the open circuit voltage. From Fig. 7, we can observe that, with an optimized Rs of 5 ohm.cm2 and Rsh of 6.32E+4 ohm.cm2 , the device performance parameters are Jsc: 1.83mA/cm2 , Voc:1.33 V, FF: 85.91%, Eff: 40%.

3.5 Effect of variation in light source and light intensity on device performance

We observed the device performance under different light sources and from Fig. 4 (a), we can observe device has best performance under LED 2700K (Warm LED) light source. We varied the light intensity of WLED from 200 lux to 1000 lux. From Fig. 4 (b), we can observe that performance reduces as we reduce light intensity. This is because on increasing light intensity, number of photons striking the cell increases. This increases the generation rate of charge carriers.

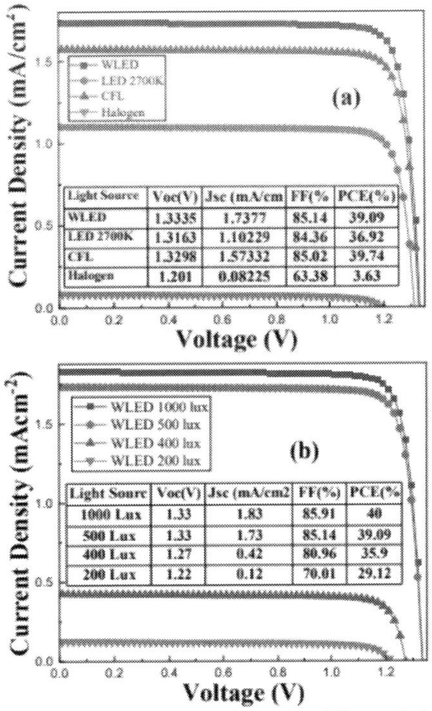

Light Source	Voc(V)	Jsc (mA/cm	FF(%	PCE(%)
WLED	1.3335	1.7377	85.14	39.09
LED 2700K	1.3163	1.10229	84.36	36.92
CFL	1.3298	1.57332	85.02	39.74
Halogen	1.201	0.08225	63.38	3.63

Light Soure	Voc(V)	Jsc (mA/cm2	FF(%)	PCE(%
1000 Lux	1.33	1.83	85.91	40
500 Lux	1.33	1.73	85.14	39.09
400 Lux	1.27	0.42	80.96	35.9
200 Lux	1.22	0.12	70.01	29.12

Fig. 4: (a) J-V characteristics under different indoor light sources (inset shows respective device perfromance parameters) (b) Variation in performance parameters due to variation in light intensity of WLED light source (inset shows respective device perfromance parameters)

4 CONCLUSIONS

In this study, we focused on one of the promising lead-free PIM, BiOI. Baseline model for the device is developed based on previous experimental works and simulation is done in SCAPS 1D under 1000 lux WLED for standard power density 521 mW/cm2 . Efficiency achieved through baseline model is 4%. The performance of the proposed device is improved by optimizing HTL, absorber and HTL/absorber interface. Final optimization is done for series and shunt resistances. After optimization, the simulated device was able to achieve PCE of 40%. At last, we observed the effect of varying light sources such as CFL, Halogen, and WLED. Moreover, with the variation in light intensity of WLED, performance reduces with reduction in light intensity.

ACKNOWLEDGEMENTS

This work is supported by seed grant [dean(R&C)/seed money/2021-22/10783 dated 07/01/2022) provided by SVNIT Surat.

REFRENCES

[1] I. Mathews, S. N. Kantareddy, T. Buonassisi, I. M. Peters, Technology and Market Perspective for Indoor Photovoltaic Cells, Joule, Volume 3, Issue 6, 2019, 1415-1426

[2] Peng, Y., Huq, T. N., Mei, J., Portilla, L., Jagt, R. A., Occhipinti, L. G., MacManus-Driscoll, J. L., Hoye, R. L. Z., Pecunia, V., Lead-Free Perovskite-Inspired Absorbers for Indoor Photovoltaics. Adv. Energy Mater. 2021, 11, 2002761

[3] Pecunia, V., Occhipinti, L. G., Hoye, R. L. Z., Emerging Indoor Photovoltaic Technologies for Sustainable Internet of Things. Adv. Energy Mater. 2021, 11, 2100698

[4] Salem, M.S.Shaker, A.Zekry, A.Abouelatta, M.Alanazi, A.Alshammari, M.T.Gontand, C. Analysis of Hybrid Hetero-Homo Junction Lead-Free Perovskite Solar Cells by SCAPS Simulator. Energies 2021, 14, 5741.

[5] H. Wang, X. Gong, D. Zhao, Y. Zhao, S. Wang, J. Zhang, L. Kong, B. Wei, R. Q. Bermudez, O. Voznyy, Y. Shang, Z. Ning, Y. Yan, E. H. Sargent, X. Yang,A Multi-functional Molecular Modifier Enabling Efficient Large Area Perovskite Light-Emitting Diodes,Joule,4,2020, 977-1987,2542-4351

[6] Zyoud SH, Zyoud AH, Ahmed NM, Prasad AR, Khan SN, Abdelkader AFI, Shahwan M. Numerical Modeling of High Conversion Efficiency FTO/ZnO/CdS/CZTS/MO Thin Film-Based Solar Cells: Using SCAPS-1D Software. Crystals. 2021; 11(12):1468.

[7] X. Wei, X. Wang, H. Jiang, Y. Huang, A. Han, Q. Gao, J. Bian, Z. Liu, Numerical simulation and experimental validation of inverted planar perovskite solar cells based on NiOx hole transport layer,Superlattices and Microstructures,112,2017;383-393,0749-6036

[8] Pecunia, V., Zhao, J., Kim, C., Tuttle, B. R., Mei, J., Li, F., Peng, Y., Huq, T. N., Hoye, R. L. Z., Kelly, N. D., Dutton, S. E., Xia, K., MacManus-Driscoll, J. L., Sirringhaus, H., Assessing the Impact of Defects on Lead-Free Perovskite-Inspired Photovoltaics via Photoinduced Current Transient Spectroscopy. Adv. Energy Mater. 2021, 11, 2003968

10.4229/EUPVSEC2023/2BV.1.20
020085-004

PHOTOVOLTAIC MODULES FOR PEDESTRIAN SURFACES USING A COMPOSITE MATERIAL WITH ENHANCED SLIP- RESISTANCE AS ENCAPSULANT

Gorka Imbuluzqueta, Naiara Yurrita, Jon Aizpurua, Juan M. Hernández, Werther Cambarau, Arantxa Pérez, Francisco J. Cano, Eduardo Román, Oihana Zubillaga
TECNALIA, Basque Research and Technology Alliance (BRTA)
Mikeletegi Pasealekua 2, 20009 Donostia-San Sebastian, Spain

ABSTRACT: Photovoltaic (PV) modules were manufactured encapsulating monocrystalline silicon cells in transparent composite material. The surface of the modules was modified with several coatings and additives to improve their slip-resistance and enable their use for pedestrian transit. The coating offering the best slip-resistance according to the pendulum friction method showed a standardized *Pendulum Test Value* of 74, meaning it can be used for wet indoor and outdoor pedestrian surfaces, like the ones existing onboard vessel decks expected in marine applications.

The slip-resistant modules were appropriately modified to improve their performance stability under ultraviolet radiation, showing an initial cell-to-module loss in short-circuit current density of 2.2%. The modules containing these coatings were exposed to accelerated aging tests to evaluate their performance stability under ultraviolet (UV) radiation, damp-heat (DH) and neutral salt spray (NSS). The modules incorporating the slip-resistant coating with UV additives showed a decrease of 2.2% in Isc after DH test, with losses below 1% for UV and NSS exposure.

The study concluded that composite based encapsulation, with proper surface modification, can be an option to obtain slip-resistant PV modules valid for marine applications as vessel decks, while retaining a suitable PV performance and stability.

Keywords: slip-resistance, composite, encapsulation, marine, stability

1 INTRODUCTION

Glass fiber reinforced composite can be used as alternative protective encapsulant material for photovoltaic (PV) cells [1], leading to lightweight multifunctional modules that can be integrated in several applications such as building or transport [2]. Focusing on transport application, incorporating PV modules to boats may reduce the negative environmental impacts and promote the transition of boats to the use of more sustainable energy sources [3].

More particularly, walkable pavements in boats are an option for PV module placement. However, these potential PV surfaces should offer safe movements to users, presenting adequate slip-resistant properties [4].

The present work deals with the development of PV modules for pedestrian transit in wet boat surfaces. These modules were composed of crystalline silicon PV cells, transparent composite as encapsulant, and slip-resistant surface modification.

2 EXPERIMENTAL

2.1 Materials and module manufacturing

The photovoltaic cells were encapsulated using linear vacuum resin infusion process. A 0/90° glass fiber fabric with a 300g/m2 areal weight was used as reinforcement. For the composite matrix, a clear epoxy resin system based on a clear bisphenol-A epoxy resin and an amine curing agent was used. The cells were monocrystalline silicon cells with 5 busbars.

For the slip-resistance (SR) study, monomodules consisting of one cell were manufactured. For the accelerated aging tests, minimodules composed of four cells were conceived.

The modules were coated with diverse solutions to improve the slip-resistance of the front surface, as presented in table I.

Further, in order to improve the module performance stability under ultraviolet (UV) radiation, two module configuration approaches were studied (figure 1). One solution consisted on incorporating UV additives in the SR coating (figure 1a). In the second one a multilayer approach was studied, with a coating stack composed of a first layer of a UV protective varnish on the composite surface, followed by the SR coating on the outer surface (figure 1b).

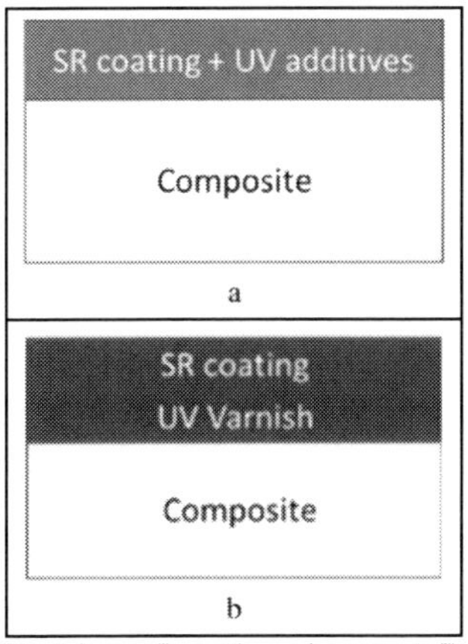

Figure 1: Module configuration schemes. a) Composite with SR coating containing UV additives (in blue), and b) Composite with a coating stack composed of UV protection varnish and a final SR coating (in red)

2.2 Module Characterization

The slip-resistance characterization of the module surfaces was carried out according to the standard UNE 41901:2017-*Surfaces for pedestrian transit. Determination of the slip resistance by pendulum friction method. Wet test.* After measurement of the Pendulum Test Value (PTV), the surfaces were classified according to the standard.

Concerning the electrical performance of the modules, external quantum efficiency (EQE) of the modules was measured using PVE300 equipment from Bentham. EQE values for the 300-1300 nm wavelength range were obtained. Short-circuit current density (Jsc) values were calculated from the EQE spectra [1], and cell-to-module

10.4229/EUPVSEC2023/2BV.1.22
020088-001

(CTM) losses due to the diverse encapsulation solutions analyzed.

2.3 Accelerated aging tests

The minimodules were exposed to accelerated indoor tests according to IEC-61215:2021 standard to evaluate their performance stability. Damp-heat (DH) and UV radiation exposure tests were carried out. The damp-heat test comprised an exposure of 1000 hours to 85ºC and 85% relative humidity. Regarding the UV radiation test, the modules were exposed during 380 hours at 60ºC to achieve an overall radiation of 15 KWh/m2.

Additionally, neutral salt-spray (NSS) exposure according to ISO9227 standard was carried out. The modules were exposed during 1000 hours at 35ºC to an atmosphere of 50g/L NaCl at pH:6.5-7.2.

The progress of the weathering was evaluated by solar simulator and analysis of current-voltage curves, obtained prior to the exposure and at the end of the aging test.

3 RESULTS AND DISCUSSION

3.1 Module manufacturing and characterization

Monolithic modules with a thickness of around 3mm where the cells and connectors were completely embedded in the composite were obtained (figure 2a). No major defects in cells, composite or cell-composite interface were detected. Further, the coatings to improve the slip-resistance feature were successfully deposited on the front surface, fully covering cell and composite edge area. A detail of the surface finish obtained by the SR coating (solution 6 in table I) can be observed in figure 2b.

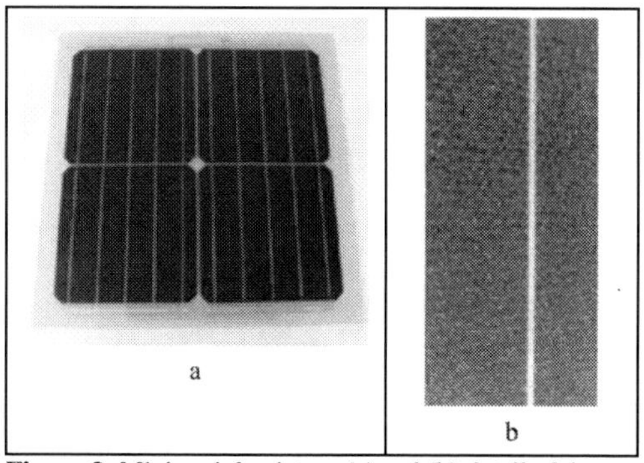

Figure 2: Minimodule picture (a) and (b) detail of the slip-resistant surface finish of the module

The PTV values measured for the evaluated surface finishes are presented in table I. The bare composite and standard varnish showed similar value, being classified as surfaces valid for indoor dry surfaces with limited slope. Modifying the standard varnish with SR additives at different contents showed slightly higher PTV values, but still remained at the same SR category. Finally, improving the PTV value up to 74 and corresponding classification was reached with a specific SR coating (number 6 in table I). The surface finish obtained with this solution would be valid for wet indoor and outdoor pedestrian surfaces, like the ones present in vessel decks in marine application.

Table I: Pendulum test values (PVT) of tested SR solutions, and classification.

Coating	PTV	Class.*
1-Bare composite	16	1
2-Standard varnish	17	1
3-Standard varnish with SR adds.	24	1
4-Standard varnish with SR adds x1.5	26	1
5-Standard vanish with SR adds. x 2	30	1
6-SR coating	74	3

**1-Valid for dry indoor surfaces with <6% slope*
2-Valid for wet indoor and outdoor pedestrian surfaces

The EQE measurements and further calculation of the Jsc for bare composite and SR coated modules, showed a reduction of 0.9% in Jsc due to the presence of the SR coating. When a varnish was incorporated together with the SR coating (figure 1b), and the SR coating was modified with additives (figure 1a), in order to reach the required UV protection, the Jsc loss increased up to 2.6% and 2.2% respectively. As can be observed in figure 3, this higher loss come mainly from the shift to higher values in the UV cut-off wavelength when comparing with the bare composite. This is due to the presence of UV absorbers incorporated in the coating, which decreases the light reaching the photovoltaic cell surface in the 300-500nm wavelength range of the spectrum. No significant influence of the coating is observed in higher wavelength values of the spectrum.

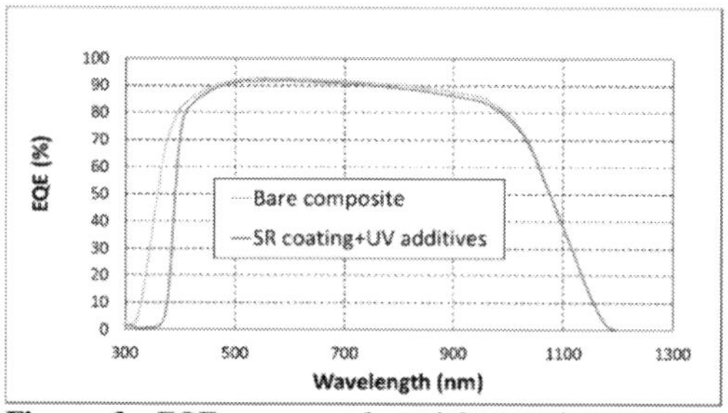

Figure 3: EQE spectra of modules made of bare composite, and composite with the SR coating with UV additives

3.2 Accelerated aging test results

The decrease in short-circuit current (Isc) presented by the modules at the end of the damp-heat, UV radiation and salt-spray aging tests is depicted in figure 4.

Overall, the major degradation is observed in the damp-heat test, and the SR coating with additives performed better than coating stack composed of an inner UV varnish and the outer SR coating.

The module incorporating the SR coating with additives showed a decrease of 2.2% in Isc after 1000 hours of DH exposure (figure 4a). The Isc loss after UV and NSS exposure radiation exposure was below 1%, in the range of the measurement error (figures 4b and 4c). Regarding the coating stack, the decrease observed in Isc reached 2.6% in damp-heat, slightly higher than the UV modified SR coating.

Thus, considering the cell-to-module losses and durability of the modules, best overall electrical performance was offered by the SR coating containing UV additives.

10.4229/EUPVSEC2023/2BV.1.22
020088-002

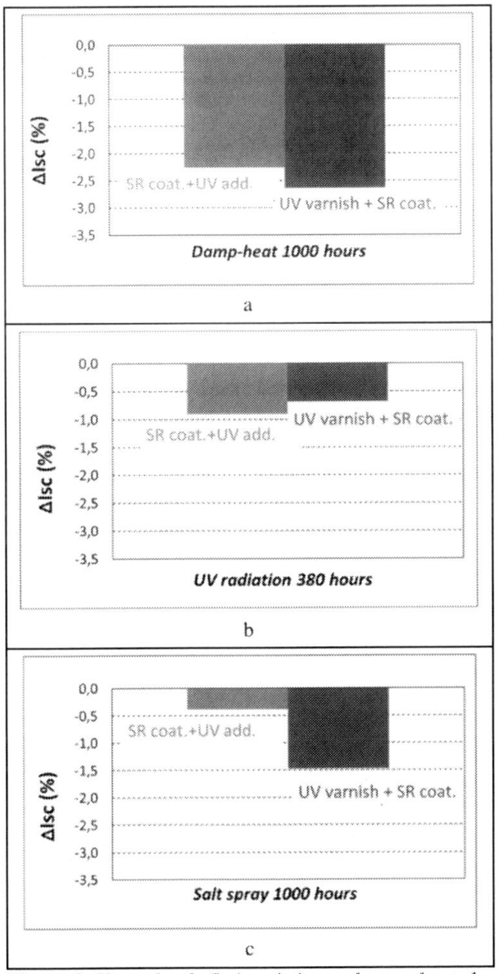

Figure 4: Short-circuit (Isc) variation under accelerated aging tests. a) After 1000h damp-heat test, b) after 380h UV radiation and c) after 1000h neutral salt-spray exposure

4 CONCLUSIONS

Photovoltaic modules using glass-fiber reinforced composite as encapsulant with further surface modification by incorporating a coating, were classified as slip-resistant surfaces valid for wet indoor and outdoor pedestrian surfaces, according to measured pendulum test value.

These wet slip-resistant photovoltaic modules, incorporating UV protection, showed low cell-to-module losses and good performance under accelerated aging tests of damp-heat, UV radiation and salt-spray.

Thus, the study concluded that the developed slip-resistant photovoltaic modules, which retain acceptable photovoltaic performance and stability in ambient conditions, could be valid for marine applications as vessel decks.

5 ACKNOWLEDGEMENTS

This work, performed within the ENGIMMONIA project, has received funding from the European Union's Horizon 2020 research and innovation program under grant agreement n° 955413.

6 REFERENCES

[1] G. Imbuluzqueta, N. Yurrita, J. Aizpurua, F.J. Cano, O. Zubillaga, Composite material with enhanced ultraviolet performance stability for photovoltaic modules, Sol. Energy Mater. Sol. Cells 200 (2019), 109947.

[2] E. Rico, I. Huerta, T. del Caño, L. Villada, Á. Gallego, V. Velasco, O. Zubillaga, J.M. Vega de Seoane, I. Arrizabalaga, N. Yurrita, J. Aizpurua, G. Imbuluzketa, F.J. Cano, PVCOM Project: Manufacture of PV Modules Encapsulated in Composite Materials for Integration in Urban Environments, Commun. Comput. Inf. Sci. 978 (2019), 38-52.

[3] S. H. Tercana, B. Eidb, M. Heidenreichc, K. Kogler, Ö. Akyürek, Financial and Technical Analyses of Solar Boats as A Means of Sustainable Transportation, Sustainable Production and Consumption 25 (2021) 404–412.

[4] International Maritime Organization, MSC-MEPC.7/Circ.3, Framework for consideration of ergonomics and work environment, 2006.

THE EXTRACTION OF THE SPATIAL EXTERNAL LUMINESCENCE EFFICIENCY FOR HIGH RESOLUTION MAPPING OF CHARGE RECOMBINATION

Tamir Yeshurun, Mor Fiegenbaum-Raz, Gideon Segev
Tel-Aviv University, Israel
gideons1@mail.tau.ac.il

ABSTRACT: Continuous effort is being put towards device optimization for high efficiency solar energy conversion systems. As various loss mechanisms of structures and materials greatly affect the performance of such devices, in-depth mapping of loss and charge transport properties can be essential for rapid device optimization. The spatial external luminescence efficiency (SELE) is defined as the probability of an electron-hole pair that is photo-generated at a certain depth to contribute to photoluminescence from the device. In this contribution, we propose a method to extract the SELE by coupling incident wavelength-dependent external luminescence efficiency measurements with optical modeling. The extraction method is demonstrated on a GaAs wafer, serving as a model system. The SELE directly maps the competition between radiative and non-radiative recombination within the device and can thus be used to study the properties of surfaces and interfaces. Furthermore, since the external luminescence efficiency is related to the obtainable photovoltage from the device, extraction of the SELE profile can be used to map the contribution of different regions in a device to the photovoltage.
Keywords: Characterization; Photoluminescence; Mapping; Losses; GaAs;

1 INTRODUCTION

The performance of solar cells is determined by the competition between charge carrier collection and various recombination and loss mechanisms. Photogenerated charge carriers that do not contribute to output current, recombine through bulk and surface defect states, or by radiative recombination, thus generating photons. These photons can either be reabsorbed within the device or emitted from the device through photoluminescence (PL). The internal luminescence yield, η_{int} is the fraction of the electron and hole pairs that recombine radiatively, and the probability for a recombination event to result in a photon emission from a device is the External Luminescence Efficiency (ELE), η_{ext} [1]–[4]. Since both the radiative recombination rate and the photovoltage are determined by the separation of electrons and holes quasi-fermi levels, the open circuit voltage of a solar cell, V_{oc}, is tightly coupled to its ELE [3], [5]:

$$qV_{oc} = qV_{oc,ideal} + kTln(\eta_{ext}) \qquad (1)$$

Where $V_{oc,ideal}$ is the open circuit voltage of an ideal solar cell, k is Boltzmann's constant, and T is the device temperature. Thus, ELE measurements, provide an estimate of the maximum open circuit voltage that can be obtained from a light absorbing material. Consequently, device optimization towards maximizing the extraction of radiative recombination photons led to a significant increase in the performance of GaAs solar cells, achieving a record efficiency of 28.8% [4].

Different regions within the device contribute differently to the photoluminescence. Thus, the photoluminescence photon flux can be described in terms of the probability for different regions to contribute to the photo-emission process:

$$\phi_{out}(\lambda) = \int_0^d \Delta G(\lambda, z) S(z) dz \qquad (2)$$

where $\phi_{out}(\lambda)$ is the measured PL photon flux in response to illumination with a monochromatic beam with a wavelength λ. $\Delta G(\lambda, z)$ is the optical generation induced by the monochromatic light, and d is the device thickness. $S(z)$ is the Spatial External Luminescence Efficiency (SELE) which is the probability for photogenerated electron hole pairs generated at a depth z from the surface to contribute to the PL photon flux. Under open circuit conditions, the external luminescence efficiency is the fraction of absorbed photons that are emitted in the PL process:

$$\eta_{ext}(\lambda) = \frac{\phi_{out}(\lambda)}{\phi_0(\lambda)A(\lambda)} = \frac{1}{\phi_0(\lambda)A(\lambda)} \int_0^d \Delta G(\lambda, z) S(z) dz \qquad (3)$$

where $\phi_0(\lambda)$ is the incident monochromatic photon flux and $A(\lambda)$ is the sample absorption. Thus, combining optical modeling to calculate $\Delta G(\lambda, z)$, with measurements of absorbed photon flux and ELE, the SELE profile can be extracted using equation (3) with regularization procedures. In this contribution, we describe the extraction method and extract the SELE profile of a GaAs wafer, serving as a model system.

2 METHOD

Tikhonov regularization was used to extract the SELE profile as previously applied for spatial collection efficiency extraction [6], [7]. First, equation (3) is discretized:

$$\bar{\phi}_{out}(\lambda) = \overline{\overline{\Delta G}}(\lambda, z_i) \cdot \bar{S}(z_i) \qquad (4)$$

Where $\overline{\overline{\Delta G}}(\lambda, z_i)$ is a matrix, describing the total optical generation within the element located at position z_i that is induced by monochromatic illumination with wavelength λ and $\bar{S}(z_i)$ is the average SELE within the element z_i. Since equation (4) has an infinite number of solutions, the candidate solutions must be constrained. In the regularization process, a vector $\bar{S}(z_i)$ is found such that the following expression is minimized:

$$min\,\{\, \| \overline{\overline{\Delta G}}(\lambda, z_i) \cdot \bar{S}(z) - \bar{\phi}_{out}(\lambda) \|_2 + \kappa^2 \,\| L(\bar{S}(z)) \|_2 \,\} \qquad (5)$$

The first term in the brackets defines the accuracy in which the candidate solution reproduces the measured ELE as described with equation (4). L is an operator such

10.4229/EUPVSEC2023/2BV.1.34
020090-001

that the second term in the brackets constrains the derivative and the second derivative of the solution. The parameter κ is a weight used to balance between a solution that reproduces the measured ELE while maintaining physical features. The Candidate solutions are SELE vectors that minimize equation (5), each with a different value of κ. From this group of solutions those that produce both a small deviation from measurements and are smooth, i.e., have small first and second derivatives, are averaged to produce the mean profile which is the extracted SELE.

3 RESULTS AND DISCUSSION

The SELE extraction method was tested on a p-type Zinc doped GaAs wafer with carrier concentration of $(1.98 \pm 0.28) \cdot 10^{19}$ $[cm^{-3}]$ that serves as a model system thanks to its well understood properties and strong PL signal. The incident wavelength-dependent ELE was obtained by combining wavelength-dependent PL measurements taken in a calibrated integrating sphere setup together with spectrophotometric absorption measurements. The inset in Figure 1a shows the measured PL spectrum. These spectra were integrated to obtain the incident wavelength-dependent ELE as shown in Figure 1a. The SELE profile of the GaAs wafer was then extracted as detailed above. Figure 1b shows the extracted SELE profile focusing on the region near the surface. The inset shows the SELE all through the device. The solid line is the average profile obtained from all the candidate solutions that can reproduce the measured ELE and are sufficiently constrained. The shaded area is the standard deviation of these solutions. Once the SELE profile is extracted, equation (3) is used to reconstruct the wavelength-dependent ELE which is compared with the measured ELE spectrum as shown in Figure 1a.

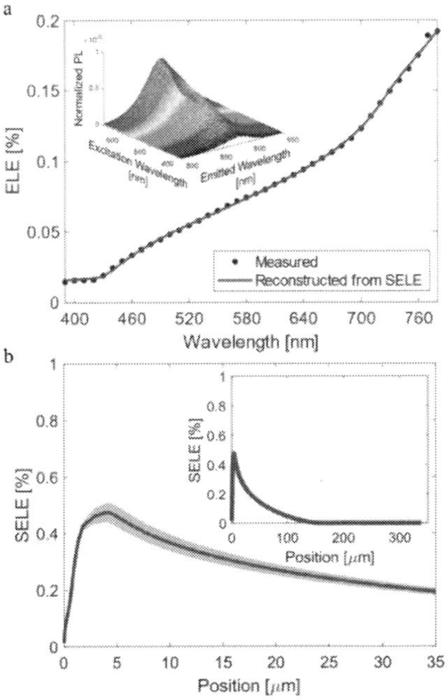

Figure 1 – ELE and SELE profiles of a p-type GaAs wafer. (a) Incident wavelength-dependent ELE measurement, and reconstruction from SELE regularization. The inset shows the wavelength-dependent PL measurements. (b) Extracted SELE profile. The shaded area corresponds to the standard deviation of possible solutions generated in the regularization process.

The SELE profile maps the competition between radiative and non-radiative recombination processes, as well as the probability for photons that are generated within the bulk of the device to be emitted. Near the surface, the low SELE is a result of surface recombination through defects, suppressing radiative recombination processes. As a result, the SELE initially increases with the depth from the surface. However, deeper within the material, radiative recombination photons are less likely to be emitted due to self-absorption, and the SELE profile decreases. A considerable portion of the photoluminescence photons are at energies below the band gap. Since these photons are hardly reabsorbed even when generated far from the surface, the decay of the SELE with depth is significantly more moderate than the decay of the optical absorption profiles. The SELE profile is also affected by the optical coupling to the ambient. Thus, regions with a high SELE suggest higher photon extraction from the device, while low SELE might be indicative of high optical losses.

4 CONCLUSION

The concept of SELE was introduced and an SELE profile of a GaAs wafer was extracted by a combination of incident wavelength-dependent PL measurements and optical modeling. The SELE profile maps the competition between radiative and non-radiative recombination mechanisms, as well as the reabsorption of photons throughout the device. The SELE profile near the surface is highly affected by surface recombination, and further away from the surface the SELE quantifies the internal luminescence efficiency and the reabsorption of photons. The SELE profile can therefore be used as a tool to quantify surface recombination velocity, and to evaluate surface passivation quality as well as optical losses in different regions within the device. Thus, extraction and analysis of SELE profiles of advanced solar cells, can help map loss mechanisms, and to quantify physical properties at the surface and interfaces while using a non-destructive method under operando conditions.

5 ACKNOWLEDGMENTS

This work was partially funded by the Israeli Ministry of Energy, and by the Israel Science Foundation.

6 REFERENCES

[1] V. Ganapati, M. A. Steiner, and E. Yablonovitch, "The Voltage Boost Enabled by Luminescence Extraction in Solar Cells," *IEEE J. Photovoltaics*, vol. 6, no. 4, pp. 801–809, 2016.

[2] M. A. Steiner *et al.*, "Effects of internal luminescence and internal optics on Voc and Jsc of III-V solar cells," *IEEE J. Photovoltaics*, vol. 3, no. 4, pp. 1437–1442, 2013.

10.4229/EUPVSEC2023/2BV.1.34
020090-002

[3] O. D. Miller, E. Yablonovitch, and S. R. Kurtz, "Strong internal and external luminescence as solar cells approach the Shockley-Queisser limit," *IEEE J. Photovoltaics*, vol. 2, no. 3, pp. 303–311, 2012.
[4] E. Yablonovitch, O. D. Miller, and S. R. Kurtz, "The opto-electronic physics that broke the efficiency limit in solar cells," *Conf. Rec. IEEE Photovolt. Spec. Conf.*, vol. 3, no. 1, pp. 1556–1559, 2012.
[5] U. Rau, "Reciprocity relation between photovoltaic quantum efficiency and electroluminescent emission of solar cells," *Phys. Rev. B*, vol. 76, no. 8, p. 085303, Aug. 2007.
[6] G. Segev *et al.*, "The Spatial Collection Efficiency of Charge Carriers in Photovoltaic and Photoelectrochemical Cells," *Joule*, vol. 2, no. 2, pp. 210–224, 2018.
[7] G. Segev, C. M. Jiang, J. K. Cooper, J. Eichhorn, F. M. Toma, and I. D. Sharp, "Quantification of the loss mechanisms in emerging water splitting photoanodes through empirical extraction of the spatial charge collection efficiency," *Energy Environ. Sci.*, vol. 11, no. 4, pp. 904–913, 2018.

10.4229/EUPVSEC2023/2BV.1.34
020090-003

PROSPECTIVE PERFORMANCE ENHANCEMENT OF $Cu_2BaSn(S,Se)_4$ BASED SOLAR CELL BY OPTIMIZING BUFFER LAYER AND METAL CONTACT

Hitarth Narsi Patel[1], Rajesh Kumar Sharma[2], Deepak Joshi[3], Vivek Garg[4]
Optoelectronics2Application (O2A) Research Group, Department of Electronics Engineering, SVNIT Surat, India
d21ec004@eced.svnit.ac.in[1], d21ec011@eced.svnit.ac.in[2], d.joshi@eced.svnit.ac.in[3],
vivekg@eced.svnit.ac.in[4]

ABSTRACT: $Cu_2BaSn(S,Se)_4$ (CBTSSe) is a prominent member of multinary chalcogenide family which provides the potential of a decreased amount of antisite defects and properties that can be tuned for photovoltaic (PV) applications. It possesses similar electrical properties as $Cu_2ZnSn(S,Se)_4$ (CZTSSe) due to the analogous connectivity of the Cu and Sn chalcogenide tetrahedra. By varying the S:Se ratio, the CBTSSe offers a wide range tunable bandgap (1.5-2.0 eV), making it an attractive choice for single and multijunction PV systems. CBTSSe has attracted the attention of many researchers due to its earth-abundant material and environmental friendliness. In this paper, we have performed numerical simulation to investigate the suitability of different buffer layers and the effect of metal contacts on performance of solar cell. Optimizing the interface defect between the absorber and buffer has significantly enhanced the Power Conversion Efficiency (PCE) from 6.17% to 13%. We also noted that the deficit in open circuit voltage (Voc) is majorly due to the poor interface between the absorber and buffer, which needs to be optimized to improve cell performance. We have demonstrated the effect of replacing the buffer layer with other potential materials, which increased PCE to 15.3%. Additionally, study on the effect of metal contact work function and incorporation of Anti-reflection coating (ARC) is performed. Finally, the performance parameters of the optimized PV device are: Jsc: 24.35 mA/cm^2, Voc: 1.08 V, FF: 88.16%, PCE: 23.32%.
Keywords: CBTSSe, Thin Films, Chalcogenide

1 INTRODUCTION

Kesterite CZTSSe has attracted significant interest because of its inherent eco-friendliness and inexpensive cost considering the raw material's toxicity and supply. In the last few years, various deposition techniques have led to PCE for CZTSSe that range from 12 to 12.6%, with only 0.2% improved performance in 8 years.[1]–[3] The progress is stalled due to low Voc, which is lower than 60% [4] of the theoretical Shockley–Queisser (SQ) limit [5] for an absorber with 1.13 eV band gap. It has been speculated that cation antisite disorder is the primary reason of Voc deficit. This disorder results from coordination similarities among component metals, which in turn promotes band tailing and deep defects [6]–[10]. A theoretical study suggests that the existing 13% PCE [11] for CZTS may already reflect the maximum feasible value due to deep defects resulting from Sn_{Zn} antisites and the restricted growth conditions available to inhibit the formation of associated defects. One reported theoretical analysis indicates that the trap-assisted PCE is inherently restricted to 20% [12] for CZTSSe, and atomic replacement and extrinsic doping may pave a path to better efficiency. CBTSSe is one of the promising materials with elements of different sizes and has proven the reduction in antisites defects.

CBTSSe PV has already achieved around 5% PCE for vacuum-deposited [12] and solution-deposited [13] fabrication techniques. Although cation antisite disorder and band tailing were successfully suppressed, the Voc of the best vacuum-based cell is 610 mV, and the Voc of the solution-based cell is 605 mV, which is below the SQ limit by a factor of 50%. Despite performing thorough optimization on the cell, the researchers achieved a Voc of SQ limit of 60%. Recently Teymur et al. [14] demonstrated over 6% efficient CBTSSe cell fabricated by the solution deposition technique. The improvement in efficiency was due to the careful selection of the buffer layer. The CdS buffer layer was replaced with a low electron affinity ZnCdS (ZCS) buffer layer.

2 SIMULATION METHODOLOGY AND DEVICE STRUCTURE

In this study, we have used Solar Cell Capacitance Simulator in One Dimension (SCAPS-1D) version 3.3.10. It offers precise predictions as well as the characterization of the electrical properties of solar cells. The continuity equation analyzes the generation and recombination of electrons and holes. The Poisson equation is used to extract electric potential. The schematic structure of CBTSSe based solar cell is shown in Fig. 1.a. To create the baseline of fabricated cell, parameters were taken from experimental data provided in [14]. The capture cross-section area of electrons and holes was assumed to be $1x10^{-15}cm^{-3}$. Flat bands were assumed at the front and back contacts for the simplicity of the analysis. Front contact reflectance was added from reflectance data provided in experimental literature[14]. All simulations are performed under standard AM1.5G spectrum with an incident light power of 1000W/m^2 at 300 K.

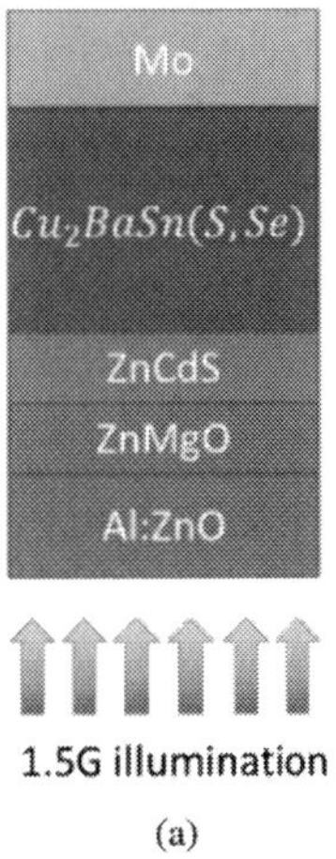

(a)

10.4229/EUPVSEC2023/2BV.1.40
020093-001

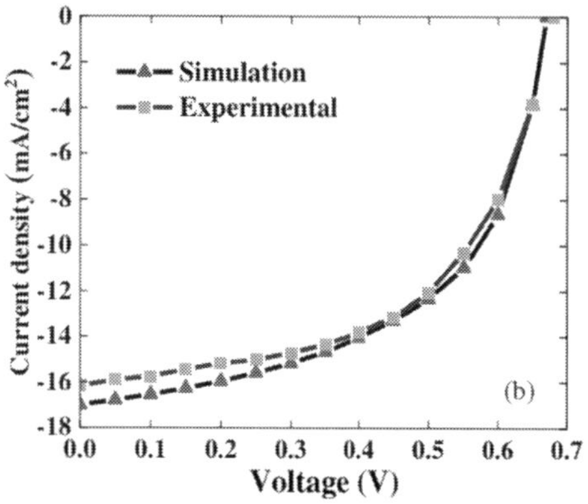

Figure 1. a: Structure of CBTSSe based solar cell. b: J-V Characteristic of Simulated cell and experimental cell

Table I: Comparison of performance parameter of simulated and experimental cell

Base model	PCE (%)	VOC (V)	JSC (mA/cm^2)	FF (%)
Exp. [14]	6.17	0.68	16.2	56.1
Simulation	6.17	0.67	16.99	53.86

Table II: Parameter used in the simulation

Parameter	CBT (S,Se) [14]	ZCS [14]	ZMO [14]	AZO [14]
Thickness (nm)	1000	50	50	250
Bandgap (eV)	1.59	2.45	3.8	3.27
Electron affinity (eV)	3.74	3.93	4.2	4.4
Dielectric permittivity (relative)	5.4	10	9	9
CB effective density of states (cm^{-3})	2.20 x10^{18}	2.20 x10^{18}	2.20 x10^{18}	2.20 x10^{18}
VB effective density of states (cm^{-3})	1.8 x10^{19}	1.8 x10^{19}	1.8 x10^{19}	1.8 x10^{19}
Electron thermal velocity (cm/s)	10^7	10^7	10^7	10^7
Hole thermal velocity (cm/s)	10^7	10^7	10^7	10^7
Electron mobility (cm^2/V. s)	100	10	10	10
Hole mobility (cm^2/V. s)	25	5	5	5
Shallow uniform donor density ND (cm^{-3})	0	2 x10^{16}	0	1 x10^{19}
Shallow uniform acceptor density NA (cm^{-3})	2 x10^{13}	0	0	0
Absorption coefficient (1/cm. eV$^{1/2}$)	Exp.	Exp.	Exp.	10^5
Defect density (cm^{-3})	2.8x 10^{17}	5.5x 10^{16}	1x10^{20}	5x 10^{19}

Apart from the parameters mentioned in Table II, defects were added at the interface of the absorber/buffer. Defects at the interface were set to neutral and single, and the defect level was 0.6 eV above the highest valance band with a defect density of 5.15x10^{12}cm^{-3}. The J-V characteristic of the simulated and experimental device is shown in Fig. 1.b, and their comparison is given in Table I.

3 BASE MODEL OPTIMIZATION

Interface defects are one of the significant reasons for a deficit in Voc. The improvement in the fabrication process can highly reduce the defect density and increase the Voc and the PCE. The defect density at the interface was varied from 1x10^{8}cm^{-3} to 1x10^{15}cm^{-3}. For defect density less than 1x10^{10}cm^{-3}, the defect density remains almost constant at around 13% (Fig. 2.a) due to the longer diffusion length of minority carriers. PCE reduces as defect density increases beyond 1x10^{10}cm^{-3}. This happens because of increased surface recombination, which reduces the number of carriers reaching the contacts. The thickness and defect density of the absorber were varied simultaneously, and other parameters were kept constant. Thickness and defect density are critical in determining the cell's performance. Thickness was varied from 100nm to 2µm, and defect density was varied from 1x10^{11}cm^{-3} to 1x10^{20}cm^{-3}. PCE improved further to 14.3%. From the contour plot in Fig. 2.b, we could conclude that for high-performance cell, the thickness should be 1µm, and the defect density should be less than 1x10^{16}cm^{-3}.

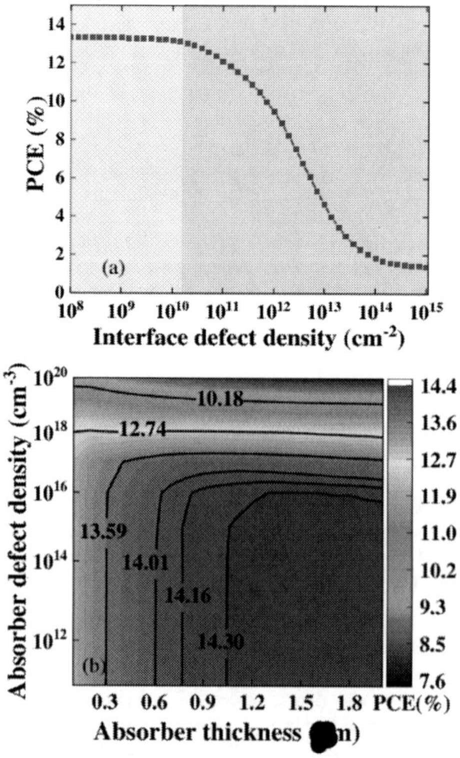

10.4229/EUPVSEC2023/2BV.1.40

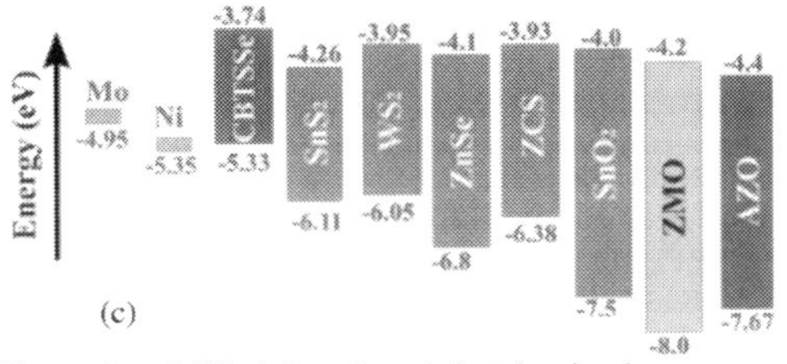

Figure 2. a: PCE v/s Interface defect density. b: Absorber defect density v/s thickness c: Band gap and electron affinity of material used for simulation

As defect density increases, the carrier's lifetime decreases, because of which they recombine quickly without reaching the contacts. CBTSSe material doesn't allow the change of the doping density easily. The doping density found in experimental literature [14]is in the order of ~10^{13}cm^{-3}. So here, we keep the doping density constant to 2x10^{13}cm^{-3} and move toward optimizing the next layer.

4 BUFFER LAYER OPTIMIZATION

The difference in electron affinity (EA) between the absorber and buffer layer is offset. The offset plays a vital role in determining the characteristic of the cell. The EA of the buffer layer (offset) was varied from 3.64 eV to 4.14 eV (0.1eV to -0.4eV), and we found a region in which PCE is greater than 13% (Fig. 3.a). This allowed us to replace ZCS with other suitable materials whose EA (offset) falls between 3.85eV to 4eV (-0.11eV to -0.26eV). We found various materials like SnO_2[16], WS_2[17], SnS_2 [18] and ZnSe [19].

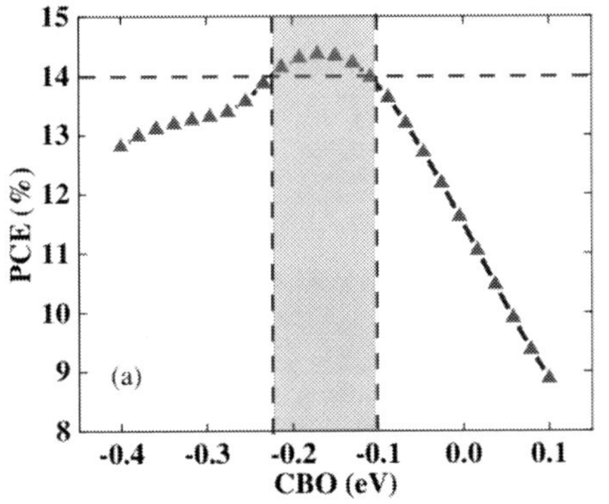

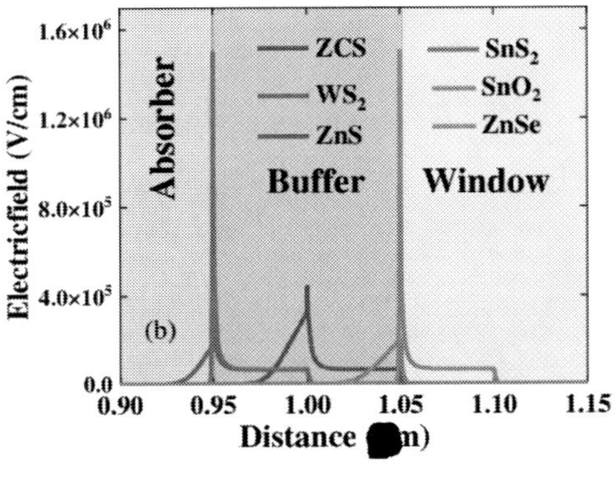

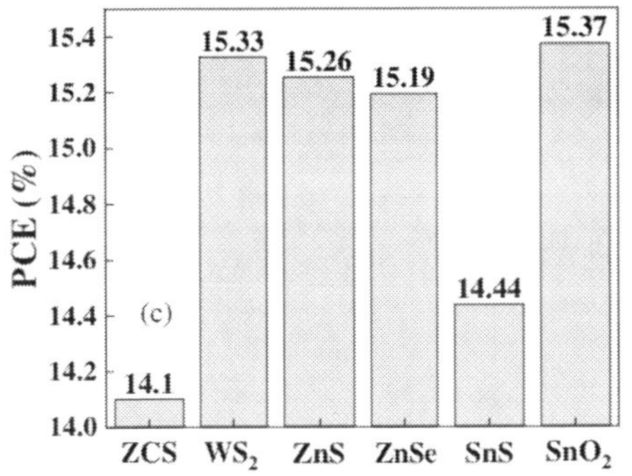

Figure 3. a: PCE v/s Absorber - Buffer offset. b: Electric field v/s distance for cells with various buffer layers. c: Comparison of PCE achieved by employing various buffer layers.

We replaced the ZCS with each of them, and to understand the effect of different buffer layers on performance, we plotted the electric field of each device with a different buffer layer (Fig. 3.b). Since the electric field is directly proportional to the doping concentration, as expected, it shoots up in the case of SnS_2 [18] and ZnSe [19]. However, the electron affinity had a higher "cliff-like" band alignment which isn't favorable for generating high PCE. So, the other material with high doping density and suitable EA was found to be WS_2 and SnO_2, with similar PCE of 15.33% and 15.37%, respectively (Fig. 3.c).

5 FRONT METAL CONTACT OPTIMIZATION

The simulation were carried out by considering the reflectance results reported in the literature. [14] The reflectance can be reduced if we put an ARC. ARC reduces the reflectance, allowing more photons to reach the subsequent layers, resulting in more free carrier generation and boosting PCE. We have used the experimentally generated reflectance data of MgF_2 ARC and simulated the cell. The application of ARC resulted in PCE of 18.6% for WS_2-based cell and 18.72% for SnO_2-based cell.

6 BACK METAL CONTACT OPTIMIZATION

Selecting the back metal contact with the appropriate work function is very important since it determines the direction of the electric field. The generated electric field must permit holes and block the electrons from reaching the back metal contact. The work function of back metal contact was varied from 4eV to 5.5eV, and we recorded the performance of cell employing WS_2 and SnO_2 buffer layer. We also analyzed the electric field for various metals with work functions 5eV and 5.5eV (Fig. 4.a-b). We noticed that the electric field is in the positive direction for the metal with a work function of 5eV, which means it is from the metal to the absorber. Electrons flow in the direction opposite to the electric field; hence, electrons would flow from absorber to metal, and holes would flow from metal to absorber, which is undesired. For the metal contact with a work function of 5.5eV, the electric field is negative, so the direction would be from the absorber to metal which would enable electrons to flow from metal to absorber and holes to flow from absorber to metal contact.

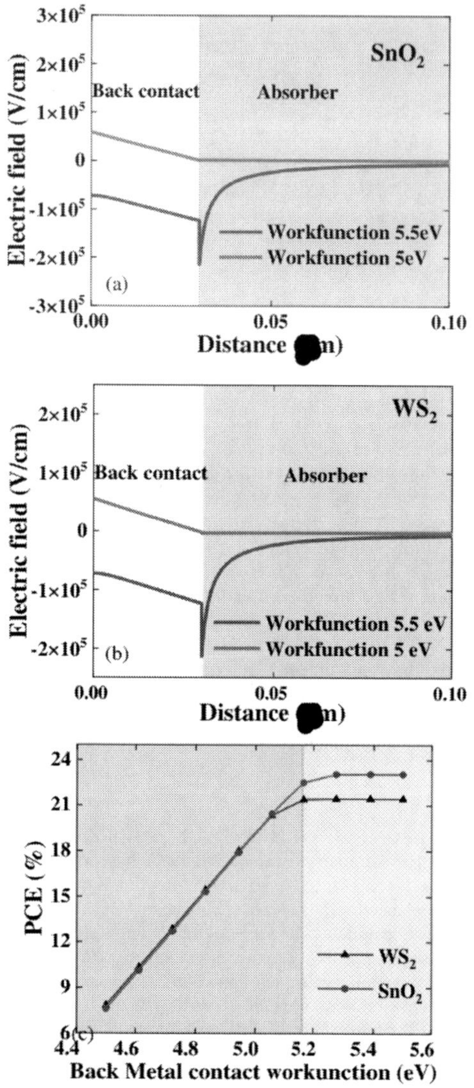

Figure 4. a: Electric field v/s distance for cell with SnO_2 buffer. b: Electric field v/s distance for cell with WS_2 buffer. c: PCE v/s back metal contact work function.

This would encourage more holes to reach the metal contact since the absorber is P-type. The base model used Mo as back contact, which resulted in low PCE. We replace Mo with other metals (Au, Pt or Ni) with a work function greater than 5.2eV (Fig. 2.c) to get maximum efficiency of ~ 23%.

7 CONCLUSION

CBTSSe is an emerging absorber material for photovoltaic applications. Experimentally the cell has achieved an efficiency of 6.17% within five years of research. Voc deficit is majorly due to high surface recombination between the absorber and buffer layer. A careful investigation in the fabrication process can help to reduce the defect. For defect density, less than $1x10^{10}cm^{-3}$ PCE of 13% can be achieved. We further proposed two buffer layers, SnO_2 and WS_2, which can replace the ZCS to improve the efficiency to ~15.3%. The coating of ARC is recommended on the front contact to reduce the reflectance. The ARC on the front layer increased the PCE to ~18.6%. Finally, optimization of the back metal contact results in the best optimized device with the photo conversion efficiency of ~23%.

8 REFERENCES

[1] K. Pal, P. Singh, A. Bhaduri, and K. B. Thapa, "Current challenges and future prospects for a highly efficient (>20%) kesterite CZTS solar cell: A review," *Sol. Energy Mater. Sol. Cells*, vol. 196, pp. 138–156, Jul. 2019, doi: 10.1016/j.solmat.2019.03.001.

[2] D.-H. Son *et al.*, "Effect of solid-H 2 S gas reactions on CZTSSe thin film growth and photovoltaic properties of a 12.62% efficiency device," *J. Mater. Chem. A*, vol. 7, no. 44, pp. 25279–25289, 2019, doi: 10.1039/C9TA08310C.

[3] Y. Gong *et al.*, "Identifying the origin of the V oc deficit of kesterite solar cells from the two grain growth mechanisms induced by Sn 2+ and Sn 4+ precursors in DMSO solution," *Energy Environ. Sci.*, vol. 14, no. 4, pp. 2369–2380, 2021, doi: 10.1039/D0EE03702H.

[4] S. Hadke *et al.*, "Emerging Chalcogenide Thin Films for Solar Energy Harvesting Devices," *Chem. Rev.*, vol. 122, no. 11, pp. 10170–10265, Jun. 2022, doi: 10.1021/acs.chemrev.1c00301.

[5] W. Shockley and H. J. Queisser, "Detailed Balance Limit of Efficiency of p-n Junction Solar Cells," *J. Appl. Phys.*, vol. 32, no. 3, pp. 510–519, Mar. 1961, doi: 10.1063/1.1736034.

[6] G. Rey *et al.*, "The band gap of Cu2ZnSnSe4: Effect of order-disorder," *Appl. Phys. Lett.*, vol. 105, no. 11, p. 112106, Sep. 2014, doi: 10.1063/1.4896315.

[7] S. Chen, A. Walsh, X.-G. Gong, and S.-H. Wei, "Classification of Lattice Defects in the Kesterite Cu 2 ZnSnS 4 and Cu 2 ZnSnSe 4 Earth-Abundant Solar Cell Absorbers," *Adv. Mater.*, vol. 25, no. 11, pp. 1522–1539, Mar. 2013, doi: 10.1002/adma.201203146.

[8] S. Chen, L.-W. Wang, A. Walsh, X. G. Gong, and S.-H. Wei, "Abundance of Cu Zn + Sn Zn and 2Cu Zn + Sn Zn defect clusters in kesterite solar cells," *Appl. Phys. Lett.*, vol. 101, no. 22, p. 223901, Nov. 2012, doi: 10.1063/1.4768215.

[9] S. Bourdais *et al.*, "Is the Cu/Zn Disorder the Main Culprit for the Voltage Deficit in Kesterite Solar Cells?," *Adv. Energy Mater.*, vol. 6, no. 12, p. 1502276, Jun. 2016, doi: 10.1002/aenm.201502276.

[10] T. Gokmen, O. Gunawan, T. K. Todorov, and D. B. Mitzi, "Band tailing and efficiency limitation in kesterite solar cells," *Appl. Phys. Lett.*, vol. 103, no. 10, p. 103506, Sep. 2013, doi: 10.1063/1.4820250.

[11] W. Chen, D. Dahliah, G.-M. Rignanese, and G. Hautier, "Origin of the low conversion efficiency in Cu 2 ZnSnS 4 kesterite solar cells: the actual role of cation disorder," *Energy Environ. Sci.*, vol. 14, no. 6, pp. 3567–3578, 2021, doi: 10.1039/D1EE00260K.

[12] D. Shin, T. Zhu, X. Huang, O. Gunawan, V. Blum, and D. B. Mitzi, "Earth-Abundant Chalcogenide Photovoltaic Devices with over 5% Efficiency Based on a Cu 2 BaSn(S,Se) 4 Absorber," *Adv. Mater.*, vol. 29, no. 24, p. 1606945, Jun. 2017, doi: 10.1002/adma.201606945.

[13] B. Teymur, L. Choubrac, H. Hempel, O. Gunawan, T. Unold, and D. B. Mitzi, "Influence of Copper Composition on Cu 2 BaSn(S,Se) 4 Solution-Deposited Films and Photovoltaic Devices with Over 5% Efficiency," *ACS Appl. Energy Mater.*, vol. 5, no. 9, pp. 10645–10656, Sep. 2022, doi: 10.1021/acsaem.2c01364.

[14] B. Teymur, Y. Kim, J. Huang, K. Sun, X. Hao, and D. B. Mitzi, "Top Stack Optimization for Cu 2 BaSn(S, Se) 4 Photovoltaic Cell Leads to Improved Device Power Conversion Efficiency beyond 6%," *Adv. Energy Mater.*, vol. 12, no. 40, p. 2201602, Oct. 2022, doi: 10.1002/aenm.202201602.

[15] Y. Zhou *et al.*, "Efficient and Stable Pt/TiO 2 /CdS/Cu 2 BaSn(S,Se) 4 Photocathode for Water Electrolysis Applications," *ACS Energy Lett.*, vol. 3, no. 1, pp. 177–183, Jan. 2018, doi: 10.1021/acsenergylett.7b01062.

[16] H. Sun *et al.*, "Efficiency Enhancement of Kesterite Cu 2 ZnSnS 4 Solar Cells via Solution-Processed Ultrathin Tin Oxide Intermediate Layer at Absorber/Buffer Interface," *ACS Appl. Energy Mater.*, vol. 1, no. 1, pp. 154–160, Jan. 2018, doi: 10.1021/acsaem.7b00044.

[17] M. K. S. Bin Rafiq *et al.*, "WS2: A New Window Layer Material

10.4229/EUPVSEC2023/2BV.1.40
020093-004

for Solar Cell Application," *Sci. Reports 2020 101*, vol. 10, no. 1, pp. 1–11, Jan. 2020, doi: 10.1038/s41598-020-57596-5.

[18] Y. K. Reddy, V. Manjunath, S. Bimli, and R. S. Devan, "Futuristic kusachiite solar cells of CuBi2O4 absorber and metal sulfide buffer Layers: Theoretical efficiency approaching 28 %," *Sol. Energy*, vol. 244, pp. 75–83, Sep. 2022, doi: 10.1016/j.solener.2022.08.034.

[19] M. Saadat, O. Amiri, and P. H. Mahmood, "Analysis and performance assessment of CuSbS2-based thin-film solar cells with different buffer layers," *Eur. Phys. J. Plus*, vol. 137, no. 5, p. 582, May 2022, doi: 10.1140/epjp/s13360-022-02804-6.

InGaAs-based MQWs PHOTOVOLTAIC UNDER CONCENTRATION

Gaurav Siddharth[1], Ruchi Singh[2], Mayank Dubey[3], Ritesh Bhardwaj[4] and Vivek Garg[5]
[1]National Institute of Technology Calicut, Kerala, 673601, India
[2]Visveswarya National Institute of Technology Nagpur, Maharashtra, 440010, India
[3]Indian Institute of Technology Indore, Madhya Pradesh, 453552, India
[4]LNM Institute of Information Technology Jaipur, Rajasthan, 302031, India
[5]Sardar Vallabhbhai National Institute of Technology Surat, Gujarat, 35007, India

ABSTRACT: This work investigates the GaAs-based multiple quantum well (MQWs) solar cell under concentrated sunlight using the developed analytical model. The variation of device performance parameters are examined for different irradiance as in time application solar cell are operated at non-standard test condition (STC). The results present that performance of solar cell can be improved by using concentrators. With increase in irradiance level, the figure of merits of GaAs-based MQWSC varies as follows: J_{sc} increases linearly, V_{oc} and conversion efficiency increases logarithmically. Significantly, the American Society for Testing and Materials (ASTM) standards data sheets and spectral irradiance are utilized for attaining photon flux density instead of black body radiation formula.
Keywords: MQW, ASTM, Concentration, Photon Flux

1 INTRODUCTION

The single junction solar cell has limitations arises by the mismatch between the fixed absorber energy bandgap and broad solar spectrum. The MQWs structures have found application in photovoltaic devices because it has overcome certain limitations of the *p-i-n* structures. The multiband structure such as MQW are used to minimize the thermalization and transmission losses offered by single junction solar cell. Also, MQW structure offers better thermal stability compared to *p-i-n* structures, as one of the transport mechanism in quantum wells is thermal activated [1]. MQWs insertion in the *p-i-n* solar cell (SC) leads to the improvement of the figure of merits of SC by individually controlling and optimizing both the current and voltage performance parameters [2]. The MQWSC have an advantage of adjustable bandgap at constant lattice parameters with energy bandgap lower than bulk material, and also have high thermal stability therefore they are good candidates for PV with concentrated sunlight [1]. The MQW based SC displayed considerably high short-circuit current density (J_{sc}), since additional charge carriers are generated because of the lower bandgap of well material, however open- circuit voltages (V_{oc}) marginally decreases but decrease is not much as it is predominantly governed by wider band gap i.e. of barrier material. The MQW structure also have potential application in the multi-junction solar cells (MJSCs) to optimize the energy bandgap so as to have maximum possible current matching of the top, middle and bottom cells [3].

The GaAs-based solar cells provide record conversion efficiency by having advantage of better physical and optical properties. The energy bandgap of the GaAs material is very much close the optimized value of the single junction absorber material. Consequently, it is important to optimize the GaAs solar cells, and then the use of multiples quantum wells in GaAs, as semiconductor host is a better option. Now to enhance the efficiency of the solar cell the InGaAs material is one candidate material for the GaAs-based MQWSC, as the introduction of In leads to the decrease in the effective bandgap of the material [4].

In this study, photon flux density (obtained from American Society for Testing and Materials (ASTM) standards data sheets and spectral irradiance) is utilized for obtaining performance parameters and *J-V* characteristics of the MQWSC device. The present work demonstrates the analytical model developed for the MQWSC to obtain the *J-V* characteristics and the device performance parameters of the GaAs-based MQWSC. The analytical study investigates the effect of solar irradiance level on the MQWSC performance parameters and its characteristics. Further, the variation of conversion efficiency with the number of QWs in MQWSC is also studied.

2 MODELLING

To gauge the performance parameters of MQWSC, an analytical study of GaAs-based MQWSC is performed using the model developed in our earlier work and is described here concisely with modification i.e. including the effect of solar irradiance.

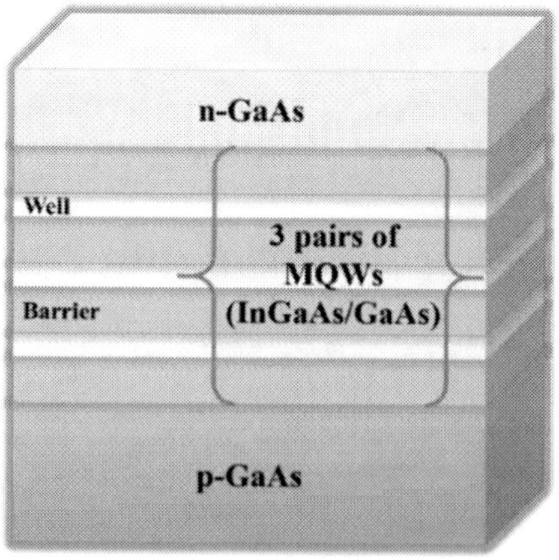

Figure 1: Schematic of CdZnO/ZnO-based MQWs structure.

Fig. 1 shows the schematic representation of the GaAs-based MQW structure used for the analytical modelling. In the schematics 3 pairs of the MQWs are shown with InGaAs as well and GaAs as barrier materials.

The spectral photon flux ($N_{ph}(\lambda)$) and irradiance (I) are given by expressions [5]

$$N_{ph}(\lambda) = P(\lambda)\frac{\lambda}{hc} = 5.034 \times 10^{14} P(\lambda)\lambda \quad (1)$$

$$I = \int P(\lambda)\, d\lambda \quad (2)$$

The spectral power density, P(λ) for AM1.5 is available from ASTM standard data sheets [6] as a function of wavelength.

The above equations are further used for obtaining the figure of merits of the SC and also to derive current density vs voltage (*J-V*), power density vs voltage (*P-V*) characteristics.

The spectral power density calculated from Planck's Law is given by [7]:

$$P(\lambda, T) = \frac{2hc^2}{\lambda^5}\frac{1}{exp\left(\frac{hc}{\lambda k_B T}\right) - 1}\pi\left(\frac{R_s}{Au - R_e}\right)^2 \quad (3)$$

where R_s and R_e are the radii of the Sun and Earth, respectively, *Au* is the distance between the Sun and Earth, and *T* is the Sun temperature.

The spectral power density available from ASTM data sheets for AM1.5 and Planck's Law are shown in Fig. 2. From comparison it can be concluded that the more accurate figure of merits of the SC are be obtained with ASTM data sheets for AM1.5 as compared to that obtained by using Planck's law [7]. Therefore, in this present model development, the ASTM data sheets for AM1.5 are utilized instead of Plank's Law.

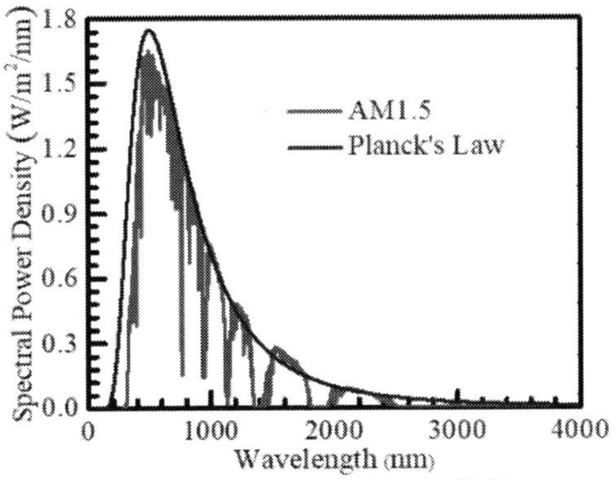

Figure 2: Spectral power density from ASTM data sheets for AM1.5 and Planck's Law.

The *J-V* characteristics and figure of merits of the MQWSC are developed in our earlier work and are briefly explained here [8-9]:

$$J(V) = J_s(1 + r\beta_1)[\exp(V/V_t) - 1] - J_{sc} \quad (4)$$

$$J_{sc} = -qI[f_w N_{ph}(E_B > E > E_W) + (1 - f_w)N_{ph}(E > E_B)] \quad (5)$$

$$V_{oc} = V_t \ln\left[\frac{J_s(1 + r\beta_1) - J_{sc}}{J_s(1 + r\beta_1)}\right] \quad (6)$$

where J_s is the reverse saturation current density, β_1 represents the radiative recombination current density, ratio *r* describes the radiative enhancement, V_t is the thermal voltage, f_w is the fraction associated with well and barrier region, and E_B and E_W are the energy bandgaps of barrier and well materials, respectively.

The maximum power density point (V_m, J_m) in active region of SCs are obtained by $\frac{\partial P}{\partial V} = 0$ and $\frac{\partial P}{\partial J} = 0$, where *P* is the power density. Finally, the iterative methods are used for the calculation of maximum power density point:

$$P = JV \quad (7)$$

$$exp\left(\frac{V_{oc}}{V_t}\right) - exp\left(\frac{V_m}{V_t}\right)\left[1 + \frac{V_m}{V_t}\right] = 0 \quad (8)$$

$$\frac{J_m}{J_m - J_{sc} + J_s(1 + r\beta_1)} = ln\left[\frac{J_s(1 + r\beta_1)}{J_m - J_{sc} + J_s(1 + r\beta_1)}\right] \quad (9)$$

3 RESULTS

The dark and light current density versus voltage (J-V) and power density versus voltage (P-V) characteristics of the GaAs-based MQWSC (5 pairs of QWs with well and barrier thicknesses of 5 and 9 nm, respectively) are depicted in Fig. 3. The figure of merits of GaAs-based MQWSC, Jsc = 23.93 mA/cm2, Voc = 951 mV, and η = 19.98%, are achieved under standard test condition (STC, irradiance level of 100 mW/cm2 at AM1.5G spectrum, 25 °C).

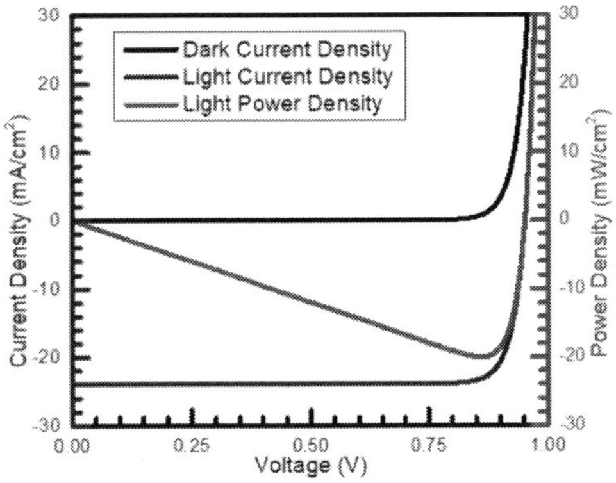

Figure 3: *J-V* and *P-V* characteristic of MQWSC at STC.

Generally, performance of the solar cells are reported at STC. However, actually the solar cells are operated in non STC, where solar irradiance level varies with location, weather and day time. Therefore, a study is done to analyse the effect of solar irradiance level (from 0.1 to 10 Suns) on the figure of merits of MQWSC. Analysis is also done for more than 1 suns, concentrators can be used for increasing the number of suns.

Fig. 4 shows the variation of figure of merits of GaAs-based MQWSC with the effect of change in the irradiance level. From Eq. 5, it can be seen that J_{sc} is linearly dependent on the solar irradiance [2], because of the additional photo generated charge carriers as irradiance level increases and similar effect is observed in the Fig. 4. However, the same linear effect is not observed with the η of the MQWSC, since the η benefits also arises from the logarithmic dependence of the V_{oc} on J_{sc}. Therefore, V_{oc} increases logarithmically with solar irradiance [2]. With increase in irradiance level from 10 to 1000 mW/cm^2, the figure of merits increases as follows: J_{sc} from 2.39 to 239.28 mA/cm^2, V_{oc} from 892 to 1011 mV and η from 18.60 to 21.37 %. It appears that η is also logarithmically increasing with the solar irradiance similar to V_{oc}, it is because of the fact that linear effect of the J_{sc} is

10.4229/EUPVSEC2023/2BV.1.43
020095-002

compromised with the linear effect of irradiance, therefore, it varies according to V_{oc} with the change in solar irradiance. So, it is concluded that performance of the solar cell decreases as solar irradiance decreases it might be due to bad weather condition, locations etc. However, performance can be improved by using concentrators.

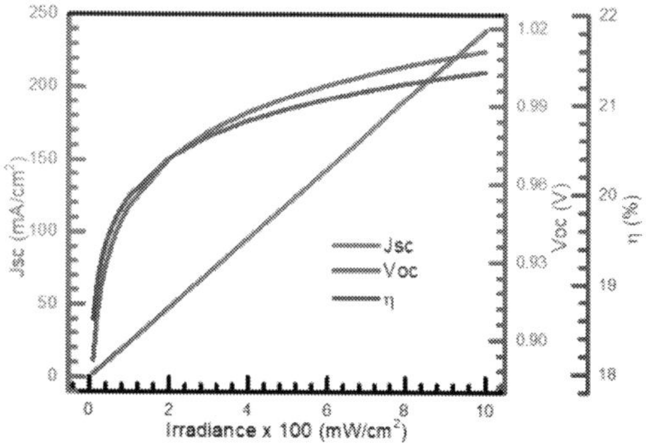

Figure 4: Figure of merits of MQWSC with variation in Solar Irradiance level (10 to 1000 mW/cm²).

4 CONCLUSION

An analytical study has been presented for the GaAs-based MQWSC. The change in the irradiance level leads to the change in all the performance parameters of the MQWSC. The conversion efficiency of the GaAs-based MQWSC increases by ~2.8% as the irradiance level increases from 10 to 1000 mW/cm². This study considerably explores the likelihood of efficient absorption of low wavelength photons and improving η by using concentrators (increasing irradiance level).

5 REFERENCES

[1] K. Y. Lai, G. J. Lin, C.-Y. Chen, Y.-L. Lai, and J. H. He, "Origin of hot carriers in InGaN-based quantum-well solar cells," IEEE Electron Device Letters, vol. 32, no. 2, pp. 179-181, Dec. 2010.

[2] R. Dahal, J. Li, K. Aryal, J. Y. Lin, and H. X. Jiang, "InGaN/GaN multiple quantum well concentrator solar cells," Appl. Phys. Lett., vol. 97, no. 7, p. 073115, Aug. 2010.

[3] B. Kim, K. Toprasertpong, A. Paszuk, O. Supplie, Y. Nakano, T. Hannappel, and M. Sugiyama, "GaAsP/Si tandem solar cells: Realistic prediction of efficiency gain by applying strain-balanced multiple quantum wells," Solar Energy Materials and Solar Cells, vol. 180, pp. 303-310, Jun. 2018.

[4] Inoue, Tomoyuki, et al., "Enhanced light trapping in multiple quantum wells by thin-film structure and backside grooves with dielectric interface," IEEE Journal of Photovoltaics, vol. 5, no. 2, pp. 697-703, Feb. 2015.

[5] R. Hulstrom, R. Bird, and C. Riordan, "Spectral solar irradiance data sets for selected terrestrial conditions," Sol. Cells, vol. 15, no. 4, pp. 365–391, Dec. 1985.

[6] "ASTM G173-03 Reference Spectra Derived from SMARTS v. 2.9.2." [Online]. Available: https://www.nrel.gov/grid/solar-resource/spectra.html.

[7] A. H. Smets, K. Jäger, O. Isabella, R. A. Swaaij, and M. Zeman, Solar Energy: The Physics and Engineering of Photovoltaic Conversion, Technologies and Systems. Cambridge, U.K.: UIT Cambridge, 2015

[8] G. Siddharth, V. Garg, B. S. Sengar, R. Bhardwaj, P.Kumar, S. Mukherjee, "Analytical Study of performance parameters of InGaN/GaN Multiple Quantum Well Solar Cell," IEEE Transactions on Electron Devices, vol. 66 no. 8, pp. 3399-3404, Aug. 2019.

[9] G. Siddharth, R. Singh, V. Garg, B. S. Sengar, M. Das, B. Mandal, M. T. Htay, M. Gupta, and S. Mukherjee, Investigation of DIBS-deposited CdZnO/ZnO-based multiple quantum well for large-area photovoltaic application, IEEE Transactions on Electron Devices, vol. 67, no. 12, pp. 5587-5592, Nov. 2020.

10.4229/EUPVSEC2023/2BV.1.43

Principles for construction of solar cells using terrestrial counter-radiation

C. Graf von Westarp

Physics Lab, Faculty Life Sciences, University of Applied Sciences Hamburg,
Ulmenliet 20, 21033 Hamburg
Carsten.Westarp@haw-hamburg.de

Abstract

Up to now climate change and ist numourous consequences such as extreme weather, threat of biodiversty and food supply has been recognizes as a major challenge for humanity, as declared in the Paris Agreement of 2016. Among different measures to limit climate change, the decarbonization of power generation is a crucial factor and among the numerous emission-free energy sources that have been developed so far, one of the most important is the use of solar radiation by solar cells. However, solar cells have the disadvantage that the usable power depends on numerous circumstances such as weather conditions, times of day and night or seasons and, therefore, fluctuates. In order to compensate these fluctuations, extensive investments in the development of transport and storage technologies are necessary.Thus, this work does not examine the radiation from the sun for power generation, but the infrared terrestrial counter-radiation from earth. After identifying said frequency range, a material must be found that is suitable for constructing solar cells active at these energies. A promising candidate is the III-V semiconductor indium antimony that is already being used for infrared detectors. By the implementation of gallium the effectiveness of the semiconductor can be improved at the desired frequencies. Based on these findings, the development of a new type of solar cells based on the absorption of the terrestrial counter-radiation can be developed.

Keywords— New materials, terrestrial counter-radiation, narrow-band-gap semiconductors

1 Introduction

Photovoltaics have been continuously expanded since the beginning of the millennium and have now become one of the most important renewable energy sources. In 2019, photovoltaics reached a share of 50% Wirth (2020) on several days in Germany. However, this technology has the fundamental disadvantage that the volume of electricity production depends on numerous boundary conditions, including the season, the coverage, the time and the temperature (Quaschning (2015)). This is mainly because the solar radiation used by the solar cell fluctuates strongly due to the conditions mentioned. These fluctuations are shown in Fig. 1 depending on the season and the weather conditions for a location in central Germany according to Koblin (1984).

This means that photovoltaics can not provide the so-called base load, which was between 20 and 40 GW in Germany in 2019 (see Insitut für Solare Energieforschung (ISE) (2019)). One solution may be to use not optical solar radiation, but thermal infrared radiation from the earth. Parameters such as the content of greenhouse gases and the temperature of the clouds and the ground are crucial, however, the fluctuations are lower in comparision to solar radiation (see Fig. 1).

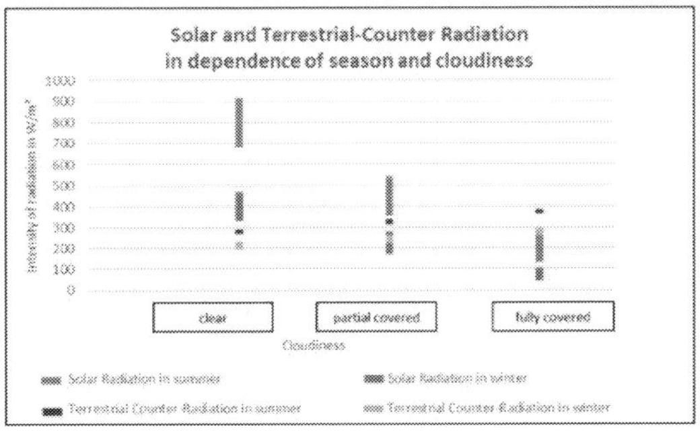

Figure 1: Radiation intensity of solar radiation and terrestrial conuter-radiation depending on season and cloudiness in central Germany.

2 Origin of the terrestrial counter-radiation

Terrestrial radiation is largely the thermal radiation of the Earth. The radiation intensity of the total emission is described by Stefan-Boltzmann-law:

$$I = \sigma T^4 \quad (1)$$

Here I describes the radiation intensity of terrestrial radiation, σ the Stefan-Boltzmann constant, and T the temperature of the emitting body. The spectral distribution gives Planck's law:

$$dI_\lambda = \frac{2\pi hc^2}{\lambda^5} \frac{1}{e^{\frac{hc}{kT\lambda}} - 1} d\lambda \quad (2)$$

Here k is the Boltzmann constant, e is the Eulerian number and dI_λ is the radiation intensity in the spectral interval $d\lambda$.
By use of a typical temperature of 20 ◦C in Equation 1, the result is a value for I of approx. 420 W/m^2, which is largely within the range of solar radiation on Earth, the so-called solar constant of 1000 W/m^2.
This emitted radiation from Earth reaches the atmosphere and is absorbed and re-emitted or transmitted (outlined in Fig. 2). On a global and spectral average, approximately 15% of the radiation is transmitted (based on results from Trenberth et al. (2009)), the rest radiates back to the Earth, the so-called terrestrial counter-radiation. This is generally referred to as 'greenhouse effect' and is dependent on the one hand on the thermal radiation emitted from Earth according to Eq. 1 and on the other hand on the composition of the atmosphere.

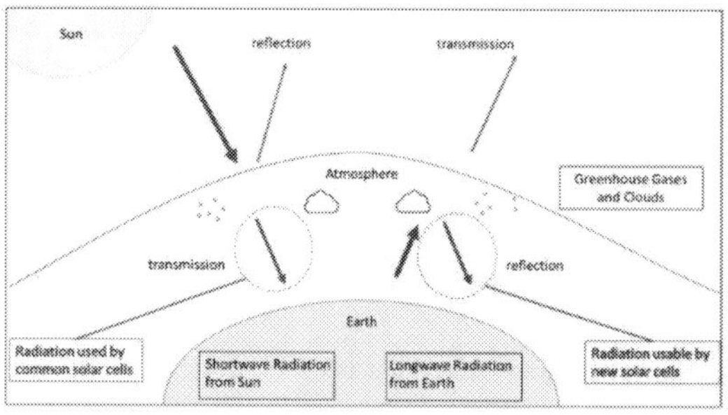

Figure 2: Effects in the atmosphere that lead to the formation of terretrial counter-radiation

The Earth's thermal radiation is largely in the mid-infrared wavelength range between 2 and 30 µm, with the maximum being 9.6 µm. Accordingly, it makes sense to to use the wavelength range between 8 and 13 µm to generate energy. However, as the terrestrial counter-radiation also depends on the composition of the atmosphere, the optical properties of the atmosphere must also be taken into account.
The atmosphere consists of more than 99.6 % of nitrogen N_2, oxygen O_2 and argon Ar (see Warnecke (1997) and Viudez-Mora (2011)), which are not optically active in the infrared radiation range are. In the mid-infrared, the dominating gases are water vapor H_2O, carbon dioxide CO_2, ozone O_3, methane CH_4, nitrous oxide N_2O, the so-called *greenhousegases*. A distinction is made between short-term volatile and long-term volatile greenhouse gases. The atmospheric content of short-term volatile gases such as H_2O, and O_3 fluctuates greatly depending on temperature and solar radiation on a daily and annual average. In addition to greenhouse gases, clouds have a significant influence on atmospheric counterradiation, although the exact relationships are still part of active research, as Viudez-Mora (2011) describes. It is known that the influence of clouds on terrestrial counter-radiation depends on cloud type, height, composition, ice content and coverage (Viudez-Mora (2011)). These properties are also volatile in the short term. Clouds have the greatest influence on the terrestrial counter-radiation in the range between 8 and 13 µm and dominate this.
In summary, this shows that the atmospheric counterradiation is constant over time in some areas and volatile in others. Since the goal of this work is to find an energy source that is constant over time, it makes sense to concentrate on an area

that is as constant as possible. In this work, measurements of atmospheric counterradiation were carried out from 2016 and 2020. A Rocket FT-IR Fourier spectrometer from ARCoptik was used, which can resolve a wide range of infrared radiation between 2.5 and 12 µm. Figure 3 shows 20 spectrally resolved measurements on different days, at different times, degrees of cloudness and temperatures. As expected, constant ranges (between 6 and 8 µm) and volatile ranges (3 to 6 µm and 8 to 12 µm) are obvoius.
The fluctuations in the range between 3 µm and 5 µm can be mainly attributed to clouds. The range between 6 and 8 µm is constant and H_2O dominates here. It may initially be surprising that the radiation intensity in this range is constant over time, although it is primarily determined by the short-term volatile H_2O. This can be explained by the absorption coefficients of H_2O is close to 1 (see Coulson (1975)) and accordingly complete absorption and emission prevails even with small amounts of water vapor. Fluctuations in the water vapor content then have no effect on the radiation intensity. If radiation from clouds occurs in this area, it does not reach the ground, so that a short-term increase in radiation, as in the volatile ranges, is not possible. In the range between 8 to 12 µm, the radiation intensity fluctuates strongly between measurements. This can clearly be attributed to the occurrence of volatile O_3 and clouds based on Viudez-Mora (2011).

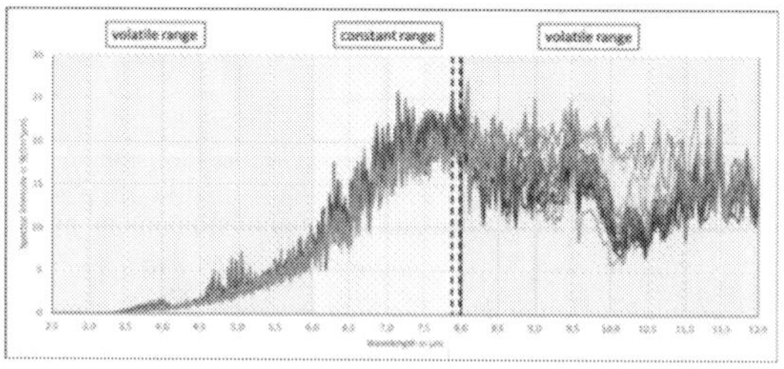

Figure 3: Spectrally resolved measurements of the terrestrial counter-radiation show the spectral radiation intensity with two volatile ranges (red) and one constant range (green).

3 The selection of a sensible semiconductor for the specific radiation

After the spectrum of terrestrial counter-radiation is presented and explained, the goal is now to find a specific wavelength at which the solar cell should work. As shown in the previous chapter, the terrestrial counter-radiation has a constant range between 6 µm and 8 µm. Since the goal of this work is to construct a constant energy source, only this ranges comes into account. In this context, it makes sense to choose a wavelength with the greatest possible spectral intensity so that the energy generated is as large as possible but also has a small fluctuation. This is the case with a wavelength of 7.8µm and 7.96 µm. The corresponding energies are calculated with the following formula:

$$E = h \cdot f = \frac{hc}{\lambda} \tag{3}$$

Here h is the Planck's constant and c the velocity of light. This results in energeries of $E_{7.8} = 0.160$ eV and $E_{7.96} = 0.157$ eV . This is far from semiconductors usually taken for solar cells like silicon (band gap of 1.11 eV) and gallium arsenide (band gap 1.43 eV). Thus, so-called narrow-band semiconductors are necessary. Among the known ones indium antimony (*InSb*) is a direct semiconductor with a band gap of 0.17 eV (Cleanroom (2020)) and is already often used as a detector in infrared sensors. By replacing Small amounts of antimony with arsenic can be reduced to the desired value of 0.16 eV. According to Cleanroom (2020), a proportion of approx. 5% arsenic is necessary.

4 Conclusion

This work presents principles that form the volatile and constant range of terrestrial counter-radiation. From this a wavelength range was determined that is suitable and *InSb* with a small amount of *As* is presented as a possible semiconductor for the construction of infrared solar cells. Further research steps should now follow, including determining useful dopants. In the long term, the construction of a prototype is planned, which can be used to optimize the conversion of thermal radiation into electrical current in a concrete and application-related manner.
In summary the principles are presented for the construction of solar cells that work at the required wavelengths and could generate power emission free permanently and constantly.

References

Cleanroom, 2020. BYU Cleanroom.

Coulson, K.L., 1975. Solar and Terrestrial Radiation. Academic Press.

Insitut für Solare Energieforschung (ISE), 2019. Energie Charts - Stromproduktion 2019.

Koblin, W., 1984. Handbuch Passive Nutzung der Sonnenergie.

Quaschning, V., 2015. Regenerative Energiesysteme. Carl Hanser Verlag M"unchen. 9. edition.

Trenberth, K.E., Fasullo, J.T., Kiehl, J., 2009. Earth's global energy budget. Bulletin of the American Meteorological Society 90, 311–323.

Viudez-Mora, A., 2011. Atmospheric downwelling longwave radiation at the surface during cloudless and overcast conditions. Measurements and modeling. Ph.D. thesis.

Warnecke, G., 1997. Meteorologie und Umwelt. Springer Verlag. 2. edition.

Wirth, H., 2020. Aktuelle Fakten zur Photovoltaik in Deutschland.

CHARACTERIZATION AND DEGRADATION OF PEROVSKITE MINI-MODULES

R. Ebner[1], A. Mittal[1], G. Ujvari[1], M. Hadjipanayi[2], V. Paraskeva[2], G. E. Georghiou[2], A. Hadipour[3], A. Aguirre[4,5,6], T. Aernouts[4,5,6], T. Fontanot[7], S. Pechmann[7], S. Christiansen[7,8], V. Zardetto[9]
[1]AIT Austrian Institute of Technology, Center for Energy, Giefinggasse 2, 1210 Vienna, Austria, T +43 50550-6628, F +43 50550-6390, rita.ebner@ait.ac.at, www.ait.ac.at
[2]University of Cyprus, 1 Panepistimiou Avenue, 2109 Aglantzia, Nicosia, Cyprus, www.foss.ucy.ac.cy
[3]Kuwait University, Department of Physics, Condensed Matter Physics Group, University City Shadadiya Campus, Kuwait
[4]Imec, imo-imomec, Thin Film PV Technology – partner in Solliance, Thor Park 8320, 3600 Genk, Belgium
[5]EnergyVille, imo-imomec, Thor Park 8320, 3600 Genk, Belgium
[6]Hasselt University, imo-imomec, Martelarenlaan 42, 3500 Hasselt, Belgium
[7]Fraunhofer Institute for Ceramic Technologies and Systems IKTS, Äußere Nürnberger Str. 62, 91301 Forchheim, Germany, www.ikts.fraunhofer.de
[8]Max Planck Institute for the Science of Light, Günther-Scharowsky-Strasse 1, 91058 Erlangen, Germany, mpl.mpg.de
[9]TNO Energy Transition, High Tech Campus, Eindhoven, The Netherlands, valerio.zardetto@tno.nl

ABSTRACT:
Organic-inorganic hybrid metal halide perovskites are poised to revolutionize the next generation of photovoltaics with their exceptional optoelectronic properties compatibility with low-cost and large-scale fabrication methods. The leap forward in the power conversion efficiency (PCE) enabled by lead halide perovskites is unprecedented, with PCEs emerging from 3.8% in its first study to a current certified value of 25.5% in single-junction and 33.7% in perovskite-silicon tandem devices [1-3]. The main challenge for the successful commercialization of perovskite solar cells is to achieve high stability at the module level. The commercially available solar modules undergo a series of characterization procedures that analyze their properties and ensure their quality. However, these procedures and protocols cannot unambiguously be applied to perovskite solar modules (PSM) due to its unpredictable degradation mechanisms. Therefore more advanced characterization methods are needed to understand the degradation mechanisms in the PSM. In this context, optical and electrical measurement methods are effectively employed in quality control and development support and are essential characterization tools in industry and research.

KEYWORDS: Perovskite, Optical and electrical characterization, Degradation

1 INTRODUCTION

In the proposed work, we employed optical and electrical characterization methods to understand the degradation of perovskite mini-modules. Optical techniques such us electroluminescence (EL), photoluminescence (PL), and dark lock-in thermography (DLIT) are non-destructive measurement techniques and provide high-resolution images showing a two-dimensional distribution of the characteristic features of PV cells and allowing the investigation of cracks, defects, shunts, and stacking faults in the cells [4]. Furthermore, electrical measurements like current-voltage (IV) characteristics and external quantum efficiency (EQE) can provide information on the power output and other device parameters, which could be used to identify the possible degradation route.

2 EXPERIMENTS

2.1 Perovskite mini-modules

A - double cation-double halide perovskite active layer with the composition $\mathbf{Cs_{0.18}FA_{0.82}PbI_{2.82}Br_{0.18}}$ was used. To make large-area devices, so-called mini-modules were produced by laser scribing, to generate 7 sub-cells connected in series. To prevent penetration of metallic particles of the top electrode into the soft perovskite layer, ITO (indium-tin-oxide) was used. ITO was also selected as a top electrode to obtain semi-transparent modules. The module stack was as follows: glass/ITO/Hole transport layer (HTL)/560nm 2C Perovskite with bandgap of 1.6 eV/Electron transport layer (ETL)/ITO/glass. Fig.1 depicts the structure of the perovskite sub-cell and Fig. 2 illustrates the cross-section of the mini-module. Fig.3 shows two images of the front and back side of one perovskite mini-module (substrate size: 3cm x 3cm and module size: 2cm x 2cm) and Fig.4 shows the EL-images (front and backside) of one perovskite mini-module.
Four perovskite mini-modules ("S9", "S10", "S11" and "S12") were produced and characterized by applying DLIT-, EL- and PL methods alongside IV and EQE measurements.
IV measurements as well as EL measurements were performed regularly from **July 2021 (07/21) to May 2023 (05/23)** to determine the **aging behavior** of the mini-modules.
The IV measurement setup at AIT was the following:
TRI-SOL Solar Simulator 1-1.6 KW:
Class AAA (IEC 060904-9), Steady State Continuous, Air Mass Filter 1.5 Global, Typical Power Output: 100mW/cm2 (1Sun) ± 20%, vary Intensity from 0 - 1.2Sun, Lamp Type and Power Xe Arc Lamp (Lamp Power:1600W), contacting with crocodile clips.

The results are presented in section 3.1.

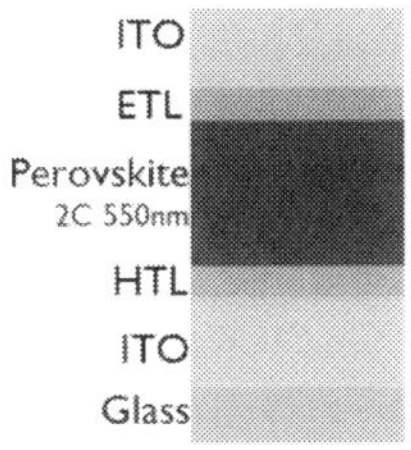

FIG.1 PIN-structure of each sub-cell.

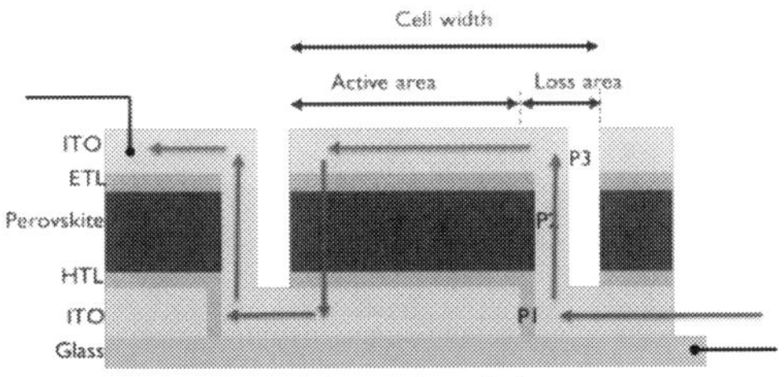

FIG.2 Cross-section of the mini-module.

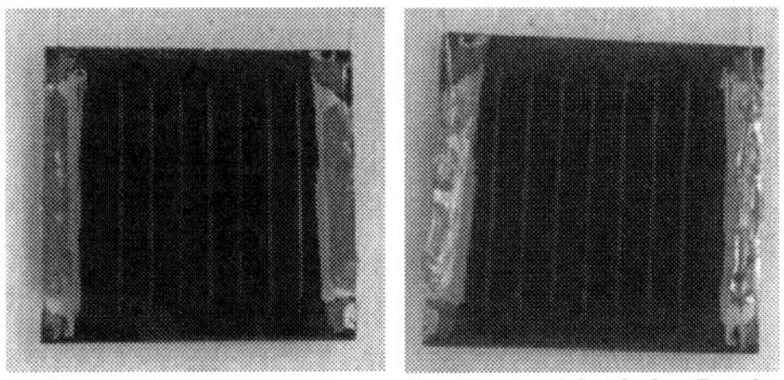

FIG.3 Perovskite mini-module: Front side (left), Back side (right).

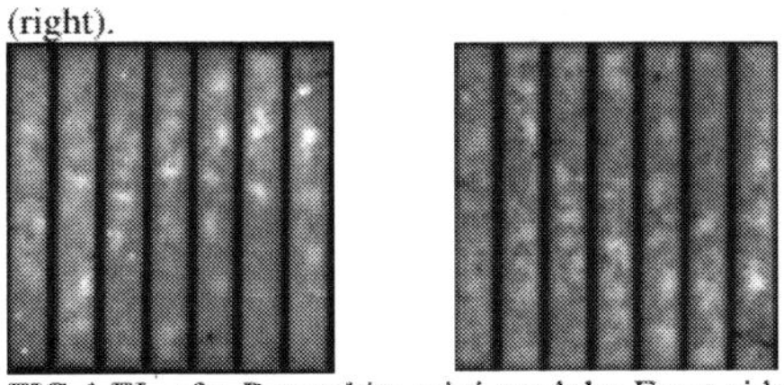

FIG.4 EL of a Perovskite mini-module: Front side (left), Back side (right).

2.2 Perovskite cells

The perovskite solar cell stack consists of a **glass/ITO/HTL/perovskite/passivation_layer/C_{60}/SnO_x /ITO/Ag finger** in p-i-n configuration. Encapsulated semitransparent perovskite solar cells, without edge sealer and with a mask used to define the aperture area taped on the front glass, were investigated (substrate size: 2.5cm x 2.5cm and cell size: 0.5mm x 0.5mm). See the image of Sample 2 in Fig. 5 and the structure of all samples in Fig. 6.

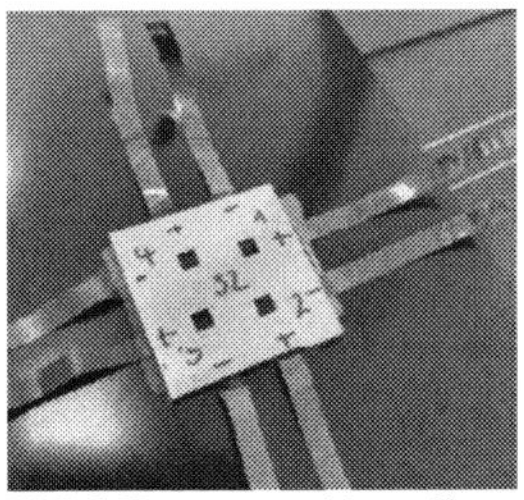

FIG.5 Four perovskite cells on Sample 2

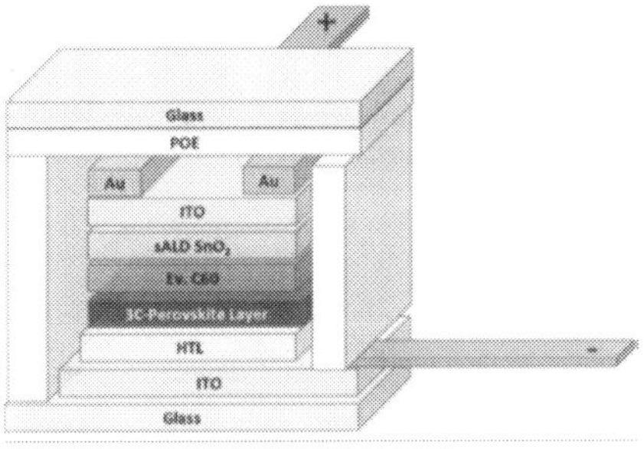

FIG.6 Structure of the perovskite sample

Four Samples with each 4 perovskite cells were produced and characterized by applying EL- and PL methods alongside IV and EQE measurements.
The results are presented in section 3.2.

3 RESULTS

3.1 Perovskite mini-modules

The perovskite mini-modules were kept indoors at room temperature and measured every month. The modules were stored in the dark between the measurements. As can be seen in **Fig.7**, the mini-modules behaved in a very stable manner over this period. Only after some measurements (e.g. DLIT, Raman) there is sometimes a clear increase or decrease in power, partly reversible and also irreversible.

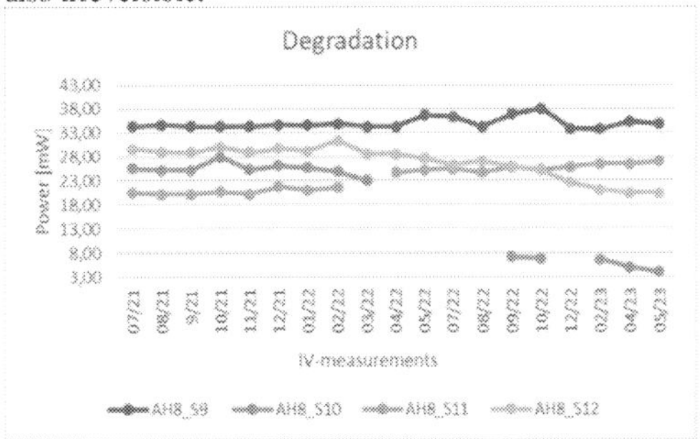

FIG.7 Aging behavior of the perovskite mini-modules

3.1.1 Mini-module "S10"

Fig. 8 shows the EL images of mini-module "S10" with two inactive cells.

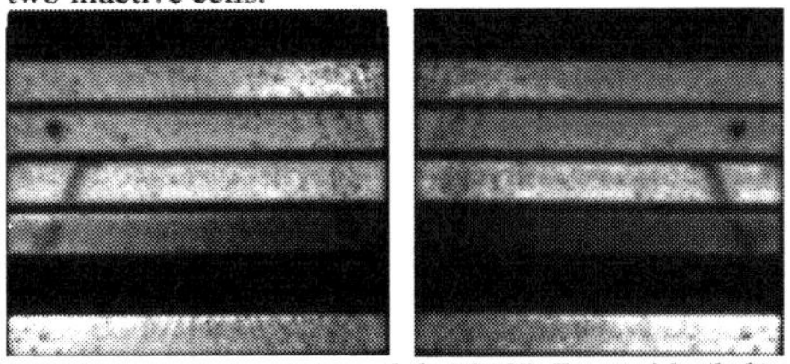

FIG.8 Perovskite mini-module "S10": Front side (left) and back side (right).

On mini-module "S10" DLIT-measurements were carried out.

3.1.1.1 DLIT-measurements

DLIT measurements were carried out on the selected mini modules. However, it soon became evident that these measurements put a great strain on the modules, with the possibility to induce degradation, interrupt electrical contact or even re-establish broken contacts.

10.4229/EUPVSEC2023/2BV.2.5
020099-002

In the case of the mini-module "S10", it can be clearly seen in the EL image (see Fig.9), that an interrupted contact was reactivated, after the DLIT measurement performed on 10/2021 was carried out. After DLIT measurement there was only one inactive subcell instead of two.

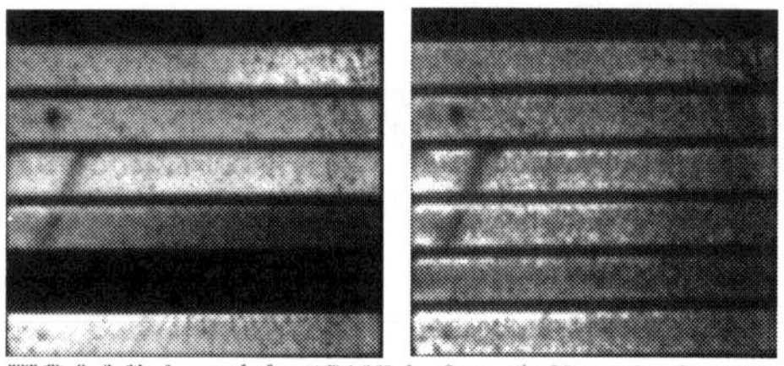

FIG.9 Mini-module "S10" before (left) and after (right) DLIT- measurement

The IV measurement results also showed a significant increase in performance after the DLIT measurement was performed (see Table 1).

Table 1: IV-measurement results of the mini-module "S10", before and after DLIT measurement (10/21).

Mini-Modul	I_{SC} [mA]	V_{OC} [V]	FF [%]	P_{MPP} [mW]	Jsc [mA/cm²]
S10	10.56	6.106	39.70	25.59	13.48
S10 DLIT	10.39	6.756	39.81	27.94	13.27

However, there was already a drop in the measurement performed on 11/2021 and a subsequent return to the initial values (see Table 2).

Table 2: IV-measurements on "S10"

Mini-module	I_{SC} [mA]	V_{OC} [V]	FF [%]	P_{MPP} [mW]	Jsc [mA/cm²]
S10 (initial value)	10.56	6.106	39.70	25.59	13.48
S10 DLIT Okt. 21)	10.39	6.756	39.81	27.94	13.27
S10 (Nov. 21)	10.48	6.748	38.18	26.99	13.38
S10 (Dez. 21)	10.41	6.737	37.31	26.18	13.30
S10 (Jan. 22)	10.43	6.764	36.95	25.78	13.17
S10 (Feb.22)	10.45	6.843	34.80	24.89	13.35
S10 (March 22)	10.38	6.397	34.73	23.07	13.26

After the performance of some optical measurements in April 2022 the minimodule "S10" was completely degraded (see Fig.7).

3.1.2 Mini-module "S11"

Fig.10 shows the EL recordings of the mini-module "S11" with three inactive cells. The IV measurement results of the mini-module "S11" are presented in Table 3.

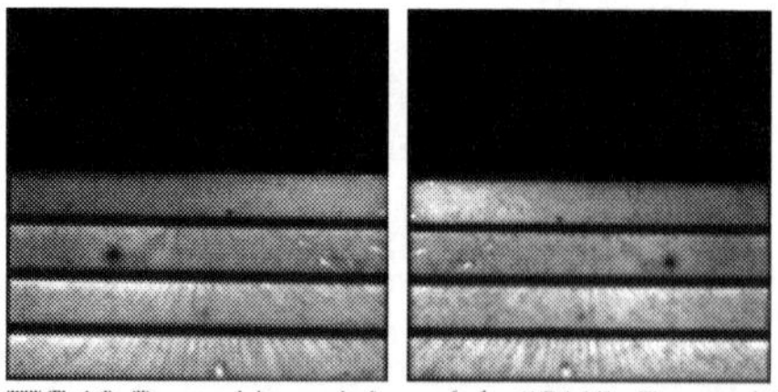

FIG.10 Perovskite mini-module "S11": Front side (left) and back side (right).

Table 3:
IV-measurement results of the mini-module "S11"

I_{SC} [mA]	V_{OC} [V]	FF [%]	P_{MPP} [mW]	Jsc [mA/cm²]
10.37	5.260	37.60	20.50	13.24

In the case of the minimodule "S11", some optical measurements were carried out in March. After the return of the minimodule, a power increase (red arrow) was also determined, which is still present and further increasing (see Fig.11).

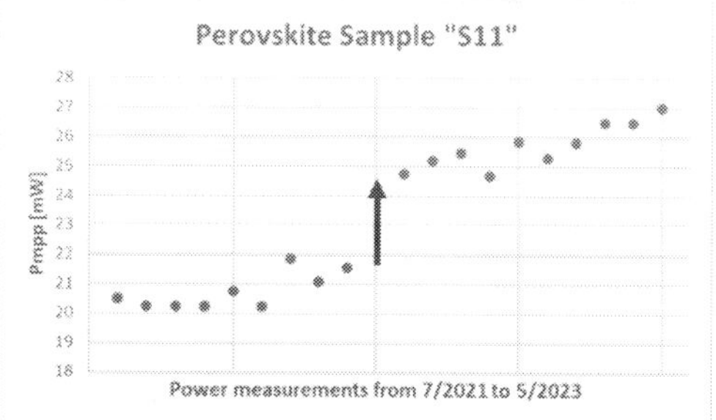

FIG.11 Performance measurements of mini-module "S11"

3.1.3 Mini-module "S12"

Fig.12 shows the EL images of the minimodule "S12"

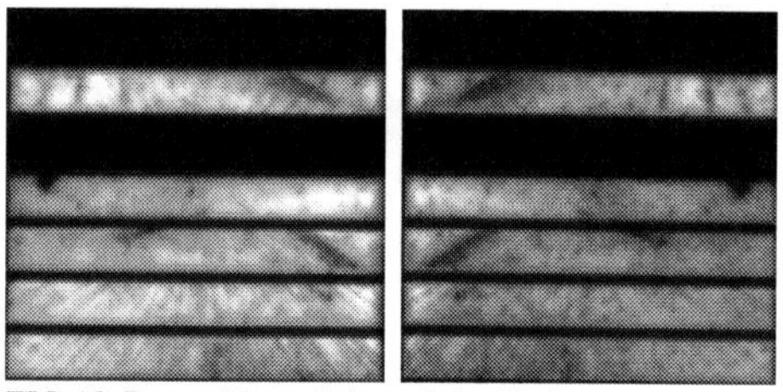

FIG.12 Perovskite mini-module "S12": Front side (left) and back side (right).

The IV measurement results of the mini-module "S12" are listed in Table 4.

Table 4:
IV-measurement results of the mini-module "S12"

I_{SC} [mA]	V_{OC} [V]	FF [%]	P_{MPP} [mW]	Jsc [mA/cm²]
10.44	6.993	40.60	29.64	13.33

A significant increase in power was measured for the minimodule "S12" in February 2022. However, there was already a decrease in power in March 2022. The reason for the increase in February 2022 and the decrease in March 2022 is unclear. Since March 2022, a slight drop in performance has been noticeable for the "S12" mini-module.

3.2 Perovskite cells

The perovskite cells were kept indoors in a glove box between the measurements.

Fig. 13 shows the **EL images** of four perovskite cells (front side) of Sample 2. The appearance of the cells is very homogeneous. Only on cell 3 a defect, a dark spot, is visible.

10.4229/EUPVSEC2023/2BV.2.5
020099-003

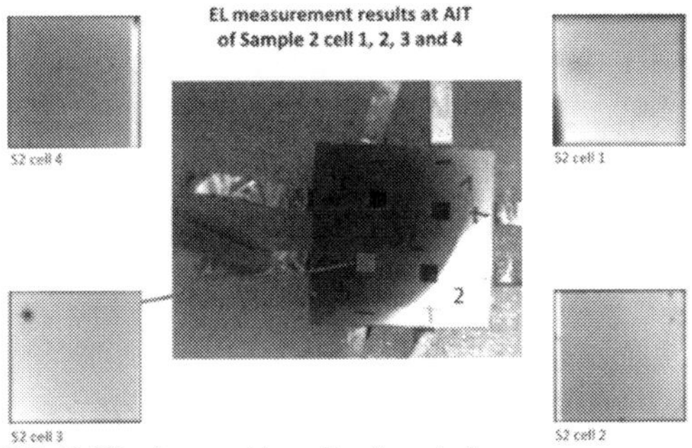

FIG.13 EL of perovskite cells, Sample 2

3.2.1 The IV measurement protocol

The IV measurement protocol for the perovskite cells was the following:

1. 1st JV scan from 1.2V to -0.2V and from -0.2V to 1.2V with a scan rate of 175mV/s; 20mV increment.
2. 2nd JV scan from 1.2V to -0.2V and from -0.2V to 1.2V with a scan rate of 175mV/s; 20mV increment.
3. MPPT of the (best) pixel for 180s.
4. 3rd JV scan from 1.2V to -0.2V and from -0.2V to 1.2V with a scan rate of 175mV/s; 20mV increment.

Table 2 shows the **IV-measurement results** of cell 1 on Sample 2.

Table 2: IV-measurement results of cell 1, Sample 2

Voc [V]	Jsc [mA.cm^{-2}]	Pmax [mW]	FF [%]
1.117	19.94	16.8	72.65

FIG. 14 and FIG. 15 show the IV curves of forward and reverse measurements of cell 3 on Sample 2, according to the IV measurement protocol (see chapter 3.2.1, initial measurement 1 and final measurement 3 after MPP tracking).

Comparing the two IV curves from the initial and final measurements in Fig. 14 and 15, an improvement of the Isc and the overall cell performance after MPP tracking for 180s is achieved. The hysteresis from forward to reverse IV trace reduced from -1.9% to 0.23% from its MPP.

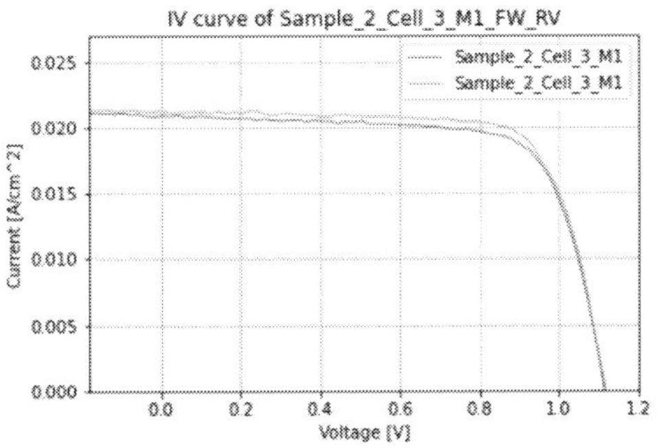

FIG.14 IV Measurement of forward and reverse curves of cell 3 on Sample 2, initial measurement

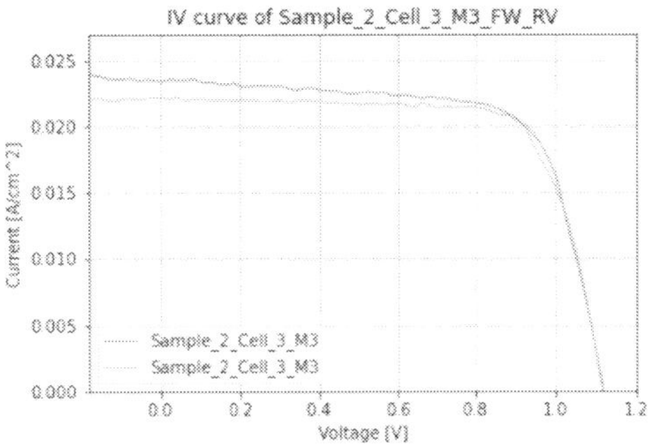

FIG.15 IV Measurement of forward and reverse curves of cell 3 on Sample 2, final measurement after the MPP tracking.

FIG. 16 presents the **EQE** measurement results of cells from different samples.

It is clear to see that the EQE-measurement results of the cells from different samples perfectly superimpose over each other.

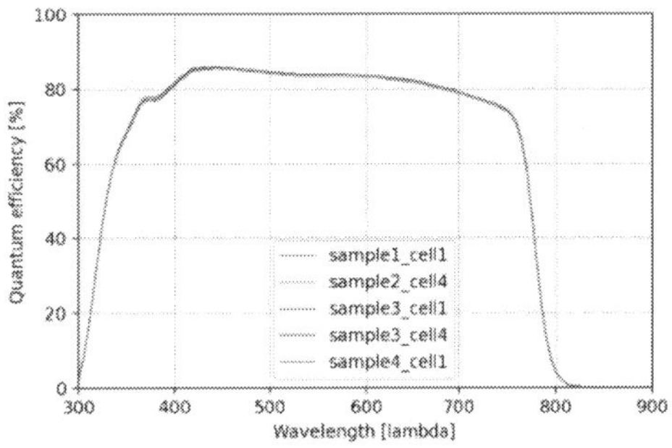

FIG.16 EQE-measurement results of 5 perovskite cells

4 CONCLUSIONS

Initial defects and shunts were identified by EL, PL and IV measurements.

It should be noted that some measurements (DLIT, Raman) led to an increase or decrease in power, partly reversible but also irreversible. Thus, perovskites are very sensitive to some optical measurements.

Further indoor tests and outdoor tests of differently structured perovskite samples and perovskite tandem cells have to be performed in order to identify more defects and thus be able to improve the perovskite cell and module structure. The aging behavior of the perovskite mini-modules will also be further analyzed. In addition, measurement protocols for indoor and outdoor tests of perovskites will be established.

5 REFERENCES

[1] Antonio Urbina 2020 J. Phys. Energy 2 022001

[2] "Oxford PV retakes tandem cell efficiency record",https://www.pv-

10.4229/EUPVSEC2023/2BV.2.5
020099-004

magazine.com/2020/12/21/oxford-pv-retakes-tandem-cell-efficiency-record/
[3] Enzheng Shi et al., "Two-dimensional halide perovskite lateral epitaxial heterostructures", *Nature*, 2020; 580 (7805): 614
[4] R. Ebner et al., "Non-destructive techniques for quality control of PV modules", 39th Annual

Acknowledgement:
This work was funded through the European Regional Development Fund and the Republic of Cyprus in the framework of the project **"DEGRADATIONLAB"** with grant number INFRASTRUCTURES/1216/0043.
http://www.foss.ucy.ac.cy/degradationlab/

"VIPERLAB" has received funding from the European Union's Horizon 2020 research and innovation programme under grant agreement N°101006715

EFFECT OF ETL, AND MAPBBRI3 QUANTUM DOTS AT HTL/ABSORBER INTERFACE ON THE PERFORMANCE OF (SN,GE) BASED PEROVSKITE SOLAR CELLS

Ravula V. N. Sai[1], Swarna S. R. Reddy[1], Anjali Mantri[1], Brajendra S. Sengar[2], Vivek Garg*[1]
[1]Department of Electronics Engineering, SVNIT, Surat-395007, India
[2]Department of Electronics and Communication Engineering, NIT Srinagar-190006, India *Email: brajendra.singh@nitsri.ac.in, +91-9179615146

ABSTRACT: Lead-free perovskite solar cells are attracting researchers to eliminate the toxic Pb. However, the photovoltaic energy conversion efficiency of Pb free devices is primarily hampered by low values of the open-circuit voltage (Voc). In this manuscript, theoretical analysis of FA0.75MA0.25Sn0.95Ge0.05I3 (Sn–Ge perovskite) has been performed to identify the potential electron transport layer (ETL) for extracting the best performance from the Sn-Ge based perovskite solar cells. The simulation results suggest the material with the electron affinity ~3.7 eV will be the suitable candidate. Based on the electron affinity, Mg0.30Zn0.70O (MZO) layer has been selected for the ETL Layer in the device simulation, and the obtained performance parameters are Voc: 0.93 V, Jsc: 25.07 mA/cm2, FF: 82.13 %, and Eff : 19.28 %. Additionally, for further improvement in the device performance, interface engineering at HTL/Absorber interface is performed using the quantum dots of CH3NH3PBr0.9I2.1, which results in the further improvement in the power conversion efficiency (PEC) by ~7 %.

1 INTRODUCTION

The Metal halide perovskites with chemical formula of ABX3 (where A = Cs+, CH3NH3+ (MA+) or CH (NH2)2 + FA+; B = Pb2+, Sn2+, or Ge2+; X = Cl-, Br-, I- or their combination) have gain interest for the PV application because of their extraordinary optoelectronic properties, such as tunable energy bandgaps, high absorption coefficient, and high carrier mobilities [1]. The lead-based perovskite solar cells have achieved the conversion efficiency of 20%, comparable to silicon solar cells. However, researchers are investigating lead-free materials to eliminate the toxic effects of Pb. Therefore, (Sn, Ge)-based perovskites have been proposed by researchers due to their nontoxic nature and abundance. But Sn-based perovskites have low stability because of the oxidation of Sn from Sn2+ to Sn4+ [2] and hence, leading to lower PCE. So, a combined Tin(Sn) and Germanium(Ge)-based perovskite material is used in this study to mitigate the above limitations. The bandgap of roughly 1.4 to 1.5 eV for the Sn–Ge perovskite materials has been found using Photoacoustic spectroscopy, which is appropriate for solar cells. The conversion efficiency of 3.31% has been achieved from tin-based perovskite solar cells with an inverted (pin) planar configuration. When evaluated in the air without encapsulation, the efficiency of a pure tin-based perovskite with 5% germanium doping rose to 4.48% (6.90% after 72 hours) [2] Showing mixed Sn–Ge perovskite displays stability and improved PCE. Several methods have been explored to have better interface properties by optimizing interfacial charge transfer processes. For this, Quantum dots (QDs) can be incorporated in various interfaces involved in PSCs, such as the inserting QDs between perovskite/ETL and perovskite/HTL [7]. Cho AnNa, et al. Have introduced PbS QDs as an inorganic HTL into the planar PSCs, [8], and this also leads to the increase in the absorption spectrum of PSCs by extending it to the infrared region [8]. Different studies of QDs incorporated in PV structures have been performed because of their tunable bandgap and higher absorption coefficients [8]. However, the band alignment study of QDs between perovskite and HTL has yet to be explored significantly. The fabrication of an optimized interface between HTL and perovskite is challenging because its processing steps should align with the beneath perovskite layer without introducing any significant distortion in the structure. In this paper, the ETL and perovskite interface is first optimized by performing a band offset study for improved performance parameters. Afterward, MgZnO as a potential ETL material has been investigated to improve the device performance. Further, we have performed the analysis of the introduction of QDs between the perovskite and HTL layers to improve the device performance.

2 DEVICE SIMULATION PARAMETERS

One-dimensional device simulation software SCAPS ver. 3.3.10 developed by Gent University was utilized for the simulation [1]. The general structure of the FA0.75MA0.25Sn0.95Ge0.05I3 perovskite solar cells, where the FA0.75MA0.25Sn0.95Ge0.05I3 is expressed as FAMASnGeI3, is composed of ITO/PEDOT:PSS (30 nm)/FAMASnSeI3 (300 nm)/PCBM (25 nm). The band levels for FAMASnGeI3 perovskite solar cells are shown in Fig. 1(a). Additionally, the details of primary input parameters used in the simulation are populated in Table 1.

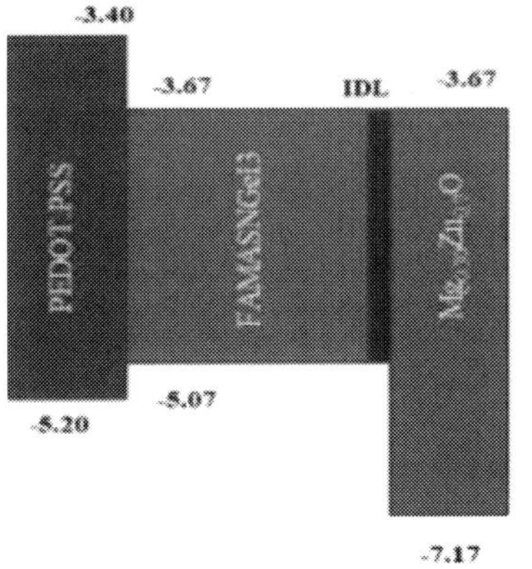

Figure 1: Energy band diagram for the simulated device structure

3 RESULTS AND DISCUSSIONS

Electron Transport Layer The baseline model for ITO/PEDOT:PSS (30 nm)/FAMASnSeI3 (300 nm)/PCBM (25 nm) device structure is developed by using the experimental data reported by Nozomi Ito, et.al. [2] Afterwards, the different band offsets have been

10.4229/EUPVSEC2023/2BV.2.6
020100-001

incorporated in the structure by varying the electron affinity of the Electron Transport Layer (ETL), and its effect on the device performance parameters has been investigated. The best optimized value of the electron affinity of the ETL layer estimated from the simulation is 3.67 eV. Next, the material with the electron affinity in the similar range is scrutinized for the implementation. Therefore, Mg0.30Zn0.70O (ETL) layer is utilized for the device implementation. The performance parameters of the simulated device structure [PEDOT:PSS (30 nm)/FAMASnSeI3 (300 nm)/MZO(25 nm)] is plotted in the Fig 2 a). The observed performance parameters of the device are Voc: 0.93 V, Jsc: 25.07 mA/cm2, FF 82.13 %, and Eff : 19.28 % . Additionally, the quantum efficiency of the simulated device is shown in Fig 2 b), which shows the efficiency of more than 80% in the wide spectral range of 300-700 nm.

Table 1: Details of primary input parameters used in the simulation

	PEDOT:PSS	$MAPbBr_{0.9}I_{2.1}$ Quantum Dots	$FAMASnGeI_3$ [4]	IDL [4]	$Mg_{0.3}Zn_{0.7}O$ (ETL) [5,6]
Thickness(μm)	0.03	0.005	0.30	0.005	0.025
Eg (eV)	1.8	2.120	1.4	1.4	3.5
χ (eV)	3.4	3.280	3.67	3.67	3.67
ε_r	18	25	8.2	8.2	9
N_C cm^{-3}	$2.2x10^{18}$	$2.2x10^{20}$	$2.2x10^{18}$	$2.2x10^{18}$	$2.2x10^{18}$
N_V cm^{-3}	$1.8x10^{19}$	$1.8x10^{20}$	$1.8x10^{19}$	$1.8x10^{19}$	$1.8x10^{19}$
N_A cm^{-3}	$1x10^{20}$	$1.0x10^{17}$	–	–	–
N_D cm^{-3}	–	--	$1x10^{13}$	$1x10^{13}$	$1x10^{17}$
μ_n cm^2/Vs	$4.5x10^{-2}$	10	2	2	100
μ_p cm^2/Vs	$4.5x10^{-2}$	10	2	2	15
E_t(eV)	0.9	0.7	0.7	0.7	1.05
N_t cm^{-3}	$1x10^{15}$	$1x10^{15}$	$2.5x10^{13}$	$2.5x10^{19}$	$1x10^{15}$
V_e & V_p (cm/s)	$1x10^7$	$1x10^7$	$1x10^7$	$1x10^7$	$1x10^7$
Δe & δp cm^2	$2.2x10^{-14}$	$2.2x10^{-14}$	$2.2x10^{-14}$	$2.2x10^{-14}$	$2.2x10^{-14}$

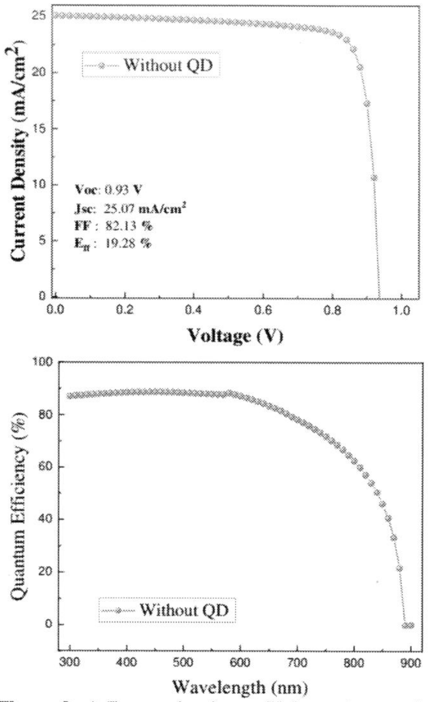

Figure 2: a) Current density vs. Voltage characteristics of simulated devices, b) Quantum Efficiency plot of optimized simulated devices.

Effect of MaPbBI3 quantum dots (QD) as an Interfacial layer: The optimized structure with the MgZnO (ETL) is further optimized. MAPbBrI QDs, have been introduced to provide a better interfacial charge transfer process between the perovskite and HTL. The appropriate band offset (conduction band offset: 0.39 eV and valence band offset: 0.33 eV) between the perovskite and MAPbBrI QDs is obtained from the Br and I compositions of 0.9 and 0.21, respectively, and the energy band diagram for the same is shown in fig. 3. The comparison of I-V characteristics and QE plots for with and without QDs incorporated solar cell (SC) structures is shown in Fig. 4 a) and b) respectively. The observed performance parameters of the device with the quantum dots at the HTL absorber interface are Voc: 1.14 V, Jsc: 25.55 mA/cm2, FF: 89.30 %, and Eff : 19.28 %. From fig. 4 (a), it can be clearly seen that efficiency is improved by ~7% by the incorporation of QDs, which is mainly because of the increase in Voc and fill factor by 210 mV and 26.23 %. The improvement in the efficiency can also be justified from the QE plot in fig. 4 (b), from which we can observe that with the inclusion of QDs, QE increases in the wavelength range from 300 to 600 nm leading to improved performance parameters of SC.

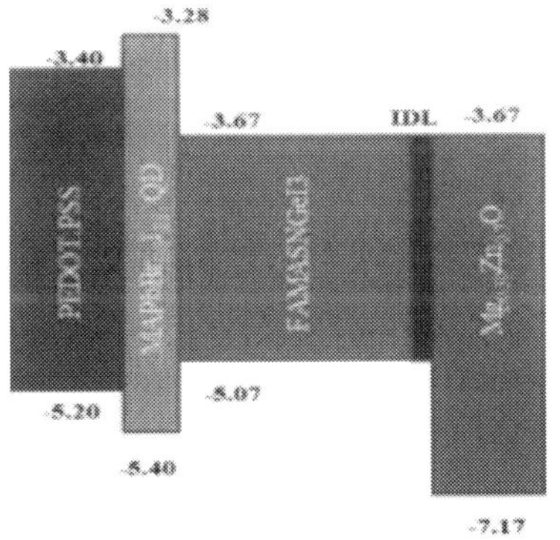

Figure: 3: Energy band diagram for the simulated device structure with the $MAPbBr_{0.9}I_{2.1}$ quantum dots-based interface layer between absorber and HTL layer.

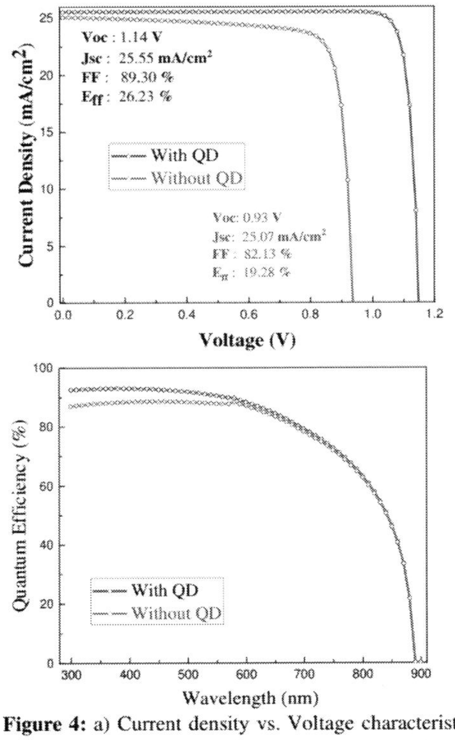

Figure 4: a) Current density vs. Voltage characteristics of simulated devices without QD interface layer and with QD interface layer, b) Quantum Efficiency plot of simulated devices without QD interface layer and with QD interface layer

4 CONCLUSION

(Sn, Ge) based perovskite materials have gained significant interest in the field of PV to remove the material's toxicity. But the performance of the SC is reduced for the lead-free perovskite materials. Therefore, a theoretical study of the interface of lead-free FA0.75MA0.25Sn0.95Ge0.05I3 (Sn–Ge perovskite) with ETL and HTL has been performed to extract the best performance. The simulation results suggest that Mg0.30Zn0.70O (MZO) layer with the electron affinity ~3.7 eV can be used as an ETL material for optimized performance. Additionally, for further improvement, quantum dots of MAPbBr0.9I2.1 are introduced at the HTL and perovskite interface, which results in the further improvement in the performance parameters, Voc increased by 210 mV, fill factor increased by ~7.17 %, and finally, the conversion efficiency increased by ~7%.

5 REFERENCES

1. Burgelman, Marc, Peter Nollet, and Stefaan Degrave. "Modelling polycrystalline semiconductor solar cells." *Thin solid films* 361 (2000): 527-532.
2. Ito, Nozomi, et al. "Mixed Sn–Ge perovskite for enhanced perovskite solar cell performance in air." *The journal of physical chemistry letters* 9.7 (2018): 1682-1688.
3. Sengar, Brajendra S., et al. "Numerical simulation: Design of high-efficiency planar pn homojunction perovskite solar cells." *IEEE Transactions on Electron Devices* 68.5 (2021): 2360-2364.
4. Minemoto, Takashi, et al. "Theoretical analysis of band alignment at back junction in Sn–Ge perovskite solar cells with inverted pin structure." *Solar Energy Materials and Solar Cells* 206 (2020): 110268.
5. Chantana, Jakapan, et al. "Micro-scale current path distributions of Zn1-xMgxO-coated SnO2: F transparent electrodes prepared by sol-gel and sputtering methods in perovskite solar cells." *Thin Solid Films* 669 (2019): 455-460.
6. Garg, Vivek, et al. "Investigation of valence plasmon excitations in GMZO thin film and their suitability for plasmon-enhanced bufferless solar cells." *Solar Energy* 178 (2019): 114-124.
7. Cha, Mingyang, et al. "Enhancing perovskite solar cell performance by interface engineering using $CH_3NH_3PbBr_{0.9}I_{2.1}$ quantum dots." *Journal of the American Chemical Society* 138.27 (2016): 8581-8587.
8. Cho, An-Na, and Nam-Gyu Park. "Impact of interfacial layers in perovskite solar cells." *ChemSusChem* 10.19 (2017): 3687-3704.

10.4229/EUPVSEC2023/2BV.2.6
020100-003

TESTING EVA, PMMA AND PVDF UNDER UV IRRADIATION FOR A NEW LOW TEMPERATURE ENCAPSULATION METHOD FOR PEROVSKITE SOLAR CELLS

L. Ocaña[1,2], C. Montes[1,2], B. González-Díaz[2], S. González-Pérez[3], E. Llarena[1].
[1]Instituto Tecnológico y de Energías Renovables, S. A. (ITER)
Pol. Industrial de Granadilla, s/n, E 38600 Granadilla de Abona, Spain.
Ph. +34 922 747 700 / Fax +34 922 747 701 / E-mail lmgonzalez@iter.es
[2]Departamento de Ingeniería Industrial. Universidad de La Laguna.
Camino San Francisco de Paula, s/n, 38206 San Cristóbal de La Laguna. S/C de Tenerife. Spain.
[3]Departamento de Didácticas Específicas. Universidad de La Laguna.
c/ Pedro Zerolo, s/n, Edificio Central Planta 2. Apartado 456, 38200 San Cristóbal de La Laguna. S/C de Tenerife. Spain.

ABSTRACT: The Solar Cell Laboratory at the Instituto Tecnológico y de Energías Renovables (ITER), in collaboration with La Laguna University (ULL), has been dedicated to advancing the development of low-temperature encapsulation methods for safeguarding Perovskite Solar Cells (PSCs). This document provides a comprehensive summary of our efforts to assess the performance of perovskite solar cells based on MAPbI3 absorbers, which were encapsulated using low-temperature techniques and subsequently subjected to UV irradiation. Our study focused on three types of polymer materials: Ethyl Vinyl Acetate (EVA), Poly(methyl methacrylate) (PMMA), and a combination of Polyvinylidene fluoride (PVDF) and EVA. The degradation of the samples was meticulously characterized through imaging microscopy and optical ellipsometry, while the electrical performance was simulated using the SCAPS-1D software. Our findings reveal that the EVA-encapsulated samples exhibited remarkable stability, maintaining a power conversion efficiency (PCE) of 15.36% after 70 minutes of exposure to UV irradiation, in contrast to the pristine samples which retained a PCE of 14.75%. This corresponds to a mere 3.99% decrease in total efficiency. In contrast, PMMA-encapsulated samples suffered a substantial loss of 62.28% in PCE, while EVA-PVDF encapsulated samples demonstrated a relatively modest reduction of 12.29% in PCE.
Keywords: Perovskite, degradation, absorption, encapsulated, Polymer Film.

1 INTRODUCTION

Prior research has emphasized the pivotal role of UV light as a significant factor in the degradation of perovskite layers [1–6]. The current study builds upon the ongoing research efforts conducted by ITER and ULL, aimed at devising a low-temperature encapsulation method to ensure the stability of perovskite solar cells when exposed to environmental factors and light.

Throughout the course of our investigation, we have continued to enhance the encapsulation method. The initial generation of encapsulated devices demonstrated resilience at temperatures up to 150°C [7] and endured 350 hours of exposure under full solar radiation [1,8]. The subsequent generation, wherein curing times of the encapsulation process were meticulously optimized, exhibited improved resistance to temperature variations, remaining stable after enduring 200 thermal cycle tests, as prescribed by the IEC-61215 standard [9]. The third generation has been evaluated with three distinct polymers, namely EVA, PMMA, and PVDF, following the protocols of the International Summit on Organic Photovoltaic Stability (ISOS-T). It was confirmed that EVA-encapsulated devices remained stable after undergoing the requisite 200 thermal cycles, maintaining a power conversion efficiency (PCE) of 15.06% in comparison to the pristine ones, which achieved a PCE of 15.96%. This represents a modest 5% reduction in total efficiency [10].

Our most recent endeavors have demonstrated that the UV-blocking characteristics of low-temperature EVA encapsulation serve as an effective protective barrier. This enabled the laminated samples to maintain stability even after exposure to 70 minutes of total UV irradiation (48 kW/m2) and 350 hours of full solar irradiation (6,051 kW/m2) [1].

To further enhance this novel low-temperature encapsulation method, this study delves into the analysis of the same three encapsulating materials: ethylene-vinyl acetate (EVA; Encapsolar PC-135 A, sourced from Stevens Urethane, Easthampton, MA, USA), poly methyl methacrylate (PMMA; Alfa Aesar, Thermo Fisher Kandel GmbH), and polyvinylidene difluoride (PVDF; X 3M 8590M Deutschland GmbH). These materials have been specifically chosen to safeguard perovskite samples from UV irradiation in ambient conditions.

To achieve this goal, we began by depositing perovskite samples composed of $MAPbI_3$ ($CH_3NH_3PbI_3$) films onto glass substrates using spin coating techniques. These samples were then effectively encapsulated using the newly developed polymer-based method described earlier.

In order to thoroughly assess the impact of UV irradiation on the aging process of the perovskite layers and its potential interactions with the hole transporter layer (HTL) and the electron transporter layer (ETL), the samples were subjected to UV lamp irradiation for a duration of 70 minutes, with an irradiance of 48 kW/m2, under standard room conditions, maintaining a relative humidity level within the range of 60-80%. Our analytical toolkit included a digital microscope, spectroscopic ellipsometer, Fourier-transform infrared spectroscopy (FTIR) system, semiconductor characterization system equipped with a solar simulator, and a spectral response system.

These comprehensive analyses allowed us to monitor and assess various aspects of the samples during the UV aging tests, including macroscopic defects, alterations in absorbance curves, changes in functional characteristic groups, variations in electrical parameters, and overall efficiency.

2 EXPERIMENTAL

To assess the performance of the various encapsulating materials, we designed three types of samples, each measuring 25 mm x 25 mm (see Table 1):

1. Samples comprised solely of a layer of the encapsulating polymer material, sandwiched between two glass substrates (referred to as G-EVA, G-PMMA, and G-PVDF).
2. Samples featuring perovskite thin films based on CH3NH3PbI3 absorbers [7], encapsulated with the polymer material and a single glass substrate (referred to as Pk-EVA, Pk-PMMA, and Pk-PVDF).
3. Simulated samples equipped with fully functional Perovskite Solar Cells (PSCs), encapsulated with the polymer material and sandwiched between glass substrates (referred to as PSC_EVA, PSC_PMMA, and PSC_PVDF).

This comprehensive set of samples enabled us to evaluate several critical factors: the transmittance and absorption coefficient of the encapsulating polymer film in the first set, the behavior of the perovskite layer in the second set, and the overall impact on the complete devices in the third set.

All sample fabrication took place within an ISO 7 cleanroom under ambient conditions [7]. The EVA and PVDF sheets were applied directly to the perovskite film in their commercial state, without any prior treatment. In contrast, PMMA solutions were prepared with a concentration of 10% PMMA by weight, achieved by adding 0.874 g of PMMA to 7 ml of chlorobenzene. This PMMA solution was then deposited onto the perovskite films using a 200 µl droplet. Subsequently, the solar cells underwent encapsulation following the low-temperature encapsulation method previously documented [9,10].

Table I: Research samples

ID	Description	Characterization
G-EVA	Glass-EVA-Glass	Transmittance
G-PMMA	Glass-PMMA-Glass	
G-EVA-PVDF	Glass-PVDF-Glass	
Pk-EVA	Glass-$CH_3NH_3PbI_3$-EVA-Glass	Absorption coefficient - Absorption coefficient variation
Pk-PMMA	Glass- $CH_3NH_3PbI_3$-PMMA-Glass	
Pk-EVA-PVDF	Glass- $CH_3NH_3PbI_3$-PVDF-Glass	
PSC-EVA	Glass-PSC-EVA-Glass	Simulation current voltage (J-V) curves
PSC-PMMA	Glass-PSC-PMMA-Glass	
PSC-EVA-PVDF	Glass-PSC-PVDF-Glass	

Building upon the parameters established in prior experiments [1], we conducted aging processes using a UV lamp rated at 24 W (365-395 nm). This lamp was positioned at a 90° angle above the samples, with variations in irradiation time ranging from 0 minutes to 70 minutes (at an intensity of 48 kW/m2), all conducted under standard room conditions with a relative humidity ranging between 60% to 80%.

The primary objective of this test was to assess the light-induced degradation of the perovskite films resulting from UV irradiation. This evaluation holds particular significance for photovoltaic technologies intended for outdoor use, as it allows us to gauge the impact of UV solar spectral irradiation [1].

3 CHARACTERIZATION

In order to assess the performance of the encapsulation materials (EVA, PMMA, and PVDF) when exposed to UV irradiation and humidity, we structured the characterization process into two distinct phases:

(i) The first phase involved characterizing the samples prior to undergoing the aging process. This step was essential to establish their pristine condition.

(ii) In the second phase, we conducted the same measurements after subjecting the samples to the aging processes. This allowed us to scrutinize the effects of UV irradiation on the perovskites encapsulated with the studied polymers. Moreover, this data was vital for informing subsequent simulations aimed at evaluating the impact on fully assembled perovskite devices.

To characterize the samples comprehensively, we conducted the following measurements:

1. We employed a digital microscope (U500X by Unimake) with a focus range spanning from 15 mm to 40 mm and a magnification range of 50X to 500X [11] to inspect the surface of the absorbing layers for any macroscopic defects.
2. Utilizing a spectroscopic ellipsometer (M-2000X) equipped with a 75 W Xenon lamp and a focused beam size of 500 µm, we examined the absorption coefficient curves. This analysis allowed us to identify variations associated with material degradation. It's worth noting that prior studies have indicated that exposure to UV irradiation affects the absorption coefficient curve within the spectral range of 600 to 800 nm, leading to a reduction in the perovskite band gap [1]. Consequently, we determined absorbance values and the band gap by conducting transmittance measurements across a spectral range spanning from 300 to 1000 nm [12].

Finally, the electrical parameters (Jsc, Voc, FF, PCE, and Pmax) were acquired for the encapsulated Perovskite Solar Cells (PSCs) throughout the aging process. We accomplished this by employing the SCAPS-1D simulation software [13]. This approach enabled us to assess the level of stability by observing any reductions in these parameters.

4 RESULTS AND DISCUSSION

4.1 Digital microscope

Figure 1 illustrates images of the samples both before and after the UV irradiation test. The first column corresponds to samples composed of a perovskite layer encapsulated with EVA, the second column shows samples with a perovskite layer encapsulated with PMMA, and the third column pertains to samples featuring a perovskite layer encapsulated with a combination of EVA and PVDF.

Upon close examination, it becomes evident that after 70 minutes under the UV lamp, the samples encapsulated with EVA retained their original appearance and exhibited no discernible macroscopic defects. This observation indicates a notably high level of stability throughout the duration of the test.

In stark contrast, the samples encapsulated with the other polymers, namely PMMA and EVA combined with PVDF, displayed numerous spots on their surfaces. This suggests that these materials are likely unsuitable for use over perovskite layers and may not offer adequate protection against UV irradiation in humid environments.

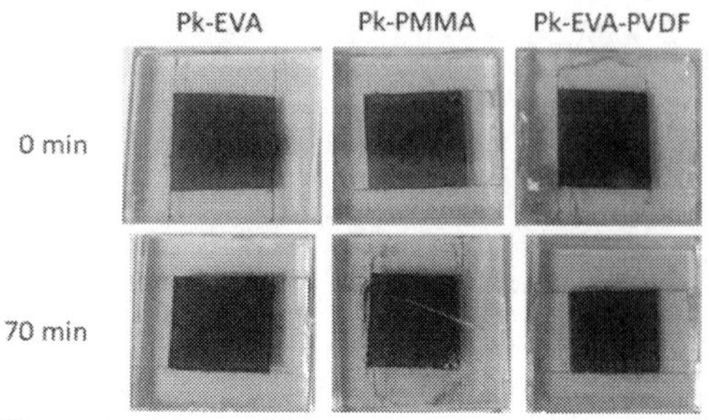

Figure 1. - Images of the samples before and after de UV irradiation test.

4.2 Spectroscope ellipsometer

To begin, we conducted an analysis of the transmittance of the three encapsulating polymeric materials using the G-EVA, G-PMMA, and G-EVA-PVDF samples (as listed in Table I). Figure 2 illustrates the results of this analysis. Notably, PMMA allowed light to pass through from 300 nm onwards, whereas the encapsulations that included an EVA layer effectively blocked irradiation above 400 nm. Consequently, the latter materials appear to be better suited for safeguarding perovskite layers from the UV portion of the spectrum.

Furthermore, the sole-EVA encapsulation exhibited higher transmittance compared to the EVA plus PVDF combination. This difference implies that the perovskite can absorb more light in the former case, potentially leading to increased efficiency [1,10]. Additionally, these findings are presented alongside the spectrum of the UV lamp employed in the degradation processes. As depicted, encapsulants composed of glass and EVA, as well as glass and EVA plus PVDF, serve as effective UV blocking agents below 400 nm.

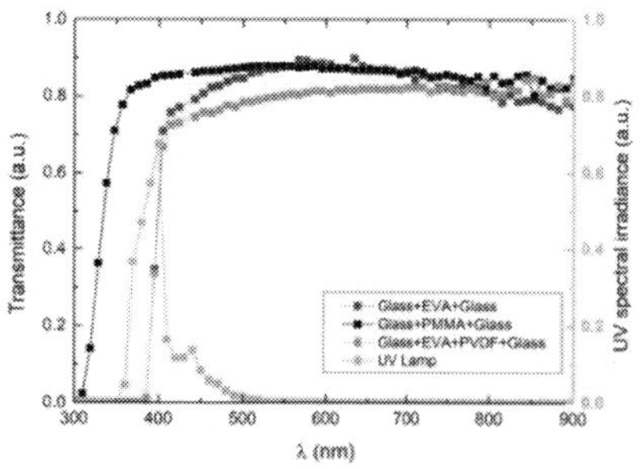

Figure 2. - Transmittance of the EVA, PMMA and EVA-PVDF after having been encapsulated (Y-axis on the left) against the irradiance of the UV lamp (Y-axis on the right).

To assess the impact of EVA, PMMA, and EVA-PVDF encapsulation on the perovskite layers during the UV irradiation test, we measured the absorption coefficient curves of the samples both before and after the degradation process.

As illustrated in Figure 3, the samples encapsulated with EVA exhibited a minimal reduction in the absorption coefficient curve, measuring less than 0.87%, within the spectral range of 400 to 760 nm (which lies below the perovskite's bandgap of 1.6 eV). In contrast, the samples encapsulated with PMMA and EVA-PVDF yielded less favorable outcomes, showing reductions in the absorption coefficient of 9.74% and 5.53%, respectively. Notably, the samples encapsulated with PMMA displayed their most significant degradation in the range of 450-550 nm, while for the EVA-PVDF combination, the decrease was relatively consistent across the entire spectrum.

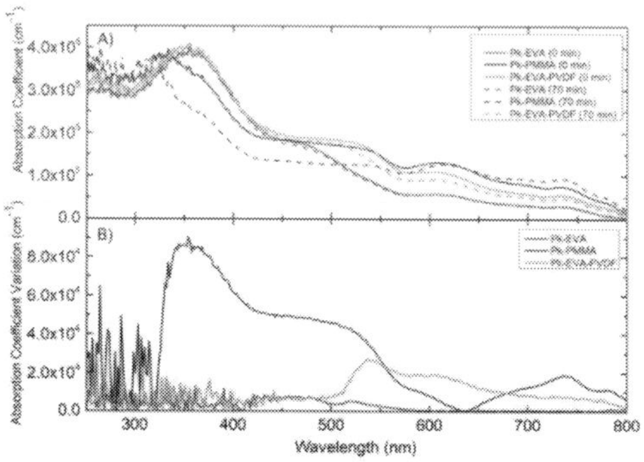

Figure 3. - (A) Absorption coefficient of the Pk-EVA, Pk-PMMA and Pk-EVA-PVDF before and after the UV irradiation test. (B) Absorption coefficient variation of the Pk-EVA, Pk-PMMA and Pk-EVA-PVDF after 70 min UV irradiation.

4.3 Encapsulated PSC simulations

In order to assess the potential impact of the aforementioned degradation of the perovskite layers on complete cells encapsulated with the studied materials, we conducted simulations using the SCAPS-1D simulation software. Developed by the Department of Electronics and Information Systems (ELIS) at the University of Ghent, Belgium, this software is widely recognized for its accuracy in the study of Perovskite Solar Cells (PSCs) [14–19].

With the thickness, band gap, and absorbance curve data for the device layers, we were able to extrapolate the degradation observed in the samples after the UV irradiation test to the SCAPS 1-D software. This allowed us to estimate their potential influence on critical parameters such as the short-circuit current (Jsc), open circuit voltage (Voc), fill factor (FF), and overall energy efficiency (PCE) [1,9].

Figure 4 presents a comparison of the IV curves for the samples encapsulated with EVA, PMMA, and EVA+PVDF (referred to as PSC-EVA, PSC-PMMA, and PSC-EVA-PVDF, respectively) both before and after undergoing the UV irradiation test.

As evident from the graph, the samples encapsulated with EVA effectively maintained their power conversion efficiency after the test, experiencing only a minimal reduction of 3.9%. In stark contrast, the samples encapsulated with PMMA saw a significant decrease from 19.81% to 7.47%. Conversely, the samples encapsulated with EVA+PVDF went from 18.68% to 16.38% (as detailed in Table III).

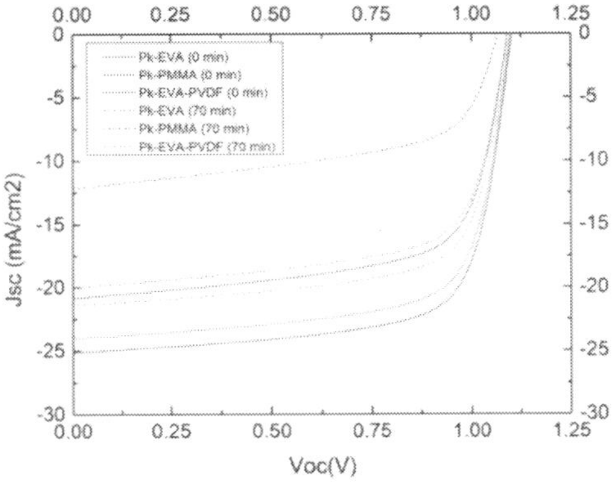

Figure 4. - Curve IV of the Pk-EVA, Pk-PMMA and Pk-EVA-PVDF before and after the UV irradiation test.

Table III: Electrical parameters

Parameters	PSC-EVA		PSC-PMMA		PSC-EVA-PVDF	
	0 min	70 min	0 min	70 min	0 min	70 min
V_{oc} (V)	1.09	1.10	1.10	1.08	1.10	1.10
J_{sc} (mA/cm^2)	20.84	20.01	25.11	12.17	24.03	21.40
FF (%)	67.55	67.00	71.62	56.85	70.73	69.61
PCE (%)	15.36	14.75	19.81	7.47	18.68	16.38
V_{mp} (V)	0.91	0.92	0.93	0.88	0.92	0.92
J_{mp} (mA/cm2)	16.79	16.03	21.40	8.49	20.23	17.81

5 CONCLUSIONS

In this study, we evaluated the performance of three different polymers, EVA, PMMA, and EVA combined with PVDF, as encapsulation materials to shield perovskite films from the effects of UV irradiation and humid conditions.

Following the completion of the tests, it became evident that encapsulation using EVA alone emerged as the superior choice. Solo EVA demonstrated significantly better results compared to PMMA and EVA combined with PVDF in terms of protection against UV radiation. EVA, by itself, effectively blocked irradiation below 400 nm, resulting in only a marginal 0.87% decrease in absorption coefficient curves. In contrast, samples encapsulated with the other polymers experienced more substantial reductions, ranging from 5.53% to 9.74%, respectively.

Using these results, we leveraged the absorbance, band gap, and thickness data of the perovskite layers to simulate the impact of subjecting the encapsulated perovskite solar cells to the aforementioned UV irradiation. The outcome revealed that the electrical parameters of EVA-encapsulated cells, including power conversion efficiency, fill factor, short-circuit current, and open-circuit voltage, remained stable after UV irradiation. Specifically, the efficiency remained at 14.75%, compared to the initial 15.36%.

6 ACKNOWLEDGEMENTS

This work was developed within the MACLAB-PV project framework, which has been co-financed by the INTERREG Madeira-Azores-Canarias Territorial Cooperation Programme (MAC) 2014-2020. 2nd Call. Axis 1. Enhancing research, technological development and innovation.

7 REFERENCES

1. Ocaña, L.; Montes, C.; González-Pérez, S.; González-Díaz, B.; Llarena, E. Characterization of a New Low Temperature Encapsulation Method with Ethylene-Vinyl Acetate under UV Irradiation for Perovskite Solar Cells. *Appl. Sci.* **2022**, *12*, 5228, doi:10.3390/app12105228.
2. Farooq, A.; Hossain, I.M.; Moghadamzadeh, S.; Schwenzer, J.A.; Abzieher, T.; Richards, B.S.; Klampaftis, E.; Paetzold, U.W. Spectral Dependence of Degradation under Ultraviolet Light in Perovskite Solar Cells. *ACS Appl. Mater. Interfaces* **2018**, *10*, 21985–21990, doi:10.1021/acsami.8b03024.
3. Ji, J.; Liu, X.; Jiang, H.; Duan, M.; Liu, B.; Huang, H.; Wei, D.; Li, Y.; Li, M. Two-Stage Ultraviolet Degradation of Perovskite Solar Cells Induced by the Oxygen Vacancy-Ti4+ States. *iScience* **2020**, *23*, 101013, doi:10.1016/j.isci.2020.101013.
4. Ye, Q.Q.; Li, M.; Shi, X.B.; Zhuo, M.P.; Wang, K.L.; Igbari, F.; Wang, Z.K.; Liao, L.S. UV-Stable and Highly Efficient Perovskite Solar Cells by Employing Wide Band Gap NaTaO3 as an Electron-Transporting Layer. *ACS Appl. Mater. Interfaces* **2020**, *12*, 21772–21778, doi:10.1021/acsami.0c04934.
5. Liu, R.; Wang, L.; Fan, Y.; Li, Z.; Pang, S. UV Degradation of the Interface between Perovskites and the Electron Transport Layer. *RSC Adv.* **2020**, *10*, 11551–11556, doi:10.1039/c9ra10960a.
6. Thilakan, A.P.; Li, J.X.; Chen, T.P.; Li, S.S.; Chen, C.W.; Osada, M.; Tsukagoshi, K.; Sasaki, T.; Yabushita, A.; Wu, K.H.; et al. Origin of Extended UV Stability of 2D Atomic Layer Titania-Based Perovskite Solar Cells Unveiled by Ultrafast Spectroscopy. *ACS Appl. Mater. Interfaces* **2019**, *11*, 21473–21480, doi:10.1021/acsami.9b02434.
7. Ocaña, L.; Montes, C.; De Sousa-Vieira, L.; Moreno-Ramírez, J.S.; González-Pérez, S.; González-Díaz, B.; Hernández-Rodríguez, C.; Friend, M.; Cendagorta, & M. On the Stability of Planar CH3NH3PbI3 Perovskite Solar Cells Produced in Ambient Conditions by Using Polymer Encapsulates. In Proceedings of the 36th European Photovoltaic Solar Energy Conference and Exhibition; Marseilles, 2019; pp. 721–727.
8. Ocaña, L.; Montes, C.; González-Pérez, S.; González-Díaz, B.; Friend, M.; Cendagorta, & M. Improvements on the Stability of MAPbI3 Thin Films against UV Radiation by a Low Temperature Encapsulation Method That Uses Commercial Ethylene-Vinyl Acetate Sheets. In Proceedings of the 37th European Photovoltaic Solar Energy Conference and Exhibition; EU PVSEC, Ed.; 2020; pp. 659–662.
9. Ocaña, L.; Montes, C.; González-Pérez, S.; González-Díaz, B.; Llarena, E. Testing Encapsulated Perovskite Solar Cells in a

Climatic Chamber by Following the IEC 61215 and IEC 61646 Standards for the Thermal Cycling Test. In Proceedings of the 38th European Photovoltaic Solar Energy Conference and Exhibition; 2021; pp. 406–410.

10. Ocaña, L.; Montes, C.; González-Díaz, B.; González-Pérez, S.; Llarena, E. Testing EVA, PMMA and PVDF Encapsulated Perovskite Solar Cells in a Climatic Chamber by Following the International Summit on Organic Photovoltaic Stability (ISOS-T) Protocols. In Proceedings of the 8th World Conference on Photovoltaic Energy Conversion; Milan, 2022; pp. 291–295.
11. Digital Microscope U500X, Focus Range 15mm - 40mm, Range 50X-500X by Unimake.
12. Montes, C.; Ocaña, L.; De Sousa-Vieira, L.; Moreno-Ramírez, J.; González-Pérez, S.; González-Díaz, B.; Hernández-Rodríguez, C.; Friend, M.; Cendagorta, M. Producing Uniform and Smooth Thin Layers of Perovskite under Ambient Conditions by Adjusting to the Existing Levels of Moisture. In Proceedings of the 35th European Photovoltaic Solar Energy Conference and Exhibition; Brussels, 2018; pp. 933–937.
13. SCAPS-1D Simulation Software, Developed by the Department of Electronics and Information Systems (ELIS) of the University of Ghent Has Been Used, Belgium.
14. Chowdhury, M.S.; Shahahmadi, S.A.; Chelvanathan, P.; Tiong, S.K.; Amin, N.; Techato, K.; Nuthammachot, N.; Chowdhury, T.; Suklueng, M. Effect of Deep-Level Defect Density of the Absorber Layer and n/i Interface in Perovskite Solar Cells by SCAPS-1D. *Results Phys.* 2020, *16*.
15. Slami, A.; Belkaid, A.B. *Numerical Study of Based Perovskite Solar Cells by SCAPS-1D*;
16. Mandadapu, U.; Vedanayakam, S.V.; Thyagarajan, K. Simulation and Analysis of Lead Based Perovskite Solar Cell Using SCAPS-1D. *Indian J. Sci. Technol.* **2017**, *10*, 1–8, doi:10.17485/ijst/2017/v10i11/110721.
17. Lin, L.; Jiang, L.; Li, P.; Fan, B.; Qiu, Y.; Yan, F. Simulation of Optimum Band Structure of HTM-Free Perovskite Solar Cells Based on ZnO Electron Transporting Layer. *Mater. Sci. Semicond. Process.* **2019**, *90*, 1–6, doi:10.1016/j.mssp.2018.10.003.
18. Bansal, S.; Aryal, P. Evaluation of New Materials for Electron and Hole Transport Layers in Perovskite-Based Solar Cells through SCAPS-1D Simulations. In Proceedings of the Conference Record of the IEEE Photovoltaic Specialists Conference; Institute of Electrical and Electronics Engineers Inc., November 18 2016; Vol. 2016-November, pp. 747–750.
19. Chakraborty, K.; Choudhury, M.G.; Paul, S. Numerical Study of Cs2TiX6 (X = Br−, I−, F− and Cl−) Based Perovskite Solar Cell Using SCAPS-1D Device Simulation. *Sol. Energy* **2019**, *194*, 886–892, doi:10.1016/j.solener.2019.11.005.

REVIEW OF A RESEARCH CARRIED OUT TO PRODUCE CONDUCTIVE INKS AND AGGLOMERATES THAT MAKE USE OF NON-PRECIOUS MATERIALS TOGETHER WITH VEHICLES COMPATIBLE WITH THIN LAYERS OF PEROVSKITE

C. Montes[1,3], L. Ocaña[1,3], B. González-Díaz[3], S. González-Pérez[2], E. Llarena[1].
[1]Instituto Tecnológico y de Energías Renovables, S. A. (ITER)
Pol. Industrial de Granadilla, s/n, E 38600 Granadilla de Abona, Spain. Ph. +34 922 747 700. Fax +34 922 747 701. E-mail: cmontes@iter.es
[2]Departamento de Didácticas Específicas.
C/ Pedro Zerolo, s/n
Edificio Central Planta 2
Apartado 456. Código postal 38200
San Cristóbal de La Laguna. S/C de Tenerife. Spain.
[3]Departamento de Ingeniería Industrial.
Universidad de La Laguna.
Camino San Francisco de Paula, s/n
38206 S/C de Tenerife, Spain

ABSTRACT: This paper summarizes the work that has been carried out at the Instituto Tecnológico y de Energías Renovables, with the aid of the University of La Laguna (Tenerife. Spain) over the last few years, in order to produce conductive inks and pastes that make use of non-precious materials, such as Graphite, Black Carbon, aluminium and nickel, together with vehicles comprised by binders such as epoxy resin, collodion, EVA and PVDF, as well as solvents used mainly as thinners for the mix, such as toluene and chlorobenzene, due to their compatibility with perovskite when deposited and cured in thin layers to produce solar cells. Contacts with volume resistivity (ρ) as low as 2.28 Ω·cm by way of screen printing were achieved. As contacts deposited via lamination techniques from the latest ink formulations, that uses graphite-carbon black and MAI as conductive powders and PVDF and Chlorobenzene as binder and thinner, respectively, do not comply with the 50:1 geometry required for the F 1896 ASTM standard, it was not possible to obtain their volume resistivity. Nevertheless, direct resistance measurements carried out on these contacts suggest that a significant reduction has been achieved.
Keywords: Perovskites, Contact, Metallization, Graphite

1 INTRODUCTION

In this paper, a brief but comprehensive description will be provided of the work carried out by the Instituto Tecnológico y de Energías Renovables, with the aid of the University of La Laguna (both located in Tenerife, Spain), in order to obtain conductive inks that were compatible with their deposition onto thin perovskite layers, initially using screen printing techniques [1]–[4] and, later on, lamination techniques, such as those commercially employed in the manufacturing of photovoltaic modules [5].

2 EXPERIMENTAL

2.1 Conductive inks with graphite and epoxy as a binding element.

Through a trial-and-error method, a preliminary recipe for conductive inks was developed, allowing them to be used for screen printing by mixing graphite (Graphite powder SGA 20 99.9 from M. Graphite Kropfmühl GmbH), epoxy (Epoxy resin SR 8500 / hardener SZ 8525 from Sicomin Epoxy System), and toluene (Toluene Puriss, 32249 Sigma-Aldrich) in a weight percentage (wt%) of 30%/58%/10%.

The mixing process for the preparation of these inks was divided into three stages: In the first stage, resin and most of the graphite were mixed for 10 minutes while constantly stirring using a stirring rod (Hei-TORQUE 400 model, from Heidolph). In the second stage, also for 10 minutes, the remaining graphite was gradually added to the mixture along with the solvent (initially toluene), all while continuing to stir. Finally, in the third stage, the hardener was incorporated into the mixture. From this point onwards, the mixture was continuously stirred, and ink prints were made every 5 minutes using a screen printing device (485 model, from HMI). The printed samples were cured for 10 minutes at 100°C in an oven (CONTERM model, from JPSelecta).

Measurements of the volumetric resistivity of the deposited contacts could not be taken due to the lack of screen prints with the necessary pattern geometry so as to use the methodology specified in the Standard Method F 1896 by the American Society for Testing Materials (ASTM) [6]. However, contacts intended for perovskite metallization were successfully printed using the specified graphite and epoxy, but by changing the type of solvent used in the mixture. Available solvents in the laboratory were used for this purpose to estimate which one produced more conductive contacts. These solvents included toluene, chlorobenzene (Chlorobenzene anhydrous, 284513 Sigma-Aldrich), DMF (N,N-Dimethylformamide anhydrous, 227056 Sigma-Aldrich), and NMP (1-Methyl-2-pyrrolidinone anhydrous, 328634 Sigma-Aldrich). Since the deposited contacts were all the same, it was possible to measure and compare their resistance, yielding the results indicated in Table 1.

Table 1: Resistances measured in the contacts deposited via screen printing..

Time (mins)	Toluene (kΩ)	Chlorobenzene (kΩ)	DMF (kΩ)	NMP (kΩ)
5	142.79	201.01	120.21	153.10
10	395.73	459.83	166.93	165.75

20	447.89	380.88	298.32	366.27
30	1632.20	947.15	273.85	364.18

Based on these results, it was possible to infer that the contacts deposited in shorter times (from the addition of the hardener to the mixture) perform better. Furthermore, solvents with higher boiling points [7], such as DMF or NMP, appeared to stabilize the inks, allowing for an extended period of use (i.e., the time from adding the hardener to the epoxy until the ink is used for printing). Unfortunately, this latter characteristic is of little use in the problem at hand, as it is not possible to use conductive inks containing polar solvents like DMF or NMP on perovskite layers, as they disrupt its crystalline structure [8].

2.2 Conductive inks with different types of conductive powders (graphite, aluminum, nickel, and copper), epoxy as a binding element, and non-polar solvents.

For this study, we focused on the use of a single non-polar solvent (toluene), the epoxy resin that was available, varying the type of conductive powders used, as shown in Table 2.

Table 2: Conductive Powders Used in the Tests.

Conductive powders	Maximum particle size (μm)	Purity (%)
Graphite powder 282863 Sigma-Aldrich	<20	n/a
Graphite powder SGA 20 OS. Graphite Kropfmühl GmbH	<50	99.9
Graphite powder SC20OS. Graphite Kropfmühl GmbH	<60	99
Graphite powder SC20OS. Graphite Kropfmühl GmbH Ferrocene F408 98%. Sigma-Aldrich	<60 / n/a	99 / 98
Graphite powder FP 99,5. Graphite Kropfmühl GmbH]	<75	99.95
Aluminium (20093), 99%, powder, 200 mesh. ACROS Organics™	<75	99
Aluminium (11010) powder fine, >93%. Sigma-Aldrich	n/a	93
Nickel (266981) powder. <50μm, 99.7% trace metals basis. Sigma-Aldrich	<50	99.7
Copper (357456) powder (dendritic). <45μm, 99.7% trace metals basis. Sigma Aldrich	<45	99.7

Furthermore, to assist in the mixing process, a three-roll mill (model ESSG62, from Shanghai Espread) was used. Lastly, custom-made screen printing screens were also produced, which were specifically designed with a 60:1 (Length:Width) pattern geometries, enabling the measurement of volumetric resistivity in the deposited contacts.

Table 3: Summary of the best results obtained in the ink tests, along with their corresponding recipes.

Conductive powder	ρ (Ω·cm)	Recipe (wt%)		
		Conductor	Binder	Solvent
Graphite powder SGA 20 OS	3.31	36.2	54.1	9.7
Aluminium 20093	3.40	70.4	25.6	4.0
Nickel 266981	6.26	68.4	26.9	4.7

As can be seen in Table 3, the introduction of metallic-type conductive powders, which consequently have higher conductivities and apparent densities than graphite, did not result in a significant improvement in the resistivity of the deposited contacts. From this, it can be deduced that the issue with the inks may lie in the choice of binder, in our case, epoxy, which does not evaporate sufficiently when the printed contacts are cured at 100°C. In fact, we conducted a test by annealing the contacts in the oven for 30 minutes at 300°C, and we obtained extraordinarily low resistivities, as shown in Table 4. Unfortunately, this curing temperature is incompatible with their use on perovskite substrates.

Table 4: Results obtained after annealing the contacts at 300°C for 30 minutes..

Conductive powder	ρ (Ω·cm)	Recipe (wt%)		
		Conductor	Binder	Solvent
Nickel 266981	0,02	68,4	26,9	4,7
Copper 357456	$1{,}02 \times 10^{-3}$	78,7	17,8	3,5

2.3 Inks with graphite and collodion as the binding element.

Collodion is a viscous solution that has been in existence since 1846 and is commonly used as a vehicle in paints and varnishes to create coatings with high adhesion and quick drying [9], [10]. Collodion can be obtained by dissolving nitrocellulose in a mixture of ethers and alcohols in a volume ratio of approximately 3:1, respectively [11], [12]. Moreover, ethers, particularly ethyl ether, are used as an anti-solvent, i.e., a reagent applied during the washing treatment stage (single-step spin coating deposition method) to produce thin perovskite layers [13]–[15]. Combining these two concepts led to the hypothesis that collodion could serve as an alternative binder to epoxy in order to produce screen printing inks compatible with perovskite.

To produce the collodion solution, it was first necessary to synthesize nitrocellulose or gun cotton. Many factors determine the required nitration reaction to produce this compound from cellulose, including the reaction temperature, the concentration of acids involved, and their water content [16].

The chemical reactions involved in the synthesis of nitrocellulose are as follows:
Synthesis of the nitrating agent NO_2+.

$$HNO_3 + 2\,H_2SO_4 \rightarrow NO_2^+ + H_3O^+ + 2HSO_4^- \quad (1)$$

Nitration of cellulose.

$$C_6H_7O_2(OH)_3 + 3NO_2^+ \rightarrow C_6H_7O_2(NO)_3 + 3H_2O \quad (2)$$

The reactions for the production of nitrocellulose were carried out in a beaker into which a mixture of concentrated sulfuric acid (H2SO4, 98%) and concentrated nitric acid (HNO3, 70%) was introduced. This mixture of acids is called the "Nitrating Mixture." The beaker was placed, uncovered (in a water bath), inside another container with water and ice. This was done to control the reaction temperature. When the nitrating mixture reached 21°C [17], a sample of 5 grams of commercial cotton, as a source of cellulose, was immersed in it. After 20 minutes, the temperature of the sample, now converted to nitrocellulose, dropped to 12°C, and it could be carefully removed from the beaker. The acid residues (HSO_4^-) attached to the nitrocellulose were neutralized using a sequential washing procedure that involved seven beakers. The first five beakers contained a 1 molar (1 M) solution of sodium carbonate (Na_2CO_3), and the last two contained deionized water (18 MΩ·cm). The pH was monitored after each wash. This phase ended when the measured pH in the washing beakers was neutral. Finally, the nitrocellulose sample was squeezed to remove excess liquid, spread as thinly as possible, and placed in a desiccator for 24 hours to obtain a highly dehydrated product.

To produce collodion compatible with perovskite, a systematic procedure was carried out to prepare solutions, following the original recipe that uses nitrocellulose (NC) (3-7% by weight) in ethyl ether (DE) (65-75% by weight) and ethanol (EtOH) (20-30% by weight), with an attempt to decrease the volume percentage (Vol.%) of DE and EtOH. Two anti-solvents used in perovskite production, chlorobenzene (CB) and toluene, as well as Butyl Diglycol (BDG), which is a glycol ether widely used to produce inks requiring a low evaporation rate to balance that of DE [18], were used.

Each collodion recipe was prepared by mixing 0.5 grams of nitrocellulose with the chosen solvent mixture, vigorously stirring the mixture in a sample vial with a lid for 2 minutes.

Table 5: Collodion recipes made based on the volume percentage (Vol. %) of the solvents used, indicating whether the mixture resulted in an appropriate solution.

Recipe #	DE	BDG	EtOH	CB	Toluene	Resultado de la disolución
Original	75.50		24.5	-	-	Ok
P2	74.36	-	25.64	-	-	Ok
P3	50.64	24.68	24.68	-	-	Ok
P4	-	75	25.00	-	-	Ok
P6	-	37.66	24.68	-	37.66	Ok
P8	-	75.00	-	25.00	-	Ok
P9	-	75	-	-	25.00	Ok
P10	-	74.03	12.98	12.99	-	Ok
P11	-	74.03	12.98	-	12.99	Ok
P12	37.66	37.66	-	24.68	-	Ok
P13	37.66	37.66	-	-	24.68	Ok
P14	14.72	14.72	9.64	-	60.92	No
P15	21.83	7.61	9.64	60.92	-	No
P16	19.91	19.91	4.63	-	55.55	No
P17	17.68	19.91	4.63	-	57.78	No
P18	19.91	19.91	4.63	-	55.55	No

Although different collodion-based ink recipes (or C-inks) were tested, the one that yielded the best results was prepared in four mixing stages:

1. In the first stage, 21.6 grams of graphite powder (Graphite powder SGA 20 OS, Graphite Kropfmühl GmbH) were mixed with 59.14 ml of collodion, prepared according to recipe P3 (see Table 4), using a stirring rod.
2. In the second stage, a three-roll mill was used, giving the mixture four passes to homogenize it.
3. In the third stage, the stirring rod was used again, and 47.04 ml of chlorobenzene and 63.93 ml of toluene were added to the mixture.
4. In the final stage, the mixture was given four more passes in the three-roll mill.

Thus, in terms of wt%, the resulting recipe was 13.0:25.4:61.6, distributed among conductive powders, binder, and solvent, respectively.

Subsequently, contacts with a 50:1 geometry were printed on a glass substrate using a screen printing machine. These contacts were then left to cure at room temperature for 24 hours, during which it was confirmed that they were uniform and well adhered to the substrate. Additionally, the volume resistivity (ρ) of the contacts deposited with the graphite ink in collodion was 0.061 Ω·cm, with a standard deviation of 0.006 Ω·cm.

Finally, upon depositing the produced ink onto a substrate coated with a perovskite layer, it was observed that it rapidly deteriorated the perovskite layer within seconds of deposition.

2.4 Inks with graphite and a mixture of epoxy and collodion as binding elements.

In this study, an attempt was made to determine the amount of collodion that could be mixed with epoxy in a way that the resulting binder would not harm the perovskite layers. After several tests, it was found that the maximum weight percentage of collodion was approximately 5% (95:5% epoxy:collodion).

This time, the ink itself was prepared with 6.175 grams of the epoxy:collodion binder and 5.161 grams of graphite powder. Initially, it was mixed by hand and then subjected to five passes through the three-roll mill. To the 8.054 grams recovered from the three-roll mill, 0.2 ml of hardener and 1.46 ml of toluene were added to complete the mixture.

It's important to note that, to improve the chances of the ink not damaging the perovskite layers, the amount of hardener used on this occasion was not the manufacturer's recommended ratio of 80:20% by weight of epoxy:hardener. Instead, the applied weight percentage was 95:5%.

This time, the resulting recipe was 38.6:41.5:19.9, in terms of its wt%, distributed among conductive powders, binder, and solvent, respectively

The printing of contacts was carried out using the screen printing device, and drying was accomplished using an oven at 100°C for 10 minutes.

The average volumetric resistivity measured in the printed contacts was 2.282 Ω·cm, with a standard deviation of 1.621 Ω·cm. this result only marginally improved upon the results obtained with the previously developed epoxy resin formulations, with the only notable enhancement being that the printed contacts exhibited more consistent resistivity values across all printed substrates. Also, as anticipated, upon inspecting the contacts printed with the developed ink on substrates coated with perovskite layers using a digital microscope, no degradation of these layers was observed following the curing process.

Anyhow, although the results obtained may be interpreted as a slight improvement compared to previous attempts, they do not appear to be significant enough to warrant further exploration of this research avenue, especially considering that this approach involves the synthesis of nitrocellulose, which adds considerable levels of complexity and risk.

2.5 Inks with graphite and a mixture of epoxy and EVA as binding elements.

As an alternative to using collodion in combination with epoxy, an approach involving the development of a binder composed of epoxy resin and a solution of ethylene-vinyl acetate (EVA) polymer in anhydrous toluene was tested. EVA was chosen because it had also been successfully used as an encapsulating agent to protect perovskite layers from degradation due to environmental exposure[19], [20].

The EVA in toluene solution, hereinafter referred to as "EVATol" for simplicity, was prepared by mixing small pieces of a commercial EVA sheet (Encapsolar PC-135 A, purchased from Stevens Urethane) with anhydrous toluene (Toluene anhydrous, 99.8%, 244511 Sigma-Aldrich) in a weight percentage (wt%) ratio of 25:75, using a heated magnetic stirrer (BasicMagMix model, from OVAN), for 24 hours at 450 rpm and 100°C. It's important to note that once this mixture was allowed to cool to room temperature, it quickly turned into a gel. However, by heating it again for 15 minutes at 100°C and stirring, it was possible to re-liquefy it.

When epoxy resin and EVATol were mixed at a 50% by weight ratio, using a spatula and a beaker, both compounds quickly merged into a gel-like ball that could only be dissolved by adding considerable amounts of toluene. Since toluene is the solvent used in EVATol, a trial-and-error procedure was carried out to obtain a stable solution (in the sense that it remained liquid) by reducing the amount of epoxy in the mixture. This goal was achieved by mixing epoxy and EVATol in a weight percentage (wt%) ratio of 25:75%.

Subsequently, several inks were prepared using the aforementioned epoxy and EVATol mixture along with graphite powder (Graphite powder SGA 20 OS, from Graphite Kropfmühl GmbH), using a method that combines manual mixing in a beaker with a laboratory spatula and five passes through the three-roll mill.

This time, the chosen recipe for producing the inks was 13.9:19.9:66.2, in terms of its wt%, distributed among conductive powders, binder, and solvent, respectively

As before, contacts were printed with a 50:1 geometry to measure their volumetric resistivity. Printing was done using a screen printer on glass substrates. Also, since no hardener was used in this formulation, the printed contacts were cured in an oven for 30 minutes at 100°C.

The average volumetric resistivity measured in the contacts printed with the best ink was 2.364 Ω·cm, with a standard deviation of 1.052 Ω·cm. This means that no significant improvements were achieved compared to previous results obtained with other formulations that also used epoxy resin.

2.6 Paste obtained from a mixture of graphite and carbon black as conductive materials, PVDF as a binding material, and toluene as a diluting material.

This study was inspired by the work of Xingzhong Zhao et al. [21], in which an electrode was developed from a commercial carbon suspension [22] that was modified for deposition onto perovskite layers using Doctor Blade techniques. Similar carbon suspensions are used in the manufacturing of electrodes for lithium batteries and can be obtained using vehicles formed by solutions consisting of binders like polyvinylidene fluoride (PVDF) [23] and solvents like N-Methyl-2-pyrrolidone (NMP) [24], [25].

In our case, we implemented a procedure to create a graphite and carbon black agglomerate with PVDF in toluene, starting from a PVDF solution in NMP, following the studies conducted by A. Bottino et al. [26], where they analyzed the solubility parameters of PVDF in 50 different solvents. We also considered the work of Mengyuan Li et al. [27], in which PVDF is dissolved in DMF at temperature.

Thus, the process began with the preparation of a solution in a 7 ml vial of PVDF (Kureha KF W#1700 powder) in DMSO (Sigma Aldrich 276855) at 15:85 % by weight (%Wt), by mixing 1.008 g of PVDF in 5 ml of DMSO. The mixing method initially involved using a spatula and then a magnetic stirrer (3D, IBX Instruments) set at 300 rpm for 5 minutes. The mixture was then transferred to a magnetic stirrer with a heating element (BasicMagMix, OVAN), where it continued to stir for 60 minutes at 75°C and 450 rpm. As soon as the PVDF comes into contact with the DMSO, it forms a whitish dispersion (resembling dissolved aspirin in water) that, when heated, thickens until it gels, acquiring a light yellowish color.

Next, in a beaker, 4.875 g of the previous solution (PVDF in DMSO), 9.098 g of graphite (Graphite powder SC 20 OS, Graphite Kropfmühl GmbH), and 3.534 g of carbon black (Conductex SC Ultra, Birla Carbon) were mixed, resulting in 18.184 g of mixture (with a 28:52:20% wt%). To accommodate such a quantity of conductive powder, an additional 20 ml of DMSO was added, bringing the final weight percentage of the mixture components to 2:66:23:9% (PVDF:DMSO:Graphite:Carbon black, wt%) before taking the mixture to a three-roll mill (ESSG65, Shanghai Espread), where it was homogenized with 5 passes, resulting in a very dark and dense paste.

The following step involved evaporating the DMSO, for which the previous paste was spread onto glass using a spatula, creating the thinnest possible layer, and then the glass was placed in an oven (CONTERM 19L, J.P. Selecta) for seven hours at 130°C. Finally, the dried product was scraped from the glass with a steel spatula and placed in a coffee grinder for pulverization.

10.4229/EUPVSEC2023/2BV.2.12

After this process, with all the DMSO evaporated from the powder, we obtained a mixture consisting of PVDF:Graphite:Carbon black with wt% of 6:68:26%, meaning that it contained 94% by weight of conductive powders, in which the small percentage of PVDF was adequately distributed.

Finally, in order to prepare the graphite, carbon black, and Methyl Ammonium Iodide (MAI) paste with PVDF in chlorobenzene, the procedure involved adding, in a small beaker, 1.043 g of the powder prepared as indicated in the previous paragraph, along with 0.241 g of MAI (Sigma Aldrich 793493), followed by 4 ml of chlorobenzene (Sigma Aldrich 284513). It was then mixed using a spatula for 5 minutes, resulting in a not very cohesive mixture, with a weight percentage of 17/6/77% (wt% of conductive powders/MAI/Chlorobenzene) in which the various materials forming the solute were adequately dispersed in the solvent. However, in terms of wt% of conductive powder, binder and solvent, the final paste was 21.7:0.9:77.4, thus producing the paste with the least amounto of vinder developed so far.

The pastes produced using this method did not turn out to be sufficiently fluid for application through the Doctor Blade method. Nevertheless, to deposit contacts with measurable electrical properties, a technique was devised. Small amounts of paste were spread onto glass substrates in the form of narrow, straight lines, with Copper tape (3M Copper tape 1181) affixed at both ends of these lines (refer to Figure 1). These contacts were then covered with Kapton tape and cured under the pressure of a heavy steel plate at 100°C for 30 minutes in an oven.

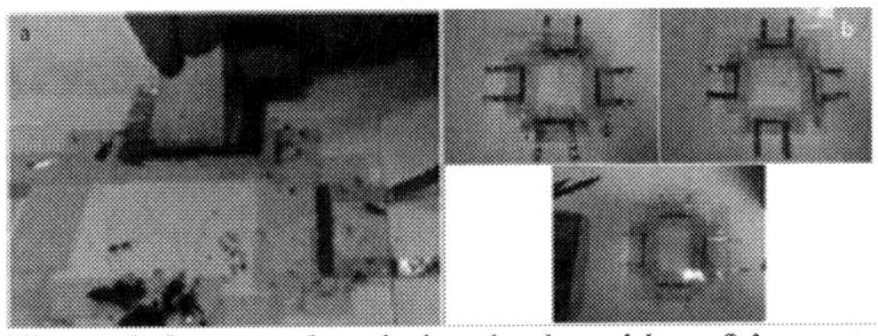

Figure 1: Images taken during the deposition of the paste (a) and of three deposited samples, each with four contacts, ready for measurement (b).

The average resistance measured in the deposited and cured contacts was 214.64 Ω, with a standard deviation of 148.79 Ω. It is important to note that the geometry of these contacts did not comply with the 50:1 pattern required by the F1896 standard, which is why it was not possible to obtain their volume resistivity, and therefore, it was also not possible to compare them under the same conditions as the contacts deposited with screen printing techniques.

3 RESULTS AND CONCLUSIONS

Table 6: Summary of the best results obtained in the ink tests conducted so far, indicating their compatibility with perovskite and their corresponding recipes.

Powder	Binder	Deposition	Compatible?	Recipe (wt%) Conductor	Binder	Solvent	ρ (Ω·cm)
Graphite	Epoxy	Screen printing	Yes	30	60	10	n/a
Graphite	Epoxy	Screen printing	Yes	36.2	54.1	9.7	3.31
Aluminium	Epoxy	Screen printing	Yes	70.4	25.6	4	3.40
Nickel	Epoxy	Screen printing	Yes	68.4	26.9	4.7	6.26
Graphite	Collodion	Screen printing	No	13	25.4	61.6	0.061
Graphite	Epoxy:Collodion	Screen printing	Yes	38.6	41.5	19.9	2.28
Graphite	Epoxy:EVATol	Screen printing	Yes	13.9	19.9	66.2	2.36
Graphite-Black Carbon-MAI	PVDF	Lamination	Yes	21.7	0.9	77.4	n/a

As observed, the inks developed using epoxy resin for screen printing deposition yielded contacts with a volume resistivity as low as 2.28 Ω·cm. Also, attempts with alternative binders, such as collodion or a mixture of collodion with epoxy, or epoxy blended with other polymers like EVA, were unsuccessful. Collodion attacked the perovskite layer, while the latter, although not harming the perovskite, failed to deliver a significant improvement in the conductivity of the produced contacts. However, a distinct approach that resulted in a paste requiring minimal binder in the form of PVDF, necessitating lamination techniques for deposition, generated contacts with a measured resistance approximately three orders of magnitude lower than those produced using epoxy and graphite via screen printing. This suggests that a promising avenue may have been identified for producing back contacts for perovskite solar cells.

4 ACKNOWLEDGEMENTS

This work has been developed within the MACLAB-PV project framework. which has been co-financed by the INTERREG Madeira-Azores-Canarias Territorial Cooperation Programme (MAC) 2014-2020. 2nd Call. Axis 1 – Enhancing research. technological development and innovation.

5 REFERENCES

[1] C. Quinto *et al.*, "Screen Printing for Perovskite Solar Cells Metallization," *EU PVSEC Proc.*, pp. 1144–1148, 2015, doi: 10.4229/EUPVSEC20152015-3BV.5.41.

[2] M. C. C. Montes, L. Ocaña, C. Quinto, S. González-Pérez, B. González-Diaz, M. Friend, "Conductive Inks with Epoxy Resin Based Vehicles for Perovskite Screen Printing Metallization as a Viable and Low-Cost Alternative to Thermal Evaporation," in *EU PVSEC Proceedings*, 2017, pp. 1158–1162. doi: 10.4229/EUPVSEC20172017-3DV.2.56.

[3] C. Montes *et al.*, "Exploring the use of collodion as a binder in screen printing vehicles for perovskite solar cells metallization," in *35 th European*

Photovoltaic Solar Energy Conference and Exhibition, WIP, Nov. 2018, pp. 938–932. doi: 10.4229/35THEUPVSEC20182018-3BV.3.34.

[4] M. C. C. Montes, L. Ocaña, S. González-Pérez, B. González-Diaz, M. Friend, "Technical Analysis of Mixed Binders Based on Epoxy Resin and Ethylene-Vinyl Acetate to Produce Conductive Inks for the Metallization of Perovskite Solar Cells via Screen Printing Technique," in *EU PVSEC Proceedings*, 2020, pp. 663–666. doi: 10.4229/EUPVSEC20202020-3BV.1.23.

[5] C. Montes, L. Ocaña, S. González-Pérez, B. González-Díaz, and E. Llarena, "Assessing the use of polyvinylidene fluoride as a binder material for producing carbon based inks suitable for HTM-free perovskite solar cells," *38th Eur. Photovolt. Sol. Energy Conf. Exhib.*, pp. 426–429, 2021, doi: 10.4229/EUPVSEC20212021-3BV.1.9.

[6] U. S. ASTM International, 100 Barr Harbor Drive, PO Box C700, West Conshohocken, PA 19428-2959, "Test Method for Determining the Electrical Resistivity of a Printed Conductve Material", doi: 10.1520/F1896-16.1.

[7] "Properties of Solvents Used in Organic Chemistry." http://murov.info/orgsolvents.htm (accessed Sep. 15, 2023).

[8] R. Liu and K. Xu, "Solvent engineering for perovskite solar cells: A review," *Micro Nano Lett.*, vol. 15, no. 6, pp. 349–353, 2020, doi: 10.1049/mnl.2019.0735.

[9] M. R. Fisch, *Liquid crystals, laptops, and life*, vol. 7, no. 9. Elsevier, 2004. doi: 10.1016/S1369-7021(04)00408-0.

[10] D. Sherrington, "Polymer Modification: Principles, Techniques, and Applications Edited by J. J. Meister. Marcel Dekker: New York. 2000. 914 pp. Price $235. ISBN 0-8247-0078-3.," *Org. Process Res. Dev.*, vol. 5, no. 2, pp. 189–190, 2001, doi: 10.1021/op0001133.

[11] R. Kohli, "Strippable Coatings for Removal of Surface Contaminants," in *Developments in Surface Contamination and Cleaning: Particle Deposition, Control and Removal*, William Andrew Publishing, 2010, pp. 177–224. doi: 10.1016/B978-1-4377-7830-4.10005-2.

[12] C. Identification, "Collodion USP, Material Safety Data Sheet," 2003.

[13] J. Troughton, K. Hooper, and T. M. Watson, "Humidity resistant fabrication of CH3NH3PbI3 perovskite solar cells and modules," *Nano Energy*, vol. 39, pp. 60–68, Sep. 2017, doi: 10.1016/J.NANOEN.2017.06.039.

[14] C. Aranda, C. Cristobal, L. Shooshtari, C. Li, S. Huettner, and A. Guerrero, "Formation criteria of high efficiency perovskite solar cells under ambient conditions," *Sustain. Energy Fuels*, vol. 1, no. 3, pp. 540–547, May 2017, doi: 10.1039/C6SE00077K.

[15] L. Contreras-Bernal *et al.*, "Homeopathic Perovskite Solar Cells: Effect of Humidity during Fabrication on the Performance and Stability of the Device," *J. Phys. Chem. C*, p. acs.jpcc.8b01558, 2018, doi: 10.1021/acs.jpcc.8b01558.

[16] W. B. Robertson, "Factors Governing the Nitration of Cellulose," Cornell University, 1946. Accessed: Jul. 27, 2018. [Online]. Available: http://hdl.handle.net/10945/6411

[17] D. Trache, K. Khimeche, A. Mezroua, and M. Benziane, "Physicochemical properties of microcrystalline nitrocellulose from Alfa grass fibres and its thermal stability," *J. Therm. Anal. Calorim.*, vol. 124, no. 3, pp. 1485–1496, Jun. 2016, doi: 10.1007/s10973-016-5293-1.

[18] "BASF Product overview Alcohols and Solvents." [Online]. Available: https://www.basf.com/global/documents/en/products-and-industries/alcohols-and-solvents/BASF_Solvents_Produktblatt_Web_02.pdf.assetdownload.pdf

[19] L. Ocaña, C. Montes, S. González-Pérez, B. González-Díaz, and E. Llarena, "Testing encapsulated perovskite solar cells in a climatic chamber by following the IEC 61215 and IEC 61646 standards for the thermal cycling test," in *38th European Photovoltaic Solar Energy Conference and Exhibition*, WIP, Nov. 2021, pp. 406–410. doi: 10.4229/EUPVSEC20212021-3BV.1.4.

[20] E. L. L. Ocaña, C. Montes, B. González-Diaz, S. González-Pérez, "Testing EVA, PMMA and PVDF Encapsulated Perovskite Solar Cells in a Climatic Chamber by Following the International Summit on Organic Photovoltaic Stability (ISOS-T) Protocols," in *8th World Conference on Photovoltaic Energy Conversion*, Milan, 2022, pp. 291–295. doi: 10.4229/WCPEC-82022-2AV.1.16.

[21] Z. Yu *et al.*, "Stable Organic–Inorganic Perovskite Solar Cells without Hole-Conductor Layer Achieved via Cell Structure Design and Contact Engineering," *Adv. Funct. Mater.*, vol. 26, no. 27, pp. 4866–4873, Jul. 2016, doi: 10.1002/ADFM.201504564.

[22] "9000 SERIES PET INK/JUJO CHEMICAL CO., LTD." http://www.jujo-chemical.co.jp/e/tec_info/yozai/9000_PET.html (accessed Jun. 12, 2021).

[23] "KF Polymer | KUREHA CORPORATION." https://www.kureha.co.jp/en/business/material/kfpolymer.html (accessed Jun. 12, 2021).

[24] Y. Wang, H. Zheng, Q. Qu, L. Zhang, V. S. Battaglia, and H. Zheng, "Enhancing electrochemical properties of graphite anode by using poly(methylmethacrylate)-poly(vinylidene fluoride) composite binder," *Carbon N. Y.*, vol. 92, pp. 318–326, Jul. 2015, doi: 10.1016/j.carbon.2015.04.084.

[25] M. Yoo and C. W. Frank, "Surface Chemistry and Morphology of Binders in Graphite Anodes of Lithium Ion Batteries," *Electrochem. Soc. 198th Meet.*, no. March, 2000.

[26] A. Bottino, G. Capannelli, S. Munari, and A. Turturro, "Solubility parameters of poly(vinylidene fluoride)," *J. Polym. Sci. Part B Polym. Phys.*, vol. 26, no. 4, pp. 785–794, Apr. 1988, doi: 10.1002/polb.1988.090260405.

[27] M. Li *et al.*, "Controlling the microstructure of poly(vinylidene-fluoride) (PVDF) thin films for microelectronics," *J. Mater. Chem. C*, vol. 1, no. 46, pp. 7695–7702, Dec. 2013, doi: 10.1039/c3tc31774a.

BOTTLENECKS IN PEROVSKITE SOLAR CELL RECYCLING

Elena S. Akulenko*, Mahboubeh Hadadian, Maryam Esmaeilzadeh, Rustem Nizamov, Kati Miettunen

Department of Mechanical and Materials Engineering, Faculty of Technology, University of Turku, Turku, FI-20014, Finland
*e-mail address: elena.akulenko@utu.fi

ABSTRACT: In the realm of recycling, the predominant emphasis has been generally on retrieving the metals. However, the highest value end product, usually with the least process energy, is realized through the revival of the device. Carbon-based perovskite solar cells (C-PSC) are a promising candidate for reviving the whole device owing to their structure (porosity) that facilitates both the flushing of the degraded perovskite as well as infiltrating new perovskite. Until now, reviving C-PSCs has been proposed by gas treatment, whereas for industrial applications liquid treatment would be preferable. This study focuses on liquid treatment by investigating how various organic solvents impact the technical feasibility of this process. Our findings highlight the key role of the mechanical stability of carbon mesoporous layer in relation to the organic solvents used in reviving. γ-Valerolactone dissolved perovskite in mesoporous layer with none or small carbon peeling off, while N,N-dimethylformamide and dimethyl sulfoxide showed in most cases high peeling off. We demonstrate a 65% revival level for PSCs using γ-Valerolactone. When employing the identical approach to revive samples that were aged under high humidity (70%) and elevated temperature (50 °C), the revival level decreased to 41%. This decline suggests that there is an irreversible degradation. These observations serve as a first step to boost a discussion about reviving C-PSCs with liquid treatment.

Keywords: Eco-design, Photovoltaics, Revival, Perovskite solar cells, Recycling

1 INTRODUCTION

In the field of perovskite photovoltaic research, in addition to investigations of efficiency and stability issues, researchers are getting increasingly concerned with recycling technology to avoid lead waste [1]. According to the eco-design concept, recycling processes can be considered under four levels depending on the processing energy inputs and value outputs: reviving device, reusing components, recycling compounds, and recovering elements [2]. In our previous publication, we examined these four levels for perovskite photovoltaics in a broader context, including carbon-based perovskite solar cells (C-PSCs) [3]. Carbon electrodes are considered as promising sustainable alternatives to metal electrodes (i.e., Au, Ag etc.) [4]. Metal electrodes are one of the main sources of degradation in perovskite solar cells (PSCs) because of their ion migration to the underneath layers, while carbon electrodes are inert to ion migration and chemically stable [5], [6].

The known methods of reviving PSCs by liquid treatment were reported for devices with metal top electrodes such as nanoporous Au [7] or mesoporous Ni [8]. These methods use N,N-dimethylformamide (DMF) for washing degraded perovskite material. Alternative reviving solvent might be a dimethyl sulfoxide (DMSO), which can be decomposed efficiently to methanesulfonic acid without releasing harmful by-products [9]. The latter is a benefit for industrial scalability. However, according to solvent survey selection guide, DMF and DMSO are considered as hazardous and problematic [10], which is a bottleneck for their industrial application. A gas treatment has been reported for reviving C-PSCs [11]. However, reviving C-PSCs by liquid treatment, which would be preferred for large scale industrial processing, is missing from the literature.

Another bottleneck in reviving by liquid treatment is the encapsulants, due to the difficulty of separating it from other components. So far, several materials (organic and inorganic) have been reported and discussed as encapsulants for perovskite photovoltaics [12]. Though, there is insufficient understanding of how encapsulation affects the recyclability of PSCs, particularly, how the encapsulants react with organic solvents supposed for revival.

This work focuses on reviving of C-PSCs by liquid treatment, utilizing DMF, DMSO, and a non-toxic alternative, γ-Valerolactone (GVL), given its recent use in the fabrication of C-PSCs [13]. We investigate behaviour of carbon electrodes in these organic solvents. The revival of C-PSCs with GVL is observed and discussed. We study the surface of C-PSCs after fabrication, aging, and revival under a scanning electron microscope. This work establishes the first step for further investigation of reviving of C-PSCs through liquid-based approaches.

2 MATERIALS AND METHODS

2.1. Materials

Zirconium (IV) oxide (ZrO_2, nanopowder, <100 nm particle size), graphite (powder, <20 μm, synthetic), ethyl cellulose (48.0-49.5% (w/w) ethoxyl basis), terpineol ((mixture of isomers) for synthesis), DMSO (anhydrous, 99.9%), DMF (anhydrous, 99.8%), and GVL (anhydrous, 99%) were purchased from Sigma Aldrich. Carbon black (CAS- 1333-86-4) was purchased from Cabot. Carbon monolithic electrode (CME), zirconia electrodes, polyimide adhesive mask, and perovskite precursor solution (consisting of PbI_2 (1.2 Molal), methylammonium iodide (1.2 Molal), 5-ammonium valeric acid iodide (5% mol) in gamma-butyrolactone (GBL) and ethanol mixture) were purchased from Solaronix.

2.2 Samples Fabrication and Revival Process

Carbon monolithic electrodes (CME) consisted of a carbon layer on top of a zirconia electrode, including a compact TiO_2 layer, a mesoporous TiO_2 layer, and a mesoporous ZrO_2 layer on fluorine-doped tin oxide (FTO) glass. The samples were heated for 30 min at 400 °C with a 30 min ramp. Once cooled to room temperature, a polyimide adhesive mask was applied on the CMEs before the perovskite precursor solution was dispersed.

Afterward, the mesoporous layers of CMEs were infiltrated with 5.74 µL of methylammonium lead iodide ($MAPbI_3$) precursor solution inside a glovebox. The samples were then heated at 50 °C for 10 min on a hot plate.

Two other carbon electrodes were prepared by applying carbon pastes of different compositions on top of the zirconia electrodes and subjected to similar processing. The first carbon paste consisted of 1.5 g graphite, 0.2 g ZrO_2, 0.2 g ethyl cellulose as a binder, and 4 ml terpineol as a solvent. The second paste had additionally 0.5 g of carbon black. After three hours of ball milling (MIXER MILL MM 400 from Retsch), the carbon pastes were deposited by doctor blading onto pre-heated zirconia electrodes and dried at 100 °C for 10 min. Next, the carbon layers were heated at 400 °C and 450 °C for 30 min after 30 min ramp correspondingly. The process of perovskite infiltration remained consistent with that utilized for CMEs.

Before conducting the revival process, several organic solvents were tested to ensure the stability of the carbon top electrode layer. Three different carbon electrodes were subjected to testing. The samples were immersed into DMF, DMSO, and GVL for 30 min and rinsed with ethanol. Next, they were dried on a hot plate at 150 °C for 10 min. The stability of carbon layer was assessed visually based on the extent of carbon layer peeling (Table I).

The revival process was carried out by immersing the samples into the GVL solvent for 30 min at room temperature. Then, the devices were washed with ethanol and then dried on a hot plate at 150 °C for 10 min. After cooling down to room temperature, the $MAPbI_3$ precursor was infiltrated into the samples by the process mentioned above. Fig. 1 shows the general process flow for reviving C-PSCs with GVL. Fresh and aged samples were washed with GVL and infiltrated with $MAPbI_3$.

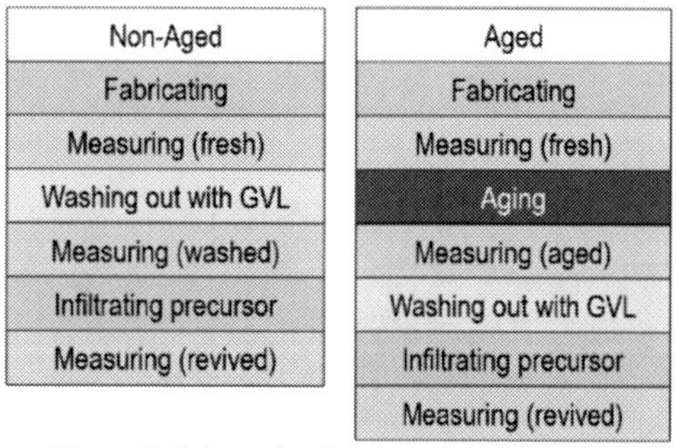

Figure 1. Schematic of process flow for reviving the samples with GVL.

2.3 Samples Characterization and Aging

The performance of solar cells was measured using a solar simulator (PEC-L01, PECCELL) comparable to AM 1.5G standard testing conditions. The J-V measurement was done with PalmSens4, which operated at an initial voltage of −0.2 V within three second holding time as preconditioning samples and was scanned up to the vortex potential set at 1 V and then back to the initial voltage. The scan rate was 5 mV/s and step size was 10 mV. A mask with a rectangular aperture 0.64 cm^2, smaller than the active area of the solar cell 12.5 x 12 mm^2, was applied on top of the cell for measuring. Furthermore, the morphology of the samples was analysed by scanning electron microscopy (SEM) (Thermo Scientific Apreo S).

The aging process was carried out according to the ISOS-D-3 [14] in a weather chamber (Aralab TESTA_e) under 50 °C temperature and 70% humidity for 19.4 h.

3 RESULTS AND DISCUSSION

3.1 Impact of Organic Solvent on Stability of Carbon Top Electrode

The high stability of the carbon top electrode is desirable for revival by liquid treatment to ensure its integrity for cyclic refurbishing and reuse. To identify the appropriate organic solvent for the revival of C-PSCs, we investigated the influence of DMF, DMSO and GVL on carbon top electrode by immersing samples into these solvents.

Our results presented in Table 1 show that GVL is a promising candidate for revival of C-PSCs compared to DMF and DMSO. GVL caused no or small peeling of carbon mesoporous layer, while DMF and DMSO destroyed most carbon layers. Only samples with carbon electrodes comprising graphite and ZrO_2 and annealed at 450 °C showed high stability during the washing-out step with all tested solvents. The proportion of ZrO_2 in these samples was 3.56 wt % which is higher than the samples with carbon black, 3.26 wt%. Therefore, one explanation might be that ZrO_2 serves as a binder for carbon particles and its sintering is improved at high temperatures so that it becomes strong enough to endure the solvent treatment. Overall, the different behaviour of varying carbon electrodes in organic solvents indicates that the fabrication method of C-PSCs plays an important role in top contact stability.

Table I: Results of six carbon electrode behaviour in different organic solvents, representing the carbon layer stability as stable, where no or small peeling off has been observed (stable), and non-stable with high peeling off (-).

Carbon Electrode	Annea-ling T °C	DMF	DMSO	GVL
CME Solaronix	400	-	-	stable
CME Solaronix	450	-	-	stable
Graphite + ZrO_2	400	-	-	stable
Graphite + ZrO_2	450	stable	stable	stable
Graphite + ZrO_2 + Carbon black	450	-	-	stable

3.2 Solar Cells' Performance Before and After Revival with GVL

Based on the solvent testing results (Table I), GVL was selected for reviving C-PSCs with CME. After immersing non-aged (fresh) samples in GVL, the solvent close to the active area of the sample immediately started to change color from colorless to yellow. After 30 min, the whole volume of GVL in a beaker was yellow. The color changing is due to yellow-colored lead iodide (PbI_2) left as a degradation compound. Yellow-colored areas were also observed in aged samples. These areas appeared along the active area of the cells. In the case of non-aged samples, GVL caused the rapid decomposition of $MAPbI_3$, being in direct contact with the fresh absorber layer. In contrast, in the case of aged samples, GVL mainly washes out the products already decomposed earlier such as PbI_2.

According to the process flow presented in Fig. 1, the photovoltaic performance of the samples was measured after fabrication, after washing out/aging, and after perovskite precursor infiltration. Measured characteristics are presented in Table II.

Table II. Average photovoltaic characteristics of six aged and five non-aged samples before and after revival with GVL.

	Samples	J_{sc} (mA/cm²)	V_{oc} (V)	FF	η (%)
Non-Aged	fresh	18.9±0.8	0.85±0.01	0.58±0.03	9.4±0.8
	washed	0.00±0.00	0.44±0.08	0.29±0.02	0.00±0.00
	revived	11.6±0.8	0.89±0.01	0.59±0.06	6.1±0.8
Aged	fresh	18.4±0.8	0.85±0.01	0.56±0.03	8.8±0.7
	aged	0.4±0.3	0.50±0.08	0.4±0.1	0.07±0.05
	revived	8±3	0.82±0.04	0.5±0.1	4±2

The average efficiency of washed samples (0%) in comparison to the ones of fresh samples indicates that GVL efficiently washed out the perovskite crystals from carbon mesoporous top layer. Both aged and washed samples showed very low performance: J_{sc} of washed samples was 0 mA/cm² while J_{sc} of aged samples was 0.4±0.3 mA/cm². The average revival rate for aged samples was 41%, while for non-aged samples it was 65% (Fig. 2). The highest revival rates were 74% for non-aged devices and 67% for aged devices. Notably, the aged samples exhibited a higher deviation in comparison to the non-aged ones. Open circuit voltage (V_{oc}) and fill factor (*FF*) of samples revived after aging were lower in comparison to those for samples revived after fabrication (non-aged). Short circuit density (J_{sc}) of samples revived after aging differs by almost 30% and have a higher standard deviation compared to samples revived after fabrications (non-aged). These results show that aged samples lost the ability to produce the original maximum photocurrent density at short circuit conditions. We also noticed that fresh samples had 20±2 Ω average series resistance. Whereas after revival, the average series resistance of non-aged samples was 30±4 Ω, and the average serial resistance of aged samples was 40±11 Ω. There is a direct correlation between the series resistance and the revival level of the samples. Lower series resistance implies that there could be some damage in the carbon layer resulting in lower conductivity.

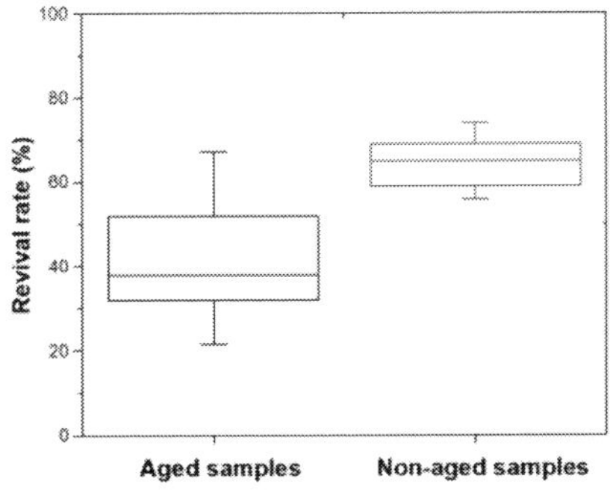

Figure 2: Revival level of aged and non-aged samples.

One possible explanation is that the aging process significantly damages the integrity of the carbon mesoporous structure, which also impedes the proper formation of the absorbing layer after the second infiltration of the perovskite precursor. The difference between revival level of non-aged and aged samples underscores the importance of further investigating the degradation mechanisms of C-PSCs to evaluate their reuse potential and to engineer their architecture for cyclic refurbishing.

3.4 Morphology of Fresh, Aged and Revived Samples

For a better understanding of changes in the morphology of carbon electrodes due to aging and revival, top-view SEM images of the samples were performed. Figure 3 shows the microscopy images of (a) C-PSCs after fabrication, (b) aging, and (c) revival. Besides the different structures of perovskite, the morphology of carbon electrodes varies as well. It appears fresh samples after fabrication (Fig. 3a) compared to both aged (Fig. 3b) and revived (Fig. 3c) samples, which seem more compact. The top view of samples after fabrication is quite uniform and there is some perovskite on top of carbon. The top view of revived samples (Fig. 3c) significantly differs from the top view of fresh samples, particularly, formation of perovskite, since those samples have aggregated perovskite crystals (stone-shape) on top of the carbon layer. One reason could be that the perovskite is partially forming on top. This formation seems more dominant for reviving process.

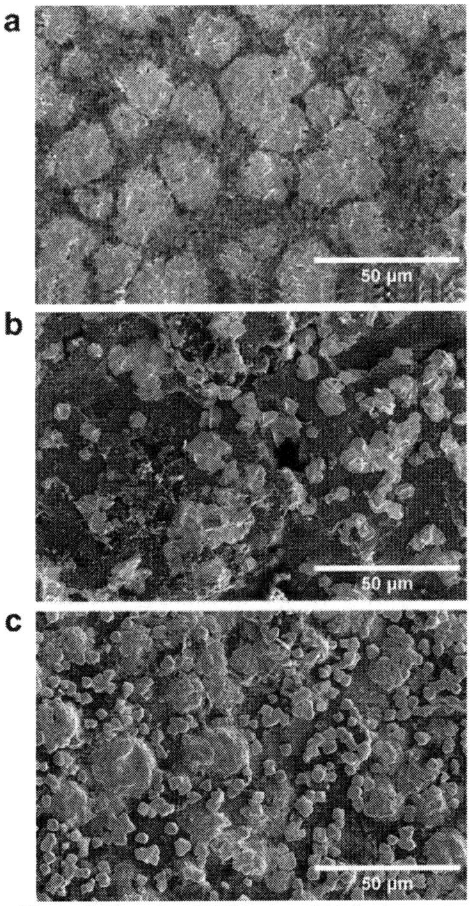

Figure 3. Top-view SEM image of (a) fresh samples with uniform distribution of perovskite, (b) aged samples with some PbI_2 aggregates, and (c) revived samples.

4 CONCLUSION

We explored the impact of various organic solvents on carbon electrodes with different compositions of carbon paste. The results show that the amount of ZrO_2 and annealing temperature have an impact on the mechanical stability of carbon electrodes when immersed in organic solvents for reviving. GVL was chosen for reviving C-PSCs due to minimal or no peeling of the carbon mesoporous layer. We demonstrated the potential of GVL solvent treatment to revive C-PSCs, achieving an average of 65% revival level for non-aged cells. However, aged samples reached only an average revival level of 41%. The difference in the revival level of non-aged and aged samples might be attributed to irreversible degradation mechanisms stemming from the aging process. Moreover, resistivity data indicates that there is a direct correlation between the series resistance and revival level of the samples. The resistance in the revived samples after aging was found to be higher compared to the revived samples after fabrication. Thus, there seems to be damage in the conductivity of the carbon layer. SEM images of the C-PSC surface after fabricating, aging, and reviving showed different distributions of perovskite in the mesoporous layer. The morphology of revived sample in comparison with fresh and aged samples, suggests that perovskite is partially forming on top. The demonstrated revival of C-PSCs with a non-toxic GVL serves as a first step to developing a scalable liquid-based recycling process for C-PSCs.

REFERENCES

[1] N. G. Park, "Perovskite Solar Cell: Research Direction for Next 10 Years," *ACS Energy Lett.*, vol. 4, no. 12, pp. 2983–2985, 2019, doi: 10.1021/acsenergylett.9b02442.

[2] K. Miettunen and A. Santasalo-Aarnio, "Eco-design for dye solar cells: From hazardous waste to profitable recovery," *J. Clean. Prod.*, p. 128743, 2021, doi: 10.1016/j.jclepro.2021.128743.

[3] E. S. Akulenko, M. Hadadian, A. Santasalo-Aarnio, and K. Miettunen, "Eco-design for perovskite solar cells to address future waste challenges and recover valuable materials," *Heliyon*, vol. 9, no. 2, p. e13584, 2023, doi: 10.1016/j.heliyon.2023.e13584.

[4] G. A. Dos Reis Benatto *et al.*, "Carbon: The Ultimate Electrode Choice for Widely Distributed Polymer Solar Cells," *Adv. Energy Mater.*, vol. 4, no. 15, 2014, doi: 10.1002/aenm.201400732.

[5] M. Hadadian, J. H. Smått, and J. P. Correa-Baena, "The role of carbon-based materials in enhancing the stability of perovskite solar cells," *Energy Environ. Sci.*, vol. 13, no. 5, pp. 1377–1407, 2020, doi: 10.1039/c9ee04030g.

[6] D. Bogachuk *et al.*, "Low-temperature carbon-based electrodes in perovskite solar cells," *Energy Environ. Sci.*, vol. 13, no. 11, pp. 3880–3916, 2020, doi: 10.1039/d0ee02175j.

[7] M. H. Li *et al.*, "Robust and Recyclable Substrate Template with an Ultrathin Nanoporous Counter Electrode for Organic-Hole-Conductor-Free Monolithic Perovskite Solar Cells," *ACS Appl. Mater. Interfaces*, vol. 9, no. 48, pp. 41845–41854, 2017, doi: 10.1021/acsami.7b12367.

[8] Z. Ku, X. Xia, H. Shen, N. H. Tiep, and H. J. Fan, "A mesoporous nickel counter electrode for printable and reusable perovskite solar cells," *Nanoscale*, vol. 7, no. 32, pp. 13363–13368, 2015, doi: 10.1039/c5nr03610k.

[9] N. G. Park, "Green solvent for perovskite solar cell production," *Nat. Sustain.*, vol. 4, no. 3, pp. 192–193, 2021, doi: 10.1038/s41893-020-00647-6.

[10] D. Prat, J. Hayler, and A. Wells, "A survey of solvent selection guides," *Green Chem.*, vol. 16, no. 10, pp. 4546–4551, 2014, doi: 10.1039/c4gc01149j.

[11] L. Hong *et al.*, "Improvement and Regeneration of Perovskite Solar Cells via Methylamine Gas Post-Treatment," *Adv. Funct. Mater.*, vol. 27, no. 43, pp. 1–8, 2017, doi: 10.1002/adfm.201703060.

[12] R. K. Raman, S. A. Gurusamy Thangavelu, S. Venkataraj, and A. Krishnamoorthy, "Materials, methods and strategies for encapsulation of perovskite solar cells: From past to present," *Renew. Sustain. Energy Rev.*, vol. 151, no. August, 2021, doi: 10.1016/j.rser.2021.111608.

[13] C. Worsley *et al.*, "γ-Valerolactone: A Nontoxic Green Solvent for Highly Stable Printed Mesoporous Perovskite Solar Cells," *Energy Technol.*, vol. 9, no. 7, pp. 1–10, 2021, doi: 10.1002/ente.202100312.

[14] M. V. Khenkin *et al.*, "Consensus statement for stability assessment and reporting for perovskite photovoltaics based on ISOS procedures," *Nat. Energy*, vol. 5, no. 1, pp. 35–49, 2020, doi: 10.1038/s41560-019-0529-5.

ACKNOWLEDGEMENTS

K.M., M.H., M.E. and R.N. acknowledge financial support from the Research Council of Finland (BioEST, project numbers 336577 and 336441, ECOSOL 347275). E.S.A. acknowledge financial support from Fortum and Neste Foundation (20210154). M.H. thanks SUSMAT profiling funding (Research Council of Finland and University of Turku). M.E. is grateful for UTUGS doctoral program funding. Strategic Materials Research Infrastructure (MARI) at University of Turku was used for this study.

LONG-TERM MONITORING SYSTEM FOR PEROVSKITE SOLAR CELLS UNDER REALISTIC INDOOR CONDITIONS

Matija Pirc, Marko Jošt, Žan Ajdič, Marko Topič
University of Ljubljana, Faculty of Electrical Engineering, Laboratory of Photovoltaics and Optoelectronics (LPVO)
Tržaška 25, SI-1000 Ljubljana, Slovenia

ABSTRACT: Indoor photovoltaics (IPV) have a tremendous market potential with the expected growth of various small electrical devices needing a perpetual energy source, notably internet of things (IoT) devices. Among the contenders for this pivotal role, perovskite solar cells (PSCs) are emerging as frontrunners, supported by a body of research attesting to their great potential. This study aims to deepen our understanding of perovskite solar cell performance in realistic indoor conditions. To this end we created an indoor monitoring system which maintains four solar cells at their maximum power points while simultaneously logging their performance and environmental conditions. Over the course of a year we monitored the behavior of three PSCs and one crystalline silicon solar cell (c-Si SC). Results for the first year show a remarkable three to one advantage in favor of the best PSC over the c-Si SC. The best of the three PSCs also retained most of its initial performance throughout the year, which may already be sufficient for some applications.
Keywords: perovskite solar cells, indoor photovoltaics, monitoring, realistic uncontrolled indoor conditions

1 INTRODUCTION

There is a myriad of small electronic devices in our homes, offices and industrial halls that could potentially be powered by indoor photovoltaics (IPV) [1]. Ambient energy harvesting like IPV is an excellent way to reduce either usage of batteries, or reduce installation complexity by removing the need for wiring, or to prolong the maintenance-free operational period of a device. The Internet of Things (IoT) is one of the areas that could greatly benefit from IPV. IoT is also a rapidly growing market, having reached an estimated 12.2 billion [2] to 200 billion [3] devices so far and still expanding. Many of these devices are expected to operate in indoor conditions therefore presenting a great opportunity for IPV.

Mature solar cell (SC) technologies, like crystal Silicon solar cells (c-Si SCs) are optimized for outdoor conditions with typical irradiances between 100 and 1000 W/m^2. However, in indoor conditions where typical irradiance is closer to 5 W/m^2, alternative SC technologies such as perovskite solar cells (PSCs) could be a better choice.

Other research groups have previously focused on this topic with efforts ranging from measuring the performance of different PV cell technology in indoor conditions [4], [5], to modelling IPV energy yield prediction based on SC parameters and indoor spectra [6], to optimizing PSC technology for indoor conditions [7]–[9]. Although a lot of measurements have been performed so far under controlled conditions simulating typical indoor conditions, we still lack an understanding of PSC performance under real-life, uncontrolled indoor conditions, which can vary considerably throughout the year.

The aim of this work is to present an in-house developed monitoring system for long-term monitoring of the performance of four small-area solar cells and their ambient conditions, along with some preliminary results.

2 EXPERIMENT

2.1 System overview

The monitoring system is designed to continuously track the maximum power point (MPP) of 4 independent small-area SCs and log their current and voltage. Along with SC operating parameters, it measures and logs basic ambiental parameters such as irradiance, temperature, humidity, lighting color and irradiance in 9 possible orientations. Every half an hour the maximum power point tracking (MPPT) is paused to perform an *I-V* scan of all 4 SCs. The system is connected to the internet via a Wi-Fi connection and stores all the measurements to a database in real time.

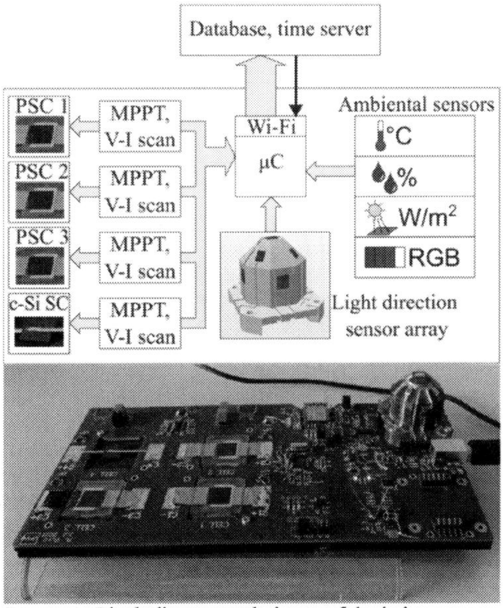

Figure 1: Block diagram and picture of the indoor monitoring system

2.2 Location

With IPV, the designer of a device usually can't pick and choose location, nor precisely optimize its orientation. Instead, the size of the IPV must be optimized for worst-case conditions. To record the performance of SCs in somewhat typical, verging on unfavorable conditions, the monitoring system was placed approximately in the center of our north-northwest facing office LPVO-1, located in Ljubljana, Slovenia (46° north latitude). Except for a few

10.4229/EUPVSEC2023/2BV.2.18
020108-001

minutes in the evenings of the summer months, natural light reaching the system consists entirely of diffuse irradiance. Natural light is sometimes supplemented with artificial light coming from fluorescent ceiling tubes.

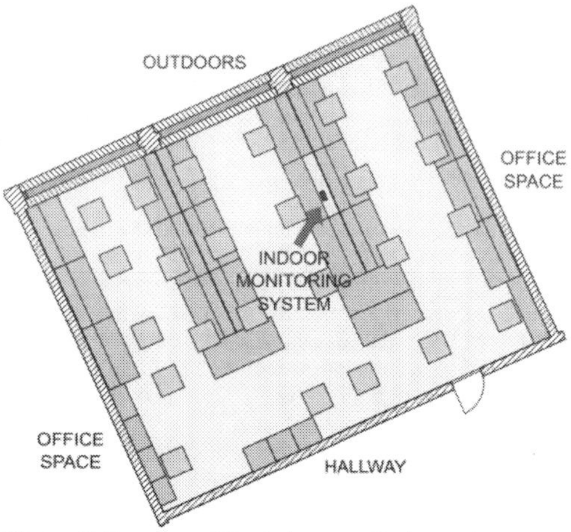

Figure 2: Location of the indoor monitoring system within the office and orientation of the office

2.3 Devices under test

The three SCs under test are PSCs and one c-Si SC was added for comparison. The performance parameters of all four SCs under STC are summarized in Table I.

Table I: Solar cell parameters under STC

Solar cell	PSC 1	PSC 2	PSC 3	c-Si SC
Area [cm^2]	1.12	1.06	1.10	0.81
J_{SC} [mA/cm^2]	22.89	22.30	20.52	38.99
V_{OC} [V]	1.11	1.13	1.12	0.60
FF [%]	78.4	78.6	79.4	78.5
PCE [%]	19.9	19.9	18.3	18.4

The PSCs have a p-i-n architecture fabricated in a superstrate configuration consisting of layers: Glass, ITO, MeO-2PACz, perovskite, C_{60}, SnO_2, Cu. A layer of MeO-2PACz was spin-coated (3000 rpm for 30 s) on a glass substrate with pre-coated and patterned ITO, and annealed at 100°C for 10 min. The perovskite layer (FACs type with the chemical formula $FA_{0.83}Cs_{0.17}Pb(I_{0.83}Br_{0.17})_3$) was spin-coated (4000 rpm for 35 s, 0.4 ml ethyl acetate applied after 25 s) and annealed at 100 °C for 30 min. Layers of C_{60} and copper were made by evaporation. The SnO_2 layer was deposited by atomic layer deposition. The thicknesses of the individual layers are about 500 nm for perovskite, 21 nm for C_{60}, 20 nm for SnO_2 and 100 nm for copper. Monolayer (MeO-2PACz) is in the sub-nanometre range. The back and front contact are connected with two copper stripes. The device is edge sealed with a second layer of glass and two-component glue. The device active areas are approximately 1 cm^2 and their PCE under STC approximately 19 %.

The c-Si SC is a back surface field (BSF) p-type silicon SC. Its nominal power conversion efficiency (PCE) is 20.1 %. To be comparable with PSCs, only 1 cm^2 square was cut out of a full sized cell with a laser. Taking into account the cell's bus-bar area, the final aperture area is slightly smaller at 0.81 cm^2. The influence of recombination centers along the laser cut lines reduced the cell PCE to 18.4 % under STC.

2.4 Monitoring system details

The core of the monitoring system is an STM32G483 microcontroller, which performs all the measurements and an MPPT. An ESP8266 Wi-Fi module is responsible for synchronization with a time server, triggering measurements at appropriate times and sending the measured results to a remote database. *I-V* scans are performed at full hours and 30 minutes past the full hour and all other measurements are performed in 5-minute intervals starting at the full hour in order to be synchronized with all other measurements performed at our outdoor test site, e.g. global horizontal irradiance, which may provide additional insights into the obtained results.

The integrated MPPT uses a perturb and observe tracking algorithm. Cell current and voltage are measured every 100 ms and the operating point of the SC is adjusted with the same frequency. The minimum, maximum, average value, standard deviation, number of measurements and duration of the last measurement interval are calculated and sent to the database every 5 minutes. Although the system has a permanent and independent power supply, the MPPT algorithm is configured to pause its operation when the cell current falls below 30 μA. This behavior was implemented to simulate a situation when the energy cost of running a MPPT algorithm would exceed its energy gains.

I-V scans start at open-circuit voltage, progress towards short-circuit, and when-short circuit is reached the direction is reversed to obtain another *I-V* scan in the opposite direction. Voltage steps are 20 mV and the time delay between the steps is 60 ms. At each step of the *I-V* scan, a measurement of the indoor irradiance is also performed and stored with each *I-V* scan data point. The addition of irradiance data enables later identification of anomalous events, like a sudden change in irradiance during an *I-V* scan, and the filtering out of such *I-V* scans. Such events are not unlikely in an uncontrolled environment.

Irradiance measurements are performed with a silicon photodiode SFH2440L with an IR-cut filter. A silicon photodiode with an IR-cut filter was selected because its EQE matches the EQE of a perovskite solar cell better than the EQE of a silicon photodiode without a filter. The photodiode for irradiance measurements was calibrated using a Newport Oriel Class A solar simulator and a calibrated reference c-Si SC. However, because the spectrum of the artificial light differs significantly from the solar simulator spectrum, which is a good match for natural light spectrum, the irradiance measurements in the presence of artificial light aren't very accurate and serve more as an indication than an actual measurement.

Temperature and humidity measurements are performed by a SHT21 digital sensor every 200 ms. Ambient light color is measured by an APDS9151 digital sensor once per second. The APDS9151 sensor measures red, green, blue and infrared components of light. Every 5 minutes, all the measurements are aggregated into a single datapoint and stored into a database, in the same way as the SC current and voltage measurements.

Irradiance in different orientations is measured only once every 5 minutes. Irradiance is measured by 9 additional photodiodes SFH2440L mounted onto a 3D printed support structure to achieve desired orientations.

3 RESULTS

Measurements started in August 2022. Fig. 3 shows the measured irradiance and P_{MPP} and V_{MPP} for a PSC and a c-Si SC during four days representing typical seasons of the year – both solstices and both equinoxes. On three of the four days, the cells are producing power, but on the winter solstice, irradiance is so low that the MPPT went into standby. This is to simulate a realistic indoor PV powered device, which would disable the MPPT when power gains would be smaller than the energy cost of running the MPPT circuitry. When the MPPT is active, the PSCs show much higher power production than the c-Si SC. The V_{MPP} of the PSC 1 rises more throughout the day than changes in irradiance would warrant, indicating the presence of a light-soaking effect.

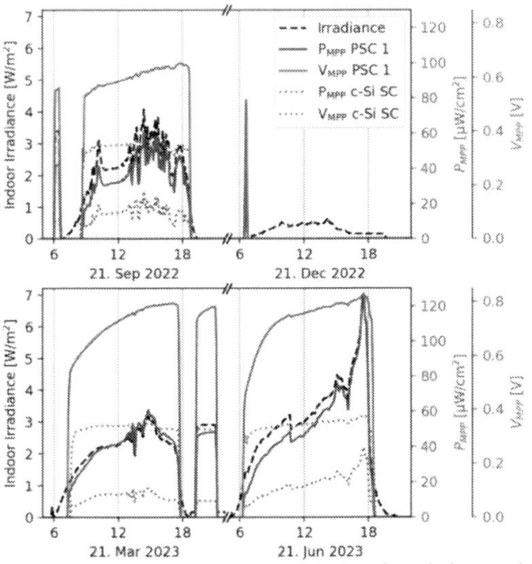

Figure 3: Solar cell performance during the solstices and equinoxes. The black dashed line shows irradiance, red lines show P_{MPP} and blue lines V_{MPP}. Solid lines are used for PSC and dotted lines for c-Si SC

The light-soaking effect is further confirmed by Fig. 4 showing the *I-V* curves recorded on 21 June 2023. It shows the *I-V* curves shifting ever more to the right during the day with the increasing irradiance, but in the evening, when the irradiance drops, the *I-V* curves remain shifted to the right, exhibiting high open circuit voltage V_{OC} and V_{MPP} even at very modest irradiance. This is important because it confirms that the light soaking effect plays an important role even at the very low irradiance levels typical for indoor conditions.

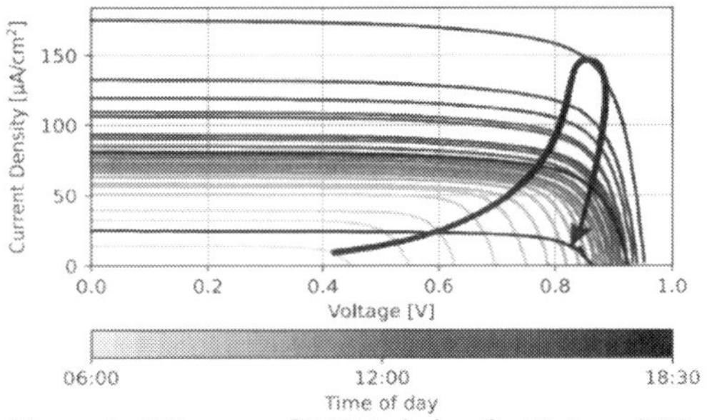

Figure 4: *I-V* curves of PSC 1 during the 21 June 2023. The brightness of each curve indicates when it was recorded, from bright in the morning to dark in the evening. The arrow indicates how the MPP was changing throughout the day

Fig. 5 shows the measured color components of light recorded on 27 December 2022 when the office lights were switched on for about half of the day. The spectrum of artificial lights is usually almost entirely within human visual range and a good match for wide bandgap solar cells like perovskite solar cells. Measuring color components of light could be a good way to detect when office lights are switched on.

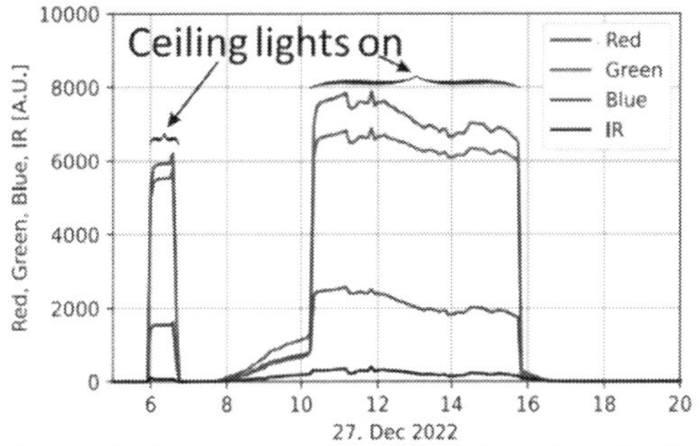

Figure 5: Color components of light throughout the 27 December 2022

Histograms of temperature and humidity in Fig. 6 show that the temperature remained between 22 and 34 °C and relative humidity between 20 and 65 %. This is a very mild environment for any solar cell, which is particularly important for more sensitive technologies like perovskite solar cells, allowing them to operate for a long time despite their inherent weakness.

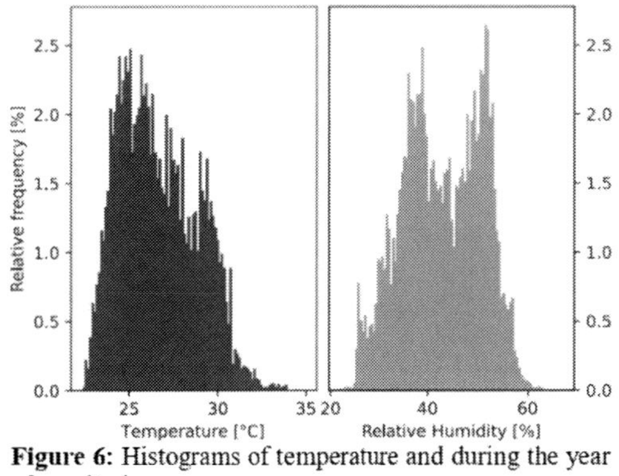

Figure 6: Histograms of temperature and during the year of monitoring

Fig. 7 shows monthly energy yields of all four solar cells and their monthly irradiation. At the beginning of the measurements, the three PSCs exhibit similar energy yields, which are between 2 and 2.5 times higher than that of the c-Si SC. In December, the energy yield of the best PSC is approximately 4 times higher than that of the c-Si SC, most probably because the fill factor of the c-Si SC falls more than that of the PSCs at low irradiance because its V_{OC} is lower in the first place. At this time, however, it is evident that the performance of PSCs 2 and 3 has already started dropping due to degradation. By the end of the first year, the performance of PSC 2 has already dropped below the performance of the c-Si SC. On the other hand, PSC 1 barely started showing signs of degradation by the end of the year and keeps well ahead of the c-Si SC, demonstrating that after ironing out some production difficulties, PSCs could already be a viable option for some applications.

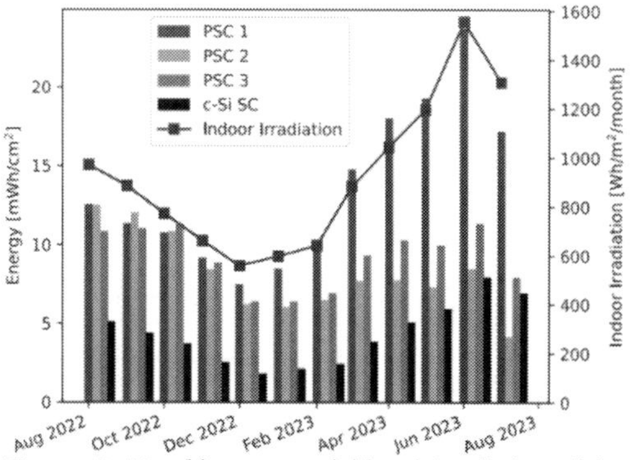

Figure 7: Monthly energy yield and irradiation of the monitored solar cells

4 CONCLUSIONS

We have monitored the performance of three PSCs and one c-Si SC in realistic uncontrolled indoor conditions for one year. During the first year of operation, the best PSC outperformed the c-Si SC by more than three times. During this time, we observed that even in the low irradiance typical of indoor environments, the light-soaking effect plays an important role in PSC PCE. However, at irradiance levels typical for indoor conditions, it takes much longer to reach saturation than under STC. The very forgiving environmental conditions found in indoor environments were also very beneficial to the PSCs enabling them to operate for much longer than in harsher outdoor environments. The best PSC was able to operate for nearly a whole year without degradation, with the first signs of possible degradation appearing only during the last month of the first year of operation.

ACKNOWLEDGEMENTS

We would like to thank D. Uršič, G. Matič and M. Jankovec for many fruitful discussions. The Research was funded by the Slovenian Research and Innovation Agency, program P2-0415.

REFERENCES

[1] K.-L. Wang, Y.-H. Zhou, Y.-H. Lou, and Z.-K. Wang, "Perovskite indoor photovoltaics: opportunity and challenges," *Chemical Science*, vol. 12, no. 36, pp. 11936–11954, 2021, doi: 10.1039/D1SC03251H.

[2] "State of IoT 2022: Number of connected IoT devices growing 18% to 14.4 billion globally," *IoT Analytics*, May 18, 2022. https://iot-analytics.com/number-connected-iot-devices/ (accessed Mar. 28, 2023).

[3] "Scaling Considerations for Organic Photovoltaics for Indoor Applications - Burwell - 2022 - Solar RRL - Wiley Online Library," 2022. https://onlinelibrary.wiley.com/doi/10.1002/solr.202200315 (accessed Mar. 27, 2023).

[4] G. Apostolou, A. Reinders, and M. Verwaal, "Comparison of the indoor performance of 12 commercial PV products by a simple model," *Energy Science & Engineering*, vol. 4, no. 1, pp. 69–85, 2016, doi: 10.1002/ese3.110.

[5] I. Raifuku, Y. Ishikawa, S. Ito, and Y. Uraoka, "Characteristics of Perovskite Solar Cells under Low-Illuminance Conditions," *J. Phys. Chem. C*, vol. 120, no. 34, pp. 18986–18990, Sep. 2016, doi: 10.1021/acs.jpcc.6b05298.

[6] B. Politi *et al.*, "Practical PV energy harvesting under real indoor lighting conditions," *Solar Energy*, vol. 224, pp. 3–9, Aug. 2021, doi: 10.1016/j.solener.2021.05.084.

[7] J. Dagar, S. Castro-Hermosa, G. Lucarelli, F. Cacialli, and T. M. Brown, "Highly efficient perovskite solar cells for light harvesting under indoor illumination via solution processed SnO2/MgO composite electron transport layers," *Nano Energy*, vol. 49, pp. 290–299, Jul. 2018, doi: 10.1016/j.nanoen.2018.04.027.

[8] J. W. Lim *et al.*, "Unprecedentedly high indoor performance (efficiency > 34 %) of perovskite photovoltaics with controlled bromine doping," *Nano Energy*, vol. 75, p. 104984, Sep. 2020, doi: 10.1016/j.nanoen.2020.104984.

[9] R. Cheng *et al.*, "Tailoring Triple-Anion Perovskite Material for Indoor Light Harvesting with Restrained Halide Segregation and Record High Efficiency Beyond 36%," *Advanced Energy Materials*, vol. 9, no. 38, p. 1901980, 2019, doi: 10.1002/aenm.201901980.

40th European Photovoltaic Solar Energy Conference and Exhibition

This presentation was selected by the Sc. Committee of the EU PVSEC 2023 for submission of a full paper to one of the EU PVSEC's collaborating peer-reviewed journals.

EFFECT OF ANTISOLVENTS IN AMBIENT AIR ON PEROVSKITE SOLAR CELLS

Takayuki NEGAMI*, Kousei SAWAKI, Abdurashid MAVLONOV, Yu KAWANO, Takashi MINEMOTO
Ritsumeikan University
1-1-1 Nojihigashi, Kusatsu, Shiga 525-8577
*Phone: +81-77-561-3065, e-mail: tnt24345@fc.ritsumei.ac.jp

ABSTRACT: We have investigated the effect of antisolvents on the performance of $MAPbI_3$ perovskite solar cells fabricated in ambient air for low-cost production of modules. We selected chlorobenzene, toluene, diethyl ether, methyl acetate, ethyl acetate, propyl acetate and butyl acetate as antisolvents. The cells using EA, PA, BA and DE showed higher efficiencies than those using CB, TL and MA. The decrease in cell performance is due to current shunting and carrier transport inhibition caused by many pinholes and aggregation of grains in the films prepared using CB and TL. Residues (presumably PbI_2) on the film surface using MA degraded the cell performance. Among the antisolvents of EA, PA, BA and DE, the cells showed almost the same Voc and Jsc, however the cell using EA showed the highest FF, resulting in the highest efficiency of 15.4 %. It is noteworthy that the cell using BA with a high flash point of 22°C showed an efficiency of 15.0%, which is comparable to the cell using EA. These results can contribute to facilitate low cost production of perovskite PV modules.
Keywords: perovskite solar cell, antisolvent, ambient air, $MAPbI_3$ film

1 INTRODUCTION

Perovskite PV modules have entered early production in recent years. Expanding the market for perovskite modules requires low-cost production and high efficiency. High efficiency perovskite solar cells are fabricated using antisolvent treatments in the laboratory under inert gas or dry air with very low dew points [1,2]. For large area module production, the equipment cost of the process in inert gas and the maintenance cost of dry room operation are important issues to reduce production costs [3]. However, the antisolvent treatment is very sensitive to moisture in the air. Meanwhile, several studies have reported that ethyl acetate as an antisolvent improves the film uniformity and cell performance of perovskite films prepared under high relative humidity [4-7]. Ethly acetate is one of the antisolvent candidate for low cost production of the perovskite PV modules. However, the low flash point -3.0°C of ethyl acetate is a serious issue for mass production of the modules. Domestic laws may limit the amount of volatile solvents stored in a facility. In this contribution, we report that different organic compounds mainly acetic acid-based solvents were used as antisolvents to prepare $MAPbI_3$ films in ambient air. We also discuss a useful antisolvent for manufacturing the perovskite PV modules on cell performance and process compatibility.

2 EXPERIMENTS

We fabricated the perovskite solar cells with a structure of glass / FTO / compact TiO_2 (c-TiO_2) / meso-porous TiO_2 (mp-TiO_2) / $MAPbI_3$ / Spiro-OMeTAD / Au. The fabrication process is described below.

2.1 Solution preparation

A precursor solution for the c-TiO_2 layer was prepared by stirring 0.3 ml titanium diisopropoxide bis(acetyl-acetonate) solution (75 wt.% in isopropanol) and 8.0 ml ethanol for 30 min. The mp-TiO_2 precursor solution was prepared by diluting 226.8 mg of titanium paste with 1.0 ml ethanol and then stirring for 1 h. To prepare $MAPbI_3$ perovskite precursor solution, 1082 mg of PbI_2 and 310.5mg of methylamine hydroiodide ($CH_3NH_2 \cdot HI$) were dissolved in 1.45 ml of DMF and DMSO at 3:1 v/v and stirred for 1 hour at 40°C. Chlorobenzene (CB), toluene (TL), diethyl ether (DE), ethyl acetate (EA), methyl acetate (MA), propyl acetate (PA) and butyl acetate (BA) were used for antisolvent treatment. The hole transport material (HTM) solution was synthesized by mixing 75 mg of spiro-OMeTAD, 28.8 μl of 4-tert-butylpyridine (tBP), and 17.5 μl of bis(trifluoromethane)-sulfonimide lithium (LiTFSI) in 1 ml of CB. The resulting solutions of $MAPbI_3$ and HTM were filtered using a 0.45 μm polytetrafluoroethylene (PTFE) syringe filter.

2.2 Cell fabrication

All layers, except Au grid contact, were prepared in an ambient air atmosphere at room temperature (RT, 25-30 °C) under relative humidity (RH) of 30-40%. A 50-nm-thick c-TiO_2 layer was deposited on the FTO/glass substrates using a spray pyrolysis method at a substrate temperature of 400°C. Subsequently, an mp-TiO_2 layer with a thickness of about 200 nm was spin-coated onto the c-TiO_2/FTO/glass substrate at 3000 rpm for 40sec. After deposition, the mp-TiO_2 layer was immediately dried at 125 °C for 5 min, followed by post annealing at 400°C for 60 min. A 100μl of $MAPbI_3$ solution was dropped and then spin-coated at 4000 rpm for 30 sec, after which the sample was preheated at 65°C for 1 min and annealed at 120°C for 10 min. For antisolvent treatment, a 200 μl amount of antisolvent compound was dropped 20 sec after the start time of the spin coating process. A 100 nm thick Spiro-OMeTAD HTM layer was then deposited by spin coating at 3000 rpm for 40 sec. Finally, an 80 nm thick Au electrode was deposited by thermal vacuum evaporation.

Here, the samples for XRD, PL, etc. characterization consist of a glass/FTO/$MAPbI_3$ structure without TiO_2 and Spiro-OMeTAD layers. The preparation method of $MAPbI_3$ films is the same as described above.

3 RESULTS AND DISCUSSION

3.1 Characterization of perovskite films

Figure 1 shows optical microscopy images of the $MAPbI_3$ films prepared using several antisolvents. Mirror-like reflection is observed in the $MAPbI_3$ films prepared

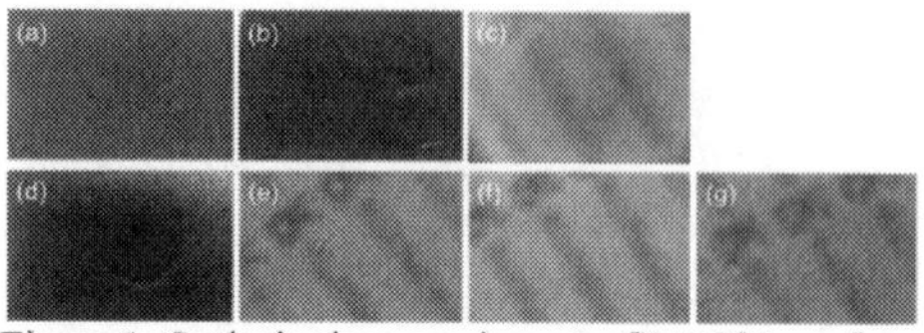

Figure 1: Optical microscoy images of $MAPbI_3$ sruface prepared using CB (a), TL (b), DE (c), MA (d), EA (e), PA (f), BA (g), respectively.

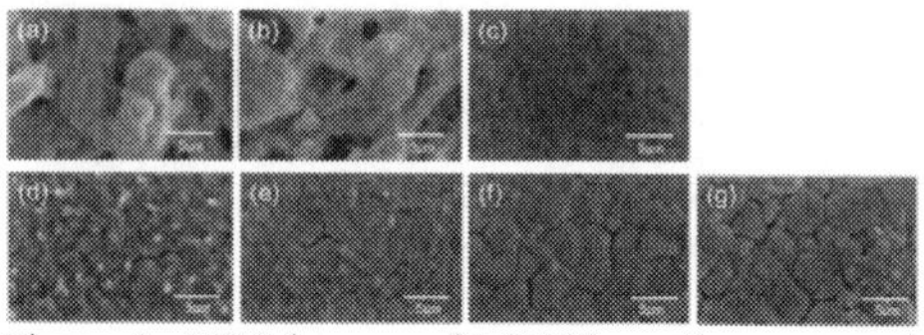

Figure 2: SEM images of $MAPbI_3$ sruface prepared using CB (a), TL (b), DE (c), MA (d), EA (e), PA (f), BA (g), respectively.

using DE, EA, PA and BA. The high reflectance of the film indicates that the layer is dense and free of pinholes. In contrast, $MAPbI_3$ films prepared using CB, TL, and MA exhibit haze due to light scattering by pinholes and rougher perovskite films.

Figure 2 shows the SEM images of the $MAPbI_3$ films prepared using several antisolvents. Pinhole-free and smooth surface morphologies were observed for $MAPbI_3$ films prepared using DE, EA, PA and BA antisolvents. The film prepared using MA has residues on their surface, indicated by small white spots. The residue seems to be PbI_2. Meanwhile, rough surface morphologies and pinholes were observed in the films prepared using CB and TL. The morphology appears similar to films prepared without using antisolvent in inert gas or dry air. This suggests that moisture inhibits continuous nucleation in the process using CB and TL as antisolvents [8,9].

Figure 3 shows XRD patterns of the $MAPbI_3$ films. All diffraction peaks can be indexed based on the tetragonal unit cell of $MAPbI_3$, except the peaks diffracted from PbI_2 and FTO. No other impurity phases are observed regardless of the antisolvent. The film prepared using EA was preferably oriented with (110) plane. The peak intensity ratio between 001 of PbI_2 and 110 of $MAPbI_3$ for the films prepared using DE and MA is higher than the other films. More unreacted PbI_2 would remain in the film using DE or MA as antisolvents.

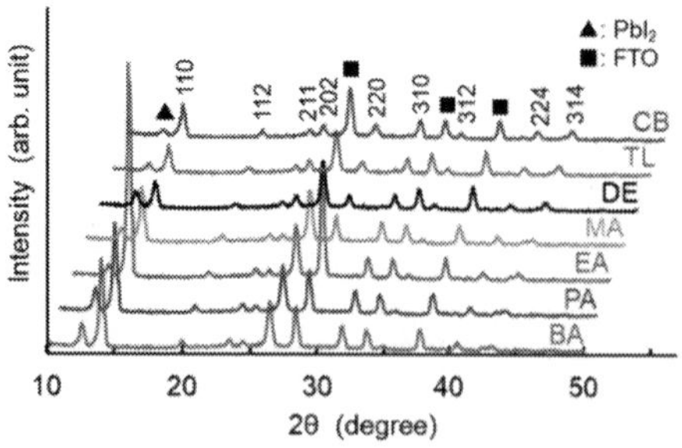

Figure 3: XED patterns of $MAPbI_3$ films
• and ■ peaks indicate the diffraction of PbI_2 (001) and FTO.

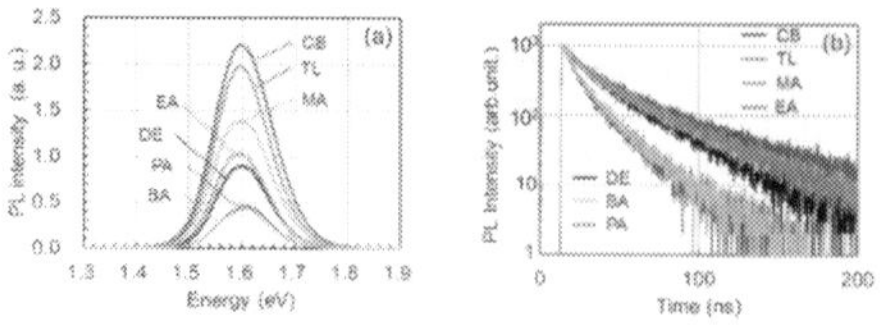

Figure 4 : PL spctra (a) and TR-PL decay curvs (b) of $MAPbI_3$ films prepared usign several antisolvnts

The steady-state photoluminescence (PL) and time-resolved PL (TRPL) measurements performed by 532nm laser incidence on the $MAPbI_3$ films. The PL spectra and TRPL decay curves are shown in Fig. 4 (a) and (b), respectively. The films prepared using CB and TL show higher PL intensity than those using DE and acetic acid-based solvents. The PL intensity of the $MAPbI_3$ film using CB and TL is enhanced by the accumulation of excited carriers within the film due to imperfect contact such as voids between the $MAPbI_3$ film and the FTO film [10]. The higher PL intensity of the film using MA compared to the films using other acetic acid-based solvents is also due to inhibition of carrier extraction by residues. The PL intensity of the films using DE, EA, PA and BA decreases due to efficient carrier transfer from the $MAPbI_3$ film to the FTO film. The TRPL decay curves shown in Fig. 4 (b) supports the results of the PL spectra. The long lifetime of the films using CB, TL and MA is due to the accumulation of excited carrier within the films. Among EA, PA and BA, the film using EA had the longest PL lifetime. This is attributed to the lower defect density of the film using EA than those using PA or BA. The $MAPbI_3$ film prepared using EA is expected to be suitable as an absorber for solar cells based on the results of SEM images and PL spectra.

3.2 Cell performance

Figure 5 shows the cell performance of the $MAPbI_3$ cells. Black and red circles indicate the performance of the I-V curves measured with forward and reverse voltage sweep, respectively. There is no significant difference in I-V hysteresis for antisolvents. All performances of the cells prepared using DE, EA, PA and BA are higher than those using CB, TL and MA. This is mainly due to the surface morphology of the $MAPbI_3$ films and contacts between TiO_2 and $MAPbI_3$ films. As shown in Fig. 2, films prepared using CB and TL have many pinholes and aggregation of crystal grains, which induce current shunting and carrier transport inhibition, resulting in a decrease in cell performance. The residues (presumably PbI_2) on the surface of the film using MA degraded the cell performances. Among the antisolvents of EA, PA, BA and DE, the cells showed almost the same Voc and Jsc, however the cell using EA showed the highest FF, resulting in the highest efficiency of 15.4 %.

The external quantum efficiency (EQE) of the cells is shown in Fig. 6. Where the cell spectra with DE and PA have been omitted for clarity of the figure. The cells using EA and BA have higher EQE than other films. The $MAPbI_3$ films using EA and BA has sufficient carrier diffusion length for light absorption. In contrast, the EQE of the films using CB and TL decreases with increasing wavelength. This indicates that carriers excited at a deep depth by long wavelengths are not efficiently extracted. Cells with MA showed lower EQE than other cells at all wavelengths. Residual PbI_2 is one cause of low EQE, but there may be other factors as well. Further analysis is required.

10.4229/EUPVSEC2023/2BV.2.27
020109-002

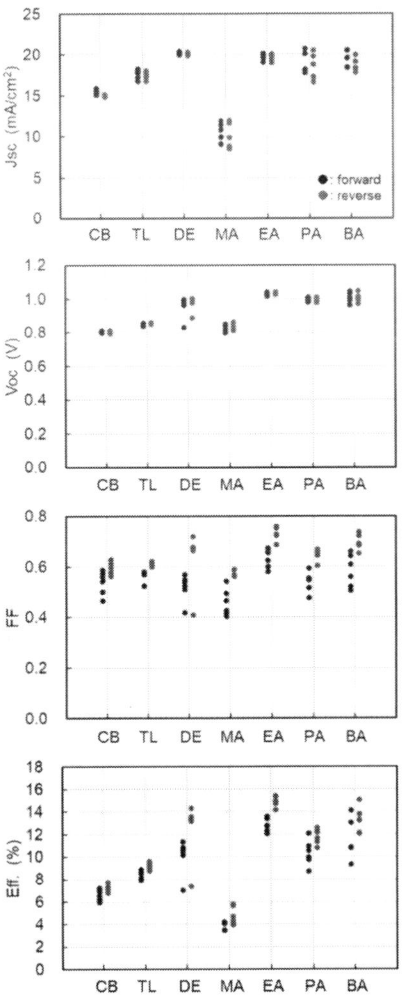

Figure 5: Performance of cells with $MAPbI_3$ films prepared using several antisolvents.
Black and red circles indicate the performance of the I-V curves measured with forward and reverse voltage sweep, respectively.

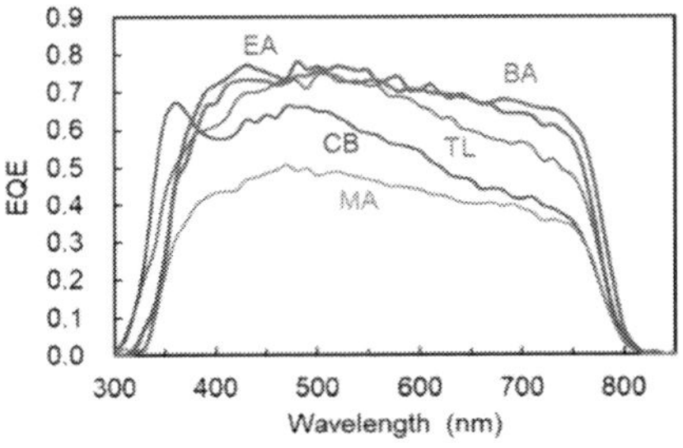

Figure 6: External quantum efficiency of $MAPbI_3$ cells

Although anti-solvent treatment is a useful method to achieve high efficiency, the flash point of antisolvents that limits storage within facilities is a key issue for module manufacturing. The flash points of DE, EA, PA, and BA, which showed high efficiency in ambient air processes, are -45, -3, 10, and 22°C, respectively. It is worth noting that the cell using BA with a higher flash point of 22°C showed an efficiency of 15.0%, which is comparable to the cell using EA with a lower flash point of -3°C.

4 CONCLUSIONS

Antisolvents suitable for production in ambient air have been studied from the perspective of module manufacturing. $MAPbI_3$ films were prepared using chlorobenzene (CB), toluene (TL), diethyl ether (DE), methyl acetate (MA), ethyl acetate (EA), propyl acetate (PA) and butyl acetate (BA) as antisolvents in ambient air. The films prepared using DE, EA, PA, and BA have dense and smooth surfaces, and reduces the PL intensity and TR-PL lifetime due to the effective extraction of excited carriers from $MAPbI_3$ to the FTO film. In contrast, the films using CB, TL have rough surfaces results from pinholes and grain aggregation, and enhanced PL intensity and TR-PL lifetime due to accumulation of excited carriers.

The cells using EA, PA, BA and DE showed higher efficiencies than those using CB, TL and MA. The decrease in cell performance is due to current shunting and carrier transport inhibition caused by many pinholes and aggregation of grains in the films prepared using CB and TL. Residues (presumably PbI_2) on the film surface using MA degraded the cell performance. Among the antisolvents of EA, PA, BA and DE, the cells showed almost the same Voc and Jsc, however the cell using EA showed the highest FF, resulting in the highest efficiency of 15.4 %. It is noteworthy that the cell using BA with a high flash point of 22°C showed an efficiency of 15.0%, which is comparable to the cell using EA.

These results can contribute to facilitate low cost production of perovskite PV modules. Further development of antisolvents suitable for processes in ambient air is required to improve efficiency.

ACKNOWLEGEMENTS

This work is partly supported by NEDO (New Energy and Industrial Technology Development Organization).

REFERENCES

[1] N. -J. Jeon et al., Nat. Mater. 13 (2014) 897.
[2] M. Xiao et al., Angew. Chem. Int. Ed. 53 (2014) 9898.
[3] S. Ahmed et al., J. Power Sources 326 (2016) 490.
[4] J. Troughton et al., Nano Energy 39 (2017) 60.
[5] T. Singh, T. Miyasaka, Adv. Energy Mater. 8 (2018) 1700677.
[6] Z. Zhang et al., J. Solid State Chem.274 (2019) 199.
[7] D.Ahmed et al., J. Mater. Sci. 56 (2021) 15205.
[8] A. R. Pascoe et al., Sci. China Mater. 60 (2017) 617.
[9] V. K. Lamer, R. H. Dinegar, J. Am. Chem. Soc. 72 (1950) 4847.
[10] Y. Gao et al., Superlattices and Microstructures 113 (2018) 761.

NUMERICAL INVESTIGATION OF LEAD FREE HALIDE PEROVSKITE WITH ALL-INORGANIC TRANSPORT LAYER

Sujit Kumar[1], V.P.Vinturaj[1], Vivek Garg[2], Sushil Kumar Pandey[1*]
[1]Department of Electronics and Communication Engineering, National Institute of Technology Karnataka, Surathkal,575025, India
[2]Department of Electronics and Communication Engineering, Sardar Vallabhbhai National Institute of Technology, Surat, 395007, India
*Author for correspondence (skpandey@nitk.edu.in)

ABSTRACT: The extensive commercial implementation of lead-based perovskite solar cells faces a major problem of stability and toxicity. Halide based Titanium perovskite having an energy band gap of 1.8 eV in longer and controlled carrier-diffusion wavelengths of more than 100 nm, appropriate energy levels, and exceptional intrinsic and environmental stability can be the best alternative for all-in-organic lead-free perovskite solar cell. Herein, we present an n-i-p all-in-organic, lead-free, stable, higher power conversion efficiency, Cs_2TiBr_6 halide-based perovskite solar cell. Scrutinizing features like hole/charge transport layers and their optimum thickness, the optimum thickness of absorber, defect densities of the absorber and interface. We introduced graphene nanoflakes and examined their effect on power conversion efficiency for a high-performance perovskite solar cell with a power conversion efficiency of 13.47 % and external quantum efficiency of 78.32 %, suggesting Cs_2TiBr_6 can play a pivotal part as an absorbing perovskite material for higher efficient all-in-organic perovskite photovoltaics (PV) technology.
Keywords: Perovskite solar cell, Cs_2TiBr_6, in-organic

1. INTRODUCTION

Energy is essential for the survival of humans in this world. Currently, projected energy demand is as high as 1GW/day [1]. It is harnessed from different renewable and non-renewable sources. However, conventional fossil-based energy sources are limited and leave behind high carbon footprints. To meet our ever-increasing energy needs, we moved towards sustainable renewable energy. One of these is photovoltaics (PV) dominated by solar cells. Jana et al. constructed the first third-generation perovskite solar cell (PSC) with a lead halide absorber that is both organic and inorganic layer [2], advancing the efficiency of perovskite solar cells from 3.8% to 25% all through the previous decade [3-7]. The benefit of this technology is its low manufacturing cost and simple fabrication. The problem with this organic-inorganic absorber layer is its toxicity, which is highly unstable because organic cations are present and reduced shelf life [8]. Due to the presence of moisture, oxygen and light electron charge layer (ETL) also becomes unstable in due course of time [9,10]. So, we are investigating an all-in-organic charge transport material. As previously reported by Chen et al., structure FTO/ TiO_2/Cs_2TiBr_6/P3HT achieved a 2.15% power conversion efficiency (PCE) [11], and Saif et al. structure FTO/SnO_2 /Cs_2TiBr_6/ MoO_3/Au reported 11.49 % PCE [12].

In this work, using SCAPS (Solar Cell Capacitance Simulator), we present a structured all-inorganic n-i-p PSC FTO/ TiO_2(GR 20%) /Cs_2TiBr_6 /MoO_3/Au. We analysed the effect of thickness on PCE. Also, we examined different interface and absorber defects effects on the performance parameters of PSC. We obtained a novel FTO/TiO_2(GR 20%)/Cs_2TiBr_6/MoO_3/Au configuration, analyzing performance parameters like Jsc, Voc, PCE, and FF.

2. METHODOLOGY

As depicted in Figure 1, an n-i-p planar PSC structure composed of various layers of TCO, ETL, Active absorber layer, HTL, and Back Contact is a linear n-i-p structured planar heterojunction. TCO is the transparent conducting oxide layer that acts as front contact, the n-type region is represented by the electron transport layer, the intrinsic (i) -region is represented by the perovskite active Layer, and the p-type region is represented by the hole Transport Layer. We have considered the work function of gold (Au) to be 5.1 eV and the work function of fluorine-doped Tin oxide (FTO) to be 4.4 eV. The provided information for each defect layer includes the following: neutral defect type, with a characteristic energy of 0.1 eV, an energetic distribution that follows a Gaussian profile, and an energy level of 0.6 eV different from the reference level. We primarily employed interface defects at the junctions between the absorber and the hole transport layer (HTL) as well as between the absorber and the electron transport layer (ETL). We conducted the simulation under the following conditions: a constant illumination of 1000 W/m^2 at AM 1.5 G, a continuous temperature of 300 K, a shunt resistance of 4200 Ω-cm^2 and a series resistance of 1 Ω-cm^2.

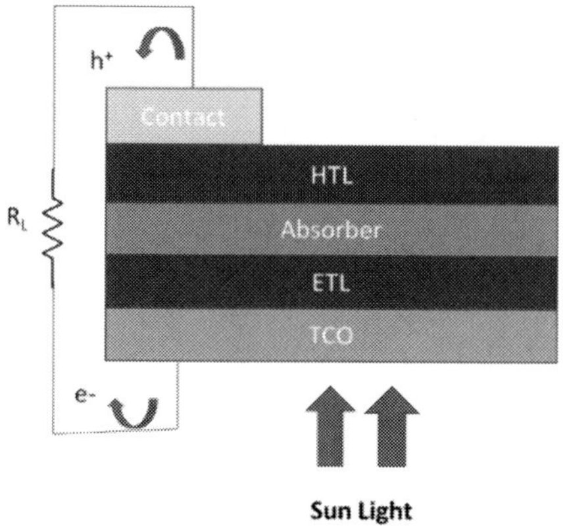

Figure 1: Schematic Perovskite Solar Cell

10.4229/EUPVSEC2023/2BV.2.28
020110-001

Table I: Input Electrical Parameters for PSC

	FTO	MoO_3	TiO_2 (20% GR)	Cs_2TiBr_6
d(nm)	500	30	50	200
χ (eV)	4	2.5	4.31	4.0
Eg (eV)	3.5	3	2.4	1.8
μn (cm^2 Vs)	20	25	350	.236
(ε_r)	9	12.5	7.8	10
Nc (cm^{-3})	2.2e18	2.2e18	1e19	6e19
N_D (cm^{-3})	1e19	0	1e19	3e19
Nv (cm^{-3})	1.8e19	1.8e19	1e19	2.14e19
μp(cm^2/Vs)	10	100	350	.171
Reference	[13]	[14]	[15]	[11]

3. RESULTS AND DISCUSSION

In this section, we have optimized the thicknesses of different layers in PSC for an efficient Novel all-inorganic perovskite solar cell. To accomplish this task, we first selected MoO_3 for the hole transport layer (HTL) and TiO_2(20 % GR) for the electron transport layer (ETL), whose electrical properties are listed in Table I. We optimized the performance of the device structure of FTO/TiO_2(20% GR)/Cs_2TiBr_6/ MoO_3/Au by determining the ideal thickness value for each layer to get an efficient, high-performance structure. We also optimized the absorber defect density and interface defect density of the proposed lead-free all-inorganic PSC.

3.1 Absorber Layer Optimization

In this section, we optimized the absorber layer thickness to achieve an optimum thickness since the absorber layer is in charge of producing electron-hole pairs., which directly influences the performance of the PSC. We simulated the absorber layer thickness from 1nm to 500 nm and observed that the PCE reaches its peak at 150 nm, and after 150nm thickness, it starts decreasing due to the recombination of electron-hole pairs, as shown in Figure 2. So, we fixed 150 nm thickness for the absorber.

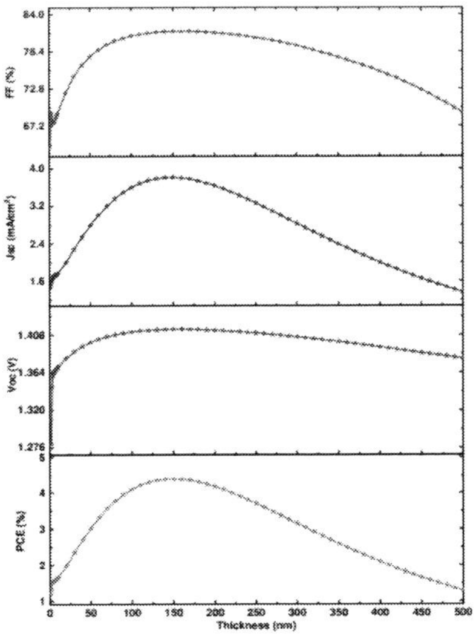

Figure 2: Thickness optimization of the Cs_2TiBr_6.

3.2 HTL Optimization

Following the thickness optimization of the active absorber layer, the thickness of the HTL was varied from 1 nm to 50 nm. We observed that as we increase the thickness, the PCE reaches the peak at 12 nm and then decreases, as shown in Figure 3. So, we kept 12 nm as the ideal thickness for the MoO_3.

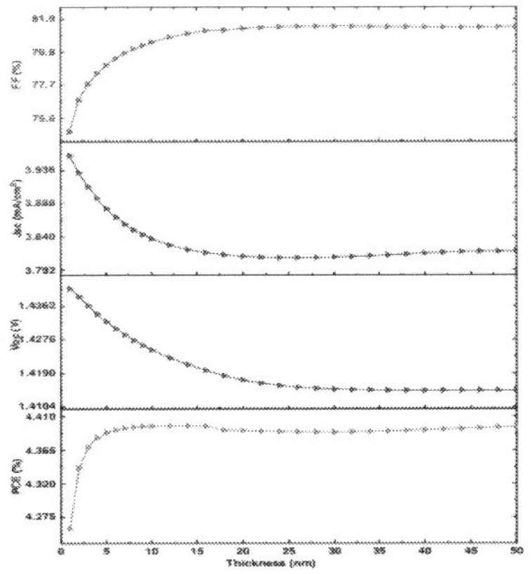

Figure 3: Thickness optimization of MoO_3

3.3 ETL Optimization

Following the optimization of the active absorber layer and hole transport layer, the thickness of ETL varied from 10 nm to 50 nm. We observed that as the thickness increases, the PCE decreases due to increased series resistance, as shown in Figure 4. So, we kept 10 nm as the ideal thickness for ETL with PCE 4.57 %.

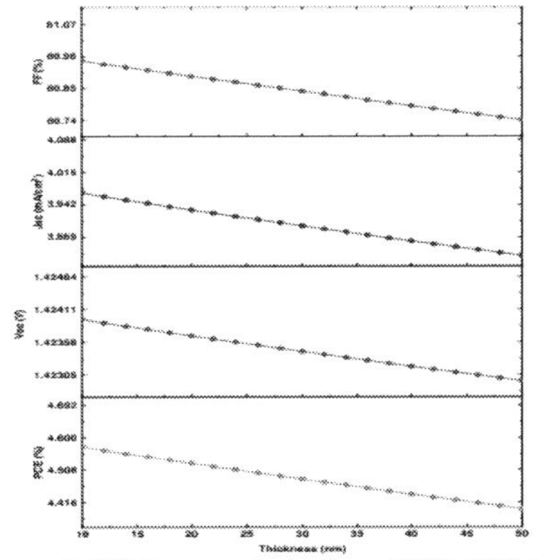

Figure 4: Thickness optimization of TiO_2(20%GR)

3.4 Absorber Defect optimization

We varied the defect density between $10^{12}cm^{-3}$ and $10^{16}cm^{-3}$ to investigate the impact of using optimized ETL, HTL, and the absorber layer. J-V characteristics for different absorber defect density is demonstrated in Figure 5. It can be observed from Figure 5 that when defect density is increased, PSC deteriorates from 8.81% at $10^{12}cm^{-3}$ to 1.54 % at $10^{16}cm^{-3}$ due to an increase in recombinational rate. So, we selected 10^{12} cm^{-3} as the optimum defect density for our device.

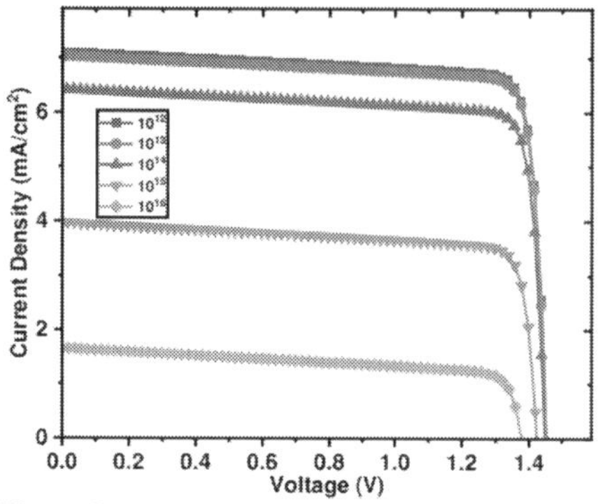

Figure 5: Absorber defect density J-V optimization

3.5 Interface defect density optimization

We varied the interface defect HTL/Absorber and Absorber/ETL interface defect from 10^8 cm^{-3} to10^{13} cm^{-3} range to investigate the interface defect impact. Figure 6 illustrates the J-V curves for various interface defects.
As we increased the interface defect density, the PCE decreased from 13.48 % at 10^8 cm^{-3} to 4.14 % at 10^{13} cm^{-3}. So, we selected 10^8 cm^{-3} as the ideal interface defect for the device.

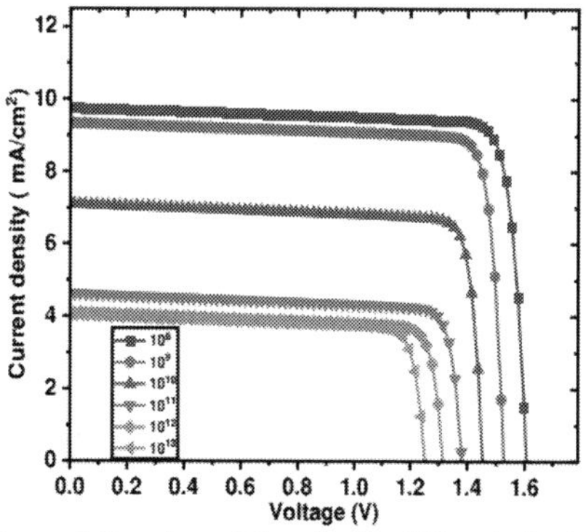

Figure 6: Interface defect density J-V optimization

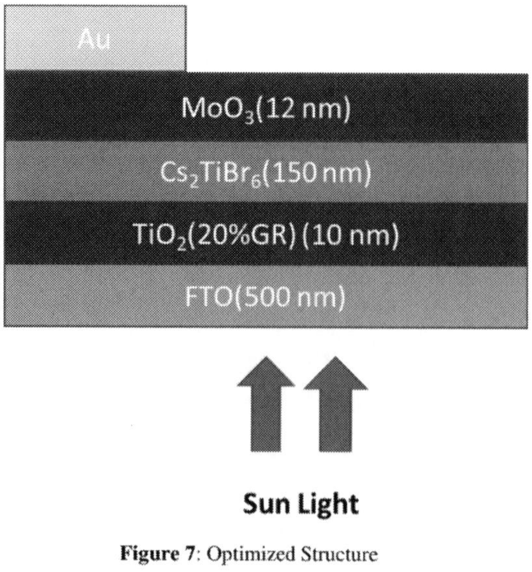

Figure 7: Optimized Structure

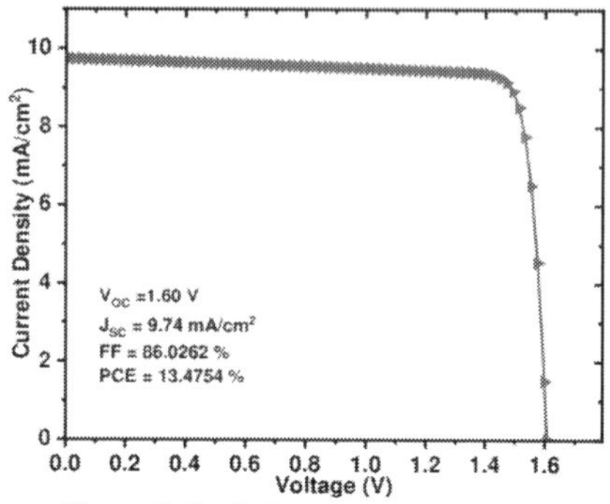

Figure 8: Optimized Structure J-V cure

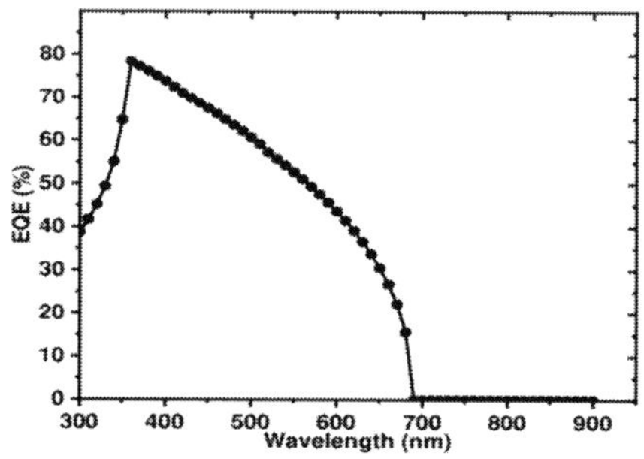

Figure 9: EQE CURVE

3.6 Optimized Device

After optimizing ETL, HTL, and the absorber layer, we identified a novel high-performance PSC with FTO/TiO_2(20%GR)/Cs_2TiBr_6/MoO_3/Au Configuration with PCE 13.48 %. Figure 7 depicts the rough novel structure's overview. Figure 8 depicts the characteristics of the J-V curve, and Figure 9 depicts the EQE curve of the optimized device.

4. CONCLUSION

In this paper, a highly efficient double halide-based perovskite solar cell is developed by analysing and optimising the impact of various layers on device performance using a unique combination of FTO/TiO_2 (20%GR)/ Cs_2TiBr_6 /MoO_3/Au. The suggested device has an exceptional PCE of 13.48%, Voc 1.6 V, Jsc 9.74 mA/cm^2, FF 86.03%, and an EQE of 78.32 % at visible spectrum, showing that our device makes the best use of solar irradiance.

5. REFERENCES

[1] N. Espinosa, M. Hösel, D. Angmo, F.C. Krebs, Energy Environ. Sci. 5 (2012) 5117–5132.
[2] A.K. Jena, A. Kulkarni, T. Miyasaka, Chem. Rev. (2019). 119, 5, 3036–3103.
[3] B. Xin, N. Alaal, S. Mitra, A. Subahi, Y. Pak, D. Almalawi, N. Alwadai, , Small 16 (43) (2020) 2004513.
[4] H. Tang, S. He, C. Peng, Nanoscale Res. Lett. 12 (1) (2017) 410.
[5] Y. Chen, L. Zhang, Y. Zhang, H. Gao, H. Yan, RSC

10.4229/EUPVSEC2023/2BV.2.28
020110-003

Adv. 8 (19) (2018) 10489–10508.
[6] J.O. Oji, N. Idusuyi, T.O. Aliu, M.O. Petinrin, O.A. Odejobi, A.R. Adetunji, Int. J. Energy Eng. 2 (2) (2012) 54–59.
[7] D. Vatansever, R.L. Hadimani, T. Shah, E. Siores Struct. 20 (5) (2011), 055019.
[8] F. Giustino, H.J. Snaith, ACS Energy Lett. 1,6 (2016), 1233–1240.
[9] B.V. Lotsch, Angewandte Chemie - International Edition, (2014).
[10] L. Calio, S. Kazim, M. Grätzel, S. Ahmad, Angewandte Chemie - International Edition, (2016).
[11] M. Chen, et al, Joule 2 (3) (2018) 558–570.
[12] S. Ahmed, F. Jannat, Md. A. K. Khan, and M. A. Alim, Optik, vol. 225, p. 165765, Jan. 2021.
[13] E. Karimi, S.M.B. Ghorashi, Optik (2017), vol 130,p 650-658.
[14] W. Li, W. Li, Y. Feng, C. Yang, Solar Energy (2019).
[15] M Dadashbeika , D Fathia, M Eskandari, Solar Energy, 207 (2020), 917-924.

INVESTIGATION OF ETL/ABSORBER HETEROJUNCTION FOR EFFICIENT CHARGE EXTRACTION FROM FORMAMIDINIUM TIN-BASED PEROVSKITE SOLAR CELL

Sakshi Gupta, Dr. Vivek Garg, Dr. Jignesh N.Sarvaiya
Optoelectronics2Application (O2A) Research Group, Department of Electronics Engineering
SVNIT, Surat-395007, India

ABSTRACT: Organic−inorganic hybrid lead (Pb) halide perovskite solar cells (PSCs) have attracted much attention in the past decades, and tremendous research efforts have rapidly propelled the power conversion efficiency (PCE) from 3.8 to 25.5%, rivaling that of crystalline silicon solar cells [1,2]. However, the use of toxic Pb elements poses a rigorous limit on their future commercialization. Thus, alternative elements in the same main group such as tin (Sn) and germanium (Ge), as well as other promising ones such as bismuth (Bi), have been explored to construct perovskite materials and pursue low toxicity as well as efficient solar cells [3]. The proposed structure ITO/PEDOT: PSS/$FASnI_3$/C_{60}/Ag has been simulated using SCAPS-1D software. The parameters from various experimental and theoretical reported works were used to simulate the proposed solar cell and it resulted in 7.30% power conversion efficiency (PCE). To investigate the impact of the active layer on the performance of the proposed device, the parameters of the active layer like thickness and defect density have been varied. Furthermore, parameters such as electron affinity, doping density, and thickness of ETL have also been varied to examine their influence on the output device parameters and optimized results deduced accordingly. Finally, the device simulation with the optimized parameters has resulted in improved output parameters with V_{OC} of 1.13V, J_{SC} of 26.69mA/cm^2, FF of 88.62%, and PCE of 26.77%, thus, paving the way for novel perovskite solar cells employing tin which are environmentally safe being lead-free and have efficiency comparable to the lead-based counterparts.
Keywords: Perovskite Solar Cells, Power Conversion Efficiency, electron affinity, doping density, etc.,

1 INTRODUCTION

Perovskite solar cells based on organometal lead halides emerged as the fourth generation of photovoltaics technology in a short period having high power conversion efficiency (PCE), thus, making it the strongest competitor of Silicon-based solar cells [4-7]. High-efficiency PSCs cluster around the region between 1.53 and 1.6 eV, which corresponds to the bandgaps for formamidinium (FA) and methylammonium (MA) lead iodide perovskites, $FAPbI_3$ and $MAPbI_3$, respectively. High photovoltage perovskite cells are enabled as bromide substitutes for iodide in the perovskite composition. Here, several devices based on the organometallic $FAPbI_3$ and $MAPbI_3$, and the inorganic $CsPbBr_3$ have already reported Voc values higher than 1.5 V with efficiencies above 10%. However, the use of toxic Pb elements poses a serious limit to their future commercialization. Thus, alternative elements of the same group such as tin (Sn) and germanium (Ge), as well as bismuth (Bi), have been explored to construct perovskite materials and obtain low-toxicity and efficient solar cells. Sn-based perovskites adopt the same general formula $ASnX_3$ [A = CH_3NH_3+, HC $(NH_2)_2^+$, or Cs^+; X = I^-, Br^-] as the Pb based analogs and exhibit similar superior photovoltaic properties such as superior light-absorbing ability, high carrier mobility, and small exciton binding energy. Although many researchers have used $CH_3NH_3SnI_3$ ($MASnI_3$) as an active layer in perovskite solar cells, highly efficient tin-based perovskite solar cells primarily employ Formamidinium tin iodide [$CH_4N_2SnI_3$ ($FASnI_3$)] as an active layer in tin-based perovskite solar cells. Milot-group has reported that $FASnI_3$ exhibits higher carrier mobility with low Auger recombination and strong radiating bi-molecular recombination rate constants as compared to $MASnI_3$ [8].
Krishnamoorthy et al. reported the research work on $FASnI_3$-based perovskite solar cells with a PCE of 1.41% in 2015 [9]. All possible techniques for boosting the efficiency of lead-based perovskite solar cells were employed in Sn-based counterparts.

2 DEVICE STRUCTURE AND ENERGY BAND DIAGRAM

The proposed structure and the band alignment of the device are shown in Fig. 1(a) and Fig. 1(b). In the present work we have investigated the simulation of $FaSnI_3$-based perovskite solar cell with PEDOT: PSS as Hole Transport Layer (HTL) and C_{60} as Electron Transport Layer (ETL) along with Indium-doped tin oxide (ITO) as back contact and Ag as front contact. The simulation study has been carried out through SCAPS (version 3.3.10) software under AM1.5G.

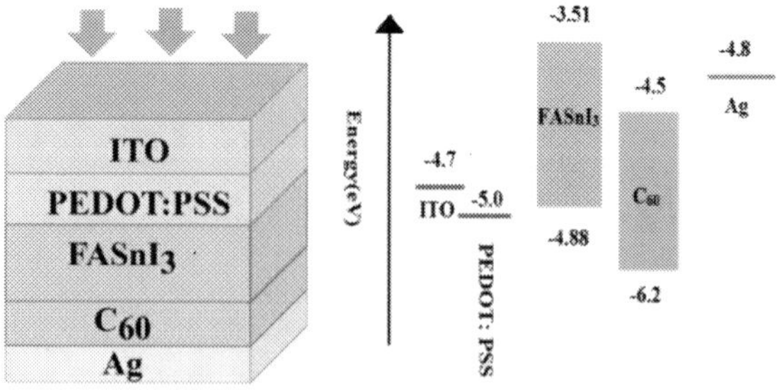

Figure 1: Proposed $FASnI_3$ perovskite a) device architecture and b) energy band alignment

Table 1: Numerical Parameters used in the Simulation.

Parameters	PEDOT: PSS	$FaSnI_3$ [10]	C_{60}
Thickness (nm)	150	220	50
Bandgap(eV)	2.2	1.41	1.7
Electron affinity (eV)	2.45	3.52	4.5

Dielectric Permittivity	3.00	8.20	4.5
Electron mobility (cm^2/V-s)	2	22	10^{-2}
Hole Mobility (cm^2/V-s)	2	22	10^{-5}
Shallow Donor Density N_D (cm^{-3})	-	-	2 x 10^{18}
Shallow Acceptor Density N_A (cm^{-3})	10^{17}	7 x 10^{16}	-

Table II: Defects of $FaSnI_3/C_{60}$ Interface

Parameters	$FaSnI_3/C_{60}$ interface
Defect Type	Neutral
Capture cross section for electrons and holes (cm^{-2})	10^{-18}
Energetic Distribution	Single
Energy with respect to Reference (eV)	0.700
Total density (integrated over all energies) ($1/cm^2$)	1×10^{10}

3 RESULTS AND DISCUSSIONS

Simulation of the device structure has been carried out at illumination of AM 1.5G spectrum. After performing the simulation following Tables I and II, the solar cell output parameters have been obtained as: V_{OC} = 0.55V, J_{SC}= 22.08 mA/cm^2, FF = 59.91%, and PCE = 7.30%. The simulation results obtained have been compared with the available experimental and theoretical studies of $FASnI_3$ perovskite-based solar cells are shown in Table III.

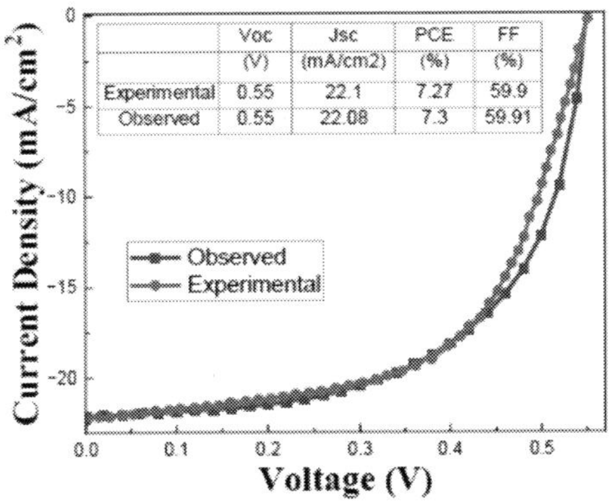

Figure 2: J-V characteristics of simulated and experimental baseline model of $FASnI_3$ device

Table III: Experimental and Simulation Comparison Results

Parameters	Experimental [10]	Simulation
Voc(V)	0.55	0.55
Jsc(mA/cm^2)	22.1	22.08
FF (%)	59.9	59.91
PCE (%)	7.27	7.3

As evident from Table III, the simulated output parameters are in close agreement with the previously reported experimental works. The J-V curve of the $FASnI_3$ perovskite solar cell with experimental and simulation results is shown in Figure. 2.

3.1 Impact of Thickness of the Active Layer

The thickness of the active layer is an important factor for determining the performance of solar cells so its impact on the output parameters of solar cells has been studied through simulation. The active layer thickness is varied from 100nm to 1μm in steps of 100nm and its impact on other parameters has been observed. It has been found that V_{OC} attains the maximum value of 0.674V at the thickness of 200nm and after that, it starts decreasing as can be seen in Fig. 3(a).

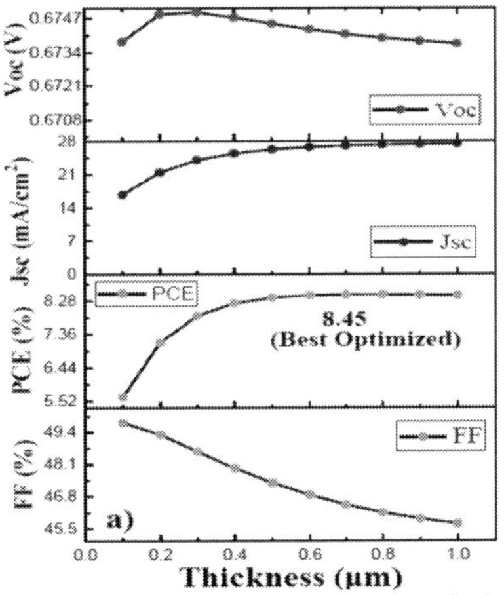

Figure 3 a): Impact of absorber thickness on device parameters

J_{SC} increases significantly with the increment in the value of thickness and attains the maximum value of 27.4 mA/cm^2 at the thickness of 1000 nm as shown in Fig. 3(a). FF has been seen decreasing with an increase in thickness from 49.8% to 45.7%. Further, PCE increases from 5.64 % at 100 nm thickness to 8.45% at 600 nm of active layer thickness.

3.2 Impact of Defect Density of Active Layer

By varying the defect density of the active layer keeping the optimized thickness at 600nm, the impact of the performance parameters of solar cells is observed as shown in Figure 3. (b). As N_t increases, diffusion length decreases which results in a decrease in the lifetime of carriers and thus, more recombination resulting in a reduction in J_{SC}, V_{OC}, and PCE. At low defect density, the carrier diffusion length is high, and the recombination process is low thereby contributing to better PV performance. As shown in Figure 4 a), the defect density of the absorber is varied from 10^9cm^{-3} to 10^{15}cm^{-3}, and the thickness of the absorber is varied from 100nm to 1μm. The efficiency is observed as 13.39% at 5.08 x 10^9cm^{-3} at 600nm thickness of the absorber. Hence, the optimized defect density and the thickness of the absorber observed are 5.08 x 10^9cm^{-3} and 600nm.

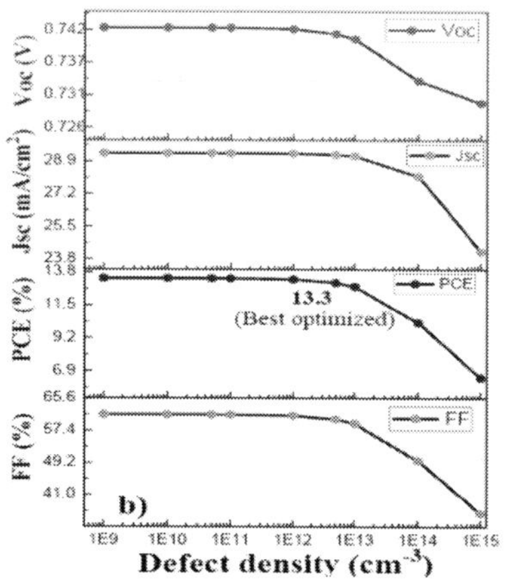

Figure 3 b): Impact of absorber defect density on device parameters

3.3 Impact of electron affinity and doping density of ETL on photovoltaic parameters

To achieve higher efficiency, there has to be less recombination at the active layer and ETL/HTL interfaces. To attain this the conduction band offset and valence band offset need to be optimized properly. In the case of the CBO of the buffer/absorber interface, electrons generated at the absorber flow to the surface side for collection. If the electron affinity of the ETL is larger than the electron affinity of the absorber, i.e., the conduction band of the ETL is lower than that of the absorber, then the energy cliff is formed at the interface and no barrier for the electron is formed.

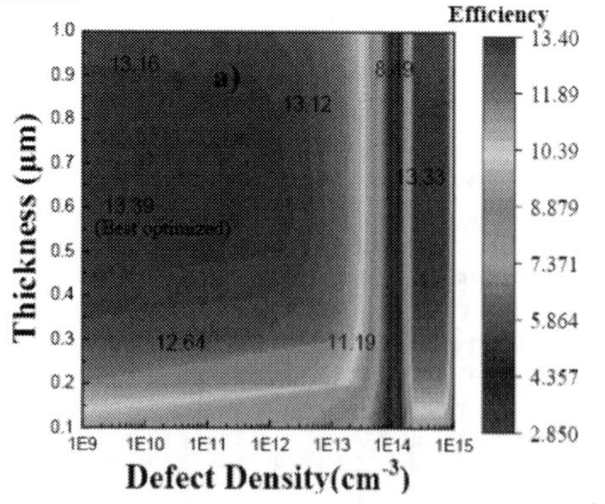

Figure 4 a): Variation in the efficiency of $FaSnI_3$ solar cell with change in absorber layer thickness and defect density

This usually happens for the negative values of CBO. However, for the positive values of CBO, an energy spike is formed at the interface, which can act as a barrier for electrons. The electron affinity of ETL i.e., C_{60} in this structure, is varied from 3.4eV to 4.8eV which gives CBO in the range of 0.12eV to -1.28eV.

$$CBO = \chi_{perovskite} - \chi_{ETL} \quad (1)$$

As doping density can be varied in ETL and HTL materials by changing the types of dopants for the efficient performance of perovskite solar cells, we have varied the doping density of ETL from 10^{15} cm^{-3} to 5 x 10^{21} cm^{-3} to analyze the impact on output solar cell parameters as shown in Figure 4. It has been found that the efficiency of the device attains the maximum value of 26.77% at a doping density of 4 x 10^{21} cm^{-3} and electron affinity of 3.71eV as shown in Fig 4. Hence the optimized doping density of ETL is chosen as 4 x 10^{21} cm^{-3}.

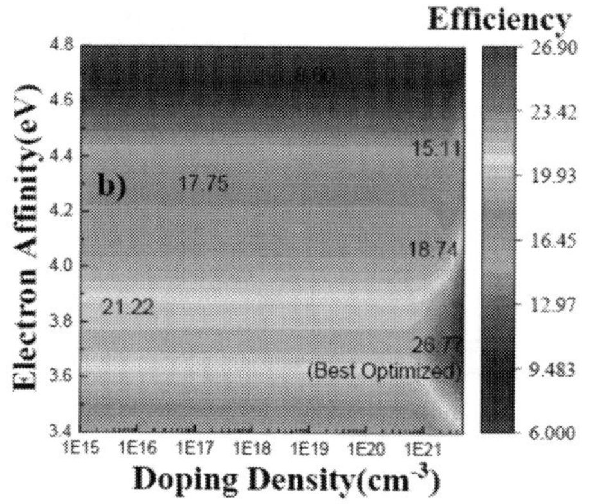

Figure 4 b): Variation in the efficiency of $FaSnI_3$ perovskite solar cell with change in ETL layer electron affinity and doping density

4 CONCLUSIONS

Although many computational and experimental studies have been undertaken on $FASnI_3$-based solar cells, the maximum PCE has remained below 20%. In this work, we proposed the solar cell device with C_{60} as ETL and PEDOT: PSS as HTL layer and simulated it using SCAPS-1D. We demonstrated the critical role of the thickness of the absorber layer on the performance parameters of Solar Cells. The thickness of $FASnI_3$ varied from 100nm to 1um. Among the different thicknesses investigated, the optimal 600-nm-thick layer-based perovskite solar cell showed the highest efficiency of about 8.45% and V_{OC}, J_{SC}, and FF of about 0.674V, 27mA/cm^2 and 46.5% respectively. Moreover, the lower value of defect density seems to be critical for the best efficiency. After optimizing the defect density of the absorber layer, doping density, and thickness of ETL, the final efficiency achieved was 26.77% where V_{OC}, J_{SC}, and FF are obtained as 1.13V, 26.69mA/cm^2, and 88.62% respectively.

4.1 References

[1] Kojima, A.; Teshima, K.; Shirai, Y.; Miyasaka, T. Organometal Halide Perovskites as Visible-Light Sensitizers for Photovoltaic Cells.J. Am. Chem. Soc. 2009, 131, 6050−6051.

10.4229/EUPVSEC2023/2BV.2.31

[2] Green, M.; Dunlop, E.; Hohl-Ebinger, J.; Yoshita, M.; Kopidakis, N.; Hao, X. Solar cell efficiency tables (version 57). Progr. Photovolt.: Res. Appl. 2021, 29, 3−15.

[3] Xiao, Z.; Song, Z.; Yan, Y. From Lead Halide Perovskites to Lead-Free Metal Halide Perovskites and Perovskite Derivatives. Adv. Mater. 2019, 31, No. e1803792.

[4] P. Tonui, S.O. Oseni, G. Sharma, Q. Yan, G.T. Mola, Perovskite photovoltaic solar cells: an overview of current status, Renew. Sustain. Energy Rev. 91 (2018) 1025–1044.

[5] H.J. Snaith, Present status and future prospects of perovskite photovoltaics, Nat. Mater. 17 (5) (2018).372:376.

[6] J.Y. Kim, J.W. Lee, H.S. Jung, H. Shin, N.G. Park, High-efficiency perovskite solar cells, Chem. Rev. 120 (15) (2020) 7867–7918

[7] K. Akhiro, T. Kenjiro, S. Yasuo, M. Tsutomu, Organometal halide perovskites as visible light sensitizers for photovoltaic solar cells, J. Am. Chem. Soc. 131 (2009) 6050–6051.

[8] R.L. Milot, G.E. Eperon, T. Green, H.J. Snaith, M.B. Johnston, L.M. Herz, Radiative monomolecular recombination boosts amplified spontaneous emission in HC (NH2)2SnI3 perovskite films, J. Phys. Chem. Lett. 7 (20) (2016) 4178–4184

[9] T.M. Koh, T. Krishnamoorthy, N. Yantara, C. Shi, W.L. Leong, P.P. Boix, A. C. Grimsdale, S.G. Mhaisalkar, N. Mathews, Formamidinium tin-based perovskite with low Eg for photovoltaic applications, J. Mater. Chem. A 3 (29) (2015) 14996–15000.

[10] Feidan Gu, Chengbo Wang, Ziran Zhao, Ge Zhan, Zhiwei Liu, Zuqiang Bian, and Chunhui Huang ACS Applied Materials & Interfaces 2021 13 (37), 44157-44164.

10.4229/EUPVSEC2023/2BV.2.31
020112-004

INVESTIGATION OF $ASnI_2Br$ PEROVSKITE/C60 HETEROJUNCTION FOR IMPROVED CHARGE EXTRACTION FROM SOLAR CELL

Rajesh Kumar Sharma[1], Hitarth Narsi Patel[2], Shivendra Yadav[3], Vivek Garg[4]
Department of Electronics Engineering, SVNIT, Surat, India
d21ec011@eced.svnit.ac.in[1], d21ec004@eced.svnit.ac.in[2], shivendra.y@eced.svnit.ac.in[3], vivekg@eced.svnit.ac.in[4]

ABSTRACT: Tin-perovskite solar cells (Sn-PSCs) are becoming popular in the current era due to their lesser toxicity than lead-perovskite solar cells (Pb-PSC). Generally, the Sn-PSCs are reported as a low bandgap but a wider-band gap Sn-PSCs possibly be achieved by X-site anion engineering. In this paper, a wide-bandgap Sn-PSC is numerically simulated, and then investigate the properties of the absorber/electron transport layer (ETL) and its interface to achieve maximum power conversion efficiency (PCE). Absorber material for this PSC is $ASnI_2Br$ with doping of 3% GeI_2, where A-site cation is considered as $EDA_{0.01}(GA_{0.06}(FA_{0.8}Cs_{0.2})_{0.94})_{0.98}$. By optimizing the absorber thickness, improvement in the PCE is achieved from 4.82% to 5.10%, followed by the optimization of absorber defects, PCE enhanced from 5.10% to 5.40%, with a 46.3 mV increment in Voc. Further, the doping of the ETL is altered from 2.6×10^{18} cm^{-3} to 1×10^{20} cm^{-3} to enhance the charge extraction, which helps to increase current density and hence efficiency from 5.40% to 6.70%. Then the offset of ETL is changed from 0.66 eV to -0.04 eV to improve PCE from 6.70% to 15.05%. Finally, the series resistance (Rs) is altered from 12.6 Ω.cm^2 to 5 Ω.cm^2, giving a PCE of 16.93%.

Keywords: wide-bandgap, tin-perovskite, interface defects, band offset, SCAPS-1D.

1 INTRODUCTION

Mixed halide perovskite gives the ability to tune the bandgap of the material by compositional engineering in the ABX_3 perovskite structure [1], [2]. Cesium (Cs), formamidinium (FA), and methylammonium (MA) have been widely used at the A-site to achieve thermal stabilities [3], while lead (Pb), tin (Sn), and germanium (Ge) are used at the B-site [4], [5]. Pb^{2+} remains the most efficient and stable among them, but it is toxic to the environment, so Sn^{2+} is used at B-site due to its lower toxicity, and similar ionic radius (Sn^{2+}-118 pm, Pb^{2+}-119 pm), and higher efficiency than other Pb-free PSCs. Researchers from all over the world are trying to maximize the PCE of lead-free PSCs above the ceiling set by single-junction solar cells. Although Sn-PSCs with a bandgap of 1.3 eV to 1.4 eV were reported frequently, Sn-PSCs with a bandgap of around 1.6 eV were rarely researched by the scientific community [6], [7]. As a result, current research focuses on designing wide-bandgap PSCs that can be used as the top layer of multijunction solar cells with silicon and other photovoltaic solar cells at the bottom layer, which has a narrow bandgap. We aim to investigate the different lead-free materials with a wide bandgap and optimize them.

2 AIM AND APPROACH

In this paper, a wide-bandgap material, 3% GeI_2 doped-$ASnI_2Br$ is considered as an absorber for the optimization [8]. Using SCAPS-1D, a planar structure of FTO/PEDOT:PSS/Absorber/C60/Ag is numerically simulated. The properties of the absorber are optimized first, that increases the PCE of the device to 5.4%. Further, the PCE of 15.05% is obtained by ETL optimization. By optimizing the absorber/C60 heterojunction, the PCE of the device enhanced up to 15.07%. The maximum PCE of 16.93% was finally achieved with the proposed structure with series resistance of 5 Ω.cm^2, while the V_{OC}, J_{SC}, and FF values were 1.29 V, 17.4 mA/cm^2, and 75.42%, respectively. J-V curve of experimental and calibrated device with performance parameters is shown in Figure 1.

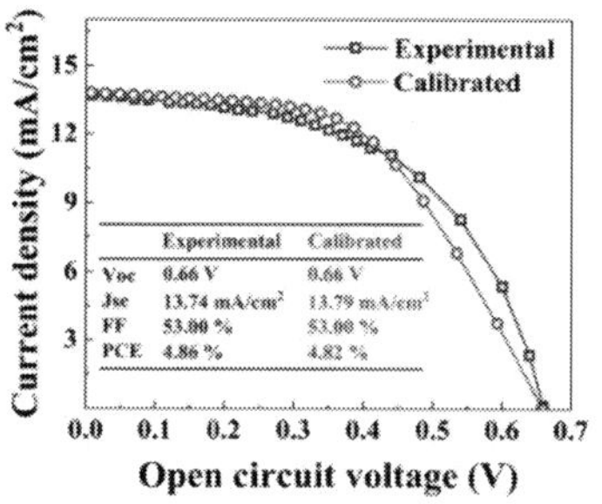

Figure 1: J-V characteristics of the experimental and simulated result. (Inset shows performance parameters of experimental and simulated device)

3 STRUCTURE AND SIMULATION PARAMETERS

The structure of the selected Sn-PSC is depicted in Figure 2. The parameters used for the simulation are selected from the literature and listed in Table I [8], [9]. The thicknesses of HTL and ETL are considered as 30 nm and 25 nm, respectively, while the absorber layer is taken as 200 nm. The light is incident from the top of fluorine-doped tin oxide (FTO). The work function of front contact (FTO) and rear contact (Ag) is 4.8 eV and 4.7 eV, respectively. Spectrum AM1.5G is used for all simulations with an irradiance power of 100 mW.cm^{-2} at 300 K and series resistance of 12.6 Ω.cm^2.

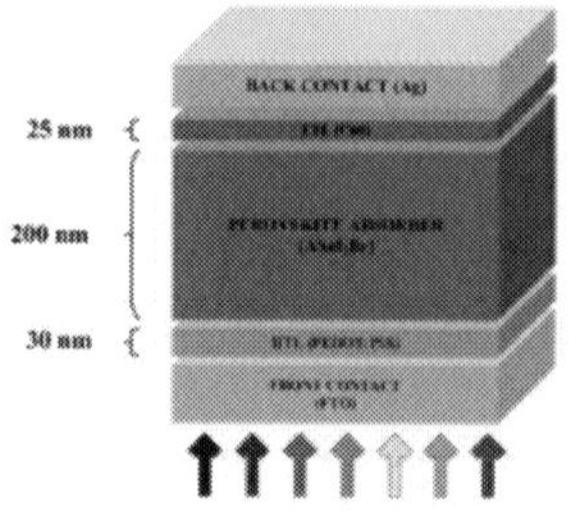

Figure 2: The structure of tin-based perovskite solar cell

10.4229/EUPVSEC2023/2BV.2.32
020114-001

Table I: Parameters considered for base model calibration

Parameters	Unit	HTL [9]	Absorber [8]	ETL [9]
Thickness, t	nm	30	200	25
Bandgap, Eg	eV	2.2	1.62	1.7
Electron affinity, χ	eV	2.9	3.84	4.5
Dielectric constant, ε_r	-	3	28	4.2
CB effective density of states, N_C	cm^{-3}	2.2×10^{15}	1×10^{19}	8×10^{19}
VB effective density of states, N_V	cm^{-3}	1.8×10^{18}	1×10^{18}	8×10^{19}
Electron thermal velocity, $v_{th,n}$	cm/s	1×10^{7}	1×10^{7}	1×10^{7}
Hole thermal velocity, $v_{th,p}$	cm/s	1×10^{7}	1×10^{7}	1×10^{7}
Electron mobility, μ_n	cm²/Vs	10	22	2.17
Hole mobility, μ_p	cm²/Vs	10	22	3.5×10^{-3}
Donor density, N_D	cm^{-3}	-	2.44	2.6×10^{18}
Acceptor density, N_A	cm^{-3}	3×10^{17}	1×10^{14}	-
Defect Density 1	cm^{-3}	1×10^{14}	4.38×10^{16}	-
Defect Density 2	cm^{-3}	-	1×10^{15}	-
Defect Density 3	cm^{-3}	-	1×10^{15}	-

4 RESULTS AND DISCUSSION

4.1 Absorber optimization

To analyze the effect of the absorber thickness variation on the performance parameter, the thickness of the absorber is varied from 0.1 µm to 1 µm. In Figure 3 (a) open circuit voltage (Voc) increases with thickness up to 0.5 µm and after that, it saturates to 0.73 V due to the increase in trap-assisted recombination, and short circuit current density (Jsc) increases rapidly with thickness due to the large absorber area and hence generation of charge carriers. Fill factor (FF) decreases with an increase in absorber thickness due to the inverse relation with Voc and Jsc. PCE first increases to 5.1% at 0.43 µm and then decreases again as indicated in Figure 3 (b). All the defect densities of the absorber are considered 1×10^{13} cm^{-3} since PCE for all the defects is constant on or below 1×10^{13} cm^{-3} after that efficiency degrades rapidly due to the trap-assisted recombination as depicted in Figure 3 (c).

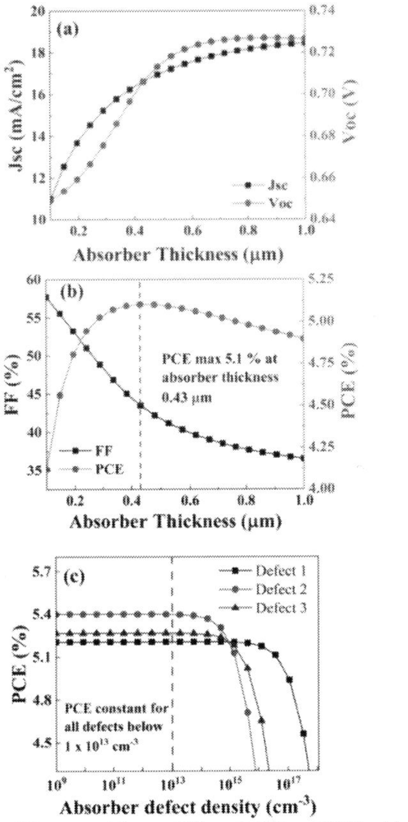

Figure 3: Effect on (a) Jsc and Voc, (b) FF and PCE with thickness, and (c) PCE with three defect densities

4.2 ETL optimization

For the optimization of ETL, the donor density is varied from 1×10^{15} cm^{-3} to 1×10^{21} cm^{-3}, it is observed from Figure 4 (a) & (b) that all the parameters increase with the increase in doping density. High doping in ETL creates a large depletion area in the absorber; hence, large amount of charge carriers are injected into the ETL from the absorber due to the high electric field created across the junction. PCE of 6.72% is obtained while considering the donor density of ETL as 1×10^{20} cm^{-3}. We proposed the Cerium (IV) oxide (CeO_2) as a new ETL material for the optimized device, due to having similar range of donor density i.e., 1×10^{20} cm^{-3} [10]. Then conduction band offset (CBO) is varied from -0.34 eV to 0.66 eV and the maximum PCE of 15.05% is achieved at -0.04 eV as represented in Figure 4 (c).

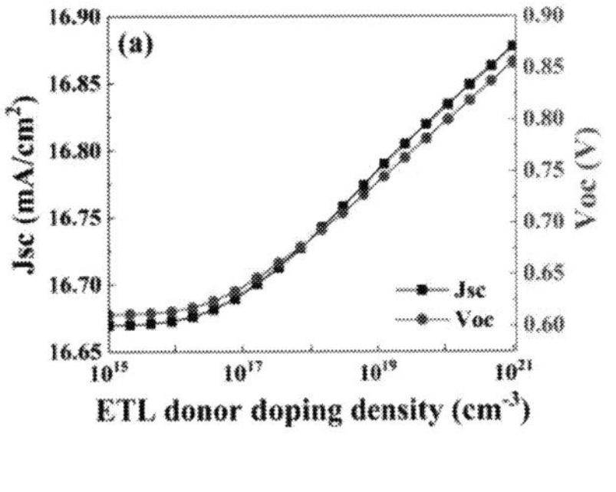

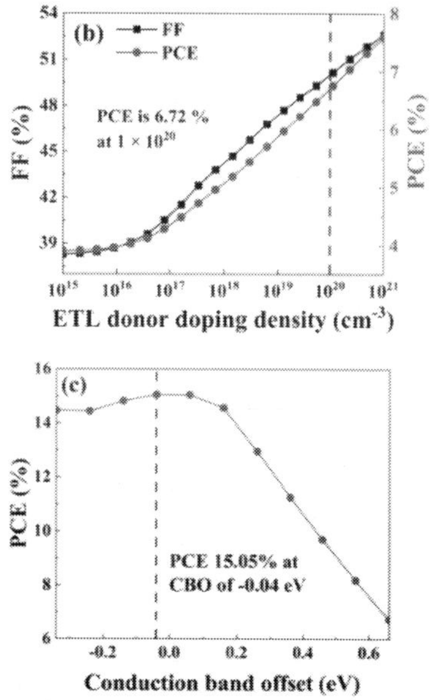

Figure 4: Effect on (a) Jsc and Voc, (b) FF and PCE with ETL donor density & (c) PCE with ETL offset variation.

4.3 Interface optimization

We further optimize the device by selecting the appropriate defect density of the interface. Figure 5 indicates the PCE is constant below 1×10^{10} cm^{-2} defect density and after that, it decreases rapidly due to the charge carrier recombination. At 1×10^{10} cm^{-2} the PCE of the device is 15.07% with a high value of Voc of 1.23 V, Jsc of 17.28 $mA.cm^{-2}$, and FF of 67.59%. Although the FF of the device is low it can be improved by reducing the series resistance.

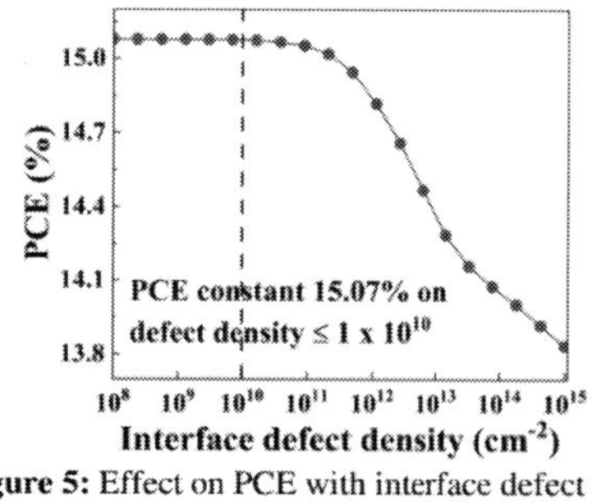

Figure 5: Effect on PCE with interface defect density

4.4 Series resistance (Rs) optimization

The series resistance is then optimized by varying the value from 1 $\Omega.cm^2$ to 15 $\Omega.cm^2$. The value of PCE continuously decreases as Rs increases since Rs limits the current density of the device hence PCE. Since the Rs of the commercial device is reported around 1 $\Omega.cm^2$, for the device optimization, we consider the Rs of 5 $\Omega.cm^2$ as per the fabrication standpoint, which can be obtained by reducing the defects in the absorber. Thus, changing the Rs from 12.6 Ω cm^2 to 5 $\Omega.cm^2$ gives the PCE of the proposed device 16.93%.

4.5 Proposed device

The optimized value of absorber thickness is 430 nm and the defect density is 1×10^{13} cm^{-3} is considered for all 3 defects, along with these parameters, the electron affinity of ETL is selected 3.8 eV, which is used to simulate the optimized device. The J-V characteristics for the optimized device is represented in Figure 6, which shows the PCE approaches 16.93% from 4.82% with Voc increased from 0.66 to 1.29 V, Jsc increased from 13.79 to 17.4 mA cm^{-2}, and FF increased from 53 to 75.42%.

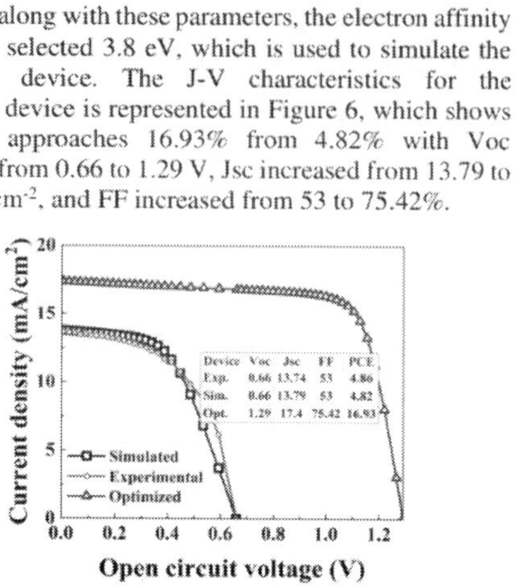

Figure 6: J-V characteristics for the optimized device. (Inset shows performance parameters of exp, sim and optimized device)

5 CONCLUSIONS

A wide-bandgap $ASnI_2Br$ (3% GeI_2 doping) absorber with a bandgap of 1.62 eV was selected for baseline simulation. The SCAPS-1D simulation tool is used to simulate the structure. For a better result, the PSC absorber layer is optimized for absorber thickness, and defect density. PCE of 5.1% achieved at optimized absorber thickness of 430 nm with increment in Voc and Jsc is 45.8 mV and 2.82 $mA.cm^{-2}$ respectively but FF is decreased by 9.47%. Considering all the defects of the absorber to 1×10^{13} cm^{-3} we achieved a PCE of 5.4% with Voc of 0.71 V, Jsc of 16.76 $mA.cm^{-2}$, and FF of 45.63%. Then donor density and electron affinity of ETL are optimized to enhance the efficiency of the device further. The optimized donor doping density of ETL is 1×10^{20} cm^{-3} hence CeO_2 can be proposed as a new ETL due to having a similar donor density. With the proposed ETL material the PCE increased by 1.32%, which majorly depends on the Voc of 0.8 V and FF of 50.11% but only a small increment is observed in Jsc of 0.08 $mA.cm^{-2}$. PCE is largely improved by CBO optimization, an increment of 8.33% is observed with Voc of 1.28 V, Jsc of 17.28 mA cm^{-2}, and FF of 67.98%. After the perovskite/C60 interface defect density optimization, PCE increases to 15.07% with Voc of 1.29 V, Jsc of 17.28 mA cm^{-2}, and FF of 67.68%. Finally, the series resistance decreases to 5 $\Omega.cm^{-2}$ and the PCE of 16.93% is achieved with an increment of FF of 7.79%, Voc of 1.29 V, and Jsc of 17.40 $mA.cm^{-2}$. It is observed that there is a minor improvement in the efficiency due to the absorber optimization but due to the ETL optimization efficiency is improved largely.

6 REFERENCES

[1] M. Saliba *et al.*, "Cesium-containing triple cation perovskite solar cells: Improved stability, reproducibility and high efficiency," *Energy Environ Sci*, vol. 9, no. 6, pp. 1989–1997, Jun. 2016, doi: 10.1039/c5ee03874j.

[2] J. H. Noh, S. H. Im, J. H. Heo, T. N. Mandal, and S. Il Seok, "Chemical management for colorful, efficient, and stable inorganic-organic hybrid

nanostructured solar cells," *Nano Lett*, vol. 13, no. 4, pp. 1764–1769, Apr. 2013, doi: 10.1021/NL400349B.

[3] G. E. Eperon, S. D. Stranks, C. Menelaou, M. B. Johnston, L. M. Herz, and H. J. Snaith, "Formamidinium lead trihalide: A broadly tunable perovskite for efficient planar heterojunction solar cells," *Energy Environ Sci*, vol. 7, no. 3, pp. 982–988, Mar. 2014, doi: 10.1039/C3EE43822H.

[4] Y. Ogomi *et al.*, "CH3NH3SnxPb(1-x)I3 perovskite solar cells covering up to 1060 nm," *Journal of Physical Chemistry Letters*, vol. 5, no. 6, pp. 1004–1011, Mar. 2014, doi: 10.1021/JZ5002117.

[5] F. Hao, C. C. Stoumpos, D. H. Cao, R. P. H. Chang, and M. G. Kanatzidis, "Lead-free solid-state organic-inorganic halide perovskite solar cells," *Nat Photonics*, vol. 8, no. 6, pp. 489–494, 2014, doi: 10.1038/NPHOTON.2014.82.

[6] T. Yokoyama, T. Bin Song, D. H. Cao, C. C. Stoumpos, S. Aramaki, and M. G. Kanatzidis, "The Origin of Lower Hole Carrier Concentration in Methylammonium Tin Halide Films Grown by a Vapor-Assisted Solution Process," *ACS Energy Lett*, vol. 2, no. 1, pp. 22–28, Jan. 2017, doi: 10.1021/ACSENERGYLETT.6B00513.

[7] D. Moghe *et al.*, "All vapor-deposited lead-free doped CsSnBr3 planar solar cells," *Nano Energy*, vol. 28, pp. 469–474, Oct. 2016.

[8] M. Chen *et al.*, "Large synergy effects of doping, a site substitution, and surface passivation in wide bandgap Pb-free ASnI2Br perovskite solar cells on efficiency and stability enhancement," *J Power Sources*, vol. 520, p. 230848, Feb. 2022, doi: 10.1016/J.JPOWSOUR.2021.230848.

[9] M. Mehrabian, E. N. Afshar, and O. Akhavan, "TiO2 and C60 transport nanolayers in optimized Pb-free CH3NH3SnI3-based perovskite solar cells," *Materials Science and Engineering: B*, vol. 287, p. 116146, Jan. 2023, doi: 10.1016/J.MSEB.2022.116146.

[10] M. K. Hossain, M. H. K. Rubel, G. F. I. Toki, I. Alam, M. F. Rahman, and H. Bencherif, "Effect of Various Electron and Hole Transport Layers on the Performance of CsPbI3-Based Perovskite Solar Cells: A Numerical Investigation in DFT, SCAPS-1D, and wxAMPS Frameworks," *ACS Omega*, vol. 7, no. 47, pp. 43210–43230, Nov. 2022, doi: 10.1021/ACSOMEGA.2C05912/ASSET/IMAGES/LARGE/AO2C05912_0013.JPEG.

INJECTION OVERMOLDED FLEXIBLE PEROVSKITE SOLAR MODULES

Riikka Suhonen, Kaisa Leena Väisänen, Marja Välimäki, Timo Kurkela, Mari Ylikunnari, and Thomas M. Kraft*
VTT Technical Research Centre of Finland Ltd., Kaitoväylä 1, Oulu 90590, Finland
*Telephone: +358 20 722 2070, Email: thomas.kraft@vtt.fi

ABSTRACT: Integration methods for perovskite solar cells (PSC) are being developed with the evolving printed electronics pilot line infrastructures enabling material processing to create tailored electronic components, hybrid systems, and further integration in structural electronics (i.e. in-mold electronics). The processing can be done by sheet based or roll-to-roll pilot lines to upscale the production. In this work, flexible roll-to-roll processed PSC modules were integrated via the plastic injection overmolding process. The PSC must be tolerant to the overmolding process conditions which proved feasible in this study leading to improved performance. Evaluation of the overmolded PSC was performed through characterization under 1 sun illumination and stability measurements. For a fully integrated energy module approach to the PSC fabrication, a double sided hybrid electronics solution was fabricated.
Keywords: perovskite solar cells, injection overmolding, in-mold electronics, energy autonomous module

1 INTRODUCTION

Advances in the performance, sustainability, and roll-to-roll (R2R) processing opportunities of flexible thin-film photovoltaics (PV) have led to the volume production of PV modules, application breakthroughs, and routes for seamless integration in systems [1]–[5]. For example, the processing work done with R2R processed flexible organic photovoltaics (OPV) over the years have now reached a level of market commercialization and are being integrated in glass and plastics for various applications. With the recent advances in the next generation solution processed perovskite solar cells this trend with integrated thin-film PV solutions continues.

Within the past decade perovskite solar cells (PSCs) have been a subject of intense research. The perovskite photo-absorptive materials offer inherent advantages leading to high power conversion efficiencies (currently 26.1% for rigid and 23.6% for flexible PSCs) [6]; long carrier diffusion lengths, high carrier mobilities, low exciton binding energies, high absorption coefficients and band gap tunability via interchanges of the precursor components [7]. One important benefit of the perovskite material is the possibility to deposit the thin films via solution-based processing, which enables high-throughput, cost-effective production of flexible and light-weight PSCs further expanding the commercialization potential of PSCs into variety of applications such as mobile electronic devices, vehicle- and building integrated PVs, and the emerging applications within the Internet of Things (IoT) market.

The PSC processing can be done by sheet based and/or R2R printing pilot lines to upscale their production [8]. The focus on fully additive solution processing of the flexible PSC stack starting from R2R wet etching of the PET/ITO substrate is seen as a viable solution and opportunity for development [2], [9], [10]. Even though the improvement in efficiency and stability of R2R processed flexible PSC remains as a very relevant research topic, the integration of such devices further into proof-of-concept lightweight systems is also required [11].

Here, we introduce the first results on the performance of flexible, R2R fabricated PSC modules integrated via a plastic injection overmolding process. Also, the characterization data of a fully autonomous energy module utilizing PSC as the energy harvester is presented.

2 PROCESSING OF PEROVSKITE PV MODULES

The architecture of the pilot processed perovskite modules was "n-i-p" consisting of flexible and transparent PET polymer substrate, transparent indium tin oxide (ITO) electrode, tin oxide (SnO_2) electron transporting layer, perovskite photoactive layer, poly(3-hexylthiophene-2,5-diyl) (P3HT) hole transporting layer and gold (Au) metallic top electrode. From these, ITO, SnO_2, perovskite and P3HT were R2R processed at the pilot printing line shown in Figure 1.

Figure 1. Photographs from the processing of flexible PSC modules on the R2R pilot printing line at VTT, Oulu, Finland.

First, the ITO layer was patterned via a previously developed R2R etching process [12]. As an electron transporting layer, commercially available aqueous tin oxide SnO_2 nanoparticle ink (15% in H_2O from Alfa Aesar) was used. The nanoparticle dispersion was modified with 2-butoxyethanol to reach a desired rheology for the R2R gravure printing process. The printing parameters for SnO_2 were the following: PET/ITO substrate was pre-treated with N_2/Ar plasma (300 W), web speed was 4 m/min, annealing temperature was 140°C (hot-air ovens), and engraving line density was 120 lines/cm in the gravure printing roll.

After SnO_2 printing and annealing, a FA-Cs-perovskite ($FA_{0.83}Cs_{0.17}Pb(I_{0.93}Br_{0.07})_3$) dissolved in dimethyl sulfoxide (DMSO) solvent was R2R gravure printed and annealed on top. A starch polymer and methyl ammonium chloride (MACl) were used in the perovskite ink as a rheological additive and as a crystallization aid, respectively [9]. The printing parameters for the perovskite were the following: PET/ITO/SnO_2 substrate was pre-treated with N_2/Ar plasma (500 W), printing was done at web speed 3 m/min with 140°C (hot-air ovens) annealing temperature, a second annealing step with 0.5 m/min web speed took place right after the printing step, and the engraving line density of 120 lines/cm in the gravure printing roll was used.

The hole transporting polymer P3HT (#4002-E, Rieke Metals) was dissolved in chlorobenzene and R2R gravure printed on top of the perovskite layer. A heated ink container was used and the P3HT ink was printed as a warm solution (ca. 50°C) at a printing speed of 6 m/min and warm air (30°C) was used to evaporate the solvent.

To finalize the module stack, samples were cut from the PET | ITO | SnO_2 | perovskite | P3HT roll, and Au metallic electrode (80nm) was thermally evaporated on top in a sheet-based process. After the evaporation, the modules were encapsulated by lamination on both sides with a transparent, commercially available barrier foil (ATCJ from AMCOR) utilizing a pressure-sensitive adhesive (PSA) (EL-92734 from Adhesives Research).

3 RESULTS AND DISCUSSION

3.1 Injection overmolding of PSCs

The plastic overmolding process used in this study utilizes a traditional injection mold with the PV component added to the mold cavity prior to plastic injection. The PSC part must therefore be tolerant to the shear force, temperature, and pressure of the overmolding process, whose plastic injection can be modelled in advance [13]. To evaluate the opportunities in wearable electronics and the overmolding process' effect on the R2R processed PSC modules, a half-wrist band mold was selected for the process. As the overmolding material, Thermoplastic Polyurethane (TPU) was used due to its transparency and the relatively mild injection overmolding processing conditions. Photographs and parameters of the overmolding process are shown in Figure 2.

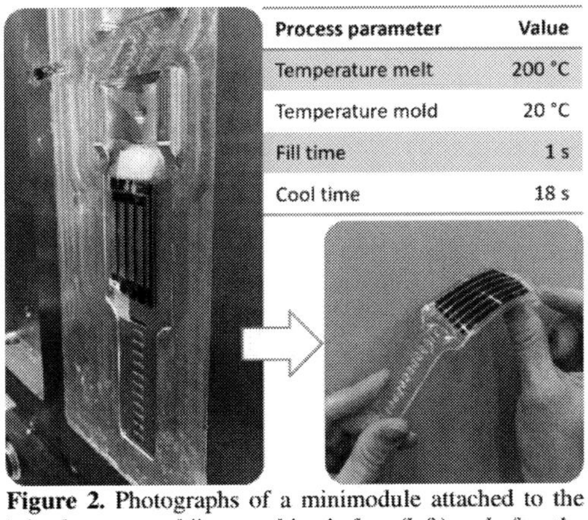

Process parameter	Value
Temperature melt	200 °C
Temperature mold	20 °C
Fill time	1 s
Cool time	18 s

Figure 2. Photographs of a minimodule attached to the injection overmolding machine before (left) and after the molding process. The applied process parameters are shown top right.

3.2 Performance of the overmolded perovskite modules

The overmolding test was repeated twice (Test I and Test II) and, in both tests, 9-12 minimodules were first encapsulated with a flexible barrier foil, (1) measured (IV-scan under AM1.5 solar simulator), (2) overmolded with TPU, and then (3) re-measured. The solar parameters short-circuit current density (J_{SC}), open-circuit voltage (V_{OC}), fill factor (FF) and power conversion efficiency (PCE) derived from the IV-scans are shown as statistical box plots in Figure 3. In Test I, the deviation in performance between the individual modules was more pronounced than in Test II but all the solar parameter values were at the same range in both tests.

In both tests, the performance of the flexible PSC minimodules was not degraded by the overmolding process. On the contrary, in Test II, the PCE of the modules is even increased after the overmolding due to enhancements in J_{SC} and V_{OC}, which is promising for future integration studies, and PSC module improvements.

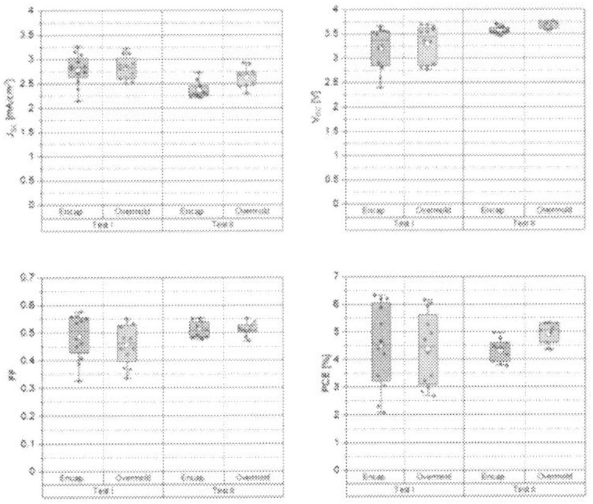

Figure 3. The solar parameters J_{SC}, V_{OC}, FF and PCE derived from the IV-curves of the encapsulated and overmolded perovskite minimodules.

3.2 Stability of the overmolded perovskite modules

After the overmolding, the performance of the minimodules (MM) over time was measured. To measure the effect of the overmolding on the stability of the MMs, the overmolded samples were compared with samples that were only encapsulated and not overmolded.

In Figure 4, the normalized PCE of the encapsulated and encapsulated & overmolded MMs under atmospheric controlled conditions of 21°C temperature and 40% relative humidity is shown. The encapsulated modules show an increase in PCE over time with the efficiency increasing for the first 30 days after which it gradually decreases until the half-lifetime (LT50) is reached after 90 days.

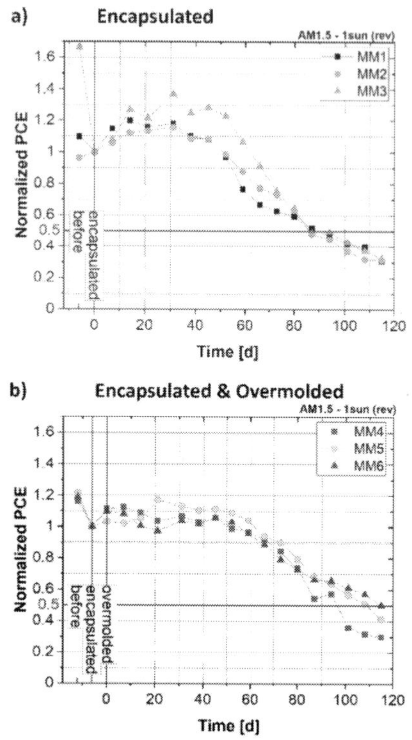

Figure 4. The normalized efficiencies of minimodules (MM) after storing under ambient conditions 21°C / 40% rH. All results are normalized to the values measured after encapsulation. MMs were periodically measured under AM1.5 solar simulator illumination.

For the overmolded modules, there was an approximately 10% increase in PCE when measured right after the overmolding process after which the modules show stable efficiency for 60 days. Thus, it has been found that the overmolding process accelerates the PCE increase phenomenon which is measured over time for the MMs without overmolding. The overmolded MMs reach their LT50 between 100 and 115 days thus showing a slight increase in the lifetime when compared with the MMs that were not overmolded.

3.3 PSC energy module

Utilizing the developed printing processing of flexible PSCs, a proof-of-concept flexible hybrid energy harvesting, and storage module was printed on both sides of a PET substrate. The double-sided system is a lightweight, flexible platform printed on both sides of the substrate using R2R manufacturing process of through-substrate vias enabling the double-sided functionality [14], [15].

The process flow in the energy module manufacturing was the following: i) ITO was R2R patterned as a negative image using rotary screen printing, ii) through-substrate via-holes were R2R laser-cut and aligned with the ITO-patterns, iii) conductive tracks were rotary screen printed on both sides of the substrate also filling the via-holes, iv) on top of the R2R patterned ITO, the perovskite PV module was fabricated by sheet-based gravure printing (SnO_2, perovskite and P3HT) and thermal evaporation processes (Au), v) the PSC module was encapsulated with barrier foils and vi) finalized with hybrid assembly of the surface mount devices (SMD) by pick and place equipment, followed by the battery attachment. From these processing steps, i-iii) were R2R and iv-vi) S2S processes.

The perovskite PV based energy module is a lightweight, flexible platform which was a winner in the OE-A Competition 2023 [16]. In addition to the PSC, the system is composed of an energy harvesting power management integrated circuit (IC), a Zener diode, a LED (to visualize when the battery is charged), a Li-polymer pouch cell battery, and a number of capacitors, inductors, and resistors. The layout of the energy module along with a photograph of it are shown in Figure 5.

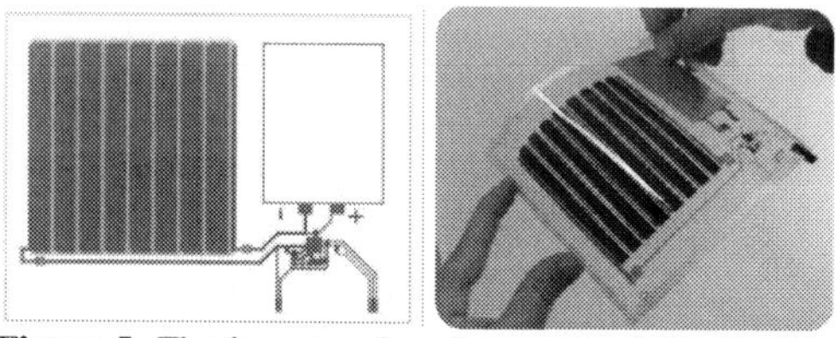

Figure 5. The layout and a photograph of the flexible double-sided perovskite PV energy module.

To provide a general context of the PSC energy module performance, the devices were measured under indoor LED light conditions at 500lux and 1000lux, and the harvesting circuit efficiency was evaluated (see Table I). 80% and 83% harvesting efficiencies were achieved at 500lux and 1000lux illumination intensities, respectively.

The harvesting efficiency losses could be improved by reducing the current loss of the conductive printed interconnects as well as tuning the maximum power point (MPP) of the energy harvesting circuit to suit the printed PSC. Nevertheless, even with the harvesting efficiency reported here, the energy module could be used to power some low power applications such as used in internet-of-things (IoT).

Table I. Harvesting circuit efficiency of the energy module under indoor lighting conditions.

Lighting condition	MPP [% of V_{OC}]	PV power [uW]	Charging power [μW]	Harvesting circuit efficiency
500 lux	50.3% (AM1.5)	86	69	80 %
1000 lux	50.3% (AM1.5)	301	251	83 %

4 SUMMARY AND CONCLUSIONS

In this study, we report the performance of injection overmolded flexible perovskite minimodules. The performance of the PSC minimodules was not degraded by the overmolding process but, for many samples, the PCE was increased after the overmolding. Furthermore, the stability of the overmolded samples was slightly increased when compared with the modules that were not overmolded. These results indicate the resilience of the flexible PSCs towards the overmolding processing conditions which opens new potential applications in structural electronics.

Also, based on the results reported here, there is clear potential for a printed perovskite energy module to be used for indoor IoT applications with a daytime harvesting

10.4229/EUPVSEC2023/2BV.2.36
020116-003

protocol. In turn, illustrating that these processing opportunities of flexible thin-film photovoltaics (PV) can enable volume production techniques for PV module integration and application breakthroughs for system integration.

7 ACKNOWLEDGMENTS

This research has received funding from the European Union's Horizon 2020 research and innovation program under grant agreement no.861985 (PeroCUBE), from Horizon Europe Research and Innovation Action programme under Grant Agreement n° 101082176 (VALHALLA) and Academy of Finland Flagship Programme, Photonics Research and Innovation (PREIN), decision 320168.

The authors acknowledge the following for their assistance: Anne Peltoniemi and Jenni Tomperi (characterization, lab-scale processing), Kyösti Heikkilä (injection overmolding), Elina Jansson (R2R via processing), Hannu Sääskilahti (R2R perovskite and via processing), Mikko Hietala (laser processing), Risto Karjalainen (SMT), Pentti Korhonen (materials), and Jari Rekilä (layout designs).

8 REFERENCES

[1] M. Ylikunnari *et al.*, "Flexible OPV modules for highly efficient indoor applications," *Flex. Print. Electron.*, vol. 5, no. 1, p. 014008, Feb. 2020, doi: 10.1088/2058-8585/ab6e73.

[2] Y. Y. Kim *et al.*, "Gravure-Printed Flexible Perovskite Solar Cells: Toward Roll-to-Roll Manufacturing," *Adv. Sci.*, vol. 6, no. 7, p. 1802094, Apr. 2019, doi: 10.1002/advs.201802094.

[3] Y. Khan, A. Thielens, S. Muin, J. Ting, C. Baumbauer, and A. C. Arias, "A New Frontier of Printed Electronics: Flexible Hybrid Electronics," *Adv. Mater.*, vol. 32, no. 15, p. 1905279, Apr. 2020, doi: 10.1002/adma.201905279.

[4] J. Park, J. Lee, S. Park, K.-H. Shin, and D. Lee, "Development of hybrid process for double-side flexible printed circuit boards using roll-to-roll gravure printing, via-hole printing, and electroless plating," *Int. J. Adv. Manuf. Technol.*, vol. 82, no. 9–12, pp. 1921–1931, Feb. 2016, doi: 10.1007/s00170-015-7507-2.

[5] M. Bakr, F. Bossuyt, and J. Vanfleteren, "The integration of electronic circuits in plastics using injection technologies: a literature review," *Flex. Print. Electron.*, vol. 7, no. 2, p. 023001, Jun. 2022, doi: 10.1088/2058-8585/ac6719.

[6] "NREL Best Research-Cell Efficiency Chart," Sep. 21, 2023. https://www.nrel.gov/pv/cell-efficiency.html

[7] M. Li *et al.*, "Multifunctional succinate additive for flexible perovskite solar cells with more than 23% power-conversion efficiency," *The Innovation*, vol. 3, no. 6, p. 100310, Nov. 2022, doi: 10.1016/j.xinn.2022.100310.

[8] M. K. Välimäki *et al.*, "Accuracy control for roll and sheet processed printed electronics on flexible plastic substrates," *Int. J. Adv. Manuf. Technol.*, vol. 119, no. 9–10, pp. 6255–6273, Apr. 2022, doi: 10.1007/s00170-022-08717-z.

[9] F. Bisconti *et al.*, "One-step polymer assisted roll-to-roll gravure-printed perovskite solar cells without using anti-solvent bathing," *Cell Rep. Phys. Sci.*, vol. 2, no. 11, p. 100639, Nov. 2021, doi: 10.1016/j.xcrp.2021.100639.

[10] Y. Y. Kim *et al.*, "Roll-to-roll gravure-printed flexible perovskite solar cells using eco-friendly antisolvent bathing with wide processing window," *Nat. Commun.*, vol. 11, no. 1, p. 5146, Oct. 2020, doi: 10.1038/s41467-020-18940-5.

[11] G. Tang and F. Yan, "Recent progress of flexible perovskite solar cells," *Nano Today*, vol. 39, p. 101155, Aug. 2021, doi: 10.1016/j.nantod.2021.101155.

[12] M. Välimäki *et al.*, "R2R-printed inverted OPV modules – towards arbitrary patterned designs," *Nanoscale*, vol. 7, no. 21, pp. 9570–9580, 2015, doi: 10.1039/C5NR00204D.

[13] T. Kololuoma *et al.*, "Adopting Hybrid Integrated Flexible Electronics in Products: Case—Personal Activity Meter," *IEEE J. Electron Devices Soc.*, vol. 7, pp. 761–768, 2019, doi: 10.1109/JEDS.2019.2903868.

[14] E. Jansson, A. Korhonen, M. Hietala, and T. Kololuoma, "Development of a full roll-to-roll manufacturing process of through-substrate vias with stretchable substrates enabling double-sided wearable electronics," *Int. J. Adv. Manuf. Technol.*, vol. 111, no. 11–12, pp. 3017–3027, Dec. 2020, doi: 10.1007/s00170-020-06324-4.

[15] M. Kujala, T. Kololuoma, J. Keskinen, D. Lupo, M. Mäntysalo, and T. M. Kraft, "Bending reliability of screen-printed vias for a flexible energy module," *Npj Flex. Electron.*, vol. 4, no. 1, p. 24, Sep. 2020, doi: 10.1038/s41528-020-00087-4.

[16] "Winner of the category 'Freestyle Demonstrator': Printed Perovskite Solar Energy Module," Mar. 30, 2022. https://oe-a.org/printed-perovskite-energy-solar-module

CORRELATIVE MICROSCOPY AND SPECTROSCOPY OF PEROVSKITE THIN FILMS: DEGRADATION ANALYSIS

V.Paraskeva[1], E. Peraticos[1], S. Pechmann[2,3], T. Fontanot[2,3], A. Aguirre[4,5,6], T. Aernouts[4,5,6], A. Krishna[4,5,6], M. Hadjipanayi[1], M. Sergides[7], A. Othonos[7], S. Christiansen[2,3], R. Ebner[8], G.E. Georghiou[1]
[1]University of Cyprus, PV Technology Laboratory, Department of Electrical & Computer Engineering, Nicosia, Cyprus
[2]Fraunhofer IKTS, Fraunhofer Institute for Ceramic Technologies and Systems, Äußere Nürnberger Str. 62, 91301 Forchheim, Germany
[3]Max-Planck-Institut für die Physik des Lichts, 91058 Erlangen, Germany
[4]Imec, imo-imomec, Thin Film PV Technology – partner in Solliance, Thor Park 8320, 3600 Genk, Belgium
[5]EnergyVille, imo-imomec, Thor Park 8320, 3600 Genk, Belgium
[6]Hasselt University, imo-imomec, Martelarenlaan 42, 3500 Hasselt, Belgium
[7]Laboratory of Ultrafast Science, Department of Physics, University of Cyprus, Nicosia, 1678, Cyprus
[8]AIT Austrian Institute of Technology, Center for Energy, Giefinggasse 2, 1210 Vienna, Austria

ABSTRACT: Perovskite materials have excellent prospects for semiconducting applications due to their desirable photoelectric properties. The morphology and the structure of the light absorption layer are crucially important for the device performance and their changes through the degradation period are of upmost importance for understanding degradation pathways. In this work, pristine perovskite thin films with and without FACl additives in the active perovskite layer have been characterized with Scanning Electron Microscopy (SEM), Atomic Force Microscopy (AFM) and Energy-Dispersive Spectroscopy (EDX) to identify changes in the perovskite grain size, grain boundaries and current densities. Moreover, femtosecond transient transmission measurements have been employed to detect changes in carrier relaxation dynamics in encapsulated perovskite thin films before and after outdoor exposure.
Keywords: perovskites, microscopy, spectroscopy

1 INTRODUCTION

Perovskite solar cells have demonstrated outstanding performance in recent years, achieving power conversion efficiencies (PCE) as high as 25.5% with theoretical efficiencies calculated up to 31% [1]. A lot of indoor characterization testing was implemented the last years at different temperatures, humidity levels, bias loading demonstrating the impact of each parameter on the perovskite lifetime and stability [2]. Understanding the degradation pathways of perovskite materials is crucial to improve their stability and pave their way to commercialization. To achieve this, understanding of the fundamental material properties is required through the utilization of advanced microscopic and spectroscopic techniques. Device engineering, morphology, and light-harvesting properties of the perovskite material are the key factors which help to improve the efficiency of perovskite solar cells [3]. Measurement results from those methods can be utilized to understand the long-term outdoor operation of devices and their degradation mechanisms.

Methods of scanning electron microscopy (SEM), and conductive Atomic Force Microscopy (c-AFM) provide information regarding the understanding of the fundamental properties of perovskite materials and solar cells at the nanoscale. One common use of the SEM method is to identify changes in the perovskite grain size and grain boundaries. Grain boundaries are considered detrimental to the performance of the perovksites and modify the charge carrier trasport pathway in the material and hence the performance of perovskites [4]. c-AFM was widely used in the past to characterize the local electrical properties of perovskite thin films [5]. c-AFM gives also valuable conductivity results of grains and grain boundaries as the SEM method which are very helpful for increasing the performance and stability of perovskite-based devices [6].

Energy-Dispersive X-Ray Spectroscopy is used to provide elemental composition details about the speciment under test. It is typically used concurrently with image acquistion in order to confirm that the specimen is composed of the expected elements and that these have the expected stoichiometric ratio.

Ultrafast pump-probe spectroscopy provides a powerful approach for deep insight into the transient photophysics of electron-hole interactions. Correlation between carrier dynamics and aging of the device was found in the past providing an insight into the factors driving degradation in perovskite solar cells [7].

In this work microscopic investigation of pristine perovskite thin films of two different compositions has been performed using Scanning Electron Microscopy (SEM), Energy Dispersive X-Ray Spectroscopy (EDX) and Atomic Force Microscopy (AFM and c-AFM) to identify differences in their grain sizes and conductive currents. At the same time, fully-fuctional devices with the same perovskite layer were exposed outdoors for investigation of their long-term operation and understanding of their outdoor behavior based on their morphology and photocurrent investigated with the above-mentioned methods. Moreover, Ultrafast spectroscopy has been applied in encapsulated perovskite thin films before and after outdoor testing to reveal changes in carrier dynamics during the outdoor exposure of samples.

2 EXPERIMENTAL APPROACH

2.1 Microscopic measurements (SEM, AFM, EDX)

Several perovskite thin films (i.e., no encapsulation, only active layer deposited on a silicon substrate) were provided to Fraunhofer IKTS for structural characterization. Two different types of samples have been utilized labelled as type A and type B perovskite. Type A perovskite has a perovskite composition $Cs_{0.18}FA_{0.82}PbI_{20.82}Br_{0.18}$ while type B has a perovskite composition of $Cs_{0.1}FA_{0.9}PbI_{2.865}Br_{0.135}$ (+ 5mol% excess of FA and Cl).

During microscopy measurements all samples were cut to smaller pieces to always examine a pristine surface that has not been altered by exposure to probes from other methods such as electrons or photons. The samples were handled with great care in inert atmosphere to avoid uncontrolled degradation from exposure to moisture and air. Sample transport between the different instruments and the glovebox was also conducted under inert conditions via a special transfer of modules (see Figure 1).

Figure 1: Sample mounted in vacuum shuttle for analysis in Zeiss SEM.

For surface and cross section SEM a Zeiss Crossbeam 550 SEM system was utilized. The system is equipped with a focused ion beam (FIB) for material ablation and cross-section preparation and an energy dispersive X-ray spectrometer (EDX) for element analysis. Images have been acquired exploiting different type of detectors (in-lens detector, secondary electrons detector) and different accelerating voltages (5, 10, 20 kV). The EDX detector (Ultim Max, Oxford Instruments, Abingdon on Thames, UK) was operated with 20 kV electron beam voltage to detect signal from heavier elements according to the perovskite composition. The AFM used was a NX20, manufactured by Park System (Suwon, South Korea), which was exploited to scan the surface of the samples acquiring 2D images to study the morphology and the local conductivity of the sample surface. The measurements were conducted in non-contact mode using a NanoWorld® Arrow™ NCR probe having a tip radius of curvature of less than 10 nm.

2.2 Ultrafast pump-probe spectroscopy

For the spectroscopic investigation of samples, several encapsulated perovskite thin films have been provided to the Laboratory of Ultrafast Science at University of Cyprus. Perovskite thin films consist by only the perovskite layer and this structure was selected for those type of measurements since complete solar cells modules consist of various layers of materials and this makes impossible to understand the complicated ultrafast dynamics in these samples. Ultrafast measurements have been conducted before and after outdoor exposure of the perovskite thin films. The experimental apparatus utilized for those measurements consisted of a Ti:Saphire ultrafast amplifier producing 100 fs pulses centred at 800 nm at a repetition rate of 1 KHz. The pump wavelength at 400 nm was generated using a BBO non-linear crystal. A small part of the fundamental energy (~2.5 μJ) was used to generate white light continuum using a 4 mm sapphire window. This white light was used in probing different energy states in the VIS region of the spectrum between 420 nm to 780 nm.

2.3 Outdoor apparatus

Fully-functional devices with the same perovskite composition used in the microscopy and spectroscopy measurements have been provided to FOSS Research Centre for Sustainable Energy for outdoor testing and investigation of performance degradation outdoors. Comparison of the outdoor degradation of devices with the information provided by the microscopy/spectroscopy techniques is then implemented. During the outdoor testing, the perovskite devices have been located at fixed plane array and current-voltage (I-V) measurements have been collected at regular intervals over several weeks of testing. Forward and reverse voltage sweeps have been applied during each IV curve. Between IV scans, the modules were kept at open-circuit voltage (V_{oc}). The voltage sweep rate was chosen to be 1 V/s while a forward-first voltage sweep strategy has been used in all instances. The outdoor testing of the samples started on the 12th of January 2021 and lasted for 16 weeks. Alongside the I-V traces from the devices, environmental sensors have been used to collect solar irradiance in the plane of array, ambient and device temperature, wind velocity and humidity/precipitation levels.

3 RESULTS & DISCUSSION

3.1 SEM measurements

Surface SEM measurement analysis yields information on the film homogeneity and morphology. All films appear densely packed and cover the whole surface homogeneously. Different grain sizes for the different perovskite types are clearly determined, with a mean diameter of 128 nm (perovskite A) versus 192 nm (perovskite B). The SEM images of the two different types of samples are depicted in Figure 2. According to the literature larger grains were found to exhibit improved photovoltaic properties.

Figure 2: Top view SEM images for perovskite type A (#1) and perovskite type B (#2).

Cross-section SEM was conducted next to have a closer look into the layer interface in order to check adhesion, layer thickness and composition variations. A fractured edge method was utilized for cross-section preparation but this method exhibited mechanical damage on the samples.

3.2 EDX measurements

The elemental composition of the samples was determined via EDX analysis performed on the sample cross-section. Therefore, the whole cross-section was mapped. The EDX elemental map of perovskite type A with composition $Cs_{0.18}FA_{0.82}PbI_{20.82}Br_{0.18}$ is depicted in

Figure 3. The perovskite films show for both samples the expected element composition (Cs, Pb, I, Br, C and N from FA), which is equal for perovskites A and B. Fluor from the added FACl and LiF is detected in type B perovskite as well as Indium and Oxygen from the ITO surface layer. Also, the underlying silicon substrate can be identified. Elemental maps have shown that elements were distributed homogeneously in the perovskite thin films, which confirms that they are of good quality.

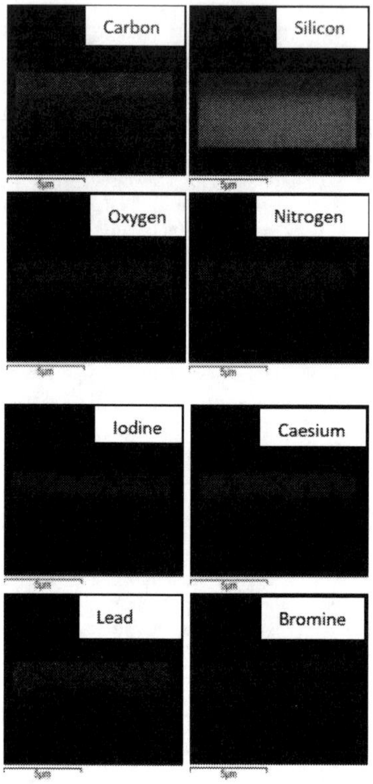

Figure 3: EDX elemental maps of perovskite type A. The perovskite composition of this sample is $Cs_{0.18}FA_{0.82}PbI_{20.82}Br_{0.18}$.

3.3 Atomic Force Microscopy

Height scans of the two different types of perovskites were implemented first (Figure 4). From the topographic images shown in Figure 4, a homogenous coverage and film morphology can be inferred. The grain size distributions fit to the observations from SEM. For perovskite B, a lot of grains apparently contain 2D-defects such as twin boundaries. The average roughness derived from the scanned area calculates to ~7nm and ~15nm for perovskite A and B, respectively.

The film thickness was measured at a film edge prepared by gently scratching with a razor blade. After image processing (slope correction), the histogram is used for thickness determination (see right part of Figure 4). The distance between the background/substrate peak and the film yields the average film thickness; ~470nm for perovskite A and ~610nm for perovskite B.

Conductive Atomic Force Microscopy (c-AFM) is a special AFM mode, which provides information on the electronic properties (local current inhomogeneities coupled to topography) by measuring in contact mode and applying

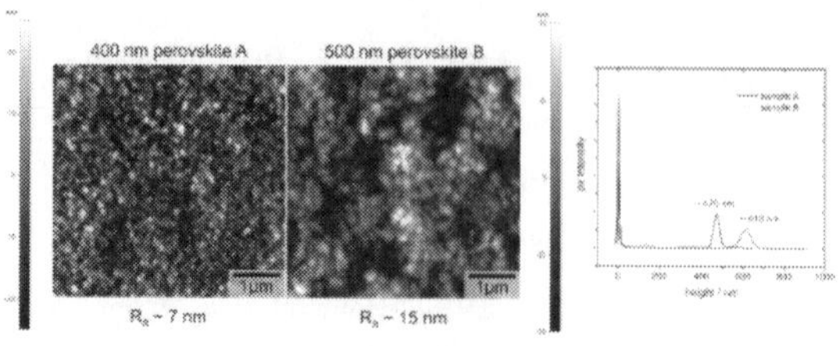

Figure 4: Left: overview AFM height scans of samples A and B in an area of 5x5μm. Right: pixel histogram derived from AFM edge scans to determine the average film thicknesses.

a bias to the sample. The signal is thereby generated by the current flow between the tip and its contact point to the sample. Additionally, the topography is recorded by deflection of the cantilever. Both images are correlated afterwards. For the investigated perovskite samples (A and B, without additional layers on top), perovskite B with its larger grains and less grain boundaries exhibits considerably higher current densities than perovskite A (mean value of 4.1 nA for perovskite B compared to 2.9 nA for perovskite A). Grain boundaries are clearly visible in the c-AFM images, hence, impact the optoelectronic properties. The c-AFM images are depicted in Figure 5.

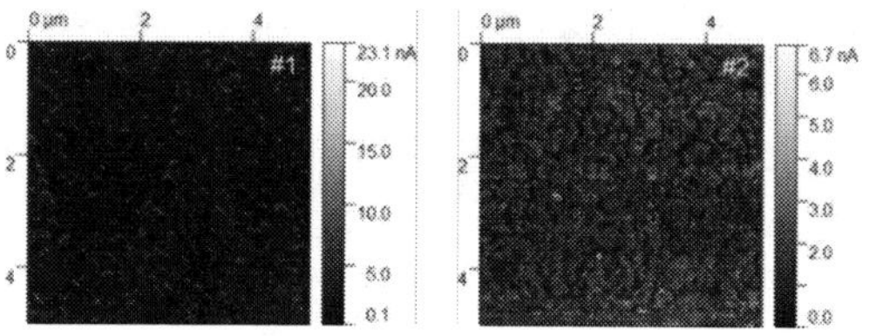

Figure 5: AFM current image of perovskites A (#1) and B (#2). The applied sample bias is -2/2 V.

3.4 Ultrafast spectroscopy

Ultrafast time-resolved differential transmission spectra were obtained for the samples using a continuous of probing wavelengths ranging from 420 nm – 780 nm at room temperature. Samples of perovskite type A were studied with this technique. Initially, femtosecond differential transmission measurements were conducted at pristine samples and at different fluence values (see Figure 6 for the spectra at pump powers 1 μJ). In particular, fluence was varied between 100 μJ/cm^2 and 400 μJ/cm^2. The spectra showed a dominant peak centred at 740 nm. Two additional peaks at 720 nm and 755 nm were also observed. A possible reason for these side-peaks is the presence of defect states energetically located close to the main 740 nm peak. Additionally, negative differential transmission values between 530 nm - 660 nm were observed indicating free carrier absorption. Carriers in this energy range were absorbed by the sample. Eventually, the system evolved to equilibrium. The behavior of the material was not affected by the different pump powers as all spectra obtained present the same features at the different pump powers studied.

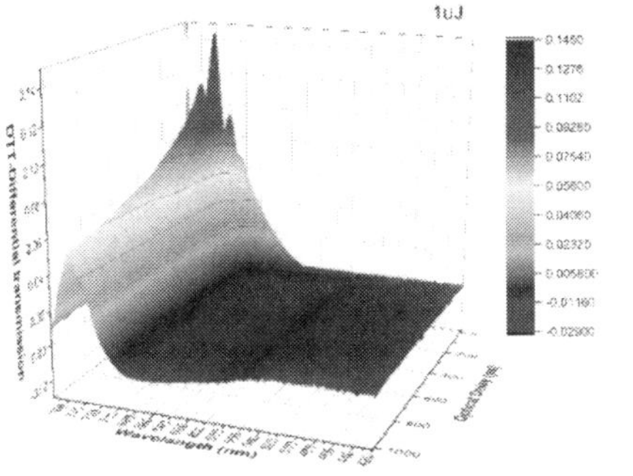

Figure 6: Differential transmission spectra of a pristine perovskite thin film at pump power of 1 μJ. Pump wavelength was set to 400 nm.

Then the differential transmission spectra were studied in encapsulated mini-modules before and after outdoor exposure. Figure 7 demonstrates the pump-probe spectra as collected on the initial date and after degradation. It is evident that the main characteristics i.e. position of the main peak (state filling) and differential transmission amplitude showed little change after this period of time.

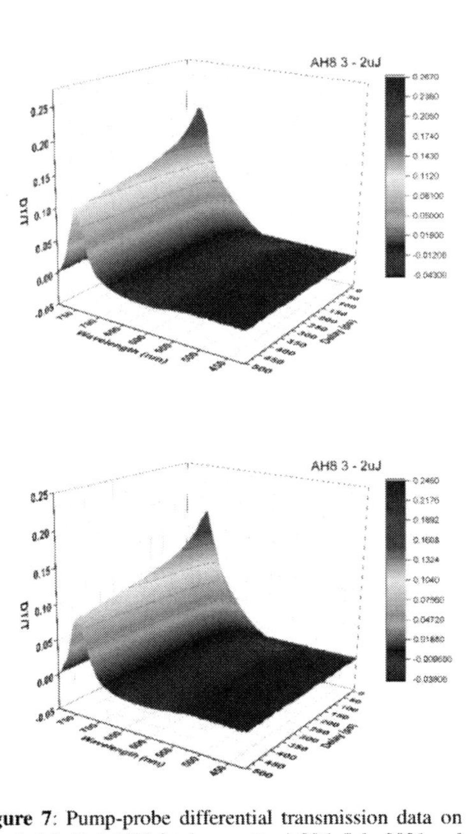

Figure 7: Pump-probe differential transmission data on sample labelled AH8 3 taken on (top) 29th July 2021 and (bottom) on 16th September 2021.

Further analysis of the ultrafast spectra was carried out to detect any differences in the decay constants. Table I presents the decay constants as derived by the exponential fittings on the main peak of the collected data. It is apparent that the degradation data revealed similar values for the peak wavelength and slight increase in relaxation times. This can be attributed to minor changes in the composition of the perovskite material. As changes in the chemical composition are becoming more significant the relaxation times obtained become noticeably increased.

Table I: Decay lifetimes before and after outdoor degradation as derived from pump probe measurements. Time constant have units of ps.

Sample	Before degradation (29/7/2021)	After degradation (16/9/2021)
AH8 3	y_0=0.54±0.01	y_0=0.49±0.01
	a_1=0.47±0.01	a_1=0.51±0.01
	τ_1=157.5±8.1	τ_1=191.5±11.4

3.5 Outdoor testing

After the microscopy and spectroscopy studies in perovskite thin films, six (6) encapsulated fully-functional devices with perovskite composition A and B have been monitored outdoors over the period from 12th of January until the 13th of May 2021 (16 weeks).

The normalized power conversion efficiency (PCE) over the sixteen weeks of outdoor exposure from all the 6 modules under test is depicted in Figure 8. Modules A1, A2, and A3 are of type A perovskite while modules B1, B2 and B3 are of type B perovskite. These measurements were taken at reverse sweeps and all parameters have been normalized to their initial values obtained outdoors. Over the entire period of exposure, type A modules presented higher PCE values than those of type B and this is attributed to the significant efficiency reduction of type B modules the first week of outdoor testing. The significant efficiency reduction in type B modules was found to be correlated to the open-circuit voltage loading between IV scans during outdoor exposure. This conclusion was found after indoor studies of both types of perovskites in the presence of Voc and MPP load [8]. Thus, even though the type B perovskites (thin film) demonstrated larger grains sizes in morphology and higher conductive currents when probed using microscopy methods, the outdoor behavior of those modules was significantly different from the results extracted in thin film level. The addition of extra layers (ETL, HTL etc.) in the full device adds more sources of degradation in devices and thus the behavior of the perovskite in a thin film level was shown to differ significantly compared to the perovskite in a full device.

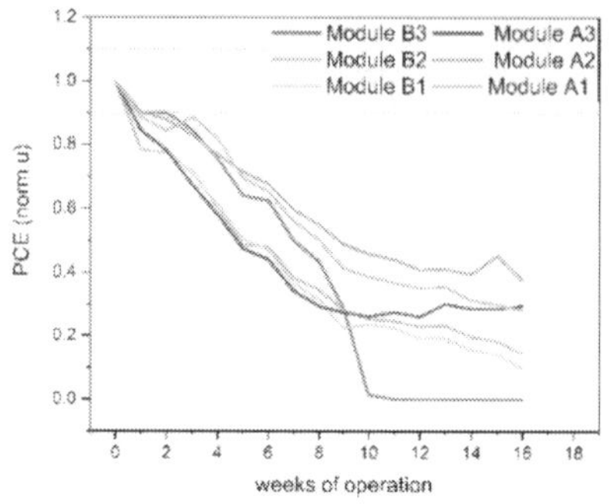

Figure 8: Weekly average performance (normalized values) of perovskites mini-modules of type A (A1, A2, A3) and B (B1, B2, B3) the first 16 weeks of outdoor exposure. Efficiency at reverse sweeps is reported on the graph.

4 CONCLUSIONS

In summary, several microscopic and spectroscopic techniques have been applied in bare and encapsulated thin film perovskite samples of two different perovskite compositions (labelled as type A and type B perovskites) and at different degradation stages (before and after their outdoor exposure). The SEM and AFM results showed different grain sizes and current densities at bare thin films. Perovskite thin films with additives (type B perovskite) exhibit larger grain size and less grain boundaries and therefore considerably higher current densities. However, during the outdoor exposure of full devices with type B perovskite composition, significant degradation was obtained due to the bias load applied and thus the beneficial effect of the additives was not detected. Therefore, the morphology results of thin perovskite films cannot be directly correlated with the outdoor behavior of full devices where all layers co-exist in the device.

Ultrafast spectroscopy was applied in only one type of perovskite (type A) before and after outdoor exposure. Pump-probe differential transmission data after degradation outdoors revealed slight changes in main peak position, differential transmission amplitude and carrier relaxation times.

5 AKNOWLEDGEMENTS

This work has been financed by the European Union through the TESTARE project (Grant ID: 101079488) and by the European Regional Development Fund and the Republic of Cyprus through the Cyprus Research and Innovation Foundation and the DegradationLab project (Grant ID: INFRASTRUCTURES/1216/0043).

6 REFERENCES

[1] C. A. Aranda, L. Caliò, and M. Salado,,Crystals 11 (2021) 1-16

[2] K. Domanski, E. Alharbi, A. Hagfeldt, M. Gratzel, and W. Tress, Nat. Energy 3 (2018) 61-67.

[3] M. Saliba *et al.*, Energy Environ. Sci., 9 (2016) 1989-1997

[4] C. S. Pathak, B. J. Chang, and S. Song, Dye. Pigment 218 (2023) 111469.

[5] A. Gomez, S. Sanchez, M. Campoy-Quiles, and A. Abate, Nano Energy 45 (2018) 94–100

[6] J. J. Li, J. Y. Ma, Q. Q. Ge, J. S. Hu, D. Wang, and L. J. Wan ACS Appl. Mater. Interfaces 7 (2015) 28518-28523.

[7] A. P. Thilakan *et al.*, ACS Appl. Mater. Interfaces 11 (2019) 21473-21480.

[8] V. Paraskeva *et al.*, Energies 16 (2023) 1-19.

40th European Photovoltaic Solar Energy Conference and Exhibition

This presentation was selected by the Sc. Committee of the EU PVSEC 2023 for submission of a full paper to one of the EU PVSEC's collaborating peer-reviewed journals.

OPTIMIZING THE ITO INTERCONNECTION LAYER FOR MONOLITHIC PEROVSKITE SILICON TANDEM SOLAR CELLS WITH HIGH SHORT-CIRCUIT CURRENTS OF 20 mA/cm^2

Özde Şeyma Kabaklı[1*], Kaitlyn McMullin[1], Leonard Tutsch[1], Martin Bivour[1], Oliver Fischer[1,2], Alexander J. Bett[1], Martin Hermle[1], Jan Christoph Goldschmidt[3], Stefan W. Glunz[1,2], Juliane Borchert[1,2], Patricia S. C. Schulze[1]

[1]Fraunhofer Institute for Solar Energy Systems, Heidenhofstr. 2, 79110 Freiburg, Germany

[2]University of Freiburg, Department of Sustainable Systems Engineering (INATECH), Emmy-Noether-Str. 2, 79108 Freiburg

[3]University of Marburg, Department of Physics, Renthof 7. 10, 35032 Marburg

*Corresponding author: Özde Şeyma Kabaklı[1], +49 761 4588-2447, oezde.seyma.kabakli@ise.fraunhofer.de

ABSTRACT: In 2-terminal perovskite silicon tandem solar cells the interconnection layer that electrically connects the top and bottoms cells plays a pivotal role for reaching highest efficiency. Typically made from a transparent conductive oxide, its electrical properties determine series resistance losses and shunt resilience, while its optical properties determine via reflection and transmission the current matching between the sub-cells. In this study, ITO interconnection layers are optimized for two-terminal monolithic p-i-n perovskite silicon tandem solar cells with a flat front side and textured rear side silicon bottom solar cell. The oxygen gas flow ratio during the sputter process of interconnection ITO is varied between 0% and 12% in oxygen and argon mixed gas. The effect of the variation on the opto-electrical properties of ITO thin films and tandem solar cell parameters is investigated. Optimization of the oxygen flow rate during sputtering increased the absolute tandem efficiency from 27.1% to 28.9%, while reaching a significantly higher short-circuit current density of 20 mA/cm^2.

Keywords: Perovskite, Silicon, Tandem, Interconnection, ITO, photovoltaics

1 INTRODUCTION

Monolithic two-terminal perovskite silicon tandem solar cells have the potential to reach 45.1% power conversion efficiency (PCE) by reducing thermalization losses[1]. In monolithic perovskite silicon tandem solar cells, indium tin oxide (ITO) is often used as the electrical interconnection layer of the two sub solar cells [2]. An interconnection ITO should (i) have high vertical conductivity to reduce series resistance losses, (ii) have low lateral conductivity to supress leakage current by shunt pathways and (iii) cause only minor optical losses[3]. An approach to increase sheet resistance (R_{Sheet}) and decrease optical absorption of the ITO layers is to increase the oxygen flow rate during sputtering [4]. Depending on the specifics of the shunts, the contribution of the bottom cell to the lateral transport and other factors, a high sheet resistance of the recombination (interconnection) junction can mitigate the impact of sub-cell shorting in the presence of local shunt paths[5–7].

The hydroxyl group concentration at the ITO surface is important as the binding mechanism for 2PACz self-assembling molecules involves binding with a covalent bond to the hydroxyl group at the surface and forming P-OH and P=O groups with H_2O as a side product. The ITO surface activity can be increased via surface cleaning methods such as hydrofluoric acid and UV-Ozone treatment[8]. Altering the ITO surface chemistry by oxygen flow might also influence the adsorption mechanism of the SAM by creating more stoichiometric oxide binding sites facilitating easy and strong adsorption.

This study investigates the effect of oxygen flow rate during the sputter process of the interconnection ITO in the tandem structure on solar cell parameters and assembly of 2PACz on ITO. The oxygen flow rate ratio during the sputter process is changed (0 sccm to 180 sccm) while keeping the thickness constant at 20 nm.

2 EXPERIMENTAL

2.1 Perovskite Silicon Tandem Solar Cells

Rear side textured, 250 µm thick p-type silicon float zone wafers (base resistivity 1 Ω cm) are used for fabrication of silicon heterojunction bottom solar cells. The rear side electrode consists of 195 nm ITO and 1000 nm Ag. Front side interconnection ITO thickness is 20 nm[9]. The ITO films are deposited in an industrial type in-line tool via DC sputtering from a rotary target. The oxygen flow rate during interconnection ITO sputtering is varied from 0% to 12% keeping the thickness constant. 2PACz is spin coated on the front ITO, followed by perovskite spin coating. 2PACz and perovskite solution preparation and deposition steps are done as described in ref [10]. LiF (1 nm) and C_{60} (15 nm) are evaporated under high vacuum. SnO_x (20 nm) is deposited using atomic layer deposition. Front side ITO (25 nm) is DC sputtered. Finally, Ag (200 nm) and MgF_2 (100 nm) are evaporated under high vacuum. Interconnection ITO, front side ITO and Ag depositions are fabricated using a shadow mask defining the individual active solar cell areas on substrates (4 solar cell of 0.25 cm^2 active area per substrate). All the steps of top solar cell processing, except the front side ITO sputtering, are carried out in inert atmosphere.

2.2 Characterization

Hall measurements are conducted to determine the electrical properties (sheet resistance, mobility, electron density) of the ITO thin films deposited on glass.

I-V measurements are done using a Wacom sun simulator consisting of a filtered Xenon (UV and visible light) and a filtered Halogen lamp (infrared light), and a Keithley 2400 source meter. Lamp intensities are adjusted using relative spectral response values per each solar cell group variation following the procedure described by Meusel et al. [10]. Measurements are performed in ambient air with the sample chuck temperature kept constant at 25 °C. EQE measurements of tandem solar cells are done using a chopped light (~130Hz frequency) of a Xenon lamp. The measurements of the perovskite and silicon sub-solar cells are done separately by applying a bias voltage and bringing the measured sub-cell to current limitation by LED bias illumination. Detailed description of how tandem cells need to be measured can be found in [11–15]

The solar cell active area is defined by shadow masks during *I-V* measurements. *I-V* measurements are done under 1 sun illumination with a solar simulator with two lamps allowing for adjustment of the spectrum.

The series resistance of the tandem solar cells (R_{series}) is calculated from the slope of the *I-V* curve at V_{OC}.

Reflection measurements are conducted on full solar cells with the LOANA tool.

3 RESULTS

The oxygen flow rate during ITO thin film sputter process is varied between 0%-12% (oxygen percentage in oxygen and argon mixed gas). Hall measurements show that the sheet resistance increased from 392 Ω/sq to 10700Ω/sq. The increase in the sheet resistance is dominated by the decrease in electron density, while highest mobility is obtained with 5% oxygen flow rate (Figure 1).

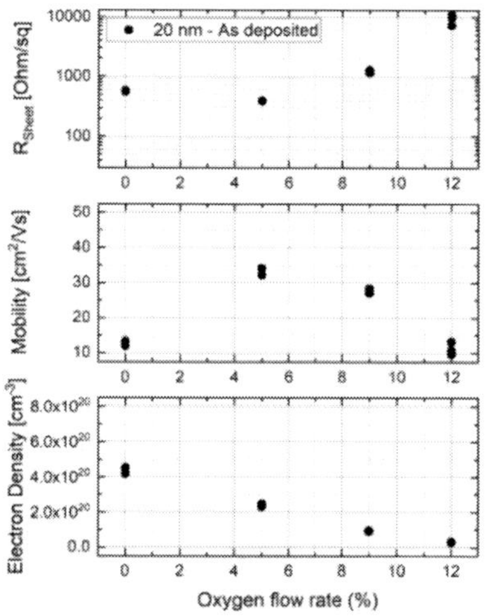

Figure 1: The flow rate of oxygen during sputtering of the as-deposited interconnection ITO on glass is varied between 0%-12%. Sheet resistance increases as the oxygen flow rate increases from 0% up to 12% Highest mobility is obtained for 5% oxygen flow rate. Highest electron density is obtained for 0%.

The 20 nm ITO thin films with 0%, 5%, 9% and 12% oxygen flow rate and R_{Sheet} of 575 Ω/sq, 392 Ω/sq, 1260 Ω/sq, 10700 Ω/sq on glass are implemented in perovskite silicon tandem solar cells, respectively (Figure 2).

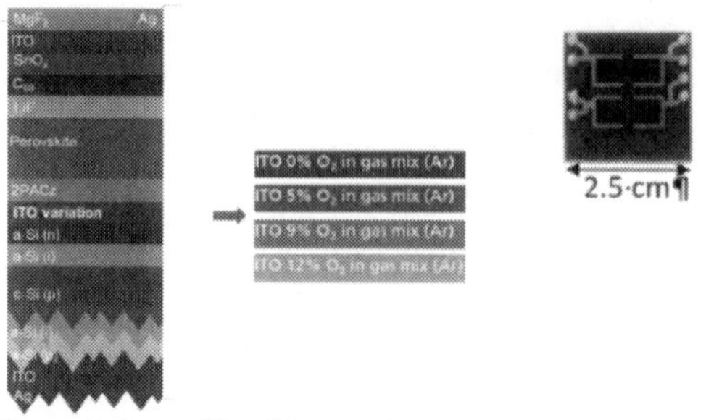

Figure 2: Perovskite silicon tandem solar cell sketch (left) with self-assembling molecular layer as hole transport layer and a thin LiF passivation at the electron transport layer-perovskite interface. Oxygen flow rate during sputtering of ITO interconnection layer is varied while keeping the rest constant. Each substrate (2.5 cm x 2.5 cm) has 4 solar cells with active areas (0.25 cm²), defined by a shadow mask that is used during processing (right).

The normalized EQEs of the highest efficiency solar cell per group show small changes for 5% and 9% oxygen flow rate ITO compared to 0% and 12% for the silicon sub-cell. In the normalized EQE graph, between 900-1100 nm wavelength range, a negligible change in the shape is observed. Reflection losses of 5% group is lower compared to other oxygen flow rate groups at 1100-1200 nm wavelength range. The perovskite silicon tandem solar cell with 9% oxygen flow rate ITO interconnection layer has the highest iV_{OC} (Figure 3).

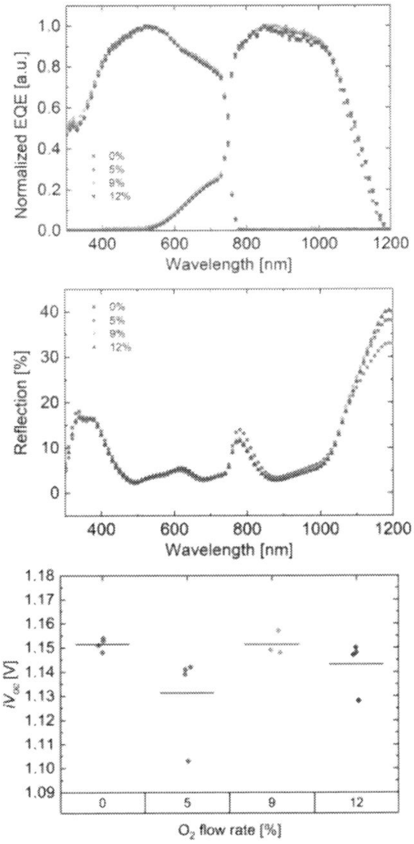

Figure 3: 2-terminal monolithic perovskite silicon tandem solar cell EQE is normalized to a maximum value of 1 (a), reflection at high wavelengths is reduced for 5% oxygen flow (b), iV_{OC} changed depending on the oxygen flow rate(c).

There is large spread within each *I-V* data group, a trend of increasing average *jsc* from 9% to 12% oxygen flow can be observed. The highest average *jsc*, 20.2 mA/cm^2 in forward scan and 20.1 mA/cm^2 in reverse scan, is obtained with the highest oxygen flow group of 12% oxygen flow. In a previous study a similar tandem solar cell with 5% oxygen flow rate for the ITO interconnection layer was shown to be almost current matched with a slightly lower current from the bottom solar cell[10]. Taking this into consideration the increase of the short circuit current might be related to higher bottom solar cell current with increased transparency of the ITO interconnection layer with higher oxygen flow rate. More efficient infrared light harvesting can be expected if the reduced parasitic absorption by increasing the oxygen flow rate is the dominating optical effect. However, the trend in reflection does not follow the oxygen flow rate (Figure 3). Even though the parasitic absorption might be lower for higher oxygen flow rates, a significant change of the parasitic absorption of the ITO layer above 5% oxygen flow rate is not expected. One other optical reason for such a trend can be increasing refractive index for higher oxygen flow rates. Spectrometric characterization is necessary to investigate if there are any changes in current matching point or if there are any other effects dominating [10–12]. Both 9% and 12% groups can achieve power conversion efficiencies above 26% (Figure 4). The absolute V_{OC} exhibits an upwards trend for higher O_2 flow rates, whereas the average is around 1.825 V for all groups due to large data spread. One of the reasons for the increasing absolute V_{OC} could be reduced non-radiative recombination, which can be observed by iV_{OC} determined by PLQY measurements on tandem solar cells. Overall, highest stabilized power conversion efficiency over 3 minutes, 28.9%, 20.03 mA/cm^2 j_{SC}, 1.85 V V_{OC}, and 78% *FF* is obtained with 9% oxygen flow rate with in-house measurements. The first two groups have less data points due to a smaller number of samples available. We are currently working on further data collection.

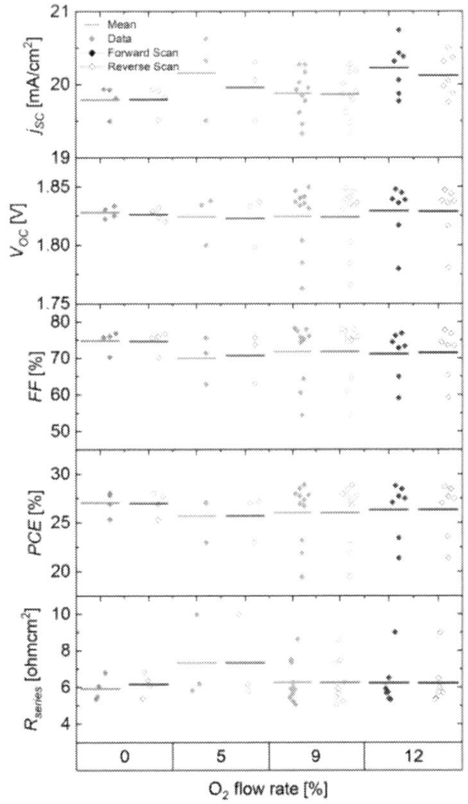

Figure 4: *I-V* data of the perovskite silicon tandem solar cells. Highest mean *jsc* is obtained for 12% xygen flow rate. High PCE above 27% is obtained for 9% and 12% oxygen flow.

As mentioned in the experimental details, the ITO interconnection layer is subjected to UV-ozone treatment before the 2PACz spin coating. The UV-Ozone treatment of the ITO surface might have a stronger influence on the surface chemistry of the 2PACz compared to the sputter gas oxygen flow rate. One possible reason for not observing a significant trend between the groups might be the dominating effect of the UV-ozone treatment on the surface chemistry of ITO.

10.4229/EUPVSEC2023/2CO.1.5
020119-003

The highest efficiency solar cell from the group with 9% oxygen flow was measured at Fraunhofer ISE CalLab PV Cells and 27.1% power conversion efficiency was certified. One of the reasons for lower power conversion efficiency result compared to in-house measurements could be LiF induced degradation of the perovskite top solar cell stack over time.

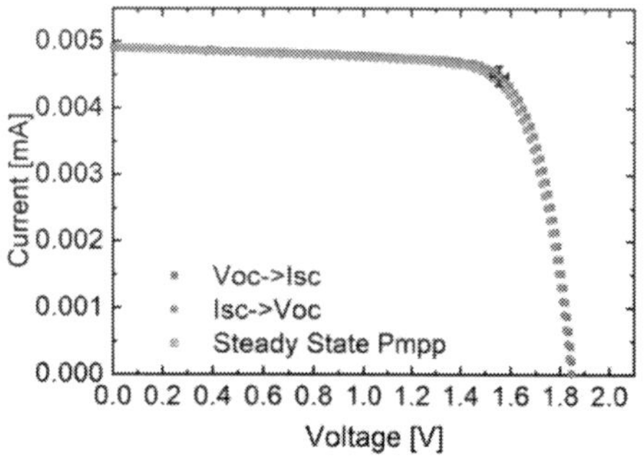

Figure 4: The highest efficiency solar cell was sent to CalLab at Fraunhofer ISE for certification three weeks later then the initial *I-V* measurements. Some degradation effect is observed over the weeks.

A summary of the certified result can be found in Table 1. The solar cell shows almost no hysteresis.

Table I: Solar cell parameters from the CalLab. The steady state PCE at maximum power point (MPP) is certified as 27.1±1.2%.

	Forward Scan		Reverse Scan	
V_{OC}	1845 ± 19	mV	1853 ± 19	mV
I_{SC}	4.900 ± 0.094	mA	4.908 ± 0.094	mA
I_{MPP}	4.46	mA	4.50	mA
V_{MPP}	1530	mV	1553	mV
P_{MPP}	6.83	mW	6.99	mW
FF	75.5	%	76.9	%
PCE				
Steady State MPP				
I_{MPP}	4.49 ± 0.15	mA		
V_{MPP}	1553 ± 35	mV		
P_{MPP}	6.97 ± 0.29	mW		
FF				
PCE	27.1 ± 1.2	%		

The highest efficiency solar cell was measured in-house over 10 days. The power conversion efficiency decreased by 1% over the 10 days (Figure 5).

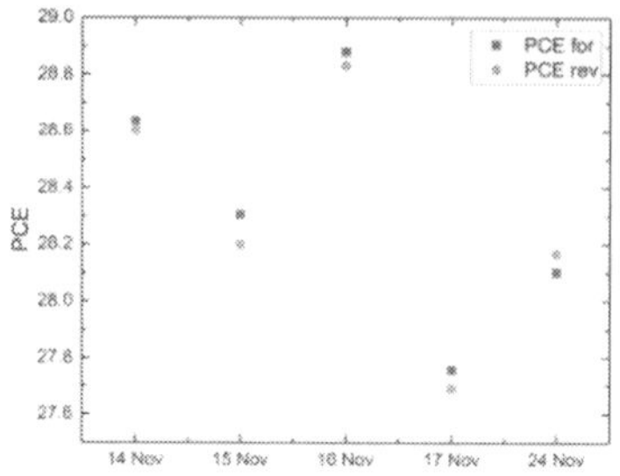

Figure 5: The in-house *I-V* measurements of the highest efficiency solar cell over 10 days. The degradation of the solar cells might be related with the LiF passivation interlayer. The migration of Li^+ and F^- ions might deteriorate the C_{60} interface [16, 17]. We are currently performing long term stability tests to understand the instability behaviour of the solar cells.

4 DISCUSSION and CONCLUSION

In this study, oxygen flow rate during sputtering is varied between 0% to 12% for the perovskite silicon tandem solar cells ITO interconnection layer. The increase from 0% to 12% resulted in a slight upward trend in j_{SC}. The slight changes in j_{SC} could be related to a change in tandem current matching condition due to a decreased absorption with higher oxygen flow rate or change in the refractive index. Implied open-circuit voltage (iV_{OC}) results suggest that the ITO interconnection layer oxygen flow rate could additionally influence the non-radiative recombination in the tandem solar cell. Also, the slope from j_{SC} to j_{MPP} could be an indication of parallel resistance problems and non-homogeneity of SAM binding over the active area. The bridging hydroxyl groups, spanning two or more metal sites, cause looser binding of the adsorbed SAMs on the surface compared to the terminal hydroxyl groups, that bond to a single metal atom site. The loosely bind SAMs can easily be disconnected from the ITO surface, which result in small pin-holes and induce current leakage.[8]. The non-homogeneity might be due to non-sufficient terminal hydroxyl groups in the ITO surface, also reducing the fabrication yield of results. However, in the 0%-12% range, no clear trend or significant influence of the oxygen flow rate can be observed in solar cell parameters. There is also no direct correlation found between higher sheet resistance of the ITO interconnection layer and shunt quenching[6]. This might indicate that either there are no dominant losses caused by local shunts or the increasing series resistance did not have significant impact on shunt quenching.

To address these questions, spectrometric characterization[11, 12] and PL imaging is planned as next steps. Moreover, the effect of higher oxygen flow rates on 2PACz assembly on ITO surface is currently under investigation. All in all, it is shown that with our optimizations, very high short-circuit current density, of 20 mA/cm^2 and high efficiency of 28.9% can be achieved. In order to reach higher efficiency and stability, the loss mechanisms at the interfaces will be addressed in further studies, such as implementing MgF_2 or Al_2O_3 as the passivation interlayer instead of LiF [17].

10.4229/EUPVSEC2023/2CO.1.5
020119-004

5 ACKNOWLEDGEMENTS

This work was supported by Fraunhofer Gesellschaft, Germany via the Fraunhofer MANITU Lighthouse Project. Ö.Ş.K. gratefully acknowledges scholarship from the Dr. Ruth Heerdt Stiftung.

6 REFERENCES

[1] M. H. Futscher and B. Ehrler, "Efficiency Limit of Perovskite/Si Tandem Solar Cells," *ACS Energy Letters*, vol. 1, no. 4, pp. 863–868, 2016, doi: 10.1021/acsenergylett.6b00405.

[2] C. Li, Y. Wang, and Choy, Wallace C. H., "Efficient Interconnection in Perovskite Tandem Solar Cells," *Small Methods*, vol. 4, no. 7, p. 2000093, 2020, doi: 10.1002/smtd.202000093.

[3] Y. Ko, H. Park, C. Lee, Y. Kang, and Y. Jun, "Recent Progress in Interconnection Layer for Hybrid Photovoltaic Tandems," *Advanced materials (Deerfield Beach, Fla.)*, vol. 32, no. 51, e2002196, 2020, doi: 10.1002/adma.202002196.

[4] A. Chen, K. Zhu, H. Zhong, Q. Shao, and G. Ge, "A new investigation of oxygen flow influence on ITO thin films by magnetron sputtering," *EuroSun2004*, vol. 120, pp. 157–162, 2014, doi: 10.1016/j.solmat.2013.08.036.

[5] K. O. Brinkmann *et al.*, "Perovskite-organic tandem solar cells with indium oxide interconnect," *Nature*, vol. 604, no. 7905, pp. 280–286, 2022, doi: 10.1038/s41586-022-04455-0.

[6] C. Blaga, G. Christmann, M. Boccard, C. Ballif, S. Nicolay, and B. Kamino, "Palliating the efficiency loss due to shunting in perovskite/silicon tandem solar cells through modifying the resistive properties of the recombination junction," *Sustainable Energy Fuels*, 2021, doi: 10.1039/D1SE00030F.

[7] F. Ventosinos, J. Klusacek, T. Finsterle, K. Kunzel, F.-J. Haug, and J. Holovsky, "Shunt Quenching and Concept of Independent Global Shunt in Multijunction Solar Cells," *IEEE Journal of Photovoltaics*, vol. 8, no. 4, pp. 1005–1010, 2018, doi: 10.1109/JPHOTOV.2018.2828850.

[8] M. Wu *et al.*, "Reconstruction of the Indium Tin Oxide Surface Enhances the Adsorption of High-Density Self-Assembled Monolayer for Perovskite/Silicon Tandem Solar Cells," *Advanced Functional Materials*, 2023, doi: 10.1002/adfm.202304708.

[9] P. S. C. Schulze *et al.*, "25.1% High-Efficient Monolithic Perovskite Silicon Tandem Solar Cell with a High Band Gap Perovskite Absorber," *Solar RRL*, vol. 4, no. 7, p. 2000152, 2020, doi: 10.1002/solr.202000152.

[10] M. Heydarian *et al.*, "Maximizing Current Density in Monolithic Perovskite Silicon Tandem Solar Cells," *Sol. RRL*, vol. 7, no. 7, p. 2200930, 2023, doi: 10.1002/solr.202200930.

[11] A. J. Bett *et al.*, "Spectrometric Characterization of Monolithic Perovskite/Silicon Tandem Solar Cells," *Sol. RRL*, 2022, doi: 10.1002/solr.202200948.

[12] M. Meusel, R. Adelhelm, F. Dimroth, A. W. Bett, and W. Warta, "Spectral mismatch correction and spectrometric characterization of monolithic III-V multi-junction solar cells," *Progress in Photovoltaics: Research and Applications*, vol. 10, no. 4, pp. 243–255, 2002, doi: 10.1002/pip.407.

[13] M. Meusel, C. Baur, G. Létay, A. W. Bett, W. Warta, and E. Fernandez, "Spectral response measurements of monolithic GaInP/Ga(In)As/Ge triple-junction solar cells: Measurement artifacts and their explanation," *Progress in Photovoltaics: Research and Applications*, vol. 11, no. 8, pp. 499–514, 2003, doi: 10.1002/pip.514.

[14] M. Saliba and L. Etgar, "Current Density Mismatch in Perovskite Solar Cells," *ACS Energy Lett.*, vol. 5, no. 9, pp. 2886–2888, 2020, doi: 10.1021/acsenergylett.0c01642.

[15] M. Mundus *et al.*, "Spectrally resolved nonlinearity and temperature dependence of perovskite solar cells," *Solar Energy Materials and Solar Cells*, vol. 172, pp. 66–73, 2017, doi: 10.1016/j.solmat.2017.07.013.

[16] A. Al-Ashouri *et al.*, "Monolithic perovskite/silicon tandem solar cell with 29% efficiency by enhanced hole extraction," *Science (New York, N.Y.)*, vol. 370, no. 6522, pp. 1300–1309, 2020, doi: 10.1126/science.abd4016.

[17] J. Liu *et al.*, "Efficient and stable perovskite-silicon tandem solar cells through contact displacement by MgFx," *Science (New York, N.Y.)*, vol. 377, no. 6603, pp. 302–306, 2022, doi: 10.1126/science.abn8910.

CP.3.4

Opportunities for Thin Film PV in the Next Decade in Europe

A. Lange First Solar, Germany
M. Glöckler First Solar, USA
M. Merfert First Solar, Netherlands

First Solar | LEADING THE WORLD'S SUSTAINABLE ENERGY FUTURE

020120-001

The Western Hemisphere's Largest Solar Manufacturer

- Founded in 1999
- Differentiated thin film technology
- Leader in R&D and manufacturing of Thin Film PV
- Industry Leader in Responsible Solar
  - Lowest carbon PV technology and lowest life cycle environmental footprint
  - Member of Responsible Business Alliance
  - EPEAT Silver Rated

020120-002

 2

40th European Photovoltaic Solar Energy Conference and Exhibition

First Solar's CdTe Thin Film: General Characteristics

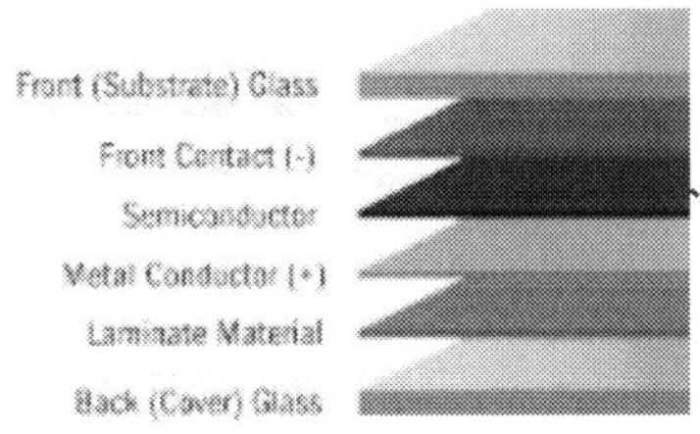

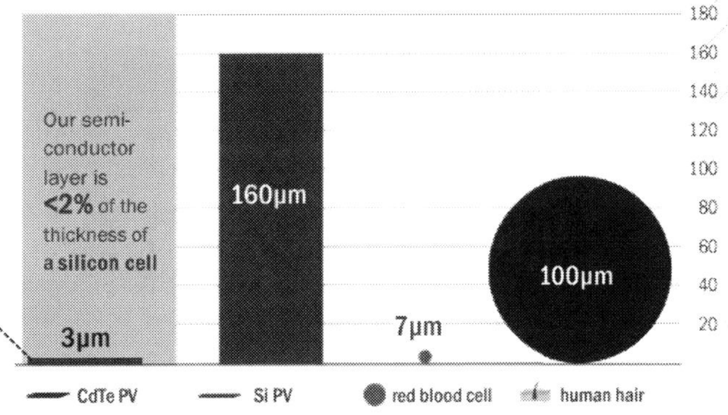

- CdTe as light absorbing material (instead of crystalline silicon)
- Thin film semiconductor: 3 micron instead of 160 micron (silicon)
- Efficiency: 19+%
  - Better quality control in production process
  - Year-1 energy: superior energy yield in hot / humid climates, CdTe does not exhibit LID
  - Lifetime energy: low annual degradation rate (-0.3%/yr warranted)

020120-004

4

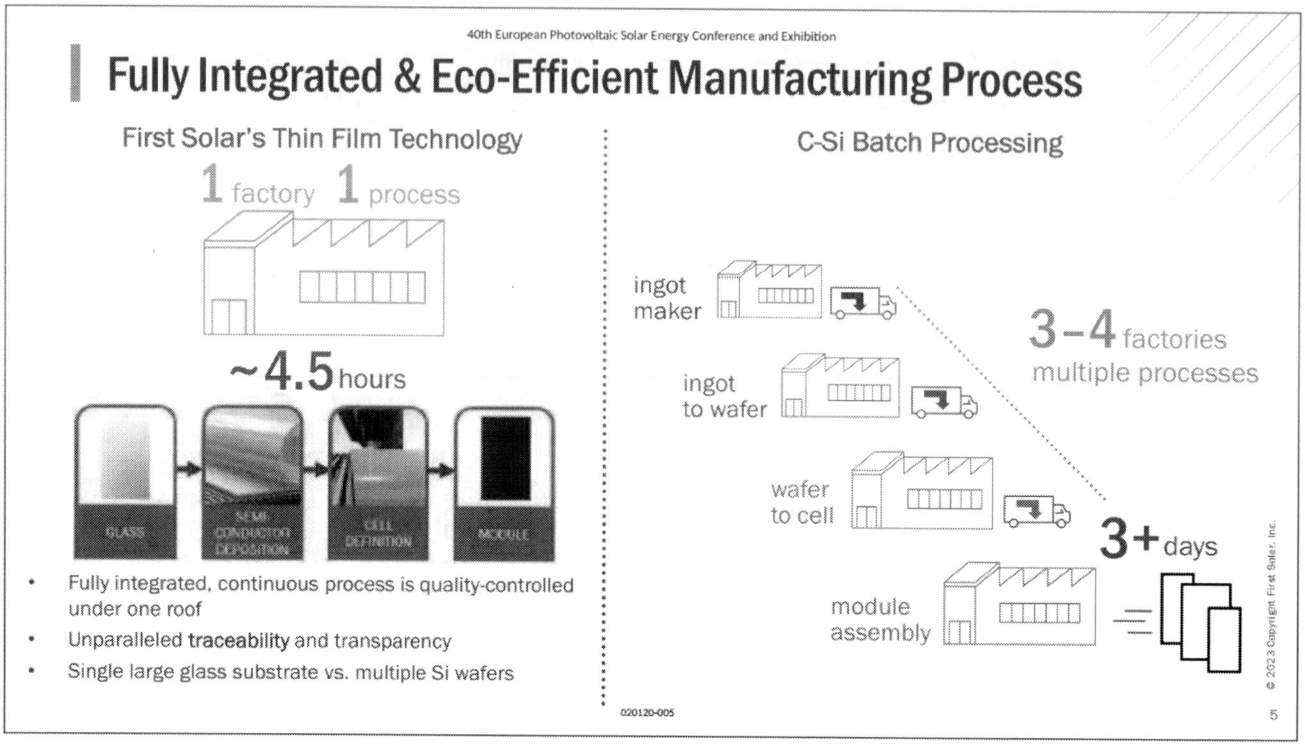
40th European Photovoltaic Solar Energy Conference and Exhibition
Fully Integrated & Eco-Efficient Manufacturing Process
First Solar's Thin Film Technology
1 factory 1 process
~4.5 hours
GLASS
SEMI CONDUCTOR DEPOSITION
CELL DEFINITION
MODULE
• Fully integrated, continuous process is quality-controlled under one roof
• Unparalleled traceability and transparency
• Single large glass substrate vs. multiple Si wafers
C-Si Batch Processing
ingot maker
ingot to wafer
wafer to cell
module assembly
3–4 factories
multiple processes
3+ days
020120-005
© 2023 Copyright First Solar, Inc.
5

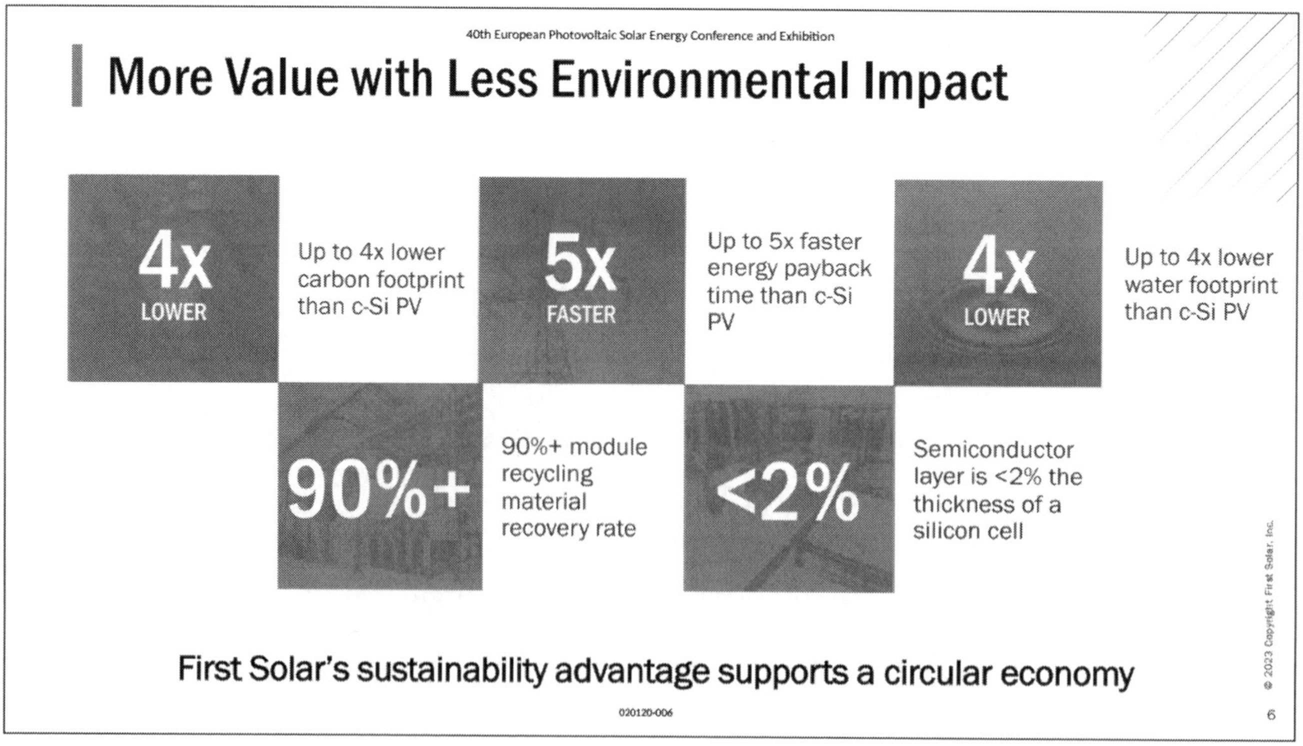
40th European Photovoltaic Solar Energy Conference and Exhibition
More Value with Less Environmental Impact
4x
LOWER
Up to 4x lower carbon footprint than c-Si PV
5x
FASTER
Up to 5x faster energy payback time than c-Si PV
4x
LOWER
Up to 4x lower water footprint than c-Si PV
90%+
90%+ module recycling material recovery rate
<2%
Semiconductor layer is <2% the thickness of a silicon cell
First Solar's sustainability advantage supports a circular economy
020120-006
© 2023 Copyright First Solar, Inc.
6

40th European Photovoltaic Solar Energy Conference and Exhibition
First Solar
Thin Film Innovation for the Future of PV
020120-007

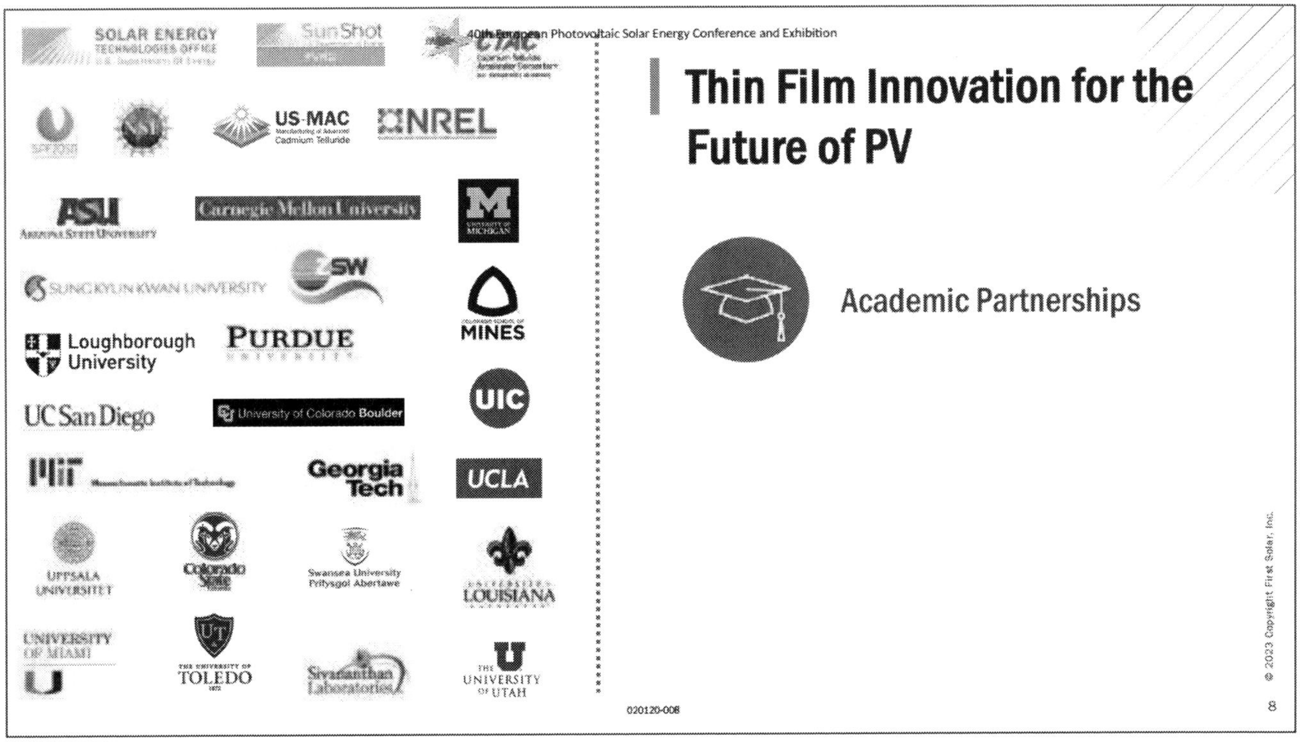
40th European Photovoltaic Solar Energy Conference and Exhibition
SOLAR ENERGY TECHNOLOGIES OFFICE
SunShot
CTAC
US-MAC
Cadmium Telluride
NREL
ASU
Carnegie Mellon University
SUNGKYUNKWAN UNIVERSITY
SW
MINES
Loughborough University
PURDUE
UC San Diego
University of Colorado Boulder
UIC
MIT
Georgia Tech
UCLA
UPPSALA UNIVERSITET
Colorado State
Swansea University Prifysgol Abertawe
LOUISIANA
UNIVERSITY OF MIAMI
THE UNIVERSITY OF TOLEDO
Sivananthan Laboratories
THE UNIVERSITY OF UTAH
Thin Film Innovation for the Future of PV
Academic Partnerships
020120-008
© 2023 Copyright First Solar, Inc.
8

EVOLAR
BOOSTING SOLAR POWER

Headquarters in Uppsala, Sweden

First Solar European Technology Center

Thin Film Innovation for the Future of PV

- Academic Partnerships
- Acquiring Know-How

020120-009

- New dedicated ~$78M perovskite development line
- Potential for full-scale pilot manufacturing line in the Jim Nolan Center for Solar Innovation

Thin Film Innovation for the Future of PV

- Academic Partnerships
- Acquiring Know-How
- Investing in Scale

020120-010

40th European Photovoltaic Solar Energy Conference and Exhibition
First Solar
Product Innovation
Roadmap

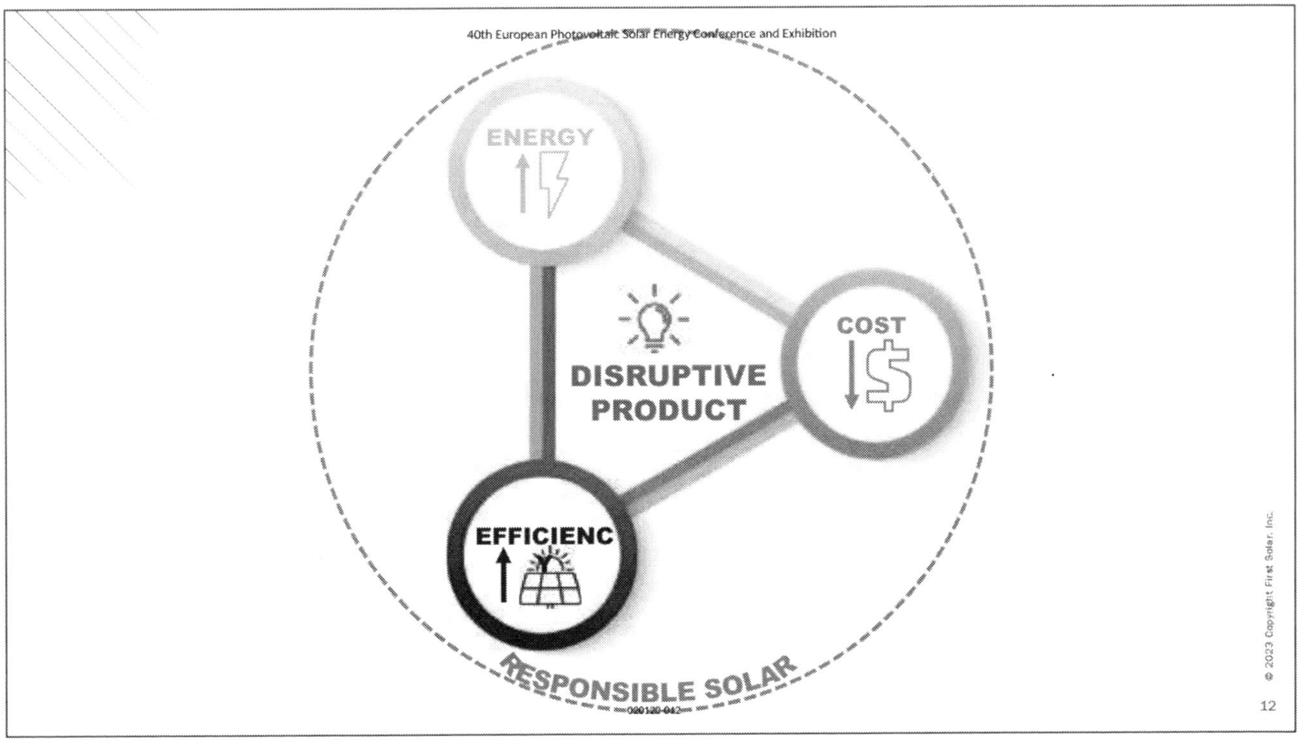
40th European Photovoltaic Solar Energy Conference and Exhibition
ENERGY
COST
DISRUPTIVE
PRODUCT
EFFICIENC
RESPONSIBLE SOLAR
© 2023 Copyright First Solar, Inc.
12

Single-Junction CdTe: 25%+ Cell Eff. Target

Ramping R&D funding and resources

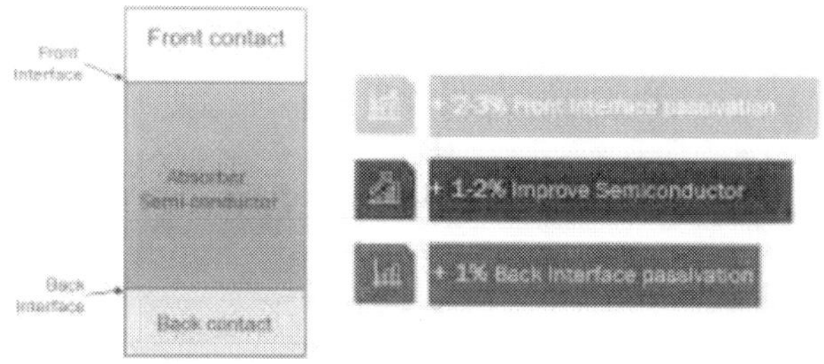

DOE Announces Goal to Cut Solar Costs by More than Half by 2030

https://www.energy.gov/articles/doe-announces-goal-cut-solar-costs-more-half-2030

Collaborators

$20MM CdTe PV Accelerator Program

- Enhance US competitiveness
- Enable ~25% cell by 2025

40th European Photovoltaic Solar Energy Conference and Exhibition

Bifacial Thin Film

15% ⇨ **25%+** **1-2%+**

Bifaciality — Bifaciality Roadmap — Real World Annual Energy

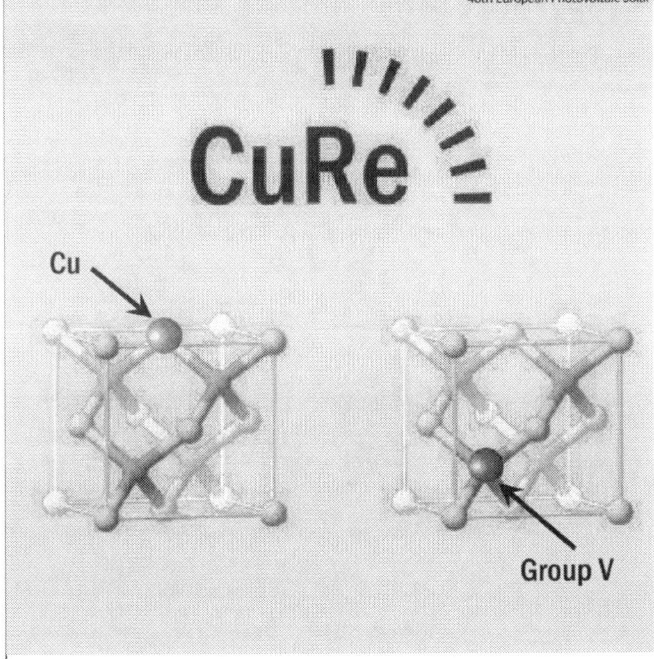

Improving the Semiconductor

- Improved Bifaciality
  15% → 25%
- Lower Degradation
  -0.3 → -0.2 %/year
  Path to -0.1 %/year
- Improved Temperature Coefficient
  -0.32 → -0.24 %/°C

Advanced Interconnect: The Opportunity

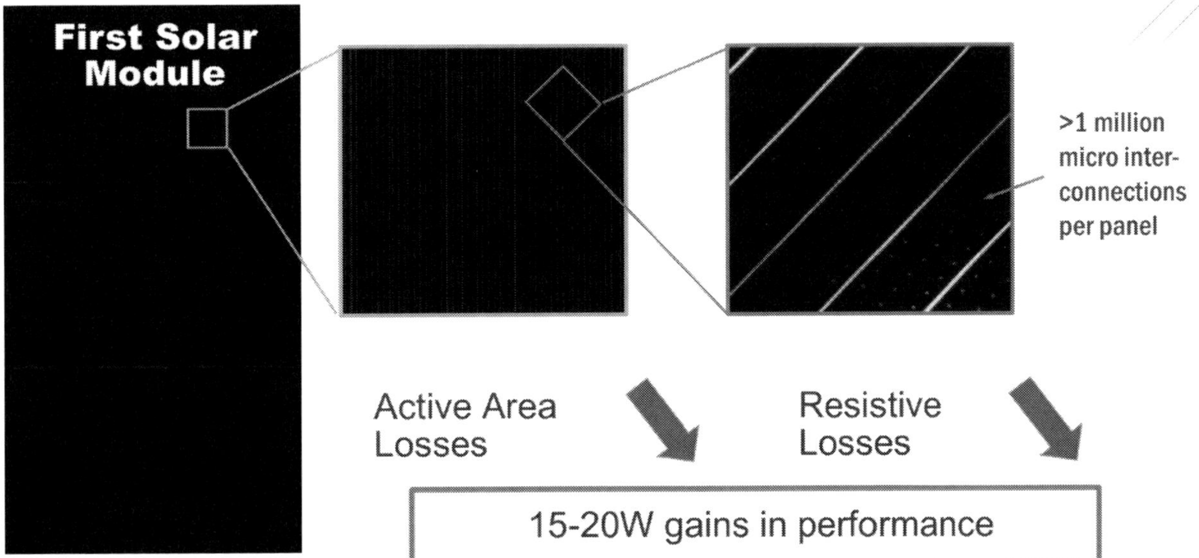

Active Area Losses

Resistive Losses

15-20W gains in performance

Cell Technology Roadmap

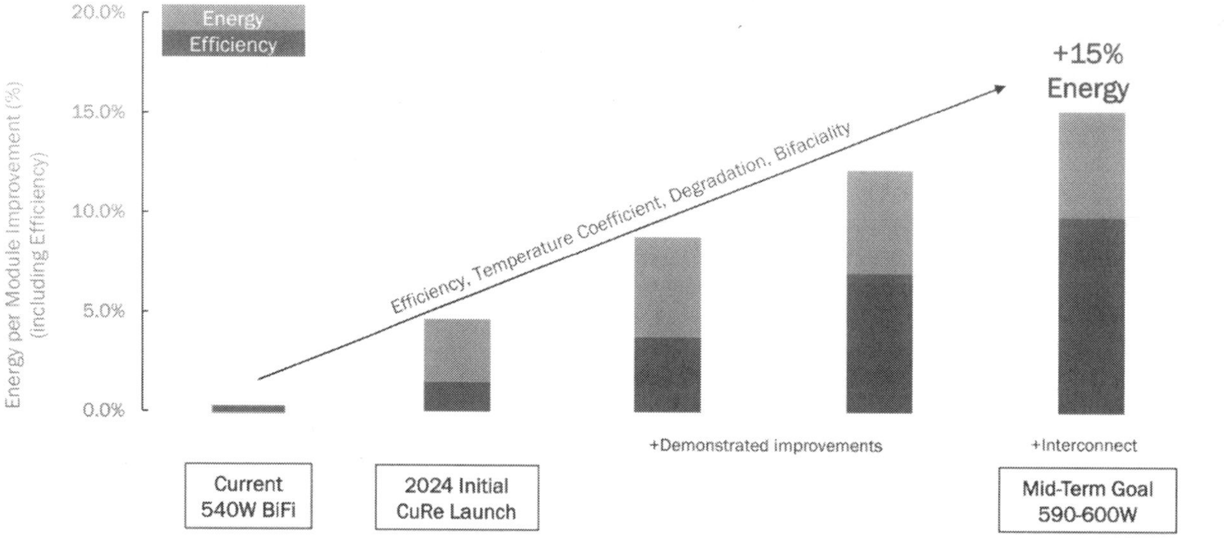

Energy represented is modelled with TMY3 weather files, NREL NSRDB albedo and PlantPredict software averaging 58 evenly distributed locations across 13 US states below 37 degrees latitude and greater than 100 degrees longitude (SE US)

ADD footnote on bins

020120-017

17

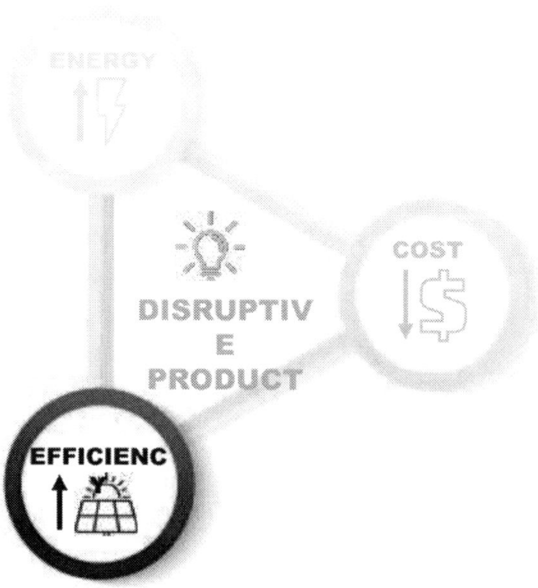

First Solar

Going Beyond: Path to a DISRUPTIVE EFFICIENCY PV Technology

020120-018

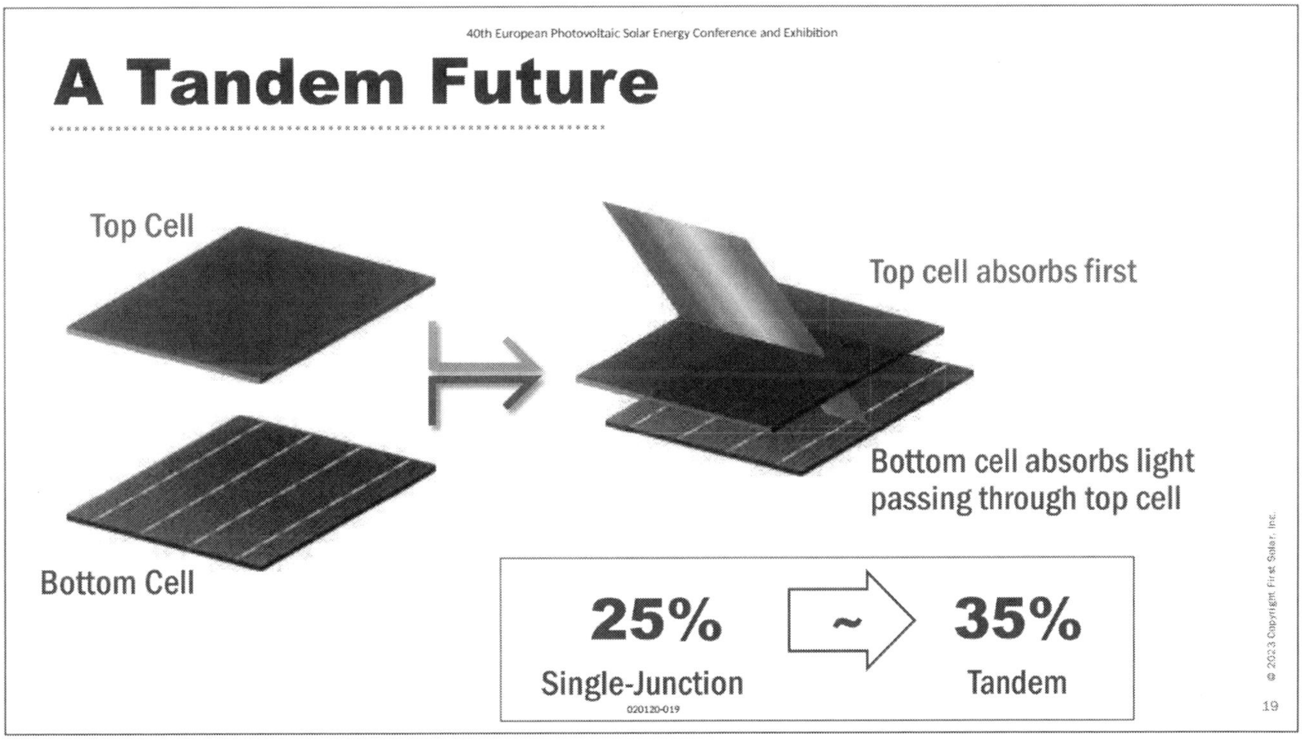
40th European Photovoltaic Solar Energy Conference and Exhibition
A Tandem Future
Top Cell
Bottom Cell
Top cell absorbs first
Bottom cell absorbs light passing through top cell
25%
Single-Junction
~
35%
Tandem
020120-019
© 2023 Copyright First Solar, Inc.
19

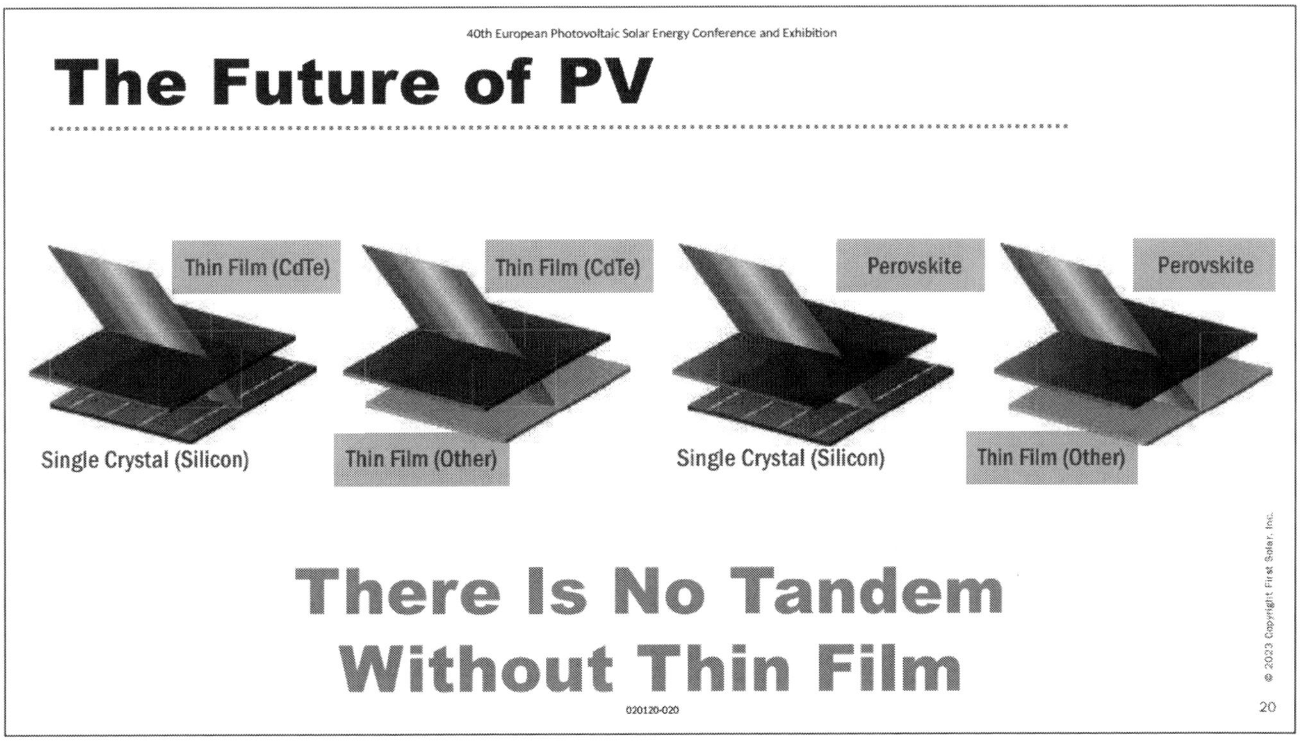
40th European Photovoltaic Solar Energy Conference and Exhibition
The Future of PV
Thin Film (CdTe)
Single Crystal (Silicon)
Thin Film (CdTe)
Thin Film (Other)
Perovskite
Single Crystal (Silicon)
Perovskite
Thin Film (Other)
There Is No Tandem
Without Thin Film
020120-020
© 2023 Copyright First Solar, Inc.
20

40th European Photovoltaic Solar Energy Conference and Exhibition

First Solar

EU Market & Policy Environment

020120-022

40th European Photovoltaic Solar Energy Conference and Exhibition

Market and Competitive Situation: A Fork in the Road

Sole-Source Dependence

- Chinese c-Si PV modules have a dominant share (85%+), sold at dumping prices
- Built with lower social and sustainability standards, subsidized
- European-made Si modules can cost more than twice as much as Chinese Si modules
- Thin film is fundamentally cost advantaged

Non-Financial Criteria can Level the Playing Field

- The **Net-Zero Industry Act** (NZIA) can overcome the cost gap with environmental, social and resilience criteria in PV auctions
- Draft **Forced Labor Ban** could enter into force by 2025
- U.S. domestic production is competitive again with UFLPA and IRA incentives and regulations

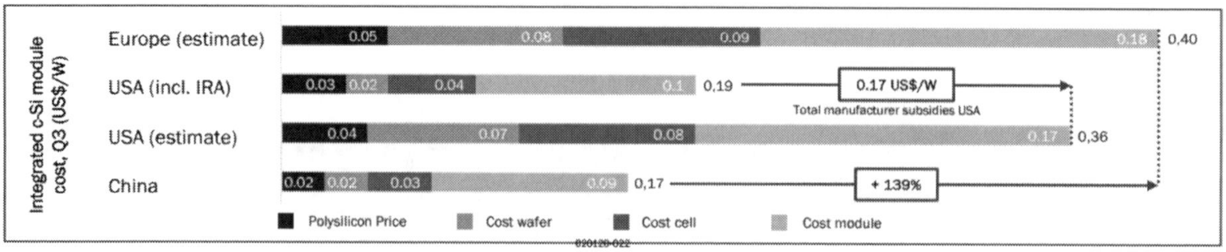

020120-022

22

First Solar

LEADING THE WORLD'S
SUSTAINABLE ENERGY FUTURE

020120-024

24

THE IMPACT OF DIFFERENT SPIN COATING ACCELERATION INDEXES ON THE PROPERTIES OF Cu_2ZnSnS_4 NANOCRYSTAL THIN FILMS

Asaad Alluhaybi [1,2], Douglas P. Halliday [2], Marek Szablewski [2]
[1] Jeddah University, Khulais College of Science and Arts, Department of Physics, Khulais, 21921, Saudi Arabia, Email: asaad.m.alluhaybi@durham.ac.uk, aalluhaybi@uj.edu.sa
[2] Durham University, Department of Physics, Centre for Materials Physics, South Road, Durham, DH1 3LE, United Kingdom

ABSTRACT: In this work, high quality copper zinc tin sulfide (CZTS) thin films were successfully synthesized via the low-cost hot-injection method and deposited on soda lime glass (SLG) substrates by the spin-coating technique. Different spin-coating acceleration indexes (001, 005, 010, 015, 020, 025 and 030) were investigated for the structural and optical properties of CZTS thin films. X-ray diffraction (XRD), Focused Ion Beam microscopy (FIB), Energy dispersive X-ray spectroscopy (EDS), Raman spectroscopy and Ultraviolet-visible spectroscopy (UV-vis) were used to characterize these films. XRD data confirms the kesterite formation structure for CZTS thin films for all samples. There were no secondary phases observed by Raman. The band gaps achieved were in the range of the expected theoretical values. It is observed that there is a general trend of increasing band gap with the increase of the spin coating acceleration index. This is becasue of increased acceleration index leading to films with smaller nanoparticles which are found to have a higher concentration of Zn. This is an indication that altering acceleration index provides a means of controlling the energy bandgap in CZTS thin films which provides additional control methods when depositing CZTS thin films. Keywords: CZTS, Hot-injection synthesis, Spin coating, Thin-films

1 Introduction

Cu_2ZnSnS_4 as a p-type quaternary semiconductor has recently considered to be a promising material for thin-film photovoltaics receiving significant interest due to its suitable bandgap and optical absorption properties [1, 2]. Its high absorption coefficient greater than 10^{-4} cm^{-1} at the fundamental energy bandgap, and the direct energy band gap of 1.5 eV makes CZTS an ideal absorber material for thin film solar cell applications [3, 4]. This value of the band gap energy is very desirable for achieving a PV device conversion efficiency close to the Shockley-Quessier theoretical limit for single junction cells about 32% [5]. However, the highest efficiency reported for CZTS to date is 10% which invites study to improve the PV device performance. Moreover, CZTS consists of earth abundant elements which might lead to CZTS being one of the most suitable thin film materials for low-cost scalable production of photovoltaics cells [6, 7]. Low toxicity is another positive feature for CZTS in comparison to the other thin-film based photovoltaic technology copper indium gallium-based selenides (CIS and CIGS), cadmium telluride (CdTe) and Gallium Arsenide (GaAs) [8-11]. A particular challenge for CZTS is the high density of native point defects and defect complexes which introduce fluctuating potentials and reduce the bandgap. Device performance is limited by a deficit in the open circuit voltage (V_{OC}) through interface recombination [12].

CZTS fabrication can be done via vacuum and non-vacuum-based methods. In this study, the non-vacuum method, hot-injection solution synthesis, was used for the production of CZTS nanoparticle thin films for the reasons of lowering the production cost of the cell, stability and uniformity of the nanoparticle inks [13-16].

The aim of this work is to study the impact of the spin-coating acceleration index on CZTS nanocrystal thin film properties. The acceleration index value is the rate of increase of the spin coating speed per second and the acceleration value for each index. For example, the lowest acceleration index 001 is equivalent to a value of 80 rpm/s, meaning it needs 32 seconds to reach to the set spin coating speed of 2500 rpm in this study.

2 Experimental details

The hot-injection method was applied here to fabricate CZTS nanocrystal inks for subsequent spin-coating with Cu-poor Zn-rich composition following procedures developed by Guo et al. and a previous work done by our group [17, 18]. The precursor molar ratios for synthesis were 1.44 mmol (0.3769 g) of copper (II) acetylacetonate, 0.94 mmol (0.2478 g) of zinc (II) acetylacetonate, 0.75 mmol (0.2909 g) of tin (IV) bis(acetylacetonate) dichloride and 4 mmol (0.1283 g) of sulphur. The CZTS nanocrystal inks were successfully synthesized with Cu-poor Zn-rich composition (Cu/Zn+Sn) = 0.85, (Zn/Sn) = 1.25 and (S/Cu+Zn+Sn) = 1.27, confirmed by elemental analysis, as described in full previously [18].

In this method, the metal salts Cu, Zn and Sn were mixed with 10 ml of oleylamine (OLA) in a 250 ml three-neck flask with stirring. The mixture was heated up to the required reaction temperatures 250 °C where sulphur was injected into the flask in 4 ml OLA and reacted for 30 minutes. The heat was then removed for 20 minutes where the mixture cooled to 50 °C. CZTS powder was extracted by centrifuging in toluene and isopropanol and the clean nanoparticles dried under vacuum overnight. 250 mg of CZTS in 1 ml of hexanethiol was used to create the ink solution.

In this work we determine that 15 seconds was enough time to coat the substrate by the Laurell technology WS-400 Lite Series spin-coating instrument used in this study. Cleaned soda lime glass (SLG) was used as substrates for spin coating the thin films. The Parameters for the spin-coater were 2500 rpm for 15 seconds with different acceleration indexes (001, 005, 010, 015, 020, 025 and 030). Two drops of volume of 40 microlitres of ink was used to make films for FIB, EDS, and UV-vis measurements and 10 drops to make thicker films for XRD and Raman measurements. To remove the solvent from the substrate, a heating mantle was used at 300 °C for 30 seconds.

10.4229/EUPVSEC2023/2CV.2.2
020121-001

3 Results and discussion

The XRD structural measurements were done using Cu-Kα x-rays in Bruker d7 diffractometers to determine CZTS nanoparticle thin film crystal structures and the size of the scattering domain.

The XRD peak positions are determined from the fitting of the spectra by using Lorentzian curves.

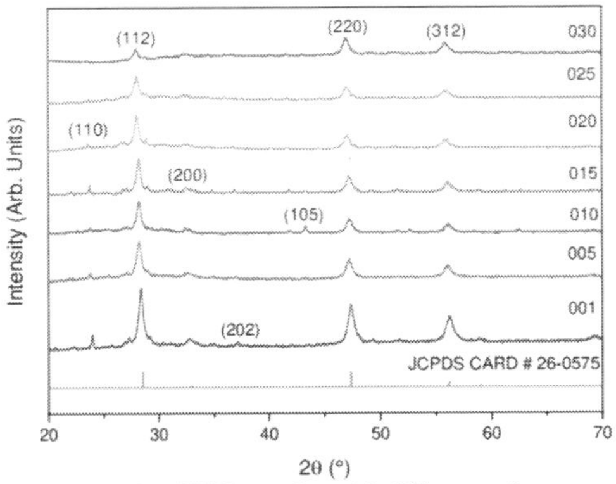

Fig. 1 XRD for CZTS samples with different spin acceleration index.

The XRD data in fig. 1 confirmed the kesterite structure formation in CZTS thin films for all samples. The four main diffraction peaks (112), (200), (220) and (312) for 005-015 acceleration indexes appeared in locations corresponding to JCPDS card number 26-0575. Moreover, (110) diffraction peak appeared in the samples 001-020 which also correspond to JCPDS card number 26-0575. (110) diffraction peaks disappeared in the high acceleration indexes 025-030 and (200) also disappeared in 020-030. It is noted that (105) diffraction peak appeared in 010 only. The peaks for the (112) structure planes were located at 28.32°, 28.18°, 28.22°, 28.05°, 27.91°, 28.12° and 27.85° for 001, 005, 010, 015, 020, 025 and 030 acceleration indexes respectively. The values of *d*-spacing were 0.3139 nm, 0.3148 nm, 0.3151 nm, 0.3166 nm, 0.3189 nm, 0.3160 nm, and 0.3190 nm respectively. The values of lattice parameter *a* were 0.5430 nm, 0.5440 nm, 0.5447 nm, 0.5441 nm, 0.5451 nm, 0.5457 nm, and 0.5461 nm respectively. The values of lattice parameter *c* were 1.091 nm, 1.096 nm, 1.092 nm, 1.114 nm, 1.135 nm, 1.101 nm, and 1.131 nm respectively.

The scattering domain size in table I was calculated from Debye-Scherrer formula:

$$L = \frac{k\lambda}{D\cos\theta} \qquad (1)$$

Where k is a constant = 0.9, λ is the x-ray wavelength (for Cu $K\alpha$, this is 0.15406 nm), D is the full width at half maximum in radians (FWHM) and θ is the Bragg angle of diffraction (peak position) [19].

Table I: Values of the size of scattering domain

Spin acceleration index	Scattering domain (nm)
001	21.70
005	20.40
010	15.85
015	14.14
020	12.98
025	12.53
030	12.22

The size of scattering domain showed a clear trend which decreases with the rise of acceleration index from 21.70 to 12.22 nm.

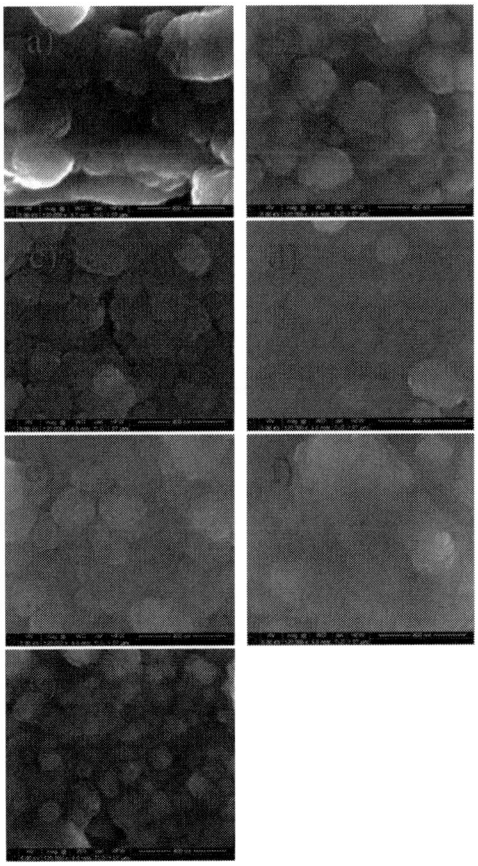

Fig. 2 FIB images for CZTS nanoparticles as a) 001, b) 005, c) 010, d) 015, e) 020, f) 025 and g) 030.

In the FIB images presented in fig. 2, the nanoparticles appear combined on top of each other creating layers of nonuniform nanoparticles except in 015 acceleration index where the grouping of nanoparticles were less. The shape of the nanoparticles was approximately spherical for all samples. An analysis of the images in fig. 2 shows that the size of the particles decreased here from 26-14 nm with the increase of spin coating acceleration index which is in good agreement with XRD data shown in table I.

Data in table II, suggests that to get the closest Cu-poor Zn-rich ratio value for efficient solar cell devices around 12%, a balance of acceleration indexes of between (010-

10.4229/EUPVSEC2023/2CV.2.2
020121-002

020) should be applied. There is a trend with increasing Zn/Sn ratio with the increase of acceleration indexes as fig. 3 below shows.

Table II: Composition ratios of CZTS nanoparticles by EDS

Acceleration index	Cu/(Zn+Sn)	Zn/Sn	S/(Cu+Zn+Sn)
001	0.76	0.67	1.27
005	0.61	0.93	1.21
010	0.78	1.02	1.12
015	1.81	1.07	1.27
020	1.80	1.13	1.22
025	0.75	1.75	1.04
030	0.24	2.15	0.50

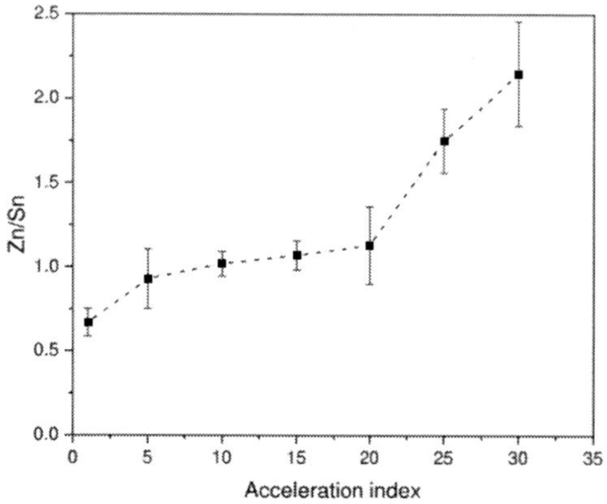

Fig. 3 The trend of Zn/Sn ratio increases with the increase of spin acceleration indexes.

Horiba JY lab RAM-HR microscope was used for the Raman spectra measurements using 532 nm excitation wavelength. The Raman peak fitting is done by using Lorentzian curves. Raman main and strong peaks of CZTS in fig. 4 were located at 340 cm^{-1} for 001, 339 cm^{-1} for 005, 338 cm^{-1} for 010, 340 cm^{-1} for 015-020 and 337 cm^{-1} for 025-030. Fig. 4 shows that there were no other prominent Raman peaks due to secondary phases detected.

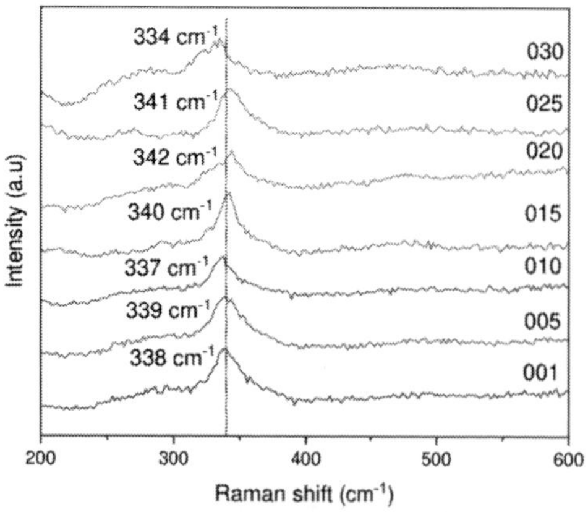

Fig. 4 Raman spectra of CZTS thin films with different spin acceleration index.

Shimadzu UV-3600 UV-vis spectrometer was used to measure the optical properties. The absorbance spectra were recorded in the range of 700-1200 nm. The energy bandgap of the thin films was determined from the Tauc plot as shown in fig. 5 where a straight line can be drawn to extrapolate the plot of absorption squared $(\alpha h\nu)^2$ in the y axis versus the energy bandgap (Eg) in the x axis using the following equation:

$$(\alpha h\nu) = A\sqrt{(h\nu - Eg)} \qquad (2)$$

where A is a constant determined by the electronic band structure, h is the Planck constant, ν is the frequency of the light, and Eg is the band gap of the material [20].

Table I-I-I: Values of the energy band gaps for different spin coating acceleration indexes

Acceleration index	Eg (eV)
001	1.34 ± 0.008
005	1.38 ± 0.010
010	1.57 ± 0.070
015	1.41 ± 0.020
020	1.43 ± 0.025
025	1.47 ± 0.030
030	1.60 ± 0.040

Samples of 010-030 correspond well with the theoretical expected bandgap values of CZTS of between 1.40-1.60 eV [21]. It is observed in table I-I-I and fig. 5 that the band gap increases with the increase of the spin coating acceleration index, but 010 sample appears to be anomalous and is out of line with overall trend. Fig. 6 shows the trend of band gap with spin acceleration index. Table I confirms that increasing acceleration index results in films with an overall smaller nanoparticle size as expected because of the increasing centrifugal forces at the commencement of spinning when the solvent is evaporating. Comparing this with figure 3 shows that there is increasing Zn in the smaller nanoparticles. This may be due to the diffusion of different atomic species during the growth stage of the CZTS nanoparticles. Fig. 6 indicates that increased acceleration index results in an increase in energy bandgap. This may be due to the increased Zn ratio. These factors indicate that altering acceleration index provides a means of controlling the energy bandgap in CZTS thin films which provides additional control methods when depositing CZTS thin films.

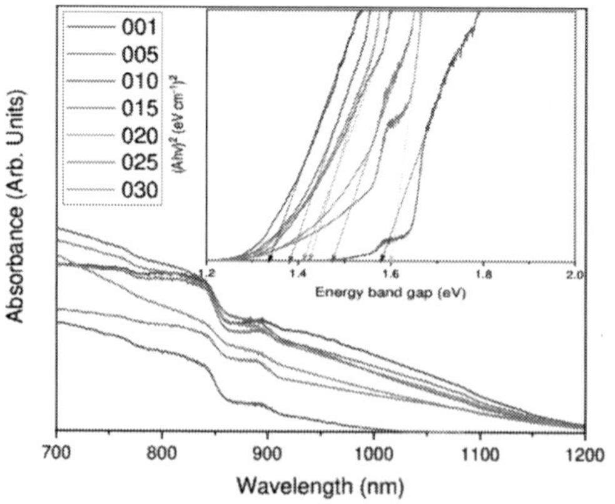

Fig. 5 Absorbance spectrum of CZTS samples with different spin coating acceleration index and the inset shows Tauc plot to determine the band gap of CZTS.

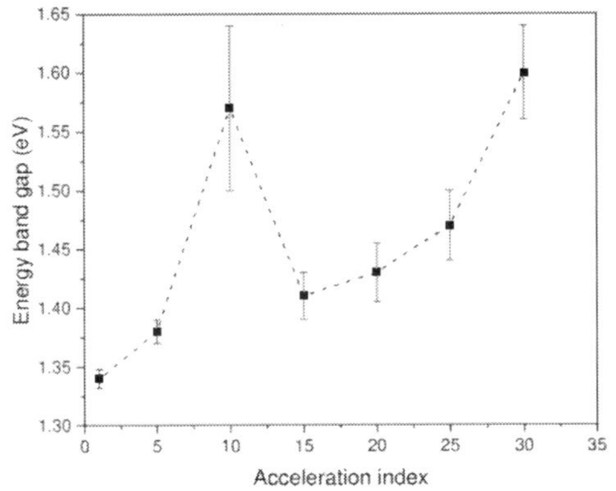

Fig. 6 The trend of the bandgap of as synthesized CZTS thin films vs different spin acceleration indexes.

4 Conclusion

In conclusion, CZTS nanoparticles inks were fabricated by the hot-injection method and deposited by spin-coating on SLG substrates. A range of different spin coating acceleration indexes (001-030) were examined to find out about the impact of them on the properties of the inks via characterization techniques: XRD, FIB, EDS, Raman, and UV-vis. XRD and Raman data confirmed the formation of the CZTS kesterite structure and showed no secondary phases in the thin films. In addition, the FIB image for 015 seems to be the most uniformed nanoparticles comparison with the other samples. The EDS values shows an excellent ratio for Cu-poor Zn-rich in sample 010, 015 and 020. The bandgap for 010-030 indicating that they are in good agreement with the expected theoretical value. We observe impact of spin coating acceleration index on the properties of CZTS resulting in smaller particles at larger acceleration index values. We find that larger acceleration index produces films with a high zinc concentration and a larger energy bandgap. This provides additional flexibility in depositing thin films for CZTS devices. From the characterisation results, 015 index of acceleration spin-coating might be the most suitable upon other indexes to improve the performance of a CZTS solar cell device.

Acknowledgments

Authors are thankful to Durham University, UK and Jeddah University, Saudi Arabia for support. The authors acknowledge Professor A. Beeby for the use of Raman facilities.

4.1 References

[1] R. Pandiyan, et al, Applied Surface Science, 2017. 396. p. 1562-1570.
[2] M. C. Naylor, et al, Faraday Discussions, 2022. 239. p. 70-84.
[3] A. Ziti, et al, Journal of Materials Science: Materials in Electronics, 2019. 30. p. 13134-13143.
[4] R. J. Deokate, et al, Journal of Materials Science: Materials in Electronics, 2019. 30. p. 3530-3538.
[5] W. Shockley, et al, Journal of Applied Physics, 2004. 32. p. 510-519.
[6] R. J. Deokate, et al, Ceramics International, 2022. 48. p. 795-802.
[7] K. S. Gour, et al, Solar RRL, 2021. 5. 2000815.
[8] A. Tumbul, et al, Journal of Alloys and Compounds, 2019. 781. p. 280-288.
[9] A. Tumbul, et al, Materials Chemistry and Physics, 2021. 258. 123997.
[10] D. K. Dwivedi, et al, Optik, 2020. 222. 165407.
[11] S. Tripathi, et al, Optik, 2021. 227. 166087.
[12] J. Park, et al, Applied Physics Letters, 2018. 113. 212103.
[13] S. Engberg, et al, American Chemical Society Omega, 2020. 5. p. 10501-10509.
[14] D. M. Romero, et al, Crystals, 2020. 10. 10070578.
[15] C. Mary, et al, Journal of Materials Science: Materials in Electronics, 2020. 31 p. 18164-18174.
[16] C. Mary, et al, Journal of Materials Science: Materials in Electronics, 2022. 33 p. 894-906.
[17] Q. Guo, et al, American Chemical Society, 2009. 131. p. 11672–11673.
[18] A. Alluhaybi, et al, EU PVSEC Conference Processing, 2022. p. 346-350.
[19] C. Nefzi, et al, Materials Science and Engineering: B, 2020. 254. 114509.
[20] R. B. Mahewar, et al, Indian Journal of Science and Technology, 2020. 13 p. 2149-2156.
[21] T. Ozdal, et al, Ceramics International, 2021. 47. p. 24841-24851

DEVELOPMENT OF FABRICATION PROCESS FOR ELECTROPLATED CZTS ABSORBER

Io Mizushima[1], Peter Torben Tang[1], Alexander Holm Ohrt[2] & Jørgen Schou[2]
1: IPU P/S, Bredevej 2B, 2830 Virum, Denmark
2: Technical University of Denmark (DTU Electro), 2800 Kgs. Lyngby, Denmark

ABSTRACT: Electroplating is a cost-effective method for deposition of copper-zinc-tin (CuZnSn) alloys which can be a precursor to fabricate CZTS absorbers for thin-film solar cells. Focus has been on development of simple and robust electroplating processes for deposition of both pure metals, copper (Cu), tin (Sn) and zinc (Zn), and alloys (CuZn, CuSn, SnZn and CuZnSn) on a molybdenum (Mo) substrate. The subsequent sulphurisation and annealing process is also discussed, and the resulting semiconductor material characterised by Raman spectroscopy. Of all the possible combinations of alloys and layers, direct deposition of brass alloy (CuZn) onto the molybdenum substrate, followed by electroplating of pure Sn, was found to be most feasible. It was also observed that tin electrodeposition in general suffers from instability of the electrolytes, resulting in large variation of the chemical composition tin alloys, as well as delamination problems. Only CuZn, deposited from an alkaline glycerol electrolyte, provided acceptable adhesion to the molybdenum substrates. Furthermore, a pure Sn electrolyte has been developed and optimized, based on pyrophosphate, which has a relatively neutral pH and good stability. Annealing experiments, with sulphur powder in combination with electroplated stacks of Sn/CuZn on a Mo substrate, show a clear CZTS phase formation.
Keywords: CZTS, electroplating, sulphurisation, molybdenum substrate

1 INTRODUCTION

Fabrication of CZTS absorbers by electroplating is an effective, fast and economical method, however, it is not straight forward to optimise and prepare the process for commercialisation. Although many articles have been published on the topic [1-2], few of them address practical issues such as delamination of deposits from the molybdenum substrate and instability of tin electrolytes. In this study several electrodeposition processes have been optimised, resulting in a stable and homogeneous CZTS formation, with focus on avoiding practical issues and prepare the method for commercialisation.

Although several electroplating methods and electrolytes exist for deposition of pure Cu, Zn or Sn metals, as well as the binary alloys CnZn and CuSn, and even the ternary alloy CuSnZn, few of them are useful for deposition of functional CZT-precursors.

Molybdenum is a typical substrate for CZTS solar cells, but it is also a difficult substrate for electroplating, since molybdenum is not stable in air or water but quickly develops an oxide layer which leads to delamination of the electroplated material – sometimes even during the deposition process. This can be verified by studying the potential-pH equilibrium diagram, also called Pourbaix diagram [3]. Especially electrolytes with very high or very low pH-values give rise to severe adhesion problems. Molybdenum also has a high bond strength to hydrogen [4], thus co-deposition or formation of hydrogen could disturb adhesion of metals deposited on molybdenum.

Tin electrolytes in general, except for alkaline cyanide Sn(IV) electrolytes, have a problem of stability due to oxidation of Sn(II) to Sn(IV). This effect will lead to instability and potentially also delamination.

Using a new almost neutral tin electrolyte, combined with direct deposition of CuZn onto the Mo substrate, 44%Cu33%Zn23%Sn electrodeposits – avoiding both challenges mentioned above – have been achieved. Furthermore, the order of the electrodeposits in the stack, has been found to be critical for formations of voids and secondary phases during annealing and sulphurisation.

2 EXPERIMENTAL

Molybdenum foil (0.5 mm in thickness) was used as substrate. The foil was cut into square shaped pieces and treated by cathodic degreasing followed by chemical activation in a solution containing 1% hydrogen peroxide and 10% hydrochloric acid.

Electroplating was performed in a 1L glass beaker, using the Mo foil as cathode (substrate) and either pure Sn or pure Cu as anode. The current density was controlled using a DC power supply being able to accurately control the overall current. The solutions were agitated with a magnetic stirrer and heated (if necessary) with the integrated hotplate. Electroplating of Sn was done at 50 °C using a tin anode and electroplating of CuZn was done at 25 °C using a copper anode. The electrolyte for CuZn electroplating contained 15 g/L copper sulphate, 30 g/L zinc sulphate, 20 g/L glycerol and 120 g/L sodium hydroxide. For Sn electroplating two different commercial solutions were used (modified cyanide based MIRALLOY® from Umicore and sulphate based SLOTOTIN 30® from Schlötter) as well as an electrolyte containing 15 g/L tin pyrophosphate and 165 g/L potassium pyrophosphate.

The chemical composition of CuZnSn electrodeposits were analysed with X-ray fluorescence (MAXXI6, Oxford). The surface topography and cross-sections were examined with a scanning electron microscope (TM3030, Hitachi).

For fabrication of CZTS absorbers, the electroplated CuZnSn precursors were annealed in a graphite box placed inside a temperature-controlled quartz tube with an inert atmosphere created by reducing the pressure and purging with nitrogen. The pressure inside the tube was 175 mbar at the start of each annealing. The temperature was raised at a rate of 20 °C/min. until a desired temperature was reached. Four different annealing schemes were tested, and they encompass different combinations of pre-annealing and sulphurisation durations (Table 1). For sulphurisation, 50 mg sulphur powder is placed inside the graphite box holding the CuZnSn precursors.

10.4229/EUPVSEC2023/2CV.2.3
020122-001

Table I: Annealing schemes

	Pre-annealing		Sulphurisation	
	Temp. [°C]	Time [min.]	Temp. [°C]	Time [min.]
1)	-	-	585	15
2)	-	-	585	10
3)	300	60	585	15
4)	300	90	585	15

After sulphurisation, the phases were identified by Raman Spectroscopy, where the samples are excited by a green (532 nm) continuous wave diode laser. The Raman setup is built by DTU Electro.

3 RESULTS

3.1 Adhesion layer on molybdenum substrate

All attempts to deposit the pure metals (Cu, Zn or Sn) from acidic or alkaline (typically cyanide based) electrolytes were not successful due to massive hydrogen evolution on the substrate. Cyanide is strong complex agent for all three metals and increases the deposition potential (a higher potential is required to release the metal ions from the cyanide complex), leading to more co-deposition of hydrogen - even in an alkaline solution. Pyrophosphate or glycerol/sodium hydroxide electrolytes facilitate electroplating on Mo because less strong complexes are formed. The use of Cu, CuZn and Sn as adhesive layer onto molybdenum substrates has been investigated. Simple tape tests demonstrated that only a brass alloy (67%Cu33%Zn) coating completely survives the test, while Sn, Sn on Cu and Sn on 90%Cu10%Zn delaminated (see figure 1).

Tin has higher electrodeposition potential than hydrogen, and hydrogen evolution could interfere with tin adhesion to the substrate. Regarding copper-molybdenum adhesion, the phase-diagrams of the Cu-Mo system [5] indicates bad affinity between copper and molybdenum, while the phase diagram of the Zn-Cr system (chromium is similar to molybdenum) [6] indicates a higher affinity for mixing. These considerations lead to the best option being fabrication of CuZnSn precursor alloys using two electroplating steps; first direct deposition of CuZn onto the Mo substrate, followed by electroplating of pure Sn.

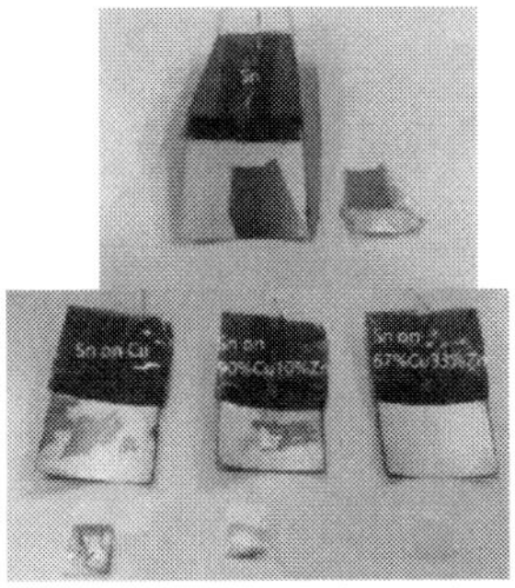

Figure 1: Photos of tape tested samples. Sn electro-deposited directly on Mo (top) and Sn coated on Cu or CuZn alloys, electrodeposited on Mo (bottom).

3.2 Optimization of electrolyte for Sn electroplating

Tin electroplating on CuZn coated molybdenum substrates was investigated. Oxalate and pyrophosphate Sn(II) based electrolytes give rise to variation in chemical composition of the alloys as a function of solution aging, while the more stable Sn(IV) cyanide electrolytes cause delamination of the already deposited CuZn from the Mo substrate. Both commercial tin electrolytes tested results in massive hydrogen evolution during electroplating. The pyrophosphate tin electrolyte is the most balanced one with regards to both stability of chemicals and appearance of the coating (see Table II). By adding sulfosalicylic acid as complexing agent for Sn(II) and dipotassium phosphate as pH buffer, the stability of this Sn electrolyte has been improved.

Table II: Features of pure tin electrolytes

Electrolyte type	Advantages	Disadvantages
Cyanide, Sn(IV)	Commercial, stable	Delamination
Acidic, sulfate Sn(II)	Commercial, relatively stable	Delamination
Pyrophosphate, Sn(II)	Good adhesion, good appearance	No commercial solution exists

3.3 Fabrication of CZTS

CuZnSn deposited samples were prepared by electroplating as a combination of CuZn and pure tin plating, both deposited at a current density of 10 mA/cm^2 and with the same time durations. The chemical compositions of the combined deposits are approximately 43 at.% Cu, 34 at.% Zn and 23 at.% Sn. There is a linear relationship between the thickness and the plating time for both electrolytes. A deposition time of 45 seconds in both electrolytes results in approximately 750 nm in total thickness. The samples are rinsed with water between the two electrolytes in order to avoid contamination of the latest (Sn) electrolyte.

Samples with the above composition were used for annealing and sulphurisation experiments, using the different annealing schemes indicated in Table I. The so-called pre-annealing step (only used in scheme 3 and 4) was introduced to allow more time for diffusion. After cooling down, the samples were removed from the oven, and characterized by Raman Spectroscopy.

In figure 2 samples with signatures only from CZTS (285, 325-330 and 367 cm^{-1}) are the top left and middle left, corresponding to 600 nm and 750 nm thick electroplated layers (30 s and 45 s of plating time respectively) that have only been annealed for 15 minutes at 585 °C. The other charts have signatures at 310 cm^{-1} (from Cu_2Sn_3) and 230 cm^{-1} (from SnS and Sn_2S_3). Signatures beyond 400 cm^{-1} are from MoS_2.

Pre-annealing improves CZTS formation for thick CZTS layer (1000 nm), illustrated by the yellow graphs in figure 2, while there is no significant influence on the 600 nm (blue graphs) or 750 nm layer (orange graphs).

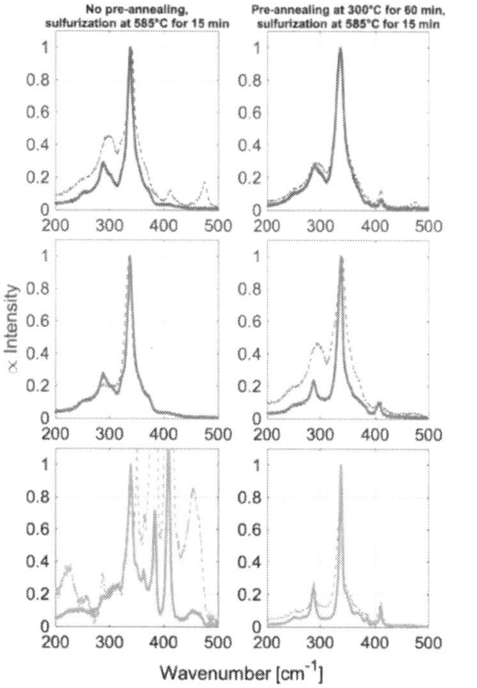

Figure 2: Raman spectra of CZTS absorber layers on Mo foil. The blue graphs are from samples where the electroplated layers have a total thickness of approximately 600 nm. The orange graphs correspond to samples with a 750 nm total thickness, and the yellow graphs correspond to a thickness of 1000 nm. The graphs to the left are samples that have been annealed according to scheme 1 in Table 1, and the graphs to the right have been annealed according to scheme 3. Each sample has been measured twice. The measurement of each sample with the sharpest features from CZTS is shown with a solid thick line, and the other measurement is shown with a thin dashed line. All graphs have been normalized to their main CZTS peak at 335-340 cm^{-1}.

Back scattering images of CZTS prepared from a 750 nm thick precursor, shows a dark and homogeneous surface, while the surface of a similar sample prepared from a 1000 nm precursor reveals several brighter phases identified as SnS/Sn_2S_3.

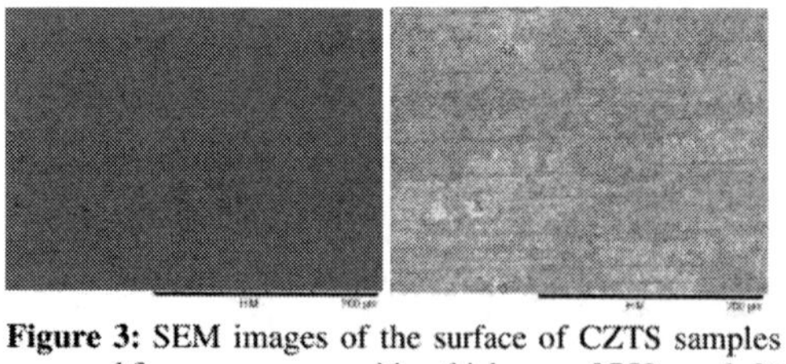

Figure 3: SEM images of the surface of CZTS samples prepared from precursors with a thickness of 750 nm (left) or 1000 nm (right) deposited on Mo foil.

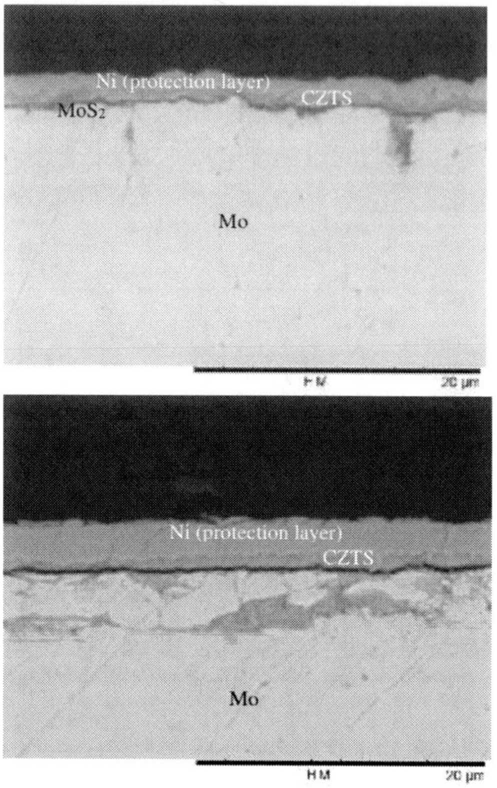

Figure 4: SEM images of cross-sections of CZTS samples prepared from precursors with thickness of 600 nm (top) or 750 nm (bottom) deposited on Mo foil. The top light-grey layer visible is a nickel protection layer, applied only to be able to prepare the polished cross-section. The black area on top of the nickel layer is the epoxy used to cast the sample after cutting and before polishing.

4 CONCLUSIONS

60-70%Cu30-40%Zn alloy electrodeposit from glycerol-NaOH electrolyte has been found as the best adhesive layer on a molybdenum substrate. Tin electroplating on the CuZn coated Mo is feasible only in a pyrophosphate-based electrolyte. By adding buffer and complex agents the stability of tin electrolyte was improved. 46%Cu-30%Zn-24%Sn was obtained by electroplating of first CuZn and second Sn in the developed electrolyte with same time duration. However, the layer thickness is crucial for secondary phase. A thin layer tends to form MoS_2. Sulphurisation of 750nm electrodeposited CuZnSn pre-cursor formed pure CZTS both with/without pre-annealing.

10.4229/EUPVSEC2023/2CV.2.3
020122-003

5 REFERENCES

[1] D. Colombara, Phys. Status Solidi A212,1, 88-102 (2015)
[2] J. Scragg, Ph.D. thesis, ISSN 2190-5053, Springer (2011)
[3] M. Nishimoto, Journal of The Electrochemical Society, 166 (11) C3081-C3089 (2019)
[4] S. Trasatti, Journal of Electroanalytical Chemistry and interfacial Electrochemistry, 39 (1) p.163-184 (1972)
[5] H. Okamoto, Desk Handbook Phase Diagrams for BinaryAlloys, Materials Park, Ohio (2000)
[6] H. Okamoto, J. phase Equilibrium and Diffusion 33, 246-247 (2012)

6 ACKNOWLEDGEMENTS

This work is part of a project called "PlateCell" financed by Danish Ministry of Energy (EUDP) under file number 64020-1088.

ELECTRICALLY CONDUCTIVE OPTICAL NOTCH FILTERS FOR COLORFUL TRANSPARENT ELECTRODE LAYERS IN THIN-FILM SOLAR CELLS

Dae-Hyung Cho[1,2], Woo-Jung Lee[1,2], Tae-Ha Hwang[1], and Yong-Duck Chung[1,2]
[1]Superintelligence Creative Research Laboratory, Electronics and Telecommunications Research Institute (ETRI), Daejeon, Korea
[2]Department of Advanced Materials and Device Engineering, Korea University of Science and Technology (UST), Daejeon, Korea

ABSTRACT: Recently, there has been an increasing industrial need to apply solar modules to various applications such as buildings, vehicles, and mobile devices for energy self-sufficiency in urban areas. Aesthetics are very important for solar modules that are easily visible to people. The solar panel should be in harmony with the surrounding environment and visually pleasing to the viewer. Therefore, there is a need to add colors that can be blended in the surroundings rather than the conventional dark black solar panels. Multi-layer optical filters with interference coatings are commonly used to realize the color of solar cells. When oxide-based thin films with high refractive index and low refractive index are alternately deposited, interferences at a specific wavelength occur, which enables obtaining a desired color. However, oxide film-based color filters usually have very low electrical conductivity, so an additional layer such as a metal film is required to improve conductivity. If the oxide itself of the color filter is conductive, it has the advantage of realizing colors by replacing the transparent electrode layer in solar cells. In this paper, the electrical conductivity, refractive index, and extinction coefficient were adjusted by changing the SnO_2 composition of the conducting indium tin oxide (ITO) thin film. The ITO thin films with different SnO_2 compositions were repeatedly deposited using sputtering to fabricate optical notch filters with high conductivity and vivid colors, and the results of applying the filters as transparent electrodes for Cu(In,Ga)Se_2 (CIGS) thin-film solar cells are reported.
Keywords: conductive color filter, indium tin oxide, thin-film solar cell

1 INTRODUCTION

For color implementation, a multilayer optical filter with an interference coating is mainly used. A multilayer optical filter is a technology widely used in various devices for implementing low or high reflection in a desired wavelength range. Mainly oxide-based thin films with high refractive index and low refractive index are alternately formed to cause constructive and destructive interference of a desired specific wavelength to realize reflection or transmission, and by using this, a desired color can be obtained. Adjusting the refractive index, thickness, and number of layers in the thin-film can control the colors. For application to a solar module, it is common to form a multilayer interference coating on a glass substrate and place it on a solar cell to realize color.

For example, the refractive indices of SiO_2 and Al_2O_3 in the visible region are about 1.47 and 1.75, respectively. Red, green, and blue colors can be realized by varying the thickness and number of layers of the thin film. A high refractive index thin film such as Nb_2O_5 can be used instead of Al_2O_3.

However, such an oxide film-based layered color filter has very low electrical conductivity and does not conduct electricity. Since materials such as SiO_2, Al_2O_3, and Nb_2O_5 have insulating properties, an additional conductive layer is required to have electrical conductivity. If the color filter has conductivity, it has the advantage of removing an additional electrode layer for various devices.

Conventionally, the technique for achieving high electrical conductivity and color implementation by inserting a very thin metal layer with a thickness of several nm between oxide films has been reported. However, since an opaque metal layer is used, a decrease in light transmittance is unavoidable, and the range of color implementation is also limited because the thickness and number of metal layers are restricted. In addition, a transparent color electrode has been reported by adjusting the porosity of the ITO thin film to change the refractive index and stacking it. However, since the refractive index is adjusted by tilting the substrate during thin film deposition, the cost increases during mass production.

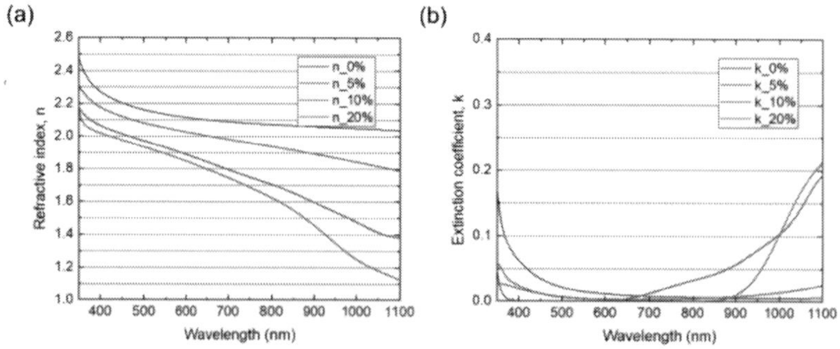

Figure 1: Refractive index (a) and extinction coefficient (b) according to SnO2 composition (0%, 5%, 10%, and 20 wt%) of ITO thin films

10.4229/EUPVSEC2023/2CV.2.11
020124-001

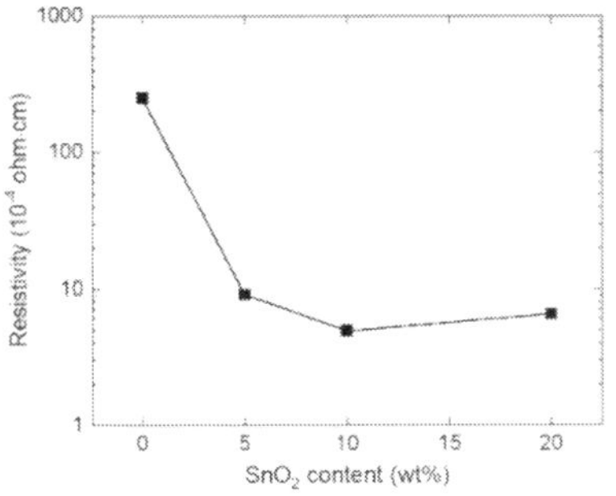

Figure 2: The resistivity of ITO thin films according to SnO_2 composition (0%, 5%, 10%, and 20 wt%)

If the color layer itself has electrical conductivity, it can replace the transparent electrode layer and lower the process cost compared to the case of forming an additional color layer on the solar cell. A colorful optical notch filter with vivid and excellent electrical conductivity was fabricated using only ITO-based thin films on CIGS solar cells without additional processes such as metal film deposition.

Table 1: Simulation results of thickness and color of ITO(0 wt%)/ITO(20 wt%) stack structures for realizing RGB colors

[0.5ITO(Sn 20%)/In_2O_3/0.5ITO(Sn 20%)]^n, n=18								
	Target wavelength (nm)	0.5ITO (Sn 20%) (nm)	In_2O_3 (Sn 0%) (nm)	0.5ITO (Sn 20%) (nm)	Color index			Color patch
					x	y	Y	
Red	600	40.6	70.9	40.6	0.5439	0.3363	24.183	
Green	500	32.3	57.9	32.3	0.2906	0.5199	28.001	
Blue	460	29.2	52.5	29.2	0.2459	0.2658	18.591	

2 RESULTS AND DISCUSSION

The ITO Thin films were deposited using sputter targets with different SnO_2 compositions of 0, 5, 10, and 20 wt%. As the SnO_2 composition of the ITO thin film increased, the refractive index in the entire visible light range decreased (Figure 1(a)). The extinction coefficient decreased at short wavelengths, but increased at long wavelengths (more than 900 nm) (Figure 1(b)).

As the SnO_2 composition increased, the resistivity decreased, showed a minimum value of 4.9×10^{-4} Ωcm at 10 wt%, and increased again at 20 wt% (Figure 2).

To realize color, a stack structure was designed in which ITO (0 wt%) and ITO (20 wt%) thin films with the largest difference in refractive index were alternately deposited 18 times. To implement red, green, and blue colors, the thickness was simulated using the Essential Macleod simulation tool. The thickness, color index, and color patch of each layer are shown in Table 1.

The color index plot and the light reflectance and light transmittance results according to the wavelength are shown in Figure 3. Both high transmittance and color implementation were satisfied by utilizing the stacked ITO thin film structure of different SnO_2 compositions.

3 CONCLUSIONS

A colorful and electrically conductive optical filter is applied to the solar cell. The optical and electrical properties are controlled by changing the SnO_2 content of the ITO thin film, which is widely used as a transparent electrode. It is intended to achieve the desired color and electrical conductivity by alternately stacking thin films with different SnO_2 contents.

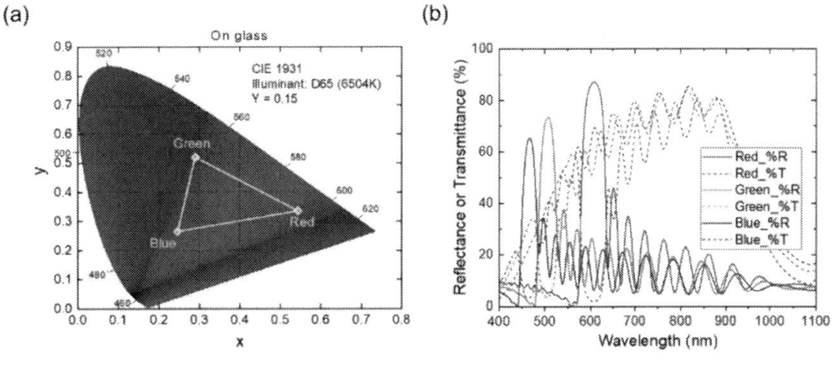

Figure 3: The Color indices (a) and the reflectance and transmittance (b) of the ITO (0 wt%)/ITO (20 wt%) stack structures deposited on a glass substrate to realize RGB color

ENHANCED PERFORMANCE OF CZTSSe SOLAR CELLS THROUGH CHLORINE-BASED IMPURITY INCLUSION PROCESS

O.K.Simya, Ikram Anefnaf, Narges Torabi, Prabeesh Punathil[1], Solidea Zanetti[2], Luca Lozzi, Elisa Artegiani, Mariyam Mukhtar and Alessandro Romeo*

LAPS-Laboratory for Photovoltaics and solid state physics, Department of Computer Science, University of Verona

Ca' Vignal 1, Strada Le Grazie 15, 37134 Verona, Italy, tel.: +39-045-8027936,

*author for correspondence, e-mail address: alessandro.romeo@univr.it

[1]Northumbria university, Newcastle, [2]also at Isopan S.p.A., via Giona 5, Trevenzuolo, Italy,

ABSTRACT: Despite its lack of rapid progress in terms of efficiency, $Cu_2ZnSn(S,Se)_4$ is still considered a viable option for solar cell absorber materials due to its relatively low cost and simple fabrication process, as well as the abundance of its constituent elements. To take advantage of this, we have adapted an easy, low-cost, non-vacuum fabrication process for $Cu_2(Zn,Sn)(S,Se)_4$ thin-film absorber layers by using spin coating of precursors and subsequent low temperature selenization. This results in efficiencies of around 5.5 %. To further improve performance, an impurity inclusion process has been introduced by treating the surface of the precursor film with chlorine-based compounds containing Na, K, or Ge. Additionally, a study has slightly modified the process by eliminating the surface cleaning step after drop casting and heating. This modification results in an efficiency of up to 7.1 %, with a significant improvement in Voc of 20 %, indicating that the use of $GeCl_4$ is a viable means of increasing efficiency by reducing defect density. The structural modifications have been analyzed using techniques such as X-ray diffraction, Raman spectroscopy, and scanning electron microscopy. In addition, current-voltage, and drive level capacitance profiling have been used to analyze the performance and defect density of the finished devices.

Keywords: CZTSSe, Selenization, Thin Film Solar Cells.

1 INTRODUCTION

$Cu_2(Zn,Sn)S_4$/$Cu_2(Zn,Sn)(S,Se)_4$, also known as CZTS/CZTSSe, is a material that is being studied for its potential use in photovoltaic applications. Despite the large amount of papers published on this type of solar cell, the highest reported efficiency is 13.2% by a non-hydrazine route with the addition of Li [1],[2]. However, CZTS/CZTSSe has a tunable band gap and a high absorption coefficient, which makes it a potentially scalable technology for photovoltaics. The main issue with CZTSSe solar efficiency is its low open circuit voltage, which has been attributed to band tailing, band misalignment of the absorber with other layers, and the presence of deep defects [3]. Researchers have found that incorporating alkali metals into CZTSSe can improve efficiency. Different methods, such as doping, alloying, and surface treatments, have been used to improve CZTSSe devices [4]. Studies have shown that incorporating elements such as Na, K, and Ge into CZTSSe can have a positive effect on device performance. However, the presence of these elements at the surface of the absorber layer mainly depends on the growth process and K inclusion can segregate at the grain boundaries [5]. The introduction of Ge as an impurity in the CZTSSe material has been found to promote the growth of the preferred (112) crystal orientation. Additionally, Ge is considered to be an effective dopant to replace Sn^{4+} due to its stable oxidation state [5],[6].

This work is mainly focused with lost-cost fabrication and doping of the chloride compounds which is used as an efficient conveyors for doping the CZTS film. The present study emphasizes a new and simpler method for doping CZTS films, which involves drop casting a solution containing an impurity element onto the CZTS film and then annealing it in Se atmosphere. The use of chlorine as an impurity conveyor has been found to increase efficiency by up to 25 %. The effects of including elements such as Na, K, and Ge on the physical and electrical properties of the CZTSSe absorber are also studied and presented in the work. This approach is different from most previous work, which relied on vacuum deposition techniques and doping the material either during deposition or through post-deposition treatments. This process is unique as it is a low-cost fabrication method with potential for large-scale production.

2 EXPERIMENTAL

2.1 Preparation of the devices

CZTS precursor films are made by depositing a Mo bilayer onto soda lime glass substrates using RF sputtering with a 99.95 % pure Mo target. The same device fabrication process is adapted from our previously reported work on CZTSSe [2]. However a slight modification of the absorber layer with the drop-casting of the dopant solution is employed. A new process is introduced in this paper, where chloride compounds are used as efficient conveyors for doping the CZTS film. Specifically, germanium chloride ($GeCl_4$), potassium chloride (KCl), and sodium chloride (NaCl) are dissolved in methanol and applied to the absorber's surface. The solution is then drop-casted onto the CZTS film and the sample is heated on a hot plate at 150 °C for 3 minutes, followed by rinsing the compound in distilled water. The films are selenized at a maximum temperature of 450 °C for 30 minutes in an argon atmosphere. A reference device without treatment solution is also made using the same fabrication process for comparison. Other studies have shown higher efficiency devices using similar non-vacuum processes, but at much higher selenization temperatures and with the addition of expensive and toxic materials. The finished devices are labeled as CZ-ref, CZ-Na, CZ-K, and CZ-Ge corresponding to untreated, Na-treated, K-treated and Ge-treated CZTSSe samples respectively.

10.4229/EUPVSEC2023/2CV.2.20
020126-001

2.2 Characterization

The crystal structure is studied by X-ray diffraction (XRD) analysis with a Thermo ARL X'TRA powder diffractometer (in Bragg-Brentano geometry, equipped with a Cu-anode X-ray source (Kα, λ=1.5418 Å) and a Peltier Si (Li) cooled solid state detector) and by Micro-Raman spectroscopy experiments in backscattering geometry with Thermo Scientific DXR2 Raman Microscope, over the Stokes-shifted region between 100 cm^{-1} to 350 cm^{-1} by a diode laser (l_{exc}=633 nm). Scanning electron microscopy (SEM) with a Zeiss Gemini500 and energy dispersive X-Ray spectroscopy (EDXS) detector were used to study the compositional analysis. Drive level capacitance profiling (DLCP) and capacitance-voltage (CV) measurements were processed with a HP4284A LCR in a vacuum of 10^{-6} mbar. Current density-voltage (JV) characteristics are measured with a Keithley Source Meter 2420, under an AM 1.5 spectrum at 100 mW/cm^2, using a LOT Quantum Design Europe solar simulator LS0306.

3 DEVICES CHARACTERIZATION

3.1 Structural properties

The XRD patterns of both treated and untreated Mo/CZTSSe films are shown in Fig. 1. According to ICDD file number 01-082-9159, all of them exhibit the typical peaks of the CZTSSe compound. The (112) peak of treated samples displays a slight shift, which we have emphasized by magnifying the pattern in the 2θ range between 26.8 ° and 28 °, and illustrated in the inset of Fig. 1. The peak for pure CZTSe is usually located at 27.16 °, but as sulfur takes the place of Se, the peak moves to a higher 2θ value of 28.44 °, which corresponds to pure CZTS [7]. The lattice parameter of the untreated absorber matches the standard values obtained from ICDD data.

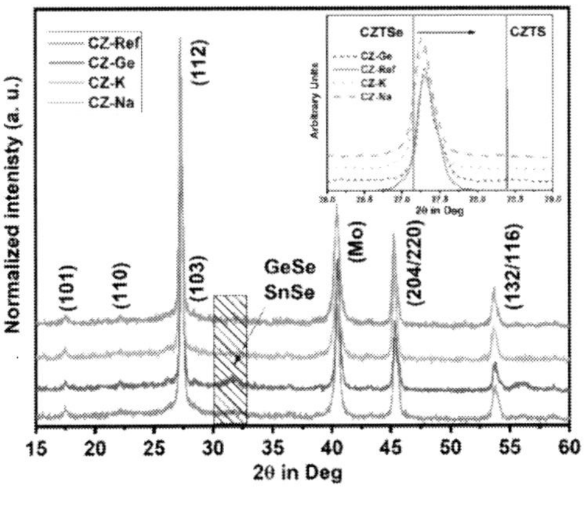

Fig 1: XRD pattern of treated and untreated samples [8].

However, when impurities are inserted, the lattice parameters increase from the standard values, with the least increase observed in the case of Ge and the greatest increase in the case of K. This suggests that each of these elements is effectively included in the lattice structure. It appears that, regardless of the type of impurity introduced, it can affect the stoichiometry, it can induce small additional phases and it can alter the lattice parameter, but not the overall structure (orientation and strain) of the absorber.

The Raman spectra of the treated and untreated films are presented in Fig. 2, obtained by using a laser with an excitation wavelength of 532 nm. All the spectra exhibit the typical peaks of CZTSSe at 56 cm^{-1}, 80–82 cm^{-1}, 174–177 cm^{-1}, 196–198 cm^{-1}, 235–250 cm^{-1}, and 329–332 cm^{-1}, which are associated with the E, A, A, B, and A vibrational modes of the CZTSSe phase respectively [9].

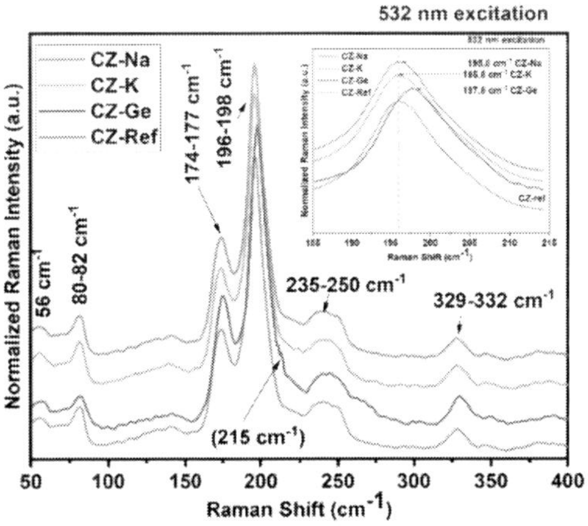

Fig.2: Raman Spectra of treated and un-treated CZTSSe absorber [8].

In the case of CZ-Ge film, a secondary phase is observed at 215 cm^{-1}, which is identified as GeSe [6], thus, supporting the XRD analysis that Ge and Se are forming a secondary phase. The peaks of the treated absorbers of CZTSSe are observed at higher wavenumbers; the shift is more prominent in the case of CZ-Ge while it is less in CZ-Na, as shown in the inset of Fig.2. This supports the observation that the S/Se ratio is higher when Ge is included, indicating a lower concentration of selenium.

3.2 Compositional Studies

The EDX analysis presented in Table 1 revealed that the impurity concentration in both sodium (Na) and potassium (K) is remarkably minimal, even falling below detectable levels. While the germanium (Ge) samples displayed only minor variations in stoichiometry with reduced selenium composition. Therefore, a comprehensive EDXS mapping was conducted for CZ-Ge films, revealing a uniform distribution of CZTSSe elements, except for a few dark grey areas, not shown here [8]. This clearly highlights the excessive amount of germanium present in this sample, preventing its inclusion in the absorber lattice. The existence of both germanium (Ge) and oxygen (O) indicates the creation of GeO_2, which acts as a passivating agent for the germanium.

Table: 1 EDXS compositional profile for treated and untreated samples

Element	CZ-Ref (at %)	CZ-K (at %)	CZ-Na (at %)	CZ-Ge (at %)	CZ-Ge without washing (at %)
Cu	24.57	25.45	25.96	24.96	20.08
Zn	14.47	13.6	13.36	13.47	12.12
Sn	12.58	12.88	12.94	13.1	9.95
S	7.02	7.82	6.56	6.87	8.69
Se	41.35	40.25	41.18	39.76	32.88
Impurity	0	0	0	1.11	8.48
O	0	0	0	0.76	7.79
Cu/Zn + Sn	0.91	0.96	0.99	0.94	0.91
Zn/Sn	1.15	1.06	1.03	1.03	1.22
S/Se	0.17	0.19	0.16	0.17	0.26

3.2 Performance parameters

Table. 2 presents the efficiency parameters of the devices. The reference devices exhibit an efficiency of approximately 5 %, with the highest conversion efficiency recorded at 5.42 % (Voc = 334 mV, Jsc = 28 mA/cm^2, and FF = 58 %). The primary limitation is the open circuit voltage, which is expected to be in the range of 600 mV.

The inclusion of Na, K, and Ge in the cells results in a slight improvement in performance, with the best cells achieving efficiencies of 5.4 %, 5.60 %, and 5.9 %, respectively. The improvement is primarily attributed to an increase in Voc, with minimal impact on Jsc. Furthermore, modifying the process by removing the surface cleaning after drop casting and heating has been studied. This approach has the benefit of increasing the incorporation of Ge but it also leaves an excess of Ge on the absorber's surface. However, this modification results in an efficiency of up to 7.1 %, with a significant improvement in Voc of 20 %, indicating that the use of $GeCl_4$ is a viable means of increasing efficiency by reducing defect density.

Table: 2 Efficiency parameters of the devices presented in Fig.4

Sample	Voc (mV)	Jsc (mA/cm^2)	FF (%)	η (%)
CZ-Ref	334	28	58	5.4
CZ-Ge	349	33	51	5.8
CZ-Ge-without washing	390	33	55	7.1
CZ-Na	360	29	52	5.4
CZ-K	351	29	55	5.6
AST-CZ-Ref	339	31	55	5.7
AST-CZ-Ge	362	32	52	6.0
AST-CZ-Ge without washing	390	32	54	6.7
AST-CZ-Na	352	30	55	5.8
AST-CZ-K	352	30	52	5.4

4 CONCLUSIONS

A simpler and novel approach of doping the absorber layers with different elements has been investigated, where a chloride solution of the elements is drop casted onto the surface of the CTZS film. This method of doping with elements such as Ge has been found to improve the efficiency by up to 7.01 % with Voc = 390 mV, Jsc = 33 mA/cm^2, and FF = 55 %. However, doping with elements such as K and Na had little effect on the efficiency of the devices. Additionally, exposing the devices to light at 80° C for 24 hours resulted in an improvement of the PCE for all devices, except for non-washed Ge treated samples.

5 ACKNOWLEDGMENTS
This project has been funded by Cariverona Foundation in the project INSOBILD Ref. 2022.0091-CUP B33C22001760007.

6 REFERENCES

[1] J. Zhou *et al.*, "Regulating crystal growth via organic lithium salt additive for efficient Kesterite solar cells," *Nano Energy*, vol. 89, no. PB, p. 106405, 2021.
[2] P. Punathil, S. Zanetti, E. Artegiani, V. Kumar, and A. Romeo, "Analysis of the drying process for precursors of Cu2ZnSn(S, Se)4 layers by low cost non vacuum fabrication technique," *Sol. Energy*, vol. 224, no. July, pp. 992–999, 2021.
[3] S. Chen *et al.*, "Classifi cation of Lattice Defects in the Kesterite Cu 2 ZnSnS 4 and Cu 2 ZnSnSe 4 Earth-Abundant Solar Cell Absorbers," 2013.
[4] Y. E. Romanyuk *et al.*, "Doping and alloying of kesterites," *JPhys Energy*, vol. 1, no. 4, 2019.
[5] B. H. Lee *et al.*, "Effect of Ge nanolayer stacking order on performance of CZTSSe thin film solar cells," *Mater. Lett.*, vol. 284, p. 128981, 2021.
[6] J. Li, H. Shen, J. Chen, Y. Li, and J. Yang, "Growth mechanism of Ge-doped CZTSSe thin film by sputtering method and solar cells," *Phys. Chem. Chem. Phys.*, vol. 18, no. 41, pp. 28829–28834, 2016.
[7] P. M. P. Salomé *et al.*, "Growth and characterization of Cu2ZnSn(S,Se)4 thin films for solar cells," *Sol. Energy Mater. Sol. Cells*, vol. 101, pp. 147–153, Jun. 2012.
[8] P. Punathil, E. Artegiani, S. Zanetti, L. Lozzi, V. Kumar, and A. Romeo, "A simple method for Ge incorporation to enhance performance of low temperature and non-vacuum based CZTSSe solar cells," *Sol. Energy*, vol. 236, no. March, pp. 599–607, 2022.
[9] M. Dimitrievska *et al.*, "Raman scattering quantitative analysis of the anion chemical composition in kesterite Cu2ZnSn(SxSe1-x)4 solid solutions," *J. Alloys Compd.*, vol. 628, pp. 464–470, 2015.

A NEW DESIGN : THREE TERMINAL BAND OFFSET BARRIER ORGANIC/SI TANDEM SOLAR CELLS

J.P. Connolly[1,2], J. Alvarez[1,2], J.P. Kleider[1,2], P. Roca i Cabarrocas[3], S. Chambon[4], L. Hirsch[4], L. Vignau[4], G. Wantz[4], M.E. Gueunier-Farret[4]

[1] *Université Paris-Saclay, CentraleSupélec, CNRS, Laboratoire de Génie Electrique et Electronique de Paris, 91192, Gif-sur-Yvette, France*
[2] *Sorbonne Université, CNRS, Laboratoire de Génie Electrique et Electronique de Paris, 75252, Paris, France*
[3] *École Polytechnique, LPICM, CNRS UMR 7647, Institut Polytechnique de Paris, 91120 Palaiseau, France*
[4] *Université de Bordeaux, IMS-CNRS, UMR 5218, Bordeaux INP, ENSCBP, F33405 Talence, France*

ABSTRACT: Organic solar cells are particularly attractive for their very low fabrication cost, and lack of use of critical raw materials. While this has to date been hobbled by low efficiency, recent achievements demonstrate organic single junction technology at 20% efficiency approaching economically viable performance.
We survey the reasons for these recent high efficiency organic solar cell achievements. They include bulk heterojunction (BHJ) concepts, composed of electron donor materials, most often a semiconductor polymer and electron accepting materials, most often a small molecule, and the novel non-fullerene acceptor materials. The range of commercially available donor and acceptor materials available offers a large design flexibility for low or high band gap devices. A natural development for this emerging thin film technology reaching competitive efficiencies is its integration in tandem solar cells.
In the context of the breakthroughs in organic solar cell efficiency, this work presents first results of the ORGANIST project (2023 – 2027) supported by the French Research National Agency. The presentation of organic materials for top cells and their efficient integration on a high efficiency silicon IBC cell is discussed with a critical focus on the interface between top organic solar cell and bottom silicon IBC cell. We present preliminary theoretical results evaluating the performance of 3T devices being developed in ORGANIST and based on identified prototype materials for the organic cell, strategies for efficient optical and electrical connection between the two subcells, aiming for low cost three terminal organic / silicon tandems with high efficiencies above the Shockley-Queisser single bandgap limit for Silicon.

Keywords: Multijunction ; Organic ; Silicon; High efficiency
Author for correspondence : james.connolly@centralesupelec.fr

1 INTRODUCTION

Multijunction cells are the most efficient solar cell designs to date. These designs have historically been fabricated only with high cost III-V materials and have been restricted to space applications. Terrestrial applications under concentration which can significantly reduce cost. However they have not, to date, led to a breakthrough in applications due to underlying materials and fabrication costs, and to the additional complexity and associated cost of light concentration.
Concerning multijunction performances, a record certified terrestrial power conversion efficiency of 38.8% performance was achieved in 2014 [1] under standard AM1.5G test conditions but without notable subsequent progress [2]. The III-V approach, therefore, while promising, has not yet delivered for terrestrial applications though work is ongoing.
In that context, the obvious attraction of tandem solar cells with silicon lower bandgap subcells have emerged, based due to the observation that the silicon gap of 1.12 eV is nearly ideally suited to the bottom gap of a tandem solar cell [3]. This too has been the focus of much research for over a decade, and has now delivered high efficiency over 33% for both III-V/Si and perovskite/Si tandems [2].
And while other materials combinations on Si have been studied, we will not list them exhaustively, only remark that none has led to more promising performance so far.
In this context, organic solar cell (OSC) performance has recently shown significant improvements, achieving 20% efficiency [4]. This makes this system a candidate for tandem solar cells, in particular in developing high bandgap organic subcells associated with Si low bandgap subcells in silicon-organic tandem devices.
This paper presents early projections of expected device performance based on a brief survey of data from the literature, which lays the basis for device performance which may be achieved within the project.

2 DEVICE DESIGN

2.1 Ideal limits and materials compromise

We first look at achievable radiative efficiencies [3] for the materials under consideration. The donor and acceptor materials considered are PBDB and GS-ISO families respectively, which yield an organic subcell bandgap of the order of 2 eV.
The radiative efficiency limit of three terminal solar cells is shown in figure 1 in order to emphasise the tolerance of the three terminal (3T) design to subcell bandgaps. With the constraint of a Si bottom subcell bandgap of 1.12 eV, the ideal top subcell has a bandgap of 1.82 eV and the corresponding tandem efficiency is at 42.2% under standard AM1.5G test conditions. If we increase the top bandgap to 2 eV, the tandem efficiency is still at 41.7%, with a drop of just 0.5% compared to the optimum case. This 2 eV material is chosen on the basis of recent trends in the field which are non-fullerene wide bandgap acceptors coupled with PBDB-TF donor [4, 5].

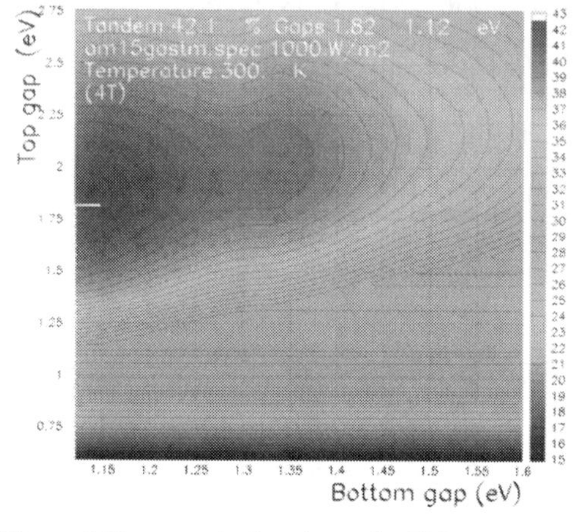

Figure 1 Three terminal tandem cell efficiency contour plot under standard AM1.5G test conditions, as a function of top and bottom subcell bandgaps in the radiative limit. Assuming a silicon bottom cell, the tandem cell maximum efficiency is 42.2% (bandgaps 1.12 and 1.82 eV).

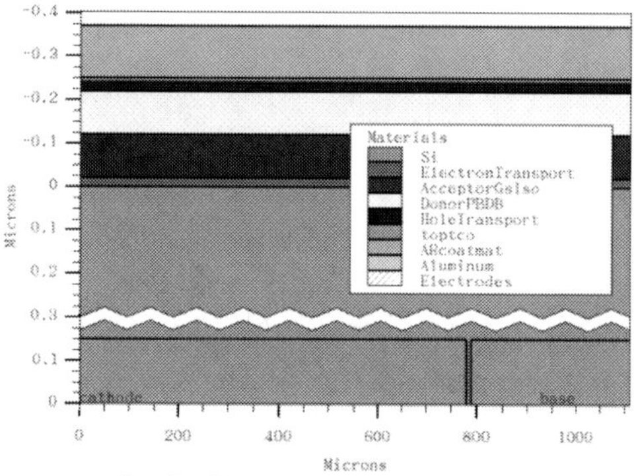

Figure 2 Sketch of the 3T OSC/Si-IBC tandem structure showing the OSC layer sequence for an n-type IBC.

2.2 Subcell design

We use a standard Si-IBC based on the ZEBRA design commercialised by ISC-Konstanz [6], partner of the ORGANIST project. We do not detail performance but note that the modelled cell efficiency is for an untextured IBC in order to facilitate OSC integration. The base Si-IBC has the following simulated performance: short circuit current density J_{SC} = 32.3 mA/cm^2, open circuit voltage V_{OC} = 0.73V, FF = 82%, and efficiency of 20%.

When shaded by the OSC top cell, however, the efficiency of the Si-IBC is naturally reduced. We note that the J_{SC} is reduced to 13.4 mA/cm^2, the V_{OC} to 0.71 V, and the fill factor slightly improved at FF = 84%, leading to an efficiency in the 3T tandem of just 8%.

The organic cell is modelled following the detailed modelling analysis of Koster et al. [7] and associated SILVACO example solarex07.in. This model includes organic semiconductor properties in particular excitonic transport and dissociation at donor/acceptor interfaces.

The modelled structure in the first instance is a bilayer heterojunction with 100 nm donor and acceptor layer thicknesses, to be developed further to enable simulation of bulk heterojunction blended organic absorbers which offer advantages for carrier collection in these materials with very short exciton lifetimes and diffusion lengths.

In the early stages of the ORGANIST project the modeling uses data obtained in the literature [8,9].

This approach yields the following simulated OSC performance : short circuit current density J_{SC} = 10.6 mA/cm^2, open circuit voltage V_{OC} = 1.48V, FF = 90%, and efficiency of 14.2%.

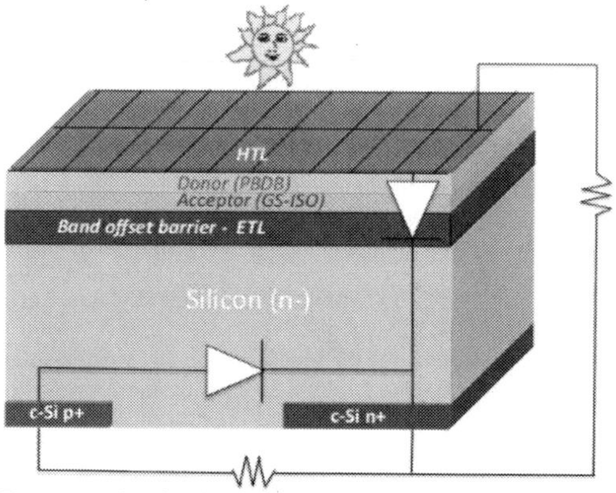

Figure 3 Sketch of the 3T OSC/Si-IBC tandem structure for an n-type IBC device.

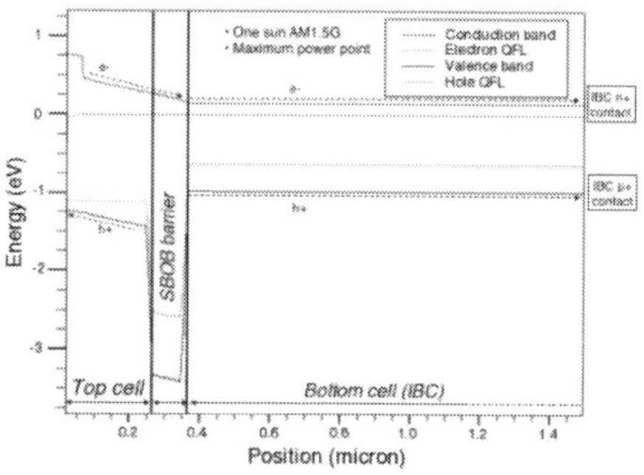

Figure 4 Schematic band diagram of the 3T-band offset barrier tandem, showing the band offset separating hole populations in top and bottom cells, and leading to different quasi-Fermi levels under one sun illumination, the simulation being at the maximum power point.

2.4 Three terminal OSC/IBC band offset barrier tandem

The 3T band offset barrier tandem device concept has been described previously [10] and is briefly summarised here. The device is sketched in figure 3 in the n-type IBC embodiment of the concept. The valence band offset barrier in this case prevents transport of holes between top and bottom subcells, such that hole populations in the two cells are decoupled, and each defined by an independent quasi-Fermi level.

Figure 4 shows the simulated band structure and quasi-Fermi levels at maximum power point voltage, under AM1.5G illumination. The different quasi-Fermi level separations in top and bottom cells allow independent operation of the two cells meaning that this structure is not subject to the constant series current constraint which applies to two terminal tandems. The maximum efficiency of the 3T tandem is the straight sum of the efficiency of both subcells operating independently at their maximum power voltages.

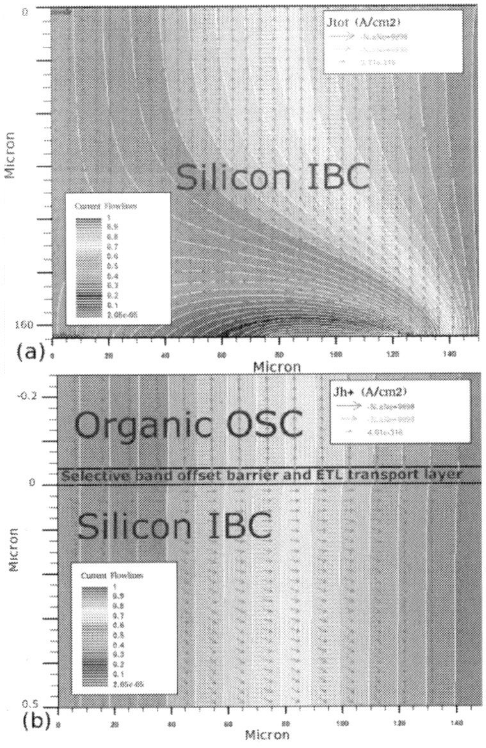

Figure 5 Numerical simulation of current flows at maximum power point showing (a) current flows in the Si-IBC between interdigitated contacts, and towards the organic cell and (b) showing hole currents only, such that the blocking behaviour of the band offset barrier can be seen.

3 CELL PERFORMANCE MODELLING

The three terminal organic / silicon tandem structure described in section 2 has been simulated with Silvaco ATLAS [11]. The current-voltage curves are simulated separately as validated by previous studies [12].

Figure 5 shows current flows in the structure at the maximum power point. Figure 5a shows current flow between the Si-IBC subcell interdigitated back contacts, and current flow into the top front surface anode. Figure 5b presents a close-up on hole transport at the front of the device, showing the organic subcell in its entirety and the top of the Si-IBC. One can see the blocking behaviour of the selective band offset barrier, such that holes in the Si-Si-IBC are blocked from flowing to the organic top cell, while organic subcell holes flow to the front surface p-type contact.

The resulting cell performance in figure 6 shows the independent power / efficiency curves of top and bottom OSC and Si-IBCs. The tandem efficiency is, in this case, 22%, which is the sum of OSC and Si-IBC subcell efficiencies. We note that there is much potential for improvement, given that the radiative limit for these materials, as we have seen earlier, is close to 42%.

On a practical level, the route to increasing this 22% efficiency starts with the organic cell. The efficiency we have evaluated with a planar heterostructure and materials parameters taken from the literature is 14%. This is far from the best OSC results in the literature which as we have seen have achieved over 20% efficiency albeit for lower bandgap materials. The next steps are therefore optimisation of the OSC which will benefit from less stringent constraints on OSC bandgap.

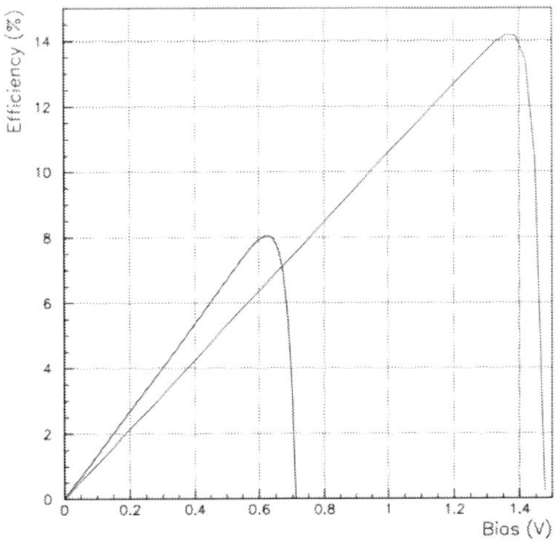

Figure 6 Performance of organic top (red) and bottom silicon (blue) cells, presented as power output relative to incident AM1.5G power, that is, subcell efficiency as a function of applied bias. The total power of the tandem is the sum of power output of each subcell.

4 CONCLUSIONS

We have presented the principles of operation of a novel tandem device, which is the three terminal selective band offset barrier tandem, based on organic and silicon materials. The performance of this device being significantly less sensitive to subcell bandgaps, the range of organic materials that can be adapted is broader. Preliminary evaluation of achievable efficiencies, using state of the art but untextured Si bottom cells, coupled with the highest efficiency organic cells but using high bandgap embodiments of donor and acceptor materials, yields a realistic efficiency potential in the short term of 22%. This being well below the radiative limit of 42.2% for these bandgaps, there remains much potential for subsequent increases in efficiency by optimising the high bandgap OSC.

10.4229/EUPVSEC2023/2CV.2.25
020127-003

Acknowledgements : The authors acknowledge the support of the ANR French national funding agency, which have funded the ORGANIST (2023-2026) project ANR-22-CE05-0025.

References

[1] P. T. Chiu et al., 2014 IEEE 40th Photovoltaic Specialist Conference (PVSC), Denver, CO, USA, 2014, https://doi.org/10.1109/PVSC.2014.6924957

[2] Green MA, Dunlop ED, Yoshita M, et al. Solar cell efficiency tables (version 62). Prog Photovolt Res Appl. 2023;31(7):651-663. https://doi.org/10.1002/pip.3726

[3] James P. Connolly *et al.*, Progress in Photovoltaics, Volume 22, Issue 7, July 2014, Pages 810-820. https://doi.org/10.1002/pip.2463

[4] Zhong Zheng *et al.,* Joule 6, 171–184, January 19, 2022, https://doi.org/10.1016/j.joule.2021.12.017

[5] Pengqing Bi *et el.*, Adv. Mater. 2022, **34**, 2108090 https://doi.org/10.1002/adma.202108090

[6] R. Kopecek *et al.*, 2020 47th IEEE Photovoltaic Specialists Conference (PVSC), Calgary, CA. https://doi.org/10.1109/PVSC45281.2020.9300503

[7] L. J. A. Koster, et al., Phys. Rev. B 72, 085205 https://dx.doi.org/10.1103/PhysRevB.72.085205

[8] Wenchao Zhao *et al.*, Advanced Materials https://doi.org/10.1002/adma.201600281

[9] GS-ISO data sheet : https://www.1-material.com/wp-content/uploads/2022/02/1M-GS-ISO-Tandem-OPV-PCE-20.pdf

[10] J.P. Connolly, et al., 36th EU PVSEC, 2019, Marseille, https://userarea.eupvsec.org/proceedings/EU-PVSEC-2019/3BV.2.60/

[11] Silvaco, Atlas User's Manual https://silvaco.com/support/technical-library/.

[12] Hiroyuki Kanda et al., Interdisciplinary Materials. 2022;1:148–156, https://doi.org/10.1002/idm2.12006

DEVELOPMENT OF MONOLITHICALLY INTEGRATED (Al)GaAs ON SILICON FOR TANDEM SOLAR CELLS

L. Dentz[1], G. Chau[1,2], G. Hallais[1], F. Hamouda[1], J.P. Connolly[2], A. Jaffré[2], L. Vincent[1], D. Bouchier[1], D. Mencaraglia[2], C. Renard[1*]
[1]C2N, Centre de Nanosciences et de Nanotechnologies, CNRS, Université Paris-Saclay, 10 Bd Thomas Gobert, 91120, Palaiseau, France
[2]GeePs, Group of Electrical Engineering Paris, CNRS, CentraleSupelec, Université Paris-Saclay, Sorbonne Université, 3&11 rue Joliot-Curie, Plateau de Moulon, 91192 Gif-sur-Yvette CEDEX, France
*Corresponding author: charles.renard@c2n.upsaclay.fr

ABSTRACT: We present here an innovative solution to heterogeneously integrated III-V materials on silicon based on the growth of micrometric GaAs crystals on silicon without any structural defects nor stress, using Epitaxal Lateral Overgrowth on Tunnel Oxide from nanoseeds (ELTOn). This epitaxial method has the advantage of being compatible with cost effective processes, while using innovative growth mechanisms with very low III-V material requirement (a micron or so deposited and no III-V or Ge substrate required) helping keep costs competitive and with low use of critical raw materials. The main focus of this work is to address the issue of the doping determination of micrometric size GaAs crystals which is a pre-requisite for developing design and technology of multijunction solar cells based on an array of GaAs crystals free of structural defects. In a previous work, local electrical characterizations by µ-PL have revealed a rather high non intentional p- type doping. Several growth solutions (effect of V/III ratio and use of alternative precursor) will be presented to lower these high non intentional p- type doping.
Keywords: Epitaxy, III-V Semiconductors, Silicon, Multijunction Solar Cell

1 INTRODUCTION

The monolithic integration on silicon of GaAs, and more generally of III-V semiconductors, would be a very promising route for the production of high efficiencies multijunctions devices in the manner of those developed for spatial applications on germanium, but at a much lower cost suitable for terrestrial PV applications. The lattice parameter mismatch leading to threading dislocations is however the major problem that many approaches have tried to resolve but to date no one has succeeded in achieving defect-free and cost-effective direct integration of the simplest GaAs binary III-V compound, on silicon.

In previous work, we have demonstrated the perfect integration on silicon of micrometric size GaAs crystals without any structural defects nor stress, using Epitaxal Lateral Overgrowth on Tunnel Oxide from nanoseeds (ELTOn). This technic also leads to a very good electrical connection not only between the GaAs microcrystals and the Si substrate, but also across the GaAs/thin SiO_2/Si sandwich (by tunneling assisted transport) surrounding the nanoholes **[1]**. The principle of the ELTOn technic is illustrated in Figure 1.

We aim now to extend this strategy to $Al_xGa_{1-x}As$ compounds, more specifically up to an Al content x=18%, which is the composition required to achieve current matching and an optimum conversion efficiency for an AlGaAs/Si tandem photovoltaic solar cell of 32% (derived from our realistic software model called SOL). The assessment of our SOL software has been fully proven by accurate modelling of the published experimental data of record multijunctions solar cells **[2]**. Nevertheless, before investigating AlGaAs alloy we will focus on the control of the residual doping of the simply binary GaAs compound.

The residual doping control in a single GaAs micro-crystal is a pre-requisite to developing the future design and technology of multijunction solar cells based on this approach. Local electrical I/V measurements of the GaAs/Si heterojunction performed with CP-AFM showed a high rectification ratio but a high p-type non intentional doping level was suspected and related to high non-intentional carbon incorporation induced by trimethylgallium (TMGa) used as Ga precursor. The precise doping level determination in a single µ-crystal by local I/V or C/V measurements is difficult because in the first case the estimation is too indirect and in the later one, the capacitance values are too small to be easily measured. To obtain a better estimation of the doping level, we have developed a contactless method based on the measurement of the evolution of the bandgap versus temperature **[3]**. The bandgap value is extracted from the micro-photoluminescence spectra of a single GaAs crystal. The bandgap narrowing due to p-type doping is taken into account and the doping density can then be estimated from the fit of the bandgap evolution with temperature.

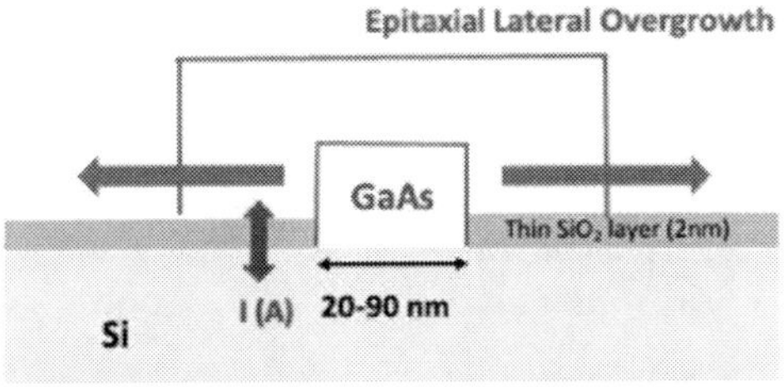

Figure 1: Schematic view of Epitaxial Lateral overgrowth on Tunnel Oxide from nanoseeds (ELTOn): the growth only begins inside the nanoholes performed in the thin silica layer covering the silicon wafer and then extend laterally on top of the SiO_2 layer. The (tunnelling assisted) electrical connection between GaAs and the Si substrate thought thin SiO_2 layer is indicated with the red bidirectional arrow.

This high carbon doping has been observed in the literature, and the concentration can be adjusted by increasing the V/III ratio during epitaxy **[4]**. Nevertheless, recent results obtained in our growth reactor not clearly seem to allow such a good control of the p-type doping level but this point will be further investigated with both higher growth temperatures and V/III ratio. However, a fallback solution is to adjust the p-type concentration by the use of TEGa (TriethylGallium) and TMGa together with TBAs. The use of TEGa with TBAs makes it possible to obtain low carbon concentrations of the order of 10^{15}-10^{16}cm^{-3} reproducibly. The simultaneous use of TEGa /

TMGa ratio in variable proportions is therefore expected to allow a good control of the level of doping over a range from 10^{14} cm^{-3} to 10^{20} cm^{-3}**[5].** These routes to a better control of the non-intentional carbon incorporation and the further p-type intentional doping will be addressed.

2 RESULTS AND DISCUSSION

2.1 Micro-scale GaAs crystals growth

The GaAs μ-crystal growth was carried out in a chemical beam epitaxy (CBE) system with the Epitaxial Lateral overgrowth on Tunnel Oxide from nano-seed (ELTOn) technique. TriMethylGallium (TMGa), TriEthylGallium (TEGa) and TertiaryButylArsine (TBAs) were used as gas sources. The (111)Si wafers were cleaned using the modified Shiraki chemical cleaning which results in a reproducible ultrathin SiO_2 layer with a thickness of 1.2 nm thick oxide layer as deduced from Auger electron spectrometry and HRTEM. The important next step is the opening of the nanoscale holes through the ultrathin silica layer prior to GaAs epitaxy.

To demonstrate the ability of the ELTOn growth for the perfect integration of micrometric size GaAs crystals, we first used a random patterning of the silica layer directly performed in situ in the growth reactor. For this purpose, the nucleation seeds must be created. Therefore, in order to open nanoscale holes through the thin silica layer, the oxidized silicon substrate was fast ramped up 750 °C, under vacuum, and maintain at such temperature during 1 min 30 s. This results in pure Si nano-areas formed via the reduction of SiO2 into volatile SiO.

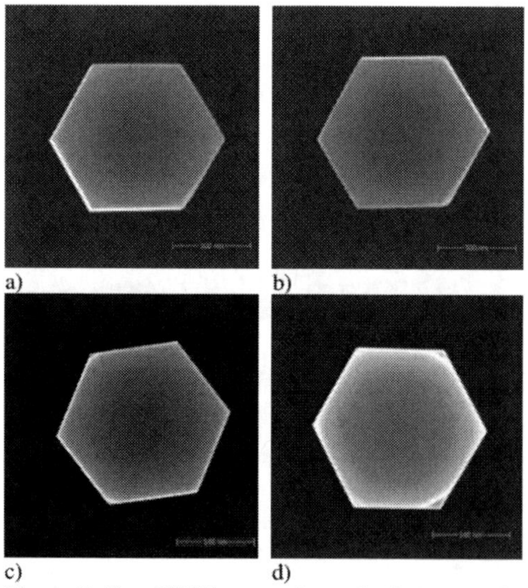

Figure 2: Top SEM images of μ-scale GaAs crystal sgrown on silicon using the ELTOn Process for different V/III ratio with TBAs and TEGa as gas precursors. a) R = 6, b) R = 10, c) R = 18 and d) R = 22.

This method has the advantage of easily providing nanoscale seeds around 50 nm in width, randomly spaced at the surface of the oxidized Si substrate [1]. After this step, the GaAs epitaxy was initiated with a two steps procedure. The first step consists in introducing solely TBAs (during 2 min) at 430 °C. In the second step TMGa, respectively TEGa, were also introduced while the growth temperature was increased up to 575 °C. After, this nucleation step that tooks 2 min, GaAs epitaxial lateral overgrowth was then continued for 45 min. In this paper we investigate particularly the effect of the As/Ga (V/III) ratio on the doping of GaAs μ-crystals. For this purpose, several V/III ratios, denoted R, will be investigated. Such as 6, 10 and 14 for [TBAs]/[TMGa]. And 6, 10, 14, 18 and 22 for [TBAs]/[TEGa]. For the sake of clarity, the V/III ratios studied in this paper are not the same for TMGa and TEGa. In fact, the results given for TMGa correspond to a previous study in which we limited ourselves to R=14. Top SEM views of μ-scale GaAs crystals grown on silicon using the ELTOn Process for different V/III ratios with TBAs and TEGa as gas precursors are showed in Figure 2.

2.2 Micro-Photoluminescence (PL) experiments

μ-PL measurements were performed on a WITec alpha-300R confocal microscope in a backscattering configuration. Samples were under vacuum at 10^{-3} mbar in a Linkam cryostat and probed at a requested temperature imposed by the temperature controller from 80K to 340 K, each 20 K. Samples were excited at 532nm through an Olympus 20x long working distance objective, with 9μW laser power (360nW/μm²). Light is collected throught a 200μm pinhole, and resolved with a 150 grooves/mm grating, with 750nm blaze wavelength. The detector is an Andor TEC cooled CCD bar at -70°C. For each spectrum, the camera thermal signal is removed.

Figures 3 and 4 show PL spectra obtained at 300 K on GaAs/Si crystals made with the TMGa or TEGa precursor respectively, for the different R values indicated in the insert. The spectrum of an undoped GaAs wafer is also shown as a reference on each of the figures. We note that in the case of TMGa (Fig. 3), the PL peaks of the crystals have an energy position lower than or very close to that of the reference peak, whereas in the case of TEGa, the energy positions of the PL peaks of the crystals have an energy position higher than that of the reference. Figure 5 summarizes these trends by showing the evolution with R of the energetic position of these peaks as a function of the ratio R.

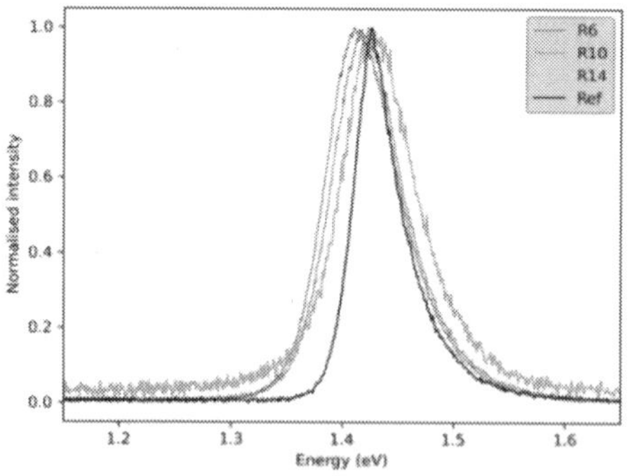

Figure 3: PL spectra at 300 K of GaAs/Si crystals made with the TMGa precursor for the different R values indicated in the insert. The black line (Ref) is the spectrum of an undoped GaAs wafer.

In the case of GaAs, it has been shown that the energetic position of the main PL peak is practically equal to the direct gap [6], to within a few meV (around 4 meV at 300 K). In Figure 5, peak energies below the reference value may therefore reflect an apparent reduction in the gap due to Band Gap Narrowing in the case of a P-type, whereas peak energies above the reference value could be explained by the Moss-Burstein effect for a degenerate N-type material. If this interpretation is correct, the different behavior of TMGa and TEGa precursors may be related to the amphoteric character of carbon with respect to GaAs doping, given the different dissociation properties of the methyl and ethyl radicals present in these precursors.

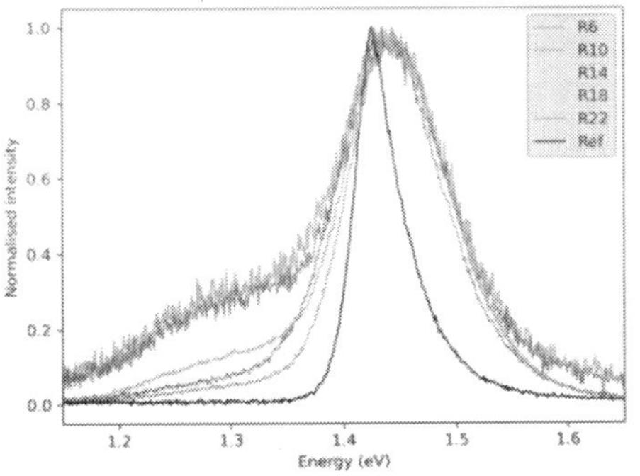

Figure 4 : PL spectra at 300 K of GaAs/Si crystals made with the TEGa precursor for the different R values indicated in the insert. The black line (Ref) is the spectrum of an undoped GaAs wafer.

Figure 5 also shows that it appears difficult to finely control doping in GaAs by using only one or other of the precursors when growing with the CBE technique. The use of a mixture of the two precursors should, in principle, provide a better answer to this problem, but this route is likely to be more tricky to implement since, with TBAs, it will require the use of three precursor gases. However, before exploring this avenue, it will be necessary to analyse in more detail our interpretation of the evolution with R of the main peak shown in Figure 5.

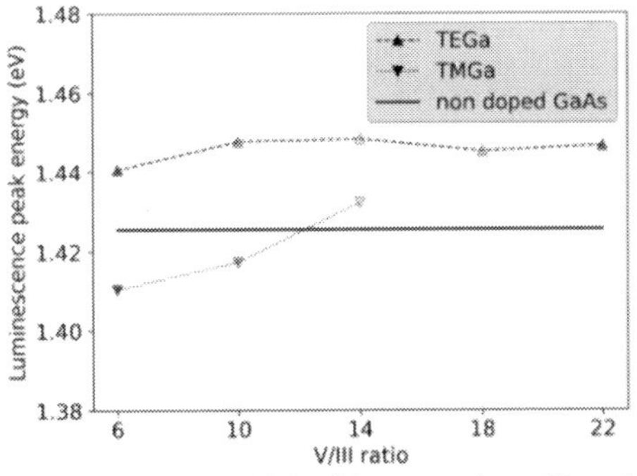

Figure 5 : Evolution with R of the energetic position of the main peak of the spectra presented in Fig. 3 and 4 as a function of the ratio R. The upper (lower) curve is for GaAs/Si crystals made with the TEGa precursor (TMGa precursor) respectively.

To this end, we will exploit photoluminescence spectra measured at different temperatures in the range 100 K to 340 K. In particular, we will analyze the evolution of the characteristics of the main peak and the secondary peak which is clearly visible at 300 K in the case of TMGa in Figure 5. We only point out here that, in a certain range of lower temperatures, this peak becomes stronger and also becomes apparent in the case of TMGa. The detailed interpretation of the evolution of the characteristics of these peaks with temperature will be one of the follow-ups to this work.

3 CONCLUSIONS

We have presented an innovative growth strategy based on Epitaxial Lateral Overgrowth On Tunnel oxide (ELTOn) in order to monolithically integrate defect-free micrometric size (Al)GaAs crystals on silicon. Integrating these crystals in a regular array to have a quasi-complete covering of the Si substrate, without coalescence of the GaAs microcrystals, is therefore a very promising route to produce in a first instance an AlGaAs/Si tandem cell at an affordable cost for terrestrial PV applications.

A fine control of the residual doping with only TMGa or TEGa as the Ga precursors during the CBE growth could not be achieved in the present work. However we have shown that mixing TMGa and TEGa precursors could be a possible route to solve this problem, taking into account the amphoteric nature of the carbon with respect to GaAs doping, and the different dissociation properties of the methyl and ethyl radicals present in these precursors.

The results already achieved have permitted the identification of the bottlenecks and discuss the possible routes for III-V on silicon multijunctions solar cells as the tandem structure presented in this paper can be the building block for further extension to triple junctions.

4 ACKNOLEGMENTS

This work was supported by the french Agence Nationale de la Recherche (ANR HELLO_PV project, Convention N° ANR-22-CE50-0011-01). The authors also thank the PIMENT platform of C2N and RENATECH for technological backup.

5 REFERENCES

[1] C. Renard, T. Molière, N. Cherkashin, J. Alvarez, L. Vincent, A. Jaffré, G. Hallais, J. Connolly, D. Mencaraglia, and D. Bouchier; *High current density GaAs/Si rectifying heterojunction by defect free Epitaxial Lateral overgrowth on Tunnel Oxide from nano-seed*; Scientific Reports (Nature) 6, 25328, 2016; doi: 10.1038/srep25328, www.nature.com/articles/srep25328

[2] J.P. Connolly, D. Mencaraglia, C. Renard, D. Bouchier, *Designing III–V multijunction solar cells on silicon*, Progress in Photovoltaics, vol. 22 (07), p. 810-820, 2014; doi: 10.1002/pip.2463

[3] T. Molière, A. Jaffré, J. Alvarez, D. Mencaraglia, J. Connolly, L. Vincent, G. Hallais, D. Mangelinck, M. Descoins, D. Bouchier, C.Renard; *GaAs micro-crystals selectively grown on silicon: Intrinsic carbon doping during chemical beam epitaxy with trimethylgallium* ; Journal of Applied Physics 121, 035704 (2017)

[4] M. Longo, R. Magnanini, A.Parisini, L.Tarricone, A.Carbognani, C.Bocchi, E.Gombi, *Controlled intrinsic carbon doping in MOVPE-grown GaAs layers by using TMGa and TBAs*, Journal of Crystal Growth, 2003, 248, 119 ; doi: 10.1016/S0022-0248(02)01846-8

[5] E. Grilli, M. Guzzi, R. Zamboni, et L. Pavesi, *High-precision determination of the temperature dependence of the fundamental energy gap in gallium arsenide*, Phys. Rev. B, vol.45, n°4, p. 1638-1644, 1992, doi: 10.1103/PhysRevB.45.1638.

COMBINED CL AND EBSD INVESTIGATION OF NON-RADIATIVE GRAIN BOUNDARY RECOMBINATION IN CdTe-CdSe THIN-FILM PHOTOVOLTAICS

Taghreed Altamimi [1,2], Jacob F Leaver [3], Jonathan D Major [3] and Budhika Mendis [1]
[1] Durham University, Physics Department, Centre for Material Physics, United Kingdom.
E-mail: Taghreed.f.altamimi@durham.ac.uk
[2] University of Hail, Physics Department, Hail 81451, Saudi Arabia.
[3] University of Liverpool, Department of Physics, United Kingdom.

ABSTRACT: This work aims to combine cathodoluminescence (CL) and electron backscatter diffraction (EBSD) to study the effect of CdTe grain boundaries on the recombination process. The grain boundary character was determined by EBSD. CL is then performed on the same area to obtain information on recombination. Recombination is monitored by the decrease in CL intensity at the grain boundary, the larger the decrease in CL intensity the stronger the non-radiative recombination at the grain boundary and therefore the more harmful to the device performance. An investigation between the non-radiative recombination and the grain misorientation in a CdTe-CdSe solar cell has been carried out. The results show that CL contrast was high within the 35°-55° range of misorientation angles, especially at 35° misorientation. This indicates that the non-radiative recombination is strong at this grain boundary misorientation. We also found CL contrast had small values around coincidence site lattice (CSL) boundaries indicating that the non-radiative recombination is low at these misorientations.
Keywords: CdTe solar cell, grain boundaries, cathodoluminescence, electron backscatter diffraction.

1 INTRODUCTION

Cadmium telluride (CdTe) solar cells are a very attractive thin-film photovoltaic technology because of low-cost manufacturing and high light absorption coefficient [1]. Electron hole pairs generated by photons absorbed in the solar cell have finite lifetime, due to radiative or non-radiative recombination processes. This has a harmful impact on solar cell efficiency [2]. One path towards increasing the efficiency of CdTe solar cells is to minimise non-radiative recombination at grain boundaries. This work combines cathodoluminescence (CL) and electron backscatter diffraction (EBSD) to study the effect of CdTe grain boundaries on the recombination process. Electron microscopy was used in this project as a valuable tool for investigating the microstructure and optical properties of solar cells at the nanoscale. It can produce high-resolution images of the solar cell microstructure and analyse the chemical composition.

2 EXPERIMENTAL METHODS

CL in the scanning electron microscope (SEM) is used to detect the radiative recombination in CdTe solar cells [3]. CL was performed on the back surface of a CdTe thin-film solar cell. EBSD is also used to provide information about the grain boundary misorientation. The electron beam hits the 70° tilted CdTe thin film solar cell backscattering to all orientations [4]. The sample used in this work has a structure of glass/fluorine-doped tin oxide (FTO)/ SnO_2/CdSe/CdTe/Au. The window layer CdSe has a thickness of 100 nm. CdTe was deposited using close-space sublimation under vacuum (i.e. 0 torr nitrogen gas pressure). The sample for scanning electron microscopy (SEM)-CL was prepared by first extracting a 2x1 mm^2 rectangular piece using a Gatan ultrasonic cutting tool. The as-deposited surface is rough due to the underlying CdTe grain structure and must therefore be flattened for CL analysis such that it is free of topographic artefacts. The sample was then imaged in a Hitachi - SU70 SEM operating at either 15 or 20 kV electron beam voltage and equipped with a GATAN MonoCL system. EBSD instrumentation was located in a different microscope, an FEI Helios 600 focused ion-beam SEM (FIB-SEM), which was operated at 30 kV electron beam voltage.

3 RESULTS

Energy dispersive X-ray (EDX) spectroscopy results (Figure 1) for the back surface showed that Cd and Te are the main components in the sample with overall weight 47.13% for Cd, 50.43% for Te and 0.64% for Se. The remaining concentration is due to aluminium (1.80 weight%), which is likely to be an impurity.

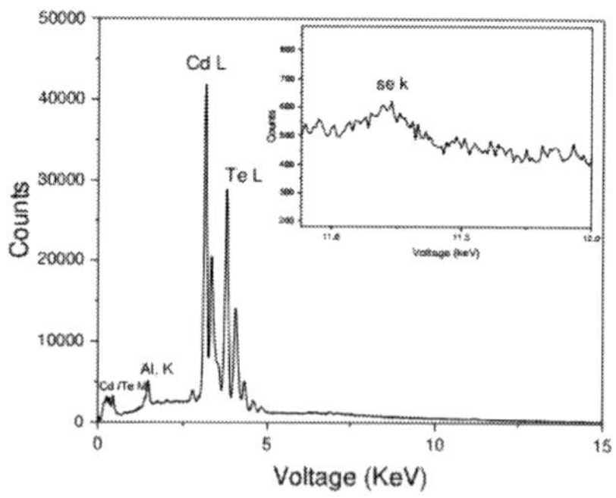

Figure 1: Energy dispersive X-ray (EDX) spectrum of the back surface of a CdTe-CdSe solar cell.

3.1 Regular grain boundaries

Approximately sixty four large grains have been picked to calculate the CL contrast and misorientation across their grain boundaries. For example, Figure 2 shows EBSD pattern quality and orientation map for some selected grains. For the same area the cathodoluminescence contrast at each individual grain boundary (GB) has been calculated from the CL intensity profile by using the intensity average at the GB and grain interior. This information is then combined with the EBSD misorientation angle for each individual GB that is provided by Aztec crystal software.

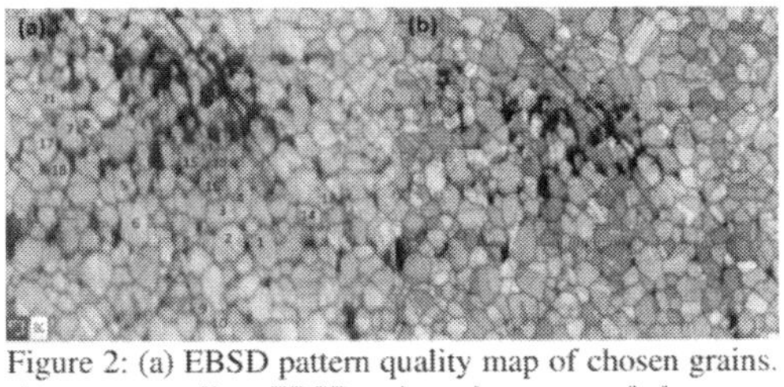

Figure 2: (a) EBSD pattern quality map of chosen grains. (b) corresponding EBSD orientation map of the same region.

Figure 3 shows the plot of CL contrast versus the grain boundary misorientation angle from our results. It reveals high contrast values with large variations from approximately 11% to 26%, depending on the misorientation angle. Overall, the range 35°-55° includes high values of CL contrast, and around the angle 35° a high peak with maximum value in CL contrast is observed. The high CL contrast means high non-radiative recombination, which is harmful to the device.

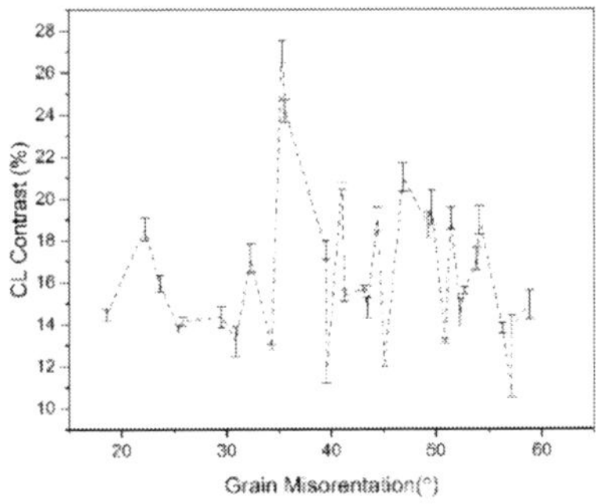

Figure 3: The CL contrast as a function of grain boundary misorientation. The line connecting the data points is a guide to the eye only.

Moseley *et al.* [5] also examined the grain boundary recombination before and after chlorine activation, by measuring the CL intensity at all GB types for both as-deposited and chlorine treated thin-films. A higher CL intensity means a low recombination rate. On the other hand, our results calculated and plotted the CL contrast, where high CL contrast denotes a high recombination rate. An increase in the average CL intensity with GB misorientation angle was observed for a given device condition (i.e. as-deposited or chlorine activated), and contrary to our results they found that the small angles are more harmful for the device. Moseley et al. assumed this might be because the deep levels were effectively passivated at the high angle GB due to chlorine activation. However, we note that the same trend is also observed for the as-deposited sample, where there is no chlorine passivation of the grain boundaries. Furthermore, we believe that a proper analysis requires a CL contrast calculation rather than the absolute CL intensity, since it is the change in CL intensity at the GB compared to the grain interior that is the relevant parameter. The absolute intensity on the other hand will be more susceptible to local variations, such as porosity, doping concentration etc.

3.2 Coincident site lattice grain boundaries

CSL grain boundaries have also been investigated. This includes the Σ3 twin boundary, which has a misorientation angle equal to 60° and appears more frequently in the sample. It has been observed that Σ3 CSL have small CL contrast values compared to the regular GBs. The CL contrast varied from 0.2% to 10%, as shown in Figure 4, where it appeared that the CL contrast for these types of GBs is smaller than the average contrast of the regular GBs. This is clear from the twin boundaries within grains 1 or 2, i.e. the red twin boundary in the EBSD map. The CL map did not show visible non-radiative recombination, indicating only a small amount of non-radiative recombination. This is not the case for the general GB between grains 1 and 2, which has strong non-radiative recombination (wider darker grain boundary). Similar observations can be made for grains 5 and 6. Therefore, Σ3 GBs are not harmful non-radiative recombination centres in the device.

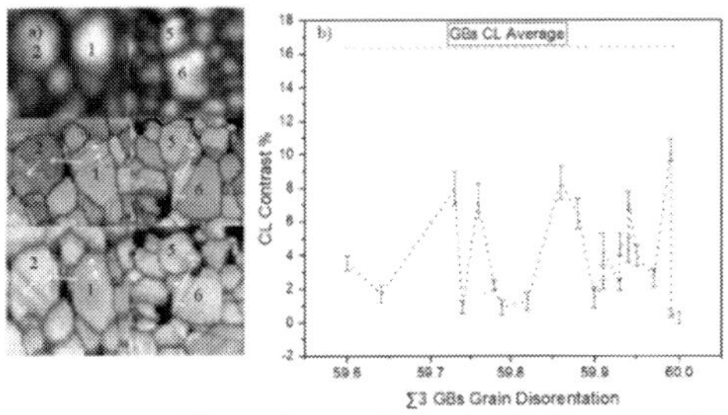

Figure 4: a) CL panchromatic (top), EBSD orientation (middle) and pattern quality (bottom) maps showing some Σ3 GBs within the grains labelled '1' and '2' as well as '5' and '6'. b) CL contrast as a function of GB misorientations close to Σ3 CSL. The line connecting the data points is a guide to the eye only. The horizontal dashed line represents the average CL contrast for the general grain boundaries in Figure 3.

Moseley *et al.* [5] also found Σ3 CSL GBs showed a high CL intensity consistent with our low CL contrast values (Figure 4b) indicating they are not harmful to the device.

3.3 Reduced recombination velocity of grain boundaries

Recombination velocity is the rate at which carriers diffuse to the grain boundary to replenish the carriers lost to recombination, thereby maintaining a steady state carrier concentration. A grain boundary with a high recombination velocity is electrically more active and has a more negative impact on the device efficiency. The recombination velocity divided by the bulk carrier diffusion velocity gives the reduced recombination velocity. Figure 5 shows the GB between two grains labelled 1 and 2, with a relatively large value (0.52) of reduced recombination velocity (Table I3.2). The CL contrast for this GB was 20%, which confirms the strong non-radiative recombination. The related misorientation at this GB was 41° as measured using EBSD (Table I). For the second GB between two different grains labelled 3 and 4, the reduced recombination velocity (0.19) was smaller than the previous GB, and the CL contrast (12.9%) and misorientation (34°) at the GB is lower than the previous one. It is clear that the nonradiative recombination is not as strong as the previous GB between grains 1 and 2. This indicates that the CL contrast analysis presented in Section 3.1 is a good indicator of the role of grain boundary misorientation on recombination activity.

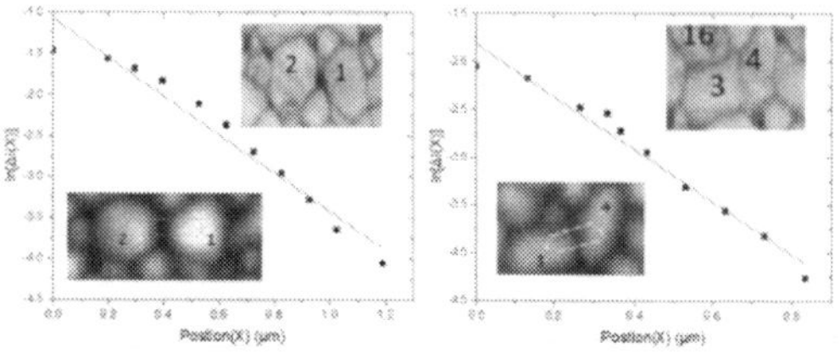

Figure 5: Reduced recombination velocity analysis across two grain boundaries. The upper inset is an EBSD pattern quality map indicating the grains that have been analysed. The lower inset is the corresponding panchromatic CL image from which data analysis is performed.

Table I: The calculated diffusion length (L), reduced recombination velocity (S), CL contrast and grain misorientation values that are related to grain boundaries 1,2 and 3,4 (Figure 5).

GB	Diffusion length(L)	Reduced velocity(S)	Mis-angle	CL Contrast
1,2	0.42±0.07	0.52±0.16	41.1°	20.62
3,4	0.36±0.02	0.19±0.01	34.3°	12.93

4 CONCLUSION

In conclusion, CL and EBSD of regular and CSL grain boundaries in CdTe-CdSe thin-film have been investigated. The grain boundaries were assumed to be pure CdTe, as EDX results showed there is no selenium appearing at the back contact surface. CL contrast has been calculated, and for the regular GBs, the contrast was significantly high in the range between 35°-55° misorientation angle. This means that those regular grain boundaries are strong non-radiative recombination centres. The highest CL contrast was observed at 35°, which might indicate this particular misorientation angle has the highest harmful effect on the device. On the other hand low index CSL such as $\Sigma3$,$\Sigma5$, $\Sigma7$ and $\Sigma9$ showed much lower contrast values which is consistent with the fact that they have high symmetry. The robustness of our results was examined by performing more detailed quantitative analysis of the reduced recombination velocity. The reduced recombination velocity results were consistent with the previous trends, i.e. the high contrast grain boundary has a larger reduced recombination velocity than the lower contrast grain boundary.

Moseley *et al* [5] used automated software to process all the grain boundaries, which is not the case with our work where the grains have been chosen and analysed by hand. Also, they used the absolute CL intensity rather than contrast, which is more susceptible to sample inhomogeneities, such as local doping concentration and porosity. Therefore, there is a need for full contrast analysis rather than only intensity analysis as we believe the contrast calculation can give more representative results that is also consistent with more detailed quantitative analyses of the reduced recombination velocity. Furthermore, Moseley *et al* [5] observed $\Sigma3$ CSL had a high intensity which is in agreement with our low contrast values, indicating that the non-radiative recombination is small at $\Sigma3$ CSL. In general, the CL contrast in CdTe-CdSe solar cells varied depending on the type of grain boundary, whether regular grain boundaries or coincidence site lattice grain boundaries.

Acknowledgments

The authors would like to thank Saudi Arabia cultural bureau in UK and Hail University in Saudi Arabia for funding and support this work. Also, Durham University for the facilities. Liverpool University for providing the sample.

4.1 References

[1] R.W. Birkmire, B.E. McCandless, CdTe thin film technology: Leading thin film PV into the future. Current Opinion in Solid State and Materials Science, (2010).

[2] H. Mohamed, Dependence of efficiency of thin-film CdS/CdTe solar cell on optical and recombination losses. Journal of applied Physics, (2013).

[3] I. Pelant, J. Valenta, Luminescence spectroscopy of semiconductors, (2012).

[4] D. Abou-Ras, T Kirchartz, Advanced characterization techniques for thin film solar cells, (2016).

[5] J. Moseley, W. K. Metzger, H. R. Moutinho, H. R, N Paudel, H. L. Guthrey, Y Yan, M. Al-Jassim, M. M. Recombination by grain-boundary type in CdTe. Journal of Applied Physics, (2015)

HEAT-TOLERANT FLEXIBLE CHALCOPYRITE SOLAR CELLS VIA ALL-SPUTTERED CD-FREE ELECTRON TRANSPORT LAYERS

Hiroki Sugimoto[1], Kazuhito Fukasawa[1], Mayo Kawahara[1], Yoshiaki Hirai[1], and Akira Yamada[2]

[1]PXP Corporation, Nishihashimoto, Sagamihara, Kanagawa, Japan
[2]Tokyo Institute of Technology, Ookayama, Meguro-ku, Tokyo, Japan

Hiroki.Sugimoto@pxpco.jp

ABSTRACT: Heat-tolerant flexible chalcopyrite Cu(In, Ga)(Se, S)$_2$ (CIGSS) solar cells were realized by introducing all-sputtered Cd-free electron transport layers (ETLs). The samples with the newly developed double ETLs showed no degradation after a long-time super heat acceleration test under the light soaking, while the samples with conventional ETLs showed large degradation. Especially fill factor drop was completely restricted, which was considered to be ascribed to the barrier effects of the new ETLs against the alkaline diffusion from the CIGSS absorber to the transparent conductive oxide through the ETLs. These results contribute to the installation of the CIGSS solar cells into the relatively higher temperature environments than the conventional ground-mounted solar cells, such as the rooftop of vehicles and the space satellites. In addition, the all-sputtered ETLs will reduce the production cost of the CIGSS solar cells due to the elimination of the high-cost conventional manufacturing processes for the ETL deposition, such as the chemical bath deposition, the chemical vapor deposition and the atomic layer deposition.
Keywords: Chalcopyrite, Cu(In, Ga)(Se, S)$_2$, CIGSS, Cd-free, All-sputter

1 INTRODUCTION

Recently, light-weight and flexible solar cells have attracted a lot of attention for the application in vehicles, drones and space satellites as well as buildings. PXP corporation (one of the small start-up company in Japan) has been developing low cost, extra light and flexible, high durability and high performance solar cells, aiming to expand thin-film photovoltaic (PV) applications into new fields, such as building integrated PV (BIPV), vehicle integrated PV (VIPV) and aerospace applications. In order to develop the next-generation solar cells applicable to these new fields, the study on durability against the higher temperature than the conventional ground-mounted PV applications is very important, because the solar cells under the anormal environments, such as the rooftop of the vehicles and the space satellites, sometimes shows much higher temperature than the ground-mounted solar cells. In these fields, the long-time heat tolerance against the temperature above 100C and up to 150C is sometimes required.

So far, a lot of papers have reported the heat stability on the chalcopyrite solar cells. According to the comprehensive review for stability tests on the chalcopyrite solar cells, the interactions among the transparent conductive oxide (TCO), the electron transport layer (ETL) and the chalcopyrite absorber were the most important key to determine the heat tolerance [1]. However, most of these papers were related to the conventional stability test around the temperature of 85C, which described that tea chalcopyrite solar cells had good heat tolerance and there was no significant difference in the degradation ratio among the variety of ETLs [2, 3]. The other papers reported the impact of the high temperature treatment above 200C within a short time (from 15min to 60min) against the variety of ETLs, such as conventional CdS [4], Cd-free In_2S_3 [5], ZnOS [6] and ZnMgO [7]. They concluded the intermixture between the ETLs and the absorbers caused degradations of the cell parameters. Not only the intermixture, but also the Na diffusion from the absorber to the TCO through the ETL was also reported to be related to the degradations [8]. Especially, the Na diffusion is considered to be the dominant factor of the degradation at the temperature range from 100C to 200C due to the low migration energy of the Na atoms in the chalcopyrite absorbers (the migration energy just about 0.3eV) [9]. There are a few reports for the investigation of the long-time heat tolerance at the temperature range from 100C to 200C [10,11]. So far, it has been difficult to prevent the Na-related degradation by the long-time heat exposure above 100C.

The purpose of this paper is to establish the heat-tolerant (up to 150C) flexible chalcopyrite Cu(In, Ga)(Se, S)$_2$ (CIGSS) solar cells by applying new ETLs. In addition, taking into account the future mass-production, we also developed the all-sputtered Cd-free ETLs, in order to reduce the production cost of the CIGSS solar cells by eliminating the high-cost conventional ETL processes, such as the chemical bath deposition (CBD), the chemical vapor deposition and the atomic layer deposition (ALD). The conventional ETL deposition method is the wet process, however it is inconvenient because both the absorber fabrication process and the TCO deposition process are the dry process. The cost impact of the introduction of all-dry ETL should be very big because it can eliminate useless repetition of vacuum brake and vacuum back as well as a plenty of toxic waste disposal.

2 EXPERIMENTAL DETAILS

In order to investigate the heat tolerance, the flexible CIGSS solar cells with the variety of ETLs were prepared. The CIGSS solar cells used for this study were specially developed for the severe space environments. Not only the high temperature tolerance but also the radiation hardness, the thermal cycle tolerance and the mechanical vibration tolerance were much improved than the conventional CIGSS solar cells [12]. The CIGSS solar cells were fabricated by highly productive sputtering-based process. The basic structure was Ag grid electrode/In_2O_3:H (IOH) TCO/ETLs/CIGSS absorber/Mo back electrode on flexible Ti film. In this study, anti-reflective (AR) coating was not deposited, however, normally relative efficiency (Eff) gain of around 8% by the AR coating has been confirmed for our cells. The thickness of the CIGSS device layer and the Ti substrate was about 3 um and 50 um, respectively. The weight of our standard CIGSS solar cells were about 250 g/m^2 (minimally 150 g/m^2 achievable).

Regarding the ETLs, at first, the conventional CdS cells with the ETL thickness of 75 nm were fabricated by the CBD method. Then, we introduced the sputtered $ZnTiO_x$ layer as the secondary wide-gap ETL on the CdS layer. The thickness of optimized $ZnTiO_x$/CdS double ETLs was 70nm and 30nm, respectively. Note that the initial efficiency (Eff) and the heat tolerance of the CdS cells and the $ZnTiO_x$/CdS cells were similar, thus the results of only the $ZnTiO_x$/CdS cells were shown in this paper. We also developed the Cd-free cells by introducing $ZnTiO_x$ single ETL with the thickness of 100nm. Finally, all-sputtered Cd-free $ZnTiO_x/In_2S_3$ double ETLs were developed as the heat tolerant and highly productive ETLs. The thickness of optimized $ZnTiO_x/In_2S_3$ double ETLs was 70nm and 30nm, respectively.

After the cell fabrication, all samples were pre-stabilized by the light soaking (LS) at 200C30min in N_2 box in order to eliminate over estimation of improvement effects by the LS. After the initial measurement of the electrical parameters, the super heat acceleration test (SHT) was performed at 150C208h under the LS in N_2 box. Then, the heat effect in dark was investigated at 150C65h in dark N_2 box. Finally, recovery effect by the LS was also investigated at 150C44h under the LS in N_2 box. The changes in the electrical parameters at respective steps were also measured.

The current–voltage (J-V) characteristics were measured by a class A solar simulator (XI-05A1V2-L, SERIC Ltd., Japan) with AM 1.5 and 100 mW/cm^2 illumination at room temperature. External quantum efficiency (EQE) curves ranging from wavelength of 300 nm to 1300 nm were measured by a spectral response measurement system (CEP25-MLT, Bunkoukeiki Co., Ltd., Japan) at room temperature without any bias. The bandgap of the CIGSS cells were extracted from the EQE curves with the derivative method [13]. Then, open circuit voltage (V_{oc}) deficit was calculated with the bandgap and V_{oc}.

3 RESULTS AND DISCUSSION

Figures 1(a)-(c) show the schematic images of the long-time SHT condition and device structures with the variety of ETLs, such as the $ZnTiO_x$/CdS double ETLs, the $ZnTiO_x$ ETL, and the $ZnTiO_x/In_2S_3$ double ETLs. The CdS layer was deposited by the CBD and the $ZnTiO_x$ layer and the In_2S_3 layer were deposited by sputtering. Before the SHT, devices were pre-stabilized by the LS at 200C30min in order to eliminate over estimation of improvement effects by the LS, then the SHT was performed at 150C208h under the LS in N_2 box.

The J-V characteristics of the $ZnTiO_x$/CdS cell, the $ZnTiO_x$ cell, and the $ZnTiO_x/In_2S_3$ cell before and after the SHT are shown in Figs 1(d)-(f). The $ZnTiO_x/In_2S_3$ double ETL cell showed no degradation after the SHT, while the $ZnTiO_x$/CdS double ETL cell and the $ZnTiO_x$ ETL cell showed large degradation. The degradation mainly caused by the fill factor (FF) drop, while the V_{oc} and the short circuit current (J_{sc}) were stable. We confirmed the conventional CdS ETL cell also showed the FF drop as same as the $ZnTiO_x$/CdS double ETL cell and the $ZnTiO_x$ ETL cell. On the other hand, the FF drop was completely restricted by the $ZnTiO_x/In_2S_3$ double ETL cell. From these results, we concluded that the important key for preventing the heat degradation should be the insertion of the In_2S_3 layer. The type of the deposition method was seemed to be less impact against the heat degradation. We also tried the In_2S_3 single ETL cell, however the initial V_{oc}

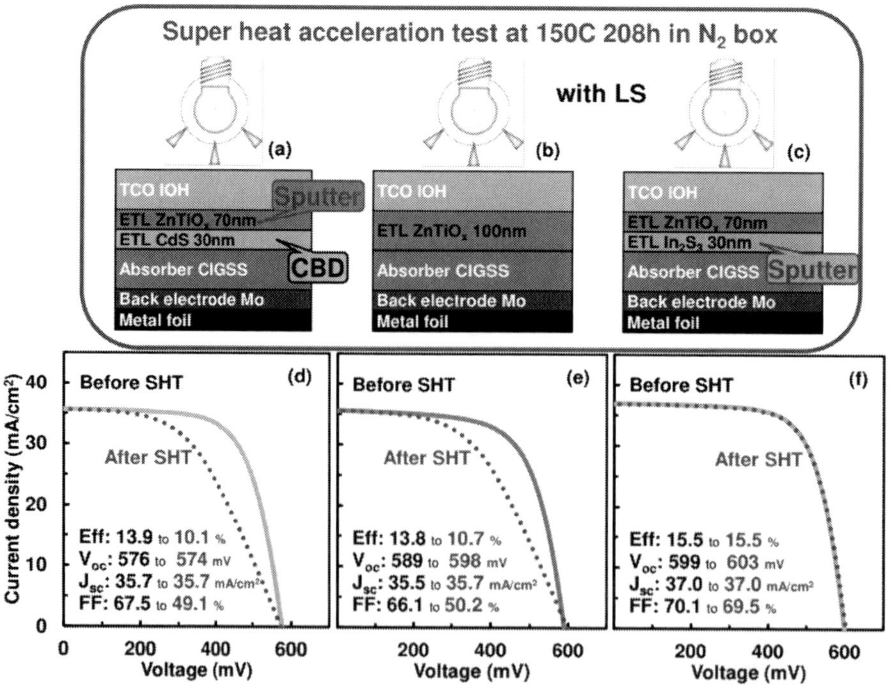

Figure 1: Schematic images of long-time SHT condition and device structures with a variety of ETLs, (a) $ZnTiO_x$/CdS double ETLs, (b) $ZnTiO_x$ ETL, and (c) $ZnTiO_x/In_2S_3$ double ETLs. CdS layer was deposited by CBD and $ZnTiO_x$ layer and In_2S_3 layer were deposited by sputtering. J-V characteristics of (a) $ZnTiO_x$/CdS cell, (b) $ZnTiO_x$ cell, and (c) $ZnTiO_x/In_2S_3$ cell before (solid lines) and after (dashed lines) SHT.

10.4229/EUPVSEC2023/2CV.2.45
020134-002

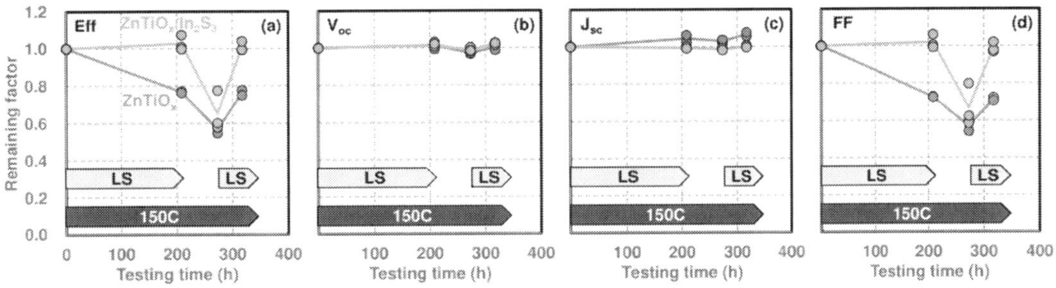

Figure 2: Changes in remaining factor of (a) Eff, (b) V_{oc}, (c) J_{sc}, and (d) FF on $ZnTiO_x$ cells (blue) and (c) $ZnTiO_x/In_2S_3$ cells (green) by SHT under LS at 150C208h, SHT in dark at 150C65h, and SHT under LS at 150C44h. Solid lines show the averaged remaining factors of respective cells.

degradation caused by the deposition damage of the TCO sputtering was not prevented. Introducing thicker In_2S_3 single ETL was one of the ways to prevent the TCO sputtering damages, however initial J_{sc} was decreased due to the less light transmission in the short wavelength region by the low bandgap of In_2S_3. This was why we introduced the wide-gap $ZnTiO_x$ layer on the In_2S_3 layer. Then we compared the ALD and the sputtering for the deposition technique of the $ZnTiO_x$. After optimization of deposition condition, fortunately the sputtered $ZnTiO_x$ cells showed comparable initial Eff with the ALD deposited $ZnTiO_x$ cells. Thus, the all-sputtered Cd-free $ZnTiO_x/In_2S_3$ double ETL structure was introduced as our basic cell structure. The strong heat tolerance of In_2S_3 layer was considered to be ascribed to the barrier effects of the ETLs against the Na diffusion from the CIGSS absorber to the TCO through the ETLs.

Figure 2 shows the changes in remaining factor of each electrical parameter on the $ZnTiO_x$ cells and the $ZnTiO_x/In_2S_3$ cells. Before the SHT, devices were also pre-stabilized by the LS at 200C30min in order to eliminate over estimation of improvement effects by the LS. At first, the SHT was performed at 150C208h under the LS in N_2 box, then in order to investigate the dark heat effects, samples were heated at 150C65h in dark N_2 box. Finally, the LS was applied again at 150C44h in N_2 box for checking the recovery effects by the LS. As same as the results shown in Fig 1(e) and (f), the $ZnTiO_x/In_2S_3$ cells also showed no degradation after the first SHT also in Fig. 2, while the $ZnTiO_x$ cells showed large degradation due to the FF drop. In this experiment, the $ZnTiO_x$ cells showed slight J_{sc} increasement. This was considered to be due to the widening of the space charge region by the decrease in the carrier density of the CIGSS absorber. We believe the alkaline diffusion in the $ZnTiO_x$ cells should decrease the carrier density. Then, after the dark heat at 150C65h, both the $ZnTiO_x/In_2S_3$ cells and the $ZnTiO_x$ cells showed the FF drop, however the FF drop was completely recovered by the successive LS at 150C44h. We considered these phenomena caused by the metastable behavior of the defect complex of Cu-vacancies and Se-vacancies (V_{Cu}-V_{Se}). The V_{Cu}-V_{Se} defects initially act as the donor-like defects, then they turn into the acceptor-like defects by the LS. However, they easily turn back to the initial state by the successive dark heating. We believe the temporary FF drop was caused by the decrease in the carrier density due to the changes in the electrical state of the V_{Cu}-V_{Se} defects by the dark heat.

Then, optimization of the Cd-free double ETLs was performed. Figure 3(a) shows the V_{oc} deficit of the $ZnTiO_x/CdS$ cells, the $ZnTiO_x$ cells, and the $ZnTiO_x/In_2S_3$ cells. We confirmed not only the heat-tolerance but also the V_{oc} deficit was improved by the $ZnTiO_x/In_2S_3$ double ETLs as compared with the conventional ETLs. Finally, Eff of 16.9% was achieve by the all-sputtered Cd-free $ZnTiO_x/In_2S_3$ cell without AR coating, as shown in Fig 3(b). These results contribute to the installation of the CIGSS solar cells in the relatively higher temperature environments than the conventional ground-mounted solar cells, such as the rooftop of vehicles and the space satellites. In addition, these achievements will reduce the production cost of the CIGSS solar cells due to eliminating the high-cost conventional manufacturing processes for the ETL depositions.

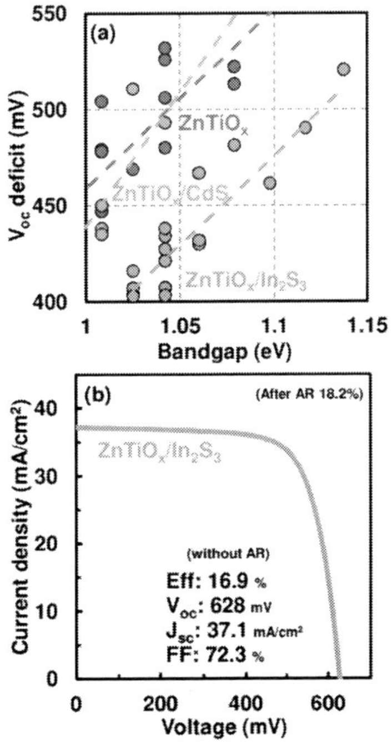

Figure 3: (a) V_{oc} deficit of $ZnTiO_x/CdS$ cells (orange), $ZnTiO_x$ cells (blue), and $ZnTiO_x/In_2S_3$ cells (green). Dashed lines show fitted lines of respective cells. (b) J-V characteristics of optimized. $ZnTiO_x/In_2S_3$ cell without AR coating.

4 SUMMARY

In this paper, we investigated the durability against the higher temperature than the conventional ground-mounted PV applications. At first, we prepared test samples with variety of ETLs and performed SHT at 150C208h in N_2 box with the LS. The samples with the $ZnTiO_x/In_2S_3$ double ETLs showed no degradation after the SHT, while the samples with conventional ETLs showed large degradation. We also investigated the heat effects in dark, then temporary FF drop was observed, however it was fully recovered after the LS. After optimization of ETLs, we confirmed not only the heat-tolerance but also the V_{oc} deficit was improved by the newly developed double ETLs as compared with the conventional ETLs. Finally, the Eff of 16.9% without the AR coating was achieved by the all-sputtered Cd-free $ZnTiO_x/In_2S_3$ double ETLs. We believe these results will contribute to the expansion of the thin-film PV into the BIPV, VIPV and aerospace applications.

REFERENCES

[1] M. Theelen, and F. Daume, Solar Energy, vol. 133 (2016) 586.

[2] N.A. Allsop, A. Hänsel, S. Visbeck, T.P. Niesen, M.C. Lux-Steiner, and Ch.-H. Fischer, Thin Solid Films, vol. 511 (2006) 55.

[3] S. Spiering, D. Hariskos, S. Schröder, and M. Powalla, Thin Solid Films, vol. 480 (2005) 195.

[4] A. Koprek, P. Zabierowski, M. Pawlowski, L. Sharma, C. Freysoldt, B. Gault, R. Wuerz, and O. Cojocaru-Mir´edin, Solar Energy Materials and Solar Cells, vol. 224 (2021) 110989.

[5] N. Naghavi, J.-F. Guillemoles, D. Lincot, B. Canava, A. Etcheberry, S. Taunier, S. Spiering, M. Powalla, M. Lamirand, and L. Legras, Proceedings 19th European Photovoltaic Solar Energy Conference (2004) 1733.

[6] J-H. Wi, T. G. Kim, J. W. Kim, W-J. Lee, D-H. Cho, W. S. Han, and Y-D. Chung, ACS Applied Materials and Interfaces, vol. 7 (2015) 17425.

[7] C-S. Lee, S. Kim, E. A. Al-Ammar, HS Kwon, and B. T. Ahn, ECS Journal of Solid State Science and Technology, vol. 3 (2014) Q99.

[8] H. A. Yetkin, T. Kodalle, T. Bertram, A. Villanueva-Tovar, M. Rusu, R. Klenk, B. Szyszka, R. Schlatmann, and C. A. Kaufmann, Progress in Photovoltaics, vol. 29 (2021) 1034.

[9] T. Maeda, A. Kawabata, and T. Wada, Japanese Journal of Applied Physics, vol. 54 (2015) 08KC20.

[10] M. Theelen, V. Hans, N. Barreau, H. Steijvers, Z. Vroon, and M. Zeman, Progress in Photovoltaics, vol. 23 (2015) 537.

[11] T. Ott, T. Walter, D. Hariskos, O. Kiowski, and R. Schaffler, IEEE Journal of Photovoltaics, vol. 3 (2013) 514.

[12] H. Sugimoto, T. Nakamura, M. Imaizumi, S. Sato, and T. Ohshima, Proceedings 50th IEEE Photovoltaic Specialist Conference (2023) unpublished.

[13] R. Carron, C. Andres, E. Avancini, T. Feurer, S. Nishiwaki, S. Pisoni, F. Fu, M. Lingg, Y. E. Romanyuk, S. Buecheler, and A. N. Tiwari, Thin Solid Films, vol. 669 (2019) 482.

STUDY OF DIFFERENT APPROACHES TO FABRICATE CdSeTe/CdTe DEVICES BY THERMAL EVAPORATION

Elisa Artegiani, Andrea Gasparotto*, Mariyam Mukhtar, Simya Olavil Karayi, Narges Torabi, Ikram Anefnaf and Alessandro Romeo*

Laboratory for Photovoltaics and Solid-State Physics (LAPS), Department of Computer Science, University of Verona
Ca' Vignal 1, Strada Le Grazie 15, 37134 Verona, Italy
* Department of Physics and Astronomy "G. Galilei", University of Padova
Via F. Marzolo 8, 35131 Padova, Italy
*author for correspondence, e-mail address: alessandro.romeo@univr.it

ABSTRACT: Introducing a CdSeTe compound at the junction of CdTe-based devices enhances photons absorption across a broad spectrum of short and long wavelengths.
CdSeTe/CdTe devices can be fabricated by sequentially depositing a CdSe and a CdTe layer. A crucial aspect of these devices is the intermixing of the layers, where they must combine to create the desired $CdSe_xTe_{1-x}$ compound at the junction by completely consuming the CdSe.
In this paper we investigate different strategies to fabricate CdSeTe/CdTe devices by sequential evaporation of a CdSe and a CdTe layer, coupled with the introduction of vacuum or air annealing either between or after the deposition of the layers.
The samples have been fabricated by low substrate temperature (LST) thermal evaporation, method that reduces energy consumption and enables the possibility to use flexible substrates. Furthermore, we have reduced the absorber thickness to 2 µm, which is half the usual thickness, aiming to minimize the materials impact in terms of cost, pollution and availability. Despite the reduction in absorber thickness, the devices exhibit remarkable current densities exceeding 28 mA/cm^2 and have achieved efficiencies approaching 15 %, setting a new record for such a thin CdTe cell produced by LST method.
Keywords: CdTe, CdSeTe, thin film, solar cells.

1 INTRODUCTION

In the last 10 years, after a long period of stagnation, the efficiency of CdTe devices has jumped from 16.5 % to 22.3 %. The breakthrough is mainly due to the introduction of selenium into the absorber layer to form a CdSeTe compound at the junction; this narrows the band gap increasing the portion of wavelengths absorbed [1]. Moreover, CdSeTe forms a good junction with wide band gap buffers, such as tin oxide and magnesium zinc oxide, allowing the complete removal of the CdS layer and therefore enhancing the absorption at shorter wavelengths [1,2]. Furthermore, CdSeTe/CdTe devices show a longer minority carrier lifetime compared to standard CdTe [3,4].

In our prior work, we introduced an innovative process to develop CdSeTe/CdTe cells. This method involves treating a thin CdTe layer at 400 °C in a selenium atmosphere (selenization), which gets subsequently mixed with a thicker CdTe layer [5–7]. However, this work describes the fabrication of CdSeTe/CdTe devices by sequentially depositing CdSe and CdTe layers, employing a low substrate temperature thermal evaporation (LST) method. This approach reduces energy consumption and enables the use of flexible substrates [8]. Moreover, we reduced the thickness of the absorber down to 2 µm, which is half the typical thickness, this in turn reduces the materials impact in terms of cost, pollution and availability [9]. Despite the thinning of the absorber, the devices exhibit impressive current densities exceeding 28 mA/cm^2 and have achieved an efficiency of 14.7 %, which is the highest ever recorded for such a thin CdTe cell using the LST method [7].

A critical point in CdSe/CdTe devices is the mixing of the layers, which must form the desired $CdSe_xTe_{1-x}$ compound at the junction by completely consuming the CdSe. In fact, CdSe in its wurtzite phase is not photoactive, and its presence at the junction implies parasitic absorption of all wavelengths [1,10]. To achieve this, during the deposition of the absorber we have investigated various annealing: of the CdSe, of the CdSe and a first thin CdTe stack, or of the completed absorber, both in air and in vacuum. The best results are achieved by annealing a first CdSe-CdTe stack at 450 °C in vacuum before finishing the deposition of the absorber.

2 EXPERIMENTAL

2.1 Device Fabrication.

In our laboratory we fabricate CdSeTe/CdTe-based devices on superstrate configuration by a LST fabrication process [8]. The front contact consists of a commercial transparent conductive oxide (TCO) supplied by NSG Pilkington, made of a fluorine-doped tin oxide (FTO) followed by a tin oxide (TO) layer. On this substrate we deposit a 300 nm thick CdSe layer through vacuum evaporation (VE) at a substrate temperature of 340 °C. After that, in the same vacuum chamber, we evaporate a 1.7 µm thick CdTe layer at the same substrate temperature. The absorber thickness is around 2 µm, thus half the usual thickness for this type of cell. To obtain a proper mixing of the CdSe and CdTe layers, during the deposition of the absorber we have investigated various annealing: 1) of the CdSe, 2) of the CdSe and a first thin CdTe stack, and 3) or of the completed absorber. An activation treatment is applied to all the devices by dropping a $CdCl_2$ solution on the surface of the absorber, followed by an annealing at 400°C in air. This process results in the re-crystallization and passivation of the grain boundaries in the absorber, as well as the blending of the CdSe and CdTe layers to form a CdSeTe compound near the junction. To remove the

$CdCl_2$ residuals from the surface, the sample is then etched in a Br-MeOH solution. This step also serves to create a surface rich in Te, which promotes the formation of Cu_xTe compounds at the junction between CdTe and back contact [11]. Subsequently, a back contact is created by depositing a 1 nm thick layer of Cu followed by a 30 nm thick layer of Au by VE. Finally, the cells are annealed at 200 °C in air, to diffuse Cu in the absorber [12] and to form the Cu_xTe compounds.

2.2 Characterization

Current density-voltage (JV) characteristics are acquired with a Keithley Source Meter 2420, under an AM 1.5 spectrum at 100 mW/cm^2, using a LOT Quantum Design Europe solar simulator LS0306.

SIMS depth profiles are obtained on a CAMECA IMS-4f using a Cs+ primary ion beam at a 10 kV accelerating voltage (corresponding to 5.5 keV impact energy) and detection of positive secondary ions.

The crystal structure is investigated by X-ray diffraction (XRD) analysis with a Thermo ARL X'TRA powder diffractometer (in Bragg-Brentano geometry, equipped with a Cu-anode X-ray source (Kα, λ=1.5418 Å) and a Peltier Si (Li) cooled solid-state detector.

3 RESULTS

3.1 Fabrication methods

Different approaches are employed and investigated for the fabrication of the absorber layer. The standard cells are produced through the sequential evaporation of CdSe and CdTe. In the case of other samples, the CdSe layer undergoes annealing either at 450 °C in a vacuum or at 400 °C or 450 °C in an air environment before the deposition of CdTe. In other devices, an initial 300 nm thin CdTe layer is deposited on the CdSe, and then the stack is subjected to annealing at 450 °C in either a vacuum or an air environment, followed by an additional 1400 nm thick CdTe layer. Finally, some samples are subjected to annealing at 450 °C in vacuum after the complete deposition of the absorber. Table I provides an overview of the various fabrication methods, whereas Table II presents the corresponding average efficiency parameters of the samples.

Table I: summary of the different fabrication processes.

Id.	CdSe (nm)	Ann (°C)	CdTe (nm)	Ann (°C)	CdTe (nm)
A-std	300		1700		
B	300	450 vacuum	1700		
C	300	450 air	1700		
D	300	400 air	1700		
E-best	300		300	450 vacuum	1400
F	300		300	450 air	1400
G	300		1700	450 vacuum	

Table II: average efficiency parameters of the samples.

Id.	FF (%)	Voc (mV)	Jsc (mA/cm^2)	Eff (%)
A-std	67 ± 2	748 ± 12	25.5 ± 0.9	12.8 ± 0.4
B	56 ± 3	720 ± 9	26.9 ± 0.8	10.9 ± 0.6
C	57 ± 1	703 ± 8	26.5 ± 0.7	10.7 ± 0.2
D	63 ± 1	765 ± 8	28 ± 0.6	13.5 ± 0.5
E-best	67 ± 3	767 ± 12	27.6 ± 1	14.2 ± 0.3
F	60 ± 2	727 ± 9	27.3 ± 0.8	11.9 ± 0.3
G	58 ± 3	689 ± 15	25.3 ± 0.7	10.1 ± 0.2

Annealing the CdSe layer after deposition could be beneficial for recrystallizing the layer and preventing excessive and uncontrolled mixing with CdTe during the activation treatment. On the other hand in samples B, where the CdSe layer is annealed at 450° in vacuum, there is a decrease in both FF and Voc. One possible explanation could be the re-evaporation of selenium during the annealing process of the CdSe in a vacuum environment. For this reason, samples C are annealed at 450°C in an air environment, however the results are similar. On the other hand, by reducing the temperature to 400°C, samples D exhibit an enhanced efficiency compared to the standard procedure, mainly attributed to an increase in current density. This suggests the formation of a compound more rich in Se at the junction. In order to increase the temperature without losing Se, the CdSe layer has been covered by a thin CdTe layer before annealing. Actually, samples E show the highest average efficiency, with increased Voc and Jsc compared to the standard samples. The same conditions are employed for sample F but with annealing in air. This strongly reduces the FF and Voc, leading to lower efficiencies. Finally, with sample G, where the annealing is performed after the deposition of the whole absorber, the FF, Voc and Jsc are reduced. Probably in this case the diffusion gradient pushes the selenium away from the junction, leading to excessive mixing with CdTe before the activation treatment. In fact, the low efficiency parameters suggest a compound poor in selenium at the junction.

3.2 Study of Se distribution

In order to investigate how the fabrication process affects the performances of the devices, the selenium distribution and the composition of the CdSeTe/CdTe absorber have been studied. In particular, we present the comparison between the standard devices (A-std) and the best cells (E-best). In this case, a bi-layer made of 300 nm of CdSe and 300 nm of CdTe has been annealed at 450 °C in vacuum, followed by the deposition of 1400 nm of CdTe.

SIMS analysis has been performed on the samples in order to explore the selenium distribution in the finished cells. SIMS provides a qualitative analysis and considering the antagonistic nature of Se and Te within the CdSeTe compound, Fig.1 illustrates the distribution of the Se/Te ratio in the absorber, presented on a logarithmic scale to enable a comparison between the cells. On the other hand, the abscissa points out the depth from the absorber surface to the TCO.

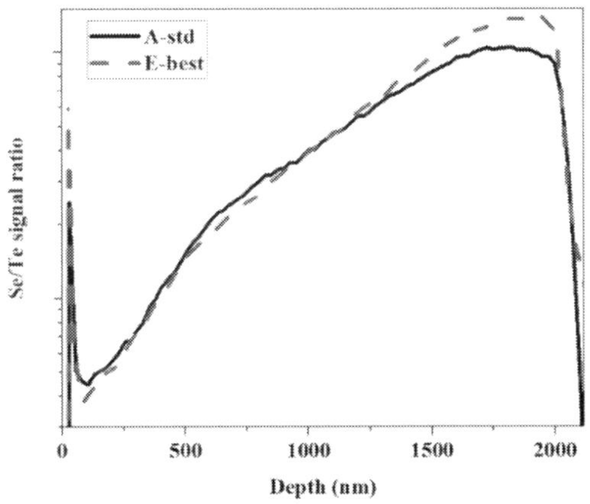

Figure 1: SIMS depth profiles showing selenium distribution through the absorber normalized on Te signal (in logarithmic scale) in A-std and E-best.

E-best shows a higher Se to Te ratio near the junction, indicating the formation of a compound richer in selenium; this accounts for the observed increase in current density of the cells. On the other hand, the standard cell displays a slightly higher amount of selenium in the region towards the back contact. SIMS analysis corroborates the hypothesis that annealing at 450 °C results in the recrystallization of the CdSe layer, preventing excessive Se diffusion away from the junction.

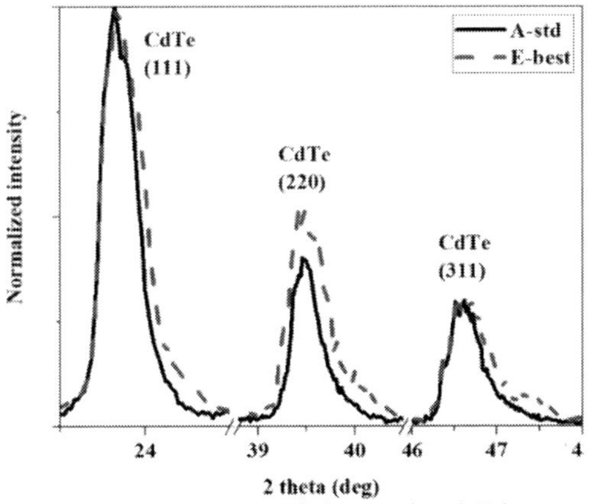

Figure 2: XRD patterns of the A-std and E-best samples.

This is further confirmed by XRD analysis. Figure 2 presents the magnification of the CdTe peaks for both A-std and E-best samples. The graph highlights a slight right-shift of the peaks for the E-best cell. This right shift indicates a higher incorporation of selenium in the compound [2], corroborating the findings from SIMS measurements.

3.3 Performance of the devices

As emphasized by the structural analysis, annealing CdSe under vacuum with a thin CdTe facilitates the recrystallization of the CdSe layer and initiates a mixing with the CdTe, thereby preventing an excessive selenium diffusion away from the junction. This fabrication method leads to an efficiency near 15 %. Figure 3 illustrates the current-voltage characteristic of one of the best-performing E-best devices.

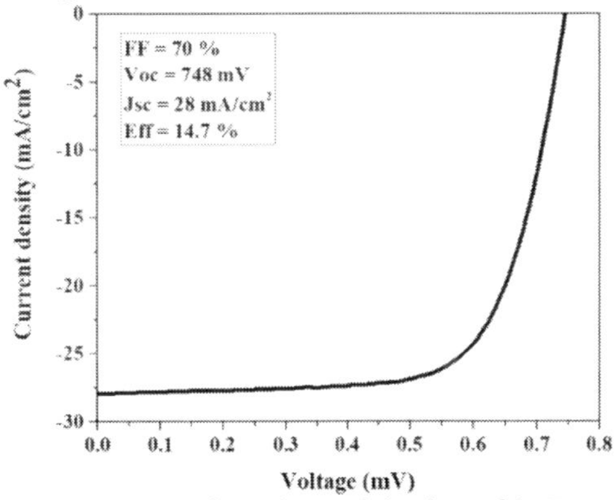

Figure 3: current-voltage characteristic of one of the E-best cells.

Indeed, the efficiency parameters of this sample are noteworthy. It achieves an efficiency of 14.7 % setting a record for CdTe devices produced at low substrate temperature with an absorber thickness reduced to 2 μm. Furthermore, these devices exhibit exceptional current densities exceeding 28 mA/cm^2, which is a remarkable result since it is actually the absolute record for such a thin absorber [7].

4 CONCLUSIONS

CdSeTe/CdTe devices have been fabricated by different processes and studied. The aim of this work is to investigate which manufacturing method is suitable to obtain the desired Se distribution in the absorber and the appropriate CdSeTe compound at the junction. Moreover, the structure of the finished cells has been analyzed in order to correlate fabrication method, performance of the cells and composition of the absorber.

The recrystallization of the CdSe layer before the activation treatment seems to be fundamental to avoid excessive diffusion of selenium away from the junction. On the other hand, this annealing is beneficial when the CdSe is covered by a thin CdTe layer to avoid Se loss. Cells fabricated by annealing at 450°C in vacuum a 300 nm thick CdSe – 300 nm CdTe bi-layer, followed by a thicker CdTe layer, show enhanced efficiency parameters compared to a standard deposition consisting of a sequential evaporation of CdSe and CdTe.

The measurements performed on the samples reveal that this 450 °C annealing favors the formation of a compound richer in Se at the junction, explaining the average current density gain of the devices.

These samples have reached an efficiency close to 15 %, with current densities exceeding 28 mA/cm^2, which is the record for such a thin CdTe-based absorber.

5 REFERENCES

[1] N.R. Paudel, Y. Yan, Enhancing the photo-currents of CdTe thin-film solar cells in both short and long wavelength regions, Appl Phys Lett. 105 (2014) 1–6. https://doi.org/10.1063/1.4901532.

[2] D.E. Swanson, J.R. Sites, W.S. Sampath, Co-sublimation of CdSexTe1−x layers for CdTe

solar cells, Solar Energy Materials and Solar Cells. 159 (2017) 389–394. https://doi.org/10.1016/j.solmat.2016.09.025.

[3] A.H. Munshi, J. Kephart, A. Abbas, J. Raguse, J.N. Beaudry, K. Barth, J. Sites, J. Walls, W. Sampath, Polycrystalline CdSeTe/CdTe absorber cells with 28 mA/cm2 short-circuit current, IEEE J Photovolt. (2018). https://doi.org/10.1109/JPHOTOV.2017.2775139.

[4] X. Zheng, D. Kuciauskas, J. Moseley, E. Colegrove, D.S. Albin, H. Moutinho, J.N. Duenow, T. Ablekim, S.P. Harvey, A. Ferguson, W.K. Metzger, Recombination and bandgap engineering in CdSeTe/CdTe solar cells, APL Mater. 7 (2019). https://doi.org/10.1063/1.5098459.

[5] E. Artegiani, A. Gasparotto, P. Punathil, V. Kumar, M. Barbato, M. Meneghini, G. Meneghesso, F. Piccinelli, A. Romeo, A new method for CdSexTe1-x band grading for high efficiency thin-absorber CdTe solar cells, Solar Energy Materials and Solar Cells. 226 (2021) 111081. https://doi.org/10.1016/j.solmat.2021.111081.

[6] E. Artegiani, P. Punathil, V. Kumar, M. Bertoncello, M. Meneghini, A. Gasparotto, A. Romeo, Effects of CdTe selenization on the electrical properties of the absorber for the fabrication of CdSexTe1-x/CdTe based solar cells, Solar Energy. 227 (2021) 8–12. https://doi.org/10.1016/j.solener.2021.08.070.

[7] E. Artegiani, A. Gasparotto, M. Meneghini, G. Meneghesso, A. Romeo, How the selenium distribution in CdTe affects the carrier properties of CdSeTe/CdTe solar cells, Solar Energy. 260 (2023). https://doi.org/10.1016/j.solener.2023.05.058.

[8] A. Romeo, E. Artegiani, D. Menossi, Low substrate temperature CdTe solar cells: A review, Solar Energy. (2018). https://doi.org/10.1016/j.solener.2018.02.038.

[9] V. Fthenakis, C. Athias, A. Blumenthal, A. Kulur, J. Magliozzo, D. Ng, Sustainability evaluation of CdTe PV: An update, Renewable and Sustainable Energy Reviews. 123 (2020). https://doi.org/10.1016/j.rser.2020.109776.

[10] E. Bastola, A.B. Phillips, G. Barros-King, M.K. Jamarkattel, D.B. Li, A. Quader, D. Pokhrel, J.M. Gibbs, Y. Yan, R.J. Ellingson, M.J. Heben, Understanding the Interplay between CdSe Thickness and Cu Doping Temperature in CdSe/CdTe Devices, in: Conference Record of the IEEE Photovoltaic Specialists Conference, 2021. https://doi.org/10.1109/PVSC43889.2021.9518832.

[11] I. Rimmaudo, A. Salavei, E. Artegiani, D. Menossi, M. Giarola, G. Mariotto, A. Gasparotto, A. Romeo, Improved stability of CdTe solar cells by absorber surface etching, Solar Energy Materials and Solar Cells. (2017). https://doi.org/10.1016/j.solmat.2016.12.044.

[12] E. Artegiani, J.D. Major, H. Shiel, V. Dhanak, C. Ferrari, A. Romeo, How the amount of copper influences the formation and stability of defects in CdTe solar cells, Solar Energy Materials and Solar Cells. 204 (2020) 110228. https://doi.org/10.1016/j.solmat.2019.110228.

PHOTON MANAGEMENT BY FUNCTIONAL BACK CONTACTS FOR ULTRATHIN CIGSE SOLAR CELLS: INSIGHTS FROM EXPERIMENT AND SIMULATION

M. Demir[1], A. Sprafke[1], B. Fuhrmann[2], T. Schneider[1], T. Hölscher[1], H. Kempa[1] and R. Scheer[1]
[1]Martin-Luther-University Halle-Wittenberg, Institute of Physics, Von-Danckelmann-Platz 3, 06120 Halle, Germany
[2]Interdisiplinary Center of Material Research, Heinrich-Damerow-Straße 4, 06120 Halle,Germany

ABSTRACT: The aim of this research is to design novel cell architectures for ultrathin CIGSe solar cells that reach or even surpass the power conversion efficiency of their conventional counterparts with absorber layer thicknesses of 2-3μm. In this contribution, ultrathin CIGSe solar cells are combined with nano-textured SiO_2 and a metal back mirror in order to increase the power conversion efficiency of ultra-thin CIGSe solar cells through enhanced absorption. To investigate the impact of the functional back contact, solar cell parameters are extracted and compared between CIGSe solar cells with a conventional back contact which is flat molybdenum, with a flat aluminum back reflector, and with a scattering SiO_2 structure below and above the back reflector. External quantum efficiency measurements reveal that by employing a functional back contact, it is possible to have optical gain in the wavelength range between 800 and 1200 nm in comparison to the flat molybdenum back contact. In addition, when comparing the solar cells with a textured back contact, a planar reflector underneath the scattering elements leads to higher absorption than a textured reflector on top of the scattering elements. Based on optical simulations, the creation of waveguide modes by the back contact texture is proposed to be responsible for the gain in absorption.
Keywords: Ultrathin, light management, absorption

1 INTRODUCTION

Recent research on Cu(In,Ga)Se2 (CIGSe) solar cells is focused on thinning down the absorber layer to enable less material consumption and cost effective large scale production. However, having ultra-thin CIGSe solar cells with absorber layer thicknesses on the sub-micron level limits absorption of the solar flux and hence leads to lower energy conversion efficiencies. So far, the highest efficiency of an ultrathin CIGSe solar cell, with 550 nm absorber layer with molybdenum back contact, is 15.2% [1]. Recently published research showed that the open circuit voltage can be increased by rear passivation of ultrathin CIGSe solar cells, and demonstrated efficiency is 13% for 640 nm CIGSe absorber thickness [2]. Considering efficiencies reached by standard thick CIGSe solar cells up to 23.6 % [3], it is clear that there are still open questions in ultrathin CIGSe solar cells research.

Poor optical absorption of the ultrathin absorber can be overcome by the cell architecture including light management techniques [4][5]. These can be macroscopic concepts, such as optimizing the refractive indices gradient or the reflectivity by adjusting the thicknesses of the front layers [6]. Furthermore, application of anti-reflection coatings is a well-known approach [7]. On the other hand, nanoscopic light management techniques can be applied as functional back contacts for the elongation of light paths within the cell and for direct enhancement of absorption due to interference [8]–[11]. Increased roughness with periodic nano-structures at the rear side can create electric field "hot" spots where the absorption is high due to multi-resonant standing waves. Moreover, the combination of a back mirror with a backside nano-texture has already been investigated [12]. Our preliminary results showed that aluminum, as a highly reflective metal, can increase the light reflection and hence provide additional double-pass absorption[13]. However, considering the high deposition temperatures of CIGSe solar cells (ca. ≈ 600 °C), a barrier layer, such as conductive oxide, is required to impede metal diffusion from the reflective metal into the CIGSe thin film. Unfortunately, such oxide layers may form metal oxides with the reflector metal as a by-product of the CIGSe process which can result in solar cells having poor electrical properties. Hence, various metals as a back reflector have been investigated extensively in the literature [14]–[17].

In this work, we study the optical gain by the implementation of metallic back reflectors, aluminum and gold, in combination with nano-textured scattering elements in comparison with typical CIGSe solar cell on molybdenum back contacts.

2 EXPERIMENTAL DETAILS

Four different types of CIGSe solar cells were prepared with varying back contacts as shown in Figure 1. For each sample, commercial soda lime glass covered by Molybdenum with sodium diffusion barrier was used as a substrate. The back reflector metal, a 100 nm thick aluminum or gold layer, was deposited by electron beam evaporation, followed by the metal diffusion barrier ITO sputtered on top. Nano-textured SiO_2 layers were realized by laser interference lithography [18]. For this, a deep UV negative photoresist (AR-N 4240, diluted with AR 300-12 1:0,75, Allresist) was applied by spin coating at 4000 rpm and exposed to a laser light with a wavelength of 266 nm through a 3-mirror Fresnel type laser interferometer. After baking on a hotplate for 5 min at 95°C and developing (AR 300-475, Allresist) of the photoresist for 45 s, a pretreatment in Oxygen plasma for 3 min was performed. SiO_2 with different thicknesses was deposited by e-beam evaporation to create the scattering elements. Back texturing was completed by a lift-off process taking place in an ultrasonic bath for 30 min in the remover solution (AR 300-76 Allresist). A cross sectional SEM micrograph of a flat reflector sample is shown in Figure 2. The resultant nano-texturing elements in hexagonal lattice have a pitch of 1.96 and 1 μm for the textured reflector and the flat reflector sample, respectively. CIGSe thin films were fabricated by physical vapor deposition in the so-called three stage process [19]. The composition and the thickness of the films were controlled by laser light scattering during the growth. Resulting GGI ([Ga]/([Ga]+[In])) and CGI ([Cu]/([Ga]+[In])) profiles were extracted by glow discharge optical emission spectroscopy (GDOES), where the spatial averaged values came out as GGI=0.3 and CGI=0.8. All samples underwent NaF post deposition treatment at 450°C. CIGSe thin films were completed to solar cells by 50 nm CdS by chemical bath deposition followed by 100 nm i-ZnO and 200 nm IT0 front layers sputtering. Ni/Al/Ni metal grids were applied by e-beam evaporation. Solar cell separation was realized by chemical etching of TCO and CdS layers with HCl solution of 5 % concentration using a fine brush. The final layer thicknesses of solar cell stacks are summarized in Table 1.

I-V measurements were performed in four-point configuration with a source meter under a class AAA solar simulator. External quantum efficiency was measured with a dedicated commercial system, also in four-point configuration. The details of the optical simulations explained in [4].

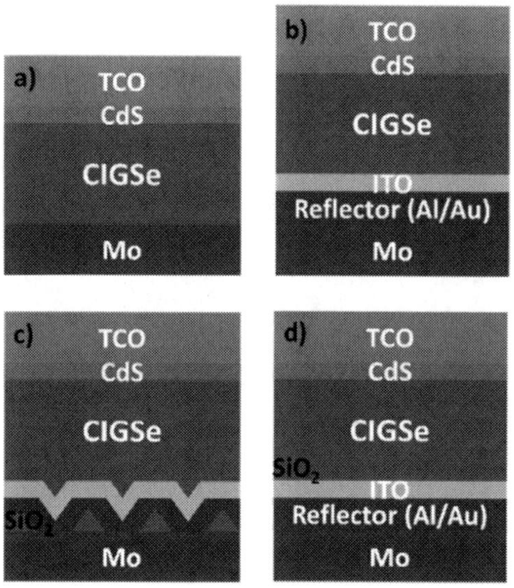

Figure 1: Solar cell layer stacks with different back contacts; a) No reflector, standard solar cell with Mo back contact b) Plain reflector, with additional flat metal layer and diffusion barrier c) Textured reflector, with metal layer and diffusion barrier on top of SiO_2 structure. d) Flat reflector, SiO_2 structure on top of flat metal layer and diffusion barrier

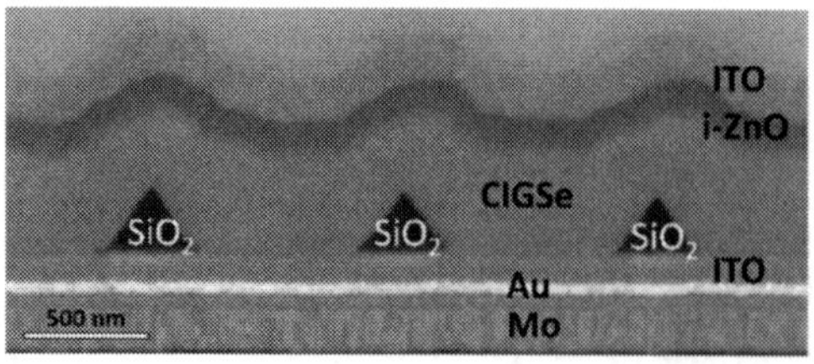

Figure 2: FIB cut crossection image of the sample flat reflector with Au back reflector

Table 1: Layer thicknesses of the solar cells

Back Contact	Thickness (nm)				
	Mo	Al	ITO	SiO_2	CIGSe
Mo	500	-	-	-	600
Plain Reflector	500	110	250	-	600
Textured Reflector	500	110	250	550	600
Flat Reflector	300	130	140	410	530

3 RESULTS & DISCUSSIONS

External quantum efficiency (EQE) curves for the different solar cell designs are presented in Figure 3 and corresponding current densities derived from the EQE spectra are shown in Figure 4.

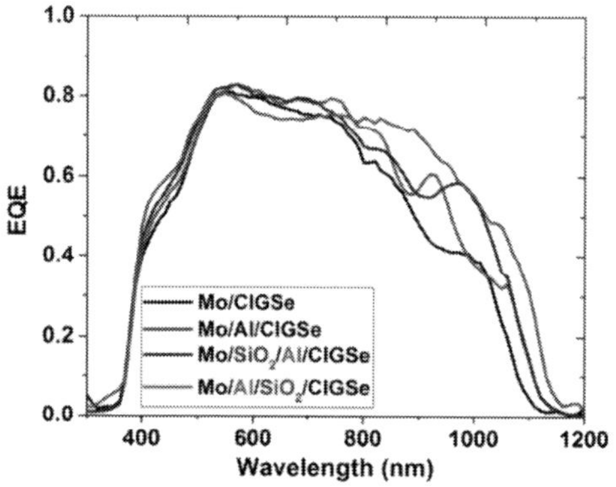

Figure 3: EQE comparison of the samples with varying functional back contacts

The differences between the curves in the short wavelength region of the spectrum, between 400 nm and 550 nm, can be explained by minor thickness variations of CdS between different batches. Likewise, the differences in the plateau region, between 550 nm and 750 nm, can be attributed to variations in thickness and optical properties of the CIGSe and window layers. The loss due to poor absorption of ultrathin CIGSe layer results in a triangle-like shape of the EQE, as can be seen in the case of the solar cell with no reflector. The positive impact of photon management techniques is visible for the other three samples in the wavelength range between 750 nm and the band gap. For the sample with plain reflector, an additional EQE peak appears at~ 920 nm resulting from the interference of the light reflected by the back mirror. This peak shifts to~970 nm when back scattering elements are introduced underneath the reflector and additional gain in current density due to double pass absorption occurs. However, the highest EQE is observed for the flat reflector sample where the scattering elements are located on top of the back reflector. The EQE curve of this sample approaches a more rectangular shape in comparison with the sample without reflector. Furthermore, the gain around the band gap must also be attributed to optical enhancement introduced by light management elements since all CIGSe films have a similar composition and band gap.

10.4229/EUPVSEC2023/2DO.6.4
020136-002

The surprising difference between the EQE curves of the samples with textured and flat reflector can be explained with the help of optical simulations. In Figure 5, color plots of the calculated local absorption at a wavelength of 940 nm are shown for solar cells with 500 nm thick CIGSe absorber layers and both types of functional back contacts. In these plots, the regions of high electric field intensity can be identified as locally enhanced absorption spots. In Figure 5, while these spots of high electric field intensity are present in both cell designs, in Figure 5.b additional spots form in the valleys between the scattering elements leading to higher charge carrier generation and to an improved EQE.

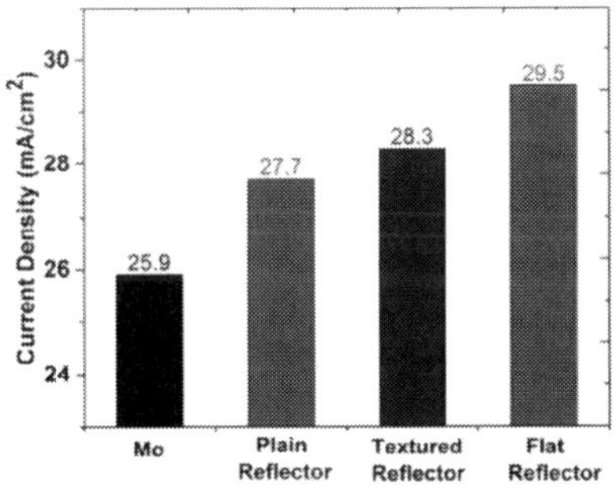

Figure 4: Current densities of the samples

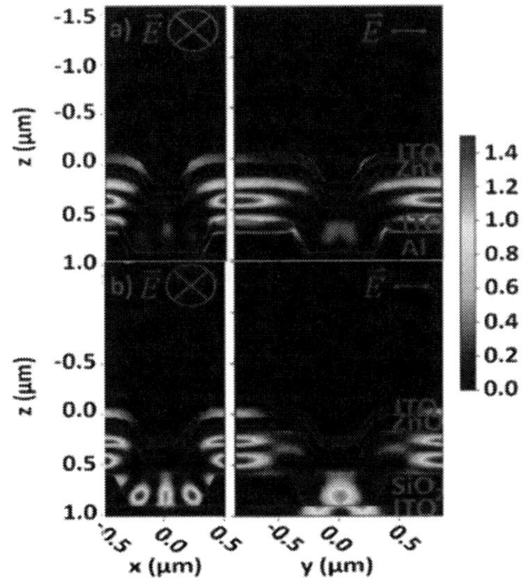

Figure 5: Calculated electric field intensity of samples at a wavelength of 940 nm with nano-texturing pitch of 1 µm a) textured reflector, b) flat reflector

It is found that not only the position of the back reflector but also the material itself has an impact especially on solar cell parameters. The effect of exchanging the Al by an Au reflector in the flat reflector configuration is presented in Figure 6, where the EQE curves and the J-V characteristics are shown. By comparing the EQE, it can be seen that the absorption is nearly identical in both cases. This is expectable since the reflectivity of Al and Au is higher than 80% especially in the near IR range of the spectrum [20]. On the other hand, the electronic characteristics of these samples are different. Solar cell parameters are summarized in Table 2. As can be derived from the inset of Figure 6 the solar cell with Al reflector has a higher series resistance which results in a poor fill factor. This can be explained by the formation of an AlO_x layer between the Al reflector and the ITO diffusion barrier during the growth of the CIGSe thin film. Depending on the thickness, this metal oxide layer can behave as a barrier at the back contact of the solar cell.

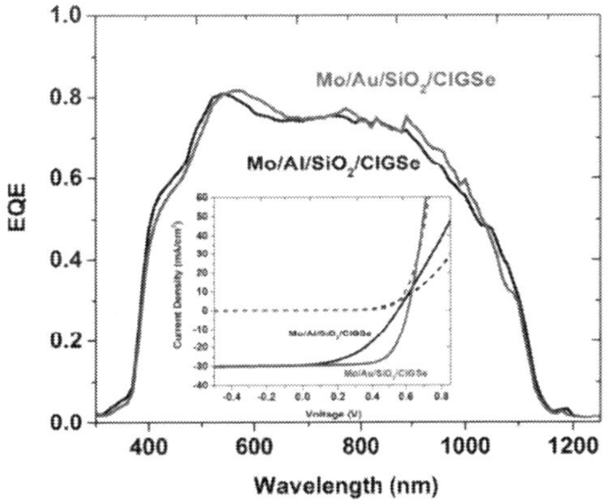

Figure 6: EQE comparison of aluminum and gold for the flat reflector case; Inset: Current-Voltage characteristics.

Table 2: Solar Cell Parameters

	Jsc (mA/cm2)	**Voc (mV)**	**FF (%)**	**η (%)**
$Mo/Al/SiO_2/CIGSe$	29.5	553	44	7.2
$Mo/Au/SiO_2/CIGSe$	29.6	611	67	12.1

4 CONCLUSION

In this ongoing work, we have shown that the light absorption in ultrathin CIGSe solar cells can be enhanced by implementing functional back contacts for light management. EQE spectra of solar cells with different back contacts showed an increased short circuit density due to double pass absorption achieved by a metallic back reflector. Moreover, the combination of a back reflector with nano-textured scattering elements enhanced the absorption especially in the long-wavelength region of the spectrum. Hereby, a planar metal reflector underneath the scattering elements results in better absorption compared to a textured reflector. With the help of optical simulations we attributed this finding to localized regions of enhanced absorption between the scattering elements. Although aluminum and gold have a similarly high reflectivity, samples with gold reflector display superior solar cell characteristics due to reduced series resistance.

5 REFERENCES

[1] L. M. Mansfield *et al.*, "Efficiency increased to 15.2%

for ultra-thin Cu(In,Ga)Se2 solar cells," *Prog. Photovoltaics Res. Appl.*, vol. 26, no. 11, pp. 949–954, 2018, doi: 10.1002/pip.3033.

[2] A. Oliveira *et al.*, "Over 100 mV VOC Improvement for Rear Passivated ACIGS Ultra-Thin Solar Cells," *Adv. Funct. Mater.*, 2023, doi: 10.1002/adfm.202303188.

[3] M. Green, E. Dunlop, J. Hohl-Ebinger, M. Yoshita, N. Kopidakis, and X. Hao, "Solar cell efficiency tables (version 57)," *Prog. Photovoltaics Res. Appl.*, vol. 29, no. 1, pp. 3–15, 2021, doi: 10.1002/pip.3371.

[4] J. Krc, M. Sever, A. Campa, Z. Lokar, B. Lipovsek, and M. Topic, "Optical confinement in chalcopyrite based solar cells," *Thin Solid Films*, vol. 633, pp. 193–201, 2017, doi: 10.1016/j.tsf.2016.08.056.

[5] M. Kovacic *et al.*, "Light management design in ultra-thin chalcopyrite photovoltaic devices by employing optical modelling," *Sol. Energy Mater. Sol. Cells*, vol. 200, no. May, 2019, doi: 10.1016/j.solmat.2019.109933.

[6] M. Schmid, "Review on light management by nanostructures in chalcopyrite solar cells," *Semicond. Sci. Technol.*, vol. 32, no. 4, p. 43003, 2017.

[7] C. Ji *et al.*, "Recent Applications of Antireflection Coatings in Solar Cells," *Photonics*, vol. 9, no. 12, 2022, doi: 10.3390/photonics9120906.

[8] G. Yin, M. W. Knight, M. C. van Lare, M. M. Solà Garcia, A. Polman, and M. Schmid, "Optoelectronic Enhancement of Ultrathin CuIn1–xGaxSe2 Solar Cells by Nanophotonic Contacts," *Adv. Opt. Mater.*, vol. 5, no. 5, 2017, doi: 10.1002/adom.201600637.

[9] M. J. Shin *et al.*, "Semitransparent and bifacial ultrathin Cu(In,Ga)Se2 solar cells via a single-stage process and light-management strategy," *Nano Energy*, vol. 82, no. October 2020, p. 105729, 2021, doi: 10.1016/j.nanoen.2020.105729.

[10] Y. Li, S. W. Tabernig, G. Yin, A. Polman, and M. Schmid, "Beyond Light-Trapping Benefits: The Effect of SiO2 Nanoparticles in Bifacial Semitransparent Ultrathin Cu(In,Ga)Se2 Solar Cells," *Sol. RRL*, vol. 6, no. 11, pp. 1–8, 2022, doi: 10.1002/solr.202200695.

[11] W. Wang, J. Zhang, X. Che, and G. Qin, "Large absorption enhancement in ultrathin solar cells patterned by metallic nanocavity arrays," *Sci. Rep.*, vol. 6, no. October, pp. 2–9, 2016, doi: 10.1038/srep34219.

[12] T. Schneider, J. Tröndle, B. Fuhrmann, F. Syrowatka, A. Sprafke, and R. Scheer, "Ultrathin CIGSe Solar Cells with Integrated Structured Back Reflector," *Sol. RRL*, vol. 4, no. 10, 2020, doi: 10.1002/solr.202000295.

[13] T. Schneider and R. Scheer, "Aluminum based back reflectors for ultrathin Cu (In, Ga) Se2 solar cells with ito diffusion barrier," in *36th European Photovoltaic Solar Energy Conference and Exhibition (EU-PVSEC), Marseille, France*, 2019, pp. 684–688.

[14] L. Gouillart *et al.*, "Reflective Back Contacts for Ultrathin Cu(In,Ga)Se2-Based Solar Cells," *IEEE J. Photovoltaics*, vol. 10, no. 1, pp. 250–254, 2020, doi: 10.1109/JPHOTOV.2019.2945196.

[15] L. Gouillart *et al.*, "Development of reflective back contacts for high-efficiency ultrathin Cu(In,Ga)Se 2 solar cells," *Thin Solid Films*, vol. 672, no. July 2018, pp. 1–6, 2019, doi: 10.1016/j.tsf.2018.12.041.

[16] L. Gouillart, I. Photovoltaïque, W. Chen, J. Keller, and M. Jubault, "Ultrathin Cu (In , Ga) Se 2 solar cells with Ag-based reflective back contacts," pp. 1481–1484, 2020.

[17] L. Gouillart *et al.*, "Interface engineering of ultrathin Cu(In,Ga)Se2 solar cells on reflective back contacts," *Prog. Photovoltaics Res. Appl.*, vol. 29, no. 2, pp. 212–221, 2021, doi: 10.1002/pip.3359.

[18] A. Facchini, "Laser Interference Lithography Technique for Back-Contact Microstructuring of CuIn(1-x)GaxSe2 Solar Cells," no. September, pp. 1–71, 2018.

[19] A. M. Gabor *et al.*, "Band-gap engineering in Cu(In,Ga) Se2 thin films grown from (In,Ga) 2Se3 precursors," *Sol. Energy Mater. Sol. Cells*, vol. 41–42, pp. 247–260, 1996, doi: 10.1016/0927-0248(95)00122-0.

[20] Z. J. Li-Kao *et al.*, "Towards ultrathin copper indium gallium diselenide solar cells: proof of concept study by chemical etching and gold back contact engineering," *Prog. Photovoltaics Res. Appl.*, vol. 20, no. 5, pp. 582–587, 2012, doi: https://doi.org/10.1002/pip.2162.

Tokyo Tech

Analysis of EBIC time variation with V_{Se} - V_{Cu} vacancy complex in $Cu(In,Ga)Se_2$ solar cells

Tokyo Institute of Technology
Ryotaro Fukuda, Takahito Nishimura, Akira Yamada

020137-001

2DO.8.3

Background

Tokyo Tech

$Cu(In,Ga)Se_2$(CIGS) solar-cell

CIGS solar-cell

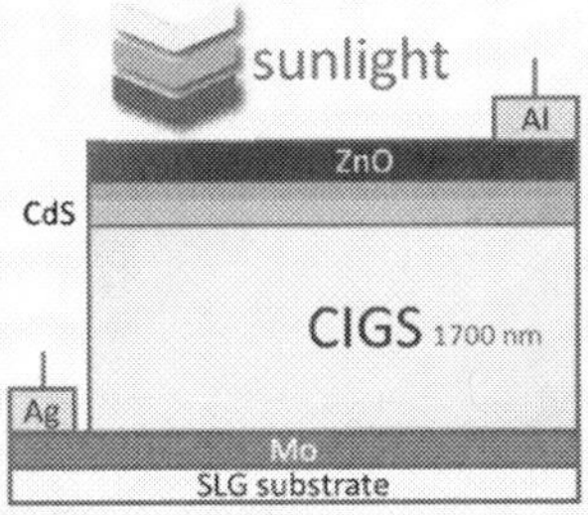

Merits

✓ **High efficiency 23.35%*.**

✓ **Long-term stability**

Problems

✓ **The difficulty of** analysis of **GB and interface.**

time variation

Heat and Light Soaking (HLS) treatment

» Heating + several hundred hours of solar irradiation

» Carrier concentration increases and conversion efficiency improves.

(>100℃)

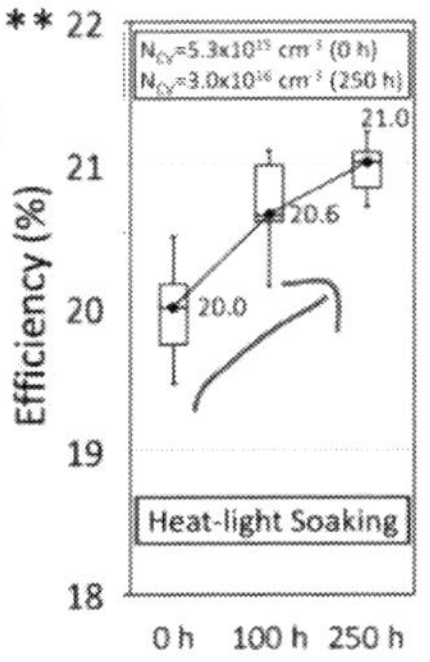

*nakamura2019 **nishinaga2017

020137-002

Electron Beam Induced Current (EBIC) Measurement

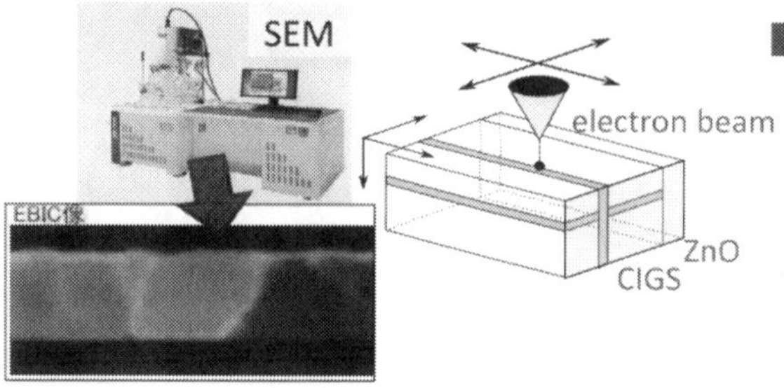

■ What is EBIC?

⇒We calculate the current from excited carriers generated by the electron beam.

→It shows the carrier collection efficiency.

⇒**For analysis of complex carrier behavior**

time variation

Time variation of EBIC

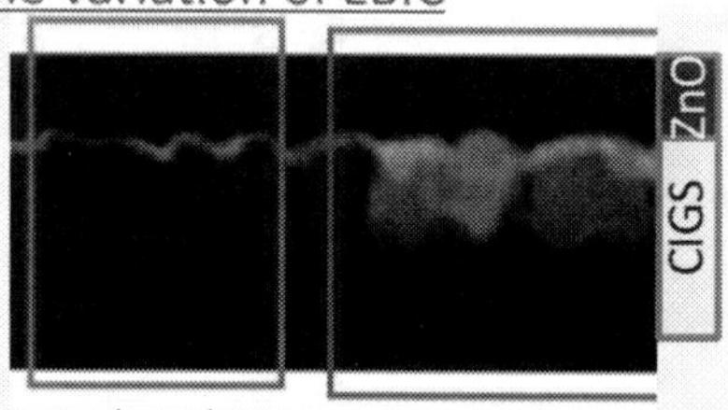

Immediately after irradiation — Long-term irradiation (300 sec.)

» After 300 seconds of electron irradiation, EBIC intensity increase overall.

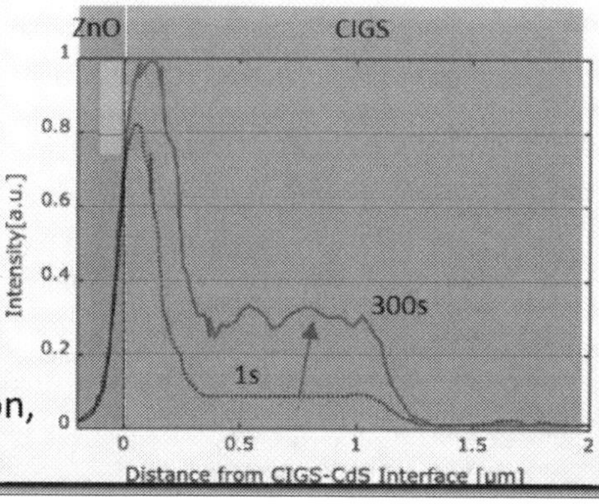

40th European Photovoltaic Solar Energy Conference and Exhibition

EBIC time variation with VV defect *

*S. Lany and A. Zunger, *J. Appl. Phys.* 100, (2006), p. 113725

V_{Se} - V_{cu} vacancy complex (VV defect)

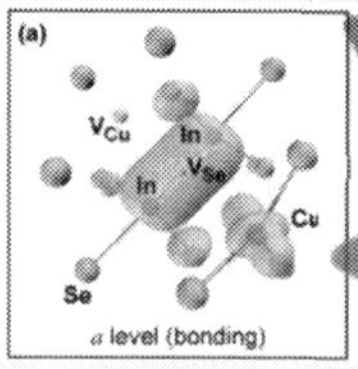

One of the **leading factors of the time variation of CIGS.**

There are few examples of **analysis (simulation).**

Defect level change with structure change

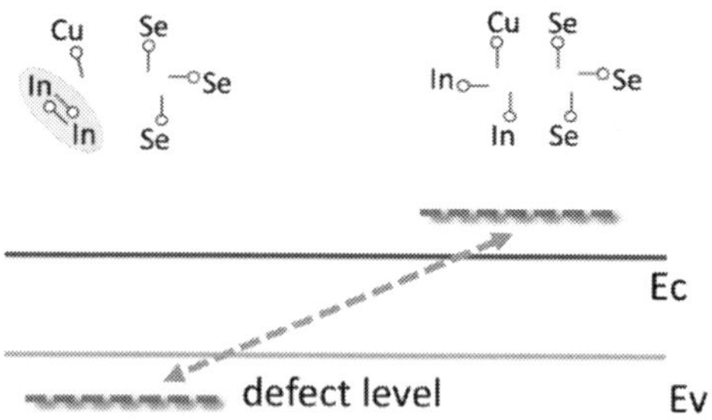

Research objectives

Understanding the behaviour of VV defects from EBIC time variation to discuss performance improvement in HLS processing.

Tokyo Tech

Construction of vacancy complex model

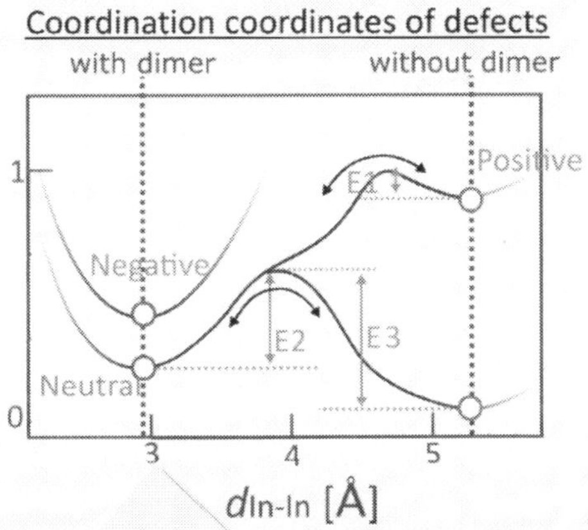

$$S = N_{e,h} \cdot \nu \cdot \exp\left(-\frac{f(E_{1\ldots3}, E_{fp,fn})}{kT}\right)$$

S: Transition rate

Transition paths

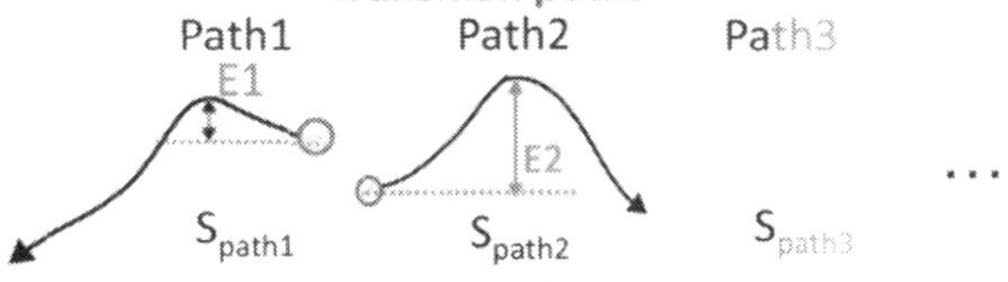

Add up all transition rates

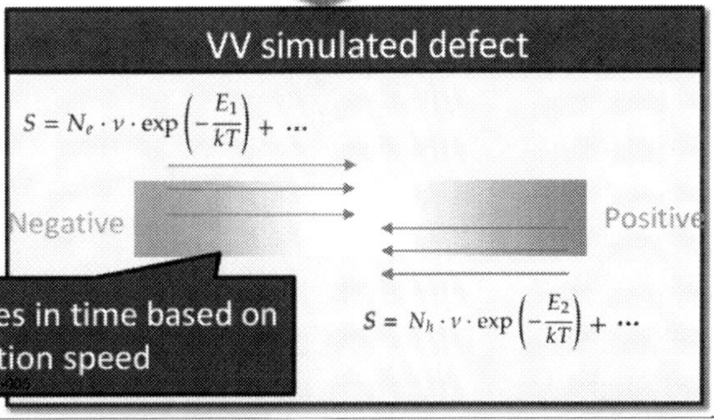

The four metastable states are connected by transition paths.

O O O Metastable state

——— Transition path

Those paths have energy barriers.

E1,E2,E3

Charge varies in time based on total transition speed

Tokyo Tech

EBIC transient response model with VV defect

✓ EBIC transient response simulation

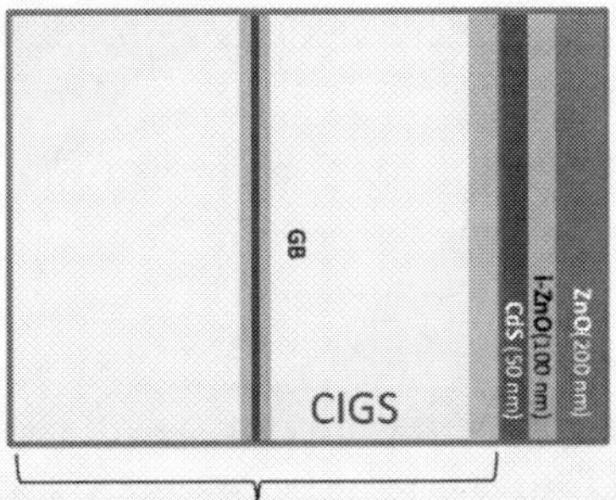

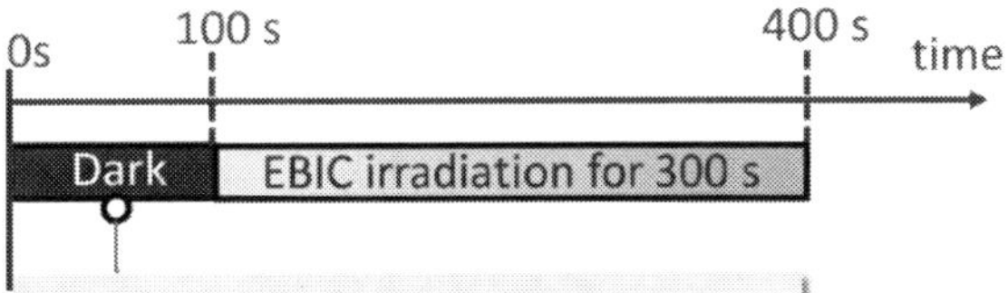

Left in darkroom for 100 seconds to stabilize defect charge state.

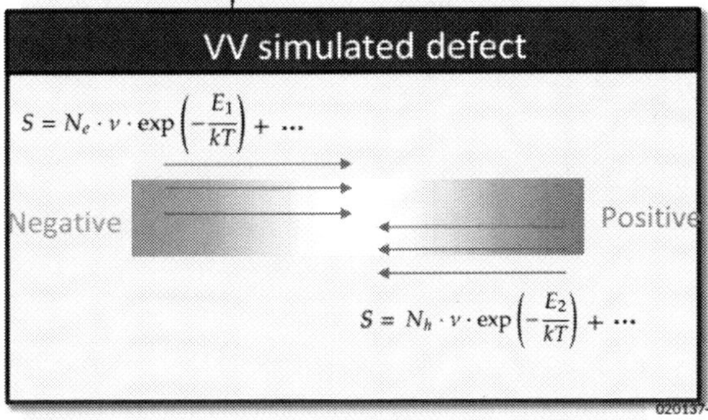

EBIC transient response model with VV defect

✓ EBIC transient response simulation

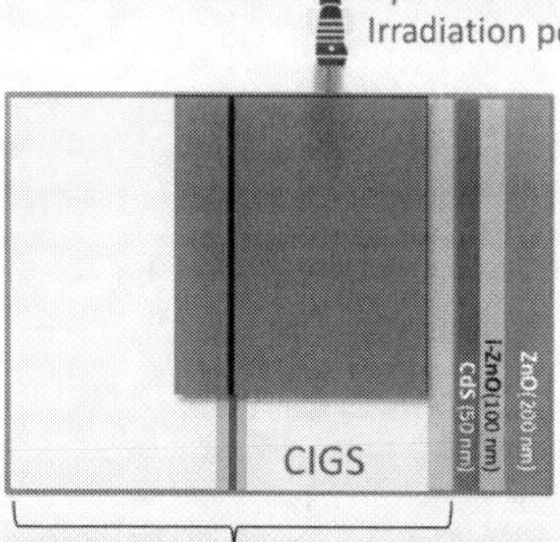

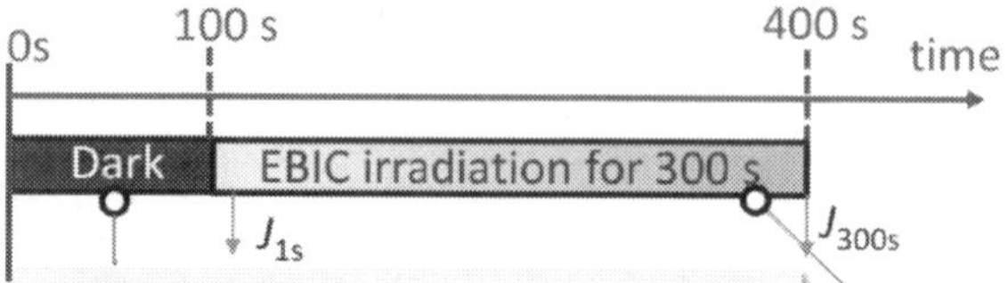

Left in darkroom for 100 seconds to stabilize defect charge state.

Calculate the increase rate (R) of EBIC current from 1 s to 300 s of irradiation. $R = J_{300s}/J_{1s}$

VV simulated defect

$S = N_e \cdot v \cdot \exp\left(-\frac{E_1}{kT}\right) + \cdots$

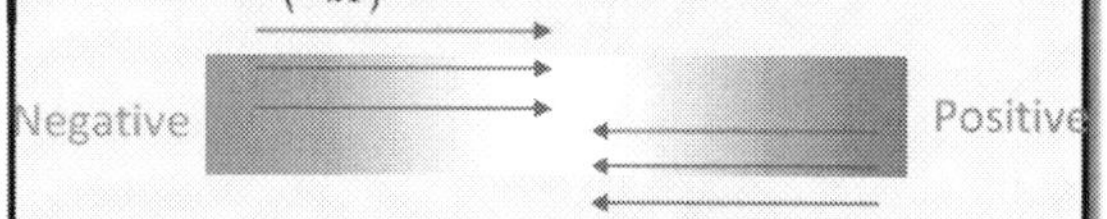

$S = N_h \cdot v \cdot \exp\left(-\frac{E_2}{kT}\right) + \cdots$

020137-007

7

40th European Photovoltaic Solar Energy Conference and Exhibition

EBIC transient response model with VV defect

✓ EBIC transient response simulation

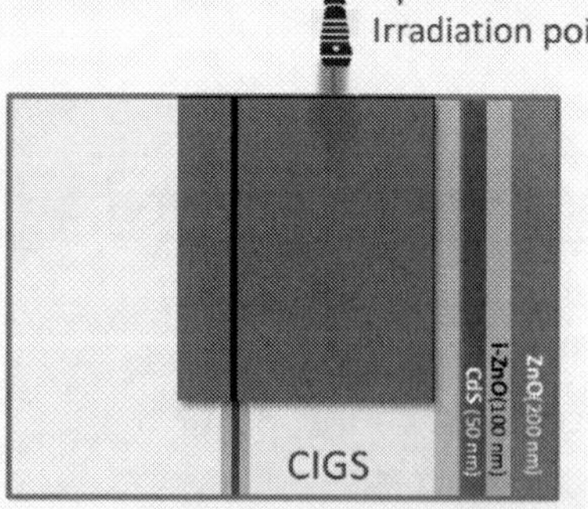

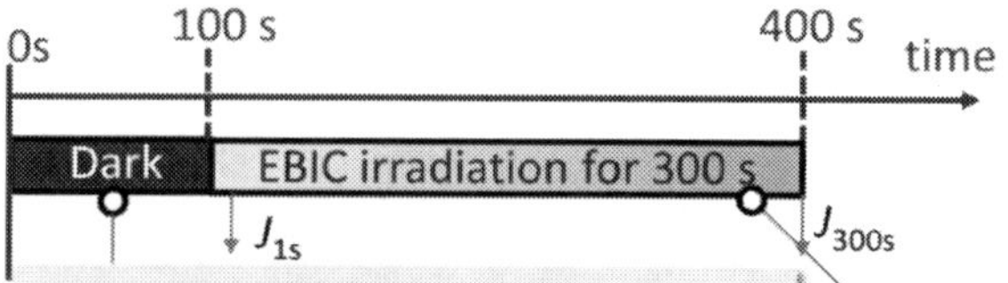

Left in darkroom for 100 seconds to stabilize defect charge state.

Calculate the increase rate (R) of EBIC current from 1 s to 300 s of irradiation. $R = J_{300s}/J_{1s}$

<u>Experimental result</u>

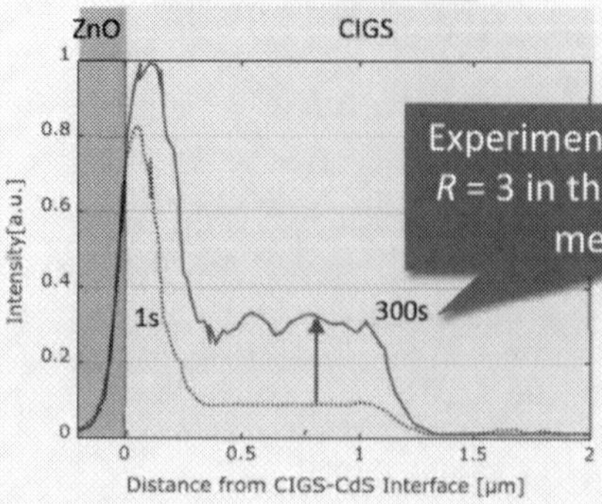

Experimental results show R = 3 in the middle of the membrane

020137-008

8

Application of Bayesian optimization

Experimental result

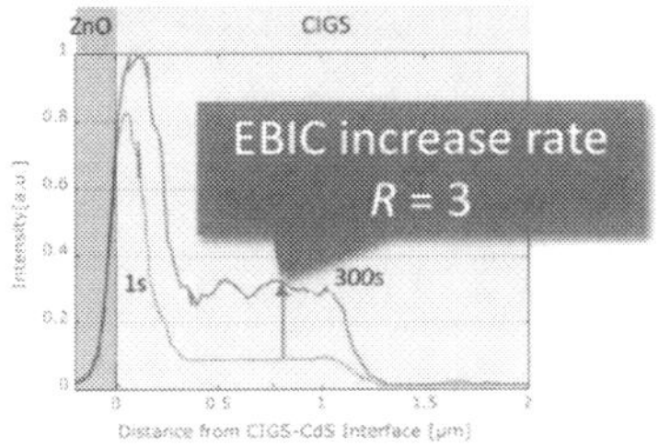

Simulated EBIC current

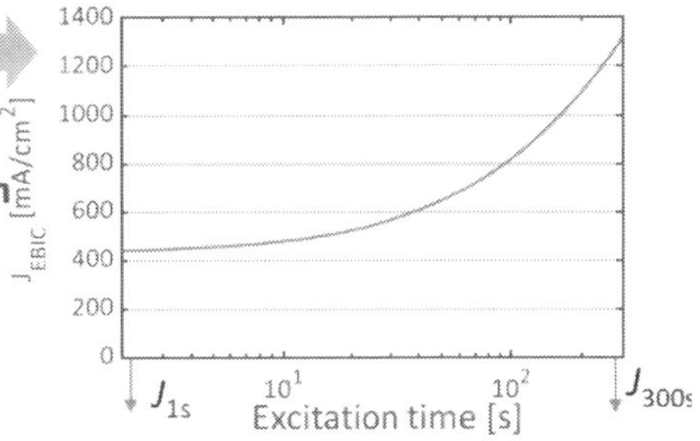

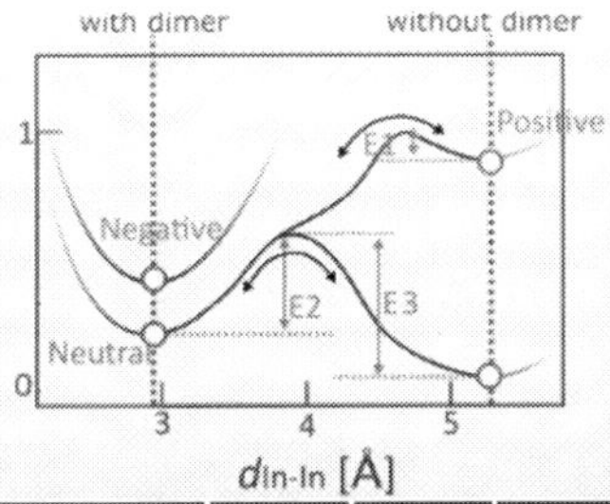

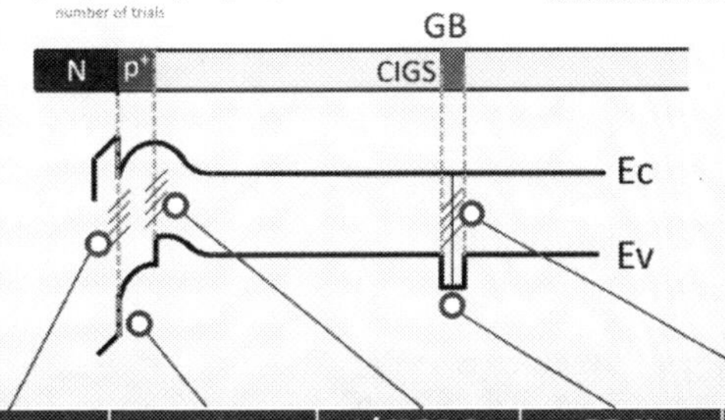

76th trial	E_1 (eV)	E_2 (eV)	E_3 (eV)	p/n N_t (cm^{-2})	p+ N_a (cm^{-3})	p+/CIGS S_{ssf} (cm/s)	GB N_a (cm^{-3})	GB S_{ssf} (cm/s)
optimal	0.91	0.51	0.67	2.67×10^{11}	2.95×10^{17}	2.92×10^4	4.56×10^{15}	69.2

020137-009

Discussion of E1, E2, and E3 values

Bayesian optimization
76th trial

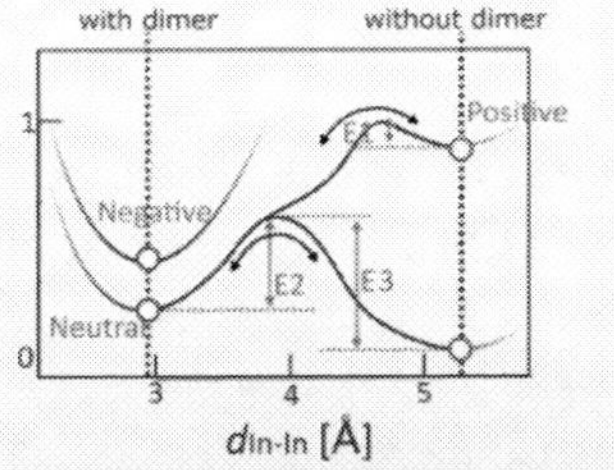

76th trial	E_1 (eV)	E_2 (eV)	E_3 (eV)	p/n N_t (cm^{-2})	p+ N_a (cm^{-3})	p+/CIGS S_{ssf} (cm/s)	GB N_a (cm^{-3})	GB S_{ssf} (cm/s)
optimal	**0.91**	0.51	0.67	2.67×10^{11}	2.95×10^{17}	2.92×10^4	4.56×10^{15}	69.2

020137-010

40th European Photovoltaic Solar Energy Conference and Exhibition

Tokyo Tech

Discussion of E1, E2, and E3 values

Values of **E2** and **E3** are relatively **close**, but our value of E1 is larger than the previous calculation.

Values from first principle calculations by Zunger et al.

76th trial	E_1 (eV)	E_2 (eV)	E_3 (eV)	p/n N_t (cm^{-2})			E_1 (eV)	E_2 (eV)	E_3 (eV)		GB S_{ssf} (cm/s)
optimal	**0.91**	0.51	0.67	2.67×10^{11}	2	Zunger's	**0.1**	0.35	0.73	15	69.2

020137-011

11

40th European Photovoltaic Solar Energy Conference and Exhibition

Tokyo Tech

Relationship between E1 value and EBIC time variation

The value of E1 and the increase rate of EBIC

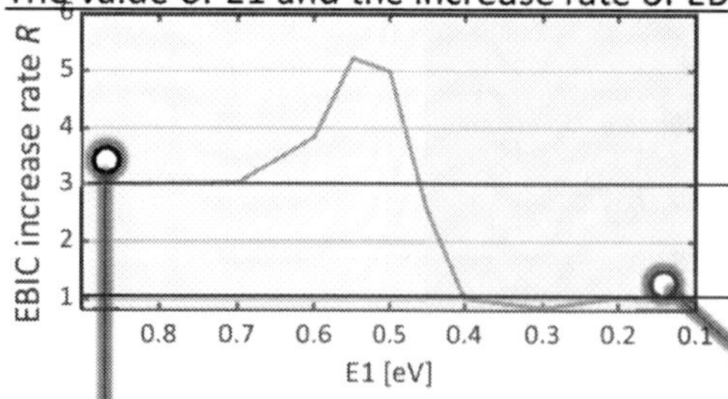

R=3: Experimental EBIC increase rate

R=1: EBIC does not change

Values of **E2** and **E3** are relatively **close**, but our value of E1 is larger than the previous calculation.

Values from first principle calculations by Zunger et al.

76th trial	E_1 (eV)	E_2 (eV)	E_3 (eV)	p/n N_t (cm^{-2})			E_1 (eV)	E_2 (eV)	E_3 (eV)		GB S_{ssf} (cm/s)
optimal	**0.91**	0.51	0.67	2.67×10^{11}	2	Zunger's	**0.1**	0.35	0.73	15	69.2

020137-012

12

40th European Photovoltaic Solar Energy Conference and Exhibition

Reasons for EBIC signal improvement

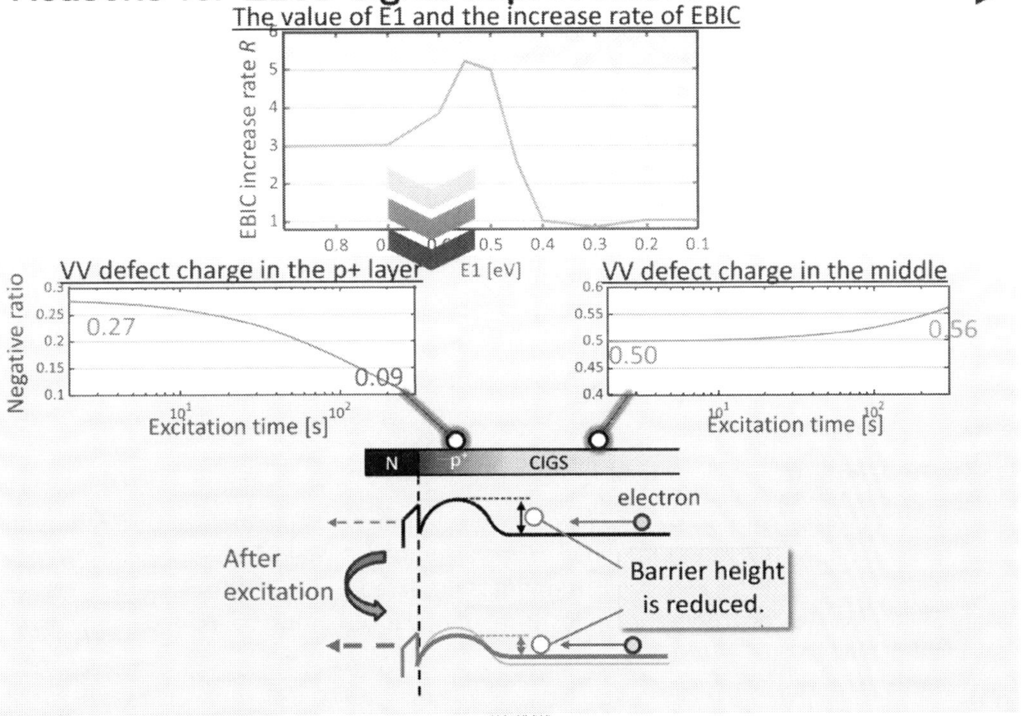

020137-013

40th European Photovoltaic Solar Energy Conference and Exhibition

E1 value and hole concentration with HLS process

→ HLS processing has been found to increase hole concentrations.

Transient response simulation of HLS process

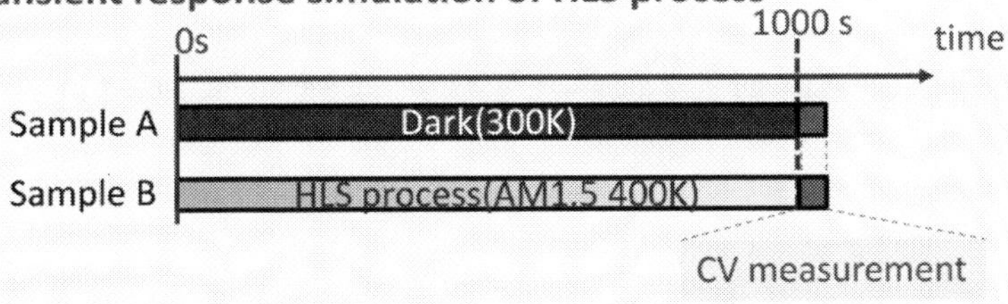

Calculated hole concentration from CV results

	E1=0.9 eV	**E1=0.5 eV**	**E1=0.1 eV**
Sample A: N_{CV} Without HLS	3.26×10^{16} cm^{-3}	3.26×10^{16} cm^{-3}	1.24×10^{17} cm^{-3}
	Increase	Increase	stable
Sample B: N_{CV} With HLS	6.69×10^{16} cm^{-3}	6.67×10^{16} cm^{-3}	1.24×10^{17} cm^{-3}

For large E1 values, we see a double increase in acceptor concentration

No increase at low E1.

020137-014

Summary

- A transient response simulation of CIGS, including charge changes in VV defects, has been successfully constructed
  - Time variation in EBIC could be explained.
- The value of the energy barrier for the VV charge transition was determined.
  - E1 must be larger than Zunger's value (more than 0.4 eV)
- HLS simulation also shows that large values of E1 are required to reproduce the CIGS time variation.

Acknowledgement

This work was supported by the New Energy and Industrial Technology Development Organization (NEDO), and JST SPRING, Grant Number JPMJSP2106.
The CIGS solar cells for EBIC measurement is prepared by Dr. S. Ishizuka and Dr. Y. Kamikawa from the National Institute of Advanced Industrial Science and Technology (AIST).

Tokyo Tech

Thank you for your attention.

Comparison with Zunger' s values

	E_1 (eV)	E_2 (eV)	E_3 (eV)
Optimal	**0.91**	0.51	0.67

	E_1 (eV)	E_2 (eV)	E_3 (eV)
Zunger's	**0.1**	0.35	0.73

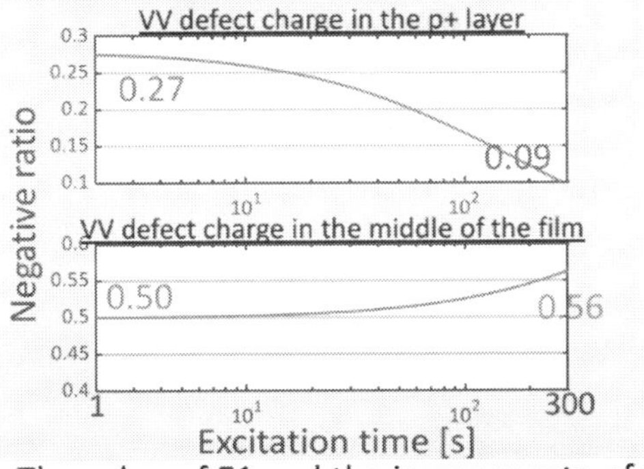

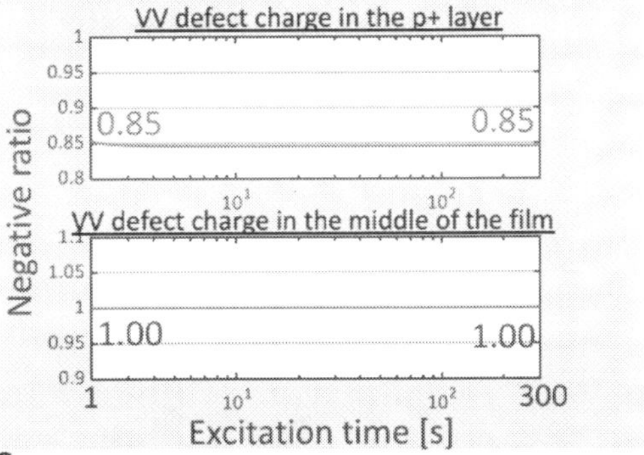

The value of E1 and the increase rate of EBIC

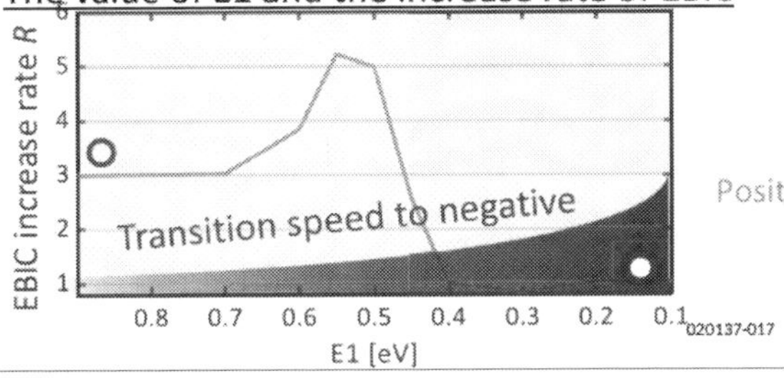

$$S = N_e \cdot \nu \cdot \exp\left(-\frac{E_1}{kT}\right)$$

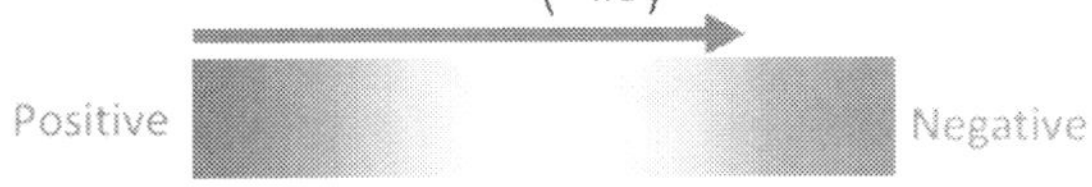

Optimal parameter to reproduce EBIC time variation.

EBIC change rate distribution in the plane of E1 and E2

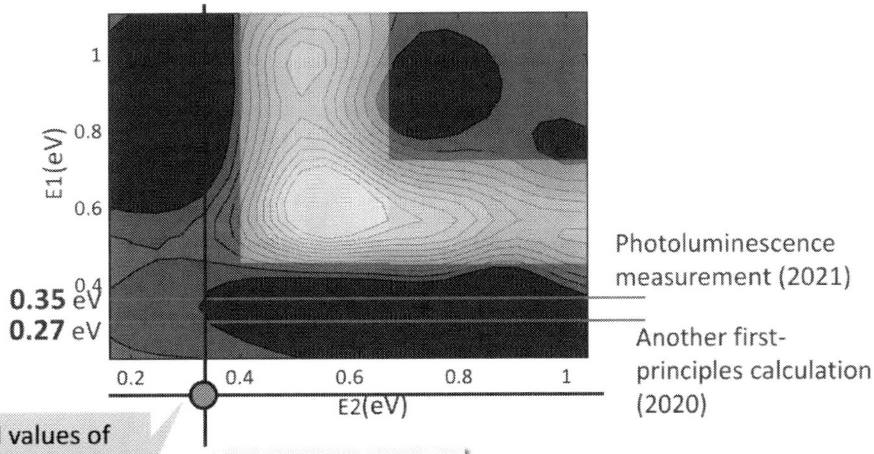

Calculated values of Zunger et al. (2006)
E1 = **0.1** eV
E2 = 0.35 eV

0.45 eV < E1
0.4 eV < E2

Tokyo Tech

Thank you for your attention.

STUDY OF UNIQUENESS OF HIGHLY FLEXIBLE CIGS SOLAR CELLS ON MICA SUBSTRATE OF ONE STEP DEPOSITION PROCESS

Maliya Syabriyana, Pin-Ying Huang, and Chih-Huang Lai
Department of Materials Science and Engineering, National Tsing Hua University
No. 101, Section 2, Kuang-Fu Road, Hsinchu, Taiwan 30013, R.O.C.
Email: syabriya@gmail.com, chlai@mx.nthu.edu.tw

ABSTRACT: The study examines the uniqueness of mica substrate for one-step Cu(In,Ga)Se2 (CIGS) deposition in the production of highly flexible solar cells. Results demonstrate that mica substrate provides unique advantages for one-step CIGS deposition, such as improved crystal growth and uniformity, leading to enhanced photovoltaic performance of flexible solar cells. The improvement mainly on short circuit current (Jsc) and Fill Factors (FF) which is resulted from better light absorption due to high quality of crystal growth of film on mica. The present of rear side of alkali metal from mica also influent the device performance. Bending test shows that CIGS on mica retaining more than 95% of the original PCE after 3000 cycles of large deformation bending. These findings contribute to the development of new, cost-effective and scalable production methods for high temperature process of flexible solar cells, with potential for widespread implementation in flexible photovoltaic devices especially in wearable devices.
Keywords: CIGS, one step sputtering, mica, flexible

1 INTRODUCTION

Nowadays, solar cells industry has focused their product not only for low cost but also for light weight, foldable, and can be installed to uneven surface [1]. The development of highly flexible and robust CIGS solar cells on mica has important implications for the future of photovoltaic technology. Flexible solar cells have the potential to revolutionize the way we generate and use solar energy, as they can be integrated into a wide range of devices and applications, including wearable devices, building-integrated photovoltaics, and portable electronics.

Several substrates have been used such as polyimide [2]–[5], polyethylene terephthalate (PET)[6], polyethylene naphtholate (PEN), paper, metal foils [7], [8]thin glass[8] [9], polyvinyl chloride (PVC), for flexible thin film, a-Si, organic and perovskite solar cells. The highest efficiency, 20.8% achieved on PI substrate[9] and followed by 19.1 % on stainless steel [1]. The challenge of having polymer-based substrate is the need of a low-temperature process in the contrary (< 450oC), meanwhile a good CIGSE quality is grown at higher temperature (around 550°C). In addition, polymer substrate has poor adhesion due to large gap of CTE [10] [11]. On the other hand, metal foils can withstand higher temperature but impurities are the limitation. This enable printable fabrication process which lead to more efficient and low cost production. In addition, the inherent issues like short duration and environmental instability have restricted their improvement over long time. The low-temperature deposition process will deteriorate device performance due to the emergence of the Schottky contact at the interface of CIGSE/Mo. Thus, flexible substrate which endure high temperature process, more than 500°C, transparent, chemically inert and cheap is highly needed to improve the solar cells performance. Therefore, finding a proper flexible substrate that can adhere to high-temperature processes and is chemically stable.

Mica seems to fulfil overmentioned requirement. This material also known as muscovite, has been widely use in electronic and electrical industries for its unique properties, named optical transparency, mechanical flexibility and high thermal stability. In addition, it withstands with oxygen and moisture. The use of mica as a substrate material offers an accessible and cost-effective solution for the large-scale and roll-able production of flexible and light weight solar cells, making renewable energy more accessible and sustainable for communities around the world.

In the present work, we focused attention on developing a low-cost, flexible, light weight and durable CIGSe thin film solar cell on chemical inert mica substrate through single steps absorber process.

2 EXPERIMENTS

2.1 Preparation of CIGSe Absorber and Solar Cells

A quaternary CIGSe target with the composition of Cu21.3In19.7Ga6.9Se52.1 was prepared by a hot-pressed sintering process from high-quality Cu2Se, In2Se3, and Ga2Se3 powders. Stainless steel (SS) foil (AISI 430) and mica were used as substrates with a thickness of 50 μm and 20 μm respectively. The 800-nm-thick Mo back contact layer was deposited by DC-magnetron sputtering under an Ar pressure of 1 mTorr on SS. On mica, before sputtering Mo, 100nm TiN was deposited as adhesion layer by DC magnetron sputtering under an Ar pressure of 5 mtorr. The CIGSe films were fabricated by using pulse DC-magnetron sputtering with an arcing monitor (Advanced Energy MDX power supply with a Sparc-le V unit), and the sputtering chamber was pumped down to the base pressure of 5 × 10−6 Torr. CIGSe absorbers were grown on Mo coated stainless steel and mica substrates with the area of 2 x 2 cm2 under a pure Ar pressure of 15 mTorr at 550 °C. Extra Na was incorporated into CIGSe films by evaporation of sodium fluoride (NaF) to 50-nm-thick and annealed in vacuum at 400°C for 60 min after CIGSe deposition. A 50-nm-thick CdS buffer layer was fabricated by chemical bath deposition at 60 °C. A 200-nm-thick ZnO:Al (Al_2O_3 2 wt %) film was deposited as a window layer in sequence by using radio frequency magnetron sputtering at room temperature. An Al grid was used as a top electrode by E-beam evaporation with a contact mask. The cells with an active area of 0.30 cm2 were isolated by photolithography.

2.2 Characterization Methods.

Microstructures were studied by using field-emission

scanning electron microscopy (FE-SEM, JSM-6500F, JEOL). X-Ray diffraction was measured using Bruker D2 diffractometer. The composition depth profile of CIGS films was also carried out by time-of-flight secondary ion mass spectroscopy (TOF-SIMS IV). The current−voltage (J−V)of solar cells were measured by a Keithley 4200-SCS semiconductor analyzer. Device performance was measured under an AM 1.5 solar simulator with a power density of 100 mW/cm2 at 25 °C. A 1000 W AAA-xenon lamp was used as the light source and the intensity was calibrated by a monocrystalline Si reference cell (certified by the Fraunhofer Institute for Solar Energy Systems ISE). The external quantum efficiency (EQE) (Enlitech) spectrum of devices was measured from 300 to 1350 nm with a 75 W xenon lamp, monochromator, optical chopper, and lock-in amplifier. Reflectance was measured by UV-Visible Horiba.

2.3 Bending Curvature and Cycle Tests

Bending radius and cycle were measured using an automatic stepper bending tool. The CIGS solar cell on mica was bent from one end to other end, and the bending curvature was then measured. Bending cycles were determined by the times when the substrate is bend to desired radius upon repeated bending at 10 and 5 mm.

3 RESULTS AND DISCUSSION

SEM images of samples prepared on different substrates are compared in Figure 1. When the film is deposited on mica and SS. Grains with two different sizes are observed. In contrast, CIGS film on mica shows less small grain size, see **Fig. 1(a),** and two different structures, smaller grain on back side and wide columnar at front side. However on SS, as shown in **Fig. 1(b)**, the smaller grain structure on back side occupied almost half of CIGS thickness. Better crystalline quality of Mo and absorber on mica is observed from XRD analysis (**Fig. 1(c) and (d)**). The smooth and weak van der wall epitaxy of the mica surface lead to higher surface diffusion which results to denser and bigger grains on mica compare to SS. Lower FWHM indicates bigger average crystallite size of CIGS on mica. Thin films with large crystallites can also exhibit reduced grain-boundary scattering, thereby contributing to an increase in Hall carrier mobility, which could reduce electrical resistivity [11].

Current density-voltage (J-V) characteristics of the CIGS are summarized in **Table I**. Device on mica has higher Jsc and FF and lower Rs, leading to PCE of 11.92%. The decreasing of electrical resistivity and improvement of FF due to larger and denser absorber's crystalline size, confirmed by XRD analysis (**Fig. 1(d)**). Since crystal quality also play important role on light absorption, therefore Jsc improved on mica, mainly on higher light absorption at long wave length region. To observe contribution of optical properties improvement, reflectance measurement also performed (**Fig. 2(a)**). It shows that CIGS on mica has lower light reflection, suggests that grain-boundary scattering is reduced due to better absorber quality, leading to Jsc enhancement.

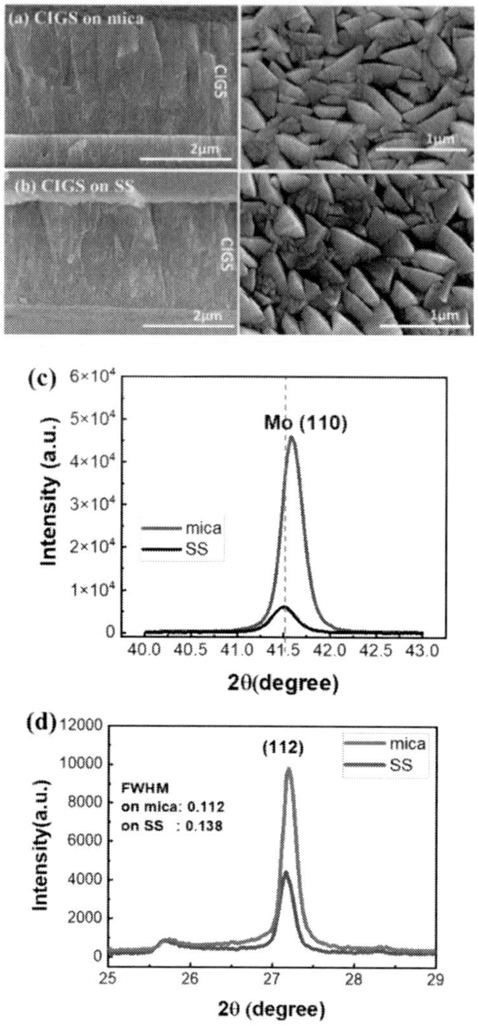

Fig. 1 SEM image of CIGS growth on (a) mica and (b) SS substrates, XRD spectra of (c) Mo and (d) CIGS deposited on mica and SS.

Table 1. CIGS solar cell performance for cells on SS and mica.

Subs	Voc (mV)	Jsc (mA cm^{-2})	FF (%)	η (%)	R_{sh}	R_s
Mica	533	32.46	68.9	11.92	1003.8	1.92
SS	542	31.64	67.2	11.52	408.71	1.76

Since mica substrate has higher content of potassium and low content of sodium, alkali metals diffusion mechanism is carefully studied using SIMS analysis, **Fig. 2(b)**. SIMS result shows that K and Na from mica diffused along the interface. After Na-PDT, thermal energy push K further to the bulk and ion exchange between Na and K at the back side resulting Na gradient. The accumulation of K at the Mo/CIGS interface reduces the Shoctky barrier at the back contact [12], which improved fill factor and resistance of device on mica.

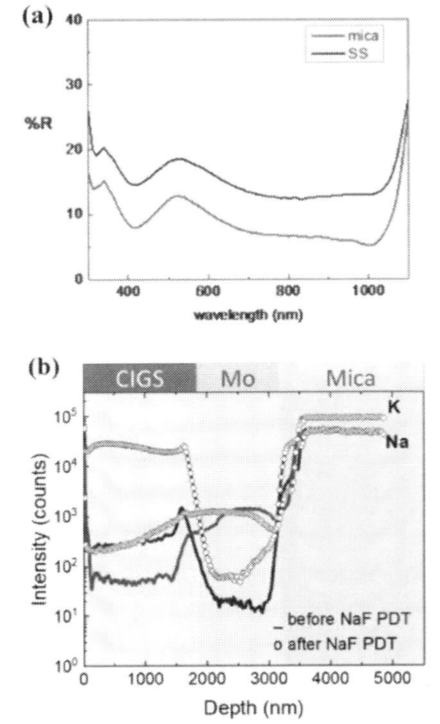

Fig. 2 (a) UV-Vis reflectance spectra of CIGS SCs on mica and SS, (b) SIMS depth profile of Na and K before NaF-PDT, after NaF-PDT on mica substrate.

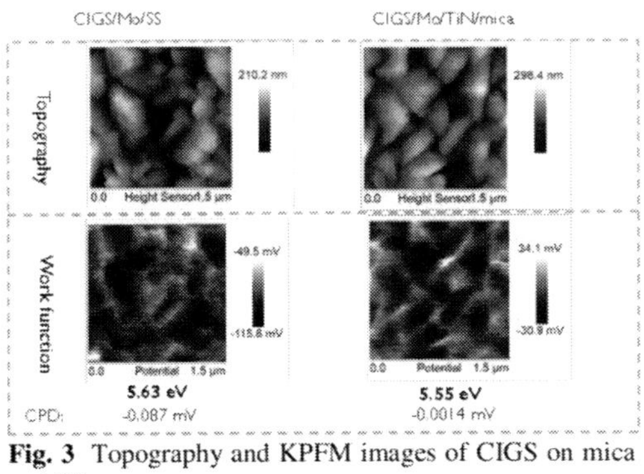

Fig. 3 Topography and KPFM images of CIGS on mica and SS

However, Voc on mica device is lower compare to on SS. Therefore, a surface potential measurement of CIGS using Kelvin Probe force microscope (KPFM) was performed to understand the influence of rear side alkali metals diffusion to the change of electrical properties of device. The measured contact potential differences (CPD), that is, the work function differences between the tip and samples (φtip -φsample). The CIGS thin film on mica exhibited higher CPD, means lower work function, indicates lower hole barrier on the surface (**Fig. 3**). Therefore recombination more probably takes place at the surface, resulting on lower Voc, unlike CIGS on SS. The presence of K from back side may reduce the p-type doping of the quaternary-sputtered CIGS, suggesting that the number of donor defects in the CIGS was indeed increase in comparison to CIGS on SS.

CIGS solar cells on mica were bended at different bending radius, 10mm and 5 mm with automatic stepping motor device to demonstrate its high flexibility. At bending radius 10 mm and 5 mm, the device kept 95% and 98%, respectively, from its original performance after 3000 cycles (**Fig. 4**). Since mica is formed by multilayers single crystal and has weak Vander Waals force, during the bending, the mechanical constraint is relaxed and further the strains is reduced due to interlayer slipping of mica. Thus makes mica as a robust substrate and able to retain good performance after bending.

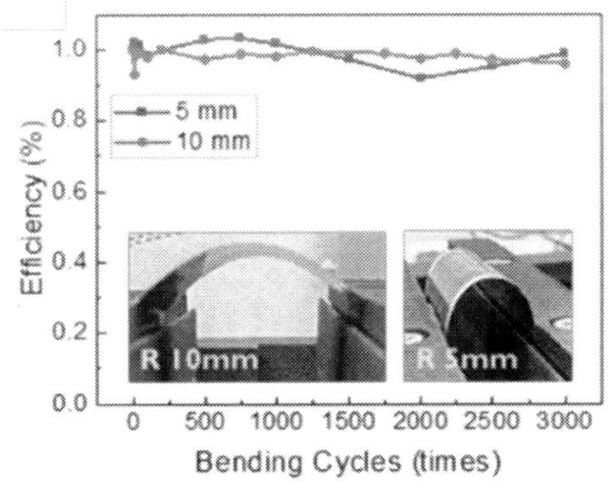

Fig. 4 Bending cycles, under radius of curvature of 10 mm and 5 mm.

Table 2 Device perfromance before and after 3000 cycle bending

Sample	Voc (mV)	Jsc (mA/cm^2)	FF (%)	PCE (%)
Before bending	507	33.76	65	11.16
After bending	506	34.07	64	11.01

4. CONCLUSIONS

In conclusion, highly flexible, robust, stable, and high temperature process CIGS SCs based on mica substrate has been successfully developed, exhibiting a power conversion almost 12.0%, retaining more than 95.8% of the original PCE after 3000 cycles of large deformation bending. The device performance outweigh the device grown on SS substrate by 0.4% efficiency due to improvement of Jsc, FF, and Rs. Better crystal quality, light absorption and potassium contributed to performance enhancement. Thus in the future better device quality can be achieved by controlling CdS thickness. Mica delivered prominent flexible, ultra-thin, chemical inert, cheap and high temperature resistance that suitable high temperature process for good quality CIGS growth.

5. ACKNOWLEDGEMENT

Ministry of Science and Technology (MOST), Taiwan for Research Funding. Mr. Wang Chiung-Chi for SIMS-TOF analysis. Prof. Yin-Hao Chu for bending test instrument. Indonesia's Endowment Fund for Education for my PhD study funding and Universitas Serambi Mekkah, Aceh, for my study support.

6. REFERENCES

[1] J. Ramanujam *et al.*, "Flexible CIGS, CdTe and a-Si:H based thin film solar cells: A review," *Progress in Materials Science*, vol. 110, p. 100619, May 2020, doi: 10.1016/j.pmatsci.2019.100619.

[2] R. Caballero, C. A. Kaufmann, T. Eisenbarth, T. Unold, R. Klenk, and H.-W. Schock, "High efficiency low temperature grown Cu(In,Ga)Se $_2$ thin film solar cells on flexible substrates using NaF precursor layers: Low T CIGSe solar cells on polyimide foil," *Prog. Photovolt: Res. Appl.*, vol. 19, no. 5, pp. 547–551, Aug. 2011, doi: 10.1002/pip.1064.

[3] K. Kim *et al.*, "A simple and robust route toward flexible CIGS photovoltaic devices on polymer substrates: Atomic level microstructural analysis and local opto-electronic investigation," *Solar Energy Materials and Solar Cells*, vol. 195, pp. 280–290, Jun. 2019, doi: 10.1016/j.solmat.2019.03.008.

[4] J. H. Shi, Z. Q. Li, D. W. Zhang, Q. Q. Liu, Z. Sun, and S. M. Huang, "Fabrication of Cu(In, Ga)Se2 thin films by sputtering from a single quaternary chalcogenide target," *Prog. Photovolt: Res. Appl.*, vol. 19, no. 2, pp. 160–164, Mar. 2011, doi: 10.1002/pip.1001.

[5] M. G. Faraj, K. Ibrahim, and A. Salhin, "Investigation of CIGS Solar Cells on Polyethylene Terephthalate Substrates," *International Journal of Polymeric Materials*, vol. 60, no. 10, pp. 817–824, Sep. 2011, doi: 10.1080/00914037.2010.551363.

[6] J.-K. Sim, S. Kang, R. Nandi, J.-Y. Jo, K.-U. Jeong, and C.-R. Lee, "Implementation of graphene as hole transport electrode in flexible CIGS solar cells fabricated on Cu foil," *Solar Energy*, vol. 162, pp. 357–363, Mar. 2018, doi: 10.1016/j.solener.2018.01.053.

[7] T. Yagioka and T. Nakada, "Cd-Free Flexible Cu(In,Ga)Se $_2$ Thin Film Solar Cells with ZnS(O,OH) Buffer Layers on Ti Foils," *Appl. Phys. Express*, vol. 2, p. 072201, Jun. 2009, doi: 10.1143/APEX.2.072201.

[8] A. Gerthoffer *et al.*, "CIGS solar cells on flexible ultra-thin glass substrates: Characterization and bending test," *Thin Solid Films*, p. 6, 2015.

[9] A. Chirilă *et al.*, "Potassium-induced surface modification of Cu(In,Ga)Se2 thin films for high-efficiency solar cells," *Nature Mater*, vol. 12, no. 12, pp. 1107–1111, Dec. 2013, doi: 10.1038/nmat3789.

[10] S. Shi *et al.*, "Recent progress in the high-temperature-resistant PI substrate with low CTE for CIGS thin-film solar cells," *Materials Today Energy*, vol. 20, p. 100640, Jun. 2021, doi: 10.1016/j.mtener.2021.100640.

[11] W.-S. Liu, W.-T. Hsieh, S.-Y. Chen, and C.-S. Huang, "Improvement of CIGS solar cells with high performance transparent conducting Ti-doped GaZnO thin films," *Solar Energy*, vol. 174, pp. 83–96, Nov. 2018, doi: 10.1016/j.solener.2018.08.050.

[12] C. P. Muzzillo, "Review of grain interior, grain boundary, and interface effects of K in CIGS solar cells: Mechanisms for performance enhancement," *Solar Energy Materials and Solar Cells*, vol. 172, pp. 18–24, Dec. 2017, doi: 10.1016/j.solmat.2017.07.006.

[13] C.-H. Hsu, W.-H. Ho, S.-Y. Wei, and C.-H. Lai, "Over 14% Efficiency of Directly Sputtered Cu(In,Ga)Se $_2$ Absorbers without Postselenization by Post-Treatment of Alkali Metals," *Adv. Energy Mater.*, vol. 7, no. 13, p. 1602571, Jul. 2017, doi: 10.1002/aenm.201602571.

WAFER-BONDED TWO-TERMINAL III-V//SI TRIPLE-JUNCTION SOLAR CELL WITH POWER CONVERSION EFFICIENCY OF 36.1 % AT AM1.5G

Patrick Schygulla[1], Ralph Müller[1], Oliver Höhn[1], Michael Schachtner[1], David Chojniak[1], Andrea Cordaro[2], Stefan Tabernig[2], Benedikt Bläsi[1], Albert Polman[2], Gerald Siefer[1], David Lackner[1], and Frank Dimroth[1]
[1]Fraunhofer Institute for Solar Energy Systems ISE, Freiburg, Germany
[2]NWO-Institute AMOLF, Amsterdam, The Netherlands
patrick.schygulla@ise.fraunhofer.de

ABSTRACT: In this work, we present the fabrication and analysis of a wafer-bonded GaInP/GaInAsP//Si triple-junction solar cell with 36.1 % conversion efficiency under AM1.5g spectral illumination. The new cell design presents an improvement over previous III-V//Si triple-junction cells by the implementation of a rear-heterojunction design for the middle cell. Furthermore, an advanced metallodielectric rear-side grating was used for light trapping enhancement in the silicon bottom cell that increased the silicon subcell current by 1.4 mA/cm². A luminescent coupling factor of 0.46 between the middle and bottom subcell was determined. The share of recombination in the middle cell space-charge region was experimentally shown to be insignificant as intended by the rear-heterojunction design. Overall, the open-circuit voltage increased by 61 mV compared to the previous generation. Given the established long-term stability of III-V and silicon based solar cells these results are promising steps towards the future employment of III-V/Si tandem solar cells.
Keywords: silicon-based tandem solar cell, III-V//Si, multi-junction, MOVPE, GaInAsP

1 INTRODUCTION

The record efficiency of silicon single-junction solar cells is close to its theoretical efficiency limit of 29.4 – 29.5 % [1,2]. Further increasing the cell efficiency [3] is an important driver for photovoltaics system cost reduction. Hence, there is the strong need for new solar cell concepts exceeding the silicon single-junction efficiency limit. Multi-junction solar cells are a promising concept to this end with demonstrated efficiencies of up to 39.5 % for a triple-junction cell fully made from III-V semiconductors [4]. Such cells are grown by epitaxial deposition on an expensive GaAs or InP substrate. Hence, we investigated the efficiency potential of III-V top structures combined with a silicon bottom cell to improve the efficiency-to-cost ratio.

Wafer-bonded III-V//Si multi-junction solar cells were the first monolithic silicon-based tandem cells to surpass a conversion efficiency of 30 %. In the past, we have shown further efficiency improvements by a) the inclusion of a TOPCon surface passivation of the silicon bottom cell together with integration of a diffractive light scattering structure on the rear side, reaching 33.3 % [5], b) the employment of an absorber material with a higher band gap in the second junction, AlGaAs instead of GaAs, reaching 34.1 % [6], and c) the employment of a material with higher minority carrier lifetimes, GaInAsP instead of AlGaAs, reaching 35.9 % [7].

Other promising tandem cell concepts involving silicon bottom cells are dual junctions employing a perovskite top cell. With this technology, the highest efficiencies exceed 30 % on a 1 cm² device [8]. Solar cells including perovskite absorbers exhibit degradation over time, for which a solution has not yet been found, though. [9]. Combining more than two junctions using perovskite absorbers is challenging due to wet-chemical processing issues and reduced availability of materials at the required higher band gaps. The highest efficiency for a monolithic triple-junction perovskite/perovskite/silicon device amounts to 20 % [10].

Here, we present a GaInP/GaInAsP//Si triple-junction solar cell with a further improved heterojunction structure of the III-V layer stack, combined with the previously demonstrated metallodielectric light trapping back-side reflector. We measured a conversion efficiency of 36.1%, the highest value observed for a Si-based solar cell. The high efficiency enables a detailed study of the carrier collection quantum efficiency and spontaneous emission processes that determine the high efficiency.

2 METHODS

2.1 Epitaxial Growth

III-V semiconductor layers were epitaxially grown lattice matched to GaAs by metalorganic vapour phase epitaxy in a commercial AIX2800G4 reactor from AIXTRON. 4" (001)-oriented GaAs wafers with an offcut of 6 ° into the [111]-B direction were used as substrates. TMAl, TMGa, TMIn, AsH_3, PH_3, DMZn, and SiH_4 were used as precursors. In both the middle and the top cell, a rear-heterojunction design was employed. The first reason for this choice is the higher minority carrier lifetime of n-GaInAsP, the only absorber material in the rear-heterojunction, compared to p-GaInAsP [11], the thickest part of the absorber in a homojunction. The second reason is the reduction of the dark current in the space charge region because of the increased band gap inside the space charge region at the rear-side of the absorber [12].

2.2 Solar Cell Processing

The middle cell was deposited first and the top cell second on the GaAs substrate to reduce the voltage drop in the top cell due to a high thermal load [7]. The upright growth fashion made it necessary to temporarily bond the epitaxial structure to a sapphire wafer as mechanical support. After that, the GaAs substrate was removed by wet chemical etching with citric acid. Then, the exposed bond layer was chemically mechanically polished by the company III/V-Reclaim. Next, a direct wafer bond to a TOPCon silicon cell was performed before the temporary bond could be released again by thermal slide at 190 °C. Solar cell processing involving photolithography, mesa etching, metal contact formation, and sputter deposition of an anti-reflection coating (ARC) followed. Twelve solar cells with a size of 2 × 2 cm² and a metal grid optimised for one-sun operation were processed on each wafer. After solar cell processing, the wafers were nano-imprinted with a diffractive grating on the rear side at AMOLF [13]. The rear-side grating is designed to enhance the optical path length in the silicon bottom cell and thus increases Si subcell current. It is designed to minimize light scattering from the back-side reflector in normal direction and to enhance large-angle scattering in the Si cell, while minimizing parasitic absorption in the metal [14]. More details on the solar cell processing can be found in [5,7,15]. The final layer stack is shown in Figure 1.

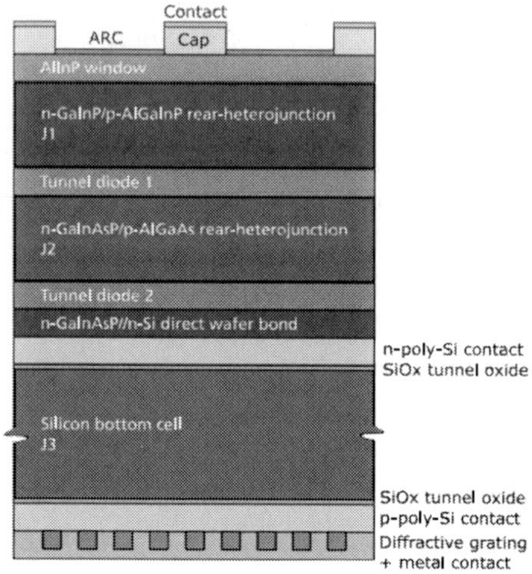

Figure 1. Layer stack of the III-V//Si triple-junction solar cell. The GaInP/GaInAsP top structure was attached to a TOPCon silicon bottom cell by surface-activated direct wafer bonding. The rear-side grating was implemented by nanoimprint lithography.

2.3 Characterisation

The solar cells were characterised by external quantum efficiency and current-voltage measurements before and after the diffractive rear-side grating was applied to evaluate the current gain. The *IV*-curves were recorded under calibrated spectral conditions of AM1.5g (IEC 60904-3:2019 (ed.4)) in the Fraunhofer ISE Calibration Laboratory (CalLab). Current-voltage characteristics were measured using a hard mask with an aperture area of 3.987 cm², which is slightly lower than the mesa area of the solar cells, in order to prevent a lateral current contribution from the silicon bottom junction outside of the mesa area.

Additionally, the sub-cell currents were analysed under different spectral conditions to determine the luminescent coupling between sub-cells and the radiative recombination current of the top cells. For this advanced characterisation method, a Wavelabs SINUS 220 LED based solar simulator equipped with 20 spectrally different LED light sources was used [16]. Using the different LED channels the injection level on each subcell was varied in a systematic manner while the remaining subcells were light biased at AM1.5g conditions. A custom software was used to control the illumination unit of the solar simulator while the spectral adjustment of the individual light sources was realized using the procedure described in [17].

3 RESULTS

3.1 Current-Voltage Characteristics

The current champion solar cell has a power conversion efficiency of 36.1 % under the AM1.5g spectrum as was determined by a calibrated current-voltage measurement that is shown in Figure 2. The main reason for the higher efficiency compared to the last generation of III-V//Si triple-junction solar cells made at Fraunhofer ISE is the increase in open-circuit voltage by 61 mV. The major difference compared to the previous generation is the inclusion of a rear-heterojunction cell configuration in place of the former homojunction cell design for the middle cell.

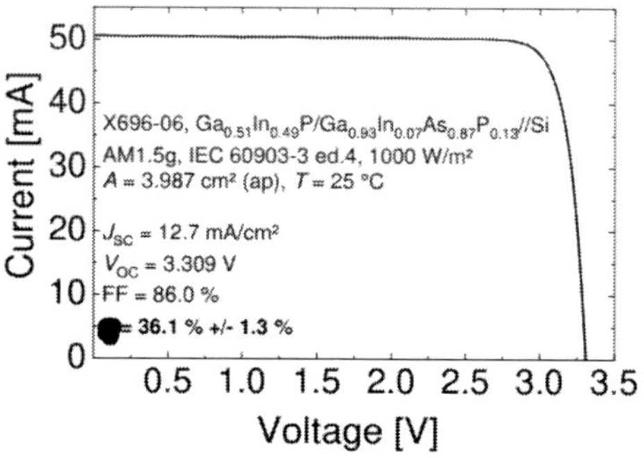

Figure 2. Current-voltage characteristics of the champion GaInP/GaInAsP//Si triple-junction solar cell measured in the Fraunhofer ISE CalLab under a calibrated AM1.5g spectrum.

3.2 External Quantum Efficiency

The external quantum efficiency (EQE) of the champion cell X696-6 is shown in Figure 3. The subcell current density is calculated from the injection current variation experiments explained below. That analysis shows that the middle cell is current limiting with 12.6 mA/cm². The sum of the subcell currents amounts to 38.9 mA/cm² which implies current matched conditions at a jsc of around 13.0 mA/cm² at the currently prevailing optical and electrical performance. Thus, the middle cell is limiting by 0.4 mA/cm². The difference of 0.1 mA/cm² in the short-circuit current density of the entire triple-junction, 12.7 mA/cm², compared to the subcell current density of the middle cell, 12.6 mA/cm², is caused by two effects. The first effect is luminescence coupling from the top to the middle cell. Considering the difference in subcell current between the top and middle cell, 0.6 mA/cm², this implies that around a fifth to a quarter of the excessively injected charge carriers recombine radiatively in the top cell and these photons are then reabsorbed in the middle cell. The second effect is a small decrease in current density from $V = 0$ towards the short-circuit point of the current limiting subcell, at $V = V_{oc} - V_{oc,j2}$ due to a finite parallel resistance.

Before the nanoimprint of the diffractive rear-side structure the short-circuit current density was limited by the silicon bottom cell with 11.7 mA/cm². Hence, the light path enhancement that leads to increased absorption in the long-wavelength regime close to the absorption threshold for silicon, around 1100 nm, resulted in an increase in current density by 1.4 mA/cm² to a total value of 13.1 mA/cm². Differently designed versions of a rear-side diffractive structure in previous experiments lead to a current density gain of 1.1 mA/cm² in the silicon bottom cell [18].

If the subcell currents of a multi-junction solar cell are perfectly matched, then the rounding of the individual IV-curves at the maximum power point adds up and the fill factor of the IV-curve of the entire device is reduced. Conversely, in strong current-mismatched conditions the fill factor is dominated by the fill factor of the current limiting subcell [19]. If the available solar resource was distributed more evenly between all subcells, while maintaining the material quality and thus the open-circuit voltage, the power conversion efficiency could still be improved by 0.4 $\%_{abs}$ to 36.5 % with mere adjustments to the absorber thicknesses.

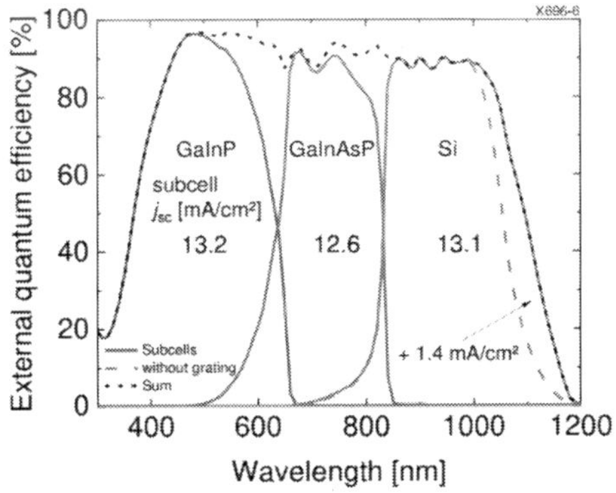

Figure 3. External quantum efficiency of the three subcells (solid line) and of the entire triple-junction solar cell (black, dotted line). The quantum efficiency of the silicon bottom cell before the imprint of the diffractive rear-side grating is shown by the orange dashed line.

3.3 Luminescence Coupling

The high amount of radiative recombination in the middle cell results in luminescence coupling from the second to the third junction. Photons that are emitted after radiative recombination of charge carriers in the middle cell have a certain probability of leaving the junction towards the bottom and be reabsorbed by the bottom cell. If the bottom cell limits the overall current, then this luminescent contribution increases the short-circuit current. Following the procedure suggested by Steiner et al. [20], the irradiance of each subcell was systematically decreased and increased with respect to AM1.5g 1-sun conditions while the other subcells were biased with light to maintain an AM1.5g injection level. The measured short-circuit currents for the different irradiance conditions are shown in Figure 4.

On the left side of the diagram where the slopes are constant, the subcell for which the photocurrent is varied is current limiting in each case. The second junction is current limiting at 1 sun. This is why the photocurrent of the second junction must be increased above 1 sun to reach current matching with the top cell at 13.2 mA/cm². Then, the top cell becomes limiting and the overall current stagnates. For the first junction, one can see that upon further increase of the photocurrent beyond 0.96 suns the overall current increases again due to luminescent coupling from the first to the second junction with the coupling factor η_{12}. If the irradiance of the third junction is decreased below around 0.97 suns, then that junction becomes limiting. From the curve shape of the injection variation of the bottom cell the luminescence coupling factor η_{23} from the second to the third junction can be extracted. The second junction is optically thick for light emitted by the first junction so the contribution of coupled light from the first directly to the third junction can be neglected.

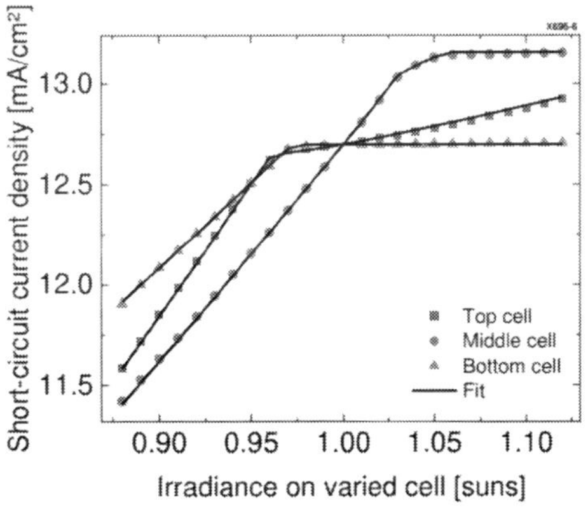

Figure 4. Measured short-circuit current of the III-V//Si 3j champion device upon selective variation of the photocurrent in the top (blue squares), middle (cyan circles), and bottom junction (orange triangles) around AM1.5g conditions at 1 sun. The other junctions are biased at the 1-sun conditions, respectively. The fitted curves to the data are shown by solid black lines.

4 CONCLUSIONS

We present a III-V//Si triple-junction solar cell with a GaInP top cell, a GaInAsP middle cell, and a silicon bottom cell with a conversion efficiency of 36.1 %, the highest efficiency reported for a Si-based multi-junction solar cell reported to date. The III-V top cell structure was wafer-bonded to the silicon bottom cell for a monolithic integration. A metallodielectric back-side reflector was integrated to optimize light trapping in the Si bottom cell. By employing a rear-heterojunction cell design in both the middle and top cell, the open-circuit voltage of the triple-junction device could be improved by 61 mV compared to previous generations. Since radiative recombination of minority carriers is dominant in this structure strong luminescent coupling between the middle and bottom subcell was observed. This was carefully considered for accurate quantification of the external quantum efficiency. An absolute determination of the external quantum efficiencies of every subcell was important for evaluating the current gain caused by the back-side reflector. Acquiring current-voltage curves at systematically varying irradiance conditions selectively for each subcell is a useful method for obtaining the luminescence coupling factors and the externally injected currents for calibration of the external quantum efficiency.

ACKNOWLEDGEMENTS

The authors would like to thank F. Sahajad, S. Stättner for help with epitaxial growth, F. Schätzle, A. Leimenstoll,

A. Lösel, J. Polzin, R. Freitas, R. Koch, and R. Neubauer for sample processing, M. Bauer from Freiburg University for a-Si deposition, F. Martin, E. Fehrenbach, J. Aulich, and A. Wekkeli for solar cell measurements, and H. Helmers and J. Ohlmann for helpful discussions. The work received funding by the Fraunhofer Gesellschaft through the ICON project MEEt. Parts of the cell design were developed in the project H2Demo (03SF0619A) funded by the German Federal Ministry of Education and Research (BMBF). The overall solar cell structure and processing was developed in the project PoTaSi (0324247) funded by the German Federal Ministry for Economic Affair and Climate Action. Work at AMOLF is part of the research program of the Dutch Research Council. P. Schygulla acknowledges his PhD scholarship from the Heinrich-Böll Foundation.

The authors have no conflict of interests to declare.

References

1. Richter A, Hermle M, Glunz SW. Reassessment of the Limiting Efficiency for Crystalline Silicon Solar Cells. *IEEE J. Photovolt.* 2013; **3(4)**: 1184–91, DOI: 10.1109/JPHOTOV.2013.2270351.
2. Veith-Wolf BA, Schäfer S, Brendel R, Schmidt J. Reassessment of intrinsic lifetime limit in n-type crystalline silicon and implication on maximum solar cell efficiency. *Sol. Energy Mater. Sol. Cells* 2018; **186**: 194–9, DOI: 10.1016/j.solmat.2018.06.029.
3. Benda V, Černá L. PV cells and modules - State of the art, limits and trends. *Heliyon* 2020; **6(12)**: e05666, DOI: 10.1016/j.heliyon.2020.e05666.
4. France RM, Geisz JF, Song T, Olavarria W, Young M, Kibbler A, Steiner MA. Triple-junction solar cells with 39.5% terrestrial and 34.2% space efficiency enabled by thick quantum well superlattices. *Joule* 2022; **6(5)**: 1121–35, DOI: 10.1016/j.joule.2022.04.024.
5. Cariou R, Benick J, Feldmann F, Höhn O, Hauser H, Beutel P, Razek N, Wimplinger M, Bläsi B, Lackner D, Hermle M, Siefer G, Glunz SW, Bett AW, Dimroth F. III–V-on-silicon solar cells reaching 33% photoconversion efficiency in two-terminal configuration. *Nat. Energy.* 2018; **3(4)**: 326–33, DOI: 10.1038/s41560-018-0125-0.
6. Lackner D, Höhn O, Müller R, Beutel P, Schygulla P, Hauser H, Predan F, Siefer G, Schachtner M, Schön J, Benick J, Hermle M, Dimroth F. Two-Terminal Direct Wafer-Bonded GaInP/AlGaAs//Si Triple-Junction Solar Cell with AM1.5g Efficiency of 34.1%. *Sol. RRL* 2020; **4(9)**: 2000210, DOI: 10.1002/solr.202000210.
7. Schygulla P, Müller R, Lackner D, Höhn O, Hauser H, Bläsi B, Predan F, Benick J, Hermle M, Glunz SW, Dimroth F. Two-terminal III–V//Si triple-junction solar cell with power conversion efficiency of 35.9 % at AM1.5g. *Prog Photovolt Res Appl.* 2022; **30(8)**: 869–79, DOI: 10.1002/pip.3503.
8. Green MA, Dunlop ED, Siefer G, Yoshita M, Kopidakis N, Bothe K, Hao X. Solar cell efficiency tables (Version 61). *Prog Photovolt Res Appl.* 2023; **31(1)**: 3–16, DOI: 10.1002/pip.3646.
9. Tockhorn P, Sutter J, Cruz A, Wagner P, Jäger K, Yoo D, Lang F, Grischek M, Li B, Li J, Shargaieva O, Unger E, Al-Ashouri A, Köhnen E, Stolterfoht M, Neher D, Schlatmann R, Rech B, Stannowski B, Albrecht S, Becker C. Nano-optical designs for high-efficiency monolithic perovskite-silicon tandem solar cells. *Nat. Nanotechnol.* 2022; **17(11)**: 1214–21, DOI: 10.1038/s41565-022-01228-8.
10. Zheng J, Wang G, Duan W, Mahmud MA, Yi H, Xu C, Lambertz A, Bremner S, Ding K, Huang S, Ho-Baillie AWY. Monolithic Perovskite–Perovskite–Silicon Triple-Junction Tandem Solar Cell with an Efficiency of over 20%. *ACS Energy Lett.* 2022; **7(9)**: 3003–5, DOI: 10.1021/acsenergylett.2c01556.
11. Schygulla P, Heinz FD, Dimroth F, Lackner D. Middle Cell Development for Wafer-Bonded III-V//Si Tandem Solar Cells. *IEEE J. Photovolt.* 2021; **11(5)**: 1264–70, DOI: 10.1109/JPHOTOV.2021.3090159.
12. Hwang S-T, Kim S, Cheun H, Lee H, Lee B, Hwang T, Lee S, Yoon W, Lee H-M, Park B. Bandgap grading and Al0.3Ga0.7As heterojunction emitter for highly efficient GaAs-based solar cells. *Sol. Energy Mater. Sol. Cells* 2016; **155**: 264–72, DOI: 10.1016/j.solmat.2016.06.009.
13. Cordaro CEA, Tucher N, Tabernig S, Hauser H, Höhn O, Müller R, Bläsi B, Polman A. Plasmonic and Mie scattering in nanopatterned back reflectors for III–V-on-silicon solar cells (Conference Presentation). In: Sprafke A, Goldschmidt JC, Pandraud G, (eds). *Photonics for Solar Energy Systems VIII: 6-10 April 2020, online only, France.* Bellingham, Washington, USA: SPIE; 2020, p. 1136607.
14. Cordaro A, Müller R, Tabernig S, Tucher N, Schygulla P, Höhn O, Bläsi B, Polman A. *Nano-patterned back-reflector with engineered near-field/far-field light scattering for enhanced light trapping in silicon-based multi-junction solar cells*; 2023.
15. Müller R, Schygulla P, Lackner D, Höhn O, Hauser H, Richter A, Fell A, Bläsi B, Predan F, Benick J, Hermle M, Dimroth F, Glunz SW. Silicon-Based Monolithic Triple-Junction Solar Cells with Conversion Efficiency >34%. In: *37th European Photovoltaic Solar Energy Conference and Exhibition.* Proceedings; 2020, pp. 574–578.
16. Scherff MLD, Nutter J, Fuss-Kailuweit P, Suthues J, Brammer T. Spectral mismatch and solar simulator quality factor in advanced LED solar simulators. *Jpn. J. Appl. Phys.* 2017; **56(8S2)**: 08MB24, DOI: 10.7567/jjap.56.08mb24.
17. Chojniak D, Bett AJ, Hohl-Ebinger J, Reichmuth SK, Schachtner M, Siefer G. LED solar simulators – A spectral adjustment procedure for tandem solar cells. *AIP Conf. Proc.* 2023; **2826(1)**: 30003, DOI: 10.1063/5.0140990.
18. Bläsi B, Höhn O, Hauser H, Cariou R, Benick J, Feldmann F, Beutel P, Lackner D, Hermle M, Siefer G, Glunz SW, Bett AW, Dimroth F, Tucher N. Photonic structures for III-V//Si multijunction solar cells with efficiency >33%. In: Wehrspohn RB, Sprafke AN, (eds). *Photonics for Solar Energy Systems VII*: SPIE; 2018, p. 2.
19. McMahon WE, Emery KE, Friedman DJ, Ottoson L, Young MS, Ward JS, Kramer CM, Duda A, Kurtz S. Fill factor as a probe of current-matching for GaInP 2 /GaAs tandem cells in a concentrator system during outdoor operation. *Prog Photovolt Res Appl.* 2008; **16(3)**: 213–24, DOI: 10.1002/pip.798.

20. Steiner M, Geisz JF, Moriarty TE, France RM, McMahon WE, Olson JM, Kurtz SR, Friedman DJ. Measuring *IV* curves and subcell photocurrents in the presence of luminescent coupling. *IEEE J. Photovolt.* 2013; **3(2)**: 879–87, DOI: 10.1109/jphotov.2012.2228298.

10.4229/EUPVSEC2023/2DO.9.2

40th European Photovoltaic Solar Energy Conference and Exhibition

This presentation was selected by the Sc. Committee of the EU PVSEC 2023 for submission of a full paper to one of the EU PVSEC's collaborating peer-reviewed journals.

Approaches for III-V/Si Tandem Solar Cells and Comparative Studies on Si Tandem Solar Cells

Masafumi Yamaguchi[1*], Kyotaro Nakamura[1], Ryo Ozaki[1], Tatsuya Takamoto[2], Taizo Masuda[3], Takashi Mabuchi[3], Kenichi Okumura[3], Nobuaki Kojima[1] and Yoshio Ohshita[1]
[1]Toyota Technological Institute, Nagoya 468-8511, Japan
[2]Sharp Corporation, Nara, 639-1186, Japan
[3]Toyota Motor Corporation, Susono 410-1193,, Japan

ABSTRACT - VIPV (Vehicle Integrated PV)-powered vehicle applications are very attractive for reducing CO_2 emission and creation of new market. Development of Si tandem solar cells is very promising for high-efficiency and low-cost solar cells. This paper presents our approaches for III-V/Si 3-junction tandem solar cells. In this paper, 24.9% efficiency Si heterojunction solar cells and mechanically stacked 4-terminal 35.8% InGaP/GaAs/ Si 3-junction tandem solar cell with an area of 23cm^2) are shown. This paper presents high efficiency potential of III-V/Si 3-junction and perovskite/Si 2-junction solar cells analyzed by using our analytical procedure and discusses about non-radiative recombination, optical and resistance losses in those Si tandem cells. Most recently, the authors have new world record efficiency (33.7%) with InGaP/GaAs/Si 3-junction solar cell module with an area of 775 cm^2.

1. Our approaches for III-V/Si 3-junction tandem solar cells

The Si tandem solar cells are very attractive for VIPV applications because of high-efficiency and low-cost potential [1]. In this paper, our recent results for the III-V/Si 3-junction solar cell and modules are presented.

As bottom Si solar cell, the authors have studied on Si hetero-junction (SHJ) solar cell which is composed of an intrinsic (i-type) a-Si layer and a p-type a-Si layer deposited on a randomly textured phosphorous-doped n-type Czochralski (CZ) crystalline silicon wafer (resistivity 5.2 Ωcm, thickness 180 μm). As shown in Fig. 1, The SHJ solar cell with an area of 48 mm × 48 mm shows 24.9% efficiency (J_{sc} = 40.13 mA/cm^2, V_{oc} = 0.729 V, FF = 0.85) under AM1.5G illumination condition. The current status of efficiencies of our Si bottom cell, upper III-V 2-junction solar cell and III-V//Si 3-junction tandem solar cell are shown to be 5.6% and 30.2% and 35.8%.

	Jsc [mA/cm^2]	Voc [V]	FF	η [%]
TTI SHJ (48mm x 48mm)	40.13	0.729	0.85	24.88

	Voc (V)	Jsc (mA/cm2)	FF	Cell efficiency (%)	Tandem cell efficiency (%)
InGaP/GaAs 2J top cell	2.546	13.81	0.860	30.23	35.81
Si bottom cell	0.691	9.96	0.811	5.58	

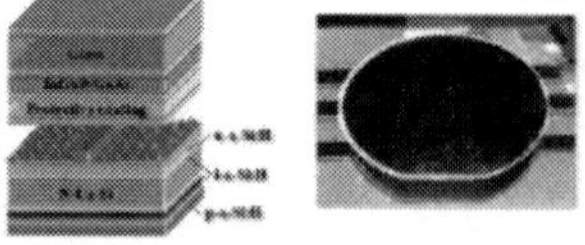

Fig. 1. A structure, photo and properties of Si hetero-junction solar cell and 4-terminal InGaP/GaAs/Si 3-junction solar cell.

2. Analysis for non-radiative recombination loss and efficiency potential of III-V single-junction and multi-junction solar cells

By using our analytical procedure [2], practical limiting efficiencies of various solar cells were estimated by assuming ERE of 10%, optical loss of 5% and resistance loss of 2%. Because 3-junction solar cells have higher potential efficiency of more than 43%, further development of e-junction solar cells includimg perovskite based 3-junction solar cells is very important.

3. Development of 3-junction Si tandem cell module

Development of high-eficiency solar cell modules is very important for various applications. For example, the high-efficincy solar cell modules with an efficiency of more than 35% will prove longer driving distance of more than 30 km/day average and more than 50 km/day on a clear day [3]. Most recently, the authours have acheved new world record efficincy (33.7%) III-V/Si 3-junction tandem slar cell module with an area of 775 cm^2 as shown in Table 1.

Table 1. The characteristic 3-junction Si tan\ndem solar cell module parameter confirmed by the AIST.

	Area (cm^2)	Isc (A)	Voc (V)	FF (%)	Pm (W)	Module Efficiency (%)	Tandem Cell Module Efficiency (%)
InGaP/GaAs 2J top cell module	775	1.245	20.29	86.5	21.85	28.17	33.66
Si bottom cell module	775	1.927	2.831	73.0	4.25	5.49	

Acknowledgments

The authors express thanks to the NEDO for support.

References

[1] M. Yamaguchi et al., J. Phys. D. Appl. Phys. **51**, 133002 (2018).
[2] M. Yamaguchi et al., Prog. Photovol. **26**, 543 (2018).
[3] M. Yamaguchi et al., Prog. Photovolt. **29**, 684 (2021).
[4] M.A. Green et al., Progress in Photovoltaics **31**, 651 (2023).

HIGHLY EFFICIENT TANDEM PEROVSKITE SOLAR CELL USING BANDGAP TUNABLE $MAPB_XSN_{1-X}I_3$ PEROVSKITES VIA SANDWICH EVAPORATION TECHNIQUE

Jia-Ci Jhou[1], Hsin-Ting Lin[2], Ching-Fuh Lin[1, 2]*

1. Graduate Institute of Photonics and Optoelectronics, National Taiwan University, No. 1, Sec. 4, Roosevelt Road, Taipei, Taiwan, 10617 R.O.C.
2. Graduate school of Advanced Technology Program in Semiconductor Devices, Materials, and Hetero-integration, National Taiwan University, No. 1, Sec. 4, Roosevelt Road, Taipei, Taiwan, 10617 R.O.C.

Email:
1. r09941004@ntu.edu.tw
2. R10k43005@ntu.edu.tw
*lincf@ntu.edu.tw

ABSTRACT: Double-Sided Sandwich Evaporation Technique is used to fabricate mixed halide-based perovskite absorbing layer. By adjusting the halide ratio and B-site doping, we successfully produced $MAPbI_xBr_{3-x}$, $MAPb_xSn_{1-x}I_3$, and formulated the triple-junction tandem solar cell. The tandem solar cell containing subcells of $MAPbI_xBr_{3-x}$ (1.96 eV)/$MAPb_xSn_{1-x}I_3$ (1.31 eV)/Si (1.12 eV) shows a high power conversion efficiency of 26.04 %.
Keywords: Perovskite, Tandem solar cells, Tunable bandgap, Power conversion efficiency.

1 INTRODUCTION

Many reports have shed light on the importance of tin doping for perovskites to lower bandgap for its practical application in multi-junction solar cells and extend the light absorption limit of the perovskite layer. As a result, we fabricated pure $MASnI_3$ perovskite and partially substituted $MAPb_xSn_{1-x}I_3$ perovskite composition by varying the co-evaporation parameter of the middle layer of the sandwich structure to control the ratio of Pb and Sn under the same sandwich evaporation technique (SET) [1] condition effectively. The tin-based perovskite fabricated using double-sided sandwich evaporation technique (DS-SET) showed an improvement in the structural morphology and exhibited a significant V_{oc} value with a maximum 0.882 V for a co-evaporation condition of PbI_2 = 35 mins, SnI_2 = 15 mins. As a result, we formulated a tandem configuration with various bandgaps of 1.96/1.31/1.12(eV) based on $MAPb_xSn_{1-x}I_3$ as an intermediate layer. Finally, a power conversion efficiency (PCE) value of 26.04% is obtained.

2 EXPERIMENTS

2.1 Manufacturing of Perovskite Solar Cell Devices

The fabrication process included cleaning an ITO-coated substrate (15 mm × 15 mm × 0.7 mm) with various solvents and UV ozone treatment. A hole-transport layer (PEDOT: PSS) was spin-coated at 4000 rpm for 30 secs and annealed at 120 °C for 10 mins. Excess MAI was rinsed with IPA at 4000 rpm for 10 secs and annealed at 120 °C for 5 mins. For chlorine-based structures, DMSO was used as a solvent via the SE-SA [2] technique. An electron transport layer (PC60BM) in chlorobenzene (20 mg/mL) was spin-coated at 3000 rpm for 60 secs and vacuum-treated for 12 hours to volatilize CB. Finally, 5 nm BCP layer and 120 nm Ag electrodes were thermally evaporated to complete the solar cell structure.

2.2 Perovskite Thin Films

Through the conventional DS-SET method, we fine-tuned the manufacturing conditions. This involved adjusting the SET duration for the initial MAI layer and selecting the inorganic components (PbI2, PbBr2, SnI2) for the middle layers of the perovskite sandwich structure. This method allowed us to vary the halide ratio and obtain perovskite materials with precise bandgap characteristics. Consequently, we successfully produced two distinct perovskite materials, as outlined in Table I.

Table I: The manufacturing parameters employed at various stages of the perovskite sandwich structure to create different compositions.

Material	MAI (1st) (SET)	Inorganic material (2nd) (Thermal Evaporator)	MAI (3rd) (SET)
$MAPbI_xBr_{3-x}$	3 torr 15-60 min	$PbBr_2$ 185 nm	3 torr 30 min
$MAPb_xSn_{1-x}I_3$	3 torr 30 min	PbI_2 25/35 min SnI_2 10-25 min	

The tandem configuration was achieved by combining the perovskite and Si cell using a six-terminal methodology. Additionally, the bandgap was determined via a Tauc plot.

3 RESULTS AND DISCUSSION

Fig. 1 shows that the incorporation of appropriate amounts of Pb^{2+} into Sn^{2+} to form Pb-Sn mixed perovskite can reduce the probability of perovskite lattice destruction. $MAPb_xSn_{1-x}I_3$ thin films with different Pb-Sn ratios were achieved by simultaneous deposition of intermediate layers of PbI_2 and SnI_2 under different co-evaporation conditions. The Tauc plot shows the $MAPb_xSn_{1-x}I_3$ bandgap increases as the deposition time of PbI_2 increases under the same SnI_2 deposition time. Moreover, at the same PbI_2 deposition time, the $MAPb_xSn_{1-x}I_3$ bandgap increases as the deposition time of SnI_2 decreases.

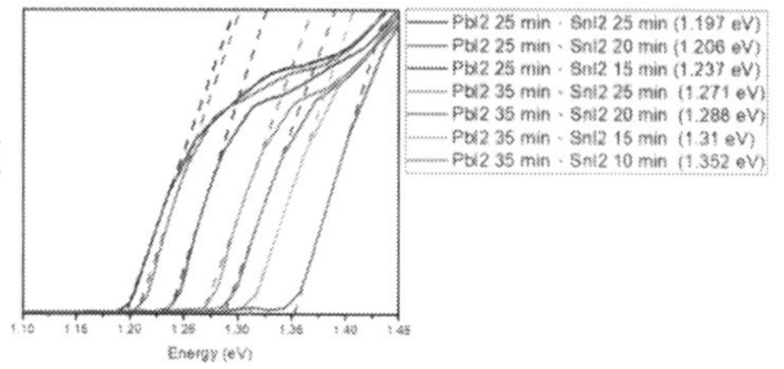

Figure 1: Tauc plot for bandgap determination of $MAPb_xSn_{1-x}I_3$ perovskite thin films under different co-evaporation conditions.

Furthermore, compared to $MASnI_3$, Fig. 2a shows that the doping of Pb in Sn conduces to reduce the surface defects of perovskite, resulting in an increase in V_{oc}. On the other hand, compared with $MAPbI_3$, changing the Pb-Sn ratio helps improve the perovskite crystallization. The optimal co-evaporation conditions are PbI_2 for 35 min and SnI_2 for 10 min, with a PCE of 17.26%.

Finally, we measured the J-V curve of $MAPb_xSn_{1-x}I_3$ as a middle cell, as shown in Fig. 2b, and combined $MAPbI_xBr_{3-x}$ as a top cell [3], a silicon solar cell as the bottom cell. The PCE of the triple-junction tandem solar cell measured through six-terminal technique, is up to 26.04%.

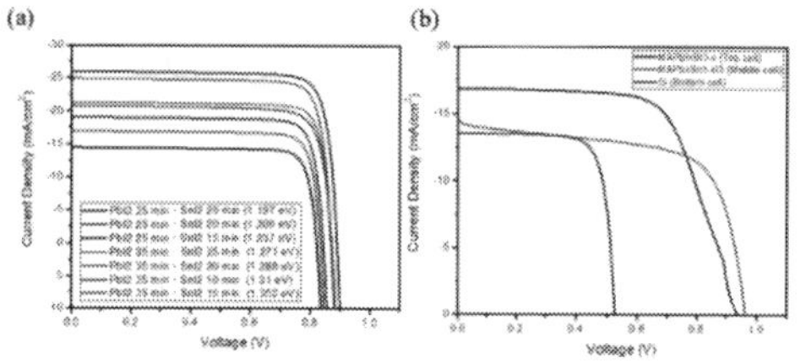

Figure 2: (a) J-V characteristic of perovskite solar cells for MAI (SET) - PbI_2/SnI_2 - MAI (SET) structure. (b) J-V characteristic of $MAPbI_xBr_{3-x}$/ $MAPb_xSn_{1-x}I_3$ /Si triple junction tandem solar cell.

4 CONCLUSIONS

In conclusion, by adjusting the halide ratio and B-site doping, we successfully produced $MAPbI_xBr_{3-x}$ and $MAPb_xSn_{1-x}I_3$, with bandgaps of 1.96 eV and 1.31 eV, respectively. The top and the middle cell can effectively separate the light absorption ratio to each sub-cell. Finally, we obtained a triple-junction tandem solar cell with a PCE as high as 26.04% by six-terminal measurement of $MAPbI_xBr_{3-x}$ / $MAPb_xSn_{1-x}I_3$ / Si.

5 ACKNOWLEDGEMENT

The authors gratefully acknowledge Ministry of Science and Technology in Taiwan for supporting the research with the contract numbers: MOST 108-2221-E-002-145-MY3, MOST 111-2221-E-002-080-MY3.

6 REFERENCE

[1] Jia-Ci Jhou, Ashish Gaurav, Chung-Han Chang, Ching-Fuh Lin." Enhanced Efficiency of Semitransparent Perovskite Solar Cells via Double-Sided Sandwich Evaporation Technique for Four Terminal Perovskite-Silicon Tandem Application." Nanomaterials 2022, 12 (9), 1569. https://doi.org/10.3390/nano12091569.

[2] Chun-Hsiao Kuan, Po-Tsun Kuo, Hui-Hung Shen, Cheng-Hung Hou, Jing-Jong Shyue, Li-Jen Huang, Ching-Fuh Lin." Sandwich Evaporation–Solvent Annealing Fabrication of Highly Crystalline MAPbIxCl3–x Perovskite Solar Cells." ACS Applied Materials & Interfaces 2021, 13 (38), 45355–45364. https://doi.org/10.1021/acsami.1c08780.

[3] Jia-Ci Jhou, Hsin-Ting Lin, Ashish Gaurav, Ching-Fuh Lin, " Bandgap tunable MAPbIxBr3-x Perovskite Thin Film Via Double-sided Sandwich Evaporation Technique For Solar Cells," Optics & Photonics Taiwan, International Conference2022 (OPTIC 2022), Paper number: 2022-SUN- P0802-P002, National Central University, Taoyuan, Taiwan, December 2-4, 2022.

40th European Photovoltaic Solar Energy Conference and Exhibition (EUPVSEC 2023)

Highly Efficient Tandem Perovskite Solar Cell Using Bandgap Tunable $MAPb_xSn_{1-x}I_3$ Perovskites via Sandwich Evaporation Technique

Jia-Ci Jhou[1], Hsin-Ting Lin[2], Ching-Fuh Lin[1,2*]

1. *Graduate Institute of Photonics and Optoelectronics, National Taiwan University*
2. *Graduate Institute of Advance Technology, National Taiwan University*

*lincf@ntu.edu.tw

Abstract

Double-Sided Sandwich Evaporation Technique is used to fabricate mixed halide-based perovskite absorbing layer. By adjusting the halide ratio and B-site doping, we successfully produced $MAPbI_xBr_{3-x}$, $MAPb_xSn_{1-x}I_3$, and formulated the triple-junction tandem solar cell. The tandem solar cell containing subcells of $MAPbI_xBr_{3-x}$ (1.96 eV)/$MAPb_xSn_{1-x}I_3$ (1.31 eV)/Si (1.12 eV) shows a high power conversion efficiency of 26.04 %.

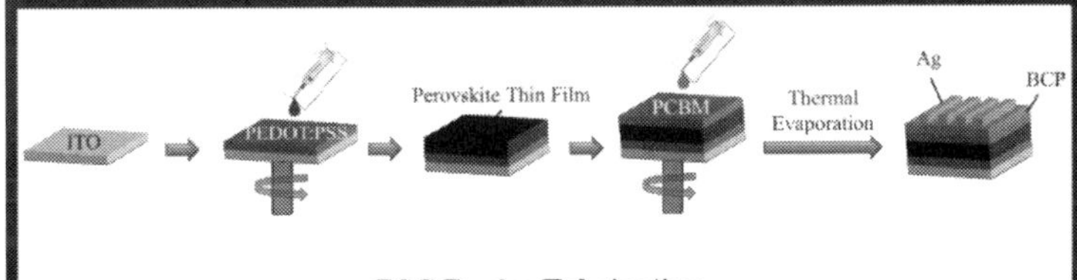

PSC Device Fabrication

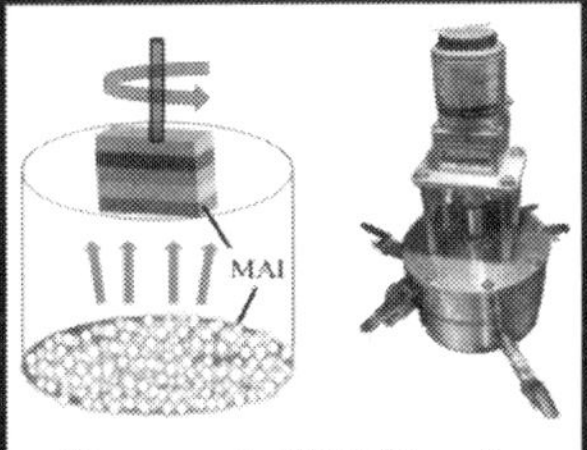

Homemade SET Chamber

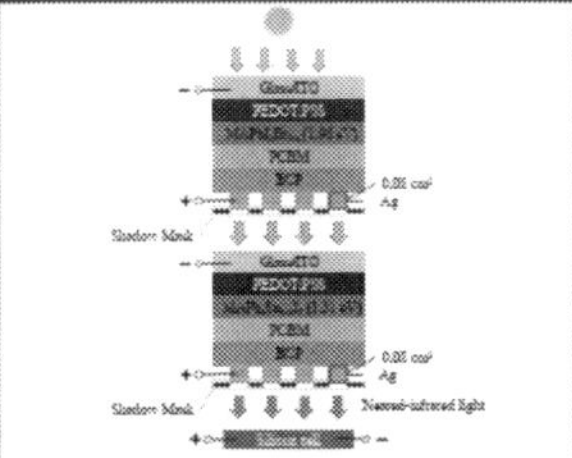

6T Tandem Configuration

Results and Discussion

Incorporating Pb^{2+} into Sn^{2+} to form Pb-Sn mixed perovskite effectively reduces lattice damage. We achieved varied Pb-Sn ratios in $MAPb_xSn_{1-x}I_3$ through simultaneous PbI_2 and SnI_2 deposition under different co-evaporation conditions, resulting in an increased bandgap (Tauc plot). Pb doping in Sn enhances V_{oc} and crystallization compared to $MASnI_3$. Optimal co-evaporation (35 mins PbI_2, 10 mins SnI_2) yields a 17.26% PCE. Furthermore, the J-V curve for the triple-junction tandem solar cell shows an impressive PCE of up to 26.04%. These findings emphasize the potential of bandgap-tunable perovskites for advancing solar cell technology.

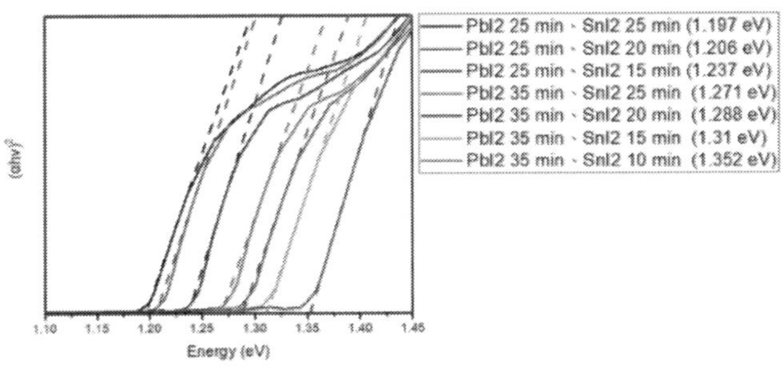

Fig 1. Tauc plot for bandgap determination of $MAPb_xSn_{1-x}I_3$ perovskite thin films under different co-evaporation conditions.

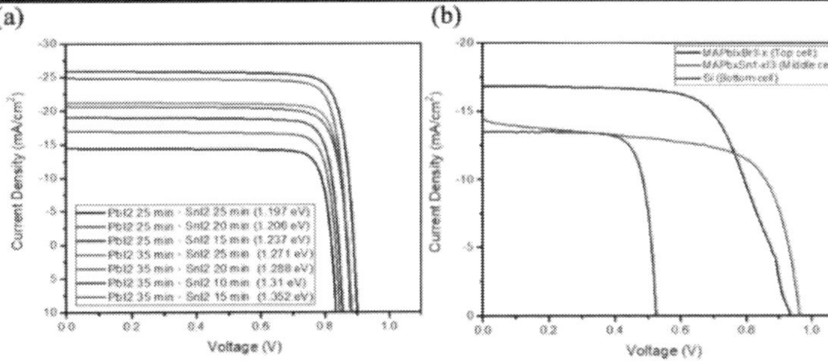

Fig 2. (a) J-V characteristic of perovskite solar cells for MAI (SET) - PbI_2/SnI_2 - MAI (SET) structure. (b) J-V characteristic of $MAPbI_xBr_{3-x}$/ $MAPb_xSn_{1-x}I_3$ /Si triple junction tandem solar cell.

Conclusion

In conclusion, by adjusting the halide ratio and B-site doping, we successfully produced $MAPbI_xBr_{3-x}$ and $MAPb_xSn_{1-x}I_3$, with bandgaps of 1.96 eV and 1.31 eV, respectively. The top and the middle cell can effectively separate the light absorption ratio to each sub-cell. Finally, we obtained a triple-junction tandem solar cell with a PCE as high as 26.04% by six-terminal measurement of $MAPbI_xBr_{3-x}$ / $MAPb_xSn_{1-x}I_3$ / Si.

References

1. Jia-Ci Jhou, Ashish Gaurav, Chung-Han Chang, Ching-Fuh Lin. " Enhanced Efficiency of Semitransparent Perovskite Solar Cells via Double-Sided Sandwich Evaporation Technique for Four Terminal Perovskite-Silicon Tandem Application. " Nanomaterials 2022, 12 (9), 1569. https://doi.org/10.3390/nano12091569.
2. Jia-Ci Jhou, Hsin-Ting Lin, Ashish Gaurav, Ching-Fuh Lin," Bandgap tunable $MAPbI_xBr_{3-x}$ Perovskite Thin Film Via Double-sided Sandwich Evaporation Technique For Solar Cells," Optics & Photonics Taiwan, International Conference2022 (OPTIC 2022), Paper number: 2022-SUN-P0802-P002, National Central University, Taoyuan, Taiwan, December 2-4, 2022

40th European Photovoltaic Solar Energy Conference and Exhibition

This presentation was selected by the Sc. Committee of the EU PVSEC 2023 for submission of a full paper to one of the EU PVSEC's collaborating peer-reviewed journals.

MONOLITHIC 3-TERMINAL PEROVSKITE/SILICON HBT-BASED TANDEM COMPATIBLE WITH INDUSTRIAL SILICON BOTTOM CELLS: A THEORETICAL STUDY

Gemma Giliberti, Matteo Cagnoni and Federica Cappelluti
Department of Electronics and Telecommunications, Politecnico di Torino
Corso Duca degli Abruzzi 24, 10129 Torino, Italy
E-mail: gemma.giliberti@polito.it, matteo.cagnoni@polito.it, federica.cappelluti@polito.it

ABSTRACT: Perovskite/silicon tandem solar cells are among the most promising solutions to overcome the efficiency bottleneck of single-gap Si cells. Multi-terminal configurations are interesting alternatives to the more studied series connected two-terminal tandem, because they allow for looser band gap requirements and higher energy yield in the field. Among three-terminal architectures, the heterostructure bipolar transistor (HBT) solar cell offers a promising approach for the integration of the top perovskite cell with both-side contacted silicon cells, which are the industry standard. Material engineering makes it possible to prevent minority carriers transport across the interface layers between the two sub-cells, allowing them to work independently. However, the introduction of the middle contact at the interface layers requires the adoption of interdigitated contact grids for current collection, whose associated optical and resistive losses need to be minimized to not jeopardize the high efficiency potential of the intrinsic device. In this work, we use physics-based and circuit-level simulations to address such issues in the perspective of developing high efficiency large-area perovskite/silicon tandems with 3T-HBT structure.

Keywords: perovskite/silicon tandems, 3-terminal tandems, heterostructure bipolar transistor, current collecting grid photovoltaics

1 INTRODUCTION

Perovskite/silicon solar cells combine high energy efficiency and low cost and are expected to enter the market soon [1], provided that their long-term stability issues can be significantly mitigated. Research has been mostly focused on 2-terminal (2T) devices, but multi-terminal 3T and 4T configurations are drawing increasing attention because of their higher flexibility in terms of perovskite band gap. Indeed, this provides higher degrees of freedom to solve stability issues [2], and less sensitivity to spectral and temperature fluctuations, and therefore higher predicted energy yield in the field [3], [4].

3T tandems reported so far are mainly based on the idea of using an interdigitated back contact (IBC) silicon bottom cell [5][6]. However, solutions compatible with both-side contacted silicon cells, such as Al-BSF, PERX, and heterojunction [7, 8] would be of great interest, because these are the preferred choice in industry [9, 10] because of their lower complexity. Possible 3T tandem approaches could be pursued either by introducing an intermediate transparent conductive layer between the two sub-cells, as done in [11], or by adopting the heterostructure bipolar transistor (HBT) architecture [12]. The HBT architecture enables the seamless connection of two sub-cells of opposite electrical polarity, by exploiting three semiconductor regions with alternating doping and each having its own electrical contact. Since the seminal work in [12], significant progress has been reported on III-V HBT solar cells [13], that proves the feasibility of this idea. In the context of perovskite/silicon (PVK/Si) technology, we have proposed a few designs of 3T-HBT with planar [14] and textured [15] Si bottom cells, predicting high theoretical efficiency on the basis of optical and transport simulations.

However, the HBT structure requires the introduction of a middle contact at the interface layers between the two sub-cells, and the adoption of interdigitated grids for current collection. These cause optical and resistive losses that need to be carefully assessed, especially in view of the rather limited (when compared to crystalline semiconductors) electrical conductivity of the materials commonly exploited as interfacial layers, such as organic electron and hole transport layers of the perovskite cell and hydrogenated amorphous silicon of heterojunction silicon cells.

In this work, we lay the groundwork for this, by studying HBT architectures with interdigitated contact layouts by means of a mixed-mode approach that combines transport and optical simulations for characterizing the intrinsic behavior of the HBT multilayer stack with circuit-level simulations for the analysis of shading and resistive loss associated to the metal grids. We study tandem architectures based on homo- and hetero-junction silicon cells to analyze to what extent the middle contact could induce resistive mediated cross-talk between the junctions and discuss prospects for the scalability of these devices to large areas.

2 DEVICE AND MODELING

Here, we consider a *p-n-p* HBT tandem made on a c-Si homojunction cell whose structure is depicted in Fig. 1(a). To realize the three-terminal connection, we introduce top interdigitated contact (TIC) grids for the emitter and base terminals, as shown in Fig. 1(b).

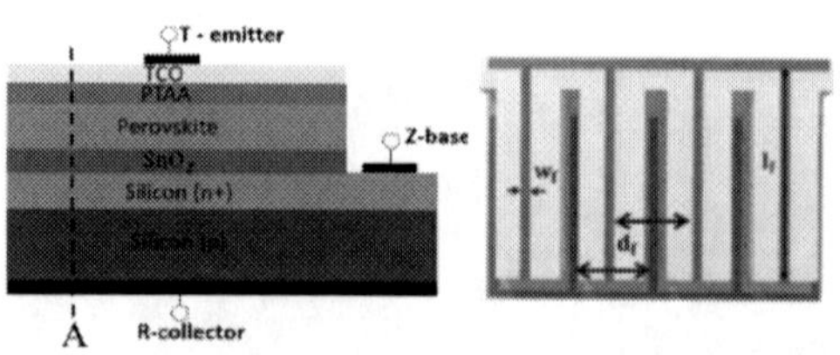

Figure 1: Example of structure of *p-n-p* 3T-HBT tandem based on *n-i-p* PVK cell combined with homojunction *n-p* Si bottom cell (a), TIC metal layout (b).

We start by analyzing the "intrinsic" device, i.e. the multilayer stack identified by cutline A in Fig. 1(a). The tandem structure is the following: from the top, thin conductive oxide (TCO) (34 nm), PTAA (11 nm) as hole transport layer (HTL), PVK (480 nm, Eg=1.55eV), SnO_2 (25 nm) electron transport layer (ETL), *n*-Si (150 nm), and 150 μm thick *p*-type c-Si. The cell architecture is completed by a 92 nm thick MgF antireflection coating. The alternate doping layers of the PVK and Si sub-cells form a structure analogous to that one of a bipolar transistor (*p*-Emitter, *n*-Base, *p*-Collector).

We use quasi-1D physics-based simulations (Transfer Matrix Method for the optical model coupled with Poisson-Drift-Diffusion for the transport model) to characterize the photovoltaic behavior of the intrinsic device.

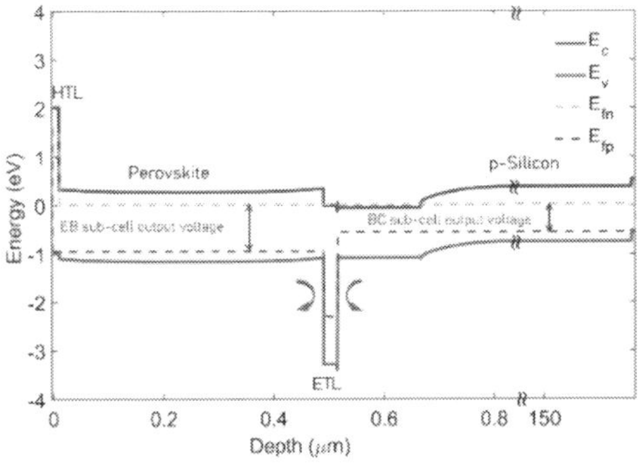

Figure 2: Energy band diagram of the *p-n-p* 3T-HBT with each junction biased at maximum power point. The dashed yellow and pink lines are the electron and hole quasi-Fermi levels, respectively.

Fig. 2 reports the device energy band diagram at maximum power point (MPP). One can see that the energy band alignment at the E/B (i.e. PVK/ETL) and B/C (i.e. n^+-Si/*p*-Si) heterointerfaces allows collection of electrons at the base contacts, whereas *repels* holes photogenerated in each sub-cell towards their respective contacts, i.e. emitter and collector, avoiding any injection from one junction to the other. This enables the perovskite and silicon layers to support a different splitting of electron and hole quasi-Fermi-levels, coherently with their own bandgap. The PVK and Si sub-cells reach open circuit voltage of 1.06 and 0.64, respectively. Owing to the independent operation of the two sub-cells, the tandem MPP takes place when each sub-cell is biased at MPP, with an overall efficiency higher than 29 % under AM1.5G.

Although the *intrinsic* performance of the HBT device is suitable for high power conversion efficiency, considering a realistic device layout as that one depicted in Fig. 1, two main effects must be carefully analyzed:

(i) the possible sub-cells cross-talk due to the parasitic resistance induced by the lateral transport across the thin base region (Fig. 1(b));

(ii) the implications of optical and resistive losses caused by the additional contact grid in terms of scalability to large areas.

To assess these effects, we resort to circuit-level simulations with the equivalent circuit model in Fig. 3. We extract the compact model of the intrinsic device by fitting the photovoltaic characteristics calculated from the physics-based simulation.

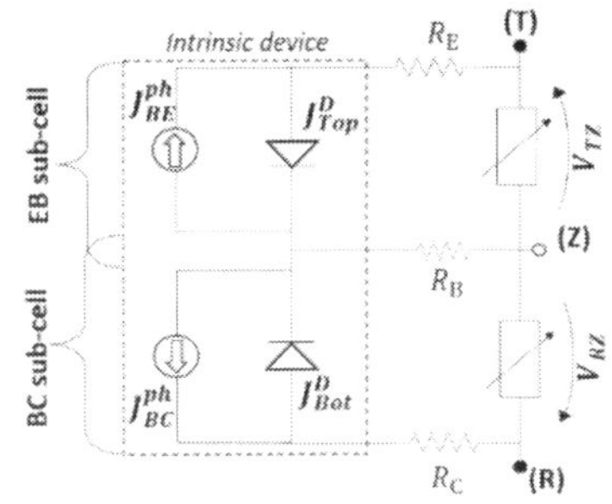

Figure 3: Equivalent circuit of the 3T-HBT tandem. Emitter, base and collector contacts are named as T, Z and R respectively, following the notation in [16].

The lumped parasitic resistances R_E R_B and R_C are estimated by analytical models of the grid unit cell [17], including lateral thin film and finger conduction, and contact resistance. In particular, with reference to the device structure in Fig. 1, the sheet resistance modeling the lateral transport across emitter and base is calculated as the parallel connection of the sheet resistances of their constituting thin film layers, e.g. ETL/n-Si for the base region of the HBT in Fig.1(a). The circuit simulation also accounts for the shadow loss due to the front metal grids. More details of the model will be discussed in the full work [18]. As seen in the equivalent circuit in Fig. 3, the impact of the base parasitic resistance, R_B, on the tandem performance is going to be more critical than the one of the emitter and collector resistances, because of the higher current flow across R_B.

3 RESULTS

To analyze the strength of the sub-cells cross-talk induced by the common base resistance, we start by considering a small-area device, with finger spacing and length of 1.5 mm and 1.5 cm, respectively.

Fig. 4 shows the interdependence of the MPP voltages of the two sub-cells and the corresponding efficiency reduction for several values of base sheet resistance, R_{sh}^{Base}.

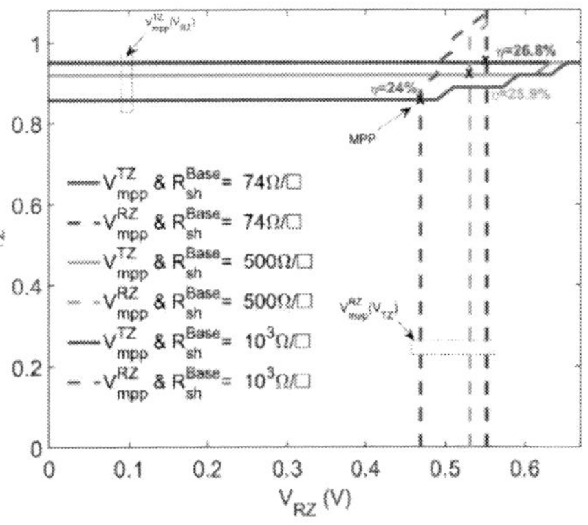

Figure 4: EB and BC MPP voltages (V^{TZ}_{mpp} *and* V^{RZ}_{mpp}) as a function of V^{RZ} and V^{TZ} voltages.

For the tandem PVK/Si-homojunction tandem of Fig. 1(a), we estimated a base sheet resistance of ≈ 74 Ω/□, limited by the lateral transport across the (n^+)c-Si layer. Higher values might be representative of other tandem architectures on heterojunction Si cells, due to the high sheet resistance of the thin a-Si:H layers (up to 10^5 Ω/□). One can observe that the two sub-cells remain independent up to the MPP, regardless of the value of R_B. The R_B mediated cross-talk onsets only at voltages higher than the MPP one and causes an increase of the open circuit (OC) voltage as the bias of the other sub-cell grows towards OC, as observed experimentally in III-V HBT tandems [19]. Therefore, the R_B mediated interdependence between the sub-cells has a marginal impact on the attainable efficiency. On the other hand, for each sub-cell, V_{mpp} becomes smaller as R_B increases, because of the series resistance effect. In this sense, it is important to assess the implications in terms of scalability of the technology.

The efficiency reduction due to shadow and resistive losses caused by the metal grids is studied in Fig. 5 and Fig. 6. In Fig. 5, we analyze the impact of R_{sh}^{Base} in terms of efficiency loss for finger distance in the range [0.5 ÷10] mm, with finger length of 1.5 cm. We find that for a R_{sh}^{Base} up to 100 Ω/□, the efficiency loss can be minimized to less than 3% with finger distance in the range [1.5 ÷ 2.9] mm, in line with typical finger distance found in Si cells. Instead, by considering higher values of $R_{sh}^{Base} \approx 1000$ Ω/□ (representative of PVS tandems on heterojunction c-Si cells), the finger spacing shall be reduced to about 1 mm, with a minimum efficiency penalty of ≈ 5%. Lower finger distance implies a higher penalty due to the dominant effect of shadow losses.

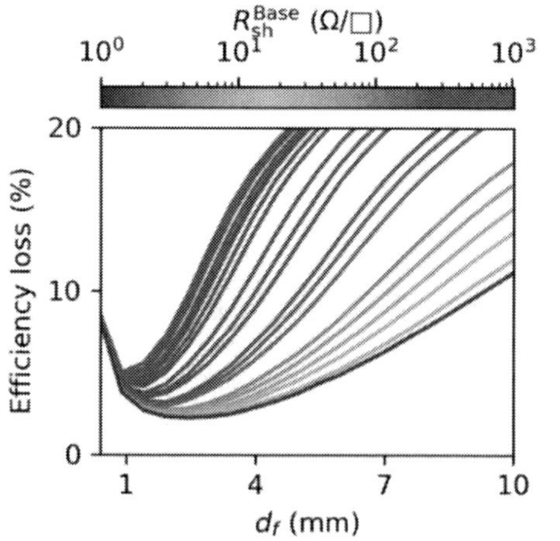

Figure 5: Efficiency loss as a function of the finger distance (d_f) and base sheet resistance (R_{sh}^{Base}).

Fig. 6 shows how the efficiency penalty scales with finger length and spacing, for an HBT with homojunction Si cell. With an optimized layout, efficiency loss remains lower than 5% for finger lengths up to about 8 cm, which is promising in the perspective of scaling up to large areas. The HBT with heterojunction Si cell is more critical, because of the low conductivity of the amorphous silicon layers, but still manageable by suitable design layouts, as discussed in the full paper [18].

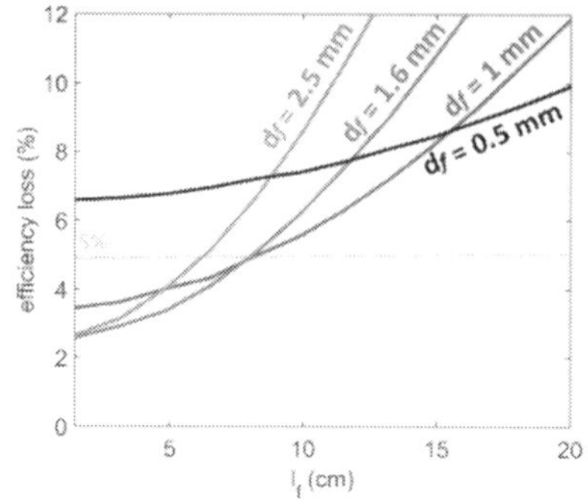

Figure 6: Efficiency loss as a function of finger length (l_f) and spacing (d_f) for a perovskite/silicon 3T-HBT on homojunction Si cell ($R_{sh}^{Base} \approx 74\ \Omega/\square$).

4 ACKNOWLEDGEMENTS

The authors acknowledge Francesco di Giacomo, Centre for Hybrid and Organic Solar Energy (CHOSE), University of Rome Tor Vergata for useful discussions. This research was partly funded by the ECSEL Joint Undertaking (JU) under grant agreement No. 101007247. The JU receives support from the European Union's Horizon 2020 research and innovation programme and Finland, Austria, Germany, Ireland, Iceland, Italy, Sweden, and Switzerland.

5 REFERENCES

[1] International technology roadmap for photovoltaic (ITRPV), 14th edition, (2023)

[2] Reddy, S. H., Di Giacomo, F., & Di Carlo, A. (2022). Low-temperature-processed stable perovskite solar cells and modules: a comprehensive review. Advanced Energy Materials, 12(13), 2103534.

[3] Gota, F., Langenhorst, M., Schmager, R., Lehr, J., & Paetzold, U. W. (2020). Energy yield advantages of three-terminal perovskite-silicon tandem photovoltaics. Joule, 4(11), 2387-2403.

[4] Warren, E. L., Deceglie, M. G., Rienäcker, M., Peibst, R., Tamboli, A. C., & Stradins, P. (2018). Maximizing tandem solar cell power extraction using a three-terminal design. Sustainable Energy & Fuels, 2(6), 1141-1147.

[5] Wagner, P., Tockhorn, P., Hall, S., Albrecht, S., & Korte, L. (2023). Performance of Monolithic Two-and Three-Terminal Perovskite/Silicon Tandem Solar Cells Under Varying Illumination Conditions. Solar RRL, 7(5), 2200954.

[6] Kanda, H., Dan Mihailetchi, V., Gueunier-Farret, M. E., Kleider, J. P., Djebbour, Z., Alvarez, J., ... & Connolly, J. P. (2022). Three-terminal perovskite/integrated back contact silicon tandem solar cells under low light intensity conditions. Interdisciplinary Materials, 1(1), 148-156.

[7] Messmer, C., Goraya, B. S., Nold, S., Schulze, P. S., Sittinger, V., Schön, J., ... & Hermle, M. (2021). The race for the best silicon bottom cell: Efficiency and cost evaluation of perovskite–silicon tandem solar cells. Progress in Photovoltaics: Research and Applications, 29(7), 744-759.

[8] Messmer, C., Schön, J., Lohmüller, S., Greulich, J., Luderer, C., Goldschmidt, J. C., ... & Hermle, M. (2022). How to make PERC suitable for perovskite–silicon

tandem solar cells: A simulation study. Progress in Photovoltaics: Research and Applications, 30(8), 1023-1037.

[9] Ballif, C., Haug, F. J., Boccard, M., Verlinden, P. J., & Hahn, G. (2022). Status and perspectives of crystalline silicon photovoltaics in research and industry. Nature Reviews Materials, 7(8), 597-616.

[10] International technology roadmap for photovoltaic (itrpv), 13th edition, april (2022)

[11] Park, I. J., Park, J. H., Ji, S. G., Park, M. A., Jang, J. H., & Kim, J. Y. (2019). A three-terminal monolithic perovskite/Si tandem solar cell characterization platform. Joule, 3(3), 807-818.

[12] Martí, A., & Luque, A. (2015). Three-terminal heterojunction bipolar transistor solar cell for high-efficiency photovoltaic conversion. Nature communications, 6(1), 6902.

[13] Antolín, E., Zehender, M. H., Svatek, S. A., Steiner, M. A., Martínez, M., García, I., ... & Martí, A. (2022). Progress in three-terminal heterojunction bipolar transistor solar cells. Progress in Photovoltaics: Research and Applications, 30(8), 843-850.

[14] Giliberti, G., Di Giacomo, F., & Cappelluti, F. (2022). Three Terminal Perovskite/Silicon Solar Cell with Bipolar Transistor Architecture. Energies, 15(21), 8146.

[15] Giliberti, G., & Cappelluti, F. (2022, March). Physical simulation of perovskite/silicon three-terminal tandems based on bipolar transistor structure. In Physics, Simulation, and Photonic Engineering of Photovoltaic Devices XI (Vol. 11996, p. 1199602). SPIE.

[16] Warren, E. L., McMahon, W. E., Rienacker, M., VanSant, K. T., Whitehead, R. C., Peibst, R., & Tamboli, A. C. (2020). A taxonomy for three-terminal tandem solar cells. ACS Energy Letters, 5(4), 1233-1242.

[17] Mette, A. (2007). New concepts for front side metallization of industrial silicon solar cells (Doctoral dissertation, Freiburg (Breisgau), Univ., Diss., 2007).

[18] Giliberti G, Cagnoni M. and Cappelluti F. (2023). Monolithic 3-terminal perovskite/silicon HBT-based tandem compatible with both-side contact silicon cells: a theoretical study. EPJ Photovoltaic. Accepted for publication. 10.1051/epjpv/2023024.

[19] Zehender, M. H. (2022). Three-terminal heterojunction bipolar transistor solar cells= Células solares de tres terminales tipo transistor bipolar de heterounión (Doctoral dissertation, Telecomunicacion).

Numerical Simulation: Design and Optimization of $CsSnI_3$ and $Cs_3Sb_2Br_9$ based Multijunction Solar Cell

Arka Samanta[1], Shivendra Yadav[1], Sushil K. Pandey[2], Gaurav Siddharth[3], Vivek Garg[1]
[1]Department of Electronics Engineering, SVNIT, Surat-395007, India
[2]Department of Electronics Engineering, NIT Surathkal-575025, India
[3]Department of Electronics Engineering, NIT Calicut-673601, India
*Email:vivekgarg0101@gmail.com

ABSTRACT: In this work, two lead free materials based multi-junction solar cell has been investigated. First the simulation model is developed from the experimental data reported on $Cs_3Sb_2Br_9$ and $CsSn_3I$ based devices. Simulated model is well matched with the experimental values of performance parameters for $Cs_3Sb_2Br_9$ (PEC~15.69 %) and $CsSn_3I$ (PEC~7.5 %). Further, the baseline models were optimized to improve the performance parameters by optimizing a) absorber, b) interface defect densities, and c) capture cross section areas. The PEC observed for the optimized $Cs_3Sb_2Br_9$ and $CsSnI_3$ are 19.15 % and 27.15 %, while Jsc of 24.8 mA/cm^2 and 14.34 mA/cm^2, respectively. As the Jsc are different, current matching is performed by reducing the thickness of the $CsSnI_3$ from 1000 nm to 350 nm to achieve the current density of 14 mA/cm^2. Using the current matched models, the design and simulation of MJSC with $Cs_3Sb_2Br_9$ as the top cell and $CsSn_3$ as the bottom cell has been performed. For the multijunction solar cell, the extracted final efficiency of 34.1% is observed.

1 INTRODUCTION

As the world moves towards green and carbon free source of energy, PV technology becomes the best choice to look upon. For solar energy to get used at a much wider range, more and more efficient solar cells are required so that higher energy can be generated per unit area. Based upon theoretical calculations done using Shockley-Quisser limit, a single junction solar cell can theoretically reach a maximum power conversion efficiency of around 33%. Practically the highest efficiency achieved by a single junction solar cell in research is around ~26 % till date. [10] Multijunction solar cells or tandem solar cells consists multiple solar cell with varying band gaps of the absorber stacked one over the other in such a way that the top cell i.e. the cell which faces the incident light is of the largest band gap, the subsequent cells have a lower band gaps and the bottom cell which faces in the direction opposite to the incident light has the lowest band gap. The photons which are having energies greater than the bandgap of the material are absorbed by the solar cell and the lower energy photons get transmitted through the solar cell unabsorbed (called as absorption loss). Now, if the energy of the photons incident on the solar cell are much higher than its band gap than the electron from the valance band jumps into the conduction band to a level much higher than the bottom of the conduction band and then eventually releases some energy to settle down to the bottom of the conduction band to increase stability resulting in thermalization loss. The working principal of MJSCs minimizes both these losses as the unabsorbed photons in the upper cells gets absorbed in the lower cells (minimizing absorption loss) and the high energy photons barely reach the lower band gap bottom cells as they mostly get absorbed in the upper cells (preventing thermalization loss) and therefore absorbing broader part of the solar spectrum effectively and hence making it possible to increase the efficiency beyond the single junction Schokley-Quisser limit. Apart from efficiency there are other advantages of MJSCs like have a greater resistance to radiation as only the top cell gets effected mostly by the incoming radiation which is another reason why they are preferred in space missions. Till date most of the solar cell producing plants run using energy produced from conventional sources of energy like thermal power plants etc. These higher efficiency solar cells would help us to make a faster transition to green energy and a time may come when energy will be made primarily from solar.

Despite the fact that MJSCs have a huge potential to generate more power per unit area than Single Junction cells still they are not used in the commercial market due to their high design and fabrication cost. Their application is limited to a very small field like space technology where higher efficiency is the primary criterion. In this work, design and optimization of a MJSC using two lead free perovskites ($Cs_3Sb_2Br_9$ and $CsSn_3I$) sub cells, is thoroughly investigated for highly efficient devices.

2 RESULTS AND DISCUSSIONS

2.1 Baseline models of $CsSnI_3$ and $Cs_3Sb_2Br_9$ Solar cells

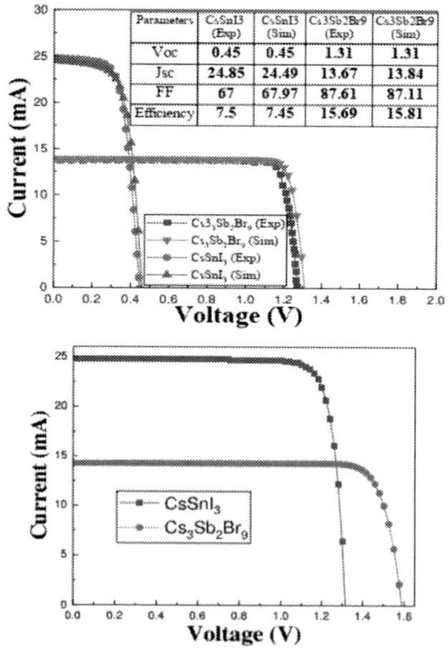

Parameters	CsSnI3 (Exp)	CsSnI3 (Sim)	Cs3Sb2Br9 (Exp)	Cs3Sb2Br9 (Sim)
Voc	0.45	0.45	1.31	1.31
Jsc	24.85	24.49	13.67	13.84
FF	67	67.97	87.61	87.11
Efficiency	7.5	7.45	15.69	15.81

Fig. 1. a.) $CsSnI_3$ and $Cs_3Sb_2Br_9$ baseline JV curve comparison (inset shows the performance parameters value for experimental and simulated devices, b) J-V characteristics of the optimized $CsSnI_3$ and $Cs_3Sb_2Br_9$ cells.

First, baseline models for both $Cs_3Sb_2Br_9$ and $CsSn_3I$ based devices has been developed using the material properties populated in Table I. [1, 7] Device structure for

Table I: Material parameters for Baseline development

	$CsSnI_3$ baseline model				$Cs_3Sb_2Br_9$ baseline model			
Parameters	**P3HT**	**$CsSnI_3$**	**TiO_2**	**FTO**	**Spiro-OMeTAD[7]**	**$Cs_3Sb_2Br_9$ [6]**	**TiO_2 [7,8]**	**FTO**
Thickness	0.03[1]	1[1]	0.1[1]	0.3[3]	150	1000	100	0.303[1]
Bandgap(eV)	2[5]	1.31[1]	2.3[2]	3.5[3]	3.1	2.0	3.2	3.5[1]
Electron affinity(eV)	3.5[5]	4.47[1]	4.2[2]	4.[3]	2.2	3.1	4	4.[3]
Dielectric (ε_r)	3.53[4]	10.59[1]	10[2]	9[3]	3.0	10	10	9[3]
DOS_{CB} (cm^{-3})	1E19	1.58E19[1]	2E17[2]	1E19[3]	2.5E+18	1E14	2E+18	1E19[3]
DOS_{VB} (cm^{-3})	1E19	1.47E18[1]	6E17[2]	1E18[3]	1.8E+19	1E14	2E+18	1E18[3]
μe(cm^2 /Vs)	2.5E-1	4.37[1]	1E1[2]	2E1[3]	2E-4	22.1	250	2E1[3]
μh(cm^2 /Vs)	1E-1	4.37[1]	2.5E1[2]	1E1[3]	2E-4	1.1	25	1E1[3]
Accep. Con. (cm^{-3})	1E10	1E16[1]	9E16[2]	1E18[3]	1E+15	1E+9	0	0
Donor Con. (cm^{-3})	1E15	1E16[1]	0	0	1E+19	1E9	1E19	1E18
Defect density(cm^{-3})		1E15				1E12	1E15	

both the materials consists of FTO/TiO_2/$CsSnI_3$/P_3HT and FTO/TiO_2/$Cs_3Sb_2Br_9$/Spiro-OmeTAD. The J-V characteristics of experimental and the simulation results for both $Cs_3Sb_2Br_9$ and $CsSn_3I$ based devices are depicted in Fig 1 a), which confirms the validity of the developed device model for the simulation. Next, the baseline models are further optimized by varying Absorber and Interface properties for improved performance. Further, as the device will be connected in the series manner, the proper current matching of the cells are required, as the cell with the lower current density value will limit the overall current of the multijunction device. Further, both sub-cell will be combined using an interconnecting layer to form the multijunction solar cell, and performance will be analyzed.

TABLE II: Device performance parameters of the optimized $CsSnI_3$ and $Cs_3Sb_2Br_9$ based solar cells

	$CsSnI_3$ solar cell		Cs_3Sb_2Br9 Solar cell	
Parameters	Baseline model	Optimized model	Baseline model	Optimized model
Voc (V)	0.45	1.316	0.733	1.58
Jsc (mA/cm^2)	24.49	24.82	3.95	14.4
FF (%)	67.97	83	82.05	83.9
PCE (%)	7.45	27.15	3.39	19.14

2.2 Sub-cell Optimization

The performance of the $CsSnI_3$ device is optimized by reducing its interface defects density and capture cross section between the $CsSnI_3$ and P_3HT layer from 9 $x10^{11}$ to 10^8 and 2.1 x 10^{-16} to 10^{-19} respectively. Further, the absorber defect density is altered from 10^{16} to 10^{10} cm^{-3} and capture cross section from 10^{15} to 10^{16}. For the optimized $CsSnI_3$ based device observed performance parameters are Voc: 1.31 V, Jsc: 24.8 $mAcm^{-2}$, FF: 83 % and PEC: 19.14 % as depicted in Fig 1 b). On the other hand, we improve the $Cs_3Sb_2Br_9$ cell by reducing its absorber ETL interface defects to 10^9 from $1.5x10^{12}$ cm^{-3}, absorber effect density from 10^{12} to 10^{10} and the thickness of absorber from 1000 nm to 1500 nm. The optimized device parameters are: Voc: 1.58V, Jsc: 14.34 $mAcm^{-2}$, FF: 83.9 %, and PEC: 15.08% as shown in Fig 1 b).

2.3 Current Matching

After optimizing the individual cells, the next step is to perform the current matching, as the current in the series connection is limited by the device with the lower current. Therefore, the current matching is achieved by varying the absorber thicknesses, and keeping the other parameters of the material constant. After repetitive simulations, it is observed that the J_{sc} top cell saturates to 14 mA/cm^2 on further increasing the thickness. Therefore, the absorber thickness of the bottom cell is reduced from 1000 nm to 350nm to achieve a J_{sc} of 14.55 mA/cm^2 as populated in Table II. Reducing the thickness of the absorber also means using less material, which eventually reduces the cost of the solar cell. After current matching the bottom cell has an individual efficiency of 15.1%. Further, we use the current matched devices to design the final Multijunction solar cell.

2.4 Multijunction Device Structure & Band Alignment

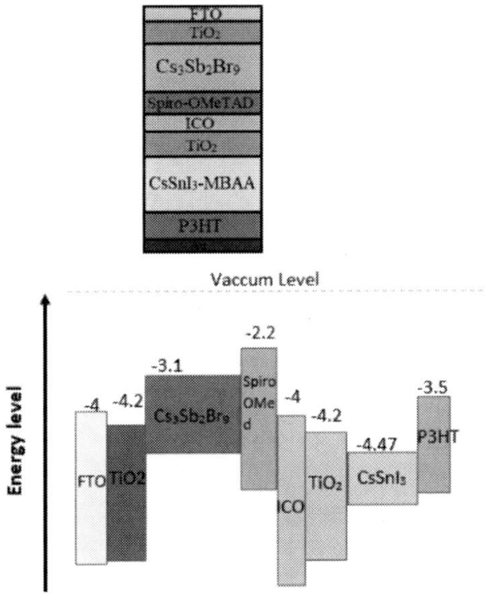

Fig. 3. a) $CsSnI_3$ - $Cs_3Sb_2Br_9$ MJSC structure, b) $CsSnI_3$ - $Cs_3Sb_2Br_9$ MJSC band alignment

The multijunction cell is designed with Cs_3Sb_2Br as the top cell as it has higher band gap (2eV), than $CsSnI_3$ which has a lower band gap of 1.31 eV, as shown in Fig 3 a). The reason behind that is the solar spectrum consist of a wide wavelength range. The higher energy photons (i.e. energy more than band gap of top cell) get absorbed in the top cell generating electron-hole pair, while the lower energy photon gets transmitted through the top cell and gets absorbed in the bottom cell. Only those photons get absorbed whose energy is less than the bandgap of the material on which its incident. Therefore, the effecive light absorbtion increases and a greater efficiency is observe in MJSCs. But if we would kept the $CsSnI_3$ as the top cell than the high energy lights would had not been able to reach the bottom cell neither they would have effectively get absorved by the narrow band gap $CsSnI_3$ cell due to thermalization cell. The band alignment at different junctions of the multijunction solar cell are shown in Fig 3 b).

2.5 Device Performance of Multijunction Solar cells

Fig. 4 depicts the J-V curve of the proposed $CsSnI_3$/ $Cs_3Sb_2B_9$ based multi junction solar cell, exhibits the Jsc of 14.33 mA/cm^2. For tandem solar cell, the observed value of open circuit voltage is the summed up values of the top (1.58 V) and the bottom (1.28) sub-cell, results in the Voc of 2.87 V. The observed FF is approx equals to the average value of the individual sub-cells FF. The individual device performance along with the multijunction solar cell is populated in the table III. The observed device power conversion efficiency of the multijunction solar cell is 34 %, which is almost equals to the summation of the individual solar cell PEC.

TABLE 3: Solar cell parameters of the current matched $CsSnI_3$ and $Cs_3Sb_2Br_9$ based solar cells.

Parameters	Current Matched $CsSnI_3$ model	Current Matched $Cs_3Sb_2Br_9$ model	MJSC
Voc (V)	1.28	1.58	**2.876**
Jsc (mA/cm^2)	14.55	14.34	**14.33**
FF (%)	80.39	83.9	**82.64**
PCE (%)	15.08	19.14	**34.07**

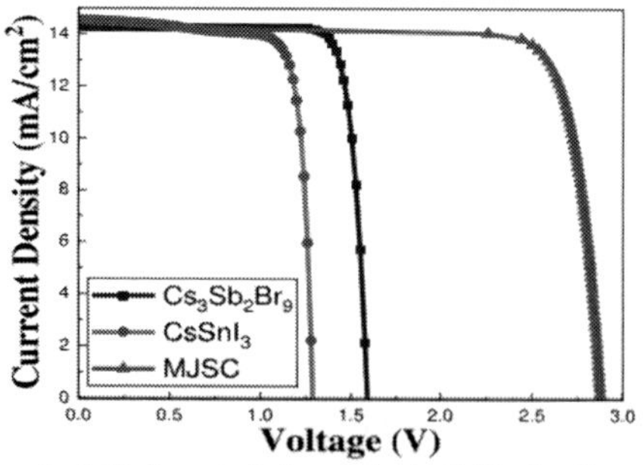

Fig. 4. J-V characteristics of the individual cells and the $CsSnI_3$-$Cs_3Sb_2Br_9$ MJSC cell

3 CONCLUSION

In this work, $CsSnI_3$ and $Cs_3Sb_2Br_9$ based multijunction solar cell is investigated for highly efficient photovoltaic device applications. In this regard, first, the baseline models for $CsSnI_3$ and $Cs_3Sb_2Br_9$ cells with the PEC of 7.5 % and 15.6 %, respectively, are developed. Next, the baseline model is further optimized by varying the properties of absorber, HTL, ETL, and respective interfaces, which results in the PEC of 27 % and 19 % for $CsSnI_3$ and $Cs_3Sb_2Br_9$ cells. The optimized bottom cell ($CsSnI_3$) cell has Voc: 1.31 V, Jsc: 24.8 mAcm^{-2}, FF: 83 % and PEC: 19.14%, while the top cell $Cs_3Sb_2Br_9$ has Voc: 1.58V, Jsc: 14.34 mAcm^{-2}, FF: 83.9 %, and PEC: 15.08%. Further, current matching is performed by varying the thickness of the absorber layer. After current matching the current density is matched at ~14 mA/cm^2. Next, the MJSC with the current matched device is examined. For the MJSC, the observed performance parameters are Voc: 2.87 V, Jsc: 14.33 mA/cm^2, FF: 82.64 %, and PEC: 34 %.

4 REFERENCES

[1] Tao Ye,* Ke Wang, Yuchen Hou, Dong Yang, Nicholas Smith, Brenden Magill, Jungjin Yoon,Rathsara R. H. H. Mudiyanselage, Giti A. Khodaparast, Kai Wang,* and Shashank Priya* "Ambient-Air-Stable Lead-Free CsSnI3 Solar Cells with Greater than7.5% Efficiency"

[2] R.T. Mouchou a, T.C. Jen a,, O.T. Laseinde a, K.O. Ukoba a Numerical simulation and optimization of p-NiO/n-TiO2 solar cell system using SCAPS

[3] Zyoud, S.H., Zyoud, A.H., Ahmed, N.M., Prasad, A.R., Khan, S.N., Abdelkader, A.F. and Shahwan, M., 2021. Numerical modeling of high conversion efficiency FTO/ZnO/CdS/CZTS/MO thin film-based solar cells: Using SCAPS-1D software. Crystals, 11(12), p.1468.

[4] Chen S. Dielectric constant measurement of P3HT, polystyrene, and polyethylene (Doctoral dissertation, Faculty of Science and Engineering).

[5] Rahmani, M.K., Khan, S.A., Khan, M.F. and Kang, M.H., 2022. Fully solution-processed organic RRAM device with highly stable butterfly-shaped hysteresis. Materials Science and Engineering: B, 282, p.115784.

[6] Y.L. Liu, C.L. Yang, M.S. Wang, X.G. Ma, Y.G. Yi, Theoretical insight into the optoelectronic properties of lead-free perovskite derivatives of Cs3Sb2X9 (X = Cl, Br, I), J. Mater. Sci. 54 (2019) 4732–4741, https://doi.org/10.1007/s10853-018- 3162-y

[7] A. Husainat, W. Ali, P. Cofie, J. Attia, J. Fuller, Simulation and analysis of methylammonium lead iodide (CH3NH3PbI3) perovskite solar cell with Au contact using SCAPS 1D simulator, Am. J. Opt. Photon. 7 (2019) 33, https://doi.org/ 10.11648/j.ajop.20190702.12.

[8] M.I. Hossain, F.H. Alharbi, N. Tabet, Copper oxide as inorganic hole transport material for lead halide perovskite based solar cells, Sol. Energy 120 (2015) 370–380, https://doi.org/10.1016/j.solener.2015.07.040.

[9] Garg, V., Kumar, A. and Sharma, P., 2021. Numerical simulation of novel lead-free Cs3Sb2Br9 absorber-based highly efficient perovskite solar cell. Optical Materials, 122, p.111715.

[10] Almora, O., Baran, D., Bazan, G.C., Cabrera, C.I., Erten-Ela, S., Forberich, K., Guo, F., Hauch, J., Ho-Baillie, A.W., Jacobsson, T.J. and Janssen, R.A., 2023. Device Performance of Emerging Photovoltaic Materials (Version 3). Advanced energy materials, 13(1), p.2203313.

10.4229/EUPVSEC2023/2DV.1.30
020148-003

STRUCTURED BACKSIDE GLASS FOR IMPROVED EFFICIENCY IN SOLAR MODULES

A SIMULATION STUDY

L. Stevens[1], B. Bläsi[1], C. Horch[2], S. Ihlow[2], M. Mittag[1]
[1] Fraunhofer ISE, Heidenhofstraße 2, 79110 Freiburg, Germany
[2] GMB Glasmanufaktur Brandenburg GmbH, Spremberger Straße 4, 03130 Tschernitz, Germany
laura.stevens@ise.fraunhofer.de

ABSTRACT: Increasing the efficiency of a photovoltaic module is still an ongoing research topic. To achieve this, we focus on the light management in the module. We perform a simulation study using Raytracing and a bottom-up multi-physic loss channel cell-to-module (CTM) analysis to study the influence of structured backside glass for bifacial glass-glass modules on the performance. The study was done to determine the module power output based on an AM1.5g spectrum including different incoming angles as an estimation of the yearly energy yield. These results are compared to modules with conventional white polymer backsheet or planar backside glass. It was found that a structured backside glass improves the light coupling into the module and internal light management through enhanced internal reflection. We study a cone structure and perform a variation of the cone angle to find an optimum in the reflection gain. We show that the highest reflection gain of 1.43% compared to a full absorbing backside for a cone structure with an angle of 35° is achievable. The theoretical cone structure was then adapted to realistic production conditions by implementing roundings. For structures with 20% rounding the reflection gain was lowered to 1.39%. The reflection gain of a white polymer backsheet was 1.47% and, thus, higher compared to the structured glass, However, compared to a planar glass with 0.11% the reflection gain of the structured glass is much higher and allows for improved bifacial modules. We also study the influence of the backside irradiance and show that for modules with structured glass a gain is obtained in comparison to modules with a white polymer backsheet. Hence, glass with optimized structures can replace backsheets without compromising the internal reflection gains.

Keywords: structured module glass, light guiding, raytracing, CTM analysis

1 INTRODUCTION

A good light management in solar modules is important to maximize the energy conversion efficiency. The front glass is typically coated with an antireflection coating (ARC) to increase the transmission of the light. Advantages of macroscopic structures in terms of yield have already been confirmed [1] for front side glasses. The advantages of a front side glass structuring can also be used for a backside glass.

The optimization in this study aims for trapping light in the module and redirecting rear side light to bifacial solar cells in the module. This can increase the module efficiency and is, therefore, relevant for the optimization of solar modules. In solar modules a white backsheet or a planar backside glass are commonly used at the rear side. Modules with a white backsheet can gain from a good scattering of transmitted light [2]. In comparison to that, solar modules with a transparent backside can make use of albedo illumination from the rear. Having a structured backside glass improves the light coupling into the module and internal light management through light steering and, thus, enhance internal reflection. In Figure 1 a schematic cross section view of the three backside types with possible light paths are shown.

The aim of this study is to design and evaluate structured backside glasses for glass-glass-modules. To this end, a simulation study was performed to find optimal structure parameters to increase reflection gains (for front side illumination) compared to a fully absorbing backside. Also, the influence on the module power with different backside illumination for a module with a white backsheet in comparison to a module with a structured backside glass was investigated.

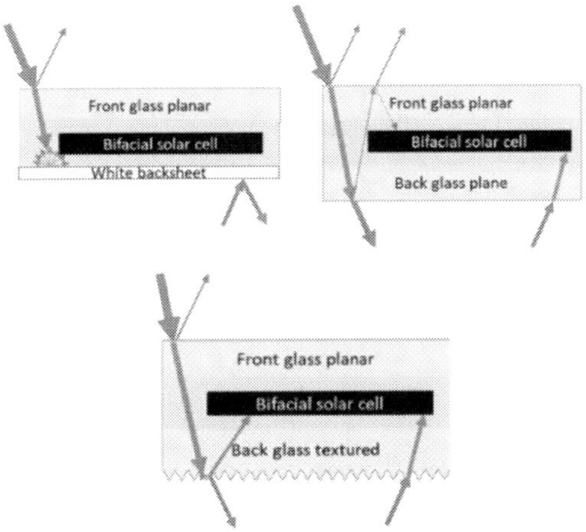

Figure 1: Cross-section views of the modules with different backsides and some possible light paths from the front and the backside: white backsheet (left), planar backside glass (middle), structured backside glass (right).

2 METHOD

The starting point for the simulation of the solar modules is the software tool SmartCalc.Module [3–5] , implementing a bottom-up multi-physics loss channel analysis of PV modules developed at Fraunhofer ISE. It allows modules to be simulated optically, electrically, and thermally. The optics of a planar encapsulation material can be described by the spectral refractive index and planar

10.4229/EUPVSEC2023/3AV.1.3
020151-001

interface reflectance and transmittance. For structured interfaces, the optical models are extended to describe feature sizes that allow to use geometric optics. Therefore, a CAD model of the structure was used in ray tracing simulations. The ray tracing framework Raytrace3D [6] developed at Fraunhofer ISE was used to investigate three-dimensional surfaces. The raytracing results along with 1000 W/m² direct normal irradiance of a AM1.5g light are then used as an input for the SmartCalc.Module analysis. The incoming light angles are varied so that the course of the sun around a year is simulated. A schematic workflow is shown in Figure 2. The single steps are described in more detail in the following.

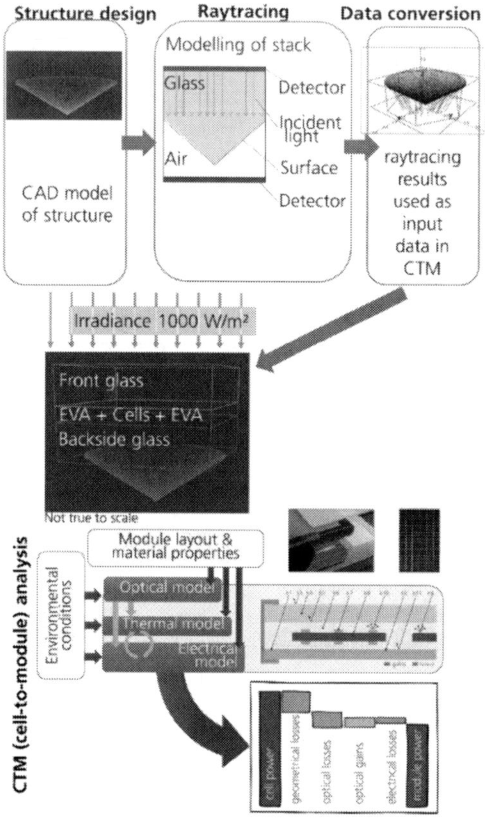

Figure 2: Schematic workflow of the simulation process to calculate the reflection gain for a structured backside glass, partly adapted from [7,8].

2.1 Raytracing analysis

The aim of the raytracing analysis is to generate a reflection or transmission distribution of incident light on a structured glass surface. The light can be emitted both inside and outside the glass, so that in the first case the internal and in the second case the external reflection at the glass can be considered.

Raytrace3D [6] includes a library of ray tracing functions written in C, which can be controlled and used through Python scripts. Various files containing the model parameters are used as input, as well as a simulation script that bundles all the information about the system to be simulated. For example, this script defines the optical properties of the materials in the model, the triangulated glass structure (from a 3D model), the shape and size of the light source, and the number and properties of the beams. Additional scripts were added, which control the running of multiple simulations with different parameters and transfer the result into a format that can be further used. Thus, multiple angles of incidence and wavelengths as well as structural variations can be run. For this simulation study zenith angles θ for 0°,30°,60°,90° and azimuth angle ϕ for angles from 180° to 147.5° in 2.5° steps have been simulated. The simulation results from raytrace3D will be used as input for the cell-to-module (CTM) analysis.

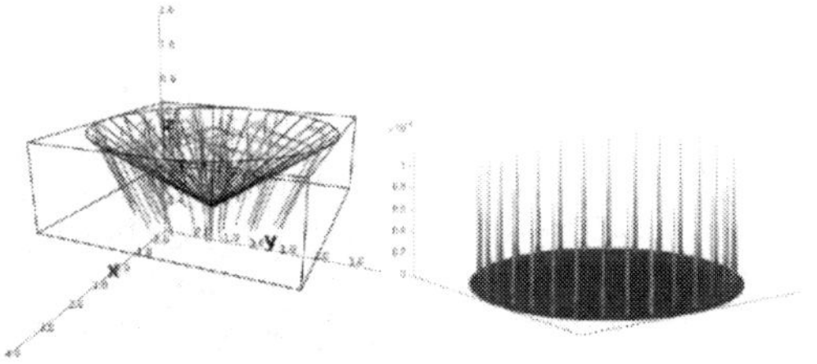

Figure 3: Ray tracing for a segmented cone at perpendicular incidence: visualization of the model (left) and the resulting reflectance distribution as polar plot (zenith angle radial, azimuth angle circular) (right).

2.2 CTM (cell-to-module) analysis

Cell-to-module loss analysis involves the consideration of various loss mechanisms that result from the encapsulation of a solar cell into a solar module. In Figure 4 the eleven different loss (orange) and gain (green) mechanisms [3] in a solar module are schematically shown.

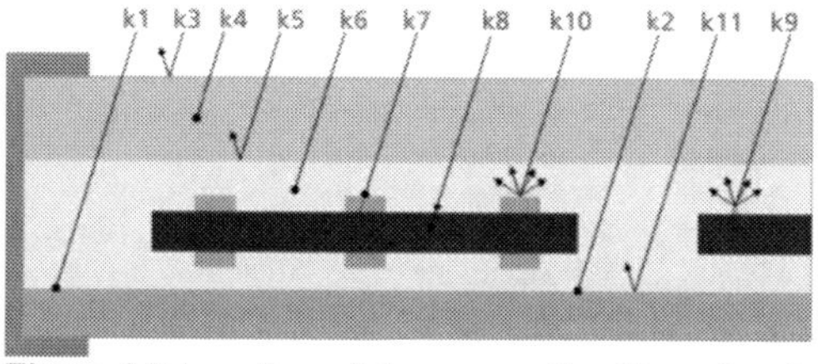

Figure 4:Schematic module cross-section illustrating the loss and gain mechanisms in the solar module.

Optical losses are primarily due to reflection from the module surface, usually a glass surface on the front of the module (k3 in Figure 4). In addition, light is absorbed in the front side encapsulation materials and is, therefore, not available to generate electricity, but only contributes to the heating of the module (k4 and k6 in Figure 4). To determine these losses, the spectral reflectance and transmittance of the glass (and encapsulation material) is needed. For a more detailed consideration, and in particular for structured surfaces, the angular distribution of the light is also relevant instead of the hemispherical values of the reflection and transmission. The bidirectional reflection or transmission distribution function (BRDF, BTDF) is then required, which describes the intensity distribution of the reflected and transmitted light in each spatial direction for a given angle of incidence. This then also provides a direct estimate of the glare effect of a surface, which is relevant for modules near airports or on building facades, for example.

An optical gain can occur in the solar module compared to the solar cell due to internal optical coupling, for example, when light impinging on "inactive" module surface, i.e. surface not occupied by cells, is reflected at the back of the module and then impinges on the front or back (for bifacial cells) of the cell (k11 in Figure 4). To calculate this coupling gain, the BRDF of the backside interface is needed. Another additional power gain occurs in bifacial modules due to light incident on the cell through the module back surface. For this, the same material information like the BRDF and material properties are needed for the module back surface as for the front surface. The results from the ray tracing study provide the relevant data for the investigated structures here.

In addition to the irradiation in the wavelength range relevant for the cell, the heat input into the module is also considered. This and the thermal properties of the module materials influence the operating temperature of the solar module under real weather conditions compared to temperature-controlled power measurement or simulation under standard test conditions (STC).

The I-V characteristic of the cell is then used to determine the power output of the module under given irradiance, temperature and wind conditions. Thus, a power simulation can be performed for various operating conditions, so that a complete annual cycle can also be mapped.

The results of the CTM analysis are then used to calculate the diffuse reflection gain. A detailed description of the method is published here [9].

3 RESULTS

We performed a simulation study of a cone structure where we changed the slope angle of the cone from 5° to 55°. This structure was then integrated into the module simulation of SmartCalc. A module with 60 cells with 4 mm cell spacing was selected for this purpose. The structured glass was selected for the rear side instead of a backsheet. We investigated the diffused reflection gain of the different structures in comparison to a fully absorbing backside. In Figure 5 the results of this simulation study are shown.

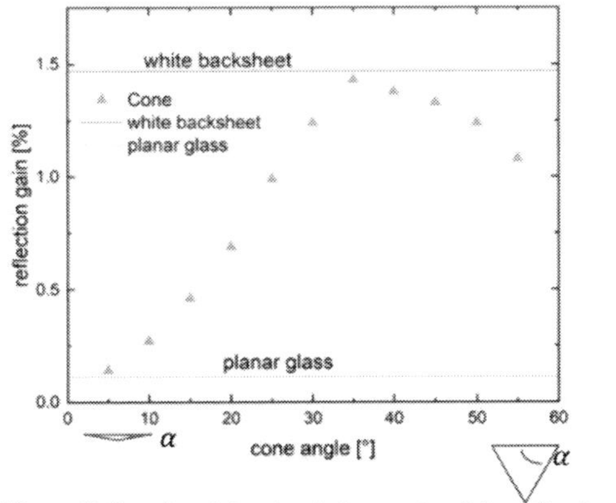

Figure 5: Results of the simulation study of the reflection gain for variating cone angle.

It is visible that the reflection gain increases with increasing slope angle of the cone until it reaches its optimum in reflection gain of 1.43% for a cone angle of 35°. The reflection gain of a white backsheet is included for comparison. The white backsheet has a very good light scattering but does not allow any yields from the back due to back illumination of the solar module.

3.1 Adaption: Roundings

The cone structure investigated before is quite theoretical. In a rolled glass production line there would be a rounding at the cone tip. Therefore, the structure was adapted to realistic production conditions by implementing roundings of 10 % or 20 % of the cone base radius. A visualization is presented in Figure 6 in the top as a sketch and below as an 3D image.

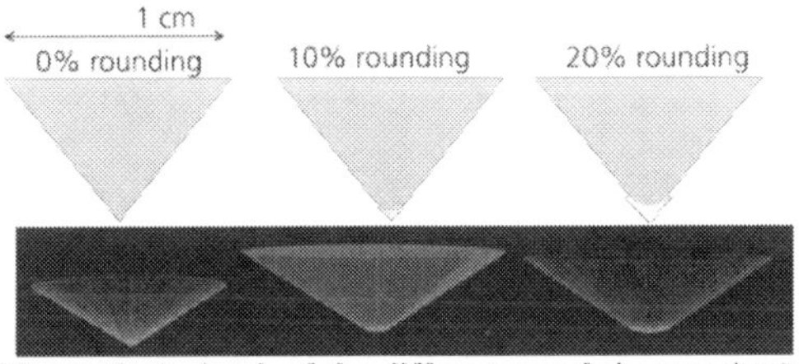

Figure 6: A sketch of the different rounded cones (top) and 3D images of the adapted cone structures with the different rounding (bottom).

The simulation process was again used for the adapted structures. The simulation results presented in Figure 7 show that by implementing roundings the reflection gains decrease slightly. The cone with 10% rounding has a reflection gain of 1.41%, the cone with 20% rounding has a reflection gain of 1.39%. The reflection gain of the white polymer backsheet is 1.47% and, thus, higher compared to a planar glass with 0.11% as the reflection gain of the structures is significantly higher.

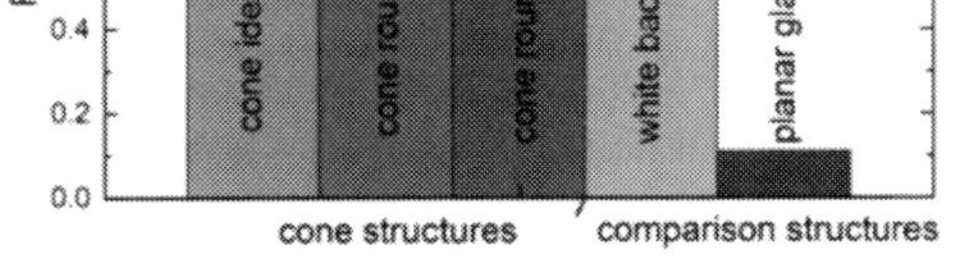

Figure 7: Reflection gain comparison for the ideal cone structure and for cones with different roundings compared to a white backsheet and a planar glass.

3.2 Adaption: Backside irradiance

The advantage of a module using backside glass is that also light from the backside can enter the module and thereby increase the power.

In Figure 8 the results of a SmartCalc.Module analysis of the module power for a module with a white backsheet and for a module with the structured glass is presented. The front side irradiance is 1000 W/m² and the backside irradiance is varied from 100 to 200 W/m².

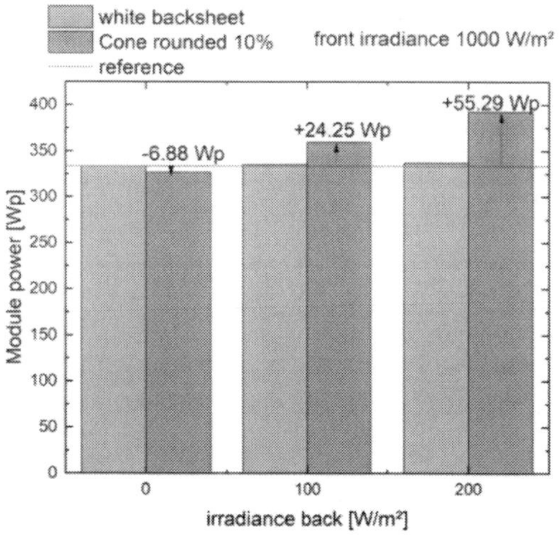

Figure 8: Module power for modules with either white backsheet or a structured glass with variating backside irradiance. The front irradiance is fixed at 1000 W/m².

It is visible that with increasing backside illumination the module with the backside glass has an advantage in the module power and shows a plus of 55.29 Wp for 200W/m² backside irradiance.

3 CONCLUSION

We used a combination of a raytracing and cell-to-module analysis to evaluate the reflection gain of a cone structure applied on the outside of a glass used on the module rear side. We could find an optimum in the reflection gain with 1.43% for a cone with a slope angle of 35°. The theoretical structure was adapted to realistic production conditions by implementing rounding at the cone tip. These roundings decrease the reflection gain down to 1.39% for a rounding of 20% of the cone radius. Compared to a standard white backsheet, which has a reflection gain of 1.47%, the cone structure features a slightly lower gain. However, the glass-glass module has the advantage of using albedo light from the backside. Our simulations could show that there is a significant module power increase with backside module illumination.

We could show that the structured backside glass helps to increase the module power output compared to a planar glass and can replace backsheets without compromising internal reflection gains.

References

[1] M. Duell, M. Ebert, M. Muller, B. Li, M. Koch, T. Christian, R. F. Perdichizzi, B. Marion, S. Kurtz, and Doble, D. M. J., "Impact of structured glass on light transmission, temperature, and power of PV modules," in *Proceedings of the 25th European Photovoltaic Solar Energy Conference and Exhibition/ 5th World Conference on Energy Conversion* (2010), pp. 3867–3872.

[2] M. Mittag, A. Grünzweig, M. Wiese, N. Mahmoud, A. Schmid, and M. Heinrich, "Analysis of backsheet and rear cover reflection gains for bifacial solar cells," 33rd European PV Solar Energy Conference and Exhibition **2017**.

[3] I. Hädrich, U. Eitner, M. Wiese, and H. Wirth, "Unified methodology for determining CTM ratios. Systematic prediction of module power," Sol Energ Mat Sol C **131**, 14–23 (2014).

[4] M. Mittag and M. Ebert, "Systematic PV module optimization with the cell-to-module (CTM) analysis software," Photovoltaics International, 97–104 (2017).

[5] Fraunhofer Institute for Solar Energy Systems ISE, "SmartCalc.Module," http://www.cell-to-module.com.

[6] P. Schöttl, G. Bern, P. Nitz, F. Torres, and L. Graf, *Raytrace3D by Fraunhofer ISE. Accurate and efficient ray tracing for concentrator optics*, 2022.

[7] A. Pfreundt, M. Mittag, and J. Shahid, "Cell-To-Module Analysis Beyond Standard Test Conditions," in *47th IEEE Photovoltaic Specialists Conference 2020* .

[8] M. Mittag, L. Vogt, C. Herzog, and D. H. Neuhaus, "Thermal Modelling of Photovoltaic Modules in Operation and Production," in *36th European Photovoltaic Solar Energy Conference and Exhibition (EUPVSEC) 2019* .

[9] A. Pfreundt, M. Mittag, M. Heinrich, and U. Eitner, "Rapid Calculation of the Backsheet Coupling Gain Using Ray Groups," in *35th European Photovoltaic Solar Energy* .

10.4229/EUPVSEC2023/3AV.1.3
020151-004

INTERMETALLIC PHASE GROWTH AND MICROHARDNESS OF SN42BI58 SOLDER JOINTS ON SILICON SOLAR CELLS

Derya Güldali[1,3], Angela De Rose[1], Sabine Oeser[2], Achim Kraft[1] and Ulrich Tetzlaff[3]
[1] Fraunhofer Institute for Solar Energy Systems ISE, Heidenhofstr. 2, 79110 Freiburg
[2] Fraunhofer Institute for Mechanics of Materials IWM, Wöhlerstr. 11, 79108 Freiburg
[3] Technische Hochschule Ingolstadt THI, Esplanade 10, 85049 Ingolstadt
Corresponding author: Derya Güldali, +49 (0)761 4588 2582, derya.gueldali@ise.fraunhofer.de

ABSTRACT: The temperature sensitivity of new solar cell technologies forces to use solder alloys with lower melting temperatures for interconnection [1].Our study explores the dynamics of intermetallic phase growth and microhardness in Sn42Bi58 solder joints applied to low-temperature silver metallization on silicon heterojunction (SHJ) solar cells. Through rigorous experimentation and analysis, a comprehension of the influence of these factors on the mechanical and material properties of the solder joint is achieved. Microstructural changes in Sn42Bi58 solder compared to conventional tin-lead solder are investigated through microscopy of cross sections, revealing enlarged intermetallic particles and phase boundary growth. These changes, attributed to lower homologous temperature of the low-melting solder, are expected to negatively impact the mechanical strength of the solder joint. For the intermetallic phase Ag_3Sn simulations predict a potential layer thickness of 20 µm after 25 years SHJ module operation. Our results show that the Ag_3Sn phase has a significant impact to the microhardness. Following the aging process, the nano hardness of the low-temperature Ag metallization experiences a twofold increase, shifting from 660 ± 53 N/mm² to 1367 ± 411 N/mm². This strengthening is primarily attributed to the prevailing influence of the Ag_3Sn intermetallic phase.
Keywords: Lead-free soldering, intermetallic compounds, microhardness, interconnection, long-term stability

1 INTRODUCTION

Research on low-temperature soldering is important to realize the industrialization of lead-free Photovoltaic-modules (PV modules) and the interconnection of temperature sensitive high efficiency cell technologies (*e.g.*, tandem solar cells). Both will be highly relevant for the PV industry in the near future. The progress in solar cell architecture to more efficient solar cells [2] leads to an adaptation of the conventional backend processes (*i.e.*, metallization and interconnection) in PV module manufacturing as the maximum soldering temperature has to be reduced to avoid damaging the solar cells. In turns, this means switching to alternative solder alloys with a lower melting point than the current standard Sn60Pb40 with $T_m \approx$ 183-190 °C.

However, due to the fact that PV is still excluded from the current European RoHS (Restriction of Hazardous Substances) regulations [3], Pb-containing solder is still dominant due to its good performance and low price. For general low-temperature (LT) soldering processes as *e.g.*, in microelectronics, several LT solder alloys could be used. For soldering in PV, the restrictions (*i.e.*, material compatibility with solar cell, long-term stability, costs etc.) drastically reduce the possible choice of LT alloy. To lower the melting point and to decrease the soldering temperature, either one of the elements Sn, In, Bi or Ga has to be used [4]. In PV, the most well-known low-melting solder alloy is the eutectic tin-bismuth alloy (Sn42Bi58). It has been under investigation by research and industry in the past decades [5–8].

SnBi solder has some advantages in terms of processability (relevant physical properties *e.g.*, melting point) and human and environmental acceptability (*e.g.*, toxicity). A comparison of the conventional tin-lead solder with the lead-free alternative regarding the most relevant criteria for low-temperature soldering is shown in Figure 1. The dominant solder alloy of current PV modules Sn60Pb40 contains 40 $\%_{wt}$ lead, which sums up to about 8-10 g/m² lead per full-size solar module [9, 10]. Lead is considered hazardous to health and environment and is included in the category of RoHS-restricted materials for electrical and electronic equipment [3]. In fact, it is only a question of time until this restriction will also cover PV products so that Sn60Pb40 has to be substituted by alternative Pb-free solder alloys. In general, Sn42Bi58 has no known harmful properties compared to Sn60Pb40 solder, making it friendlier to the environment and health. Compared to tin-lead solder, bismuth-based alloys have a lower electrical conductivity and a higher tendency for oxidation during soldering, which leads to reduced wettability. Sn42Bi58 is a promising and cost-effective alternative solder alloy for the industry, but research in this area is not yet complete and more understanding is needed in terms of fundamentals and long-term behavior of the solder.

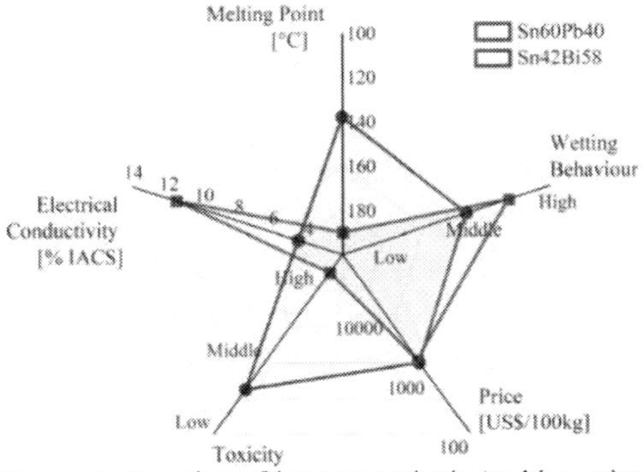

Figure 1: Overview of important criteria (melting point, electrical conductivity according to IACS, toxicity, price and wetting behavior) for the application of the most prominent solder alloys in PV: Sn60Pb40 and the lead-free alternative Sn42Bi58. The further out the alloy is located on the respective axis, the more suitable it is for low-temperature interconnection for Si solar cells. Data taken from [11–15].

2 THEORETICAL BACKROUND

2.1 Solder joint of silicon solar cells

Solder joints of Si solar cells have an important task as they mechanically and electrically contact the cell's electrodes for collection of the generated current. This is achieved by establishing a reliable connection between the cell ribbon and the metallization of the solar cell (see Figure 2).

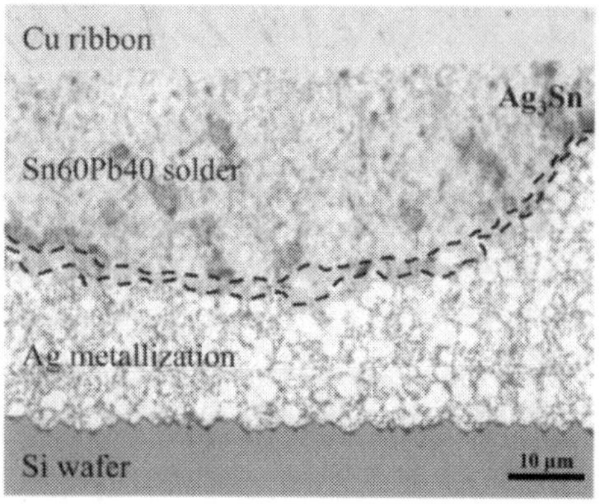

Figure 2: Microscopic image (cross section) showing the solder joint with Sn60Pb40 solder and Cu-ribbon on LT Ag metallization of an SHJ solar cell. The formation of the intermetallic phase Ag_3Sn (highlighted by black dashed line) implies successful wetting and solder joint formation.

To achieve a large contacted area between solder and metallization, sufficient and continuous wetting of the busbars by the solder is necessary (see chapter 2.2). Additionally, the proportion of non-electrically conductive areas, such as cracks and voids (*e.g.*, air inclusions), should be minimized. Considering that the module is subjected to various thermomechanical and mechanical stresses, the solder joint should exhibit stability against these loads.

The quality and design of solder joints have a significant impact on several aspects, including mechanical stress during string production, transport of the electric current of photoconversion and long-term stability of the photovoltaic module. Cases of coarsening of the solder and fatigue cracks in the solder joint have been reported for photovoltaic modules that have been exposed to outdoor conditions for an extended period of time [16, 17].

Therefore, fundamental understanding of solder joints as well as detailed analysis is evident to evaluate modifications such as the solder coating of the ribbons, in order to ensure the continued stability of the solder joints.

2.2 Intermetallic compounds

As mentioned in chapter 2.1, contact formation between the cell ribbon and the busbar is essential for the power output of the PV module. This process is known as joining. According to DIN 8580, this involves the joining of two or more components or materials (here: cell ribbon and metallization) with a formless solid (here: solder) [18].

In PV, mostly infrared soldering is used with solder-coated Cu ribbons (see Figure 2). Due to the soldering temperatures during the process, atomic exchange processes lead to interatomic diffusion at the interfaces between Cu ribbon and solder and between solder and metallization, resulting in the formation of diffusion zones. As a result, solid solutions and/or intermetallic compounds/phases (IMC) can form at this interface. The formation of these phases depends on factors such as the mutual solubility of the materials, the crystal structures of the materials to be joined and the temperatures prevailing during the soldering and the subsequent cooling process. The existence of a diffusion zone is a prerequisite for forming a reliable electrical contact.

When joining Sn- and Ag-containing components, an IMC of Sn-Ag-phases is formed at the interface. According to the binary phase diagram (see Figure 3), this phase consists of 75 $\%_{at}$ Ag and 25 $\%_{at}$ Sn (Ag_3Sn) and is stable at room temperature. On conventional Si solar cells, this phase can be found at the interface between the Ag metallization and the Sn-containing solder (see Figure 4).

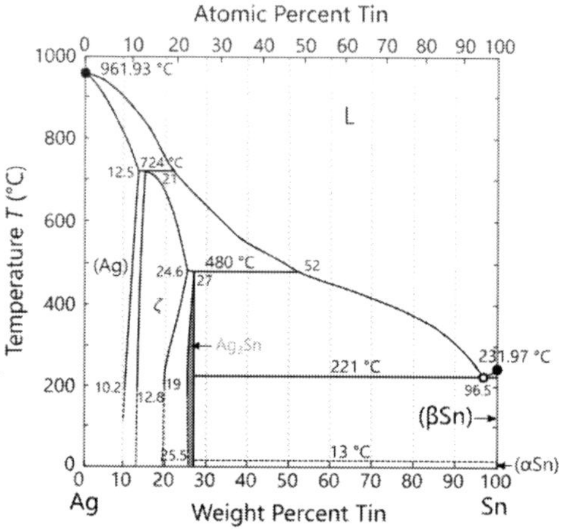

Figure 3: Binary tin-silver phase diagram. The intermetallic phase Ag_3Sn is highlighted in orange. Adopted from [11].

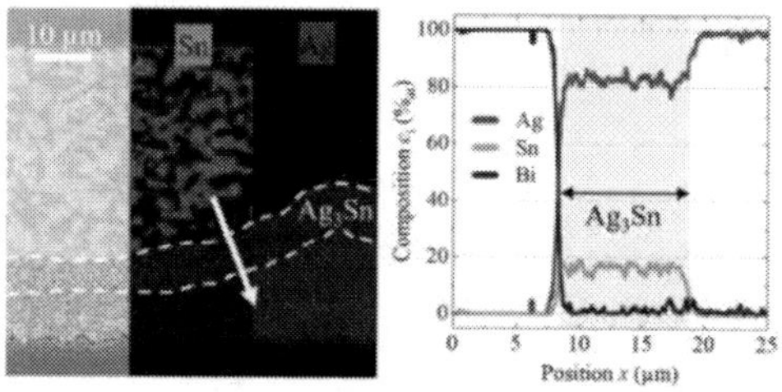

Figure 4: Analysis to identify Ag_3Sn in the solder joint, shown for a lead-free joint. Left: SEM image and EDX mapping of the solder joint. Right: EDX line scan across the interface solder/metallization.

With this intermetallic phase being part of the solder joint, the mechanical as well as electrical properties of the joint change. When the solder joint is in operation and exposed to various environmental influences (*e.g.*, high outdoor temperatures and temperature changes), a change in the microstructure of the solder joints is expected, which consequently changes the properties of the solder joint. The thickness as well as hardness of the IMC is an important parameter for the quality of the whole solder joint. The growth behavior of various IMC under isothermal aging has already attracted much attention in other electronic fields [19, 20] as well as in PV [5, 21–24]. Studies have also shown that the failure mechanisms of solder joints are often associated with the presence and growth of intermetallic compounds at the substrate-solder interface [25–27]. IMCs often exhibit very high hardness and lower electrical conductivity due to their complex crystal structure, among other factors. As a result, an increase of the thickness of this layer would lead to embrittlement of the solder joint and crack initiation or propagation.

The understanding of the microstructure of the solder joint and especially of the IMC as well as its effect on the stability of PV modules is currently not entirely developed and needs more investigations.

2.3 Modelling the growth characteristics of IMCs

In the joint of the solar cell, there is a special solid-state diffusion process between the Sn in the solder and the Ag in a metallization. Since it is not a pure metal-metal diffusion, describing this diffusion is complex. Therefore, for the subsequent calculations, a simple approximation with a known solution of Fick's law is chosen in the first step. This approach has already been used in other studies investigating intermetallic phase growth [23, 26, 28].

The growth behavior of the intermetallic layer Ag_3Sn during solid state aging can be described by the diffusion of Sn and Ag atoms to each other. As the thickness of the IMC layer increases over time, the flux of atoms during diffusion decreases. This behavior can be represented by the parabolic time law (Eq. 1). The equation can be interpreted as a special one-dimensional solution of the second Fick's law, where x^2 is defined as the mean square diffusion distance of the atoms over the time t [29]. For all following calculations, it is assumed that the root of the mean square diffusion distance x^2 is equal to the intermetallic layer thickness x. By transforming the equation, the obtained relationship for the thickness x of the intermetallic phase is achieved:

$$x^2 = 2Dt \qquad \text{Eq. 1}$$

$$x = \sqrt{2Dt} \qquad \text{Eq. 2}$$

The Arrhenius relationship, given in Eq. 3, describes the temperature-dependent behavior of the diffusion coefficient D in materials [29, 30]. It states that the diffusion coefficient is exponentially proportional to the reciprocal of temperature T, where an increase in temperature leads to an acceleration of diffusion processes:

$$D = D_0 e^{-\frac{Q}{RT}} \qquad \text{Eq. 3}$$

The equation involves the diffusion constant as a thermally independent constant D_0, activation energy Q, the gas constant R and the absolute temperature T as components, expressing the temperature dependence of diffusion behavior. D_0 and Q are material-specific constants, both depending on the considered diffusion mechanism and system. Combining Eqs. 2 and 3 leads to Eq. 4:

$$x = \sqrt{2\,D_0 e^{-\frac{Q}{RT}}\,t} \qquad \text{Eq. 4}$$

To determine the activation energy Q, the Eq. 4 is linearized through the application of the natural logarithm, leading to the formulation of Eq. 5:

$$\ln(x) = \frac{1}{2}\ln(2D_0 t) - \frac{Q}{2R}\frac{1}{T} \qquad \text{Eq. 5}$$

By simplifying the function ln(x) with respect to $1/T$, a linear relationship between Q and R is obtained, which is characterized by the slope labeled "a":

$$Q = -2aR \qquad \text{Eq. 6}$$

There are two possible approaches for calculating the diffusion constant D_0: For the first option, the relation ln x and $1/T$ is used as already done for the calculation of the activation energy (Eq. 6).

The second approach uses the Eq. 2 and the relation $x \sim t^{1/2}$. The slope a_* of this equation allows to calculate the diffusion coefficient, resulting in the diffusion constant using the Arrhenius relationship (Eq. 3):

$$D_0 = \frac{D}{e^{-\frac{Q}{RT}}} = \frac{a_*^2}{e^{-\frac{Q}{RT}}} \qquad \text{Eq. 7}$$

Since the first approach generates a large error, the second approach (Eq. 7) is used for the calculations in this work.

Utilizing these two quantities, it becomes possible to forecast the increase in layer thicknesses at various temperatures and durations as by Eq. 4. This equation can also be employed to infer aging parameters, including time and temperature, to achieve a specific layer thickness:

$$t = \frac{x^2}{2 \cdot D_0 \cdot e^{-\frac{Q}{RT}}} \qquad \text{Eq. 8}$$

$$T = \frac{Q}{R} \cdot \frac{1}{\ln(D_0 t) - 2\ln(x)} \qquad \text{Eq. 9}$$

These calculations allow to estimate the growth of the Ag_3Sn intermetallic phase and to investigate the relevance of the influence of the phase on the module stability. The aim is to predict the changes of critical phases in the solder joint between Cu ribbon and solar cell over the lifetime of a PV module and to study its impact.

3 MATERIALS AND METHODS

3.1 Materials

Using a screen-printing process, the metallization is realized at Fraunhofer ISE PV-TEC laboratory in a five busbar (BB) layout on the front and rear side, with a constant BB width of 0.8 mm on industrial silicon heterojunction wafer (M6, n-type, Cz-Si, textured, bifacial). Table 1 lists the cell ribbons used for the interconnection of the SHJ solar cells.

Table I: Compilation of the used cell ribbons with their respective geometric and material-specific properties.

Leaded Solder (Reference)
Cu ribbon (0.90 × 0.22 mm²) coated with 20-25 µm **Sn60Pb40** (near-eutectic, T_m: 183-190 °C)
Lead-Free Solder
Cu ribbon (0.90 × 0.22 mm²) coated with 15-25 µm **Sn42Bi58** (eutectic, T_m: 139 °C)

3.2 Solar cell interconnection

Soldering is realized on a manual laboratory setup to be more flexible for process parameter and material variations. We used manual soldering with a solder iron supported by heating plates for preheating of the solar cells, using the following set-temperatures:

- $Sn_{60}Pb_{40}$: Preheating 120 °C, Soldering 240 °C
- $Sn_{42}Bi_{58}$: Preheating 100 °C, Soldering 180 °C

Prior to soldering, a conventional no-clean flux with a low solid content is applied locally to the solar cell busbars to facilitate the reduction of oxides and support wetting.

3.3 Metallography

For the evaluation of the solder joint microstructure, metallographic cross sections are prepared out of the soldered SHJ solar cells. After laser cutting of small pieces along the soldered busbar, the samples are embedded into epoxy resin and polished perpendicular to the cross section of the Cu ribbon.

3.4 Isothermal aging

To comprehend the metallurgical behavior within the solder joints, we conduct isothermal aging experiments to assess how microstructural changes evolve concerning temperature and time. The aging process is conducted in an atmospheric air-drying oven, exposing prepared samples to aging at 65 °C, 85 °C, 105 °C and 125 °C for a total duration of 168 hours (7 days). Isothermal aging is performed similarly to the standard DIN EN 60068-2-2, a norm for a dry heat test method [31]. Optical microscopy images are captured in the initial state and at specific time intervals (24 h, 96 h, 168 h).

3.5 Characterization of microstructure

Both scanning electron microscopy (SEM) and energy diffractive X-ray (EDX) spectroscopy are employed to analyze the samples in their initial state and after 168 hours. These techniques provide comprehensive insights into diffusion zones and IMC formation at the joints' interfaces.

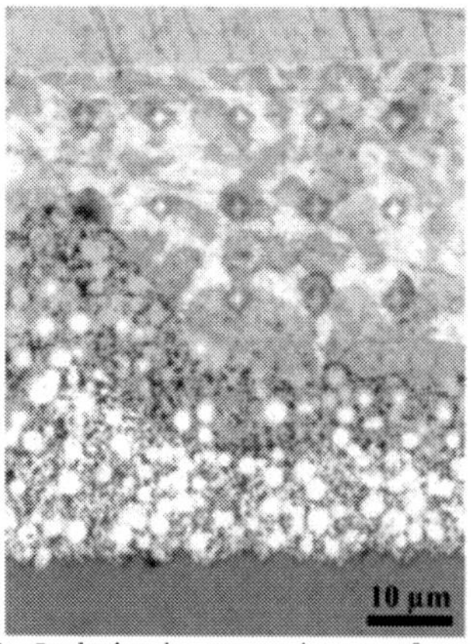

Figure 5: Optical microscope image of a solder joint (cross section view) after hardness measurement using nanoindentation [32]. The hardness indentations of the Berkovich pyramid in the solder are clearly visible. Due to the higher hardness of the other solder joint components, the indents in the Cu ribbon, the Ag metallization and the Si wafer are hard to identify in the image.

Resulting images will provide the basis for thickness measurements of the IMC for the calculation of the growth parameters given in chapter 2.3. In addition, an instrumented indentation test in the nano range according to DIN EN ISO 14577-1 [32] is employed to determine the microhardness in the solder joint. The hardness measurements were performed at Fraunhofer Institute for Mechanics of Materials (IWM). Multiple indentations are created per cross section sample using a standardized Berkovich diamond tip with an applied force of 2 mN and a dwell time of 10 s to ascertain the hardness values. To achieve this, a grid (50 × 70 µm) with 5 × 9 indentations was defined to cover each phase within the solder joint (see example in Figure 5).

4 RESULTS AND DISCUSSION

4.1 Evaluation of microstructural changes after isothermal aging

As grain coarsening and IMC growth primarily result from diffusion processes, which are significantly temperature-dependent, it's reasonable to investigate these phenomena exclusively through isothermal aging experiments. External factors like humidity are negligible and have no impact on the quantitative outcomes of our study. Specifically, comprehending the growth kinetics of IMCs and deriving practically applicable model parameters necessitates systematic aging experiments conducted at various isothermal temperature.

At each time interval and at specific aging temperatures, the thickness of the IMC layer is measured using an optical microscope. In Figure 6, the resulting IMC layer thickness x, determined by 20 measurements per cross section, is plotted against the square root of time $t^{1/2}$. Note that the chosen approach from Eq. 2 correlates well with the results. A linear regression is necessary to read out the y-axis intercept c according to Eq. 7, the diffusion constants D_0 for the Ag_3Sn phase can then be calculated using "c".

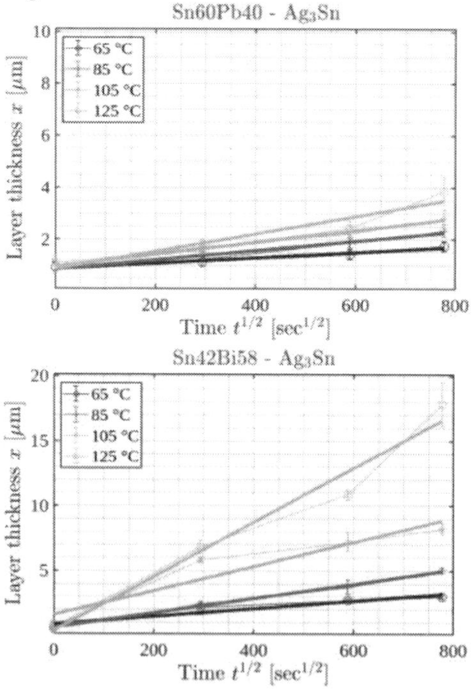

Figure 6: The plots represent the growth of the Ag_3Sn phase in the bismuth-containing (top) and lead-containing (bottom) solder joints and the linear fit of each temperature. The layer thickness x is plotted over the square root of time $t^{1/2}$. Note that different scaling is applied to each diagram for clarity.

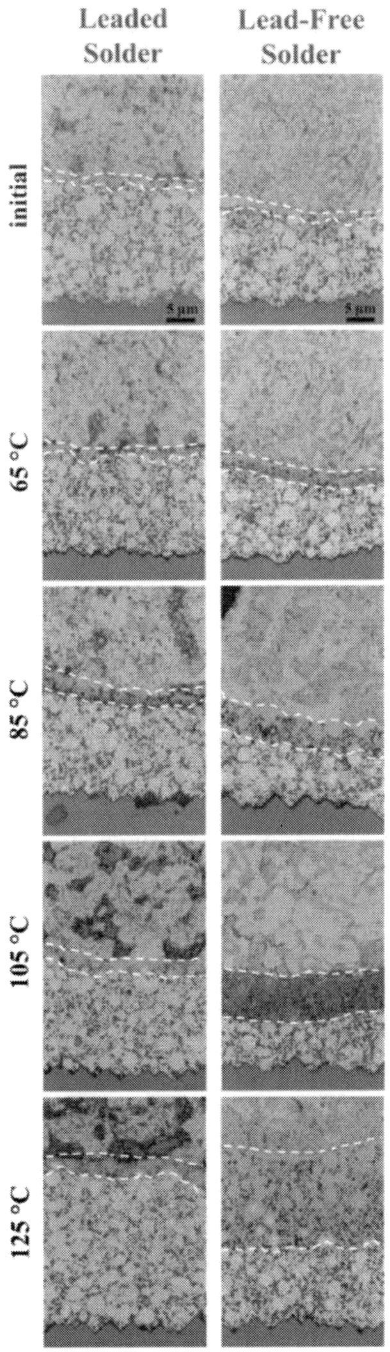

Figure 7: Microscopic images (cross sections) showing joints on SHJ solar cells with soldered ribbons (blue: lead-free solder, red: leaded solder) in the initial state, at 65 °C, 85 °C, 105 °C and 125 °C after 168 h aging in atmospheric air. After aging, the intermetallic phase Ag_3Sn (in white) grows into the Ag metallization, which is a consequence of the diffusion of Sn from Sn42Bi58 and Sn60Pb40 solder.

The analysis of the cross section by microscopy reveals a different microstructure and aging behavior of Bi-containing solder joints compared to conventional Pb-containing solder joints (see Figure 7). For both, Sn diffusion into the Ag metallization is observed, forming the IMC Ag_3Sn at the interface. Due to the lower melting point of SnBi and an increased atom mobility within the crystal lattice, the Sn diffusion is more pronounced within the SnBi solder joint [33–36]. The Sn atoms at the interface diffuse into the metallization to form the Ag_3Sn phase. As a result, the Bi-rich phases remaining at the interface come together to form larger grains, similar to the process of Ostwald ripening [35].The closer the aging temperature to the melting point of the respective solder alloy, the stronger becomes the diffusion velocities in the crystal lattice due to the increasing homologous temperature.

The plots in Figure 8 show the natural logarithm of the Ag_3Sn layer thickness x plotted against $1/T$ for the three different aging times. From the arithmetic mean of the three fits, the activation energy Q can be determined in accordance with Eq. 6. from whose slope a and intercept c (intersection with y-axis), the activation energy Q can be determined in analogy to Eq. 6.

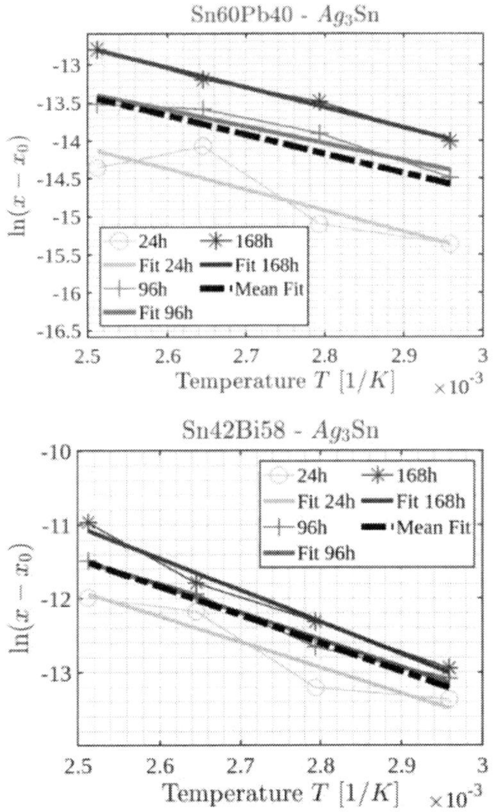

Figure 8: Arrhenius plots of Ag_3Sn IMC growth in both solder alloys for different aging times. The natural logarithm of IMC phase thickness $\ln(x\text{-}x_0)$ is plotted over the inverse of the temperature $1/T$.

As explained in chapter 2.3, the diffusion constant D_0 is calculated using the slope a of the diagrams in Figure 6.

10.4229/EUPVSEC2023/3AV.1.4

The mean value of all 4 temperature graphs is calculated and listed in Table II.

Table II: Activation energies Q and diffusion coefficients D_0 resulting from the slope and the ordinate intercept of the data plotted in Figure 7, using Equations 6 & 7.

Solder alloy	Q [kJ/mol]	D_0 [m²/s]
Sn60Pb40	42.1 ± 4.8	$(1.9 \pm 0.2) \times 10^{-12}$
Sn42Bi58	63.5 ± 7.4	$(3.3 \pm 1.0) \times 10^{-8}$

The calculated values for Q and D_0 clearly illustrate why there is a stronger growth of the Ag_3Sn phase in the Sn42Bi58 solder. With increasing temperature, a higher activation energy leads to a steeper rise in the diffusion coefficient D, considering the Arrhenius relationship from Eq. 3. Furthermore, from the comparison of the activation energy of the two solders, it can be interpreted that there is a correlation with the Sn content in the solder. The lead-containing solder has a higher Sn content (60 $\%_{wt}$) in comparison to the bismuth-containing solder (42 $\%_{wt}$). Since only Sn contributes to the formation of the Ag_3Sn phase, it is logical that a higher Sn content in the solder lowers the energy barrier for the formation of this intermetallic phase, as more Sn is available for diffusion [26]. However, the Sn60Pb40 solder still exhibits slower phase growth, as its homologous temperature (for 125 °C: Sn60Pb40 0.87-0.89 and Sn42Bi58 0.97) is higher than that of the lead-free, low-temperature solder. This factor also significantly influences the growth rate. Based on the measured layer thicknesses and the resulting calculated material properties, a trend towards almost 5-fold stronger growth of the intermetallic phase in the Sn42Bi58 solder on this low-temperature metallization can be observed.

For both solder alloys, a growth of the Ag_3Sn phase is observed, which is driven by the segregation of the base materials. This process is based on the effort to reduce the total energy in the system and to establish an equilibrium [37]. Looking into the alloy phase diagrams (not shown here), it is obvious that the respective elements have a very low solubility (for the temperature range used here) in each other, resulting in precipitates/phases which subsequently agglomerate into lager grains to reduce the surface energy. As a result, the increase of Bi-rich grains can lead to a reduction in fatigue strength, ending in a lower resilience to mechanical stress, during the lifetime of photovoltaic modules. Additionally, the dominance of grain coarsening as a failure mode in solder joints was a significant reliability issue in electronics [38, 39]]. This phenomenon was clearly associated with fatigue cracking in PV modules after extended outdoor exposure [16, 40–42].

4.2 Modeling of Ag_3Sn intermetallic phase growth

Utilizing these material attributes, it becomes feasible to predict the phase growth under the specified reliability test conditions, which are commonly employed for the certification of PV modules. An established and important test for heat exposure is the damp heat test according to IEC 61215:2021 [43]. For 1000 h, the specimens are aged in the simulation at a constant temperature of 85 °C under 85 % relative humidity. Based on Eq. 4, the calculated results are shown in Figure 9, illustrating that the Ag_3Sn phase in low-melting solders, such as Sn42Bi58, is expected to grow excessively up to 12 µm layer thickness x after 1000 h at 85 °C. Note that the influence of the humidity is neglected here. Based on the calculated material parameters, the leaded solder joint would reach a thickness of about 4 µm after 1000 h damp heat test, ~1/3 of the phase thickness in the lead-free solder. It should be noted that in the simulations, there is no upper limit for the layer thickness of intermetallic phases. However, under real conditions, growth is constrained by the thickness of the metallization. Additionally, the depletion of Sn in the solder layer also leads to a slowdown of Sn diffusion towards the Ag metallization.

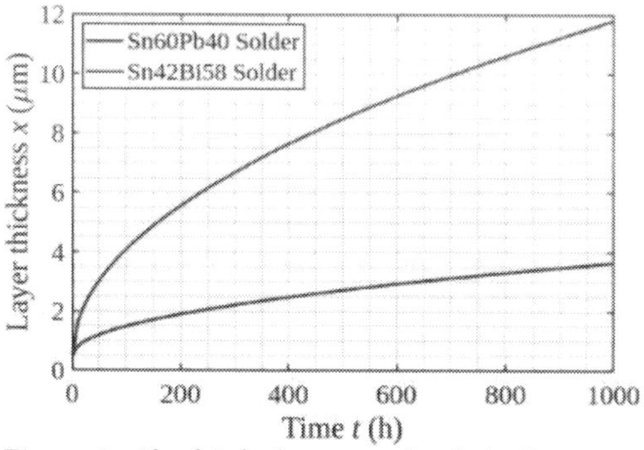

Figure 9: Simulated phase growth of Ag_3Sn over a duration of 1000 hours at 85 °C, reproducing the temperature conditions of the damp heat tests as defined in IEC 61215:2021 [43].

Usually, manufacturers provide a 25-year warranty on their PV modules. Using the numerical simulations and the corresponding calculations, the growth of the Ag_3Sn phase in LT metallization over 25 years was determined (see Figure 10). For this purpose, an annual average ambient temperature in Germany of 10.5 °C (2022) [44] and an initial layer thickness of $x_0 = 0.5$ µm after interconnection are assumed. It turns out that the IMC reaches a layer thickness of ~ 11 µm for the low-melting solder (blue line in Figure 10). The thickness for the conventional tin-lead solder is ~ 7.8 µm.

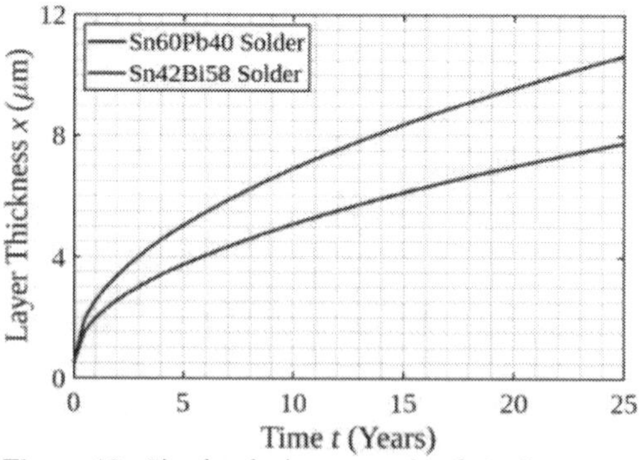

Figure 10: Simulated phase growth of Ag_3Sn over a duration of 25 years at 10.5 °C with an initial layer thickness of $x_0 = 0.5$ µm.

Nevertheless, the simulation misses validation by comparison with experimentally measured IMC growth after long-term exposure to humid heat, thermal cycling or extended outdoor conditions, which will be addressed in future work. The simulations help to understand and predict the behavior of intermetallic phase growth in LT Ag metallization, as it is used for SHJ or perovskite Si tandem solar cells. It is important to mention that the

10.4229/EUPVSEC2023/3AV.1.4
020152-006

calculated characteristic values for the activation energy Q and diffusion constant D_0 are valid for this joining system. Changes in the metallization paste lead to changes in the diffusion process, including the presence of more or less vacancies and diffusion paths, which would accelerate or slow down Sn diffusion.

4.3 Effect of IMC growth on microhardness

The clear Ag_3Sn phase growth influences the microstructure of the solder joint. Beside electrical changes, this causes an increase in microhardness in particular regions. Figure 11 shows the Martens hardness HM plotted against the position within the solder joint for the initial state (top) and after isothermal aging at 125 °C for 168 h (bottom). Post aging, the hardness of the Ag metallization witnesses a 2-fold rise, increasing from 660 ± 53 N/mm² to 1367 ± 411 N/mm², with the presence of the Ag_3Sn intermetallic phase predominantly influencing this enhancement. Similarly, the hardness of the Ag_3Sn phase in the lead-containing solder displays comparability at 1366 ± 237 N/mm² (not shown here). The hardness of the other components (Si wafer, pure Ag metallization, solder and Cu ribbon) is only slightly affected by the temperature exposure The hardness of the other components (Si wafer, pure Ag metallization, solder and Cu ribbon) is only slightly affected by the temperature exposure.

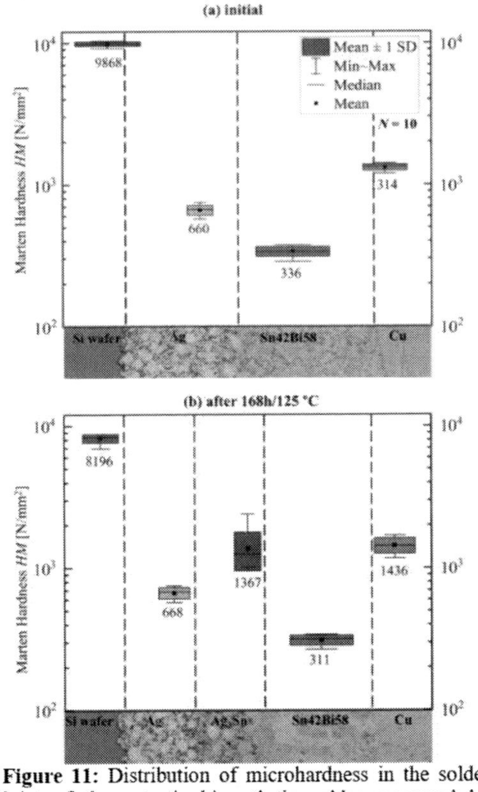

Figure 11: Distribution of microhardness in the solder joint of the eutectic bismuth-tin solder measured by nanoindentation [32] in the initial state and after 168 h at a constant aging temperature of 125 °C in atmospheric air (10 measurements values per group). The effect of the observed high growth of the intermetallic phase Ag_3Sn in Figure 5 is reflected here in the increase in hardness within the Ag metallization.

However, the significant expansion of the IMC in the Bi solder joint indicates that due to the lower homologous temperature, the overall solder joint hardness grows progressively over time. During temperature cycling, these components with significantly higher hardness than the surrounding materials, can lead to thermal stress during temperature cycling. This thermal stress can result in solder joint cracking or detachment, especially if the thermal expansion coefficients of the materials are mismatched. The higher hardness of the IMC can be more prone to brittle fracture, especially if the solder joint experiences rapid temperature changes during day-night cycle or mechanical loading by *e.g.*, snow. Brittle fractures can lead to solder joint failure [39, 45]. Consequently, the occurrence of a joint crack could lead to elevated series resistances and a subsequent reduction in current delivery.

5 CONCLUSIONS AND OUTLOOK

In conclusion, this study explored intermetallic phase growth and microhardness in Sn42Bi58 solder joints on silicon heterojunction solar cells. Through rigorous experimentation and analysis, we gained a comprehensive understanding of how these factors affect solder joint mechanical and material properties.

Microstructural changes, observed through microscopy, revealed significant alterations in the Sn42Bi58 solder matrix compared to traditional tin-lead solder. These changes included thicker Ag_3Sn intermetallic compound layers and increased grain boundary growth between phases, potentially compromising solder joint mechanical strength. Simulations, based on an upper limit scenario, indicated that this could lead to a layer thickness of approximately 11 μm over 25 years at 10.5°C. These microstructural variations also influenced microhardness, with an observable increase from 660 N/mm² to 1367 N/mm², primarily due to the Ag_3Sn phase.

Comparing these two solder compositions highlighted distinct kinetics and behaviors of intermetallic phases. The prevalence of the Ag3Sn phase in Sn42Bi58 solder joints on low-temperature Ag metallization emphasized its potential impact on joint mechanical properties. The interplay between intermetallic compounds and different solder materials offered insights for optimizing solder joint reliability and performance in various applications.

These findings provide a foundation for enhancing soldering processes and material selection across industries. Further exploration is needed to understand the effects of intermetallic layer growth, either alongside or compared to grain coarsening in the solder matrix. This exploration should include failure analysis of field-aged PV modules and correlating IMC growth with solder joint mechanical properties.

Moreover, comprehensive understanding of processes and materials, along with stringent control, is vital to establish meaningful correlations between simulation and experimentation. Our simulation results serve as comprehensive predictions within the given conditions.

This study underscores the importance of intermetallic phases in solder joint behavior, particularly their role in governing microhardness and, consequently, the overall structural integrity of solar cell assemblies. These insights offer valuable groundwork for enhancing photovoltaic module reliability and performance, contributing to advancements in solar energy technology. The study focused on Sn42Bi58 solder on low-temperature metallization. Future research should investigate whether other low-temperature solders exhibit similar behavior. Our findings suggest that the growth of the brittle Ag_3Sn phase depends on factors like solder homologous temperature, Sn content, element solubility, and metallization characteristics. Low-temperature solders with different properties may have different phase growth behaviors, emphasizing the complexity of solder joint behavior and the need for comprehensive analysis.

ACKNOWLEDGEMENTS

This work was supported by the German Federal Ministry for Economic Affairs and Climate Action within the project Utility4Indium (contract no. 020E-100487483).

REFERENCES

[1] B. A. Korevaar, J. A. Fronheiser, X. Zhang, L. M. Fedor, and T. R. Tolliver, "Influence of annealing on performance for hetero-junction a-Si/c-Si devices," in *Proceedings of the 23rd European Photovoltaic Solar Energy Conference and Exhibition*, Valencia, Spain, 2008, pp. 1859–1862.

[2] ITRPV, "International Technology Roadmap for Photovoltaic (ITRPV): 2022 Results," 2023.

[3] *RoHS-Richtlinie 2011/65/EU: Beschränkung der Verwendung bestimmter gefährlicher Stoffe in Elektro- und Elektronikgeräten*, 2011.

[4] G. Humpston and D. M. Jacobson, *Principles of soldering*. Materials Park, Ohio: ASM International, 2004.

[5] T. Geipel, M. Moeller, J. Walter, A. Kraft, and U. Eitner, "Intermetallic compounds in solar cell interconnections: Microstructure and growth kinetics," *Sol Energ Mat Sol C*, vol. 159, pp. 370–388, 2017, doi: 10.1016/j.solmat.2016.08.039.

[6] B. Lalaguna, P. Sánchez-Friera, I. J. Bennett, D. Sánchez, L. J. Caballero, and Alonso J., "Evaluation of bismuth-based solder alloys for low-stress interconnection of industrial crystalline silicon PV cells," in *Proceedings of the 22nd European Photovoltaic Solar Energy Conference and Exhibition (EUPVSEC)*, Milan, Italy, 2007, pp. 2712–2715.

[7] M. Nowottnick and A. Novikov, "Possibilities and Limits of Bismuth Solders," in *Proc. of SMTA International*, Rosemont, USA, 2017, pp. 195–200.

[8] D. Güldali and A. De Rose, "Material Joint Analysis of Lead-Free Interconnection Technologies for Silicon Photovoltaics," in *45th International Spring Seminar on Electronics Technology (ISSE)*, 2022.

[9] Paul Gebhardt *et al.*, "Lead-Free PV Modules: Industrial Realization and Evaluation of Environmental Impact," in *8th World Conference on Photovoltaic Energy Conversion*, Milano, 2022, pp. 930–934.

[10] Dr. Harry Wirth | Fraunhofer ISE, "Recent Facts about Photovoltaics in Germany," 2021. [Online]. Available: https://www.ise.fraunhofer.de/en/publications/studies/recent-facts-about-pv-in-germany.html

[11] H. Baker, *Alloy Phase Diagrams: ASM Handbook*, 3rd ed.: ASM International, 1992.

[12] DERA and BGR, "Preismonitor September 2022," [Online]. Available: https://www.deutsche-rohstoffagentur.de/DERA/DE/Aktuelles/Monitore/2022/09-22/2022-09-preismonitor.pdf?__blob=publicationFile&v=2

[13] J. R. Rumble, *CRC handbook of chemistry and physics*, 98th ed. Boca Raton, Florida: CRC Press, 2017. [Online]. Available: http://hbcponline.com/

[14] S. Zhou, C. Yang, Y.-A. Shen, S. Lin, and H. Nishikawa, "The newly developed Sn–Bi–Zn alloy with a low melting point, improved ductility, and high ultimate tensile strength," *Materialia*, vol. 6, p. 100300, 2019, doi: 10.1016/j.mtla.2019.100300.

[15] D. C. Adriano, *Trace Elements in Terrestrial Environments: Biogeochemistry, Bioavailability, and Risks of Metals*, 2nd ed.: Springer New York, 2001.

[16] D. L. King, M. A. Quintana, J. A. Kratochvil, D. E. Ellibee, and B. R. Hansen, "Photovoltaic module performance and durability following long-term field exposure," (in English), *Prog. Photovolt: Res. Appl.*, vol. 8, no. 2, pp. 241–256, 2000.

[17] J.-S. Jeong, N. Park, and C. Han, "Field failure mechanism study of solder interconnection for crystalline silicon photovoltaic module," *Microelectronics Reliability*, vol. 52, 9-10, pp. 2326–2330, 2012, doi: 10.1016/j.microrel.2012.06.027.

[18] *DIN 8580:2003-09, Fertigungsverfahren_-Begriffe, Einteilung*, DIN Deutsches Institut für Normung e. V. (DIN), Berlin, 2003.

[19] M. Schaefer, R. A. Fournelle, and J. Liang, "Theory for intermetallic phase growth between cu and liquid Sn-Pb solder based on grain boundary diffusion control," (in English), *Journal of Elec Materi*, vol. 27, no. 11, pp. 1167–1176, 1998, doi: 10.1007/s11664-998-0066-7.

[20] T. An and F. Qin, "Effects of the intermetallic compound microstructure on the tensile behavior of Sn3.0Ag0.5Cu/Cu solder joint under various strain rates," *ELTE 2010 & IMAPS/CPMT Poland 2010*, vol. 54, no. 5, pp. 932–938, 2014, doi: 10.1016/j.microrel.2014.01.008.

[21] T. Geipel, M. Moeller, A. Kraft, and U. Eitner, "A comprehensive study of intermetallic compounds in solar cell interconnections and their growth kinetics," *Enrgy Proced*, vol. 98, pp. 86–97, 2016.

[22] P. Schmitt, P. Kaiser, C. Savio, M. Tranitz, and U. Eitner, "Intermetallic phase growth and reliability of Sn-Ag-soldered solar cell joints," *Enrgy Proced*, vol. 27, pp. 664–669, 2012, doi: 10.1016/j.egypro.2012.07.126.

[23] Y.-J. Jeon, M.-S. Kang, and Y.-E. Shin, "Growth of an Ag3Sn Intermetallic Compound Layer Within Photovoltaic Module Ribbon Solder Joints," *Int. J. of Precis. Eng. and Manuf.-Green Tech.*, vol. 7, no. 1, pp. 89–96, 2020, doi: 10.1007/s40684-019-00028-1.

[24] H. Gopalakrishna *et al.*, "Activation Energy for End-of-Life Solder Bond Degradation: Thermal Cycling of Field-Aged PV Modules," *IEEE Journal of Photovoltaics*, vol. 10, no. 6, pp. 1762–1771, 2020, doi: 10.1109/JPHOTOV.2020.3025726.

[25] P. L. Tu, Y. Chan, and J. Lai, "Effect of intermetallic compounds on the thermal fatigue of surface mount solder joints," *IEEE Trans. Compon. Packag. Manuf. Technol. B*, vol. 20, no. 1, pp. 87–93, 1997, doi: 10.1109/96.554534.

[26] D. R. Frear and P. T. Vianco, "Intermetallic growth and mechanical behavior of low and high melting temperature solder alloys," *Metall and Mat Trans A*, vol. 25, no. 7, pp. 1509–1523, 1994, doi: 10.1007/BF02665483.

[27] G. Zeng, S. Xue, L. Zhang, L. Gao, W. Dai, and J. Luo, "A review on the interfacial intermetallic compounds between Sn–Ag–Cu based solders and substrates," *Journal of Materials Science: Materials in Electronics*, vol. 21, no. 5, pp. 421–440, 2010, doi: 10.1007/s10854-010-0086-y.

[28] M. H. Poech, "Schädigungsmechanismen in Lötverbindungen bei erhöhter Temperatur," *Verfahrenstechnik der Energiewandlung (VTE)*, vol. 14, no. 1, pp. 12–18, 2002.

[29] R. J. Borg and G. J. Dienes, *An Introduction to Solid State Diffusion*. San Diego: Academic Press, 1988.

[30] H. L. Harald Ibach, *Solid-State Physics: An Introduction to Principles of Materials Science,* 4th ed. Berlin, Heidelberg: Springer Berlin Heidelberg, 2009.

[31] *Umgebungseinflüsse - Teil 2-2: Prüfverfahren - Prüfung B: Trockene Wärme*, DIN EN 60068-2-2, Deutsches Institut für Normung e.V. (DIN), 2008.

[32] *DIN EN ISO 14577-1:2015-11, Metallische Werkstoffe_- Instrumentierte Eindringprüfung zur Bestimmung der Härte und anderer Werkstoffparameter_- Teil_1: Prüfverfahren (ISO_14577-1:2015); Deutsche Fassung EN_ISO_14577-1:2015*, Berlin.

[33] D. Hull and D. J. Bacon, *Introduction to Dislocations,* 5th ed.: Elsevier, 2011.

[34] G. Gottstein, *Materialwissenschaft und Werkstofftechnik,* 4th ed.: Springer-Verlag, 2014.

[35] J. A. Marqusee and J. Ross, "Theory of Ostwald ripening: Competitive growth and its dependence on volume fraction," *The Journal of Chemical Physics*, vol. 80, no. 1, pp. 536–543, 1984, doi: 10.1063/1.446427.

[36] J. Rösler, H. Harders, and M. Bäker, *Mechanisches Verhalten der Werkstoffe,* 4th ed. Wiesbaden: Springer Fachmedien Wiesbaden, 2012.

[37] P. G. Harris, K. S. Chaggar, and M. A. Whitmore, "The effect of ageing on the microstructure of 60:40 tin-lead solders," *Soldering & Surface Mount Tech*, vol. 3, no. 1, pp. 20–33, 1991, doi: 10.1108/eb037741.

[38] H. Ma and J. C. Suhling, "A review of mechanical properties of lead-free solders for electronic packaging," *J Mater Sci*, vol. 44, no. 5, pp. 1141–1158, 2009, doi: 10.1007/s10853-008-3125-9.

[39] Y. Tian, W. Liu, R. An, W. Zhang, L. Niu, and C. Wang, "Effect of intermetallic compounds on fracture behaviors of Sn3.0Ag0.5Cu lead-free solder joints during in situ tensile test," *J Mater Sci: Mater Electron*, vol. 23, no. 1, pp. 136–147, 2012, doi: 10.1007/s10854-011-0538-z.

[40] M. A. Quintana, D. L. King, T. J. McMahon, and C. R. Osterwald, "Commonly observed degradation in field-aged photovoltaic modules," in *Proceedings of the 29th IEEE Photovoltaic Specialists Conference*, New Orleans, Louisiana, USA, 2002, pp. 1436–1439.

[41] U. Itoh, M. Yoshida, H. Tokuhisa, K. Takeuchi, and Y. Takemura, "Solder joint failure modes in the conventional crystalline Si module," *Energy Procedia*, vol. 55, pp. 464–468, 2014.

[42] G. Cuddalorepatta, A. Dasgupta, S. Sealing, J. Moyer, and Tolliver, T.Loman, J., "Durability of Pb-free solder between copper interconnect and silicon in photovoltaic cells," *Prog. Photovolt: Res. Appl.*, vol. 18, no. 3, pp. 168–182, 2010, doi: 10.1002/pip.944.

[43] *Terrestrial photovoltaic (PV) modules – Design qualification and type approval – Part 1-3: Special requirements for testing of thin-film amorphous silicon based photovoltaic (PV) modules*, IEC 61215-1-3:2016, International Electrotechnical Commission (IEC), Geneva, Switzerland, 2016.

[44] Deutscher Wetterdienst, "Klimastatusbericht Deutschland Jahr 2022," Geschäftsbereich Klima und Umwelt, Offenbach, 2022. [Online]. Available: www.dwd.de/DE/derdwd/bibliothek/fachpublikationen/selbstverlag/selbstverlag_node.html

[45] T. An and F. Qin, "Relationship Between the Intermetallic Compounds Growth and the Microcracking Behavior of Lead-Free Solder Joints," *J. Electron. Packag.*, vol. 138, no. 1, 2016, doi: 10.1115/1.4032349.

TEMPERATURE DISTRIBUTION DURING THE LAMINATION PROCESS OF PV MODULES AND ITS INFLUENCE ON THE DEGREE OF CROSSLINKING FOR EVA - SIMULATION VS TEST RESULTS

Aksel Kaan Öz, Japan Vasani, Christian Reichel, Christine Wellens, Dirk Holger Neuhaus
Fraunhofer Institute for Solar Energy Systems ISE, Heidenhofstraße 2, 79110 Freiburg, Germany
Corresponding Author: Aksel Kaan Öz | +49 (0) 761 4588 5025 | e-mail: aksel.kaan.oez@ise.fraunhofer.de

ABSTRACT: The lamination process of photovoltaic (PV) modules significantly influences their long-term reliability. One way to control the quality of the lamination process is measuring the degree of crosslinking of the modules, reflecting sufficiency of process parameters like lamination temperature and lamination duration. In this study, we conducted thermocouple measurements across module layers to observe temperature profiles and gauge the degree of crosslinking in glass-backsheet (GB) and glass-glass (GG) modules, both with and without cells. These were compared with the results of a simulation tool that can model the temperature profile and the degree of crosslinking during the lamination. The measurements were compared with the simulations, yielding promising matches with observed values. The simulation effectively captured temperature trends during lamination. Notably, inadequate processing led up to a 6.5% discrepancy in crosslinking between frond and back side of modules, a concern addressed by using a plate-plate chamber for GG modules or extending the process duration in the plate-membrane chamber. The study underscored the pronounced influence of cells on crosslinking; their inclusion decreased it by up to 6.5%. Furthermore, we validated the simulation tool's crosslinking prediction against Soxhlet extraction results.

Keywords: PV Modules, Lamination, Degree of Crosslinking, Simulation, Module Manufacturing

1 INTRODUCTION

In recent years, the photovoltaic (PV) industry has witnessed remarkable advancements in cell technologies, sizes, and materials, driven by the pursuit of cost reduction, environmental friendliness, and enhanced module efficiency while ensuring prolonged operational lifetimes with minimal power degradation [1,2]. The lamination process is an important key when it comes to achieving long-term reliability [3]. A robust adhesion among module layers and a good degree of encapsulant crosslinking are essential shields against external stressors, humidity, and other factors that could compromise cell integrity.

To ensure a good lamination process, the optimization of key parameters such as temperature, time, and pressure according to the specific material combinations is essential. Furthermore, it is imperative to achieve a consistent temperature distribution within the module, both laterally but also in the different layers of the module [4].

This paper scrutinizes temperature homogeneity within the module during the lamination process through two distinct methods. Firstly, by comparing the degree of crosslinking in the encapsulant on the front and rear side of the cell, and secondly, by conducting thermocouple measurements to capture the temperature profile within the module during lamination. Additionally, the measurements are compared with simulation from a tool that is capable of modeling the lamination process and is providing temperature data for both the front and rear sides of the cell. The tool may simplify the assessment of new bills of materials (BOM) without the need for time-consuming tests, ultimately leading to valuable time and cost savings.

The study is primarily focused on ethylene-co-vinyl acetate (EVA), the dominant encapsulant in the PV market with approximately 70% market share [5]. To validate our findings, we compare the measured degree of crosslinking and temperature profiles from the lamination process with the simulation results.

In summary, this research sheds light on the critical role of the lamination process in achieving PV module longevity and quality. By investigating temperature homogeneity and introducing a simulation tool, we aim to contribute to streamlined evaluation processes within the PV industry.

2 MATERIALS AND METHODS

In our initial phase, we fabricated six mini modules in both glass-glass (GG) and glass-backsheet (GB) configurations, with three modules for each setup. The modules were laminated with an EVA encapsulant, featuring a size of 680 mm x 370 mm. For both lay-up types (GB, GG) the EVA-GB process was used, which took place in the plate-membrane chamber (HP1) of the Ypsator of Bürkle.

In our extensive study (module matrix shown in Figure 1) we analyzed twelve different conditions and laminated 24 modules. To have a better understanding and reliable results, 2 modules were manufactured for each case and tested twice for their degree of crosslinking. To simplify the sample preparation for the determination of the degree of crosslinking an additional layer of Teflon was introduced between the encapsulant and the glass/rear side layer. Additionally, the lamination process for GB and GG modules was segregated. GB modules underwent lamination in HP1, while GG modules were pre-processed in HP1 and subsequently transferred to the plate-plate chamber (HP2) for crosslinking. The main difference between HP1 and HP2 is that, in HP2 the module is heated actively from both sides and is also in direct contact to the heating plates, whereas in HP1, the module is only in direct contact to the bottom heat plate and gets less heat from the top side.

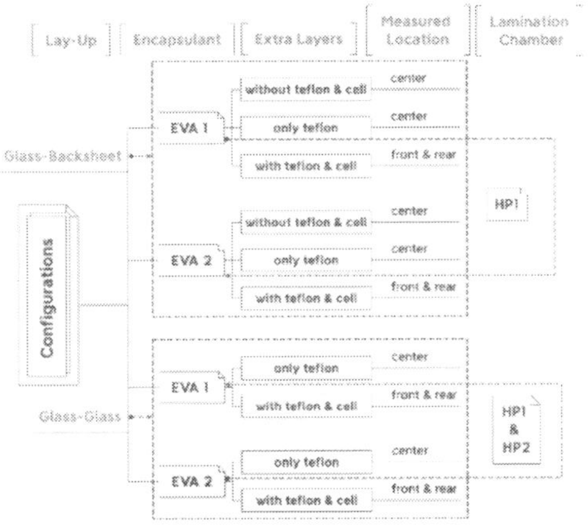

Figure 1: Module Matrix

For the different lay-ups the samples for the Soxhlet extraction were taken from the locations mentioned in Figure 1. The modules including cells were measured on both sides of the cell. The samples of the modules with no solar cells, called "laminates", were taken from their center. The study also compared the degree of crosslinking of laminates and modules to investigate the influence of cells and also checks for the influence of the Teflon sheet. The specifications of the glasses, encapsulants and other materials used are shown in Table 1.

Table 1: List of used encapsulation materials and their specifications

Layer	Material	Thickness [mm]
glass 1	solar glass	3
glass 2	solar glass	2
backsheet	PET/Primer	0.220
encapsulant 1	EVA	0.500
encapsulant 2	EVA	0.500
teflon sheet	Polytetrafluoroethylene (PTFE)	0.08

2.1 Soxhlet Extraction

To assess the lamination process, the degree of crosslinking values measured on the front and rear side of the PV modules are compared. For this purpose, we adopted Soxhlet extraction, a straightforward time-intensive technique widely used for measuring crosslinking levels in PV modules [6,7].

The Soxhlet degree of crosslinking was calculated by the following formula [6]:

$$X\ (Soxhlet)[\%] = \left(\frac{M_2 - M_0}{M_1 - M_0}\right)\ x\ 100,\ M_2 \leq M_1$$

where "M_0" indicates the pre-determined weight of the cylindrical stainless steel mesh tube, "M_1" the total weight of the sample in its tube before the process and "M_2" the weight of the sample and its tube after being extracted and dried in the vacuum oven.

The measurements were carried out by using the Behrotest apparatus in accordance with the standard outlined in IEC 62788-1-6. Each module was subjected to at least two sample measurements, the results are presenting the mean value and standard deviation of these measurements.

2.2 Thermocouple Measurements

To precisely monitor the temperature dynamics during lamination, we employed type K thermocouples to measure the actual temperature within a standard-sized module (1960 mm x 1010 mm). Simultaneously, the temperature progression throughout the process was logged using the Datapaq Q18 datalogger. To evaluate the uniformity of heat distribution across the module and comparing the heat variations between its front and rear sides, we strategically positioned four sensors within the module, as depicted in Figure 1.

It is important to note that the PV module lay-up utilized for the thermocouple measurements slightly differs from the standard industry configuration. In our measurement setup, an additional Teflon sheet was inserted between the sensors and the encapsulant. This design modification allows the sensors to be reused for different lay-ups (GB, GG) and distinct processes (EVA-GB, EVA-GG). In Figure 2, the sensors are placed between the inner side of the glass and the additional Teflon sheet, for the GB lay-up, the rear sensors are placed between the backsheet and the Teflon sheet.

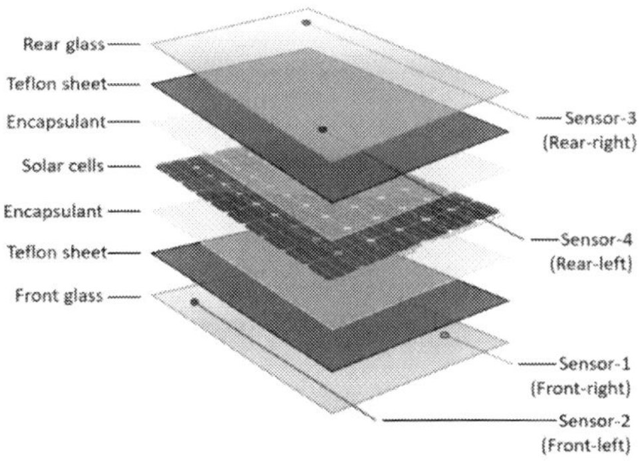

Figure 2: Module lay-up and sensor locations for the thermocouple measurements

2.3 Simulation

The lamination and the crosslinking process of the front and rear EVA in a GB and GG module as shown in Figure 1 is modelled to verify the temperature differences and the experimentally observed degree of crosslinking. A thermal model for photovoltaic modules is developed that determines the one-dimensional temperature distribution within the different module layers, provided by a heating plate in the lamination process [8]. Based on the modelled temperature profile of the EVA in the PV module, a model for the cure kinetics which is used to predict the degree of crosslinking during the lamination process was developed.

In order to achieve the degree of crosslinking of the encapsulation material in the lamination model, the chemical reactions of the EVA is determined by Differential Scanning Calorimetry (DSC), allowing to obtain the crosslinking reaction, which assists in determining the activation energy. The activation energy is calculated using the Friedman kinetic analysis method to see the changes in the degree of crosslinking via the activation energy. To solve the kinetic equation, the kinetic parameters are required with the slope formula to get the activation energy. For the calculation of the kinetic parameters, DSC measurements of the cured sample are taken into account such as reaction that occurs with respect to temperature and time at different heating rates. The activation energy and temperature profile of the encapsulation material are used to acquire the final results for the degree of crosslinking.

The investigated temperature profile is characterized taking into account the material values of the materials used at the module and laminator level. Due to the flexible design of the boundary conditions on both sides of the module, the possibility to extend the module structure (adding or removing any layers from the module) and manufacturing processes (laminators with one heating plate or both side heating plates) is given.

3 RESULTS AND DISCUSSIONS

In this section the degree of crosslinking for the different conditions is analyzed and the results of the simulation are introduced under four main headlines. These are the comparison of the degree of crosslinking on the front and rear side of the cell, the change in degree of crosslinking after including cells into the laminate, the temperature profile during lamination versus the simulation and the degree of crosslinking curve, respectively.

3.1 Front versus Rear Crosslinking

In the initial trial, all laminates and modules underwent the EVA 1 GB process in the plate-membrane chamber. In the extensive study, we employed different EVAs and alternated between the GB process and GG process based on module lay-up. The results obtained are shown in Figure 3.

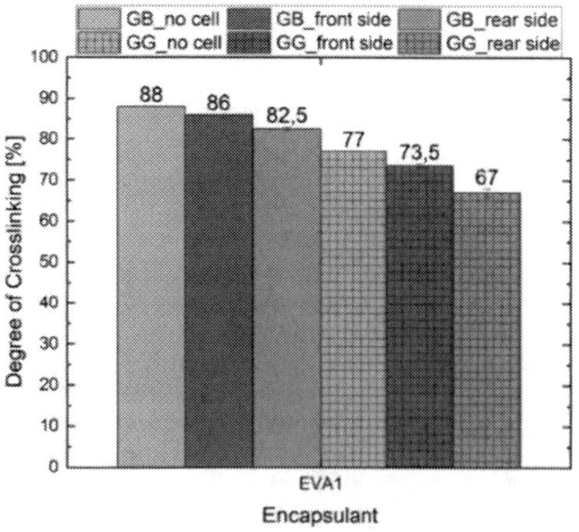

(a)

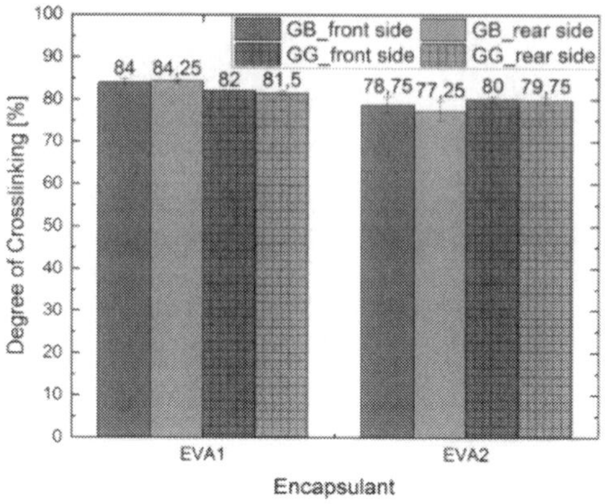

(b)

Figure 3: Comparison of the degree of crosslinking on the front/rear side of the module, (a) for initial trial, (b) for extensive study

The results, illustrated in Figure 3 (a), highlight a noteworthy reduction in crosslinking for all GG modules in the initial trial. This decrease, particularly evident in the 6.5% difference between rear and front side crosslinking in the GG module, suggests a temperature disparity between both sides. This difference is attributed to the lamination process occurring primarily in the GB process. For improved temperature transmission and crosslinking uniformity in GG modules, an extended process duration in HP1 or transitioning to HP2 (plate-plate chamber) for GG modules is recommended (Figure 6).

In Figure 3 (b), a reduced difference in crosslinking between front vs. rear side is observed compared to Figure 3 (a). This improvement is attributed to GG modules being laminated in a combination of HP1 and HP2, facilitating better temperature transmission and crosslinking uniformity. Notably, EVA2 consistently exhibited lower crosslinking compared to EVA1. A potential reason might be differences in the amount of curing agents in the encapsulant or that the used process was more suitable for EVA1 [9].

3.2 Laminate versus Module Crosslinking

Given the lower crosslinking degrees observed for modules (with cells) compared to laminates (without cells) in the initial trial, we examined the influence of cells on crosslinking. While laminates were sampled from the center, module sampling was constrained by solar cells. The degree of crosslinking of the GB module and GG module in Figure 4 shows the mean values measured on the front and rear side of the modules.

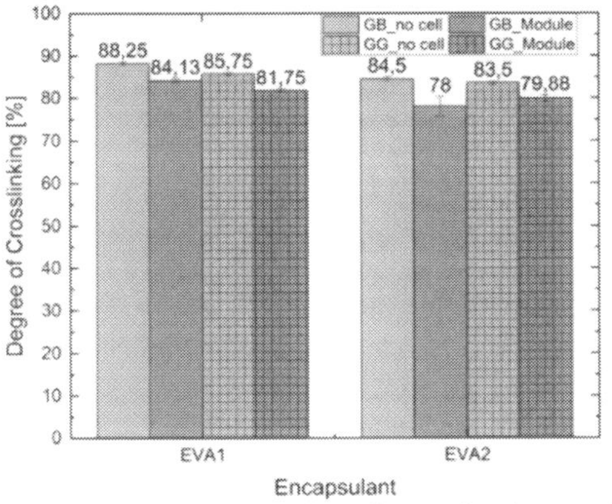

Figure 4: Influence of solar cells on the degree of crosslinking

Figure 4 presents the crosslinking degrees for the GB and GG modules, revealing differences ranging from 3.5% to 6.5%, based on module lay-up and encapsulant. The graph highlights a reduction in crosslinking after inserting cells for all cases. Notably, the difference remained relatively stable (around 4%) with EVA1, while EVA2 was more influenced by the lay-up.

3.3 Temperature Profile during Lamination

To investigate the crosslinking disparities between the front and rear side of the module, we studied the temperature profile during lamination. The exact locations of the sensors are shown in Figure 2. The results of the thermocouple measurements and the simulations for the GB and GG processes together with their lamination parameters are compared in and Figure 6.

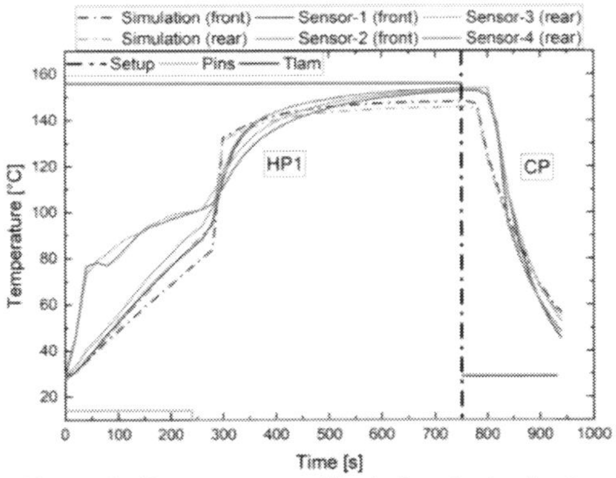

Figure 5: Temperature profile during the lamination process for a glass-backsheet module and lamination process parameters of the glass-backsheet process used in this study

The simulated temperature process follows the thermocouple measurements, demonstrating consistency in temperature evolution. In both a certain temperature raises out to 240 seconds is observed, after this point, both show initially a faster increase in temperature followed by a smoother curve. At this point the pins go down and the module gets in direct contact with the heating plate, which provides a faster temperature transfer.

After 750 seconds the module goes from HP1 to the cooling press CP where it is cooled down. In the measurement we can see a small delay compared to the simulation, because in reality it takes some seconds while the module is transferred from one chamber to the other.

The results indicate that the length of the process is almost optimal. The maximum temperatures reached by the sensors on the front and rear side are 153.9 °C and 153.4 °C, respectively. For the simulation these values are 148.2 °C and 145.7 °C whereas the set temperature off the process was 155 °C.

An interesting observation was the temperature profile at the beginning of the process, where the rear side of the module had a higher temperature compared to the front side. This phenomenon was also observed in the simulation. The reason for that might be, that at that point there is no direct contact on both sides, the module lies on the pins and the heat transfer through the backsheet is going faster compared to the glass on the front side. However, for a reliable statement further research needs to be done.

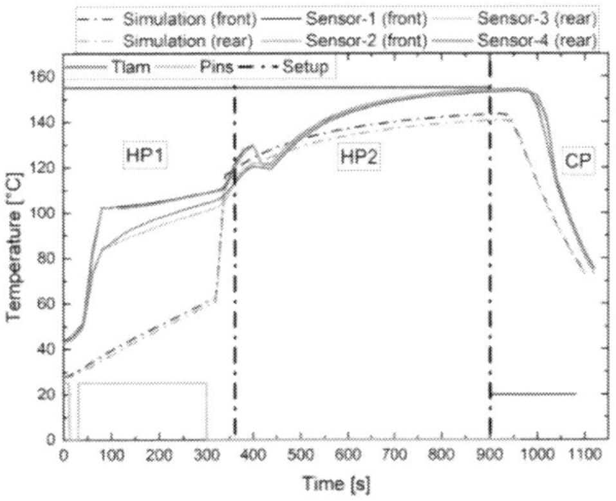

Figure 6: Temperature profile during the lamination process for a glass-glass module and lamination process parameters of the glass-glass process used in this study

Figure 6 shows that our simulation tool also simulated promising results in comparison to the outcomes of the measurements. This time the measurements show a higher temperature on the front side of the module compared the rear side, different as observed in the GB module lay-up. There are two main reasons for this. One is the short direct contact (when the pins are down) at the beginning of the process. The other factor is that this time we have a GG module with same thermal characteristics on both sides.

Same as in , the influence of the pins and the delay in cooling while the transfer to the CP can be easily read from Figure 6. In addition to these, the GG module is also going through the phase where it enters the HP2, here it is heated actively from both sides. That provides a better heat transfer and helps that a homogeneous temperature is reached on both sides of the module. In this phase the temperature gap from HP1 closes and the temperature on both sides approaches each other.

The first thing that attracts the attention when comparing the simulation and the measurement is the temperature profile in the first 300 seconds. In the GG process shown in Figure 6 the pins are going down at the beginning and then go up again. This causes a direct contact with the heating plate, resulting in a fast temperature transfer. However, after the pins go up again, the temperature curve becomes flatter. The influence of this small step was not considered and therefore neglected in the simulation. This might also be the explanation for the

10.4229/EUPVSEC2023/3AV.1.5

difference in peak temperatures. The peak temperatures on the front and rear side measured are 154.4 °C and 154.1 °C, in the simulation, however, these are 143.6 °C and 140 °C, respectively.

Another point is the measured decrease in temperature between 350 and 400 seconds. Here the module is transferred from HP1 to HP2. The contact with the room temperature during this time is causing this small decrease. In the simulation this was not considered.

3.4 Degree of Crosslinking Curve

Our simulation tool also enables crosslinking degree predictions, potentially avoiding the need for costly and time-consuming prototype production and testing.

The results of the simulation are compared with the measurements conducted in our previous study, where we analyzed the degree of crosslinking over time for different curing durations at 150° C (see Figure 7) [2]. The simulation parameters were adopted. The glass thickness was changed to 3.2 mm and the lamination temperature was set to 150 °C, instead of 155 °C.

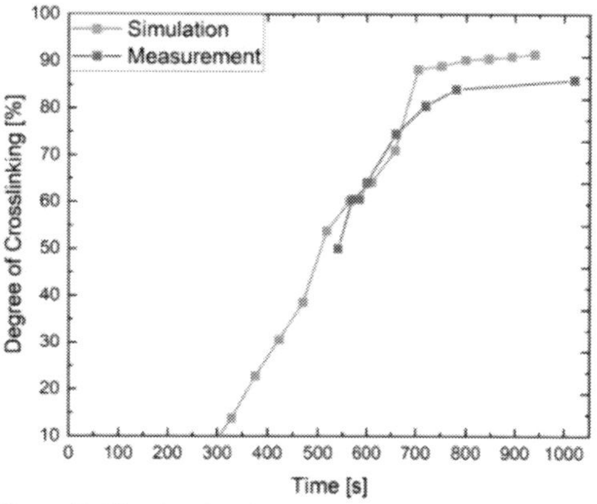

Figure 7: Simulated and measured degree of crosslinking over time for an EVA1, glass-backsheet module

Comparing simulation results with measurements from our prior study (Figure 7), we observed good agreement for most of the process. However, a divergence emerges after the 660 seconds, reaching a 5% difference at the end of the process (91% simulation versus 86% measurement). Therefore, we are investigating improvements on the simulation tool.

In summary, the results emphasize the significance of module lay-up, encapsulant choice, and temperature profiles in determining crosslinking uniformity and stability in photovoltaic modules.

4 CONCLUSION AND OUTLOOK

In conclusion, this study reveals crucial aspects of the lamination process for photovoltaic modules and its impact on crosslinking behavior. Through a systematic analysis of various module lay-ups, encapsulant materials, and process parameters, several significant findings have emerged.

Firstly, our initial investigation into the comparison of crosslinking on the front and rear sides of the module revealed that the degree of crosslinking was 6.5% lower on the rear side of glass-glass (GG) modules compared to the front side, indicating a temperature difference between the sides. This difference is attributed to the lower temperature transfer at the plate-membrane chamber or the lamination duration. To mitigate this, active heating on both sides during lamination, or extending the process duration in HP1 is proposed. Notably, the encapsulant type significantly influenced the crosslinking outcome, with EVA2 consistently exhibiting lower degrees of crosslinking compared to EVA1 when processed at the same temperature.

Moreover, incorporating cells into the laminate led to an important observation, the degree of crosslinking decreased by 3.5% to 6.5%. This shows the influence of cells on the degree of crosslinking and that this factor should not be neglected.

The temperature profile during lamination, analyzed through thermocouple measurements and simulation, exhibited a high degree of conformity between the two methods. Notably, the simulated temperature peaks closely matched the measured values, reaching 148.2 °C on the front and 145.7 °C on the rear side for GB modules. Additionally, the degree of crosslinking curve, simulated to save time and resources, exhibited a commendable alignment with measurements. Although a slight discrepancy was noted after 660 seconds, the simulation maintained a satisfactory correspondence, underscoring its practical value.

These results collectively illuminate key factors influencing the lamination process and crosslinking behavior in photovoltaic modules. Our findings not only contribute to enhancing the understanding of module manufacturing but also offer a validated simulation tool for predicting crosslinking behavior. With implications for module quality, reliability, and operational stability, this study provides valuable insights that could help to improve manufacturing processes.

In prospect, further research can delve deeper into optimizing process parameters, exploring advanced encapsulant materials, and extending the applicability of the simulation tool to various scenarios.

5 ACKNOWLEDGEMENT

This work was supported by the German Federal Ministry for Economic Affairs and Energy (BMWi) under the contract number 0324287C, acronym GEPARD.

6 REFERENCES

[1] Cattaneo, Gianluca & Faes, Antonin & Li, Heng-Yu & Galliano, Federico & Gragert, Maria & Yao, Yu & Grischke, Rainer & Söderström, Thomas & Despeisse, Matthieu & Ballif, Christophe & Perret-Aebi, Laure-Emmanuelle, Lamination process and encapsulation materials for glass–glass PV module design, Photovoltaics International (2015).

[2] A.K. Öz, C. Wellens, M. Mittag, M. Wiese, S. Sraisth, D. Klaus et al., 2.5 Minutes Lamination Process and the Influences of the Degree of Cross-Linking and the Moisture Ingress on the Degradation of PV Modules, 7 pages / 8th World Conference on Photovoltaic Energy Conversion; 816-822 (2022).

[3] A.K. Öz, C. Herzog, C. Wellens, D.E. Mansour, M. Heinrich, A. Kraft, The Impact of the Lamination Process on the Adhesion Properties at the Glass-Encapsulant Interface and Damp Heat Stability of PV Modules, 7 pages / 38th European Photovoltaic

Solar Energy Conference and Exhibition; 708-714 (2021).
[4] S. Sraisth, A.K. Öz, D. Klaus, C. Wellens, M. Heinrich, Influence of the Lamination Pressure on the Adhesion, Degree of Cross-Linking, and Bubble Formation of PV Modules, 7 pages / 8th World Conference on Photovoltaic Energy Conversion; 823-829 (2022).
[5] M. Fischer, M. Woodhouse, P. Baliozian, J. Trube, International Technology Roadmap for Photovoltaic (ITRPV) 14 (2023).
[6] C. Hirschl, L. Neumaier, S. Puchberger, W. Mühleisen, G. Oreski, G.C. Eder et al., Determination of the degree of ethylene vinyl acetate crosslinking via Soxhlet extraction: Gold standard or pitfall?, Solar Energy Materials and Solar Cells 143 (2015) 494–502.
[7] C. Lux, U. Blieske, E. Malguth, N. Bogdanski, Variations in Cross-Link Properties of EVA of Un-Aged and Aged PV-Modules, 5 pages / 29th European Photovoltaic Solar Energy Conference and Exhibition; 2462-2466 / 29th European Photovoltaic Solar Energy Conference and Exhibition; 2462-2466 (2014).
[8] M. Mittag, L. Vogt, C. Herzog, A. Pfreundt, J. Shahid, D.H. Neuhaus et al., Thermal Modelling of Photovoltaic Modules in Operation and Production, 9 pages / 36th European Photovoltaic Solar Energy Conference and Exhibition; 892-900 (2019).
[9] G.M. Wallner, B. Adothu, R. Pugstaller, F.R. Costa, S. Mallick, Comparison of Crosslinking Kinetics of UV-Transparent Ethylene-Vinyl Acetate Copolymer and Polyolefin Elastomer Encapsulants, Polymers 14 (2022).

10.4229/EUPVSEC2023/3AV.1.5
020153-006

ANALYSIS OF OPTICAL COUPLING GAINS FROM CELL INTERCONNECTION FOR THE ENERGY RATING OF PV MODULES

Alexander Protti, Anmol Welpulwar, Jibran Shahid, Max Mittag, Ammar Tummalieh, Christian Reichel
Fraunhofer Institute for Solar Energy Systems ISE, Heidenhofstr. 2, 79110 Freiburg, Germany
alexander.aguilar.protti@ise.fraunhofer.de

ABSTRACT: This paper discusses how the geometric design, and the optical properties of the cell interconnection ribbons and wires (CIR) could be adapted for improving the rated yield of PV modules. For such purpose, a comprehensive study of the shading losses and the coupling gains is done for rectangular, round, and triangular CIR cross-section geometries. A sensitivity analysis of the CIR reflectance and geometrical parametrization of each geometry is done with the aid of spectrally resolved raytracing. The impact of the effective width (EW) of the CIR on the rated power, the rated energy yield, and the climate specific energy rating (CSER) of a TOPCon module is then evaluated following the methods for a simulation-based Energy Rating. Results show that improved ribbon reflectance or geometry can enhance module performance by 0.7%. However, the combined effect of better reflectance and geometry can increase the power of a module by 1.6% and the energy yield per area by 1.5%. Furthermore, it is shown that the optical coupling gains can reduce the material consumption of CIR by 15% with still better module performance.
Keywords: Cell Interconnector Ribbons, Effective Width, Optical Coupling Gains, Energy Rating, CTM analysis, Ray Tracing

1 INTRODUCTION

The international standard IEC-61853 series [1] provides the guidelines for determining a rated energy yield [kWh/m²] and a Climate Specific Energy Rating (CSER) of photovoltaic (PV) modules. In the frame of the Ecodesign Directive and the Energy Labelling Regulations, the European Commission [2,3] proposed that manufacturers declare the performance of their modules using the procedures from this standard (IEC-61853), enabling the comparison of PV modules based on the rated energy yield [1]. This motivates manufacturers to look beyond the rated power [Wp] at standard test conditions (STC) by including the angular, spectral, low-light, and thermal response in the optimization of PV modules.

In a PV module, the cell interconnectors transport the energy from one solar cell to another, enabling the aggregation of voltage and power. The architecture of most solar cells [4] requires the use of cell interconnector ribbons and wires (CIR) which are soldered on top of the busbars and fingers, casting a shadow that reduces the rated power and energy yield from a module [5–7].

For any given number of CIR, decreasing their cross-section area, drops their optical footprint, but increases their electric resistance [7], leading to a classical optimization task [8,9].

The trend to reduce the finger width and increment the number of ribbons (i.e. busbars) [4] impacts both module reliability and the cell-to-module (CTM) efficiency, due to a redistribution of the ohmic losses in the string connectors [10]. However, a reduction in the cross-section area of each ribbon must follow, so that the sum of all cross-sections remains constant. The resulting spread of the optical footprint from CIR but does not decrease its magnitude.

Therefore, reduction of the shade from CIR requires alternative solutions, such as reflecting the light towards the cell surface (i.e. generating optical coupling gains) [11], shingling the cells [12], or completely removing the metallization from the cell surface (i.e. IBC cells). Still, shingled and IBC cells entail challenges that restrict their effectiveness and utilization. Thus, it is expected that CIR be the predominant technology in the market for at least the next decade [4], which creates the urge for maximizing the optical coupling gains from CIR.

Hence, this work explores how the geometry and optical properties of CIR can be adapted for improving the performance of a module. In this search, three main objectives are pursued. First, describe a formal mathematical parametrization of the net shading caused by cell metallization. Second, analyze how the net shading from CIR changes for different geometries, materials, and angles of incidence. Third, estimate the impact of these changes on the module performance, rated power, rated energy yield and Climate Specific Energy Rating.

2 METHODOLOGY

The coupling gains from the CIR can be experimentally determined either by reflection [11] or by light beam induced current (LBIC) [6] measurements. However, none of these methods are used, as both require preparation of samples (i.e., mini modules), complementary measurements and/or post processing that might increase the time, cost, and uncertainty of the results. Furthermore, the experimental methods in general do not offer better understanding of how the ribbons improve the performance of the modules, which is of relevance for this study.

Therefore, a ray tracer model has been developed and optimized for making spectrally resolved simulations for a wide range of CIR geometries. The ray tracer has been integrated into the cell-to-module (CTM) analysis model SmartCalc.Module by Fraunhofer ISE [13].

Other models have been used in studies to analyze the effect of the optical gains from CIR on module power and yield [5,9,14,15]. However, these analysis focus on the net effect (Wp, kWh) and offer limited knowledge about the mechanisms inside the module for overcoming the limitations of current ribbon designs. In the following sections a brief explanation complements existing literature [16–18] and helps to better understand the results of the present study. The equations presented in the next section are compatible with the CTM methodology

10.4229/EUPVSEC2023/3AV.1.9
020155-001

described in [6].

2.1 Optical properties of the CIR

Reflection of light from metals can be defined as specular, with a reflection coefficient described by the Fresnel optics [19]. Models that describe the effect of roughness on the material surface can be found elsewhere [20]. However, for materials with relatively low melting point (such as soldering allows) the coating of the ribbons melts during soldering, leading to a relatively low roughness [21,22]. The fact that most ribbon coatings are produced with a hot-dip process, contributes to this assumption, in case electrically conductive adhesives (ECA) are used. Hence, in the present analysis the roughness of the CIR is not considered, although other studies have made different assumptions [14].

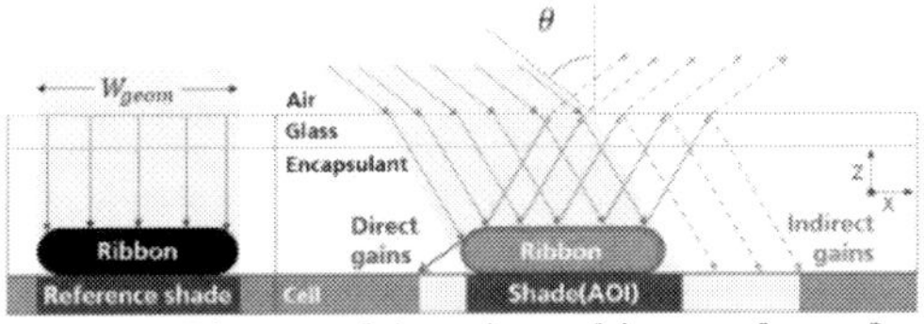

Figure 1: Diagram of the gains and losses of a perfect absorber (left) and a specular-reflecting CIR (right) for a given zenith angle (θ). The geometrical width (W_{geom}) is the width of the shade casted by the perfect absorber, when illuminated from the zenith (θ=0°). Dimensions are not to scale.

2.2 Mechanisms for generating coupling gains

The coupling gains can be classified in direct or indirect (Fig.1). In an unencapsulated cell, a fraction of the light that hits the metallization is either reflected towards the cell (i.e. direct gains) or absorbed by the metal, while most of the light is reflected away from the cell and lost [18]. Nevertheless, when the solar cell is embedded in a module, a fraction of the upwards reflected light is reflected downwards at the glass-air interphase, which increases the probability that it is used to generate electricity (i.e. indirect gains) [16,17]. This reduces the net shading effect of the cell metallization. How much of these indirect gains are achieved by partial or total internal reflection will depend on the angle of incidence (AOI) and the incident light spectrum, as well as the geometry and the optical properties of glass, encapsulant, CIR and cell (reflection, absorption, roughness, texture, etc.).

2.3 Analysis of the coupling gains

In the field of CTM analysis, the level of optical coupling caused by a "shading object" (busbar, finger or ribbon) can be parametrized in terms of the "effective width" (EW) [6,11], which is the complement of the "transparency" [14]:

$$EW = 1 - Transparency \tag{1}$$

Although, most readers might find the concept of "transparency" more intuitive to understand than the EW, the latter is easy to use in CTM analysis calculations and is always a positive value. Hence, it is explained in further detail in the next section.

2.4 The effective width (EW)

The EW indicates the normalized net shading loss after accounting for any reflection gains. Mathematically, this is the difference between the shaded flux (losses) [W] and the recovered flux (gains) [W] normalized by the shade caused by a perfect absorber (zero reflectance) with the same geometry, when illuminated from the zenith ("reference shade") (Fig.1). For a given pair of zenith (θ) and azimuth (ϕ) angles, the EW is the difference between normalized shade and normalized gains:

$$EW(\theta,\phi) = \frac{net\ shade(\theta,\phi)}{reference\ shade} \tag{2a}$$

$$EW(\theta,\phi) = \frac{shade(\theta,\phi) - gains(\theta,\phi)}{shade(0°,\phi)} \tag{2b}$$

$$EW(\theta,\phi) = normalized\ shade - normalized\ gains \tag{2c}$$

An ideal "coupling object" would have an EW=0%, as all the light would be always optimally redirected to the cell. The opposite would be an EW=100% at an AOI=0°, which then increases for AOI>0, as the shade covers a wider cell area, but no shaded light is recovered ($W_{shade} > W_{geom}$).

The angular shade and gains, depend not only on the AOI, but also on the spectral distribution of the incoming light, the geometry, and optical properties (refractive index, extinction coefficient) of the shading object and the module layers. The optical properties of the cell (reflectance, spectral response, texture, angular response, etc.) affect the EW as well.

2.5 Effective shade

In the case of a full-squared solar cell, the effective shade (ES) casted by a shading object with a given geometrical width (W_{geom}) and length (L_{geom}), can be defined as:

$$ES_i(\theta,\phi) = EW_i(\theta,\phi) * W_{i,geom} * L_{i,geom} * N \tag{3}$$

Here, N is the number of elements of one type of shading object i which can cause optical couplings, such as fingers (F), busbars (BB), or CIR.

2.6 Net shading loss from CIR

The following equation exemplifies the use of the ES for obtaining the "net fractional reduction in cell illumination caused by the CIR after encapsulation" (k_{CIR}):

$$k_{CIR} = \frac{A_{cell,geom} - ES_{F,encaps} - ES_{CIR/BB,encaps}}{A_{cell,geom} - ES_{F,encaps} - ES_{BB,encaps}} \tag{4}$$

Here, the subindex "CIR/BB" indicates that the CIR is placed on top of the busbar and both objects are accounted as one. The subindices "*encaps*" and "*air*" indicate whether the cell is encapsulated or not. In the case of a full square cell, the geometrical area of the cell can be obtained by the product of its length and width ($A_{cell,geom} = W_{geom,cell} * L_{geom,cell}$).

2.7 Incidence angle modifier

The incidence angle modifier describes the angular dependency of the short circuit current (Isc) of a module at given zenith (θ) and azimuth (ϕ) angles with respect to the Isc at zenithal irradiance:

$$IAM(\theta,\phi) = \frac{Isc(\theta,\phi)}{Isc(0,\phi)} \quad (5)$$

The EW behaves differently for each azimuth angle [14], which impacts the IAM. In such case, as specified in [1], the IAM of the module should be measured at two perpendicular azimuth angles ϕ_1 and $\phi_2 = \phi_1 + 90$. In the current analysis, the average of those values is considered for the virtual energy rating calculations:

$$IAM(\theta) = \frac{IAM(\theta,\phi_1) + IAM(\theta,\phi_2)}{Isc(0,\phi)} \quad (6)$$

3 VARIATION OF THE EW WITH RESPECT TO THE ANGLE OF INCIDENCE

In this section, the AOI is defined as the zenith angle of the incidence light on the glass, before refraction takes place, when the azimuth is parallel to the x coordinate (Fig.1). This direction is where the net shade is most sensitive to changes in the cross-section geometry of the shading object.

Since most used CIR nowadays are round wires, the angular dependency of the EW is exemplified for unitary height-to-width ratio (H/W=1). This is displayed in Fig. 2, 3 and 4 for square, round, and triangular CIR, where the black line represents the normalized shade, and the green line, the sum of both direct and indirect gains. The difference between the green and black line is the EW.

In the case of the square ribbon (Fig.2), the EW has a value of 100% at AOI=0° because all the light that hits the ribbon is reflected outside of the module. The direct gains increase proportional to the shaded area, which depends on the AOI (notice the symmetry in the gray shade and the red gains in Fig.2 upper right). The angular dependency of the indirect gains is described by the angular reflection on the glass-air surface (Fresnel's equations). So, light that hits the glass at an AOI close to 90° will be refracted inside the module almost at the critical angle. Since the horizontal surface of the ribbon is parallel to the glass surface, when the AOI is close to 90°, the indirect gains approach total internal reflection, reducing the EW. From this analysis the following conclusion can be derived: increasing the H/W ratio increases both normalized shade and direct gains but reduces the indirect gains.

For the round ribbon (Fig.3), there is an overlapping of indirect and indirect gains on the cell surface, because light is reflected in all directions. For an AOI=0°, the upper half of the ribbon surface is responsible for indirect reflection. However, the parts of the CIR surface with tilts greater than 45° (both left and right) reflect the light downwards, causing direct gains. As the AOI increases, the ratio between direct and indirect gains, as well as the EW remain practically constant. Still, changing the H/W ratio has a similar effect as described for the rectangular ribbon.

In the case of the triangular ribbon whose tilt is greater than 45° (Fig.4), there is a transition from direct to indirect coupling gains when the refracted angle (AOI_{inner}) is greater than a certain threshold ($AOI_{inner} > 2 * tilt - 90°$). For monochromatic light, this transition would cause a discontinuity in the EW. However, for polychromatic light, the transition takes place gradually (i.e. for a given AOI only a fraction of the spectrum is reflected close to horizontally). Furthermore, the shade of the ribbon increases when $AOI_{inner} > 90° - tilt$. For extremely flat (H/W<<1) or wide (H/W>>1) ribbons, the tendencies described for rectangular ribbons with respect to the H/W apply. In all cases, both direct and indirect gains decrease proportional to the reflectance of the CIR coating.

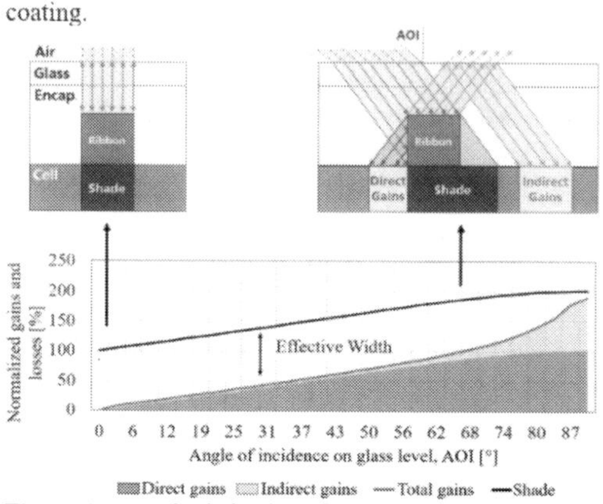

Figure 2: Angular behavior of the coupling gains, losses and effective width for a perfectly specular reflecting cell interconnector with a square cross-section area.

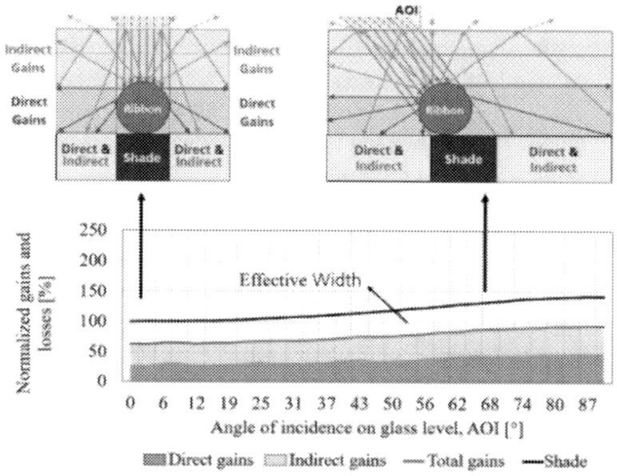

Figure 3: Angular behavior of the coupling gains, losses and effective width for a perfectly specular reflecting cell interconnector with a round cross-section area.

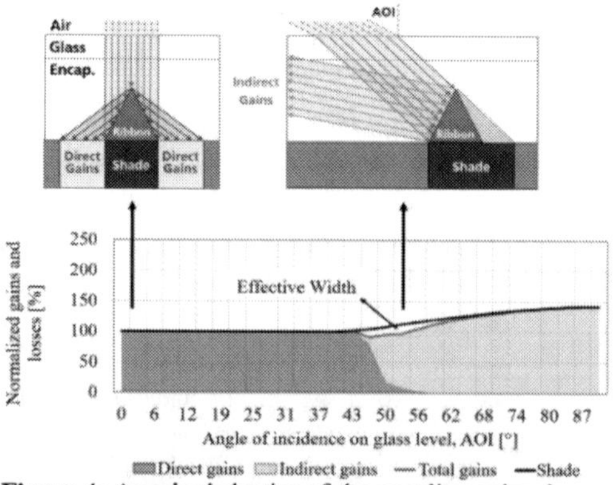

Figure 4: Angular behavior of the coupling gains, losses and effective width for a perfectly specular reflecting cell interconnector with a triangular cross-section area, with a unitary heigth-width ratio (H/W=1) and a tilt of 63.4°.

10.4229/EUPVSEC2023/3AV.1.9
020155-003

4 IMPACT OF THE EW ON MODULE PERFORMANCE

To evaluate the impact on module performance of different CIR designs, the module characteristics listed in the Table below are considered. In all cases, the reference ribbon is a tin-lead coated (Sn63Pb37, abbreviated as SnPb for simplicity) round wire, with a copper core and a diameter of 300 µm, which is widely used in current PV modules [4].

Table I: Characteristics of the simulated module

Layout	Cell spacing: 2 mm String spacing: 2 mm Margins: 21mm Number of cells: 120 Module length = 2.18 m Module width = 1.31 m Module area = 2.856 m²
Cell	Cell type: half-cut M12 TOPCon Efficiency = 24.4% Bifaciality = 83% $W_{F,geom}$ width = 0.05mm $W_{BB,geom}$ width = 0.05mm N_F = 60 (front) / 69 (rear) N_{BB} = 18
Cover	Front: 2.0 mm clear glass with ARC Rear: 2.0 mm clear glass with ARC
Encapsulant	Front: UV transmitting EVA Rear: UV blocking EVA Thickness = 600 µm
Cell interconnectors	Resistivity = 1.8 µΩ-cm Dimensions: see table II
String interconnectors	Resistivity = 1.8 µΩ-cm Width = 5.072 mm Thickness = 0.472 mm
Junction box	Number: 2 Internal resistance = 0.03 mΩ
Cables	Length = 0.5 m Cross section = 4 mm² Specific resistance = 0.02 Ω-mm²/m Contact resistance plug/jack=0.3mΩ

4.1 Variation of the ribbon reflectance

The effect of the reflecting properties of the coating is studied by using the refractive index and extinction coefficient of the reference ribbon (tin-lead- solder alloy (Sn63Pb37)) to that of copper (Cu), aluminum (Al) and silver (Ag). In all cases, a width of 300 µm and an area of 0.071 mm² is used (see Table II).

The results are summarized in Figure 5. Here, for all geometries, the reference ribbon (blue) has the lowest power, energy yield and energy rating, while silver (green bar) presents consistently the highest values. In the case of the round ribbon, a silver coating can increase the power and yield of the module by 0.7%. That is 4.2 Wp or 2.2 kWh/m². Results for coating materials such as Al and Cu are between the limits of SnPb and Ag.

This relative behavior between the coatings is due to the differences in the spectral and angular reflectance of the different materials (see Fig.6). Here the spectral response is included to highlight the relative importance of the reflectance in each wavelength. The tin-lead solder alloy has overall the lowest reflectance, which is similar to copper for wavelengths below 600 nm. At greater wavelengths, copper has a remarkably better reflectance. In the case of aluminum, it is overall a better reflector than copper, but at around 600 nm is outperformed by copper. Silver behaves almost as a perfect reflector for wavelengths greater than 500 nm.

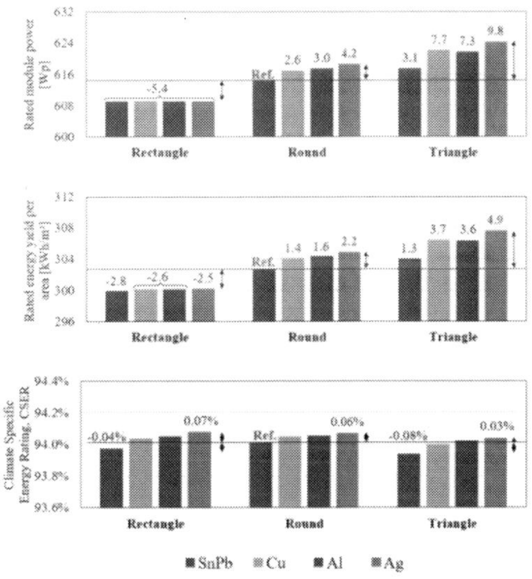

Figure 5: Effect of the coating and shape of the ribbon on the module performance

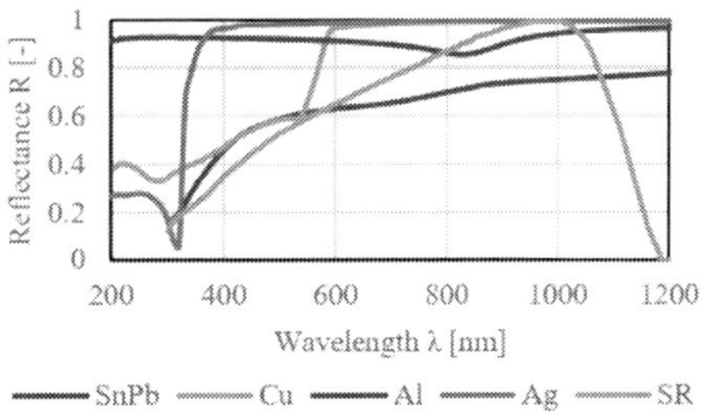

Figure 6: Spectral reflectance of tin-lead (SnPb), copper (Cu) aluminum (Al) and silver (Ag) coatings different coating materials for a normal incident angle. The normalized spectral response (SR) of the cell is included to illustrate the relative relevance of each reflected wavelength on the optical gains.

4.2 Variation of the ribbon shape

Figure 5 shows that rectangular ribbons cause a reduction of 5.4 Wp (-0.88%) and 2.8 kWh/m² (-0.92%) in module performance. On the other hand, changing the shape from round to triangular, causes an increment of 3.1 Wp (0.5%) and 1.3 kWh/m² (0.4%) in module power and yield per area, respectively, without changing the ribbon coating. However, a combination of using silver on triangular-shaped ribbons causes an increment of almost 9.8 Wp (1.6%) and 4.9 kWh/m² (1.5%) with respect to the reference.

Whenever the shape or the coating are changed, the CSER remains practically constant because power and yield are increased in a similar proportion.

10.4229/EUPVSEC2023/3AV.1.9
020155-004

Table II: Area, height and height-to-width ratio (H/W) of analyzed ribbon shapes with a width of 300 µm

Area, mm²	Height, µm			H/W (Tilt, °)		
	■	▲	●	■	▲	●
0.045	150	**300**	191	0.50	**1.00 (63.4°)**	0.64
0.060	200	400	255	0.67	1.33 (69.4°)	0.85
0.071	236	471	**300**	0.79	1.57 (72.3°)	**1.00**
0.090	**300**	600	382	**1.00**	2.00 (76.0°)	1.27

4.3 Variation of the ribbon cross-section area and aspect ratio

The variations in the ribbon cross-section were done using the dimension showed in Table II. Figure 7 shows that power and yield improve with greater ribbon cross-section area, because the ohmic resistance decreases. This is not that important under low-light irradiance because the resistive losses decrease with the square of the current ($\Delta P=R*I^2$). Therefore, changes in module power due to reduced series resistance are relatively greater than the improvements in module yield, which explain the decreasing trend in the CSER with respect to the cross-section area. Still, the coupling gains of triangular ribbons can improve module power, yield and CSER, even when the cross-section area of the ribbon is reduced by 15% (from 0,071 to 0,06 mm²).

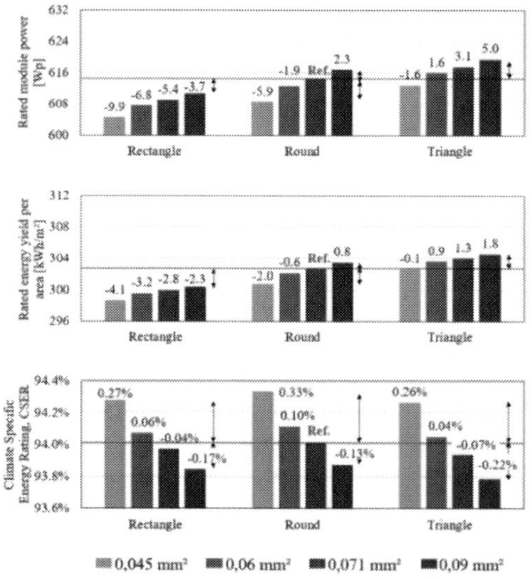

Figure 7: Effect of the cross section area and shape of the ribbon on module performance

5 CONCLUSION

A formal mathematical description of the optical coupling gains of cell interconnector ribbons was presented in this study. An analytical explanation of the angular behavior of the EW for different ribbon shapes elucidated the benefits of using triangular ribbons with H/W>1. Furthermore, the values presented here work as first approximations when performing cell-to-module (CTM) analysis under STC and non-STC conditions, simplifying the acquisition of data for CTM analysis.

The impact of different ribbons geometries was studied for different performance indicators, including the energy rating. It was shown that the optical coupling gains not only can improve module power, rated energy yield and energy rating, but also allow saving of ribbon material.

6 ACKNOWLEDGEMENT

This work was supported by the Federal Ministry for Economics and Climate Action (BMWK) under contract number 03EE1045B, with the acronym SMART.

7 REFERENCES

[1] International Electrotechnical Commission, Photovoltaic (PV) module performance testing and energy rating – Part 4: Standard reference climatic profiles, 1st ed., IEC Central Office, Geneva, Switzerland, 2018.

[2] N. Dodd, N. Espinosa, Preparatory study for solar photovoltaic modules, inverters and systems, 2019.

[3] A. Wade, et al., The Product Environmental Footprint (PEF) of photovoltaic modules-Lessons learned from the environmental footprint pilot phase on the way to a single market for green products in the European Union, Prog Photovolt Res Appl 26 (2018) 553–564. https://doi.org/10.1002/pip.2956.

[4] ITRPV, International Technology Roadmap for Photovoltaic (ITRPV): 2022 Results, fourteenth ed., 2023.

[5] M. Mittag, et al., Triangular ribbons for improved module efficiency, in: Proceedings of the 32nd European Photovoltaic Solar Energy Conference and Exhibition, Munich, Germany, 2016, pp. 169–172.

[6] I. Hädrich, Unified methodology for determining CTM ratios: Systematic prediction of module power, in: Proceedings of the 4th International Conference on Crystalline Silicon Photovoltaics, 's-Hertogenbosch, Netherlands, 2014.

[7] H. Hanifi, et al., Investigation of cell-to-module (CTM) ratios of PV modules by analysis of loss and gain mechanisms, Photovoltaics International (2016) 89–99.

[8] A. Protti, J. Shahid, M. Mittag, D. H. Neuhaus, U. Kräling, M. Kaiser, Virtual Energy Rating: a method for optimizin module performance through cell-to-module analysis.

[9] H. Hanifi, et al., Optimized tab width in half-cell modules, Enrgy Proced 92 (2016) 52–59. https://doi.org/10.1016/j.egypro.2016.07.009.

[10] Student Awards Finalist Presentation: Impact of String Connection and Contact Degradation on Electrical Current Distribution on Solar Cell and Photovoltaic Module Level: a Magnetic Field Imaging Validated Model, 2023.

[11] I. Hädrich, et al., Finger and ribbon optics for increasing module power, in: Proceedings of the 31st European Photovoltaic Solar Energy Conference and Exhibition, Hamburg, Germany, 2015.

[12] M. Mittag, et al., Cell-to-Module (CTM) Analysis for Photovoltaic Modules with Shingled Solar Cells, in: 2017 IEEE 44th Photovoltaic Specialist

Conference (PVSC), Washington, DC, IEEE, 25.06.2017 - 30.06.2017, pp. 1531–1536.
[13] Fraunhofer Institute for Solar Energy Systems ISE, SmartCalc.Module. http://www.cell-to-module.com (accessed 1 September 2023).
[14] I. Haedrich, M. Ernst, Impact of angular irradiance distributions on coupling gains and energy yield of cell interconnection designs in silicon solar modules in tracking and fixed systems, J. Phys. D: Appl. Phys. 54 (2021) 224003. https://doi.org/10.1088/1361-6463/abe967.
[15] I. Haedrich, M. Ernst, Impact of Optimizing Cell Metallization for Local Conditions on the Module Energy Yield, in: 2021 IEEE 48th Photovoltaic Specialists Conference (PVSC), Fort Lauderdale, FL, USA, IEEE, 2021, pp. 2104–2110.
[16] A.W. Blakers, Shading losses of solar-cell metal grids, Journal of Applied Physics 71 (1992) 5237–5241. https://doi.org/10.1063/1.350580.
[17] M.F. Stuckings, A.W. Blakers, A study of shading and resistive loss from the fingers of encapsulated solar cells, Sol Energ Mat Sol C 59 (1999) 233–242. https://doi.org/10.1016/S0927-0248(99)00024-0.
[18] R. Woehl, M. Hoerteis, S.W. Glunz, Analysis of the Optical Properties of Screen-Printed and Aerosol-Printed and Plated Fingers of Silicon Solar Cells, Advances in OptoElectronics 2008 (2008).
[19] Frank L. Pedrotti Leno M. Pedrotti Leno S. Pedrotti, Introduction to Optics 3/E.
[20] Matt Pharr, Wenzel Jakob, Greg Humphreys, Physically Based Rendering: From Theory To Implementation, 2023. https://pbr-book.org/.
[21] R. Meier, et al., Microstructural Optimization Approach of Solar Cell Interconnectors Fatigue Behavior for Enhanced Module Lifetime in Extreme Climates, Energy Procedia 92 (2016) 560–568. https://doi.org/10.1016/j.egypro.2016.07.020.
[22] Tae-Sik Cho, Chul-Sik Cho, Enhancement of the Surface Smoothness of Cu Ribbon for Solar Cell Modules 16 (2015) 20–24.

SIMULATION OF HEAT FLOW AND NOVEL MANUFACTURING DESIGN OF PVT MODULES

R. Koepge, P. Sheela, U. Zeller, H. Schwabe
Fraunhofer Center for Silicon Photovoltaics CSP
Otto-Eissfeldt-Strasse 12, 06120 Halle, Germany

ringo.koepge@csp.fraunhofer.de

ABSTRACT: Nowadays PVT modules (Photovoltaic Thermal modules) consist of a standard PV module and a separated heat absorber. The aim of such devices is to generate heat and electricity simultaneously in a compact unit. Justified by the fact that standard PV panels are connected to a common heat exchanger, this investigation faces two major topics of a new PVT designs. The technical and scientific innovation is based on the attempt of joint manufacturing of both components (PV module and heat exchanger) to decrease production costs and increase the PVT module performance. A novel module design is determined by replacing the polymer back sheet with a thin aluminum plate. That plate represents one surface of a full area contact heat exchanger. Manufacturing challenges caused by different thermal strain were solved. In addition, several heat exchanger designs have been simulated to improve PVT module weight by decreasing the volume of the water body as well as to ensure an almost homogeneous temperature distribution. Furthermore, the module design should cause a small drop of water pressure when the heat exchanger is working to ensure an economic use of the PVT module. The presented PVT modules operate at ambient temperatures to achieve an optimum of the electrical yield at the expense of low fluid temperatures. But, for that new PVT approach the lower thermal level is sufficient for low temperature heat cycles or heat pump applications. Finally, a PVT module and heat exchanger design will be presented having the potential to be transferred into production.

Keywords: PVT, Lamination, Manufacturing, Thermal Expansion, Simulation, Heat Flow, Heat Exchanger

1 INTRODUCTION

Applications for solar energy usage can be subdivided into two major categories, photovoltaics and solar thermal modules. In photovoltaics the solar energy is transformed into electrical energy using semiconductor materials like silicon-based cells, whereas in solar thermal modules the radiation is converted into heat. By combining these two methods known as Photovoltaic Thermal collector (PVT) systems electrical and thermal efficiencies can be improved. One of the main parameters reducing the electrical efficiency is the increase of cell temperatures in the module. Currently, the efficiency of the commercially available solar modules is in the range of 15 to 20 percent. This amount of the incoming solar irradiation is used by the cells for generating electricity, the rest of the energy is either reflected or absorbed by the cells leading to increased temperatures of the cells and module. PVT systems are using forced convection of water or air as the working medium. Heat absorbers are installed alongside the PV modules influencing the cell operating temperature. Since the cell temperatures are reduced the electrical yield is increased. Numerous investigations of established PVT modules have already been carried out [1-7]. Since standard PV panels are usually combined with a separate heat exchanger, a major challenge arises for a new design of PVT modules. It would be advantageous to design the module as a single compact unit.

This investigation faces at first a novel module design that is focused on replacing the polymer back sheet with a thin aluminum sheet. That sheet acts simultaneously as one surface of a full area contact heat exchanger. Second, a novel heat exchanger design is simulated to improve PVT properties. In detail, the drop of fluid pressure, the weight of a PVT module and its temperature distribution are considered for improvement. The new designed PVT module in general operates at ambient temperatures to achieve an optimum of the electrical yield at the expense of outcoming coolant temperatures. But, for this PVT approach the thermal properties of heat absorber and coolant are suited for low temperature heat cycles or heat pump applications. In summary, the PVT module market is at its beginning and new module designs and concepts are required for market penetration and dedicated applications.

2 METHODS

The development of a novel PVT design was subdivided into two stages. Stage one faces the manufacturing part, how to solve the challenges caused by the lamination process. The second stage deals with the topic of finding a heat exchanger design that can be applied to a clean flat aluminum back sheet. The fabrication of the heat exchanger and its application is part of future investigations outside of this current study.

2.1 Manufacturing

The lamination of materials with a huge difference in thermal expansion coefficient will cause mechanical stress inside the module. The following Table 1 shows a general overview of the coefficients. Standard modules approximatively have a length of 1.7 m. In that case the difference in thermal expansion between aluminum and glass parallel to the length of the module is in a range of 3 mm. The expansion of the encapsulant is neglectable because of the very low Youngs Modulus. The silicon cell also shows a thermal expansion but, the length of one cell in comparison to the length of the glass and aluminum is quite short. In addition, each cell is surrounded by encapsulant material, and the cell position is almost in the middle (neutral axis). That´s why the thermal expansion of the silicon cell is also not important for module manufacturing.

Table 1 Thermal expansion coefficients [8-10]

No.	Material	Thermal Expansion Coefficient α
1	glass	3,3 to 10 x 10^{-6}/K
2	aluminum	23,1 x 10^{-6}/K

The cross linking of the encapsulant EVA (ethylene–vinyl acetate) that was used, takes place at approximatively 150 °C. At this moment an irreversible connection between glass and aluminum is made. By cooling down the module to room temperature the higher contraction of the aluminum generates residual stresses inside the module. These stresses are causing a module deformation that means a module bending.

The approach to get rid of the deformation is to apply a counter bend during the lamination. Therefore, the expected bending that is caused by the thermal expansion needs to be quantified and a lamination tool needs to be constructed. The module bending was determined by measuring the distance between the back plate surface and the ground. Three measurements for each module are taken. The schematic side view can be seen in Figure 1.

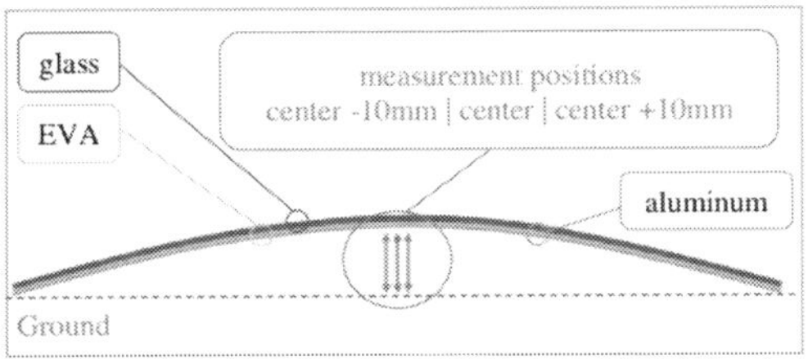

Figure 1 Schematic side view of solar module laminate without solar cells after lamination process. The residual stresses cause a module bending that was quantified by three measurements in the center of each module.

The lamination process was performed at Meyer Solar Solutions GmbH ICOLAM 28/26 at a temperature of 155 °C to ensure the full cross linking of the encapsulant. As mentioned, an inverse bending of the module is needed to manufacture a plane module. That bending needs to take place in two dimensions, parallel to the long edge and parallel to the short edge of the glass plate. The bending along the long edge has the major impact. Thus, a set up for a one-dimensional bending was applied. Five PV modules were manufactured with a counter bending lamination process,

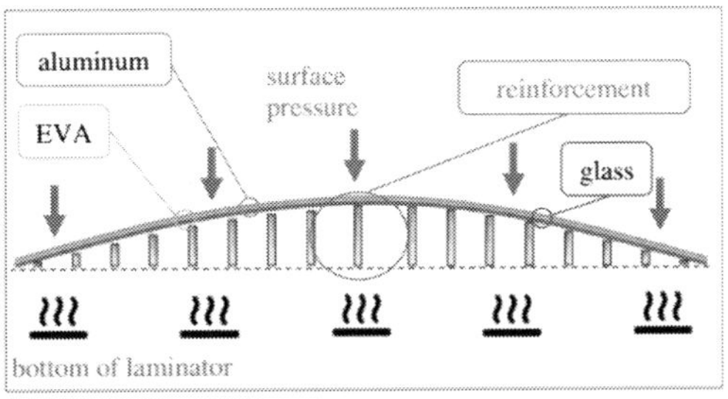

Figure 2 Schematic side view of solar module laminate in counter bend status during the lamination process.

The reinforcement of the deformed module was done by duromer polymer foam body placed at the same distance under the module.

2.2 Simulation | Approach and targets

Standard PV module design was drawn using FUSION360. FUSION360 is a commercial computer-aided design (CAD) software application by Autodesk. The module consists of a front glass, front EVA encapsulant, 120 half cells (161x80.5 mm²) in butterfly configuration, a back EVA encapsulant and an aluminum back sheet (Figure 3). The dimension of the module is 1000 x 1680 mm², the water body of the heat exchanger is neglected in this figure.

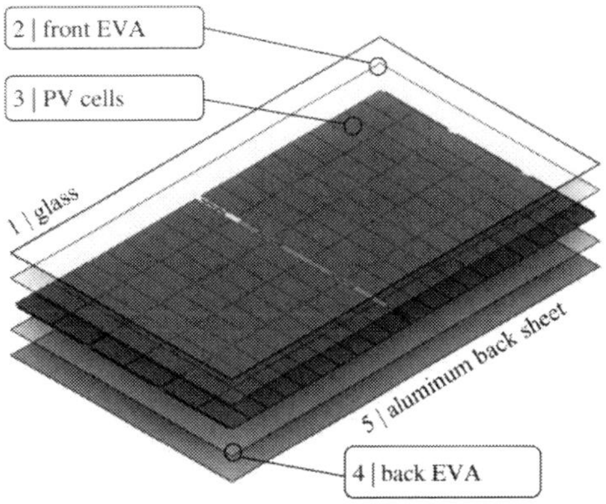

Figure 3 Exploded view of designed 3D CAD model of PV module without heat exchanger water body.

Two heat exchanger layouts are exemplary used for simulation, see Figure 4. Layout 1 (Figure 4a) is defined by a constant thickness of the fluid water within the heat exchanger (7.5 mm) with 18 circle imprints with a depth of 4 mm (edge fillet 5 mm) and a diameter of 150 mm. The second layout (Figure 4b) has a wedge shape, the thickness of the fluid water within the heat exchanger increases linear from the top (water inlet) to the bottom (water outlet) from 5 mm to 9.5 mm. This so-called radiator model has 6 imprints with a depth of 4 mm.

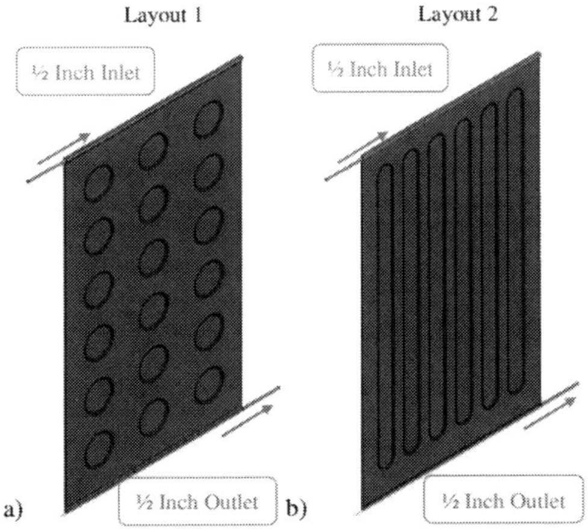

Figure 4 Heat exchanger design, a) Layout 1: circle imprints and constant thickness of the heat exchanger, b) Layout 2: ellipse imprints and linear change of heat exchanger thickness from top to the bottom (wedge shape)

The following cross-section view (Figure 5) shows all materials that were considered for simulation as well as the following table (Table 2) shows general material and geometric data.

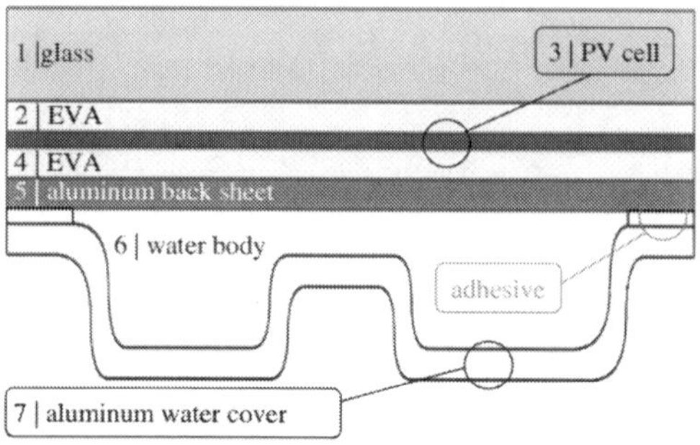

Figure 5 Cross-section is showing how PVT system was designed.

The cross-section view shows all components of Figure 3 extended by the heat exchanger (water body and an aluminum water cover sheet metal. The connection of the aluminum water cover was done by adhesive bonding. These material properties were not considered for simulation.

Table 2 Overview of PVT module materials

No.	Layer [ ]	Thickness [mm]	Specific Heat Capacity [J/Kg K]	Density [kg/m³]	Thermal Conductivity [W/m K]
1	glass	3.00	500	3000	1.80
2	EVA	0.45	2090	960	0.35
3	cells	0.20	677	2330	148
4	EVA	0.45	2090	960	0.35
5	aluminum (back sheet)	0.30	691	2400	32
6	Water body		4200	998	0.598
7	aluminum (water cover)	3.00	691	2400	32

The major targets for simulation are the reduction of the total PVT module weight (including water) below 30 kg, a maximum of temperature difference across the module of 5 K and a pressure drop between water in and water outlet less than 0.05 MPa. Ansys 2022 R1 steady state thermal is utilized to simulate the 3D model. Various mass flow rates and water inlet temperatures were used throughout the entire setup and process. The following Table 3 shows an overview of the simulation DoE. The water inlet temperature was set constant to 5°C.

Table 3 Overview of simulation runs DoE parameter; water inlet temperature was set constant to 5°C.

Simulation Run No.	Layout No.	Mass Flow Rate	Ambient Temperature
1	1	100 l/h	5°C
2	1	200 l/h	5°C
3	1	100 l/h	20°C
4	1	200 l/h	20°C
5	2	100 l/h	5°C
6	2	200 l/h	5°C
7	2	100 l/h	20°C
8	2	200 l/h	20°C

2.3 Simulation | Conditions and assumptions

In the present simulation there are two domains - solids and fluids. A conjugate heat transfer mechanism was considered to assess the heat flow through fluid-solid interactions. To achieve this, the entire model is split into two domains depending on how they physically appear: the water constitutes a fluid domain, while the solar cells, EVA, glass, and aluminum back sheet constitute a solid domain. All fluid-solid barriers were subject to a no-slip criterion. To resolve and examine the turbulence fluid pipe, the k-two-transport-equation model with an improved wall treatment was used. The technical data module defines the material properties and characteristics for each solar module layer, including density, thermal conductivity and specific heat capacity.

The PVT developed in the present system is based on the following assumptions:

1. The flow is considered as k epsilon turbulence under steady state condition.
2. All the properties of the materials are considered independent of temperature.
3. No shading and dust formation on the PVT module.
4. The heat transfer coefficient to the atmosphere is assumed as per ambient temperature.
5. No formation of air bubbles in the fluid flow.
6. Transmissivity of the EVA is considered as 100%.
7. Arrangement of the junction boxes is neglected

The 3d-CAD model (from FUSION360) is imported into the modelling component of ANSYS. For the steady state thermal simulation, the fluid domains are neglected/suppressed. The remaining solid components are assigned the corresponding material data/parameters.

Thermal boundary conditions are considered with standard irradiation of 1000 W/m² perpendicular to the glass plane of the module and film coefficient to the atmosphere of 5 W/m²K for free convection. The ambient temperature to the atmosphere is 5 °C and 20 °C according to the cases.

3 RESULTS

Manufacturing issues were solved. First modules to demonstrate a successful fabrication of a PV module with an aluminum back sheet were manufactured and characterized. The calculated weight including the PV module, the fluid water and aluminum water cover sheet below 30kg. The simulation analysis was focused on the temperature distribution. The simulation model has been established so far, but still requires additional validation calculations. Since these were not intended within the project framework, they are part of planned follow-up research tasks. Consequently, the final results are taking into account temperature driven electrical and hydraulic performance are still pending.

3.1 Demo Modules

The module bending after the lamination process was determined (see Figure 1), when changing the standard polymer back sheet versus an aluminum sheet. The following Table 4 shows the results of measuring the module deformation at three points in the center of each

10.4229/EUPVSEC2023/3AV.1.19
020160-003

module that were manufactured under different process conditions. No process window was observed that can avoid or significantly reduce the module deformation. Thus, a manufacturing setup was established that realizes a counter bending (see Figure 2).

Table 4 Overview of module deformation for different Process condition

Process No.	Position 1 Center -10 mm	Position 2 Center	Position 3 Center +10 mm
1	49,1 mm	46,7 mm	46,7 mm
2	48,7 mm	46,5 mm	46,5 mm
3	48,0 mm	47,4 mm	47,4 mm
4	51,2 mm	48,3 mm	48,3 mm
5	50,7 mm	48,2 mm	48,2 mm
6	50,5 mm	48,6 mm	48,6 mm
7	49,3 mm	47,8 mm	47,8 mm
8	48,4 mm	47,5 mm	47,5 mm
9	47,3 mm	45,1 mm	45,1 mm

With the help of the counter bending tool a manufacturing of a solar module with an aluminum back sheet was possible. To ensure a short circuit free PV module an ETFE layer was applied between the aluminum back sheet and the solar cells. In detail the layer stack beginning from back side consists of an aluminum back sheet, EVA layer, ETFE layer, EVA layer, PV cells, EVA layer and a front glass. The resulting demo module can be seen in Figure 6a. After manufacturing all modules are checked by electroluminescence photography for cell cracks and short circuit areas. All five modules passed the quality check. Exemplary an electroluminescence image of one of the five modules is shown in Figure 6b.

At this point the demo modules are ready for the next step of new PVT design development. The heat exchanger needs to be designed and applied to the demo module. Thus, a fist simulation was performed to evaluate the performance of a full surface heat exchanger concept.

a) b)

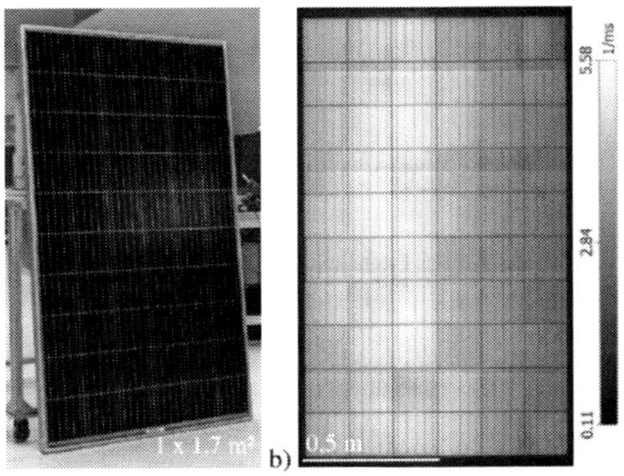

Figure 6 One of five 60-cells-modules with an aluminum back sheet and an optimized light absorbing surface coating, a) light photography, b) Electroluminescence image, no defects are detected after manufacturing.

3.2 Temperature distribution

The simulation result will help to design a new type of full surface heat exchanger. Two layouts were analyzed regarding their temperature distribution, especially the temperature of the solar cells are under investigation. Module temperature correlates to the module performance. Temperature coefficient of short circuit current (I_{SC}), temperature coefficient of open circuit voltage (V_{OC}) and temperature coefficient of maximum output power (P_{max}) define the correlation between temperature variation and resulting performance drop or increase of the module performance in comparison to operation at standard conditions. Therefore, it is important to take care of a module warming in general as well as temperature differences between the cells on the same module because it will also cause a performance drop. Figure 7 shows the temperature distribution of two modules, in this case the cell temperatures of simulation run No. 2 and simulation run No. 6. Both simulations run with the following parameters, 200 l/h mass flow rate and 5 °C ambient temperature.

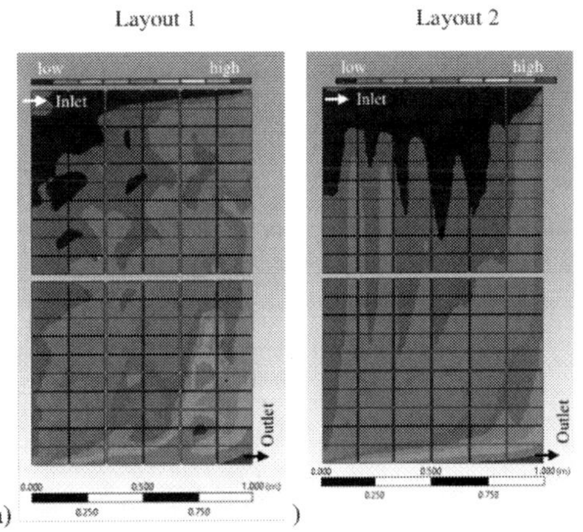

a))

Figure 7 Simulation results of cell temperature distribution caused by sunlight irradiation and water cooling of the backside at 200 l/h mass flow rate and 5 °C ambient temperature, a) Layout 1, circle imprints and plane-parallel water body shape b) Layout 2, ellipse imprints and wedge shape of water body.

Temperature differences between the inlet and outlet area were analyzed in all simulation runs. The best cooling performance was observed at 5 °C ambient temperature and a mass flow rate of 200 l/h (simulation run No. 2 and 6). A systematic approach for heat exchanger layouts is ongoing.

4 SUMMARY AND CONCLUSION

A novel manufacturing design of a PVT module successfully was implemented. Challenges regarding the lamination of two materials with a large difference of the thermal expansion coefficient were solved and five demonstration modules were fabricated without any defects. The demo modules are ready for the next steps of PVT development. Thus, a first simulation of two heat exchanger layouts was exemplary performed. Both layouts show an almost equal cooling performance, the major influencing parameters are ambient temperature and mass flow rate. Simulation results still need to be validated. I.e., the boundary conditions and results need to be checked and reviewed. Based on the first module prototypes with an aluminum back sheet and an accompanying simulation of two first full surface heat exchanger the next development will focus on the fabrication and application of the heat exchanger.

5 ACKNOLEGEMENT

Financial support by the "Investitionsbank Sachsen-Anhalt" of the funded project "Spekab" (Grant notice number 2204/00016) as well as the Federal Ministry for Economic Affairs and Climate Action funded project "AluPV" (FKZ: 03EN1069B) is gratefully acknowledged.

6 REFERENCES

[1] Arvind Tiwari, M.S. Sodha, Performance evaluation of hybrid PV/thermal water/air heating system: A Parametric study, Renewable Energy, Volume 31, Issue 15, 2006, Pages 2460-2474, ISSN 0960-1481

[2] Ulrich Fritzsche, Markus Schweiger, Florian Reil, Performance Prediction of PVT Modules – the Link between Thermal and Electrical Operation, EU PVSEC, 2017

[3] T.T. Chow, A review on photovoltaic/thermal hybrid solar technology, Applied Energy, Volume 87, Issue 2, 2010, Pages 365-379, ISSN 0306-2619

[4] A. Mellor, D. Alonso Alvarez, I. Guarracino, A. Ramos, A. Riverola Lacasta, L. Ferre Llin, A.J. Murrell, D.J. Paul, D. Chemisana, C.N. Markides, N.J. Ekins-Daukes, Roadmap for the next-generation of hybrid photovoltaic-thermal solar energy collectors, Solar Energy, Volume 174, 2018, Pages 386-398, ISSN 0038-092X

[5] H.A. Zondag, Flat-plate PV-Thermal collectors and systems: A review, Renewable and Sustainable Energy Reviews, Volume 12, Issue 4, 2008, Pages 891-959, ISSN 1364-0321

[6] Saeed Abdul-Ganiyu, David A Quansah, Emmanuel W Ramde, Razak Seidu and Muyiwa S. Adaramola, Investigation of Solar Photovoltaic-Thermal (PVT) and Solar Photovoltaic (PV) Performance: A Case Study in Ghana. Energies 2020, 13(11), 2701

[7] S. Diwania1, S. Agrawal, A.S. Siddiqui, S. Singh. Photovoltaic–thermal (PV/T) technology: A comprehensive review on applications and its advancement. International Journal of Energy and Environmental Engineering (2020) 33–54

[8] Schott Technische Gläser - Physikalische und chemische Eigenschaften, Mainz im Mai 1999

[9] Paetec GmbH: Formeln und Tabellen für die Sekundarstufen I u. II. Berlin 1996, Ausdehnungskoeffizienten von Feststoffen und Flüssigkeiten.

[10] William M. Haynes (Hrsg.): CRC Handbook of Chemistry and Physics. A ready-reference Book of chemical and physical Data. 92. Auflage. CRC Press, Boca Raton FL u. a. 2011, ISBN 978-1-4398-5511-9

10.4229/EUPVSEC2023/3AV.1.19
020160-005

A "NEW" PV GENERATOR AT THE IES-UPM WITH BIFACIAL PV MODULES MANUFACTURED IN 1987

F. Martínez-Moreno*, P. Merodio, E. Lorenzo

Instituto de Energía Solar – Universidad Politécnica de Madrid. Grupo de Sistemas Fotovoltaicos (IES-UPM).
Campus Sur UPM. Ctra. Valencia km. 7. ETSI Sistemas de Telecomunicación. 28031 Madrid, Spain.

*Corresponding author: francisco.martinez@ies.upm.es

ABSTRACT: This work introduces the new PV generator installed on the rooftop of the IES headquarters at the Campus Sur of the Universidad Politécnica de Madrid. The peculiarity of this generator lies in its modules, the oldest commercially manufactured bifacial modules from the 1980s. 59 of these modules have been recovered from real installations after at least 20 years of outdoor exposure. Prior to connection, the modules were cleaned, repaired and tested for reliability. On May 5, 2021, 56 of the 59 modules were connected constituting a 3.79 kW PV generator (nominal power of the front side) which is injecting power into the grid. The results obtained in the power characterization of the modules, as well as the yearly bifacial performance ratio of the generator, are a clear evidence of their good health. In addition, a comparison between different versions of the performance ratio to evaluate a bifacial generator is shown.

Keywords: bifacial PV modules, bifacial PV generator, ageing, reliability, performance ratio.

1 INTRODUCTION

Bifacial technology seems to be called to lead the global PV market in the coming years. Nowadays, 30% of the manufactured PV modules are entirely bifacial, but about 25% of the monofacial modules are also manufactured with bifacial cells. In other words, more than 50% of the modules manufactured today include bifacial cells. Moreover, it is expected that 60% of the modules to be manufactured during the first years of the next decade will be fully bifacial [1].

It can be though that bifacial technology is a novelty which has appeared in the last 5 – 10 years. Nothing could be further from the truth: although the major production of bifacial modules started in the second half of 2010s, the first commercialization of bifacial PV modules was developed by ISOFOTON in 1980s [2].

Researchers from the IES-UPM have recovered 59 modules "ISO-90" manufactured in 1987 from real installations after about 20 years of outdoor exposure (Fig. 1).

Figure 1: Modules ISO-90 manufactured by ISOFOTON in Spain (1987).

These modules give us the opportunity to evaluate the degradation of the oldest bifacial PV modules working so long time to answer two key questions:

- Is bifacial technology as reliable as traditional monofacial technology?
- Is the commitment to bifacial technology the right one?

2 THE PV MODULES

The ISO-90 PV modules have 72 circular bifacial cells based on n-PERT technology from 100 mm diameter wafers. They are connected in a series/parallel array (36 in series/2 in parallel) and are encapsulated inside the module with EVA. The rear side has a polymeric sheet to protect the module (TEDLAR).

This configuration leads to a rated front power at Standard Test Conditions (STC are defined as $G^* = 1000\text{W/m}^2$ and $T_C^* = 25°\text{C}$) of 67.7 W for the whole PV module (0.8 m^2), with a bifaciality close to 70%. So, the total power rating expected for these modules according ISOFOTON procedures is 90 W.

Once the modules had been recovered after at least 20 years of outdoor exposure, they were cleaned to characterize them according to international standards [3] to determine their state of health.

2.1 Visual inspection.

First, we noticed that these modules have open-air contacts, which represent a potential risk of direct contact with people. Therefore, they must be installed in a low DC voltage configuration to avoid electrical hazards.

In addition, this feature means that the contacts could be prone to develop corrosion, as can be seen in Fig. 2a. In most of the modules, we were able to replace the rusted screw to restore the functionality of the module (Fig. 2b). But in 7 out of the 59 modules the corrosion was so severe that we were forced us to break the terminal and directly solder a cable on the terminal place (Fig. 2c). Once these contacts were repaired, the modules were ready to be electrically characterized.

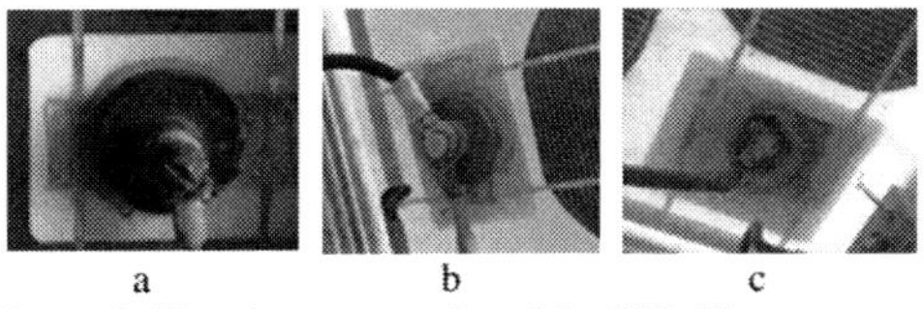

Figure 2: Electric contacts of modules ISO-90.

10.4229/EUPVSEC2023/3AV.1.26
020162-001

Secondly, the back-sheet does not show any visual symptoms of degradation, such as bubbles, delamination etc. However, the EVA shows a noticeable browning in the central part of the front side of the solar cells (Fig. 3). Nevertheless, the power measurements presented below suggest the impact of this browning on performance is rather modest, as reported by other authors [4].

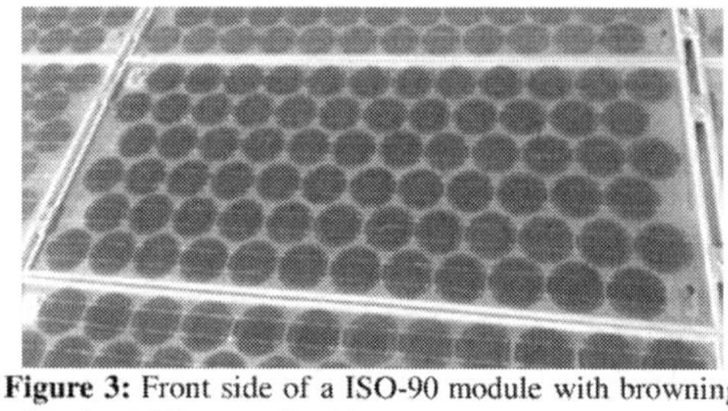

Figure 3: Front side of a ISO-90 module with browning after about 20 years of outdoor exposure.

The rear part does not present this phenomenon because the UV radiation reaching this side has been significantly lower (Fig. 4).

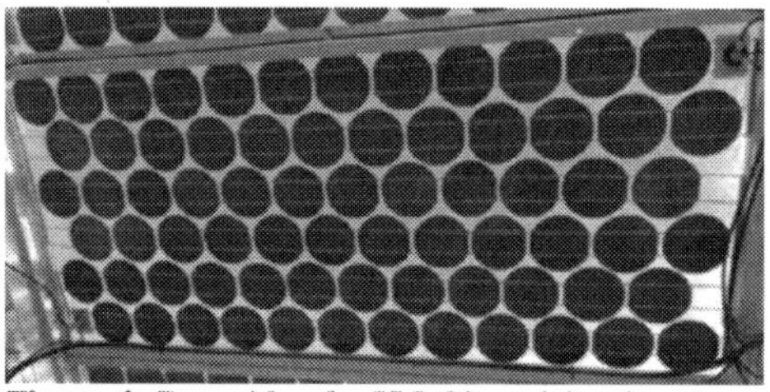

Figure 4: Rear side of a ISO-90 module after about 20 years of outdoor exposure (all without browning).

Finally, we have detected some misalignments in the intercell connection ribbons of the modules (Fig. 5a) and some cells with marks (Fig. 5b). Both features are typical of the manual assembling and they have no negative impact on performance (the marks are related to the tapes that were used for assuring the solar cells before module lamination).

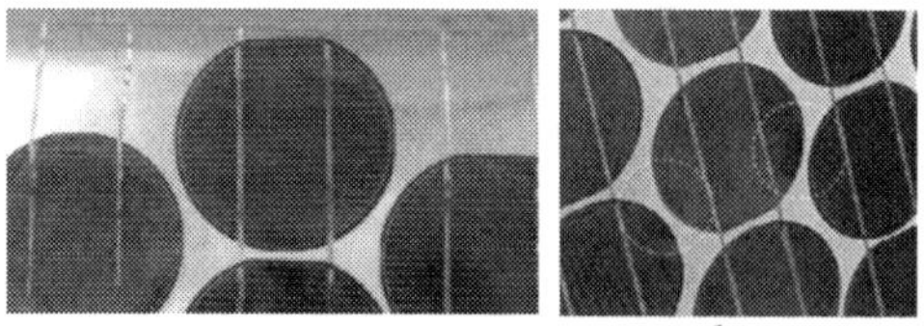

a b

Figure 5: Signs of manual module assembling.

2.2 Electrical isolation.

In order to check if the modules are electrically safe, we have performed the electrical isolation test (dry conditions) according to the procedure explained in the international standards [3]. Although the modules are to be installed in a low DC voltage configuration (< 60 V_{DC}), we have considered a maximum voltage system of 600 V_{DC} to perform the test. The minimum isolation resistance that a 0.8 m^2 module must have is 51.3 MΩ. The module with the lower isolation resistance has 700 MΩ. Therefore, the use of these modules will not pose any electrical risk.

2.3 Power characterization.

All the PV modules have been tested at STC with a standard monofacial solar simulator. Both front and rear I-V curves have been obtained independently (the untested side has been covered with an opaque sheet to avoid its electrical contribution). Fig. 6 and Fig. 7 show their respective power histograms. Note that the front power results, with differences of 19% between maximum and minimum values, are fully consistent with the typical tolerance of ±10% reported by manufacturers at that time [5].

As can be seen, the front power has a normal distribution, with a mean value at STC of 62 W. That is, the power losses related to nominal STC power are 8.4%. This means that modules show an aging of -0.42% or lower (these losses result from a combination of original underrating –an extended practice at that time– and aging, so actual aging should be lower; moreover, it is assumed that these modules have been exposed for at least 20 years; if a figure higher than 20 years is considered, the yearly losses decrease). Thus, the modules show a really good state of electrical health, with degradation rates similar that those currently guaranteed by the manufacturers (36 years later) and better than those reported at that time (between -0.8%/year and -1%/year).

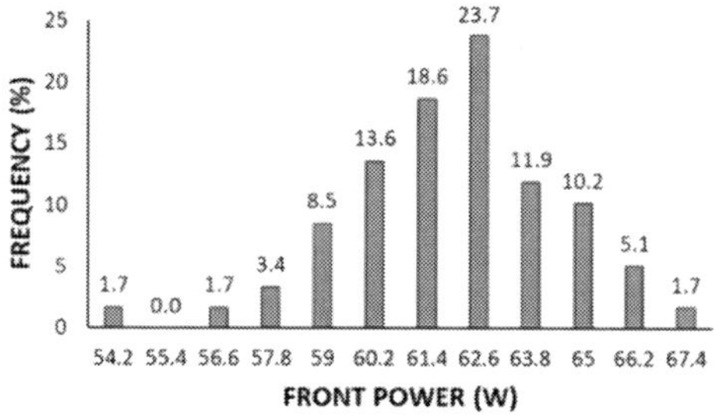

Figure 6: Histogram of P^*_{FRONT}.

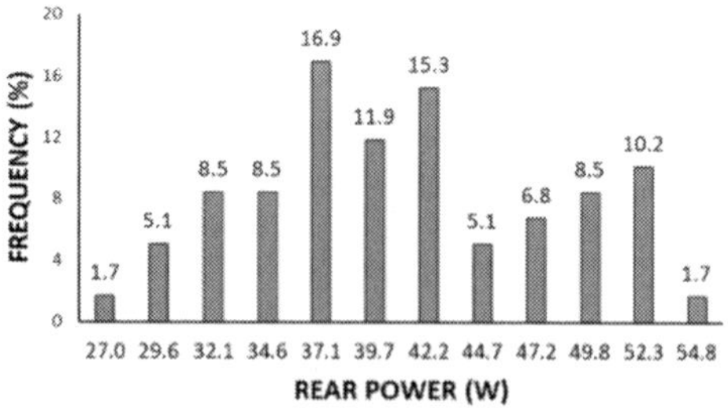

Figure 7: Histogram of P^*_{REAR}.

The actual mean power at STC of the rear side, is 41 W. Therefore, the actual power bifaciality is $\varphi_P = 66\%$, close to the nominal value. It is clear that dispersion of rear power (standard deviation of 6.9 W, leading to a coefficient of variation[1] of 16.8%) is much larger than the one related to front power (standard deviation of 2.5 W,

[1] The coefficient of variation is defined as the ratio between the standard deviation and the mean value.

with a coefficient of variation of 4%). This is also a common practice in current bifacial PV modules [6]: the manufacturing of the front side of the cells is done more carefully and homogeneously compared to the rear side one. This is because the front side is the main responsible for power generation as it receives most of the solar irradiance (in sunny conditions, usually 80% or higher).

3 THE PV GENERATOR

Once the modules were cleaned, repaired and tested for reliability, 56 of the 59 modules were connected constituting a 3.79 kW PV generator (nominal power of the front side) which is injecting power into the grid. To minimize electrical risk, as well as mismatch losses, we have connected them in a low DC voltage configuration. The modules have been arranged in 2 strings in parallel, each consisting of 2 modules in series and connected to the grid through Huayu HY-1200-Plus microinverters with four different MPP inputs at 30 V_{DC}. Each one of these sets of 4 modules has been connected to each of these inverter inputs, which convert DC power into AC power with a maximum efficiency of 95% (i.e. the minimum measured value of conversion losses is 5%). Besides, seeking to minimize mismatch losses, modules in series have similar current at the MPP.

The system has been in routine operation since May 5, 2021. In its first two years of operation, the generator has injected into the grid 11480 kWh (5722 kWh in its first year and 5758 kWh in its second year). Fig. 8 and Fig. 9 show the front and the rear sides of this generator.

Figure 8: Front view of the PV generator.

Figure 9: Rear view of the PV generator.

In order to evaluate the quality of the installation we calculated the performance ratio, PR. Fig. 10 shows the weekly PR values (this explains the subscript "w") during its first year of operation:

- PR_W^F (orange squares) represents the traditional performance ratio, which accounts only for the irradiation received on the front side:

$$PR_W^F = \frac{E_{AC,W}}{P_{FRONT}^{*} \int_W G'_{FRONT}\, dt}$$

In this expression, $E_{AC,W}$ is the energy injected by the generator into the grid during the considered week, P_{FRONT}^{*} is the front power of the generator (3.79 kW) and G'_{FRONT} is the irradiance reaching the front side of the generator, G_{FRONT}, divided by the reference irradiance at STC, G^{*}:

$$G'_{FRONT} = \frac{G_{FRONT}}{G^{*}}$$

- PR_W^B (blue circles) represents the bifacial performance ratio [7], which accounts all the irradiation received on both sides, front and rear:

$$PR_W^B = \frac{E_{AC,W}}{P_{FRONT}^{*} \int_W G'_{FRONT}(1 + \varphi_P \cdot GBR)\, dt}$$

In this expression, GBR is the so-called Irradiance Bifacial Ratio, defined as the ratio between the rear and front irradiances:

$$GBR = \frac{G_{REAR}}{G_{FRONT}}$$

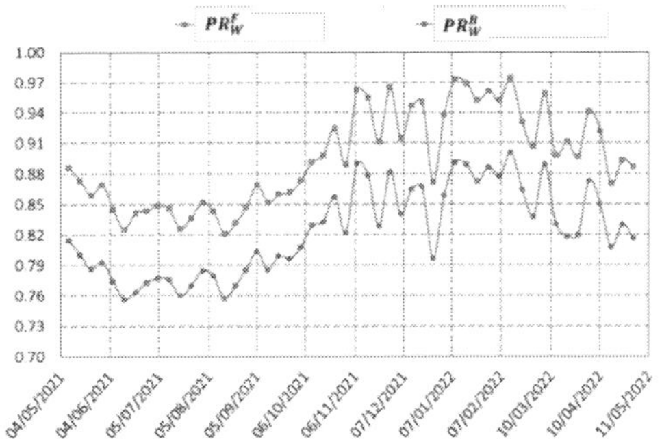

Figure 10: Weekly performance ratio (front and bifacial) of the PV generator during its first year of operation.

It is noticeable that PR_W^B values are lower than those of PR_W^F. This is because PR_W^F disregards the rear irradiation contribution to the energy yield. In fact, the physical meaning of PR_W^F is doubtful: some values for winter weeks (from November to March) are higher than 0.95. Thus, if losses due to inverter conversion are considered (5% or higher), the new performance value obtained will be higher than 1, which is not consistent (in addition, there are other losses that have not been considered yet: temperature, module mismatch, soiling, wiring, etc.).

Thus, although the front performance ratio (typical of monofacial installations) is often used for bifacial PV plants [8] [9], its use to properly evaluate the performance of PV installations with bifacial PV modules seems erroneous. It is more reasonable to take into account the rear irradiation and calculate the bifacial performance ratio, a variable with more consistent values (now leaving enough space to include the rest of losses which are present in the system but are not yet considered in this variable).

Furthermore, as can be seen both graphs have the same shape, with performance ratios values changing throughout the year: the lowest values are related to the summer weeks (from June to September) while the highest values belong to the winter weeks (from November to March). This is mainly due to the different temperature losses affecting the PV modules in the different seasons. Therefore, in order to correctly assess the quality of a PV installation it is necessary to consider the entire year of operation. Otherwise, erroneous conclusions could be drawn. In this case, the yearly value for both indicators are $PR_Y^F = 0.88$ and $PR_Y^B = 0.81$.

But, what if a quick quality control is needed (one week or one month)? In this case, depending on the season of the year, the result of the performance evaluation using any of these performance ratios can be very different. Thus, the bifacial performance ratio corrected in temperature, $PR25_W^B$, which has been plotted in Fig. 11, is even better than the bifacial performance ratio:

- $PR25_W^B$ (yellow triangles) represents the weekly bifacial performance ratio which also accounts for the temperature losses:

$$PR25_W^B = \frac{E_{AC,W}}{P^*_{FRONT}\int_W G'_{TOTAL}\left[1+\frac{\gamma}{100}\cdot\left(T_C - T_C^*\right)\right]dt}$$

In this expression, γ is the temperature coefficient of power (%/°C), T_C is the cell temperature and

$$G'_{TOTAL} = G'_{FRONT}(1+\varphi_P\cdot GBR)$$

It is now evident that in predominantly sunny weeks, regardless the season, the values of $PR25_W^B$ show lower variations (between 0.87 and 0.93) than the corresponding to PR_W^B (between 0.76 and 0.89). In addition, the weekly values are closer to the yearly one, $PR25_Y^B = 0.90$.

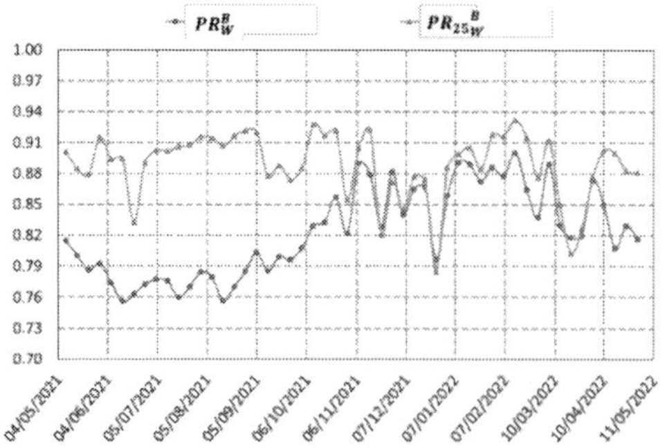

Figure 11: Weekly bifacial performance ratio (with and without temperature correction) of the PV generator during its first year of operation.

5 SUMMARY

A 3.79 kW PV generator (front side) consisting of 56 of the oldest commercially manufactured bifacial modules from the 1980s has been installed on the rooftop of the IES headquarters at the Campus Sur of the Universidad Politécnica de Madrid. This generator is in operation injecting power into the grid from May 5, 2021.

The PV modules have been recovered from real installations after at least 20 years of outdoor exposure. Once cleaned, they were characterized to determine their state of health.

A visual inspection revealed that some of them had had broken or severely corroded contacts, but have been repaired to recover its electrical functionality. In addition, several modules show noticeable browning on their front side. As alarming as this may seem, it does not affect their electrical insulation or electrical response: all the modules are electrically safe and their average front power is at most an 8.4 lower than the nominal value. Thus, their degradation rate -0.42%/year is in line with the current degradation rates guaranteed by manufacturers 36 years later: -0.45%/year.

Another evidence of the good health of these modules is the yearly bifacial performance ratio of the generator: taking into account both frontal and rear irradiation, its bifacial performance ratio is 81%. It is advisable to use this parameter to evaluate the quality of a bifacial PV installation, instead of the traditional performance ratio used in monofacial installations, which only considers the frontal irradiation. Otherwise, inconsistent results could be achieved. Moreover, when quality tests are performed in short time periods lower than a year (one week, one month), the bifacial performance ratio corrected in temperature is preferred to avoid seasonal dependences: in the sunny periods, the weekly values of this bifacial performance ratio are more similar to the yearly one (90%).

Thus, despite these modules have been in operation for decades, they are still safe and maintain a remarkable capacity of energy production. Therefore, bifacial technology appears to be as reliable and durable as the traditional monofacial technology. The results obtained in this work support the current commitment to bifacial technology, which is based on manufacturing techniques already used more than 30 years ago.

6 ACKNOWLEDGEMENTS

This work was supported by the Spanish MCIN/AEI/10.13039/501100011033 through Project GREASE (Proyectos de I + D + i Retos Investigación 2020), with grant number PID2020-113533RB-C31.

7 REFERENCES

[1] International Technology Roadmap for Photovoltaic (ITRPV). 2021 results. 13th edition. March 2022

[2] Eguren J., Martínez-Moreno F., Merodio P., Lorenzo E. First bifacial PV modules early 1983. Solar Energy 243, 327-335 (2022). https://doi.org/10.1016/j.solener.2022.08.002

[3] IEC 61215-2: 2021. Terrestrial photovoltaic (PV) modules - Design qualification and type approval - Part 2: Test procedures

[4] Jordan D.C., Silverman T.J., Wohlgemuth J.H, Kurtz S.R., VanSant K.T. Photovoltaic failure and degradation modes. Progress in Photovoltaics, 25, 318-326 (2017). https://doi.org/10.1002/pip.2866

[5] TamizhMani M.G., Mikonowicz A., Kuitche J. Nameplate, datasheet and sampling requirements for PV modules. A Solar ABCs Proposed Standard.

10.4229/EUPVSEC2023/3AV.1.26
020162-004

Stakeholder Webinar 3 November 2011. http://www.solarabcs.org/about/publications/meeting_presentations_minutes/2011/11/pdfs/webinar-Nameplate-rating-tolerance-3Nov2011-v2.pdf (consulted on 17th July 2022)

[6] Parra V., Gómez J.C., Vázquez J.C., Álvarez F. Tackling inconsistencies in bifacial PV technology. PV Tech Power (24) 29-34 (2020)

[7] Ledesma J.R., Almeida R.H., Martínez-Moreno F., Rossa C., Martín-Rueda J., Narvarte L., Lorenzo E. A simulation model of the irradiation and energy yield of large bifacial photovoltaic plants. Solar Energy 206, 522-538 (2020). https://doi.org/10.1016/j.solener.2020.05.108

[8] Ayala P., Muñoz C., Osorio N., Hernández C., Zurita F., Gutierrez V., Ramirez G., Mancilla F., Valdivia P., Cuevas F., Ferrada P. Bifacial technology performance compared with three commercial monofacial PV technologies under outdoor high irradiance conditions at the Atacama Desert. IEEE 7th WCPEC (a joint conference of 45th IEEE PVSC, 28th PVSEC & 34th EU PVSEC), 672–675 (2018).

[9] Singh J.P., Walsh T.M., Aberle A.G. Performance investigation of bifacial PV modules in the Tropics. 27th EUPVSEC, 3262-3266 (2012). https://doi.org/10.4229/27thEUPVSEC2012-4BV.2.15

INFLUENCE OF BACK SHEET THERMAL INSULATOR ON OPERATIONAL TEMPERATURE OF PHOTOVOLTAIC–THERMAL (PVT) MODULES

R.R. Vardanyan, N.K. Badalyan, M.G. Travajyan
National Polytechnic University of Armenia
105 Teryan Str. Yerevan 0009, Armenia
Corresponding author: ruben.vardanyan@polytechnic.am

ABSTRACT: The hybrid PVT system proved to be much more energy efficient with improved electrical efficiency due to the cooling of PV cells and gained thermal energy due to heat absorbed from the heated PV cells. The glazed, unglazed, and different thermal absorber configurations of PVT modules are investigated by different research teams. In this work, the influence of the thermal insulating back sheet on the operational temperatures of PVT modules theoretically and experimentally is investigated. The PVT hybrid modules with the thermal insulating back sheet and without the thermal insulating back sheet are considered. The comparative analysis of operational temperatures of PVT modules and general PV modules is conducted. It is shown that at low ambient temperatures (below some "critical" value) the temperature of the PVT module can be higher than the temperature of the general PV module ("reverse" process). It is also shown that the PVT hybrid module without a thermal insulating back sheet can be used in a broader range of ambient temperatures than the PVT module with the thermal insulating back sheet. The results of this work can be used by scientists and engineers for optimal design and application of PVT hybrid modules and systems.
Keywords: photovoltaic, thermal, hybrid, module

1 INTRODUCTION

Solar photovoltaic and thermal (PVT) is a hybrid technology that combines photovoltaic (PV) solar cells, which converts solar energy into electricity, and a solar thermal collector, which transfers the unused excess heat from the PV module into the useful thermal energy in the form of heated water or air. The hybrid PVT system proved to be much more energy efficient with improved electrical efficiency due to the cooling of PV cells and gained thermal energy due to heat absorbed from the heated PV cells.

Numerous research works have been conducted for the investigation of the physical parameters and properties of PVT hybrid systems [1- 4]. Various design configurations for hybrid PVT collectors are compiled and assessed [5]. In most cases, the glazed (covered) and non-glazed (non-covered) PVT modules are considered. The PVT systems with different absorber configurations, various materials, and geometry also are investigated [6]. Unfortunately, there is a lack of investigations on the role and importance of the application of thermal insulation back sheets.

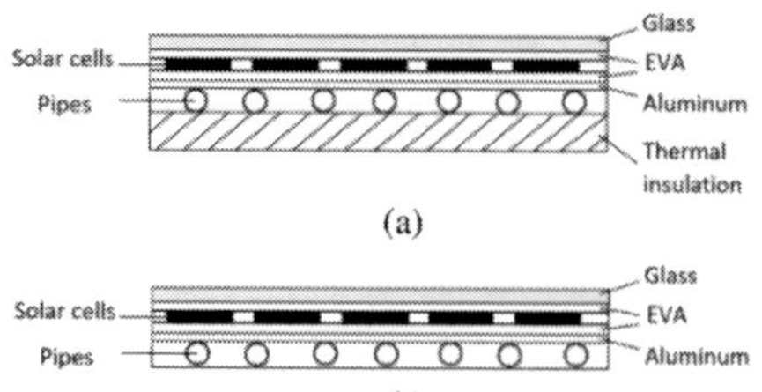

Figure 1: PVT collectors with the thermal insulating back sheet (a) and without the thermal insulating back sheet (b)

At present, the configuration of hybrid PVT modules, consisting of a general PV module integrated with a thermal absorber and back sheet thermal insulator is widely used (Fig. 1a). The PVT hybrid modules with back sheet thermal insulators are produced by different companies. In parallel with this, there are some designs without a back sheet thermal insulator (Fig. 1b), or it is stated by the producer that the thermal insulating back sheet is optional. In this work, the influence of the thermal insulating back sheet on the operational temperatures of PVT modules is investigated.

2 THERMAL INSULATING BACK SHEET INFLUENCE ON PARAMETERS OF PVT MODULE

The operational temperature of the PVT hybrid module can be defined with the use of the expression [7]:

$$T_{cell.PVT} = \frac{1}{K1+K2}[F(\tau\alpha) + K1\ Tin + K2\ Ta], \qquad (1)$$

where $K1 = 2\frac{mC}{AS}$, $K2 = \frac{FU}{G}$, m is the mass of fluid in the storage tank (g), C is the heat capacity of fluid (J/g ^{0}C), A is the area of the PVT solar collector receiving sun rays, S is the solar radiation energy per square meter, F is the PVT collector's heat removal factor, $(\tau\alpha)$ is the effective transmittance of the cover (τ) and absorbance of collector (α) factor, U is the overall heat loss coefficient of the collector, G is the intensity of solar radiation on the unit of the surface (W/m^2), Tin is inlet temperatures of PVT collector (^{0}C), T_a is the ambient temperature (^{0}C). In accordance with formula (1), the overall heat loss coefficient of the collector U depends on the application or not of the thermal insulating back sheet (Fig. 1).

The main purpose of the application of a PVT hybrid system is the reduction of the operating temperature of solar cells of PV modules. Therefore, it is reasonable to compare the temperature of the PVT hybrid collector with the temperature of the general PV module in different weather conditions, i.e. under the influence of different ambient temperatures.

The temperature of the solar cells of PV modules can be calculated on the base of the NOCT (Nominal Operating Cell Temperature). The NOCT is defined as the temperature reached by open-circuited cells in a module under the standard reference environment defined as follows: solar irradiation 800 W/m^2, ambient temperature 20 ^{0}C, wind velocity 1m/s, and mounting type - open backside. The best module operated at a NOCT of 33 °C, the worst at 58 °C, and the typical module at 48 °C respectively. An approximate expression for calculating the cell temperature is given by [8]

$$T_{PV} = Ta + G\,\frac{NOCT-20}{800}\,. \qquad (2)$$

Thus, the temperature of the solar cell in the PV module T_{pv} increases linearly with an increase in ambient temperature (Fig. 2).

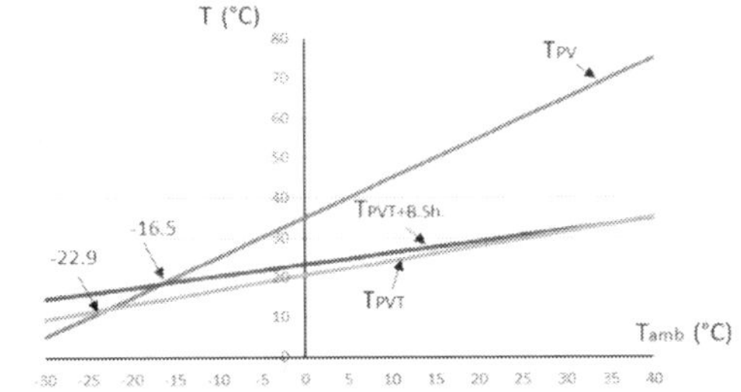

Figure 2: The temperatures of the PVT hybrid modules without thermal insulating back sheet (T_{PVT}) and with the thermal insulating back sheet ($T_{PVT+B.Sh.}$), as well as the temperature of general PV module (T_{PV}), vs ambient temperature T_{amb} (G = 1000 W/m^2, NOCT = 48 ^{0}C, PVT system tank capacity is 100 l, A = 2 m^2, Tin = 15 ^{0}C, for unglazed flat without thermal insulator $F(\tau\alpha) = 0.45$, FU = 15 W/m^2 ^{0}C and for unglazed flat PVT+B.Sh. with a thermal insulator $F(\tau\alpha) = 0.45$, FU = 10.15 W/m^2 ^{0}C [9])

In Fig. 2 the dependences of operational temperatures of PVT hybrid modules without a thermal insulating back sheet (T_{PVT}) and with a back sheet ($T_{PVT+B.Sh.}$) from the ambient temperature plotted by formula (1) are presented. It can be seen from Fig. 2 that the operational temperatures T_{PVT} of the PVT hybrid modules without thermal insulating back sheet (Fig. 1b) at different ambient temperatures T_{amb} is less than the temperatures $T_{PVT+B.Sh.}$ of the PVT hybrid modules with the thermal insulating back sheet (Fig. 1a). Thus, the application of the insulating back sheet increases the temperature of PVT hybrid module and consequently, the thermal output of the PVT system is increasing, but the electrical efficiency of solar cells is decreasing.

The interception points of lines T_{PVT} (without thermal insulating back sheet) and $T_{PVT+B.Sh.}$ (with the thermal insulating back sheet), with the characteristic of general PV module T_{PV}, are at T_{amb} = - 22.9 ^{0}C and T_{amb} = - 16.5 ^{0}C respectively. These temperatures are called "critical" [7]. Below these temperatures, the PVT hybrid modules are warmer than the general PV module ("reverse" process). As a result of this process, the electrical efficiency of the PVT module is becoming smaller than the efficiency of the PV module. Consequently, the application of PVT hybrid systems in these cold conditions is not reasonable, since the main purpose of a PVT system is the cooling of the PV module. It can be concluded also that the PVT module without a thermal insulating back sheet can be used in a wider range of ambient temperatures than the PVT module with a thermal insulating back sheet.

3 EXPERIMENTAL SETUP AND METHODOLOGY

The influence of the back sheet thermal insulator on the operational temperature of the PVT hybrid module experimentally is investigated. For comparison, the temperature of a general PV module in the same conditions is measured as well. The experiments are conducted in outdoor conditions of the research laboratory for "Semiconductor photovoltaic devices" at the National Polytechnic University of Armenia (Fig.3).

The PVT module is prepared by attaching the thermal collector to the rear side of a general PV module (unglazed type). For this purpose as well as for the reference PV module the polycrystalline PV modules with an output power of 275 W, produced in Armenia are used. The hybrid PVT and PV modules are placed close to each other and tilted by 30° from horizon (Fig. 3).

The measurements were conducted for the PVT system with and without thermal insulating back sheet on the same PVT module. The circulation of water in the PVT module was realized by using the thermo-siphon system with the thermally insulated water tank as well as by forced circulation.

Figure 3: PVT hybrid module (from the left) and reference PV module (from the right)

For measurements of temperatures at different points of the experimental setup, an automated measurement system was developed (Fig. 4). It consists of an Arduino microcontroller, DS18B20-type temperature sensors, a Micro SD card module for saving the measured data, and a power supply unit.

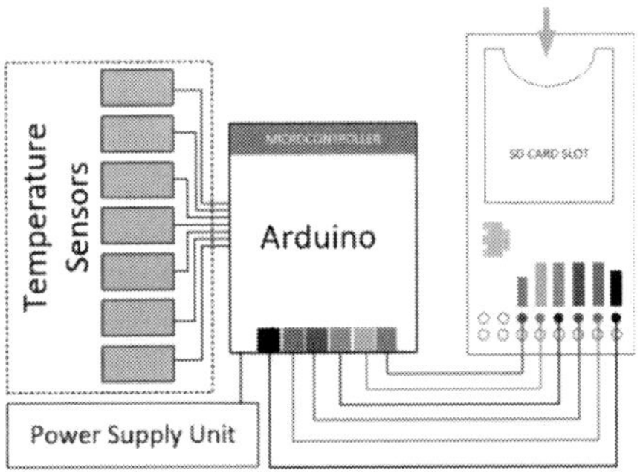

Figure 4: Block diagram of the automated system for the measurements of temperature at different points

The measurements were performed according to the pre-written program code for the microcontroller. The measurements were conducted at 7 points at the same time with the time intervals of 20 seconds. The measured data is stored in a Micro SD memory card from which the data is imported to a computer for processing and analysis.

The block diagram of the experimental setup with the

connections of temperature sensors is presented in Fig. 5.

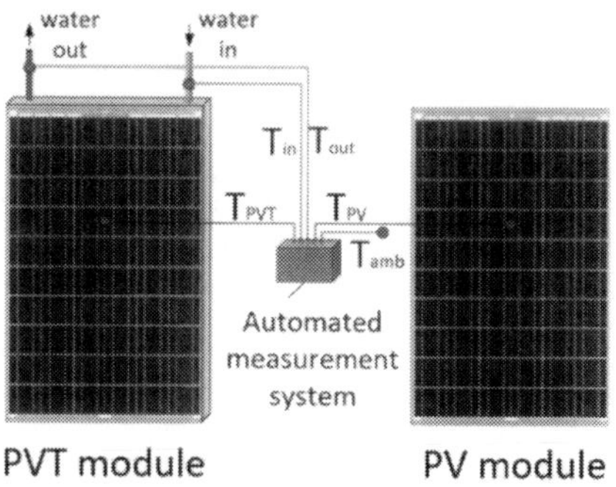

Figure 5. Block diagram of the experimental setup

4 RESULTS AND DISCUSSION

As mentioned above, the measurements were conducted for the thermo-siphon and forced-type water circulation in the PVT module. The results for each type of water circulation are presented below.

4.1 Measurements of temperatures in thermo-siphon system.

The thermo-siphon circulation of water through the PVT module was realized by connection to the PVT module of a thermally insulated water tank with a capacity of 100 l. Measurements were conducted on 02.05.2023 and 10.05.2023 between 10:00 and 15:00. Following temperatures were measured.

- T_{amb} - ambient temperature,
- T_{PV} – reference PV module temperature,
- T_{PVT} - temperature of the PVT hybrid module without an insulating back sheet,
- $T_{PVT+B.Sh.}$ - temperature of the PVT hybrid module with the thermal insulation back sheet,
- T_{in} - inlet water temperature into the PVT module,
- T_{out} - outlet water temperature from the PVT module,
- T_{tank} - temperature of the water inside the tank.

The 700 measurements of temperatures of the PVT module without the thermally insulating back sheet as well as the temperatures of the reference PV module and other components of the thermo-siphon system were conducted on 10.05.2023 from 10:00 to 15:00. Obtained data are presented in Fig. 6.

As could be expected, the temperatures of PVT and PV modules are higher than the ambient temperature, and this difference is about 25-30 °C. It can be seen from Fig. 6 that the temperature of the PVT module T_{PVT} in most cases is smaller than the temperature of the PV module T_{PV}. The higher temperature values of the PVT module at the beginning of measurements ("reverse" process) can be described by the slow cooling ability of the thermo-siphon system, and at the end of measurements – by increasing of temperature of water in the tank, due to the limited capacity of the tank.

Thus, it can be concluded that the PVT module without the thermally insulating back sheet in the thermo-siphon system operates normally and cools the solar cells, while the solar cells in the reference general PV module operate under higher temperatures.

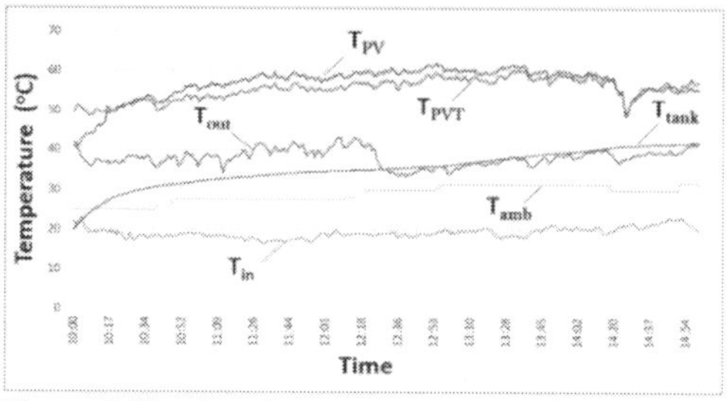

Figure 6: The data of the temperatures on the surfaces of the PVT module without a thermally insulating back sheet (T_{PVT}), PV module (T_{PV}), T_{tank}, T_{in}, T_{out} and T_{amb} measured in 10.05.2023 from 10:00 to 15:00

The other set of experiments with 700 measurements of temperatures was conducted when the PVT module was covered by the thermal insulating back sheet. The temperatures of the reference PV module and other components of the thermo-siphon system were measured as well on 02.05.2023 from 10:00 to 15:00. Obtained data are presented in Fig. 7.

In this case, also, the temperatures of PVT and PV modules are higher than the ambient temperature, and this difference is about 25 °C. It can be seen that the temperature of the PVT module is $T_{PVT+B.Sh.}$ is higher than the temperature of the reference PV module T_{PV} ("reverse" process). The higher temperatures of the PVT module in comparison with the reference general PV module can be described by the reduced convection cooling of the PVT module due to the covering it by a thermal insulating back sheet as well as by the low cooling possibility of the given thermo-siphon system.

Thus, it can be concluded that the PVT module with the thermal insulating back sheet in the given thermo-siphon system does not reduce the operation temperature of solar cells, moreover, the temperature of solar cells in the PVT module is increased and the electrical efficiency is decreased ("reverse" process) in comparison with reference general PV module. To overcome this problem, an appropriate cooling system for PVT modules must be applied.

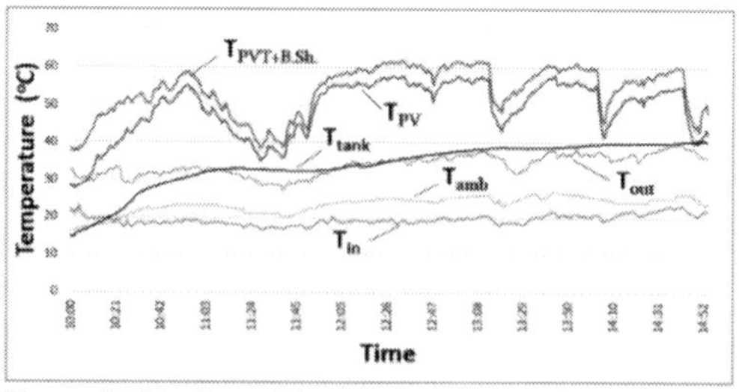

Figure 7: The data of the temperatures on the surfaces of the PVT module with a thermal insulating back sheet ($T_{PVT+B.Sh}$), PV module (T_{PV}), T_{tank}, T_{in}, T_{out}, and T_{amb} measured on 02.05.2023 from 10:00 to 15:00

4.2 Measurements of temperatures in a forced water circulation system.

The temperature measurements of the PVT module at a water flow of 1.5 liters/minute were conducted on 28.04.2023 and 08.05.2023 between 10:00 and 15:00. Following temperatures were measured: T_{PVT}, $T_{PVT+B.Sh}$,

T_{PV}, and T_{amb}.

The 715 measurements of temperatures of the PVT module without the thermal insulating back sheet as well as the temperatures of the reference PV module and ambient temperature were conducted on 08.05.2023 from 10:00 to 15:00. Using this data, the dependences of temperature of the PVT module without the back sheet and the reference PV module from the ambient temperature are plotted (Fig. 8).

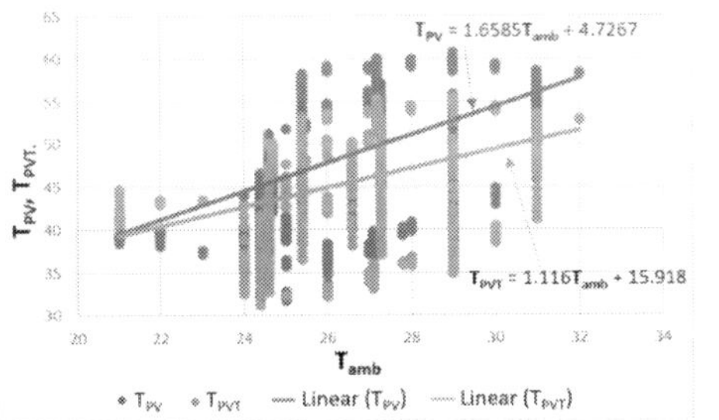

Figure 8: Dependences of temperatures of the PVT module without thermal insulating back sheet (T_{PVT}), and PV module (T_{PV}), from the ambient temperature T_{amb}.

The linear approximation of measured T_{PVT} and T_{PV} shows that the average temperature of the PVT hybrid module is smaller in comparison with the temperature of the reference PV module. The angular coefficient of the line for T_{PVT} is smaller (1.116) than the angular coefficient of the line plotted for T_{PV} (1.6585). Note that the same type of characteristics are obtained by the above-presented theory (Fig. 2).

Thus, it can be concluded that in a PVT module with a given cooling regime (forced water flow of 1.5 liters/minute) the average temperature of solar cells is smaller than the temperature of the reference general PV module. But at low ambient temperatures, below about 21 °C ("critical" ambient temperature), due to the different angular coefficients of the plotted dependences, the temperature of the cells in the PVT module can be higher than in the PV module ("reverse" process). To overcome this negative effect, the appropriate cooling regime for the PVT module must be applied.

The other set of experiments with 740 measurements of temperatures was conducted when the PVT module was covered by the thermal insulating back sheet. The temperatures of the reference PV module and the ambient temperature were measured as well on 28.04.2023 from 10:00 to 15:00. In this case also, using the obtained data, the dependences of the temperature of the PVT module with the back sheet $T_{PVT+B.Sh.}$ and the reference PV module T_{PV} from the ambient temperature are plotted (Fig. 9).

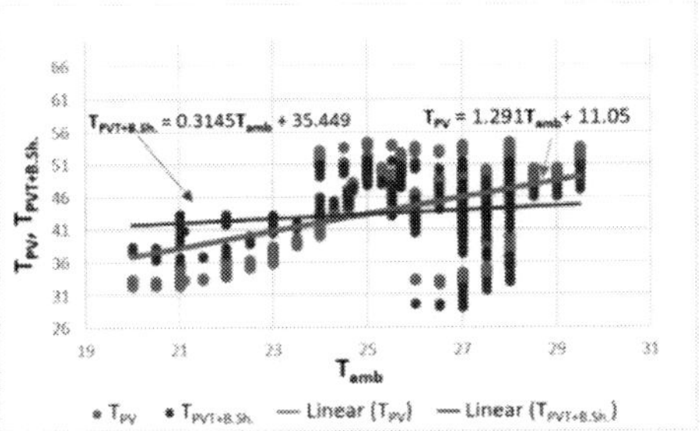

Figure 9: Dependences of temperatures of the PVT module with a thermal insulating back sheet ($T_{PVT+B.Sh}$) and PV module (T_{PV}) from the ambient temperature T_{amb}.

The linear approximation of measured $T_{PVT+B.Sh}$ and T_{PV} show that in this case the angular coefficient of the line for hybrid module $T_{PVT+B.Sh.}$ is smaller (0.3145) than the angular coefficient of the line plotted line for reference PV module TPV (1.291). This is also coherent with the theory (Fig. 2).

In comparison with the PVT module without a back sheet (Fig. 8) it can be seen that the interception of linear approximations ("critical" ambient temperature), acquires a higher value of ambient temperature (T_{amb} = 25 °C), as it was in theoretically plotted diagram (Fig. 2). It means that at high ambient temperatures, above 25 °C, the average temperature of PVT module with the back sheet is smaller than the temperature of reference PV module. But below the 25 °C of ambient temperature the PVT module is hotter than the reference PV module ("reverse" process). This negative effect must be considered during the development of PVT modules and the PVT hybrid systems.

5 CONCLUSIONS

- The influence of the back sheet thermal insulator on the operational temperature of PVT modules theoretically and experimentally is investigated. The experimental results are in good correlation with the theory.
- The application of the thermally insulating back sheet results in an increase in the temperature of the PVT hybrid module, and consequently, the thermal output of the PVT system increases, and the electrical efficiency of solar cells decreases.
- Below the so-called "critical" ambient temperature the PVT hybrid modules are warmer than the general PV module ("reverse" process), and the application of a PVT system below the "critical" temperature (cold conditions) is not reasonable.
- The PVT module without a thermal insulating back sheet can be used in a broader range of ambient temperatures than the PVT module with a thermally insulating back sheet.
- To overcome the negative effect of heating the PVT modules ("reverse" process), the appropriate structures of the PVT module and PVT system must be considered during the design of PVT hybrid systems.

6 REFERENCES

[1] C.A. Ramos, A.N. Alcaso, A.J. Cardoso. Photovoltaic-thermal (PVT) technology: Review and case study. IOP Conf. Series: Earth and Environmental Science 354 (2019) 1- 12.

[2] S.H. Bandaru, et all. A Review of Photovoltaic Thermal (PVT) Technology for Residental Applications: Performance Indicators, Progress, and Opportunities. Energies 2021, 14, 1-48.

[3] M.G. Noxpanco, J. Wilkins, S. Riffat. A Review of the Recent Development of Photovoltaic/Thermal (PV/T) Systems and Their Applications. Future Cities and Environment. 6(1) 9. 2020, 1-16.

[4] S. Diwania, S. Agrawal, A. Siddiqui, S. Singh. Photovoltaic-thermal (PV/T) technology: a comprehensive review on applications and its advancement. International Journal of Energy and Environmental Engineering (2020) 33-54.
[5] F.A. Sachit, et all. Current progress on flat-plate water collector design in photovoltaic thermal (PV/T) systems: A review. Jour of Adv Res. in Dynamical & Control Systems, V. 10, 2018 680-694.
[6] V. Tirupati, Y. Sekhar. Hybrid Photovoltaic/Thermal (PVT) Collector Systems Wit Different Absorber Configurations For Thermal Management – A Review. Energy & Environment 22. 2021 https://doi.org/10.1177/0958305X211065575
[7] R.R.Vardanyan, V.K. Dallaqyan, M.G. Travajyan. Temperature Dependent Performance of Solar Photovoltaic and Thermal Hybrid Systems. Springer, Journal of Contemporary Physics, V. 55(4), (2020) 339-344. http://link.springer.com/article/10.3103/S1068337220040180
[8] R. G. Ross. "Flat-Plate Photovoltaic Array Design Optimization", 14th IEEE Photovoltaic Specialists Conference. San Diego, CA, (1980) 1126-1132.
[9] J.H. Kim, J.T. Kim. Comparison of Electrical and Thermal Performances of Glazed and Unglazed PVT Collectors. International Journal of Photoenergy. Volume 2012, (2012) p.1-7.

ANTI-GLARE BIFACIAL MODULE DESIGNS WITH STRUCTURED GLASS FOR FAÇADE APPLICATION

Marc Hofmann[1], Janina Willmann[1], Lasse Bienkowski[1,2], Benedikt Bläsi[1], Martin Zimmer[1], Dirk Reinwand[1], Frank Ensslen[1], Stephan Riepe[1], Sebastian Nold[1], Andreas Brand[1], Tilmann E. Kuhn[1]
[1]Fraunhofer Institute for Solar Energy Systems ISE, Heidenhofstrasse 2, 79110 Freiburg, Germany
[2]Institut für Ionenphysik und Angewandte Physik, Universität Innsbruck, 6020 Innsbruck, Austria
Tel.: +49 761 4588-5051, email: marc.hofmann@ise.fraunhofer.de

ABSTRACT: Façades of buildings can be an important part of the upcoming energy system if photovoltaic (PV) modules are installed. In this context, we propose the integration of PV module lamellae mounted perpendicularly to the original façade optimizing the compromise between PV yield, shading and interior lighting of the building. Mitigating glare caused by sun reflection on the cover glass of a PV module is of paramount importance to ensure the effective operation of a PV system integrated into façades. This is achieved by glass surface structuring using laser processing in combination with wet chemical treatment. Bidirectional reflectance distribution function (BRDF) measurements revealed a reduction by more than three orders of magnitude for the treated glass surface.

Keywords: Building-integrated photovoltaics, glass structuring, laser, wet etching

1 INTRODUCTION

Mankind has already occupied large portions of earth's surface and built streets, buildings, parking lots, etc.. To omit further negative impact on nature (biodiversity, water management, ...), the multifunctional use of already sealed surfaces is wise. Roof areas are more and more not only a protective skin of the building's inside anymore but also produce electricity by photovoltaic module integration. It was shown that façades of buildings can also provide significant amounts of electricity by implementing PV modules [1]. However, among others, one has to take care of a few aspects:

- Find architectural appealing and aesthetic design opportunities.
- Find a concept that allows applicable cost structures for a widespread application.
- Omit glare/ reflection effects that could create limited sight during passing (e.g., driving) by or could reduce life quality inside or outside the house.
- Find a good compromise between electricity generation by PV and illumination of the living/working space inside adjacent houses.

Additional aspects include electrotechnical, mechanical, and legal topics.

Currently, the mass market for PV module producers does not take façade applications into account. Besides façades, also other perpendicular structures like noise protection walls along highways or similar are relevant examples for the application of such modules. According to an estimation of Eggers et al. [1], one could install – with current PV module technology efficiencies – about 400 GW_P of PV power at existing German façades of buildings with a minimum of 500 $kWh/m^2/a$ of solar illumination. Using a significant part of this potential would provide an important share of the needed PV power capacity in the German grid. Our developments of anti-glare bifacial module technologies help to overcome hurdles in using these potentials.

In this work, we present innovative façade design concepts that combine PV lamellae with structured glass to achieve a harmonious blend of high electrical performance, abundant natural lighting inside the building, and comfortable shading to protect occupants from excessive heat and glare. The structured glass is realized through a two-step approach combining laser irradiation and a subsequent wet chemical treatment.

2 MOTIVATION / APPLICATION CASE

Façades show a huge potential for PV integration, allowing for a significant share of the needed PV integration in existing buildings to meet the electricity needs of our societies. Thus, good concepts for the transformation of passive into electricity generating façades are very useful now and in the future.

In Figure 1, façade sketches are shown with perpendicular thin PV modules placed in front of a freely designable glass window/concrete/wood/steel/ or other type of wall. The PV modules (lamellae) have a front and a back side, each absorbing light and generating electricity (bifacial solar cells and modules). The lamellae can be placed near to each other resulting in greater shading of the building's inside (i.e., cooling effect in summer and less blinding by direct sunlight). A larger distance between the lamellae would allow a clearer view from inside the building to the outside. More lamellae would obviously typically lead to higher cost of the structure.

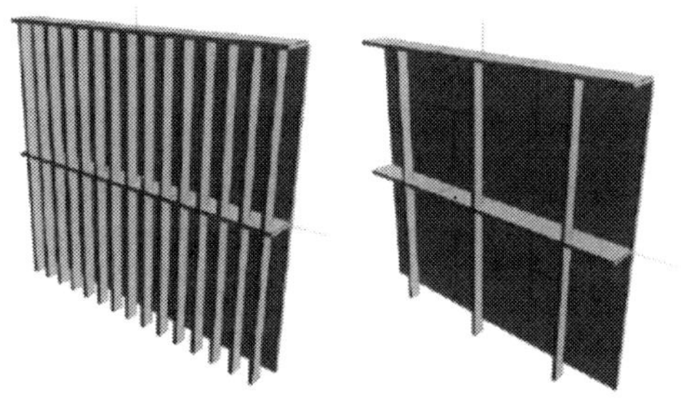

Figure 1: Sketches of façades with differently distanced perpendicularly arranged lean PV modules (lamellae).

Preliminary simulations were performed with a similar context to investigate the electricity generation potential of such installations. Here, the case of a façade facing south with the lamellae facing east and west is considered. Figure 2 shows the results of the simulation. The upper image shows a false colour image of an exemplary design and visualizes the sum of the light hitting either side of the lamella over the course of a year – up to 1000 kWh/m2/a of illumination can be achieved. However, the light energy is not equally distributed onto the lamella surface portions. Closer to the inner building and going higher towards the horizontal stabilizing part, the amount of light is decreasing due to shading of the supporting structure.

Thus, a smart module concept would be beneficial for the electricity output.
Additionally, in case the lamellae would be made with flat glass on the outside, light reflection effects would be an important drawback of the integration of such installations into a façade. Thus, a technically viable way to reduce glare very effectively is essential. It would allow a naturally illuminated living and working area inside the building without glare-depending disturbance and without the need of sun-blinds for most of the daytime.

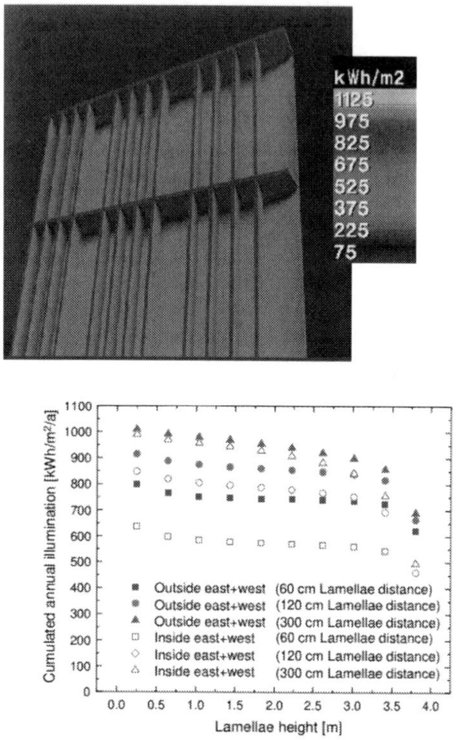

Figure 2: *Top:* Simulation results showing the cumulated annular light energy hitting different surface portions of a façade with differently spaced perpendicular PV lamellae in Germany.
Bottom: Simulated cumulated annual light energy hitting front and back of a PV lamella in Germany. The amount of light is depending on the distance of the lamellae and on the height position on the lamella since shading occurs due to horizontal elements in the simulated case.

3 STRUCTURING OF GLASS

In our research, we have learned that a significant degree of freedom in altering the size and shape of structure features on the glass surfaces can be achieved by a two-step process which combines a laser pre-process and a wet chemical post-process. The laser process prepares the glass surface in such a way that wet chemical etching for example results in very steep features. In Figure 3 and Figure 4, exemplary measurement results of such treated glass surfaces are presented. Figure 5 shows an example of a glass sample that was partly treated by a laser and wet etching process sequence.

Please refer to a paper of L. Bienkowski et al. on this topic which will be submitted shortly [2].

Figure 3: Glass surfaces were treated in a two-step process sequence using laser processing and wet chemical etching. An exemplary false-color image of the height profile of a periodic structure measured applying confocal microscopy is shown.

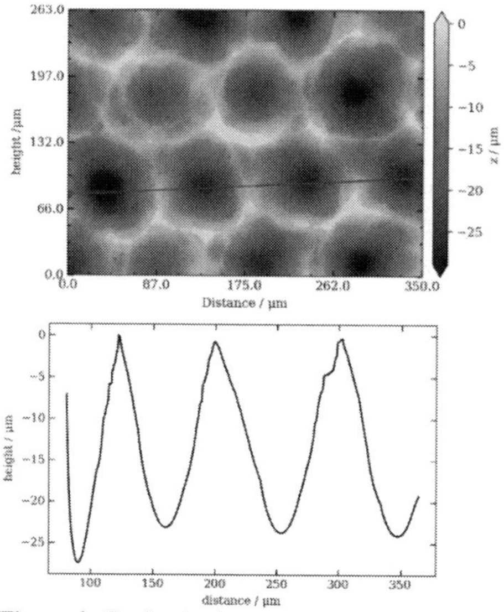

Figure 4: Confocal microscopy results: top: false-color height profile, bottom: line scan result following the red line in the top image. Surface scan results show steep structures at the glass surface for a process with periodically distributed laser spots. This presents another process example, not directly comparable to Figure 3.

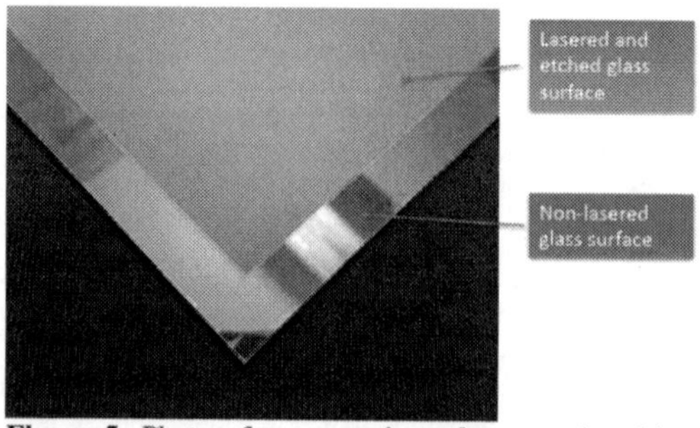

Figure 5: Photo of an exemplary glass sample with a portion of its surface sequentially lasered and etched. Along the edge of the sample, non-lasered areas are visible and can be used as visual reference.

10.4229/EUPVSEC2023/3AV.1.34
020166-002

Mapping a larger portion of such a treated glass surface by confocal microscopy and plotting the first derivative of the resulting data as a slope angle histogram resulted in Figure 6.

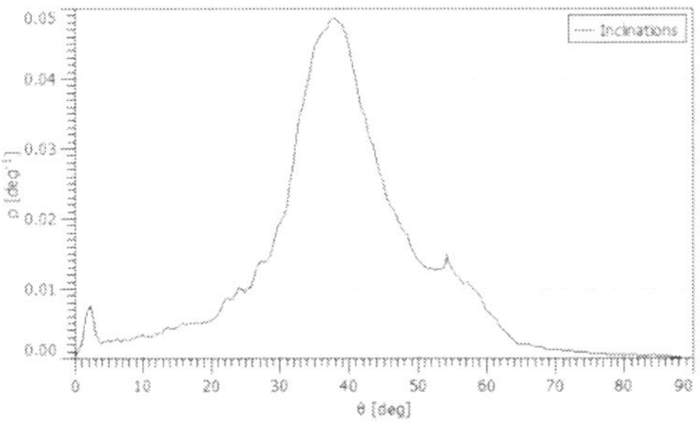

Figure 6: This histogram of the surface angle distribution of the treated glass sample shows a steep and widely distributed structure with the peak of distribution in the range 35-40°.

4 ANTI-GLARE EFFECT

In order to reduce the luminance of the sunlight reflected from the surface to such an extent that glare no longer occurs, a strongly diffuse reflection behavior is aimed for. One measure of this is the surface angle distribution: the wider the surface angle distribution, the more diffuse the reflection behavior. Figure 6 shows a wide surface angle distribution.

For a better quantification, the luminance of the reflected sun light was estimated for an incidence angle of the sun of 10° based on the surface angle distribution. When comparing a non-treated flat glass showing a luminance of more than 300,000 cd/m^2 to the lasered and etched sample presented above with less than 5,000 cd/m^2, we see that a large reduction of luminance can be achieved while still keeping the transmission through the glass (and towards the solar cells) high.

The angle-dependent reflection behavior of a material is described by its Bidirectional Reflectance Distribution Function (BRDF). It is defined as the quotient between the radiance of the incident light and the reflected irradiance as a four-dimensional function over the angle of incidence and reflection. Figure 7 shows the BRDF of an untreated glass (top) and the lasered and etched glass (bottom) measured with a 3D scanning goniophotometer (PG II Gonio-Photometer, pab advanced technologies Ltd, Freiburg, Germany). The untreated glass shows a sharp peak and thus a strong specular reflection. In comparison, the lasered and etched glass reflects light diffusely over a wide solid angle.

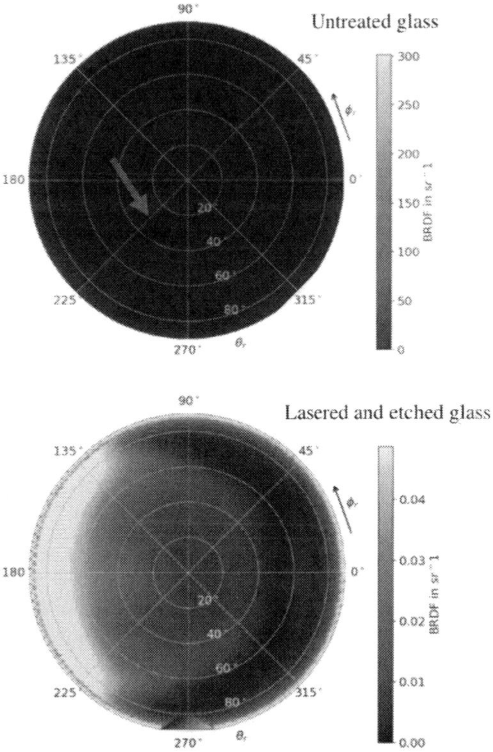

Figure 7: Bidirectional reflectance distribution functions (BRDF) for a polar angle of incidence of 30° and an azimuthal angle of incidence of 0°.

Top: Untreated glass showing a very high peak at the direct reflection angle of 30°. The red arrow points at the small but very bright dot that represents the reflection position.

Bottom: BRDF of a lasered and wet etched glass sample with very low BRDF value range showing a large distribution of the reflected light (large scattering).

Please note that the color scale differs by almost 4 orders of magnitude.

5 STUDY ON GLARE ASSESSMENT

Glare can be dangerous and/or life quality decreasing in different situations. One very important segment influenced by glare is the mobility sector. We plan to perform studies on the assessment of glare in the case of driving in a car and passing by glare effects. Two study parts are currently under preparation, a first investigation will look at different lighting and glare situations in a lab environment. In a second step, field tests will be performed. The studies shall also help to generate a proposal for standardization work and will also take the pre-work of Fraunhofer ISE into account [3-6].

Figure 8: Example for a glare situation (new library of the university of Freiburg); image is courtesy of Fraunhofer ISE, taken from the website bipv-bw.de

8 SUMMARY

Structuring of glass surfaces with the combination of laser treatment and a subsequent wet etching process offers a large variety of surface morphologies. The creation of structures that lead to large scattering of reflected light resulting in a strong suppression of glare was possible and presented. This helps to outline technological paths for a large application of PV also for facades, noise barriers, etc.

Acknowledgements

This work was funded by the German Federal Ministry for Economic Affairs and Climate Action within the research project "HighFas" (contract no. 03EE1105A).

Johannes Eisenlohr is thanked for illumination intensity distribution simulations.

References

[1] J.-B. Eggers, M. Behnisch, J. Eisenlohr, H. Poglitsch, W.-F. Phung, M. Münzinger, C. Ferrara, T.E. Kuhn , „PV-Ausbauerfordernisse versus Gebäudepotenzial: Ergebnis einer gebäudescharfen Analyse für ganz Deutschland", 2020, 35th PV-Symposium, Bad Staffelstein, pp. 837-856

[2] L. Bienkowski, M. Zimmer, B.-U. Sander, J. Willmann, A. Brand, M. Hofmann, B. Bläsi, J. Nekarda, „Etching-Assisted Laser Surface Functionalization Process for an Anti-Glare Structure on PV Module Glass", 2023, to be published

[3] J. Wienold, T. Iwata, M. Sarey Khanie, E. Erell, E. Kaftan, R. G. Rodriguez, J. A. Yamin Garreton, T. Tzempelikos, I. Konstantzos, J. Christoffersen, T. E. Kuhn, C. Pierson and M. Andersen, "Cross-validation and robustness of daylight glare metrics", Lighting Res. Technol. 2019; 1–31

[4] J. Wienold, T. E. Kuhn, J. Christoffersen, M. Sarey Khanie, M. Andersen, "Comparison of Luminance Based Metrics in Different Lighting Conditions", Proceedings of the CIE 2017, Midterm Meeting, Jeju Island, Korea

[5] J. Wienold, J. Christoffersen: "Evaluation methods and development of a new glare protection model for daylight environments with the use of CCD cameras", Energy and Buildings, vol. 38 (2006), p. 743-757.

[6] H. R. Wilson, B. Bueno, J. Eisenlohr, T. E. Kuhn, "Evaluating BIPV modules with respect to textural appearance, glare, colour and electricity yield", 35. PV-Symposium 2020

IMPROVING ALLBLACK MODULE EFFICIENCIES

B. Jaeckel[1], M. Pander[1], M. Evertz[2], M. Brandenburg[2], M. Robertz[2], A. Linsenmeyer[3]
[1] Fraunhofer Center for Silicon Photovoltaics CSP, Otto-Eissfeldt-Str. 12, 06120 Halle (Saale), Germany
[2] Aluminium Féron GmbH & Co. KG (Feron), Hammerbenden 1 – 13, 52355 Düren, Germany
[3] Sunset Energietechnik GmbH, Industriestrasse 8 – 22, 91325 Adelsdorf
Corresponding author: bengt.jaeckel@csp.fraunhofer.de

ABSTRACT:

To further increase the acceptance of PV in urban environments it is essential to enhance the optical appearance of the PV modules. Costs came down to the point that a few Euros more to get aesthetically pleasant modules increase the system costs negligible. Most straight forward are homogeneous black colored modules, including the frame and the use of black clamps for mounting. Unfortunately, the black color typically comes with a reduced efficiency. However, there are technical possibilities to reduce this drawback. Herein we explore the potential of using bifacial cells in a mono-facial module design while optimizing the reflectance of the used black backsheet.

A detailed study was performed to analyze different bifacial cells (PERC and HJT). Differences in rear side EQE were determined that can cause some current mismatch on the rear side of the module, especially for bifacial module constructions. After determination of solar cells' EQE, transmittance and reflectance of different backsheets a set of theoretical backsheets was studied to investigate possible current gains from behind a bifacial cell in a mono-facial PV module design. The potential was determined to be in the range of 0.6% Isc gain for a PERC cell. HJT will be higher due to typically higher bifaciality coefficients.

The promising results were used by Feron to manufacture black backsheets with increased reflectance in the range higher than 750nm. The optimized backsheets achieve also values of up to 0.6% Isc gain. Currently full-size modules are built to verify the calculations and first modules power measurements agree well with the calculated Isc gains.

Keywords: module efficiency, color, bifacial, CTM

1. INTRODUCTION and MOTIVATION

Within the "Energiewende" it is crucial to further enhance efficiency of PV cells and modules. Besides the efficiency of the PV modules the aesthetics of a PV system is getting more and more important. High reflective white backsheets were developed in the past to increase photon collection from between the solar cells mainly with respect to mono-facial cells in a mono-facial module design. This is well known and a major part of CTM-gain [1]-[3] as illustrated in **Figure *1*** with the coefficients k1 to k11 for mono facial cells and mono facial module designs. For bifacial modules the figure is updated to include rear-side effects.

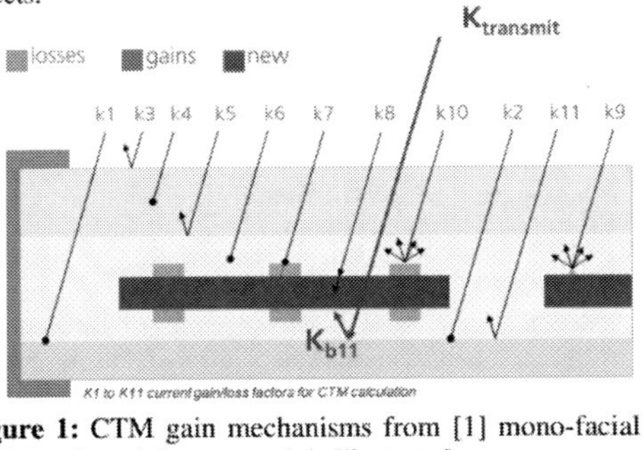

Figure 1: CTM gain mechanisms from [1] mono-facial modules. In red the new path is illustrated.

However, the potential of current collection does not stop underneath the solar cells. Today most commercially available H-grid cells are bifacial and extra current gains can be achieved from the rear side. Typical bifacial modules use therefore transparent polymer backsheet or a glass cover. For open-rack mounted modules, including fixed and 1x-tracking systems, this is the optimal module design to choose when significant amounts of light can be directly brought into the module from the rear side. However, this construction is not necessarily optimal for close roof attached or building integrated PV modules as typically no photon collection can be achieved from the rear side. With the additional need for improved aesthetically pleasant roof top installation it is necessary to enhance photon collection from within the module.

Highly reflective white backsheets are fundamentally best to gain current (see example in Figure 2), but typically the white areas between the solar cells are non-aesthetic. Therefore, often black backsheets are used that are typically non-reflective and absorb all light (see Figure 2) resulting in typically lower output power while using same cell efficiency and also lead to higher operating temperatures for black modules.

The motivation now is to take the best out of both worlds. First, use bifacial solar cells instead of mono-facial cells and second, develop and use a backsheet that appears black to the human eye, but is highly reflective in the regions higher than 750nm to around 1200nm.

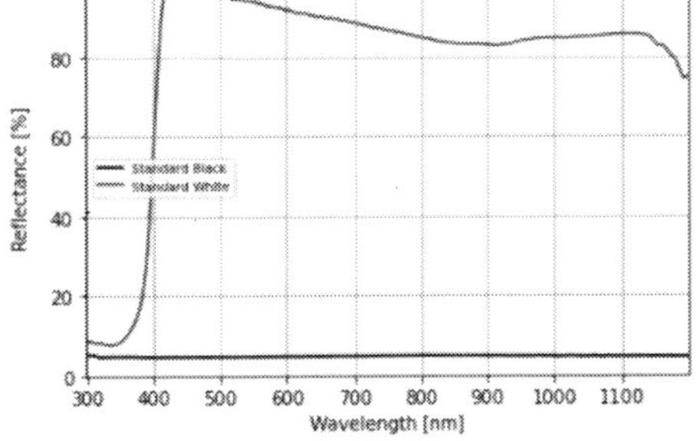

Figure 2: Typical reflectance curves of a white and a black backsheet

The study was divided into several steps. First, the potential for efficiency (current gain) was estimated based on impact on cell technology (PERC; HJT) and its bifaciality coefficient, and the possible impact on cell sorting. The second and last step was the development of an optically black, but for the relevant solar spectrum reflective backsheet.

2. STUDY of POTENTIAL Isc GAINS

Herein we estimate the potential of current gain due to the use of bifacial cells in a mono facial module. The focus lays on the large area available underneath the solar cell. Cell-to-cell gaps can be handled by normal CTM approach but those are getting less important as cell spacing is currently limited to a technical minimum (typically 2mm) to increase module efficiency (W/m²). Some designs basically reduce those to only string-to-string gaps [4]-[8] or even to zero [5]-[8].

To understand the potential of bifacial cells in a mono-facial module design the amount of transmitted light through the solar cell is important. Thinner wafers and only partly metalized rear sides of the cells allow that some light passes through the cell.

The AM 1.5 spectrum (incoming light – brown)), the average transmittance of a PERC cell (green) and the resulting potentially available spectrum (orange) for reflection and re-absorption in the cell is given in Figure 3. The transmittance of 9 cells was measured and averaged. There is no significant difference between the cells and for the following calculations only the average is used.

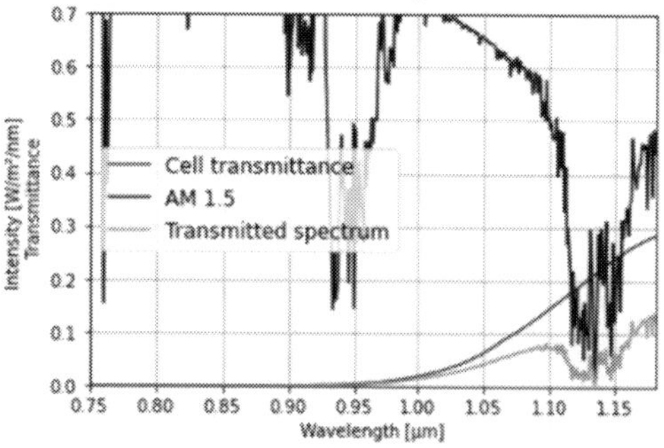

Figure 3: AM 1.5(brown), PERC cell transmittance (green) and calculated remaining spectrum (organge) for reflection at backsheet.

With the knowledge of the amount of transmitted light, the reflectance of the backsheet and the EQE of the solar cell from the back a potential current gain can be calculated.

In the first step front- and rear side EQEs were measured from different batches of cells (PERC and HJT). Figure 4 shows the measured dataset for the PERC cell. The EQEs on the front side are very similar, as to be expected for color sorted cells within a cell box. However, the rear side shows especially in the region between 400-600nm significant differences. For a bifacial module with assumed direct front and rear side illumination current mismatches will occur. The corresponding boxplot analyze is given in the right in Figure 4. Isc(front) is less effected and is a direct result of cell front side sorting (power and color). Rear side sorting is typically not done resulting in potentially extra rear side current mismatch.

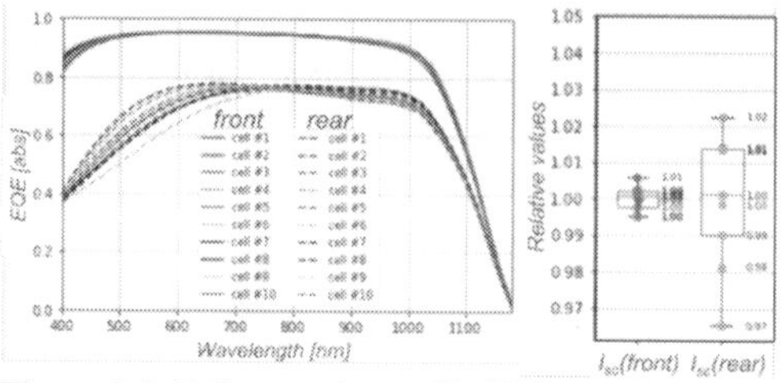

Figure 4: left) Front and rear side EQE measurement of commercially available 22% bifacial PERC cells. right) Boxplot analyze of resulting parameters affecting PV module performance.

For a detailed mismatch calculation this may need to be considered. For the potential study herein an average front and rear side EQE are used as shown in Figure 5.

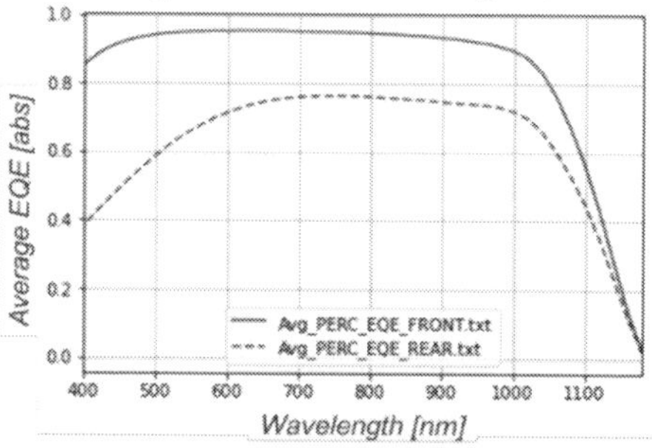

Figure 5: Average EQE from Figure 4 for further calculations, bifacially coefficient is around 74%, as expected for this cell technology.

For the potential gain study for an optically black module a standard black (black), a standard white (green) and three different reflective "theoretical" backsheets (blue, red and turquoise) were considered. The used reflectance curves are given in Figure 6.

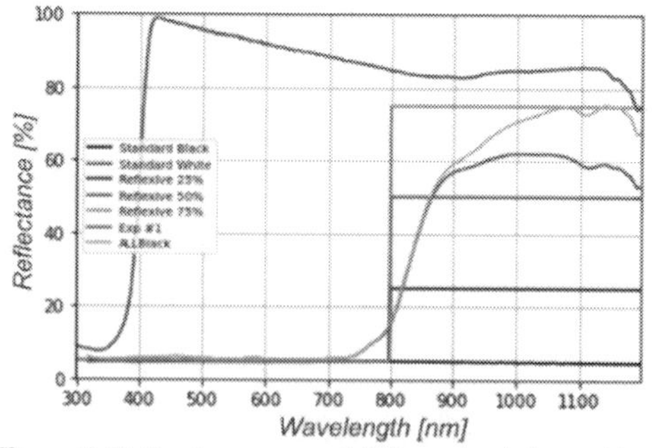

Figure 6: Reflection curves of different backsheets: black (black) and white (green) are standard commercially available backsheets. Indicated with blue, red and turquoise color are theoretical black appearing backsheets with different reflectance in the near infrared. Purple and ochre curves are black, but reflective manufactured backsheets.

The results of the calculations are given in Table 1. As expected, a black backsheet does not contribute to a current gain from the back side. The gain by using a white backsheet in combination with bifacial cells will lead to a current gain of more than 0.6%, excluding the CTM-gains between the cells. CTM gains can be determined e. g. by SmartCalc.CTM and the results are in overall agreement with previously determined CTM-values [9][10].

Depending on the assumed theoretical reflectance of the optically appearing black backsheet a current gain behind the bifacial cell of almost 0.6% for the 75% reflective backsheet can be calculated.

3. Real AllBlack backsheet

With the understanding of the fundamental potential to gain rear-side currents a black appearing backsheet was developed by Feron. After some experimental trails the first promising results were achieve and displayed in Figure 6 (Exp #1 – purple) with a significant increased reflectance. With process optimization a first very promising AllBlack backsheet could be manufactured and scaled up to be able to manufacture prototype full-size PV modules.

Using the same calculation procedure as for the theoretical backsheets the newly developed backsheet shows a current gain of around 0.55% for a PERC cell (see Table 1) and even more for cells with a higher bifaciality such as HJT-cells can be expected.

Table 1: Summary of resulting current gains from rear side of a bifacial solar cell.

Backsheet	Absolute ISC gain in [A] for M6 half cut bifacial cell	%-gain for PERC cell
Black	0.014	0.034
White	0.253	0.628
25pers	0.076	0.189
50pers	0.150	0.373
75pers	0.223	0.554
AllBlack	0.219	0.543

Currently a larger batch of modules are manufactured to proof the calculation and to directly evaluate the impact on current and module performance of such black reflective backsheets.

Overall it is planned to manufacture 12 different bill-of-materials (BOMs) to differentiate between cell shape (full vs. half cell), impact on reflectance of backsheet and the impact on color cell sorting of the rear side of the cell to experimentally see impact of the findings of Figure 4.

The module manufacturing took place at Sunset Solar under standard manufacturing conditions. This was done to utilize as close as possible industry machinery to also evaluate the processibility of the AllBlack backsheet in comparison to standard materials.

All modules are made out of the same batch of 22.2% efficient bifacial PERC cells with the same cell sorting (color code), same front glass, same encapsulation material, frame and contacting protocol. The main modification is the backsheet. For most BOMs at least 10 modules are planned to be build. Due to machinery difficulties, it was not possible to build all module types in time for the conference. The detailed analyze will therefore be published elsewhere.

Results from the first batch of modules is presented in Figure 7. Mod-types #1 and #2 are two white backsheets with a minor difference in reflectance where the overall reflectance of the backsheet in Mod #1 is slightly higher compared to Mod #2. Mod #3 is a standard, non-reflective black backsheet. A drop of approx. 3% in current is measurable. Mod #5 within this figure represents the newly developed AllBlack by Feron. About 2% current can be regained by using the new backsheet.

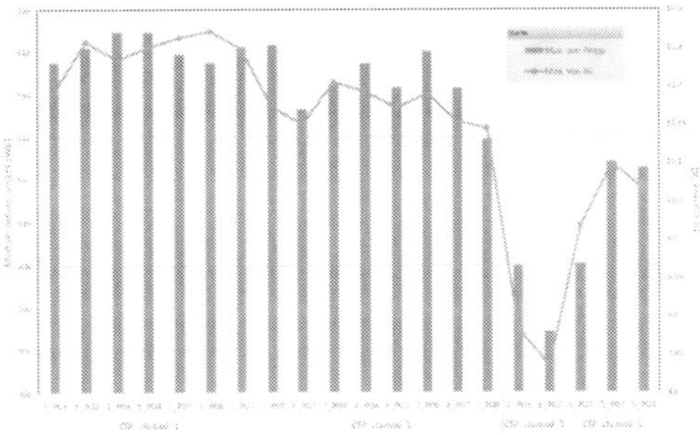

Figure 7: Module power and short circuit current of modules from first batch of modules manufactured to evaluate the gain of the AllBlack backsheet.

More modules are currently in the manufacturing process. This is important to better quantify the gains on module level. As no production is perfect and transportation also can cause damage there are some modules with some cracked cells. An example is given in Figure 8 from P01 of Mod #5 and a crack free module P07 from Mod #2.

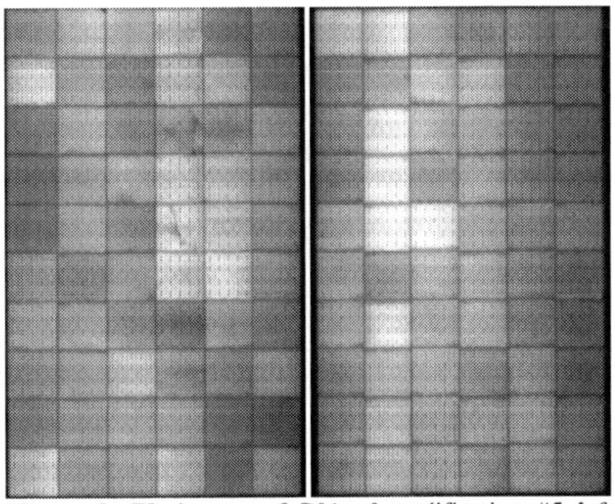

Figure 8: EL images of P01 of modification #5 left (AllBlack) and a reference module without cell cracks P07 of modification #2 (right).

4. SUMMARY

The study investigated the potential Isc current gains of visually black, but reflective in the range above 750nm backsheets using bifacial solar cells. The focus lays on the gains underneath the solar cell, not in the cell gaps.

The theoretical evaluation was done by creating simple rectangular reflectance curves of 25%, 50% and 75% and compared to standard white and black backsheet to estimate the Isc gain potential. It is demonstrated that more 0.6% can be gained by using standard bifacial PERC c-Si cells and more can be expected for cells with higher bifaciality and similar transmittance.

The trend from calculations was possible to verify within a first batch of full-size PV modules. The current development prototype of the AllBlack backsheet showed a possible gain for PERC solar cells close to 0.55% from cell's back side and further improvements are ongoing. Higher gains can be expected while using cells with higher bifaciality coefficients such as HJT. The modules with the AllBlack backsheet including black frames create nice and aesthetically attractive PV modules.

The longevity, the gains in output power and the optical appearance will be studied next. Different modified

10.4229/EUPVSEC2023/3AV.1.36
020168-003

modules will be exposed outdoors to evaluate real outdoor performance and check for operating temperatures in comparison to standard black backsheets. The stability of the black color will be further studied with accelerated aging tests in analogy with IEC-protocols for backsheets and PV modules.

5. ACKNOWLEDGEMENTS

The authors gratefully acknowledge the financial support by the German Federal Ministry for Economic Affairs and Climate Action for financial support of the project "CTS1000plus" under grant no. 03EE1028A.

6. REFERENCES

[1] SmartCalc.CTM & Module: Cell to Module Analysis by Fraunhofer ISE: https://www.cell-to-module.com
[2] H. Hanifi, "Investigation of cell-to-module (CTM) ratios of PV modules by analysis of loss and gain mechanisms," Photovoltaic international, 2016.
[3] H. Hanifi, "Optimized tab width in half-cell modules," 6th International Conference on Silicon Photovoltaics, vol. 92, pp. 52–59, 2016, doi: 10.1016/j.egypro.2016.07.009
[4] B. Jaeckel, "Filling in the gaps: The evolution of high-density module design - Technical possibilities and Challenges", PV magazine, June 2020
[5] A. Mondon, „Comparison of Layouts for Shingled Bifacial PV Modules in Terms of Power Output, Cell-to-Module Ratio and Bifaciality," 35th European Photovoltaic Solar Energy Conference - Brussels, 2018
[6] N. Klasen, „Quantitative Evaluation of the Shading Resilience of PV Modules", European Photovoltaic Solar Energy Conference, 2021
[7] N. Klasen, „Performance of shingled solar modules under partial shading", Progress in Photovoltaics Research and Application. 2021, 1-14, DOI: 10.1002/pip.3486
[8] D. v. Kutzleben, "Development of Shingle Matrix Technology for Integrated PV Applications", WCPEC-8, 2022
[9] M. Mittag, "Electrical and Thermal Modeling of Junction Boxes" 33rd European Photovoltaic Solar Energy Conference - Amsterdam, 2017
[10] M. Mittag, "Analysis of Backsheet and Rear Cover Reflection Gains for Bifacial solar cells", 33rd European Photovoltaic Solar Energy Conference - Amsterdam, 2017

VALIDATION OF DECLARED EFFICIENCY IN COMMERCIAL PV INVERTERS' DATASHEETS

Geyciane P. de Lima[1†], Andrei C. Ribeiro[1], Paulo R. D. R. da Silva[1], Guilherme C. S. Prym[1], Mendelsson R. M. Neves[1], João Lucas de S. Silva[1], Hugo da S. Alvarez[2], Rodrigo M. Garcia[2‡], Tárcio A. S. Barros[1], and Marcelo G. Villalva[1]

[1]School of Electrical and Computer Engineering, Dept. of Systems and Energy - DSE, Campinas, Brazil,
†geycianedelima@gmail.com

[2]Build Your Dreams Company - BYD, Dept. Research and Development - R&D, Campinas, Brazil,
‡rodrigo.garcia@byd.com

ABSTRACT: This study aims to ensure the efficiency of commercial photovoltaic inverters in Brazil following Ordinance 140/2022. Our study evaluates the overall efficiency of two PV inverters sold in Brazil. We found a significant discrepancy between declared and measured efficiency values on one commercial inverter, violating regulations. Brazil's commitment to the energy transition process has led to reforming the Ordinance governing tests to ensure safe and quality equipment sales. The certification from the National Institute of Metrology, Quality, and Technology (INMETRO) is mandatory for PV inverters connected to the grid. The new Ordinance 140/2022 introduces a broader scope of testing, including the efficiency test of PV inverters, referencing the central efficiency standard IEC 62891:2020. Our laboratory tests, conducted under controlled conditions, revealed that while Inverter 1 met the required standards, Inverter 2 showed a 2.80% discrepancy between declared and measured values. Such findings underscore the importance of accurate efficiency measurements for designers and consumers, ensuring a fair energy transition in Brazil..

1 SUMMARY OF THE ABSTRACT

Brazil is committed to the energy transition process, within all the advances proposed by the country, the Ordinance that governs the tests was recently reformulated so that the equipment is sold in a safe and quality way.

For the photovoltaic (PV) inverter connected to the grid to be sold in Brazil, it is necessary to obtain certification from the National Institute of Metrology, Quality and Technology (INMETRO). The tests are governed by Ordinance 004/2011 [1] until March 2024, as after that period, the new Ordinance 140/2022 [2] will mandatorily come into force for the certification of photovoltaic inverters.

INMETRO Ordinance 004/2011 aims at energy efficiency and adequate safety level of photovoltaic systems. Tests such as anti-islanding, grounding, isolation devices and short circuits are required. Also, such issues as direct current (DC) injection limitations, harmonics, flicker, voltage, reconnection lag, and reconnection time [3].

Therefore, to evaluate not only the quality of the photovoltaic inverter with the electrical grid but also the quality of the equipment under test, Ordinance 140/2022 has a more extended scope of testing that includes the efficiency test of photovoltaic inverters.

In this context, evaluating the quality of the photovoltaic inverter through the efficiency parameter is a good indication for both the designer and the consumer. In this way, it is possible to identify the quality of the equipment being used and improve the quality of the project so that it is more consistent with reality.

Ordinance 140/2022 refers to the central efficiency standard IEC 62891:2020 [4] for the test to be carried out. To verify whether the equipment would be approved following the new Ordinance 140/2022, a trial will be carried out on seven photovoltaic inverters connected to the grid in order to compare whether the values reported in the data sheets are the same as those measured in the laboratory according to the IEC 62891:2020 standard. In this way, it is possible to raise the quality of equipment sold in the country to achieve a fair energy transition.

2 METHODOLOGY

For the test to be carried out in the laboratory, it is normatively necessary that the environment is controlled at 25°±5°C and that the setup is by Figure 1. Where 1 is the data acquisition and collection equipment, 2 are the photovoltaic simulator sources, and 3 is the alternative current (AC) network emulator source.

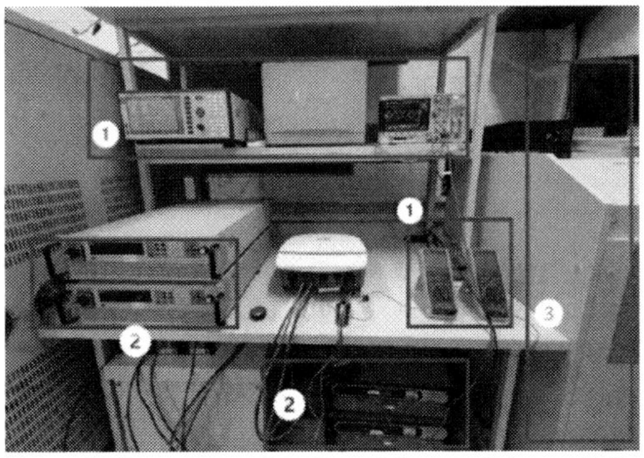

Figure 1: Setup for testing the efficiency of photovoltaic inverters

The test is performed through IEC 62891:2020, in which the regulation mentions European and Californian efficiency. Ordinance 140/2022 requires only European efficiency, also known as EURO efficiency.

To calculate the EURO efficiency, it is necessary to understand that the IEC 62891:2020 standard presents

10.4229/EUPVSEC2023/3AV.1.51
020171-001

three efficiencies: the conversion efficiency, the Maximum Power Point Tracking (MPPT) efficiency, and the total efficiency. Conversion efficiency is the ratio of output alternating current power to direct current input power. MPPT efficiency is the ratio of the DC input power to the theoretical maximum power point value. Finally, the total efficiency is the product of conversion and MPPT efficiency. Or the ratio of the AC output power to the theoretical value of the maximum power point.

Defined as efficiencies, the standard presents EURO efficiency according to equation 1.

$$\eta_{EURO} = 0,03\eta_{t_{5\%}} + 0,06\eta_{t_{10\%}} + 0,13\,\eta_{t_{20\%}} + 0,10\,\eta_{t_{30\%}} + 0,48\,\eta_{tl_{50\%}} + 0,20\,\eta_{t_{100\%}} \quad (1)$$

The equation 1 is explained through equation 2.

$$\eta_{EURO} = weight * \eta_{t_{(power\ inverter\ ex,:\ 5\%)}} + weight * \eta_{t_{(power\ inverter,\ ex,:\ 10\%)}} + \ldots + \quad (2)$$

In which, weight is built up over the amount of time the inverter operates at a specific level from each site's irradiance database. The total conversion efficiency (t) is the ratio between the AC output power and the theoretical maximum power point value. Finally, power inverter is the load that is representative of the power of the inverter. Therefore, the test will occur in all these loads to calculate the equation.

That said, for data to be collected and EURO efficiency to be continued, in addition to the power loads that the inverter needs to comply with, it is necessary that the equipment is tested at minimum, rated and maximum voltages.

Having defined the test method, IEC 62891:2020 does not specify tolerance of the value indicated in the datasheet with the value measured in the laboratory. However, Ordinance 140/2022 acquires as a technical requirement that on-grid inverters must present a measurement of energy efficiency equal to or greater than the value declared by the manufacturer on the datasheet or in the product manual, with a tolerance of -1.00 percentage point.

3 RESULTS

The two inverters tested are 8 kW, called inverter 1 and inverter 2, respectively. As mentioned in the methodology of this article, the equipment is approved to be sold in Brazil if the difference between the measured and declared values in the datasheets is at most 1%. As presented in Table 1, inverter 1 would be approved according to the new ordinance, and invert 2 would not. In the case of inverter 2, the manufacturer would have to modify the datasheet value to the value measured in the laboratory, respecting the tolerance margin.

As well as, Fig. 2 to 7 present efficiency curves (conversion, MPPT and total efficiency) of two photovoltaic inverters sold in Brazil. Inverter 1 is approved for sale in Brazil in the efficiency test of Ordinance 140/2022. While inverter 2 has a difference of 2.80% between the measured value and the value declared in the datasheet.

Table 1: Comparison between the values measured in the laboratory and declared in the datasheets.

Description	Power (kW)	Eff. EURO (Datasheet)	Eff. EURO (Measure)	Percentage Difference
Inverter 1	8	97.00%	96.34%	0.66%
Inverter 2	8	98.00%	95.20%	2.80%

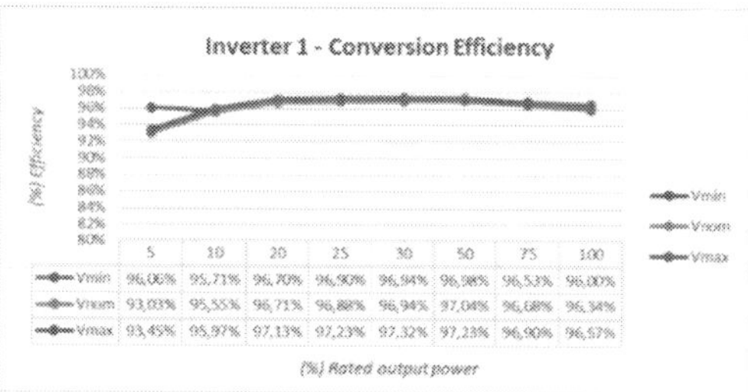

Figure 2: Inverter 1- Conversion Efficiency Curve

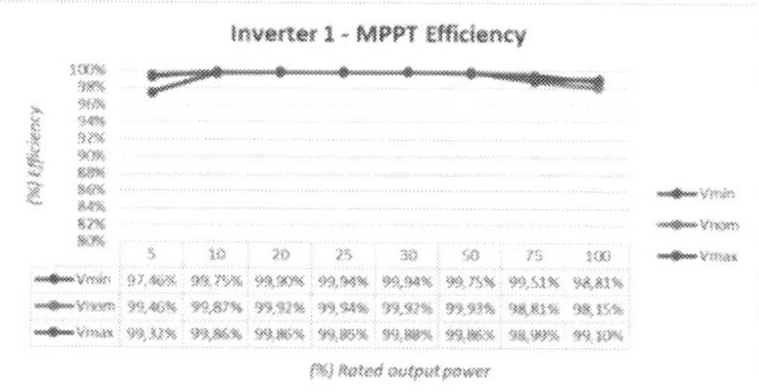

Figure 3: Inverter 1- MPPT Efficiency Curve

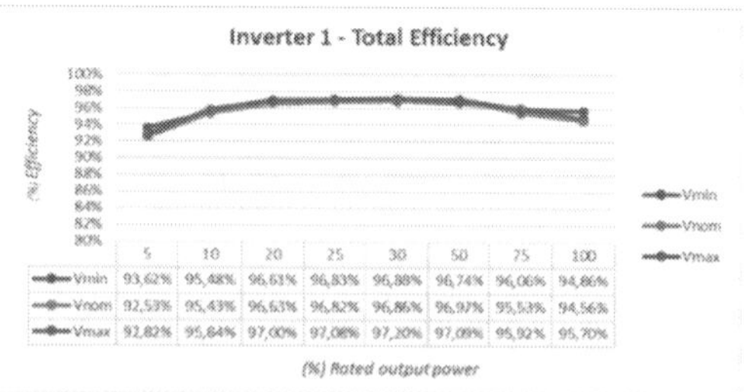

Figure 4: Inverter 1- Total Efficiency Curve

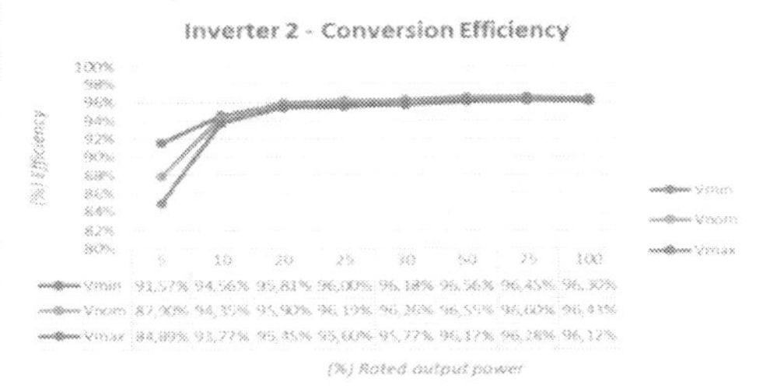

Figure 5: Inverter 2 - Conversion Efficiency Curve

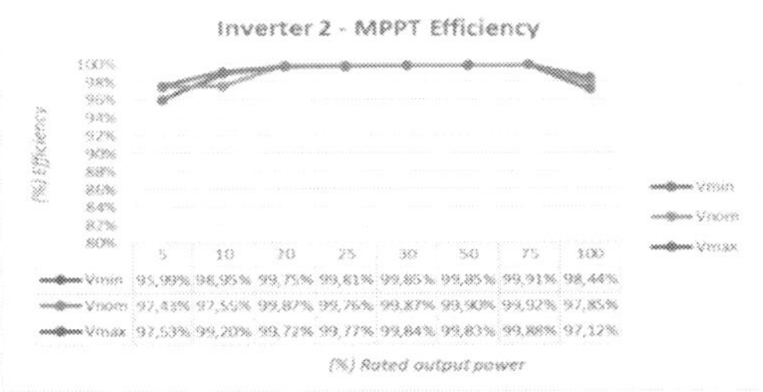

Figure 6: Inverter 2 - MPPT Efficiency Curve

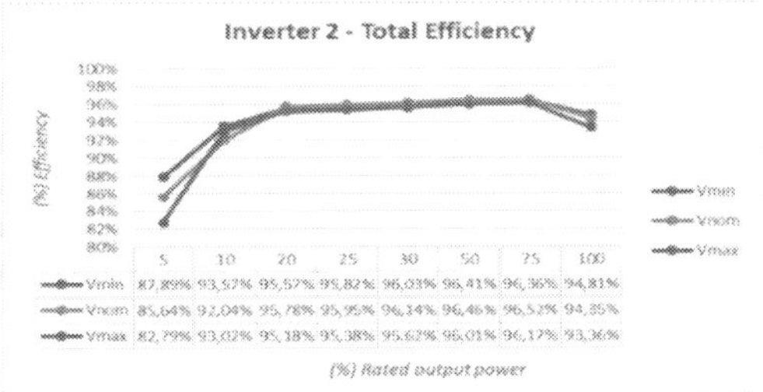

Figure 7: Inverter 2 - Total Efficiency Curve

4 CONCLUSION

This work revealed the importance of Ordinance 140/2022 in Brazil for improving the quality of photovoltaic inverters. It was observed that the efficiency of some PV inverters sold in Brazil does not correspond to the value declared in their technical sheets, with emphasis on Inverter 2, which presented a discrepancy of 2.80%. This difference is significant, as it directly influences the return on investment and the system's energy production. Furthermore, efficiency curves (conversion, MPPT, and total efficiency) are fundamental to optimizing equipment performance. For future work, it is suggested to test more samples of commercial photovoltaic inverters and carry out reliability tests to verify the quality of the converters. Additionally, PVsyst software can be a valuable tool for determining the impact of measured and declared efficiency on photovoltaic generation.

5 REFERENCES

[1] Instituto Nacional de Metrologia, Qualidade e Tecnologia. Portaria N°.004 Rio de Janeiro, Brasil, 2011. Available: <http://www.inmetro.gov.br/legislacao/rtac/pdf/rtac001652.pdf>

[2] Instituto Nacional de Metrologia, Qualidade e Tecnologia. Portaria N°.140 Rio de Janeiro, Brasil,2011.Available:<https://in.gov.br/en/web/dou/-/portaria-n-140-de-21-de-marco-de-2022-389587680>

[3] Figueira, H. H., Hey, H. L., Schuch, L., Rech, C., & Michels, L. (2015, June). Brazilian grid-connected photovoltaic inverters standards: A comparison with IEC and IEEE. In 2015 IEEE 24th International Symposium on Industrial Electronics (ISIE) (pp. 1104-1109). IEEE.

[4] Maximum power point tracking efficiency of grid connected photovoltaic inverters, IEC 62891:2020, International Standard, International Electrotechnical Commission, July 2020.

CENTRAL INVERTER TESTING UNDER REAL OUTDOOR CONDITIONS. A CONTROLLABLE ANALYSIS UNDER NON-CONTROLLABLE CONDITIONS USING STATISTICS. A REAL CASE STUDY

S. Suárez*, D. Villoslada, G. Navas, J. Vilela, I. Fernández, S. Rodríguez-Conde
Enertis Applus+, www.enertisapplus.com
Parque Empresarial Las Mercedes, C/ de Campezo, 1, 28022 Madrid, Spain.
*sergio.suarez@enertisapplus.com

ABSTRACT: Central inverters are usually rigorously tested in lab settings before field deployment [1]. Despite this, post-installation evaluations pose challenges due to electrical-topological complexities and measurement accuracy. Breaking away from traditional lab-based assessments, this study introduces a novel on-site testing methodology, effectively addressing these limitations. The on-site approach is particularly innovative for evaluating key parameters like the European Efficiency (η_{EU}), a weighted-average metric that is critical for real-time performance and quality assurance. Utilizing this metric and conforming to the UNE 50530 standard [2], the methodology successfully generated efficiency curves for various DC input voltages and assessed harmonic distortion dependence on irradiance levels. Our findings underscore the significant influence of central inverters on the field performance of photovoltaic (PV) strings. This directly impacts electrical signal quality, thereby emphasizing the value of our on-site testing approach for more accurate and relevant performance metrics
Keywords: Monitoring, Balance of System, PV Central inverter, European Efficiency, Testing On-Site

1 INTRODUCTION

Photovoltaic (PV) systems have undergone significant advancements over the last few decades, with inverters playing a central role in their effectiveness and efficiency. Traditionally, central inverters are evaluated in controlled laboratory settings to ensure they meet technical specifications before being deployed in the field [1]. While this is a standard industry practice, it is insufficient for assessing long-term performance and efficiency in real-world operational conditions.

Haeberlin's seminal work on the evolution of inverters outlines the substantial changes and improvements in inverter technologies from 1989 to 2000[1]. More recent studies have delved into specific performance metrics like the DC voltage dependency of inverter efficiency [2] and a more comprehensive understanding of total efficiency [3]. However, these studies primarily focus on laboratory-based testing methodologies.

A significant research gap exists in understanding the long-term performance and efficiency of central inverters in real-world, on-site conditions. On-site evaluations are challenging due to the complexities associated with electrical topology and the difficulties in maintaining measurement accuracy over time. This study aims to bridge this gap by introducing a novel on-site testing methodology, offering a more context-relevant evaluation of central inverters' performance metrics, including European Efficiency (η_{EU}).

The primary contribution of this paper is the development and application of an on-site testing methodology for assessing central inverters. Building upon the foundational work by Baumgartner et al. on the DC voltage dependency of inverter efficiency [2], our methodology innovatively adapts this understanding to real-world operational settings. This approach not only addresses the limitations of laboratory testing but also emphasizes the crucial role of on-site evaluations for more accurate and relevant performance metrics.

2 EUROPEAN EFFICIENCY AND LABORATORY-BASED PERFORMANCE METRICS

2.1 European Efficiency (η_{EU})

One of the most widely accepted metrics for evaluating the efficiency of central inverters is the European Efficiency (η_{EU}). This metric is standardized according to UNE 50530 & IEC 61830 and provides a weighted-average efficiency calculated at constant input voltages (V_{DC}) [2]. η_{EU} serves as a comprehensive representation of the inverter's overall performance, allowing for a standardized comparison across different products and technologies.

2.2 UNE 50530 & IEC 61830 Standards

UNE 50530 and IEC 61830 are critical standards that guide the performance testing of PV inverters. Specifically, UNE 50530 outlines the procedures for evaluating the performance of inverters under controlled conditions, emphasizing constant V_{DC} input. Similarly, IEC 61830 focuses on the general requirements and methodologies for assessing PV inverters. Both standards advocate for testing under stable V_{DC} conditions, thereby serving as a benchmark for laboratory testing.

2.3 V_{DC} instability

While laboratory-based testing benefits from controlled settings, on-field evaluations present a distinct set of challenges. A key issue is the inherent instability of V_{DC} during on-field testing. Unlike in laboratory conditions where V_{DC} is stabilized, field conditions are subject to fluctuations due to varying environmental factors like solar irradiance and temperature.

Given the unstable nature of V_{DC}, direct application of η_{EU} can yield misleading results. To address this, η_{EU} overall is calculated during the on-field tests, and statistical filters are applied to obtain η_{EU} (V_{DC}) across the entire value range. This adjustment aims to account for the variability in V_{DC}, providing a more accurate depiction of the inverter's field performance.

10.4229/EUPVSEC2023/3AV.1.52
020172-001

3 EXPERIMENTAL DESCRIPTION

This section outlines the design and conditions of the experiment aimed at evaluating the on-field performance of central inverters using a novel on-site methodology. The focus was on assessing vital performance parameters such as European Efficiency (η_{EU}) under real-world conditions.

3.1 Data Acquisition and Sampling

Within a single photovoltaic project, five central inverters (Fig. 1) of the model "Power Electronic HEMK 660V FS3510K" were inspected. Data recording was carried out for a minimum of one full day for each inverter, ensuring that 100% load was reached in every case. Additionally, the tests were designed to include at least one day with entirely clear skies to provide a robust set of data.

Figure 1: Inverter Power Electronic HEMK 660V FS3510K tested during this work.

The measurement equipment comprised a high-precision Zimmer LMG power meter, supplemented with current and voltage measurement accessories for both AC and DC. The device boasted an uncertainty level below 0.4%. Data was collected with a temporal resolution of 200ms, allowing for a highly detailed analysis of the inverter performance under varying conditions.

3.2 Climatic Conditions During Tests

The impact of environmental variables is crucial for a comprehensive interpretation of the results. Testing was carried out under diverse weather conditions, ranging from sunny days with high levels of solar irradiance to overcast scenarios with reduced irradiance. Temperature, wind speed, and humidity were also recorded to offer a more holistic view of the environmental factors influencing inverter efficiency.

4 EXPERIMENTAL RESULTS

This section discusses the findings derived from the on-field testing of the five central inverters ("Power Electronic HEMK 660V FS3510K") previously mentioned. Two critical facets are explored: the behavior of V_{DC} in an uncontrolled setting and the statistical techniques applied to ascertain V_{DC} profiles.

4.1 V_{DC} in an Uncontrolled Analysis.

Tests were conducted in the normal operating conditions of a photovoltaic plant, where V_{DC} is inherently uncontrollable. Consequently, the European Efficiency (η_{EU}) exhibited significant fluctuations across different load levels of the inverters. These variations were especially pronounced when V_{DC} showed greater instability within the operational range of the inverter.

The Fig. 2 provides a graphical representation of the average efficiency, independent of V_{DC}, highlighting both its variability and its dependency on V_{DC} fluctuations.

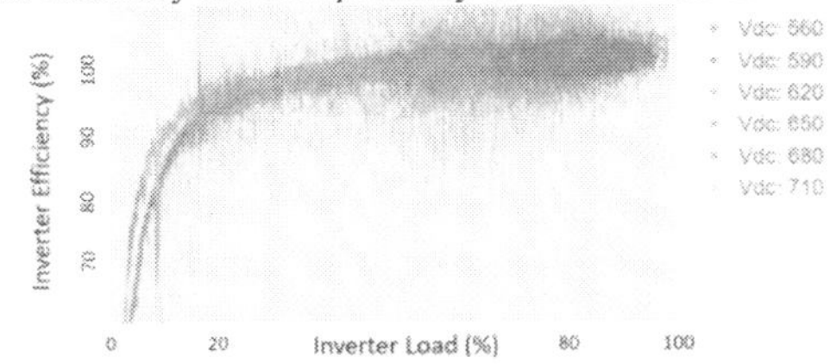

Figure 2: Noise measurement along 'efficiency curves vs. load' under varying V_{DC} conditions.

The inverter's efficiency curve illustrates the device's performance across various load conditions. Generally, the efficiency increases asymptotically up to 15% of the load, after which it follows a gentle exponential trend. This trend approximates a near-linear behavior, with efficiency continuing to rise to the maximum operational loads.

The Fig. 3 employs a Box Plot to visually depict the voltage variability, represented through the Standard Error of the Mean (SEM). Each box plot in Fig. 3 represents the mean efficiency value (and its standard deviation) for a 5% load range of the inverter, spanning from 20% to 100% of the inverter's total operating load.

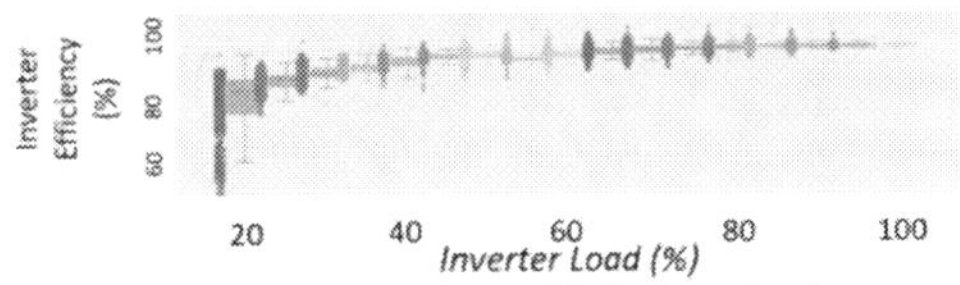

Figure 3. Boxplot of Efficiency Vs Inverter load

This graphical representation allows for a more nuanced understanding of the voltage variability and its impact on the overall efficiency.

To mathematically represent the average efficiency value over all possible inputs $\eta_{EU\,overall}$, the following equation is introduced:

$$\eta_{EU\,overall} = \int \eta_{EU}\,(V_{DC})\delta\,V_{DC}$$

In this equation, $\eta_{EU}(V_{DC})$ represents the European Efficiency function respect V_{DC} at constant inverter load and δV_{DC} represent infinitesimal variation on the functional V_{DC}.

The formula serves to quantify the average efficiency, η_{EU}, over varying V_{DC} values, offering a robust metric that complements our on-field experimental findings.

4.2 Obtaining V_{DC} Profiles through Statistical Analysis

To achieve a detailed understanding of η_{EU} as it relates to varying V_{DC}, statistical filtering techniques were employed [1]. The filtered data allowed for the calculation of η_{EU} for each specific V_{DC} value. Nonlinear regression was subsequently applied to model this data, and the Maximum Power Point Tracker (MPPT) was validated.

To enhance the clarity of the inverter efficiency curves and isolate the effects of V_{DC} voltage variability, Fig. 4 is presented. This figure exclusively reports the mean efficiency values, thereby omitting the associated dispersion for each inverter load level. This graphical representation reveals how the inverter's efficiency improves at specific V_{DC} voltage ranges, which correspond to the optimal operating region of the MPPT.

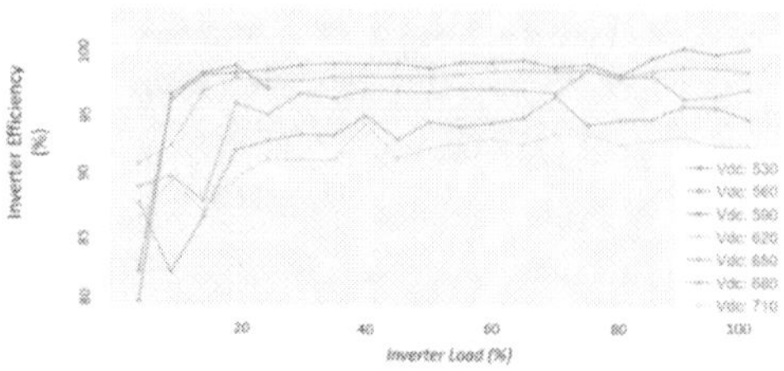

Figure 4. Average efficiency values at different V_{DC} values using filtering data.

The specific voltage at which η_{EU} peaks was identified and denoted as η_{EU} (MPPT) [3] [4]. This corresponds to V_{DC} where the MPPT should be working.

To mathematically calculus the MPPT value over all voltage rage the following equations is introduced:

$$\frac{\partial \eta_{EU}}{\partial V_{DC}} = 0$$

This offers a clear explanation of the practical implications of the formula, linking it directly to the Maximum Power Point Tracking (MPPT) and its operating range.

By employing these empirical data both pre and post-statistical filtering, nonlinear regression methods can be applied to model the inverter's efficiency curves at various V_{DC} voltage levels, as depicted in the subsequent Fig. 5

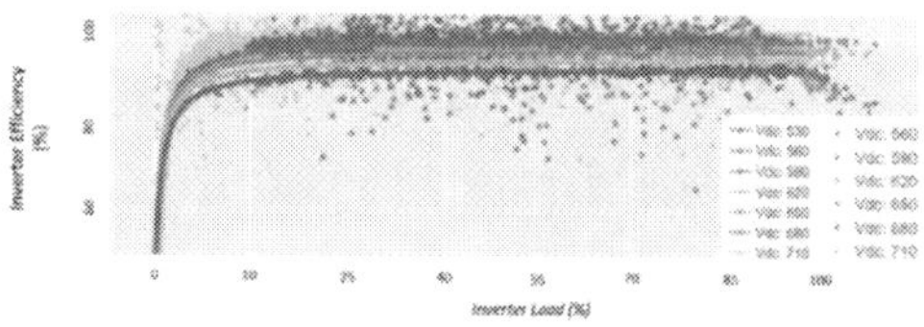

Figure 5. Non-linear regression model fitting Efficiency vs load

In conclusion, it is evident that it is possible to measure the inverter's efficiency without controlling the V_{DC} variable. Subsequent statistical filtering can then be applied to the voltage values, which can be utilized to construct a nonlinear mathematical model. This approach enables the assessment of a central inverter's quality status across multiple operational ranges

4.3 Experimental results on five central inverters

Five central inverters were rigorously tested using the mathematical method delineated earlier in this paper. For each inverter, η_{EU} values were computed for varying V_{DC} inputs, and the operational range of the MPPT was validated.

Table 1 summarizes these findings, reporting the overall η_{EU} and indicating the V_{DC} at which maximum efficiency is achieved for each inverter.

Table I: Summary of inverter efficiency tests conducted on-field for five central inverters.

Sample	$\eta_{EU\ overall}$	$\eta_{EU\ max}$	V_{DC}
Inverter 1	98.7±0.8	98.8±0.3	753.2V
Inverter 2	99.1±0.4	99.1±0.2	749.4V
Inverter 3	97.8±0.7	98.4±0.5	747.1V
Inverter 4	96.9±1.1	97.3±0.7	739.9V
Inverter 5	98.1±0.9	98.2±0.5	755.1V

The parameters listed in Table 1 include the overall European Efficiency ($\eta_{EU\ overall}$) cited before in the paper and the $\eta_{EU\ maximum}$ corresponds to the maximum European Efficiency achieve by the inverter by the V_{DC} which maximum efficiency is attained.

Each parameter has been assessed against the manufacturer's recommended thresholds, showing that all the tested inverters exceed these limits, thereby affirming their optimal condition and effective operation.

5 CONCLUSIONS

Efficiency tests for inverters were successfully conducted, demonstrating reproducible results. While these results align with tests performed in controlled environments.

They also remain consistent under real-world, uncontrolled field conditions. A new domain in high-power electronics testing has been unveiled in field applications.

6 REFERENCES

[1] H. Haeberlin, "Evolution of Inverters for Grid connected PV-Systems from 1989 to 2000," 17th Eur. Photovolt. Sol. Energy Conf., p. 5, 2001.

[2] F. P. Baumgartner, H. Schmidt, B. Burger, R. Brundlinger, H. Haberlin, and M. Zehner, "Status and relevance of the DC voltage dependency of the inverter efficiency," 22nd Eur. Photovolt. Sol. Energy Conf. Milano, vol. 1999, no. September, pp. 3–7, 2007.

[3] H. Haeberlin, "Evolution of Inverters for Grid H. Haeberlin, L. Borgna, M. Kaempfer, and U. Zwahlen, "Total Efficiency Htot – a New Quantity for Better Characterisatic of Grid-Connected Pv Inverters," no. June, pp. 1–4, 2005.

[4] H. Haeberlin and P. Schaerf, "New Procedure for Measuring Dynamic MPP-Tracking Efficiency at Grid-Connected PV Inverters," Sol. Energy, no. September, pp. 21–25, 2009.

DESIGN AND IMPLEMENTATION OF A POWER DC/DC CONVERTER FOR APPLICATION IN PHOTOVOLTAIC PANELS

Patricio Valdivia-Lefort*, Antonio Sanchez*, Rodrigo Barraza**
*Electrical Engineering Department, Federico Santa Maria Technique University
**Mechanical Engineering Department, Federico Santa Maria Technique University
Av Vicuña Mackenna 3939, San Joaquin, Santiago - Chile
Telephone: +56 22 3037238, Email address: patricio.valdivial@usm.cl

ABSTRACT: In this work it is Design and implementation of a power dc/dc converter for application in photovoltaic panels by redefining a new relationship between electronics of power and photovoltaic systems, opening the possibility of testing different types of control schemes. To accomplish this, it is proposed a management control of a photovoltaic system in a stable manner over time throughout its operating range. To achieve this, were evaluated different topologies that are adapted to this objective, once the topology is chosen, different control schemes were evaluated through simulations and discriminated based on to the main objective. The implementation was thought to simulate a photovoltaic system generic system, which includes a dc link at the output of the power converter. The results obtained show that a good performance of the converter and its control scheme control, being able to operate at different points of operation and showing good stability in the face of changes in environmental conditions.

Keywords: DC-DC converter, Photovoltaic Module, PV panel operation points

1 INTRODUCTION

The growing development of energy management systems for photovoltaic systems has turned to the academy to satisfy the demand that the industry demands. In this sense, consistent progress has been made in power converters driving panels photovoltaics in order to obtain the maximum power from them. And while this development has It has been fruitful to a few problems that have been limited to only one edge of investigation.

Such is this case, where the control scheme of the converters is restricted to the control of tension of the panel and the points of operation only to those where it is obtained maximum power. However, little has been evaluated the performance of these systems in suboptimal, or their behavior at different points of operation and their rejection to disturbances or even failures.

Then arises the need to develop technology that allows to monitor the state of these DC generation sources, evaluating their performance at different points of operation. This will also allow, as a last point, to distinguish the status of operation of a photovoltaic solar panel remotely, being essential that its implementation is validated in a topology as close as possible to an electrical microgrid.

In the present work its seeks to design and implement a dc/dc converter and its control strategy that manages to take a photovoltaic solar panel to different points of operation in a stable way.

The Scientific innovation and relevance of this work is justified by the fact that a new relationship has been proposed and a new dc/dc converter has been designed and implemented to operate a photovoltaic solar panel at different points of operation in a stable way.

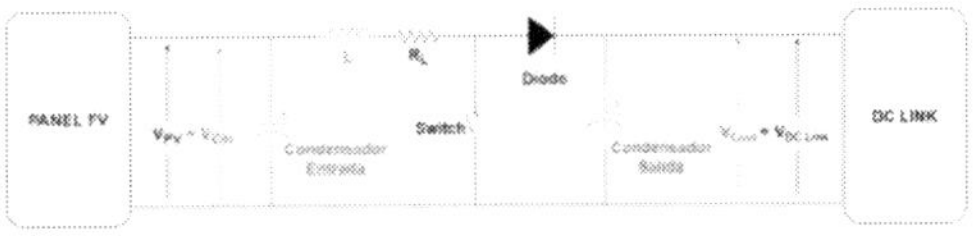

2 SETUP:

The implemented setup is depicted bellow, where it can be seen a photovoltaic module connected to the proposed DC-DC converter, a battery, and a control system.

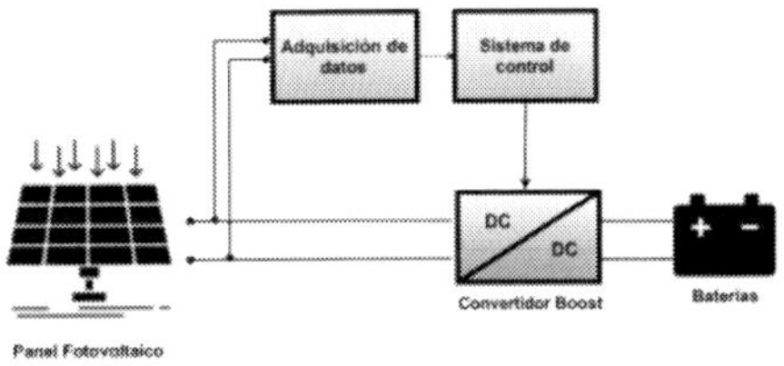

Figure 1 implemented setup

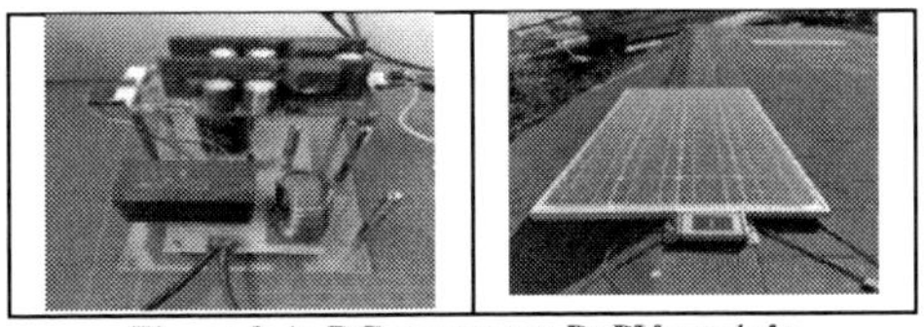

Figure 2 A: DC converter; B: PV module

3 RESULTS OBTAINED FOR THE DC – DC CONVERTER

A DC-DC converter was proposed and implemented with three different control strategies; the results obtained from the DC-DC converter proposed are shown as follows.

10.4229/EUPVSEC2023/3AV.1.53
020173-001

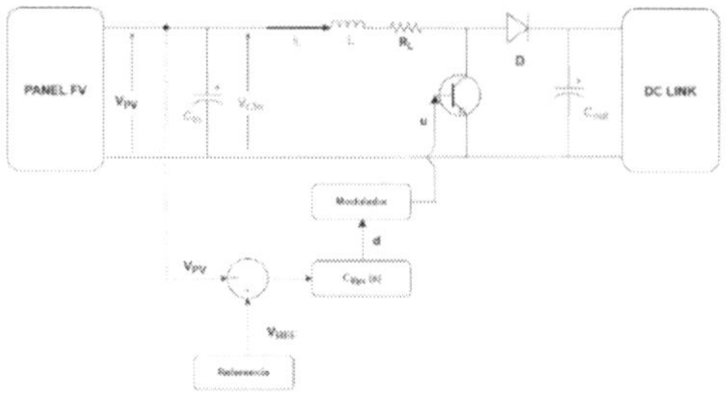

Figure 3 1: DC -DC converter proposed.

3.1.- PID Voltage Control:

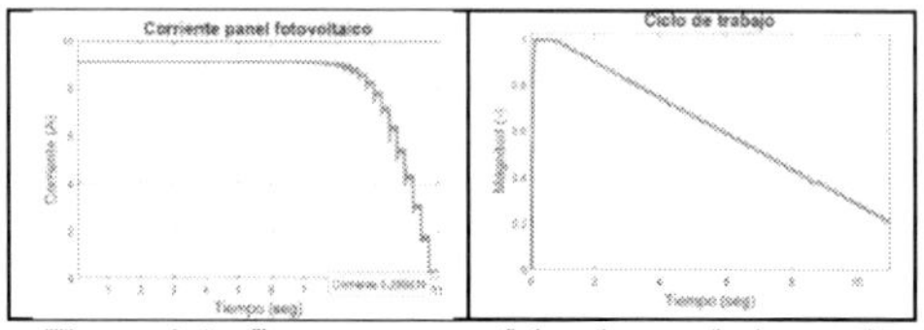

Figure 4 A: Current curve of the photovoltaic panel with a PID voltage control strategy. B: Duty cycle for PID voltage control.

3.2.- SMC+PI Voltage Control:

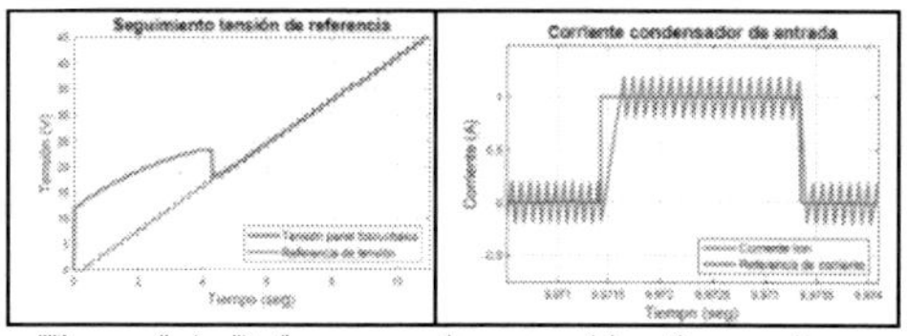

Figure 5 A: Reference voltage tracking for a voltage control PI + SMC current control. B: Photovoltaic panel with a PI voltage control + SMC current control.

3.3.- IP Current Control:

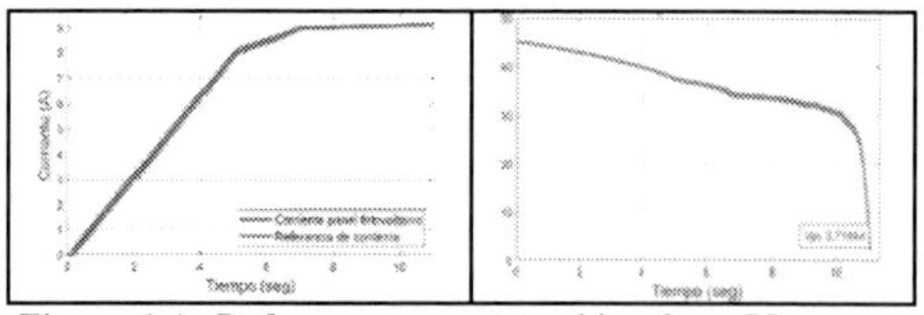

Figure 6 A: Reference current tracking for a PI current control. B: Voltage curve of the PV module with a PI current control.

As a summary of the three control schemes utilized, the following chart shows their performances:

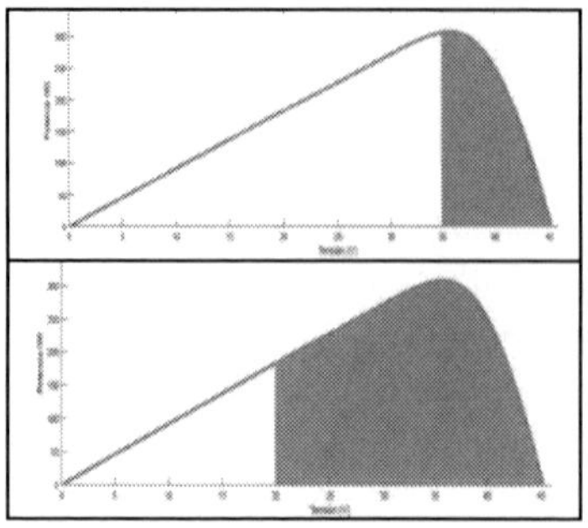

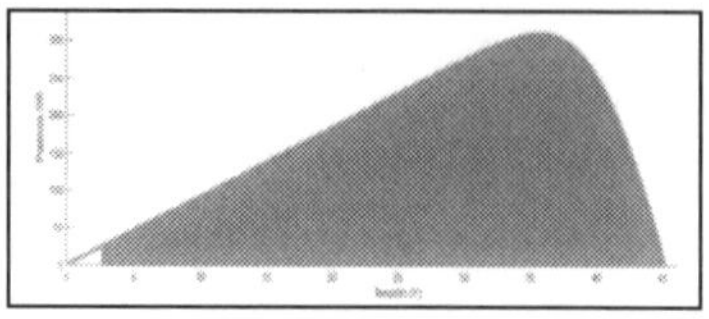

Figure 7: (a) PV curve for PID voltage control. (b) PV curve for PI+SMC control. (C): PV curve for PI current control.

It can be seen from the figures that the control scheme which allowed to obtain a greater number of operating points was the linear control of PI current. In contrast, the control schemes of PID voltage and PI voltage + SMC current, they presented better performances but in more limited areas of operation, with a minor ripple and faster responses.

CONCLUSIONS

It was possible to design a boost type power converter that allowed to control a photovoltaic panel of 100 W in different points of operation, even under different environmental conditions of irradiance and temperature.

Regarding the definition of a topology, the choice of the boost type converter demonstrated that this is still a very robust and versatile topology, since it allows to build relatively simple control schemes and obtain performances that meet the implementation goals. In this particular case, it was sought that the topology would allow operate in the entire range of operation area of the photovoltaic panel, which although it was not a 100% achieved, a performance very close to the ideal was obtained.

The validation of the topology and control scheme was achieved through simulations in Simulink®. These simulations allowed to see differences obtained by using different control approaches. The robustness of the system in the face of irradiance variations is highlighted, since the results were consistent with what was obtained in the preliminary tests.

The results obtained show that a good performance of the converter and its control scheme control, being able to operate at different points of operation and showing good stability in the face of changes in environmental conditions.

REFERENCES

[1] Shaw, Priyabrata & Garanayak , Priyabrat . (2018). Analysis , design and implementation of analog circuitry based maximum power point tracking for photovoltaic boost DC/DC converter . Transactions of the Institute of Measurement and Control.

[2] Giovanni Petrone; Carlos Andres Ramos Paja; Giovanni Spagnuolo, Modeling the PV Power Conversion Chain ,"in Photovoltaic Sources Modeling , IEEE, 2016, pp.127 164, doi : 10.1002/9781118755877.

[3] Giovanni Petrone; Carlos Andres Ramos Paja; Giovanni Spagnuolo , "Control of the Power Conversion Chain ," in Photovoltaic Sources Modeling , IEEE, 2016, pp.165 184, doi : 10.1002/9781118755877.

[4] Texas Instruments, "TMS320F2837xD Dual Core Microcontrollers " in Technical Reference Manual, December 2013 Revised 2019.

[5] Texas Instruments, "LAUNCHXL F28379D Overview ", August 2016 Revised March 2019.

[6] Texas Instruments, "TMS320F2837xD Dual Core MCUs " in Silicon Revisions C, B, A, 0 , December 2013 Revised June 2020.

[7] Di Piazza, Maria Carmela & Vitale , G.. (2013). Photovoltaic sources : Modeling and emulation . 10.1007/978 1 4471 4378 9.

[8] Muhammad H. Rashid, "ELECTRONICA DE POTENCIA, Circuitos, dispositivos y aplicaciones". México. 3ª Edición.

SIMULATION OF ELECTRICAL LOSSES IN BIFACIAL PV AND VALIDATION WITH REAL VERTICAL SYSTEM IN GERMANY

*Dennis Castillo[1], *Odbayar Boldbaatar[1], Eva-Maria Grommes[2], Ulf Blieske[2]
Cologne Institute for Renewable Energy
University of Applied Sciences Cologne
Betzdorferstraße 2, 50679 Cologne, Germany
odbayar.boldbaatar@smail.th-koeln.de, dennis.castillo@smail.th-koeln.de, eva-maria.grommes@th-koeln.de, ulf.blieske@th-koeln.de

ABSTRACT: There has been significant progress in the simulation of bifacial photovoltaic (PV) systems in the past. Despite this, the electrical losses of bifacial PV systems have yet to be thoroughly investigated in terms of simulation. This study focuses on the cables and the inverters by quantifying and validating the electrical losses caused by the resistivity of the wires and the efficiency of the inverters. For the simulation, a script is implemented into the code of BifacialSimu [1]. Furthermore, individual monthly simulations were compared with data measured from a PV plant in southern Germany and with simulations made in PVsyst [2] to validate their accuracy. On average, the simulation for the cable losses was 0.137% in BifacialSimu and 0.135% in PVsyst for the months in consideration. On the other hand, the average simulated losses for the inverter were 2.07%, while the losses measured on-site were 2.09%. These results demonstrate the high accuracy of the models applied.
Keywords: Photovoltaic, Bifacial Photovoltaic, Electrical losses, Inverter losses, Cable losses

1 INTRODUCTION

According to the International Energy Agency (IEA), in the year 2022, there was a remarkable increase of 270 TWh in solar PV production, accounting for 26% growth. It is considered to become the lowest-cost option for new electricity generation in most of the world [3]. Despite all the advantages of PV technology, one of the biggest concerns for solar arrays with standard fixed tilt or tracking systems is the enormous spatial demand required for their installation. For this reason, bifacial technology is attracting attention all over the world. The main advantage of this technology is that bifacial cells can use both sides of the cell to convert solar energy to electricity, unlike the conventional monofacial ones, leading to higher efficiency [4].

As the demand for these types of systems continues to grow, more accurate and reliable models for simulating them to optimize their performance are extremely important. Several models to predict the performance of bifacial PV systems have been developed. Typically, these models can have an optical, thermal, or electrical approach [4]. In this paper, we will consider the optical approach, given that the software used for the simulation is based on it. For this model, the front side irradiance is measured similarly as in monofacial systems, considering the direct, diffuse, and reflected irradiance. The rear-side irradiance results are a bit more complex, given that it is caused by a combination of diffuse and direct irradiance reflected from the ground, which the PV modules and external bodies can also shade [4]. Two main concepts have gained popularity among the scientific community for the development of these models. The state-of-the-art models are based on ray tracing (RT) and view factor (VF) concepts. Ray tracing models allow the simulation with more complex geometries, such as the shading of the module's rear sides by external bodies like the mounting structure, but view factors are preferred for simple system configurations, such as PV systems, in which installation is optimized for bifacial modules [5].

Despite the advances in these models, the electrical losses of bifacial PV systems have yet to be thoroughly investigated in terms of simulation. Identifying and minimizing the losses in electrical devices is one of the most significant challenges facing solar system developers [6]. The most critical components of a solar power system are the inverter, the cables, and the solar panels. The losses caused by these components' operation directly affect the solar system's performance, hence their analysis's importance.

This study focuses on the cables and the inverters by quantifying and validating the electrical losses caused by the resistivity of the wires and the efficiency of the inverters. For the simulation, a script is implemented into the code of BifacialSimu [1], an open-source software for simulating the energy yield of Bifacial PV systems. The validation is done by comparing the simulation results with measured data from a utility-scale PV plant installed in southern Germany [7], as well as with similar simulations made with PVsyst [2], a commercially available software for PV systems simulations. The system used for the calculation is comprised of a 60kW inverter with an input of 240 vertically installed bifacial solar modules connected in 12 equal strings. Weather data of the location is used in the software to calculate the output of the modules, and consequently, the ohmic losses are calculated and compared with a simulation made with the same parameters in PVsyst [2]. Similarly, the output of the inverter is calculated by reducing the calculated losses due to its efficiency and then compared to the energy yield measured on-site. Simulations are made using the different efficiency values provided by the manufacturer: European efficiency (η_{EU}), California Energy Commission efficiency (η_{CEC}), maximum efficiency (η_{max}), and the weighted efficiency curve.

2 METHODOLOGY

In this study, a simulation-based approach is applied to calculate the electrical losses in PV systems, which are further compared with simulation results from PVsyst and actual data measured in a utility-scale PV plant to validate its accuracy.

2.1 BifacialSimu

The tool is a Python package developed by the Cologne Institute for Renewable Energy that utilizes a variety of simulation models to estimate the energy yield of bifacial PV systems [1]. As with many other similar

10.4229/EUPVSEC2023/3AV.1.55
020175-001

tools, a set of parameters is required to run the simulation. The user can enter these parameters through a very intuitive graphical user interface (GUI). First, a scene with the collector field has to be defined by selecting a fixed tilt or tracking system, the corresponding azimuth, the arrangement of the modules by rows, and the albedo characteristics. Location and weather information are needed to estimate the energy yield. This information can be either downloaded for a specific set of coordinates or a weather file containing information like the radiation, pressure, and temperature with the appropriate format can be imported. Furthermore, a time frame and a simulation mode have to be selected. For estimating the radiation, the software allows the user to select between view factors, ray tracing, or a combination of both models for the most accurate result. Finally, the equipment has to be selected. A set of predefined modules is available for selection, and additionally, it is possible to add a specific module by entering its corresponding parameters.

All this allows the user to estimate the energy yield delivered by the PV system at the end of the array. However, this study aims to estimate the electrical losses between the array of modules and the inverter output. Therefore, the main focus is on the selection of the wire and inverter parameters that will be necessary to estimate the losses due to resistivity and efficiency. For calculating the cable losses, the required parameters are the wire's cross-sectional diameter, its material, and the average length from the array(s) to the inverter(s). For estimating the inverter's losses, there is a tab to select the corresponding parameters, in which, like the process of selecting a PV module, a pre-established inverter can be selected, or the relevant parameters can be entered manually to create a generic model. In this case, given that the losses of the inverter can be summarized in the efficiency (refer to section 2.2.2), the user can select between the different efficiency models η_{max}, η_{EU} or η_{CEC} or can manually input the weighted efficiency values for each corresponding level of power or voltage. By entering individual efficiency values, the program constructs the inverter's efficiency curve by the process of interpolation, which is then used to estimate the losses for every single input.

2.2 Electrical Losses

In this paper, we classify the electrical losses as the energy lost in the cables and the inverter, which are caused by the resistivity and the efficiency respectively.

2.2.1 Cable Losses

Typically, the selection of the cross-sectional area of a cable is based on the maximum permissible voltage drop from the source to the load, which is usually between 1 – 2% for the current under Standard Test Conditions (STC) [8]. The power losses $P_{W,Loss}$ in a cable are dependent on the equivalent resistance R [Ω], which is shown in Equation 1. Where ρ represents the resistivity of the wire [Ωm], which is dependent on the material of the conductor. a is the cross-sectional area of the conductor [m²], and l is the length [m] of the cable from the source to the load [9].

$$R = \frac{\rho * l}{a} \tag{1}$$

Furthermore, $P_{W,Loss}$ [W] is calculated with Equation 2, where I is the current flowing through the circuit [A] [10].

$$P_{W,Loss} = R * I^2 \tag{2}$$

2.2.2 Inverter Losses

The overall losses of a PV inverter are represented by its efficiency (η) [11]. The efficiency is defined as the ratio between the energy delivered at the AC terminals of the inverter and the energy delivered by the PV array at the DC terminals of this device [12]. The inverter's efficiency is not fixed, and it changes depending mainly on the input power and the voltage of the PV array [12, 13]. Based on the manufacturer, the data sheet might include a fixed value for Maximum Efficiency, European Efficiency, and California Efficiency, or it can provide an efficiency curve or table with a weighted efficiency [12]. An example of the latter is shown in Figure 1. This is the efficiency curve of the inverters used in this case study.

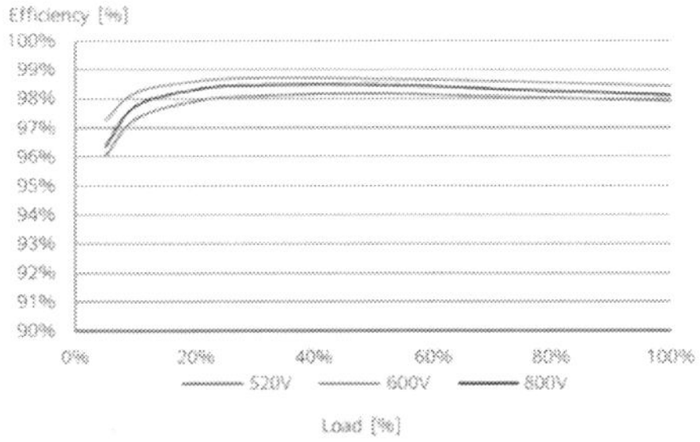

Figure 1: Efficiency curve of Huawei Smart Inverter SUN2000-60KTL [14]

The value for the Maximum Efficiency is obtained from this graph as the highest achievable efficiency in the range of operation of the inverter. But, as expected, the inverter will not be operating at this point all the time, and therefore, there is a need to adjust this value. For this purpose, two models have been developed that consider the distribution of irradiation over a year at the cities of Trier, Germany, for η_{EU} and California, US, for η_{CEC}; both of which prioritize the ranges in different operating points at these specific locations [12].

Currently, the European Efficiency η_{EU} is the most used weighted efficiency [15]. This is usually given by the manufacturer of the inverter and it is calculated with equation 3, where $\eta_{x\%}$ represents the efficiency of the inverter when it operates at x% of its nominal power.

$$\eta_{EU} = 0.03\eta_{5\%} + 0.6\eta_{10\%} + 0.13\eta_{20\%} + 0.10\eta_{30\%} + 0.48\eta_{50\%} + 0.20\eta_{100\%} \tag{3}$$

Similarly, equation 4 is used for calculating the California Efficiency (η_{CEC}). [13]

$$\eta_{CEC} = 0.04\eta_{10\%} + 0.05\eta_{20\%} + 0.12\eta_{30\%} + 0.21\eta_{50\%} + 0.53\eta_{75\%} + 0.05\eta_{100\%} \tag{4}$$

To calculate the losses of the inverter $P_{I,Loss}$ [W], Equation 5 is used, where P_{in} is the input power [W] at the DC terminals of the device and η is the selected efficiency of the inverter [unitless].

$$P_{I,Loss} = \eta * P_{in} \tag{5}$$

10.4229/EUPVSEC2023/3AV.1.55
020175-002

2.3 Simulation

As stated in section 2.1, when using BifacialSimu, the user can enter the data and select between the available options through the GUI. Here, the user has the option to enable or disable the calculations for the electrical losses. If these are enabled, the software will first calculate the equivalent resistance of the cables by using Equation 1 with the corresponding parameters.

Consequently, to estimate the power output, the code must calculate the output current of the field of modules. This DC current I_{dc} is used in Equation 2 to estimate $P_{I,Loss}$. Since this value depends on the information from the weather file, the results will vary for every time stamp and therefore, an iteration must be done for every hour.

After calculating, this is then subtracted from the power output of the modules P_{bi} to obtain the power that is being delivered to the input of the inverter P_{in}. This is then used in Equation 5 along with the selected efficiency to calculate $P_{I,Loss}$. Finally, a graph is computed showing the electrical losses of the system. A brief description of the steps of the simulation can be seen in Figure 2.

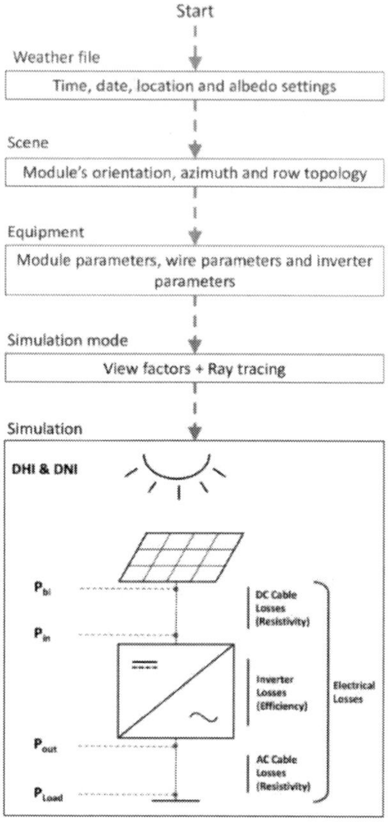

Figure 2: Flow chart diagram of the simulation model

2.4 Site Description

The simulation results are compared with a utility-scale Agro-PV plant located in Aasen, Germany. The plant was built by Next2Sun Technology GmbH and is now operated by Solverde Bürgerkraftwerke Energiegenossenschaft [7]. It is comprised of 11,000 bifacial modules of 380 and 385W (Jolywood JW-D72N Series) connected in strings of 20 modules each. These modules are vertically installed, meaning that they have a fixed tilt of 90°. Each set of 12 strings is connected to a 60kW string inverter (Huawei SUN2000-60KTL Series).

2.5 Data Acquisition

Given that there is no on-site irradiance measurement system, the weather data was obtained from a climate data center from the German Weather Service (DWD), which is located 15km from Aasen; therefore, there is a slight deviation from the weather conditions in the site.

The data provided from the solar park includes information about the installed equipment and an electric plan with the corresponding connections. The scale in the plan, along with information from the publicly available press release report, was used to calculate the average distance of the cables. The energy yield data of the plant was also provided for April, July, August, and September 2021. This data contains information like the energy output for every single string in DC and every inverter in AC. To validate the simulation results, only one inverter with 12 strings, or 240 modules, is considered to reduce the simulation times in the software.

Finally, a simulation was made in PVsyst to obtain the cable losses to be further compared. Both BifacialSimu and PVsyst simulations were made using the same weather files and equipment parameters.

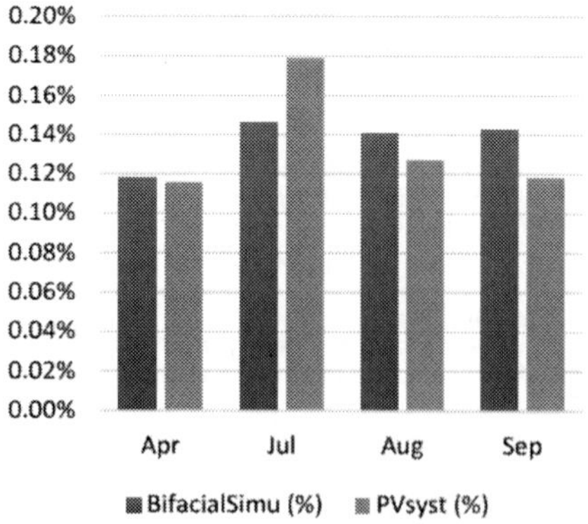

Figure 3: Percentage of energy lost due to resistivity in the cables

3 DATA VALIDATION

The comparison of the results of the energy lost in the cables between both simulations is shown in Figure 3. Overall, the results were very similar; PVsyst showed average losses of 0.135% during the four months, while BifacialSimu showed 0.137%. It was not possible to measure the cable losses during plant operation.

Consequently, the energy yield from the PV field was simulated to be compared with the values measured on-site, shown in Figure 4.

10.4229/EUPVSEC2023/3AV.1.55
020175-003

The deviation between the simulations and the measured data is, on average, 1.92%. In August, the results from BifacialSimu show the lowest deviation with 0.54%. While there is a discrepancy of around 6.37% in September.

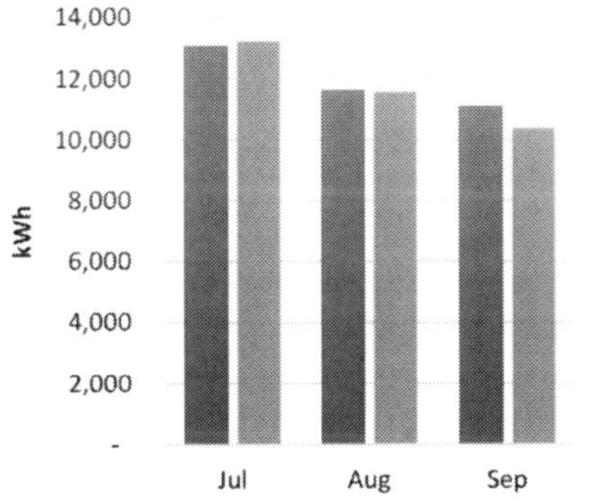

Figure 4: Monthly energy yield comparison between measurement and simulation

To get a better comprehension of these variations, an analysis was computed for short-time periods, which is shown in Fig. 5. As seen from this graph, there are some variabilities in the magnitude and in the time of the results. This might be because the station that provided the weather file is not exactly in the location of the PV plant, as stated in section 2.5. Given that this paper focuses on analyzing only the electrical losses, it was decided to consider the percentage of energy lost instead of the actual energy in kWh.

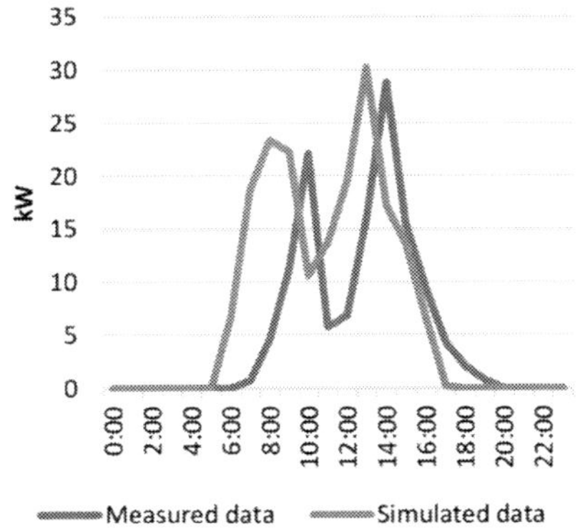

Figure 5: Energy yield comparison on September 16th, 2021

The inverter losses as percentages are shown in Figure 6. In this case, if we consider the month of September, which is the month with the highest accuracy, the simulation results had a deviation of 3.88%. On average, the simulated inverter losses were 2.04%, while the losses recorded on-site were 2.15% in these months.

Overall, the deviation between measured and simulated losses induced by the inverter is approximately 5.47%.

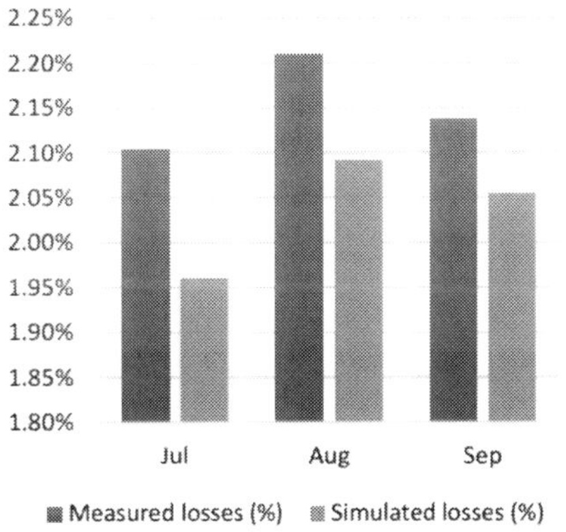

Figure 6: Percentage of the energy lost due to inverter losses

September was selected to compare the results with the different efficiency values summarized in Table I. As a consequence, the most accurate result was obtained using the weighted efficiency curve. The result was 2.05%, which was slightly lower than the measured losses of 2.28%. In comparison, the maximum and European efficiency values are 1.10% and 1.30%, respectively.

Table 1: Comparison of different types of efficiency for inverter losses in September

Measured losses (%)	2.28%
Weighted efficiency loss (%)	2.05%
Maximum efficiency loss (%)	1.10%
Euro efficiency loss (%)	1.30%

4 DISCUSSION

The comparison between the simulated results and the measured data showed that doing simulations for short-term or daily periods may result in unwanted deviations; therefore, trying more extended periods is recommended. Given that the results are strongly dependent on the weather information, it is crucial to simulate data from a location as accurately as possible. The shift in the generation curves that can be observed in Figure 5 is due to the differences in the weather file used for the simulation and the actual weather conditions on-site. It is also essential to consider that every software has different parameters and models for the simulation, and these might lead to different results among them. Furthermore, other external factors might affect the measured data on-site, like system maintenance, defective equipment, or changes in the grid parameters. Despite all these differences, the overall results were remarkably similar between the simulated and measured data.

Further analysis can be made to improve the precision of the calculation of the cable losses by considering the changes in the material's resistivity due to the temperature.

10.4229/EUPVSEC2023/3AV.1.55
020175-004

5 CONCLUSIONS

This study provides valuable insights into the electrical losses in bifacial PV systems. On average, when comparing the results of the cable losses with other simulation software using similar parameters, they were almost identical with PVsyst, a commercial software, showing 0.135%, while BifacialSimu, an open-source software, resulted in 0.137% for the four months period considered.

When comparing the inverter losses, the simulated results were 914 kWh, while the measured ones were 993 kWh. Nevertheless, it is essential to note that there is also a variation between measured and simulated energy yield. Therefore, when comparing only the percentage of electricity lost in the inverter, the average simulated losses are 2.07%, while the losses measured on-site were 2.09%.

As expected, the analysis of the different efficiency models showed that the most accurate results are provided by following the efficiency curve, with a difference of only 0.23% from the measured data during the month of September. These findings can help improve the design and optimization of bifacial PV systems to reduce electrical losses and enhance the energy generation potential.

6 ACKNOWLEDGMENTS

The authors thank Peter Bendix from Next2sun for supplying the experimental data from the plant in Aasen, Germany. Furthermore, one of the authors, Eva-Maria Grommes, has been financed by the Mathilde-von-Mevissen scholarship by TH Köln.

7 REFERENCES

[1] E.-M. Grommes and U. Blieske, "BifacialSimu: Holistic Simulation of large-scale Bifacial Photovoltaic Systems," *JOSS*, vol. 7, no. 78, p. 4443, 2022, doi: 10.21105/joss.04443.

[2] PVsyst SA, *PVsyst*. Accessed: Mar. 6 2023. [Online]. Available: https://www.pvsyst.com/

[3] IEA, *Solar PV*. [Online]. Available: https://www.iea.org/energy-system/renewables/solar-pv (accessed: Sep. 1 2023).

[4] E. Mouhib, L. Micheli, F. M. Almonacid, and E. F. Fernández, "Overview of the Fundamentals and Applications of Bifacial Photovoltaic Technology: Agrivoltaics and Aquavoltaics," *Energies*, vol. 15, no. 23, p. 8777, 2022, doi: 10.3390/en15238777.

[5] R. Kopecek and J. Libal, "Bifacial Photovoltaics 2021: Status, Opportunities and Challenges," *Energies*, vol. 14, no. 8, p. 2076, 2021, doi: 10.3390/en14082076.

[6] S. Maniktala, "Conduction and Switching Losses," in *Switching Power Supplies A - Z*: Elsevier, 2012, pp. 311–341.

[7] N. Riedel-Lyngskær *et al.*, "Validation of Bifacial Photovoltaic Simulation Software against Monitoring Data from Large-Scale Single-Axis Trackers and Fixed Tilt Systems in Denmark," *Applied Sciences*, vol. 10, no. 23, p. 8487, 2020, doi: 10.3390/app10238487.

[8] Deutsche Gesellschaft für Sonnenenergie, *Planning and installing photovoltaic systems: A guide for installers, architects and engineers*. London: Earthscan, 2013.

[9] J. G. Webster, *Electrical Measurement, Signal Processing, and Displays*: CRC Press, 2003.

[10] S. Ekici and M. A. Kopru, "Investigation of PV System Cable Losses," *IJRER*, v7i2, 2017, doi: 10.20508/ijrer.v7i2.5660.g7062.

[11] Z. Salam and A. A. Rahman, "Efficiency for photovoltaic inverter: A technological review," in *2014 IEEE Conference on Energy Conversion (CENCON)*, Johor Bahru, Malaysia, 2014, pp. 175–180.

[12] A. Harish, A. Ramanan, P. K. Das, G. Giridhar, K. Balaraman, and T. Prabu, "Formulation of Efficiency of Inverters for Solar Photovoltaic Power Plants - Indian Case Study," in *2019 2nd International Conference on Power Energy, Environment and Intelligent Control (PEEIC)*, Greater Noida, India, 2019, pp. 196–201.

[13] N. Ketjoy, W. Chamsa-ard, and P. Mensin, "Analysis of factors affecting efficiency of inverters: Case study grid-connected PV systems in lower northern region of Thailand," *Energy Reports*, vol. 7, pp. 3857–3868, 2021, doi: 10.1016/j.egyr.2021.06.075.

[14] HUAWEI, *SUN2000-60KTL-M0 Smart String Inverter*. [Online]. Available: https://digitalpower.huawei.com/power-marketing/#/en/material/fusionsolar (accessed: 2023).

[15] A. Ab. Rahman, Z. Salam, S. Shaari, and M. Z. Ramli, "Methodology to Determine Photovoltaic Inverter Conversion Efficiency for the Equatorial Region," *Applied Sciences*, vol. 10, no. 1, p. 201, 2020, doi: 10.3390/app10010201.

10.4229/EUPVSEC2023/3AV.1.55
020175-005

COMPARISON OF TRACKING ALGORITHMS FOR PHOTOVOLTAIC SYSTEMS ON IRREGULAR TERRAINS

J. Robledo[1], P. Sánchez-Friera[2], J. Leloux[1], R. Fernández[3], A.I. Suárez[3], L. Pérez[3]
[1] LuciSun, Rue Saint-Jean 29, 1495 Sart-Dames-Avelines, Belgium
[2] Solkeys, Mar Cantábrico 16, 33204 Gijón, Spain
[3] Gonvarri MS R&D, PI Cancienes, 33470 Corvera de Asturias, Spain

ABSTRACT: New solar tracking algorithms which account for sloped terrains are necessary to avoid significant energy generation losses, as well as potential reliability issues associated with partial shading, and enable a wider solar photovoltaic (PV) deployment including irregular terrains. To develop and validate these novel tracking algorithms, advanced system simulations are essential. For the first time, a tool which enables accurate estimations of the components of the irradiance with spatial resolution at cell level or higher has been employed to assess the benefits of employing advanced tracking algorithms. A novel tracking algorithm which considers the irregularities of the terrain has been compared to a reference backtracking algorithm, and generation gains of up to 2.5% have been estimated for specific days in the year.
Keywords: single-axis trackers, tracking algorithms, energy yield simulation

1 INTRODUCTION

The deployment of tracking photovoltaic (PV) systems on terrains with irregular topology is recently becoming a common practice. Conventional backtracking algorithms work well on flat and horizontal terrains. However, when the terrain is uneven, they present limitations. They are either unable to completely avoid row-to-row shading or else, they overcorrect the tracking angle and lead to misalignment losses. Both situations can take place for independent trackers at the same time, or for the same tracker at different times in the day. For example, if we consider a typical setting with horizontal single-axis trackers (HSAT) whose rotation axis are in the north-south direction, on a terrain sloping downwards towards the east, conventional backtracking will not be able to avoid row-to-row shades in the evening. On the other hand, in the morning, the shadows will be shorter than expected on a horizontal terrain, and therefore the trackers will be inefficiently oriented if a standard algorithm is used, leading to generation losses.

Previous works have shown that significant energy generation losses can be associated with sloped terrains, in comparison to flat terrains, when conventional back-tracking algorithms are employed [1, 2, 3, 4, 5]. The exact gain associated with an optimised tracking algorithm depends on the plant conditions and will generally be higher the steeper the terrain. High ground cover ratios (GCR) and high share of the direct irradiance component also make the system more sensitive to the tracking algorithm employed. Terrain sloping in the N-S direction can also cause significant additional energy losses due to shading between trackers [6].

New solar tracking algorithms which account for sloped terrains are necessary to avoid significant generation losses, as well as potential reliability issues associated to partial shading, and enable a wider solar PV deployment including terrains which are not necessarily flat and/or horizontal. Gonvarri MS R&D is aware of this market need and has been working in recent years in the development of advanced backtracking algorithms adapted to sloped terrains.

An essential tool in the development of new tracking algorithms is the correct validation of their impact on the energy generation. This requires advanced and flexible simulation tools in which the new tracking algorithms can be implemented, and that enable an accurate estimation of the irradiance on the PV generator plane at any given time. The aim of this work is to assess the performance of a newly developed tracking algorithm proposed by Gonvarri for irregular terrains, which has been compared to a standard backtracking algorithm [7].

2 METHODOLOGY

2.1 Software package

To evaluate the performance of the proposed algorithm, we have used the simulation package LuSim [8, 9] developed by LuciSun. Lusim is designed to perform customized solar energy simulations, particularly adapted for complex shading scenarios. It is a web-based tool running a proprietary GPU (Graphic Processor Unit) based simulation engine that allows to parallelize the shading simulation and obtain results at cell level for large solar photovoltaic installations.

As opposed to other advanced simulation tools based on 2D view factors or raytracing, LuSim is based on high-resolution 3D view factors. The tool uses 3D graphic libraries to provide a comprehensive visual environment that allows to observe the details of the shading phenomena. Additionally, it considers the electrical architecture of the modules as well as the electrical interconnection between them to capture the electrical mismatch that occurs for a particular shading distribution between the cells and irradiance conditions. LuSim offers as advantages high accuracy, low resource requirements and a dynamic and visual simulation environment which can be very useful for in-depth analysis of the plant operation.

The tool allows the setup of 3D scenarios with terrain-following solar trackers, in which each individual tracker has a specific axis orientation and height. In addition, it is possible to associate individual kinematics (tracking algorithms) to each tracker in the system, in order to evaluate customized backtracking strategies.

2.2 System description

The simulation has been performed by modelling a real PV system located in the province of Badajoz (Spain), which is represented in Figure 1.

10.4229/EUPVSEC2023/3AV.1.56
020177-001

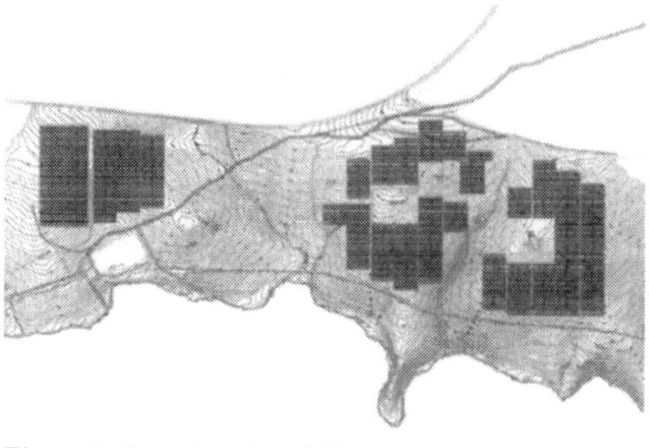

Figure 1: Overview of the PV system.

The plant includes 705 TracSmarT+® 1V solar trackers, adding up to a total capacity of 22 MW. Each tracker has a length of 68.3 m and a collector height of 2.26 m. The terrain presents slopes of up to 5% approximately with different local features across the PV plant. The average pitch distance between two tracker rows is 6.5 m (trackers are twin type, with 6 m pitch between rows connected to the same motor and 7 m between twin rows), which corresponds to a GCR of 34.7%. For the simulation we have employed a generic 72-cell PV module with nameplate peak power 540 Wp and dimensions 2.25 m x 1.13 m. Each tracker allocates 58 PV modules.

2.3 Tracking algorithms

We have evaluated the energy yield of the system using two different tracking algorithms. The reference algorithm A is a standard backtracking algorithm, which would eliminate the shadows between trackers if the terrain was flat and horizontal. This algorithm defines a constant tracking angle for all the trackers in the plant, which depends on the system ground cover ratio.

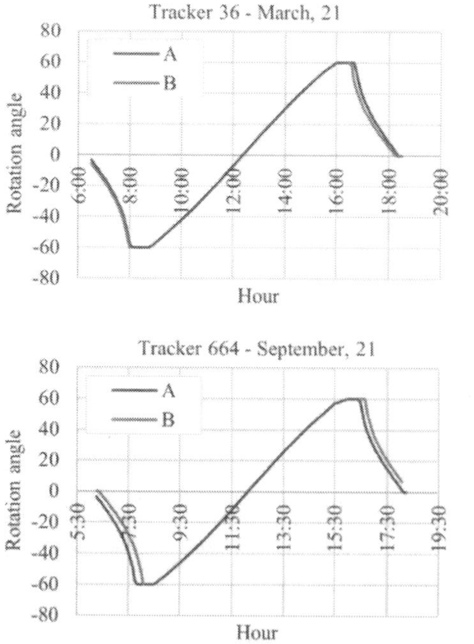

Figure 2: Tracking laws A and B for two specific trackers in the plant belonging to different slopes orientations.

The novel backtracking algorithm B has been developed by Gonvarri MS R&D specifically for irregular terrains, and it takes partially into consideration some of the terrain local features.

Examples of the tracking laws for two representative trackers on slopes are shown in Figure 2.

2.4 Approach

We have modeled the full PV system using Lusim, employing the two tracking algorithms A and B, for four representative days in the year, which correspond to the equinoxes and solstices (21st of March, 21st of June, 21st of September and 21st of December). A typical meteorological year (TMY) for the location was selected to estimate the irradiance values in these 4 days.

A variable time resolution was used with 5 min time step in the periods where there is backtracking (in the morning and evening), and 30 min in the central hours when there is no backtracking.

In order to quantify the shading that occurs at cell level, the PV module is subdivided in cells, and each cell is subdivided in a set of evaluation points (25 for this particular study) where the shading condition is obtained through a depth analysis comparison from an orthographic projection obtained from the Sun's location. For each of the solar trackers, 58 modules x 72 cells x 25 detection points = 104400 locations were evaluated at each of the time steps requested (283 timesteps over the four representative days considered).

The shading average for the 25 points at each cell is calculated. These values are then used to estimate the mismatch conditions that will occur depending on the irradiance situation.

The shading conditions will consequently take into account the relative location between the trackers, as well as their individual orientation and kinematics. This analysis allows to identify which areas present an improvement and which ones present a performance reduction when applying a specific tracking algorithm.

From the shading conditions and considering the direct irradiance situation for a particular instant, it is possible to define a concept of "effective" shading, which considers the electrical architecture and connection between the modules. A more detailed description of this concept can be found in [8]. The energy generated by the system can be readily obtained in the subsequent steps.

3 RESULTS

3.1 Scene representations

The simulation enables a detailed analysis of each subsystem and tracker in the PV plant, representing the shadows on the trackers with a cell-level resolution, for any given time.

Figure 3 and Figure 4 show two examples of the shading improvement when implementing algorithm B over A on two different slopes at morning and evening. The small strip of shaded modules has an impact on the energy generated due to the 1V configuration and bypass diode activation that it is not initially avoided by the conventional backtracking law used by algorithm A.

Figure 3: Scene showing the shadows on a particular tracker (tracker ID 36) on the 21st of March at 17:00, employing algorithm A (top) and algorithm B (bottom).

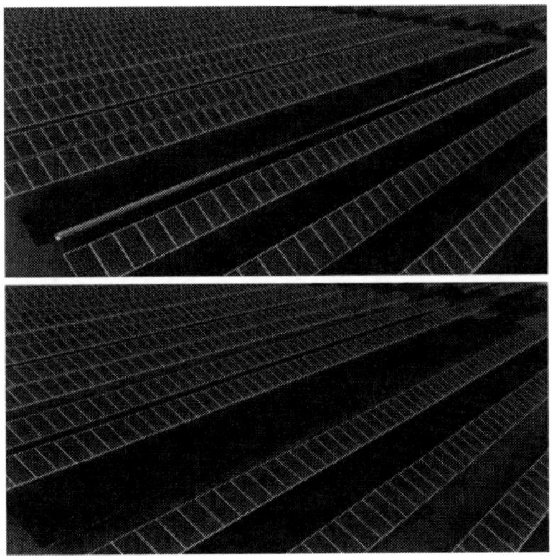

Figure 4: Scene showing the shadows on a particular tracker placed (tracker ID 664) on the 21st of September at 7:30, employing algorithm A (top) and algorithm B (bottom).

3.2 Daily power generation

The generation gains are highly dependent on the local terrain variations, and can take place at different times of the day for each particular tracker. The power generation profiles for the trackers selected above (ID 36 and ID 664) are shown in Figure 5 and Figure 6. It is observed that the new algorithm increases the generation of both trackers, as the shadows are avoided due to an increased backtracking in a customized way depending on the installation conditions.

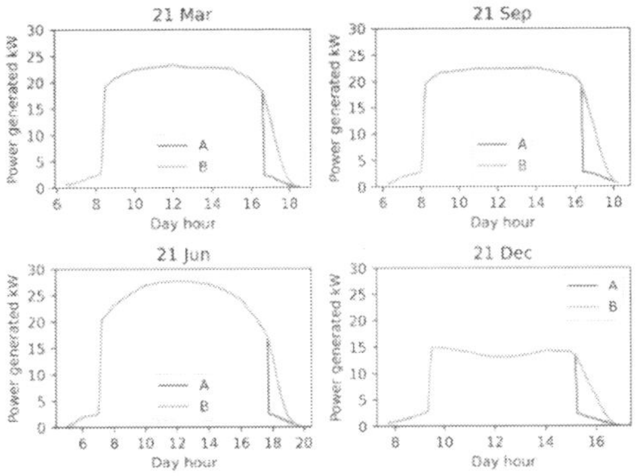

Figure 5. Power generated by a selected tracker (ID 36) during the four reference days using algorithms A (reference) and B (novel).

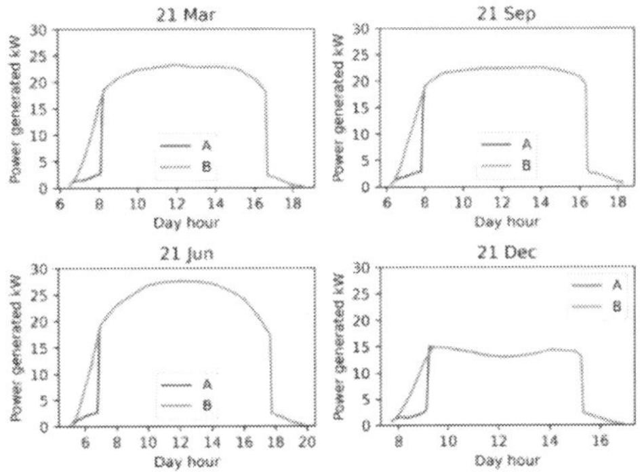

Figure 6. Power generated by a selected tracker (ID 664) during the four reference days using algorithms A (reference) and B (novel).

3.3 Total DC energy delivered by the system

The total DC energy generated by the system is shown in Table I for each representative day and the two employed algorithms.

Table I. DC energy generated (MWh) by the system for each representative day using algorithms A (reference) and B (novel).

Day \ Algorithm	21-mar	21-jun	21-sep	21-dec
A	138.87	198.3	139.76	65.33
B	140.35	199.38	141.42	66.95
Gain	1.1%	0.5%	1.2%	2.5%

It is observed that the novel algorithm leads to a systematic improvement of the energy generation across the PV plant, for all four selected days. As expected, the gain is larger in the winter. This is because the sun position is lower and the inter-tracker shadows play a larger role, considering that the location is characterized by a high proportion of clear sky days all through the year. The new algorithm is effectively reducing the shadows between trackers, and this is the main cause for the increased energy generation. For the sum of the 4 selected days, the overall gain is 1.1%.

4 SUMMARY AND CONCLUSIONS

We have analyzed the benefits of employing a newly developed algorithm for irregular terrains, applying it to a real PV system of 22 MW peak power located in Spain. The system has been modeled in detail using the Lusim simulation package, which enables the introduction individual tracking laws for each tracker in the plant and an estimation of the irradiance with a cell-level resolution. The new algorithm has been shown to reduce the shadows between trackers located on slopes, with respect to a conventional backtracking algorithm. This is translated into an overall energy gain of 1.1%, which goes up to 2.5% in the winter.

It has been shown that advanced simulations are key to develop and validate novel tracking algorithms. Their optimization, and their comparison with reference algorithms cannot be implemented using standard tools.

Considering the enormous growth projections of PV deployment worldwide in the near future [10], in particular for tracking utility-scale PV systems, it is expected that PV modules and trackers will be increasingly installed on irregular topographies. New tracking algorithms such as the one developed by Gonvarri MS R&D are essential to enable this growth and maximize the energy generation of any given PV system.

REFERENCES

[1] B. Nascimento, D. Albuquerque, M. Lima and P. Sousa, "Backtracking Algorithm for Single-Axis Solar Trackers installed in a sloped field," *Int. Journal of Engineering Research and Applications*, vol. 5, pp. 100-103, 2015.

[2] K. Passow, K. Lee, S. Shah, D. Fusaro, J. Sharp and L. Creasy, "Strategies to Optimize and Validate Backtracking Performance of Single-Axis Trackers on Sloped Sites," *Proc. of 48th IEEE Photovoltaic Specialists Conference (PVSC)*, pp. 1960-1964, 2021.

[3] M. A. Mikofski and P. J. Rainey, "Tracker Terrain Losses," *Proc. of 47th IEEE Photovoltaic Specialists Conference (PVSC)*, pp. 1859-1862, 2020.

[4] M. Leung, M. A. Mikofski, M. Hamer, A. Neubert, A. Parikh, P. Rainey and R. Kharait, "Tracker Terrain Loss Part Two," *IEEE Journal of Photovoltaics*, vol. 12, no. 1, pp. 127 - 132, 2022.

[5] K. Anderson, "Maximizing Yield with Improved Single-Axis Backtracking on Cross-Axis Slopes," *Proc. of 47th IEEE Photovoltaic Specialists Conference (PVSC)*, pp. 1466-1471, 2020.

[6] A. A. Shishavan, V. R. Abbaraju, A. Dobos and F. Borrelli, "Terrain-following single-axis tracking PV systems: advantages and performance analysis," *Proc. of 38th European Photovoltaic Solar Energy Conference and Exhibition (EUPVSEC)*, pp. 1060 - 1064, 2021.

[7] E. Lorenzo, L. Narvarte and J. Muñoz, "Tracking and back-tracking," *Progress in Photovoltaics: Research and Applications*, vol. 19, pp. 747-753, 2011.

[8] J. Robledo, J. Leloux and E. Lorenzo, "3D simulation of complex shading affecting PV systems taking benefit from the power of graphics cards developed for the video game industry," *Proc. of 29th European Photovoltaic Solar Energy Conference and Exhibition (EUPVSEC)*, pp. 2310 - 2315, 2014.

[9] J. Robledo, J. Leloux, B. Sarr, C. A. Gueymard and A. Driesse, "Dynamic and visual simulation of bifacial energy gain for photovoltaic plants," *Proc. 38th European Photovoltaic Solar Energy Conference and Exhibition (EU PVSEC)*, pp. 1313 - 1320, 2021.

[10] Solar Power Europe, "Global Market Outlook for Solar Power 2023-2027," 2023.

10.4229/EUPVSEC2023/3AV.1.56

INFLUENCE OF THE MATERIAL COMBINATION ON THE UV FLUORESCENCE OF FILMS FOR PHOTOVOLTAIC MODULES

Zonghan Jiang(1), Patrick Wessel(1,2), Sebastian Dittmann(1,2), Carlos Meza(1), Ralph Gottschalg(1,2)
Hochschule Anhalt University of Applied Sciences, Bernburger Str. 55, 06366, Köthen, Germany (1)
Fraunhofer Center for Crystalline Silicon Photovoltaics CSP, Otto-Eißfeldt-Straße 12, 06120, Halle (Saale), Germany (2)
Email: zonghan.jiang@hs-anhalt.de

ABSTRACT: The lifetime of photovoltaic (PV) modules is influenced by many factors. Aging and degradation of the PV module encapsulation are important factors. The most commonly used material for encapsulation today is ethylene vinyl acetate (EVA), and in the recent, not only EVA but also other polymers are also commonly used for PV module encapsulation, and the degradation behavior varies from polymer to polymer. The ultraviolet fluorescence (UVF) inspection, as a non-destructive inspections method, is ideally suited for monitoring the aging and degradation behavior of polymers. In this research, a quantitative non-destructive method to characterize the fluorescence's changes of different material combinations during the aging process is presented. Furthermore, the suitability of UVF inspection for different material combinations and the behavior and properties of the different material combinations during aging are investigated using UVF. In addition, the spectral changes of the fluorescence of various material combination during degradation will be characterized by an RGB-based UVF analysis method.
Keywords: Ultraviolet fluorescence, degradation, material combination, encapsulation, photovoltaic

1 INTRODUCTION

Aging and degradation of the PV module encapsulation are important factors for the lifetime of the PV modules. Polymeric materials are used for the encapsulation of solar cells to protect them from environmental influences and to absorb mechanical stresses caused by daily temperature fluctuations and the different thermal expansion coefficients of the module components, mainly silicon and glass. The most common encapsulation material is EVA (ethylene vinyl acetate) [1]. However, not only EVA, but also other polymers are used for PV module encapsulation in the recent years. What these polymers have in common is that they fluoresce under ultraviolet fluorescence inspection during the aging process.

Ultraviolet fluorescence (UVF) is an inspection technology that can be used to detect cell cracks and other defects in most silicon PV modules [2]. According to other studies, the fluorescence intensity in UVF images can also be used to determine the degradation and aging of PV encapsulation materials [3-7]. Thus, UVF technology is a promising tool for quality and condition control of PV module encapsulation. However, as mentioned above, the aging behavior of different polymer combinations is expected to be different. Therefore, it is important to investigate and research the aging behavior of different polymers by UV fluorescence.

In this paper, the relationship between fluorescence of different material combination and their age is investigated. The aim of the paper is to investigate long-term information on the formation of various defects and to realize material-based indicators of module quality. In this study, a workflow for quantitative field analysis of encapsulations is also proposed and a method for non-destructive evaluation of material condition for module aging is developed. Different material combinations are measured by ultraviolet fluorescence at different time points of the aging process and the aging of different material combinations is quantified by fluorescence intensity(pixel value). Through the results, the applicability of UV fluorescence to different material classes and the different aging behavior of different material combinations is demonstrated.

2 MATERIAL CHANGES AND FLUOROSCOPIC PHENOMENA DURING AGING

Many studies have shown that the photoluminescence of polymers during degradation is due to the substances formed during degradation. In general, carbonyl and polyene groups are held responsible for the luminescence of polymers. In the case of polyvinyl chloride, it is known that polyenes are formed during degradation [8]. In the case of EVA, the discoloration is due to various degradation processes leading to the formation of chromophores in the polymer. In the thermal and photochemical degradation processes of EVA, the formation of polyenes and α, β-unsaturated carbonyl groups have been shown to lead to discoloration and fluorescence. Additives used for curing or stabilization degrade and also cause fluorescence [9]. Due to this phenomenon, the aging behavior of EVA can be characterized and quantified by UV fluorescence. And because of the different degradation processes of different polymers, the number of substances formed during degradation that cause the fluorescence phenomenon may be different.

3 AIM AND EXPERIMENTAL APPROACH

In this study, the following tasks are set and investigated.

a. Trend and change in fluorescence area for the same encapsulation material over the aging period.
b. Investigation of the data contained in the different color layers of a UVF image.
c. Comparison the degradation behavior of different material combinations through reconstructed mathematical function.

To achieve these objectives, the experimental approach will be divided in the following steps.

1) Development and optimization of a UVF measurement prototype.
2) Different encapsulation materials are aged in climatic chambers (DH-UV) for 1250h in total:
3) Regular measurements with the UVF prototype during the aging process.

4) Calculation of the average pixel value of a polymer fluorescence image.
5) Analysis and comparison of differences between color layers of RGB images to improve the evaluation quality.
6) Evaluation of fluorescence trends in polymers by reconstructed mathematical function.
7) Comparison of UVF images of different encapsulation material combination at the same aging stage.

3.1 Design of the UVF prototype
The UVF prototype consists of the following components: Camera, UV LED, heat sink, band pass filter for LED, long pass filter for camera and external structure.
The entire frame of the device is made of aluminum profiles. The reason for using aluminum profiles is that the prototype can later be equipped with other additional devices. The material of the top cover is also made of aluminum. Since the light source generates more heat when the device is used, this affects the results of the experiment. For this reason, aluminum was chosen as the material for the top cover because of its high heat transfer capability. Another reason is the relative inexpensiveness of aluminum. Six 365nm UV LEDs are mounted on the top cover of the prototype as the light source. The rated power of each LED is 1W. A bandpass filter (Hoya color glass U360 bandpass filter) is installed in front of each group of LEDs. The bandpass filter is used to block the visible light spectrum emitted by the LEDs, thus reducing the effects of external influences on fluorescence detection [10]. Each group of LEDs is equipped with a heat sink whose function is to accelerate the heat dissipation from the LED. The interior of the prototype will be covered with black paper to shield external light sources and reduce the effect of fluorescence from other substances. The front camera of iPhone11 is used as the camera. And the exposure time is set to 0.
This prototype is currently being converted into a more permanent setup, the components of which unfortunately could not yet be fully obtained due to various supply problems.

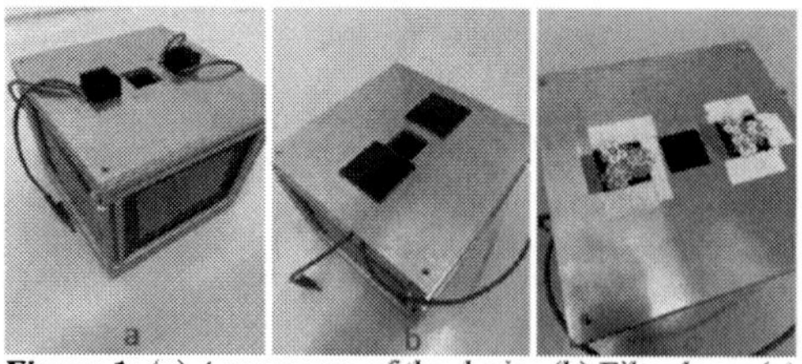

Figure 1: (a) Appearance of the device (b) Filter layer (c) Light source layer

3.2 Sample preparation and aging process
In this study, 3 polymers of backsheet and 2 polymers of encapsulation were used as samples and assembled by lamination in different combinations. To control the number of variables in the experiment, all samples were laminated simultaneously for 1 hour.
Polymers for encapsulation:
1. ENC1 (film material: BPO),
2. ENC2 (film material: EVA),
3. ENC4 (film material: POE)
Polymers for backsheet:
1. BS1 (base material: PPE).
2. BS4 (base material: FFC//PET//FFC)
The combinations between all the different samples can be seen in Table 1.

Table I: Type of the Material Combination

ID	Combination's type
1	BS1 + ENC1
2	BS1 + ENC2
3	BS1 + ENC4
4	BS4 + ENC1
5	BS4 + ENC2
6	BS4 + ENC4

All specimens were initially classified into two distinct categories. The primary category pertained to individual polymers, while the secondary category encompassed combinations of diverse polymers. (Refer to Table 1.) An identical replica was created for each specimen and combination using the same procedure. After fabrication, all specimen samples underwent lamination. Following the lamination process, all specimens were subjected to cooling and aging within a controlled environment chamber. For this study, DH-UV (85°C, 60% relative humidity, irradiation ranging between 170 and 300 W/m² at wavelengths of 280 - 400nm) was utilized as the specified aging condition.
The total aging duration is 1250 hours, after each 250 hours, the specimens will undergo ultraviolet fluorescence (UVF) testing, and the data will be meticulously documented.

3.3 UVF test setting and analyze method.
The UVF test is performed in a low-light environment. The UVF images of all samples are recorded only after the samples have been prepared and have completely cooled. After recording, all samples are aged in the climatic chamber. All samples are taken out of the climatic chamber at different times and photographed.
During all the tests, the exposure of the cell phone camera and the flash were turned off and no zoom function was used. The supply voltage and current of the led lamps were kept at the same value in all the tests to control the luminous flux of the emitted UV light. Since the temperature of the led lamp also affects the efficiency of the UV LEDs, in order to minimize this effect, all photography will be done within 10s and the system will be turned off after each test to avoid flux changes due to high temperature.
In order to analyze the changes in the fluorescence intensity of the samples during the aging process, the average pixel value of the UVF images of all the samples will be used as the basic parameter to quantify the changes in fluorescence intensity. Due to the destruction as well as degradation of the sample during the aging process, calculating the average pixel value of the entire image would result in a bias. For this reason, a small part of the UVF image (200*200) will be extracted from the whole picture and used in the analysis and calculation of the average pixel value.
Since some of the samples had many different patterns and cracks on their surfaces during the aging process, the areas to be analyzed were selected using standard areas wherever possible.

4 RESULTS

In this study, the results of the experiment will be presented in following three sections.

- Visual inspection
- UVF image analysis based on grayscale.
- Analysis based on different RGB (red, green, blue) layers.

Furthermore, the results of these three different analytical methods will be compared and evaluated.

4.1 Visual inspection

Visual inspection is one of the most traditional ways of UVF analysis. The UVF images of all samples at different aging stages are first observed by visual inspection. Visual inspection reveals that the fluorescence of most samples shows a tendency to become brighter during the aging process. However, the amount of fluorescence increase for each sample is different at different stages of aging. For example, the fluorescence of single-substance samples ENC1, ENC2, and ENC4 is more rapid than that of other samples. In some samples, the change in the middle aging stage is so small that it is impossible to observe the difference with the naked eye. In terms of the fluorescence change phenomenon, there are also some samples (ENC1, ENC2) that change to transparent first, and then produce a brighter blue light at the aging stage. In terms of the damage range, most of the single-substance samples also showed cracks and degradation on the surface during the aging process or even the samples were completely damaged (ENC4 was completely damaged after 1000h aging). Meanwhile, composite samples showed greater resistance to damage. In summary, it can be confirmed that during the aging process and after the aging process, different materials and combinations show different changes under UVF testing. This further reveal that the UVF inspection is applicable to all material combinations used in the sample.

Figure 2: UVF images of ENC2, BS1 with aging time 0h, 250h, 500h, 750h, 1000h, 1250h (top: ENC2, down: BS1) under visual inspection

4.2 Grayscale-based analysis of UVF Images

In order to have a basic understanding of how the UVF images of different combinations of substances change during the aging process, all pictures were first converted into grayscale maps and the average pixel values of the samples in the UVF pictures were calculated using the analysis mentioned above. Figure 3 shows the changes of average pixel values of UVF images of all samples in the grayscale map during the aging process and the trend.

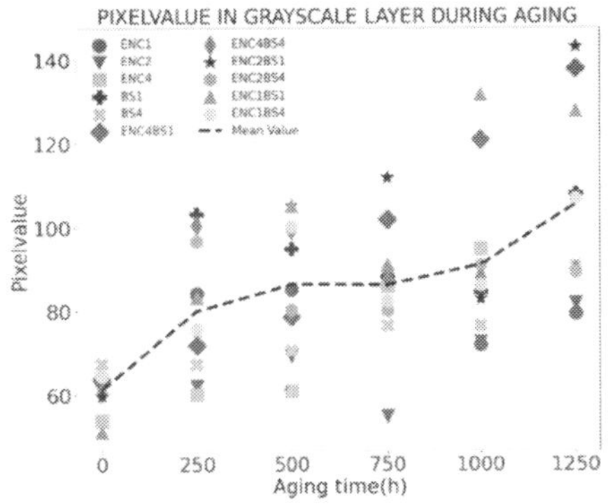

Figure 3: Average pixel value of different material combinations and mean pixel value of all samples in grayscale

In order to better observe the fluorescence changes of all samples during the aging process, the mean value was adopted to describe the trend of all samples. As shown in the Figure 3, the mean values of the pixel values of all samples showed a growing trend.

To provide a more accurate description of the fluorescence trend of each sample during the aging process, polynomial mathematical functions for pixel value change from different samples will be created. Some of the anomalous pixel values that dropped significantly during the aging phase were excluded to minimize the effect of measurement errors on the results. After evaluating the systematic error as well as the fit of the function, a polynomial mathematical function of the following form was adopted.

$$y = ax^3 + bx^2 + cx + SV \qquad (1)$$

In this functional expression, y represents the average pixel value of the sample, x represents the aging time, and SV represents the average pixel value of the sample at 0 h. The three parameters (a, b, and c) will be optimized so that the equation fits the change in the average pixel value of the respective samples during the aging period (0-1250h).

Table II: Coefficients of the reconstructed mathematical function for all samples

Sample type	a	b	c
ENC1	1.1611e-07	-2.4358e-04	1.3552e-01
ENC2	1.3433e-08	-1.6387e-05	1.5268e-02
ENC4	-1.2280e-08	3.9015e-05	3.7933e-03
BS1	2.3537e-07	-4.6002e-04	2.4412e-01
BS4	2.9281e-08	-3.1181e-05	1.1715e-02
ENC1BS1	-1.5821e-08	-3.8001e-05	1.3367e-01
ENC1BS4	9.9815e-08	-2.0083e-04	1.2741e-01

ENC2BS1	2.5654e-07	-4.4358e-04	2.1783e-01
ENC2BS4	8.7945e-08	-1.9587e-04	1.2887e-01
ENC4BS1	-3.0050e-08	8.4537e-05	1.0326e-03
ENC4BS4	1.5215e-07	-3.4863e-04	2.2217e-01

From the different mathematical functions for the different samples, it can be concluded that the rate of change of fluorescence varies from sample to sample. Furthermore, all the change trends can be categorized into two types altogether, one is rising first and then falling and finally continuing to rise, and the other is continuously rising.

However, the fluorescence intensity in the mathematical function of some samples produced a downward trend during some aging stages, but a comparison of the results of the visual inspection leads to the conclusion that the fluorescence intensity of the sample has increased, but the color of the fluorescence has changed. The change in color may be due to the production of certain substances during degradation that emit fluorescence at different wavelengths.

For this reason, it can be guessed whether some changes also occur at the spectral level of the substance and the grayscale images of the UVF images cannot contain this information, because the grayscale image shows only the changes in the total light intensity.

4.3 RGB-based analysis of UVF Images

Due to the assumptions mentioned in the previous section, an RGB-based analysis is introduced. Since the RGB layers of the different images each represent the sum of light intensities at different wavelength ranges on the spectrum, the changes in the spectral level of the sample under UVF detection during aging can also be roughly estimated by analyzing the changes in the different RGB layers of the images.

The difference between the RGB-based analysis method and the grayscale map-based analysis method is that instead of converting the UVF image into a grayscale map, the average pixel values of all color layers of the sample's UVF image are sampled and calculated in the same way. As shown in the figure 4, 5, 6, the average pixel value of all the sample during aging exhibits very different changes in the different RGB layers.

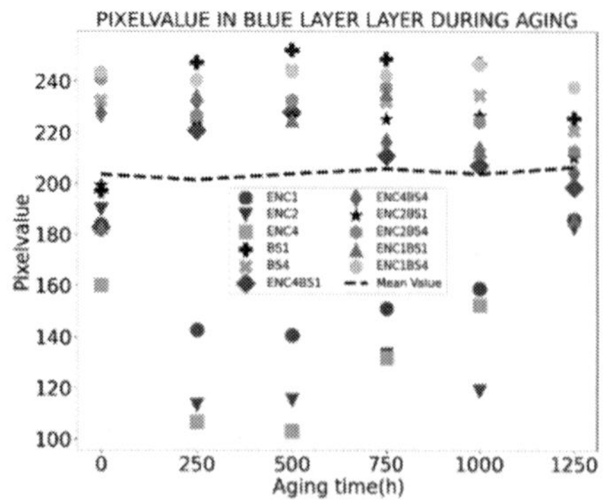

Figure 4: Average pixel value of different material combinations and mean pixel value of all samples in blue layer

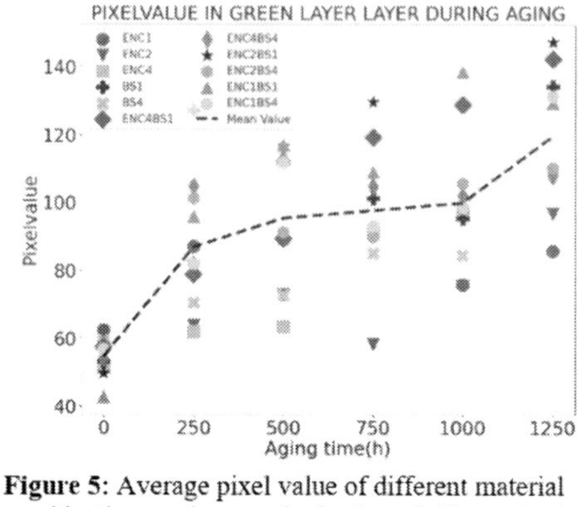

Figure 5: Average pixel value of different material combinations and mean pixel value of all samples in green layer

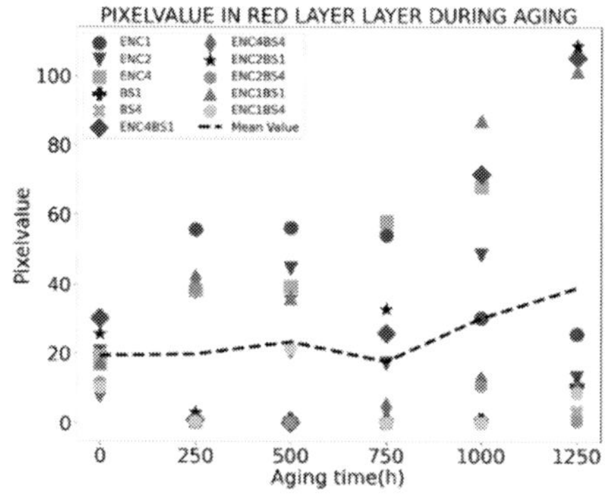

Figure 6: Average pixel value of different material combinations and mean pixel value of all samples in green layer

The result can be analyzed on several levels. First of all, the change in the mean value of each color layer need to be observed. In the green layer it can be noticed that the trend of the pixel values is basically similar to the grayscale map analysis. The only difference lies in the aging period of 500h-1000h, where the mean value in the grayscale map has a slight decreasing trend compared to the green layer. In contrast, the mean values in the blue as well as in the red layer have a relatively large difference.

The second is the pattern of change in the average pixel value of each sample in different layers. In the blue layer, the change in the mean value is completely unrepresentative of the pattern of change in some of the samples. For example, the pixel value of sample ENC1, ENC2, ENC4 decrease rapidly and finally rise during the aging process, and by comparing the pictures in the visual inspection, it was found that the blue fluorescence of these three samples firstly decreases rapidly but during the period, and this blue fluorescence rises rapidly in the final aging stage. But at the same time, the total intensity of the light didn't change too much. Through the different fluorescence colors, a hypothesis was also raised whether different polymers produced during the degradation process possess different wavelengths of excitation fluorescence light.

The most noteworthy point is in the red layer. After the first stage of aging, the average pixel values of most of the samples, except for a few (ENC1, ENC2, ENC4, ENC2BS4, ENC4BS4), gradually converge to 0 and then most of the samples show a fluctuating trend or stay stable. In the following aging phase, most of the samples showed fluctuating trends or remained stable, except for ENC4, ENC1BS1, ENC2BS1, and ENC4BS1, which showed a rapid increase in the later stages of aging. (ENC4 has been completely damaged, so there is no 1250h data for this sample).

By comparison, it can be concluded that the RGB-based analysis method provides more information about spectral changes of material combinations during aging compared to the grayscale-based analysis. Grayscale-based analysis essentially only represents changes in fluorescence in the green layer.

And it can be drawn from the results that the excitation fluorescence of different combinations of substances has significantly different spectral changes during the degradation process, furthermore, many of differences are likely to be located in the spectral range of the near infrared. Based on this phenomenon, spectral analysis of UVF inspection is crucial and needed in the future.

5 CONCLUSIONS AND FUTURE WORKS

From the above evidence, the following conclusions can be summarized.

1. And it was confirmed that UV fluorescence detection is applicable to all combinations of materials used in the samples, and the change of the fluorescence of different substances and their combinations has different trends in the aging process.
2. the RGB-based analysis method provides more information as well as details of the spectral changes compared to the grayscale-based analysis method.
3. In order to analyze detail of the aging behavior of photovoltaic encapsulations by UVF inspection, the spectral analysis of the samples under UVF inspection need to be applied in the future research.

For future work, the following points could be improved.

1. Reduction of systematic errors and more advanced testing equipment.
2. Investigation of the effect of different environmental factors on the results.
3. Spectral changes (For example, in the near-infrared region) in the fluorescence of aging polymers under UVF detection will be measured as well as analyzed in the future.

6 REFERENCES

[1] Kontges, Marc; Morlier, Arnaud; Eder, Gabriele; Fleis, Eckhard; Kubicek, Bernhard; Lin, Jay (2020): Review: Ultraviolet Fluorescence as Assessment Tool for Photovoltaic Modules. In IEEE J. Photovoltaics 10 (2), pp. 616–633. DOI: 10.1109/JPHOTOV.2019.2961781.

[2] Grabmayer, Klemens; Wallner, Gernot M.; Beißmann, Susanne; Schlothauer, Jan; Steffen, Ronald; Nitsche, David et al. (2014): Characterization of the aging behavior of polyethylene by photoluminescence spectroscopy. In Polymer Degradation and Stability 107, pp. 28–36. DOI: 10.1016/j.polymdegradstab.2014.04.030.

[3] Kim, Jae Hyun; Lyu, Yadong; Miller, David C.; Gu, Xiaohong (2017): Characterizations of aged Glass/Ethylene Vinyl Acetate/Glass using fluorescence spectroscopy and instrumented indentation. In: 2017 IEEE 44th Photovoltaic Specialist Conference (PVSC). 2017 IEEE 44th Photovoltaic Specialists Conference (PVSC). Washington, DC, 2017/6/25 - 2017/6/30: IEEE, S. 3195–3199.

[4] Roeder, Beate; Schlothauer, Jan; Koehl, Michael (2011): Fluorescence imaging for analysis of the degradation of PV-modules. In: 2011 37th IEEE Photovoltaic Specialists Conference. 2011 37th IEEE Photovoltaic Specialists Conference (PVSC). Seattle, WA, USA, 2011/6/19 - 2011/6/24: IEEE, S. 3606–3608.

[5] Schlothauer, Jan; Jungwirth, Sebastian; Köhl, Michael; Röder, Beate (2012): Degradation of the encapsulant polymer in outdoor weathered photovoltaic modules: Spatially resolved inspection of EVA ageing by fluorescence and correlation to electroluminescence. In: Solar Energy Materials and Solar Cells 102, S. 75–85. DOI: 10.1016/j.solmat.2012.03.022.

[6] Gopalakrishna, Hamsini; Arularasu, Pooja; Dolia, Kshitiz; SINHA, Archana; TamizhMani, GovindaSamy (2019): Characterization of Encapsulant Degradation in Accelerated UV Stressed Mini-Modules with UV-cut and UV-pass EVA. In: 2019 IEEE 46th Photovoltaic Specialists Conference (PVSC). 2019 IEEE 46th Photovoltaic Specialists Conference (PVSC). Chicago, IL, USA, 2019/6/16 - 2019/6/21: IEEE, pp. 1961–1964.

[7] Grabmayer, Klemens; Wallner, Gernot M.; Röder, Beate (2016): Correlation of spatially resolved photoluminescence and viscoelastic mechanical properties of encapsulating EVA in differently aged PV modules. In Prog. Photovolt: Res. Appl. 24 (6), pp. 855–870. DOI: 10.1002/pip.2734.

[8] Schlothauer, Jan C.; Grabmayer, Klemens; Hintersteiner, Ingrid; Wallner, Gernot M.; Röder, Beate (2017): Non-destructive 2D-luminescence detection of EVA in aged PV modules: Correlation to calorimetric properties, additive distribution and a clue to aging parameters. In Solar Energy Materials and Solar Cells 159, pp. 307–317. DOI: 10.1016/j.solmat.2016.09.011.

[9] Schlothauer, Jan C.; Röder, Beate (2018): Prospects of 2D-Luminescence Spectroscopy for Aging Investigations of the Embedding EVA Polymer in PV Modules: Revealing DLO Conditions. In: Service Life Prediction of Polymers and Plastics Exposed to Outdoor Weathering: Elsevier, pp. 161–184.

[10] Schlothauer, Jan Caspar; Peter, Clea; Hirschl, Christina; Oreski, Gernot; Röder, Beate (2017): Non-destructive monitoring of ethylene vinyl acetate crosslinking in PV-modules by luminescence spectroscopy. In J Polym Res 24 (12). DOI: 10.1007/s10965-017-1409-y.

BACKSHEET DEGRADATION IN MODERATE AND DESERT CLIMATE

[(1)]A.A. Abdallah*, [(4)]O. Stroyuk, [(1)]H.M. Abufares, [(1,3)]M. Abdelrahim, [(1,3)]M. Elgaili, [(2)]K. Mroue, [(2)]A. Samara, [(2)]M. Pasha, [(4)]C. Buerhop-Lutz
[1]*Qatar Environment and Energy Research Institute (QEERI), Hamad Bin Khalifa University (HBKU), Qatar Foundation, P.O. Box 34110, Doha, Qatar*
[2]*Core Laboratories, Hamad Bin Khalifa University (HBKU), Qatar Foundation, Doha, Qatar*
[3]*Bin Omran Trading & Telecommunications, P.O. Box 288, Doha, Qatar*
[4]*Forschungszentrum Juelich GmbH, Helmholtz Institute Erlangen-Nuernberg for Renewable Energy (HI ERN), 91058 Erlangen, Germany*
*Contact: +974 445 46682, aabdallah@hbku.edu.qa

ABSTRACT: In this paper, we compared PV module backsheet (BS) degradation observed in moderate and desert climates, represented by Germany and Qatar, respectively. The results reveal various types of polyethylene terephthalate (PET) and polyamide (PA)-based BS degradation. Visual inspection and various characterization techniques were used to gain more insight into the degradation mechanism of BS materials. Cross-sectioning method, used in combination with Raman, was found to provide details about the initiation of BS cracking. The results showed a similarity in the failure pattern of the BSs, but occurring in a relatively shorter time in the desert environment. The desert climatic conditions have shown an earlier degradation of PA after 4 years of installation; while in moderate climate obvious, back sheet degradation was observed after 6-7 years of installation. Degradation of the PET-based BS starts to initiate later when compared with PA BS, followed by crack propagation. Yellowing of the PET BSs is an indication of the change in the polymer surface properties and subsequent cracking, while PA-based sheets show broad open and closed cracks running under the cell interconnect and the solar cell. Embrittlement of the BS caused by UV exposure and subsequent chemical aging is expected to initiate the BS degradation. The outcome of this paper can be used as a guideline for PV module manufacturers to improve module material durability for different climatic conditions.

Keywords: back sheet; cracking; aging, durability

1 INTRODUCTION

The deployment of solar photovoltaic (PV) technology has grown in the last decades reaching 1.2 TWp in 2022 [1] to continue playing a crucial role in meeting the growing demand for sustainable electricity supply. In Qatar, the first 800 MWp silicon PV plant has been inaugurated last year covering approximately 10% of the local peak electricity demand. To further accelerate the deployment of PV in the country, reliable operating PV systems and reliable, productive and long-lasting PV modules are required.

The desert regions present unique challenges for solar PV performance and reliability due to harsh environmental conditions, such as high ultraviolet (UV) and visible solar radiation, high operating temperatures, and frequent occurrence of dust storms. The backsheet (BS) material, which is typically made of polymer-based materials, plays a critical role in protecting the PV module and maintaining its performance safely over the module's lifespan. BS degradation is a known failure mode observed in different PV modules causing loss of the electrical insulating properties, hence a safety problem. Furthermore, the degradation of layered BSs may not be directly observed in terms of power degradation; however, it could be an indication of a later PV system failure. Understanding the impact of these environmental conditions and failure mechanisms of PV module materials is essential for long-term PV system performance and degradation.

The impact of climatic conditions on PV module material degradation has been reported in the literature [2]–[5]. Exposure to UV radiation and high operating temperatures can have a significant impact on the durability and reliability of PV backsheet materials. UV radiation can cause photodegradation, leading to e.g. cracking, embrittlement, delamination, and discoloration of the backsheet material [6].

Physical aging due to high operating temperature has also being considered as an important accelerating factor for BS cracking of PA and PVDF in the field, as reported by different authors [4], [7]. However, to ensure a reliable PV module installation in desert climate (with high UV radiation and high operating temperatures), an in-depth understanding of the PV module material crack initiation and propagation mechanism is essential. Additionally, the development of different solar cell configurations, such as Tunnel Oxide Contact (TOPCon), silicon heterojunction (HJT) and interdigitated back contact (IBC), along with the different PV applications (e.g., floating PV), requires highly improved and superior BS properties. For instance, TOPCon, HJT and IBC solar cells are highly sensitive to moisture ingress through the BS compared to today's standard cell technology of passivated emitter and rear contact (PERC), thus requiring a BS with enhanced barrier properties [8]. The impact of UV radiation on different solar cell technologies has been studied by Sinha et al. [9]. Their results indicate that HJT, IBC and PERC solar cells are more sensitive to UV-induced degradation, resulting in a larger power drop of about 3.6% compared with the conventional aluminum back surface field (Al-BSF) solar cells that show less than 1% drop. Recent publications reporting BS failures [10], [11] have also underlined the need for a better understanding of the main causes of BS degradation in the field.

The aim of this paper is to compare the aging behavior of PV module BS materials in moderate (Germany) and desert climatic (Qatar) conditions in order to gain an accurate understanding of the main causes of BS degradation. The findings of this work and the proposed workflow will provide a better understanding of the impact of UV irradiance, module temperature, and relative humidity on the durability of PV module BSs. Further identification of the causes of BS degradation in the field will lead to improving BS material selection and design to ensure reliable PV systems. The development of PV module material will support long-term lifetime beyond 30 years.

2 METHODOLOGY

In this paper, ground-mounted silicon PV system installed in Europe (Germany) and in the Middle East (Qatar) will be presented. The failure observed on the PV module BSs will be reported. The BS materials under study are multi-layer non-fluoropolymer-based Polyethylene terephthalate (PET) and Polyamide (PA).

Firstly, visual inspection of PV modules installed in two PV sites has been conducted to assess PV module material degradation. An overview of the BS failure modes observed will be presented. Secondly, the meteorological data of the two PV sites were compared. Lastly, for a better understanding of the BS degradation mechanisms (physical, chemical, structural) occurring in the field during the testing period, cross sectioning of samples was performed for different advanced characterization methods. Near-infrared absorption (NIRA), Raman and UV fluorescence (UVF) spectroscopy will be used to visualize the different failure modes and to identify the composition and layer structure of the BS material and the main cause of BS degradation/cracking.

3 RESULTS

The results report on the impact of different environmental parameters on PA and PET BS material durability.

3.1 Meteolorgical data analysis

According to Köppen-Geiger climate classification [12], Qatar is located in a hot desert climate (BWh) with typically high irradiance, high temperature and high relative humidity. On the other hand, Germany shows two main distinct climates, namely the temperate continental climate (Dfb) and the temperate oceanic climate (Cfa). Fig. 1 shows the annual UV irradiance, the maximal, mean, and minimal values of daily ambient temperature, and relative humidity measured at the two locations under study (Germany and Qatar) [13]. The annual UV irradiance in Qatar is higher by a factor of 1.6 than in Germany. The maximum ambient air temperature in Qatar is about 38°C, while in Germany is about 25°C. Note that the module temperature may reach 70°C during the summer months in desert climate. Additionally, the desert climate has high thermal cycles during day and seasonal variation. Negative temperatures are not stated for Qatar, different from Germany, where the temperatures can fall below 0°C. Relative humidity in Germany is higher than in Qatar; dry days with relative humidity below 40% are seldom.

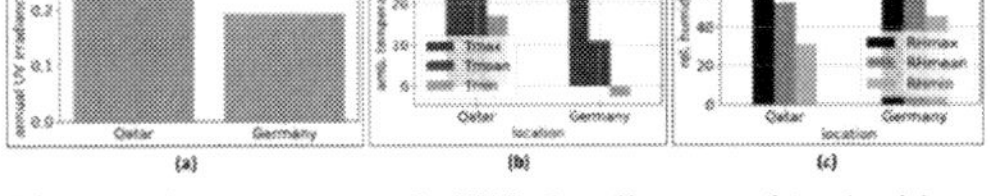

Figure 1: (a) Annual UV irradiance, (b) Ambient temperature, and (c) relative humidity for Germany (moderate climate) and Qatar (desert climate)

3.2 Backsheet degradation

Fig. 2 shows representative photographic images of degraded polyamide (PA) BS with different failure patterns observed in the two ground-mounted silicon PV sites in moderate climate (Germany) and desert climate (Qatar). Fig. 2a shows BS cracking running under the solar cell interconnect, where cracks initiated from the airside layer and the core layer followed by crack propagation and delamination. This type of BS cracking is running under the busbar of the solar cells. BS chalking (deposition of rutile Titania pigment), green spots and burn-marks were also observed (indicated by arrows in Fig. 2b). In our previous publication [14], we have shown that white BS chalking is related to formation of organic chunks covered with titania (TiO_2) nanoparticles, while the greenish chalking deposit is composed mainly of copper hydroxo sulfate $Cu_3(SO_4)(OH)_4$. The latter suggested that chalking is essentially chemical degradation of the PA BS and copper (Cu) originally from the cell interconnections. Chalking has also been reported by other authors [15], [16].

Fig. 2c shows another PV module type with BS cracking surrounding the solar cell. This type of BS cracking has been observed in different climatic conditions, often 3-6 years after initial deployment [7]. Eder et al. suggested that the lack of UV-absorber in the encapsulant material is the main reason for BS cracking around the cells in the cell gaps. We have also observed that BS cracking starts earlier in desert climate as compared with moderate climate. In the desert climate, PA cracking was observed after 4 years from the installation, while in moderate climate PA degradation was observed after 6-7 years of the module commissioning. This suggests that BS degradation depends on the exposure time and the climatic conditions, in particular the UV and the amplitude of temperature cycling. In addition, PA BS degraded in Qatar (desert climate) showed a larger cracking and yellowing, while in Germany (moderate climate) typically a lower extentr of BS cracking and yellowing was observed.

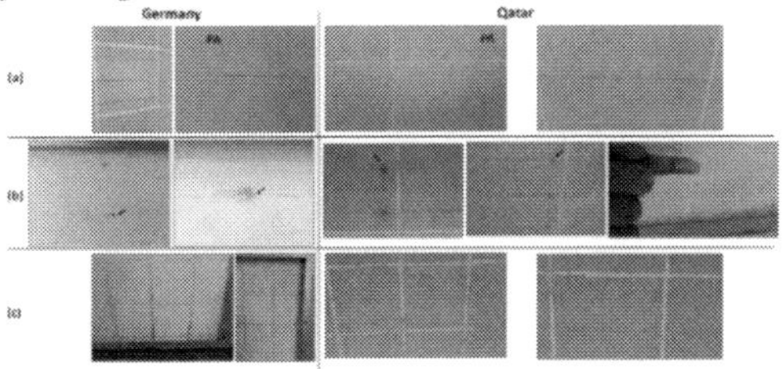

Figure 2: Photographic images of polyamide (PA) BS degradation patterns observed by visual inspection of ground-mounted silicon PV modules installed in Germany (left) and Qatar (right). Similar PA BS failure-modes were observed at two different sites. (a) PA BS cracking observed along the ribbon connection, (b) Burn-marks and green spots, and (c) BS cracking surrounding the solar cell.

Fig. 3 shows selected photographic images of degraded PET-based BS observed at the PV site in Qatar. Field data in Germany is not available for the PET BS. Outdoor UV exposure of the PET BS in desert climate showed yellowing, cracking and delamination. Yellowing of the PET BS is an indication of the change in the polymer surface properties and may lead to subsequent cracking. Yellowing typically represents degradation of the polymer that can lead to backsheet cracking. Cracks in PET-based BS were observed to propagate in two directions, one along the width direction and the other perpendicular to it (along the module length direction). The surface morphology of the polyester showed local delamination of the BS layered-structure (Fig. 3b). Similar PET BS failure modes were reported in the literature for different climatic zones [17], [18].

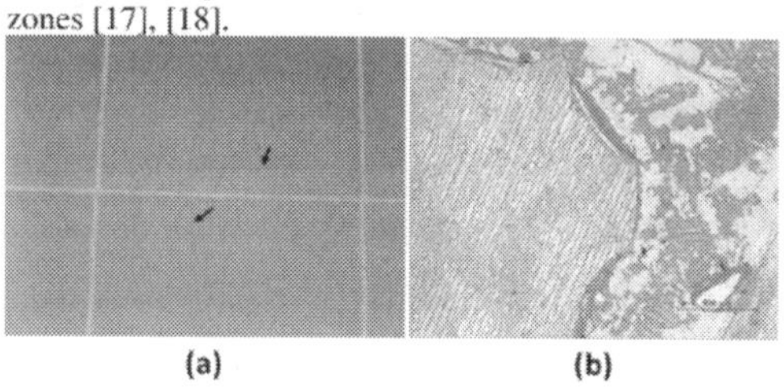

Figure 3: (a) Photographic images of degraded PET-based BS on silicon PV system installed in desert climate; (b) Optical back scattered field image of PET showing partial delamination of the top layer of PET-based BS.

From a more detailed microscopic analysis, plastic deformation of the PA and PET BS cracking is clear in the SEM images. Fig. 4a shows an example of cracking in PA BS with clearly visible partially delaminated top layer (left image) as well as a magnified section showing the surface microstructure near the crack region (right image, also indicated by a red square on left image). Fig. 4b shows the example of PET BS cracking with multiple small cracks surrounding the main crack. Comparing the PA and PET BS cracking shown in Fig. 4a and 4b, using similar magnification, reveals that the crack in the PA BS is much deeper when compared with the crack in PET, which could be an indication of a higher UV resistivity of the top BS layer. This also agrees with the usage of the PET-based BS as a core layer to provide the mechanical strength of the BS.

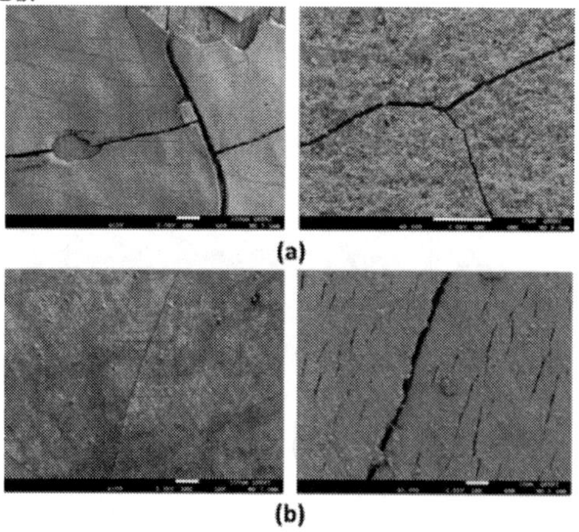

Figure 4: SEM images showing the details of the BS cracking for: (a) polyamide BS and (b) PET BS. The red square indicates the area around the cracks for which a magnified image is provided.

A cross-section of the BS was investigated by Raman micro-spectroscopy. Fig. 5a shows the PET BS sample from Qatar showing the presence of three layers. A strongly degraded and partially delaminated air side and a core layer (totally 140-150 µm) were observed. The thicker inner layer (25 µm) was identified by Raman spectroscopy as rutile-filled PET; however, no meaningful Raman spectra can be registered from delaminated air layer and inner thinner layer due to strong fluorescence background. Unfortunately, the particular BS type used in PV installations in Qatar was not found on PV plants in Germany, making a direct comparison impossible. In general, PET-based Bss deployed in milder climate of Germany do not show such drastic degradation of the air and core layers, the formastion of cracks typically observed mostly in inner rutile-filled polypropylene layers. Also, yellowing of PET-based BSs inGermany is much lower occurring mostly locally on the hot spots due to thermal degradation.

Furthermore, quite high carbonyl indices CI between 0.19 and 0.25 for EVA were measured by NIRA spectroscopy, indicating degradation of the encapsulant polymer (in Germany CI = 0.14 – 0.15).

Fig. 5b shows the PA BS sample from Qatar with one layer (320-330 µm) with cracking on the air side. Very similar images and cracking patterns were also collected from cracked BSs from PV fields in Germany.

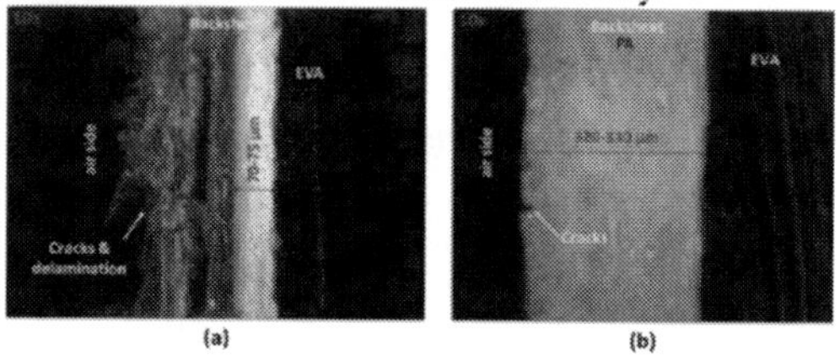

Figure 5: Cross-sectional microscopic images of the samples subjected to Raman study: (a) degraded PET BS sample (TiO_2/PET) with internal cracks and delamination. The TiO2-filled front layer is missing due to complete delamination, (b) degraded PA BS sample (TiO_2/PA/TiO_2) with cracking

Fig. 6 shows the UV-excited fluorescence (UVF) spectra collected for the degraded PET and polyamide BS samples from Qatar. Both PA and PET BSs showed rather intense UVF indicating a strong degradation, while no UVF can be detected for original BS materials. Comparing the UVF intensity, PET has a higher UVF intensity than polyamide. In contrast to the samples from the desert climate, imilarly aged and cracked PA BS samples collected in Germany show very weak (typically below detection limit) UVF signals. In the case of PET-based BSs aged in Germany noticeable UVF signals can only be detected for very old modules (ca. 30 years) or in the spots of high thermal impact (burn marks or hot spots) while the rest of BS area remains not fluorescent. The UVF observations, therefore, show a striking contrast between the BS samples collected in the desert (Qatar) and moderate (Germany) climates. The UVF in BS materials is assumed to originate from carbonyl-based fluorophors forming as a result of oxidative degradation of polymers. In this view, the intensity of UVF can be taken as a direct measure of the oxidative degradation of BS materials.

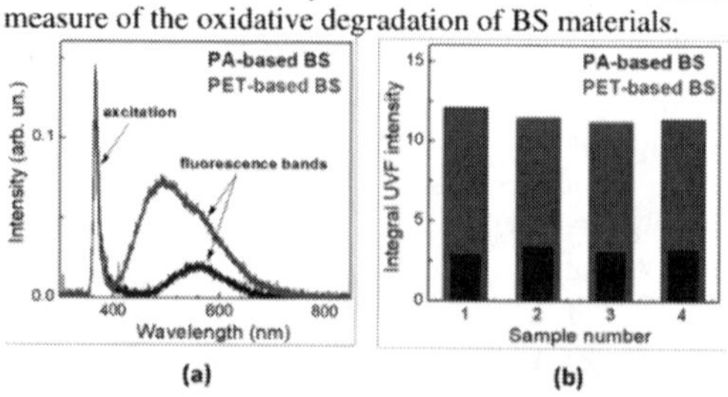

Figure 6: (a) UV fluorescence spectra for polyamide and PET BS samples, excitation beam around 200 nm, (b) UVF intensity averaged for a series of four degraded PA and PET BS samples.

Inner and outer layer of the BS were investigated after years of operation in the field. The stacks of the BS has been identified using Near-infrared absorption (NIRA). Fig. 7 shows typical NIRA spectra registsred for PET-based (a) and PA-based (b) BS samples from Qatar. The spectra taken in two different spots of each sample were found to be identical. NIRA spectrum of the PET-based BS reveals characteristic signals at 1660 nm and 2100-2200 nm originating from =C-H vibrations in PET. Pronounced signals at 1700-1800 nm stem from multiple aliphatic C-H vibrations in methylene and methyl groups of PET and EVA. A pronounced band at 1910 nm is related to water. In general, the shape NIRA spectrum of PET-based BSs from Qatar is the same as for the samples from Germany, because NIRA has a large penetration depth and provides a cumulative signal from all polymer components present in the BS and encapsulant. At the same time, the intensity of water-related band for BS samples from the desert climate is much higher than for the corresponding samples from Germany. The latter observation indicates on a considerably strongler degradation extent of the Qatar samples as compared to those from Germany. Oxidative degradation of PET layer makes is more hydrophilic and enhanced absorption of water from atmosphere during the NIRA measurements, the adsorbed water serving as an indicator of the degradation degree of the BS sample.

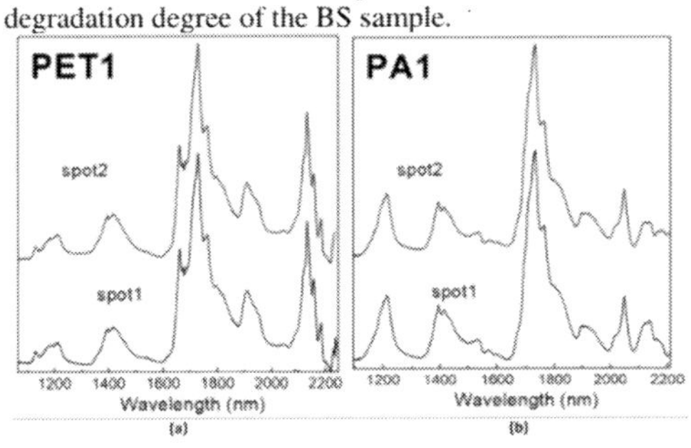

Figure 7: NIRA-spectra for PET and PA of two spots on the degraded samples

NIRA spectra of PA BS show characteristic amide signal at 2050 nm related to C-N-H vobrations, as well as low-intensity characteristic N-H vibrations at ca. 1520-1550 nm. The PA BS samples from Qatar show a very prominent water band, while for the PA samples from Germany water-related signal is typically very low. This difference is also indicative of the much deeper state of oxidative degradation of PA Bss from the desert climate as compared to the samples from moderate European climate, in accordance with the above discussed microscopic observations and UVF measurements.

4 CONCLUSIONS

This paper aims at comparing backsheet degradation observed in moderate and desert climates for two different BS types. Typical desert climatic condition has shown to cause earlier degradation of PA. In desert climate, cracking, chalking and delamination of PA BS observed earlier after 4 years (Qatar) compared with 6 years in moderate climate (Germany). Our results showed that outdoor exposure of the PV module in an high UV irradiance and high operating temperature initiates and accelerates degradation of the backsheet material. Yellowing, embrittlement, cracking, delamination and subsequent chemical aging were observed for the samples from Qatar to be expressed much stronger as for the corresponding samples from Germany. These results suggest that in order to support the deployment of more PV intallations in desert climates, PV modules should possess more resistance to the high UV radiation when compared with BS material for moderate climates.

A high rate of backside yellowing was observed in polyester-based backsheets. PA back sheets showed open and close crack running under the cell interconnect and appeared to be consistent throughout the module for both locations. On the other hand, less excessive degradation of the PET starts to initiate later, followed by excesssive crack propagation throughout the BS, compared with PA.

4 ACKNOWLEDGEMENTS

This publication was made possible by NPRP Grant # NPRP11S-1220-170110 from the Qatar National Research Fund (a member of Qatar Foundation). CB and OS acknowledge financial support from the German Federal Ministry for Economic Affairs and Energy (project "dig4morE", FKZ: 03EE1090B), The findings herein reflect the work, and are solely the responsibility, of the authors.

5 REFERENCES

[1] D. Feldman, K. Dummit, J. Zuboy, and R. Margolis, National Renewable Energy Laboratory (NREL), Golden, CO (United States), NREL/PR-7A40-86215, May 2023.
[2] D. C. Jordan, J. H. Wohlgemuth, and S. R. Kurtz, National Renewable Energy Lab. (NREL), Golden, CO (United States), NREL/CP-5200-56485, Oct. 2012.
[3] D. E. Mansour, C. Barretta, L.P. Bauermann,.G. Oreski, A. Schueler, D. Philipp, and P. Gebhardt, Sustainability, vol. 12, no. 12, Art. no. 12, 2020.
[4] Y. Lyu, A. Fairbrother, M. Gong, J.H. Kim, X. Gu, M. Kempe, S. Julien, K.-Tak Wan, S. Napoli, A. Hauser, G. O'Brien, Y. Wang, R. French, L. Bruckman, L. Ji, K. Boyce, Sol. Energy, vol. 199 (2020) 425.
[5] Y. Lyu, A. Fairbrother, M. Gong, J.H. Kim, A. Hauser, G. O'Brien, X. Gu, Prog. Photovolt. Res. Appl. (2020) ; 28:704.
[6] Y. Voronko, G. C. Eder, M. Knausz, G. Oreski, T. Koch, and K. A. Berger, Prog. Photovolt. Res. Appl., vol. 23, no. 11 (2015) 1501.
[7] G. C. Eder, Y. Voronko, G. Oreski, W. Mühleisen, M. Knausz, A. Omazic, A. Rainer, C. Hirschl, H. Sonnleitner, Sol. Energy Mater. Sol. Cells, vol. 203 (2019) 110194.
[8] O. A. Arruti, L. Gnocchi, Q. Jeangros, A. Virtuani, and C. Ballif, Proceedings IEEE 48th Photovoltaic Specialists Conference (PVSC), Fort Lauderdale, FL, USA: IEEE, (2021) 2032.
[9] A. Sinha, J. Qian, S.L. Moffitt, K. Hurst, K. Terwilliger, D.C. Miller, L.T. Schelhas, P. Hacke, Prog Photovolt Res Appl. 31 (2023) 36.
[10] S. Ulicna, A. Sinha, M. Springer, D.C. Miller, P. Hacke, L.T. Schelhas, M. Owen-Bellini, IEEE J. Photovolt., VOL. 11, NO. 5 (2021).
[11] G. Oreski, C. Barretta, A. Macher, G. Eder, L. Neumaier, M. Feichtner, M. Aarnio-Winterhof, Sol. Energy Mater. Sol. Cells, vol. 259 (2023) 112438.
[12] J. Ascencio-Vásquez,K. Brecl, M. Topič, Solar Energy, Volume 191, (2019) 672.
[13] "Global Modeling and Assimilation Office - NASA, MERRA-2."

https://gmao.gsfc.nasa.gov/reanalysis/MERRA-2/ (accessed Aug. 17, 2023).
[14] A.A. Abdallah, K. Mroue, A. Samara, B. Aissa, M.M. Kivambe, B. Figgis, J. Lopez-Garcia, S. Grosser, M. Pander, B. Jaeckel, C. Hagendorf, M. Ebert, K. Ilse, Proceedings the 8th World Conference on Photovoltaic Energy Conversion WCPEC-8, 2022.
[15] P. Gebhardt, L.P. Bauermann, D. Philipp, Proceedings the European Photovoltaic Solar Energy Conference and Exhibition 2018.
[16] J. Markert, S. Kotterer, D. E. Mansour, D. Philipp, and P. Gebhardt, EPJ Photovolt., vol. 12 (2021) 7.
[17] W. Gambogi Y. Heta, K. Hashimoto, J. Kopchick, T. Felder, S. MacMaster, A. Bradley, B. Hamzavytehrany, L. Garreau-Iles, T. Aoki, K. Stika, T. J. Trout, and T. Sample, IEEE J. Photovolt., vol. 4, no. 3 (2014) 935.
[18] A. Omazic, G. Oreski, M. Edler, G.C. Eder, C. Hirschl, G. Pinter, M. Erceg, J. Appl. Polym. Sci., vol. 137, no. 30 (2020) 48899.

EVALUATION OF CELL CRACKS IN CRYSTALLINE SILICON PHOTOVOLTAIC MODULES USING IMAGING TECHNIQUES – FIRST STEPS TOWARDS PREPARING AN EVALUATION GUIDELINE

C. Buerhop-Lutz[1], T. Winkler[1], S. Rupp[2], M.B. Koentopp[2], A. Linsenmeyer[3], B. Jaeckel[4], A. Shirgupe[5], R. Gottschalg[4,5], G. Kleiss[6], I.M. Peters[1]
[1] Forschungszentrum Jülich GmbH, HI ERN, Erlangen, Germany
[2] Hanwha Q CELLS GmbH, Bitterfeld-Wolfen, Germany
[3] Sunset Energietechnik GmbH, Adelsdorf, Germany
[4] Fraunhofer CSP, Halle, Germany
[5] Hochschule Anhalt – APOLLO, Köthen, Germany
[6] 8.2 Arp & Kleiss GmbH, Berlin, Germany
c.Buerhop-lutz@fz-juelich.de

ABSTRACT: The question whether cell cracks in PV modules are harmful and will reduce performance or present a safety risk, or are stable and without long-term risks, is currently discussed controversially. This discussion is also leading to legal disputes. An interdisciplinary consortium came together in the project "PV-Riss" to develop a proposal for guidelines for the evaluation of cell cracks in PV modules. Three steps of cell crack assessment are distinguished: 1. crack detection, i.e. data acquisition using imaging methods, 2. crack evaluation, i.e. data processing and power impact estimation, 3. crack prognosis, i.e. statements on crack and impact evolution over time or during the service life.

A survey of reported results based on literature is given amended with our own studies covering thousands of PV modules tested in the lab and investigated outdoors in PV power plants. The dataset includes footage of electro-luminescence (EL), UV-fluorescence (UV) and visual (VIS) images. For quantification of the resulting power, IV-curves and monitoring data are evaluated. Cell cracks visible in EL do not cause power reduction unless the cracks disconnect cell fragments electrically from the busbars. Various studies demonstrate that most cracks do not change or become cell fractures with time using mechanical loading test sequences in the lab. Field experiments seem to confirm the lab findings but are too scarce to be more than indications. Today, the majority of published findings are based on two- to four-busbar module technologies and AL-BSF cell technologies. Very first insight gathered from testing modern module technologies shows that the significance of cell cracks decreases with increasing number of busbars, but that some aspects of module design such as cell-to-cell distance can have an impact on crack tolerance depending on manufacturing process quality. Here, more tests and guidelines for testing are necessary.

1 INTRODUCTION

Cracks in PV-modules have been reported manifold. They are either directly visible due to snail tracks or easily visualizable through e. g. electroluminescence (EL) measurements. Thus, they are generally identifiable and are a cause for concern with respect to the value of the system when ownership is about to change or with respect to the long-term performance and safety of the system. Often technical advisors (TA) are called in to assess and either advice on severity and impact of these issues.

There is no standard of how to assess the impact cracks will have on performance. There are some observational assessments which correlate number of cracks with potential power loss. The best assumption to date is that cracks will not cause any power loss unless an area of the cell is electrically separated from the remainder of the cell (= black areas in EL images). The current of this particular cell will then, as a first approximation, be reduced by the fraction of the cell which is isolated. This results in an equal current loss in the entire string in the module. This superficial description demonstrates why simplistic assessments will not work. Furthermore, a crack may behave different to another depending on shape and length and position within the cell. Each manufacturer has, e. g., its own classification system and definition of what is acceptable and what not. A standard approach to evaluate cell cracks is required as the TA needs this to deliver a judicially valid statement and ideally mediate between different stakeholders. PV-Riss aims to develop the foundation for this evaluation process.

The situation is further complicated by technological developments. Most literature reports are on Al-Back-Surface-Field (Al-BSF) cells with two or three busbars. Recent years have seen a progression in technologies from this technology to first PERC and now towards TopCon technologies with 9-16 busbars depending on cell size. Alternatively, the market for HJT technologies as well as innovative contacting schemes is growing rapidly. Each cell technology will behave slightly different with respect to performance and safety of the module. Different contact technologies will affect the impact on the module as well as represent new ageing paths. The guidelines to be developed must include an update to current technologies.

An item not being addressed by today's assessments is the prognosis of the system, i.e. progression of cracks and associated production losses. [1] PV-Riss will deliver further insights and guidelines on how to assess this objectively.

SCIENTIFIC INNOVATION AND RELEVANCE

The relevance of cell cracking in PV-modules is controversially discussed in the industry. It is cause for protracted legal disputes between (international) contracting parties. The discussion is based on data with historic cell types and connection approaches. Typically, studies were conducted on 2- or 3-bus bar Al-BSF cells, which have been substituted by multi-bus-bar PERC or TOPCon cells. The literature gives some indication on the relevance of cracks, but technical advisors have no scientifically verified assessment protocol which enables them to deliver an objective, legally accepted evaluation of observed cell cracks. This will be addressed by PV-Riss.

We want to

- investigate cell cracking,
- collate data to assess impact on energy generation
- Develop a method of prognosis for affected systems
- develop a standardization proposal.

Previous results and available criteria will be taken into account, as well as current findings on new module technologies. Thus, PV-Riss will deliver an advance on a topic which has an unfortunate relevance in terms of legal disputes and provide an update for current cell technologies.

2 APPROACH

The evaluation of cell cracks in PV modules, whether they are harmful and reduce performance or stable and without long-term changes, is currently controversial and leads to disputes. In an interdisciplinary consortium, we are working on a proposal for a guideline for the evaluation of cell cracks in PV modules in the project "PV-Riss". We distinguish between three steps in cell crack assessment: 1. crack detection, i. e. data acquisition using imaging methods, 2. crack evaluation, i. e. data processing and power determination, 3. crack prognosis, i. e. statements on crack evolution over time or during the service life.

We present reported results on cell cracks from literature as well as own studies covering thousands of PV modules tested in the lab and investigated at real operating conditions in PV power stations. The dataset includes footage of electroluminescence (EL), UV-fluorescence (UV) and visual (VIS) images. For quantification of the resulting power, IV-curves and monitoring data are evaluated. Most experience exists for two- to four-busbar module technologies.

3 RESULTS

The results address awareness to cell crack detection, cell crack evaluation and cell drack prognosis.

3.1 Crack detection

In an experiment conducted in 2022, looking at large-scale PV power stations, 25.4% of modules with cell cracks (snail trails) are identified by visual inspection in PV system A (7,937 inspected modules). By EL-inspection 76.6% of modules with cell cracks are detected within 2,133 inspected modules in PV-system B. Table 1 shows the most frequent crack patterns and their frequency. Cracks parallel to busbar are the most dominating ones followed by cracks at the cell corners and point-like, local structures. Less often occur cracks perpendicular to Busbars, diagonal orientated and branched cracks.

10.4229/EUPVSEC2023/3AV.2.8
020181-002

Table 1: typical crack pattern detected by visual imaging in two PV power stations A and B. Multiple annotations were possible, since many modules have cells with different crack structures.

	cracks - parallel to busb	cracks - perpendicular busbar	cracks - circle at the ce corner	cracks - pointlike, local	cracks - (strongly) branched	cracks - diagonal	Total cracked modules	PV-system
Visual image			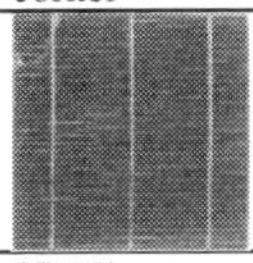	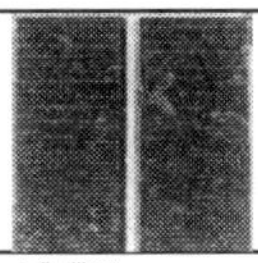		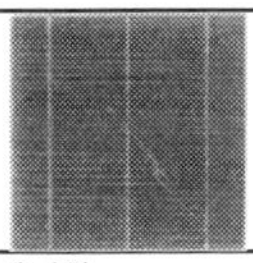		
Freq. VIS	83.4%	7.8%	27.1%	13.7%	8.8%	2.4%	2,016	A
Freq. EL	86.0%	12.4%	30.3%	23.8%	7.0%	12.4%	1,634	B

Statistics in Figure 1 show that the probability of a module having a cracked cell is high (40%); however, more than 20 cracked cells are very rare. The lower rows are most affected, regardless of the module orientation (landscape or portrait) and number of rows per mounting rack (2, 3, 4, or 5).

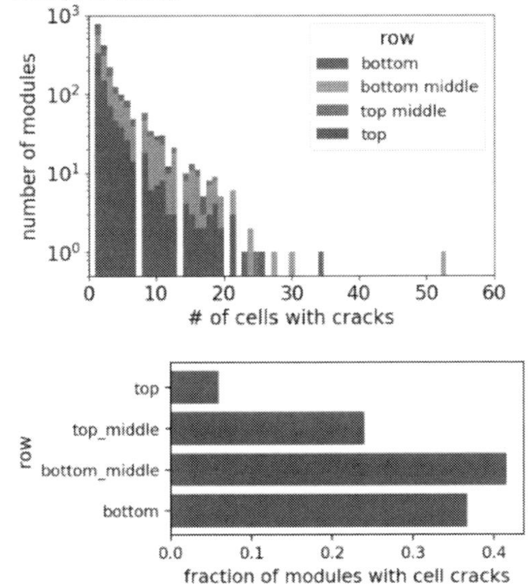

Figure 1: Distribution of cell cracks by visual inspection of 2,016 modules in field for snail trails (25.4% of inspected modules in the field have snail trails), top) module counts versus number of cells with cracks, bottom) distribution of modules with cell cracks with respect to installed row in mounting rack

3.2 Crack evaluation

Important for operation and evaluation is the effect of cracks on power production. In EL images, it is easy to distinguish between cracks without impact on power reduction and cell fractures with resulting power loss. Cell fractures are dark to black in EL images due to decontacting of cell fragments from the current carrying busbars.[1, 2]

The resulting power can be measured either in the laboratory with a solar simulator at defined conditions or in the field at operating conditions with a mobile IV-tracer. This results in advantages and disadvantages for both measurements, e. g. transport. Currently, research is being conducted to predict the IV curve and thus the module and string power from EL images using machine learning methods.[3] It is found that cell fractures can be associated with power loss, depending on the size of the fracture area and the remaining quality of the electrical contact; cell cracks can NOT be associated with power reduction, so far.

3.3 Crack prognosis

Both, crack detection by means of imaging methods and punctual power measurement of modules with cell cracks, are snapshots. The measurements give no indication of perspective crack evolution. Therefore, numerous tests have been developed to predict further evolution of existing crack structures and potential power losses. Field studies are hardly known, but the performance under operating conditions is relevant for predicting resulting power and yield losses. In addition to tests according to standard IEC61215-2 [4] with a test load of 5400 Pa, e.g. executed with sandbags and stamps, advanced test devices were set up by different research groups, in which the load is homogeneously adjusted by air pressure. Furthermore, EL images and IV curves can be recorded at the same time. [5-8] The aim was to simulate realistic load scenarios, e. g. snow loads. The tests showed that under load the module power decreases; unloaded it reaches the original power again. This happens because under load the cracks open, dark appearing fracture with electrically decontacted fragments are created, when relieved, the cracks close, the electrical contact and the power are restored. Cyclic test series up to 100,000 cycles (ZAE), 1,000,000 cycles (NREL) at approximately 1000 Pa were performed without significant changes in crack structures and power production.[9] Even temperature changes did not lead to power decreases in the laboratory. [10, 11] [10, 11] [10, 11]

Reports on how modules with existing cell cracks perform in the field are rare. Due to the installation situation, the modules sag a few millimeters. In the field, it should be noted that the modules sag due to the installation situation, that they operate at elevated temperatures, that they are exposed to fluctuating environmental conditions, e. g. irradiance, wind and daily and seasonal temperature cycles. A module is not operated by its own, a single module is in series connection with about 20 other modules, so the operating point may be shifted. Evaluating power and power changes in the field are complex. For the PV system, the power production of the string is relevant. Alternatives to a snapshot are time series, e. g. from monitoring or measurement repetitions.

For module monitoring, a one-year field study shows that power and voltage of modules with cell fractures are reduced in summer compared to winter, analyzing relative values.[3] This phenomena was attributed to temperature and expansion of the polymers.

So far single modules have been evaluated. For operation, however, it is important which impact a faulty module has on the string performance.

Furthermore, we want to sensitize for the influence of the measurement conditions and the choice of the evaluation basis, e. g. datasheet, flash list, reference module, for the crack assessment. Here, further test series and evaluation criteria are necessary to quantify the effects of cracks in a module on the string yield in the field.

In general, most experience is available for modules with full cells with 2-4 busbars. Newer technologies have hardly been investigated or evaluations published.

First tests of half-cell technologies were carried out at the mechanical stress test setup in the laboratory. Figure 2 shows EL images of a "zerogap" module after loading test with a pressure of p = 2000 Pa and p = 2600 Pa. In contrast to the experience with loads of 2- to 4-busbar technologies, the cracks do not close AFTER the loading (recognizable by the dark cells). For this specific module, this is associated with permanent power decreases of 3% and respectively 10%. The role of module assembly, architecture and material selection needs to be investigated in further experiments.

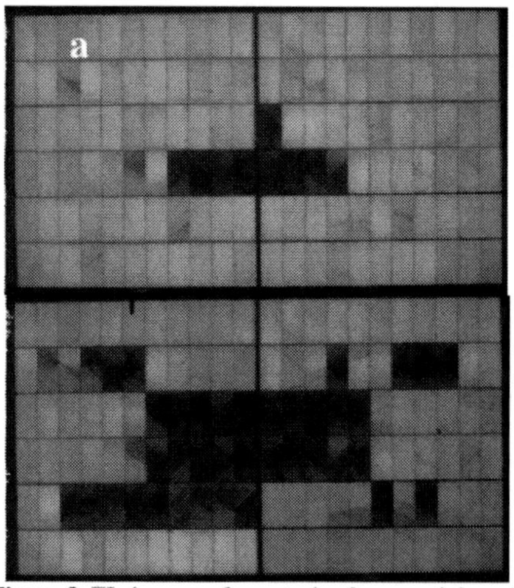

Figure 2: EL-images of new technology modules with cell cracks after mechanical loading tests in the lab, a) EL **after** load p=2000 Pa, power loss ΔP = -3%, b) EL **after** load p=2600 Pa, power loss ΔP = -10%

4 SUMMARY

We are working on a guideline for the assessment of cell cracks, taking into account the interests of different stakeholders of PV power stations. There are three aspects: 1) image acquisition (e. g. resolution, excitation current, location (lab or field measurement)), 2) crack evaluation (e. g. quantifying potential power loss) and 3) crack prognosis (studying potential propagation of cracks during field operation in the lab and in the field). Important is to extent existing studies and experience on 3- or 4 busbar module technologies to modern multi-wire and half-cut cell module technologies.

5 ACKNOWLEGDEMENT

We gratefully acknowledge the German Federal Ministry for Economic Affairs and Climate Action (BMWK) for funding *PV-Riss* (FKZ: 03TN0033A-H).

6 REFERENCES

[1] M. Köntges, et al. (2014).
[2] M. Köntges, et al.; IEEE Journal of Photovoltaics 3 (2013) 95.
[3] C. Buerhop-Lutz, et al., 36th EU-PVSEC (2019) 1336.
[4] Standard IEC 61215 second edition 10.15 2005 Crystalline silicon terrestrial photovoltaic (PV) modules - Design qualification and type approval wet leakage current test.
[5] A. M. Gabor, et al., 43rd IEEE, Photovoltaic Specialists Conference (PVSC) (2016).
[6] E. J. Schneller, et al., 44th IEEE (2017).
[7] C. Buerhop-Lutz, et al., 33rd EU-PVSEC (2017) 1451 .
[8] C. Buerhop, et al.; Progress in Photovoltaics 26 (2018) 261.
[9] C. Buerhop-Lutz, et al., 35th EU-PVSEC (2018) 1554.
[10] T. J. Silverman, et al., NREL reliability workshops (2018).
[11] C. Buerhop-Lutz, et al., 35th EU-PVSEC (2018) 1147.

10.4229/EUPVSEC2023/3AV.2.8
020181-004

RELIABILITY OF LIGHTWEIGHT PV LAMINATIONS UNDER HARSH ENVIRONMENTAL CONDITIONS

Alicia Buceta, Ana B. Cueli and Jaione Bengoechea
National Renewable Energy Center (CENER)
Avenida Ciudad de la Innovación, 7, 31621 Sarriguren (Navarra) SPAIN
Tel.: +34 948 25 28 00 Fax: +34 948 27 07 74

ABSTRACT: Based on previous results, promising lightweight modules' reliability under harsh environmental conditions was assessed. The outer layer materials used in this design were **polypropylene honeycomb** (two different thicknesses) with non-woven polyester skin as **backsheet** and a **1 mm thick glass** as **frontsheet**. These modules already proved their capacity to withstand the hail impact test; herein, their resistance to damp heat conditions and ultraviolet is presented. The modules were subjected to 2,550 hours of damp heat conditions, more than 2.5 times the severity of IEC 61215 standard. During the intermediate diagnosis tests conducted no major visual defect was detected and the modules remained fully operative. The electrical performance characterization shows a decrease of maximum power below 10% after 2,215 hours of damp heat which might be caused by slight oxidation on the contacts. Alternative solutions to reduce moisture permeation were studied; same basic configurations but different encapsulants and seals were exposed to condensation conditions. The use of POE encapsulant proved to be the best performing one. On the other hand, the susceptibility of the HC material to ultraviolet radiation has been investigated. The yellowing effect and the influence on the mechanical strenght after high doses of ultraviolet radiation have been tested.
Keywords: c-Si, lightweight module, honeycomb, stability tests

1 INTRODUCTION

The global energy situation showcases the need of new solutions that ease the use of renewable energies. Photovoltaic applications such as BAPV (Building Applied Photovoltaics), BIPV (Building Integrated Photovoltaics), VIPV (Vehicle Integrated Photovoltaic) and IIPV (Infrastructure-Integrated Photovoltaic) can be part of the solution.[1] For instance, in the coming years the rooftop solar Photovoltaic (PV) installation market is expected to increase but is still limited by the conventional PV modules' weight.[2, 3] Therefore, finding a lightweight solution for c-Si modules will increase their applications, especially considering light roofs or historical buildings with weight restrictions.[4, 5]

Most of the commercial c-Si PV modules use glass as frontsheet which impose roughly 60-70% of their total weight. PV industry is exploring new solutions to protect the modules by the front side as thinner glasses or polymeric films in order to achieve lighter modules that meet the technical requirements.[6] The critical issue when conventional PV glass is replaced by lighter frontsheets is to maintain the mechanical integrity of the modules. Therefore, it becomes necessary to reinforce the rest of the layers comprised in the PV laminate (encapsulants, backsheet) to prevent cells from breaking. The approach applied in this work is the use of honeycomb (HC) structures as backsheet due to their high strength and lightweight.

In a previous work the lamination process to obtain lightweight PV modules using alternative materials with HC as backsheet was presented.[7] It was also validated their capability to withstand the hail impact test described in IEC 61215 standard without causing any cracks on the solar cells. Once that the mechanical properties of the new developed products have been checked, the next step is to verify their behaviour under demanding climatic conditions as damp heat and high ultraviolet (UV) radiation.

The processes and equipment used in the present study for the fabrication of the mini-modules are the ones currently used in the industry to manufacture conventional c-Si PV modules. Therefore, these new low weight PV designs can be easily adopted by producers in order to achieve high degree of automation.

2 METHODOLOGY

The tested module configurations were based on previous results obtained in the frame of AiSoVol2 project and this work is the next progress phase.[8]

The UV weathering and climatic tests performed on the lab scale modules are based on IEC PV module qualification standards,[9] although, they will be tested beyond the standards requirements.

2.1 Materials

Two polypropylene (PP) HCs with non-woven polyester skin capable of withstanding punctually temperatures of 140 ºC were used as backsheets. They are opaque white materials, rigid and very light. One was 10 mm thick (HC10) with 8 mm hexagonal cell size and the other sample was 14 mm thick (HC14) with 5 mm hexagonal cell size (Figure 1).

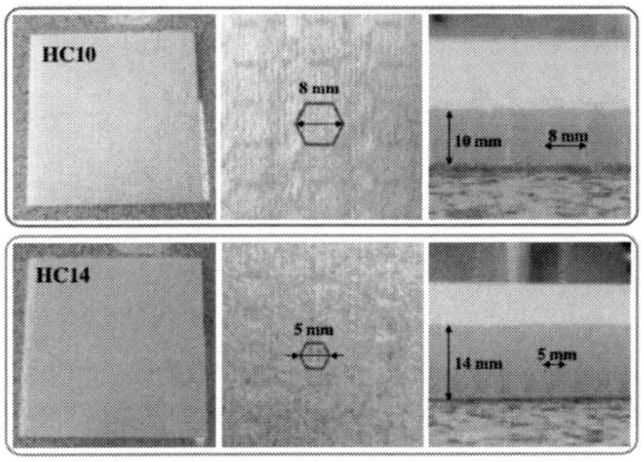

Figure 1. Characteristics of the used HC.

As frontsheet a thin glass, 1 mm thick, was used with around 90% transmittance.

10.4229/EUPVSEC2023/3AV.2.9
020182-001

An ultra-fast cure EVA (Ethylene-vinyl acetate) was chosen as the encapsulant. As alternatives, POE and TPO were tested.

All the tested materials are commercially available.

2.2 Lamination

The laminator used to obtain the PV mini-modules was a PEnergy Lab-Laminator L036LAB, which is an automated module laminator for samples up to 60 x 60 cm and 4 cm thickness maximum. It has a membrane on the upper chamber and the hot plate on the lower chamber.

2.3 Climatic chamber

Their resistance to temperature and humidity was tested in a C-70/200 climate chamber from CTS. The damp heat test (MQT 13) was performed following the IEC 61215-2 standard.

2.4. Ultraviolet exposure

The UV irradiation was applied in a weathering chamber using as light source Heraeus-OHC402S lamps with a special filter to match the spectral distribution stated in the qualification standard of PV modules. The temperature of the samples during UV irradiation was fixed at 60ºC.

2.5 I-V curves measurements, electroluminescence and optical measurements

All the IV curves were measured in a Pasan flash Sun Simulator. The electroluminescence (EL) tests were performed with a digital CMOS camera Orca-flash from Hamamatsu and a CAS140CT spectroradiometer from Instrument Systems was used for the optical measurements. The wavelength range goes from 300 nm to 1600 nm.

3 RESULTS AND DISCUSSION

The main goal of this work was to assess the behaviour of the lightweight PV module solution previously selected for withstanding the hail impact under harsh environmental conditions.

The tested modules size was 40 x 37 cm, with 4 polycrystalline silicon solar cells serially connected. The sandwich structure was ThinGlass/2xEVA/SolarCell/EVA/PPHoneyComb, showed in Figure 2.

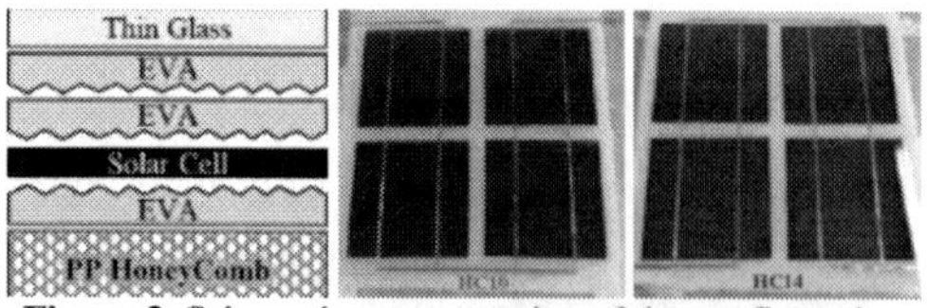

Figure 2. Schematic representation of the configuration tested, where the HC can be 10 mm or 14 mm thick.

In our previous work, with this configuration a clear presence of humidity was observed after the first 140 hours at 85 ºC and 85% of relative humidity (RH), although, it disappeared after a few days out of the climatic chamber and the performance was not jeopardized by it.[7] Therefore, a more extensive study of the effect of damp heat was carried out, showing a clear effect of humidity after 2,215 h. To reduce their water permeation, alternative solutions were laminated, where POE seemed to be a better encapsulant than EVA for this purpose.

Also, given the fact that the polypropylene HCs with non-woven polyester skin were new materials to be used as backsheet, they were exposed to high UV irradiation doses up to 70 kWh/m^2, where not much yellowing appears to occur but their mechanical properties drop drastically.

3.1 Climatic chamber

The modules were tested in the climatic chamber under damp heat conditions (85 ºC and 85% RH) a total of 2,550 h and new configurations to avoid water permeation were tested in condensation conditions (85 ºC and 95% RH) for 160 h.

3.1.1 Damp heat

To study the evolution of the modules and their potential degradation, the experiment is stopped after different exposure times and the samples are characterized before continuing with the test. The characterisation consists on detailed visual inspection, I-V curve measurement, electroluminescence test and optical characterization. When comparing the maximum power (P_{MAX}) obtained by the I-V curve measurement, it was observed that after 1,050 h under damp heat condition the lost was around 7% for both samples, (Figure 2). When continuing the process up to 2,215 h the drop in P_{MAX} was around 10%, while after 2,550 h decreases down to around 30%.

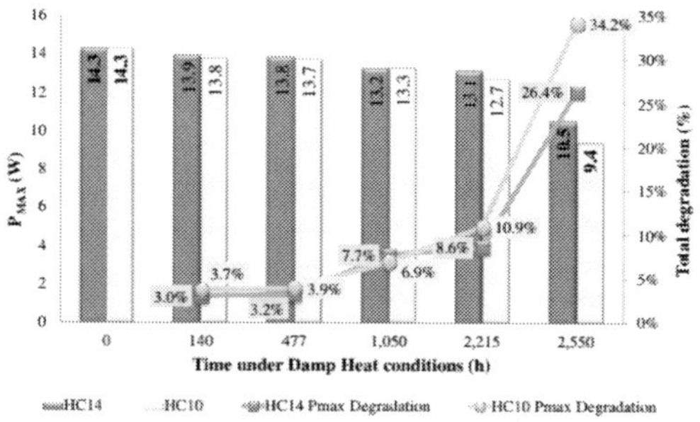

Figure 3. Representation of P_{MAX} after different time exposure under damp heat (column graph) and its degradation.

This decrease in P_{MAX} can be explained with the EL, were clear water damage is observed after 2,215 h of exposure, which gets significantly worst after 2,550 h (Figure 4).

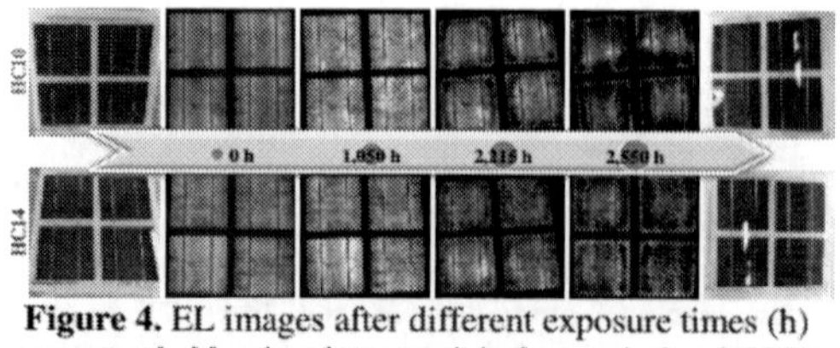

Figure 4. EL images after different exposure times (h) accompanied by the photograph before and after 2,550 h in the climatic chamber.

The behaviour can be also corroborated in the visual inspection, where the ribbons oxidation started to appear after 2,215 h exposed (Figure 5).

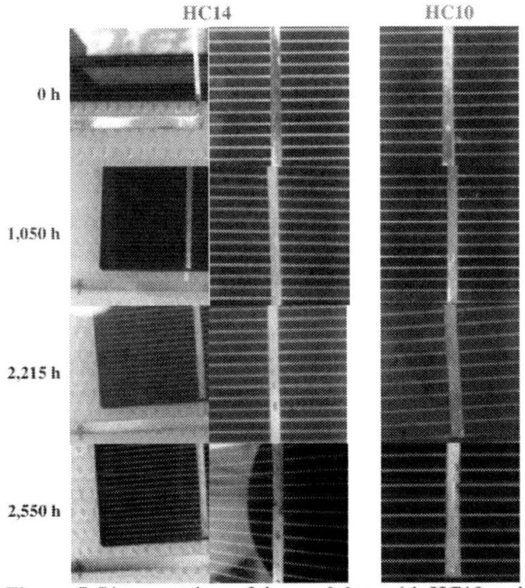

Figure 5. Pictures taken of the modules, with HC10 and HC14, after different exposure times.

On the other hand, the optical measurements can give us information about the yellowing index (YI) of the PV module. By using the reflection measurements, the graph showed in Figure 6 indicates a correlation between the increasing of the YI and the increase of the exposure time.

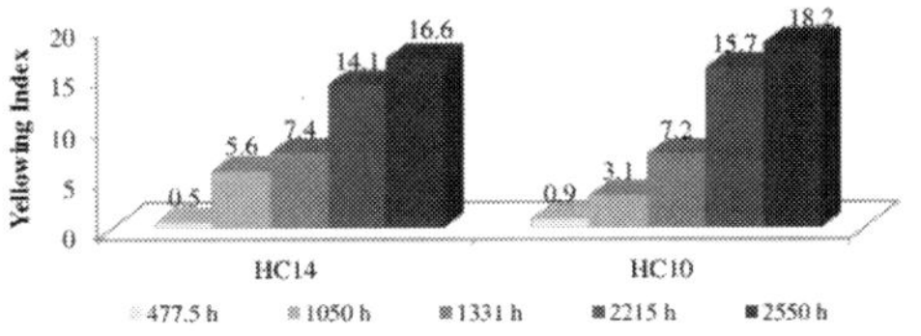

Figure 6. Yellowing Index calculations using reflectance measurements of the two tested modules using HC10 and HC14 after different exposure time under damp heat.

It seems clear that some moisture penetration occurs during the damp heat, whose effects are especially noticeable after 2,215 h of time exposure.

3.1.2 Condensation conditions

Given the previous results, small samples with new configurations are tested to reduce water ingress. In Figure 7 a schematic representation of the laminated sandwiches is presented. In order to better determine the barrier effect provided by each encapsulant, transparent tedlar is used instead of thin glass in all samples. The use of tedlar polymeric film with higher WVTR (Water Vapor Transmision Rate) than glass accelerates the effects of humidity over the samples. The used encapsulants were EVA to compare with our original samples, POE and TPO as alternatives.

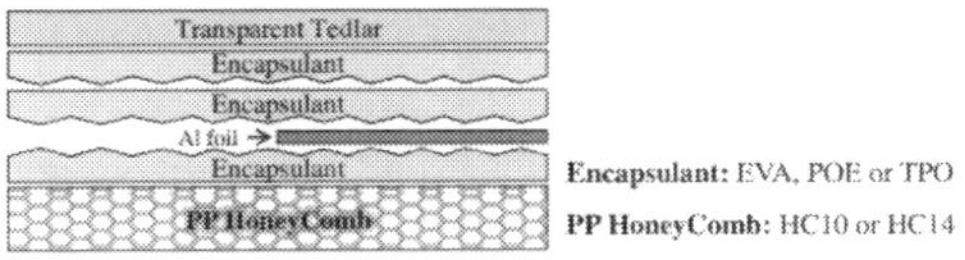

Figure 7. Schematic of the laminated sandwiches to study the influence of the encapsulant in water permeation.

Additionally, two sealed samples were prepared with EVA and HC14, one was only sealed at the edges with a waterproof tape while the other one was also sealed at the back of the HC14 with white Tedlar® (Figure 8).

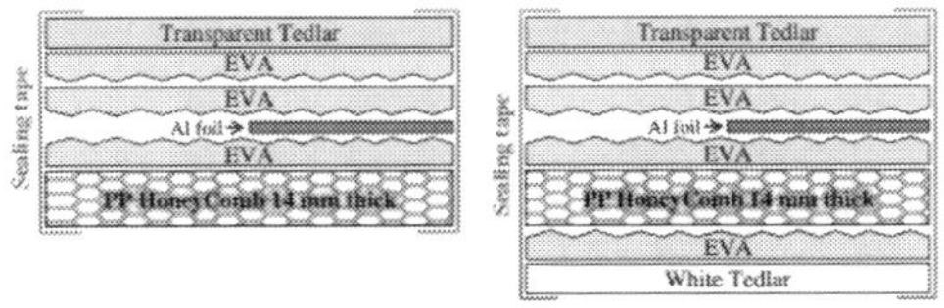

Figure 8. Schematics of sealed samples tested.

The moisture ingress was measured indirectly by comparing the reflectance before and after being exposed to 85 °C and 95% RH for 160 hours. No solar cell was included in these first trials, although, half of each sample had an aluminium foil where the solar cell should be to facilitate the reflectance measurements used to compare the samples. The reflectance was measured in two points as showed in Figure 9, one with the HC and the other one with the aluminium foil in the background.

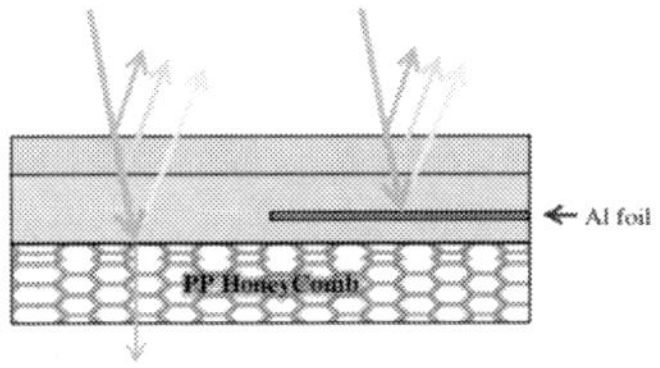

Figure 9. Representation of the points where the reflectance was measured.

The results from the optical measurements with the HC in the background showed that when using POE the variation in the average reflectance of the sample before entering the climatic chamber and afterwards was minimal. Since the reflectance remains similar it could be inferred that no presence of humidity is detected inside the sample. This effect was also observed visually, the EVA and TPO samples had more haze than POE. In Figure 10 the absolute value of the average reflectance variation before and after the climatic chamber is presented. The reflectance growths after the climatic chamber due to the haze increasing.

10.4229/EUPVSEC2023/3AV.2.9

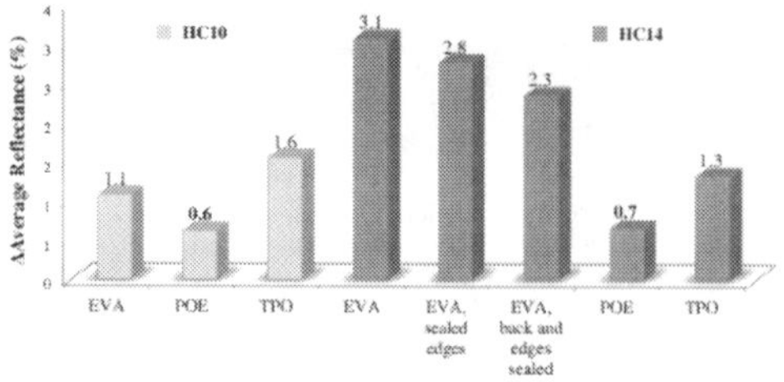

Figure 10. Absolute value of the variation of the average reflectance before and after the climatic chamber. Measurements done with the HC in the background.

When the measurement takes place on top of the aluminium foil, the behaviour is slightly different. In this case the measurement will give us information only about the fronsheet and first encapsulant sheets, not the HC. The aluminium foil has specular reflection which is greater than the reflectance measured at the HC, and in this case, the haze that appears due to humidity presence (diffuse reflectance) decreases the reflection of the sample, the opposite of what it is observed when the measurements is made over the sample area with HC in the background. POE was still showing the best behaviour but the difference was not as prominent. Also, the sealed edges and back of the sample seems to be detrimental, this might indicate that the principal path for the humidity penetration is through the tedlar sheet, and sealing the sample reduces the possibility to evacuate it (Figure 11).

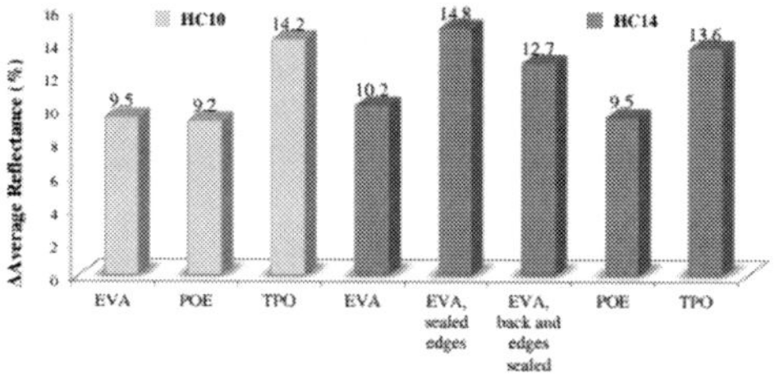

Figure 11. Absolute value of the variation of the average reflectance before and after the climatic chamber. Measurements done with the aluminium foil in the background.

These are preliminary results that need further testing and study.

3.2 UV Irradiation

The UV exposure test was conducted to assess the susceptibility of the honeycomb to this type of irradiation as it is known that polymers might be degraded by UV light. Therefore, samples of HC10 and HC14 sheets were exposed a total UV dose of 70 kWh/m^2, although optical measurements were performed at intermediate exposure times, their mechanical properties where just tested at the end of the experiment.

The reflectance was measured to calculate the yellowing index of the samples after each exposure to UV. As it can be observed in Figure 12 a slight yellowing occurs over time, although it is barely noticeable to the naked eye.

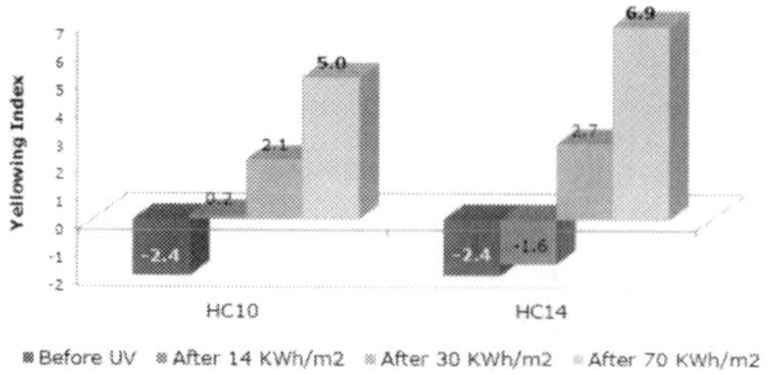

Figure 12. Yellowing index of HC10 and HC14 after different time exposure under UV light. Negative values indicate a bluish colour.

On the other hand, during the inspections conducted after each irradiation period, it was detected some embrittlement of the honeycombs. Therefore, the mechanical properties were tested following the UL1703 standard,[10] using a 535 g and diameter of 5.1 cm steel ball that was dropped from a height of 1.3 m. The crater formed in the samples not exposed to UV light was compared to the one obtained on the weathered samples. As it can be observed in Figure 13, the mechanical properties of the HC are severely affected.

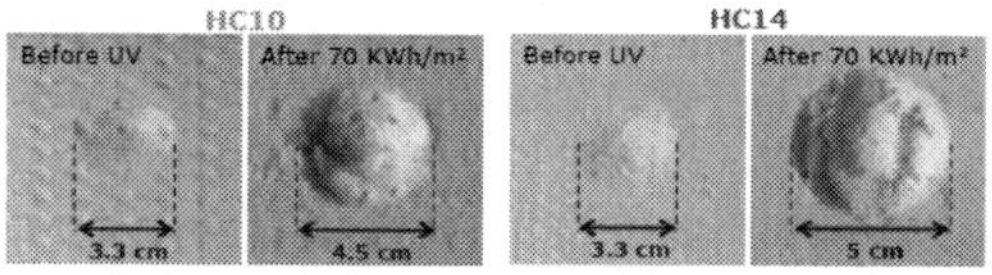

Figure 13. Aspect of the honeycombs after the mechanical test, before and after being exposed to UV light.

The total UV dose applied to the honeycomb samples is more than 4 times the dose required by IEC 61215-2 standard for the front side of the module. Taking into account that this material will be used as backsheet, only a portion of the front UV irradiation will reach it. The amount of stress on the backside of the module will depend on site specific variables; considering the representative conditions for rack mounted modules given in IEC TS 62788-7-2 the total dose applied in this experiment would be equivalent to 7 years for arid climates and more than 11 years in a temperate one.[11]

4 CONCLUSIONS

The lightweight solution proposed withstands, with minimal effect, twice the severity of the IEC 61215 requirements for damp heat.

A better solution was found that would help control the humidity filtration. The water permeation can be reduced by the use of POE instead of EVA.

When the HC is tested under UV irradiation no significant yellowing is observed although the mechanical properties are affected.

As future research, the behavior of the mini-modules with POE encapsulant should be tested under damp heat to validate the reduction of the humidity permeation. On the other hand, further UV protection to polypropylene HC could be provided, although it would depend on the final application.

10.4229/EUPVSEC2023/3AV.2.9
020182-004

5 ACKNOWLEDGEMENTS

These results are part of project AiSoVol2, extension of AiSoVol (photovoltaic generation solution as an alternative construction material). It is a project funded by the Ministry of Economy and Competitiveness, within the State Program for R&D&I oriented to meet societal challenge 3: Secure, Clean and Efficient Energy Challenge. The project, which began on 1st June 2020 and will end on 31st December 2023, is composed by the partnership of the Technological Institute of Renewable Energies, SA (ITER) as coordinator and the National Renewable Energy Centre, CENER Foundation.

Special thanks to the entire technical team for their invaluable help in carrying out the tests.

6 REFERENCES

[1] Singh, D.; Chaudhary, R.; Karthick, A. *Review on the Progress of Building-Applied/Integrated Photovoltaic System*; Environmental Science and Pollution Research, 2021; Vol. 28. https://doi.org/10.1007/s11356-021-15349-5.

[2] Huld, T.; Bodis, K.; Pinedo Pascua, I.; Dunlop, E.; Taylor, N.; Jäger-Waldau, A. The Rooftop Potential for PV Systems in the European Union to Deliver the Paris Agreement. *european energy innovation*. 2018, pp 12–15.

[3] Philipps, S. (Fraunhofer I.; Warmuth, W. (PSE P. G. *Photovoltaics Report (Updated 24 February 2022)*; Freiburg, 2022.

[4] Kuhn, T. E.; Erban, C.; Heinrich, M.; Eisenlohr, J.; Ensslen, F.; Neuhaus, D. H. Review of Technological Design Options for Building Integrated Photovoltaics (BIPV). *Energy Build.*, **2021**, *231*, 110381. https://doi.org/10.1016/j.enbuild.2020.110381.

[5] Schmid, C. Lightweight module durability: Working to address a market barrier https://www.pv-tech.org/lightweight_module_durability_working_to_address_a_market_barrier/ (accessed Aug 22, 2022).

[6] Menéndez, A.; Martínez, A.; Santos, A.; Ruiz, B.; Moritz, K.; Klein, I.; Díaz, J.; Lagunas, A. R.; Sauermann, T.; Gómez, D. A Multifunctional ETFE Module for Sustainable Façade Lighting: Design, Manufacturing and Monitoring. *Energy Build.*, **2018**, *161*, 10–21. https://doi.org/10.1016/j.enbuild.2017.12.023.

[7] Buceta, A.; Aguirre, M.; Cueli, A. B.; Ezquer, M.; Llarena, E.; Cal, S.; Linares, A.; Castelo, R. Lightweight PV Modules Compatible with One Step Lamination Process. In *8th World Conference on Photovoltaic Energy Conversion*; 2022; pp 847–850.

[8] CENER. AiSoVol2 https://www.cener.com/en/areas/photovoltaic-solar-energy-department/outstanding-projects/aisovol-2-adaptive-photovoltaic-generation-solution-for-use-in-building-and-distributed-generation/.

[9] IEC. *IEC 61215-2 Terrestrial Photovoltaic (PV) Modules – Design Qualification and Type Approval – Part 2: Test Procedures*, 2nd ed.; Geneva, 2021.

[10] UL Standard. *UL 1703 Ed. 3-2002 Standard For Flat-Plate Photovoltaic Modules And Panels*, Standard 1.; 2002.

[11] IEC. *82/1212/DTS*; 2017.

ANALISYS OF THE EFFECTS OF LEAKAGE CURRENT PRODUCED BY POTENTIAL INDUCEDDEGRDATION (PID) IN PHOTOVOLTAIC MODULES WITH PERC AND TOPCON TECHNOLOGY

Mendelsson R. M. Neves[1], Allan M. C. Silveira[1], João Frederico S. de Paula[1], Harison F. Santos2, Thais Crestani2, Lia S. Costa2, Rodrigo M. Garcia2, Marcelo G. Villalva1 and Tárcio A. dos S. Barros[1]
[1] School of Electrical and Computer Engineering, UNICAMP, Campinas, Brazil.
[2]BYD Energy R&D from Brazil, Campinas, Brazil.
m264655@dac.unicamp.br,harison.santos@byd.com, thais.crestani@byd.com, lia.costa@byd.com, rodrigo.garcia@byd.com, villalva@unicamp.br

ABSTRACT: Durability and reliability are essential factors to consider when evaluating PV modules. Durable and rugged photovoltaic modules can withstand environmental factors such as high temperatures, humidity, and extreme weather conditions and continue to produce electricity throughout their expected lifetime. Several factors have resulted in the durability and reliability of photovoltaic modules, including the quality of the materials used in the module, the module design, and the manufacturing process. Potential-induced degradation (PID) and leakage current testing are essential methods used to evaluate the performance and durability of photovoltaic modules. In this article, the relationship between leakage current and the effect of PID degradation on the performance of mini photovoltaic modules is studied. The minimodules were manufactured with solar cells with p-type PERC technology and TOPCon with n-type cells to study the influence of the module encapsulants in this type of degradation. The study seeks to correlate the leakage current with the encapsulating materials, and minimodules were manufactured with a variation of the encapsulant using EVA (Ethylene Vinyl Acetate), EPE (Expanded Polyethylene), and POE (Polyolefin Elastomers).

Keywords: leakage current, modules photovoltaics

1 INTRODUCTION

The durability and reliability of photovoltaic (PV) modules are two variables that the photovoltaic industry seeks to extend, as it is a critical factor in reducing the costs of PV systems in the long term. Thus, there is some concern about how long the module will perform at its maximum and continue to generate energy. For this, the photovoltaic industry invests in introducing new materials, processes, and technologies for solar cells. The primary solar cell technologies used in the market are PERC (Passivated Emitter and Rear Contact) and TOPCon (Tunnel Oxide Passivated Contact).

PERC solar cells are silicon solar cells with an additional dielectric layer on the back of the cell, which improves the efficiency of the cell by reflecting backlight that would otherwise be absorbed by the substrate [1], [2]. TOPCon solar cells are a variation of the PERC cell that uses a different type of dielectric material and structure to improve efficiency further [3], [4]. Both technologies are used in high-efficiency solar panels and are becoming increasingly popular in the solar industry. TOPCon cells offer higher efficiency, better-shading tolerance, and better performance under non-uniform light conditions than PERC cells [5], [6].

Potential-induced degradation (PID) is a degradation mechanism associated with system polarization in which the performance of the photovoltaic module is reduced due to the electrical potential difference between the various components of the module [7], [8], which causes allow leakage currents to flow from the module frame to the cells or from the cell to the frame, causing the affected cells to lose part of their output power [7], [9]. Leakage current can contribute to PID by creating a potential difference between the front and back contacts of the module, which can result in loss of power and reduction in the overall efficiency of the module [10], [11]. As a result, it is essential to minimize leakage current to prevent or reduce PID effects.

When a solar cell is subjected to a high negative voltage, a notable voltage disparity arises between the cell and the module's structure. The module structure typically maintains a zero potential as it is usually grounded. Because of the short distance between the solar cells and the structure, along with potential impurities in the encapsulating material, this can give rise to a current flow between the cells and the structure, resulting in current leakage across the entire photovoltaic module[12].

Figure 6 illustrates a cross-section of a PV module, displaying various routes for leakage currents from the cell circuit to the frame. In this study, all modules utilized a soda-lime glass top cover, while the back cover was made of a polymer backsheet. On the front side, the bulk resistance of the encapsulant, bulk resistance of the front glass, and surface resistance of the glass are connected in series. On the back side, due to the notably high bulk resistance of the backsheet, minimal current passes through this path. Consequently, the two primary routes for leakage currents are through the bulk of the front glass and laterally along the interfaces between the glass-encapsulant and backsheet-encapsulant [9], [11], [12].

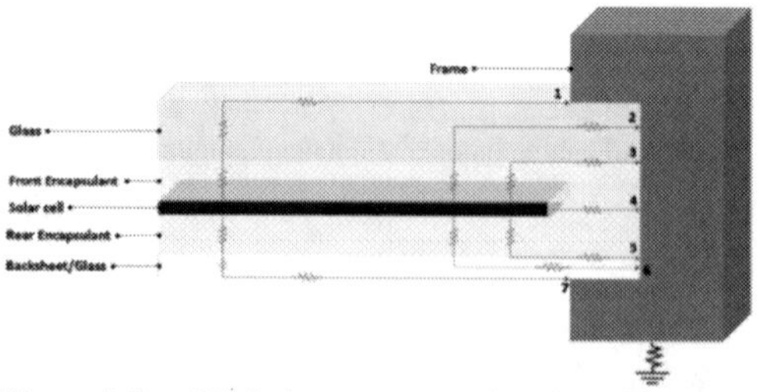

Figure 1: Possible leakage current path. Adapted from [6].

The main objectives of this study were to evaluate new encapsulating materials with low ionic migration rates, which can help prevent PID, typically based on silicone or

10.4229/EUPVSEC2023/3AV.2.12
020184-001

other polymers with high resistance to ionic migration. Likewise, it was possible to observe the effects of leakage current by using new materials and design modules that can reduce the susceptibility of the module to leakage current. For the product engineering area of the photovoltaic industry, it allowed evaluation of the different module designs that can reduce the risk of PID by incorporating an adequate grounding system and

designing the module with improved thermal management. In addition, it allowed the evaluation of degradation mechanisms to identify the different mechanisms that can compromise the durability of photovoltaic modules, such as electromigration, and develop a better understanding of this phenomenon.

2 EXPERIMENTAL DETAILS

The PID test methodology for PV modules evaluates the performance and durability of PV minimodules under conditions that can cause PID. PID can occur when a PV module is exposed to high voltage potentials and high humidity conditions, which can cause a decrease in the module's power output. In the test procedure, a high voltage potential is applied to the mini photovoltaic modules for a certain period, usually 96 hours, corresponding to one cycle. The output power of the minimodules is measured before and after the voltage stress test to determine the decrease in output power due to the PID by measuring the I-V parameters.

The performance of the minimodule is evaluated, and any problems are identified, such as hot spots or cracks in the cells. This is done by visually inspecting the minimodule and using the electroluminescence camera to detect any problems before and after the degradation. The test is performed under determined humidity and temperature levels, with varying materials (encapsulants and backsheets), and cell types (PERC and TOPCon). The results are compared with electrical parameters to determine the PID level. Mono crystalline silicon solar modules of 30x30 cm2 area were used for the study as shown in Fig. 2.

Figure 2: Sample minimodule.

Encapsulants of different brands and types, EVA, EPE and POE, were chosen to reflect their relevance in the photovoltaic industry. Furthermore, two main solar cell technologies have been incorporated, including PERC and TOPCon cells. The comprehensive approach of combining minimodule assembly with different types of encapsulants and cell technologies provided a detailed assessment of performance and efficiency characteristics in a wide variety of scenarios. Table I shows how the combination between the components of the minimodules was made.

Table I: Composition of the minimodules samples

Sample	Solar Cell	Encapsulant					
		Front			Rear		
		EVA	EPE	POE	EVA	EPE	POE
A1	PERC	✓	---	---	✓	---	---
A2	TOPCON	✓	---	---	✓	---	---
B1	PERC	---	---	✓	---	---	✓
B2	TOPCON	---	---	✓	---	---	✓
C1	PERC	---	✓	---	---	✓	---
C2	TOPCON	---	✓	---	---	✓	---
D1	PERC	---	---	✓	✓	---	---
D2	TOPCON	---	---	✓	✓	---	---

This test methodology was based on the IEC 61215-2:2021 MQT 21 standard, which provides a consistent and reliable method to evaluate PV modules PID resistance, helping to select modules with high PID resistance. In this standard, the degradation experiment is carried out in a climatic chamber applying a DC voltage of -1500V on the photovoltaic modules to the module structure, with a temperature of 85 ± 2°C and relative humidity of 85 ± 2% for different durations of time.

The PID test setup was set up to evaluate the solar module components and the leakage current present in the mini modules, the experiment setup is shown in Fig 3. A DC voltage of 1500 V was applied using a DC power supply for a period of time of 192 hours and the leakage current was measured using the HOTOTECH model HTPV-16 control module.

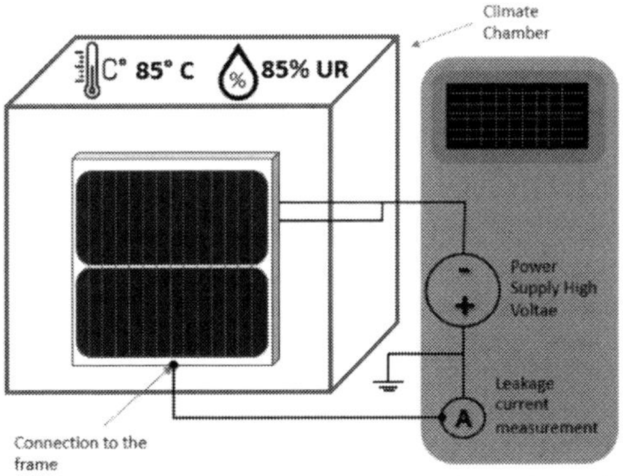

Figure 3: Possible leakage current path. Adapted from [6].

The characterization processes involve electroluminescence (EL) imaging and current-voltage (I-V) measurements as show in Fig. 4. An analysis of the leakage currents and the load transferred to the ground of the minimodules is carried out in each combination of materials used in the function of the ambient conditions, including temperature and relative humidity. The results of this study demonstrate the differences that the materials can present in the PID effect and how much the cells can suffer from this degradation. It is essential to mention that PID tests are performed under laboratory conditions, and

10.4229/EUPVSEC2023/3AV.2.12

results may differ under actual operating conditions.

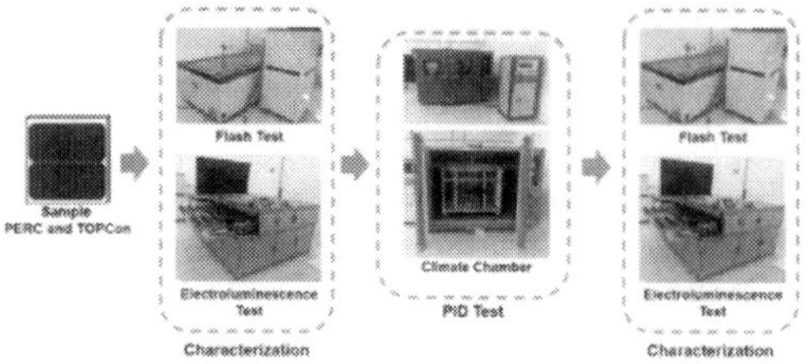

Figure 4: Details of the steps of the experimental procedure

3 RESULTS AND DISCUSSION

3.1 Sample PERC

During the analysis of minimodules assembled with various solar cell technologies and encapsulant samples, it was observed that PERC (Passivated Emitter Rear Cell) cells consistently exhibited higher levels of leakage current compared to other solar cell technologies. This intriguing observation led to a more in-depth investigation into the influence of different encapsulant samples on this increase in leakage current as show in Fig. 5.

It became evident that the different encapsulant samples used in the minimodule configurations significantly contributed to the increase in leakage current in PERC cells, with a particular focus on EVA (Ethylene Vinyl Acetate). While all encapsulant samples were selected based on their technical specifications and compatibility with photovoltaic modules, a notable variation was observed in terms of electrical resistivity and isolation behavior when subjected to high reverse voltages.

Minimodules using EVA as the encapsulant showed particularly high sensitivity to leakage current, especially under high-stress testing conditions. In contrast, encapsulant samples such as EPE (Ethylene Propylene Ethylene) and POE (Polyolefin Elastomer) demonstrated more effective electrical resistance and a lesser contribution to the increase in leakage current in PERC cells.

These results underscore the critical importance of the specific encapsulant choice when designing photovoltaic modules using PERC cells. In addition to considering the solar cell technology, the electrical compatibility of the encapsulant with the cell and its performance under different operational conditions should also be carefully assessed. Encapsulant selection strategies, such as preferring EPE or POE over EVA, and potential modifications in the encapsulation process, may be necessary to minimize the impact of differences in encapsulant samples on leakage current and optimize the overall performance of photovoltaic modules.

3.2 Sample TOPCon

Firstly, it became evident that TOPCON (Tunnel Oxide Passivated Contact) cells consistently exhibited significantly lower levels of leakage current compared to other solar cell technologies as show in Fig. 6. This noteworthy observation was especially pronounced when the minimodules were subjected to high-stress testing conditions, such as exposure to elevated temperatures and intense sunlight.

Furthermore, the investigation revealed that the choice of encapsulant had a substantial impact on leakage current. Different encapsulant samples played a crucial role in influencing the overall behavior of leakage current in the minimodules. Notably, minimodules utilizing POE (Polyolefin Elastomer) as the encapsulant demonstrated greater resistance to leakage current compared to those using EVA (Ethylene Vinyl Acetate) as the encapsulant.

These findings highlight the importance not only of the solar cell technology but also the choice of encapsulant when designing photovoltaic modules. While TOPCON

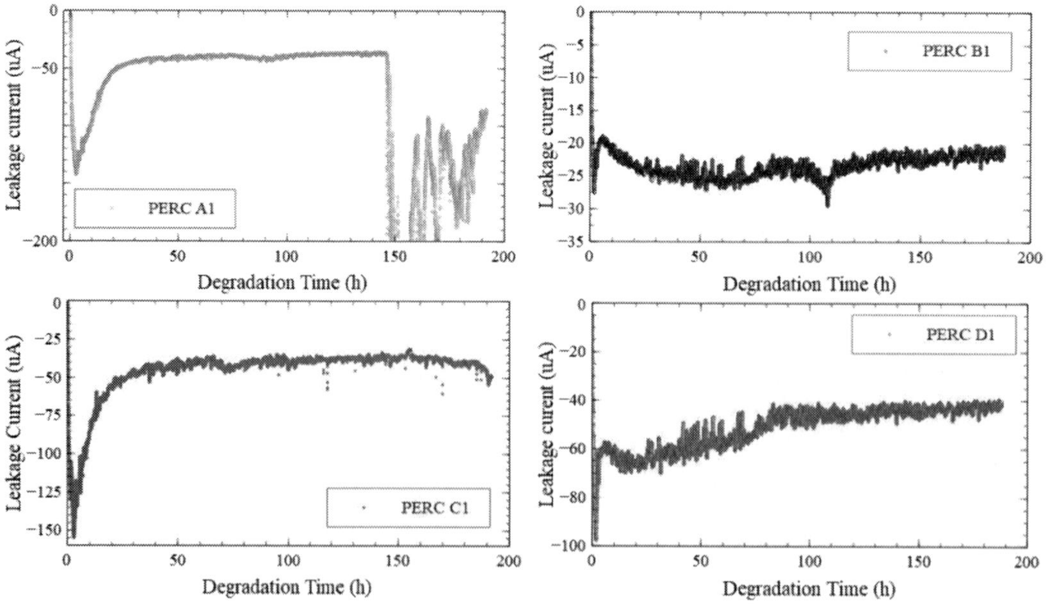

Figure 5: Measurement of leakage current in samples of PERC minimodules.

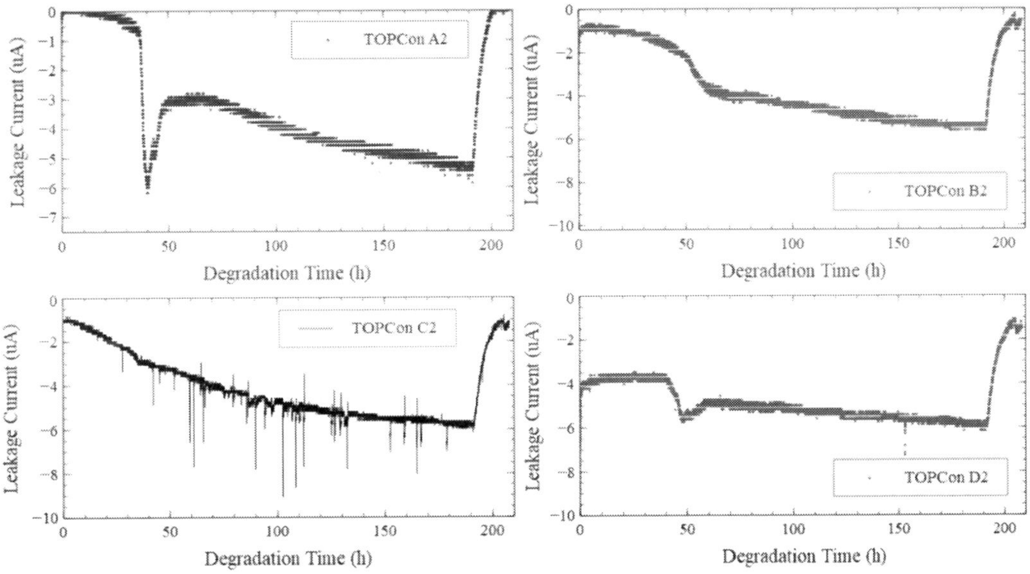

Figure 6: Measurement of leakage current in samples of TOPCon minimodules.

cells inherently displayed lower levels of leakage current, selecting an appropriate encapsulant, such as POE instead of EVA, further improved resistance to leakage current and contributed to superior module performance.

4 CONCLUSIONS

The study can conclude that the choice of encapsulant directly impacts the module's susceptibility to PID, with some encapsulants shown to be more effective in preventing potential-caused transmission than others. A switch to more effective encapsulants can increase the durability of photovoltaic modules, potentially resulting in longer system life and significantly reducing handling over time. For PV system owners, the key takeaway may be that careful selection of encapsulants can improve return on investment (ROI) due to greater efficiency and more effective reduction of handling over time. Regarding the specific PID resistance of the indicated encapsulants, EPE (Ethylene Propylene Ethylene) is often considered the most resistant encapsulant among the three. EPE has superior dielectric and chemical properties that make it less susceptible to manipulation caused by potential. Its molecular structure is less permeable to moisture and more resistant to high temperatures, providing greater electrical insulation and chemical resistance, contributing to its ability to prevent PID. However, it should be noted that EPE can be more expensive compared to EVA and POE. The choice between these encapsulants must be based on the specific needs of a photovoltaic project, considering factors such as cost, PID resistance, and environmental conditions.

ACKNOWLEDGEMENTS

This work was supported by BYD Energy Brazil (through the PADIS/MCTI program, Funcamp project N.5779)

REFERENCES

[1] S. Kashyap, J. Madan, R. Pandey, e R. Sharma, "Comprehensive Study on the Recent Development of PERC Solar Cell", em *2020 47th IEEE Photovoltaic Specialists Conference (PVSC)*, jun. 2020, p. 2542–2546. doi: 10.1109/PVSC45281.2020.9300985.

[2] J. Kim, S. Lee, S. Chowdhury, e J. Yi, "A Brief Review of Passivation Materials and Process for High Efficiency PERC Solar Cell", *Trans. Electr. Electron. Mater.*, vol. 23, nº 1, p. 1–5, fev. 2022, doi: 10.1007/s42341-021-00366-5.

[3] H. Yousuf *et al.*, "A Review on TOPCon Solar Cell Technology", *Current Photovoltaic Research*, vol. 9, nº 3, p. 75–83, set. 2021, doi: 10.21218/CPR.2021.9.3.075.

[4] D. K. Ghosh *et al.*, "Fundamentals, present status and future perspective of TOPCon solar cells: A comprehensive review", *Surfaces and Interfaces*, vol. 30, p. 101917, jun. 2022, doi: 10.1016/j.surfin.2022.101917.

[5] B. Kafle, B. S. Goraya, S. Mack, F. Feldmann, S. Nold, e J. Rentsch, "TOPCon – Technology options for cost efficient industrial manufacturing", *Solar Energy Materials and Solar Cells*, vol. 227, p. 111100, ago. 2021, doi: 10.1016/j.solmat.2021.111100.

[6] L. Wang, Y. Tang, S. Zhang, F. Wang, e J. Wang, "Energy yield analysis of different bifacial PV (photovoltaic) technologies: TOPCon, HJT, PERC in Hainan", *Solar Energy*, vol. 238, p. 258–263, maio 2022, doi: 10.1016/j.solener.2022.03.038.

[7] I. M. Kwembur, J. L. Crozier McCleland, E. E. van Dyk, e F. J. Vorster, "Detection of Potential Induced Degradation in mono and multi-crystalline silicon photovoltaic modules", *Physica B: Condensed Matter*, vol. 581, p. 411938, mar. 2020, doi: 10.1016/j.physb.2019.411938.

[8] M. R. M. Neves, G. C. S. Prym, e M. G. Villalva, "DEGRADAÇÃO INDUZIDA POR POTENCIAL (PID) EM MÓDULOS FOTOVOLTAICOS: UMA BREVE REVISÃO", *Congresso Brasileiro de Energia Solar - CBENS*, p. 1–10, ago. 2022.

[9] V. Naumann, C. Hagendorf, e K. Ilse, "On the Discrepancy between Leakage Currents and Potential-Induced Degradation of Crystalline Silicon Modules", 2014, doi: 10.4229/28thEUPVSEC2013-4DO.3.2.

[10] V. Kumari *et al.*, "Estimation of potential induced degradation in solar Mini-modules", *Materials Today: Proceedings*, vol. 30, p. 229–233, 2020, doi: 10.1016/j.matpr.2020.06.258.

[11] Wei Luo *et al.*, "Potential-induced degradation in photovoltaic modules: a critical review", *Energy & Environmental Science*, vol. 10, nº 1, p. 43–68, jan. 2017, doi: 10.1039/C6EE02271E.

[12] N. G. Dhere, N. S. Shiradkar, e E. Schneller, "Evolution of leakage current paths in MC-Si PV modules from leading manufacturers undergoing high-voltage bias testing", *IEEE Journal of Photovoltaics*, vol. 4, nº 2, p. 654–658, mar. 2014, doi: 10.1109/JPHOTOV.2013.2294764.

THE INFLUENCE OF SALINITY ON POTENTIAL-INDUCED DEGRADATION (PID) IN MONOCRYSTALLINE SILICON PHOTOVOLTAIC MODULES

Crestani, T.[1]; Santos, H. F.[1]; Mendes, E.[1]; Carvalho, G.S[1]; Oliveira, M.S[1]; Neves, M. R. M.[2]; Garcia, R.M[1]; Miranda, C. G[1].

[1]Build Your Dreams Company - BYD, Dept. Research and Development - R&D, Campinas, Brazil
[2]School of Electrical and Computer Engineering, Dept. of Systems and Energy - DSE, Campinas, Brazil
thais.crestani@byd.com

ABSTRACT: Photovoltaic modules, when installed in a tropical climate like Brazil, suffer specific degradation due to the hot and humid climate we have here throughout the year. Over time, photovoltaic modules, due to a series of degradation mechanisms, reduce their energy conversion capacity. Knowing the degradation mechanisms of photovoltaic modules, and knowing the influence of environmental factors on the degradation rate of these devices, this work was carried out with the aim of investigating the correlation between salinity, photovoltaic modules installed in coastal regions, and resistance to the PID (potential induced degradation) effect of photovoltaic modules constructed with full-cell crystalline silicon cells. The photovoltaic module exposed to salinity (called conditioned) was tested in a salt spray chamber for seven days, the module (called unconditioned) was not exposed to salinity and both were subjected to the PID degradation test for 96h, electrically characterized and then exposed. for another 96h, totaling 192h of tests. It is concluded that the time of exposure to salinity of the conditioned sample did not influence the degradation induced by potency.
Keywords: PV modules, Solar Energy, Potential-Induced Degradation (PID), Salinity.

1 INTRODUCTION

Photovoltaic modules, also called solar panels, are devices that convert light into electrical energy in the form of direct current through the photovoltaic effect. In general, photovoltaic devices are assembled using aluminum structures, tempered glass, encapsulating materials, photovoltaic cells and backsheet. The main element is the cells, they effectively convert energy, the other elements have the function of protecting and enabling the passage of electrical current.

Photovoltaic modules installed in tropical regions face different environmental conditions. The tropical climate presents several challenges for photovoltaic modules, high humidity throughout the year is the biggest threat to their durability, while high temperatures and highly diffused light conditions can negatively affect their output power [1].

Over time, photovoltaic modules suffer degradation and reduce their energy conversion capacity. In the most common devices constructed from c-Si crystalline silicon cells, the main effects that cause efficiency reduction involve Light Induced Degradation (LID) as well as Potential Induced Degradation – PID). The first mechanism is associated with excess illumination above the bandgap range, causing a decrease in short circuit current (I_{SC}) and open circuit voltage (V_{OC}) due to minority carrier recombination [2]. In the case of the second, the reduction in performance is strongly related to the difference in electrical potential between the modules, causing leakage currents to flow from the frame to the solar cells, or vice versa [3].

Many investigations were conducted in order to understand the influence of environmental conditions such as temperature, humidity, condensation, exposure to light. In this context, results were obtained that demonstrate a strong correlation to higher PID degradation rates with increasing temperature and humidity in modules using crystalline silicon cells [4]. Degradation profiles confirmed in the study of [5] show that PID is accelerated by certain types of salt spray preconditioning. For PID acceleration, sodium (Na) ions penetrating ambient PV modules can also cause degradation according to proposed PID mechanisms, because sodium ions existing in the front glass cause PID. Furthermore, this type of degradation may indicate the phenomenon of wear after long exposure in the field (mainly close to the coast).

Locations close to the ocean or marine environments have a lot of salt and water vapor mixed with acidic and basic substances (atmosphere), acting directly on the surface of the module. These chemical agents can eventually degrade some components of the photovoltaic module (corrosion of metallic parts and deterioration of non-metallic components due to the assimilation of salts), causing degradation of the module's power and even the module's electrical safety, impairing its operation. Figure 1 shows how salt deposition can occur on the module surface. In this study [6], corrosion tests were carried out in a closed laboratory climatic chamber, as they allow the simulation of an aggressive environment under which the module will be exposed in a short period, compared to when it will be in the field. With this method it is possible to obtain results similar to real conditions.

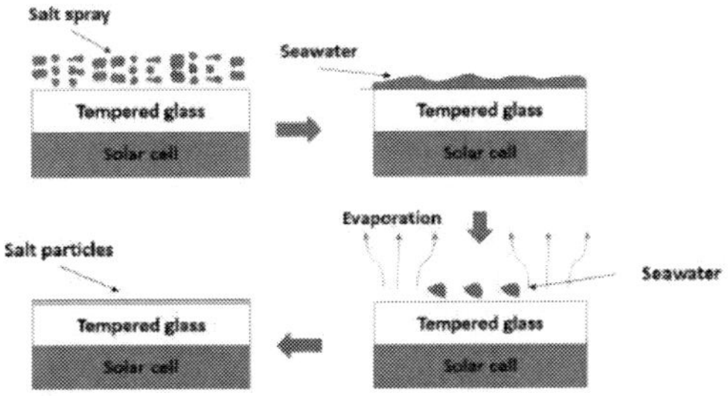

Figure 1: Salinity effect test process [6].

For p-type solar cells, the charged silicon nitride (SiNx) layers are expected to be negatively charged to reach degradation saturation. However, for n-type cells, degradation saturation is reached when positively charged by the SiNx layers. When comparing the polarization of the PID type in c-Si cells with layers of silicon dioxide

(SiO_2) it was found that in cells of this type degradation is greater and that it saturates in less time than in cells of type p. PID degradation depends on several factors, such as the quality of chemical passivation, the depth of the junction and the structure of the solar cell used in the photovoltaic module [7].

P-type silicon solar cells with PERC structure (passivated emitter and rear cell) in bifacial photovoltaic modules show potential-induced degradation (PID) especially at the rear. The study of the impacts of different layers of back passivation with the intention of generating anti-PID and altering the electrical properties of bifacial PERC cells carried out by Ma et al. (2022). Both simulation and characterization showed that the introduction of SiNx films with high refractive index and additive oxide films favored the reduction of the rear PID effect. Along with the well-resolved front anti-PID technique in the industry, the new rear passivation layer presented lays a foundation for the efficient applications of bifacial p-Si PERCs [8].

Knowing the degradation mechanisms of photovoltaic modules, and knowing the influence of environmental factors on the degradation rate of these devices, the objective of this work was to investigate the correlation between salinity and resistance to the PID effect in photovoltaic modules built with full crystalline silicon full-cell.

2 METHODS

The testing methodology was carried out with samples, "conditioned" exposed to salinity and unconditioned not exposed to salinity. Both samples were electrically characterized in the solar simulator model PASAN Highlight AAA+++, provided by the company Meyer Burger, Figure 2, calibrated with a Golden. The so-called "conditioned" sample was subjected to the salt spray test, according to method 3 of the IEC 60068-2-52 standard, with a 7-day cycle, using a concentration of 50 g/L in a saline chamber provided by the company Equilam, model SS3000, Figure 3.

Figure 2: Solar simulator model PASAN Highlight AAA+++, Meyer Burger.

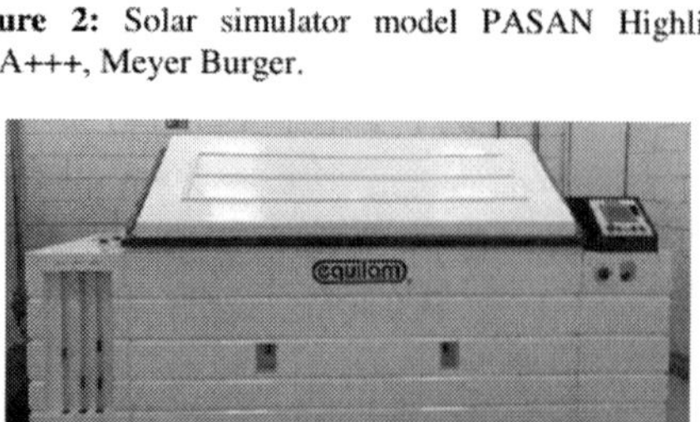

Figure 3: Saline chamber supplied by Equilam, model SS3000.

The two samples, conditioned and unconditioned, were subjected to the Induced Potential Degradation test - PID according to item MQT 21 of the IEC 61215-2021 standard using a climatic conditioning chamber provided by the company Hototech, model HTPV-16, Figure 4, with temperature of 85°C and relative humidity of 85%, -1500 VDC, for an uninterrupted period of 96 hours, followed by another cycle of 96 hours, totaling the test period of 192 hours.

Figure 4: Climate conditioning chamber, Hototech, model HTPV-16.

3 RESULTS AND DISCUSSIONS

Samples of photovoltaic modules with a nominal power of 400 W/m², built from MonoPERC M3 cells, 3.2 mm front glass, with ethylene vinyl acetate (EVA) encapsulation, 430 and 410 g/m² of grammature - front and rear, respectively, "full-cell" architecture. The equipment, initially, was also exposed to environmental conditions in external exposure (light, heat and humidity) during a period of 10 days to stabilize the output power. The samples were electrically characterized using a PASAN Highlight AAA+++ solar simulator, supplied by Meyer Burger, calibrated with Golden GDP221010-2. The so-called "conditioned" sample was subjected to the saline mist test, according to method 3 of the IEC 60068-2-52 standard [9], with a 7-day cycle, using a concentration of 50 g/L in a saline chamber provided by the company Equilam, model SS3000. Then, both samples were submitted to the PID test, according to item MQT 21 of the IEC 61215:2021 standard [10] using a climate conditioning chamber provided by the company Hototech, model HTPV-16, with a temperature of 85 ° C and relative humidity of 85%, -1500 VDC, for an uninterrupted period of 96 h, followed by another cycle of 96 h, totaling the test period of 192 h.

The PID test was carried out to simulate the climatic conditions that can cause this type of degradation in monocrystalline silicon photovoltaic modules. The test was performed by applying a high voltage potential to the module and exposing it to high humidity conditions. You can then measure the decrease in the module's power output and identify any other problems such as hot spots or cracks in the cells. PID degradation can occur when a PV module is exposed to high voltage potentials and high humidity conditions, which can cause a decrease in module output power, with the salinity, test the influence of sodium chloride concentration (conditioning) in photovoltaic modules. For the test, the output power of the module was measured before and after conditioning in exposure to sodium chloride saline solution (NaCl). Module performance was evaluated by measuring electroluminescence in order to identify any problems such as hot spots or cracks in the cells.

Figure 5 shows the evolution of the normalized power for the tested samples. In this context, it is possible to notice the drastic reduction in the performance of both samples, greater for the sample conditioned at 96h, but equal at 192h. The normalized power reduction of both items is directly associated with PID degradation.

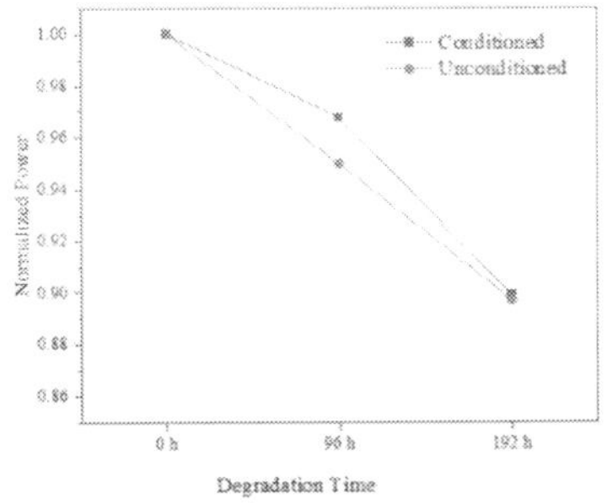

Figure 5: Evolution of the normalized potency of the samples after the degradation test.

In the electroluminescence image in Figure 6 in (a) the image of the conditioned sample, that is, exposed to the salinity test and in (b) the image of the unconditioned sample, that is, exposed only to the degradation chamber. Cells at the edges of modules are more prone to potential-induced degradation due to the ease of migration of ions from the glass into the cell. It is noted that in both samples there is evidence of the beginning of degradation.

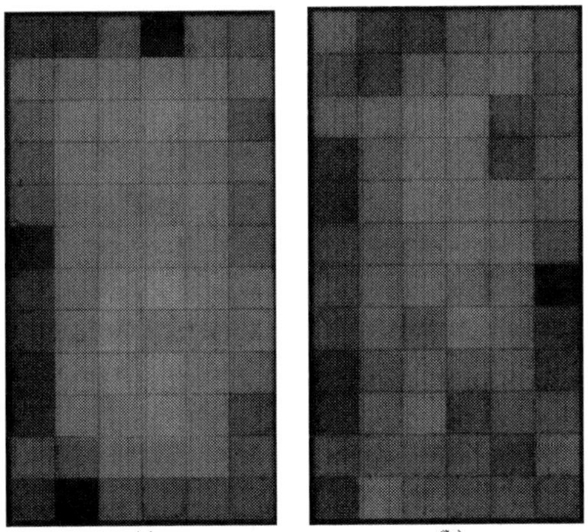

(a) (b)

Figure 6: Electroluminescence images in (a) conditioned and (b) unconditioned.

When analyzing the electrical parameters of the conditioned and unconditioned photovoltaic module, it is noted that in Figure 7 (a) the normalized short-circuit current (Isc) for the unconditioned module was greater than the conditioned one. Figure 7 (b) presents the normalized open-circuit voltage (V_{OC}) for the samples throughout the tests. Note that there was no significant variation in V_{OC} after exposure to salinity and the performance of the PID test for 96h and 192h. In Figure 7 (c) it is observed that the resistance in series of both samples had a significant increase.

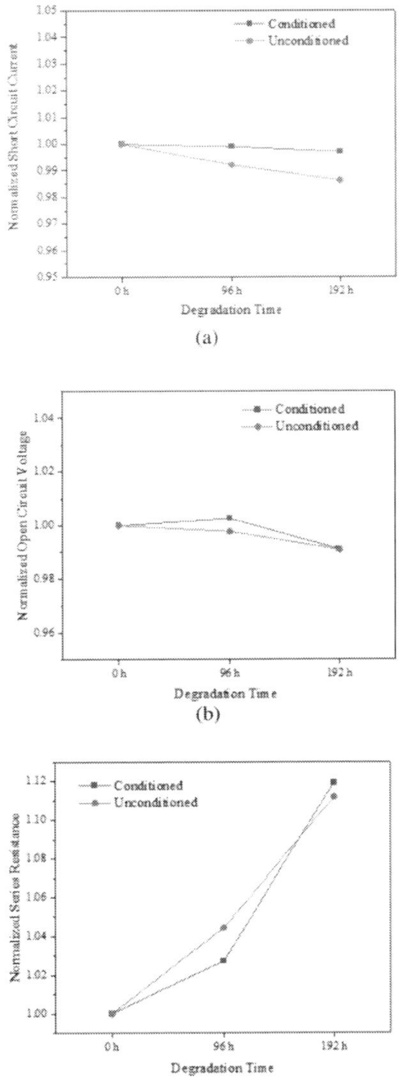

Figure 7: Graphs of short-circuit current (a), open-circuit voltage (b) and series resistance (c) normalized as a function of degradation time.

4 CONCLUSION

The influence of salinity on potential-induced degradation (PID) in monocrystalline silicon photovoltaic modules was analyzed with respect to the electrical parameters and the electroluminescence image of the modules, conditioned (exposed to salinity) and unconditioned. The so-called "conditioned" sample was subjected to the saline mist test, with a 7-day cycle, using a concentration of 50 g/L in a saline chamber. Then, both samples were submitted to the PID test, using a climate conditioning chamber, with a temperature of 85 °C and relative humidity of 85%, -1500 VDC, for a period of 96 uninterrupted hours, followed by another cycle of 96 h, totaling the test period of 192 h. It is concluded that the time of exposure to salinity of the conditioned sample did not influence the degradation induced by potency.

5 ACKNOWLEDGMENT

The authors are grateful for the support of BYD (Build Your Dreams) Energy Brazil for developing this type of research in its R&D department in partnership with MCTI (Ministry of Science, Technology and Innovation) through incentives from PADIS (Brazilian Semiconductor Industry Technological Development Program).

6 REFERENCES

[1] Walsh, TM, Xiong, Z., Khoo, YS, Tay, AAO, Aberle, AG (2012). Singapore Modules - Optimized PV Modules for the Tropics. Energy Procedia, 15, pp. 388-395.

[2] Lindroos, J. and Savin, H. (2016). Review of light-induced degradation in crystalline silicon solar cells. Solar energy Materials and Solar Cells. See 147, pp. 115-126.

[3] Luo, W., Khoo, YS, Hacke, P., Naumann, V., Lausch, D., Harvey, SP, Prakash, J., Chai, J., Wang, Y., Aberle, AG, Ramakrishna S. (2017). Potential-induced degradation in photovoltaic modules: a critical review. Energy & Environmental Science, vol. 10. DOI: 10.1039/c6ee02271e.

[4] Hoffmann, S. and Koehl, M. (2012). Effect of humidity and temperature on the potential-induced degradation. Progress in photovoltaics: research and applications. DOI: 10.1002/pip.2238.

[5] Suzuki, S., Nishiyama, N., Yoshino, S., Ujiro, T., Watanabe, S., Doi, T., Masuda, A. and Tanahashi, T. (2015) Acceleration of potential-induced degradation by salt-mist preconditioning in crystalline silicon photovoltaic modules. Japanese Journal of Applied Physics 54. DOI: 10.7567/JJAP.54.08KG08.

[6] Neves, M., Silveira, A., Garcia, R. Alvarez, H. Marques, F., and Villalva, M. "Effects of salt spray on csi photovoltaic modules in the Brazilian region," in 2023 IEEE Photovoltaic Specialists Conference (PVSC 50). IEEE, 2023, pp. 1–6.

[7] Yamaguchi S., Jonai S., Nakamura K., Marumoto K., Ohshita Y., Masuda A. (2022). Polarization-Type Potential-Induced Degradation in Front-Emitter p-Type and n-Type Crystalline Silicon Solar Cells. ACS Omega 2022, 7, pp. 36277 – 36285. DOI: 10.1021/acsomega.2c03866.

[8] Ma, S., Tong, R., Wu, XY, Li, ZP, Kong, XY, Shen, WZ (2022). Effective way to reduce rear-side potential-induced degradation of bifacial PERC solar cells. Solar Energy Materials and Solar Cells, v. 239. DOI:10.1016/j.solmat.2022.111687.

[9] IEC 60068-2-52:2018 Environmental testing Part 2: Tests – Test Kb: Salt mist, cyclic (sodium chloride solution) (IEC 60068-2-52:2017). BS EN IEC 60068-2-52:2018.

[10] IEC 61215-2: 2021 Terrestrial photovoltaic (PV) modules - Design qualification and type approval. Part 2: Test procedures MQT 21. BS EN IEC 61215-2:2021.

NUMERICAL DESIGN STUDY FOR DESERT CLIMATE APPLICATIONS OF BIFACIAL PV MODULES

Matthias Pander*(1), Bengt Jaeckel(1), Stephan Großer(1), Matthias Ebert(1), Amir A. Abdallah(2)
(1)Fraunhofer-Center for Silicon-Photovoltaics (CSP), Halle (Saale), Germany
(2)Qatar Environment and Energy Research Institute (QEERI), Hamad bin Khalifa University (HBKU), Qatar Foundation, P.O. Box 34110, Doha, Qatar
*Contact: +49 345 5589 5215, matthias.pander@csp.fraunhofer.de

ABSTRACT: Standard crystalline silicon solar modules have shown to be robust under moderate climate conditions but with the need for sustainable energy production, both the quantity of PV installations and the application in other climatic zones must be investigated. Due to the high annual amount of solar irradiation (2 to 3 times as much as in Central European climates), desert regions are actually ideal for solar applications in terms of potential energy yield. However, there are various challenges for the long-term reliability of the modules and components. In addition to well-known short-term effects, such as, soiling losses, the thermo-mechanical stress in desert regions is very challenging due to, from one hand, the higher ambient temperatures that could reach on average 40°C, while the module temperature could reach 70°C. From the other hand, the large temperature fluctuations between day and night that occur in desert climate induce thermo-mechanical stress.

In parallel, a significant increase in cell size and thus change in module size is taking place. The long-established M2 wafer format is already being replaced by M6 and represents only a transition size to M10 and M12 wafers. In desert applications, bifacial cells have a particularly high potential for maximum utilization of the available irradiation due to the high albedo.

In this paper, different cell sizes, cell spacings and cell connector thicknesses are analyzed and compared under thermo-mechanical loading using a finite element simulation. The aim is to find an optimum between the possible power gains by improving the cell to module (CTM) ratio, the area efficiency and module reliability in desert climate.

Keywords: Finite Element Simulation, FEM, Fatigue, Crystalline Silicon

1 INTRODUCTION

Standard crystalline silicon solar modules have proven reliable in temperate climates, but given the need for sustainable energy production, both the quantity of PV installations and the application in other climates need to be further increased. Due to the high annual solar radiation (two to three times more than in central European climates), desert regions are ideal for solar applications in terms of potential energy yield. However, there are several challenges for the long-term reliability of the modules and components. In addition to the well-known short-term effects, such as soiling losses, the thermomechanical stress in desert regions is very demanding, on the one hand because of the higher ambient temperatures, which can reach 40°C on average, while the module temperature can reach 70°C. On the other hand, the large temperature fluctuations between day and night that occur in desert climates lead to thermo-mechanical stress.

According to ITRPV a significant increase in cell size and thus change in module size is taking place [1]. The M2 and M3 cell sizes on the market are increasingly losing relevance and there will be predominantly M6, M10 and M12 formats, with mainly half cells being used up to M10. For M12, third cells will also become relevant.

A large proportion of current cell technologies already have bifacial properties and the bifaciality factors will further increase with upcoming technologies [1], which offer additional advantages in desert regions due to the potentially high albedo. The module structure for bifacial applications is assumed to be a glass-glass structure with 2 mm glass. It is known from previous studies that cell spacing plays an important role for reliability and is also important for characteristic values, such as, CTM [2]. Another factor is the number of busbars or interconnector ribbons or wires used. The number depends on cell size and cross section of the interconnector. For M10, the prediction is at least 9 busbars [1].

In this paper the finite element method (FEM) is used to analyze the thermo-mechanical loading of the cell connections for different cell formats, cell spacing and connector thicknesses. The scope is to find an optimum between the possible power gains by improving the cell to module (CTM) ratio, the area efficiency and module reliability.

2 MODEL SETUP AND ASSUMPTIONS

2.1 Simulation of interconnector stress

A parameterized finite element model was set up for the investigations. Since a large number of variants are to be investigated, a simplified 2D cross-section is analyzed. Current and future wafer and cell sizes are taken from the ITRPV [1]. The interconnection technology studied here is flat standard cell interconnectors. The connector cross-section depends on the intended connector number for current transport and the allowable electrical losses [3]. To simplify the analysis, only the thickness will be changed and the width will be adjusted accordingly. A larger width may have an effect on fatigue crack growth, but this is neglected here. An unrestricted transfer to round wires is not possible with the current state of results, but it can be assumed that the relationships are comparable. However, round wires can also buckle out of the plane of observation during loading, so that a 3D model would be required. The following basic configuration is defined for analyses and comparisons:

- cell thickness: 180 μm
- front encapsulant thickness: 0.4 mm
- back encapsulant thickness: 0.4 mm
- glass thickness: 2 mm
- backside glass thickness: 2 mm

Five cells in a row are simulated in the global model which is sufficient to accurately calculate the cell shift and stress on the ribbon.

The selected parameters result in a simulation matrix with 126 variants:

- Wafer sizes: M2, M3, M6, M6_HC, M10_HC, M12_HC, M12_TC (HC = half-cell, TC = third cell) → 7 variants
- Cell gaps (distance between cells): 1, 2, 3, 5, 10, 20 mm → 6 variants
- Thickness of ribbon: 0.16 mm, 0.2 mm, 0.24 mm → 3 variants

For the comparison of designs the temperature cycle test (TCT) according to IEC 61215 MQT 11 is used as the load case. The applied temperature profile is shown in Figure 1.

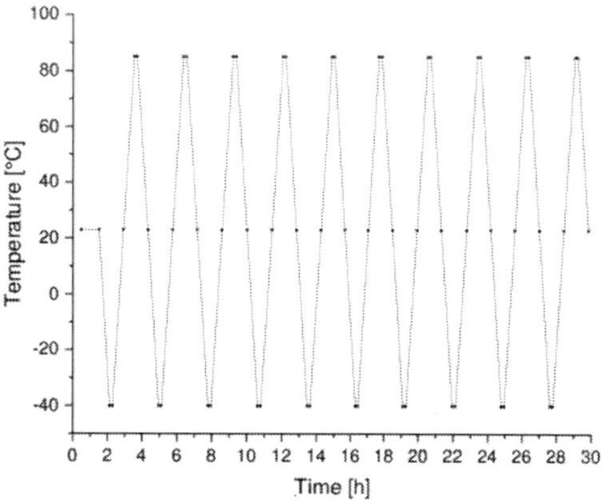

Figure 1: Defined temperature profile according to the IEC 61215 MQT 11

Material data is taken from literature and internal measurements [4]. The calculation is performed in two steps. First, the cell displacements are calculated in a global model and then the local stresses at the connectors are calculated in a detailed submodel. Plastic strain energy density is evaluated as an indicator of damage ([4], [5]). The mesh is adapted according to the required details. Especially in the detailed model, an increased mesh density is required to correctly determine the plastic strain from the simulation. The meshed cell interconnections of different cell gaps are shown in Figure 2.

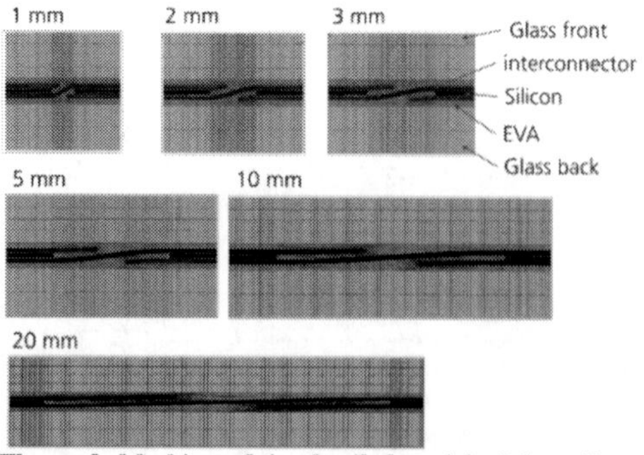

Figure 2: Meshing of the detailed model of the cell gap with the interconnector ribbon.

The result for a calculation of 10 cycles is shown in Figure 3. The maximum load occurs at the inflection points of the buckled connector. One can determine an average increase in plastic strain energy density from the calculation and thus extrapolate to any number of temperature cycles. For the following evaluation, 200 cycles are considered.

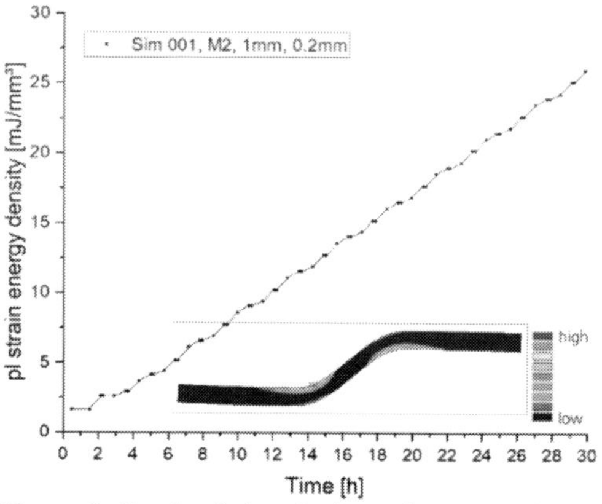

Figure 3: Result of the plastic strain energy density (plsed) within the interconnector at the M2 cell size, 1 mm cell gap and 0.2 mm ribbon thickness for 10 cycles

2.2 CTM and module efficiency calculation

The estimation of module power and efficiency of designs is based on the CTM model from [6]. In this model, thirteen factors k_j are defined that describe gains and losses when the cell is transferred to the module. Based on the sum of the power of the individual cells, the module power is determined.

$$P_{mod} = P_{[0]} \prod_{j=3}^{13} k_j$$

For better comparability the following parameters are fixed for the CTM calculation:

- 22.2 % cell efficiency regardless of format
- Pseudo square cells
- Number of interconnectors (n_{bb}) is fixed per format:
  - M2: $n_{bb}=3$
  - M3: $n_{bb}=5$
  - M6: $n_{bb}=9$
  - M10: $n_{bb}=12$
  - M12: $n_{bb}=15$
- Number of cells per module: 60 FC, 120 HC, 180 TC
- Module border top and bottom: 27 mm
- Module border left and right: 25 mm
- Module area in the middle for HC/TC: 15 mm

Further assumptions and simplifications are applied regarding the k-factors. Since only changes in cell spacing and connector geometry take place, only the following factors in the CTM model [3] are affected:

- k_1 – inactive module border area:

$$k_1 = 1 - \frac{A_{border}}{A_{mod}}$$

- k_2 – inactive area between cells and strings

$$k_2 = 1 - \frac{A_{cell,spacing}}{A_{mod}}$$

- k_7 – Shading of the cell by interconnector ribbons

$$k_7 = 1 - \frac{A_{ribbon} - A_{front,pad}}{A_{cell}}$$

- k_{11} – Coupling gain from surrounding, two variations are considered here:
    - transparent backside, resulting in $k_{11} = 1$
    - white grid where optical gain is depending on cell distance and gain factors are assumed to follow data from Fig. 8 in [6]
- k_{12} – Cell interconnection losses:

$$\Delta P_{[12]} = n \cdot \left(\Delta P_{ribbon,front} + \Delta P_{ribbon,back} + \Delta P_{cellgap}\right)$$

$$\Delta P_{ribbon} = \sum_{k=1}^{n_{bb}} \frac{1}{3} \cdot I_k^2 \cdot \overline{\rho_{bb+ribbon}} \cdot l_{ribbon}$$

Nomenclature

A_{mod} – total module area
A_{border} – area from module edge to cell edge
$A_{cell,spacing}$ – area between cells
A_{cell} – total cell area
A_{ribbon} – total area of the ribbon
$A_{front,pad}$ – area of the front metallization of the busbars
$\overline{\rho_{bb+ribbon}}$ – effective resistivity of busbar metallization and ribbon
l_{ribbon} – length of ribbon excluding cell distance area
$\Delta P_{ribbon,front/back}$ – electrical losses generated in busbar and ribbon on cell front / backside
n – amount of cells in module
I_k – current collection from one busbar from one cell side to the other
n_{bb} – amount of busbars

The other factors are assumed to be unaffected by the changes made or the change in the factor is neglectable:

- k_3 - Reflection of light at the front interface of the module
- k_4 - Absorption of light in the front cover
- k_5 - Reflection of light at the interface between front cover and encapsulation material
- k_6 - Absorption of light in the encapsulation material
- k_8 - Reduced reflection of the cell due to encapsulation
- k_9 - Reflection of light from the cell metallization on the active cell area
- k_{10} - Reflection of light from the interconnector ribbons on the active cell area
- k_{13} - string interconnection losses

Based on these assumptions, an optimal ribbon width (wribbon) can be determined for the different variants with the CTM model, which leads to the maximum module power. From this power, the module efficiency can be determined with the resulting module area.

3 ANALYSIS OF INTERCONNECTOR STRESS

Figure 4 shows the plastic strain energy density calculated for a TC200. It can be readily seen that cutting of the cells reduces the interconnector stress. Increasing the cell size increases the stress again, but an M12 HC (half-cell) is still better than M2 FC (full cell). Narrow cell gaps lead to higher stresses on the interconnector and will reduce the interconnector lifetime. For large cell distances (10 and 20 mm), it was found that only elastic stresses occur in the interconnector and therefore cause almost no damage. Thinner ribbons (here 0.16 mm) show especially for narrow cells gaps (1 to 3 mm) lower stresses compared to 0.2 mm thickness. This advantage vanishes for ≥ 5 mm cell distance. At high cell spacings, the stresses are actually higher, but at a generally low level. For the thick ribbon (0.24 mm) the stress is higher for very small cell gap (1 mm). Surprisingly there are not much difference for the moderate cell distances compared to 0.2 mm. For high cell gaps (>10 mm) the thick ribbon has small advantages.

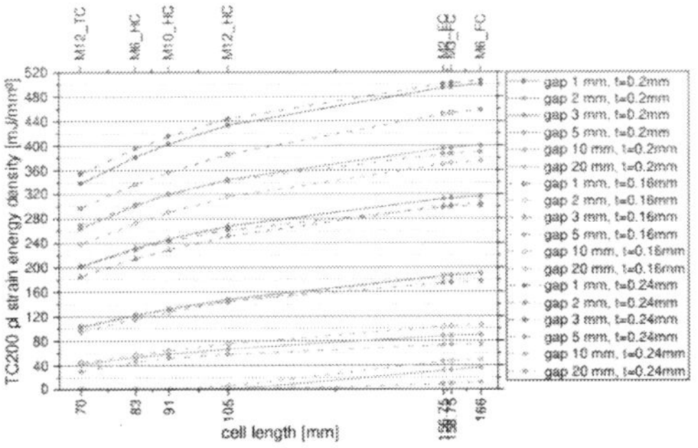

Figure 4: simulated plastic strain energy density after TC200 for the selected parameters.

4 EVALUATION OF DESIGNS

The determination of the optimum ribbon width for a combination of cell gap and ribbon thickness is shown in Figure 5 for modules with M6 full and half-cells. There is an optimum at which the lowest possible electrical losses and the lowest possible shading (optical gains) occur. Without taking internal reflections into account, the module power remains approximately constant as the module spacing increases. Consequently, the area efficiency decreases (Figure 6). The optical gains due to reflections increase the module power with increasing cell spacing. Gains up to 3 mm spacing are particularly clear. The required ribbon width must increase slightly since cell spacing increases the electrical resistance between the cells. Comparing M6 half-cell and M6 full-cell designs reveals the well-known fact that cell cutting halves the current in the substring, thereby quartering electrical losses [3]. This overcompensates for the losses due to inactive areas and results in a higher area efficiency despite the larger module area. It should be noted that the range for optimum ribbon width is smaller for half-cells compared to full cells.

The maximum power at optimum ribbon width was determined and the module efficiency was calculated. The results for selected cell sizes are shown in Figure 6. The larger cells lead to a lower proportion of inactive area in the module, which consequently increases the module efficiency. In terms of module efficiency, the thick ribbons (t=0.24 mm) lead to the best results, since the required width is the smallest and thus the least shading occurs. It was also found that the required width sometimes is less than the height, which consequently leads to the use of round wires, as it is already common in the industry. Larger cell spacing without taking optical losses into account quickly leads to a reduction in module efficiency (~0.16 %/mm for the half cells). For the M12 third cell, the drop is even steeper (~0.23 %/mm). The optical gains can only compensate the losses by reducing the active area up to about 2 mm. After that, a steady reduction also takes place.

10.4229/EUPVSEC2023/3AV.2.18
020188-003

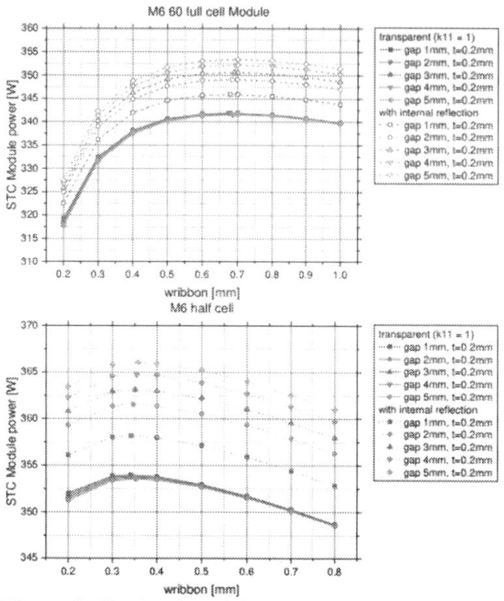

Figure 5: Simulated module power for M6 FC and HC (for t=0.2 mm, nbb=9) for transparent backside and reflective backside, determination of optimal ribbon width

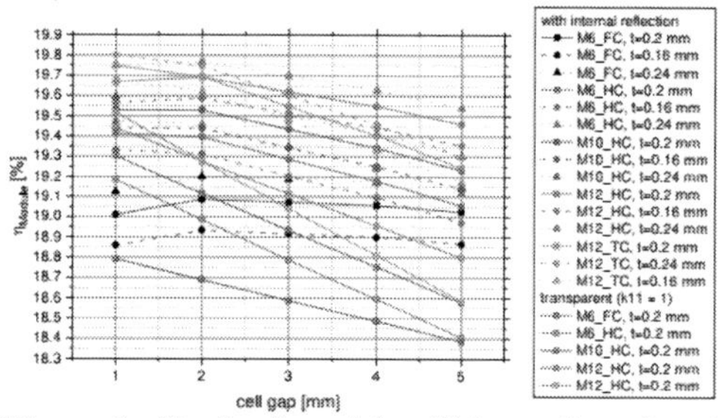

Figure 6: Simulated module efficiency for selected designs with optimal ribbon width (transparent only given for one width)

5 DISCUSSION

With respect to reliability the best design would be a high cell gap (5 mm and more), thin ribbon and a small cut cell (M6_HC or M12_TC). But this setup results in comparably low active area (respectively low module area efficiency), additional shading of active cell area because a thin ribbon needs a higher width, to keep electrical losses remain at the same level. However, the difference between the ribbon thickness above 3 mm is less significant compared to the cell gap. Comparing 1 mm and 3 mm cell gap results in almost 30 - 40 % less stress for the interconnector. The utilization of optical gains can contribute to increase the profitability of the more reliable designs in terms of performance. Highly reflective coatings of glass, so called white grid, are already available for this purpose. The consideration of the ribbon cross-section has been made only for 1000 W/m² irradiation. In desert regions, higher irradiation and a high albedo are partly present, so that resulting currents for 1200 to 1300 W/m² must be considered. Increasing the module current would result in slightly wider ribbons and thus reduce STC performance. Therefore, it is possible that modules optimized for maximum yield in the desert may have lower STC performance.

6 SUMMARY AND OUTLOOK

In this paper interconnector stress was simulated for diverse cell formats, ribbon geometries, and cell spacing. Modules with high power density due to low cell distance (<2 mm) show highest ribbon stress which can result in fatigue-related reliability issues especially in regions with high daily temperature changes, especially desert climates. As the size of the solar cells increases, the load on the cell interconnectors increases. Dividing the cells is beneficial for reducing the electrical losses and the load on the cell interconnectors. Increasing the cell spacing can significantly reduce the ribbon stress compared to other geometry changes but reduces module efficiency. The use of a reflective backside can compensate for losses in area efficiency due to increased cell spacing and increase reliability.

In further steps, simulations with specific real climate data can be performed to determine the load on the interconnectors of the identified design candidates as well as specific yields under the irradiation conditions with the goal of achieving the maximum yield with the highest possible long-term reliability.

7 REFERENCES

[1] ITRPV, "International Technology Roadmap for Photovoltaic (ITRPV)", 2022

[2] M. Pander et al., "Digital Prototyping – Application of Numerical Methods in Module Development", 36th EU PVSEC, 2019, 10.4229/EUPVSEC20192019-4BO.11.2

[3] H. Hanifi, "Comparison of optical gains and electrical losses in modules with different designs of partial cells in desert regions", 35th EU-PVSEC, 2017, 10.4229/EUPVSEC20172017-5BV.4.49

[4] M. Pander et al., "Lifetime estimation for solar cell interconnectors", 28th EU PVSEC, 2013, 10.4229/28thEUPVSEC2013-4CO.10.3

[5] D. Kujawski, Fatigue Failure Criterion based on strain energy density, Theoretical and Applied Mechanics, Vol. 7, 1, 1989, pp.15-22

[6] I. Haedrich et al., "Unified methodology for determining CTM ratios: Systematic prediction of module power", Solar Energy Materials & Solar Cells 131, 2014

[7] R. Meier et al., "Microstructural optimization approach of solar cell interconnectors fatigue behavior for enhanced module lifetime in extreme climates", Energy Procedia, Vol. 92, 2016, 10.1016/j.egypro.2016.07.020

8 ACKNOWLEDGEMENT

This publication was made possible by NPRP grant # NPRP11S-1220-170110 from the Qatar National Research Fund (a member of Qatar Foundation). Financial support by the Federal Ministry for Economic Affairs and Climate Action (BMWK) for funding the project "CTS1000plus" (FKZ: 03EE1028A) is gratefully acknowledged. The findings herein reflect the work, and are solely the responsibility, of the authors.

QUANTIFICATION AND MITIGATION OF THE PV-SOILING IN BIFACIAL MODULES CONFIGURATION

Brahim Aïssa*, Amir. A. Abdallah, Benjamin W. Figgis, Maulid M. Kivambe, Juan Lopez-Garcia
Qatar Environment & Energy Research institute (QEERI), Hamad Bin Khalifa University (HBKU), Qatar Foundation, P.O. Box 34110 Doha, Qatar.
Corresponding author: E-mail (B. Aïssa): baissa@hbku.edu.qa, Phone: + 974 44547475

ABSTRACT: Bifacial PV modules technology is emerging as a reliable solution to enhance the output power energy compared to traditional monofacial modules since both sides of the PV module absorb solar radiation. In addition, bifacial PV modules reduce the LCOE and might be a natural soiling mitigation approach, with small or no soiling in the rear side of the module. In this work, dust accumulation was measured for bifacial configuration based on uncoated and silica-coated glass coupons. The outdoor testing was carried out in the harsh desert conditions of the state of Qatar. Samples were characterized every two months to assess the soiling-rate and the silica coating degradation. Silica films were deposited by a reactive RF magnetron sputtering of a SiO_2 target under different oxygen to argon flow ratios [r(O2) = O2/Ar, ranging from 0 to 25%]. The refractive index (n), extinction coefficient, total transmission and reflectance were systematically investigated, and three-dimensional average roughness data construction was carried out to probe the surface morphology. Contact angle measurements were performed to assess the wettability of these silica coatings as a function of r(O2). The dust accumulation (soiling) was expressed in terms of the surface coverage (SC) by the dust particles and then correlated to the r(O2) and the tilt angle (θ)of these samples. Both front and rear sides of these coupons were considered and empirical equations representing the SC were deduced for each r(O2) as a function θ. The degradation of these silica coatings was assessed through the micro hardness (H) analysis and has shown that after 24 months of outdoor exposure, the films hardness decreased by only 5 %. Finally, a computational analysis based on the measured experimental optical properties were performed to simulate the photovoltaic (PV) device performance and a clear improvement in the photoconversion efficiency was observed with respect to r(O2). Several optimizations and simulations are currently ongoing to see the effects of soiling rate, albedo, rain, cleaning costs and frequency on the bifacial PV system.

1 AIM AND APPROACH

The aim is to develop and antireflection and antidust silica coatings with optimized properties to suit desert climate conditions and to assess this coating in real world conditions characterizing the desert environment of the state of Qatar. Subsequently, the developed coatings have been used to quantify the accumulated dust in bifacial configuration.

Material deposition and outdoor testing: In this work, the optical, microstructural, surface and mechanical properties of 100 nm silica thin films grown on a glass substrate by the radiofrequency (RF) sputtering technique at 200° C under different oxygen to total flow ratios [r(O2) = O2/Ar, ranging from 0 to 25%] were investigated. Once the growth of these silica layers was optimized, the coated glass coupons (in monofacial and bifacial configurations) were placed at different tilt angles in outdoor conditions at the outdoor test facility (OTF). OTF is a 35,000 m2 outdoor site with various plug & Play PV testing systems. The extreme environmental conditions at OTF (high irradiation, temperature, albedo, soiling, humidity leading to cementation, coastal marine aerosols, etc.) are unique and help to evaluate the performance of different solar technologies and devices and determine how to maximize their energy production and reliability [1,2]. After each period of 2 months, these coupons are taken back to assess the dust accumulation (in both front and rear sides) and to measure their stability/degradation through their mechanical properties by means of micro indentation.

2 SCIENTIFIC INNOVATION AND RELEVANCE

Anti-soiling coatings (ASC) is one of the leading soiling mitigation techniques thoroughly researched in the community. Ideally, ASCs have to be optically transparent, with anti-reflection and self-cleaning ability, non-toxic, stable, durable, cost-effective, and deployable at an industrial level. This rends ASC somehow the "Holy Grail" of the soiling community [3]. On the other hand, it was reported that bifacial PV harvests sunlight from both sides of the module, which can significantly increase the energy yield per module by up to 30% compared to monofacial and therefore, the LCOE, and studying the soiling effects in bifacial modules is critical since dust accumulation is one of the most impacting factors for PV power loss, and hence a relevant parameter to consider when designing the bifacial system. Soiling on bifacial PV systems is a complex problem and requires an in-depth investigation of installation parameters: such as tilt angle, mounting height, and rear/front side dust deposition rates [4]. In principle, all PV systems (bifacial or monofacial), experience a soiling effect; however, the effect on bifacial PV is less pronounced [5,6]. This is primarily due to a smaller amount of dust exposure on the rear side of the module in tilted systems or due to the vertical installation practice of bifacial technology. Quantifying the soiling in bifacial configuration was rarely investigated in the relevant literature and this work addresses these two key issues: ASC and soiling quantification in bifacial configuration.

3 RESULTS AND DISCUSSIONS

Figure 1a shows a photo of the OTF area, where dust caused PV energy to decrease by 0.49%/day on average each day without cleaning has been measured. Fig. 1b shows a summary of the tested samples, namely, uncoated and silica coated glass coupons, with different oxygen content r(O2), placed at different tilt angles. These silica coatings were first optimized before being assessed as anti-dust coating in outdoor conditions, where film thickness and r(O2) were investigated. Figures 2a-b show the representative SEM micrographs of the limit oxygen content values under study, namely, silica without the additional oxygen content (i.e., r(O2) = 0%) and silica with the richest oxygen content (i.e., r(O2) = 25%). In both cases, films seem dense, homogenous, and pinhole- and cracks-free with no voids observed and cover the entire substrate surface uniformly.

Figure 2c shows the multiple correlations between the refractive index and hydrophobicity properties, suggesting that adding oxygen to the silica film plays a role not only in terms of optical properties (i.e., films become more transparent) but also in their morphological and surface properties. Hence, a significant gain in terms of optical transmission and consequently self-cleaning capability is expected under specific weather conditions. Furthermore, coated coupons were placed in OTF to undergo real word conditions. Their stability/degradation was assessed through their mechanical properties using Vickers micro-indentations. Results are shown in Figure 2d. The mechanical testing demonstrated the following: (i) porous silica was found to be rather a hard material, which is only 10 % less hard than a tempered glass; (ii) more oxygen into silica led to a decrease in its hardness (i.e., likely becoming more porous), however, more hydrophilic; (iii) almost no degradation has been witnessed after 2 months exposure, and (iv) only 2.5 % of hardness-decrease has been measured after 12 months and about 5 % of hardness-decrease of 24 months outdoor exposure. Figure 4a presents selected SEM micrographs of the front and rear sides of the soiled silica coated coupons, after 2 months of outdoor exposure in OTF.

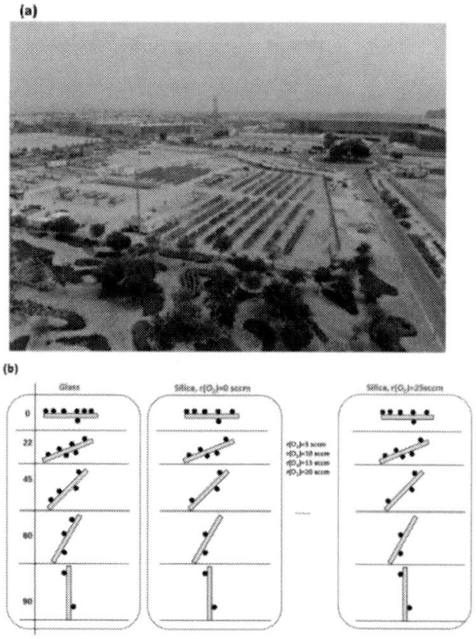

Figure 1: (a) Photo showing an overview of the OTF area located in the desert environment of the state of Qatar. (b) schematic depicting the different configurations of the tested coupons: (i) uncoated vs. coated, (ii) silica with different oxygen content, and (iii) various tilt angles.

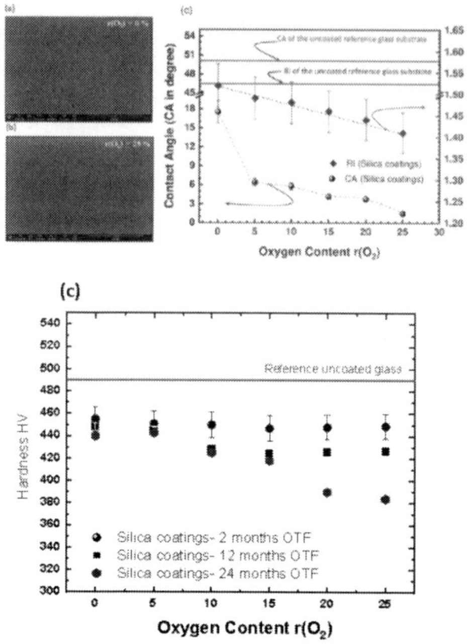

Figure 2: (a-b) FESEM of the magnetron sputtering deposited silica coatings with $r(O_2)$ = 0 and 25%, respectively. (c) Correlations between $r(O_2)$, contact angle and the refractive index of these silica. (d) Harness calculated onto an average of 10 micro indentations after 2 months, 12 months and 24 months in OTF.

Photos are showing representative samples with r(O2) = 5 %. Soiling was assessed by means of the SEM surface coverage (SC) by dust and was taken as a key-parameter to compare samples under different conditions (i.e., r(O2), tilt angle, front vs. rear side). As expected, higher tilt angle led to accumulating less dust, and the soiling was found to be proportional to the cosine of the tilt angle θ. Silica with higher oxygen content, showed higher hydrophilicity and collected less dust (without any cleaning). On the other hand, the quantification of the bifacial vs. monfacial soiling was found to satisfy the following empirical equation:

$$SC\ (Rear) = 15\%\ SC\ (Front) \times \cos(\theta) + 1.6 \quad \text{(Eq. 1)}$$

Additional tests are now in progress to expose these samples to longer outdoor periods to verify repeatedly this empirical equation. Finally, these silica coatings were simulated to be applied to solar cells as anti-reflection coatings (ARC) using SCAPS and OptiLayer software. Both the absorption coefficient and band gap of the silica layers were deduced from the experimental data (Results not shown in this abstract and will be thoroughly discussed in the full paper). A set of thicknesses between 20 and 200 nm was launched to optimize the ARC layer thickness. As calculated, adding an oxygen-rich $r(O_2)$ = 25%, ARC layer showed an improvement in the photoconversion efficiency (PCE) by ~ 27%, and was attributed to the reduction in light reflection (i.e., increase of photon absorption). The calculated values confirm that silica layers developed with a high oxygen concentration (i.e., r(O2) = 25%) result in a higher Jsc (47 ma/cm2) and Voc (0.750 V). Such performance is mainly due to the improvement of optical

10.4229/EUPVSEC2023/3AV.2.20
020190-002

properties and related to improving Jsc, as expected, due to less reflection, resulting in higher solar spectrum absorption in the Si absorber layer. In addition, OptiLayer software has been used to find out the minimum achievable reflectance for stacking of oxide layers. A two-layer antireflection design is simulated using this software. The desired wavelength range has been set from 420 to 1200 nm during this analysis. Our simulations pointed out that silica grown at r(O2) = 10% with a refractive index of 1.415 can be used with another optimized oxide layer of 2.35 refractive index to have a minimum reflectance of 0.4%.

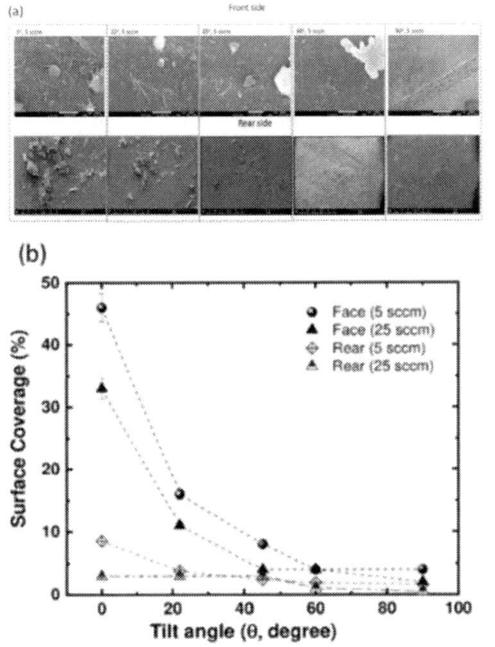

Figure 3: (a) Typical SEM micrographs of the front and rear sides of the soiled silica coated glass coupons, after 2 months of outdoor exposure in OTF. (b) Soiling as measured by the surface coverage with dust as a function of the tilt angle.

4 SUMMARY

In this work, a detailed study has been conducted on the optical, microsctructural, surface, mechanical and morphological properties of silica thin layers deposited by reactive RF magnetron sputtering. Multiple correlations have been pointed out between the optical properties, the surface wettability and the mechanical characteristics of the films. Outdoor measurements have permitted to assess the anti-soiling capability of such coatings in addition to quantify the dust accumulation in bifacial configuration, where an empirical correlation have been highlighted between the dust accumulation in the two faces of the bifacial coupons and the tilt angle. Comprehensive discussions of these results will be provided in the full version of this paper.

ACKNOWLEDGEMENT:
Authors thank the Qatar National Research Fund (QNRF, Member of Qatar Foundation) for the NPRP Grant #11S-1220-170110 which is funding this research.

5 REFERENCES

[1] B. Figgis "QEERI Outdoor Test Facility Technical Details" DOI: 10.13140/RG.2.2.29594.98244. Document available at: file:///C:/Users/baissa/Downloads/QEERIOTFTechnical Detailsv2021-03.pdf

[2] Figgis, B. QEERI Solar Test Facility 5-Year Report, 2018. DOI: 10.13140/RG.2.2.12062.23361

[3] M. Istiaque Hossain, A. Ali, V. Bermudez Benito, B. Figgis and B. Aïssa "Anti-Soiling Coatings for Enhancement of PV Panel Performance in Desert Environment: A Critical Review and Market Overview" Materials 2022, 15(20), 7139 (2022).

[4] Gu, W., Ma, T., Ahmed, S., Zhang, Y. & Peng, J. A comprehensive review and outlook of bifacial photovoltaic (bPV) technology. Energy Convers. Manag. 223, 113283 (2020).

[5] Ullah, A., Amin, A., Kazmi, T., Saleem, M. & Butt, N. Investigation of soiling effects, dust chemistry and optimum cleaning schedule for PV modules in Lahore, Pakistan. Renew. Energy 150, 456–468 (2020).

[6] M.I. Hossain, B. Aïssa, A. Samara, S. Mansour, C. Broussillou, V. Bermudez "Hydrophilic Antireflection and Antidust Silica Coatings" ACS Omega, 2021, 6, 8, 5276–5286, (2021)

PERFORMANCE AND RELIABILITY OF PV MODULES IN INDONESIA

Oo Abdul Rosyid[1]*, Hendrik Straeter[2], Nelly Malik Lande[1], Andrianshah Priyadi[1], Hartadhi1, Anita Faradilla[1], Fariz Maulana Riza Nulhaq[1], Lily Sapinah[1], Adjat Sudrajat[1], A. Taradini[1], M. Ridlo Rosyid[1]

[1]National Research and Innovation Agency (BRIN), Kawasan Sains Teknologi BJ Habibie, Jl. Raya Puspiptek, Tangerang Selatan, 15314 Banten, Indonesia
[2]Physikalisch-Technische Bundesanstalt (PTB), Bundesallee 100, 38116 Braunschweig, Germany

*Corresponding author, e-mail to: abdul65.rosyid@gmail.com

ABSTRACT: A solar power photovoltaic (PV) plant generates sustainable and clean electricity with minimal environmental impact. However, the reliability and performance of the PV plant depends significantly on the quality of the components used. PV modules play a crucial role in terms of performance and reliability of the PV plants. This research discusses the methodology and analysis of a 2-year qualification test experience of PV modules at the BRIN PV Laboratory in Indonesia. The study tested and analyzed more than 14 types of PV module designs (a total of 168 pieces) from both local and imported industries. The results showed that approximately 43% of these PV modules failed to meet the qualification criteria, with local products contributed to the main failures (about 83.3%) compared to the imports. The identified failures were mainly related to wet leakage current (59%), followed by cell metallization burn (12%), glass breakage (8%), back sheet delamination (8%), and other (8%). Besides, some modules experienced degradation of maximum power by more than 5% at the end of the test. Careful selection of high-quality raw materials, ensuring quality assurance in the manufacturing process, and proper handling and transportation of PV modules are key factors in improving the quality of PV module products.
Keywords: photovoltaics, PV module, crystalline silicon, qualification testing, reliability, and performance

1 INTRODUCTION

In response to growing concern over global warming and its connection to carbon emissions, there is a widespread recognition of the need to transition away from fossil fuels and increase the share of renewable energy sources in the global energy mix [1]. Solar photovoltaic (PV) energy, viewed as a solution to address the energy crisis, particularly as a substitute for fossil fuels, has emerged as a paramount global priority for numerous countries [2]. In addition to the unlimited resources, operating technology is also non-polluting and environmentally friendly in facing the challenges of energy issues in the future. The photovoltaic module industry is one sector that is expected to contribute to the development of the national electricity system as an alternative to the use of fossil energy [3].

Indonesia's experience with PV technology has been marked by remarkable progress and achievements, particularly in rural electrification and large-scale solar projects [4]. PV modules used in Indonesia generally come from local industry or imported products. According to the Indonesian Ministry of Energy and Energy Resources (ESDM), currently there are more than 15 national solar module manufacturers, 11 of which are part of the Indonesian Solar Module Manufacturers Association (APAMSI) with an annual production capacity of around 500 MWp. However, challenges related to reliability and performance continue to be significant focal points.

The quality factor of PV module is one of the important issues in ensuring the performance and reliability of the PV power system [5]. According to data reported by the National Renewable Energy Laboratory (NREL) in 2017, the most common degradation modes in modules for the last 10 years were hot spots (33%) followed by ribbon discoloration (20%), glass breakage (12%), encapsulant discoloration (10%), cell breakage (9%), and potential-induced degradation (PID, 8%) [6]. Other study conducted by TamizhMani et al. [7] found that failures in PV systems are generally caused by 55% installation errors, followed by 25% planning and documentation errors, 9% product defects, 5% environmental factors, 5% maintenance, and others 1%. Köntges et al. [8] also examined that typical failures of PV module products are divided into three categories, i.e.: infant failures, midlife-failures, and wear-out-failures. The infant-failures happen at the start of PV module's working life due to defective construction, faults in production and non-conforming materials. The major failures include loose frame, glass breakage, failures in junction box, cell-interconnection and delamination are the most significant failures for PV modules in the field. In this phase transportation damage contributes to 5 % of all failures.

Design qualification testing using tests such as IEC 61215 has been key for achieving high reliability [9]. Therefore, this study assessed the qualification method for PV modules and presented the results of two years of experience at the BRIN PV Laboratory, focusing on terrestrial photovoltaic devices utilizing mono- and polycrystalline silicon technology in Indonesia. This comprehensive investigation encompassed over 14 design variations of PV modules, sourced from both local and international manufacturers.

2 MATERIALS AND METHODS

2.1 Test samples

The study focused on Si-based PV modules, which represent the dominant technology with over 94% of the market share and of cumulative installations [10]. The core device inside a Si-based PV module is the solar cell, a semiconductor material able to convert the energy of light into electricity by the photovoltaic effect. A PV module is the assembly of several solar cells that are (i) electrically interconnected (typically in series, to increase the voltage output), and (ii) encapsulated between some materials that protect them from the environmental stress factors and provide electrical isolation of the electrical circuit. The study considered 14 design types of PV module with the total number of 168 pieces, as listed in Table I.

Table I: Technical specifications of the photovoltaic module test samples.

ID	Type	Voc [V]	Isc [A]	Vmp [V]	Imp [A]	Pmp [W]
01	M-Si	49.2±3%	10.1±3%	40.3	9.68	390 ± 3%
02	M-Si	44.7±5%	9.57±5%	37.26	8.86	330 ± 5%
03	M-Si	48.0±3%	7 ± 3%	40	6.7	270 ± 3%
04	M-Si	48.8±3%	9.9 ±3%	41.6	9.62	400 ± 3%
05	M-Si	46.49	9.32	37.8	8.99	340 ± 5%
06	M-Si	47.9±3%	9.9 ±3%	40.71	9.33	380 ±3%
07	M-Si	47.7	9.91	39.5	9.37	370 +3%
08	P-Si	45.3± 3%	9.1 ± 3%	38.68	8.53	330 ± 3%
09	M-Si	49.2± 3%	10.1± 3%	40,30	9.68	390 ± 3%
10	M-Si	49.6± 3%	13.0± 3%	40,90	12,47	510 ± 5%
11	M-Si	50.3± 3%	10.6± 3%	42.3	10.17	430 ± 3%
12	M-Si	457± 3%	16.1± 3%	38.4	15.07	575 ± 3%
13	M-Si	46 ± 3%	15.9± 3%	38.4	14.84	570 ± 3%
14	M-Si	42.5± 3%	14.2± 3%	35.21	13.49	475 ± 3%

2.2 Methods

These tests are done in different sequences in accordance with IEC 61215 standards [11]–[14]. outlined in the standard as shown in Figure 1. After the initial tests, the modules are separated and subjected to one of the different sequence tests. The sequences are called A, B, C, D or E, and afterwards tested again for efficiency and safety. Each single sequence contains stress tests which specifically aim at clarifying one of the identified main degradations causes that are commonly faced in the field.

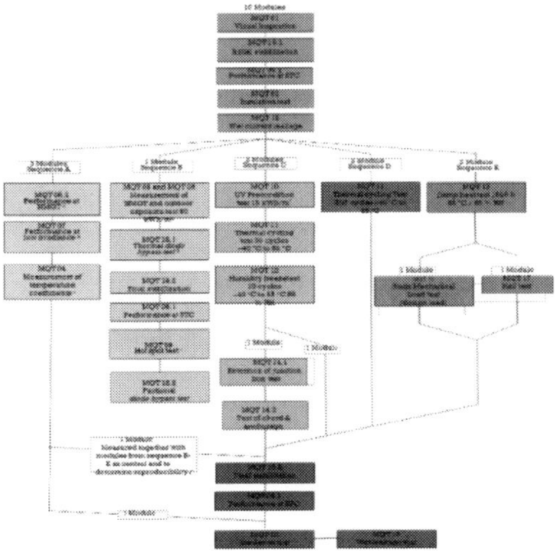

Figure 1: Test flow diagram of the IEC 61215:2016 [12].

2.3 Pass-fail criteria

Each of the modules must successfully pass the so-called test criteria [12]:

a) There is no visual evidence of a major defect, as defined in the IEC 61215-1.
b) The measured insulation resistance of the modules shall not be less than 400 MΩ (for modules with an area of less than 0,1 m^2), and 40 MΩ-m^2 (for modules with an area larger than 0,1 m^2)
c) Verification of rated label values (Gate1) and maximum power degradation during type approval testing (Gate2) requirements are met after the tests.

After stabilization, the measured maximum power, Pm(Lab) the modules shall be within the power rating of the name plate Pm(NP) including stated measurement uncertainty m_1, called Gate1, as stated in Eq. (1).

$$P_m(Lab)\left[1+\frac{|m_1|(\%)}{100}\right] \geq P_m(NP)\left[1-\frac{|t_1|(\%)}{100}\right] \quad (1)$$

where, m_1 is the measurement uncertainty in % of laboratory for Pm (expanded combined uncertainty (k=2), ISO/IEC Guide 98-3), and t_1 is the manufacturer's rated lower production tolerance in % for Pm.

For the arithmetic average of the measured maximum, Type equation here.Pm(Lab) shall be within the power rating of the name plate Pm(NP), as stated in the Eq.(2).

$$\bar{P}_m(Lab)\left[1+\frac{|m_1|(\%)}{100}\right] \geq P_m(NP) \quad (2)$$

The measured open-circuit voltage, Voc(Lab) of each module must meet criteria, as stated in Eq. (3).

$$V_{oc}(Lab)\left[1+\frac{|m_2|(\%)}{100}\right] \leq V_{oc}(NP)\left[1+\frac{|t_2|(\%)}{100}\right] \quad (3)$$

where, Voc(Lab) is a measured maximum Voc of each module under steady state, Voc(NP) is the maximum Voc stated on the nameplate of each module with no tolerance, m_2 is the measurement uncertainty in % of laboratories for Voc, and t_2 is the manufacturer's rated upper production tolerance in % for Voc.

Measured short-circuit current, Isc(Lab) of each module must meet the criteria, as stated in Eq. (4):

$$I_{sc}(Lab)\left[1+\frac{|m_3|(\%)}{100}\right] \leq I_{sc}(NP)\left[1+\frac{|t_3|(\%)}{100}\right] \quad (4)$$

where, Isc(Lab) is the maximum measured Isc of each module in stable conditions, Isc(NP) is the maximum Isc stated on the nameplate of each module with no tolerances; m_3 is the measurement uncertainty in % of laboratories for Isc, and t_3 is the manufacturer's rated upper production tolerance in % for Isc.

At the end of each test sequence or for sequence B after bypass diode test, the maximum power output drop of each module Pm(LabGate2) shall be less than 5%, referenced to the module's initial measured output power Pm(LabGate1). Each test sample shall meet the criterion, as stated in Eq. (5).

$$P_m(Lab_{Gate2}) \geq 0.95 P_m(Lab_{Gate1})\left[1-\frac{r(\%)}{100}\right] \quad (5)$$

The reproducibility r shall be less than stated in the technology specific parts of this standard. The reproducibility r is verified by comparing the control module(s) from sequence A after initial stabilization (beginning of the test) and after final stabilization (end of tests from sequence B to E).

2.4 The BRIN PV Laboratory

The BRIN PV laboratory was built and put in

operation in 2019. The ISO 17025 accredited laboratory tests and certifies PV modules according to national and international standards, including IEC 61215. Figure 2 and Figure 3 show the PV module test facilities at the BRIN PV Laboratory.

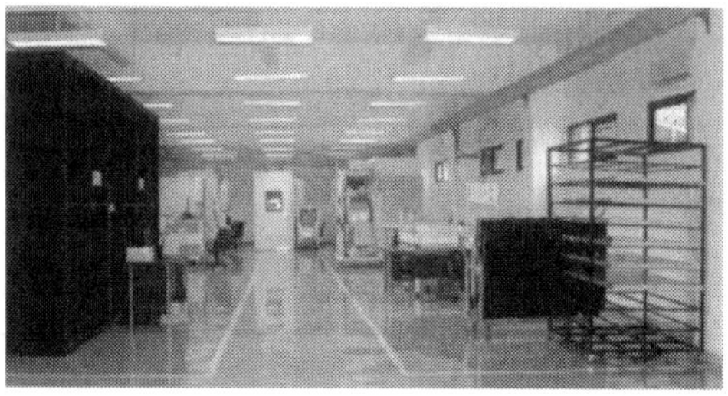

Figure 2: PV laboratory (indoor) at the BRIN KST Serpong, Indonesia.

Figure 3: Outdoor facility of the PV laboratory at the BRIN KST Serpong, Indonesia.

The BRIN PV laboratory is equipped with the main equipment units, as shown in Tabel II.

Table II: List of the main testing equipment of the BRIN PV Laboratory

No	Equipment	Descriptions
1	Insulation tester	To evaluate the electrical insulation and wet leakage current test for MQT 03,15 in the module
2	Pulsed solar simulator	To determine the maximum module power at STC, low irradiance, and NMOT for MQT 02, 04, 06, 07
3	Steady-state solar simulator	To assess module to withstand heat effects caused by hots pots for MQT 09, 19
4	Climate chamber (3 units)	To test the module to withstand thermal mismatch, fatigue, and other stresses; high-temperature and humidity effects (MQT 11, 12, 13)
5	UV precondition chamber	To identify materials and bonds vulnerable to UV-induced degradation for MQT 10
6	Visual inspection and EL tester	To detect visual defects in the module for MQT 01
7	Outdoor exposure test equipment	To measures module operating temperature for MQT 01 and 08
8	Static mechanical load tester	To verify the module ability to withstand the minimum static load specified by the manufacturer for MQT 16
9	Hail impact tester	To verifies that the module can withstand hail impacts for MQT 17
10	Bypass diode testers	To evaluate the thermal design feasibility and long-term reliability of the bypass diode for MQT 18
11	Robustness of terminations tester	To determine module terminations and cable attachment to the module can withstand possible mechanical pressure during assembly, installation, or normal module handling for MQT 14

3 RESULTS AND DISCUSSION

The study tested more than 14 module type design or in total of 168 pieces of C-Si PV modules from different local manufactures as well as imported products. Several PV module types failed to meet the pass criteria. Figure 4 displays the test results of 57% PV modules passed, and 43% failed. The 83% failed samples (5 out of 6) were from local manufacturer, highlighting performance and manufacturing concerns.

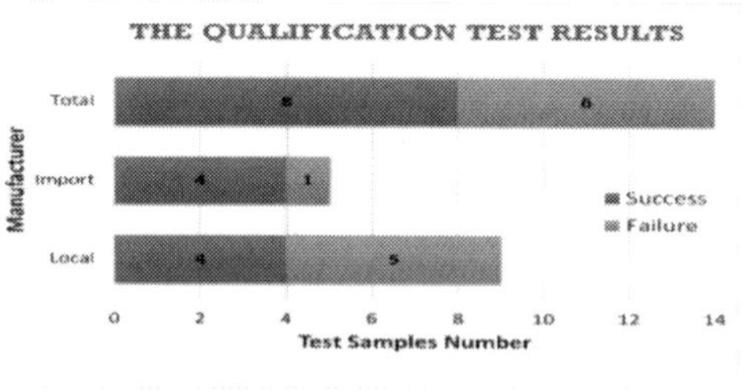

Figure 4: Qualification test results of 14 PV module design.

3.1 Types of failure

Figure 5 shows the main issues with the PV modules were wet leakage current (59%), followed by cell metallization burn (12%), glass breakage, and frame looseness (8%). Additionally, back sheet delamination

(8%) and other problems (8%). The observed visual defects of the modules observed are shown in Figure 6.

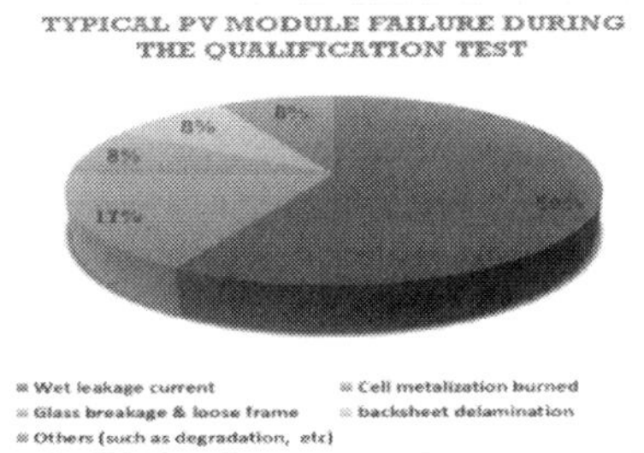

Figure 5: Type of failure from the test results.

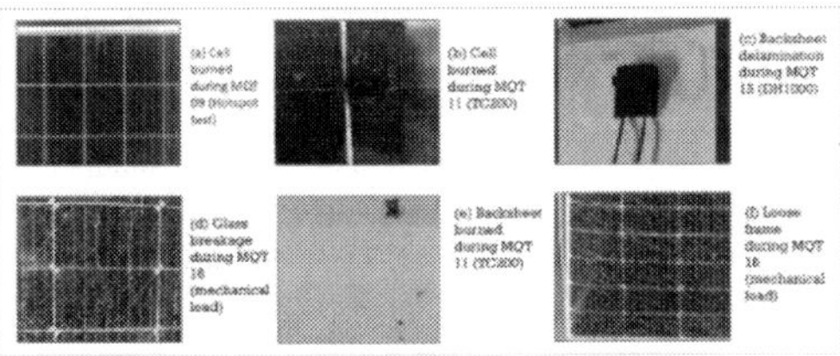

Figure 6: Visual failure from the test results.

3.1.1 Visual failures

The burning or damage to solar cells during the qualification test of PV modules can occur due to several reasons, and it is an indication of potential issues in the module's design, manufacturing, or materials. The qualification test is intended to evaluate the performance, reliability, and safety of the PV module under extreme conditions. Figure 6(a) shows cell burning or damage, during hotspot (MQT09) and Figure 6(b) cell burning during thermal cycling (TC200, MQT-11) tests. Some possible reasons for solar cell burning during the qualification test, such as manufacturing defects, substandard materials, etc. To prevent solar cell burning during qualification testing, manufacturers must adhere to stringent quality control measures, use high-quality materials, and ensure proper module design and assembly. Thorough testing and continuous improvement based on test results are essential to enhance the reliability and safety of PV modules in the field.

Backsheet delamination is a common issue that occurs due to the combination of moisture and heat. Figure 6(c) and and 6(e) show backsheet delamination occurred during DH and TC200. The damp heat test is a reliability test used to assess the long-term durability and performance of PV modules under high humidity and elevated temperature conditions. During this test, the PV modules are subjected to high temperatures (typically around 85°C) and high humidity (typically around 85% relative humidity) for an extended period, often 1000 hours or more. Several factors contribute to backsheet delamination during the damp heat test, such as: moisture ingress, thermal cycling, backsheet material, manufacturing defects, etc. To prevent backsheet delamination during damp heat testing, manufacturers need to ensure high-quality backsheet materials, suitable adhesives, and robust lamination processes. Regular testing and improvements in backsheet design and materials can help enhance the reliability and longevity of PV modules in damp and hot environmental conditions.

Structural failure during the qualification test of PV modules can occur due to a variety of reasons related to design, manufacturing, and environmental factors. The qualification test is intended to evaluate the module's ability to withstand harsh conditions and ensure its long-term reliability. Figure 6(d) shows structural failure, i.e. glass breakage and loose frame. These failures occurred during the mechanical load test (MQT-16) in the sequence E of IEC 61215:20016. Some possible reasons for structural failure during the qualification test, such as: adhesive bond failure, manufacturing defects, material incompatibility, etc. To prevent structural failure during the qualification test, manufacturers need to ensure robust design and construction, proper material selection, and adherence to industry standards and testing protocols. Regular testing and quality control throughout the manufacturing process are essential to identify and address potential issues before modules are deployed in the field.

3.1.2 Wet leakage failure

The wet leakage current, also known as moisture-induced leakage current, can occur during the qualification test of PV modules due to the presence of moisture or water ingress into the module's components. This phenomenon is a concern because it can lead to performance degradation and potential safety hazards. The wet leakage testing is carried out after the insulation test and repeated at the end of the quality tests after all other tests have been passed. The insulation resistance shall not be more than 40 MΩ per each square meter of the modules which have an area bigger than 0.1 m^2. The wet leakage current failure is the highest ranked failure during the PV modules test at the laboratory, as seen in the Figure 7.

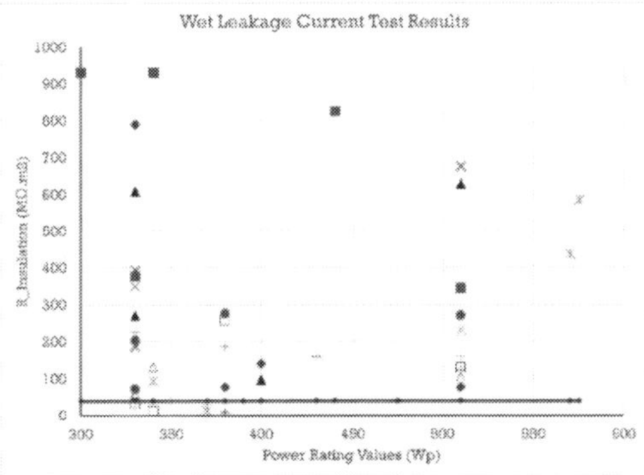

Figure 7: Wet current leakage failure for different PV modules.

The wet leakage failure mainly occurred after the humidity freeze test or damp heat test for PV modules which may fail due to poor lamination or edge sealing defect during manufacturing, cracks or microcracks, delamination, etc. To prevent wet leakage current during the qualification test, it is essential for manufacturers to employ robust design and manufacturing processes. Proper encapsulation with high-quality materials, thorough sealing of all components, and rigorous quality control measures are crucial to ensure the module's resistance to moisture ingress. Regular testing and continuous improvement based on test results help enhance the reliability and performance of PV modules in damp environments.

3.2. Verification of the rated values

Maximum power verification according to IEC 61215 consists of verification of maximum power according to nameplate value (Gate1) and maximum degradation after final test (Gate2). In the Gate1 analysis, the results of the maximum power measurement of MQT 6.1 in the initial test are compared with the nominal value on the nameplate according to Eq. (1). Apart from that, the value of open-circuit voltage and short-circuit current is also compared with the nominal value of the nameplate, according to Eq. (2) and Eq. (3), respectively.

3.2.1 Maximum power values verification

Verification of the nameplate value for the maximum power (Pm) of the test sample is done in according to Eq. (1) and (2). Figure 8 and Figure 9 show that the measured (Pm) values of maximum power and the average (Pm_avg) values of PV modules at the initial test are larger than the nameplate values. Two samples fail to meet the requirement due to their Pm values are larger than 0%, in the range of 0.87% to 6.59% than their nameplate.

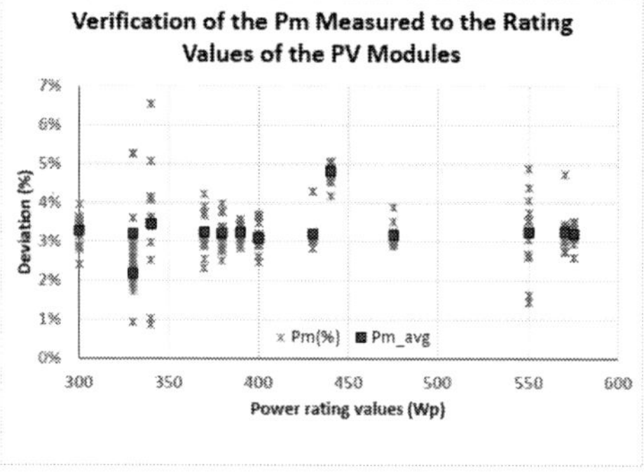

Figure 8: Deviation of the measured Pm and Pm_avg to the rated values.

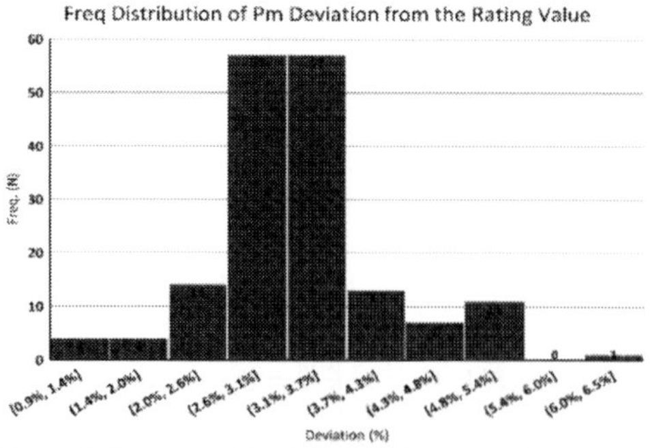

Figure 9: Frequency distribution of the measured Pm to the rated values.

3.2.2 Open circuit voltage values verification

Verification of the nameplate value for the open circuit voltage (Voc) of the test sample is done in according to Eq. (3). The measured values for open circuit voltage (Voc) of the module are in general lower than the nameplate values. Figure 10 shows that deviation of the measured open circuit voltage (Voc) values of the PV modules in the range of -12.07% to 3.33%. Several modules fail to meet the passing criteria due to their Voc values are bigger than the rated values. The higher measured Voc might be danger for the design of PV system, especially for the input of MPPTs.

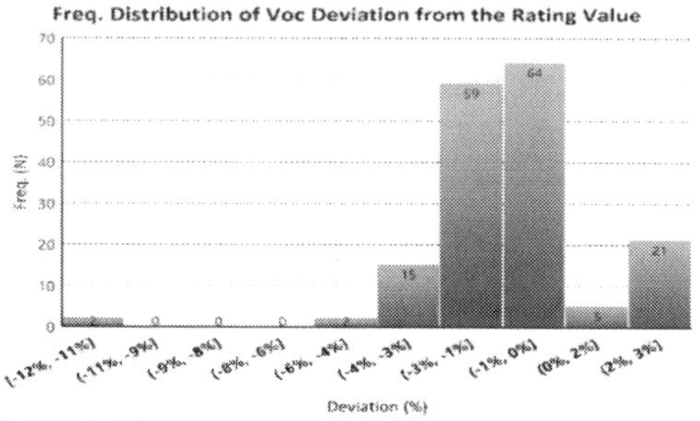

Figure 10: Frequency distribution of the measured Voc to the rated values.

3.2.3 Short-circuit values verification.

Verification of the nameplate value for the short-circuit current (Isc) of the test sample is done in according to Eq. 4. Figure 11 shows that the measured short-circuit current Isc(Lab) values of the PV modules vary between -5% to 20% from the name plant values, Isc(NP). Around 30% of the test samples fails to meet the requirement due to the Isc(Lab) values are larger than the Isc(NP). The higher measured Isc might be danger for the design of PV system, especially for the grid-tie inverter.

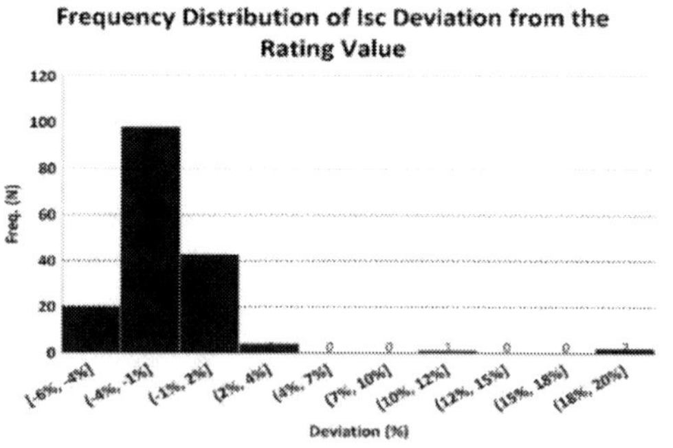

Figure 11: Frequency distribution of the measured Isc to the rated values.

3.3. Maximum power degradation (Gate-2)

In general, the maximum power of silicon PV modules will experience degradation during operation at a rate of approximately -1% per year [15]. The maximum power degradation of of a PV module during the qualification test can be verified according to the Eq. (5). To pass this test, the observed degration should be less than -5%. Figure 12 and Figure 13 show that the maximum power degradation of the PV modules is mainly lower than -5%, with the average value of -0.9%. Only 4 out of 168 test samples (2.38%) fail to meet the requirement.

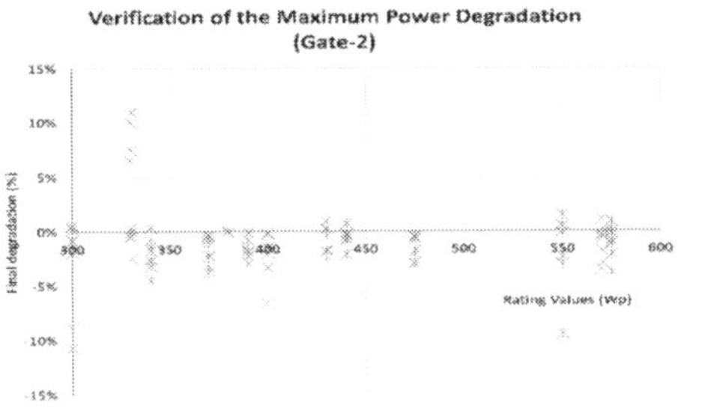

Figure 12: Deviation of the maximum power degradation.

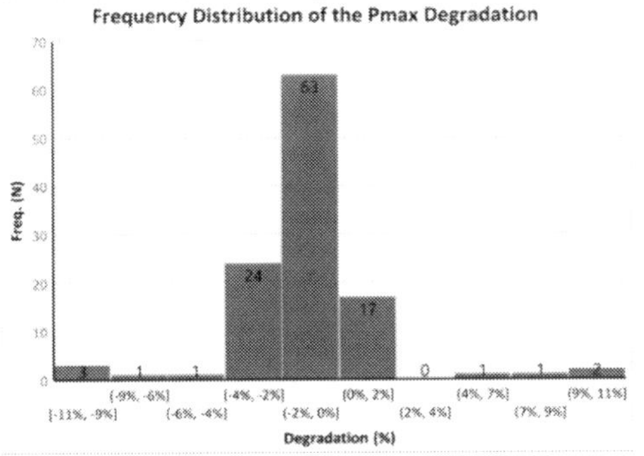

Figure 13: Frequency distribution of the maximum power degradation.

4. CONCLUSION

PV module performance and reliability issues is one of the challenges in solar PV development. This study has effectively identified, evaluated, and analyzed the performance and reliability of PV modules available in Indonesia, encompassing products from both local and imported manufacturers, in accordance with the guidelines outlined in IEC 61215:2016. In general, It is evident that PV modules produced locally, require enhancements in product quality, particularly concerning material selection, manufacturing processes, as well as the handling and transportation procedures for these modules.

5. ACKNOWLEDGEMENTS

I would like to express our heartfelt gratitude to the BRIN PV Laboratory for their valuable facilitation and support in our research endeavors. Furthermore, we extend our sincere appreciation to PTB Germany for their kind facilitation, which has allowed us to participate in the esteemed EU PVSEC conference. PTB Germany's support and encouragement have been crucial in making our participation in this prestigious event possible.

6 REFERENCES

[1] B. N. Stram, "Key challenges to expanding renewable energy," Energy Policy, vol. 96, pp. 728–734, Sep. 2016, doi: 10.1016/j.enpol.2016.05.034.

[2] O. A. Al-Shahri et al., "Solar photovoltaic energy optimization methods, challenges and issues: A comprehensive review," Journal of Cleaner Production, vol. 284. Elsevier Ltd, Feb. 15, 2021. doi: 10.1016/j.jclepro.2020.125465.

[3] N. S. M. N. Izam, Z. Itam, W. L. Sing, and A. Syamsir, "Sustainable Development Perspectives of Solar Energy Technologies with Focus on Solar Photovoltaic - A Review," Energies, vol. 15, no. 8. MDPI, Apr. 01, 2022. doi: 10.3390/en15082790.

[4] B. K. Sovacool, "Success and failure in the political economy of solar electrification: Lessons from World Bank Solar Home System (SHS) projects in Sri Lanka and Indonesia," Energy Policy, vol. 123, pp. 482–493, Dec. 2018, doi: 10.1016/j.enpol.2018.09.024.

[5] G. Petrone, G. Spagnuolo, R. Teodorescu, M. Veerachary, and M. Vitelli, "Reliability issues in photovoltaic power processing systems," IEEE Transactions on Industrial Electronics, vol. 55, no. 7, pp. 2569–2580, Jul. 2008, doi: 10.1109/TIE.2008.924016.

[6] D. C. Jordan, T. J. Silverman, J. H. Wohlgemuth, S. R. Kurtz, and K. T. VanSant, "Photovoltaic failure and degradation modes," Progress in Photovoltaics: Research and Applications, vol. 25, no. 4, pp. 318–326, Apr. 2017, doi: 10.1002/pip.2866.

[7] G. TamizhMani, B. Li, C. L. S. K. C.Dreier, and Shisler, Regional Influence on Module Design Quality: Qualification Testing Failure Rate Results from Six Regional Labs of TUV Rheinland around the World. IEEE, 2014.

[8] Köntges et al., Performance and reliability of photovoltaic systems subtask 3.2: Review of failures of photovoltaic modules : IEA PVPS task 13 : external final report IEA-PVPS. 2014.

[9] S. Kurtz et al., "Photovoltaic Module Qualification Plus Testing", 2013. [Online]. Available: www.nrel.gov/publications.

[10]U. Gangopadhyay, S. Jana, and S. Das, "State of Art of Solar Photovoltaic Technology," Conference Papers in Energy, vol. 2013, pp. 1–9, May 2013, doi: 10.1155/2013/764132.

[11]A. Firman, M. Cáceres, A. R. González Mayans, and L. H. Vera, "Photovoltaic Qualification and Approval Tests," Standards, vol. 2, no. 2, pp. 136–156, Apr. 2022, doi: 10.3390/standards2020011.

[12]IEC 61215-1, "Terrestrial photovoltaic (PV) modules : design qualification and type approval. Part 1, Test requirements", 2016.

[13]IEC 61215-2, "Terrestrial photovoltaic (PV) modules : design qualification and type approval. Part 2, Test procedures", 2016.

[14]A. Firman, M. Cáceres, A. R. González Mayans, and L. H. Vera, "Photovoltaic Qualification and Approval Tests," Standards, vol. 2, no. 2, pp. 136–156, Apr. 2022, doi: 10.3390/standards2020011.

[15]J. Y. Ye, T. Reindl, A. G. Aberle, and T. M. Walsh, "Performance degradation of various PV module technologies in tropical Singapore," IEEE J Photovolt, vol. 4, no. 5, pp. 1288–1294, 2014, doi: 10.1109/JPHOTOV.2014.2338051.

Defects and degradations in photovoltaic modules from hot Middle East deserts

S. Kumar[1,*,#], B. Adothu[1,#], B. Jäckel[2], Z. Shekason[1], D. Daßler[2], K. Chapaneri[1], A. Baloch[1], P. Gabbadi[1], Y. Kumar[1], A. Alheloo[1], A. Almheiri[1], J. J. John[1], G. Mathiak[1], V. Alberts[1], and R. Gottschalg[2]

[1] DEWA Research & Development Center, P.O.Box: 564 Dubai, United Arab Emirates
[2] Fraunhofer Center for Silicon Photovoltaics CSP, Otto-Eissfeldt-Str. 12, 06120 Halle (Saale), Germany
* corresponding author: phone +971 58 539 5739 | sagarika505@gmail.com
# both authors have equally contributed to this manuscript

Abstract— **The Middle East is witnessing rapid solar energy growth, exemplified by the Mohammed Bin Rashid (MBR) Solar Park in the UAE. With 1000 MW operational capacity, and plans to reach 5 GW by 2030, harsh desert conditions pose challenges. High solar irradiation, intense UV exposure, extreme temperatures, and sand-dust accumulation lead to premature defects and power loss in PV modules. Existing testing lacks desert-specific standards. This study establishes hot desert environmental guidelines by analyzing common defects and degradation in MBR Solar Park modules. An outdoor facility monitored 120 modules for 7 years, employing various indoor techniques. The research identifies degradation modes, conducts root cause analysis, and assesses long-term performance impact, instrumental in guiding solar installations in desert environments.**

Keywords—desert climate, crystalline PV modules, thermo-mechanical fatigue, finger defects

I. Introduction

The Middle Eastern region is seeing rapid growth in solar installations. The Mohammed Bin Rashid (MBR) Solar Park, located in a hot desert in United Arab Emirates (UAE), recently reached an operational solar PV capacity of about 1000 MW in April 2021, and plans to install 5 GW by 2030 [1]. However, such desert regions pose harsh climate conditions, owing to a very high solar resource. This leads to early defects leading to power degradations in the installed PV modules, causing long-term reliability and durability issues [2]. The major environmental contributors to such degradations are high solar irradiation with strong UV dose, high ambient temperature, large variations between day and night temperatures, and sand-dust accumulation, as shown in Fig. 1 (a). Among these, the dominant stressor is high solar irradiance and UV dosage, which is more than double that of a city in Germany.

The existing IEC testing protocols do not cater to any specific climate conditions, especially the harsh climate conditions of this region, and are not sufficient to replicate the observed degradation modes. The standard development project is working towards developing hot desert environmental standards for photovoltaic modules being installed in such harsh conditions [3]. As a precursor, it is important to identify and analyze different types of commonly observed defects and degradations from the field. For this purpose, an in-house outdoor testing facility (OTF-A) was set up within the desert environment, as shown in Fig. 1(b). It consists of a total of 120 modules (30 types) installed in 4 rows at 3 angles of installation (0°, 25°, and 90°) for a period of 7 years. Periodic investigations and analyses have been conducted on these modules using outdoor and indoor techniques [4]. The outdoor performance and weather data are also being constantly monitored and analyzed. The modules are also tested in the in-house testing facility for a more in-depth analysis of the observed defects and degradation. This paper is focused on findings from the indoor testing of these modules and presents the use of illuminated current-voltage (I-V) analysis, electroluminescence (EL) imaging, ultra-violet fluorescence (UV-f) imaging, module-external quantum efficiency (EQE) analysis, Raman spectroscopy, and colorimetry. The findings include the major degradation modes observed in the field, including their root cause analysis and assessment of their impact on long-term performance. The purpose of this backyard study is to provide feedback and benchmarking for developing hot desert standards based on the findings. The first draft of the hot desert standard has been presented in [3]. A study based on material analysis on modules from different manufacturers will also lead to selecting hot desert-appropriate PV module packaging materials.

II. Methodology

This study presents findings and learnings from 7-year-old exposed modules from a hot desert outdoor testing facility at DEWA R&D center, Dubai, UAE. The site presents harsh weather conditions. Such harsh weather conditions have led to defects and degradations within different components of the PV modules during the 7 years of exposure. Fig. 1 shows the visual defects observed in all modules, binned based on the defect severity.

10.4229/EUPVSEC2023/3AV.2.27
020197-001

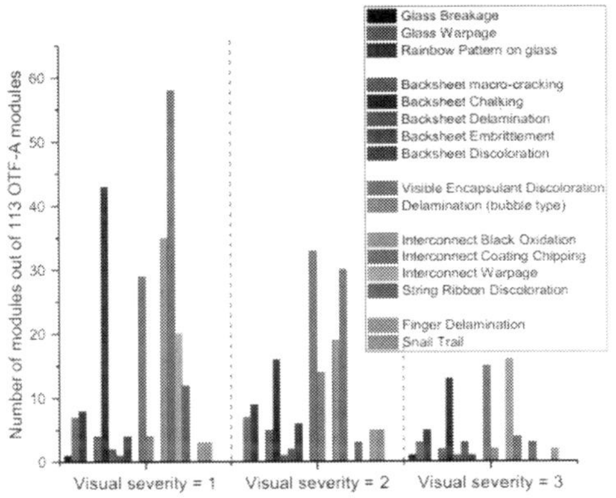

Fig. 1. Visual defects observed in the OTF under 7 years of exposure, binned according to the visible severity of defects.

Visible discoloration, interconnect oxidation, and back sheet chalking has been commonly observed in the modules. More details on observed visual defects are presented in [5]. Such defects have been studied in detail using indoor characterization techniques (I-V, EL, UV-F imaging, module-EQE, Raman spectrometry, and colorimetry). This paper focuses on short circuit current (Isc) losses and investigates the corresponding defects leading to such losses.

III. RESULTS AND DISCUSSION

The overall degradation scenario for the OTF modules can be seen in Fig. 2, which presents the annual degradation rate of the modules calculated using indoor I-V data. Two major degradation modes have been observed, where Isc losses were most dominant. EL imaging showed various structural defects at the cell level, including finger breakages and cracks.

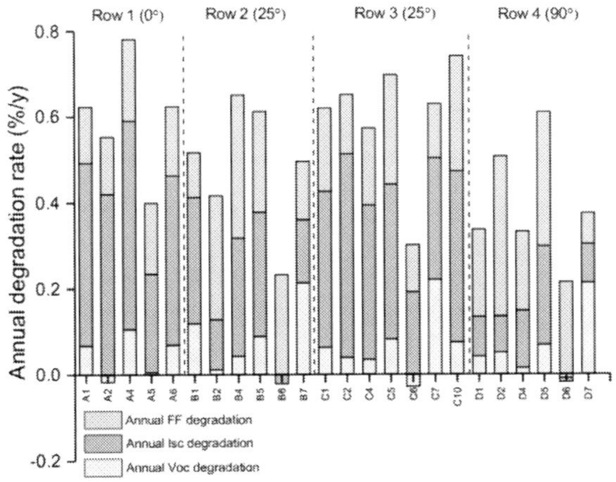

Fig. 2. EL image (with intensity in blue line) of a cell with different type of finger defects observed in a outdoor exposed PV cell.

Material deterioration (fluorophore formation and genesis of photobleaching) has been observed in all modules from the OTF using UV-f imaging (365 nm source). Two major categories of patterns are observed viz., framed and filled pattern. For the purpose of demonstration, findings from 4 representative c-Si glass-backsheet modules (B1, B2, B4, and B6) have been presented. These modules belong to 4 different manufacturers. Fig. 3 shows UV-F images of one cell from these modules. These modules showed uniform (bright) fluorophore formation (visible discoloration in some modules), barring at region of cracks or near busbar in a few cells. The non-bright region is proposed to be due to the photobleaching of fluorophores.

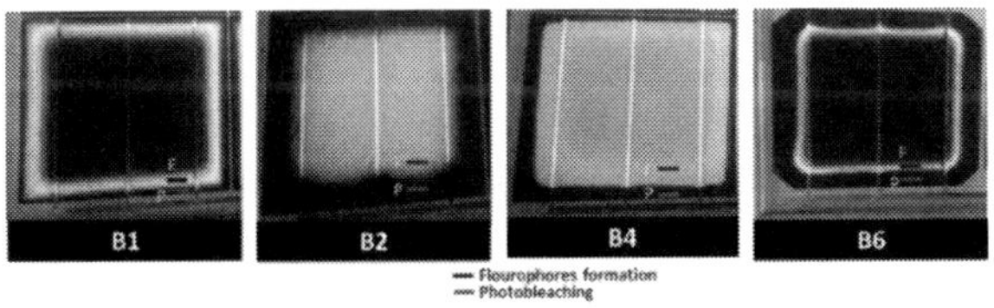

Fig. 3. UV-f images of 4 representative modules (B1, B2, B4, and B6; c-Si, glass-backsheet) showing material deterioration (fluorophore formation and photobleaching)

Correspondingly, the degradation in Isc (from installation) and change in yellowness index (YI) at a bright region from a non-bright region in UV-F image has been shown in Fig. 4. The framed patterns showed negative changes in YI, whereas, filled pattern showed positive. The same trend has been observed for all other modules from the OTF. To be able to comment on the degree of fluorophore formation or discoloration, Raman spectra have been obtained while focusing on the encapsulant of the modules through the glass side, using appropriate spacers, as shown in Fig. 5. A one-to-one trend has been observed between the loss in Isc and Raman intensity of the 4 modules, showing potential cause-effect relation between observed material deterioration and Isc losses. This also indicates the sensitivity of Raman analysis for fluorophore detection and its relative quantification.

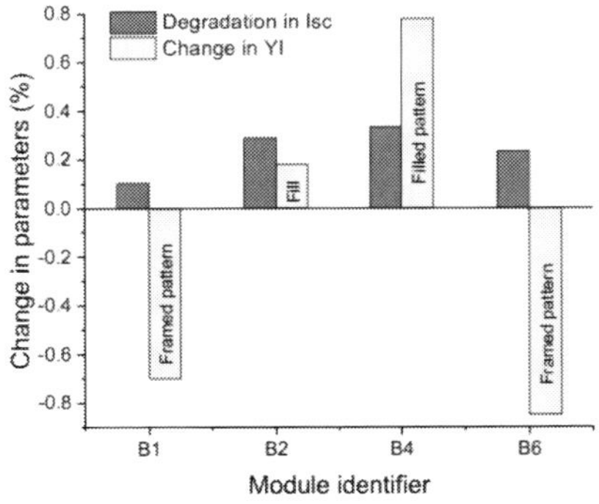

Fig. 4. Degradation in Isc (from installation) and change in yellowness index (YI) at the bright region from a non-bright region in UV-F image

10.4229/EUPVSEC2023/3AV.2.27
020197-002

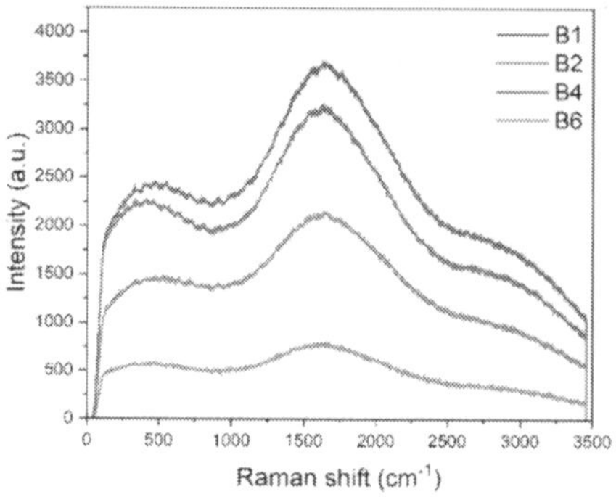

Fig. 5. Raman spectra at the corresponding fluorophore appearing regions, as seen in UV-f images

To be able to obtain more information on the impact of such observed defects, EQE has been obtained at the regions of interest indicated as F and P in the UV-f images shown in Fig. 6. The analysis has been presented for modules B4 and B6, as examples of cases with filled and framed patterns, respectively. It can be observed for module B6 (frame) that the region with fluorophores shows lower EQE in comparison to the photobleached region, as lower wavelengths lie in the UV region. On the other hand, module B4 (filled) shows the opposite behavior indicating a degradation product in the photobleached region for framed modules, which may not be visible at the given UV light source. Corresponding impact, when translated in terms of Isc losses (calculated under AM1.5G spectrum) shows a much higher reduction of 15.6% at the photobleached region for module B4 compared to a lower Isc reduction of 6.1% at the region visible as fluorophores in the module B6.

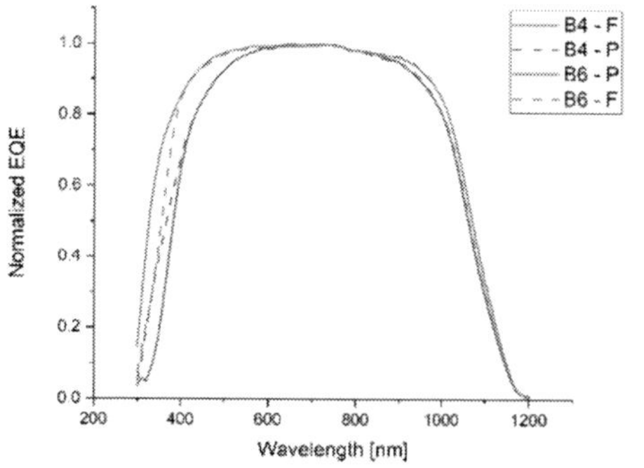

Fig. 6. EQE at the indicated points P and F (UV-f image) for modules B4 and B6.

References

[1] MESIA, Solar Outlook Report 2020. Dubai: MESIA, 2020.

[2] J. J. John, A. Elnosh, A. Safieh, and A. Alnuaimi, "Investigation of degradation observed on recent polycrystalline silicon PV modules installed in desert region," In IEEE 46th Photovoltaic Specialists Conference (PVSC), pp. 0870-0873, June 2019.

[3] J. J. John, B. Jäckel, S. Kumar, B. Adothu, G. Mathiak, and V. Alberts, "Hot desert standard development for improved PV module reliability and durability in Middle-East region", EUPVSEC-2023 (abstract submitted)

[4] S. Kumar, A. Alheloo, S. Hassan, H. Hanifi, H. Alhamadani, H. Alshanqiti, J. J. John, G. Mathiak, and V. Alberts, "Investigation and Analysis of Defects and Degradations in Desert Fielded Photovoltaic Modules" In EEE 48th Photovoltaic Specialists Conference (PVSC), Jun. 2021, pp. 0788-0792, June 2021.

[5] B. Adothu, S. Kumar, B. Jäckel, N. L. Muttumthala, Z. Shekason, D. Daßler, K. Chapaneri, P. Gabbadi, Y. Kumar, A. Alheloo, A. Almheiri, J. J. John, G. Mathiak, V. Alberts, and R. Gottschalg, "Identification and Investigation of Materials Degradation in Photovoltaic Modules from Middle East Hot Desert", EUPVSEC-2023 (abstract submitted)

[6] S. Kumar, B. Adothu1, B. Jäckel, Z. Shekason, A. Moosawi, A. Seentakath, D. Daßler, K. Chapaneri, J. J. John, G. Mathiak, V. Alberts, and R. Gottschalg, "Identification and analysis of metallization defects in hot Middle East desert-operated photovoltaic modules", EUPVSEC-2023

IDENTIFICATION AND INVESTIGATION OF MATERIALS DEGRADATION IN PHOTOVOLTAIC MODULES FROM MIDDLE EAST HOT DESERT

B. Adothu[1,#], S. Kumar[1,#], B. Jaeckel[2], N. L. Muttumthala[1], Z. Shekason[1], D. Daßler[2], K. Chapaneri[1], P. Gabbadi[1], Y. Kumar[1], A. Alheloo[1], A. Almheiri[1], J. J. John[1], G. Mathiak[1], V. Alberts[1], and R. Gottschalg[2]

[1] DEWA Research & Development Center, P.O.Box: 564 Dubai, United Arab Emirates
[2] Fraunhofer Center for Silicon Photovoltaics CSP, Otto-Eissfeldt-Str. 12, 06120 Halle (Saale), Germany
* corresponding author: phone +971 522 578980 | balojiadothu@gmail.com
# both authors have equally contributed to this manuscript

ABSTRACT: The desert climate is the best solar climatic zone on Earth for harvesting solar power because of its higher irradiance resources. The severity of material degradation depends upon the photovoltaic (PV) modules installed in which type of climate zones, weather conditions, and module tilt angles. The paper's main objective is to show the desert's harsh weather conditions and material degradation modes. The encapsulant discoloration and chalking in the backsheet were two dominant material degradation modes observed in desert-operated PV modules at different tilt angles. Majorly, short circuit current (Isc) and fill factor (FF) losses were observed almost in all outdoor teste facility (OTF) modules. Fill and frame discoloration patterns were observed in fluorescence images of OTF PV modules. Encapsulant type is detected with Raman spectroscopy. Ethylene vinyl acetate is identified as an encapsulant material used in c-Si PV modules. The annual Isc degradation rate is 0.28%/year and 0.41 %/year for frame and filled discoloration patterns. Backsheet chalking is analyzed with scanning electron microscopy (SEM) and Energy dispersive X-ray spectroscopy. Chalking is due to the titanium dioxide formation on the outer layer of the backsheet. This paper motivates the researcher and PV community to study the PV materials degradation behavior and defects impact on PV modules operated in desert climates.

Keywords— Desert climate, PV component defects, discoloration patterns, encapsulant, back sheet.

1 INTRODUCTION

Solar energy is becoming a promising candidate in the renewable energy technology world, especially in the Middle East (ME) hot desert, where there is an abundance of solar irradiation [1], [2]. Therefore, the ME regions are adopting PV modules power plants at both residential and utility-scale. In this view, Dubai has launched a Clean Energy Strategy that aims to produce 75 % of its energy requirements from clean energy sources by 2050 [3]. This includes initiatives such as Mohammed Bin Rashid Al Maktoum Solar Park (launched in 2012 in Dubai). According to J. Ascencio-Vásquez, et al. and J.-F. Leli`evre et al., a hot desert climate is the best solar climatic zone on Earth for harvesting solar power because of its higher irradiance resources [4], [5]. However, there is no indication for selecting packaging materials for PV modules for the desert climate application. A PV module is typically made of multiple components and together called the 'bill of materials (BOMs) [6] namely solar cells, encapsulant film, glass, backsheet film, connecting ribbons, protective frame, junction box, sealants, and external cable. Encapsulant discoloration/yellowing, delamination, corrosion, snail trails, chalking, hot spots, interconnects and finger breakages & discoloration, and cell cracks are the common defects observed in PV modules. When a PV module operates under different climatic conditions shows different degradation modes and defects [4,7–11]. The material degradation in each PV component is different. The severity of material degradation depends upon the PV modules installed in which type of climate zones, weather conditions, and module tilt angles. The formation of defects shows a significant impact on the performance of the PV modules. PV modules in the desert experience multiple stressors such as large temperature differences in day and night temperatures, high UV radiation, medium humidity, windspeed, and soiling[11]. It is necessary to understand the defect formation and their impact on PV modules especially in desert climates.

The IEC 61215 standard series is recognized for approval type and designed for moderate climates [12]. This standard is not sufficient to represent the desert conditions and desert-induced defects and degradation modes. The reliability and durability of PV modules are of great importance for the development of desert standards for all PV utility industry segments.

The paper's main objective is to show the PV component defects and their identification in PV modules exposed to ME hot desert regions. As is the case with the ME hot desert, which is characterized by harsh weather conditions, it can raise concerns about the PV modules' long-term performance. Collecting harsh weather conditions, and surveys of PV module defects motivate the researcher and PV community to focus on hot desert standard development for representing the desert-induced defects and degradation modes in the PV modules exposed in the ME region.

2 METHODOLOGY

The crystalline silicon PV modules were installed at the Dubai Electricity and Water Authority (DEWA) R&D Center within the MBR Solar Park, Dubai, UAE, and continuously monitored for the last seven years. Our outdoor test facility (OTF) has continuously recorded weather data. This study investigated 120 c-Si PV modules for visual defect identification at different tilt angles. 25 modules were investigated for power degradation. Weather data, visual defects survey, material degradation, and the annual degradation rates have been used for defining the desert-induced materials defects in the PV modules.

Visual images have been used to identify the component defects in PV modules. Fluorescence images were used to identify discoloration patterns, whereas Raman spectroscopy was used to identify encapsulant types. HALM simulator was used for indoor I-V and degradation analysis. Backsheet chalking morphology is analyzed with scanning electron microscopy (SEM) and elemental compositions are analyzed with energy-dispersive X-ray spectroscopy.

3 RESULTS AND DISCUSSIONS

Desert climatic parameters such as irradiance, UV dose, maximum ambient temperature, large variation in the day & night temperatures, relative humidity, wind speed, and high soiling rates in Dubai's hot desert region are shown in Fig. 1.

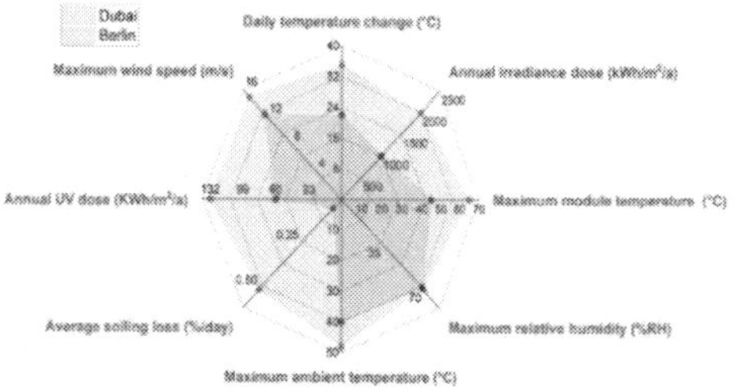

Fig. 1. Dubai's hot desert weather conditions.

Fig. 2 shows the desert-induced defects in PV components. The warpage in solar glass and snail trials in solar cells are shown in Fig. 2 (I and II).)). Due to the heavy weight of glass and thermal mismatches, the PV module bends towards the center side in frameless modules. Snail tracks are visible discolored solar cell defects caused by silver contact discoloration and corrosion. Encapsulant shows different visual discoloration patterns at the center and edges of the solar cells (Fig. 2 III. (a & b). The combination of dryness and thermal cycling can cause severe rusting, chipping, and cracks of busbars/interconnects in the PV modules shown in Fig. 2 IV. (a, b, & c). Backsheet chalking, & embrittlement are shown in Fig. 2 V. (a & b). Chalking is a white powder that can be removed by wiping or washing on the air side layer of the backsheet. During the field operation, polymer films are directly exposed to weather conditions. This reduces the molecular weight and mechanical strength and increases the crystallinity, leading to embrittlement failure of the backsheet.

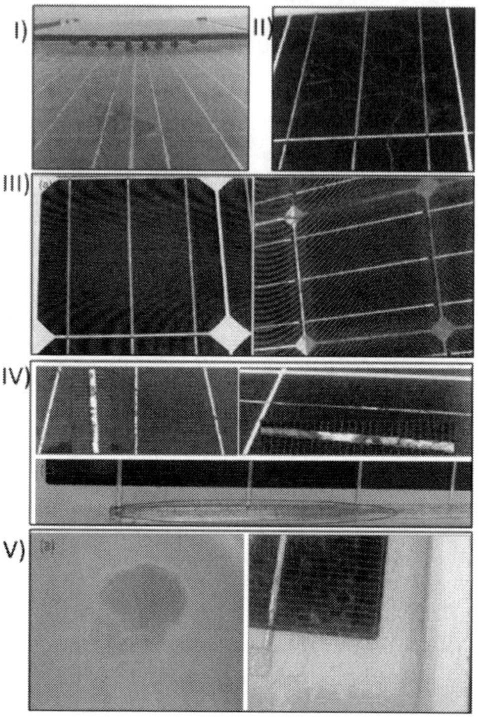

Fig. 2. Visual defects: I). Glass warpage, II). Solar cell snail trials, III). encapsulant discolorations, (a) fill pattern and (b) frame pattern, IV). Interconnects defects, (a) rusting, (b) chipping and (c) cracks, V). (a) & (b) Backsheet chalking, & embrittlement.

Fig. 3 shows the distribution of visual defects of individual components collected from the desert-exposed PV modules at different tilt angles (5°, 25°, and 90°). Visible encapsulant discoloration is the most dominant degradation mode in all the modules tilted at different angles. Backsheet chalking and interconnect oxidations are the two other dominant visually observed material defects in the ME hot desert region. Higher defects and significant discoloration are found at the 25° tilted modules.

The overall annual degradation rate of PV modules shown in Fig. 4 was calculated from the indoor I-V analysis. All the defects shown in Fig. 2 contribute to the annual degradation of PV modules. Majorly Isc and FF losses have been observed almost in all the OTF modules. It has been observed that Isc loss is the most dominant degradation in all electrical parameters. Encapsulant discoloration and chalking are further analyzed as these are dominant degradation modes observed in field-operated modules.

Encapsulant discoloration is the most dominating degradation mode in desert-exposed PV modules. PV module shows two discoloration patterns at the center of the cell called filled patterns and at the edge of the cell called frame patterns shown in the UV fluorescence image in Fig. 5 (a) & (b). Ethylene Vinyl acetate (EVA) based reference module Raman spectra compared with 7 years aged modules.

10.4229/EUPVSEC2023/3AV.2.29
020198-002

The acetate vibrational group (C=O) belongs to the vinyl acetate, and the methylene vibrational group (CHx) belongs to the polyethylene unit in EVA. Both fill and prame pattern modules are identified as EVA encapsulant-based modules shown in Fig. 6. Highly discolored modules show a strong fluorescence background in Raman spectra (Modules such as A, F, and I). For both frame and filled discoloration patterns annual Isc degradation is shown in Fig. 7.

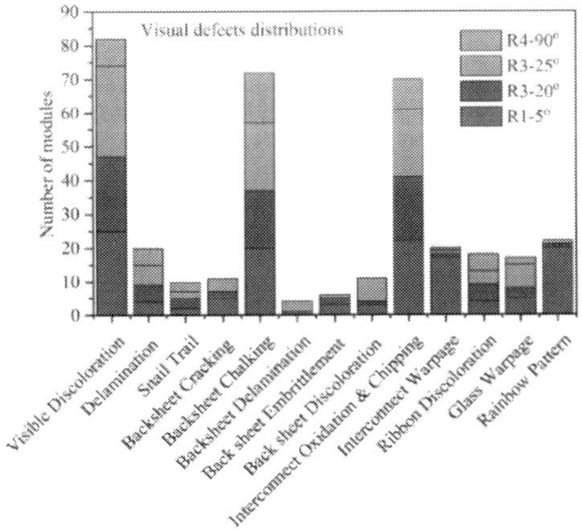

Fig. 3. Dominant defects in PV components at different tilt angles.

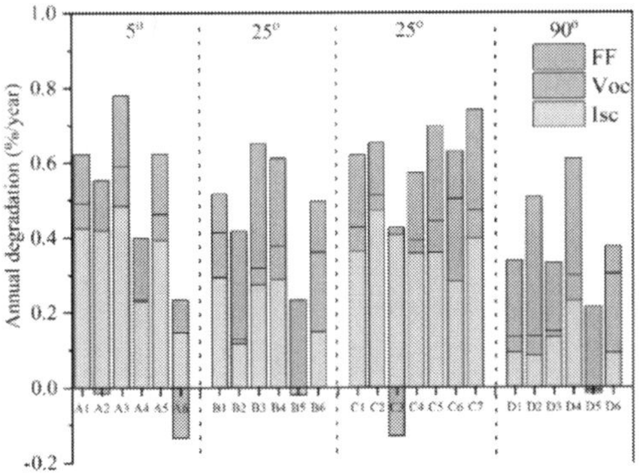

Fig. 4. Annual degradation of Isc, Voc, and FF for seven years field aged PV modules at different tilt angles.

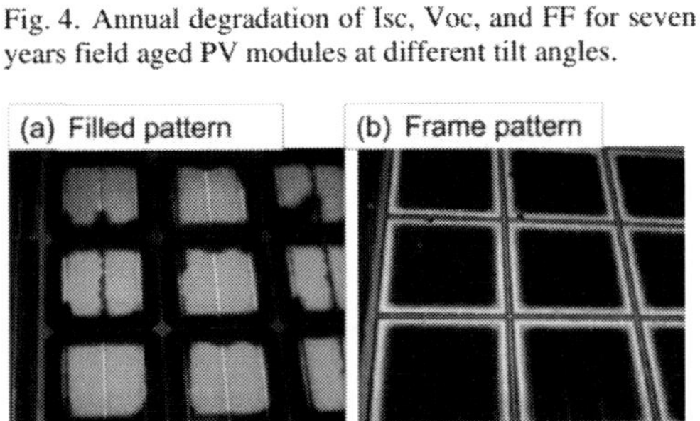

Fig. 5. UV-FL image of encapsulant filled and frame discoloration patterns

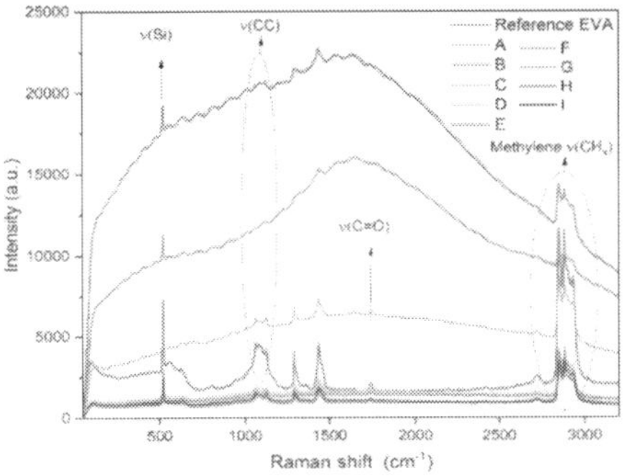

Fig. 6. Raman spectra of nine (A to I)) field-aged modules.

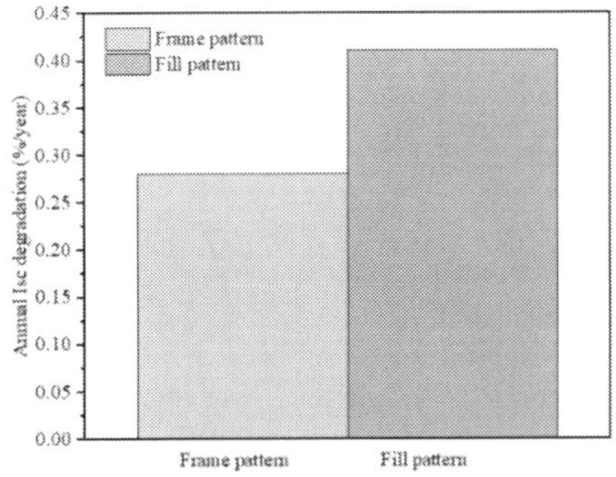

Fig. 7. Annual Isc degradation rate for fill and frame pattern.

Backsheet chalking is the second dominating degradation mode in the ME hot desert region. Almost all the modules showed white powder coming out after wiping the back side of the PV modules, which are exposed in a hot desert climate. The additives that were added to the backsheet (during the fabrication process) appeared in the form of white powder over the aging in the field conditions. The backsheet chalking powder morphology is shown in scanning electron microscopy (SEM) image Fig. 8 (a). The powder particles are distributed with less than 3 um. Chalking powder analysis has been done at seven different positions as shown in Fig. 8 (b). The chalking powder elemental spectra show that Titanium is the main element that has oxidized in the desert region. The chalking powder elemental spectra and their weight percentage distribution have been shown in Fig. 8 (c). Spectra indicates that the TiO_2 is the main product that causes the chalking of the backsheet in the PV module. Table 1 shows the elemental composition of the chalking powder. The chalking is an early stage of backsheet material degradation and does not immediately affect the module performance. Nevertheless, it can cause the potential development of backsheet failure in the form of moisture ingress, delamination of inner layers, and insulation failures.

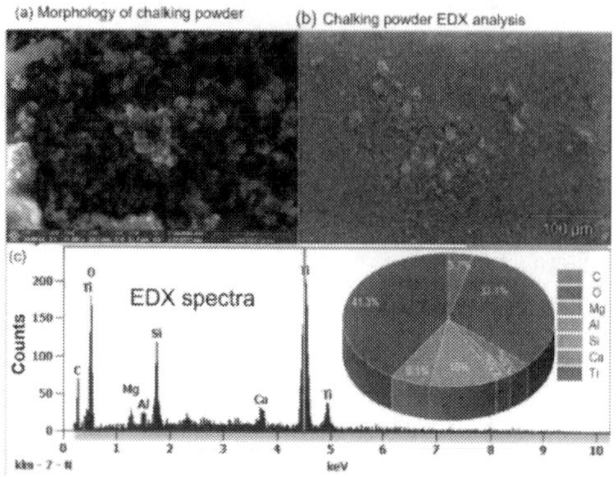

Fig. 8 Chalking powder SEM analysis: (a) morphology of powder, (b) EDX analysis at seven points in SEM image, (c) EDX Elemental spectra and insert Pic char shows the elemental weight distributions.

Table 1. Chalking elemental atomic weight percentage at 7 points.

C	O	Mg	Al	Si	Ca	Ti
1.0	13	0.8	0.8	4.7	3.2	76.4
7.6	45	7.3	1.8	21.6	11.3	3.8
2.1	37	1.7	1.7	5.5	2.2	43.7
0.9	12	0.4	0.5	1.9	3.5	80.3
5.0	24	2.8	3.2	15.4	4.2	5.8
5.9	55	8.1	5.2	8.41	15.3	0
2.4	40	2.5	3.4	11.13	2.45	33.2
3.6	2	3.4	2.4	9.80	6.01	40.5

4 SUMMARY OF THESE NOTES

Middle East's hot desert region is covered with harsh weather conditions. These harsh conditions are the sources that cause the major materials and power degradations in the PV modules. The discoloration in the encapsulants, chalking in the backsheet, rusting, coating chipping and cracks in interconnects, snail train in solar cells, and warpage in solar glass are the dominant visual defects in PV module components. Three modes of material degradation were studied and investigated in the desert such as encapsulant discoloration, backsheet chalking, and interconnect oxidations. EVA is identified as an encapsulant material used in all PV modules and shows two different discoloration patterns. The filled pattern shows higher fluorescence and higher current reduction in the PV modules than the frame pattern. Backsheet chalking is the second dominating degradation mode in the PV modules. The chalking powder shows that TiO_2 is the main component that comes out in the form of white powder from the backsheet. Chalking does not show an immediate effect on the PV module performance but can be a potential cause for other degradation modes.

5 REFERENCES

[1] S. Griffiths, "Strategic considerations for deployment of solar photovoltaics in the Middle East and North Africa," *Energy Strategy Reviews*, vol. 2, no. 1, 2013, doi: 10.1016/j.esr.2012.11.001.

[2] M. Zubair and A. B. Awan, "Economic viability of solar energy export from the Middle East and North Africa to Europe and South Asia," *Environ Dev Sustain*, vol. 23, no. 12, 2021, doi: 10.1007/s10668-021-01424-x.

[3] J. F. Lelièvre *et al.*, "Desert label development for improved reliability and durability of photovoltaic modules in harsh desert conditions," *Solar Energy Materials and Solar Cells*, vol. 236, 2022, doi: 10.1016/j.solmat.2021.111508.

[4] J. F. Lelièvre *et al.*, "Desert label development for improved reliability and durability of photovoltaic modules in harsh desert conditions," *Solar Energy Materials and Solar Cells*, vol. 236, Mar. 2022, doi: 10.1016/j.solmat.2021.111508.

[5] J. Ascencio-Vásquez, K. Brecl, and M. Topič, "Methodology of Köppen-Geiger-Photovoltaic climate classification and implications to worldwide mapping of PV system performance," *Solar Energy*, vol. 191, pp. 672–685, Oct. 2019, doi: 10.1016/j.solener.2019.08.072.

[6] B. Adothu, P. Bhatt, S. Zele, J. Oderkerk, F. R. Costa, and S. Mallick, "Investigation of newly developed thermoplastic polyolefin encapsulant principle properties for the c-Si PV module application," *Mater Chem Phys*, vol. 243, p. 122660, Mar. 2020, doi: 10.1016/J.MATCHEMPHYS.2020.122660.

[7] B. Adothu, A. K. Singh, S. Kumar, S. Zele, and S. Mallick, "Effect of curing temperature on properties of Ethylene vinyl-acetate used for crytalline silicon solar module encapsulation," in *36th European Photovoltaic Solar Energy Conference and Exhibition*, 2019, pp. 1204–1207.

[8] B. Adothu, S. Chattopadhyay, P. Bhatt, P. Hui, F. R. Costa, and S. Mallick, "Early-stage identification of encapsulants photobleaching and discoloration in crystalline silicon photovoltaic module laminates," *Progress in Photovoltaics: Research and Applications*, 2020, doi: 10.1002/pip.3269.

[9] B. Adothu *et al.*, "Newly developed thermoplastic polyolefin encapsulant–A potential candidate for crystalline silicon photovoltaic modules encapsulation," *Solar Energy*, vol. 194, pp. 581–588, Dec. 2019, doi: 10.1016/j.solener.2019.11.018.

[10] S. Chattopadhyay, R. Dubey, and V. Kuthanazhi, *All-India Survey of Photovoltaic Module Reliability : 2016*, no. May. 2016.

[11] S. Kumar *et al.*, "Investigation and Analysis of Defects and Degradations in Desert Fielded Photovoltaic Modules," in *Conference Record of the IEEE Photovoltaic Specialists Conference*, Institute of Electrical and Electronics Engineers Inc., Jun. 2021, pp. 788–792. doi: 10.1109/PVSC43889.2021.9518896.

[12] International Electrotechnical Commission. Technical Committee 82, *IECTS62804-1, Photovoltaic (PV) modules : Test methods for the detection of potential-induced degradation*. 2015.

10.4229/EUPVSEC2023/3AV.2.29
020198-004

OPEN FAULT DETECTION TECHNOLOGY OF BYPASS CIRCUIT OF PV MODULE WITH IR CAMERA -REDUCING OF REFLECTION EFFECT BY CLOUDS-

Shogo. NISHIKAWA & Hiro KUBOTA
Nihon University
1-8-14, Surugadai Kanda Chiyoda-ku, Tokyo, Japan

ABSTRACT: The one of the existing most popular technology for finding open bypass circuit to prevent hotspot is the measurement of surface temperature with an IR camera. However, this solution has defects. For example, thermal image is affected by surrounding structure such as anntena and building, so on, and it is difficult to measure the true surface temperature of PV modules with IR camera. To solve the problems mentioned previously, we developed new detection technology for open fault of bypass circuit. The reflection effect is deleted and the position of open fault part is identified exactly by the development technology. However, the effect of reflection of moving clouds is not deleted. Therefore, we studied the advanced detection technology to delete the reflection effect of moving clouds. In this paper, the outline and effect of new propsed technology is described in this paper.

Keywords: open fault of bypass circuit, detection technology of fault position, IR camera, reflection effect of moving clouds

1 INTRODUCTION AND OBJECTIVE

Bypass circuits of PV module is very important protective circuit to avoid hotspot of module. If bypass circuit becomes open fault, it is possible that hotspots occur in the PV cells without protection of bypass circuit. In the worst case, hotspot is a cause of fire, and we have to find hotspots and open fault of bypass circuit as soon as possible. The one of the existing most popular technology for finding hotspot is the measurement of surface temperature with an IR camera. However, this solution has defects. The typical defect of existing technology with IR camera is that we cannot find hotspot under irradiance less than 600 W/m^2 **[1]**. Because the surface temperature of PV module does not enough increase. And since thermal image is affected by anntena and surrounding structure, it is often difficult to measure the true surface temperature of PV module with IR camera.

To solve the problems mentioned previously, we developed new detection technology for open fault of bypass circuit **[2]**.

By using this developed technology, temperature of only open fault part changes in accordance with change of applying voltage. Therefore, the effect of reflection is deleted and the position of open fault part is identified exactly.

On the other hand, the effect of reflection of moving clouds is not deleted. When one of the components of change rate equals to the frequency of applied voltage, the application of periodic voltage with only one frequency is not always useful. In this case, the temperature change of normal part also increases and we can not identify the position of open fault part.

The objective of this study is to develop the new detection technology to reduce the effect of reflection of moving clouds.

2 EFFECT OF REFLECTION OF MOVING CLOUDS

Figure 1 shows optcal images of PV modules. The upper one is image of no reflection of clouds during sunny day. On the other hand, the lower one is image of reflection of clouds during cloudy day.

a) no reflection of clouds

b) reflection of clouds on PV modules
Figure 1: Optical image of PV modules

Figure 2 shows the thermal images of PV modules with reflection of moving clouds. As shown in these thermal images, surface temperature of PV modules are affected by reflection of clouds and true temperature can not be measured. Moreover, temperarure distributuon of PV modules changes in accordance with move of clouds.

Figure 2: The reflection of sunlight by moving clouds (thermal image)

3. Principle of detection technology

The principle of proposed detection technology for confirming the open fault location is described here. Normal cluster with normal bypass circuit and abnormal cluster with open bypass circuit are connected in series as shown in Figure 3. We apply a periodic voltage Vr across

10.4229/EUPVSEC2023/3AV.2.32
020199-001

the string. The polarity of applied voltage is the direction that bypass diodes(Db) turns on. When we applied voltage across the string shown in Figure 3, the voltage of normal cluster is limited some 1 V because of Db turn-on. On the other hand, the voltage of abnormal cluster is very big more than that of normal cluster.

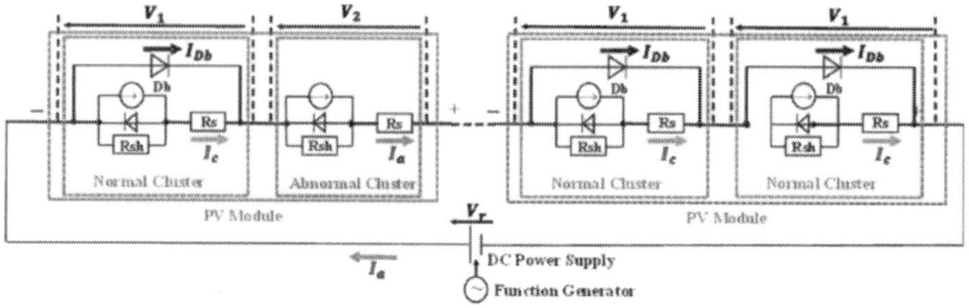

Figure 3: Test circuit for identifying the open fault of bypass circuit in a string

The normal cluster voltage V_1 is very small compared to abnormal cluster voltage V_2 because of on state of bypass diode of normal cluster. Therefore, dissipated energy in the abnormal cluster is bigger than that in the normal cluster, and temperature change in the abnormal cluster is also bigger than that in the normal cluster. Moreover, applied voltage *Vr* including AC components with specified frequency is useful to avoid the effect by surrounding structure and so on.

When reflection of movong clouds does not affect on surface of PV modules, we can detect open fault cluster with square wave or sine wave. However, when reflection of movong clouds affect on surface of PV modules, we can not detect open fault cluster with square wave or sine wave. Because, if the speed of moving clouds includes the component of frequency eqauls to frequency of applied voltage, the amplitude of temperature change of normal cluster can become bigger than threshold[3]. Therefore, we apply the voltage waveform indicated in Figure 4. This waveform is associated wave with square wave whose frequency is 0.01Hz and 0.005Hz, duty ratio is 0.5.

However, the frequency spectrum of surface temparature change of PV modules also includes 0.0075Hz besides 0.01Hz and 0.005Hz.

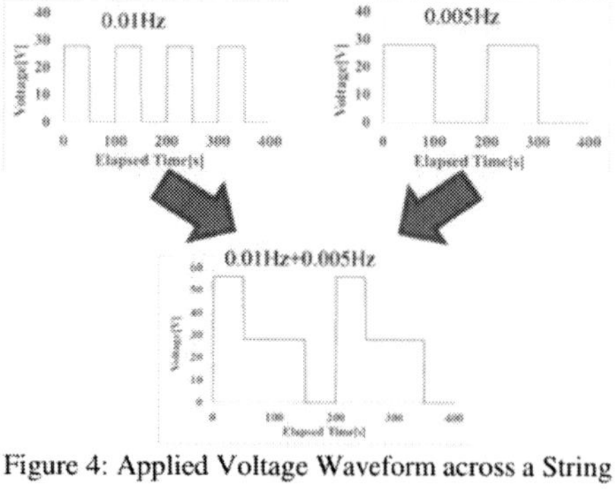

Figure 4: Applied Voltage Waveform across a String

4. ANALYSIS METHOD OF MEASUREMENT DATA

The amplitude of temperature change at all pixeles of thermal image which frequency is 0.01Hz, 0.0075Hz and 0.005Hz are calculated. And binarized image at three types of frequency is made from amplitude of temperature change at all pixeles The threshold of magunitude of temperature change for binarized image is decided based on reference [3]. The red pixel shows that temperature change of a pixel is bigger than that threshold. Moreover, the synthetic binarzed image is made by the procedure as shown in Figure 5.

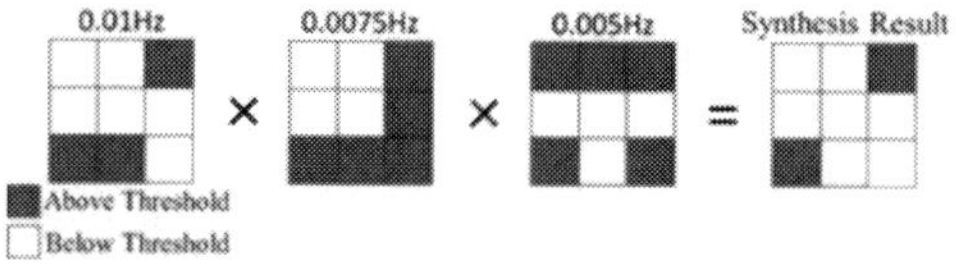

Figure 5. The procedure of synthetic binarized image

The ratio of red pixels RP at every cluster is calculated from equation (1). The RP of open fault cluster is higher than that of normal cluster very much.

$$RP[\%] = \frac{total\ number\ of\ red\ pixels\ in\ a\ cluster}{total\ number\ of\ all\ pixels\ in\ a\ cluster} \times 100 \quad \cdots (1)$$

5. RESULTS AND DISCUSSION

5.1 In the case of no reflection effect

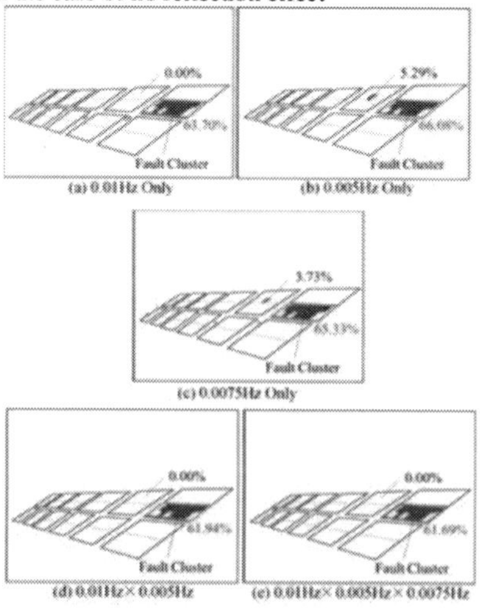

Figure 6: Example of Binarization Result by two types and three types (Detectable, No reflection effect)

10.4229/EUPVSEC2023/3AV.2.32
020199-002

An example of detectable of open fault position is shown in Figure 6. In this case, no reflection effec does not occur in every frequency. The value in the figure is RP for cluster. The RP of open fault cluster decreases a little by synthesizing binarized image with different frequency comonent. And RP of all normal cluster is zero.

The frequency spectrum of temperature changes in normal and open fault cluster with no reflection effect is shown in Figure 7. In this case, the amplitude of temperature change of open fault cluster at every frequency is bigger than threshold. On the other hand, the amplitude of normal cluster at every fequency is smaller than threshold.

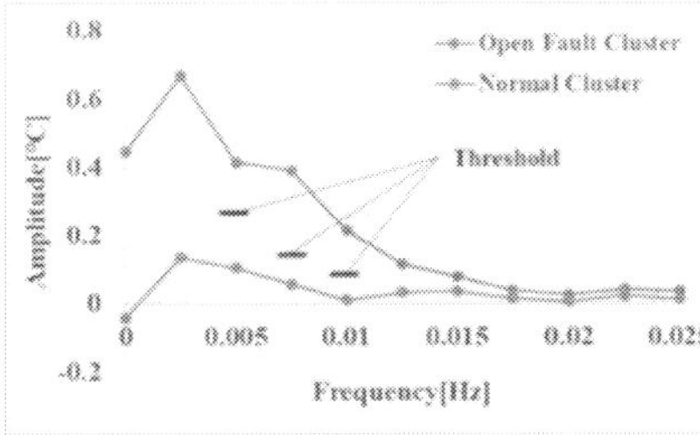

Figure.7: Frequency Spectrum of Temperature Changes in Normal / Open Fault Cluster (No Reflection Effect, Detectable)

5.2 In the case of detectable with reflection effect

An example of detectable of open fault position is shown in Figure 8. In this case, reflection effect occurs at 0.005Hz. However, the RP of open fault cluster decreases by synthesizing binarized image with different frequency comonent. And RP of all normal cluster is almost zero.

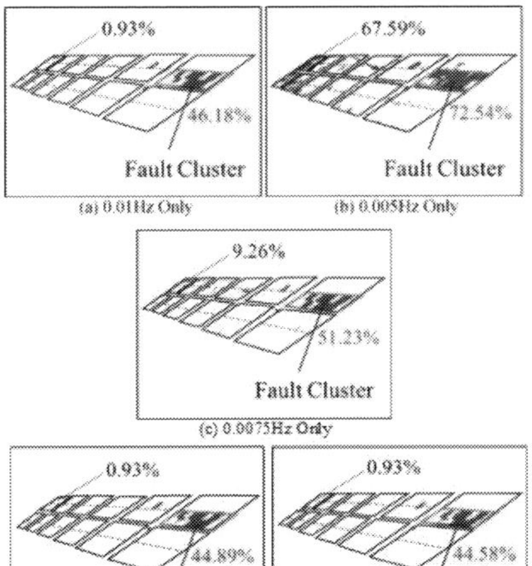

Figure.8: Example of Binarization Image by Two Types and Three Types of Frequency (Affected by Reflection, Detectable)

The frequency spectrum of temperature changes in normal and open fault cluster with reflection effect is shown in Figure 9. In this case, the amplitude of temperature change of open fault cluster at 0.005Hz only is bigger than threshold.

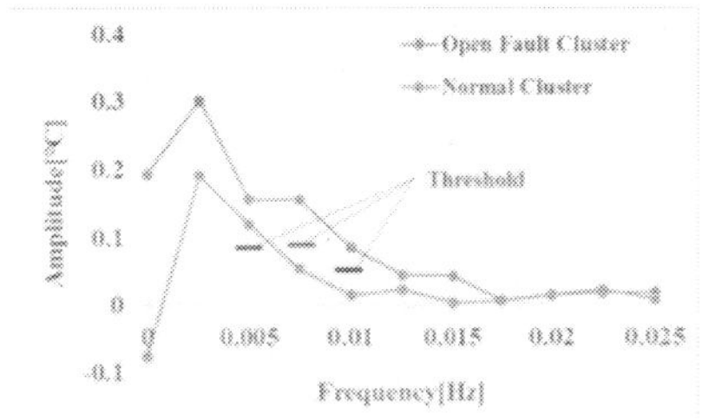

Figure.9: Frequency Spectrum of Temperature Changes (Affected by Reflection Effect, Undetectable)

5.3 In the case of undetectable with reflection effect

An example of undetectable of open fault position is shown in Figure 10. In this case, reflection effec occur at all frequency. Therefore, the RP of normal cluster does not decrease by synthesizing binarized image with different frequency component.

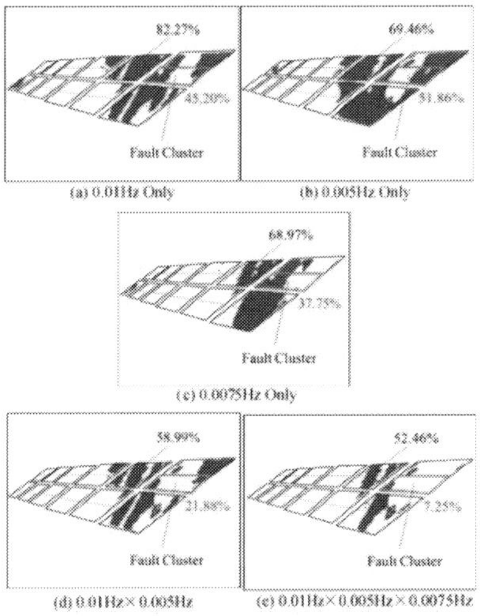

Figure.10: Example of Binarization Image by Two Types and Three Types of Frequency (Affected by Reflection, Undetectable)

The other frequency spectrum of temperature changes in normal and open fault cluster with reflection effect is shown in Figure 11. In this case, the amplitude of temperature change of normal cluster at every frequency is bigger than threshold.

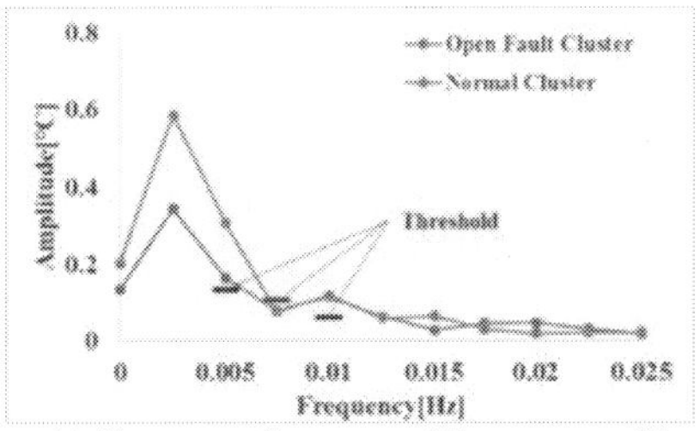

Figure.11: Frequency Spectrum of Temperature Changes (Affected by Reflection Effect, Undetectable)

10.4229/EUPVSEC2023/3AV.2.32
020199-003

6. CONCLUSION

We proposed new technology to detect open fault position of bypass circuit by deleting the reflection effect of moving clouds. Though this technology is useful for some cases, it is not imperfect.

7. REFERENCES

[1] IEC 62446-3 TS: Photovoltaic (PV) systems – Requirements for testing, documentation and maintenance – Part 3: Outdoor infrared thermography of photovoltaic modules and plants

[2] S. Nishikawa, N. Fujita, and H. Kuroda, "Fundamental Study on Open Fault Detection Technology of Bypass Circuit of PV Module with IR Camera", in 37th European Photovoltaic Solar Energy Conference and Exhibition, 2020, pp.914-918

[3] N.Otsu, "An Automatic Threshold Selection Method Based on Discriminant and Least Squares Criteria", IEICE trans. D, Vol.J63-D, No.4,pp.349-356(1980-4) (in Japanese)

10.4229/EUPVSEC2023/3AV.2.32
020199-004

ACCELERATED AGING STUDY OF BACKSHEET REPAIR WITH FLOWABLE SILICONE SEALANT

Guy Beaucarne[1], Jorge Lima Garcia[1], Emmanuel Jadot[1], Kayla Kenney[2], Anika Gassner[3], Gabriele Eder[3]

[1]Dow Silicones Belgium SRL, Rue Jules Bordet Parc Industriel Zone C, B-7180 Seneffe, Belgium
[2]Dow Silicones Corporation, 2200 W Salzburg Road, Auburn, MI 48611, USA
[3]OFI - Austrian Research Institute for Chemistry and Technology, Franz-Grill-Straße 1, 1030 Wien, Austria

ABSTRACT: We investigate the repair of polyamide AAA backsheets on test samples and on complete PV modules that had been degraded during years of operation in a solar park, and study the durability of the repair through further climate chamber aging. We find that coating backsheets with a flowable silicone sealant layer reduces further damage of backsheets, and that increasing the layer thickness improves that effect. Experiments in which PV modules with badly degraded AAA backsheets were repaired with a layer of silicone repair coating material of 400 µm average thickness showed restored electrical insulation as evidenced by insulation resistance measurement in wet leakage conditions. The repair effectiveness was maintained after applying Sequence B of IEC 61730. The extent of further backsheet degradation upon further aging after repair appears to depend on the initial pre-repair degradation. It is therefore recommended to apply the repair coating at the earliest stage as possible, before extensive cracking has occurred.

Keywords: Backsheet, degradation, repair, silicone, insulation

1 INTRODUCTION

In many solar parks installed between 2010 and 2016, severe degradation of some PV modules has recently been observed through formation of cracks in the backsheet resulting in loss of electrical insulation and inverter tripping. In many cases, this phenomenon has been associated with a specific type of backsheet, namely coextruded polyamide backsheets ("AAA backsheets") [1-3], which was unfortunately widely used at the time due to inadequate testing standards. PV modules affected by this problem sometimes get replaced but this is challenging in terms of cost and some practical aspects, such as the different size and specifications of modern modules compared to the original modules. An alternative is to repair the modules, for which different solutions have been developed, either based on tapes or on repair coatings [4]. A promising type of coating-based approach is to use a flowable silicone sealant [5], because of the well-known excellent durability and electrical insulation properties of silicones. There is a need to determine the durability of repair solutions, and their ability to prevent further backsheet degradation. In this work we investigated AAA backsheet degradation in accelerated aging with and without a flowable silicone protective coating.

2 EXPERIMENTAL DETAILS

2.1 Tests on coupons

Test coupons were made consisting of float glass, EVA (0.9 mm) and AAA backsheet. The sample size was approximately 8 x 25 cm^2. Cracks were simulated by cutting the backsheet over 5 cm with a sharp knife. Then some coupons were covered with layers of a white silicone coating material of type DOWSIL™ PV-9001 Backsheet Coating with various thicknesses by doctor blade coating. All samples were subjected to a backsheet durability test sequences in climate chambers and assessed visually between the different steps. The test sequence was based on the PVEL backsheet durability test sequence [6], designed to be applied on new modules to identify problematic backsheets.

DH1000
Back UV 65 kWh/m2
TC50 + HF 10
Back UV 65 kWh/m2
TC50 + HF 10
Back UV 65 kWh/m2
TC50 + HF 10
Back UV 65 kWh/m2

Figure 1: PVEL backsheet durability test sequence

To assess the impact of damp heat, some samples underwent the sequence without the initial damp heat step.

2.2 Tests on field-aged modules

Six PV modules showing severe degradation were obtained. These modules had been originally manufactured by a tier 1 PV manufacturer. They featured 60 multicrystalline silicon solar cells and had a nameplate power of 245W. AAA backsheets had been used in the bill of material. The panels had operated in a solar park in Europe for an estimated period of 8 to 10 years before being decommissioned. Severe degradation could be observed, with deep cracks in a checkered pattern ("squared cracks"[1]), visible encapsulant degradation and some signs of corrosion in some places on the bussing ribbons. The question whether such highly degraded modules should be considered for repair is debatable, but for the sake of the study and as a sort of worse case scenario, repair was attempted nevertheless.

Figure 2: Photo of one of the PV modules used for study

The back of the modules was cleaned by wiping it with a moist cloth, after which it was dried with a dry cloth. The backsheet was then covered by a thin layer of the flowable silicone. For each module, the target average thickness was 400 µm. Modules 1 to 5 were coated by pouring the material and spreading with a squeegee until a uniform layer was obtained. For module 6, the coating was sprayed on using a high-pressure hydraulic pump. After coating and curing, the modules were shipped to certified testing lab AIT in Austria. There the insulation resistance (R_{iso}) of the modules was measured under wet leakage conditions (IEC 61730 MST 17). Subsequently, three of the modules were subjected to a climate chamber aging sequence, namely sequence B from IEC 61730-2 (sequence introduced in the standard in 2016 to simulate conditions that cause backsheet cracking). Three modules were selected for those tests. Two modules (1, 6) were selected for the full aging sequence, while module 4 underwent the sequence without the UV exposure steps. Wet leakage tests were performed after the first humidity freeze (HF) cycle and at the end of the sequence. The modules were observed closely between each step, in particular the backsheet and the repair coating.

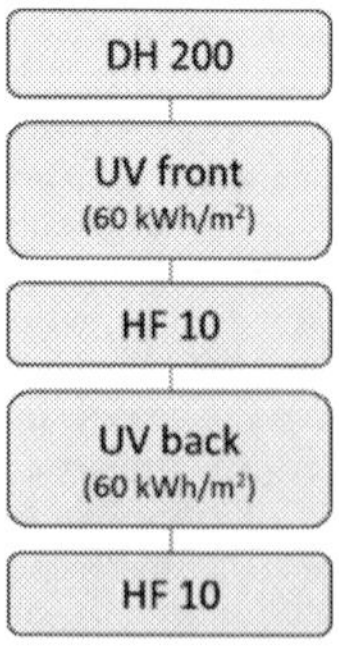

Figure 3: Sequence B from IEC 61730-2:2016

3 RESULTS

3.1 Tests on coupons

Figure 4 shows photos of the samples after the complete ageing cycle including damp heat.

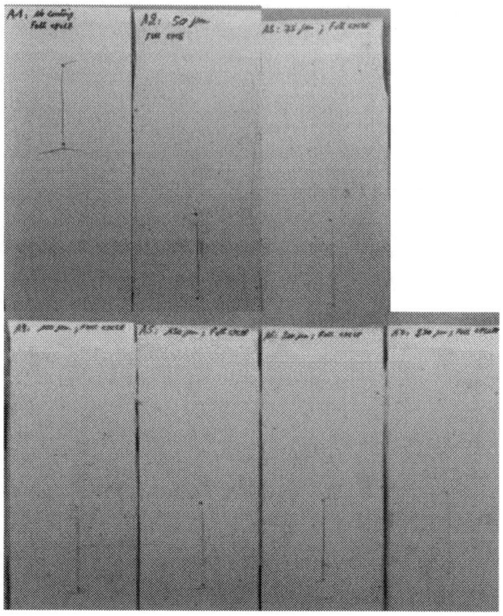

Figure 4: Photos of glass – EVA - AAA backsheet samples with an initial cut, taken after the complete ageing test sequence. Silicone layer thickness ranges from 0 µm (no coating) to 250 µm

As can be seen in Figure 4, the aging results in the initiation of cracks at the extremities of the cuts, perpendicular to the cut direction. This is because, in this set of experiments, the cuts were made in a direction perpendicular to the 'machine direction', the direction of film extrusion during backsheet manufacturing. In hindsight this was the wrong direction to choose if the cut aimed to be an 'artificial crack'. The strength of the material is the lowest in the transverse direction, and therefore the cracks appear first in the machine direction upon thermal cycling [6].

A close examination of the photos shows that very thin silicone coatings above the 'artificial cracks' appear to come open upon thermal cycling whereas thicker layers stay intact (Figure 5).

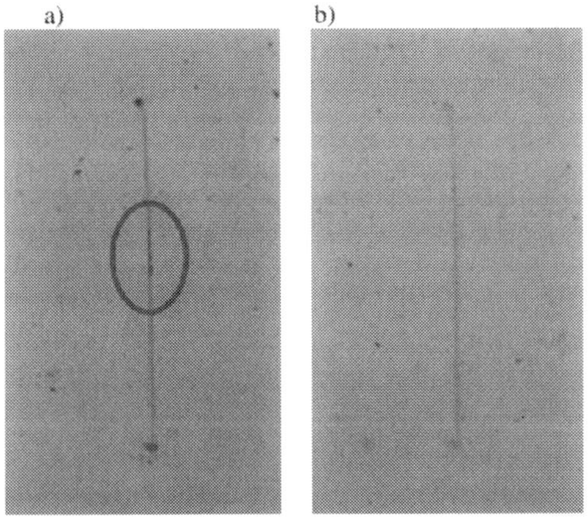

Figure 5: Close-up photos of two samples glass-EVA-AAA backsheet shown in Figure 1, illustrating that the silicone layer local appears to break open for very thin coatings. a) 75 µm thick silicone layer, b) 250 µm layer

The initial damp heat treatment in the backsheet ageing sequence turns out to have a large impact on backsheet degradation. Table 1 compares samples that underwent the accelerated ageing sequence according to Figure 1, but with and without the initial damp heat treatment. It is clear that the absence of initial damp heat leads to a lower level of degradation. This is very important to keep in mind when comparing aging results of different backsheet durability testing sequences, because of some of them do not include a damp heat step [7].

Table 1: glass – EVA - AAA backsheet laminate samples with cut, with or without silicone coating, after aging sequence including or not initial damp heat treatment

	0 µm:no coating	100 µm coating	250 µm coating
with DH	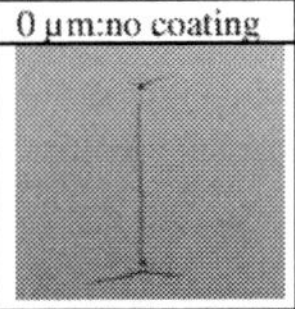	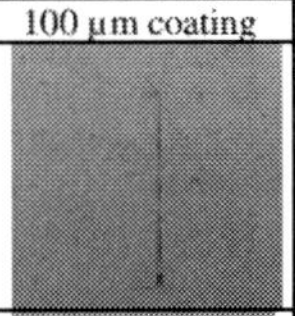	
No DH	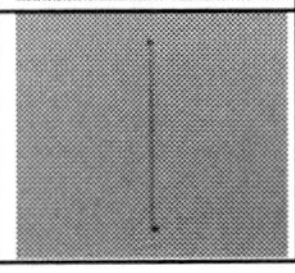		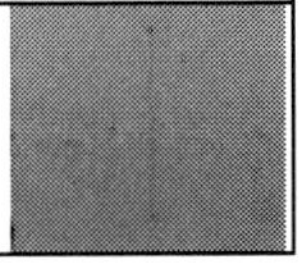

One of the reasons why a thicker layer protects better from degradation is that it allows less UV light to reach the backsheet. The flowable sealant used contains a mineral filler that scatters and reflects lights. With increasing film thickness, the transmitted light through the film decreases. This is illustrated in the transmittance measurements in Figure 6. These measurements were made on samples consisting of the silicone coating material between to pieces of optical glass. Transmittance is expected to be even lower for the case of air/silicone/backsheet structure, as a lot of light will be reflected at the air/silicone interface.

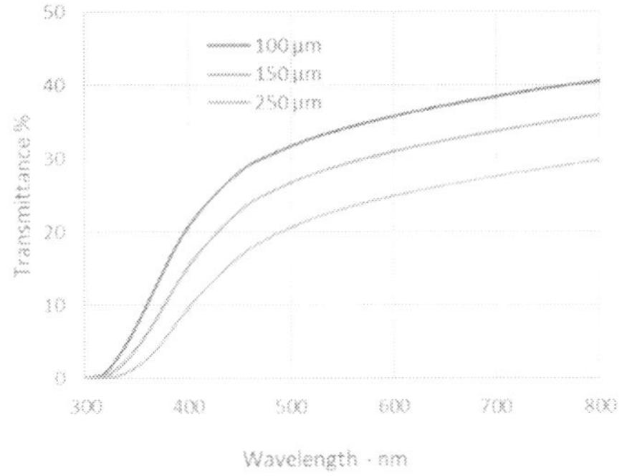

Figure 6: Total transmittance curves for glass-silicone sealant-glass samples with different silicone layer thicknesses (100 µm, 150 µm, 200 µm).

3.2 Tests on field aged modules

Table 2 shows the measurement results of the insulation resistance of the 6 repaired modules under wet leakage conditions according to IEC 61730 MST 17. It can be seen that R_{iso} has been restored to high values (much higher than the minimum required 25 MΩ), and this for both for the squeegee- and spray-applied repair coatings.

Table 2: Insulation resistance measurement in wet leakage conditions for the repaired field-aged modules

Sample	Coating method	R_{iso} (MΩ)
1	Squeegee	360
2	Squeegee	287
3	Squeegee	311
4	Squeegee	340
5	Squeegee	381
6	Spray	350

Figure 7 shows the back of module 1 in the same spot at different stages of the aging sequence.

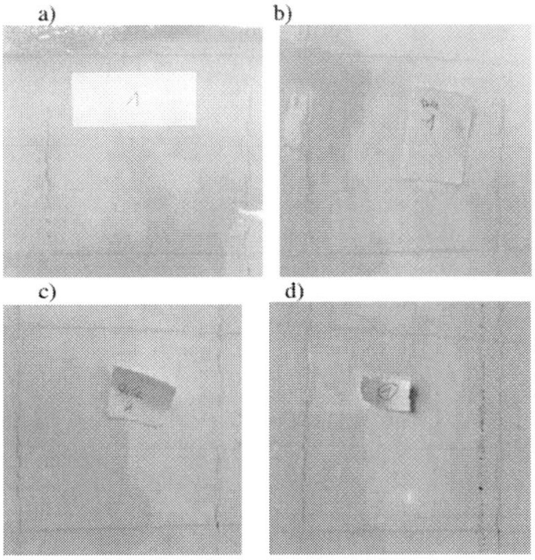

Figure 7: Photo of the back of module 1 at different stages of the aging sequence: a)initial stage, b)after DH

and front UV, c) after DH, front UVand first HF, d) after complete sequence

The appearance of the backsheet (at the rear) and coating after damp heat and front UV steps does not change compared to the initial situation. The humidity freeze cycle in contrast turned out to have a significant impact on the backsheet appearance. The silicone coating above the cracks in the backsheet appear to open in some places, see for instance in the right bottom corner of Figure 7 (c).

In order to assess the impact of the apparent coating failure on electrical insulation, an intermediate R_{iso} measurement under wet leakage was carried out after the first HF10 cycle of sequence B. As can be seen in Table 3, last but one column, electrical insulation has in fact remained high, well above the minimum value to pass the wet leakage test.

Table 3: Insulation resistance measurement in wet leakage conditions for the repaired field-aged modules before, during and after sequence B aging

Sam-ple	Comment	**Initial R_{iso} (MΩ)**	**R_{iso} after first HF10 (MΩ)**	**Final R_{iso} (MΩ)**
1	Squeegee applied	360	318	344
4	Squeege applied, no UV	340	272	236
6	Spray applied	350	302	307

Sequence B was subsequently completed with the steps of back UV exposure and the final HF10 cycle. Further changes to the backsheet appearance could be observed. While the openings in the silicone coating did not evolve much (Figure 7(d)), new cracks in the backsheet were observed in places where there was no crack initially, including some behind interconnecting ribbons. These cracks were well covered by the silicone coating.

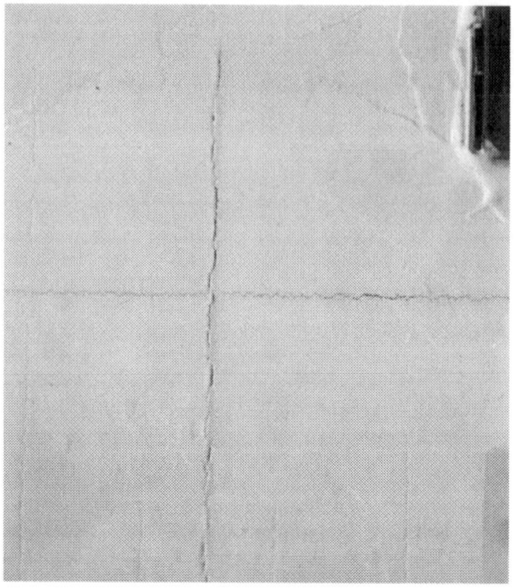

Figure 8: Photo of the back of module 1 after completion of sequence B. Some newly formed cracks are visible below the silicone coating

Insulation resistance under wet leakage remained high, as can be seen in the last column of Table 3.

4 DISCUSSION

The results presented above show that the degradation history prior to repair (extent of material properties degradation of the backsheet, presence of already formed cracks, etc.) has an impact on how the backsheet and the laminate as a whole is going to degrade further. The worse the degradation that has already taken place, the worse further degradation will be. The presence of the silicone coating can have a preventive action on further degradation, as can for instance be seen Figure 1 and Table 1, where the coating clearly prevents crack formation in the machine direction. However, in case of already bad degradation, further degradation is not stopped. It is therefore recommended to apply the repair coating at the earliest stage as possible, before extensive cracking has occurred.

The repair coating nevertheless can restore electrical insulation even on laminates with badly cracked backsheets, and maintain it upon further aging even when the backsheet degrades further under the coating. The observation that insulation resistance in wet leakage was maintained even when the silicone layer appeared to open upon extensive aging, is both striking and intriguing. Several hypotheses can be put forward. Possibly the silicone does not open through the whole thickness of the layer. The low surface tension of the silicone coating might also play a role. Further study of these modules will shed light on this surprising aspect of backsheet repair with silicone coating.

5 CONCLUSION

We investigated the repair of AAA polyamide

backsheets on test samples and on complete PV modules that had been degraded during years of operation in a solar park, and studied the durability of the repair through further climate chamber aging. We find that coating backsheets with a silicone layer reduces further damage of backsheets upon accelerated aging. The extent of further backsheet degradation upon further aging after repair appears to depend on the initial pre-repair degradation, suggesting that carrying out repair in an early stage of degradation, prior to extensive backsheet damage, is preferable.

It was shown that coating PV modules with degraded AAA backsheets with a layer of a silicone repair coating with average thickness of 400 µm restores electrical insulation as evidenced by insulation resistance in wet leakage conditions, even for modules with extensive backsheet cracking. The repair effectiveness was maintained after applying Sequence B of IEC 61730-2:2016, which includes multiple environmental stresses.

7 ACKNOWLEDGEMENTS

We thank Aimé Tuyishimire from Dow Silicones Belgium for the transmittance measurements. The authors are also grateful to Rirae srl for providing the field-aged modules.

6 REFERENCES

[1] G. Eder, Y. Voronko, G. Oreski, W. Mühleisen, M. Knausz, A. Omazic, A. Rainer, C. Hirschl, H. Sonnleitner, Error analysis of aged modules with cracked polyamide backsheets, Solar Energy Materials and Solar Cells, 203 (2019) 110194.
[2] J. Markert, S. Kotterer, D.E. Mansour, D. Philipp, P. Gebhardt, Avanced analysis of backsheet failures from 26 power plants, 38th European PV Solar Energy Conference (2021).
[3] J. Pascual, M. García, J. Marcos, L. Marroyo, Analysis of polyamide and fluoropolymer backsheets: Degradation and insulation failure in field-aged photovoltaic modules, Progress in Photovoltaics: Research and Applications, (2022).
[4] Y. Voronko, G. Eder, C. Breitwieser, W. Mühleisen, L. Neumaier, S. Feldbacher, G. Oreski, N. Lenck, Repair options for PV modules with cracked backsheets, Energy Science & Engineering, (2021) 1-13.
[5] G. Beaucarne, G. Eder, E. Jadot, Y. Voronko, W. Mühleisen, Repair and preventive maintenance of photovoltaic modules with degrading backsheets using flowable silicone sealant, Progress in Photovoltaics: Research and Applications, 30 (2022) 1045-1053.
[6] P. Lechner, H. Wirth, J. Schnepf, S. Hummel, D. Geyer, B. Weinreich, R. Haas, Analysis of Fielded PV-Modules with Backsheet Issues, 36th European Photovoltaic Solar Energy Conference, (2019).
[7] M.D. Kempe, T. Lockman, J. Morse, Development of Testing Methods to Predict Cracking in Photovoltaic Backsheets, in: 2019 IEEE 46th Photovoltaic Specialists Conference (PVSC), 2019, pp. 2411-2416.

The complete results of the tests carried out at AIT can be found in the AIT final report, available upon request to this paper's first author.

Relevance of Pre-exponential constant and Activation energy in the degradation of encapsulant (EVA/TPO) in c-Si photovoltaic modules

Maruthi Avula[1], Baloji Adothu[1,2], Krishan Kumar Patel[1], Siddharth Singh[1], Sudhanshu Mallick[1]

[1]Metallurgical Engineering & Materials Science Department, Indian Institute of Technology Bombay, Mumbai, 400076, India

[2]Dubai Electricity and Water Authority, Research & Development Center, Al Qudra, MBR Solar Park, Dubai, UAE

Corresponding author Email id: 204110019@iitb.ac.in

Abstract: Encapsulant discoloration is a significant cause of degradation in the Power conversion efficiency of solar modules. Evaluating the encapsulant's degradation under accelerated aging conditions and predicting the lifetime is critical. The current work reports the calculation of Activation energy to understand the discoloration of Ethylene Vinyl Acetate (EVA) and Thermoplastic Polyolefin (TPO) in C-Si Photovoltaic (PV) solar modules encapsulants using an indoor accelerated aging ultraviolet (UV) stress test of 244 days. Multiple modules are aged simultaneously at different temperatures in a single weathering chamber. The percentage change in short circuit current (Isc) per hour, Yellowness Index (del. YI), and Diffusion Coefficient(D) were used as the primary responses in the Arrhenius equation for the calculation of the activation energy. UV-365 illumination and Fluorescence imaging are used to identify the fluorophore generation in the encapsulant of PV modules. The values we got here for both EVA and TPO revealed the importance of pre-exponential constant along with activation energy and the methodologies reported here are unique and straightforward methods to evaluate the reliability and study the encapsulant degradation mechanisms using relatively less time and resources. Understanding the degradation characteristics can lead to the development of appropriate formulations/additives in EVA/TPO for better performance in different climatic conditions.

Keywords: Photovoltaic solar cell; Activation energy; Inverted grayscale; Fluorescence imaging;

Introduction:

Solar energy is a prominent source of renewable energy because of the availability of insolation throughout the year in some geographical locations. Rapid technological advances and industries make this area the most promising renewable energy technology. Despite technological advances, there are many hurdles affecting the reliability of PV modules. A major challenge is the degradation and delamination of module encapsulation which depletes the lifetime of solar modules [1]. Encapsulation is a crucial element of the PV module. Generally, a module consists of a top layer, typically glass, followed by solar cells sandwiched between two encapsulant layers, the back sheet, and the junction box. The life of a PV module is maintained by choosing the right components, and the encapsulant is a major part that needs to be carefully chosen due to its susceptibility to degradation [2]. Ethyl vinyl acetate (EVA) is a material that is used predominantly as an encapsulant. EVA undergoes thermal degradation and UV aging in the presence of weathering elements, affecting the PV module's overall efficiency [3]. Compared with other layers, this encapsulant has a high chance of degradation as soon as it is exposed to humidity, weather conditions, and UV radiation. Various additives such as UV absorbers, UV light stabilizers, and antioxidants are added to the polymer composition of EVA to protect the solar cell and increase its service life [4]. PV modules have an expected life of 25-30 years, meaning they must perform at 80% of their efficiency till the end of life [5]. In a study conducted by Gagliardi et al., modules set up in different regions of the world for 20 years were collected and studied. It was noticed that modules from hotter, warmer, and more irradiated zones suffered more EVA degradation than modules from other parts of the world [6]. While both high temperature and UV exposure play a role in the degradation of EVA, the extent of their contribution might not be the same. Kshitij Dolia [7] experimented with UV-cut and UV-pass EVA to determine the extent of discoloration under the same conditions for both. UV-cut EVA contains additives that absorb UV light below 360 nm, while UV-pass EVA allows the transmission of UV light to the cell. UV-cut EVA showed a high amount of discoloration. Although no discoloration was observed in UV-pass EVA, the complete delamination of the encapsulant was most likely due to the absence of adhesion-promoting additives in EVA.

Indoor accelerated activation energy testing involves putting the modules under various stresses and theoretically simulating the effect of years of usage. The activation energy calculated using this method is used in various empirical models that estimate the time to failure in outdoor conditions considering the weathering elements. David Miller et al. performed accelerated UV stress testing on various EVA-based encapsulants under different steady-state conditions to calculate the activation energy for encapsulant discoloration. The data were fitted by an Arrhenius relation, and the activation energy was estimated to be 30 to 60 kJ/mol (0.31 to 0.62 eV) [8]. TPO is gaining a lot of interest in the PV industry compared to EVA due to the lack of vinyl acetate groups, restricting any formation of acetic acid - which is linked to many failure mechanisms in EVA PV Modules. Its higher volume resistivity and lower water transmission rates are other advantages. Baloji et al. performed thermal analysis and reported that the thermal stability of TPO is better than EVA, and the crystallinity remains the same or shows a minor increase for TPO [9]. The present work determines the activation energy and the pre-exponential factor for the degradation of TPO and White TPO (Thermoplastic Polyolefins) along with EVA to compare the stability of the various encapsulants.

2. Experimentation

2.1 PV Module Fabrication Materials: For this experiment, three types of modules were used: i) TC Module - Top is UV-pass EVA; Bottom is UV-cut EVA ii) TT Module - Top and Bottom are both UV-pass TPO iii) TW Module – Top is UV-pass TPO.; Bottom is white TPO. White TPO is milky white and has higher reflectivity. This higher reflectivity in the white TPO (90%) as compared to transparent EVA (75%) results in a higher module power output when used as the encapsulant on the back side of the solar cell. However, the peel strength between the white TPO/Back sheet interface is lower than the transparent EVA/Back sheet interface, which may result in early delamination and failure, and the schematic representation of this arrangement is shown in Fig .1.

2.2 UV Accelerated Aging Test: A T-type thermocouple is attached using aluminum tape to the backside center of each sample, for monitoring sample temperature. The samples are kept in a 12-cm glass wool insulation box and tightly packed with aluminum tape and the whole system is kept beside the black panel temperature sensor inside the Xenon accelerated test chamber. The Xenon test chamber is operated with a total UV intensity of 72 W/m^2 at a black panel temperature of 60°C and a chamber air temperature of 40°C at 20% chamber relative humidity. The samples were kept in a chamber for 244 days, and photobleaching and discoloration patterns were observed and identified using nondestructive techniques.

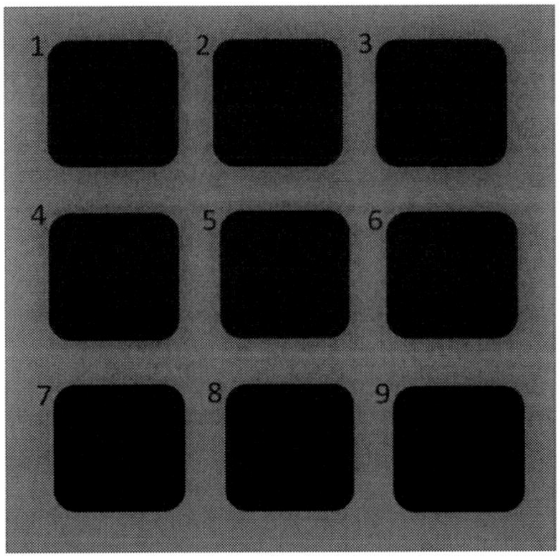

Figure 1. Schematic representation of PV cells (9 cells in each module encapsulated with different materials)

Calculation of Activation Energy: A simple Arrhenius equation is used to calculate the activation energy for the discoloration or photobleaching of the PV modules [10,11].

$$R = Ae^{-\frac{Ea}{KT}} \qquad \textbf{Equation.............(1)}$$

$$\ln R = \ln A + (-\frac{Ea}{K})\frac{1}{T} \qquad \textbf{Equation.........(2)}$$

Where 'A' is a pre-exponential factor; Ea = Activation Energy (eV); k = Boltzmann constant (8.615 * 10-5 eV/K); T= Temperature (K), and 'R' is substituted as – % change in short-circuit current (Isc) per hour or Change in Yellowness Index (del. YI) or Diffusion Coefficient. Plotting the natural logarithm of each of these against (1/T) for each of the mini-PV modules and employing a best-fit line to the corresponding points allows us to calculate the activation energy and the preexponential factor by equating the slope to (-Ea/k) and the intercept to (A), according to equation (2).

2.3 Nondestructive Characterizations: Fluorescence imaging and its grayscale image are used to characterize discoloration and photobleaching in the PV mini-modules. Fluorescence images were captured by using a digital camera under UV-365 light illumination. The optical transmittance and absorbance of glass-to-glass EVA laminates (unexposed samples) have been measured with UV/V is spectrophotometer from 200- to 900-nm spectral range with a 1-nm step size at an average integrating time of 0.1s. Yellowness Index (YI) is a dimensionless number calculated from spectrophotometric data that determines the degree to which a sample's color shifts away from an ideal white. It is measured with a portable colorimeter at a 10°observer angle and a D65 light source. Current-voltage (I-V) measurement has been recorded on a Class AAA Spire Solar Simulator

10.4229/EUPVSEC2023/3AV.2.36

3. Results and Discussion: -

3.1. Fluorescence Imaging: Fluorescence imaging is a technique used to identify the fluorophore generation in the encapsulant of PV modules. Fluorophores are generated as a result of the degradation of polymers or additives. Generally, fluorophores are excited with UV light and emit radiation at higher wavelengths (visible region). Fluorescence imaging shows the fluorescence of fluorophores in the degraded sample. There are no fluorophores present in all the prepared samples, which is clearly shown in the initial fluorescence image and its inverted grayscale images. Degradation of mini-modules was assessed after 244 days of UV accelerated aging test. All three types of mini-modules (TC, TT, and TW) show fluorescence under UV-365 illumination.

3.2. Inverted Grayscale Profile: The individual mini-module fluorescence images are converted into an inverted grayscale image by using ImageJ software (Fig.2.) The white areas are the photobleached regions (at the edges of the modules), while the dark black color indicates discoloration. The conversion to an inverted grayscale image is important to find the amount of photobleaching that occurred by calculating oxygen diffusion length (ODL). This is calculated by measuring the length of the photobleached region from all sides of the mini-module and then averaging it (Red lines in Fig .2(d)). This gives us the ODL for all the mini-modules.

Then equation 3 is used to calculate the Diffusion Coefficient,

$$L = \sqrt{Dt} \qquad (3)$$

In the above equation where 'L' is ODL, 'D' is the diffusion coefficient, and 't' is the time in seconds.

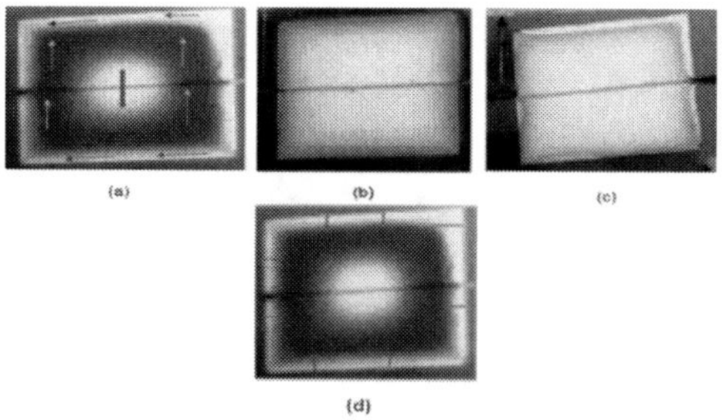

Figure 2: (a) TC (b)TT (c)TW, (For TC) Red - Photobleached; Orange - Discolored; Blue – Undegraded and (d)TC (Red color lines indicate the ODL)

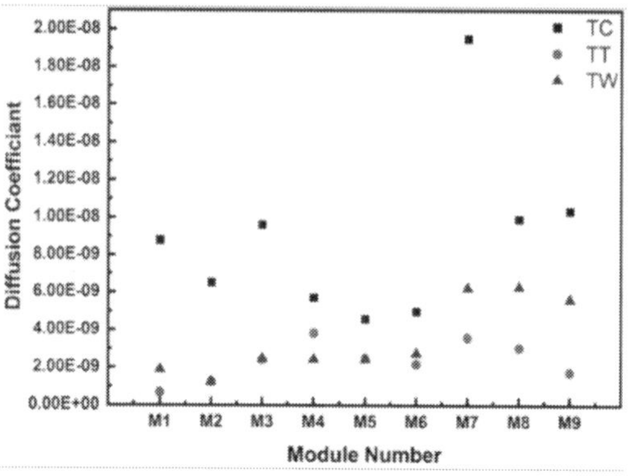

Figure 3: Diffusion Coefficient of PV modules encapsulated with EVA, TPO, and BPO

The Diffusion Coefficient was calculated by using equation 3 and the values were plotted against the number of modules the plot was shown in (fig.3.). This diffusion coefficient will be used for the calculation of Activation Energy for photobleaching, while a loss in short-circuit current (Isc) and change in Yellowness Index (del. YI) will be used to calculate Activation Energy for discoloration by using the Arrhenius equation.

3.3. Calculation of activation energy by using Isc, YI, and D

3.3.1. Activation Energy using Isc: Initial Isc (before UV aging) and final Isc (after UV aging) are recorded for each of the mini-modules. Percentage drop In Isc per hour is calculated for each mini-PV Module, and its natural log is plotted against the inverse of temperature (1/T). Using a best-fit line, the slope is calculated for TC, TT, and TW modules. Using that, the Activation energy and the Pre exponential factor for discoloration are calculated using the Arrhenius equation (equation (2)). The respective plots along with the best-fit line, are shown in fig (4).

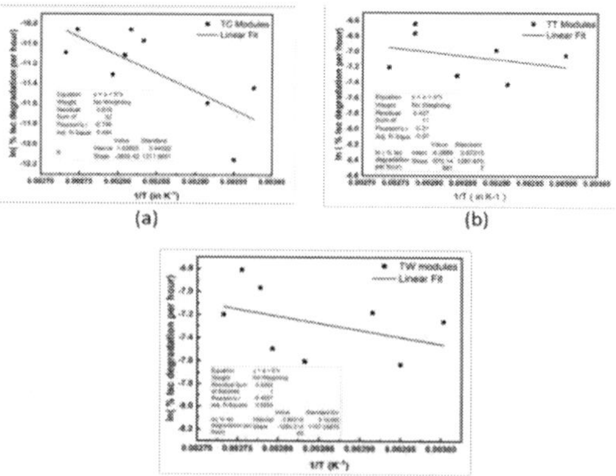

Figure 4: *(a) TC (b)TT (c)TW, ln (% Isc deg rate) vs 1/T (K^{-1})*

10.4229/EUPVSEC2023/3AV.2.36
020203-003

Table 1: *Activation energy and Pre-exponential factor using Isc*

Module	Slope	Ea	A
TC	-3602.42	0.31 eV	0.358
TT	-970.14	0.08 eV	0.014
TW	-1204.21	0.10 eV	0.021

3.3.2. Activation Energy using Yellowness Index: Similar to the Isc calculation, we have used the Arrhenius equation to calculate activation energy by using a change in yellowness index (YI). The initial and final Yellowness Index (YI) values are recorded, and the natural logarithm of the change in YI is plotted against 1/T. Activation energy is calculated according to equation 2, and the respective plot, along with the best-fit line, is shown in Fig (5).

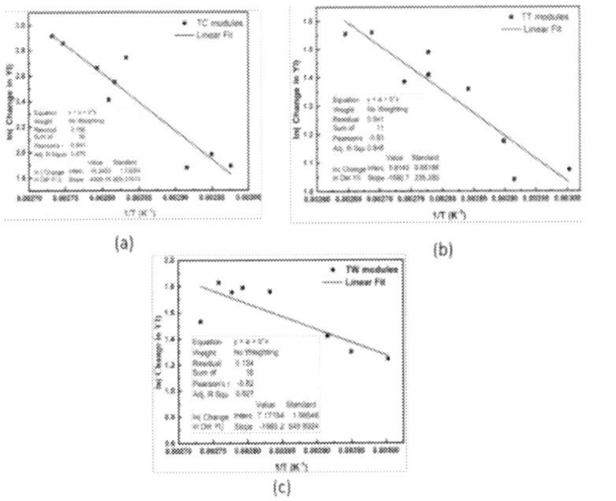

Figure 5: (a) TC (b)TT (c)TW, ln (Change in YI) vs 1/T (K^{-1})

Table 2: Activation energy and Pre-exponential factor using YI

Module	Slope	Ea	A
TC	-4509.16	0.39 eV	4.15*106
TT	-1592.71	0.14eV	333.61
TW	-1965.27	0.17eV	1299.84

3.3.3. Activation Energy using Diffusion Coefficient (D): Similar to the previous methodology, the activation energy for photobleaching is calculated using the diffusion coefficient as the parameter in the Arrhenius relation.. The respective plot, along with the best-fit line, is shown in fig (6). In the plot, one can easily notice the wild spread of the data, and several data points need to be excluded to get the Arhennius fit, so the measurement technique by diffusion length is not reliable. But at the same time, the inverted grayscale image, method easily shows visible signs of degradation and photobleaching.

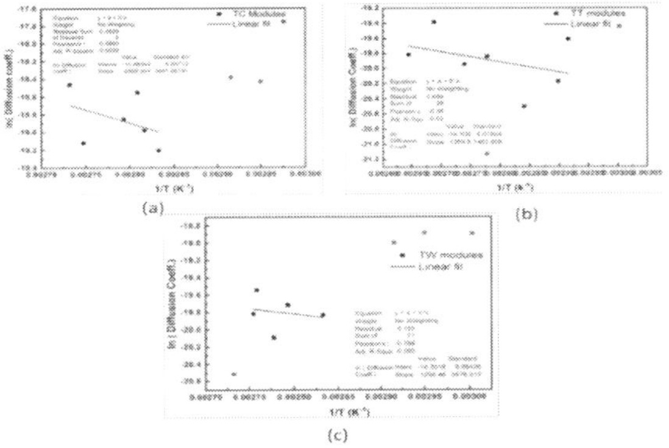

Figure 6: (a) TC (b)TT (c)TW, ln (Diffusion coefficient (D)) vs 1/T (K^{-1})

Table 3: Activation Energy and Pre-exponential factor using Diffusion Coefficient

Module	Slope	Ea	A
TC	-2968.30	0.26 eV	$2.54*10^{-5}$
TT	-1359.84	0.12eV	$1.01*10^{-7}$
TW	-1256.46	0.11eV	$8.33*10^{-8}$

Summary

An indoor accelerated UV testing was performed in a Xenon weathering chamber replicating the field-observed encapsulant discoloration in the glass/back sheet laminated PV modules. Three different modules, a total of nine in number, were tested in two different temperatures in a single chamber using insulation of different thicknesses on the module back sheet. The activation energy was estimated to be in the range of 26 to 39 kJ/mol (0.27 to 0.39 eV) reported for EVA. The values obtained are also in this range, and all three methods have given approximately equal values, so the methods are reliable. The activation energy and the pre-exponential constant values for EVA TPO and white TPO were presented in Table (4). We have tried to establish the conditions and effect of various parameters on the discoloration of PV modules encapsulant. This helps determine the service life of a PV module at a particular geographical location. The activation energy required for discoloration and photobleaching calculations is done for different PV modules using various parameters. This is crucial in identifying the appropriate module type for any geographical location

Table 4: Activation Energy and Pre-exponential factors for the Discoloration and Photobleaching of PV modules

Module	Ea (using Isc) eV	A (using Isc) hr^{-1}	Ea (using YI) eV	A (using YI) a.u.	Ea (using D) eV	A (using D) $mm^2 s^{-1}$
TC	0.31	0.358	0.39	$4.15x10^{6}$	0.26	$2.54x10^{-5}$
TT	0.08	0.014	0.14	333.61	0.12	$1.01x10^{-7}$
TW	0.10	0.021	0.17	1299.84	0.11	$8.33x10^{-8}$

Conclusion

The activation energy we have calculated by the Arrhenius equation matches the reported work. From the fluorescence imaging and inverted grayscale images, one can clearly notice that EVA-encapsulated cells discolor faster than TPO-encapsulated PV cells. From the Arrhenius equation, the change in the yellowness index (YI) is equal to the product of the pre-exponential factor and the activation energy. While the value of activation energy obtained for the degradation of TPO is lower compared to EVA, their pre-exponential factors are several orders of magnitude lower, resulting in a lower overall degradation rate for TPO as compared to PV cells encapsulated with EVA.

Declaration of interest: All authors declared that no conflict of interest in this work. Furthermore, the authors are grateful to IIT Bombay for providing research funds and facilities.

References:

[1] B. Adothu, P. Bhatt, S. Zele, J. Oderkerk, F.R. Costa, S. Mallick, Investigation of newly developed thermoplastic polyolefin encapsulant principle properties for the c-Si PV module application, Mater Chem Phys. 243 (2020) 122660. https://doi.org/10.1016/j.matchemphys.2020.122660.

[2] J. Tracy, N. Bosco, F. Novoa, R. Dauskardt, Encapsulation and backsheet adhesion metrology for photovoltaic modules, Progress in Photovoltaics: Research and Applications. 25 (2017) 87–96. https://doi.org/10.1002/pip.2817.

[3] F. PERN, A. CZANDERNA, Characterization of ethylene vinyl acetate (EVA) encapsulant: Effects of thermal processing and weathering degradation on its discoloration, Solar Energy Materials and Solar Cells. 25 (1992) 3–23. https://doi.org/10.1016/0927-0248(92)90013-F.

[4] M.C.C. de Oliveira, A.S.A. Diniz Cardoso, M.M. Viana, V. de F.C. Lins, The causes and effects of degradation of encapsulant ethylene vinyl acetate copolymer (EVA) in crystalline silicon photovoltaic modules: A review, Renewable and Sustainable Energy Reviews. 81 (2018) 2299–2317. https://doi.org/10.1016/j.rser.2017.06.039.

[5] V. Sharma, S.S. Chandel, Performance and degradation analysis for long term reliability of solar photovoltaic systems: A review, Renewable and Sustainable Energy Reviews. 27 (2013) 753–767. https://doi.org/10.1016/j.rser.2013.07.046.

[6] M. Gagliardi, M. Paggi, Long-term EVA degradation simulation: Climatic zones comparison and possible revision of accelerated tests, Solar Energy. 159 (2018) 882–897. https://doi.org/10.1016/j.solener.2017.10.081.

[7] K. Dolia, Accelerated UV Testing and Characterization of PV Modules with UV-cut and UV-pass EVA Encapsulants, 2018.

[8] D.C. Miller, E. Annigoni, A. Ballion, J.G. Bokria, L.S. Bruckman, D.M. Burns, Xinxin Chen, L. Elliott, Jiangtao Feng, R.H. French, S. Fowler, Xiaohong Gu, P.L. Hacke, C.C. Honeker, M.D. Kempe, H. Khonkar, M. Kohl, L.-E. Perret-Aebi, N.H. Phillips, K.P. Scott, F. Sculati-Meillaud, T. Shioda, S. Suga, S. Watanabe, J.H. Wohlgemuth, Degradation in PV encapsulation transmittance: An interlaboratory study towards a climate-specific test, in: 2015 IEEE 42nd Photovoltaic Specialist Conference (PVSC), IEEE, 2015: pp. 1–6. https://doi.org/10.1109/PVSC.2015.7355607.

[9] B. Adothu, P. Bhatt, P. Kartikay, S. Zele, F.R. Costa, J. Oderkerk, S. Mallick, Determination of Crystallinity and Thermal Stability of Newly Developed Thermoplastic Polyolefin Encapsulant for the c-Si PV Module Application, in: 2019 IEEE 46th Photovoltaic Specialists Conference (PVSC), IEEE, 2019: pp. 0487–0490. https://doi.org/10.1109/PVSC40753.2019.8980526.

[10] A. Sinha, D. Kumar, M. Kempe, D. Jordan, G. TamizhMani, Activation Energy Determination for Photovoltaic Encapsulant Discoloration by Indoor Accelerated UV Testing, in: 2018 IEEE 7th World

Conference on Photovoltaic Energy Conversion (WCPEC) (A Joint Conference of 45th IEEE PVSC, 28th PVSEC & 34th EU PVSEC), IEEE, 2018: pp. 1336–1341. https://doi.org/10.1109/PVSC.2018.8547726.

[11] M. Naskar, H.M. Dharmendra, K.P. Meena, Thermal Lifetime Estimation of EVA Encapsulants from Activation Energy based Method, Power Research - A Journal of CPRI. 17 (2022) 13. https://doi.org/10.33686/pwj.v17i1.167354.

10.4229/EUPVSEC2023/3AV.2.36
020203-006

An Analysis of the Operational Efficiency of Solar Modules through Empirical Research

Jeong-Eon Lee*, Han-Yeol Kim**, Jae-Hyeok Hur**, Myoung-Geon Seo***, Bo-Woon Han***, Min-Seok Kim***, Sol-Yi Lee***, Yeon-Bum An***, Su-Mi Yang***
Dept. of Green Energy Engineering Far East University
1523-8@hanmail.net, klhan3091@gmail.com, hgh9848@naver.com, kkk10108@naver.com, onlyyou558@gmail.com, rlaalstjryes@naver.com, thftka0513@naver.com, dksdusqja@naver.com, esther4853@kdu.ac.kr

ABSTRACT

In the 2000s, investment in solar power plants and power generation facilities increased rapidly with the implementation of korea's policy to supply solar modules in earnest, and the generation of aged solar modules is expected to increase rapidly. Accordingly, Most of the aged solar modules in korea are buried without professional pretreatment and reuse technologies, and furthermore, they are leading to social and environmental problems. Therefore, in order to solve this problem, research and development of technology related to resource recovery through recycling are being carried out, but the system for recycling old and waste modules and judgment procedures for building databases are still insufficient. In order to solve this problem, it is necessary to establish a database for reuse of old and discarded modules and to analyze the operational efficiency of obsolete modules. Accordingly, this study analyzed the operation efficiency and performance according to the aging degree of each solar module along with the measurement of hot spot using a thermal imaging camera, and Through this, established a database necessary for the development of reuse technology. This study analyzed the operational efficiency according to the degree of aging based on the empirical study of old solar modules. Accordingly, this paper intends to present a procedure for determining whether it is reusable based on a database through empirical reaearch for the reuse of old solar modules and criteria for whether it is reusable according to operational efficiency.

Keywords—Solar Module, Hot Spot, Photovoltaic, Thermal image, Deteriorated module

I. Introduction

The world has come to an era of climate crisis due to the acceleration of climate change, and research on alternative energy has been actively conducted over the past few decades [1]. Accordingly, as interest in eco-friendly energy increases due to climate change issues, Korea aims to convert to renewable energy, and is actively pursuing renewable energy 3020 policy to increase the proportion of renewable energy sources to 20% of all energy sources by 2030. Moreover, investment and facilities in solar power plants have increased rapidly as the policy of distributing solar modules has been implemented in earnest, and the amount of aging solar modules is expected to increase rapidly. However, the domestic old solar module is buried without expertise and reuse technology, and environmental problems are also continuous, and social problems. Therefore, research and development of technologies related to resource recovery through recycling is being carried out to solve this problem, but the system for reuse of old and waste modules and the judgment procedure for establishing a database are still insufficient.

Solar modules, a key facility for solar power plants, are technically changing rapidly as the solar industry advances. The resulting technology for solar waste module reuse should also be able to respond to these changes, and standards and standards should be established to accurately distinguish which modules should be reused or discarded(recycled).

Accordingly, this study measured hot spots using drones and thermal imaging cameras and analyzed operational efficiency and performance according to the degree of aging for each solar module. In addition, for the reuse of aging solar modules, we propose a judgment procedure and criteria for reuse based on a database through empirical research.

II. Subject

This study obtained data from comparative analyses of four place installed in Eumseong-gun, Chungcheongbuk-do, 10[kW], 49.4[kW], 13.6[kW], and 20.16[kW]. The empirical research and analysis period of the solar power generation system analyzed characteristics and efficiency from November 17, 2022 to December 16, 2022, and the solar cell module used single crystal silicon products of 250W to 380W in Korea.

It was composed of a total of four solar power generation systems, and the solar cell modules used two companies' products, and the module aging level was 4, 6, and 7 years, forming a comparative group.

The factors affecting photovoltaic power generation for empirical research were largely divided into meteorological, geographical, and facility elements, and shown in <Table I> [2].

<TABLE I> Factors Affecting Solar Power Generation

Sortation	Factors Affecting Photovoltaic
Meteorological Element	Meteorological Element Affecting Photovoltaic (insolation, sunshine duration, cloud cover etc.)
	Meteorological Element Affecting the Efficiency of Photovoltaic (temperature, wind velocity etc.)
Geographical Element	Location of installation of Photovoltaic installation (latitude, longitude etc.)
	Installation height of photovoltaic installations (height above sea level etc.)
Facilities Element	Installation type of module (fixed type, fixed variable, Tracking formula)
	Efficiency of power generation facilities (transformer, inverter, module etc.)
	Track loss and Voltage dip

A. Standard of Solar Power Generation System

The PV system is located at a longitude of 127.69 latitude and 36.9 latitude, and has a relatively high solar radiation distribution compared to other regions.

And An empirical study was conducted to compare and analyze the operation characteristics of PV systems such as power generation performance, reliability, and stability according to changes in environmental conditions [1].

[Fig.1] is a diagram showing where the PV system is installed on a map, and <TABLE II> shows the specifications for each module applied to the demonstration experiment as a table.

[Fig. 1] PV System Installation Location

<TABLE II> Photovoltaic System Module Specification

	Site1	Site2	Site3	Site4
P max(W)	250	380	340	320
V oc(A)	37.7	48.72	46.30	45.90
I sc(A)	8.73	10.06	9.34	9.17
V mp(V)	30.20	39.63	38.10	36.7
I mp(A)	8.28	9.59	8.93	8.71
Cell Type	Single crystal Si	Single crystal Si	Single crystal Si	Single crystal Si

B. An Analysis of the Operating Efficiency of Solar Power Generation System

This study collected data based on the specifications of modules manufactured between 2016 and 2019 through empirical research. In addition, the manufacturing year, installation year, and the degree of capacity and aging for each module are shown in <TABLE 1>. Among them, operational efficiency was analyzed for four power plants of 10kW, 13.6kW, 49.4kW, and 20.16kW.

<TABLE III> Target module information applied to the empirical study

Place	Production	Installation	Capacity	Deterioration
A	2016.04	2016.06	10kW	7year
B	2016.05	2016.05	10kW	7year
C	2017.05	2020.07	49.4kW	6year
D	2017.05	2020.07	15kW	6year
E	2017.05	2017.08	20.16kW	6year
F	2017.11	2018.04	13.6kW	6year
G	2019.05	2020.03	35.52kW	4year
H	2019.06	2020.04	51kW	4year

10.4229/EUPVSEC2023/3AV.2.38

[Fig.2] shows the panoramic view of the 10kWclass solar power generation system in which 10 250W modules applied in this study are combined.

[Fig. 2] View of 10kw Class Solar Power System

[Fig.3] and [Fig.4] is a detailed photograph of the front and rear of the module with a thermal image detector to check whether there is a hot spot phenomenon per module of a 10kW solar power system. The front and rear of the module were photographed with a thermal image detector to infer temperature changes, hot spot phenomena, and the possibility of ignition. Considering that the module manufactured in 2016 was aging for about 7 to 8 years, 0 to 1 hot spot phenomenon appeared, and the temperature change test showed that the normal temperature was maintained, and the possibility of ignition was close to zero.

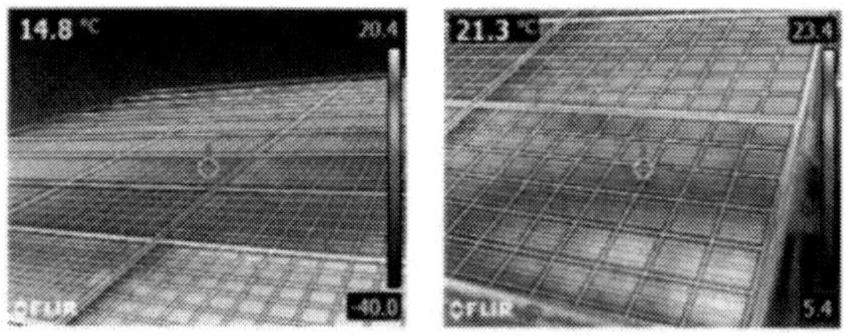

[Fig. 3] 10kW Solar Module Front Hot Spot

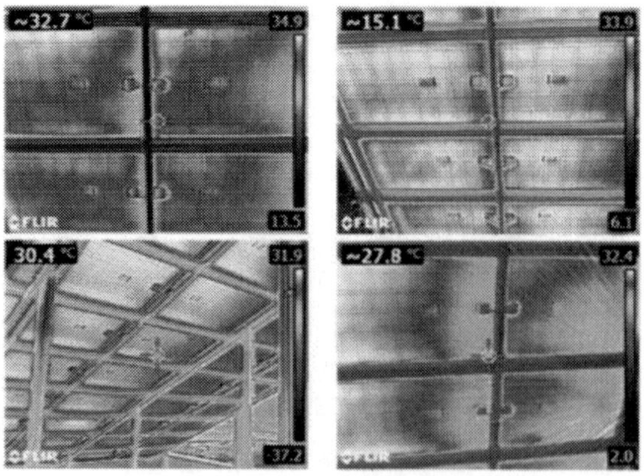

[Fig. 4] 10kW Solar Module Rear Hot Spot

[Fig.5] is a panoramic view of the 49.4kW photovoltaic power generation system that combines 130 380W modules using drones.

[Fig. 5] View of 49.4kW Class Solar Power System

[Fig.6] is a detailed photograph of the front and rear of the module with a drone and a thermal imaging detector to check the presence or absence of spot phenomenon per module of the 49.4kW solar power system. There are two to three hot spot phenomena per module, and considering the degree of aging, it was confirmed that the operation efficiency and performance were lower than the average aging module.

[Fig.7] and [Fig.8] are photographs of the outside and inside of the inverter and switchboard taken with a thermal imaging detector. As a result of the temperature change test, it was confirmed that certain factors important for power generation showed the possibility of ignition and relatively high temperature values.

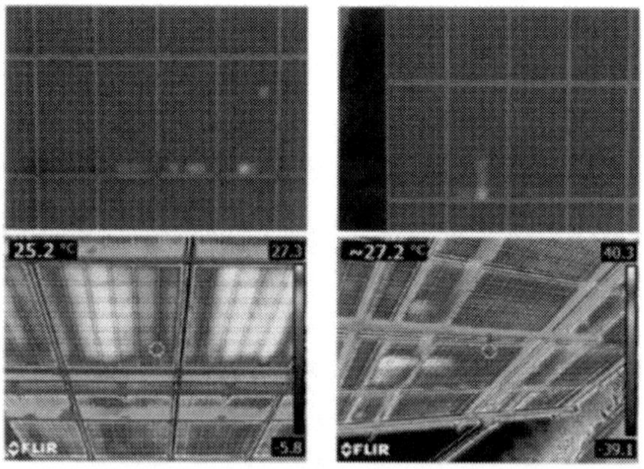

[Fig. 6] 49.4kW Solar Module Front and Rear Hot Spot

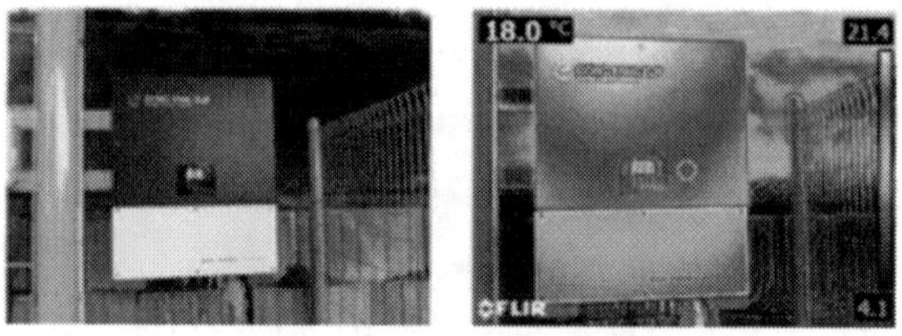

[Fig. 7] Inverter Thermal Imaging

10.4229/EUPVSEC2023/3AV.2.38

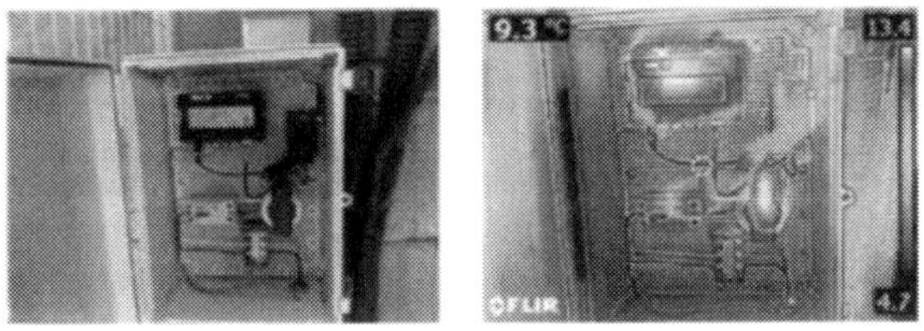

[Fig. 8] Switch Board Internal Thermal Imaging

[Fig.9] is a panoramic view of the 13.6kW photovoltaic power generation system that combines 40 340W modules using drones and thermal imaging detectors.

[Fig. 9] View of 13.6kW Class Solar Power System

[Fig.10] shows the measurement of the front and rear of the module of the 13.6kW photovoltaic power generation system with a thermal image detector. Measurements of the front and rear of the module showed that there were one or two hot spot per module, and it can be seen that dead cells may occur if this state persists.

[Fig.11] is a photograph of the inverter and the connection board measured with a thermal imaging detector. The areas with severe heat generation could not be confirmed by thermal imaging detectors, and compared to new models, it was confirmed that they were in relatively normal condition and operating normally.

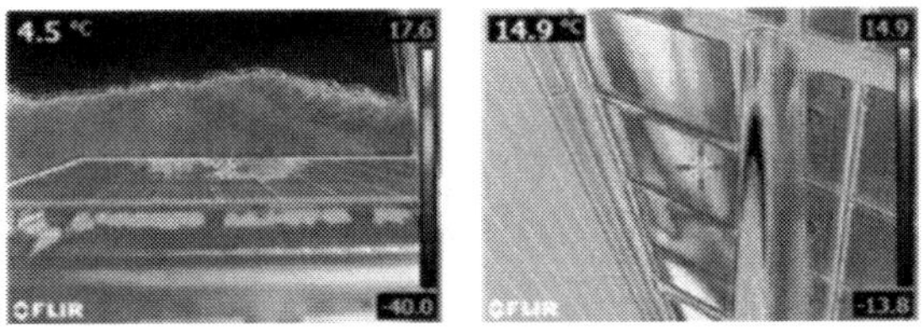

[Fig. 10] 13.6kW Solar Module Front and Rear Hot Spot

[Fig. 11] Inverter and Connector Band Thermal Imaging

[Fig.12] is a panoramic view of the 20.16kW solar power generation system that combines 63 320W modules using drones and thermal imaging detectors.

[Fig. 12] View of 20.16kW Class Solar Power System

[Fig.13] and [Fig.14] are photographs of the front and rear of the module with thermal imaging detectors.

As a result of measuring the front and rear of the 20.16Kw photovoltaic module with a thermal imaging detector, it was confirmed whether the module temperature, hot spot, dead cell, were likely to occur. Hot Spot was confirmed to be 0 to 1 per module, and the temperature change test was normal and the possibility of ignition was analyzed to be low.

As a result, it was confirmed that there was no possibility of a Dead Cell occurring, and that the operation efficiency in power generation such as module state, temperature, and amount of power generation was normal.

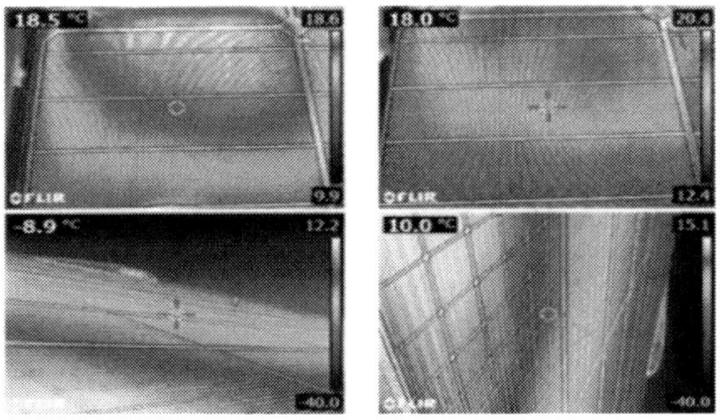

[Fig. 13] 20.16kW Solar Module Front Hot Spot

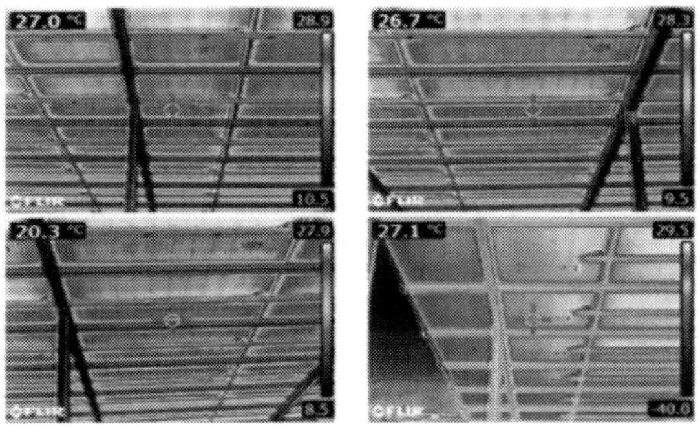

[Fig. 14] 20.16kW Solar Module Rear Hot Spot

III. Conclusion

This study analyzed the operational efficiency evaluation and performance of old solar modules through empirical research on four 10[kW], 49.4[kW], 13.6[kW], and 20.16[kW] solar power generation systems installed in Eumseong-gun, Chungcheongbuk-do. In the case of site1, it was confirmed that there were 0 to 1 hot spot phenomenon when the drone and thermal image were measured, and the temperature was relatively normal when the module temperature test was conducted. It is judged that there will be no possibility of ignition or dead cell occurrence over the next 5 to 10 years. In the case of Site2, 2 to 3 hot spot phenomena were identified as a result of thermal image measurement, and it is judged that the operation efficiency was significantly reduced compared to the degree of aging. In addition, as a result of the module temperature test, the module temperature was measured at a relatively high temperature, and the possibility of ignition was confirmed to be high in the future, and it is judged that Dead Cell will occur accordingly. Site3 was confirmed 1 to 2 hot spot phenomena when thermal imaging was judged that this can affect power generation, module efficiency, and performance. Site4 confirmed that the number of hot spot phenomena was 0 to 1, and both the temperature change test and the possibility of firing were normal.

As nature of the PV system, the main loss factors were inverter loss, loss due to solar radiation variation, loss due to shade generation, and efficiency change due to temperature change [1].

In addition, it is judged that detailed analysis is needed using I-V Test and EL equipment for output, voltage, current, Crack, Dead Cell, pollution, etc. which cannot be checked in detail with the measurement equipment applied in this study, Through this, it is judged that standards for reuse and waste treatment(recycling) of old modules can be established.

In further analysis, Based on the operation efficiency analysis of old modules, we will study to establish optimal solar power system reuse certification and reusable standards by checking the Pmax value calculated from the graph value and the crack and dead cell of old modules through EL equipment.

Reference

[1] Wang-Cheol Cha, Joung-ho Park, Uk-rae Cho, and Jae-Chul Kim. (2015). A Study on solar Power Generation Efficiency Empirical Analysis according to Temperature and Wind speed. KIEEP Insight, Vol. 64, No. 1, pp. 1~6, 2015

[2] Jeong-Min Park, Seong-Kyeol Kim, Byung-Ho Jung, Geum-Bae CHO, Hyung-Rae Baek. (2003). Using Demonstration PV System for Operation Efficiency Anal

10.4229/EUPVSEC2023/3AV.2.38
020205-005

A Study on Efficiency and Comparative Analysis of 100kw Eco Module Test-field Construction

Jae-Hyeok Hur*, Jeong-Eon Lee**, Han-Yeol Kim**, Myoung-Geon Seo**, Bo-Woon Han**, Su-Mi Yang**
Dept. of Green Energy Engineering Far East University /Eco-friendly Future Energy Lab
hgh9848@naver.com, 1523-8@hanmail.net, klhan3091@gmail.com, kkk10108@naver.com, onlyyou558@gmail.com, esther4853@kdu.ac.kr

ABSTRACT

As of 2018, with the rapid increase in investment and power generation facilities for domestic solar power plants in South Korea, the amount of aging solar modules has also significantly increased. Furthermore, there are various environmental issues arising from the lack of clear policies and guidelines for the disposal of end-of-life solar modules and the reuse of aging modules. However, in South Korea, there is still a need for standardized measures and solutions for the reuse of aging modules, ensuring their safety, durability, and performance.
In this study, to establish the reliability of reused modules, we constructed a 100 kW (50 kW: Eco-Modules, 50 kW: New Modules) power generation demonstration site. Through efficiency research using the Eco-Modules for the first time in South Korea, we established the reliability of reused modules. Additionally, through the demonstration research, we compared and analyzed the efficiency of aging solar modules, and built a necessary database for the certification standards for reuse. Therefore, in this paper, we propose the establishment of standards through the construction of a 100 kW (50 kW: Eco-Modules, 50 kW: New Modules) power generation demonstration site, and the comparison and analysis of operational efficiency for the certification standards of reused modules.

Keywords— solar cells, ECO modules, reuse panels,

I. Introduction

The need for environmentally friendly energy has been rapidly increasing due to the rise in international oil prices and the phenomenon of global warming. In particular, in South Korea, there is a focus on expanding the renewable energy industry to reduce greenhouse gas emissions, with solar energy being highly regarded as an alternative energy source. As a result, investments and power generation facilities for solar power plants have been rapidly increasing, leading to a significant increase in the amount of aging solar modules. Consequently, the disposal of end-of-life solar modules has become a pressing issue, as there are no clear guidelines for their proper handling. As a result, most of these modules are being landfilled without proper recycling or reuse, leading to additional environmental problems.

To address this issue, it is crucial to establish policy regulations for the disposal of end-of-life solar modules and provide clear guidelines for their reuse. However, in South Korea, there is still a lack of standardized measures and solutions for the reuse of aging modules, as well as a lack of certification standards ensuring their safety, durability, and performance. Therefore, there is a need to establish certification standards for reused modules and develop a database for the comparison and analysis of efficiency between aging and new modules, as well as for standardization.

In this study, we aimed to establish certification standards and ensure the reliability of reused modules by constructing a 100 kW (50 kW: Eco-Modules, 50 kW: New Modules) Eco-Module power generation demonstration site, which is the first of its kind in South

Korea. Through efficiency research using Eco-Modules, we provided a basis for standardizing renewable modules and their performance. Additionally, through demonstration research using EL, IV-Test, and module temperature sensor equipment, we compared and analyzed the operational efficiency between aging solar modules and new models, and established a database and standardization for reliability verification.

Overall, this research contributes to the establishment of certification standards for reused modules, the standardization of performance, and the development of a database necessary for certification standards by conducting efficiency research using Eco-Modules and comparing the operational efficiency between aging and new solar modules.

II. Subject

In this paper, we constructed a 100 kW-scale Eco-Module demonstration site to demonstrate efficiency and reliability, and presented the research findings in a data-driven manner.

The study focused on performance evaluation of aging solar modules installed in Eumseong-gun, Chungcheongbuk-do, South Korea. We collected modules that were manufactured after 2010 and had been in operation for over 10 years. The performance evaluation was conducted at the Solar Research Center of Chungbuk Technopark, using EL equipment and I-V Test equipment to analyze the module's performance. Additionally, we built a power generation system using Eco-Modules and new modules to collect data and conducted research on the efficiency of Eco-Modules.

III. ECO module condition check and grade analysis

A. ECO module grade analysis

The classification and evaluation of ECO modules were conducted based on the criteria listed in <Table 1> to assess the module's condition. The decision on whether to reuse or recycle the module was determined based on the total deduction points.

< TABLE I> Details in Criterion of Process

Category	Failure Types	Failure Details	Subtract Points
Visual & Thermal Inspection	Tempered Glass	Glass Sealing Deterioration	-10
		Contamination	
	Back Sheet	Back sheet Scratch, Dent, Discoloration	-10
	Encapsulate Material	EVA Discoloration	-5
		EVA Delamination	-20
	Corrosion	Ribbons, Metal Corrosion	-10
	Thermal	Hotspots on Junction Box Area	-5
		Under 3 Hotspots	
EL Inspection	Crack	Fine Cracks Through EL Images	-10
	Contamination	Belt Print, Picker Print	
	Diode Defect	Diode Defect Through EL Images	
	PID	PID Through EL Images	-20
Output Test	Electrical Characteristics Test (Pmax)	Above 90% Efficiency	0
		Above 80% Under 90%	-10
		Under 80%	Recycle

< TABLE II> Rating Based on Deduction

Rate	Subtract Points	Process
A	Under 0 ~ 20	Maintenance work and Reuse
B	Over -20 ~ Under -30	
F	over –30	Recycle

[Fig. 1] shows the appearance of F-grade modules and Rework modules that occur during the process of determining the grade using the module classification criteria mentioned above. Modules with open circuits or those that can be repaired easily undergo rework, while the remaining F-grade modules proceed to the recycling process.

[Fig. 1] Separation by module grade.

B. Collection and use of ECO modules equipment

This study conducted experiments on old modules that have been used for over 10 years in order to establish a 50KW ECO module power generation system. [Fig. 2] shows the research process for grade analysis and the appearance of modules classified by grade.

[Fig. 2] Collection photo of modules manufactured in 2010 and used for over 10 years.

The measurements were conducted using both I-V Test and EL, and [Fig. 3] shows the photo of the equipment used for the measurements and the process of conducting the measurements and classification. The modules under measurement were installed inside the Solar Simulator Chamber for the measurements.

Through the I-V Test, the maximum efficiency of the module under STC conditions and its current efficiency were graphically displayed for comparison and analysis. EL measurements were used to identify various EL defects such as micro-cracks.

[Fig. 3] Analysis photo using EL and IV-Test equipment.

IV. Efficiency analysis of ECO modules and new modules

A. Analysis of weekly average module power generation ef ficiency

A 100KW power generation site was established using both ECO modules and new modules. An efficiency analysis of the solar power generation output was conducted for the months of July and August after installation, resulting in the findings shown in [Fig. 4] and [Fig. 5].

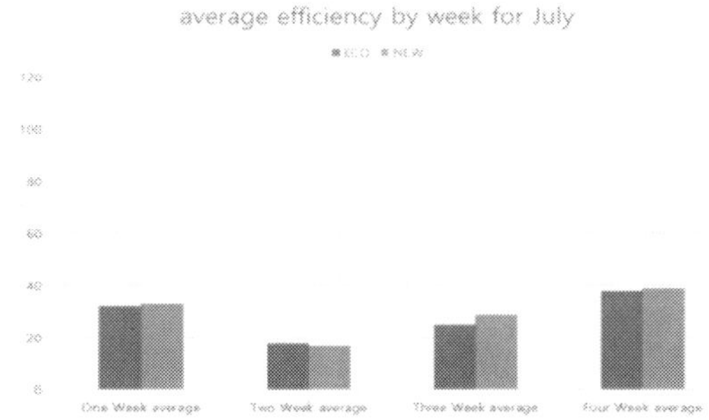

[Fig. 4] Average by week in July

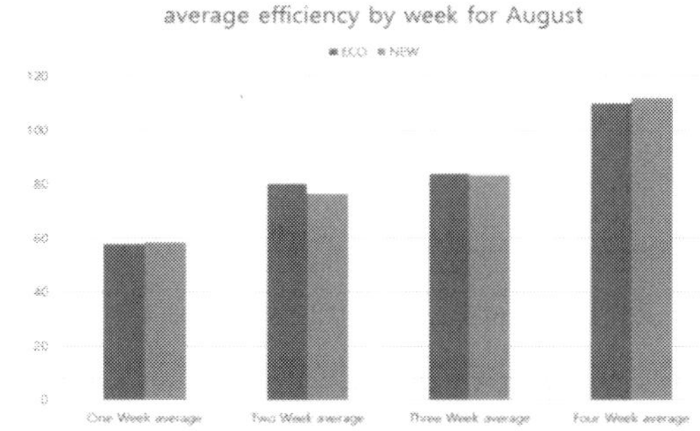

[Fig. 5] Average by week in August

[Fig 4] represents the average power generation output for the month of July, while [Fig 5] shows the average power generation output for the month of August. The difference in average power generation output between the two graphs is attributed to factors such as typhoons or monsoon seasons, which caused a decrease in power generation output in July overall. Additionally, when comparing the power generation outputs, it was observed that there is a minimal difference of less than 5% between the ECO modules and the new modules. Furthermore, it was also noted that in certain weeks, the power generation output of the ECO modules performed better than the new modules.

10.4229/EUPVSEC2023/3AV.2.39
020207-003

B. Analysis of degradation and reliability of ECO modules after the construction of the power generation site

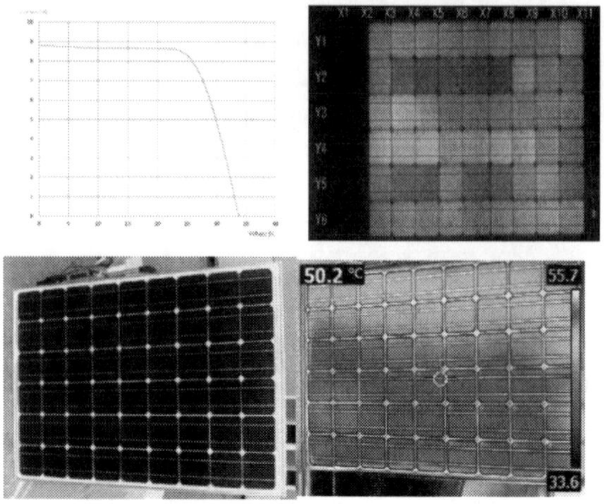

[Fig. 6] Re-measurement results of modules used for the demonstration study.

To verify the degradation and reliability of the ECO modules used in the experimental site over the course of two months, re-measurements were conducted. EL, I-V Test, and thermal imaging measurements were performed. When compared to the measurements taken before the establishment of the experimental site, there were no significant observations of deterioration in EL defects or changes in efficiency. [Fig. 6] represents the re-measurement results of the modules used in the experimental study, and no changes in the I-V curve, deterioration of EL defects, or generation of hotspots were observed.

V. conclusion

In this paper, a 100 kW ECO module demonstration site (50 kW: ECO module, 50 kW: new module) was constructed to demonstrate efficiency and reliability, and the research was dataized and presented. The modules were classified by grade, including aging modules, and ECO modules were used for the construction of the ECO module power generation site. Based on the measurement data after the construction of the demonstration site, the average power generation was analyzed, and the power generation between July and August was averaged and compared. The difference in power generation between the ECO module and the new module was less than 5%, indicating a small deviation. It was also observed that in certain weeks, the power generation of the ECO module was higher than that of the new module. Despite the decrease in power generation due to the rainy season in July, the difference in power generation was not significant. This indicates that although aging modules may deteriorate due to long-term power generation and moisture infiltration, there was no impact from typhoons or heavy rain. To confirm the degradation of the ECO modules, a re-measurement was conducted on the ECO modules that were deployed in the demonstration site for two months. As a result, no deterioration of EL defects, efficiency changes, or hotspot generation was observed, which increased the reliability of the ECO modules. It is determined that further verification research using the ECO module power generation site is necessary to collect measurement data for durability and stability, and additional research on the reliability of the ECO modules is needed. In addition, the ECO module power generation site can be utilized for collaborative research in the field of renewable energy and can be a representative case for various reuse studies of renewable energy technologies.

DAMP-HEAT ENDURANCE INVESTIGATION OF PV MODULES BASED ON N-TYPE BIFACIAL PASSIVATED CONTACT CELLS

Yanfang Zhou, Daoyuan Chen, Yuqiu Ye, Haipeng Yin, Xinwei Niu
JA Solar Technology Co., Ltd
Yangzhou, Jiangsu Province, P. R. China, (86)13222698735, zhouyf@jasolar.com

ABSTRACT: N-type bifacial passivated contact cells have higher efficiency and considerable manufacturing cost than PERC cells, but they are more susceptible to degradation under acidic humid environment. Therefore, the module damp-heat endurance requires close investigation for long-term reliability concerns. In this study, we investigate the effect of module structures and encapsulation materials on the damp-heat endurance of single and double glass modules by examining power loss and electroluminescence imaging following extended damp-heat aging. For double glass modules, the moisture transmission is restricted with limited water vapor entry path from the module edges, different module structures and encapsulation materials result in modest reliability performance deviation. For single glass modules, extended DH aging reveals significant differences in the power loss and electroluminescence imaging depending on used module structures and encapsulation materials due to enlarged water infiltration. The white EVA and polymer/metal composite backsheets also demonstrate the ability to prevent metal corrosion with lower power losses. The effective means of enhancing the damp-heat endurance of the module is to obstruct water penetration into the front surface of the cell. This can be achieved by either upgrading the module structures or opting for materials with lower vinyl acetate content and increased water resistance.
Keywords: n-type cells, damp-heat endurance, module structure, encapsulation material

1 INTRODUCTION

Currently, the mainstream PV modules are based on the p-type passivated emitter rear contact (PERC) structure, but its future improvement is limited, and n-type bifacial passivated contact (Tunnel Oxide Passivated Contact, TOPCon) cells have become a leading candidate to replace PERC as the next generation of mass production solar cells with greater than 1% increased efficiency and a considerably lower manufacturing cost per watt. However, a direct comparison of TOPCon and PERC metallization with respect to corrosion, as reported by N. Iqbal et al. [1], showed that TOPCon metallization on the front side is more susceptible to acid hydrolysis from encapsulation materials upon exposure to moisture. Additionally, Sommeling et al. [2] studied the corrosion process of TOPCon in an acidic environment and demonstrated that the deterioration of TOPCon metallization is heavily influenced by the glass frits. The glass frits are crucial constituents of metallization pastes that aid in dissolving the silicon nitride coating, resulting in improved performance of the metallization pastes for contact formation [3]. However, they can have a significant detrimental impact on TOPCon metallization when exposed to moisture where acid-induced corrosion occurs [4]. Obviously, the reliability of photovoltaic (PV) modules based on TOPCon cells faces significant challenges. The behavior of corrosion is complex, and its mechanism is not yet fully understood.

For PERC cell modules, ethylene vinyl acetate (EVA) is a cost-effective encapsulating material with high reliability and process feasibility. However, EVA is susceptible to moisture and can release acetic acid through hydrolysis, which may cause corrosive stress on the metallization [5]. Therefore, for PV modules based on TOPCon cells, polyolefin elastomers (POE) without acetate groups are a viable alternative to EVA [6]. However, POE, despite its reliability, is not a manufacturing-friendly material due to higher costs and limited process windows. On the other hand, EPE (EVA/POE/EVA) is designed to strike a balance between reliability requirements and manufacturing efficiency. This study also compares the reliability performance of PV modules based on TOPCon cells using various encapsulation materials.

Typically, commercial photovoltaic (PV) modules are categorized into two types based on their encapsulation solutions: double glass (DG) modules and single glass (SG) modules. DG modules feature a glass encapsulation backsheet material, whereas SG modules feature a polymer encapsulation backsheet material. G. Oreski et al.[6] thought that a DG module can build up higher acid concentrations than a SG module once the degradation of EVA has started. On the other hand, A. Sinha et al.[7] thought that the polymer backsheet of SG modules allows higher moisture permeation leading to more acetic acid and more terrible corrosion. This study makes a complete assessment of the damp-heat endurance of both DG and SG modules with different encapsulation solutions to reveal the differences of corrosion in two types of modules.

Module reliability performance is an essential consideration for cell and module technology for real application up to 25 or more years. Through our previous studies and comprehensive evaluations, we have determined that TOPCon cells are particularly vulnerable to degradation in acidic and humid environments. Consequently, it is imperative to conduct thorough investigations of TOPCon cell-based module damp-heat (DH) endurance to address long-term reliability concerns. In this paper, we investigate the impact of module structures and encapsulation materials on the power loss (PL) and electroluminescence (EL) imaging of n-type PV modules following extended damp-heat aging. Module performance is evaluated using different wafer sizes and cell gaps. We compare the application differences in DG and SG modules by evaluating the encapsulation materials EVA, POE, and EPE. Additionally, we compare EVA with white filler and a polymer-metal composite backsheets.

2 EXPERIMENTAL

10.4229/EUPVSEC2023/3AV.2.41
020210-001

Commercially available PERC and TOPCon cells were connected in series with 0.26mm circular interconnection ribbons to form cell strings. Different wafer sizes and cells gaps are also designed to verify the impact of module structures on moisture transmission. As shown in Table I, various encapsulation materials with different water vapor transmission rates (WVTR) are also commercially available and are used to make double glass and single glass laminates, which undergo framing, J-box installation and curing to become the final modules for DH aging. W-EVA(W-E) means a kind of EVA filled with TiO_2 powder with the size of 200-300nm. And the polymer/metal composite backsheet (P/ M) means a kind of backsheet including a metal layer of 40-50μm thickness between two polymer layers. Different modules with the same size and the same number of cells in each module were prepared for performance evaluation. The encapsulation solutions of the DG and SG modules are shown in Table II and Table III, respectively. The extended damp-heat tests(85℃/85% RH) up to 3000 hours were conducted to investigate the corrosion behavior of various metallic components and delamination properties. Optoelectrical parameters, EL images and appearance before and after DH aging with an interval of 1000 hours are collected to reveal the effect of water vapor on the module performance.

Table I: Encapsulation materials used to make modules

Encapsulation materials		Thickness (mm)	WVTR (g/m^2.day)
Glass	SG	3.2	<0.01
	DG	2.0	<0.01
Film	POE	0.5	3.86
	EPE	0.5	8.50
	EVA	0.5	23.71
	W-E	0.5	22.39
Backsheet	Polymer	0.3	2.07
	P/ M	0.3	0.11

Table II: DG modules with different encapsulation solutions prepared for DH aging

DG	Cell Type	Cells Gap (mm)	Front Film	Rear Film
1	TOPCon	1.8	POE	POE
2	TOPCon	1.8	EPE	EPE
3	TOPCon	1.8	EPE	EVA
4	TOPCon	-0.3	EVA	EVA
5	TOPCon	-0.3	EPE	EPE

Table III: SG modules with different encapsulation solutions prepared for DH aging

SG	Cell Type	Cell Gap (mm)	Front Film	Rear Film	Backsheet
1	PERC	-0.3	EVA	EVA	Polymer
2	TOPCon	-0.3	POE	POE	Polymer
3	TOPCon	-0.3	EPE	EPE	Polymer
4	TOPCon	-0.3	POE	EVA	Polymer
5	TOPCon	-0.3	POE	W-E	Polymer
6	TOPCon	1.8	EPE	EPE	P/ M
7	TOPCon	1.8	POE	POE	Polymer

3 RESULTS AND DISCUSSION

3.1 The effect of module structure on DG module damp-heat endurance

Modules (DG-2 and DG-5 in Table II) containing identical encapsulation materials but varying in wafer size and cell gap were prepared and subjected to extended damp-heat aging to investigate the impact of module structure on the endurance of DG modules. Fig.1 demonstrates that, after DH3000 aging tests, DG-2 and DG-5 modules each experienced power degradation of 2.29% and 1.95%, respectively. These results suggest a negligible influence of cell gap on the reliability performance of DG modules. The losses were primarily observed in the Isc variation, which is expected for modules tested under harsh DH conditions. Typically, as the DH aging process extends, more moisture may infiltrate the modules and lead to the continuous hydrolysis of EVA. This causes acid to reach a critical concentration and corrode the metallization. The corrosion will deteriorate the contact between the metal grids and silicon, resulting in increased resistance of the metal grids and interconnection ribbons. Finally, the module Rs increases together with the decreased module power [8]. For DG-2 and DG-5 modules, little change in Rs values is observed before and after the DH aging, indicating minimal corrosion occurred after the harsh tests.

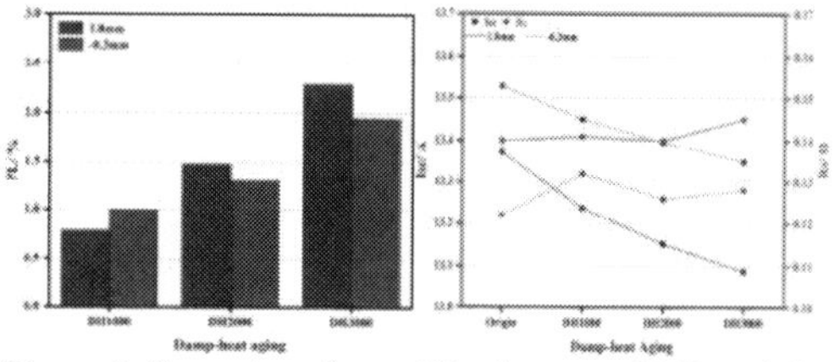

Figure 1: Power loss, Isc and Rs changes of DG modules with different cells gaps after DH aging

Fig. 2 displays EL images of DG modules with various cell gaps following DH3000: (a) 1.8mm; (b) -0.3mm. It is evident that no notable dissimilarities were observed, barring sharper cell boundaries on the 1.8mm cell gap module. This can be attributed to lower photoelectric activity on the gaps compared to the cell body. The slightly blackened areas on the module edges appeared similar and suggested possible corrosion. As observed in Fig. 3, severe corrosion is typically found at the edge of DG modules encapsulated with dual EVA, which provides further evidence for the edge corrosion phenomenon of DG modules.

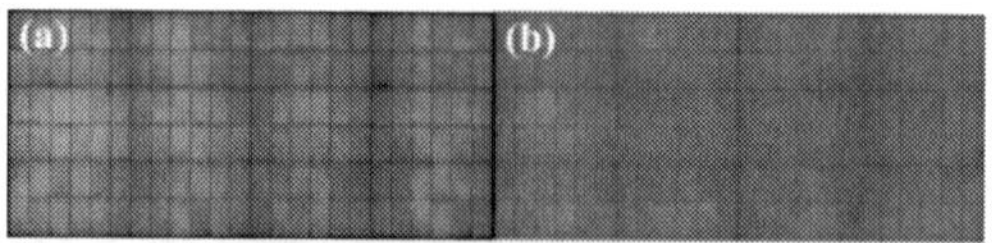

Figure 2: EL images of DG modules with different cells gaps after DH3000: (a) 1.8mm; (b) -0.3mm

Figure 3: The severe corrosion typically found at the edge

of DG modules encapsulated with dual EVA

3.2 The effect of encapsulation materials on the damp-heat endurance of DG modules

Modules (DG-1, DG-2, DG-3, and DG-4 in Table II) were prepared with various encapsulation materials and subjected to extended damp-heat (DH) tests to assess the effect of these materials on the endurance of DG modules. As depicted in Figure 4, there is little difference in power loss (PL) among the modules that underwent DH2000 aging, and all PL values were less than 2%. However, when extending the damp-heat aging test to 3000 hours, the power loss (PL) of the module encapsulated by dual ethylene-vinyl acetate (EVA) increases to 8.28%, while other groups show less than 3%. It is demonstrated that the modules with TOPCon cells, even DG modules, face the failure of damp-heat endurance when dual EVA is adopted as encapsulation material. What's more, we see that the modules PLs increase with the increase of EVA content in the encapsulation material due to the release of more vinyl acetate after hydrolysis[9].

Usually, with the extension of DH tests, the metallization corrosion becomes severe, contributing to the increased Rs and decreased FF. In particular, the FF of the DG module encapsulated with dual EVA sharply decreases by 6.8% (from 81.01% to 75.50%) after DH3000 aging, resulting in serious module power loss.

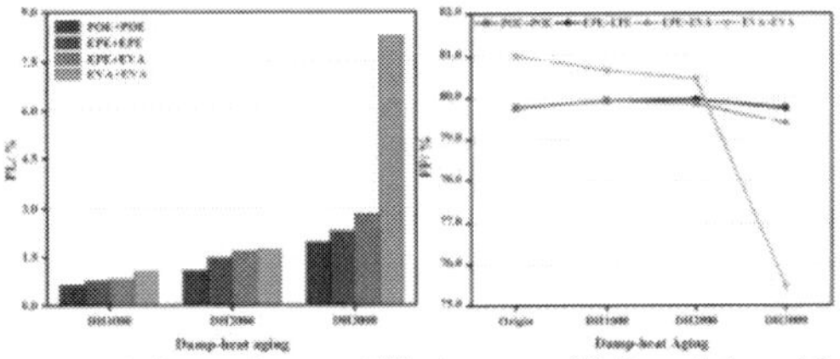

Figure 4: Power loss and FF changes of DG modules with different encapsulation materials after DH aging

Fig. 5 shows EL images of DG modules with varying encapsulation materials after the DH3000 test. The images illustrate that black inactive regions are present at the edges of all modules, suggesting corrosion occurred following the harsh DH3000 tests. Moreover, as the EVA content increases in the encapsulation material, the black regions progressively expand towards the center of the module. The module with dual EVA exhibits the most pronounced corrosion, aligning with the PL result and being as predicted.

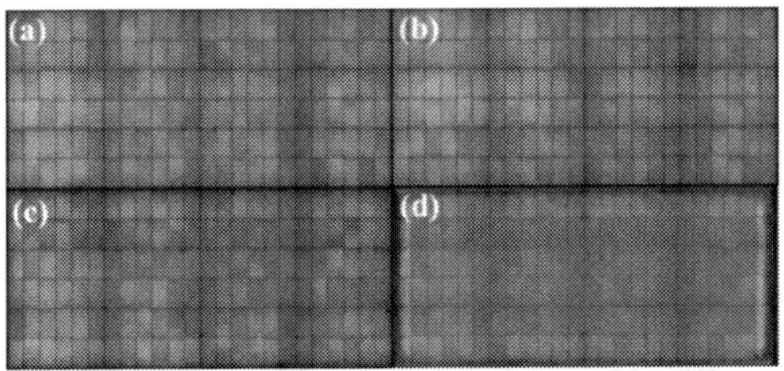

Figure 5: EL images of DG modules with different encapsulation materials post DH3000: (a) POE+POE; (b) EPE+EPE; (c) EPE+EVA; (d) EVA+EVA

For DG modules, water vapor primarily enters at the module edges. While different encapsulation materials and structures can result in some reliability performance deviation, it is not significant. Encapsulation materials with a double EPE structure (EVA/POE/EVA) are sufficient to pass the 3000-hour damp-heat endurance test. Furthermore, when replacing the rear film from EPE to EVA on the DG module, it still maintains high reliability with <3% power loss after 3000 hours of DH testing, as depicted in Table IV. However, if the front film is also swapped from EPE to EVA, the PL of the DG module after DH3000 aging exhibits a sharp incline, due to the increased acid concentration caused by more EVA, which results in direct corrosion to the front metallization of TOPCon cells. This finding suggests that the back metallization of TOPCon cells is less susceptible to acid corrosion compared to the front metallization. Additionally, DG modules that use dual EVA as an encapsulation material are at higher risk of potential induced degradation (PID) due to the lower volume resistivity of EVA [10].

Table IV: Damp-heat endurance comparison of different DG modules

Sample	Power Loss after DH aging (%)		
	1000h	2000h	3000
DG-1	0.63	1.11	1.95
DG-2	0.80	1.48	2.29
DG-3	0.84	1.68	2.80
DG-4	1.08	1.73	8.28
DG-5	1.00	1.31	1.95

3.3 The effect of module structure on damp-heat endurance of SG modules

Modules (SG-2 and SG-7 in Table III) with the same encapsulation materials but different wafer sizes and cell gaps were prepared and subjected to extended DH aging to evaluate the effect of module structure on damp-heat endurance of SG modules. As shown in Fig. 6, after DH2000 aging, the power losses of SG modules with -0.3mm and 1.8mm cells gaps are 4.21% and 6.35%, respectively. In addition, as the increase of DH duration, the FF gaps between 2 modules become larger, indicating that the corrosion on the 1.8mm gap module is more severe than that on the -0.3mm gap module. This assumption can be further verified by the EL images presented in Fig. 7. A closer inspection of the EL image of the 1.8mm SG module reveals inactive black areas distributed along the interconnection ribbons. However, the EL image of the -0.3mm SG module shows a much more uniform performance. The phenomenon differs from that of DG modules because of the varying entry path of water vapor in the two modules.

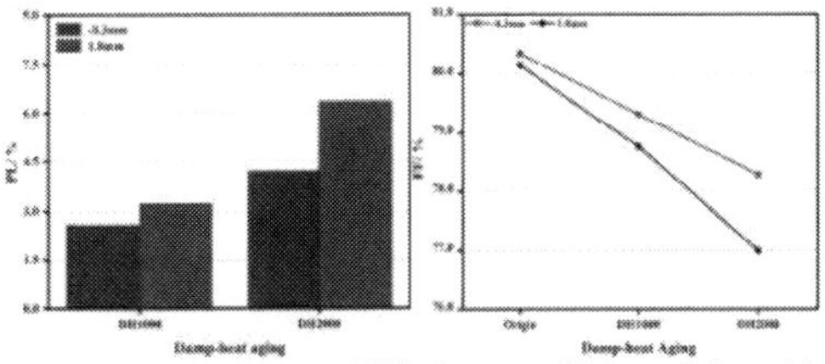

Figure 6: Power loss and FF changes of SG modules with different cell gaps after DH aging

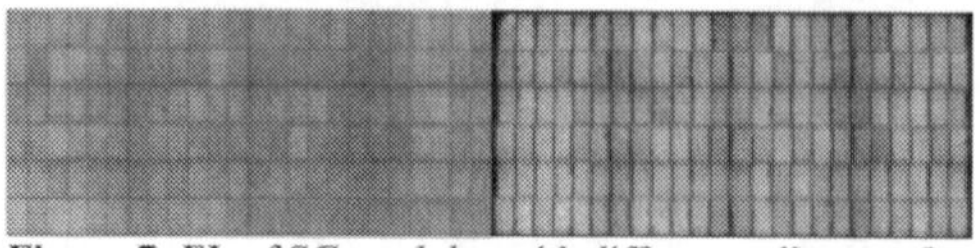

Figure 7: EL of SG modules with different cell gaps after DH1000: (a) -0.3mm; (b) 1.8mm

A model is proposed in Fig. 8 to describe the possible corrosion process in the 1.8mm SG module. In general, the water vapor enters into the module through the full polymer backsheet with inadequate water resistance, penetrates the encapsulation film to reach the back surface of the cell as well as the interconnection ribbons, then spreads along the ribbon to the front surface of the cell, resulting in corrosion of the front metallization, which is more susceptible to moisture and acid. The films typically struggle to wrap the interconnection ribbons entirely because lamination is difficult or due to expansion mismatches during aging, leaving gaps between the ribbon and the film. As depicted in Fig. 9, moisture is then able to spread along the ribbons through these gaps. For the standard SG module with a 1.8 mm cell gap, moisture may directly reach the gaps between the front film and ribbons through the rear film. This shortens the distance of moisture to the front metallization, leading to more severe corrosion. Alternatively, reducing the cell gap to -0.3 mm replaces the original cell gap with the wafer, which blocks moisture efficiently, greatly reducing the corrosion situation.

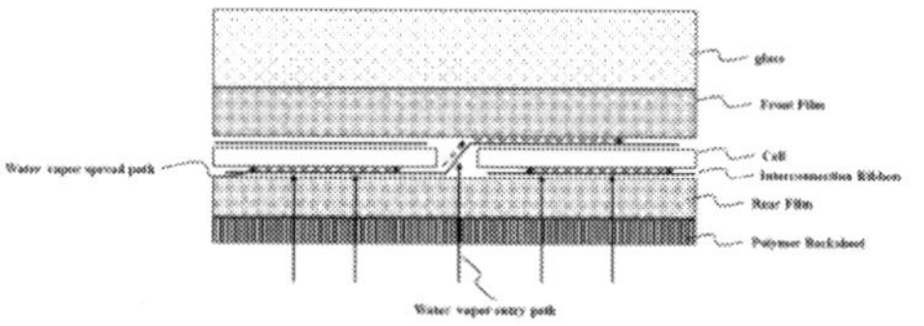

Figure 8: Diagram of water vapor entry and spread path in the SG module

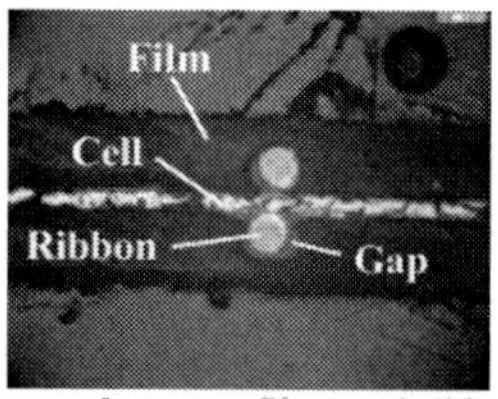

Figure 9: The gaps between films and ribbons

3.4 The effect of encapsulation materials on the damp-heat endurance of SG modules

Modules (SG-1, SG-2, SG-3, and SG-4 in Table III) were prepared using various cells and encapsulation materials. These modules were then tested for extended DH to evaluate the impact of cells and encapsulation materials on SG module damp-heat endurance. As depicted in Figure 10, the endurance of DH in SG PERC module does not appear to be influenced by the encapsulation material used. Even when dual EVA encapsulation is employed, the module's power loss remains below 3% following DH2000 aging. Furthermore, as shown in Figure 11(a), there are no noticeable inactive regions in the EL image. Conversely, the DH endurance of TOPCon-based SG modules is greatly influenced by the encapsulation materials used. After the DH2000 test, the PL of the dual POE solution is measured at 4.21%. However, when the rear POE is replaced with EVA, the module PL post DH2000 increases to 5.84%. Furthermore, replacing dual POE with dual EPE causes the power loss to rise to 7.28%, indicating an increase in module corrosion with the increase in EVA content in the encapsulation material. Among all the splits, the front side is more critical on the EVA application.

Fig.10 demonstrates the changes in FFs of SG modules with varying cells and encapsulation materials following DH aging. Results show that FFs decrease with extended DH durations, more significant corrosion, steeper FF changes, and increased power losses.

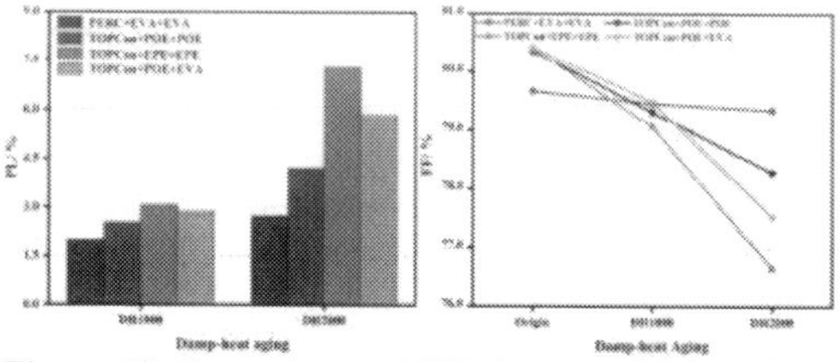

Figure 10: Power loss and FF changes of SG modules with different cells and encapsulation materials after DH aging

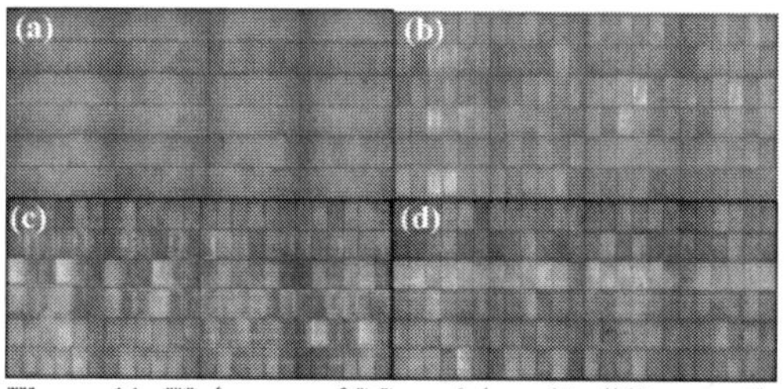

Figure 11: EL images of SG modules with different cells and encapsulation materials after DH2000: (a) PERC+ EVA+EVA; (b) TOPCon+POE+POE; (c) TOPCon+ EPE+EPE; (d) TOPCon+POE+EVA

Modules (SG-5 and SG-6) were prepared to evaluate the effect of W-EVA and polymer/metal composite backsheet on the performance the damp-heat endurance,

As shown in Figure 12, the PL of modules using W-EVA encapsulation and regular transparent EVA after DH1000 is similar, with a difference of only 0.05%. However, after DH2000 tests, the PL difference between the two modules increases to 0.92%, indicating that the SG module encapsulated with W-EVA has better resistance to damp-heat aging than the transparent EVA under extended testing. W-EVA is an EVA film uniformly dispersed with nanoparticles of TiO_2 measuring between 20 and 300nm in size. The presence of TiO_2 particles is believed to help adsorb or block acidic moisture, resulting in a slower corrosion process, as depicted in Fig. 14(a).

With an incredibly low WVTR value, the polymer/metal composite backsheet significantly contributes to the DH durability of the SG module. Even with a 1.8mm cell gap, the PLs after DH1000 and DH2000 tests are 1.53% and 1.8%, respectively, behaving similarly to DG modules. As depicted in Figure 14(b), the backsheet structure of the polymer/metal composite features a metal layer sandwiched between two polymer films, effectively preventing water penetration and eliminating the potential risk of metallization corrosion.

10.4229/EUPVSEC2023/3AV.2.41
020210-004

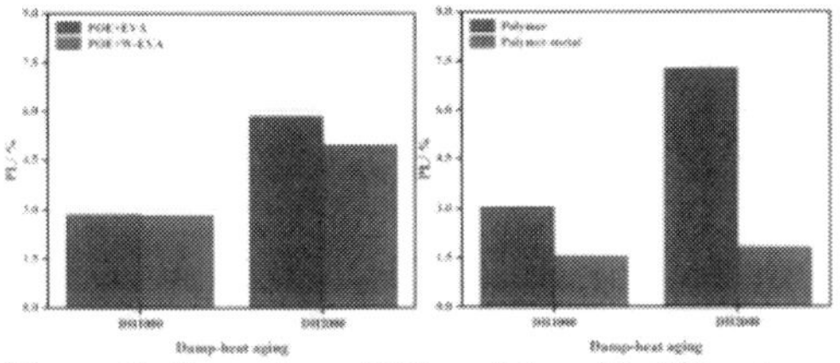

Figure 12: Power loss of SG modules with different rear films and backsheets after DH aging

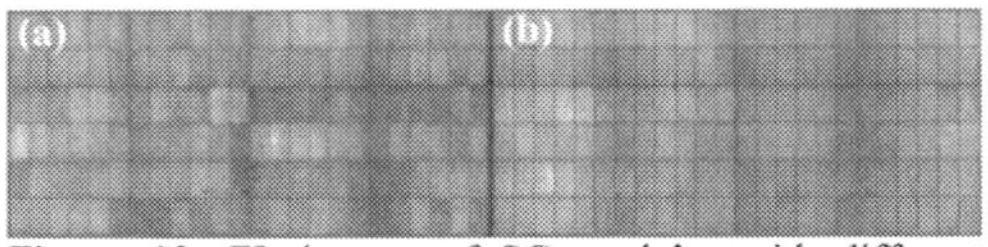

Figure 13: EL images of SG modules with different encapsulation materials after DH2000: (a) -0.3mm cell gap+ POE+ W-EVA+ polymer backsheet; (b) 1.8mm cell gap+ EPE+ EPE+ polymer/ metal backsheet

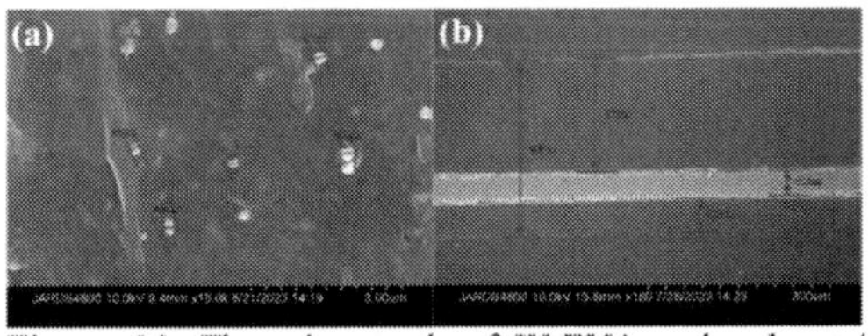

Figure 14: The micrograph of W-EVA and polymer/ metal composite backsheet: (a) W-EVA; (b) Polymer/ metal composite backsheet

In summary, the damp-heat endurance of SG modules employing TOPCon cells encounters a significant challenge as demonstrated in Table V. In contrast to DG modules, the module structure of SG modules plays a crucial role on their damp-heat durability. When the cell gap is reduced to -0.3mm, the initial gap is filled by wafer rather than polymer materials, which offers superior moisture resistance, resulting in a noteworthy decline in power loss. Reducing the amount of EVA used in the encapsulation material and enhancing the moisture resistance of the backsheet can effectively decelerate the corrosion process, thus enhancing the damp-heat endurance of SG modules.

Table V: Damp-heat endurance comparison of different SG modules

Sample	Power Loss after DH aging (%)	
	1000h	2000h
SG-1	1.98	2.74
SG-2	2.54	4.21
SG-3	3.06	7.28
SG-4	2.87	5.84
SG-5	2.82	4.92
SG-6	1.53	1.80
SG-7	3.24	6.35

4 CONCLUSIONS

In this paper, we investigate the impact of module structures and encapsulation materials on the endurance of n-type PV modules under damp-heat aging. To achieve this, we test the power loss (PL) and electroluminescence (EL) imaging following an extended damp-heat (DH) aging period. For DG modules, limited water vapor entry occurs at the module edges, thereby restricting moisture transmission. We observe modest reliability performance deviation linked with varying module structures and encapsulation materials. Encapsulation material with dual EPE is adequate to pass the 3000-hour damp-heat endurance test. Moreover, switching the rear side film from EPE to EVA still preserves high reliability with less than 3% power loss after DH3000 aging. For SG modules, the situation changed, different encapsulation solutions lead to distinct power losses and electroluminescence imaging after DH aging. With a regular module structure featuring a 1.8mm cells gap, the power loss after DH 1000 exceeds 5%, even with dual POE encapsulation. This loss is due to high water vapor transmission through the polymer backsheet. However, after decreasing the cells gap to -0.3mm, module power loss significantly decreases after DH1000 aging. This indicates that the corrosion mainly occurred at the front side of n-type cells. White EVA with TiO_2 filler added has been proven to help resist metallization corrosion with less power loss. Additionally, SG modules with a polymer/metal composite backsheet have similar damp-heat endurance to DG modules. In summary, the degradation post-damp heat can be adequately explained by corrosion of the front-side paste due to water penetration. The acidic environment resulting from EVA and additives can exacerbate the degradation, resulting in severe module performance deterioration. The effective way to enhance the module's damp-heat endurance is to prevent water penetration into the cell front surface. This can be achieved by using encapsulation materials containing less vinyl acetate and a lower water vapor transmission rate (WVTR). Additionally, improving the module's structure and implementing corrosion preventive metallization of n-type cells can also be beneficial.

REFERENCES

[1] N. Iqbal, Solar Energy Materials and Solar Cells (2023) 250.
[2] P.M. Sommeling, Solar Energy Materials and Solar Cells (2023) 256.
[3] M. Hoerteis, Advanced Functional Materials 20 (2010) 20 476-484.
[4] C. Peike, Solar Energy Materials and Solar Cells 116 (2013) 49-54.
[5] N. Kyranaki, Progress in Photovoltaics 30 (2022) 1061-1071.
[6] G. Oreski, Progress in Photovoltaics 28 (2020) 1277-1288.
[7] A. Sinha, Journal of Physics D-Applied Physics 54 (2021).
[8] A. Kraft, Ieee Journal of Photovoltaics 5 (2015) 736-743.
[9] A.W. Czanderna, Solar Energy Materials and Solar Cells 43 (1996) 101-181.
[10] K. Hara, Solar Energy Materials and Solar Cells 140 (2015) 361-365.

COMPARISON OF THE FLOW BEHAVIOR AND ADHESION OF POLYOLEFIN REAR INSLUATORS WITH EVA FOR BACK CONTACT SOLAR MODULES

Meijers Guido, Proost Kristof, Duijnhoven van Frank, Thellen Chris, Maesen Franc, Pasmans Peter, Schoot Hugo
Endurans Solar
Mauritslaan 49 6129 EL Urmond

ABSTRACT: Interdigitated Back Contacted (IBC) solar modules have the highest power density to date and are now widely commercially available. The ideal configuration to maximize module power and energy output is the combination of back-contact solar cells, conductive backsheet (CBS), low temperature solder (LTS) for inter-connection between the cells and the conductive metal foil and a Rear Perforated Insulator (RPI) to act as an insulator between the solar cells and the conductive backsheet. Low temperature solder is placed in the punched holes within the insulator.

This study shows that the punched holes in a standard EVA encapsulant close during module lamination due to which the interconnection between the solar cell and the conductive backsheet can be negatively influenced. The holes close because EVA is already melting and flowing before it is fully cross-linked. Specially designed polyolefin RPIs do not show this flow due to which the holes remain open, and the interconnection remains good.

Furthermore, a comparison is made among various commercially available RPIs, showing that not all RPIs exhibit the adhesive properties needed to meet long-term ageing requirements. It is demonstrated that newly developed polyolefin RPIs combine hole integrity with high adhesion, which is essential for passing the stringent ageing requirements.

Keywords—Back-contact modules, IBC, encapsulant, series resistance, accelerated testing, rheology.

1 BACKGROUND AND INTRODUCTION

Back-contacted solar technology has been used for many years in premium PV modules, and is now seeing a rapidly growing market share, mainly sparked by increased developments and availability of IBC cells. The most used cell types for back-contacted modules are metal wrap through (MWT) and interdigitated back-contacted (IBC) solar cells. These back-contacted solar cells have demonstrated exceptional energy output and are among the cells with the highest peak power density in the market [1], [2]. The ideal configuration to maximize module power and energy output is the combination of back-contact solar cells, conductive backsheet (CBS), low temperature solder for inter-connection and a perforated rear insulator sheet (RPI sheet) between the solar cells and the conductive backsheet (see schematic drawing 1). This combination holds great value for both the module maker and end-users.

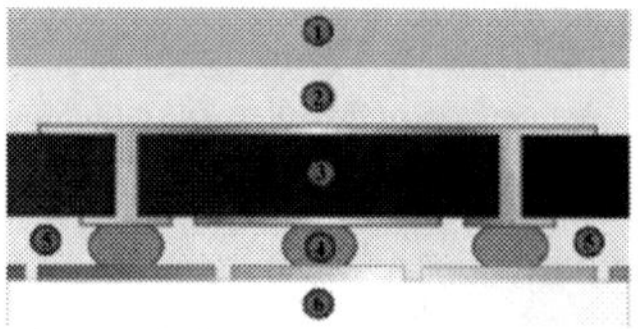

Schematic drawing 1: Schematics of the built-up of a back contact module: 1-glass; 2- front encapsulant; 3- solar cell; 4-low-temperature solder; 5-rear perforated insulator; 6-conductive backsheet.

The combination of CBS with back-contacted cells results in the highest module efficiency due to very low cell-to-module (CTM) losses. This is primarily attributed to reduced resistive losses and a smaller module area using the same number of cells [3-5]. Furthermore, the relative energy output is higher compared to more traditional module configurations, thanks to a lower nominal operating cell temperature (NOCT) and good low-irradiance performance of the cells [6-8].

During the development of back contacted modules, initially, standard available encapsulants were used to separate the cells from the conductive backsheet. However, it became apparent that these materials could not be easily used, requiring special properties. For example, holes need to be punched in the encapsulant to insert the conductive adhesive. Most standard encapsulant sheets expand and shrink after punching. Furthermore, during module lamination, more traditional encapsulants like EVA and POE sheets lack dimensional stability due to snap-back below the melting point.

Above the melting point of the encapsulant, shrinkage and polymer flow complicate the size and position of the holes, resulting in poor connections between the cell and conductive backsheet. Additionally, most standard encapsulants available have a low viscosity at low melting temperatures, causing the punched holes to close before a proper connection is established between the CBS and the solar cells. Therefore, there is a strong need for encapsulants with a wide processing window and dimensional stability during lamination for back-contact modules.

In this work, the authors present a comparison between a standard encapsulant, and a specially designed polyolefin rear perforated insulator (RPI) as back-encapsulant for back-contact technology. The results include evaluations of lamination, rheology measurements, and adhesion characterization.

2 Experimental

2.1 Flow behavior during module lamination

In establishing the design criteria for a new RPI sheet, a comprehensive study of the dimensional stability and flow characteristics of the specific back-encapsulants was conducted. The newly developed RPI sheet was compared to a standard commercially available EVA encapsulant, with both materials having 4 mm punched holes. During module lamination in an SM Innotech Profilam laminator, the temperature profile within the encapsulant was recorded. Dummy modules were created by stacking front glass, back-encapsulant and CBS. A glass plate with a

10.4229/EUPVSEC2023/3AV.2.42
020212-001

thermocouple gage was placed adjacent to this stack on the laminator heating plate. The temperature profile during the lamination process was recorded using a Datapaq temperature recorder from Fluke (Q18). Two different stacks were assembled: one using a commercial fast-curing EVA encapsulant and the other utilizing with the newly developed polyolefin RPI sheet. The lamination cycle was interrupted at specific points, both before and after the melting of the EVA, with the exact interruption time determined through multiple experiments. Table I provides a summary of the lamination process parameters and Figure 1 illustrates the corresponding temperature profiles in the RPI sheet and EVA back-encapsulants.

Table I: Lamination protocol used in the SMInnotech Profilam laminator

profile	T. (°C)	Vacuum Pins up (s)	To Press. Atmosph (s)	Time (s/ min)
Full* lamin.(A)	150	240	60 pins down	1080/ 18
Before flow(B)	150	180	60 pins up	240/ 4
After flow (C)	150	210	60 pins down	270/ 4.5

*Between "Vacuum Pins up" and "To pressure atmosphere" extra cycles were added to reach full lamination. What was added was 120 seconds at Vacuum with pins down, pressure to 800 mbar for 60 seconds with the pins down and crosslinking occurred during 600 seconds with the pins down.

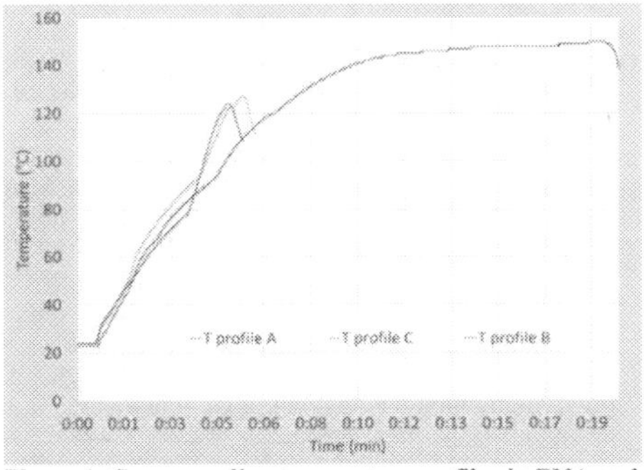

Figure 1: Corresponding temperature profiles in EVA and RPI back-encapsulants.

2.2 Gel content measurements

It is well-known that most EVA-encapsulants contain peroxide, which initiates cross-linking. Cross-linking is aimed at preventing flow after module lamination (use phase). Consequently, it is expected that cross-linking will occur during module lamination, preventing melting or limiting flow, and ensuring that the punched holes in the encapsulant remain open.

Performing gel content measurements will teach us at which point during module lamination, cross-linking is at an appropriate level. For this investigation, samples were prepared according to lamination profiles A, B and C (see Table I), aiming to determine the gel-content.

2.3 Rheology measurements

The RPI has a different composition compared to EVA which should prevent any flow during module lamination and the punched holes will not close. However, the RPI should exhibit good adhesion to a conductive backsheet and the solar cells. Rheological properties were therefore measured using an Anton Paar MCR301 rheometer (SN80108634) equipped with a set-up with a Peltier heated bottom plate and hood (P-PTD200-H-PTD200 SN81926411-81901487). For the measurements a disposable geometry was used (D-PP25, SN12151). Measurements on RPI samples were performed in oscillation to study their behavior.
Heating and cooling curves were made from 20 to 150°C with a rate of 11 °C/min and a frequency of 10 rad/s.

2.4 Adhesion measurement

The peel strength of different laminates was tested to investigate the difference in adhesion between EVA and different (commercially available) RPIs. A stack was made of glass, EVA, EVA-or-RPI and conductive backsheet after which the stack was placed in a module laminator for 18 minutes at 150°C to reach full lamination. Laminated stacks were placed in a thermal cycling (TC) and a damp heat (DH) chamber. The TC and DH tests are described in IEC 61215.
Peel strength measurements were performed between the EVA-or-RPI and the conductive backsheet before and after TC- and DH-ageing.

3 Results and Discussion

3.1 Flow behavior during module lamination

Picture 1 depicts the encapsulant evolution over lamination time. In the case of the punched EVA, the hole positions remained the same, but they completely disappear after melting. As a comparison the RPI did not show any visual flow or melting and as a result the holes did not close. This observation highlights that the primary factor contributing to hole distortions is not so much the shrinkage or snap back below the melting point of the polymer matrix's melting point, but rather the flow of the polymer above the melting temperature, leading to complete closure of the holes. In an actual back-contact module, such closure would severely impact the conductive adhesive or solder connecting the cells and the CBS, potentially resulting in a catastrophic loss of power output of the module.

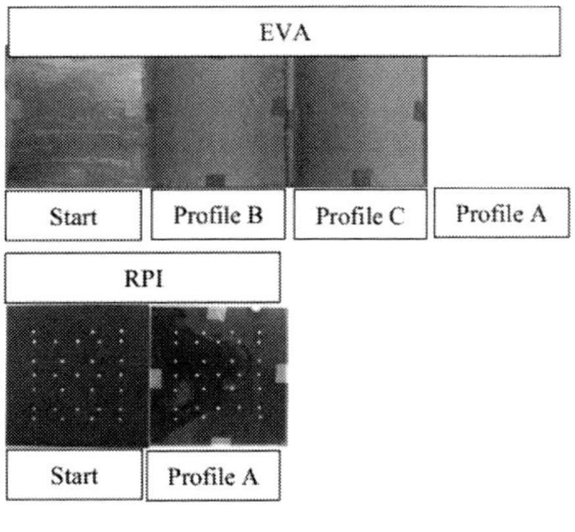

10.4229/EUPVSEC2023/3AV.2.42
020212-002

Picture 1: Integrity of the holes using different cycles.

3.2 Gel content measurements

It was observed that the holes did close during module lamination, prompting an investigation into the cause by conducting gel-content measurements on EVA. The results are given in Table II.

The gel-content measurements clearly demonstrate that no cross-linking occurs after T-profile B and C, whereas sufficient cross-linking occurred only after Full lamination (T-profile A). EVA typically starts melting around 70-75°C, and after 4.5 minutes of lamination the melting temperature is already above 100°C explaining why the holes are closed.

Table II: Gel content measurement

Setting	lamination at 150°C (s/min)	Gel content EVA (%)
Full lamination (A)	1080 (18)	87.8
Before flow point EVA (B)	240 (4)	0
After flow point EVA (C)	270 (4.5)	0

3.3 Rheology measurements

Rheology measurements were executed on two RPIs, RPI A and RPI Black. RPI Black has been developed by Endurans Solar and RPI A is a competitive product. The results revealed that the viscosity of both RPI sheets only experienced a slight decrease at a temperature of 150°C, see table III. Furthermore, the viscosity remained high at this temperature of 150°C and can therefore be explained why the holes of both RPIs will not close during module lamination.

Table III: Rheology measurements: Viscosity

	RPI A	RPI black
Viscosity (Pa.s)		
20°C	2.42E+06	1.79E+06
105°C	5.45E+05	1.86E+05
150°C	6.90E+03	9.66E+03

From the measurements, also the phase shift angle (tan delta) was taken, which is shown in table IV. The phase shift angle (δ) is a measure of the elastic or viscous behavior of the material. The lower this phase shift angle the more rigid the sample will behave. When the phase shift angle is larger than 45° the material will behave more like a fluid. It can be seen that both materials are converted from a rigid material at low temperature to a more mobile material at high temperatures as the phase shift angle increases. Moreover, the tan delta results indicated that the tested RPI Black maintained a solid state at the highest module lamination temperature of 150°C. The result demonstrates that this RPI sheet retains its solid characteristics during the module lamination process, providing an explanation for the non-closure of the punched holes. In comparison RPI A shows a more fluid behavior as the phase shift angle is 58° at a temperature of 150°C. This does not immediately result in closing of the punched holes but can result in possible movement of the solar cells which are on top of this rear insulator.

Table IV: Rheology measurements: Phase shift angle (δ)

	RPI A	RPI black
δ		
20°C	5	5
105°C	8	17
150°C	58	44

In conclusion, it can be affirmed that ideal RPI sheets should possess a melting point higher than the maximum lamination temperature employed in module production. Nonetheless, achieving satisfactory adhesion levels to other surfaces, such as CBS, cells, and front encapsulant, can be challenging when the material is not in a molten state. Another option considered was completing crosslinking before the polymer reaches a molten state, but it was found to be unattainable.

3.4 Adhesion measurements

Peel strength measurements were performed between the conductive backsheet and the RPI or the EVA, and the results are shown in Figures 2 and 3.

Results illustrate that both EVA as the RPI sheet from Endurans solar exhibit initial adhesion values of 40 to 50 N/cm, while other commercial RPI sheets demonstrate significantly lower initial adhesion below 10 N/cm and a few even below 1 N/cm. After undergoing 1000 hours of damp heat aging according to IEC61215 standards, the peel adhesion force of most encapsulants, including EVA, decreases to below 5 N/cm. In contrast, the adhesion of the Endurans RPI encapsulant remains high, in the range of 30 N/cm.

During thermal cycling, both EVA and the Endurans solar RPI sheet maintain a high level of adhesion (> 35 N/cm), while the adhesion of the commercial RPIs remains below 5 N/cm.

Low adhesion can lead to delamination between the encapsulant and the RPI sheet, particularly during the module's service. Outdoor exposure to temperature fluctuations between night and day, as well as seasonal variations between summer and winter, coupled with water ingress, can initiate delamination. A weak interlayer adhesion will further exacerbate this issue. Delamination between the conductive backsheet and the RPI sheet will increase the pressure in the laminate structure, which subsequently can cause the conductive paste to delaminate. This results in a loss of electrical connection, leading to reduced module efficiency or even complete module power output failure.

10.4229/EUPVSEC2023/3AV.2.42
020212-003

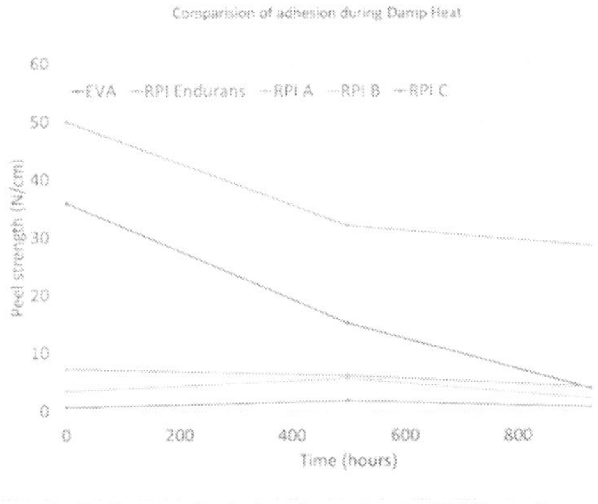

Figure 2: Peel strength during damp heat of EVA and different RPIs

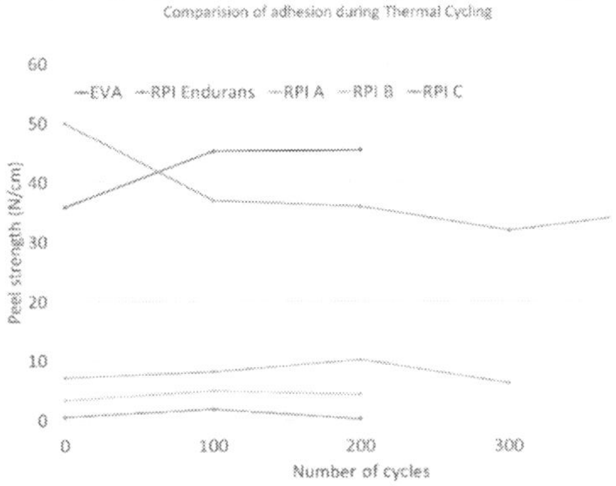

Figure 3: Peel strength during thermal cycling of EVA and different RPIs

4 Summary

In back-contact modules, a perforated rear insulator is required to separate the back contacts of the cells from the conductive backsheet. Standard encapsulants like EVA and POE pose challenges in fulfilling this purpose due to the loss of structural integrity and polymer melt flow that occurs during lamination. It was demonstrated that the flow of the polymer melt is the primary complication with the use of more traditional materials, rather than snap-back below the polymer's melting point or shrinkage above it. A comparison between the rheological behavior of traditional encapsulant and a newly developed polyolefin RPI sheet provided insights into the melting behavior of these materials during the module lamination process. While other commercially available RPI sheets exhibit low adhesion to other surfaces, this is not the case for the newly developed polyolefin RPI sheet. The newly developed RPI sheet combines hole integrity during lamination with high adhesion to other surfaces in the module, which results in superior service life stability of back-contacted photovoltaic modules.

5 References

[1] ITRPV Eighth Edition, "International Technology Roadmap for Photovoltaic Results 2016", 2017.
[2] M. A. Green, Y. Hishikawa, E. D. Dunlop, D. H. Levi, J. Hohl-Ebinger, and A. W. Ho-Baillie, "Solar Cell Efficiency Tables (version 51)", Progress in Photovoltaics, 26(1), 3, 2018
[3] H. Schoot, Intersolar 2022, BC Workshop
[4] I. J. Bennett, E. Bende, M. J. A. A. Goris and W. Eerenstein, "An Overview of Developments in Foil-Based Back-Contact Modules" presented at 29th European Photovoltaic Solar Energy Conference and Exhibition, Amsterdam, The Netherlands, 22-26 September 2014
[5] H. Schulte-Huxel, et al., High-efficiency modules with passivated emitter and rear solar Cells? An analysis of electrical and optical losses, IEEE J. Photovolt., 7 (1) (Jan. 2017), pp. 25-31
[6] Presented at "Energie Vakbeurs 2015 6-8 October 2015" by Tulipps Solar systems solutions
[7] PES Solar Issue 30 2017 Pages 51 – 100
[8] https://exasun.com/
[9] Effect of Novel Encapsulants and Backsheets on Short Circuit Current in Interdigitated Back Contact Solar cells based PV Modules; Asma Shamim et al; July 2020; Optik - International Journal for Light and Electron Optics

10.4229/EUPVSEC2023/3AV.2.42
020212-004

MEASUREMENT UNCERTAINTY ANALYSIS FOR LARGE AREA HIGH-EFFICIENCY MODULES

K. Ramspeck, C. Böhmer and M. Meixner
halm elektronik GmbH
Friesstraße 20, 60388 Frankfurt

ABSTRACT: In this contribution we present an analysis of measurement uncertainty for a new cetisPV-Moduletest4 system for large-size high-efficiency silicon photovoltaic modules. In our analysis we follow the approach used by Dirnberger et al. [1] evaluating all contributions with special focus on the main driving quantities irradiation inhomogeneity, spectral mismatch and transient effects in *IV*-curve measurements which became increasingly important with n-type cell technologies gaining market share. Using state of the art procedures and equipment we reach a measurement uncertainty of only 1.5 % for a system with 2.7 m x 1.6 m illumination area at a projection distance of 4 m only. From our analysis we derive optimization potential on which further development of the system will focus in near future, trying to reduce the measurement uncertainty towards 1 %.
Keywords: power output measurement, measurement uncertainty, large area modules

1 INTRODUCTION

Measurement uncertainty (MU) is the main differentiator between calibration laboratories as lowering the uncertainty increases confidence about output power of the PV industry and reduces financial risks in PV projects very directly. Besides this, lower measurement uncertainty helps to distinguish technological advantages and is thus a driver of innovation, too. Round robin experiments between different test laboratories [2] still show significant differences between the P_{MPP} values measured. Since then, through the development of new, larger modules, and upcoming cell technologies, the properties of the device under test (DUT) have undergone significant changes in the recent past. In order to catch up with these changes, measurement systems were redesigned to meet the demands for larger measurement areas. These new measurement conditions require a revisit of measurement uncertainty, especially as the larger modules with new cell designs lead to an increase in measurement uncertainty values if no counter measures are taken. We demonstrate how careful system design and analysis of the systems properties allows to keep measurement uncertainty at low levels even for latest module technologies. By performing a measurement uncertainty analysis, we identified, that main drivers remain lateral irradiance inhomogeneity and spectral mismatch. Therefore, within this paper we focus on the determination of their contributions together with analysis of the impact of transient effects which have gained increasing attention in latest cell technologies today. Please note, that the number resulting from the measurement uncertainty analysis is very specific to assumptions, methods and instrumentation used in its derivation and cannot be seen as a simple system property. It thus remains an important task of each calibration laboratory to assess this quantity carefully and minimize it as much as possible by careful system analysis and optimized procedures.

2 INSTRUMENTATION

2.1 Flasher system

We analyze the measurement uncertainty achievable with our cetisPV-Moduletest4 system providing an illumination area of 270 cm x 160 cm with irradiance inhomogeneity below 0.5 %, excellent spectral homogeneity and an A+ spectral match according to IEC 60904-9:2020. While achieving such large illumination area, the system design is kept compact with only 4 m projection distance between light source and measurement plane. Figure 1 shows an image of the system (a), a close-up of the light source (b) and the irradiance distribution of a fully homogenized setup (c). Please note, that the picture shown in (a) includes the cetisPV-Therm-package-M which allows to control the module temperature within a range from 15 °C – 75 °C.

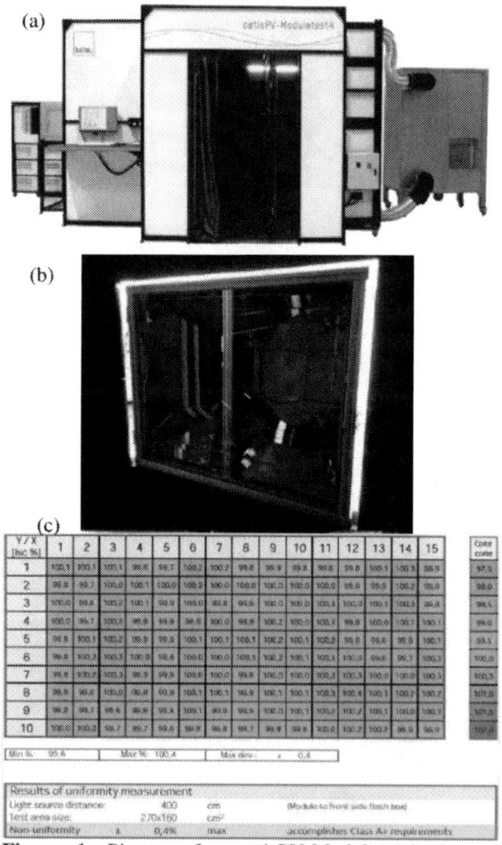

Figure 1: Picture of a cetisPV-Moduletest4 system including the thermal control option (a), close-up of the light source (b) and irradiance homogeneity map (c) of the system.

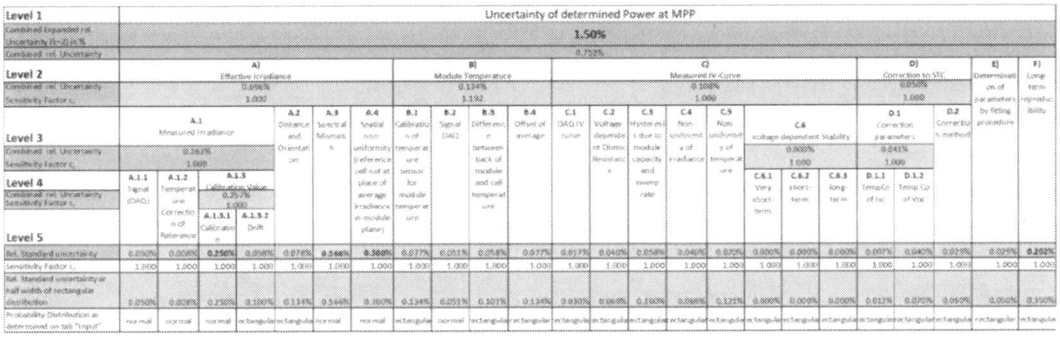

	A.1.1	A.1.2	A.1.3.1	A.1.3.2	A.2	A.3	A.4	B.1	B.2	B.3	B.4	C.1	C.2	C.3	C.4	C.5	C.6.1	C.6.2	C.6.3	D.1.1	D.1.2	D.2	E)	F)
Level 1	Uncertainty of determined Power at MPP																							
Combined Expanded rel. Uncertainty (k=2) in %	**1.50%**																							
Combined rel. Uncertainty	0.752%																							
Level 2	A) Effective Irradiance							B) Module Temperature				C) Measured IV-Curve								D) Correction to STC			E) Determination of parameters by fitting procedure	F) Long-term reproducibility
Combined rel. Uncertainty	0.696%							0.134%				0.108%								0.050%				
Sensitivity Factor c_i	1.000							1.192				1.000								1.000				
Level 3	A.1 Measured Irradiance				A.2 Distance and Orientation	A.3 Spectral Mismatch	A.4 Spatial non-uniformity (reference cell not at place of average irradiance in module plane)	B.1 Calibration of temperature sensor for module temperature	B.2 Signal (DAQ)	B.3 Difference between back of module and cell temperature	B.4 Offset of average	C.1 DAQ IV curve	C.2 Voltage dependent Ohmic Resistance	C.3 Hysteresis due to module capacity and sweep rate	C.4 Non-uniformity of irradiance	C.5 Non uniformity of temperature	C.6 Voltage-dependent Stability			D.1 Correction parameters		D.2 Correction method		
Combined rel. Uncertainty	0.262%																0.000%			0.041%				
Sensitivity Factor c_i	1.000																1.000			1.000				
Level 4	A.1.1 Signal (DAQ)	A.1.2 Temperature Correction of Reference	A.1.3 Calibration Value														C.6.1 Very short-term	C.6.2 short-term	C.6.3 long-term	D.1.1 TempCo of Isc	D.1.2 TempCo of Voc			
Combined rel. Uncertainty			0.257%																					
Sensitivity Factor c_i			1.000																					
Level 5			A.1.3.1 Calibration	A.1.3.2 Drift																				
Rel. Standard uncertainty	0.050%	0.008%	**0.250%**	0.058%	0.078%	**0.566%**	**0.300%**	0.077%	0.051%	0.058%	0.077%	0.017%	0.040%	0.058%	0.040%	0.070%	0.000%	0.000%	0.000%	0.007%	0.040%	0.029%	0.029%	**0.202%**
Sensitivity Factor c_i	1.000	1.000	1.000	1.000	1.000	1.000	1.000	1.000	1.000	1.000	1.000	1.000	1.000	1.000	1.000	1.000	1.000	1.000	1.000	1.000	1.000	1.000	1.000	1.000
Rel. Standard uncertainty or half width of rectangular distribution	0.050%	0.008%	0.250%	0.100%	0.134%	0.566%	0.300%	0.134%	0.051%	0.101%	0.134%	0.030%	0.069%	0.100%	0.069%	0.121%	0.000%	0.000%	0.000%	0.012%	0.070%	0.050%	0.050%	0.350%
Probability Distribution as determined on tab "Input"	normal	normal	normal	rectangular	rectangular	normal	normal	rectangular	normal	rectangular	rectangular	rectangular	rectangular	rectangular	rectangular	rectangular	rectangular	rectangular	rectangular	rectangular	rectangular	rectangular	rectangular	rectangular

Figure 3: Results of measurement uncertainty analysis and contributions to the same.

2.2 Analysis instruments

The system described briefly in 2.1 has been examined thouroughly using spectroradiometric measurements, homogeneity measurements, *IV*-measurements and examining temperature and geometric effects. For spectrum measurements an Avantes AvaSpec-Hero Spectroradiometer is used for λ<1020 nm and an Avantes NIR256512-1.7-HSC-EVO spectroradiometer is used in the near infrared. Both spectroradiometers are regularly calibrated on our own inhouse spectroradiometer calibration setup with a primary light source calibrated directly from the national german metrology institute, PTB. For homogeneity measurements, an encapsulated 6-inch monocrystalline PERC cell mounted on an xy-stage is used. Additionally, measurements are performed with a WPVS-like cell mounted on the same stage and with a self-build 144 channel homogeneity measurement tool which is displayed in Figure 2. This tool reads out 144 channels simultaneously 50 times during one single flash and provides precise high resolution (2 cm x 2 cm) irradiance data. It is calibrated in a large homogeneous light field of a Moduletest-system in which any obstructing elements have been removed and projection distance increased to ensure local homogeneous irradiance.

Figure 2: Self-built 144 channel homogeneity measurement tool with 2 cm x 2 cm resolution.

IV curves are measured using a cetisPV-EL3-M and cetisPV-CT-L2 3-Quadrant measurement system. Temperatures are measured using Pt100 sensors. As modules are at ambient conditions, temperature inhomogenenity and temperature differences between junction and module surface are very small. We use flash lengths with a maximum of 90 ms in this study, temperature increase during a single flash is thus kept below 0.2 K.

3 MEASUREMENT UNCERTAINTY ANALYSIS

For the general measurement uncertainty analysis, we follow the procedure proposed by Dirnberger et al. [1]. Examination of many different quantities contributing to the total measurement uncertainty of P_{MPP} leads us to the result shown in Figure 3. Most contributions proof to be rather small and contribute little to the total MU due to the quadratic addition of independent errors. Spectral mismatch, inhomogeneity of irradiation, primary calibration and long-term effects tend to be the quantities with the largest impacts. The first two of them, which can be analyzed and worked on by flasher design will be treated in more detail in this contribution. In addition, we will show how to reduce the contribution of transient effects to negligible levels using advanced hysteresis evaluation [3]. Alltogether an expanded measurement uncertainty of 1.5 % (k=2) is reached with the system. This value depends however strongly on assumptions in the derivation of the input quantities as well as on each labs procedures and cannot be seen as a simple system property.

3.1 Transient effects

In order to examine the contribution of transient effects in the cells *IV*-measurement to the total MU, we examine a commercially available HJT module with respect to different measurement times. In this examination we use single sweep measurements, advanced hysteresis evaluation with linear voltage ramping and advanced hysteresis evaluation with adapted voltage ramping. For very long single sweep measurement simulation, we use 84 ms sweep time in a reduced voltage range around P_{MPP} (prior determined from advanced hysteresis) spanning different voltage ranges. This avoids heating up the module during very long illumination times or large numbers of flashes. The results, displayed in Figure 4 show strong deviations from the steady-state P_{MPP} value for sweep times below 500 ms if a linear single sweep is used. Measurements which are in agreement within 0.1 %.rel are only achieved for single sweep measurements with strongly reduced voltage range which correspond to above 1 sec total measurement time for a full *IV*-curve. The same precision is reached for advanced hysteresis measurements with > 30 ms sweep times,

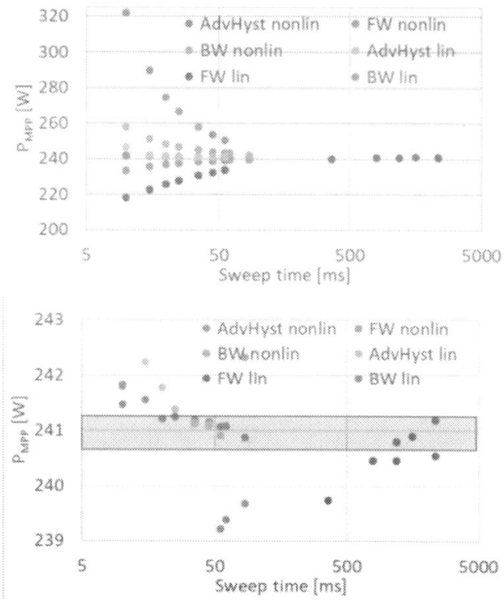

Figure 4: P_{MPP} results of sweep time variations on a commercial HJT solar module with strong transient effects. The upper graph shows the full data, the lower graph a restricted view of +/-2 W around the steady-state value for this specific module.

especially if optimized voltage ramping is used. This reduces the total required measurement time to the range of 60 ms – a value which can is reached within one single flash with the system.

Both methods help to minimize transient effect related MU to a value below 0.1 % and thus make its contribution to the total MU nearly negligible.

3.2 Irradiance inhomogeneity

For analysis of the contribution of irradiance inhomogeneity, we performed homogeneity measurements with different measurement resolution and different detectors. For these measurements, a system without final homogenization was used, thus a slightly higher inhomogeneity close to 1% is present during this examination. Inhomogeneity was measured on this system using a 6-inch encapsulated cell mounted on an automated xy-stage, using the 144-channel homogenization measurement tool mounted on the same xy-stage and using a WPVS-like cell mounted again on the same xy-stage. While the measurement with the full-size cell and the homogeneity measurement tool were performed on the full area, the measurements with the WPVS-like cell were performed only in locations which were found to be suitable for calibration from the measurement with the 6-inch cell. Figure 5 shows the high resolution (2 cm x 2 cm) irradiance map measured with the homogeneity tool on a total area of 288 cm x 192 cm. A reduced intensity at the edges outside of the homogeneous illumination area of 270 cm x 160 cm is visible on this map. Besides this, a slight high resolution inhomogeneity is visible which is of importance for calibration with a small sized WPVS cell. To evaluate the general precision of the measurement with the high resolution tool, we compare averaged values from this map with the direct homogeneity measurement with a full sized cell in Figure 6.

Figure 5: High resolution (2 cm x 2 cm) irradiance map of the system examined in this study.

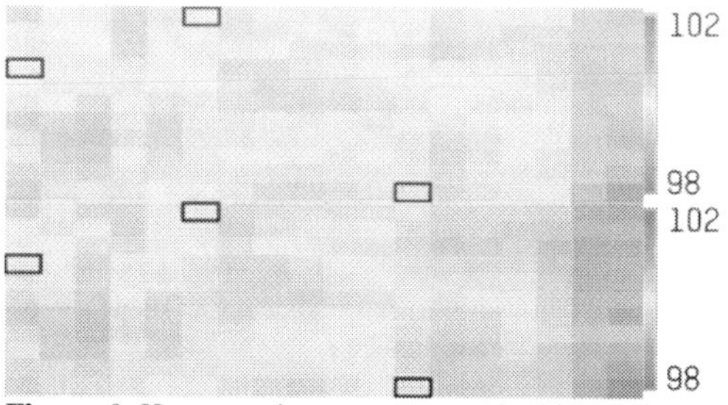

Figure 6: Homogeneity map measured with a full sized cell (top) and averaged from the high resolution image (bottom).

The images show, that both measurements are in close agreement, with only slight differences in the range of +/-0.2 %. For further evaluation of measurement uncertainty caused by local inhomogeneity we examine three positions more closely, which are marked by black boxes in Figure 6. They are chosen such, that the irradiance measured in these positions with the 6-inch encapsulated cell equals the average value of the irradiance measured in an anticipated module area of 2.4 m x 1.3 m, which is 1000 W/m². These areas are chosen at the edges of the measurement area in order to allow mounting a WPVS-like solar cell besides a large sized module during calibration and are used further as calibration sites for a WPVS-like cell. In order to assess calibration uncertainty caused by local intensity variations in these positions, scans with the WPVS-like cell are performed to cover the whole area of the 6-inch cell with

99.8	100.1	100.0	100.0	100.1	100.1	100.1	100.0
99.8	100.0	100.0	100.0	100.1	100.0	100.0	100.0
100.0	100.1	100.2	100.1	100.1	100.1	100.0	100.1
100.1	100.3	100.2	100.1	100.1	100.1	100.0	100.1
99.9	100.2	100.3	100.2	100.1	100.1	99.9	100.0
100.0	100.1	100.2	100.1	100.1	100.1	100.0	100.2
100.0	100.2	100.2	100.2	100.3	100.3	100.2	100.3
100.0	100.1	100.2	100.1	100.3	100.4	100.3	100.3
100.1	100.2	100.1	100.1	100.1	100.2	100.3	100.3
100.0	100.3	100.2	100.2	100.1	100.1	100.1	100.2
99.9	100.1	100.0	100.1	100.0	100.1	100.1	100.2
99.8	100.0	99.9	99.9	100.0	100.0	100.0	100.1
99.9	100.2	100.0	99.9	100.0	100.1	100.1	100.2
100.0	100.2	100.1	100.1	100.2	100.3	100.2	100.4
100.1	100.3	100.1	100.2	100.2	100.4	100.3	100.5
100.2	100.4	100.2	100.3	100.3	100.5	100.4	100.6
99.8	99.8	99.8	99.7	99.7	99.7	99.6	99.4
99.8	99.8	99.7	99.6	99.6	99.7	99.6	99.4
99.7	99.7	99.7	99.6	99.6	99.7	99.5	99.4
99.7	99.7	99.6	99.5	99.5	99.6	99.5	99.3
99.6	99.6	99.5	99.5	99.5	99.6	99.6	99.4
99.7	99.7	99.6	99.6	99.6	99.6	99.5	99.3
99.8	99.8	99.7	99.7	99.8	99.9	99.7	99.4
99.8	99.8	99.7	99.7	99.6	99.7	99.6	99.5

Figure 7: local irradiance inhomogeneity in the three fields marked in Figure 6 by black boxes.

10.4229/EUPVSEC2023/3AV.3.7
020213-003

the small reference device. These scans, displayed in Figure 7, deliver 192 intensity values from which the standard deviation is derived and used as a measure for the MU contribution from irradiation inhomogeneity.
The resulting value of the standard deviation for the data shown in Figure 7 of 0.3 % (k=1) is used as measurement uncertainty caused by irradiation inhomogeneity in this study. Further reduction of this value is possible by choosing a position with average irradiation inside one of the marked 6-inch cell areas for calibration. However, this optimization potential is still to be realized through thorough analysis and optimization of the comparability of different local irradiation measurement tools to ensure the effectiveness of this precise positioning.

3.3 Spectral mismatch

Uncertainty in the determination of spectral mismatch (MM) contributes significantly to the total MU of P_{MPP}. Spectral mismatch is calculated according to equation (1),

$$MM = \frac{\frac{\int SR_{DUT}(\lambda)E_{SIM}(\lambda)\partial\lambda}{\int SR_{DUT}(\lambda)E_{STC}(\lambda)\partial\lambda}}{\frac{\int SR_{RC}(\lambda)E_{SIM}(\lambda)\partial\lambda}{\int SR_{RC}(\lambda)E_{STC}(\lambda)\partial\lambda}} \quad (1),$$

where $SR_{DUT}(\lambda)$ is the spectral response of the device under test, $E_{SIM}(\lambda)$ is the spectral irradiance of the simulator, $SR_{RC}(\lambda)$ is the spectral response of the reference cell and $E_{STC}(\lambda)$ the spectral irradiance defined for AM1.5G.

Three of the four quantities used for calculation of the MM value are measurement results, bearing their own uncertainty. They contribute through the rather complex form of equation (1), to the uncertainty of the MM value which in turn is a contribution to the total MU of P_{MPP}. Hohl-Ebinger et al. proposed a random walk analysis [4], choosing spetrum and spectral response data within the ranges of their measurement uncertainties and deriving many possible mismatch values from which, then, a standard deviation can be calculated to assess the value of MM uncertainty. Such a random walk analysis accounts well for correlations between neighbouring wavelengths. Correlations we observe in recalibration of different spectroradiometers however show correlated deviations over broad ranges of wavelengths, often shifting the UV part upwards and the IR part downwards or vice versa. Such long range correlations will not be assessed by random walks typically which is why in this study we have chosen a different approach. Assuming two different measurement uncertainty distributions for spectral response measurement and spectrum measurement we calculated boundary external quantum efficiency (EQE) and spectrum curves. These curves were used together with measured ones to calculate mismatch values using all possible combinations. From the resulting mismatch values, different subsets of data can be used to derive standard deviations to be used for measurement uncertainty calculation. Figure 8 shows the ingoing SR curves and Figure 9 the blue part of the ingoing spectra. The measurement uncertainties assumed are +/-10 % absolute in EQE where either 10 % are added or subtracted for wavelengths below 850 nm and vice versa for wavelengths above 850 nm to create a large impact on spectral mismatch. In a second approach, up to 30 % absolute measurement uncertainty for the wavelengths of 300 nm and 1200 nm were assumed, reducing the uncertainty by 1 % every 10 nm. For the center part between 600 nm and 900 nm, no uncertainty is assumed in this approach, as EQE values typically are nearly constant in this spectral range. Assumptions on measurement uncertainty of spectra were set similarly. Either a 10 % relative error was assumed, switching its sign at 850 nm wavelength, or up to 25 % relative error at 300 nm and 1200 nm were assumed, decreasing by 1 % every 10 nm. Altogether 9 spectra were used in this contribution the measured one, the ones with changes only in the infrared of blue wavelength region (with varying measurement uncertainty of up to 25 %) and spectra were blue and infrared values were shifted by 10 % relative into different directions for below and above 850 nm wavelength. Figure 8 displays external quantum efficiency (EQE) curves used for the generation of mismatch values.

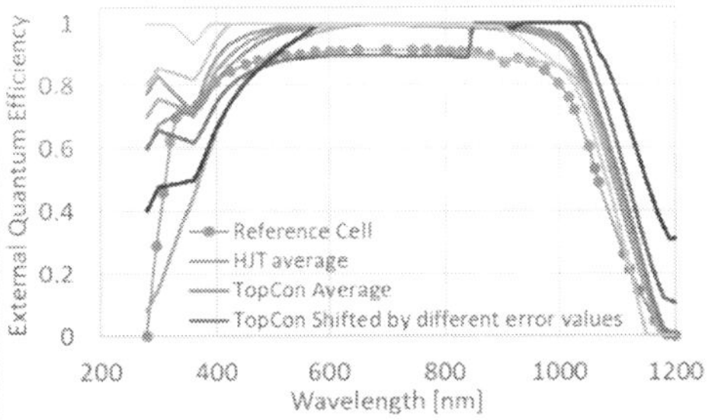

Figure 8: External quantum efficiency curves used for the mismatch evaluation.

The values of external quantum efficiency were limited at 1 and 0 to remain physically possible. TopCon and HJT data of bare cells was used and TopCon curves shifted by the stated values of measurement uncertainty were generated. Besides the EQE curves displayed in figure 8, curves measured on 11 TopCon and 4 HJT cells were used directly. Together with the spectra 5 of which are shown in the blue range in Figure 9, altogether 189 mismatch values were calculated.

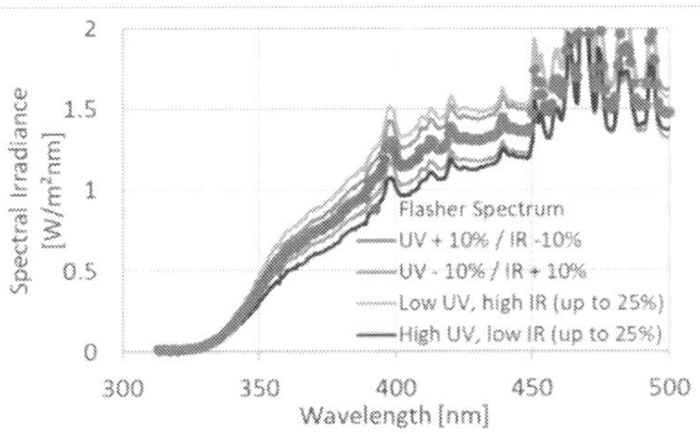

Figure 9: Spectral irradiance data in the blue range, used for evaluation of mismatch values.

Our approach assumes very high correlation over a large wavelength range, avoiding up and down shifting of EQE curves and being in agreement with possible shifts we observe in spectroradiometer recalibration. Assuming rather high measurement uncertainties in combination with this approach we make sure not to underestimate the MU contribution from spectral mismatch.

Figure 10 shows mismatch values calculated for 4 of the 9 spectra used, displaying all 21 values for the EQE curves available for each spectrum. Ovals mark the results for the HJT-cell EQE curves, rectangular boxes the results for the EQE curves shifted by error assumptions. The mismatch values in the boxes show the largest scatter, due to the large variations in blue and infrared response assumed. A different level of mismatch values is observed for the different spectra. Changing the spectrum shifts the rather constant mismatch calculated with measured TopCon EQE curves by about +/-0.3 %.

Figure 11 displays standard deviations of spectral mismatch for several subsets of the total data. Using all 189 values of MM, the standard deviation is 0.42 %. Using only the spectral response data shown in Figure 10, a higher value of 0.56 % results, as data with added or subtracted measurement uncertainty has a larger share in this data subset. Assuming only measured TopCon quantum efficiencies (7 different processes and 11 different cells alltogether) the standard deviation reduces drastically to 0.22 %. Similar values result if the data subset is restricted to one single spectrum thus eliminating the spectrum MU.

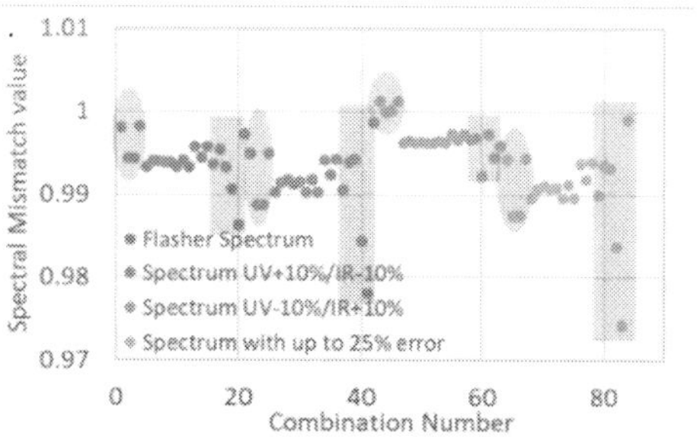

Figure 10: Spectral mismatch values calculated for 4 spectra and the 21 spectral response curves used in this study.

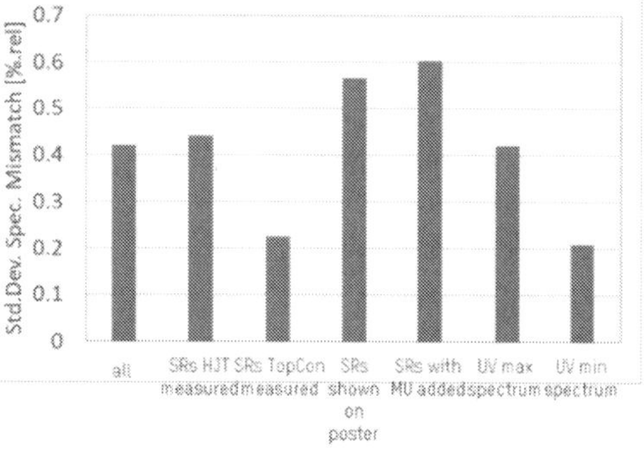

Figure 11: Standard deviation calculated for different subsets of spectral mismatch values.

In this case, different spectra still may show significant differences in the resulting MU. For evaluation of the total MU we chose the high value of 0.56 % as standard deviation of spectral mismatch in order to fully account for measurement uncertainties in spectral response and spectrum measurements. Further work will be necessary to reduce this value by optimization of evaluation procedures on the one hand, and spectral irradiance distribution on the other hand.

3.4 Total results

Bringing all ingoing quantities together we derive a value of 1.5 % for the expanded measurement uncertainty of P_{MPP}, as displayed in Figure 3. This value is mainly dominated by contributions from spectral mismatch, irradiation inhomogeneity, primary calibration of reference cell and long-term reproducibility. It is important to understand, that this value is not a property of the system but strongly depends on methods for - and uncertainties of measurements of the optical properties of the system and other values contributing to the uncertainty. Through these measurements it is a value which is governed by the quality system and instrumentation of a laboratory performing a MU assessment and thus a value specific for a laboratory and even a module type and size examined in combination with the reference cell used and all measures taken to make sure that the system is operated correctly. In this regard, our analysis can only give guidance on ways to calculate the uncertainty and give an exemplary value reachable under the circumstances described here.

4 CONCLUSION

Concluding we have shown how a measurement uncertainty for P_{MPP} measurements of 1.5 % is achievable for large sized capacitive modules using a cetisPV-Moduletest4 system. Further optimization potential, especially in the field of spectral mismatch uncertainty has been demonstrated. With further work being dedicated to system optimization and optimized procedures for measurement uncertainty determination we are confident that a further reduction of measurement uncertainty towards a value of 1 % will be possible in near future. However, strict quality measures and precise measurement equipment for spectrum and homogeneity measurement will be required to achieve lowest measurement uncertainty.

References

[1] D. Dirnberger and U. Kräling in IEEE J. of Photovoltaics, vol. **3**, no. 3, pp. 1016 – 1026, 2013
[2] C. Monokroussos, E. Salis, D. Etienne, X. Zhang, S. Dittmann, G. Friesen, K. Morita, J. Stang, T. Herbrecht, V. Fakhfouri, N. Rebaud, D. Pavanello and H. Müllejans in Prog Photovolt Res Appl. 2019, **27**, pp. 603-622
[3] K. Ramspeck, S. Schenk, L. Komp, A. Metz and M. Meixner in Proc. 29th EUPVSEC, p. 1253, 2014
[4] J. Hohl-Ebinger and W. Warta in Prog Photovolt Res Appl. 2011, **19**, pp. 573 - 579

10.4229/EUPVSEC2023/3AV.3.7
020213-005

SHORT TERM OUTDOOR ENERGY RATING VS CLIMATE SPECIFIC ENERGY RATING (CSER)

Mariella Rivera, Christian Reise
Fraunhofer Institute for Solar Energy Systems ISE
Heidenhofstrasse 2, 79110 Freiburg, Germany
mariella.josefina.rivera.aguilar@ise.fraunhofer.de

ABSTRACT: This study validates the Site Specific Energy Rating (SSER) of PV modules. The SSER follows the Climate Specific Energy Rating (CSER) according to the IEC 61853 series of standards, but uses a site specific meteorological profile. At first, the laboratory characterization is performed using the energy rating methodology on monocrystalline silicon heterojunction (SHJ) PV modules. Subsequently, a site specific climate profile for Freiburg, Germany, is generated using ground-based irradiance measurements. The Site-Specific Energy Rating (SSER) is then calculated based on this profile and compared with CSER results from the IEC 61853-4 data sets. For validation purposes, short-term outdoor characterizations of the same modules are conducted, capturing real-world performance. The results demonstrate a good correlation between the SSER derived from outdoor measurements and the one based on the standard methodology, with deviations ranging from 0.23% to 1.27%. However, we find that the IEC 61853-4 standard profiles fail to accurately represent locations like Freiburg, as the differences between CSER and SSER exceed the expected uncertainty. This study emphasizes the importance of incorporating site-specific climates profiles and validates, with short-term outdoor exposures, the seasonal behavior of the SSER. The findings contribute to refining the Energy Rating approach and improving the accuracy of PV module performance assessments.

Keywords: Energy Rating, Performance Ratio, Outdoor validation

1 INTRODUCTION

The accurate assessment of PV module performance is crucial for optimizing and ensuring the reliable operation of solar energy systems. The IEC 61853 1-4 series of standards offers a standardized approach to determine the energy output of PV modules under specific climate conditions, known as Climate Specific Energy Rating (CSER). Part 1 describes the test procedures for evaluating the behavior of PV modules in relation to irradiance and temperature. It involves power rating measurements conducted over a range of different irradiances and temperatures levels. Part 2 provides detailed procedures for considering various factors that impact module performance, including angle of incidence (AOI), module operating temperature, and spectral effects. Part 3 presents the energy rating calculation procedure using the defined climatic profiles in Part 4. Collectively, these standards offer a comprehensive framework for assessing the energy performance of PV modules in accordance with standardized reference climatic profiles [1].

Previous studies have compared PV module performance using the standard methodology, primarily conducted before the full publication of the IEC 61853 standard. For instance, simulation tools like PVsyst have shown agreements within 0.5% [2, 3], while methodologies from the European Union Joint Research Center (JRC), known as ESTI-ER, have shown comparable accuracies with differences of approximately 1.8% [4]. A modified IEC 61853 methodology was also compared to outdoor measurements, resulting in deviations of around 2% for different time periods [4]. Furthermore, a round-robin study comparing different energy rating methods, including a combination with IEC 61853, yielded deviations of +/- 3% with short and long term outdoor data (without considering spectral or reflection effects) [5]. In the latest publication on the comparison of measured data with the standard methodology, differences of 5% were estimated for SHJ PV modules, which improved to 3.5% by incorporating degradation rates into the procedure [6].

Therefore, this paper aims to contribute to the validation of the accuracy of the now complete standard procedure calculations by comparing them with outdoor measurements of DC Performance Ratio (PR_{DC}). The goal is to determine if effects like light reflection, real module temperature, and irradiance dependent behavior are well represented by the methodology. Furthermore, the study aims to demonstrate that the seasonal behavior of CSER can be correlated with short-term outdoor measurements, providing a comprehensive view of PV module performance, as specified in the standard, under a range of representative weather conditions.

Our outdoor measurements are carried out in Freiburg, Germany. However, previous studies have shown that the standard climate profile, *temperate continental*, does not adequately represent the conditions in southern Germany. The performance behavior of monocrystalline-silicon modules exhibited deviations (up to 3.32%) larger than the expected uncertainty in energy rating calculations of around 2% [7, 8]. Therefore, we utilize site-specific climate data for Freiburg and calculate the Site Specific Energy Rating (SSER) [9], which is directly compared and validated with the performance in outdoor conditions during the same time period.

2 METHODOLOGY

The energy rating (ER) of a PV module is defined as the ratio of the energy produced by the module under specific climate conditions over a period (1 year according to IEC 61853-3) to the energy that would have been produced if the module had worked at standard test conditions (STC). In this paper, the ER will be calculated for different periods of time (daily, weekly) to assess its seasonal variability (Equation 1).

$$SSER_T = \frac{G_{STC} \sum_n E_n}{P_{STC} \sum_n H_n} = PR_{DC,T} \qquad (1)$$

Where $SSER_T$ [dimensionless] is the Site Specific Energy Rating over a time period T; G_{STC} [W/m²] is the irradiance at STC; P_{STC} [W] is the PV module's maximum power at STC; $E_T = \sum_n E_n = \sum_n (P_{MPP,n} \times 1h)$ [Wh] is the total energy produced by the PV module during the time period T over hourly time steps according to IEC 61853-4; $\sum_n H_n$ [Wh/m²] is the total irradiation during the time period.

The SSER was calculated according to the IEC 61853-3 procedure ($SSER_{MODEL}$) for Freiburg, Germany, using hourly values of ground-based broadband and spectral irradiance, as well as weather monitoring parameters (ambient temperature, wind speed) from our monitoring station for the same time period of outdoor exposure, following the format of the climate data sets from IEC 61853-4. For the $SSER_{REAL}$, the accumulated power output and irradiance in plane of array (POA) (Gpoa) measured in outdoor conditions are used to calculate the PR_{DC} based on Equation 1 (Fig. 1).

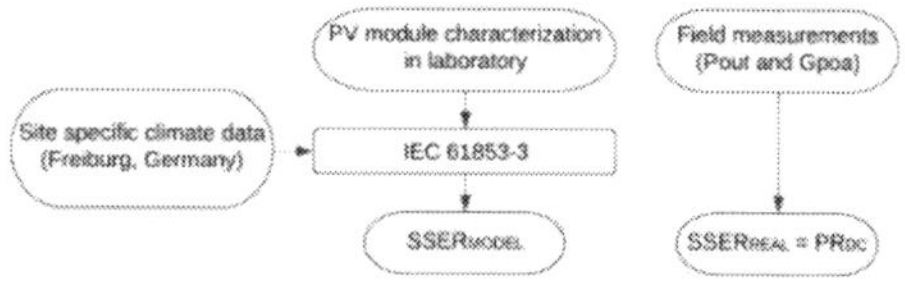

Figure 1: Methodology to calculate the modelled (left) and real (right) SSER.

2.1 Module characterization

We assessed the performance of four SHJ PV modules from two manufacturers (M01 and M02). To ensure the confidentiality of all parties involved, the PV modules are reported here anonymously. Our laboratory, CalLab PV Modules at Fraunhofer ISE, conducted tests to determine their angular, spectral, and thermal response, as well as their power matrix (according to IEC 61853-1). Table I and Figures 2 and 3 present their characteristics. From this characterization, it is possible to assess that M02 exhibits a higher power output at higher temperatures and a narrower spectral response compared to M01.

Measurement uncertainties for the IEC 61853 methodology have been evaluated in our laboratory for various technologies, including c-Si, a-Si, CdTe and CIGS, as documented in previous studies [7]. The uncertainty (k=1) in measuring the STC power range from 1.1% to 2%, depending on the specific technology. Additionally, for southern region of Germany, we estimated the uncertainty associated with factors influencing the energy rating as follows: 0.9%...1.8% for spectral effects, 0.6%...1.1% for irradiance effects, 0.1%...0.3% for temperature effects, and roughly 1% for angular effects. Among these factors, STC power measurements contribute the most to the overall uncertainty in the methodology. The characterization measurements in the laboratory are dependent on the PV module's previous exposure to light and temperature (metastability effects), especially for new cell technologies. Regarding the effects of light-induced degradation and dark storage, an uncertainty of 0.7% was estimated for c-Si modules and around 1% for thin film types.

For our study, these uncertainty values will also be considered for SHJ modules since no reference data has been found. Furthermore, in a recent round robin study on the metastability effects of SHJ, they shown lower deviation (2%) of their initial maximum power after exposure, compared to thin film technologies (13%). [10]

Table I: PV module characterization parameters acc. to IEC 61853-1 from CalLab PV Modules: angular response (*ar*), Pmpp temperature coefficient (TkPmpp) and thermal coefficients (*u0* & *u1*).

Module	**Area [m²]**	***ar***	**TkPmpp [%/°C]**	***u0 & u1***
M01	1.89	0.10596	-0.34	$u0 = 30.1$
M02	1.84	0.14720	-0.27	$u1 = 4.4$

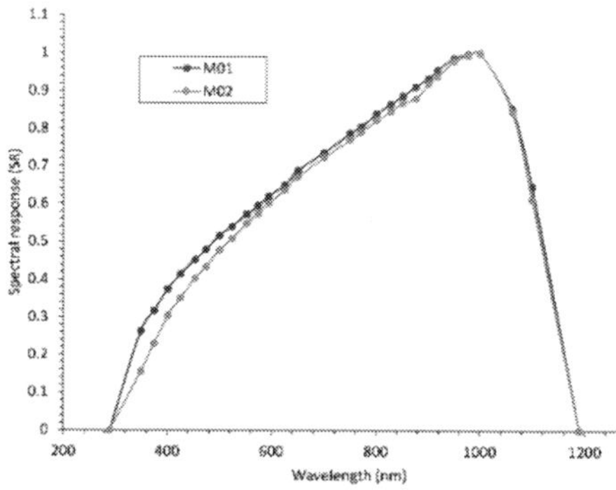

Figure 2: Spectral response of tested modules from manufacturers M01 and M02.

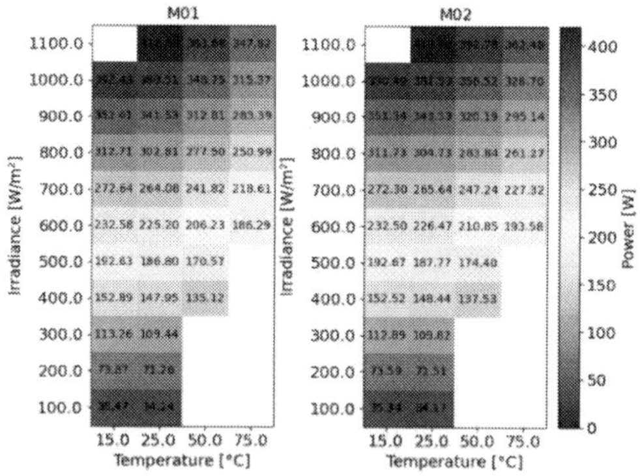

Figure 3: Power Matrices for both manufacturers M01 and M02.

2.2 Outdoor measurements

The PV modules were installed in Freiburg, Germany, on a rooftop, facing south with a tilt angle of 15°. Alongside the modules, a pyranometer and a silicon reference cell were placed in the same POA (Fig. 4). Additionally, a complete weather monitoring station and sun tracker were utilized to record the conditions during exposure time, ensuring representative site-specific conditions (Table II). To achieve data accuracy, we filtered the collected data for a time frame from 9:00 to 15:00, considering solar elevations larger than 5° and Gpoa larger than 10 W/m². The filtering process helped to avoid any erroneous values due to non-ideal shading conditions or high AOI.

Table II: Outdoor conditions in time of exposure. Minimum – maximum and mean values.

	Global irradiance in POA [W/m²]	Diffuse irradiance [W/m²]	Ambient temperature [°C]	Wind speed [m/s]
Autumn (28.10-21.12.2022)	(10.2 – 951.5 W/m²) 245.5 W/m²	(11.7 – 258.8 W/m²) 100.0 W/m²	(-7.5 – 30.5 °C) 8.9 °C	(0.02 – 5.8 m/s) 1.1 m/s
Winter (22.12-20.03.2023)	(10.4 – 1200 W/m²) 306.9 W/m²	(10.3 – 441.7 W/m²) 123.1 W/m²	(-3.0 – 23.7 °C) 8.8 °C	(0.03 – 10.7 m/s) 1.8 m/s
Spring (21.03-20.06.2023)	(10.1 – 1368.9 W/m²) 555.5 W/m²	(10.5 – 577.9 W/m²) 214.2 W/m²	(4.8– 35.9 °C) 19.5 °C	(0.02 – 7.9 m/s) 1.5 m/s
Summer (21.06-28.08.2023)	(15.8 – 1276.4 W/m²) 564.6 W/m²	(14.3 – 587.9 W/m²) 234.4 W/m²	(13.9 – 38.7 °C) 26.8 °C	(0.03 – 8.3 m/s) 1.7 m/s

Figure 4: Four PV Modules, with one pair for each manufacturer M01 and M02, installed on a rooftop in Freiburg, Germany.

During the time of outdoor exposure from autumn 2022 until summer 2023, we observed larger soiling depositions on the modules in the spring season, which was quantified during two cleaning periods. It's important to note that this soiling is attributed to occasional occurrences of soiling due to pollen and not something regular throughout the entire period of exposure at this site. The soiling ratio was calculated by comparing the performance ratio (PR), as the average power output divided by the Gpoa, over a period of 1 week when the module was soiled and 1 week after cleaning procedure. A soiling ratio (PRsoiled /PRclean) of 1 indicates no significant performance difference between both periods. For M01, the module presented a soiling ratio of 0.96, while for M02, it was 0.95.

Regarding degradation rates over this exposure period, we filtered the power output of the modules for conditions with Gpoa ranging from 800 to 1000 W/m² and with a percentage of fluctuation of less than 5%. Subsequently, we corrected the power for a Gpoa of 1000 W/m² and a more normal operating temperature condition of 45°C, rather than the STC condition of 25°C [11]. The linear fit to this power data revealed degradation rates of -0.19 %/year for M01 and -0.16 %/year for M02. It's essential to further analyze these degradation rates calculated with a shorter amount of data to understand any correlation with long-term calculations and datasheet estimations.

3 EVALUATION

3.1 Module behavior

In the initial phase of our evaluation, we present the differences in behavior between the two type of modules in this study by conducting a loss analysis in accordance with reference [9]. The quantification of losses attributed to each influencing parameter is derived from the correction procedures outlined in the ER methodology as specified in IEC 61853-3. Fig. 5 illustrates the percentage of losses in SSER, from the exposure time in Freiburg. It considers the effects of reflected irradiance resulting from variations in the AOI, the gains or losses from fluctuations in ambient temperature and irradiance levels, as well as the impact of spectral variations. Our findings reveal that during the fall and winter seasons, significant losses primarily result from low irradiance levels and losses related to angular response. Notably, M02 exhibits a greater susceptibility to AOI variations in comparison to M01. Conversely, during spring and summer, the primary cause of losses for both module types is the ambient temperature, with M02 displaying greater resilience in these conditions.

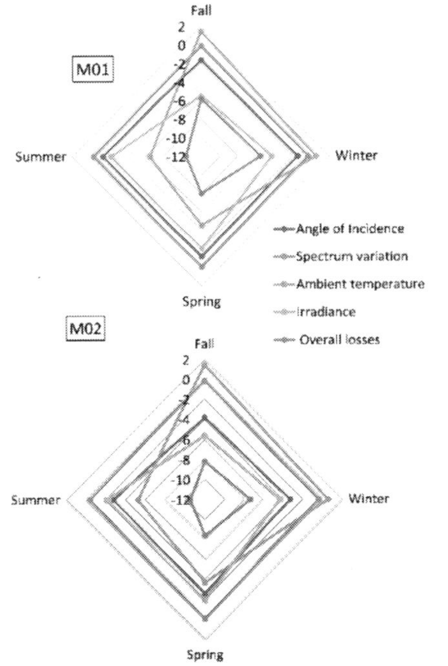

Figure 5: Seasonal ER loss analysis for module M01 (upper graph) and M02 (lower graph). Losses are expressed in percent with the higher values in inner curves (negative values) and include the overall losses.

3.2 CSER vs SSER

The SSER for the complete year 2022, modeled using the standard methodology in the specific location of Freiburg, Germany, was 0.93 for both type of modules. When modeling for a shorter period of 10 months, between October 2022 and August 2023, the results were very similar at 0.92. (Table III). However, whit the extension of the ER climate data sets to specific sites, it became evident that locations like Freiburg are not accurately represented by the standard profiles, such as the *temperate continental*, where modules present higher CSER of 0.99 (M01) and 0.98 (M02), resulting in differences between SSER and CSER larger than the expected uncertainty in the methodology [8, 9]. Notably, SSER values derived from direct field measurements (PR_{DC}) closely matched the modeled results for Freiburg, underscoring the need for region-specific modeling for energy rating assessments.

Table III: SSER results for Freiburg in years 2022 and short-term exposure between 2022-2023. CSER for the profile "temperate continental".

Profile	M01	M02
CSER *"temperate continental"*	0.99	0.98
SSER Freiburg (2022)	0.93	0.93
$SSER_{MODEL}$ Freiburg (Oct. 2022 – Aug. 2023)	0.92	0.92
$SSER_{REAL}$ Freiburg (Oct. 2022 – Aug. 2023)	0.92	0.89

3.3 SSER validation

To further validate our site-specific modeled ($SSER_{MODEL}$) outcomes, we conducted calculations on a daily and weekly basis (Equation 1), enabling a comparative assessment with outdoor measurements ($SSER_{REAL}$). This approach allowed us to understand any seasonal deviations and the impact of weather conditions. The weekly analysis revealed deviations between the *temperate continental* CSER and Freiburg SSER, in both modeled and real values, larger than 10% during the winter period, while lower deviations of around 5% in summer (Fig. 6). Further examination of the weekly $SSER_{MODEL}$ and $SSER_{REAL}$ values in Freiburg highlighted a consistent seasonal pattern. Some of the disparity points are attributed to differences in the sources of irradiance in POA between the modeled and real datasets used for the calculations. The IEC 61853-3 method necessarily derives G_{POA} from DNI and DIF, while the measured PR_{DC} is based on measured G_{POA}.

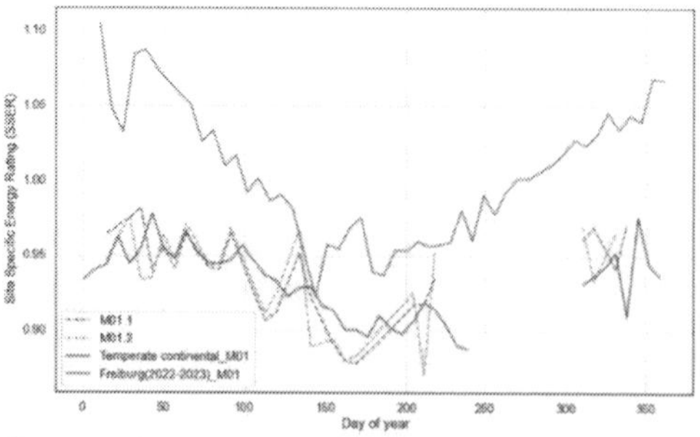

Figure 6: Weekly CSER acc. to IEC 61853 in *temperate continental* profile (green). Weekly $SSER_{MODEL}$ in Freiburg (red) and $SSER_{REAL}$ (dashed lines) in 2022-2023 for M01 (M01.1 and M01.2).

The daily analysis showed average deviations between *modelled* and *real* SSER for M01 of 1.27% (M01.1 and M01.2), whereas for M02, these deviations were 0.23% (M02.1) and 1.12% (M02.2). These small deviations indicate a good correlation between short term outdoor ER and the standard methodology. Fig. 7 and 8 show the daily SSER values with the corresponding deviations between modeled and real. Over the course of exposure is visible that warmer months with higher levels of irradiance result on smaller range of deviations.

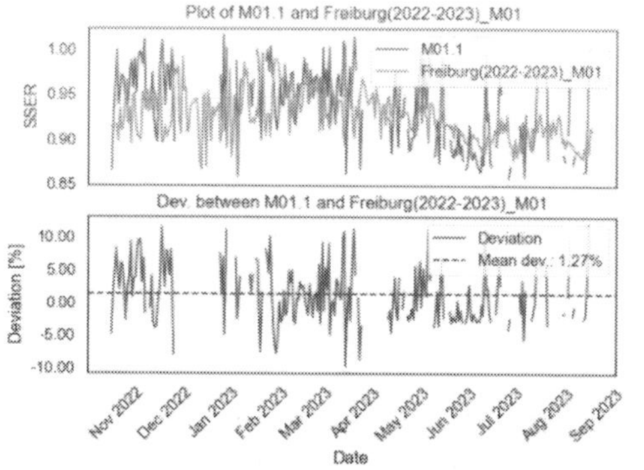

Figure 7: Upper graph shows the daily $SSER_{REAL}$ (blue) and $SSER_{MODEL}$ (orange) in Freiburg from November 2022 to August 2023 for M01 (M01.1). Lower graph shows the daily and average deviation (red line) between modeled and real SSER.

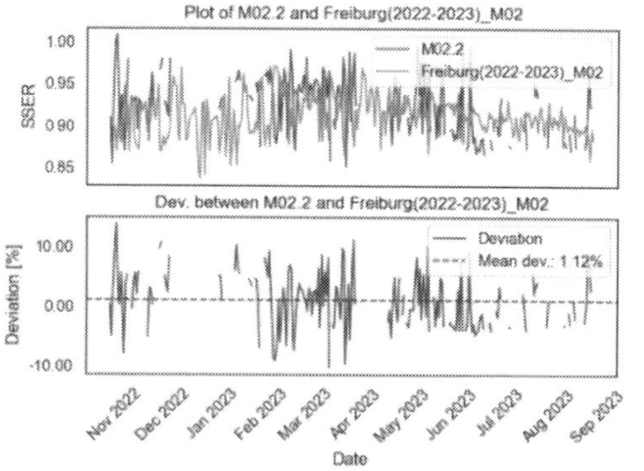

Figure 8: Upper graph shows the daily $SSER_{REAL}$ (blue) and $SSER_{MODEL}$ (orange) in Freiburg from November 2022 to August 2023 for M02 (M02.2). Lower graph shows the daily and average deviation (red line) between modeled and real SSER.

4 CONCLUSIONS

The study involved conducting short-term outdoor measurements of PV module performance in Freiburg, Germany, with the aim of validating the ER methodology. The focus was particularly on validating the modeling of SSER for this specific location and identifying seasonal variations, along with determining the primary sources of losses. The short-term behavior of SSER was assessed on weekly and daily basis. Weekly results provided insight into the variation of SSER across the different seasons of the year. These values tend to be higher during winter and lower in summer. The representation of SSER through weekly values also highlighted the significant deviations when compared to standard profiles (CSER) like *"temperate continental"*. This finding reinforces the necessity of employing site-specific data, as stated in previous studies [9].

The daily SSER derived from outdoor measurements ($SSER_{REAL}$) effectively validated site-specific modeled results ($SSER_{MODEL}$), exhibiting average deviations ranging from 0.23% to 1.27%, for both type of modules. These deviations fall within the anticipated uncertainty in the methodology, which was estimated on our laboratories' assessments and findings from previous studies [7]. The seasonal analysis revealed that the lowest range of deviation between real and modeled SSER occurs during spring, making it an optimal season for evaluating short-term outdoor measurements and validating the ER methodology. Notably, the occurrences of soiling in spring and the expected degradation rate did not have significant effects on the modules' ER.

The ER losses analysis aligned well with the indoor characterization of the modules, demonstrating its effectiveness not only in comparing different climates profiles but also in analyzing seasonal effects. These findings emphasize the significance of incorporating local climate data and acknowledging performance variations across the year. Such insights have the potential to enhance the testing and evaluation of modules, optimizing energy output in different seasons. Furthermore, this type of analysis can result in more precise energy yield predictions and well-informed decisions for solar energy systems situated in specific geographical locations.

5 FURTHER WORK

To deepen comprehension of the energy rating methodology, future work will involve the utilization of an expanded outdoor dataset to conduct both uncertainty and sensitivity analyses. Furthermore, we aim to undertake a comparative study involving indoor characterization and outdoor procedures, which could offer valuable insights into the advantages of adopting site-specific analysis and outdoor characterization for emerging technologies. For example, this could involve evaluating thermal coefficients and power matrices within both controlled indoor settings and actual outdoor conditions, according to the procedures outlined in IEC 61853 1-2. This approach has the potential to yield more accurate assessments of PV module performance in real-world applications.

6 REFERENCES

[1] *IEC 61853-3 Photovoltaic (PV) module performance testing and energy rating— Part 3: Energy rating of PV modules*, 2018.

[2] C. Monokroussos, Y. Zhang, E. W. Lee, F. Xu, A. Zhou, Y. Zhang, and W. Herrmann, "Energy performance of commercial c-Si PV modules in accordance with IEC 61853-1, -2 and impact on the annual specific yield," *EPJ Photovolt.*, vol. 14, p. 6, https://www.scienceopen.com/document_file/936f3a53-84e3-42d3-88eb-3d8989c8b208/API/pv220042.pdf, 2023.

[3] C. Monokroussos *et al., Eds., Energy Rating of c-Si and mc-Si commercial PV-modules in accordance with IEC 61853-1,-2,-3 and impact on the annual yield*, 2017.

[4] Y. Poissant, S. Pelland, D. Turcotte, Ed., *A Comparison of Energy Rating Methodologies Using Field Test Measurements*, 2008.

[5] S. Dittmann *et al., Eds., Results of the 3rd Modelling Round Robin within the European Project „PERFORMANCE"– Comparison of Module Energy Rating Methods*: WIP-Munich, 2010.

[6] B. Bora, R. Kumar, O. S. Sastry, B. Prasad, S. Mondal, and A. K. Tripathi, "Energy rating estimation of PV module technologies for different climatic conditions," *Solar Energy*, vol. 174, pp. 901–911, https://www.sciencedirect.com/science/article/pii/S0038092X18309630, 2018.

[7] D. Dirnberger, B. Müller, and C. Reise, "PV module energy rating: opportunities and limitations," (en), *Prog. Photovolt: Res. Appl.*, vol. 23, no. 12, pp. 1754–1770, https://onlinelibrary.wiley.com/doi/full/10.1002/pip.2618, 2015.

[8] T. Huld, E. Salis, A. Pozza, W. Herrmann, and H. Müllejans, "Photovoltaic energy rating data sets for Europe," *Solar Energy*, vol. 133, pp. 349–362, https://www.sciencedirect.com/science/article/pii/S0038092X16300391, 2016.

[9] M. Rivera, C. Reise, and U. Kräling, "Representativeness of Energy Rating acc. to IEC 61853 for Different Locations in Middle and South Europe," *Energy Tech*, 2023.

[10] M. Pravettoni and A. S. Rajput, "On the metastability of silicon heterojunction solar photovoltaic modules," (En;en), *MRS Bulletin*, pp. 1–9, https://link.springer.com/article/10.1557/s43577-023-00475-x, 2023.

[11] R. M. Smith, D. C. Jordan, and S. R. Kurtz, "Outdoor PV Module Degradation of Current-Voltage Parameters," Boulder, CO: American Solar Energy Society (ASES), National Renewable Energy Lab. (NREL), Golden, CO (United States), 2012.

LIMITS OF THE SINGLE DIODE MODEL IN VIEW OF ITS APPLICATION TO THE LATEST PV CELL TECHNOLOGIES

Agnes Bridel-Bertomeu, Michele Oliosi, André Mermoud, Bruno Wittmer,
PVsyst SA
Route de la Maison-Carrée 30 - 1242 Satigny – Switzerland, support@pvsyst.com

ABSTRACT: The single diode model is a fundamental element in photovoltaic (PV) performance simulation tools, playing a pivotal role in characterizing PV module behavior as a function of irradiance and temperature.

This study is underscored by advancements in PV cell technology. Recent developments have yielded cells with exceptionally high fill factor, particularly prominent in TOPCon and Heterojunction cells. This brings the methods that are currently used to derive the single diode model parameters to their limit and can lead to situations where in order to describe accurately the MPP behavior, the precision for describing the open circuit voltage V_{oc} is compromised.

The primary objective of this study is to explore the boundaries of the single diode model as a means to describe the I-V characteristics of PV cells. The focus lies in investigating the implications of these high fill factors that inherently result in an elevated $\frac{V_{MPP}}{V_{OC}}$ ratio and how the single diode model can effectively encapsulate such behavior. By addressing this challenge, we aim to advance our understanding of the model's limits and its potential adaptations to accommodate the evolving landscape of PV cell technologies.
Keywords: Single diode model, Advanced PV cell technology, Modelling, Simulation, Limits of parameters

1 SINGLE DIODE MODEL

The single diode model [1] is an equivalent circuit that can be used for an individual PV cell, a module consisting of several cells or an array consisting of several modules. The concept of the single diode model finds its root in the history of solar cell research with foundational work dating back to several decades.

1.1 Description of the single diode model

The equivalent circuit to describe a PV cell is illustrated Figure 1 and the expression describing the general single-diode model for a given set of references conditions (STC: $1000\ \mathrm{W/m^2}, 25°\mathrm{C}, \mathrm{AM} = 1.5$) reads,

$$I = I_{ph} - I_0\left(e^{\left(\frac{V+I\cdot R_s}{N_{cells}\cdot n\cdot V_{th}}\right)} - 1\right) - \frac{V + I\cdot R_s}{R_{sh}} \tag{1}$$

with, two unknowns,

- I: output current of the PV module [A],
- V: output voltage of the module [V],

five parameters to be determined,

- I_{ph}: photocurrent generated by the module[A],
- I_0: diode saturation current [A],
- R_s: series resistance [Ω],
- R_{sh}: shunt resistance [Ω],
- n: diode ideality factor,

and physical constants,

- V_{th}: thermal voltage (given by $\frac{kT}{q}$, where $k = 1.381 \times 10^{-23}$ J/K is Boltzmann's constant, T is the temperature in Kelvin, and $q = 1.302 \times 10^{-19}$ Coulomb is the electronic charge).

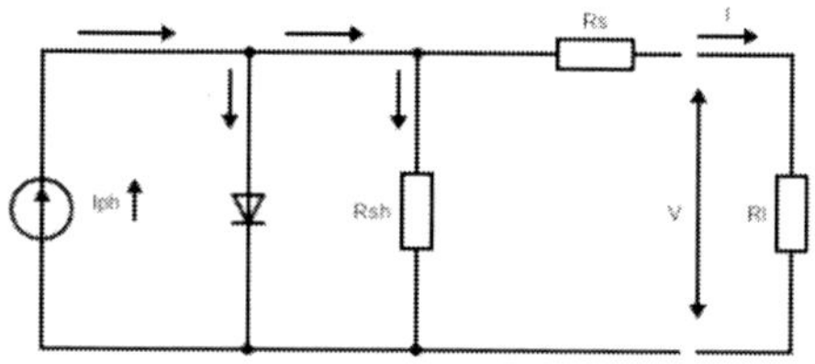

Figure 1: Diagram of the equivalent circuit to describe a PV cell.

The model parameters can depend on the irradiance and temperature. In the simplest approach, I_{ph} is proportional to the irradiance G, and I_0 is a function of the temperature, T.

The single diode model in PVsyst adds two small modifications to this basic description:

1. The shunt resistance, R_{sh} is assumed to increase exponentially with falling irradiance. This is described by two additional parameters. The value for R_{sh} at $G = 0$ and the exponential factor for the decrease.
2. A small temperature dependance of the ideality factor, n which allows to tune the MPP temperature coefficient at standard test conditions (STC) to a given measured value.

In this work, we restrict out considerations to the STC and to the behavior at $200\ \mathrm{W/m^2}$ and $25\ °\mathrm{C}$. Therefore, the temperature dependences will have no impact on the discussion and the results.

To solve equation (1) , we will first have to determine the five parameters based on the STC data of the modules commonly provided by manufacturers. The details on the determination of these parameters are not explained [2] in this paper, and for the interested reader, please refer to [1] [2].

1.1 Output of the single diode model

After obtaining the five parameters, the single diode model can be solved either by using a successive approximation approach to predict the I-V characteristics of a PV cell, or by explicitly expressing I(V) with the help of the Lambert W function[2]. The typical I-V and P-V curves of solar cells are shown in Figure 2 along with its characteristic points, including the (i) short-circuit point $(I_{sc}, 0)$, (ii) open-circuit point $(0, V_{OC})$ and (iii) maximum power point (I_{MPP}, V_{MPP}). These characteristic points are normally known from the data sheet of the module.

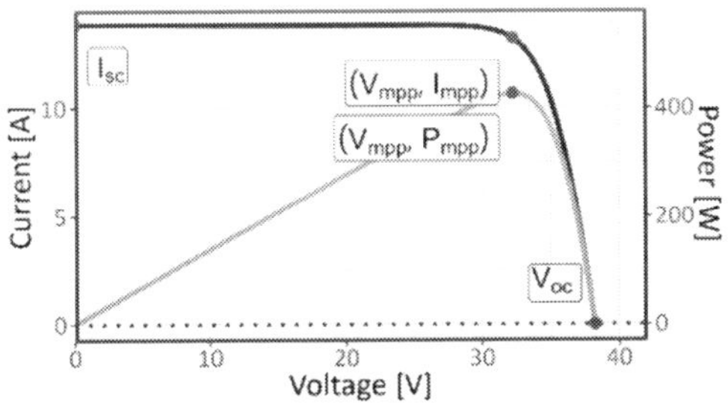

Figure 2: I-V and P-V curve a PV cell with its characteristic points $I_{sc}, V_{oc}, (V_{MPP}, I_{MPP})$ and (V_{MPP}, P_{MPP}).

The single diode model allows to describe the PV module behavior at all temperatures and irradiances as illustrated in Figure 3. These curves show the efficiency as function of irradiance for different temperatures. They exemplify the behavior of a typical state of the art monocrystalline PV module. The detailed parameters that were used for these curves are given in section 3.

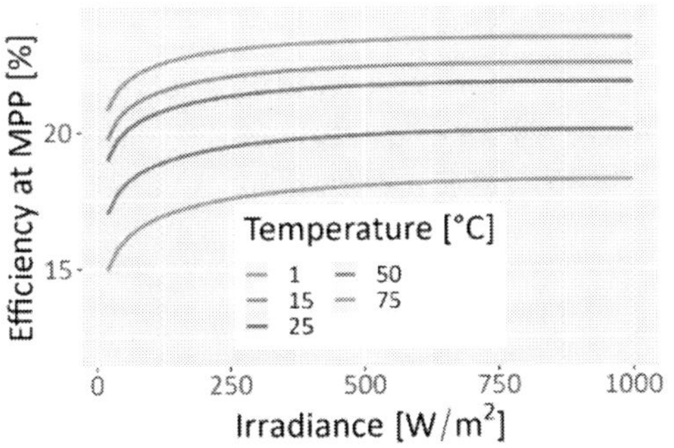

Figure 3: Behavior of a PV module at all temperatures as a function of irradiance.

2 HISTORICAL TRENDS OF: V_{MPP}/N_{CELLS}

Photovoltaic technology encompasses a range of technologies and materials. The preeminent technology is centered on silicon-based PV modules where silicon solar cells offer an optimal balance between cost-effectiveness and efficiency. However, driven by the quest for enhanced efficiency and improved performance, the technology of solar cells has been evolving. This evolution has a profound impact on the fill factor (FF). in the context of solar cells, FF represents a key parameter that characterizes the overall performance and efficiency of a PV module. This dimensionless quantity is represented as

$$FF = \frac{I_{MPP} \cdot V_{MPP}}{I_{SC} \cdot V_{OC}} \quad (2)$$

Technologies such as TOPCon and HIT are advanced silicon variants that serves at reducing recombination losses to enhance FF and efficiency. The trend to larger FF goes along with an increase of the cell voltage $\frac{V_{MPP}}{N_{cells}}$. The V_{oc} shows a less pronounced increase, leading to ever higher ratios of $\frac{V_{MPP}}{V_{OC}}$. Figure 4 illustrates the evolution of $\frac{V_{MPP}}{N_{cells}}$ for Si-mono, Si-poly and HIT PV modules in the PVsyst database and demonstrates an upward trend of the $\frac{V_{MPP}}{N_{cells}}$ between year 2000 and 2023.

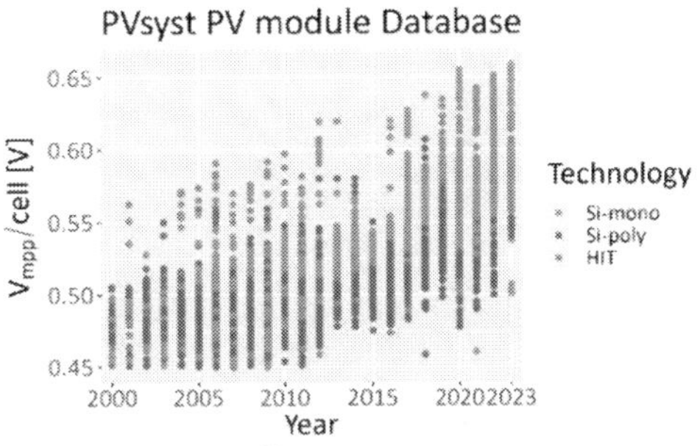

Figure 4: Trends of $\frac{V_{Mpp}}{N_{cells}}$ of PV modules with different technologies, i.e., Si-mono, Si-poly and HIT, in the PVsyst database.

In the next section, we will study the interdependence of different outputs of the single diode model. We will look especially at the ratio $\frac{V_{MPP}}{V_{OC}}$, and how it is linked to the ratio $\frac{V_{Mpp}}{N_{cells}}$ and the behavior of the model at low irradiance. This is motivated by the fact that the trend of increasing $\frac{V_{MPP}}{V_{OC}}$ is reaching a limit, where the single diode model cannot be mathematically established.

3 PARAMETER RANGES AND CONSTRAINTS

In order to analyze the behavior of the single diode model, we will start with a set of parameters that describes a typical monocrystalline 426 Wp module. The parameters for the single diode model for this PV module are presented in Table 1. The parameters I_{ph} and N_{cells} set the global power of the PV module and can be changed without changing the following discussion. The other parameters will have an impact on the detailed shape of the I-V and P-V curves which will change the values like $\frac{V_{MPP}}{V_{OC}}$ or $\frac{V_{Mpp}}{N_{cells}}$.

Example PV module	
I_{ph} [A]	13.84
I_0 [pA]	15
N_{cells}	54
n	1
R_s[mΩ]	120
R_{sh}[Ω]	800

Table 1: Parameters of a monocrystalline 456 Wp PV module.

We first have a look at the parameter $\frac{V_{MPP}}{V_{OC}}$ and its dependence on R_s and I_0. The value of $\frac{V_{MPP}}{V_{OC}}$ increases as R_s and I_0 decreases. This behavior is illustrated in Figure 5 and Figure 6 where an almost linear dependency for R_s and a logarithmic dependency for I_0 is observed respectively. We can see that by changing these two parameters a large range of $\frac{V_{MPP}}{V_{OC}}$ ratios are covered, ranging from 0.80 to 0.88.

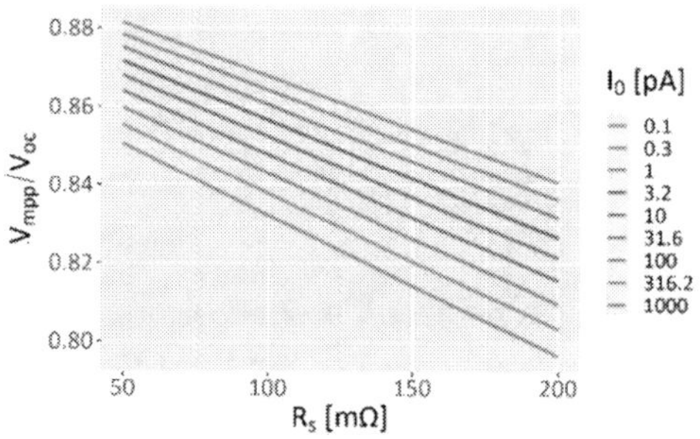

Figure 5: $\frac{V_{MPP}}{V_{OC}}$ in terms of R_s for different I_0.

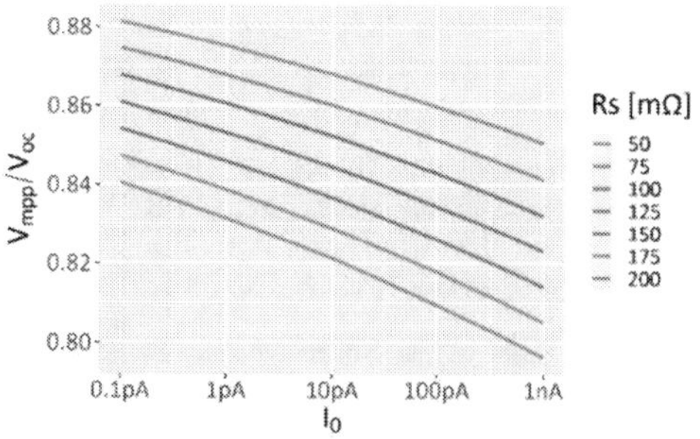

Figure 6: $\frac{V_{MPP}}{V_{OC}}$ in terms of I_0 for different R_s.

However, changing R_s and I_0 will have also an impact on other results of the single diode model, such as the $\frac{V_{MPP}}{N_{cells}}$ and the low light efficiency. A realistic modeling of a PV module will require that these quantities stay within certain limits, and as we will see, this will limit the $\frac{V_{MPP}}{V_{OC}}$ to a much smaller range.

Now, if we analyze the V_{MPP} of PV cells with different technology, it is observed that in general, the $\frac{V_{MPP}}{N_{cells}}$ does not change much. The histogram in Figure 7 shows this ratio for the PV modules that were introduced in the PVsyst database over the past five years. One can see that this value lies between 0.55V and 0.6V for Si-mono modules and between 0.60V and 0.65V for HIT modules. The medians for these distributions lie at 0.57V and 0.63V respectively. We will take the range from 0.55V to 0.65V as the range of realistic values for $\frac{V_{MPP}}{N_{cells}}$.

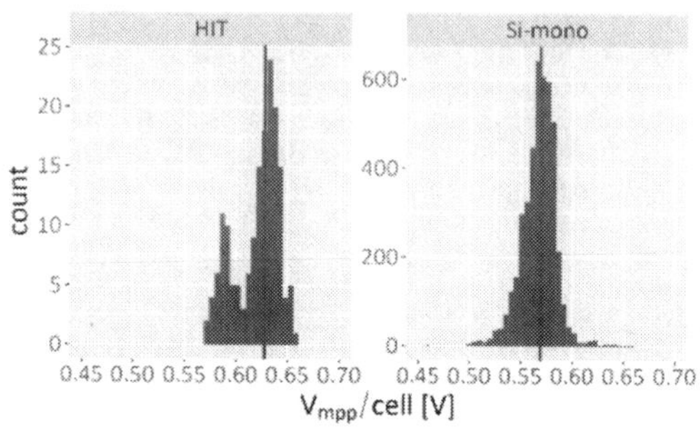

Figure 7: Histogram of $\frac{V_{MPP}}{N_{cells}}$ for HIT and Si-mono PV modules.

Another property of the single diode model that can be used as a criterion for a realistic description of crystalline Si PV modules is the low light behavior. As a measure for this behavior, we take the relative efficiency loss of the PV module at 200 W/m^2 and 25°C with respect to STC. Based on measurements for many silicon PV modules, we have observed that this relative drop in efficiency lies always very close to -3%. As shown in Figure 8, we can see that limiting the loss to values close to -3% puts a strong limit on the range of possible values for R_s. To calculate the relative efficiency loss in Figure 8, we assumed that R_{sh} at 200 W/m^2 is approximately 4 times the values of R_{sh} at STC. This assumption is based on the default values in Pvsyst for R_{sh} at G=0 and the exponential factor of 5.5.

In the following section, we will see how the restriction to these conditions, which we consider a realistic description of a crystalline Si module, will limit the possible ranges for $\frac{V_{MPP}}{V_{OC}}$ and $\frac{I_{MPP}}{I_{SC}}$, which together determine the Fill Factor of the PV module.

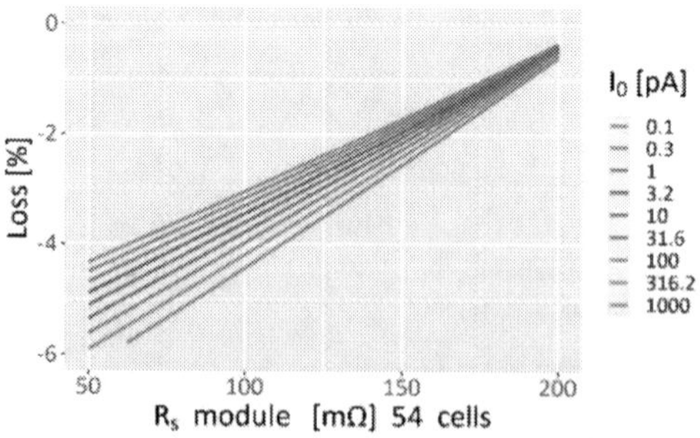

Figure 8: The low light efficiency in terms of R_s for different I_0.

4 LIMITS FOR THE RATIOS: VMPP/VOC AND IMPP/ISC

Now that we have seen the dependence of $\frac{V_{MPP}}{V_{OC}}$ on different parameters, we are in position to analyze the

10.4229/EUPVSEC2023/3AV.3.14
020216-003

limits of ratios, specifically the $\frac{V_{MPP}}{V_{OC}}$ and $\frac{I_{MPP}}{I_{SC}}$.

The main findings show that limiting parameters $\frac{V_{MPP}}{N_{cells}}$ and low-light efficiency to typically observed values leads also to a limited range for $\frac{V_{MPP}}{V_{OC}}$ and $\frac{I_{MPP}}{I_{SC}}$. When the voltage at the maximum power point per cell, $\frac{V_{MPP}}{N_{cells}}$ is confined within the range of 0.55V to 0.65V and a relative efficiency loss of 3% is observed at an irradiance level of $200\,W/m^2$, an intriguing limitation emerges concerning the $\frac{V_{MPP}}{V_{OC}}$ ratio. This ratio is constrained to values encompassing the range of 83% to approximately 85%. The upper limit of 85% signifies the necessity for exceptionally diminutive saturation currents, I_0, that is typically on the order of a few pA at the standard temperature of 25°C. However, this constraint implied the emergence of even more minuscule saturation currents at lower temperatures, such as around -10°C, thereby emphasizing the profound sensitivity of diode behavior to temperature fluctuations within the PV system. Figure 9 shows the $\frac{V_{MPP}}{V_{OC}}$ ratio as a two-dimensional contour plot along with the ranges for $\frac{V_{MPP}}{N_{cells}}$ and the relative efficiency and the $\frac{V_{MPP}}{V_{OC}}$ for both Si-mono and HIT technology PV modules are included and represented with dotted lines.

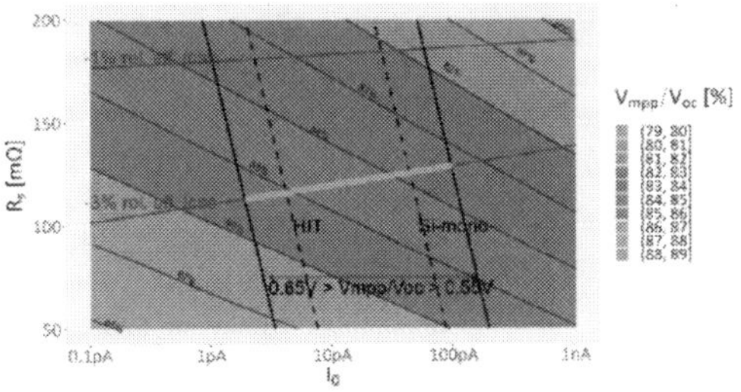

Figure 9: 2D contour plot of the $\frac{V_{MPP}}{V_{OC}}$ ratio - for $0.55V < \frac{V_{MPP}}{N_{cells}} < 0.65V$ and relative efficiency loss of 3% at $G=200\,W/m^2$, $\frac{V_{MPP}}{N_{cells}}$ is constrained between 83% to 85%.

As for the $\frac{I_{MPP}}{I_{SC}}$ ratio, the relationship between the short-circuit current I_{sc} and the current at the maximum power point I_{mpp} is primarily influenced by $\frac{V_{MPP}}{N_{cells}}$. In practical scenarios with typical values for $\frac{V_{MPP}}{N_{cells}}$, the ratio of $\frac{I_{MPP}}{I_{SC}}$ tends to remain within a relatively narrow range, generally between 95% to 96%. This observation implies that variations in I_{mpp} and I_{sc} have only a marginal impact on increasing the fill factor of the PV system. An illustration of the ranges of $\frac{I_{MPP}}{I_{SC}}$ along with the ranges for $\frac{V_{MPP}}{N_{cells}}$ and the relative efficiency for both Si-mono and HIT PV modules is presented in Figure 10.

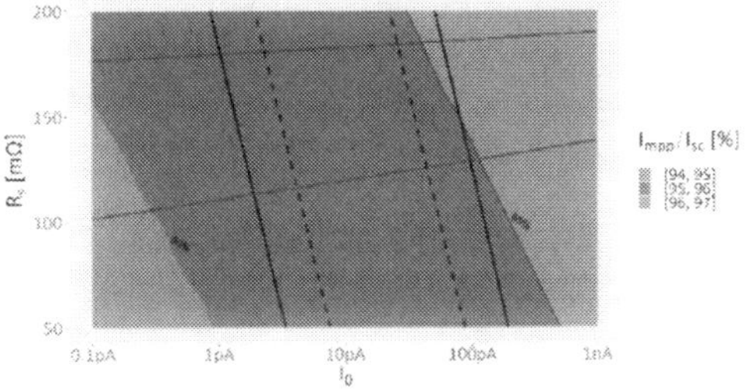

Figure 10: 2D contour plot of the $\frac{I_{MPP}}{I_{SC}}$. ratio - for $0.55V < \frac{V_{MPP}}{N_{cells}} < 0.65V$ and relative efficiency loss of 3% at $G=200\,W/m^2$, $\frac{V_{MPP}}{N_{cells}}$ is constrained between 83% to 85%.

5 DISCUSSION AND PERSPECTIVES

In the preceding section, we examined the intricated interplay of several parameters in PV modules and their influence on the $\frac{V_{MPP}}{V_{OC}}$ and $\frac{I_{MPP}}{I_{SC}}$ ratios. Several insights can be deduced from this analysis.

The key takeaway is that R_s should be meticulously determined to strike a balance between maximizing voltage ratios and minimizing efficiency losses. We observed that increasing R_{sh} marginally extends the range towards higher $\frac{V_{MPP}}{V_{OC}}$ and $\frac{I_{MPP}}{I_{SC}}$ ratios. However, it is crucial to note that a smaller R_s results in larger $\frac{V_{MPP}}{V_{OC}}$, yet this advantage comes at the cost of higher efficiency losses when the module operated under low irradiance conditions.

In PVsyst, the R_s is fixed by the required relative efficiency at $200\,W/m^2$. Furthermore, the ideality factor, n and reverse saturation current, I_0 play intricate roles in the behavior of the diode within PV modules. While it is understood that selecting values of $n < 1$ and reducing I_0 can widen the range of possible $\frac{V_{MPP}}{V_{OC}}$ ratios, a physical interpretation of these values remains elusive. A deeper understanding of semiconductor physics governing PV module operation is required in order to harness the benefits of adjusting n and I_0 for optimizing $\frac{V_{MPP}}{V_{OC}}$ ratios.

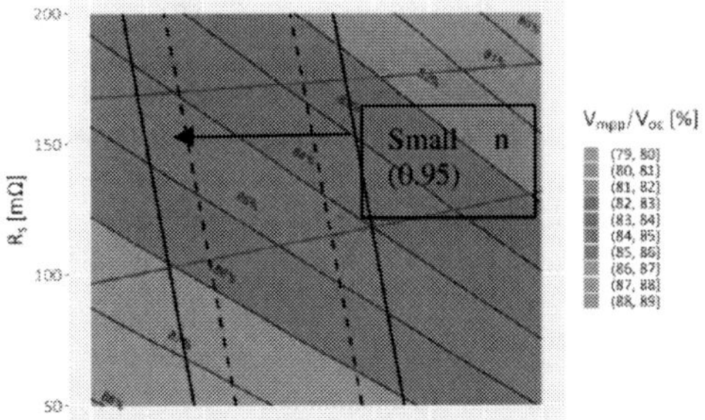

Figure 11: 2D contour plot of the $\frac{V_{MPP}}{V_{OC}}$ ratio with a smaller ideality factor n.

It occurs that in the PVsyst database, the $\frac{I_{MPP}}{I_{SC}}$ ratio has to be limited by increasing the values of V_{OC}. This ensures a description of the MPP that is consistent with the data sheets. [2]

6 CONCLUSION

The modules of new technologies (TOPCon, HJT) exhibit a very high Fill Factor. The main contribution to this increase is the $\frac{V_{MPP}}{V_{OC}}$ ratio; the $\frac{I_{MPP}}{I_{SC}}$ is rather stable and does not have a big importance in the Fill Factor.

The problem is that the one-diode model (especially the Shockley diode behaviour) is no longer applicable with such values. This study aims to identify the limits on the parameters, which mathematically allow to apply the one-diode model in satisfactory conditions, corresponding to the physical reality.

Some of these physical realities are derived from the PV modules characteristics, as present in the PVsyst database for recent modules (5 years). At STC, the voltage $\frac{V_{MPP}}{N_{cells}}$ is distributed between 0.55 and 0.65 V (0.63V average for HIT). And in most measurements that we have received from testing laboratories, the relative efficiency is close to -3% at $200\ W/m^2$. The diode saturation current should not become lower than some few pA at STC, as it has a huge decrease according to the lower temperatures.

Now the low-light efficiency is mainly related to the R_s value. Therefore, we have evaluated the $\frac{V_{MPP}}{V_{OC}}$ value, as a function of the R_s and the diode saturation current. We found that the low-light efficiency limit of 3% imposes a maximum $\frac{V_{MPP}}{V_{OC}}$ ratio of the order of 0.86, or more likely 0.84-0.85 for reasonable Io and $\frac{V_{MPP}}{N_{cells}}$ values. The R_{sh} value has a marginal effect on these limitations.

The main objective of the one-diode model is to evaluate the P_{mpp} value during the simulations, as a function of the irradiance and the temperature. The I-V curve is only used in special conditions, i.e., limitations of the inverter (overload) or electrical shadings calculations. In practice, we can "save" the one-diode model for modules of extreme fill factors, by artificially increasing the V value at STC (usually by some fractions of percent). This will not modify the P_{mpp} behaviour, and very slightly modify the I-V curve at low currents.

6 REFERENCES

[1] J. A. Duffie et W. A. Beckman , Solar Engineering of Thermal Process, John Wiley and Sons, 2nd edition, 1991.

[2] A. Jain et A. Kapoor, «Exact analytical solutions of the parameters of real solar cells using Lambert W-function,» Solar Energy Materials and Solar Cells, pp. 269-277, 2004.

[3] A. Mermoud et T. Lejeune , «Performance Assessment of a Simulation Model for PV Modules of Any Available Technology,» 25th European Photovoltaic Solar Energy Conference and Exhibition / 5th World Conference on Photovoltaic Energy Conversion, pp. 4786--4791, 2010.

10.4229/EUPVSEC2023/3AV.3.14
020216-005

DO WE CORRECTLY DEFINE THE NOMINAL POWER OF THIN-FILM MODULES? DETERMINATION OF POWER STABILISATION COEFFICIENTS FOR THIN-FILM PV MODULES TO REFLECT THE LOCATION DEPENDENT INFLUENCE

Thomas Weber[*,1], Muktaparna Boruah[1,2], Sandra Lust[1], Nattapark Pongthanacharoenkul[1], Nithin Murali[1], Benjamin Lippke[1], Steven Xuereb[1]

[1]PI Photovoltaik-Institut Berlin AG, Wrangelstraße 100, 10997 Berlin, Germany
[2]Carl von Ossietzky University of Oldenburg, Ammerländer Heerstraße 114-118, 26129 Oldenburg, Germany

[*]weber@pi-berlin.com; phone +49 30 8145264 - 111

ABSTRACT: Thin-film modules exhibit metastability due to light exposure which is reversible at high temperatures. This phenomenon demonstrates a seasonal trend and poses a challenge for accurate and reproducible module power measurements. This study has developed an approach to quantify the seasonal deviation in terms of "seasonal coefficients". Data was taken from an installed thin-film plant. The dc power was adjusted to STC and plotted over the evaluation period (2013 to 15). The adjusted power showed a seasonal sinusoidal trend – values gradually increased from winter to summer and then decreased from summer to winter. A sine curve was fitted to this data to determine the seasonal coefficients. The coefficients were incorporated into the output from a simulated model of the plant. The effects of integrating the coefficients were analysed using the ASTM E-2848 standard. Results showed an improvement of 20 % in accuracy of simulated data incorporated with seasonal coefficients. Therefore, these coefficients can be used to determine a variable power to which the module's actual performance in the field can be compared. The seasonal coefficients can also be used to improve the results of predicted energy yield obtained from simulation software.
Keywords: thin-film, stabilization, power determination, seasonal deviation, location dependence

1 INTRODUCTION

The motivation behind this study is the lack of standard procedures to evaluate seasonal deviations in thin-film module performance that occur due to its metastability. This study has developed an approach to quantify this seasonal deviation in terms of "power stabilisation coefficients" or "seasonal coefficients", with a focus on microcrystalline-amorphous (micromorph, a-Si/μc-Si) modules. However, previous work has shown that other thin-film technologies are also sensitive to this phenomenon [1].

Studies have shown that thin-film modules exhibit reversible metastable behaviour when exposed to light due to the Staebler-Wronski effect [2]. This behaviour of thin-film modules has been shown to exhibit a seasonal trend [3–5]. However, this seasonality is different from temperature coefficients given in datasheets. The performance of the modules would also vary from place to place based on the site climate. Performance is higher in warmer and lower in colder climates, e.g. for micromorph Si modules [6]. The seasonal variation in power of micromorph modules is depicted in **Figure 1**, in terms of time and location dependency. A warm (red) and cold (blue) power curve for the same technology at different locations is shown.

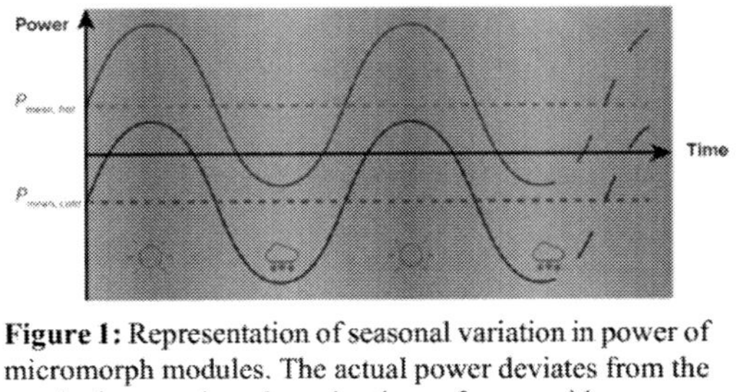

Figure 1: Representation of seasonal variation in power of micromorph modules. The actual power deviates from the nominal power based on the time of year, with summer represented as a sun and winter represented by a cloud. The power is also location dependent. A warm (red) and cold (blue) power curve for the same technology at different locations is shown.

The seasonal deviation in power affects the power stability of the module, and consequently poses a challenge for accurate and reproducible power measurements of modules [7]. This is why procedures for stabilisation are required to stabilise the module's electrical characteristics before measuring its power. The metastable behaviour of thin-film modules differs according to the cell technology and therefore each type of thin-film module requires its own set of stabilisation procedures [6].

The objective of this study is to develop an approach to quantify the seasonal deviations in actual power of thin-film modules installed in the field. This could be used to determine a variable power to which the module's actual performance in the field can be compared. Subsequently, it would also enable simulated models to produce results closer to the actual performance of thin-film PV systems.

2 METHODOLOGY

A thin-film PV plant installed in Jastrabá, Slovakia was chosen for investigation. The plant consists of micromorph modules (Sharp NA-F128) with a capacity of 111 kWp. It was grid-connected in 2012. The methodology comprises of four steps. Step 1 was to gather the measured data of the plant and use this data to quantify seasonal deviations in the power of the **measured data** in terms of the 'seasonal coefficient'. Step 2 was to model the as-built design using the measured meteorological data, which resulted in the **simulated data**. Step 3 was to incorporate the seasonal coefficients (correction function) with the simulated data, resulting in **modified data.** Step 4 was to analyse the results of all three datasets using a

10.4229/EUPVSEC2023/3AV.3.16
020217-001

regression analysis method called the Capacity Test, incorporated in the ASTM-E2848 standard [8]. The methodology overview is depicted in **Figure 2**.

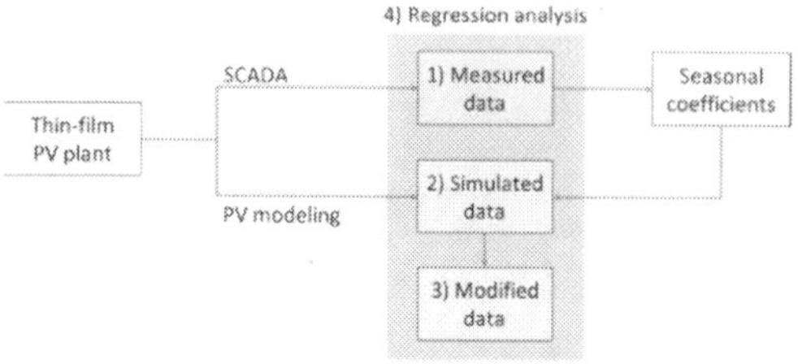

Figure 2: Graphical representation of methodology

2.1 Step 1: Measured Data

Data was collected for the evaluation period January 2013 to December 2015, with a temporal resolution of 15 minutes. The parameters for which data was collected were plane-of-array (POA) irradiance, module temperature, DC current and DC voltage. Then, this data was filtered using three criteria taken from the Capacity Test to improve data quality. As per these criteria, the following data were filtered out:

- Datapoints corresponding to POA irradiance less than 200 W/m²
- Datapoints clipped by the inverter
- Datapoints that did not follow the linear relationship (outliers) of power vs irradiance

After the data was filtered, the power values were adjusted to STC (irradiance of 1000 W/m² and module temperature of 25°C) [9] and the STC adjusted power was plotted over time. A sine curve (P_{sine}) was fitted to the adjusted power. The seasonal coefficients were calculated using the following formula:

$$Seasonal\ Coeff\ (t) = \frac{P_{sine}(t)}{P_{installed}},$$

where $P_{installed}$ is the installed capacity of the plant, with a value of 111 kWp.

2.2 Step 2: Simulated Data

The simulated power (P_{sim}) was obtained by modelling the PV plant in the simulation software SAM, from NREL [10]. Similar to Step 1, the data was filtered, power was adjusted to STC, and adjusted power was plotted over time.

2.3 Step 3: Modified Data

The modified power (P_{mod}) was calculated using the following formula:

$$P_{mod} = P_{sim} \cdot Seasonal\ Coeff\ (t)$$

Similar to Step 1, the data was filtered, power was adjusted to STC, and adjusted power was plotted over time.

2.4 Step 4: Regression Analysis

This approach is taken from ASTM E-2848 and is based on the following multiple linear regression equation[1]:

$$P = G(a + b \cdot G + c \cdot T),$$

where P is the system output power [W], G is the POA irradiance [W/m²], T is the ambient temperature [°C], and a, b, c are the regression coefficients.

The data was filtered as given in Steps 1, 2 and 3. Regression was run monthly for the three-year evaluation period. Each month had a set of reporting conditions (RC) that represented the data for that month. G_{RC} is the 60th percentile and T_{RC} is the mean value of each month. The datasets are further filtered for regression to be within ±20 % of G_{RC}. After running the regression, a P_{RC} was obtained for each month from 2013 to 15.

[1]The original equation has a third variable (wind speed) which has not been included to reduce uncertainty in results [11]

3 RESULTS

3.1 Seasonality in Measured Power

Figure 3 shows the measured power values adjusted to STC plotted over the evaluation period and fitted with the sine curve (according to step 1) Measured data). A clear seasonal trend can be seen. Power values gradually increase from winter to summer and then decrease from summer to winter. Despite the power values for the measured dataset being adjusted to STC, very few datapoints reach the installed capacity power level. The sine mean deviates from installed power by -16 %. If the annual degradation of 1 % was considered, the expected capacity would stand at around 109 kW in 2015, and the sine mean would deviate from this value by -14 %.

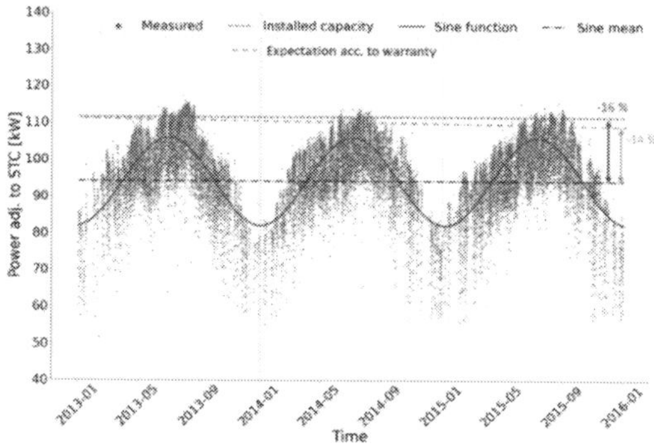

Figure 3: Measured power adjusted to STC shows a sinusoidal trend. The sine mean (dashed blue line) deviates from installed power (dashed red line) by -16 %. Considering annual module degradation rate, the expected installed power would be 109 kW in 2015 (dashed green line), and the sine mean would deviate by -14 %

3.2 Seasonality in Measured, Simulated, and Modified Power

Figure 4 shows a comparison of adjusted power to STC values for all three datasets (step 1 to 3). The measured and modified data show similar trends in seasonal deviation whereas the simulated data shows a slight sinusoidal trend and sits right on top of both measured and modified. Also, the measured and modified powers reach the installed capacity only in summer months while the simulated data is predominantly higher than this value.

The energy yield from the measured, simulated, and modified datasets were also compared to see if there was any influence of the seasonal deviation on the results. The energy yield for the measured data was taken directly from the SCADA platform, while the energy yield for the simulated and modified data was calculated from the

10.4229/EUPVSEC2023/3AV.3.16
020217-002

obtained simulation results. The results are in Table 1:

Table 1: Total energy yield in MWh from 2013-15

	Measured	Simulated	Modified
Total MWh	360	425	370

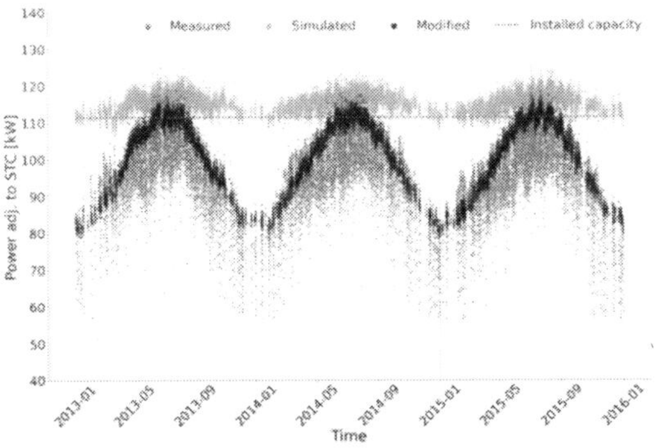

Figure 4: Simulated power (in pink) shows a very small sinusoidal trend, while modified power (in dark blue) and measured power (in light blue) show a sinusoidal trend with a 4-time higher amplitude. Measured and modified powers only reach the installed capacity in the summer months, while simulated power values are above the installed capacity throughout the evaluation period

3.3 Regression Analysis

Figure 5 shows the percentage difference in power at reporting conditions (P_{RC}) between simulated vs measured and modified vs measured datasets (according to step 4). The percent difference of simulated is highest in the winter months and lowest in the summer months, with an average difference of ~25 %. Meanwhile the percent difference of modified dataset is relatively lower and consistent, with an average difference of ~5 %. Therefore, there was an overall 20 % improvement in the accuracy of simulated power results, after being incorporated with the seasonal coefficients.

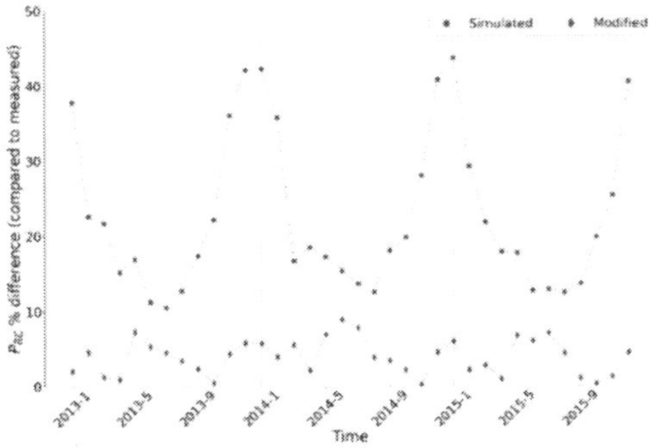

Figure 5: Percent difference in P_{RC} between simulated vs measured (blue squares), and modified vs measured datasets (purple diamonds)

4 CONCLUSION

Through the methodology developed in this study, the seasonal deviation of power was quantified in the form of seasonal coefficients. A seasonal trend observed in the measured power adjusted to STC during the evaluation period from 2013 to 2015. This result validates reviewed literature which state that modules exhibit maximum performance output in the summer when the average temperature conditions are highest. It is also known from literature that heat treatment, e.g., 90 °C for 48 hours stabilize Si cell technology on its maximum [?] and is used in the industry [5]. This is also known as annealing. From this it is deduced that the power would vary with location, with higher powers in warmer climates and lower powers in colder climates.

However, very few power values actually reach the installed capacity level - the rated capacity is reached only in the summer months. The mean of the sine curve of measured data is around 95 kW, whereas the installed capacity is 111 kW_p. Even if the annual degradation of 1 % would have been considered, the expected capacity would be around 109 kW, which is still significantly larger than the sine mean. Moreover, the simulated yield is higher than the measured yield by 16 %. The sinusoidal trend shows up but is underestimated by a factor of 4.

In summary, the nominal (fixed) value of the module does not represent the mean value of the seasonal deviation. For which location/climate does that nominal value apply? At least a reference location/climate should be mentioned in the warranty and standards.

The methodology developed in this study can be used to determine a variable nominal power opposed to the fixed value given in module datasheets. This paves the way to enable constructive warranty discussions for low-performing systems. Finally, results from the regression analysis showed that there was an overall improvement of 20 % in accuracy of simulated data incorporated with the seasonal coefficients.

5 OUTLOOK

The developed methodology should be replicated to check seasonality for all other thin-film and c-Si technologies. Especially Pero/Si tandem and high efficiency cell technologies should be investigated, because they contain thin-films. The developed methodology should also be performed on existing plants at different locations. This would lead to a (type or technology) matrix with a combination of mean irradiance and temperature values, related to a seasonal coefficient. Also, the effect of spectral mismatch on seasonal deviation should be considered. Finally, a laboratory procedure can be performed to determine the temperature coefficients depending on the stabilization status of the module.

6 REFERENCES

[1] T. Weber *et al.*, "Do We Correctly Determine the Power of Thin-Film Modules? Investigation of Stabilization Procedures for Power Determination of Thin-Film Modules," in *38th European Photovoltaic Solar Energy Conference and Exhibition*, Lisbon, Portugal, 2021.

[2] M. Nikolaeva-Dimitrova, R. P. Kenny, E. D. Dunlop, and M. Pravettoni, "Seasonal variations on energy yield of a-Si, hybrid, and crystalline Si PV modules," *Prog. Photovolt: Res. Appl.*, vol. 18, no. 5, pp. 311–320, 2010, doi: 10.1002/pip.918.

[3] M. Gostein and L. Dunn, "Light soaking effects on photovoltaic modules: Overview and literature review," in *37th IEEE Photovoltaic Specialists Conference*, Seattle, USA, 2011, pp. 3126–3131.

[4] R. Gottschalg, J. del Cueto, T. R. Betts, S. R. Williams, and D. G. Infield, "Investigating the seasonal performance of amorphous silicon single- and multi-junction modules," in *3rd World Conference onPhotovoltaic Energy Conversion, 2003. Proceedings of*, 2003, 2078-2081 Vol.2.

[5] Sharp Corporation, "Maximum Power Measurement Guideline for Thin-Film PV Module,"

[6] T. Weber *et al.*, "Investigation of Stabilization Procedures for Power Determination of Thin-Film Modules," in *36th European Photovoltaic Solar Energy Conference and Exhibition*, Marseille, France, 2019.

[7] M. G. Deceglie, T. J. Silverman, B. Marion, and S. R. Kurtz, "Temperature-dependent light-stabilized states in thin-film PV modules," in *2015 IEEE 42nd Photovoltaic Specialist Conference (PVSC)*, 2015, pp. 1–3.

[8] *Test Method for Reporting Photovoltaic Non-Concentrator System Performance,* E44 Committee, West Conshohocken, PA.

[9] H. K. Singh *et al.*, "Modified STC Correction Procedure for Assessing PV Module Degradation in Field Surveys," in *2017 IEEE 44th Photovoltaic Specialist Conference (PVSC)*, 2017, pp. 1995–1999.

[10] NREL System Advisor Model (SAM). [Online]. Available: https://sam.nrel.gov/

[11] K. Anderson, J. Kemnitz, and M. Boyd, "Evaluating Cell Temperature Models and the Effect of Wind Speed in PV System Capacity Testing: Preprint," National Renewable Energy Lab. (NREL), Golden, CO (United States) NREL/CP-5K00-78950. [Online]. Available: https://www.osti.gov/biblio/1820579

DETERMINING THE SPECTRAL RESPONSIVITY OF PV MODULES WITH A TUNABLE LASER LIGHT SOURCE

Konstantin Ladner, Hendrik Sträter, Stefan Winter
Physikalisch-Technische Bundesanstalt (PTB), Bundesallee 100, 38116 Braunschweig, Germany
Konstantin.Ladner@ptb.de

ABSTRACT: The spectral responsivity (SR) of photovoltaic (PV) devices is important for accurate electrical parameter measurements. Determining SR for PV modules is challenging due to their size. This study presents an improved and non-destructive method for measuring the SR for PV modules. The setup uses a tunable laser system with a wavelength range spanning 280 nm to 1320 nm with customizable step width. It integrates a chopper and a laser beam homogenizer (LBH) to generate a uniform light field, enabling simultaneous irradiation of the entire PV module and a WPVS reference cell at the same time. Simultaneously, an LED-based solar simulator emits an AM 1.5 like spectrum, used as bias light to irradiate the PV module under realistic current-generating conditions. We confirmed our results by employing a large area reference cell (LARC) and a WPVS reference cell of know spectral responsivity. The laser's properties were analyzed, and the beam's various aspects are presented. Our methodology enhances the precision of a PV module's IV curve corrections by reducing uncertainties associated with the spectral mismatch factor and the determined SR can be used as input data for Energy Rating of solar modules.
Keywords: PV module, Spectral responsivity, DSR method, spectral mismatch, solar simulator.

1. INTRODUCTION

The advancement of photovoltaic (PV) technology depends on the precise understanding and measurement of its inherent properties. A critical property is the spectral responsivity (SR), which is essential in determining the spectral mismatch - a key factor for the accurate measurement of a PV device's electrical parameters. The method introduced here is fundamentally derived from the DSR method established at PTB [1] and has been further refined, culminating in a laser-DSR facility [2]. However, the challenge becomes more pronounced when dealing with PV modules, mainly due to their larger size compared to individual solar cells. Traditional methods for determining this responsivity often involve examining single cells within a module or are using a limited number of filters with a high bandwidth [3].

An alternative approach involves an LED-based solar simulator with modulated LED channels, providing the SR of solar modules limited by the range and width of the spectrum emitted by the LEDs [6].

We therefore developed a method based on a tunable laser system, which is capable of measuring relative Spectral Responsivity (SR) of a solar module. In order to illuminate the whole module, a laser beam homogenizer is used to provide a uniform large scale light field. The properties of the light field are discussed as they represent the limiting factor in the measurement. This innovative approach improves the precision of SR measurements for PV-modules.

Moreover, SR measurements, involving world photovoltaic scale (WPVS) reference cells with an active area of 2x2 cm² and large area reference cells (LARC) with and active area of 16x16 cm², have validated the efficacy of this method, highlighting its potential in enhancing PV calibration processes.

2. SETUP

The setup utilizes a constant bias light from an LED based solar simulator and a chopped laser beam to irradiate the Device Under Test (DUT). Both the DUT and a WPVS reference cell with a known spectral responsivity (SR) are measured simultaneously. Both devices are connected in each case to an impedance converter, which maintains the devices in a short-circuit current state (Isc) state. The current is then separated to DC and AC and converted into a proportional voltage. The applied lock-in technology detects the AC signal from the chopped laser light, allowing for the determination of the SR of the DUT (PV module and a LARC). A multimeter measure the DC voltage originating from the bias current.

2.1. The Laser System

The primary source of the optical power originates from a tunable laser, which emits irradiance between 660 nm and 1320 nm, as depicted in Figure 1. The laser paths were established using flip-mirrors.

This irradiance can be adjusted further through either Second Harmonic Generation (SHG: 330 nm to 659 nm) or Third Harmonic Generation (THG: 280 nm to 329 nm). Depending on the wavelength, the laser beam is guided into one of two distinct optical fibers and is subsequently chopped by two choppers. Suitable collimators are used to direct the beam into the glass fiber.

To monitor variations and drifts in the power of the laser system, monitor diodes are utilized. The mirrors and other optical instruments are optimized for the spectral range of wavelengths, and each wavelength has a path via flip mirror mounts. Power loss due to the 20 m glass fiber length and optical losses from all optical components must be considered and had been optimized.

Consequently, our system enables the span of accessible wavelengths to extend from 280 nm to 1320 nm, covering the wavelength range of silicon-based photovoltaic devices.

To monitor variations and drifts in the power of the laser system, monitor diodes equipped with a beam splitter are utilized. Consequently, this enables the span of accessible wavelengths to extend from 280 nm to 1320 nm, a range pertinent to silicon-based photovoltaic devices. The mirrors and other optical instruments are optimized for the spectral range of the laser light.

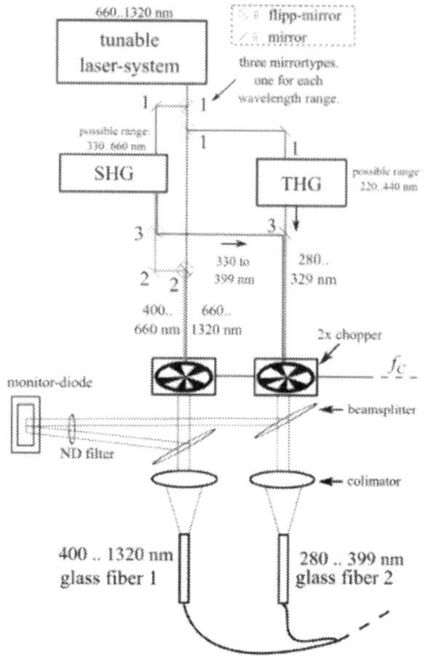

Figure 1: Schematic of the setup including the laser and its harmonic generator (SHG, THG), (flip-) mirrors, chopper, beam splitter with monitor diode, collimator, and glass fiber. To irradiate the DUT and reference cell with this large wavelength interval, all light paths must be optimized to its specific optical power throughput. Finally, the 20 m glass fiber couples into a laser beam homogenizer (LBH), which irradiates the DUT and reference cell.

2.2. Irradiance with bias and the LBH

Both, the constant light source (E_{bias}) and the chopped monochromatic light (E_{mono}) are directed onto the DUT and the WPVS reference cell, see Figure 2 and Figure 3.

The LBH composes of two micro lens arrays and a Fourier-lens to enlarge the laser beam to a size large enough to measure a whole PV module. The LBH has been specifically engineered to meet certain criteria: high power, extended distance, and broad homogenic irradiance. Each micro lens array employed is a cylindrical lens with a focal point less than 1 mm and a columnar distance of approximately 0.3 mm and 0.4 mm respectively. The secondary cylindrical micro lens array is oriented perpendicularly. The Fourier-lens pattern is finalized with an approximate ratio of 1.0 m to 0.75 m at a distance of 5.0 m. Due to constraints, the irradiated area proved to be too inhomogeneous for larger PV modules.

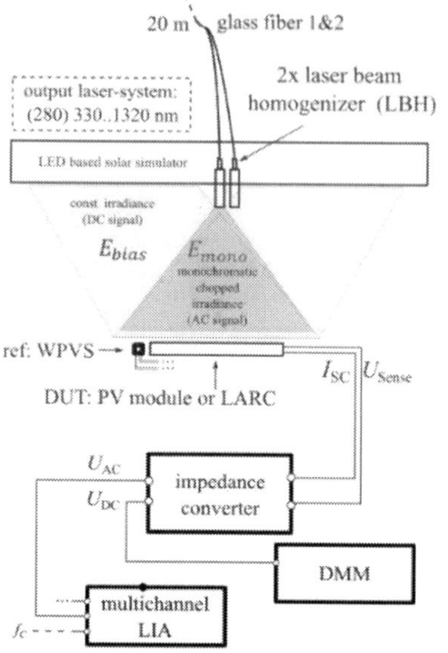

Figure 2: Schematic of the setup including irradiance of the DUT and reference cell. The exposed devices are connected each to an impedance converter which finally measure the separated signal of the irradiation.

The synchronal transversion of the I_{SC} of both devices can be detected the multichannel Lock-In Amplifier (LIA) and the direct relation of both signals leads to a relative SR expressed by the equation:

$$s_{dut}(\lambda) = \frac{I_{dut}(\lambda)}{I_{ref}(\lambda)} \cdot s_{ref}(\lambda)$$

Here I_{dut} (λ) denotes the measured short-circuit current of the device under evaluation, I_{ref} (λ) represents the concurrently measured short-circuit current of the reference device, and s_{ref} (λ) is the established spectral responsivity of the reference device.

A depiction of the irradiance of all devices are presented in Figure 3. Evidently, the irradiance of the LBH fully irradiates all devices. A scan of the irradiated area is shown in Figure 6.

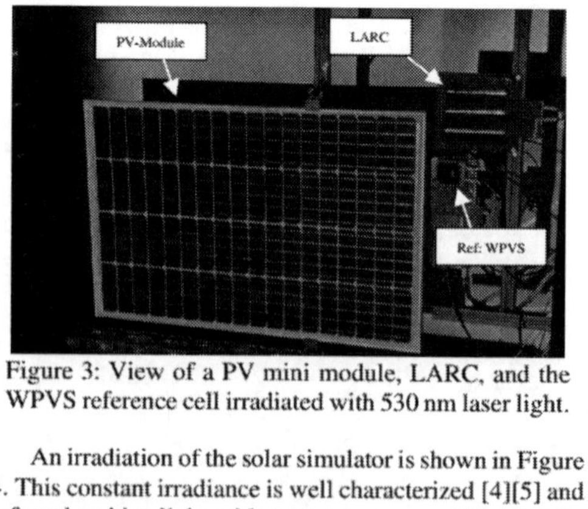

Figure 3: View of a PV mini module, LARC, and the WPVS reference cell irradiated with 530 nm laser light.

An irradiation of the solar simulator is shown in Figure 4. This constant irradiance is well characterized [4][5] and referred as bias light with a spectrum near AM 1.5. The shift in the short-circuit current induced using bias light can be corrected if the SR is well-known, by employing the spectral mismatch factor (MMF).

Figure 4: Image of the solar simulator used as bias light source. It produces a rectangular light field of 2 m x 1 m with a non-uniformity of less than 2 %.

2.3. The Laser Beam

To reliably determine the SR, the spectral properties and optical power of the laser beam has been examined to understand the type of illumination being dealt with. First, the power was assessed, followed by the non-uniformity, and then the full width at half maximum (FWHM). Subsequently, the SR measurement were addressed with the generated irradiance.

2.3.1. Power of the Laser System

Figure 5 illustrates the output power of the laser system as function of wavelength with a peak power of 3.5 W at 900 nm, displaying noticeable fluctuations over the spectral range. A significant decrease of power occurs below 780 nm.

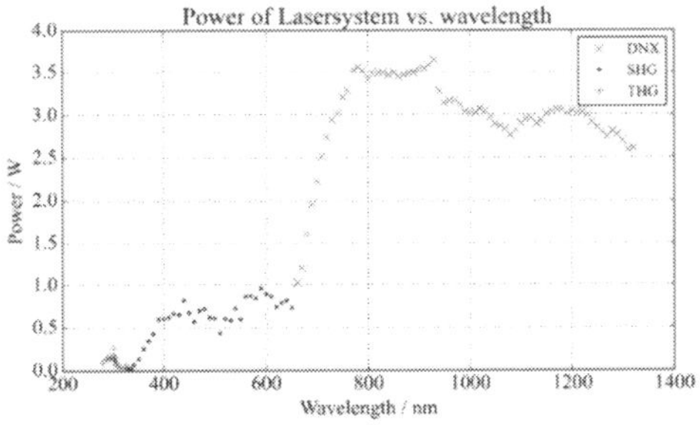

Figure 5: Optical power of the laser system determined directly at the output of the tunable laser system, SHG, and THG, respectively.

The tunable laser system provides the most power output while the SHG yields power from 330 nm to 660 nm. It is important to consider the transition at 400 nm and 660 nm, where signal intensity changes and requires corrections through overlap measurements and fittings.

2.3.2. Non-Uniformity

Figure 6 illustrates the NU scan of the first irradiance area produced by the LBH. By positioning the LBH central to reference cells (WPVS), a scan was conducted an area used by the LARC and the reference cell. Furthermore, a second scan area was done to determine a non-uniformity used by a PV module and a reference cell.

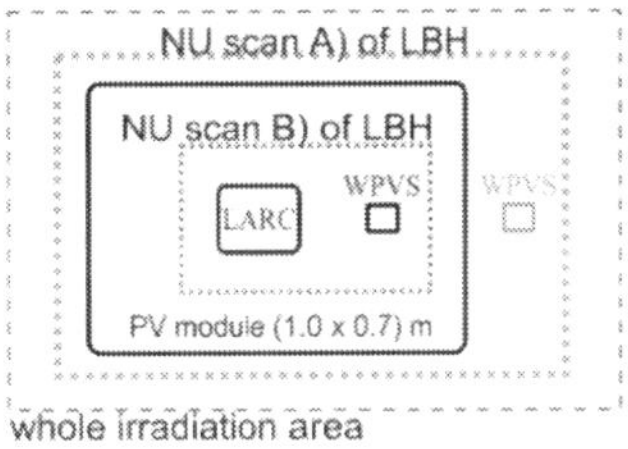

Figure 6: The initial irradiance area is demarcated with scanned regions to assess NU high-resolution scans are indicated in red, with a NU of less than 5%, while low-resolution scans are shown in green and exhibit approximately 10% NU.

Upon defining the scan regions, multiple scans were performed, as illustrated in Figure 7. The non-uniformity (NU) exhibits variability across different wavelengths, prompting questions regarding the irradiated area's stability. Regardless of this concerns later SR measurements seems appropriate.

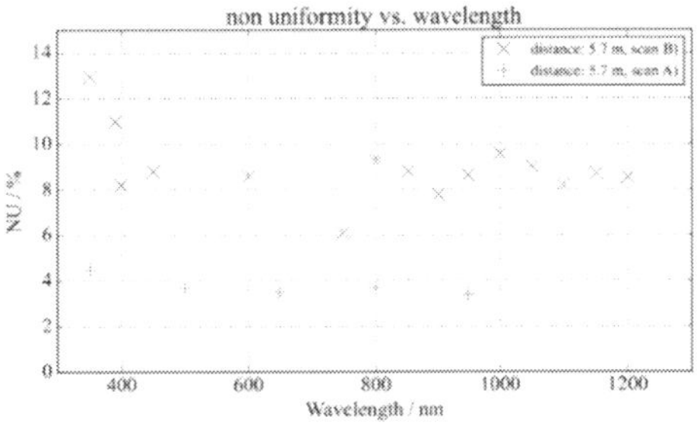

Figure 7: Resulting NU values for selected wavelengths and two sizes of the measurement area. The NU is still wavelength-dependent and exhibits significantly smaller values in for a smaller measurement area.

The NU of the LBH for a selected wavelength is presented in Figure 8. The shown NU is for the small area, referred to as Scan A).

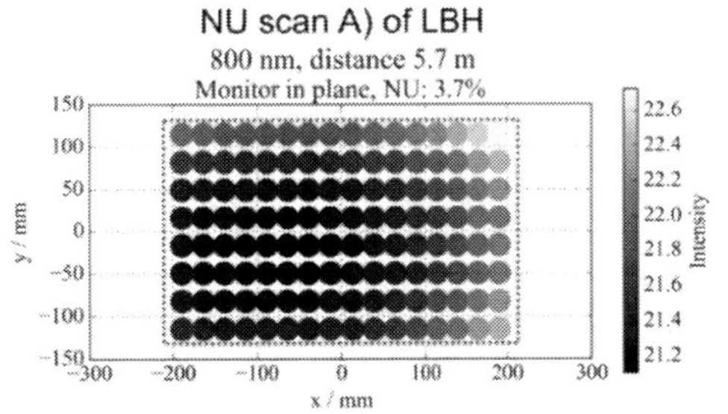

Figure 8: NU Scan for the smaller region (Scan A) resulting in a NU of 3.7%. A distinct spot of heightened intensity is observed slightly to the left of the center of about 50 mm. which arise from a small LBH misalignment.

Achieving a low and same NU for all wavelengths is crucial for an accurate SR measurement to ensure all cells generate the same current. When determining the SR, the cell receiving the least irradiance becomes the limiting factor. With the non-uniformity scan, it becomes evident that the central solar cell receives greater irradiation than the surrounding cells, leading to higher current generation.

In essence, the SR of this minimally irradiated solar cell is what we predominantly measure. While a certain degree of current mismatch is unavoidable, consistent deviations can still yield acceptable results.

2.3.3. Spectral Properties

A narrow FWHM denotes a more monochromatic light source, essential for determining the spectral responsivity of solar cells at specific wavelengths. It ensures the cell predominantly receives the intended wavelength, allowing for precise measurements.

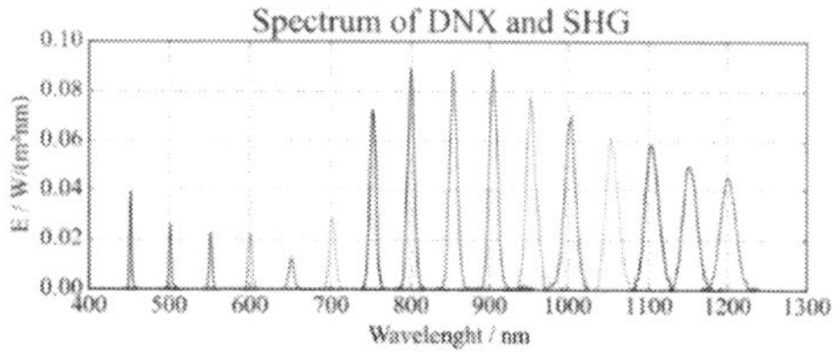

Figure 9: Spectra of selected wavelengths from the laser system in the measurement plane. The full width at half maximum (FWHM) increases with increasing wavelength.

Moreover, a reduced FWHM minimizes interference from adjacent wavelengths, ensuring the generated current is mainly due to the desired wavelength. Our observations underscored the light source's consistency and reliability, with no FWHM-related issues identified. It's noteworthy that an increasing FWHM is evident, as illustrated in Figure 10.

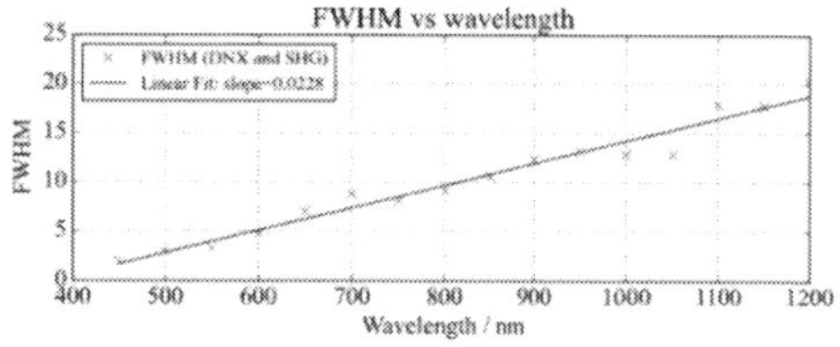

Figure 10: Full Width at Half Maximum (FWHM) of selected spectra increases with increasing wavelength. A fit is depicted to approximate a slope of 2.3 nm for each 100 nm change of the output wavelength.

3. RESULTS

To demonstrate, that the method of measuring an SR of a single solar cell can be transferred to a PV module, a two-step procedure was used to validate the setup. Firstly, two WPVS reference cell SR curves are presented.

Secondly a Large Area Reference Cell (LARC) was measured and depicted with its well-known SR curve. Finally, a small PV-module shows a smooth and reasonable SR curve as we would expect from a PV module using this solar cell technology. valid results. Additionally, preliminary results from a standard-sized PV module also indicated a favorable SR curve.

3.1. Validation with a WPVS reference cell

In Figure 11 one can see a WPVS reference cell, which is used to validate the setup with evident satisfactory alignment. The peak of the SR exhibits the most significant deviation, approximately 0.7 %. This discrepancy is most likely due to the suboptimal use of the impedance converter, which is designed for high currents (around 9A). However, connecting a WPVS reference cell results in a weak, noisy signal (700 mA). Connecting a device with a higher current will eliminate this noise, as will be demonstrated in the subsequent section. Overall, the measurement is considered a significant success.

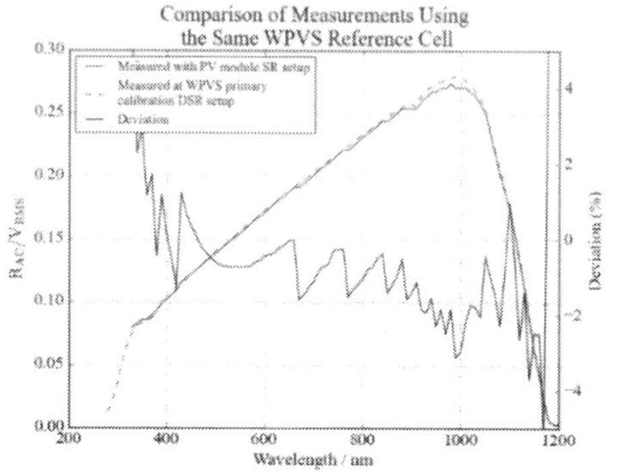

Figure 11: Direct comparison of a WPVS reference cell measured with the setup for PV modules compared with the known data from our primary calibration laboratory for WPVS reference cells. indicating a deviation across the entire wavelength range of approximately 3 %.

3.2. Large Reference Cell (LARC)

Figure 12 shows several SR curves. The WPVS reference cell is represented in red ich was used to calculate the SR curve of a LARC in green, without applied bias irradiance. Additionally, a second SR of the same LARC curve is depicted. Both SR curves are closely aligned. Notable deviations exist below 400 nm, attributed to differences in the steep slope of the curve. Deviations between 600 nm and 1000 nm are less than 1 %, and deviations less than 5 % are observable. Significant deviations are present between 380 nm and 620 nm. Despite these variations, system verification is feasible, particularly when the size of the LBH's irradiance is not a concern.

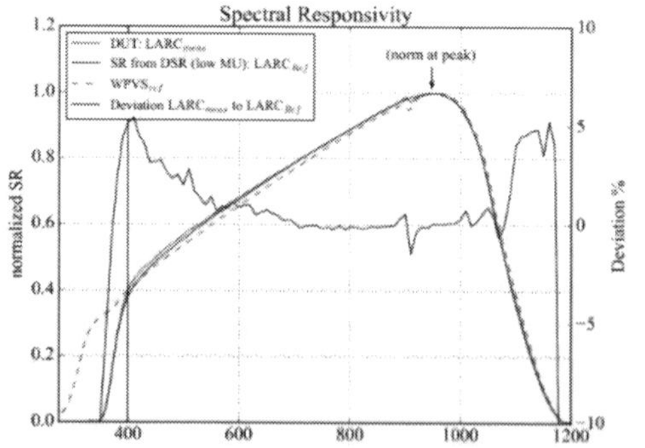

Figure 12: SR with a reference cell (WPVS). For a comparison the norm was set to the peak values of both measurements. Deviations of the DUT (green) and the reference cell (LARC, well know SR) are below 5 % and replicable, which leads to a verification of monochromatic light beam. Given the minimal deviations observed, the

setup is deemed to be reliable. Subsequent steps will involve the calculation of measurement uncertainties.

3.3. Small PV Module

To demonstrate the suitability of our methods for PV modules, which consists of solar cells connected in series, we determined the SR of a small PV module, as shown in Figure 13. It shows a slight decrease in SR above 800 nm. The SR curve appears smooth and reasonable. Measurements with bias light had been conducted and are very similar and results in a deviation of less than 0.5 %. This PV module was also turned by 180° and shows a very similar SR curve with a deviation of again less than 0.5 %, indicating consistent and reliable measurements.

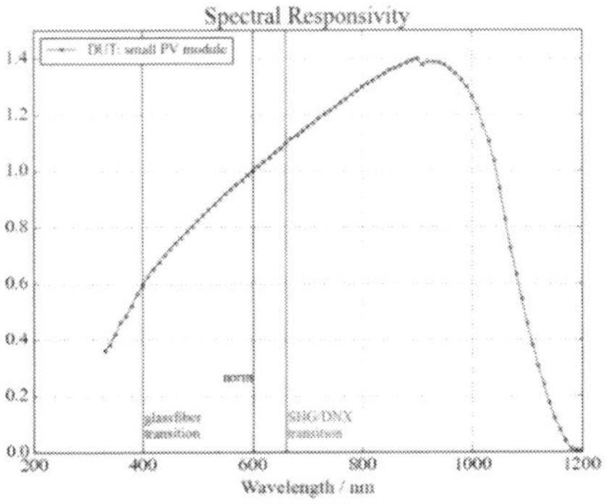

Figure 13: SR curve of a small PV module with solar cells connected in series.

3.4. Industrial Size PV Module

One variation in this examination was the size of the PV module and the distance (7 m) of DUT and LBH. Bias light was set to 300 W/m². All measurements were normalized to 600 nm. Limitations in the solar simulator setup suggest that this distance may cause high deviations. While the curves fit well, they don't match the quality of initial measurements (see Figure 14). Deviations, most pronounced above 900 nm, are likely due to noise and equipment misalignment. The right curve shows more IR deviations due to higher spectral responsivity and temperature effects. Given the measurement duration of about half an hour, temperature increases over this period.

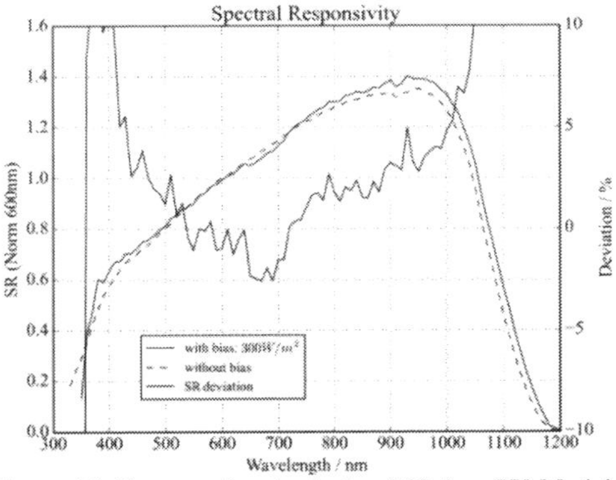

Figure 14: Comparative Analysis of SR in a PV Module with and without bias. Notable deviations are evident, particularly in the near infrared region. The LBH was positioned 7 m away, primarily due to the substantial size of the standard PV module. Alterations in distance resulted in NU variations, yet the measurements presented remain valid.

3.5. Discussion

Further research and methodological adjustments can reduce these deviations. One approach is to alter the order of set wavelengths, starting at 1200 nm and ending at 330 nm. Initial observations indicate fewer deviations in the infrared. Furthermore, the size of the DUT and distances indicated the challenges and demands further optimization, regarding LBH and maybe also every other aspect of irradiation.

4. CONCLUSIONS

We demonstrated the successful transition of the DSR method from individual solar cells to a photovoltaic (PV) module. The module exhibited satisfactory alignment and yielded plausible results, thereby underscoring the adaptability of the method. A systematic process was employed to validate the setup, which included measuring a LARC (Large Area Reference Cell) alongside a WPVS reference cell, both of which exhibited favorable alignment. Further validation using two designable reference cells revealed even superior alignment. While the adaptation to a larger PV module presented challenges, mainly due to setup modifications not meeting specific requirements, the initial success with the smaller module offers promise. The endeavor to utilize an LBH for broadening the light beam and further adapting the DSR Method to larger PV modules remains a challenge. Our findings indicate potential for improvement through various adjustments, with the challenges posed by the DUT size being an active area of research to optimize the setup for industrial PV modules. Importantly, the resulting curves from our study underscore the feasibility of calculating the MMF for small PV modules.

5. ACKNOWLEDGEMENT

The DSR setup is developed within the project „MetroKomPV“, which is funded by the Federal Ministry for Economic Affairs (BMWi), Germany

6. REFERENCES

[1] J. Metzdorf, “Calibration of solar cells. 1: The differential spectral responsivity method”, Applied Optics, vol. 26, No. 9, 1987.

[2] S. Winter, “Design, realization and uncertainty analysis of a laser-based primary calibration facility for solar cells at PTB”, Measurement, vol. 51, pp. 457–463, 2014.

[3] Clara Edmonds and Harald Müllejans, Improving Spectral Responsivity Measurements by Correcting for Filter Bandwidth, IEEE Journal of Photovoltaics, Vol. 13, No. 1, 2023

[4] S. Riechelmann, H. Sträter, S. Winter, Determination of a PV Module Power Matrix with an LED Solar Simulator, Proceedings 37th European Photovoltaic Solar Energy Conference (2020)

[5] S. Riechelmann, Characterizing the Angular Distribution of an LED-Based Solar Simulator for PV Modules, 36th European Photovoltaic Solar Energy

Conference (2019)
[6] Sträter, H.; Riechelmann, S.; Neuberger, F.; Winter, S. (2019): LED-Based Differential Spectral Responsivity Measurements of PV Modules. 36th European Photovoltaic Solar Energy Conference and Exhibition; 1121-1126 (2019)

THERMAL ISSUES ON HALF- CELL BIFACIAL MODULES. A WAY THROUGH ALBEDO AND MISMATCH VOLTAGE

S. Suárez*, J. M. Álvarez, D. Villoslada, G. Navas, J. Vilela, I. Fernández, S. Rodríguez-Conde
Enertis Applus+, www.enertisapplus.com
Parque Empresarial Las Mercedes, C/ de Campezo, 1, 28022 Madrid, Spain.
*sergio.suarez@enertisapplus.com

ABSTRACT: This study presents a comprehensive thermal analysis of half-cell bifacial photovoltaic modules, emphasizing the role of ground albedo and voltage mismatch in hot spot formation. Contrary to prior simulation-based findings suggesting thermal advantages of half-cell designs, the on-field data in this work offer a nuanced perspective. Specifically, half-cell bifacial modules exhibit a greater number of hot spots compared to monofacial modules, cited in this work as Hotspot mirroring (H_{mirr}) Additionally, the study reveals that under specific partial shading scenarios and limited thermal dissipation, the thermal gradients in H_{mirr} can be greater and comparable to normal hotspot produced by shadows. The study also establishes a direct correlation between the characteristics of the hot spots and the ground albedo. The research concludes by highlighting the need for revised technical standards and considerations for bifacial solar technology.

Keywords: Bifacial, Half-cell, Hotspot, Albedo Testing On-Site.

1 INTRODUCTION

The accelerated adoption of photovoltaic (PV) technologies serves as a cornerstone in global efforts to mitigate climate change and to provide sustainable, renewable energy. However, the operational efficiency and performance of PV modules are influenced by a myriad of factors that can adversely affect their longevity and effectiveness. Among these, the thermal management of photovoltaic modules has garnered significant attention in scientific literature.

Half-cell designs and bifacial technology have emerged as potential solutions for enhancing the operational efficiency of PV modules. Specifically, half-cell bifacial modules have shown promise in terms of improved efficiency and operational flexibility. Despite these advantages, there exists a notable paucity of comprehensive studies addressing the thermal behavior of these modules, particularly in the context of ground albedo and voltage mismatch.

This study aims to fill this research gap by investigating the influence of ground albedo and voltage mismatch on hot spot formation in half-cell bifacial PV modules. Employing an empirical approach based on on-field data collection, this work offers a nuanced perspective that diverges from prior simulation-based findings. Through this inquiry, the concept of Hotspot Mirroring (H_{mirr})) is introduced to describe a unique characteristic of half-cell modules, and its implications on thermal management are discussed.

The ultimate objective of this research is to provide insights that could inform future technical guidelines and standards for the design and deployment of half-cell bifacial photovoltaic modules.

2 HOTSPOT MIRRORING- VOLTAGE MISMATCH CASE

The phenomenon of hotspot mirroring (H_{mirr}) is particularly noteworthy in our study and is a consequence of voltage mismatch between the upper and lower substrings in half-cell modules. This mismatch occurs when a cell with forced shading begins to generate a negative voltage.

The repercussion of this is a disruption in the normal functioning of cells located on the opposite side of the shaded cell. Specifically, thermal patterns resembling a checkerboard emerge, affecting the weakest cells of the adjacent substring.

Our experimental findings on (H_{mirr}) corroborate those from prior simulation-based research [1]. To illustrate this effect, we include an on-field infrared (IR) image that clearly shows the checkerboard thermal patterns across the affected cells on the Fig. 1.

Figure 1: Element inducing an upper-position hotspot shadow, and a corresponding mirrored hot spot in the lower position.

This confirmation of the hotspot mirroring phenomenon through experimental data underscores the importance of considering such effects in the thermal management of bifacial half-cell solar technologies.

3 EXPERIMENTAL SETUP

To carry out an exhaustive analysis of thermal behavior in half-cell bifacial and monofacial photovoltaic modules, a comprehensive experimental setup was devised. Both types of devices were equipped with a monitoring system measuring irradiance and temperature in different points of the devices under test.

3.1 Instrumentation

Each module was fitted with five thermocouples for the purpose of temperature measurement. To assess the irradiance conditions, two irradiance sensors were installed on each module, denoted as G_{front} and G_{rear} to measure the frontal and rear irradiance respectively.

A schematic representation of the sensor arrangement is provided in Fig. 2 for further clarity.

10.4229/EUPVSEC2023/3AV.3.19
020221-001

Figure 2: Irradiance and thermal sensor located on monofacial (left) and bifacial (right) half-cell devices.

These instruments enabled the accurate quantification of the incident irradiance conditions affecting the PV modules current production and the temperature achieve by different points in the modules.

3.2 Hotspot forcing and measurement methodology.

Intentional shadowing was employed to both monofacial and bifacial half-cell modules to induce what we refer to as hotspot mirroring (H_{mirr}). The existence and distribution of these hotspots on the module were previously verified through thermographic camera imaging (IR).

To provide an in-depth understanding of the thermal behavior of the modules under study, five thermocouples were strategically positioned at the following locations:

A) Shadow-induced Hotspot (H_{shad})
B) Hotspot Mirroring (H_{mirr})
C) Upper Substring Region with non-thermal abnormalities
D) Lower Substring Region with non-thermal abnormalities
E) Central position with non-thermal abnormalities

In the Fig. 3, we present the designated locations for the thermocouples as well as the area subjected to intentional shadowing.

Figure 3: Designated locations of the thermocouples (left) and the area subjected to intentional shadowing (right)

This setup ensured a comprehensive thermal map covering different areas with thermal effects and not, including those instigated by intentional shadowing and mirroring effect over a 30-days recording period.

4 RESULTS

4.1 Voltage mismatch analysis

To quantify and compare voltage mismatch in both configurations, temperature gradients were calculated using thermocouples placed strategically across the modules. Over a 30-day period, a strong correlation was found between temperature gradients and the level of irradiance received by each module—most pronounced on clear days with maximum irradiance. Temperature differences were calculated using the central thermocouple (point E) as a reference, and by comparing shadow-induced hotspots with hotspot mirroring.

As indicated in the subsequent Fig. 4, temperature differences were higher for H_{mir} than for H_{shad}.

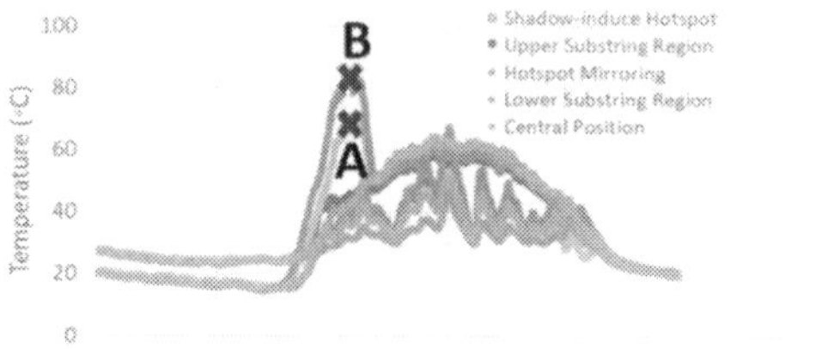

Figure 4: Temperature profiles in two day-types (clear and cloudly) with A) H_{shad} and B) H_{mirr}

Maximum values were computed, where the table following this figure reports maximum deviations of 20.2°C for monofacial modules and 31.1°C for bifacial modules. Maximum values for other measurements were also provided and compared with the central point of the module.

Table I: Summary of Temperature deviation on both devices.

Sample	Max A-B (∘C)	Average A-B (∘C)	A-E (∘C)	B-E (∘C)
Monofacial	20.2	1.81	6.52	4.70
Bifacial	31.1	2.70	2.42	0.40

4.2 Albedo effects

Albedo's influence is another area where bifacial and monofacial modules exhibited diverging characteristics. While albedo's influence on temperature was subtle in monofacial modules, bifacial modules showed a more pronounced sensitivity to variations in ground reflection, particularly in hot spots' characteristics under different shading scenarios.

The study analyzes the maximum irradiance levels at which the protection diodes in both monofacial and bifacial modules are activated. The activation of the diode can be identified by a discernible decrease in the temperature of the hotspots. Unlike monofacial modules, bifacial modules are exposed to rear irradiance (G_{rear}), which contributes to the circulating current and consequently results in greater thermal gradients.

Comparative analysis, as illustrated in the Fig. 5, indicates that the monofacial module activates its protection diode at an irradiance level of 770 (W/m^2).

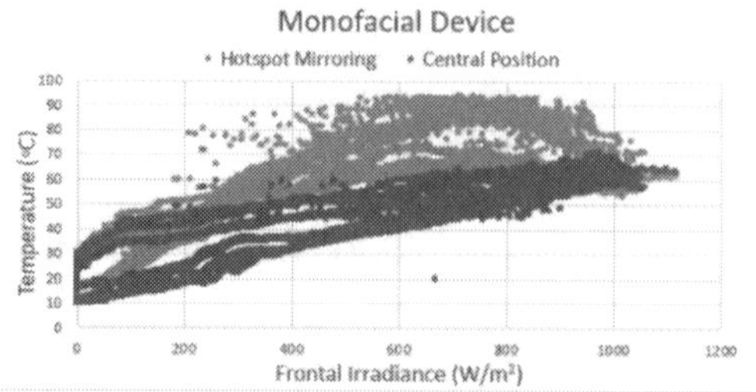

Figure 5: H_{mirr} vs Central position temperature's profile in monofacial device.

10.4229/EUPVSEC2023/3AV.3.19
020221-002

Comparing the central position with Hotspot mirroring profiles can be used to identify when the hotspot temperature cools enough to achieve thermal equilibrium with the rest of the module.

The deviation in the slope of the profile indicates diode activation due to dangerous heat production by the affected cells and negative voltage in the shadowed area.

As seen in Fig. 6, values are significantly lower for the monofacial case compared to the bifacial case, while the bifacial module activates its diode at 630 (W/m^2)

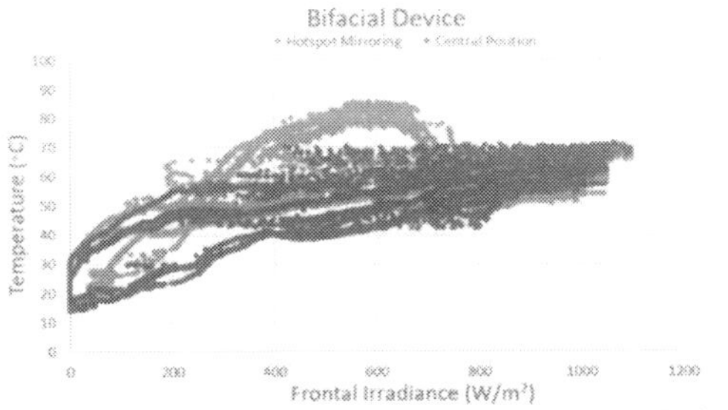

Figure 6: H_{mirr} vs Central position temperature's profile in bifacial device

5 IEC62446-3 ANALYSIS

Upon completion of the study, a review of the existing IEC 62446-3 standard, which outlines the steps for conducting and analyzing on-field thermography, was undertaken. The results of this work reveal subtle discrepancies that suggest a potential need for updates to the regulation.

Notably, the phenomenon of hotspot mirroring, which is a relatively new occurrence associated with half-cell technology, is not currently accounted for in the standard. This oversight is particularly concerning, given that half-cell technology has been present in utility-scale markets for several years.

Additionally, criteria for defect identification based on thermal gradients, as well as minimum irradiance levels for inspection, warrant reconsideration for bifacial modules. The role of rear irradiance (G_{rear}) in these modules is substantial, exacerbating the severity of hotspots and lowering the frontal irradiance levels at which diode activation occurs.

One crucial aspect emphasized in the standard's guidelines is the extrapolation of thermal gradients, as illustrated in the forthcoming Fig. 7. Here, gradients calculated at different irradiance levels relative to Standard Test Conditions (STC-1000W/m^2) are intended to be extrapolated due to the linear relationship between incident irradiance and the thermal gradient of defects.

$$\Delta T_2 = \left(\frac{G_2}{G_1}\right)^x \Delta T_1$$

Figure 7: Equation for extrapolating thermal gradients to STC according to IEC 62446-3.

In this equation ΔT_2 corresponds to the thermal gradient at STC (G_2), while ΔT_1 represents the thermal gradient measured at the irradiance G_1, typically lower than STC. The factor x is 1 for modules and varies for other components.

The extrapolation guidelines in IEC 62446-3 currently do not account for the G_{rear} effect, which, as demonstrated in this study, is a key consideration.

6 CONCLUSIONS

Diverse thermal behaviors were observed on Half-Cell technology, including significant new hotspots in shadow-adjacent areas called in this work as Hotspot Mirroring (H_{mirr}).

A linear relationship was found between thermal gradient and ground albedo, suggesting behavior modifications due to bifacial gain (G_{front} & G_{rear}) during thermal inspections.

The IEC 62446-3 standard provides guidelines for normalizing thermal gradients to Standard Test Conditions (STC) of 1000 W/m^2, due to the dependency between thermal gradients and irradiance levels. For bifacial systems, normalization should also account for the bifacial gain.

7 REFERENCES

[1] J. Quian, "Comparison of Half-Cell and Full-Cell Module Hotspot-Induced Temperature by Simulation" IEEE JOURNAL OF PHOTOVOLTAICS, VOL. 8, NO. 3, May 2018

Experimental study of temperature and performance in BIPV and comparison with numerical simulation

Paul Schenk*, Jens Froebel, Sebastian Schindler, Matthias Pander, Bengt Jaeckel
Fraunhofer-Center for Silicon-Photovoltaics (CSP), Otto-Eissfeldt-Str. 12, 06120 Halle (Saale), Germany
Fraunhofer Institute for Microstructure of Materials and Systems IMWS, Walter-Huelse-Str. 1, 06120 Halle, Germany
*=corresponding author: phone: +49-345-5589 5220, fax: +49-345-5589 5999, paul.schenk@csp.fraunhofer.de

ABSTRACT: Numerical simulations have become indispensable in modern product development and design optimization. They offer a fast possibility to compare different variants under various aspects. The validity and correctness of a simulation depends mainly on experience and already known data sets. In the case of new challenges and issues, these data sets are often missing. Especially in complex simulations, in which real effects can only be reproduced to a limited extent, the results are difficult to classify. To confirm the accuracy of simulations, they must be backed up by validating experiments.

Concrete façade elements with integrated photovoltaic modules were developed within the SOLARcon project. These were investigated and evaluated by simulations regarding their thermomechanical behaviour. To better understand the results and the model, a demonstrator was built and equipped with sensors for temperature measurement and an I-V tracing device. This demonstrator for instance provides temperature values for the modules and the concrete body as well as current yield data of the PV modules. The data from the demonstrator is in well alignment with the simulation results.

Keywords: Simulation, Building Integrated PV (BIPV), Thermal Performance, FEM

1 INTRODUCTION

Numerical simulations have become indispensable tools in modern product development, enabling the evaluation of different design variants and predicting system performance. However, accurately simulating the thermal behavior of Building-Integrated Photovoltaic (BIPV) systems poses unique challenges due to complex interactions between materials and environmental conditions. In the SOLARcon project, BIPV systems with concrete façade elements integrated with PV modules were investigated. This paper presents an experimental study aimed at validating a numerical simulation model to predict the thermal behavior of BIPV systems.

The thermal behavior of free-standing PV modules is relatively well understood and mathematically described ([1]-[4]). The small thickness of the component means that there are no large temperature differences, and the thermal inertia is so small that simple steady-state models can be applied without major limitations. In a building integration with a solid concrete body, larger temperature differences are expected. Thermal inertia is also relevant and more complex models are required. Measured values are required to verify the simulation models.

For the further development and the build-up of know-how in simulation models for BIPV issues, the simulation results already achieved [5] are to be evaluated with the help of an experiment. The goal is also to collect more data to improve the models in the future and to find a workaround that can find the most accurate statements with fast simulations. In addition, the power of the installed modules was measured to be able to make statements about the yield, but also about the aging/damage of the modules.

A demonstrator was built from the concrete elements designed in the project and manufactured by the partners, which was equipped with PV modules. Two of the four modules were equipped with temperature sensors on the back. In addition, the concrete body was also equipped with two temperature sensors to determine the difference between the module and the concrete. The entire structure was installed on the open field in November 2022, so that there are at least six months of data available to compare measurement and simulation. This covers a wide range of weather conditions to investigate the influence of different boundary conditions.

2 EXPERIMENTAL SETUP

2.1 Demonstrator

The demonstrator setup is shown in Figure 1. The concrete bodies were supplied by the project partner Hering Bau. The PV modules were manufactured at the Fraunhofer CSP, they consist of two 4 mm glasses, of which the rear glass was printed in black. The encapsulation material is EVA and there are two half cells per module. The four modules are connected in series. The PV modules were bonded into the concrete elements with Sikasil SG-20 adhesive. The module temperature sensors are attached on the backside of the laminate and the concrete sensor is in the core of the concrete (Figure 2).

Figure 1: SOLARcon demonstrator at Fraunhofer CSP test field

2.2 Measurement equipment

The continuously recorded measurement data includes in particular the environmental values such as temperature at the back of the solar module, the temperature of the concrete element, the ambient temperature, the irradiation in the plane of the solar cells and the wind speed. As a data logger a Raspberry PI with a Raspbian distribution was used. Temperature measurement is performed by digital sensors DS18B20 from Dallas/Maxim. The sensors are factory calibrated with an accuracy of +/-0.5°K. The connection is made via the so-called 1-wire protocol which is natively supported by the data logger. The measured values are measured every minute and written to a sqlite3 database.

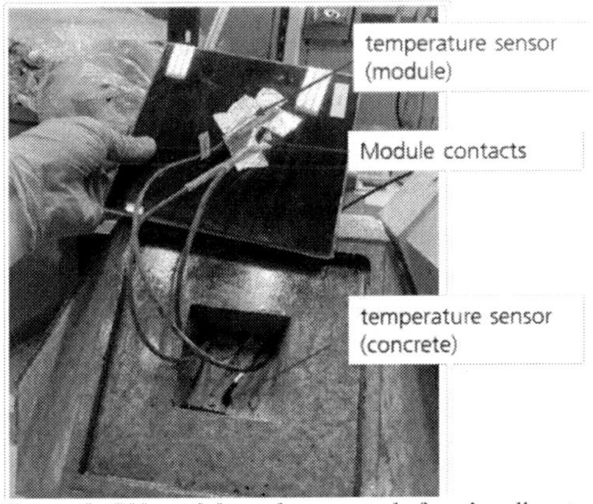

Figure 2: PV module and concrete before bonding together with temperature sensors (black cables)

An anemometer with pulsed output is used to measure the wind speed. The pulses are interpreted into measurement data by a microprocessor board Arduino-Uno with a shield and program code made for the purpose. The irradiation values of the sensor used are measured by the same measuring board measuring the I-V-curves.

The modules used consist of 2 halved solar cells each, which are connected in series. All modules are connected in series and connected to the MPP characteristic curve measuring system. The measuring board for determining the current-voltage values of the solar cells regularly determines the complete I-V-curve of the series connection. The usual characteristic data points are extracted from the curves: short circuit current, open circuit voltage, fill factor and peak power as well as current and voltage at maximum power point. To keep the solar cells in optimal load between the recording of the measurement curves, the last determined maximum power point is set as the operating point at the electronic load. The generated energy is dissipated as heat via a heat sink. In the present setup, the measurement data is saved to the memory once a minute.

2.3 Measurement data

To optimize computational efficiency and reduce the complexity of the simulation, the recorded experimental data are processed. Initially collected at minute intervals, these data points are averaged to half-hourly values. This data reduction allows for effective modeling of temperature trends and fluctuations without an extensive number of simulation steps.

Figure 3 shows the irradiances averaged over half an hour for two selected days. Figure 4 gives the I-V curves at the time of maximum irradiation on these days. These data can later be used for a yield evaluation, which is not part of this paper. The data of the open circuit voltage were used for the plausibility check of the temperature measurement of the modules.

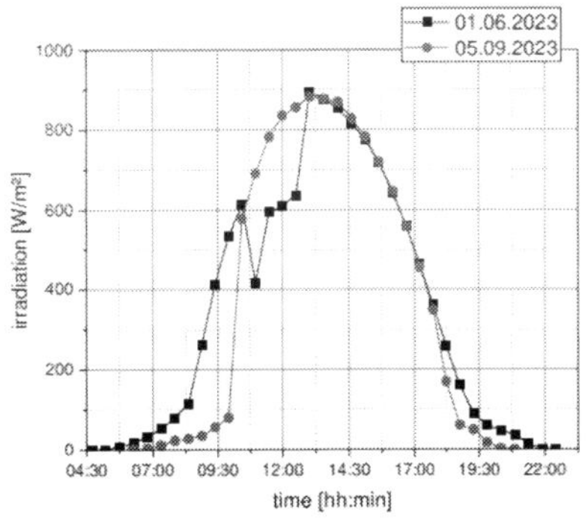

Figure 3: Irradiation curves on the days selected for the simulation

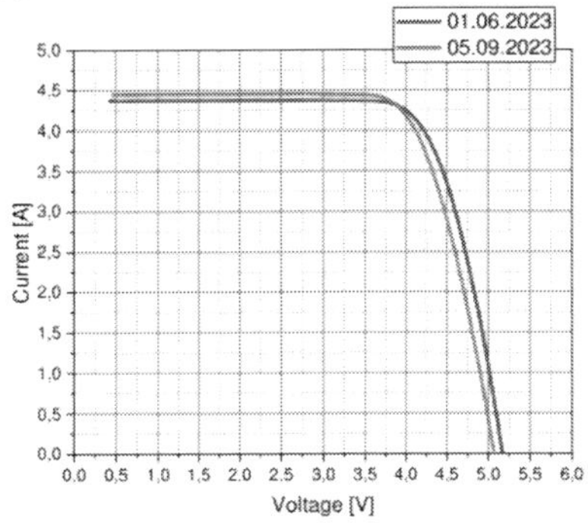

Figure 4: I-V curves at the time of maximum irradiation in Figure 3

3 SIMULATION

Accurately simulating the interaction between concrete, PV modules, and various environmental factors is a complex task. This section outlines the key components and processes of the simulation, providing insight into how the model captures the dynamic behavior of the BIPV system.

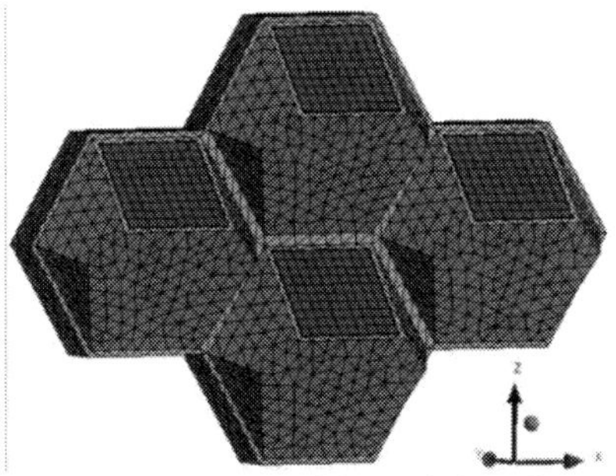

Figure 5: Meshed simulation model, consisting of concrete, adhesive, PV module (2x4 mm glass, encapsulation material and cell)

3.1 Geometric model

The meshed 3D-model of investigated structure is shown in Figure 5. Since the temperature simulation does not have such high demands on the mesh compared to structural simulation, a comparatively coarse mesh was used to optimize the computing time. Ideal thermal contact between the individual components is assumed, i.e. high adhesion and wetting of concrete and module with the silicone.

3.2 Boundary conditions

The primary objective of the simulation is to replicate the thermal dynamics of the BIPV system throughout a typical day. To achieve this, a transient simulation approach is employed. Instead of considering a full 24-hour period, the simulation focuses on the specific time frame during which the irradiation sensor records the first and last values. This time frame is crucial for capturing the critical variations in temperature and performance.

A steady state starting point is required for the transient simulation. This point uses the temperatures at the beginning of the day when there is an almost homogeneous temperature distribution.

One of the key components of the simulation is modeling the heat generated within the solar cells due to electrical efficiency. As it is known solar cells convert only a portion of the incident sunlight into electrical energy, with the rest being dissipated as heat. In the model the generated heat is a function of irradiance and temperature in the previous step, to account for temperature dependence of energy conversion. Approximately 80 % of the incident energy is converted to heat [6]. This heat source is introduced into the simulation to account for the thermal effects of solar energy conversion.

The absorbed radiation on the concrete elements must also be considered to simulate the temperature behavior within the BIPV system. As sunlight strikes the concrete surface, it imparts thermal energy, leading to temperature increases.

In addition to heat generation, the simulation considers heat transfer through convection. Convection is the process by which heat is exchanged between the BIPV system and its surroundings, primarily influenced by ambient temperature and wind speed. To avoid time-consuming flow simulations, the heat transfer coefficient is estimated based on models from the literature [7].

3.3 Material data

The material data is another important component of the simulation, without a correct material behavior it is not possible to achieve reliable results with the simulation. Most of the material data comes from the project partners or suppliers, as well as from Fraunhofer CSP's own material databases. Some of the material data could not be provided by the partners, especially for concrete and adhesives. For this purpose, online material sources were used and checked with inhouse data. Table 1 lists the material parameters relevant for the thermal simulation for all materials used.

Table 1: Material data for the materials used in the simulation

Material	Density [kg/m³]	Thermal conductivity [W/m*K]	Specific heat [J/kg*K]
Concrete [8]	2300	1.35	840
Glass	2500	1	720
Adhesive (Sikasil SG-20) [9]	1370	0.29	1175
EVA	950	0.22	2250
Silicon	2329	150	703

3.4 Evaluation of simulation results

To validate the simulation results against real-world data (see measurement setup in section 2), the locations where temperature sensors were installed in the experimental setup are replicated within the simulation model. These sensors include two embedded in the concrete and two placed on the back of PV modules. The temperatures recorded in the experiment are summarized and averaged so that one temperature value is presented for PV module and concrete. The temperatures evaluated for the simulation and shown in the diagram are taken at the exact same points as in the experiment.

4 RESULTS

Two different days were selected for simulation: one in early summer with significant cloud coverage in the morning and another in late summer with clear skies. In Figure 6 the comparison of simulated and measured temperature curves is shown. The simulation results effectively captured the daily temperature trends of the BIPV system for both days. On both days, the module temperature increases as soon as the sun shines on the structure. On the clear day the trend of temperature is monotonous rising and falling. As expected, the temperature development in the concrete is slower and the maximum temperature is reached about 2.5 h later. On the other day during periods of cloud cover, noticeable drops in temperature were observed, indicating the impact of weather conditions on the thermal behavior of the system. It can be seen, that the module reacts much faster to the change in environmental conditions than the concrete. This confirms the assumption of a comparatively high thermal inertia of the system. Figure 7 shows the simulation result at the time of the maximum module temperature (approx. 2 p.m.) as a contour plot over the geometry. Here it can be seen very clearly that the temperature decreases significantly starting from the module over the concrete.

While the simulations generally aligned with the experimental data, slight temperature differences were observed. Specifically, the PV modules tended to be warmer in the simulation, while the concrete exhibited slightly lower temperatures. These discrepancies could be attributed to differences in material properties and real-world behavior between the simulated and actual systems. Nevertheless, the presented simulation model demonstrated the ability to accurately simulate the thermal behavior of BIPV systems based solely on available weather data, material parameters, and electrical characteristics.

The findings from the experimental validation of the simulation model highlight its accuracy and versatility for various BIPV applications. The model's non-geometry-specific nature allows for its applicability to different BIPV system configurations and geometries. This makes it a valuable tool for product development, performance evaluation, and optimization in the field of Building-Integrated BI and Vehicle Integrated VI Photovoltaics.

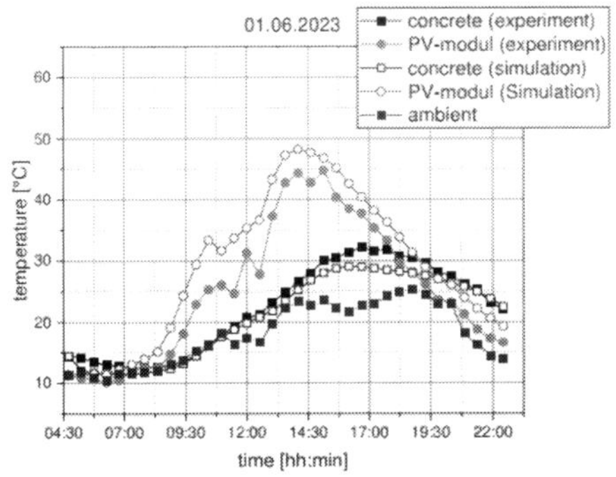

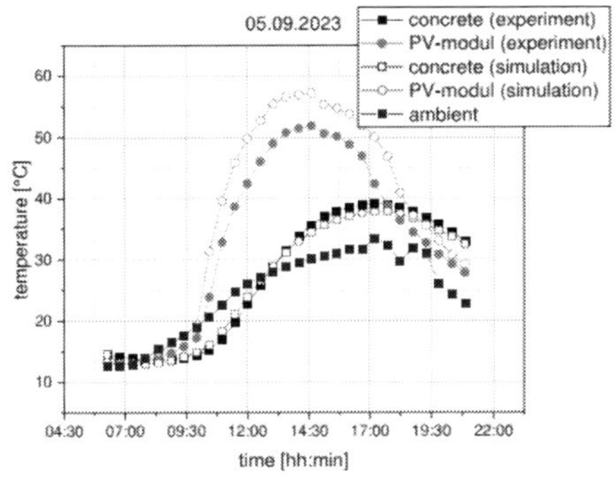

Figure 6: Comparison of the simulated temperatures for the concrete body and the PV module (outlined) with the temperatures recorded in the demonstrator (filled) and the ambient temperatures for 01.06.23 (top) and 05.09.23 (bottom)

Figure 7: simulated temperature contour plot on 01.06.2023, at the time of maximum temperature (approx. 2 p.m.)

5 CONCLUSIONS AND OUTLOOK

In conclusion, the experimental study conducted as part of the SOLARcon project successfully validated a numerical simulation model for predicting the thermal behavior of the demonstrator BIPV system. The simulations captured the daily temperature trends and demonstrated the impact of weather conditions on the system's thermal behavior. While slight temperature differences were observed between the simulations and experimental data, the model proved to be accurate and versatile for various BIPV applications.

One of the strengths of this simulation model is its versatility. It is not restricted to a specific BIPV system geometry or configuration. Instead, it can be applied to a range of BI or VIPV applications, making it a valuable tool for predicting the thermal behavior of various integrated photovoltaic systems.

While the simulation methodology presented here is a significant step forward in accurately predicting BIPV system behavior, it is not without limitations. Future research can focus on refining material parameters and incorporating more detailed models of electrical efficiency and heat transfer. Other material stacks, including all polymer systems, can be assessed, and optimized in geometry, size, mounting configuration. Furthermore, it can be reversely used to better predict the energy yield of new applications.

6 ACKNOWLEDGEMENTS

The authors thank the Federal Ministry of Economics and Energy for the financial support of the work in the project "SOLARcon" (grant no. 03EE1016B).

7 REFERENCES

[1] E. Skoplaki, A.G. Boudouvis, J.A. Palyvos, Solar Energy Materials & Solar Cells 92 (2008) 1393
[2] Ulli Zeller, The influence of air flow on the temperature distribution within crystalline solar modules, 28th EU-PVSEC, 2013
[3] M. Pander et al., Lifetime estimation for solar cell interconnectors, 28th EU-PVSEC, 2013
[4] M. Pander et al., Digital Prototyping – application of numerical methods in module development, 36th EU-PVSEC, 2019
[5] P. Schenk et al., Analysis of the thermomechanical behaviour of concrete façade elements with integrated photovoltaic modul[es], 38th EU-PVSEC, 2021
[6] Peter Würfel, Physik der Solarzelle, Spektrum Akademischer Verlag / Springer Spektrum, 2000
[7] VDI Gesellschaft: VDI-Wärmeatlas. Wiesbaden: Springer Berlin Heidelberg, 2005.
[8] EN, 1992-1-2: design of concrete structures. Part 1-2: general rules—structural fire design," Eurocode 2, European Committee for Standardization, Brussels, Belgium, 2004
[9] Sika AG, https://deu.sika.com/
[10] Goodfellow GmbH, https://www.goodfellow.com/de/p/si30-rd-000120/silicone-elastomer-rod

IMPACT OF SOILING ON TRANSPARENCY OF DIFFERENT GLASS TYPES: A COMPARATIVE STUDY

Sara El Hassani, Donat Hess, Christof Bucher, Matthias Burri
Bern University of Applied Sciences (BFH), Jlcoweg 1, 3400 Burgdorf, Switzerland
sara.elhassani@bfh.ch, donat.hess@bfh.ch, matthias.burri@bfh.ch, christof.bucher@bfh.ch

To improve the optical properties of photovoltaic (PV) modules, many manufacturers use new glass surface structures that reduce glare. However, these modifications can increase susceptibility to soiling. This research presents a new method for assessing soiling by positioning glasses without solar cells on a reference PV module. With this method, the soiling of the glasses can be measured without the systematic measurement uncertainties from the power or irradiance measurement negatively affecting the measurement accuracy. To demonstrate the method, five glass types are tested at four tilt angles at three locations. In this paper, the preliminary results are presented together with the soiling data after two months at a selected site.

1 INTRODUCTION

Photovoltaic (PV) modules are a widely utilised method for converting solar energy into electricity, making solar energy a highly promising renewable energy source. However, the accumulation of dust and other pollutants on the surface of the glass is a major challenge for PV systems. This accumulation significantly reduces the transmissivity of the glass cover, resulting in a decrease in the amount of solar irradiation received by the solar cells. As a result, the power production performance of the solar PV modules is degraded, reducing their efficiency. Large differences in soiling effects are reported depending on the type of system and location. In some regions, depending on the local environmental conditions and the design of the PV module, this soiling can cause energy losses of up to 8.41% in its peak power output compared to a cleaned module [1]. This problem is particularly prevalent in dusty countries, where weather conditions can negatively affect the productivity of photovoltaic cells [2]. In desert areas, where airborne dust is prevalent, the productivity of PV modules is further reduced, despite their high irradiation and ample space for construction [1].

Several previous studies have investigated the effects of soiling on solar panels and the optimisation of solar panel performance in different environments. For example, the study conducted by Hassan Z. Al Garni [3] examined the impact of soiling on solar photovoltaic (PV) performance in Saudi Arabia. The study evaluated the performance of solar PV systems under soiling in different regions of Saudi Arabia, finding that PV performance can be reduced by 2% to 50%, and even up to 20% during a single sandstorm. The study highlighted several mitigation techniques, including regular cleaning, the use of bi-facial PV solar panels with solar trackers, and robotic cleaning systems. The study also characterised the size and composition of dust particles in different locations in Saudi Arabia and concluded that dust accumulation and cleaning costs are not significant barriers to large-scale, cost-effective solar PV deployments in Saudi Arabia, particularly in the central region.

Ramadan J. Mustafa et al [4] conducted a study to examine the impact of four environmental factors (dust accumulation, water droplets, birds' droppings, and partial shading) on the performance of photovoltaic (PV) systems. This is the first study to investigate all four factors simultaneously. The study found that water droplets had a positive impact on PV performance, reducing the temperature of the PV panel and improving power output by at least 5.6%. On the other hand, dust accumulation reduced power output by 8.80% and efficiency by 11.86%. The study highlights the importance of considering all environmental factors that can impact PV performance when designing and operating PV systems.

J.K. Kaldellis et al [5] conducted a study on the effect of air pollution on the performance of photovoltaic (PV) panels. They analysed the impact of three representative air pollutants - red soil, limestone, and carbonaceous fly-ash particles - on the energy performance of PV installations. They found that the deposition of solid particles on PV panels due to atmospheric air pollution resulted in a considerable reduction in PV energy performance, with the extent of reduction being dependent on the particles' composition and source. The study developed a theoretical model to predict the impact of regional air pollution on PV performance, which was validated using experimental results obtained in an urban environment with aggravated air pollution. The findings of this study highlight the need to consider air pollution as a factor that affects the performance and maintenance of PV systems, especially in urban areas.

Bouchra Laarabi et al [6] conducted a study on the characterisation of soiling on glass samples exposed to real conditions in the Rabat-Sale-Kenitra region in Morocco. The study found that soiling is a localised phenomenon, with deposition densities ranging from 0.37 g/m2/month to 2.86 g/m2/month. The effect of soiling on glass transmittance showed a reduction in the range of 1.05%/month to 10.04%/month, depending on the tilt angle and exposure period. There is a positive and linear correlation between the mass density of the deposit and the loss of transmittance, which is influenced by the period and location of exposure. The type of dust may also affect the degree of transmittance reduction. The particle size distribution was found to be prevailing in the range of 3-14 μm, with most particles having a regular shape. The study also observed enhanced cementation on the samples located near the ocean due to the presence of salt crystals on the surface.

The contemporary realm of photovoltaic research underscores a pressing imperative to delve deeper into the ramifications of soiling on solar panels across a spectrum of environmental paradigms, and more pertinently, the interplay between glass surface attributes and soiling dynamics. In the quest to optimise optical characteristics for Building Integrated PV (BIPV) and concurrently

10.4229/EUPVSEC2023/3AV.3.25
020226-001

mitigate glare-associated challenges, numerous manufacturers are pioneering innovative glass surface architectures. However, the effects of these changes on soiling, particularly the accumulation of dust an on the glass facades and roofs, are still insufficiently understood.

To fill this gap, the present study aims to investigate the relationship between glass surface properties and the propensity for particle accumulation on PV modules. A series of five different glass variants have been positioned at different tilt angles in three different locations.

Subsequent evaluations include a assessment of the effectiveness of the PV glass, measured using a mini-module to assess glass transmittance. In soiling research, where two sites are located next to each other, it is common for researchers to name geographically distinct sites, each symbolic of unique climatic peculiarities [[7], [8], [9]].

2 OBJECTIVE

The results of the study are aimed in particular at specialists and companies who want to use new glass on photovoltaic modules and would like to know more about their tendency to soil. Two goals are set in the study:

1. A measurement procedure is to be developed, set up and tested with which the soiling of different types of glass can be tested under various conditions.
2. The first results shall quantify how much the tilt angle and the glass surface of a PV module influences the soiling.

The second objective is to be investigated on a larger scale in a later publication. This document only deals with the measurement system and first results as a proof of concept.

3 METHODOLOGY

3.1 Basic concept

The study process follows the following steps:

1. The transmission of all glass samples is initially measured (clean glasses)
2. The glasses are installed in different locations, inclinations and weathered.
3. The transmission is measured again periodically.
4. The degree of soiling (power reduction) is measured in percent from the loss of transmission.

The disadvantage of this concept is that the energy yield, which is proportional to the integral of soiling over time, cannot be determined. For this purpose, the decrease in transmission can be determined more precisely and attributed to the soiling with little error tolerance. Especially in the case of differently oriented glasses, it would not be possible to make an exact statement about soiling solely by monitoring the energy yield.

3.2 Transmission measurement

The measurement of irradiance (W/m^2) and irradiation (kWh/ m^2) is associated with large uncertainties even with high-quality measuring instruments. Similarly, the reduced yield or reduced power of a PV module cannot be clearly assigned to a cause using simple measurement and interpretation methods. For this reason, only relative measurements are used in this paper. A single PV mini-module is used as a sensor, which is covered with a cleaned reference glass for each measurement location and then operated for a short time (approx. 15 minutes) under stable and constant irradiation and temperature conditions. Thereby, spectral shifts due to weather conditions might influence the result, but the influence is proportional to the expected energy yield of a real PV module.

Between each individual power measurement of the mini-module with a glass sample, the power measurement of the mini-module with the reference glass is repeated. Thus, 7 x 4 + 1 = 29 reference measurements are obtained per measurement session. The irradiation-corrected variation of these basically identical measurements is considered as stochastic measurement error (measurement noise).

3.3 Preparation of measurements

In order to obtain measurement results that are as precise as possible, the following points are taken into account:

- The glass samples are aligned perpendicularly to the solar radiation for the measurement
- Measurements are only carried out with an irradiation of at least 900 W/m2.
- Measurements are only taken when the sky is cloudless.
- Measurements are only taken into account if the solar irradiation does not vary by more than +/- 20 W/m^2 during the entire measurement.
- The reference glass is cleaned on the front side before each measurement campaign. The back side is not cleaned or is cleaned as well for all sample glasses.

3.4 Initial measurements

Before installing the glass samples, initial transmission measurements are carried out on the glasses. The subsequent soiling of the glass is determined based on these initial measurements.

To determine the transmission of a given glass sample S1 relative to the transmission of a reference glass Ref1, the following initial measurements are carried out:

- the power $P_{Ref1}(t-1)$ of the mini-module covered with the reference glass Ref1 are measured.
- The power $P_{S1}(t)$ of the mini-module covered with sample S1 are measured.
- The power $P_{Ref1}(t+1)$ of the mini-module covered with the reference glass Ref1 are measured again.

Where t denotes the time step or the number of measurements done.

Subsequently, the weighted mean value of the reference measurement is determined in order to compensate for any changes in irradiation between measurement 1 and measurement 2:

$$P_{Ref1}(t) = \left(\frac{1}{\Delta t_1} P_{Ref1}(t-1) + \frac{1}{\Delta t_2} P_{Ref1}(t+1)\right) \cdot \frac{1}{\frac{1}{\Delta t_1}+\frac{1}{\Delta t_2}}$$

Δt_1 and Δt_2 are the time difference between the $P_{Ref1}(t-1)$ and $P_{Ref1}(t+1)$ respectively and the $P_{S1}(t)$ measurement. They are used as weighting factors to take into account the fact that the reference measurement that is made closer in time to the sample measurement is closer to the conditions under which $P_{S1}(t)$ is measured. This means that changing irradiation conditions can be partially compensated for.

The relative transmission of the glass sample is then calculated as follows:

$$T_{S1}(t) = P_{S1}(t) / P_{Ref1}(t)$$

With this measurement procedure, the relative transmission before installation compared to the reference glass is determined for each glass sample. For future soiling measurements, these transmission values will be used for comparison.

Because all 7x4 = 28 glass samples are measured directly one after the other, all reference measurements can be used for the glass sample measured immediately before as well as for the glass sample measured immediately after.

3.5 Soiling measurements

The procedure for measuring soiling is identical to the initial measurements. The only difference is that the reference glass Ref1 is cleaned before the measurements start. Because the reference glass is weathered like all other glass, any changes in the glass (e.g. turbidity) are not erroneously evaluated as soiling, but are compensated for with the help of the relative measurements.

By comparing the transmission of the samples measured in the soiling measurement with the initial transmission measurements $T_{S1}(t)$, the extent of soiling on each glass sample can be determined. The difference between the initial transmission values and the soiling measurements indicates the level of soiling caused by soiling on the glass samples.

3.6 Soiling calculation

Based on the measurements, the degree of soiling of the glass samples S_{S1} can be determined.

$$S_{S1} = 1 - T_{S1,soiled} / T_{S1,initial}$$

3.7 Estimation of measurement error

The uncertainty of the relative transmission ΔT_s is estimated by the standard deviation of repeated measurements of an identical glass. For this purpose, the reference glass $Ref1$ is used, since the most measurements are available for this glass. $T_{ref}(t)$ is determined as:

$$T_{ref1}(t) = P_{ref1}(t) / P'_{ref1}(t)$$

where $P'_{ref1}(t)$ is the time-weighted average of $P_{ref1}(t-2)$ and $P_{ref1}(t+2)$. The difference of four-time steps ($t-2$ to $t+2$) is since the measurements in between $P_s(t-1)$ and $P_s(t+1)$ are sample measurements and not reference measurements. Due to the larger time intervals, it can be assumed that this procedure leads to a conservative estimation of the uncertainty.

A disadvantage of this error estimation is that it cannot be assumed that the obtained T_{ref1} values are normal distributed, since T_{ref1} is not a measured value but a calculated value. To be able to estimate to what extent T_{ref1} is normal distributed, the distribution is visualised in a QQ plot Figure 4.

The uncertainty of soiling ΔS_{s1} is calculated based on the Gaussian error propagation:

$$\Delta S_{s1}^2 = \frac{\partial S_{s1}(T_{s1,soiled}, T_{s1,initial})}{\partial T_{s1,soiled}}^2 \Delta T_{s1,soiled}^2 + \frac{\partial S_s(T_{s1,soiled}, T_{s1,initial})}{\partial T_{s1,initial}}^2 \Delta T_{s1,initial}^2 = \left(-\frac{1}{T_{S1,initial}}\right)^2 \Delta T_{s1,soiled}^2 + \left(\frac{T_{S1,soiled}}{T_{S1,initial}^2}\right)^2 \Delta T_{s1,initial}^2$$

This error propagation calculation is used to compute the error bars of the soiling losses in **Fehler! Verweisquelle konnte nicht gefunden werden.**.

3.8 Used glass samples

In this study, a total of five glass types or glass surfaces are to be examined. Table I shows the different glass types.

Table I: Glass types used

Glass type (Abbreviation)	**Description**
Standard Solar Glass (SSG1, SSG2, SSG3)	Standard solar glass is used in most PV modules. The reflection has some beam spread. The glass is highly transparent and has a relatively smooth surface. For reference and comparison reasons, 3 samples of this glass type are used.
Reference Glass (Ref1)	The reference glass is a SSG which is cleaned before each measurement session.
Float Glass (FG)	Float glass looks like window glass. The surface is perfectly flat. The reflection does not show any beam spread.
Satinated Glass (SatG)	This glass has a matt surface that strongly diffuses the reflected light. It is used especially in BIPV to achieve a matt surface.
Sandblasted Glass (SBG1, SBG2)	Float glass was sandblasted with two different grain types (glass pearls and Korund Biloxit) in the PV laboratory of the BFH. The surface of the glass is roughened and becomes matt.
Coated Glass (CG)	In this work, glass with a special anti-glare coating is to be used. However, at the time of going to press, the coated glass could not yet be installed.

While sandblasted glass is hardly used in practice, the use of standard solar glass, float glass and satinated glass in particular is common today. Figure 1 shows four example modules with similar glass as used in the study.

10.4229/EUPVSEC2023/3AV.3.25
020226-003

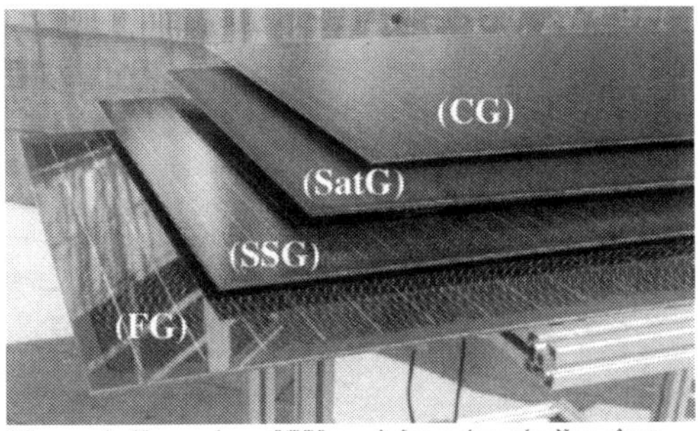

Figure 1: Examples of PV modules using similar glass

3.9 Mounting system

The glass samples used in the study were mounted on a specially designed fixture that ensured that the samples were tilted at the optimal angles of 1.5°, 10°, 30° and 75° for the tests. The mounting structure also prevented water from flowing from one glass sample to the other, which could potentially lead to additional cleaning of the lower glass samples.

To ensure the stability of the fixture and reliable results throughout the testing process, the fixture was installed on a stable foundation that consisted of either garden slabs or concrete slabs, depending on the location. This foundation provided a secure base for the experiment and prevented potential disturbances that could affect the testing process.

Figure 2 shows the planned setup of the glass samples mounted on the specially designed structure and illustrates how the structure was installed to maintain the angle of inclination and prevent water from flowing between the samples. These measures were taken to ensure the accuracy of the study results and minimise potential errors that could result from unstable or improperly mounted samples. Figure 3 shows the first installation made in Hindelbank.

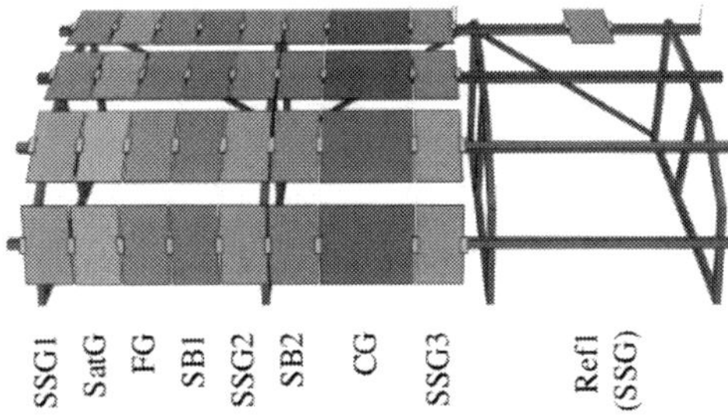

Figure 2: Mounting structure design

Figure 3: Test bench installed in Hindelbank

3.10 Locations

The installations are made at three different locations in Switzerland. In Table II, the different locations are described and classified.

Table II: Locations for the test bench installation

Location	Description
Burgdorf, Gsteig	Burgdorf is a small town in a rural setting. The test stand is located at around 570 m above sea level on the flat roof of a building belonging to the Bern University of Applied Sciences. The soiling level is estimated to be rather low. 
Hindelbank	The test stand is installed at 540 m above sea level on a farm in a rural area. The soiling level is very high due to the farming operation. 
Thun, Gwatt	The test stand is installed on an industrial building at 580 m above sea level in Thun. The soiling level is considered to be low. 

4 RESULTS

4.1 Uncertainty

The measurement uncertainty is determined by examining the deviation between measurements of the identical glass Ref1. Figure 4 shows the QQ plot for the initial measurements and the first soiling measurements in Hindelbank.

It can be seen from the QQ plots that the calculated deviations largely follow a standard normal distribution. However, very large and very small deviations are no longer exactly in the expected normally distributed range. Also, the measurement error for the soiling measurements in Hindelbank are larger than the measurement errors for the initial values.

From this it can be deduced that the error calculation is fundamentally correct for most values, but for large and small measurement values, calculated soiling becomes inaccurate. In particular, the outlier in Figure 7 cannot be explained by the calculated and shown error range. For illustration purposes, the individual calculated values are also listed in Figure 5.

The measurement uncertainty calculated on the basis of the values shown in the QQ plots are shown in Table III. The large difference in measurement uncertainty between the two measurements can be explained by the less constant weather conditions in Hindelbank.

Table III: Measurement uncertainty

Measurements	Measurement uncertainty
All initial measurements, all sites	σ = 0.00269 2σ =0.005392
Soiling measurements, Hindelbank only	σ = 0.00864 2σ =0.01728

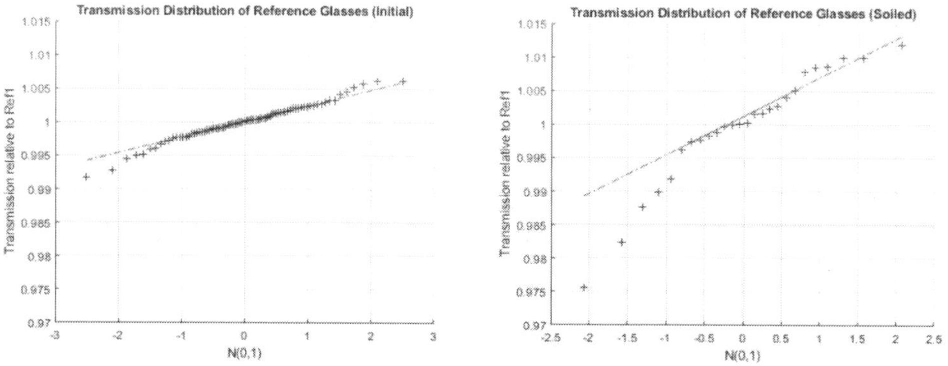

Figure 4: QQ plots for initial measurements (left) and soiling measurements in Hindelbank (right).

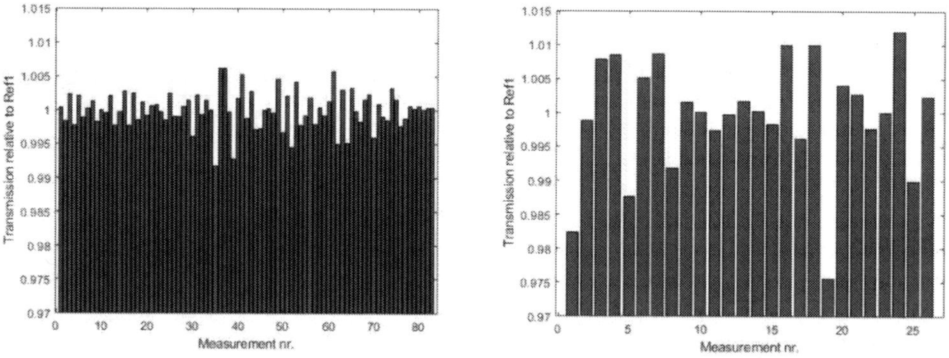

Figure 5: Calculated transmission values of the first measurements (left) and the first measurements in Hindelbank (right). The baseline values include the reference module measurements from all three locations, whereas the Hindelbank measurements only include the measurements from the glasses installed in Hindelbank.

4.2 Initial measurements

The initial measurements show the differences in the transmission of the tested glasses when new (Figure 6 and Table IV). The specified range refers to the spread of the measured values. The measurement uncertainty can be stated as 2σ =0.005392 according to Table III. The following findings can be made

- The standard solar glass SSG has the highest transmission when new. The SSG is given as a reference lens with a value of 1.0.
- The frosted SatG glass is just behind the reference glass with a transmission of 0.978. The variation between the frosted glasses is small.
- The float glass FG has an even lower transmission of 0.949.
- The transmission of the sandblasted glasses is significantly lower than that of the other glasses at 0.876 and 0.855. In addition, the dispersion between the individual sandblasted glasses is two to four times as high as the dispersion of the reference glass.

Table IV: Initial relative transmission measurement results (measurement precision: 2σ =0.005392)

Glass type	Transmission relative to SSG
SSG	1.000 ± 0.005
SatG	0.978 ± 0.007
FG	0.949 ± 0.008
SB1	0.876 ± 0.014
SB2	0.855 ± 0.024

10.4229/EUPVSEC2023/3AV.3.25
020226-005

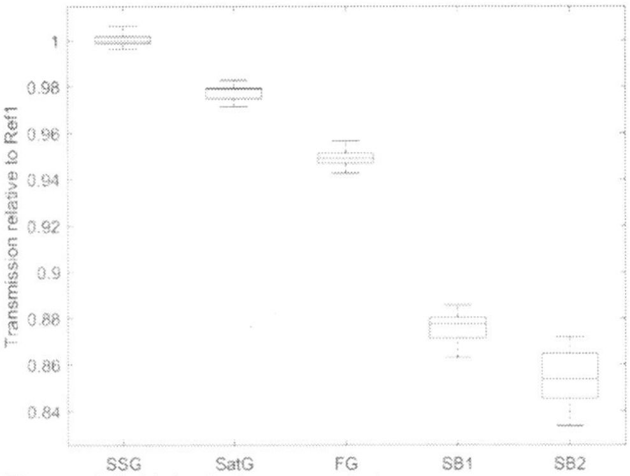

Figure 6: Initial relative transmission measurement results (measurement precision: 2σ =0.005392, not indicated in the box plot).

4.3 Soiling measurements

The aim of this project is to carry out regular soiling measurements at all three locations. The results presented here refer only to the first, initial soiling measurements at the Hindelbank site. Because the soiling pressure is very high due to the agricultural operation, soiling levels of up to 6% can be measured after around two months of exposure.

Figure 7 and Table VI show the results of the soiling measurements. Table V summarises all values of a specific inclination angle. The value at 1.5° inclination of the SSG3 glass was not taken into account because it is an outlier that arose due to a measurement error. Table VI shows the average soiling across all angles of inclination for each glass type.

The following observations can be made:

- The steeper the glasses are installed, the less soiling there is.
- The differences are particularly high between standard solar glass and frosted glass.
- Sandblasted glass, but also float glass, shows less dependence on the angle of inclination when it comes to soiling than other glasses.
- The error range is large relative to the soiling. There is therefore hardly any trend to be identified, particularly for float glass and sandblasted glasses.

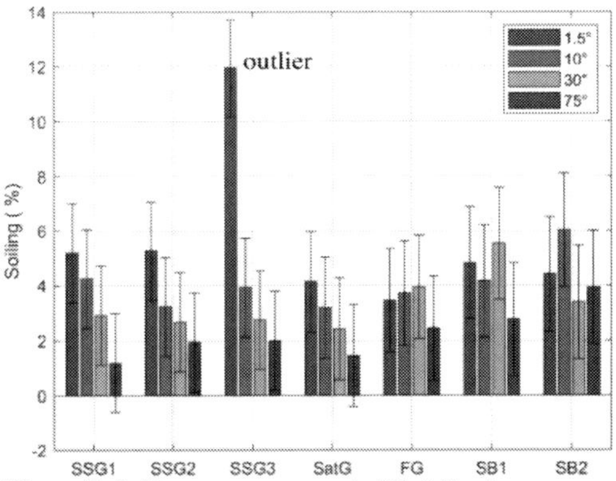

Figure 7: Soiling measurements in Hindelbank (measurement precision of transmission measurements: 2σ =0.0185—0.0209, calculated for the soiling ratio and indicated in the error bars)

Table V: Soiling measurements in Hindelbank by tilt angle (measurement precision: 2σ =0.01728)

Tilt angle	Mean soiling (%)
1.5°	4.55 ± 0.78 (without SSG3,)
10°	4.08 ± 0.72
30°	3.37± 0.72
75°	2.24 ± 0.72

Table VI: Soiling measurements in Hindelbank by glass type (measurement precision: 2σ =0.01728)

Glass type	Mean soiling (%)
SSG	3.37 ± 0.55 (without SSG3, 1.5°)
SatG	2.80 ± 0.93
FG	3.39 ± 0.95
SB1	4.33 ±1.02
SB2	4.44 ± 1.04

The errors shown in Table V and Table VI is calculated using the Gaussian error propagation.

5 CONCLUSION

In this study, a method for the systematic evaluation of the soiling tendency of different solar glass types was presented and demonstrated. Although the method is based on relative measurements, it should be noted that the measurement uncertainties, which range from 1.85 % to 2.09 %, exceed the soiling values in some cases. Despite this limitation, the data show a clear trend. In particular, it was found that after two months of soiling, glasses with a strong tilt accumulate up to three times less dirt than glasses with a slight tilt. This effect is more pronounced for glasses with a smooth surface than for those with a matt surface. For glasses with a matt surface, the soiling tendency seems to be less influenced by the inclination angle and corresponds more to the soiling behaviour of glasses with a less smooth surface.

6 OUTLOOK

This project is scheduled to continue in 2024. The focus is then on repeatedly measuring the degree of soiling of the glasses. The preliminary results shown here should be validated and generalised.

7 ACKNOWLEDGEMENT

This publication was written as part of the AURORA project funded by Innosuisse and with the support of the Swiss Government Excellence Scholarships program ESKAS. The authors thank you for the support.

8 REFERENCES

[1] Dida, M., Boughali, S., Bechki, D., Bouguettaia, H., 2020. Output power loss of crystalline silicon photovoltaic modules due to dust accumulation in Saharan environment. Renew. Sustain. Energy Rev. 124, 109787. https://doi.org/10.1016/j.rser.2020.109787

[2] Chaichan, M.T., Mohammed, B.A., Kazem, H.A., 2015. Effect of pollution and cleaning on photovoltaic performance based on experimental study. Int. J. Sci. Eng. Res. 6, 594–601. https://doi.org/10.13140/RG.2.1.2928.2725

[3] Al Garni, H. Z. (2022). The Impact of Soiling on PV Module Performance in Saudi Arabia. Energies, 15(21), 8033.

[4] Mustafa, R. J., Gomaa, M. R., Al-Dhaifallah, M., & Rezk, H. (2020). Environmental impacts on the performance of solar photovoltaic systems. Sustainability, 12(2), 608.

[5] Kaldellis, J. K., Fragos, P., & Kapsali, M. (2011). Systematic experimental study of the pollution deposition impact on the energy yield of photovoltaic installations. Renewable energy, 36(10), 2717-2724.

[6] Laarabi, B., El Baqqal, Y., Dahrouch, A., & Barhdadi, A. (2020). Deep analysis of soiling effect on glass transmittance of PV modules in seven sites in Morocco. Energy, 213, 118811.

[7] Herrmann, W., Schweiger, M., Tamizhmani, G., Shisler, B., & Kamalaksha, C. (2015). Soiling and self-cleaning of PV modules under the weather conditions of two locations in Arizona and South-East India. IEEE 42nd Photovoltaic Specialist Conference (PVSC 2015).

[8] Dahlioui, D., Laarabi, B., & Barhdadi, A. (2019). Investigation of soiling impact on PV modules performance in semi-arid and hyper-arid climates in Morocco. Energy & Sustainable Development, 51, 32-39. https://doi.org/10.1016/j.esd.2019.05.001

[9] Alami Merrouni, A., Lopes, D., Alae, A., Silva, H.G., Bennouna, E.G., et al. (2019). A comparative study of soiling on solar mirrors in Portugal and Morocco: Preliminary results for the dry season. AIP Conference Proceedings, 2126, 220001.

[10] Kipp & Zonen, Pyranometers v. Reference Cells for PV Installations, solar and direct solar radiation, https://www.kippzonen.de/, downloaded September 2023

MODULE EXTERNAL QUANTUM EFFICIENCY MEASUREMENTS AS A METHOD FOR DEFECTS ANALYSIS IN BIFACIAL PV MODULES.

Ahmad Alheloo, Ali Almheiri, Sagarika Kumar, Baloji Adothu, Gerhard Mathiak, Vivian Alberts
Dubai Electricity and Water Authority
ahmad.ibrahim@dewa.gov.ae, ali.almheiri@dewa.gov.ae, sagarika.kumar@dewa.gov.ae
DEWA R&D Center, Dubai, UAE.

ABSTRACT: This paper aims to further expand on defects analysis of PV modules by introducing a module-level External Quantum Efficiency (m-EQE) measurement technique onto bifacial solar modules exposed to harsh desert conditions for 7 years. The technique is used to measure the EQE of a cell within the module with a spatial resolution of 5 cm x 2.5 cm, on degraded bifacial modules. Results show a clear difference in EQE between the rear and front surfaces, with the rear surfaces showing lower EQE in all four cases (two cells per module) at lower wavelengths, and higher EQE in the rear surface at higher wavelengths and the infrared range. Results are correlated with other testing methods, including current-voltage (IV), Electroluminescence (EL), and Ultraviolet Fluorescence (UV-F), which showed a current loss that was attributed to encapsulant degradation seen in UV-F patterns and EQE loss in lower wavelengths. Module-EQE is a non-destructive technique that provides critical insight into the spectral behavior of the modules. Adding module EQE as part of the intermediate testing technique for defect analysis test cycles is suggested, especially to progressively observe changes under a desert-based test cycle.
Keywords: Module-EQE, defects analysis, bifacial EQE, spectral response, encapsulant degradation.

1 INTRODUCTION

Solar energy, especially in the form of photovoltaics (PV), is slowly becoming a dominant source of energy production globally. This rise in solar energy production, combined with the continued decline in the cost of solar energy, makes PV an essential aspect of meeting global electricity demand. As such, it is crucial to thoroughly understand the nature of the various defects and degradation mechanisms that negatively impact solar module performance, especially in the harsh desert conditions where the solar resource is most abundant.

Of the various available means to analyze solar module performance and reliability, the measure of External Quantum Efficiency (EQE) is considered to be an excellent method for determining defects and performance optically and on a cell level, even being able to identify and locate such losses within the specific layers of the solar cells making up the module. Such defects include the likes of encapsulant discoloration, surface glass abrasions, and transmittance losses [1,2]. Hence, this method would prove especially useful when analyzing bifacial solar modules due to the differences between the front and back surfaces in material composition and layers. However, spectral response measurement, and in extension EQE, is typically performed on separate solar cells, and there isn't a very extensive amount of literature on module-level.

This paper aims to introduce a module-level EQE measurement technique and equipment where the EQE of any cell within the module is measured at a spatial resolution and area of 5 cm x 2.5 cm. These measurements were done on bifacial solar modules which have been exposed to 7 years of degradation in desert conditions and compared to conventional testing techniques in order to further understand how EQE testing can be implemented in PV defects analysis, especially for bifacial modules.

2 LITERATURE REVIEW

Despite the proven reliability of EQE measurements as an accurate and useful means for analysis in solar cells of all types, its use on a module level is not commonly used or entirely well-understood yet, which is especially true regarding bifacial modules and their unique design.

However, various authors have attempted to tackle this issue by designing and utilizing systems for module-level EQE [3,4]. Some of these designs are unique in implementation, including Perulli et al.'s utilization of a technique that combines photocurrent (PC) and photoluminescent (PL) mapping to calculate EQE, while using fluorescence maps to differentiate the data points based on different layers within each solar cell [5].

Moreover, a few examples exists where such techniques were used to test bifacial modules specifically. For instance, a test on PERC bifacial modules was carried out by Luo et al., where degradation due to PID was correlated to EQE loss at higher wavelengths (800-12000 nm) [6]. The authors conclude that this degradation occurred at the rear side, recommending the reinforcement of these areas in the module. Another study by Sulas-Kern et al. tests bifacial mini-modules, where the EQE loss was attributed to rear-side surface recombination in EVA modules, since POE modules did not exhibit such losses [7].

3 METHODOLOGY

3.1 Equipment Setup

The module-EQE measurements are carried out using a custom-made Bunkoukeiki device, shown in Fig. 1. The equipment is able to hold up a PV module on a test bench and expose it to a monochromatic light that can simulate a range of wavelengths. The device also provides bias light to each cell within the module to ensure homogeneity throughout the cells. The setup also consists of power sources for voltage injection and a data acquisition setup. The device then starts taking spectral response readings for the cell as it cycles through the wavelength range provided, creating a spectral response curve that is translated into an EQE curve.

This data is then compared to other various testing techniques commonly used for solar PV degradation analysis. This includes our use of current-voltage IV curves, Ultraviolet Fluorescence (UV-F) imaging, and Electroluminescence (EL) imaging.

10.4229/EUPVSEC2023/3AV.3.28
020228-001

Figure 1: Customized equipment setup for module-EQE measurements.

3.2 Experimental Procedure

The goal of this paper's procedure is to obtain enough initial information to allow the understanding of how and where module-EQE testing could be utilized within the PV module defects analysis field. Hence, the flowchart seen in Fig. 2 shows the procedure followed in this paper.

Figure 2: A flowchart of the experimental procedure.

To achieve both tests, modules are brought in from the nearby Outdoor Testing Facility (OTF), located within the Mohammed Bin Rashid (MBR) Solar Park in Dubai, which has nearly 2GW of solar power. The park is exposed to harsh desert conditions and large amounts of UV dosage, which are suitable for such a degradation analysis experiment. Moreover, two modules are used for each of the two test processes for a total of four modules, shown in Fig. 3.

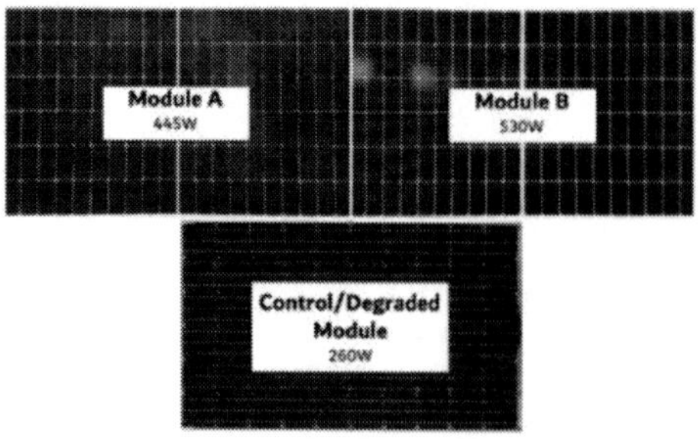

Figure 3: Details on the modules selected for the study.

For the first test, it consists of testing the other two modules by measuring their module EQE, IV curves, UV-F, and EL. These modules consist of a clean module untouched by the outdoors, and a degraded module which has been exposed to seven years of harsh conditions, named "Control" and "Degraded", respectively.

On the other hand, the second test utilizes two modules, named Module A and Module B, have their module-EQE measured in various cells within the module in addition to both front and back readings. Both modules have half-cut cells of differing makes. The EQE curves are then compared.

4 RESULTS & DISCUSSION

Given that the focus is on two different experiments with results that are to be combined, the section is split into two segments below for each of the two experiments described.

4.1 EQE Degradation Analysis

When the Control and Degraded modules are compared, the results through use of both conventional techniques and module-EQE show various outcomes. The analysis of IV curves, shown in Fig. 4, between the degraded and control modules reveals a significant overall loss in current throughout the entire curve. This unmistakably indicates a substantial reduction in output resulting from the module's degradation. Through examination of Electroluminescence (EL) images, shown in Fig. 5, where it is clear that some finger breakages were propagated, the module remains nearly unchanged in EL, possibly signifying that the power loss and degradation is caused by issues that lie elsewhere.

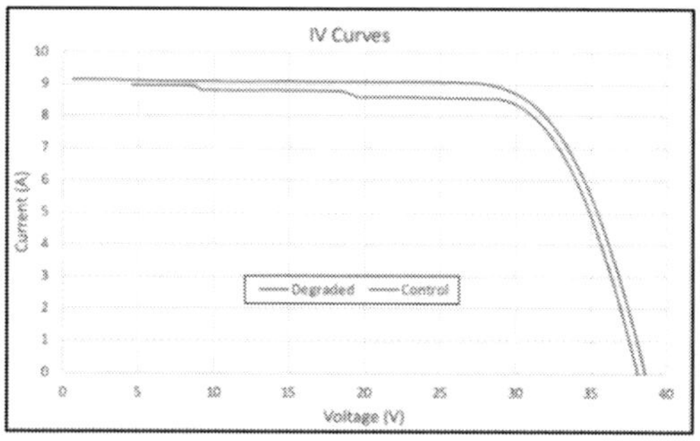

Figure 4: IV curves of both the degraded and clean modules.

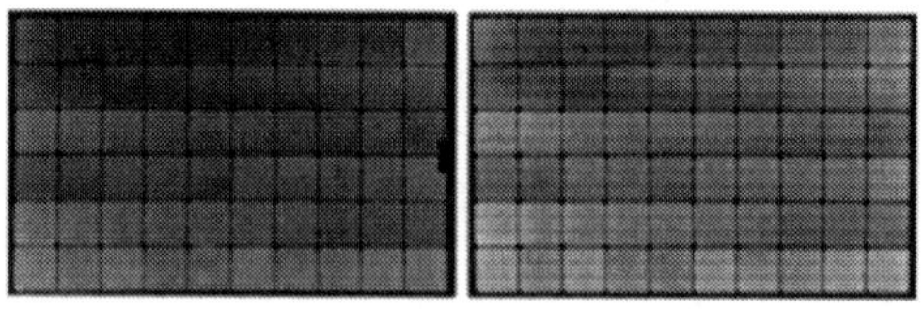

Figure 5: EL images of the front and back of the modules.

Hence, to further understand the underlying factors contributing to this power loss, we turn our attention to the observed white edge patterns in the UV-F images, as shown in Fig. 6. These patterns appear to be intricately connected to the loss in External Quantum Efficiency (EQE) at the edges of the photovoltaic cell within the UV wavelength range (300-500 nm), shown in Fig. 7. This means that EQE is able to detect and quantify the defects seen in UV-F and could provide further insight.

10.4229/EUPVSEC2023/3AV.3.28

Figure 6: UV-F image of the degraded module, showing the pattern of a white frame on the edges.

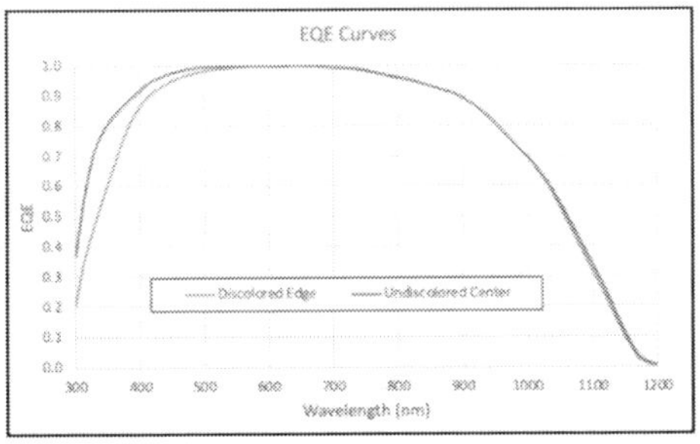

Figure 7: EQE measurements for the edge and center of the solar cell.

4.2 EQE Front-Back Comparison:

Fig. 8 and 9 show the EQE and IV, respectively, of both tested modules under different cell locations and front/back surface testing, where A and B denote the Module A and B respectively. Both modules consistently demonstrate a significant EQE gap between their rear and front surfaces, with rear surfaces consistently showing lower EQE, especially at lower UV wavelengths due to surface recombination-related losses. Conversely, the front surface EQE remains relatively stable across different cells. In contrast, rear surfaces exhibit EQE variations when comparing different cells within the same module. This EQE discrepancy aligns with lower current observed in the rear surface, corroborating the link between EQE disparities and rear surface performance.

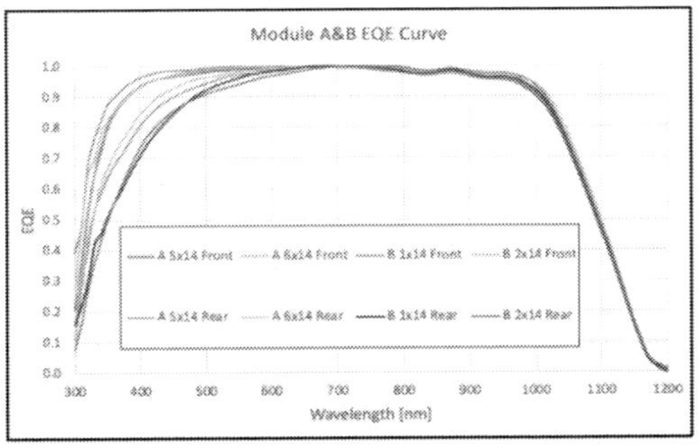

Figure 8: Module-EQE for both modules at different locations/cases.

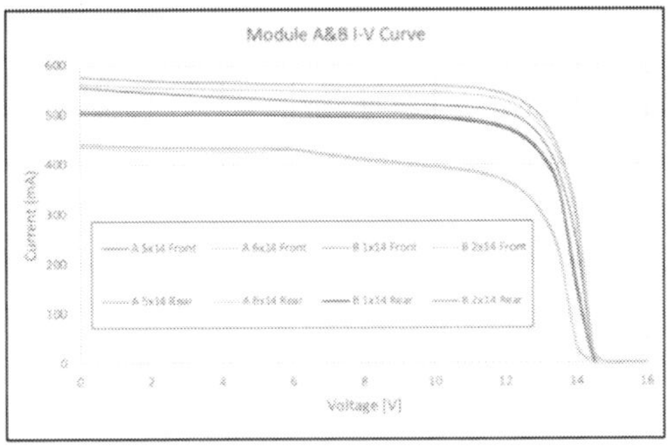

Figure 9: IV curves for both modules at different locations/cases.

This highlights the importance of understanding and addressing the unique performance characteristics of both front and rear surfaces in bifacial modules for optimizing durability. Hence, its implementation as part of a unique UV testing cycle for desert conditions would be critical, where such a cycle is highly recommended [8].

7 CONCLUSION

This study was carried out on four PV bifacial modules split into two tests using a unique module-EQE analysis method. The first was on examining both the front and rear surfaces, where it was observed that there was a consistent reduction in EQE at lower wavelengths (300-600 nm) in the rear surfaces of these modules. This EQE decrease appears to possibly be correlated with defects in the rear surface or degradation mechanisms occurring in this region. These findings suggest the potential for further comprehensive testing of rear surface EQE, particularly in the context of harsh desert conditions.

Furthermore, the second test consisted of a comparative analysis was conducted between two bifacial modules of the same type, revealing that the degraded module exhibited a loss in current. This loss was attributed to UV patterns, which were subsequently correlated with a reduction in EQE at lower wavelengths.

In light of these observations, our study highlights the significance of combining module-EQE testing with accelerated aging processes designed to simulate harsh desert conditions. This integrated approach enables a detailed exploration of defect mechanisms affecting bifacial modules, offering valuable insights into the intricate interactions within the module and cell layers, especially in harsh desert environments.

8 REFERENCES

[1]: B. Adothu, S. Kumar, B. Jäckel, N. L. Muttumthala, Z. Shekason, D. Daßler, K. Chapaneri, P. Gabbadi, Y. Kumar, A. Alheloo, A. Almheiri, J. J. John, G. Mathiak, V. Alberts, and R. Gottschalg, "Visual Defects in Photovoltaic Modules from Middle East Hot Desert", EUPVSEC-2023.

[2]: J. J. John, B. Jäckel, S. Kumar, B. Adothu, G. Mathiak, V. Alberts, "Hot desert standard development for improved PV module reliability and durability in Middle-East region", EUPVSEC-2023.

[3]: Jyothi, K. N. B., Shiradkar, N., Arora, B. M., &

Narasimhan, K. L. (2019, June). Non-Destructive Technique for Measurement of Quantum Efficiency of Encapsulated Solar Cells in PV Modules. In 2019 IEEE 46th Photovoltaic Specialists Conference (PVSC) (pp. 0388-0391). IEEE.

[4]: Siefer, G., Gandy, T., Schachtner, M., Wekkeli, A., & Bett, A. W. (2013, June). Improved grating monochromator set-up for EQE measurements of multi-junction solar cells. In 2013 IEEE 39th Photovoltaic Specialists Conference (PVSC) (pp. 0086-0089). IEEE.

[5]: Perulli, A., Lattante, S., Persano, A., Cola, A., & Anni, M. (2019). On the homogeneity of the external quantum efficiency in a free OPV roll-to-roll flexible solar module. Synthetic Metals, 247.

[6]: Luo, W., Hacke, P., Terwilliger, K., Liang, T. S., Wang, Y., Ramakrishna, S., ... & Khoo, Y. S. (2018). Elucidating potential-induced degradation in bifacial PERC silicon photovoltaic modules. Progress in Photovoltaics: Research and Applications, 26(10), 859-867.

[7]: Sulas-Kern, D. B., Owen-Bellini, M., Ndione, P., Spinella, L., Sinha, A., Uličná, S., ... & Schelhas, L. T. (2022). Electrochemical degradation modes in bifacial silicon photovoltaic modules. Progress in Photovoltaics: Research and Applications, 30(8), 948-958.

[8]: S. Kumar, B. Adothu, B. Jäckel, Z. Shekason, D. Daßler, K. Chapaneri, P. Gabbadi, Y. Kumar, A. Alheloo, A. Almheiri, J. J. John, G. Mathiak, V. Alberts, and R. Gottschalg, "Defects and degradations in photovoltaic modules from hot Middle East deserts", EUPVSEC-2023.

ENCAPSULANT-FREE FULL-SIZE N.I.C.E. MODULES IN OUTDOOR PERFORMANCE TEST: FIRST YEAR RESULTS AND ROADMAP TOWARDS HIGHER EFFICIENCY

Christoph Pönisch, Lukas Schanz, Jesus da Costa Fernandes, Michael Schmidt, Daniel Kray
Institute for Sustainable Energy Systems (INES), Offenburg University of Applied Sciences, Germany
Badstraße 24, 77652 Offenburg, Germany

ABSTRACT: Encapsulant-free N.I.C.E. modules have strong ecological advantages compared to conventional laminated modules but suffer generally from lower electrical performance. Via long-term outdoor monitoring of full-size industrial modules of both types with identical solar cells, we investigated if the performance difference remains constant over time and which parameters influence its value. After assessing about a full year's data, two obvious levers for N.I.C.E. optimization are identified: The usage of textured glass and transparent adhesives on the module rear side. Also, the performance loss could be alleviated using tracking systems due to lower AOI values. Our measurements show additionally that N.I.C.E. module surfaces are in average about 2.5°C cooler compared to laminated modules. With these findings, we lay out a roadmap to reduce today's LIV gap of about 5%$_{rel}$ by different optimizations.

Keywords: encapsulant-free modules, outdoor degradation, bifacial solar modules, N.I.C.E., temperature behavior.

1 INTRODUCTION

Encapsulant-free modules are one of the most promising candidates for "terawatt-ready" PV technologies. Their recyclability and general sustainability are outstanding. However, the lack of encapsulant induces an additional front side reflection thus reducing electrical conversion efficiency. If the BOM savings and general production cost reductions overcompensate this loss, then the LCOE of encapsulant-free modules is competitive, nevertheless. This must be shown for different applications worldwide.

2 AIM AND METHODS

Industrial bifacial glass-glass 60-cells modules were provided by Apollon Solar (N.I.C.E. [1]) and SoliTek (reference: Conventional lamination). Identical monocrystalline PERC solar cells with 5 busbars were used (M2 wafer size). While the complete BOM was unknown to us, two main differences between both module types were visible:

- Frame: Black (ref) vs. plain Al (N.I.C.E.)
- Front glass: Textured (ref) vs. float glass (N.I.C.E.)

The modules were initially characterized via LIV in a Wavelabs LED flasher, mounted on a south-west facing fixed-tilt monitoring setup for approximately one year duration, afterwards unmounted and LIV flashed again. One reference module and 2 N.I.C.E. modules were stored inside in the dark and flashed at the same times.

3 INITIAL CHARACTERIZATION

A lab-type Wavelabs LED flasher was used to record LIV data under STC. P_{mpp} was adjusted for temperature using experimentally determined temperature coefficients during the outdoor measurements (cf. section 6). The measurements are summarized in Table I.

From these initial measurements we concluded that the average MPP STC power of N.I.C.E. modules was 6.2%$_{rel}$ lower compared to the reference. That value is consistent with literature [2]. Clearly, this effect is dominated by lower I_{sc}. As some of the modules are stored in the dark for degradation references, we calculated the average MPP STC power difference of the 4 modules used for outdoor testing to be just **4.4%$_{rel}$**. This number was used later as baseline to evaluate differences in long-term DC yield.

Table I: Initial LIV data at STC.

Module	I_{sc} / A	V_{oc} / V	FF / %	R_s / Ω	P_{mpp} / W	Bifi / %
Reference (n=2)	9.66	40.99	78.4	0.234	310.3	77
N.I.C.E. (n=5)	9.10	40.95	78.1	0.234	291.2	66

Bifaciality differed by 11%$_{abs}$ which is due to the usage of black adhesives on the rear side of N.I.C.E. modules, cf. Fig. 1. The usage of less and possibly transparent adhesive is an obvious approach to tackle this issue.

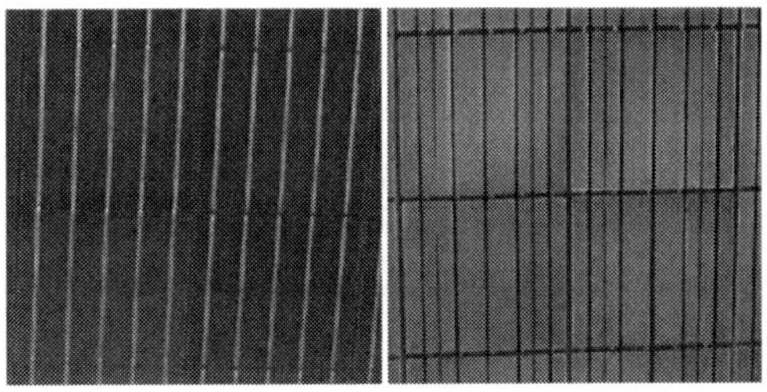

Figure 1: Comparison of module rear side configurations. Left: Reference, right: N.I.C.E. with black adhesive lines.

4 OUTDOOR MONITORING SETUP

In the next step, the modules were installed on a fixed-tilt (26°) rack on the institute's roof terrace, grid-connected with individual micro inverters and monitored (DC and AC data) together with weather and irradiation conditions (pyranometer and reference solar cells for front and rear irradiance, anemometer for wind measurement, temperature sensors for every module and the ambient temperature) in 1-minute intervals. The setup is shown in Fig. 2. While the location is sub-optimum in terms of daily partial shading and orientation (facing south-west), we verified via PV*SOL [3] simulations that all 4 modules experience the same cumulative irradiance and shading during a full year.

The collected data set was then reduced taking the following steps to increase the data quality:

- Irradiance in plane-of-array (POA) > 50 W/m^2
- Module powers > 10 W
- Exclude maintenance sequences

Figure 2: Outdoor monitoring rack on INES roof with 3 N.I.C.E. modules and 1 reference module.

The reduced data set was then analyzed using Python [4]. In this paper we present data between September 2nd 2022 and August 10th 2023.

5 DATA EVALUATION AFTER ABOUT 1 YEAR

First, all modules were removed from the rack, cleaned and flashed for STC LIV again. This procedure was also performed for the modules stored in the dark that were not exposed to outdoor conditions. The results are presented in Table II.

Table II: STC LIV data from flasher measurements at the end of the study. *Delta* = difference between initial (09/22) and final (08/23) measurements.

Module	*Delta* P_{mpp} / W	*Delta Pmpp* / %$_{rel}$
Reference outdoor (n=1)	-2.4	-0.78
N.I.C.E. outdoor (n=3)	-2.2	-0.76
Reference dark-stored (n=1)	+0.6	+0.21
N.I.C.E. dark stored (n=2)	-2.0	-0.68

As the relative differences between initial and final measurements are well below 1%$_{rel}$, we concluded that no significant degradation was visible on none of the module types. Next, we evaluated the cumulative DC energy production over the testing period. The results are shown in Table III for the full period and wintertime (December through February).

Table III: Measured DC energy production data.

Module	DC energy production/kWh	DC energy production Dec-Feb/kWh
Reference (n=1)	203.08	23.29
N.I.C.E. avg (n=3)	191.59	20.71
Relative difference N.I.C.E. vs. reference	**-5.7%**	-11.1%

The recorded difference of 5.7% over the whole period was slightly higher than the 4.4% from the initial flasher measurement in section 3. In the following, we investigated the reasons for the relatively lower outdoor performance. For that, we plotted the ratio of the average N.I.C.E. P_{mpp} and the reference P_{mpp} versus different parameters like angle-of-incidence (AOI), front and rear irradiance, cf. Fig. 3 to 5 for clear sky data [5]. While we did not see significant influence of front / rear irradiance or ambient temperature on this ratio, the effect of AOI in Fig. 3 was striking. For AOI smaller than approx. 35°, N.I.C.E. performed slightly better compared to the flasher data, while for AOI larger than 40°, the difference increased quickly. We attribute this behavior to the textured glass of the reference module that is known to outperform flat front glasses especially for larger AOI [6].

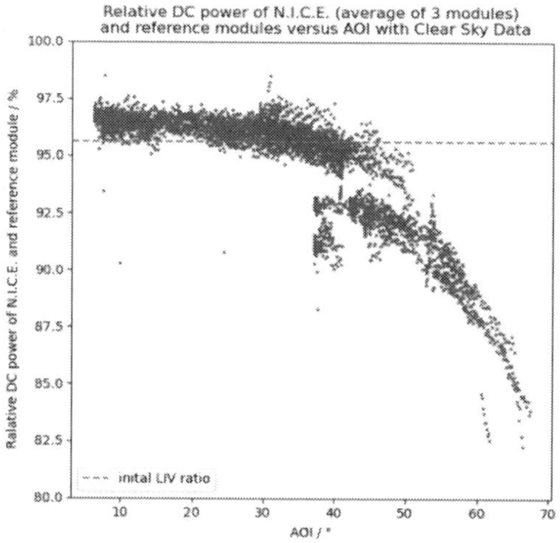

Figure 3: Ratio of N.I.C.E. and reference DC power versus AOI for clear sky conditions. Dashed line: Ratio from initial flasher tests at STC.

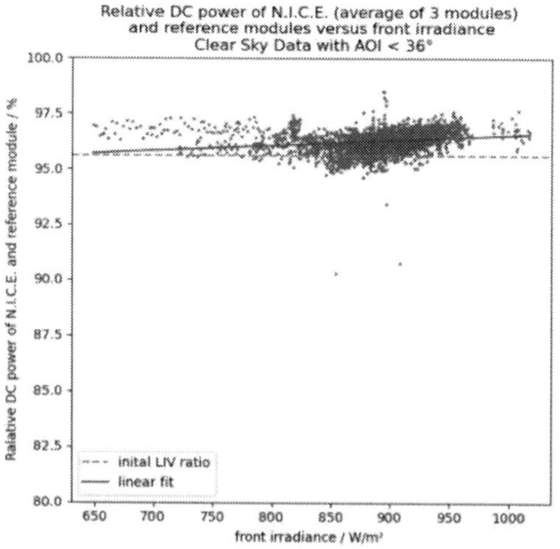

Figure 4: Ratio of N.I.C.E. and reference DC power versus front irradiance for clear sky conditions and near-normal incidence (AOI < 36°). Dashed line: Ratio from initial flasher tests at STC.

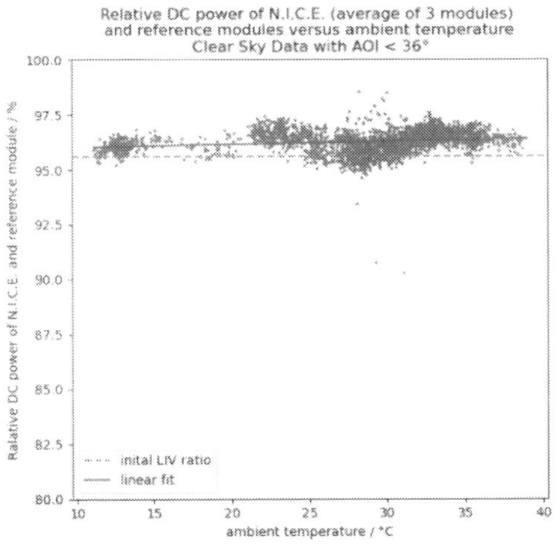

Figure 5: Ratio of N.I.C.E. and reference DC power versus ambient temperature for clear sky conditions and near-normal incidence (AOI < 36°). Dashed line: Ratio from initial flasher tests at STC.

6 TEMPERATURE BEHAVIOR

To assess the relatively higher performance of N.I.C.E. modules at low AOI compared to initial flasher data (cf. Fig. 4 and 5), we investigated the recorded temperature data. Therefore, we experimentally determined the MPP power temperature coefficients of all four modules: We filtered the clear sky data to effective irradiance values of 1000 +/- 10 W/m^2. These data sets were linearly fitted so that the sought coefficients could be identified from the respective slopes. The average coefficient for the three N.I.C.E. modules was about 10% lower than for the reference module (-0.63 W/K compared to -0.69 W/K).

This first indication of better temperature behavior was supported by comparing the average module rear surface temperature between N.I.C.E. and the reference module, cf. Fig. 6. Clearly, an average difference of about 2.5°C cooler N.I.C.E. modules can be seen for clear sky conditions, independent of the reference module temperature.

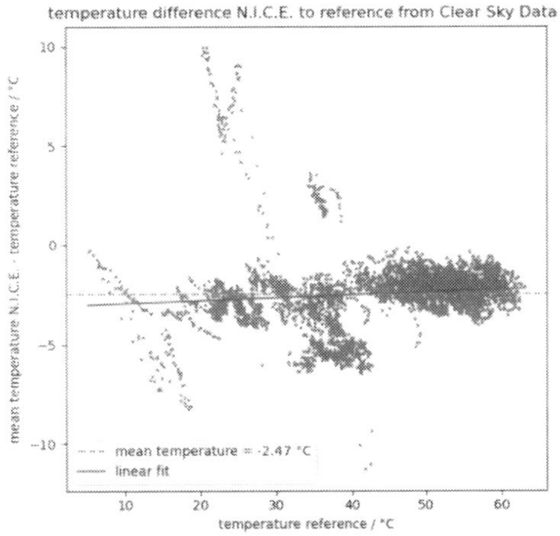

Figure 6: Average mean rear surface temperature difference between N.I.C.E. and reference module. Negative values indicate that N.I.C.E. modules are cooler.

7 ROADMAP TO HIGHER EFFICIENCY

While STC LIV measurements reproduce the well-known lower optical performance of N.I.C.E. compared to conventional laminated modules, the long-term outdoor monitoring paints a more differentiated picture: The overall DC energy production is slightly lower than expected while the N.I.C.E. performance over the year is strongly depending on AOI. This allows to set up a clear roadmap for higher efficiencies of encapsulant-free modules:

(1) Use textured front glass
(2) Improve bifaciality by using transparent rear-side adhesives.
(3) Use preferably with tracking systems.
(4) Use suitable optical material for better optical coupling.

The Apollon modules suffer strongly from the front float glass and the rear shading by black adhesives. Steps (1) and (2) are relatively easy to implement. In tracking applications, the comparison will be more in favor of N.I.C.E. as AOI will be constantly lower. Finally, step (4) would fully eliminate any optical loss of encapsulant-free modules. However, this is still just a theoretical concept.

8 OUTLOOK AND SUMMARY

Our study showed that encapsulant-free N.I.C.E. modules from Apollon Solar suffer from lower optical performance, especially at higher AOIs. However, there are hints that the temperature behavior is superior as average rear surface temperature is about 2.5°C lower compared to the laminated reference module with identical solar cells.

In further work, we will conduct a simulation study to quantify the potential of optimized encapsulant-free modules. Using textured front glass and transparent rear adhesives are two simple steps to optimize performance. Also, tracking applications will be beneficial for relative N.I.C.E. performance as AOI is constantly lower. We expect that the relative performance difference can be reduced to very low values that will make encapsulant-free modules attractive for terawatt-scale production.

10.4229/EUPVSEC2023/3AV.3.32
020229-003

Bibliography

[1] J. Dupuis, E. Saint-Sernin, O. Nichiporuk, P. Lefillastre, D. Bussery, and R. Einhaus, "NICE module technology - From the concept to mass production: A 10 years review," in *38th IEEE Photovoltaic Specialists Conference*, Austin, TX, USA, 2012, doi: 10.1109/PVSC.2012.6318254.

[2] R. Einhaus, K. Bamberg, R. De Franclieu, and H. Lauvray, "Recent Progress with NICE Module Technology," *Proceedings of the 20th European PVSEC, Barcelona,* pp. 1993-1996, 2005 2005.

[3] V. Software. "PV*SOL." https://valentin-software.com/produkte/pvsol/ (accessed 08/22/2023, 2023).

[4] P. S. Foundation. "Python." https://www.python.org/ (accessed 08/22/2023, 2023).

[5] P. Ineichen and R. Perez, "A New Airmass Independent Formulation For The Linke Turbidity Coefficient," *Solar Energy,* vol. 73, no. 3, pp. 151-157, 2002.

[6] P. J. Sánchez-Illescas, P. Carpena, P. Bernaola-Galván, M. Sidrach-de-Cardona, A. V. Coronado, and J. L. Álvarez, "An analysis of geometrical shapes for PV module glass encapsulation," *Solar Energy Materials and Solar Cells,* vol. 92, no. 3, pp. 323-331, 2008/03/01/ 2008, doi: https://doi.org/10.1016/j.solmat.2007.09.008.

Assessing uncertainties from reflected irradiance in bifacial PV simulations through a 3D view factor model and rear sensor measurements

Stéphane Mollier, John (Ioannis) A. Tsanakas

Univ. Grenoble Alpes, CEA, Liten, Campus Ines, 73375 Le Bourget du Lac
stephane.mollier@cea.fr
ioannis.tsanakas@cea.fr

ABSTRACT: Emerging PV applications such as agrivoltaics (agriPV) or bifacial PV present design particularities and are sensitive to irregular surroundings. Realistic and accurate modeling of these applications is both essential to minimize the perceived risk for their financing but is also challenging and requires a very careful model definition and calibration. In this study, we describe a simulation method based on 3D view factors. Using on site measurements, we show how details can influence the irradiance estimation of reference cells on the backside of the same PV module array. Intermediate simulation results such as the evolution of shading over time can provide a first insight of the differences in reflected irradiance measurements. Similarly, we present how the view factors can help to understand the impact of obstacles. Finally, we validate preliminary results of optical simulations by comparing them to on-site measurements.

Keywords: PV modelling, energy yield simulations, bifacial PV, view factors, reflected irradiance.

1. SCIENTIFIC RELEVANCE AND AIM

1.1 Context – State of play

Today's commercial solutions for photovoltaic (PV) energy yield simulations are widely used for modelling standard PV systems. Yet, due to their "black-box" nature, such modelling tools present considerable limitations in addressing design- and technology-specific parameters of PV modules and systems [1,2]. On the other hand, PV systems based on recent and emerging PV technologies, notably bifacial PV, require tailored modelling/simulation approaches (and better understanding) to ensure highly accurate, realistic and reliable assessments of their PV energy yield. In turn, from a grid penetration and investment perspective, realistic and accurate PV energy simulations are essential to minimize the perceived technical (and economic) risk of bifacial PV projects and their financing. Indeed, with the increasing market uptake and deployment of bifacial PV systems, consequent research efforts have been focusing on developing physics-based "customized" models for simulating and assessing the PV energy yield, under bifacial PV-specific conditions (e.g. albedo dependence, rear-side illumination and mismatches, etc).

On the latter, Chudinzow et al. [3] introduced a bifacial PV model, with which they investigated and quantified experimentally the influence of ground size, cast ground shadows as well as ground reflectivity on the energy yield of bifacial PV system. Sensitivity analysis-based results of their model, showed that the extension of ground area, which contributes to ground-reflected irradiation, resulted in a slight, asymptotical increase of the energy yield. With the same model, it was possible to show, through ground shadow-free simulations, that the presence of ground shadows can reduce the annual bifacial PV yield by almost 4%, while the influence (and contribution) of ground reflectivity on the energy yield was modelled and validated for five different ground surfaces (dry asphalt, grass, dry grassland, white gravel and white membrane). Durković et al. in [4] presented a model for more accurate calculation of irradiation for large bifacial PV power plants, i.e. PV power plants characterized by significantly bigger lengths than heights of PV rows. The model is claimed convenient for hourly meteorological data, which enables calculations of typical daily and annual electricity production of PV power plants. Unlike other existing similar models, the model proposed in [4] considered the fact that less diffuse irradiation incident onto the surface between PV rows which are in the shade of previous PV rows, than entire horizontal diffuse irradiation; thus view factors (VFs) are calculated by simple algebraic relations instead of solving of a double integral. Similarly, Ledesma et al. also developed in 2019 and presented in [5] their VF based two-dimensional model for calculating the rear irradiance in large bifacial PV plants has been developed now, then integrated in SISIFO (an free simulation tool developed at IES-UPM) for static structures and also for horizontal single-axis trackers. In either modelling approaches above, uncertainties (errors) in corresponding energy yield estimation derive primarily from the intrinsic non-uniformities of rear-side irradiance. Finally, in an effort focusing on modelling bifacial PV systems with single-axis tracking, Pelaez et al. [6] evolved the Radiance bifacial PV model [7], including additional modelling steps, i.e. for calculating the array tilt, ground clearance, and row-to-row spacing for each time step. Unlike to the conventional fixed-tilt simulation workflow (annual average bifacial gain calculation based on a single annual sky source), for or a tracked system, multiple scene geometries were considered (for each tracker tilt in 5° increments), along with the solar resource (cumulative hourly values) corresponding to each tracker tilt angle.

1.2 Motivation and aim

The deployment of PV installations, especially in urban, peri-urban and agricultural areas, faces the inevitable constraint of limited space, as available land is (or becomes) scarce in such areas, or has to be shared with other sectors (e.g. agriculture, infrastructure, buildings). Emerging PV applications and configurations, notably agrivoltaics (agriPV), building-integrated PV (BIPV), floating PV and vertical bifacial PV, offer a solution to such constraint, promoting a more rational use of available land or spaces for an optimal so-called landscape integration of PV. However, these new PV applications present design particularities and complex layouts, which render their modeling and simulation of their energy yield a challenging task, in the same sense as discussed in Section 1.1. PV systems in shared spaces, with complex or irregular surroundings and layouts, require therefore a very careful model definition, yield simulation and comparative assessment with on-site measurements.

An example of how details can influence a measurement of irradiance data (and consequently a PV energy yield simulation) is illustrated in Fig. 1; where we show (on left) the rear side of the first row of monitored PV arrays installed at INES test site, and (on right) the irradiance (W/m²) data readings, with a 5-minute time step, during an indicative (random) day in winter. The irradiance sensors, i.e. six reference cells denoted A to F from the front to the background respectively are mounted at the same height and with the same tilt angle (30°) as the modules plane and are oriented to the ground. We can see that six sensors on the very same row of modules that (seem to) measure the same data (irradiance in this case), with the same inclination can give significantly different results.

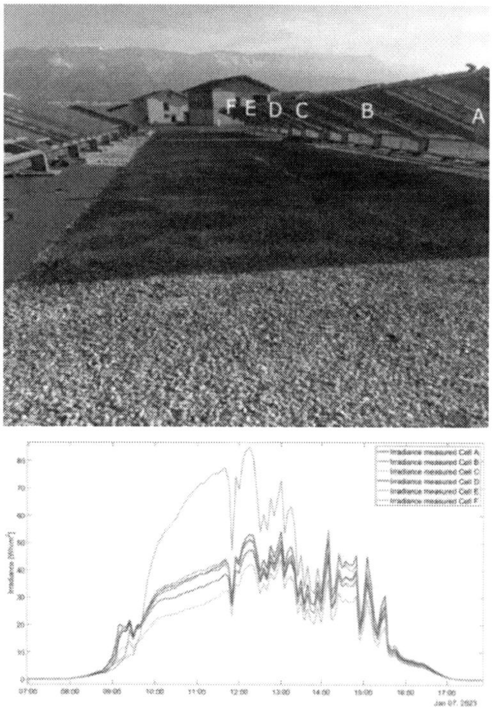

Fig. 1: Observed deviations of in-plane irradiance readings from six identical sensors (reference cells) at the same tilt and orientation (right figure), installed on the rear side of PV modules, monitored at INES test facilities (left figure).

The objectives of this research work is to be able to explain and reproduce these differences by modeling. The modeling tools being sometimes complex, they require more advanced validation and calibration which is sometimes difficult with a limited number of sensors. By understanding the main contribution of irradiance, the model can be fine-tuned for increased accuracy, especially for complex modeling scenes and PV sites-scenarios.

2. APPROACH METHODOLOGY

2.1 Modeling aspects

The simulation tool developed for this study is based on TriFactors, a simulation software developed by CEA to model bifacial PV plants. This new version, also based on 3D view factors, allows removing some limitations on the definition of the scene such as taking into account obstacles or structures. The simulation method is globally similar and have the following main steps:

- Define the 3D scene (module, structure, obstacle) with a meshing of the ground
- Estimate the view factors between every elements of the scene.
- For each time step:
    1. Compute the complex shadow of the 3D scene
    2. Transpose global/diffuse irradiance to the surfaces of interest
    3. Combine the contributions of irradiance (direct, diffuse and reflected) to obtain the global irradiance

To simulate the reflected irradiance, the two main approaches are the Ray-Tracing (RT) and the View Factors (VFs). RT is generally very accurate but have a high computation cost. In this abstract, we consider the VFs approaches for his ability of giving contextual information that can help to understand the contribution of irradiance. An important hypothesis of VFs is that it assumes the surface to be Lambertian, i.e. to reflected the light equally in every direction. This hypothesis might not be realistic in all type of ground especially when the sun is low in the horizon. The ideal way to take into account the direction of reflection is to design the BRDF (Bidirectional Reflectance Distribution Function). However, these functions are difficult to estimate especially with a limited number of sensors available and are not considered in this study.

The reflected irradiance on a surface of interest is computed by summing the contribution of all surfaces as follows:

$$I^i_{reflected} = \frac{1}{A_i} \sum_{k}^{all\,surface} VF_{k\rightarrow i} * Albedo_k * A_k I^k$$

where I^k is the irradiances received by the surface k (ground or object) including the impact of shadowing, the $Albedo_k$ corresponds to the proportion of light that is reflected, and the view factor $VF_{k\rightarrow i}$ is the proportion of this reflected light reaching the surface of interest. All the view factors are estimated by numerical integration. The contribution through multiple reflection are neglected.

2.2 Data aspects

The input data needed for the model are first the Global Horizontal Irradiance (GHI), the outside temperature and a general representation of the PV plant. To increase the accuracy, we also have collected with sensors additional information such as the albedo and the Diffuse Horizontal Irradiance (DHI). For validation, the site has been equipped by irradiance sensors (cell ref) with the same tilt than the module on both the rear and the front side.

3. SIMULATION RESULTS & DISCUSSION

3.1 Understanding sensors difference through modeling

As we have shown in the example of Figure 1, we are assessing a bifacial PV array installed at INES field testing platform, for which we witness significant deviations between the readings from rear-side irradiance sensors, for the same row of PV modules. In this Section, we try to understand and discuss these observations. We also use a modelling approach to give an additional insight into the source of such deviations and their (potential) contribution in uncertainties in PV energy yield simulations/assessments.

A first observation is that the reference cell F gives irradiance readings relatively lower first during a short period in the morning; and then significantly higher compared to all the rest of reference cells. A most plausible explanation is the presence of transient shading on this row, i.e. during a short period in the morning, due to presence of a building in proximity (Fig. 2). In turn, the shading decreases the irradiance incident on the surrounding ground and therefore the light (irradiance) reflected to the sensor in question.

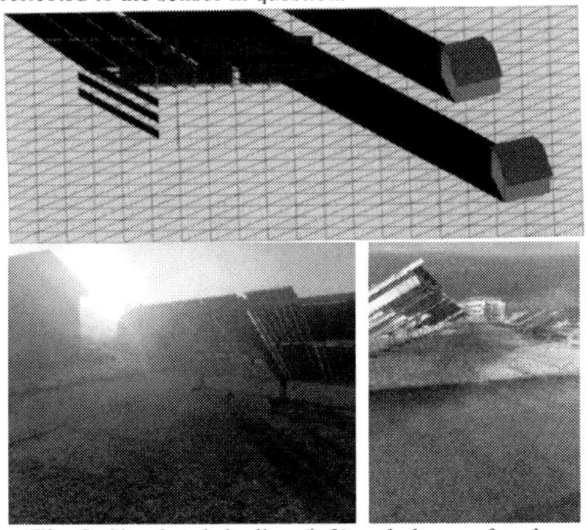

Fig. 2: Simulated shading (left) and photos after the shading moved (right)

To validate the ground the shading, data from camera on INES bench sites are used for a given time comparison. The Figure 3 illustrates a comparison between simulation (top) and image from camera (bottom), for the same time (10:27) and day (May, 11th 2023), on the East of the PV row, near the sensors cell E and F.

We can noticed that some gaps between the PV modules along the mounting structure allow more light incident to (and reflected from) the ground behind the rest of the modules in the same row, which explains higher rear-side irradiance values recorded by the corresponding sensors.

Besides, the fact that this sensor positioned at the side of the row closest to the east, explains its significantly higher irradiance readings in the morning, compared to the afternoon, as the shading of the modules spreads towards west in the morning.

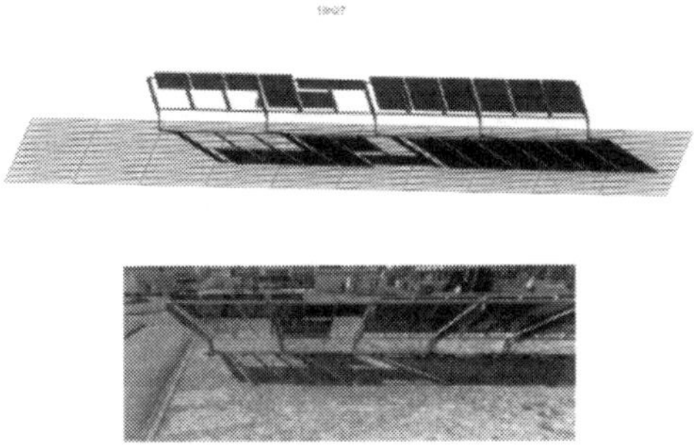

Fig. 3: Comparison of ground shadow near the cell E and F from simulation (top) and camera image (bottom)

Another interesting observation is that the overall irradiance readings measured by the reference cell C are lower than for the rest of the reference cells. Near this sensor, we can notice that there is a combiner box (also present near the other sensors), but also a small structure on the ground, with a metal drain cover. Through View Factor (VF) modelling, we can describe which portion that such obstacles represent in sight of the reference cell (Fig. 4).

Fig. 4: Understanding the impact of small obstacle using view factors computation.

We can see that the calculated VF of the reference cell to the drain cover is equal to 15.8% which implies a strong influence of this structure to the overall modelled irradiance – and, consequently, to the resulting (simulated) bifacial PV energy yield. On the other hand, the modelled VF from the cell to the adjacent combiner box is equal to 4.2%. As all the sensors have a box at proximity, we can fairly assume that the monitored irradiance data to be slightly underestimated for all sensors. The VF of the obstacle is not directly equivalent to a loss in irradiance but only represents the portion of an obstacle in the sight of the reference cell. However, this could give a rough idea of the obstacle's impact on the irradiance. Eventually, this type of analysis can be replicated and upscaled at a PV string(s) level, for instance, to understand the impact of the overall structure.

3.2 Preliminary findings in simulations and comparative analysis

In this section, we present results from a comparative analysis between the on-site measurements (monitored) data and the simulations. An essential point to validate is the optical modelling results, especially for the rear-side of the modules, as these may be influenced by considerable inhomogeneities of the incident and reflected irradiance, as discussed in the previous section. Fig. 5 gives an overview of the measured and simulated rear-side irradiance, for the studied bifacial PV array, indicatively for the reference cells B, C and F.

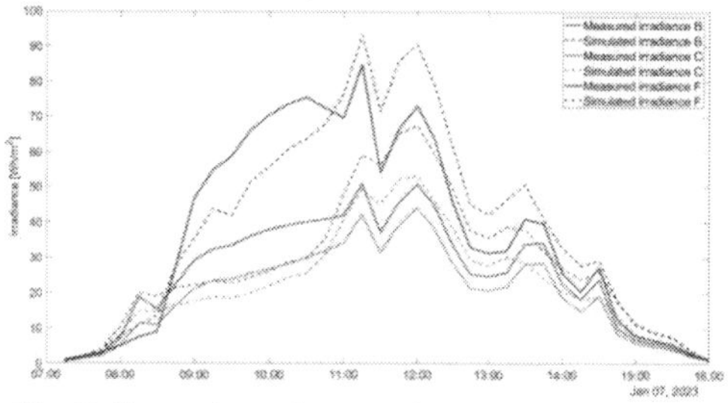

Fig. 5: Comparison of rear irradiance measurement and simulation with a 15mn time step. The reference cells are located in the same array of bifacial module as described in Fig. 1.

We observe that the simulated irradiance for the sensor F is lower than for the two others (B and C), between 8:00 and 8:30, in the morning. This effect has been already observed in the (real) on-site measurements and – as discussed in the previous section, it is explained by the transient near-shading from buildings in proximity, that reduced the amount of light incident to and reflected from the surrounding ground. Yet, this effect is generally underestimated by the model, when compared to the actual measurement. Similarly, the higher simulated irradiance for the sensor F is in agreement with the conclusions also discussed in the previous section: gaps (missing modules) in the mounting structure of the PV array allow more incident and reflected light to the ground below and eventually to the corresponding reference cell. We also observe that the simulated irradiance for the cell C is lower than the other cells as it was expected due to the presence of the drain blocking a part of the ground reflection.

We can notice that there is still some important difference between the measurement and the model that can partially be explained by a lack of calibration of the model. Indeed, the complexity of the model required numerous parameters that are not necessarily known. More effort will be invested in calibration in ongoing and near-future work, which will also include the results from our comparison between simulated and monitored (actual) bifacial PV power outputs. Other possibilities to explain these differences could be spectral effect or bidirectional reflectance distribution function (BRDF) that are not included yet in the tool.

Acknowledgments

The authors acknowledge the valuable contribution of all SERENDI-PV project partners. This project has received funding from the European Union's Horizon 2020 research and innovation programme under grant agreement No. 953016. For CEA team, part of this work was also supported by the French National Program "Programme d'Investissements d'Avenir - INES.2S" under Grant Agreement ANR ANR-10-IEED-0014 0014-01.

References

[1] Horvath, I.T., et al. (2018). Next Generation Tools for Accurate Energy Yield Estimation of Bifacial PV Systems – Best Practices, Improvements and Challenges, 36th European Photovoltaic Solar Energy Conference and Exhibition (EU-PVSEC), Marseille, France.

[2] Stein, J.S., Farnung, B. (2017). PV Performance Modeling Methods and Practices. Report IEA-PVPS T13-06:2017. ISBN 978-3-906042-50

[3] Chudinzow, D., et al. (2019). Simulating the energy yield of a bifacial photovoltaic power plant. Solar Energy 183, pp. 812–822.

[4] Durković, V. and Durković, Ž. (2019). Extended model for irradiation suitable for large bifacial PV power plants. Solar Energy 191, pp. 272–290.

[5] Ledesma, J.R., et al. (2019). A simulation model of the irradiation and energy yield of large bifacial photovoltaic plants. Solar Energy 206, pp. 522-538.

[6] Pelaez, S.A., et al. (2019). Model and Validation of Single-Axis Tracking With Bifacial PV, IEEE Journal of Photovoltaics 9(3), pp. 715-722.

[7] Ward, G.J. (1994). The RADIANCE lighting simulation and rendering system. In: Proc. 21st Annual Conference on Computer Graphics and Interactive Techniques, pp. 459–472.

AUTHOR INDEX

40th European Photovoltaic Solar Energy Conference and Exhibition (EU PVSEC 2023)

Lisbon, Portugal
18-22 September 2023

Volume 2 of 3

ISBN: 979-8-3313-1553-5

Printed from e-media with permission by:

Curran Associates, Inc.
57 Morehouse Lane
Red Hook, NY 12571

Some format issues inherent in the e-media version may also appear in this print version.

Printed with permission by Curran Associates, Inc. (2025)

For permission requests, please contact WIP – Renewable Energies
at the address below.

WIP – Renewable Energies
Sylvensteinstr. 2
81369 Munchen
Germany

Phone: +49 89 72012735
Fax: +49 89 72012791

wip@wip-munich.de

Additional copies of this publication are available from:

Curran Associates, Inc.
57 Morehouse Lane
Red Hook, NY 12571 USA
Phone: 845-758-0400
Fax: 845-758-2633
Email: curran@proceedings.com
Web: www.proceedings.com

TABLE OF CONTENTS

VOLUME 1

VOLUME 2

VOLUME 3

Mask for Routine Testing Using Time Domain Reflectometry on Solar Modules: Measures to Investigate Non-visible Failures.

A.M.C. Silveira[1] Student Member IEEE, M.R.M. Neves[2], R. Garcia[4], H. Alvarez[4],
🎗 In Memory M.G. Villalva[2] Member IEEE, F.C. Marques[3] and L.C. Kretly[1] Member IEEE
[1]School of Electrical and Computer Engineering, Communications Department - DECOM, Campinas, Brazil
[2]School of Electrical and Computer Engineering, Dept. of Systems and Energy - DSE, Campinas, Brazil
[3]Institute of Physics Gleb Wataghin, Dept. of Applied Physics - DFA, Campinas, Brazil
[4]Build Your Dreams Company - BYD, Dept. Research and Development - R&D, Campinas, Brazil

ABSTRACT: This work investigated the application of the TDR technique in solar modules to explore its practical applications in the module certification laboratory, commissioning of photovoltaic systems and also in the assembly line itself. The objective of this work is to establish masks of routine tests through the electronic signature measured by the technique reflectometry in the time domain - TDR, in this way it is possible to investigate non-visual defects in the solar module. Currently, it is common to use the electroluminescence technique to detect visual and solar cell degradation caused by PID effect – Power Induced Barrier. However, this work seeks to make the TDR technique viable as an auxiliary inspection to the electroluminescence technique in the failure of non-visual defects in solar modules.

Keywords: Solar Module, TDR, Non-visual Defects, Visual Inspection, Electroluminescence, Routine Test, Electronic Signature Mask.

1 Introduction

In photovoltaic - PV context, the main method to detect defects which decrease their energy efficiency in solar module [1] consists of visual inspection assisted by Electroluminescence - EL technique [2]. In Fig. 1 is illustrated the EL technique.

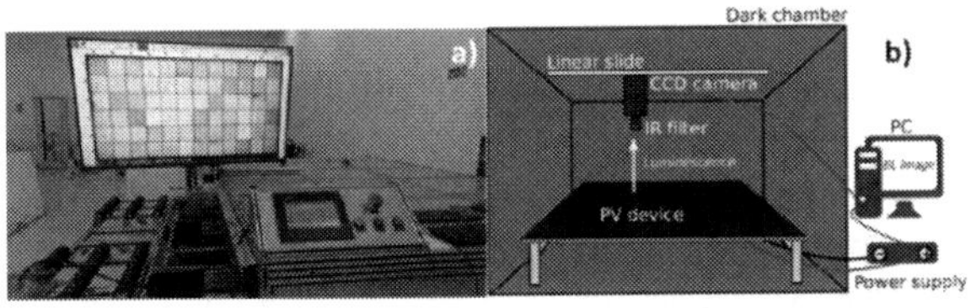

Figure 1. Electroluminescence technique. a) EL Setup in laboratory and b) Schematic of the EL technique . [adapted from 3]

This technique is used during the assembly of solar modules, commissioning of photovoltaic systems and also certification process and makes visual defects easier to find, such as: cracked, bent, broken cells, cracked cells, faulty interconnections or joints, failure of adhesive bonds, sticky surfaces of plastics, defective terminations, solar cell degradation caused by PID effect and others conditions that may affect performance, these defects are also described in IEC 61215 standard [4] dealing with solar module safety and peformance.

2 Time Domain Reflectometry Technique

The Time Domain Reflectometry - TDR technique consists of sending a reference signal along a system or transmission line and the reflection is measured along the entire length [5], [6]. The analysis of the reflected waveform makes it possible to determine the location and nature of any anomalies or defects in the transmission line. Fig. 2 shows the schematic of TDR measurement in solar modules.

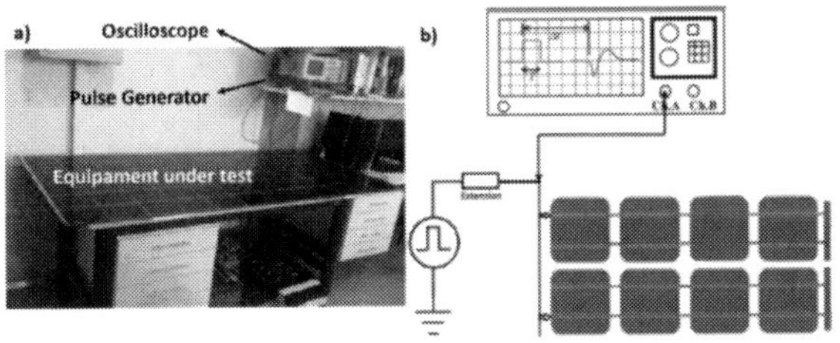

Figure 2. TDR technique. a) TDR setup in the laboratory and b) schematic of the TDR technique. [adapted from 5]

In the context of solar module such as transmission line and equipment under test (EUT), non-visible defects like cold solder joint [7], busbar interruption or solar cell degradation can be easily detected using the TDR technique.

The Fig. 3 shows the non-visible defect that could appear on the front grid of the solar cells.

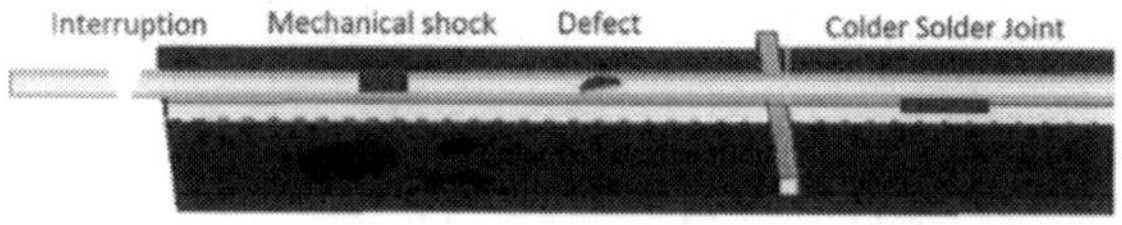

Figure 3. Illustration of the non-visible defect on the front grid degradation of the solar cell. [adapted from 8]

The TDR technique quickly reveals the characteristic impedance of the transmission line and shows its position and nature of each discontinuity or load [9]. Since the reflection coefficient ρ is represented in the following expression in 1.

$$\rho = \frac{E_r}{E_i} = \frac{Z_L - Z_0}{Z_L + Z_0} \tag{1}$$

Where E_R and E_i are reflected and incident energy, respectively Z_0 characteristic impedance of the line Z_L impedance of the load or equipment under test.

The distance D from the fault or anomaly can be defined by the following expression.

$$D = \frac{c}{\sqrt{\epsilon_r}}\frac{\Delta T}{2} = V_P \frac{\Delta T}{2} \tag{2}$$

Where V_P is the propagation velocity of the pulse in the material, ΔT is the time interval to be monitored, c is the speed of light in vacuum and ϵ_r is the specific electrical permittivity of the material.

2.1 *Solar module modeled as transmission lines*

The TDR technique does not analyze the DC resistance of the solar module, but its AC impedance. The TDR pulse has a spectral component with respect to the rise time of the TDR pulse, which takes into account the skin effect of the solar module connection busbars, acting as a transmission line [10]–[12].

This paper addresses the solar module as transmission lines, the propagation of the TDR signal is strongly linked to the substrate material and the radio aspect of the solar module [13]–[15], as shown in the equation 3 and 4,

$$\epsilon_{eff} = \frac{\epsilon_r + 1}{2} + \left[\frac{\epsilon_r - 1}{2\sqrt{1 + 12\left(\frac{H}{W}\right)}}\right] \tag{3}$$

$$Z_0 = \frac{120\pi}{\sqrt{\epsilon_{eff}\left[\frac{W}{H} + 1.393 + \frac{2}{3}\ln\left(\frac{W}{H} + 1.444\right)\right]}} \tag{4}$$

Where ϵ_{eff} is effective electrical permittivity, ϵ_r is relative permittivity, H is substrate thickness, W is connection width and Z_0 is microstrip impedance.

Due to this issue, the characterization of the solar module is approached as a eletronic signature derived from busbar grid reflection coefficient as a function of time.

3 Setup of the TDR Measurements on Solar Module

This work carried out the TDR measurement in a solar module without the by-pass diode and without the lamination process, only the solar cells and the busbar connections, thus reproducing the assembly process of the solar module in the factory.

3.1 *Assembly Setup and Precautions.*

The TDR measurement for solar modules is not a conventional measurement, because the sample is an active element. Therefore, additional care must be taken when assembling the TDR setup.

The Fig. 4 shows the manual assembly process of the solar module in the laboratory and table I presents the equipment and configurations used in the TDR measurement.

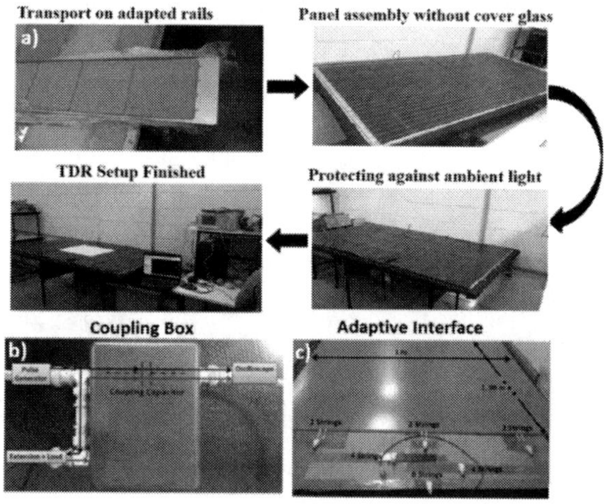

Figure 4. Assembly Setup in laboratory. a) Illustration of the manual process of assembling the solar module in the laboratory, b) Coupling box used in TDR measurement and c) Illustration of the adaptive interface used in TDR measurements for different combinations of strings.

10.4229/EUPVSEC2023/3AV.3.36
020231-002

Table I. Equipment and configurations used

Equipment	Configurations
Digital Oscilloscope	200MHz 1Gsa/s
Bench Multimeter	Autoranging
Pulse Generator	1kHz, 500ns, 1V, 50 ps
Coupling Box	0.18μF and 200 V
Photovoltaic module	320Wp - Doble Glass Polycrystalline
Control Software	3000 Series Scope Connect Software
4 meter extension	Connections
Adaptive Interface	6 - BNC connectors

The coupling of the TDR pulse coming from the signal generator to the solar module under test is performed by a capacitor in series in the test circuit and by the same way the signal reflected back is measured by the oscilloscope, as shown in fig. 4 b) thus avoiding damage caused by some DC current from the solar module under test that can feed back to the pulse generator and/or oscilloscope. The identification of the electrical signature of the solar module was carried out under shading condition and temperature conditions, as the equipment under test is active in nature, as shown in Fig. 4 a).

However, for the measurements to be perfectly executed, it is necessary to develop an interface, in which the TDR pulse can be easily applied and repeated, maintaining the accuracy of the results. The adaptive interface must be matched to 50Ω so that there are minimal reflections in the measurement system, as shown in fig. 4 d).

3.2 Measurement System Calibration.

Calibration of the measurement system used the open, short, load technique, in addition to using a 4-meter extension cable as a reference standard. Listed below is the calibration method and Fig. 5 illustrates the oscilloscope screen for each check.

- In open-circuit calibration, check the maximum extension of the reference extension cable, before the reflection tends to infinity.
- In short-circuit calibration, after the pulse runs through the entire reference extender, the reflection tends to zero, the minimum reflections observed on the oscilloscope concern the measurement system itself.
- Finally, knowing well the minimum and maximum reflections of the system and also the standard of the extender used, the measurement can be carried out with the load or equipment under test.

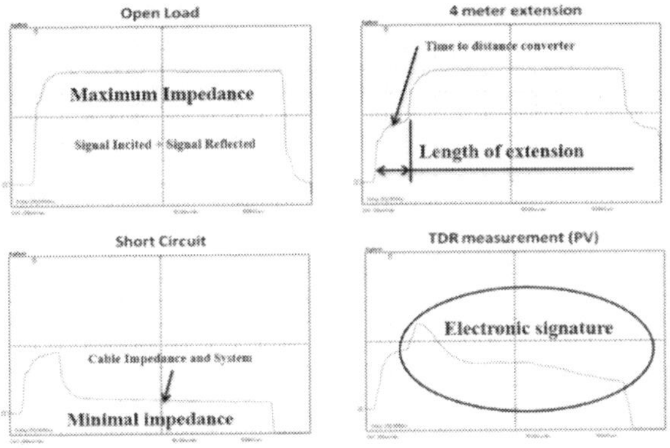

Figure 5. Oscilloscope screens referring to the Open, Short, Load calibration of the TDR setup.

3.3 TDR Spatial Resolution

Spatial resolution refers to the ability of TDR systems to precisely locate two closely spaced impedance discontinuities [16]. The shape and rise-time of the step stimulus is important for accurate TDR measurements [17], [18]. Spatial resolution is a function of the risetime of the incident TDR pulse and the effective dielectric constant of the transmission line being measured [19]. The equation 5 shows the minimum two-event resolution of a TDR measurement setup.

$$l = \frac{t_r c}{2\sqrt{\epsilon_{eff}}} \tag{5}$$

Where l is the length of the minimum resolved feature size, c is the speed of light in air, t_r is the rise-time of the incident TDR pulse, ϵ_{eff} is the effective dielectric constant.

However, the resolution of a TDR is directly related to the system rise-time, which is actually the overall rise-time of the measurement system, as shown in equation 6.

$$tr_{system} = \sqrt{tr_{step}^2 + tr_{scope}^2} \tag{6}$$

Where tr_{system} is the overall system rise-time, tr_{step} is the risetime of the incident pulse, tr_{scope} is the rise-time of the sampling oscilloscope.

10.4229/EUPVSEC2023/3AV.3.36
020231-003

4 Measurements Results

The results obtained from the TDR measurements are shown in Fig. 6 and 7.

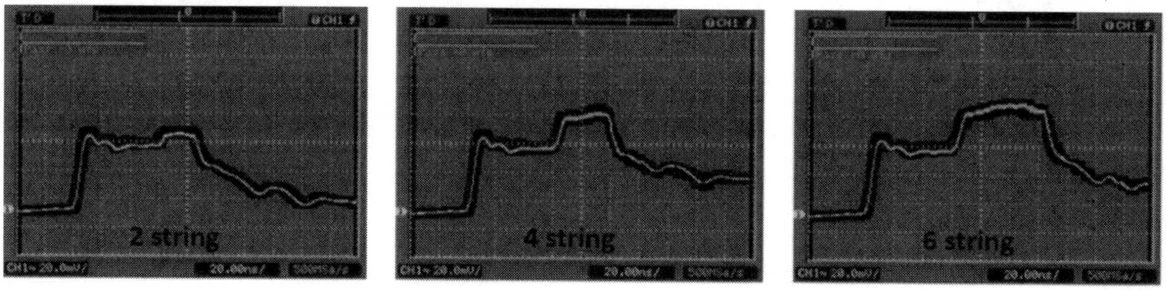

Figure 6. Routine test masks for solar module in manufacture in 2 string, 4 string and 6 string configuration to check for non-visible defects

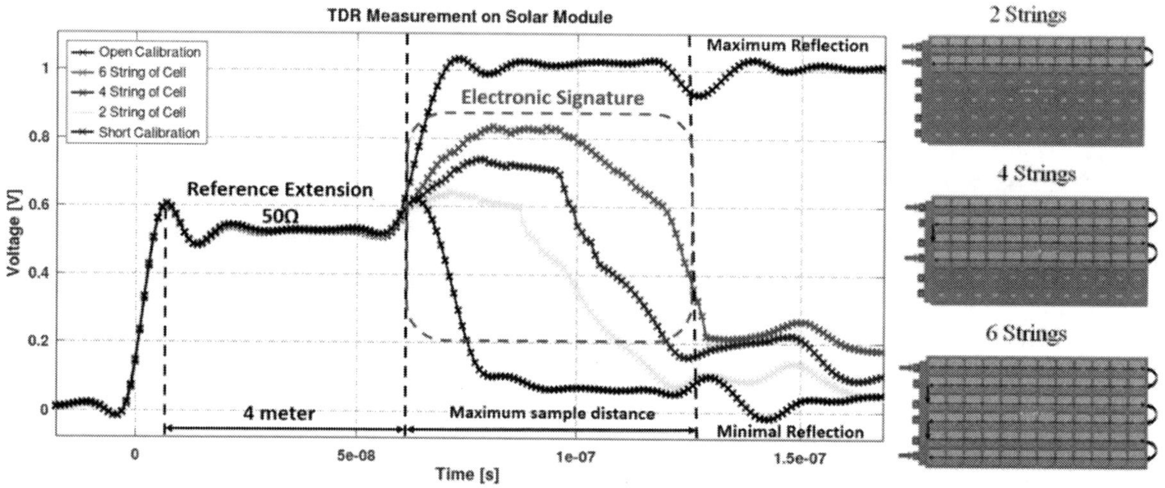

Figure 7. Parameterized measurement result in 2 string, 4 string and full configurations and Illustration of settings used in the TDR measurement in the solar module.

The TDR measurement was carried out in 3 different configurations: 2 strings, 4 strings and 6 strings or complete solar module, which presented an electronic signature for each configuration and consequently a routine test mask, as observed in Fig. 6.

5 Discussion of results

Note the TDR measurement in Fig. 7, that the reference extension with impedance close to 50Ω propagates up to approximately 650 milisecond in the graph, from there the electronic signature of the solar module is obtained.

The distance traveled by the TDR pulse and the reflection of the signal in the 6-string configuration is greater than the other configurations, as the TDR pulse travels a greater distance and consequently presents greater impedance due to the greater number of solar cells in series.

As observed in the 3 and 4 equations, any degradation that occurs in the silicon substrate of the solar cells that impacts the ϵ_r parameter will affect the electronic signature of the previously characterized solar module, the error being captured by the electronic signature mask.

6 Conclusion

The TDR electronic signature mask can be used for routine pass-no-pass type test purposes, due small variations in the TDR electronic signature mask allow identification of errors in the solar module assembly, commissioning of photovoltaic systems and certification process.

In this case, the signature of the 320Wp - Double Glass Polycrystalline model was obtained, however each model of solar module without the bypass diode will have a particular electronic signature and consequently a TDR mask for that signature for pass-no-pass test.

Using the TDR technique for inspection of non-visible defects it may be possible to detect defects such as: busbar connection problems, cold solder joint and solar cell degradation caused by PID effect. The TDR mask is presented as a complementary technique for the inspection of defects by electroluminescence in solar modules.

7 Acknowledgment

This work was partially supported by the Research Project Unicamp-Funcamp-BYD Energy Brazil under contract nº 84406-22 and PADIS (Brazilian Program for the Technological Development of the Semiconductor Industry) of MCTI, Brazil. Also thanks to CNPq (National Council for Scientific and Technological Development) for granting of scholarship to the author with the process number 445074/2020-5.

References

[1] E. L. Meyer and E. E. Van Dyk, "Assessing the reliability and degradation of photovoltaic module performance parameters," *IEEE Transactions on reliability*, vol. 53, no. 1, pp. 83–92, 2004.

[2] T. Potthoff, K. Bothe, U. Eitner, D. Hinken, and M. Köntges, "Detection of the voltage distribution in photovoltaic modules by electroluminescence imaging," *Progress in Photovoltaics: Research and Applications*, vol. 18, no. 2, pp. 100–106, 2010.

[3] K. G. Bedrich, "Quantitative electroluminescence measurements of PV devices," Ph.D. dissertation, Loughborough University, 2017.

[4] I. E. Commission *et al.*, "Iec 61215-1: 2021 terrestrial photovoltaic (pv) modules—design qualification and type approval—part 1: Test requirements," *International Electrotechnical Commission: Geneva, Switzerland*, 2021.

[5] D. Smolyansky, "Electronic package fault isolation using tdr," *ASM International*, pp. 289–302, 2004.

[6] N. Ida, "Engineering electromagnetics," in *New York etc: Springer, 2000*. Springer, 2000, pp. 1–3.

[7] N. Bosco, T. J. Silverman, and S. Kurtz, "Climate specific thermomechanical fatigue of flat plate photovoltaic module solder joints," *Microelectronics Reliability*, vol. 62, pp. 124–129, 2016.

[8] A. Silveira, S. Barbin, and L. Kretly, "Using tdr-time domain reflectometry measurements to compare ribbon busbar versus wire busbar connections in polycrystalline solar cells: The signature approach," in *2018 IEEE-APS Topical Conference on Antennas and Propagation in Wireless Communications (APWC)*. IEEE, 2018, pp. 1–3.

[9] T. A. Note, "Tdr impedance measurements: A foundation for signal integrity," 2004.

[10] M. Buzuayene, "Rise time vs. bandwidth and applications," *Interference Technol. Available online: https://interferencetechnology. com/rise-time-vs-bandwidth-and-applications/(accessed on 30 October 2019)*, 2008.

[11] H. Casimir and J. Ubbink, "The skin effect," *Philips Technical Review*, vol. 28, no. 9, pp. P275–276, 1967.

[12] A. Silveira, M. Neves, R. Garcia, H. Alvarez, M. Villalva, F. Marques, and L. Kretly, "Characterization of solar cell busbar grid for different technologies by time domain reflectometry simulation: Transmission line approach," in *2023 IEEE Photovoltaic Specialists Conference (PVSC 50)*. IEEE, 2023, pp. 1–5.

[13] N. M. Din, C. K. Chakrabarty, A. B. Ismail, K. K. A. Devi, and W.-Y. Chen, "Design of rf energy harvesting system for energizing low power devices," *Progress In Electromagnetics Research*, vol. 132, pp. 49–69, 2012.

[14] R. Garg, I. Bahl, and M. Bozzi, *Microstrip lines and slotlines*. Artech house, 2013.

[15] M. Neves, A. Silveira, R. Garcia, H. Alvarez, F. Marques, and M. Villalva, "Effects of salt spray on csi photovoltaic modules in the brazilian region," in *2023 IEEE Photovoltaic Specialists Conference (PVSC 50)*. IEEE, 2023, pp. 1–6.

[16] D. Antonovici, "Advances in time domain reflectometry characterisation for high speed interconnects," in *2015 IEEE 21st International Symposium for Design and Technology in Electronic Packaging (SIITME)*. IEEE, 2015, pp. 37–40.

[17] M. D. Tilden, "Measuring controlled-impedance boards with tdr," *Printed Circuit Fabrication*, vol. 14, no. 2, 1992.

[18] K. Schoen, "Sub-10 picosecond differential tdr and tdt pulse source," in *Conference, 2003. Fall 2003. 62nd ARFTG Microwave Measurements*. IEEE, 2003, pp. 25–34.

[19] H. Packard, "Time domain reflectometry," *Application Note*, 1998.

A Study on the Quality Analysis of Aged Solar Modules for Residential Use

:Analysis of defect type of deteriorated solar module

Han-Yeol Kim*, Jeong-Eon Lee**, Jae-Hyeok Hur**, Myoung-Geon Seo***, Bo-Woon Han***, Min-Seok Kim***, Sol-Yi Lee***, Yeon-Bum An***, Su-Mi Yang***
Dept. of Green Energy Engineering Far East University
klhan3091@gmail.com, 1523-8@hanmail.net, hgh9848@naver.com, kkk10108@naver.com, onlyyou558@gmail.com, rlaalstjryes@naver.com, thftka0513@naver.com, dksdusqja@naver.com, esther4853@kdu.ac.kr

ABSTRACT

Since the solar industry in Korea was fully activated in the 2000s, as the installation of solar modules has increased rapidly, the number of defective modules has also been increasing due to the increase in old modules. In particular, due to the nature of solar power systems, which are heavily influenced by the natural environment, degradation and failure can occur due to various causes. In addition, modules that have been exposed to the natural environment for a long time may change some of the properties of the materials that make up the module or develop electrical characteristics that were not a problem when the module was manufactured. In this study, we analyzed the causes of solar module failures and the types of failures by module type, as there are different factors that cause failures between modules. Old solar modules from small and large solar power plants were measured using EL meters, I-V tracers and thermal imaging cameras, and the results were analyzed to create a database of defective solar module types. In particular, in the case of crystalline silicon modules, the rated power of the modules has been reduced due to both failures caused by the modules themselves and failures caused by external factors that develop after installation, and in the case of modules manufactured before the 2010s, the quality standards for clear module performance have not been established, resulting in problems caused by raw material degradation in addition to electrical failures. Therefore, this paper analyses the types of defects in PV modules more than 10 years old and proposes grading criteria based on the types of defects, such as electrical characteristics and performance.

Keywords—PV Module, Defect Type, EL Inspection, Thermal Image

I. Introduction

As the core equipment of a solar power plant, solar modules are undergoing rapid technological change as the solar industry becomes more sophisticated. The technology for reusing waste solar panels must also be able to respond to these changes. As a leader in renewable energy, Europe is already preparing for this change. According to the KEA(Korea Energy Agency), research is underway in Europe, Japan and China not only to repair defects to extend the life of solar modules, but also to create improved recycling plans and technologies for existing solar modules, develop materials for sustainable solar modules and design new solar modules[1]. For these government efforts to be effective, the industry needs to establish standards to accurately identify which panels should be recycled, reused or scrapped.

The general reuse process is to first assess the manufacturer, model number, application situation and photographs, then carry out cleaning, insulation test, IV curve test and EL test to assess the quality as a reuse module, and then decide whether to supply it as a reuse panel or a recycled panel. For this to be widely implemented, it is important to establish criteria for the classification phase, technical guidelines and standards detailing the classification and testing steps. This is essential to ensure the high quality of reusable modules and can make them more competitive.

In this paper, the requirements for the selection of reusable modules and the basic classification steps of identification of defect types that can be analyzed by visual inspection and analysis of defect types using EL cameras are presented, and the grade evaluation criteria for establishing quality standards for reusable modules are divided into A, B and F grades by reference to IEA-PVPS. Each type of failure is rated with a different number of points and the criteria is based on the impact of the failure type on the performance and lifetime of the solar module. In addition, the electrical characteristics were evaluated and the output tested using an I-V tester, and a rating was calculated based on the output efficiency.

II. Analysis of the type of failure and process for re-use of obsolete modules

For crystalline silicon module products, degradation of rated power may occur as the module evolves after installation, due to failures caused by the module itself and failures caused by external factors. Modules that have been exposed to the natural environment for a long period of time may experience some changes in the properties of the materials that make up the module, or electrical characteristics that were not a problem when the module was manufactured may evolve and cause failure. The latest modules are installed and developed by minimising defects through material selection and strict quality control during the manufacturing process. However, there are no established quality standards for clear module performance for modules manufactured before the 2010s, leading to problems with electrical failures and material degradation. Among them, various defects observable with the naked eye are listed in <Table I>.

A. Visual inspection process for reusable modules

Sorting aged solar modules by visual inspection is one of the most important steps in removing bad modules from further processing. <Table I> shows the typical defects listed in IEA-PVPS, which are easy to identify because they are obvious on the module. There are three main types of damage caused by external forces or pressures, such as glass cracks and frame damage. Types of moisture ingress include delamination, which is the separation of different interfaces, and snail traces, which are caused by chemical degradation due to moisture ingress into microcracks in the cell. The last type is occurs during power generation, such as EVA discolouration, where the packaging material changes colour due to the heat of the cell, burning of the EVA and backsheet as it heats up along the busbar, and hot spots due to bypass diode failure, leaving burn marks. These sources of failure can, in the worst case, cause a fire and should be checked regularly.

<Table I> Typical defects of Visual inspection

PV module component	PV module failures
Front Surface of PV	Air Bubble, Peel, Yellowing, browning
PV Cell	Broken cell, cracked cell, AR discoloration, snail track
Electrodes in PV Cell	Burned, Oxidation
Back Sheet of PV	Air Bubble, Peel, Yellowing, Scratch, Burned
Frame	Crooked, Broken, Scratch, Warped
Junction Box	Loose, Oxidation, Corrosion
Wire	Separated, Fragile, Component Exposed

If an old module has visible defects and the inspection shows that the damage is severe enough to affect performance, such as broken modules or dislodged frames, it is classified as Grade F, which means it cannot be reused, and is sent to the recycling process.

B. EL defects of PV module and incidence rate by type

In this paper, solar modules from small and large photovoltaic power plants were examined, and modules that were not abnormal under visual observation were cleaned for reuse and subjected to EL testing and I-V curve testing. EL measurements can reveal whether a module is shunted, PID, defective cells, bad diodes, broken cells, cracked cells, etc. <Table II> shows the approximate information of the deteriorated solar modules examined, and [Fig. 1] shows the four most common EL defects of the aged modules.

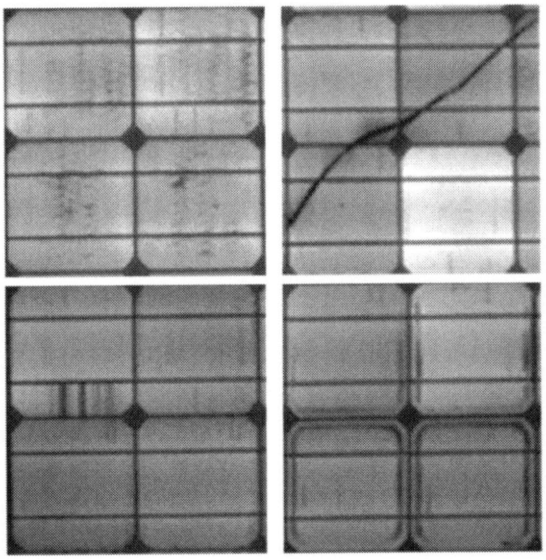

[Fig.1] Types of errors commonly found in EL measurements

[Fig.2] Front view of the PV module under test

<Table II> Information of PV Module

Manufacturer	Shinsung
Module Output	215 [W]
Cell Type	single crystalline
Busbar	2
Year Of use	11~12 Years

[Fig. 3] is a graph of EL defect incidence by module type examined. The study was carried out on 80 old solar modules that had been visually inspected and washed. According to the data collected, the most common failure is finger interruptions, which is attributed to poor screen printing during solar cell manufacture. The second most common defect was Belt Print, a contamination caused by the lack of regular belt cleaning during the solar cell manufacturing process.

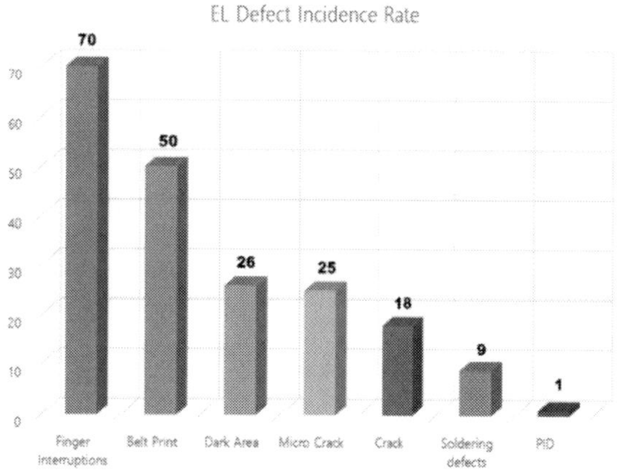

[Fig. 3] EL Defect Incidence Rate by Type

In addition to the experimental data, the EL photographs in [Fig. 1] show that the most common type of defect is a failure in the cell manufacturing process. This could be due to a lack of control at the cell manufacturer's end of the cell manufacturing process, a specific recipe or an equipment problem. In the case of cell cracks, some are due to expansion of microcracks in the cell itself and durability, such as external impact, but it is speculated that cracks often occurred during demolition, so they were a common occurrence.

C. Correlation between EL defects and thermal images

In some cases, the types of defects identified by EL testing do not appear when measured with a thermal imaging camera during actual power generation. In this study, we compared the thermal images of normal and defective modules with EL images and confirmed the relationship between EL and thermal images. Thermal imaging measurements were taken using a Flir EX5T.

[Fig. 4], [Fig. 5] show thermal images of normal and defective modules in actual power generation and EL, I-V test results.

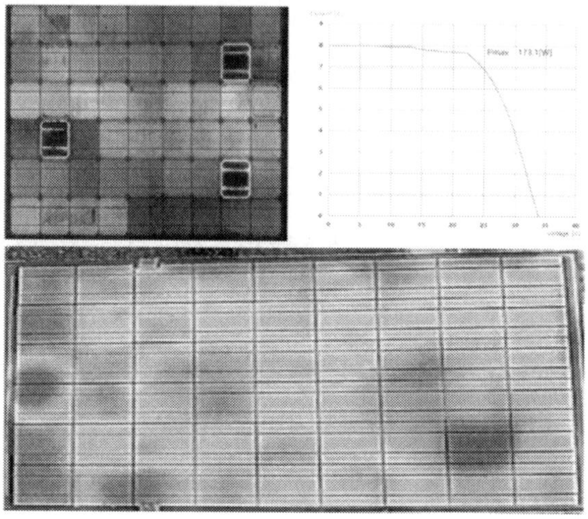

[Fig. 4] Test results of defective module

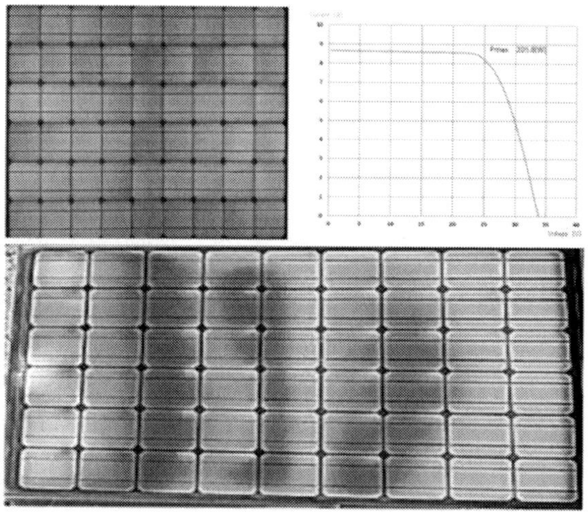

[Fig. 5] Test results of normal module

[Fig. 4] shows the results of a F Grade module. The EL results showed multiple defects in the cells, but the thermal images did not show all of the hot spots at the defect sites. [Fig. 5] is the result of a normal module, and while the EL result was clean with no defects, the thermal imaging measurement showed a hot spot near the diode. This confirms that not all defects in EL results can be detected by thermal imaging and, conversely, not all defects in EL can be detected by thermal imaging.

III. Suggestions for PV module grading criteria for re-use

In this study, the types of defects and their incidence rates were databased through research on the reuse of deteriorated photovoltaic modules, and based on this, the reusability

10.4229/EUPVSEC2023/3AV.3.37
020232-003

assessment procedure and criteria were proposed.

<Table III> Details in Criterion of Process

Category	Failure Types	Failure Details	Subtract Points
Visual & Thermal Inspection	Tempered Glass	Glass Sealing Deterioration	-10
		Contamination	
	Backsheet	Backsheet Scratch, Dent, Discoloration	-10
	Encapsulant Material	EVA Discoloration	-5
		EVA Delamination	-20
	Corrosion	Libbons, Metal Corrosion	-10
	Thermal	Hotspots on Junction Box Area	-5
		Under 3 Hotspots	
EL Inspection	Crack	Fine Cracks Through EL Images	-10
	Contamination	Belt Print, Picker Print	
	Diode Defect	Diode Defect Through EL Images	
	PID	PID Through EL Images	-20
Output Test	Electrical Characteristics Test (Pmax)	Above 90% Effciency	0
		Above 80% Under 90%	-10
		Under 80%	Recycle

<Table IV> Rating Based on Deduction

Rate	Subtract Points	Process
A	Under 0 ~ -20	Maintenence work and Reuse
B	Over -20 ~ Under -30	
F	Over -30	Recycle

When evaluating deteriorated modules, the status of the module was checked according to the checklist in <Table III> so that the total number of points could be used to decide whether the module should be reused or recycled. <Table IV> is the treatment standard according to the total number of points, if the total number of points of old solar modules exceeds -30, they are subject to the recycling process as grade F, and if it is below -30, they are classified into grades A to B to determine the required reuse process or treatment direction after detailed inspection. After the EL test, the module is subjected to a maximum power test using an I-V tester. The measurement is performed at STC conditions and the efficiency is measured by measuring the final output value against the module specification.

The following equation is for measuring the efficiency of the module.

$$Eff(\%) = \frac{Actual\ Output\ \ Value}{Output\ \ Value\ on\ Spec} \times 100$$

If the module's efficiency was below 80%, it was immediately classified as F and sent for recycling.

IV. Conclusion

In this study, a classification and criteria were developed to analyze the failure types of deteriorated solar modules and to proceed with the subsequent treatment process such as recycling and reuse. EL measurements were used to analyze the types of EL defects in old solar modules. The incidence rates were plotted on a graph. We conducted a study of 80 solar modules of the same specification that were visually and thermally inspected, and found that finger interruptions and belt prints accounted for the majority of failures. While this causes a long-term output degradation, it's a small enough defect that it can be reused, so it's not considered a significant penalty. We also correlated the EL test results with the thermal images and found that the defect areas did not match up completely.

The criteria for grading in the reuse process were presented, and a grade was set according to the total number of points deducted, and subsequent processing was carried out accordingly. This study will contribute to the establishment of performance evaluation standards for domestic old photovoltaic modules. Further research should be needed to develop appropriate systems for managing deteriorated PV modules.

Reference

[1] Song Myung-kyu, "The abandoned solar power, Industrial Zone Rebirth, Today Energy, January 1, 2022, http://www.todayenergy.kr/news/articleView.html?idxno=243987

10.4229/EUPVSEC2023/3AV.3.37
020232-004

EVALUATION AND PREDICTION OF SOILING ON PV SITES AND ADAPTATION OF CLEANING STRATEGIES

Yaman Al-Riyalat*[1], Jan de Leeuw[2], Martin Dennenmoser[1], Kai Saegebarth[2], Michael Schmidt[3]
[1] BayWa r.e. Solar Projects GmbH, Kaiser-Joseph-Str. 263, 79098 Freiburg im Breisgau
[2] BayWa r.e. Solar Projects GmbH, Arabellastr. 4, 81925 Munich
[3] Offenburg University, Institute for Sustainable Energy Systems (INES), Badstr. 24, 77652 Offenburg
*Phone: +49 151 70563308 ; Email: Yaman.Al-Riyalat@baywa-re.com

ABSTRACT: Soiling is an important issue in the renewable energy sector since it can result in significant yield losses, especially in regions with higher pollution or dust levels. To mitigate the impact of soiling on photovoltaic (PV) plants, it is essential to regularly monitor and clean the panels, as well as develop accurate soiling predictions that can affect cleaning strategies and enhance the overall performance of PV power plants. This research focuses on the problem of soiling loss in photovoltaic power plants and the potential to improve the accuracy of soiling predictions. The study examines how soiling can affect the efficiency and productivity of the modules and how to measure and predict soiling using machine learning (ML) algorithms. The research includes analyzing real data from large-scale ground-mounted PV sites and comparing different soiling measurement methods. It was observed that there were some deviations in the real soiling loss values compared to the expected values for some projects in southern Spain, thus, the main goal of this work is to develop machine learning models that could predict the soiling more accurately. The developed models have a low mean square error (MSE), indicating the accuracy and suitability of the models to predict the soiling rates. The study also investigates the impact of different cleaning strategies on the performance of PV power plants and provides a powerful application to predict both the soiling and the number of cleaning cycles.
Keywords: Predictions, Monitoring, Performance, Cleaning, Machine Learning, Decision-making

1 INTRODUCTION

The main factors determining how to place PV systems for the best yield are the geographic location (latitude and available solar radiation) and installation design (tilt, orientation, and altitude). However, other parameters could affect how well the system performs (system efficiency and performance) such as dust and bird droppings which are both often unmarked and underestimated [1] [2].

Soiling loss occurs when dust and dirt build up on PV surfaces over time, reducing the amount of sunlight that can reach the solar panels. Soiling loss plays a significant role in determining PV system losses since a PV system's efficiency is directly related to how much sunlight can be converted into electrical energy. The quantity of wind, dirt, local traffic, and the frequency and intensity of rain all significantly impact yield loss due to soiling.

Since the solar panel is sensitive to the density of dust, it is crucial to study the impact of soiling on solar panels and have a cleaning strategy to remove dust from the panel's surface [3].

Cleaning the panel surface reduces the impact of soiling on PV performance. However, frequent cleaning drives up maintenance and operation costs. As a result, the frequency of panel cleaning must be evaluated in relation to the cost of cleaning, location, PV design, soiling rate, and anticipated power output [4].

The spatial variability of soiling deposition is a crucial issue since it can economically be more advantageous to clean only a part of the strings rather than the entire PV system. The soiling rates and the regularity of natural cleanings can also shift with the seasons and from year to year [5].

This study aims to investigate the issue of soiling loss in PV plants and to improve the accuracy of soiling predictions. This includes understanding how soiling can affect the efficiency and productivity of the panels, measuring and predicting soiling rate with machine learning algorithms, and evaluating the impact of cleaning strategies on the performance of PV power plants. In other words, the research questions are:

1. How to measure soiling?
2. How to predict it effectively?
3. How to find the optimum cleaning strategy?

2 METHODOLOGY

Machine learning models will be used to predict the soiling in different locations. Therefore, input (weather data) and output (soiling rate) data from several locations should be collected so that the models can be trained on these data. Historical weather data and other performance-related data from yield studies and a monitoring portal [6] were collected for different projects in Spain and France.

2.1 Photovoltaic Plants

To understand the effect of soiling under different conditions, five PV plants, four in Spain and one in France with different soiling rates and loss have been analyzed in this study.

2.2 Collected Dataset

The energy generation values, radiation data, specific yield, and data from sensors which are installed in each project have been collected to calculate the performance ratio (PR) of the project and the soiling rates and losses.

Furthermore, to understand the impact of weather on dust accumulation and to study the effect of rain on cleaning the modules, several historical weather datasets (humidity, dew, precipitation, wind speed, and wind direction) have been collected from a publicly available database (Visual Crossing) [7] for each project.

10.4229/EUPVSEC2023/3AV.3.38
020233-001

2.3 Performance Difference Method

A performance difference (PD) method was developed to determine the PR values. The tool's objective is to determine the soiling losses based on the deviation between the actual PR and the adjusted PR, which is the budget/expected PR in yield studies minus the module degradation value which is 0.25% each year.

Yield studies provide a monthly PR (budget PR) and specific yield which could be compared to the actual monthly PR (calculated PR) and specific yield to indicate the soiling loss if the PR is not affected by different factors such as plant outages or power limitations. By taking the monthly budget PR and the specific yield, temperature losses would be excluded (from budget values). This ensures that the difference between the budget and the actual PR can be mainly accounted for soiling and dust accumulation.

There are some uncertainties in this approach, as there is still some influence of temperature loss or other losses on the calculated PR. However, for locations with high soiling rates, this method can provide a good soiling estimation.

This method is used to calculate the annual soiling losses which depend on the PR difference and how much this difference affects the total yield of the PV plant. In this step and after cleaning the data, the daily specific yield in [kWh/kWp] and the PR for each inverter are calculated by the following equations:

$$PR = \frac{Daily\ production}{Radiation\ yield * Inverter\ capacity} \quad (1)$$

$$Specific\ energy\ yield = \frac{Daily\ energy\ production}{Inverter\ capacity} \quad (2)$$

After performing the calculations above, the estimated daily specific yield loss due to soiling (YL) in [kWh/kWp] can be calculated as follows:

$$YL = \frac{Budget\ PR - Actual\ PR}{Actual\ PR} x\ Specific\ energy\ yield \quad (3)$$

2.4 Artificial Neural Network

The ANN model was developed in Python using the Scikit-learn library which is an open-source Python library [8] that provides an implementation of the best hyperparameters for training the model. Hyperparameters are the settings that control a model's behavior. These hyperparameters are defined outside of the model, however, they are directly related to the model's performance.

The following five parameters are used as inputs: dew, relative humidity, precipitation, wind speed, and wind direction, with just one parameter predicted as output: the daily soiling rate. Before the training process, daily data was collected for three different fixed-tilt projects with different weather conditions across Spain. Both input and output data sets were normalized to the range between 0 and +1.

Normalization aims to increase the model's accuracy by converting the values of numeric columns in a dataset to a common scale without changing differences in value ranges [9].

After normalization, the entire dataset was divided into two groups, one for artificial neural network model training (80%) and another one for testing (20%).

It is extremely difficult to choose the number of neurons in a hidden layer and the number of hidden layers for any artificial neural network model. Therefore, two functions have been developed to overcome this difficulty. The first one will find the number of nodes for the hidden layers by linearly changing the number of nodes between predefined random values for the hidden layers. Then, another function dynamically creates new models based on given hyperparameters. Finally, the "GridSearchCV" algorithm of the Scikit-Learn library in Python was used to find the optimal model performance score and the optimal parameters.

Based on the above discussion, one hidden layer and 16 neurons were found to be the best structure to predict the daily soiling rate with maximum accuracy. Figure 1 shows the schematic diagram of the artificial neural network model used in this study.

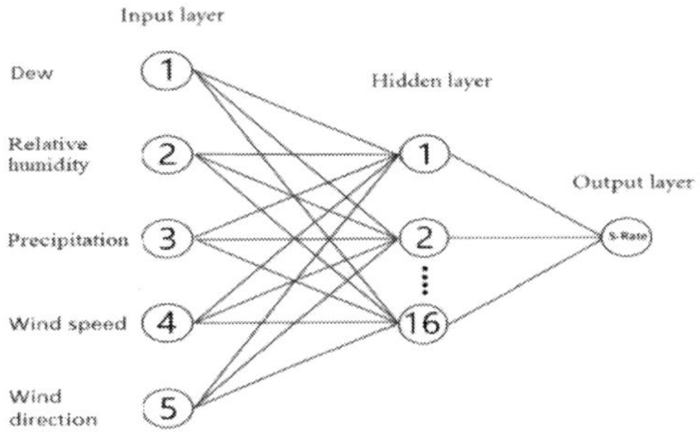

Figure 1: The schematic diagram of the artificial neural network model used in this study.

2.5 Random Forest

Unlike ANN, the random forest (RF) algorithm has a straightforward training procedure. When compared to artificial intelligence machine-based models, the training procedure is much faster and data normalization is not required.

In this study, the RF model was developed in Python using the Scikit-learn library. The same dataset and input parameters used in the ANN model are used as inputs in this model.

To optimize the best hyperparameters for the model, a random search goes through a predefined distribution of hyperparameters to find the best set of parameters for a given model.

Finally, a prediction is made using the random forest regressor model testing data value to compare the predicted value to the actual value of the testing data.

3 ANALYSIS, RESULTS, AND VALIDATION

3.1 Analysis and Results of the PD Method and Soiling Sensors

Figure 2 illustrates the soiling loss estimation analysis for one of the projects in southern Spain for the year 2021. The blue points represent the daily actual PR, while the budget PR (from yield studies) is represented by the black dashed line. The gray columns show the cleaning events and the orange columns represent rain events of 2mm/ day or higher.

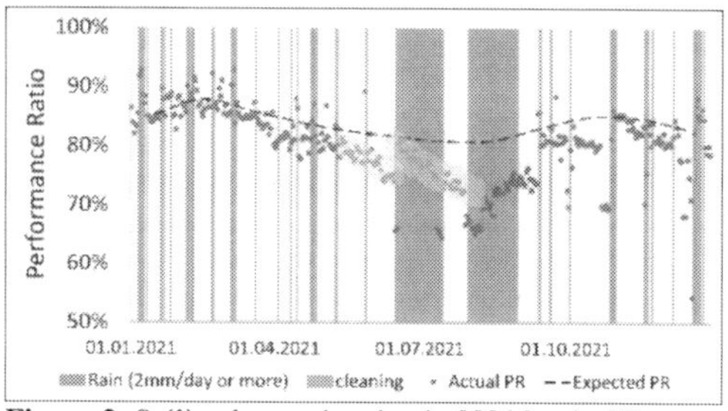

Figure 2: Soiling loss estimation in 2021 by the PD method for a project in southern Spain.

A gradual but marked decrease in system performance through the dry season (summer months) is observed in this project in southern Spain. System performance returns to normal levels after a period of rain following the dry season. Two cleaning rounds were done each year, the first cleaning round started on 15.06, and the second started on 01.08. Each cleaning round lasted up to 6-7 weeks, which explains the gradual increase in the performance ratio in these two periods.

The same analysis was repeated for the same years to compare the results obtained from the soiling sensors with those obtained through the PD method. Figure 3 shows the soiling loss for the year 2021 as determined by the PD method and the soiling sensor in the same project.

A significant increase in soiling losses can be observed during the summer period for both the sensors and the PD method, with the two being in close range for dry periods. However, the PD method estimated lower soiling losses during wet periods for all years. Moreover, significant variability in the estimation of soiling losses using the PD method can be observed, particularly during months with reduced soiling accumulation.

Additionally, rainfall events in the amount of ≥ 2mm/day were found to be sufficient to enhance the performance of the modules.

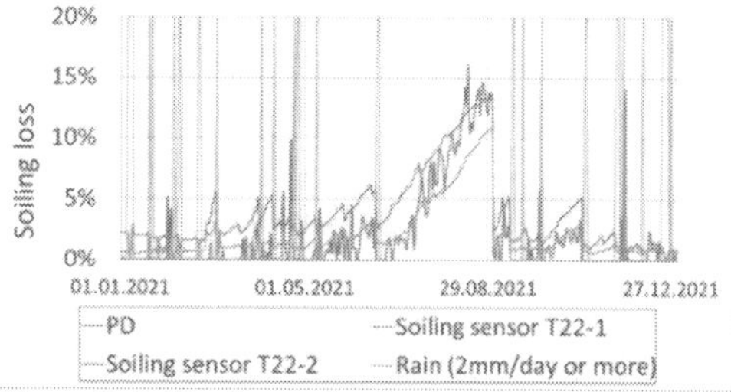

Figure 3: Soiling loss estimation in 2021 for transformer 22.

The rate of soiling, which quantifies the daily loss in system performance, can be estimated by determining the slope of the decrease in performance. The measurement must be taken over an extended period, preferably during a dry season, when no rain events or cleaning activities occur. The resulting soiling rate was determined to be approximately 0.27% per day.

The same procedure has been done for other projects to find the soiling rate that will be used for the training of the machine learning models.

3.2 Machine Learning Models Analysis and Results

In this study, a database includes historical weather data for the same three projects, as well as the monthly soiling rates was constructed.

To gain insight into the relationship between the weather data and the soiling rate, a correlation matrix was generated and presented in Figure 4. This matrix displays the strength and direction of the relationships between the different weather variables and the soiling rate (SR), providing valuable information for further analysis and understanding of the impact of weather on soiling.

The results indicated that dew and humidity had a significant impact on the soiling rate, with dew having a positive effect of 0.56, and humidity having a negative effect of -0.66. This means that locations with higher humidity are likely to experience lower soiling rates. The impact of wind speed and direction was also positive on soiling. On the other hand, precipitation had a low negative correlation with the soiling rate. However, precipitation events with sufficient rain effectively clean the modules and prevent dust accumulation, resulting in a near-zero soiling loss. As a result, precipitation, despite having a low impact on the soiling rate, has a high impact on the annual soiling loss.

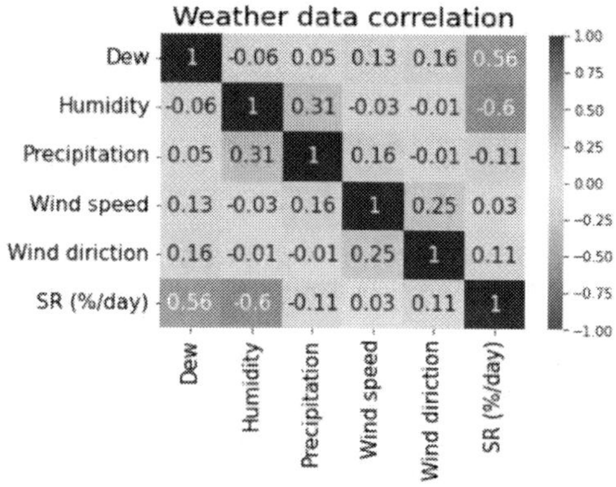

Figure 4: Weather data with the soiling rate correlation matrix.

The performance of both models was evaluated using the mean square error (MSE) and the results showed a high accuracy with an error of less than 2%. Figure 5 presents the results of the average of both models, in terms of monthly soiling rates and annual soiling loss for several locations in Europe. The annual soiling loss calculation was made under the assumption that the soiling loss would be reset to zero if a precipitation event with a minimum of 5mm/day occurred, which is defined as the precipitation threshold.

The figure clearly shows the difference between the soiling rate and the soiling loss. Several locations could have almost the same daily or monthly soiling rate that is affected by weather conditions (as explained before in the correlation matrix) but the annual soiling loss could vary a lot while it depends on the amount of precipitation per year.

	Month	Madrid	Tordesillas	Rome	Munich
Soiling rate (%/day)	Jan	0.09	0.08	0.1	0.08
	Feb	0.11	0.1	0.11	0.09
	Mrz	0.14	0.11	0.12	0.11
	Apr	0.14	0.11	0.13	0.12
	Mai	0.18	0.14	0.16	0.12
	Jun	0.22	0.17	0.25	0.18
	Jul	0.24	0.19	0.31	0.17
	Aug	0.24	0.2	0.28	0.17
	Sep	0.2	0.16	0.24	0.14
	Okt	0.16	0.13	0.15	0.1
	Nov	0.11	0.09	0.13	0.08
	Dez	0.1	0.09	0.11	0.08
Soiling loss (%)	Annual soiling losses **without cleaning** Rain threshold 5mm/day	2.1%	2.5%	4.6%	0.7%

Figure 5: Soiling predictions using the average of ML models for different locations.

3.3 Validation of Machine Learning Models vs. PD Method and DustIQ Sensors

The accuracy of the machine learning models developed in the previous section will be validated against the results obtained from the PD method and DustIQ sensors for two projects, one in southern Spain (Project 1) and another one in northern Spain (Project 2). The results of this comparison are presented in Table 1 and show that the machine learning models predict the soiling rates in a range that is comparable to the results obtained from both the PD method and DustIQ sensors.

Table 1: Comparison of soiling loss results.

Project name	Year	PD	DustIQ	Machine learning models
Project 1	2021	4.5%	3.7%	4.1%
	2022	7.8%	6.2%	5.6%
Project 2	2022	2.8%	1.6%	2.0%

Furthermore, comprehensive evaluations were carried out to evaluate the performance of the developed ML models with real soiling data obtained from various BayWa r.e. projects as well as an external company. Remarkably, the obtained results consistently demonstrated the predictive capabilities of the developed models and validated the accuracy of soiling loss predictions.

4 CLEANING STRATEGIES

A tool has recently been developed to determine the optimal cleaning strategy for each project. The tool also calculates the energy and revenue gain, or loss associated with different amounts of cleaning cycles, as well as the best time to clean the modules to minimize the soiling loss.

It uses a combination of precipitation data, monthly specific yield, installed DC capacity from yield studies, predicted soiling rates from the ML models, cleaning costs, and a fixed energy price per kWh as inputs. Through simulations, it calculates the annual soiling loss and the optimal number of cleanings that result in the lowest revenue loss.

The result is determined by averaging the best cleaning strategies from each year in the simulation, considering the number of cleanings and the soiling loss threshold for the initiation of cleaning.

A simulation has been performed for Project 1 with the assumption of a 2mm/day precipitation threshold. Figure 6 presents the annual soiling loss for different cleaning cycles and soiling loss thresholds at which cleaning starts. The results indicate that 3 cleaning cycles, starting at a 2% soiling loss, result in the lowest annual energy production loss of 3.4%.

Annual loss of energy production in %

Number of cleanings	Soiling threshold 2%	4%	6%	8%
0	7.6	7.6	7.6	7.6
1	6.1	5.4	4.9	4.7
2	3.8	3.6	3.7	3.8
3	3.4	3.5	3.6	3.8

Annual money loss [thousand €]

Number of cleanings	Soiling threshold 2%	4%	6%	8%
0	1375	1375	1375	1375
1	1177	1052	965	936
2	852	818	835	866
3	872	884	899	883

Figure 6: Impact of different cleaning strategies on soiling and money loss.

However, since each cleaning cycle is associated with additional costs, the same simulation but in terms of revenue loss has been done. The results suggest that 3 cleaning rounds will lead to a higher revenue loss. The lowest annual revenue loss is achieved with 2 cleaning rounds that start at a 4% soiling loss.

Figure 7 shows the difference between an optimized cleaning strategy and no cleaning, for the years between 2018 till 2023.

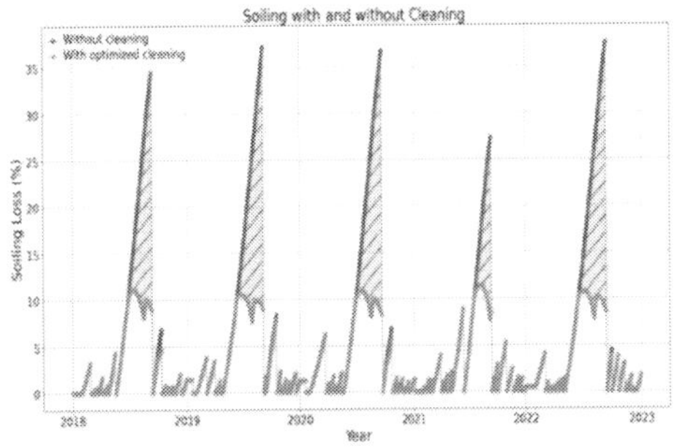

Figure 7: Project 1 soiling loss simulations with and without cleaning.

The diagram demonstrates the significant difference with and without cleaning, and in a location like Project 1 in southern Spain, the soiling could reach a high level of around 40% without cleaning and can be reduced to a maximum of around 12% with optimized cleaning cycles.

5 CONCLUSION

The usage of machine learning models has been investigated throughout this research to improve the prediction of soiling rates for new projects in different locations. With the enhanced accuracy of the soiling predictions, the best cleaning strategy can be identified, and the PV plants' yield can be optimized.

The study started by comparing two different methods for measuring soiling loss, including DustIQ soiling sensors and the Performance Difference (PD) method. In general, the results showed that the PD method provided lower soiling loss values in winter because of the effect of several other factors on the performance of the PV modules, and some outliers that affected the calculations, while the DustIQ sensor results were more realistic.

The results showed that soiling is a complicated function of environmental variables. The main significant variables are dew and relative humidity, which determine the rate at which dust accumulates on module surfaces, affecting their performance.

Additionally, two machine learning models were developed, including Random Forest (RF), and Artificial Neural Networks (ANN). The models were trained to predict the daily soiling rate by considering the impact of weather variables. The added value of these models was analyzed, and it was found that both RF and ANN models were suitable for capturing the complex relationship between PV soiling loss and environmental variables and could accurately predict soiling rates with a Mean Square Error (MSE) of less than 2%.

The soiling profile of each site is impacted by both the soiling rate and the precipitation amount, with no clear threshold for rainfall cleaning impact being identified. However, it was generally found that 2mm/day was good to enhance the performance of PV modules.

Furthermore, soiling can vary dramatically from year to year and therefore the average soiling loss over several years should be taken to estimate the soiling loss for each site.

Finally, this study highlights the importance of accurate soiling predictions and effective cleaning strategies in optimizing the performance and productivity of PV power plants.

The soiling predictions make the business case for BayWa r.e. and the investors more economically safe. This research demonstrates that machine learning algorithms can predict the soiling more accurately and help optimize the cleaning strategy.

6 REFERENCES

[1] M. Mani and P. Rohit, "Impact of dust on solar photovoltaic (PV) performance: Research status, challenges and recommendations," *ELSEVIER*, vol. 14, no. 9, pp. 3124-3131, December 2010.

[2] A. Ali, "Photovoltaic Module Dust Cleaning Techniques," in *The Effects of Dust and Heat on Photovoltaic Modules: Impacts and Solutions. Green Energy and Technology*, Springer, Cham, 2022.

[3] P. Patil, J. Bagi and M. Wagh, "A review on cleaning mechanism of solar photovoltaic panel," *IEEE*, 2017.

[4] L. Dorobantu, M. Popescu, C. Popescu and A. Craciunescu, "The effect of surface impurities on photovoltaic panels," *Renewable Energy and Power Quality*, 2011.

[5] D. L. Alvarez, A. S. Al-Sumaiti and S. R. Rivera, "Estimation of an Optimal PV Panel Cleaning Strategy Based on Both Annual Radiation Profile and Module Degradation," *IEEE*, vol. 8, pp. 63832-63839, 2020.

[6] be4energy, "be4energy," 2022. [Online]. Available: https://www.be4energy.com/index.php?option=com_content&view=category&layout=blog&id=8&Itemid=166.

[7] visualcrossing, "visualcrossing," 2022. [Online]. Available: https://www.visualcrossing.com/weather/weather-data-services#. [Accessed 2022].

[8] F. Pedregosa, G. Varoquaux and A. Gramfort, "Scikit-learn: Machine Learning in Python," *Journal of Machine Learning Research*, 2011.

[9] U. Jaitley, "Medium," 2018. [Online]. Available: https://medium.com/@urvashilluniya/why-data-normalization-is-necessary-for-machine-learning-models-681b65a05029#:~:text=Normalization%20is%20a%20technique%20often%20applied%20as%20part,machine%20learning%2C%20every%20dataset%20does%20not%20require%20norma. [Accessed 2023].

This presentation was selected by the Sc. Committee of the EU PVSEC 2023 for submission of a full paper to one of the EU PVSEC's collaborating peer-reviewed journals.

Identification and analysis of metallization defects in desert-operated photovoltaic modules

S. Kumar[1], B. Adothu[1], B. Jäckel[2], Z. Shekason[1], A. Moosawi[1], A. Seentakath[1], D. Daßler[1], K. Chapaneri[1], J. J. John[1], G. Mathiak[1], V. Alberts[1], and R. Gottschalg[2]

1 DEWA Research & Development Center, P.O.Box: 564 Dubai, United Arab Emirates
2 Fraunhofer Center for Silicon Photovoltaics CSP, Otto-Eissfeldt-Str. 12, 06120 Halle (Saale), Germany
* Corresponding author: phone +971 58 539 5739 | sagarika505@gmail.com

Abstract—**The day and night temperature differences in desert regions are relatively extreme compared to moderate conditions, causing thermo-mechanical fatigue within installed photovoltaic (PV) modules. This leads to breakages in cells and its metallization, which causes resistive losses. Herein, finger metallization are major current-carrying components, which are reported to fail under such conditions. There are various types of finger defects depending on the mode of degradation, source of breakage and degradation severity. This paper discusses and analyses different types of finger defects observed in 5-year desert-aged modules. The work also proposes methods to differentiate between aged finger defects from manufactured-based ones.**

Keywords—desert climate, crystalline PV modules, thermo-mechanical fatigue, finger defects

I. Introduction

The Middle East and North Africa (MENA) region has been witnessing exponential growth in utility solar plants, with expected solar capacities reaching up to almost 35GW by 2024 [1]. However, PV modules in desert climate are exposed to environmental stressors such as elevated temperature and ultraviolet (UV) radiation, sand abrasion and large temperature cycles [2] [3] [4] [5], which can degrade different components of the module and eventually lead to defects such as encapsulant discoloration and delamination, cell cracks, hot spots, finger breakages, discoloration of metallization, solder-bond failure and glass abrasion and breakages [2] [3] [4] [5] [6]. These defects and degradations reduce the power output of the system, as well as the lifetime of PV modules, affecting system reliability.

Many studies conducted in hot and dry climates identified series resistance as a major mode of degradation [2] [7] [8]. Emitter resistance, fingers, busbars, solder, and interconnects are major contributors to series resistance. During operation, high-temperature variation between day and night in desert climates subject the PV module components to thermo-mechanical loads [2] due to the difference in their thermal expansion coefficients [9]. Finger-solder interfaces are more susceptible to such conditions in comparison to other components, owing to the small size of fingers and material properties of solder. Solder thermal fatigue failures results in finger breakages, which increases series resistance and reduce fill factor, causing power losses. In addition, finger and busbars can also corrode under chemical modes, which arises due to the ingression of moisture and gas to the PV cell through the encapsulant and back sheet, resulting in metallization corrosion and increased series resistance and power losses [10]. Thermo-mechanical-based finger defects are more commonly observed in the desert region due to the high-temperature variation between day and night [2]. Moreover, another mode of finger defect is manufacturer-induced due to insufficient silver pasting during the screen-printing process of metallization or non-optimal soldering conditions. In addition, fingers defect may be introduced as a result of accidental damages during transportation or installation [10]. This type of defect is referred to as finger interruption and can occur anywhere along the finger length [10]. Finger interruptions are often mistaken for finger breakages as they both look very similar in electroluminescence (EL) imaging; presented as dark elongated box-like features, which can lead to misanalyses of data in reliability studies.

Thus, current-carrying metallization is important in reliability studies and requires detailed studies on the identification, investigation and analysis of different finger defects commonly found in desert climates. This paper contains a study on different modes of finger defects and degradation modes observed in 5-year-old desert fielded PV modules installed in DEWA R&D center, Dubai, UAE. It utilizes non-destructive techniques such as electroluminescence (EL) imaging and microscopic visual inspection for structural defect detection, identification, analysis and differentiation between different types of finger defects. This would be instrumental in early detection of such defects and differentiation between them based on their respective reason of breakage which could be either induced during manufacturing process, during transportation/installation or field operation. Furthermore, this paper could assist in the mitigation of such defects in the design of desert-resistant PV modules, as well as contribute to the development of PV standards specific to desert conditions.

II. Methodology

The crystalline silicon PV modules were installed at Dubai Electricity and Water Authority (DEWA) R&D Center within the MBR Solar Park, Dubai, UAE for a period of 5 years. For this study, 20 crystalline silicon PV modules were investigated, which were installed at an angle of 25° and were manually cleaned weekly.

The study utilizes electroluminescence (EL) imaging and microscopic visual inspection for detection, analysis and differentiation of different finger defects of the cell. Cell EL imaging has been performed using an InGaS type detector at constant short-circuit current conditions and integration time of 3.5s, to identify the structural and electrical defects of the cell including finger defects type, severity and location. For a comparative analysis, the EL images were used to analyse the modules pre- and post- field exposure. In addition, for detailed analysis the EL intensity data of the cell was also extracted. For further investigation of source of origin of the different finger defects, microscopic inspection has been conducted using a custom-made system consisting of a visual camera with a zoom lens moving along X-Y channels.

III. RESULTS AND DISCUSSION

The sample modules showed different defects and degradations wherein encapsulant discoloration was dominantly observed. The second dominant defect observed were finger breakages seen in 70% of the modules in varying severity. Correspondingly, fill factor losses were also observed in these modules. Fig. 1 shows the EL image of a representative cell with different types of finger defects.

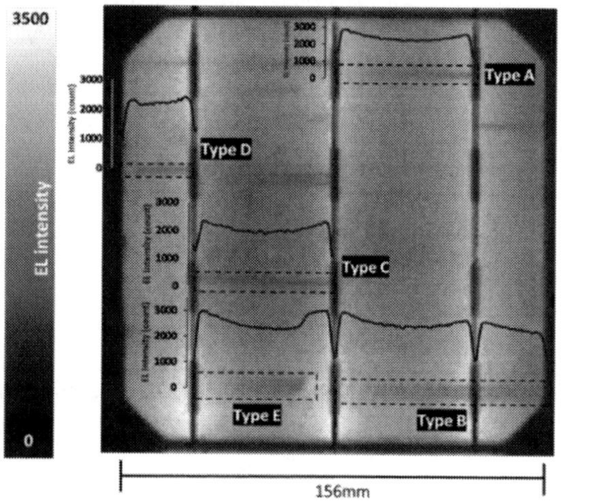

Fig. 1. EL image (with intensity in blue line) of a cell with different type of finger defects observed in a outdoor exposed PV cell.

The EL image was used to identify different types of finger defects shown as type A to type E. These types have been investigated further individually to identify their origin of breakage, mode of degradation and severity in comparison to other types. Table 1 shows these different types of defects observed in the modules after 5-year exposure. These types have been explained in detail in the subsequent subsections.

TABLE 1. SUMMARY OF FINGER DEFECTS SEEN IN 5-YEAR EXPOSED PV MODULES

A. *Electrical characterization of module defects*

I-V+EL+Isc from EQE

Trying to do IQE if it works should do:

1. Localized current comparison between all the types

	Type A and Type B	Type C and Type D	Type E
Type of defect	Finger breakage	Finger interruption and breakage	Finger shorting
Mode of degradation	Outdoor exposure (thermal fatigue)	Manufacturing process induced (improper screen printing) + outdoor exposure (thermal fatigue)	Manufacturing process induced (excessive silver paste spillage)
Origin of breakage	Finger- solder interface	Any point along the finger and/ or finger-solder interface	Any point along the finger

B. *Type A and Type B: Finger breakage under outdoor exposure*

Type A and type B dark rectangular defects visible in the EL image are identified as a finger breakage observed under outdoor exposure, owing to its characteristic origin of breakage from the busbar region, as visible in Fig. 1. These defects were not visible in the initial EL images. The exact location can be attributed finger-solder junction breakage due to thermo-mechanical fatigue within cell, which disrupts the current path [11]. However, it is difficult to detect these breakages with only microscopic visual inspection since, the point of breakage can also be under the busbar at the finger-solder junction.

Furthermore, to be able to comment on the severity of this breakage, the EL intensity curve above the type A defect was evaluated. It can be seen that there is almost no gradient with a plateau formed due to the single source of the busbar. To be able to understand the impact on the operation of solar cell, a current conduction diagram is shown in Fig. 2, based on the EL image.

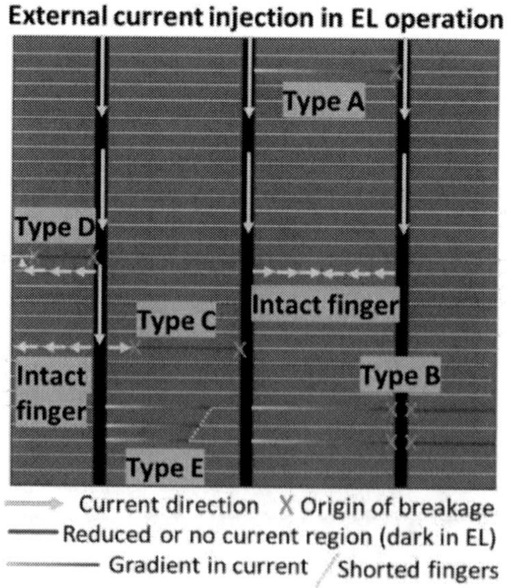

Fig. 2. Current conduction diagram for different types of finger defects in PV cell under EL operation, where current is forced through the interconnects into the cell.

C. Type C and Type D: Finger interruption with finger breakage

Type C and type D finger defects (Fig. 1) are cases where we can demonstrate the difference between a finger breakage and finger interruption effectively. Finger interruptions are formed due to an improper screen-printing process [10]. It can be seen in the initial EL image that this type of defect is restricted to the first busbar (Fig. 3) or near the edge of the cell (Fig. 4). However, after the 5-year exposure, a new type of finger breakage has appeared from the second busbar, along with the same finger. More information on the current conduction and severity comparison of defects has been added in the longer version of the manuscript.

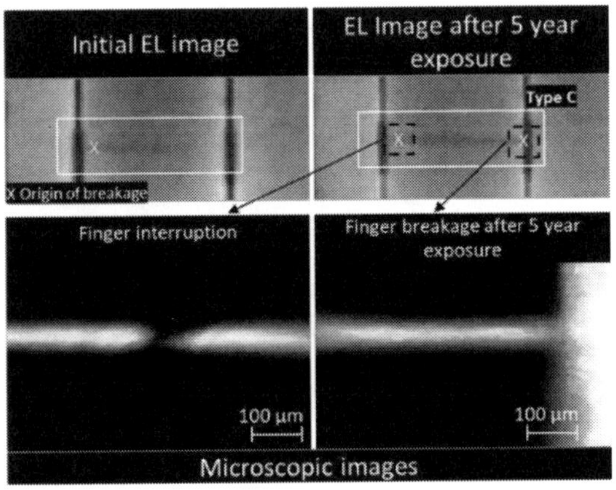

Fig. 3. Intial and final EL image after 5-year exposure in desert climate indicating two different types of breakages in a single finger, with corresponding microscopic images of type C finger.

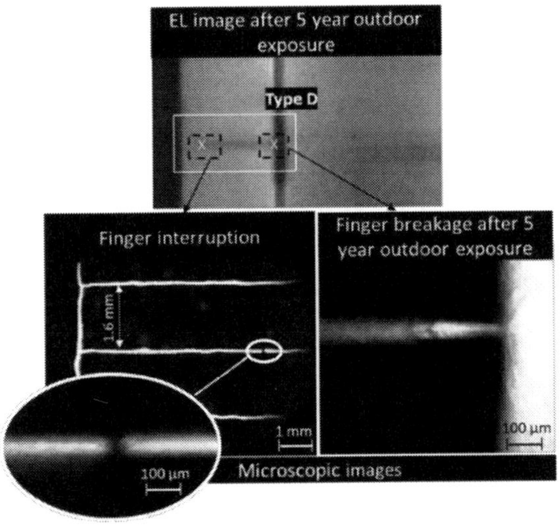

Fig. 4. Initial and final EL image after 5-year exposure in desert climate indicating two different types of breakages in a single finger, with corrosponding microscopic images of type D defect.

D. Type E: Finger shorting during screen-printing process

Type E finger defect was observed in the initial EL images, as shown in Fig. 5. However, since the source of breakage is away from the solder-busbar junction, the defect is formed before the outdoor exposure since there is no vulnerability of breakage under thermomechanical fatigue. For further investigation, the region of interest has been observed under the microscope, as shown in Fig. 5. It can be seen that the spilling of excess silver paste across the fingers, created a shorting effect, which is visualized in the EL image as the accumulation of current on the right-hand side of the shorted region in Fig. 5. More information on the current conduction and severity comparison of defects has been added in the extended version of the manuscript.

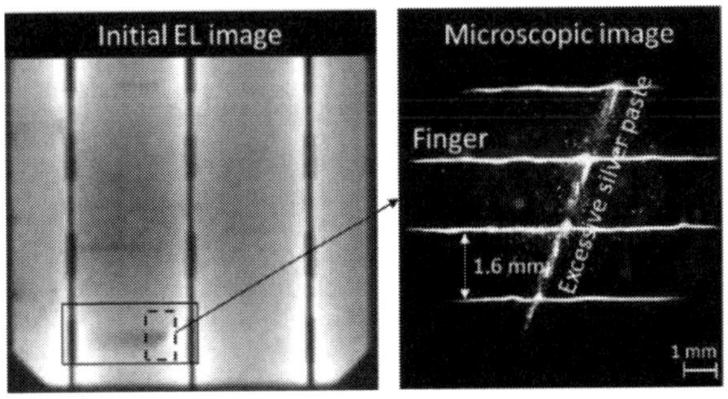

Fig. 5. Initial EL image indicating type C finger breakage, where finger shorting is observed in the corresponding microscopic visual image.

E. Cracks propagation and generation

IV. Conclusion

This paper discussed different types of finger defects in modules exposed for 5 years to a desert region. The defects have been differentiated on the basis of the origin of breakage, mode of degradation, and severity using EL imaging and microscopic visual inspection imaging. The information can be used to differentiate between manufactured and aged defects using EL imaging. Furthermore, this paper could assist in the mitigation of such defects in the design of desert-resistant PV modules, as well as contribute to the development of PV standards specific to desert conditions.

References

[1] Middle East Solar Industry Association MESIA, Solar Outlook Report 20201. Dubai, January 2021

[2] S. Kumar, A. Alheloo, S. Hassan, H. Hanifi, H. Alhamadani, H. Alshanqiti, J. John, G. Mathiak, and V. Alberts, "Investigation and analysis of defects and degradations in desert fielded photovoltaic modules," IEEE 48th Photovoltaic Specialists Conference (PVSC), pp. 0788-0792, 2021.

[3] S. Kumar, H. Alhamadani, S. Hassan, A. Alheloo and J. John, "Comparative investigation and analysis of discoloration and glass abrasions in desert exposed photovoltaic modules," IEEE 48th Photovoltaic Specialists Conference (PVSC), 2021.

[4] Z. Kherici, H. Cheghib, N. Kahoul, M. Youness, B. C. Affari, and T. Hazem, "Failure modes of standard photovoltaic modules in Sahara

10.4229/EUPVSEC2023/3BO.11.1
020235-003

Desert," IEEE 11th International Renewable Energy Congress (IREC), pp. 1-4, October 2020.

[5] A. Omazic, G. Oreski, M. Halwachs, G.C. Eder, C. Hirschl, L. Neumaier, G. Pinter and M. Erceg, "Relation between degradation of polymeric components in crystalline silicon PV module and climatic conditions: A literature review," Solar Energy Material and Solar Cells, vol. 192, pp. 123-133, April 2019.

[6] A. Bouraiou, M. Hamouda, A. Chaker, A. Neçaibia, M. Mostefaoui, N. Boutasseta, A. Ziane, R. Dabou, N. Sahouane and S. Lachtar, "Experimental investigation of observed defects in crystalline silicon PV modules under outdoor hot dry climatic conditions in Algeria," Solar Energy, vol. 159, pp. 475-487, January 2018.

[7] S.V. Janakeeraman, J. Singh, J. Kuitche, J.K. Mallineni and G. TamizhMani, "A statistical analysis on the cell parameters responsible for power degradation of fielded PV modules in a hot-dry climate," IEEE 40th Photovoltaic Specialist Conference (PVSC), pp. 3234-3238, June 2014.

[8] N. Kahoul, H. Cheghib, M. Sidrach-de-Cardona, B.C. Affari, M. Younes and Z. Kherici, "Performance degradation analysis of crystalline silicon solar cells in desert climates," Energy for Sustainable Development, Vol. 65, pp. 189-193, December 2021.

[9] L.C. Rendler, P. Romer, A.J. Beinert, J. Walter, S. Stecklum, A. Kraft, U. Eitner and S. Wiese, "Thermomechanical stress in solar cells: Contact pad modeling and reliability analysis," Solar Energy Materials and Solar Cells, vol. 196, pp. 167-177, July 2019

[10] S. Kumar, R. Meena, and R. Gupta, "Finger and interconnect degradations in crystalline silicon photovoltaic modules: A review," Solar Energy Materials and Solar Cells, vol. 230, Sep 2021.

[11] S. Kumar and R. Gupta, "Investigation and analysis of thermo-mechanical degradation of fingers in a photovoltaic module under thermal cyclic stress conditions," Solar Energy, vol. 174, pp. 1044–1052, Nov 2018.

40th European Photovoltaic Solar Energy Conference and Exhibition

This presentation was selected by the Sc. Committee of the EU PVSEC 2023 for submission of a full paper to one of the EU PVSEC's collaborating peer-reviewed journals.

A NEW INTRINSIC MODULE PARAMETER APPROACH FOR DEGRADATION ASSESSMENT OF PHOTOVOLTAIC MODULES: RESULTS BASED ON DESERT DATA

[1]Dhanup S. Pillai*, [1]Amir Abdallah, [1,2]Mohamed Abdelrahim, [1,2]Mohamed Elgaili
[1]Qatar Environment and Energy Research Institute (QEERI), Hamad bin Khalifa University (HBKU), Qatar Foundation. P.O. Box 34110, Doha, Qatar
[2]Bin Omran Trading and Telecommunication W.L.L, Qatar
dsomasekharanpillai@hbku.edu.qa, aabdallah@hbku.edu.qa, momohamed@hbku.edu.qa, mohammohamed@hbku.edu.qa

ABSTRACT: Degradation of PV modules is usually studied and understood pertinent to the drop in any of the following performance indicators after a certain period of outdoor exposure: 1. maximum power, 2. performance ratio and 3. energy yield, assessed either based on module or system level analysis. The most common assessment method for individual PV module degradation is the indoor testing using a dedicated solar simulator as per the IEC guidelines Predominantly, by virtue of indoor testing, degradation of PV modules is quantified in terms of annual degradation rates, indicative of an average percentage reduction in maximum power over the tested period. Therefore, existing studies only provide degradation rates and trends relevant to the aforesaid performance indicators for different PV module technologies. Distinct to these approaches, this article for the first time presents a different dimension for degradation by analyzing the impact of intrinsic module parameters (shunt resistance, R_{sh} and series resistance, R_s) on percentage drop in the maximum power of PV modules. Indoor test results obtained for a set of multi-crystalline PV modules after 8 years of outdoor operation in Qatar's desert climate, and its interpretations based on modelled intrinsic parameter values are presented in this article. The results indicate that the proposed findings would be pivotal for the global solar community to better assess PV module degradation, indicative of the expected future degradation.

Keywords: Photovoltaic (PV), degradation, diode model, parameter estimation, module reliability

1 INTRODUCTION

Photovoltaic (PV) module degradation is usually assessed based on the annual percentage reduction in maximum power (P_{max}) over a certain period of on-field operation [1]. Hence, it has often been considered as the key performance criterion for determining and explaining module performance [2]. However, calculating the annual power drop alone might not be informative to interpret the impact of present degradation mechanisms and failure modes on the expected future performance of PV modules. As it is, from a predictive maintenance perspective, this prevents the end-user/asset-owner from making informed decisions on module replacements, which is important for optimal system performance. Thus, the motivation of this work is to improve the overall framework in which PV module degradation is assessed and propose key parameters that might be helpful for predicting the imminent performance of PV modules.

In the above context, literature studies [3], [4] have showcased that most (almost all) of the defect mechanisms contributing to degradation would eventually affect the intrinsic parameters of PV modules, primarily the shunt (R_{sh}) and the series resistances (R_s). Therefore, evaluating the change in these parameters after certain years of field operation is likely to provide better insights on module performance and main causes of degradation. Subsequently, it also helps to interpret the current state of health and future expected degradation of PV modules. While these intrinsic parameters in-principle evolve from the fundamental single diode (SD) and double diode (DD) models, the hypothesis of this research is to utilize these models to identify the impact of these intrinsic parameters on maximum power degradation of PV modules. Thus, the objective is to provide new insights to PV module degradation by co-relating the indoor test measurement results with the acquired knowledge of benchmark intrinsic parameter values.

In this communication, 8 multi-crystalline silicon modules returned from the field (after 8 years of operation) are sequentially studied utilizing the following procedures: 1. Visual inspection to identify the visual defects, 2. Electroluminescence (EL) imaging to identify the presence of micro-cracks, and 3. IV curve measurement to test P_{max} at different irradiance and temperature conditions. Finally, the results are modelled to assess its implications on module parameters and co-related with the experimental findings to evaluate module performance.

Note that the idea to evaluate model parameters triggered from the indoor test results of one multi-crystalline module that recorded nearly the STC P_{max} value (among the 8 tested) even with a significant drop (69%) in R_{sh}. Thus, the key scientific contribution is the new findings pertinent to the variations in intrinsic module parameters, its degradation trends and peculiar impact on P_{max}. Furthermore, degradation of each module tested is expressed in terms of quantitative benchmarks set for R_{sh} and R_s values that corresponds to 20% drop in initial P_{max}. The results reveal that for multi-crystalline modules, R_{sh} and R_s have contrasting degradation trends, which could be very reliable predictors of early degradation.

In general, the findings in this work are representative of different module technologies, and similar analysis has been performed for: 1. multi-crystalline, 2. monocrystalline, and 3. heterojunction modules (see Figure 1). More importantly, it is worthwhile to mention that the results obtained for the intrinsic parameter effect are identical for all the 3 technologies. Therefore, for brevity, discussions pertinent to monocrystalline and heterojunction modules are not presented in this communication.

2 INDOOR TEST FINDINGS

The objective of this work is to propose a modelling-based approach to evaluate PV module degradation that not only expresses the annual power degradation rate, but also is indicative of the expected future degradation. An attempt has been made in this article to achieve this objective by virtue of intrinsic parameter modelling of PV modules. For this study, as explained earlier, a set of 8 multi-crystalline modules (having a P_{max} value of 235W at STC) returned from similar outdoor testing conditions in typical Qatar desert climate after 8 years of on-filed operation are selected. The desert climate in Qatar is unique and is typically characterized by high UV irradiance with annual relative humidity (RH) of 75.2 mm in combination with high module operating temperatures that may reach 70°C during summer.

As indicated above, the first phase of assessment includes visual inspection, EL imaging and indoor flash testing of the selected PV modules as per the IEC guidelines. First, screening for visual defects is performed, followed by EL imaging pertinent to different module short circuit currents to identify the presence of micro-cracks, shunts, inactive cells, and other internal defects. For clarity on the broadness of the proposed study, the imaging results obtained for one module representing the three module technologies are shown in Figure 1. However, further evaluations and results for heterojunction and monocrystalline technologies are not included as a part of this proceeding.

In this context, the set of multi-crystalline modules under study showed no significant visual defects on visual inspection, while EL imaging showed the presence of dark areas (indicative of shunts) in module 1 (numbered in chronological order). Then, flash testing was carried out using an A+A+A+ pulsed solar simulator to determine the I-V curves at various irradiance and temperature conditions. For each test case, the resultant values of P_{max}, R_{sh} and R_s are recorded at all tested conditions for comparison with modelled values. The results obtained at STC for the 8 multi-crystalline modules are illustrated in Table I. From the results, following conclusions and interpretations can be derived.

1. No significant power drop is witnessed after 8 years of operation
2. The annual degradation rates with respect to P_{max} lie within the expected limits, ranging from 0.22%/year (module 1) to 1.06%/year (module 7)
3. Interestingly, module 1 that recorded the highest P_{max} value seems to be having a very low shunt resistance
4. Unfortunately, the initial values for R_{sh} and R_s are unknown for the tested modules
5. The impact of intrinsic parameters on the percentage drop in P_{max} is not clear and require further investigations.

Table I: Flash test results obtained for multi-crystalline modules at STC

Module	Measured Values at STC		
	P_{max} (W)	R_s (Ω)	R_{sh} (Ω)
Module 1	**231.1**	**0.477**	**70.773**
Module 2	218.87	0.703	349.188
Module 3	218.8	0.713	247.97
Module 4	222.08	0.822	317.8
Module 5	218.71	0.769	217.6
Module 6	217.47	0.741	322.2
Module 7	214.86	0.826	400.56
Module 8	220.9	0.778	255.81

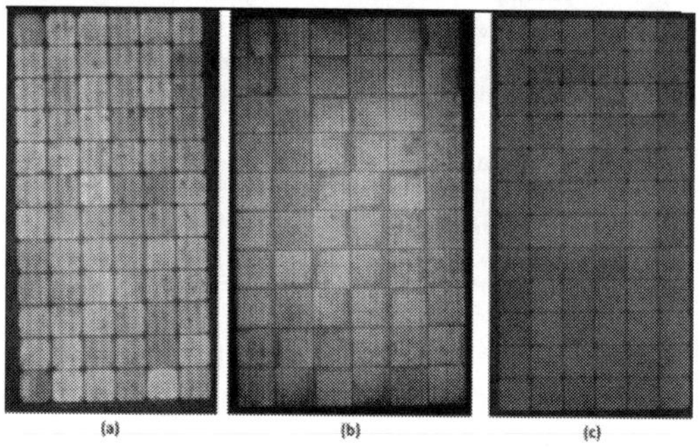

Figure 1: EL images representative of different PV module technologies under study: (a) mono-crystalline silicon, (b) multi-crystalline silicon and (c) silicon heterojunction (Note: For brevity, no further evaluations are presented for technologies (a) and (c))

3 PV MODELLING

Even though the change in P_{max} can be assessed based on a comparison with the name plate data, the variations in R_{sh} and R_s cannot be evaluated if the initial test values are unknown. This is because benchmark values for R_{sh} and R_s are usually not available in module data sheet. Unfortunately, the initial intrinsic parameters of the modules considered in this study are unknown for a realistic comparison with the test data. Therefore, the equivalent diode model of the same module is realized in MATLAB using the data sheet parameters and manufacturer's data (PAN file) available in the PVsyst software database. Among SD and DD models, SD model (shown in Figure 2) has been utilized in this work owing to its simplicity and the well-known current equation derived from the SD model is detailed in Equation (1).

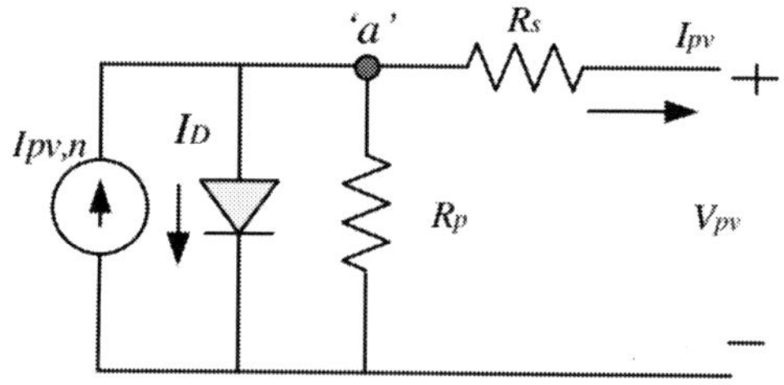

Figure 2: Equivalent SD model developed in MATLAB for estimating R_{sh} and R_s values

$$I = I_{ph} - I_o\left[exp\left(\frac{V+IR_s}{a+V_{T1}}\right) - 1\right] - \left(\frac{V+IR_s}{R_p}\right) \quad (1)$$

While the intrinsic parameters are inherently unknown in the model, it's important to mention that the ideal values for the parameters at various operating conditions (varying irradiance and temperature) need to be estimated using a suitable parameter estimation approach. To achieve this task, an optimization-based curve-fit approach utilizing the root mean square error (RMSE) objective function is utilized. For optimization, recently proposed Wind Driven Optimization (WDO) algorithm is programmed in MATLAB and the results obtained are validated using the literature data [5, 6]. The curve-fit results obtained for two different test conditions are shown in Figure 3 (a) and

Figure 3 (b), respectively to highlight the accuracy of the parameters obtained at different irradiance levels. It can be observed that the RMSE values typically in the range of E^{-3} and E^{-4} are in good agreement with the reference curves. It is worth mentioning that the experiments are also extended for all indoor test conditions to record parameter values. However, for brevity, those results are not shown.

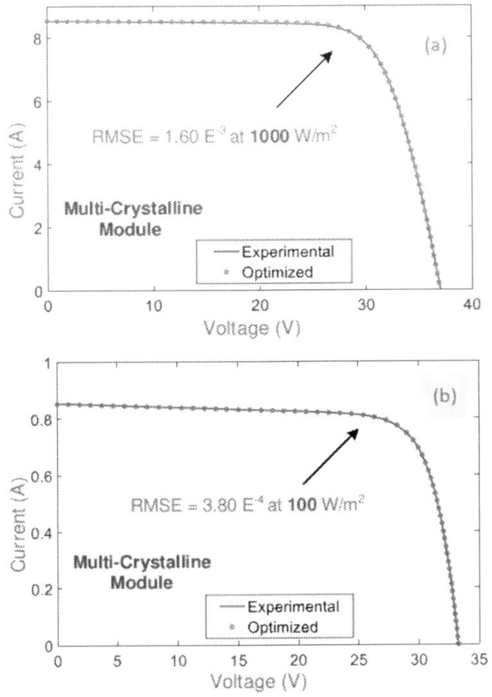

Figure 3: WDO optimization results for obtaining initial intrinsic module parameters: (a) STC and (b) 100 W/m^2

4 RESULTS

Once the parameters are estimated using the above curve-fit approach, the tested values are compared with the ideal ones to determine the parameter variations over time. In the proposed study, the results obtained highlight quite interesting findings indicative of a unique dependency between module parameters and P_{max} degradation. For instance, the module recorded the highest P_{max} (module 1, 231.1W) is found to have extremely low R_{sh} value (70 Ω) during indoor testing, which is 69% less than its initial value at STC, 223 Ω (modelled). More importantly, this module has substantially better R_s (0.570 Ω) compared to the other 7 modules (R_s in the range of 0.8 – 0.9 Ω and R_{sh} above the modelled values). These results are suggestive of the fact that except for module 1, the witnessed drop in maximum power is only due to increased R_s values. Another interesting observation is that, for module 1, even though the R_{sh} is degraded to nearly 70% of its initial value, the inherent degradation mechanism does not seem to be contributing much to the reduction in P_{max}, which eventually raises the research question on the impact of intrinsic parameters on P_{max}.

Therefore, to better understand the impact of intrinsic parameters on the percentage drop in P_{max}, the intrinsic parameter values (i.e., R_{sh} and R_s) expected for 1%, 5%, 10% and 15% drop in P_{max} has also been identified using the proposed model. These values are further utilized to plot a graph in order to verify the relation between the percentage drop in intrinsic parameters and the percentage drop in P_{max}. Perhaps, this analysis determines the overall degradation trends for R_{sh} and R_s with respect to the percentage drop in P_{max}. Based on the obtained values, the impact of R_{sh} and R_s on percentage reduction in P_{max} is shown in Figure 4 and Figure 5 respectively. Some conclusive remarks based on this assessment are that illustrates the impact of these parameters on the reduction in P_{max} are as follows:

1. The overall degradation trend in multi-crystalline modules due to R_{sh} is unique as observed in Figure 4. It can be observed that the initial degradation is usually unnoticeable since the drop in R_{sh} required for even 1% reduction in P_{max} is usually around 50% of its initial value at STC. Thus, early degradation due to R_{sh} (up to 5% P_{max} loss) can be very slow compared with the later rapid degradation phases (trend becoming more linear after 10% drop in P_{max}). This is a crucial parameter for predicting module performance.
2. In contrary, PV module degradation due to R_s shows almost a negative exponential characteristic, where all the degradation phases are almost constant. i.e., the percentage drop in R_s corresponding 0-5% degradation in P_{max} of the PV module is identical to that required for 5-10%, 10-15%, and 15-20% drop in P_{max}.
3. On the other hand, compared to the trend in R_{sh}, the degradation in P_{max} by virtue of a degradation mechanism that affects the R_s of a PV module is expected to be rather slow as it requires nearly 250% increase in its initial value to reach 20% degradation in P_{max} at STC (see Figure 5).
4. The overall trend is identical in all operating conditions irrespective of the irradiance and temperature changes for both R_{sh} and R_s (as can be seen in Figure 4 and 5).
5. The findings illustrate that both R_{sh} and R_s are contrastingly sensitive to degradation, and hence are strong predictors of early degradation in PV modules.

In the final phase, the developed model has been utilized to set benchmarks values (for R_{sh} and R_s) for 20% degradation in P_{max}, which would be the maximum allowable limit to guarantee reliability over 25 years operation. The present values of R_{sh} and R_s are then co-related with the benchmark values and the overall degradation trend (see Figures 4 and 5) to interpret the state of health and predict the future performance of tested modules. The interpretation of the future performance of the tested PV modules based on the benchmarks set are graphically illustrated in Figure 6, which shows that module 1 is already in the rapid degradation phase in terms of R_{sh} and is a likely candidate for replacement soon. On the other hand, all other modules are well below the threshold limits of R_{sh}. While in terms of R_s, though modules 4,7, and 8 are near to its threshold limits, it can still perform further owing to the constant degradation expected due to R_s. Note that these findings are well illustrative of the indoor test results in Table I.

10.4229/EUPVSEC2023/3BO.11.2
020236-003

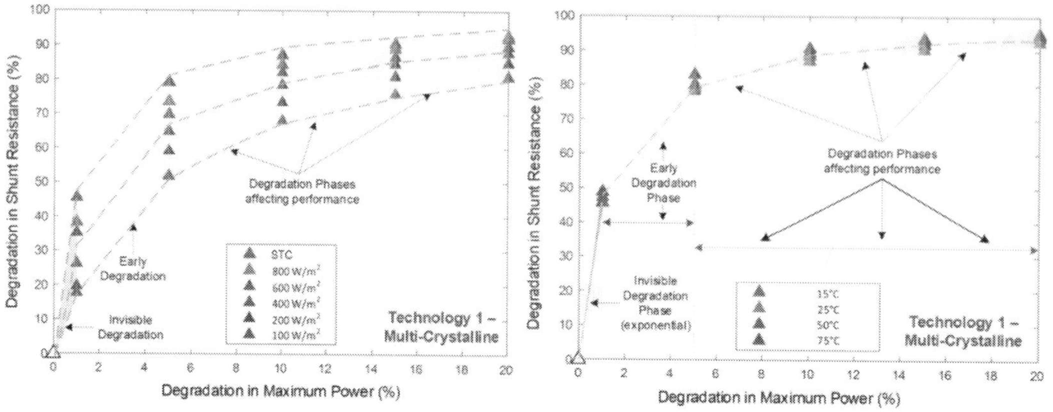

Figure 4: Impact and overall degradation trend in P_{max} due to R_{sh}: varying irradiance (left) and temperature (right)

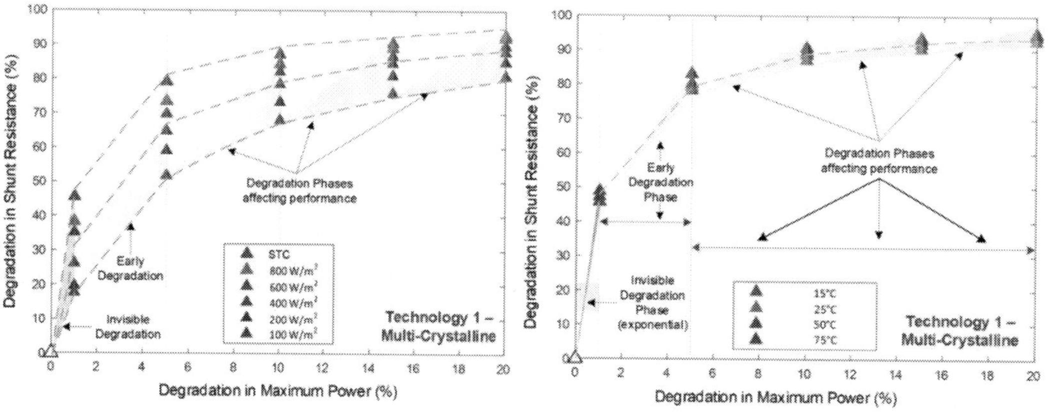

Figure 5: Impact and overall degradation trend in P_{max} due to R_s: varying irradiance (left) and temperature (right)

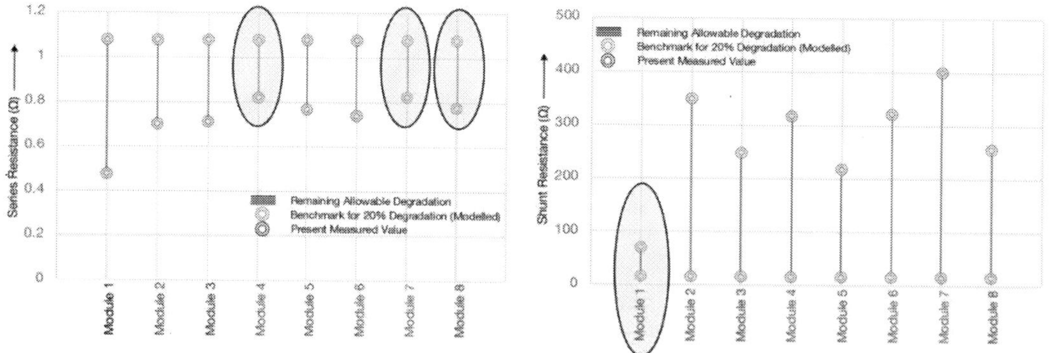

Figure 6: Estimated degradation based on module parameter benchmarks and predicted future degradation

5 SUMMARY

This article proposed a new approach for assessing degradation that could be helpful for making informed decisions on module replacements. The key take away is that while degradation is conventionally assessed based on the drop in the energy yield, performance ratio and/or P_{max}, this study shows that monitoring R_{sh} and R_s values can provide additional information on the degradation thresholds and expected future performance. Furthermore, the uniqueness in the contrasting degradation trends representative of the impact of intrinsic parameters can be used as reliable predictors for early degradation. The future scope of this work would include the implementation of the proposed approach for on-field measurements in order to assess module as well string level degradation in PV systems.

REFERENCES

[1] Y. R. Golive et al., "Analysis of Field Degradation Rates Observed in All-India Survey of Photovoltaic Module Reliability 2018," IEEE J Photovolt, vol. 10,

no. 2, pp. 560–567, Mar. 2020, doi: 10.1109/JPHOTOV.2019.2954777.

[2] J. A. Clavijo-Blanco, G. Álvarez-Tey, N. Saborido-Barba, J. L. Barberá-González, C. García-López, and R. Jiménez-Castañeda, "Laboratory tests for the evaluation of the degradation of a photovoltaic plant of 2.85 MWp with different classes of PV modules," *Renew Energy*, vol. 174, pp. 262–277, Aug. 2021, doi: 10.1016/j.renene.2021.04.024.

[3] M. 1972- Köntges et al., Performance and reliability of photovoltaic systems subtask 3.2: Review of failures of photovoltaic modules : IEA PVPS task 13 : external final report IEA-PVPS.

[4] Z. Liu et al., "Quantitative analysis of degradation mechanisms in 30-year-old PV modules," Solar Energy Materials and Solar Cells, vol. 200, Sep. 2019, doi: 10.1016/j.solmat.2019.110019.

[5] C. S. Ruschel, F. P. Gasparin, and A. Krenzinger, "Experimental analysis of the single diode model parameters dependence on irradiance and temperature," Solar Energy, vol. 217, pp. 134–144, Mar. 2021, doi: 10.1016/j.solener.2021.01.067.

[6] Mathew D, Ram JP, Pillai DS, Kim YJ, Elangovan D, Laudani A, Mahmud A, "Parameter Estimation of Organic Photovoltaic Cells–A Three-Diode Approach Using Wind-Driven Optimization Algorithm," IEEE Journal of Photovoltaics. vol. 12, pp. 327-36, Dec. 2021.

Abrasion of Anti-Reflective Coatings by Robot Cleaning

Sep. 2023

Ben Figgis, Amir Abdallah, Maulid Kivambe, Brahim Aissa, Ayman Samara, Juan Lopez Garcia, Veronica Bermudez

QEERI
Qatar Environment & Energy Research Institute
HAMAD BIN KHALIFA UNIVERSITY

40th European Photovoltaic Solar Energy Conference and Exhibition

BACKGROUND

- Use of cleaning machines & robots now the norm in desert PV plants
- Cleaning robot research @ QEERI —
  - What % of dust do they remove?
  - What is the optimal run schedule?
  - Wet or dry cleaning method?
  - Do they micro-crack PV cells?
  - Moving shadow effect on PV output?
  - How rapidly do they wear out?
  - Do robots abrade modules' anti-reflective coating (ARC)?

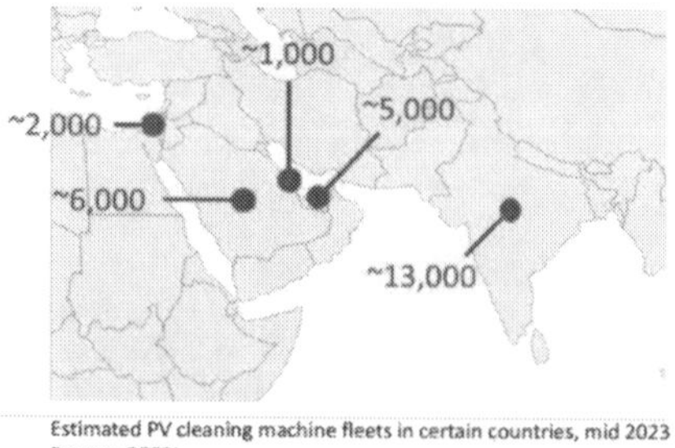

Estimated PV cleaning machine fleets in certain countries, mid 2023
Source: QEERI

MEASUREMENT CHALLENGE

- How to test ARCs, modules and robots?
- Methods and advantages —

Lab Bench	Field
Fast	Real-world environment
Standardized, repeatable	Real-world products
Simple versions are low cost	
Good for prototyping & comparing ARCs	Good for validating ARC durability

Accelerated	Normal
Fast	Real-world dust and humidity
Good for comparing ARCs	Good for validating an ARC

Glass coupons	Modules
Fast & easy to prepare	Industrial ARC properties
Detailed lab analysis	Direct Pmax results
	Non-uniformity effects
Good for prototyping & comparing ARCs	Good for validating ARC durability

MEASUREMENT CHALLENGE

~All published studies so far have been accelerated tests, mostly in the lab

Lab Bench	Field
Fast	Real-world environment
Low cost	Real-world products
Standardized/ repeatable	
Good for prototyping & comparing ARCs	Good for validating ARC durability

Accelerated	Normal
Fast	Real-world dust and humidity
Good for comparing ARCs	Good for validating an ARC

Glass coupons	Modules
Fast & easy to prepare	Industrial ARC properties
Detailed lab analysis	Direct Pmax results
	Non-uniformity effects
Good for prototyping & comparing ARCs	Good for validating ARC durability

Lange et al, Solar Energy Materials & Solar Cells 240 (2022) 111732

MEASUREMENT CHALLENGE

This study focused on real-world modules, robot and conditions

Lab Bench	Field
Fast	Real-world environment
Low cost	Real-world products
Standardized/ repeatable	
Good for prototyping & comparing ARCs	Good for validating ARC durability

Accelerated	Normal
Fast	Real-world dust and humidity
Good for comparing ARCs	Good for validating ARC

QEERI Outdoor Test Facility

Glass coupons	Modules
Fast & easy to prepare	Industrial ARC properties
Detailed lab analysis	Direct Pmax results
	Non-uniformity effects
Good for prototyping & comparing ARCs	Good for validating ARC durability

40th European Photovoltaic Solar Energy Conference and Exhibition

TEST SET-UP

- Samples
  - 6 types of PV modules
  - Glass coupons with 6 types of ARC
  - Dry-brush cleaning robot (Nomadd)
- Robot cleaning schedules
  - Daily
  - Weekly
  - None (weathering only)
- 2 years of testing
- Hot desert climate @ QEERI Outdoor Test Facility, Qatar
- Study conducted by Solar Consortium

MODULE Pmax VS. CLEANING FREQUENCY

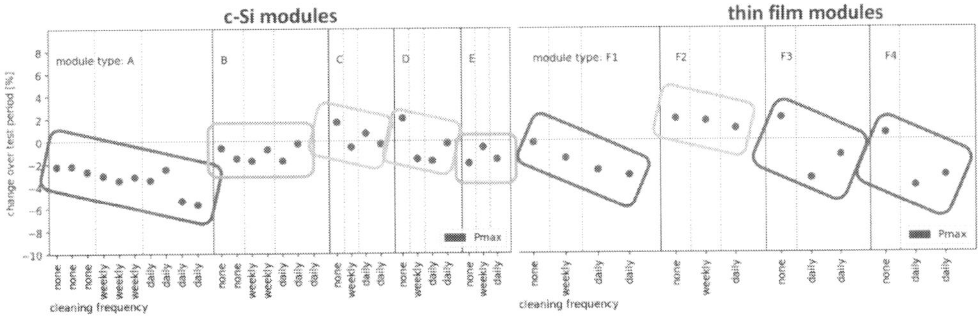

➡ more frequent robot cleaning

→ SOME MODULES' Pmax IMPACTED BY DRY ROBOT CLEANING, OTHER MODULES NOT AFFECTED

→ NEED TO TEST SPECIFIC ROBOT/MODULE COMBINATIONS, NO "RULE OF THUMB"

020237-007

ROOT CAUSE OF Pmax VARIATION

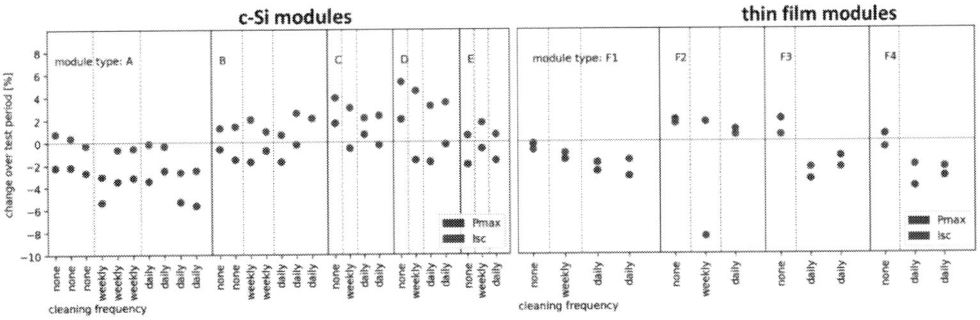

→ Pmax CHANGES CORRELATE WITH Isc CHANGES

→ INDICATES Pmax LOSSES ARE MAINLY OPTICAL

→ SUGGESTS ROOT CAUSE IS ARC ABRASION

020237-008

REFLECTIVITY CHANGES - MODULES

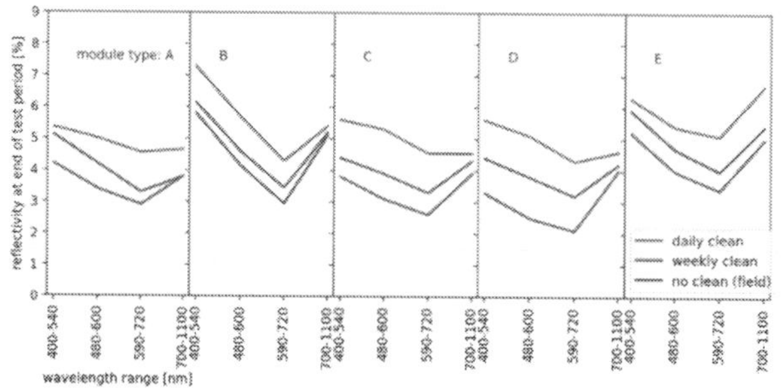

→ REFLECTIVITY INCREASED WITH MORE ROBOT CLEANING, FOR ALL c-SI MODULES

020237-009

40th European Photovoltaic Solar Energy Conference and Exhibition

REFLECTIVITY CHANGES - MODULES

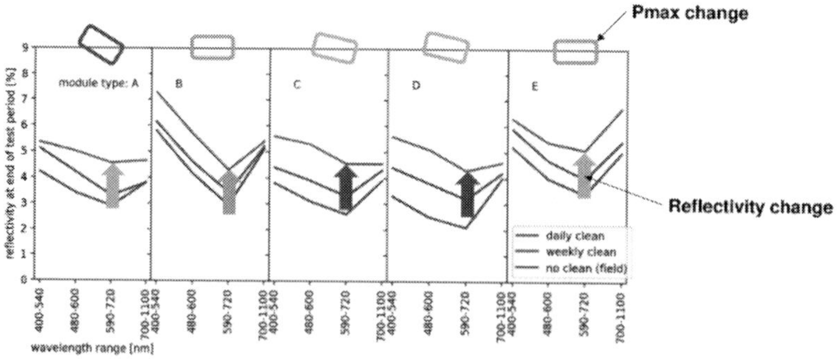

→ REFLECTIVITY INCREASED WITH MORE ROBOT CLEANING, FOR ALL c-SI MODULES

→ ...BUT ΔREFLECTIVITY HAS LOW CORRELATION WITH ΔPmax

020237-010

COUPONS — REFLECTIVITY

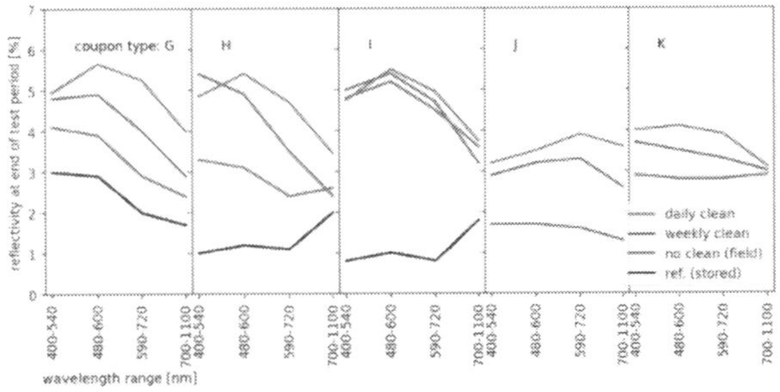

→ TESTED ARCs HAD GREATLY DIFFERENT ABRASION RESISTANCES
...BUT SOME WERE EXPERIMENTAL

→ ARCs DEGRADED WITH OUTDOOR EXPOSURE ALONE (NO ROBOT CLEANING)

COUPONS (TYPE J) — SCRATCHES

daily clean

weekly clean

no clean

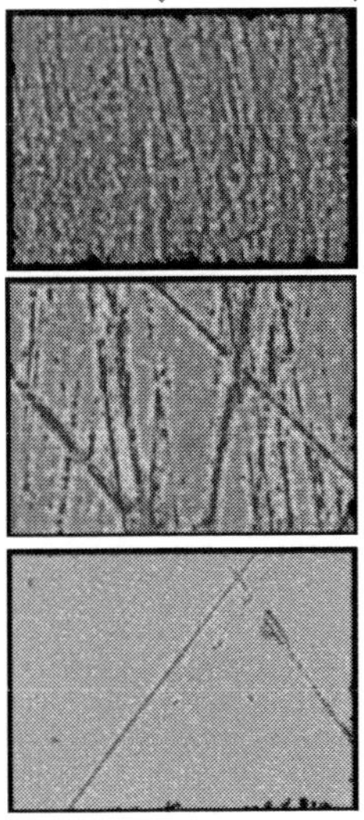

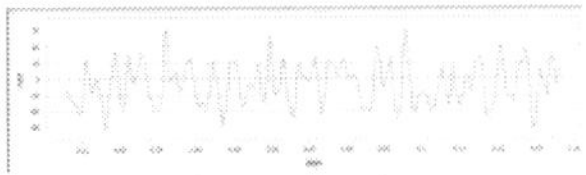

Extensive scratches
Deepest ~90 µm

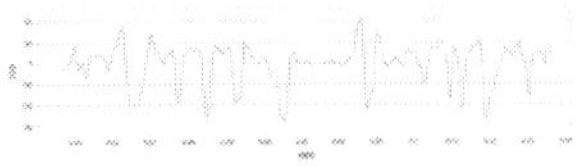

Moderate scratches
Deepest ~60 µm

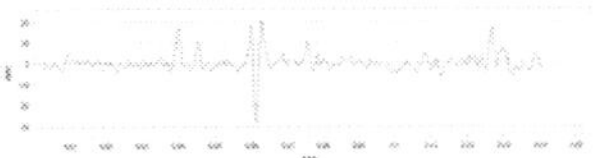

Very few scratches
Deepest ~20 µm

→ INCREASED SCRATCHES QUANTITY AND DEPTH WITH MORE ROBOT CLEANING

SCRATCH OBSERVATIONS NOT EXTENDABLE TO MODULES, DUE TO MANUFACTURING DIFFERENCES

DOES 18 MONTHS OF DRY ROBOT CLEANING ABRADE PV ANTI-REFLECTIVE COATING?

"YES" RESULTS

For some modules, more cleaning → larger Pmax loss

ΔPmax ≈ ΔIsc, suggesting optical loss, suggesting ARC abrasion

For all c-Si modules and coupons, more cleaning → greater reflectivity

For typical coupon, more cleaning → more severe scratches

"NO/MAYBE" RESULTS

For some modules, no correlation between Pmax loss and cleaning cycles

Low correlation between ΔPmax/Isc and Δreflectivity → root cause ambiguous

Reflectivity results for thin-film modules not understood

MODULE VIBRATION BY CLEANING ROBOT

- Module center vibration during robot cleaning:
  - c-Si modules (2x1 m): <1 mm
  - TF modules (1.2x0.6 m): ~0 mm
- Strong wind: ~2mm ...more than robot cleaning
- → Robot cleaning vibration unlikely to damage PV cells

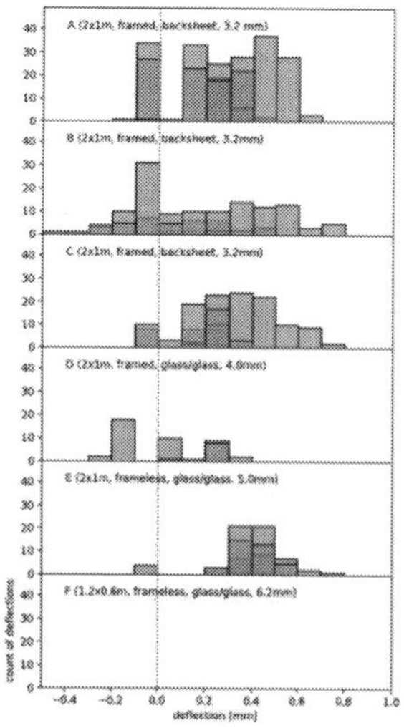

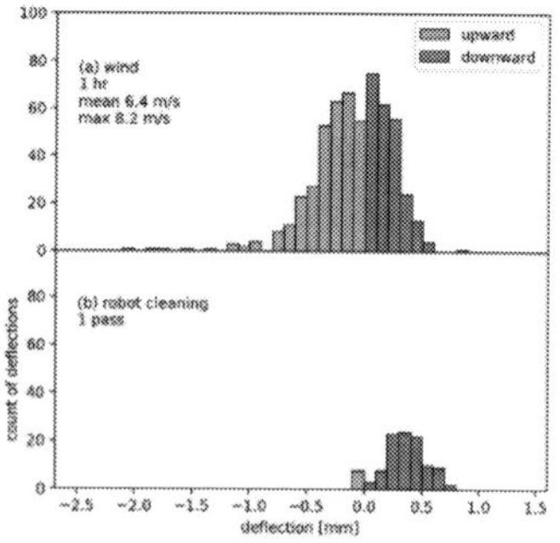

KEY TAKE-AWAYS

- Dry robot cleaning **can** abrade PV anti-reflective coatings within 18 months.
  Depends on the module (and probably robot)
- Reflectivity data and coupon results don't necessarily translate to module Pmax
- For high confidence of PV plant performance, need to test actual robot/module pair and measure Pmax
- Further research needed to translate accelerated lab tests to real-world results

THANK YOU

Fraunhofer CSP

3BO.11.5

PV Module Mission Profile from Middle East Deserts vs temperate European climate

Bengt Jaeckel, David Daßler, Matthias Pander
Fraunhofer Center for Silicon Photovoltaics CSP, Otto-Eissfeldt-Str. 12, 06120 Halle (Saale), Germany

Jim Joseph John, Sagarika Kumar, Baloji Adothu
DEWA Research & Development Center, P.O. Box: 564 Dubai, United Arab Emirates

Mission Profile
Outline

1. **Definition of Mission Profile (MP)**
2. **Use of MPs**
3. **Single-Stress MPs**
   - UV, amb. T, rH
   - Temperature change
4. **Mulitple-Stress MPs**
   - rH-T-pairs
   - UV-T-pairs and UV-module T pairs
5. **Summary**

Image: https://enterfea.com/thermal-loads/

Mission Profile

Definition

Motivation:

The concept of Mission Profiles (MP) is aimed to make an assessment for PV module durability **more location and installation specific**.

The concept:

A Mission Profile is a simplified representation of all relevant **static and dynamic load profiles** that a population of **electrical, mechanical, electronic, and electro-mechanical components** is **exposed during their entire life cycle** (expected service life)[1].

The term "Load" describes **environmental loads or a functional load** that applies to a component within an operating mode. Generally, loads for a PV module are **mainly mechanical, thermal, electrical, chemical type and fundamentally most relevant radiation**. The latter interacts with all other load types in one or the other way such as **extra heat, high voltages** or as an extra driving force for electro-chemical reactions.

[1] http://www.mpfo.org/mpfo-0.7/mpfo_ch_introduction.html (modified)

Fraunhofer CSP

020238-003

40th European Photovoltaic Solar Energy Conference and Exhibition

Mission Profile

Starting point

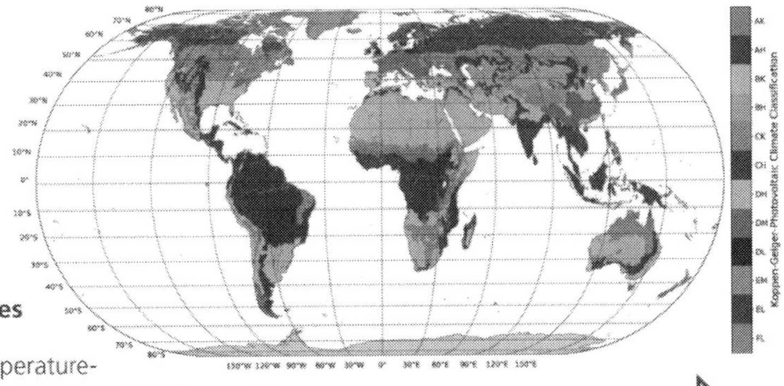

Köppen-Geiger-Photovoltaic climate classification with the **12 most relevant climate zones**

The first letter indicates the Temperature-Precipitation (TP)-zones: A-Tropical, B-Desert, C-Steppe, D-Temperate, E-Cold and F- Polar.

The second letter indicates the **4 Irradiation (I)-zones**: K-Very High, H-High, M-Medium and L-Low irradiation.

Great base to get global overview, but still difficult to compare local stress profiles

J. Ascencio-Vasquez, "Methodology of Koppen-Geiger-Photovoltaic climate classification and implications to worldwide mapping of PV system performance", Solar Energy 191 (2019) 672–685

Fraunhofer CSP

020238-004

Mission Profile

Why MPs?

Historic understanding:

- Main source of large scale systems is for temperature / Mediterian climates; not tropic or deserts
- Challenges for temperate climates are known (to some extend)
- High resolution weather data is available PV systems, including module temperatures and energy yield data

Haves for new sites and climate challenges:

- Climate data (at least from satellite data) is available globally
- Energy yield estimations (quite good!)

Needs:

- Simple procedure to assess possible risks by the new climatic conditions
- Fast priority setting for relevant extended stress tests

Ambient weather data
↓
+e.g. device temperatures or approximations
↓
Mission Profiles
↓
Use of acceleration protocols e.g. Arrhenius
↓
Compare stress for individual reactions and degradation modes
↓
Choose most relevant stress tests

See EPJ paper

Mission Profile

Single Stressor: Examples: UV, relative humidity and temperature ambient profiles

One full year of weather data (30-60sec)

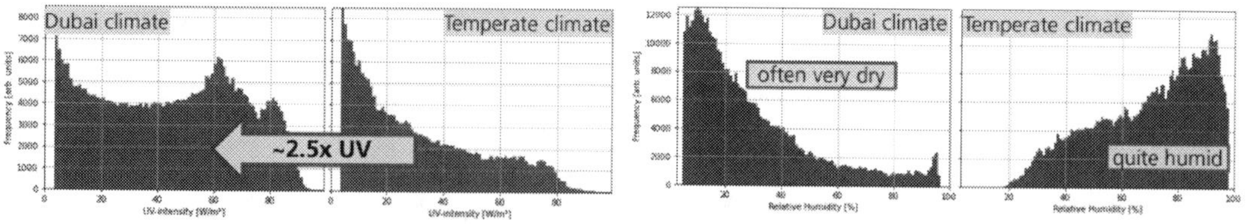

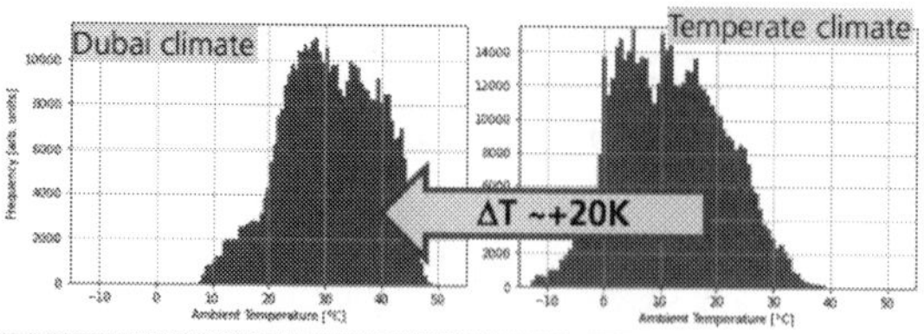

Good for some aspects (**UV-dose**), but to understand degradation in more detail, typical the **combination e.g. of T and rH** is of highest relevance

Mission Profile

Single Stressor: PV module temperature change

- Temperature change is most critical for all interconnection as daily thermo-mechanical stress is applied
- Example: Daily Min-Max module DTs for a temperate and the hot climate
  - Wide spread for the temperate site
  - Most days a temperature change of 30-50K can be observed, at elevated temperatures compared to the temperate site

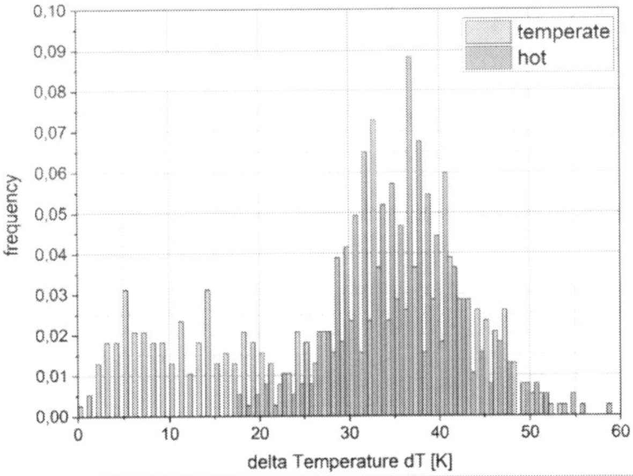

40th European Photovoltaic Solar Energy Conference and Exhibition

Mission Profile

Multiple Stressor: **rH-T** and UV-T pairs

Most chemical reactions and diffusion processes are highly temperature accelerated and are often related to moisture increases or corrosion reactions.

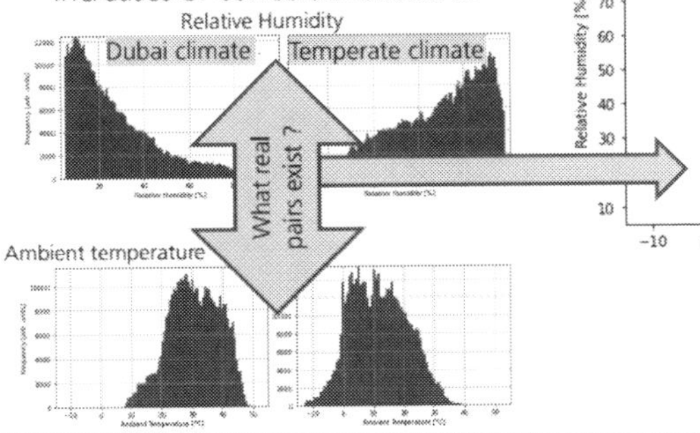

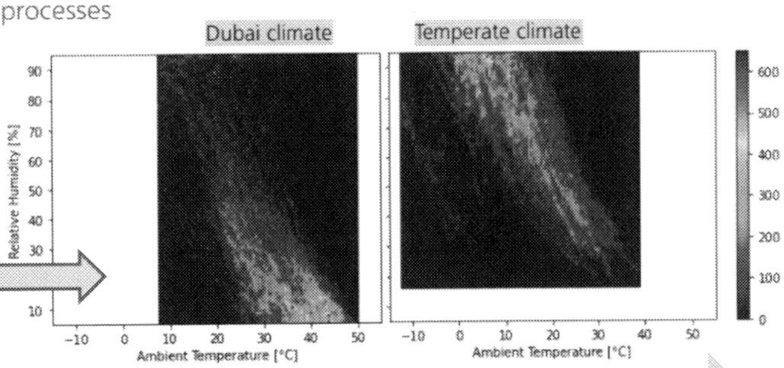

Dubai: typically warm and dry, but also events at very high humidity at temperatures above 10°C
Temperate: typically "cold" and more humid

Mission Profile

Multiple Stressor: rH-T and **UV-T pairs**

UV-degradation is dependent on (device) temperature.

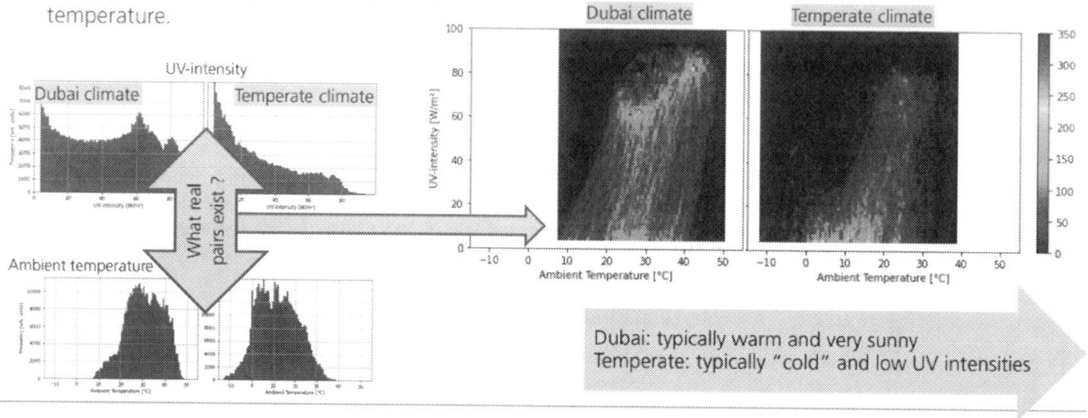

40th European Photovoltaic Solar Energy Conference and Exhibition

Mission Profile

Multiple Stressor: rH-T and **UV- PV module T pairs**

UV-degradation is dependent on (device) temperature.

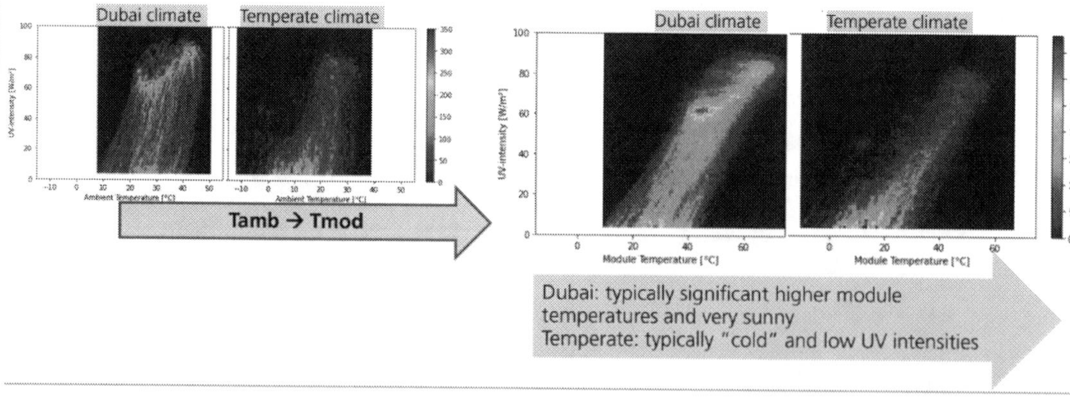

Mission Profile

Summary

Ambient weather data
↓
+e.g. device temperatures or approximations
↓
Mission Profiles
↓
Use of acceleration protocols e.g. Arrhenius
↓
Compare stress for individual reactions and degradation modes

1. Mission profiles represent expected loads of a certain device. E.g. temperature, temperature change, UV-dose
2. Multi-stress MPs are great to look at combintions of stress factors such as rH and T
3. MPs can serve as a base to benchmark new locations in respect to known places

Fraunhofer CSP

Fraunhofer CSP

Thank you for your attention!

Contact
Dr. Bengt Jaeckel
Group manager PV Modules, Components and Manufacturing
Bengt.jaeckel@csp.fraunhofer.de

Fraunhofer CSP
Otto-Eißfeldt- Straße 12
06120 Halle (Saale)
www.csp.fraunhofer.de

Fraunhofer
CSP

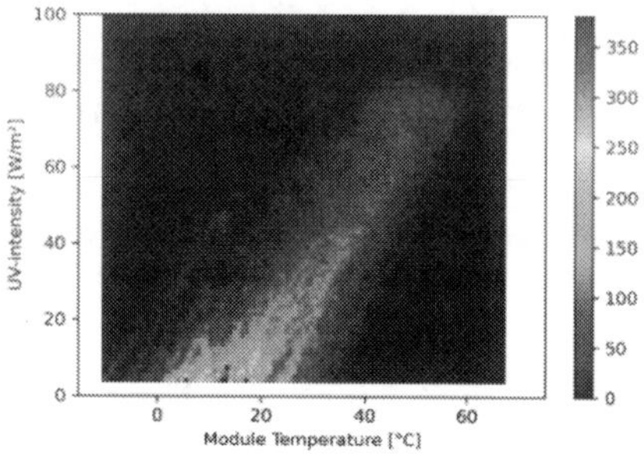

Publication note:

A detailed Paper on Mission Profiles will be published in EPJ as invited paper of EU Pvsec later this year. Please check journal webpage or contact Bengt Jäckel

Title:
„Mission Profile concept for PV modules: use case - Middle East Deserts vs temperate European climate

020238-013

40th European Photovoltaic Solar Energy Conference and Exhibition

This presentation was selected by the Sc. Committee of the EU PVSEC 2023 for submission of a full paper to one of the EU PVSEC's collaborating peer-reviewed journals.

NEAR-INFRARED SPECTROSCOPY OF EVA ENCAPSULANTS AS A TOOL FOR PROBING DEGRADATION STATUS OF SILICON PV MODULES

Oleksandr Stroyuk, Claudia Buerhop, Jens Hauch, Ian Marius Peters
High Throughput Methods in Photovoltaics, Forschungszentrum Jülich GmbH, Helmholtz Institute Erlangen-Nürnberg for Renewable Energy (HI ERN), Immerwahrstraße 2, 91058 Erlangen, Germany

ABSTRACT: We developed a non-invasive field-ready approach for the estimation of the relative degradation status of PV modules in the field using near-infrared absorption (NIRA) spectroscopy of ethylene vinyl acetate (EVA) encapsulants. The proposed methodology allows the general degradation of a PV module to be evaluated from the spectral parameters of the frontal EVA encapsulant layer without any major intervention in the module operation. NIRA spectra of EVA encapsulant provide several quantitative markers of the degradation status of PV modules, including carbonyl index and water content, both representative of the oxidation state of the encapsulant due to environment-driven degradative processes. The approach was tested in a series of field measurements. At that, pairs of NIRA spectra of BSs and EVA encapsulants were collected for several hundred modules on five PV power stations in Germany and Southern Europe in 2021-2022. NIRA spectroscopy of BSs allowed four major BS types to be identified in the collected dataset and to track the dependence of the degradation markers in EVA on the composition of BSs. The carbonyl index of EVA was found to depend on the total, cumulative exposure of tested PV modules to solar irradiation, increasing with the age of the PV field.

Keywords: spectroscopic analysis; backsheets; encapsulants; polymer degradation

1 INTRODUCTION

Environmental degradation of polymer components (backsheets (BSs), encapsulants, etc.) of silicon PV modules has recently been recognized as a major reason for failures resulting in considerable safety concerns, the inefficiency of the inverter operation, and, eventually, a loss in the yield of PV fields [1,2]. In this view, new field-capable approaches allowing us to evaluate the current state of degradation of polymer components and the module in general in a fast and non-invasive way are of paramount importance, spectroscopy and spectro-imaging being among the most natural options.

Near-infrared absorption (NIRA) spectroscopy allows organic polymer components of PV modules to be selectively probed while being non-sensitive to inorganic compounds (silicon, glass, etc.) and surface contaminations [3,4]. This feature of NIRA provides unique opportunities, in particular, it allows ethylene vinyl acetate (EVA) co-polymer encapsulants to be probed non-invasively through the protective glass and to collect various information on degradation events inside PV modules without the interference of operation of the PV power station.

In the present work, we developed a non-invasive field-ready approach for the estimation of the relative degradation status of PV modules in the field using NIRA spectroscopy of EVA encapsulants. The proposed methodology allows the general degradation of a PV module to be evaluated from the spectral parameters of the frontal EVA encapsulant layer without any major intervention in the module operation. The method is selective toward the type of backsheet and can be potentially upscaled to a high-throughput regime.

2 METHODS

The NIRA spectra were registered using a fiber-coupled FT-NIR Rocket spectrometer (ArcOptix, Switzerland) in a spectral range of 900-2600 nm. The samples were illuminated with a broad-emitting stabilized fiber-coupled NIR lamp. The spectrometer and the lamp were connected to a probe block at an angle of 90° via 1000 μm-thick optical fibers transmitting in the range of 400-2600 nm. The probe was placed on the inspected surface and held manually, and a series of 10 NIR reflectance spectra were collected and averaged. The original NIR reflectance spectra were then converted into NIRA spectra by dividing the reflectance spectrum of the lamp and the reflectance spectrum of the sample.

During field measurements, a couple of NIRA spectra of BS and EVA encapsulant were taken from each inspected module. The NIRA spectra of BSs were taken from the air side behind one of the lowest Si cells of the module, regardless of the module orientation. The NIRA spectra of the EVA encapsulant were taken above metal interconnects close to the points where BS spectra were registered. Typically, EVA was probed over a section of metal interconnect located above a Si cell in the left upper corner of the PV module when looked at from the frontal side. Only the lowest rows of modules were used for the spectral data collection. The spectra were collected on clear days with no rain at ambient temperatures of 22-28oC from dry modules with a rate of 30-40 pairs of BS/EVA spectra per hour allowing the data for each particular field to be collected in approximately steady ambient conditions.

3 RESULTS AND DISCUSSION

The quality of NIR reflectance (NIRR) spectra and derivative NIRA spectra of EVA encapsulant was found to depend strongly on the measurement location (Fig. 1a). The EVA NIRR spectra measured over busbars (spot 2) show a much higher signal-to-noise ratio as compared to those registered above the Si cells (spot 1), this effect becomes even more pronounced for measurements over broad metal interconnects (spot 3), which deliver spectra with the highest quality (compare curves 1-3 in Fig. 1b).

A NIRA EVA spectrum registered in this mode shows a set of C-H vibration bands at 1700-1800 nm stemming from the EVA hydrocarbon backbone, a characteristic carbonyl vibration band at 2140 nm, and a broad absorption band of water centered at 1900-1910 nm (Fig.

1c). The latter two bands can be used to evaluate the content of C=O and water in EVA, both indicative of the current state of oxidative degradation of the encapsulant.

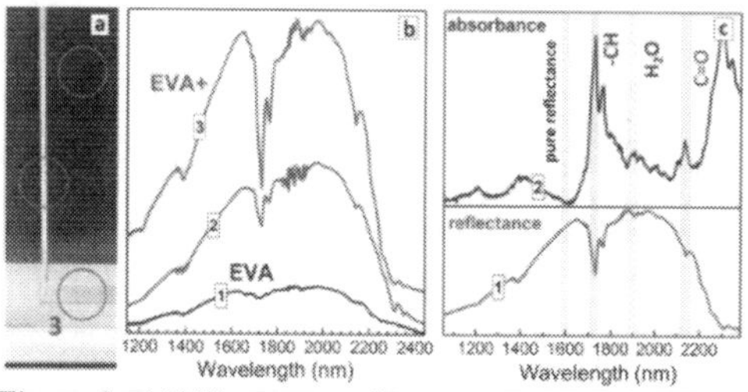

Figure 1: a) Different spots for measurements of NIRA spectra of EVA encapsulant. (b) NIRR spectra acquired in spots 1-3 are indicated in (a). (c) Exemplary NIRR spectrum (curve 1) and NIRA spectrum (curve 2) were used for PV module analysis.

In the following, we use the carbonyl index CI and the water index WI as quantitative measures of the C=O and H2O content, calculated as ratios of the intensities 2140-nm and 1910-nm bands, respectively, to the intensity of a reference C-H band at 1730 nm. Additionally, EVA is characterized by zero NIR absorption at 1600 nm, allowing this spectral range to be used to evaluate pure metal reflectivity from the metal interconnect and in this way evaluate the current state of metal corrosion. Similar to CI and WI, reflectivity was calculated as a ratio of reflectance at 1600 nm to the reference reflectance signal at 1730 nm.

To illustrate the relationship between the spectral parameters extracted from the NIRA EVA spectra and the state of degradation of the corresponding PV module we compared the CI, WI, and metal reflectivity of a series of 12 PV modules with their isolation resistance R_{iso} measured by the "wet" leakage test [5]. It was found that both CI and WI increase with a decrease in R_{iso}, clearly indicating that more degraded PV modules with lower isolation resistance exhibit higher carbonyl and water contents in the EVA encapsulant (Fig. 2a,b). On the opposite, the metal reflectivity was found to increase with an increase in R_{iso} (Fig. 2c), showing that the interconnects of less degraded modules are less affected by the corrosion and formation of non-reflective corrosion products.

The feasibility and perspectives of NIRA spectroscopy of the EVA encapsulant were tested during a field measurement campaign in 2021-2022 carried out on a series of five PV plants in Germany (North – NG, South – SG) and Southern Europe (SE). The age of installations varied from 5 to 10 years as noted in Table 1.

Table 1: Summary of the PV modules tested during the field measurements

	Name	BS	N	Age, years
1	FC	a-FC/PET/PP	167	10
	PA	a-PA/PA/PA	71	
2	PET-1	a-PET/PET/PE	77	10
3	PET-2	a-PET/PET/PE	26	5
4	PVF-1	a-PVF/PET/PVF	171	10
5	PVF-2	a-PVF/PET/PVF	65	8

During field measurements, pairs of NIRA spectra were collected from PV modules in the lowest rows of PV installations, including a NIRA spectrum of EVA taken on the module front side over a metal interconnect, and a NIRA spectrum of the module BS taken from the back side approximately at the same spot as the front-side EVA spectrum. The NIRA spectrum of BS was then used to identify the BS composition [3,4], while the NIRA spectra of EVA were processed to extract values of CI and WI. In total, four BS types were identified with a different BOM (Table 1) designated by the polymer type in the air layer as FC (fluorinated coating), PA (polyamide), PET (polyethylene terephthalate), and PVF (polyvinyl fluoride). In the case of field #4, the modules were delivered to the lab and kept at low relative humidity (RH) for 3 weeks before the measurements, while other modules were tested in the field under the same climatic conditions.

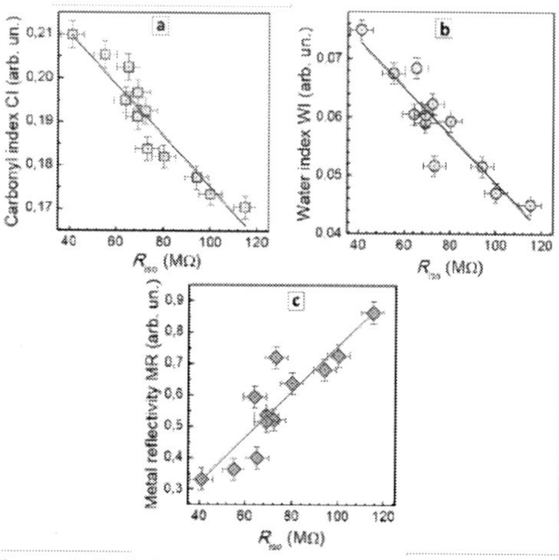

Figure 2: Relationships between carbonyl index CI (a), water index WI (b), and metal reflectivity (c) determined from NIRA spectra of EVA encapsulant and the insulation resistance of a series of PV modules with PA BSs. Pearson correlation coefficients are -0.95 (a), -0.92 (b), and 0.90 (c). Solid lines represent linear fits of the data. b) for PV strings with BS-types quantified by the amount of FC1-BS within the strings

The carbonyl index of EVA encapsulant of the field-aged PV modules was found to depend considerably on the BS material and, for the same BS type, - on the age of the PV power station (Fig. 3). For the same module age, the lowest CIs were found for PV modules with PVF-type BSs (median CI of ca. 0.17), followed by PET- (0.18-0.19), PA- (0.19-0.20), and FC-type BSs (ca. 0.21). This progression correlates with the insulation behavior of modules with these BS types evaluated earlier in the lab [5], with the BSs with lower R_{iso} showing higher CIs.

The lowest R_{iso} and the highest median CI was found for FC-based BSs which systematically showed inferior isolation properties both in the lab [5] and in field measurements on module, string, and inverter levels [6,7]. For the same BS type, the CI was found to depend on the age of the field (compare PET-1 and PET-2, Fig. 3), indicating a cumulative effect of the field aging time on the degradation status of BSs and PV modules in general.

To assess the effect of age we calculated the total exposures of the tested PV plants to solar irradiation using open data on cumulated annual irradiation doses for different climates [8]. A distinct correlation was found between the median CI and total solar exposure, with the depth of EVA oxidation (CI) increasing steadily with the solar illumination dose (Fig. 4).

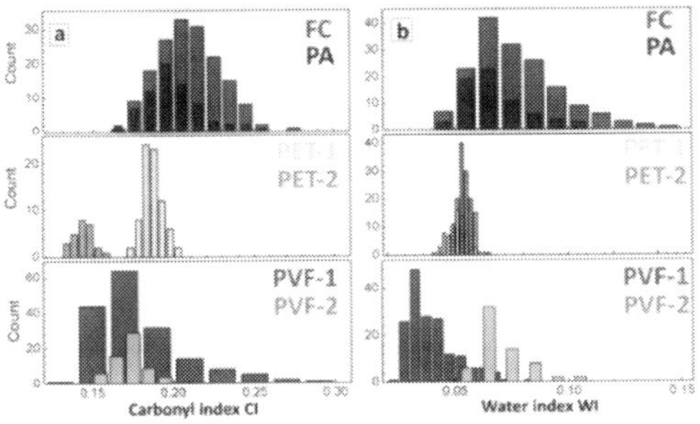

Figure 3: (a,b) Distributions of carbonyl index CI (a) and water index WI (b) collected for PV modules with different BS types (FC, PA, PVF, PET) and at different locations (PET-1,2 and PVF-1,2).

This correlation is entangled with the above-discussed dependence of the median CI on the composition of tested BSs. The combination of age and BOM can be clearly seen for FC- and PA-based BSs having identical exposure but differing significantly in the median CI. The water content was also found to be dependent on the BS composition, increasing from PET- to PA- to FC-based BSs (Fig. 3).

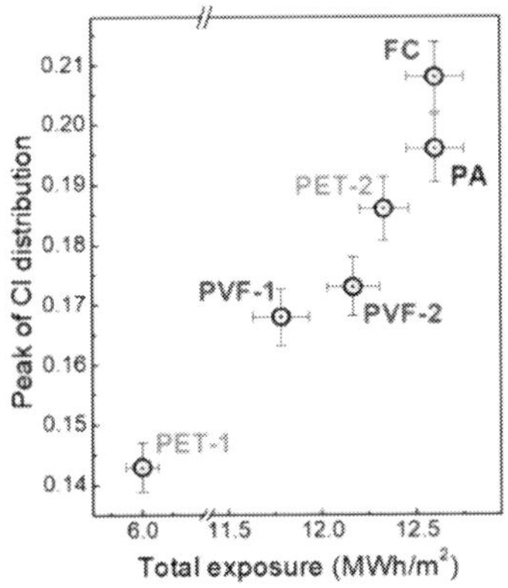

Figure 4: Relationship between the peak of CI distributions collected on different PV fields and the total exposure of the tested PV modules to solar irradiation.

The PVF-based BSs fall out of this progression because PVF-1 samples were measured in a dry lab long after the delivery from the field, while PVF-2 samples were tested in a strongly different climate (SE) as compared to PV fields tested in Germany. These observations indicate that WI values are strongly dependent on the measurement conditions and should be treated cautiously when comparing datasets collected in different climates, while the CI values are much more climate-independent and representative of the general degradation status of PV modules.

4 SUMMARY

In conclusion, we show that NIRA spectra of EVA encapsulant provide quantitative markers of the degradation status of PV modules, which can be collected in the field in a relatively fast and non-destructive way with no major intervention in system operation. We found that carbonyl content in EVA can be used to evaluate encapsulant oxidation and depends strongly on the BS type and the total exposure to solar irradiation. Water content in EVA is also indicative of the degradation state of the module but can be affected by climatic conditions and the history of module storage.

The approach was tested in a series of field measurements. At that, pairs of NIRA spectra of BSs and EVA encapsulants were collected for several hundred modules on five PV power stations in Germany and Southern Europe in 2021-2022. NIRA spectroscopy of BSs allowed four major BS types to be identified in the collected dataset and to track the dependence of the degradation markers in EVA on the composition of BSs. The depth of degradation as shown by NIRA spectroscopy of EVA was found to depend on the insulation properties of BS polymers. Along with this, the carbonyl index of EVA was found to depend on the total, cumulative exposure of tested PV modules to solar irradiation, increasing with the age of the PV field. Water content in EVA is also indicative of the degradation state of the module. However, this marker was found to be affected by climatic conditions and the history of module storage before the measurements.

5 ACKNOWLEDGEMENT

This work is supported by the German Federal Ministry for Economic Affairs and Energy (project "COSIMA", FKZ: 0324291A and "dig4morE", FKZ: 03EE1090B), and Zentrales Innovationsprogramm Mittelstand (ZIM, project "PolymerCHECK", No. 16KN083038).

6 REFERENCES

[1] L. Hernández-Callejoa, S. Gallardo-Saavedraa, V. Alonso-Gómez, Sol. Energy 2019, 188, 426-440.

[2] J. Denz, J. Hepp, C. Buerhop, B. Doll, J. Hauch, C.J. Brabec, I. M. Peters, Energy Environ. Sci., 2022,15, 2180-2199.

[3]. O. Stroyuk, C. Buerhop-Lutz, A. Vetter, J. Hauch, C.J. Brabec, Sol. Energy Mater. Sol. Cells 2020, 216, 110702.

[4]. O. Stroyuk, C. Buerhop-Lutz, A. Vetter, J. Hepp, J. Hauch, I.M. Peters, C.J. Brabec, Progr Photovoltaics, 2022, 30, 859-868.

[5]. C. Buerhop-Lutz, O. Stroyuk, J. Zöcklein, T. Pickel, J. Hauch, I.M. Peters, Progr. Photovoltaics, 2022, 30, 938-947.

[6]. C. Buerhop-Lutz, T. Pickel, O. Stroyuk, J. Hauch, I.M. Peters, Solar Energy Mater. Solar Cells 2022, 246, 111913.

[7]. C. Buerhop-Lutz, O. Stroyuk, T. Pickel, J. Hauch, and I.M. Peters, Progr. Photovoltaics, 2022, http://doi.org/10.1002/pip.3621.

[8]. Global Modeling and Assimilation Office. MERRA-2 tavg1_2d_slv_Nx: 2d,1-Hourly, Time-Averaged, Single-Level, Assimilation, Single-Level Diagnostics V5.12.4, Greenbelt, MD, USA, Goddard Earth Sciences Data and Information Services Center (GES DISC). 10.5067/VJAFPLI1CSIV.

PCCL
Polymer Competence Center Leoben

Design of Experiment for degradation modelling of PV modules

G. Oreski, E. Helfer, K.A. Berger, G.C. Eder, Y. Voronko, B. Brune, I. Ortner, K. Knöbl, L. Neumaier, M. Feichtner

Polymer Competence Center Leoben GmbH,
Roseggerstraße 12, 8700 Leoben
+43 664 88679331
gernot.oreski@pccl.at

INFINITY

European Photovoltaic Solar Energy Conference
Lisboa, 19.09.2023

ADVANCE!

Introduction

PCCL
Polymer Competence Center Leoben

INFINITY – Climate sensitive long-time reliability Photovoltaics (2015-2018)

INFINITY

- WP3 - Climate sensitive accelerated aging tests
  - ✓ *Identical modules for exposure to 10 different accelerated aging tests simulating different climatic conditions*
  - ✓ *100 six cell modules & 100 single cell modules*
- WP4 - Advanced high efficiency PV module components and materials for predefined environmental conditions
  - ✓ *Modules with combinations of different components: 3 different encapsulants & 2 different backsheets*
  - ✓ *72 six cell modules & 72 single cell modules*
- Available data
  - ✓ *> 1300 images (EL, UV-F)*
  - ✓ *> 2000 spectra and characteristic curves (IV, FTIR, UVF, UV/VIS/NIR)*

Main issue: Project ended with a lot of data not sufficiently analyzed

Solution: Application for a follow up project dealing with data mining and degradation modelling

ADVANCE!

2

Objectives

PCCL Polymer Competence Center Leoben

Main Objective: to derive an optimum design of experiment for degradation modelling of PV modules based on the experiences made in the Austrian flagship projects Infinity and ADVANCE!

Inventory and cataloging of existing data & setup of database → Feature extraction from images and spectra → Degradation modelling → Analysis of experimental design → Derivation of optimum design of experiment

See also at #EUPVSEC2023:

- 3BO.15.3 - E. Helfer - Influence of External Stresses on Degradation Pathways of Photovoltaic Modules

Papers:

- B. Brune et al., Connecting material degradation and power loss of PV modules using advanced statistical methodology, https://doi.org/10.1016/j.solmat.2023.112485
- B. Brune et al., Quantifying the influence of encapsulant and backsheet composition on PV-power and electrical degradation, https://doi.org/10.1002/pip.3679
- L. Neumaier et al., Advanced UV-fluorescence image analysis for early detection of PV-power degradation, https://doi.org/10.1051/epjpv/2023001

- **What surprising results or insights did we obtain?**
- **What could we do better in planning and executing our project?**

Design of Experiment

PCCL Polymer Competence Center Leoben

Expectation | Reality

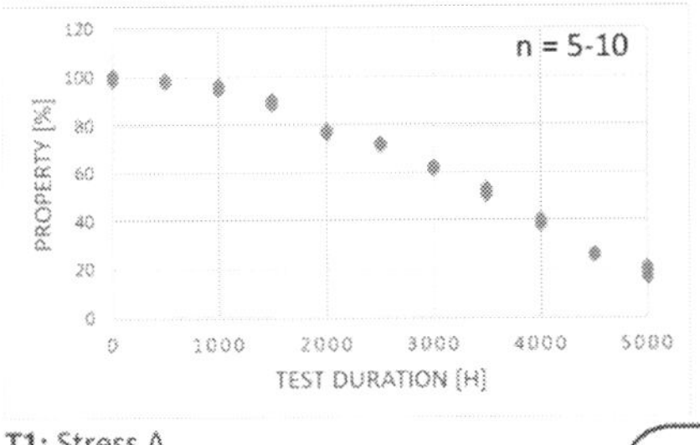

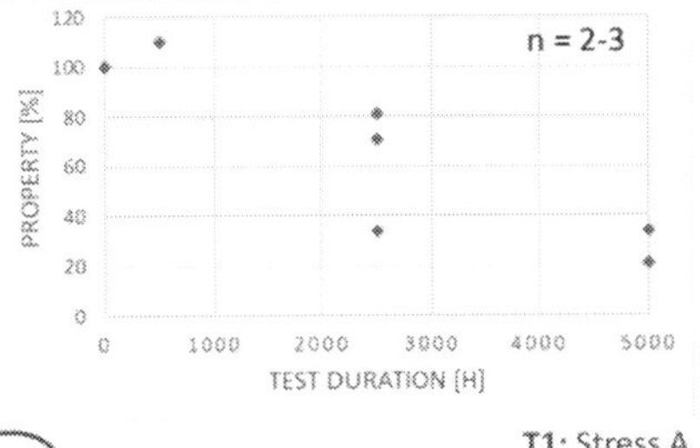

T1: Stress A
T2: Stress A + Stress B
T3: Stress A + Stress C
T4: Stress A + Stress B + Stress C
Identical number of steps and test duration

- Not enough materials / broken specimen
- Availability and limits of aging chambers
- Project ends
- Repurposed samples

T1: Stress A
T2: Stress B + Stress C
T3: Stress C + Stress D
T4: Stress A + Stress B + Stress D
Varying number of steps and test durations

4

Data availability

PCCL
Polymer Competence Center Leoben

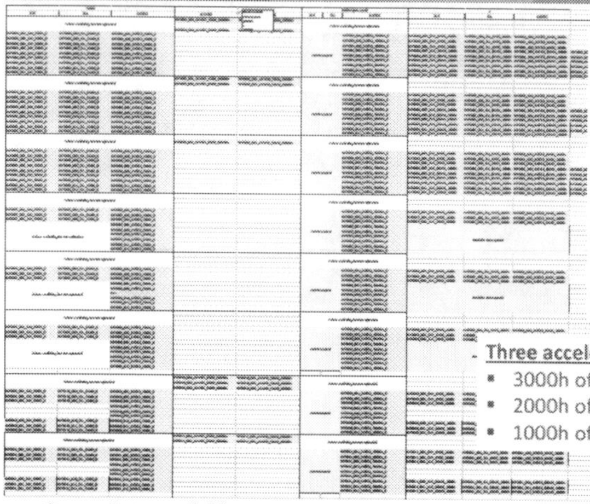

TABLE 1. List of test modules (12 identical modules named 001 … 012 were manufactured for each type).

Module name	Encapsulant	Backsheet
INFS004_001 … 012	EVA	PPF
INFS005_001 … 012	TPO	PPF
INFS006_001 … 012	POE	PPF
INFS007_001 … 012	EVA	CPO
INFS008_001 … 012	POE	CPO
INFS009_001 … 012	TPO	CPO

Three accelerated aging tests

- 3000h of DH (85°C / 85% RH, 1000h steps)
- 2000h of irradiance (1000W/m², 50°C, > 40% RH, 1000h steps)
- 1000h of sequential test (200h DH + 40 cycles of Irradiance + DH)

- **Green:** Data available
- **Yellow:** Modules not exposed or data not measured

How to model degradation as a factor of different stressors with these incomplete data sets?

B. Brune et al., Quantifying the influence of encapsulant and backsheet composition on PV-power and electrical degradation, https://doi.org/10.1002/pip.3679

020241-005

Specimen design

PCCL
Polymer Competence Center Leoben

Material coupons
- Destructive & non-destructive material testing

Single cell modules
- Non-destructive material characterization
- Electrical characterization

Six cell modules
- Electrical characterization
- Non-destructive material characterization

Full size modules
- Validation

Comparability of different specimen sizes

- UV cutoff of glazing material needs to be identical - UV radiation between 300 and 320 nm most damaging for polymers
- Different diffusion pathways and diffusion lengths - overestimated degradation for smaller samples

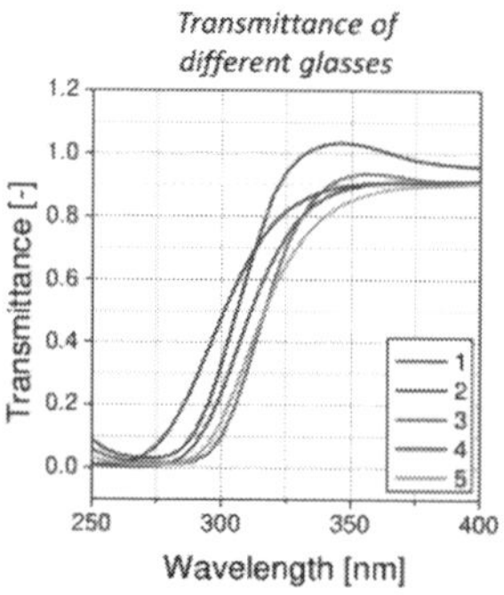

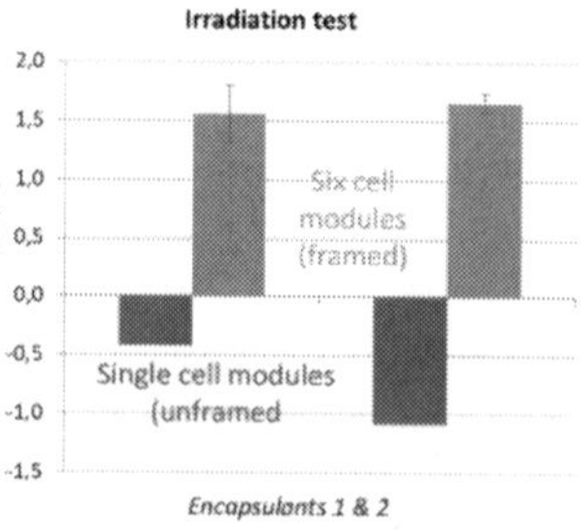

6

020241-006

Sample preparation

PCCL

Manual work during production of test modules as well as different production dates may cause higher scattering in the data or in the worst-case misleading results

- Typical sources of error caused by manual work
  - ✓ *Soldering of the interconnects*
  - ✓ *Handling & alignment of cells and cell strings*
  - ✓ *Cutting and alignment of the polymer films*
- Problem of different production dates
  - ✓ *Shelf life of components: e.g. encapsulant films & adhesion promoters; adhesives;*
  - ✓ *New production batches of components*
  - ✓ *Differences in production equipment, e.g. thermal behavior of laminator*
  - ✓ *Different operator / handler*

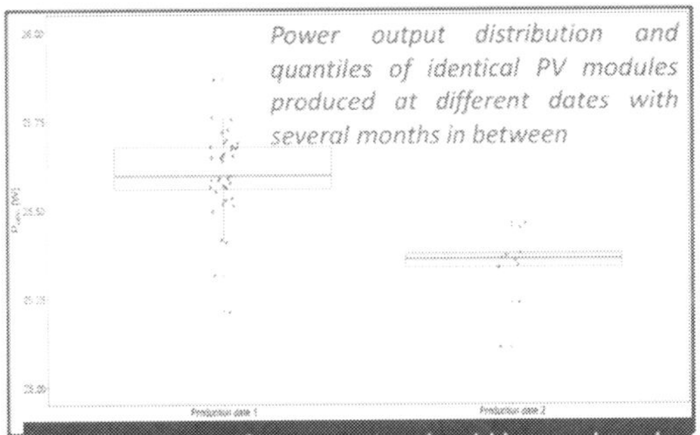

Power output distribution and quantiles of identical PV modules produced at different dates with several months in between

- All samples of a test series should be produced at the same time with the same material and component batches and production parameters and facilities
- Precise sample preparation is the basis for reliable and reproducible measurement data, both, at module and material level

020241-007

Data availability

PCCL

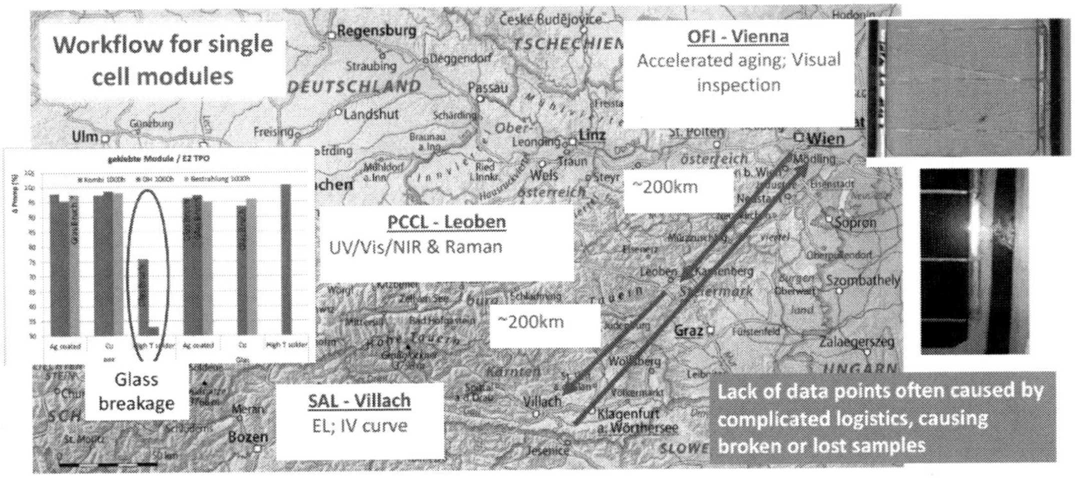

© https://commons.wikimedia.org/wiki/File:%C3%96sterreich_%C3%9Cbersichtskarte.png

8

020241-008

Data accuracy

PCCL

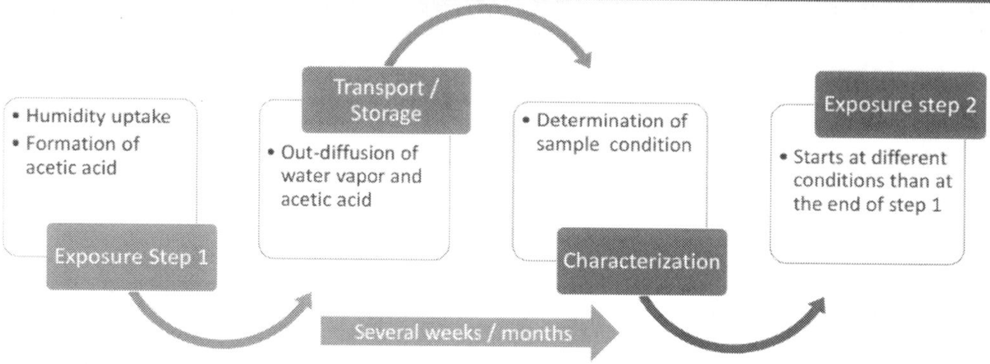

- Long transport or storage times between exposure and sample characterization can affect measurement results due to diffusion processes, especially for small samples / mini modules
- Re-exposing of samples does not start at the same conditions → usually underestimation of degradation processes

9

020241-009

40th European Photovoltaic Solar Energy Conference and Exhibition

Material characterization methods

PCCL

Obtaining material characteristics of all layers/components integrated into a test module´s multi-material stack can be challenging

- UV/Vis/NIR spectroscopy (back)
- FTIR-ATR spectroscopy
- Raman

- Nanoindentation

How do material changes correlate with module power?

- NIR spectroscopy
- Fluorescence spectroscopy
- UV/Vis/NIR spectroscopy (front)
- Raman

- Thermal analysis (DSC, TGA, EGA)
- Additive analysis (GC/MS)

See also 3BO.14.4 and 4AO.7.3

020241-010

Material changes vs module power?

PCCL
Polymer Competence Center Leoben

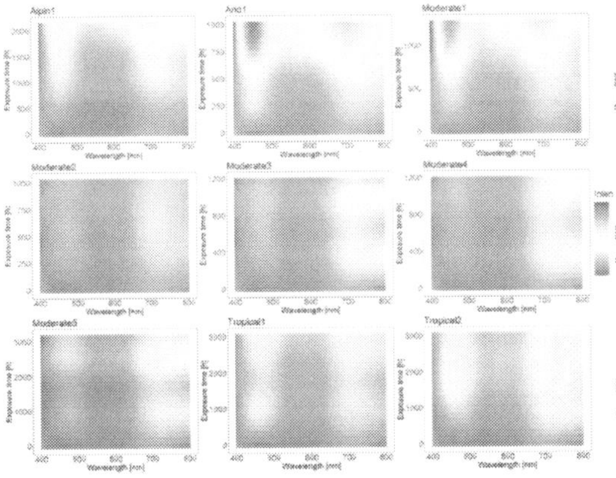

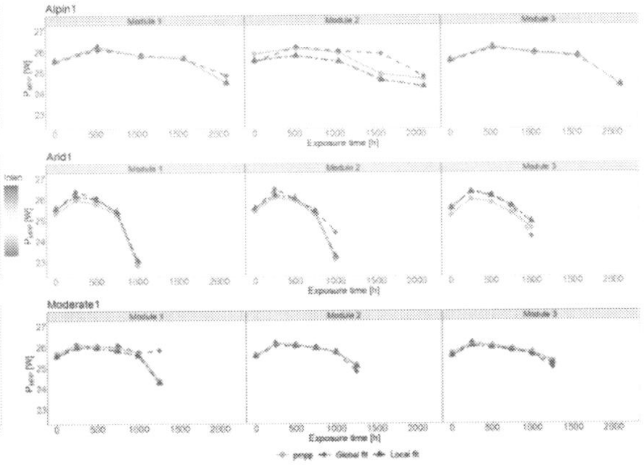

Spectral intensity (400 to 800 nm) of the UVF spectra

UV fluorescence spectra of encapsulants can explain power loss of PV modules using advanced statistical modelling

B. Brune et al., Connecting material degradation and power loss of PV modules using advanced statistical methodology, https://doi.org/10.1016/j.solmat.2023.112485

11

020241-011

40th European Photovoltaic Solar Energy Conference and Exhibition

Automated data processing

PCCL
Polymer Competence Center Leoben

Several measures enable automated data processing and data reduction:

- ✓ Consistent data naming convention
- ✓ Uniform data sets and data files (and folder structure)
- ✓ Uniform settings for recording of images (particularly uniform illumination)

12_42xPVP_2x3_Module_4BB_Mono
19 Glas-Glas Module
☑ wp4 6 cell modules_E2
WP4_6cell modules LCA

Module title Advance!	Module group Infinity	# Infinity	No of cells	Module #	Encapsulant	Backsheet	Aging Test	Remark
INFS004_001	INFS-B1-M1	1a	6	1	EVA	PPF	REF	
INFS004_002	INFS-B1-M1	1b	6	2	EVA	PPF	REF	
INFS004_003	INFS-B1-M1	1c	6	3	EVA	PPF	REF	
INFS004_004	INFS-B1-M1	2a	6	4	EVA	PPF	DH85	
INFS004_005	INFS-B1-M1	2b	6	5	EVA	PPF	DH85	
INFS004_006	INFS-B1-M1	2c	6	6	EVA	PPF	DH85	
INFS004_007	INFS-B1-M1	3a	6	7	EVA	PPF	ISUN1	
INFS004_008	INFS-B1-M1	3b	6	8	EVA	PPF	ISUN1	
INFS004_009	INFS-B1-M1	3c	6	9	EVA	PPF	ISUN1	
INFS004_010	INFS-B1-M1	4a	6	10	EVA	PPF	COMB	
INFS004_011	INFS-B1-M1	4b	6	11	EVA	PPF	COMB	
INFS004_012	INFS-B1-M1	4c	6	12	EVA	PPF	COMB	

Identifying, cataloging and renaming of the files took several weeks!

12

B. Brune et al., Quantifying the influence of encapsulant and backsheet composition on PV-power and electrical degradation, https://doi.org/10.1002/pip.3679

020241-012

Data processing - feature extraction

PCCL

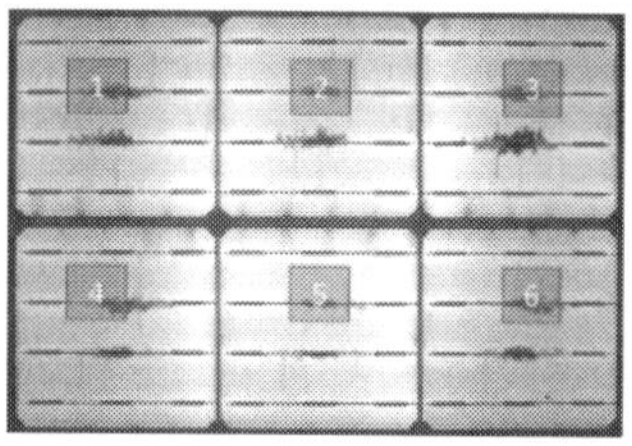

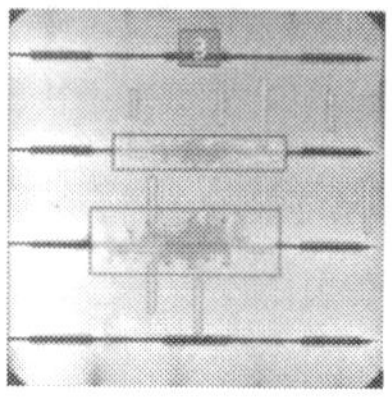

Automatic evaluation of
- ✓ *Number and length of broken grid fingers*
- ✓ *Discolored, corroded area next to the busbars*

Deep learning based automated feature selection in EL images

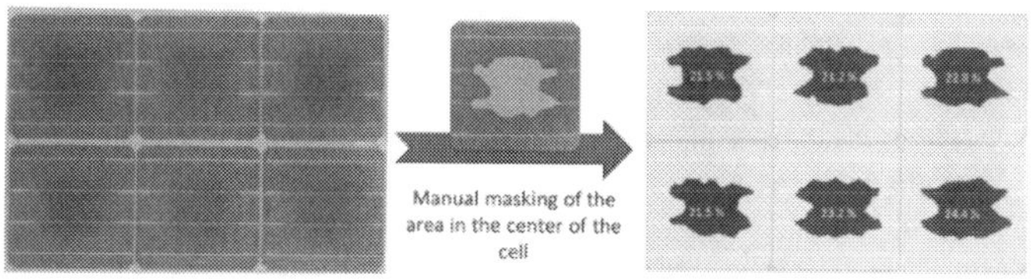

Manual data extraction from UV fluorescence images

Main reason for manual feature extraction:
Differences in the intensity caused by changing lighting situations during image recording

13

020241-013

40th European Photovoltaic Solar Energy Conference and Exhibition

Summary & Conclusions

PCCL

An optimum design of experiment which generates the necessary input data for meaningful degradation modelling of PV modules needs to focus on

Reproducible sample preparation

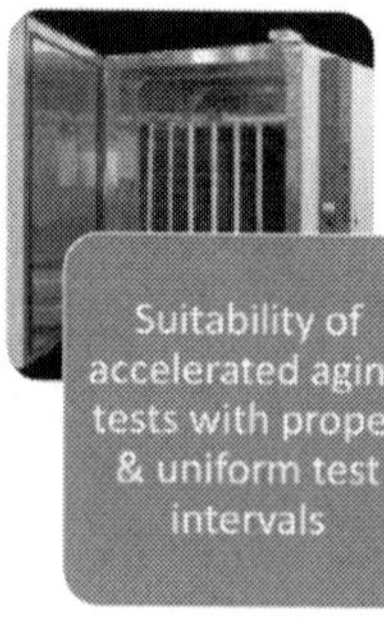

Suitability of accelerated aging tests with proper & uniform test intervals

Suitability of characterization methods

Appropriate data preparation allowing for automated feature extraction

14

020241-014

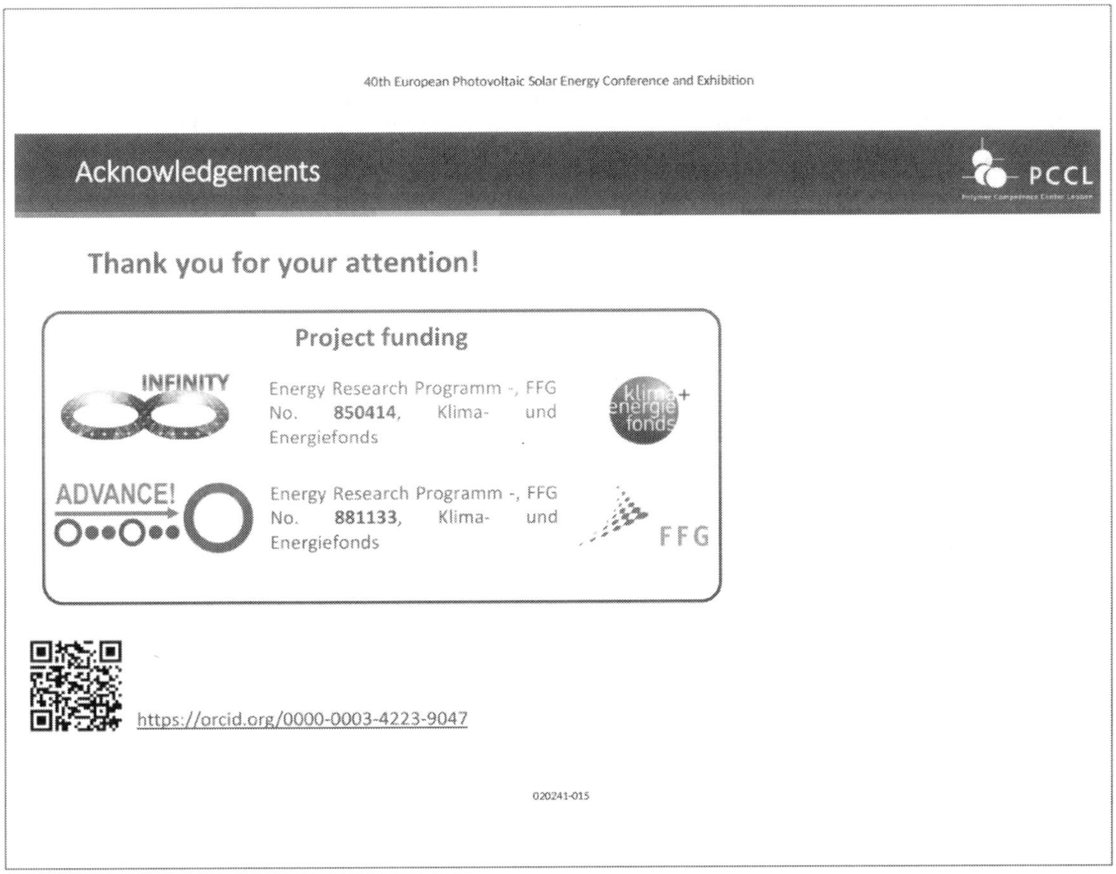
40th European Photovoltaic Solar Energy Conference and Exhibition
Acknowledgements
PCCL
Thank you for your attention!
Project funding
INFINITY
Energy Research Programm -, FFG No. 850414, Klima- und Energiefonds
klima+ energie fonds
ADVANCE!
Energy Research Programm -, FFG No. 881133, Klima- und Energiefonds
FFG
https://orcid.org/0000-0003-4223-9047
020241-015

ADVANCE!

Influence of external stresses on degradation pathways of photovoltaic modules

E. Helfer[1*], G. C. Eder[2], Y. Voronko[2], B. Brune[3], I. Ortner[3], K. A. Berger[4], K. Knöbl[5], L. Neumaier[6] & G. Oreski[1]

1) Polymer Competence Center Leoben, Leoben, Austria
2) OFI Austrian Research Institute for Chemistry and Technology, Vienna, Austria
3) TÜV Austria Data Intelligence, Vienna, Austria
4) AIT Austrian Institute of Technology, Vienna, Austria
5) University of Applied Sciences Technikum Wien, Vienna, Austria
6) Silicon Austria Labs, Villach, Austria

EU PVSEC 2023

18.09.-22.09.2023, Lisbon, Portugal

*) Roseggerstraße 12, 8700 Leoben, Austria
+43 3842 42962 88
eric.helfer@pccl.at

www.pccl.at

020242-001

40th European Photovoltaic Solar Energy Conference and Exhibition

Introduction

Previous project "INFINITY"

- The goal was to find solutions to design modules for specific climate zones.
- Hundreds of sample modules were subjected to precisely defined accelerated aging scenarios.
- PV modules
  - 6-cell sample modules
  - mono-crystalline Si cells with SnPb ribbons interconnections
  - front glass
  - polyethylene terephthalate (PET) backsheet (fluoro-coating on the inner side)
  - ethylene-vinyl acetate (EVA) encapsulant

INFINITY

https://energieforschung.at/projekt/climate-sensitive-long-time-reliability-of-photovoltaics/

- Many measurements have been carried out:
  - electrical characterization (e.g. I-V curves, EL…)
  - material characterization data (e.g. IR spectra and color measurements of backsheets, UVF-spectra of encapsulants)

"Overflow" of data

Need for additional data evaluation → follow-up project ADVANCE!

www.pccl.at 2

020242-002

40th European Photovoltaic Solar Energy Conference and Exhibition

Objectives

Objectives of ADVANCE!

Develop

- an exploratory data analysis tool for automated modeling/prediction of photovoltaic (PV) module reliability under predefined real-world climatic conditions.
- algorithms for automatic data preparation with the aim of generating high-quality data sets that are suitable for analysis.
- a methodology to derive single point values as degradation indicators.
- validated models to identify correlations between:
  - electrical performance degradation
  - specific property changes of the materials
  - impact of the climatic stress conditions

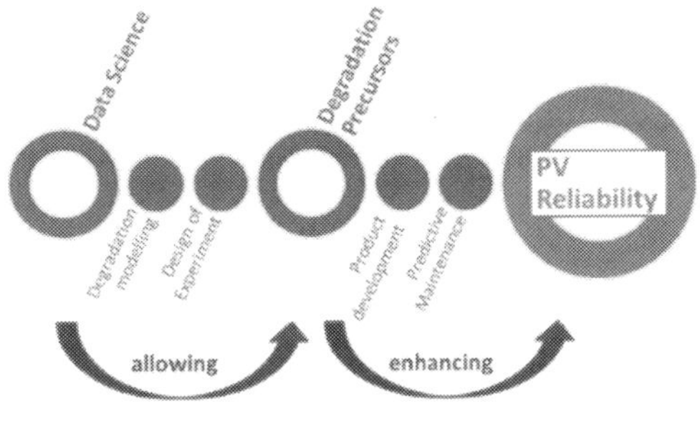

- Derive an optimal design of experiments in terms of:
  - aging tests
  - characterization methods
  - data preparation to avoid data gaps

40th European Photovoltaic Solar Energy Conference and Exhibition

Experimental: Accelerated aging tests

- Reference accelerated aging action according to standard IEC 61215, MQT 1311 (most frequently used in PV community)
- Variation in:
  - temperature
  - humidity
  - irradiation
  - salt exposure
  - temperature cycles
  - dynamic mechanical load

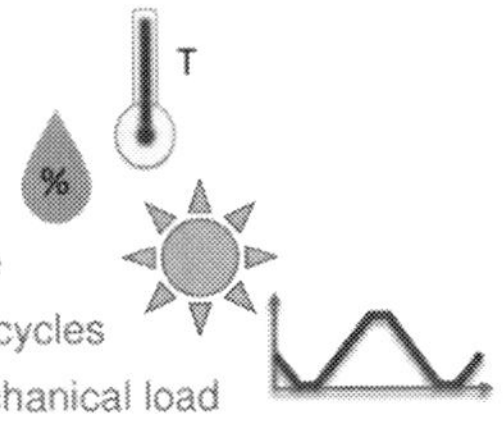

Aging action	Duration		Temperature T Module	Chamber	Rel. Humidity H Module	Chamber	Irradiance I	DML	Salt	TC	Intervals
Reference	1000h	1	-	85°C	-	0.85	-	-	-	-	constant
Moderate 1	1000h	1	113°C	85°C	0.304	0.85	1000 W/m²	-	-	-	constant
Moderate 2	1000h	1	78 °C	60°C	0.181	0.4	1000 W/m²	-	-	-	48h
	≈ 7 cycles	2	-	85°C	-	0.85	-	-	-	-	96h
Moderate 3	1000h	1	78°C	60°C	0.181	0.4	1000 W/m²	-	-	-	48h
	≈ 7 cycles	2	-	85°C	-	0.85	-	-	-	-	96h
		3	-	-	-	-	-	1000 c	-	-	24h
Moderate 4	1000h	1	78°C	60°C	0.181	0.4	1000 W/m²	-	-	-	48h
	≈ 7 cycles	2	-	85°C	-	0.85	-	-	-	-	96h
		3	-	60°C	-	-	-	-	Salt-mist	-	24h
Moderate 5	3000h	1	78°C	60°C	0.181	0.4	1000 W/m²	-	-	-	48h
	≈ 7 cycles	2	-	85°C	-	0.85	-	-	-	-	96h
		3	-	-40/+85°	-	-	-	-	-	50 c	300h
Alpin 1	2000h	1	-	85°C	-	0.85	-	-	-	-	250h
	≈ 4 cycles	2	119°C	85°C	0.249	0.85	1200 W/m²	-	-	-	250h
		3	-	-	-	-	-	1000 c	-	-	24h
Tropical 1	3000h	1	-	85°C	-	0.85	-	-	-	-	constant
Tropical 2	3000h	1	-	90°C	-	0.9	-	-	-	-	constant
Arid 1	1000h	1	129°C	95°C	0.157	0.5	1200 W/m²	-	-	-	constant

[1] Eder, G. C. et al. (2019): Climate specific accelerated ageing tests and evaluation of ageing induced electrical, physical, and chemical changes. In: Prog Photovolt Res Appl 27 (11), S. 934–949. DOI: 10.1002/pip.3090.

Data extraction and modelling

Data extraction

Data modelling

www.pccl.at 5

020242-005

40th European Photovoltaic Solar Energy Conference and Exhibition

Data extraction and modelling

Data extraction

www.pccl.at 6

020242-006

Results: Data extraction

Electrical measurement data:

- Electroluminescence (EL) images
- I-V curves

- Length of the cracks in the module
- "Shading" area due to corrosion of the busbars
- Area of the finger defects

- P_{MPP}
- U_{OC}, I_{SC}
- R_S, R_P

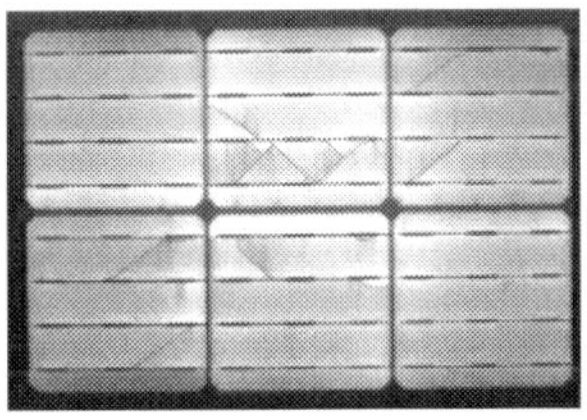

Cracks

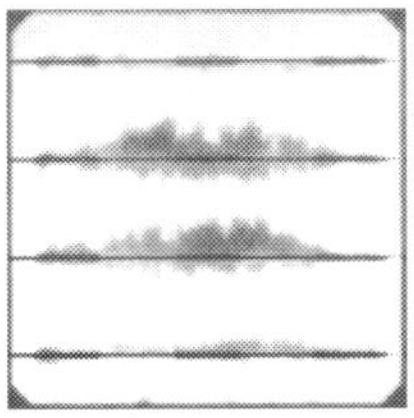

Shading area

Results: Data extraction

Material characterization data:

- ATR infrared spectroscopy (IR) spectra of the backsheet surface
- Color measurements of the backsheet material
- Ultra violet fluorescence (UVF) images and spectra of the encapsulant material

- 28 IR bands evaluated
- band maximum
- height, area

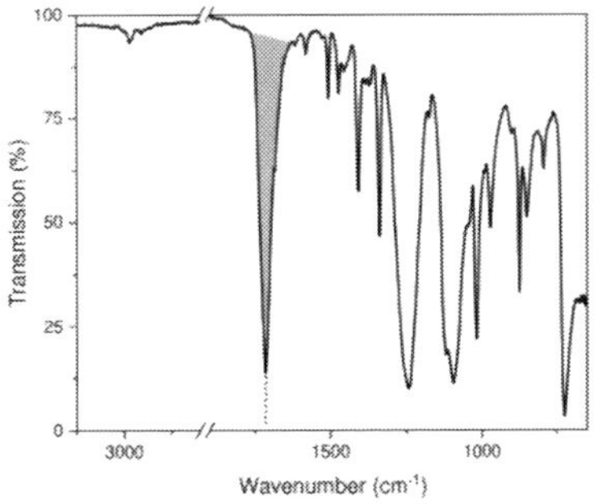

Important IR bands

Decrease due to chain scission

- 1240 cm^{-1} (ester group)
- 1715 cm^{-1} (carbonyl group)

Increase due to chain scission

- 1265 cm^{-1} (acid group)
- 1675 cm^{-1} (carbonyl group)

→ Chemical change of backsheet surface

Results: Data extraction

Material characterization data:

- ATR infrared spectroscopy (IR) spectra of the backsheet surface
- Color measurements of the backsheet material
- Ultra violet fluorescence (UVF) images and spectra of the encapsulant material

- L*, a*, b*
- ΔE*

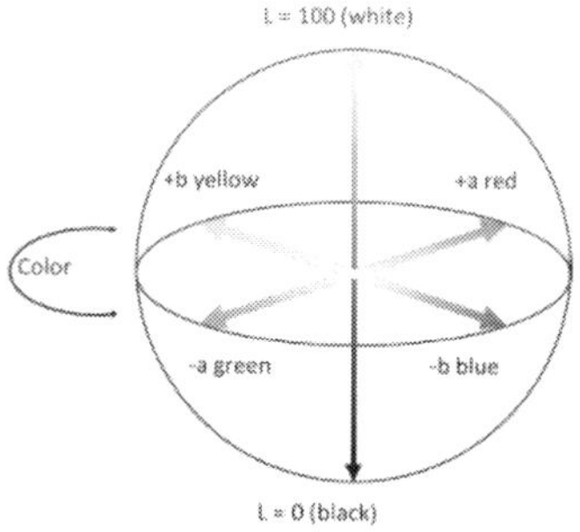

→ Chemical change of backsheet surface

$$\Delta E^* = \sqrt{\Delta L^{*2} + \Delta a^{*2} + \Delta b^{*2}}$$

40th European Photovoltaic Solar Energy Conference and Exhibition

Results: Data extraction

Material characterization data:

- ATR infrared spectroscopy (IR) spectra of the backsheet surface
- Color measurements of the backsheet material
- Ultra violet fluorescence (UVF) images and spectra of the encapsulant material

- 3 peaks
- peak maximum
- height, area

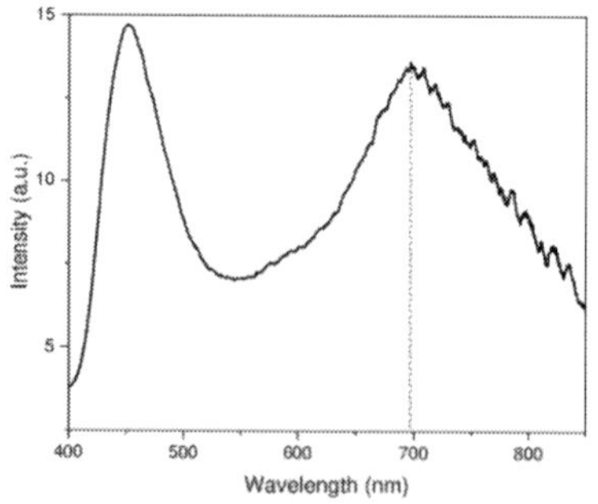

Important peaks

- 450 – 500 nm (high humidity)
- ~550 nm (natural weathering)
- ~700 nm (high temperature)

→ Chemical change of encapsulant

Data extraction and modelling

ADVANCE! PCCL Polymer Competence Center Leoben

Data extraction

Data modelling

www.pccl.at 11

020242-011

40th European Photovoltaic Solar Energy Conference and Exhibition

Data extraction and modelling

ADVANCE! PCCL Polymer Competence Center Leoben

Data modelling

www.pccl.at 12

020242-012

Approach: Data modelling

ADVANCE! PCCL Polymer Competence Center Leoben

Network Structural Equation Modelling (netSEM) [2, 3]

- Package in R (https://cran.r-project.org/web/packages/netSEM/index.html)
- Methodology for determining degradation mechanisms and rates
- Understand how mechanisms interact
- Extract the combined effects on performance loss due to material degradation

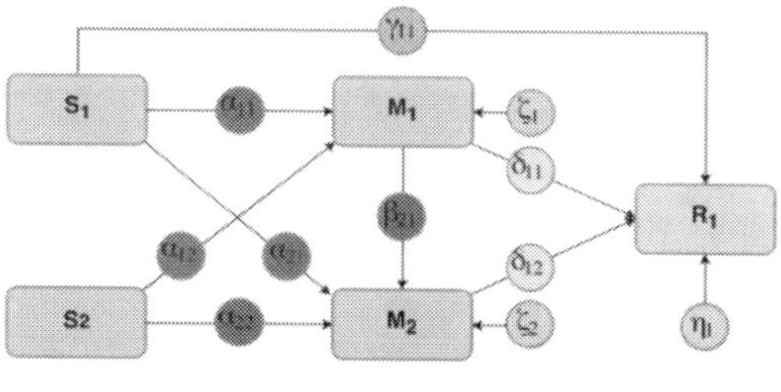

$$M_1 = x + \alpha_{11}S_1 + \alpha_{12}S_2 + \zeta_1$$
$$M_2 = y + \alpha_{21}S_1 + \alpha_{22}S_2 + \beta_{21}M_1 + \zeta_2$$
$$R_1 = z + \gamma_{11}S_1 + \delta_{11}M_1 + \delta_{12}M_2 + \eta_1$$ [4]

netSEM pathway diagram

[2] Bruckman, L. S. et al. (2013): "Statistical and Domain Analytics Applied to PV Module Lifetime and Degradation Science." IEEE Access 1 (2013): 384-403. doi:10.1109/ACCESS.2013.2267611

[3] Bruckman, L. S. et al. (2013): "Photovoltaic Lifetime and Degradation Science Statistical Pathway Development: Acrylic Degradation." In SPIE Solar Energy+ Technology, 8825:88250D-8. International Society for Optics and Photonics, 2013. doi:10.1117/12.2024717

[4] Gok, A. et al. (2019): "Degradation Science and Pathways in PV Systems." In Durability and Reliability of Polymers and Other Materials in Photovoltaic Modules, pp. 47–93. DOI: 10.1016/B978-0-12-811545-9.00003-3

40th European Photovoltaic Solar Energy Conference and Exhibition

Approach: Data modelling

ADVANCE! PCCL Polymer Competence Center Leoben

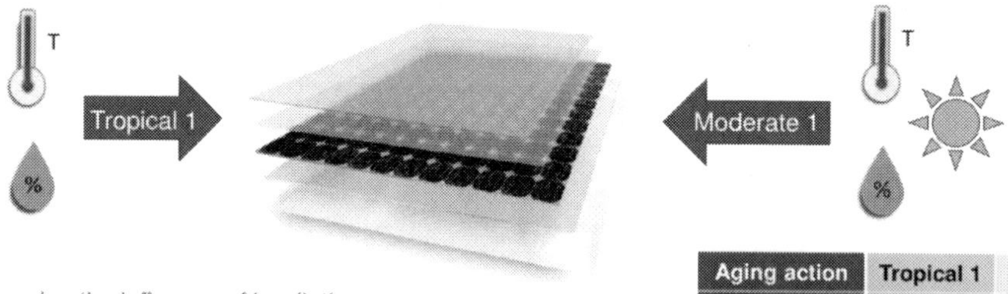

- Determine the influence of irradiation on degradation pathways of PV modules.
  - aging time as main stressor
  - P_{MPP} as main response
  - variables from each analysis method
  - Tropical 1 and Moderate 1

➡ **correlate material and electrical properties**

Aging action	Tropical 1	Moderate 1
Duration	3000 h	1000 h
Temperature (Chamber)	85 °C	85 °C
Rel. Humidity (Chamber)	0.85	0.85
Irradiance	-	1000 W/m²

- FTIR and UVF variables: area information for each peak was used.

Results: Data modelling – Tropical 1

Tropical 1 – damp heat

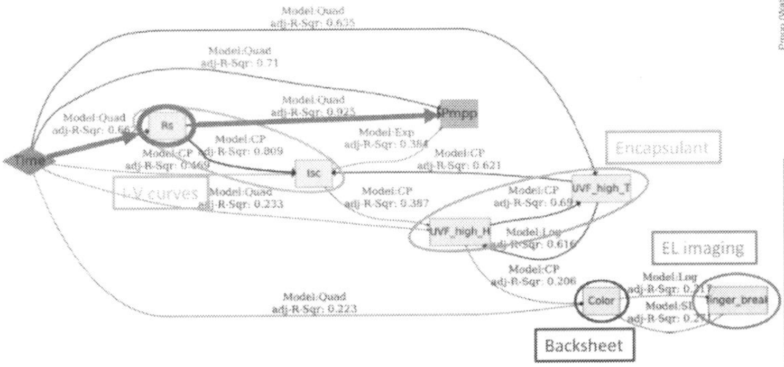

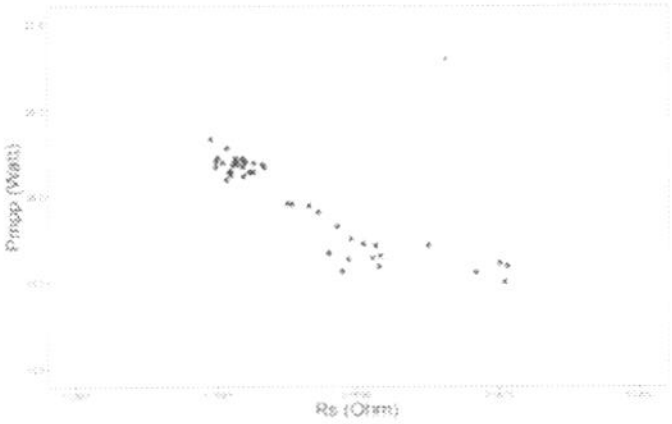

Main degradation path through R_S (corrosion).

Encapsulant fluorescence (UVF peaks) and backsheet degradation (change of color) do not seem to have a big influence on the power loss.

Time: aging time
finger_break: area of finger breaks
UVF_high_H: UVF peak at 450 – 500 nm
Color: total change of color (ΔE*)
UVF_high_T: UVF peak at ~700 nm
www.pccl.at

15

020242-015

40th European Photovoltaic Solar Energy Conference and Exhibition

Results: Data modelling – Moderate 1

Moderate 1 – damp heat + irradiation

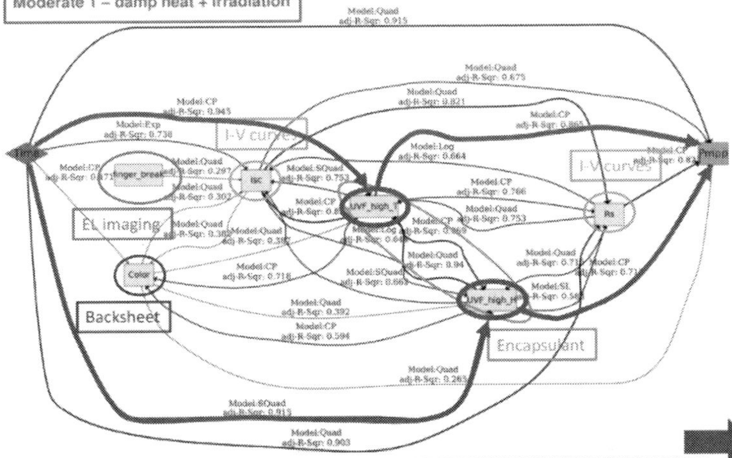

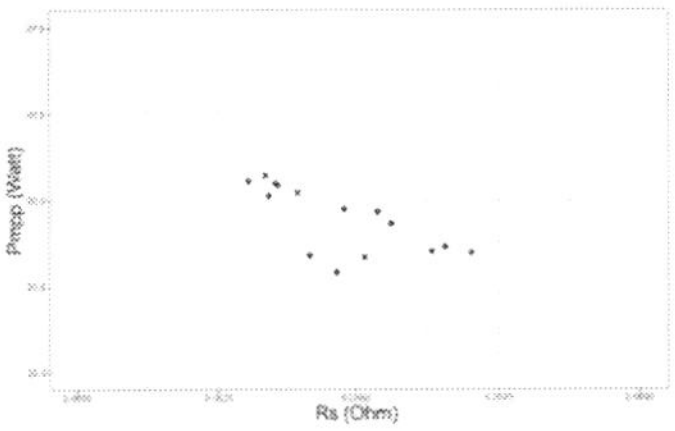

Degradation at the backsheet surface (change of color) does not show a significant correlation with decrease in P_{MPP}.

R_S still shows a good correlation to P_{MPP}.

With additional irradiance, encapsulant degradation (UVF) shows increased impact on P_{MPP}.

Time: aging time
finger_break: area of finger breaks
UVF_high_H: UVF peak at 450 – 500 nm
Color: total change of color (ΔE*)
UVF_high_T: UVF peak at ~700 nm
www.pccl.at

16

020242-016

Summary

More characteristic values were extracted from existing data

- characteristic values from spectra (FTIR, UVF)
- single point data from EL and UVF images

Temperature and humidity load (DH–storage)

- corrosion effects (measurable as change in the serial resistance) were found to be the main degradation pathway

Testing conditions without irradiation useful?

Additional irradiance leads to a change of the degradation pathway of PV modules.

- degradation of encapsulant influences the electrical performance

40th European Photovoltaic Solar Energy Conference and Exhibition

Outlook

!Full database will be published soon!

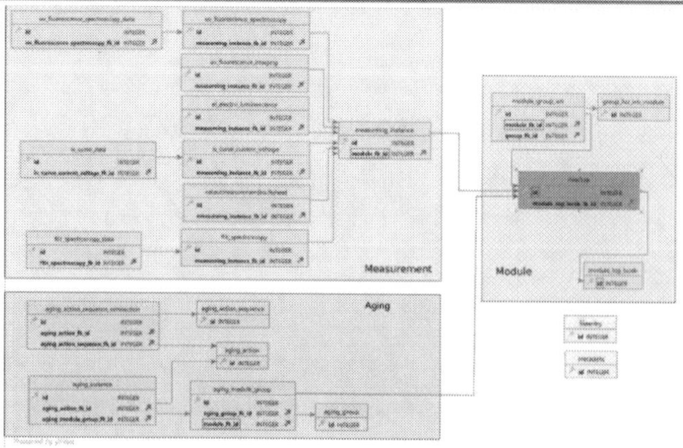

Thank you for your attention!

ADVANCE! receives funding by the Austrian government, by means of the Austrian Federal Ministry for Climate Action, Environment, Energy, Mobility, Innovation and Technology (bmk.gv.at), represented by the Austrian Research Promotion Agency (FFG, https://www.ffg.at/), under contract No. 881133, within the 6. e!MISSION call.

Eric Helfer
Roseggerstraße 12,
8700 Leoben, Austria
+43 3842 42962 88
eric.helfer@pccl.at

www.pccl.at

020242-019

PCCL

Design and Testing of PV Modules Based on Glass/Glass Configuration to Achieve Extended Lifetime

C. Barretta[1*], L. Meinhart[1], H. Krebs[2], J. Wittfoth[2], A. Brandstätter[3], D. Geyer[4], R. Einhaus[4], A. Gok[5], G. Oreski[1]

1) Polymer Competence Center Leoben GmbH (PCCL), Leoben, Austria
2) CS Wismar GmbH Sonnenstromfabrik, Wismar, Germany
3) Lenzing Plastics GmbH, Lenzing, Austria
4) Zentrum für Sonnnenenergie- und Wasserstoff-Forschung Baden-Württemberg (ZSW), Stuttgart, Germany
5) Gebze Technical University, Gebze/Kocaeli, Turkey

40th EU PVSEC, 19.07.2023, Lisbon

*) Saurauagasse 1, 8700 Leoben, Austria
+43 3842 42962 722
chiara.barretta@pccl.at

www.pccl.at

020243-001

40th European Photovoltaic Solar Energy Conference and Exhibition

How can we extend PV module's lifetime?

PCCL

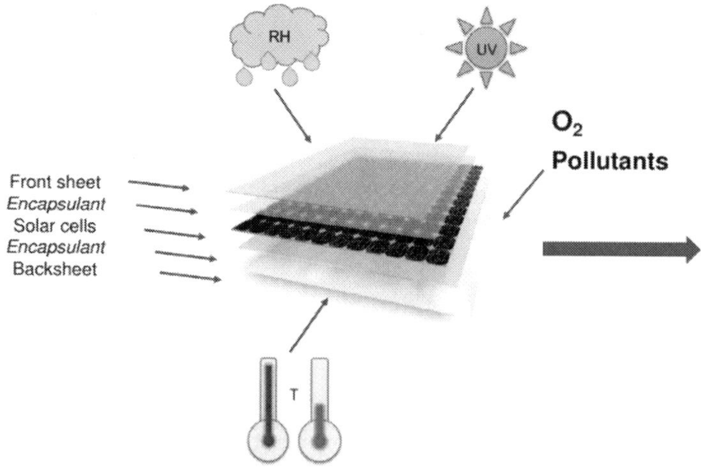

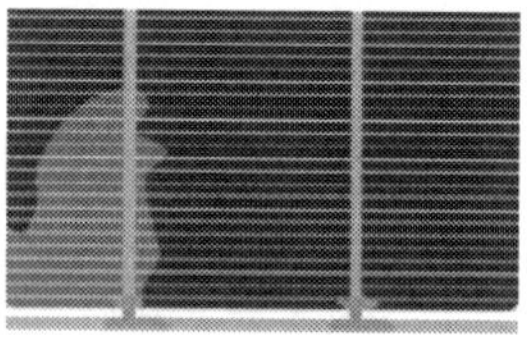

Degradation modes

- Discoloration (yellowing/browning) of encapsulant
- Delamination
- Corrosion of metallization and interconnections
- Potential induced degradation (PID)
- Backsheet cracks

www.pccl.at 2

020243-002

How can we extend PV module's lifetime?

PCCL
Polymer Competence Center Leoben

PV module technology → Bill of materials → Appropriate accelerated aging procedures

- Glass/glass
- Bifacial cells (PERC)

Encapsulant	Chemical structure
EVA	Ethylene vinyl acetate
TPO	Ethylene α-olefin
POE	Ethylene α-olefin
ION	Ethylene acrylic acid

Edge seal – polyisobutylene-based

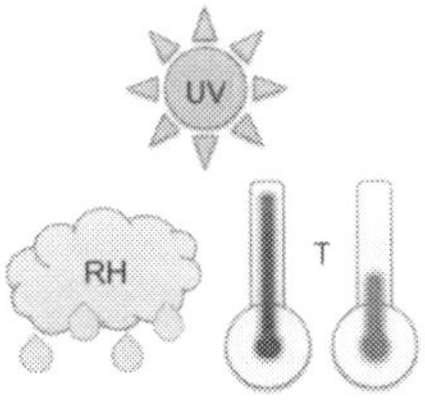

85 °C, 85% r.H. up to 5000 hours [1, 2]

[1] Kempe, M. D., Panchagade, D., Reese, M. O., and Dameron, A. A. (2015), Modeling moisture ingress through polyisobutylene-based edge-seals. *Prog. Photovolt: Res. Appl.*, 23: 570–581. doi: 10.1002/pip.2465

[2] Koehl, M., Hoffmann, S., and Wiesmeier, S. (2017) Evaluation of damp-heat testing of photovoltaic modules. Prog. Photovolt: Res. Appl., 25: 175–183. doi: 10.1002/pip.2842

How can we extend PV module's lifetime?

PCCL
Polymer Competence Center Leoben

Bill of materials

3AV.1.35 Oreski G.
Polyethylene copolymers as Solar Cell Encapsulants: a Critical Overview

Encapsulant	Chemical structure	Crosslinking during lamination	Melting temperature (°C)
EVA	Ethylene vinyl acetate	yes	64
TPO	Ethylene α-olefin	no	120
POE	Ethylene α-olefin	yes	70
ION	Ethylene acrylic acid	no	93

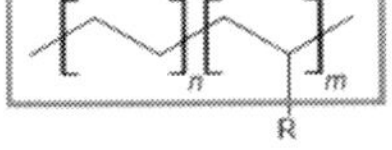

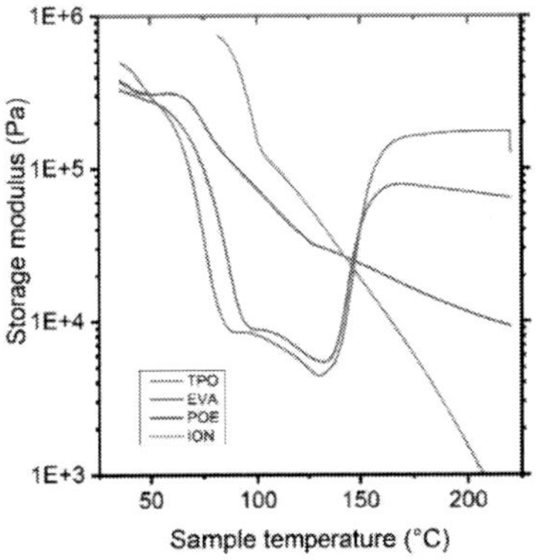

How can we extend PV module's lifetime?

PCCL
Polymer Competence Center Leoben

Appropriate accelerated aging procedures

Coupons / Test laminates
- Material compatibility
- Encapsulant degradation

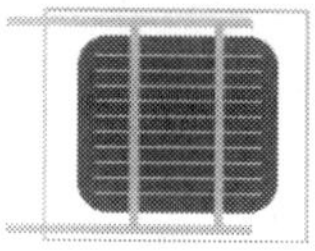

Test mini modules
- Information about power / electrical performances

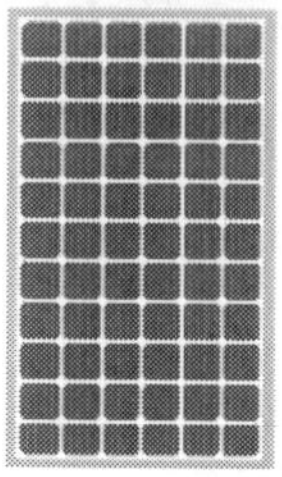

Full-scale modules
- Field performances considering actual microclimate

Characterization method	Properties of interest
UV-Vis-NIR spectroscopy	Optical properties, discoloration
Thermo-desorption Gas Chromatography coupled to Mass Spectrometry (TD-GC/MS)	Qualitative additive analysis
Fourier Transform Infrared Spectroscopy – Attenuated Total Reflectance (FTIR-ATR)	Chemical aging
Differential Scanning Calorimetry (DSC)	Thermal properties, chemical aging and morphology
Thermogravimetric Analysis (TGA)	Thermal stability, additive consumption

3BO.14.3 Einhaus R. *Test and Evaluation of Combinations of Encapsulant Materials towards a Long Service Lifetime of PV Modules: PV 40 Plus*

www.pccl.at 5

020243-005

Results
Visual inspection

PCCL
Polymer Competence Center Leoben

EVA DH 5000 h

- Discoloration
- Adhesion issues

POE DH 5000 h

- Slight discoloration

ION DH 5000 h

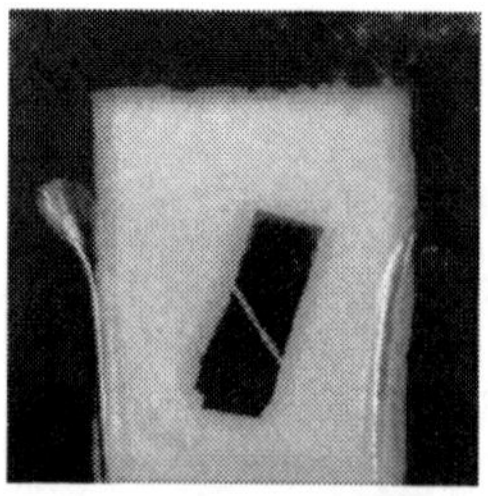

- Adhesion issues at different interfaces

TPO DH 5000 h

- Discoloration

www.pccl.at 6

020243-006

Results
UV-Vis-NIR spectroscopy: yellowness index

PCCL

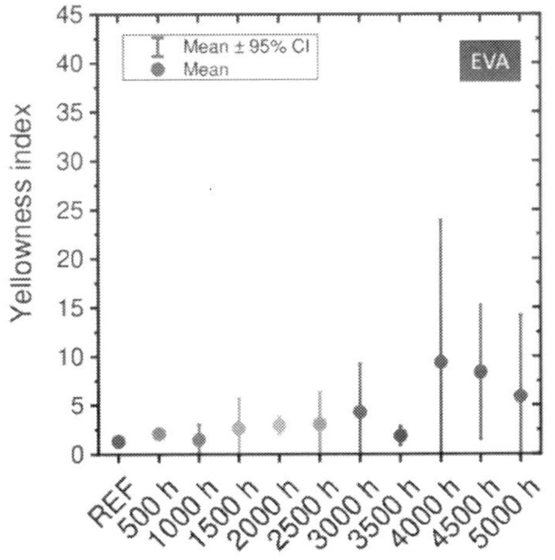

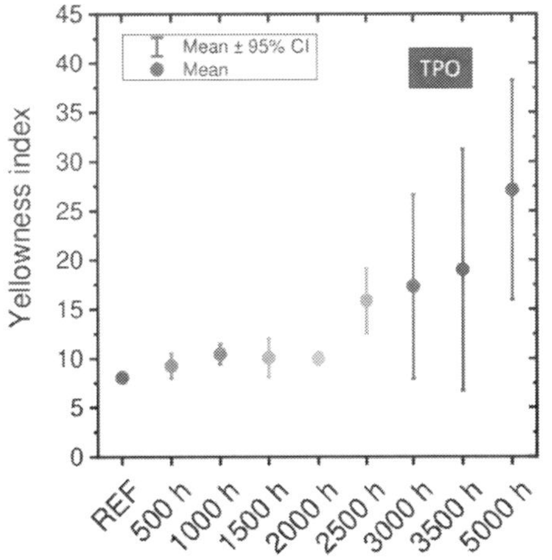

www.pccl.at 7

020243-007

Results
UV-Vis-NIR spectroscopy: yellowness index

PCCL

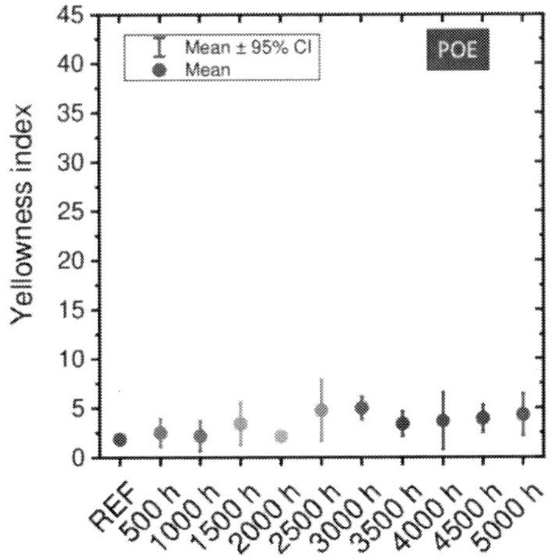

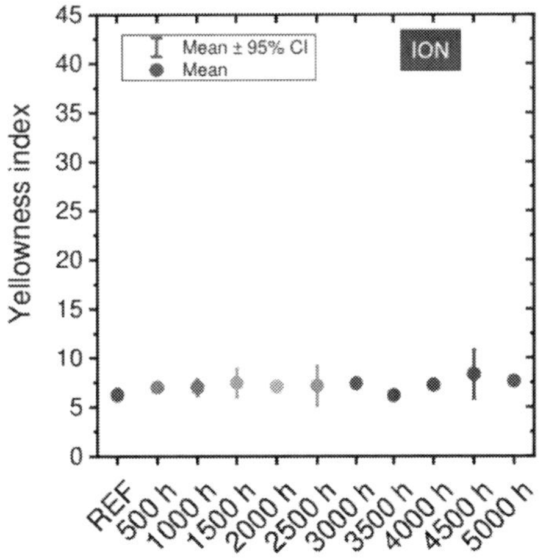

www.pccl.at 8

020243-008

Results
TD-GC/MS: Qualitative additive analysis

PCCL

Stabilizers are still present after the exposure to DH

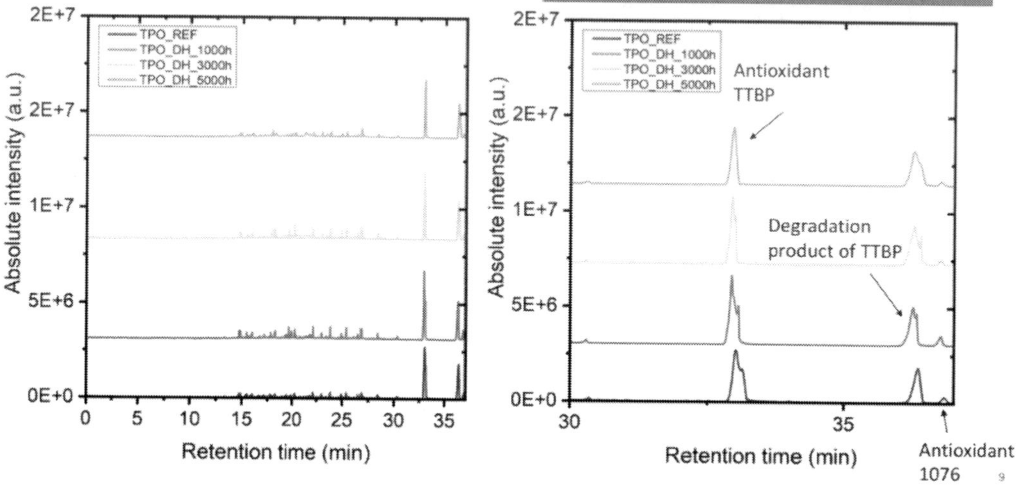

9

020243-009

40th European Photovoltaic Solar Energy Conference and Exhibition

Results
TD-GC/MS: Qualitative additive analysis

PCCL

Encapsulant	DH testing time (hours)	Antioxidant TTBP	Antioxidant 1076	UV Absorber	HALS
EVA	0	–	Yes	Yes	Yes
EVA	5000	–	Yes	Yes	Yes
POE	0	–	–	Yes	Yes
POE	5000	–	–	Yes	Yes
TPO	0	Yes	Yes	–	–
TPO	5000	Yes	Yes	–	–
ION	0	–	–	–	–
ION	5000	–	–	–	–

www.pccl.at

10

020243-010

Results
FTIR-ATR: Chemical aging

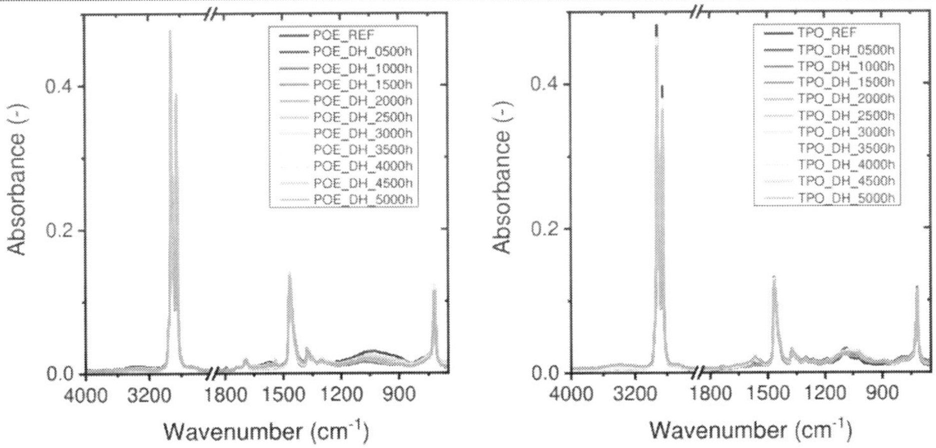

No significant signs of polymer chemical degradation

40th European Photovoltaic Solar Energy Conference and Exhibition

Results
TGA: Thermal stability

Encapsulant	5% weight loss temperature, T5 (°C)
EVA	333
TPO	420
POE	414
ION	414

- EVA shows two decompositions steps
- TPO, ION and POE show only one decomposition step
- Temperature values where 5% weight loss takes place (T5) do not change significantly over DH exposure time

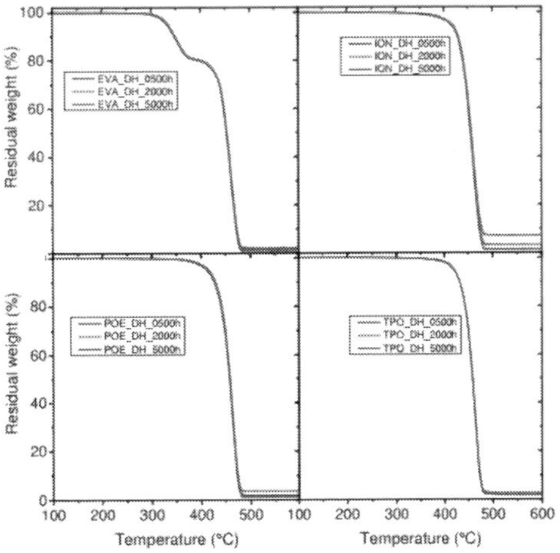

Results
DSC: Thermal properties and morphology

PCCL

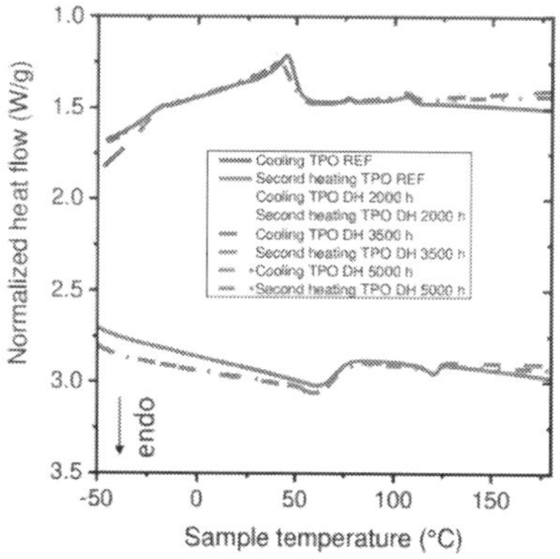

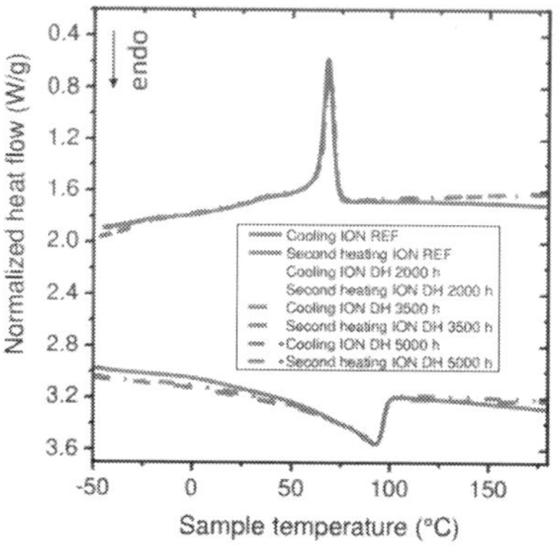

- TPO presents two melting peaks, possibly due to presence of base resins with different densities
- POE presents only one melting peak (as well as EVA and POE, not shown)
- No significant changes in melting peak temperature over DH exposure → **no chemical aging**

Summary and outlook

PCCL

Observed degradation:
- **Discoloration** (yellowing) → might only an artifact due to DH and be **reversible** upon exposure to UV
- **Delamination** (especially for ION) → might be prevented by **optimizing processing parameters**

- All materials showed generally **good resistance** to DH
- Significant depletion of stabilizers was not observed
- **Chemical characteristics and thermal properties** remained substantially **unchanged** upon **DH exposure**

- Testing of coupons under UV irradiation is currently ongoing

Thank you for your attention!

Chiara Barretta
Sauraugasse 1, 8700 Leoben, Austria
+43 3842 42962 722
chiara.barretta@pccl.at

This work was conducted as part of the Austrian "e!MISSION.at - Energy Mission Austria" project "PV40+" (FFG No. 881868) funded by the Austrian Climate and Energy Fund and the Austrian Research Promotion Agency (FFG), and by The Scientific and Technological Research Council of Turkey (TUBITAK) under the Grant No: 120N520.

3BO.14.3 Einhaus R.
Test and Evaluation of Combinations of Encapsulant Materials towards a Long Service Lifetime of PV Modules: PV 40 Plus

3AV.2.10 Gok A.
Equivalent Damp Heat Testing Time Map for service lifetime of Photovoltaic Modules

www.pccl.at 15

020243-015

TWO-DIMENSIONAL REPRESENTATION OF THE BIDIRECTIONAL REFLECTANCE DISTRIBUTION FUNCTION OF PHOTOVOLTAIC MODULES

Christof Bucher[1], Andreas Bohren[2], Donat Hess[1], Sara El Hassani[1], Matthias Hügi[1]
[1]Bern University of Applied Sciences (BFH), Switzerland; [2]Ostschweizer Fachhochschule (OST)
christof.bucher@bfh.ch, andreas.bohren@ost.ch, donat.hess@bfh.ch, sara.elhassani@bfh.ch, matthias.huegi@bfh.ch

The optical reflectance properties of PV modules can be described using the bidirectional reflectance distribution function (BRDF). In this paper, a simplified two-dimensional approach to represent the BRDF is proposed. The shape of the solar reflectance image on the PV module is described by comparing a mathematical model and two measurement approaches. For standard PV modules using anti-reflective solar glass, the shape of the solar reflection image is regular and typically elliptical and depends on the angle of solar incidence α. The elliptical shape is described using a diagram for each horizontal and vertical axis based on the constant luminance zone. The dimensions of the axes are approximated using an empirical mathematical function. The result can be used to include glare information in the data sheet of a PV module and to correctly implement the phenomenon of beam spread in tools for glare hazard analysis.

1 INTRODUCTION

Glare hazards on PV modules are an increasing concern in the build-over area, especially when eastern, western and northern roofs in the northern hemisphere are covered with PV modules [1], [2]. Glare analysis tools such as Sandias Solar Glare Hazard Analysis Tool (SGHAT, [3], [4]) typically model the beam spread as a symmetric phenomenon that is independent of the angle of incidence α. This leads to overestimation or underestimation of the effective glare duration. The correct modelling of the shape of the reflection beam is seen to be complex, because the structure of the glass surface is irregular and leads to irregular reflection cones[5], [6].

Although PV modules are typically coated with an anti reflective coating (ARC) and/or manufactured using a textured surface to maximise the light transmission through the solar glass they don't provide reliable protection against glare hazards[7].

Due to geometrical reasons tilted PV modules facing south in the northern hemisphere do normally not cause any glare hazards to the neighbouring area. However, due to cost reduction of photovoltaics, more and more rooftops are equipped with PV independent of the roof orientation. Thereby it was found that rooftops facing north are critical regarding glare hazards for observation points north of the building.

The most relevant factor to quantify glare is the luminance in cd/m^2, describing the "brightness" of a surface. Because the disturbance caused by glare depends on a large number of factors, no guideline or limit values have been published yet. Depending on the source, so-called absolute glare (saturation of the eyes) occurs from about 100,000 cd/m^2. However, disturbing effects that lead to an afterimage can already occur at significantly lower values [7], see also the next section "definition of glare".

Originally, glare reports were prepared for PV systems at airports. However, at least in Switzerland, more and more glare reports are being requested to clarify glare in neighbourhood disputes. Glare reports typically calculate at what time of day and year the physical law of reflection "angle of incidence = angle of emergence" is observed. The beam spreading is either neglected or represented with a highly simplified procedure. The free tool for calculating glare in Switzerland (www.blendtool.ch [8]) as well as the SGHAT, for example, add the angle of beam spreading as a constant and angle-independent number to the reflection diagram. This is critical for two reasons:

1. The angle of beam spreading is not defined and not known for most glasses.
2. The beam spread is not constant but dependent on the angle of incidence α.

Information on reflections is sometimes reproduced on data sheets of PV modules. This information refers either to the angle-dependent reflection coefficient (Figure 1) or to the maximum expected angle-dependent luminance (Figure 2). While the reflection coefficient cannot be used to draw conclusions about the glare properties because it does not take beam spreading into account, the luminance provides the most important information about glare. However, luminance can only be used to determine how bright a PV module can be from which angle. The information on how large this bright spot is is missing, and it is therefore not possible to determine how long a glare can last. However, because the duration of glare is one of the main factors for assessing glare, the current data sheets of glare-reduced PV modules are only suitable for assessing possible glare to a limited extent.

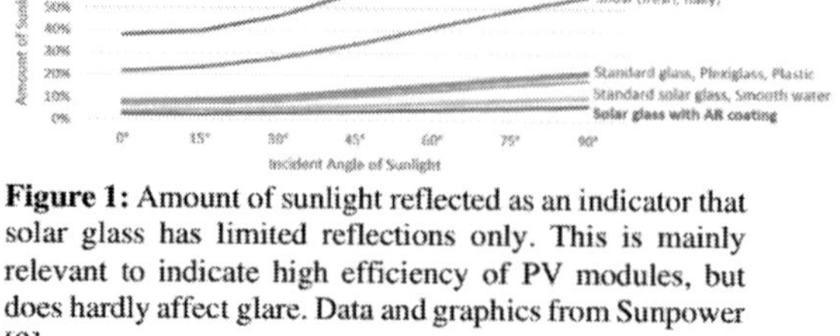

Figure 1: Amount of sunlight reflected as an indicator that solar glass has limited reflections only. This is mainly relevant to indicate high efficiency of PV modules, but does hardly affect glare. Data and graphics from Sunpower [9].

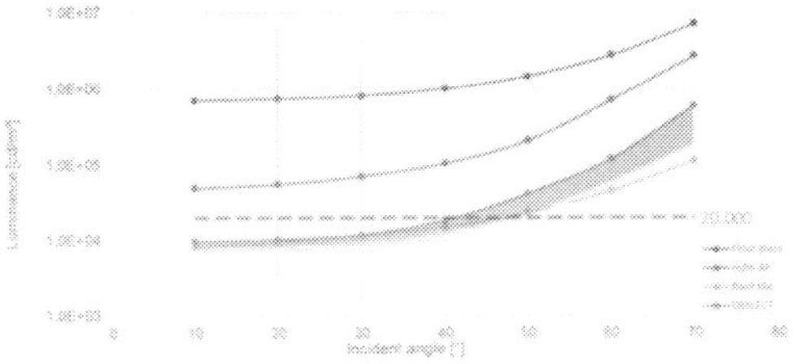

Figure 2: Glare representation of a PV module equipped with glare-reduced Deflect ® glass. Source: Data sheet [10].

2 DEFINITIONS

2.1. Definition of Glare

Depending on the situation different definitions of glare are found. The European standard EN 12665 "Light and lighting Basic concepts and criteria for specifying lighting requirements" [11] defines glare as an "unpleasant visual condition caused by unfavourable luminance distribution or excessive contrast". This very general definition touches on a wide range of different phenomena. Specifically, a disturbance that is subjectively perceived as unpleasant is called psychological glare ("discomfort glare"). Physiological glare or absolute glare, on the other hand, is a measurable impairment of the visual functions that can also lead to permanent damage to the retina.

Absolute glare is the situation in which the eye can no longer adapt to the brightness. Depending on the area of the light source and the aperture, luminance levels of $1.0 \cdot 10^4$ to $1.6 \cdot 10^5$ cd/m² lead to absolute glare [12]. With absolute glare, the eye can no longer adapt and above this limit, differences in brightness can no longer be perceived. From a medical point of view, it is advisable to avoid long-lasting situations with absolute glare. However, many common materials such as white plaster, glass or wet surfaces easily cause absolute glare in full sunlight. This means that absolute glare sensations are common in a normal environment.

Psychological glare (discomfort glare) is considered to induce a significant impairment of general well-being, work performance, safety, ability for concentration, etc. However, depending on the person and the situation, it is possible that even absolute glare does not necessarily lead to discomfort glare.

Different methods and standards have been developed to quantify glare. However, the best developed methods are for indoor use only and are applicable for the evaluation of illumination concepts.

In addition to luminance, there are various other quantities for determining glare. One of the best known is the "Unified Glare Rating" (UGR) developed by the "Commission International de l'Eclairage" (CIE). However, UGR was developed for small, point light sources in front of a dark background and is not suitable for evaluating the glare emitted by larger surfaces [13], [14].

However, in order to assess the glare on a PV module in reality, these definitions are only of limited use. In particular, these definitions do not take into account that virtually all surfaces can be glare under certain conditions, and yet are not considered to be glare. For this reason, this paper compared luminance measurements of different materials that are not associated with glare. A key factor in evaluating these materials is the angle of light incidence α. For virtually any surface, the flatter the angle α, the greater the luminance of the surface. To evaluate whether a surface is perceived as glare or not, the measured materials are therefore compared with each other.

The maximum value of a material not associated with glare is proposed as the limit value for a PV module to be considered "glare-free" in this paper (Figure 3). This is angle-dependent and is only defined up to $\alpha = 80°$ (flat incidence angle). In reality, flatter angles α lead to double glare, i.e. the sunbeam and the glare beam come from practically the same direction. The glare beam is thus drowned out by the sun beam. Figure 3 shows the luminance levels of various surfaces measured by SPF. It also shows the resulting proposed limit value for assessing a PV module as a "glare-free PV module". The values for assessing the absence of glare are set at approx. 90% of the measured values in order to maintain a safety margin and are rounded generously for better usability.

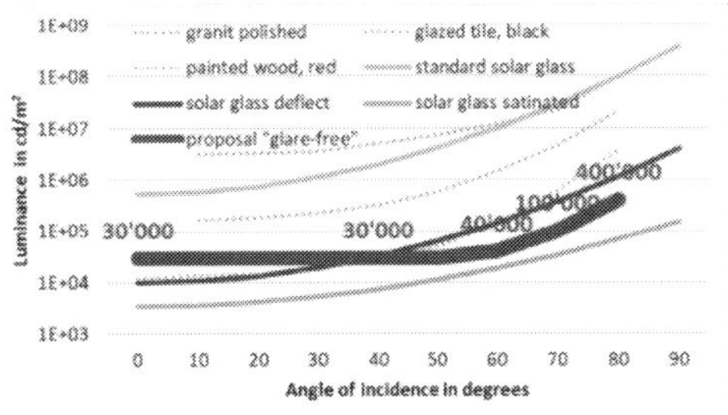

Figure 3: Luminance of different surfaces and proposed limit value for "glare-free PV modules".

2.2 Description of the Shape of Glare

From our daily experience, we know that glare on a PV module usually has an elliptical appearance (Figure 4). It is assumed, that the horizontal and vertical axis of this ellipse depend on the type of solar glass. The aim of this paper is to characterise and mathematically describe both the size and the shape of this elliptical appearance.

Figure 4: The appearance of glare is usually elliptical.

The exact size of the glare ellipse depends on the definition of the glare and the beam spread respectively. This is defined in the following section.

2.3. Definition of Beam Spread

Beam spread is a common concept for describing the shape of glare on a PV module. The following definition is used[3], [4], [17]:

$$\beta = 2 \cdot (4.65 + \sigma)\ mrad$$

Where 4.65 $mrad$ is the subtended half-angle of the sun and σ is the slope error of the reflective surface. The factor of 2 accounts for the resulting reflected angle from the surface. As this equation assumes round (circular) reflection beams and does not depend on the incident

10.4229/EUPVSEC2023/3CO.10.1
020244-002

angle, it is not suitable for accurately model the shape of the glare.

Therefore, a new definition of beam spread is proposed in this paper. It is based on the assumption that the shape of glare is elliptic and consists of a horizontal and vertical axis. To calculate the horizontal and vertical axis of the reflection ellipse, two approaches are proposed.

- The contour line at 30'000 cd/m^2
- The Full With Half Maximum (FWHM) of the cone (see also Figure 12)

These two approaches are presented in the following sections.

Contour line at 30'000 cd/m^2

In this approach the glare ellipse is defined as the part of the image which exceeds 30'000 cd/m^2. In contrast to Figure 3, a constant threshold value is proposed in this paper. This is done to avoid arbitrary curves in the results. However, the threshold value is set to the value found for most angles in Figure 3, which is 30'000 cd/m^2. This corresponds approximately to the white areas in the pictures of Figure 4.

Full With Half Maxima (FWHM)

Instead of defining the area with a luminance larger than 30'000 cd/m^2 as the glare area, the glaring area can be described with the FWHM. However, this underestimates the size of the glare ellipse for standard solar glass and underestimates it for glasses with luminance values close to the glare limit.

3 METHODS TO DESCRIBE THE ELLIPSE

3.1 Geometrical model to calculate the horizontal and vertical axis of a glare ellipse

The approach of describing a glare as an ellipse is more phenomenological than scientific. The results are determined on the basis of calibrated measurements, not on the basis of a physical model. Nevertheless, a physical model is proposed in this chapter. This is intended to give the phenomenological observations a physical legitimation.

The physical model assumes that the glass surface is not flat but slightly modelled with high and low points. Accordingly, the areas between the high and low points do not lie in the supposed module plane, but at a certain flat angle to it. The modelling in this paper assumes that there are five surface areas on each surface element: One of them lies in the module plane, the other four are each inclined at constant angles in a cardinal direction. A glare ray is thus divided into five reflection rays (see Figure 5) neglecting the fact that in reality the surface does not have five discrete zones, but that these zones merge smoothly into one another. The five reflection rays symbolise the ends of the axes and the centre of the ellipse.

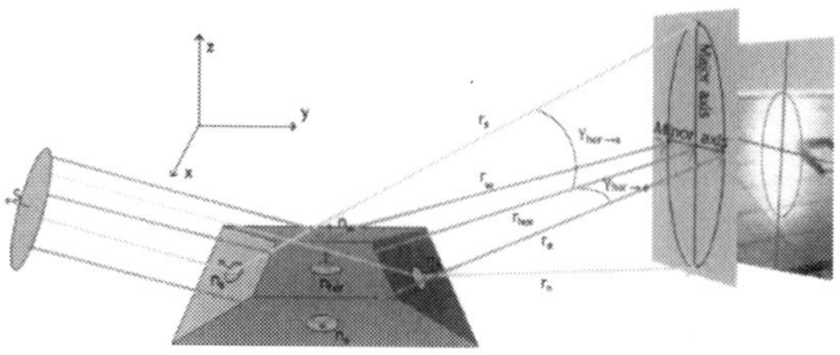

Figure 5: Geometrical approach to model the shape of the reflection

To calculate the reflection beams, the following definitions are made. For simplified representation, the module is placed in the plane of the coordinate system and the azimuth angle of the sun is defined as zero:

- Elevation angle of the sun: α
- Sun vector: $\vec{s} = \begin{pmatrix} 0 \\ cos\alpha \\ sin\alpha \end{pmatrix}$
- Angle difference between the plane of module and the maximum deviation: β
- Normal vector of the horizontal plane of module $\overrightarrow{n_{hor}} = \begin{pmatrix} 0 \\ 0 \\ 1 \end{pmatrix}$
- Normal vector of the east west deviated surface: $\overrightarrow{n_{ew}} = \begin{pmatrix} sin\beta \\ 0 \\ cos\beta \end{pmatrix}$
- Normal vector of the north south deviated surface: $\overrightarrow{n_{ns}} = \begin{pmatrix} 0 \\ sin\beta \\ cos\beta \end{pmatrix}$

The angles and vectors are shown in Figure 5 and Figure 6.

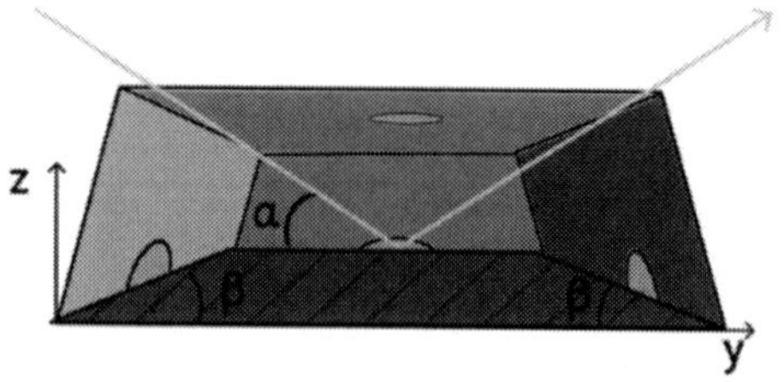

Figure 6: Definition of the angles used in this paper. The incidence angle α is used throughout the paper.

Based on the optical law "angle of incidence = angle of reflection", the reflection beam is calculated as $\vec{r} = \vec{s} - 2 \cdot (\vec{s} \cdot \vec{n}) \cdot \vec{n}$ for a normal vector $\vec{n}$ of the plane of reflection. For the three normal vectors $\overrightarrow{n_{hor}}$, $\overrightarrow{n_{ew}}$, and $\overrightarrow{n_{ns}}$ this results in the following reflection vectors $\overrightarrow{r_{hor}}$, $\overrightarrow{r_{ew}}$ and $\overrightarrow{r_{ns}}$:

$$\overrightarrow{r_{hor}} = \vec{s} - 2 \cdot \underbrace{(\vec{s} \cdot \overrightarrow{n_{hor}})}_{sin\alpha} \cdot \overrightarrow{n_{hor}} = \begin{pmatrix} 0 \\ cos\alpha \\ sin\alpha \end{pmatrix} - 2 \cdot sin\alpha \cdot \begin{pmatrix} 0 \\ 0 \\ 1 \end{pmatrix} = \begin{pmatrix} 0 \\ cos\alpha \\ -sin\alpha \end{pmatrix}$$

$$\overrightarrow{r_{ew}} = \vec{s} - 2 \cdot \underbrace{(\vec{s} \cdot \overrightarrow{n_{ew}})}_{sin\alpha \cdot cos\beta} \cdot \overrightarrow{n_{ew}}$$

$$= \begin{pmatrix} 0 \\ cos\alpha \\ sin\alpha \end{pmatrix} - 2 \cdot sin\alpha \cdot cos\beta \cdot \begin{pmatrix} sin\beta \\ 0 \\ cos\beta \end{pmatrix}$$

$$= \begin{pmatrix} -2 \cdot sin\alpha \cdot cos\beta \cdot sin\beta \\ cos\alpha \\ sin\alpha - 2 \cdot sin\alpha \cdot cos\beta \cdot cos\beta \end{pmatrix}$$

$$\overrightarrow{r_{ns}} = \vec{s} - 2 \cdot \underbrace{(\vec{s} \cdot \overrightarrow{n_{ns}})}_{\substack{cos\alpha \cdot sin\beta + sin\alpha \cdot cos\beta \\ = sin(\alpha+\beta)}} \cdot \overrightarrow{n_{ns}}$$

$$= \begin{pmatrix} 0 \\ cos\alpha \\ sin\alpha \end{pmatrix} - 2 \cdot sin(\alpha+\beta) \cdot \begin{pmatrix} 0 \\ sin\beta \\ cos\beta \end{pmatrix}$$

$$= \begin{pmatrix} 0 \\ cos\alpha - 2 \cdot sin(\alpha+\beta) \cdot sin\beta \\ sin\alpha - 2 \cdot sin(\alpha+\beta) \cdot cos\beta \end{pmatrix}$$

However, it is not the reflection vectors themselves that are of interest, but the angles $\gamma_{hor\to ew}$ and $\gamma_{hor\to ns}$ between the reflection vector at the module plane and the other reflection vectors. Because all vectors are normalised in length, the angle between two vectors is calculated with the help of the scalar product as follows:

$$cos\gamma_{hor\to ew}(\alpha,\beta) = \frac{\overrightarrow{r_{hor}} \cdot \overrightarrow{r_{ew}}}{\underbrace{|\overrightarrow{r_{hor}}|}_{1} \cdot \underbrace{|\overrightarrow{r_{ew}}|}_{1}} = \overrightarrow{r_{hor}} \cdot \overrightarrow{r_{ew}}$$

$$= \begin{pmatrix} 0 \\ cos\alpha \\ -sin\alpha \end{pmatrix} \cdot \begin{pmatrix} -2 \cdot sin\alpha \cdot cos\beta \cdot sin\beta \\ cos\alpha \\ sin\alpha - 2 \cdot sin\alpha \cdot cos\beta \cdot cos\beta \end{pmatrix}$$
$$= (cos\alpha)^2 - (sin\alpha)^2 + 2 \cdot (sin\alpha)^2 \cdot (cos\beta)^2$$

$$cos\gamma_{hor\to ns}(\alpha,\beta) = \frac{\overrightarrow{r_{hor}} \cdot \overrightarrow{r_{ns}}}{\underbrace{|\overrightarrow{r_{hor}}|}_{1} \cdot \underbrace{|\overrightarrow{r_{ns}}|}_{1}} = \overrightarrow{r_{hor}} \cdot \overrightarrow{r_{ns}}$$

$$= \begin{pmatrix} 0 \\ cos\alpha \\ -sin\alpha \end{pmatrix} \cdot \begin{pmatrix} 0 \\ cos\alpha - 2 \cdot sin(\alpha+\beta) \cdot sin\beta \\ sin\alpha - 2 \cdot sin(\alpha+\beta) \cdot cos\beta \end{pmatrix}$$
$$= (cos\alpha)^2 - 2 \cdot cos\alpha \cdot sin(\alpha+\beta) \cdot sin\beta - (sin\alpha)^2 + 2 \cdot sin\alpha \cdot sin(\alpha+\beta) \cdot cos\beta$$
$$= cos(2\beta)$$

The last two equations thus give the horizontal and vertical axes of the presumed reflection ellipse. Figure 7 shows the two angles graphically. With this representation alone, the elliptical shape can be justified, but not the observed increase in the major axis of the ellipse at flat angles.

The angle of the minor axis is defined by $\gamma_{hor\to ew}(\alpha)$. Like the angle of the major axis, it starts at $\gamma_{hor\to ew}(0°) = 2\beta$ for $\alpha = 0°$ (perpendicular to the module surface), but converges towards $\gamma_{hor\to ew}(90°) = 0°$ for $\alpha = 90°$. The angle of the main axis, on the other hand, is independent of α and is $\gamma_{hor\to ns}(\alpha) = 2\beta$ for all incident angles. This consideration only applies for $\alpha \geq \beta$ (Figure 5)

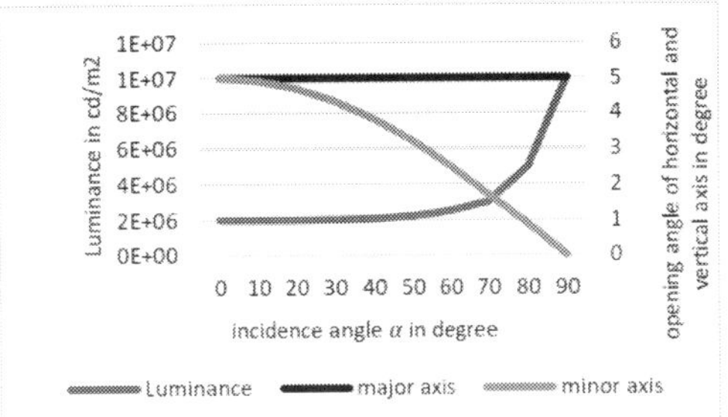

Figure 7: Numeric result of horizontal and vertical axis for the angle β =1.25°

Figure 7 suggests that an ellipse can be observed. However, this ellipse would become smaller and smaller as the angle α increases. This does not correspond to the observations. In order to describe the phenomenon of the ellipse becoming larger, the model from Figure 5 must be modified.

Figure 5 assumes that the horizontal and vertical axes of the ellipse are given by a clearly defined angle, caused by the discretely divided surface structure. In reality, however, the five discrete zones merge smoothly into one another and therefore the ellipse does not have clear dimensions but becomes faded with increasing distance from the centre. The photographed ellipse results from the saturation of the camera sensor. At the same settings, the sensor saturates at the same luminance each time. Because the luminance of the blurred ellipse increases with increasing α, the glare limit shifts outwards.

To illustrate this observation, it is assumed in the following that the angle β increases proportionally to the logarithm of the luminance. The increase in β is argued by the fact that as α increases, reflections greater than 30'000 cd/m² are possible for surface areas with larger southward tilt β. The proportionality factor is not known and is arbitrarily set at 1. Thus, if the luminance increases by a factor of 10 (i.e. by 1 power of ten), β increases by a factor of 1. Figure 9 shows the resulting axis sizes.

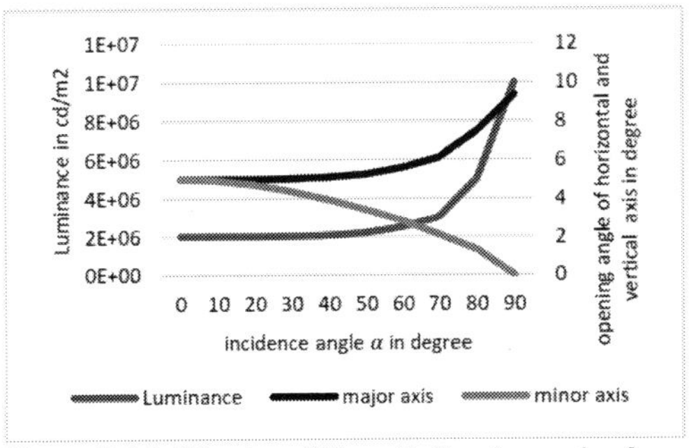

Figure 8: Modification of Figure 7: Opening angle of horizontal and vertical axis, corrected by the Luminance.

The reduction of the minor axis size with increasing α does not agree with all observations. This can be explained by the fact that our model only considers planes with north/south and east/west orientation. Such planes cannot deflect irradiation with large α laterally. A diagonal surface with, for example, south/west or south/east orientation could. In order to make the model more accurate, such planes would therefore also have to be taken into account.

With this model, the observed phenomena can be explained partially, but not completely. In the following of this paper, we will therefore not work with the geometric model, but with the metrological investigations.

3.2. Phenomenological approach to the representation of the ellipse

In this approach, the glare on a standard photovoltaic module is photographed at different observer distances, different orientations of the PV module and for a variation of observation angles. The camera is set so that the sensor goes into saturation at approx. 30'000 cd/m². Thus, the white areas in the photos are a appropriate visualisation of the glare.

In addition, the software iQ-Luminance from the company Image Engineering GmbH & Co. KG is used to convert the photos into luminance images. Since one camera setting only allows for reliable luminance values within a decade, each luminance image is based on six serial pictures, each with decreasing exposure times or increasing aperture values. The six luminance images were then merged into one image, considering only those areas of each image where the pixel sensors were neither oversaturated nor undersaturated. This allows to create a luminance picture of the hole glare.

Figure 9 serves as example of such a luminance image. The angles 1°, 5°, 10°, 20° are shown as concentric circles around the glare maximum to allow qualitative estimation of the glare size. The white contour line represents the iso line of 30'000 cd/m² and the red line its elliptical approximation. The good correspondence between the 30'000 cd/m² iso line and the elliptic fit confirms the assumption that the glare has an elliptic shape. Therefore, only the fit curve is shown in the subsequent plots. The horizontal and vertical axes of the fit allow the size of the glare to be quantified.

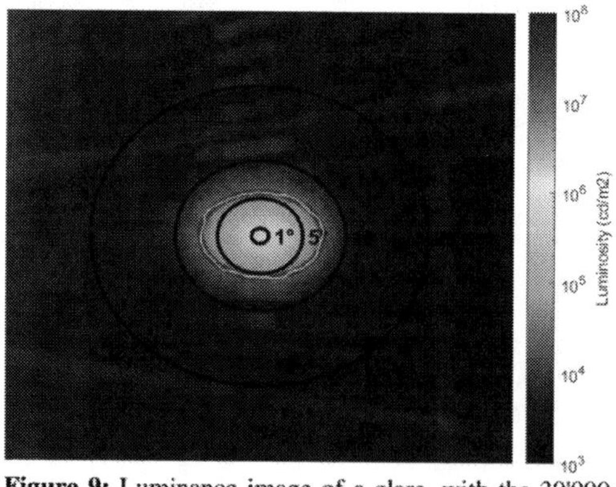

Figure 9: Luminance image of a glare, with the 30'000 cd/m² iso line (thin white) and its elliptic approximation (red). Concentric angles (1°, 5°, 10° and 20°) around the center of the glare are shown in black.

Photos of PV modules from different distances show that the size of the glare is independent of the distance of the observer for standard and float glasses (see Figure 10).

Figure 10: Phenomenological approach to show the independence of horizontal and vertical axis of the glare ellipse from the distance of the observer for standard (upper row) and float module (lower row)

Glare images of differently oriented modules (horizontal and vertical), under constant irradiation angle, show no difference in the case of the float glass but do show differences in the case of the standard glass module Figure 11. In this case, the difference in the length of the major axis is noticeable.

Figure 11: Phenomenological approach to investigate the influence of PV module orientation on the horizontal and vertical axes of the glare ellipse for standard (top row) and float (bottom row) modules.

To investigate the dependence of the glare of a standard module on α, photos (saturation at ca. 30'000 cd/m²) and luminance images were generated for nine different angles between 5° and 80°. Based on the luminance images, the maximum luminance as well as the horizontal and vertical length of the glare ellipse were determined (Table I).

Table I: Photographed glare and the corresponding luminance image for different α.

Optical picture		**Luminance image**	
α (°)	**Luminance max cd/m²**	**Vertical axis (°)**	**Horizonal axis (°)**
$\alpha = 5°$	594'000	5.0	5.7
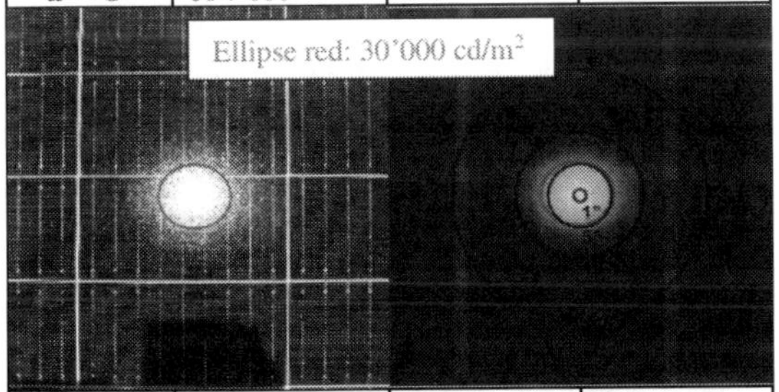			
$\alpha = 10°$	577'000	4.7	5.9
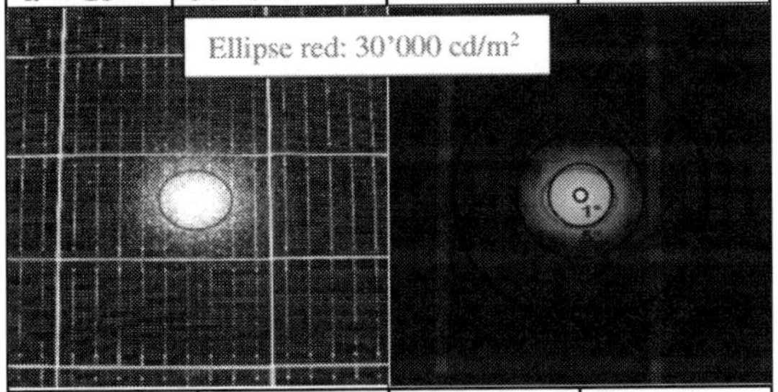			
$\alpha = 20°$	539'000	4.9	5.4
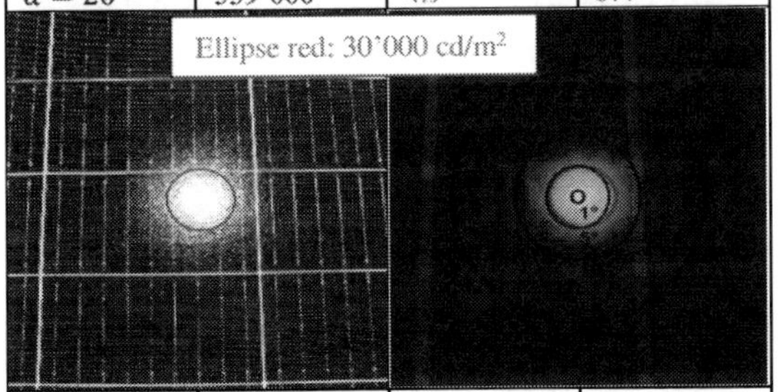			
$\alpha = 30°$	587'000	4.9	5.2
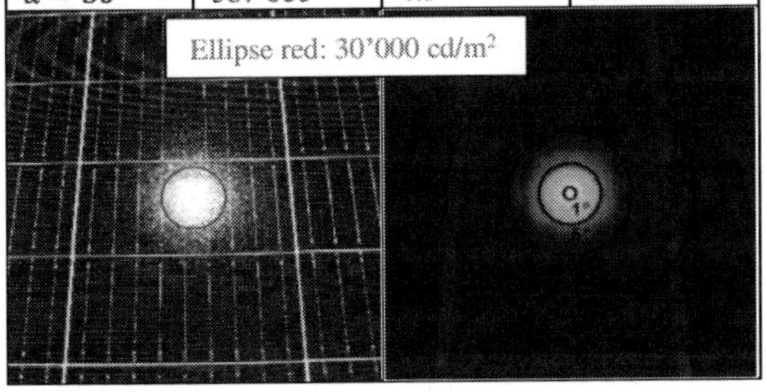			
$\alpha = 40°$	495'000	5.1	5.0
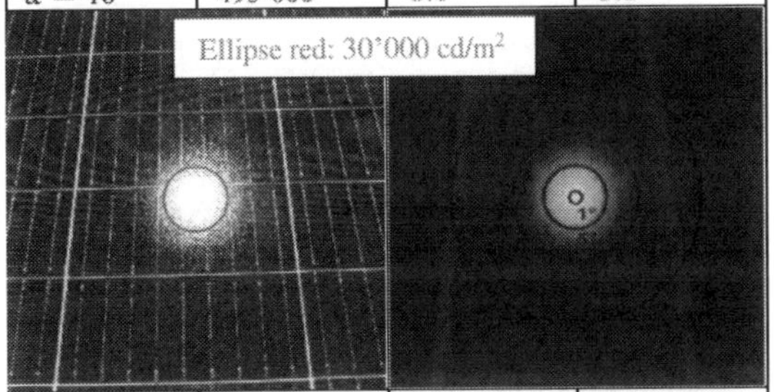			
$\alpha = 50°$	839'000	6.3	5.0
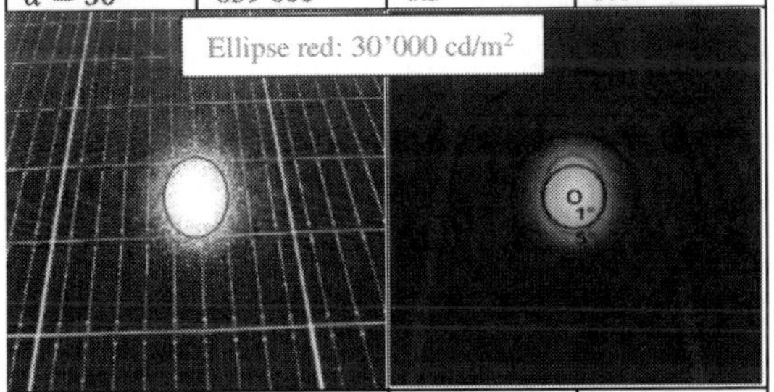			
$\alpha = 60°$	2'400'000	7.8 (7.4)	5.0 (4.7)
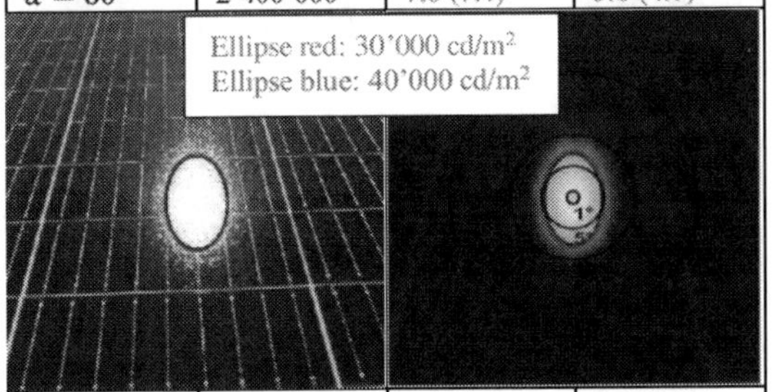			
$\alpha = 70°$	11'900'000	9.8 (8.2)	4.4 (3.6)
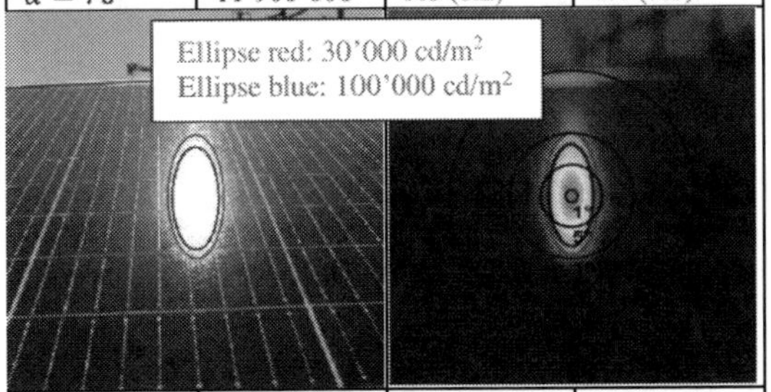			
$\alpha = 80°$	47'500'000	10.7 (8.2)	3.8 (2.3)
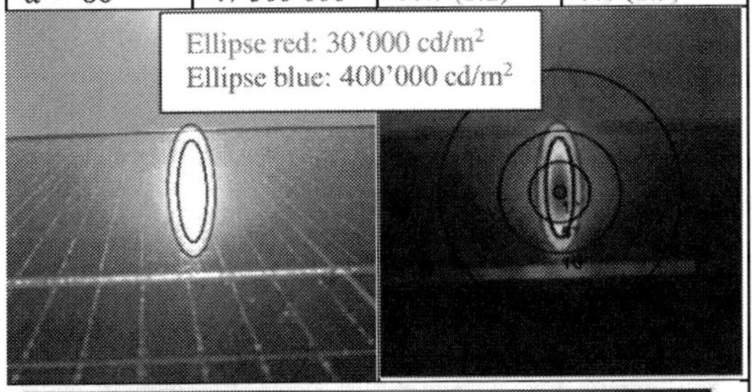			
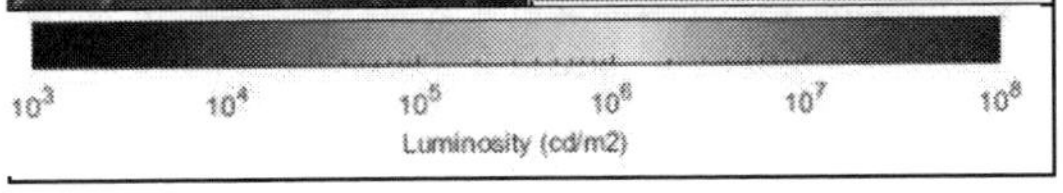			

3.3 BRDF Approach

In a third, phenological approach using calibrated indoor laboratory measurements, the shape of the glare beam and thus the size of the horizontal and vertical axis are calculated using the bidirectional reflectance distribution function (BRDF). In a generic approach this can be done by approximating the BRDF conical shape with a two-dimensional Gaussian curve. As described in Chapter 2, both the FWHM and the contour line approach can be used to describe the elliptic shape of the cross section of the cone in Figure 12.

In contrast to the measurement method from Chapter 3.2, the BRDF measurement in the laboratory has the advantage that it is independent of the weather and can therefore be repeated at any time. The disadvantage, however, is that only a small area of the PV modules is measured. If there are inhomogeneities on the PV module, which can be caused, for example, by the conductor tracks in the modules, the BRDF measurement can no longer be used. During the measurement, it must be defined which point on the module is measured, which has a major influence on the result.

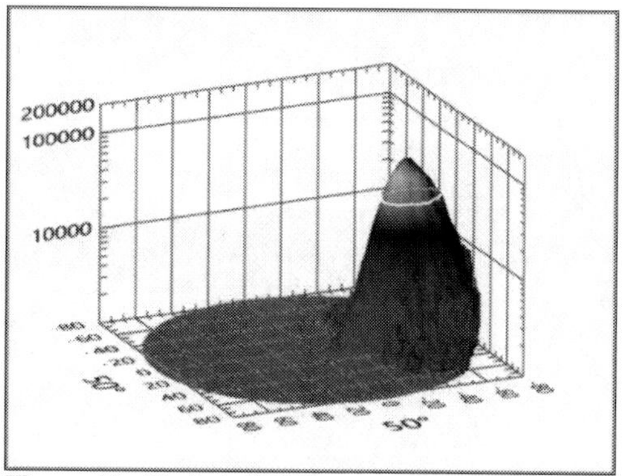

Figure 12: Typical BRDF measurement of a glare-reduced PV module (Figure 15) at an incidence angle of α =50°. The white line indicates the contour line for 30'000 cd/m^2.

The measured contour line is shown in Figure 13. An ellipse with a major and a minor axis is placed on top. This shows that the reflection surface that reflects with at least a certain luminance can be represented as an ellipse.

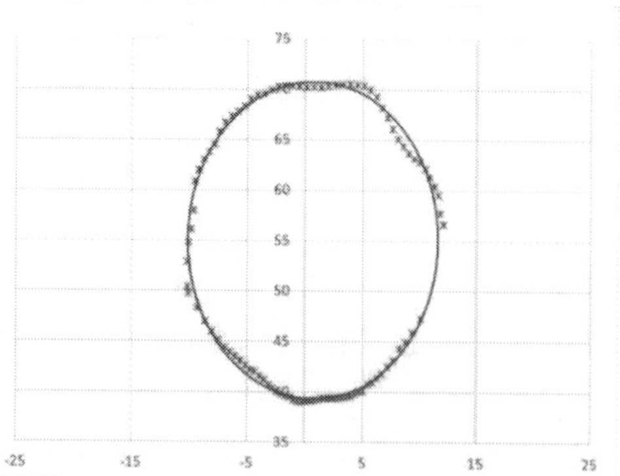

Figure 13: Elliptical approximation of the contour line at 30'000 cd/m^2 based on the example in Figure 12.

For each α in 10° steps, the luminance of the reflection is determined using the BRDF. The corresponding measurement results are shown in the form of elliptical contour lines in Figure 14. It can be seen that the ellipses hardly change for steep angles α. However, when α is flat, the reflecting surfaces become larger. For α starting from 50°, in accordance with Figure 3 in Chapter 2, the 30'000 cd/m^2 contour line is no longer specified, but instead the higher limit values.

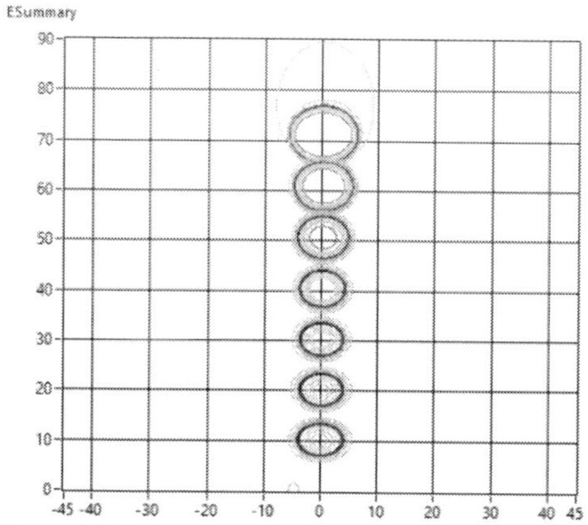

Figure 14: BRDF measurements shown as contour lines for different incident angles. Each set of circles refers to one specific incident angle with different colours for different luminance levels. The red circles show the contour line for a luminance of 30'000 cd/m^2.

For technical reasons the instrument that is used to measure the BSDF functions is not made for α of 80° or more. The measurements are therefore restricted to a maximum of α = 70°. Even these measurements show, that or many glasses with only slight textures surfaces the 30'000 cd/m^2 contour line is passing over the 90° reflection angle.

Assuming a luminance of 30'000 cd/m^2 being critical for glare hazards, a contour line at 30'000 cd/m^2 can be formed (see chapter 2 of this paper). Adopting the approach that was developed in Chapter 2, just means that the contour line is adjusted for higher angles α.

In analogy to Table I, all measurement results for α between 0° and 70° are shown in Table II. Because the measured solar glass differs from that in Chapter 3.2, the results are not directly comparable. However, both tables show that the approach can be presented in the same way regardless of the implementation and regardless of the product being measured.

Figure 15: Glass sample used to demonstrate the BRDF approach.

Table II: Left: Typical BRDF measurements of a glare-reduced PV module at different α. The z-axis for the luminance in cd/m^2 is logarithmic. The flat ground is caused by the lower cut at 1000 cd/m^2. Reflections below this value are irrelevant. Right: Luminance contour lines (approximated ellipses) for a typical PV surface. The red line indicates the contour line for 30'000 cd/m^2 and the blue line indicates the contour line when following the glare criteria defined in chapter 2.

Optical picture		**Luminance image**	
α (°)	**Luminance max cd/m^2**	**Vertical axis (°)**	**Horizonal axis (°)**
$\alpha = 0°$	22'000	7.1	7.5
Ellipse red: 30'000 cd/m^2			
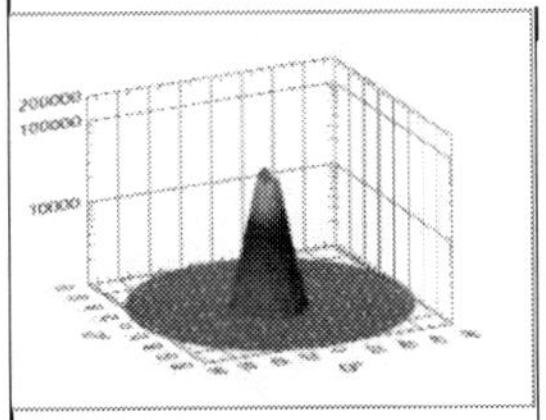		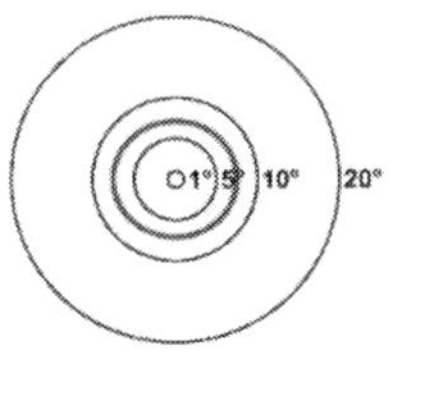	
$\alpha = 10°$	23'900	7.0	8.5
Ellipse red: 30'000 cd/m^2			
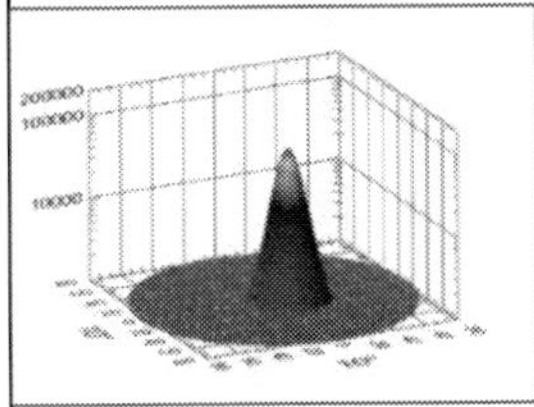		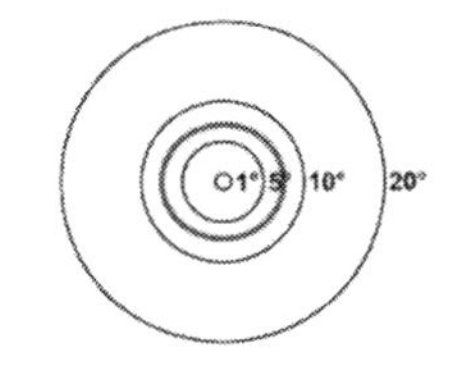	
$\alpha = 20°$	26'400	7.5	8.4
Ellipse red: 30'000 cd/m^2			
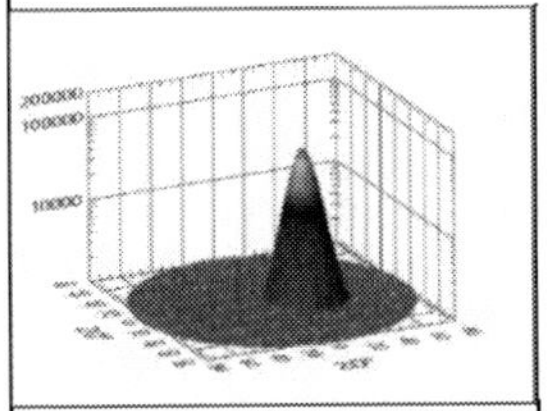		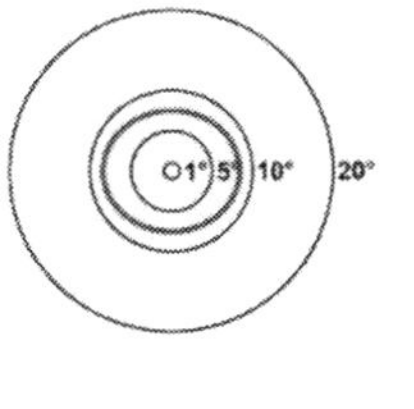	
$\alpha = 30°$	31'700	8.4	8.5
Ellipse red: 30'000 cd/m^2			
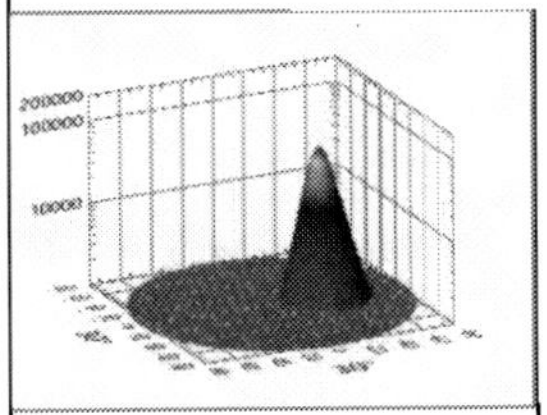		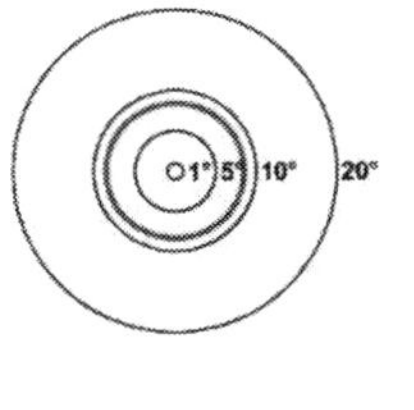	
$\alpha = 40°$	43'400	10.2	9.6
Ellipse red: 30'000 cd/m^2			
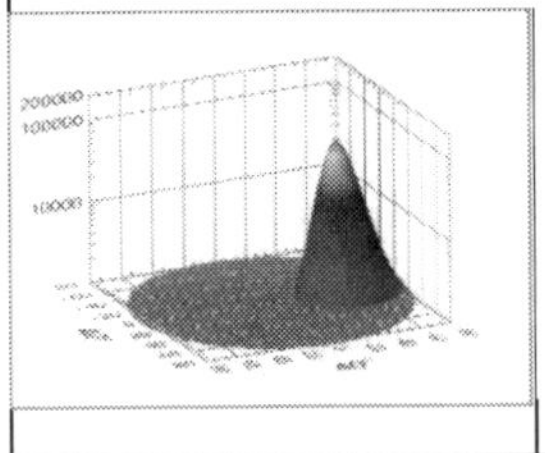		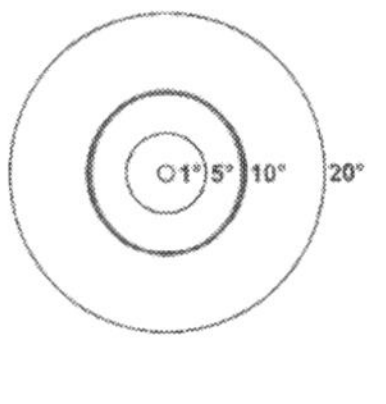	
$\alpha = 50°$	75'700	14.9	11.7
Ellipse red: 30'000 cd/m^2			
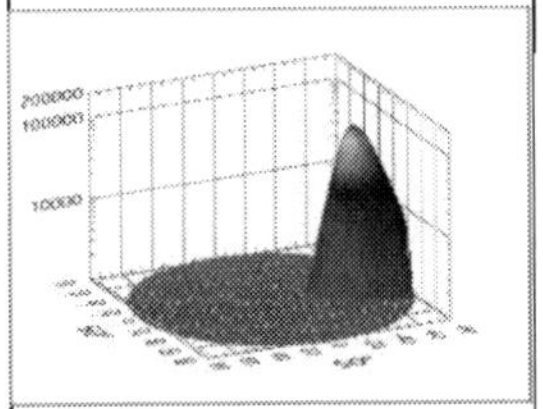		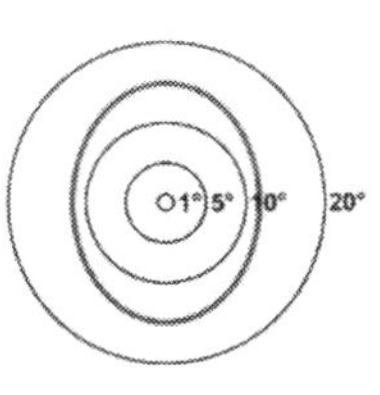	
$\alpha = 60°$	182'500	23.3 (17.7)	15.7 (15.2)
Ellipse red: 30'000 cd/m^2 Ellipse blue: 40'000 cd/m^2			
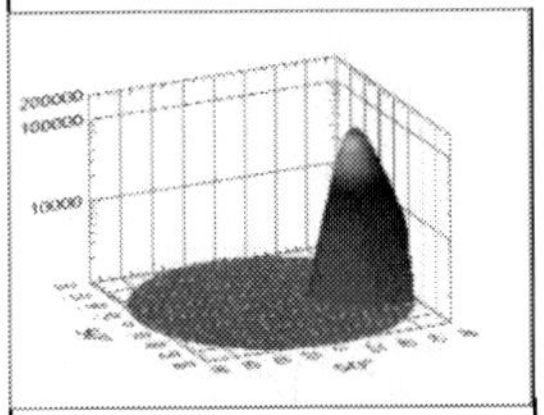		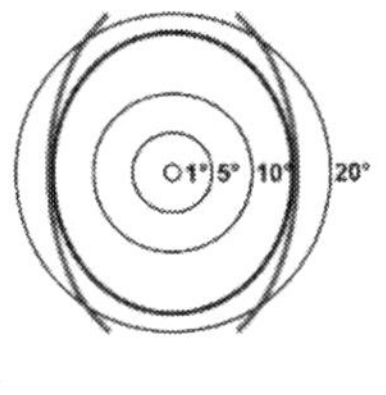	
$\alpha = 70°$	342'900	34.0 (18.0)	23.0 (12.9)
Ellipse red: 30'000 cd/m^2 Ellipse blue: 100'000 cd/m^2			
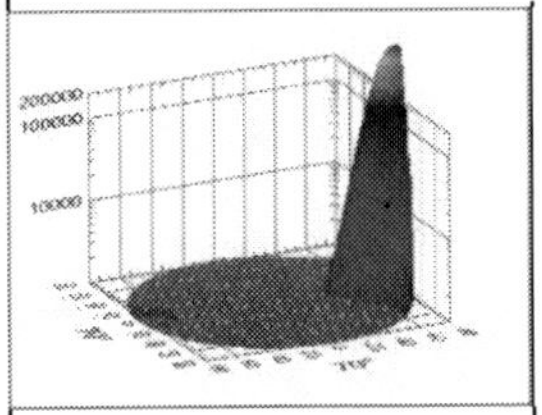		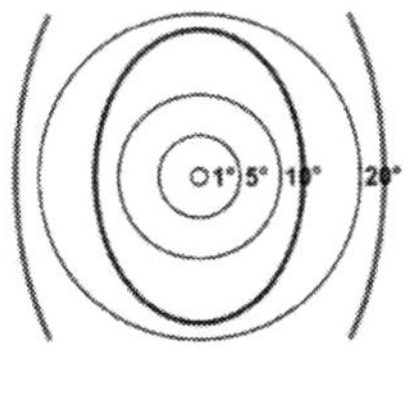	

In comparison to Table I which is based on a standard PV module, the luminance of the glare-reduced PV module measured in Table II is much lower, but the beam spread is greater. Overall, this leads to a larger ellipse of possible glare. Luminance values for α between 0° to 20° are below 30'000 cd/m^2 and do therefore not cause glare according to Chapter 2.

10.4229/EUPVSEC2023/3CO.10.1
020244-008

4 INTERPRETATION

Table III: Identified glare properties of standard module, based on luminance images of Chapter 3.2. Values in blue (in brackets) refer to the increased glare limit for angles of a =60° and larger.

α (°)	Limit (cd/m²)	Lmax (cd/m²)	Vertical Axis (°)	Horizontal Axis (°)
5°	30'000	594'000	5.0	5.7
10°	30'000	577'000	4.7	5.9
20°	30'000	539'000	4.9	5.4
30°	30'000	587'000	4.9	5.2
40°	30'000	495'000	5.1	5.0
50°	30'000	839'000	6.3	5.0
60°	40'000	2'400'000	7.4 (7.8)	4.7 (5.0)
70°	100'000	11'900'000	8.2 (9.8)	3.6 (4.4)
80°	400'000	47'500'000	8.2 (10.7)	2.3 (3.8)

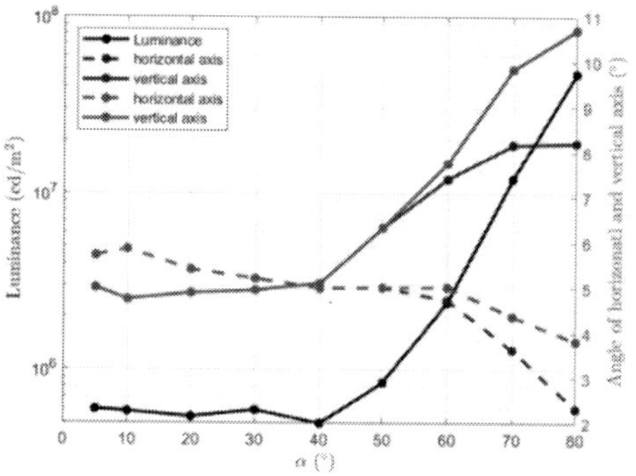

Figure 16: Plotted glare properties of a standard module, based on luminance images (module presented in Chapter 3.2).

Table IV: Identified glare properties of the glare-reduced module, based on BRDF approach of Chapter 3.3. Values in blue (in brackets) refer to the increased glare limit for angles of a =60° and larger.

α (°)	Limit (cd/m²)	Lmax (cd/m²)	Vertical Axis (°)	Horizontal Axis (°)
0°	30'000	22'300	7.1	7.5
10°	30'000	24'100	7.0	8.5
20°	30'000	26'700	7.5	8.4
30°	30'000	32'000	8.4	8.5
40°	30'000	44'000	10.2	9.6
50°	30'000	76'100	14.9	11.6
60°	40'000	185'000	17.7 (23.4)	15.2 (15.8)
70°	100'000	347'000	18.0 (34.0)	13.0 (23.0)
80°	400'000	na	na	na

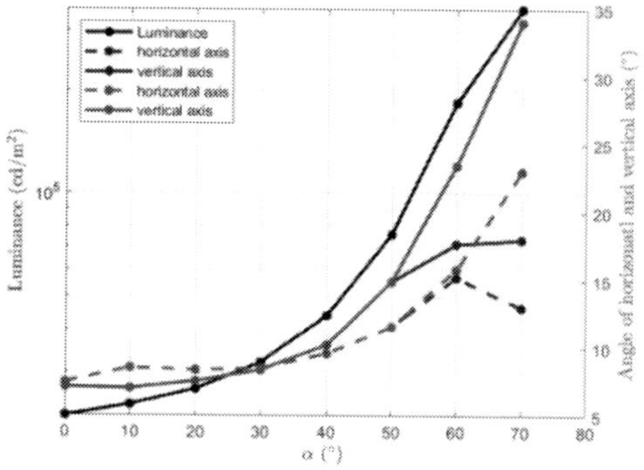

Figure 17: Plotted glare properties of a glare-reduced module, based on BRDF approach. Luminance values below the curve given in Figure 3 are not considered to cause glare (module presented in Chapter 3.3).

The angular dependence of the glares studied in Chapter 3.2 are summarised in Table III and Figure 16; the ones for Chapter 3.3. in Table IV and Figure 17, respectively. The results serve as an example of how the angular dependence of glare can be characterised in a simple two-dimensional graph.

The graphical representation (Figure 16 and Figure 17) allows to identify the glare characteristics of the studied modules and to compare them. Since the methods in Chapter 3.2. and 3.3. examine different PV modules, different glare characteristics can be seen. While the luminance for angles $\alpha < 50°$ is independent for the standard module investigated in chapter 3.2 (Figure 16), the luminance of the glare-reduced module investigated in chapter 3.3 (Figure 17) already increases for low α. Complementary behaviour is also seen in the horizontal axis of the glare ellipse. While the horizontal axis of glare ellipse of the standard module decreases with higher alpha, the same axis of the glare-reduced module rather increases with higher alpha.

6 PRESENTATION OF THE RESULTS AS A FUNCTION

The luminance as well as the horizontal and vertical axes can be represented as a function. Due to the various influences that lead to the characteristics of the two axes, empirical functions are sought. It seems clear that the best candidate can vary depending on the type of glass. It also makes a big difference whether the function should display the contour line with a constant 30'000 cd/m² or the one with higher limit values for flatter angles as introduced in Chapter 2. In this paper, the latter is chosen, as it has more relevance in practice.

Since the human eye sees logarithmically, it makes more sense to fit the logarithm of the luminance instead of the luminance itself. In this way, the fit for smaller luminance values deviates less from the measured values in absolute terms than for large luminance values, which is necessary to adequately represent the luminance for small alpha.

Two functions are tested to fit the logarithm of luminance. First, since the logarithm of the luminance (mostly) increases more with a larger angle of incidence α, i.e. its second derivative is greater than or equal to 0, the following cosine function is tested:

$$ln(y) = a + \frac{b}{cos(\alpha)^c}$$

Here a, b, and c are parameters to be defined and α denotes the angle of incidence. Second, to better account for the fact that the luminance values can remain constant at low luminance values, α up to 50°, the biquadratic function, which has been investigated in [1] is tested:

$$ln(y) = a \cdot \alpha^4 + b \cdot \alpha^2 + c$$

Table V: Parameters identified for the standard PV module measured in Chapter 3.2.

	Luminance		**Vertical axis**	**Hor. axis**
y	cos	biquadratic	Logistic	
	$e^{a+\frac{b}{cos(\alpha)^c}}$	$e^{a\cdot\alpha^4+b\cdot\alpha^2+c}$	$\frac{a}{1+e^{-(\alpha-b)\cdot d}}+c$	
a	2.173	$1.235\cdot 10^{-7}$	3.411	-1395
b	10.82	$-5.504\cdot 10^{-5}$	52.46	190.5
c	0.211	13.2	4.854	5.621
d	na	na	0.167	0.05465
R^2	0.958	0.983	0.99516	0.955

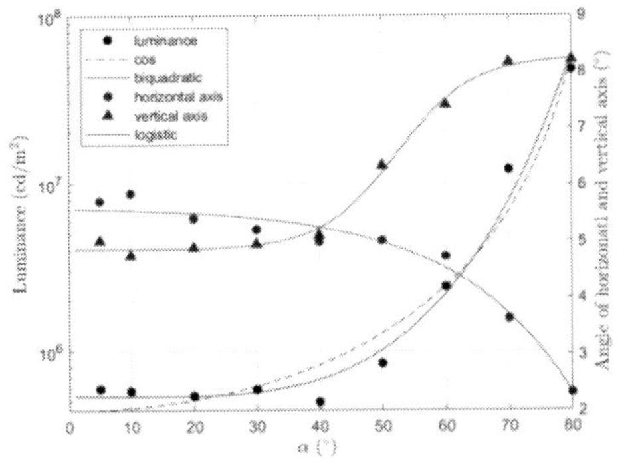

Figure 18: Fit of luminance image-based measurements of the standard PV module in Chapter 3.2.

The dimensions of the vertical and horizontal axes of the ellipse, on the other hand, are almost constant at steep angles of incidence, then increase and stagnate again at shallow angles α. For this reason, an S-curve (sigmoid function, logistic function) is chosen as the fit function.

$$y = \frac{a}{1 + e^{-(\alpha-b)\cdot d}} + C$$

For the results in Table IV and Table III, the parameters in Table V and Table VI were found. The fits to the respective parameters are shown in Figure 18 and Figure 19

The biquadratic function seems to be more suitable for luminance values that remain constant during small incident angles (α <50°), which is the case for the standard solar glass of Chapter 3.2 (R2 = 0.983 vs. R2 = 0.958, see Figure 18). This may be because the biquadratic function, unlike the cos function, is generally not a monotonically increasing function (between 0° and 90°).

The logarithm of the luminance values measured in Chapter 3.3. is monotonically increasing (see Figure 19) and can be fitted equally adequately by the biquadratic function (R2 = 0.993) as well as by the cos function (R2 = 0.993).

How well the logistic function agree with the measured axis values depends in particular on how the limit values for flat angles of incidence (α > 50°) are chosen. If these are greatly increased as suggested in this paper, sigmoid functions suitable. However, if the limit values were left at 30'000 cd/m², the biquadratic or cosine function would maybe be more suitable.

Table VI: Parameters identified for the glare-reduced PV module measured in Chapter 3.3.

	Luminance		**Vertical axis**	**Hor. axis**
y	cos	biquadratic	Logistic	
	$e^{a+\frac{b}{cos(\alpha)^c}}$	$e^{a\cdot\alpha^4+b\cdot\alpha^2+c}$	$\frac{a}{1+e^{-(\alpha-b)\cdot d}}+c$	
a	18.5	$2.733\cdot 10^{-8}$	11.11	5.882
b	-8.52	$4.439\cdot 10^{-4}$	45.31	47.03
c	-0.3774	9.999	7.23	8.214
d	na	na	0.169	0.2229
R^2	0.992	0.993	0.998	0.918

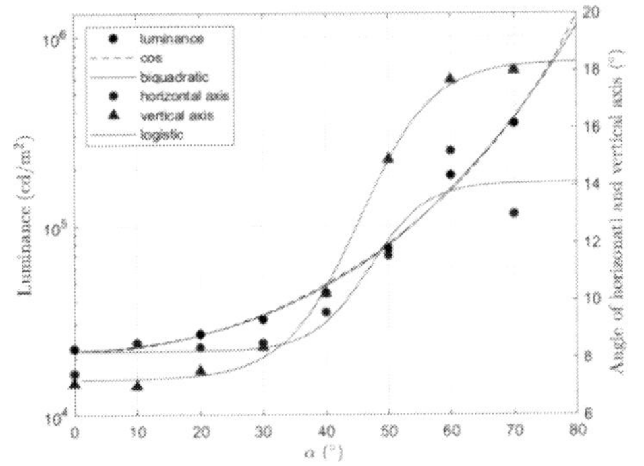

Figure 19: Fit of BRDF approach-based measurements of the glare-reduced PV module in Chapter 3.3.

7 CONCLUSION

In this paper, a new definition and form of representation of the beam spreading of reflections from solar glasses was introduced and discussed. The proposal made is intended to describe properties regarding glare on PV modules in a simple and yet practically relevant way.

The proposal is based on the fact that glare on solar glasses is mostly elliptical and can therefore be characterised with a horizontal and vertical axis as well as the maximum luminance of the glare surface.

The method presented here works for glasses whose glare is elliptical. However, it does not work for characterising a reflection in general, especially when a PV module is produced with a glare-free glass. The method is particularly designed for implementation in software tools that are intended to calculate the possible glare on an optically slightly scattering surface in a simple way.

10.4229/EUPVSEC2023/3CO.10.1

Although this paper examines various aspects of the form of glare, not all questions have been answered. Chapter 8 therefore lists which points should be further investigated.

8 OUTLOOK

The method presented here has not yet been used in a glare calculation tool and thus has not yet been tested for its practicality. This is proposed as the next step. Also, no product screening has been done yet. It is proposed to create a database of typical solar glasses and to record their luminance, as well as horizontal and vertical axis of the glare ellipse.

Furthermore, the paper has some inconsistencies that should be further investigated in the future:

1. It is assumed that the specular reflection for standard solar glasses is independent of the glass orientation. Figure 11 shows that this is not true or only partially true. To what extent this renders the statements of the paper invalid needs to be checked.
2. The validity restriction of the method presented in this paper refers to reflections that have an elliptical appearance. This restriction needs to be defined more precisely.
3. The methods in chapter 3.2 and chapter 3.3 should lead to identical results. The extent of the differences between the methods still needs to be validated.
4. The angle of incidence-dependent threshold value of 30,000 cd/m^2 to 400,000 cd/m^2 is proclaimed in this paper, but has not yet been confirmed. Further work and plausibility checks are necessary.
5. The geometric model in Chapter 3.1 does not fully correspond to reality and should be improved if this approach is to be pursued further.
6. The fit functions suggested in Chapter 6 should be tested, optimized and validated using different PV modules or surfaces.

9 ACKNOWLEDGEMENT

This publication was written as part of the project “Grundlagen zu "reflexionsarm" gem.Art. 32a RPV”. This project is being implemented by the Swiss photovoltaic association Swissolar in collaboration with the OST University of Applied Sciences and the Bern University of Applied Sciences. The project is financed by the Federal Office of Energy.

10 REFERENCES

[1] Bucher, Christof; Danaci, Sirin; Wandel, Jasmin (25 Oktober 2021). Glare Hazard Analysis of Novel BIPV Module Technologies In: Solar World Congress. Monaco (virtual). 25.-29. Oktober 2021.

[2] Andreas Bohren, 2015. Blendung von Solaranlagen Übersicht zur aktuellen Rechtslage. 25. OTTI Symposium Thermische Solarenergie, 6.-8. Mai 2015, Kloster Banz, Bad Staffelstein.

[3] Clifford K. Ho, 2011, Methodology to Assess Potential Glint and Glare Hazards From Concentrating Solar Power Plants: Analytical Models and Experimental Validation, Journal of Solar Energy Engineering, Concentrating Solar Technologies Department, Sandia National Laboratories, Albuquerque, USA.

[4] Clifford K. Ho, 2015, Solar Glare Hazard Analysis Tool (SGHAT) Technical Reference Manual, Sandia National Laboratories, Albuquerque, USA.

[5] Florian Ruesch et al., 2015, Quantification of Glare from Reflected Sunlight of Solar Installations, SPF Institute for solar technology, HSR Rapperswil university of applied science, SHC 2015, International Conference on Solar Heating and Cooling for Buildings and Industry.

[6] Florian Ruesch et al., 2016, Methode zur Quantifizierung der Blendung durch Solaranlagen - Vergleich mit anderen Materialien der Gebäudehülle, Institut für Solartechnik SPF, Hochschule für Technik Rapperswil HSR, 26. Symposium Thermische Solarenergie, 20.-22.04.2016 in Bad Staffelstein, Germany.

[7] Referenz: Christof Bucher, Reflexionen an Photovoltaikanlagen, Ursachen und Lösungen, Bulletin electrosuisse, September 2021 [https://www.bulletin.ch/de/news-detail/reflexionen-an-photovoltaikanlagen.html].

[8] Kanton Bern / Meteotest, Blendtool V1.2.0, www.blendtool.ch, Switzerland, accessed 11th of September 2023.

[9] Photovoltaic Systems: Lower Levels of Glare and Reflectance vs. Surrounding Environment, Sunpower Corporation, Document # s107905 Rev A, 1st of October 2019

[10] Data sheet, Kioto Solar, Sonnenkraft, Glas-Folien Solarmodul, Power-66 HC, 405Wp / 410Wp HC, 400Wp HC black, Blendreduziertes DEFLECT Glas, June 2022

[11] EN 12665, Light and lighting - Basic terms and criteria for specifying lighting requirements, 2018

[12] Marc Wittlich, 2010, Blendung – Theoretischer Hintergrund, Institut für Arbeitsschutz der Deutschen, Gesetzlichen Unfallversicherung (IFA), Sankt Augustin, Germany, May 2010

[13] EN 12464-1, Light and lighting - Lighting of work places - Part 1: Indoor work places

[14] Leonie Geerdinck: Glare perception in terms of acceptance and comfort. Graduation Report an: Faculty Industrial Engineering & Innovation Sciences, Technische Universiteit Eindhoven. http://alexandria.tue.nl/extra2/afstversl/tm/Geerdinck_2012.pdf, August 2012

[15] Robert D Clear, Discomfort glare: What do we actually know? In: Lighting Research and Technology. vol. 45 no. 2 141-158 doi:10.1177/1477153512444527, April 2013

[16] Hiroshi TAKAHASHI, Yoshinori KOBAYASHI, Shou ONDA, Takashi IRIKURA: Position Index for the Matrix Light Source. In: Journal of Light & Visual Environment. Vol. 31 (2007) No. 3 S. 128–133., https://www.jstage.jst.go.jp/article/jlve/31/3/31_3_128/_article

[17] Julius Yellowhair and Clifford K. Ho, Assessment of Photovoltaic Surface Texturing on Transmittance Effects and Glint/Glare Impacts, Proceedings of the ASME 2015 Power and Energy Conversion.

Test requirements for angular response measurement of encapsulated solar cell coupons

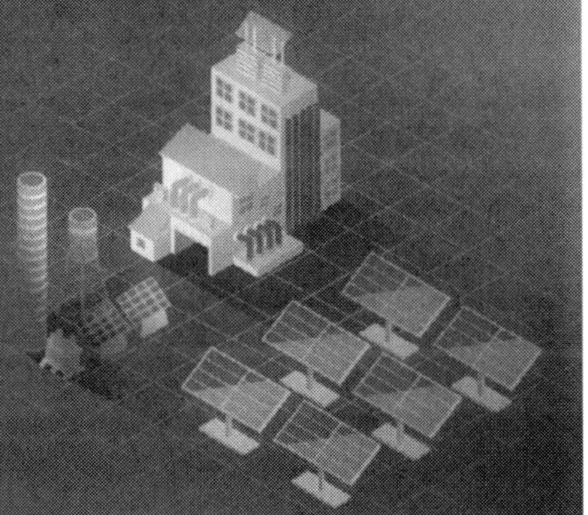

Werner Herrmann[1], Giorgio Bardizza[1], Lorenz Rimmelspacher[1], Min Hsian Saw[2], Mauro Pravettoni[2], Nicholas Riedel-Lyngskær[3], Markus Babin[3], Ingo Kröger[4]

[1] TÜV Rheinland, Cologne, Germany,
[2] National University of Singapore, SERIS, Singapore
[3] DTU Electro, Roskilde, Denmark,
[4] PTB, Braunschweig, Germany

40th EUPVSEC, Lisbon, 18 – 22 September 2023

NUS National University of Singapore | SERIS Solar Energy Research Institute of Singapore — PTB — DTU — TÜVRheinland® Precisely Right.

Motivation

- **Single cell coupons** are commonly used for comparative angular response measurements in the laboratory, but their use is not covered by the IEC 61853-2 test standard.
- **Previous test work:** Variety of indoor measurements test methods with full illumination or partial illumination of the test cell
- Lack of test experience for **back-contact cells and bifacial cells**

▼

- Study the IAM comparability between **partial illumination and full illumination** of cell coupons
- Study the impact of **masking methods** for full illumination of cell coupons
- Study **diffuse irradiance impacts** for IAM measurement with large area solar simulators
- **Define test requirements** for angular response measurement of encapsulated solar cell coupons

Introduction

Climate Specific Energy Rating (CSER) of PV modules defined in IEC 61853-3

$$CSER = \frac{E_{MOD,Year}/H_{Year}}{P_{MAX,STC}/G_{REF,STC}}$$

- CSER is the DC performance ratio of a PV module.
- CSER=1 means that the annual average operational efficiency corresponds to the STC efficiency.

$E_{MOD,Year}$	Simulated PV module annual energy yield (IEC 61853-3) for six reference climates (IEC 61853-4)
H_{Year}	Annual in-plane solar radiation for selected reference climate (IEC 61853-4)
$P_{MAX,STC}$	PV module nominal output power at STC according to manufacturers specification
$G_{REF,STC}$	STC reference irradiance = 1000 W/m²

PV module performance parameters resulting from laboratory measurements:

1) G-T power matrix acc. to IEC 61853-1: 100 W/m² to 1100 W/m², 15°C to 75 °C
2) Modelling parameters u_0, u_1 for PV module operating temperature acc. to IEC 61853-2
3) Spectral response curve acc. to IEC 61853-2/IEC 61853-7
4) Angular response parameter a_r acc. to IEC 61853-2

Angular responsivity measurement of PV devices acc. to IEC 61853-2 [1]

Step 1: Measure the short circuit current $Isc(\theta)$ of the PV device in the angular range -80° to +80°

Step 2: Calculate the Incident Angle Modifier *IAM(θ)*:

$$IAM(\theta) = \frac{Isc(\theta)/Isc(0°)}{\cos(\theta)}$$

Step 3: Plot the *IAM(θ)* data points in a diagram and determine the fitting parameter a_r of the angular response curve *AR(θ)* [2].

$$AR(\theta) = \frac{1 - \exp(-\cos\theta/a_r)}{1 - \exp(-1/a_r)}$$

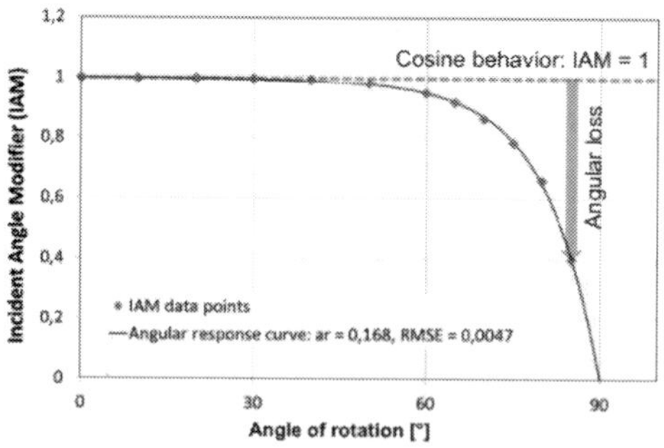

The greater the a_r value, the higher the angular loss (a_r range for PV modules: 0.13 .. 0.18)

[1] IEC 61853-2: Photovoltaic (PV) module performance testing and energy rating – Part 2: Spectral response, incidence angle and module operating temperature measurements
[2] N. Martin et al.: Calculation of the PV modules angular losses under field conditions by means of an analytical model, Solar Energy Materials & Solar Cells 70 (2001)

Indoor measurement practices

Indoor IAM test procedure ⇨ Use of an optically matched mini-module:

- Identical in materials and design
- Equivalent refraction and reflection processes

Intercomparison test work [3] [4] ⇨ Use of encapsulated single cell coupons

Test methods for solar cell coupons:

a) Full illumination of the test cell
   - Indoor measurement with a solar simulator
   - Outdoor measurement under natural sunlight

b) Partial illumination of the test cell
   A small beam size illuminates just a sub-area of the cell

Need for a standardized test procedure for single cell coupons!

Optically matched mini-module

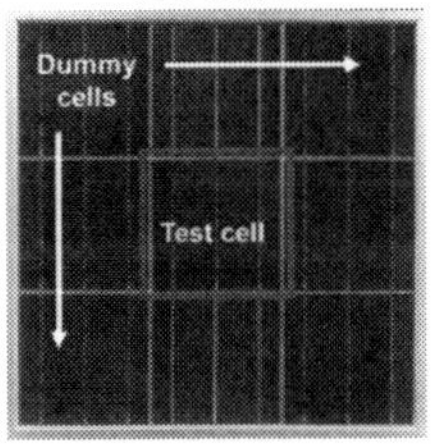

Coupon of an encapsulated single cell

[3] N. Riedel-Lyngskær et al.: Interlaboratory comparison of angular-dependent photovoltaic device measurements: Results and impact on energy rating. doi.org/10.1002/pip.3385
[4] N. Riedel-Lyngskær et al.: Interlaboratory Comparison of Methodologies for Measuring the Angle of Incidence Dependence of Solar Cells, 35th EU PVSEC 2018

Full illumination vs. partial illumination of the test cell

Full illumination of the cell coupon

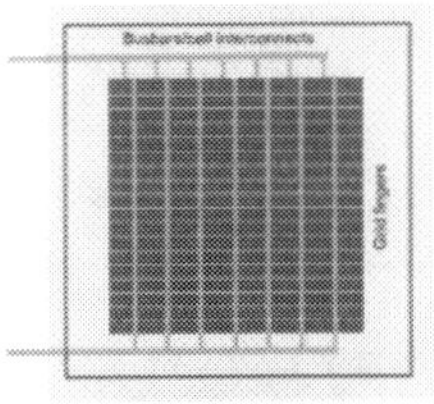

- Test method closer to real operating conditions of PV modules
- Masking of cells is required to avoid stray light effects to enter into the encapsulation around the cell and through the edges of the package
- Bifacial sample requires additional measures to avoid irradiation of rear side

Partial illumination of the cell coupon

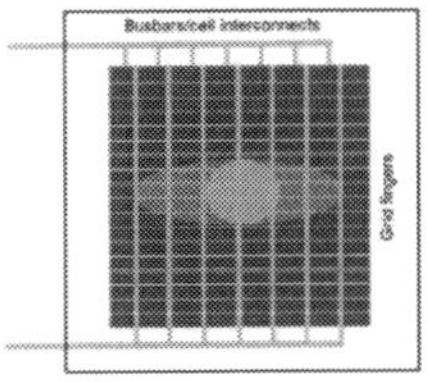

- Size of test area increases as a function of AoI (elliptical shape):
   - Circular beam at normal incidence ⇨ Tripling of the illuminated area at AoI=80°
   - Conflict with busbars ⇨ Variation in the percentage illuminated active area
- Multiple reflections are mostly not considered
- Low photocurrent: Bias light and lock-in technique will be required to avoid non-linearity effects

Test cell coupons

Cell type: **Cell dimension:** **Coupon dimension:** **Optical structure:**	HJT (multiwire) 16.2 cm x 8.1 cm 30 cm x 30 cm Glass/EVA/BS (black)	IBC 15.6 cm x 15.6 cm 20 cm x 20 cm Glass/EVA/BS (white)	Bifacial (front) 15.6 cm x 15.6 cm 30 cm x 30 cm Glass/EVA/Glass
			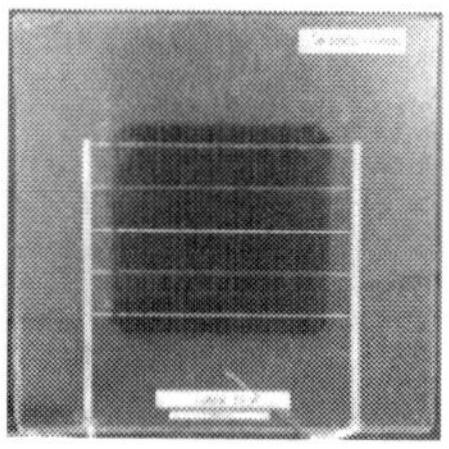

Masking methods for full illumination:

- Black tape around sample edges
- Bifacial cell: Black cloth on rear side and around sample edges

Test equipment – Full illumination of the test cell

TUV Rheinland:

- Pasan 3b pulsed solar simulator with 4 Xenon arc lamps
- Distance of lamps to axis of rotation: 8 m
- Reduction of diffuse irradiance in the test area with three baffles in the beam of light:
- Max. AoI variation for M12 wafer due to light divergence: 2°

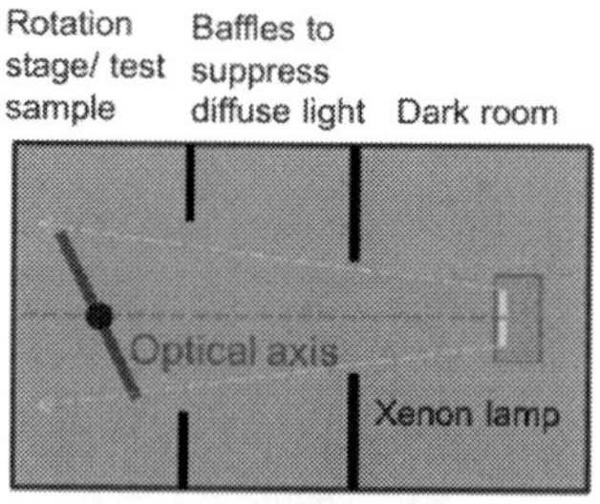

PTB:

- Laser-DSR facility with 2-axis goniometer
- Modulated monochromatic light (Step width: 50 nm, spot size: 300 x 300 mm², Lock-In technique)
- Bias light of test cell
- Measurement of spectrally resolved angular response
- Calculation of integral AM1.5 weighted angular dependence

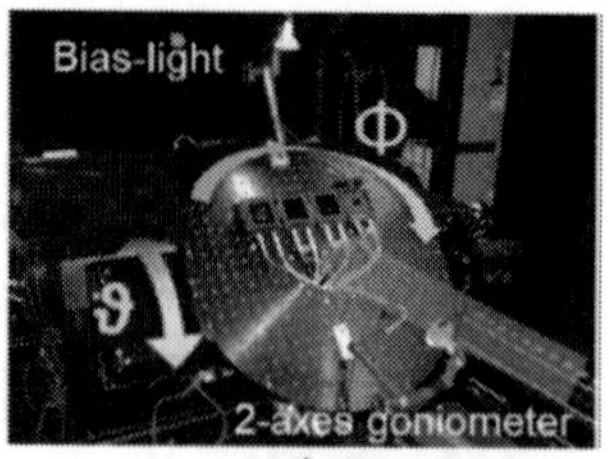

Test equipment – Partial illumination of test cell

SERIS:

a) SERIS customized system
  - Chopped xenon light with lock-in technique
  - Optical fiber attached to an angle probe holder
  - Bias light of test cell or PV module
  - Spot size: 10 mm

b) PASAN 3b pulsed solar simulator

DTU:

- Laser driven light source (LDLS),
- Continuous spectral irradiance: 300 nm to 1400 nm
- Spot size: 12 mm diameter

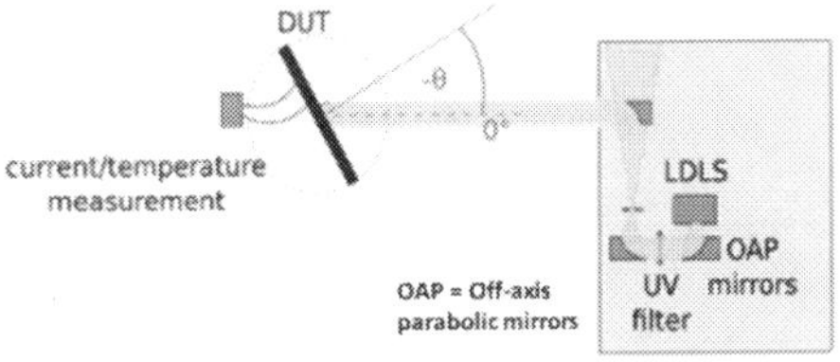

TÜVRheinland®
Precisely Right.

Test results for full and partial illumination

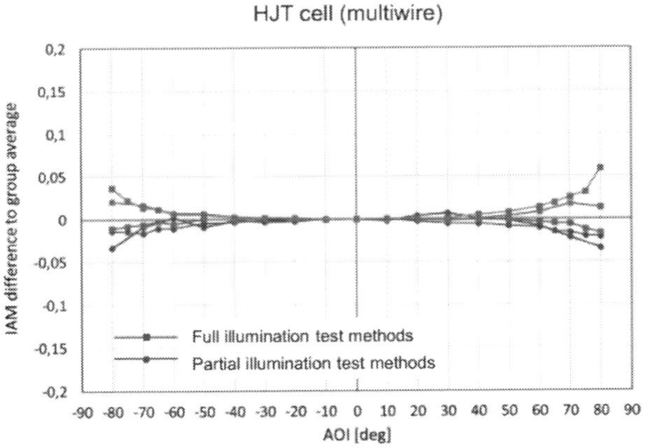

	ar	RMSE
Seris (partial)	0,187	0,0036
DTU (partial)	0,185	0,0029
Seris (full)	0,164	0,0004
TÜV (full)	0,157	0,0059
PTB (full)	0,179	0,0008

$\Delta a_r = 0{,}03$

Temperate coastal climate (IEC 61853-4)
ΔCSER = 0,016

TÜVRheinland®
Precisely Right.

Test results for full and partial illumination

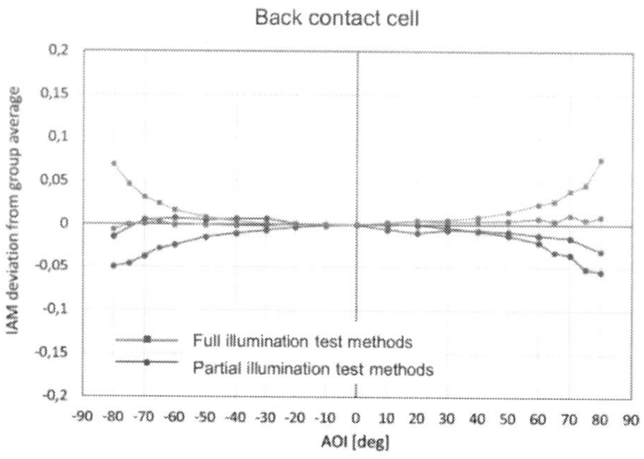

	ar	RMSE
Seris (partial)	0,173	0,0037
TUV (full)	0,140	0,0085
PTB (full)	0,161	0,0056
DTU (partial)	0,193	0,0038

Δa_r = 0,053

Temperate coastal climate (IEC 61853-4)
ΔCSER = 0,027

40th European Photovoltaic Solar Energy Conference and Exhibition

Test results for full and partial illumination

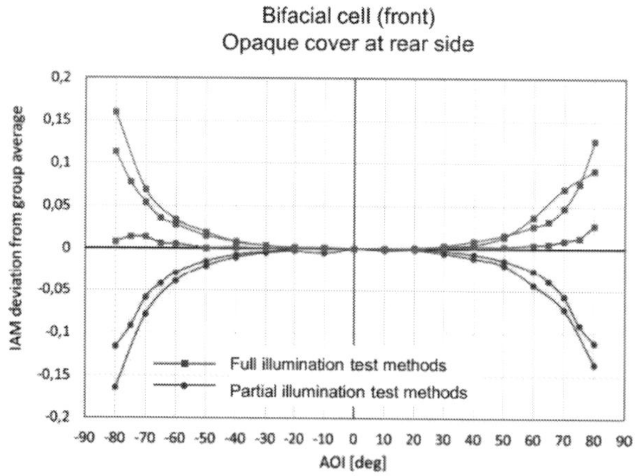

	ar	RMSE
Seris (partial)	0,179	0,0019
DTU (partial)	0,164	0,0026
Seris (full)	0,074	0,0094
TÜV (full)	0,113	0,0043
PTB (full)	0,081	0,0075

Δa_r = 0,105

Temperate coastal climate (IEC 61853-4)
ΔCSER = 0,046

Impact of different masking methods for full illumination

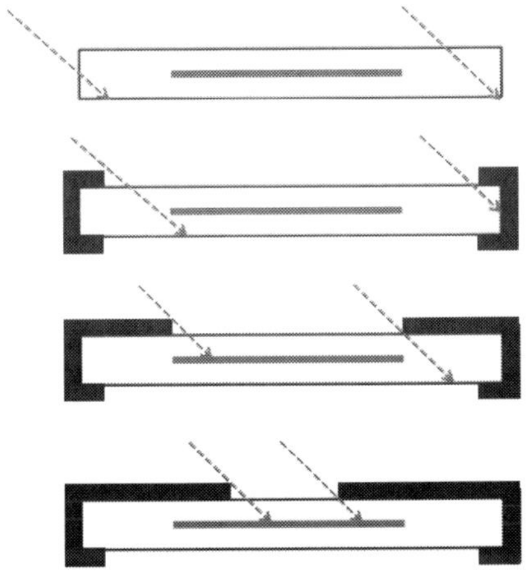

No masking:
- Stray light effects from coupon edges
- Multiple reflection inside package

Opaque tape around the sample edges
- Multiple reflection inside package

Opaque mask reaching the cell edges:
- Partial shading of cell caused by the mask. Shade length for AoI=80°: Mask/glass/encapsulation thickness
- Multiple reflection on the opposite side

Partial illumination of the test cell:
- Shift and reduction of the illuminated area as a function of AoI and mask thickness

TÜVRheinland®
Precisely Right.

Impact of different masking methods for full illumination

Back contact cell

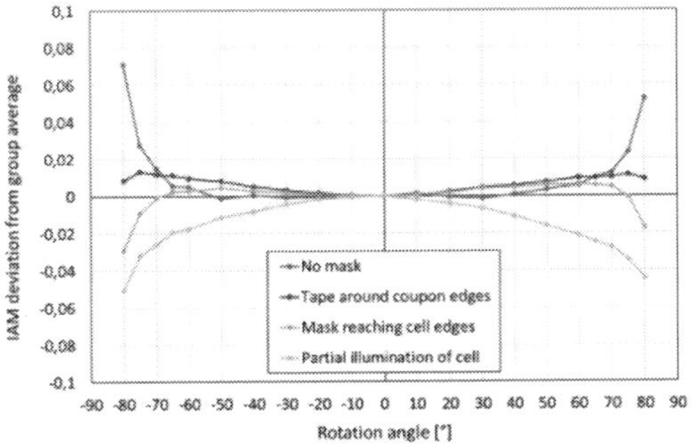

Bifacial cell (front)

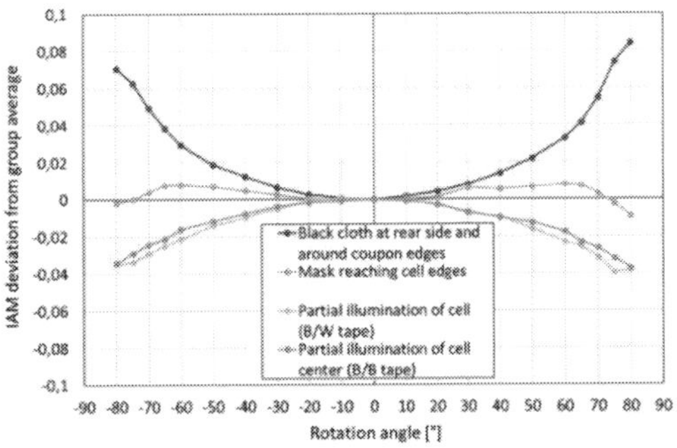

- Significant impact of masking the sample edges
- Mask should not reach cell edges to avoid self-shading

- Significant impact of masking up to the cell edges
- Low impact of tape type

TÜVRheinland®
Precisely Right.

Impact of diffuse irradiance in the test room of solar simulators

Solar simulator type: Pasan 3b solar simulator, tunnel dimensions: 3 m x 3 m x 8 m

Technique: Opaque paddle in the beam of light ⇨ Shading of test cell for all rotation angles

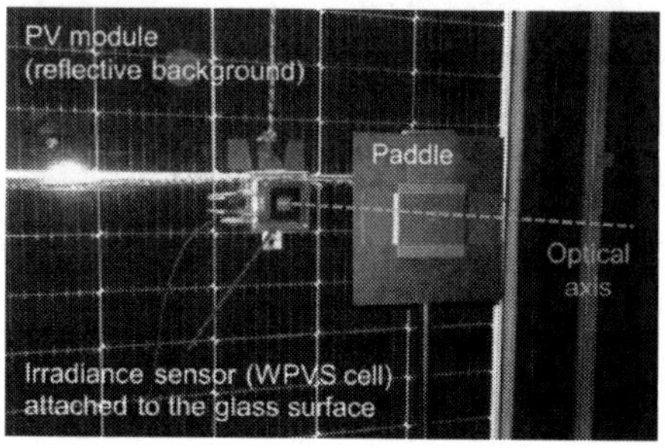

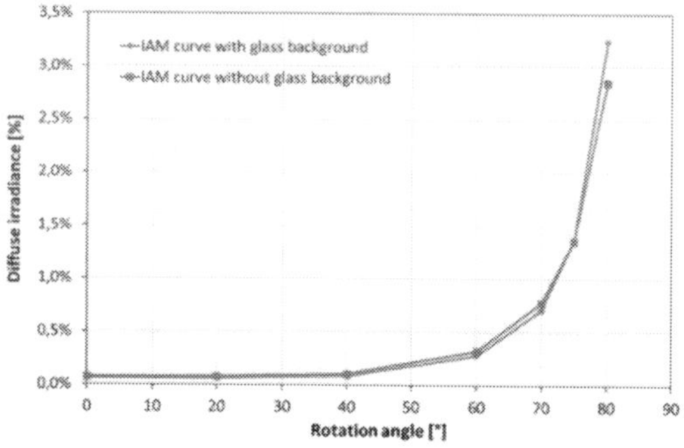

Results:

- Diffuse light inside the test room of a solar simulator is not negligible
- Diffuse irradiance is AoI dependent, but is independent on surface reflections from a PV module

TÜVRheinland® Precisely Right.

40th European Photovoltaic Solar Energy Conference and Exhibition

Comparability of test methods

- Partial illumination (Series, DTU)
- Full illumination + masking the cell + diffuse irradiance correction (TUV)

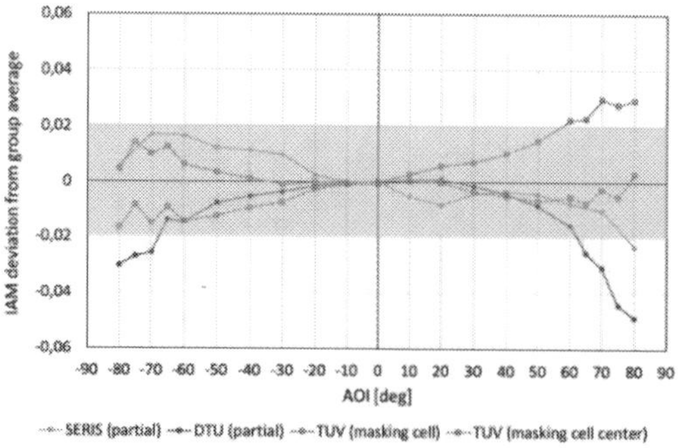

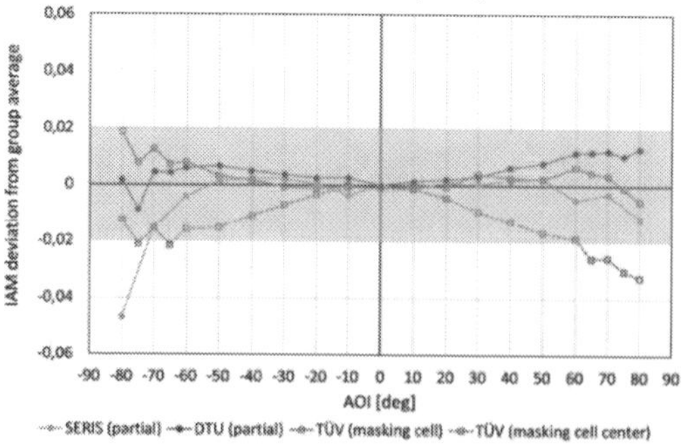

TÜVRheinland® Precisely Right.

Conclusions and outlook

- **IAM measurement on cell coupons with simulated sunlight**:
  - Partial illumination vs. full illumination: Apply suitable masking method to achieve optimal comparability
- IAM measurement with **large area solar simulator**: Diffuse irradiance correction required
- **Recommended masking method for full illumination**:
  - Use thin opaque tape as masking material
  - Opaque mask around the test cell, but leave space to cell edges to avoid self-shading (spacing dimension ≈ thickness of tape/glass/encapsulation layers)
- Further results also for **outdoor IAM measurement** expected
- **Provide input** for revision of the IAM section of IEC 61853-2

40th European Photovoltaic Solar Energy Conference and Exhibition

Thank you for your attention!

Contact:

Dr. Werner Herrmann
TÜV Rheinland Solar GmbH
51101 Cologne, Germany
werner.herrmann@de.tuv.com

TÜVRheinland®
Precisely Right.

CHARACTERIZATION OF CURVED MODULES FOR VIPV

I. Antón[1], R. Herrero[1], F.J. Martín[1,2], S.Askins[1], J. Macías[1], L. J. San José[1], G. Vallerotto[1], R. Núñez[1], C. Domínguez1
[1]Instituto de Energía Solar-Universidad Politécnica de Madrid (IES-UPM), Madrid, Spain
2Solar Added Value (SAV), Madrid, Spain

The vehicle integrated photovoltaic (VIPV) technology demands 3D curved PV surfaces to meet the specific design constraints in the automotive industry. The proper characterization of three dimensional PV surfaces requires specific methods and equipment that must be developed. This paper describes the procedures carried out for both outdoor and indoor characterization of modules with curvatures between 1 and 3 meter. IV curves at STC and the angular response of the modules are compared, showing remarkable resemblance. A unique collimated solar simulator has been adapted for indoor characterization while a 2-axis tracker with programmable deviation of the pointing to the sun has been used outdoors.
Keywords: solar simulator, vehicle integrated photovoltaics, characterization

1 INTRODUCTION

The deployment of the EV arose the interest of vehicle integrated photovoltaics (VIPV), which consists of integrating PV solar panels in car surfaces such as roof, hood, doors, and even windows [1]. In this market, 3-dimensional (3D) shapes are demanded for the PV surfaces depending on the body car. Thus, VIPV surfaces performance would be dependent on the intrinsic lack of irradiance uniformity proper to the curved shape, and also on the non-stationary working conditions producing, for example, irradiance patterns affected by shadows or fast changing of angle of incident (AOI).

Suitable standards must be developed considering these VIPV particularities not addresed in current ones such as flat IEC 60904 series [2]. In this regard, the standardization body for photovoltaic technology (IEC TC82) has organized an international team (PT600) formed by scientists, engineers, and experts of different disciplines and organizations aimed to developed fair and scientific testing and rating [3]. Among all the activities carried out in the working group, it is worth mentioning the development of an international Round Robin (RR), supported by METI and JET/JEMA (Japan), in which photovoltaic curved samples have been measured by different laboratories and institutions.

Conventional flat PV modules are reliably characterized indoors using solar simulators, which are devices capable of providing irradiance conditions similar to the sunlight in a laboratory environment or production line. Most widely used solar simulators for PV panels are based on a point light source placed far enough from the panel, producing a light pulse long enough to assure a steady-state condition during the IV sweep. This type of solar simulator has demonstrated to achieve the highest quality in terms of spatial non-uniformity of irradiance, spectral mismatch and temporal stability accordingly to [4]. VIPV imposes new constraints for a solar simulator, being the most important related to the curved shape of the PV surfaces. This characteristic causes an inherent non-uniform solar irradiance throughout the surface due to the shape but also dependent on the light source angular distribution. Fig. 1 shows the scheme of collimated solar simulator and conventional one for flat modules. For the case of conventional simulators, the non-uniform irradiance over the curved surface is boosted by the increasing angle of incidence ($\delta > \theta$) throughout the aperture area caused by the divergence of the beam (β). Nevertheless, the collimated solar simulator, with the divergence of the beam ($\pm\alpha$) close to $\pm 0.27°$, reproduces the curved module performance as under sunlight in terms of irradiance over the surface.

This study presents the work carried in the context of the international RR by the Institute of Solar Energy (IES-UPM) focusing on revealing specific considerations in the outdoor and indoor testing of curved surfaces, in particular, analyzing the convenience of a collimated solar simulator for this purpose.

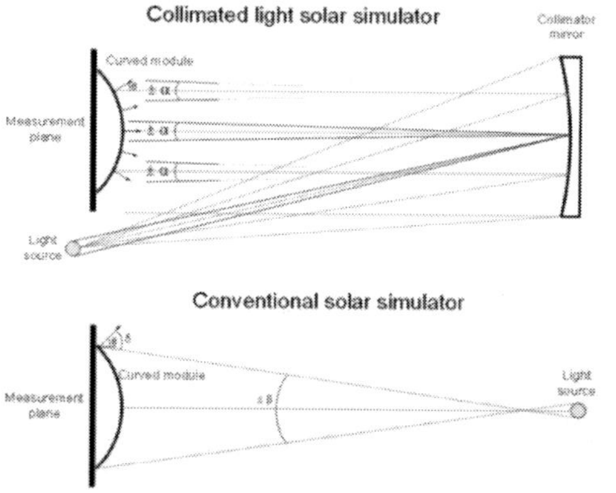

Figure 1: Scheme of collimated solar simulator compared to a conventional one.

2 EXPERIMENTAL

The measured samples are cilindrical curved modules (1380 mm x 700 mm) manufactured by AGG Energy with different radious of curvature (1 m, 2 m, 3 m, and flat), each module is formed by 8 strings of 4 monocrystalline silicon cells (MWT) series connected, with the inclusion of a bypass diode per string.

The international RR was organized in several routes, one was specifically devoted to research activities for the development of meaningful test procedures for curved PV modules. IES-UPM has participated in this branch, the main goal was the comparison of outdoors and indoors power rating at Standard Test Conditions (GNI=1000 W/m^2, module temperature of 25°C, and air mass (AM) of 1.5).

10.4229/EUPVSEC2023/3CO.10.4
020247-001

2.1 Outdoor measurements

The modules under test were installed on a two axis tracker (azimuth-elevation), all in landscape configuration and arranged in the same column avoinding any reflective surfaces at the sides of the modules (see Fig. 2). The IV curve of the 4 modules were measured simultaneously together with temperature probes attached to the rear side of the modules, the GNI irradiance given by a reference monocrystalline silicon cell (calibratred by Fraunhofer ISE CalLab PV Cells) and meteo station data (DNI, GNI, diffuse, ambient temperature, wind, humidity, spectrally resolved DNI 300-1100 nm, etc).

A measurement campaign has been carried out at winter conditions in Madrid at which cold and clear sky days are quite common in this time of the year. The modules performance have been evaluated by means of IV curves measured at steadystate conditions, IV curved measured under thermal transient (to obtain STC) and IV curves measured during 2-axis tracking angular sweeps by following the procedure of [5] (to evaluate the angular performance).

Figure 2: Curved modules installed in two-axis tracker of different radius of curvature (from top to bottom: 1 m, 2 m, 3m and flat).

These are the steps of the measurement procedure carried out to optain the power rating at Standard Test Condition:

- Calibration of the Isc of each module against the reference cell (GNI) at clear sky conditions.
- Evaluation of modules during thermal transient. First, modules are covered with cardboard, assuring they are at room temperature. After this, the modules are uncovered and IV curves are measured together with the module temperature during warming up, aiming at STC conditions.

2.2 Indoor measurements

IES-UPM developed the Helios 3198, a collimated solar simulator devoted for indoor characterization of CPV modules [6-10]. The illumination system consists of a round collimator with a diameter of 2 meters and a focal distance of 6 meters, and a small toroid flash lamp of 6.5 cm in diameter. A black tunnel for stray light rejection has been designed and built, including the baffles and light trap box and chambers as detailed in Fig. 3.

The modules have been measured at STC (1000W/m2 and module temperature of 25°C), at normal irradiance and with different angle of incidence to determine the rated power of the modules and the angular response.

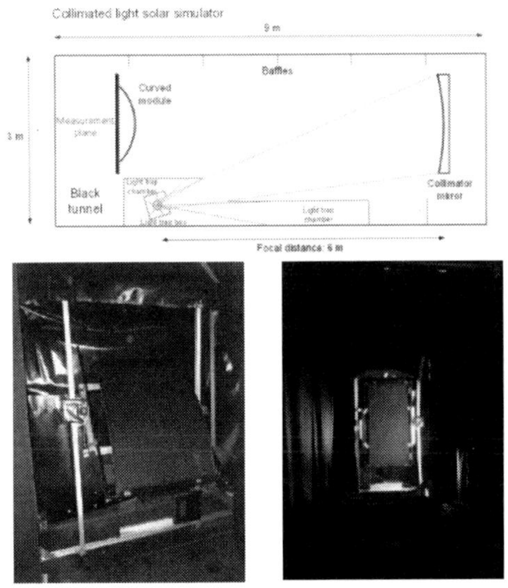

Figure 3: (up) Scheme of Helios 3198 solar simulator adapted to VIP; (left) module and rotating structure to evaluate angular response: (right) collimator mirror reflecting VIPV modules under test.

3 RESULTS AND CONCLUSIONS

3.1 Power rating at STC

The IV curves of the 4 modules at STC conditions were obtained both indoors and outdoors. Fig. 4 presents the maximum power (Pmp) measured for the 4 modules under inspection. These Pmp values are related to IV curves measured at steady conditions and IV curve measured during thermal transient. Outdoor STC Pmp values (red circles), and IV curves, are obtained by filtering to achieve measurement conditions as close as possible to STC.

In this study, direct comparison of raw data (IV curves measured at STC both indoors and outdoors) has been carried out. Translation equations and procedures not suitable for curved surfaces due to the steps-shaped IV curve are avoided as much as possible, only minimum irradiance correction has been applied.

Fig. 5 shows the IV curves of the modules in both indoor and outdoor. Relative errors between indoor and outdoor measured Pmp values are lower than 1.5% for all cases. To compare this curves, minimun correction has been performed. For example, for the case of R=1000, the outdoor IV curve was measured at 994 W/m2 and module temperature of 24°C. The outdoor IV curve has been corrected to 1000W/m^2 (only current values were corrected, voltage correction has considered negligible). The rated power, measured in the solar simulator was 127.4 Wp, while outdoors was 126.5 Wp at 994W/m2 (127.3 Wp if corrected to 1000W/m^2).

10.4229/EUPVSEC2023/3CO.10.4

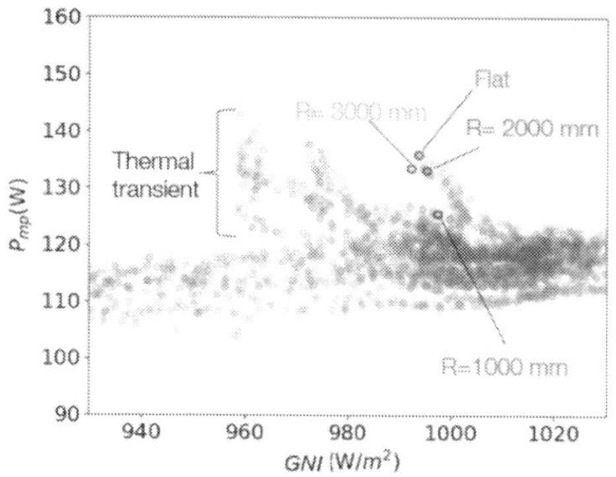

Figure 4: Pmp vs. GNI for the modules with different radious of curvature (1000 mm, 2000 mm, 3000mm and flat).

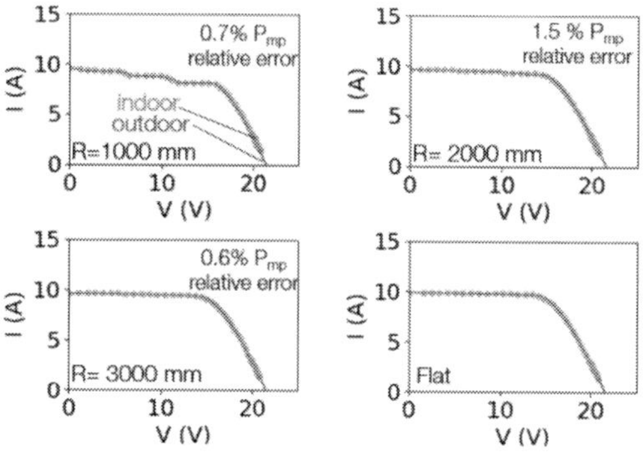

Figure 5: Measured IV curves (indoor and outdoor) for curved and flat modules.

3.2 Angular performance

The angular response of VIPV modules is particularly relevant due to both curvature and non-stationary working conditions due to vehicle movement. It must be pointed out that this response depends not only on the curvature, but also on the interconnection topology of the cells comprising the module (cell size, series/parallel connection, bypass diodes…).

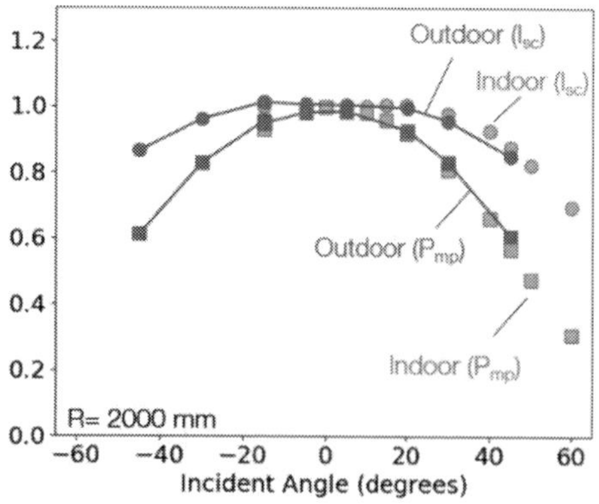

Figure 6: Measured Pmp and Isc (normalized) vs. Incident angle for the modules with R=2000 mm.

Fig. 6 shows the results achieved for the module with 2 m of radius of curvature both indoors and outdoors and at two bias conditions (Isc and Pmp). When cells are connected in series and the module integrates a high number of by-pass diodes, the angular response depends on the bias point. This is a very well known effect in CPV [11], where there is typically one by-pass diode per cell [12]. For the case of curved modules, the inherent non-uniformity caused by the curvature makes it convenient to integrate a greater number of diodes than in conventional flat modules [13]. The most relevant angular response is obviously the one obtained at the maximum power point. For this reason, a set of IV curves has been measured indoor and outdoor at different angular positions and the angular response is deduced based on Isc and Pmp values.

ACKNOLEDGMENTS

This work has been supported by the Grant PID2021-128853OB-I00 funded by MCIN/AEI/ 10.13039/501100011033 and by "ERDF A way of making Europe". The authors express sincere thanks to the AGG Energy for the manufacturing of the module, and JEMA and JET program for sponsoring the activities with those modules, entrusted by METI (Japan) and carried out under the umbrella of the IEC TC82/PT600 group devoted to Vehicle Integrated PV Systems.

REFERENCES

[1] "State-of-the-Art and Expected Benefits of PV-Powered Vehicles" IEA-PVPS, ISBN 978-3-907281-15-4

[2] K. Araki, et. Al."To Do List for Research and Development and International Standardization to Achieve the Goal of Running a Majority of Electric Vehicles", *Solar Energy Coatings*, vol.8, no. 7, pp, doi: 10.3390/coatings8070251

[3] K. Araki, et. Al. "International collaboration for standardization (testing and rating) for VIPV" , in 33rd International Photovoltaic Science and Engineering Conference, 2022.

[4] C. Droz, et Al. "Evaluation of Commercial Large Area Solar Simulator: Features Exceeding the IEC Standard Class AAA" in 25th European Photovoltaic Solar Energy Conference and Exhibition, 2010, pp. 3884 – 3888.

[5] D. Riley et Al. ASME. J. Sol. Energy Eng. June 2015; 137(3): 031008, doi:10.1115/1.4029379

[6] C. Domínguez et Al. Optics Express, vol. 16, no. 19, pp. 14 894–14 901, doi: 10.1364/OE.16.014894

[7] C. Domínguez et Al., "Characterization of five CPV module technologies with the Helios 3198 solar simulator," in *34th IEEE Photovoltaic Specialists Conference* 2009, pp. 001004-001008, doi: 10.1109/PVSC.2009.5411192

[8] C. Domínguez et Al., "Solar simulator for indoor characterization of large area high-concentration PV modules", in *33rd IEEE Photovoltaic Specialists*

10.4229/EUPVSEC2023/3CO.10.4
020247-003

Conference, 2008, pp. 1-5, doi: 10.1109/PVSC.2008.4922739.

[9] I. Antón et Al., "Characterization Capabilities of Solar Simulators for Concentrator Photovoltaic Modules," *Japanese Journal of Applied Physics*, vol. 51, p. 10ND12, doi: 10.1143/JJAP.51.10ND12

[10] C. Domínguez et Al., "Indoor Characterization of CPV Modules Using the Helios 3198 Solar Simulator," in *24th European Photovoltaic Solar Energy Conference*, 2009, pp. 165-169, doi: 10.4229/24thEUPVSEC2009-1BO.7.6

[11] I. Antón y G. Sala, «Losses caused by dispersion of optical parameters and misalignments in PV concentrators», Prog. Photovolt. Res. Appl., vol. 13, n.o 4, pp. 341-352, jun. 2005, doi: 10.1002/pip.582.

[12] M. Wiesenfarth, I. Anton, and A. W. Bett, "Challenges in the design of concentrator photovoltaic (CPV) modules to achieve highest efficiencies," Applied Physics Reviews, vol. 5, no. 4, p. 041601, Dec. 2018, doi: 10.1063/1.5046752

[13] J. Macias, R. Herrero, R. Nunez, y I. Anton, «On the effect of cell interconnection in Vehicle Integrated Photovoltaics: modelling energy under different scenarios», en 2021 IEEE 48th Photovoltaic Specialists Conference (PVSC), Fort Lauderdale, FL, USA: IEEE, jun. 2021, pp. 1336-1339. doi: 10.1109/PVSC43889.2021.9518935.

THE EFFECT OF IRRADIANCE NON-UNIFORMITY ON THE PERFORMANCE OF STATE-OF-THE-ART PV-MODULE DESIGNS

Y.Zhang[1], I.Margaris[2], Q.Gao[1], W.Xu[1], G. Bardizza[3], W. Herrmann[3], C. Monokroussos[1]
[1] TÜV Rheinland (Shanghai) Co., Ltd, Shanghai, P.R. China
[2] Department of Electrical and Computer Engineering, University of Thessaly, Volos, Greece
[3] TÜV Rheinland Energy GmbH, Cologne, Germany
Corresponding author: Christos Monokroussos
Telephone: +86 21 6108 1188
Email: Christos.Monokroussos@tuv.com

ABSTRACT: The market share of innovative PV module designs with features like cut cells and unconventional topologies is growing rapidly. These features make the power rating measurement under standard testing conditions of innovative PV modules different from that of conventional PV modules. This work aims to investigate the effect of irradiance non-uniformity on the power rating of state-of-the-art PV module designs. The current-voltage characteristics of three different PV-module designs (half-cell, third-cell, and shingled PV module) under different irradiance non-uniformity were simulated in MATLAB utilizing the Bishop model and Lambert W function. The results show that under random non-uniform irradiance conditions, the short-circuit current and maximum power errors were approximately equivalent between conventional and innovative PV module designs. However, under non-uniform irradiance exhibiting specific profiles, the half-cut, third-cut, and shingled PV modules consistently exhibited lower maximum power losses compared to the conventional design.
Keywords: non-uniform irradiance, half-cut cell, shingled, power rating

1 INTRODUCTION

Significant progress in the development of cell-cutting and interconnection technologies in recent years has introduced innovative PV-module designs in the photovoltaic (PV) market. Unlike conventional PV module designs with only series-connected strings of 60 or 72 solar cells, state-of-the-art PV module designs, including cut-cell and shingled solar cells employ new circuit topology and geometric layout. Two common features of these innovative PV-module designs are cut-cells and the series–parallel-series (SPS) topology. The global market share of these PV-module designs has risen rapidly, exceeding 90% by the end of 2022 [1]. One major advantage of these innovative PV modules is their improved tolerance to partial shading [2-4]. However, the effect of non-uniform irradiance on power rating under standard test conditions (STC) has not been studied for these designs. In our previous research, irradiance non-uniformity has been identified as a primary source of measurement uncertainty for conventional PV-module designs [5]. Investigating the impact of non-uniform irradiance on power rating will pave the way for the power rating and assessment of measurement uncertainty (MU) for innovative PV modules [6].

This work aims to quantitatively determine the effect of irradiance non-uniformity on the performance of state-of-the-art PV-module designs. The current-voltage (I-V) characteristics of three different PV-module designs (half-cell, third-cell, and shingled PV module) under different irradiance non-uniformity were simulated and compared to a conventional PV module. The simulation methodology and results of this work provide a theoretical basis for the power rating and MU calculation of indoor PV module measurement.

2 METHODOLOGY

This work analyzed four PV module architectures, as illustrated in Figure 2. The I-V characteristics of PV modules were simulated in MATLAB based on the Bishop one-diode circuit model (as shown in Figure 1) which considers the avalanche breakdown behavior of solar cells [5]. Thus, both the forward and reverse characteristics of the solar cells were included in the analysis. Bypass diodes were modeled as ideal Schottky diodes. For shingled PV module, the parameters of the Bishop one-diode circuit model were extracted from measured I-V curves under unshaded and shaded conditions using a Differential Evolution Algorithm [7]. For better comparability, the model parameters (I_{ph}, R_s, R_{sh}, I_0, and n) for the conventional, half-cut, and third-cut PV module designs were scaled based on the model parameters of the shingled PV module and Ohm's law. The reverse bias coefficients of m and a in the Bishop model are assumed constant for all the PV modules.

By applying Kirchhoff's circuit laws, the module I-V characterization was obtained by integrating the I-V curves of all solar cells and bypass diodes within the module. To achieve both the accuracy and efficiency of the simulation, the Lambert W function was employed in solving the I-V implicit equation. The I-V curves simulated using the Lambert W function showed good agreement with the numerical simulation results obtained under varying irradiance. Moreover, the computational time decreased 10x when applying the Lambert W function for solving the implicit I-V equation, demonstrating its advantage in terms of simulation efficiency and speed.

The study examined the effect of irradiance non-uniformity by simulating different non-uniform distribution patterns, including random, horizontal asymmetry, and vertical asymmetry, as illustrated in Figure 3. For each irradiance distribution profile, the simulations were performed with irradiance non-uniformity magnitudes of 1%, 2%, and 5%, corresponding to solar simulator classifications A+, A, and B defined in IEC 60904-9:2020 [8]. A Monte-Carlo simulation approach was adopted to stochastically model various irradiance non-uniformity scenarios and analyze the distribution of PV module performance metrics statistically. With an average irradiance fixed at 1000 W/m2, 1000 random non-uniformity patterns were assessed per non-uniformity case.

10.4229/EUPVSEC2023/3CO.10.5
020248-001

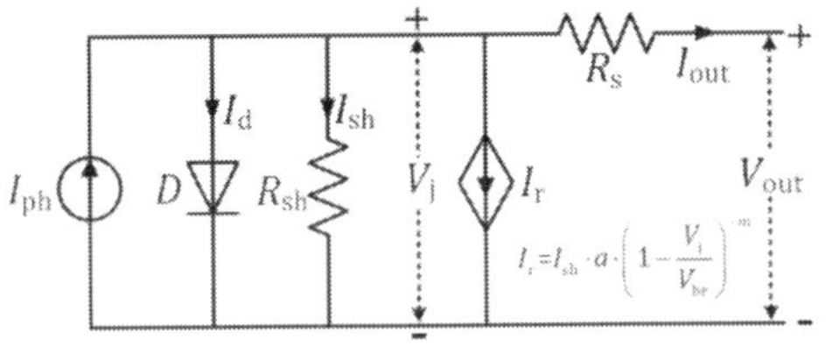

Figure 1: Bishop one-diode circuit model considering reverse characteristics of PV device

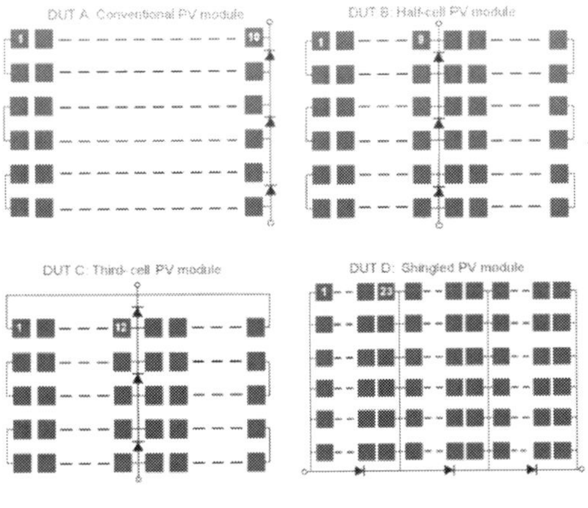

Figure 2: Schematic diagram of the PV-module designs of four PV module designs

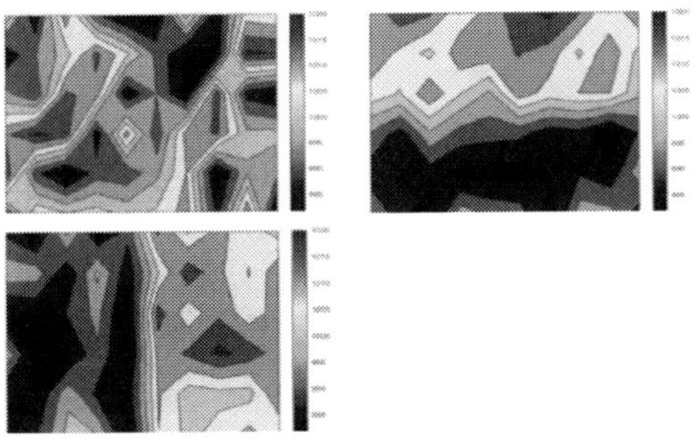

Figure 3: Schematic diagram of the irradiance profiles: (a) random (b) horizontal asymmetry (c) vertical asymmetry

3 RESULTS AND DISCUSSIONS

In the simulation, the irradiance distributions of different irradiance non-uniformity profiles and magnitudes were generated via the Monte Carlo method. Boxplot diagrams were utilized to visualize the distributions of simulated I_{sc} and P_{max} errors under different conditions.

3.1 PV-module performance under random non-uniform irradiance

Figures 4 and 5 show the simulated I_{sc} and P_{max} error distributions for random non-uniformity ranging from 1%, 2% to 5% of four PV-module designs. Overall, the influence of irradiance non-uniformity on I_{sc} is larger than that of P_{max} for all the DUTs, and the error profile of I_{sc} and P_{max} increases with the non-uniformity for each DUT. As shown in Figure 6, for conventional and shingled PV modules under 5% irradiance non-uniformity, the segment between 0 V and V_{mpp} on output characteristic curves exhibits significant deformation. The extent of curve deformation depends on the reverse characteristics of PV devices, interconnect configurations, and the irradiance mismatch level. Conventional, half-cut, and third-cut PV module designs exhibited comparable errors in I_{sc} under random non-uniform irradiance. In contrast, the shingled PV module design demonstrated minimal I_{sc} variation compared to other module types. For irradiance non-uniformity levels of 1%, 2%, and 5%, the maximum absolute deviation of the median of I_{sc} error was 0.035%, 0.071%, and 0.162% respectively among four PV module designs. Compared to conventional PV module design, the maximum deviation of I_{sc} error values for the innovative module designs (half-cut, third-cut, shingled) reaches up to 0.412%, 0.854%, and 2.166% under irradiance non-uniformity of 1%, 2%, and 5%, respectively. The standard deviation of the distribution is progressively decreasing from conventional, half-cut, third-cut, and shingled modules.

Compared to the conventional PV module, the half-cut, third-cut, and shingled PV module designs exhibited relatively lower errors in P_{max} under random non-uniform irradiance. Specifically, the maximum absolute deviation of the median of P_{max} error was 0.004%, 0.016%, and 0.07% with irradiance non-uniformity of 1%, 2%, and 5%, respectively. Relative to the conventional PV module, the maximum deviation found in P_{max} error values for the innovative module designs (half-cut, third-cut, shingled) could reach up to 0.021%, 0.093%, and 0.541% under irradiance non-uniformity of 1%, 2%, and 5%, respectively. The standard deviation of the error is progressively decreasing, consistently with the findings for I_{sc} In summary, when subjected to conditions of random irradiance non-uniformity, the simulation results indicate that the I_{sc} and P_{max} error can be considered approximately equivalent for both the conventional PV module design and the innovative module types (half-cut, third-cut, shingled).

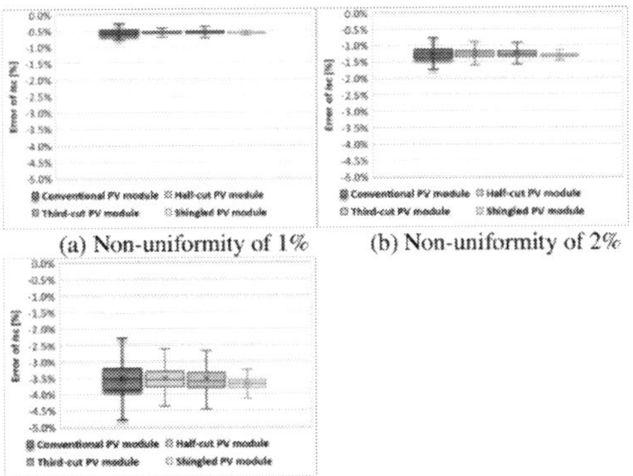

(a) Non-uniformity of 1% (b) Non-uniformity of 2%

(c) Non-uniformity of 5%

Figure 4: Simulation results of I_{sc} error distributions for random non-uniformity varying within 1-5% of four PV-module designs

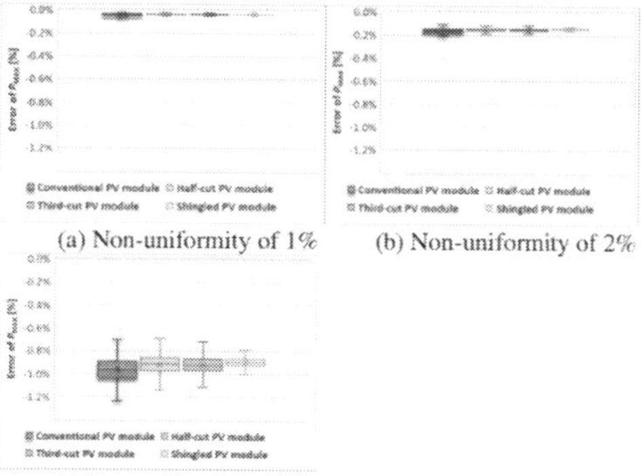

(a) Non-uniformity of 1% (b) Non-uniformity of 2%

(c) Non-uniformity of 5%

Figure 5: Simulation results of P_{max} error distributions for random non-uniformity varying within 1-5% of four PV-module designs.

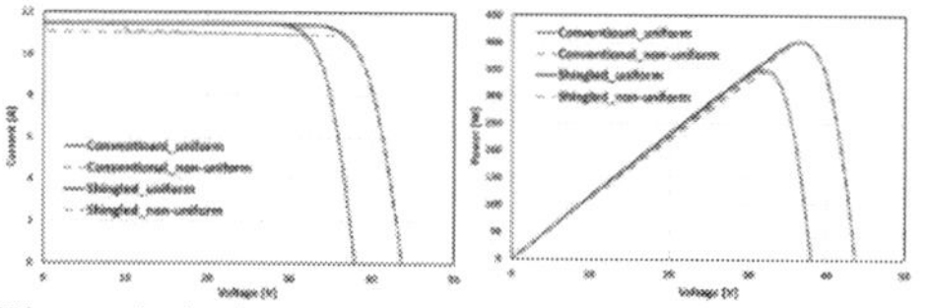

Figure 6: An instance of I-V and P-V characterization of conventional and shingled PV modules with non-uniformity of 5% and random distribution

3.3 PV-module performance under non-uniform irradiance with horizontal asymmetry distribution

It should be noted that in real situations, the distribution of irradiance non-uniformity on a module surface may not be purely random, but rather exhibit certain profiles. For example, solar simulators may produce non-random distributions due to differing lamp positions or degradation rates of lamps. When a horizontal asymmetry irradiance distribution was specifically modeled, the simulated I_{sc} and P_{max} error distributions differed from the random case, as shown in Figures 7 and 8. The simulation results showed that irradiance non-uniformity led to an underestimation of I_{sc} for the shingled PV module design, but overestimation for the other designs, which is highly related to the layout of the bypass diodes. Figure 9 shows the I-V and P-V curves of conventional and shingled PV modules under a horizontal asymmetry irradiance distribution profile. For conventional PV modules, a small step appears on the I-V curve, which is due to the overall irradiance differs for each substring, and the output current of each substring would be affected by the bypass diode. However, the step was so minimal that would not manifest as a visible feature on the smooth P-V curve. For shingled PV modules, the irradiance experienced by each substring group is relatively uniform under the horizontal-asymmetry irradiance pattern.

Consistent with the findings under random non-uniform irradiance, conventional, half-cut and third-cut PV modules performed similarly on I_{sc}, while the shingled PV module exhibited a greater absolute error in I_{sc} For relative magnitudes of P_{max} error, the conventional PV module exhibited the largest, followed by the half-cut module, then the third-cut module, with the shingled module demonstrating the lowest. The maximum absolute deviation of the median of P_{max} error was 0.033%, 0.124%, and 0.724% for the irradiance non-uniformity of 1%, 2%, and 5%. Under the worst-case scenario, compared to a conventional design, the maximum potential deviations of P_{max} error observed for innovative PV module designs reached up to 0.041%, 0.175%, and 1.005% under 1%, 2%, and 5% non-uniformity irradiance, respectively.

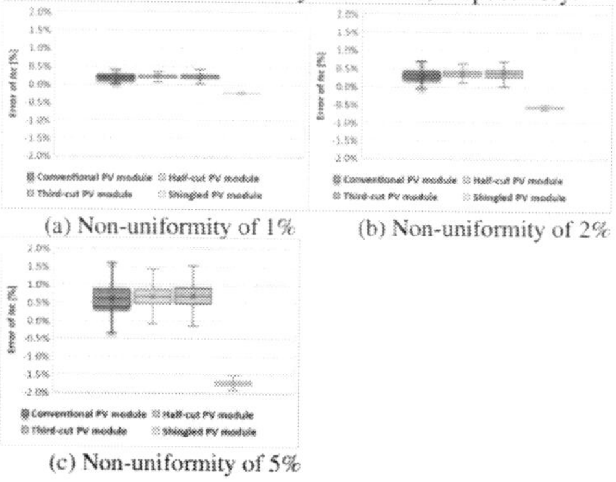

(a) Non-uniformity of 1% (b) Non-uniformity of 2%

(c) Non-uniformity of 5%

Figure 7: Simulation results of I_{sc} error distributions for horizontal asymmetry non-uniformity varying within 1-5% of four PV-module designs

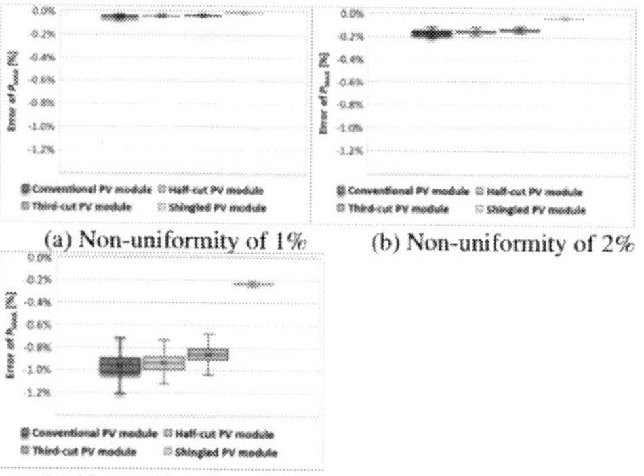

(a) Non-uniformity of 1% (b) Non-uniformity of 2%

(c) Non-uniformity of 5%

Figure 8: Simulation results of P_{max} error distributions for horizontal asymmetry non-uniformity varying within 1-5% of four PV-module designs

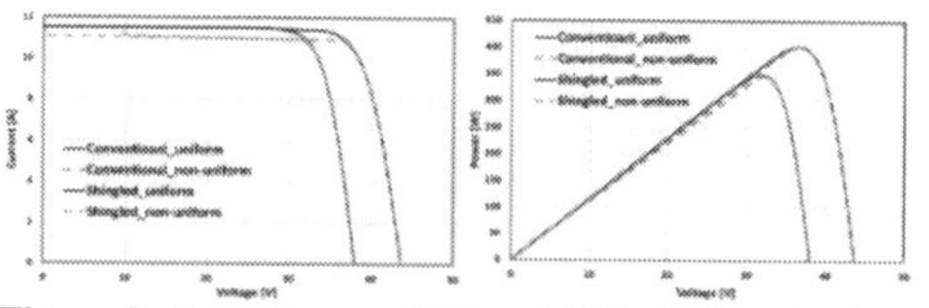

Figure 9: An instance of I-V and P-V characterization of conventional and shingled PV modules with non-uniformity of 5% and horizontal asymmetry distribution

3.3 PV-module performance under non-uniform irradiance with vertical asymmetry distribution

With the non-uniform irradiance with vertical asymmetry, the simulated I_{sc} and P_{max} errors are shown in Figures 10 and 11. In contrast to the case under a horizontal asymmetry irradiance distribution, the shingled PV module design demonstrated a positive I_{sc} error, whereas the other PV module designs exhibited negative I_{sc} error values. When the non-uniform irradiance distribution occurred in a left-to-right configuration, the irradiance became uniformly distributed across each substring for the half-cut, third-cut, and shingled PV module designs. However, for the shingled module configuration, the situation was reversed. As Figure 12 shows, a small step was observed on the I-V curve of the shingled PV module. Among four PV module designs, the conventional design demonstrated the greatest magnitude of absolute I_{sc} error.

For P_{max} error across four photovoltaic module designs, the conventional PV module exhibited the highest, followed by the shingled PV module design, then half-cut and third-cut PV module designs demonstrated the lowest. The maximum absolute deviation of the median of P_{max} error is 0.029%, 0.121%, and 0.71% for the irradiance non-uniformity of 1%, 2%, and 5%. Under the worst-case scenario, the maximum potential deviations of P_{max} error observed for innovative photovoltaic module designs, relative to a conventional design, were up to 0.042%, 0.175%, and 1.018% under 1%, 2%, and 5% non-uniformity irradiance, respectively.

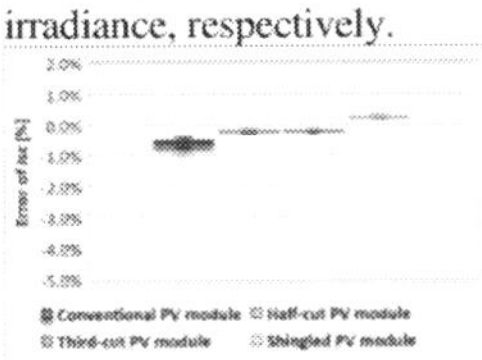

(a) Non-uniformity of 1%

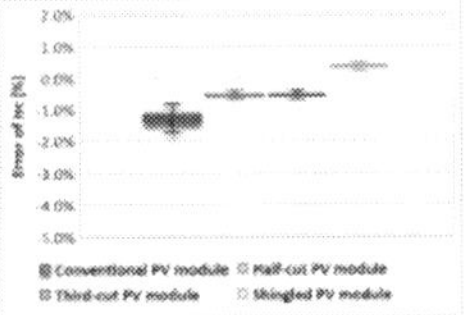

(b) Non-uniformity of 2%

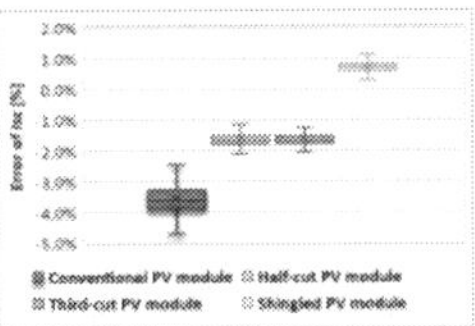

(c) Non-uniformity of 5%

Figure 10: Simulation results of I_{sc} error distributions for vertical asymmetry non-uniformity varying within 1-5% of four PV-module designs

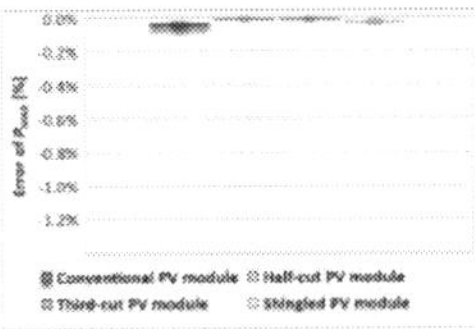

(a) Non-uniformity of 1%

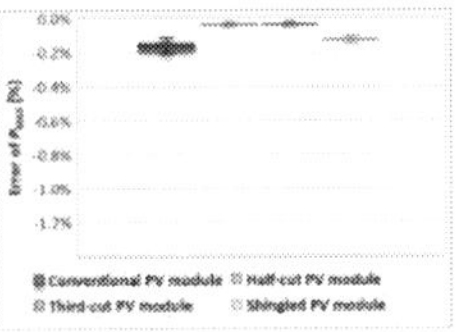

(b) Non-uniformity of 2%

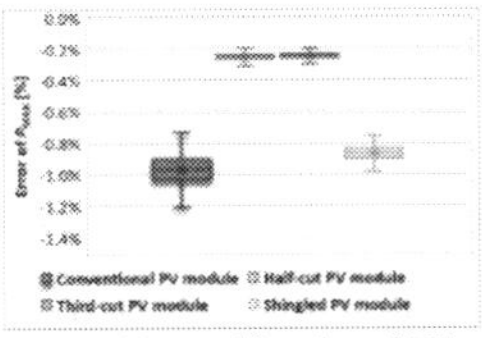

(c) Non-uniformity of 5%

Figure 11: Simulation results of P_{max} error distributions for vertical asymmetry non-uniformity varying within 1-5% of four PV-module designs

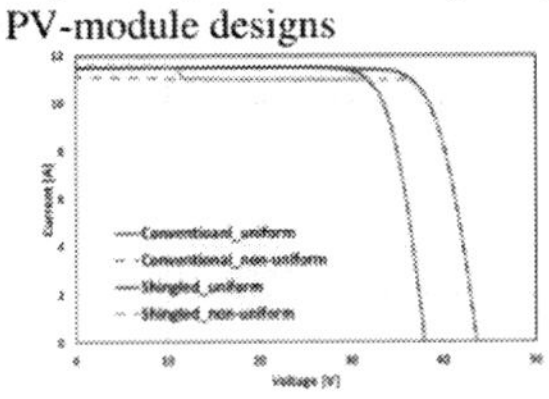

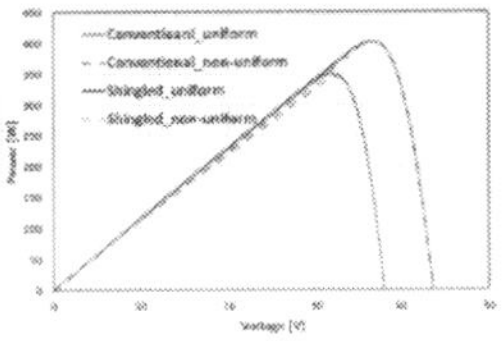

Figure 12: An instance of I-V and P-V characterization of conventional and shingled PV modules with non-uniformity of 5% and vertical asymmetry distribution

5 CONCLUSION

In this study, simulations were conducted to evaluate and compare the performance of conventional and innovative photovoltaic module designs under non-uniform irradiance conditions with varying non-uniformity magnitudes and irradiation profiles. Multiple factors were found to potentially influence the observed performance of photovoltaic modules, including the solar cell characteristics, module geometric layout, irradiance non-uniformity magnitudes, and irradiation profiles.

Under random non-uniform irradiance conditions, the I_{sc} and P_{max} errors in conventional PV module designs were approximately equivalent to those of innovative designs. The largest deviation of medians of I_{sc} and P_{max} errors across all designs was under 0.2% at irradiance non-uniformity of 5%.

With non-uniform irradiance exhibiting specific profiles, the observed deformation of I-V curves was found to be primarily dependent on the magnitude of irradiance variation across individual substrings within the PV modules. Under all irradiance conditions tested, the maximum power losses were consistently lower for the half-cut, third-cut, and shingled PV module designs compared to the conventional design. The maximum absolute deviation of the median of P_{max} error across the four PV modules is around 0.03%, 0.12%, and 0.70% for the irradiance non-uniformity of 1%, 2%, and 5%. Comparing the single P_{max} error between innovative and conventional PV modules, the worst case would be approximately 0.04%, 0.18%, and 1.02% given irradiance non-uniformity levels of 1%, 2%, and 5% respectively.

The reductions in I_{sc} and P_{max} error caused by irradiance non-uniformity should further minimize measurement uncertainty resulting from irradiance non-uniformity. This work assumed that the relationship in single-diode model parameters for different PV module architectures follows Ohm's law, neglecting losses caused by cell cutting and interconnecting. Given this precondition, the qualitative I_{sc} and P_{max} error distributions should be similar for PV module designs irrespective of the cell number.

6 REFERENCES

[1] ITRPV, International Technology Roadmap for Photovoltaic (ITRPV): Results 2022, VDMA e.V., Frankfurt 2022.

[2] Klasen, N, Weisser, D, Rößler, T, Neuhaus, DH, Kraft, A. Performance of shingled solar modules under partial shading. Prog Photovolt Res Appl. 2022; 30(4): 325-338.

[3] Oliver Kunz, Rhett J. Evans, Mattias K. Juhl, Thorsten Trupke, Understanding partial shading effects in shingled PV modules, Energy,Volume 202,2020,Pages 420-428.

[4] Hanifi, Hamed & Schneider, Jens & Bagdahn, Jörg.

(2015). Reduced shading effect on half-cell modules – measurement and simulation.
[5] Monokroussos C, Etienne D, Morita K, et al. Impact of calibration methodology into the power rating of c-Si PV modules under industrial conditions. 28th EUPVSEC, Paris, 2013: 2926-2934.
[6] W. Herrmann, W. Wiesner. Modelling of PV modules - the effects of non-uniform irradiance on performance measurements with solar simulators. 16th EUPVSEC, Glascow,2000.
[7] Ishaque K, Salam Z. An improved modeling method to determine the model parameters of photovoltaic (PV) modules using differential evolution (DE)[J]. Solar energy, 2011, 85(9): 2349-2359.
[8] IEC 60904-9: 2020 Photovoltaic devices-Part 9: Classification of solar simulator characteristics. International Electrochemical Committee, Geneva, Switzerland, 2020.

10.4229/EUPVSEC2023/3CO.10.5
020248-005

| Sebastian Dittmann

40th European Photovoltaic Solar Energy Conference and Exhibition

Advanced analysis of spectral and spatial non-uniformity of an LED sun simulator with double-side illumination for bifacial PV Module power rating

Sebastian Dittmann[1]
Giuliano L. Martins[2]
Hugo Sanchez[1]
Zonghan Jiang[1]
Ralph Gottschalg[1]

[2]Statkraft UK Ltd
[3]Fraunhofer CSP

[1]Anhalt Photovoltaic Performance and Lifetime Laboratory (APOLLO)

Photovoltaic research

T: +49 3496 67 2354
M: sebastian.dittmann@hs-anhalt.de

| Sebastian Dittmann

40th European Photovoltaic Solar Energy Conference and Exhibition

Outline

- Anhalt Photovoltaic Performance and Lifetime Laboratory (APOLLO)
- Motivation
- LED-Sun Simulator specifications
- Uncertainty sources
  - Spatial non-uniformity
  - Spectral non-uniformity
- Conclusion

| Sebastian Dittmann

40th European Photovoltaic Solar Energy Conference and Exhibition

Apollo Anhalt Photovoltaic Performance and Lifetime Laboratory

HOCHSCHULE ANHALT University of Applied Sciences

PHOTOVOLTAIC RESEARCH

APOLLO (Anhalt Photovoltaic Performance and Lifetime Laboratory)

Outdoor PV Laboratory:

- **PV module monitoring** to measure maximum power output in real operating conditions and different applications
- **30 kWp grid-connected PV system** consists of mono-polycrystalline and thin-film technologies [1]
- **Bifacial PV systems** with different module technologies and different orientations
- **AgriPVplus** demonstrator with vertical east/west orientation
- **High-precision** meteorological equipment for on-site measurement of environmental parameters

Indoor PV Laboratory:

- **LED sun simulator** with two independent light sources for **mono** and **bifacial** module characterization
- **Mobile LED sun simulator** to characterize PV modules in the field
- Tools for material characterization, e.g., **Raman Spectrometer**

PVCAMPER

TruePower™ Alliance

agriPV+

[1] one of the four test fields of the TruePower™ Alliance of SERIS

20.09.2023 40th European Photovoltaic Solar Energy Conference & Exhibition, Lisbon, Portugal 3

| Sebastian Dittmann

40th European Photovoltaic Solar Energy Conference and Exhibition

Apollo Anhalt Photovoltaic Performance and Lifetime Laboratory

HOCHSCHULE ANHALT University of Applied Sciences

Motivation

- Energy yield estimation and bankability of large PV systems rely on precise input parameters → low measurement uncertainty reduce financial risk
- Standard sun simulators, e.g., Xenon pulsed based were continuously improved (uncertainties of 1.7% with MMF, 2.5% w/o MMF)
- **New PV module technologies**, e.g., bifacial PV modules requires new or improved measurement methods or devices
  - IEC 6904-1-2, two different measurement methods (one side or Ge method, double-side)
    - Are they comparable? Measurement uncertainties?
  - **Novel measurement devices**, e.g., sun simulator based on light-emitting-diodes (LED)
    - No proper uncertainty calculation for LED sun simulators (light with high defuse light, multiple controlled LED channels)
    - No uncertainty calculation for the use of **two** simultaneously light sources

Aim

- Analyze and identify technology-specific uncertainty sources of a LED based sun simulator

20.09.2023 40th European Photovoltaic Solar Energy Conference & Exhibition, Lisbon, Portugal 4

| Sebastian Dittmann

40th European Photovoltaic Solar Energy Conference and Exhibition

LED-Sun Simulator

Light source and modifications

	Front-side	Rear-side
Dimensions of light source	1150 x 2100 mm	
Light source	7704 LEDs	5328 LEDs
Spectral range [nm]	350 – 1100 nm	400 – 1000 nm
LED-boards	18 with 428 LEDs each	18 with 296 LEDs each
No. of LED-channels	**each board has 36 individual powered LED channels**	
Cooling system	active, 20°C	passiv
Edge of light source	reflective mirrors	
Distance to PV module	25 cm	

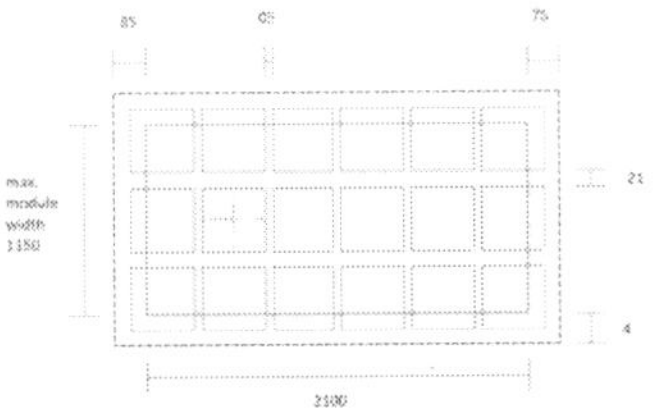

Dimension of the front- and rear-side light source, 18 LED-boards divided in four sub boards.

Modifications:

- Irradiance measurement for both light sources with calibrated WPVS reference cell connected via a trans-impedance amplifier
- Irradiance correction of the I-V curve
- I-V sweep-time: **500 ms** for monofacial and **80 ms** for bifacial modules*
- Module temperature 25 ± 2°C
- Participation in an international RR for bifacial PV Module measurement [6, G. Koutsourakis et. al]

*limitation for bifacial PV module measurement due to passive cooling of rear-side light source

| Sebastian Dittmann

40th European Photovoltaic Solar Energy Conference and Exhibition

Uncertainty sources

Two light-source – bifacial measurement

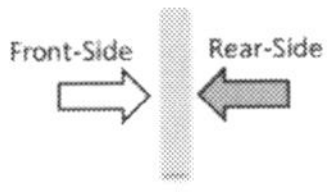

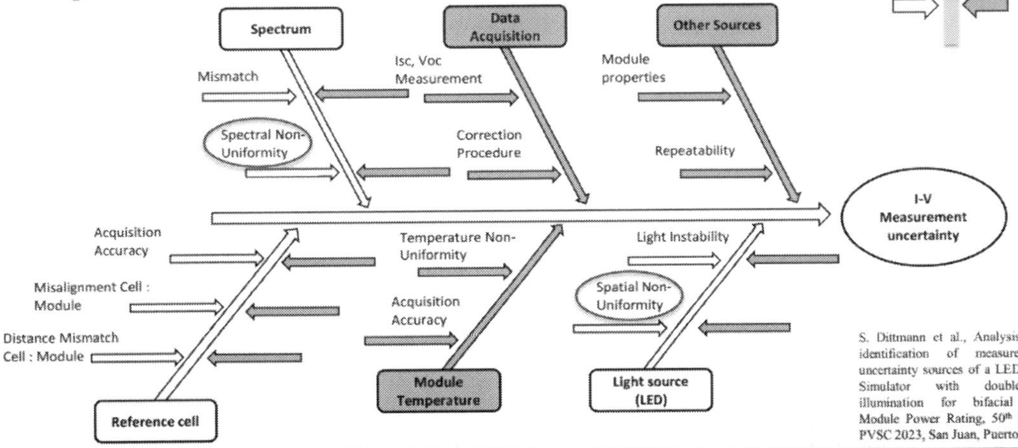

S. Dittmann et al., Analysis and identification of measurement uncertainty sources of a LED Sun Simulator with double-side illumination for bifacial PV Module Power Rating, 50th IEEE PVSC 2023, San Juan, Puerto Rico

Spatial Non-Uniformity
Methodology

1. Measurements with a **uniformity PV module**
   - 72 cells (6') with independent Isc measurements
   - Two different modules of the same type and calibration institution
     - Manufacturer
     - APOLLO

2. Measurement with a WPVS **reference cell** (smaller detector)
   - calibrated WPVS reference (2x2cm)
   - Measurement points:
     - a) 18 point, each in the center of the board
     - b) 220 points, one measurement for each sub-board across module plane

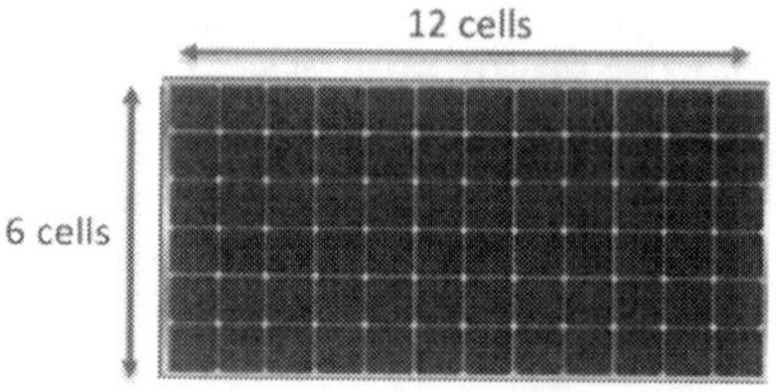

Uniformity module with independent Isc measurement

Uniformity measurement with a WPVS reference cell, 18 and 220 points

Spatial Non-Uniformity
Uniformity module

MEASURED SPATIAL UNIFORMITY WITH A UNIFORMITY MODULE, VALUES IN %

Irradiance [W/m²]	Front side		Rear side
	Manufacture	APOLLO	APOLLO
1000	1.72	2.13	4.02
800	1.59	2.41	3.02
600	1.63	2.66	3.58
400	1.54	2.44	3.98
200	2.33	4.64	3.89
100	2.33	12.46	6.66

→ Same measurement method leads to different results

Front-side at 1000 W/m²

Manufacturer

APOLLO

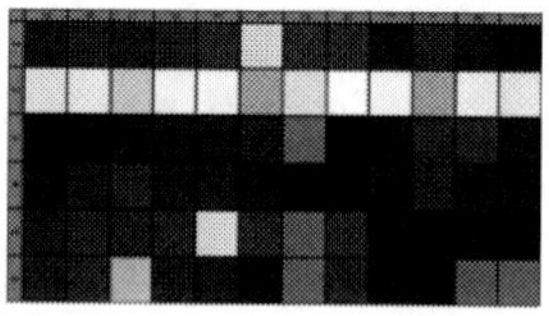

Rear-side at 1000 W/m²

Manufacturer

N/A

APOLLO

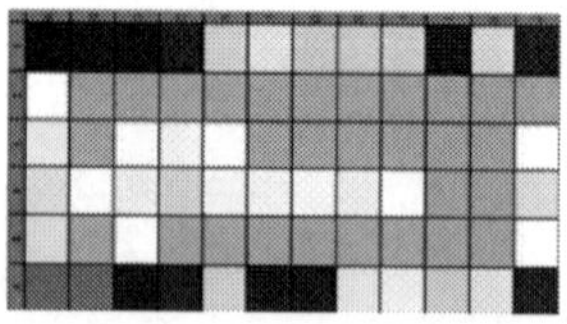

| Sebastian Dittmann

40th European Photovoltaic Solar Energy Conference and Exhibition

Spatial Non-Uniformity

WPVS reference cell - 18 measurement points

LED board with sub-boards

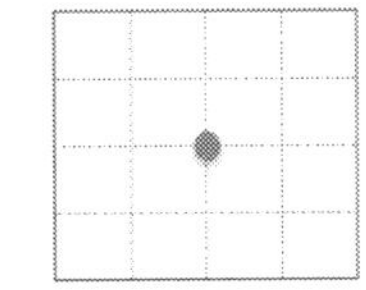

- Detector 2cm x 2cm
- Measurements at the center of the LED boards
- 18 measurement in total

Front-side at 1000 W/m²

	1	2	3	4	5	6
1	1034.8	1028.5	1051	1033.9	1041.9	1035.5
2	987.7	988.6	997.2	978.1	1001.3	984.8
3	1019.5	1017.4	1023.2	1020.9	995.1	1002.7

Non-Uniformity at 1000W/m²

- Front-side: 3.42%
- Rear-side: 5.98%

Non-Uniformity at 200W/m²

- Front-side: 5.20%
- Rear-side: 5.85%

→ Increase of non-uniformity with smaller detector

Rear-side at 1000 W/m²

	1	2	3	4	5	6
1	1030.7	1006.2	1006.5	997.4	987.2	973.2
2	1039.9	1025.7	1009.6	992.8	1007.9	970.5
3	993.2	992.2	979.9	976.4	956.4	914.6

20.09.2023 40th European Photovoltaic Solar Energy Conference & Exhibition, Lisbon, Portugal 9

| Sebastian Dittmann

40th European Photovoltaic Solar Energy Conference and Exhibition

Spatial Non-Uniformity

WPVS reference cell - 220 measurement points

LED board with sub-boards

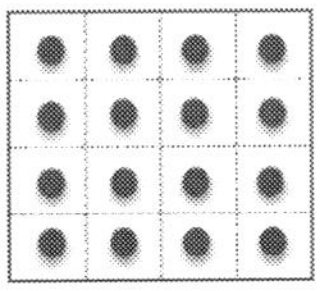

- Detector 2cm x 2cm
- Measurements at center of each LED sub-board
- 220 measurements in total

Front-side at 1000 W/m²

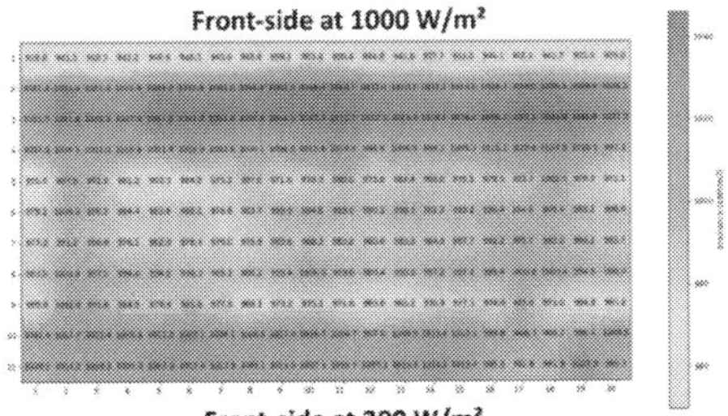

Front-side at 200 W/m²

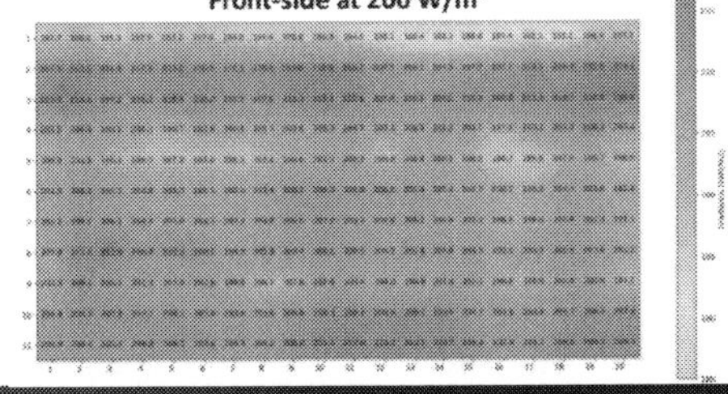

Non-Uniformity at 1000W/m²

- Front-side: 6.95%
- Rear-side: 6.77%

Non-Uniformity at 200W/m²

- Front-side: 6.98%
- Rear-side: 6.85%

→ Increase of non-uniformity with smaller distance

20.09.2023 40th European Photovoltaic Solar Energy Conference & Exhibition, Lisbon, Portugal 10

| Sebastian Dittmann

40th European Photovoltaic Solar Energy Conference and Exhibition

Apollo Anhalt Photovoltaic Performance and Lifetime Laboratory

HOCHSCHULE ANHALT University of Applied Sciences

Spatial Non-Uniformity
Summary

MEASURED SPATIAL UNIFORMITY WITH A UNIFORMITY MODULE, VALUES IN %

Irradiance [W/m²]	Uniformity module manufacturer	Uniformity module APOLLO	WPVS cell 18 pionts	WPVS cell 220 pionts
	Front side			
1000	1.72	2.13	3.42	6.95
200	2.33	4.64	5.20	6.77
	Rear side			
1000	-	4.02	5.98	6.98
200	-	3.89	5.85	6.85

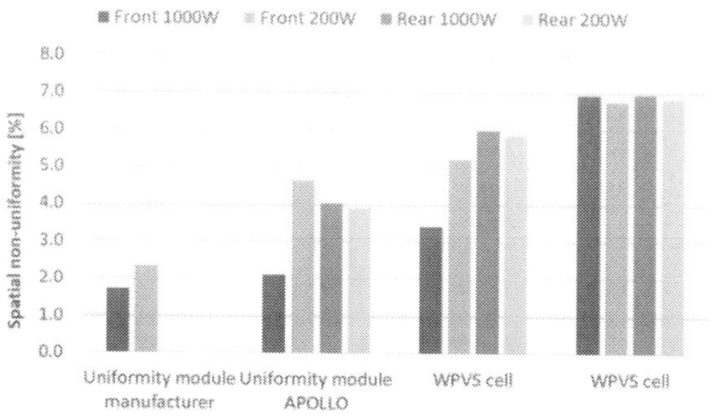

→ Size of used detector influence the measurement, large detector might not be sufficient for LED sun simulators

→ Uncertainty of measurement method needs to be considering in overall uncertainty budget

| Sebastian Dittmann

40th European Photovoltaic Solar Energy Conference and Exhibition

Apollo Anhalt Photovoltaic Performance and Lifetime Laboratory

HOCHSCHULE ANHALT University of Applied Sciences

Spectral Non-Uniformity
Classification

MEASURED SPECTRAL MATCH AND CORRESPONDING IEC CLASSIFICATION

Spectral range [nm]	Front side		Rear side	
	Spectral match	IEC Class	Spectral match	IEC Class
400-500	0.70	B	0.88	A
500-600	0.79	A	0.82	A
600-700	0.83	A	0.84	A
700-800	0.74	B	0.79	A
800-900	0.56	C	0.72	B
900-1100	0.45	C	0.99	A

- **Differences in Mismatch Factor between Class A Xe and the LED based simulator**
- Front-side: 0.5% or Rear-side: 4 %

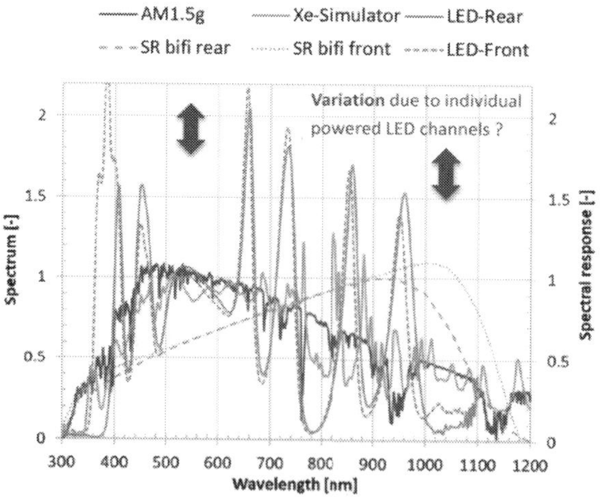

| Sebastian Dittmann

40th European Photovoltaic Solar Energy Conference and Exhibition

Spectral Non-Uniformity

Spectrum at 220 measurement points

LED board with sub-boards

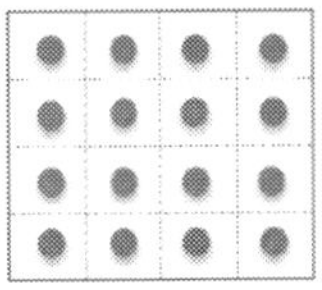

- Measurements at center of each LED sub-board
- 220 measurements in total

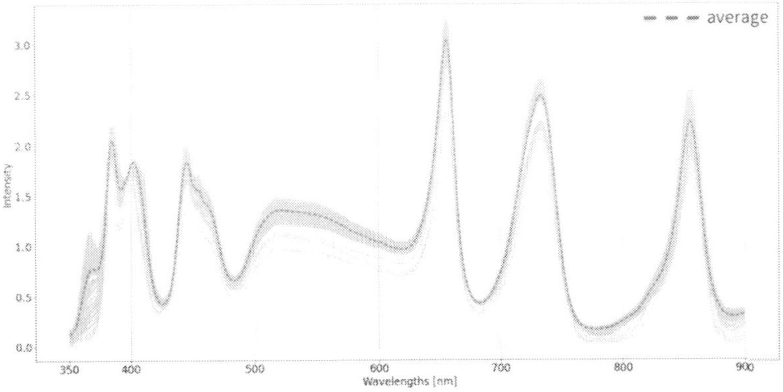

| Sebastian Dittmann

40th European Photovoltaic Solar Energy Conference and Exhibition

Spectral Non-Uniformity

Spectrum at 220 measurement points

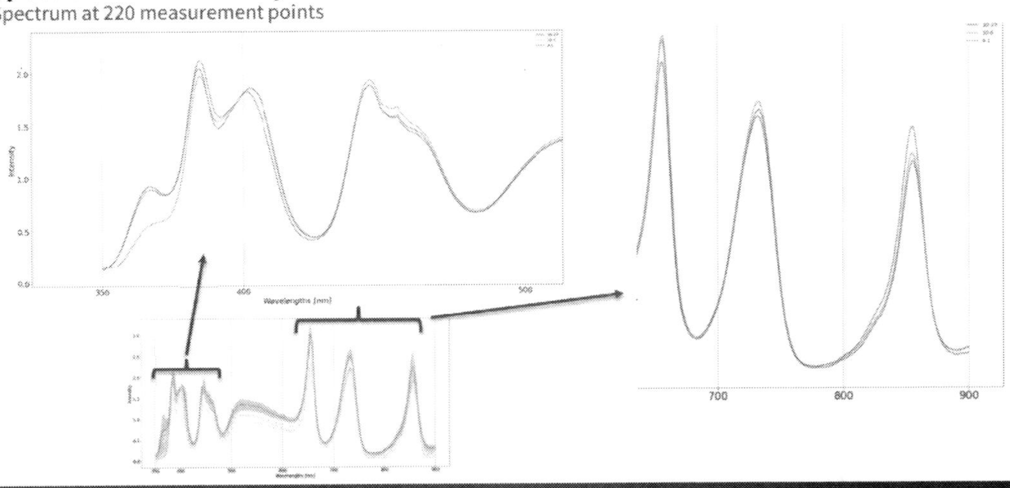

Spectral Non-Uniformity

Spectrum at 220 measurement points

350-400nm 23%

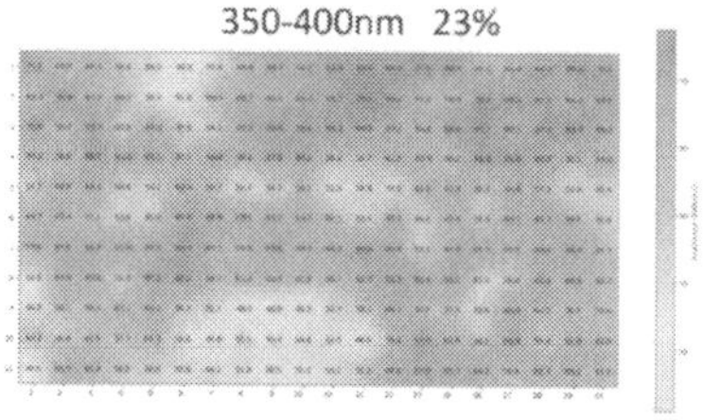

400-500nm 8%

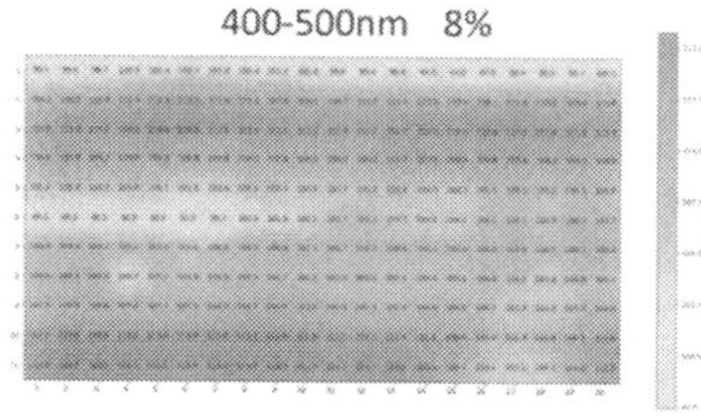

500-600nm 9%

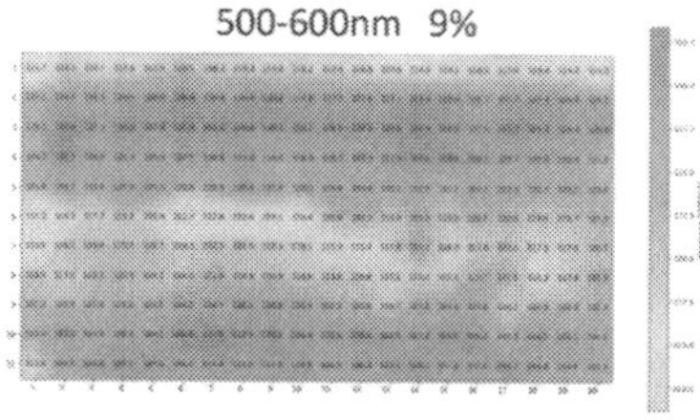

700-800nm 11%

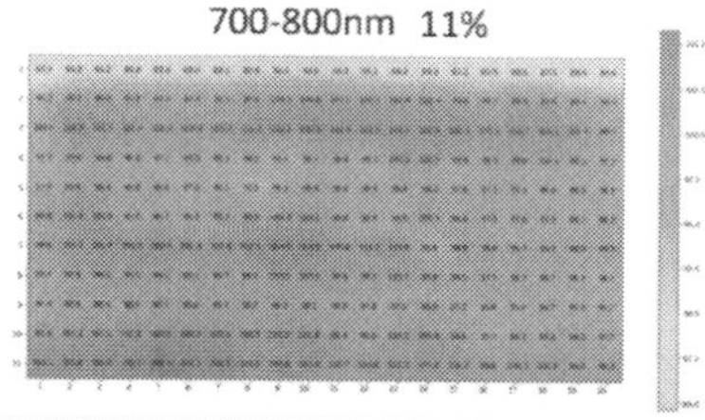

Overview of uncertainties

Front-, rear- and double side

Standard uncertainty component	Source	Distribution	Front side @1000 W/m²	Rear side @1000 W/m²	Double-sided @1000/200 W/m²
Solar Simulator - Data Acquisition					
Data acquisition error - Voltage	Manufacturer	Gaussian	± 0.03 %	± 0.03 %	± 0.03 %
Data acquisition error - Current	Manufacturer	Gaussian	± 0.03 %	± 0.03 %	± 0.03 %
Light Source (LED)					
Front side - non-Stability	Measurement	Rectangular	± 1.15 %	-	± 2.08 %
Front side - Spatial Non-Uniformity	Measurement	Rectangular	± 2.42 %	-	± 3.64 %
Rear side - non-Stability	Measurement	Rectangular	-	± 0.43 %	± 0.43 %
Rear side - Spatial Non-Uniformity	Measurement	Rectangular	-	± 5.00 %	± 5.00 %
Module Temperature					
Acquisition Accuracy	Manufacturer	Rectangular	± 0.34 %	± 0.34 %	± 0.34 %
Temperature Non-Uniformity[a]	Literature	Rectangular	± 0.67 %	± 0.67 %	± 0.67 %
Measurement Temperature	Manufacturer	Rectangular	± 0.67 %	± 0.67 %	± 0.67 %
Spectrum					
Spectral mismatch - Reference Cell		Gaussian	± 1 %	-	± 1 %
Spectral mismatch - Mini Module	Measurement	Gaussian	-	± 3%	± 3 %
Reference cells					
Acquisition Accuracy (USB 4704)	Manufacturer	Gaussian	± 0.10 %	± 0.10 %	± 0.10 %
Calibration - Mini Module	Measurement	Gaussian	-	± 1.00 %	± 1.00 %
Calibration - ISE Reference Cell	Manufacturer	Gaussian	± 0.10 %	-	± 0.10 %
Angle misalignment - Module and reference device	Literature	Rectangular	± 3.00 °	± 3.00 °	± 3.00 °
Distance Mismatch - Module and reference device	Measurement	Rectangular	± 1.00 %	± 1.00 %	± 1.00 %

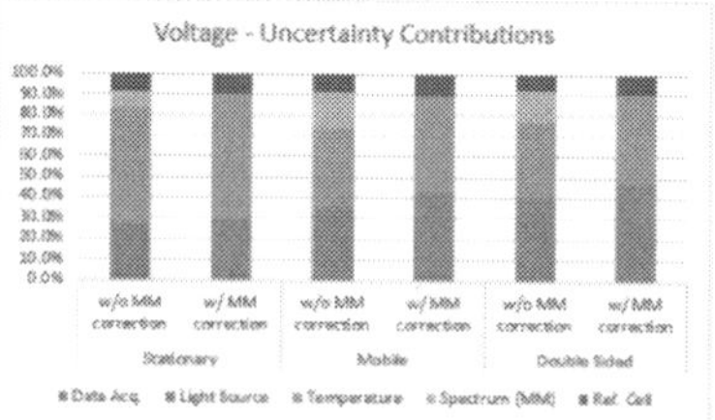

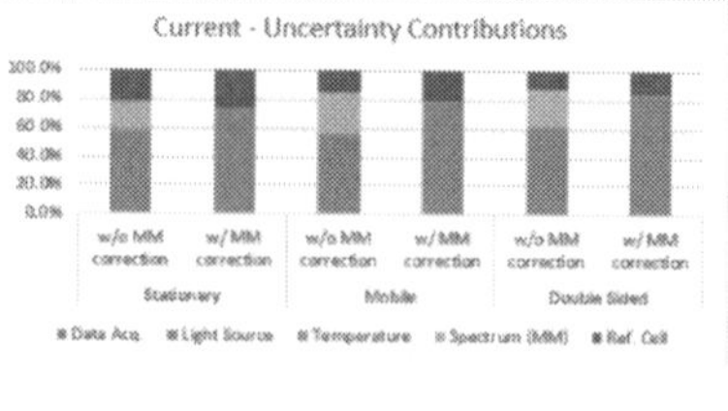

Expanded uncertainty

Front-, rear- and double side

EXPANDED UNCERTAINTY (K=2) WITHOUT MISMATCH CORRECTION

Component	Front side @1000 W/m²	Rear side @1000 W/m²	Double-sided @1000/200W/m²
Current (I)	± 3.5 %	± 6.7 %	± 8.3 %
Voltage (V)	± 0.4 %	± 0.5 %	± 0.6 %
Power (P)	± 3.5 %	± 6.7 %	± 8.4 %

IEC 60904-9 classification

Irradiance [W/m²]	front-side	rear-side
1000	CBB	BBB
800	CBB	BBB
600	CBB	BBC
400	CBB	BBC
200	CBC	BBC
100	CCC	BCC

IEC classification at 1000 W/m²:

- Front-side light source: class CBB
- Rear-side light source: class BBB
- increase in uncertainty for low irradiances below 400 W/m²,
  - important for bifacial PV modules, where rear-irradiance conditions are mainly below 200 W/m²

Conclusion

- Detail analysis of the spatial and spectral non-uniformity and compared different measurement methods
- **spectral-** and **spatial non-uniformity** are critical uncertainty sources due to independent powered LED channels
- Size of used detector influence the measurement, large detector might not be sufficient for LED sun simulators
- Uncertainty of non-uniformity measurement method needs to be considering in overall uncertainty budget

Future work

- What are the influences of the spectral non-uniformity on I-V measurements?

| Sebastian Dittmann

40th European Photovoltaic Solar Energy Conference and Exhibition

Apollo Anhalt Photovoltaic Performance and Lifetime Laboratory

HOCHSCHULE ANHALT University of Applied Sciences

References

[1] D. Dirnberger, U. Kräling, "Uncertainty in PV Modules Measurement – Part I: Calibration of Crystalline and Thin-Film Modules, IEEE Journal of Photovoltaics, Vol. 3, No. 3, 2013

[2] H. Müllejans, W. Zaaiman, and R. Galleano, "Analysis and mitigation of measurement uncertainties in the traceability chain for the calibration of photovoltaic devices," Meas. Sci. Technol., vol. 20, no. 7, p. 75101, 2009.

[3] K. Emery, "Uncertainty analysis of certified photovoltaic measurements at the national renewable energy laboratory," Nat. Renewable Energy Lab. (NREL), Golden, CO, USA, NREL/TP-520-45299, Aug. 2009.

[4] Photovoltaic Devices— Part 1-2: Measurement of current-voltage characteristics of bifacial photovoltaic (PV) devices, Int. Electrotech. Comm. (IEC), Geneva, Switzerland, IEC 60904-1 Ed. 1.0, 2019

[5] C. Monokroussos, Harald Müllejans, S. Dittmann, et al., "Electrical characterization intercomparison of high-efficiency c-Si modules within Asian and European laboratories", Progress in Photovoltaics, 2019

[6] G. Koutsourakis, M. Rauer, T. R. Betts, J. C. Blakesley, S. Dittmann, W. Herrmann, S. Riechelmann, et al., "Results of the Bifacial PV Cells and PV Modules Power Measurement Round Robin Activity of the PV-Enerate Project,", 2020

[7] S. Dittmann, G. Luchetta, R. Gottschalg, et al., "Advanced analysis of spectral and spatial non-uniformity of an LED sun simulator with double-side illumination for bifacial PV Module power rating", 2023, 40th EUPVSEC, Lisbon

| Sebastian Dittmann

40th European Photovoltaic Solar Energy Conference and Exhibition

Apollo Anhalt Photovoltaic Performance and Lifetime Laboratory

HOCHSCHULE ANHALT University of Applied Sciences

Thank you for your attention

Photovoltaic Research:
sebastian.dittmann@hs-anhalt.de

Anhalt Photovoltaic Performance and Lifetime Laboratory
Bernburger Str. 55
06366 Köthen (Anhalt)

Follow us on LinkedIn

Apollo Anhalt Photovoltaic Performance and Lifetime Laboratory

agriPV+

DTU

EU PVSEC

Impact of Coloured Interlayers on Modelling Module Temperature of Vertically Mounted Glass-Glass BIPV Modules

Markus Babin, Ingvar H. Jóhannsson, Michael L. Jakobsen, Sune Thorsteinsson

EUPVSEC 2023 – 3CO.11.1

DTU

Introduction

Coloured PV:

- Technology variations
  - Absorptive colours
  - Structural colours
- Performance impact:
  - Transmission losses
  - Angular-dependent reflection losses (IAM)
  - **Module temperature**

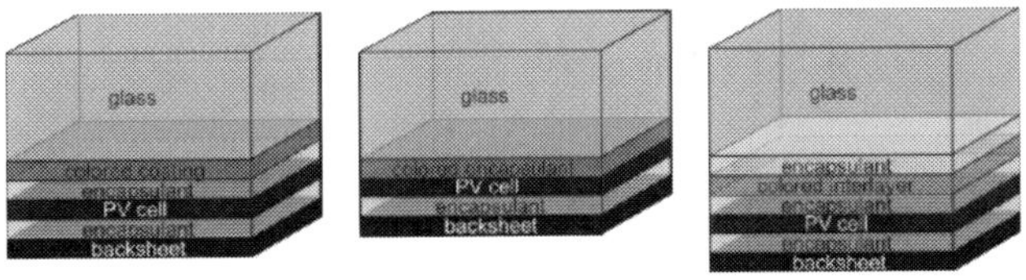

Agenda:

- BIPV test site at DTU
- Temperature measurements
- Recommendations for temperature modelling

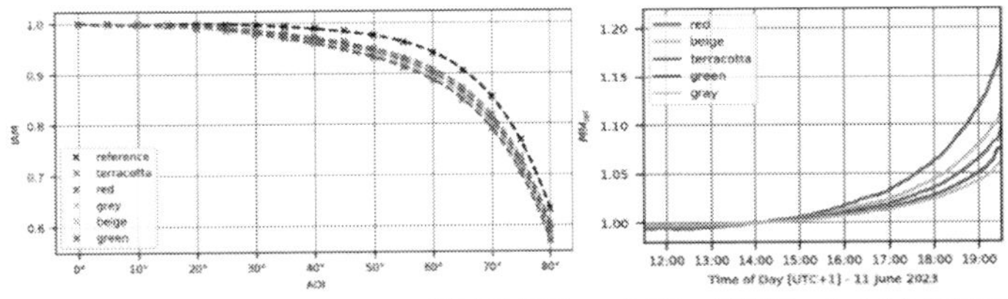

4BO.16.6: IAM Losses in Colored BIPV: From the Lab to the Field

Modelling PV performance:

- Sequential modelling chain
- Multiple options for each modelling step
  - Varying complexity and accuracy
- Coloured BIPV
  - Optical losses ← Colouration
  - Temperature models ← Mounting
- Effective irradiance
  - Different definitions in different models/tools
  - Irradiance on PV cell surface
  - Accounts for shading, soiling, reflectance, IAM, spectral mismatch, …
- Temperature models
  - Depend on irradiance and wind speed
  - Steady-state models more common than transient models

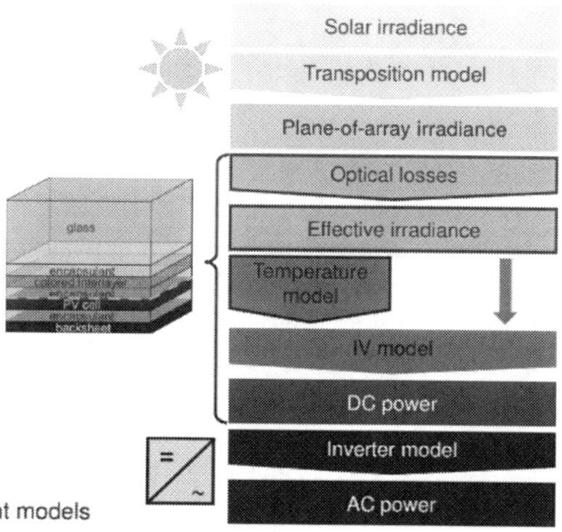

BIPV test site at DTU Risø campus

- Curtain wall system
  - Adapted office container
  - Glass-glass PV modules
  - Satinated glass
  - Coloured interlayers
  - No ventilation on back
- Continuous monitoring
  - G_{POA} pyranometer (10 sec)
  - Power optimizers (5 min)
  - Back-of-module (10 sec) temperature sensors
  - Nearby weather station for DNI, DHI, etc (10 sec)

DTU

BIPV test site at DTU Risø campus

Single-cell coupon test site
- Similar sample construction to modules in façade
  - Standard PV glass (instead of satinated)
  - Integrated thermocouples (back of cell)
- Vertical airflow behind modules (instead of insulated)
- Samples in short-circuit (instead of at MPP)

STC performance characterisation

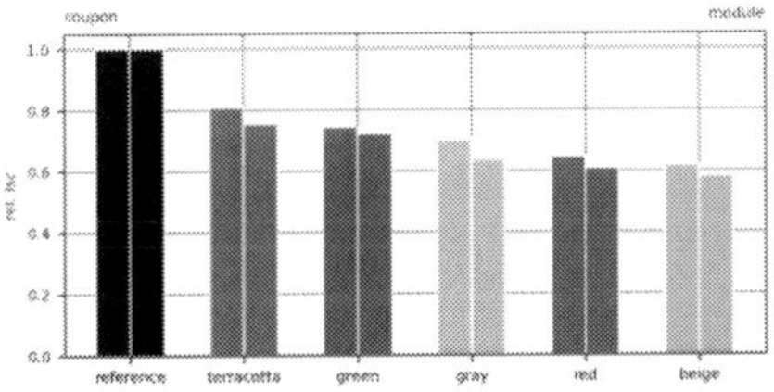

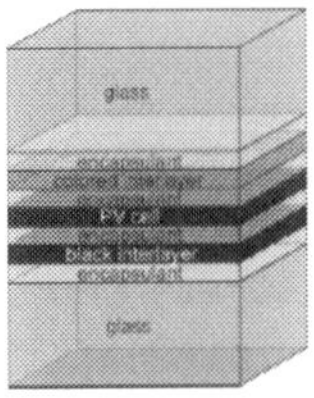

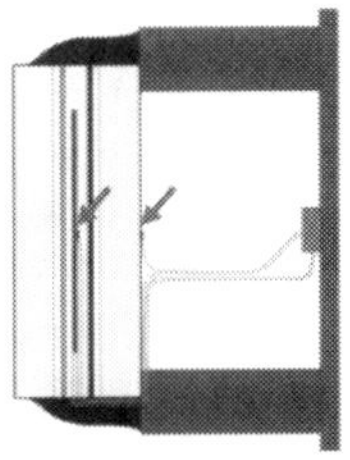

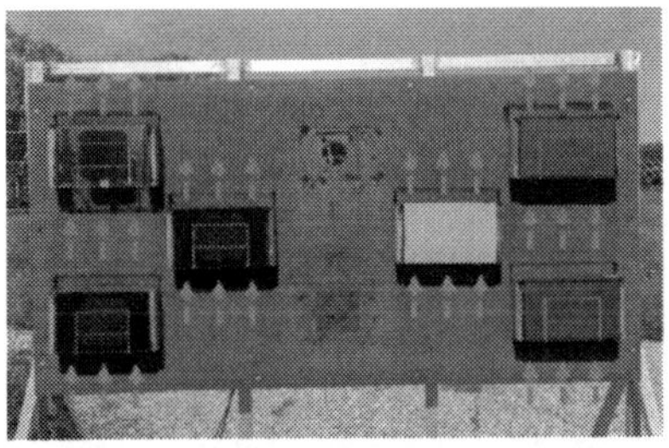

DTU

Monitored temperature – clear sky day

Time delay in heating/cooling
- Higher in façade
- High thermal capacity
- Reduced ventilation

Colour dependency
- Lower temperatures for light colours
- Similar temperatures for dark colours despite optical losses

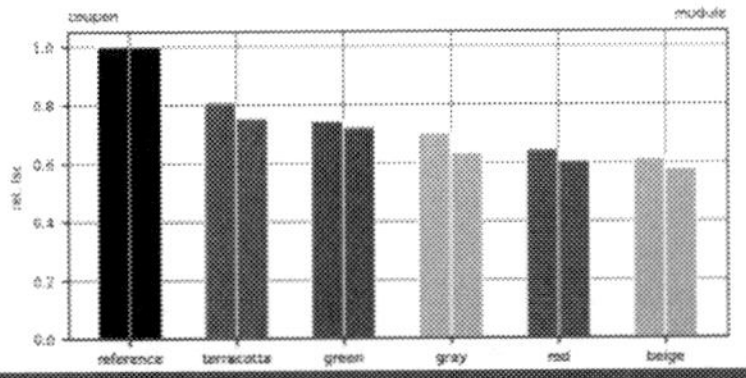

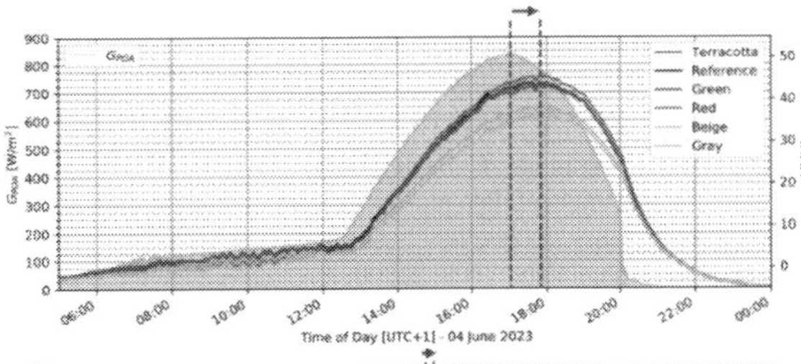

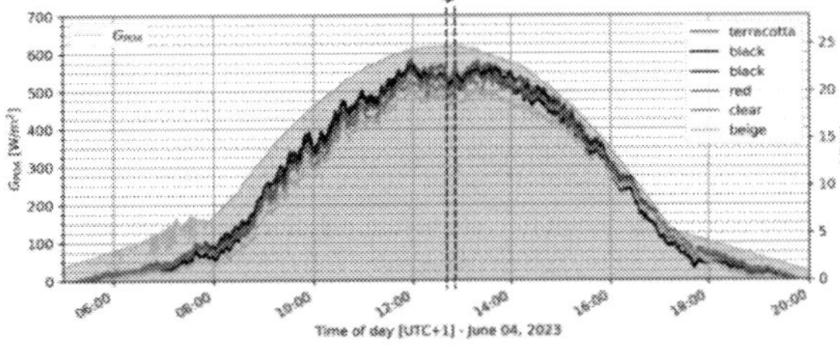

Modelling temperature

Ross temperature coefficients

- Does not consider wind
- Mainly influenced by mounting configuration
- Differences according to colour:
  - Dark colours: 43.8 – 46.2 $\left[K/\frac{kW}{m^2}\right]$
  - Light colours: 36.6 – 37.0 $\left[K/\frac{kW}{m^2}\right]$

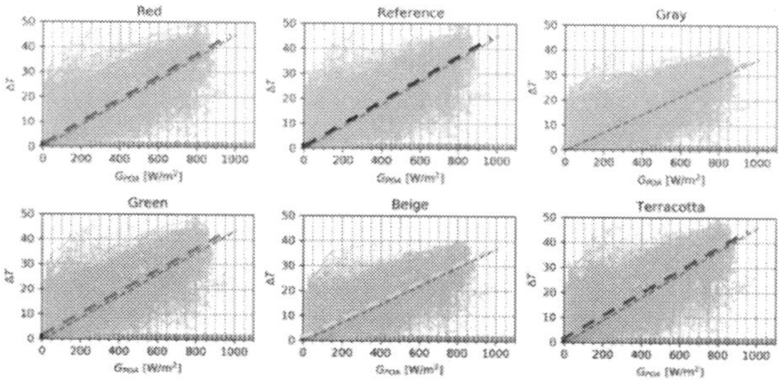

40th European Photovoltaic Solar Energy Conference and Exhibition

Modelling temperature

Relative temperature difference to ambient

- Light colours ~15% lower than reference and dark colours
- No good correlation with total optical losses

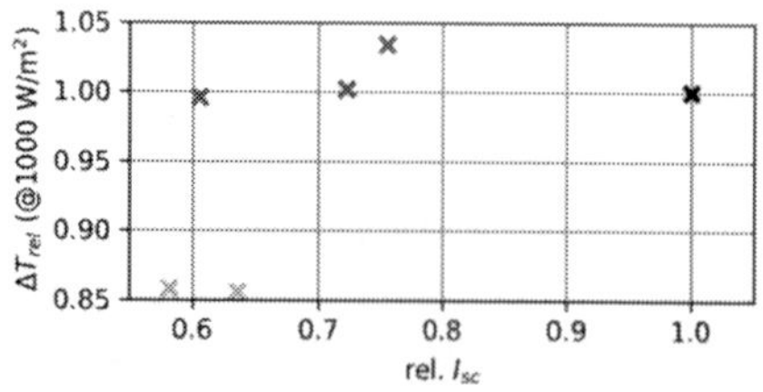

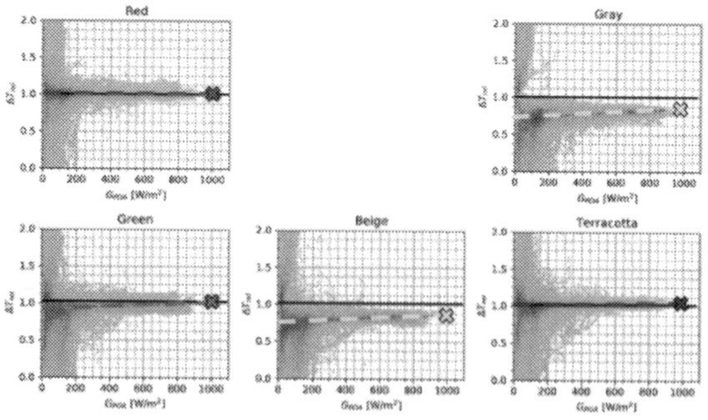

DTU

Spectral reflectance

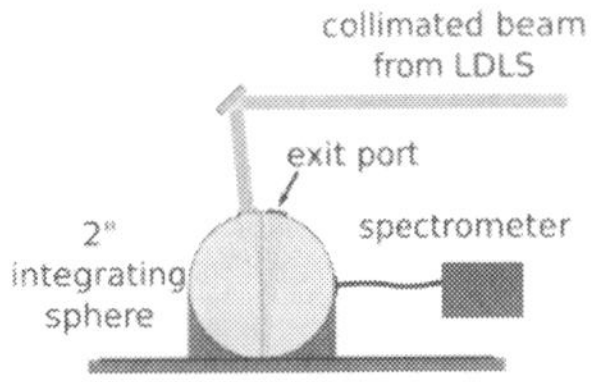

Spectral total reflectance measurements
- Significantly higher for gray and beige samples

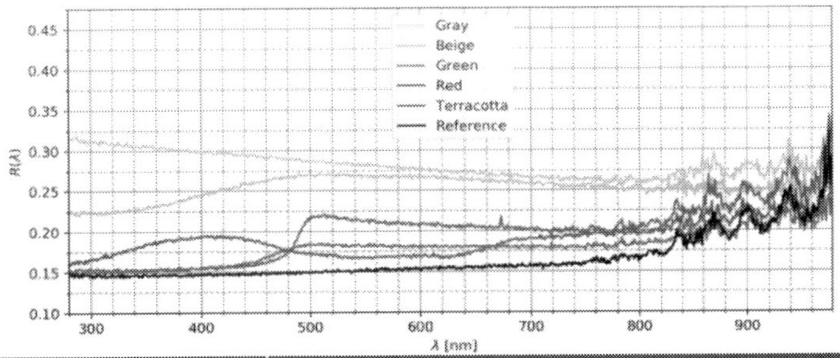

DTU

Temperature correlation with reflectance

Weighted total reflectance:

$$R' = \frac{\int R(\lambda)\cdot G_{AM1.5}(\lambda)\cdot SR(\lambda)\, d\lambda}{\int G_{AM1.5}(\lambda)\cdot SR(\lambda)\, d\lambda}$$

Relative effective irradiance:

$$G'_{rel} = \frac{1-R'}{1-R'(ref)}$$

Light absorbed in coloured layer contributes significantly to module heating

Light reflected in coloured layers does not contribute to module heating

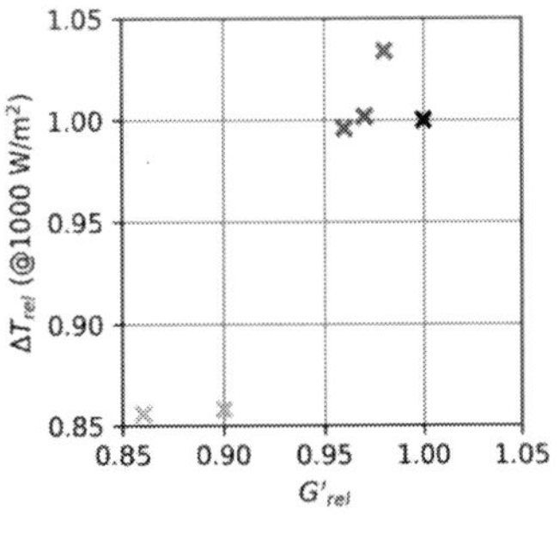

DTU

Modelling recommendations:

Include absorbed irradiance in temperature modelling:
- Effective irradiance (after reflections)

Exclude absorbed irradiance in electrical modelling:
- Effective irradiance (at cell level)

Alternative:
- Ross temperature model with module specific coefficients
- Combination of mounting configuration and colour
- INOCT, eff. NOCT

No simple way to handle thermal capacitance of poorly ventilated BIPV façades
- Transient models instead of steady-state models

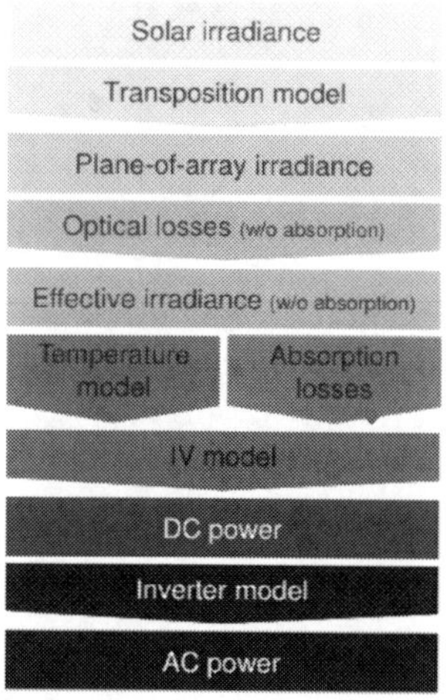

DTU

Conclusions

BIPV test installation with coloured interlayers

Insulated façade shows significant thermal delay

Light absorption in coloured BIPV affect thermal models significantly
- Not directly correlated with total optical losses
- Recommended to separate reflection and absorption losses in coloured BIPV modelling

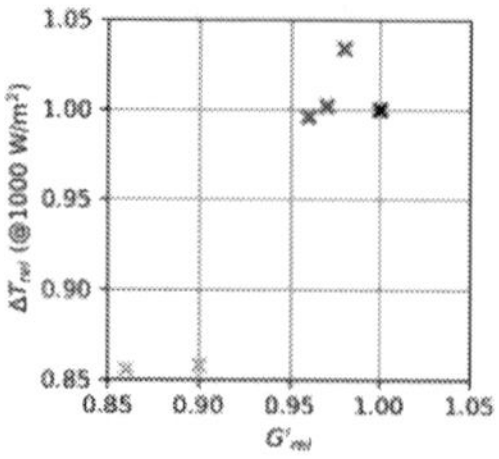

Supported by:
EUDP
The Energy Technology Development and Demonstration Programme
64021-1079 UnitSun is supported by the Energy Technology development and Demonstration Program (EUDP)

Thank you!
Ingvar H. Jóhannsson
Michael L. Jakobsen
Sune Thorsteinsson
SPVS group
UnitSun project partners

Contact Information
Markus Babin
marbab@dtu.dk
DTU Electro
Frederiksborgvej 399
4000 Roskilde
Denmark

Article expected be published in EPJ Photovoltaics later this year.

DTU

020250-013

020251-001

40th European Photovoltaic Solar Energy Conference and Exhibition

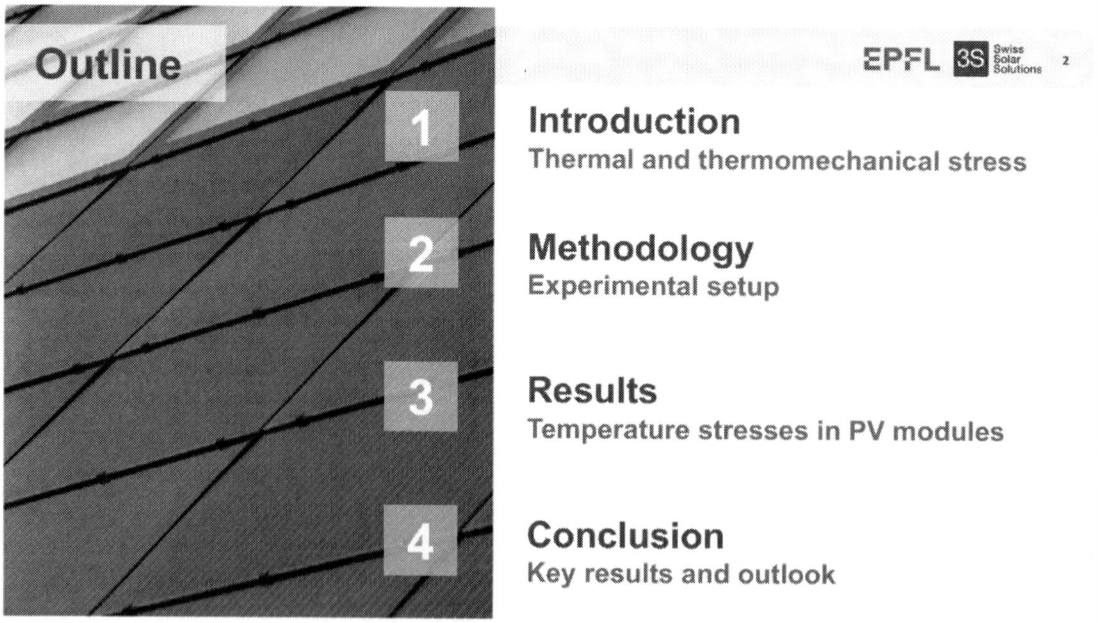

020251-002

Outline

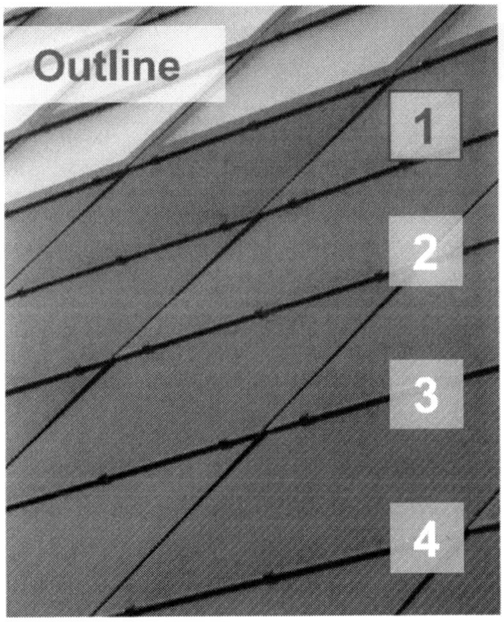

EPFL 3S Swiss Solar Solutions 3

1 **Introduction**
Thermal and thermomechanical stress

2 **Methodology**
Experimental setup

3 **Results**
Temperature stresses in PV modules

4 **Conclusion**
Key results and outlook

020251-003

40th European Photovoltaic Solar Energy Conference and Exhibition

1 | Introduction

Literature on thermal performance of PV modules

EPFL 3S Swiss Solar Solutions 4

Building-integrated PV (with reduced ventilation) are exposed to **higher temperatures**.

Harsher operating conditions can translate to **accelerated degradation mechanisms**.

What about temperature variations ?

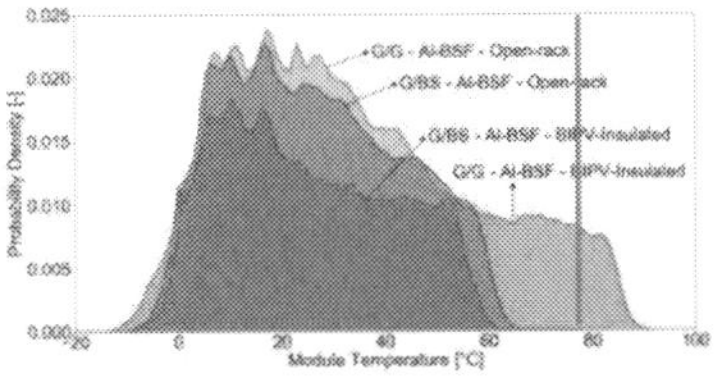

Solar Energy

Modelling the performance of amorphous and crystalline silicon in different typologies of building-integrated photovoltaic (BIPV) conditions

PROGRESS IN PHOTOVOLTAICS

Research Article | Full Access

Evaluation of high-temperature exposure of photovoltaic modules

Operating Temperatures and Diurnal Temperature Variations of Modules Installed in Open-Rack and Typical BIPV Configurations

Applied Energy

BIPV-temp: A demonstrative Building Integrated Photovoltaic installation

IEC TS 63126

020251-004

1 | Introduction

Thermal vs. thermomechanical stress

EPFL 3S Swiss Solar Solutions 5

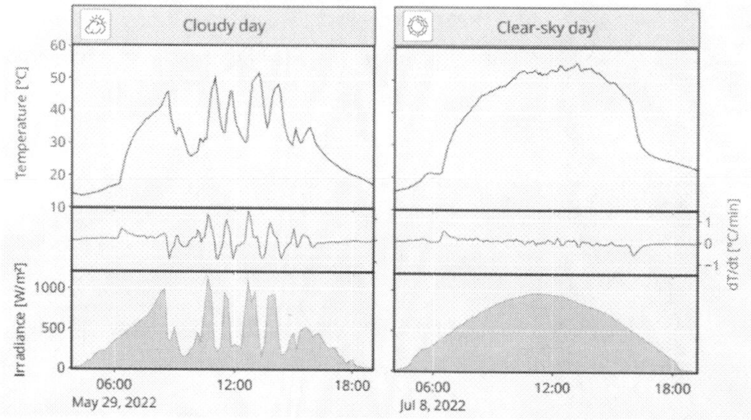

Thermal stress

exposure to high temperature

- Encapsulant degradation
- Cell damage
- Safety hazards

Thermomechanical stress

Thermal coefficient mismatch during temp. variations

- Solder joint fatigue
- Interconnect damage
- Cell cracking

020251-005

Outline

EPFL 3S Swiss Solar Solutions 6

1 Introduction
Thermal and thermomechanical stress

2 **Methodology**
Experimental setup

3 Results
Temperature stresses in PV modules

4 Conclusion
Key results and outlook

020251-006

2 | Methodology
Experimental setup

BE SMART

EPFL 3S Swiss Solar Solutions 7

- Single-cell mini modules
- 36 configurations
- One full year of data (Dec. 2021 – Dec. 2022)
- Neuchâtel, CH

Sensor location	Glass thickness	Insulation	Inclination
Back-of-cell	3.2 mm	Insulated	0°
Back-of-module	6 mm	Close-roof	45°
		Open-rack	90°

020251-007

Outline

EPFL 3S Swiss Solar Solutions 8

1 Introduction
Thermal and thermomechanical stress

2 Methodology
Experimental setup

3 **Results**
Temperature stresses in PV modules

4 Conclusion
Key results and outlook

020251-008

3 | Results
Thermal stress

98th percentile temperature (T98)

EPFL 3S Swiss Solar Solutions 9

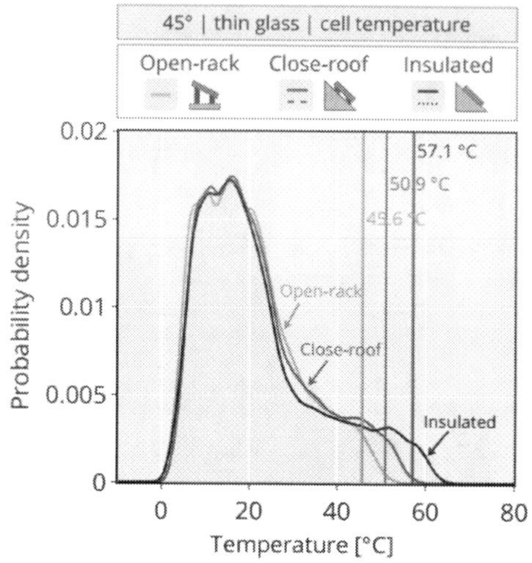

020251-009

40th European Photovoltaic Solar Energy Conference and Exhibition

3 | Results
Thermal stress indicators

98th percentile temperature (T98)

EPFL 3S Swiss Solar Solutions 10

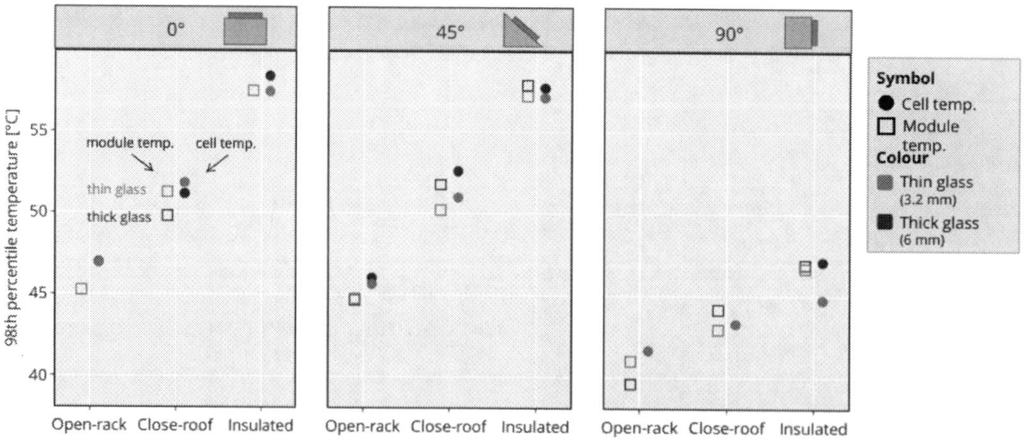

020251-010

3 | Results
Thermomechanical stress

Temporal temperature gradient

EPFL 3S Swiss Solar Solutions 11

Total temperature travelled

Thermomechanical stress $\sigma \propto \Delta T$

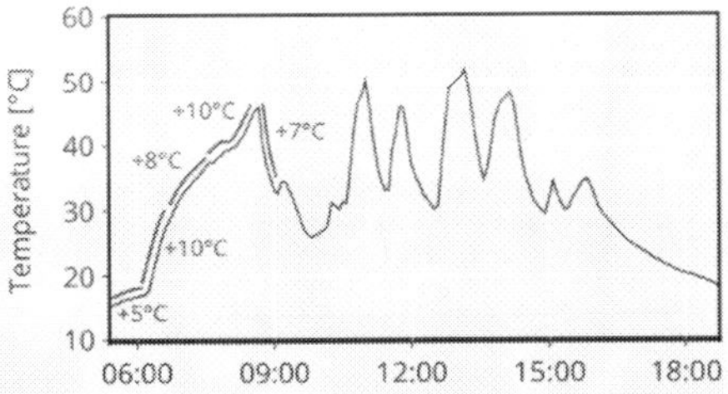

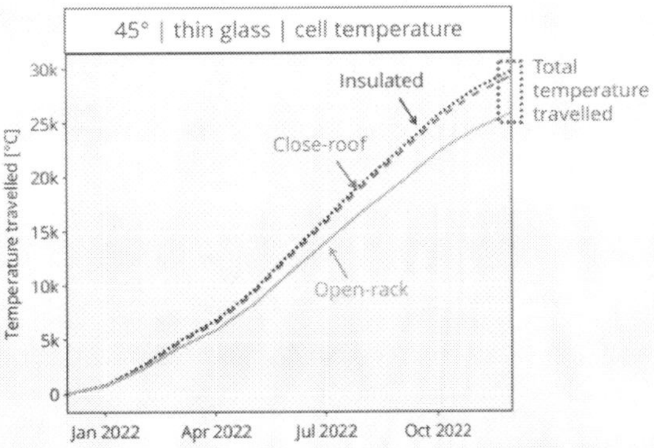

020251-011

40th European Photovoltaic Solar Energy Conference and Exhibition

3 | Results
Thermomechanical stress

Temporal temperature gradient

EPFL 3S Swiss Solar Solutions 12

Total temperature travelled

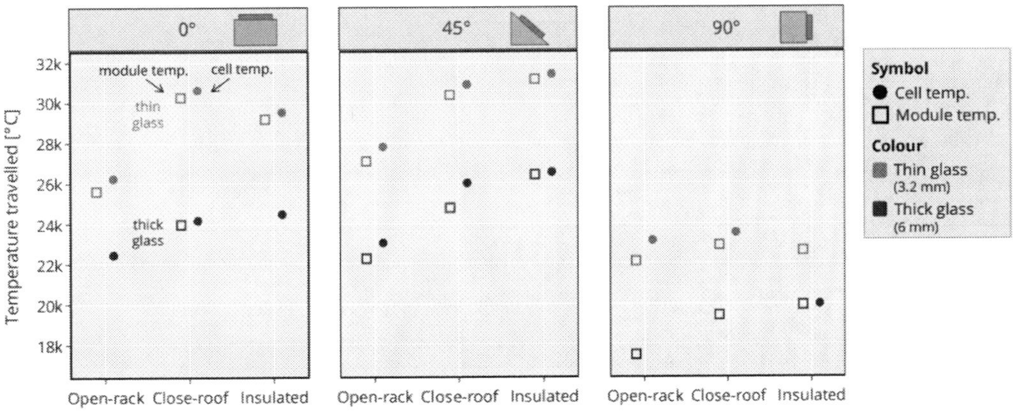

020251-012

40th European Photovoltaic Solar Energy Conference and Exhibition

3 | Results
Thermomechanical stress

Spatial temperature gradient

EPFL 3S Swiss Solar Solutions 13

Cell vs. Module temp. difference

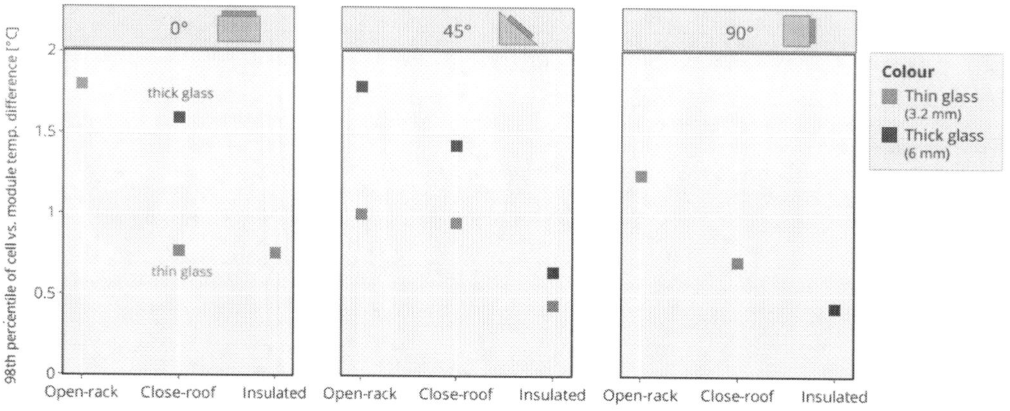

020251-013

40th European Photovoltaic Solar Energy Conference and Exhibition

Outline

EPFL 3S Swiss Solar Solutions 14

1 **Introduction**
Thermal and thermomechanical stress

2 **Methodology**
Experimental setup

3 **Results**
Temperature stresses in PV modules

4 **Conclusion**
Key results and outlook

020251-014

4 | Key results and outlook

EPFL 3S Swiss Solar Solutions 15

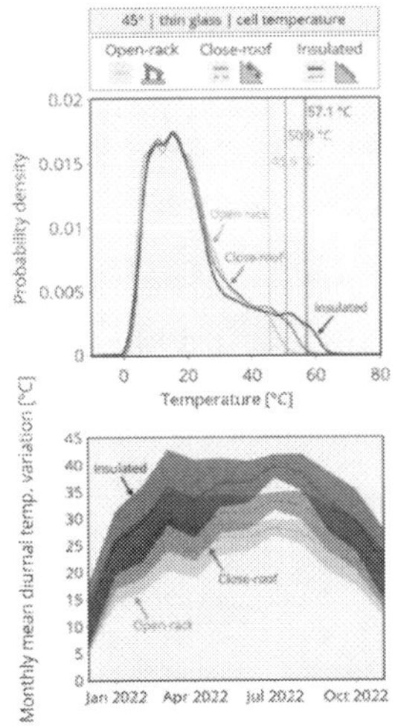

Thermal stress indicators

Higher insulation levels lead to higher T98

Higher T98 translates to higher diurnal temperature variation.

Insulation in PV installation leads to increased thermal stress, but **intraday temperature variations are not captured**.

020251-015

40th European Photovoltaic Solar Energy Conference and Exhibition

4 | Key results and outlook

EPFL 3S Swiss Solar Solutions 16

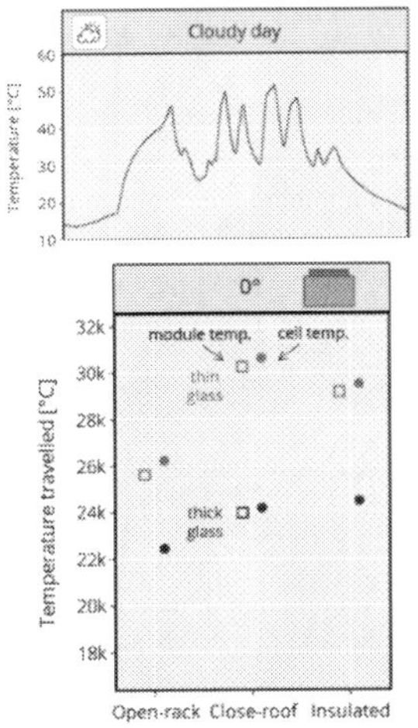

Thermomechanical stress indicators

Close-roof configurations are found to experience larger temporal temperature gradients

Insulation reduces the spatial temperature gradient, reducing material stress due to thermal coefficient mismatch

Mechanical stress due to temperature variations may **unexpectedly be higher for close-roof mounting** than for insulated modules

020251-016

020251-017

40th European Photovoltaic Solar Energy Conference and Exhibition

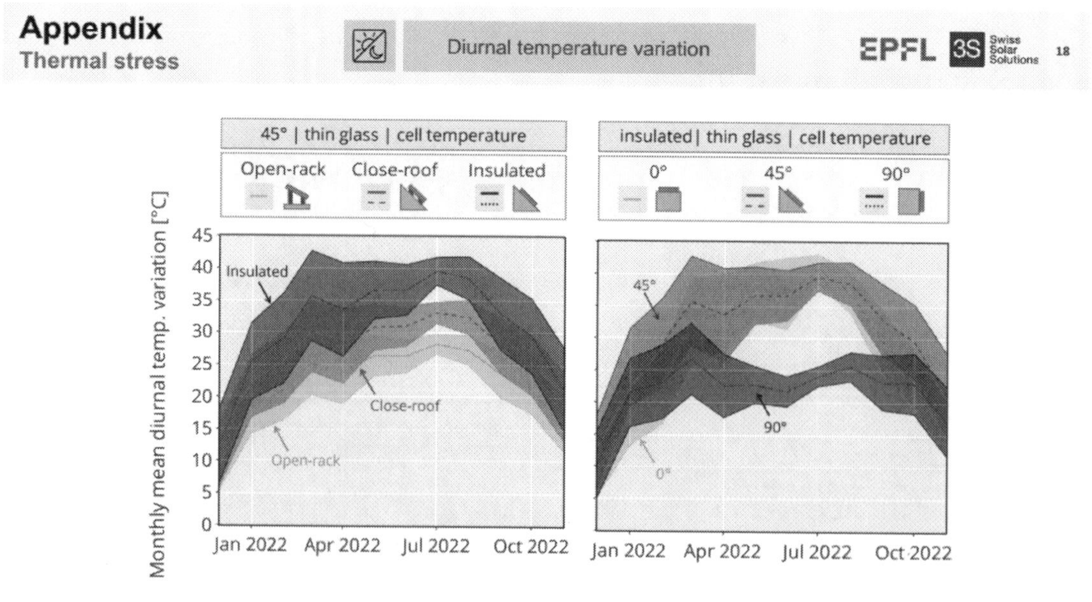

020251-018

Appendix

Thermomechanical stress

Temporal temperature gradient

EPFL 3S Swiss Solar Solutions 19

98th percentile temp. gradient

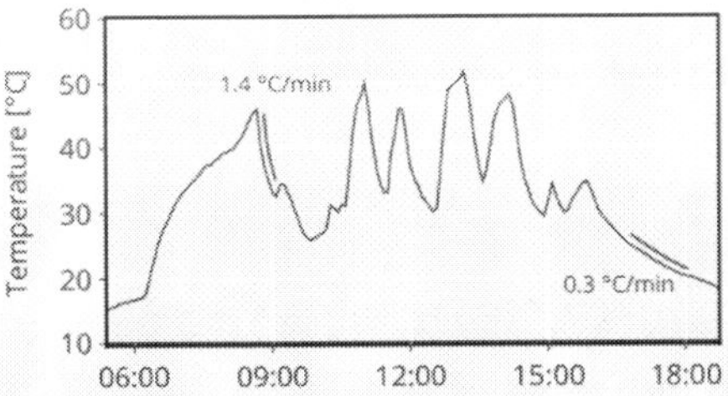

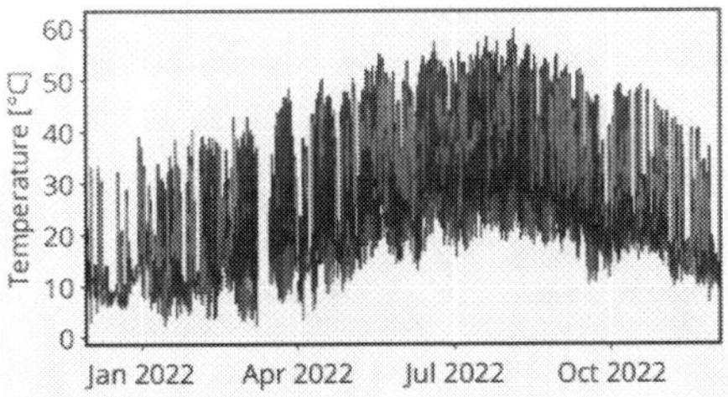

Appendix

Thermomechanical stress

Temporal temperature gradient

EPFL 3S Swiss Solar Solutions 20

98th percentile temp. gradient

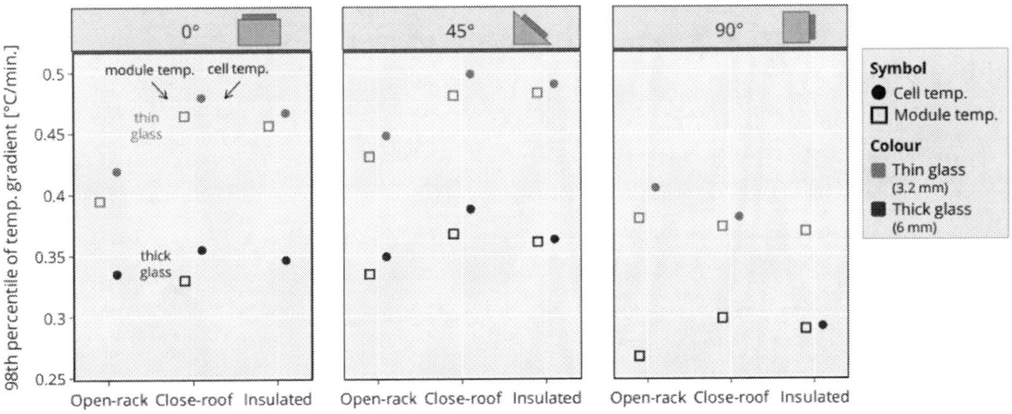

Appendix

Thermomechanical stress

EPFL 3S Swiss Solar Solutions 21

Temperature gradient (absolute value)

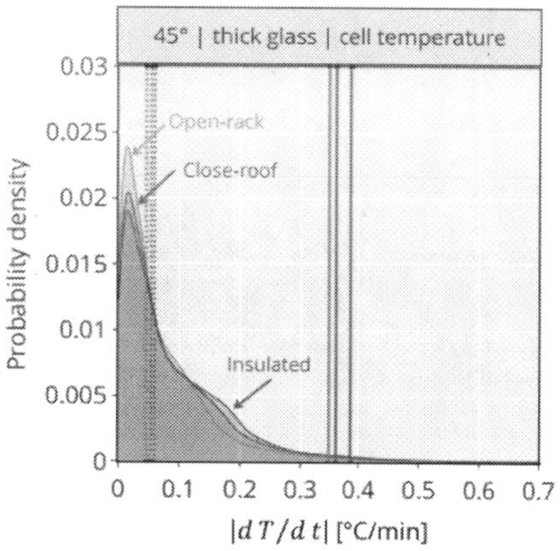

Cell vs. module temp. difference

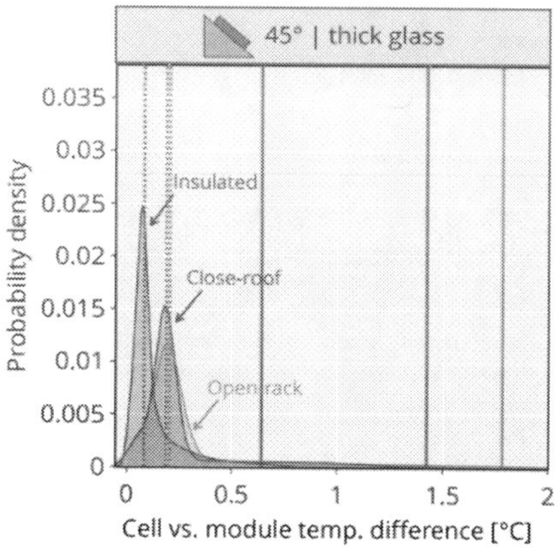

020251-021

40th European Photovoltaic Solar Energy Conference and Exhibition

Appendix

Thermal cycling and ramping literature

EPFL 3S Swiss Solar Solutions 22

Bosco et al. (2016), *'Climate specific thermomechanical fatigue of flat plate photovoltaic module solder joints'*

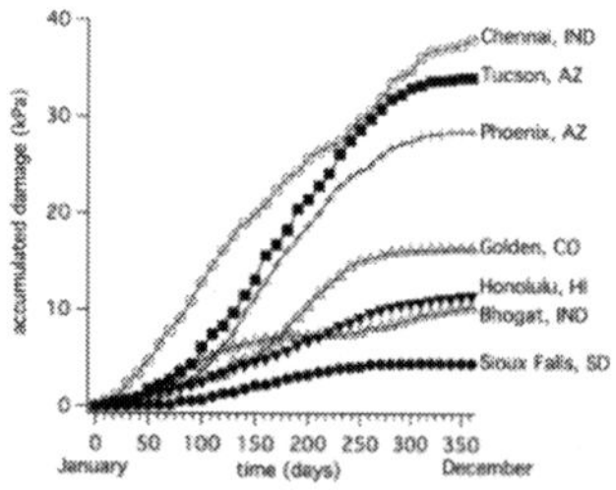

These loading cycles manifest as temperature reversals of the module while at a relatively high temperature of 56.4 °C, and therefore are likely due to clouding events on a hot, sunny day. Comparing the two hottest cities, Chennai and Phoenix: while their mean daily maximum temperature change is roughly the same Chennai has almost twice the number of reversals as Phoenix and therefore accumulates ~35% more damage. Phoenix and Tucson have roughly the same number of reversals and almost identical mean daily maximum temperatures. However, Tucson's larger mean temperature change results in it accumulating almost 20% more damage than Phoenix.

Bosco et al. (2012), *'On the Effect of Ramp Rate in Damage Accumulation of the CPV Die-Attach'*

Preliminary simulations indicated that the rate of inelastic energy (damage) accumulation dramatically increases with increasing temperature ramp rates in accelerated thermal cycling. Thermal cycling experiments with temperature ramp rates up to 140°C/min were explored. The resulting experimentally measured crack growth was found to be consistent with FEM simulations. This agreement between simulation and experiment provides confidence for using this simulated energy approach to develop efficient thermal cycling test sequences. The prospect of this work is a significant decrease in overall accelerated test time to reach an equivalent amount of thermal fatigue damage.

Lau & Dauksher (2005), *'Effects of Ramp-Time on the Thermal-Fatigue Life of SnAgCu Lead-Free Solder Joints'*

For both temperature ranges considered, the shorter (or faster) ramp times (rates) produce larger stress ranges in the lead-free solder joints.

020251-022

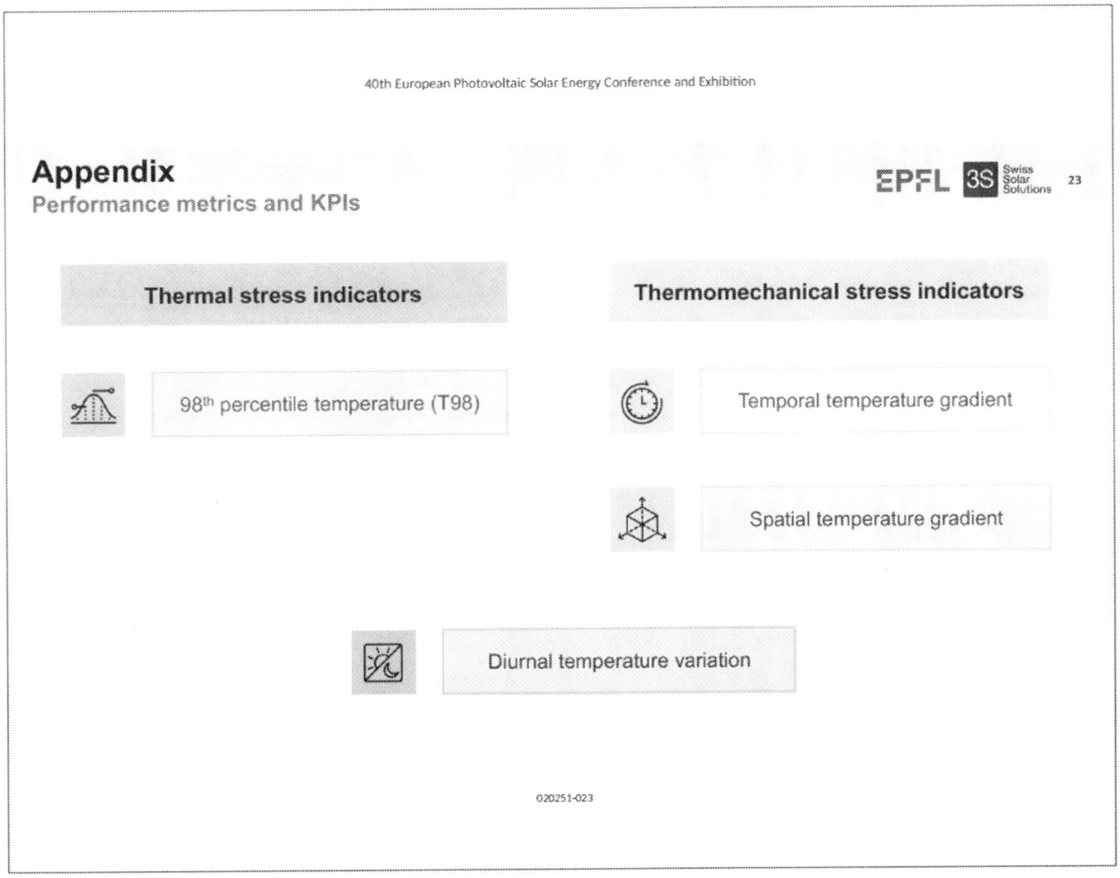
40th European Photovoltaic Solar Energy Conference and Exhibition
Appendix
Performance metrics and KPIs
EPFL
3S
Swiss Solar Solutions
23
Thermal stress indicators
98th percentile temperature (T98)
Thermomechanical stress indicators
Temporal temperature gradient
Spatial temperature gradient
Diurnal temperature variation
020251-023

PREDICTING OUTDOOR PERFORMANCE OF PEROVSKITE MODULES WITH MODIFIED RELIABILITY TESTING

P. Manshanden, G. Coletti, M.W. Ronchetti, M.J. Jansen, V. Rosca, M.Späth, V. Zardetto, W.J.H. Verhees, I. Dogan, H. Fledderus, S.C. Veenstra
TNO Energy & Material Transition
Westerduinweg 3, 1755 LE Petten, the Netherlands, High Tech Campus 21, 5656 AE Eindhoven, the Netherlands

ABSTRACT: Perovskite based photovoltaics still has some reliability problems to overcome before it can be commercially successful. At the current development stage, besides (adapted) IEC 61215 tests, specific reliability tests for perovskites should be devised, based on real outdoor parameters.
Two points regarding reliability testing are raised in this paper. First of all: we observed that our perovskite samples all (temporarily) regain efficiency after high temperature/high irradiance stress. This is an issue for monitoring tests outdoor where degradation is measured at STC indoor, as the power during operation is one under elevated temperature and irradiance. Remedying this problem will require adaptions to IEC 61215 tests, as well as careful attention for newly developed tests.
Secondly, during in-situ monitoring, we observed that using maximum power point tracking to obtain power data instead of optimizer behaviour leads to less stress on the samples than happens in field operation. Also, as the stress seems to induce meta-stable instead of permanent defects, the stress type and induced defects also appears to be not identical.

Keywords: perovskite, reliability, outdoor performance

1 INTRODUCTION

As currently the module costs comprises less than half of the total installation costs for utility scale and less than a quarter for residential photovoltaics[1], the most efficient way to reduce the levelized cost at device level is to increase the efficiency at the same price. This increase in efficiency cannot come at the price of device or module lifetime, as that would negate any financial advantages.

One of the ways to improve module efficiency is to employ tandem structures, for instance by stacking a perovskite layer on top of a silicon solar cell. As the added cost of the perovskite layer is modest [2], and the efficiency of the perovskite is rising steeply, this option currently attracts much interest[3][4][5].

As perovskite and perovskite-silicon modules come closer to commercial field implementation, the question of how one assures a device lifetime in decades in real outdoor circumstances arises. As long as devices are at technological readiness levels below 5, one can assume that reliability tests as currently described in the IEC 61215 can be used as initial tests. With rising technological readiness levels comes the requirement to match the reliability tests to the devices.

The procedure to arrive at device matched reliability tests has been described in the nineties for silicon solar modules: the failure mechanisms as seen outdoor should be mimicked indoor. A critical look at what happens to outdoor exposure in the field is essential in order to identify all possible failure mechanisms.

In this paper, an often overlooked feature of commercial deployment is investigated for implications on the reliability of perovskites and perovskite-silicon devices, namely the global maximum power point finding routines for irregular irradiation (shade): regular full IV sweeps. In addition, we have a look at the influence of a pause between the reliability stressor and measurement.

2 EXPERMIMENTAL SETUP

As the core idea of the paper is to compare outdoor performance to indoor reliability tests, the experimental setup consists of both indoor and outdoor testing facilities. The samples are both perovskite and perovskite-silicon tandem structures.

2.1 Experimental samples

The perovskite samples are semi-transparent triple cation mixed halide pin structure pixel cells arranged with four electrically isolated 4x4 mm cells on a 3x3 cm glass plate. Although the samples are in lab-scale dimensions, they were made using scalable S2S coating methods such as sputter coatings and slot die coatings.

Figure 1: schematic of pixel cell plates

The glass plate with four cells is encapsulated between two 7x7 cm glass plates with commercial TPO, with a black EPE full rear side and front mask. Because the glass plates are thicker than the TPO, TPO is also used to the four sides of the perovskite plate. Connection to the cells is made with copper foil and conductive glue, which in turn are soldered to standard tabs leading outside. The edges are sealed with black edge sealant.

2.2 Outdoor testing facilities

Our outdoor facility is located in Petten, the Netherlands, located within 200 m of the North Sea (latitude 52.78, longitude 4.67). The prevailing climate is moderate maritime. As such, temperature extremes (lower than 0 or above 30 °C) are rare, the wind chill is significant and the chemical conditions are especially harsh

The outdoor test facility is oriented towards the full south on a flat roof , with the devices at an angle of 30 degrees. The devices are enveloped in an opaque plastic

10.4229/EUPVSEC2023/3CO.11.3
020252-001

holder which fully covers the rear side to reduce wind cooling. The temperature of the devices is measured by a PT100 at the rear of the devices (inside the holder). Irradiance is measured by silicon devices, filtered silicon devices spectrally matched to perovskites and calibrated spectrometers. The samples are measured every ten minutes, with full synchronization of the irradiance, spectrometer, temperature and IV measurements, and kept at maximum power point between measurements. The measurement setup used has measurement ranges from 5 mA to 20 A
and 1 to 100 V.

Figure 2: photograph of outdoor test location showing samples and measurement devices, with the North sea in the background

2.3 Indoor testing facilities

Our indoor testing facilities include a class A solar simulator with suitable spectrally matched reference cells where perovskite samples can be measured at standard test conditions (STC).

For special reliability testing, a setup with adaptable heating and lighting was engineered attached to a hill climbing maximum power point tracer and a tracer with regular IV sweeps. The illumination consisted of white LED light. Class C irradiance can be used to for light soaking purposes. An advantage to LED illumination is that the lamps cause very little additional heat, allowing temperature setpoints to be reached within minutes. To aid the temperature stability of the samples, the setup was heated before the samples were entered.

Resistors were used to set superfluous cells in the test environment to approximately maximum power point, to avoid detrimental effects of setting to open circuit voltage.

3 OUTDOOR RESULTS

All samples were subjected to short damp heat testing prior to outdoor placement to ensure the edge sealing is water resistant.

3.1 Perovskite samples

The perovskite pixel cells were all individually contacted and driven, for a total of eight cells. One of these cells was broken at the start, and another one broke after a couple of days. The other cells were still actively producing energy after 76 days. The samples were numbered S1 to S7 (skipping the initially broken one).

The direct analysis of IV data from outdoor measurements is complicated by the variability of outdoor conditions. It is obviously not possible to use any measurement data taken during which period the irradiance varied, and this data is fully excluded from all consideration. Further limiting the variability of the environment by data selection is a straightforward method of reducing the variance, provided the remaining range contains sufficient data points.

In the case of our outdoor site, one of the more common irradiance levels is approximately 500 W/m², occurring every day for at least ten time stamps, more in case of overcast weather. The temperature is more variable, as this concerns both morning, evening and overcast times. Therefore we took only measurements between 400-600 W/m² and corrected for specific irradiance at the time. Temperature correction is problematic for perovskite, as consensus on correct values has not yet been reached [6].

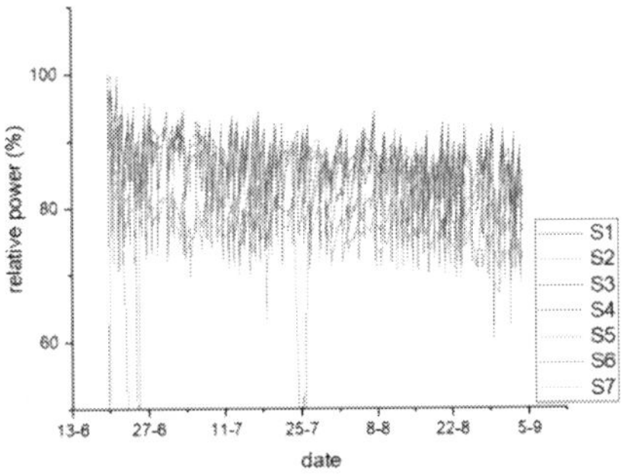

Figure 3: relative power evolution of outdoor samples. First days degradation is about 10%, with subsequent minor degradation. Spike of S7 is soiling, recovery after soiling removal.

The maximum power point evolution is shown in Figure 3. The maximum power point is normalized to the maximum value. S4 and S7 show an initial dip after the first full day outdoors, followed by a more steady decline. The others declined in a nearly constant rate. This decline is more easily observed in the linear fit to the data as demonstrated in Figure 4. The initial power of STC conforms more closely to the start of the line than to the 100% level.

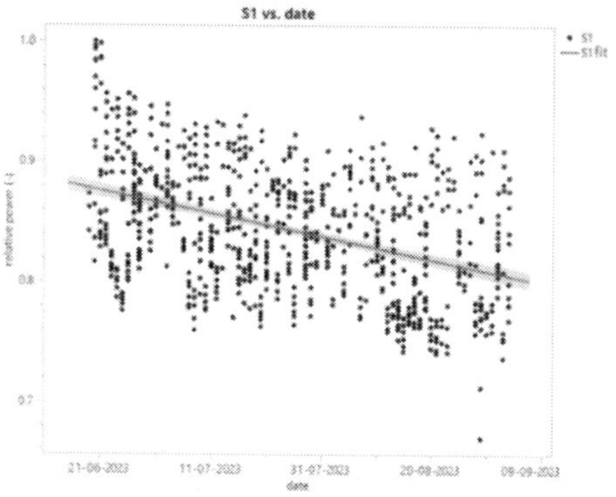

Figure 4: linear fit to the outdoor data of S1. The total decline is approximately 8%.

The degradation cause can be analyzed by looking at the IV curves themselves as shown in Figure 5 where the IV curves of S1 are compared with two months between. The times were selected such that the irradiance is 1000 W ± 5 W/m², the temperature is 35 ± 3°C and the average photon energy is 1.66 ± 0.01.

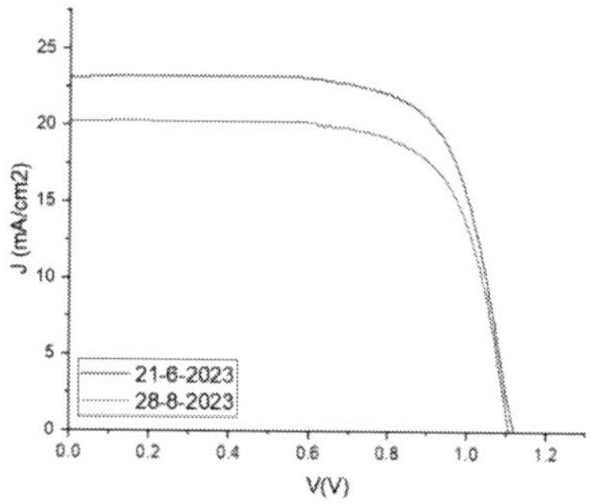

Figure 5: JV curves of S1 with two months in-between, taken at times with G=1000 W/m² and the sample temperature approximately 35 °C

The degraded component is the current, with voltage only decreased due to less current. Note that this pattern is constant for all our perovskite samples. The decrease in current unlikely to be due to loss of transparency of any layer on top of the photoactive layer, as the transparency of the entire stack is constant. This decrease is hypothesized to be due to degradation of the photoactive layer – the perovskite itself – as observed by others [7].

3.2 STC measurements of outdoor samples

The indoor solar simulator measurements before and after the indoor period of the current samples are missing because we did not choose to take the samples from the roof. The reason for this omission will be explained in the next paragraph.

Previously we have taken perovskite samples indoors. These samples were somewhat larger than the current samples, but the outdoor setup was the same. Also, the data selection was identical, but the time frame is different. The results of both the indoor and outdoor measurements are summarized in Figure 6.

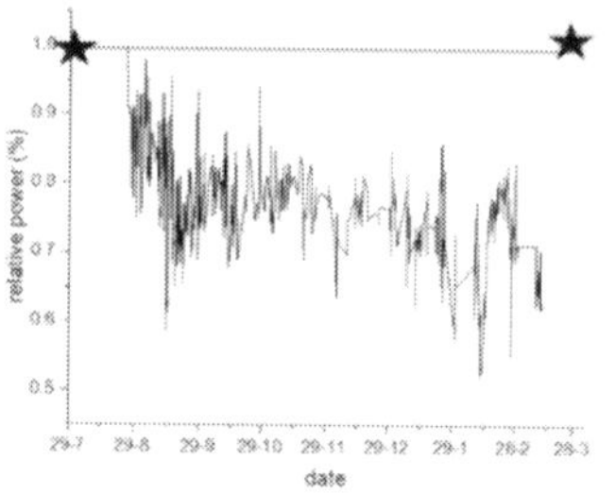

Figure 6: outdoor (line) and off-line indoor (star) measurements of the same sample compared. Time line is half a year. The period between removal outdoor and indoor test is approximately one week during which the sample was stored in the dark.

The correspondence of indoor and outdoor measurements is less than ideal, probably due to the very long time period where the sample was indoors. Please note that the correspondence for simultaneously measured silicon samples is easily within the error of the measurements.

4 INDOOR RESULTS

4.1 Standard tests preformed

Several standard tests were performed prior to the following investigation: damp-heat, UV exposure and thermal shock.

In the damp-heat test we saw 5% degradation occurring between 100 and 300 h, followed by stabilization until the end of the testing period.

UV exposure and thermal shock measurements gave more scattered results around 100% of initial power which proved difficult to analyze.

4.1 Recovery of degradation

Many methods of reliability testing assume that the sample is invariant between stress application and measurement of the effect.

As part of our investigation concentrated on effects during active maximum power point tracking, which was registered, we noticed an interesting effect which challenges this basic assumption, which is illustrated in Figure 7.

10.4229/EUPVSEC2023/3CO.11.3
020252-003

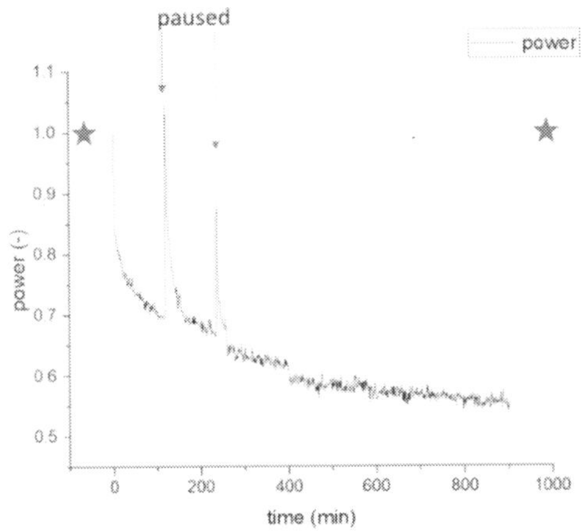

Figure 7: reliability testing at 80C and 1600 W/m². Twice testing was paused, and IV was measured at STC before and after. Note that the light was off during the measurement pause, but the heat remained on.

The test under consideration consisted of a temperature of 80 °C and 1600 W/m² During the test, testing was paused twice for a couple of minutes. On resume, the sample recovered, but resumed the degradation path. After the test, the sample was measured at STC. At that time, the sample showed nearly the same IV characteristics as initial.

This finding caused some issues with damp heat testing and thermal shock testing where IV measurements during testing are not possible as they occur in the dark. In the usual scheme of removing the samples for IV we saw no effect for 800 h of damp heat testing at 85 °C/85 rel. hum. And 500 thermal shock cycles.

4.2 Degradation at maximum power point

We used two schemes for maximum power point tracking: the simple hill climbing algorithm and one interspaced with regular IV sweeps to ensure the power point stays at the global maximum. In the field, these sweeps occur with a frequency of anything between 1 second and 10 minutes. We used 1.5 minutes in the indoor situation. Note that sweeps cannot have a much shorter interval, as perovskite samples should be measured in seconds rather than milliseconds.

During the sweeps the samples were heated or 80 °C with 1600 W/m² illumination. During stressing, two samples of each plate were monitored, one with maximum power point tracking according to a hill climbing algorithm and the other by IV sweeps every 1.5 min in addition.

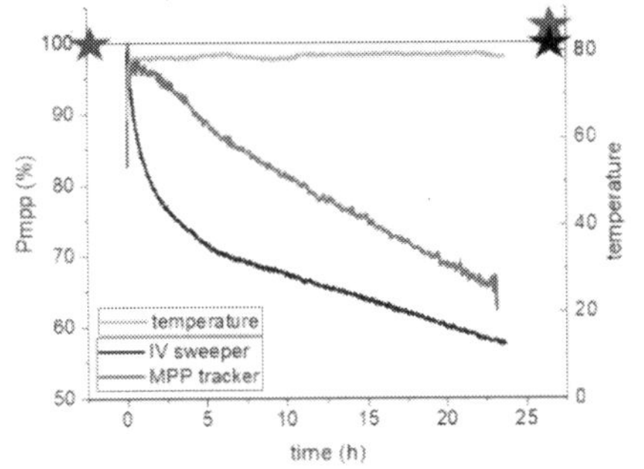

Figure 8: power evolution for maximum power point tracking and IV sweeping 80 °C and 1600 W/m², with stars denoting STC measurements before and after.

In Figure 8 the stars denote STC measurements before and after the stressor. As pointed out in paragraph 4.1, the STC results do not correspond to the full degradation. The hill climbing algorithm samples show no degradation at all when not under any form of stress. As for the IV sweeps, those mostly do show regeneration, but for less than the full amount. This shows a fundamental different end state of the samples. Also during degradation, the hill climbing algorithm tends to be better than the more realistic outdoor IV sweep procedure.

5 DISCUSSION

5.1 Recovery of degradation

The finding that stressor factors may show recovery in the period between removal from the stressor and off-line measurement is a complicating factor in the analysis of dark reliability tests and outdoor experiments. While we have not proven recovery will occur in all circumstances, our findings indicate a proof of non-recovery is necessary in order to rely on results like these.

For a generalized example, many of the dark reliability tests of IEC 61215 are to be measured off-line. In-situ tests are typically dark IV, but given that the perovskite signal is primarily on the current, extracting a perovskite degradation from the dark IV is not trivial.

For a specific example, the increasingly popular heat/irradiance stressing at V_{oc} conditions is troublesome, because tracking the voltage will not meaningfully inform of degradation as the perovskite degradation is mainly on current and off-line measuring will give flattered outcomes. Interspaced IV measurements online would of course be valid.

Also, outdoor measurements have to be analyzed in-situ, which means a fairly large amount of noisy data needs to be cleaned up in analysis. Small level degradation is even so quite difficult to prove. Researchers working on outdoor data have published extensively on analysis [8].

5.2 IV sweeps versus mpp tracking

Adding regular IV sweeps to mpp tracking causes irreversible damage to the perovskite samples under consideration. This damage is less visible while the samples are actively stressed.

Because the mpp algorithm is a simplification of the field situation, it is required to verify this simplification is a good approximation of reality. The current results cast the validity of said simplification in doubt.

5.3 Indoor and outdoor results

The outdoor degradation progresses at a much slower rate than indoor degradation, which is in principle a design feature of accelerated indoor testing.

The cause of the rate difference is twofold. First of all, all degradation parameters show comprise of extreme outdoor conditions. In the Netherlands, in-roof systems may reach 80 °C for about an hour each year. Irradiance tops at approximately 1300 W/m², so the indoor test conditions are on both counts more severe than outdoor. In addition, the outdoor sweeps had a frequency of 10 minutes, while we used 90 seconds indoors.

Secondly, the regeneration aspect also comes into play at night, granting the samples a period to recover, as previously observed [9]. Degradation is faster after recovery, but nevertheless that path has to be traversed every day again. The older outdoor results shows that eventually one night is insufficient for full recovery, which at that point can still be achieved by a longer rest period.

6 SUMMARY

Perovskites still have some reliability problems to overcome before commercialization can become successful. At the current stage, besides (adapted) IEC 61215 tests, specific reliability tests for perovskites should be devised, always based on real outdoor parameters.

One key problem has been identified in this paper: the tendency of perovskite to (temporarily) regain efficiency after stresses. This problem will come into play both for adapting existing IEC 61215 tests, as well as new types of testing and regular outdoor tests.

As a smaller problem, we identified that the simplification to use maximum power point tracking instead of optimizer behaviour leads to applying less stress on the samples than would happen in field situations. Also, as the stress seems to induce meta-stable instead of permanent defects, the stress also appears to be not identical.

7 ACKNOWLEDGEMENTS

This work was funded by the Horizon2020 project Triumph.

8 REFERENCES

[1] V. Ramasamy, "US Solar Photovotaic system and energy storage cost benchmarks, with minimum sustainable price analysis: Q1 2022", NREL/TP-7A40-83586

[2] P. Culik et al., "Design and Cost Analysis of 100 MW Perovskite Solar Panel Manufacturing Process in Different Locations", ACS Energy Letters 2022 7(9), pg 3039

[3] NREL Campion Photovoltaic Module Efficiency Chart, www.nrel.gov/pv/module-efficiency.html

[4] A. Mahapatra et al, "Recent progress in perovskite solar cells: challenges from efficiency to stability", Materials today chemistry 23 (2022), 100686

[5] E. Raza, Z. Ahmad, "Review on two-terminal and four-terminal crystalline silicon/perovskite tandem solar cells; progress, challenges and future perspectives", Energy Reports 8 (2022), pg 5820

[6] L. Lin and N.M. Ravindra, "Temperature dependence of CIGS and perovskite solar cell performance: an overview", SN Applied Sciences (2020) 2:1361

[7] J. Bisquert, E.J. Juarez-Perez, "The causes of degradation of perovskite solar cells", J. Phys. Chem. Lett. 2019, 10, pg 5889

[8] N.J.J. Dekker et al, "Accurate yearly yield calculation using PV module fingerprint method", 32th EUPVSEC (2015)

[9] V. Stoichkov et al, "Outdoor performance monitoring of perovskite solar cell mini-modules: diurnal performance, observance of reversible degradation and variation with climatic performance", Solar Energy 170 (2018), pg 549.

INCIDENCE ANGLE MODIFIER ISSUES FOR PHOTOVOLTAIC ENERGY YIELD ESTIMATIONS

Laura Barrutia[1], Ulyses Bueno[1], Mikel Muñoz[2] and Eduardo Lorenzo[1]
[1]Instituto de Energía Solar – Universidad Politécnica de Madrid, ETSIS de Telecomunicación, C/Nikola Tesla, s/n, 28031, Madrid, Spain
[2]ACCIONA-ENERGIA, Avenida Ciudad de la Innovación, 5, 31621, Sarriguren (Navarra), Spain

ABSTRACT: The aim of this work is to highlight the importance of considering not only the reflective losses governed by changes in the refractive indices of the photovoltaic module, but also the losses governed by the effect of the accumulation of dust particles on the surface of the modules under real operating conditions, known as soiling. The Martin-Ruiz Model which derives from the mathematical handling of the Fresnel equations and includes analytical solutions for diffuse and reflected irradiance components, is presented as a robust, accurate and reliable analytical model that contemplates these two factors. With the implementation of the Martin-Ruiz Model in SISIFO, a PV simulation software developed by the IES-UPM, optical losses at a 100 MW one-axis tracked PV plant in the Kalahari Desert were simulated. Four different soiling scenarios were considered. The results suggest that the incident angle modifier losses are between 44 to 64% larger than those predicted when considering only the effect of refractive index change. Simulation software integrating models that consider all of these optical losses will improve the energy yield prediction of any photovoltaic plant with more reliable estimations.
Keywords: angle of incidence, refractive index, soiling, PV modules

1 INTRODUCTION

An incidence Angle Modifier, IAM, refers to optical losses that occur when the angle between the incident ray on a surface and the normal to the surface, θ, is other than zero. Depending on both the geometry of the photovoltaic (PV) array and the prevailing solar radiation and soiling conditions, IAM losses represent between 1 to 5% of the yearly energy yield of commercial PV plants; large enough to account for these losses when modelling this yield. In fact, most, if not all of today´s PV software, includes some model for accounting for these losses. For example, PVsyst relies on the classic Fresnel equations, SAM (Solar Advised Model developed by NREL) [1] which relies on the Sandia Module Model [2], which uses a set of empirically derived coefficients (a_0... a_5) and a fifth-order polynomial equation to account for the losses. SISIFO, an open software developed by the IES-UPM, relies on the Martin-Ruiz [3] model. This model diversity can be understood as a symptom of the confusion that affects this issue. One of the objectives of the work is to highlight Martin-Ruiz model as a robust and reliable model.

Relevant aspects to be considered when analyzing the convenience of a particular IAM model are:

First, the PV module consist of a stack of different materials, so that light passes through several interfaces (air-AR coating glass; AR coating glass-glass; glass-encapsulant; encapsulant – AR coating cell; AR coating cell- cell). In principle, all these materials have different refraction index, n, so that IAM models must be aware of the multiplicity of interfaces. The great mathematical complexity of applying the classic Fresnel equations to all these interfaces has given rise to the search for other alternatives, either purely empirical ones, such as the ASHRAE and the Sandia models, or those derived from the mathematical handling of the Fresnel equations, such as the Martin-Ruiz model. However, because all these interfaces are designed to minimize reflection (which is why PV modules look very dark), considering just the first interface (air-glass or air- AR coating glass) generally suffices in practical terms.

Second, the incident solar radiation is made up of a beam component, characterized by a concrete θ value, and diffuse and reflected components characterized by their corresponding angular extensions. IAM for the first, can be modeled using an equation relating the transmittance at the air- module interface with the angle of incidence, $IAM(\theta)$. However, IAM for the diffuse and reflected components requires integration into these extensions. $IAM(\theta)$ models capable of analytical integration are preferred to those requiring numerical integration.

Third, IAM is affected by changes in the refraction index. But, despite the less known, IAM is also affected by soiling. This is due to the volumetric properties of soil and makes actual IAM energy loses which are about 30% to 100% higher than those estimated from clean PV modules. Therefore, $IAM(\theta)$ models must be capable of considering this soiling dependence.

Fourth, the model must be capable of reproducing the experimental observations. In this regard it must be remembered that the IEC 61853-2 norm [4] describes methods for measuring $IAM(\theta)$for totally clean modules. Corresponding $IAM(\theta)$ results are available for different modules, either from independent laboratories or directly from the PV module manufacturers. Moreover, this paper presents the results of IAM measurements on soiled modules.

2 MARTIN-RUIZ MODEL

The electrical characteristics of a PV module are supplied in reference to normal incidence radiation. Since ray incidence is not always perpendicular, transmittance losses at different angles must be considered.

IAM modelling can be described by Fresnel laws as the ratio between the transmission coefficient at a particular angle $T(\theta)$ and when normal to the module´s surface $T(0)$.

10.4229/EUPVSEC2023/3CO.11.6
020253-001

eq. 1 $$IAM(\theta) = \frac{T(\theta)}{T(0)}$$

In the search for an easy mathematical description of $T(\theta)/T(0)$ experimental results, alternative formulations other than Fresnel laws have been proposed [1-3]. A brief description of Martín-Ruiz model will be presented since it is the model used in this work. The model was proposed by Martin and Ruiz in 2001, given by:

eq. 2 $$IAM(\theta_i) = 1 - \frac{exp\left(-\frac{cos\theta_i}{a_r}\right) - exp\left(-\frac{1}{a_r}\right)}{1 - exp\left(-\frac{1}{a_r}\right)}$$

This model has been widely referred in the literature (364 references in google scholar on the 11th of August of 2022) and is specifically recommended in IEC 61853-2 for describing the experimental results. The angular losses coefficient a_r is an empirical dimensionless parameter. Recommended default values for the adjusting coefficient are a_r= 0.17 for bare glass and a_r= 0.14 for modules with an anti-reflection coating.

From previous work in our group, it can be claimed that for a clean module and a beam of light incident in one direction, the Martín-Ruiz model leads to results that are practically indistinguishable from those given by the Fresnel equations.

Martin-Ruiz model also considers other important issues when thinking about PV energy yield simulations:

On the one hand, simulating the transmittance of diffuse radiation requires integrating the transmitted beam irradiance into the angular extension of the corresponding components (the sky, the ground, and the horizontal band), which are functions of the tilt angle and the Ground Cover Ratio, GCR. Because of the complexity of the Fresnel Equations, it is extremely challenging to derive an analytical solution to the diffuse transmittance. Thus, it is common to resort to numerical solutions [5-7]. With the aim of reducing the computational burden, a polynomial fit to precomputed results was proposed [8]. More recently, an analytical solution called FEDIS [9] was also proposed. We believe the model proposed by Martin and Ruiz 20 years ago can also be categorized as an alternate form of the Fresnel equations whose integration has analytical solutions [3].

On the other hand, IAM in real modules is not only affected by changes in the refraction index but also by soiling. Hence, the ability of considering this additional effect also matters. Martín-Ruiz is the only model capable of stablishing the soiling effect simply by modifying the adjusting parameter a_r.

As a summary of Martin-Ruiz model, it is highlighted as a robust model with one adjusting parameter, it leads to results that are practically indistinguishable from those given by Fresnel equation, it includes analytical solutions for diffused and reflected irradiance components and it is the only model that considers effect of dirt. All these arguments are what we have considered when choosing this model in our work.

3 IAM TESTING ON CLEAN MODULES

Figure 1 shows the IAM function for three clean PV commercial modules as given in the pan.file provided by their respective manufacturers. Interestingly, the three modules seem to be otherwise equal (PERC, monocrystal, Si-n, cell size, power temperature coefficient, etc.). So that, all things being equal (price, guarantees, etc.) observed differences at IAM curves leads to preferring some modules to others. Thus, the convenience of independent IAM testing is undoubted.

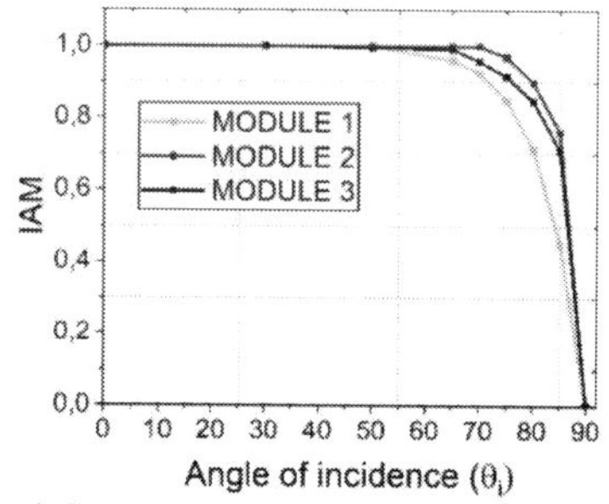

Figure 1: IAM curve comparison between three different PV commercial modules. IAM data as given in the pan.file provided by their respective manufacturers.

IEC 618532-2 establishes requirements for both indoor and outdoor IAM testing procedures. The indoor one requires electrical access to an individual string of cells, which is difficult with double glass encapsulated modules. The outdoor procedure is based on collecting the I_{SC} data of the test module throughout a wide range of θ_i. This requires the test to be carried out on very clear days (the ratio between Direct Normal Irradiance and Global Normal Irradiance must be larger than 85%) and using a highly accurate tracker (better than ± 0.5°) for varying θ_i. The latter is particularly difficult to achieve in practice, which helps to explain why only few laboratories have reported on the implementation of this test [10-12].

However, the same standard states: "For modules with a flat uncoated front glass plate made of standard solar glass, the relative light transmission into the module is primarily influenced by the first glass-air interface. In this case, the test does not need to be performed; rather, the data of a flat glass air interface can be used". This opens the door for alternative IAM testing based on comparing the I_{SC} responses of the module under test and of another coplanar module with an uncoated front glass (in practice, most PV modules produced before 2107) serving as reference.

In this work, an alternative outdoor testing is proposed. The experiment is based on a relative measurement and by using a static set-up. Figure 2 shows such an experiment performed at the IES-UPM. The experiment is based on collecting I_{sc} data of two coplanar modules, bare glass reference module and AR coated glass module. To match the most IAM sensitive θ_i range (let us say, 50°≤ θ_i ≤ 80°) with the time windows at around noon, when the diffuse fraction of the global irradiance is lowest, the modules are facing East and tilted close to the latitude of the site minus the declination of the day, the set-up is static, and the test is extended from 10:00 to 12:00. Note that for a fixed orientation and tilt, θ_i it is easily calculated by means of Sun´s position equations found in basic solar radiation literature. Thus, an accurate tracker is unnecessary.

A)

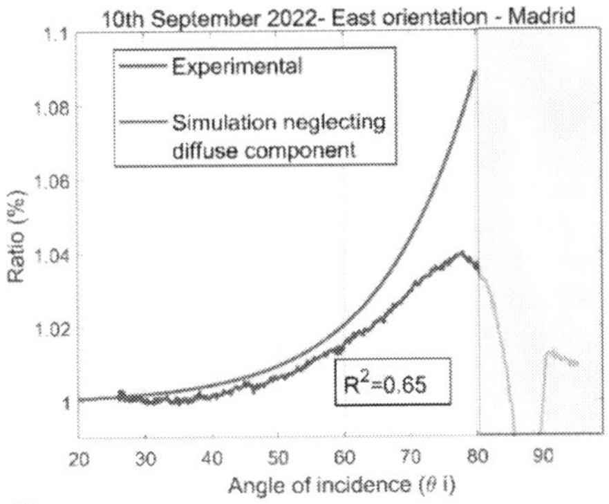

B)

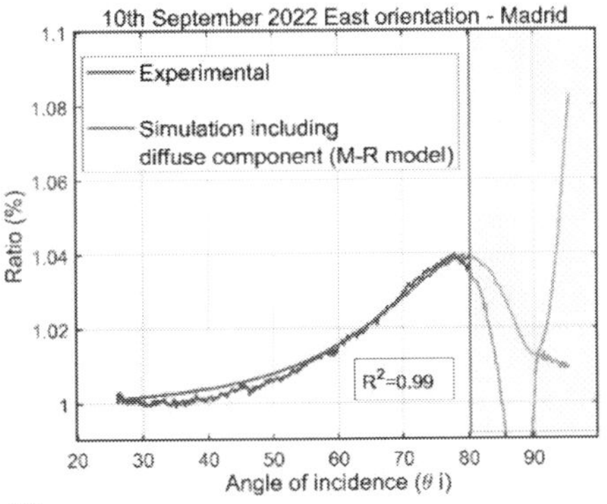

C)

Figure 2: A) IAM testing performed at IES-UPM based on relative measurements. Experimental behavior of the ratio (blue line) and simulation fit with Martin-Ruiz model (pink line) between the reference and the undertest module as a function of the angle of incidence for a clear day (10th of September 2022, Madrid). Comparison between the Martin-Ruiz model without considering diffuse radiation (B) and with its implementation (C). The grey rectangle (θi>80°) corresponds to the fitting regions where the contribution of annual energy distribution as function of the angle of incidence is less important. The Martín-Ruiz coefficient of $ar = 0.17$ remained fixed as a reference for a clean module with bare glass surface.

Figure 2 plots the evolution of the ratio between the I_{SC} of the test module (with AR coating layer) to the I_{SC} of the reference module (bare glass), both normalized to the respective STC values, which equals the ratio of the corresponding IAM values. Martin-Ruiz model allows the experimental result to be fitted. Considering the diffuse irradiance component led to an improvement in the fitting and, hence, the accuracy of the result experiment. This is clearly seen when comparing Figure 2 B (fitting without diffuse component) and Figure 2 C (fitting considering diffuse component). Diffuse irradiance was estimated from the readings of a standard meteorological station dotted with a pyrheliometer and a horizontal pyranometer. An $a_r = 0.17$ has been selected for modeling the IAM of the reference module and the best fit is achieved when modelling IAM of the tested module with $a_r = 0.145$. Note that this value is precisely the output result of the experiment, which can be directly used as input data for energy yield simulation.

4 SOILING EFFECTS ON IAM

Soiling is an IAM additional cause to changes in the refraction index. This is a direct consequence of the volumetric properties of dust particles.

Figure 3 A) shows a visual demonstration of this effect. Despite all of the modules in the PV array involved being equally soiled, they seem to be dirtier the farther away they are, because the angle of view from the camera increases. In a PV array, this angle θ_i is the angle of incidence of the sun.

Figure 3B) shows this effect for two hypothetical particles sphered and parallel piped-shaped, respectively. The key is to realize that (disregarding interference phenomena) the received shaded area and, therefore, the apparent dirtiness varies with θ_i, it being minimal with normal incidence.

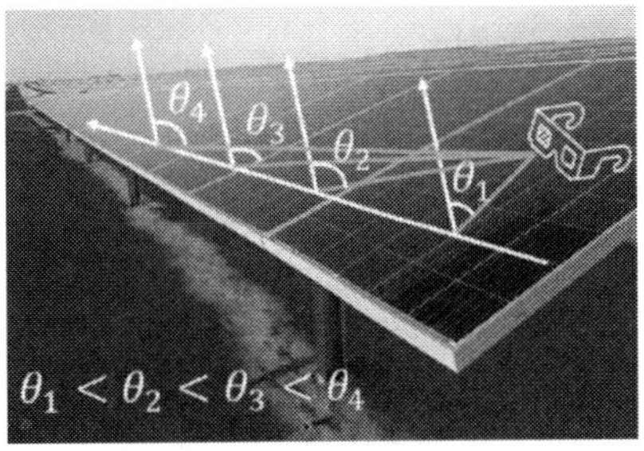

A)

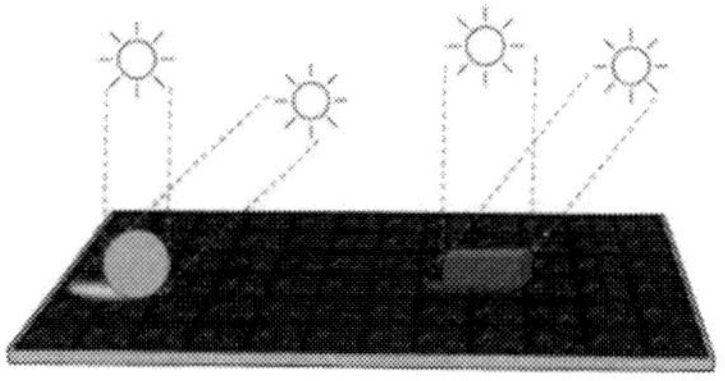

B)

Figure 3: A) the fact dust particles have volume, makes their impact in terms of shading dependence on the angle of incidence. Due to this effect, the more sideways that the

10.4229/EUPVSEC2023/3CO.11.6
020253-003

observer looks at the module, the dirtier it will seem. B) sketch example of the effect of a spherical (green) and parallel piped-shaped (purple) dust particle with the angle of incidence.

Thus, the final IAM can be understood as the product of two components:

eq. 3 $$IAM = IAM^{\Delta n} \cdot IAM^{s}$$

Where the superscripts "Δn" and "s", respectively, mean a change of refraction index and soiling. The total optical losses, TOL, are given by:

eq. 4 $$TOL = 1 - (1 - NSR) \cdot IAM$$

Where *NSR* is the normal soling ratio, understood as the reduction in transmittance for normal incidence.

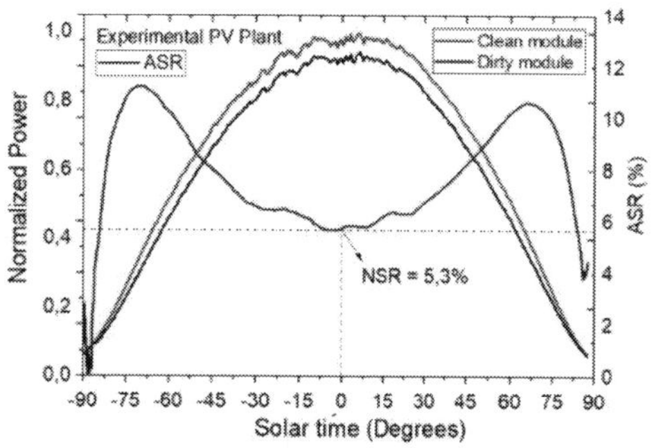

Figure 4: Evolution of the normalized powers of a clean module, p^C (red line) and a dirty module, p^S (black line) for a particular day (29th of July). The apparent soiling ratio, ASR (blue line) is also plotted. Losses go from ~ 6% at noon, to more than 11% in the early morning and late afternoon. Data taken from a real operation PV Plant in Murcia (Spain).

Figure 4 shows the result of an experiment dedicated to soiling carried out at a PV plant in Murcia (Spain). The MPP power of two equals, coplanar, south oriented and 20º tilted PV modules has been recorded throughout a clear day. One of the modules was soiled while the other was cleaned. The figure shows the daily evolution of the respective powers, normalized to the corresponding STC value, p^S (black line) and p^C (red line), and of the apparent soiling ratio, ASR, given by:

eq. 5 $$ASR = 1 - p^s/p^c$$

Similar equation to eq. 5 but expressed in terms of daily integrals, gives a good indication of the daily energy losses due to soiling, ASRd. That is:

eq. 6 $$ASR_d = 1 - \frac{\int_{day} p^S dt}{\int_{day} p^C dt}$$

In this experiment, $ASR_d = 7.5\%$, which is 31% greater than $NSR = 5.3\%$. This outlines the importance of considering the effect of soiling in IAM when estimating the energy yield of PV plants. This integration can be applied to any other period (month, year, etc.).

5 SIMULATIONS WITH SISIFO

At this point, we think there are enough arguments to select the Martin-Ruiz model as the most convenient *IAM* model for energy yield simulation purposes. It is as accurate as the Fresnel equations, it provides analytical solutions for all the radiation components reaching the PV generators´ surfaces, it reproduces well the experimental observations well in clean modules and, finally, it offers a simple way of accommodating the impact of soiling. All of these reasons led us to implement this model in SISIFO, a PV simulation software (freely accessible at https://sisifo.info) developed by the IES-UPM. Default a_r coefficient values for clean single glass and glass with an AR coating layer are 0.17 and 0.14 respectively. SISIFO assumes that *NSR* is kept constant throughout the day and accepts a set of 365 values as input for defining the yearly evolution of soiling. An interesting observation for simulation purposes is that soiling does not detectably affect the temperature of the PV modules [13].

A main difficulty for the simulation to be representative is establishing a proper scenario for the yearly evolution of *NSR*. Despite the relevant overall effort to make data on the effects of PV module dust soiling available, which is advancing towards something like the establishment of an encyclopedia of dust soiling rates [14], the huge importance of local effects (from nearby sources of dust, that affect dust accumulation rates, to even the price of electricity, which affects the cleaning schedule) makes it very difficult to develop general representative soiling evolution models. According to the IES-UPM experience in the quality control of large commercial PV plants, baseline scenarios for analyzing the bankability aspects of PV plants more often rely on assuming an *NSR* constant throughout the year, typically around 2%. However, this is frankly unrealistic, because soiling is in fact determined by seasonal effects such as atmospheric particle content, raining, cleaning. Hence, we decided to go along with the recommendation that soiling scenarios be supported by real measurements at a specific region for at least one year [15].

We have simulated the energy yield of an Acciona's 100 MW PV plant in Sishen (South Africa) in the Kalahari Desert. The Kalaharian climate is subtropical and semi-arid with a rainy season, from November to April, and a dry season from May to September. Annual precipitation is 240 mm. During the rainy season natural cleaning suffices to keep *NSR* below 3%. During the dry season artificial cleaning is required. To promote local employment, that cleaning is carried out manually, and a full cleaning campaign takes about one month. During 2020 only one cleaning campaign was carried out and *NSR* reached 10% on some days. During 2021, following the results of a dedicated study to optimize the cleaning schedule, three cleaning campaigns were carried out. Figure. 5 shows the yearly evolution of *NSR* during these years.

10.4229/EUPVSEC2023/3CO.11.6
020253-004

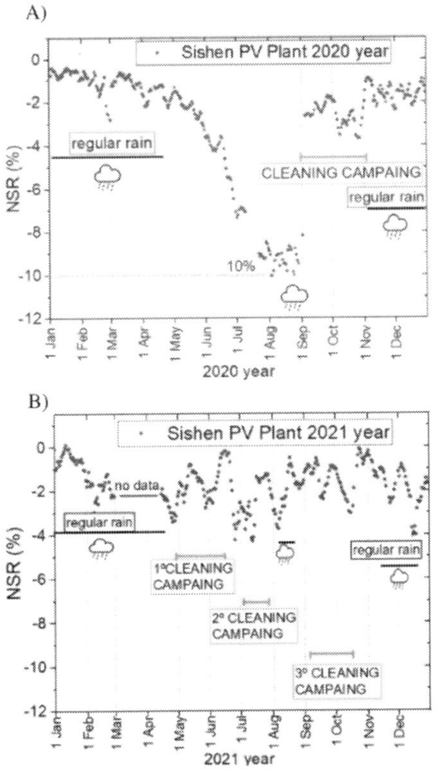

Figure 5: Real evolution ratio of *NSR* of the Sishen PV Plant (South Africa) during 2020 A) and 2021 B) where different cleaning campaigns were carried out (one during 2020 and three during 2021).

The Sishen PV plant is horizontal axis tracked with a GCR of 0.39 and Si-x PV modules covered with uncoated glass. Simulations have been carried out with the TMY as given by PVGIS, and for four different *NSR* scenarios: constant throughout the year with 0% and 2% values, as observed during 2020 and as observed during 2021. Note that in all of these scenarios the average *NSR* keeps below 3%, so that they are representative of reasonably well-maintained PV plants. Table I gives the corresponding yearly results concerning the effective irradiation, understood as the irradiation corrected by soiling and IAM effects. That is:

eq. 7 $$G_{ef,y} = G_y(1 - NSR)IAM_y$$

where G is the global in-plane irradiance and the subindex "ef" and "y" means effective and yearly, respectively. The breakdown of optical losses is also given. The outstanding conclusion is that real IAM losses are between 44 to 64% larger than those predicted by IAM models that only consider the effect of changes in the refractive index.

Table I: Effective irradiation in terms of annual results ($G_{ef,y}$ (kWh/m^2) obtained from the simulation exercise of Sishen Plant, South Africa, with an horizontal single axis system. Different NSR values have been considered in the simulations (a constant 0% throughout the year, a constant 2% and the annual average of 2020 and 2021). Optical losses and distinction between IAM$^{\Delta n}$ and IAM^{S} are also given.

	IAM Sishen Plant			
NSR scenario	0%	2%	2020	2021
$G_{ef,y}$ (kWh/m^2)	2,183	2,119	2,069	2,128
	Yearly Soiling and IAM losses (%)			
NSR	0	2	2.9	1.6
(1-IAM)	1.7	2.6	4.1	2.6
(1- IAM $^{\Delta n}$)	1.7	1.7	1.7	1.7
(1- IAM ^{S})	0	0.9	2.4	0.9
(TOL)	1.7	4.5	6.8	4.1

6 CONCLUSIONS

IAM refers to losses due to PV module transmittance reductions with the incidence angle. They typically represent 2-5% of the yearly energy yield of commercial PV plants, so that they must be conveniently modeled in corresponding simulation exercises. Despite PV modules being made up of a stack of several interfaces (air-AR coated or uncoated glass - encapsulant; – AR coating cell; AR coating cell- cell), the experimental observations are very well described by considering only the first interface.

In principle, the classic Fresnel equations applied to the first interface provide a good description for a clean module and a beam of light incident in one direction. However, other considerations arise when thinking about PV energy yield simulations. On the one hand, incident irradiance includes not only beam but also diffused and reflected components, both characterized by their respective angular extensions. Hence, easy integration (analytical better than numerical) matters. On the other hand, IAM in real modules is not only affected by changes in the refraction index but also by soiling. Hence, the ability of considering this additional effect also matters.

Martín-Ruiz model which derives from the mathematical handling of the Fresnel equations and have one adjusting parameter, includes analytical solutions for diffused and reflected irradiance components and it can establish the soiling effect simply by modifying the adjusting parameter.

Practically all the current PV modules use AR-coated glass as a frontal cover, and the module manufacturers provide the corresponding IAM information by means of a set of relative transmittances versus incidence angle values. According to this information, some modules are better than others. Thus, independent testing is required. The IEC norm [4] prescribes requirements for outdoor IAM testing, addressing only absolute measurements, that is, the relative transmittance result is established in relation with the irradiance observed by a pyranometer, whose domed glass cover makes it insensitive to IAM. This leads to considerable implementation difficulties mainly due to tracking requirements. We suggest an easier alternative consisting of a relative measurement. The module being test is static and placed side-by-side with another module covered by uncoated bare glass, whose IAM response is well known, and the relative transmittance of the tested

10.4229/EUPVSEC2023/3CO.11.6
020253-005

module is established in relation that corresponding to the latter. Interestingly, the Martin-Ruiz analytical equations for the diffuse irradiance allows for significantly improving the accuracy of the results.

All of these reasons lead us to select the Martin-Ruiz model to be implemented in SISIFO, an open-source software developed by the IES-UPM and put at public disposal from [16]. That has been used to analyze the optical losses at a 100 MW one-axis tracked PV plant in the Kalahari Desert, using daily measurements of the soiling rate throughout two years with different cleaning schedules as input. The results suggests that the IAM soiling component is between 40 and 50% of that corresponding to the change in the refractive index. Hence, taking it into consideration will increase the certainty of energy yield forecast and thus, a better PV plant project evaluation.

7 ACKNOLEDGMENTS

The authors are very grateful with Francisco Martinez-Moreno and Pedro Moreno for the assistance related to the experimental work of Figure 2A and with Javier Ledesma for the valuable support with SISIFO software. The authors also want to express their gratitude to ACCIONA for providing experimental data of their PV Plants. This work has received financial support from the Spanish MCIN/AEI/10.13039/501100011033 through Project GREASE (Proyectos de I + D + i Retos Investigación 2020), with grant number PID2020-113533RB-C31 and from an internal project with ACCIONA GENERACIÓN RENOVABLE, S.A

8 REFERENCES

[1] P. Gilman, N. Blair, M. Mehos, C. Christensen, S. Janzou, and C. Cameron, "Solar Advisor Model User Guide for Version 2.0," United States, 2008. [Online]. Available: https://www.osti.gov/biblio/937349

[2] D.L. King, W.E. Boyson, and J. Kratochvil, "Photovoltaic array performance model," United States, 2004. [Online]. Available: https://www.osti.gov/biblio/919131

[3] N. Martin and J.M. Ruiz, "Calculation of the PV modules angular losses under field conditions by means of an analytical model," *Solar Energy Materials and Solar Cells,* vol. 70, no. 1, pp. 25-38, (2001), doi: https://doi.org/10.1016/S0927-0248(00)00408-6.

[4] IEC-61853-2, "Photovoltaic (PV) module performance testing and energy rating-Part 2: Spectral response, incidence angle and module operating temperature measurements," ed: International Electrotechnical Commission Geneva, Switzerland, (2012).

[5] R. Preu, G. Kleiss, K. Reiche, and K. Bücher, "PV-module reflection losses: measurement, simulation and influence on energy yield and performance ratio," in *Proc. 13th European PVSEC,* (1995), vol. 1465, pp. 23-27.

[6] E.A. Sjerps-Koomen, E.A. Alsema, and W.C. Turkenburg, "A simple model for PV module reflection losses under field conditions," *Solar energy,* vol. 57, no. 6, pp. 421-432, (1996).

[7] A. Mermoud and B. Wittmer, "PVSYST user's manual," *Switzerland, January,* (2014).

[8] B. Marion, "Numerical method for angle-of-incidence correction factors for diffuse radiation incident photovoltaic modules," *Solar Energy,* vol. 147, pp. 344-348, (2017).

[9] Y. Xie, M Sengupta, A Habte, and A Andreas, "The "Fresnel Equations" for Diffuse radiation on Inclined photovoltaic Surfaces (FEDIS)," *Renewable and Sustainable Energy Reviews,* vol. 161, p. 112362, 2022/06/01/ (2022), doi: https://doi.org/10.1016/j.rser.2022.112362.

[10] M. García, L. Marroyo, E. Lorenzo, and M. Pérez, "Soiling and other optical losses in solar-tracking PV plants in navarra," *Progress in Photovoltaics: Research and Applications,* vol. 19, no. 2, pp. 211-217, (2011).

[11] D.L. King, J.A. Kratochvil, and W.E. Boyson, "Measuring solar spectral and angle-of-incidence effects on photovoltaic modules and solar irradiance sensors," in *Conference Record of the Twenty Sixth IEEE Photovoltaic Specialists Conference-1997,* (1997): IEEE, pp. 1113-1116.

[12] M. Ezquer, "Determination of incidence angle modifiers for different photovoltaic technologies through outdoor measurements," in *Proc. 29th Eur. Photovoltaic Sol. Energy Conf. Exhib.,* (2014), pp. 2822-2828.

[13] M.K. Smith, C.C. Wamser, K.E. James, S. Moody, D.J. Sailor, and T.N. Rosenstiel, "Effects of natural and manual cleaning on photovoltaic output," *Journal of solar energy engineering,* vol. 135, no. 3, (2013).

[14] A.A. Du Plessis, J.M. Strauss, and A.J. Rix, "Application of dust mitigation strategies to single-axis-tracking photovoltaic modules in the semi-arid areas of South Africa," *IET Renewable Power Generation,* vol. 14, no. 15, pp. 2781-2790, (2020).

[15] R. Conceição, J. González-Aguilar, A.A. Merrouni, and M. Romero, "Soiling effect in solar energy conversion systems: A review," *Renewable and Sustainable Energy Reviews,* vol. 162, p. 112434, (2022).

[16] IES-UPM. "SISIFO Simulation Software." https://sisifo.info.

PERFORMANCE GAIN OF SHADING TOLERANT PV MODULES IN DIFFERENT ELECTRICAL PV SYSTEM SETUPS

Franz Baumgartner, Cyril Allenspach
ZHAW, Zurich University of Applied Sciences, School of Engineering, IEFE and IMES
www.zhaw.ch/~bauf, Technikumstr. 9, CH-8401 Winterthur, Switzerland
Email: bauf@zhaw.ch, Website: zhaw.ch/=bauf

ABSTRACT: With the increasing number of PV power plants on roofs, less than optimal locations are also being utilised with PV modules that are partially shaded. Recently shading tolerant PV modules equipped with a higher number of bypass diodes entering the markets proposing tens of percent yield gain compared to regular PV modules. In addition to the duration and intensity of the shading, the actual losses in the bypass device itself also play a decisive role in the annual electricity yield. The PVshade tool from the ZHAW takes into account the electrical losses at the respective operating points, e.g. of each individual DC/DC optimiser, and can thus calculate the differences in the annual solar yields of different systems using geometric shadow analysis. As a result, for light shading with a chimney, a yield gain of about 2% was determined for a shade-tolerant module that uses a bypass diode in parallel with each solar cell, compared to a standard module with 3 bypass diodes when using string inverters. Whereas the additional use of individual DC/DC optimisers behind a few shading tolerant modules in conjunction with a conventional string inverter can achieve an additional yield boost of up to 1%. More robust bypass diodes in the PV modules helps a lot to reduce module hot spots and increase reliability. This reduces the risk of having to replace an optimiser or a module on the roof, which can result in considerable craftsman costs of several thousand euros per replacement.
Keywords: Shading, PV Module Design, Energy Yield, Energy Yield Simulation

1 INTRODUCTION

Shading may change solar cell current by a few tens of percent and voltage by only a few percent. Therefore, the solar cells in the PV generator would have to be connected in parallel, but this is ruled out because of the module currents of a few 100 amps if all full cells are in parallel. Consequently, over the last half century, PV modules have been produced primarily as a series connection of solar cells.

The unequal module currents in the series connection are commonly bridged by Bypass Devices (BD). Robust diodes, single FET's or elaborate DC/DC converters are used as BD. Low number of cells in a string protected by a DB leads to less power losses e.g., at partial shading and lower hot spot temperatures in the solar cell which is operated in the string not as a generator anymore. Too small numbers of serials connected cells and high forward voltage of BD generates additional losses [1,2]. To prevent reverse current in parallel connecting strings blocking DB used, first introduced in space application in the first decade after the invention of the solar cell [3].

Recently different electrical wiring layouts of shading tolerant utility-scale photovoltaic modules are investigated and compared in terms of, shading tolerance, and costs [4]. An optimum number of 12 serial connected silicon half-cells was found, which again was connected in parallels with similar substrings and protected by four bypass diodes in the PV module.

Nowadays PV module products starts on the markets which consists also of in parallel connected strings of in serial connected small size solar cells without blocking diodes included. In their advertising information's, these manufacturers give the impression of over 60% additional yield compared to other products but do not mention the measurement setup and thus deceive the customer into believing that they have an annual additional yield of this magnitude [5].

Similar marketing approaches were used by DC/DC power electronic optimizer manufacturers in the last decade, where customers were promised up to 30% additional yield, but this is usually wrong by a factor of ten for the annual balance of typical systems [6].

Commercial PV planning tools map the shadow cast on the PV generator, but today they often only use an inadequate loss model of the DC/DC optimiser, resulting in deviations of up to 10% in the annual analysis when these are compared with conventional string inverters [7]. ZHAW has determined target-oriented loss models with the efficiency measurements in the laboratory on commercial optimizer products during regular emulated operation in the string [7-13].

2 ZHAW ANNUAL SIMULATION TOOL

The PVshade tool was used to assess the annual yield of partial shaded PV generators including suitable power loss models of DC/DC optimizer and inverter and simulation of the operating point of each optimizer in the string with an interval of less than one hour. PV Shade was developed at ZHAW and the method and results of annual yield for string and optimizer system were reported [13].

2.1 Analysed PV Modul cell and BD wiring

The following four PV module layouts including standard bypass diodes used her (Fig. 1 – 4) and a 5th standard full cell module also with 3 bypass diodes.

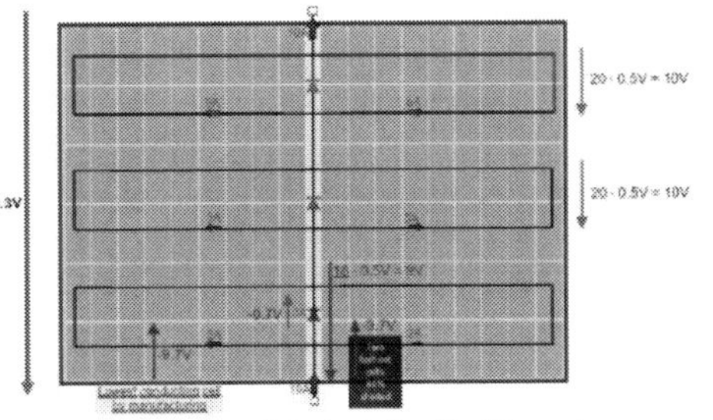

Figure 1: A 60% partial shaded cell affects the current flow within half-cut cell of butterfly PV modules equipped with 3 bypass diodes.

10.4229/EUPVSEC2023/3DO.16.1
020254-001

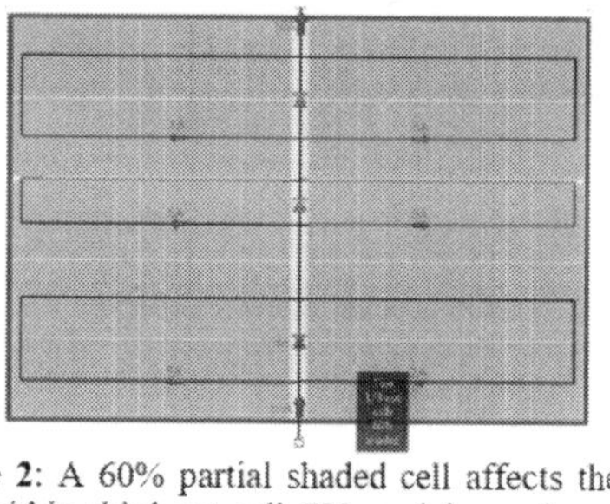

Figure 2: A 60% partial shaded cell affects the current flow within third-cut cell PV module equipped with 3 bypass diodes.

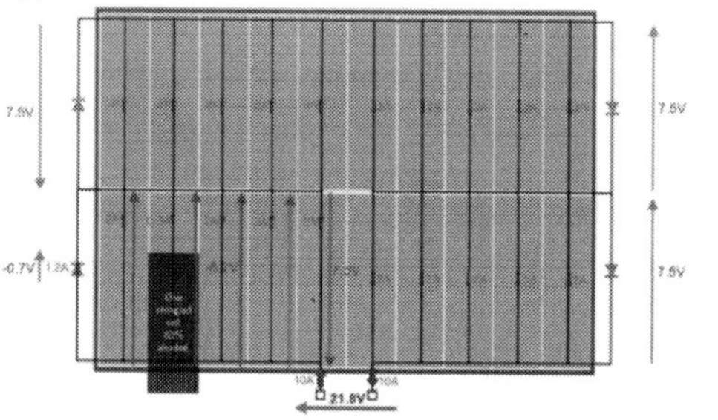

Figure 3: PV module consisting of four series-connected segments of five shingled solar strips in parallel, whereby a bypass diode (in green) is installed to each quadrant [5].

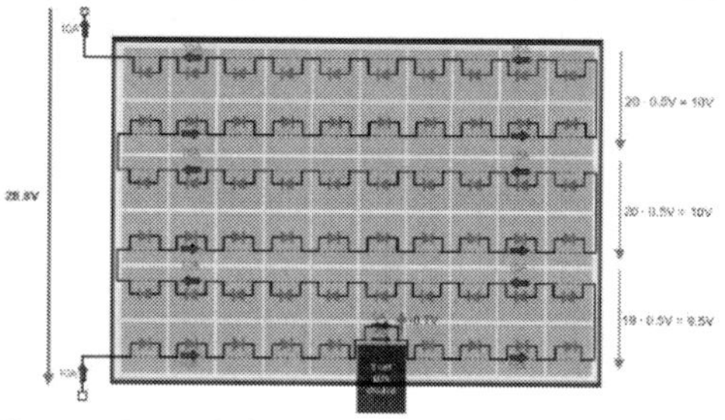

Figure 4: Internal circuitry of a (hot-spot free) PV module with a bypass diode (in green) installed at each PV cell (AESolar).

For the different DB wiring Fig. 1 to 3 a forward voltage of 0.7V was used in the simulations.

2.1 Shading Simulation and Electronic Losses

The electrical PV module output is connected in this study by a regular String Inverter (SINV) as the product from Huawei, SUN2000 3.68KTL, the use of two or four independent optimizer (from Huawei, SUN2000-450W-P2) behind the most shaded PV modules and the all DC/DC optimizer solution from Solaredge S5000B (Fig. 5).

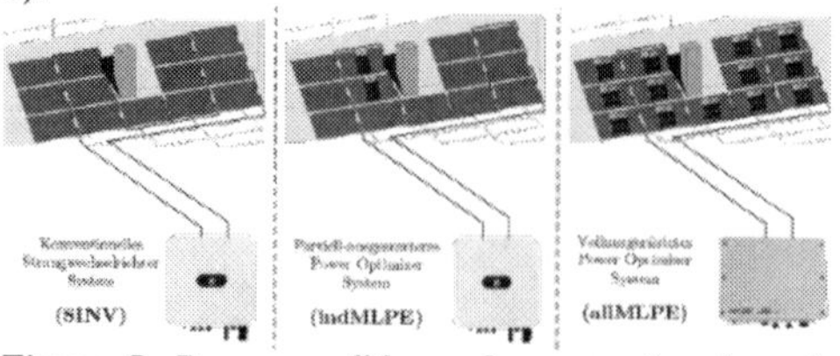

Figure 5: Power conditioner for processing the solar power from the different PV module circuits.

For die solar irradiance the data from Metosuisse for Zurich Kloten are used.

3 RESULTS AND DISCUSSION

The different shading scenarios are described by the Shading Index (SI) as defined in [16]. Light shading by a chimney is shown in Fig. 6 and 7 with the only difference in the number of used PV modules in the PV plant. The results of medium shading with SI values between 3 and 6% (Fig. 8 and 9) are given by a dormer as also often seen as shading object on tilted roofs with a smaller and large gap to PV modules beside the dormer.

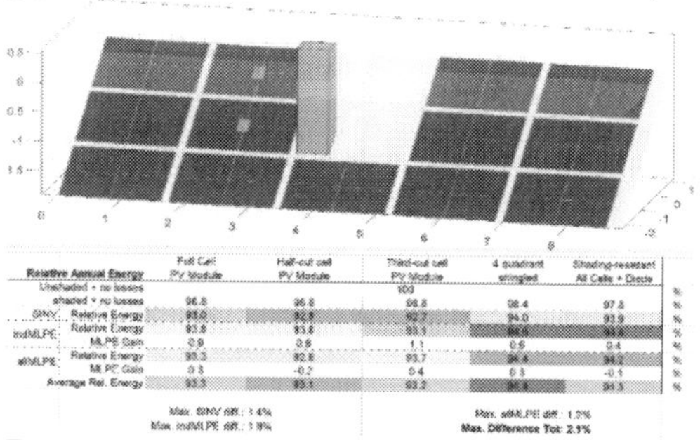

Figure 6: ZHAW PVShade annual yield comparison of different of shading scenario 1 – chimney with 13 PV modules with a shading index of about 2%. The orange dops show the location of the independent optimizer.

For the 13 modules in the string (Fig. 6) due to shading of the chimney not 100% but only 96.8% of DC electricity is available at the MPP of all PV full cell modules together. Connecting the standard full cell PV modules with STC power of 5.2kW to the standard SINV the losses in the DC/AC converter leads to only 93.0% of AC energy feed into the grid (cabling losses are neglected here).

The shading PV modules 4-quadrant shingled (Fig. 3) and bypass diodes for all cells (Fig. 4) are performing by about 1.5% better on the AC output using two independent optimisers as shown in Fig. 6. There is no relevant gain in AC output found, if all PV modules would be equipped by an optimizer due to the additional losses of these additional devices special in the unshaded operation points [13].

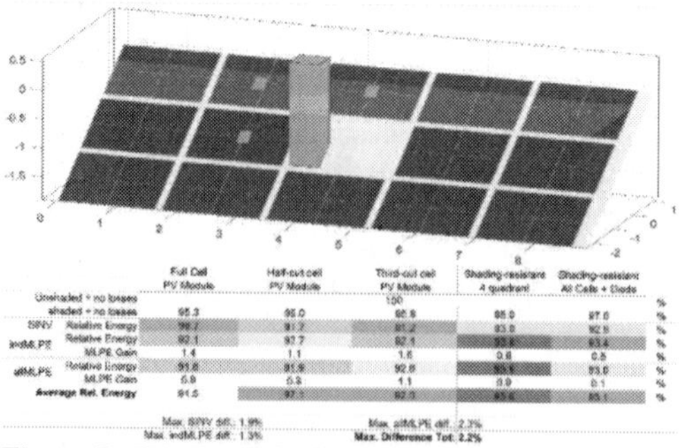

Figure 7: ZHAW PVshade annual yield comparison of different of shading scenario 2 – chimney with 14 PV modules with a shading index of about 3%.

Adding another module behind the chimney (Fig.7) increase the total AC output power by 1/13 but decrease the relative energy efficiency of the whole system by about 1% due to additional shading behind the chimney. Again, the same shading tolerant PV module than before perform in combination with independent optimizer best. Now the difference from that optimum to the full cell PV module with regular SINV increase to about 3%.

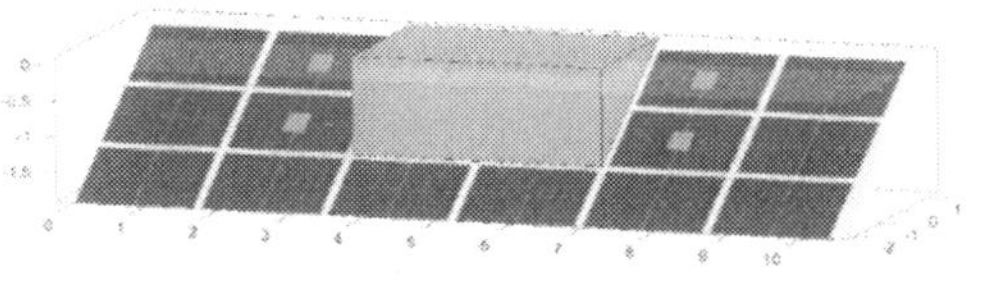

		Full Cell PV Module	Half-cut cell PV Module	Third-cut cell PV Module	Shading-resistant 4 quadrant	Shading-resistant All Cells + Diode	
Unshaded + no losses				100			%
shaded + no losses		94.3	93.5	93.5	95.5	95.6	%
SINV	Relative Energy	88.4	87.4	87.1	90.8	92.3	%
indMLPE	Relative Energy	89.8	88.7	88.7	91.6	92.7	%
	MLPE Gain	1.4	1.4	1.7	0.8	0.3	%
aMLPE	Relative Energy	88.8	89.4	89.7	92.5	92.7	%
	MLPE Gain	2.4	2.2	2.9	1.8	0.4	%
Average Rel. Energy		89.5	88.5	88.5	91.7	92.5	%

Max. SINV diff: 5.1% Max. aMLPE diff: 3.0%
Max. indMLPE diff: 3.9% **Max. Difference Tot: 8.6%**

Figure 8: ZHAW PVshade annual yield comparison of different of shading scenario 2 – dormer with 14 PV modules with a shading index between 3% and for the shading tolerant PV modules up to 6% with a smaller gap to the modules beside the dormer.

Relative Annual Energy		Full Cell PV Module	Half-cut cell PV Module	Third-cut cell PV Module	Shading-resistant 4 quadrant	Shading-resistant All Cells + Diode	
Unshaded + no losses				100			%
shaded + no losses		95.2	94.4	94.8	97.3	97.1	%
SINV	Relative Energy	90.1	89.1	88.9	92.1	92.9	%
indMLPE	Relative Energy	91.0	90.9	90.1	92.6	93.1	%
	MLPE Gain	1.0	1.1	1.4	0.8	0.3	%
aMLPE	Relative Energy	91.4	90.4	90.8	93.2	93.1	%
	MLPE Gain	1.5	1.5	2.1	1.2	0.3	%
Average Rel. Energy		90.8	89.8	90.0	92.6	93.0	%

Max. SINV diff: 3.9% Max. aMLPE diff: 2.4%
Max. indMLPE diff: 3.0% **Max. Difference Tot: 4.3%**

Figure 9: ZHAW PVshade annual yield comparison of different of shading scenario 2 – dormer with 14 PV modules and a little more distance to the modules beside with a shading index between 3% and for the shading tolerant PV modules up to 5% with a smaller gap to the modules beside the dormer.

The results for the higher degree of shading with SI up to 6% due to the placement of the modules close to the dormer, the benefit of shading tolerant PV modules is 4% in annual PV yield gain even if the cheaper SINV solution is realised.

5 CONCLUSIONS and OUTLOOK

Under light shading conditions with a chimney or medium shading of a dormer on the roof top PV system, the use of shading tolerant PV modules plus installing an independent DC/DC Optimizer behind a few shaded PV modules achieves the highest annual yields. Thus, this advanced system setup benefits in 3% higher yield, while 2% belongs to the shading tolerant module, equipped with additional built in bypass diodes. For medium shaded condition of a dormer the results similar with 3% higher output, if not the smallest gab is applied by placing the PV module beside the dormer.

However, the use of more robust BD reduces the maximum hot spot load of PV modules and thus leads to a higher reliability of the overall system. This reduces the risk of having to replace an optimiser or a module on the roof, which can result in considerable craftsman costs of several thousand euros per replacement.

Finally, the banal recommendation to install as few modules as possible in the shade and possibly to remove shading objects (chimneys) must not be missed.

For PV planners and installers, comprehensive results such as yield comparisons for different shading and system variants and recommendations, including the use of shade-tolerant PV modules, will be published in the IEA PVPS T13 report next year [19]. On the beginning of 2024 ZHAW will present the yield results of PVshade on an interactive website, for selected cases of partial shading and real electronic product losses from string inverter to types of optimizer [20].

ACKNOWLEDGEMENT

The module level power electronic performance estimation of the ZHAW is funded by the Swiss Federal Office of Energy, with Project Number: SI/502247-01. Accordingly, the ZHAW is involved in the analysis of shading scenarios, to identify characteristic, benchmark shading situations, as a part of the collaboration by the International Energy Agency (IEA) PVPS Task 13 ST2.5. To support the development of a future technical specification, benchmark shading adaption efficiency cases are developed by ZHAW together with the IEC TC 82/WG 6.

REFERENCES

[1] Viera, R.G. et al.. A Comprehensive Review on Bypass Diode Application on Photovoltaic Modules. Energies 2020, 13(10), 2472, https://doi.org/10.3390/ en13102472

[2] patents on BD,
G. Wolff et al., 1973, US 3 952 324;
R. Hollaus et. al., 1986, US 4 567 316

[3] Baron, W.R.; Virobik, P.F.. Effect of shadows on solar array output. In Proceedings of the IEEE 4th Photovoltaic Specialists Conference, Cleveland, OH, USA, 2–3 June 1964

[4] Li C. Rendler, C. Reichel, M. F. Berwind, A. Heimsath, D. H. Neuhaus, *"Innovative Layouts for Utility-Scale PV Modules: Module Characteristics, Shading Tolerance, and Electricity Costs"*, IEEE US PVSC June 2023.

[5] manufacturer of shading tolerant PV module calming their products producing more than 63% and 82% energy compared to competitor products https://www.solaria.com/powerxt-400-solar-panels [website visited 2023-09-18]

[6] SolarEdge Technologies Inc., Solaredge Fact Sheet, www.solaredge.com/sites/default/files/se-fact-sheet-na.pdf, [Online: accessed 15.05.2022].

[7] C. Allenspach, F. Baumgartner, *Lab measuments of power optimizer efficiency and performance simulation,* In Proceedings of the 40th European Photovoltaic Solar Energy Conference and Exhibition (EUPVSEC), Online. WIP Renewable Energies, Lisbon, Portugal, 2023

[8] F. Baumgartner, R. Vogt, C. Allenspach, F. Carigiet, In Proceedings of the 38th European Photovoltaic Solar Energy Conference and Exhibition (EUPVSEC), Online. WIP Renewable Energies, Munich, Germany, 2021 650-654.

[9] C. Allenspach, V. G. de Echavarri Castro, S. Richter, C. Meier, F. Carigiet, F. Baumgartner, In Proceedings of the 37th European Photovoltaic Solar Energy Conference and Exhibition (EUPVSEC). WIP Renewable Energies, Munich, Germany, 2020 1188-1194.

[10] C. Allenspach, A. Bänziger, A. Schneider, F. Carigiet, F. Baumgartner, Performance comparison of a P370 power optimizer system and a string inverter system (poster), In 20th Swiss National Photovoltaic Conference, 2022.

[11] C. Allenspach, F. Carigiet, A. Bänziger, A. Schneider, F. Baumgartner; *Power Conditioner Efficiencies and Annual Performance Analyses with Partially Shaded Photovoltaic Generators Using Indoor Measurements and Shading Simulations*; Sol. RRL 2022, 2200596.

[12] F. Baumgartner, C. Allenspach in Chapter 3.1.4 of the IEA Report, *Performance of New Photovoltaic System Designs*, Report IEA-PVPS T13-15:2021, April 2021, ISBN 978-3-907281-04.

[13] R. Bründlinger, N. Henze, G. Lauss, J. Liu, In Proceedings of the 26th European Photovoltaic Solar Energy Conference and Exhibition (EUPVSEC). WIP Renewable Energies, Munich, Germany, 2011, 3204-3211.

[14] C. Deline, B. Marion, J. Granata, S. Gonzalez, A performance and economic analysis of distributed power electronics in photovoltaic systems, Technical report, National Renewable Energy Laboratory (NREL) and Sandia National Laboratories, 2011.

[15] C. Deline, S. MacAlpine, In Proceedings of the 2013 IEEE Energy Conversion Congress and Exposition. 2013 4801-4807.

[16] A. J. Hanson, C. Deline, S. M. MacAlpine, J. T. Stauth, C. R. Sullivan, IEEE Journal of Photovoltaics 2014, 4, 6 1618.

[17] K. Sinapis, C. Tzikas, G. Litjens, M. van den Donker, W. Folkerts, W. van Sark, A. Smets, Solar Energy 2016, 135 731.

[18] C. Deline, J. Meydbray, M. Donovan, Photovoltaic shading testbed for module-level power electronics: 2016 performance data update, Technical report, National Renewable Energy Laboratory (NREL) and PV Evolution Labs / DNV GL, 2016.

[19] F. Baumgartner, C. Allenspach, R. Bründlinger, U. Jan.; *Performance of partial shaded PV generators operated by optimized power electronics*, will be published in Aug. 2024, IEA PVPS Task 13.

[20] pvshade.engineegring.zhaw.ch
Web PVShade online results available; run by ZHAW School of Engineering (ZHAW), Institute of Energy systems and Fluid-Engineering (IEFE) [Online: accessible in Jan 2024].

10.4229/EUPVSEC2023/3DO.16.1
020254-004

COMAPRISON OF INDOOR ELECTRICAL MEASUREMENT AND OUTDOOR ENERGY YIELD EVALUATION OF SHADE-RESISTANT PV MODULES UNDER SHADING CONDITIONS

Sebastian Dittman[1], Tales Siqueira[1,2], Hugo Sanchez[1], Afshin Bakhtiari[2],
Waldemar Meier[2], Carlos Meza[1], Ralph Gottschalg[1], Hamed Hanifi[2]

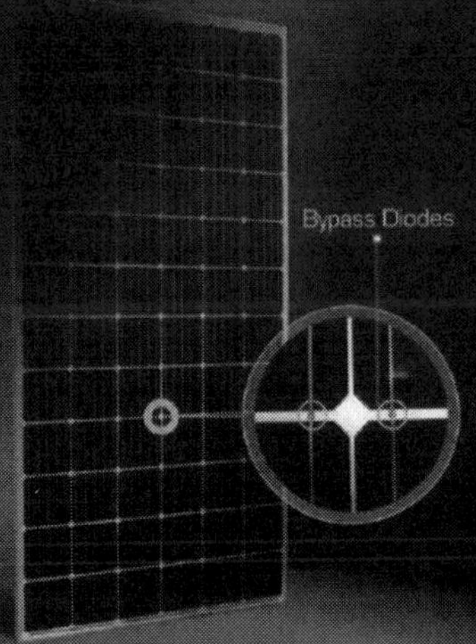

EUPVSEC 2023, Lisbon

1 Anhalt University of Applied Sciences, Koethen, Germany
2 AE Solar, Messerschmittring 54, Koenigsbrunn, Germany

AESOLAR HOCHSCHULE ANHALT University of Applied Sciences Apollo

CONTENT

Apollo Anhalt Photovoltaic Performance and Lifetime Laboratory HOCHSCHULE ANHALT University of Applied Sciences AESOLAR

MOTIVATION

An overview on market segments

Two biggest PV markets in Europe

- Germany is the biggest European PV market
- 60% rooftop[1]

- Netherlands is the biggest market per capita
- Over 80% rooftop

Rooftop installations are one of the major markets which secures the distrbuted and renewable energy

Dr. Hamed Hanifi [1]: Fraunhofer ISE 3

40th European Photovoltaic Solar Energy Conference and Exhibition

MOTIVATION

New markets

- Rooftop market is the king, but ...
- High demand to renewable energy, especially after gas crisis in Europe, has enhanced the niche PV markets
  - Example: PV modules on balconies

Dr. Hamed Hanifi 4

MOTIVATION

Challenges of rooftop PV

- Rooftops are not always 100% exposed to the sun
- Building elements can cast shadow on PV modules
  - Chimneys
  - Trees
  - Antennas
  - Other building structures

40th European Photovoltaic Solar Energy Conference and Exhibition

MOTIVATION

Challenges of rooftop PV

- Climate and land specifications increase the probability of partial shading:
  - Limited area (more obstacles and shadings + more modules per area)
  - snow
  - bird drops
- PV modules are very sensitive to partial shading and lose power drastically

MOTIVATION

Challenges of rooftop PV

- Continuous partial shading might lead to hot-spots or in the worst cases fire

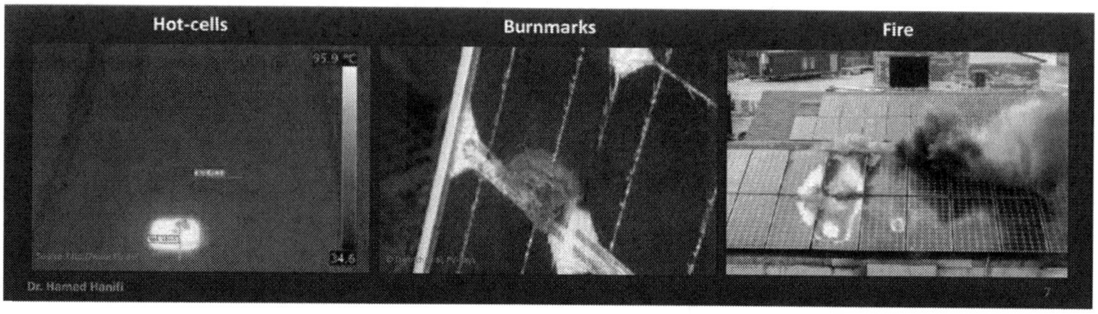

020256-007

40th European Photovoltaic Solar Energy Conference and Exhibition

SHADE-RESISTANT PV MODULE

Smart HSF (Hot-spot-free) PV module of AESOLAR

- One bypass diode for every individual solar cell
- Integrated bypass diode technology

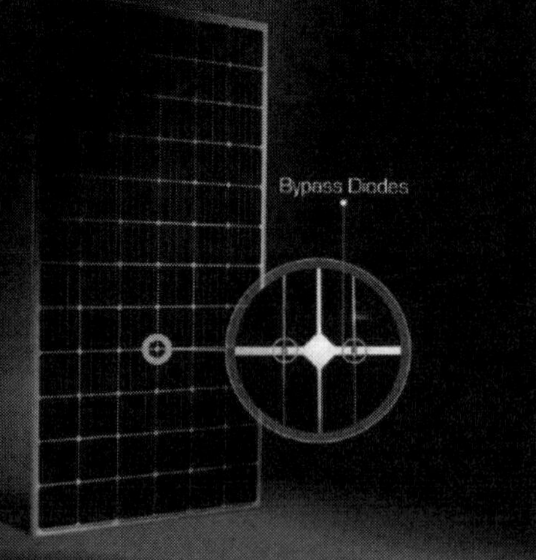

Dr. Hamed Hanifi

020256-008

METHODOLOGY

Module deisgn

- Two modules with similar bill of materials are tested under partial shading conditions
  - Shade-resistant (HSF)
  - Standard

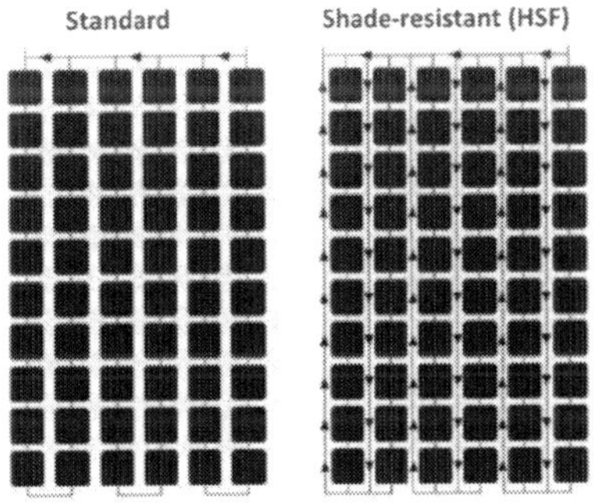

Module Properties	AE Standard Module	AE HSF Module
Manufacturer	AE Solar	AE Solar
Power rating [W]	335	335
Number of Cells	60	60
Cell Type	Monocrystalline PERC	Monocrystalline PERC
Cell Dimensions	158.75 x 158.75 mm²	158.75 x 158.75 mm²
Glass	3.2 mm tempered glass w/ AR coating	3.2 mm tempered glass w/ AR coating
Encapsulant	Ethylene Vinyl Acetate	Ethylene Vinyl Acetate
Backsheet	White multilayer backsheet	White multilayer backsheet
Frame	Anodized aluminum alloy	Anodized aluminum alloy

Dr. Hamed Hanifi

3

020256-009

METHODOLOGY

Indoor partial shading test

Measurement tool and setup

- Sun simulator based on LED
- Measurement under STC and low light conditions

Shading scenarios

- Shading the center
- Shading the bottom row
- Shading the side row
- Shading one cell
- Shading three rows
- Shading diagonal

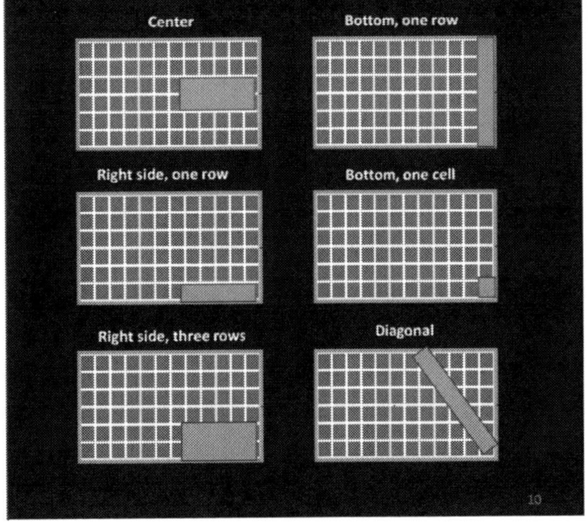

Dr. Hamed Hanifi

020256-010

METHODOLOGY

Outdoor test

Mounting in outdoor test field

- AE shade-resistant (shaded, mono-facial)
- AE standard (shaded, mono-facial)
- Reference module (unshaded, mono-facial)

Shading

- Wooden planks of 210 cm x 25 cm.

Measurement period

- 6 weeks between Oct. and Nov. 2022

Measurement tools

- All modules are connected to a SOL.Connect I-V tracer each.
- I-V curve, irradiance and backside temp. measurements are taken every 10 seconds.

Dr. Hamed Hanifi

Apollo HOCHSCHULE ANHALT University of Applied Sciences AESOLAR

Energy yield measurement setup at the Anhalt Photovoltaic Performance and Lifetime Laboratory (APOLLO) in Bernburg, Germany

RESULTS

Indoor measurement

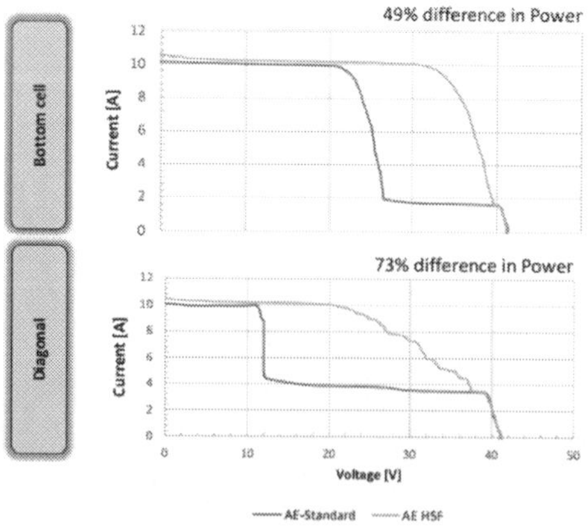

Dr. Hamed Hanifi

Apollo HOCHSCHULE ANHALT University of Applied Sciences AESOLAR

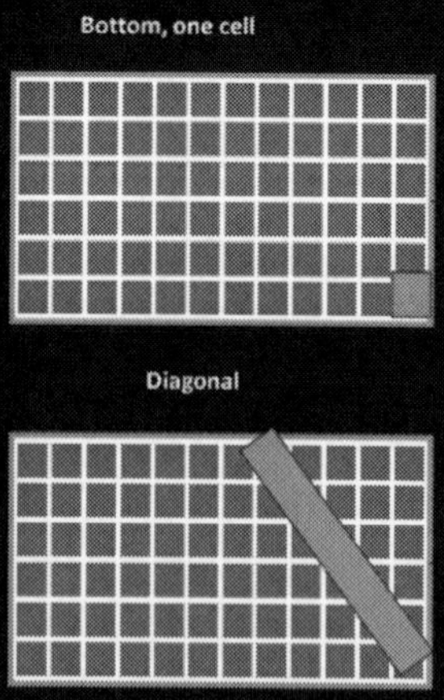

40th European Photovoltaic Solar Energy Conference and Exhibition

Apollo HOCHSCHULE ANHALT University of Applied Sciences AESOLAR

RESULTS

Outdoor measurement

- Evaluation of IV curves in real-life scenario shows a significant advantage of HSF module over equivalent standard module

Differences of 60% in Pm between AE HSF and AE Standard

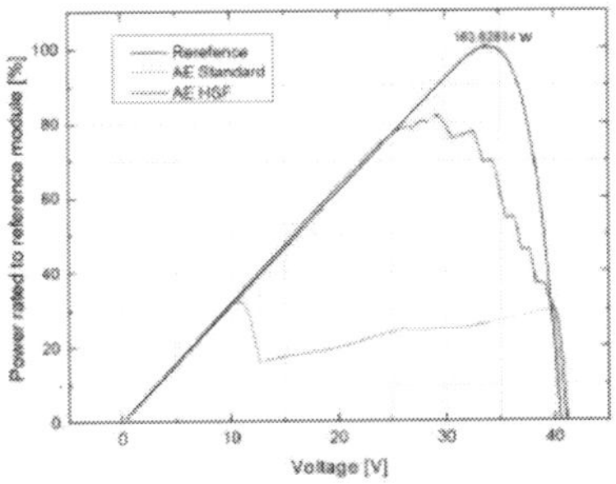

Dr. Hamed Hanifi

14

RESULTS

Outdoor measurement

Gain in energy yield of +49% compared to AE Standard

Clear day without shading

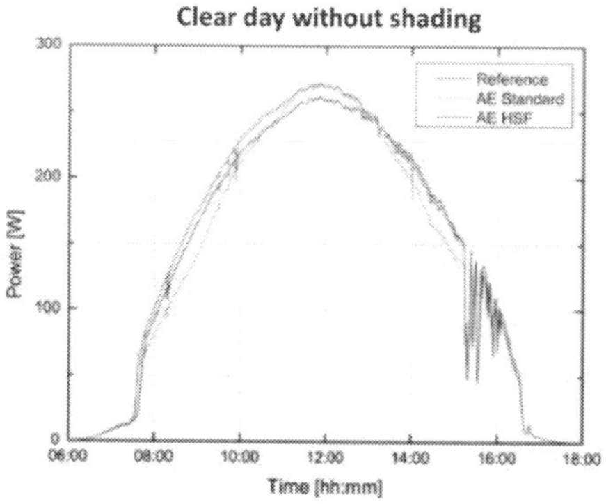

Clear day with shading

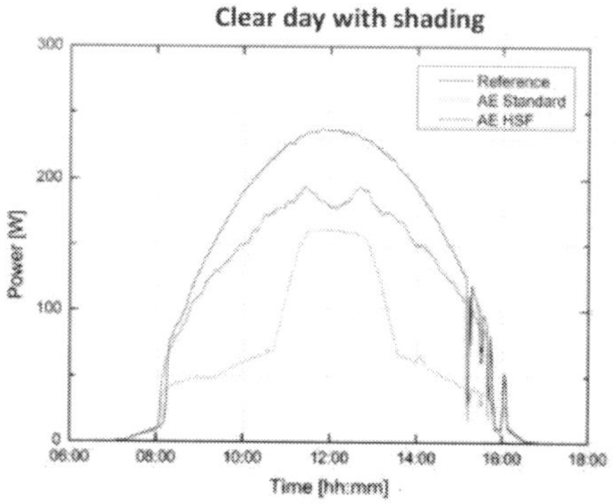

40th European Photovoltaic Solar Energy Conference and Exhibition

RESULTS

Outdoor measurement

- Spezific Energy yield for different day typs

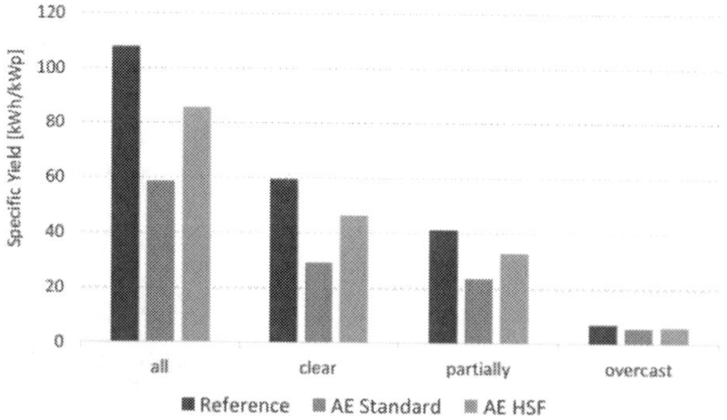

RELIABILITY

Reliability of diodes under stress test

- a test to evaluate the durability of bypass diodes under partial shading conditions
- The device switches the bias every 60 seconds for 10,000 cycles → 25 years

Forward bias

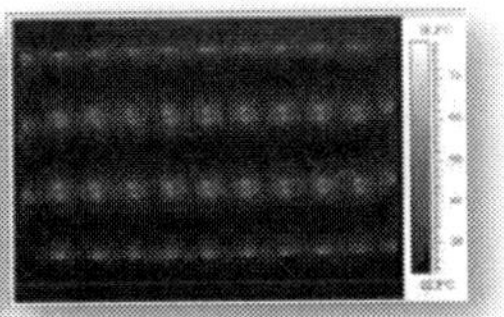

Reverse bias

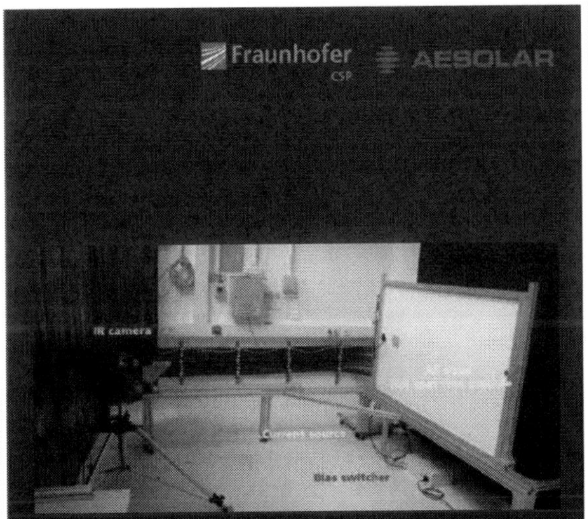

Test setup with bias switcher and IR camera to monitor module temperature on each bias mode

Dr. Hamed Hanifi   Hanifi et al, Solar Energy, 2019

40th European Photovoltaic Solar Energy Conference and Exhibition

RELIABILITY

Test results - IV

- The module is measured by flasher and electroluminescence before and after 10,000 cycles of stress test
- I-V flash results
  - no power loss
- EL results
  - No defects

The diodes are durable enough to work for 25 years

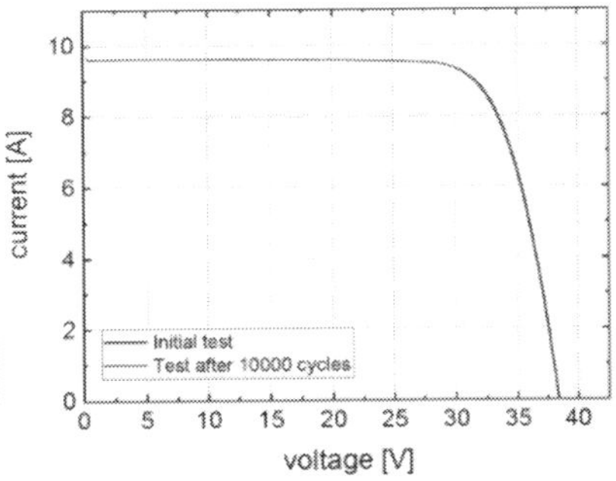

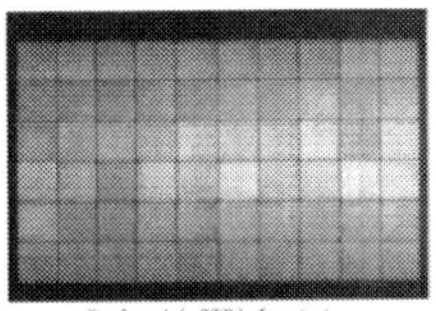

EL of module P07 before test

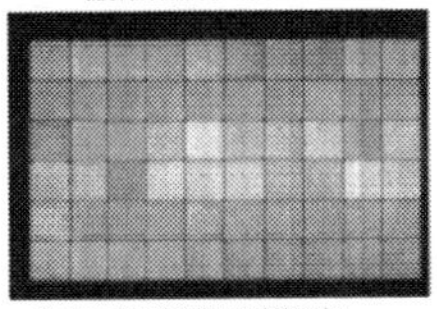

EL of module P07 after 10.000 cycles

Dr. Hamed Hanifi   Hanifi et al, Solar Energy, 2019

DISCUSSION

Economics

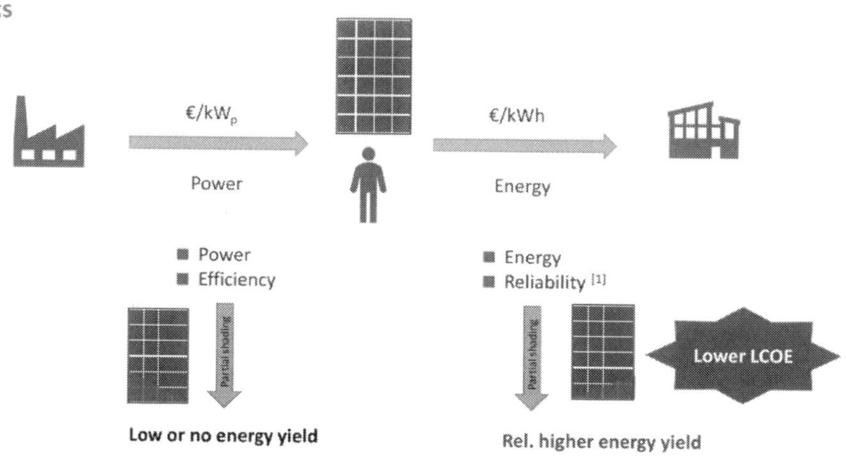

Dr. Hamed Hanifi   Hanifi et al, Solar Energy, 2019

DISCUSSION AND OUTLOOK

Power optimizers vs shade-resistant modules

Power optimizers

- String power of module is lost
- Regulates the voltage and current to avoid mismatch
- Higher capex

Shade-resistant panels

- The string is in operation
- No need for power optimizer for small to medium size shades

Lower system costs and higher yield → Lower LCOE

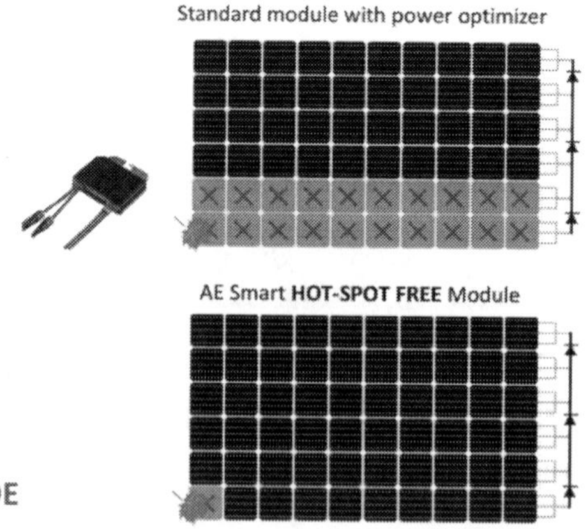

Dr. Hamed Hanifi

OUTLOOK

Smart HSF module 2.0

- First prototype was presented in Intersolar 2023
- Half-cells replaced the full-size solar cells
- Invisible bypass diodes
- Less bypass diodes and better functionality

SegmentPV project

Comparison of a standard half-cell and shade-resistant half-cell design of AESOLAR under partial shading conditions

21

Dr. Hamed Hanifi

020256-021

40th European Photovoltaic Solar Energy Conference and Exhibition

TAKEAWAY MESSAGE

SHADE-RESISTANT PV MODULES

- Customers should learn that kWh is important and not Wp!
- Project owners pay for efficiency and power [€/Wp] but they need energy [€/kWh]!

- Flexible with direction of shading

- Different shading scenarios has been testing indoors
- Indoor measurements show an **extra energy gain** up to 49% for single cell shading and up to 73% with diagonal shading

- Energy yield of unshaded modules show comparable result
- HSF module leads to 30% higher yield under shading conditions

HSF Shade-resistant module **promisses a higher energy yield to achieve a** lower LCOE **under partial shading conditions**

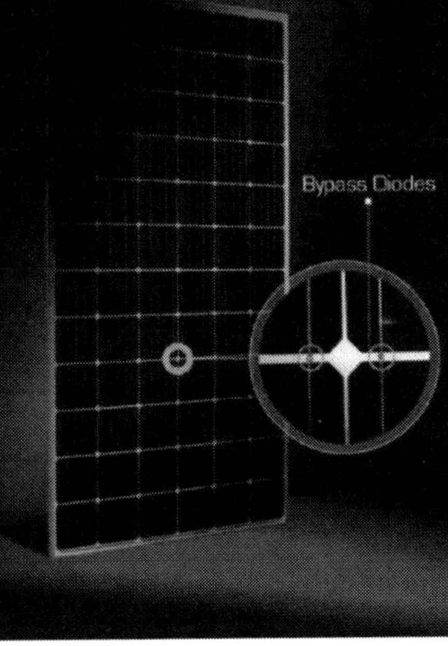

22

Dr. Hamed Hanifi

020256-022

AESOLAR | Hochschule Anhalt | Apollo

THANK YOU VERY MUCH FOR YOUR ATTENTION!

This work is supported by AESOLAR and Federal Ministry of Economics and Climate protection under project SEGMENTPV

Contact:
Dr. Hamed Hanifi
Head of Product Development and Technical Marketing

Email.: h.hanifi@ae-solar.com
Tel: +49 8231 97 82 68-2

Contact:
Sebastian Dittmann
Scientific staff Hochschule Anhalt
Team manager APOLLO

Email.: sebastian.dittmann@hs-anhalt.de
Tel: +49 3496 67 2354

A MEASURE FOR PV MODULE PERFORMANCE UNDER PARTIAL SHADING AND ITS APPLICATION TO CIGS AND CRYSTALLINE SILICON TECHNOLOGIES FOR REALISTIC SHADING SCENARIOS

Janina Moereke, Peter Borowski, Stefan Grünsteidl, and Thomas Dalibor
Avancis GmbH, Otto-Hahn-Ring 6, 81739 München, Germany
Phone: +49(0) 89 219620 442, Email: janina.moereke@avancis.de

ABSTRACT: Shading opacity and geometry of realistic shades, particularly on building façades, were measured and observed. These measurements determined that realistic shading scenarios include shading opacities between 60% to 90%, depending on the type and distance of the object casting the shadow. Lab-based experiments on the power output of commercially available PV modules (wafer-based and monolithic thin-film) were therefore performed at (partial) shading opacities in the stated range. The shading loss indicator (SLI) is introduced as a measure to benchmark the shading tolerance of a PV module for particular shading scenarios. The SLI compares power loss due to shading relative to ideal power loss under such conditions (i.e., when percentage of power lost equals the percentage of active module area shaded) and it is shown how the different interconnection and bypass diode placement strategies for different cSi and CIGS technologies result in different SLI. CIGS module performance is found to be superior at shades perpendicular and at an 45° angle to the long module edge compared to any cSi technology tested. For the least favourable geometry, where shading is applied parallel to the long module edge, performance of CIGS reached comparable levels to cSi for partially translucent shading.

Keywords: partial shading, shading opacity, power loss, BIPV, CIGS

1 INTRODUCTION

Whether in free field installations or integrated into urban structures, during operation shading will affect the performance of a photovoltaic (PV) module and system. Surrounding structures or vegetation as well as any material accumulation on the module surface will cause partial shading of modules and can hardly be completely avoided.

Earlier studies have looked into how to mitigate the effects by proposing different module interconnection strategies and the use of bypass diodes [1][2][3]. They were concerned mainly with avoiding module defects due to prolonged partial shading. This has been shown to create very high localised temperatures leading to semiconductor material failure in the module through hotspots [4][5]. As a consequence, standardised procedures exist to test for the susceptibility of a module for such permanent damage [6][7][8]. Monolithically integrated thin-film Copper Indium Gallium Selenide (CIGS) PV modules were shown to not degrade under typical partial shading scenarios [9].

Other studies have tried to accurately quantify the effect of various shading patterns [10][11][12][13]. These focused mainly on estimating the power reduction due to temporary shading. While it may not lead to permanent damage of the module, it still has an impact on the module or system performance. No standardised test exists currently to evaluate this power loss. Some efforts were made to develop such a test by looking at shading patterns typically found in roof-top installations [14][15][16]. As the possibilities of integrating photovoltaic modules into urban environments increase, shading tolerance becomes a crucial parameter for evaluating the module's performance for a variety of different shading patterns. Specifically, when integrating PV modules into the façade of a building (building-integrated PV (BIPV)), shading from protruding building features or adjacent structures will become a regular occurrence. Simulation models have been created for some PV technologies in order to predict the shading behaviour of a module [17][18], but need experimental validation.

Understanding how the performance of an individual module is affected by a certain shading pattern is important in order to provide a realistic assessment of the planned system. Ideally, power output is reduced by the same percentage as the active area that is being shaded. In reality, however, power reduction curves of photovoltaic modules in response to increasing partial shading are non-linear and the shading will affect types of PV modules differently. On system-level, analyses and development have been performed to determine how power loss from shading can be reduced through string planning and the use of module level electronics [19][20].

The present study is concerned with demonstrating how shading tolerance varies for different types of PV modules and proposing a measure to enable a fair comparison between those.

Typical shading scenarios for BIPV systems have been analysed by measuring the remaining irradiance on the shaded parts of a building façade in section 2. This follows the introduction of a measure for shading tolerance, the shading loss indicator (SLI), in section 3. The measurement setup for lab-based measurements in order to calculate the SLI is introduced in section 4 before presenting results and their interpretation in section 5 for the following PV module types and topologies: CIGS monolithic thin-film, crystalline silicon (cSi) full cell, half cell and shingled cells.

2 TYPICAL SHADING SCENARIOS IN (BI)PV SYSTEMS

For ground installed PV systems, typical shading scenarios include the accumulation of material from the environment on the usually inclined module surface, such as leaves, dust, snow or bird droppings. The resilience to partial shading of this kind is tested in the hot spot endurance test of IEC 61730-2 [7]. In these scenarios as well as in the IEC test, all of the light is blocked by the mentioned material lying directly on the module surface. Particularly (but not only) in BIPV systems, it is not uncommon that modules are shaded by objects further

Figure 1: Typical scenarios of partial shading in (BI)PV systems on sunny clear sky days: protruding building features (a and b), poles (c), trees (d and f) and adjacent building walls (e). All images show PV façades built with CIGS thin-film modules by AVANCIS. See also Figure 4.

away (protruding building elements, neighbouring buildings, poles) or with different opacity grades (trees). Figure 1 displays a few such scenarios of partial shading on photovoltaic façades. Rather than all of the light being

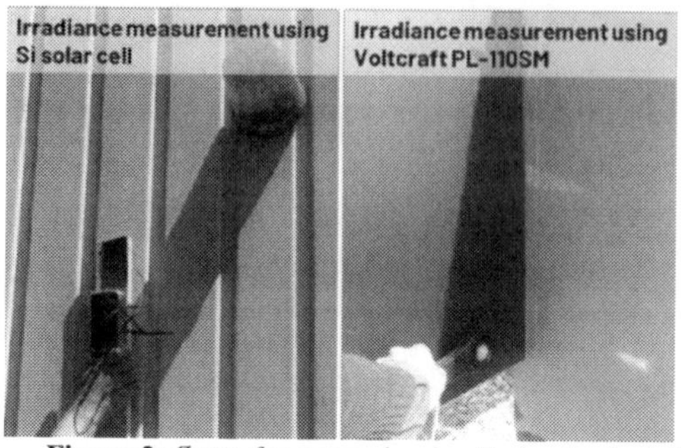

Figure 2: Setup for measuring irradiation in shaded and unshaded area on a façade from protruding building features with a Si solar cell (left) and the Voltcraft PL 110SM (right).

blocked, in these scenarios diffuse or reflected irradiation reaches the shaded area. Due to the nonlinearity of a PV cell's IV curve it is important to consider different opacity grades of shading for a realistic assessment of the performance of a PV module or system under partial shading. In order to be able to design a lab-based experiment that reflects these varying shading conditions, irradiance measurements were conducted on differently shaded building façades. Irradiance was measured using a VOLTCRAFT PL-110SM solar radiation measuring instrument with an accuracy of ±50 W/m². These measurements were validated by measuring the generated current of a Si solar cell connected to a multimeter (see Figure 2 for both setups). For each considered scenario, the ratio of irradiance measurements inside the shaded area to the unshaded area was calculated. Subtracted from 1 this then expressed the shade opacity p (see Table I). Other than the type of shading object, considered factors on shaded irradiance include:

- Weather (sunny clear sky or cloudy days)
- Façade orientation
- Time of day

Irradiance was measured in the vertical plane and in cardinal directions of south, east and west.

Table I: Measured irradiance values for realistic scenarios of partial shading. All measurements were conducted in Munich, Germany. Weather conditions were clear sky in March and light veil clouds in July with clearly visible shadows. Particularly with the latter conditions, irradiance measurements varied significantly (±50 W/m²) on a timescale of seconds.

Month and time	orientation	Shading object	Irradiance [W/m²] unshaded	Irradiance [W/m²] shaded	**Shade opacity p**
March, around noon	vertical, facing S	nearby wall	980	100	**0.90**
March, around noon	vertical, facing S	tree (without leaves), ~5m distance	880	250	**0.72**
March, around 14:00	vertical, facing S	nearby wall	690	90	**0.87**
March, around 14:00	vertical, facing S	pole, ~2m distance	1090	150	**0.86**
March, around 14:00	vertical, facing S	building S, ~30m distance	790	50	**0.94**
July, around 9:00	vertical, facing E	building, 90° to facade, ~20m long	640	120	**0.81**
July, around 9:00	vertical, facing E	broadleaf tree E, ~3m distance	690	110	**0.84**
July, around 9:00	vertical, facing E	conifer E, ~1-4m distance	710	80	**0.89**
July, around 9:00	vertical, facing E	pipe (diameter ~15cm), ~40cm distance	780	135	**0.83**
July, around 15:00	vertical, facing S	building projection (light grey)	350	120	**0.66**
July, around 15:00	vertical, facing W	building projection (grey)	500	210	**0.58**
July, around 15:00	vertical, facing W	broadleaf tree W, ~4m distance	510	220	**0.57**
July, around 15:00	30° tilt, facing S	broadleaf tree S, ~10m distance	750	310	**0.59**

10.4229/EUPVSEC2023/3DO.16.4
020257-002

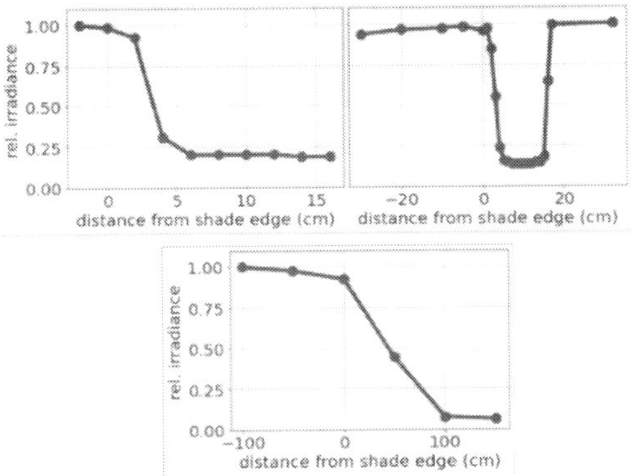

Figure 3: The transition from unshaded to shaded area for a vertically installed PV module and shading cast by an adjacent building in approximately 30m distance (bottom), by a lamp pole (top right) as well as a nearby wall (top left).

Measurement results are displayed in Table I. The values were read when the irradiance sensor was well within the respective shaded or unshaded area. These results show, that irradiance in the shaded area is reduced by 60 to 90% compared to the unshaded area right next to the shadow.

Furthermore, realistic shades are rarely uniform in shading opacity. Figure 3 illustrates this by showing the transition from unshaded to shaded area of a building wall or pole. Measurements were taken with a spatial resolution of a few centimetres. While in the main shaded area only around 10-20% of light arrives, in the transition area 30-50% of unshaded irradiation was recorded. In conclusion, any lab-based experiment would need to include shading measurements not just for an opacity grade of 100% (i.e., no light reaches the module in the shaded area), but also opacity grades between 50 and 90% (i.e., 50% to only 10% of light reaching the module in the shaded area) to be able to estimate how well a module performs under realistic partially shaded conditions.

3 DEFINING THE SHADING LOSS INDICATOR (SLI)

3.1 Aim

Similar to the temperature coefficient, with which the effect of temperature on the power output of a PV module is characterised by a single number, it is desirable to find a measure that characterises the effect of partial shading of a PV module on its power output. Ideally, the performance of a PV module under partial shading should also be characterised by a single number. Since partial shading occurs in infinitely many types (cf. section 2), this study identifies a small set of relevant scenarios of increasing partial shading with a given orientation with respect to the module that can be reproduced in the lab. With these proposed measurements, the performance of a PV module for each of these scenarios is characterised by a single number, the shading loss indicator. The increasing ratio of shaded module area, captured by the definition of the integral in the SLI, reproduces typical partial shading scenarios on building façades as they may occur on time scales of several minutes by the apparent movement of the sun (cf. Figure 4).

Figure 4: Movement of a shadow across a window on a building façade (red arrow) over the course of 20 minutes, motivating a shading measurement at increasing shaded area as defined in the SLI.

3.2 The Shading Loss Indicator (SLI)

The performance of a PV module under partial shading conditions is proposed to be characterised by a set $\{\mathrm{SLI}_\Pi\}$ of numbers for a standard set of shading scenarios Π. Each shading scenario Π includes the incremental increase in shaded area in a certain direction relative to the module's orientation, e.g., increasing shading along the short/long edge of the module or with a certain angle with respect to the long edge. For each such scenario, the SLI is defined as:

$$\mathrm{SLI}_\Pi = \frac{\int P_{\mathrm{sh}}(q)\mathrm{d}q}{\int P_{\mathrm{sh,ideal}}(q)\mathrm{d}q} - 1 \qquad (1)$$

where q denotes the fraction (from 0 to 1) of the area of the module that is shaded, $P_{\mathrm{sh}}(q)$ the power output of the partially shaded module in its maximum power point (MPP) and $P_{\mathrm{sh,ideal}}(q)$ the power output of a module with ideal power conversion under partial shading conditions. For the latter case, the power loss curve (cf. Figure 7) follows a straight line as the same percentage of power is lost as percentage of area shaded.

Depending on the performance of the module under test and the specific shading scenario Π, the SLI as defined in equation (1) can vary between 0 for a PV module with ideal response to partial shading and close to -1 for a PV module whose power decreases to zero when only a tiny part of its area is shaded.

Assuming a PV module with area A, spatially homogeneous photon conversion efficiency η, and a shaded portion of its area illuminated with a reduced irradiance $G_1 = (1-p)G_0$, where p denotes the shade opacity, the denominator in the definition of the SLI of equation (1) can be expressed as

$$A\eta \int_0^1 [G_1 q + G_0(1-q)]\mathrm{d}q = \frac{1}{2} P_{\text{unshaded}}(2-p) \quad (2)$$

For shading scenarios with only two values of irradiance on the surface of the PV module (unshaded and shaded with opacity p), the SLI can therefore be stated as

$$\mathrm{SLI}_{\Pi}^{p} = \frac{2}{P_{\text{unshaded}}\,(2-p)} \int_0^1 P_{\text{sh}}^{p}(q)\mathrm{d}q - 1 \quad (3)$$

where P_{unshaded} is the power of the fully illuminated PV module in MPP.

For completely dark shades (p=1, G_1=0W/m²) this then turns into:

$$\mathrm{SLI}_{\Pi}^{1} = \frac{2}{P_{\text{unshaded}}} \int_0^1 P_{\text{sh}}^{1}(q)\mathrm{d}q - 1 \quad (4)$$

Similarly, the SLI for the opacities later considered in lab measurements can be expressed as:

$$\mathrm{SLI}_{\Pi}^{0.5} = \frac{2}{3}\frac{2}{P_{\text{unshaded}}} \int_0^1 P_{\text{sh}}^{0.5}(q)\mathrm{d}q - 1 \quad (5)$$

$$\mathrm{SLI}_{\Pi}^{0.8} = \frac{5}{6}\frac{2}{P_{\text{unshaded}}} \int_0^1 P_{\text{sh}}^{0.8}(q)\mathrm{d}q - 1 \quad (6)$$

$$\mathrm{SLI}_{\Pi}^{0.9} = \frac{10}{11}\frac{2}{P_{\text{unshaded}}} \int_0^1 P_{\text{sh}}^{0.9}(q)\mathrm{d}q - 1 \quad (7)$$

By determining the SLI, a PV module's shading performance can be compared to that of other modules and technologies for the particular shading scenario investigated. SLI is reported as a percentage, varying between 0 % and -100%. The higher the absolute value of SLI, the more power is lost on top of ideal shading loss.

4 MEASURING THE SHADING LOSS INDICATOR

The shading loss indicator (SLI) was measured for commercially available PV modules with a variation of cell interconnection of wafer-based and thin-film technologies:

- cSi full cell
- cSi half cell
- cSi shingled
- monolithic thin-film CIGS

While the monolithically interconnected CIGS module had one bypass diode, based on the measurement results it is deduced wafer-based cSi modules include up to 3 bypass diodes each. Figure 5 illustrates the deduced current flow and bypass diode placements for all module types in this study. All modules investigated were monofacial. For bifacial modules the definition of the measurement procedure and SLI (see section 3) may have to be adjusted.

The IV curves were measured using a flash lamp sun simulator by halm in the lab environment by sweeping in reverse direction (from open circuit to short circuit). After measuring the IV characteristics of the unshaded, fully illuminated module, a shading material was placed over part of the module for partial shading and IV curves were measured again. After each measurement the shaded area was increased and the measurement repeated until all of the module was covered by the shading material. A final unshaded measurement confirmed that no permanent damage had been induced by the set of measurements with IV curves returning to their initial shape.

This set of measurements was repeated for three different shading scenarios:

1) Shading perpendicular to the long module edge
2) Shading parallel to the long module edge
3) Shading at a 45° angle to the long module edge

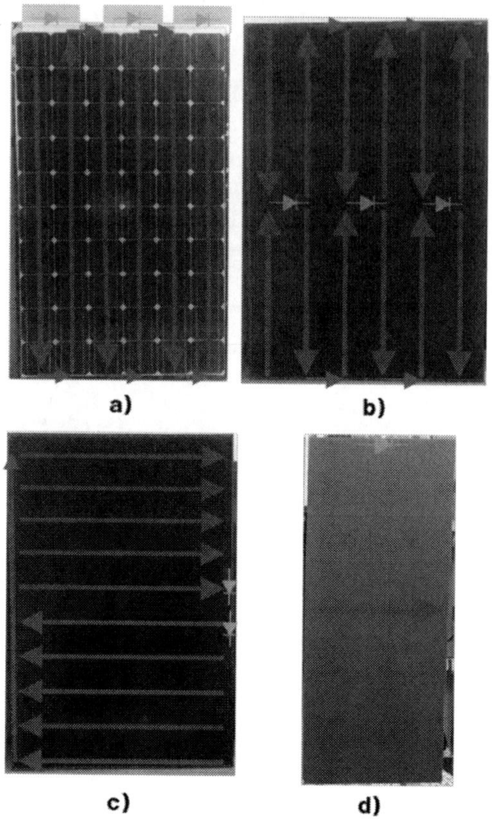

Figure 5: Deduced current flow (red arrows) and bypass diode placement (orange) of a) cSi full cell module, b) cSi half cell module, c) cSi shingled module and d) monolithic thin-film CIGS module.

Additionally, these measurements were repeated using shading material of four different opacity grades (see Table II). Varying the opacity of the shading material was included in this study in order to represent as much as possible realistic shading conditions such as established in section 2. Available materials were not able to reproduce an opacity of 60%. Therefore, measurements were performed at an opacity of 50%, which was easily produced using available large sheets of translucent paper. Spectrometer measurements comparing outdoor shade and lab measurements showed that, in the visible range for any one wavelength, transmission may differ by +/-5% while the difference can reach up to 10% for wavelengths in the UV and NIR region of the spectrum.

Table II: Opacity grades of shading materials used

	Material	opacity
1	Cardboard	100%
2	Flipchart paper	80%
3	Translucent white paper	50%
2+3	Flipchart + translucent paper	90%

10.4229/EUPVSEC2023/3DO.16.4
020257-004

From the IV characteristics the maximum power point was determined and set in relation to the maximum power point for unshaded conditions. Results were then plotted as relative maximum power against percentage of active module area shaded. From these curves the shading loss indicator as described in section 3 was calculated. Bar charts of SLI for different technologies thus show how much additional power is lost during shading compared to ideal shading loss (in the ideal case, percentage of power lost equals the percentage of active module area shaded for fully opaque shades).

Module design using parallel cell connection or bypass diodes is expected to create IV curves with more than one current plateau. The position of the maximum power point may therefore shift abruptly as shaded area is increased. Similar to a single cell as part of a series connection within a module, a change in IV curve of a module will affect the global power maximum of a string of modules. This challenges the MPP tracking algorithm of the inverter when trying to identify the global power maximum of the PV system and generally extends the shading loss to interconnected modules. From string planning, improved tracking algorithms to the use of module-level electronics, several methods have been introduced to minimise this challenge on system level. Thus, splitting longer strings into in parallel connected strings of modules with similar shading situation, adding sweeps or perturbations into the tracking algorithm, or using cell string or module optimisers will all mitigate the shading losses on generator level. Yet all these methods have their limitations and drawbacks, and generally limit the negative shading influences to smaller parts of the PV generator. Therefore, a generally better shading behaviour on generator level is reached with a better SLI on module level.

5 RESULTS AND INTERPRETATION

IV measurements have been recorded for the four module technologies mentioned in section 4 for three shading scenarios. Two examples of the evolution of typical IV curves with increasing shaded area (opacity 50%) perpendicular to the long module edge are shown in Figure 6. For the monolithic CIGS module (the BIPV module SKALA by AVANCIS) this shading geometry affects all cells equally, since they are all connected in series in one string and placed next to each other parallel to the long edge of the module. For shade advancing perpendicular to the long module edge across the module, current is gradually decreased with increasing shaded area. The top graph in Figure 6 shows this with the gradual decrease of the current plateau in the IV curve, but general shape unchanged.

For cSi shingled modules, the bottom panel of Figure 6 shows a very different behaviour under the same shading scenario (semi-transparent shade perpendicular to the long edge of the module). Here, the current plateau of the IV curve is split into two parts starting already at low area fractions shaded, leading to the emergence of a second maximum of the power-voltage curve (not shown). For this module technology, two blocks of parallel connected cell strings are connected in series with one bypass diode protecting each block (cf. Figure 5). Gradually increasing the shaded area leads to a gradual decrease of one of the current plateaus until about half the module is covered by the semi-transparent shade. Moving the shadow further and into the second block of cells decreases the second

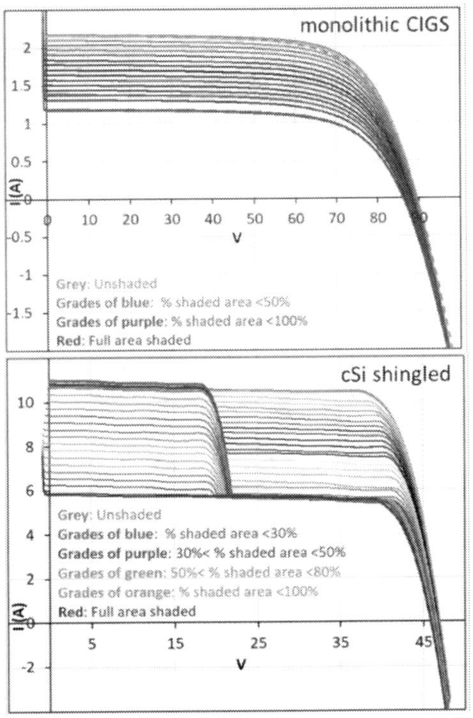

Figure 6: Examples of IV curves for shading perpendicular to the long module edge using material opacity of 50% for monolithic CIGS (top) and cSi shingled technology (bottom).

current plateau but has no significant effect on the global maximum power point due to the comparatively high remaining irradiance on the shaded part of the module. While the shape of the IV curve continues to change, the recorded power loss will remain constant when further increasing shaded area at this point. This example with multiple (local) maximum power points emerging during partial shading scenarios shows the importance of superior MPP tracking algorithms for extracting maximum electrical yield from a PV module.

5.1 Power loss for shading with 50% opacity

Figure 7 shows the relative power as a function of active area shaded for three shading scenarios for a shading material of 50% opacity.

In the first scenario, where shading material is placed perpendicular to the long module edge, CIGS modules show near-ideal power loss as power is reduced by the same percentage as active area shaded, as has been explained in the discussion of IV curves for Figure 6.

For wafer-based technologies, interconnection differs resulting in a less favourable power loss curve for this shading scenario. By gradually increasing the shaded area, whole cells are shaded within one string of series connected cells (see Figure 5 for an illustration of current flow in the module).

For the cSi module with full cells, series connection of cells runs across the whole length of the module. Bypass diodes are placed at three points dividing this series connection into three segments. As cells in all three segments are shaded in the scenario perpendicular to the

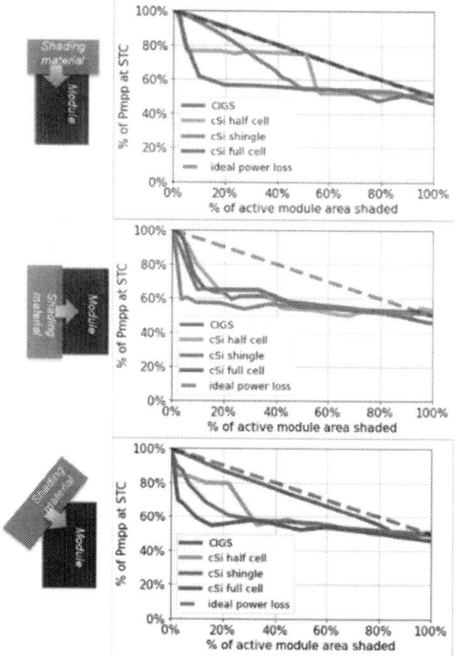

Figure 7: Power loss curves for cSi and CIGS technologies for three shading geometries and 50% opacity of the shading material. The SLI is calculated from these curves by relating the area under the curves to that under the curve for the dashed line.

long module side, power is reduced very quickly early on and drops to the minimum of 50%, once a whole row of cells is shaded, thus limiting the current available from the module to that which is being produced by the area which only sees half of the irradiation (due to the shading material with 50% opacity).

For cSi half cell technology, based on the IV curves recorded, three bypass diodes are deduced in the middle of the module and strings of cells in series are connected in parallel. As shaded area is increased from the short module side, power is reduced very quickly as contribution to module power from whole cell strings is reduced. A power plateau is reached once bypass diodes take over for one side of the parallel interconnection. From the middle of the module, as the other side of the parallel connection is shaded, power is reduced very quickly again, so that at just short of 60% of the active area shaded, power drops to the minimum of 50%.

For cSi shingled interconnection, strings of cells are also connected in parallel, but with a different layout. As shading material is moved across the module, only cell area is reduced at first for part of a series connection. This results in a gradual power reduction. Eventually whole strings of cells are deactivated before the bypass diodes, placed in the middle of the module, take over for one half of the module, creating a plateau. While for fully opaque shades, power would start to be reduced again as from the middle of the module onwards, this does not occur for partially translucent shades such as displayed here in Figure 7. IV curves in Figure 6 showed the appearance of two current plateaus. The resultant power curve as a function of voltage will show two power maxima.

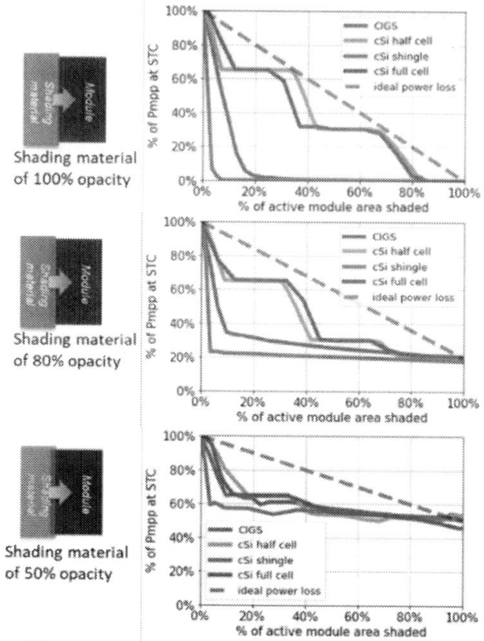

Figure 8: Power loss curves for the shading scenario under which CIGS performs weakest. This illustrates how lab-based measurements with 100% shading opacity (top) overestimate the additional power loss particularly for CIGS compared to realistic conditions with opacities of 80% (middle) or 50% (bottom).

However, in the case of partially translucent shade, the first maximum will remain the global maximum and remain at a stable position, while in the case of 100% opacity, the maximum power point will move to the second maximum and will reduce to zero as shaded area is increased.

Also in Figure 7 are shown power loss curves for shading parallel to the long module side (middle graph).

For the CIGS module, since each cell extends across the whole length of the active area on the long side, power drops very quickly to the minimum of 50% as a whole cell is shaded in the series connection. Similarly, for cSi shingled technology, power is reduced to its minimum at an even smaller fraction of the active area.

For cSi half and full cell modules, the placement of the bypass diodes prevents this fast drop. Power plateaus are created as soon as a bypass diode takes over for a string of cells. However, these plateaus appear at near minimum power, so that a significant advantage is not necessarily reached with such module designs over CIGS and cSi shingled technology at these more realistic shades with reduced opacity. Section 5.2 will discuss the differences of power loss with varying shading opacity further.

For the third shading scenario at 45° to the long edge of the module (bottom graph in Figure 7), features of both, the power loss curves for shading perpendicular and parallel to the long module edge, can be identified. In this scenario, CIGS outperforms all other cSi technologies with near ideal power loss behaviour.

5.2 Power loss for shading with opacities between 50% and 100%

When repeating shading measurements using material with 100% opacity, the behaviour resulting from cell interconnection in series and in parallel as well as the placement of bypass diodes becomes more pronounced as power loss minimum for all curves moves to 0% for fully shaded modules. Figure 8 illustrates the least favourable shading scenario for CIGS modules, where shading is applied parallel to the long module side.

The fast drop in power for both cSi shingled and CIGS technology is clearly visible as are the power plateaus for cSi half and full cell technologies. As shading opacity is reduced to 80%, the minimum power of the power loss curve is elevated to 20%. The advantage of cSi half cell technology shrinks as the second plateau sits only slightly above the power minimum. Once shading opacity is reduced to 50%, power loss curves of all technologies lie close together as power drops very soon for all technologies for the first 10% of shaded active area. This comparison shows, that judging shading loss from lab-based measurements with a black shading material (opacity 100%) may cause large overestimation for some technologies as in realistic shading scenarios, particularly for BIPV installations, there is always a certain residual amount of irradiance on the shaded part of the module.

5.3 SLI results

The different behaviour between shading scenario as well as opacity grades, such as explained in sections 5.1 and 5.2, are reflected in the calculated additional power loss compared to perfectly linear loss. This quantity has been defined as the shading loss indicator, or SLI, in section 3. In Figure 9 the SLI for all shading scenarios and shading opacities investigated in this study are shown as bar charts. Measured power loss as well as the ideal power loss curves such as shown in Figure 7 have been integrated. Using these integrals, the SLI could be calculated from equation (1).

The superior behaviour of CIGS technology for scenarios where shading occurs perpendicular to the long module side as well as at 45°, is shown by the much smaller SLI at all shading opacities compared to any cSi technologies, including the most innovative ones. For its most unfavourable shading geometry, parallel to the long module edge, CIGS modules equal the performance of cSi modules at opacity grades of 50%. Figure 9 illustrates further how important considering realistic shading opacities is in assessing a module's performance under partial shading. SLI may vary significantly depending on the shading opacity.

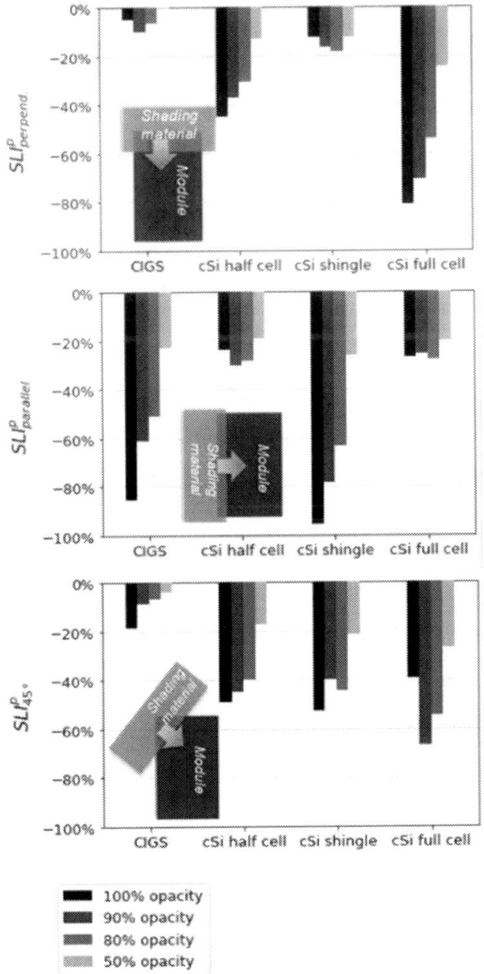

Figure 9: Shading Loss Indicator (SLI) calculated for the three shading scenarios considered (insets) and four opacity grades for CIGS as well as cSi technologies. The SLI expresses the additional power loss compared to ideal power loss such that a module with ideal power loss would yield an SLI of 0% (cf. section 3).

6 CONCLUSIONS

Outdoor measurements showed that lab-based measurements using fully opaque shading materials do not accurately represent real shading situations where diffuse and reflected light results in shading opacities between 60 to 90%. The analysis of the shading loss indicator (SLI), which has been proposed in this paper as a measure to compare PV modules according to their performance under partial shading, showed, that advantages or disadvantages in shading scenarios concluded on in such lab-based measurements may disappear for different shading opacities. CIGS modules show superior performance in all shading opacities for shading perpendicular or at 45° to the long module edge as compared to any of the tested cSi technologies. While cSi half and full cell technologies perform strongly in shading parallel to the long module edge, this advantage over CIGS modules shrinks at realistic shading opacities of 90% and 80% and disappears altogether at a shading opacity of 50%. In the designing stage of a PV system, not only the choice of module technology is therefore important when trying to minimise effects of shading, but also the module orientation relative to the reoccurring shade.

For comparison of the performance of PV modules under partial shading, it is proposed to state the SLI for a small set of standardized shading scenarios and fixed opacity grades in a 'score card' such as shown in Tables III and IV for CIGS and cSi shingled modules. It is expected that the way in which the individual solar cells and bypass diodes are arranged and connected within a PV module is the biggest influencer for the SLI and the score card. Differences between PV module manufacturers, particularly when based on the same underlying cell technology, are expected to be small for equal module technologies.

The proposed score card can complement established quantities influencing the electrical yield of a PV module or system as, e.g., the temperature coefficient and the performance at low irradiance [6]. Particularly for BIPV systems, it can be an additional factor for the energy rating that is currently not considered in the respective standard [21].

Table III: Score card for shading tolerance of CIGS module AVANCIS SKALA for three shading scenarios Π and three degrees of opacity p. The data shown was obtained from an AVANCIS SKALA 4.6 7003 (blue) and is valid for all current modules by AVANCIS of this size.

SLI_{Π}^{p} **AVANCIS SKALA**	 Π : parallel to long edge	 Π : perpend. to long edge	 Π : 45° to long edge
p=1	**-85%**	**-5%**	**-19%**
p=0.9	**-61%**	**-10%**	**-8%**
p=0.5	**-22%**	**-0.1%**	**-4%**

Table IV: As of Table III but for a module with shingled crystalline silicon (cSi) cells.

SLI_{Π}^{p} **cSi shingled**	 Π : parallel to long edge	 Π : perpend. to long edge	 Π : 45° to long edge
p=1	**-96%**	**-12%**	**-52%**
p=0.9	**-78%**	**-16%**	**-40%**
p=0.5	**-26%**	**-12%**	**-21%**

7 ACKNOWLEDGEMENTS

The authors would like to thank Jonathan Kindorf, Jasna Maksimovic, Markus Micossi and Kristian Herath for their support with performing the measurements.

8 REFERENCES

[1] H. Hanifi et al., "A novel electrical approach to protect PV modules under various partial shading situations", Solar Energy 193 (2019), 814-819.
[2] G. Sai Krishna et al., "Reconfiguration strategies for reducing partial shading effects in photovoltaic systems", Solar Energy 182 (2019), 429-452.
[3] K. Osmani et al., "Mitigating the effects of partial shading on PV system's performance through PV array reconfiguration: A review", Thermal Science and Engineering Progress 31 (2022), 101280.
[4] H. Chu et al., "Study of Hot Spots in Partially Shaded PV Modules with Large-Area Wafers", Proc. of the WCPEC-8 (2022), 917-921, 10.4229/WCPEC-82022-3DV.3.22.
[5] M. Khodapanah et al., "Partial shading detection and hotspot prediction in photovoltaic systems based on numerical differentiation and integration of the P-V curves", IET Renewable Power Generation 17 (2023), 279-295.
[6] IEC 61215-2:2021
[7] IEC 61730-2:2016
[8] IEC TS 63140:2021
[9] M. Kitze et al., "The effect of shadows from people and cleaning tools on the STC power of CIGS thin-film PV modules", Proc. of the 36th EUPVSEC (2019), 1062-1069, 10.4229/EUPVSEC20192019-4AV.1.33.
[10] J. Robledo et al., "Dynamic simulation of the shading cast by a wind farm on an adjacent photovoltaic plant", Proc. of the 38th EUPVSEC (2021), 1307-1313, 10.4229/EUPVSEC20212021-5CV.3.8.
[11] R. Simplicio et al., "Shading impact on 10 kWp rooftop grid connected photovoltaic system", Proc. of the 33rd EUPVSEC (2017), 2194-2201, 10.4229/EUPVSEC20172017-6BV.1.5.
[12] G. Van den Broeck et al., "Experimental analysis of the performance of façade-integrated BIPV in different configurations", Proc. of the 32nd EUPVSEC (2016), 2750-2755, 10.4229/EUPVSEC20162016-6AV.5.13.
[13] O. Kunz et al., "Understanding partial shading effects in shingled PV modules", Solar Energy 202 (2020), 420-428.
[14] H. Sträter et al., "An approach for a shading resistance classification of PV modules", Proc. of the WCPEC-8 (2022), 492-498, 10.4229/WCPEC-82022-3BO.12.5.
[15] C. Tzikas et al., "Do thin film PV modules offer an advantage under partial shading conditions?", Proc. of the 33rd EUPVSEC (2017), 1593-1597, 10.4229/EUPVSEC20172017-5BV.4.21.
[16] A. Gutiérrez Galeano et al., "Shading ratio impact on photovoltaic modules and correlation with shading patterns", Energies 11 (2018), 852-878.
[17] S. Gallardo-Saavedra et al., "Simulation, validation and analysis of shading effects on a PV system", Solar Energy 170 (2018), 828-839.
[18] N. Klasen et al., "Quantitative evaluation of the shading resilience of PV modules", Proc. of the 38th EUPVSEC (2021), 560-568, 10.4229/EUPVSEC20212021-4BO.2.6.
[19] I. Mehedi et al., "Critical evaluation and review of partial shading mitigation methods for grid-connected PV systems using hardware solutions: The module-level and array-level approaches", Renewable and Sustainable Energy Reviews 146 (2021), 111138.
[20] K. Sinapis et al., "A comprehensive study on partial shading response of c-Si modules and yield modelling of string inverter and module level power electronics", Solar Energy 135 (2016), 731-741.
[21] IEC 61853

IMPROVEMENTS IN THE ACQUISITION OF DAYLIGHT ELECTROLUMINESCENCE IMAGES USING HIGH SPEED CAMERAS: COMPARISON OF SQUARE AND SINUSOIDAL WAVES EXCITATIONS

C. Terrados, D. González-Francés, J. Anaya, K.P. Sulca, V. Gómez-Alonso, M.A. González, O. Martínez*
GdS-Optronlab group, Dpto. Física de la Materia Condensada. Universidad de Valladolid
Edificio LUCIA. Paseo de Belén, 19. 47011 Valladolid (Spain)
*oscar.martinez@uva.es

ABSTRACT: Electroluminescence (EL) imaging is a very powerful technique for the identification of defects in Silicon solar cells and panels. Daylight Electroluminescence (dEL) has emerged recently, aiming to avoid the need to perform the EL images on dark environments, which complicates the measurements and limits the number of panels to be inspected. In this work we present a procedure to obtain dEL images by using sinusoidal or square excitations and high speed InGaAs cameras, working in asynchronous mode between excitation and signal capture. We subsequently apply post-processing techniques based on both frequency and time domain analyses to assess the quality of the images by quantifying the signal-to-noise ratio arising from both analyses. We captured a stack of dEL images using two InGaAs cameras capable of acquiring at speeds of up to 60 and 600 frames per second. Our results demonstrate the promising potential of the proposed dEL method, as the post-processed image stack exhibits the same features as dark EL images obtained using the same InGaAs cameras. Notably, this stack of dEL images can be obtained within a very short acquisition time, typically ranging from 100 to 200 ms, while maintaining good image quality.
Keywords: Electroluminescence, defect characterization, solar plants

1 INTRODUCTION

Electroluminescence (EL) characterization has become a widely accepted method for obtaining detailed information about the defects in Si solar cells and modules. This technique offers high spatial resolution, enabling precise detection of faults such as cracks, broken cells, interconnections, shunts, potential and light-induced degradation (PID/LID), among others [1, 2]. EL imaging is now a very well-established technique [2], providing complementary, and often more comprehensive, information compared to infrared thermography (IRT) and I-V characterization [3, 4]. However, two bottlenecks appear in performing the EL characterization, especially in the case of a large number of solar panels. First, EL is usually performed in dark scenarios. Secondly, EL requires the injection of current into the solar modules, which when performed on-site requires large power sources or frequent movement of a small power source from module to module.

Given the significant number of solar modules in a photovoltaic plant, the industry requires fast inspection techniques [5]. However, a massive inspection is currently unavoidable if the EL image is taken in strictly dark conditions, disassembling the modules and sending them to a lab, or also using specialized dark containers or mobile vans. Performing on-site EL during the night is also possible, but the risk of electrical manipulation damage is high and the potential number of modules to be inspected is low. For this reason, it would be very beneficial to perform on-site EL at the solar plant without disassembling the modules and during the day. This will reduce the risk of electrical manipulation damage, as well as to suppress the risk during assembly and disassembly of the modules, allowing a faster inspection of a large number of panels.

Daylight EL (dEL) has gained rapid development and significant attention in recent years [6-10], although it is still challenging to be applied in-situ in photovoltaic plants for a large number of modules. Enabling dEL requires IR based cameras – such as those based on InGaAs sensors–, in conjunction with sophisticated filtering procedures, capable distinguish the weak luminescence emission coming from the solar module from the more intense ambient light.

Several approaches for the obtention of dEL images can be used [10-12]. In our first approach [10], we used an InGaAs camera, a power source, and a mechanism for the injection of the current in a switching mode, with "on" and "off" periods. In this mode, there is a synchronization between the current injection and the acquisition of the images, in both "on" and "off" periods. The subtraction of the images obtained in both periods, and the accumulation of periods, give enough signal to obtain final good images, even at high irradiation conditions [10]. However, relatively long total acquisition times (of the order of tens of seconds) would be needed, especially for hard conditions, such as high irradiation conditions, low emitting panels, etc.

On the other hand, specific power sources able to produce square, sinusoidal, and other kinds of wave forms, can now being applied for the current injection of the solar modules. This can be used for the decoupling of the EL image acquisition and the power injection process (non-synchronized way). To do this, InGaAs cameras capable of capturing a large number of frames per second (fps) are needed, as well as robust post-processing analysis of the collected images.

Our proposed asynchronous dEL method enables the inspection of a large number of Si modules of a PV solar plant in very short times, which is becoming a requirement for O&M activities of a medium-to-large PV solar plant. We achieve this thanks to the combination of a 15 kW power source with square and sinusoidal wave forms, used for the injection of current into the strings, and a very fast acquisition of images. In this work we show the capability of two cameras from different manufacturers, with maximum acquisition speeds of 60 and 600 fps. Finally, the use of advanced filtering techniques to discard the non-EL related radiation is shown by comparing two methods to filtrate the EL signal from the ambient light: a time-domain method using square waves for the injection of current, and a frequency-domain method using both square or sinusoidal waves for the injection of current.

2 EXPERIMENTAL DESCRIPTION

For the current injection, a 15 kW power source was used (EA-PS 91500-30 3U 19" 3U 15000W model), allowing for the injection of current with I_{sc} values for a whole solar PV string. The power source allows for both square and sinusoidal wave form injection schemes for the excitation of Si solar modules.

Two InGaAs cameras, with maximum of 60 and 600 frames per second (fps), respectively, were used and compared in this work. A Hamamatsu C12741-03 InGaAs camara, with 640 x 512 pixels and pixel pitch of 20 x 20 µm, with 14 bits' quantization and maximum speed of 60 fps, and a First Ligh C-RED 2 Lite InGaAs camera, with 640 x 512 pixels and pixel pitch of 15 x 15 µm, with 14-bit quantization and 16-bit dynamical range with maximum speed of 600 fps, were used. We use a Kowa short wave infrared (SWIR) optical system with 16 mm focal length for image acquisition. This allows an entire module per image to be viewed with the camera placed 2.5 meters away from the module. On the other hand, in order to suppress as much ambient light as possible, a SWIR bandpass filter, centered around 1160 nm with a bandwidth of 150 nm and a transmittance close to 90%, is used.

For the wave excitation, various frequencies (f) were tested while keeping the maximum intensity around I_{sc}. Additionally, exposure times (t_{exp}) were varied from 1.2 to 10 ms, depending on the selected acquisition velocity (number of fps) of the camera and irradiation conditions.

To conduct this analysis both a multi-Si Al-BSF (Sharp, ND-AR330H 330 W, V_{oc}=45.5 V, I_{sc}=9.40 A) and a multi-Si PERC module (Canadiansolar, CS6U 345 W, V_{oc}=46 V, I_{sc}=9.69 A) were used as test specimens.

3 DATA ANALYSIS

3.1 Square and sinusoidal wave excitation

Opposite to our previous synchronous method to obtain the dEL images of the modules [10], here we describe the use of an asynchronous scheme in which the excitation of the module and the acquisition of the images are decoupled. To do this, a periodic oscillating signal is needed, for which both square and sinusoidal schemes have been used for the current injection. The true EL signal coming from the module is discriminated from the ambient light (noise) by filtering the non-modulated component of the light acquired by the camera.

It is worth noting that the EL emission can be very small when compared with the background light, particularly for high irradiation conditions and older technology panels. This makes it challenging to distinguish signal from noise. Fig.1 illustrates this phenomenon for a multi-Si Al-BSF solar panel exposed to 750 W/m^2 of solar irradiation. The panel was excited with a sinusoidal wave of 10 A in amplitude, a frequency of 4 Hz with 10 ms of exposure time and recorded at 32 fps. In the image the average of the intensity recorded for each frame by the First Light camera vs the number of acquired frames is shown. Under these conditions the EL signal is only a very small fraction of the recorded intensity, with a modulated signal of less than 20 units over a background of ~11500 units. To make things worse, the background is not constant due to changes in the ambient light during the acquisition of the images which lead to a strong modulation of the background (see for instance Figure 1, the initial jump in the first 100 frames). It is worth noting that the longer the time required to record the stack of images the worse this modulated background can be, and more difficult is to extract the information from the panel. Consequently, a robust baseline correction step is imperative for each pixel in the image stack before differentiating modulated light from background light.

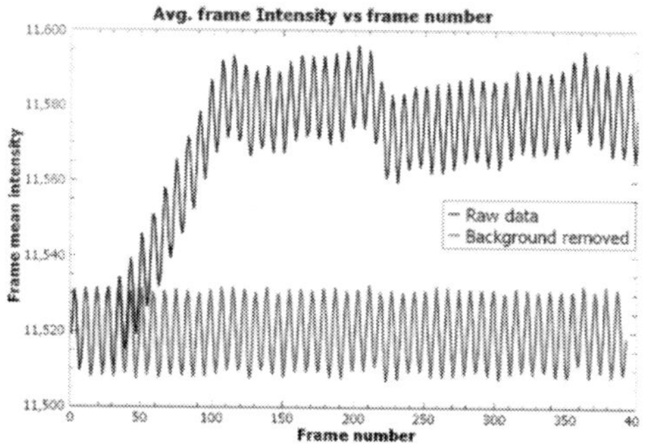

Figure 1: Intensity levels of the signal captured with the First Light camera (400 images) for the multi-Si Al-BSF module with a sinusoidal function. G = 750 W/m^2, f = 4 Hz, 32 fps, t_{exp} = 10 ms.

3.2 Frequency domains analysis

In the frequency domain, both sinusoidal and square waves can be processed by calculating the real valued Fast Fourier Transform (FFT) of each pixel in the image stack. In this process, we extract the signal to noise ratio (SNR) of the excitation frequency against the background defined in the region $[f/2, 2f]\backslash\{f\}$. The SNR is calculated from the squared amplitude at frequency f and the variance of the noise level in the specified region of the frequency spectrum signal analysis:

$$SNR = \begin{cases} 10log_{10}\left(\frac{(Peak\ amplitude)^2}{Var(Noise)}\right) & if > 10 \\ 0 & if \leq 10 \end{cases}$$

All pixels with SNR lower than 10 dB (i.e., equivalent to a signal three times higher than the standard deviation of the noise) are filtered, as they may not be distinguishable from noise with sufficient certainty as per the standard practice in signal analysis in frequency domain. As a quality metric the average SNR of the pixels in the lower quartile is defined (SNR_{25}=$Avg(Q1$(SNR)); i.e., the average SNR of those pixels with the lowest values which are associated with the regions with weakest EL emission. Therefore, the higher is this value, the better we can distinguish the weakest part of the image from noise. Finally, all frequencies but the excitation frequency f for sinusoidal waves, and odd multiples of f for square waves, [f, $3f$, $5f$, ...], are set to 0. Then the image stack is transformed to time domain to recover the image in real space containing only the EL component of the light. The last step is integrating the new image stack and normalizing to the number of periods to recover a single image with all the EL information. After this process, a final assessment quantifies the information recovered in the filtered image, typically represented as the number of bits needed to represent the bandwidth of the image. This measurement is based on the number of levels needed to represent the average of the most intense decile of pixels in the image to weight for potential spurious pixels (labelled here as Resolution).

3.3 Time domain analysis

In this process the *On* states are separated from the *Off* states, and therefore only square waves can be handled. After separating the *On* and *Off* states their averages are computed, and their difference is calculated for each pixel. The SNR is calculated using the same principles as described for frequency domain, but in this case the signal is the square of the average difference between the On and *Off* sates, and the noise is the variance of the *Off* levels. The variance is determined by averaging the variance of the *Off* states per period if there are more than 4 data points in an *Off* state during each period. Otherwise, the total variance of the *Off* states in the image stack is used. This is done to mitigate the effect of the residual background discussed in 3.1. For this analysis all pixels with less than 3 dB (i.e., equivalent to approximately 92% certainty of the pixel not being noise if noise is normally distributed) are filtered:

$$SNR = \begin{cases} 10\,log_{10}\left(\frac{\left(On_{avg} - Off_{avg}\right)^2}{Var(Noise)}\right) & if > 3 \\ 0 & if \leq 3 \end{cases}$$

The quality metric is identical to the one defined for frequency domain, the average SNR of the pixels in the lower quartile, SNR_{25}=*Avg*(*Q1*(SNR)). It is worth noting even being the same SNR_{25} for both types of analysis the numbers are not directly comparable because their detection threshold differs. This is because in time domain we are working with averages of the whole image stack, which are less sensitive to detect modulation in the signal than the frequency analysis which uses information of all individual frames in the image stack.

4 RESULTS DISCUSSION

4.1 Square and sinusoidal wave excitation

Figure 2 shows the dark EL image obtained with the Hamamatsu camera for the PERC module, which is the image against we benchmark the dEL images. In this case, the *Off* signal has been first collected, then the *On* signal for a constant current injection of 9 A (I_{sc}) has been collected. An artificial square image has been constructed using both *On* and *Off* signals, in order to create a synthetic image stack of 400 images like those obtained when exciting with a square signal with *f*=2.5 Hz, 50 fps, and t_{exp} = 15 ms.

A time domain analysis as described in 3.3 has been performed in this synthetic dataset, which result is the final dark EL imaged showed in Figure 2. The SNR_{25} marker in this case is ~63.6 dB, due to the completely dark conditions. The information recovered from this image stack has a resolution of ~11.2 bits, namely 2353 levels are needed to represent all the EL information from the image stack.

Figure 3(a, b) shows the processed dEL images for both sinusoidal and square excitations schemes on the same PERC module shown in Figure 2. Image stacks of 400 frames were acquired using the same camera, with *f* = 10 Hz, 46 fps and t_{exp} = 2 ms (G = 875 W/m²). SNR_{25} values are in the order of 25 dB, indicating that even the weakest part of the image is filtered with a signal 8 times above the variance of the noise level, resulting in images where it is possible to clearly distinguish the state of all the cells of the module and their defective areas. Indeed, the visual information is almost indistinguishable from the image obtained in dark conditions. It is worth noting that the SNR_{25} is slightly better for the sinusoidal wave, while the square wave conveys slightly more information. Nonetheless, both types of excitations produce nearly indistinguishable results. When comparing the amount of information in these images, only approximately 8 out of the original 14 bits have been recovered. In other words, approximately 256 levels are recovered from the original 16384 levels of the camera sensor, which is nearly a tenfold reduction compared to the dark image. Therefore, both types of panel excitation, square and sinusoidal, when properly generated, will result in similar image quality and there is no clear reason to prefer one over the other. However, by exciting with square waves we allow for a time processing of the image, which is not currently achievable with the sinusoidal wave.

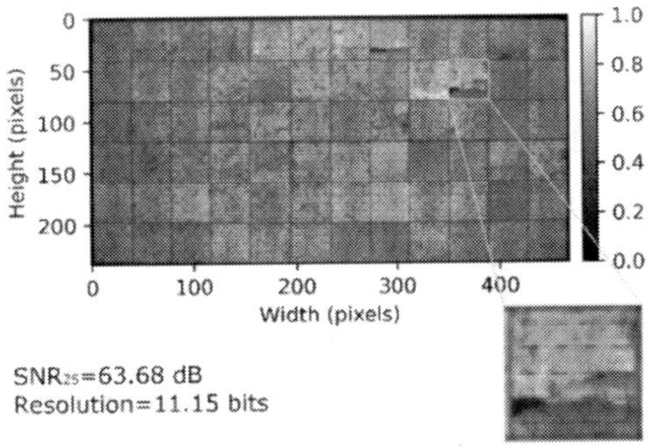

Figure 2: Dark EL image of the PERC module obtained with the Hamamatsu camera working with an acquisition velocity of 50 fps and t_{exp} = 15 ms. In order to compare the quality of the image with the dEL images shown in figure 3, a virtual square wave has been generated, with *f* = 2.5 Hz, and a post-process analysis has been carried out in the time domain. The inset shows the zoom for a single cell with dark areas and a crack.

Figure 3(c) shows the processed image of the PERC module with square excitation and time domain analysis. Once again, the image quality is sufficient to clearly distinguish the same features as in dark conditions and nearly indistinguishable from the frequency domain analysis. It's worth noting that the difference in the SNR_{25} metric between time and frequency analysis is attributed to the distinct mathematical procedures used in the analysis as described in 3.2 and 3.3. In any case, we can perform a direct comparison with the image shown in Figure 2 since both are processed with the same method. By doing this, it is observed that, although the dEL image is visually comparable with the dark EL image, the quality markers are very different. In dEL we are recovering only ~222 levels to represent the image, i.e. we have 10 times less resolution than in dark EL. However, even with this loose in recovered information both dark EL and dEL allow to recognise the same defective regions and even small features in the panel. This confirms the capability of the dEL process in the non-synchronized way carried out in this work.

4.2 Comparison between cameras with different acquisition velocities

Very recently, a new generation of InGaAs cameras with very high acquisition velocities have appeared in the market. This paves the way for a much faster acquisition of the dEL images. In order to test the capabilities of such cameras, we have tested the First Light C-RED 2 Lite InGaAs camera, with a maximum acquisition velocity 10 times higher than the one used in Figures 2 and 3 (600 fps vs 60 fps). This increase in the acquisition velocity means also a decrease of the maximum exposure time, and therefore potentially to a more challenging filtering of the background light if the sensors are not able to detect more efficiently the photons from the panel.

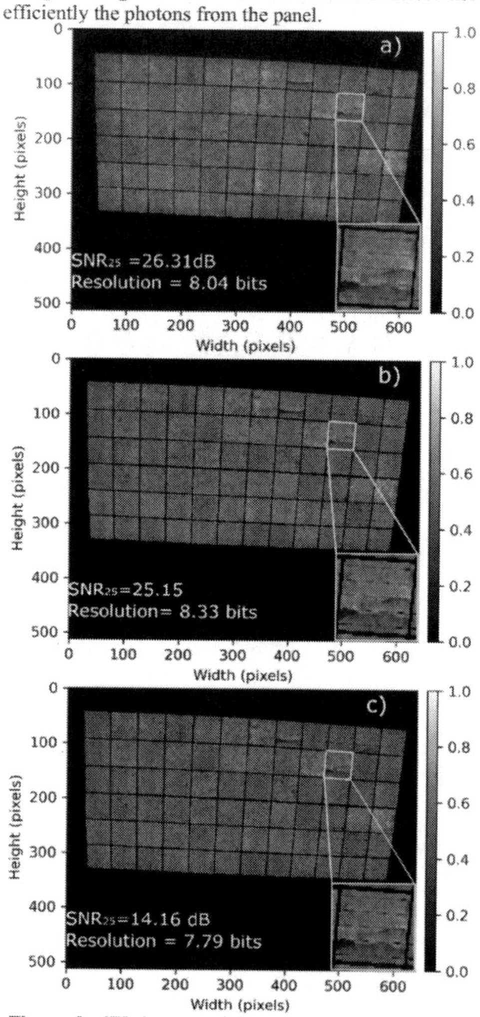

Figure 3: dEL images of the PERC module obtained with the Hamamatsu camera and with sinusoidal (a) and square (b, c) wave excitations. The post-process analysis has been done in the frequency domain (a, b) or in the time domain (c). G = 875 W/m², f = 10 Hz, 46 fps, t_{exp} = 2 ms. The inset in each image shows the zoom for a single cell with dark areas and a crack.

Figure 4 (a, b) shows the dEL images obtained with both the Hamamatsu and the First Light cameras, respectively. A square wave excitation was used in both cases. The acquisition conditions were f = 2 Hz, 46 fps, t_{exp} = 2 ms to obtain 400 images with the Hamamatsu camera, and f = 62.5 Hz, 600 fps, t_{exp} = 1.2 ms to obtain 594 images with the First Light camera. It should be noted that the measurements with both cameras were taken in different days, and unfortunately the irradiance conditions were different, G = 780 W/m² and 340 W/m² for the Hamamatsu and First Light, respectively. To somewhat compensate for this low irradiance, we have selected for this comparison a more challenging panel which has a much weaker EL response than the previously described PERC panel. Therefore, an older multi-Si Al-BSF module was selected for this comparison. A post-process analysis in the frequency domain was performed in both cases, yielding SNR_{25} and Resolution values of 20.36 dB and 9.06 bits for the Hamamatsu camera and 18.02 and 9.53 bits for the First Light camera. In both cases the images yield similar level of information and allow to clearly detect all the features shown by the dark EL image from that module, Fig. 4 c, which was obtained with the Hamamatsu camera for comparison. It should be noted that the images were obtained at different temporal moments, and the module has degraded in the last images (obtained with the Hamamatsu camera) respect to the image obtained with the First Light camera.

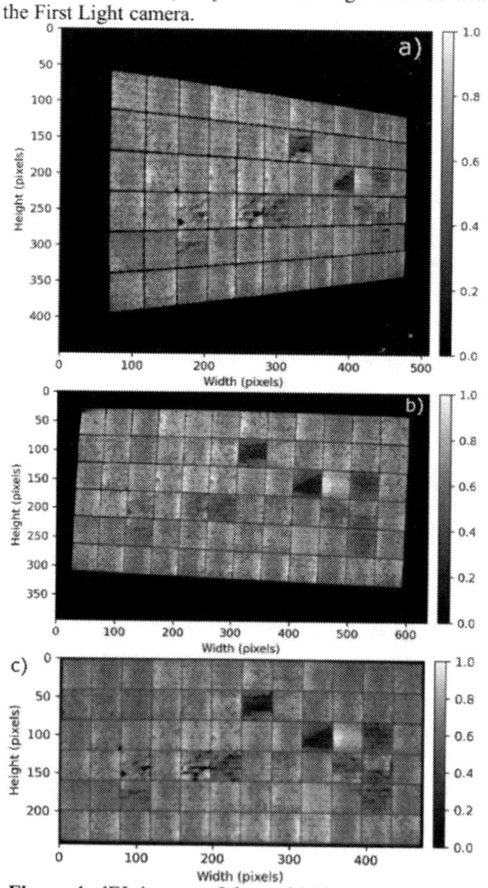

Figure 4: dEL images of the multi-Si Al-BSF module obtained with both the Hamamatsu camera (a) and the First Light camera (b) with square wave excitations. The post-process analysis has been done in the frequency domain. In a) G = 780 W/m², f = 2 Hz, 46 fps, t_{exp} = 2 ms. In b) G = 340 W/m², f = 62.5 Hz, 600 fps, t_{exp} = 1.2 ms. (c) Dark EL image of the same module obtained with the Hamamatsu camera working with an acquisition velocity of 32 fps and t_{exp} = 15 ms. An artificial square wave has been generated, with f = 2 Hz, and a post-process analysis has been done in the frequency domain.

Figure 4 shows that working with short exposure times and fast acquisition times is not linked to a worsening of the dEL image quality, and therefore it opens the door to perform the analysis in much shorter times. Note that the Hamamatsu camera needed ~8.7 s to collect all the images while the First Light used slightly less than 1 s even collecting 50% more images.

To test how much we can reduce the time needed to obtain a final image with enough quality we have removed frames from the 594 image stack taken with the First Light camera. For this we have removed full periods of data from the end of the stack and extracted the SNR_{25} marker from the resultant image stack. This analysis is shown in Fig. 5 for both frequency and time domains. Here it is observed that for frequency domain analysis by increasing the number of periods from 5 up to 20 the SNR_{25} value increases, then reaching a plateau until 40 periods, and then degrades for 50 and 60 cycles. This behaviour occurs because the signal degrades in frequency over time, likely due to fluctuations in the background light that introduce a small phase shift when subtracting the baseline. This figure demonstrates that there is little improvement in image quality when extending the measurement time from ~100 ms to ~600 ms, and in some cases, extending the measurement time beyond that limit can result in worse dEL images. On the other hand, for the time domain analysis the SNR_{25} slightly increases monotonically with the number of cycles/frames in the stack (Figure 5-a, left axis). This improvement occurs because the difference between the average and noise level becomes more significant as more data is added, while it is less sensitive to phase degradation. When evaluating the resolution of the final images, a pattern akin to that observed for the SNR_{25} marker in frequency analysis becomes evident. Specifically, the resolution plateaus at 10 periods and exhibits a slight degradation when exceeding 40 periods. In contrast, in the time-dependent analysis, resolution displays a nearly constant trend regardless of the number of periods. As previously discussed, this behaviour is expected since the addition of more periods primarily enhances the ability to discriminate signal from noise rather than amplifying the signal level itself.

This finding illustrates that remarkably short acquisition times (~160 ms) are sufficient for acquiring dEL images in very short timeframes, paving the way to the implementation of dEL in the field using cameras installed on drones for efficient inspection of a large number of modules.

5 CONCLUSIONS

Large improvements for the obtention of dEL images have been proven, with the combined use of a 15 kW power source with square and sinusoidal waves excitation modes and a very high acquisition speed camera. Two methods for analyse the dEL signal from Si solar panels have been disclosed: one in the frequency domain and one in the time domain. Both methods have been applied to images taken from multi-Si low emitting (Al-BSF) and high emitting (PERC) panels. It is demonstrated that, with proper data filtering, the information recovered from dEL images can show a similar amount of visual information than images taken in dark conditions. A metric (SNR_{25}) has been proposed to quantify the quality of the images. dEL images with good quality can be obtained in very short timeframes (approximately 160 ms). These improvements in the acquisition of the images can open the door for the massive inspection of Si solar plants by a very powerful technique such as EL imaging.

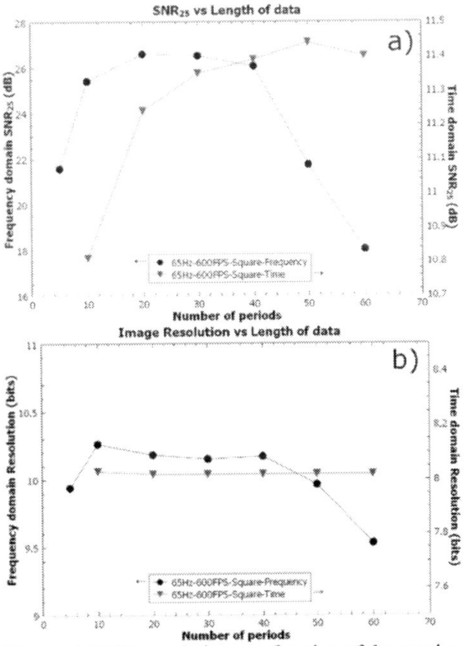

Figure 5: SNR_{25} evolution as a function of the number of analysed cycles (a). The data correspond to the dEL images of the multi-Si Al-BSF module obtained with the First Light camera with square wave excitation. The post-process analysis has been done both in the frequency and time domains. G = 340 W/m^2, f = 62.5 Hz, 600 fps, t_{exp} = 1.2 ms. Also, the evolution of the resolution of the images with the number of periods is shown (b).

6 ACKNOWLEDGMENTS

This work has been financed by the Spanish Ministry of Science and Innovation, under project PID2020-113533RB-C33, and by the Regional Government of Castilla y León (Junta de Castilla y León) and by the Ministry of Science and Innovation and the European Union NextGenerationEU / PRTR under the project "*Programa Complementario de Materiales Avanzados*".

7 REFERENCES

[1] L. Koester, S. Lindig, A. Louwen, A. Astigarraga, G. Manzolini, and D. Moser, "Review of photovoltaic module degradation, field inspection techniques and techno-economic assessment," Renew. Sustain. Energy Rev., vol. 165, no. November 2021, p. 112616, 2022, doi: 10.1016/j.rser.2022.112616.

[2] IEC, "IEC TS 60904-13 Photovoltaic Devices - Part 13: Electroluminescence of Photovoltaic Modules," http://www.iec.ch/. [Online]. Available: http://www.iec.ch/

[3] S. Gallardo-Saavedra, L. Hernández-Callejo, M.C. Alonso-García, J.D. Santos, J.I. Morales-Aragones, V.

Alonso-Gómez, A.M. Moretón-Fernández, M.A. González-Rebollo, O. Martínez, "Nondestructive characterization of solar PV cells defects by means of electroluminescence, infrared thermography, I-V curves and visual tests: Experimental study and comparison". Energy 205, 117930 (2020).

[4] M. Aghaei, A. Fairbrother, A. Gok, S. Ahmad, S. Kazim, K. Lobato, G. Oreski, A. Reinders, J. Schmitz, M. Theelen, P. Yilmaz, J. Kettle, "Review of degradation and failure phenomena in photovoltaic modules". Renew. Sustain. Energy Rev. 159, 112160 (2022).

[5] I. Høiaas, K. Grujic, A. Gerd, I. Burud, E. Olsen, N. Belbachir, "Inspection and condition monitoring of large-scale photovoltaic power plants: A review of imaging technologies". Renew. Sustain. Energy Rev. 161, 112353 (2022).

[6] L. Stoicescu, M. Reuter, J.H. Werner, "Daysy: luminescence imaging of PV modules in daylight", in 29th Eur. Photovolt. Sol. Energy Conf. Exhib. (2014), p. 2553.

[7] J. Adams et al., "Non-Stationary Outdoor EL-Measurements with a Fast and Highly Sensitive InGaAs Camera," 32nd Eur. Photovolt. Sol. Energy Conf. Exhib., 2016.

[8] T. J. Silverman, M. G. Deceglie, K. Vansant, S. Johnston, and I. Repins, "Illuminated Outdoor Luminescence Imaging of Photovoltaic Modules," IEEE Photovolt. Spec. Conf., 2017.

[9] G. Alves Dos Reis Benatto et al., "Drone-Based Daylight Electroluminescence Imaging of PV Modules," IEEE J. Photovoltaics, vol. 10, no. 3, pp. 872–877, 2020, doi: 10.1109/JPHOTOV.2020.2978068.

[10] M. Guada et al., "Daylight luminescence system for silicon solar panels based on a bias switching method," Energy Sci. Eng., no. June, pp. 1–15, 2020, doi: 10.1002/ese3.781.

[11] C. Mantel et al., "SNR Study of Outdoor Electroluminescence Images under High Sun Irradiation," in 2018 IEEE 7th World Conference on Photovoltaic Energy Conversion (WCPEC) (A Joint Conference of 45th IEEE PVSC, 28th PVSEC & 34th EU PVSEC), Jun. 2018, no. 1, pp. 3285–3289. doi: 10.1109/PVSC.2018.8548264.

[12] G. A. dos Reis Benatto et al., "Daylight Electroluminescence of PV Modules in Field Installations: When Electrical Signal Modulation is Required?," in 8th World Conference on Photovoltaic Energy Conversion, 2022, pp. 735–739. doi: 10.4229/WCPEC-82022-3BV.3.44.

10.4229/EUPVSEC2023/3DO.16.5
020258-006

Fraunhofer ISE

Impact of String Connection and Contact Degradation on Electrical Current Distribution on Solar Cell and Photovoltaic Module Level: a Magnetic Field Imaging Validated Model

Ammar Tummalieh[1,2], M. Mittag[1], J. Weber[1], D. Yucebas[1], L. Schäfer[1], R. Quay[2,3], C. Reichel[1], D. Neuhaus[1]
[1]Fraunhofer Institute for Solar Energy Systems (ISE), Heidenhofstrasse 2, 79110 Freiburg, Germany
[2] University of Freiburg, Department of Microsystems Engineering (IMTEK), Georges-Koehler-Allee 103, 79110 Freiburg, Germany
[3] Fraunhofer Institute for Applied Solid State Physics IAF, Tullastrasse 72, 79108 Freiburg, Germany

40th EU PVSEC 2023
Lisbon, Portugal
September 21st, 2023
www.ise.fraunhofer.de

This work will be published at Progress in Photovoltaics: Research and Applications

PROGRESS IN PHOTOVOLTAICS RESEARCH AND APPLICATIONS

40th European Photovoltaic Solar Energy Conference and Exhibition

Motivation

Position of string connector terminal

- Placement of the terminal at the middle in research solar cells (single-cell modules)

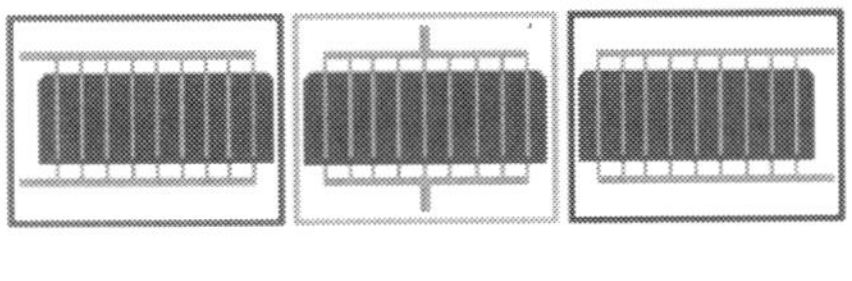

Contact failure

- Misallignment or contact failure between ribbon and metallization

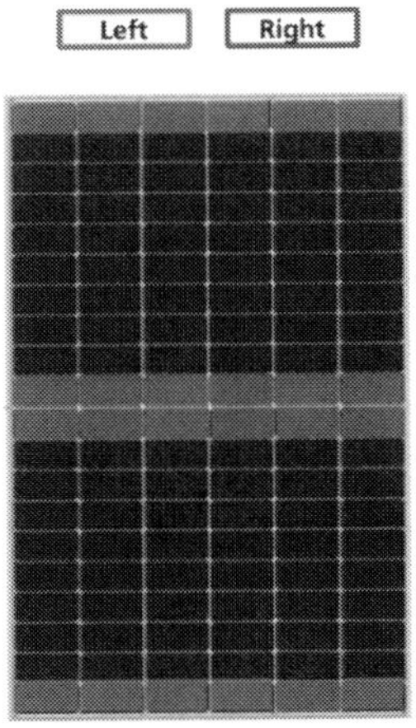

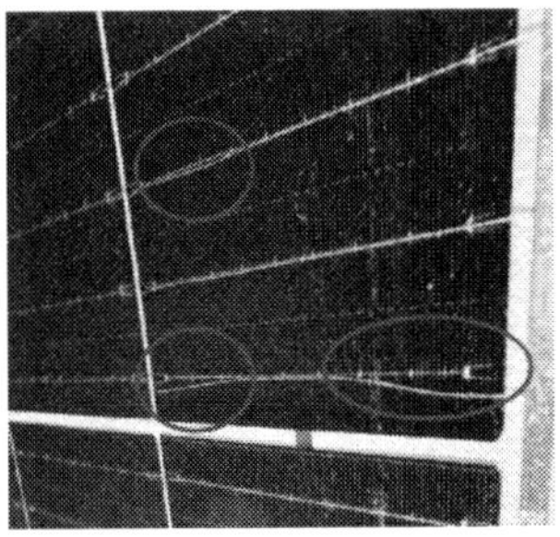

[1]

2

[1] D. Mesquita et al., "A review and analysis of technologies applied in PV modules" in Proceedings of the IEEE PES Innovative Smart Grid Technologies Conference, 2019

universität freiburg Fraunhofer ISE

Motivation

Open questions

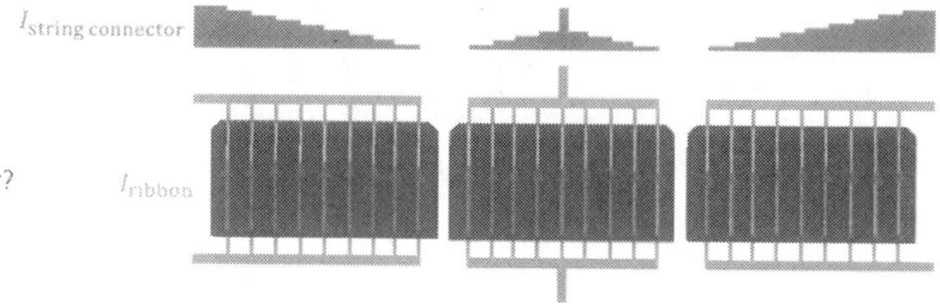

Does the position of the string connector terminal matter? And how?

How does that impact the current distribution and the power of solar cell and PV module?

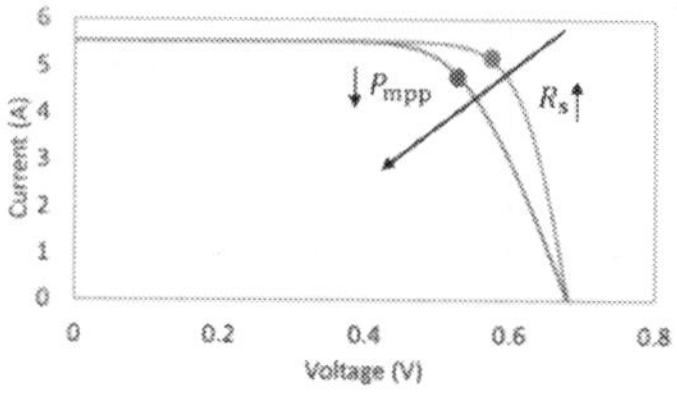

How do different contact failure scenarios impact cell output?

universität freiburg Fraunhofer ISE

Introduction

State of the art

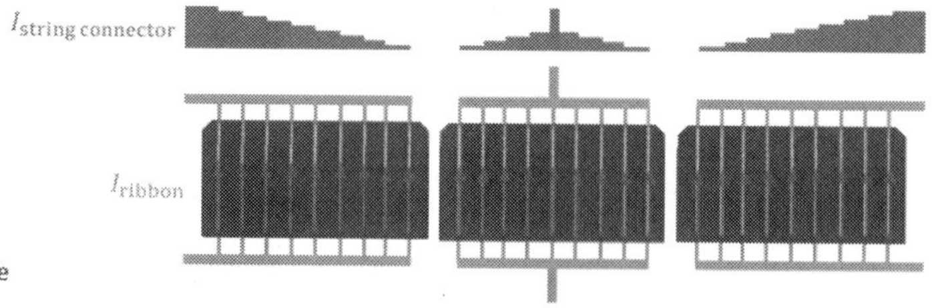

Terminal position: **Left** **Middle** **Right**

- Electrical current at all ribbons is equal, assuming:
  - Even distribution of cell metallization
  - Identical ribbon and contact resistance

$\rightarrow I_{\text{ribbon}} = \frac{I_{\text{cell}}}{n}$, n: number of busbars

- Power loss in ribbons $\rightarrow P_{\text{loss, ribbons}} = n \times I^2_{\text{ribbon}} \times R$
- Power loss in string connector $\rightarrow P_{\text{loss, string connector}} = \begin{cases} \sum_{k=0}^{k=\frac{n-1}{2}} \left(\left(2 \times \left((n-k)^2 \times I^2_{\text{ribbon}} \times R_{\text{string connector}}\right)\right)\right) + n^2 \times I^2_{\text{ribbon}} \times R_{\text{string connector}}, \text{Terminal at the middle} \\ \sum_{k=0}^{k=n-1} ((n-k)^2 \times I^2_{\text{ribbon}} \times R_{\text{string connector}}, \text{Terminal at right or left} \end{cases}$

universität freiburg Fraunhofer ISE

Methodology

Applying string connector

- Developing a resistance network model in MATLAB
- It has an ohmic resistance
- Voltage drop occurs across it
- Definition of the ohmic resistance of each current path depending on ribbon position relative to the string connector terminal

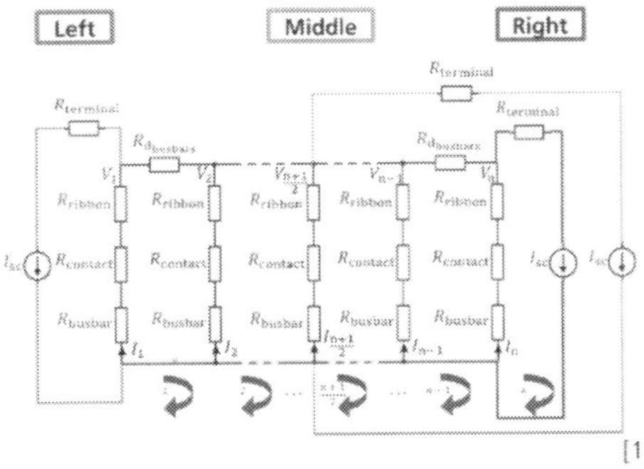

[1]

$$I_1(R_{busbar} + R_{contact} + R_{ribbon}) + I_1\left(R_{d_{busbars}}\right) + (I_1 - I_2)(R_{ribbon} + R_{contact} + R_{busbar}) = 0$$

5

[1] A. Tummalieh et al., "Impact of String Connection and Contact Degradation on Electrical Current Distribution on Solar Cell and Photovoltaic Module Level: A Magnetic Field Imaging Validated Model" Progress in Photovoltaics: Research and Applications (tbp).

universität freiburg Fraunhofer ISE

020259-005

40th European Photovoltaic Solar Energy Conference and Exhibition

Methodology

Applying string connector

- Developing a resistance network model in MATLAB
- It has an ohmic resistance
- Voltage drop occurs across it
- Definition of the ohmic resistance of each current path depending on ribbon position relative to the string connector terminal
- Smaller string connector → higher ohmic resistance → higher voltage drop → more inhomogeneous currents

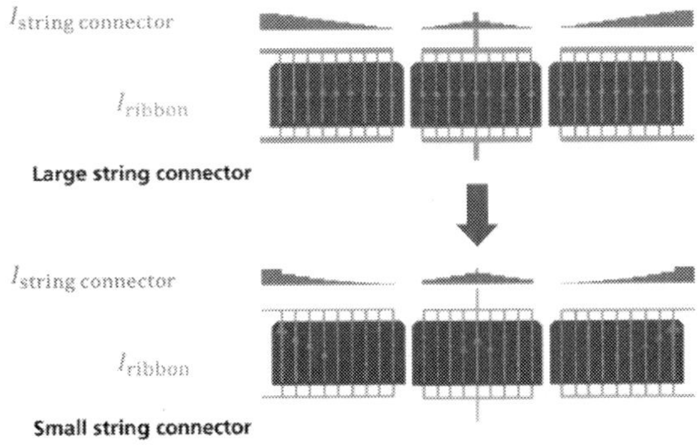

6

universität freiburg Fraunhofer ISE

020259-006

Methodology

Contact defects

- Contact resistance is calculated based on
  - Area of metallization element
  - Area of solder/ECA
  - Resistivities of the different elements

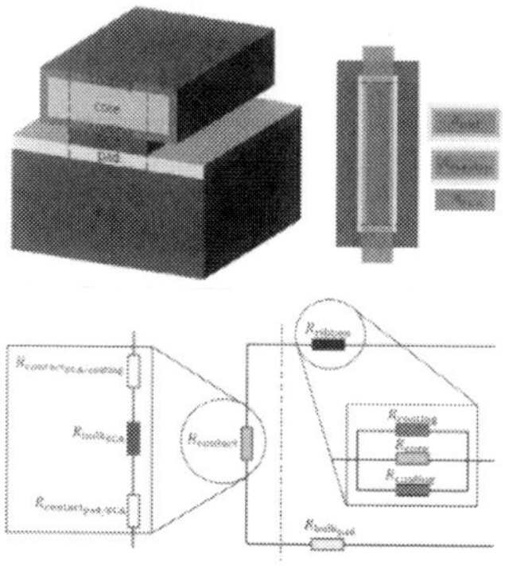

$$R_{\mathrm{ECA\,(bulk)}} = \frac{\rho_{\mathrm{ECA}}\, t_{\mathrm{ECA}}}{A_{\mathrm{ECA}}}$$

$$A_{\mathrm{ECA}} = \frac{f_{\mathrm{share}}\, A_{\mathrm{overlap}}}{100\%}$$

$$R_{\mathrm{contact}} = \frac{\rho_{\mathrm{contact}}}{A_{\mathrm{ECA}}}$$

[1]

[1] A. Tummalieh et al., "Interconnection Technology in PV Modules: Impact of Ribbons, Tab Connectors and Electrically Conductive Backsheet on Module Performance," in Proceedings of the 8th World Conference on Photovoltaic Energy Conversion, Milan, Italy, 2022.

universität freiburg Fraunhofer ISE

020259-007

40th European Photovoltaic Solar Energy Conference and Exhibition

Methodology

Contact defects

- Contact resistance is calculated based on
  - Area of metallization element
  - Area of solder/ECA
  - Resistivities of the different elements
- Cell connector failure
  - Calculation of the area of interest AOI
  - Current is distributed on other ribbons based on the distance
- String connector failure
  - Same approach is applied, considering a failure of whole ribbons
  - Number of ribbons depends on the position of the failure in the string connector

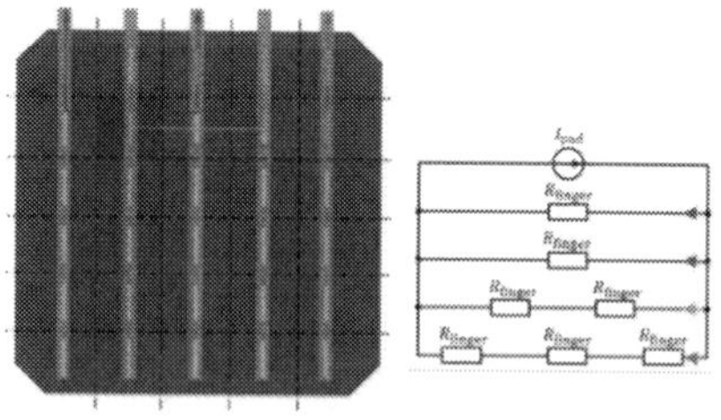

[1]

$$I = I_{\mathrm{pad}} \frac{R_{\mathrm{eq}}}{3\, R_{\mathrm{finger}}}$$

[1] A. Tummalieh et al., "Impact of String Connection and Contact Degradation on Electrical Current Distribution on Solar Cell and Photovoltaic Module Level. A Magnetic Field Imaging Validated Model" Progress in Photovoltaics: Research and Applications (tbp).

universität freiburg Fraunhofer ISE

020259-008

Methodology

Magnetic Field Imaging (MFI)

- MFI can reveal relative current strength and direction [1]
- A 3-dimensional axes system uses a magnetic sensor array to scan the sample in all possible directions with millimeter accuracy. [1,2]

[1] J. Weber et al., "Magnetic Field Imaging (MFI) of Shingle Solar Modules," in Proceedings of the 49th IEEE PVSC 2022 (tbp).

[2] U. Zelleret al., "Comparison of Magnetic Field Imaging (MFI) and Magnetic Field Simulation of Silicon Solar Cells," SiliconPV, 2019.

40th European Photovoltaic Solar Energy Conference and Exhibition

Methodology

Magnetic Field Imaging (MFI)

- MFI can reveal relative current strength and direction [1]
- A 3-dimensional axes system uses a magnetic sensor array to scan the sample in all possible directions with millimeter accuracy. [1,2]
- Magnetic flux density (B_x, B_y, B_z) induced by the currents flowing in the samle is measured [1]
  - Each peak in the line plot corresponds to a ribbon current
- Backgroud measurement is essential [1,2]

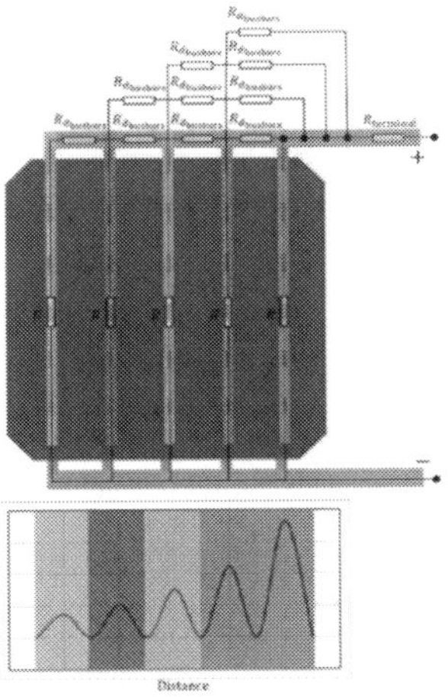

[1] J. Weber et al., "Magnetic Field Imaging (MFI) of Shingle Solar Modules," in Proceedings of the 49th IEEE PVSC 2022 (tbp).

[2] U. Zelleret al., "Comparison of Magnetic Field Imaging (MFI) and Magnetic Field Simulation of Silicon Solar Cells," SiliconPV, 2019.

Results

—

Position of string connector terminal

- Applying a 5x0.3 mm² string connector
- Trend in current distribution is recognizable in the simulation
- Higher currents at ribbons near to the string connector terminal
- ~ 2.5% maximum difference in current share

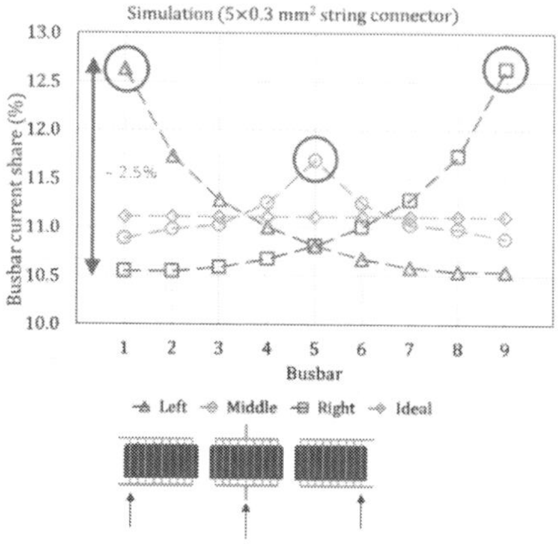

40th European Photovoltaic Solar Energy Conference and Exhibition

Results

—

Position of string connector terminal

- Applying a 5x0.3 mm² string connector
- Trend in current distribution is recognizable in the simulation
- Higher currents at ribbons near to the string connector terminal
- ~ 2.5% maximum difference in current share
- No recognized trend in the MFI measurement

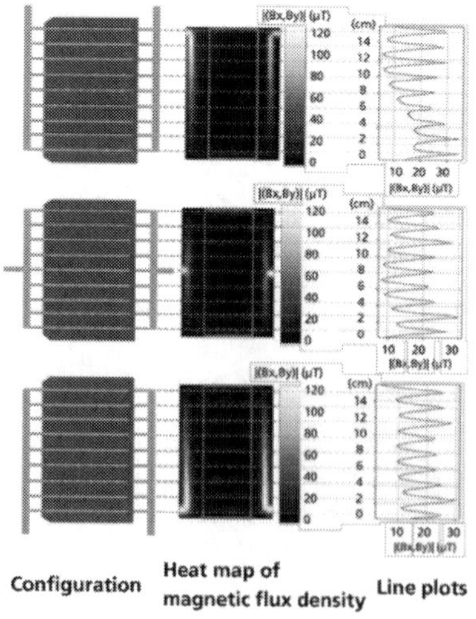

Results

Position of string connector terminal

- Applying a 1x0.08 mm² string connector
- Similar trend in current distribution
- ~ 18% maximum difference in simulated current share

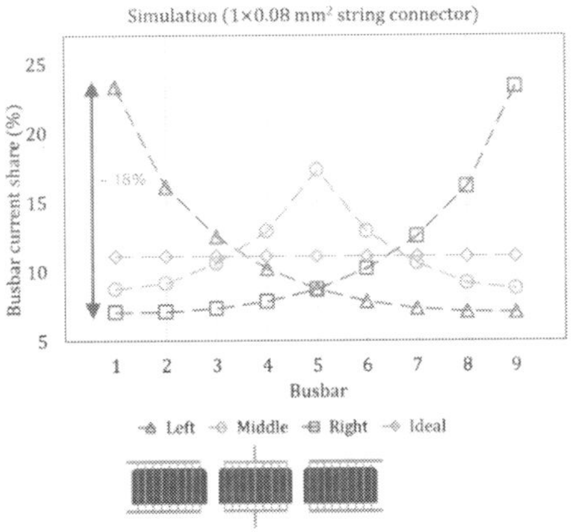

40th European Photovoltaic Solar Energy Conference and Exhibition

Results

Position of string connector terminal

- Applying a 1x0.08 mm² string connector
- Similar trend in current distribution
- ~ 18% maximum difference in simulated current share
- A similar trend in MFI measurement is recognized
- ~ 17% maximum difference in measured current share

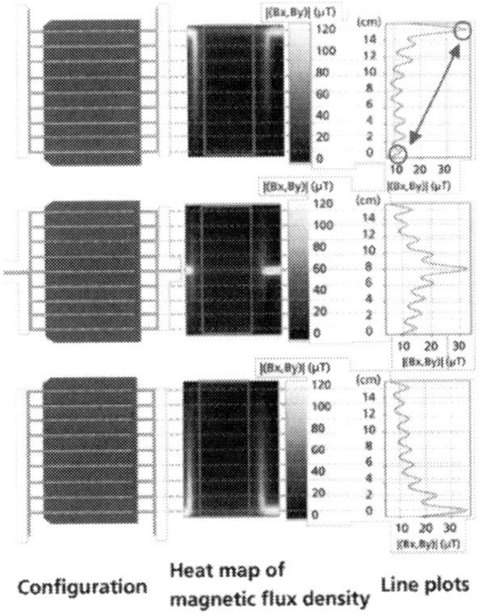

Results

Additional ohmic loss

- Applying a 1x0.08 mm² string connector
- Simulation of ohmic losses including
  - Losses in ribbons
  - Losses in string connector
- ~ 46% higher ohmic losses by positioning the terminal at the rightmost or leftmost ribbon compared to a terminal at the middle

→ ~ 0.35 W lower P_{mpp} of solar cell

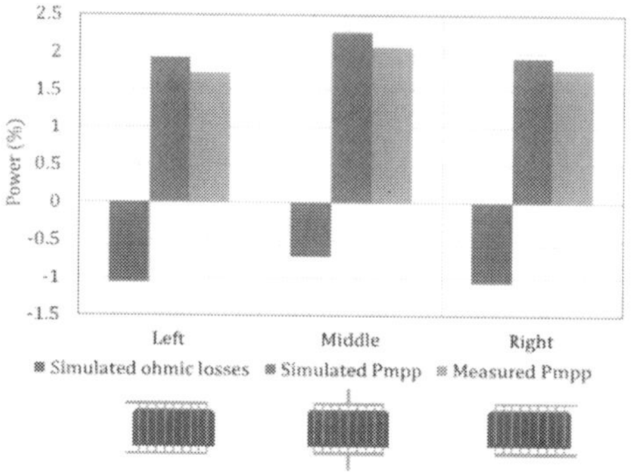

universität freiburg Fraunhofer ISE

020259-015

40th European Photovoltaic Solar Energy Conference and Exhibition

Results

Impact on module level

- Considering a 120-half-cell module with 12 substrings
- 24 cells connected directly to the string connectors
- < 0.1 Wp less module power (large string connector)
- ~ 3.2 Wp less module power (small string connector)
→ ~ 0.9% of 359 Wp module

Comparison with the case of equally distributed current

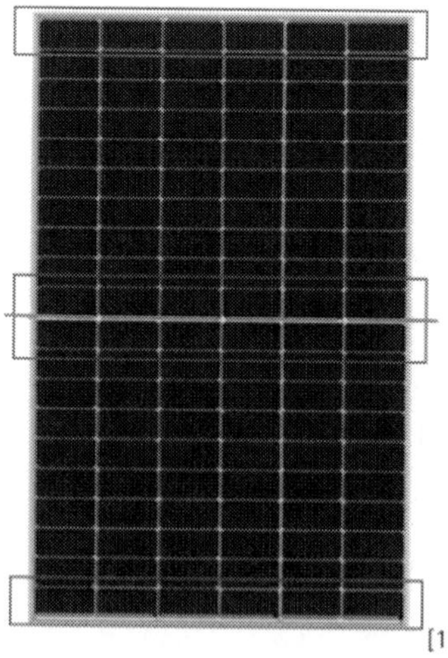
[1]

[1] D. Mesquita et al., "A review and analysis of technologies applied in PV modules" in Proceedings of the IEEE PES Innovative Smart Grid Technologies Conference, 2019

universität freiburg Fraunhofer ISE

020259-016

Results

Contact defects

- Detaching ribbon from contact pads
- Detaching complete ribbons
- Breaking string connector

- MFI results show redistribution of ribbon current

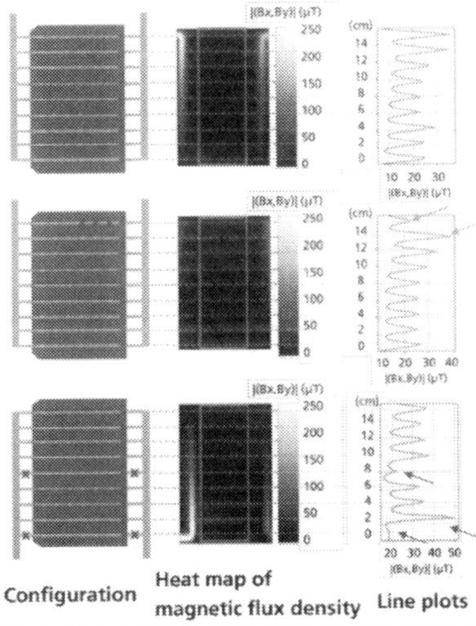

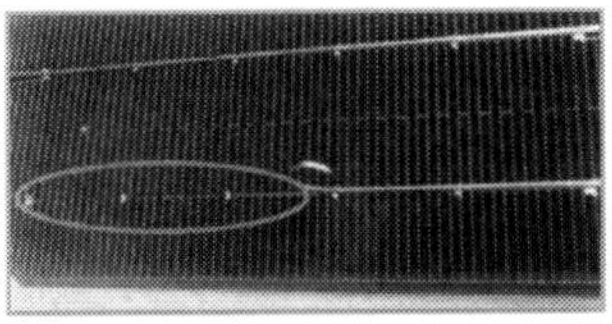

universität freiburg Fraunhofer ISE

020259-017

40th European Photovoltaic Solar Energy Conference and Exhibition

Results

Contact defects

- Detaching ribbon from contact pads
- Detaching complete ribbons
- Breaking string connector

- MFI results show redistribution of ribbon current

- Simulation results exhibit similar current redistribution

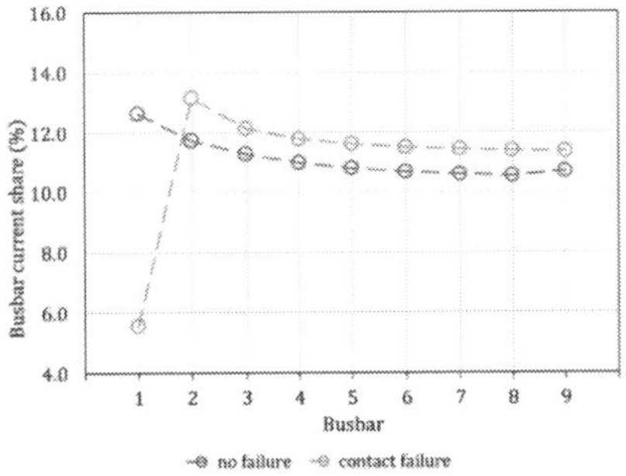

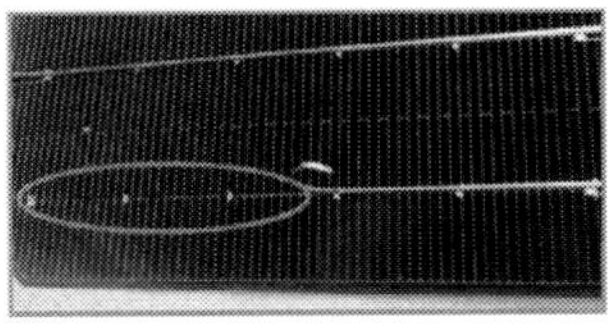

universität freiburg Fraunhofer ISE

020259-018

Results

Contact defects

- Simulation of different scenarios of contact defects
  - Detaching ribbon from contact pads
    - Higher impact for ribbons near string connector terminal

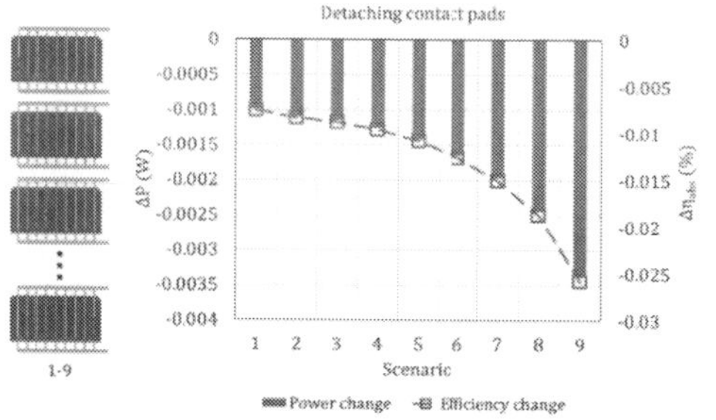

Results

Contact defects

- IV-curve measurement of a cell with broken ribbons
- ~1 W in Pmpp after breaking half of the ribbons
- ~ 65% reduction in Pmpp after breaking 8 ribbons

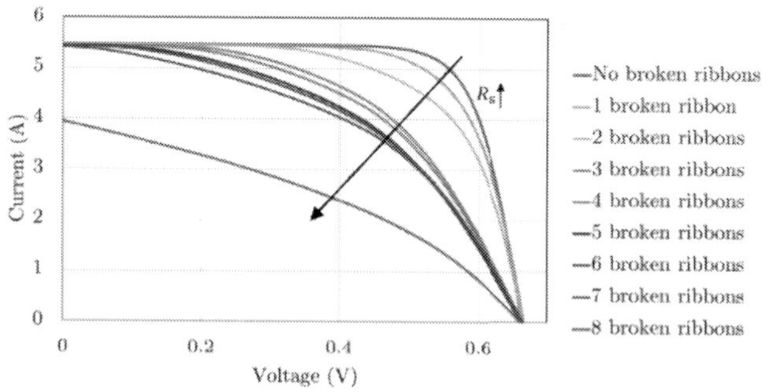

Results

Contact defects

- Breaking 1 ribbon of 1 cell
- Simulating 120-half-cell module with butterfly topology
- ~ 0.3 Wp reduction in module Pmpp

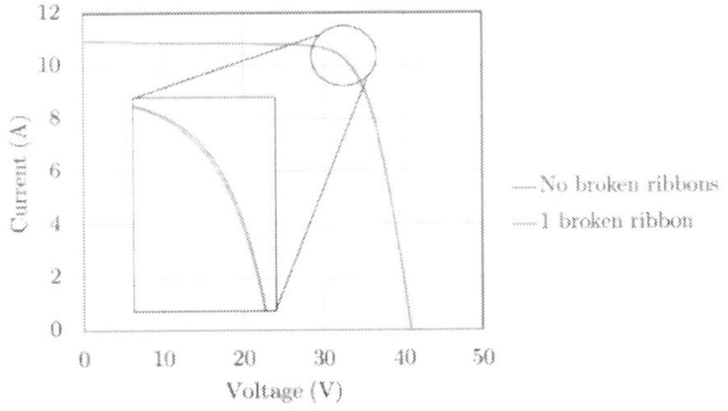

universität freiburg Fraunhofer ISE

Summary

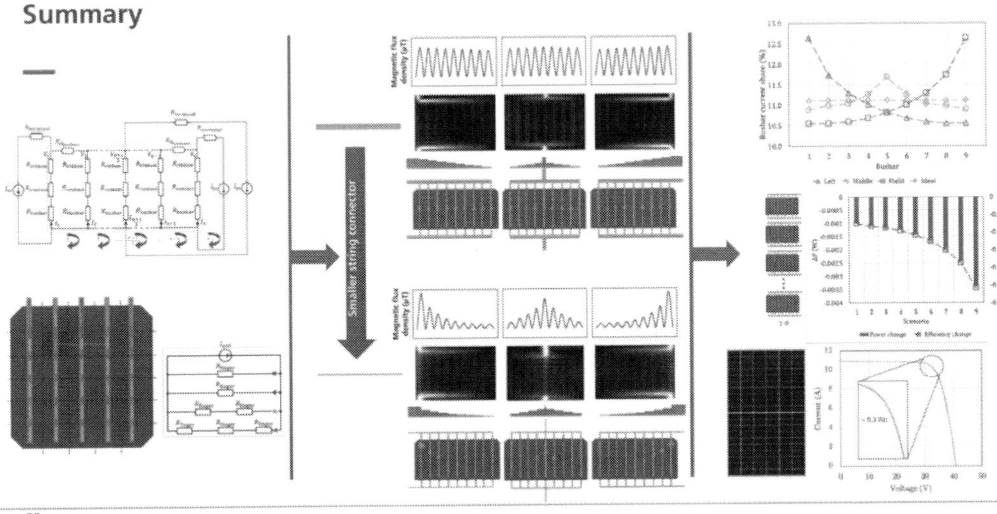

universität freiburg Fraunhofer ISE

Fraunhofer ISE

Thank You for Your Attention!

Contact

Fraunhofer ISE

Ammar Tummalieh
Team „Module Assessment"
ammar.tummalieh@ise.fraunhofer.de

Fraunhofer ISE
Heidenhofstrasse 2
79110 Freiburg
Germany
www.ise.fraunhofer.de

Supported by:

Federal Ministry for Economic Affairs and Climate Action

on the basis of a decision by the German Bundestag

Fraunhofer ISE contributions of the 40th EU PVSEC

This work was financially supported the German Federal Ministry for Economic Affairs and Climate Action within the research project "CTS1000plus" under grant number 03EE1028A.

This work will be published at Progress in Photovoltaics: Research and Applications

PROGRESS IN PHOTOVOLTAICS RESEARCH AND APPLICATIONS

Fraunhofer CSP

40th European Photovoltaic Solar Energy Conference, Lisbon, 21.09.2023

Local resolution of currents through electrical joints consisting of materials with different conductivity

Stephan Großer, Matthias Pander, Ulli Zeller, Bengt Jäckel

3DO.17.6

Motivation

Upcoming interconnection technologies

Lead free interconnection

Conductive adhesive and lead free soldering (market share up to 40% till 2033) [1]

Low temperature contact formation

Advantage of low thermomechanical stress, stress absorbance [2] and low thermal budget

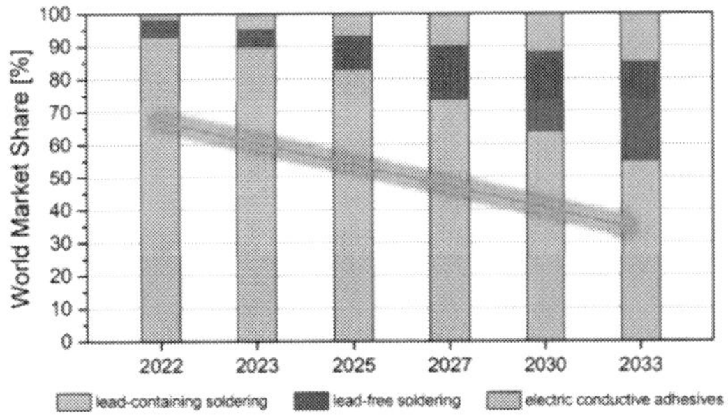

Different technologies for cell interconnection[1]

[1] Redrawn from ITRPV, "International Technology Roadmap for Photovoltaic (ITRPV) - 2022 Results," (2023)
[2] D. Tune et al., Photovoltaic international (2021), pp. 57-77.

Fraunhofer CSP

Electrically conductive adhesives (ECA)

State of the technology for PV

Electrically conductive adhesive (ECA)

Adhesive matrix + conductive filler → Very flexible functionalization!

Application

Shingle cells, cell interconnection and conductive backsheets

Reliability

Similar to solder contacts – BOM screening!

Characterization / Assessment

Behavior, performance, reliability and damage mechanism

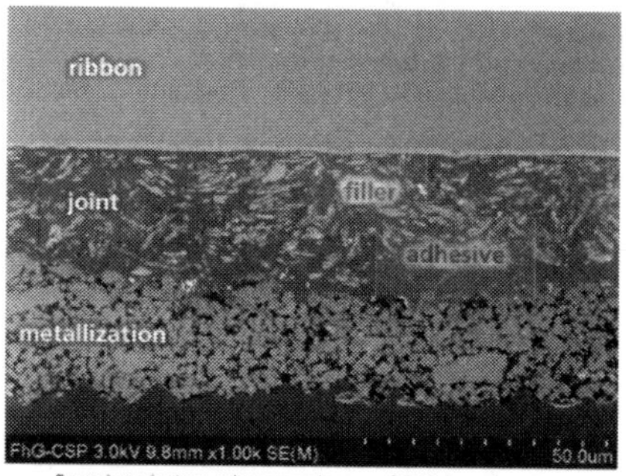

Scanning electron microscopy image of ECA-based contact (colored)

T. Geipel, et al., Energy Procedia 55 (2014) 336 – 341
T. Geipel, PhD thesis, Gottfried Wilhelm Leibniz University Hannover (2017)
D. Tune et al., Photovoltaic international (2021), pp. 57-77.
G. Oreski et al., in Photovoltaics International. (2019), pp. 38-45.
T. Timofte et al., 8th WCPEC, Milano (2022), pp. 551-558.
J. Weber, T. Roessler, Progress in Photovoltaics: Research and Applications, (2023) 10.1002/pip.3699
M. I. Devoto Acevedo et al., SOLMAT, 260, (2023) 10.1016/j.solmat.2023.112490
S. Großer et al., 8th WCPEC, Milano (2022), pp. 907-911.
B. Hartweg et al., 8th Metallization & Interconnection Workshop. (2019)
M. Springer et al., IEEE J. Photovoltaics 11 (2021) pp. 1040-1047

Fraunhofer CSP

020260-003

40th European Photovoltaic Solar Energy Conference and Exhibition

Scope

Insight in contact function

Main challenges for ECA development[1-3]

Cost: Ag-reduction

Performance: Good electrical properties

Reliability: Maximum adhesion strength and stability

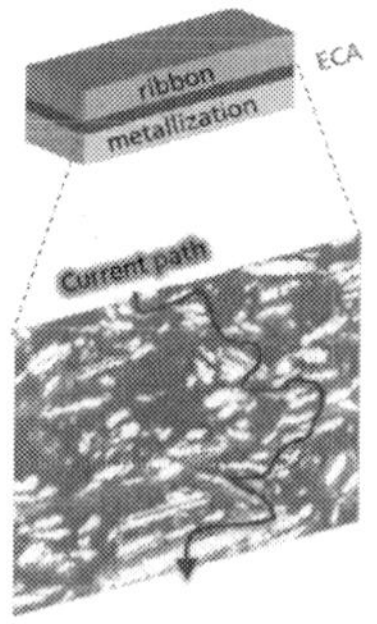

Conduction at the percolation threshold (illustration)

Scope: Assess the joint material impact on the current distribution

- **Approach: Visualize existing current transport**
- **Application: Joint materials with different volume resistivity**

1) D. Tune et al., in Photovoltaic international. (Solar Media Ltd., 2021), pp. 57-77
2) Y. Sakaniwa et al., ICEP-IAAC Proceedings (2015). Pp. 252-257
3) S. Großer et al., 8th WCPEC (2022), pp. 907-911.

Fraunhofer CSP

020260-004

Approach
Design of experiment - sample

Simple symmetric sample design

Interface – contact material to ribbon relevant

0.3 x 5mm²

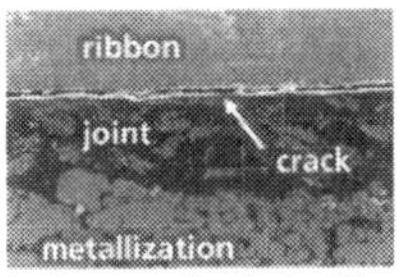

Aged contact with adhesion failure

Joint between 2 copper ribbons

Copper ribbon, $Sn_{60}Pb_{40}$ coated ($\rho_V = 1.7 \cdot 10^{-6}\ \Omega \cdot cm$)

Low-resistive joint – $Sn_{60}Pb_{40}$ ($\rho_V = 1.5 \cdot 10^{-5}\ \Omega \cdot cm$)

High-resistive joint – ECA, low Ag content ($\rho_V = 1 \cdot 10^{-2}\ \Omega \cdot cm$)

- Scraper coated
- Curing @ 150 °C / 10 min / 10 kPa

Test structures

40th European Photovoltaic Solar Energy Conference and Exhibition

Approach
Design of experiment - setup

e^- drift generates magnetic field

Biot-Savart-law

Spatially resolved detection

Magnetic flux density (vector) measured

1. Current density
2. Geometric distance

$$|B|^* \sim \frac{1}{d} \cdot j$$

Symmetric sample layout

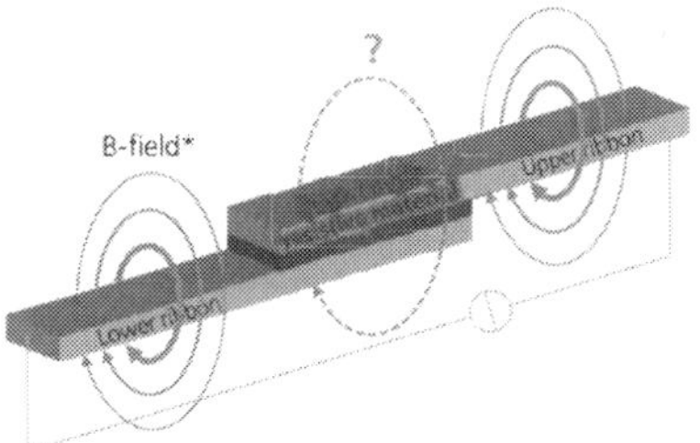

Configuration
(cross-section view)

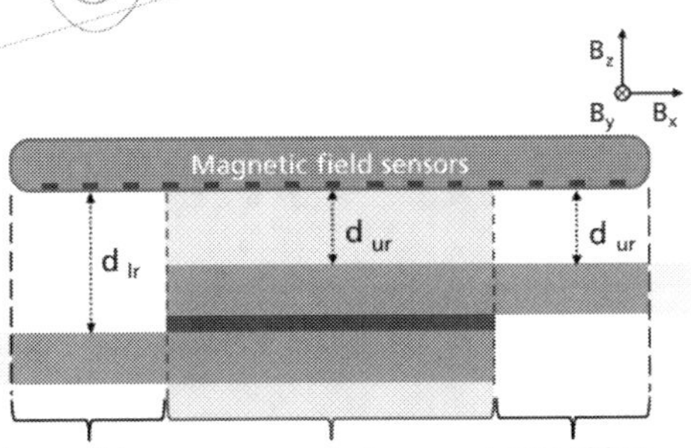

*) simplification for infinite straight wire

Current path
Low-resistive joint to Cu-ribbon ($Sn_{60}Pb_{40}$ solder)

Flux density profile → current distribution in the contact

Kirchhoff's law

Borders: localized current injection into contact

Center: equal current density in both ribbons

Result

Current flow imaging method demonstrated

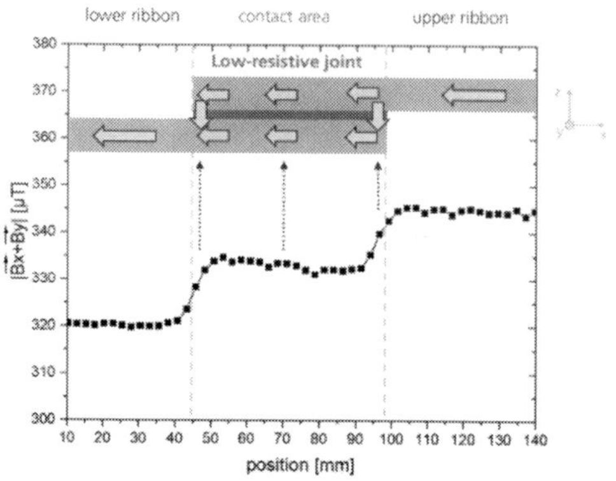

- Joint length = 50 mm
- Volume resistivity: joint ~ 10 times higher than copper

40th European Photovoltaic Solar Energy Conference and Exhibition

Current path
High-resistive joint to Cu-ribbon (ECA)

Current distribution in the high-resistive contact

Homogeneous current transport (from ribbon into joint)

Full contact area contributes to current transport

Joint material resistivity influence with current distribution

High-resistive contact material leads to extended current distribution in the contact material

Is the profile characteristic for solder and ECA? → NO

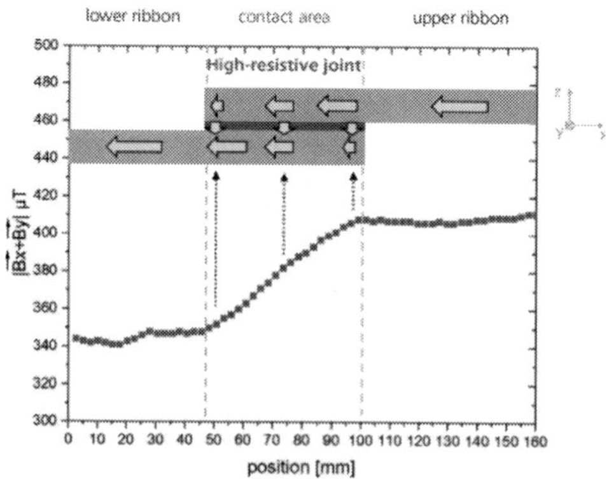

- Joint length = 50 mm
- Volume resistivity: joint ~ 6000 times higher than copper

Finite element simulation

Current densities and magnetic flux density

.

Modelling and finite-element analysis with ANSYS

Mimic experiment setup

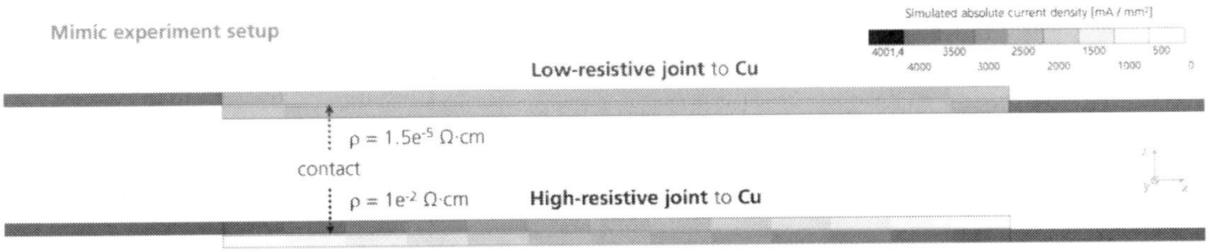

Modell confirmed local current distribution: basis for simulation of magnetic flux density B

Low-resistive joint → Rapid bisection of current

High-resistive joint → Extended length of current transport

Fraunhofer CSP

40th European Photovoltaic Solar Energy Conference and Exhibition

Finite element simulation

Current densities and magnetic flux density

B-field profile shape is independent from material type

Calculation of magnetic flux density confirms the measurement

Reveal impact of contact resistance

Resistivities and contact area determine the current distribution

For smaller contacts length

→ Current distribution for high- and low-resistive materials gets similar

→ High resistive contact with increased power dissipation!

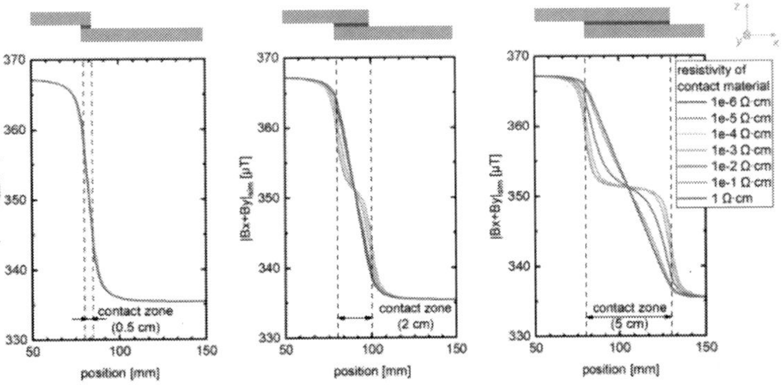

Fraunhofer CSP

Summary and conclusion

- Determination of current distribution in contacts by magnetic field imaging is feasible

- Local current distribution in contacts depends on
  - Volume and contact resistivity
  - Material combination (wire / joint)
  - Contact area formed by joint

- **Application potential**
  - Contact homogeneity → manufacturing
  - Understand material (analysis precisely at region of interest) → Ag-reduction
  - Contact stability → degradation, redundancy function / mitigation of failures

Fraunhofer CSP

40th European Photovoltaic Solar Energy Conference and Exhibition

Fraunhofer CSP

Publication in review → **EPJ Photovoltaics**

Acknowledgments

My colleagues from Fraunhofer CSP

Project "Zquadrat" and the financial support by BMWK funding (FKZ: 03EE1005B)

Supported by:

Federal Ministry for Economic Affairs and Climate Action

on the basis of a decision by the German Bundestag

Thank you for your attention!

Fraunhofer CSP

Contact

Dr. Stephan Großer
PV Modules, Components and Manufacturing
Tel. +49 345 5589-5112
Fax +49 345 5589-5999
Stephan.grosser@csp.fraunhofer.de

DESIGN, MANUFACTURING AND RELIABILITY OF SHINGLE MATRIX MODULES

Najwa Abdel Latif[1], Jochen Markert[1], Daniel von Kutzleben[1], Sebastian Birnkammer[1], Torsten Rößler[1], Julian Weber[1], Achim Kraft[1], Holger Neuhaus[1]
[1] Fraunhofer Institute for Solar Energy Systems ISE
Heidenhofstraße 2, 79110 Freiburg im Breisgau, Germany

ABSTRACT: Fraunhofer ISE, in collaboration with M10 Solar Equipment GmbH, addresses the demand for efficient, visually appealing and shading resilient solar panels by introducing and developing the shingle matrix technology. The matrix technology also ensures efficiencies surpassing those for ribbon-connected modules due to the increased cell packing density. This study demonstrates significant progress in advancing shingle matrix technology towards commercial viability. Optimization of shingle cutting, shingle placement precision, and electrically conductive adhesives (ECAs) curing conditions were successfully realized in this study to establish a reliable automated interconnection. Eleven full size shingle matrix modules were produced by industrial scale processes and underwent IEC 61215-2:2021-based climate chamber tests and tests addressing the electrical safety under combined environmental stresses according to IEC 61730. The modules showed performance degradations < −3.4% aligning with IEC 61215-2 criteria. These efforts ensure matrix technology's stability and safety, facilitating broader industry acceptance and demonstrate the commercial potential of this innovative technology.
Keywords: Photovoltaic Modules, Shingle Matrix Technology, Shingling, Module Reliability, Accelerated Ageing, Electrically Conductive Adhesives

1 INTRODUCTION

Shingle principle of the matrix technology was introduced already in 1990 [1]. Fraunhofer ISE developed full scale modules based on this concept to address the demand for PV-modules combining high efficiency with an excellent visual appearance and a superior shading resilience [2, 3, 4, 5].

Previously, Fraunhofer ISE research focused on understanding cell-to-module losses of linear shingled string modules and investigated critical processing steps for shingle module reliability [6]. The same methodology was applied to shingle matrix technology due to its significance in advancing photovoltaic module efficiency and aesthetics. This paper focuses on the scientific evaluation of shingle matrix technology, particularly its reliability both generally and in comparison, to historical module certification data [7], which highlights its progression towards commercialization.

While matrix shingling has been acknowledged for years, attaining commercial viability has remained elusive. Recent presentations from Fraunhofer ISE have illuminated matrix shingling potential [2]. This study highlights the technology's readiness for mass production through the successful performance of all important reliability tests.

The following sections will delve into the optimization process details, module production challenges, and an in-depth discussion of testing results. These discussions aim to validate the robustness and potential widespread adoption of shingle matrix modules.

2 MATERIALS AND METHODS

Key aspects of the shingle matrix technology were optimized, including shingle cutting, precise shingle placement, and the curing conditions of electrically conductive adhesive (ECA). To demonstrate technological maturity, eleven full-format shingle matrix modules underwent several reliability and safety tests following IEC 61215-2:2021 and IEC 61730-2:2016 standards. This testing encompassed thermal cycling (TC), damp-heat (DH), humidity-freeze (HF), mechanical load (ML) and hot-spot endurance from IEC 61215, as well as additional material creep and sequential testing using the sequence B1 from IEC 61730.

2.1 Shingle Separation Processes

The cells used for shingle matrix production are p-type mono-facial PERC cells in M6 format. To create shingle cells, M6 PERC host wafers are divided into 1/6th using laser cutting. In the industry, two common laser cutting techniques are utilized: thermal laser separation (TLS) and laser scribing combined with mechanical cleaving (LSMC) [8]. However, both methods, to varying degrees, lead to undesired charge carrier recombination at the cut edges, resulting in reduced module efficiencies.

LSMC is recognized for inducing more efficiency losses due to recombination when separating intact host wafers, unlike TLS. This differentiation can be identified by a distinct decrease in power output resulting from a lower fill factor (*FF*). The resulting shingles are 166.00×27.67 mm² and come in pseudo-square (PSQ) and full-square (FSQ) configurations. PSQ shingles consist of the two outer shingles of a host cell, while FSQ shingles have two laser-separated edges. In this study, only FSQ shingles separated using both laser techniques are used to create the shingle matrices and a comparison in the performance is carried out.

2.2 Electrically Conductive Adhesive (ECA)

To establish a robust interconnection between the shingles facilitating efficient current flow and mechanical stability, an epoxy based ECA containing silver particles, developed by Polytec PT, was utilized. This ECA is commercially known under the name SB 1242.

2.3 Matrix Shingling

The application of ECA, shingle assembly, and ECA curing were carried out using a shingle matrix stringer developed by M10 Solar Equipment GmbH. The process utilizes shingle cells as input materials. First, the ECA is applied via dispensing units onto the backside busbar of each shingle. These cells are then arranged in a matrix configuration controlled by a vision system and organized into rows and columns. Next, the shingles are positioned on an assembling tray with precise alignment. Finally, an infrared lamp curing unit is employed to facilitate the curing of the matrices.

10.4229/EUPVSEC2023/3DO.18.1
020261-001

2.4 Module Production

The glass-backsheet modules underwent lamination using a Bürkle laminator, employing ethylene-vinyl acetate (EVA) as encapsulant and a polyethylene terephthalate (PET) based backsheet. The glass has a 3.2 mm thickness and is coated with an antireflection layer. These materials were chosen as these are industry-standard options frequently employed in module manufacturing. The produced framed modules are connected with 3 bypass diodes and after the lamination process, they attain dimensions of 1761 × 1035 mm².

2.5 Module Reliability Testing

To assess the stability and electrical safety of the matrix shingle interconnection technology compared to standard modules with soldered ribbon interconnection, different environmental stress scenarios were employed. Accelerated ageing tests and sequences defined in IEC 61215-2:2021 [9] and IEC 61730-2:2016 [10] (as indicated in Table I) were utilized according to IEC 17025 [11] certified in the PV Modules TestLab at Fraunhofer ISE for this purpose.

Table I: Tests/sequences applied to the respective number of modules; the specifications for each test and sequence are defined in the following paragraphs 2.5.1 – 2.5.7.

Test / Sequence	Standard	No. of Modules
Hot-spot endurance test	IEC 61215	1
TC200	IEC 61215	4
DH1000 + ML	IEC 61215	3
TC50 + HF10	IEC 61215	1
Material creep test	IEC 61730	1
Sequence B1	IEC 61730	1

The results were evaluated in accordance with the defined acceptance criteria. To measure the impact of the tests, a series of characterization steps were taken both initially and after each test:

- Visual inspection acc. to IEC61215-2:2021, MQT 01
- Electroluminescence imaging (EL) acc. to IEC TS 60904-13:2018
- Wet leakage current test acc. to IEC 61215-2:2021, MQT 15
- Insulation test acc. to IEC 61215-2:2021, MQT 03
- Power measurements at standard conditions acc. to IEC60904 (-1 and -3)

Prior to the reliability tests defined in IEC 61215, a preconditioning step involving light soaking (30 kWh/m²) was implemented to stabilize any boron-oxygen light-induced degradation (BO LID) of the modules. In the discussions that follow, the term "initial" pertains to results obtained from the performance measurements after the initial stabilization phase.

2.5.1 Hot-Spot Endurance Test

The test assesses PV module's resilience against reverse bias hot-spot heating effects, simulating worst-case shading. The module is exposed to 1000 W/m² solar radiation in a sun simulator with homogeneity class B, which states an arial irradiation inhomogeneity of < 5 % [9]. The shingle matrix technology represents a combination of series and parallel cell connections and is not yet considered in the specifications for the test procedure defined in IEC 61215-2, in which a worst-case shading scenario was assumed, and therefore each cell row was treated as one full cell. Sequential shading of rows identifies the lowest shunt resistance row. Shading with different extents (25%, 50%, 75% and 100%) identified the worst-case condition, with the maximum temperature measured after 1 h using an infrared camera (model 882) from the manufacturer Testo.

2.5.2 Thermal Cycling

During TC200 testing, 200 cycles of temperature variations between 85 °C and −40 °C were applied. This evaluation aimed to gauge the module's resilience against potential mismatches in thermal expansion coefficients (CTE), which can lead to thermomechanical stresses. These stresses are more likely to occur in areas where cells are interconnected. The dwell time at the maximum temperatures was 10 min and the heating/cooling rate was restricted to 100 K/h.

2.5.3 Damp Heat

In order to evaluate matrix module resilience against increased humidity levels, the modules underwent exposure to a temperature of 85 °C at a relative humidity of 85% for a duration of 1000 hours in a climate chamber.

2.5.4 Static Mechanical Load Test

This test assesses the module's capability to endure wind, snow, static, or ice loads. It involves applying a dynamic load with three cycles alternating pressure and suction of a minimum ± 2400 Pa, each held for one hour, to both the front and rear sides of the module. For this study, various load profiles were employed for each module, including ± 2400 Pa, ± 3600 Pa, and + 5400 Pa/−3600 Pa. The modules were mounted using four clamps at one quarter of the long module side, with cross braces running below.

2.5.5 Humidity Freeze

In the HF10 test (10 cycles), the resistance of a module against increased humidity levels at high temperatures followed by sub-zero temperatures is determined. In contrast to the TC test, the dwell times are increased to 20 h at 85 °C and 30 min at −40 °C and the humidity level during the high temperature phase is set to 85 % r.h.

2.5.6 Material Creep Test

The material creep test defined in IEC 61730-2 aims to validate that materials used in the PV module will not show creep or lose adhesion when operated at the highest temperatures typically experience in the field [10]. Therefore, a frameless module is exposed to 105 °C for 200 hours in a climate chamber.

2.5.7 Sequence B1

Testing in accordance with sequence B1 entails a combination of environmental stresses with a cold conditioning phase (−40 °C for 48 hours), succeeded by dry heat conditioning (105 °C for 200 hours), followed by HF10 exposure, another cold conditioning, and a final HF10 exposure.

10.4229/EUPVSEC2023/3DO.18.1

3 RESULTS

The performance of the modules before and after each reliability test was evaluated, and comparisons were carried out to assess the eventually developing degradation effects.

Negligible changes of I_{SC} ($\Delta P_{MPP} < 0.4\%$) and V_{OC} ($\Delta V_{OC} \leq 0.1\%$) during the initial stabilization indicated no significant levels of BO LID present in the modules.

3.1 Hot-Spot Endurance Test

PV module susceptibility to hot-spot formation during partial shading varies based on cell-interconnection technology and number of cells per string. The unique combination of series and parallel cells in matrix shingle technology adds complexity to the assessment of hot-spots.

When the test was performed as described above, the module exhibited a low susceptibility for critically increased temperatures due to partial shading. The highest temperature occurred during 75% shading of a single cell row after one hour, finally reaching a temperature of 104.3 °C, which was measured near the module edge (Figure 1).

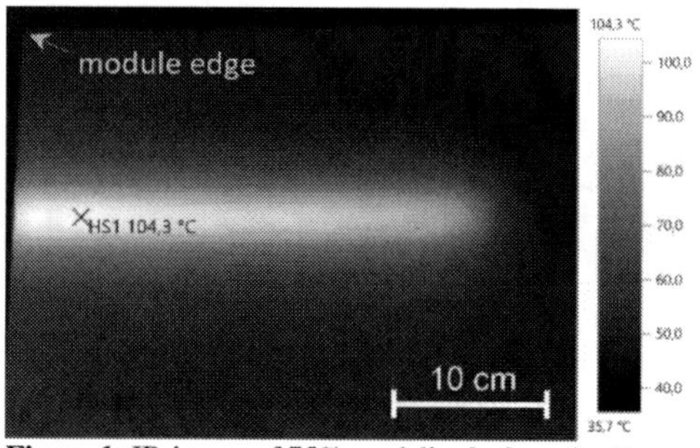

Figure 1: IR-image of 75% partially shadowed cell row after one hour of equilibration under illumination.

Temperatures in such ranges can be considered non-critical, as the thermal degradation of materials commonly used in module manufacturing is typically observed at temperatures beyond 160 °C. However, the specific materials used during module production can strongly influence this threshold [12].

This assessment is supported by the absence of visual damages, negligible performance losses ($\Delta P_{MPP} = -0.3\%$), and intact electrical insulation properties ($\geq$ 5000 MΩ) after the test.

3.2 TC200

To demonstrate the stability of the modules against CTE mismatches and resulting thermomechanical stress, four modules underwent a TC200 test according to IEC 61215-2:2021, MQT 11.

With an average P_{MPP} degradation of −0.6% (max. −1.3%, Figure 2), mainly caused by losses in *FF*, the modules showed an average performance degradation comparable to typical c-Si modules.

No visual defects could be observed after the test. Electroluminescence (EL) images before and after the test (Figure 3) revealed similar, and only negligible changes in brightness for all modules, indicating no severe degradation potential arising from thermal mismatch of the materials used for cell interconnection. Furthermore, the electrical insulation properties of the modules remained intact ($\geq$ 3180 MΩ) without any critical approximation to the pass/fail criterion according to IEC 61215-2:2021 (= 22 MΩ).

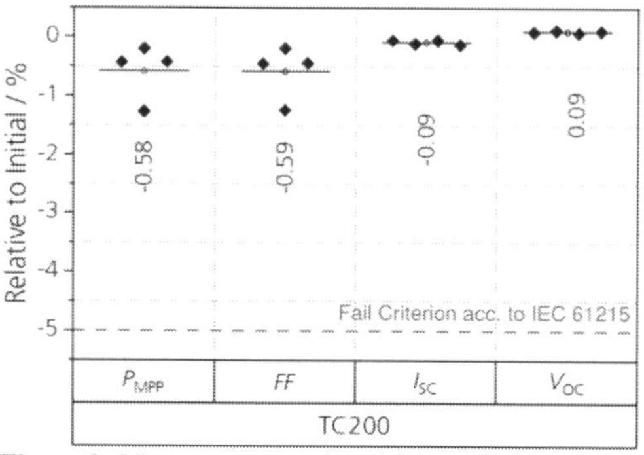

Figure 2: Measurement results of P_{MPP}, *FF*, I_{SC} and V_{OC} after the TC200 test relative to the initial measurement; Dashed line: 5% loss criteria according to IEC 61215-2:2021.

Figure 3: Exemplary EL image of one module before (left) and after (right) the TC200 test.

3.3 DH1000 and ML Tests

Three modules underwent DH1000 followed by ML tests (IEC 61215-2:2021, sequence E). Post-DH1000, the modules were subjected to different load profiles (Table II). No visible defects were observed after the tests. The EL images after the DH 1000 test exhibited darker patterns on shingle cells (Figure 4, red square), likely due to moisture accumulation under EVA stickers used at some positions for module production. It is hypothesized that increased moisture content in the EVA tape led to localized cell degradation. However, no major impact on module reliability is expected, and the issue can be addressed through material changes. After ML tests, additional cell cracks appeared (Figure 4, right: dark spots in EL images), with the number of cracks correlating with higher load profiles (Table II).

During DH followed by ML tests, two modules exhibited significantly brighter spots in EL near their centre, close to the top and bottom edges. This is caused by a known degradation of the module's terminal connections and has been fixed in the current technology revision. DH1000 resulted in an average performance degradation of −2.1% (max. = −2.3%) due to increased series resistance and resulting losses in *FF* combined with losses in I_{SC} (Figure 5). Degradation effects during DH ageing are typically caused by humidity ingress and subsequent hydrolysis reactions with the encapsulant leading to corrosion at cell connectors [13]. Average results from previous module certification data (median = −0.28%) suggest an under performance. However, looking at the distribution of the previous results, it appears that 14% of the modules showed a positive change of P_{MPP}, which can be attributed to meta stability effects that are assumed to play a negligible role in the current investigation [7].

Table II: Load profile of respective ML test, number of additional cells cracks after the test and corresponding performance loss.

Load Profile	No. of cells with additional cracks	Relative performance loss / %
2400 / −2400 Pa	3	−2.1
3600 / −3600 Pa	9	−2.9
5400 / −3600 Pa	14	−2.3

ML tests introduced additional performance losses, which, however, could not be correlated with the number of cracked cells (Table II and Figure 5). Inconsistencies between number of cracked cells and performance loss likely arise from manufacturing uncertainties and material inhomogeneities. Compared to certification data, results appear above average (median = −1.76%). However, comparing these test results lack differentiation based on load profiles, mounting specs, cell tech, materials, frame, and glass thickness parameters, which can all impact module stability and performance degradation.

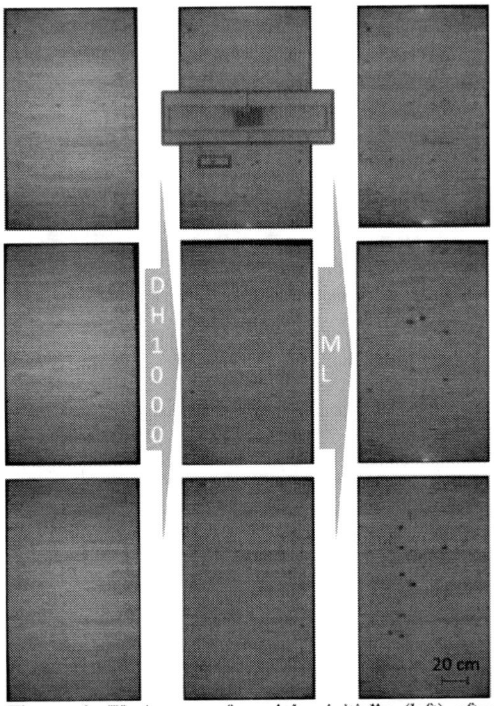

Figure 4: EL images of modules initially (left) after DH1000 (centre) and additional ML (right); Red square: Darker appearing patterns developing during DH ageing caused by EVA tapes used for module production.

Despite this limitation, the comparison to the −5% criteria defined in IEC 61215 suggests fundamental stability in module design, even when subjected to elevated load levels. The electrical insulation remained unaltered, with a minimum value of 2500 MΩ after the test at ±3600 Pa.

3.4 TC50 and HF10

Another module underwent testing involving TC50 followed by HF10. Similar to observation after the TC200 tests, the modules did not show any indications of degradation in the EL images due to thermomechanical stresses or increased humidity ingress (Figure 6). The cause of a single cell appearing brighter at the lower edge, like modules in DH/ML, is briefly discussed in Section 3.3.

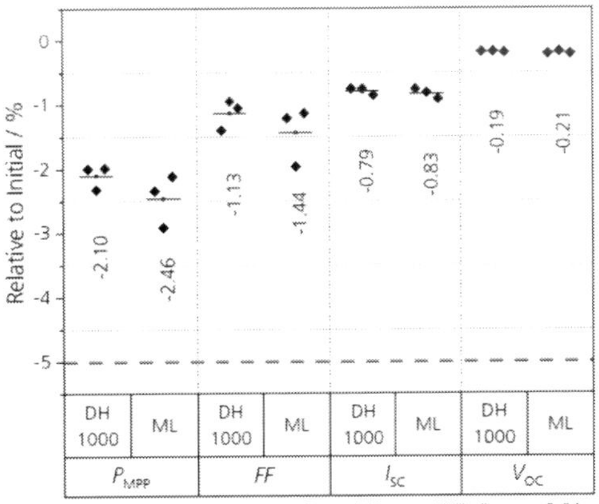

Figure 5: Measurement results of P_{MPP}, FF, I_{SC} and V_{OC} after DH1000 followed by ML testing relative to the initial measurements; Dashed line: 5% loss criteria according to IEC 61215-2:2021.

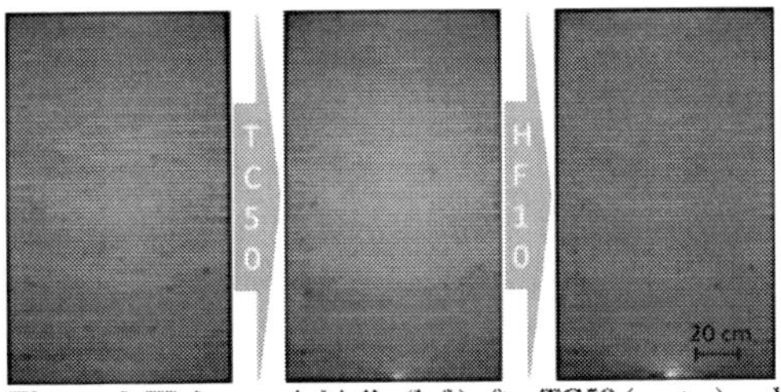

Figure 6: EL images initially (left) after TC50 (centre) and after additional HF10 ageing (left).

Figure 7 shows the electrical parameters after the TC50/HF10 sequence relative to the initial module's performance. After TC50, the impact on the module's power was in a similar range to that after TC200 (−0.7%), suggesting that the first temperature cycles have a predominant influence on the FF degradation.

Despite this, the electrical insulation remained unaffected, with a final value of 2860 MΩ.

A severe power loss down to −3.3% was detected after HF10. The cause of this deterioration was traced to an increased series resistance that developed during HF10 ageing. This points towards an added effect of humidity ingress during the elevated temperature phases (also observed in modules in DH) and subsequent sub-zero temperatures.

3.5 Material Creep Test

The frameless module showed intact electrical insulation properties with 2230 MΩ after the material creep test. No visual degradation or shifting of module components could be observed on the laminate. A similar brighter-appearing area, as already observed in the modules in TC and DH, could be observed in the EL image and is assumed to be responsible for a reduced *FF* and resulting performance loss of −2%.

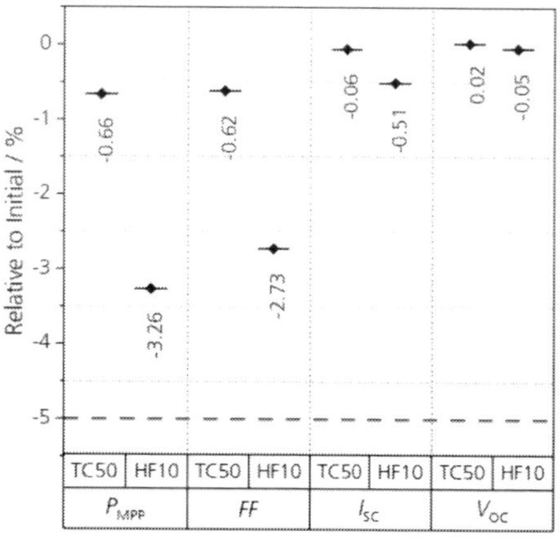

Figure 7: Measurement results of P_{MPP}, *FF*, I_{SC} and V_{OC} after TC50 followed by HF10 testing relative to the initial measurement; Dashed line: 5% loss criteria according to IEC 61215-2:2021.

3.6 Sequence B1

The module showed intact electrical insulation properties after the combined stress testing with the sequence B1 following IEC 61730-2:2016 [10] but with a significantly reduced value 161 MΩ (pass/fail = 22 MΩ). This reduction developed in conjunction with a −4.4% performance loss during the final HF10 ageing cycle (final: −7.5%), suggesting a common root cause. A bright area, similar to the ones that were found after other reliability tests developed in EL images, is suspected to be responsible for at least a part of the *FF* and subsequently performance loss. No other defects could be observed during the characterization.

Generally, the performance of the modules after ageing tests was highly dependent on factors such as the laser separation method used for shingle production, the dispensing parameters of ECA, and the precision of shingle placement.

Reliability tests showed acceptable performance during TC, DH, and ML tests. TC50 with subsequent HF10 testing showed the maximum power reduction with −3.3%, however, still meeting the −5 % power loss requirement in IEC 61215-2:2021. Furthermore, electrical insulation properties remained intact, and no visual defects could be observed after each of the tests.

4 CONCLUSIONS

In the scope of this investigation, the general stability of matrix shingle technology against stress scenarios outlined in IEC 61215 and IEC 61730 and therefore the readiness for module certification could be demonstrated. This was successfully achieved by satisfying all requirements stipulated by the standard concerning power loss and electrical safety for all conducted test scenarios. Furthermore, for all tests, except of the DH test, the performance degradation after climate chamber testing was comparable or lower than the median values from historical module certification data reported in literature [7]. The shingle matrix modules showed a sufficient gap to the 5% loss criteria defined in IEC 61215. Small irregularities observed in EL are assumed to originate from inaccuracies during module production and can be easily avoided in future processes. No visual defects or loss of electrical insulation properties were observed. A maximum temperature of approximately 104 °C in the hot spot test suggested a reduction in the risk of formation of hot-spots during partial shading scenarios, but given the use of an internal approach, these findings warrant further validation by subsequent experiments. The results after material creep test and sequence B1 according to IEC 61730 indicate no major electrical safety concerns, even due to combined environmental stresses. It can be concluded that these results demonstrate the high potential and readiness for commercialization of shingle matrix modules regarding certification. The results from reliability tests that can be reached by modules from an optimized industrial production line are expected to be even better because of small material and handling issues that can be solved easily. However, the understanding for long-term reliability and electrical safety of matrix shingle modules will be extended by additional investigations in the future.

5 ACKNOLEDGMENT

We would like to thank all the colleagues and students who participated in this work as well as the Federal Ministry for Economic Affairs and Climate Action for funding this study within the scope of the project "Shirkan" (contract no. 03EE1026A).

6 REFERENCES

[1] W. Schmidt and K.-D. Rasch, "New interconnection technology for enhanced module efficiency," IEEE Trans. Electron Devices, vol. 37, no. 2, pp. 355–357, 1990, doi: 10.1109/16.46366.

[2] Daniel von Kutzleben et al., "Development of shingle matrix technology for integrated PV applications," WCPEC-8 Conference Proceedings, 2022.

[3] A. Mondon, N. Klasen, E. Fokuhl, M. Mittag, M. Heinrich, and H. Wirth, "Comparison of Layouts for Shingled Bifacial PV Modules in Terms of Power Output, Cell-to-Module Ratio and Bifaciality," in Proceedings of the 35th European Photovoltaic Solar Energy Conference and Exhibition, 2018, pp. 1006–1010.

[4] N. Klasen, D. Weißer, T. Geipel, D. H. Neuhaus, and A. Kraft, "Performance of shingled solar modules under partial shading," Prog. Photovolt: Res. Appl., pp. 1–14, 2021, doi: 10.1002/pip.3486.

[5] N. Klasen, F. Lux, J. Weber, T. Roessler, and A. Kraft, "A Comprehensive Study of Module Layouts for Silicon Solar Cells Under Partial Shading," IEEE J. Photovoltaics, vol. 12, no. 2, pp. 546–556, 2022, doi: 10.1109/JPHOTOV.2022.3144635.

[6] T. Rößler et al., "Progress in shingle interconnection based on electrically conductive adhesives at Fraunhofer ISE," Aip Conf Proc, 2022.

[7] P. Gebhardt, G. Mülhöfer, A. Roth, and D. Philipp, "Statistical analysis of 12 years of standardized accelerated aging in photovoltaic-module

certification tests," Prog Photovolt Res Appl, vol. 29, no. 12, pp. 1252–1261, 2021, doi: 10.1002/pip.3450.
[8] A. Münzer et al., "Laser Assisted Separation Processes for Bifacial pSPEER Shingle Solar Cells," in Proceedings of the 37th European Photovoltaic Solar Energy Conference and Exhibition, online, 2020.
[9] IEC 61215: Terrestrial photovoltaic (PV) modules: Design qualification and type approval: Part 2: Test Procedures, Edition 2.0, International Electrotechnical Commission, Geneva, Switzerland, 2021.
[10] IEC 61730: Photovoltaic (PV) module safety qualification: Part 2: Requirements for testing, Edition 2.0, International Electrotechnical Commission, Geneva, Switzerland, 2016.
[11] DIN EN ISO/IEC 17025-2018: General requirements for the competence of testing and calibration laboratories, DIN Deutsches Institut für Normung e. V., Mar. 2018.
[12] D. Philipp, H. Manuel, E. Fokuhl, and G. Mülhöfer, "Hot-Spot Endurance Test: Modifications for Bifacial Photovoltaic Modules," 35th European Photovoltaic Solar Energy Conference and Exhibition, pp. 1029–1033, 2018.
[13] Nikoleta Kyranaki et al., "Damp-heat induced degradation in photovoltaic modules manufactured with passivated emitter and rear contact solar cells," Damp-heat induced degradation in photovoltaic modules manufactured with passivated emitter and rear contact solar cells. In: Prog Photovolt Res Appl 30 (9), S. 1061–1071. DOI: 10.1002/pip.3556.

IMPROVEMENT OPTIONS FOR PV MODULES BY GLASS STRUCTURING

Marc Hofmann[1], Laura Stevens[1], Patrick Hör[1], Philipp Barth[1], Benedikt Bläsi[1], Stephan Riepe[1], Svenja Kalthoff[1], Bishal Kafle[1], Martin Zimmer[1], Max Mittag[1], Sebastian Nold[1], Ilker Sen[2], Johanna Reck[3], Nikolas Schröer[4], Laurent Clochard[5], Sebastian Ihlow[6], Carsten Horch[6]
[1]Fraunhofer Institute for Solar Energy Systems ISE, Heidenhofstrasse 2, 79110 Freiburg, Germany
[2]Gebr. Schmid, Robert-Bosch-Str. 32-36, 72250 Freudenstadt, Germany
[3]SENTECH Instruments, Schwarzschildstr. 2, 12489 Berlin, Germany
[4]ICB, Ludwig-Erhard-Ring 21, 15827 Dahlewitz, Germany
[5]Nines Photovoltaics, Synergy Centre TU Dublin- Tallaght Campus, Dublin 24 D24 A386, Ireland
[6]GMB Glasmanufaktur Brandenburg, Spremberger Str. 4, 03130 Tschernitz, Germany
Tel.: +49 761 4588-5051, email: marc.hofmann@ise.fraunhofer.de

ABSTRACT: The structuring of glass surfaces offers a wide area of application for photovoltaics: Increasing the energy yield and decreasing glare are achievable and become important factors for applications to building surfaces like roofs facing north, façades or walls along streets (e.g., noise-barrier). We investigated ways to reach specific glass surface morphologies and optical behaviors using wet and dry etching, combinations of blasting and etching, and imprinting into hot glass. We found that when a structured glass surface is present at the solar module's front, an increase in electricity yield can be achieved, with the largest gains under angles of incidence above 60°. Also, the application of structured glass to the solar module's rear can have a positive impact on the performance; we saw an increase in energy yield with illumination from the front especially for angles of incidence of 40° and above. The glass surface properties were investigated using a newly setup laser-based measurement tool that helped understanding light diffraction.

Keywords: module glass structuring, glass imprinting, glass etching, module performance improvement.

1 INTRODUCTION

Photovoltaic module glass surface structuring offers the chance to engineer the optical properties of reflection and transmission of light at and through the glass. Such treated glasses could provide higher electricity generation of the encapsulated solar cells and reduce glare effects. This would allow the extensive use of building envelopes that are not in consideration yet due to either lower electrical performance (e.g., roofs facing northern directions) or that would lead to significant glare issues (e.g., façades near public streets, …). It was shown that, in Germany, the relevant shares of existing building envelopes would allow for about 1 TW of PV installation [1]. Figure 1 brings together the findings of Wirth [2] with building-integrated photovoltaics' (BIPV) potential being in the range of 1 TW. Additionally, other already-mankind-treated areas in Germany hold significant PV-integration potential.

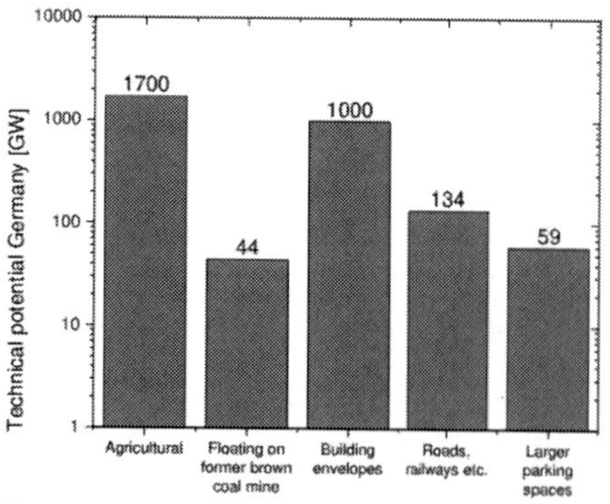

Figure 1: Technical potential for integration of PV in already existing human land use in Germany. Data taken from [2].

Performance increase and cost considerations allow the determination of the impact a certain improvement or cost change might have. In Figure 2, we used ISE's cost calculation tool SCost [3]. For example, increasing the electricity yield by 0.15%$_{abs}$ would allow a decrease in levelized cost of electricity (LCOE) by about 0.01 €ct/kWh. In other words, higher area-related costs of 0.8 €/m2 would theoretically be possible leading to the same LCOE.

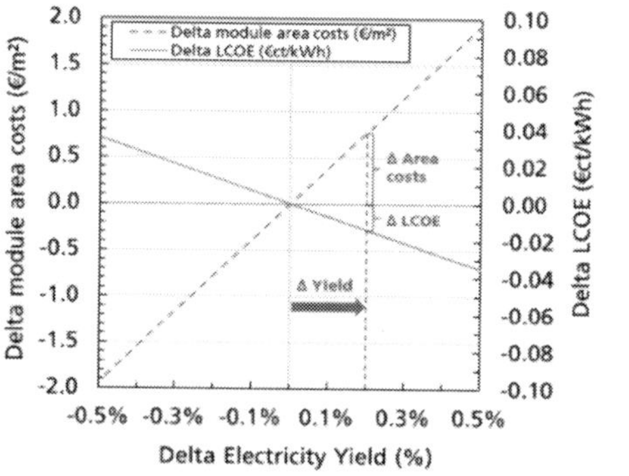

Figure 2: Calculation of interdependence of total cost, electricity yield and LCOE.

The structuring of glass surfaces is a promising way to reduce glare, increase performance and, as a result, enlarge the application possibilities of PV modules.

Glass structuring was investigated before. The company Saint-Gobain has offered structured glasses under the product name Albarino [4]. Such glasses were successfully tested in indoor and outdoor conditions as presented by Grunow et al. [5] who showed efficiency gains for the modules.

In the approach presented here, we are working on different technologies to achieve structured glass surfaces that facilitate optical reflection and transmission engineering in a solar PV module. Etching of (rather) cold glass is performed. Etching was carried out using two different approaches: wet chemicals or dry gases. Besides or prior to the etching, the effect of an additional blasting treatment was evaluated. Surface structure results are characterized with a newly designed measurement tool that analyses the diffusivity of the light transmission through a treated glass.

The current state of the art to improve light absorption is to texture the surface of the solar cell and apply an anti-reflective layer on both solar cell and glass of the module. Structuring the glass itself could generate further electric gains, increasing efficiency, but also further facilitate the application of PV modules by reducing angular dependence and glare. Huge potential for PV incorporation is found in building surfaces. Mostly, houses are currently only using roof areas facing south directions due to highest PV electricity harvesting. Roofs facing north and façades could provide a large share of electricity generated by BIPV, but the glare effects affecting the vision of e.g., car drivers passing by are an issue. Moreover, for Agrivoltaic (APV) more and more upright installations (fence-style) are considered. Additionally, for all application surfaces and tilting angles of modules, an increase of light absorption in the solar cells due to glass structuring would be a benefit, especially for shallow angles of incidence.

2 SIMULATION OF GLASS SURFACES

Optical simulations of glass surfaces included into photovoltaic modules were performed.

Here, one focus was put on the structuring of the front glass surfaces of glass-glass modules. Significant yield gain can be expected from a structured glass surface, according to the simulation results, presented in Figure 3.

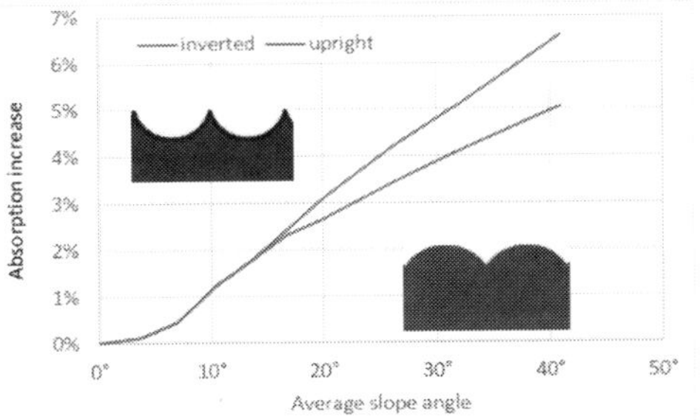

Figure 3: Estimation of the yield enhancement that can be achieved by texturing the front surface of the module glass. As a simple model, the increased optical in-coupling caused by spherical caps (inverted and upright) was modeled using the SunSolve raytracer of PV Lighthouse [6]. The simulation based on diffusely incident radiation gives a good first estimate for the enhanced annual yield. Especially the inverted spherical caps with their "bowl-like" shape are beneficial, since their steep sidewalls couple in light incident at shallow angles.

A simulation study on structured glass surfaces for the application as rear module surface is presented by L. Stevens et al. within this conference [7].

3 ETCHING OF GLASS SURFACES

For the structuring of glass, wet and dry etching technologies were evaluated.

First, we want to report on the wet etching process development. Here, one important goal was to increase the process speed. Currently, we achieved a process duration of about 3 min to end up with the surface structure presented in Figure 4. The etching process bath includes hydrofluoric acid plus additives and is performed in the acidic regime. Next steps in our work will include the upscaling of the etching process to larger glass surfaces.

Figure 4: Example results of a wet-etched glass surface. Top: Photo of a treated glass.
Bottom: Confocal microscopy image of a treated glass surface.

Dry technologies can also be applied to etch glass surfaces. NinesPV and Fraunhofer ISE worked on the development of a plasma-free dry etching process. Figure 5 gives an example of glass surfaces after dry etching. Here, the thermally driven etching process based on climate-friendly F_2 gas led to the creation of a nanoporous layer at the glass surface. Transmission can be improved with an antireflective layer created by this etching. The etching depth corresponds to the wavelength showing the peak in transmission (optical depth $\approx \lambda/4$). It was found that the structures we could achieve applying the dry etching approach differed significantly from the structures reached with the wet-chemical processes within the investigated parameter ranges.

10.4229/EUPVSEC2023/3DO.18.3

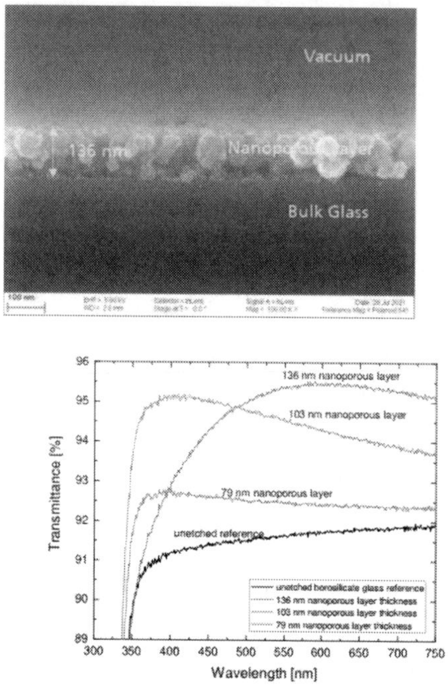

Figure 5: Results after dry etching of glass using a plasma-less process applying climate-friendly F_2 gas.
Top: Scanning electron microscopy (SEM) image of nanoporous layer etched into the surface of a glass sample.
Bottom: Transmittance data showing the increase by etching a glass surface. The transmittance peak shifts with the depth of the porous layer. The peak for sample exhibiting the 79 nm nanoporous layer is expected to be in the wavelength range dominated by the lower transmission of the glass.

4 BLASTING PLUS ETCHING

Besides solely etching glass surfaces, Schmid and Fraunhofer ISE have evaluated the combination of blasting a glass surface with subsequent wet etching. This approach allows a larger degree of freedom for the engineering of glass surface shapes. Figure 6 shows the surface angle distributions of 3 "blast plus etch" processes and an "only etch" sample for comparison in a histogram-style representation. The more area portions with higher surface angles are detected in this measurement, the more light incident from shallow angels can be in-coupled (compare to Figure 3).

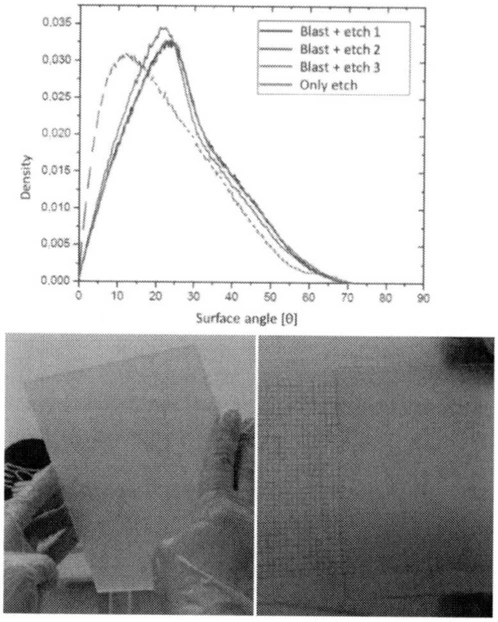

Figure 6: Results of blasted plus wet-etched glass.
Top: Surface angle distribution of wet-etched and blasted plus wet-etched glass.
Bottom: Photographs of blasted plus wet-etched glass sample with homogenous structuring.

5 IMPRINTING IN HOT GLASS

The production of flat glass starts with the mixture of the ingredients. Then, the heating-up to transfer into the liquid phase takes place, followed by a cooling-down. While the glass is still hot, the imprinting of structures into the glass surface can be facilitated. By these means, the company GMB is already manufacturing different products. We have investigated novel structures to improve the optical performance of solar modules. In cooperation of GMB and Fraunhofer ISE, a back-sheet solar glass with specialized optical properties was developed by raytracing simulations and optical measurements of the incident lights. In order to implement the proposed glass structures into the glass production process and produce the narrow and small glass structures, a significant amount of pre-work was performed. In a duplex iteration process GMB explored the necessarily needed hot end parameters as glass temperature, roller pace, annealing lehr's conditions etc. for the best outcome of the glass structures' dimension values in order to produce the solar glass as close to the simulated glass structures (around hundredths of a millimeter) in terms of dimensions as possible on an industrial scale. This produced glass has unique optical properties as a better light incident from the backside of the backsheet over all incident angles, which leads to a higher luminous efficacy and yield of the solar panels.

6 MEASUREMENT OF OPTICAL PROPERTIES

Within our work, a novel measurement tool was developed by Sentech, analyzing the light scattering taking place while light transmits through a structured glass sample. A quantification of the scattering that is achieved can be made and thus a comparison of samples on this basis is possible. Figure 7 (left) shows the principal setup of the measurement tool and an exemplary 2D diffraction image.

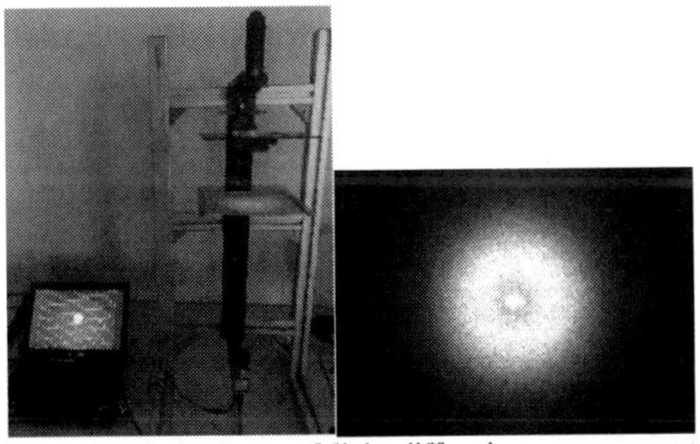

Figure 7: Left: Setup of light diffraction measurement tool. Right: Example 2D diffraction image achieved using the novel tool.

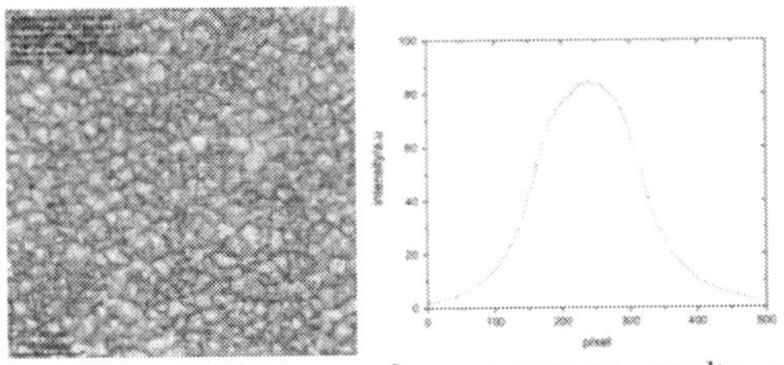

Figure 8: Right: example measurement results of diffracted light achieved using the novel measurement tool. A line scan of the light diffraction distribution is shown. Here: relatively homogeneous structuring was achieved. Left: confocal microscopy image of the same sample.

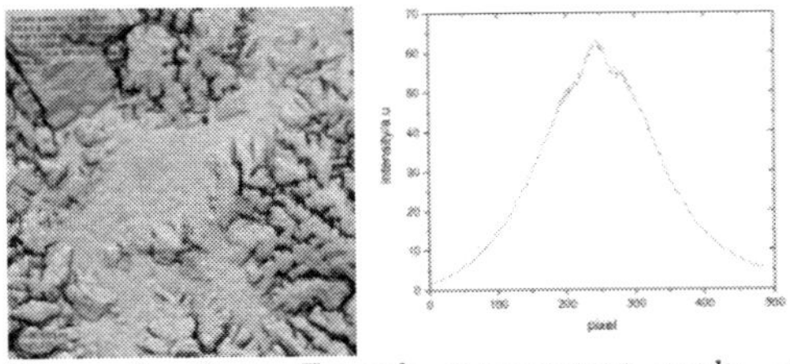

Figure 9: Right: Example measurement results of diffracted light achieved using the novel measurement tool. Here: relatively inhomogeneous structuring was achieved with significant area shares being rather flat leading to a peak in the measurement signal. Left: confocal microscopy image of the same sample.

Figure 8 and Figure 9 show exemplary measurement results of structured glasses applying the novel measurement tool in comparison to confocal microscopy images. One can see that the line scans of the diffraction pattern depend on the homogeneity of the glass structure distribution.

7 IMPLEMENTATIONS INTO SOLAR MODULES

Structured glasses were implemented into small solar modules to investigate the related electrical performance ($200x200\ mm^2$). In the characterization of the solar modules, special care was taken of the performance under different angels of incidence θ. To compare the results, the "incidence angle modifier" IAM was calculated from the short-circuit current densities of the modules and the correction factor cos(θ):

$$IAM = \frac{Isc(\theta)}{\cos(\theta) * Isc(0)}$$

Figure 10 shows IAM values for differently structured small solar modules. The top graph was created using solar modules with differently structured module front glass and with illumination from the front. We can see that the unstructured reference glass performs about the same as the structured counterparts up to an angle of 50°. Then, with increasing angle of incidence, structured glasses show benefits, with the blasted-plus-etched glasses leading to the highest gains under high angles, outperforming the "only-etched" samples. Imprinted glasses were not tested here.

Figure 10 (bottom) shows results of the application of a flat front with a variation of the rear glass, applying an illumination from the front. Thus, investigating the effect on the internal rear reflectance. Here, the structured rear glass showed improved module performance already starting at an angle of ~40°. In contrast to Figure 10 (top) is the bottom graph looking at average values of 3 identically prepared mini modules of each kind. Here, the highest short-circuit current density (Jsc) at perpendicular incidence of each kind of module was used as reference (100%).

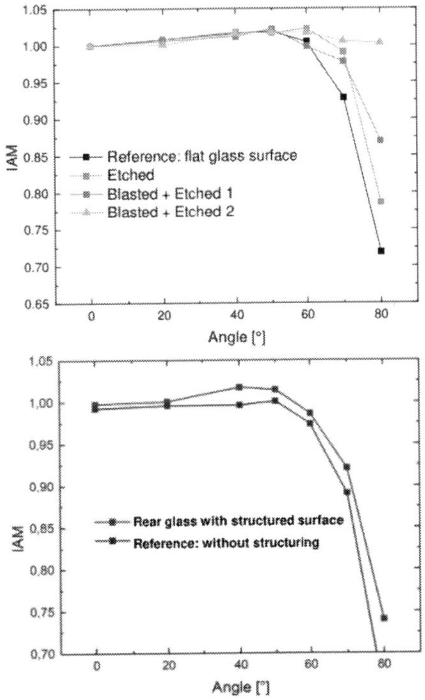

Figure 10: Top: IAM of differently structured module front glasses. Bottom: IAM of module rear glass with or without structuring.

10.4229/EUPVSEC2023/3DO.18.3

The benefits of structured glass result from a reduced reflection especially at higher angles of incidence (Figure 11, a, Figure 10) of the glasses as well as an improved internal light management (Figure 11, d). Due to the expectable non-perpendicular irradiance of light on the rear side of the module the additional gains are likely for bifacial modules (Figure 11, c). An increase of reflection at perpendicular irradiance (Figure 11, b) might reduce the power of the module at Standard Testing Conditions (STC) thus increasing the necessity to assess the module regarding performance i.e. using extended simulation [8], energy rating [9] or other advanced yield calculation methods. However, using the investigated glass structures, no decrease of module power was measured at perpendicular irradiance for most of the investigated samples.

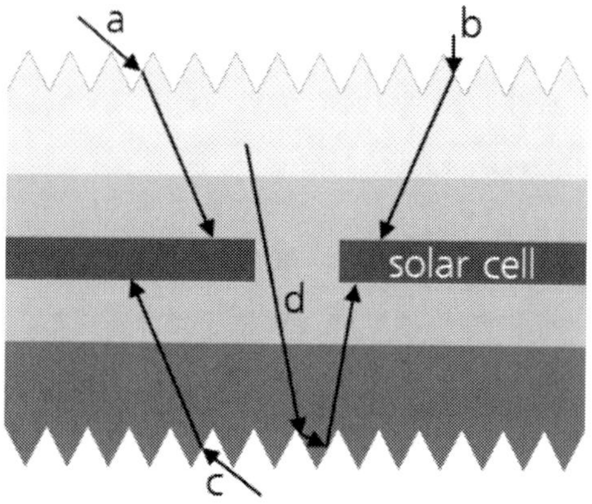

Figure 11: Schematic of light paths impacted by structured glass covers.

8 SUMMARY

Glass structuring by different means is capable of leading to benefits in the solar module performance. It was shown that imprinting into hot glass, etching the glass or blasting plus etching glass can show improvements in the energy yield. This is especially due to improvements in the reflection properties of the glass for angles in the range of ~60° or higher, a region where the incidence angle modifier of the flat reference glass typically drops sharply.

Acknowledgements

This work was funded by the German Federal Ministry for Economic Affairs and Climate Action within the research project "OPTOMOD" (contract no. 03EE1079A).

References

[1] Eggers J-B, Behnisch M, Eisenlohr J, Poglitsch H, Phung W-F, Münzinger M, Ferrara C, Kuhn T, PV-Ausbauerfordernisse versus Gebäudepotenzial: Ergebnis einer gebäudescharfen Analyse für ganz Deutschland, 2020, 35th PV-Symposium, Bad Staffelstein, pp. 837-856

[2] Wirth H, Recent Facts about Photovoltaics in Germany, https://www.ise.fraunhofer.de/content/dam/ise/en/documents/publications/studies/recent-facts-about-photovoltaics-in-germany.pdf, accessed on 26 January 2023.

[3] Nold S, Techno-ökonomische Bewertung neuer Produktionstechnologien entlang der Photovoltaik-Wertschöpfungskette: Modell zur Analyse der Total Cost of Ownership von Photovoltaik-Technologien. Dissertation: Fraunhofer Verlag, 2019.

[4] Blieske U, Doege T, Gayout P, Neander M, Neumann D and Prat A, "Light-trapping in solar modules using extra-white textured glass," 3rd World Conference on Photovoltaic Energy Conversion, 2003., pp. 188-191.

[5] Grunow, P., Sauter, D., Hoffmann, V., Huljic, D., Litzenburger, B., & Podlowski, L. (2005). The influence of textured surfaces of solar cells and modules on the energy rating of PV systems. 20th EPVSEC.

[6] https://www.pvlighthouse.com.au/sunsolve

[7] Stevens, L., Bläsi, B., Horch, C., Ihlow, S., Mittag, M., „Structured Backside Glass for Improved Efficiency in Solar Modules – A Simulation Study", 40th EUPVSEC, 2023.

[8] Pfreundt, A., et al., "Cell-to-module Analysis beyond Standard Test Conditions," *2020 47th IEEE Photovoltaic Specialists Conference (PVSC)*, Canada

[9] Protti, A., et al. "Virtual Energy Rating: A Method For Optimizing Module Performance Through Cell-To-Module Analysis", 8th WCPEC, 2022

10.4229/EUPVSEC2023/3DO.18.3
020263-005

40th European Photovoltaic Solar Energy Conference and Exhibition

This presentation was selected by the Sc. Committee of the EU PVSEC 2023 for submission of a full paper to one of the EU PVSEC's collaborating peer-reviewed journals.

CHALLENGES FOR QUALITY CONTROL POSED BY NEW PV ENCAPSULATION MATERIALS

Sandra Lust[1], Nola Schnitzler[1], Alexander Brendler[1], Stefan Wendlandt[1], Thomas Weber[1], Shiva Ram Kuntamukkula[1], Anton Mordvinkin[2], Michael Wendt[2]
[1]PI Photovoltaik-Institut Berlin AG, Wrangelstr. 100, 10997 Berlin, Germany
[2]Fraunhofer-Center für Silizium-Photovoltaik CSP, Otto-Eißfeldt-Straße 12, 06120 Halle
Corresponding author: Sandra Lust | +49 30 814 52 64 -202 | lust@pi-berlin.com

ABSTRACT: In this paper we investigate to what extent the standard GC tests for quality assurance of solar modules have to be adapted to the special requirements of combinations of encapsulation materials newly come out on the market. To investigate possible interactions when EVA and POE materials come into direct contact, GC tests were carried out on laminated samples and mini-modules of different EVA and POE material combinations, as well as on commercially available modules. Current test standards do not specify that front and back should be tested; in practice, often only one encapsulation film in the module is tested, usually the one on the back. In the case of material combinations, however, both films must be tested. When using a combination of EVA and POE films, the process time window covering recommended lamination durations becomes more narrow, making the lamination process more susceptible to process variations. Since cross-linking degrees of 70% and higher have been required for EVA films, the area of low GC values has not been the focus of scientific investigations so far. However, for certain POE materials, the required minimum GC is exactly in this range. Direct contact of EVA and POE films may have an influence on the GC values of the individual films, as a slight decrease of the GC values in the case of the EVA/POE combinations were observed. The significance of GC testing for POE films is limited because international standards give different results at low GC levels.

Keywords: PV modules, encapsulant, POE, gel content, reliability testing

1 INTRODUCTION

Although ethylene vinyl acetate copolymer (EVA) is still the dominant PV encapsulation material, polyolefins (PO) have gained market share in recent years [1,2]. Like EVA, PO can be mixed with a cross-linking agent and chemically react during the lamination process to form polyolefin elastomers (POE). Furthermore, EVA-POE-EVA (EPE) multilayer films are increasingly coming onto the market. All the materials mentioned can be used on the front and/or the rear of the module. POE films do not contain corrosive acetic acid as a degradation product; they are less permeable to vapour and have a lower conductivity than EVA, thus preventing PID.

The cross-linking reaction initiated by the thermal process in the laminator is essential for the material properties of the encapsulation films and thus for the long-term stability of the module [3-10]. The cross-linked and thereby insoluble fraction of the encapsulation can be determined by extracting the soluble part of the material with solvent. The degree of cross-linking or gel content (GC) results from the mass ratio after and before extraction of the samples.

This article examines how the use of EVA and POE material combinations on the front and rear of the solar module affects the measured GC values.

2 INDUSTRIAL FRAMEWORK CONDITIONS

Manufacturer recommended GC values compiled from publicly available data sheets of commercially available POE, EVA and EPE encapsulation materials are shown in table I [1,2,11-13]. The recommended GC values for POE and EPE are systematically and substantially lower than for EVA, varying from 60% for the POE encapsulant TF4 (First) to 80% for the EVA encapsulant S201MT1 (HIUV) and B601HP (Better). Without knowledge of the BOM and thus the encapsulation materials used, it is therefore no longer possible to make a statement about the quality of the cross-linking for newer solar modules. This is a big issue for quality control at module customer site.

Table I: Manufacturer recommended GC values of various POE, EPE and EVA encapsulants [1,2,11-13].

Encapsulant	Supplier	GC/ %
POE		
CONSERV E 360	Renewsys	50-80
TF4, TF8	First	>60
TR02BA-45T VR01BA-45T	Mitsui Chemicals	60-80
PO8110	3M	65-80
CybrightT22	Cybrid	75-95
EPE		
B602M	Better	>70
P507-T1	HIUV	>70
EVA		
F806, F406P	First	>75
SV-15296P	Sveck	>75
CybrightT11	Cybrid	75-95
B601HP	Better	80-90
S201MT1	HIUV	80-95
EVA9110T	3M	
CONSERV P 360 - 14 FC, CONSERV P UVT - 14 FC	Renewsys	≥80

For various currently produced solar modules, information about the encapsulant material was extracted from the BOMs. Lamination temperatures and holding times recommended by the manufacturer were compared for the pair of encapsulant material within one module type. Recommended lamination temperatures for POE tend to be higher than for EVA and holding times longer, thus resulting in a narrow process window for combinations of the two materials. It was even found that for a certain EVA/POE combination, the holding times for the two materials vary so much, that the two process windows no longer overlap (see figure 1). Even though it is likely that satisfactory lamination results can be achieved outside the manufacturer's recommended parameters, this example shows that in the case of EVA/POE combinations the encapsulation material deserves an increased quality assurance attention.

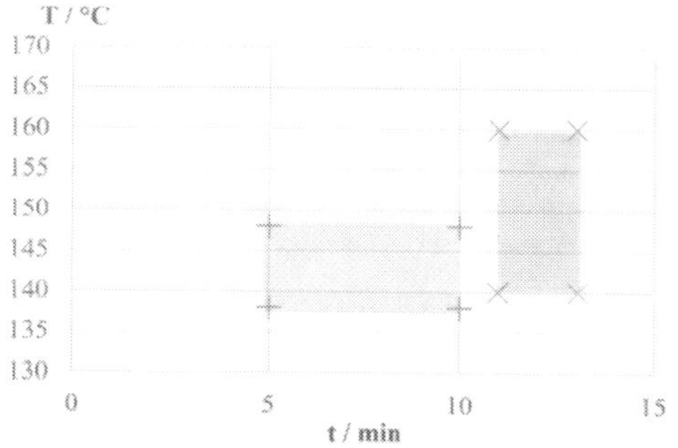

Figure 1: Lamination process windows recommended for the two encapsulant materials (front and rear) of a commercially available solar module.

The analytical methods for determining the GC value are described in ASTM D2765, IEC 62788-1-6 and the Chinese GB/T29848-2018 standard, although these standards do not specify at which positions in the PV module the encapsulation material should be extracted for further testing. In mass production, regular lamination tests with subsequent analysis of the GC provide the module manufacturer with information on the quality of the front and back encapsulant used. For determination of the GC of processed modules, the encapsulant material has to be mechanically extracted from the module. In practice, the GC of only one of the two encapsulation films is measured, usually the encapsulant at the back of the solar module. This article examines the extent to which the GC values of the two encapsulation materials in the module differ from each other.

To investigate possible interactions when EVA and POE materials are in direct contact, GC tests with laminates, mini-modules and commercial modules with combinations of different encapsulation materials were carried out.

3 MATERIAL AND METHODS

3.1 Sample preparation

To test possible interactions of EVA and POE materials in direct contact with each other, two-layer laminates were produced from four POE (P_x, x=1-4) and four EVA (E_y, y=1-4) sheets, all commercially available. 5cmx10cm pieces were cut from each sheet, two of them were placed on top of each other and laminated between two teflon films. Each encapsulation material was laminated with itself and to each of the seven other encapsulation materials, resulting in a total of 36 material combinations. The lamination parameters and gel content values recommended by the manufacturer can be found in table II. A laminator S1815 from S3 Swiss Solar System AG was used for lamination of the foils and mini-modules. The evacuation time was 5 minutes (min) for all samples. The lamination temperatures and holding times for the laminates were varied:

L1: 130 °C and 15 min,
L2: 130 °C and 25 min,
L3: 150 °C and 15 min and
L4: 140 °C and 25 min.

Using different combinations of encapsulants E_1, E_2, P_1 and P_2, single-cell TOPCon glass-backsheet mini-modules were laminated using the parameter set L3.

Table II: Lamination parameters and gel content values recommended by the encapsulant manufacturer.

		T/ °C	t/ min	GC / %
E_1	3M	145-155	8-13	
E_2	Renewsys	145-150	8-10	≥80
E_3	Renewsys	145-150	8-11	≥80
E_4	Renewsys	unknown	unknown	≥80
P_1	3M	160	11	65-80
P_2	Renewsys	150-155	12-16	50-80
P_3	Mitsui Chemicals	150 (160)	10-20 (7-11)	60-80
P_4	Mitsui Chemicals	150 (160)	10-20 (7-11)	60-80

For the round robin test described in section 3.4, laminates were produced using a POE film (TF4, First, see table I). The evacuation time was 5 min, the curing time was 15 min and the temperature was changed from 120°C to 170°C in 10-degree Celsius steps. The samples were cut into approximately (5x5) mm^2 pieces, which were then thoroughly mixed to produce samples with identical properties. In the case of in-house testing according to GB/T29848-2018, the particle size was then further reduced to approximately (1x1) mm^2.

3.2 Gel content tests

If not mentioned otherwise, the cross-linking tests were carried out according to IEC 62788-1-6 standard using a Soxhlet-like extraction apparatus, in which the extraction cycles are carried out by heating and cooling the solvent (THF). One cycle of heating and cooling lasts 2 min, 150 cycles were performed. Three extractions were performed for each parameter set and the mean GC value was calculated.

For the round robin test described in section 4.4, GC values were mesured at PI Berlin and at two external laboratories, according to the Chinese GB/T29848-2018 and the IEC 62788-1-6 standard.

4 RESULTS

4.1 GC of homogeneous *laminates* of EVA and POE

Figure 2 shows on the left the measured GC values of the two-layered homogeneous EVA and POE laminates (E_xE_x, P_yP_y) at the four different lamination parameter sets. The corresponding GC values recommended by the manufacturer can be found in table II. At 150°C (L3 and L4), all encapsulants meet the GC value recommended by the manufacturer. The EVA films reach values of 90%, the values for the POE films are significantly lower. For P_2, the required value of 50% cross-linking is only just achieved at a holding time of 25 min (L4). All EVA materials tested achieve cross-linking degrees of well over 50% even at 130°C (L1 and L2). Encapsulants E_2 and E_3 achieve the required cross-linking degree of over 80% at 130°C and 25 min holding time (L2). Three of the four POEs tested (P_2, P_3, P_4), on the other hand, show no crosslinking at 130°C even with a holding time of 25 min (L2). Only in the case of P_1 cross-linking is detected, while the range of variation is considerable. These results underline that for EVA/POE combinations in solar modules, the lamination parameters must be chosen with special care. At values below 60 %, the reproducibility of the GC test is no longer satisfactory.

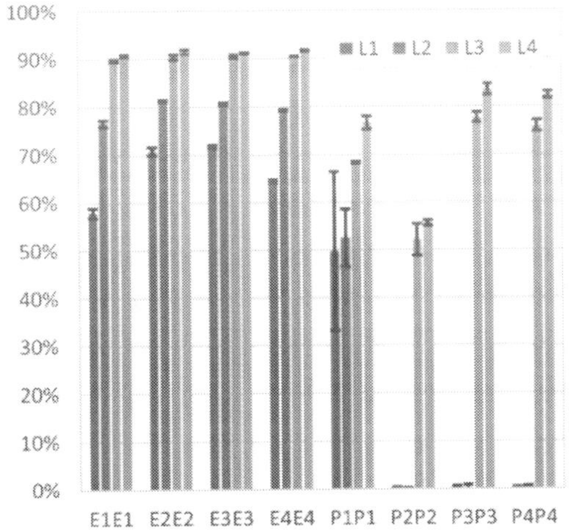

Figure 2: Measured GC values of the two-layered homogeneous laminates.

4.2 GC of heterogeneous *laminates* of EVA and/or POE

When combining two different encapsulation materials A_x and B_y, the resulting GC value $GC_{mes(AxBy)}$ is expected to equal the mean value of the GC values of the individual films:

$$GC_{calc(AxBy)} = [GC_{mes(AxAx)} + GC_{mes(ByBy)}]/2.$$

Figure 3 shows the difference between the measured and the calculated GC values:

$$\Delta GC_{AxBy} = GC_{mes(AxBy)} - GC_{calc(AxBy)}$$

of the two-layered combinations of POE (P_xP_y) and of EVA (E_xE_y).

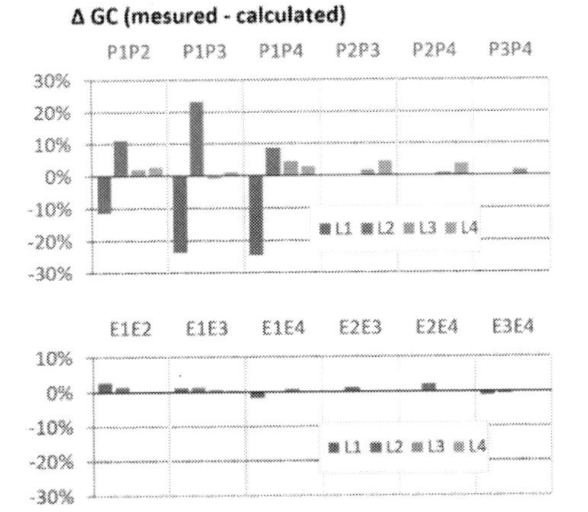

Figure 3: Difference between measured and calculated GC values of the two-layered combinations of POE (top) and of EVA (bottom).

For combinations of EVA films (E_xE_y), as well as for combinations of POE films (P_xP_y), the calculated mean values of the GC of the single sheets equals the experimentally determined GC values of the heterogeneous samples. Deviations occur for combinations including material P_1 (P_1P_y) at lower temperature (130°C, L1 and L2), which could be due to the high error in determining the GC value for this material.

Differences between measured and calculated GC values of the two-layered combinations of POE with EVA are shown in figure 4. Again, large differences for combinations with the POE encapsulation film P_1 (E_xP_1) could be observed. But also for P_3 and P_4, values determined for the combination of EVA with POE films (E_xP_3, E_xP_4) were slightly lower than the mean values of the GC of the individual films would have suggested.

It can be provisionally stated that the lamination of combinations of EVA (E_xE_y) and POE (P_xP_y) encapsulants in contact with each other does not influence the GC values of the individual films. However, the slight decrease of the GC value in the case of EVA/POE material combinations (E_xP_y) requires a more detailed examination.

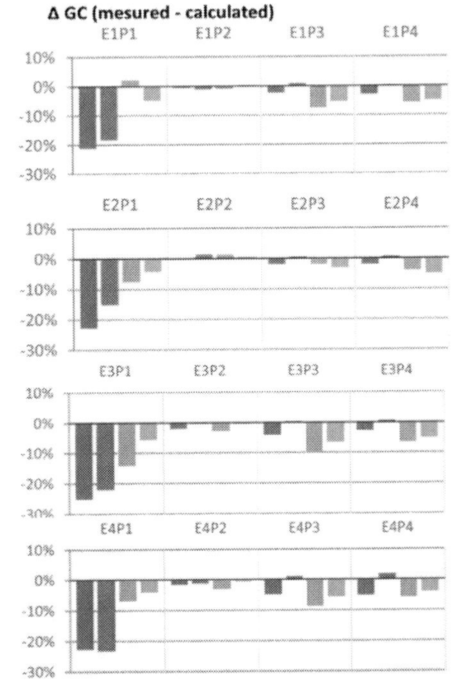

Figure 4: Difference between measured and calculated GC values of combinations of POE with EVA.

4.3 GC of *modules* with EVA/POE combinations

Figure 5 shows the GC of the encapsulation materials E_1, E_2, P_1 and P_2 extracted from the front and back of the four mini-modules, made on the basis of different combinations of these encapsulants. For comparison, the GC values of the encapsulation laminates are also given.

10.4229/EUPVSEC2023/3DO.19.3
020264-003

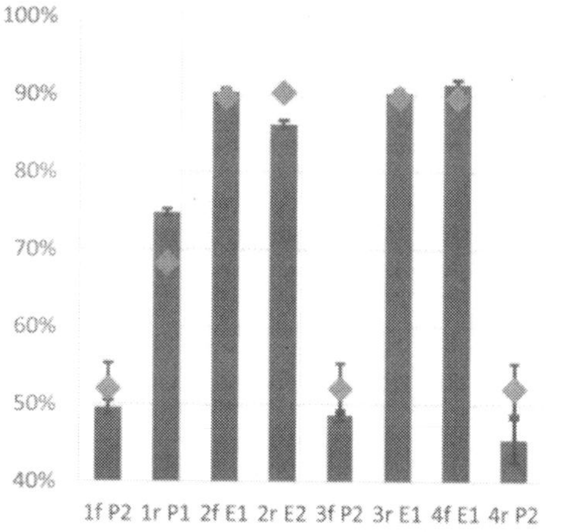

Figure 5: GC values of front (f) and rear (r) side of mini-modules (1 to 4) with POE (P_1, P_2) and EVA (E_1, E_2) (blue columns); GC values of laminates of P_1, P_2, E_1, E_2 (orange rhombs).

For the encapsulant E1, the GC of the mini-modules match that of the laminate. For encapsulants P1, P2 and E2 there are slight deviations, which in the case of P2 lead to a GC remaining below the manufacturer's recommended limit of 50%. For the materials E2 and P1 deviations exceeding the standard deviations were found. E2 and P2 show a higher value, P1 a lower value in the module than in the laminates. Inhomogeneities in the chemical composition of the film could be a reason for the deviation. Possible interactions between the encapsulation materials cannot be detected in these measurements, as the samples were taken both in front of and behind the solar cell, so that the encapsulation foils were not in direct contact.
For P2 on the rear of the module, the values were slightly lower than for the same material on the front, this also applies to E1.

Figure 6 shows the GC values of a test series on commercially available modules with EVA on front and rear side.

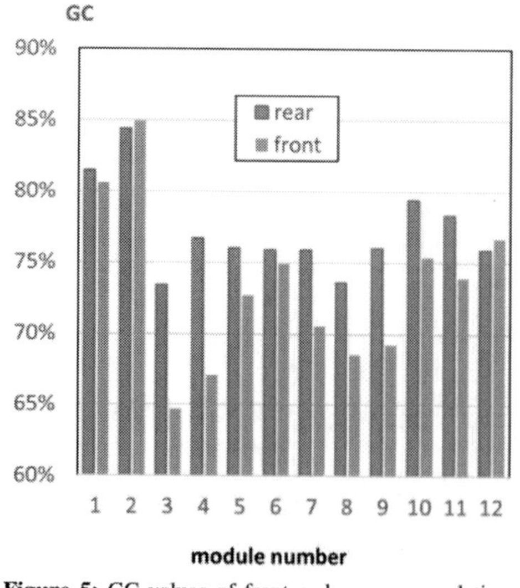

Figure 5: GC values of front and rear encapsulation of commercial modules with EVA on front and rear side.

The material of the rear side has a GC below 75% in only two cases, the material of the front side in eight out of twelve cases, in four cases the value is even below 70%. Given the GC values of 75% and higher, recommended by the manufacturer for EVA encapsulation, this example shows the necessity of the lamination control of both encapsulants of the module.

4.4 Applicability of the standard GC tests to POE

In order to test the applicability of the GC tests to low GC values and therefore to POE materials, a lamination series was prepared in which the temperature was varied from 120 to 170°C in 10°C increments. The resulting samples were tested at PI Berlin and sent to two external laboratories to measure the gel content. A Chinese laboratory used the method as described in the Chinese national standard GB/T29848 and a German laboratory - the Fraunhofer CSP in Halle – as described in the IEC 62788-1-6 standard (using toluene). At PI Berlin, we tested both methods.

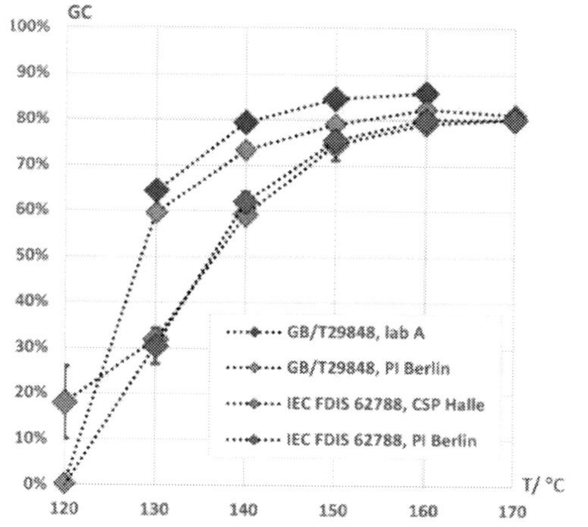

Figure 7: GC values of POE (TF4) samples laminated at various temperatures and tested per GB/T29848 and IEC 62788-1-6.

The results of the extraction show the dependence of the degree of cross-linking on the lamination temperature, which is typical for chemically cross-linked films, with the degree of cross-linking increasing sharply at first and then reaching saturation. The GB/T29848 extractions show consistently higher values than the two IEC 62788-1-6 extractions. Whilst similar values are obtained in the plateau region with the different methods, there are large differences in the lower temperature regions. As the encapsulant suppliers recommend GC values of 60% and above, these differences cannot be neglected and are the subject of further investigations to be published in a EPJPV paper in a paper in the framework of this conference [14].

5 CONCLUSIONS

Material combinations of EVA with POE pose the following challenges for quality control: (i) The lower limits for GCs recommended by manufacturers vary for POE and are often significantly lower than the values for EVA. Without BOM, it is therefore no longer possible to make a statement about the quality of the cross-linking. (ii) Since cross-linking degrees of 70% and higher have been required for EVA films, the area of low GC values has not been the focus of scientific investigations so far. However, for certain POE materials, the required minimum GC is exactly in this range. (iii) Current test standards do not specify that front and back should be tested; in practice, often only one encapsulation film in the module is tested, usually the one on the back. In the case of material combinations, however, both films must be tested. (iv) When using a combination of EVA and POE film, the process time window covering recommended lamination durations becomes more narrow, making the lamination process more susceptible to process variations. (v) The significance of GC testing for POE films is limited because international standards give different results at low GC levels.

5. ACKNOWLEDGEMENT

This work was supported by the German Federal Ministry for Economic Affairs and Energy (BMWi) under the contract number 03EE1112D, acronym EVAplus.

6. REFERENCES

[1] S. K. Chunduri, M. Schmela, TaiyangNews Market Survey Backsheets & Encapsulation (2021), TAIYANGNEWS. Düsseldorf 2021. https://taiyangnews.info/reports/taiyangnewsbacksheets-encapsulation/
[2] S. K. Chunduri, M. Schmela, TaiyangNews Market Survey Backsheets & Encapsulation (2020), TAIYANGNEWS. Düsseldorf 2020. https://taiyangnews.info/reports/marketsurvey-backsheets-encapsulation-2020/
[3] M. Aghaei, A. Fairbrother, A. Gok, S. Ahmad, S. Kazim, K. Lobato, G. Oreski, A. Reinders, J. Schmitz, M. Theelen, P. Yilmaz, J. Kettle. Review of degradation and failure phenomena in photovoltaic modules. Renewable and Sustainable Energy Reviews 159 (2022) 112160.
[4] A. K. Öz, C. Herzog, C. Wellens, D. E. Mansour, M. Heinrich, A. Kraft. The Impact of the Lamination Process on the Adhesion Properties at the Glass-Encapsulant Interface and Damp Heat Stability of PV Modules. 38th EUPVSEC (2021).
[5] G. Oreski, A. Omazic, G. C. Eder, Y. Voronko, L. Neumaier, W. Mühleisen, C. Hirschl, G. Ujvari, R. Ebner, M. Edler. Properties and degradation behaviour of polyolefin encapsulants for photovoltaic modules. Prg. Photovolt. Res. Appl. 28 (2020) 1277-1288.
[6] S. Pingel, S. Janke, B. Stannowski, S. Fechner, L. Podlowski. Advanced testing of PV Module encapsulants. 35th EUPVSEC (2018).
[7] G. Oreski, A. Rauschenbach, C. Hirschl, M. Kraft, G.C. Eder, G. Pinter. Crosslinking and post-crosslinking of ethylene vinyl acetate in photovoltaic modules. J. Appl. Polym. Sci. (2017) 134.
[8] G. Oreski, A. Rauschenbach, C. Hirschl, M. Kraft, G. Eder, G. Pinter. Crosslinking and post-crosslinking of ethylene vinyl acetate in photovoltaic modules. Journal of Applied Polymer Science 134 (2017) 44912.
[9] C. Hirschl, M. Biebl-Rydlo, M. DeBiasio, W. Mühleisen, L. Neumaier, W. Scherf, G. Oreski, G. Eder, B. Chernev, W. Schwab, M. Kraft. Determining the degree of crosslinking of ethylene vinyl acetate photovoltaic module encapsulants – a comparative study. Solar Energy Materials & Solar Cells 116 (2013) 203-218.
[10] A. W. Czanderna, F.J. Pern, Encapsulation of PV modules using ethylene vinyl acetate copolymer as a pottant: a critical review. Solar Energy Materials and Solar Cells 43 (1996) 101-181.
[11] https://www.firstpvm.com/en/pro_cate.asp?cid=203#auto
[12] https://www.cybrid.com.cn/product/329.html
[13] https://multimedia.3m.com/mws/media/1216185O/3m-solar-tech-data-sheetpo8110.pdf
[14] S. Lust, N. Schnitzler, A. Brendler, S. Wendlandt, T. Weber, S. R. Kuntamukkula, A. Mordvinkin, M. Wendt, to be published.

LIGHT AS HEAVEN, STRONG AS HELL(?): TESTING HONEYCOMB-BASED LAMINATES FOR LIGHT-WEIGHT C-SI PV APPLICATIONS

Jonathan Govaerts[1], Paul Dufke[1], Bin Luo[1], Matthias Bonnart[1], Rik Van Dyck[1], Tom Borgers[1], Jef Poortmans[1], Arne Derluyn[2], Jarne Saelens[2], Wouter Winant[2], Meric Caliskan Arslan[3], Umang Desai[4], Fabiana Lisco[4], Antonin Faes[4] Christophe Ballif[4], Gernot Oreski[5], Nikolina Pervan[5], Gabriele Eder[6]

[1]imec-EnergyVille-UHasselt, Genk, Belgium
[2]EconCore NV, Leuven, Belgium
[3]KalyonPV, Turkey
[4]École Polytechnique Fédérale de Lausanne (EPFL), Institute of Electrical and Micro Engineering (IEM), Photovoltaics and Thin-Film Electronics Laboratory (PV-lab), Rue de la Maladière 71b, 2000 Neuchâtel, Switzerland
[5]Polymer Competence Center Leoben, Leoben, Austria
[6]OFI, Austrian Research Institute for Chemistry and Technology, Vienna, Austria.

ABSTRACT: In this paper we discuss evaluation results of light-weight PV laminates using sandwich panels based on honeycomb cores and different types of (fibre-reinforced) thermoplastics as sandwich skins. First, we discuss the fabrication and resulting performance of the PV modules, and then the results of reliability testing that has been carried out (thermal cycling, damp heat and humidity freeze). During these tests, an increased brittleness was observed. We investigated the underlying mechanism and, by using differential scanning calorimetry, could pinpoint it to a hydrolysis effect. Finally, after failing initial hail impact testing with the honeycomb laminates, we indicate how this could be resolved by tuning the skin and frontside encapsulant buildup.

Keywords: light-weight, reliability, c-Si PV

1 BACKGROUND

Light-weight PV modules can address different markets and applications, and can reduce glass consumption - and therefore the weight - compared to standard modules with a glass-glass or glass-backsheet buildup.

As the weight in standard modules is dominated by the glass, different approaches target reducing the amount of glass by implementing thinner glass layers, and/or replacing them with fibre-reinforced polymers [1]. Of course, the functionalities of the glass (thermal, mechanical, optical, electrical and hydrostatic stability) then have to be taken over by these alternatives.

One option to provide mechanical stability at a reduced weight is by using a sandwich panel buildup as a back cover for PV modules [2]. In this work, we evaluate such sandwich panels based on a honeycomb core and (fibre-reinforced) polymeric skins. These have the additional benefit of an increased sound insulation, which could be beneficial for VIPV and BIPV applications. The sandwich structures are fabricated by EconCore, using different materials and a patented method for making honeycomb cores [3], schematically shown in Figure 1. These are then used as backsheets in PV laminates, which can be evaluated in reliability testing.

This work is carried out in the frame of the Delight project, targeting BIPV applications with light-weight PV modules, and looking at in-depth characterization and modification of polymer materials to address specific requirements for this application (such as aesthetics, colour, performance, reliability, fire safety, shade tolerance, ...) [4].

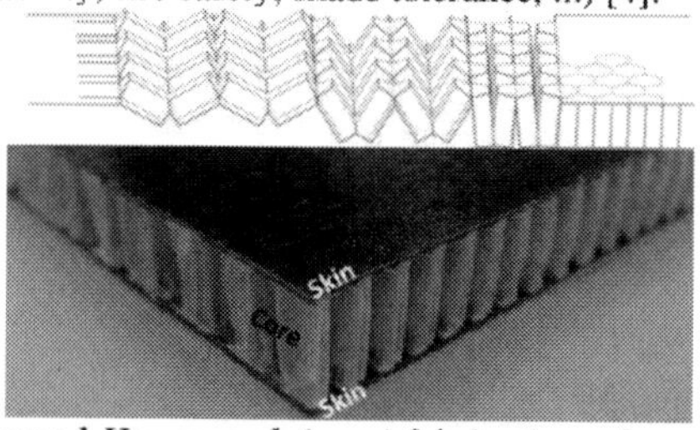

Figure 1 Honeycomb (core) fabrication schematic (top) and sandwich panel buildup (bottom)

2 LAMINATE BUILDUP AND FABRICATION

In first instance, a set of honeycomb panel materials was chosen (cf. Table 1) for an extensive test matrix, where PP stands for polypropylene, GF for glass fiber, and BOPET for biaxially oriented polyethylene terephthalate. As reference, a glass-glass buildup was included. For the buildup with the honeycomb panels, each laminate has identical (commercial) encapsulant (polyolefin-based) and polymer frontsheet (PET-based) materials. Furthermore, either full-size M2 PERC cells were incorporated in a 2-cell 5BB configuration or half-cut M10 PERC cells in a 4-cell 10BB configuration.

Table 1 Different honeycomb panels implemented as back cover for PV laminates

Type	Core	Skins	Thickness

10.4229/EUPVSEC2023/3DO.19.5
020266-001

PPGF8	White PP core	white PP/GF	12 mm
PPGF6	White PP core	black PP/GF	12 mm
BOPET	Black PET core	transparent BOPET	10 mm
PETGF	Black PET core	greyish PET/GF	11 mm

Lamination was carried out in a standard membrane laminator at a temperature of 135 °C and a pressure of 300 mbar. The parameters were chosen based on a previous screening exercise with the implemented materials. A frame all-around was included during lamination to minimize pinching at the edges.

In second instance, a set of modules was prepared to evaluate identical half-cut M10 PERC strings in a reference light-weight buildup, without honeycomb but glass- resp. carbon-fibre reinforced polypropylene backsheets (GFPP resp. CFPP). [1]

3 LAMINATE PERFORMANCE

With the above materials and process, a set of 32 laminates (mini-modules) was prepared. Figures 2 (frontside view) and 3 (rearside view) give an impression of the samples.

Figure 2 Frontside view on test laminates: glass-glass 5BB and 10BB reference (left), 10BB (middle) and 5BB (right) on (from front to back) of PPGF8, PPGF6, PETGF and BOPET

Figure 3 Backside view on samples from Figure 2 (perspective 90 degrees turned)

The fabricated samples were then characterized with Electroluminescence (EL) and current-voltage (IV)-measurements. Tables 2 and 3 compile the resulting averages and standard deviations of 4 (for glass-glass and 10BB laminates) or 3 (for 5BB laminates) identical samples.

Table 2 Weight [g] of the fabricated laminates, averages of at least 3 samples

Weight [g]	10BB	5BB
GLASS	2024	
PPGF8	487	483
PETGF	426	421
BOPET	339	334
PPGF6	456	451

By implementing the honeycomb panels, the weight can obviously be reduced compared to glass-glass configurations from 12.7 kg/m^2 (glass-glass) down to 3.4 kg/m^2 (BOPET). On a smaller scale, there is a consistent difference in weight of roughly 5g between the 10BB (4x M10 halfcut) and 5BB (2x M2 fullsize) samples.

Table 3 Initial IV parameters of the fabricated laminates: averages and standard deviations of at least 3 samples

	I_{sc} [A]		V_{oc} [V]		FF [%]		P_{mpp} [W]	
10BB_GLASS	**6.90**	±0.13	**2.79**	±0.01	**80.6**	±0.4	**15.51**	±0.24
10BB_PPGF8	**6.74**	±0.07	**2.78**	±0.01	**81.1**	±0.3	**15.22**	±0.16
10BB_PETGF	**6.60**	±0.08	**2.78**	±0.00	**81.3**	±0.4	**14.92**	±0.16
10BB_BOPET	**6.52**	±0.08	**2.78**	±0.01	**81.7**	±0.3	**14.80**	±0.16
10BB_PPGF6	**6.51**	±0.08	**2.78**	±0.01	**81.3**	±0.2	**14.70**	±0.20
5BB_GLASS	**9.92**	±0.31	**1.34**	±0.01	**78.0**	±0.4	**10.39**	±0.29
5BB_PPGF8	**9.91**	±0.17	**1.34**	±0.01	**78.3**	±0.1	**10.41**	±0.20
5BB_PETGF	**9.74**	±0.12	**1.34**	±0.01	**78.0**	±0.1	**10.20**	±0.19
5BB_BOPET	**9.66**	±0.13	**1.35**	±0.00	**78.3**	±0.2	**10.18**	±0.13
5BB_PPGF6	**9.56**	±0.11	**1.34**	±0.00	**78.5**	±0.2	**10.05**	±0.12

For open-circuit voltage (V_{oc}) and fill factor (FF), there are no significant differences for samples using the same (5BB or 10BB) solar cell strings. Variations in power (P_{mpp}) are due to differences in short-circuit current (I_{sc}) that may be attributed to different absorption coefficients of the polymer and glass front cover and differences in optical reflectance between the different honeycomb panels behind the cells.

4 RELIABILITY TESTING

In preliminary testing, we submitted each type of laminate to IEC61215 test schemes, including thermal cycling (TC), damp heat (DH) and humidity freeze (HF) testing.

Figure 4 indicates the performance degradation during TC in terms of FF. The performance at maximum power point (P_{mpp}) could also be shown but due to the limited degradation so far (in the range of percentages) this would distort the graphs due to variations in I_{sc} related to the calibration accuracy of the solar simulator (as these are also in the lower %-range).

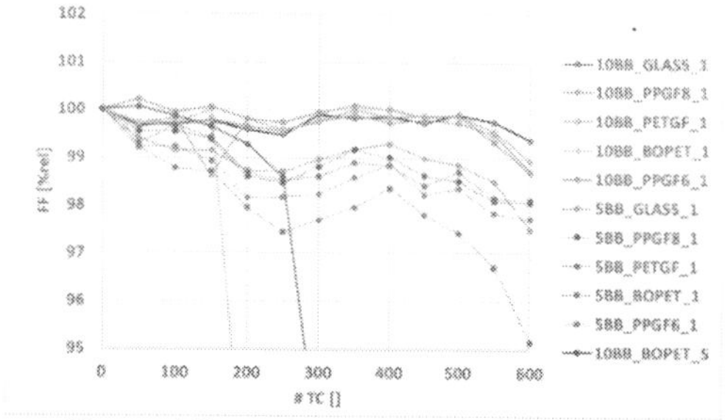

Figure 4 Degradation of FF throughout TC

Very limited degradation is observed for all 10BB samples, except for the BOPET samples (an additional one, 10BB_BOPET_5, was made as verification after the first one failed early). This sample showed a significant drop in FF and consequently P_{mpp} beyond 150 thermal cycles. EL imaging in Figure 5 indicates that this is due to an interconnect failure. Though it could be that 1 or more wire got fully detached from the cell metallization, it is more likely that 1 or more wires was broken, a failure mode that has been reported earlier (albeit with ribbons instead of wires) [1].

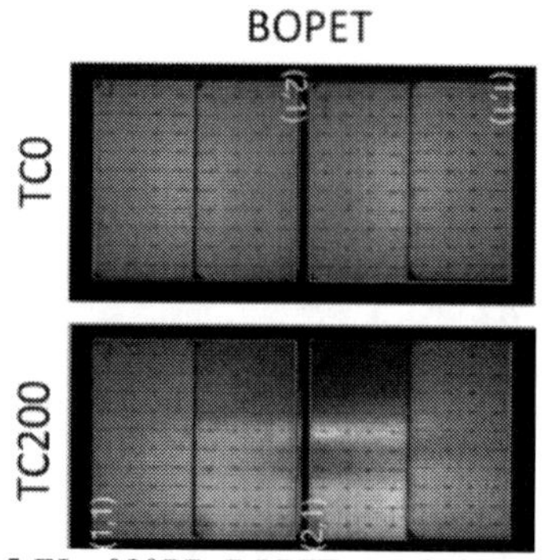

Figure 5 EL of 10BB_BOPET sample before (top) and after (bottom) 200 TC

Figure 4 also indicates a slight but consistent degradation of all 5BB samples compared to the 10BB ones. This might be attributed to a pressure contact at the edges of the cells, which is formed during lamination but is released in subsequent TC. Figure 6 shows the EL footprint of this mechanism for the BOPET sample as a reduced intensity at the edges of the cells, which is also observed for all other 5BB samples. Similarly as for the 10BB laminates, the BOPET version of the 5BB laminates is the first to fail.

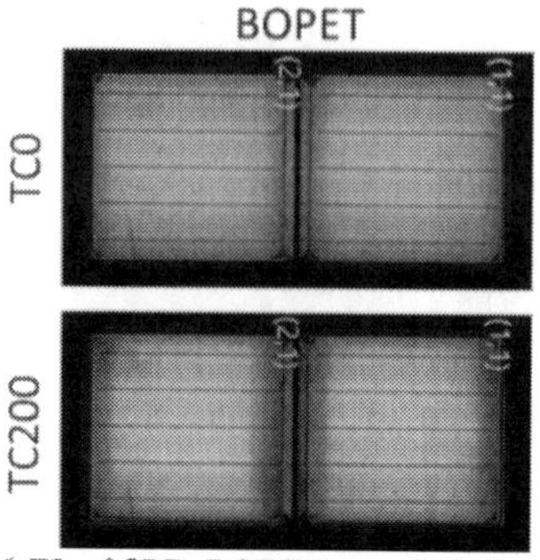

Figure 6 EL of 5BB_BOPET sample before (top) and after (bottom) 200 TC indicating degradation at the edges of the cells

During DH testing, for which the results are shown in Figure 7, no significant degradation is observed except for the 10BB_PETGF sample.

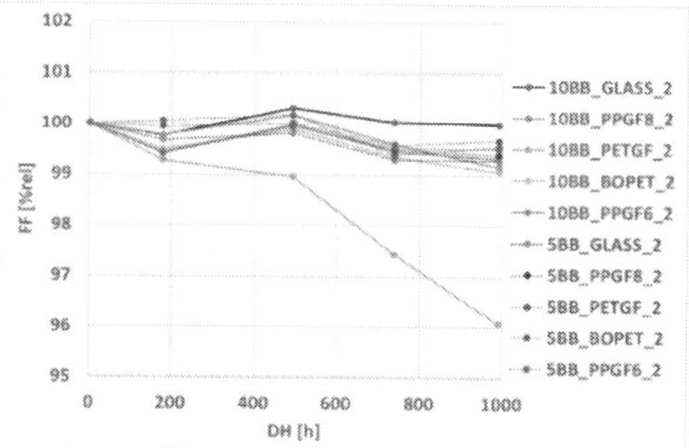

Figure 7 Degradation of FF throughout DH

Looking into the EL images before and after DH as shown in Figure 8, some degradation is observed throughout the different cells in the module. It is clear that the degradation is not specifically coming from the edges, and not all cells are affected equally. Unlike with glass-glass encapsulation, where moisture can only gradually get to the cells from the sides, the light-weight (glass fibre reinforced) polymer encapsulation will be less of a barrier due to the higher WVTRs of the polymers. The edges of the honeycomb are also not fully sealed.

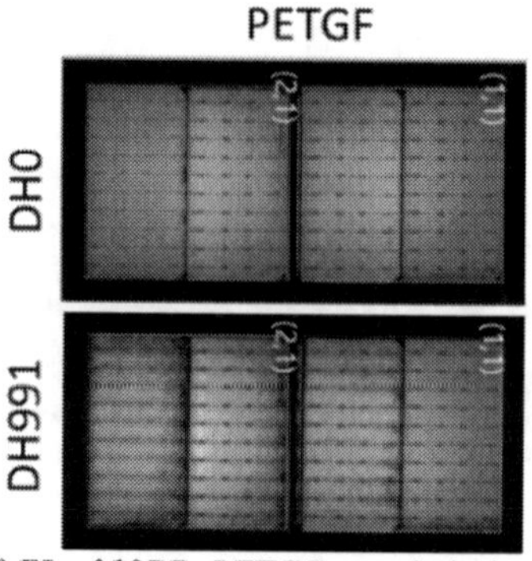

Figure 8 EL of 10BB_PETGF sample before (top)

10.4229/EUPVSEC2023/3DO.19.5
020266-003

and after (bottom) 991 hours of DH

HF testing results are given in Figure 9. With the basic test comprising 10 HF cycles, no significant degradation was observed. However, with extended testing to 20 HF cycles, some degradation appears for the 10BB_PETGF sample and to a lesser extent for the 10BB_BOPET sample.

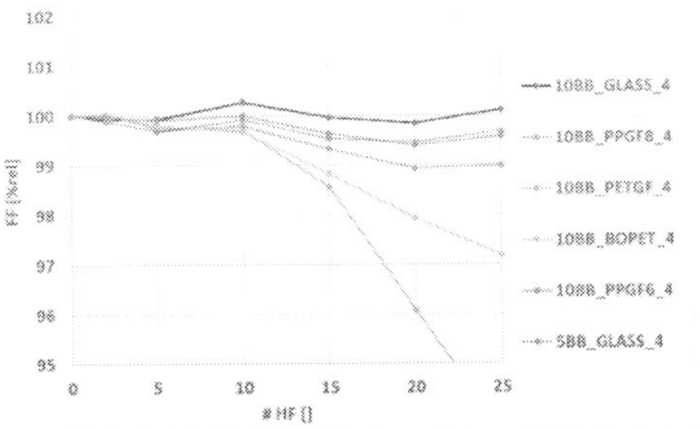

Figure 9 Degradation of FF throughout HF

At first glance, EL imaging after HF (Figure 10) indicates a similar degradation footprint as for DH (Figure 8).

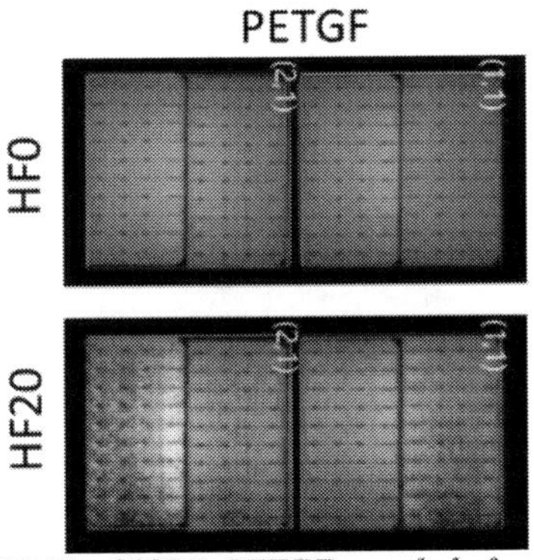

Figure 10 EL of 10BB_PETGF sample before (top) and after (bottom) 20 HF cycles

While the reliability testing in terms of electrical performance is quite promising for all types in the above results, the mechanical integrity of the PET-based sandwich panels was compromised. The increased brittleness of the PET honeycomb observed in those samples would be detrimental during mechanical testing (as e.g. impact resistance tests). Most likely hydrolysis of the PET-based materials is the main reason for this loss in performance. One way to confirm this is to analyse the samples before and after these aging trials using Differential Scanning Calorimetry (DSC). Hydrolysis causes chain scission in the polymer which results in a loss in molecular weight, and a reduced impact resistance of the structure. This causes the material to behave extremely brittle. This hydrolysis process is sped up tremendously at elevated temperatures. The DSC investigation is performed on the PP and PET honeycomb cores and on the BOPET skins.

Only the cooling curve is presented as this gives the most information in this context where loss of molecular weight/IV is analysed. The cooling rate of the DSC test was 10°C/min. A higher solidification crystallization temperature indicates a lower molecular weight, which would be visible as sharper exothermic peaks at higher temperatures in a DSC graph [5]. An overlay was made of the cooling curves of the PET core samples before and after the different aging trials.

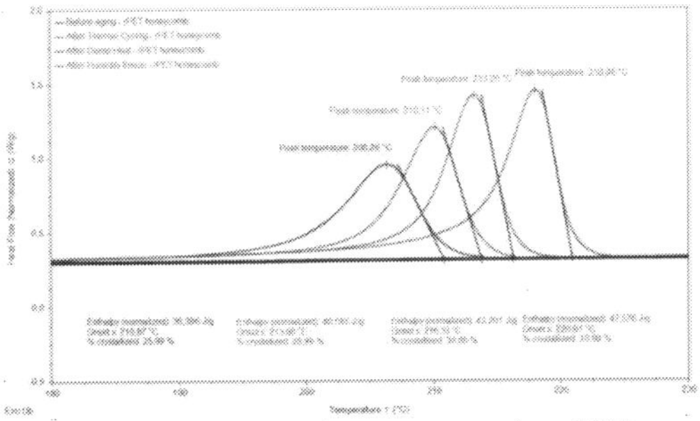

Figure 11: DSC cooling curve overlay: PET honeycomb before and after aging

As seen in the DSC overlay in Figure 11, all three aging trials affected the PET material by lowering the molecular weight. TC has the smallest effect on the molecular weight of the PET material, and also no severe increase in brittleness was observed for this material after TC was completed. This is due to the lower humidity present during this aging test. For DH and HF tests a severe increase in brittleness is observed, and is confirmed in the overlay with a difference in crystallisation speed much larger before and after aging for these two samples compared to TC. As expected, due to the nature of the test, the DH aging test caused the most damage to the material. The long exposure (1000h) to high temperature (85°C) at a high humidity (85%RH) caused severe hydrolysis which decreased the molecular weight of the polymer (chain scission). As this test is the most severe of the three aging tests performed on the material, only samples after this aging trial were tested for the BOPET skin and the PP honeycomb core.

While for the BOPET skin no changes were noted during initial (visual and mechanical) inspection, the DH storage does affect the BOPET skin in terms of material degradation (decrease in molecular weight) as indicated by the cooling curve DSC (graph not included here). Further mechanical evaluation (tensile testing/bending testing) could prove interesting to see what the aging's effect is on the materials performance.

For the PP honeycomb core, the effect on the material is minimal, both in visual and mechanical

10.4229/EUPVSEC2023/3DO.19.5
020266-004

inspection and in DSC. This makes sense as PP is not prone to hydrolysis.

To limit the material degradation of PET (during simulated aging and in actual outdoor conditions), both edge-closing methods and additives can be considered to increase the material's resistance to hydrolysis.

Additionally, on a similar set of samples, also hail testing was performed. While the reference glass-glass modules passed the tests, both for 5BB and 10BB strings, the honeycomb samples all suffered from significant cracking and performance degradation as shown in Figure 12.

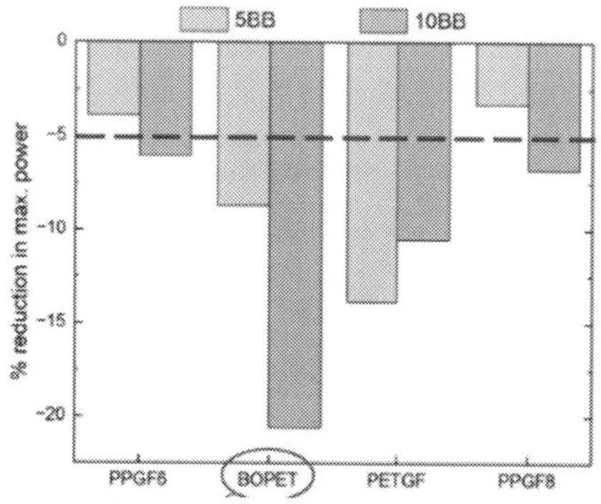

Figure 12 Degradation in Pmpp after hail testing of PV laminates of the different honeycomb types

Figure 13 illustrates for one sample (10BB-PETGF), reproduced in the other types, how each hail stone in the active area results in cell cracks at the impact location.

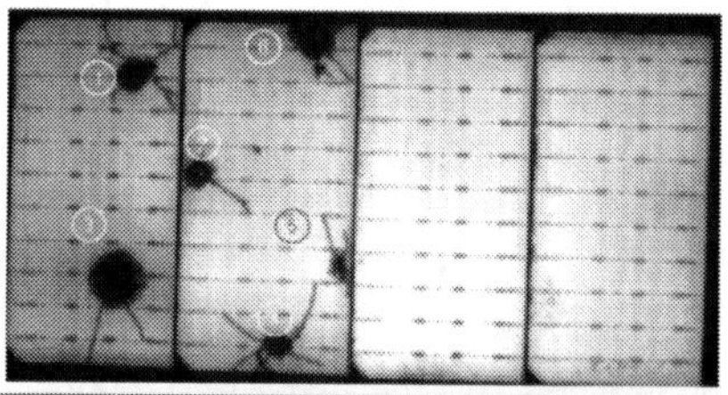

Spot #	Spot	Mass of the ice ball (mg)	Speed of the ice ball (m/s)
1	On the busbar	7.36	23.21
2	Close to interconnect	7.34	21.82
3	On the busbar	7.56	23.29
4	On the busbar	6.91	22.7
5	Close to interconnect	6.83	18.7
6	Close to edge	7.2	21.56
7	Edge of a module	6.78	22.24
8	Corner of a module	6.94	21.84

Figure 13 Conditions and resulting EL picture of hail testing on the 10BB-PETGF sample

To resolve this and be able to pass hail testing, previous work [1] has indicated that the skin layer closest to the cell and encapsulant layers in front of the cell have to be adapted. In order to demonstrate the effectiveness of such reinforced skin layers also with the 4-cell half-cut M10 strings, we evaluated GFPP and CFPP as reference light-weight buildups without honeycomb, submitting them to the same reliability testing as above. The results are indicated in Figures 14 and 15.

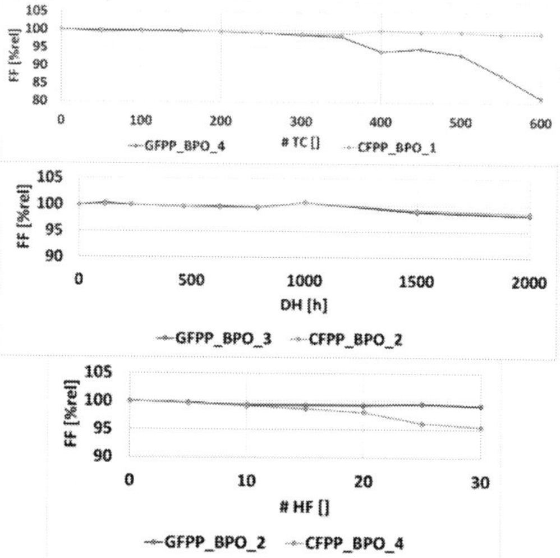

Figure 14: Degradation of FF throughout TC (top), DH (middle) and HF (bottom) testing of the GFPP and CFPP lightweight references

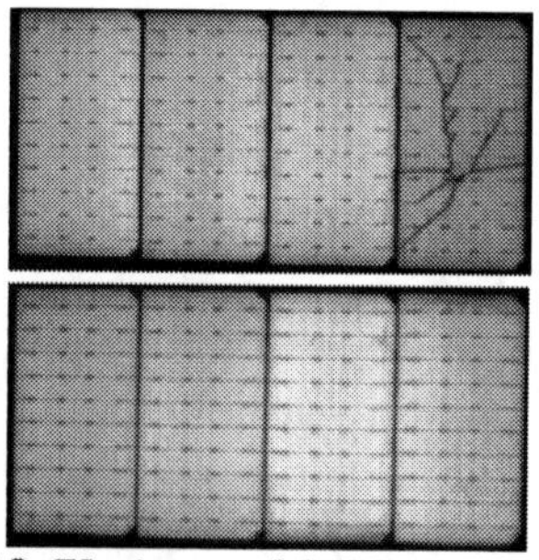

Figure 15: EL picture of CFPP (top) and GFPP (bottom) laminate after hail testing (5 stones)

Throughout the testing, only limited degradation is observed. A slight drop in FF for the GFPP laminate after 400 cycles can likely be attributed by EL to a broken interconnect wire. In hail testing, one hail stone impact (out of 5 for each laminate) results in a cell crack (visible in the EL picture of the CFPP laminate).

SUMMARY

In this work, we report on the preliminary testing of different types of honeycomb-based sandwich panels acting as PV module backsheet (and references without honeycombs). While environmental testing in terms of electrical

10.4229/EUPVSEC2023/3DO.19.5
020266-005

performance after thermal cycling, damp heat exposure or humidity freeze cycling can be readily passed, the mechanical behaviour of the PET-based sandwich panel materials are compromised. The effect is likely caused by hydrolysis and can be reduced by eliminating, protecting or modifying the PET materials. For passing the hail test, it is important to tune the skin layer immediately behind the cells.

ACKNOWLEDGMENT

The authors gratefully acknowledge the Solar-eranet project "Delight", its consortium partners and the financial support by the Flemish and other involved governments, as well as the European Union's Horizon 2020 Programme for research, technological development and demonstration for funding part of this work under Grant Agreement no. 857793 (HighLite). The authors also would like to acknowledge the partial funding by the Kuwait Foundation for the advancement of Sciences under project number P115-15EE-01.

REFERENCES

[1] J. Govaerts et al. "Development and testing of light-weight PV modules based on glass-fibre reinforcement", EPJ Photovoltaics 13, 13 (2022).
[2] J. Rion et al. "Ultra-light asymmetric photovoltaic sandwich structures", Composites: Part A 40 (2009) 1167–1173.
[3] J. Pflug, I. Verpoest, "Half closed thermoplastic honeycomb, their production process and equipment to produce", International Patent Nr WO2006/053407 A1, Nov. 21, 2005
[4] N. Pervan et al., "Honeycomb structures as backsheets for light weight PV modules", EUPVSEC (submitted)
[5] X. Chen et al., "Effect of molecular weight on crystallization, melting behavior and morphology of poly(trimethylene terephalate)", Polymer Testing, Volume 26, Issue 2, 2007, Pages 144-153 ISSN 0142-9418, https://doi.org/10.1016/j.polymertesting.2006.08.011

10.4229/EUPVSEC2023/3DO.19.5
020266-006

PV MICROINVERTERS: LATEST EFFICIENCY RANKINGS, ENERGY YIELD ASSESSMENTS, FIRMWARE ISSUES

Stefan Krauter, Jörg Bendfeld
Paderborn University, EET-NEK, Pohlweg 55, D-33098 Paderborn, Germany, E-mail: Stefan.Krauter@upb.de

ABSTRACT: The market for microinverters is growing, especially in Europe. Driven by rising electricity prices, eased legislation, many mini photovoltaic energy systems are being installed. Since 2014, microinverters have been studied indoors and outdoors at Paderborn University. In the indoor lab, conversion efficiencies as a function of load have been measured with high accuracy and ranked according to Euro- and CEC weightings: The latest rankings of 2023 are included in this paper. In the outdoor lab, energy yields have been measured using identical and calibrated crystalline silicon PV modules: Until 2020, measurements were carried out by 215 W_p modules. Because of increasing module power nowadays, 360 W_p modules are now being used. A simple formula method to determine actual energy yield has been developed, so yield for any module & inverter configuration can be determined by just two coefficients *a* and *b*. In 2023, new inverters (Anker MI 60, APSystems DS3-S/600W, Deye SUN-300-G3, Ecoflow Powerstream 60, Hoymiles HM 700, NEP BDM-600, Parkside PBKW-300-A1, Technaxx TX204) entered the market, they have been measured and ranked according to their EU- and CEC-efficiencies. Presently, they are undergoing yield assessments which will take at least one year to be valid and comparable. Interesting observations have been made: Inverters of the same brand, type, and date have been equipped with different firmware versions, resulting in different MPPT-accuracies and approach speeds. On the other hand, inverters of different brands have been equipped with the same hardware and firmware. Outlook: While grid connection is increasingly carried out via household plugs with non-insulated contacts, switch-off time after grid-disconnection (e.g., according to VDE-AR-N 4105) will also be measured and ranked in upcoming publications.
Keywords: Microinverter, System Performance, Energy Rating, Yield, Balcony Power Plant.

1 INTRODUCTION

Microinverters are inverters that are connected mostly to a single PV module (occasionally to two modules, as indicated in the tables; few are available for four modules, but these are not considered here), so each module–inverter combination acts as an independent power plant. The microinverter consists of a maximum power point tracker (MPPT), the DC-AC inverter, and an islanding protection unit (see e.g., [1]). For higher power requirements, several module-inverter combinations are interconnected in parallel on the AC output side. This configuration offers various advantages: Easier planning and installation, easy up- and downscaling of a plant, including extensions or repair that could be carried out even during power plant operation. Logistics is simplified. Effect of shadowing is very limited, and due to low system voltages, potential induced degradation (PID) does not occur. An excellent overview of the development and the advantages of microinverters has been compiled by H. Oldenkamp [2]. However, costs of power plants based on micro inverters are about 10–20% higher. Some of the inverters cannot be operated by themselves and require a control unit (often combined with a remote shutdown option and a monitoring system), or a protective device for grid interfacing (depending on national regulations), thus adding extra costs. Also, conversion efficiency may not be as high as for central inverters. Due to smart master–slave concepts centralized solutions with multiple but relatively large inverters may offer higher yields under weak light conditions. [3] is giving a performance comparison of a microinverter, a power-optimizer, and a central inverter.

2 MEASUREMENTS

2.1 Indoor tests for conversion efficiency measurements

Due to the reproducible test conditions in the indoor lab, the inverters have been examined individually with predefined and controlled input data. Input has been a PV module simulator with data being set corresponding to the modules used in the outdoor test. The main output data being recorded is the delivered AC power of the inverters. Besides input power, output is also a function of input voltage. If input voltage is getting too low, the inverters even stop operating. The following examinations are based on the possible range of input data (including voltage) given the specific PV module also used for the outdoor investigation.

Peak efficiency is often reached close to the maximum load of the inverter. Peak efficiency (often promoted in data sheets) is not a helpful value since most of the time the inverters operate in the range of 20% to 40% of their rated power – at least under non-arid conditions. Consequently, an adequately weighted efficiency is a more adequate value to rate conversion devices. One type of weighted efficiency is the so-called "European Efficiency" η_{Euro} which it is calculated according to:

$$\eta_{Euro} = 0.03 \cdot \eta_{5\%} + 0.06 \cdot \eta_{10\%} + 0.13 \cdot \eta_{20\%} + 0.1 \cdot \eta_{30\%} + 0.48 \cdot \eta_{50\%} + 0.2 \cdot \eta_{100\%} \quad (1)$$

The other is the "CEC efficiency" by the California Energy Commission (CEC). CEC efficiency is computed as an average value of DC–AC conversion efficiencies at six pre-defined relative output values between 10% and 100% of its rated power (with an emphasis on higher irradiance levels) is determined by:

$$\eta_{CEC} = 0.04 \cdot \eta_{10\%} + 0.05 \cdot \eta_{20\%} + 0.12 \cdot \eta_{30\%} + 0.21 \cdot \eta_{50\%} + 0.53 \cdot \eta_{75\%} + 0.05 \cdot \eta_{100\%} \quad (2)$$

For the "European Efficiency", weighting factors for high relative power values are lower.

The output power values used for the inverters (adjusted by controlling the DC input current) are continuously increased in 1024 steps from 0 to maximum. Each step takes eight seconds while the measurement

10.4229/EUPVSEC2023/3EO.1.2
020267-001

duration is 500 ms. Figure 1 shows an example for the measuring procedure.

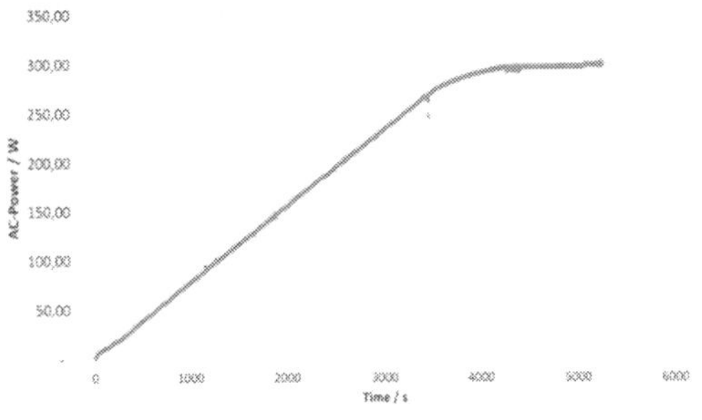

Figure 1: Example for measured AC power output (in Watt) as a function of measurement duration (in seconds) for linear increasing DC input current.

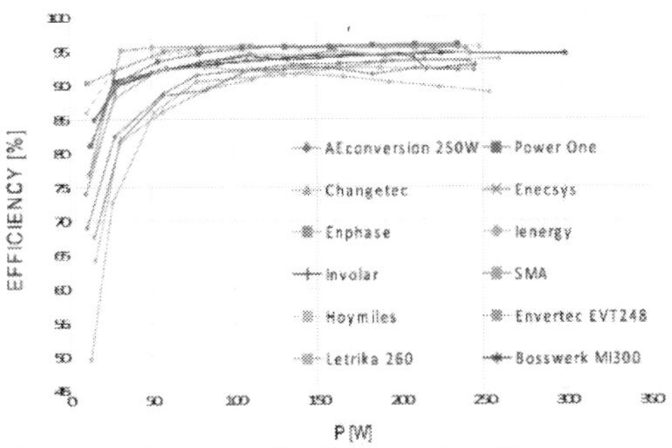

Figure 2: Measured DC-AC conversion efficiencies as a function of power output, for twelve microinverters with single PV module inputs (updated for 2022).

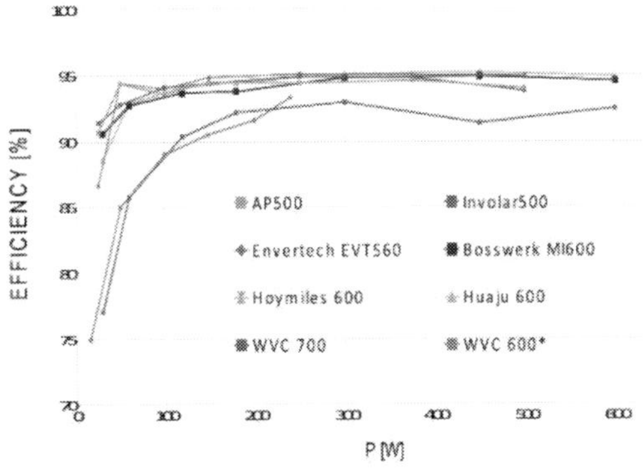

Figure 3: Measured DC-AC conversion efficiencies as a function of power output, for eight microinverters with two PV module inputs, including the latest models of Hoymiles, Huaju, Bosswerk.

The measured DC-AC conversion efficiencies of all inverters are shown in Fig. 2 and Fig. 3. Based on those measurements, the European (EU) efficiencies and the CEC efficiencies for the micro inverters have been calculated according to (1) and (2), eleven micro-inverters are designed for single modules, 18 inverters have inputs for two PV modules: Anker Solix MI 60; APSystems YC 500, DS3-S; Involar MAC 500; Deye Sun 600 G3; Ecoflow Powerstream 600; Envertech EVT-560; Hoymiles MI 500, 600, 700; Huaju HY 600; NEP BDM 600; Bosswerk Mi 600; Parkside PBKW-300-A1; Technaxx TX 204; Tsun TSOL-MS600; WVC 600, 700.

The ranking considering the European (EU) conversion efficiency is shown in Table I.

Comments: Envertech EVT 560 and PowerOne/ABB Micro-0.25-i have the same conversion efficiency, thus sharing rank number 5, so do DeyeSun G3 and Huaju HY600, sharing rank 7; Involar MAC 500 and Bosswerk Mi600, thus sharing rank 9 as well as AEconversion INV 250-45 and Enecsys SMI-S-240W, thus sharing rank 24.

Table I: Ranking of all tested microinverters by "European Conversion Efficiency", according to (1).

Rank No.	Manufacturer Model, Type	European Efficiency	Relative eff. of #1
1	SMA Sunnyboy 240	95.4%	100.0%
2	Enphase M 215	95.2%	99.8%
3	Hoymiles MI 500	95.0%	99.5%
4	Hoymiles MI 600	94.7%	99.3%
5	Envertech EVT-560	94.6%	99.2%
5	PowerOne/ABB Micro-0.25-i	94.6%	99.2%
7	Deye Sun 600 G3*	94.5%	99.0%
7	Huaju HY 600	94.5%	99.0%
9	Involar MAC 500	94.3%	98.8%
9	Bosswerk Mi 600	94.3%	98.8%
11	Technaxx TX 204*	94.2%	98.7%
12	APSystems YC 500	94.1%	98.6%
13	Anker Solix MI 60*	93.6%	98.1%
14	Bosswerk Mi 300	93.5%	98.0%
15	Envertech EVT-248	93.2%	97.7%
16	APSystems DS3-S*	93.0%	97.5%
17	Ecoflow Powerstream 600*	92.7%	97.2%
18	Involar MAC 250	92.7%	97.2%
19	Hoymiles HM 700*	92.5%	97.0%
19	NEP BDM 600**	92.5%	97.0%
21	Tsun TSOL-MS600*	92.4%	96.9%
22	WVC 700 (at 600 W)	91.6%	96.0%
23	Changetech ELV 300-25	90.9%	95.3%
24	AEconversion INV 250-45	90.4%	94.7%
24	Enecsys SMI-S-240W	90.4%	94.7%
26	Ienergy GT 260	89.9%	94.3%
27	Parkside PBKW-300-A1*	88.9%	93.2%
28	Letrika 260	88.7%	93.0%
29	WVC 700 (at 700 W)	73.3%	76.8%
30	WVC 600 (failed)	0.0%	0.0%

* new in 2023 ** switched off for powers above 500 W

Comment: WVC 600 stopped operating at a measured power of 250 W. After a test run at higher temperatures, the inverter failed constantly. Since the documentation of WVC 600 and WVC 700 inverter has been extremely poor, its rated power has been assumed. For this reason, WVC 700 is shown first with the assumed rated power of 600 W, then with 700 W. The maximum measured power of the WVC 700 inverter was 600 W only.

10.4229/EUPVSEC2023/3EO.1.2

Table II shows the same, but with the CEC-efficiency formula (2) applied.

Table II: Ranking of all microinverters by "CEC Efficiency", calculated according to (2).

Rank No.	Manufacturer Model, Type	CEC-Efficiency	Relative eff. of #1
1	Enphase M 215	95.6%	100.0%
2	PowerOne/ABB 0.25-i	95.5%	99.9%
3	Hoymiles MI 500	95.4%	99.8%
4	SMA Sunnyboy 240	95.1%	99.5%
5	Hoymiles MI 600	95.0%	99.4%
6	Huaju HY 600	94.9%	99.3%
7	Technaxx TX 204*	94.8%	99.2%
8	Envertech ENV-560	94.6%	99.0%
8	Involar MAC 500	94.6%	99.0%
8	Bosswerk Mi 600	94.6%	99.0%
11	APSystems YC 500	94.5%	98.9%
12	Deye Sun 600 G3*	94.4%	98.8%
13	Bosswerk Mi 300	94.1%	98.5%
14	Envertech EVT-248	94.1%	98.4%
15	Anker Solix MI 60*	93.9%	98.2%
15	Involar MAC 250	93.9%	98.2%
17	Ecoflow Powerstream 600*	92.9%	97.2%
17	NEP BDM 600**	92.9%	97.2%
19	Tsun TSOL-MS 600*	92.8%	97.1%
20	APSystems DS3-S*	92.7%	97.0%
21	Enecsys SMI-S-240W	92.0%	96.3%
22	WVC 700 (at 600 W)	91.6%	95.9%
23	Hoymiles HM 700*	91.5%	95.8%
23	Letrika 260	91.5%	95.8%
25	Ienergy GT 260	91.4%	95.7%
26	AEconversion 250	91.2%	95.5%
27	Changetech ELV 300-25	90.9%	95.1%
28	Parkside PBKW-300-A1*	89.7%	93.9%
29	WVC 700 (at 700 W)	87.5%	91.6%
30	WVC 600 (failed)	0.0%	0.0%

* new in 2023 ** switched off for powers above 500 W

Comment: Envertech EVT-560, Involar MAC 500, and Bosswerk Mi600 have the same conversion efficiency, thus sharing rank 8. The same applies to Anker MI 60 and Involar MAC 250, sharing rank 15; as well as for Ecoflow Powerstream 600 and NEP BDM 600, sharing rank 17; also Hoymiles HM 700 and Letrika 260S-C60-P260 sharing rank 23.

2.2 Thermal issues

In 2021/22, we started also to investigate thermal behavior of newer inverters: Some of the inverters failed when operated close to their nominal power (WVC 600, 700), independent of temperature (see Table I and II). Others (e.g., the Huaju HY 600) reduce their power output at 60°C, but operate normally at 25°C (see Fig. 4), but after cooling down, the inverters operated normally, without any damage. Possibly, that overheating issue was due a relatively compact inverter size.

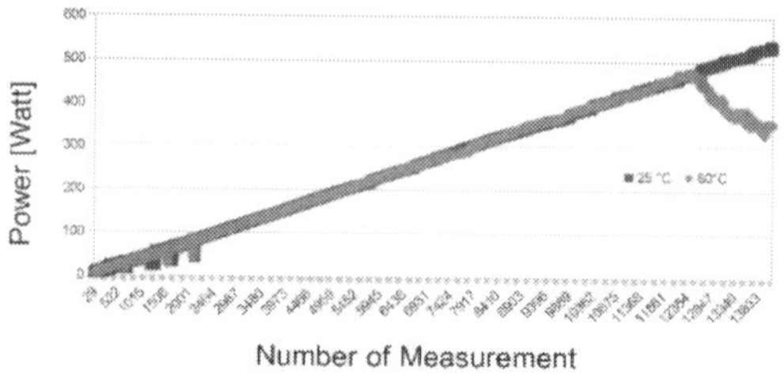

Figure 4: AC power output of the Huaju HY 600 inverter at 25°C (blue line) and at 60°C (red line) ambient temperature (heat chamber in the laboratory).

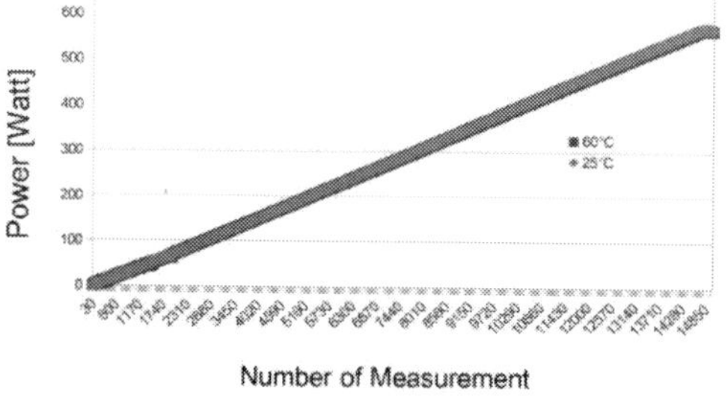

Figure 5: AC power output of inverters at 25°C (blue line) and at 60°C (red line) ambient temperature of Bosswerk Mi600 and Hoymiles HM 600.

Most other inverters did not show any power reduction (e.g., Bosswerk Mi600, Mi300, and Hoymiles HM600), even at elevated temperatures (see Fig. 5).

2.3 Outdoor measurements for yield assessments

The new configuration for the tests, using ten 360 W_p modules (lower row, from left), is shown in Figure 6. Modules have been manufactured by Solarwatt®, the power output at STC of each module has been measured in the factory in Dresden (Germany). Additionally, one module has been sent for a precision measurement to the testing laboratory ISFH in Hameln (Germany). It turned out that the factory measurements have been very accurate (362 W_P vs. 359.34 W_P ±3% at ISFH in July 2021).

Figure 6: Configuration of PV modules of PV outdoor laboratory for electrical energy yield comparison of microinverters using eight equal, calibrated PV modules (of 360 W_p each) as inputs.

Besides the effects already observed with the 215 W_p modules, such as distinct conversion efficiencies at different irradiance levels, speed of MPPT algorithms, minimum thresholds for initiating operation; additionally, temporal saturation effects are observed at some inverters with the new 360 W_p modules applied.

The resulting electrical energy yields during the course of a day for the different microinverters and module configurations are shown in Figure 7 for a daily course and in Table III (see [4]) over a longer period of time for the 215 W_p modules. To some extent, the above-mentioned effects can be observed.

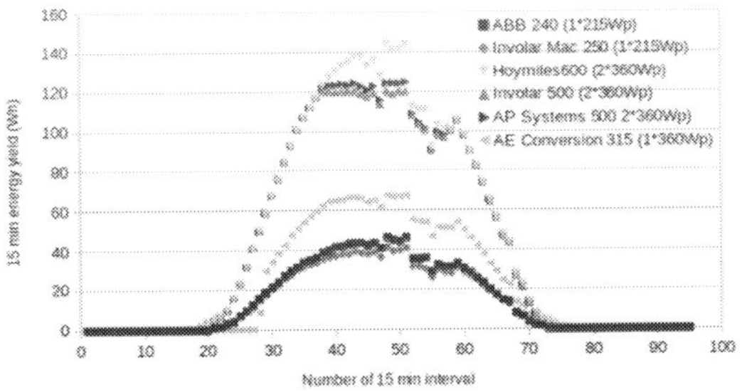

Figure 7: Example of electrical energy yield measurements (during an interval of 15 minutes) of different inverters and two different PV module sizes during a mostly clear day (some clouds in the afternoon).

Table III: Former ranking of microinverters by relative energy yield measured, using the 215 W_p modules (1 or 2 of them, as number of inputs), referenced to former #1 (Power One/Aurora/ABB, type Micro-0.25-i) [4].

Rank No.	Manufacturer Model, Type	Relative yield vs. ABB
1	Involar MAC 500	100.7 %
2	Power One/Aurora/ABB Micro-0.25-i	100.0 %
3	APSystems YC 500	99.3 %
4	Hoymiles MI 600	97.4 %
5	SMA Sunnyboy 240	95.2 %
5	Enphase M215	95.2 %
7	Involar MAC 250	94.2 %
8	Envertech EVT 300	94.0 %
9	WVC 700 (at 600 W)	91.7 %
10	AEconversion/Aptronic INV 250-45	92.5 %
11	Envertech EVT 248	92.1 %
12	Ienergy GT 260	91.5 %
13	Enecsys SMI-S-240W	88.7 %
14	Hoymiles MI 250	78.4 %
15	Changetec ELV 300-25	75.6 %

While the different types of effects make it quite cumbersome to predict an energy yield for a certain configuration at a certain location, a more consumer-friendly yield-predicting method has been elaborated by performing some yield data analysis.

Each microinverter has been directly connected to a calibrated electrical energy meter with a S_0-interface. To secure an accurate yield measurement, the calibrated electrical energy meters are replaced on a regular base with new freshly calibrated ones. All S_0-interfaces have been connected to a server-based data acquisition system.

3 UNIVERSAL YIELD ASSESSMENT

To ease the characterization of a specific combination of PV module & microinverter, a linear equation has been applied to a well investigated reference characteristics of a very good inverter without issues for low irradiance, MPPT, and saturation. The inverter chosen as a reference has been the Enphase M 215, which ranked #1 at the CEC-efficiency rankings, see [4].

Plotting a function of the actual yield (y) over the reference yield (x), that function would be $y = a\,x + b$ with the trivial coefficients $a = 1$ and $b = 0$ for the reference configuration (Enphase M 215 with the Q-cells 215 W_p module). Figure 8 shows the original configuration with the inverters for single modules and the 215 W_p modules attached.

The coefficients of the different inverters for the relative yield equation $y = ax + b$ have been elaborated in Table IV: It can be observed that for low daily yields Involar MAC 250 is performing a little bit better than the reference, so b is above 0, for high yields its performance is decreasing (relative to the reference), so a is above 1. For the Envertech EVT 300 the characteristics is vice versa: Performance at low yields is worse than the reference, so b is negative; relative performance is increasing towards high reference yields, so steepness of curve is higher, resulting in an $a > 1$.

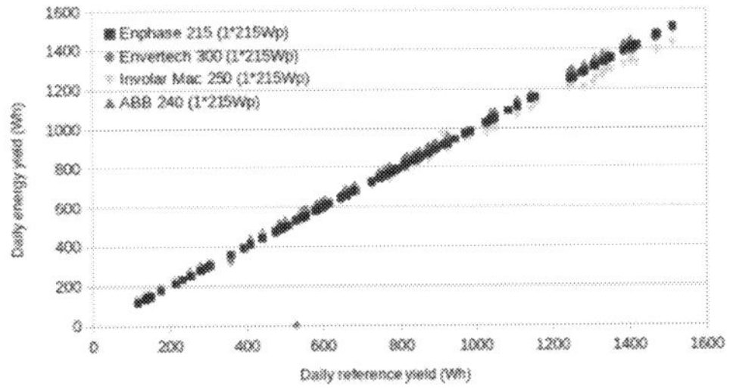

Figure 8: Electrical energy yields of different inverters for single modules with a 215 W_p module attached. Daily reference yield (x-axis) is the energy yield (AC) achieved by an Enphase M 215 inverter with a single 215 W_p module applied.

Table IV: Coefficients for relative daily yield $y = a\,x + b$ (referenced to Enphase M 215, all with single 215 W_p modules), yield is given in AC electrical energy

Manufacturer	Type	a	b (Wh)
Involar	MAC 500	0.923	+ 43.35
Power One /Aurora/ABB	Micro-0.25-i	1.011	+ 25.90
Envertech	EVT 300	1.020	- 33.45
Enphase	M 215	1.000	± 0.00
Bosswerk	Mi 300	0.969	+ 4.58

Figure 9 shows the characteristics of different microinverters that can serve two modules, either with two 215 W_p (older measurements) or two 360 W_p modules (latest measurements). Table V shows the corresponding coefficients a (for "steepness") and b (for "offset") of the relative daily yield curve.

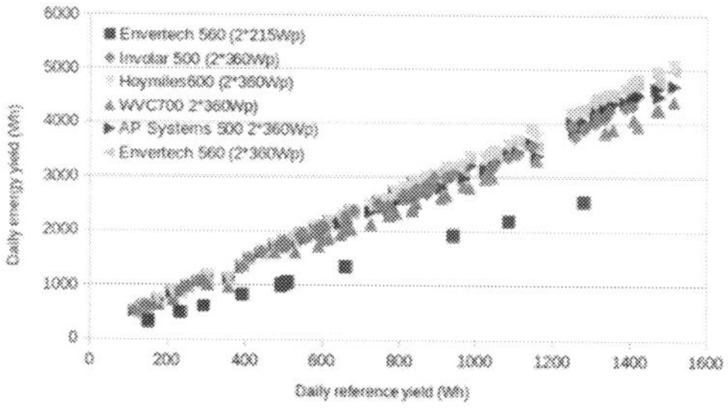

Figure 9: Daily energy yields (AC) of different inverters for two modules with two 215 W_p or 360 W_p modules attached. Reference yield (x-axis) is the yield achieved by an Enphase M215 with a single 215 W_p module applied.

Table V: Coefficients for relative daily yield $y = a\,x + b$ for microinverters serving two modules, either 215 W_p or 360 W_p types (referred to Enphase M 215 with a 215 W_p module), yield is given in electrical AC energy, order is alphabetic.

Manu-facturer	Type (module power)	a	b (Wh)
APSystems	DS-S* (2 x 360 W_p)	3.201	+ 63.00
APSystems	YC 500 (2 x 360 W_p)	2.953	+ 254.77
Bosswerk	Mi 600 (2 x 360 W_p)	3.122	+ 112.17
Deye	Sun 600 G3* (2 x 215 W_p)	1.876	+ 62.00
Deye	Sun 600 G3* (2 x 360 W_p)	3.124	+ 92.00
Envertech	EVT 560 (2 x 215 W_p)	1.983	+ 37.80
Envertech	EVT 560 (2 x 360 W_p)	3.227	+ 109.97
Hoymiles	MI 600 (2 x 360 W_p)	3.189	+ 168.32
Hoymiles	MI 700 * (2 x 360 W_p)	3.247	+ 133.00
Huaju	HY 600 (2 x 360 W_p)	3.141	+ 153.90
Involar	MAC 500 (2 x 360 W_p)	2.889	+ 180.70
Lidl Parkside	PBKW-300-A1* (1 x 160 W_p)	0.673	- 41.00
NEP	BDM 600 * (2 x 360 W_p)	2.700	+ 276.00
Technaxx	TX 204* (2 x 360 W_p)	3.156	+ 190.00
WVC	WVC 700 (2 x 360 W_p)	2.750	+ 172.39

* new in 2023 ** switched off for powers above 500 W

The coefficients of determination R^2 for all regressions of the measurement values to determine the coefficients a and b have been in the vicinity of 0.99 or above.

4 CONCLUSIONS

Efficiency: Interestingly, newer inverters did not make it to the top of the efficiency rankings. Apparently, efficiency is not a top priority for the manufacturers (any more).

Yield: The use of a reference configuration together with the two coefficients of a linear equation enables a simple method to describe quite accurately the daily yield performance of any microinverter in combination with any PV module, even with under- or oversized ones. While prices of PV modules are decreasing at a higher pace than prices for microinverters, we will see more configurations with oversized modules and more saturated microinverters more often in the future. This underlines the necessity of a method (e.g., as described) to extrapolate energy yield.

5 OUTLOOK

While grid connection of small PV systems is increasingly carried out via household plugs with non-insulated contacts within so called "balcony power plants", switch-off time after grid-disconnection (e.g., according to VDE-AR-N 4105) will also be measured and ranked in upcoming publications. In many countries, even a galvanic separation of the contacts after switch-off is required, which will be verified. Also, yield measurements for these balcony power plants shall be carried out, including modules supplied in a kit with typically vertical mounting.

6 REFERENCES

[1] W-F. Lai, S-M. Chen, T-J. Liang, K-W. Lee, A. Ioinovici: "Design and Implementation of Grid Connection Photovoltaic Inverter", Proceedings IEEE Energy Conversion Congress and Exposition (2012) 2426.

[2] H. Oldenkamp, I.J. de Jong: "The return of the AC-module inverter", Proceedings 24th European Photovoltaic Solar Energy Conference (2009) 3101.

[3] D. Stellbogen, P. Lechner, M. Senger: "Field and laboratory performance characterisation of microinverter and Power optimizer systems", Proceedings 32nd European Photovoltaic Solar Energy Conference (2016) 1654.

[4] S. Krauter, J. Bendfeld: "Micro-Inverters: An Update and Comparison of Conversion Efficiencies and Energy Yields". Proceedings of 37th European Photovoltaic Solar Energy Conference (2020) 935.

MEDIUM VOLTAGE STRING INVERTER FOR FUTURE UTILITY-SCALE PV POWER PLANTS

Michael Geiss, David Derix, Andreas Hensel, Jürgen Thoma, Dirk Kranzer
Fraunhofer Institute for Solar Energy Systems ISE
Heidenhofstr. 2, 79110 Freiburg, Germany

ABSTRACT: State of the art utility-scale PV power plants use AC-Voltages up to 880 V_{AC}. This voltage increased steadily in the past, because with higher system voltage the currents decrease. Due to the smaller currents, cable cross-sections can also be decreased. At the same time the subsystem power in a PV plant can be increased. This results in a reduced number of transformers for the same power plant size. These points lead to significant savings at cost for materials as well as for installation.
In present there is no further increase in the system voltage despite further increasing output powers of the string inverters. The reasons can be found in the now necessary step beyond the low voltage range.
At Fraunhofer ISE a new system concept for future utility-scale power plants was developed that is based on this step beyond the low voltage range with $V_{AC} \geq 1{,}000$ V. Further a medium voltage string inverter was developed that perfectly fits into the new concept. It has a line to line output voltage of 1,500 V and a output power of 250 kVA. The new concept and the developed inverter are shown in this paper.
Keywords: medium voltage, SiC, utility-scale, string inverter

1 MOTIVATION

Large quantities of raw materials are needed for the planned expansion of photovoltaics to achieve climate targets. These include copper and aluminum in cables and transformers. In the MS-LeiKra project (FKZ: 03EI4014A), funded by the German Federal Ministry of Economics and Climate Protection (BMWK), a string inverter was developed that can drastically reduce the required cable cross-sections by using a higher voltage level and the resulting lower currents.

The extreme savings potential that could be exploited by further increasing the voltage is shown in **Figure *1***. With an output line to line voltage of around 800 V_{AC} possible today, a minimum cable cross-section of 120 mm² is required according to DIN VDE 0298-4, Table 9.2 when assuming a string inverter with a power of 250 kVA. If the voltage is increased to 1,500 V_{AC}, the cable cross-section drops to 35 mm². When thinking about 5 kV_{AC}, only 4 mm² of required copper cross-section remains. With double-digit cable lengths in km for a large-scale PV power plant, the savings potential here is enormous.

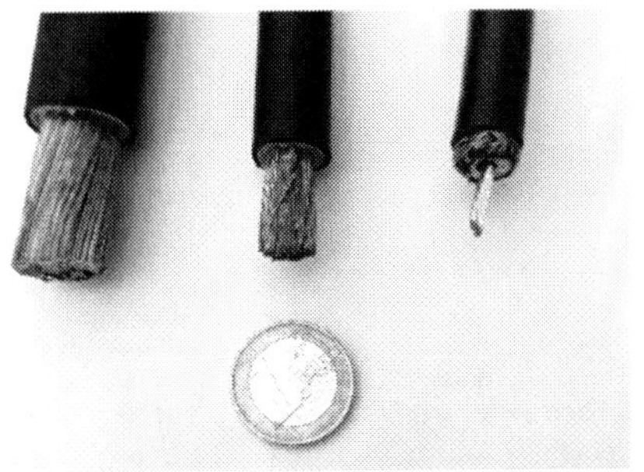

Figure 1: Minimum cable cross sections for 250 kW at different voltage levels. Left to right: 120 mm² @ 800 V_{AC}, 35 mm² @ 1,500 V_{AC} und 4 mm² @ 5,000 V_{AC}

Because of the missing standards and the difficulties to build a high efficient medium voltage string inverter based on silicon semiconductors this potential is not used yet. This paper shows that with the use of silicon carbide semiconductors the technical side is not a problem any more.

2 NEW POWER PLANT CONCEPT

The new string inverter is designed for use in large-scale PV power plants. Due to the large expansion and the associated long cable lengths, there is considerable potential for savings here when using higher voltage and thus lower currents. By focusing on large-scale PV power plants and excluding the residential sector, there are more degrees of freedom in the possible power plant concepts.

The power plant concept is based on the use of string inverters. An overview of the new concept is given in **Figure *2***. The individual inverters are connected to the central station of a sub-station via stub lines with fuse disconnectors. Circuit breakers are used only for larger blocks because they are very expensive and otherwise poorly utilized. However, by the use of fuse disconnectors, the advantage of high availability of the string inverter concept remains intact.

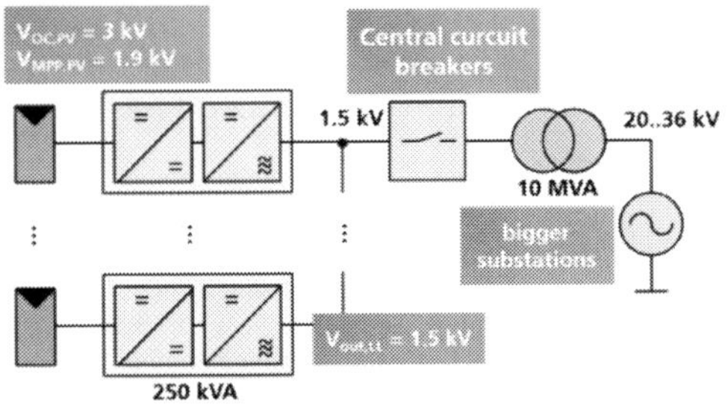

Figure 2: Plant concept for future utility scale pv power plants

The open-circuit voltage of the PV generator was set to a maximum of 3 kV. This voltage should be feasible with glass-glass modules in the near future. The minimum MPP voltage is thus 1.7 kV. In the course of the project, a variant for today's 1500 V modules was also developed, but this will not be discussed in detail here.

Since in large power plants the alignment of the modules is very homogeneous and individual shadings occur only very rarely, the string inverter is equipped with only one MPPT.

10.4229/EUPVSEC2023/3EO.1.3
020268-001

The power of the individual string inverters was set at 250 kVA. Various aspects were taken into account to set this power, such as the cable lengths and cross-sections on the AC and DC sides, but also suitable semiconductors for the inverters.

The individual inverters are connected to the central station of a sub-station via spur lines. The line to line voltage here is 1,500 V. In the central stations they are collected at one circuit breaker for 5 to 10 inverters. This is where the advantage of excluding the residential can be used. Since there is always an internal grid in large power plants and a switchgear at the connection point to the public grid, the relays in the inverter can be omitted. Fuse disconnectors are used to protect the stubs between the inverter and the central station. In this way, individual stubs can be disconnected in the event of a fault or for maintenance purposes, and the high availability of the string inverter concept is maintained.

The higher voltages within the power plant result in a large savings potential for the cables. With the same cross-section, losses can be greatly reduced and thus a higher yield can be achieved. Optionally, cable cross-sections can be reduced while losses remain the same. This reduces material and installation costs. This effect will have a even stronger impact if medium-voltage cables are increasingly used and the resulting increase in the number of units reduces the difference in costs between low- and medium-voltage cables.

The subsystem size should be between 10 MVA and 12 MVA. This Power corresponds to the maximum power that can still be realized with a transformer with a 1,500 V secondary winding. Today's concepts use subsystems in the size between 3 and 5 MVA. A larger power is difficult to realize in a low-voltage transformer with only one secondary winding. The larger subsystems result in further savings because fewer transformers, switchgear, and associated civil work are required.

An approach with central inverters would also be conceivable. The connection between the central inverter and the DC/DC stations would then be made via a medium voltage DC bus. However, since components for medium voltage DC were rare and very expensive at the beginning of the project, this approach was not pursued.

3 MEDIUM VOLTAGE STRING INVERTER

3.1 General Data

Adapted to the new power plant concept, a medium voltage string inverter was developed. In order to achieve a reasonable AC voltage and power of the individual inverters, a cost study was carried out based on material costs. A subsystem size of 12 MW was assumed. The cost study resulted in an AC output line to line voltage of 1,500 V with an inverter power of 250 kVA. The most important specifications of the two-stage inverter consisting of DC/DC converter and DC/AC inverter can be found in Table I.

Table I: Electrical parameters of the string inverter

Parameter	Value
Power	250 kW
DC-Link voltage	2.4 kV
AC output voltage	1.5 $kV_{linetoline}$
AC output current	96 A_{RMS}
Inverter topology	hybride ANPC with 3rd harmonic injection
Switching frequency inverter	24 kHz
PV open circuit voltage	3 kV
PV MPP voltage	1.7-2.4 kV
DC/DC topology	3 times interleaved boost
Switching frequency DC/DC	24 kHz

3.2 DC/DC Stage

A triple interleaved boost converter is used for the DC/DC converter, as can be seen in **Figure *3***. The total resulting current ripple is reduced by a phase-shifted control of the individual legs. This helps for a smaller dimensioning of the input and output capacitors.

The semiconductors used are 3.3 kV SiC MOSFETs in a half bridge configuration. The switching frequency could be set to 24 kHz despite the high reverse voltage. By using half-bridge modules, synchronous rectification can be achieved and thus an improvement in losses.

The three boosters are connected in parallel on the input and output side. Thus a single MPP tracking takes place.

The DC link with center point serves as the connection point between the DC/DC converter and the inverter. The nominal voltage here is 2.4 kV.

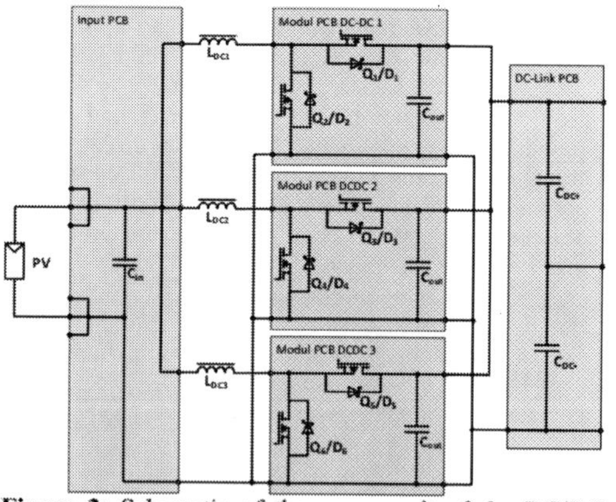

Figure 3: Schematic of the power unit of the DC/DC converter

3.3 Inverter Stage

On the other side of the DC link is the inverter section. Here a hybrid version of the active neutral point clamped (ANPC) topology is used. It consists of one module per phase with two SiC MOSFETs and four Si-IGBTs with a blocking voltage of 1.7 kV. The hybrid approach allows the advantages of Si and SiC to be used simultaneously. The Si IGBTs only have to be switched at the mains frequency, while the SiC MOSFETs switch at a high switching frequency of 48 kHz. This significantly reduces the filtering effort. At the same time, by using only two instead of six SiC switches, the cost of the semiconductor modules can be adapted to the enormous price pressure in the PV sector.

10.4229/EUPVSEC2023/3EO.1.3
020268-002

To obtain the maximum possible output voltage, a modulation with injection of the 3rd harmonic in the line to center voltage is used.

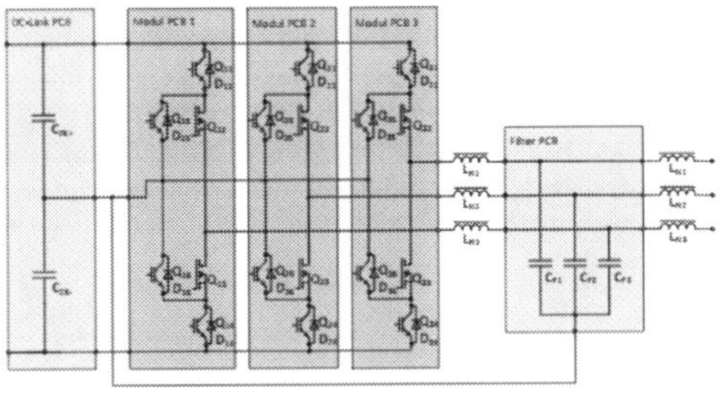

Figure 4: Schematic of the power unit of the inverter

3.4 Mechanical Design

Figure 5 and **Figure 6** show pictures of the front- and back view of the developed string inverter. All components are arranged around two heat sinks. The part in the left of the front view is the DC/DC converter, the part in the right the inverter.

On the lower left side one can see the input PCB. This carries the DC terminals, the current measurement and the input filter capacitors. The three PCBs above are the module PCBs for the three DC/DC phases. These are used to connect the power modules and carry the gate drivers and their supplies, among other things. On the back side of the heat sink the tank with the three potted boost chokes is mounted. Below the heat sink there is space for the input EMI filter.

Figure 5: Front view of the developed string inverter

Between the two heat sinks the DC link capacitors and the corresponding DC link board is mounted. The control board is mounted in front of the DC link board. It carries the DSP controllers for the DC/DC converter and the inverter.

Figure 6: Rear view of the developed string inverter

The right part belongs mainly to the inverter. On the top right site of the front view the three module boards with the gate drivers and the corresponding supplies are mounted. The PCB below is the self-supply for the complete inverter. At the bottom end is the output board. It carries the mains voltage measurement and the AC connection terminals. The LCL filter components are mounted on the back of the heat sink with the filter board next to them. It carries the measurements for the choke currents and the filter capacitor voltages and is used to connect all LCL components

Active air cooling is used in both stages. Below the semiconductor modules heatpipes are embedded into the heatsink for a better heat spreading and therefore a better utilization of the heatsink.

The overall size of the demonstrator is (75 x 72 x 35) cm^3 (h x w x d) leading to a volume of 189 l. This is a comparable size with todays string inverters on the market.

4 MEASUREMENT RESULTS

The demonstrator has already been successfully put into operation. In a first step only the inverter stage was put into operation in open loop control. A medium voltage source was used to power the DC-Link. On the AC side a resistor was connected as load. The test was carried out without 3rd harmonic injection. Due to the used resistor the power was limited to 72.5 kW. **Figure 7** shows the voltage and current waveforms of this measurement. The DC-Link was fully charged to 2.4 kV resulting in an AC line to midpoint voltage of 745 V_{RMS} due to the missing 3rd harmonic. This equals to a line to line voltage of 1290 V_{RMS}. At this operating point an efficiency of 99 % could be reached for the inverter stage.

Besides this tests always one single phase of the inverter and also of the DC/DC stage was tested under full voltage and full power conditions in open loop control.

10.4229/EUPVSEC2023/3EO.1.3

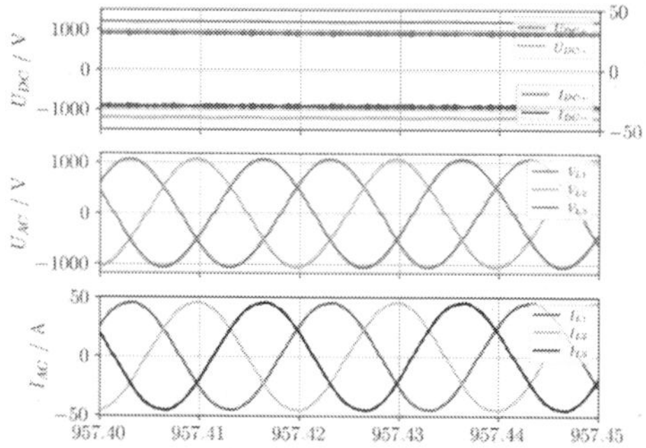

Figure 7: Voltage and current waveforms of the inverter stage at 72.5 kW in open loop at an ohmic load.

After the successful passing of all open loop tests, the complete string inverter was put into closed loop operation. For these tests a DC-Source was used that simulates the PV generator. On the AC side the inverter was connected to the 20 kV medium voltage grid via a 20 kV to 1.5 kV transformer. **Figure *8*** shows the voltage and current waveforms at an operating point of 13 kW. The inverter output line to line voltage at this point was 1,512 V_{RMS} (873 V_{RMS} line to midpoint)

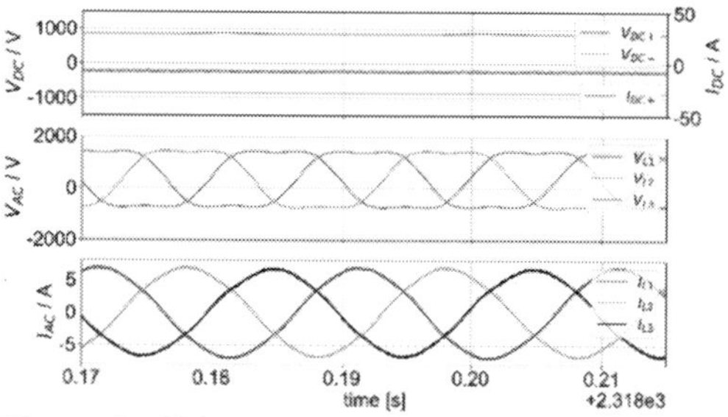

Figure 8: Voltage and current waveforms of the stringinverter when feeding 13 kW into the grid.

After the successful connection to the medium voltage grid, the next step will be to increase the power up to the full power and to do efficiency tests.

5 CONCLUSION AND OUTLINE

It is shown, that there is a tremendous potential of saving resources when stepping up the voltages of an PV inverter beyond the low voltage range.

During the project MS-LeiKra a medium voltage string inverter has been developed, and successfully put into operation. It has an input open circuit voltage of 3 kV and an output voltage of 1.5 kV line to line. The 250 kVA power stack is based on 3.3 kV and 1.7 kV SiC semiconductors to reach the high voltage.

A power plant concept for future utility-scale PV power plants is presented. It is based on the step into medium voltage range and the use of the presented inverter with a focus on only large power plants. The concept is explained in and possible cost saving potentials are presented.

With the demonstrator Fraunhofer ISE has shown, that from the inverter side there are no technical limits for introducing medium voltage in PV. As a next step, it is planned to build a pilot plant to prove feasibility on system level.

Also, the PV standards must be adapted to the higher voltages. Here a range from 1 kV_{AC} to 5 kV_{AC} would be really interesting as there will be semiconductors available in the close future for this voltage range. This range is not only interesting for the PV sector. In Wind turbines, EV-charging parks and industrial grids the similar saving potentials can be used.

10.4229/EUPVSEC2023/3EO.1.3
020268-004

40th European Photovoltaic Solar Energy Conference and Exhibition

This presentation was selected by the Sc. Committee of the EU PVSEC 2023 for submission of a full paper to one of the EU PVSEC's collaborating peer-reviewed journals.

EVALUATION OF HORIZONTAL SINGLE-AXIS SOLAR TRACKER ALGORITHMS IN TERMS OF ENERGY PRODUCTION AND OPERATIONAL PERFORMANCE

Ildefonso Muñoz; Adrián Guinda, Luis Casajus, Gregorio Olivares, Sara Díaz, Ana Gracia
CENER (National Renewable Energy Centre of Spain) - Department of Solar Energy Technologies and Storage
c/Ciudad de la Innovación, 7 – 31621 Sarriguren, Navarra, Spain (+34 948 252 800)
E-mail: imunoz@cener.com, aguinda@cener.com, lcasajus@cener.com, golivares@cener.com, sdreal@cener.com, agracia@cener.com

ABSTRACT: Horizontal single-axis solar tracking systems have become more widely used than other solar tracking systems, due to their optimum trade-off between rates of energy production, installation cost (CAPEX) and maintenance (OPEX). It is well known that the most widely used solar tracking algorithm is the so-called Astronomical algorithm with backtracking capability. However, this solar tracking strategy is not always optimal from the point of view of energy production, durability and reliability of the solar tracker. This paper compares the performance of three tracking algorithms to the Astronomical one. Two algorithms aim at optimizing the received irradiance focusing on the diffuse component. The third one, called Analytical, is presented as a new development in this study, defining the optimal inclination angle depending on all radiation components. The comparison focuses on different parameters: in-plane irradiance, DC power output from a monofacial PV system and operational aspects like the performed number of movements. High time resolution (1 minute) simulations have been performed with proprietary software of CENER, based on open code libraries (PVLIB Python), taking into account several effects on the effective irradiance and power output estimation, such as shading effects or PV configuration. Six locations with different climatic conditions have been analysed.
Keywords: horizontal solar tracker, tracking algorithms, PV performance, high resolution

1 INTRODUCTION

Currently, the most widely used solar tracking system in large utility-scale PV installations is horizontal single-axis tracking, based on a north-south rotation axis parallel to the ground, on which the PV modules are placed. Compared with fixed structures PV plants, this type suppose an increase of the operation and maintenance costs due to the inclusion of new elements in the installation, but this cost is compensated with an increase of the energy generated by the installation. A mechanical drive provides an east-west rotation of the plane of array (POA) throughout the day, according to a tracking algorithm implemented in an electronic control that commands the mechanical drive.

The most commonly used solar tracking strategy in single-axis horizontal trackers is the so-called Astronomical algorithm, which in order to maximize the received in-plane irradiance, aims at all times at following the solar trajectory with a simple geometrical approach. The position of the PV modules is established aiming at minimizing the angle of incidence between the normal vector of the POA and the solar vector. [1] The problem of self-shading or shading generated on immediately contiguous rows is also well known, using backtracking strategies as a modification of Astronomical algorithm in order to minimize or in some cases even altogether avoid the problem of mutual shading between nearby trackers.

However, the aforementioned solar tracking strategies, despite being the most commonly used, are not always optimal from the point of view of energy production, durability and reliability of the solar tracker. Due to the fact that this new approaches can be easily implemented in the electronic control system of the solar tracker, this provides the opportunity to explore possible improvements in PV plant performance without using additional equipment or incurring further costs, through the implementation of new solar tracking strategies.

This paper presents an in-depth comparative and evaluation studies of three tracking algorithms for horizontal single axis solar tracker, taking account the location of solar tracker, atmospheric conditions (radiation and temperature), energy production and reliability factors based on high temporal resolution data (1 minute). This paper compares the performance of three tracking algorithms to the Astronomical one, being one of them, called Analytical algorithm, a new algorithm developed by CENER.

2 METHODOLOGY

This section describes the proposed solar tracking algorithms, the methodology applied for obtaining the solar radiation and climatological data used as input in the simulations, the selection of locations to be considered and, lastly, the PV simulation software developed by CENER for the simulation and estimation of the energy production of PV systems at high spatial and temporal resolution.

2.1 Proposed solar tracking algorithms

Diffuse algorithm. It's based on the fact that on cloudy days the beam component of the radiation is secondary to the diffuse component in their contribution to the G_{POA}. In addition to the irradiance at the tilted angle imposed by the Astronomical tracking algorithm at each time step, the global horizontal irradiance (GHI) is needed, either measured or estimated by radiation transposition calculations [2]. Once both irradiances are known and compared, if GHI > G_{POA}, the tracker is instructed to be positioned horizontally. If this condition is not met, the command sent to the tracker is to perform Astronomical tracking.

Diffuse+Nowcasting algorithm. The Diffuse algorithm follows a simple premise, but certain problems can arise in its practical implementation. On days with cloudy intervals, cloud passing may occur frequently, so that the relation between GHI and G_{POA} varies with the same frequency. In that case, there may be situations where the position of the solar tracker is constantly changing between the horizontal position and the position determined by the Astronomical algorithm, increasing the consumption of the tracker drive, the operation and maintenance costs and reducing the lifetime of the installation components. It is even possible that the scenario in which moving the modules to the horizontal position is not justified because the gain in irradiance obtained at that moment is lower than the extra consumption due to that movement.

To solve the problems described above, the use of the data immediately after the instant when the tracker position was calculated has been considered. This is nowadays possible thanks to the forecasting techniques of atmospheric conditions at a very short period of time (nowcasting), which are sufficiently developed, with a reasonably reliable prediction [3-5]. The defined Diffuse+Nowcasting algorithm considers a fixed and predicted at a window of fifteen minutes. At the starting point of each predicted data window, the total radiation that will be present in the horizontal plane (GHI) and in the plane of array (G_{POA}) is calculated. The mode of operation of the solar tracker during this fixed window will be determined according to the following situations defined in Equation (1) and (2):

If currently in Astronomical tracking mode, change to horizontal position if relationship in Equation 1 is satisfied, otherwise maintain the Astronomical tracking:

$$\sum_t GHI - \sum_t G_{POA} \geq G_{min} \quad (1)$$

If currently in horizontal position, change to Astronomical tracking if relationship in Equation 2 is satisfied, otherwise maintain the horizontal position:

$$\sum_t G_{POA} - \sum_t GHI \geq G_{min} \quad (2)$$

In Equation (1) and (2), G_{min} represents the minimum radiation gain that must be obtained in order to introduce changes in the tracking mode so that the power generation of the PV generator field is higher, taking into account the additional consumption of the solar tracker. The parameter G_{min} is a function of the PV generator (PV module efficiency and performance ratio, PR), the mechanical drive of the solar tracker (power consumption) and the selected calculation window (parameter t). The parameter G_{min} can also be increased depending on the design criteria of the installation, to minimize the number of tracker movements and considering only relatively large radiation increments. For the present paper, the parameter G_{min} is defined as 0 and the parameter t as fifteen minutes. These imply that if the irradiation received by the PV module with the actual algorithm being applied during the next fifteen minutes is smaller than the irradiation received with the other algorithm, the solar tracker change the actual algorithm, keeping this change during the next fifteen minutes. Meaning that if the algorithm that maximizes the received irradiation during this time is the Astronomical, the tracker will apply it during these fifteen minutes, with the respective movements, but if the optimal algorithm is the horizontal, the tracker will remain in the horizontal position during these fifteen minutes, after which a new evaluation process will be applied by the algorithm.

Analytical algorithm. The algorithms described above are based on the assumption that the optimal positions of the POA of a single-axis horizontal solar tracker could be defined according to the Astronomical tracking or the horizontal position. This is correct provided the beam radiation or diffuse radiation component is very large relative to the global radiation on the POA, in the first case for the Astronomical tracking (completely clear days) and in the second case for the Diffuse Radiation algorithm (completely cloudy days).

However, depending on the location where the solar tracker is placed, the fraction of beam and diffuse radiation of the global radiation may not be so different. For example, days with the presence of less dense and less dark clouds allow some beam radiation component, but not of a higher order than the diffuse radiation. This may result in the optimum angle being in an intermediate position between the horizontal plane and the angle set by the Astronomical algorithm. Moreover, on days with cloudy intervals, an intermediate position makes it possible to be closer to the optimum position in the event of a clear or cloudy day, reducing the number of turns and the consumption of the drive.

Based on the above, a new solar tracking algorithm for single-axis horizontal trackers is proposed. The global irradiance on the plane of array and its components can be estimated according to the considered isotropic model [6] from Equations (3) to (7):

$$G_{POA} = G_B + G_D + G_G \quad (3)$$

$$G_B = DNI * cos(AOI) \quad (4)$$

$$cos(AOI) = cos\,\theta_s * cos\,\beta + sin\,\theta_s * sin\,\beta * cos(\gamma_s - \gamma) \quad (5)$$

$$G_D = DHI * \frac{1+cos\,\beta}{2} \quad (6)$$

$$G_G = GHI * a * \frac{1-cos\,\beta}{2} \quad (7)$$

Where:

G_{POA}:	global irradiance in the Plane of Array
G_B:	beam irradiance of G_{POA}
G_D:	sky diffuse irradiance of G_{POA}
G_G:	ground (reflected) irradiance of G_{POA}
DNI:	direct normal irradiance
DHI:	sky diffuse horizontal irradiance
GHI:	global horizontal irradiance
AOI:	angle of incidence of the sun vector and POA
θs:	solar zenith angle
γs:	solar azimuth angle
β:	tilt angle of the POA
γ:	azimuth angle of the POA
a:	albedo of the ground

The maximum irradiance condition as a function of the angle of the POA of the solar tracker can be mathematically expressed as the partial derivative of the expression of the global irradiance as a function of the β angle equal to zero, as shown in Equation (8):

$$\frac{\partial G_{POA}}{\partial \beta} = \frac{\partial G_B}{\partial \beta} + \frac{\partial G_D}{\partial \beta} + \frac{\partial G_G}{\partial \beta} = 0 \rightarrow Max\ irradiance \quad (8)$$

Calculating appropriately, we obtain the expression shown in Equation (9) for the β angle that maximizes the irradiance over the plane of array as a function of GHI, DNI, DHI, location and design parameters of PV installation.

$$\beta_{opt} = arctan\left(\frac{DNI * sin\,\theta_s * cos(\gamma_s - \gamma)}{\frac{1}{2}(DHI - GHI \cdot a) + DNI * cos\,\theta_s}\right) \quad (9)$$

The calculation of the optimum angle β is completed

in this new algorithm with the considerations of the mechanical limits of the solar tracker and the implementation of the backtracking strategy (as has been done in the others) in a similar way as detailed above. [7]

2.2 Solar Radiation and Meteorological data

In order to analyse and compare the performance of the considered tracking algorithms not only with regard to the in-plane irradiance but also the performance of the PV modules, various sites within the spatial coverage of the considered datasets have been identified. The selection process aimed at identifying locations with different ranges of global horizontal irradiance, and diffuse and direct fraction, as well as ambient temperature profiles as the main variables affecting the PV performance. Therefore, not only the yearly irradiation values or the long-term average temperatures were analysed, but long-term datasets were studied in order to take into account the inter- and intra-annual variability of both irradiance and temperature.

The standard IEC 61853-4 describes six reference climates representative of the climatic conditions most PV installations worldwide will be subjected to. The six climates are defined as: Tropical humid (TH), Subtropical arid (SA), Subtropical coastal (SC), Temperate coastal (TC), High elevation (HE) and Temperate continental (TN).

For the present study six different locations have been selected, one for each of these reference climates. Details of these locations are available in Table 1.

Table I: Selected locations and yearly values of global and diffuse horizontal irradiation H_{GHI}, H_{DHI} (kWh·m^{-2}), diffuse ratio DR (%) and temperature ambient T_{amb} (°C).

Site (climate)	Lat./Lon.	H_{GHI}	H_{DHI}	DR	T_{amb}
Afghanistan (HE)	33.7/63.9	2113	544	25.7	13.4
Congo (TH)	-3.5/13.0	1591	974	61.2	26.4
Belarus (TN)	54.0/27.5	1049	527	50.2	7.1
Namibia (SA)	20.7/14.0	2455	499	20.3	25.8
Spain (SC)	42.8/-1.6	1527	600	39.2	14.5
Scotland (TC)	57.0/-3.6	808	528	65.3	6.6

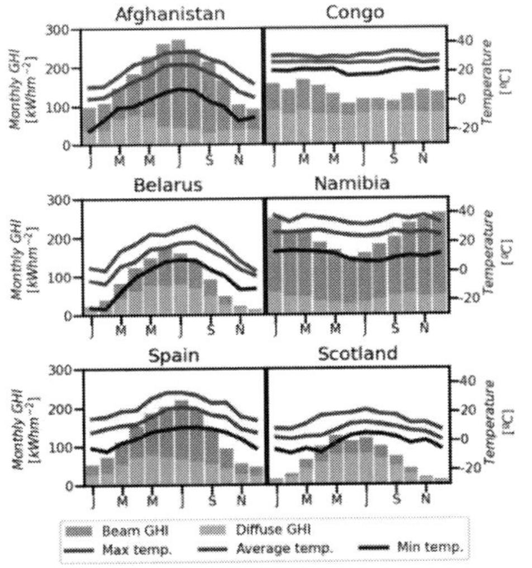

Figure 1: Monthly GHI and T_{amb} in selected locations

The selected sites show significant variability both in terms of the total global horizontal irradiation levels and the diffuse and direct fraction of the said irradiation levels. Sites like for example Afghanistan and Namibia with similar global irradiation levels show significant differences in the temperature profiles which could affect the performance of the trackers and the PV installation.

The characterization of the operating conditions of the solar trackers in terms of solar radiation and climatological variables has been performed by means of the Typical Meteorological Year (TMY). The TMY used in the present analysis is an annual time series of minute data defined based on the irradiance values so that the TMY values correspond to the 50th percentile of the annual values of the long-term database used to define the TMY. Therefore, the probability of exceeding the corresponding annual values is 50%.

The methodology followed for the generation of the TMY at minute resolution can be divided into two main steps: extraction of modelled data from databases and long-term analysis and TMY generation. Modelled solar radiation and climatological minute data over a period of 19 consecutive years (February 2004 - January 2023) have been obtained through Solar Radiation Data (SODA) platform [8] which provides access to the following databases: Copernicus Atmosphere Monitoring Service (CAMS) for solar radiation variables and Modern-Era Retrospective analysis for Research and Applications, Version 2 (MERRA-2) for the meteorological ones. SODA pre-processes CAMS and MERRA-2 original data in order to provide data at various time resolutions, from monthly to minute data.

For the present study, time series of minute data from the long-term time period specified above, have been retrieved for the following variables: ambient temperature, relative humidity, atmospheric pressure at 2 m above ground level, wind speed & direction at 10 m above ground level and rainfall.

Once the modelled series were obtained, the long-term analysis of the GHI and DNI irradiance values was carried out following the UNE 206013:2017 standard [9]. This standard was established in 2014 under the auspices of the Spanish Association for Standardization and Certification (AENOR), with CENER participation.

2.3 Estimation of energy production

The evaluation of the different algorithms for single-axis horizontal solar trackers proposed in this article has been carried out with the simulation software developed by CENER called SIMPV. This software, based on the Python PVLIB library [10] estimates the energy produced by a PV installation using meteorological data as input and taking into account different effects which affect the performance and the power generation of PV installations. The methodology applied for the estimation of the energy produced by any PV string implemented in SIMPV can be divided in two steps: the estimation of the effective irradiance on the POA and the electrical output.

The estimation of the effective irradiance takes into account the effect of the angle of inclination of the solar tracker imposed by the tracking algorithm, the calculation of the global, diffuse, direct and reflected incident irradiance on the module's surface for each simulation step, using the three component anisotropic model defined by Perez et al. [11], the effect of self-shading or shadings between contiguous rows of PV modules and the IAM factor.

The effective irradiance captured by the PV modules enables to transform the energy of the received photons to electric energy due to the photoelectric effect. However, there are various effects that limit the efficiency of the said transformation modifying the ability of the module to produce electricity, like the irradiance applied to the module, the temperature of the solar cells or the electrical layout. The effects on the electrical output considered on the simulation program SIMPV are estimated based on the calculation of IV curves of the PV modules (considering effects of temperature, shadings, etc.) and the sum of this curves according to the electrical connection of PV modules.

The main advantage of the CENER proprietary software, SIMPV, is that there is no limitation in the layout of the PV installation or in the temporal resolution of the meteorological data to be used as input data and the subsequent power output estimation. These capabilities make SIMPV suitable for the detailed evaluation and comparison of the performance of the different tracking algorithms analysed in the present study.

2.4 Study scenarios

In order to analyse the effect of the mounting configuration on the PV power output and analyse in detail the electrical impact of the shading effect, the performance of the different tracking algorithms has been evaluated considering two different mounting configurations. The estimation of the electrical production has been performed considering a single solar tracker located in the central part of a PV plant (surrounded by other identical solar trackers), containing a string of 30 PV modules of 575 Wp of monofacial "half-cell" monocrystalline Si technology with three bypass diodes (size of 2.4x1.1 m), all of them connected in series. The ground albedo has been defined as 0.25. The two different configurations regarding the physical disposition of the photovoltaic modules are:

- 1P disposition. PV modules placed in portrait position (vertically) along the axis of rotation in one single row of 30 modules. A pitch between trackers of 6.0 m has been considered.
- 2L disposition. PV modules placed in landscape position (horizontally) along the axis of rotation in two parallel rows of 15 modules each. A pitch between trackers of 5.5 m has been considered.

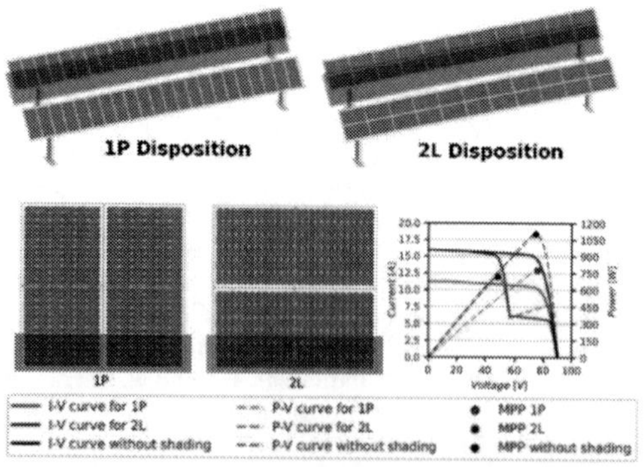

Figure 2: Schematic representation of the effect of self-shading on two single-axis solar trackers with 1P and 2L configurations, respectively (top). Diagram of two half-cell PV modules in 1P and 2L disposition with the same shading condition and their corresponding IV and PV curves, showing the different electrical behaviour with the same shading (bottom).

In this way, the two physical arrangements of the PV modules have the same active area and the same GCR (0.4), so the electrical production results are only dependent on the tracking algorithm and the physical arrangement of the modules, as shown in Fig.3. Related to the solar tracker mechanical drive, all values of irradiance estimated on the POA have been obtained with the consideration of the real movement of a horizontal single-axis solar tracker. For the present study a step movement of 1.5° and a maximum turning speed of 10° per minute have been considered, similar to the applied angle shown in Fig.3.

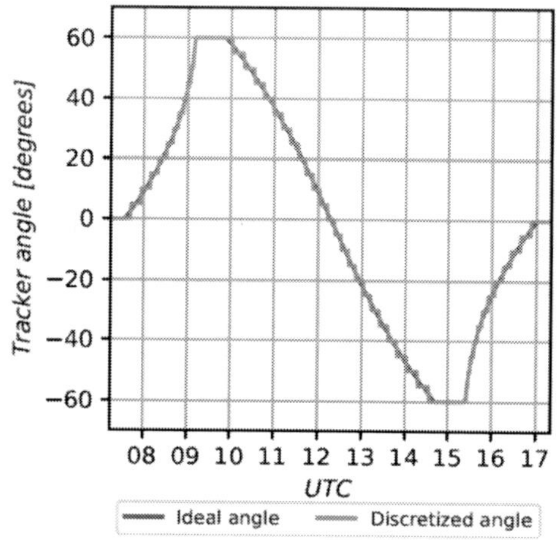

Figure 3: Ideal and real (discretized) movement of a typical solar tracker mechanical drive.

3 RESULTS

The first results regard the comparison of the irradiance values on the POA obtained through the implementation of the proposed algorithms, with and without the application of the backtracking strategy (BT). As shown in Fig.4, in all sites with any PV module disposition and any solar tracking algorithm, the no application of backtracking causes DC shading losses compared with the same algorithms with backtracking, varying between 1.2% and 3.3% in 1P disposition, and 2.4% and 5.7% in 2L disposition.

Based on the result exposed above, in all the considered tracking algorithms have been included the backtracking strategy (therefore without electrical shading losses). From the point of view of PV performance, Fig.4 shows the gain on the generated DC energy of the proposed solar tracking algorithms versus the Astronomical algorithm in the studied locations for both 1P and 2L configurations.

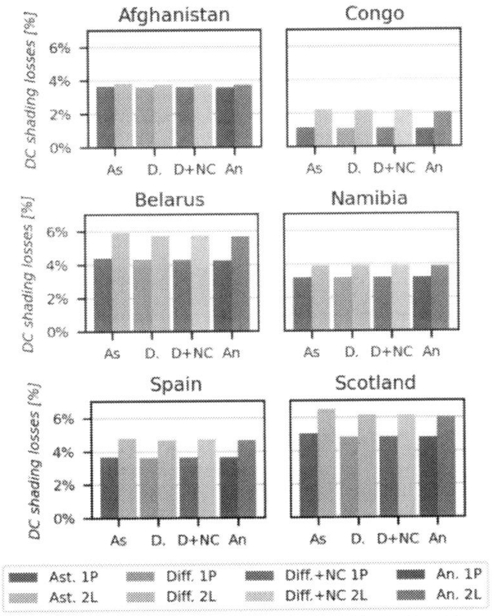

Figure 4: DC shading losses of tracking algorithms without backtracking in the considered sites.

The gain obtained on the DC energy output, compared to the Astronomical algorithm, is higher in locations where the diffuse component is predominant, like Scotland. The lower the diffuse fraction, the lower the gain from the considered tracking algorithms compared to the Astronomical one, like in the location in Namibia, as shown in Fig.5.

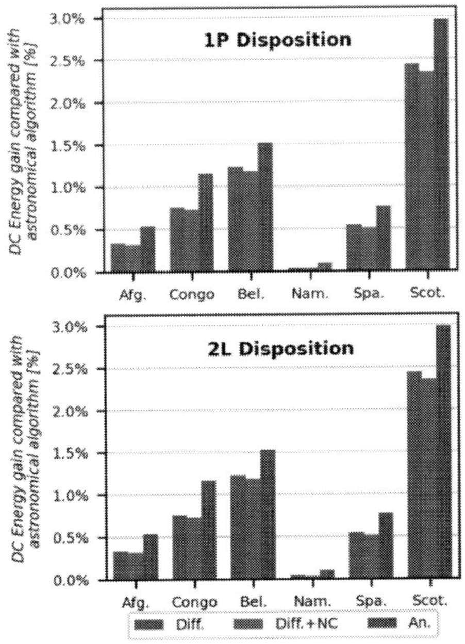

Figure 5: Yearly DC energy gain of solar tracking algorithms compared with Astronomical algorithm with backtracking strategy for the 1P disposition (top) and 2L disposition (bottom).

From the three algorithms analysed, the Analytical one tends to perform better than the ones based on the diffuse irradiance analysis for both 1P and 2L dispositions. For the 1P disposition, the PV output gained from the Analytical algorithm compared to the Astronomical one varies between 0.09% obtained in Namibia to 2.96% in Scotland. The corresponding values for the Diffuse algorithm are 0.03% and 2.43%, while the values for Diffuse+Nowcasting algorithm are 0.03% and 2.34%. Similar values are observed in the 2P configuration. The DC output gain ranging between 0.04% and 2.44% for the Diffuse algorithm, 0.03% and 2.35% for the Diffuse+Nowcasting algorithm, and 0.10% and 2.99% for the Analytical algorithm. These two algorithms based on the diffuse irradiance show similar results on all sites, with an average lower gain obtained by the one considering nowcasting. Even though, on average for both configurations (1P and 2L) and locations, the Analytical algorithm results in DC output gains compared to the Astronomical algorithm, 70% higher than the other two algorithms.

Finally, the last comparison between the performances of the analysed tracking algorithms regards the performed movements is exposed in Fig.6. Due to the considered set up used for the 1P and 2L PV modules disposition and the same GCR obtained, both configurations perform the same number of movements. In all considered sites, the Analytical algorithm performs the highest number of movements with the subsequent higher risk of failure of the solar tracker. On the contrary, the algorithm using forecasting needs the lowest number of movements, which could increase the lifetime of the tracker. In some locations, the number of movements from this algorithm could be as few as half of those required by the Analytical algorithm, especially at locations with high diffuse fraction. While in locations like Namibia with high irradiation levels, all algorithms require similar number of movements, very similar as well to the Astronomical one.

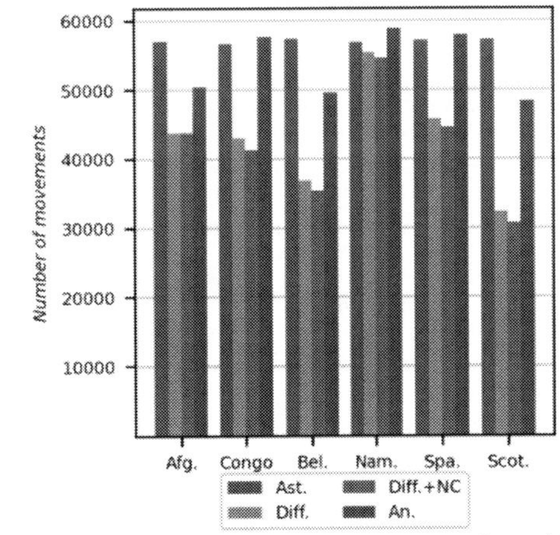

Figure 6: Annual number of movements performed by the solar tracker depending on location and solar tracking algorithm with backtracking.

Even though the Analytical algorithm requires the highest number of movements, it is not the algorithm that shows the highest accumulated angular movement. In opposition to this, the Diffuse algorithm is the one that accumulates the highest angular movement and presents the highest average angular movement per movement performed by

10.4229/EUPVSEC2023/3EO.1.4

the tracker, as shown in Fig.7.

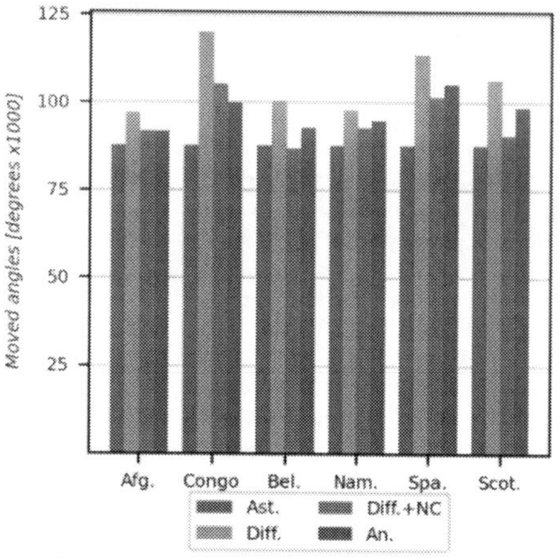

Figure 7: Accumulated angular movement in a year.

4 CONCLUSIONS

Current horizontal single axis solar tracker presents a chance for designing and implementing new tracking algorithms in an easy way, which could increase de performance of the PV plant without adding new elements or technologies.

The performance of the tracking algorithms depends on the climatological conditions, especially on the diffuse fraction. In general, the higher the diffuse component, the higher benefit obtained from implementing other algorithms like the Diffuse Radiation or the Analytical one, compared to the Astronomical one. While, the in-plane irradiation gain obtained by the three algorithms compared to the Astronomical one at locations with predominantly clear sky conditions is below 0.13%, for locations with high diffuse fraction values, the new Analytical algorithm presented in this paper derives on almost 3% higher values than the Astronomical one, higher gain values than the other two. If the DC power output values are analysed, the performance of the three algorithms in comparison to the Astronomical one also depends on the climatological conditions. For example, in locations with 60% diffuse fraction the DC gain could reach almost 3% by the Analytical algorithm, which performs better than the other two algorithms. In locations with predominant clear sky conditions, all algorithms perform similarly and require, as well, similar number of movements. However, as overcast skies become more frequent, the Analytical algorithm performs higher number of movements than the other two diffuse irradiance based algorithms, and shown similar numbers to the Astronomical algorithm. Nevertheless, if the total accumulated angular movement is considered, the Diffuse Radiation algorithm is the one that, in most considered locations, shows the higher numbers.

5 REFERENCES

[1] W. F. Marion, A. P. Dobos, Technical Report NREL/TP-6A20-5881, USA 2013.
[2] T. Muneer, Building Serv. Eng. Res. Technol 1990, 11, 153.
[3] C.A. Gueymard, J.A. Ruiz-Arias, Sol Energy 2016, 128, 1.
[4] A. Sanfilippo, L. Martin-Pomares, N. Mohandes, D. Perez-Astudillo, D. Bachour, Sol Energy 2016, 125, 77.
[5] T. Schmidt, J. Kalisch, E. Lorenz, D. Heinemann, Atmos. Chem. Physics 2016, 16, 3399.
[6] H. Hottel, B. Woertz, Trans. Am. Soc. Mech. Eng 1942, 64.
[7] K. Anderson, M. Mikofski, Technical Report NREL/TP-5K00-76626, USA 2020.
[8] SODA (https://www.soda-pro.com/web-services)
[9] UNE 206013:2017. Solar thermal electric plants. Procedure for the generation of solar radiation percentiles years. UNE Normalización Española 2017.
[10] W.F. Holmgren, C.W. Hansen, M.A. Mikofski, J. Open Source Softw. 2018, 3, 884.
[11] R. Perez, P. Ineichen, R. Seals, J. Michalsky, R. Stewart, Sol Energy 1990, 44, 271.

10.4229/EUPVSEC2023/3EO.1.4
020270-006

PCCL
Polymer Competence Center Leoben

Analysis of Performance of PV Systems and Insights on Packaging Degradation Using NIR Spectroscopy

C. Barretta[1*], S. Lindig[2], M. Bredács[1], A. Astigarraga[2], E. Helfer[1], G. Oreski[1]

1) Polymer Competence Center Leoben GmbH (PCCL), Leoben, Austria
2) EURAC Research, Bolzano, Italy

40th EU PVSEC, 18.09.2023, Lisbon

*) Sauraugasse 1, 8700 Leoben, Austria
+43 3842 42962 722
chiara.barretta@pccl.at
www.pccl.at

eurac research

Introduction

PCCL
Polymer Competence Center Leoben

Oreski, G, Omazic, A, Eder, GC, et al. Properties and degradation behaviour of polyolefin encapsulants for photovoltaic modules. *Prog Photovolt Res Appl.* 2020; 28: 1277– 1288. https://doi.org/10.1002/pip.3323

Visual inspection

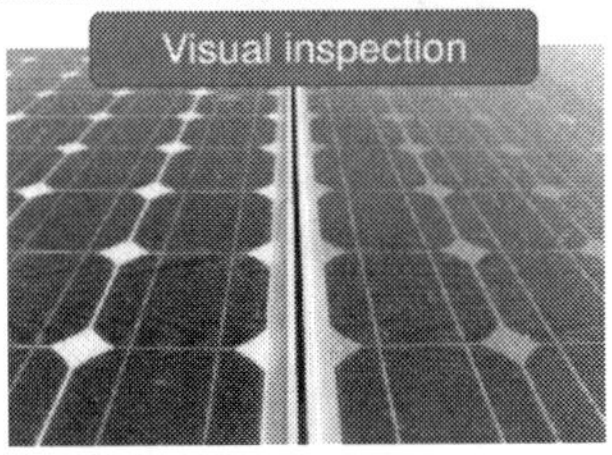

Field inspection methods

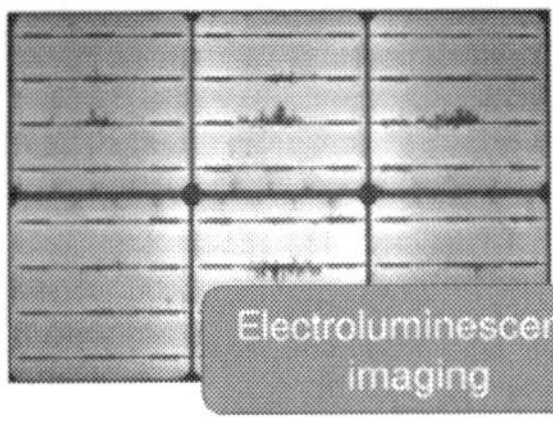

Electroluminescence imaging

IV curve measurements

UV fluorescence imaging

IR imaging

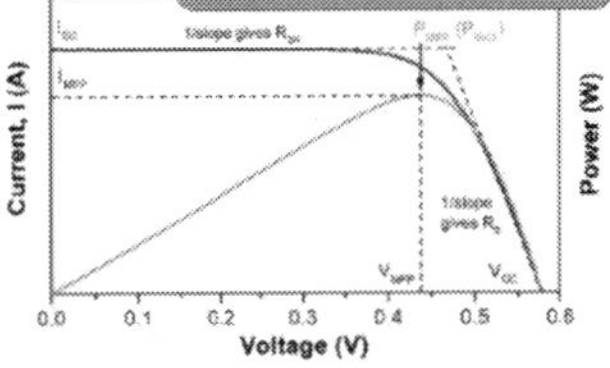

M. Köntges, A. Morlier, G. Eder, E. Fleiß, B. Kubicek and J. Lin, "Review: Ultraviolet Fluorescence as Assessment Tool for Photovoltaic Modules," in IEEE Journal of Photovoltaics, vol. 10, no. 2, pp. 616-633, March 2020, doi: 10.1109/JPHOTOV.2019.2961781.

www.pccl.at

2

Experimental
Near Infrared (NIR) spectroscopy in the field

PCCL

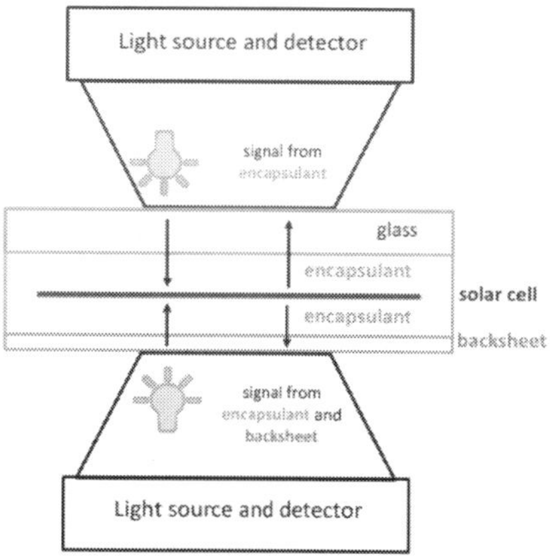

Measurement range:
1450 – 2450 nm
(7000 – 4000 cm^{-1})

www.pccl.at 3

020273-003

40th European Photovoltaic Solar Energy Conference and Exhibition

Experimental
NIR spectroscopy in the field: data processing and evaluation

PCCL

NIR spectroscopy

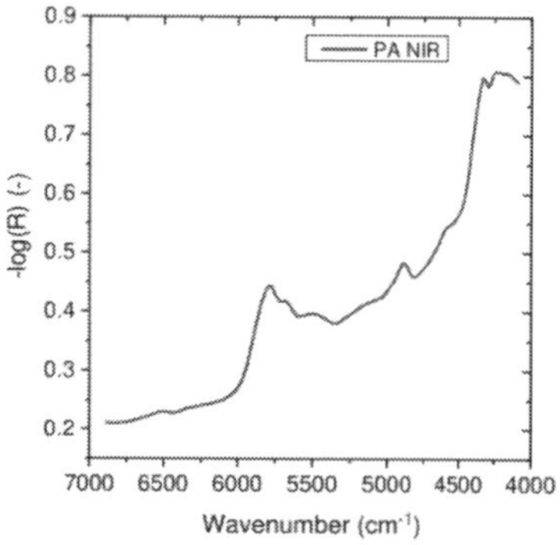

FTIR-ATR spectroscopy

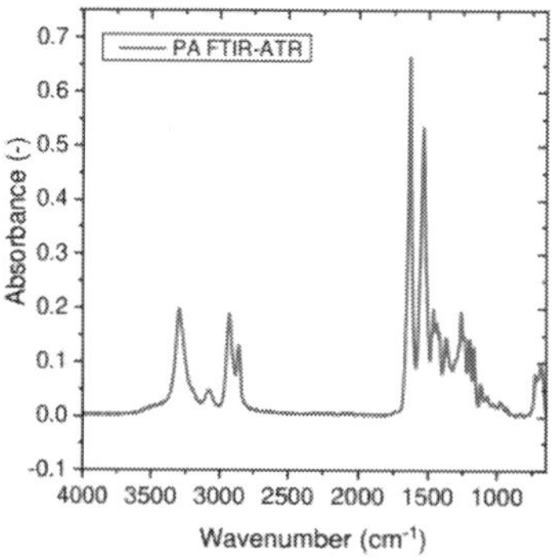

Multivariate Data Analysis (MVDA)

Principal Component Analysis (PCA)

- **Unsupervised classification method** (qualitative)
- **Separation** or sorting of groups **of objects** into one or more **classes** based on distinctive features in the objects
- **Unlimited variables** (dimensions) that create a multidimensional environment
- Identification of **principal components** that describe differences between the groups (e.g. wavenumbers)

www.pccl.at 4

020273-004

Results
Performance loss rates

PCCL

- **19 modules** from **6 PV systems**
- PV modules based on c-Si technology with glass/encapsulant/backsheet configuration
- Exposed in the field for about **12 years** in northern Italy (temperate climate with medium irradiation)
- **Performance loss rates** between –0.89 %/year and –1.56 %/year
- The **reference modules** for each system were available and **stored under lab conditions**

PV system	Performance loss rate [%/year]
mc-Si A	–1.13
pc-Si A	–1.56
pc-Si B	–0.89
pc-Si C	–1.10
pc-Si D	–1.23
pc-Si E	–1.5

I. Kaaya, S. Lindig, K-A. Weiss, A. Virtuani, M. Sidrach de Cardona Ortin, D. Moser: "Photovoltaic lifetime forecast model based on degradation patterns". Prog Photovolt Res Appl. 2020; 28: 979–992. https://doi.org/10.1002/pip.3280

www.pccl.at

020273-005

40th European Photovoltaic Solar Energy Conference and Exhibition

Results
BOM identification

PCCL

PV system	Encapsulant	Backsheet
mc-Si A	EVA	PET core + F + EVA
pc-Si A	EVA	PA
pc-Si B	EVA	PA
pc-Si C	EVA	PET core + F
pc-Si D	EVA	PET core + F
pc-Si E	EVA	PET core + PVDF

EVA – Ethylene vinyl acetate
PET – Polyethylene terephthalate
PA – Polyamide
F – Fluorine-containing polymer
PVDF – Polyvinylidene fluoride

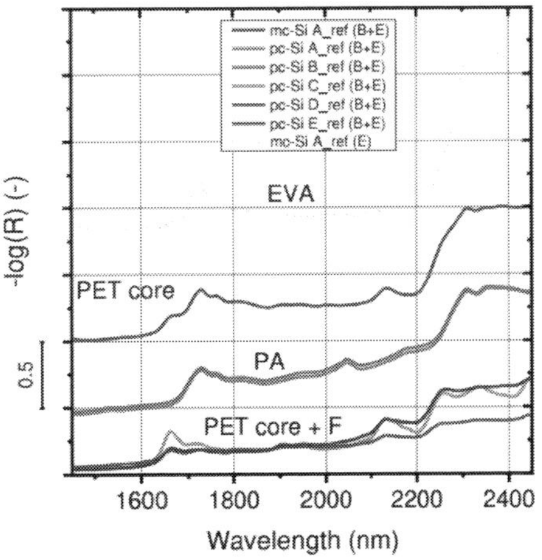

www.pccl.at 6

020273-006

40th European Photovoltaic Solar Energy Conference and Exhibition

Results
Case Study mc-Si A

PV system	Encapsulant	Backsheet	Performance loss rate [%/year]
mc-Si A	EVA	PET core + F + EVA	–1.13

- Encapsulant discoloration on the cells
- No observable backsheet degradation

www.pccl.at 7

020273-007

40th European Photovoltaic Solar Energy Conference and Exhibition

Results
Case Study mc-Si A

Front side – information about encapsulant (EVA)

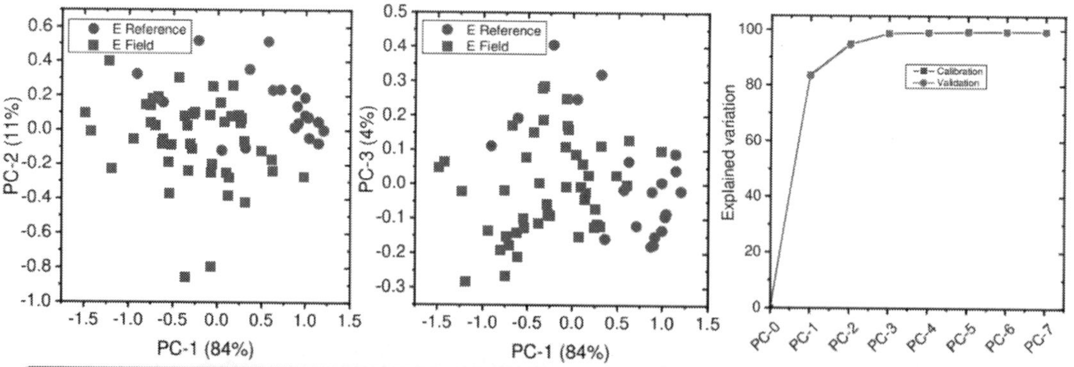

- Weak model
- PC1 describes the difference between measurements referred to exposed and unexposed PV modules

www.pccl.at 8

020273-008

Results
Case Study mc-Si A

PCCL

Back side – information about back encapsulant and backsheet (PET core + F + EVA)

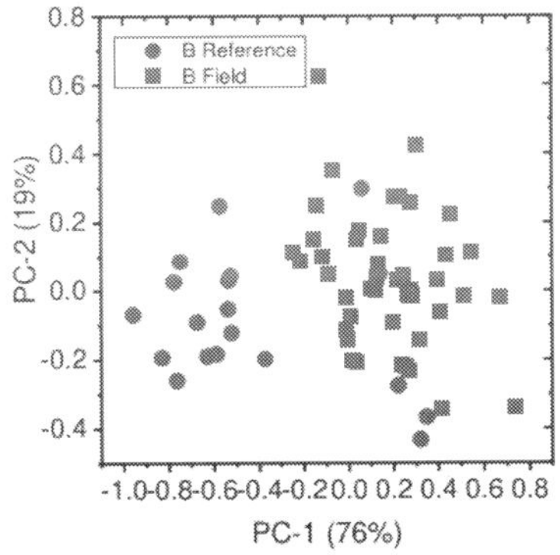

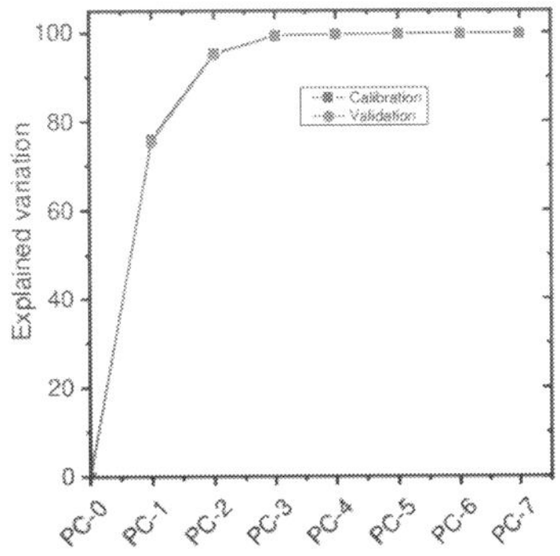

- Good model
- PC1 (76%) describes the aged samples
- Outliers due to inhomogeneity within the PV modules

Results
Case Study pc-Si B

PCCL

PV system	Encapsulant	Backsheet	Performance loss rate [%/year]
pc-Si B	EVA	PA	–0.89

- Encapsulant discoloration (not homogeneous)
- Humidity ingress
- Cell cracks
- Corrosion
- Backsheet chalking
- Backsheet cracks

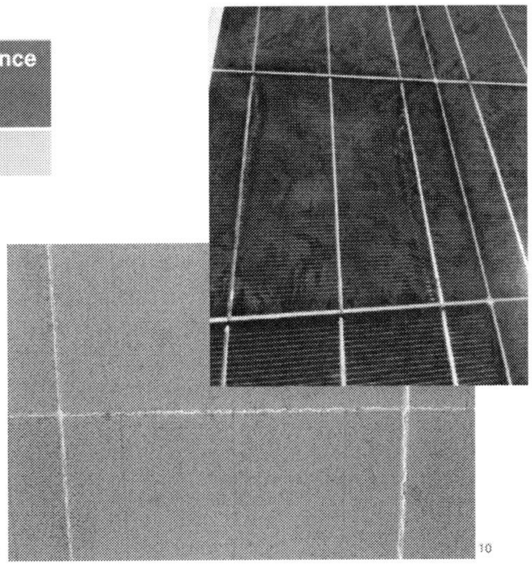

Results
Case Study pc-Si B

PCCL

Front side – information about front encapsulant (EVA)

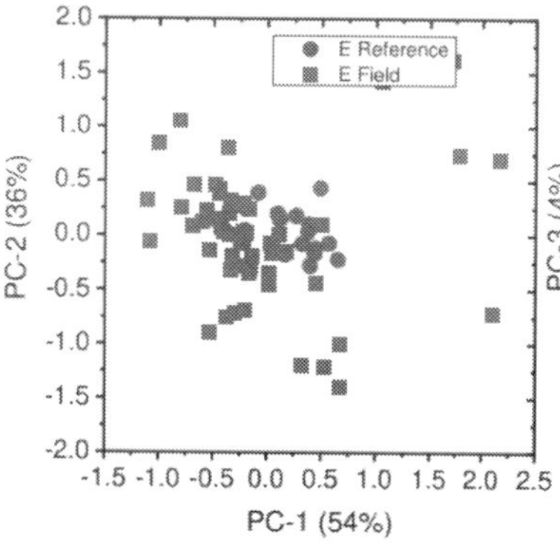

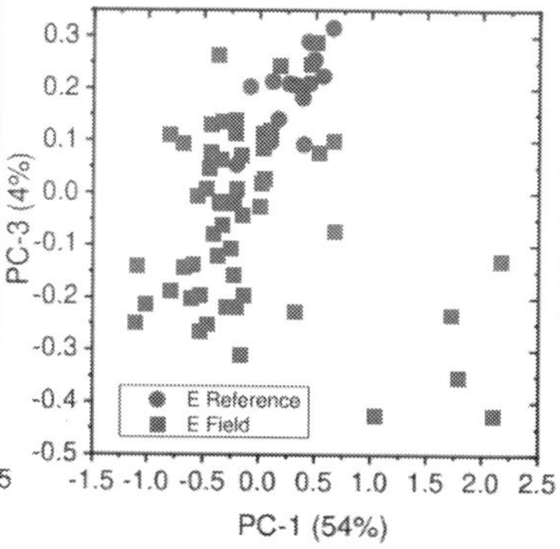

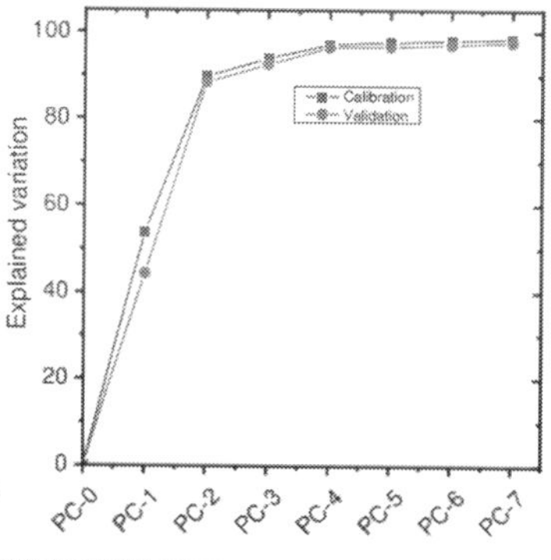

- The difference between the measurements on exposed and unexposed modules is unclear
- Changes due to additives cannot be detected
- Not reliable model

Results
Case Study pc-Si B

PCCL

Back side – information about back encapsulant and backsheet (PA)

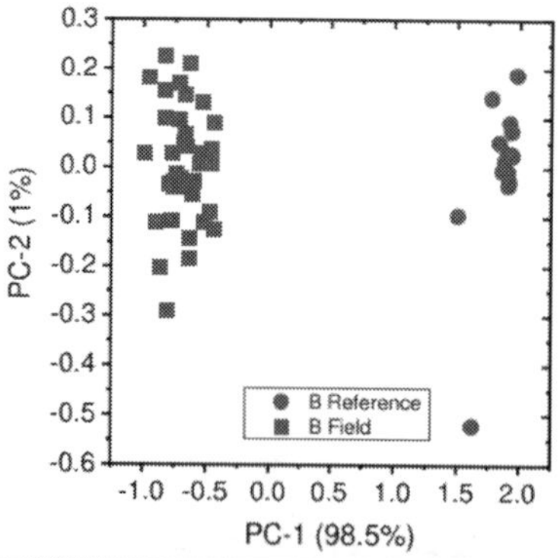

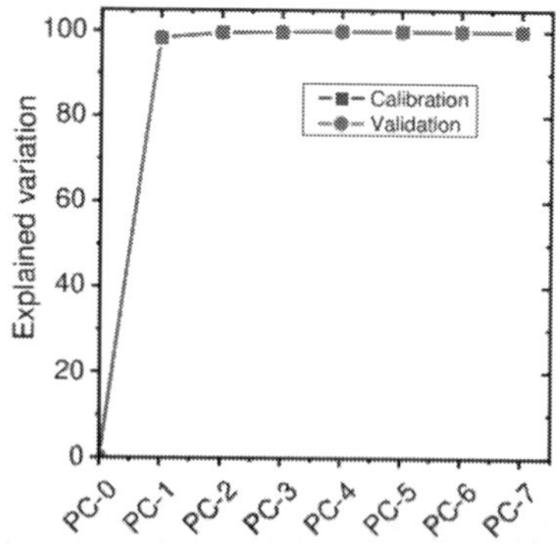

- Very robust model
- PC1 (98%) negative peaks describe the aged samples
- Difference due to loss of gloss → chalking

Summary and outlook

PCCL
Polymer Competence Center Leoben

Advantages
- ✓ Fast measurement method
- ✓ Allows BOM identification
- ✓ Applicable for automated condition monitoring
- ✓ Backsheet degradation can be detected

Challenges
- ✓ Advanced data modelling is necessary
- ✓ Contact measurement
- ✓ Encapsulant degradation difficult to detect

Current status

Proof of concept
- NIR can detect polymer aging

Future work

Modelling of polymer degradation over exposure time
- Time series upon accelerated aging tests

Correlation between **performance losses** and **material degradation**
- Combined model for performance losses and polymer degradation

Condition monitoring
Estimation of **remaining lifetime**

www.pccl.at 13

020273-013

40th European Photovoltaic Solar Energy Conference and Exhibition

Thank you for your attention!

PCCL
Polymer Competence Center Leoben

Chiara Barretta
Sauraugasse 1, 8700 Leoben, Austria
+43 3842 42962 722
chiara.barretta@pccl.at

eurac research

www.pccl.at 14

020273-014

PV PERFORMANCE IN THE YEARS AFTER EXTREME WEATHER EVENTS

Dirk C. Jordan*, Kirsten Perry, Robert White, Chris Deline
National Renewable Energy Laboratory, 15013 Denver West Parkway, Golden, CO 80401

ABSTRACT: The impact of extreme weather events on photovoltaic (PV) performance was studied by comparing the National Oceanic and Atmospheric Administration (NOAA) database on severe weather with the National Renewable Energy Laboratory (NREL) PV Fleet database on continuous fielded PV performance. We identified 170 systems that were immediately impacted by weather events. Those systems exhibited a small, typical (median) loss of 1% of total production during the calendar year of the extreme weather event. 12 systems out of more than 6400 systems (0.2 %) were more heavily impacted with losses of 10% or more indicating the continued need to engineer systems to withstand the extreme weather that may occur over the decades-long expected lifespan of the panels. At the median, in 2-3 days the PV systems show production again, suggesting that profound, permanent damage did not occur in most systems—in some cases it is possible only the monitoring communication equipment was damaged. Additionally, we found that hail size of 25mm (1 in.) or larger is associated with higher long-term degradation (despite typical qualification for 25mm with traditional IEC 61215 testing)--utilization of newer, more stringent hail technical specification (IEC TS 63397) is recommended. For wind speeds we found the threshold of higher PLR to be around 90 km/hour. Detailed examination of O&M records revealed that module damage may occur at sites from lightning when unprotected by lightning rods. Furthermore, increased tracker maintenance can be expected when tracked systems are exposed to high wind events. Properly built distributed energy generation including PV solar and energy storage systems have demonstrated usefulness in providing power through extreme weather events and their aftermath, even during grid blackouts. Keywords: Photovoltaics, performance, degradation, reliability, extreme weather

1 INTRODUCTION

PV modules have been tested for mechanical load and hail because of field exposure to wind, snow and hail since the Jet Propulsion Laboratory (JPL) Block Buy program during the late 1970s and early 1980s [1]. The developed tests were added to the module qualification test standard IEC 61215 because of its wide application and was most likely a major contributor to the increased lifetimes observed during the Block Buy program [2,3]. Hail resistance is one of the more dominating field module failures even today [4]. When the front glass is cracked the damage is easily visible, yet internal damage when the front glass is not broken needs to be detected expediently [5]. Depending on location, hail size and mounting configuration the damage can range from minor to multi-dollar insurance claims [6,7]. In addition, the front glass thickness of PV modules is a key module component with thicknesses of 3.2 to 4 millimeter substantially decreasing hail-induced damage [8].

Resilience against extreme weather conditions is key for future power generation, as the frequency and intensity of extreme weather has been increasing. To address this question, we cross-compared National Oceanic and Atmospheric Administration (NOAA) database on extreme weather with one of the largest databases on time series-based PV from the PV Fleet Performance Data initiative [9,10]. The associated PV Fleet database contains more than 24,000 inverter data streams from almost 2,500 commercial and utility PV sites, totaling more than 8 Gigawatts in the USA, with a mean site age of more than five years [11]. Additionally, in this research we obtained operations & maintenance (O&M) records to provide insights obtained from our performance investigations. O&M records by automated tickets or manual entries can provide valuable insights of weak points in PV systems and provide an overview of the industry as general [12,13]. Harmonizing records and automation are a key ingredient to prioritize OM procedures to reduce lifecycle costs [14]. Transitioning to predictive maintenance from a reactive approach will further reduce costs [15]. Regarding O&M records and extreme weather, snow events and hurricanes have been shown to be two major contributors [16].

2 DATA SET CURATION

In this analysis, the authors examined both PV and extreme weather time series data, as well as storm-related operations & maintenance (O&M) logs, to gain insights. For the time series analysis, the Storm Events Database, which is publicly available from NOAA, was leveraged to identify extreme weather events in the United States. The Storm Events Database contains reported extreme weather events across the US, including hurricanes, floods, hail and windstorms, and tornadoes, among others. Data in the storm events database includes the start and end date of the event, event type, starting and ending latitude-longitude coordinates, as well as event severity, when applicable. It is important to note that multiple storm event types can occur simultaneously in the same geographic area; for example, a location may experience lightning and a windstorm simultaneously.

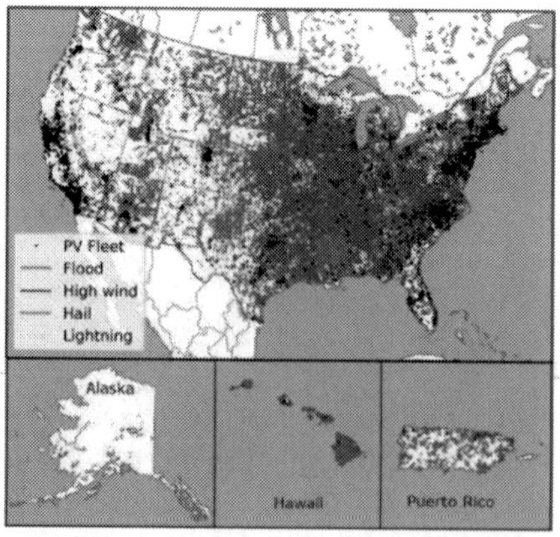

Figure 1: Map of the USA including the territory of Puerto Rico with extreme weather events (flood, high wind, hail and lightning) and the PV systems from PV Fleet.

Figure 1 shows a map of extreme weather events, including flood, high wind, hail, and lightning events, and PV Fleets systems. To link these PV systems to specific weather events, storms occurring within 10 km of each system's latitude-longitude coordinates during a period where PV AC power or AC energy time series data was available were identified. Once this subset of linked weather-system events was identified, each individual linked storm event was further compared against the associated system's time series data to determine system impacts (i.e., outages). System outages occurring within one calendar day of an associated event were recorded, and length of the outage was calculated. Overall, a total of 199 storm events, associated with 170 distinct crystalline silicon systems, led to PV system outages over a geographical diverse region commensurate with the PV Fleet distribution.

3 SHORT-TERM IMPACT—OUTAGES

Following the data set compilation, additional variables associated with each outage were determined, including classifying if the outage was full (all power streams affected) or partial (not all power streams affected), and calculating the length of the system outage in hours. Full and partial outages immediately impact the power production following the weather event and are summarized in the histograms of Fig. 2 with additional details provided in Table I. We calculated the percentage of production loss in two ways. First, we calculated the energy loss by the number of days the system was offline as a percentage of the calendar years. This obviously assumes equal production throughout the year. Secondly, we obtained annual and monthly production for each affected system through PVWatts simulation [17]. Then we estimated the loss in the occurring months proportional to the outage length. The overall median loss between the two methods agreed within 0.05 % and for individual storm categories within 0.1 % providing a minimum estimate of the uncertainty. At the median, all of the weather events have an impact of around 1 % annual loss or between 2 and 4 days, which is relatively small, as shown in Table I. However, a few systems are much more severely impacted, as can be seen in the tail of Fig. 2 with ca. 0.2 % losing more than a month of. This tail is especially pronounced for floods and high wind events.

Table I: Outage summary table

Event	Systems impacted (%)	Lost production, median (%)	Lost days, median
Flood & rain	3.0	1.1	4.2
Hail	1.6	0.8	3.1
High wind	3.2	0.7	2.7
Lightning	1.4	1.0	3.6
Total	2.7	0.9	3.5

Therefore, from a fleet perspective the impact by these extreme events is relatively minor, yet the risk is exemplified in the long tail of the distributions. Because we did not have full operations and maintenance (O&M) tickets for these specific systems, it is not clear if the loss was caused by damage to the system and possible associated safety aspects or merely a communications issue or perhaps due to wider grid outages.

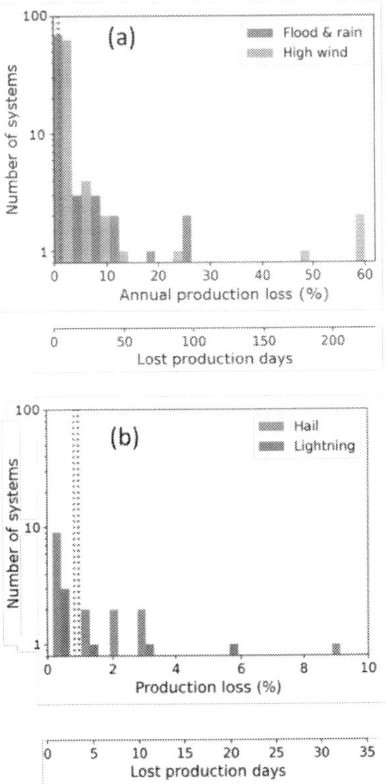

Figure 2: Histograms of the production impact caused by extreme flooding and high wind events (a) and hail and lightning (b). The secondary horizontal axis displays the lost production days and the medians by vertical dashed lines.

4 LONG-TERM IMPACT—DEGRADATION

Beyond the immediate impacts of discrete weather events, such as system outages as defined above, long-term impacts in the form of performance loss rate (PLR) could be affecting systems for many years following an extreme weather event. Hail impacts or mechanical loads—dynamic or static—from wind or snow can lead to cell cracking within modules or may lead to other degradation modes [18]. Although how cell cracking leads to power loss is still being actively researched [19,20]. To determine PLRs, we use RdTools, an open-source software package, co-developed by NREL that uses the year-on-year methodology, in which the normalized performance is compared to the same day year-over-year assuming linearity[21]. Because this method requires at least two years of data, we investigated all impacted systems that had at least 2 years of data prior and following a weather event. For irradiance data we used satellite data from the National Solar Radiation Database (NSRDB) to avoid potential confounding effects because of irradiance sensor drift [17]. To reduce noise in the RdTools analysis, we further filtered for clear sky conditions and used a 7-day moving Hampel outlier filter, as detailed in [22]. These added steps allowed us to detect smaller changes in PLR when systems were exposed to hail wind, or extreme snow events. A summary of PLRs on systems impacted by hail of various sizes from peanut to walnut size and larger regardless of outage detection following the storm is presented in Fig. 3. Boxplots of the distributions in each

10.4229/EUPVSEC2023/4AO.7.5

category are shown using blue for PLRs before the storm (pre) and orange after the storm (post). Individual data points, each representing an inverter, are overlaid on the boxplots with a representative uncertainty bar displayed by gray vertical bars. The representative uncertainty bar is calculated as the median of all uncertainty bars in each category with additional details provided in summary Table II.

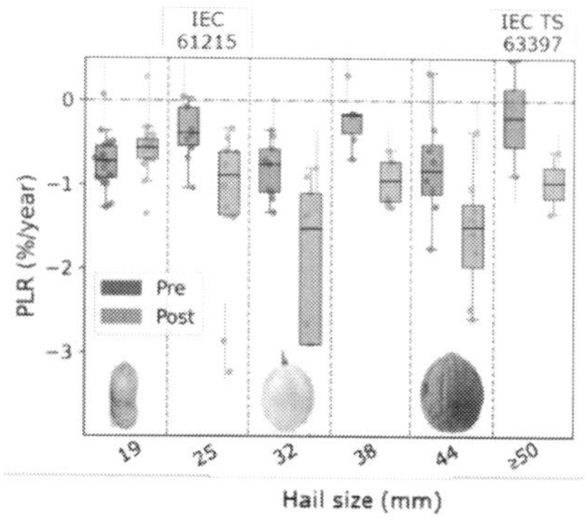

Figure 3: RdTools analysis of performance loss rates pre- (blue) and post-storm (orange) for hail events on all systems. The extent of the boxplot represents the interquartile range with the median displayed by a horizontal line within each box. For size comparison, a peanut, grape and walnut are shown.

Table II: Summary table for hail impacted systems.

Hail size (mm)	Median pre-storm (%/yr)	Median post-storm (%/year)	Interquartile range (%/yr)
19	-0.72	-0.56	0.40
25	-0.38	-0.89	0.54
32	-0.76	-1.52	0.62
38	-0.17	-0.94	0.47
44	-0.83	-1.49	0.80
≥50	-0.20	-0.97	1.37

For hail approximately the size of peanuts, no statistically significant higher PLRs after hail exposure was detected. On the contrary, despite PLR uncertainties and without exception larger hail categories exhibited higher PLRs after hail impact. However, it appears that larger hail sizes may not necessarily lead to higher PLRs, rather it appears that higher PLRs will occur if a minimum damage threshold is exceeded. Of particular note is the 25 mm diameter size hail category because the module qualification test IEC 61215 uses 25 mm diameter sized hail [2]. Yet, systems exposed to 25 mm hail under field conditions displayed higher PLR despite being qualified to this specific hail size. Several possibilities exist that could explain the discrepancy. In naturally occurring hailstorms multiple strikes may impinge on a module below the maximum determined hail size, yet potentially deliver more kinetic energy to a module [23]. In addition, actual hail is often not perfectly round and may not resemble indoor testing. Differences in mounting may also exist between indoor testing and actual field installation. Regardless of the underlying cause, the need for more stringent hail testing is evident from these data and is currently under development and published as technical specification (TS) IEC TS 63397 in 2022 [24].

5 OPERATIONS & MAINTENANCE

The storm-related O&M data set described in the Methodology section above was analyzed to determine the primary failure mechanism/issue associated with each different storm type. The results of this analysis are provided in this section. Hurricanes and other high wind events were grouped together and further categorized for fixed tilt and tracked systems, as shown in Fig. 4. The most dominant issues regardless of mounting type were ground faults and utility recloser trips. A difference can be seen in the subsequent category; for fixed tilt systems module and inverter damage are next while for tracked systems the tracker itself is a major cause for maintenance.

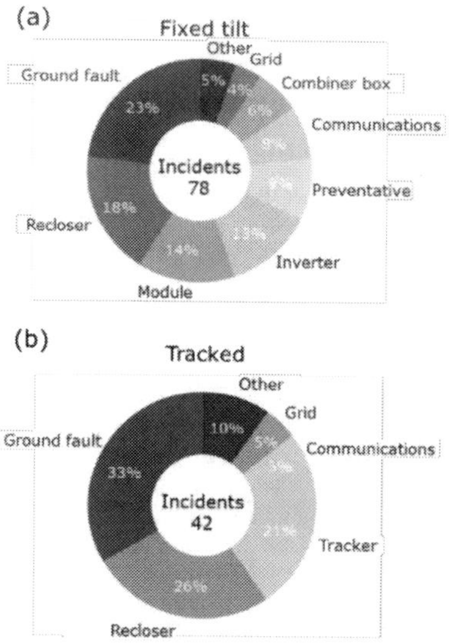

Figure 4: Itemization of O&M tickets for fixed tilt (a) and tracked systems (b) impacted by high wind events.

Besides the mounting configuration, installation practices and quality can have a substantial influence on the outcome when PV systems are exposed to extreme weather events. The outcome of a system affected by wind gusts of around 100 km/hour gusts is shown in Fig. 5. Modules were damaged because of installation practices not suited for regions susceptible to high winds. In high wind areas modules should be mounted using through-bolting and sturdy brackets because double clamps often get bent and pose the threat that modules may be destroyed and lead to safety issues [25] Several O&M best practices guidelines exist and even guidelines in preparation for extreme weather that help with installation best practices [26,27]. Thes guidelines may not eliminate entirely the risk of damage to a PV plant, but it can increase the survival probability.

Figure 5: Example of damage caused by wind gust. Double module clamp loosened (a) and bent mounting bracket (b) because of insufficient bracketing strength.

6 CONCLUSION

Long lifetime for PV systems is a win-win for all stakeholders, however with increased lifetime an enhanced probability exists to be exposed to extreme weather events during that time. To investigate how these events impact PV performance from a fleet perspective, we compared two large-scale databases for PV time series and extreme weather to get initial, quantifiable insights into the effects of extreme weather events of PV systems. We determined that median short-term outages led to production losses of only approximately 1 % of annual production per event. Yet, a few systems were more substantially impacted. In addition, long-term consequences expressed in higher PLRs that may go undetected can lead to increased financial losses. For hail we found hail size of 25 mm or above led to increased performance losses providing a reason for more stringent hail testing currently already under development. System modules qualified through IEC 61215 for 25 mm hail showed higher PLRs when exposed to that same size hail in natural settings. Furthermore, increased tracker maintenance can be expected when tracked systems are exposed to high wind events.

We demonstrated that from module qualification to installation best practices all play important roles in the resilience of PV systems. Although a serendipitous element with extreme weather may not be eliminated, adhering to strict quality guidelines from module manufacturing to installation and O&M practices may increase the probability to survive these extraordinary events.

7 ACKNOWLEDGMENT

We would like to thank our corporate partners who provided the PV system data in this document. We would also like to thank the NREL PV reliability group, and Katherine Jordan. This work was authored in part by Alliance for Sustainable Energy, LLC, the manager and operator of the National Renewable Energy Laboratory for the U.S. Department of Energy (DOE) under Contract No. DE-AC36-08GO28308. Funding provided by the U.S. Department of Energy's Office of Energy Efficiency and Renewable Energy (EERE) under Solar Energy Technologies Office (SETO) Agreement Number 30295. The views expressed in the article do not necessarily represent the views of the DOE or the U.S. Government. The U.S. Government retains and the publisher, by accepting the article for publication, acknowledges that the U.S. Government retains a nonexclusive, paid-up, irrevocable, worldwide license to publish or reproduce the published form of this work, or allow others to do so, for U.S. Government purposes.

8 REFERENCES

[1] J.H. Wohlgemuth, Photovoltaic Module Reliability, John Wiley & Sons, NJ, USA, 2020.

[2] International Electrotechnical Commission 61215-1 2021 Terrestrial photovoltaic (PV) modules—design qualification and type approval.

[3] R.G. Ross, Jr., M.1. Smokier, ,JPL Publication 86-31, 1986.

[4] M. Aghaei, A. Fairbrother, A. Gok, S. Ahmad, S. Kazim, K. Lobato, G. Oreski, A. Reinders, J. Schmitz, M. Theelen, P. Yilmaz, J. Kettle, Renewable and Sustainable Energy Reviews, 159, 112160, (2022).

[5] W. Muehleisen, G.C. Eder, Y. Voronko, M. Spielberger, H. Sonnleitner, K. Knoebl, R. Ebner, G. Ujvari, C. Hirschl, Renewable Energy 118, 138—145, (2018).

[6] S. Johnston, D.B. Sulas-Kern, D. Jordan, 48th IEEE Photovoltaic Specialists Conference, virtual, 2021.

[7] https://www.pv-magazine.com/2021/03/19/storm-season-has-the-us-solar-industry-looking-to-protect-assets-from-costly-hail-damage/

[8] S. Chakraborty, A. Kumar Haldkar, N.M. Kumar, Renewable Energy, 203, 345–356, (2023).

[9] NOAA National Centers for Environmental Information (NCEI) U.S. Billion-Dollar Weather and Climate Disasters (2023).

[10] C. Deline, K. Anderson, D. Jordan, A. Walker, J. Desai, K. Perry, M. Muller, B. Marion, R. White, in 48th IEEE Photovoltaic Specialist Conference, virtual (2021).

[11] D.C. Jordan, K. Anderson, K. Perry, M. Muller, M. Deceglie, R. White, C. Deline, Progress in Photovoltaics, 30 (10), 1166 (2022).

[12] A. Golnas A, IEEE Journal of Photovoltaics, 3(1), 416-421, (2013).

[13] D.C. Jordan, B. Marion, C. Deline, T. Barnes, Mark Bolinger, Progress in Photovoltaics, 28(8), (2020)

[14] S. Lindig, Sandra Gallmetzer, M. Herz, A. Louwen, E. Koumpli, P.S.E. Paez, D. Moser, Progress in Photovoltaics:1–20, 2022.

[15] G. Guerra, P. Mercade, L. Landberg, Predictive maintenance of solar PV plants: The time is now, DNV, 2021.

[16] N.D. Jackson, T. Gunda, Applied Energy, 302, 117508 (2021).

[17] M. Sengupta, Y. Xie, A. Lopez, A. Habte, G. Maclaurin, J. Shelby, Renewable and Sustainable Energy Reviews 89, 51 (2018).

[18] M. Köntges, S. Kurtz, C. Packard, U. Jahn, K.A. Berger, K. Kato, T. Friesen, H. Liu, M. Van Iseghem, International Energy Agency, Report IEA-PVPS, T13-01:2014, 2014.

[19] C. Buerhop, S. Wirsching, A. Bemm, T. Pickel, P. Hohmann, M. Nieß, C. Vodermayer, A. Huber, B. Glück, J. Mergheim, C. Camus, J. Hauch, C. J. Brabec, Progress in Photovoltaics, 26, 261 (2018).

[20] T.J Silverman, N. Bosco, M. Owen-Bellini, C. Libby, M.G. Deceglie, IEEE Journal of Photovoltaics 12 (4), 1090 (2022).

[21] M. Deceglie, A. Nag, C. Deline, A. Shinn, G. Kimball, D. Ruth, D. Jordan, J. Yan, K. Anderson, K. Perry, M. Mikofski, M. Muller, W. Vining, RdTools v2.1.0 beta.1. Computer Software. https://github.com/NREL/rdtools. DOI: 10.5281/zenodo.4307010.

[22] D.C. Jordan, C. Hansen, Renewable Energy, 209, 393 (2023).

[23] P. Bostock, PV Reliability Workshop, Golden, CO, USA, 2023.

[24] IEC TS 63397:2022, "Photovoltaic (PV) modules – Qualifying guidelines for increased hail resistance", 2022.

[25] C. Burgess., S. Detweiler, C. Needham, F. Oudheusden, Clinton Foundation, FCX Solar & Rocky Mountain Institute (2020).

[26] National Renewable Energy Laboratory, NREL/TP-7A40-73822 (2018).

[27] E. Hotchkiss et al., Preparing Solar Photovoltaic, Systems Against Storms, NREL/FS-5R00-81968 (2022).

SOLARGIS

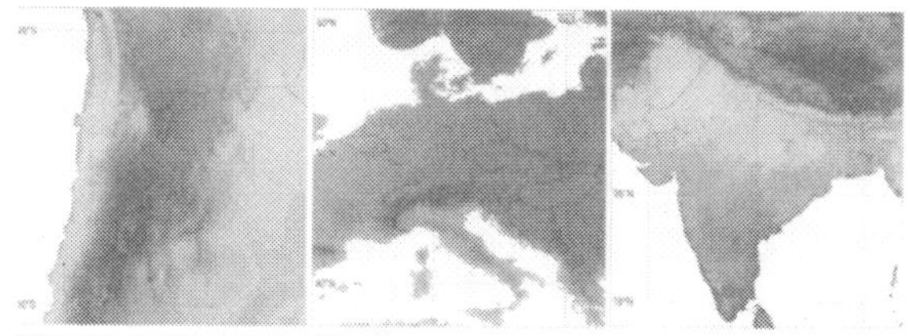

Global High-Resolution Map of the Lowest Expected Operating Temperature for Optimum Sizing of PV Arrays

Artur Skoczek[1], Juraj Betak[1], Jose Antino Ruiz-Arias[1,2], Jan Veres[1], Tomas Cebecauer[1]
[1] Solargis, Slovakia
[2] University of Malaga, Malaga, Spain

EU PVSEC **18 — 22 September** 2023

Definition based on IEC 62738:
Ground-mounted photovoltaic power plants - Design guidelines and recommendations

The maximum PV string open circuit voltage is calculated from the module STC V_{oc} rating corrected for lowest expected operating temperature, per **IEC 62548** Photovoltaic arrays - Design requirements.

The **lowest expected operating temperature shall be the mean of annual extreme low temperatures**; use of the record low temperature shall not be required.

If sufficient data is available the mean data can be limited to sunlight hours, using a very low threshold for irradiance, i.e. 10 W/m² or less.

PV array maxumum voltage based on IEC 62548

Table 4 – Voltage correction factors for crystalline and multi-crystalline silicon PV modules

Lowest expected operating temperature °C	Correction factor
24 to 20	1,02
19 to 15	1,04
14 to 10	1,06
9 to 5	1,08
4 to 0	1,10
–1 to –5	1,12
–6 to –10	1,14
–11 to –15	1,16
–16 to –20	1,18
–21 to –25	1,20
–26 to –30	1,21
–31 to –35	1,23
–36 to –40	1,25

40th European Photovoltaic Solar Energy Conference and Exhibition

Lowest expected operating temperature (TLEO), calculation methodology

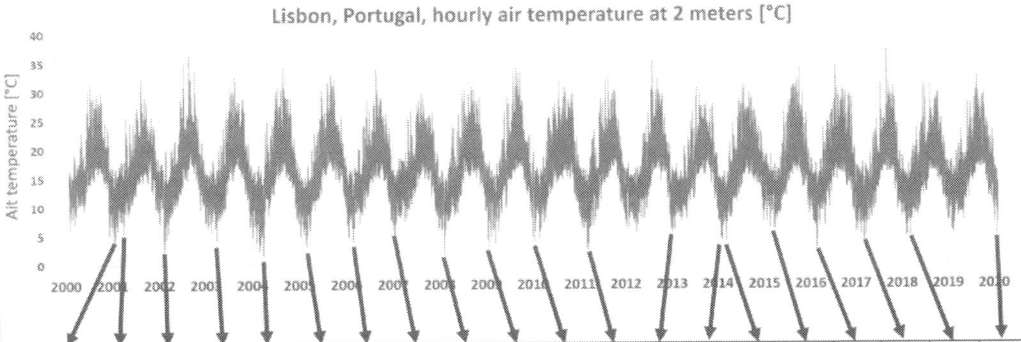

2001	2002	2003	2004	2005	2006	2007	2008	2009	2010	2011	2012	2013	2014	2015	2016	2017	2018	2019	2020
4.1	5.5	2.3	4.2	1.6	3.2	4.0	6.1	1.8	3.7	4.6	2.6	5.5	4.4	4.0	6.1	2.5	4.2	4.7	5.1

Lowest expected operating temperature (TLEO) = 4.0°C

020275-003

40th European Photovoltaic Solar Energy Conference and Exhibition

Lowest expected operating temperature, the source data: ERA5

Climate reanalysis combine past observations with models to generate consistent time series of multiple climate variables. Reanalyses are among the most-used datasets in the geophysical sciences. They provide a comprehensive description of the observed climate as it has evolved during recent decades, on 3D grids at sub-daily intervals.

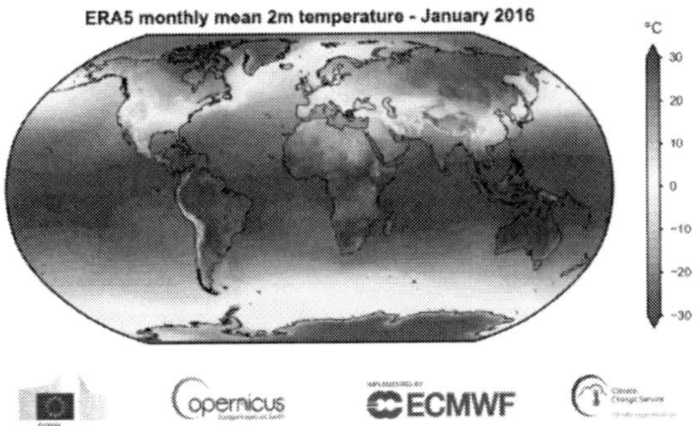

Modern Era Reanalysis v5 (ERA5) is the latest climate reanalysis produced by **ECMWF**, providing hourly data on many atmospheric, land-surface and sea-state parameters together with estimates of uncertainty.
ERA5 data are available at 0.25° x 0.25° resolution, with atmospheric parameters on 37 pressure levels.
ERA5 is available from 1940 and continues to be extended forward in time, with daily updates being made available 5-7 days behind real time

020275-004

Enhancing the resolution of meteorological model using digital elevation model and lapse rate

Dry Adiabatic Lapse Rate
As long as an air parcel is unsaturated (RH< 100%), the rate at which its temperature will change will be constant. A decrease in temperature with height is called a lapse rate. This drop in temperature is due to adiabatic expansion and a decrease in internal energy.

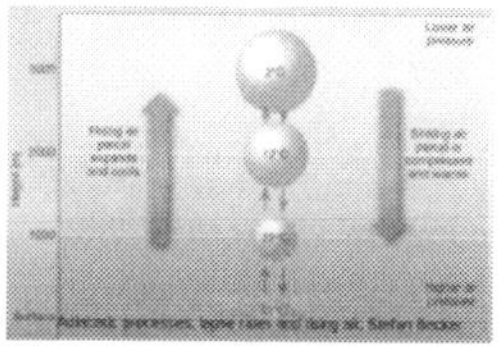

Dynamic Lapse Rate, derived from the model vertical pressure levels (hourly data)

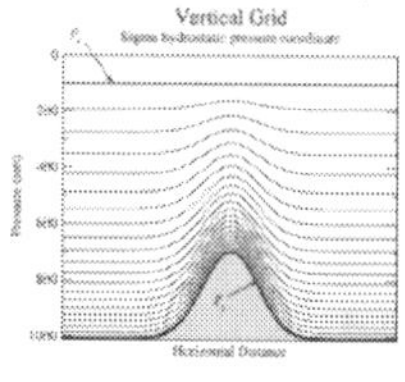

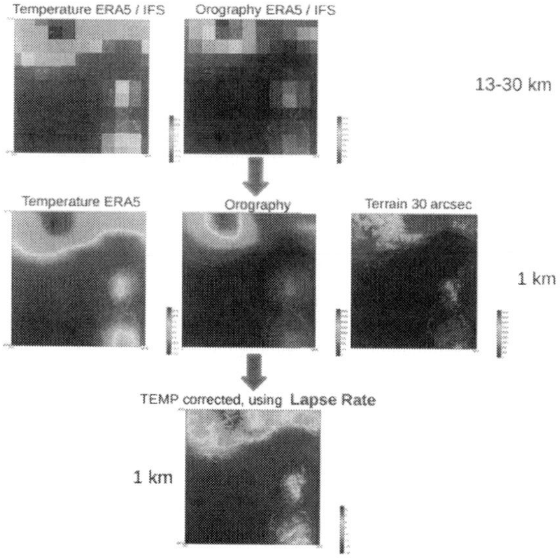

40th European Photovoltaic Solar Energy Conference and Exhibition

Enhancing the resolution of meteorological model using digital elevation model and lapse rate

Terrain 30 arcsec

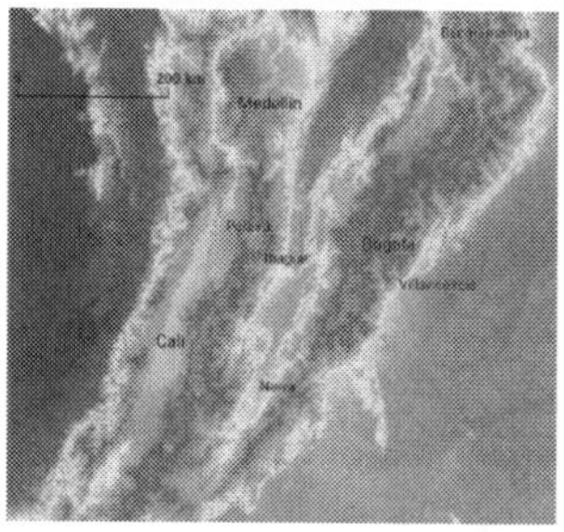

Orography ERA5 0.25°

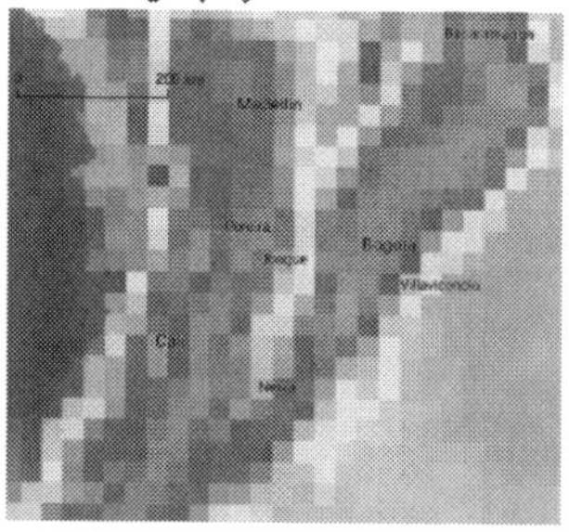

Orography diff (+/- 1000m)

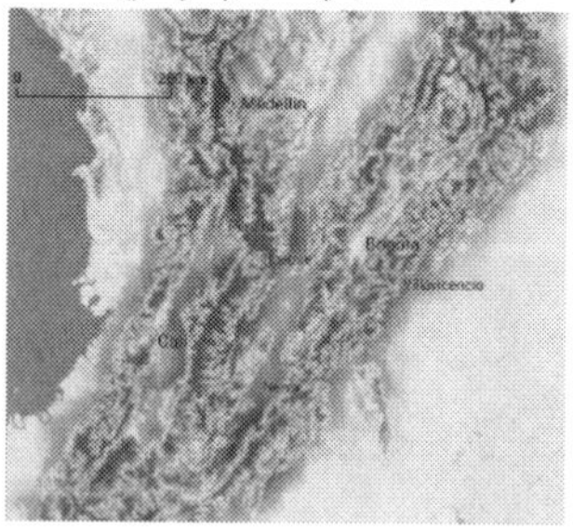

$$T(z) = T_{model} + \Delta T(L_{p1}, L_{p2}; \Delta z)$$

$$\Delta T(L_{p1}, L_{p2}; \Delta z) = sign(\Delta z) \times \begin{cases} L_{p1}|\Delta z|, if\ |\Delta z| < 500m \\ L_{p1} \times 500 + L_{p2}(|\Delta z - 500|), if\ |\Delta z| \geq 500m \end{cases}$$

Where:
T – air temerature
L_{p1} and L_{p2} are the lapse rates below and above 500 m, respectively.
$\Delta z = z - z_{model}$ (z and z_{model} are the elevations of digital terrain model 30 arcsec and the weather model, respectively, and given in meters)

Validation of the model TLEO with NOAA Integrated Surface Database (ISD)

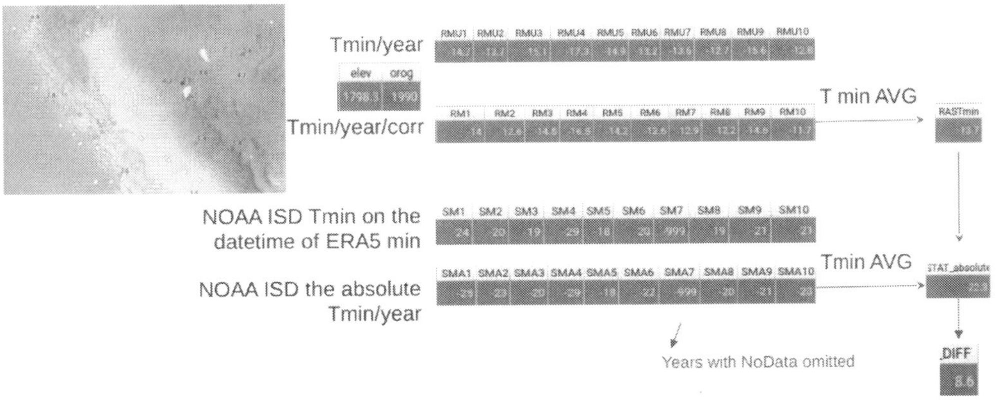

The Integrated Surface Database (ISD) is a global database maintained by US NOAA that consists of hourly and synoptic surface observations compiled from numerous sources into a single common ASCII format and common data model.
ISD includes numerous parameters such as wind speed and direction, wind gust, temperature, dew point, cloud data, sea level pressure, altimeter setting, station pressure, present weather, visibility, precipitation, snow depth, and various other elements as observed by each station.

The validation results

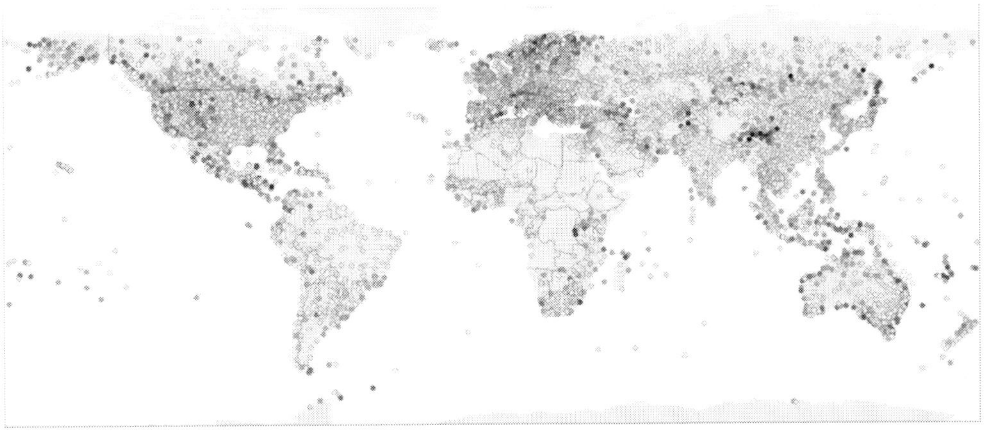

Multiple station's filtering criteria: 8507 meteorological stations from NOAA ISD database.
Validation years: 2001 to 2020

The overall bias, calculated as the difference between the modeled and measured TLEO is 1.2°C, standard deviation 2.4°C. The most extreme values (P5 and P95) drop below -2.6°C and exceed 5.1°C, respectively.

The validation results - examples

Italy, alpine region | **Australia/Tasmania coastal** | **Norway, mountainous region**

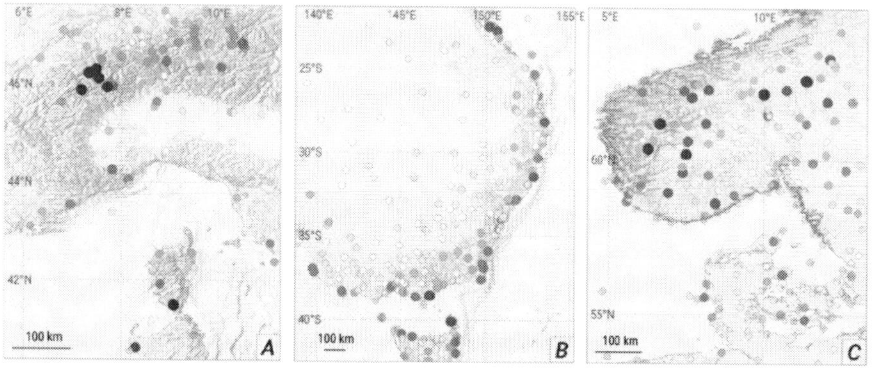

Difference [°C] ● <-9 ● -9--7 ● -7--5 ● -5--3 ● -3--1 ● -1-1 ● 1-3 ● 3-5 ● 5-7 ● 7-9 ● >9

Selected areas with significant differences between TLEO calculated by Solargis and TLEO derived from NOAA ISD meteorological stations

The validation results - examples

Spatial validity of stations
Simplified Cost Distance Analysis Based on Terrain Complexity (Voronoi Polygons)

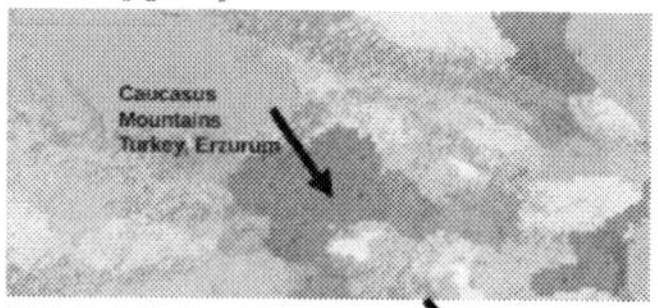

Common Problems: ERA5 too large pixel for such orography.

- Meteo-sites mostly in the valleys; ERA5 orography height is typically much higher, therefore the values colder.
- Even if the lapse-rate correction works in a right direction, mostly it's not sufficient.
- Temperature inversion in "cold" valleys

examples: Central Alps, Tibet, East (and also West) Himalayas, Iran West, Turkey SE, and many others

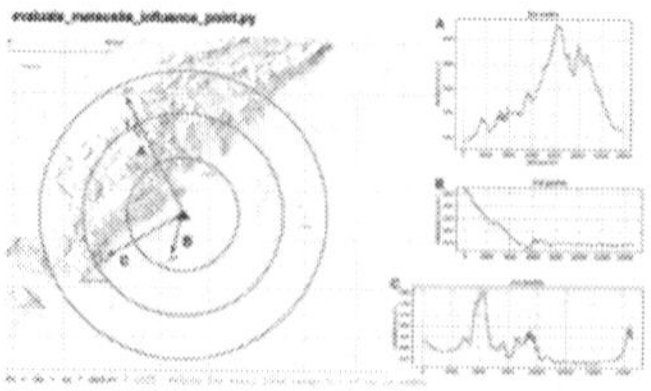

The validation results - examples

Common Problems

Underestimated continentality influence, mainly on the coasts (ERA5 overestimates sea/ocean influence) / cold air falling from the mountains toward the water bodies.

- Affects many mountainous islands
- Continental effect

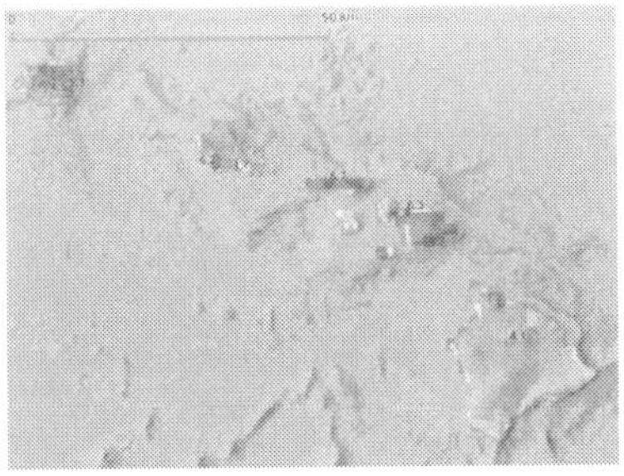

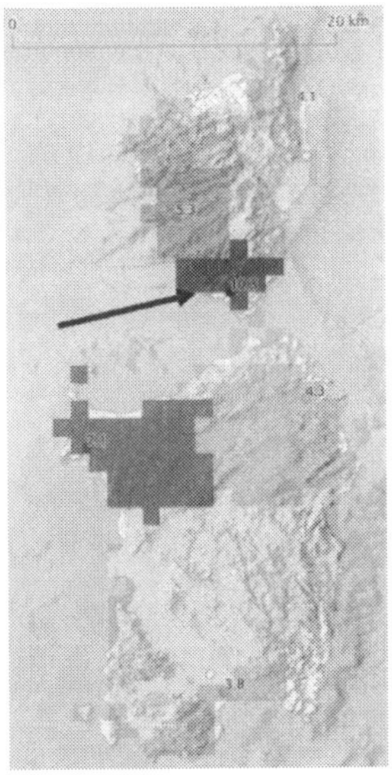

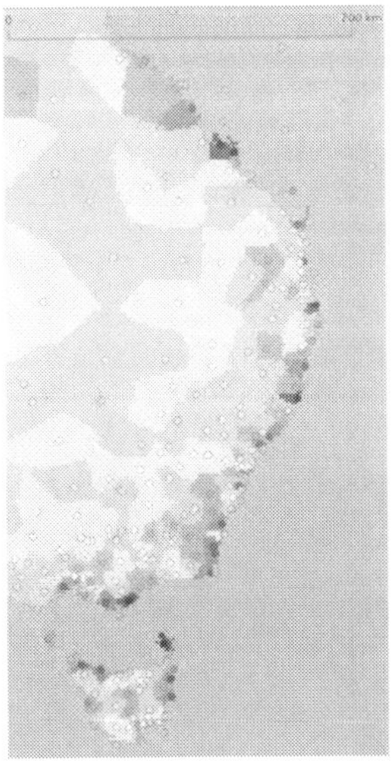

Attempts to estimate the uncertainty of TLEO map

Locally, microclimate effects may cause significant differences when comparing measured to modeled data.
The classification is based on statistical evaluation of several factors (stations density, stations distance, variability of differences) with the two most influential ones:

- terrain dissection
- the proximity to large water bodies.

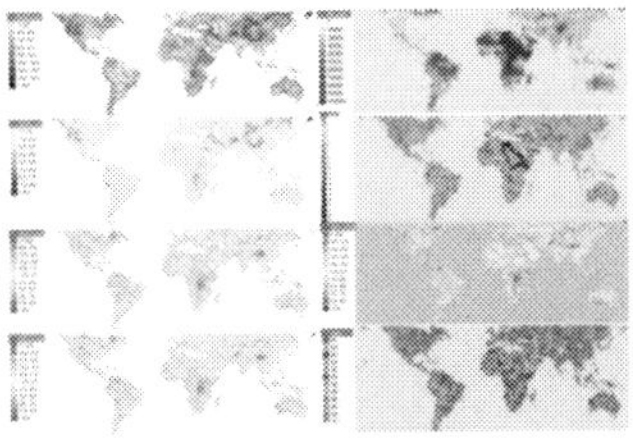

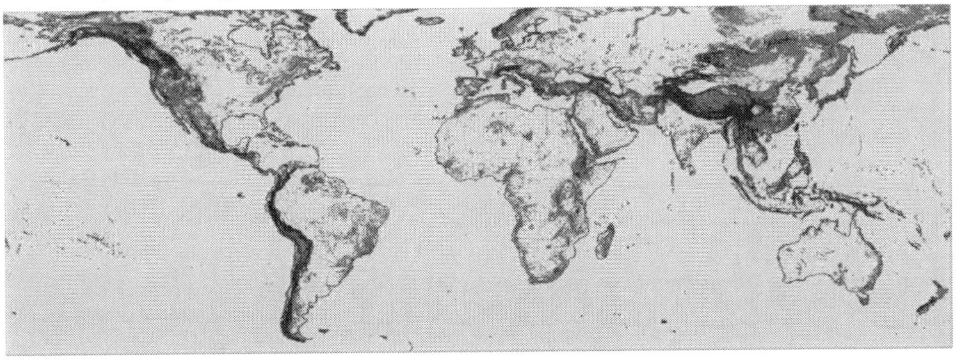

Probability of regional microclimate occurence Low Moderate High

40th European Photovoltaic Solar Energy Conference and Exhibition

Raster data calculation simplification

In order to speed up the calculations of global high resolution TLEO map a simplification in the elevation correction was introduced: monthly average lapse rate profile instead of a full time series of lapse rate.

The impact of this simplification was found to be significantly smaller when compared against the distribution of standard deviations of annual TLEO and the distribution of the absolute difference between record high and record low annual TLEO.

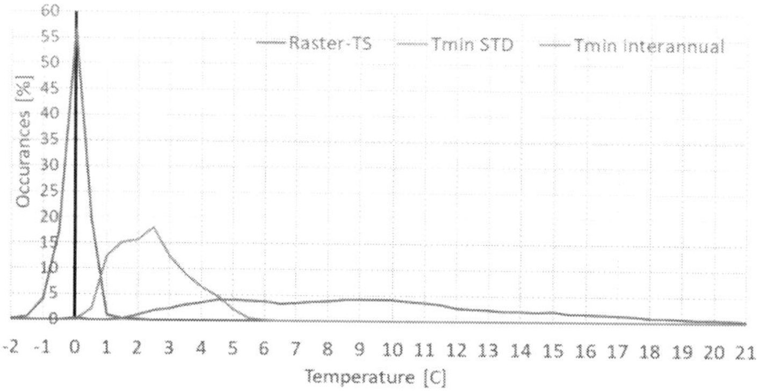

020275-013

40th European Photovoltaic Solar Energy Conference and Exhibition

Absolute TLEO against TLEO as defined in IEC 62738

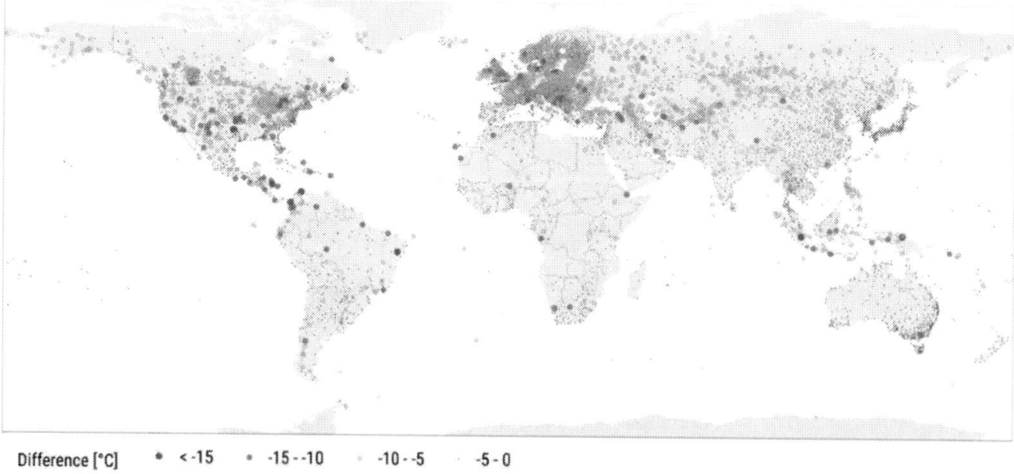

The difference between the absolute TLEO (single year) and the average TLEO (IEC 62738) calculated for a period 2001-2020, derived from NOAA ISD database. In many cases the values exceed 5°C, occasionally 10°C.

020275-014

Global map of TLEO

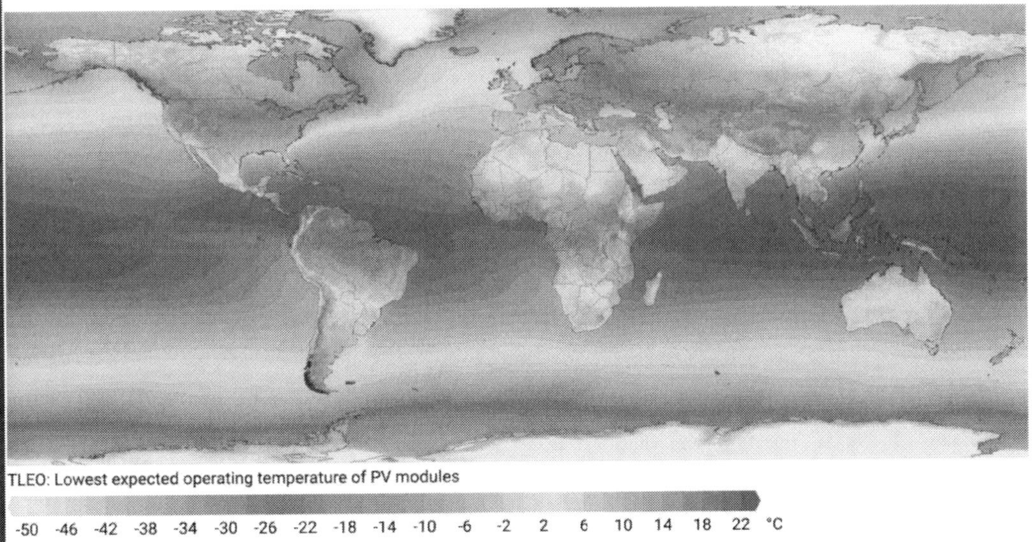

020275-015

40th European Photovoltaic Solar Energy Conference and Exhibition

Focus on the regions with high TLEO gradients:
Pacific coast in South America (left), Central Europe (middle) and South Asia (right)

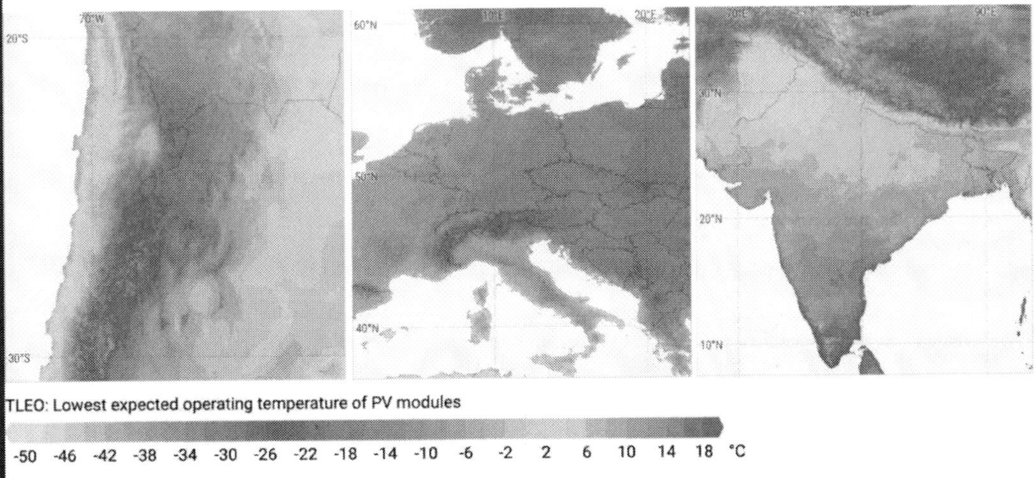

020275-016

This presentation was selected by the Sc. Committee of the EU PVSEC 2023 for submission of a full paper to one of the EU PVSEC's collaborating peer-reviewed journals.

MODELING THE MOISTURE-INDUCED DEGRADATION IN PV MOUDLES WITH SILICON HETEROJUNCTION CELLS

Suraj Ravindrababu[1], Paul Gebhardt[1], Daniel Philipp[1], Ingrid Hädrich[1]
[1]Fraunhofer Institute for Solar Energy Systems
Heidenhofstrasse 2, 79110 Freiburg

ABSTRACT: Moisture ingress into photovoltaic (PV) modules initiate and propagate several degradation modes, affecting the maximum power. This paper proposes a model to predict the degradation of power exemplarily for one type of encapsulated heterojunction solar cells by correlating the instantaneous power as a function of water content near the vicinity of the cell. Through this the temporally varying stressors i.e., temperature and relative humidity is modeled implicitly. The work combines experimental results, conducted on single/half-cell mini-modules, and a computational model of the same to propose an empirical model which predicts the power as a function of the water content. The proposed model is fit to the data with a R^2-value of 0.9928 and the empirical parameters are determined for the cell. The capability of extending the model to predict the performance of large modules under the influence of moisture is illustrated.
Keywords: degradation model, moisture ingress, heterojunction solar cells, finite element method

1 INTRODUCTION

Moisture is one of the critical stressors which affects the reliability of PV modules [1]. Modeling the ingress of moisture is critical but there are several challenges to this. The diffusion properties and their dependency on the environmental stresses (temperature and relative humidity) are unknown or not quantified for all BOMs [2]. Diffusion properties and their relation to varying stressors is imperative for modeling the ingress of moisture in modules installed outdoor. Majority of the degradation models are modelled with data from a handful of specific locations, affecting their applicability to versatile climatic conditions [3]. If the degradation model must be valid for various combinations of the stressors, then they must be trained with data obtained from a complex experimental design which might not be feasible in a realistic time frame.

With the case of silicon heterojunction (HJT) modules, the assessment of reliability is further challenging. The unavailability of long-term field performance data and degradation models specific the HJT modules prove to be the major challenge [4]. Although the modeling of moisture ingress in PV modules has been simulated before using several methods, including the finite element method (FEM) [5–7], the effect of the collected moisture inside the module on the overall electrical performance needs to be described effectively as a mathematical model. This model should be capable of adapting to different degradation trend shapes (response by different cell types to temperature and humidity) and to climate data at any given location. An empirical model is proposed in this article to address this which enables the prediction of a cell's power when the water content near the cell is known. An outlook on how the proposed model could be used to predict field performance is mentioned briefly. Here, the experimental and numerical analysis is conducted on encapsulated HJT cells, but the approach can be adopted to study and model degradation in any moisture sensitive cell technology.

The paper is divided into the following sections. Section 2 mentions the approach adopted in the paper and the assumptions in the approach. Section 3 shows the materials, module design, experiments and measurements performed. Section 4 shows the construction of the FEA model in COMSOL. Section 5 illustrates the empirical model in detail followed by a note on future work and a summary, listed in Section 6.

2 APPROACH

The approach adopted in this article is illustrated in Figure 1. Experiments are conducted on single or half-cell modules to obtain power measurements. Simultaneously, a computational model of the single/half-cell model is created in a FEA environment. The boundary conditions are defined according to the environment the modules are subjected to during the experiments. The computational model provides us with the water content in the vicinity of the cell. The accumulation of water in modules, especially above and below the cells, for outdoor and IEC conditions has been recorded in [8]. The effect on cells due to this accumulated moisture needs to be quantified and a mathematical model has to be defined to easily predict the degradation in larger modules. To this end, an empirical model equating the instantaneous maximum power (P_{mpp}) as a function of the water content (W) in the volume above and below the cell marked by the cell surface area is proposed. Using the experimental and numerical results of the single/half-cell modules, the parameters of the empirical model (α, β) is obtained using the least squares method [9]. The extension of this model to predict the performance of larger modules will be done in the future.

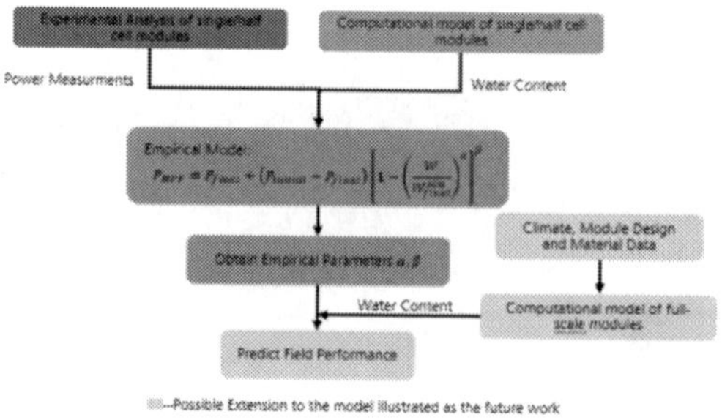

Figure 1: Workflow for the formulation of the empirical model

10.4229/EUPVSEC2023/4AO.8.2
020276-001

Please note, it is assumed that temperature does not directly degrade the cell, but it only assists in the ingress of moisture and catalyzes the degradation process that is initiated in the presence of moisture. However, the proposed model could be extended to fit the effect of temperature on degradation by repeating the single/half-cell experiments at different temperatures.

3 EXPERIMENTAL INVESTIGATION

Experimental analyses were done on single/half-cell modules with two different wafer sizes. Three single cell modules with M2 HJT wafers are encapsulated in EVA and sandwiched between two glass plates (size 20 x 20 x 0.3 cm³). This set of modules are named 'Set-M2'. Four half-cell modules with M6 HJT wafers are encapsulated using POE and are sandwiched between glass plates of same size as seen in Set-M2. Two of the modules in Set-M6 were embedded with moisture strips to visually observe the ingress of moisture. The modules with M6 wafers will be referred as 'Set-M6'. All modules were exposed to a damp-heat chamber with constant 85°C and 85% rH . Electroluminiscence (EL), visual inspection and power measurements (IV) were performed every 1000h for Set-M2 and every 500h for Set-M6. The results are displayed in Figure 2 and Figure 3.

From the EL images, it is seen that Set-M6 shows little to no degradation while Set-M2 shows high degradation. From the power measurements, the cell degrades by 20% after 1000 h of damp heat exposure. For illustrating the empirical degradation model, the data from Set-M2 is taken as it shows considerable degradation.

Set-M2 shows extreme degradation arising from the busbars near the edges of the cell and the degradation propagates towards the centre of the cell. In Set-M6, after 1000h, the EL images show random dark spots which seem to evolve as the ageing time increases. A deep discussion on the various degradation modes is done in [10]. Since we focus more on formulating an approach for degradation modeling, discussion on the degradation mechanisms and analysis on how they initiate and propagate are not discussed here.

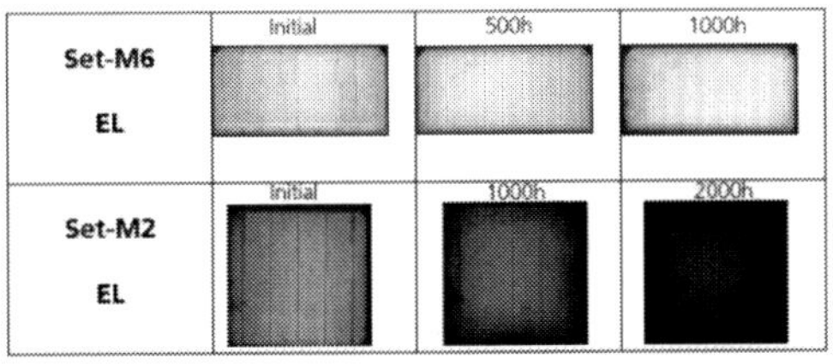

Figure 2: EL Images on Set-M2 and Set-M6 Modules.

4 COMPUTATIONAL MODELING

The computational model of Set-M2 is done in COMSOL Multiphysics 6.0. Owing to the symmetry in the module design of Set-M2, a quarter-model of the module is modeled in COMSOL to improve computational time. The developed model is shown in Figure 4.

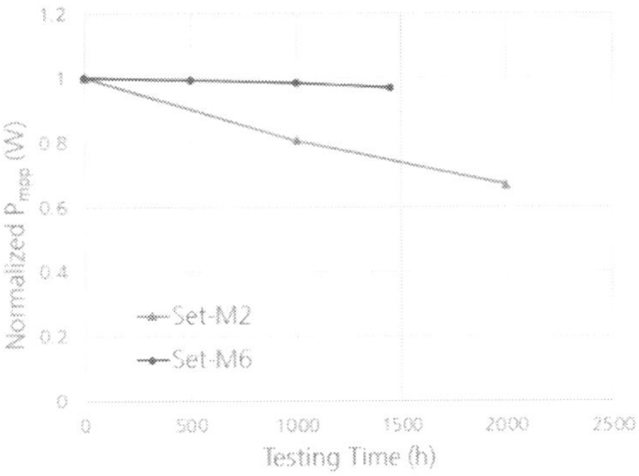

Figure 3: Power measurements as a function of damp heat exposure time for Set-M2 and Set-M6

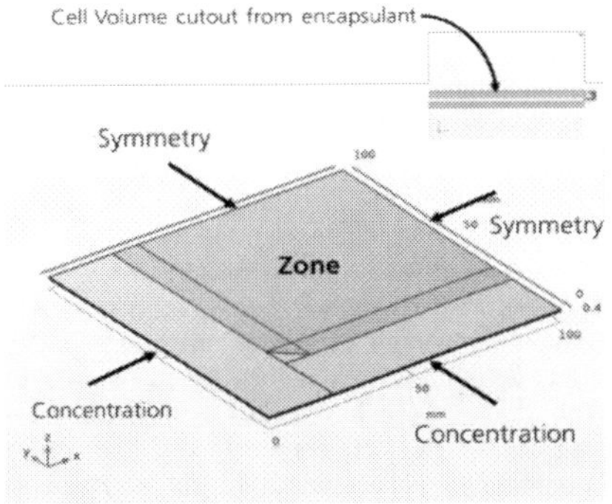

Figure 4: FEA quarter model of Set-M2

The volume of the cell (thickness 200um) is treated as a 'empty area' in the model. Therefore, the remaining continuum can be regarded as encapsulant, and the corresponding diffusion properties are given to this region. Diffusion coefficient (*D*) and the saturation concentration (*C*) of EVA at 85°C and 85% rH is calculated from [11–13]. To compare the moisture ingress in EVA and POE, a half-model of Set-M6 is also constructed with a similar approach and simulated in COMSOL. Since the POE encapsulant used here and in [14] are similar, the diffusion properties for POE at 85°C and 85% rH are calculated from the cited article. The values used in the simulation are shown in Table 1.

Table I: Diffusion properties of EVA and POE

Material	Diffusion Coefficient, D (m²/s)	Concentration, C (mol/m³)
EVA	8e-10	177.63
POE	1.854e-10	47.21

In Figure 4,, the area near the vicinity of the cell is marked in blue. This region we now call 'zone' in this paper. The amount of water (in mol) entering the zone is computed through volume integration for all time steps. The response is the saturation water content in the zone (W_{final}^{sim}). The water content vs. the simulation time plot is shown in Figure 5 and the zone saturation water content is marked.

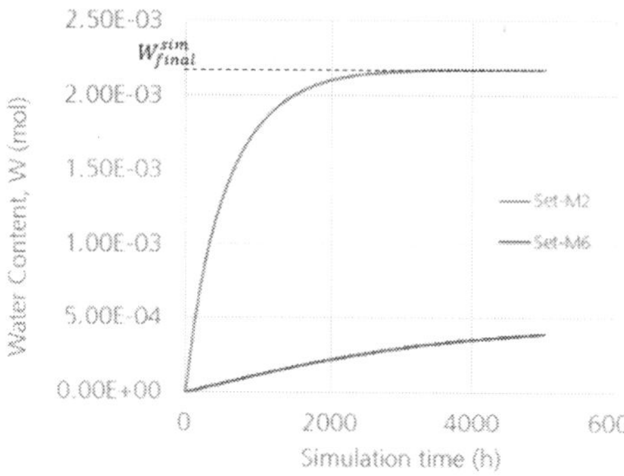

Figure 5: Water content (mol) vs. simulation time (h)calculated in COMSOL for Set-M2 and Set-M6

5 THE EMPIRICAL MODEL

The degradation shape of P_{mpp} in a full-scale PV module is known to follow a hyperbolic tangent function as described in [15]. The degradation of this metric attains a saturation after a certain testing time. The authors explain that this saturation in P_{mpp} occurs due to the equilibrium which is reached within the module. This stops moisture ingress into the module, avoiding further degradation of the cells. The formulated hyperbolic tangent function which evaluates the power as a function of testing time for the whole module is used to predict the failure time for two different combinations of temperature and humidity. It is seen that in both cases, the formulated empirical model does not predict the failure times exactly. This proves that formulating a mathematical model to explain failure times for a complete module could be futile. It gives more control and insight in modeling degradation when we assess the failure at a smaller scale (mini module/microclimate assessment) as adopted here.

Modeling the performance metric as a function of time is problematic, as seen in [15]. In damp heat conditions, the time to failure is a function of both temperature and humidity. With three levels for each stressor, we need eight experiments to compute the time to failure of the modules at different temperature and relative humidity. This can be costly and time consuming. By modeling the degradation as a function of a local diffusion parameter, such as water content, the effect of moisture at the cell level could be understood. The effect of temperature on the cell level could be differentiated by conducting the single/half-cell damp heat experiments at two different temperatures at the same humidity level.

The measured cell power as a function of water content (mol) in the zone is shown in Fig 6. The orange and black data points are extrapolated values from the obtain experimental data. It is seen that at 80h the cell has undergone around 2% degradation and at around 5000h, the cell had degraded by 60%. The data points are chosen in order to select the appropriate empirical relation which provides a good fit to the data. The chosen empirical formula is of the form shown below [16]:

$$P_{MPP} = P_{final} + (P_{initial} - P_{final})\left[1 - \left(\frac{W}{W_{final}^{sim}}\right)^{\alpha}\right]^{\beta} \quad (1)$$

In equation 1, $P_{initial}$ is the P_{mpp} value recorded before the start of the ageing experiments. In other words, this is the power of the 'dry' module. P_{final} is the final power when the zone has reached the saturation. P_{mpp} can further decline even after the zone has reached saturation. It is best to do the experiment till the cell is seen to have fully deteriorated (either through EL or power measurements) or a saturation is observed as seen in [15]. The empirical model (eq. 1) is taken from [16] and an extra shape parameter β. is added to the model to fit the sharp curvature as the zone reaches saturation.

The model is fitted to the data using the least squares method and the R^2 value of 0.99928 shows a good fit between measurement data and model.

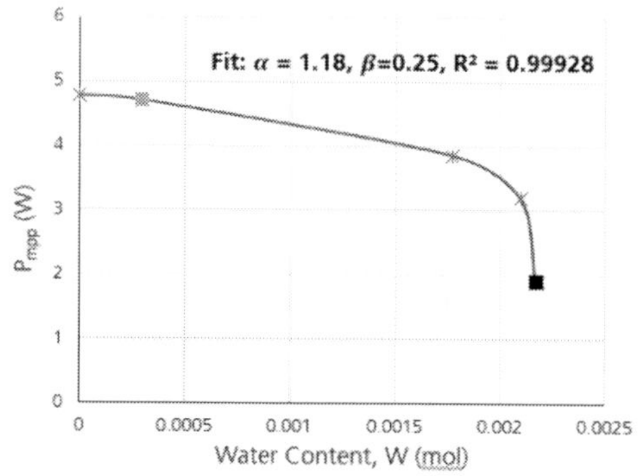

Figure 6: Pmpp vs. zone water content. Eq 1. fitted to the data.

6 EXTENSION OF THE EMPIRICAL MODEL TO LARGER MODULES

The formulated empirical model could be extended to predict the power degradation in larger modules at different locations. An illustration is show in Figure 7. This can be accomplished by creating a computational model of the larger module to simulate the ingress of moisture at a given climatic condition. When an array of cells is present, a zone for each of these cells could be created in the computational model as shown earlier. Experiments on single/half-cell modules should be conducted at an elevated temperature or humidity. A similar BOM and module configuration should be used to construct the mini modules. The use of single or half-cell depends on the wafer size and BOM availability. The modules should be designed in a way that the zone gets saturated quickly. Then, from the power – water content curve (or respectively FF, I_{sc} etc.) the empirical model can be fit and the parameters α, β can be determined. Further experiments at two different temperatures could be conducted to identify the effect of temperature on the performance metric. Using these additional experiments, α, β could be modeled as a function of temperature.

The empirical model with the fitted parameters could be used in conjunction with the FEA simulation of water ingress in larger modules. When $W^{i,j}$ in Eq. 1 is substituted with the water content from simulation of larger module, the corresponding power of the cell is obtained. From this the power output of the whole module could be calculated.

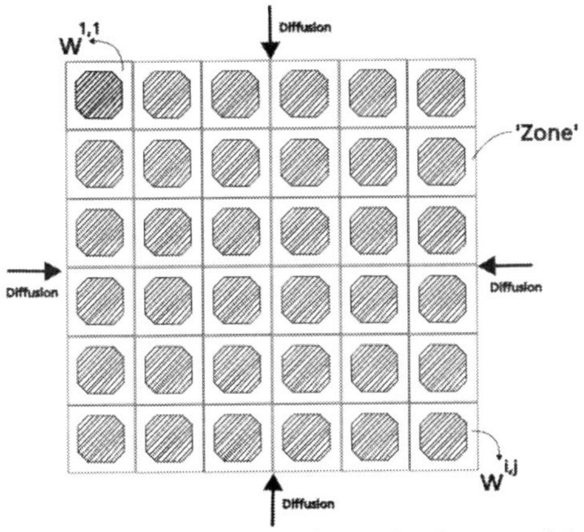

Figure 7: Figure shows several zones in a larger module. $W^{i,j}$ is the water content at a zone of a cell in ith row and jth column.

7 SUMMARY

An empirical model relating the electrical performance of a HJT cell as a function of a local diffusion parameter has been set up. The model is derived from experimental and numerical investigation of water accumulation near the vicinity of the cell, which we termed as 'zone' in here. The model can be applied to cells which show different degradation shapes to the here investigated HJT cells by changing the value of shape parameters α & β. Possibilities how to extend the model to predict field degradation in large modules has been discussed. By conducting the experiments at two different temperature settings (apart from the prescribed 85°C), the effect of temperature on the degradation could also be calculated by hierarchically modelling α & β as a function of temperature.

REFERENCES

[1] Segbefia O. K., Imenes A. G., and Sætre T. O., 2021,"Moisture ingress in photovoltaic modules: A review," Solar Energy, **224**, pp. 889–906.

[2] van der Wel G. K., and Adan O., 1999,"Moisture in organic coatings — a review," Progress in Organic Coatings, **37**(1), pp. 1–14.

[3] I. Kaaya, M. Koehl, A. P. Mehilli, S. de Cardona Mariano, and K. A. Weiss, 2019,"Modeling Outdoor Service Lifetime Prediction of PV Modules: Effects of Combined Climatic Stressors on PV Module Power Degradation," IEEE Journal of Photovoltaics, **9**(4), pp. 1105–1112.

[4] Olatz Arriaga Arruti, Alessandro Virtuani, and Christophe Ballif,"Long-term performance and reliability of silicon heterojunction solar modules,"

[5] S. Mitterhofer, C. Barretta, L. F. Castillon, G. Oreski, M. Topič, and M. Jankovec, 2020,"A Dual-Transport Model of Moisture Diffusion in PV Encapsulants for Finite-Element Simulations," IEEE Journal of Photovoltaics, **10**(1), pp. 94–102.

[6] M. Jankovec, S. Mitterhofer, J. Slapšak, and M. Topič, 2020,"Monitoring of lateral moisture ingress across the surface of a solar cell in PV modules," *2020 47th IEEE Photovoltaic Specialists Conference (PVSC)*, pp. 1308–1312.

[7] J. Slapšak, S. Mitterhofer, M. Topič, and M. Jankovec, 2019,"Wireless System for In Situ Monitoring of Moisture Ingress in PV Modules," IEEE Journal of Photovoltaics, **9**(5), pp. 1316–1323.

[8] Hülsmann P., and Weiss K.-A., 2015,"Simulation of water ingress into PV-modules: IEC-testing versus outdoor exposure," Solar Energy, **115**, pp. 347–353.

[9] Box G. E. P., Hunter J. S., and Hunter W. G., 2005, Statistics for experimenters, *Wiley series in probability and statistics*, Wiley Hoboken, NJ.

[10] Sen C., Wang H., Wu X., Khan M. U., Chan C., Abbott M., and Hoex B., 2023,"Four failure modes in silicon heterojunction glass-backsheet modules," Solar Energy Materials and Solar Cells, **257**, p. 112358.

[11] Dadaniya A., and Datla N. V., 2019,"Water diffusion simulation in photovoltaic module based on the characterization of encapsulant material using in-situ gravimetric technique," Solar Energy Materials and Solar Cells, **201**, p. 110063.

[12] Liu Z., Reinoso J., and Paggi M., 2022,"Hygro-thermo-mechanical modeling of thin-walled photovoltaic laminates with polymeric interfaces," Journal of the Mechanics and Physics of Solids, **169**, p. 105056.

[13] M. Jankovec, F. Galliano, E. Annigoni, H. Y. Li, F. Sculati-Meillaud, L. -E. Perret-Aebi, C. Ballif, and M. Topič, 2016,"In-Situ Monitoring of Moisture Ingress in PV Modules Using Digital Humidity Sensors," IEEE Journal of Photovoltaics, **6**(5), pp. 1152–1159.

[14] R. E. Kumar, G. v. Gastrow, J. Leslie, R. Meier, M. I. Bertoni, and D. P. Fenning, 2019,"Quantitative Determination of Moisture Content in Solar Modules by Short-Wave Infrared Reflectometry," IEEE Journal of Photovoltaics, **9**(6), pp. 1748–1753.

[15] Koehl M., Hoffmann S., and Wiesmeier S., 2017,"Evaluation of damp-heat testing of photovoltaic modules," Progress in Photovoltaics: Research and Applications, **25**(2), pp. 175–183.

[16] Schaff J. R., and Davidson B. D., 1997,"Life Prediction Methodology for Composite Structures. Part I—Constant Amplitude and Two-Stress Level Fatigue," Journal of Composite Materials, **31**(2), pp. 128–157.

DNV

GreenPowerMonitor
a DNV company

Long-term PV System Modelling and Degradation using Neural Networks

Gerardo Guerra[1], Pau Mercade-Ruiz[1], Gaetana Anamiati[1], Lars Landberg[2]
[1]GreenPowerMonitor a DNV Company
[2]DNV

18 September 2023

020278-001

Content

2 DNV © 18 SEPTEMBER 2023

DNV

020278-002